2025 IEEE Applied Power Electronics Conference and Exposition (APEC 2025)

Atlanta, Georgia, USA
16-20 March 2025

Pages 1-585

IEEE Catalog Number: CFP25APE-POD
ISBN: 979-8-3315-1612-3

**Copyright © 2025 by the Institute of Electrical and Electronics Engineers, Inc.
All Rights Reserved**

Copyright and Reprint Permissions: Abstracting is permitted with credit to the source. Libraries are permitted to photocopy beyond the limit of U.S. copyright law for private use of patrons those articles in this volume that carry a code at the bottom of the first page, provided the per-copy fee indicated in the code is paid through Copyright Clearance Center, 222 Rosewood Drive, Danvers, MA 01923.

For other copying, reprint or republication permission, write to IEEE Copyrights Manager, IEEE Service Center, 445 Hoes Lane, Piscataway, NJ 08854. All rights reserved.

*** This is a print representation of what appears in the IEEE Digital Library. Some format issues inherent in the e-media version may also appear in this print version.

IEEE Catalog Number: CFP25APE-POD
ISBN (Print-On-Demand):   979-8-3315-1612-3
ISBN (Online):   979-8-3315-1611-6
ISSN: 1048-2334

Additional Copies of This Publication Are Available From:

Curran Associates, Inc
57 Morehouse Lane
Red Hook, NY 12571 USA
Phone: (845) 758-0400
Fax: (845) 758-2633
E-mail: curran@proceedings.com
Web: www.proceedings.com

TABLE OF CONTENTS

Versatile Controller Architecture for a Universal DC Fast Charging Front-End 1
Anurag Singh, Sayan Paul, Tejas Bhuse, Trent Martin, Hien Nguyen, Inder Vedula, Nikola Milivojeviæ, Dragan Maksimoviæ, Luca Corradini

A 10 kV SiC MOSFET based Three-Phase Single-Stage Isolated MVAC/LVDC Converter for Solid State Transformer Applications 9
Anup Anurag, Chi Zhang, Rudy Wang, Peter Barbosa

Direct Digital Control Applied to T-Type Vienna Rectifiers for Power Factor Correction 16
Jun-Yang Chang, Tsai-Fu Wu, Chien-Chih Hung, Jui-Yang Chiu

Active Power Decoupling Method based on Dual Active Bridge Converter without Additional Components 21
Kosuke Takeuchi, Takashi Ohno, Hiroki Watanabe, Yuki Nakata, Jun-Ichi Itoh

An ANPC-Based Building Block for Medium-Voltage Applications 27
Ahmed Rahouma, Hui Cao, David A. Porras, Zhuxuan Ma, Yue Zhao, Juan C. Balda

Analog Control of a 2.5 kW GaN based CRM PFC with Input Filter Optimization 34
Naveed Ishraq, Ayan Mallik

An iTHD and Efficiency Optimized Control Method for Triangular Conduction Mode Totem Pole Bridgeless PFC with Zero Current Detection 42
Brent McDonald, Sheng-Yang Yu

Resonance Current Suppression for AC-DC Active-Clamp Flyback Converter by Triangular Current Mode 48
Yasuo Uchida, Hiroki Watanabe, Jun-Ichi Itoh

A Universal DC Fast Charging Front-End with Optimized Film Capacitor Design 54
Sayan Paul, Anurag Singh, Tejas Bhuse, Trent Martin, Hien Nguyen, Inder Vedula, Nikola Milivojeviæ, Dragan Maksimoviæ, Luca Corradini

Power Characterization of a 1200-V/800-V 22-kVA 30-kHz Unity-Gain Dual-Active-Bridge Converter Prototype 62
Radhika Sarda, Abishek Sethupandi, Madasamy Palavesha Thevar, Howe Li Yeo, Praveenkumar Palani, Vaisambhayana B. Sriram, Anshuman Tripathi

Design of Fully Soft-Switched Semi-Dual Active DC-DC Converter for Battery Charging Application 69
Siva Prabhakar, Shiladri Chakraborty, Sandeep Anand

A ZCS-ZVS Strategy for Low Impedance Dual Active Bridges in MHz Range 77
Pushkar Saraf, Michael Solomentsev, Alex Hanson

A 6.6 kW Highly Efficient Reconfigurable Dual Active Bridge Converter Designed using Planar Transformer, SiC-Fets and Monolithic Bidirectional Devices 90
Reza Barzegarkhoo, Fabian Groon, Arkadeb Sengupta, Marco Liserre

Interleaved Switched-Inductor-Based SIPO Partial Power Converter Module for Battery Management Systems 98
Fengwang Lu, Henry Shu-Hung Chung

Single Sensor-Based Fault Localization and Detection in GaN Three-Phase Dual Active Bridge Converters 103

Satyam Sa, Yi Han, Cheng Feng Wang, Olivier Trescases

Enhanced Cocharge Operation Scheme in Bidirectional PhaseShift Full-Bridge Converters with Eliminated Voltage Overshoot and Reduced Freewheeling Current 111

Tien-Sheng Li, Minh Ngo, Rolando Burgos, Dong Dong

DC Bias Elimination in Isolated DC-DC Converters using Fundamental-Frequency Ripple 118

Arkadeb Sengupta, Thiago Antonio Pereira, Marco Liserre

Tunable Matching Network with Dual Phase-Switched Impedance Modulation Actuators 124

Alexander Jurkov, David Perreault

Soft-Switched Pulsed Bias Plasma Supply System 132

Julia Estrin, Alexander Jurkov, David J. Perreault

Analysis and Design of a Cyclo-Active-Bridge Inverter for Single-Stage Three-Phase Grid Interface 139

Mian Liao, Tanuj Sen, Yang Wu, Minjie Chen

Modular Nanosecond Pulse Generator Leveraging GaN and SiC for Versatility and Performance 147

P. Briz, H. Sarnago, O. Lucía

A Variable Frequency Technique for EMI and Efficiency Improvements in High-Level Count Flying Capacitor Multilevel Converters 151

Francesca Giardine, Sahana Krishnan, Logan Horowitz, Robert C. N. Pilawa-Podgurski

Analysis and Implementation of Minimum-Sensor Capacitor Voltage Estimators for Flying Capacitor Multilevel Converters 157

S. Tahmid Mahbub, Rahul K. Iyer, Ivan Z. Petriæ, Robert C. N. Pilawa-Podgurski

Single-Stage Bidirectional High-Frequency Link DC to Three-Phase AC (4-Wire) Grid-Tied Microinverter 164

Aniruddh Marellapudi, Satish Belkhode, Joseph Benzaquen Sune, Deepak Divan

Analysis and Design of a Constant Current LCC Class-E Inverter 171

Ju Gao, Ziheng Liu, Jiayin He, Hongjie Peng, Chengkang Ao, Jinyan Wang

Series Connected Class-E Push-Pull Converters using GaN HEMT for High-Efficiency RF Generators in Float Zone Silicon Production 175

Faheem Ahmad, Thore Stig Aunsborg, Jannick Kjær Jørgensen, Stig Munk-Nielsen

State of the Art 1.7kV Lateral GaN HEMTs, an Alternative to SiC 180

Karthick Murukesan, Robert Yang, Kamal Varadarajan, Sorin Georgescu, Doug Kang

Modeling and Characterization of Current and Future 1.2 kV Wide Bandgap Semiconductor-Based MOSFETs 185

Sushanta Gautam, Austin M. Szczublewski, Samuel K. Atwimah, Aidan P. Fox, William M. Collings, Tolen Nelson, Daniel G. Georgiev, Raghav Khanna, Andrew D. Koehler, Karl D. Hobart

2.5-kV 6.4-ns 100-kHz Repetitive GaN Marx Generator 192

Ruize Sun, Ci Pan, Wanjun Chen, Bo Zhang

Novel Dual Output LDO Architecture in 650-V GaN Technology for Power ICs 195

Plinio Bau, Thanh Hai Phung, Deniz Aygun, Bart Coomans, Mike Wens

Impact of Substrate Bias on the Stability of Bidirectional GaN HEMT in Hard- and Soft-Switching 202
Qihao Song, Hongchang Cui, Qiang Li, Yuhao Zhang

Characterization of LED Driven GaN-Based Photoconductive Switches 207
Samuel K. Atwimah, Tolen M. Nelson, Geoffrey M. Foster, Daniel G. Georgiev, Andrew D. Koehler, Alan G. Jacobs, Karl D. Hobart, Micheal R. Hontz, Raghav Khanna

Development and Validation of Repetitive Transient Gate Overvoltage Rating for GaN HEMTs 214
Ricardo Garcia, Angel Espinoza, Siddhesh Gajare, Shengke Zhang

Junction Temperature Monitoring of GaN HEMT by using On-Resistance with Voltage Clamp and
Current Shunt 219
Xiao Wang, Mingrui Zou, Jiakun Gong, Yulei Wang, Zheng Zeng

False Turn-On Failure and Protection of p-Gate GaN HEMT in MHz Class-E Resonant Inverter 225
Ziheng Liu, Ju Gao, Hongjie Peng, Jiayin He, Jinyan Wang, Maojun Wang

Heat Extraction from Ferrite Cores using Metallic Laminations 231
Alyssa Brown, Duy T. Nguyen, Alex J. Hanson

Folded Flex-PCB Winding Planar Transformer for High-Frequency Isolated DC-DC Converters 238
Soundhariya G. Soundararajan, Hans Wouters, Wout Vanderwegen, Wilmar Martinez

Winding Strategy Analysis and Optimization for High-Current Matrix Transformer 246
Bima Nugraha Sanusi, Pinhe Wang, Michael A. E. Andersen, Ziwei Ouyang

Investigation on Impact of Transformer Parasitic Capacitance on Standby Power Consumption in
Power Converters 252
Kamran Kamran, Andrea Russo, Federica Cammarata, Claudia Malannino, S. Yuri Ciardo,
Ziwei Ouyang

PCB-Winding Integrated Transformer for 800-V Dual Active Bridge Converter using 1.2-kV GaN
Devices 258
Hans Wouters, Wei-Ren Lin, Nicolas Pirson, Thomas Jochmans, Yu Zuo, Wilmar Martinez

Comparative Assessment of Inductance Modeling for PCB-Based Circular Spiral Coils in Inductive
Power Transfer Systems 266
Gaia Petrillo, Drazen Dujic

Compact Air-Core Inductors for Variable Frequency Soft-Switching in 3 Phase Inverters 272
Youssef A. Fahmy, Matthias Preindl

Simulation and Experimental Research on Cooling Performance of Fully-Immersed Evaporative
Cooling High-Frequency Transformer 278
Zhanlei Liu, Lingyu Zhu, Yuntian Gao, Yongliang Dang, Cao Zhan, Shengchang Ji

High-Efficiency PCB-Embeddable Inductor for Vertical Power IVR Applications 285
Youssef Kandeel, Liang Ye, John Flannery, Cian Ó Mathúna, Ranajit Sai, Seamus O'Driscoll,
Takayuki Tsuchida, Naoya Terauchi, Sumiaki Kishimoto, Toshio Hiraoka, Masanori Nagano

An Adaptive Zero Current Switching Control Technique for Multi-Resonant Switched-Capacitor
Converters 291
Haifah B. Sambo, Rose A. Abramson, Sahana Krishnan, Robert C. N. Pilawa-Podgurski

Small-Signal Analysis and External Ramp Design for Multiphase Current-Mode Constant On-Time
Control with Phase Overlapping 299
Sundaramoorthy Sridhar, Qiang Li

Multiphase Constant-on-Time Minimum-Deviation Controller for Modern Processors.................................. 307
  Duo Li, Gianluca Roberts, Aleksandar Prodić, Alan Wu

Closed-Loop Control of a Dual-Side Series/Parallel Piezoelectric-Resonator-Based DC-DC
Converter... 315
  Wen-Chin B. Liu, Gaël Pillonnet, Patrick P. Mercier

High-Bandwidth Embedded Rogowski Coil on Multilayer Substrate with Minimal Contribution to
Power Loop Inductance.. 321
  Takahiro Okamoto, Masataka Ishihara, Kazuhiro Umetani, Eiji Hiraki

Operating and Switching Frequency Circulating Current Control in Paralleled High Power
Adjustable Speed Drives with Common DC Link... 327
  Kevin Lee, Zhihao Song, Wenxi Yao, Bo Wei

Mixed-Signal Sliding Mode Controller for Non-Inverting Buck-Boost Photovoltaic DC Optimizers............. 334
  Anurag Singh, Sayan Paul, Dragan Maksimović, Luca Corradini

A Current Sensorless Output Voltage Tracking Controller-Observer for a Boost Inverter using
Feedback Linearization 342
  Ion Leandro Dos Santos, Tailan Orlando, Yohannes Amilcar Tekle Scherer, Telles Brunelli
  Lazzarin, Hector Bessa Silveira

Modeling and Control of a Cyclo-Active-Bridge Inverter for Single-Stage Three-Phase Grid
Interface... 349
  Tanuj Sen, Mian Liao, Yang Wu, Minjie Chen

Turn-On Transient Modeling of 10 kV SiC MOSFET Half-Bridge Power Module in LTspice 357
  Nianzun Qi, Jannick Kjær Jørgensen, Gao Liu, Zhixing Yan, Morten Rahr Nielsen, Asger
  Bjørn Jørgensen, Hongbo Zhao, Stig Munk-Nielsen

A Compact, Automated Sawyer-Tower System for Characterization of the High-Frequency, Soft-
Switching C_{oss} Loss of Wide Bandgap Devices.. 363
  Katherine Liang, Malachi Hornbuckle, Juan Rivas-Davila

Enhancing Behind-the-Meter Visibility of Grid Edge PV Systems and Electric Vehicle Charging
Loads Through Integration of Compact Low-Cost Sensors 370
  Mehrnaz Madadi, Paul Ohodnicki, Subhashish Bhattacharya

Supercapacitor based TMS Pulse Generator Design-Experimental Results Versus MATLAB
MOSFET Simulation Model 378
  Soniya Raju, Nihal Kularatna, Marcus Wilson, Alistair Steyn-Ross

Application of Artificial Intelligence for Modeling SiC Power MOSFETs.. 385
  Fredo Chavez, Danial Bavi, Sourabh Khandelwal

Multi-Objective Design Automation in Power Electronics using Bayesian Optimization Techniques 389
  Tung-Tan Nguyen, Man-Hay Pong, Huang-Jen Chiu

Reduced Order Thermal Modelling of Multi-Chip Silicon Carbide Power Modules.................................... 395
  Aamir Rafiq, Blake Nelson, Marshal Olimmah

Design and Evaluation of Dual-Resolver Emulation for Control System Verification in Aerospace
Actuation Applications 401
  Tomas Sadilek, Julian Opificius, Jason Wright, Alec Leslie, Jeremie Tuzizila, Cesar Alzate,
  Hunter Burnett, Joshua Atkinson, Justin Stricula

Un-Terminated Blackbox Modeling for Electric Machines.. 409
  Xinliang Yang, Vladimir Mitrovic, Qing Lin, Rolando Burgos

7.2 kW GaN-Based DAB Converter with 37 kW/L Power Density and High Efficiency............................... 416
  Esmaeil Jalalabadi, Xiaoyu Wang, Jaksa Rubinic, Yang Jiao, Lucas Lu

A Novel Interleaving Method for High Power Integrated Electric Vehicle Charger with Three-Phase
Permanent Magnet Synchronous Motor 423
  Ryota Tanaka, Toshihiro Kai, Kenta Takishima, Yoshiyuki Nagai, Tetsuya Hayashi, Kantaro
  Yoshimoto

A Three-Phase CLLC Resonant Converter with Integrated Planar Magnetics for 22-kW On-Board
Chargers.. 429
  Tianlong Yuan, Zhangwei Xiang, Abdelrahman Ali, Feng Jin, Qiang Li, Wendell Da-Cunha-
  Alves, Xiaoshan Liu

Reconfigurable LLC Resonant Converter for Wide Voltage Range and Reduced Voltage Stress in
DC-Connected EV Charging Stations 436
  Yu Zuo, Xiaobing Shen, Bangli Du, Qingcheng Sui, Tim Geboers, Wilmar Martinez

Design and Control of GaN based Three-Phase / Single-Phase Combo Three-Level Flying
Capacitor PFC for OBC Applications.. 442
  Nidhi Haryani, Laszlo Huber, Anup Anurag, Juan Ruiz, Peter Barbosa

Optimization Strategy for Battery Electric Vehicle (BEV) DC Fast Charging (FC) in Cold
Environments.. 449
  Seif Sarofim, Cheng Feng Wang, Satyam Sa, Avram Kachura, Isaac Muscat, Olivier Trescases

DC-Link Voltage Reduction with Synergetic Common-Mode Voltage Control of Single-Phase Two-
Stage Non-Isolated EV Chargers 457
  Dongsu Lee, Juwon Lee, Jung-Ik Ha

DC-DC Converter Architecture for Fast Electric Vehicle (EV) Battery Charging Applications 464
  Shibaji Basu, Arjun Ivimey, Praveen Jain

Fast Simulator for the Estimation of Inverter DC-Link Temperature in e-Drives Subjected to Highly
Variable Working Cycles 472
  Simone Giuffrida, Fabio Mandrile, Radu Bojoi

A Monolithic Regulated 160 MHz Resonant DC-DC Converter 479
  Giacomo Ripamonti, Stefano Michelis, Georgios Bantemits, Pablo Daniel Antoszczuk, Khalil
  Khalife, Nils Hans Van Der Blij, Sokratis Koseoglou, Mattia Balutto, Francesco Driussi,
  Stefano Saggini

Reconfigurable Trans-Inductor Voltage Regulator with Improved Light Load Efficiency in Data
Center Applications 485
  Ziyao Wang, Zehui Li, Haoyu Wang

Fully Integrated Voltage Regulators (FIVRs) with Package In-Situ Coupled CoaxMIL Inductor for
High Power Density Microprocessor Applications 491
  Jaeil Baek, Beomseok Choi, Siddharth Kulasekaran, Huong Do, Brandon Marin, Jose
  Chavarria, Leigh Wojewoda, Kaladhar Radhakrishnan

Multiphase Lateral Flux Indirect Coupled Inductor for Vertical Power Delivery Voltage Regulator
Module 498
  Adhistira M. Naradhipa, Qiong Wang, Qiang Li

A High Density Three-Level Quadratic Buck Hybrid Converter for 48V-to-PoL Conversion.........................505
  Kejia Wang, Si Yuan Sim, Yin Quen Choong, Xin Zhang, Sriharsh Pakala, Cheng Huang

Air-LEGO: A Magnetic-Free Ultra-Thin 24V-to-1V 120A VRM with Air-Coupled Inductors.........................510
  Haoran Li, Wenliang Zeng, Youssef Elasser, Minjie Chen

A 15A 48V-Input Dual-Path Hybrid Dickson Converter with 6 mm³ Low Saturation Current
Inductors for Point-of-Load Conversion 518
  Hua Chen, Young-Seok Noh, Minxiang Gong, Vivek De, Arijit Raychowdhury

An Ultra-Fast Control Strategy and Pre-Current-Balancing Measures Prepared for Rapid Transients
in Constant On-Time Controllers 524
  Yijie Qian, Yuan Gao, Wenze Shu, Lingyun Li, Shen Xu, Weifeng Sun

Loosely Coupled Trans-Inductor Voltage Regulator (LC-TLVR) Inductor as Compensation Inductor
(Lc) 530
  Pavan Kumar, Arturo Sanchez Hernandez

Novel Complex Permeability Model of Powder Magnetic Materials...538
  Lukas Mueller, James Cox, Jun Wang, Enrique Garcia

Design Study Evaluating Impact of Gap Loss on Nanocrystalline Inductor Cores with Experimental
Validation 544
  Maurice Sturdivant, Brandon Grainger, Christopher Bracken, Paul R. Ohodnicki

A Permanent Magnet Variable Inductor for DC Fault Current Limiting Applications 552
  Mark Nations, Subhashish Bhattacharya

Design-Oriented Modeling and Multi-Objective Optimization of Two-Phase Coupled Inductors in
Multiphase PWM Converters 558
  Yicheng Zhu, Jiarui Zou, Robert C. N. Pilawa-Podgurski

MagNetX: Extending the Magnet Database for Modeling Power Magnetics in Transient 566
  *Hyukjae Kwon, Shukai Wang, Haoran Li, Youssef Elasser, Gyeong-Gu Kang, Daniel Zhou,
  Davit Grigoryan, Minjie Chen*

Non-Monotonic Influence of DC Bias on Ferrite Core Loss Up to 10 MHz with Sine Wave
Excitation 573
  Bohua Zhang, Martin Pfost

Comprehensive SPICE Model for Inductors Considering Magnetic Losses Under DC Bias Current 579
  Yuki Sato, Hirokazu Matsumoto, Junichi Kotani, Shohei Tomioka, Kenichiro Tanaka

Indented Core to Reduce and Desensitize Inductor's Fringing Losses without Increasing Volume.................586
  Rajaie Nassar, Promit Datta, Guo-Quan Lu, Christina DiMarino, Khai Ngo

Coupled Inductor Analysis and Finite Element Modeling Assisted Design for Boost Extender
Topology..594
  Vikas Kumar Rathore, Michael Evzelman, Mor Mordechai Peretz

Stability Analysis of Current-Limited Grid-Forming Inverters with Frequency Stabilization: An
Equivalent Impedance Approach..602
  Bowen Yang, Gab-Su Seo

Revisit Active Power Oscillation in Multi-Virtual Synchronous Generators Gride 609
  Junjie Xiao, Pavol Bauer, Zian Qin

A Novel Current Control Technique for Off-Grid Single-Phase Inverters 616
Arpan Laha, Abirami Kalathy, Praveen Jain, Majid Pahlevani

Intelligent Low-Bandwidth Frequency Controller for VSGs at Economic Dispatch in Islanded
Microgrid.. 622
Shraf Eldin Sati, Ahmed Al-Durra, Hatem H. Zeineldin, Tarek H.M. El-Fouly, Ehab F. El-Saadany

Hardware-in-the-Loop of a Grid Forming Control Strategy Applied to a DC Off-Grid Green
Hydrogen Production System.. 629
Diego Montoya-Acevedo, René Contreras-Barrios, Ángel Maureira-Riquelme, Esteban Ibáñez-Muñoz, Catalina Gonzalez-Castaño, Carlos Restrepo

Experimental Validation of a 40kW, 480V Point-to-Point DC Interlinks for Controller-Agnostic,
Interoperable Networked Microgrids 637
Maximiliano Ferrari, Michael Starke, John Smith, Joao Pereira, Misael Montejano

Andronov-Hopf Oscillator-Based Grid-Forming Converters with Embedded Disturbance Rejection
for Non-Ideal Loading Condition 645
Vikram Roy Chowdhury, Gab-Su Seo, Barry Mather

Estimation of Rectifier Output Current of the LLC Converter.. 651
Xin Wu, Yi Zhou, Haihong Long, Dehong Xu

A 100kHz Digitally Controlled 10kW, 2-Channel Solar MPPT Converter using 3-Level Topology
with >75W/in³ Power Density and >98.5% Peak Efficiency.. 658
Ranajay Mallik, Akshat Jain

A Bootstrapless KY-S-Hybrid Buck-Boost Converter with Full Range iLs Reduction and 400%
Line Transient Response Acceleration for AI-Mobile Application.. 664
Chuan-En Chang, Cheng-Ta Chuang, Hao-Ran Huang, Chieh-Ju Tsai, Ching-Jan Chen

Digital Control of a 600-V to 28-V 20-kW Two-Stage DC-DC Converter 670
Shreyas B. Shah, Rachit Pradhan, Jiaqi Yuan, Mohamed Ibrahim, Ahmed Elezab, Samuel Hemming, Giorgio Pietrini, Piranavan Suntharalingam, Mario F. Cruz, Ali Emadi

Self-Calibrated Digital Current Emulation for High-Frequency Hysteretic Current-Mode Control in
GaN PFC Converters... 676
Mohammad Shawkat Zaman, Olivier Trescases

High-Frequency Flying Capacitor Four-Level Drain Supply Modulator 682
Audrey Cheshire, Paul Flaten, Zoya Popoviœ, Dragan Maksimoviœ

Discontinuous Modulation Strategy for Voltage and Temperature Balancing of MMCs 689
Davide D'Amato, Stayner Nóbrega Barros, Jun-Hyung Jung, Marco Liserre

Damping Control and Improvement of Grid-Forming Inverter from a Wideband Stability
Perspective... 696
Rui Kong, Subham Sahoo, Yubo Song, Frede Blaabjerg

A Grid-Forming Split-Phase Three-Leg Inverter with Unbalanced Loading and Active Power
Decoupling 703
Namwon Kim, Renata Kimpara, Michael Starke

Completely Decentralized Active and Reactive Power Control of Grid-Connected Cascaded H-Bridge Inverters with Integrated Battery Storage 711
Soham Dutta, Brian Johnson

Small-Signal Modeling and Damping Design of Unfolding-Based Single Stage AC-DC Converter using the Extra Element Theorem 719
Dakota Goodrich, Aditya Zade, Shubhangi Gurudiwan, Mahmoud Mansour, Regan Zane, Hongjie Wang

Methods to Enhance Cybersecurity of Multiple Inverters in Large Grid Connected PV / Battery Energy Storage Systems 727
Hasan Ibrahim, Jaewon Kim, Peng-Hao Huang, Vishwam Raval, Prasad Enjeti

Optimal DC-DC Converter Topology and Control Algorithm for Fuel Cell Electric Vehicle with Series-Connected Supercapacitor 733
Hyeon Soo Kim, Yun Seong Hwang, Seung Hyun Kang, Man Jae Kwon, Byoung Kuk Lee

Reliability-Constrained Design of a High-Gain Power Optimizer based on a Real Mission Profile 738
Stefano Cerutti, Francesco Iannuzzo, Ariya Sangwongwanich, Tamás Kerekes, Mario Giuseppe Pavone, Francesco Gennaro, Natale Aiello, Francesco Musolino, Paolo Stefano Crovetti

Submodule Voltage Balancing Technique of Solar MMC for Firing the Switches using Integrated PWM Modules 746
Ahmed Elsanabary, Saad Mekhilef, Mokhtar Aly, José Rodriguez

Single-Stage High-Frequency-Link Split-Phase Microinverter with High Voltage Gain based on Buck-Boost AC Chopper 751
Xuewen Li, Jia Liu, Jinjun Liu

Fault Diagnosis and Tolerant Strategy for Triple-Port Hydrogen Converter using SSA-Optimized Random Forest Algorithm 757
Shiqi Zhang, Yiyina Teng, Naizhe Diao, Xiaoqiang Guo, Vladimir Terzija, Lichong Wang

Resilient Operation for Grid-Connected Cascaded H-Bridge Multilevel Inverter with Improving PV Source Stress 761
Jinli Zhu, Yuan Li, Hector Akuta, Jeonghun Kim, Uthandi Selvarasu, Shumeng Wang, Vikram Roy Chowdhury, Brad Lehman, Fang Z. Peng

A Medium Voltage Grid-Connected PV Inverter with a New Modular High Voltage Gain Converter Featuring Internal Modified Voltage Doubling Balancers 768
Kajanan Kanathipan, Muhammad Ali Masood Cheema, John Lam

Split-Source Common-Ground Inverter for Photovoltaic Applications 775
Mahmoud A. Gaafar, Mohamed Orabi, Samir Kouro, Ahmed Ibrahim, Eltaib Abdeen D. Ibrahim

Comprehensive Investigation and Proposal of a New Wireless Charging Road Structure using Low-Environmental-Impact Magnetic Concrete 782
Shuntaro Inoue, Yuko Kano, Shin Tajima

Design of a Bidirectional High Power Inductive Power Transfer System with Auxiliary Winding for Automotive Applications 788
Luis Ruiz Chamorro, Nikola Mirkoviœ, Alberto Delgado Expósito, Pedro Alou Cervera, Miroslav Vasiœ

Mutual Inductance and Load Identification Method based on the Voltage Transients of WPT Systems 795
Xiaosheng Wang, C.Q. Jiang, Yibo Wang, Liping Mo

Digitally Controlled Misalignment-Tolerant Inductive Power Transfer System with Adaptive Hybrid Compensation for CC/CV Charging of E-Scooter 801

Niranjan Shrestha, V.S.R.Varaprasad Oruganti, Sheldon Williamson

On/Off Control of Modular Inductive Power Transfer System 809

Kunxiao Zhou, Guangdong Ning, Heyuan Li, Xinlin Wang, Minfan Fu

Receiver Side Regulation of LCC Wireless Power Transfer System with Variable Notch Filter 815

Hsin-Che Hsieh, Jih-Sheng Lai

84.7 Percent Peak Efficiency Stress Tolerant DC DC Buck Converter for Li Ion Battery Driven Standby Circuits in 18nm FDSOI............... 821

Gautam Dey Kanungo, Pijush Kanti Panja, Vikas Bugade, Kallol Chatterjee

Leveraging Ultrasound and Neural Networks for Non-Invasive Power Converter Efficiency Estimation............... 828

Youssof Fassi, Vincent Heiries, Jérôme Boutet, Julien Marianne, Sébastien Martin, Mathilde Chareyron, Clément Chambon, Sébastien Boisseau

A Load-Independent Multi-Relays Wireless Power Transfer with Self-Regulation and Single Compensation Network............... 834

Jong-Hun Kim, Najam Ul Hassan, Seogyong Jeong, Myeong-Ho Kim, Min-Sik Kim, Jee-Hoon Jung, Byunghun Lee, Se-Un Shin

A GaN-Based Single-Stage Solid-State Transformer Replacement for 40 VA Class 2 Line-Frequency Transformers............... 840

Allen T. Nguyen, Charles R. Sullivan

Survey of Components and Topologies for High-Efficiency and High-Power Density 48V DC-DC Converters 848

Joseph Winkler, Niklas Deneke, Bernhard Wicht

A Novel Solid-State Circuit Breaker using B-TRAN™ 854

Mudit Khanna, Ruiyang Yu, Milad Tayebi, Jiankang Bu, Jeffrey Knapp

Development of a Supercritical Fluid-Insulated Fast Mechanical Switch for MVDC Hybrid Circuit Breakers............... 860

Zhiyang Jin, Qichen Yang, Alfonso Cruz, Lukas Graber

Dynamic Impedance Matching for a Variable Reluctance Energy Harvesting Application with Constrained Space 868

Fernando Pérez, Alejandro Redondo, Airán Francés, Gabriel Mujica

Renewable Energy-Powered DC-Converted Refrigerator based on a Supercapacitor-Assisted Technique 874

Nirashi Polwaththa Gallage, Nihal Kularatna, Alistair Steyn-Ross, Dulsha Kularatna-Abeywardana

Design and Evaluation of Flexible Inductors for Wearable Power Electronics............... 880

Sean Logi, F. Selin Bagci, Katherine A. Kim

Design of Boost Power Factor Corrector and Asymmetrical Half-Bridge Flyback Converter for USB-PD Applications............... 887

Yun-Keng Cheng, Tsorng-Juu Liang, Kai-Hui Chen, Ming-Chang Tsou

Computationally Efficient Current Sensorless Predictive Control for PMSM Drive Fed by a Matrix Converter with CMV-Free Operation 895

Ali Sarajian, Ibrahim Harbi, Quanxue Guan, Davood Arab Khaburi, Ralph Kennel, José Rodriguez, Patrick Wheeler, Mokhtar Aly

PMSM Motor Drive with Current Direct Digital Control and Near 1st-Order Speed Control 900

Po-Chang Lee, Tsai-Fu Wu, Han Ku, Chien-Chih Hung, Jui-Yang Chiu

Fault-Tolerant Multilevel Converter for Multiphase Switched Reluctance Motor Drives based on q+2 Converter 906

Mahmoud A. Gaafar, Mohamed Orabi, Hao Chen, Mostafa Dardeer

Uncertainty-Aware Artificial Intelligence for Gear Fault Diagnosis in Motor Drives 912

Subham Sahoo, Huai Wang, Frede Blaabjerg

Neural Network based Digital Twin Health Monitoring of BLDC Motor Drives for Robots 919

Mohamed Y. Metwly, Benjamin Luckett, Landon Clark, JiangBiao He, Biyun Xie

MTPA Control using Predictive P&O Method for Dual Parallel Surface-Mounted Permanent Magnet Synchronous Motor Drives Fed by a Single Inverter 925

Jae-Seong Kim, Kyo-Beum Lee

A Novel I-f Startup Strategy with Smooth Transition to Sensorless Control for CSI-Fed PMSM Drives used in Submersible Pumps 930

Milad Bahrami-Fard, Majid Ghasemi Korrani, Babak Fahimi

Simulation-Assisted Design and Implementation of an Electrically Excited Synchronous Motor Drive System 938

Shih-Gang Chen, Jun-Ming Hsu, Chun-Yen Chen, Ming-Shi Huang

Implementation and Analysis of Direct Torque Control on High-Speed PMSMs: A Comparative Study of Commercial and Laboratory-Developed Motors 943

Md Moniruzzaman, Kishor Joshi, Md Rashedur Rahman, Md Khurshedul Islam, Seungdeog Choi, Masoud Karimi Ghartemani

A Ferrite based Carbon Reinforced Composite Wrapped IPM Rotor Design for High-Speed Traction Applications 951

Md Rashedur Rahman, Md Khurshedul Islam, Md Moniruzzaman, Seungdeog Choi, Han-Gyu Kim, Andrew Walters

A Novel Phase-Mode Controller for Resonant Converters 958

Claudio Adragna, Daniele Cazzaniga, Stefano Manzoni

A Regulated 36V-60V-Input VIN-Insensitive Resonant Switched-Capacitor Converter with Large Voltage Conversion Ratio 966

Yichao Ji, Jingyi Yuan, Lin Cheng

A Hybrid Switched Capacitor Converter Enabling Capacitive-Based Wireless Power Transfer for Battery Charging Applications 971

Jade Sund, Samantha Coday

A 48V to 50-110V Resonant Power-Bus Charger with Reduced Conduction Loss for MHz-Frequency Long-Range LiDAR Driver 978

Hangxiao Ma, Xuchu Mu, Yang Jiang, Weihang Zhang, Jincheng Zhang, Rui P. Martins, Pui-In Mak

A Trajectory Controlled 48-to-24 V Resonant Switched Capacitor Converter with 98.7% Efficiency and Ultrafast Dynamic Response 983
Hélène T.W. Ma Yang, Liang Wang, Haoyu Wang, Wai Tung Ng

Low Power, Non-Isolated, Extremely-High Step-Up, Quasi-Resonant Hybrid DC–DC Converter 990
Kumar Joy Nag, Aleksandar Prodiœ

Isolated Soft-Switching Flying-Capacitor based Quasi-Resonant Step-Up Converter 997
Kumar Joy Nag, Aleksandar Prodiœ

Accurate Small-Signal Phasor Transformation-Based Modeling of Secondary-Side Diode-Bridge Rectifiers for Battery Charging Applications 1004
Aditya Zade, Regan Zane

High-Efficiency Isolated Piezoelectric Transformers for Magnetic-Less DC-DC Power Conversion 1012
Sourav Naval, Wentao Xu, Mustapha Touhami, Jessica D. Boles

First Characterization of GaN Power Device and IC at Deep Cryogenic Temperatures Down to 100 mK 1020
Xin Yang, Matthew Porter, Zineng Yang, Zichen Xi, Liyang Jin, Liyan Zhu, Linbo Shao, Yuhao Zhang

Dynamic Environment-Aware Lifetime Prediction of SiC MOSFET Modules Through LSTM 1026
Md Zakir Hasan, Seungdeog Choi, Youssef Aider, Prashant Singh, Chun-Hung Liu

Guarding-Based C-V Characterization of 10 kV SiC MOSFET in Half-Bridge Module Configuration 1034
Nianzun Qi, Gao Liu, Zhixing Yan, Shaokang Luan, Pawel Piotr Kubulus, Yuan Gao, Stefan Meyer, Hongbo Zhao, Asger Bjørn Jørgensen, Stig Munk-Nielsen

Automated Characterization Platform for Comprehensive Dynamic R_{dson} Assessment of GaN HEMTs from 50 K to 400 K 1040
Tian Qiu, Zheyu Zhang, Purushottam Khadka, Ahmed Siraj, Dilip Rana

A Gate Driving Scheme for GaN Git with Enhanced Short Circuit Capability for Motor Drive Application 1047
Zongjie Zhou, Yan Cheng, Kevin J. Chen

Online Detection and Reduction of the Influence of Parameter Tolerance of Paralleled SiC MOSFETs in an EV Inverter Environment 1051
Hadiuzzaman Syed, Jochen Streit, Robert Kragl, Muhammad Muneeb Alam, Alberto Martinez-Limia, Karl Oberdieck, Ertuðrul Sönmez

Dynamic Current Sharing Issues with Paralleling SiC Power MOSFETs 1058
Ching-Yao Liu, Chen-Chan Lee, Jih-Sheng Lai

Integrated Short-Circuit Protection Design based on Dual-Channel Gate Driver for Series Connected Medium-Voltage SiC MOSFETs 1063
Rui Wang, Drazen Dujic

Long-Term High-Temperature Dynamic Gate Stress Reliability of a Last-Generation, Automotive-Grade, Planar 1200 V SiC MOSFET 1070
Giuseppe Mauromicale, Alessandro Sitta, Michele Fiore, Michele Calabretta, Francesco Iannuzzo

Innovative Gate Driver Structure Achieving Low Time Skew Across Isolation Barrier for Parallel Connected SiC Modules 1076
Louison Gouy, Anne-Sophie Descamps, Nicolas Ginot, Christophe Batard

Fully Integrated Closed-Loop Active Gate Driver IC with Real-Time Control of Gate Current Change Timing by Gate Current Sensing 1084
Yaogan Liang, Katsuhiro Hata, Makoto Takamiya

Analyze and Design of Digitally Load Current Modulated Active Gate Driver for GaN HEMTs based Buck DC-DC 1090
Wentao Liu, Zhina Lian, Taotao Wu, Xiaochuan Peng, Hao Min

Impact of Real-Time Variable Gate-Drive Strength on Drive Cycle Efficiency in SiC Inverter-Fed PMSM Traction Drives 1096
Matteo Pizzuto, Aiswarya Balamurali, Aniket Anand, Narayan C. Kar

Demonstration of Efficiency Increase of 350 V-to-13.3 V Isolated DC-DC Converters for Electric Vehicles by Active Gate Driving 1102
Yohei Sukita, Katsuhiro Hata, Hiroki Kondo, Kenichi Watanabe, Kenichi Nagayoshi, Makoto Takamiya

A Multi-Level Active Gate Driver for Achieving Thermal Balance in Parallel Connected Power MOSFETs 1108
Jingyuan Liang, Lingwei Sun, Wen Tao Cui, Wai Tung Ng, Motomitsu Iwamoto, Haruhiko Nishio

A Fast Short-Circuit Protection Method for Ohmic Gate P-GaN HEMT based on Gate Charge 1114
Yue Wu, Xi Jiang, Song Yuan, Xiaowu Gong, Zhaoheng Yan, Jiahong Chen, Yun Xu, Jinjie Liu

Comparison of Ultrafast-Rise-Time Gate Drivers for Wide-Bandgap Devices in Sub-Microsecond Pulsed Power Applications 1121
Soham Roy, Duy T. Nguyen, Neeraj Anantha, Alex J. Hanson

A Discrete Multilevel Active Gate Driver for GaN HEMTs to Optimize the Switching Behavior 1129
Celine Lawniczak, Martin Pfost

Attenuation of Fundamental Component of Differential Mode Noise using Active EMI Filter 1135
Guru Abhilash Mulumudi, Naveed Ishraq, Ayan Mallik

Graph Neural Network based Performance Modeling for the Dual Active Bridge Converter with Operational Generalization 1143
Weihao Lei, Fanfan Lin, Xinze Li, Xiaokun Bao, Xin Zhang

An Augmented State Space Modelling Approach for DC-DC Converter Start-Up in Closed Loop 1148
Waseah Anjum, Arkadeb Sengupta, Marco Liserre

The Utilization of a Parallel Computing Algorithm for Accelerating Switching-Level Modeling of Power Electronics Simulations in a T-Type PV Inverter 1153
Buck F. Brown III, Liwei Wang, Zheyu Zhang, Johan Enslin, Yi Li

A New Reduced Order Analytical Switching Model for eGaN HEMTs 1159
Ruqi Li, Douglas Arduini, Phen Lumod, Shobhana Punjabi, River Lin, Harold Gutierrez

Proposal of an Alternative Reverse Recovery Calculation Method 1167
Brian Deboi, Blake Nelson, Austin Curbow

Improvement of CM EMI Attenuation Ability of Transformer with Negative Capacitor 1173
    Qinghui Huang, Yiming Li, Yirui Yang, Shuo Wang, Yanwen Lai, Zhedong Ma

Damping Factor based PCB Parasitic Inductance Value Optimization to Minimize Voltage
Overshoot and Settling Time of Semiconductors 1179
    Reza Shahbazi, Yunting Liu

Hardware Implementation of Virtual Resistance based FRT Logic in Programmable 3-Level ANPC
Inverters 1184
    Mohammad Safayet Hossain, Shuvangkar Chandra Das, Paychuda Kritprajun, Amin Banaie,
    Tapas Barik, Deepak Ramasubramanian, Aboutaleb Haddadi, Evangelos Farantatos, Ulrich
    Muenz

Rad-Hard PSFB Controller for High-Voltage Space Applications 1190
    Reynaldo S. Gonzalez, Robert E. Bolaños

Modeling, Control and Digital Implementation of a Buck Converter Operating in Triangular
Current Mode for a Wide Output Voltage Range Space Application 1197
    Regina Ramos, Sara Pérez, Guillermo Núñez, Pedro Alou, Javier Torres

Thermal Model and Optimization of a Multi-Winding Transformer for Lunar Surface Power
Transmission 1203
    Zhining Zhang, Yuzhou Yao, Junchong Fan, Juchen Yang, Robert Guenther, Pengyu Fu, Jin
    Wang

Active Gate Driver Power Supply for High-Reliability Applications 1211
    Joseph P. Kozak, Juan Ramirez, Jesse Lin, Allison Orr, Alexander Martin, Hala Tomey

A Hybrid Energy Storage System for eVTOL Unmanned Aerial Vehicles using Supercapacitors 1217
    Ali Alenezi, PengHao Huang, Prasad Enjeti

Evaluation of Retired Lithium-Ion Batteries for Second-Life Applications Through Electrochemical
Impedance Spectroscopy 1224
    Latha Anekal, Sheldon Williamson

Uninterruptable Non-Isolated Integrated Power Electronics Converter (UNIPEC) for Commercial
Truck Auxiliary Power Unit 1230
    Pouya Zolfi, Ahmad Alzahrani, Ayman El-Refaie

Investigation of Electrical Safety for Non-Isolated Single-Phase On-Board Chargers used in
BEV/PHEV 1237
    Soya Kataoka, Shohei Funatsu, Hiroaki Matsumori, Takashi Kosaka, Keisuke Nakamura,
    Subrata Saha

An 8-Level Flying Capacitor Multilevel Converter for Electric Aircraft Pulse Deicing 1242
    Nicole Stokowski, Andrew Freeman, Aidan Rodgers, Aria Delmar, Jonathan Sengstock, Alex
    Solecki, Andrew Stillwell

Impact of Position Measurement Delay Angle on Performance of PMSM Drives for Electric Power
Steering in a Wide Speed Range 1248
    Yingzhe Wu, Hengbin Zhang, Yuxiang Xue, Lisheng Wang, Hui Li, Shan Yin

Physical Parameter Estimation for a Two-Level VSI Three-Phase PMSM Electric Drivetrain 1255
    Bernard Steyaert, Ananda Tjakra Adisurja, Matthias Preindl

A Novel Two-Dimensional Random Switching Frequency PWM Method for Variable Frequency Drives 1261

Mostafa Abarzadeh, Kevin Lee

Optimized Maximum Torque and Minimum Loss Fault-Tolerant Control Schemes for Dual Three-Phase PMSM 1267

Syed Mohammad Maaz, Dong-Choon Lee

Wireless Actuation of Magnetic Robots with a Modular 60 mT 3-D Helmholtz Coil System 1274

Konstantinos Manos, Yifan Rao, Tuo Zhao, Kevin Liu, Daniel Zhou, Calvin Nguyen, Eric Chen, Glaucio H. Paulino, Minjie Chen

A Versatile PHIL based Motor Emulator Testbench using a High-Performance Power Amplifier Testbench 1279

Seyedeh Nazanin Afrasiabi, Rajendra Thike, Mathews Boby, K. S. Amitkumar

A 450V Three Phase GaN IPM Achieving 99.1% Efficiency in Smallest 12mm x12mm Package for 250W Power Delivery without Heatsink 1286

Maik Kaufmann, Manu Balakrishnan, Stefan Herzer, Anand Chellamuthu, Hely Zhang

FEA based High-Frequency Synthesis for the Design and Optimization of GaN-Based Dual Three Phase Motor Drive System 1294

Syed Imam Hasan, Alper Uzum, Ashraf Siddiquee, Yilmaz Sozer, Krishna Namburi

Evaluation of Passive Common-Noise Canceller Considering Both of Thermal Equilibrium and Common-Mode Noise Cancellation 1299

Koji Mitsui, Kenshiro Katsura, Koki Notake, Koji Yamaguchi

Performance Evaluation of Isolated DC/DC Converters in Modularized Bridge Rectifier Solid-State Transformer 1305

Zhenchao Li, Andrea Cervone, Drazen Dujic

Active and Reactive Power Flow Control of the Dual Active Bridge Converter 1311

Lauryn Morris, Thomas W. Francois, Jonathan Saelens, Oroghene Oboreh-Snapps, Arnold Fernandes, Praneeth Uddarraju, Sophia A. Strathman, Jonathan W. Kimball

Comparative Analysis of Carbon Footprints and Material Usage of Solid-State Transformers and Low-Frequency-Transformer-Based MVac-LVdc Interfaces for High-Power EV Charging 1318

Luc Imperiali, Rudy Wang, Anup Anurag, Peter Barbosa, Johann W. Kolar, Jonas Huber

Trade Study of Isolation Requirements and Magnetic Core Selection for Medium Frequency-Medium Voltage Transformers 1326

Mohendro Kumar Ghosh, Mark A. Juds, Brandon Grainger, Ahmad El Shafei, Bogdan S. Borowy, Paul Ohodnicki

Comparative Evaluation of a Multilevel LLC Resonant Converter for a Modular DC/DC Stage in a Electrolyzer Power Supply 1334

Samuel S. Queiroz, Levy F. Costa

Cost-Effectiveness Assessment of SiC MOSFET and Si IGBT Semiconductors in a Three-Level Resonant Converter for Solid-State Transformer 1341

Samuel S. Queiroz, Levy F. Costa

Comparative Performance Analysis of Medium Voltage 3L-ANPC and 3L-DNPC Pole Enabled by Series-Connection of 10kV SiC MOSFETs and 10kV SiC JBS Diodes for Sine Triangle PWM Operation 1347
  Sanket Parashar, Shubham Rawat, Nithin Kolli, Raj Kumar Kokkonda, Subhashish Bhattacharya

A Zero Harmonic Distortion Master Converter for Medium Voltage Microgrids 1355
  Gabriel V. Ramos, Dener A. de L. Brandão, Thiago M. Parreiras, Danilo I. Brandão, Braz de J.C. Filho

An MILP Approach for Modeling and Analyzing the BESS for Smoothing Renewable Fluctuations Considering BESS Capacity Attenuation in the Bulk Power System with High Inverter-Based Resource Penetration 1363
  Hualong Liu, Wenyuan Tang

Thermal and Efficiency Characterization of Immersion Cooled SiC Traction Inverter 1368
  Yiju Wang, Reza Ilka, JiangBiao He

FPGA-Based Hybrid Simulator for Real-Time 3-D Temperature Monitoring of Power Converters 1375
  Xianghao Mo, Daniel Ríos Linares, Regina Ramos, Miroslav Vasiœ

A New Subassembly Concept for Enhanced Heat Dissipation and Reliability of Power Module 1383
  Yosuke Nakata, Yuji Sato, Shin Uegaki, Jun Fujita, Akihiko Furukawa, Masayoshi Tarutani

Stand-Alone R_{DS-ON} Sensor for In-Situ Prognostic, Protection and Reliability Enhancement of Power Converters 1388
  Zaheen Mustakin, Qiang Mu, Lucas Pereira, Jiale Zhou, Tiefu Zhao, Babak Parkhideh

Electrical Evaluation of a Modular High Voltage 3D Power Module using Direct Dielectric Liquid Cooling 1396
  Omar Sanjakdar, Yvan Avenas, Rachelle Hanna, Guillaume Piquet Boisson, Emmanuel Marcault, Antoine Philippe

Board Level Reliability of Gull-Wing, Micro-Leaded and Lead-Less Packaged MOSFETs in Automotive Environments 1403
  Christopher Liu, Vijayakrishna Satyamsetti, Xuanjing Wei, Christian Radici, Peter Vines, Wayne Lawson

Cost Effective and High Noise Immunity Methodology for Aging Evaluation of DC-Link Capacitors in Traction Inverters 1408
  Seyed Hossein Aleyasin, Fausto Stella, Radu Bojoi, Enrico Vico

A 3D Structure of Single-Sided Cooling Power Module with Low Thermal Resistance and Low Inductance 1414
  Hirofumi Hisamochi, Koki Notake, Yoshiaki Takahashi, Koji Yamaguchi

Aging of Y-Capacitor in an EMI Filter and Its Impact on Common-Mode Noises 1420
  Tahmid Ibne Mannan, Seungdeog Choi, Subarto Kumar Ghosh, Md Moniruzzaman

2200A/48V-to-1V Low-Profile Direct Power Converter with Standard PCB Transformer 1427
  Alejandro Figueroa, Pablo Mazariegos, Álvaro Cobos, Javier Goicoechea, Alejandro Castro, José A. Cobos

Single-Stage 48V-to-1V Regulator with a Half-Turn Transformer and Current-Doubler Rectifier.............. 1433
  Xinmiao Xu, Qiang Li

Ultra-Low-Profile Single-Stage Voltage Regulator Module (VRM) for Next-Generation AI Accelerators.. 1439
Xufu Ren, Jinfeng Zhang, Zhenshuai Rong, Borong Hu, Teng Long

Novel TLVR Operation in Multi-Stage Voltage Regulator Module with Current Multipliers....................... 1444
Kevin Zufferli, Roberto Rizzolatti, Mario Ursino, Simone Mazzer, Gerald Deboy, Stefano Saggini

Interphase LC-Oscillation Suppression with Fast Line-Transient Response in 48-V Series-Capacitor Buck Converters for Automotive Applications 1451
W.L. Jiang, Y. Liu, N. Khan, J. Pigott, H.J. Bergveld, V. Chaturvedi, O. Trescases

An Approach to Compensate for Low Frequency DC-Link Voltage Ripple in High Power ANPC Inverter 1459
Shaozhe Wang, Ankit Vivek Deshpande, Rolando Sandoval, Erick Pool-Mazun, Enrique Garza-Arias, Prasad Enjeti

A Cascaded Multilevel Inverter System with Hot-Swapping and Fault Isolation Capability for Improved Resiliency.. 1465
Uthandi Selvarasu, Vikram Roy Chowdhury, Shumeng Wang, Jinli Zhu, Mahshid Amirabadi, Yuan Li, Brad Lehman

Layout Optimization for Parasitic Inductance Reduction of GaN-Based NPL.X Multilevel Inverter 1473
Ali Halawa, Jinyeong Moon, Woongkul Lee

Topology Selection and Design Methodology for SiC based Solar Photovoltaic Medium Voltage Direct Grid Connect Inverters 1481
Jenson Joseph C. Attukadavil, Baylon G. Fernandes

EMI Modeling of PCB-Based Three-Level Active Neutral-Point-Clamped GaN Converter....................... 1489
Mohammad Hassan Adeli, Necmi Altin, Erkan Deniz, Adel Nasiri

A Novel Layout for Improving Current Sharing of Paralleled SiC MOSFETs with TO-247 Package........... 1495
Che-Wei Chang, Matthias Spieler, Rolando Burgos, Ayman El-Refaie, Renato Amorim Torres, Dong Dong

A Sensor-Less IGBT On-State Voltage Estimation Method using Inverter Control Variables 1501
Shuyu Ou, Subham Sahoo, Ariya Sangwongwanich, Yongjie Liu, Frede Blaabjerg

A Novel Non-Intrusive Online Monitoring Method for Diagnosing the Lift-Off of Bonding Wires in SiC MOSFETs... 1507
Keqi Song, Henry Shu-Hung Chung, Ho-Tin Tang

Optimizing MOSFET Selection for EMC-Critical Automotive Applications 1512
Sacha J. Cazzitti, Christian Radici, Andrew J. Forsyth, Cheng Zhang, Peter Vines

Improving Dynamic Current Sharing Between Parallel MOSFETs by Optimizing Device Parameters 1519
Kunal Jha, Kapil Kelkar, Marina Hedenik, David Penof

A 21.6 kW/L Two-Phase Immersion-Cooled Isolated DC-DC Converter... 1529
Aleksandar Ristic-Smith, Kawsar Ali, Daniel Rogers

Extraction of Common Mode Parasitic Capacitance in Balance Filter for the Prediction of EMI Noise Suppression 1537
Qiuzhe Yang, Xingyu Chen, Zijian Wang, Qiang Li

A 660W, 96% Efficiency 3D Heterogeneously Integrated Digital DC/DC Power Module for Vertical Power Delivery 1544
    Haoyu Wang, Xuliang Wang, Yan Wang, Xiaosen Liu

Planar Rogowski Coil-Based Switch Current Measurement for a 1.2 kV SiC MOSFET Embedded Die PCB... 1551
    Matthias Spieler, Che-Wei Chang, Ayman M. El-Refaie, Dong Dong, Rolando Burgos

Effect of Magnetic Couplings on Conducted EMI of GaN-Based PFC Converter 1557
    Tyler McGrew, Qiang Li

Optically-Controlled 3.3 kV SiC MOSFET with Fast Switching Speed and Low Optical Power 1564
    Xin Yang, Guannan Shi, Liyang Jin, Yuan Qin, Matthew Porter, Che-Wei Chang, Xiaoting Jia, Dong Dong, Linbo Shao, Yuhao Zhang

Optimization Techniques for Parallel-Connected Devices in IPMs for Consumer Use 1569
    Keisuke Kawamoto, Haruhiko Murakami, Teruaki Nagahara, Michael Rogers, Akiko Goto, Shoji Saito, Koichiro Noguchi

Investigating the Temperature Dependency and Operating Parameters of a Self-Driving Active Gate Driver 1576
    Vin Loong Choo, Martin Pfost

Use of Switched-Capacitor Circuit to Generate Negative Gate-Source Voltage Pulses 1582
    Ho-Tin Tang, Henry Shu-Hung Chung

An Optically Isolated Gate Driver with Simultaneous Data and Power Transmission Through a Miniaturized, Efficient Photonic Platform... 1590
    Jiajun Li, Mariia Klymenko, Yanqiao Li, William Scheideler, Jason T. Stauth

Optimal Shared Energy Storage Capacity Configuration in Multi-Energy Microgrids Considering Battery Lifetime Loss based on Relaxation Techniques... 1597
    Hualong Liu, Wenyuan Tang

Virtual Resistance Control for an Active Battery Management System 1602
    Alastair P. Thurlbeck, Ashraf Siddiquee, Mithat John Kisacikoglu, Yilmaz Sozer

Internal Voltage Source Saturation Impact on Stability Limits of Grid Forming Converter 1610
    Divyanshu Bansal, Aravind G., L. Umanand

A Zero Harmonic Distortion Grid-Connected Grid-Forming Converter for Battery Energy Storage System Applications 1615
    Gabriel V. Ramos, Thiago M. Parreiras, Fangzhou Zhao, Xiongfei Wang, Braz de J.C. Filho

Single Cell Energy Router Justification for Three Phase Near Zero Energy Buildings 1622
    Hossein Nourollahi Hokmabad, Tala Hemmati Shahsavar, Oleksandr Matiushkin, Tanel Jalakas, Oleksandr Husev, Juri Belikov

A Multi-UAV Charging Station Enabling Free Landing by Grid Pattern Transmitter.................... 1629
    Jungho Kim, Hyunkyeong Jo, Seoktae Seo, Bonyoung Lee, Hyungki Min, Franklin Bien

Capacitor Design for Self-Resonant Coils for Long-Distance Wireless Power Transfer System.................. 1635
    Mostak Mohammad, Vandana Rallabandi, Omer C. Onar, Gui-Jia Su

A 10.4-kW High-Power-Transfer-Density Multi-MHz Capacitive Wireless Power Transfer System for EV Charging Utilizing Stacked-Inverter Stacked-Rectifier Architecture 1640
Dheeraj Etta, Miguel Alvarez Dominguez, Sounak Maji, Syed Saeed Rashid, Khurram K. Afridi

Reduced-Fringing-Field Multi-MHz Capacitive Wireless Power Transfer System using Metasurface-Based Couplers with Active Field Cancellation 1646
Syed Saeed Rashid, Dheeraj Etta, Matteo Ciabattoni, Francesco Monticone, Khurram K. Afridi

Living Object Detection in Wireless Power Transfer Systems using Remote Capacitive Bio-Signals Monitoring... 1653
Bruno M.G. Rosa, Paul D. Mitcheson

Modified N:1 Switched Capacitor Converter with Reduced Capacitor DC Bias Voltage for High Power Density 1659
Taewoo Lee, Dam Yun, Sunghyuk Choi, Jung-Ik Ha

Wide Range Digital Control for Three-Level Buck Converters with Sensorless Flying-Cap Voltage Balancing... 1666
Hossein Hajisadeghian, Giovanni Bonanno

A Comparative Investigation of a New Continuous Voltage Conversion Ratio Approach in a Zero-Inductor Voltage Converter... 1673
Sina Salehi Dobakhshari, Aamna Nasir Hameed, Binghui He, Mojtaba Forouzesh, Yan-Fei Liu

A 96.1% Peak Efficiency, 6.8 kW/in³, 48V-to-6V On-Package Intermediate Bus Converter with LV-GaN Power Transistors... 1681
Mausamjeet Khatua, Nachiket Desai, Harish Krishnamurthy, Sheldon Weng, Jingshu Yu, Huong Do, Samuel Bader, Han Wui Then, Krishnan Ravichandran, James Tschanz, Kaladhar Radhakrishnan, Vivek De

A 48V to 2.4V-5V 95.8%-Peak-Efficiency 869W/in³-Power-Density Fibonacci Dual-Path Hybrid DC-DC Converter with Inductor Current Reduction and Low Output Resistance..................................... 1687
Yichao Ji, Zeguo Liu, Lin Cheng

An Ultra-Fast Very Large Scale Interleaved Li-Fi Transmitter 1693
Daniel H. Zhou, Konstantinos Manos, Minjie Chen

Isolated PWM DC-DC Converter with Single Magnetic Component, ZVS and Self-Balanced Switched-Capacitor Voltage 1701
Pablo M. Gil, Juan Rodríguez, Diego G. Lamar

Analysis and Design of a Low-Complexity ZVS Buck-Boost Converter 1707
Burkhard Ulrich

A High Conversion-Ratio Hybrid Series-Parallel DC-DC Converter with Pseudo-Soft-Charging and Inductor Current Frequency Multiplication.. 1715
Avinash Maddela, Kishalay Datta, Jason T. Stauth

A Real-Time Variation Control of Deadtime in GaN-Based Bidirectional Buck-Boost Converter for Lithium-Ion Battery Formation System... 1723
Jong-Hun Lim, Go Woon Heo, Je-Yeong Lim, Dong Hwan Kim, Byoung Kuk Lee

A Space Vector PWM Strategy for Charging of Bootstrap Capacitor in Three-Level Neutral-Point-Clamped Inverter 1728
Anantha Hegde, Asamira Suzuki, Hirokazu Nakamura, Takamune Kabashima, Koji Higashiyama, Keiji Akamatsu

A Complementary Carrier based PWM Strategy for Average Current Sampling of Three-Phase Inverter using Single Current Sensor 1734
Byeong-Il Kim, Joon-Seok Kim, Yeongsu Bak, June-Seok Lee

Short-Circuit Ride-Through for a CRM-Based Soft-Switching Three-Phase Inverter 1741
Xingyu Chen, Gibong Son, Qiang Li

Modified Space Vector Modulation with Low Bandwidth Sensor to Reduce Losses in Soft Switching Three-Phase Inverters 1746
Md Didarul Alam, Nazmul Hassan, Iqbal Husain, Liming Liu, Hongrae Kim

A Feedforward Ripple Reduction Control Strategy based on a Hybrid GaN/Si Interleaved Inverter 1754
Mowei Lu, Jurgis Reinotas, Xiaoyang Tian, Stefan M. Goetz

IGBT Comparison for Optimized Switching Behavior in the SiC/Si-Hybrid Switch 1759
Adrian Amler, Thomas Heckel, Daniel Ruppert, Cornelius Rettner, Martin März

Forward Recovery and its Mitigation in Hybrid Si/SiC-Based DC–AC Converters 1767
Yun Zhou, Thomas Lehmeier, Adrian Amler, Martin März

Real-Time IGBT Module Ageing Characterization Through Temperature Monitoring 1774
Quirc Perez-Farre, Luis F. Gomez-Rivera, Carlos Lopez-Torres, Kai Dannehl, Antoni García-Espinosa, Alejandro Paredes-Camacho

Experimental Validation of Triangular SOA via Infrared Thermography of a MOSFET Die Operating in the Thermally Unstable Linear-Mode for Automotive Applications 1781
Yacine Ayachi Amor, Christian Radici, Kerry J. Abrams, Philip Ellis, Peter Vines, Wayne Lawson

Feasibility Study of the SuperIGBT: A Series-Connected High Voltage IGBT with a Single Gate 1786
Junhong Tong, Alex Q. Huang, Huanghaohe Zou, Zhiyuan Ma

Low Profile, Laminated Nife Transformers for Flyback Converters 1791
Xuan Wang, Reza Mounesi, Matthew Catanoso, Matthew Fox, Adel Nasiri, Mark G. Allen

Comprehensive Demonstration of New Magnetic Designs Utilizing Magnetic Anisotropy of the Cores for Integrated Magnetics 1797
Yota Takamura, Honami Nitta, Tatsuya Miyazaki, Kimito Yamanaka, Ryosuke Ishido, Akira Namba, Keisuke Fujisaki, Shigeki Nakagawa

A Two – Stage Artificial Neural Network (ANN) – based Design and Optimization of High Frequency Transformers for Dual Active Bridge Converter 1803
Lufan Zhou, Alberto Delgado Expósito, Adam Ruszczyk, Simon Round, Miroslav Vasiœ

Modeling and Optimizing Winding Arrangement for Gapped Planar Magnetics based on Artificial Neural Network 1810
Hanqing Cao, Bima Nugraha Sanusi, Ziwei Ouyang

Free-Shape Optimization of VHF Air-Core Inductors using a Constraint-Aware Genetic Algorithm 1816
Thomas Guillod, Charles R. Sullivan

Organic Direct Bonded Copper-Based Rapid Prototyping for Silicon Carbide Power Module Packaging 1824
  Shuofeng Zhao, Joshua Major, Douglas DeVoto, Sarwar Islam, Xiaoling Li, Mike Tant, Faisal Khan, Sreekant Narumanchi

Discrete Power Device Packaging with Integrated Direct Two-Phase Cooling 1832
  Jinpeng Cheng, Jinxiao Wei, Hao Feng, Li Ran

Investigation of Die Top-Side Re-Metallization for SiC-Based Double-Side Cooled Power Modules 1836
  Narayanan Rajagopal, Christina DiMarino

Design of Low Parasitic Inductance GaN HEMT Flip-Chip Power Module 1844
  Mohammad Dehan Rahman, Tanzila Akter, Abu Shahir Md Khalid Hasan, H. Alan Mantooth, Xiaoqing Song

A Scalable Dual-Orthogonal-Cooling Packaging Concept for Parallel-Series SiC Chips 1850
  Ekaterina Muravleva, Youssef Abotaleb, Blake Anderson, Zichen Zhang, Boyi Zhang, Jerry Hudgins, Jun Wang

Parasitic Impact Analysis and Design of Hybrid EMI Filter for Active Clamp Flyback SMPS 1858
  Tahmid Ibne Mannan, Seungdeog Choi, Masoud Karimi-Ghartemani

Overview of Dynamic Characterization of Switches for Three Phase Voltage Source, Current Source, and Matrix Converter Applications 1866
  Sneha Narasimhan, Sathya Rupan Thirumoorthi, Subhashish Bhattacharya

Advanced Modeling Technique of Class-E Inverter Considering Low R_{on} of eGaN FETs and Different Design Procedures 1874
  Manas Palmal, Jungwon Choi

PiezoNet and Data-Driven Models for Time-Domain Characterization of Piezoelectric Resonators 1882
  Davit Grigoryan, Mian Liao, Haoran Li, Shukai Wang, Tanuj Sen, Matthew Tan, Minjie Chen

A New Gate Charge De-Embedding Method for Accurate On-Wafer Characterization of HV MOSFET Devices 1889
  João R.R.O. Martins, Rachid Hamani, Vincent Quenette, Joerg Gessner

4 kW Auxiliary Power Module for Electric Vehicles Utilizing a Dual-Phase LLC DC-DC Converter 1892
  Mojtaba Forouzesh, Xiang Yu, Yan-Fei Liu, Paresh C. Sen

New Reverse Mode Control Method of Phase-Shift Full-Bridge Converter for Bidirectional Auxiliary Power Module 1899
  Jongyoon Chae, Dongmin Kim, Dongmin Choi, Gun-Woo Moon

In-Situ EV EIS with a High-Density Flying Capacitor Multi-Level Converter Supercapacitor System 1905
  Avram Kachura, Gaël Vergès, Samantha K. Murray, Olivier Trescases

A Novel 500-kHz LLC-T Resonant Converter with Wide Output Range 1913
  Zhengming Hou, Dong Jiao, Jih-Sheng Lai

High Efficiency Traction Drive Operation with a Partial Load Three-Phase Triangular Current Mode Modulation Concept 1919
  Bhaskar Chatterjee, Jan Allgeier, Thomas Plum, Marc Hiller

Analysis of Maximum Power Transfer Limit for Linear Operation of Dual-Active-Bridge Converters 1927
Radhika Sarda, Ezequiel Ramos Rodriguez, Gaowen Liang, Glen G. Farivar, Josep Pou, Vaisambhayana B. Sriram, Anshuman Tripathi

Enhanced Control for Integrated Active Power Decoupling in Single-Phase Three-Level Flying Capacitor PFC Converter 1935
Gleisson Balen, Cristian Blanco, Ángel Navarro-Rodríguez, Pablo García, Rafael Peña-Alzola

Improving Transient Stability of PLL-Synchronized Grid-Following Inverters 1940
Surya Prakash, Kalpana Beura, Mohamed Alkhatib, Omar Al Zaabi, Khalifa Al Hosani, Utkal Ranjan Muduli

Online Impedance-Based Analysis for Power System Stability Assessment using Transformer-Less and Filter-Less Switch-Mode Perturbation Generator 1946
Tomoya Ide, Yuko Hirase, Cheng Huang, Takanori Isobe

PIR-R Control for Three-Phase Grid-Connected Inverter with Unbalanced Grid Current Correction 1953
Haneen Ghanayem, Xingyu Yang, Mohammad Alathamneh, R.M. Nelms

Design and Placement of a Passive Clamp Snubber for Isolated SEPIC and Cuk Converters Working as Automatic Power Factor Correctors 1959
Abraham López, Juan Rodríguez, Duberney Murillo-Yarce, Javier Sebastián, Diego G. Lamar

Current Sensorless Control Strategy for Single-Phase T-Type PFC Converter 1967
Che-Yu Lu, Jia-En Zeng

Three-Phase Single-Stage Multiport AC-DC Converter with Integrated DC-DC Conversion Stages 1972
Asad Hameed, Gerry Moschopoulos

High Efficiency AC-Adapter Realized by Voltage-Clamper with Mid-Voltage AHB Converter using Synchronous Rectification 1977
Shuichiro Motoori, Akihiro Kawano, Toshiyuki Zaitsu, Riku Tatetsu, Kohei Sebata, Kazuki Miyanjou, Kimihiro Nishijima

Active Soft Switching Technique for Single Phase Series Capacitive Link Universal Rectifier 1983
Anran Wei, Brad Lehman, Mahshid Amirabadi

A Multi Mode Control Algorithm for Totem-Pole Bridgeless PFC 1990
Bosheng Sun, Sheng-Yang Yu, Amir Hussain

Protection Strategy for Flying Capacitor Totem-Pole PFC Under the AC Drop Transient 1995
Yanqing Wu, Wending Zhao, Zhenhai Zhu, Xinke Wu

Three-Phase with Three Single-Phase Single-Stage Isolated AC-DC Converters for EV Charging Station Applications 2002
Misha Kumar, Peter M. Barbosa, Juan M. Ruiz

400V SiC in Next-Generation 3-Level Flying Capacitor Bridgeless Totem-Pole PFC 2009
Rytis Beinarys, Seamus O'Driscoll

Extended Smart-Link Quasi-Single-Stage 3-Phase AC-DC Power Supply Module for AI-Driving Data Centers 2014
D. Biadene, J. Huber, J.W. Kolar, P. Mattavelli

A New Three-Phase Multi-Mode AC/DC LLC Converter with Output-Controlled Active Rectifier (with V2G and G2V Functions) for Fast DC Charging Application.. 2022
  Xiaoyi Xia, John Lam

Capacitorless Notch Resonant Converters for Miniaturized LLCLC Resonant Converters in Electric Vehicle Charging Applications 2029
  Haitham Kanakri, Euzeli Cipriano Dos Santos Jr., Maher Rizkalla

Multiple-Core Transformer Design based on Half-Turn Structure in Two-Stage DC-DC Converter for Battery Storage System.. 2035
  Yilei Li, Bima Nugraha Sanusi, Pinhe Wang, Tianming Luo

Bidirectional DC-DC Converter Utilizing Coupled Inductors for Energy Storage System.......................... 2043
  Wen-Hsuan Lee, Jiann-Fuh Chen, Hsuan Liao, Kuo Fu Liao

Comparison of 2-Level and Quasi-2-Level Topologies in a Bidirectional Isolated DC-DC Converter for MVDC Networks.. 2051
  José Andrés Aguilar Croston, Jean-Yves Gauthier, Cyril Buttay, Maryam Saeedifard, Besar Asllani, Piotr Dworakowski

Sling Forward Converter for Offline Operation: Achieving High Efficiency and Wide Voltage Range Performance 2059
  Nasherul Islam, Guozhu Chen, Honglei Miao, Fuxing Zhang

A Pulse Width Alternating Modulation Strategy for Three-Level Buck-Boost Converter 2066
  Xinlong He, Caifeng Liu, Xudong Zou, Jiaao Zou, Tianyi Zhang, Yong Kang

ZVT Circuit Applied for Wide Input Range Isolated Converters 2070
  Linguo Wang, Zhongyin Guo, Junjie Zhu, Bing Zhang, Zhiling Zuo, Xiaoguang Gao, Guangji Ma

Impact of Asymmetrical Leakage Inductance on a 380 V-12 V LLC Converter with Synchronous Rectifier for DCX Application 2075
  Jinshu Lin, Shan Yin, Chen Song, Honglang Zhang, Minhai Dong, Limei Xu, Hui Li

Start-Up Techniques and Universal Closed Loop Control of Immittance Network based Resonant Converter... 2082
  Ripun Phukan, Misha Kumar, Randy Beckemeyer, Juan Ruiz, Peter Barbosa

Multi-Objective Efficiency-Oriented Optimization for DAB Converters Minimizing Current Stress and Backflow Power with Soft-Switching Assurance 2088
  Kun Wang, Ian Laird, Jun Wang

An ISOP-PSFB PWM Converter based on Coupled Output Inductors and Phase-Shifted Modulation with Full ZVZCS Range 2096
  Kang Hong, Guo Xu, Guangfu Ning, Mei Su

Design and Implementation of a GaN-Based Soft-Switched Series-Capacitor Buck Converter Operating at the CCM-DCM Boundary for High-Performance Computing Systems 2101
  Ramin Rahimzadeh Khorasani, Kolman Puterman Ghitelman, Madhavan Swaminathan

Intrinsic Feedback Model for Coupled-Damped Self-Balancing of General Multiphase Hybrid Converters 2109
  Haoran Xu, Weijia Hao, Desheng Zhang, Run Min, Qiaoling Tong, Xuecheng Zou

A High-Efficiency Switching Oscillation Suppression Strategy based on Damped Oscillation for Synchronous DC-DC Converter 2117
    Hao Yuan, Chuan Ni, Zhengyu Ye, Wei Lu, Hui Xue, Ting Qian

Efficient and Streamlined Demodulation Strategy for High-Frequency Talkative DC-DC Converters 2125
    Abdelmoumin Allioua, Hendrik Gockel, Gerd Griepentrog

A 90.9% Peak Efficiency KY Single-Inductor Bipolar-Output Converter with Conductance Modulation Controller for Active-Matrix Organic Light Emitting Diode Power Supply............................ 2131
    Sheng-Han Yu, Chieh-Ju Tsai, Hao-Ran Huang, Ching-Jan Chen

Constant-on-Time Control for Zero-Bias Trans-Inductor Voltage Regulators................................. 2138
    Hank Zeng, Justin Lee, Rixin Lai, Hang Shao

An Improved PFM Control Scheme for Three-Level Buck Converter based on Ton Extension Achieving an 810% Frequency Reduction 2143
    Yi-Chun Chang, Chieh-Ju Tsai, Ting-Lun Lee, Ching-Jan Chen

A Concept for Current Ripple or Transient Improvements in Multiphase Converters 2149
    Alexandr Ikriannikov, Alex Gao

System Solutions and Design Trade-Offs to the Input Filter Interactions with Battery Chargers 2157
    Xigen Zhou, Dan Mavencamp, Kuang-Yao Cheng

Modeling and Implementation of a Zero Bias TLVR 2162
    Lei Wang, Travis Guthrie, Peyman Asadi, Mark Alexander, Kunrong Wang, Brandon Howell

cGANET-Enhanced Voltage Gain Modeling: Elevating CLLC Converter Accuracy.................................. 2167
    Yu Zuo, Xiaobing Shen, Fanghao Tian, Jiaze Kong, Hans Wouters, Wilmar Martinez

Capacitive vs Inductive Coupling based DC-DC Converter Operating in MHz Switching Frequency Range... 2173
    Saeid Pourjafar, Parham Mohseni, Oleksandr Husev, Ryszard Strzelecki, Oleksandr Matiushkin

LLC Converter Main Transformer Losses: Eliminating Air Gaps and Integrating Parallel External Inductors.. 2179
    Yu-Chen Liu, Shang-Syun Wu

Small-Signal Phasor Modeling of T-Type Bridge-Based Single-Sided and Double-Sided LCC Resonant Converters for WPT Applications... 2194
    Aditya Zade, Shubhangi Gurudiwan, Regan Zane

A Hybrid Three-Port Topology for Urban Charging Stations... 2202
    Mohammadreza Khodaparast Klidbari, Naser Souri, Zahra Sadat Habibolahi, Hamid Montazeri Hedeshi

Reconfigurable H5-Bridge based LLC-DAB Sigma Converter for EV Fast Charging Stations 2207
    Huangsheng Xu, Mingde Zhou, Qishan Pan, Haoyu Wang

A Resonant Reset Forward Converter with Ultra-High Conversion Gain using Differential Transformation Technique (DTT)... 2213
    Shubham Srivastava, Mandeep S. Rana, Santanu K. Mishra

Full-Range ZVS Modulation of Switched Capacitor Converter for Sensorless Voltage Balancing 2220
    Md Tanvir Ahammed, Wensong Yu

Dimensional Parasitics Absorption in Capacitively-Isolated Ćuk Converter for Medium-Voltage High Step-Down Converters 2228
Aakash Kamalapur, Jung-Soo Bae, Mark Cairnie, Rajaie Nassar, Jack Knoll, Dushan Boroyevich, Guo-Quan Lu, Christina DiMarino, Qiang Li, Khai D. T. Ngo

A 36-to-60V Input Dual-Phase 2MHz 93%-Efficiency ZVT Series-Parallel Hybrid Buck Converter using Single Auxiliary Inductor and Adaptive Time Multiplexing Control............ 2236
Qi Cheng, Hoi Lee

Improved Efficiency in a 10 W Class-Φ_2 Converter Utilizing a Resonant Gate Drive 2241
Malachi Hornbuckle, Katherine Liang, Juan Rivas-Davila

The Analysis and Design of a Resonant Capacitively-Isolated Cockcroft-Walton Converter............ 2249
Elizabeth Rabenold, Raiphy Jerez, Samantha Coday

SHSC: Non-Isolated High-Density 4:1 IBC for 48 V Applications 2254
Mario Ursino, Roberto Rizzolatti, Simone Mazzer

High-Performance Current Multiplier: A Hybrid Switched Capacitor Solution for High-Current Applications............ 2260
Kevin Zufferli, Roberto Rizzolatti, Mario Ursino, Simone Mazzer, Gerald Deboy, Stefano Saggini

Representation and Design Methodology for Generalized Switched-Capacitor Converter Topologies 2268
Seokwon Choi, Dam Yun, Jung-Ik Ha

A 48-V-to-1-V Gallium Nitride Switching Bus Converter for Processor Vertical Power Delivery with 2.7 mm Thickness and 3048 W/in³ Power Density 2276
Jiarui Zou, Yicheng Zhu, Nathan M. Ellis, Logan Horowitz, Robert C. N. Pilawa-Podgurski

Ripple Reduction and Efficiency Improvement of Always-Dual-Path Hybrid DC-DC Converter based on Phase Shift Operation............ 2284
Katsuhiro Hata, Shinsaku Tanaka, Toru Ashikaga, Yasuhiro Rikiishi

Ultralocal PQ Theory: A New Approach for Model-Free Predictive Direct Power Control of Shunt Active Power Filters 2290
Mahdi S. Mousavi, Abolfazl Nassaji, Ibrahim Harbi, Behnam Nikmaram, S. Alireza Davari, Mokhtar Aly, José Rodriguez

Symmetrical Balanced Circuit for Common-Mode Noise Mitigation in LCL-T Resonant Converter............ 2296
Ripun Phukan, Boyi Zhang, Juan Ruiz, Peter Barbosa

A Single-Phase Soft-Switching Buck-Boost Inverter............ 2303
Lukas Wipprecht, Burkhard Ulrich

Low-Complexity Model Predictive Control Method for Driving Dual Induction Motors Fed by Five-Leg Inverter............ 2311
Jun Young Lee, Eun Woo Lee, Dongho Choi, June-Seok Lee

Overvoltage Mitigation Filter using High-Frequency Cable Modeling in Long Transmission Lines for Silicon Carbide Inverter Systems............ 2317
Yun-Jin Lee, Kyo-Beum Lee

Power Delivery Network (PDN) Design and Analysis to Achieve Low Impedance in Fast Edge Rate DC-DC Converters for EMI Compliance............ 2322
Manraj Singh Ladhar, Sheldon Williamson

Enhancing the Performance of Dual Input Split Source Inverters using an Advanced Modulation Strategy............2327
  Mustafa Abu-Zaher, Fang Zhuo, Mokhtar Aly, Mahmoud A. Gaafar, Mohamed Orabi, José Rodriguez, Alaaeldien Hassan, Jiachen Tian, Samir Kouro

A Novel GaN-HEMT Single-Phase Single-Stage Buck-Boost Micro-Inverter Topology for PV Applications............2332
  Pengwei Li, Uiliam Kutrolli, Ali Bazzi

A Dynamic Current Sharing Method using Novel Clip Considering Mutual Inductance Coupling............2343
  Zexiang Zheng, Jianwei Lv, Yiyang Yan, Baihan Liu, Yifan Zhang, Linhao Ren, Jiaxin Liu, Cai Chen, Yong Kang, Xiong Zhang, Hao Yu, Wei Jiang

Application-Oriented Test Setup for Measuring Dynamic Output and Transfer Characteristics of GaN-HEMTs............2348
  Philipp Swoboda, Martin Fein, Simon Frank, Andreas Liske, Marc Hiller

Mitigating Gate Voltage Oscillation in Parallel SiC Power Modules for xEV............2356
  Hideo Komo, Michael Rogers, Mark Steiner, Eric Motto, Koichi Taguchi, Chihiro Kawahara, Junichi Nakashima, Yasushige Mukunoki, Seiichiro Inokuchi, Rei Yoneyama

Switching Performance Comparison of Low-Voltage GaN and Si Devices............2361
  Tianxiao Chen, Haoyang Liu, Pedro A.M. Bezerra, Eckart Hoene, Sibylle Dieckerhoff

Modeling of Switching Transients for Frequency-Domain CM EMI Analysis in Double Sided Cooling Power Modules............2369
  Sijia Liu, Liu Yang, Heng Zhang, Yifan Zhang, Zexiang Zheng, Jianwei Lv, Jiaxin Liu, Cai Chen, Yong Kang, Yuebin Zhou, Daming Wang, Shuang Zhao

Leakage Current Detection Scheme for Aging Test of 10kV SiC MOSFET Power Module............2375
  Peiyang Ding, Hong Zhang, Tianshu Yuan, Qiling Chen, Jiacheng Guo, Dingkun Ma, Peiyuan Sun, Ting Hou, Laili Wang

Physics-Informed Neural Network Approach for Early Degradation Trajectory Prediction of Power Semiconductor Modules............2380
  Jie Kong, Yi Zhang, Yichi Zhang, Lukas Wick, Frederik Lillebæk Hansen, Dao Zhou, Huai Wang

Nonlinear Output Capacitance of Bidirectional Gallium Nitride Power Switches............2387
  Michael Bosch, Jeremy Nuzzo, Dominik Koch, Mathias C.J. Weiser, Ingmar Kallfass

Novel Approach of Determining and Predicting SiC MOSFET's on Resistance from Device Case Temperature using Machine Learning............2393
  Paul Bradford, Conner Deppe, Hongjie Wang

Comparison of Static Characteristics in GaN HEMTs Across 50K to 400K Considering Diverse Techniques and Statistical Variation............2400
  Purushottam Khadka, Saumil Shivdikar, Zheyu Zhang, Tian Qiu, Ahmed Siraj

Compact Model of β-Ga$_2$O$_3$ Schottky Barrier Diode............2407
  Abu Shahir Md Khalid Hasan, Mohammad Dehan Rahman, Tanzila Akter, Md Majharul Islam, Md Maksudul Hossain, Xiaoqing Song, H. Alan Mantooth

DC-Link Capacitor Board Design for Low Parasitic Inductance............2413
  Mikayla Benson, Lifang Yi, Kangbeen Lee, Jinyeong Moon, Woongkul Lee

First Demonstration of a Gallium Oxide Power Converter 2419
Joshua J. Piel, Elizabeth A. Sowers, Daniel M. Dryden, Thaddeus J. Asel, Adam T. Neal,
Brenton A. Noesges, Shin Mou, Andrew J. Green

Optimized Integrated EMI Filter Design in SiC Power Modules with Terminal Inductor for Better
High-Frequency EMI Suppression 2426
Yifan Zhang, Wenzhe Xu, Jianwei Lv, Yiyang Yan, Baihan Liu, Sijia Liu, Jiaxin Liu, Cai Chen,
Yong Kang, Xiong Zhang, Hao Yu, Wei Jiang

Balanced Technique using Integrated Winding Coupled Inductor for High-Power Density Two-
Phase Interleaved Boost Converter 2431
Yuta Imaeda, Jun Imaoka, Masayoshi Yamamoto, Hiroyuki Onishi

MagNetX: Foundation Neural Network Models for Simulating Power Magnetics in Transient 2438
Shukai Wang, Hyukjae Kwon, Haoran Li, Youssef Elasser, Gyeong-Gu Kang, Daniel Zhou,
Davit Grigoryan, Minjie Chen

Revisiting Models of Common Mode Inductors to Include the Magnetized Capacitance Effect 2446
Rafael Bogo Portal Chagas, Marcelo Lobo Heldwein

A High Frequency Coupled Inductor with Distributed Air Gap for High Power DC-DC Converters 2453
Muhammad Fasih Uddin, Ahmed H. Ismail, Peyman Darvish, Baher Abu Sba, Yue Zhao

High-Power Planar Transformer Design for Four-Port Converters 2461
Arya Sadasivan, Behrooz Mirafzal

Optimal Design of Inductors with Aluminum Litz Wire for Inductive Power Transfer Systems 2468
Jesús Acero, Claudio Carretero, Ignacio Lope, Óscar Lahuerta, José-Miguel Burdío

Analytic Design of Flat-Wire Inductors for High-Current and Compact DC-DC Converters 2474
Sajjad Mohammadi, James L. Kirtley, Alireza Namadmalan

Insulation Dielectric Loss of High-Frequency Transformer Under Square Voltage Excitation with
Edge Oscillation 2482
Zhanlei Liu, Lingyu Zhu, Yuntian Gao, Yongliang Dang, Cao Zhan, Shengchang Ji

Improved High-Speed Thermal Analysis based on Two-Step Simulation for High-Frequency
Transformers 2488
Zheyuan Yi, Kai Sun, Qiang Li, Zengyang Liu

Core Material Characterization Under DC Bias Conditions 2495
Jonas Mühlethaler, Fabrice Locher, Frédéric Mathieu, Edward Herbert

A Low-Cost Setup and Procedure for Measuring Losses in Inductors 2502
Burkhard Ulrich

Effect of Temperature of Additively Manufactured Cores 2510
Ken Johnson, Ali Bazzi

Extreme Temperature Permeability Engineered Soft Magnetics 2516
Tyler W. Paplham, Alex M. Leary, Paul R. Ohodnicki Jr.

An Isolated RF Power Combining Approach with Multiple Decoupled Input Coils 2521
Ziyang Xu, Yifan Zhao, Zhan Liu, Alex J. Hanson, Ming Liu

Simulation of a Custom Core, 15kV Isolated Gap Transformer Optimized for High Power Density 2527
Andrew Galamb, Fei Teng, Srdjan Lukic

Low Interwinding Capacitance Design for PCB-Winding based Transformer in Self-Powered Gate Drive Power Supply for High-Voltage SiC MOSFET 2535
    Yuan Zhou, Li Zhang, Yilun Chen, Tianxiang Yin, Lei Lin

Integrated 4-Level Dual-Phase Superimposed Quadratic Power Converter for High-Density Direct 48V/1V Conversion 2541
    Prosenjit Ghosh, Jin Woong Kwak, Fei Zhou, D. Brian Ma

Compensation Method for Unbalance of the Multi-Channel Class E Power Amplifier using the Closed Loop Frequency Control 2547
    Kyungmin Lee, Sungku Yeo

High Temperature Operation of Digital Gate Driver Integrated Into a Power Module 2551
    Kazuma Saiga, Shohei Zaizen, Satoshi Nakano, Shigeru Kusunoki, Kiyoto Watabe, Katsuhiro Hata, Makoto Takamiya, Shin-Ichi Nishizawa, Wataru Saito

Evaluation Index-Based Multiphysics Coupling Model and Analysis Methodology for High-Reliable Power Supply Module 2556
    Haoyu Wang, Xuliang Wang, Yan Wang, Xiaosen Liu

Electrical Characterization of Modular 3D Packaging Assembled with Compressed Metal Foams 2562
    Paul Bruyere, Alexis Derbey, Betina Zynger-Capaverde, Yvan Avenas, Eric Vagnon, Jean-Luc Schanen, Jean-Michel Guichon, Omar Sanjakdar

Improvement in Short-Circuit Robustness of SiC-MOSFETs based Power Modules using Two-Level Turn-On (2LTO) 2569
    Muhammad Muneeb Alam, Saad Khalid, Nisar Ahmed Khan, Ngoc Ho Tran, Sebastian Strache

GaN-Based Two Stage Point-of-Load (PoL) Converter with 2.5D Embedded Substrate Implementation 2576
    Samuel Defaz, Yang Li, Fang Luo

Near-Field Coupling Mitigation of the Noise from High Voltage DC-Link Decoupling Capacitors in Voltage Source Converters 2582
    Yuxuan Wu, Kushan Choksi, Samuel Defaz, Fang Luo

Advantages of Paralleling SiC MOSFETs in High-Performance Power Modules 2589
    Steffen Beushausen, Dominik Alexander Ruoff, Wenqi Zhou, Karl Oberdieck

A SiC Half-Bridge Power Module based on Liquid Metal Packaging for High Performance and Low Thermal Stress 2597
    Wei Mu, Ameer Janabi, Luke Shillaber, Borong Hu, Teng Long

Analysis and Modeling of Radiated EMI Considering Coupling Between Power Converter and Power Cable with LC-Type EMI Filter 2603
    Qinghui Huang, Yingjie Zhang, Shuo Wang, Yirui Yang, Zhedong Ma, Yanwen Lai

Simple Prediction Method for Impacts of Switching Characteristics on EMI Noise of a Three-Phase PWM Inverter 2610
    Shinobu Nagasawa, Toshiya Tadakuma, Keita Takahashi

Coaxially Nested 3.3 kV SiC MOSFET Packages with Uniform Interpackage Electric Field Distribution 2616
    Jack Knoll, Mark Cairnie, Christina DiMarino

Thermal Modeling and Performance of a Bare-Die Embedded PCB for High Power Density Converters Design 2624

Shahid Aziz Khan, Feng Zhou, Mengqi Wang, DucDung Le, Shivam Chaturvedi

Research on the Voltage Fluctuation Suppression Strategy in Weak Grid Under Pulsed Power Load Integration 2628

Xi Chen, Jiazheng Zhang, Mingjun Bao

An Optimized Firmware-Based Cycle-by-Cycle Current Limiting Method for Power Electronic Converters in UPS 2634

Teng Wu, Hong Liu

Frequency Stop-Band Management System for DC-DC Converters 2640

Alessandro Bertolini, Alberto Cattani, Claudio Luise, Alessandro Gasparini

Multi-Stage Model Predictive Control with Enhanced Discrete-Time Models for Multilevel Inverters 2647

Hoang Le, Apparao Dekka, Deepak Ronanki, Abdul R. Beig

Direct Effective Power Control (D-EPC) for LLC Resonant Converters Operating in Boost Mode using Event-Driven-Timer based Digital Controller 2654

Yuto Yoshimura, Kenji Funatani, Kazuhiro Umetani, Toshiyuki Zaitsu, Akito Nakagaki, Masataka Ishihara, Eiji Hiraki

Mitigation Method of Resonance Between Paralleled On-Line UPS 2660

Teng Wu, Zhenguo Huo, Shangxian Ning

An Extra-Element Small-Signal Model for a Current-Fed Resonant Dual-Active-Bridge Converter 2667

Paolo Sbabo, Paolo Mattavelli, Giorgio Spiazzi, Andrea Petucco

Concurrent Charge Distribution and Time-Optimal Control for Unordered Single-Inductor Dual-Output Converter 2675

Xuliang Wang, Haoyu Wang, Yang Liu, Yunxin Wang, Boran Zhang, Hongru Liu, Yan Wang, Xiaosen Liu

Circulating Current Control with Loss Reduction for Parallel Connected Inverters 2681

Shun Endo, Takae Shimada, Masato Ando, Yuuichi Mabuchi, Masaki Miyamae, Naoki Takayama, Yohei Matsumoto, Naoto Onuma

Analysis of Power and Power Spectral Density for Quaternary Random Pulse Position Modulation 2687

Hung-Chi Chen, Hsiang-Kai Wu, Chih-Chiang Wu

Bidirectional CLLC Converter using a Hybrid Control Method for Wide Voltage Range Applications 2692

Jhih-Cheng Hu, Hong-Xuan Liao, Chien-Lung Liu, Wei Wang, Ming-Shi Huang

Design and Control of a High-Bandwidth Dual Active Bridge DC-DC Converter 2698

Alper Uzum, Syed Imam Hasan, Yilmaz Sozer, Kenneth A. Loparo

Unified Model Predictive Control for DC-DC Buck Converters: From Start-Up to Steady-State Operation 2703

Zhengchen Guo, R.M. Nelms

A Novel IPPC Method for Precise Overload Protection and Burst Mode Operation in LLC Resonant Converters 2708

Manikanta Pallantla, Ramkumar S

An Improved Current-Sensorless Model Predictive Voltage Control for Four-Leg Voltage Source Inverters.. 2713
  Heng Guo, Yuxin Wei, Mengmeng Jing, Wenlong Ding, Bin Duan, Chenghui Zhang

A Highly Integrable, Modular and Multi-Functional Fault Monitoring Active Gate Driver with Parallel Buffers for a Global Enhanced Reliability of Gen. 3 SiC Power MOSFETs 2718
  Mathis Picot-Digoix, Léo Seugnet, Frédéric Richardeau, Jean-Marc Blaquière, Sébastien Vinnac, Thanh-Long Le, Stéphane Azzopardi

A 24 – 16 V to 0.8 – 1.2 V Merged 4-Stage Hybrid-SC-SL Converter with 96.5% Peak Efficiency and Larger Than 50% iL Reduction.. 2725
  Chien-Hao Tseng, Cheng-Ta Chuang, Chieh-Ju Tsai, Ching-Jan Chen

Innovation Active Gate Drive Method (Named TriC3™) for MOSFET Heat Reduction and EMI 2730
  Hisashi Sugie

A KY Buck-Boost Converter with Extended Ramp Control Achieving 1500% Output Variation Reduction for Smooth Mode Transition 2735
  Yu-Ting Hung, Chieh-Ju Tsai, Ching-Jan Chen, Chun-Yu Hsieh

An USB Cable based Extended Conversion Range L-First Hybrid-Converter using Valley-Virtual-Inductor-Current-Mode Control with Auto-Tracking Slope Compensation Against ±50% Inductance Variation 2741
  Chun-I Li, Chieh-Ju Tsai, Ching-Jan Chen

Impact of Gate Resistor Configurations on Current Balancing in Paralleled SiC MOSFETs 2746
  Yifu Zhang, Shashank Karanth, Emanuel Eni

Exploring the Potential of FPGA in High-Frequency Switching DC-DC Boost Converters using Model Predictive Control 2752
  Qingcheng Sui, Bangli Du, Yu Zuo, Wilmar Martinez

A 7 Bit 5A 6.7 GHz Gate-Shaping Digital Gate Driver with Burst-Sampling ADC for Iterative Switching Optimization of SiC Power MOSFETs 2757
  Tobias Zekorn, Kenny Vohl, Erik Wehr, Leon Weihs, Michael Hanhart, Ralf Wunderlich, Stefan Heinen

Decentralized Interleaving of Series-Stacked DC-DC Converters via Extremum-Seeking Control 2764
  Ivan Petrić, Vignesh Iyer, Shoudong Hu, Chirayu Rajpurohit, Bailey Sauter, Milan Ilić, Luca Corradini, Dragan Maksimović

Online Dead-Time Control for Half Bridges without Preliminary Training based on Switching Transient Steepness 2772
  Lukas Knappstein, Niklas Falkenberg, Martin Pfost

Impedance-Based State-of-Health Estimation for Lithium-Ion Battery Management Systems 2779
  Mohammad K. Al-Smadi, Jaber A. Abu Qahouq

Stability Analysis and Resonance Damping of LC Filter-Based Voltage Source Converter with Single-Loop Voltage Control... 2785
  Aravind G., Divyanshu Bansal, L. Umanand

Finite Control Set Model Predictive Control Combined with Online Junction Temperature Estimation for Reliability Enhancement of Voltage Source Inverters 2790
  Qiang Mu, Jiale Zhou, Zaheen Mustakin, Lucas Pereira, Babak Parkhideh, Tiefu Zhao

Framework for Dynamic Control and Operation of Power Electronics Interfaces...2797
Radha Sree Krishna Moorthy, Steven Campbell

Achieving Soft-Charging and Over 20% Input Current Ripple Reduction in a 48-to-6 V Dickson
Converter using 3-Phase Split-Phase Control...2805
Nagesh Patle, Rose A. Abramson, Sahana Krishnan, Jiarui Zou, Robert C. N. Pilawa-Podgurski

Experimental Verification of Circuit-Losses Analysis-Model of DC-Output Converter Developed
using Approximated Equations from Measurement Data and Datasheet Data...2813
Ryota Kondo, Tsuyoshi Funaki

Scattering Parameter Measurement System using Probes for Surface Mount Devices Operating in
the Frequency Range from 50 kHz to 1 GHz..2821
*Ryoko Kishikawa, Masahiro Horibe, Tomokazu Shoji, Shigenori Yabuta, Toshi Ohi, Ryo
Takeda, Takamasa Arai*

Optical Transformer Design with Additional Common-Mode Noise Reduction Winding for Flyback
DC-DC Converters 2828
*Yusuke Irie, Shinichiro Eguchi, Yoichi Ishizuka, Toshiro Takeuchi, Akio Iwabuchi, Takahiro
Koga, Toshiyuki Tanaka*

Enhanced Bus Voltage Stability Through Digital Twin-Enabled Adaptive Controller Tuning.......................2833
Matthew Belanger, Andy Wong, Kerry Sado, Enrico Santi

Modeling and Performance Characterization of Lithium-Ion Capacitor at Different Temperature
and Voltage Values..2840
Mohammad K. Al-Smadi, Jaber A. Abu Qahouq, Sajad Saberi

Conveniently Identify Coils in Inductive Power Transfer System using Machine Learning.........................2846
*Yifan Zhao, Mowei Lu, Ting Chen, Heyuan Li, Xiang Gao, Zhenbin Zhang, Minfan Fu, Stefan
M. Goetz*

Accurate Modeling of LLC Resonant Converters with Enhanced Analytical Approach Considering
of Parasitic Capacitance 2851
Dong Jiao, Zhengming Hou, Jih-Sheng Lai

High-Frequency Conditioning Circuits for Power-Related Information Extraction in Non-
Sinusoidal Power Electronic Systems 2857
Haoyu Wang, Yuanxin Zhang, Di Mou, Alex Hanson, Shiqi Ji

Transconductance Model of the Dual Active Bridge Converter Under Single and Dual Phase Shift
Control..2865
Jared Cronin, Andrew Wunderlich, Enrico Santi

Lumped Parameter Modeling for Real-Time Thermal Regulation of Li-Ion Battery Packs.........................2871
Utkal Ranjan Muduli, Mohamed Shawky El Moursi, Khalifa Al Hosani, Ahmed Al-Durra

A Physics-Based Temperature Dependent Analytical Model for 2DEG Density in AlGaN/GaN
HEMT Devices 2877
Kashfia Tajmim Nabila, Jerry L. Hudgins

Comparative Analysis of Stator-PM Machines: Design Optimization and Electromagnetic
Performance Evaluation 2883
Maryam Salehi, Madhav Manjrekar

Elimination of Deadtime Effect on Resolver Offset Estimation using the Pulsating Current Command for Electric Vehicle Application 2889
Yingfeng Ji, Nurani Chandrasekhar

A Generic Load Emulator for Testing Motor Drives of E-Mobility 2894
Qingzheng Zhang, Kaiyuan Feng, Changsheng Hu, Dehong Xu

Design and Implementation of Power Assisted Control System for E-Bikes 2900
Che-Yu Lu, Tzu-Ping Cheng

A Hybrid PWM Strategy with Reduced Common-Mode Voltage and Extended Output Voltage Linearity for Adjustable Speed Drives 2907
Zhe Zhang, Kevin Lee

Single-Phase Open-Circuit Fault-Tolerant Control of Three-Phase PMSM Drives 2913
Yuichiro Minato, Yuki Nakata, Jun-Ichi Itoh

Multi-Vendor Encoder Position Sensing Interface using Programmable IP based Solution.......................... 2920
Rajul Bhambay, Dhaval Khandla, Pratheesh Gangadhar, Thomas Leyrer, Achala Ram, Manoj Koppolu, Archit Dev

Sensorless Control Method at Low-Speed Range using High-Frequency Voltage Injection for Synchronous Reluctance Motors Considering to Nonlinear Characteristic Due to Magnetic Saturation 2924
Sota Takizawa, Sari Maekawa

Hybrid Control Scheme for Permanent Magnet Gear Motor.. 2932
Bing Li, Takayoshi Matsuo, Ahmed Sayed-Ahmed, Yujia Cui, Jiangang Hu

Cost-Effective Fault Diagnosis for Motor and Inverter using Bootstrap Charging and Single DC Link Current Sensor 2937
Gyu Cheol Lim, Won Hyo Jeong, Kahyun Lee, Jung-Ik Ha

Improved PWM to Suppress Motor Overvoltage Caused by Voltage Reflection...................................... 2943
Sung-Oh Kim, Kyo-Beum Lee

Analysis of Double Pulsing Effect in Motor Drives based on Vector Diagram.. 2948
Byeong-Woo Kang, Kyo-Beum Lee

A Novel Speed Sensor-Less Control of a Solar-Powered PMSM Drive 2953
Abirami Kalathy, Arpan Laha, Praveen Jain, Majid Pahlevani

Design of a Compact Low-Loss MMC Double Submodule for MVDC and HVDC Applications 2960
Ali Sharaf Addin, Rainer Marquardt, Thomas Brückner

A Series-Type Dynamic Voltage Restorer Control Strategy to Cope with Voltage Swell........................... 2968
Jiazheng Zhang, Hongyu Chen, Xi Chen, Mingjun Bao

Machine Learning Approach for Accurate Lithium-Ion Battery Temperature Prediction using Electrochemical Features Independent of Battery SOC and SOH.. 2973
Vincent Masabiar Tingbari, Oluwaseun Isaiah Ekuewa, Anshul Nagar, Asad Abbas, Jamil Umar, Yuxin Zhang, Woonki Na, Jonghoon Kim

A Battery Strings Circulating Current Blocking Method for Battery Energy Storage Systems 2981
Haihong Long, Ziang Sun, Yucheng Fan, Xin Wu, Dehong Xu

A Hybrid Multilevel Converter-Based High-Gain Isolated DC/DC Converter for Grid-Tied Energy Storage Applications.. 2986

Pengyu Fu, Yizhou Cong, Jin Wang, Anant Agarwal

LCL Filter Parameter Selection using Graphical Method for a 13.8 kV ac 1.1 MVA 7-Level Flying Capacitor Grid-Connected Converter Utilizing Variable Switching Frequency 2992

Arthur Mendes, David Nam, Mingze Gao, Thimothy Thacker, Dong Dong, Rolando Burgos

Online Extraction of Electrochemical Impedance Spectroscopy Pattern based on EV Load Profile and Short Time Fourier Transform for Diagnosis of Lithium-Ion Battery Safety 3000

Miyoung Lee, Dongcheol Lee, Youngmin Bae, Jongchan An, Garam Yang, Woonki Na, Jonghoon Kim

Enhanced Incremental Capacity Analysis for Evaluating Battery Degradation Mechanisms of Optimized Fast Charging Methods 3006

Taehyeon Gong, Jaehyeong Lee, Sungjun Lee, Yura Kim, Bomyeong Ko, Woonki Na, Sungjin Choi, Jonghoon Kim

Co-Estimation of SOC and SOT in Lithium-Ion Batteries using an RLS-Based Heat Generation Model 3012

Seongkyu Lee, Eunjin Kang, Minhyeok Kim, Seunghyun Lee, Minwoo Song, Jaea Lee, Woonki Na, Jonghoon Kim

Three-Stage Adaptive Control Strategy for Stability Improvement of Grid-Connected Inverter in Weak Grid.. 3018

Longxiang You, Sicong Jin, Xin Zhang, Zuoshuai Wang, Sunqing Wang

Degradation Analysis of Offshore Bifacial PV Modules Under Multiple Climatic Stressors 3024

Aidha Muhammad Ajmal, Yongheng Yang

A Flexible Energy Management System for Solar Powered Electric-Bus Charging Stations 3030

Supun Amarathunga, Pasan Gunawardena, Xiaoting Wang, Yunwei Li

A Vienna Rectifier based Grid-Connected Powertrain for Hydrokinetic Turbine Systems 3036

Peidong Li, Md Tariquzzaman, Yue Cao

Condition Monitoring for DC-Link Capacitors and PV Arrays based on the Start-Up Process of the PV System 3042

Yongjie Liu, Ariya Sangwongwanich, Chen Liu, Xing Wei, Shuyu Ou, Tamás Kerekes, Jiahong Liu, Huai Wang

Electrically and Thermally Efficient Reliable Power Converter Design for Micro–Hydrokinetic Turbine 3048

Md Tariquzzaman, Peidong Li, Yue Cao

Comprehensive Evaluation of Cyber Attacks on Grid-Connected Smart Inverters 3054

Rishabh Singla, Vishwam Raval, Hasan Ibrahim, Jaewon Kim, Prasad Enjeti, Narsimha Reddy

Parallel Operation of Grid-Forming Converters based on Kuramoto Oscillators with Virtual Cable Emulation for Improved Power Sharing.. 3059

Vikram Roy Chowdhury, Gab-Su Seo, Barry Mather

Enhancing Hydrogen Production in Hybrid Standalone Microgrids 3064

Utkal Ranjan Muduli, Mohamed Shawky El Moursi, Khalifa Al Hosani, Ahmed Al-Durra

LSTM-Based Sub-Synchronous Oscillation Detection Scheme for Type 4 Wind Farm Interfaced with Weak AC Grid 3071
Omar Abu-Rub, Muhammad F. Umar, Jana A. Sheikh Ali, Yazan Qiblawey, Abdulrahman Alassi, Maryam Saeedifard, Mohammad B. Shadmand

A Study of Module Design Method to Suppress the Oscillation Occurs Between Parallel-Connected Power Devices 3077
Shinji Yato, Hiroto Sakai, Hideo Araki, Shumei Shimosako

A High-Efficient Hybrid Traction Inverter in Electric Vehicle Applications 3083
Yousefreza Jafarian, Omid Salari, Praveen Jain, Alireza Bakhshai, Mohamed Z. Youssef

Dual-Use of Onboard Chargers to Achieve Controllable DC Bus Voltage for Electric Vehicles 3089
Anuj Maheshwari, Elie Libbos, Arijit Banerjee

Isolated Single-Phase Onboard Chargers for BEV/PHEV using Active Power Decoupling Technology 3096
Yoshiki Amano, Keigo Nishimura, Hiroaki Matsumori, Takashi Kosaka, Kenichi Nagayoshi, Kenichi Watanabe

A Practical Use of xEVCap: The Modular and Standard DC-Link Capacitor Solution for the Main EV Powertrain Inverter............ 3100
David Olalla, Tomas Wagner, Fernando Rodriguez, Alberto Espinar

Optimized Bidirectional On-Board Charger using a Novel Unfolder-DAB Topology 3109
Héctor Sarnago, Ignacio Álvarez, Pablo Briz, Óscar Lucía

Critical Thermal Characterization of Next-Generation Solid-State Batteries for Automotive Battery Management Systems 3114
Chandan Chetri, Sheldon Williamson

Nanocrystalline CMC Inductors for EV Charging: Trade Studies and Testing Standardization 3119
Christopher Bracken, Mark A. Juds, Paul R. Ohodnicki, Bharadwaj Reddy Andapally, Jose Gato

Predicting Efficiency of On-Board and Off-Board EV Charging Systems using Machine Learning 3124
Mohamed Yasko, Fanghao Tian, Wilmar Martinez, Johan Driesen

High-Power and High-Speed Multi-Channel VCSEL Arrays with GaN Driver for Automotive LiDAR 3129
Yifu Liu, Sichao Li, Junlei He, Changyu Hu, Bill He, Karthik Krishnamurthy, Andy Shen

Double Pulse Test Platform for Hybrid SiC-IGBT Switch Characterization and Optimal Gate Control Strategy for EV Traction Inverters 3133
Rosario Attanasio, Harsha Ademane, Ryan Satterlee, Gianni Vitale

Critical Role of Individual Cell Temperature Monitoring in Mitigating Thermal Runaway and Reducing Accelerated Degradation in Lithium-Ion Batteries............ 3141
Mohit Sharma, Akash Samanta, William Locke, Sheldon Williamson

Loss-Optimized Design of a Triple Active Bridge DC-DC Converter for an Electric Vehicle Application 3147
Sreejith Chakkalakkal, Kyle Kozielski, Wesam Taha, Yicheng Wang, Aniket Anand, Ali Emadi

A Magnetic-Less DC/DC Converter with Pulse Charging for 800 V Powertrains from 400 V DC Fast Chargers 3155
Duc Dung Le, Shivam Chaturvedi, Shahid Aziz Khan, Mengqi Wang, Mohamed Elshaer

Boosting Charger Efficiency: A GaN-Based Flyback Converter with Energy Recycling 3160
Ahmad Nabizadah, Majid Ghasemi Korrani, Babak Fahimi

A Hybrid Three-Level Buck Converter with Flying Supercapacitor for High Load Current Surge
Capability using Peak Current Mode Control 3167
*Finlay Lodge, Rafael Peña-Alzola, Martin MacFadyen, Patrick Norman, Mark Sweet,
Graeme Burt*

Supercritical Carbon Dioxide (sCO.)-Cooled Current Source Inverter-based Integrated Motor Drive
for MW-Scale Electric Aviation Applications 3174
*Hang Dai, John Yagielski, Thomas Jahns, Kum-Kang Huh, Vandana Rallabandi, Libing
Wang, Tarak Saha, Wenda Feng, Bulent Sarlioglu*

The Challenge of Thermal Runaway in Soft Magnetic Materials for Inductive Power Transfer 3181
Yibo Wang, Ben Zhang, Weisheng Guo, Tianlu Ma, Sheng Ren, C.Q. Jiang

A Capacitively Coupled Alternative Electric Field Control for Freeze-Free based High Quality
Food Preservation 3187
Jaeyong Cho, Junhyeong Park, Sung-Bum Park, Daehyun Kim, Jinsoo Choi

The Characteristics of the Long Length Primary Loop and the Power Supply for the SCMaglev's
DWPT System 3194
Keisuke Yamamoto, Jun Enomoto, Shunsaku Koga, Junichi Kitano

A Wireless EV Charging System with a Double-Sided LCC Network using Variable Switching
Frequency and DC-Link Voltage Control... 3200
Chae-Lyn Kim, Hyeonu Jo, Ju-A Lee, Dong Hyeon Sim, Byoung Kuk Lee

Class E/EF Inductive Power Transfer to Achieve Stable Output Under Variable Low Coupling................. 3206
Yifan Zhao, Mowei Lu, Heyuan Li, Zhenbin Zhang, Minfan Fu, Stefan M. Goetz

A Motorized Air-Core Variable Inductance Winding Structure 3212
Xindong Li, Sampath Jayalath, Cheng Zhang

Wireless Power Transfer System with Automatic Tuning Capability in Metallic Environment.................... 3220
*Renjie Zhang, Yue Wu, Delin Zhao, Yaohua Li, Yongbin Jiang, Yi Tang, Huan Yuan, Xiaohua
Wang, Mingzhe Rong*

Design of Wireless Power Transmitters for Enhanced Transmission Distance and Output Power 3227
Kaiyuan Wang, Shuang Zhao, Shuye Shang, Eric Ka-Wai Cheng, Siew-Chong Tan, Yun Yang

Optimization of Wireless Power Transfer Waveforms and In-Vivo Receivers for Implantable
Medical Devices 3232
Hanbing Liu, Xin Zan

Comparison of Compact Power Amplifier Designs for High Frequency Resonant Wireless Power
Transfer Systems at 6.78 MHz using High-Q Resonators.. 3241
Manuel Rueß, Kilian Müller, Mathias C.J. Weiser, Ingmar Kallfass

Analysis and Design of Capacitive Coupling Wireless Power Transfer System using Load-
Independent Class-EF Inverter 3248
*Takumi Kobayashi, Yutaro Komiyama, Akihiro Konishi, Hiroaki Ota, Yuki Ito, Taichi Mishima,
Takeshi Uematsu, Kien Nguyen, Hiroo Sekiya*

Design and Optimization of a 600 W Wireless Drone Charger for High Gravimetric Power Density 3253
Arka Basu, Daniel Costinett

Stabilization Method for DC-Bus Oscillation in Dynamic Wireless Power Transfer Systems...... 3261
Yuki Ochiai, Keisuke Kusaka

Unveiling Aliasing Effect on Resonant Pole Locations in Wireless Battery Chargers 3267
Anwesha Mukhopadhyay, Daniel Costinett

Integrated Hybrid Inductive and Capacitive Power Transfer System with Asymmetrical PCB Self-Resonator............ 3275
Yao Wang, Zhen Sun, Xiangrong Zhang, Yun Yang, Shu Yuen Ron Hui

High Frequency Noise Reduction Method of the Class E Power Amplifier................................. 3281
Kyungmin Lee, Sungku Yeo

Single-Stage Three-Phase Buck-Matrix Rectifier with Series-Parallel Connected Transformers for High-Power 48 V Data Center Power Supplies............ 3285
Yuki Ishikura, Chinmay Bhagat

Sector Transition PWM Modulation Scheme for a Three-Phase Isolated Buck-Matrix Rectifier 3291
Chinmay Bhagat, Yuki Ishikura

Adaptive Capacitance Circuit for Optimal Dynamic Impedance Matching in Variable Reluctance Energy Harvesting Applications 3298
Alejandro Redondo, Fernando Pérez, Sofía García, Gabriel Mujica, Airán Francés

Gallium Nitride (GaN) based Topology Comparison for Low Power Battery Charging Applications 3304
Jai Aditya Chaudhary, Rosario Attanasio, Gianni Vitale

Server Motherboard Power Performance Study Under Immersion Cooling Environment............ 3312
Meng Wang, Haiyan Wang, Pavan Kumar, Haijin Zhang, Xiang Li, Fengwei Bian, Jianting Deng, Jiaqi Zhu, Yiming Lei

Practical PCB Design Considerations for GaN HEMTs based Isolated DC-DC Converter............ 3316
Gaureej Gauttam, Harish S. Krishnamoorthy, Sai Sushma Pasupuleti

Data-Driven Characterization and Forecasting of Metal-Oxide Varistor Degradation in DC Circuit Breakers............ 3321
Zhi Jin Zhang, Yang Liu, Lukas Graber, Maryam Saeedifard

A Thyristor-Based Fault Current Bypass Solid-State Circuit Breaker for DC Microgrid Applications 3328
Jiale Zhou, Xiuhu Sun, Qiang Mu, Tiefu Zhao

Single-Stage Three-Phase AC-AC Isolated Inertialess Converter (IIC) for Industrial Drives............ 3334
Brad Houska, Decheng Yan, Aniruddh Marellapudi, Satish Belkhode, Joseph Benzaquen Sune, Deepak Divan

Author Index

APEC 2026, March 22 - 26, Henry B. Gonzalez Convention Center, San Antonio, TX

Announcement and Call for Technical Session Papers

APEC 2026 continues the long-standing tradition of addressing issues of immediate and long-term interest to the practicing power electronic engineer. Outstanding technical content is provided at one of the lowest registration costs of any IEEE conference.

APEC 2026 will provide:

a) The best power electronics exposition;
b) Professional development courses taught by world-class experts;
c) Lecture and Dialogue Presentations of peer-reviewed technical papers covering a wide range of topics, and
d) A venue to network and enjoy the company of fellow power electronics professionals in a beautiful setting.

Important Milestones *(subject to change and posted at www.apec-conf.org)*:

May 29, 2025	**Digest submission site opens**
August 15, 2025	**Deadline for submission of digests**
October 14, 2025	**Author Decision Notification**
December 1, 2025	**Final papers and author registrations are due**

Submission Requirements: Prospective authors are asked to submit a digest explaining the problem that will be addressed by the paper, the major results, and how this is different from the closest existing literature. Please refer to the APEC 2026 Topics of Interest when planning and submitting your digest. Papers presented at APEC must be original material and not have been previously presented or published. The principal criteria in selecting digests will be the usefulness of the work to the practicing power electronic professional. Reviewers value evidence of completed experimental work. Authors should obtain any necessary company and governmental clearance prior to submission of digests.

Please visit *http://apec-conf.org/conference/sessions/technical/* for all details on digest and final manuscript format and how to submit your paper. APEC2026 provides extended submission windows to the authors for both digest and final paper submissions compared to prior years in order to facilitate higher quality submissions. Take advantage of this upgrade by publishing and presenting your research at APEC2026.

If a digest is accepted, authors must submit a final manuscript before the deadline, or the manuscript cannot be published in the Proceedings or presented at the conference. The paper will be scheduled for presentation in either a lecture or a dialogue format, based on the submitter choice and other qualification criteria. Submission of your digest serves as your implicit agreement and consent to electronic distribution of your presentation. Final manuscripts may be subject to charges if their papers are over the page or file-size limit. **At least one of the authors listed on a paper must be registered for either Full Registration or for Technical and Industry Sessions registration.**

Become an APEC Paper Reviewer: APEC relies upon a peer review process to ensure the quality of the technical content. To help maintain the high quality of the program, please contribute a few hours to review digests in your area of expertise by registering at www.apec-conf.org (available summer 2025 under "Paper Reviewer Sign-up").

Calls for Industry Session presentations, Professional Education Seminars, and Exhibitor Seminars will be posted at www.apec-conf.org.

Website: www.apec-conf.org	**APEC**	**APEC Sponsors**
Email: apec@apec-conf.org	3644 Wright Ter NE	Power Sources Manufacturers Association
Phone: +1-202-624-1762	Washington, DC 20018	IEEE Industry Applications Society
Facsimile: +1-202-624-1766		IEEE Power Electronics Society

APEC 2026 Announcement and Call for Technical Session Papers

APEC 2026 Topics of Interest

1. **AC-DC Converters**
   a. Single-Phase and Three-Phase Input
   b. Power Factor Correction: CCM, DCM, CRM/BCM Control, Bridgeless
   c. Embedded AC-DC Power Supplies
   d. External AC-DC Adapters
   e. Bidirectional AC-DC Converters

2. **DC-DC Converters**
   a. Hard- and Soft-Switched
   b. Resonant Converters
   c. Point-of-Load (POL) and Multi-Phase Converters
   d. Voltage Regulator Modules (VRM)
   e. Bidirectional DC-DC Converters

3. **DC-AC Inverters**
   a. Single and Multi-Phase Inverters
   b. Multilevel Inverters
   c. PWM Strategies
   d. Power Quality and EMI

4. **Devices and Components**
   a. Power Silicon MOSFETs, BJTs, IGBTs, etc.
   b. GaN and SiC Devices and Modules
   c. Ultra-Wide Bandgap Devices
   d. Capacitors, Supercapacitors
   e. Interconnects, Busbars and Fuses

5. **Magnetics**
   a. Advanced Magnetic Materials and Geometries
   b. Winding Techniques
   c. High-Frequency Magnetics
   d. Additive Manufacturing for Magnetic Materials
   e. Magnetics Modeling and Simulations

6. **Power Electronics Integration and Manufacturing**
   a. Power Electronics Packaging
   b. Power Modules / High Density Design
   c. Thermal Management
   d. Quality and System Reliability Including EMI/EMC
   e. Embedded Technologies, 3D Packaging, and Additive Manufacturing
   f. Production Processes and Design for Manufacturability

7. **Control**
   a. Control of Power Electronic Converters
   b. Current-Mode and Voltage-Mode Control
   c. Digital Control-MCUs, DSPs, FPGAs, ASICs
   d. Sensor and Sensor-less Control
   e. Gate Drive Circuits and Fault Protection
   f. Control ICs

8. **Modeling and Simulation**
   a. Circuits and Systems
   b. Device and Component Modeling
   c. Parasitics Extraction and Optimization
   d. Software Tools
   e. Hardware-in-the-Loop and Rapid Prototyping

9. **Motor Drives**
   a. AC, DC, BLDC Motor Drives
   b. Actuators
   c. Integrated Motor Drives
   d. Modeling and Control Techniques for Motor Drives
   e. Power Quality and EMI for Motor Drives
   f. Sensor Integration

10. **Power Electronics for Utility Applications**
    a. FACTs Devices and HVDC
    b. Solid-State Transformers
    c. Energy Storage Systems
    d. Distributed Energy Systems
    e. Microgrid Systems
    f. Power Quality, UPS, Active Power Filters
    g. Smart Grid and Metering

11. **Renewable Energy Systems**
    a. Photovoltaic (PV) Inverters and Micro Inverters
    b. Maximum Power Point Tracking (MPPT)
    c. Wind Energy Conversion Systems
    d. Fuel Cells and Other Emerging Renewable Energy Systems

12. **Transportation Power Electronics**
    a. Vehicular Power Electronic Circuits and Systems
    b. Power Electronics for Hybrid and Electric Vehicles
    c. On-board and Off-board Charging Systems
    d. Power Electronics for Aerospace
    e. Power Electronics for Shipboard and Other Transportation Applications

13. **Wireless Power Transfer**
    a. Wireless Charging
    b. Safety and Reliability
    c. Power for IoT
    d. Non-contact Sensors for Power Electronics

14. **Power Electronics Applications**
    a. Datacenter/Telecom Power Architecture and System Considerations
    b. Solid State and Hybrid Circuit Breakers
    c. Defense and Military Power Electronics
    d. AC-DC-AC Applications and Matrix Converters
    e. Lamp Ballasts and LED Lighting
    f. Portable Power
    g. Energy Harvesting

APEC Conference Committee

Aung Thet Tu
Tu Tech
2025 General Chair

Jin Wang
The Ohio State University
Program Chair

Stephanie Watts Butler
WattsButler, LLC
Plenary Session Chair

Laszlo Balogh
Texas Instruments
Seminar Session Co-Chair

Deepak Veereddy
Infineon
Industry Session Co-Chair

Aengus Murray
Consultant
Industry Session Co-Chair

Indumini Ranmuthu
Texas Instruments
Debate (RAP) Sessions Chair

John Vigars
Allegro MicroSystems
Exposition Chair,
Finance Co-Chair

Benjamin Benchimol
Würth Elektronik
Exposition Co-Chair

Greg Evans
Welcomm, Inc.
Publicity Co-Chair

Bob White
Embedded Power Labs
Web Chair

Frank Cirolia
Advanced Energy
Social Media Chair

Siamak Abedinpour
Publications Chair

Sonia Soin
Guest Hospitality Co-Chair

Alireza Safaee
Apple Inc.
Student Job Fair Chair

Tim McDonald
Infineon
Past General Chair 2024

Dhaval Dalal
ACP Technologies
Assistant Program Chair

Jaume Roig
onsemi
Plenary Session Co-Chair

Conor Quinn
Advanced Energy
Seminar Session Co-Chair

David Chen
Power Integrations
Industry Session Co-Chair

Eric Persson
Infineon
AV Systems Chair

Dheeraj Jain
Würth Elektronik
Debate (RAP) Sessions Co-Chair

Van Niemela
OmniOn Power
Exposition Co-Chair

Mark Nelms
Auburn University
Finance Chair

Kathy Naraghi
Welcomm, Inc.
Publicity Co-Chair

Harish Krishnamoorthy
University of Houston
Mobile App Chair

Bridget O'Gorman
PESC, Inc.
Social Media Co-Chair

Olivier Trescases
University of Toronto
Education Analytics and Awards Chair

Madeline White
Guest Hospitality Co-Chair

David Otten
Massachusetts Institute of Technology
MicroMouse Chair

APEC Conference Management

David McKennon
MMS Meetings
Strategic Account Director

Caryn Pepper
MMS Meetings
Logistics Manager

Rachel Epstein
MMS Meetings
Marketing & Logistics Manager

Catalina DeMassi
MMS Meetings
Education Manager

Ed Lonsinger
MMS Meetings
Exhibits Director

Melissa Benowitz
MMS Meetings
Food & Beverage Manager

Rebecca Armely
MMS Meetings
Registration Manager

Tom Wehner
Epapers, LLC
Abstracts Management

Kelly Laurie
MMS Meetings
Sr. Events Director

Christen Denson
MMS Meetings
Logistics Manager

Katia McKennon
MMS Meetings
Education Director

Nate Knauer
MMS Meetings
Education and Mobile App Manager

Lexie Teeter
MMS Meetings
Exhibits Manager

Adam Lewis
MMS Meetings
Registration Director

Mike Schultz
MMS Meetings
Registration Manager

Tia Fulmer
Printed and Digital
The Printing House

APEC 2025 Steering Committee

Jin Wang
The Ohio State University
APEC 2026 General Chair
Representing PELS

Aung Thet Tu
TuTech
APEC 2025 General Chair
Representing IAS

Tim McDonald
Infineon
Steering Committee Chair,
APEC 2024 General Chair
Representing PSMA

Pradeep Shenoy
Texas Instruments
APEC 2023 General Chair
Representing PELS

Omer C. Onar
Oak Ridge National Laboratory
APEC 2022 General Chair
Representing IAS

Conor Quinn
Advanced Energy
APEC 2021 General Chair
Representing PSMA

José A. Cobos
Universidad Politécnica de Madrid
APEC 2020 General Chair
Representing PELS

Ernie Parker
Crane Aerospace & Electronics
APEC 2019 General Chair
Representing IAS

Eric Persson
Infineon
APEC 2018 General Chair
Representing PSMA

Dhaval Dalal
ACP Technologies
APEC 2027 General Chair
(Non-voting Member)
Representing PSMA

APEC 2025 Program Committee Track Chairs

AC-DC Converters

Gerry Moschopoulos

Mehdi Narimani

Arijit Banerjee

DC-DC Converters

Olivier Trescases

Cahit Gezgin

Xin Zhang

Sombuddha Chakraborty

DC-AC Inverters

Matt Woongkul Lee

Yunting Liu

Karun Potty

Devices and Components

Minjie Chen

Fei Yang

Raghav Khanna

Hengzhao Yang

Magnetics

Matt Wilkowski

Ed Herbert

George Slama

Power Electronics Integration and Manufacturing

Liming Liu

Jim Marinos

Ayman Fayed

Control

Seungdeog Choi

Xiaonan Lu

Dorin Neacsu

Jaber Abu Qahouq

Modeling and Simulation

James Victory

Adam Skorek

Phani Marthi

Motor Drives

Ziaur Rahman

Dinesh Kumar

Ali Safayet

Power Electronics for Utility Applications

Jonathan Kimball

Mohammed Agamy

Ali Khajehoddin

Renewable Energy Systems

Haoyu Wang

Tiefu Zhao

Fei Gao

Transportation Power Electronics

Dong Cao

Tao Yang

Rasoul Hosseini

Harish Krishnamoorthy

Wireless Power Transfer

Erdem Asa

Khurram Afridi

Mehdi Farasat

Power Electronics Applications

Manuel Arias Perez De Azpeitia

Jeff Nilles

Foreword

40th Annual IEEE Applied Power Electronics Conference and Exposition

March 16-20, 2025, Georgia World Congress, Atlanta, GA

Welcome!

On behalf of the APEC 2025 Organizing Committee, and in memory of Tony O'Gorman, the original General Chair of APEC 2025 who tragically passed away last year, I welcome you to the 40th annual IEEE Applied Power Electronics Conference and Exposition (APEC) at the Georgia World Congress Center in the historic city of Atlanta.

Yes, APEC is forty years young. The first APEC was in April of 1986 at the Fairmont Hotel in New Orleans, Louisiana with 34 technical papers, 6 seminars, a panel discussion on the future of power electronics, 20 exhibitors, and had 250 attendees. Forty years later, APEC has grown to over 6,000 attendees, with over 750 technical presentations, 18 professional education seminars, and a three-day exposition with over 300 exhibitors.

There are some program changes for APEC 2025 I would like to highlight. On Monday, before the Plenary Session, there will now be a new Dialogue Preview Session of selected technical papers from the Thursday Dialogue Sessions. This will be a good opportunity for the attendees to "taste" the 14 different flavors of APEC's power electronic tracks and interact with the authors. In honor of Tony O'Gorman, who will forever be a friend of APEC, this session for APEC 2025 will be dedicated to him. The opening Plenary Session has also been shortened to four invited speakers. On Thursday, there will no longer be any afternoon sessions.

The rest of the traditional program remains. The 18 professional education seminars starting on Sunday are a "must" for those seeking a better in-depth understanding of a variety of power electronic topics. The Plenary Session on Monday afternoon will cover important power electronics topics by distinguished invited speakers followed by the can't-miss Welcome Reception in the Exhibit Hall. On Tuesday through Thursday, over 750 presentations will be given in lecture and dialogue portions of the Technical Sessions and in the application-oriented Industry Sessions. The 3 Debate Sessions (formerly called Rap Sessions) on Tuesday evening will offer lively and entertaining interactive discussions on current hot topics. There will also be about 50 Exhibitor Presentations showcasing the latest products and technologies from our exhibiting companies. *The challenge at APEC is not about whether you can find a topic in your area of interest. Rather, it is about choosing which session among many that covers your interests best.*

In addition to the Technical Program, we also have the FIRST Robotic demonstration and the MicroMouse competition on Monday, the Student Job Fair on Tuesday afternoon which is free of charge to attend for registered students, the Guest/Spouse program, and a special Wednesday Night Social event commemorating the 40th Anniversary of APEC. I encourage you to use the improved mobile app which does an excellent job of allowing you to plan your total APEC experience by previewing the content, customizing your calendar and receiving notifications.

APEC is a conference organized by power electronics professionals like us, for the global power electronics community. I am grateful to our exhibitors, partners, and our three sponsors (PSMA, IEEE PELS, and IEEE IAS). There are hundreds of volunteers who make this the premier event that it is. I thank the dedication, passion, knowledge, and guidance of the Organizing and Steering Committees, the Technical Program's Track Chairs, Session Chairs, our army of reviewers, and the authors and presenters. Finally, I thank the tireless support of our professional conference management company Meeting Management Services (MMS) in our second year of our partnership together.

Sincerely,

Aung Thet Tu
General Chair
2025 IEEE Applied Power Electronics Conference and Exposition

APEC History

Year	Site	Dates	General Chair	Program Chair
1986	Fairmont Hotel New Orleans, Louisiana	April 28 – May 1	John G. Kassakian	R. David Middlebrook
1987	Town and Country Hotel San Diego, California	March 2 – 6	John G. Kassakian	R. David Middlebrook
1988	Fairmont Hotel New Orleans, Louisiana	February 1 – 5	William W. Burns, III	William W. Burns, III
1989	Baltimore Convention Center Baltimore, Maryland	March 13 – 17	William W. Burns, III	Robert V. White
1990	Biltmore Hotel Los Angeles, California	March 11 – 16	Robert V. White	Charles Harm
1991	Hyatt Regency Reunion Hotel Dallas, Texas	March 10 – 15	Charles Harm	Thomas M. Jahns
1992	Weston Copley Plaza Hotel Boston, Massachusetts	February 23 – 27	Thomas M. Jahns	Kevin J. Fellhoelter
1993	Town and Country Hotel San Diego, California	March 7 – 11	Kevin J. Fellhoelter	Douglas McIlvoy
1994	Disney Contemporary Hotel Orlando, Florida	February 13 – 17	Douglas McIlvoy	Thomas Latos
1995	Hyatt Regency Reunion Hotel Dallas, Texas	March 5 – 9	Thomas Latos	Charles E. Mullett
1996	Fairmont Hotel San Jose, California	March 3 – 7	Charles E. Mullett	Thomas G. Wilson, Jr.
1997	Weston Peachtree Hotel Atlanta, Georgia	February 23 – 27	Thomas G. Wilson, Jr.	David Torrey
1998	Disneyland Hotel Anaheim, California	February 15 – 19	David Torrey	F. Dong Tan
1999	Adams' Mark Hotel Dallas, Texas	March 14 – 18	F. Dong Tan	Robert V. White
2000	Fairmont Hotel New Orleans, Louisiana	February 6 – 10	Robert V. White	R. Mark Nelms
2001	Disneyland Hotel Anaheim, California	March 4 – 8	R. Mark Nelms	V. Joseph Thottuvelil
2002	Adams' Mark Hotel Dallas, Texas	March 10 – 14	V. Joseph Thottuvelil	Bruce Miller
2003	Fontainebleau Hotel Miami Beach, Florida	February 9 – 13	Bruce Miller	Jim Kokernak
2004	Disneyland Hotel Anaheim, California	February 22 – 26	Jim Kokernak	Jason Lai
2005	Hilton Austin Austin, Texas	March 6 – 10	Jason Lai	Van Niemela
2006	Hyatt Regency Hotel Dallas, Texas	March 19 – 23	Van Niemela	Russ Spyker

Year	Site	Dates	General Chair	Program Chair
2007	Disneyland Hotel Anaheim, California	February 25 – March 1	Russ Spyker	Steve Pekarek
2008	Austin Convention Center Austin, Texas	February 24 – 28	Steve Pekarek	Kevin Parmenter
2009	Marriott Wardman Park Hotel Washington, District of Columbia	February 15 – 19	Kevin Parmenter	Babak Fahimi
2010	Palm Springs Convention Center Palm Springs, California	February 21 – 25	Babak Fahimi	Patrick Chapman
2011	Fort Worth Convention Center Fort Worth, Texas	March 6 – 10	Patrick Chapman	Frank Cirolia
2012	Coronado Springs, Disney World Orlando, Florida	February 5 – 9	Frank Cirolia	Siamak Abedinpour
2013	Long Beach Convention Center Long Beach, California	March 17 – 21	Siamak Abedinpour	Haidong Yu
2014	Fort Worth Convention Center Fort Worth, Texas	March 16 – 20	Haidong Yu	Aung Thet Tu
2015	Charlotte Convention Center Charlotte, North Carolina	March 15 – 19	Aung Thet Tu	Alireza Khaligh
2016	Long Beach Convention Center Long Beach, California	March 20 – 24	Alireza Khaligh	Jonathan Kimball
2017	Tampa Convention Center Tampa, Florida	March 26 – 30	Jonathan Kimball	Eric Persson
2018	Henry B. Gonzalez Convention Center, San Antonio, Texas	March 4 – 8	Eric Persson	Ernie Parker
2019	Anaheim Convention Center Anaheim, California	March 17-21	Ernie Parker	José A. Cobos
2020	Virtual Conference	March 15-19	José A. Cobos	Conor Quinn
2021	Virtual Conference	June 14-17	Conor Quinn	Omer C. Onar
2022	George R. Brown Convention Center, Houston, Texas	March 20-25	Omer C. Onar	Pradeep Shenoy
2023	Orange County Convention Center, Orlando, Florida	March 19-23	Pradeep Shenoy	Tim McDonald
2024	Long Beach Convention Center Long Beach, California	February 25 – 29	Tim McDonald	Tony O'Gorman

The Sponsors

Power Sources Manufacturers Association

Trifon Liakopoulos
Chairman of the Board

David Chen
Vice President

Renee Yawger
President

Tim McDonald
Secretary/Treasurer

IEEE Power Electronics Society

Johan Enslin
President

Brad Lehman
Past President

Pradeep Shenoy
Vice President (Conferences)

Joseph Kozak
Treasurer

IEEE Industry Applications Society

Ayman El-Refaie
President

Pericle Zanchetta
Vice President

Avoki Omekanda
President-Elect

Tamas Ruzsanyi
Treasurer

APEC 2025 Reviewers

Aalam, Mohammad Nair
Abbasi, Ayesha
Abdelkader, Mohamed
Abdullah, Yousef
Abramson, Rose
Acero, Jesus
Ademane, Harsha
Agarwal, Nitin
Ahluwalia, Urvi
Ahmed, Jamshed
Ajegbemika, Eniola
Al Sakka, Mustapha
Alam, Md Didarul
Alaql, Fahad
Alhoor, Wisam
Alizadeh, Rayna
Alshammari, Sulaiman
Alsmadi, Mohammad
Alzawaideh, Ayman
Amin, Ashik
Amin, Nazmul
An, Zheng
Andersen, Thomas
Ankam, Karthik
Ansari, Md Irshad
Anumakonda, Gangadhar Rao
Arias, Manuel
Arogunjo, Ezekiel
Arvind, P
Attukadavil, Jenson Joseph
Atwimah, Samuel
Azadeh, Yalda
Babaie, Mohammad
Babazadeh Dizaji, Ramin
Bai, Yijie
Banerjee, Soham
Bao, Mingjun
Barreto, Luiz Henrique
Bayliss III, Rod
Beechner, Troy
Behnia, Puya
Beig, Balanthi
Beohar, Navankur
Bhagat, Chinmay

Bhandarkar, Santosh
Bharathan, Aarranon
Bhogaraju, Sri Krishna
Bilakanti, Nishant
Biswas, Kausik
Bouderdaben, Omar
Briz, Pablo
Brooks, Nathan
Brown, Alyssa
Budiwicaksana, Lukas Antonio
Burugula, Vasishta
Byrajanda Naniappa, Aiyappa
Cao, Jijun
Carlson, Nathan
Carslake, Cameron
Celikovic, Janko
Cen, Siye
Cerutti, Stefano
Chae, Jong Yoon
Chakraborty, Sombuddha
Chandio, Rashid Hussain
Chang, Yun-Hsiang
Chaudhary, Jai Aditya
Chawda, Gajendra Singh
Chen, Chasel
Chen, Chen
Chen, Hao
Chen, Hung-Chi
Chen, Kai-Hui
Chen, Shiying
Chen, Wei
Chen, Xingyu
Chen, Yenan
Chen, Yu
Chen, Yuming
Cheng, Hsin
Cheng, Qi
Cheng, Zeyu
Cheshire, Audrey
Choi, Sihoon
Chokshi, Kunvar
Cirolia, Frank
Coday, Samantha

Croft, Nathan
Cui, Junwei
Cui, Xiaofan
Cui, Yujia
Curi Busarello, Tiago Davi
Dai, Xize
Das, Annoy Kumar
Das, Uppal
Datta, Kishalay
De Leon, Alberto
Delmar, Aria
Demirkutlu, Eyyup
Deshpande, Ankit Vivek
Dhima, Shubham
Ding, John
Ditze, Stefan
Dornala, Avinash
Dou, Yi
Dr. Schlenk, Manfred
Du, Lixiong
Dubey, Ashwini Kumar
Duffy, Maeve
Dunford, William
Duppalli, Veda Samhitha
Dutta, Oindrilla
Dutta, Soham
Ejury, Jens
Elias, Sergio
Emmett, Paul
Erturk, Mete
Esmaeili Rad, Sadegh
Estrin, Julia
Etta, Dheeraj
F. Costales, Miguel
Fahmy, Youssef
Fakunle, Samuel
Farag, Mohamed
Farasat, Mehdi
Farooq, Maida
Farsi, Homayoon
Fernández, Jose Antonio
Fernandez Miaja, Pablo
Filchev, Todor
Fishbune, Rick

Foelkel, Lorandt
Freitas, Nayara
Friedrichs, Peter
Fu, Pengyu
Fu, Zhenda
Gadelrab, Rimon
Galiano Zurbriggen, Ignacio
Gamboa, Gustavo
Gandluru, Veera Bharath
Gao, Bo
Gao, Fengqi
Gao, Hang
Gao, Mingze
Gao, Zihan
Gauttam, Gaureej
Gawhade, Pragya
Geda, Million Gerado
Gennaro, Francesco
Gerfer, Alexander
Get, Teddy
Gezgin, Cahit
Ghaffari, Abolfazl
Ghavaminejad, Mahdi
Ghimire, Dilip
Gholamian, Mahzad
Ghosh, Sumana
Giardine, Francesca
Gill, Lee
Gomez, Jonathan
Gomez, Roderick
Gómez, Alexis Anselmo
Gonçalves, Odiglei Hess
Grigoryan, Davit
Guillot, Laurent
Guleria, Ashwani
Guo, Desheng
Guo, Haidong
Guo, Zhehui
Guo, Zhengchen
Gupta, Ashish
Gupta, Chanchal
Gupta, Mahima
Gupta, Prof. Dr. Suryakant
Gupta, Shantanu

Gurudiwan, Shubhangi
Hadifar, Navid
Haji Ali Biglo, Ali
Haji Ali Biglo, Ali
Halawa, Ali
Hameed, Asad
Hameed, Muhammad Furqan
Hammer, Jan
Han, Lei
Han, Weijie
Haniyur, Sheshagiri Shey
Haque, Moinul Shahidul
Harikumaran, Jayakrishnan
Harris, Michael
Harrye, Yasen
Hasan, Md Zakir
Hassan, Faridul
Hassan, Mustafeez
Havugimana, Emmanuel
Haynes, Geoff
Henspeter, Justin
Hentschke, Theyllor
Hernandes, Daniela
Hesener, Alfred
Hiesmayr, Alfred
Hodge, Stuart
Hofmann, Viktor
Honea, Jim
Hosseinabadi, Farzad
Hou, Yuetao
Hou, Zhengming
Hsu, Fu-Jen
Hsu, Mason
Hu, Borong
Hu, Jhih-Cheng
Hu, Jiangang
Hu, Shoudong
Huang, Jiasheng
Huang, Xingxuan
Huang, Yuanqing
Hubert, Florian
Husev, Oleksandr
Hussain, Aqarib
Hwang, Inhwi
Ibrahim, Mohammad
Intriago, Raul

Iyer, Rahul
Jackson, Amanda
Jadhav, Naman
Jalalabadi, Esmaeil
Jang, Won-Yong
Jappe, Tiago
Jayaraman, Karthik
Jia, Niu
Jiang, Xi
Jiao, Dong
Jimenez, Sergio
Joisher, Mansi
Jolly, Nitish
Joo, Dongmyoung
Joshi, Piyush
Joshi, Rahul
Josipovic, Ksenija
Jung, Woohyuk
Kącki, Marcin
Kamalapur, Aakash
Kanale, Ajit
Kanaparthi, Vijaya
Karimi, Masoud
Karneddi, Harish
Kazem, Kamran
Khajehoddin, Sayedali
Khajueezadeh, Mohammadsadegh
Khanna, Mudit
Kharrich, Mohammed
Khodabakhsh, Javad
Kieu, Huu-Phuc
Kim, Jonghoon
Kim, Jong-Woo
Kim, Namwon
Kim, Seho
Kim, Yeonwoo
Kimball, Jonathan
Kitai, Hidenori
Kokkonda, Raj Kumar
Kolli, Nithin
Komeda, Shohei
Kozielski, Kyle
Krishnan, Sahana
Kuang, Kaining
Kumar, Ashish
Kumar, Manish
Kumar, Prince
Kumar, Ujjwal
Kwak, Jin Woong
Kwon, Hyukjae

L, G
La Rosa, Manuela
Lan, Yuhao
Lazzarin, Telles
Lee, Dong-Choon
Lee, Kangbeen
Lee, Sangwhee
Lee, Sangwon
Lee, Yongduk
Lessner, Philip
Li, Bin
Li, Bing
Li, Chenxi
Li, Jiacheng
Li, Ruqi
Li, Shengming
Li, Tianchen
Li, Xiang
Li, Xiaoling
Li, Yanchao
Li, Yanqiao
Li, Yaohua
Li, Zehui
Li, Zhenchao
Li, Zilin
Liang, Jiawei
Liao, Mian
Lin, Calvin
Lin, Jackman
Lin, Min
Liu, Chen
Liu, Guanjiang
Liu, Hanbing
Liu, Hualong
Liu, Jiahong
Liu, Jingcun
Liu, Lehan
Liu, Wenchin
Liu, Wenzhao
Liu, Xudan
Liu, Yang
Liu, Yongjie
Liu, Yu Chen
Liu, Zeyuan
Liu, Zhanlei
Liu, Ziheng
Lnu, Abdull Wasay
Lou, William
Lough, Ben
Lu, Che-Yu
Lu, Y
Ma, Kwokwai Ma
Ma, Rui
Mabetha, Bahlakoana

Mahajan, Sagar Bhaskar
Mahmodicherati, Sam
Mahmud, Sadab
Mahmud, Tahmin
Maji, Sounak
Maji, Sounak
Malik, Muhammad Shehroz
Mannan, Tahmid Ibne
Manos, Konstantinos
Marinos, Jim
Martínez, Alfonso
Maru, Siddharth
Masoumi, Amir Hossein
Matallana, Asier
Mateos Gil, Pablo
Mattavelli, Paolo
Maynard, Xavier
McDonough, Matthew
Mehta, Rajesh
Mei, Yunhui
Meng, Haoran
Milic, Aleksandar
Milner, Jordan
Mirdoddi, Kaushik
Mirza, Abdul Basit
Mishra, Manash
Mishra, Rahul
Mohamed, Mohamed
Mohammad, Mostak
Moniruzzaman, Md
Morand, Julien
Moreira, William
Morsali, Payam
Mosa, Mostafa
Moschopoulos, Gerry
Motoori, Shuichiro
Motto, Eric
Mu, Qiang
Mu, Wei
Mueller, Lukas
Mukhopadhyay, Anwesha
Mulkern, Joe
Mulpuri, Sai Krishna
Mulpuri, Vamsi

Muneeb, Abdul
Murray, Samantha
Murukesan, Karthick
Musetti, Alex
Mweene, Haachitaba
Nagarale, Tanvi
Naghash, Reza
Nakao, Hiroshi
Naligama, Chamila Anuradha
Namburi, Krishna Mpk
Nassar, Rajaie
Naval, Sourav
Nawafleh, Yannal
Nazir, Moazzam
Neely, Jason
New, Christopher
Ng, Edmond James
Ngo, Khai
Nguyen, Minh-Khai
Nhat Truong, Phan
Ni, Shusen
Niraula, Manish
Noon, James
Noon, John
Novak, Mateja
Nowakowski, Richard
Nuli, Prathima
Nunez, Juan
Oh, Mingi
Olayiwola, Tofopefun Nifise
Otuboah, Francis Yaw
Ou, Shuyu
Pallantla, Manikanta
Palma, Marco
Pan, Qishan
Pan, Tao
Pandey, Pankaj
Parashar, Sanket
Paredes-Camacho, Alejandro
Parida, Nibedita
Parida, Nibedita
Park, Jaeyeon
Parker, Mason
Parreiras, Thiago
Parsa Sirat, Ali
Patle, Nagesh

Paturi, Satish Kumar
Paul, Sayan
Peng, Chuhan
Peng, Yuchong
Pereira Monteiro, Amanda
Pervaiz, Saad
Phuong-Ha, La
Pinarello Scalcon, Filipe
Pong, Bryan M.H.
Pourmohammadi, Mehdi
Pradhan, Rachit
Prakash, Pranav Raj
Pramod, Prerit
Pulvirenti, Mario
Qi, Nianzun
Qian, Yijie
Qiu, Maohang
Quilici, James
Rabenold, Ellie
Rahimian, Mohsen
Rahman, Mohammad Arifur
Rahman, Mohammad Mahinur
Rajagopal, Narayanan
Ramamurthy, Anand
Ramanath, Anushree
Ranjram, Mike
Rasheed, Marium
Rashid, Syed Saeed
Rashkin, Lee
Rehman, Hamood
Ren, Xufu
Renjie, Zhang
Resalayyan, Rakesh
Ribeiro, Pedro
Rice, Ritch
Rios Linares, Daniel
Rivera, Miguel
Rizzolatti, Roberto
Robert Tertese, Yongo
Rohilla, Yogesh
Rosahl, Thoralf
Rossi, Mattia
Roy, Soham
Roy, Sukanta

Roy Chowdhury, Vikram
Rylko, Marek
Sabate, Juan
Sachan, Shailu
Sadeghi, Ayoub
Sadilek, Tomas
Sado, Kerry
Sagar, Jawahar
Sahin, Ilker
Sai, Toru
Saito, Katsuaki
Saito, Wataru
Salehi, Maryam
Salehi Dobakhshari, Sina
Sambo, Haifah
Sancar, Senol
Saraf, Pushkar
Sareen, Puneet
Sarma, Bhaskarjyoti
Sati, Shraf Eldin
Sato, Yuki
Satyamsetti, Vijayakrishna
Savulak, Stephen
Scelba, Giacomo
Schumann, Christian
Selvarasu, Uthandi
Sen, Tanuj
Sengupta, Arkadeb
Shah, Shreyas
Shahbazi, Reza
Shahzad, Khurram
Shan, Shengming
Shao, Jianwen
Sharifi, Reza
Sharma, Angshuman
Sharma, Anuj
Sharma, Rishabh Kumar
Sharma, Shrivatsal
Shin, Jongwon
Shu, Ji
Singh, Anurag
Singh, Sukhjit
Singh, Vinay Kumar
Sinha, Shivangi
Sinha, Sourish S.
Sinha, Sreyam
Som, Cem
Sondharangalla, Madhura

Song, He
Song, Qihao
Song, Yubo
Souri, Naser
Sridhar, Sundaramoorthy
Srikanta Murthy, Puneeth
Srinivasan, Kishan
Steinberger, Mike
Steyaert, Bernard
Stokowski, Nicole
Stolt, Eric
Stupar, Andrija
Sun, Min
Sun, Ruize
Sun, Xiuhu
Sund, Jade
Sur, Debotrinya
Suzuki, Asamira
Swaminathan, Niraja
Tadakuma, Toshiya
Taghavi, Masoud
Tajfar, Alireza
Takamori, Taro
Takamura, Yota
Talebzadeh, Sarah
Tariquzzaman, Md
Tasnin, Fariha
Telefus, Mark
Tenghoff, Olof
Tian, Fanghao
Tien, Kevin
Tiwari, Gyanendra
Toops, Patrick
Torrico, Grover
Touhami, Mustapha
Tourloukis, Mihalis
Tsai, Chieh-Ju
Tu, Cong
Vaghasiya, Kamal
Vechalapu, Kasunaidu
Veeramraju, Kartikeya Jayadurga
Veilleux, Elias
Venkatraman, Prasad
Villamor Baliarda, Ana
Wa, Terry Haoyu
Wang, Bixuan

Wang, Haoyu
Wang, Rui
Wang, Sofia
Wang, Tianhong
Wang, Xiaodan
Wang, Yafeng
Wang, Yafeng
Wang, Yibo
Wang, Yilin
Wang, Yuchen
Wang, Ziyao
Wei, Mengxuan
Wei, Ruizhi
Weldesamual,
    Gebrehiwet
Wen, Eric
Wen, Luowei
Wilkowski, Matt
Williams, David
Wondrak, Wolfgang
Wright, Jason
Wu, Di
Wu, Fanfu
Wu, Jingjie
Wu, Xuesong

Wu, Yifan
Wu, Yihan
Wu, Yihao
Wu, Yue
Wu, Yuxuan
Wu, Zhuoqun
Wu, Zhuoqun
Xiao, Ziheng
Xie, Tianshi
Xie, Tianshi
Xu, Jinming
Xu, Mark Dehong
Xu, Wentao
Xu, Xinmiao
Xu, Zixuan
Xu, Ziyang
Yamauchi,
    Yoshitaka
Yan, Yiyang
Yang, Bowen
Yang, Le
Yang, Qiuzhe
Yang, Xinliang
Yasko, Mohamed
Yehia, Rami

Yılmaz, Kadir
Yonezawa, Yu
Yoo, Inpil
Yousefzadeh Fard,
    Amin
Yu, Qinghong
Yu, Qixue
Yu, Sean
Yu, Xipei
Zade, Aditya
Zan, Xin
Zare, Alireza
Zeltser, Ilya
Zeng, Jianwu
Zguir, Isra
Zhan, Cao
Zhang, Alpha J.
Zhang, Cheng
Zhang, Chi
Zhang, Daifei
Zhang, Haoyu
Zhang, Kaichen
Zhang, Lei
Zhang, Ruizhe
Zhang, Shuyu

Zhang, Xueshen
Zhang, Yi
Zhang, Yingjie
Zhang, Yuanzheng
Zhang, Zichen
Zhao, Delin
Zhao, Yifan
Zheng, Sanbao
Zheng, Yifei
Zhong, Kailun
Zhou, Bokang
Zhou, Fei
Zhou, Jiale
Zhou, Linke
Zhou, Mingde
Zhu, Gangwei
Zhu, Yinxiao
Zhuge, Yingjian
Zimmerman, Alan
Zolfi, Pouya
Zou, Huanghaohe
Zou, Jiarui
Zuo, Yu

APEC 2025 Partners

As of 2/7/2025 the companies below were confirmed partners at APEC 2025. The actual partners at APEC 2025 may differ slightly from this list.

EMERALD

DIAMOND

PLATINUM

GOLD

SILVER

TAIYO YUDEN

APEC 2025 Exhibitors

As of 2/7/2025, the companies below were planning to exhibit at APEC 2025. The actual exhibitors at APEC 2025 may differ slightly from this list.

3D-Micromac AG
ABC Taiwan Electronics Corp.
ABIS CIRCUITS CO.,LTD
AC Power Corp.(Preen)
ACME Electronics Corporation
Acopian Power Supplies
Advanced Energy
Advanced Test Equipment Rentals
Aehr Test Systems
Aishi Capacitors
Aismalibar North America
AL TRANSFO TECHNOLOGY LIMITED
AllSpice.io
Allstar Magnetics
Alltronics Tech Manufacturing Limited
Alpha & Omega Semiconductor
Altair
AmberSemi
AMETEK Programmable Power
AMETHERM, INC
Analog Devices
Anbon Semi
Anhui Tiger Co., Ltd
Animato Electronics, Inc.
Apex Microtechnology
Asahi Diamond America
ATM MATERIAL.CO.LTD
AVL
B&K Precision
Batten & Allen
Berkeley Power and Energy Center (BPEC)
BH Electronics, Inc.
Bosch Semiconductors
Boschman Advanced Packaging Technology
Bourns, Inc.
Broadcom Inc.
BTCOIL USA LLC

Caerus Power Technology
CalRamic Technologies, LLC
Cambridge GaN Devices
Captor Corporation
Capxon Electronic Technology Co., Ltd.
Center for Power Electronics Systems
Central Semiconductor
Centrotherm International AG
Chang Sung Corporation
Chip-GaN Power Semiconductor
Chroma
Cleverscope
Coil Winding Specialist, Inc.
Coilcraft, Inc.
Conquer Electronics Co., Ltd.
CoolCAD Electronics
Cramer Magnetics
CuNex GmbH
Danfysik A/S
Datatronics
Dean Technology, Inc.
Delta Electronics (Americas) Ltd.
DEMAK Group
DEWESoft LLC
DEWETRON Inc.
DigiKey
DIOTEC Semiconductor America
dSPACE Inc.
Ducati Energia SPA
E&B TECHNOLOGY
EFC/Wesco
Efficient Power Conversion Corporation (EPC)
EGSTON Power Electronics
Electro Technik Industries
Electrocube, Inc.
Electronic Concepts, Inc.
Electronicon Kondensatoren GmbH

Elektrisola Inc.
ELKO EP North America LLC
Elna Magnetics
ELYTONE ELECTRONIC CO., LTD.
EMcoretech
Empower Semiconductor, Inc.
ERD
EXATRON, INC.
EXXELIA SAS
Fair-Rite Products
Ferric Inc
Ferrotec (USA) Corporation
Ferroxcube
Focused Test, Inc.
Frenetic Electronics, S.L.
Fuji Electric Corp. of America
GaNPower International Inc.
GE Aerospace
GMW Associates
Goldenbamboo Electronics (Zhuhai) Co.,Ltd
Good-Ark Semiconductor
Greenconn Technology (Shenzhen) Co. Ltd.
Hengdian Group DMEGC Magnetics Co., Ltd
Heraeus Electronics
Hesse Mechatronics
Hind Rectifiers Limited
Hioki USA Corp.
HMI
Holy Stone International
Hotland International Corp
HVM Technology
HVR Advanced Power Components
IAS (IEEE Industry Applications Society)
ICE Components, Inc.
Ideal Power
iDRC
Impedyme Inc.
Indium Corporation
Infineon Technologies Americas Corp
Innoscience Technology Co., Ltd.
Innovative Thermal Solutions, Inc
INPACK
iNRCORE, LLC
INSTEK America Corp.
Inter Outstanding Electronics Inc (IOE)

ITECH ELECTRONICS
ITELCOND SRL
ITG Electronics, Inc.
IWATSU
JARO Thermal
Nantong Jianghai Capacitor Co., LTD
Johanson Dielectrics, Inc
Jovil Universal LLC
Kendeil srl
Kewell Technology Co., Ltd
Keysight
Kikusui America, Inc.
Knowles Precision Devices
KYOCERA AVX
Laser Thermal
LinkCom Manufacturing Co., Ltd
Lodestone Pacific
Luminus Devices, Inc., featuring APC-E and Sanan
Semiconductor
MacDermid Alpha Electronics Solutions
Maditronics Inc
Magna-Power Electronics
Magnetec
Magnetic Metals Corporation
Magnetics
Magnetika, Inc.
Malico Inc.
Manutech
Marel Power Solutions
Max Echo Technology Corp
MaxLinear Corporation
Menlo Microsystems, Inc.
Mentech
Mersen
Metalor Technologies USA
Methode Electronics
MH&W International Corp
Miba Resistors
Microchip Technology
Microgate Technology Co. LTD
Micrometals, Inc.
Minebea Power Semiconductor Device
Mission Power
Mitsubishi Electric US, Inc
MK Magnetics Inc

Monolithic Power Systems, Inc.
Mouser Electronics
MS Power GmbH
Murata
NAECO
Nagase America LLC
Nanjing New Conda Magnetic Industrial Co., LTD.
National Magnetics Group
Navitas
Nayak Corporation, Inc.
New England Wire Technologies
Nexperia
Nexustest
Nichicon (America) Corp.
NINGGUO YUHUA ELECTRICAL PRODUCTS CO.,LTD
Nisshinbo Micro Devices
NoMIS Power
NORWE Inc.
Nuvoton Technology Corp America
OMICRON Lab
Onics Resistors Private Limited
onsemi
OPAL-RT TECHNOLOGIES
P. Leo & Co., Ltd.
Pacific Power Source
Pacific Sowa Corporation; C/O Epson Atmix Corporat
Parker Overseas Pvt. Ltd.
Payton America Inc.
PCIM
PCT INDUSTRIES LTD.
Peak Nano Films
Pearson Electronics
PELS (IEEE Power Electronics Society)
Pentamaster Instrumentation Sdn. Bhd.
PIN SHINE INDUSTRIAL CO., LTD
Plexim
PMK Mess- und Kommunikationstechnik GmbH
POCO Holding Co., LTD
PolyCharge America, Inc.
Power Electronic Measurements Ltd
Power Integrations
Power Management Integration Center (PMIC)
PowerAmerica
PowerELab Ltd.
Powerex, INC

Poweronics
Premier Magnetics
Premo USA, Inc
PRISOURCE ELECTRONICS CO.,LTD.
Proterial America, Ltd.
Providence Electronics Corp.
PSMA
Pulsiv Limited
Quanding Magneto-Electric Material Co.,Ltd
Quantic Capacitors and Magnetics
Reed Semiconductor Corp.
Regatron by Ohmini
REMTEC, Inc.
Renesas Electronics America Inc.
Resonac America, Inc.
RFMW
Richardson Electronics, Ltd.
Richardson RFPD
Rico Products Inc.
Rohde & Schwarz USA, Inc.
ROHM Semiconductor
Rubadue Wire
S.C.O.M.E.S. SRL
SABIC
Sager Electronics
Samwha USA
SanRex Corporation
Saras Micro Devices
Semikron Danfoss
SemiQ
Semitel International Ltd
Shaanxi Shinhom Enterprise Co.,Ltd
SHANGHAI MAGWAY MAGNETIC CO.,LTD.
Shenzhen Cenker Technology Group Co., Ltd.
Shenzhen CODACA Electronic Co., Ltd
ShenZhen East-Win Technology Co.,Ltd.
Shenzhen Liron Electronics Co., Ltd.
Shenzhen Sunlord Electronics Co.,Ltd.
Shenzhen Zeasset Electronic Technology Co., Ltd.
Shin-Etsu Silicones of America
Shiv Om Precision USA LLC.
Sichuan Zhongxing Electronic Co., Ltd
Siemens
Signal Transformer
SIMPLIS Technologies

Sinomag Technology CO., LTD
Soitec
Speedgoat
Standex Electronics
Stellar Industries Corp.
STMicroelectronics
Storm Power Components
STS Spezial-Transformatoren-Stockach GmbH & Co.KG
SUITA Electric Corporation
Sumida America Inc.
Sumitomo Chemical Co., Ltd.
SUPERWORLD ELECTRONICS (S) PTE LTD
Synopsys
TAI-TECH Advanced Electronics
Taiwan Semiconductor
Taiyo Kogyo Co., Ltd.
TAIYO YUDEN USA INC.
Talema Group LLC
Tamura Corp. of America,Tamura Japan
TCLAD Inc.
TDG Holding Co., Ltd
TDK Corporation
Tektronix Inc. + EA Elektro-Automatik
Teledyne LeCroy
Tesec, Inc.
Texas Instruments
T-Global Technology Co., Ltd.
TJ Assemblies / United Technical Products
TME US, LCC
Toshiba America Electronic Components, Inc.
Tower Semiconductor

Tran-Tec LLC
Tyndall National Institute
Typhoon HIL, Inc.
United Chemi-Con
VAC Magnetics, LLC
Vincotech GmbH
Vishay Intertechnology
Vision Technologies Co., Ltd
Vitrek-High Voltage Test & Measurement
Voltage Multipliers, Inc.
Wakefield Thermal
Wellascent Electronic (Ganzhou) Co.,Ltd
West Coast Magnetics
WIMA Capacitors GmbH & Co.KG
WISE INTEGRATION
Wolfspeed, Inc.
WorldMicro
Wurth Elektronik
Wuxi CRE New Energy Technology Co., Ltd.
Wuxi Leapers Semiconductor Co., Ltd.
X-FAB Global Services GmbH
Xiamen Faratronic Co. Ltd.
Xiinergy Systems Inc
YAGEO Group
Yokogawa Test&Measurement
YUEQING DAHE ELECTRIC CO., LTD
ZES ZIMMER Electronic Systems GmbH
Zhejiang Zuoao Technology Co.,Ltd.
ZHONGSHAN COMPETENT AUTOMATION EQUIPMENT CO.,LTD
ZHUHAI WEIHAN WIRE CO., LTD.

TECHNICAL PAPERS

Tuesday, March 18, 2025

SESSION T01: AC-DC Converters I

8:30 - 12:00
TRACK: AC-DC Converters

SESSION CHAIR(S)
Mike Ranjram, *Arizona State University*
Xiaofan Cui, *University of California at Los Angeles*

T01.1 Versatile Controller Architecture for a Universal DC Fast Charging Front-End 1
Anurag Singh[1], Sayan Paul[1], Tejas Bhuse[2], Trent Martin[2], Hien Nguyen[2], Inder Vedula[2],
Nikola Milivojević[2], Dragan Maksimović[1], Luca Corradini[1]
[1]University of Colorado Boulder, United States; [2]FreeWire Technologies, United States

> PRESENTATION   TOPIC: **Single-Phase and Three-Phase Input**

T01.2 A 10 kV SiC MOSFET based Three-Phase Single-Stage Isolated MVAC/LVDC
Converter for Solid State Transformer Applications 9
Anup Anurag, Chi Zhang, Rudy Wang, Peter Barbosa
Delta Electronics, Inc., United States

> PRESENTATION   TOPIC: **Power Factor Correction: CCM, DCM, CRM/BCM Control, Bridgeless**

T01.3 Direct Digital Control Applied to T-Type Vienna Rectifiers for Power Factor Correction 16
Jun-Yang Chang[1], Tsai-Fu Wu[2], Chien-Chih Hung[2], Jui-Yang Chiu[2]
[1]Delta Electronics Inc., Taiwan; [2]National Tsing Hua University, Taiwan

> PRESENTATION   TOPIC: **Power Factor Correction: CCM, DCM, CRM/BCM Control, Bridgeless**

T01.4 Active Power Decoupling Method based on Dual Active Bridge Converter
without Additional Components 21
Kosuke Takeuchi, Takashi Ohno, Hiroki Watanabe, Yuki Nakata, Jun-Ichi Itoh
Nagaoka University of Technology, Japan

> PRESENTATION   TOPIC: **Bidirectional AC-DC Converters**

T01.5 An ANPC-Based Building Block for Medium-Voltage Applications 27
Ahmed Rahouma, Hui Cao, David A. Porras, Zhuxuan Ma, Yue Zhao, Juan C. Balda
University of Arkansas, United States

> PRESENTATION   TOPIC: **Bidirectional AC-DC Converters**

T01.6 **Analog Control of a 2.5 kW GaN based CRM PFC with Input Filter Optimization** 34
Naveed Ishraq, Ayan Mallik
Arizona State University, United States

PRESENTATION   TOPIC: **Single-Phase and Three-Phase Input**

T01.7 **An iTHD and Efficiency Optimized Control Method for Triangular Conduction Mode Totem Pole Bridgeless PFC with Zero Current Detection** 42
Brent McDonald, Sheng-Yang Yu
Texas Instruments Incorporated, United States

PRESENTATION   TOPIC: **Power Factor Correction: CCM, DCM, CRM/BCM Control, Bridgeless**

T01.8 **Resonance Current Suppression for AC-DC Active-Clamp Flyback Converter by Triangular Current Mode** 48
Yasuo Uchida, Hiroki Watanabe, Jun-Ichi Itoh
Nagaoka University of Technology, Japan

PRESENTATION   TOPIC: **External AC-DC Adapters**

T01.9 **A Universal DC Fast Charging Front-End with Optimized Film Capacitor Design** 54
Sayan Paul[1], Anurag Singh[1], Tejas Bhuse[2], Trent Martin[2], Hien Nguyen[2], Inder Vedula[2], Nikola Milivojević[2], Dragan Maksimović[1], Luca Corradini[1]
[1]University of Colorado Boulder, United States; [2]FreeWire Technologies, United States

PRESENTATION   TOPIC: **Single-Phase and Three-Phase Input**

SESSION T02: Bi-Directional DC-DC Converters

8:30 - 12:00
TRACK: DC-DC Converters

SESSION CHAIR(S)
Olivier Trescases, *University of Toronto*
David Reusch, *Texas Instruments*

T02.1 **Power Characterization of a 1200-V/800-V 22-kVA 30-kHz Unity-Gain Dual-Active-Bridge Converter Prototype** 62
Radhika Sarda[1], Abishek Sethupandi[2], Madasamy Palavesha Thevar[2], Howe Li Yeo[2], Praveenkumar Palani[2], Vaisambhayana B. Sriram[1], Anshuman Tripathi[1]
[1]Nanyang Technological University, Singapore; [2]Amperesand, Singapore

PRESENTATION   TOPIC: **Bidirectional DC-DC Converters**

T02.2 **Design of Fully Soft-Switched Semi-Dual Active DC-DC Converter for Battery Charging Application** 69
Siva Prabhakar, Shiladri Chakraborty, Sandeep Anand
Indian Institute of Technology Bombay, India

PRESENTATION   TOPIC: **Hard- and Soft-Switched**

T02.3 **A ZCS-ZVS Strategy for Low Impedance Dual Active Bridges in MHz Range** 77
Pushkar Saraf, Michael Solomentsev, Alex Hanson
The University of Texas at Austin, United States

PRESENTATION   TOPIC: Bidirectional DC-DC Converters

T02.4 **T-Type Multilevel DAB Converter with Reduced Backflow Power** .... N/A
Piyali Pal[1], Ranjan Kumar Behera[2], Khalifa Al Hosani[1], Utkal Ranjan Muduli[1]
[1]Khalifa University, U.A.E.; [2]Indian Institute of Technology Patna, India

PRESENTATION   TOPIC: Bidirectional DC-DC Converters

T02.5 **A 6.6 kW Highly Efficient Reconfigurable Dual Active Bridge Converter Designed using Planar Transformer, SiC-Fets and Monolithic Bidirectional Devices** 90
Reza Barzegarkhoo, Fabian Groon, Arkadeb Sengupta, Marco Liserre
Christian-Albrechts-Universität zu Kiel, Germany

PRESENTATION   TOPIC: Bidirectional DC-DC Converters

T02.6 **Interleaved Switched-Inductor-Based SIPO Partial Power Converter Module for Battery Management Systems** 98
Fengwang Lu, Henry Shu-Hung Chung
City University of Hong Kong, China

PRESENTATION   TOPIC: Bidirectional DC-DC Converters

T02.7 **Single Sensor-Based Fault Localization and Detection in GaN Three-Phase Dual Active Bridge Converters** 103
Satyam Sa, Yi Han, Cheng Feng Wang, Olivier Trescases
University of Toronto, Canada

PRESENTATION   TOPIC: Bidirectional DC-DC Converters

T02.8 **Enhanced Cocharge Operation Scheme in Bidirectional PhaseShift Full-Bridge Converters with Eliminated Voltage Overshoot and Reduced Freewheeling Current** 111
Tien-Sheng Li, Minh Ngo, Rolando Burgos, Dong Dong
Virginia Polytechnic Institute and State University, United States

PRESENTATION   TOPIC: Bidirectional DC-DC Converters

T02.9 **DC Bias Elimination in Isolated DC-DC Converters using Fundamental-Frequency Ripple** 118
Arkadeb Sengupta, Thiago Antonio Pereira, Marco Liserre
Christian-Albrechts-Universität zu Kiel, Germany

PRESENTATION   TOPIC: Bidirectional DC-DC Converters

SESSION T03: Single and Multiphase Inverters

8:30 - 12:00
TRACK: DC-AC Inverters

SESSION CHAIR(S)
Woongkul Lee, *Purdue University*
Laszlo Huber, *Delta Electronics, Inc.*

T03.1 Tunable Matching Network with Dual Phase-Switched Impedance Modulation Actuators 124
Alexander Jurkov[1], David Perreault[2]
[1]MKS Instruments, Inc., United States; [2]Massachusetts Institute of Technology, United States

PRESENTATION   **TOPIC: Single and Multi-Phase Inverters**

T03.2 Soft-Switched Pulsed Bias Plasma Supply System 132
Julia Estrin[1], Alexander Jurkov[1,2], David J. Perreault[1]
[1]Massachusetts Institute of Technology, United States; [2]MKS Instruments, Inc., United States

PRESENTATION   **TOPIC: Single and Multi-Phase Inverters**

T03.3 Analysis and Design of a Cyclo-Active-Bridge Inverter for Single-Stage Three-Phase Grid Interface 139
Mian Liao[1], Tanuj Sen[1], Yang Wu[2], Minjie Chen[1]
[1]Princeton University, United States; [2]ABB Corporate Research, Sweden

PRESENTATION   **TOPIC: Single and Multi-Phase Inverters**

T03.4 Modular Nanosecond Pulse Generator Leveraging GaN and SiC for Versatility and Performance 147
P. Briz, H. Sarnago, O. Lucía
Universidad de Zaragoza, Spain

PRESENTATION   **TOPIC: Multilevel Inverters**

T03.5 A Variable Frequency Technique for EMI and Efficiency Improvements in High-Level Count Flying Capacitor Multilevel Converters 151
Francesca Giardine, Sahana Krishnan, Logan Horowitz, Robert C. N. Pilawa-Podgurski
University of California, Berkeley, United States

PRESENTATION   **TOPIC: Power Quality and EMI**

T03.6 Analysis and Implementation of Minimum-Sensor Capacitor Voltage Estimators for Flying Capacitor Multilevel Converters 157
S. Tahmid Mahbub, Rahul K. Iyer, Ivan Z. Petrić, Robert C. N. Pilawa-Podgurski
University of California, Berkeley, United States

PRESENTATION   **TOPIC: Multilevel Inverters**

T03.7 Single-Stage Bidirectional High-Frequency Link DC to Three-Phase AC (4-Wire) Grid-Tied Microinverter 164
Aniruddh Marellapudi[1], Satish Belkhode[2], Joseph Benzaquen Sune[1], Deepak Divan[1]
[1]Georgia Institute of Technology, United States; [2]Indian Institute of Technology Roorkee, India

PRESENTATION   **TOPIC: Single and Multi-Phase Inverters**

T03.8 **Analysis and Design of a Constant Current LCC Class-E Inverter** 171
Ju Gao, Ziheng Liu, Jiayin He, Hongjie Peng, Chengkang Ao, Jinyan Wang
Peking University, China

PRESENTATION

TOPIC: **Single and Multi-Phase Inverters**

T03.9 **Series Connected Class-E Push-Pull Converters using GaN HEMT for High-Efficiency RF Generators in Float Zone Silicon Production** 175
Faheem Ahmad, Thore Stig Aunsborg, Jannick Kjær Jørgensen, Stig Munk-Nielsen
Aalborg University, Denmark

PRESENTATION

TOPIC: **Single and Multi-Phase Inverters**

SESSION T04: GaN Power Devices

8:30 - 12:00
TRACK: Devices and Components

SESSION CHAIR(S)
Joseph Kozak, *Johns Hopkins University Applied Physics Laboratory*
Aaron Brovont, *Purdue*

T04.1 **State of the Art 1.7kV Lateral GaN HEMTs, an Alternative to SiC** 180
Karthick Murukesan, Robert Yang, Kamal Varadarajan, Sorin Georgescu, Doug Kang
Power Integrations, United States

PRESENTATION

TOPIC: **GaN and SiC Devices and Modules**

T04.2 **Modeling and Characterization of Current and Future 1.2 kV Wide Bandgap Semiconductor-Based MOSFETs** 185
Sushanta Gautam[1], Austin M. Szczublewski[1], Samuel K. Atwimah[1], Aidan P. Fox[1], William M. Collings[1],
Tolen Nelson[1], Daniel G. Georgiev[1], Raghav Khanna[1], Andrew D. Koehler[2], Karl D. Hobart[2]
[1]The University of Toledo, United States; [2]U.S. Naval Research Laboratory, United States

PRESENTATION

TOPIC: **GaN and SiC Devices and Modules**

T04.3 **2.5-kV 6.4-ns 100-kHz Repetitive GaN Marx Generator** 192
Ruize Sun, Ci Pan, Wanjun Chen, Bo Zhang
University of Electronic Science and Technology of China, China

PRESENTATION

TOPIC: **GaN and SiC Devices and Modules**

T04.4 **Novel Dual Output LDO Architecture in 650-V GaN Technology for Power ICs** 195
Plinio Bau[1], Thanh Hai Phung[1], Deniz Aygun[2], Bart Coomans[2], Mike Wens[2]
[1]Wise-integration, France; [2]MinDCET NV, Belgium

PRESENTATION

TOPIC: **GaN and SiC Devices and Modules**

T04.5 **Impact of Substrate Bias on the Stability of Bidirectional GaN HEMT in Hard- and Soft-Switching** 202
Qihao Song[1], Hongchang Cui[1], Qiang Li[1], Yuhao Zhang[2]
[1]Virginia Polytechnic Institute and State University, United States; [2]The University of Hong Kong, China

PRESENTATION

TOPIC: **GaN and SiC Devices and Modules**

T04.6 Characterization of LED Driven GaN-Based Photoconductive Switches 207

Samuel K. Atwimah[1], Tolen M. Nelson[1], Geoffrey M. Foster[2], Daniel G. Georgiev[1], Andrew D. Koehler[2], Alan G. Jacobs[2], Karl D. Hobart[2], Micheal R. Hontz[3], Raghav Khanna[1]
[1]The University of Toledo, United States; [2]U.S. Naval Research Laboratory, United States; [3]Naval Surface Warfare Center, United States

PRESENTATION   TOPIC: GaN and SiC Devices and Modules

T04.7 Development and Validation of Repetitive Transient Gate Overvoltage Rating for GaN HEMTs 214

Ricardo Garcia, Angel Espinoza, Siddhesh Gajare, Shengke Zhang
Efficient Power Conversion Corporation, United States

PRESENTATION   TOPIC: GaN and SiC Devices and Modules

T04.8 Junction Temperature Monitoring of GaN HEMT by using On-Resistance with Voltage Clamp and Current Shunt 219

Xiao Wang, Mingrui Zou, Jiakun Gong, Yulei Wang, Zheng Zeng
Chonqqing University, China

PRESENTATION   TOPIC: GaN and SiC Devices and Modules

T04.9 False Turn-On Failure and Protection of p-Gate GaN HEMT in MHz Class-E Resonant Inverter 225

Ziheng Liu, Ju Gao, Hongjie Peng, Jiayin He, Jinyan Wang, Maojun Wang
Peking University, China

PRESENTATION   TOPIC: GaN and SiC Devices and Modules

SESSION T05: Magnetics Applications I

8:30 - 12:00
TRACK: Magnetics

SESSION CHAIR(S)
Matt Wilkowski, Wurth Elektronik
George Slama, Wurth Elektronik

T05.1 Heat Extraction from Ferrite Cores using Metallic Laminations 231

Alyssa Brown, Duy T. Nguyen, Alex J. Hanson
The University of Texas at Austin, United States

PRESENTATION   TOPIC: Advanced Magnetic Materials and Geometries

T05.2 Folded Flex-PCB Winding Planar Transformer for High-Frequency Isolated DC-DC Converters 238

Soundhariya G. Soundararajan, Hans Wouters, Wout Vanderwegen, Wilmar Martinez
KU Leuven, Belgium

PRESENTATION   TOPIC: Winding Techniques

T05.3 Winding Strategy Analysis and Optimization for High-Current Matrix Transformer 246
Bima Nugraha Sanusi, Pinhe Wang, Michael A. E. Andersen, Ziwei Ouyang
Technical University of Denmark, Denmark

PRESENTATION

TOPIC: **Winding Techniques**

**T05.4 Investigation on Impact of Transformer Parasitic Capacitance on
Standby Power Consumption in Power Converters** 252
Kamran Kamran[1,2], Andrea Russo[1], Federica Cammarata[1], Claudia Malannino[1],
S. Yuri Ciardo[1], Ziwei Ouyang[2]
[1]STMicroelectronics, Italy; [2]Technical University of Denmark, Denmark

PRESENTATION

TOPIC: **Winding Techniques**

**T05.5 PCB-Winding Integrated Transformer for 800-V Dual Active Bridge Converter using
1.2-kV GaN Devices** 258
Hans Wouters, Wei-Ren Lin, Nicolas Pirson, Thomas Jochmans, Yu Zuo, Wilmar Martinez
KU Leuven, Belgium

PRESENTATION

TOPIC: **High-Frequency Magnetics**

**T05.6 Comparative Assessment of Inductance Modeling for PCB-Based Circular
Spiral Coils in Inductive Power Transfer Systems** 266
Gaia Petrillo, Drazen Dujic
École Polytechnique Fédérale de Lausanne, Switzerland

PRESENTATION

TOPIC: **Winding Techniques**

T05.7 Compact Air-Core Inductors for Variable Frequency Soft-Switching in 3 Phase Inverters 272
Youssef A. Fahmy, Matthias Preindl
Columbia University, United States

PRESENTATION

TOPIC: **High-Frequency Magnetics**

**T05.8 Simulation and Experimental Research on Cooling Performance of Fully-Immersed
Evaporative Cooling High-Frequency Transformer** 278
Zhanlei Liu[1], Lingyu Zhu[1], Yuntian Gao[1], Yongliang Dang[1], Cao Zhan[2], Shengchang Ji[1]
[1]Xi'an Jiaotong University, China; [2]Virginia Polytechnic Institute and State University, United States

PRESENTATION

TOPIC: **High-Frequency Magnetics**

T05.9 High-Efficiency PCB-Embeddable Inductor for Vertical Power IVR Applications 285
Youssef Kandeel[1], Liang Ye[1], John Flannery[1], Cian Ó Mathúna[1], Ranajit Sai[1], Seamus O'Driscoll[1],
Takayuki Tsuchida[2], Naoya Terauchi[2], Sumiaki Kishimoto[2], Toshio Hiraoka[2], Masanori Nagano[2]
[1]Tyndall National Institute, Ireland; [2]Taiyo Yuden Co., Ltd., Japan

PRESENTATION

TOPIC: **High-Frequency Magnetics**

SESSION T06: Control of Power Electronic Converters I

8:30 - 12:00
TRACK: Control

SESSION CHAIR(S)
Masoud Karimi, *Mississippi State University*
Misha Kumar, *Delta Electronics, Inc.*

T06.1   **An Adaptive Zero Current Switching Control Technique for Multi-Resonant Switched-Capacitor Converters** 291
Haifah B. Sambo, Rose A. Abramson, Sahana Krishnan, Robert C. N. Pilawa-Podgurski
University of California, Berkeley, United States

PRESENTATION   TOPIC: Control of Power Electronic Converters

T06.2   **Small-Signal Analysis and External Ramp Design for Multiphase Current-Mode Constant On-Time Control with Phase Overlapping** 299
Sundaramoorthy Sridhar, Qiang Li
Virginia Polytechnic Institute and State University, United States

PRESENTATION   TOPIC: Current-Mode and Voltage-Mode Control

T06.3   **Multiphase Constant-on-Time Minimum-Deviation Controller for Modern Processors** 307
Duo Li[1], Gianluca Roberts[1], Aleksandar Prodić[1], Alan Wu[2]
[1]*University of Toronto, Canada;* [2]*Intel Corporation, Canada*

PRESENTATION   TOPIC: Current-Mode and Voltage-Mode Control

T06.4   **Closed-Loop Control of a Dual-Side Series/Parallel Piezoelectric-Resonator-Based DC-DC Converter** 315
Wen-Chin B. Liu[1], Gaël Pillonnet[2], Patrick P. Mercier[1]
[1]*University of California, San Diego, United States;* [2]*CEA-Leti, Université Grenoble Alpes, France*

PRESENTATION   TOPIC: Control of Power Electronic Converters

T06.5   **High-Bandwidth Embedded Rogowski Coil on Multilayer Substrate with Minimal Contribution to Power Loop Inductance** 321
Takahiro Okamoto, Masataka Ishihara, Kazuhiro Umetani, Eiji Hiraki
Okayama University, Japan

PRESENTATION   TOPIC: Sensor and Sensor-Less Control

T06.6   **Operating and Switching Frequency Circulating Current Control in Paralleled High Power Adjustable Speed Drives with Common DC Link** 327
Kevin Lee[1], Zhihao Song[2], Wenxi Yao[2], Bo Wei[2]
[1]*Eaton Corporation, United States;* [2]*Zhejiang University, China*

PRESENTATION   TOPIC: Control of Power Electronic Converters

T06.7   **Mixed-Signal Sliding Mode Controller for Non-Inverting Buck-Boost Photovoltaic DC Optimizers** 334
Anurag Singh, Sayan Paul, Dragan Maksimović, Luca Corradini
University of Colorado Boulder, United States

PRESENTATION   TOPIC: Digital Control-MCUs, DSPs, FPGAs, ASICs

T06.8 A Current Sensorless Output Voltage Tracking Controller-Observer for a
Boost Inverter using Feedback Linearization 342
Ion Leandro Dos Santos, Tailan Orlando, Yohannes Amilcar Tekle Scherer,
Telles Brunelli Lazzarin, Hector Bessa Silveira
Universidade Federal de Santa Catarina, Brazil

PRESENTATION   TOPIC: **Control of Power Electronic Converters**

T06.9 Modeling and Control of a Cyclo-Active-Bridge Inverter for Single-Stage
Three-Phase Grid Interface 349
Tanuj Sen[1], Mian Liao[1], Yang Wu[2], Minjie Chen[1]
[1]Princeton University, United States; [2]ABB Corporate Research, Sweden

PRESENTATION   TOPIC: **Control of Power Electronic Converters**

SESSION T07: Modeling and Simulation I

8:30 - 12:00
TRACK: Modeling and Simulation

SESSION CHAIR(S)
Adam Skorek, *University of Québec at Trois-Rivières*
Yunting Liu, *Pennsylvania State University*

T07.1 Turn-On Transient Modeling of 10 kV SiC MOSFET Half-Bridge Power Module in LTspice 357
Nianzun Qi, Jannick Kjær Jørgensen, Gao Liu, Zhixing Yan, Morten Rahr Nielsen, Asger Bjørn
Jørgensen, Hongbo Zhao, Stig Munk-Nielsen
Aalborg University, Denmark

PRESENTATION   TOPIC: **Device and Component Modeling**

T07.2 A Compact, Automated Sawyer-Tower System for Characterization of the
High-Frequency, Soft-Switching C_{oss} Loss of Wide Bandgap Devices 363
Katherine Liang, Malachi Hornbuckle, Juan Rivas-Davila
Stanford University, United States

PRESENTATION   TOPIC: **Parasitics Extraction and Optimization**

T07.3 Enhancing Behind-the-Meter Visibility of Grid Edge PV Systems and Electric Vehicle
Charging Loads Through Integration of Compact Low-Cost Sensors 370
Mehrnaz Madadi[1], Paul Ohodnicki[2], Subhashish Bhattacharya[1]
[1]North Carolina State University, United States; [2]University of Pittsburgh, United States

PRESENTATION   TOPIC: **Hardware-in-the-Loop and Rapid Prototyping**

T07.4 Supercapacitor based TMS Pulse Generator Design-Experimental Results
Versus MATLAB MOSFET Simulation Model 378
Soniya Raju, Nihal Kularatna, Marcus Wilson, Alistair Steyn-Ross
University of Waikato, New Zealand

PRESENTATION   TOPIC: **Device and Component Modeling**

T07.5 **Application of Artificial Intelligence for Modeling SiC Power MOSFETs** 385
Fredo Chavez, Danial Bavi, Sourabh Khandelwal
Macquarie University, Australia

> PRESENTATION

TOPIC: **Device and Component Modeling**

T07.6 **Multi-Objective Design Automation in Power Electronics using Bayesian Optimization Techniques** 389
Tung-Tan Nguyen, Man-Hay Pong, Huang-Jen Chiu
National Taiwan University of Science and Technology, Taiwan

> PRESENTATION

TOPIC: **Parasitics Extraction and Optimization**

T07.7 **Reduced Order Thermal Modelling of Multi-Chip Silicon Carbide Power Modules** 395
Aamir Rafiq[1], Blake Nelson[2], Marshal Olimmah[2]
[1]*Wolfspeed, Inc., United Kingdom;* [2]*Wolfspeed, Inc., United States*

> PRESENTATION

TOPIC: **Device and Component Modeling**

T07.8 **Design and Evaluation of Dual-Resolver Emulation for Control System Verification in Aerospace Actuation Applications** 401
Tomas Sadilek, Julian Opificius, Jason Wright, Alec Leslie, Jeremie Tuzizila, Cesar Alzate, Hunter Burnett, Joshua Atkinson, Justin Stricula
Curtiss-Wright Corporation, United States

> PRESENTATION

TOPIC: **Hardware-in-the-Loop and Rapid Prototyping**

T07.9 **Un-Terminated Blackbox Modeling for Electric Machines** 409
Xinliang Yang, Vladimir Mitrovic, Qing Lin, Rolando Burgos
Virginia Polytechnic Institute and State University, United States

> PRESENTATION

TOPIC: **Circuits and Systems**

SESSION T08: EV Charging Applications

8:30 - 12:00
TRACK: Transportation Power Electronics

SESSION CHAIR(S)
Dong Cao, *University of Dayton*
Sheldon Williamson, *Ontario Tech University*

T08.1 **7.2 kW GaN-Based DAB Converter with 37 kW/L Power Density and High Efficiency** 416
Esmaeil Jalalabadi[1], Xiaoyu Wang[1], Jaksa Rubinic[2], Yang Jiao[2], Lucas Lu[2]
[1]*Carleton University, Canada;* [2]*Infineon Technologies AG, Canada*

> PRESENTATION

TOPIC: **On-Board and Off-Board Charging Systems**

T08.2 **A Novel Interleaving Method for High Power Integrated Electric Vehicle Charger with Three-Phase Permanent Magnet Synchronous Motor** 423
Ryota Tanaka[1], Toshihiro Kai[1], Kenta Takishima[1], Yoshiyuki Nagai[1], Tetsuya Hayashi[1], Kantaro Yoshimoto[2]
[1]*Nissan Motor Co., Ltd., Japan;* [2]*Tokyo Denki University, Japan*

> PRESENTATION

TOPIC: **Power Electronics for Hybrid and Electric Vehicles**

T08.3 **A Three-Phase CLLC Resonant Converter with Integrated Planar Magnetics for 22-kW On-Board Chargers** 429

Tianlong Yuan[1], Zhangwei Xiang[1], Abdelrahman Ali[1], Feng Jin[2], Qiang Li[1], Wendell Da-Cunha-Alves[3], Xiaoshan Liu[3]
[1]Virginia Polytechnic Institute and State University, United States;
[2]Delta Electronics, Inc., United States; [3]Valeo, France

PRESENTATION

TOPIC: On-Board and Off-Board Charging Systems

T08.4 **Reconfigurable LLC Resonant Converter for Wide Voltage Range and Reduced Voltage Stress in DC-Connected EV Charging Stations** 436

Yu Zuo, Xiaobing Shen, Bangli Du, Qingcheng Sui, Tim Geboers, Wilmar Martinez
KU Leuven, Belgium

PRESENTATION

TOPIC: Power Electronics for Hybrid and Electric Vehicles

T08.5 **Design and Control of GaN based Three-Phase / Single-Phase Combo Three-Level Flying Capacitor PFC for OBC Applications** 442

Nidhi Haryani, Laszlo Huber, Anup Anurag, Juan Ruiz, Peter Barbosa
Delta Electronics, Inc., United States

PRESENTATION

TOPIC: On-Board and Off-Board Charging Systems

T08.6 **Optimization Strategy for Battery Electric Vehicle (BEV) DC Fast Charging (FC) in Cold Environments** 449

Seif Sarofim, Cheng Feng Wang, Satyam Sa, Avram Kachura, Isaac Muscat, Olivier Trescases
University of Toronto, Canada

PRESENTATION

TOPIC: On-Board and Off-Board Charging Systems

T08.7 **DC-Link Voltage Reduction with Synergetic Common-Mode Voltage Control of Single-Phase Two-Stage Non-Isolated EV Chargers** 457

Dongsu Lee, Juwon Lee, Jung-Ik Ha
Seoul National University, Korea

PRESENTATION

TOPIC: On-Board and Off-Board Charging Systems

T08.8 **DC-DC Converter Architecture for Fast Electric Vehicle (EV) Battery Charging Applications** 464

Shibaji Basu, Arjun Ivimey, Praveen Jain
Queen's University, Canada

PRESENTATION

TOPIC: On-Board and Off-Board Charging Systems

T08.9 **Fast Simulator for the Estimation of Inverter DC-Link Temperature in e-Drives Subjected to Highly Variable Working Cycles** 472

Simone Giuffrida, Fabio Mandrile, Radu Bojoi
Politecnico di Torino, Italy

PRESENTATION

TOPIC: Power Electronics for Hybrid and Electric Vehicles

Wednesday, March 19, 2025

SESSION T09: Point-of-Load Converters

8:30 - 12:00
TRACK: DC-DC Converters

SESSION CHAIR(S)
Peyman Asadi, *Renesas Electronics*
Cahit Gezgin, *Infineon Technologies*

T09.1 A Monolithic Regulated 160 MHz Resonant DC-DC Converter 479
Giacomo Ripamonti[1], Stefano Michelis[1], Georgios Bantemits[1], Pablo Daniel Antoszczuk[1], Khalil Khalife[1],
Nils Hans Van Der Blij[1], Sokratis Koseoglou[1], Mattia Balutto[2], Francesco Driussi[3], Stefano Saggini[3]
*[1]CERN - European Organization for Nuclear Research, Switzerland; [2]Politecnico di Milano, Italy;
[3]Università degli Studi di Udine, Italy*

PRESENTATION

TOPIC: **Resonant Converters**

**T09.2 Reconfigurable Trans-Inductor Voltage Regulator with Improved Light Load
Efficiency in Data Center Applications** 485
Ziyao Wang, Zehui Li, Haoyu Wang
Shanghai Tech University, China

PRESENTATION

TOPIC: **Point-of-Load (POL) and Multi-Phase Converters**

**T09.3 Fully Integrated Voltage Regulators (FIVRs) with Package In-Situ Coupled
CoaxMIL Inductor for High Power Density Microprocessor Applications** 491
Jaeil Baek[1,2], Beomseok Choi[2], Siddharth Kulasekaran[2], Huong Do[2], Brandon Marin[2],
Jose Chavarria[2], Leigh Wojewoda[2], Kaladhar Radhakrishnan[2]
[1]Korea Advanced Institute of Science & Technology, Korea; [2]Intel Corporation, United States

PRESENTATION

TOPIC: **Point-of-Load (POL) and Multi-Phase Converters**

**T09.4 Multiphase Lateral Flux Indirect Coupled Inductor for Vertical Power
Delivery Voltage Regulator Module** 498
Adhistira M. Naradhipa[1], Qiong Wang[2], Qiang Li[1]
[1]Virginia Polytechnic Institute and State University, United States; [2]Google LLC, United States

PRESENTATION

TOPIC: **Point-of-Load (POL) and Multi-Phase Converters**

T09.5 A High Density Three-Level Quadratic Buck Hybrid Converter for 48V-to-PoL Conversion 505
Kejia Wang[1], Si Yuan Sim[1], Yin Quen Choong[1], Xin Zhang[2], Sriharsh Pakala[3], Cheng Huang[1]
*[1]Iowa State University, United States; [2]IBM T. J. Watson Research Lab, United States; [3]NXP
Semiconductors, United States*

PRESENTATION

TOPIC: **Point-of-Load (POL) and Multi-Phase Converters**

T09.6 Air-LEGO: A Magnetic-Free Ultra-Thin 24V-to-1V 120A VRM with Air-Coupled Inductors 510
Haoran Li[1], Wenliang Zeng[1], Youssef Elasser[2], Minjie Chen[1]
[1]Princeton University, United States, [2]NVIDIA Corporation, United States

PRESENTATION

TOPIC: **Voltage Regulator Modules (VRM)**

T09.7 A 15A 48V-Input Dual-Path Hybrid Dickson Converter with 6 mm³ Low Saturation Current Inductors for Point-of-Load Conversion 518

Hua Chen[1], Young-Seok Noh[1], Minxiang Gong[2], Vivek De[2], Arijit Raychowdhury[1]
[1]Georgia Institute of Technology, United States; [2]Intel Corporation, United States

PRESENTATION

TOPIC: Point-of-Load (POL) and Multi-Phase Converters

T09.8 An Ultra-Fast Control Strategy and Pre-Current-Balancing Measures Prepared for Rapid Transients in Constant On-Time Controllers 524

Yijie Qian, Yuan Gao, Wenze Shu, Lingyun Li, Shen Xu, Weifeng Sun
Southeast University, China

PRESENTATION

TOPIC: Voltage Regulator Modules (VRM)

T09.9 Loosely Coupled Trans-Inductor Voltage Regulator (LC-TLVR) Inductor as Compensation Inductor (Lc) 530

Pavan Kumar[1], Arturo Sanchez Hernandez[2]
[1]Intel Corporation, United States; [2]Intel Corporation, Mexico

PRESENTATION

TOPIC: Point-of-Load (POL) and Multi-Phase Converters

SESSION T10: Magnetics Modelling

8:30 - 12:00
TRACK: Magnetics

SESSION CHAIR(S)
Adam Skorek, *University of Québec at Trois-Rivières*
Matt Wilkowski, *Wurth Elektronik*

T10.1 Novel Complex Permeability Model of Powder Magnetic Materials 538

Lukas Mueller[1], James Cox[2], Jun Wang[1], Enrique Garcia[2]
[1]University of Nebraska-Lincoln, United States; [2]Micrometals, Inc, United States

PRESENTATION

TOPIC: Magnetics Modeling and Simulations

T10.2 Design Study Evaluating Impact of Gap Loss on Nanocrystalline Inductor Cores with Experimental Validation 544

Maurice Sturdivant, Brandon Grainger, Christopher Bracken, Paul R. Ohodnicki
University of Pittsburgh, United States

PRESENTATION

TOPIC: Magnetics Modeling and Simulations

T10.3 A Permanent Magnet Variable Inductor for DC Fault Current Limiting Applications 552

Mark Nations, Subhashish Bhattacharya
North Carolina State University, United States

PRESENTATION

TOPIC: Magnetics Modeling and Simulations

T10.4 **Design-Oriented Modeling and Multi-Objective Optimization of Two-Phase Coupled Inductors in Multiphase PWM Converters** 558
Yicheng Zhu, Jiarui Zou, Robert C. N. Pilawa-Podgurski
University of California, Berkeley, United States

PRESENTATION
TOPIC: **Magnetics Modeling and Simulations**

T10.5 **MagNetX: Extending the Magnet Database for Modeling Power Magnetics in Transient** 566
Hyukjae Kwon[1], Shukai Wang[1], Haoran Li[1], Youssef Elasser[2], Gyeong-Gu Kang[1], Daniel Zhou[1], Davit Grigoryan[1], Minjie Chen[1]
[1]Princeton University, United States; [2]NVIDIA Corporation, United States

PRESENTATION
TOPIC: **Magnetics Modeling and Simulations**

T10.6 **Non-Monotonic Influence of DC Bias on Ferrite Core Loss Up to 10 MHz with Sine Wave Excitation** 573
Bohua Zhang, Martin Pfost
Technical University Dortmund, Germany

PRESENTATION
TOPIC: **High-Frequency Magnetics**

T10.7 **Comprehensive SPICE Model for Inductors Considering Magnetic Losses Under DC Bias Current** 579
Yuki Sato[1], Hirokazu Matsumoto[1], Junichi Kotani[2], Shohei Tomioka[2], Kenichiro Tanaka[2]
[1]Aoyama Gakuin University, Japan; [2]Panasonic Industry Co., Ltd., Japan

PRESENTATION
TOPIC: **Magnetics Modeling and Simulations**

T10.8 **Indented Core to Reduce and Desensitize Inductor's Fringing Losses without Increasing Volume** 586
Rajaie Nassar, Promit Datta, Guo-Quan Lu, Christina DiMarino, Khai Ngo
Virginia Polytechnic Institute and State University, United States

PRESENTATION
TOPIC: **Advanced Magnetic Materials and Geometries**

T10.9 **Coupled Inductor Analysis and Finite Element Modeling Assisted Design for Boost Extender Topology** 594
Vikas Kumar Rathore, Michael Evzelman, Mor Mordechai Peretz
Ben-Gurion University of the Negev, Israel

PRESENTATION
TOPIC: **Magnetics Modeling and Simulations**

SESSION T11: Power Conversion for Microgrids

8:30 - 12:00
TRACK: Power Electronics for Utility Applications

SESSION CHAIR(S)
Ali Khajehoddin, *University of Alberta*
Jacob Mueller, *Sandia National Laboratories*

T11.1 Stability Analysis of Current-Limited Grid-Forming Inverters with Frequency Stabilization: An Equivalent Impedance Approach 602
Bowen Yang, Gab-Su Seo
National Renewable Energy Laboratory, United States

PRESENTATION   **TOPIC: Microgrid Systems**

T11.2 Revisit Active Power Oscillation in Multi-Virtual Synchronous Generators Gride 609
Junjie Xiao, Pavol Bauer, Zian Qin
Delft University of Technology, Netherlands

PRESENTATION   **TOPIC: Microgrid Systems**

T11.3 A Novel Current Control Technique for Off-Grid Single-Phase Inverters 616
Arpan Laha, Abirami Kalathy, Praveen Jain, Majid Pahlevani
Queen's University, Canada

PRESENTATION   **TOPIC: Microgrid Systems**

T11.4 Intelligent Low-Bandwidth Frequency Controller for VSGs at Economic Dispatch in Islanded Microgrid 622
Shraf Eldin Sati, Ahmed Al-Durra, Hatem H. Zeineldin, Tarek H.M. El-Fouly, Ehab F. El-Saadany
Khalifa University of Science and Technology, U.A.E.

PRESENTATION   **TOPIC: Microgrid Systems**

T11.5 Hardware-in-the-Loop of a Grid Forming Control Strategy Applied to a DC Off-Grid Green Hydrogen Production System 629
Diego Montoya-Acevedo[1], René Contreras-Barrios[1], Ángel Maureira-Riquelme[1], Esteban Ibáñez-Muñoz[1], Catalina Gonzalez-Castaño[2], Carlos Restrepo[1]
[1]Universidad de Talca, Chile; [2]Universidad Andrés Bello, Chile

PRESENTATION   **TOPIC: Microgrid Systems**

T11.6 Experimental Validation of a 40kW, 480V Point-to-Point DC Interlinks for Controller-Agnostic, Interoperable Networked Microgrids 637
Maximiliano Ferrari, Michael Starke, John Smith, Joao Pereira, Misael Montejano
Oak Ridge National Laboratory, United States

PRESENTATION   **TOPIC: Microgrid Systems**

T11.7 Andronov-Hopf Oscillator-Based Grid-Forming Converters with Embedded Disturbance Rejection for Non-Ideal Loading Condition 645
Vikram Roy Chowdhury, Gab-Su Seo, Barry Mather
National Renewable Energy Laboratory, United States

PRESENTATION   **TOPIC: Microgrid Systems**

T11.8 Estimation of Rectifier Output Current of the LLC Converter 651
Xin Wu, Yi Zhou, Haihong Long, Dehong Xu
Zhejiang University, China

PRESENTATION

TOPIC: Solid-State Transformers

T11.9 A 100kHz Digitally Controlled 10kW, 2-Channel Solar MPPT Converter using 3-Level
Topology with >75W/in^3 Power Density and >98.5% Peak Efficiency 658
Ranajay Mallik, Akshat Jain
STMicroelectronics, India

PRESENTATION

TOPIC: Energy Storage Systems

SESSION T12: Control of Power Electronic Converters II

8:30 - 12:00
TRACK: Control

SESSION CHAIR(S)
Jaber Abu Qahouq, *The University of Alabama (UA)*
Paolo Mattavelli, *University of Padua*

T12.1 A Bootstrapless KY-S-Hybrid Buck-Boost Converter with Full Range iLs Reduction
and 400% Line Transient Response Acceleration for AI-Mobile Application 664
Chuan-En Chang, Cheng-Ta Chuang, Hao-Ran Huang, Chieh-Ju Tsai, Ching-Jan Chen
National Taiwan University, Taiwan

PRESENTATION

TOPIC: Control ICs

T12.2 Digital Control of a 600-V to 28-V 20-kW Two-Stage DC-DC Converter 670
Shreyas B. Shah[1], Rachit Pradhan[1], Jiaqi Yuan[1], Mohamed Ibrahim[2], Ahmed Elezab[2],
Samuel Hemming[1], Giorgio Pietrini[1], Piranavan Suntharalingam[2], Mario F. Cruz[2], Ali Emadi[1]
[1]McMaster University, Canada; [2]Eaton Corporation, United States

PRESENTATION

TOPIC: Digital Control-MCUs, DSPs, FPGAs, ASICs

T12.3 Self-Calibrated Digital Current Emulation for High-Frequency Hysteretic
Current-Mode Control in GaN PFC Converters 676
Mohammad Shawkat Zaman, Olivier Trescases
University of Toronto, Canada

PRESENTATION

TOPIC: Current-Mode and Voltage-Mode Control

T12.4 High-Frequency Flying Capacitor Four-Level Drain Supply Modulator 682
Audrey Cheshire, Paul Flaten, Zoya Popović, Dragan Maksimović
University of Colorado Boulder, United States

PRESENTATION

TOPIC: Digital Control-MCUs, DSPs, FPGAs, ASICs

T12.5 Discontinuous Modulation Strategy for Voltage and Temperature Balancing of MMCs 689
Davide D'Amato[1], Stayner Nóbrega Barros[2], Jun-Hyung Jung[1], Marco Liserre[1]
[1]Fraunhofer Institute for Silicon Technology ISIT, Germany; [2]Christian-Albrechts-Universität zu Kiel, Germany

PRESENTATION

TOPIC: Control of Power Electronic Converters

T12.6 Damping Control and Improvement of Grid-Forming Inverter from a
Wideband Stability Perspective 696
Rui Kong, Subham Sahoo, Yubo Song, Frede Blaabjerg
Aalborg University, Denmark

PRESENTATION   TOPIC: Control of Power Electronic Converters

T12.7 A Grid-Forming Split-Phase Three-Leg Inverter with Unbalanced Loading and
Active Power Decoupling 703
Namwon Kim, Renata Kimpara, Michael Starke
Oak Ridge National Laboratory, United States

PRESENTATION   TOPIC: Control of Power Electronic Converters

T12.8 Completely Decentralized Active and Reactive Power Control of Grid-Connected
Cascaded H-Bridge Inverters with Integrated Battery Storage 711
Soham Dutta, Brian Johnson
The University of Texas at Austin, United States

PRESENTATION   TOPIC: Control of Power Electronic Converters

T12.9 Small-Signal Modeling and Damping Design of Unfolding-Based Single Stage
AC-DC Converter using the Extra Element Theorem 719
Dakota Goodrich, Aditya Zade, Shubhangi Gurudiwan, Mahmoud Mansour, Regan Zane, Hongjie Wang
Utah State University, United States

PRESENTATION   TOPIC: Control of Power Electronic Converters

SESSION T13: Renewable Energy Systems

8:30 - 12:00
TRACK: Renewable Energy Systems

SESSION CHAIR(S)
Yongheng Yang, *Zhejiang University*
Haoyu Wang, *Shanghai Tech University*

T13.1 Methods to Enhance Cybersecurity of Multiple Inverters in Large Grid Connected
PV / Battery Energy Storage Systems 727
Hasan Ibrahim, Jaewon Kim, Peng-Hao Huang, Vishwam Raval, Prasad Enjeti
Texas A&M University, United States

PRESENTATION   TOPIC: Photovoltaic (PV) Inverters and Micro Inverters

T13.2 Optimal DC-DC Converter Topology and Control Algorithm for Fuel Cell Electric
Vehicle with Series-Connected Supercapacitor 733
Hyeon Soo Kim, Yun Seong Hwang, Seung Hyun Kang, Man Jae Kwon, Byoung Kuk Lee
Sungkyunkwan University, Korea

PRESENTATION   TOPIC: Fuel Cells and Other Emerging Renewable Energy Systems

T13.3 **Reliability-Constrained Design of a High-Gain Power Optimizer based on a Real Mission Profile** 738

Stefano Cerutti[1], Francesco Iannuzzo[2], Ariya Sangwongwanich[2], Tamás Kerekes[2], Mario Giuseppe Pavone[3], Francesco Gennaro[3], Natale Aiello[3], Francesco Musolino[1], Paolo Stefano Crovetti[1]
[1]Politecnico di Torino, Italy; [2]Aalborg University, Denmark; [3]STMicroelectronics, Italy

PRESENTATION   TOPIC: **Photovoltaic (PV) Inverters and Micro Inverters**

T13.4 **Submodule Voltage Balancing Technique of Solar MMC for Firing the Switches using Integrated PWM Modules** 746

Ahmed Elsanabary[1], Saad Mekhilef[2], Mokhtar Aly[3], José Rodriguez[3]
[1]Port Said University, Egypt; [2]Swinburne University, Australia; [3]Universidad San Sebastián, Chile

PRESENTATION   TOPIC: **Photovoltaic (PV) Inverters and Micro Inverters**

T13.5 **Single-Stage High-Frequency-Link Split-Phase Microinverter with High Voltage Gain based on Buck-Boost AC Chopper** 751

Xuewen Li, Jia Liu, Jinjun Liu
Xi'an Jiaotong University, China

PRESENTATION   TOPIC: **Photovoltaic (PV) Inverters and Micro Inverters**

T13.6 **Fault Diagnosis and Tolerant Strategy for Triple-Port Hydrogen Converter using SSA-Optimized Random Forest Algorithm** 757

Shiqi Zhang[1], Yiyina Teng[1], Naizhe Diao[1], Xiaoqiang Guo[1], Vladimir Terzija[2], Lichong Wang[3]
[1]Yanshan University, China; [1]Newcastle University, United Kingdom; [3]Shenke Electronic Co., Ltd., China

PRESENTATION   TOPIC: **Fuel Cells and Other Emerging Renewable Energy Systems**

T13.7 **Resilient Operation for Grid-Connected Cascaded H-Bridge Multilevel Inverter with Improving PV Source Stress** 761

Jinli Zhu[1], Yuan Li[1], Hector Akuta[1], Jeonghun Kim[1], Uthandi Selvarasu[2], Shumeng Wang[2], Vikram Roy Chowdhury[3], Brad Lehman[2], Fang Z. Peng[1]
[1]University of Pittsburgh, United States; [2]Northeastern University, United States; [3]National Renewable Energy Laboratory, United States

PRESENTATION   TOPIC: **Photovoltaic (PV) Inverters and Micro Inverters**

T13.8 **A Medium Voltage Grid-Connected PV Inverter with a New Modular High Voltage Gain Converter Featuring Internal Modified Voltage Doubling Balancers** 768

Kajanan Kanathipan[1], Muhammad Ali Masood Cheema[2], John Lam[1]
[1]York University, Canada; [2]Northern Transformer Corporation, Canada

PRESENTATION   TOPIC: **Photovoltaic (PV) Inverters and Micro Inverters**

T13.9 **Split-Source Common-Ground Inverter for Photovoltaic Applications** 775

Mahmoud A. Gaafar[1], Mohamed Orabi[1], Samir Kouro[2], Ahmed Ibrahim[3], Eltaib Abdeen D. Ibrahim[4]
[1]Aswan University, Egypt; [2]Universidad Tecnica Federico Santa Maria, Chile; [3]Sohag University, Egypt; [4]High Institute for Engineering and Technology, Egypt

PRESENTATION   TOPIC: **Photovoltaic (PV) Inverters and Micro Inverters**

SESSION T14: Wireless Power Transfer: Design and Control

8:30 - 12:00
TRACK: Wireless Power Transfer

SESSION CHAIR(S)
Zeljko Pantic, *North Carolina State University*
Weijin Qiu, *John Deere*

T14.1 **Comprehensive Investigation and Proposal of a New Wireless Charging Road Structure using Low-Environmental-Impact Magnetic Concrete** 782
Shuntaro Inoue, Yuko Kano, Shin Tajima
Toyota Central R&D Labs., Inc., Japan

PRESENTATION   TOPIC: **Wireless Charging**

T14.2 **Design of a Bidirectional High Power Inductive Power Transfer System with Auxiliary Winding for Automotive Applications** 788
Luis Ruiz Chamorro, Nikola Mirković, Alberto Delgado Expósito, Pedro Alou Cervera, Miroslav Vasić
Universidad Politécnica de Madrid, Spain

PRESENTATION   TOPIC: **Wireless Charging**

T14.3 **Mutual Inductance and Load Identification Method based on the Voltage Transients of WPT Systems** 795
Xiaosheng Wang, C.Q. Jiang, Yibo Wang, Liping Mo
City University of Hong Kong, China

PRESENTATION   TOPIC: **Wireless Charging**

T14.4 **Digitally Controlled Misalignment-Tolerant Inductive Power Transfer System with Adaptive Hybrid Compensation for CC/CV Charging of E-Scooter** 801
Niranjan Shrestha, V.S.R.Varaprasad Oruganti, Sheldon Williamson
Ontario Tech University, Canada

PRESENTATION   TOPIC: **Wireless Charging**

T14.5 **On/Off Control of Modular Inductive Power Transfer System** 809
Kunxiao Zhou, Guangdong Ning, Heyuan Li, Xinlin Wang, Minfan Fu
Shanghai Tech University, China

PRESENTATION   TOPIC: **Wireless Charging**

T14.6 **Receiver Side Regulation of LCC Wireless Power Transfer System with Variable Notch Filter** 815
Hsin-Che Hsieh, Jih-Sheng Lai
Virginia Polytechnic Institute and State University, United States

PRESENTATION   TOPIC: **Wireless Charging**

T14.7 **84.7 Percent Peak Efficiency Stress Tolerant DC DC Buck Converter for Li Ion Battery Driven Standby Circuits in 18nm FDSOI** 821
Gautam Dey Kanungo[1], Pijush Kanti Panja[2], Vikas Bugade[1], Kallol Chatterjee[1]
[1]STMicroelectronics, India; [2]SiTime B.V., Netherlands

PRESENTATION   TOPIC: **Power for IoT**

T14.8   Leveraging Ultrasound and Neural Networks for Non-Invasive Power Converter Efficiency Estimation 828

Youssof Fassi[1], Vincent Heiries[1], Jérôme Boutet[1], Julien Marianne[2], Sébastien Martin[1], Mathilde Chareyron[1], Clément Chambon[1], Sébastien Boisseau[1]
[1]CEA-Leti, Université Grenoble-Alpes, France; [2]SERMA INGENIERIE, France

PRESENTATION

TOPIC: **Non-contact Sensors for Power Electronics**

T14.9   A Load-Independent Multi-Relays Wireless Power Transfer with Self-Regulation and Single Compensation Network 834

Jong-Hun Kim[1], Najam Ul Hassan[2], Seogyong Jeong[3], Myeong-Ho Kim[1], Min-Sik Kim[1], Jee-Hoon Jung[4], Byunghun Lee[5], Se-Un Shin[1]
[1]Pohang University of Science and Technology, Korea; [2]University of Michigan-Dearborn, United States; [3]Samsung Electronics, Korea; [4]Ulsan National Institute of Science and Technology, Korea; [5]Hanyang University, Korea

PRESENTATION

TOPIC: **Wireless Charging**

SESSION T15: Power Electronics Applications I

8:30 - 12:00
TRACK: Power Electronics Applications

SESSION CHAIR(S)
Jeffery Nilles, *Alpha&Omega Semiconductors*
Raj Kumar Kokkonda, *North Carolina State University*

T15.1   A GaN-Based Single-Stage Solid-State Transformer Replacement for 40 VA Class 2 Line-Frequency Transformers 840

Allen T. Nguyen, Charles R. Sullivan
Dartmouth College, United States

PRESENTATION

TOPIC: **AC-DC-AC Applications and Matrix Converters**

T15.2   Survey of Components and Topologies for High-Efficiency and High-Power Density 48V DC-DC Converters 848

Joseph Winkler, Niklas Deneke, Bernhard Wicht
Leibniz University Hannover, Germany

PRESENTATION

TOPIC: **Datacenter/Telecom Power Architecture and System Considerations**

T15.3   A Novel Solid-State Circuit Breaker using B-TRAN™ 854

Mudit Khanna, Ruiyang Yu, Milad Tayebi, Jiankang Bu, Jeffrey Knapp
Ideal Power Inc., United States

PRESENTATION

TOPIC: **Solid State and Hybrid Circuit Breakers**

T15.4   Development of a Supercritical Fluid-Insulated Fast Mechanical Switch for MVDC Hybrid Circuit Breakers 860

Zhiyang Jin[1], Qichen Yang[2], Alfonso Cruz[1], Lukas Graber[1]
[1]Georgia Institute of Technology, United States; [2]University of Central Florida, United States

PRESENTATION

TOPIC: **Solid State and Hybrid Circuit Breakers**

T15.5 Dynamic Impedance Matching for a Variable Reluctance Energy Harvesting Application with Constrained Space 868

Fernando Pérez, Alejandro Redondo, Airán Francés, Gabriel Mujica
Universidad Politécnica de Madrid, Spain

PRESENTATION

TOPIC: **Energy Harvesting**

T15.6 Renewable Energy-Powered DC-Converted Refrigerator based on a Supercapacitor-Assisted Technique 874

Nirashi Polwaththa Gallage[1], Nihal Kularatna[1], Alistair Steyn-Ross[1], Dulsha Kularatna-Abeywardana[2]
[1]University of Waikato, New Zealand; [2]University of Auckland, New Zealand

PRESENTATION

TOPIC: **Energy Harvesting**

T15.7 Design and Evaluation of Flexible Inductors for Wearable Power Electronics 880

Sean Logi, F. Selin Bagci, Katherine A. Kim
National Taiwan University, Taiwan

PRESENTATION

TOPIC: **Portable Power**

T15.8 Design of Boost Power Factor Corrector and Asymmetrical Half-Bridge Flyback Converter for USB-PD Applications 887

Yun-Keng Cheng[1], Tsorng-Juu Liang[1], Kai Hui Chen[1], Ming Chang Tsou[2]
[1]National Cheng Kung University, Taiwan; [2]Leadtrend Technology Corporation, Taiwan

PRESENTATION

TOPIC: **Portable Power**

T15.9 Computationally Efficient Current Sensorless Predictive Control for PMSM Drive Fed by a Matrix Converter with CMV-Free Operation 895

Ali Sarajian[1], Ibrahim Harbi[1], Quanxue Guan[2], Davood Arab Khaburi[3], Ralph Kennel[1], José Rodriguez[4], Patrick Wheeler[5], Mokhtar Aly[4]
[1]Technical University of Munich, Germany; [2]Sun Yat-Sen University, China; [3]Iran University of Science and Technology, Iran; [4]Universidad San Sebastián, Chile; [5]University of Nottingham, United Kingdom

PRESENTATION

TOPIC: **AC-DC-AC Applications and Matrix Converters**

SESSION T16: Motor Drives I

8:30 - 12:00
TRACK: Motor Drives

SESSION CHAIR(S)
Ali Safayet, *HL Mechatronics*
Ziaur Rahman, *Booz Allen Hamilton*

T16.1 PMSM Motor Drive with Current Direct Digital Control and Near 1st-Order Speed Control 900

Po-Chang Lee[1], Tsai-Fu Wu[2], Han Ku[2], Chien-Chih Hung[2], Jui-Yang Chiu[2]
[1]Delta Electronics Inc., Taiwan; [2]National Tsing Hua University, Taiwan

PRESENTATION

TOPIC: **AC, DC, BLDC Motor Drives**

T16.2 Fault-Tolerant Multilevel Converter for Multiphase Switched Reluctance Motor Drives based on q+2 Converter 906

Mahmoud A. Gaafar[1], Mohamed Orabi[1], Hao Chen[2], Mostafa Dardeer[1]
[1]Aswan University, Egypt; [2]China University of Mining and Technology, China

PRESENTATION

TOPIC: AC, DC, BLDC Motor Drives

T16.3 Uncertainty-Aware Artificial Intelligence for Gear Fault Diagnosis in Motor Drives 912

Subham Sahoo, Huai Wang, Frede Blaabjerg
Aalborg University, Denmark

PRESENTATION

TOPIC: AC, DC, BLDC Motor Drives

T16.4 Neural Network based Digital Twin Health Monitoring of BLDC Motor Drives for Robots 919

Mohamed Y. Metwly[1], Benjamin Luckett[2], Landon Clark[2], JiangBiao He[1], Biyun Xie[2]
[1]The University of Tennessee, United States; [2]University of Kentucky, United States

PRESENTATION

TOPIC: AC, DC, BLDC Motor Drives

T16.5 MTPA Control using Predictive P&O Method for Dual Parallel Surface-Mounted Permanent Magnet Synchronous Motor Drives Fed by a Single Inverter 925

Jae-Seong Kim, Kyo-Beum Lee
Ajou University, Korea

PRESENTATION

TOPIC: AC, DC, BLDC Motor Drives

T16.6 A Novel I-f Startup Strategy with Smooth Transition to Sensorless Control for CSI-Fed PMSM Drives used in Submersible Pumps 930

Milad Bahrami-Fard, Majid Ghasemi Korrani, Babak Fahimi
The University of Texas at Dallas, United States

PRESENTATION

TOPIC: AC, DC, BLDC Motor Drives

T16.7 Simulation-Assisted Design and Implementation of an Electrically Excited Synchronous Motor Drive System 938

Shih-Gang Chen, Jun-Ming Hsu, Chun-Yen Chen, Ming-Shi Huang
National Taipei University of Technology, Taiwan

PRESENTATION

TOPIC: AC, DC, BLDC Motor Drives

T16.8 Implementation and Analysis of Direct Torque Control on High-Speed PMSMs: A Comparative Study of Commercial and Laboratory-Developed Motors 943

Md Moniruzzaman[1], Kishor Joshi[1], Md Rashedur Rahman[1], Md Khurshedul Islam[2],
Seungdeog Choi[1], Masoud Karimi Ghartemani[1]
[1]Mississippi State University, United States; [2]ASML, United States

PRESENTATION

TOPIC: AC, DC, BLDC Motor Drives

T16.9 A Ferrite based Carbon Reinforced Composite Wrapped IPM Rotor Design for High-Speed Traction Applications 951

Md Rashedur Rahman[1], Md Khurshedul Islam[2], Md Moniruzzaman[1], Seungdeog Choi[1],
Han-Gyu Kim[1], Andrew Walters[1]
[1]Mississippi State University, United States; [2]ASML, United States

PRESENTATION

TOPIC: AC, DC, BLDC Motor Drives

SESSION T17: Resonant and Quasi-Resonant DC-DC Converters

13:30 - 17:00
TRACK: DC-DC Converters

SESSION CHAIR(S)
Mladen Ivankovic, *Infineon Technologies*
Sombudha Chakraborty, *Texas Instruments*

T17.1   **A Novel Phase-Mode Controller for Resonant Converters** 958
Claudio Adragna, Daniele Cazzaniga, Stefano Manzoni
STMicroelectronics, Italy

[PRESENTATION]   **TOPIC: Resonant Converters**

T17.2   **A Regulated 36V-60V-Input V_{IN}-Insensitive Resonant Switched-Capacitor Converter with Large Voltage Conversion Ratio** 966
Yichao Ji, Jingyi Yuan, Lin Cheng
University of Science and Technology of China, China

[PRESENTATION]   **TOPIC: Resonant Converters**

T17.3   **A Hybrid Switched Capacitor Converter Enabling Capacitive-Based Wireless Power Transfer for Battery Charging Applications** 971
Jade Sund, Samantha Coday
Massachusetts Institute of Technology, United States

[PRESENTATION]   **TOPIC: Resonant Converters**

T17.4   **A 48V to 50-110V Resonant Power-Bus Charger with Reduced Conduction Loss for MHz-Frequency Long-Range LiDAR Driver** 978
Hangxiao Ma[1,2], Xuchu Mu[1], Yang Jiang[1], Weihang Zhang[3], Jincheng Zhang[3], Rui P. Martins[1,4], Pui-In Mak[1]
[1]*University of Macau, China;* [2]*UM Hetao IC Research Institute, Shenzhen, China;* [3]*Xidian University, China;*
[4]*Universidade de Lisboa, Portugal*

[PRESENTATION]   **TOPIC: Resonant Converters**

T17.5   **A Trajectory Controlled 48-to-24 V Resonant Switched Capacitor Converter with 98.7% Efficiency and Ultrafast Dynamic Response** 983
Hélène T.W. Ma Yang[1], Liang Wang[2], Haoyu Wang[2], Wai Tung Ng[1]
[1]*University of Toronto, Canada;* [2]*Shanghai Tech University, China*

[PRESENTATION]   **TOPIC: Resonant Converters**

T17.6   **Low Power, Non-Isolated, Extremely-High Step-Up, Quasi-Resonant Hybrid DC–DC Converter** 990
Kumar Joy Nag, Aleksandar Prodić
University of Toronto, Canada

[PRESENTATION]   **TOPIC: Hard- and Soft-Switched**

T17.7   **Isolated Soft-Switching Flying-Capacitor based Quasi-Resonant Step-Up Converter** 997
Kumar Joy Nag, Aleksandar Prodić
University of Toronto, Canada

[PRESENTATION]   **TOPIC: Resonant Converters**

T17.8 Accurate Small-Signal Phasor Transformation-Based Modeling of Secondary-Side Diode-Bridge Rectifiers for Battery Charging Applications 1004

Aditya Zade, Regan Zane

Utah State University, United States

PRESENTATION

TOPIC: Resonant Converters

T17.9 High-Efficiency Isolated Piezoelectric Transformers for Magnetic-Less DC-DC Power Conversion 1012

Sourav Naval, Wentao Xu, Mustapha Touhami, Jessica D. Boles

University of California, Berkeley, United States

PRESENTATION

TOPIC: Resonant Converters

SESSION T18: GaN and SiC Power Devices

13:30 - 17:00
TRACK: Devices and Components

SESSION CHAIR(S)

Jaeil Baek, *KAIST*
Zhiguo Pan, *Schneider Electric*

T18.1 First Characterization of GaN Power Device and IC at Deep Cryogenic Temperatures Down to 100 mK 1020

Xin Yang[1], Matthew Porter[1], Zineng Yang[1], Zichen Xi[1], Liyang Jin[1], Liyan Zhu[1], Linbo Shao[1], Yuhao Zhang[2]

[1]Virginia Polytechnic Institute and State University, United States; [2]The University of Hong Kong, China

PRESENTATION

TOPIC: GaN and SiC Devices and Modules

T18.2 Dynamic Environment-Aware Lifetime Prediction of SiC MOSFET Modules Through LSTM 1026

Md Zakir Hasan[1], Seungdeog Choi[1], Youssef Aider[2], Prashant Singh[2], Chun-Hung Liu[1]

[1]Mississippi State University, United States; [2]The University of Tennessee, United States

PRESENTATION

TOPIC: GaN and SiC Devices and Modules

T18.3 Guarding-Based C-V Characterization of 10 kV SiC MOSFET in Half-Bridge Module Configuration 1034

Nianzun Qi, Gao Liu, Zhixing Yan, Shaokang Luan, Pawel Piotr Kubulus, Yuan Gao, Stefan Meyer, Hongbo Zhao, Asger Bjørn Jørgensen, Stig Munk-Nielsen

Aalborg University, Denmark

PRESENTATION

TOPIC: GaN and SiC Devices and Modules

T18.4 Automated Characterization Platform for Comprehensive Dynamic R_{dson} Assessment of GaN HEMTs from 50 K to 400 K 1040

Tian Qiu, Zheyu Zhang, Purushottam Khadka, Ahmed Siraj, Dilip Rana

Rensselaer Polytechnic Institute, United States

PRESENTATION

TOPIC: GaN and SiC Devices and Modules

T18.5 A Gate Driving Scheme for GaN Git with Enhanced Short Circuit Capability for Motor Drive Application 1047
Zongjie Zhou, Yan Cheng, Kevin J. Chen
The Hong Kong University of Science and Technology, China

PRESENTATION   TOPIC: GaN and SiC Devices and Modules

T18.6 Online Detection and Reduction of the Influence of Parameter Tolerance of Paralleled SiC MOSFETs in an EV Inverter Environment 1051
Hadiuzzaman Syed[1], Jochen Streit[1], Robert Kragl[1], Muhammad Muneeb Alam[1],
Alberto Martinez-Limia[1], Karl Oberdieck[1], Ertuğrul Sönmez[2]
[1]Robert Bosch GmbH, Germany; [2]Reutlingen University, Germany

PRESENTATION   TOPIC: GaN and SiC Devices and Modules

T18.7 Dynamic Current Sharing Issues with Paralleling SiC Power MOSFETs 1058
Ching-Yao Liu[1], Chen-Chan Lee[2], Jih-Sheng Lai[3]
*[1]National Yang Ming Chiao Tung University, Taiwan; [2]National Taiwan University, Taiwan;
[3]Virginia Polytechnic Institute and State University, United States*

PRESENTATION   TOPIC: GaN and SiC Devices and Modules

T18.8 Integrated Short-Circuit Protection Design based on Dual-Channel Gate Driver for Series Connected Medium-Voltage SiC MOSFETs 1063
Rui Wang, Drazen Dujic
École Polytechnique Fédérale de Lausanne, Switzerland

PRESENTATION   TOPIC: GaN and SiC Devices and Modules

T18.9 Long-Term High-Temperature Dynamic Gate Stress Reliability of a Last-Generation, Automotive-Grade, Planar 1200 V SiC MOSFET 1070
Giuseppe Mauromicale[1], Alessandro Sitta[1,2], Michele Fiore[1], Michele Calabretta[1], Francesco Iannuzzo[2]
[1]STMicroelectronics, Italy; [2]Aalborg University, Denmark

PRESENTATION   TOPIC: GaN and SiC Devices and Modules

SESSION T19: Gate Drive Circuits I

13:30 - 17:00
TRACK: Control

SESSION CHAIR(S)
Zahra Saadatizadeh, *Mississippi State University*
Davide Giacomini, *Infineon Technologies*

T19.1 Innovative Gate Driver Structure Achieving Low Time Skew Across Isolation Barrier for Parallel Connected SiC Modules 1076
Louison Gouy, Anne-Sophie Descamps, Nicolas Ginot, Christophe Batard
Nantes Université, France

PRESENTATION   TOPIC: Gate Drive Circuits and Fault Protection

T19.2 **Fully Integrated Closed-Loop Active Gate Driver IC with Real-Time Control of Gate Current Change Timing by Gate Current Sensing** 1084
Yaogan Liang[1], Katsuhiro Hata[2], Makoto Takamiya[1]
[1]The University of Tokyo, Japan; [2]Shibaura Institute of Technology, Japan

PRESENTATION
TOPIC: Gate Drive Circuits and Fault Protection

T19.3 **Analyze and Design of Digitally Load Current Modulated Active Gate Driver for GaN HEMTs based Buck DC-DC** 1090
Wentao Liu, Zhina Lian, Taotao Wu, Xiaochuan Peng, Hao Min
Fudan University, China

PRESENTATION
TOPIC: Gate Drive Circuits and Fault Protection

T19.4 **Impact of Real-Time Variable Gate-Drive Strength on Drive Cycle Efficiency in SiC Inverter-Fed PMSM Traction Drives** 1096
Matteo Pizzuto[1], Aiswarya Balamurali[2], Aniket Anand[2], Narayan C. Kar[1]
[1]University of Windsor, Canada; [2]R&D Americas, Schaeffler, Canada

PRESENTATION
TOPIC: Gate Drive Circuits and Fault Protection

T19.5 **Demonstration of Efficiency Increase of 350 V-to-13.3 V Isolated DC-DC Converters for Electric Vehicles by Active Gate Driving** 1102
Yohei Sukita[1], Katsuhiro Hata[2], Hiroki Kondo[3], Kenichi Watanabe[3], Kenichi Nagayoshi[3], Makoto Takamiya[1]
[1]The University of Tokyo, Japan; [2]Shibaura Institute of Technology, Japan; [3]Toyota Industries Corporation, Japan

PRESENTATION
TOPIC: Gate Drive Circuits and Fault Protection

T19.6 **A Multi-Level Active Gate Driver for Achieving Thermal Balance in Parallel Connected Power MOSFETs** 1108
Jingyuan Liang[1], Lingwei Sun[1], Wen Tao Cui[1], Wai Tung Ng[1], Motomitsu Iwamoto[2], Haruhiko Nishio[2]
[1]University of Toronto, Canada; [2]Fuji Electric Co. Ltd., Japan

PRESENTATION
TOPIC: Gate Drive Circuits and Fault Protection

T19.7 **A Fast Short-Circuit Protection Method for Ohmic Gate P-GaN HEMT based on Gate Charge** 1114
Yue Wu, Xi Jiang, Song Yuan, Xiaowu Gong, Zhaoheng Yan, Jiahong Chen, Yun Xu, Jinjie Liu
Xidian University, China

PRESENTATION
TOPIC: Gate Drive Circuits and Fault Protection

T19.8 **Comparison of Ultrafast-Rise-Time Gate Drivers for Wide-Bandgap Devices in Sub-Microsecond Pulsed Power Applications** 1121
Soham Roy, Duy T. Nguyen, Neeraj Anantha, Alex J. Hanson
The University of Texas at Austin, United States

PRESENTATION
TOPIC: Gate Drive Circuits and Fault Protection

T19.9 **A Discrete Multilevel Active Gate Driver for GaN HEMTs to Optimize the Switching Behavior** 1129
Celine Lawniczak, Martin Pfost
Technical University Dortmund, Germany

PRESENTATION

TOPIC: **Gate Drive Circuits and Fault Protection**

SESSION T20: Modeling and Simulation II

13:30 - 17:00
TRACK: Modeling and Simulation

SESSION CHAIR(S)
Jason Neely, *Sandia National Laboratories*
Jingbo Liu, *Eaton*

T20.1 **Attenuation of Fundamental Component of Differential Mode Noise using Active EMI Filter** 1135
Guru Abhilash Mulumudi, Naveed Ishraq, Ayan Mallik
Arizona State University, United States

PRESENTATION

TOPIC: **Circuits and Systems**

T20.2 **Graph Neural Network based Performance Modeling for the Dual Active Bridge Converter with Operational Generalization** 1143
Weihao Lei[1], Fanfan Lin[1], Xinze Li[2], Xiaokun Bao[1], Xin Zhang[1]
[1]Zhejiang University, China; [2]University of Arkansas, United States

PRESENTATION

TOPIC: **Circuits and Systems**

T20.3 **An Augmented State Space Modelling Approach for DC-DC Converter Start-Up in Closed Loop** 1148
Waseah Anjum, Arkadeb Sengupta, Marco Liserre
Christian-Albrechts-Universität zu Kiel, Germany

PRESENTATION

TOPIC: **Circuits and Systems**

T20.4 **The Utilization of a Parallel Computing Algorithm for Accelerating Switching-Level Modeling of Power Electronics Simulations in a T-Type PV Inverter** 1153
Buck F. Brown III[1], Liwei Wang[1], Zheyu Zhang[2], Johan Enslin[1], Yi Li[1]
[1]Clemson University, United States; [2]Rensselaer Polytechnic Institute, United States

PRESENTATION

TOPIC: **Circuits and Systems**

T20.5 **A New Reduced Order Analytical Switching Model for eGaN HEMTs** 1159
Ruqi Li, Douglas Arduini, Phen Lumod, Shobhana Punjabi, River Lin, Harold Gutierrez
Cisco Systems, Inc., United States

PRESENTATION

TOPIC: **Device and Component Modeling**

T20.6 **Proposal of an Alternative Reverse Recovery Calculation Method** 1167
Brian Deboi, Blake Nelson, Austin Curbow
Wolfspeed, Inc., United States

PRESENTATION

TOPIC: **Device and Component Modeling**

T20.7 Improvement of CM EMI Attenuation Ability of Transformer with Negative Capacitor 1173

Qinghui Huang, Yiming Li, Yirui Yang, Shuo Wang, Yanwen Lai, Zhedong Ma
University of Florida, United States

PRESENTATION

TOPIC: **Parasitics Extraction and Optimization**

T20.8 Damping Factor based PCB Parasitic Inductance Value Optimization to Minimize Voltage Overshoot and Settling Time of Semiconductors 1179

Reza Shahbazi, Yunting Liu
The Pennsylvania State University, United States

PRESENTATION

TOPIC: **Parasitics Extraction and Optimization**

T20.9 Hardware Implementation of Virtual Resistance based FRT Logic in Programmable 3-Level ANPC Inverters 1184

Mohammad Safayet Hossain[1], Shuvangkar Chandra Das[2], Paychuda Kritprajun[3], Amin Banaie[4], Tapas Barik[4], Deepak Ramasubramanian[4], Aboutaleb Haddadi[4], Evangelos Farantatos[4], Ulrich Muenz[5]
[1]University of Central Florida, United States; [2]Clarkson University, United States; [3]The University of Tennessee, United States; [4]Electric Power Reseach Institute, United States; [5]Siemens, United States

PRESENTATION

TOPIC: **Hardware-in-the-Loop and Rapid Prototyping**

SESSION T21: Critical Applications in Space and Transportation

13:30 - 17:00
TRACK: Transportation Power Electronics

SESSION CHAIR(S)
Tao Yang, *University of Nottingham*
Jean Marcos Lobo da Fonseca, *CAT*

T21.1 Rad-Hard PSFB Controller for High-Voltage Space Applications 1190

Reynaldo S. Gonzalez, Robert E. Bolaños
Southwest Research Institute, United States

PRESENTATION

TOPIC: **Power Electronics for Aerospace**

T21.2 Modeling, Control and Digital Implementation of a Buck Converter Operating in Triangular Current Mode for a Wide Output Voltage Range Space Application 1197

Regina Ramos[1], Sara Pérez[1], Guillermo Núñez[1], Pedro Alou[1], Javier Torres[2]
[1]Universidad Politécnica de Madrid, Spain; [2]Airbus Crisa, Spain

PRESENTATION

TOPIC: **Power Electronics for Aerospace**

T21.3 Thermal Model and Optimization of a Multi-Winding Transformer for Lunar Surface Power Transmission 1203

Zhining Zhang[1], Yuzhou Yao[1], Junchong Fan[1], Juchen Yang[1], Robert Guenther[2], Pengyu Fu[1], Jin Wang[1]
[1]The Ohio State University, United States; [2]GPEM LLC, United States

PRESENTATION

TOPIC: **Power Electronics for Aerospace**

T21.4 Active Gate Driver Power Supply for High-Reliability Applications 1211
Joseph P. Kozak, Juan Ramirez, Jesse Lin, Allison Orr, Alexander Martin, Hala Tomey
Johns Hopkins University Applied Physics Laboratory, United States

PRESENTATION   TOPIC: **Power Electronics for Aerospace**

T21.5 A Hybrid Energy Storage System for eVTOL Unmanned Aerial Vehicles using Supercapacitors 1217
Ali Alenezi, PengHao Huang, Prasad Enjeti
Texas A&M University, United States

PRESENTATION   TOPIC: **Power Electronics for Aerospace**

T21.6 Evaluation of Retired Lithium-Ion Batteries for Second-Life Applications Through Electrochemical Impedance Spectroscopy 1224
Latha Anekal, Sheldon Williamson
Ontario Tech University, Canada

PRESENTATION   TOPIC: **Vehicular Power Electronic Circuits and Systems**

T21.7 Uninterruptable Non-Isolated Integrated Power Electronics Converter (UNIPEC) for Commercial Truck Auxiliary Power Unit 1230
Pouya Zolfi, Ahmad Alzahrani, Ayman EL Refaie
Marquette University, United States

PRESENTATION   TOPIC: **Power Electronics for Shipboard and Other Transportation Applications**

T21.8 Investigation of Electrical Safety for Non-Isolated Single-Phase On-Board Chargers used in BEV/PHEV 1237
Soya Kataoka[1], Shohei Funatsu[1], Hiroaki Matsumori[1], Takashi Kosaka[1],
Keisuke Nakamura[2], Subrata Saha[2]
[1]Nagoya Institute of Technology, Japan; [2]Aisin Corporation, Japan

PRESENTATION   TOPIC: **On-Board and Off-Board Charging Systems**

T21.9 An 8-Level Flying Capacitor Multilevel Converter for Electric Aircraft Pulse Deicing 1242
Nicole Stokowski, Andrew Freeman, Aidan Rodgers, Aria Delmar, Jonathan Sengstock,
Alex Solecki, Andrew Stillwell
University of Illinois Urbana-Champaign, United States

PRESENTATION   TOPIC: **Power Electronics for Aerospace**

SESSION T22: Motor Drives II

13:30 - 17:00
TRACK: Motor Drives

SESSION CHAIR(S)
Ziaur Rahman, *Booz Allen Hamilton*
Ali Safayet, *HL Mechatronics*

T22.1 **Impact of Position Measurement Delay Angle on Performance of PMSM Drives for Electric Power Steering in a Wide Speed Range** 1248

Yingzhe Wu[1], Hengbin Zhang[2], Yuxiang Xue[2], Lisheng Wang[1], Hui Li[2], Shan Yin[2]
[1]Shanghai Gatek Automotive Electronics Company Ltd., China; [2]University of Electronic Science and Technology of China, China

PRESENTATION

TOPIC: **Modeling and Control Techniques for Motor Drives**

T22.2 **Physical Parameter Estimation for a Two-Level VSI Three-Phase PMSM Electric Drivetrain** 1255

Bernard Steyaert, Ananda Tjakra Adisurja, Matthias Preindl
Columbia University, United States

PRESENTATION

TOPIC: **Modeling and Control Techniques for Motor Drives**

T22.3 **A Novel Two-Dimensional Random Switching Frequency PWM Method for Variable Frequency Drives** 1261

Mostafa Abarzadeh, Kevin Lee
Eaton Corporation, United States

PRESENTATION

TOPIC: **Power Quality and EMI for Motor Drives**

T22.4 **Optimized Maximum Torque and Minimum Loss Fault-Tolerant Control Schemes for Dual Three-Phase PMSM** 1267

Syed Mohammad Maaz, Dong-Choon Lee
Yeungnam University, Korea

PRESENTATION

TOPIC: **Modeling and Control Techniques for Motor Drives**

T22.5 **Wireless Actuation of Magnetic Robots with a Modular 60 mT 3-D Helmholtz Coil System** 1274

Konstantinos Manos, Yifan Rao, Tuo Zhao, Kevin Liu, Daniel Zhou, Calvin Nguyen, Eric Chen, Glaucio H. Paulino, Minjie Chen
Princeton University, United States

PRESENTATION

TOPIC: **Actuators**

T22.6 **A Versatile PHIL based Motor Emulator Testbench using a High-Performance Power Amplifier Testbench** 1279

Seyedeh Nazanin Afrasiabi[1], Rajendra Thike[2], Mathews Boby[2], K. S. Amitkumar[2]
[1]Concordia University, Canada; [2]Opal-RT Technologies Inc., Canada

PRESENTATION

TOPIC: **AC, DC, BLDC Motor Drives**

T22.7 **A 450V Three Phase GaN IPM Achieving 99.1% Efficiency in Smallest 12mm x12mm Package for 250W Power Delivery without Heatsink** 1286
Maik Kaufmann[1], Manu Balakrishnan[2], Stefan Herzer[1], Anand Chellamuthu[3], Hely Zhang[4]
[1]Texas Instruments Incorporated, Germany; [2]Texas Instruments Incorporated, India;
[3]Texas Instruments Incorporated, United States; [4]Texas Instruments Incorporated, China

PRESENTATION

TOPIC: **AC, DC, BLDC Motor Drives**

T22.8 **FEA based High-Frequency Synthesis for the Design and Optimization of GaN-Based Dual Three Phase Motor Drive System** 1294
Syed Imam Hasan[1], Alper Uzum[1], Ashraf Siddiquee[1], Yilmaz Sozer[1], Krishna Namburi[2]
[1]The University of Akron, United States; [2]Nexteer Automotive, United States

PRESENTATION

TOPIC: **Integrated Motor Drives**

T22.9 **Evaluation of Passive Common-Noise Canceller Considering Both of Thermal Equilibrium and Common-Mode Noise Cancellation** 1299
Koji Mitsui, Kenshiro Katsura, Koki Notake, Koji Yamaguchi
IHI Corporation, Japan

PRESENTATION

TOPIC: **Power Quality and EMI for Motor Drives**

SESSION T23: Solid State Transformer Design and Control

13:30 - 17:00
TRACK: Power Electronics for Utility Applications

SESSION CHAIR(S)
Ali Khajehoddin, *University of Alberta*
Hang Dai, *GE Aerospace Research*

T23.1 **Performance Evaluation of Isolated DC/DC Converters in Modularized Bridge Rectifier Solid-State Transformer** 1305
Zhenchao Li, Andrea Cervone, Drazen Dujic
École Polytechnique Fédérale de Lausanne, Switzerland

PRESENTATION

TOPIC: **Solid-State Transformers**

T23.2 **Active and Reactive Power Flow Control of the Dual Active Bridge Converter** 1311
Lauryn Morris, Thomas W. Francois, Jonathan Saelens, Oroghene Oboreh-Snapps,
Arnold Fernandes, Praneeth Uddarraju, Sophia A. Strathman, Jonathan W. Kimball
Missouri University of Science and Technology, United States

PRESENTATION

TOPIC: **Solid-State Transformers**

T23.3 **Comparative Analysis of Carbon Footprints and Material Usage of Solid-State Transformers and Low-Frequency-Transformer-Based MVac-LVdc Interfaces for High-Power EV Charging** 1318
Luc Imperiali[1], Rudy Wang[2], Anup Anurag[2], Peter Barbosa[2], Johann W. Kolar[1], Jonas Huber[1]
[1]ETH Zürich, Switzerland; [2]Delta Electronics, Inc., United States

PRESENTATION

TOPIC: **Solid-State Transformers**

T23.4 **Trade Study of Isolation Requirements and Magnetic Core Selection for Medium Frequency-Medium Voltage Transformers** 1326
Mohendro Kumar Ghosh[1], Mark A. Juds[1], Brandon Grainger[1], Ahmad El Shafei[2],
Bogdan S. Borowy[2], Paul Ohodnicki[1]
[1]University of Pittsburgh, United States; [2]Eaton Corporation, United States

PRESENTATION

TOPIC: Solid-State Transformers

T23.5 **Comparative Evaluation of a Multilevel LLC Resonant Converter for a Modular DC/DC Stage in a Electrolyzer Power Supply** 1334
Samuel S. Queiroz, Levy F. Costa
Eindhoven University of Technology, Netherlands

PRESENTATION

TOPIC: Solid-State Transformers

T23.6 **Cost-Effectiveness Assessment of SiC MOSFET and Si IGBT Semiconductors in a Three-Level Resonant Converter for Solid-State Transformer** 1341
Samuel S. Queiroz, Levy F. Costa
Eindhoven University of Technology, Netherlands

PRESENTATION

TOPIC: Solid-State Transformers

T23.7 **Comparative Performance Analysis of Medium Voltage 3L-ANPC and 3L-DNPC Pole Enabled by Series-Connection of 10kV SiC MOSFETs and 10kV SiC JBS Diodes for Sine Triangle PWM Operation** 1347
Sanket Parashar, Shubham Rawat, Nithin Kolli, Raj Kumar Kokkonda, Subhashish Bhattacharya
North Carolina State University, United States

PRESENTATION

TOPIC: Solid-State Transformers

T23.8 **A Zero Harmonic Distortion Master Converter for Medium Voltage Microgrids** 1355
Gabriel V. Ramos[1], Dener A. de L. Brandão[1], Thiago M. Parreiras[2], Danilo I. Brandão[1], Braz de J.C. Filho[1]
[1]Universidade Federal de Minas Gerais, Brazil; [2]Centro Federal de Educação Tecnológica de Minas Gerais, Brazil

PRESENTATION

TOPIC: Microgrid Systems

T23.9 **An MILP Approach for Modeling and Analyzing the BESS for Smoothing Renewable Fluctuations Considering BESS Capacity Attenuation in the Bulk Power System with High Inverter-Based Resource Penetration** 1363
Hualong Liu, Wenyuan Tang
North Carolina State University, United States

PRESENTATION

TOPIC: Energy Storage Systems

SESSION T24: Reliability, Efficiency, and Thermal Performance of Power Modules and Components

13:30 - 17:00
TRACK: Power Electronics Integration and Manufacturing

SESSION CHAIR(S)
Vidhi Patel, *ABB*
Zhou Dong, *ABB*

T24.1 **Thermal and Efficiency Characterization of Immersion Cooled SiC Traction Inverter** 1368
Yiju Wang[1], Reza Ilka[2], JiangBiao He[1]
[1]The University of Tennessee, United States; [2]University of Kentucky, United States

> PRESENTATION

TOPIC: **Power Electronics Packaging**

T24.2 **FPGA-Based Hybrid Simulator for Real-Time 3-D Temperature Monitoring of Power Converters** 1375
Xianghao Mo, Daniel Ríos Linares, Regina Ramos, Miroslav Vasić
Universidad Politécnica de Madrid, Spain

> PRESENTATION

TOPIC: **Thermal Management**

T24.3 **A New Subassembly Concept for Enhanced Heat Dissipation and Reliability of Power Module** 1383
Yosuke Nakata, Yuji Sato, Shin Uegaki, Jun Fujita, Akihiko Furukawa, Masayoshi Tarutani
Mitsubishi Electric Corporation, Japan

> PRESENTATION

TOPIC: **Power Modules / High Density Design**

T24.4 **Stand-Alone R_{DS-ON} Sensor for In-Situ Prognostic, Protection and Reliability Enhancement of Power Converters** 1388
Zaheen Mustakin, Qiang Mu, Lucas Pereira, Jiale Zhou, Tiefu Zhao, Babak Parkhideh
University of North Carolina at Charlotte, United States

> PRESENTATION

TOPIC: **Quality and System Reliability Including EMI/EMC**

T24.5 **Electrical Evaluation of a Modular High Voltage 3D Power Module using Direct Dielectric Liquid Cooling** 1396
Omar Sanjakdar[1], Yvan Avenas[2], Rachelle Hanna[2], Guillaume Piquet Boisson[1], Emmanuel Marcault[3], Antoine Philippe[1]
[1]Université Grenoble Alpes, CEA-Liten, INES, France; [2]Université Grenoble Alpes, CNRS, Grenoble INP, G2Elab, France; [3]CEA, France

> PRESENTATION

TOPIC: **Power Electronics Packaging**

T24.6 **Board Level Reliability of Gull-Wing, Micro-Leaded and Lead-Less Packaged MOSFETs in Automotive Environments** 1403
Christopher Liu[1], Vijayakrishna Satyamsetti[1], Xuanjing Wei[2], Christian Radici[1], Peter Vines[1], Wayne Lawson[1]
[1]Nexperia, United Kingdom; [2]Nexperia, China

> PRESENTATION

TOPIC: **Power Electronics Packaging**

T24.7 Cost Effective and High Noise Immunity Methodology for Aging Evaluation of DC-Link Capacitors in Traction Inverters 1408
Seyed Hossein Aleyasin, Fausto Stella, Radu Bojoi, Enrico Vico
Politecnico di Torino, Italy

PRESENTATION

TOPIC: **Quality and System Reliability Including EMI/EMC**

T24.8 A 3D Structure of Single-Sided Cooling Power Module with Low Thermal Resistance and Low Inductance 1414
Hirofumi Hisamochi, Koki Notake, Yoshiaki Takahashi, Koji Yamaguchi
IHI Corporation, Japan

PRESENTATION

TOPIC: **Power Modules / High Density Design**

T24.9 Aging of Y-Capacitor in an EMI Filter and Its Impact on Common-Mode Noises 1420
Tahmid Ibne Mannan, Seungdeog Choi, Subarto Kumar Ghosh, Md Moniruzzaman
Mississippi State University, United States

PRESENTATION

TOPIC: **Quality and System Reliability Including EMI/EMC**

Thursday, March 20, 2025

Session T25: 48V-to-1V Direct DC-DC Converters

8:00 - 9:40
TRACK: DC-DC Converters

SESSION CHAIR(S)
Cahit Gezgin, *Infineon Technologies*
Mark DeMarie, *IBM*

T25.1 **2200A/48V-to-1V Low-Profile Direct Power Converter with Standard PCB Transformer** 1427
Alejandro Figueroa, Pablo Mazariegos, Álvaro Cobos, Javier Goicoechea, Alejandro Castro, José A. Cobos
Differential Power S.L., Spain

PRESENTATION

TOPIC: **Point-of-Load (POL) and Multi-Phase Converters**

T25.2 Single-Stage 48V-to-1V Regulator with a Half-Turn Transformer and Current-Doubler Rectifier 1433
Xinmiao Xu, Qiang Li
Virginia Polytechnic Institute and State University, United States

PRESENTATION

TOPIC: **Voltage Regulator Modules (VRM)**

T25.3 Ultra-Low-Profile Single-Stage Voltage Regulator Module (VRM) for Next-Generation AI Accelerators 1439
Xufu Ren, Jinfeng Zhang, Zhenshuai Rong, Borong Hu, Teng Long
University of Cambridge, United Kingdom

PRESENTATION

TOPIC: **Voltage Regulator Modules (VRM)**

T25.4 Novel TLVR Operation in Multi-Stage Voltage Regulator Module with Current Multipliers 1444

Kevin Zufferli[1], Roberto Rizzolatti[1], Mario Ursino[1], Simone Mazzer[1], Gerald Deboy[1], Stefano Saggini[2]
[1]Infineon Technologies AG, Austria; [2]Università degli Studi di Udine, Italy

PRESENTATION   Topic: **Voltage Regulator Modules (VRM)**

T25.5 Interphase LC-Oscillation Suppression with Fast Line-Transient Response in 48-V Series-Capacitor Buck Converters for Automotive Applications 1451

W.L. Jiang[1], Y. Liu[1], N. Khan[1], J. Pigott[2], H.J. Bergveld[2], V. Chaturvedi[2], O. Trescases[1]
[1]University of Toronto, Canada; [2]NXP Semiconductors, Netherlands

PRESENTATION   Topic: **Point-of-Load (POL) and Multi-Phase Converters**

Session T26: Multilevel Inverters

8:00 - 9:40
TRACK: DC-AC Inverters

SESSION CHAIR(S)
Yunting Liu, *Pennsylvania State University*
Dingrui Li, *Clemson University*

T26.1 An Approach to Compensate for Low Frequency DC-Link Voltage Ripple in High Power ANPC Inverter 1459

Shaozhe Wang[1], Ankit Vivek Deshpande[1], Rolando Sandoval[1], Erick Pool-Mazun[1],
Enrique Garza-Arias[2], Prasad Enjeti[1]
[1]Texas A&M University, United States; [2]Tecnológico de Monterrey, Mexico

PRESENTATION   Topic: **Multilevel Inverters**

T26.2 A Cascaded Multilevel Inverter System with Hot-Swapping and Fault Isolation Capability for Improved Resiliency 1465

Uthandi Selvarasu[1], Vikram Roy Chowdhury[2], Shumeng Wang[1], Jinli Zhu[3],
Mahshid Amirabadi[1], Yuan Li[3], Brad Lehman[1]
[1]Northeastern University, United States; [2]National Renewable Energy Laboratory, United States;
[3]University of Pittsburgh, United States

PRESENTATION   Topic: **Multilevel Inverters**

T26.3 Layout Optimization for Parasitic Inductance Reduction of GaN-Based NPL.X Multilevel Inverter 1473

Ali Halawa[1], Jinyeong Moon[2], Woongkul Lee[1]
[1]Purdue University, United States; [2]Florida State University, United States

PRESENTATION   Topic: **Multilevel Inverters**

T26.4 Topology Selection and Design Methodology for SiC based Solar Photovoltaic Medium Voltage Direct Grid Connect Inverters 1481

Jenson Joseph C. Attukadavil, Baylon G. Fernandes
Indian Institute of Technology Bombay, India

PRESENTATION   Topic: **Multilevel Inverters**

T26.5 **EMI Modeling of PCB-Based Three-Level Active Neutral-Point-Clamped GaN Converter** 1489
Mohammad Hassan Adeli, Necmi Altin, Erkan Deniz, Adel Nasiri
University of South Carolina, United States

PRESENTATION

TOPIC: Power Quality and EMI

SESSION T27: Power MOSFETs and IGBTs I

8:00 - 9:40
TRACK: Devices and Components

SESSION CHAIR(S)
Ming Liu, *Shanghai Jiaotong University*
Benjamin Lough, *Texas Instrument*

T27.1 **A Novel Layout for Improving Current Sharing of Paralleled SiC MOSFETs with TO-247 Package** 1495
Che-Wei Chang[1], Matthias Spieler[1], Rolando Burgos[1], Ayman EL-Refaie[2],
Renato Amorim Torres[3], Dong Dong[1]
[1]*Virginia Polytechnic Institute and State University, United States;* [2]*Marquette University, United States;*
[3]*General Motors, United States*

PRESENTATION

TOPIC: Power Silicon MOSFETs, BJTs, IGBTs, etc.

T27.2 **A Sensor-Less IGBT On-State Voltage Estimation Method using Inverter Control Variables** 1501
Shuyu Ou, Subham Sahoo, Ariya Sangwongwanich, Yongjie Liu, Frede Blaabjerg
Aalborg University, Denmark

PRESENTATION

TOPIC: Power Silicon MOSFETs, BJTs, IGBTs, etc.

T27.3 **A Novel Non-Intrusive Online Monitoring Method for Diagnosing the Lift-Off of Bonding Wires in SiC MOSFETs** 1507
Keqi Song, Henry Shu-Hung Chung, Ho-Tin Tang
City University of Hong Kong, China

PRESENTATION

TOPIC: Power Silicon MOSFETs, BJTs, IGBTs, etc.

T27.4 **Optimizing MOSFET Selection for EMC-Critical Automotive Applications** 1512
Sacha J. Cazzitti[1], Christian Radici[2], Andrew J. Forsyth[1], Cheng Zhang[1], Peter Vines[2]
[1]*The University of Manchester, United Kingdom;* [2]*Nexperia, United Kingdom*

PRESENTATION

TOPIC: Power Silicon MOSFETs, BJTs, IGBTs, etc.

T27.5 **Improving Dynamic Current Sharing Between Parallel MOSFETs by Optimizing Device Parameters** 1519
Kunal Jha[1], Kapil Kelkar[1], Marina Hedenik[2], David Penof[2]
[1]*Infineon Technologies AG, United States;* [2]*Infineon Technologies AG, Austria*

PRESENTATION

TOPIC: Power Silicon MOSFETs, BJTs, IGBTs, etc.

SESSION T28: Design Techniques for Power Modules

8:00 - 9:40

TRACK: Power Electronics Integration and Manufacturing

SESSION CHAIR(S)
Vidhi Patel, *ABB*
Weiqiang Chen, *ABB*

T28.1 **A 21.6 kW/L Two-Phase Immersion-Cooled Isolated DC-DC Converter** 1529
Aleksandar Ristic-Smith, Kawsar Ali, Daniel Rogers
University of Oxford, United Kingdom

PRESENTATION   TOPIC: **Power Electronics Packaging**

T28.2 **Extraction of Common Mode Parasitic Capacitance in Balance Filter for the
Prediction of EMI Noise Suppression** 1537
Qiuzhe Yang, Xingyu Chen, Zijian Wang, Qiang Li
Virginia Polytechnic Institute and State University, United States

PRESENTATION   TOPIC: **Quality and System Reliability Including EMI/EMC**

T28.3 **A 660W, 96% Efficiency 3D Heterogeneously Integrated Digital DC/DC
Power Module for Vertical Power Delivery** 1544
Haoyu Wang, Xuliang Wang, Yan Wang, Xiaosen Liu
Tsinghua University, China

PRESENTATION   TOPIC: **Power Electronics Packaging**

T28.4 **Planar Rogowski Coil-Based Switch Current Measurement for a
1.2 kV SiC MOSFET Embedded Die PCB** 1551
Matthias Spieler[1], Che-Wei Chang[1], Ayman M. EL-Refaie[2], Dong Dong[1], Rolando Burgos[1]
[1]*Virginia Polytechnic Institute and State University, United States;* [2]*Marquette University, United States*

PRESENTATION   TOPIC: **Embedded Technologies, 3D Packaging, and Additive Manufacturing**

T28.5 **Effect of Magnetic Couplings on Conducted EMI of GaN-Based PFC Converter** 1557
Tyler McGrew, Qiang Li
Virginia Polytechnic Institute and State University, United States

PRESENTATION   TOPIC: **Quality and System Reliability Including EMI/EMC**

SESSION T29: Gate Drive Circuits II

8:00 - 9:40

TRACK: Control

SESSION CHAIR(S)

Seungdeog Choi, *Mississippi State University*
Kang Wei, *Texas Instrument*

T29.1 **Optically-Controlled 3.3 kV SiC MOSFET with Fast Switching Speed and Low Optical Power** ... 1564
Xin Yang[1], Guannan Shi[1], Liyang Jin[1], Yuan Qin[1], Matthew Porter[1], Che-Wei Chang[1],
Xiaoting Jia[1], Dong Dong[1], Linbo Shao[1], Yuhao Zhang[2]
[1]Virginia Polytechnic Institute and State University, United States; [2]The University of Hong Kong, China

PRESENTATION   **TOPIC: Gate Drive Circuits and Fault Protection**

T29.2 **Optimization Techniques for Parallel-Connected Devices in IPMs for Consumer Use** 1569
Keisuke Kawamoto[1], Haruhiko Murakami[1], Teruaki Nagahara[1], Michael Rogers[2],
Akiko Goto[1], Shoji Saito[1], Koichiro Noguchi[1]
[1]Mitsubishi Electric Corporation, Japan; [2]Mitsubishi Electric US, Inc., United States

PRESENTATION   **TOPIC: Gate Drive Circuits and Fault Protection**

T29.3 **Investigating the Temperature Dependency and Operating Parameters of a
Self-Driving Active Gate Driver** 1576
Vin Loong Choo, Martin Pfost
Technical University Dortmund, Germany

PRESENTATION   **TOPIC: Gate Drive Circuits and Fault Protection**

T29.4 **Use of Switched-Capacitor Circuit to Generate Negative Gate-Source Voltage Pulses** 1582
Ho-Tin Tang, Henry Shu-Hung Chung
City University of Hong Kong, China

PRESENTATION   **TOPIC: Gate Drive Circuits and Fault Protection**

T29.5 **An Optically Isolated Gate Driver with Simultaneous Data and Power Transmission
Through a Miniaturized, Efficient Photonic Platform** 1590
Jiajun Li, Mariia Klymenko, Yanqiao Li, William Scheideler, Jason T. Stauth
Dartmouth College, United States

PRESENTATION   **TOPIC: Gate Drive Circuits and Fault Protection**

Session T30: Energy Flow and Battery Management Systems

8:00 - 9:40
TRACK: Power Electronics for Utility Applications

SESSION CHAIR(S)
Jonathan Kimball, *Missouri University of Science and Technology*
Stanley Atcitty, *Sandia National Laboratories*

T30.1 Optimal Shared Energy Storage Capacity Configuration in Multi-Energy Microgrids Considering Battery Lifetime Loss based on Relaxation Techniques 1597
Hualong Liu, Wenyuan Tang
North Carolina State University, United States

PRESENTATION

TOPIC: **Microgrid Systems**

T30.2 Virtual Resistance Control for an Active Battery Management System 1602
Alastair P. Thurlbeck[1], Ashraf Siddiquee[2], Mithat John Kisacikoglu[1], Yilmaz Sozer[2]
[1]National Renewable Energy Laboratory, United States; [2]The University of Akron, United States

PRESENTATION

TOPIC: **Energy Storage Systems**

T30.3 Internal Voltage Source Saturation Impact on Stability Limits of Grid Forming Converter 1610
Divyanshu Bansal, Aravind G., L. Umanand
Indian Institute of Science, India

PRESENTATION

TOPIC: **Distributed Energy Systems**

T30.4 A Zero Harmonic Distortion Grid-Connected Grid-Forming Converter for Battery Energy Storage System Applications 1615
Gabriel V. Ramos[1], Thiago M. Parreiras[2], Fangzhou Zhao[3], Xiongfei Wang[3,4], Braz de J.C. Filho[1]
[1]Universidade Federal de Minas Gerais, Brazil; [2]Centro Federal de Educação Tecnológica de Minas Gerais, Brazil; [3]Aalborg University, Denmark; [4]KTH Royal Institute of Technology, Sweden

PRESENTATION

TOPIC: **Energy Storage Systems**

T30.5 Single Cell Energy Router Justification for Three Phase Near Zero Energy Buildings 1622
Hossein Nourollahi Hokmabad[1], Tala Hemmati Shahsavar[1], Oleksandr Matiushkin[1], Tanel Jalakas[1], Oleksandr Husev[2], Juri Belikov[1]
[1]Tallinn University of Technology, Estonia; [2]Gdansk University of Technology, Poland

PRESENTATION

TOPIC: **Distributed Energy Systems**

SESSION T31: MHz Frequency Wireless Power Transfer

8:00 - 9:40

TRACK: Wireless Power Transfer

SESSION CHAIR(S)

Jungwon Choi, *University of Washington*
Gui-Jia Su, *Oak Ridge National Lab*

T31.1 **A Multi-UAV Charging Station Enabling Free Landing by Grid Pattern Transmitter** 1629
Jungho Kim, Hyunkyeong Jo, Seoktae Seo, Bonyoung Lee, Hyungki Min, Franklin Bien
Ulsan National Institute of Science and Technology, Korea

PRESENTATION   **TOPIC: Wireless Charging**

T31.2 **Capacitor Design for Self-Resonant Coils for Long-Distance
Wireless Power Transfer System** 1635
Mostak Mohammad, Vandana Rallabandi, Omer C. Onar, Gui-Jia Su
Oak Ridge National Laboratory, United States

PRESENTATION   **TOPIC: Wireless Charging**

T31.3 **A 10.4-kW High-Power-Transfer-Density Multi-MHz Capacitive Wireless Power Transfer
System for EV Charging Utilizing Stacked-Inverter Stacked-Rectifier Architecture** 1640
Dheeraj Etta, Miguel Alvarez Dominguez, Sounak Maji, Syed Saeed Rashid, Khurram K. Afridi
Cornell University, United States

PRESENTATION   **TOPIC: Wireless Charging**

T31.4 **Reduced-Fringing-Field Multi-MHz Capacitive Wireless Power Transfer System using
Metasurface-Based Couplers with Active Field Cancellation** 1646
Syed Saeed Rashid, Dheeraj Etta, Matteo Ciabattoni, Francesco Monticone, Khurram K. Afridi
Cornell University, United States

PRESENTATION   **TOPIC: Wireless Charging**

T31.5 **Living Object Detection in Wireless Power Transfer Systems using
Remote Capacitive Bio-Signals Monitoring** 1653
Bruno M.G. Rosa, Paul D. Mitcheson
Imperial College London, United Kingdom

PRESENTATION   **TOPIC: Non-contact Sensors for Power Electronics**

SESSION T32: 48V Intermediate Bus Converters

8:00 - 9:40
TRACK: DC-DC Converters

SESSION CHAIR(S)
Xin Zhang, *IBM*
Jason Stauth, *Dartmouth College*

T32.1 **Modified N:1 Switched Capacitor Converter with Reduced Capacitor DC Bias Voltage for High Power Density** 1659
Taewoo Lee, Dam Yun, Sunghyuk Choi, Jung-Ik Ha
Seoul National University, Korea

PRESENTATION | TOPIC: **Hard- and Soft-Switched**

T32.2 **Wide Range Digital Control for Three-Level Buck Converters with Sensorless Flying-Cap Voltage Balancing** 1666
Hossein Hajisadeghian, Giovanni Bonanno
Università degli Studi di Padova, Italy

PRESENTATION | TOPIC: **Point-of-Load (POL) and Multi-Phase Converters**

T32.3 **A Comparative Investigation of a New Continuous Voltage Conversion Ratio Approach in a Zero-Inductor Voltage Converter** 1673
Sina Salehi Dobakhshari, Aamna Nasir Hameed, Binghui He, Mojtaba Forouzesh, Yan-Fei Liu
Queen's University, Canada

PRESENTATION | TOPIC: **Point-of-Load (POL) and Multi-Phase Converters**

T32.4 **A 96.1% Peak Efficiency, 6.8 kW/in³, 48V-to-6V On-Package Intermediate Bus Converter with LV-GaN Power Transistors** 1681
Mausamjeet Khatua, Nachiket Desai, Harish Krishnamurthy, Sheldon Weng, Jingshu Yu, Huong Do, Samuel Bader, Han Wui Then, Krishnan Ravichandran, James Tschanz, Kaladhar Radhakrishnan, Vivek De
Intel Corporation, United States

PRESENTATION | TOPIC: **Point-of-Load (POL) and Multi-Phase Converters**

T32.5 **A 48V to 2.4V-5V 95.8%-Peak-Efficiency 869W/in³-Power-Density Fibonacci Dual-Path Hybrid DC-DC Converter with Inductor Current Reduction and Low Output Resistance** 1687
Yichao Ji, Zeguo Liu, Lin Cheng
University of Science and Technology of China, China

PRESENTATION | TOPIC: **Voltage Regulator Modules (VRM)**

SESSION T33: DC-DC Converter Applications

10:10 - 11:50
TRACK: DC-DC Converters

SESSION CHAIR(S)
Robert Mascia, *Hewlett Packard Enterprise*
Hoi Lee, *University of Texas at Dallas*

T33.1 **An Ultra-Fast Very Large Scale Interleaved Li-Fi Transmitter** 1693
Daniel H. Zhou, Konstantinos Manos, Minjie Chen
Princeton University, United States

PRESENTATION   TOPIC: Point-of-Load (POL) and Multi-Phase Converters

T33.2 **Isolated PWM DC-DC Converter with Single Magnetic Component, ZVS and Self-Balanced Switched-Capacitor Voltage** 1701
Pablo M. Gil, Juan Rodríguez, Diego G. Lamar
Universidad de Oviedo, Spain

PRESENTATION   TOPIC: Hard- and Soft-Switched

T33.3 **Analysis and Design of a Low-Complexity ZVS Buck-Boost Converter** 1707
Burkhard Ulrich
Reutlingen University, Germany

PRESENTATION   TOPIC: Hard- and Soft-Switched

T33.4 **A High Conversion-Ratio Hybrid Series-Parallel DC-DC Converter with Pseudo-Soft-Charging and Inductor Current Frequency Multiplication** 1715
Avinash Maddela, Kishalay Datta, Jason T. Stauth
Dartmouth College, United States

PRESENTATION   TOPIC: Point-of-Load (POL) and Multi-Phase Converters

T33.5 **A Real-Time Variation Control of Deadtime in GaN-Based Bidirectional Buck-Boost Converter for Lithium-Ion Battery Formation System** 1723
Jong-Hun Lim, Go Woon Heo, Je-Yeong Lim, Dong Hwan Kim, Byoung Kuk Lee
Sungkyunkwan University, Korea

PRESENTATION   TOPIC: Bidirectional DC-DC Converters

SESSION T34: Inverter Modulation and Control Strategies

10:10 - 11:50
TRACK: DC-AC Inverters

SESSION CHAIR(S)
Karun Arjun Potty, *Lucid Motors*
Diego Raffo, *Infineon Technologies*

T34.1 **A Space Vector PWM Strategy for Charging of Bootstrap Capacitor in Three-Level Neutral-Point-Clamped Inverter** 1728
Anantha Hegde, Asamira Suzuki, Hirokazu Nakamura, Takamune Kabashima,
Koji Higashiyama, Keiji Akamatsu
Panasonic Industry Co., Ltd., Japan

PRESENTATION
TOPIC: PWM Strategies

T34.2 **A Complementary Carrier based PWM Strategy for Average Current Sampling of Three-Phase Inverter using Single Current Sensor** 1734
Byeong-Il Kim[1], Joon-Seok Kim[1], Yeongsu Bak[2], June-Seok Lee[1]
[1]Dankook University, Korea; [2]Keimyung University, Korea

PRESENTATION
TOPIC: PWM Strategies

T34.3 **Short-Circuit Ride-Through for a CRM-Based Soft-Switching Three-Phase Inverter** 1741
Xingyu Chen[1], Gibong Son[2], Qiang Li[1]
[1]Virginia Polytechnic Institute and State University, United States; [2]Tesla, United States

PRESENTATION
TOPIC: PWM Strategies

T34.4 **Modified Space Vector Modulation with Low Bandwidth Sensor to Reduce Losses in Soft Switching Three-Phase Inverters** 1746
Md Didarul Alam[1], Nazmul Hassan[1], Iqbal Husain[1], Liming Liu[2], Hongrae Kim[2]
[1]North Carolina State University, United States; [2]Eaton Corporation, United States

PRESENTATION
TOPIC: Single and Multi-Phase Inverters

T34.5 **A Feedforward Ripple Reduction Control Strategy based on a Hybrid GaN/Si Interleaved Inverter** 1754
Mowei Lu, Jurgis Reinotas, Xiaoyang Tian, Stefan M. Goetz
University of Cambridge, United Kingdom

PRESENTATION
TOPIC: PWM Strategies

SESSION T35: Power MOSFETs and IGBTs II

10:10 - 11:50

TRACK: Devices and Components

SESSION CHAIR(S)

Marie Lawson, *Hungtington Ingalls Industries*
Robert Scibilia, *Texas Instrument*

T35.1 **IGBT Comparison for Optimized Switching Behavior in the SiC/Si-Hybrid Switch** 1759
Adrian Amler[1], Thomas Heckel[2], Daniel Ruppert[3], Cornelius Rettner[4], Martin März[1]
[1]Friedrich-Alexander-Universität Erlangen-Nürnberg, Germany; [2]Fraunhofer Institute for Integrated Systems and Device Technology, Germany; [3]Audi AG, Germany; [4]Volkswagen AG, Germany

PRESENTATION   TOPIC: Power Silicon MOSFETs, BJTs, IGBTs, etc.

T35.2 **Forward Recovery and its Mitigation in Hybrid Si/SiC-Based DC–AC Converters** 1767
Yan Zhou, Thomas Lehmeier, Adrian Amler, Martin März
Friedrich-Alexander-Universität Erlangen-Nürnberg, Germany

PRESENTATION   TOPIC: Power Silicon MOSFETs, BJTs, IGBTs, etc.

T35.3 **Real-Time IGBT Module Ageing Characterization Through Temperature Monitoring** 1774
Quirc Perez-Farre[1], Luis F. Gomez-Rivera[1], Carlos Lopez-Torres[2], Kai Dannehl[2],
Antoni García-Espinosa[1], Alejandro Paredes-Camacho[1]
[1]Universitat Politecnica de Catalunya, Spain; [2]SEAT S.A., Spain

PRESENTATION   TOPIC: Power Silicon MOSFETs, BJTs, IGBTs, etc.

T35.4 **Experimental Validation of Triangular SOA via Infrared Thermography of a MOSFET
Die Operating in the Thermally Unstable Linear-Mode for Automotive Applications** 1781
Yacine Ayachi Amor, Christian Radici, Kerry J. Abrams, Philip Ellis, Peter Vines, Wayne Lawson
Nexperia, United Kingdom

PRESENTATION   TOPIC: Power Silicon MOSFETs, BJTs, IGBTs, etc.

T35.5 **Feasibility Study of the SuperIGBT: A Series-Connected High Voltage
IGBT with a Single Gate** 1786
Junhong Tong, Alex Q. Huang, Huanghaohe Zou, Zhiyuan Ma
The University of Texas at Austin, United States

PRESENTATION   TOPIC: Power Silicon MOSFETs, BJTs, IGBTs, etc.

SESSION T36: Magnetics Design and Modelling

10:10 - 11:50

TRACK: Magnetics

SESSION CHAIR(S)

Matt Wilkowski, *Wurth Elektronik*
George Slama, *Wurth Elektronik*

T36.1 Low Profile, Laminated Nife Transformers for Flyback Converters 1791

Xuan Wang[1], Reza Mounesi[2], Matthew Catanoso[1], Matthew Fox[1], Adel Nasiri[2], Mark G. Allen[1]
[1]University of Pennsylvania, United States; [2]University of South Carolina, United States

PRESENTATION

TOPIC: **Advanced Magnetic Materials and Geometries**

T36.2 Comprehensive Demonstration of New Magnetic Designs Utilizing Magnetic Anisotropy of the Cores for Integrated Magnetics 1797

Yota Takamura[1], Honami Nitta[1], Tatsuya Miyazaki[2], Kimito Yamanaka[1], Ryosuke Ishido[2], Akira Namba[2], Keisuke Fujisaki[3], Shigeki Nakagawa[3]
[1]Institute of Science Tokyo, Japan; [2]ROHM Co., Ltd., Japan; [3]Toyota Technological Institute, Japan

PRESENTATION

TOPIC: **High-Frequency Magnetics**

T36.3 A Two – Stage Artificial Neural Network (ANN) – based Design and Optimization of High Frequency Transformers for Dual Active Bridge Converter 1803

Lufan Zhou[1], Alberto Delgado Expósito[1], Adam Ruszczyk[2], Simon Round[3], Miroslav Vasić[1]
[1]Universidad Politécnica de Madrid, Spain; [2]Hitachi Energy, Poland; [3]Hitachi Energy, Switzerland

PRESENTATION

TOPIC: **Magnetics Modeling and Simulations**

T36.4 Modeling and Optimizing Winding Arrangement for Gapped Planar Magnetics based on Artificial Neural Network 1810

Hanqing Cao, Bima Nugraha Sanusi, Ziwei Ouyang
Technical University of Denmark, Denmark

PRESENTATION

TOPIC: **Magnetics Modeling and Simulations**

T36.5 Free-Shape Optimization of VHF Air-Core Inductors using a Constraint-Aware Genetic Algorithm 1816

Thomas Guillod, Charles R. Sullivan
Dartmouth College, United States

PRESENTATION

TOPIC: **Magnetics Modeling and Simulations**

SESSION T37: Packaging of Power Devices and Modules

10:10 - 11:50

TRACK: Power Electronics Integration and Manufacturing

SESSION CHAIR(S)

Lee Gill, *Sandia National Laboratories*
Joshua Stewart, *Aerospace Corporation*

T37.1 **Organic Direct Bonded Copper-Based Rapid Prototyping for Silicon Carbide Power Module Packaging** 1824
Shuofeng Zhao, Joshua Major, Douglas DeVoto, Sarwar Islam, Xiaoling Li, Mike Tant, Faisal Khan, Sreekant Narumanchi
National Renewable Energy Laboratory, United States

PRESENTATION   TOPIC: **Power Electronics Packaging**

T37.2 **Discrete Power Device Packaging with Integrated Direct Two-Phase Cooling** 1832
Jinpeng Cheng[1], Jinxiao Wei[1], Hao Feng[1], Li Ran[2]
[1]Chongqing University, China; [2]University of Warwick, United Kingdom

PRESENTATION   TOPIC: **Power Electronics Packaging**

T37.3 **Investigation of Die Top-Side Re-Metallization for SiC-Based Double-Side Cooled Power Modules** 1836
Narayanan Rajagopal, Christina DiMarino
Virginia Polytechnic Institute and State University, United States

PRESENTATION   TOPIC: **Power Electronics Packaging**

T37.4 **Design of Low Parasitic Inductance GaN HEMT Flip-Chip Power Module** 1844
Mohammad Dehan Rahman, Tanzila Akter, Abu Shahir Md Khalid Hasan, H. Alan Mantooth, Xiaoqing Song
University of Arkansas, United States

PRESENTATION   TOPIC: **Power Electronics Packaging**

T37.5 **A Scalable Dual-Orthogonal-Cooling Packaging Concept for Parallel-Series SiC Chips** 1850
Ekaterina Muravleva[1], Youssef Abotaleb[1], Blake Anderson[1], Zichen Zhang[2], Boyi Zhang[3], Jerry Hudgins[1], Jun Wang[1]
[1]University of Nebraska-Lincoln, United States; [2]Virginia Polytechnic Institute and State University, United States; [3]Delta Electronics, Inc., United States

PRESENTATION   TOPIC: **Power Modules / High Density Design**

SESSION T38: Modeling and Simulation III

10:10 - 11:50
TRACK: Modeling and Simulation

SESSION CHAIR(S)
Rafal Wojda, *Oak Ridge National Laboratory*
Jacob Mueller, *Sandia National Laboratories*

T38.1 **Parasitic Impact Analysis and Design of Hybrid EMI Filter for Active Clamp Flyback SMPS** 1858
Tahmid Ibne Mannan, Seungdeog Choi, Masoud Karimi-Ghartemani
Mississippi State University, United States

PRESENTATION   TOPIC: **Parasitics Extraction and Optimization**

T38.2 **Overview of Dynamic Characterization of Switches for Three Phase Voltage Source, Current Source, and Matrix Converter Applications** 1866
Sneha Narasimhan, Sathya Rupan Thirumoorthi, Subhashish Bhattacharya
North Carolina State University, United States

PRESENTATION   TOPIC: **Circuits and Systems**

T38.3 **Advanced Modeling Technique of Class-E Inverter Considering Low R_{on} of eGaN FETs and Different Design Procedures** 1874
Manas Palmal, Jungwon Choi
University of Washington, United States

PRESENTATION   TOPIC: **Circuits and Systems**

T38.4 **PiezoNet and Data-Driven Models for Time-Domain Characterization of Piezoelectric Resonators** 1882
Davit Grigoryan, Mian Liao, Haoran Li, Shukai Wang, Tanuj Sen, Matthew Tan, Minjie Chen
Princeton University, United States

PRESENTATION   TOPIC: **Device and Component Modeling**

T38.5 **A New Gate Charge De-Embedding Method for Accurate On-Wafer Characterization of HV MOSFET Devices** 1889
João R.R.O. Martins[1], Rachid Hamani[1], Vincent Quenette[1], Joerg Gessner[2]
[1]*X-FAB Silicon Foundries SE, France;* [2]*X-FAB Silicon Foundries SE, Germany*

PRESENTATION   TOPIC: **Parasitics Extraction and Optimization**

SESSION T39: Auxiliary Systems and Applications in EVs

10:10 - 11:50

TRACK: Transportation Power Electronics

SESSION CHAIR(S)

Rasoul Hosseini, *General Motors*
Yateendra Deshpande, *Conifer Systems*

T39.1 **4 kW Auxiliary Power Module for Electric Vehicles Utilizing a Dual-Phase LLC DC-DC Converter** 1892
Mojtaba Forouzesh, Xiang Yu, Yan-Fei Liu, Paresh C. Sen
Queen's University, Canada

PRESENTATION   TOPIC: Power Electronics for Hybrid and Electric Vehicles

T39.2 **New Reverse Mode Control Method of Phase-Shift Full-Bridge Converter for Bidirectional Auxiliary Power Module** 1899
Jongyoon Chae[1], Dongmin Kim[2], Dongmin Choi[1], Gun-Woo Moon[1]
[1]Korea Advanced Institute of Science and Technology, Korea; [2]Electronics and Telecommunications Research Institute, Korea

PRESENTATION   TOPIC: Power Electronics for Hybrid and Electric Vehicles

T39.3 **In-Situ EV EIS with a High-Density Flying Capacitor Multi-Level Converter Supercapacitor System** 1905
Avram Kachura, Gaël Vergès, Samantha K. Murray, Olivier Trescases
University of Toronto, Canada

PRESENTATION   TOPIC: Power Electronics for Hybrid and Electric Vehicles

T39.4 **A Novel 500-kHz LLC-T Resonant Converter with Wide Output Range** 1913
Zhengming Hou, Dong Jiao, Jih-Sheng Lai
Virginia Polytechnic Institute and State University, United States

PRESENTATION   TOPIC: Power Electronics for Hybrid and Electric Vehicles

T39.5 **High Efficiency Traction Drive Operation with a Partial Load Three-Phase Triangular Current Mode Modulation Concept** 1919
Bhaskar Chatterjee[1,2], Jan Allgeier[1], Thomas Plum[1], Marc Hiller[2]
[1]Robert Bosch GmbH, Germany; [2]Karlsruhe Institute of Technology, Germany

PRESENTATION   TOPIC: Power Electronics for Hybrid and Electric Vehicles

SESSION T40: Grid-Tied Inverter Control

10:10 - 11:50
TRACK: Power Electronics for Utility Applications

SESSION CHAIR(S)
Jonathan Kimball, *Missouri University of Science and Technology*
Raj Kumar Kokkonda, *North Carolina State University*

T40.1 Analysis of Maximum Power Transfer Limit for Linear Operation of Dual-Active-Bridge Converters 1927

Radhika Sarda[1], Ezequiel Ramos Rodriguez[1], Gaowen Liang[1], Glen G. Farivar[2], Josep Pou[3], Vaisambhayana B. Sriram[1], Anshuman Tripathi[1]
[1]Nanyang Technological University, Singapore; [2]University of Melbourne, Australia; [3]City University of Hong Kong, China

PRESENTATION

TOPIC: **Solid-State Transformers**

T40.2 Enhanced Control for Integrated Active Power Decoupling in Single-Phase Three-Level Flying Capacitor PFC Converter 1935

Gleisson Balen[1], Cristian Blanco[1], Ángel Navarro-Rodríguez[1], Pablo García[1], Rafael Peña-Alzola[2]
[1]Universidad de Oviedo, Spain; [2]University of Strathclyde, United Kingdom

PRESENTATION

TOPIC: **Power Quality, UPS, Active Power Filters**

T40.3 Improving Transient Stability of PLL-Synchronized Grid-Following Inverters 1940

Surya Prakash[1], Kalpana Beura[2], Mohamed Alkhatib[2], Omar Al Zaabi[3], Khalifa Al Hosani[3], Utkal Ranjan Muduli[3]
[1]National Institute of Technology Jamshedpur, India; [2]United Arab Emirates University, U.A.E.; [3]Khalifa University, U.A.E.

PRESENTATION

TOPIC: **Power Quality, UPS, Active Power Filters**

T40.4 Online Impedance-Based Analysis for Power System Stability Assessment using Transformer-Less and Filter-Less Switch-Mode Perturbation Generator 1946

Tomoya Ide[1], Yuko Hirase[1], Cheng Huang[2], Takanori Isobe[2]
[1]Toyo University, Japan; [2]University of Tsukuba, Japan

PRESENTATION

TOPIC: **Smart Grid and Metering**

T40.5 PIR-R Control for Three-Phase Grid-Connected Inverter with Unbalanced Grid Current Correction 1953

Haneen Ghanayem[1], Xingyu Yang[2], Mohammad Alathamneh[1], R.M. Nelms[2]
[1]Al-Balqa Applied University, Jordan; [2]Auburn University, United States

PRESENTATION

TOPIC: **Power Quality, UPS, Active Power Filters**

SESSION D01: AC-DC Converters II

11:30 - 13:30
TRACK: AC-DC Converters

SESSION CHAIR(S)
Jinia Roy, *University of Wisconsin-Madison*
Jim Ching-Jan Chen, *National Taiwan University*

D01.1 Design and Placement of a Passive Clamp Snubber for Isolated SEPIC and Cuk Converters Working as Automatic Power Factor Correctors 1959
Abraham López, Juan Rodríguez, Duberney Murillo-Yarce, Javier Sebastián, Diego G. Lamar
Universidad de Oviedo, Spain

PRESENTATION   TOPIC: Power Factor Correction: CCM, DCM, CRM/BCM Control, Bridgeless

D01.2 Current Sensorless Control Strategy for Single-Phase T-Type PFC Converter 1967
Che-Yu Lu, Jia-En Zeng
National United University, Taiwan

PRESENTATION   TOPIC: Power Factor Correction: CCM, DCM, CRM/BCM Control, Bridgeless

D01.3 Three-Phase Single-Stage Multiport AC-DC Converter with Integrated DC-DC Conversion Stages 1972
Asad Hameed, Gerry Moschopoulos
Western University, Canada

PRESENTATION   TOPIC: Bidirectional AC-DC Converters

D01.4 High Efficiency AC-Adapter Realized by Voltage-Clamper with Mid-Voltage AHB Converter using Synchronous Rectification 1977
Shuichiro Motoori[1], Akihiro Kawano[1], Toshiyuki Zaitsu[1], Riku Tatetsu[1], Kohei Sebata[1], Kazuki Miyanjou[2], Kimihiro Nishijima[2]
[1]ROHM Co., Ltd., Japan; [2]Sojo University, Japan

PRESENTATION   TOPIC: External AC-DC Adapters

D01.5 Active Soft Switching Technique for Single Phase Series Capacitive Link Universal Rectifier 1983
Anran Wei, Brad Lehman, Mahshid Amirabadi
Northeastern University, United States

PRESENTATION   TOPIC: Single-Phase and Three-Phase Input

D01.6 A Multi Mode Control Algorithm for Totem-Pole Bridgeless PFC 1990
Bosheng Sun, Sheng-Yang Yu, Amir Hussain
Texas Instruments Incorporated, United States

PRESENTATION   TOPIC: Power Factor Correction: CCM, DCM, CRM/BCM Control, Bridgeless

D01.8 Protection Strategy for Flying Capacitor Totem-Pole PFC Under the AC Drop Transient 1995
Yanqing Wu[1], Wending Zhao[1], Zhenhai Zhu[2], Xinke Wu[1]
[1]Zhejiang University, China; [2]Fuzhou University, China

PRESENTATION   TOPIC: Power Factor Correction: CCM, DCM, CRM/BCM Control, Bridgeless

D01.10 **Three-Phase with Three Single-Phase Single-Stage Isolated AC-DC Converters for EV Charging Station Applications** 2002
Misha Kumar, Peter M. Barbosa, Juan M. Ruiz
Delta Electronics, Inc., United States

> PRESENTATION

TOPIC: Single-Phase and Three-Phase Input

D01.11 **400V SiC in Next-Generation 3-Level Flying Capacitor Bridgeless Totem-Pole PFC** 2009
Rytis Beinarys[1], Seamus O'Driscoll[2]
[1]*ICERGi Limited, Ireland;* [2]*Tyndall National Institute, Ireland*

> PRESENTATION

TOPIC: Power Factor Correction: CCM, DCM, CRM/BCM Control, Bridgeless

D01.12 **Extended Smart-Link Quasi-Single-Stage 3-Phase AC-DC Power Supply Module for AI-Driving Data Centers** 2014
D. Biadene[1], J. Huber[2], J.W. Kolar[2], P. Mattavelli[1]
[1]*Università degli Studi di Padova, Italy;* [2]*ETH Zürich, Switzerland*

> PRESENTATION

TOPIC: Embedded AC-DC Power Supplies

D01.13 **A New Three-Phase Multi-Mode AC/DC LLC Converter with Output-Controlled Active Rectifier (with V2G and G2V Functions) for Fast DC Charging Application** 2022
Xiaoyi Xia, John Lam
York University, Canada

> PRESENTATION

TOPIC: Single-Phase and Three-Phase Input

SESSION D02: High Voltage DC-DC Converters I

11:30 - 13:30
TRACK: DC-DC Converters

SESSION CHAIR(S)
Saad Pervaiz, *Texas Instruments*
Boyi Zhang, *Delta Electronics, Inc.*

D02.1 **Capacitorless Notch Resonant Converters for Miniaturized LLCLC Resonant Converters in Electric Vehicle Charging Applications** 2029
Haitham Kanakri, Euzeli Cipriano Dos Santos Jr., Maher Rizkalla
Purdue University, United States

> PRESENTATION

TOPIC: Resonant Converters

D02.2 **Multiple-Core Transformer Design based on Half-Turn Structure in Two-Stage DC-DC Converter for Battery Storage System** 2035
Yilei Li, Bima Nugraha Sanusi, Pinhe Wang, Tianming Luo
Technical University of Denmark, Denmark

> PRESENTATION

TOPIC: Resonant Converters

D02.3 Bidirectional DC-DC Converter Utilizing Coupled Inductors for Energy Storage System 2043

Wen-Hsuan Lee, Jiann-Fuh Chen, Hsuan Liao, Kuo Fu Liao
National Cheng Kung University, Taiwan

PRESENTATION

TOPIC: **Bidirectional DC-DC Converters**

D02.4 Comparison of 2-Level and Quasi-2-Level Topologies in a Bidirectional Isolated DC-DC Converter for MVDC Networks 2051

José Andrés Aguilar Croston[1], Jean-Yves Gauthier[2], Cyril Buttay[2], Maryam Saeedifard[3], Besar Asllani[1], Piotr Dworakowski[1]
[1]*SuperGrid Institute, France;* [2]*Université de Lyon, INSA Lyon, Laboratoire Ampère, France;* [3]*Georgia Institute of Technology, United States*

PRESENTATION

TOPIC: **Bidirectional DC-DC Converters**

D02.5 Sling Forward Converter for Offline Operation: Achieving High Efficiency and Wide Voltage Range Performance 2059

Nasherul Islam[1], Guozhu Chen[1], Honglei Miao[2], Fuxing Zhang[2]
[1]*Zhejiang University, China;* [2]*HNAC Technology Co., Ltd, China*

PRESENTATION

TOPIC: **Hard- and Soft-Switched**

D02.6 A Pulse Width Alternating Modulation Strategy for Three-Level Buck-Boost Converter 2066

Xinlong He, Caifeng Liu, Xudong Zou, Jiaao Zou, Tianyi Zhang, Yong Kang
Huazhong University of Science and Technology, China

PRESENTATION

TOPIC: **Bidirectional DC-DC Converters**

D02.7 ZVT Circuit Applied for Wide Input Range Isolated Converters 2070

Linguo Wang, Zhongyin Guo, Junjie Zhu, Bing Zhang, Zhiling Zuo, Xiaoguang Gao, Guangji Ma
ZTE Corporation, China

PRESENTATION

TOPIC: **Hard- and Soft-Switched**

D02.8 Impact of Asymmetrical Leakage Inductance on a 380 V-12 V LLC Converter with Synchronous Rectifier for DCX Application 2075

Jinshu Lin[1], Shan Yin[2], Chen Song[1], Honglang Zhang[1], Minhai Dong[1], Limei Xu[1], Hui Li[1]
[1]*University of Electronic Science and Technology of China, China;* [2]*Huawei Technologies Co., Ltd., China*

PRESENTATION

TOPIC: **Resonant Converters**

D02.9 Start-Up Techniques and Universal Closed Loop Control of Immittance Network based Resonant Converter 2082

Ripun Phukan, Misha Kumar, Randy Beckemeyer, Juan Ruiz, Peter Barbosa
Delta Electronics, Inc., United States

PRESENTATION

TOPIC: **Resonant Converters**

D02.10 Multi-Objective Efficiency-Oriented Optimization for DAB Converters Minimizing Current Stress and Backflow Power with Soft-Switching Assurance 2088

Kun Wang, Ian Laird, Jun Wang
University of Bristol, United Kingdom

PRESENTATION

TOPIC: **Bidirectional DC-DC Converters**

D02.12 An ISOP-PSFB PWM Converter based on Coupled Output Inductors and Phase-Shifted Modulation with Full ZVZCS Range 2096

Kang Hong, Guo Xu, Guangfu Ning, Mei Su
Central South University, China

PRESENTATION

TOPIC: Hard- and Soft-Switched

SESSION D03: Low Voltage DC-DC Converters I

11:30 - 13:30
TRACK: DC-DC Converters

SESSION CHAIR(S)
David Reusch, *Texas Instruments*
Justin Henspeter, *IBM*

D03.1 Design and Implementation of a GaN-Based Soft-Switched Series-Capacitor Buck Converter Operating at the CCM-DCM Boundary for High-Performance Computing Systems 2101

Ramin Rahimzadeh Khorasani, Kolman Puterman Ghitelman, Madhavan Swaminathan
The Pennsylvania State University, United States

PRESENTATION

TOPIC: Voltage Regulator Modules (VRM)

D03.2 Intrinsic Feedback Model for Coupled-Damped Self-Balancing of General Multiphase Hybrid Converters 2109

Haoran Xu[1], Weijia Hao[1], Desheng Zhang[2], Run Min[1], Qiaoling Tong[1], Xuecheng Zou[3]
[1]*Huazhong University of Science and Technology, China;* [2]*Wuhan University of Technology, China;*
[3]*Henan Academy of Sciences, China*

PRESENTATION

TOPIC: Point-of-Load (POL) and Multi-Phase Converters

D03.3 A High-Efficiency Switching Oscillation Suppression Strategy based on Damped Oscillation for Synchronous DC-DC Converter 2117

Hao Yuan[1], Chuan Ni[2], Zhengyu Ye[2], Wei Lu[2], Hui Xue[2], Ting Qian[1]
[1]*Tongji University, China;* [2]*LEN Technology Ltd., China*

PRESENTATION

TOPIC: Hard- and Soft-Switched

D03.4 Efficient and Streamlined Demodulation Strategy for High-Frequency Talkative DC-DC Converters 2125

Abdelmoumin Allioua, Hendrik Gockel, Gerd Griepentrog
Technical University of Darmstadt, Germany

PRESENTATION

TOPIC: Hard- and Soft-Switched

D03.5 A 90.9% Peak Efficiency KY Single-Inductor Bipolar-Output Converter with Conductance Modulation Controller for Active-Matrix Organic Light Emitting Diode Power Supply 2131

Sheng-Han Yu, Chieh-Ju Tsai, Hao-Ran Huang, Ching-Jan Chen
National Taiwan University, Taiwan

PRESENTATION

TOPIC: Voltage Regulator Modules (VRM)

D03.6 Constant-on-Time Control for Zero-Bias Trans-Inductor Voltage Regulators 2138
Hank Zeng, Justin Lee, Rixin Lai, Hang Shao
Monolithic Power Systems, Inc., United States

PRESENTATION TOPIC: Point-of-Load (POL) and Multi-Phase Converters

D03.7 An Improved PFM Control Scheme for Three-Level Buck Converter based on
Ton Extension Achieving an 810% Frequency Reduction 2143
Yi-Chun Chang, Chieh-Ju Tsai, Ting-Lun Lee, Ching-Jan Chen
National Taiwan University, Taiwan

PRESENTATION TOPIC: Voltage Regulator Modules (VRM)

D03.8 A Concept for Current Ripple or Transient Improvements in Multiphase Converters 2149
Alexandr Ikriannikov, Alex Gao
Analog Devices, Inc., United States

PRESENTATION TOPIC: Point-of-Load (POL) and Multi-Phase Converters

D03.9 System Solutions and Design Trade-Offs to the Input Filter Interactions with
Battery Chargers 2157
Xigen Zhou, Dan Mavencamp, Kuang-Yao Cheng
Texas Instruments Incorporated, United States

PRESENTATION TOPIC: Bidirectional DC-DC Converters

D03.10 Modeling and Implementation of a Zero Bias TLVR 2162
Lei Wang, Travis Guthrie, Peyman Asadi, Mark Alexander, Kunrong Wang, Brandon Howell
Renesas Electronics Corporation, United States

PRESENTATION TOPIC: Point-of-Load (POL) and Multi-Phase Converters

SESSION D04: High Voltage DC-DC Converters II

11:30 - 13:30
TRACK: DC-DC Converters

SESSION CHAIR(S)
Branko Majmunovic, *Texas Instruments*
Shuang Zhao, *Hefei University of Technology*

D04.2 cGANET-Enhanced Voltage Gain Modeling: Elevating CLLC Converter Accuracy 2167
Yu Zuo, Xiaobing Shen, Fanghao Tian, Jiaze Kong, Hans Wouters, Wilmar Martinez
KU Leuven, Belgium

PRESENTATION TOPIC: Resonant Converters

D04.3 Capacitive vs Inductive Coupling based DC-DC Converter Operating in
MHz Switching Frequency Range 2173
Saeid Pourjafar[1], Parham Mohseni[1], Oleksandr Husev[2], Ryszard Strzelecki[2], Oleksandr Matiushkin[3]
[1]Tallinn University of Technology, Estonia; [2]Gdansk University of Technology, Poland;
[3]University of Extremadura, Spain

PRESENTATION TOPIC: Resonant Converters

D04.4 **LLC Converter Main Transformer Losses: Eliminating Air Gaps and Integrating Parallel External Inductors** 2179
Yu-Chen Liu, Shang-Syun Wu
National Taipei University of Technology, Taiwan

PRESENTATION   TOPIC: Resonant Converters

D04.5 Comparative Analysis of Modulation Techniques for a Wide Input Voltage Range
Dual-Active-Bridge DC-DC Converter **WITHDRAWN** 2187
Alper Soysal[1,2], Enes Çatlıoğlu[1], Eyyup ... dır El[1], Kübra Uludağ[1]
[1]Aselsan Inc., Türkiye; [2]Istanbul Technical University, Türkiye

PRESENTATION   TOPIC: Bidirectional DC-DC Converters

D04.6 **Small-Signal Phasor Modeling of T-Type Bridge-Based Single-Sided and Double-Sided LCC Resonant Converters for WPT Applications** 2194
Aditya Zade, Shubhangi Gurudiwan, Regan Zane
Utah State University, United States

PRESENTATION   TOPIC: Resonant Converters

D04.7 **A Hybrid Three-Port Topology for Urban Charging Stations** 2202
Mohammadreza Khodaparast Klidbari[1], Naser Souri[2], Zahra Sadat Habibolahi[3], Hamid Montazeri Hedeshi[4]
[1]K.N. Toosi University of Technology, Iran; [2]Virginia Polytechnic Institute and State University, United States; [3]Iran University of Science and Technology, Iran; [4]University of Tabriz, Iran

PRESENTATION   TOPIC: Resonant Converters

D04.8 **Reconfigurable H5-Bridge based LLC-DAB Sigma Converter for EV Fast Charging Stations** 2207
Huangsheng Xu, Mingde Zhou, Qishan Pan, Haoyu Wang
Shanghai Tech University, China

PRESENTATION   TOPIC: Bidirectional DC-DC Converters

D04.9 **A Resonant Reset Forward Converter with Ultra-High Conversion Gain using Differential Transformation Technique (DTT)** 2213
Shubham Srivastava, Mandeep S. Rana, Santanu K. Mishra
Indian Institute of Technology Delhi, India

PRESENTATION   TOPIC: Resonant Converters

D04.10 **Full-Range ZVS Modulation of Switched Capacitor Converter for Sensorless Voltage Balancing** 2220
Md Tanvir Ahammed, Wensong Yu
North Carolina State University, United States

PRESENTATION   TOPIC: Resonant Converters

D04.12 Dimensional Parasitics Absorption in Capacitively-Isolated Ćuk Converter for Medium-Voltage High Step-Down Converters 2228
Aakash Kamalapur, Jung-Soo Bae, Mark Cairnie, Rajaie Nassar, Jack Knoll, Dushan Boroyevich,
Guo-Quan Lu, Christina DiMarino, Qiang Li, Khai D. T. Ngo
Virginia Polytechnic Institute and State University, United States

PRESENTATION

TOPIC: Hard- and Soft-Switched

SESSION D05: Low Voltage DC-DC Converters II

11:30 - 13:30
TRACK: DC-DC Converters

SESSION CHAIR(S)
Yicheng Zhu, *University of California at Berkeley*
Peyman Asadi, *Renesas Electronics*

D05.1 A 36-to-60V Input Dual-Phase 2MHz 93%-Efficiency ZVT Series-Parallel Hybrid Buck Converter using Single Auxiliary Inductor and Adaptive Time Multiplexing Control 2236
Qi Cheng[1,2], Hoi Lee[1]
[1]University of Texas at Dallas, United States; [2]Texas Instruments Incorporated, United States

PRESENTATION

TOPIC: Hard- and Soft-Switched

D05.2 Improved Efficiency in a 10 W Class-Φ_2 Converter Utilizing a Resonant Gate Drive 2241
Malachi Hornbuckle, Katherine Liang, Juan Rivas-Davila
Stanford University, United States

PRESENTATION

TOPIC: Resonant Converters

D05.3 The Analysis and Design of a Resonant Capacitively-Isolated Cockcroft-Walton Converter 2249
Elizabeth Rabenold, Raiphy Jerez, Samantha Coday
Massachusetts Institute of Technology, United States

PRESENTATION

TOPIC: Resonant Converters

D05.4 SHSC: Non-Isolated High-Density 4:1 IBC for 48 V Applications 2254
Mario Ursino, Roberto Rizzolatti, Simone Mazzer
Infineon Technologies AG, Austria

PRESENTATION

TOPIC: Resonant Converters

D05.5 High-Performance Current Multiplier: A Hybrid Switched Capacitor Solution for High-Current Applications 2260
Kevin Zufferli[1], Roberto Rizzolatti[1], Mario Ursino[1], Simone Mazzer[1], Gerald Deboy[1], Stefano Saggini[2]
[1]Infineon Technologies AG, Austria; [2]Università degli Studi di Udine, Italy

PRESENTATION

TOPIC: Resonant Converters

D05.6 **Representation and Design Methodology for Generalized Switched-Capacitor Converter Topologies** 2268
Seokwon Choi, Dam Yun, Jung-Ik Ha
Seoul National University, Korea

PRESENTATION

Topic: **Hard- and Soft-Switched**

D05.8 **A 48-V-to-1-V Gallium Nitride Switching Bus Converter for Processor Vertical Power Delivery with 2.7 mm Thickness and 3048 W/in³ Power Density** 2276
Jiarui Zou[1], Yicheng Zhu[1], Nathan M. Ellis[2], Logan Horowitz[1], Robert C. N. Pilawa-Podgurski[1]
[1]*University of California, Berkeley, United States;* [2]*University of California, Santa Cruz, United States*

PRESENTATION

Topic: **Point-of-Load (POL) and Multi-Phase Converters**

D05.9 **Ripple Reduction and Efficiency Improvement of Always-Dual-Path Hybrid DC-DC Converter based on Phase Shift Operation** 2284
Katsuhiro Hata[1], Shinsaku Tanaka[2], Toru Ashikaga[2], Yasuhiro Rikiishi[2]
[1]*Shibaura Institute of Technology, Japan;* [2]*Sanken Electric Co., Ltd., Japan*

PRESENTATION

Topic: **Voltage Regulator Modules (VRM)**

SESSION D06: DC-AC Inverters

11:30 - 13:30
TRACK: DC-AC Inverters

SESSION CHAIR(S)
Woongkul Lee, *Purdue University*
Karun Arjun Potty, *Lucid Motors*

D06.1 **Ultralocal PQ Theory: A New Approach for Model-Free Predictive Direct Power Control of Shunt Active Power Filters** 2290
Mahdi S. Mousavi[1], Abolfazl Nassaji[2], Ibrahim Harbi[3], Behnam Nikmaram[1],
S. Alireza Davari[1], Mokhtar Aly[4], José Rodriguez[4]
[1]*Shahid Rajaee Teacher Training University, Iran;* [2]*University of Science and Culture, Iran;*
[3]*Technical University of Munich, Germany;* [4]*Universidad San Sebastián, Chile*

PRESENTATION

Topic: **Power Quality and EMI**

D06.3 **Symmetrical Balanced Circuit for Common-Mode Noise Mitigation in LCL-T Resonant Converter** 2296
Ripun Phukan, Boyi Zhang, Juan Ruiz, Peter Barbosa
Delta Electronics, Inc., United States

PRESENTATION

Topic: **Power Quality and EMI**

D06.4 **A Single-Phase Soft-Switching Buck-Boost Inverter** 2303
Lukas Wipprecht, Burkhard Ulrich
Reutlingen University, Germany

PRESENTATION

Topic: **Single and Multi-Phase Inverters**

D06.5 Low-Complexity Model Predictive Control Method for Driving Dual Induction Motors Fed by Five-Leg Inverter 2311

Jun Young Lee, Eun Woo Lee, Dongho Choi, June-Seok Lee
Dankook University, Korea

PRESENTATION

TOPIC: Single and Multi-Phase Inverters

D06.6 Overvoltage Mitigation Filter using High-Frequency Cable Modeling in Long Transmission Lines for Silicon Carbide Inverter Systems 2317

Yun-Jin Lee, Kyo-Beum Lee
Ajou University, Korea

PRESENTATION

TOPIC: Power Quality and EMI

D06.7 Power Delivery Network (PDN) Design and Analysis to Achieve Low Impedance in Fast Edge Rate DC-DC Converters for EMI Compliance 2322

Manraj Singh Ladhar, Sheldon Williamson
Ontario Tech University, Canada

PRESENTATION

TOPIC: Power Quality and EMI

D06.9 Enhancing the Performance of Dual Input Split Source Inverters using an Advanced Modulation Strategy 2327

Mustafa Abu-Zaher[1,2], Fang Zhuo[1], Mokhtar Aly[3], Mahmoud A. Gaafar[4], Mohamed Orabi[4], José Rodriguez[3], Alaaeldien Hassan[5], Jiachen Tian[1], Samir Kouro[6]
[1]Xi'an Jiaotong University, China; [2]Sohag University, Egpyt; [3]Universidad San Sebastián, Chile; [4]Aswan University, Egypt; [5]South Valley University, Egypt; [6]Universidad Tecnica Federico Santa Maria, Chile

PRESENTATION

TOPIC: Single and Multi-Phase Inverters

D06.12 A Novel GaN-HEMT Single-Phase Single-Stage Buck-Boost Micro-Inverter Topology for PV Applications 2332

Pengwei Li, Uiliam Kutrolli, Ali Bazzi
University of Connecticut, United States

PRESENTATION

TOPIC: Single and Multi-Phase Inverters

SESSION D07: Power Devices and Components I

11:30 - 13:30
TRACK: Devices and Components

SESSION CHAIR(S)
Raghav Khanna, *University of Toledo*
Sushanta Gautam, *University of Toledo*

D07.1 Influence of Threshold Voltage Temperature Dependency on Junction Temperature Imbalance in Paralleled SiC MOSFETs .... N/A

Andrea Piccioni, Thomas Aichinger
Infineon Technologies AG, Austria

PRESENTATION

TOPIC: GaN and SiC Devices and Modules

D07.2 A Dynamic Current Sharing Method using Novel Clip Considering Mutual Inductance Coupling 2343

Zexiang Zheng[1], Jianwei Lv[1], Yiyang Yan[1], Baihan Liu[1], Yifan Zhang[1], Linhao Ren[1], Jiaxin Liu[1], Cai Chen[1], Yong Kang[1], Xiong Zhang[2], Hao Yu[2], Wei Jiang[2]
[1]Huazhong University of Science and Technology, China; [2]GAC Aion New Energy Automobile Co., Ltd., China

PRESENTATION   TOPIC: GaN and SiC Devices and Modules

D07.3 Application-Oriented Test Setup for Measuring Dynamic Output and Transfer Characteristics of GaN-HEMTs 2348

Philipp Swoboda, Martin Fein, Simon Frank, Andreas Liske, Marc Hiller
Karlsruhe Institute of Technology, Germany

PRESENTATION   TOPIC: GaN and SiC Devices and Modules

D07.4 Mitigating Gate Voltage Oscillation in Parallel SiC Power Modules for xEV 2356

Hideo Komo[1], Michael Rogers[1], Mark Steiner[1], Eric Motto[1], Koichi Taguchi[2], Chihiro Kawahara[2], Junichi Nakashima[2], Yasushige Mukunoki[2], Seiichiro Inokuchi[2], Rei Yoneyama[2]
[1]Mitsubishi Electric US, Inc., United States; [2]Mitsubishi Electric Corporation, Japan

PRESENTATION   TOPIC: GaN and SiC Devices and Modules

D07.5 Switching Performance Comparison of Low-Voltage GaN and Si Devices 2361

Tianxiao Chen[1,3], Haoyang Liu[3], Pedro A.M. Bezerra[3], Eckart Hoene[2], Sibylle Dieckerhoff[1]
[1]Technische Universität Berlin, Germany; [2]Fraunhofer IZM, Germany;
[3]Huawei Technologies Duesseldorf GmbH, Germany

PRESENTATION   TOPIC: GaN and SiC Devices and Modules

D07.6 Modeling of Switching Transients for Frequency-Domain CM EMI Analysis in Double Sided Cooling Power Modules 2369

Sijia Liu[1], Liu Yang[2], Heng Zhang[1], Yifan Zhang[1], Zexiang Zheng[1], Jianwei Lv[1], Jiaxin Liu[1], Cai Chen[1], Yong Kang[1], Yuebin Zhou[2], Daming Wang[2], Shuang Zhao[3]
[1]Huazhong University of Science and Technology, China; [2]State Key Laboratory of HVDC Electric Power Research Institute of China Southern Power Grid, China; [3]Hefei University of Technology, China

PRESENTATION   TOPIC: GaN and SiC Devices and Modules

D07.7 Leakage Current Detection Scheme for Aging Test of 10kV SiC MOSFET Power Module 2375

Peiyang Ding[1], Hong Zhang[1], Tianshu Yuan[1], Qiling Chen[1], Jiacheng Guo[1], Dingkun Ma[1], Peiyuan Sun[1], Ting Hou[2], Laili Wang[2]
[1]Xi'an Jiaotong University, China; [2]China Southern Power Grid Electric Power Research Institute, China

PRESENTATION   TOPIC: GaN and SiC Devices and Modules

D07.9 Physics-Informed Neural Network Approach for Early Degradation Trajectory Prediction of Power Semiconductor Modules 2380

Jie Kong[1], Yi Zhang[1], Yichi Zhang[1], Lukas Wick[2], Frederik Lillebæk Hansen[2], Dao Zhou[1], Huai Wang[1]
[1]Aalborg University, Denmark; [2]Neurospace, Denmark

PRESENTATION   TOPIC: Power Silicon MOSFETs, BJTs, IGBTs, etc.

D07.10 **Nonlinear Output Capacitance of Bidirectional Gallium Nitride Power Switches** 2387

Michael Bosch, Jeremy Nuzzo, Dominik Koch, Mathias C.J. Weiser, Ingmar Kallfass
Universität Stuttgart, Germany

PRESENTATION

TOPIC: **GaN and SiC Devices and Modules**

SESSION D08: Power Devices and Components II

11:30 - 13:30
TRACK: Devices and Components

SESSION CHAIR(S)
Raghav Khanna, *University of Toledo*
Samuel Atwimah, *University of Toledo*

D08.3 **Novel Approach of Determining and Predicting SiC MOSFET's on Resistance from Device Case Temperature using Machine Learning** 2393

Paul Bradford, Conner Deppe, Hongjie Wang
Utah State University, United States

PRESENTATION

TOPIC: **Power Silicon MOSFETs, BJTs, IGBTs, etc.**

D08.4 **Comparison of Static Characteristics in GaN HEMTs Across 50K to 400K Considering Diverse Techniques and Statistical Variation** 2400

Purushottam Khadka, Saumil Shivdikar, Zheyu Zhang, Tian Qiu, Ahmed Siraj
Rensselaer Polytechnic Institute, United States

PRESENTATION

TOPIC: **GaN and SiC Devices and Modules**

D08.5 **Compact Model of β-Ga$_2$O$_3$ Schottky Barrier Diode** 2407

Abu Shahir Md Khalid Hasan, Mohammad Dehan Rahman, Tanzila Akter,
Md Majharul Islam, Md Maksudul Hossain, Xiaoqing Song, H. Alan Mantooth
University of Arkansas, United States

PRESENTATION

TOPIC: **Ultra-Wide Bandgap Devices**

D08.6 **DC-Link Capacitor Board Design for Low Parasitic Inductance** 2413

Mikayla Benson[1], Lifang Yi[2], Kangbeen Lee[3], Jinyeong Moon[2], Woongkul Lee[3]
[1]Michigan State University, United States; [2]Florida State University, United States;
[3]Purdue University, United States

PRESENTATION

TOPIC: **Capacitors, Supercapacitors**

D08.7 **First Demonstration of a Gallium Oxide Power Converter** 2419

Joshua J. Piel, Elizabeth A. Sowers, Daniel M. Dryden, Thaddeus J. Asel, Adam T. Neal,
Brenton A. Noesges, Shin Mou, Andrew J. Green
Air Force Research Laboratory, United States

PRESENTATION

TOPIC: **Ultra-Wide Bandgap Devices**

D08.8 **Optimized Integrated EMI Filter Design in SiC Power Modules with Terminal Inductor for Better High-Frequency EMI Suppression** 2426

Yifan Zhang[1], Wenzhe Xu[1], Jianwei Lv[1], Yiyang Yan[1], Baihan Liu[1], Sijia Liu[1], Jiaxin Liu[1], Cai Chen[1], Yong Kang[1], Xiong Zhang[2], Hao Yu[2], Wei Jiang[2]

[1]Huazhong University of Science and Technology, China; [2]GAC Aion New Energy Automobile Co., Ltd., China

PRESENTATION

TOPIC: **GaN and SiC Devices and Modules**

SESSION D09: Magnetics Characterization and Designs

11:30 - 13:30
TRACK: Magnetics

SESSION CHAIR(S)
Edward Herbert, *PSMA*
George Slama, *Wurth Elektronik*

D09.1 **Balanced Technique using Integrated Winding Coupled Inductor for High-Power Density Two-Phase Interleaved Boost Converter** 2431

Yuta Imaeda[1], Jun Imaoka[1], Masayoshi Yamamoto[1], Hiroyuki Onishi[2]

[1]Nagoya University, Japan; [2]ROHM Co., Ltd., Japan

PRESENTATION

TOPIC: **Winding Techniques**

D09.2 **MagNetX: Foundation Neural Network Models for Simulating Power Magnetics in Transient** 2438

Shukai Wang[1], Hyukjae Kwon[1], Haoran Li[1], Youssef Elasser[2], Gyeong-Gu Kang[1], Daniel Zhou[1], Davit Grigoryan[1], Minjie Chen[1]

[1]Princeton University, United States; [2]NVIDIA Corporation, United States

PRESENTATION

TOPIC: **Magnetics Modeling and Simulations**

D09.3 **Revisiting Models of Common Mode Inductors to Include the Magnetized Capacitance Effect** 2446

Rafael Bogo Portal Chagas[1], Marcelo Lobo Heldwein[2]

[1]Nidec Global Appliance, Brazil; [2]Technical University of Munich, Germany

PRESENTATION

TOPIC: **Magnetics Modeling and Simulations**

D09.4 **A High Frequency Coupled Inductor with Distributed Air Gap for High Power DC-DC Converters** 2453

Muhammad Fasih Uddin, Ahmed H. Ismail, Peyman Darvish, Baher Abu Sba, Yue Zhao

University of Arkansas, United States

PRESENTATION

TOPIC: **High-Frequency Magnetics**

D09.5 **High-Power Planar Transformer Design for Four-Port Converters** 2461

Arya Sadasivan, Behrooz Mirafzal

Kansas State University, United States

PRESENTATION

TOPIC: **High-Frequency Magnetics**

D09.6 **Optimal Design of Inductors with Aluminum Litz Wire for Inductive Power Transfer Systems** 2468
Jesús Acero, Claudio Carretero, Ignacio Lope, Óscar Lahuerta, José-Miguel Burdío
University of Zaragoza, Spain

> PRESENTATION

TOPIC: **Winding Techniques**

D09.7 **Analytic Design of Flat-Wire Inductors for High-Current and Compact DC-DC Converters** 2474
Sajjad Mohammadi[1], James L. Kirtley[1], Alireza Namadmalan[2]
[1]Massachusetts Institute of Technology, United States; [2]Bourns Electronics Ireland, Ireland

> PRESENTATION

TOPIC: **Magnetics Modeling and Simulations**

D09.8 **Insulation Dielectric Loss of High-Frequency Transformer Under Square Voltage Excitation with Edge Oscillation** 2482
Zhanlei Liu[1], Lingyu Zhu[1], Yuntian Gao[1], Yongliang Dang[1], Cao Zhan[2], Shengchang Ji[1]
[1]Xi'an Jiaotong University, China; [2]Virginia Polytechnic Institute and State University, United States

> PRESENTATION

TOPIC: **High-Frequency Magnetics**

SESSION D10: Magnetics Applications II

11:30 - 13:30
TRACK: Magnetics

SESSION CHAIR(S)
Matt Wilkowski, *Wurth Elektronik*
George Slama, *Wurth Elektronik*

D10.1 **Improved High-Speed Thermal Analysis based on Two-Step Simulation for High-Frequency Transformers** 2488
Zheyuan Yi[1,2], Kai Sun[1], Qiang Li[3], Zengyang Liu[1]
[1]Tsinghua University, China; [2]Delta Electronics Shanghai Co Ltd, China;
[3]Virginia Polytechnic Institute and State University, United States

> PRESENTATION

TOPIC: **Magnetics Modeling and Simulations**

D10.2 **Core Material Characterization Under DC Bias Conditions** 2495
Jonas Mühlethaler[1,3], Fabrice Locher[1], Frédéric Mathieu[1], Edward Herbert[2]
[1]Lucerne University of Applied Sciences and Arts, Switzerland; [2]Power Sources Manufacturers Association, United States; [3]Frenetic Switzerland GmbH, Switzerland

> PRESENTATION

TOPIC: **Magnetics Modeling and Simulations**

D10.3 **A Low-Cost Setup and Procedure for Measuring Losses in Inductors** 2502
Burkhard Ulrich
Reutlingen University, Germany

> PRESENTATION

TOPIC: **High-Frequency Magnetics**

D10.4 **Effect of Temperature of Additively Manufactured Cores** 2510
Ken Johnson, Ali Bazzi
University of Connecticut, United States

PRESENTATION

TOPIC: **Additive Manufacturing for Magnetic Materials**

D10.6 **Extreme Temperature Permeability Engineered Soft Magnetics** 2516
Tyler W. Paplham[1], Alex M. Leary[2], Paul R. Ohodnicki, Jr.[1]
[1]*University of Pittsburgh, United States;* [2]*NASA Glenn Research Center, United States*

PRESENTATION

TOPIC: **Advanced Magnetic Materials and Geometries**

D10.7 **An Isolated RF Power Combining Approach with Multiple Decoupled Input Coils** 2521
Ziyang Xu[1], Yifan Zhao[2], Zhan Liu[3], Alex J. Hanson[1], Ming Liu[3]
[1]*The University of Texas at Austin, United States;* [2]*Shandong University, China;*
[3]*Shanghai Jiao Tong University, China*

PRESENTATION

TOPIC: **High-Frequency Magnetics**

D10.8 **Simulation of a Custom Core, 15kV Isolated Gap Transformer Optimized for**
High Power Density 2527
Andrew Galamb, Fei Teng, Srdjan Lukic
North Carolina State University, United States

PRESENTATION

TOPIC: **Magnetics Modeling and Simulations**

D10.9 **Low Interwinding Capacitance Design for PCB-Winding based Transformer in**
Self-Powered Gate Drive Power Supply for High-Voltage SiC MOSFET 2535
Yuan Zhou, Li Zhang, Yilun Chen, Tianxiang Yin, Lei Lin
Huazhong University of Science and Technology, China

PRESENTATION

TOPIC: **High-Frequency Magnetics**

SESSION D11: Power Converter Design, Packaging and Integration

11:30 - 13:30
TRACK: Power Electronics Integration and Manufacturing

SESSION CHAIR(S)
Rafal Wojda, *Oak Ridge National Laboratory*
Jason Neely, *Sandia National Laboratories*

D11.1 **Integrated 4-Level Dual-Phase Superimposed Quadratic Power Converter for**
High-Density Direct 48V/1V Conversion 2541
Prosenjit Ghosh, Jin Woong Kwak, Fei Zhou, D. Brian Ma
The University of Texas at Dallas, United States

PRESENTATION

TOPIC: **Power Modules / High Density Design**

D11.2 **Compensation Method for Unbalance of the Multi-Channel Class E Power Amplifier using the Closed Loop Frequency Control** 2547
Kyungmin Lee, Sungku Yeo
Samsung Electronics, Korea

PRESENTATION   TOPIC: **Quality and System Reliability Including EMI/EMC**

D11.3 **High Temperature Operation of Digital Gate Driver Integrated Into a Power Module** 2551
Kazuma Saiga[1], Shohei Zaizen[1], Satoshi Nakano[1], Shigeru Kusunoki[1], Kiyoto Watabe[1], Katsuhiro Hata[2], Makoto Takamiya[3], Shin-Ichi Nishizawa[1], Wataru Saito[1]
[1]Kyushu University, Japan; [2]Shibaura Institute of Technology, Japan; [3]The University of Tokyo, Japan

PRESENTATION   TOPIC: **Power Modules / High Density Design**

D11.5 **Evaluation Index-Based Multiphysics Coupling Model and Analysis Methodology for High-Reliable Power Supply Module** 2556
Haoyu Wang, Xuliang Wang, Yan Wang, Xiaosen Liu
Tsinghua University, China

PRESENTATION   TOPIC: **Quality and System Reliability Including EMI/EMC**

D11.6 **Electrical Characterization of Modular 3D Packaging Assembled with Compressed Metal Foams** 2562
Paul Bruyere[1,2], Alexis Derbey[1], Betina Zynger-Capaverde[1], Yvan Avenas[1], Eric Vagnon[2], Jean-Luc Schanen[1], Jean-Michel Guichon[1], Omar Sanjakdar[3]
[1]Université Grenoble Alpes, CNRS, Grenoble INP, G2Elab, France; [2]Ecole Centrale de Lyon, CNRS, Ampère, France; [3]Université Grenoble Alpes, CEA-Liten, INES, France

PRESENTATION   TOPIC: **Power Electronics Packaging**

D11.7 **Improvement in Short-Circuit Robustness of SiC-MOSFETs based Power Modules using Two-Level Turn-On (2LTO)** 2569
Muhammad Muneeb Alam, Saad Khalid, Nisar Ahmed Khan, Ngoc Ho Tran, Sebastian Strache
Robert Bosch GmbH, Germany

PRESENTATION   TOPIC: **Power Modules / High Density Design**

D11.8 **GaN-Based Two Stage Point-of-Load (PoL) Converter with 2.5D Embedded Substrate Implementation** 2576
Samuel Defaz, Yang Li, Fang Luo
Stony Brook University, United States

PRESENTATION   TOPIC: **Embedded Technologies, 3D Packaging, and Additive Manufacturing**

D11.9 **Near-Field Coupling Mitigation of the Noise from High Voltage DC-Link Decoupling Capacitors in Voltage Source Converters** 2582
Yuxuan Wu, Kushan Choksi, Samuel Defaz, Fang Luo
Stony Brook University, United States

PRESENTATION   TOPIC: **Quality and System Reliability Including EMI/EMC**

D11.10 Advantages of Paralleling SiC MOSFETs in High-Performance Power Modules 2589
Steffen Beushausen, Dominik Alexander Ruoff, Wenqi Zhou, Karl Oberdieck
Robert Bosch GmbH, Germany

PRESENTATION   TOPIC: **Power Modules / High Density Design**

**D11.11 A SiC Half-Bridge Power Module based on Liquid Metal Packaging for
High Performance and Low Thermal Stress** 2597
Wei Mu, Ameer Janabi, Luke Shillaber, Borong Hu, Teng Long
University of Cambridge, United Kingdom

PRESENTATION   TOPIC: **Power Electronics Packaging**

**D11.12 Analysis and Modeling of Radiated EMI Considering Coupling Between
Power Converter and Power Cable with LC-Type EMI Filter** 2603
Qinghui Huang, Yingjie Zhang, Shuo Wang, Yirui Yang, Zhedong Ma, Yanwen Lai
University of Florida, United States

PRESENTATION   TOPIC: **Quality and System Reliability Including EMI/EMC**

**D11.13 Simple Prediction Method for Impacts of Switching Characteristics on
EMI Noise of a Three-Phase PWM Inverter** 2610
Shinobu Nagasawa, Toshiya Tadakuma, Keita Takahashi
Mitsubishi Electric Corporation, Japan

PRESENTATION   TOPIC: **Quality and System Reliability Including EMI/EMC**

**D11.14 Coaxially Nested 3.3 kV SiC MOSFET Packages with Uniform Interpackage
Electric Field Distribution** 2616
Jack Knoll, Mark Cairnie, Christina DiMarino
Virginia Polytechnic Institute and State University, United States

PRESENTATION   TOPIC: **Power Electronics Packaging**

**D11.15 Thermal Modeling and Performance of a Bare-Die Embedded PCB for
High Power Density Converters Design** 2624
Shahid Aziz Khan[1], Feng Zhou[2], Mengqi Wang[1], DucDung Le[1], Shivam Chaturvedi[1]
[1]University of Michigan-Dearborn, United States; [2]Toyota Research Institute of North America, United States

PRESENTATION   TOPIC: **Power Electronics Packaging**

SESSION D12: Control I

11:30 - 13:30
TRACK: Control

SESSION CHAIR(S)
Jaber Abu Qahouq, *The University of Alabama (UA)*
Nathan Weise, *Marquette University*

D12.1 Research on the Voltage Fluctuation Suppression Strategy in Weak Grid Under Pulsed Power Load Integration 2628

Xi Chen, Jiazheng Zhang, Mingjun Bao
Tianjin University, China

PRESENTATION
TOPIC: Control of Power Electronic Converters

D12.2 An Optimized Firmware-Based Cycle-by-Cycle Current Limiting Method for Power Electronic Converters in UPS 2634

Teng Wu, Hong Liu
Schneider Electric, China

PRESENTATION
TOPIC: Control of Power Electronic Converters

D12.3 Frequency Stop-Band Management System for DC-DC Converters 2640

Alessandro Bertolini, Alberto Cattani, Claudio Luise, Alessandro Gasparini
STMicroelectronics, Italy

PRESENTATION
TOPIC: Control of Power Electronic Converters

D12.4 Multi-Stage Model Predictive Control with Enhanced Discrete-Time Models for Multilevel Inverters 2647

Hoang Le[1], Apparao Dekka[1], Deepak Ronanki[2], Abdul R. Beig[3]
[1]*Lakehead University, Canada;* [2]*Indian Institute of Technology Madras, India;* [3]*Khalifa University, U.A.E.*

PRESENTATION
TOPIC: Control of Power Electronic Converters

D12.5 Direct Effective Power Control (D-EPC) for LLC Resonant Converters Operating in Boost Mode using Event-Driven-Timer based Digital Controller 2654

Yuto Yoshimura[1], Kenji Funatani[2], Kazuhiro Umetani[1], Toshiyuki Zaitsu[2], Akito Nakagaki[1], Masataka Ishihara[1], Eiji Hiraki[1]
[1]*Okayama University, Japan;* [2]*ROHM Co., Ltd., Japan*

PRESENTATION
TOPIC: Control of Power Electronic Converters

D12.6 Mitigation Method of Resonance Between Paralleled On-Line UPS 2660

Teng Wu, Zhenguo Huo, Shangxian Ning
Schneider Electric, China

PRESENTATION
TOPIC: Control of Power Electronic Converters

D12.7 An Extra-Element Small-Signal Model for a Current-Fed Resonant Dual-Active-Bridge Converter 2667

Paolo Sbabo, Paolo Mattavelli, Giorgio Spiazzi, Andrea Petucco
Università degli Studi di Padova, Italy

PRESENTATION
TOPIC: Control of Power Electronic Converters

D12.8 **Concurrent Charge Distribution and Time-Optimal Control for Unordered Single-Inductor Dual-Output Converter** 2675

Xuliang Wang[1], Haoyu Wang[1], Yang Liu[2], Yunxin Wang[1], Boran Zhang[1], Hongru Liu[1], Yan Wang[1], Xiaosen Liu[1]

[1]Tsinghua University, China; [2]The Hong Kong University of Science and Technology, China

PRESENTATION   TOPIC: Control of Power Electronic Converters

SESSION D13: Control II

11:30 - 13:30
TRACK: Control

SESSION CHAIR(S)
Mehdi Farasat, *Louisiana State University*
Narayan C. Kar, *University of Windsor*

D13.1 **Circulating Current Control with Loss Reduction for Parallel Connected Inverters** 2681

Shun Endo[1], Takae Shimada[1], Masato Ando[1], Yuuichi Mabuchi[1], Masaki Miyamae[1], Naoki Takayama[1], Yohei Matsumoto[2], Naoto Onuma[2]

[1]Hitachi Ltd., Japan; [2]Hitachi Building Systems Co., Ltd., Japan

PRESENTATION   TOPIC: Control of Power Electronic Converters

D13.2 **Analysis of Power and Power Spectral Density for Quaternary Random Pulse Position Modulation** 2687

Hung-Chi Chen[1], Hsiang-Kai Wu[1], Chih-Chiang Wu[2]

[1]National Yang Ming Chiao Tung University, Taiwan; [2]Industrial Technology Research Institute, Taiwan

PRESENTATION   TOPIC: Control of Power Electronic Converters

D13.3 **Bidirectional CLLC Converter using a Hybrid Control Method for Wide Voltage Range Applications** 2692

Jhih-Cheng Hu[1], Hong-Xuan Liao[1], Chien-Lung Liu[2], Wei Wang[1], Ming-Shi Huang[1]

[1]National Taipei University of Technology, Taiwan; [2]Delta Electronics Inc., Taiwan

PRESENTATION   TOPIC: Control of Power Electronic Converters

D13.4 **Design and Control of a High-Bandwidth Dual Active Bridge DC-DC Converter** 2698

Alper Uzum[1], Syed Imam Hasan[1], Yilmaz Sozer[1], Kenneth A. Loparo[2]

[1]The University of Akron, United States; [2]Case Western Reserve University, United States

PRESENTATION   TOPIC: Control of Power Electronic Converters

D13.5 **Unified Model Predictive Control for DC-DC Buck Converters: From Start-Up to Steady-State Operation** 2703

Zhengchen Guo, R.M. Nelms
Auburn University, United States

PRESENTATION   TOPIC: Control of Power Electronic Converters

D13.7 **A Novel IPPC Method for Precise Overload Protection and Burst Mode Operation in LLC Resonant Converters** 2708
Manikanta Pallantla, Ramkumar S
Texas Instruments Incorporated, India

PRESENTATION

TOPIC: Control of Power Electronic Converters

D13.8 **An Improved Current-Sensorless Model Predictive Voltage Control for Four-Leg Voltage Source Inverters** 2713
Heng Guo, Yuxin Wei, Mengmeng Jing, Wenlong Ding, Bin Duan, Chenghui Zhang
Shandong University, China

PRESENTATION

TOPIC: Control of Power Electronic Converters

SESSION D14: Control IV

11:30 - 13:30
TRACK: Control

SESSION CHAIR(S)
Mohamed Gamal Hussien, *Tanta University*
Nidhi Haryani, *Delta Electronics, Inc.*

D14.1 **A Highly Integrable, Modular and Multi-Functional Fault Monitoring Active Gate Driver with Parallel Buffers for a Global Enhanced Reliability of Gen. 3 SiC Power MOSFETs** 2718
Mathis Picot-Digoix[1], Léo Seugnet[2], Frédéric Richardeau[2], Jean-Marc Blaquière[2], Sébastien Vinnac[2], Thanh-Long Le[1], Stéphane Azzopardi[1]
[1]*Safran Tech, France;* [2]*LAPLACE Laboratory, Université de Toulouse, CNRS, INPT, UPS, France*

PRESENTATION

TOPIC: Gate Drive Circuits and Fault Protection

D14.2 **A 24 – 16 V to 0.8 – 1.2 V Merged 4-Stage Hybrid-SC-SL Converter with 96.5% Peak Efficiency and Larger Than 50% iL Reduction** 2725
Chien-Hao Tseng, Cheng-Ta Chuang, Chieh-Ju Tsai, Ching-Jan Chen
National Taiwan University, Taiwan

PRESENTATION

TOPIC: Control ICs

D14.3 **Innovation Active Gate Drive Method (Named TriC3™) for MOSFET Heat Reduction and EMI** 2730
Hisashi Sugie
ROHM Co., Ltd., Japan

PRESENTATION

TOPIC: Gate Drive Circuits and Fault Protection

D14.4 **A KY Buck-Boost Converter with Extended Ramp Control Achieving 1500% Output Variation Reduction for Smooth Mode Transition** 2735
Yu-Ting Hung[1], Chieh-Ju Tsai[1], Ching-Jan Chen[1], Chun-Yu Hsieh[2]
[1]*National Taiwan University, Taiwan;* [2]*Novatek Microelectronics Corp., Taiwan*

PRESENTATION

TOPIC: Control ICs

D14.5 An USB Cable based Extended Conversion Range L-First Hybrid-Converter using Valley-Virtual-Inductor-Current-Mode Control with Auto-Tracking Slope Compensation Against ±50% Inductance Variation 2741
Chun-I Li, Chieh-Ju Tsai, Ching-Jan Chen
National Taiwan University, Taiwan

PRESENTATION

TOPIC: **Control ICs**

D14.6 Impact of Gate Resistor Configurations on Current Balancing in Paralleled SiC MOSFETs 2746
Yifu Zhang, Shashank Karanth, Emanuel Eni
Infineon Technologies AG, Germany

PRESENTATION

TOPIC: **Gate Drive Circuits and Fault Protection**

D14.7 Exploring the Potential of FPGA in High-Frequency Switching DC-DC Boost Converters using Model Predictive Control 2752
Qingcheng Sui, Bangli Du, Yu Zuo, Wilmar Martinez
KU Leuven, Belgium

PRESENTATION

TOPIC: **Digital Control-MCUs, DSPs, FPGAs, ASICs**

D14.8 A 7 Bit 5A 6.7 GHz Gate-Shaping Digital Gate Driver with Burst-Sampling ADC for Iterative Switching Optimization of SiC Power MOSFETs 2757
Tobias Zekorn, Kenny Vohl, Erik Wehr, Leon Weihs, Michael Hanhart, Ralf Wunderlich, Stefan Heinen
RWTH Aachen University, Germany

PRESENTATION

TOPIC: **Digital Control-MCUs, DSPs, FPGAs, ASICs**

SESSION D15: Control III

11:30 - 13:30
TRACK: Control

SESSION CHAIR(S)
Qing Ye, *Caterpillar Inc.*
Brian Johnson, *University of Texas at Austin*

D15.1 Decentralized Interleaving of Series-Stacked DC-DC Converters via Extremum-Seeking Control 2764
Ivan Petrić[1], Vignesh Iyer[2], Shoudong Hu[2], Chirayu Rajpurohit[2], Bailey Sauter[2], Milan Ilić[1], Luca Corradini[2], Dragan Maksimović[2]
[1]Hanwha Qcells Technologies, United States; [2]University of Colorado Boulder, United States

PRESENTATION

TOPIC: **Control of Power Electronic Converters**

D15.2 Online Dead-Time Control for Half Bridges without Preliminary Training based on Switching Transient Steepness 2772
Lukas Knappstein, Niklas Falkenberg, Martin Pfost
Technical University Dortmund, Germany

PRESENTATION

TOPIC: **Gate Drive Circuits and Fault Protection**

D15.3 **Impedance-Based State-of-Health Estimation for Lithium-Ion Battery Management Systems** 2779
Mohammad K. Al-Smadi, Jaber A. Abu Qahouq
The University of Alabama, United States

PRESENTATION   **TOPIC: Control of Power Electronic Converters**

D15.4 **Stability Analysis and Resonance Damping of LC Filter-Based Voltage Source Converter with Single-Loop Voltage Control** 2785
Aravind G., Divyanshu Bansal, L. Umanand
Indian Institute of Science, India

PRESENTATION   **TOPIC: Control of Power Electronic Converters**

D15.5 **Finite Control Set Model Predictive Control Combined with Online Junction Temperature Estimation for Reliability Enhancement of Voltage Source Inverters** 2790
Qiang Mu, Jiale Zhou, Zaheen Mustakin, Lucas Pereira, Babak Parkhideh, Tiefu Zhao
University of North Carolina at Charlotte, United States

PRESENTATION   **TOPIC: Control of Power Electronic Converters**

D15.7 **Framework for Dynamic Control and Operation of Power Electronics Interfaces** 2797
Radha Sree Krishna Moorthy, Steven Campbell
Oak Ridge National Laboratory, United States

PRESENTATION   **TOPIC: Control of Power Electronic Converters**

D15.8 **Achieving Soft-Charging and Over 20% Input Current Ripple Reduction in a 48-to-6 V Dickson Converter using 3-Phase Split-Phase Control** 2805
Nagesh Patle, Rose A. Abramson, Sahana Krishnan, Jiarui Zou, Robert C. N. Pilawa-Podgurski
University of California, Berkeley, United States

PRESENTATION   **TOPIC: Control of Power Electronic Converters**

SESSION D16: Modeling and Simulation IV

11:30 - 13:30
TRACK: Modeling and Simulation

SESSION CHAIR(S)
Adam Skorek, *University of Québec at Trois-Rivières*
Cao Zhan, *Virginia Tech*

D16.1 **Experimental Verification of Circuit-Losses Analysis-Model of DC-Output Converter Developed using Approximated Equations from Measurement Data and Datasheet Data** 2813
Ryota Kondo[1], Tsuyoshi Funaki[2]
[1]Mitsubishi Electric Corporation, Japan; [2]Osaka University, Japan

PRESENTATION   **TOPIC: Circuits and Systems**

D16.2 **Scattering Parameter Measurement System using Probes for Surface Mount Devices Operating in the Frequency Range from 50 kHz to 1 GHz** 2821
Ryoko Kishikawa[1], Masahiro Horibe[1], Tomokazu Shoji[2], Shigenori Yabuta[2], Toshi Ohi[2], Ryo Takeda[3], Takamasa Arai[3]
[1]National Institute of Advanced Industrial Science and Technology, Japan; [2]T Plus Co. Ltd., Japan; [3]Keysight Technologies, Japan

PRESENTATION

TOPIC: **Device and Component Modeling**

D16.3 **Optical Transformer Design with Additional Common-Mode Noise Reduction Winding for Flyback DC-DC Converters** 2828
Yusuke Irie[1], Shinichiro Eguchi[1], Yoichi Ishizuka[1], Toshiro Takeuchi[2], Akio Iwabuchi[3], Takahiro Koga[4], Toshiyuki Tanaka[1]
[1]Nagasaki University, Japan; [2]Sanken Electric Co., Ltd., Japan; [3]Smart Power Semi, Japan; [4]ANSYS, Inc., Japan

PRESENTATION

TOPIC: **Parasitics Extraction and Optimization**

D16.4 **Enhanced Bus Voltage Stability Through Digital Twin-Enabled Adaptive Controller Tuning** 2833
Matthew Belanger, Andy Wong, Kerry Sado, Enrico Santi
University of South Carolina, United States

PRESENTATION

TOPIC: **Hardware-in-the-Loop and Rapid Prototyping**

D16.7 **Modeling and Performance Characterization of Lithium-Ion Capacitor at Different Temperature and Voltage Values** 2840
Mohammad K. Al-Smadi, Jaber A. Abu Qahouq, Sajad Saberi
The University of Alabama, United States

PRESENTATION

TOPIC: **Device and Component Modeling**

SESSION D17: Modeling and Simulation V

TRACK: Modeling and Simulation

SESSION CHAIR(S)
Sheldon Williamson, *Ontario Tech University*
Abraham Gebregergis, *Drexel University*

D17.1 **Conveniently Identify Coils in Inductive Power Transfer System using Machine Learning** 2846
Yifan Zhao[1], Mowei Lu[1], Ting Chen[2], Heyuan Li[3], Xiang Gao[3], Zhenbin Zhang[2], Minfan Fu[3], Stefan M. Goetz[1]
[1]University of Cambridge, United Kingdom; [2]Shandong University, China; [3]Shanghai Tech University, China

PRESENTATION

TOPIC: **Parasitics Extraction and Optimization**

D17.2 **Accurate Modeling of LLC Resonant Converters with Enhanced Analytical Approach Considering of Parasitic Capacitance** 2851
Dong Jiao, Zhengming Hou, Jih-Sheng Lai
Virginia Polytechnic Institute and State University, United States

PRESENTATION

TOPIC: **Circuits and Systems**

D17.3 **High-Frequency Conditioning Circuits for Power-Related Information Extraction in Non-Sinusoidal Power Electronic Systems** 2857

Haoyu Wang[1], Yuanxin Zhang[2], Di Mou[2], Alex Hanson[1], Shiqi Ji[2]
[1]The University of Texas at Austin, United States; [2]Tsinghua University, China

PRESENTATION

TOPIC: **Circuits and Systems**

D17.4 **Transconductance Model of the Dual Active Bridge Converter Under Single and Dual Phase Shift Control** 2865

Jared Cronin[1], Andrew Wunderlich[2], Enrico Santi[1]
[1]University of South Carolina, United States; [2]Integer Technologies LLC, United States

PRESENTATION

TOPIC: **Circuits and Systems**

D17.5 **Lumped Parameter Modeling for Real-Time Thermal Regulation of Li-Ion Battery Packs** 2871

Utkal Ranjan Muduli, Mohamed Shawky El Moursi, Khalifa Al Hosani, Ahmed Al-Durra
Khalifa University, U.A.E.

PRESENTATION

TOPIC: **Parasitics Extraction and Optimization**

D17.6 **A Physics-Based Temperature Dependent Analytical Model for 2DEG Density in AlGaN/GaN HEMT Devices** 2877

Kashfia Tajmim Nabila, Jerry L. Hudgins
University of Nebraska-Lincoln, United States

PRESENTATION

TOPIC: **Device and Component Modeling**

D17.7 **Comparative Analysis of Stator-PM Machines: Design Optimization and Electromagnetic Performance Evaluation** 2883

Maryam Salehi, Madhav Manjrekar
University of North Carolina at Charlotte, United States

PRESENTATION

TOPIC: **Device and Component Modeling**

SESSION D18: Motor Drives III

11:30 - 13:30
TRACK: Motor Drives

SESSION CHAIR(S)
Ali Safayet, *HL Mechatronics*
Ziaur Rahman, *Booz Allen Hamilton*

D18.1 **Elimination of Deadtime Effect on Resolver Offset Estimation using the Pulsating Current Command for Electric Vehicle Application** 2889

Yingfeng Ji, Nurani Chandrasekhar
Ford Motor Company, United States

PRESENTATION

TOPIC: **Modeling and Control Techniques for Motor Drives**

D18.2 **A Generic Load Emulator for Testing Motor Drives of E-Mobility** 2894
Qingzheng Zhang, Kaiyuan Feng, Changsheng Hu, Dehong Xu
Zhejiang University, China

PRESENTATION   TOPIC: **AC, DC, BLDC Motor Drives**

D18.3 **Design and Implementation of Power Assisted Control System for E-Bikes** 2900
Che-Yu Lu, Tzu-Ping Cheng
National United University, Taiwan

PRESENTATION   TOPIC: **Modeling and Control Techniques for Motor Drives**

D18.4 **A Hybrid PWM Strategy with Reduced Common-Mode Voltage and
Extended Output Voltage Linearity for Adjustable Speed Drives** 2907
Zhe Zhang, Kevin Lee
Eaton Corporation, United States

PRESENTATION   TOPIC: **Power Quality and EMI for Motor Drives**

D18.5 **Single-Phase Open-Circuit Fault-Tolerant Control of Three-Phase PMSM Drives** 2913
Yuichiro Minato[1], Yuki Nakata[2], Jun-Ichi Itoh[2]
[1]Murata Machinery, Ltd., Japan; [2]Nagaoka University of Technology, Japan

PRESENTATION   TOPIC: **Modeling and Control Techniques for Motor Drives**

D18.6 **Multi-Vendor Encoder Position Sensing Interface using Programmable IP based Solution** 2920
Rajul Bhambay[1], Dhaval Khandla[1], Pratheesh Gangadhar[1], Thomas Leyrer[2],
Achala Ram[1], Manoj Koppolu[1], Archit Dev[1]
[1]Texas Instruments Incorporated, India; [2]Texas Instruments Incorporated, Germany

PRESENTATION   TOPIC: **Sensor Integration**

D18.7 **Sensorless Control Method at Low-Speed Range using High-Frequency Voltage
Injection for Synchronous Reluctance Motors Considering to Nonlinear
Characteristic Due to Magnetic Saturation** 2924
Sota Takizawa, Sari Maekawa
Meiji University, Japan

PRESENTATION   TOPIC: **Modeling and Control Techniques for Motor Drives**

D18.8 **Hybrid Control Scheme for Permanent Magnet Gear Motor** 2932
Bing Li, Takayoshi Matsuo, Ahmed Sayed-Ahmed, Yujia Cui, Jiangang Hu
Rockwell Automation, United States

PRESENTATION   TOPIC: **AC, DC, BLDC Motor Drives**

D18.9 **Cost-Effective Fault Diagnosis for Motor and Inverter using
Bootstrap Charging and Single DC Link Current Sensor** 2937
Gyu Cheol Lim[1], Won Hyo Jeong[1], Kahyun Lee[2], Jung-Ik Ha[1]
[1]Seoul National University, Korea; [2]Ewha Womans University, Korea

PRESENTATION   TOPIC: **AC, DC, BLDC Motor Drives**

D18.10 Improved PWM to Suppress Motor Overvoltage Caused by Voltage Reflection 2943
Sung-Oh Kim, Kyo-Beum Lee
Ajou University, Korea

PRESENTATION   TOPIC: **Modeling and Control Techniques for Motor Drives**

D18.11 Analysis of Double Pulsing Effect in Motor Drives based on Vector Diagram 2948
Byeong-Woo Kang, Kyo-Beum Lee
Ajou University, Korea

PRESENTATION   TOPIC: **Modeling and Control Techniques for Motor Drives**

D18.12 A Novel Speed Sensor-Less Control of a Solar-Powered PMSM Drive 2953
Abirami Kalathy, Arpan Laha, Praveen Jain, Majid Pahlevani
Queen's University, Canada

PRESENTATION   TOPIC: **Modeling and Control Techniques for Motor Drives**

SESSION D19: Utility Applications I

11:30 - 13:30
TRACK: Power Electronics for Utility Applications

SESSION CHAIR(S)
Stanley Atcitty, *Sandia National Laboratories*
Lingyu Zhu, *Xi'an Jiaotong University*

**D19.2 Design of a Compact Low-Loss MMC Double Submodule for
MVDC and HVDC Applications** 2960
Ali Sharaf Addin, Rainer Marquardt, Thomas Brückner
Universität der Bundeswehr München, Germany

PRESENTATION   TOPIC: **FACTs Devices and HVDC**

D19.7 A Series-Type Dynamic Voltage Restorer Control Strategy to Cope with Voltage Swell 2968
Jiazheng Zhang, Hongyu Chen, Xi Chen, Mingjun Bao
Tianjin University, China

PRESENTATION   TOPIC: **Power Quality, UPS, Active Power Filters**

**D19.9 Machine Learning Approach for Accurate Lithium-Ion Battery Temperature Prediction
using Electrochemical Features Independent of Battery SOC and SOH** 2973
Vincent Masabiar Tingbari[1], Oluwaseun Isaiah Ekuewa[1], Anshul Nagar[1], Asad Abbas[1],
Jamil Umar[1], Yuxin Zhang[1], Woonki Na[2], Jonghoon Kim[1]
[1]Chungnam National University, Korea; [2]California State University, United States

PRESENTATION   TOPIC: **Energy Storage Systems**

SESSION D20: Utility Applications II

11:30 - 13:30
TRACK: Power Electronics for Utility Applications

SESSION CHAIR(S)
Yuheng Wu, *John Deere*
Athar Hanif, *Ohio State University*

D20.1 **A Battery Strings Circulating Current Blocking Method for Battery Energy Storage Systems** 2981
Haihong Long, Ziang Sun, Yucheng Fan, Xin Wu, Dehong Xu
Zhejiang University, China

PRESENTATION   TOPIC: **Energy Storage Systems**

D20.6 **A Hybrid Multilevel Converter-Based High-Gain Isolated DC/DC Converter for Grid-Tied Energy Storage Applications** 2986
Pengyu Fu, Yizhou Cong, Jin Wang, Anant Agarwal
The Ohio State University, United States

PRESENTATION   TOPIC: **Energy Storage Systems**

D20.7 **LCL Filter Parameter Selection using Graphical Method for a 13.8 kV ac 1.1 MVA 7-Level Flying Capacitor Grid-Connected Converter Utilizing Variable Switching Frequency** 2992
Arthur Mendes, David Nam, Mingze Gao, Thimothy Thacker, Dong Dong, Rolando Burgos
Virginia Polytechnic Institute and State University, United States

PRESENTATION   TOPIC: **Distributed Energy Systems**

D20.8 **Online Extraction of Electrochemical Impedance Spectroscopy Pattern based on EV Load Profile and Short Time Fourier Transform for Diagnosis of Lithium-Ion Battery Safety** 3000
Miyoung Lee[1], Dongcheol Lee[1], Youngmin Bae[1], Jongchan An[1], Garam Yang[1], Woonki Na[2], Jonghoon Kim[1]
[1]Chungnam National University, Korea; [2]California State University, United States

PRESENTATION   TOPIC: **Energy Storage Systems**

D20.9 **Enhanced Incremental Capacity Analysis for Evaluating Battery Degradation Mechanisms of Optimized Fast Charging Methods** 3006
Taehyeon Gong[1], Jaehyeong Lee[1], Sungjun Lee[1], Yura Kim[1], Bomyeong Ko[1], Woonki Na[2], Sungjin Choi[3], Jonghoon Kim[1]
[1]Chungnam National University, Korea; [2]California State University, United States; [3]University of Ulsan, Korea

PRESENTATION   TOPIC: **Energy Storage Systems**

D20.10 **Co-Estimation of SOC and SOT in Lithium-Ion Batteries using an RLS-Based Heat Generation Model** 3012
Seongkyu Lee[1], Eunjin Kang[1], Minhyeok Kim[1], Seunghyun Lee[1], Minwoo Song[1], Jaea Lee[1], Woonki Na[2], Jonghoon Kim[1]
[1]Chungnam National University, Korea; [2]California State University, United States

PRESENTATION   TOPIC: **Energy Storage Systems**

SESSION D21: Power Electronics for Renewable Energy

11:30 - 13:30
TRACK: Renewable Energy Systems

SESSION CHAIR(S)
Tao Yang, *University of Nottingham*
Jingbo Liu, *Eaton, US*

D21.1 **Three-Stage Adaptive Control Strategy for Stability Improvement of Grid-Connected Inverter in Weak Grid** 3018
Longxiang You[1], Sicong Jin[1], Xin Zhang[1], Zuoshuai Wang[2], Sunqing Wang[3]
[1]Zhejiang University, China; [2]Wuhan Second Ship Design and Research Institute, China; [3]China Ship Scientific Research Center, China

PRESENTATION   TOPIC: Photovoltaic (PV) Inverters and Micro Inverters

---

D21.2 **Degradation Analysis of Offshore Bifacial PV Modules Under Multiple Climatic Stressors** 3024
Aidha Muhammad Ajmal, Yongheng Yang
Zhejiang University, China

PRESENTATION   TOPIC: Fuel Cells and Other Emerging Renewable Energy Systems

---

D21.3 **A Flexible Energy Management System for Solar Powered Electric-Bus Charging Stations** 3030
Supun Amarathunga, Pasan Gunawardena, Xiaoting Wang, Yunwei Li
University of Alberta, Canada

PRESENTATION   TOPIC: Fuel Cells and Other Emerging Renewable Energy Systems

---

D21.4 **A Vienna Rectifier based Grid-Connected Powertrain for Hydrokinetic Turbine Systems** 3036
Peidong Li, Md Tariquzzaman, Yue Cao
Oregon State University, United States

PRESENTATION   TOPIC: Fuel Cells and Other Emerging Renewable Energy Systems

---

D21.5 **Condition Monitoring for DC-Link Capacitors and PV Arrays based on the Start-Up Process of the PV System** 3042
Yongjie Liu, Ariya Sangwongwanich, Chen Liu, Xing Wei, Shuyu Ou, Tamás Kerekes, Jiahong Liu, Huai Wang
Aalborg University, Denmark

PRESENTATION   TOPIC: Photovoltaic (PV) Inverters and Micro Inverters

---

D21.6 **Electrically and Thermally Efficient Reliable Power Converter Design for Micro–Hydrokinetic Turbine** 3048
Md Tariquzzaman, Peidong Li, Yue Cao
Oregon State University, United States

PRESENTATION   TOPIC: Fuel Cells and Other Emerging Renewable Energy Systems

---

D21.7 **Comprehensive Evaluation of Cyber Attacks on Grid-Connected Smart Inverters** 3054
Rishabh Singla, Vishwam Raval, Hasan Ibrahim, Jaewon Kim, Prasad Enjeti, Narsimha Reddy
Texas A&M University, United States

PRESENTATION   TOPIC: Photovoltaic (PV) Inverters and Micro Inverters

D21.8 Parallel Operation of Grid-Forming Converters based on Kuramoto Oscillators with Virtual Cable Emulation for Improved Power Sharing 3059

Vikram Roy Chowdhury, Gab-Su Seo, Barry Mather
National Renewable Energy Laboratory, United States

PRESENTATION

TOPIC: Photovoltaic (PV) Inverters and Micro Inverters

D21.9 Enhancing Hydrogen Production in Hybrid Standalone Microgrids 3064

Utkal Ranjan Muduli, Mohamed Shawky El Moursi, Khalifa Al Hosani, Ahmed Al-Durra
Khalifa University, U.A.E.

PRESENTATION

TOPIC: Fuel Cells and Other Emerging Renewable Energy Systems

D21.10 LSTM-Based Sub-Synchronous Oscillation Detection Scheme for Type 4 Wind Farm Interfaced with Weak AC Grid 3071

Omar Abu-Rub[1], Muhammad F. Umar[2], Jana A. Sheikh Ali[3], Yazan Qiblawey[4], Abdulrahman Alassi[4], Maryam Saeedifard[1], Mohammad B. Shadmand[5]
[1]*Georgia Institute of Technology, United States;* [2]*Texas A&M University at Qatar, Qatar;* [3]*Qatar University, Qatar;* [4]*Iberdrola Innovation Middle East, Qatar;* [5]*University of Illinois Chicago, United States*

PRESENTATION

TOPIC: Wind Energy Conversion Systems

SESSION D22: Transportation Electrification I

11:30 - 13:30
TRACK: Transportation Power Electronics

SESSION CHAIR(S)
Rasoul Hosseini, *General Motors*
Antonio J. Marques Cardoso, *University of Beira Interior*

D22.1 A Study of Module Design Method to Suppress the Oscillation Occurs Between Parallel-Connected Power Devices 3077

Shinji Yato, Hiroto Sakai, Hideo Araki, Shumei Shimosako
ROHM Co., Ltd., Japan

PRESENTATION

TOPIC: Power Electronics for Hybrid and Electric Vehicles

D22.2 A High-Efficient Hybrid Traction Inverter in Electric Vehicle Applications 3083

Yousefreza Jafarian[1], Omid Salari[1], Praveen Jain[1], Alireza Bakhshai[1], Mohamed Z. Youssef[2]
[1]*Queen's University, Canada;* [2]*Ontario Tech University, Canada*

PRESENTATION

TOPIC: Power Electronics for Hybrid and Electric Vehicles

D22.3 Dual-Use of Onboard Chargers to Achieve Controllable DC Bus Voltage for Electric Vehicles 3089

Anuj Maheshwari[1], Elie Libbos[2], Arijit Banerjee[1]
[1]*University of Illinois Urbana-Champaign, United States;* [2]*Texas Instruments Incorporated, United States*

PRESENTATION

TOPIC: Power Electronics for Hybrid and Electric Vehicles

D22.4 **Isolated Single-Phase Onboard Chargers for BEV/PHEV using Active Power Decoupling Technology** 3096

Yoshiki Amano[1], Keigo Nishimura[1], Hiroaki Matsumori[1], Takashi Kosaka[1], Kenichi Nagayoshi[2], Kenichi Watanabe[2]
[1]Nagoya Institute of Technology, Japan; [2]Toyota Industries Corporation, Japan

PRESENTATION

TOPIC: On-Board and Off-Board Charging Systems

D22.5 **A Practical Use of xEVCap: The Modular and Standard DC-Link Capacitor Solution for the Main EV Powertrain Inverter** 3100

David Olalla[1], Tomas Wagner[2], Fernando Rodriguez[2], Alberto Espinar[2]
[1]TDK Electronics AG, Germany; [2]TDK Electronics SAU, Spain

PRESENTATION

TOPIC: Power Electronics for Hybrid and Electric Vehicles

D22.6 **Optimized Bidirectional On-Board Charger using a Novel Unfolder-DAB Topology** 3109

Héctor Sarnago, Ignacio Álvarez, Pablo Briz, Óscar Lucía
I3A, Universidad de Zaragoza, Spain

PRESENTATION

TOPIC: On-Board and Off-Board Charging Systems

SESSION D22: Transportation Electrification I

11:30 - 13:30
TRACK: Power Electronics for Utility Applications

SESSION CHAIR(S)
Rasoul Hosseini, *General Motors*
Antonio J. Marques Cardoso, *University of Beira Interior*

D22.7 **Critical Thermal Characterization of Next-Generation Solid-State Batteries for Automotive Battery Management Systems** 3114

Chandan Chetri, Sheldon Williamson
Ontario Tech University, Canada

PRESENTATION

TOPIC: Energy Storage Systems

SESSION D23: Transportation Electrification II

11:30 - 13:30
TRACK: Transportation Power Electronics

SESSION CHAIR(S)
Dong Cao, *University of Dayton*
Manish Niraula, *The University of Texas at Dallas*

D23.1 **Nanocrystalline CMC Inductors for EV Charging: Trade Studies and Testing Standardization** 3119

Christopher Bracken[1], Mark A. Juds[1], Paul R. Ohodnicki[1], Bharadwaj Reddy Andapally[2], Jose Gato[3]
[1]University of Pittsburgh, United States; [2]CBMM, Netherlands; [3]Vacuumschmelze GmbH & Co. KG, Germany

PRESENTATION

TOPIC: Vehicular Power Electronic Circuits and Systems

D23.2 Predicting Efficiency of On-Board and Off-Board EV Charging Systems using Machine Learning 3124
Mohamed Yasko, Fanghao Tian, Wilmar Martinez, Johan Driesen
KU Leuven, Belgium

PRESENTATION
TOPIC: On-Board and Off-Board Charging Systems

D23.3 High-Power and High-Speed Multi-Channel VCSEL Arrays with GaN Driver for Automotive LiDAR 3129
Yifu Liu[1], Sichao Li[1], Junlei He[1], Changyu Hu[1], Bill He[2], Karthik Krishnamurthy[1], Andy Shen[1]
[1]JFS Laboratory, China; [2]Vertilite Co., Ltd., China

PRESENTATION
TOPIC: Vehicular Power Electronic Circuits and Systems

D23.4 Double Pulse Test Platform for Hybrid SiC-IGBT Switch Characterization and Optimal Gate Control Strategy for EV Traction Inverters 3133
Rosario Attanasio, Harsha Ademane, Ryan Satterlee, Gianni Vitale
STMicroelectronics, United States

PRESENTATION
TOPIC: Power Electronics for Hybrid and Electric Vehicles

D23.5 Critical Role of Individual Cell Temperature Monitoring in Mitigating Thermal Runaway and Reducing Accelerated Degradation in Lithium-Ion Batteries 3141
Mohit Sharma, Akash Samanta, William Locke, Sheldon Williamson
Ontario Tech University, Canada

PRESENTATION
TOPIC: Vehicular Power Electronic Circuits and Systems

D23.6 Loss-Optimized Design of a Triple Active Bridge DC-DC Converter for an Electric Vehicle Application 3147
Sreejith Chakkalakkal[1], Kyle Kozielski[1], Wesam Taha[2], Yicheng Wang[2], Aniket Anand[2], Ali Emadi[1]
[1]McMaster University, Canada; [2]R&D Americas, Schaeffler, Canada

PRESENTATION
TOPIC: Vehicular Power Electronic Circuits and Systems

D23.7 A Magnetic-Less DC/DC Converter with Pulse Charging for 800 V Powertrains from 400 V DC Fast Chargers 3155
Duc Dung Le[1], Shivam Chaturvedi[1], Shahid Aziz Khan[1], Mengqi Wang[1], Mohamed Elshaer[2]
[1]University of Michigan-Dearborn, United States; [2]Ford Motor Company, United States

PRESENTATION
TOPIC: Vehicular Power Electronic Circuits and Systems

D23.8 Boosting Charger Efficiency: A GaN-Based Flyback Converter with Energy Recycling 3160
Ahmad Nabizadah, Majid Ghasemi Korrani, Babak Fahimi
The University of Texas at Dallas, United States

PRESENTATION
TOPIC: On-Board and Off-Board Charging Systems

D23.9 A Hybrid Three-Level Buck Converter with Flying Supercapacitor for High Load Current Surge Capability using Peak Current Mode Control 3167
Finlay Lodge[1], Rafael Peña-Alzola[1], Martin MacFadyen[1], Patrick Norman[1], Mark Sweet[2], Graeme Burt[1]
[1]University of Strathclyde, United Kingdom; [2]Rolls-Royce plc, United Kingdom

PRESENTATION
TOPIC: Power Electronics for Aerospace

D23.10 Supercritical Carbon Dioxide (sCO₂)-Cooled Current Source Inverter-based Integrated Motor Drive for MW-Scale Electric Aviation Applications 3174

Hang Dai[1], John Yagielski[1], Thomas Jahns[2], Kum-Kang Huh[1], Vandana Rallabandi[3],
Libing Wang[1], Tarak Saha[1], Wenda Feng[2], Bulent Sarlioglu[2]
[1]GE Aerospace Research, United States; [2]University of Wisconsin-Madison, United States;
[3]Oak Ridge National Laboratory, United States

PRESENTATION

TOPIC: Power Electronics for Hybrid and Electric Vehicles

SESSION D24: Wireless Power Transfer I

11:30 - 13:30
TRACK: Wireless Power Transfer

SESSION CHAIR(S)

Reza Tavakoli, *Rivian*
Mostak Mohammad, *Oak Ridge National Laboratory*

D24.1 The Challenge of Thermal Runaway in Soft Magnetic Materials for Inductive Power Transfer 3181

Yibo Wang, Ben Zhang, Weisheng Guo, Tianlu Ma, Sheng Ren, C.Q. Jiang
City University of Hong Kong, China

PRESENTATION

TOPIC: Safety and Reliability

D24.2 A Capacitively Coupled Alternative Electric Field Control for Freeze-Free based High Quality Food Preservation 3187

Jaeyong Cho, Junhyeong Park, Sung-Bum Park, Daehyun Kim, Jinsoo Choi
Samsung Electronics, Korea

PRESENTATION

TOPIC: Wireless Charging

D24.3 The Characteristics of the Long Length Primary Loop and the Power Supply for the SCMaglev's DWPT System 3194

Keisuke Yamamoto, Jun Enomoto, Shunsaku Koga, Junichi Kitano
Central Japan Railway Company, Japan

PRESENTATION

TOPIC: Wireless Charging

D24.4 A Wireless EV Charging System with a Double-Sided LCC Network using Variable Switching Frequency and DC-Link Voltage Control 3200

Chae-Lyn Kim, Hyeonu Jo, Ju-A Lee, Dong Hyeon Sim, Byoung Kuk Lee
Sungkyunkwan University, Korea

PRESENTATION

TOPIC: Wireless Charging

D24.6 Class E/EF Inductive Power Transfer to Achieve Stable Output Under Variable Low Coupling 3206

Yifan Zhao[1], Mowei Lu[1], Heyuan Li[2], Zhenbin Zhang[3], Minfan Fu[2], Stefan M. Goetz[1]
[1]University of Cambridge, United Kingdom; [2]Shanghai Tech University, China; [3]Shandong University, China

PRESENTATION

TOPIC: Safety and Reliability

D24.7 A Motorized Air-Core Variable Inductance Winding Structure 3212

Xindong Li[1], Sampath Jayalath[2], Cheng Zhang[1]

[1]The University of Manchester, United Kingdom; [2]University of Cape Town, South Africa

PRESENTATION   TOPIC: Wireless Charging

D24.8 Wireless Power Transfer System with Automatic Tuning Capability in Metallic Environment 3220

Renjie Zhang[1], Yue Wu[1], Delin Zhao[2], Yaohua Li[2], Yongbin Jiang[2], Yi Tang[2], Huan Yuan[1], Xiaohua Wang[1], Mingzhe Rong[1]

[1]Xi'an Jiaotong University, China; [2]Nanyang Technological University, Singapore

PRESENTATION   TOPIC: Wireless Charging

D24.9 Design of Wireless Power Transmitters for Enhanced Transmission Distance and Output Power 3227

Kaiyuan Wang[1], Shuang Zhao[2], Shuye Shang[3], Eric Ka-Wai Cheng[3], Siew-Chong Tan[4], Yun Yang[1]

[1]Nanyang Technological University, Singapore; [2]Hefei University of Technology, China; [3]University of California, Merced, United States; [4]The University of Hong Kong, China

PRESENTATION   TOPIC: Wireless Charging

SESSION D25: Wireless Power Transfer II

11:30 - 13:30
TRACK: Wireless Power Transfer

SESSION CHAIR(S)
Khurram Afridi, *Cornell University*
Weijing Qiu, *John Deere*

D25.1 Optimization of Wireless Power Transfer Waveforms and In-Vivo Receivers for Implantable Medical Devices 3232

Hanbing Liu, Xin Zan
University of Maryland, United States

PRESENTATION   TOPIC: Wireless Charging

D25.3 Comparison of Compact Power Amplifier Designs for High Frequency Resonant Wireless Power Transfer Systems at 6.78 MHz using High-Q Resonators 3241

Manuel Rueß, Kilian Müller, Mathias C.J. Weiser, Ingmar Kallfass
Universität Stuttgart, Germany

PRESENTATION   TOPIC: Wireless Charging

D25.4 Analysis and Design of Capacitive Coupling Wireless Power Transfer System using Load-Independent Class-EF Inverter 3248

Takumi Kobayashi[1], Yutaro Komiyama[1], Akihiro Konishi[1], Hiroaki Ota[2], Yuki Ito[2], Taichi Mishima[2], Takeshi Uematsu[2], Kien Nguyen[1], Hiroo Sekiya[1]

[1]Chiba University, Japan; [2]OMRON Corporation, Japan

PRESENTATION   TOPIC: Wireless Charging

D25.5 **Design and Optimization of a 600 W Wireless Drone Charger for High Gravimetric Power Density** 3253
Arka Basu, Daniel Costinett
The University of Tennessee, United States

PRESENTATION

TOPIC: Wireless Charging

D25.6 **Stabilization Method for DC-Bus Oscillation in Dynamic Wireless Power Transfer Systems** 3261
Yuki Ochiai, Keisuke Kusaka
Nagaoka University of Technology, Japan

PRESENTATION

TOPIC: Wireless Charging

D25.7 **Unveiling Aliasing Effect on Resonant Pole Locations in Wireless Battery Chargers** 3267
Anwesha Mukhopadhyay, Daniel Costinett
The University of Tennessee, United States

PRESENTATION

TOPIC: Wireless Charging

D25.8 **Integrated Hybrid Inductive and Capacitive Power Transfer System with Asymmetrical PCB Self-Resonator** 3275
Yao Wang[1], Zhen Sun[1], Xiangrong Zhang[1], Yun Yang[1], Shu Yuen Ron Hui[2]
[1]Nanyang Technological University, Singapore; [2]The University of Hong Kong, China

PRESENTATION

TOPIC: Wireless Charging

SESSION D26: Power Electronics Applications II

11:30 - 13:30
TRACK: Power Electronics Applications

SESSION CHAIR(S)
Jeffery Nilles, *Alpha&Omega Semiconductors*
Nil Patel, *RMIT University, Melbourne*

D26.1 **High Frequency Noise Reduction Method of the Class E Power Amplifier** 3281
Kyungmin Lee, Sungku Yeo
Samsung Electronics, Korea

PRESENTATION

TOPIC: Energy Harvesting

D26.2 **Single-Stage Three-Phase Buck-Matrix Rectifier with Series-Parallel Connected Transformers for High-Power 48 V Data Center Power Supplies** 3285
Yuki Ishikura, Chinmay Bhagat
Murata Manufacturing Co., Ltd., Japan

PRESENTATION

TOPIC: Datacenter/Telecom Power Architecture and System Considerations

D26.3 **Sector Transition PWM Modulation Scheme for a Three-Phase Isolated Buck-Matrix Rectifier** 3291
Chinmay Bhagat, Yuki Ishikura
Murata Manufacturing Co., Ltd., Japan

PRESENTATION

TOPIC: Datacenter/Telecom Power Architecture and System Considerations

D26.4 **Adaptive Capacitance Circuit for Optimal Dynamic Impedance Matching in Variable Reluctance Energy Harvesting Applications** 3298

Alejandro Redondo, Fernando Pérez, Sofía García, Gabriel Mujica, Airán Francés
Universidad Politécnica de Madrid, Spain

PRESENTATION

TOPIC: **Energy Harvesting**

D26.5 **Gallium Nitride (GaN) based Topology Comparison for Low Power Battery Charging Applications** 3304

Jai Aditya Chaudhary, Rosario Attanasio, Gianni Vitale
STMicroelectronics, United States

PRESENTATION

TOPIC: **Portable Power**

D26.6 **Server Motherboard Power Performance Study Under Immersion Cooling Environment** 3312

Meng Wang[1], Haiyan Wang[2], Pavan Kumar[3], Haijin Zhang[1], Xiang Li[1], Fengwei Bian[1], Jianting Deng[2], Jiaqi Zhu[2], Yiming Lei[2]
[1]Intel Corporation, China; [2]Nettrix Information Industry Corporation, China; [3]Intel Corporation, United States

PRESENTATION

TOPIC: **Datacenter/Telecom Power Architecture and System Considerations**

D26.7 **Practical PCB Design Considerations for GaN HEMTs based Isolated DC-DC Converter** 3316

Gaureej Gauttam, Harish S. Krishnamoorthy, Sai Sushma Pasupuleti
University of Houston, United States

PRESENTATION

TOPIC: **Defense and Military Power Electronics**

D26.8 **Data-Driven Characterization and Forecasting of Metal-Oxide Varistor Degradation in DC Circuit Breakers** 3321

Zhi Jin Zhang, Yang Liu, Lukas Graber, Maryam Saeedifard
Georgia Institute of Technology, United States

PRESENTATION

TOPIC: **Solid State and Hybrid Circuit Breakers**

D26.10 **A Thyristor-Based Fault Current Bypass Solid-State Circuit Breaker for DC Microgrid Applications** 3328

Jiale Zhou, Xiuhu Sun, Qiang Mu, Tiefu Zhao
University of North Carolina at Charlotte, United States

PRESENTATION

TOPIC: **Solid State and Hybrid Circuit Breakers**

D26.11 **Single-Stage Three-Phase AC-AC Isolated Inertialess Converter (IIC) for Industrial Drives** 3334

Brad Houska[1], Decheng Yan[1], Aniruddh Marellapudi[1], Satish Belkhode[2], Joseph Benzaquen Sune[1], Deepak Divan[1]
[1]Georgia Institute of Technology, United States; [2]Indian Institute of Technology Roorkee, United States

PRESENTATION

TOPIC: **AC-DC-AC Applications and Matrix Converters**

Versatile Controller Architecture for a Universal DC Fast Charging Front-End

Anurag Singh*, Sayan Paul*, Tejas Bhuse†, Trent Martin†, Hien Nguyen†, Inder Vedula†,
Nikola Milivojević†, Dragan Maksimović* and Luca Corradini*

*Colorado Power Electronics Center
Department of Electrical, Computer and Energy Engineering
University of Colorado, Boulder, CO 80309, USA
email: {sayan.paul, anurag.singh-1, maksimov, luca.corradini}@colorado.edu

† Freewire Technologies
1811 Pike Rd #2-D, Longmont, CO 80501
email: {tbhuse, tmartin, hnguyen, ikumar, nmilivojevic}@freewiretech.com

Abstract—**Universal dc fast chargers, capable of operating from either single-phase (1ϕ) or three-phase (3ϕ) grids, require a configurable, high power density ac-dc front-end which not only involves a high degree of hardware re-utilization between 1ϕ and 3ϕ operation, but a corresponding level of control versatility as well. This work develops a general control architecture for a universal ac-dc front-end based on a three-leg topology, which can operate the system as a unity power factor corrector in both 3ϕ and 1ϕ configurations. A simple and effective control solution is employed in 1ϕ operation to re-utilize one of the system legs as an active power filter and remove the $2\omega_0$ fluctuating power without needing dedicated harmonic suppression loops. The proposed control operation is validated through simulations and Hardware-in-the-Loop (HIL) testing for a $10\,\mathrm{kW}/6.6\,\mathrm{kW}$ design, with experimental results demonstrated for 1ϕ closed loop operation at reduced power.**

Index Terms—**Grid-Connected Converters, Power Factor Correction, d-q Control, Active Power Filters (APF)**

I. INTRODUCTION

Driven by the ever-increasing adoption of electric vehicles, the need for ubiquitous and versatile charging stations has surged. Off-board *universal* chargers, compatible with both 1ϕ and 3ϕ grids, can significantly improve EV integration in the current utility distribution infrastructure, especially in the U.S., where both grid types are prevalent [1]. Incorporating integrated energy storage [2] allows these chargers to be deployed in various locations, including areas affected by weak grid characteristics [3], while maintaining a limited power rating. On-board universal chargers further promote EV adoption by enabling seamless Level 2 charging compatibility in different regions [4]. Additionally, integrating bidirectional capabilities facilitates vehicle-to-grid (V2G) functionality, supporting grid stability and interactivity.

Achieving high power density in universal chargers is often constrained by the size of the dc link capacitance C_{dc} in single-phase (1ϕ) configurations. To address this issue, part of the front-end circuit can be reconfigured to operate as an Active Power Filter (APF), which buffers the $2\omega_0$ fluctuating power into a buffer capacitance, C_{buf}, thereby relaxing the requirements on C_{dc} to levels comparable to the 3ϕ case [4]. While this additional capacitance, C_{buf}, can be derived from the input LCL filter [5], this strategy necessitates the use of large X-type film capacitors to meet the minimum C_{buf} requirement and increases the current stress on the main rectifier legs leading to inefficient operation.

Fig. 1 illustrates the $3\phi/1\phi$ front-end stage of a universal charger considered in this work. The selectors configure the system to operate either as a conventional 3ϕ PFC rectifier or as a 1ϕ PFC rectifier with the unused leg modulated as an active power filter. The control structure for the $3\phi/1\phi$ rectifier is designed to achieve two primary objectives: (1) ensuring Unity Power Factor (UPF) operation of the rectifier and (2) maintaining regulation of the dc link voltage. These objectives are accomplished using a cascaded control architecture, comprising an inner loop for grid current control (i_g) and an outer loop for dc-link voltage regulation (v_{dc}). The APF leg controller operates leg C as the buck-type APF [6]–[8], re-routing the $2\omega_0$ power component to the capacitive energy buffer C_{buf}, placed as a separate unit. This work presents a versatile controller architecture for a universal charging front-end, designed to maximize software re-usability and streamline firmware development. The proposed architecture facilitates modular and reusable code, simplifying integration and testing across both configurations. The 1ϕ control is implemented in the d-q reference frame, leveraging the same core rectifier control blocks and design methodology as the 3ϕ configuration. The APF control utilizes a novel and simple duty cycle feed-forward technique that uses run-time parameters derived from sensed values, effectively eliminating the $2\omega_0$ power component without the need for dedicated harmonic suppression loops or intrusive current sensing. The architecture enables decoupled control of active and reactive components of power flow in both configurations, thus enabling grid-supporting features such as vehicle-to-grid (V2G) functionality [9], [10], enhancing grid stability and integrating renewable energy sources.

979-8-3315-1612-3/25 $31.00 © 2025 IEEE

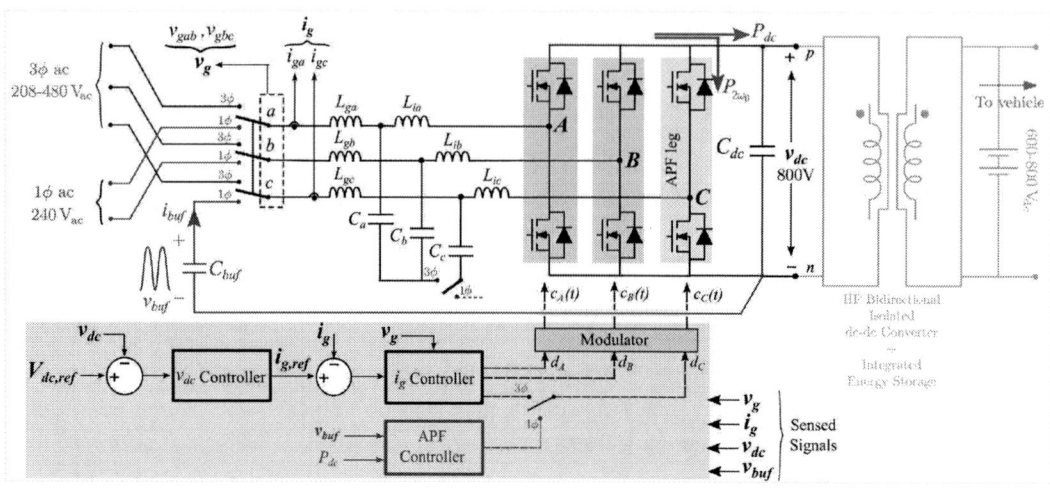

Fig. 1: High-level overview of the universal charger front-end and controller architecture.

The remainder of the paper is organized as follows: Section II discusses the proposed control architecture, providing detailed insights into the design and implementation of each sub-module. Section III validates the proposed approach using simulation and Hardware-in-the-Loop (HIL) setup, showcasing closed-loop performance under rated conditions.

II. CONTROLLER ARCHITECTURE

The complete block diagram of the controller architecture is presented in Fig. 2, illustrating the unified rectifier control framework implemented in the d-q reference frame, which is shared by both configurations. In the figure, components common to both the 1ϕ and 3ϕ configurations are depicted in black, while red and blue blocks represent elements specific to the 3ϕ and 1ϕ configurations, respectively. Irrespective of the configuration mode, the current loop compensator operates in the rotating d-q reference frame, employing simple PI controllers for each axis. This approach provides superior tracking performance, increased control bandwidths, and, most importantly, a simpler design and implementation on digital platforms compared to the traditionally used Proportional Resonant (PR) controllers. The d-axis current reference $i_{d,ref}$ is provided by the dc link voltage loop compensator, whereas the q-axis reference is set to zero for UPF operation. The core d-q current controller interfaces with the sensed signals and PWM modulator through input and output processing blocks whose function changes with the selected configuration. Notably, in the 1ϕ configuration, the duty cycle for leg C is

TABLE I: Specifications for the Universal Charger Front-end

Parameter	Value
$P_{\text{dc},3\phi}$ / $P_{\text{dc},1\phi}$	10 kW / 6.6 kW
V_{dc}	800 V
$I_{\text{ph,RMS}}$	28 A
ω_o	$2\pi \times 60$ rad s^{-1}
f_{sw}	25 kHz

modulated using a feed-forward duty cycle, $d_{buf}(t)$, derived from the sensed quantities utilized in the main rectifier control.

A. Grid Synchronization

The Phase-Locked Loop (PLL) structure overview is shown in Fig. 2. In 3ϕ structure, the Clarke transformation is applied on line-to-line voltages $v_{g,ab}$, $v_{g,bc}$, and $v_{g,ca}$ to transform them into corresponding α-β components v_α and v_β. In 1ϕ operation, a cascaded Second Order Generalized Integrator (C-SOGI) [11] stage (Fig. 3a) is used to generate an in-phase and an in-quadrature version of the sensed grid voltage, which become v_α and v_β, respectively. Such *fictitious axis emulation* [12] of the β-axis component allows the use of a unified current control architecture for both 3ϕ and 1ϕ operations. The α-β components are then used for phase estimation by the following Synchronous Reference Frame (SRF) PLL stage depicted in Fig. 3b.

In C-SOGI, the transfer functions for the α-β transformation paths are defined as follows:

$$\frac{v_\alpha(s)}{v_g(s)} = G_\alpha(s) = \frac{(K_e\omega_0 s)^2}{(s^2 + K_e\omega_0 s + \omega_0^2)^2}, \quad (1a)$$

$$\frac{v_\beta(s)}{v_g(s)} = G_\beta(s) = \frac{K_e^2\omega_0^3 s}{(s^2 + K_e\omega_0 s + \omega_0^2)^2}. \quad (1b)$$

The parameter K_e is selected based on the minimum settling time for a step response, which can be evaluated numerically. With a minimum 2% settling time of approximately 22 ms, the optimal gain value selected is $K_e \approx 1.6$. The SRF-PLL design involves the selection of PI controller parameters that will ensure that the PLL's estimated phase θ_{PLL} matches the actual phase of the sensed grid voltage $\theta = \omega_0 t$. The d-q reference frame grid voltage can be written as:

$$v_d = V_m \cos(\theta_{PLL} - \theta_a), \quad v_q = V_m \sin(\theta_{PLL} - \theta_a), \quad (2)$$

where θ_{PLL} and θ_a are the PLL's estimation and actual phase angles respectively. For small differences between the angles $\theta_{PLL} - \theta_a$, this can be approximated as $v_q \approx V_m(\theta_{PLL} - \theta_a)$.

979-8-3315-1612-3/25 $31.00 © 2025 IEEE

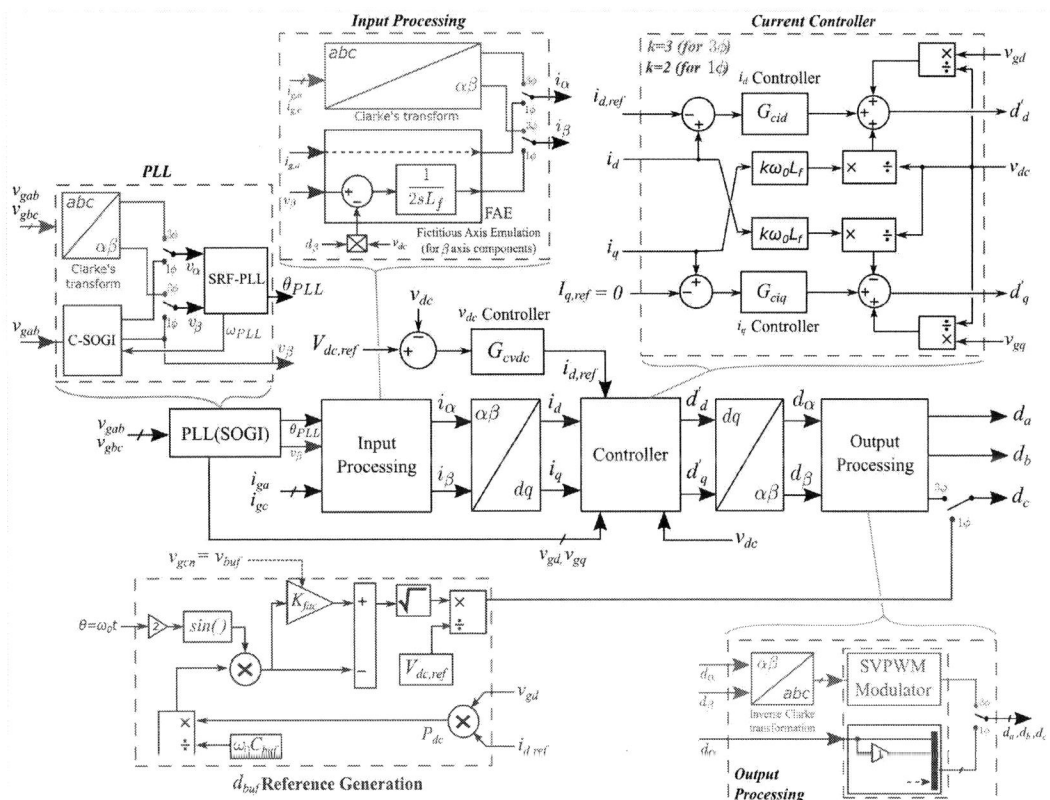

Fig. 2: Proposed controller architecture, with (b) and (c) showing the input and output processing blocks, respectively. In the diagram, $L_f = L_{fr} + L_{fg}$ is the total series inductance in 3ϕ configuration.

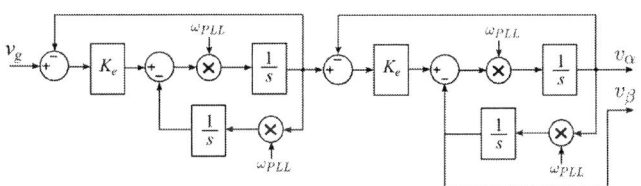

(a) Cascaded SOGI (C-SOGI).

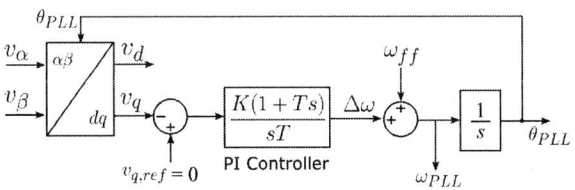

(b) Synchronous Reference Frame (SRF) PLL.

Fig. 3: PLL structures used for grid synchronization: (a) C-SOGI and (b) SRF-PLL block diagram.

The PI controller for the SRF-PLL can then be designed using the symmetric optimum method. The parameters for the PI controller (defined by (3)) are determined as shown in (4), yielding a bandwidth of $1/\alpha_B T_s$ (in rad/s) and provide a phase margin given by (5).

$$G_{c,SRF}(s) = \frac{K(1 + sT)}{sT} \quad (3)$$

$$T = \alpha_B^2 T_s, \quad K = \frac{1}{V_m \alpha_B T_s} \quad (4)$$

$$\text{Phase Margin} = \tan^{-1}\left(\frac{\alpha_B - 1/\alpha_B}{2}\right) \quad (5)$$

When selecting the bandwidth for the SRF-PLL, considerations must be made for various sources of disturbance, including DC offsets, grid unbalance, and grid harmonics, as these factors can distort the unit vectors generated by the PLL (i.e., $\sin(\theta_{\text{PLL}})$ and $\cos(\theta_{\text{PLL}})$) [13]. In three-phase ($3\phi$) systems, DC offsets can be mitigated using a band-pass filter (BPF) placed upstream of the SRF-PLL, whereas in the single-phase (1ϕ) system, this filtering action is inherently achieved through the cascaded SOGI structure. Furthermore, unbalanced grid voltages in 3ϕ systems introduce oscillations in the estimated θ_{PLL} at a frequency of $2\omega_0$, while third harmonic distortion in the grid voltage of positive sequence may also induce oscillations at the same frequency. Consequently, the PI controller's bandwidth should be carefully selected to suppress disturbances, particularly at $2\omega_0$. In this design, a bandwidth of 40 Hz has been identified as a suitable choice to achieve effective disturbance rejection.

TABLE II: SRF-PLL PI Controller Parameters

Parameter	Value
Proportional Gain (K)	0.3023
Integral Gain (K/T_s)	0.764
Bandwidth (Hz)	40
Phase Margin (degrees)	88.85

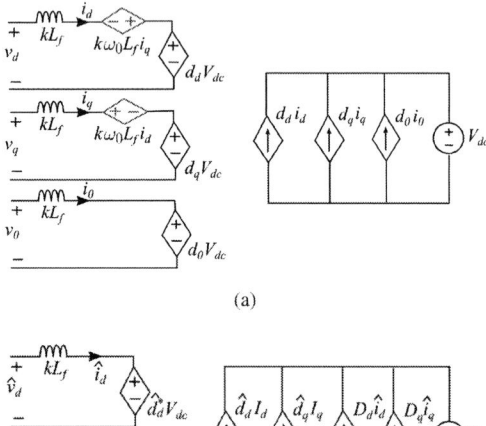

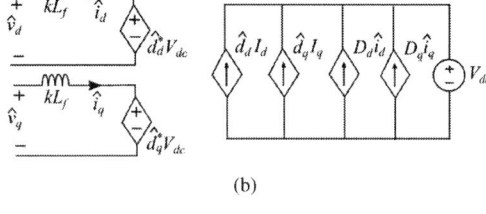

(b)

Fig. 4: Equivalent circuit for current loop design assuming output voltage is fixed at V_{dc}: (a) large-signal equivalent model in the rotating d-q frame; (b) small-signal ac equivalent with cross-coupling terms canceled out.

B. Current Controller Design

The converter small signal model for both operations is shown in Fig. 4. The cross-coupling terms arising from the d-q transformation are eliminated using proper adjustment of the d-axis and q-axis duty cycles [14]. The decoupled d-axis and q-axis small signal model become effectively similar to the dc-dc boost converter case and can be compensated using a simple PI controller. Additionally, grid voltage feed-forward can be added for improved tracking performance and disturbance rejection, as illustrated in Fig. 2.

The plant transfer function can be evaluated as

$$G_{id}(s) = \frac{\hat{i}_d}{\hat{d}_d} = \frac{V_{dc}}{ksL_f}, \quad G_{iq}(s) = \frac{\hat{i}_q}{\hat{d}_q} = \frac{V_{dc}}{ksL_f}, \quad (6)$$

where k is 3 for 3ϕ operation and 2 for 1ϕ operation.

C. DC Link Voltage Control Loop

The dc link voltage is maintained with a bandwidth lower than the line cycle to prevent distortion of the grid current. The average power balance equation for 3ϕ and 1ϕ operation is given by

$$P_{dc} = \epsilon \cdot v_{gd} \cdot i_d = v_{dc} \cdot i_{dc}, \quad (7)$$

where $\epsilon = 1$ for 3ϕ operation and $\epsilon = 1/2$ for 1ϕ operation. Assuming unity power factor (UPF) operation, the power

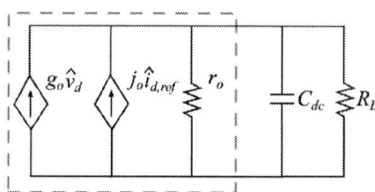

Fig. 5: Low-frequency small-signal equivalent circuit for rectifier output with resistive load R_L.

processed by the q-axis component is zero. The relationship for i_{dc} can be expressed as

$$i_{dc} = \frac{\epsilon \cdot v_{gd} i_d}{v_{dc}}. \quad (8)$$

Linearizing the model around the dc operating point (V_{gd}, I_d, V_{dc}) and noting that current loop compensator ensures $i_d \approx i_{d,ref}$, yields small signal equation listed in (9).

$$
\begin{aligned}
\hat{i}_{dc} &= \frac{\epsilon I_{d,ref}}{V_{dc}}\hat{v}_{gd} + \frac{\epsilon V_{gd}}{V_{dc}}\hat{i}_{d,ref} - \frac{\epsilon V_{gd} I_{d,ref}}{V_{dc}^2}\hat{v}_{dc} \\
&= \underbrace{\frac{\epsilon I_{d,ref}}{V_{dc}}}_{g_o}\hat{v}_{gd} + \underbrace{\frac{\epsilon V_{gd}}{V_{dc}}}_{j_o}\hat{i}_{d,ref} - \underbrace{\frac{I_{dc}}{V_{dc}}}_{1/r_o}\hat{v}_{dc}
\end{aligned} \quad (9)
$$

The transfer function from $\hat{i}_{d,ref}$ to $\hat{v}_{dc}$ for the case of resistive load (R_L) is then given in (10).

$$G_{vi}(s) = \frac{\hat{v}_{dc}}{\hat{i}_d} = \frac{\epsilon V_{gd}}{V_{dc}} \cdot \frac{R_L}{sR_L C_{dc} + 2} \quad (10)$$

The final designed current and voltage compensators and the accompanying loop gain specifications are listed in Table III.

TABLE III: Controller Design Parameters

Parameter	3ϕ Control	1ϕ Control
DQ-Axis Current Controller		
Proportional Gain ($K_{p,dq}$)	0.007	0.0047
Integral Gain ($K_{i,dq}$)	8.84	5.89
Bandwidth (Hz)	2000	2000
Phase Margin (degrees)	84.3	84.3
DC Link Voltage Controller		
Proportional Gain ($K_{p,dc}$)	0.0052	0.0148
Integral Gain ($K_{i,dc}$)	6.54	12.22
Bandwidth (Hz)	20	20
Phase Margin (degrees)	90	90

D. Active Power Filtering in 1ϕ Operation

In 1ϕ operation, the fluctuating $2\omega_0$ power component produces a ripple on the dc link capacitor, which must be sized accordingly. An Active Power Filter (APF) circuit employs a separate buffer capacitance C_{buf} to exchange this ripple power with the grid. This eliminates the $2\omega_0$ component from the C_{dc}, which now can be sized only for the switching frequency ripple. Leg C of the rectifier circuit, which, after interconnection, functions as an APF circuit along with the filter inductance L_f and C_{buf}, is now fed with a duty cycle that depends on the average power P_{dc} processed by the converter and the phase information $\theta_{PLL} = \omega_0 t$ coming from the PLL.

The APF acts as a buck circuit as viewed from the dc link side, where the voltage on the buffer capacitor C_{buf} is given by $v_{buf}(t) = d_{buf}(t) \cdot V_{dc}$, with $d_{buf}(t) \in [0, 1]$. Assuming unity power factor (UPF) operation, and expressing the grid voltage and current as $v_g = V_m \sin(\omega_0 t)$ and $i_g = I_m \sin(\omega_0 t)$, respectively, the instantaneous input power is

$$p_{in} = v_g i_g = \frac{V_m I_m}{2} - \frac{V_m I_m}{2} \cos(2\omega_0 t) = P_{dc} + p_r, \quad (11)$$

where p_r represents the fluctuating power. Ideally, the APF is designed to absorb p_r, allowing only P_{dc} to pass to the dc bus, thereby eliminating the $2\omega_0$ ripple. For the power absorbed by the buffer capacitor, one has

$$\frac{1}{2} C_{buf} \frac{dv_{buf}^2}{dt} = -\frac{V_m I_m}{2} \cos(2\omega_0 t), \quad (12)$$

which leads to

$$v_{buf}^2 = K - \frac{V_m I_m}{2\omega_0 C_{buf}} \sin(2\omega_0 t) = K - \frac{P_{dc}}{\omega_0 C_{buf}} \sin(2\omega_0 t), \quad (13)$$

where K is the constant of integration. The RMS value of the buffer voltage is related to K as $V_{buf,rms} = \sqrt{K}$ [15]. Substituting this into the expression for $v_{buf}(t)$ yields

$$v_{buf}(t) = \sqrt{V_{buf,rms}^2 - \frac{P_{dc}}{\omega_0 C_{buf}} \sin(2\omega_0 t)}. \quad (14)$$

The required duty cycle $d_{buf}(t)$ for the APF leg can then be expressed as shown in (15).

$$d_{buf}(t) = \frac{1}{V_{dc}} \sqrt{V_{buf,rms}^2 - \frac{P_{dc}}{\omega_0 C_{buf}} \sin(2\omega_0 t)} \quad (15)$$

For practical implementation, the duty cycle has to be used in a slightly modified form. To this end, we identify the parameters required to reconstruct the $d_{buf}(t)$ expression shown in (15). The reference dc link voltage $V_{dc,ref}$ is provided by the voltage control loop, while $I_{d,ref}$, the reference for the d-axis current loop, is derived from the dc link voltage loop compensator. Additionally, V_{gd} (d-axis grid voltage), ω_{PLL} (grid angular frequency), and θ_{PLL} (instantaneous phase of the grid voltage) are estimated from the PLL. Using these parameters, the duty cycle can be expressed as

$$d_{buf}(t) = \sqrt{\left(\frac{V_{gd} I_{d,ref}}{2\omega_{PLL} C_{buf} V_{dc,ref}^2} \right) (K_{fac} - \sin(2\theta_{PLL}))}. \quad (16)$$

The factor K_{fac}, which does not affect the $2\omega_0$ ripple cancellation, can be chosen within the allowable range as in (17).

$$1 < K_{fac} < \left(\frac{\omega_0 C_{buf} V_{dc}^2}{P_{dc}} - 1 \right) \quad (17)$$

This factor can be designed to ensure the peak value of $v_{buf}(t)$ remains slightly below $V_{dc,ref}$, limiting RMS current in C_{buf} while maintaining the APF leg duty cycle within acceptable limits. This feed-forward technique ensures effective ripple cancellation across all operating points while maintaining the ripple on the dc link voltage within desired constraints.

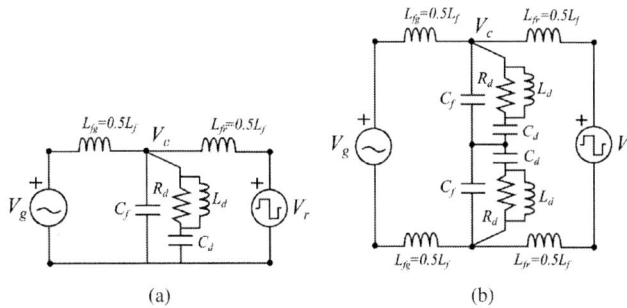

Fig. 6: Equivalent circuit of the rectifier with LCL filter in (a) 3ϕ and (b) 1ϕ configurations.

TABLE IV: Component Values for the Universal ac-dc Front-end

Component	Value
Rectifier Side Filter Inductance (L_{fr})	75 μH
Grid Side Filter Inductance (L_{fg})	75 μH
Filter Capacitor (C_f)	15 μF
Damping Capacitor (C_d)	15 μF
Damping Inductance (L_d)	150 μH
Damping Resistance (R_d)	2.74 Ω
DC Link Capacitance (C_{dc})	25 μF
Buffer Capacitance (C_{buf})	150 μF

E. Comments on Input Filter Design

Standards such as IEEE 519-2014, IEEE 1547-2018, and IEC 61000-3-2 establish permissible limits for harmonic distortion introduced into the power grid, thereby defining the maximum allowable Total Harmonic Distortion (THD) that can be injected by equipment and distributed energy resources. The input LCL filter (Fig. 6) ensures the worst-case THD is below 5% in the grid current while maintaining stable closed loop operation. Since damping the input filter is essential to ensure the rectifier control loop's stability, split-capacitor RL (SC-RL) damping is selected for the LCL filter. The design of the damping branch involves the selection of component values for C_d, R_d, and L_d to get minimum peaking at the filter resonant frequency. Inductor L_d bypasses R_d at line frequency, thus alleviating losses usually introduced by the damping branch.

The designed values are summarized in Table IV. The quality factor (QF) of the LCL filter can be judged from the frequency response of the transfer function (v_c/v_r), where v_c is the voltage on filter capacitor C_f and v_r represents the voltage on the rectifier side. The selected value of $R_d = R_d^*$ yields near-optimal damping, i.e., reduced peak in the (v_c/v_r) frequency response as illustrated in Fig. 7.

III. SIMULATIONS, HARDWARE-IN-THE-LOOP TESTS AND EXPERIMENTAL RESULTS

The components used for the 10 kW/6.6 kW design of the $3\phi/1\phi$ rectifier circuit are listed in Table IV. Simulation results from MATLAB®/PLECS® for the 3ϕ and 1ϕ configurations are presented in Fig. 8 and Fig. 9, respectively, based on the specifications listed in Table I. In the 1ϕ simulation results,

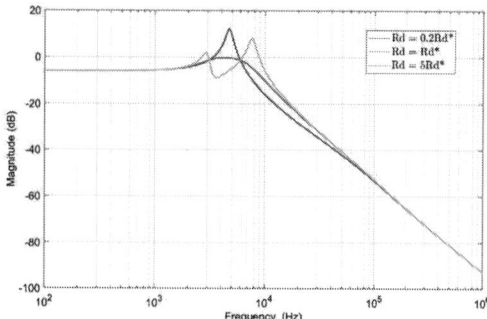

Fig. 7: Frequency response of transfer function v_c/v_r showing the change in filter Quality Factor (QF) for different R_d.

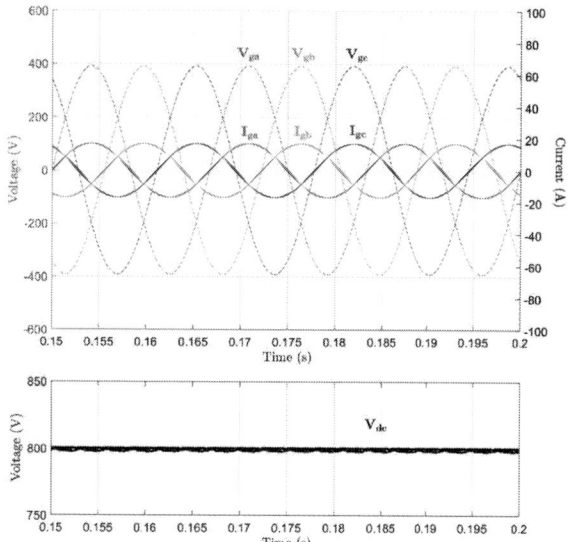

Fig. 8: Simulation results for 3ϕ Rectifier operation for $P_{dc} = 10\,\mathrm{kW}$, $V_{g,LL(rms)} = 480\,\mathrm{V}$, and $V_{dc} = 800\,\mathrm{V}$.

note that the APF adjusts well to step change in load (by 20%) and 30% sag/swell in grid voltage as illustrated in Fig. 9a and Fig. 9b respectively.

The proposed controller is then implemented on Texas Instruments® F28379D LaunchPad™. Hardware-in-the-loop (HIL) test results using Typhoon HIL® 604 for the rated specifications listed in Table I are reported in Fig. 10, confirming that UPF operation and the capability of the APF provision to limit the dc bus voltage ripple.

A rectifier prototype operating at reduced power and bus voltage levels with a resistive load is used to validate the proposed control scheme for 1ϕ configuration. Experimental results at $1.5\,\mathrm{kW}$, $240\,\mathrm{V}$ grid voltage, and $400\,\mathrm{V}$ dc output voltage with and without APF enabled are presented in Fig. 11. It is observed that the $2\omega_0$ ripple on the dc-link voltage (v_{dc}) is significantly reduced as a result of APF operation.

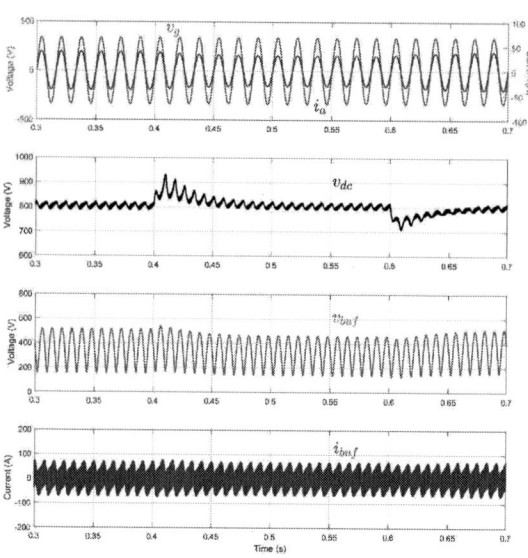

(a) System response to a 20% step change in load.

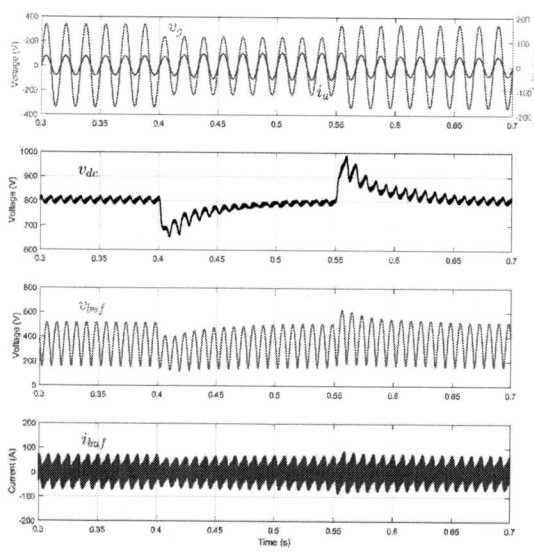

(b) Performance under 30% step change in $V_{g(\mathrm{rms})}$.

Fig. 9: Simulation results for closed-loop operation as 1ϕ rectifier at the rated conditions of $P_{dc} = 6.6\,\mathrm{kW}$, $V_{g(\mathrm{rms})} = 240\,\mathrm{V}$, and $V_{dc} = 800\,\mathrm{V}$.

IV. CONCLUSION

A controller architecture is developed for a universal $3\phi/1\phi$ front-end rectifier. The core rectifier control remains common to both 3ϕ and 1ϕ configurations, simplifying embedded firmware development and enabling efficient implementation. The APF control, employing a simple duty cycle feed-forward technique, effectively eliminates the double line frequency

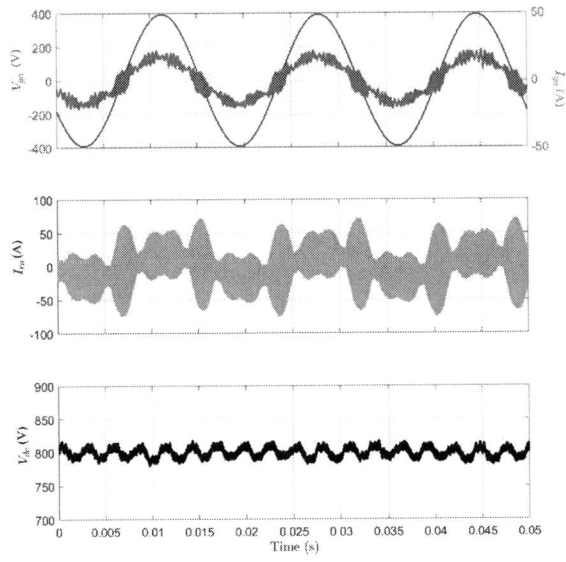

(a) 3ϕ Operation: $P = 10\,\mathrm{kW}$, $V_{g,rms} = 480\,\mathrm{V}$, and $V_{dc} = 800\,\mathrm{V}$.

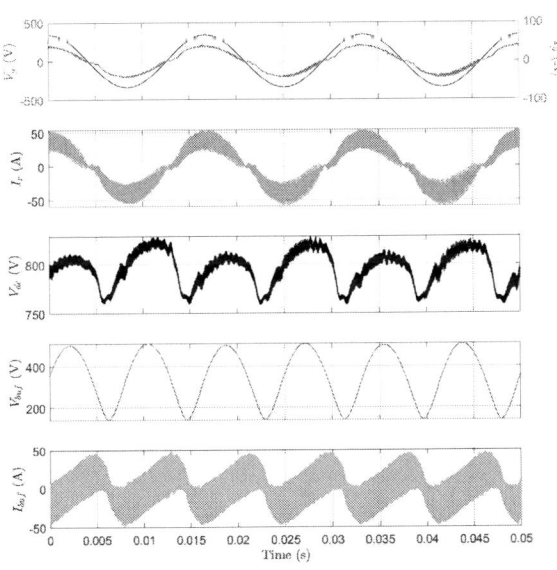

(b) 1ϕ Operation: $P = 6.6\,\mathrm{kW}$, $V_{g,rms} = 240\,\mathrm{V}$, and $V_{dc} = 800\,\mathrm{V}$

Fig. 10: HIL testing results for closed-loop operation in (a) 3ϕ and (b) 1ϕ configurations.

ripple from the dc link, thereby reducing the required dc link capacitance in 1ϕ operation. The proposed control scheme is validated for a $10\,\mathrm{kW}/6.6\,\mathrm{kW}$ system through simulations as well as HIL testing. Experimental results at reduced power and bus voltage levels are also reported, demonstrating the effectiveness of the APF control scheme in 1ϕ operation.

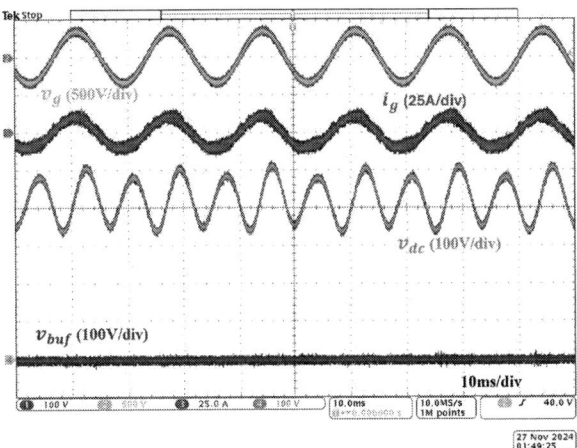

(a) APF disabled

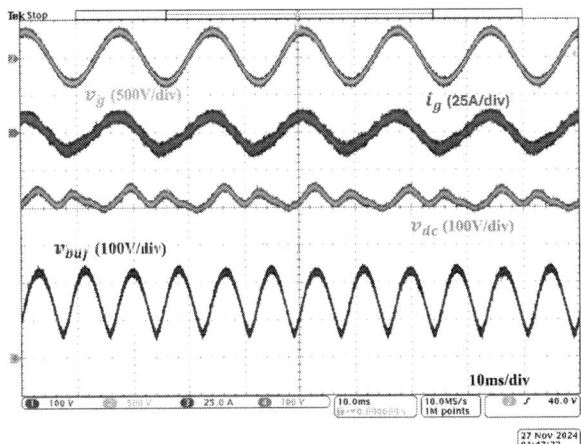

(b) APF enabled ($K_{fac} = 1.2$)

Fig. 11: Experimental results in 1ϕ operation with PFC current loop closed at $P = 1500\,\mathrm{W}$, $V_{dc} = 400\,\mathrm{V}$, and $V_{g,rms} = 240\,\mathrm{V}$.

References

[1] M. Yilmaz and P. T. Krein, "Review of battery charger topologies, charging power levels, and infrastructure for plug-in electric and hybrid vehicles," *IEEE Trans. Power Electron.*, vol. 28, no. 5, pp. 2151–2169, 2013.

[2] H. Tu, H. Feng, S. Srdic, and S. Lukic, "Extreme fast charging of electric vehicles: A technology overview," *IEEE Trans. Transport. Electrific.*, vol. 5, no. 4, pp. 861–878, 2019.

[3] M. M. Mahfouz and R. Iravani, "A supervisory control for resilient operation of the battery-enabled dc fast charging station and the grid," *IEEE Transactions on Power Delivery*, vol. 36, no. 4, pp. 2532–2541, 2021.

[4] D. Menzi, S. Weihe, J. A. Anderson, M. Kasper, J. Huber, and J. W. Kolar, "Novel s-link enabling ultra-compact and ultra-efficient three-phase and single-phase operable on-board ev chargers," in *Proc. 24th IEEE Workshop on Control and Modeling for Power Electronics (COMPEL)*, 2023, pp. 1–7.

[5] H. Zhao, Y. Shen, W. Ying, S. S. Ghosh, M. R. Ahmed, and T. Long, "A single- and three-phase grid compatible converter for electric vehicle on-board chargers," *IEEE Trans. Power Electron.*, vol. 35, no. 7, pp. 7545–7562, 2020.

[6] D. Neumayr, D. Bortis, and J. W. Kolar, "Ultra-compact power pulsation buffer for single-phase dc/ac converter systems," in *Proc. 8th IEEE International Power Electronics and Motion Control Conference (IPEMC-ECCE Asia)*, 2016, pp. 2732–2741.

979-8-3315-1612-3/25 $31.00 © 2025 IEEE

[7] R. Wang, F. Wang, R. Lai, P. Ning, R. Burgos, and D. Boroyevich, "Study of energy storage capacitor reduction for single phase pwm rectifier," in *Proc. 24th Annual IEEE Applied Power Electronics Conference and Exposition (APEC)*, 2009, pp. 1177–1183.

[8] Y. Liu, W. Zhang, Y. Sun, M. Su, G. Xu, and H. Dan, "Review and comparison of control strategies in active power decoupling," *IEEE Trans. Power Electron.*, vol. 36, no. 12, pp. 14 436–14 455, 2021.

[9] Y. Zhao, J. Wang, L. Bao, S. Dai, C. Ye, and Y. Wang, "A control strategy for building ev chargers to provide reactive power compensation for urban power systems," in *2022 IEEE/IAS Industrial and Commercial Power System Asia (I&CPS Asia)*, 2022, pp. 2031–2035.

[10] R. K. Lenka, A. K. Panda, A. R. Dash, N. N. Venkataramana, and N. Tiwary, "Reactive power compensation using vehicle-to-grid enabled bidirectional off-board ev battery charger," in *2021 1st International Conference on Power Electronics and Energy (ICPEE)*, 2021, pp. 1–6.

[11] A. Kulkarni and V. John, "Design of a fast response time single-phase pll with dc offset rejection capability," in *Proc. 31st IEEE Applied Power Electronics Conference and Exposition (APEC)*, 2016, pp. 2200–2206.

[12] A. Rufer, B. Bahrani, S. Kenzelmann, and L. Lopes, "Vector control of single-phase voltage source converters based on fictive axis emulation," in *Proc. 1st IEEE Energy Conversion Congress and Exposition (ECCE)*, 2009, pp. 2689–2695.

[13] A. Kulkarni and V. John, "Analysis of bandwidth–unit-vector-distortion tradeoff in pll during abnormal grid conditions," *IEEE Transactions on Industrial Electronics*, vol. 60, no. 12, pp. 5820–5829, 2013.

[14] S. Buso and P. Mattavelli, "Digital control in power electronics, 2nd edition," *Synthesis Lectures on Power Electronics*, Morgan & Claypool Publishers, USA, 2015.

[15] H. Sarnago and O. Lucía, "High power density on-board charger featuring power pulsating buffer," *IEEE Open Journal of Power Electronics*, vol. 5, pp. 162–170, 2024.

A 10 kV SiC MOSFET based Three-Phase Single-Stage Isolated MVAC/LVDC Converter for Solid State Transformer Applications

Anup Anurag, Chi Zhang, Rudy Wang and Peter Barbosa

Milan M. Jovanović Power Electronics Lab, Delta Electronics, Raleigh, NC 27709

Email: anup.anurag@deltaww.com

Abstract—**Due to the recent strides in developing medium voltage 10 kV Silicon Carbide (SiC) based MOSFETs, it has become possible to design MV systems without complicated multilevel converter topologies. Using these devices opens up a lot of potential to reduce the system size and complexity while preserving high efficiency. Solid-state transformer (SST) applications have recently gained a lot of traction due to the rise of artificial intelligence which fuels the demand for data centers. The realization of these SSTs using a direct single-stage AC/DC power conversion architecture is lucrative due to its low cost and component count. In this paper, a single-stage, isolated AC/DC converter using 10 kV SiC MOSFETs, connected to a 4.16 kV grid, is introduced. The topology combines the power factor correction (PFC) stage and the DC/DC stage which enables the conversion using only a single 10 kV half-bridge module. Further, the MV inductors operate in discontinuous conduction mode (DCM) which results in a simple open loop control for the switches. This single-stage converter uses a single medium voltage transformer which streamlines the complexity and reduces the system cost. An experimental hardware is built and tested up to 5 kV DC link voltage and a maximum power of 20 kW. A peak efficiency of 98.4% is observed for the laboratory prototype.**

Keywords—AC/DC converters, medium voltage, single-stage, taipei rectifier, zero voltage switching (ZVS)

I. INTRODUCTION

With the ever-changing requirements of making the grid smarter, the concept of solid-state transformers (SSTs) has garnered a lot of interest since they form the backbone of an intelligent grid system. MVAC/LVDC SSTs have become a promising AC/DC interface owing to their numerous advantages including lower number of conversion stages, smaller size, fault current limiting capacity, and renewable energy integration. Typically, MVAC/LVDC voltage conversion is realized as a two-state solution where the AC/DC stage provides the PFC function and the DC/DC stage provides the galvanic isolation while controlling the output voltage [1]–[3]. To realize these systems, an Input-Series Output-Parallel (ISOP) configuration is most commonly used [4]. The advantage of the ISOP system stems from the fact that the system uses reliable and cost-effective LV semiconductor technology to build up the voltage. Further, the inherent nature of the system makes it highly modular and the system can offer redundancy in the event of a failure. The main downside of such a system lies in the MV galvanic isolation. Due to the ISOP structure, each converter needs a transformer to provide the MV isolation which may result in a large number of lower power transformers rated for MV isolation [4]. Also, since each of the MV phases operates individually, the double-line frequency ripple needs to be suppressed within each modular

converter. Three-phase single-stage conversion, which offers a direct MVAC/LVDC conversion, has gained a lot of interest since it offers a low cost, low component count, and convenient solution where the intermediate DC bus is not necessary.

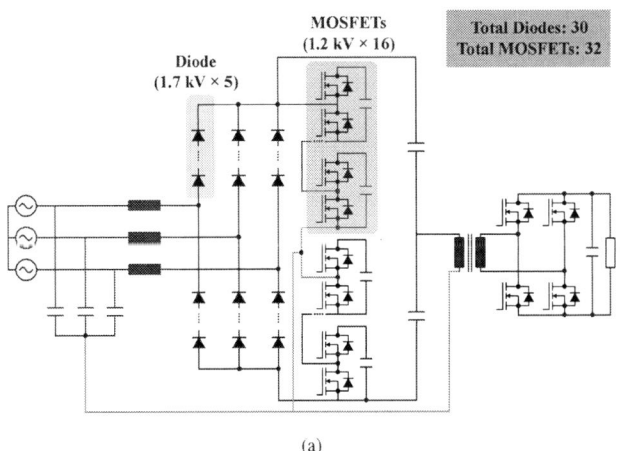

(a)

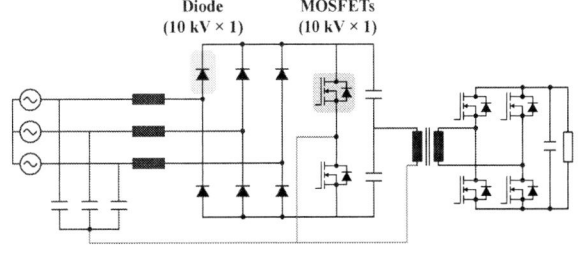

(b)

Fig. 1: (a) Realization of the single-stage MVAC/LVDC (4.16 kV MV to 800 V LV) converter using (a) 1.2 kV SiC MOSFETs in a super-switch configuration, and (b) 10 kV SiC MOSFETs.

To mitigate the challenges of the ISOP structure, while maintaining a simple low cost solution, a super-switch-based single-stage SST is developed in [5]. The schematic of the converter is shown in Fig. 1(a). This converter uses LV half-bridge cells connected to build up the medium voltage. This structure can achieve a quasi-2-level operation which reduces the dv/dt stress at the MV AC node. However, the concept still

979-8-3315-1612-3/25 $31.00 © 2025 IEEE

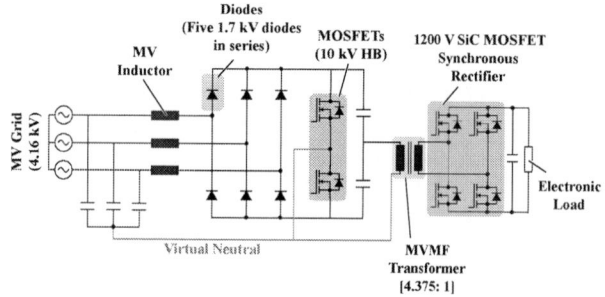

Fig. 2: Schematic of the single-stage three phase 10 kV SiC MOSFET based AC/DC SST system.

uses the LV devices as its building block which increases the complexity of the structure, especially in terms of component count as well as communication between the individual cells, which is typically carried out through optical fibers. Recent developments in MV power semiconductor devices have made it possible to have devices with blocking voltage capability of up to 10 kV [6]. The development of these MV devices has opened up opportunities to reduce the system complexity and interface simple power electronics converters directly to the MV grid. Fig. 1 shows the comparison between the usage of LV 1.2 kV devices and 10 kV SiC MOSFETs based on the structure used in [5].

The advantages offered by the 10 kV SiC MOSFETs are multifold. A reduction in the number of cells, as well as device count, is observed which vastly reduces the complexity of the system. The number of communication nodes is decreased significantly which can be translated to a huge cost saving. However, the use of MV devices also brings along its challenges in terms of the high dv/dt seen at the MVAC node. The quasi-2-level operation, which is directly dependent on the number of cells, is also affected since the number of steps is reduced. Lastly, the high costs of the current generation of the 10 kV SiC MOSFETs make it challenging to implement this on a larger scale.

This paper explores the possibility of using 10 kV SiC MOSFETs in a single-stage solution, which can convert MVAC (4.16 kV grid) to LVDC (800 V) using only two active devices. The inherent complexity of connecting multiple LV devices is eliminated, and a comparable efficiency to the LV based counterpart is observed. A detailed design of the MV system needed to implement the converter, including the MV busbar design, magnetics design, and the MV gate driver, is provided in this paper. The MV system is designed and tested in the laboratory for a 20 kW system operating at a maximum of 5 kV DC-link voltage, and a peak efficiency of 98.4% is observed. The demonstrated topology uses only a single 10 kV half-bridge module to integrate the MV SST system into a 4.16 kV MV grid. It is however worthwhile to note that for MV grid connections higher than 4.16 kV, voltage scaling-up concepts for the 10 kV SiC MOSFETs such as the super-switch solution [7] are necessary to handle the required blocking voltage.

II. HARDWARE DESIGN FOR THE MV CONVERTER SYSTEM

As shown in Fig. 2, the converter consists of different medium voltage components, which work together to realize the total converter operation. The LV side which converts the high-frequency AC to LVDC can be realized by diodes, but a synchronous rectifier is used to improve the efficiency of the system. The synchronous rectifier is built using commercially available 1.2 kV SiC MOSFETs (CAB400M12XM3) and gate drivers from Wolfspeed (CGD12HBXMP) [8]. The parameters of the system are shown in Table I. A description of the design

TABLE I: Parameters of the system

Parameter	Value
Rated power	30 kW
MV DC-link voltage	7 kV
LV DC-link voltage	800 V
Switching Frequency	44 kHz
Resonant Frequency	39 kHz

and testing of the individual MV components is necessary to understand the realization of the entire system.

A. Litz wire based MV inductors

The MV inductors for all the phases are designed using a conventional manner where spacing is provided between the winding and the core to achieve the required insulation. The core uses a low-loss ferrite N87 material and the windings are realized using litz wires and a single layer winding structure. A spacing of 5 mm is provided between the winding and the core. An optimization is carried out to decide the core size and the number of turns based on the efficiency and power density. While there is an inherent limit in scaling up the voltage levels of such a design, this design is used to demonstrate the 4.16 kV grid-connected SST application. A partial discharge test on the inductor to ensure a reliable operation in the system. A 5 kV RMS Partial discharge inception voltage (PDIV) between the winding and the core is observed which is higher than the system requirement. Fig. 3 shows the photograph of the designed inductor.

B. Medium Voltage PCB Busbars

With the use of fast switching devices, busbars are the preferred way of interconnection between the devices and form one of the critical parts of the medium voltage system. Apart from carrying the current, the busbar needs to handle the required medium voltage insulation. Conventional busbars which are made up of stacked copper plates and insulators are prone to internal air voids during the manufacturing process which can lead to partial discharge at medium voltage levels. One of the solutions to increase the PDIV level is to increase the insulation thickness. However, increasing insulation thickness leads to over-designing of the system while increasing the cost, weight, and inductance of the whole structure. To avoid these challenges, PCBs provide an elegant way to make the busbar structure due to the mature technology to manufacture a defect-free product.

To reduce the total thickness of the PCB structure while maintaining a high PDIV, a design process similar to [9] is

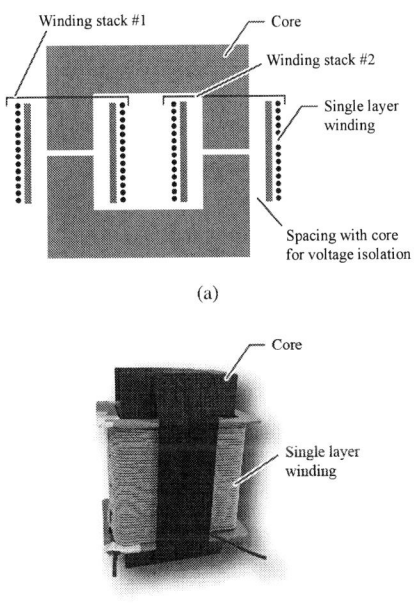

(a)

(b)

Fig. 3: (a) Schematic and (b) prototype of the MV inductor.

used. The busbar is designed for a three-phase system with positions for mounting three 10 kV SiC MOSFETs. There are three critical points to verify for the electric fields in the PCB namely, the parallel plane, the plated through holes, and the net terminations. This is shown in Fig. 4. A field grading method is applied to ensure that the electric field in the air is less than <2 kV/mm. The structure as well as the developed prototype of the board is shown in Fig. 5.

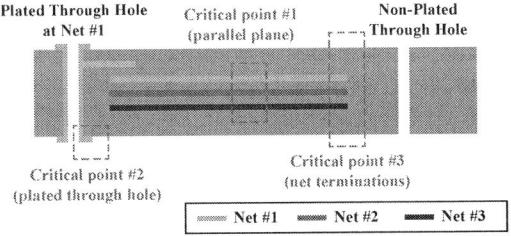

Fig. 4: Cross section of the PCB structure showcasing the critical electric field points.

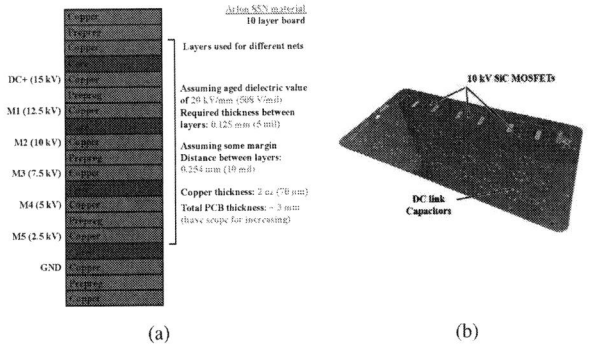

(a)   (b)

Fig. 5: (a) Schematic and (b) prototype of the MV busbar.

The electric field is simulated for the critical points and serves as a guideline for the design process. The simulation results are shown in Fig. 6. In Fig. 6(a), the effect of field grading is demonstrated. On the left side, field grading is carried out, and on the right side, there is no field grading. It is seen that the electric field in the air exceeds 2 kV/mm without any field grading. In Fig. 6(b), the design rules for a non-

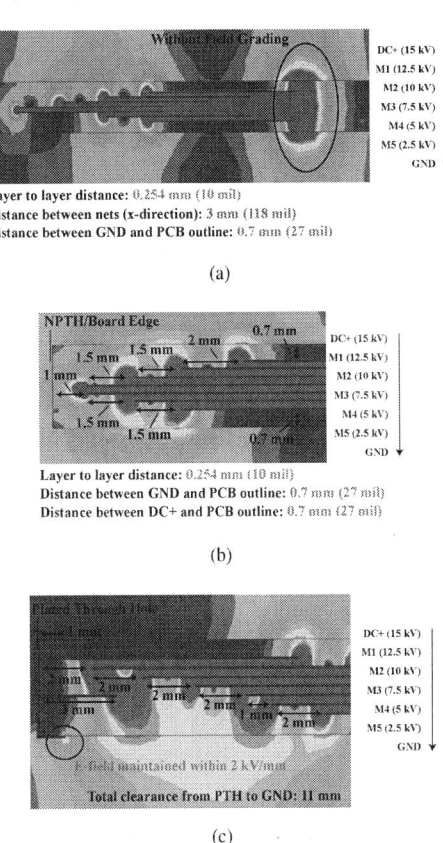

(a)

(b)

(c)

Fig. 6: (a) Schematic and (b) prototype of the MV busbar.

plated through hole (NPTH)/ or a board edge are established for a DC voltage of 15 kV. The design rules for plated through holes (PTH) are established in Fig. 6(c). In this, an embedded finger/shield is provided to divert the electric field from the boundary of the board. With the established guidelines from the simulations, the MV busbar is designed.

A partial discharge test of the busbar reveals a PDIV voltage of 16 kV RMS, which exceeds the design goals and is deemed suitable to be used in the system.

C. MV Devices and Gate Drivers

A single 10 kV SiC MOSFET half-bridge module is used for the system. A building block is created where the 10 kV SiC MOSFET is mounted to a heatsink, and Rogowski coils are added on the current path for overcurrent protection, as shown in Fig. 7(a). Two of these building blocks are used for running the converter system as described later. The MV gate drivers used for driving the 10 kV SiC MOSFETs are the ones used in [10] and are shown in Fig. 7(b). They offer a high PD voltage of 11 kV RMS and a low coupling capacitance of 6 pF and are suitable for the converter operation.

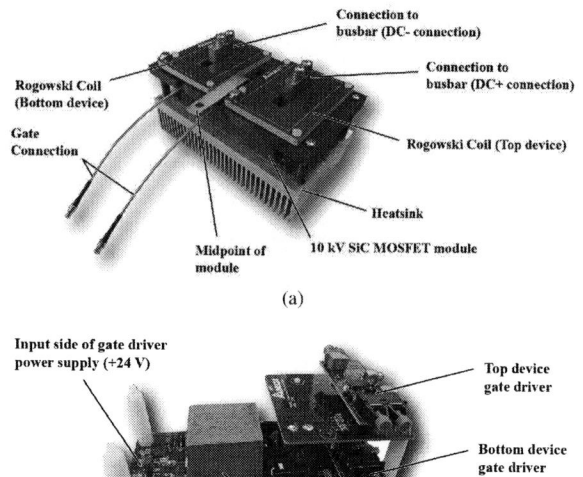

(a)

(b)

Fig. 7: (a) 10 kV SiC MOSFET half-bridge module, and (b) medium voltage gate driver used for driving the 10 kV SiC MOSFET module.

D. MV Transformer

One of the key components of the SST is the medium voltage medium frequency (MVMF) transformer which provides the galvanic isolation as well as the voltage step down between the MV and the LV side. Designing an MVMF transformer is one of the most challenging aspects due to the high insulation requirements. Due to the medium frequency operation, it is necessary to use litz wire to form the winding structure, and typically ferrite is used as the core material due to its high-frequency performance. Ideally for low power levels, a dry-type transformer is preferred. However, encapsulating litz wire structures is a complicated and unreliable process since there is a high possibility of air bubbles being trapped inside the litz wire surface. A vacuum-based encapsulation process can be carried out to ensure that the air bubbles are removed, but it still leads to unreliable results, especially for PD testing where even small air bubbles inside the structure can lead to a reduced PDIV causing insulation failure over a period of time. In [11], a dry-type transformer is designed for 50 kHz switching frequency operation at 25 kW rated power. However, for the insulation testing, only a DC test of up to 20 kV is carried out and no PD test results are shown.

To achieve a power-dense and efficient transformer, an optimization based on conventional techniques has been carried out to determine the parameters of the oil-insulated transformer. The optimization determines the core shape and size, number of turns, and maximum flux density based on the total volume and losses. The core material is fixed to N95 material and an isolation distance of 6 mm is also fixed. Based on the optimization, the closest commercially available core (B67345B0003X095) is used for the design. The schematic of the transformer, and the developed prototype is shown in Fig. 8. An impedance analyzer is used to measure the leakage inductance and the magnetizing inductance of the transformer structure. An effective leakage inductance of 311 μH and

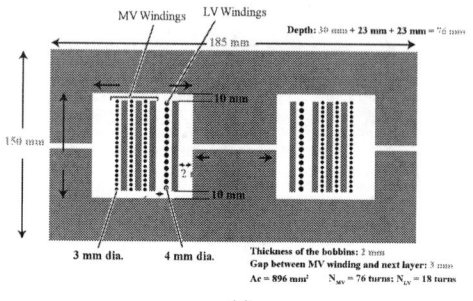

(a)

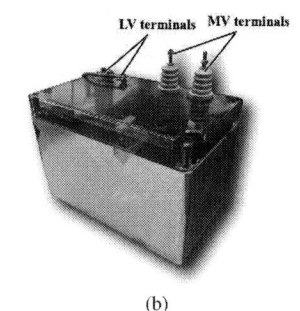

(b)

Fig. 8: (a) Schematic of the transformer, and (b) photograph of the experimental prototype.

a magnetizing inductance of 10.25 mH is measured. The resonant capacitor can be designed based on the same. A commercially available container is used for the tank structure and medium voltage bushings are provided for the insulation. The transformer is filled with oil and is put under vacuum to remove all the internal air bubbles. Fig. 9 shows the transformer structure during the vacuum process and just after removing it from the vacuum chamber. It is seen that, even if the bubbles are not visible to the naked eyes, as soon as the structure experiences low pressure, the bubbles start to flow up to the surface and escape. The transformer is kept inside a vacuum for more than 24 hours to allow for all bubbles to escape. A partial discharge test, as well as a voltage breakdown test, is carried out for the MVMF transformer. The partial discharge test setup, along with the experimental results are shown in Fig. 10. The PDIV is found to be 22 kV RMS, as shown in Fig. 10(b) which is well above the operating range for the transformer. An insulation breakdown test is also carried out to test the withstand capacity of the transformer and to identify the voltage breakdown point. A voltage withstand capacity of around 32 kV is observed as shown in Fig. 10(c). The tank is connected to the ground potential and a medium voltage is applied to the terminals. The critical weak point is identified as the distance between the terminal and the metal screw connecting the lid to the tank structure. Further improvements to the structure can be made by increasing this distance or using insulating materials between the two points. The parameters of the transformer are summarized in Table II.

III. EXPERIMENTAL PROTOTYPE DESIGN AND HARDWARE RESULTS

Fig. 11 shows the complete hardware prototype for the single-stage single-switch SST structure. To ensure DCM operation, and consequently achieve PFC, the parameters have been designed accordingly [12]. In the current setup, testing is carried out for up to 5 kV DC-link voltage and 20 kW of power. Two different switching frequencies (44 kHz and 52 kHz) are tested to compare the efficiency differences between the operations. The system is run for 15 minutes to ensure thermal stability and the efficiency measurements are carried out. The efficiency measurements are carried out using the Yokogawa WT5000

TABLE II: Specifications of the transformer

Parameter	Value
Rated power	30 kW
MV voltage	7 kV
LV voltage	800 V
Turns-ratio	4.375:1
Leakage/magnetizing inductance	311 μH/10.25 mH
PD voltage	22 kV RMS
Withstand voltage	32 kV RMS

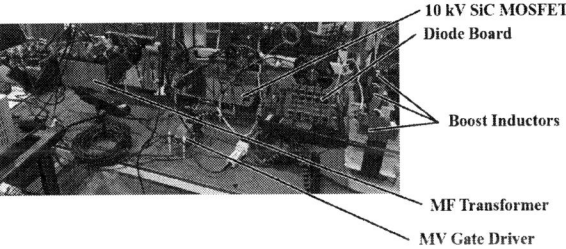

Fig. 11: Experimental setup of the single-stage SST.

series power analyzer. For the medium voltage measurements, HV CIC Research probes, which can measure up to 20 kV RMS voltage, are used. The current measurements are carried out using a Danisense 50 A current sensor. The sensors are interfaced with the power analyzer. Measurements are carried out at 4 kV DC-link voltage and 5 kV DC-link voltage. In Fig. 12, the critical experimental waveforms are shown for 4 kV and 5 kV DC-link voltage at a switching frequency of 44 kHz. It can be seen that the system achieves PFC in an open loop operation and the current is sinusoidal. A THD of around 3% is observed for the current waveforms. The devices achieve soft-switching since the resonant tank is operated above the resonant frequency. The boost inductor current as well as the resonant current both contribute to the soft-switching operation of the devices. Similar tests are repeated for a switching frequency of 52 kHz (which is further higher than the resonant frequency) and the efficiency is recorded. Some of the critical efficiency points are shown in Fig. 13. A peak efficiency of 98.4% was observed at 4 kV DC link with a power transfer of 13 kW. For the rated power of 20 kW, at 5 kV DC link voltage, an efficiency of 98.15% was observed.

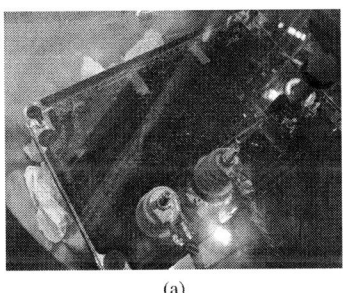

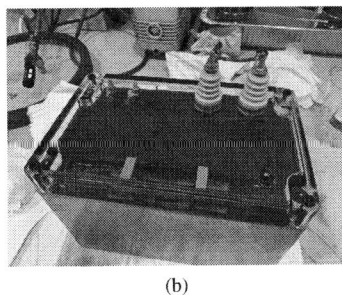

Fig. 9: Photograph of the oil transformer (a) just after starting the vacuum process, and (b) after the vacuum process is completed.

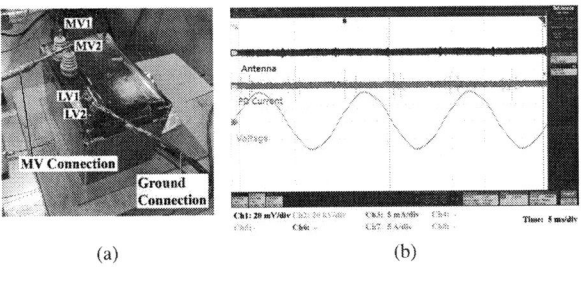

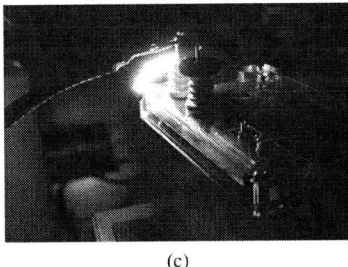

Fig. 10: (a) Developed prototype of the transformer being tested in the partial discharge chamber, and (b) partial discharge testing results of the transformer showcasing a 22 kV RMS PDIV (Ch1: Voltage across the antennae, Ch2: Current through the ground, Ch3: Applied voltage across the transformer) (c) Long exposure photograph for the AC withstand testing of the transformer. AC withstand capability of 32 kV RMS is observed.

IV. CONCLUSIONS

This paper demonstrates the operation of a single-stage, single-phase medium voltage AC/DC SST system using only a single half-bridge module. The topology offers a low-cost, low-complexity solution for direct integration to the MV grid. An efficiency of 98.4% has been demonstrated which is promising as compared to two-stage solutions. Details and design considerations regarding developing different parts of the system are also provided. A partial discharge free busbar is designed and tested. A MVMF transformer is designed which can withstand a 22 kV RMS partial discharge voltage. The current system is aimed at a 4.16 kV grid connection which needs only a single 10 kV SiC MOSFET for the required DC-link of 7 kV. For scaling up to higher voltages such as 13.8 kV, where the DC-link is around 22 kV, scale-up strategies such as series-connection, multilevel converters, or super-switch concepts need to be implemented. It is envisaged

979-8-3315-1612-3/25 $31.00 © 2025 IEEE

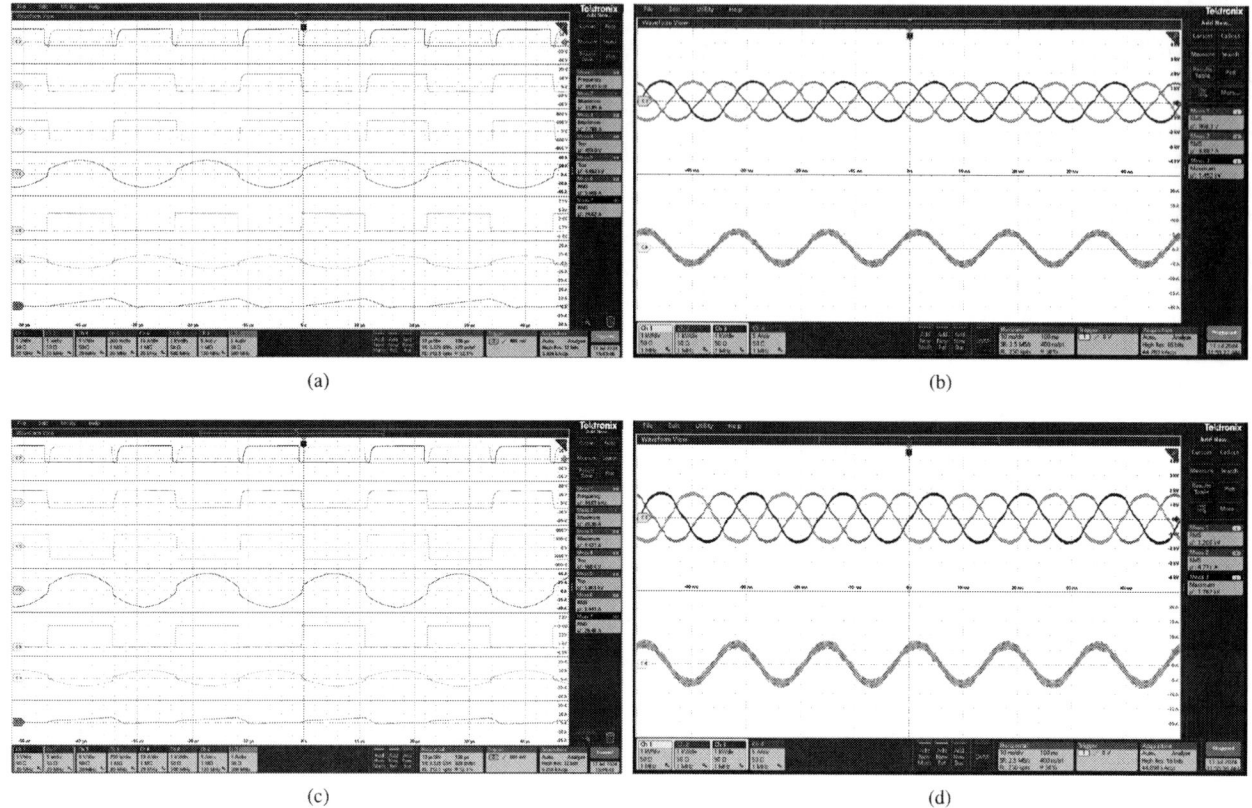

(a)   (b)

(c)   (d)

Fig. 12: Experimental waveforms showcasing the operation of the single-stage SST converter under different voltage and frequency conditions (a,b) 4 kV; 44 kHz, and (c,d) 5 kV; 44 kHz.(a,c- Ch1: SR gate voltage top device (5 V/div), Ch2: MV gate voltage top device (5 V/div), Ch3: MV gate voltage bottom device (5 V/div), Ch4: MV side resonant current (5 A/div), Ch5: Switching node voltage on LV side (200 V/div), Ch6: LV side resonant current (10 A/div), Ch7: DCM inductor through boost current (5 A/div), Ch8: Voltage across bottom device on MV side (1 kV/div) b,d - Ch1/Ch2/Ch3: Grid phase voltages (1 kV/div), Ch4: B-phase grid current (5 A/div)).

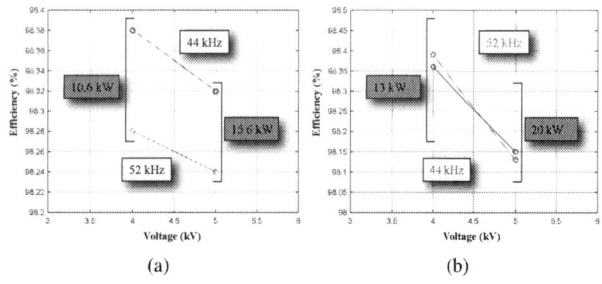

(a)   (b)

Fig. 13: Measured efficiency of the system at a (a) constant load of $R = 20\ \Omega$, and (b) $R = 16\ \Omega$. The LV side voltage is based on the turns-ratio of the transformer. The peak efficiency of the single-stage converter system is found to be 98.4%.

that the use of 10 kV SiC MOSFETS can simplify the system complexity and provide a simplified and cost-effective solution for a MVDC to LVDC connection.

ACKNOWLEDGEMENTS

The authors would like to thank the support engineers at Milan Power Electronics Lab, Mr. Ted Pappas, and Mr. Juan Ruiz for help in the PCB layout, soldering and hardware implementation of the magnetic components. The authors would also like to thank the MV team for discussions and idea development.

REFERENCES

[1] J. E. Huber and J. W. Kolar, "Solid-State Transformers: On the Origins and Evolution of Key Concepts," *IEEE Industrial Electronics Magazine*, vol. 10, no. 3, pp. 19–28, 2016.

[2] A. Q. Huang, "Medium-Voltage Solid-State Transformer: Technology for a Smarter and Resilient Grid," *IEEE Industrial Electronics Magazine*, vol. 10, no. 3, pp. 29–42, 2016.

[3] X. She, A. Q. Huang, and R. Burgos, "Review of Solid-State Transformer Technologies and Their Application in Power Distribution Systems," *IEEE Journal of Emerging and Selected Topics in Power Electronics*, vol. 1, no. 3, pp. 186–198, 2013.

[4] R. L. Da Silva, V. L. F. Borges, C. E. Possamai, and I. Barbi, "Solid-State Transformer for Power Distribution Grid Based on a Hybrid Switched-Capacitor LLC-SRC Converter: Analysis, Design, and Experimentation," *IEEE Access*, vol. 8, pp. 141 182–141 207, 2020.

[5] C. Zhang, R. Wang, Z. Shen, T. Sadilek, A. Anurag, and P. Barbosa, "A Single-Stage Three-Phase Isolated AC-DC Converter for Medium Voltage Solid State Transformer Applications," in *2023 IEEE Applied Power Electronics Conference and Exposition (APEC)*, 2023, pp. 1503–1509.

[6] Y. Li, M. ul Hassan, A. B. Mirza, Y. Xie, S. Deng, S. S. Vala, F. Luo, X. Feng, S. Narumanchi, and J. D. Flicker, "State-of-the-Art Medium- and High-Voltage Silicon Carbide Power Modules, Challenges and Mitigation Techniques: A Review," *IEEE Transactions on Components, Packaging and Manufacturing Technology*, pp. 1–1, 2024.

[7] R. Wang, C. Zhang, T. Sadilek, Z. Shen, and P. Barbosa, "Quasi-Two-Level (Q2L) Half Bridge Cascaded (HBC) Super Switch (SS) Concept for Medium Voltage Applications," in *2022 IEEE Energy Conversion Congress and Exposition (ECCE)*, 2022, pp. 1–6.

979-8-3315-1612-3/25 $31.00 © 2025 IEEE

[8] Wolfspeed. [Online]. Available: https://www.wolfspeed.com/

[9] L. Ravi, X. Lin, D. Dong, and R. Burgos, "A 16 kV PCB-Based DC-Bus Distributed Capacitor Array with Integrated Power-AC-Terminal for 10 kV SiC MOSFET Modules in Medium-Voltage Inverter Applications," in *2020 IEEE Energy Conversion Congress and Exposition (ECCE)*, 2020, pp. 3998–4005.

[10] A. Anurag and P. Barbosa, "High-Voltage Isolated Power Supply Structure for Gate Drivers of Medium-Voltage SiC Devices," *IEEE Transactions on Power Electronics*, vol. 38, no. 6, pp. 6907–6911, 2023.

[11] D. Rothmund, T. Guillod, D. Bortis, and J. W. Kolar, "99% Efficient 10 kV SiC-Based 7 kV/400 V DC Transformer for Future Data Centers," *IEEE Journal of Emerging and Selected Topics in Power Electronics*, vol. 7, no. 2, pp. 753–767, 2019.

[12] Y. Jang, M. M. Jovanović, M. Kumar, Y. Chang, Y.-W. Lin, and C.-L. Liu, "A Two-Switch, Isolated, Three-Phase AC–DC Converter," *IEEE Transactions on Power Electronics*, vol. 34, no. 11, pp. 10 874–10 886, 2019.

Direct Digital Control Applied to T-type Vienna Rectifiers for Power Factor Correction

Jun-Yang Chang
Advanced Research Section in
Photovoltaic Inverter Business Unit
Delta Electronics Inc. (Taoyuan Plant)
line 4: Taoyuan, Taiwan
Email: a0983812545@gmail.com

Tsai-Fu Wu
Elegant Power Electronics Applied
Research Laboratory in Department of
Electrical Engineering
National Tsing Hua University
Hsinchu, Taiwan
Email: tfwu@ee.nthu.edu.tw

Chien-Chih Hung
Elegant Power Electronics Applied
Research Laboratory in Department of
Electrical Engineer
National Tsing Hua University
Hsinchu, Taiwan
Email: jeff80121@gmail.com

Jui-Yang Chiu
Elegant Power Electronics Applied
Research Laboratory in Department of
Electrical Engineer
National Tsing Hua University
Hsinchu, Taiwan
Email: s110061802@m110.nthu.edu.tw

Abstract—**This paper develops a three-phase three-wire T-type Vienna rectifier with direct digital control (DDC) for power factor correction (PFC). It features lower component voltage stress, reduced input current harmonics, boost-type topology, and higher power density. The DDC strategy is adopted to simplify the derivation of control laws through a division-summation process. This control method does not need abc-to-dq frame transformations; thus, it results in a simple derivation and programming. Furthermore, not only can the control laws handle the normal input voltage, but they can handle imbalance and distortion conditions. The successful control of three-phase decoupling is achieved, making the rectifier more effective. Experimental results from a 6 kW prototype verify the analysis and discussion.**

Keywords—***Three-phase three-wire Vienna rectifier, Direct Digital Control, Grid Voltage Imbalance and Distortion Control***

I. INTRODUCTION

Vienna rectifiers are used to transfer AC to DC power, popular for charging stations. In various load applications, harmonic currents deteriorate in power quality. Thus, PFC circuits are used in electronic products to improve the power factor (PF) and stability of the products. There are two types of PFC circuits: passive and active types. Passive PFC circuits typically achieve power factors between 0.7 and 0.8 using capacitive and inductive compensation, which is bulky and heavy. Active PFC circuits use Pulse-Width Modulation to modulate the input current and feature smaller capacitors and inductors. Furthermore, it maintains a PF of higher than 0.95, suitable for high-power applications.

A. Vienna Topology

Vienna rectifiers [1]-[5] are characterized by a three-level output, which reduces the number of switches compared to the traditional three-level rectifiers, simplifying hardware design. Since there is no bridge-arm switch conduction issue, dead time is not a concern. However, the connection of DC link midpoint to each phase input inductor via switches introduces a midpoint imbalance issue, which must be addressed in controller design. Replacing all diodes in a six-switch Vienna rectifier with active switches can further reduce switching losses and improve system efficiency,

while increase control complexity. For lower cost, better efficiency and easier to control, this study adopts the six-switch Vienna rectifier with DDC.

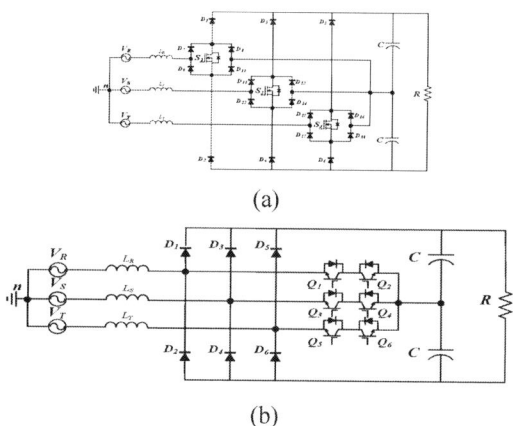

(a)

(b)

Fig. 1. Vienna rectifier Topology: (a) Three-swiitch Vienna rectifier and (b) Six-switch Vienna rectifier.

B. Control Strategy

Recent advancements in Vienna rectifier control methods are with abc-to-dq [6] based Proportional-Integral-Derivative [7]-[9] (PID) control. The commonly used PI controller is effective in correcting low-frequency errors and noise rejection, making it suitable for high-noise environments. However it has slower response and the parameters need to be tuned. The Proportional-Resonant (PR) [10]-[12] control was then adopted, but it only tracks a specific frequency to achieve good performance while suppressing harmonic interference. However, its high sensitivity to specific frequencies requires careful design. Three-phase decoupled (D^3C) [13]-[16] can derive the control law using a division-summation process, avoiding the traditional frame transformation (abc to dq). It can leverage system parameters to create an adaptive control law, as shown in Fig. 2. The benefits include better adaptability to system parameter changes, improved robustness, simpler design, and good dynamic response. Therefore, it is selected for the control in this research.

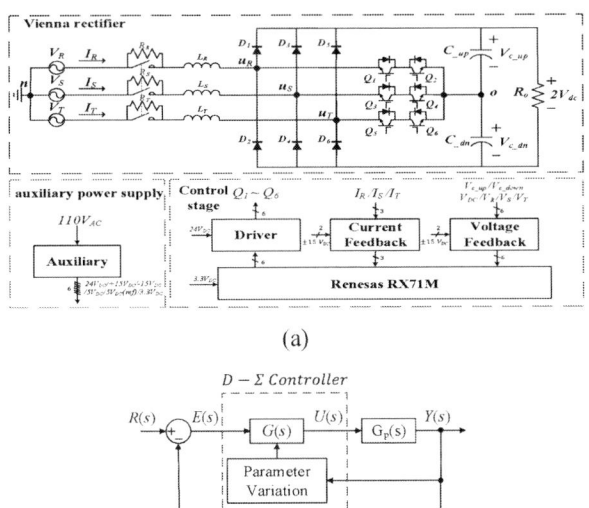

(a)

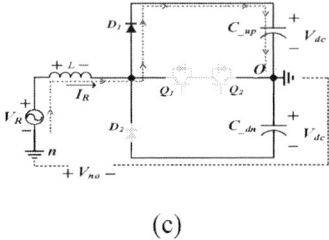

(c)

Fig. 3. Equivalent circuit diagram under $V_R > 0$: (a) Single-phase (b) Magnetization and (c) Demagnetization modes.

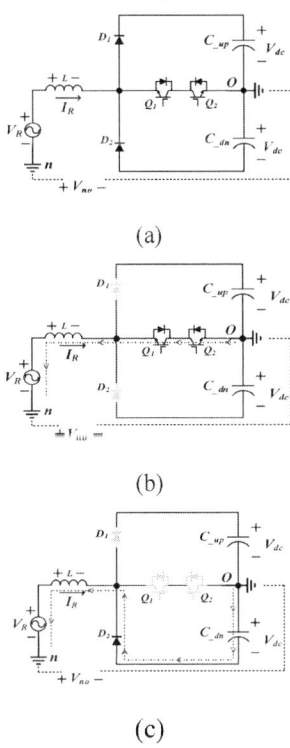

(a)

(b)

(c)

Fig. 4. Equivalent circuit diagram under $V_R < 0$: (a) Single-phase (b) Magnetization and (c) Demagnetization modes.

(b)

Fig. 2. (a) System architecture and (b) DDDC block diagram.

II. SYSTEM ARCHITECTURE AND CONTOL SCHEME

A. Circuit Operation

The three phases have been decoupled through the adaptation process [10]. They can be analyzed as three sets of single-phase rectifiers. The single-phase equivalent circuit is shown in Fig. 3(a). When the switches Q_1 and Q_2 are turned on, the input power source will store energy in the inductor, as shown in Fig. 3(b). When the Q_1 and Q_2 are turned off, either D_1 conducts if the input voltage $V_R > 0$, or D_2 conducts if $V_R < 0$, as shown in Fig. 3(c). Then, the inductor transfers energy to the dc-link capacitor and supplies the load.

B. Decoupled Direct Digital Control(DDDC)

The three-phase system is decoupled into three single-phase systems by utilizing the common-mode voltage (v_{no}) as a reference loop. The single-phase equivalent circuit is shown in Fig. 4(a), and it can be divided into two states by the different polarity of input voltage. Therefore, it has totally four sets and each set includes: magnetization mode and demagnetization mode, as shown in Fig. 4(b) and 4(c), respectively.

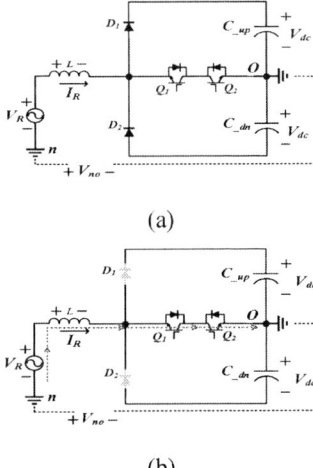

(a)

(b)

In the division process, as $V_R > 0$ and from Fig. 4(b) and (c), ΔI_{Rmag} occurs when switches Q_1 and Q_2 are turned on, and D_1 and D_2 are turned off, while ΔI_{Rdem} occurs when Q_1, Q_2 and D_2 are turned off, but D_1 is turned on. The related expressions are as follows:

Magnetization mode(Q_1 and Q_2: on, D_1 and D_2: off):

$$\Delta I_{Rmag} = \frac{V_{no} + V_R - (R_R \times I_R)}{L_R} \times D_R T_{SW} \qquad (1)$$

Demagnetization mode(D_1: on, the others: off):

$$\Delta I_{Rdem} = \frac{V_{no} + V_R - (R_R \times I_R) - V_{dc}}{L_R} \times (1 - D_R) T_{SW} \qquad (2)$$

As $V_R < 0$, from Fig. 4(b) and (c), ΔI_{Rmag} occurs when switches Q_1 and Q_2 are turned on, and D_1 and D_2 are turned off, while ΔI_{Rdem} occurs when Q_1, Q_2 and D_1 are turned off, but D_2 is turned on. The related expressions are as follows:

Magnetization mode(Q_1 and Q_2 are on, D_1 and D_2 are off):

$$\Delta I_{Rmag} = \frac{V_{no} + V_R - (R_R \times I_R)}{L_R} \times D_R T_{SW} \qquad (3)$$

979-8-3315-1612-3/25 $31.00 © 2025 IEEE   17

Demagnetization mode(D_2 is on, the others are off):

$$\Delta I_{Rdem} = \frac{V_{no} + V_R - (R_R \times I_R) + V_{DC}}{L_R} \times (1 - D_R)T_{SW} \qquad (4)$$

In summation process, as $V_R > 0$ where D_R indicates the duty ratio of phase R and T_{sw} is the switching period. Adding (1) and (2) together to obtain the current change in one cycle can be expressed as (5), and then, the duty ratio can be derived as (6):

$$\Delta I_R = \Delta I_{Rmag} + \Delta I_{Rdem} \qquad (5)$$

$$D_R = \frac{L_R \times \Delta i_R}{T_{SW} \times V_{dc}} - \frac{V_R}{V_{dc}} + \frac{R_R \times I_R}{V_{dc}} + 1 - \frac{\overline{V_{no}}}{V_{dc}} \qquad (6)$$

where $\overline{V_{no}}$ is the average of the common-mode voltage. In $V_R < 0$, adding (3) and (4) together to obtain the current change in one switching cycle can be expressed as (5), and then, the duty ratio can be derived as (7):

$$D_R = -\frac{L_R \times \Delta i_R}{T_{SW} \times V_{dc}} + \frac{V_R}{V_{dc}} - \frac{R_R \times I_R}{V_{dc}} + 1 + \frac{\overline{V_{no}}}{V_{dc}} \qquad (7)$$

Similarly, the derivation for phases S and T follows the same methodology. To achieve the D³C, the duty ratio needs to be discretized, forming the basis of the three-phase decoupled direct digital controller. The control laws are summarized in the following:

$$\begin{cases} D_R[k+1] = \frac{L_R \times (I_{Rref}[k+1] - (i_{balance}[k+1] + I_R[k]))}{T_{SW} \times V_{dc}} - \frac{V_R[k]}{V_{dc}} + 1 - \frac{\overline{V_{no}}}{V_{dc}} \\ D_S[k+1] = \frac{L_S \times (I_{Sref}[k+1] - (i_{balance}[k+1] + I_S[k]))}{T_{SW} \times V_{dc}} - \frac{V_S[k]}{V_{dc}} + 1 - \frac{\overline{V_{no}}}{V_{dc}} \\ D_T[k+1] = \frac{L_T \times (I_{Tref}[k+1] - (i_{balance}[k+1] + I_T[k]))}{T_{SW} \times V_{dc}} - \frac{V_T[k]}{V_{dc}} + 1 - \frac{\overline{V_{no}}}{V_{dc}} \end{cases} \qquad (8)$$

$$\begin{cases} D_R[k+1] = \frac{-L_R \times (I_{Rref}[k+1] - (i_{balance}[k+1] + I_R[k]))}{T_{SW} \times V_{dc}} + \frac{V_R[k]}{V_{dc}} + 1 + \frac{\overline{V_{no}}}{V_{dc}} \\ D_S[k+1] = \frac{-L_S \times (I_{Sref}[k+1] - (i_{balance}[k+1] + I_S[k]))}{T_{SW} \times V_{dc}} + \frac{V_S[k]}{V_{dc}} + 1 + \frac{\overline{V_{no}}}{V_{dc}} \\ D_T[k+1] = \frac{-L_T \times (I_{Tref}[k+1] - (i_{balance}[k+1] + I_T[k]))}{T_{SW} \times V_{dc}} + \frac{V_T[k]}{V_{dc}} + 1 + \frac{\overline{V_{no}}}{V_{dc}} \end{cases} \qquad (9)$$

where (8) is for $V_R > 0$ and (9) is for $V_R < 0$.

The dc-link voltage regulation is based on the PI control, formulated as follows:

$$i_{iref}[k+1] = \left[\left(\frac{T_{SW}}{2} K_i + K_p \right) \times (2V_{dc,ref}[k+1] - V_{c\_up}[k+1] - V_{c\_dn}[k+1]) \right. \\ \left. + \left(\frac{T_{SW}}{2} K_i - K_p \right) \times (V_{dc,ref}[k] - V_{c\_up}[k] - V_{c\_dn}[k]) + i_{ref}[k] \right] \times \sin(\theta_V[k+1]) \qquad (10)$$

where K_p indicates the proportional gain, K_i indicates the integral gain and T_{sw} is the switching period.

The balance-regulation of voltage imbalance is based on the PI control, formulated as follows:

$$i_{balance}[k+1] = \left(\frac{T_{SW}}{2} K_{i\_balance} + K_{p\_balance} \right) \times \left(V_{c\_up}[k+1] - V_{c\_dn}[k+1] \right) + \\ \left(\frac{T_{SW}}{2} K_{i\_balance} - K_{p\_balance} \right) \times \left(V_{c\_up}[k] - V_{c\_dn}[k] \right) + i_{balance}[k] \qquad (11)$$

where $K_{p\_balance}$ indicates the proportional gain, and $K_{i\_balance}$ indicates the integral gain.

Finally, based on (8) and (9), the system control block diagram is organized and depicted in Fig. 5.

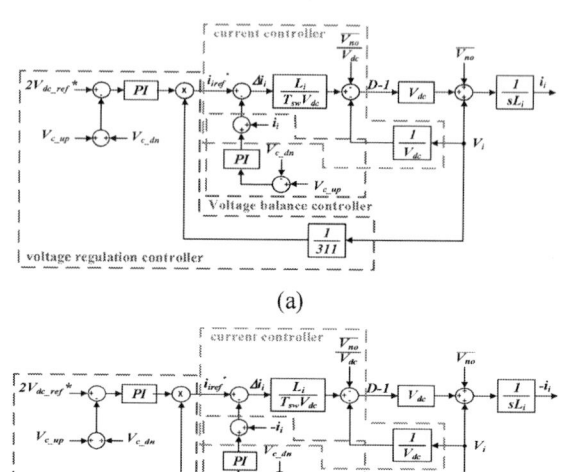

(a)

(b)

Fig. 5. Control block diagram based on the input voltage (a) positive half cycle and (b) negative half cycle.

III. SYSTEM SIMULATION AND EXPERIMENTAL RESULTS

This section presents the waveforms using D³C method applied to the Vienna rectifier. The system hardware parameters are listed in TABLE I.

TABLE I. System specifications and design parameters.

Parameter	Value
Grid Voltage(V_{LL})	380 V$_{rms}$
Inductor Current (I_{AC_line})	9.116 A$_{rms}$
DC Bus Voltage (V_{dc})	760 V
DC Bus Current (I_{dc})	7.89 A
Grid Frequency f_{AC}	60 Hz
Rated Power (P_{rate})	6 kW
Switching Frequency (f_{sw})	200 kHz
DC Bus Capacitance ($C_{\_up}$、$C_{\_dn}$)	500 µF
Inductor (L_i)	5,000 µH
Inductor core	CK740060
Switch	RJH60F7
Diode	HFA60PA120C

Fig. 6 shows the simulated and experimental results at 6 kW, including the waveforms when the input voltage is normal, unbalanced and distorted with odd harmonics.

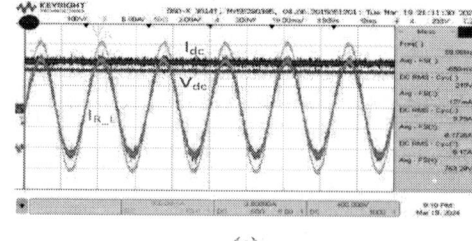

(a)

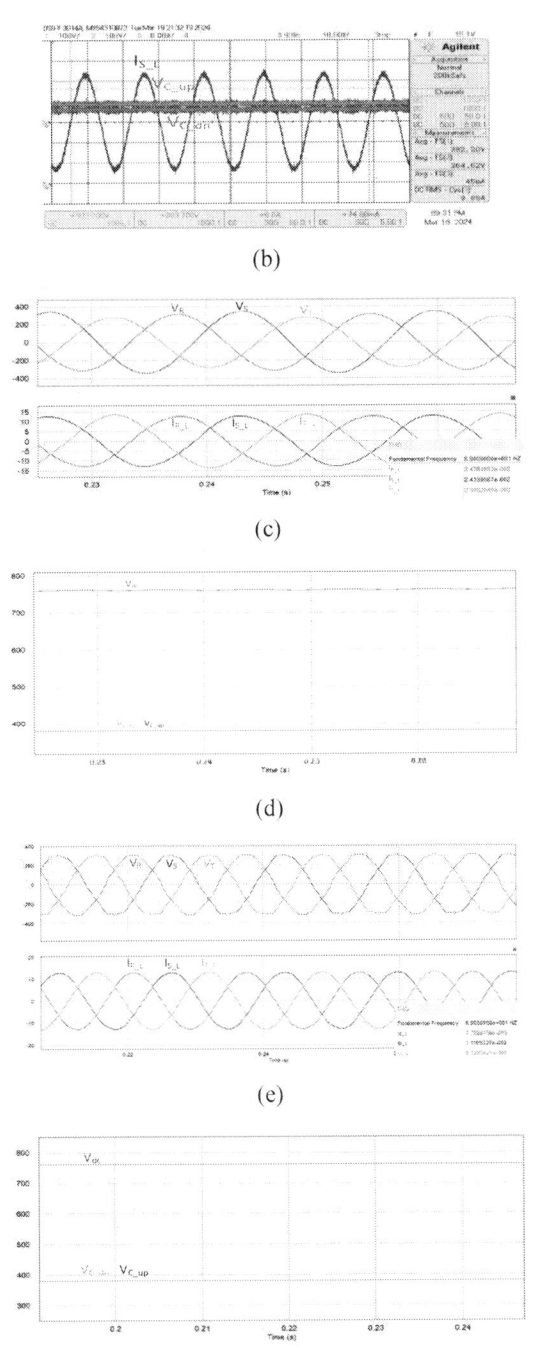

(b)

(c)

(d)

(e)

(f)

Fig. 6. Simulated and experimental results at 6 kW (a) and (b) input voltage are normal, (c) and (d) input voltage is unbalanced, and (e) and (f) input voltage are normal containing fundamental harmonics

Fig. 6(a) and (b) show phase R current I_{R_L} and voltage V_R, dc-link voltage V_{dc} and current I_{dc}. S-phase current I_{S_L}, the voltage of upper capacitor V_{c_up}, and the voltage of lower capacitor V_{c_dn} under the normal input voltage. I_{R_L} and V_R are in phase, and the PFC performance has been achieved. V_{dc} is stable, and the voltage error between V_{c_up} and V_{c_dn} is less than 5%. Fig. 6(c) and (d) show the dc-link voltage, input current and the voltage under the unbalanced input voltage with 10% difference. Fig. 6(e) and (f) show the dc-link voltage, input current and the voltage under the

distortion input voltage. The inductor current remains balanced, achieving PFC and low THD of the inductor current corresponding to the harmonic control standards for power systems. The output voltage is stable at 760V, and the upper and lower arm capacitor voltages are balanced. Fig. 7 shows the THD of phase R. The THD of phase R under full-load inductor current is 3.67%, satisfying to the harmonic standards IEEE-519.

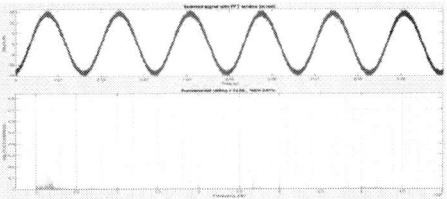

Fig. 7. The THD of the phase R inductor current

IV. CONCLUSION

This paper has used decoupled direct digital control applied to a three-phase three-wire six-switch Vienna rectifier. By utilizing the division-summation process, suitable control laws for the system have been derived, addressing the decoupling issues among the three phases. This method allows direct adaptation and control of the rectifier without the need of a complex controller tuning process. Compared to the traditional frame transformation system (abc to dq), it has significantly simplified the design complexity, making it faster and easier to implement.

REFERENCES

[1] J. W. Kolar, H. Sree, U. Drofenik, N. Mohan, and F. C. Zach, "A novel three-phase three-switch three-level high power factor SEPIC-type AC-to-DC converter," in *Proceedings of APEC 97 - Applied Power Electronics Conference*, 27-27 Feb. 1997.

[2] J. W. Kolar and F. C. Zach, "A novel three-phase utility interface minimizing line current harmonics of high-power telecommunications rectifier modules," *IEEE Trans. Ind. Electron.*, vol. 44, no. 4, pp. 456-467, 1997.

[3] J. W. Kolar and T. Friedli, "The Essence of Three-Phase PFC Rectifier Systems-Part I," *IEEE Trans. Power Electron.*, vol. 28, no. 1, pp. 176-198, Jan 2013.

[4] T. Friedli, M. Hartmann, and J. W. Kolar, "The Essence of Three-Phase PFC Rectifier Systems-Part II," *IEEE Trans. Power Electron.*, vol. 29, no. 2, pp. 543-560, Feb 2014.

[5] T. Wang, C. Chen, T. Liu, Z. Chao, and S. Duan, "Current Ripple Analysis of Three-Phase Vienna Rectifier Considering Inductance Variation of Powder Core Inductor," IEEE Trans. Power Electron., vol. 35, no. 5, pp. 4568-4578, 2020, doi: 10.1109/TPEL.2019.2944853.

[6] S. Ahmad, S. Mekhilef, and H. Mokhlis, "DQ-axis Synchronous Reference Frame based P-Q Control of Grid Connected AC Microgrid," in *2020 IEEE International Conference on Computing, Power and Communication Technologies (GUCON)*, 2-4 Oct. 2020.

[7] L. Osório, J. Mendes, R. Araújo, and T. Matias, "A comparison of adaptive PID methodologies controlling a DC motor with a varying load," in *2013 IEEE 18th Conference on Emerging Technologies & Factory Automation (ETFA)*, 10-13 Sept. 2013, pp. 1-6.

[8] J. He and X. Zhang, "Comparison of the back-stepping and PID control of the three-phase inverter with fully consideration of implementation cost and performance," *Chinese Journal of Electrical Engineering*, vol. 4, no. 2, pp. 82-89, 2018.

[9] X. Yangxu, Z. Danhong, Z. Huaiun, W. Lianshun, Q. Yue, and L. Zhiwen, "Neural network- fuzzy adaptive PID controller based on VIENNA rectifier," in *2018 Chinese Automation Congress (CAC)*, 30 Nov.-2 Dec. 2018, pp. 583-588.

[10] S. Nirmal, K. N. Sivarajan, E. A. Jasmin, M. Nandakumar, and B. Jayanand, "Steady state error elimination and harmonic compensation using proportional resonant current controller in grid-tied DC microgrids," in *2018 International Conference on Power, Instrumentation, Control and Computing (PICC)*, 18-20 Jan. 2018 2018, pp. 1-5.

[11] M. Hui, Y. Shengyang, W. Wei, and S. Wang, "The PR Control with Load Current Feed-forward for Vienna Rectifier," in 2019 IEEE International Conference on Power, Intelligent Computing and Systems (ICPICS), 12-14 July 2019 2019, pp. 508-512, doi: 10.1109/ICPICS47731.2019.8942550.

[12] T. Liu, C. Chen, T. Wang, S. Duan, and H. Cheng, "Proportional-Resonant Current Control for VIENNA Rectifier in Stationary $\alpha\beta$ Frame," in 2018 IEEE International Power Electronics and Application Conference and Exposition (PEAC), 4-7 Nov. 2018 2018, pp. 1-7, doi: 10.1109/PEAC.2018.8589971.

[13] T.-F. Wu, Y.-H. Huang, Y.-T. Liu, and M. Misra, "Decoupled Direct Digital Control with D-Σ Process and Average Common-Mode Voltage Model for 3Φ3W LCL Converters," in *2019 IEEE Applied Power Electronics Conference and Exposition (APEC)*, 17-21 March 2019, pp. 601-606.

[14] T. F. Wu, C. H. Chang, L. C. Lin, Y. C. Chang, and Y. R. Chang, "Two-Phase Modulated Digital Control for Three-Phase Bidirectional Inverter With Wide Inductance Variation," *IEEE Trans. Power Electron.*, vol. 28, no. 4, pp. 1598-1607, 2013.

[15] T. F. Wu, C. H. Chang, L. C. Lin, G. R. Yu, and Y. R. Chang, "A D-Σ Digital Control for Three-Phase Inverter to Achieve Active and Reactive Power Injection," *IEEE Trans. Ind. Electron.*, vol. 61, no. 8, pp. 3879-3890, 2014.

[16] T. F. Wu, H. C. Hsieh, C. H. Chang, L. C. Lin, and Y. R. Chang, "Improvement of Control Law Derivation and Region Selection for D-Σ Digital Control," *IEEE Trans. Ind. Electron.*, vol. 62, no. 10, pp. 6042-6050, 2015.

Active Power Decoupling Method based on Dual Active Bridge Converter without additional components

Kosuke Takeuchi
Dept. of Electrical, Electronics and Information Engineering
Nagaoka University of Technology
Nagaoka, Niigata, Japan
s191047@stn.nagaokaut.ac.jp

Takashi Ohno
Dept. of Science of Technology Innovation
Nagaoka University of Technology
Nagaoka, Niigata, Japan
s225058@stn.nagaokaut.ac.jp

Hiroki Watanabe
Dept. of Electrical, Electronics and Information Engineering
Nagaoka University of Technology
Nagaoka, Niigata, Japan
hwatanabe@vos.nagaokaut.ac.jp

Yuki Nakata
Dept. of Electrical, Electronics and Information Engineering
Nagaoka University of Technology
Nagaoka, Niigata, Japan
nakata@vos.nagaokaut.ac.jp

Jun-ichi Itoh
Dept. of Science of Technology Innovation
Nagaoka University of Technology
Nagaoka, Niigata, Japan
itoh@vos.nagaokaut.ac.jp

Abstract— **This paper proposed an active power decoupling method using a Dual Active Bridge (DAB) converter to minimize the energy buffer in single-phase AC-DC converter applications. Single-phase converters commonly require bulky electrolytic capacitors due to their power pulsation at twice the grid frequency. However, the electrolytic capacitor often decreases the power density and lifetime of the power converter. Active power decoupling is one of the solutions to this problem by using a firm or ceramic capacitor as the energy buffer. However, the additional circuit decreases the conversion efficiency and power density.**

This paper proposes the active power decoupling method using a DAB converter without additional components. The Proposed method applies the feed-forward control to determine the phase shift angle for the power decoupling capability. In addition, the ZVS condition with the proposed method is considered to clarify the design of the small DC-link capacitor. The validity of the proposed method is demonstrated by the experimental result. As the experimental result, it was confirmed that the second harmonic component of the output voltage was reduced by 91.2%.

Keywords—Dual Active Bridge converter, Active power decoupling, carrier phase shift

I. INTRODUCTION

Electric Vehicles have been widely developed to reduce the carbon emissions of traditional gasoline vehicles. An onboard charger (OBC) is necessary to realize the battery charge from the AC grid, and it requires high conversion efficiency, small volume, and lightweight to implement on EVs.

OBC typically consists of an isolated AC-DC converter with a PWM rectifier and an isolated DC-DC converter. Candidates for isolated DC-DC converters include LLC converters, full bridge converters, and Dual Active Bridge (DAB) converters. DAB converters have been extensively studied due to their bi-directional power transfer and high-efficiency operation with Zero Voltage Switching (ZVS) owing to low switching losses.

OBC connects to a single-phase or three-phase AC grid. In single-phase applications, the single-phase power pulsation with twice the grid frequency should be eliminated from the output DC port because it may become a cause of the lifetime limitation of batteries [1]-[5]. Generally, a bulky electrolytic capacitor

connects to the DC-link for the energy buffer of the single-phase power pulsation. However, electrolytic capacitors limit the reliability of the power converter. Moreover, this capacitor decreases the power density due to making a large capacitor bank by the low allowable ripple current.

An active power decoupling (APD) method is one of the solutions to replace the electrolytic capacitor in the single-phase converter [6]-[9]. This method eliminates the single-phase power pulsation from the DC port by a small capacitor such as a firm or ceramic capacitor. The simple APD implementation adds a boost or buck-type decoupling port to the DC link [10], [11]. However, the additional circuit may increase converter losses and circuit volume due to an increase of circuit components. The method of decoupling power without adding an active switch instead requires an additional capacitor across the bridge arm [12]. Therefore, a method that completely eliminates the need for additional components is required.

This paper proposes a power decoupling method by applying carrier phase shift control without extra components. This proposed method actively varied the phase shift angle according to the DC link voltage variation to obtain the constant DC power. This contribution of paper is that improvement of the power density owing to removing the bulky electrolytic capacitor without an extra APD circuit.

The remainder of the paper is organized as follows. Section II explains the circuit topology of the single-phase AC to DC and the principle of single-phase power pulsation. In section III, the proposed power pulsation compensation method is described. In addition, The ZVS range of DAB converters is presented. Finally, in section IV, the experimental results are demonstrated.

II. CIRCUIT TOPOLOGY AND PRINCIPLE OF SINGLE-PHASE POWER PULSATION

Fig. 1 shows a diagram of the isolated AC-DC converter under consideration. The isolated AC-DC converter consists of a circuit combining a PWM rectifier and a DAB converter. The

PWM rectifier has the DC-link voltage control and PFC capability for grid current. In contrast, the DAB converter has the transmission power control by the phase shift carrier control.

This circuit applies the small capacitors for C_{dc} and C_{out} to eliminate electrolytic capacitors. However, the second-order harmonics appear at DC-link voltage due to the small capacitance of C_{dc}. As a result, the transmission power fluctuates at the double-line frequency.

Fig. 2 shows the compensation principle of single-phase power pulsation. As shown in Fig. 2, the input power of a single-phase AC-DC converter pulsates at twice the frequency. In contrast, the output power is controlled constantly by charging and discharging the energy in the buffer capacitor C_{dc} to counteract input power pulsations. When both the Input voltage and current waveforms are sinusoidal, the relationship between instantaneous input power p_{in}, instantaneous output power p_{out}, and instantaneous power of the buffer p_{buf} is expressed as

$$p_{out} = p_{in} + p_{buf} = \frac{1}{2}V_{in}I_{in} - \frac{1}{2}V_{in}I_{in}\cos(2\omega t) + p_{buf} \quad (1),$$

where V_{in} is the peak voltage, I_{in} is the peak current and ω is the angular frequency of the input voltage. The second term in (1) shows the power pulsation component, which pulsates twice the input frequency. This power pulsation is superimposed on the DC link current as a ripple. When the instantaneous power is kept constant, the buffer instantaneous power p_{buf} to compensate for the input power pulsation is expressed as

$$p_{buf} = \frac{1}{2}V_{in}I_{in}\cos(2\omega t) \quad (2),$$

where the polarity of p_{buf} is positive when C_{buf} is charged. The first term in (1) is the DC amount, and the second term is the power pulsation component. The instantaneous power p_{buf} of the buffer is controlled to cancel the second term and is expressed as

$$p_{out} = \frac{1}{2}V_{in}I_{in} \quad (3).$$

The second and third terms in (1) cancel so that the output power is only the DC value shown in (3). Since the output power is constant, the output voltage is also constant when a load resistor is connected.

III. PROPOSED POWER DECOUPLING METHOD.

A. Proposed control method

The proposed method controls the transmission power of the DAB converter to compensate for the twice-frequency power pulsations that occur in the DC link. First, the relationship between the phase difference and the transmission power of a typical DAB converter is explained. The relationship between the phase difference and output power P_{out} when the transformer voltage is set to two levels is expressed as

$$P_{out} = \frac{NV_{dc}V_{out}}{\omega L_2}\delta\left(1 - \frac{\delta}{\pi}\right) \quad (4),$$

where V_{dc} and V_{out} are the DC link and output voltages, respectively, N is the turn ratio of the high-frequency transformer, and L_2 is the inductance value. It is the phase

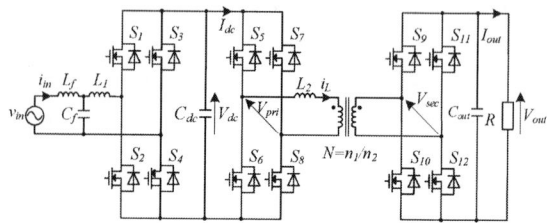

Fig. 1 Circuit configuration of isolated AC-DC converter using DAB converter.

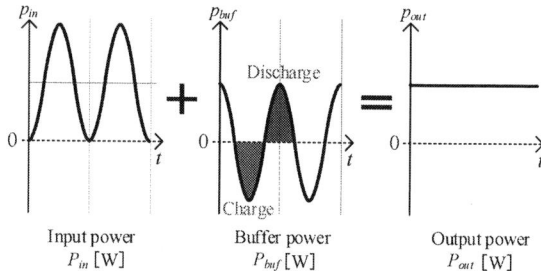

Fig. 2 Compensation principle of power ripple

difference between the primary and secondary of the DAB converter.

In the proposed circuit, the input voltage of the DAB converter pulsates at twice the frequency of the single-phase system. The DC link voltage V_{dc}, including the pulsation, is expressed as

$$V_{dc} = V_{avg} + \Delta V_c \sin(2\omega t) \quad (5),$$

V_{avg} is the average voltage of the DC link voltage, and ΔV_c is the DC link voltage pulsation amplitude. From (4) and (5), the output power P_{out}, including single-phase pulsation, is expressed as

$$P_{out} = \frac{NV_{out}\{V_{avg} + \Delta V_c \sin(2\omega t)\}}{\omega L_2}\delta\left(1 - \frac{\delta}{\pi}\right) \quad (6).$$

When the phase difference d is constant, the output power of the DAB converter pulsates at twice the frequency due to input voltage fluctuations in (6). Therefore, the proposed control actively varies the phase difference d to keep the output power constant. The phase difference of the DAB converter is expressed as

$$\delta = \frac{\pi}{2}\left(1 - \sqrt{1 - \frac{8P_{out}f_{sw}L_2}{NV_{dc}V_{out}}}\right) \quad (7),$$

where, the DC link voltage uses the instantaneous value to calculate the phase difference d of the DAB converter to compensate for single-phase power pulsations. Therefore, output power is controlled on an instantaneous power basis, which allows constant control of output power. The phase difference d to compensate for single-phase power pulsation is expressed as

$$\delta = \frac{\pi}{2}\left\{1 - \sqrt{1 - \frac{8P_{out}f_{sw}L_2}{NV_{out}\{V_{avg} + \Delta V_c \sin(2\omega t)\}}}\right\} \quad (8).$$

B. Relationship between buffer capacitance and ripple voltage

The amount of power to be compensated and the allowable fluctuation voltage of the capacitor determine the capacitor capacity for compensating single-phase power pulsation. The amount of power required to compensate for power pulsation W_c is expressed as

$$W_c = \frac{1}{2} V_{in} I_{in} \int_0^{\frac{1}{4f}} \sin(2\omega t)\, dt = \frac{V_{in} I_{in}}{2\omega} = \frac{p_{out}}{\omega} \tag{9}.$$

From the relationship between capacitor power and voltage, the amount of capacitor power is expressed as

$$W_c = \frac{1}{2} C V_{cmax}^2 - \frac{1}{2} C V_{cmin}^2 \tag{10},$$

where V_{cmax} is the maximum fluctuating voltage, and V_{cmin} is the minimum fluctuating voltage. From (9) and (10), the capacitance C_{buf} of the buffer capacitor used for single-phase power pulsation compensation is expressed as

$$C_{buf} = \frac{2P_{out}}{\omega \left(V_{cmax}^2 - V_{cmin}^2 \right)} = \frac{P_{out}}{2\omega V_{avg} \Delta V_c} \tag{11}.$$

The output power and the capacitance of the buffer capacitor determine V_{cmax} and V_{cmin}. The addition of the average voltage V_{avg} and the fluctuating voltage ΔV_c of the voltage pulsation is V_{cmax}, and the subtraction is V_{cmin}. From (11), the fluctuating voltage ΔV_c variation width and the buffer capacitor capacitance are inversely proportional.

C. ZVS range of DAB converter

This chapter investigates the ZVS range when the proposed single-phase power pulsation compensation is applied in a two-level DAB converter. To achieve ZVS over the entire operating range, even when the input voltage of the DAB converter pulsates due to single-phase power pulsation, the phase difference variation during APD operation should operate within the ZVS condition. Therefore, the ZVS range and circulating current in the proposed method are explained.

Fig. 3 shows the waveforms of the primary voltage v_{pri}, secondary voltage v_{sec}, and inductor i_L of the DAB converter. The condition for soft switching to be established in the DAB converter depends on the positive and negative inductor current i_L. Fig. 3 (a) shows the inductor current waveform when $i_L > 0$ in the interval from t_2 - t_3. Fig. 3 (b) shows that $i_L < 0$ at time t_2, and soft switching is not established when the mode switches. In Fig. 3(c), $i_L < 0$ at time t_3, and, as in Fig. 3(b), soft switching does not take place. Therefore, soft switching is established when both conditions $i_L > 0$ in the interval t_2 - t_3 and ZVS can be achieved.

The ZVS conditions for the inductor currents $i(t_2)$ and $i(t_3)$ are expressed as

$$i(t_2) = \frac{(2\delta - \pi) V_{dc} + \pi N V_{out}}{2\omega L} > 0 \tag{12},$$

$$i(t_3) = \frac{\pi V_{dc} + (2\delta - \pi) N V_{out}}{2\omega L} > 0 \tag{13},$$

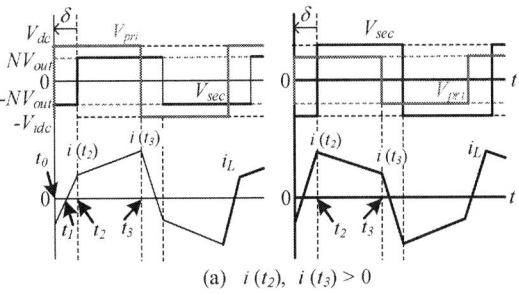

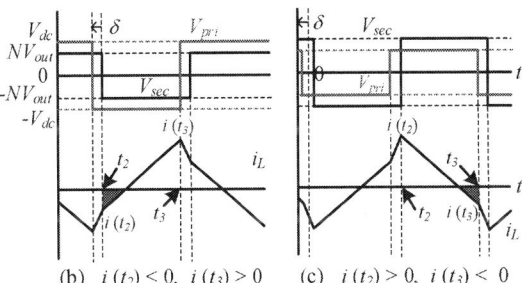

(a) $i(t_2),\ i(t_3) > 0$

(b) $i(t_2) < 0,\ i(t_3) > 0$ (c) $i(t_2) > 0,\ i(t_3) < 0$

Fig. 3 Inductor current and soft switching success or failure

where the condition for a phase difference that can achieve ZVS is expressed as

$$\delta > \frac{\pi}{2} \left(1 - \frac{N V_{out}}{V_{dc}} \right) \tag{14},$$

$$\delta > \frac{\pi}{2} \left(1 - \frac{V_{dc}}{N V_{out}} \right) \tag{15}.$$

Therefore, the ZVS condition of the DAB converter when the input voltage pulsates is expressed as

$$\delta > \frac{\pi}{2} \left\{ 1 - \frac{N V_{out}}{V_{avg} + \Delta V_c \sin(2\omega t - \pi)} \right\} \tag{16},$$

$$\delta > \frac{\pi}{2} \left\{ 1 - \frac{V_{avg} + \Delta V_c \sin(2\omega t - \pi)}{N V_{out}} \right\} \tag{17}.$$

This section describes the phase difference due to the input voltage pulsation of the DAB converter during APD. From equation (8), the expressions for the phase difference when the DC link voltage is at its maximum and minimum values are expressed as

At maximum DC link voltage

$$\delta = \frac{\pi}{2} \left\{ 1 - \sqrt{1 - \frac{8 I_{out} f_{sw} L_2}{N(V_{avg} + \Delta V_c)}} \right\} \tag{18},$$

At minimum DC link voltage

$$\delta = \frac{\pi}{2} \left\{ 1 - \sqrt{1 - \frac{8 I_{out} f_{sw} L_2}{N(V_{avg} - \Delta V_c)}} \right\} \tag{19}.$$

The phase difference in (18) and (19) must follow the ZVS condition in (16) and (17).

Fig. 4 shows the ZVS range of the DAB converter and the operating point of the phase difference during APD. In this case, the average voltage V_{avg} of the DC link voltage is 1p.u. From (11), the DC link voltage fluctuation voltage increases when the buffer capacitor is made small. With a buffer capacitor of 150μF, ZVS operates in the entire range. However, with a buffer capacitor of 100μF, the fluctuating voltage of the DC link voltage increases, resulting in sections of hard switching. Therefore, a trade-off exists between smaller buffer capacitor capacitance and ZVS operation over the entire region. The maximum and minimum DC link voltages must be within the ZVS range to achieve both APD and ZVS operations over the entire range. The boundary conditions of the ZVS range are expressed as

ZVS boundary conditions at maximum input voltage

$$\sqrt{1 - \frac{8I_{out}f_{sw}L_2}{N\left(V_{avg} + \Delta V_c\right)}} < \frac{NV_{out}}{V_{avg} + \Delta V_c} \qquad (20),$$

ZVS boundary conditions at minimum input voltage

$$\sqrt{1 - \frac{8I_{out}f_{sw}L_2}{N\left(V_{avg} - \Delta V_c\right)}} < \frac{V_{avg} - \Delta V_c}{NV_{out}} \qquad (21).$$

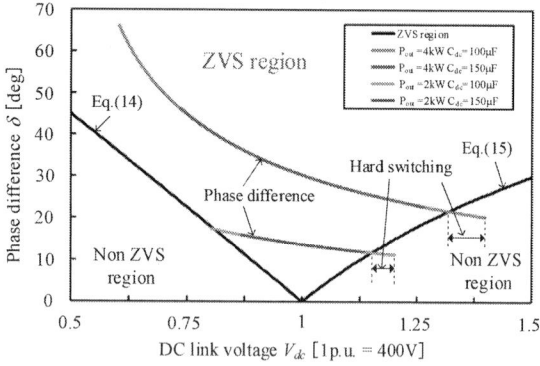

Fig. 4 Relationship between the ZVS range of DAB converter and the maximum and minimum DC link voltage pulsations.

IV. EXPERIMENTAL RESULTS

In this chapter, the effectiveness of the proposed method is verified using a prototype with a rating of 4 kW under the experimental conditions shown in Table 1. The input voltage V_{in} is 200 V_{rms}, the output voltage V_{out} is 400 V, and the grid frequency f_g is 50 Hz.

Fig. 5 shows the relationship between the maximum possible ZVS variation voltage ΔV_c and buffer capacitor capacitance C_{buf} when output power P_{out} is varied. The range in which ZVS is possible at all operating points must satisfy ZVS boundary conditions in (20) and (21). As a result, the maximum fluctuating voltage for the DAB converter to operate ZVS in the full range at 4 kW output power is 125 V. From (11), the buffer capacitance is calculated to be 125 μF for a maximum fluctuating voltage of 125 V. In this experiment, the buffer capacitor C_{buf} was designed to be 150μF to keep the DC link

Table.1 Experimental parameters

Parameter		
Rated Power	P_{out}	4kW
Grid voltage	v_{in}	200V_{rms}
Grid Frequency	f_g	50Hz
DC Link voltage	V_{dc}	400V
Output voltage	V_{out}	400V
Switching Frequency	f_{sw}	50kHz
Turn Ratio	N	n_1/n_2=1
Inductor (LC rectifer)	L_f	800μH
Inductor (PWM rectifer)	L_1	800μH
Inductor (DAB converter)	L_2	56μH
Capacitor (LC rectifer)	C_f	3nF
Buffer Capacitor	C_{dc}	150μF
Output Capacitor	C_{out}	60μF
Output Resistance	R	40 Ω
Cutoff Freq. of Current Control	f_{c_acr}	1kHz
Cutoff Freq. of Voltage Control	f_{c_avr}	10Hz

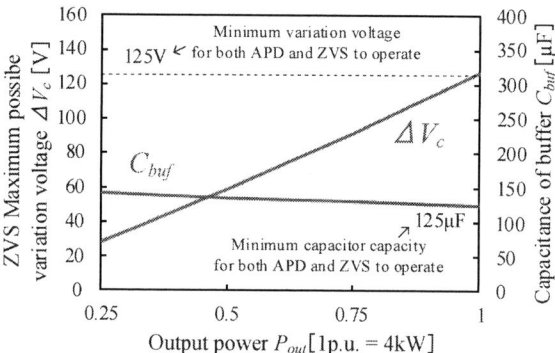

Fig.5 Maximum fluctuating voltage and buffer capacitor capacitance that can be ZVS over the entire range due to output power change.

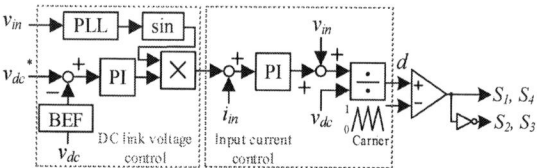

(a) Control block diagram of PWM rectifier

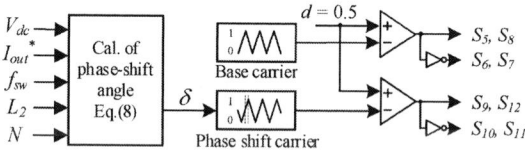

(b) Control block diagram of DAB converter

Fig. 6 Control block diagram for the isolated AC-DC converter.

voltage fluctuation voltage below 125V for ZVS operation in all regions.

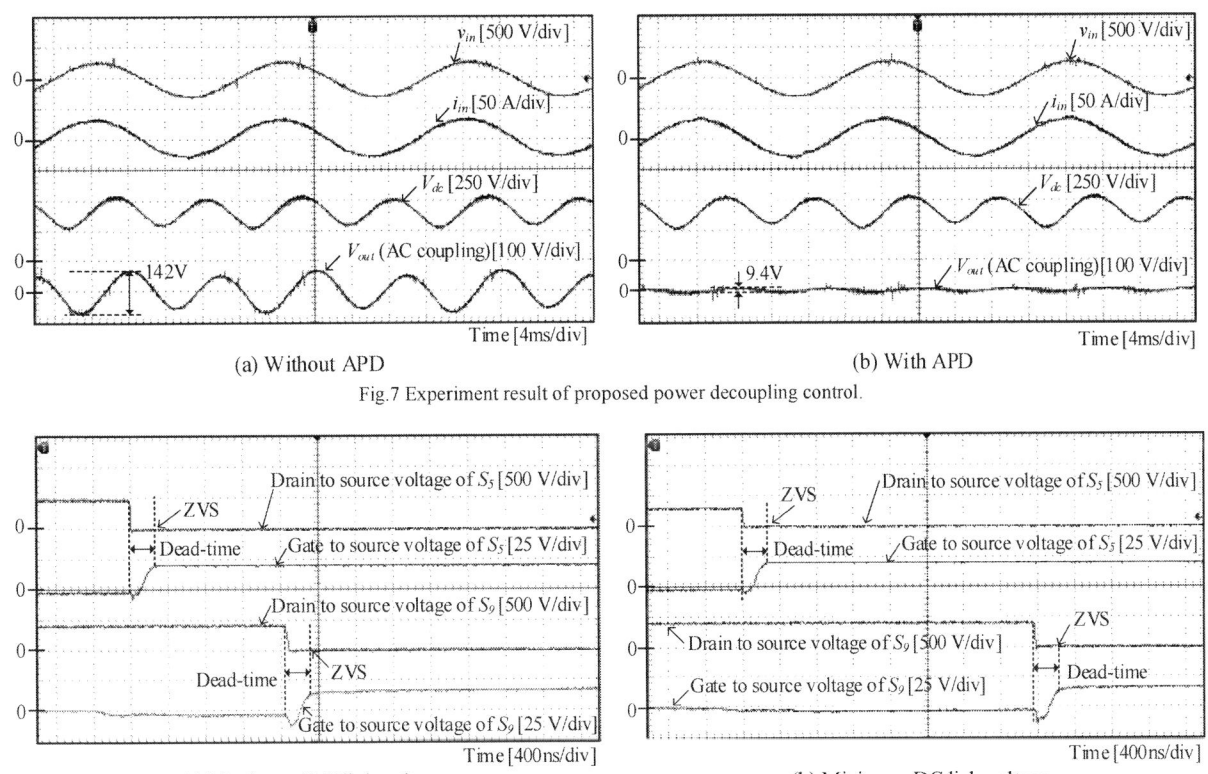

(a) Without APD　　　　　　　　　　　　(b) With APD

Fig.7 Experiment result of proposed power decoupling control.

(a) Maximum DC link voltage　　　　　　　(b) Minimum DC link voltage

Fig.8 Gate to source voltage and drain to source voltage of S_5 and S_9 at the DAB converter

Fig.6 shows the control block diagram of the proposed circuit. Fig. 6 (a) and Fig. 6 (b) show the control block diagrams of the PWM rectifier and DAB converter. The PWM rectifier controls the average value of the DC link voltage. The DAB converter performs carrier phase shift control using the phase difference calculated from (8). where using the detected value in the DC link voltage term of (8) allows feed-forward control even when the load conditions change.

Fig.7 shows the experimental waveforms to confirm the validity of the proposed APD method. The output voltage was measured with AC decoupling to verify the effect of reducing the second harmonic component. According to Fig.7 (a), the DC voltage fluctuation of 142V occurs due to the power ripple. On the other hand, The proposed APD method eliminates the low-frequency harmonics on the DC voltage and reduces the load voltage fluctuation by 93.4%, as shown in Fig. 7 (b). Moreover, the sinusoidal input current is obtained by the PFC control.

Fig.8 shows the drain-source voltage and gate-source voltage of the DAB converter. Fig.8 (a) and Fig.8 (b) show the switching waveforms at the maximum and minimum DC link voltages, respectively. The phase difference was operated based on the ZVS range to achieve both APD and ZVS. in Fig.4.

Fig.9 shows the harmonic analysis result of the DC output voltage. The ratio of each harmonic to the DC component with and without the proposed APD method is shown. According to Fig.9, Compensation for power pulsation reduced the second harmonic component by 91.2%. Therefore, power decoupling

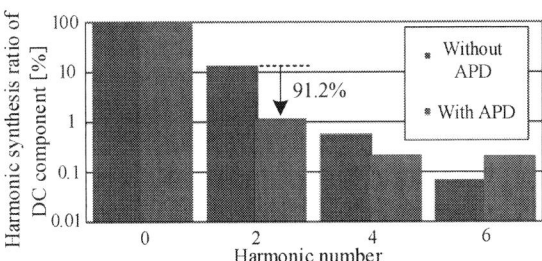

Fig.9 Harmonic analysis result of output voltage

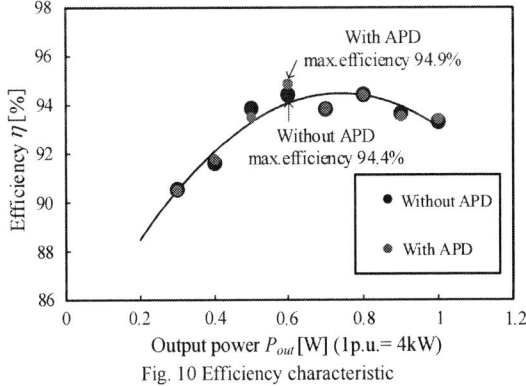

Fig. 10 Efficiency characteristic

979-8-3315-1612-3/25 $31.00 © 2025 IEEE

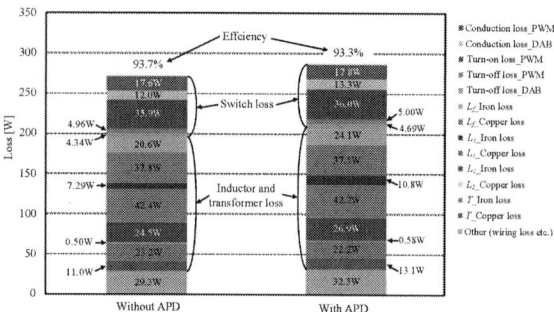

Fig.11 Loss analysis result with and without APD control.

was achieved by controlling the phase difference of the DAB converter.

Fig.10 shows the efficiency curves of the isolated AC-DC converter. According to Fig.10, the efficiency with the proposed APD is the same as the efficiency without APD. This is because the DAB converter operates in ZVS even during APD operation, which reduces switching losses.

The efficiency of this prototype is low at 93.3%. This could be due to pulsating DC link voltage. The DAB converter is based on the condition that the input-to-output voltage ratio and the transformer turn ratio match, and the effective value of the transformer current increases as the voltage varies. Therefore, under the parameter conditions of this paper, the large fluctuations in DC link voltage with and without APD increase the RMS value of transformer current and reduce efficiency.

Fig. 11 shows the results of loss separation with and without the proposed APD. The overall loss is divided into switch loss, inductor loss, and transformer loss, respectively. The losses that occur in a switch are divided into conduction losses and switching losses. The conduction loss was derived by the product of the square of the current RMS value and the winding resistance. The switching losses were derived from the turn-on and turn-off losses from the data sheets of the MOS-FET (SCT4018KR : Rohm), used in the experiments. At that time, the turn-on loss is assumed to be zero because the DAB converter operates ZVS on all switches. Separate inductor and transformer losses into iron and copper losses. The copper loss was derived by the product of the square of the current RMS value and the inductor or transformer wiring resistance. The iron loss was derived by subtracting the copper loss from the total loss of each inductor and transformer. Wiring losses and reactive power losses are considered as other losses.

The conduction, switching, iron, and copper loss with APD increased more than without APD in Fig.11. However, the increase in loss due to APD adaptation is only 0.28% of the total power and does not significantly affect the overall circuit loss. The iron and copper losses in the inductors and transformers account for about 61.9% of the total loss. In addition, the inductor on the input side of the PWM rectifier has a large ratio of iron loss to copper loss. Therefore, the optimal design of the inductor of the PWM rectifier may improve the overall efficiency of the circuit.

V. CONCLUSION

This paper proposed the APD method for DAB converters based on carrier phase shift control, and it does not require additional circuits. The proposed method compensates for the power pulsation by controlling the phase difference of the DAB converter to cancel the double-frequency pulsation occurring in the single-phase system. This method maintains constant output power with a small-capacity DC link capacitor. The conditions for ZVS corresponding to power pulsation compensation were also studied, and boundary conditions that allow ZVS in all regions were derived. The conditions for ZVS that are compatible with power pulsation compensation were also investigated. The experimental result showed that the pulsation of the output voltage was reduced by 93.4%. The operation of the proposed APD method was compatible with the ZVS operation of the DAB converter in its entire operating range. The efficiency improvement will be considered based on loss analysis in future work.

REFERENCES

[1] J. Lee and J. Won, "Multifunctional Onboard Charger for Electric Vehicles Integrating a Low-Voltage DC-DC Converter and Solar Roof," in IEEE Journal of Emerging and Selected Topics in Power Electronics, (2024)

[2] Y. Park, S. Chakraborty and A. Khaligh, "DAB Converter for EV Onboard Chargers Using Bare-Die SiC MOSFETs and Leakage-Integrated Planar Transformer," in IEEE Transactions on Transportation Electrification, vol. 8, no. 1, pp. 209-224, (2022)

[3] S. G. Barbosa, L. H. S. C. Barreto and D. d. S. Oliveira, "A Single-Stage Bidirectional AC–DC Converter Feasible for Onboard Battery Chargers," IEEE Journal of Emerging and Selected Topics in Power Electronics, (2022)

[4] X. Wang, X. Wei, Q. Chen and H. Dai, "A Novel System for Measuring Alternating Current Impedance Spectra of Series-Connected Lithium-Ion Batteries With a High-Power Dual Active Bridge Converter and Distributed Sampling Units," IEEE Trans. Ind. Electronics, (2021)

[5] H. V. Nguyen, D. -D. To and D. -C. Lee, "Onboard Battery Chargers for Plug-in Electric Vehicles With Dual Functional Circuit for Low-Voltage Battery Charging and Active Power Decoupling," IEEE Access, (2018)

[6] S. Chakraborty and S. Chattopadhyay, "A Dual-Active-Bridge-Based Fully ZVS HF-Isolated Inverter With Low Decoupling Capacitance," IEEE Trans. Power Electronics, (2020)

[7] Y.Ohnuma, J.Itoh : "A Single-phase-to-three-phase Power Converter with an Active Buffer and a Charge Circuit", IEEJ Journal of Industry Applications, (2012)

[8] L. Jin, B. Liu and S. Duan, "ZVS operation range analysis of three-level dual active bridge DC-DC converter with phase-shift control," 2017 IEEE Applied Power Electronics Conference and Exposition (APEC), Tampa, FL, USA, pp. 362-366 (2017)

[9] H. Li, Z. Gao and F. Wang, "A PWM Strategy for Cascaded H-bridges to Reduce the Loss Caused by Parasitic Capacitances of Medium Voltage Dual Active Bridge Transformers," 2022 IEEE Energy Conversion Congress and Exposition (ECCE), Detroit, MI, USA, (2022)

[10] Z. Qin, Y. Tang, P. C. Loh and F. Blaabjerg, "Benchmark of AC and DC Active Power Decoupling Circuits for Second-Order Harmonic Mitigation in Kilowatt-Scale Single-Phase Inverters," in IEEE Journal of Emerging and Selected Topics in Power Electronics, vol. 4, no. 1, pp. 15-25 (2016)

[11] H. Li, K. Zhang, H. Zhao, S. Fan and J. Xiong, "Active Power Decoupling for High-Power Single-Phase PWM Rectifiers," in IEEE Transactions on Power Electronics, vol. 28, no. 3, pp. 1308-1319, (2013)

[12] Y. Sun, Y. Liu, M. Su, X. Li and J. Yang, "Active Power Decoupling Method for Single-Phase Current-Source Rectifier With No Additional Active Switches," in IEEE Transactions on Power Electronics, vol. 31, no. 8, pp. 5644-5654 (2016)

An ANPC-Based Building Block for Medium-Voltage Applications

Ahmed Rahouma, Hui Cao, David A. Porras, Zhuxuan Ma, Yue Zhao, and Juan C. Balda
Dept. of Electrical Engineering
Univeristy of Arkansas
Fayetteville, AR, USA
Email: {arrahoum, hcao, daporras, zm009, yuezhao, jbalda}@uark.edu

Abstract—**The cascaded H-bridge (CHB) converter has shown significant potential for integrating several high-power applications into medium-voltage (MV) power grids. While high-voltage (HV) switching devices offer the advantage of reducing the number of MV building blocks (MV-BBs) required in CHB converters, they pose technical and economic challenges. These include low switching frequency, limited current capability, and high cost. This article introduces an innovative MV-BB design based on active neutral point clamped (ANPC) legs. The proposed MV-BB enables the use of switching devices with lower voltage ratings while maintaining the same number of cascaded MV-BBs. Detailed analysis of the configuration and modulation technique highlights the operational characteristics and performance benefits of this approach. Compared to the conventional MV-BB configuration, the proposed one significantly decreases system power losses, volume, and weight. For instance, the power losses of switching devices are reduced by 34%, while the size, volume, and weight of the LCL filter are decreased by at least 50%. To validate the feasibility and performance of the proposed ANPC-based MV-BB, a prototype was fabricated and tested at 15 kW.**

Keywords—*AC-DC converter, active neutral point clamped converter, cascaded H-bridge converter, dc-dc converter.*

I. INTRODUCTION

Conventional medium-voltage power conditioning systems (MV-PCSs), which are used for integrating high-power (HP) applications, commonly include line-frequency transformers (LFTs) to step up the ac voltage and provide galvanic isolation. These LFTs lead to increased system sizes and weights; imposing limits on space-constrained applications. In contrast, multilevel converters (MLCs) offer stepping up the ac voltage while significantly improving both volumetric and gravimetric power densities; that is, ease of scalability. Among various MLC topologies, the cascaded H-bridge (CHB) converter stands out as a particularly suitable choice for integrating HP applications; for example, grid-connected battery energy storage systems (G-BESSs) and STATCOMs [1]–[3].

In the context of a MV-PCS utilized for HP G-BESS, a CHB converter is employed in the MV ac/dc stage as illustrated in Fig. 1(a). The CHB converter's versatility allows it to be adapted to various MVAC systems by modifying the number of cascaded medium-voltage building blocks (MV-BBs); one is shown in Fig. 1(b). However, the complexity of the converter escalates as the number of MV-BBs increases. The integration of high-voltage (HV) semiconductor devices can effectively mitigate this complexity by reducing the number of MV-BBs required.

The voltage rating of HV SiC switching devices typically spans from 1.7 kV to 15 kV. Recently, 1.7 kV and 3.3 kV SiC MOSFET devices have been commercialized [4]–[7]. R&D efforts have yielded engineering samples with higher ratings, such as 6.5 kV and 10 kV [8]–[11]. The 15-kV devices, are still in their early stages of research [12]–[15].

A straightforward strategy to reduce the number of cascaded MV-BBs without resorting to (non-commercial) HV devices involves connecting low voltage (LV) devices in series to produce a higher voltage switching position. For instance, in [16]–[18], a 3.3 kV SiC MOSFET device was constructed by connecting two 1.7 kV SiC MOSFETs in series. Similarly, in [19], [20], a 6.5 kV SiC MOSFET device was implemented by utilizing four 1.7 kV devices, and in [21], a 15 kV SiC MOSFET device was formed by connecting ten 1.7 kV devices in series. This approach presents challenges related to common-mode current issues, unequal static and dynamic voltage sharing among the devices, the need for additional RC snubber circuits, and the requirement for special isolated gate driver power supplies [22]–[24].

An alternative approach to the series connection of devices involves employing multilevel MV-BBs based on a few MLC topologies, such as a flying capacitor (FC) and neutral point clamped (NPC) converters. In these MV-BBs, the HV device is substituted with several LV devices not requiring a series connection. For example, recent research explored utilizing an FC MV-BB [25]–[27]. The FC MV-BB configuration comprises four three-level FC leges, which replace the conventional half bridges. Each one of these legs contains a flying capacitor unit and four switches. While the FC MV-BB offers several advantages, it presents several challenges, stemming from the presence of flying capacitors. These challenges include the necessity for a pre-charge process and startup procedures, complex voltage balancing control, and additional size and weight of the capacitors [28]–[30].

An NPC-based MV-BB, as extensively discussed in [31]–[34], consists of four three-level NPC legs, with each leg comprising two clamping diodes and four switches. While the NPC MV-BB offers several advantages, it does pose certain challenges that deserve attention, including issues related to capacitor voltage imbalance, increased control complexity, operational limitations stemming from variations in clamping diode blocking voltage, and the presence of high conduction losses in the diodes, as highlighted in [35]–[37].

In summary, achieving the benefits of a reduced number of MV-BBs without encountering the mentioned drawbacks can be accomplished through utilizing an active neutral point clamped (ANPC)-based MV-BB configuration. The main contribution of this article is the introduction of an ANPC-based MV-BB, which consists of four ANPC legs, each

---

The information, data, or work presented herein was funded by PowerAmerica Institute and its members.

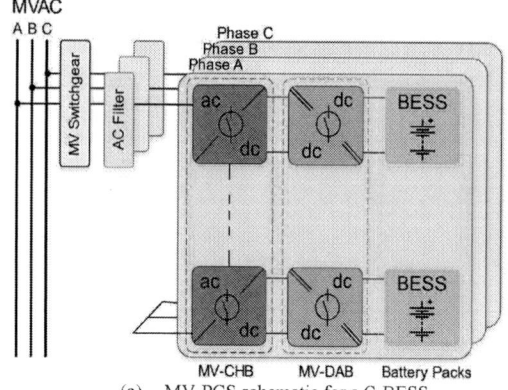

(a) MV-PCS schematic for a G-BESS

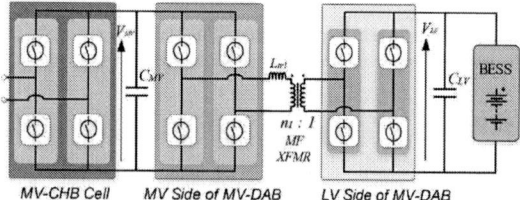

(b) Conventional HB-based MV-BB

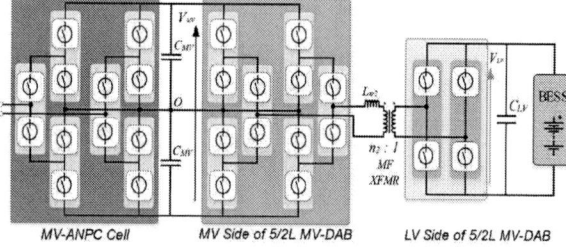

(c) Proposed 3L-ANPC-based MV-BB

Fig. 1. Structure of MV-PCS and its different MV-BBs.

equipped with six switches, as illustrated in Fig. 1(c). Two of these legs are allocated for the ac/dc stage, referred to as the MV-ANPC cell, while the other two serve the MV side of the 5/2-level MV-DAB. The LV side of the MV-DAB remains consistent for both the H-bridge MV-BB and the ANPC MV-BB, facilitating the interface with a single BESS.

The subsequent sections of this article are structured as follows: Section II discusses the configuration and benefits of the proposed ANPC-based MV-BB. Section III explains the operating principles and modulation technique of the proposed MV-BB, supported by electro-thermal simulations. Section IV analyzes the experimental results of an ANPC MV-BB prototype. Finally, Section V presents the conclusions and recommendations for future work.ng the applicable criteria that follow.

II. COMPARISON OF SEVERAL MV-BB CONFIGURATIONS

Various configurations can be employed to implement MV-BBs using lower voltage-rated switches like FC and NPC. This section delves into the investigation of an ANPC-based MV-BB, offering in-depth technical insights into its advantages compared to a traditional HB-based MV-BB.

A. Comparison with Other MV-BB Configurations

Considering that the LV side of the MV-BB remains consistent, as depicted in Figs. 1(b) and (c), Table I compares various configurations of the MV side in terms of their main components. Although these configurations utilize an equal number of legs, they differ in other aspects. The HB-based configuration stands out for having the fewest number of switching positions (SPs), despite employing SPs with higher voltage ratings, twice those of the other configurations. The ANPC-based configuration features the highest number of SPs. However, it does not require the clamping diodes needed in the NPC-based configuration or the clamping capacitors of the FC-based configuration.

Despite the ANPC-based MV-BB requiring three times the number of SPs compared to the HB-based counterpart, its overall cost is lower when considering the combined expenses of the SPs, gate driver boards, and cold plates. This cost advantage stems from the economic efficiency of using cheaper lower voltage-rated SPs. For example, a Si IGBT 1.7 kV [38] can deliver four times the current rating of a Si IGBT 3.3 kV [39]. Consequently, the SPs cost of an ANPC-based MV-BB utilizing Si IGBT 1.7 kV modules is approximately 18% less than that of an HB-based one using Si IGBT 3.3 kV modules. Furthermore, connecting four modules in parallel introduces additional complexities, particularly, managing current sharing among them. Additionally, while the SPs in the ANPC-based MV-BB occupy 25% less volume, their weight is 8.5% higher than those in the HB-based MV-BB.

B. Maintaining the Integrity of the Specifications

The SP power losses in both HB-based and ANPC-based MV-BBs can be analytically expressed as follows:

$$P_{c-HB} = 2N_p \times R_{ds-on(HV)} \times I_{D-rms}^2 \qquad (1)$$

$$P_{c-ANPC} = 4 R_{ds-on(LV)} \times I_{D-rms}^2 \qquad (2)$$

$$P_{sw-HB} = 2N_p \times f_{sw} \times (E_{on(HV)} + E_{off(HV)}) \qquad (3)$$

$$P_{sw-ANPC} = 4 f_{sw} \times (E_{on(LV)} + E_{off(LV)}) \qquad (4)$$

where P_{c-HB} denotes the SP conduction losses and P_{sw-HB} represents the SP switching losses of an HB-based MV-BB, while P_{c-ANPC} signifies the SP conduction losses and $P_{sw-ANPC}$ denotes the SP switching losses of an ANPC-based MV-BB. Additionally, N_p represents the number of paralleled HV modules required to match the current rating of the LV module.

The parameters $R_{ds-on(HV)}$ and $R_{ds-on(LV)}$ refer to the on-state resistance of HV and LV modules, respectively. Bolotnikov et al. [40] proved that replacing an HV module with two LV modules results in lower on-state resistance, i.e, $R_{ds-on(HV)} > 2 R_{ds-on(LV)}$. $E_{on(HV)}$ and $E_{on(LV)}$ represent the turn-on energy losses of HV and LV modules,

TABLE I   COMPONENTS OF THE MV SIDE OF MV-BBS

Topology Parameter	HB-based	Three Level-based		
		FC	NPC	ANPC
No. of Bridges	4	4	4	4
Switching positions	8	16	16	24
Voltage of a switch	2x	x	x	x
Clamping diodes	0	0	8	0
Clamping capacitors	0	4	0	0

respectively, while $E_{off(HV)}$ and $E_{off(LV)}$ denote the turn-off energy losses of HV and LV modules, respectively. Lastly, I_{D-rms} refers to the RMS current and f_{sw} represents the employed switching frequency.

To evaluate the effects on SP power losses in each MV-BB alternative, CAB650M17HM3, a 1.7 kV, 650 A SiC MOSFET module, is selected, while the 3.3 kV, 400 A SiC module reported in [41], [42] is considered. two 3.3 kV modules are required for the application considered here (i.e., $N_p = 2$). Considering the following: $T_{vj} = 175\ °C$, $V_{dc} = 2.4$ kV, $I_{D-rms} = 350$ A, the SP power losses in the HB-based MV-BB are $P_{c-HB} = 4.97$ kW and $P_{sw-HB} = 8.4$ kW, while those in the ANPC-based MV-BB are $P_{c-ANPC} = 4.69$ kW and $P_{sw-ANPC} = 4.18$ kW. Thus, the ANPC-based MV-BB reduces the total SP power losses by around 34 %.

C. Effect of the Proposed MV-BB on the Output Filter

The LCL filter stands out as the most preferred choice for grid-connected applications because it provides various benefits considering weight, cost, and size [35], [43], [44]. The inductance of the converter-side inductor (L_c) is determined as follows:

$$L_c = \frac{V_{dc}}{4\, I_{ac} \cdot \Delta i_{rpp} \cdot f_{eff}} \tag{5}$$

where I_{ac} denotes the output ac current of the MV-PCS, Δi_{rpp} represents the selected output current ripple percentage (i.e., 30%), f_{eff} refers to the effective switching frequency. In the case of HB-based MV-PCS, the effective switching frequency (f_{eff-HB}) can be calculated as follows:

$$f_{eff-HB} = 2\, n_{BB} \cdot f_{sw} \tag{6}$$

where n_{BB} represents the number of cascaded MV-BB. The effective switching frequency of the ANPC-based MV-PCS ($f_{eff-ANPC}$) can be determined as follows:

$$f_{eff-ANPC} = 4\, n_{BB} \cdot f_{sw} \tag{7}$$

From (6) and (7), $f_{eff-ANPC}$ is twice as f_{eff-HB} for the same device switching frequency, which leads to size reduction of the inductors of the LCL filter by 50 %. However, the maximum achievable switching frequency of the LV modules is nearly twice that of the HV ones [45], which boosts the $f_{eff-ANPC}$ up to four times as f_{eff-HB} and decreases the output inductors by 75 %.

III. ANALYSIS OF ANPC-BASED MV-BB

The proposed topology is analyzed by investigating the switching states of the ANPC-based MV-BB first. Then, the proposed topology is simulated using PLECS software to investigate the required modulation technique.

A. Operating Principles of the Proposed ac/dc Side

The proposed MV-BB comprises two 3L-ANPC legs for the ac/dc stage, each providing three different voltage levels: +V_{dc}/2, 0, -V_{dc}/2, where Vdc is the dc link total voltage. By combining the output voltage of both legs, it becomes possible to achieve five different levels of output voltage: +V_{dc}, +V_{dc}/2, 0, -V_{dc}/2, -V_{dc}. Fig. 2 illustrates the switching pattern for this stage across varying output voltages. The conduction path for each pattern is highlighted in red. In Fig. 3, the switches'

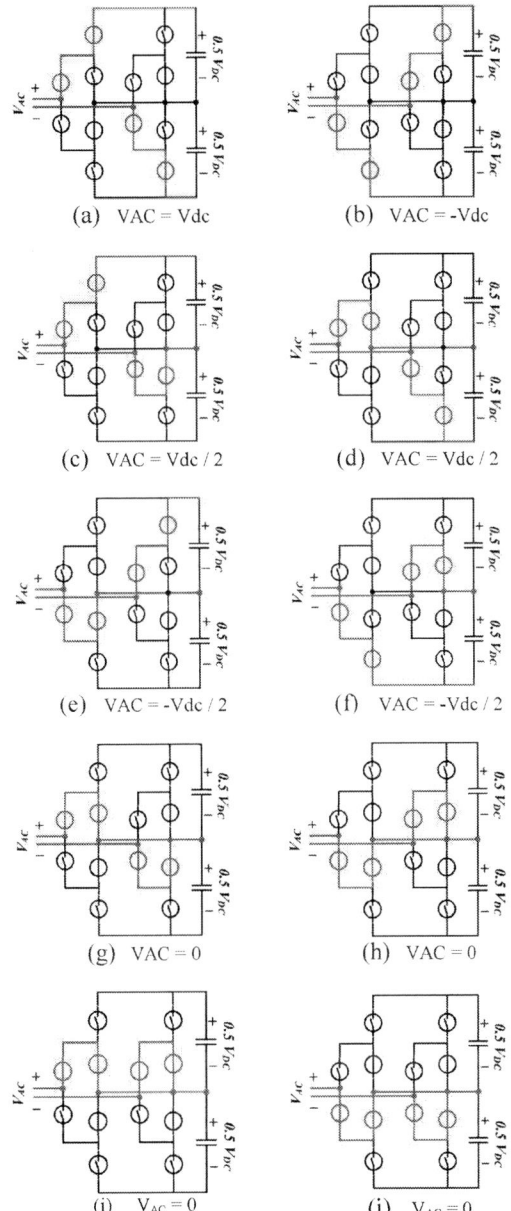

Fig. 2. Possible switching patterns of the proposed MV-BB.

names and the rails of the dc bus for each ANPC leg are presented in their corresponding positions.

As shown in Fig. 2(a), to generate a +V_{dc} at the ac output of the MV-BB, one ANPC leg connects the positive terminal of the AC output to the positive rail of the dc bus via switches S1, S2, and Sn, while another leg connects the negative terminal to the negative rail through switches S3, S4, and Sp. This configuration adheres to the gate sequence for the ANPC modulation outlined in [46], as referenced in Table II and Fig. 3, ensuring compliance with both the gate sequence and the negative output requirement.

For an ac output of +V_{dc}/2, two configurations are feasible. The first connects the positive terminal of the ac output to the positive rail of the dc bus and the negative terminal to the neutral rail as illustrated in Fig. 2(c). The second configuration

TABLE II. GATE SEQUENCE FOR ANPC MODULATION [46]

State	Sp	S1	S2	S3	S4	Sn
Positive	0	1	1	0	0	1
Neutral 1	1	0	1	0	0	1
Neutral 2	1	0	0	1	0	1
Negative	1	0	0	1	1	0

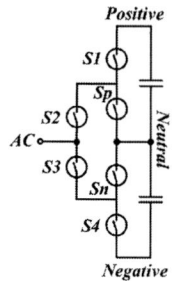

Fig. 3. Illustration of each ANPC leg components and rails.

reverses this connection, linking the positive terminal to the neutral rail and the negative terminal to the negative rail as presented in Fig. 2(d). Both configurations align with the SPWM modulation techniques and gate sequences outlined in Table II.

To achieve a zero output, both terminals of the ac/dc stage are linked to the neutral rail, utilizing the states Neutral 1 and Neutral 2 for the gate sequence of the two ANPC leges. Four configurations are available for this output voltage, as illustrated in Fig. 2(g-j). For the output of -Vdc/2, the gate sequence of the two leges is either in the Neutral 2 and Positive state or Negative and Neutral 1 state, as depicted in Fig. 2(e) and (f). Finally, to attain a -Vdc output, the gate sequence for the two legs is set to Negative & Positive, as shown in Fig. 2(b).

B. The Modulation Technique for the Proposed MV-BB

The proposed 3L-ANPC-based MV-BB is simulated to analyze its performance and study the modulation technique. The paper employs the principle of decomposition, treating each bridge independently to produce a three-level output. To achieve this output, phase disposition (PD) modulation is utilized. The used modulation technique includes two reference sine waveforms Vref_A and Vref_B, where the phase shift φ between them is π. Moreover, two triangular carriers Vcr_top and Vcr_bott are in phase and vertically disposed as depicted in Fig. 4. Noteworthy, this technique provides wo ways to control the generated voltage through

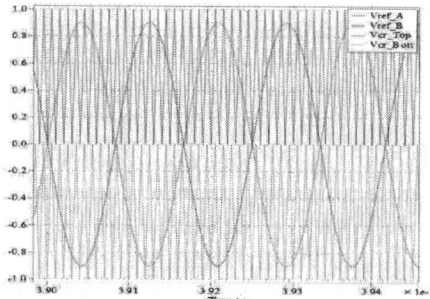

Fig. 4. Modulation waveforms of proposed MV-BB.

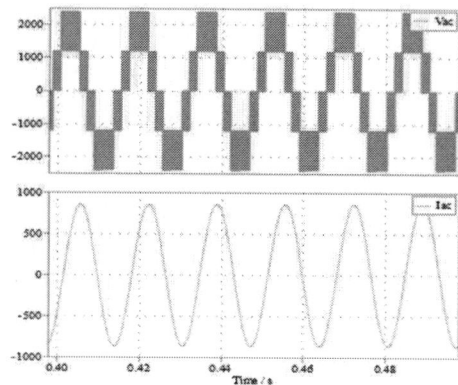

Fig. 5. Modulation waveforms of proposed MV-BB.

adjusting the modulation index and/or the phase shift between the reference waveforms.

Using PLECS software, the ANPC-based MV-BB was simulated considering 1.7 kV switching devices. These devices had an operating voltage of 1200 V. So, the dc link voltage was 2400 V. the output current and voltage are presented in Fig. 5.

IV. EXPERIMENTAL RESULTS

While the preceding sections delved into the operational principles and simulations of the proposed ANPC-based MV-BB, this section showcases the experimental results of its prototype validating the provided analysis. As illustrated in Fig. 6, the MV side of this prototype comprises four ANPC legs, each equipped with three 1.7 kV SiC MOSFET half-bridge power modules from Wolfspeed [47]. For the sake of comparison, another prototype of an HB-based MV-BB was successfully crafted, as depicted in Fig. 7. This prototype's MV side features four 3.3 kV, 200 A SiC MOSFET power half-bridge modules produced at HiDEC facilities, University of Arkansas. The static and dynamic characteristics of these modules were previously documented and analyzed [48]. Both prototypes have the same LV side of the MV-DAB which contains two 1.2 kV, 425 A SiC MOSFET half-bridge power modules from Wolfspeed [49]. The turns ratio of the MV transformer was precisely calculated to achieve the required voltage conversion between the MV and LV windings. The MV inductor was constructed with extra insulation layers to endure the high induced voltage across its terminals.

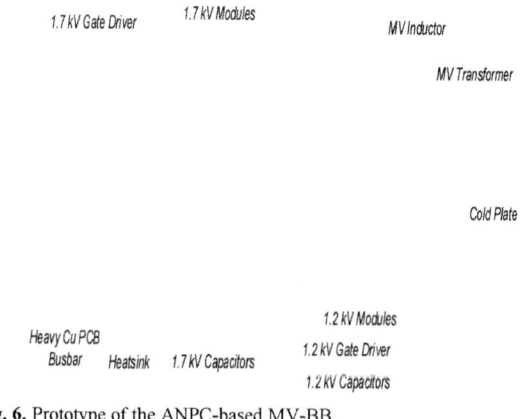

Fig. 6. Prototype of the ANPC-based MV-BB.

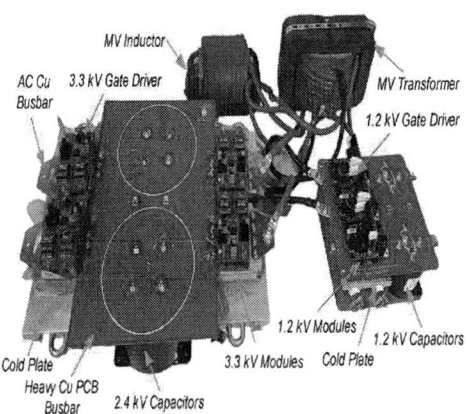

Fig. 7. Prototype of the HB-based MV-BB.

Both prototypes underwent testing considering the discharge mode of the G-BESS. To simulate this case, a Magna Power Supply was connected at the dc input terminals of the LV side of the MV-DAB, while a resistive load was attached to the ac output terminals of the MV-BBs. Table III shows the selected testing conditions for both prototypes. For comparison purposes, the switching frequency of the MV-DAB is 20 kHz limited by the 3.3 kV SiC MOSFET modules. While the MV transformer and inductor were designed at this frequency, the weight and volume of these two components could be reduced by utilizing the maximum feasible switching frequency of the 1.7 kV SiC MOSFET modules.

The ANPC-based MV-BB was tested successfully, and the captured experimental waveforms are presented in Fig. 8. In Fig. 8(a), the first part shows the output of $+1.2$ kV, while the second presents the output current. Thirdly, a 2.4 kV dc link was measured as shown. The 5/2L MV-DAB waveforms are illustrated, where the fourth part shows the 5L waveform of the MV side and the 2L one of the LV side. Therefore, the current of the 5/2L MV-DAB tends to be more sinusoidal than that of the HB-based MV-BB. The five-level output voltage, including 2.4 kV, 1.2 kV, 0, -1.2 kV, and -2.4 kV levels, was achieved successfully as shown in Fig. 8(b). Lastly, the load was not purely resistive as noted in the current waveform.

On the other side, Fig. 9 exhibits the experimental waveforms captured during testing of the HB-based MV-BB. Fig. 9(a) presents the typical waveforms of the MV-DAB, where the MV side has two levels, i.e., 2.4 kV and -2.4 kV. The LV side has the same 2L waveform as of the previous prototype. Additionally, the current waveform includes the sharp transitions between the switching patterns. Fig. 9(b) shows the three-level output current and voltage waveforms of the HB-based MV-BB.

TABLE III. TESTING CONDITIONS OF BOTH PROTOTYPES

Parameter	Symbol	Value
Voltage of the LV dc-link	V_{LV}	800 V
Voltage of the MV dc-link	V_{MV}	2400 V
Switching frequency of the MV-DAB	f_{sw_DAB}	20 kHz
Switching frequency of the ac/dc stage	f_{sw_inv}	10 kHz
Load active power	P_{load}	15 kW
Output frequency	f_g	60 Hz
Modulation index	m_a	0.8

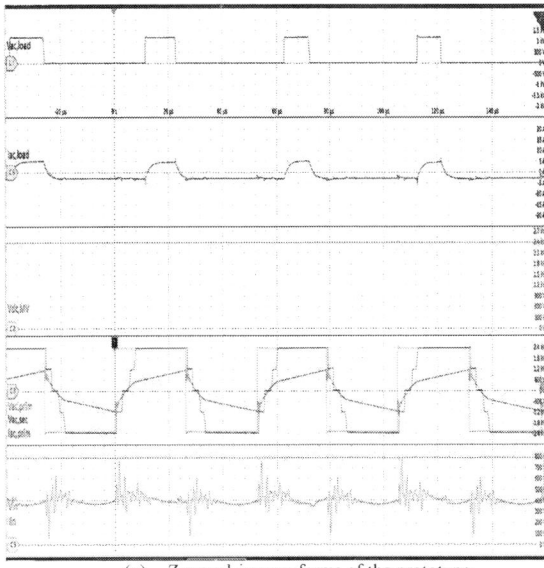

(a) Zoomed-in waveforms of the prototype.

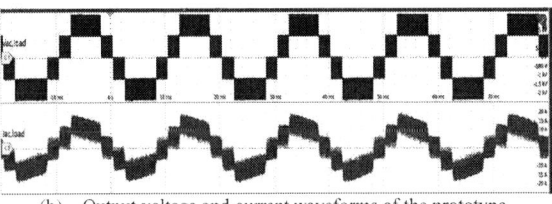

(b) Output voltage and current waveforms of the prototype.

Fig. 8. Experimental waveforms of the ANPC-based MV-BB prototype.

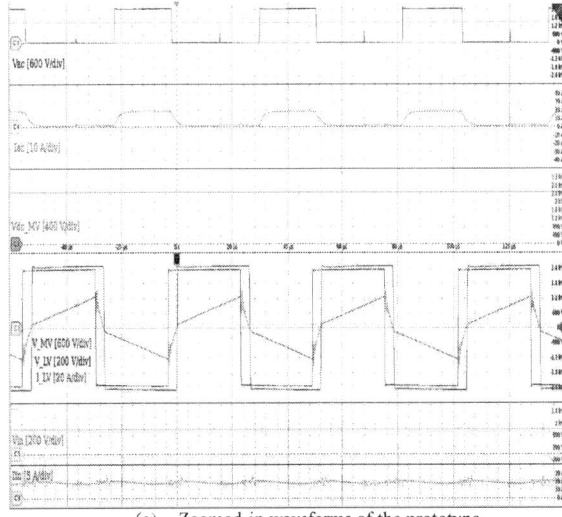

(c) Zoomed-in waveforms of the prototype.

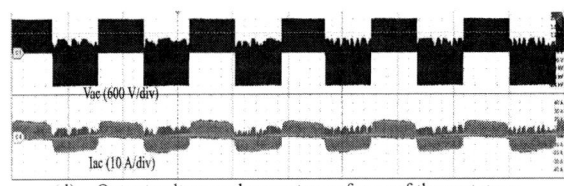

(d) Output voltage and current waveforms of the prototype.

Fig. 9 Experimental waveforms of the ANPC-based MV-BB prototype.

V. CONCLUSIONS AND FUTURE WORK

While HV modules can reduce the number of series-connected MV-BBs, they pose other challenges to the MV-PCS. A novel ANPC-based MV-BB design was investigated to demonstrate its advantages over the traditional H-based MV-BB. The proposed MV-BB involved ANPC legs, allowing the use of lower-rated voltage modules while maintaining the same dc link voltage.

Analysis of the proposed MV-BB showed approximately 34% lower power losses in its switching devices compared to the H-based one. Additionally, the volume, weight, and power losses of the inductors required for the LCL filter are reduced by at least 50%. Consequently, the proposed MV-BB is considered a more cost-effective solution than the conventional H-based one. Furthermore, a phase disposition modulation technique was employed to generate the required control signals, as demonstrated by simulations of the proposed MV-BB.

Finally, experimental results and captured waveforms validated the feasibility and advantages of the ANPC-based MV-BB compared to the H-bridge-based design.

After proving the benefits of the ANPC-based MV-BB, an MLC topology based on cascading the proposed MV-BB will be investigated to analysis the required modulation technique and control strategies.

ACKNOWLEDGMENT

The authors would like to thank Dr. Alan Mantooth and Dr. Yuxiang Chen for providing the 3.3 kV SiC MOSFET modules. Testing for this project was conducted at the National Center for Reliable Electric Power Transmission (NCREPT), the University of Arkansas' High-Power Test Facility.

REFERENCES

[1] H. Akagi, "Multilevel Converters: Fundamental Circuits and Systems," Proc. IEEE, vol. 105, no. 11, pp. 2048–2065, Nov. 2017, doi: 10.1109/JPROC.2017.2682105.

[2] A. Delavari, I. Kamwa, and A. Zabihinejad, "A comparative study of different multilevel converter topologies for Battery Energy Storage application," Can. Conf. Electr. Comput. Eng., pp. 17–21, 2017, doi: 10.1109/CCECE.2017.7946773.

[3] G. Wang et al., "A review of power electronics for grid connection of utility-scale battery energy storage systems," IEEE Trans. Sustain. Energy, vol. 7, no. 4, pp. 1778–1790, 2016, doi: 10.1109/TSTE.2016.2586941.

[4] "CAB650M17HM3 Product Information," 2022. https://assets.wolfspeed.com/uploads/dlm_uploads/2022/02/CAB650 M17HM3.pdf (accessed Jan. 20, 2023).

[5] "BSM250D17P2E004 - 1700V, 250A, Half bridge, Silicon-carbide (SiC) Power Module | ROHM Semiconductor - ROHM Co., Ltd." https://www.rohm.com/products/sic-power-devices/sic-power-module/bsm250d17p2e004-product (accessed Aug. 27, 2021).

[6] "3300 V LM Half-Bridge SiC Power Modules | Wolfspeed." https://www.wolfspeed.com/products/power/sic-power-modules/3300v-lm-silicon-carbide-half-bridge-power-modules/ (accessed Aug. 02, 2023).

[7] "MSC025SMA330 | Microchip Technology." https://www.microchip.com/en-us/product/MSC025SMA330 (accessed Oct. 28, 2023).

[8] B. Deboi, A. Lemmon, B. Nelson, C. New, and D. Hudson, "Modeling and Validation of Medium Voltage SiC Power Modules," in Conference Proceedings - IEEE Applied Power Electronics Conference and Exposition - APEC, Mar. 2020, vol. 2020-March, pp. 1964–1971, doi: 10.1109/APEC39645.2020.9124285.

[9] L. Sang et al., "Development of high-voltage SiC Power Electronic Devices," in 2021 18th China International Forum on Solid State Lighting and 2021 7th International Forum on Wide Bandgap Semiconductors, SSLChina: IFWS 2021, Dec. 2021, pp. 17–24, doi: 10.1109/SSLCHINAIFWS54608.2021.9675164.

[10] B. Passmore et al., "The next generation of high voltage (10 kV) silicon carbide power modules," in WiPDA 2016 - 4th IEEE Workshop on Wide Bandgap Power Devices and Applications, Dec. 2016, pp. 1–4, doi: 10.1109/WIPDA.2016.7799900.

[11] G. G. Oggier, R. G. Jimenez, Y. Zhao, and J. C. Balda, "Modeling and Characterization of 10-kV SiC MOSFET Modules for Medium-Voltage Distribution Systems," 2020 IEEE 11th Int. Symp. Power Electron. Distrib. Gener. Syst. PEDG 2020, pp. 583–590, Sep. 2020, doi: 10.1109/PEDG48541.2020.9244411.

[12] K. Vechalapu, S. Bhattacharya, E. Van Brunt, S. H. Ryu, D. Grider, and J. W. Palmour, "Comparative Evaluation of 15-kV SiC MOSFET and 15-kV SiC IGBT for Medium-Voltage Converter under the Same dv/dt Conditions," IEEE J. Emerg. Sel. Top. Power Electron., vol. 5, no. 1, pp. 469–489, Mar. 2017, doi: 10.1109/JESTPE.2016.2620991.

[13] Z. Zhang et al., "Packaging of a 15-kV Silicon Carbide MOSFET With Insulation Enhanced by a Nonlinear Resistive Polymer-Nanoparticle Coating," Oct. 2022, doi: 10.1109/ECCE50734.2022.9947881.

[14] Q. Zhu, L. Wang, A. Q. Huang, K. Booth, and L. Zhang, "7.2-kV Single-Stage Solid-State Transformer Based on the Current-Fed Series Resonant Converter and 15-kV SiC mosfets," IEEE Trans. Power Electron., vol. 34, no. 2, pp. 1099–1112, Feb. 2019, doi: 10.1109/TPEL.2018.2829174.

[15] S. Madhusoodhanan et al., "Three-phase 4.16 kV medium voltage grid tied AC-DC converter based on 15 kV/40 a SiC IGBTs," in 2015 IEEE Energy Conversion Congress and Exposition, ECCE 2015, Oct. 2015, pp. 6675–6682, doi: 10.1109/ECCE.2015.7310594.

[16] V. Jones, R. Fantino, A. Rahouma, J. C. Balda, and R. Adapa, "Construction and testing of a 13.8 kV, 750 kVA 3-Phase current compensator using modular switching positions," Conf. Proc. - IEEE Appl. Power Electron. Conf. Expo. - APEC, pp. 2050–2057, Jun. 2021, doi: 10.1109/APEC42165.2021.9487422.

[17] P. Trochimiuk, R. Kopacz, G. Wrona, and J. Rabkowski, "Active Voltage Balancing of Series-Connected 1.7 kV/325 A SiC MOSFETs Enabling Continuous Operation at Medium Voltage," IEEE Access, vol. 9, pp. 8604–8614, 2021, doi: 10.1109/ACCESS.2021.3049606.

[18] P. Trochimiuk, R. Kopacz, G. Wrona, and J. Rabkowski, "Medium voltage power switch based on 1.7 kV SiC MOSFETs connected in series inside power modules," Sep. 2019, doi: 10.23919/EPE.2019.8915468.

[19] K. Vechalapu, S. Hazra, U. Raheja, A. Negi, and S. Bhattacharya, "High-Speed medium voltage (MV) drive applications enabled by series connection of 1.7 kV SiC MOSFET devices," in 2017 IEEE Energy Conversion Congress and Exposition, ECCE 2017, Oct. 2017, vol. 2017-January, pp. 808–815, doi: 10.1109/ECCE.2017.8095868.

[20] E. Raszmann et al., "Design and Test of a 6 kV Phase-Leg using Four Stacked 1.7 kV SiC MOSFET High-Current Modules," in Conference Proceedings - IEEE Applied Power Electronics Conference and Exposition - APEC, Mar. 2020, vol. 2020-March, pp. 1604–1610, doi: 10.1109/APEC39645.2020.9124298.

[21] W. Xu and A. Q. Huang, "15-KV/50-A SiC AC Switch Based on Series Connection of 1.7-KV MOSFETs," IEEE Trans. Ind. Appl., Sep. 2023, doi: 10.1109/TIA.2023.3284808.

[22] J. W. Baek, D. W. Yoo, and H. G. Kim, "High-voltage switch using series-connected IGBTs with simple auxiliary circuit," IEEE Trans. Ind. Appl., vol. 37, no. 6, pp. 1832–1839, Nov. 2001, doi: 10.1109/28.968198.

[23] V. S. Nguyen, P. Lefranc, and J. C. Crebier, "Gate Driver Supply Architectures for Common Mode Conducted EMI Reduction in Series Connection of Multiple Power Devices," IEEE Trans. Power Electron., vol. 33, no. 12, pp. 10265–10276, Dec. 2018, doi: 10.1109/TPEL.2018.2802204.

[24] V. Jones, R. A. Fantino, and J. C. Balda, "A Modular Switching Position with Voltage-Balancing and Self-Powering for Series Device Connection," IEEE J. Emerg. Sel. Top. Power Electron., vol. 9, no. 3, pp. 3501–3516, Jun. 2021, doi: 10.1109/JESTPE.2020.3003796.

[25] A. K. Sadigh, V. Dargahi, and K. A. Corzine, "New Multilevel Converter Based on Cascade Connection of Double Flying Capacitor

Multicell Converters and Its Improved Modulation Technique," IEEE Trans. Power Electron., vol. 30, no. 12, pp. 6568–6580, Dec. 2015, doi: 10.1109/TPEL.2014.2387066.

[26] H. Huang, L. Zhang, O. Oghorada, and M. Mao, "Analysis and Control of a Modular Multilevel Cascaded Converter-Based Unified Power Flow Controller," IEEE Trans. Ind. Appl., vol. 57, no. 3, pp. 3202–3213, May 2021, doi: 10.1109/TIA.2020.3029546.

[27] S. H. Kim, Y. H. Jang, and R. Y. Kim, "Modeling and Hierarchical Structure Based Model Predictive Control of Cascaded Flying Capacitor Bridge Multilevel Converter for Active Front-End Rectifier in Solid-State Transformer," IEEE Trans. Ind. Electron., vol. 66, no. 8, pp. 6560–6569, Aug. 2019, doi: 10.1109/TIE.2018.2871789.

[28] M. M. Da Silva and H. Pinheiro, "Voltage balancing in flying capacitor converter multilevel using space vector modulation," Jul. 2017, doi: 10.1109/PEDG.2017.7972477.

[29] S. C. Mersche, N. Katzenburg, P. Kiehnle, B. Schmitz-Rode, D. Schulz, and M. Hiller, "Comparison of Quasi-Two-Level Operation of a Flying Capacitor Converter with Quasi-Two-Level Operation of a Modular Multilevel Converter," Mar. 2022, doi: 10.1109/SPEC55080.2022.10058325.

[30] G. V. Bharath and P. T. Balsara, "A New Carrier based Precharging Technique for Single-Phase Odd-Level Flying Capacitor Multilevel Converters (FCMCs)," in Proceedings of the Energy Conversion Congress and Exposition - Asia, ECCE Asia 2021, May 2021, pp. 1262–1267, doi: 10.1109/ECCE-ASIA49820.2021.9479443.

[31] B. Ge, F. Z. Peng, A. T. De Almeida, and H. Abu-Rub, "An effective control technique for medium-voltage high-power induction motor fed by cascaded neutral-point-clamped inverter," IEEE Trans. Ind. Electron., vol. 57, no. 8, pp. 2659–2668, Aug. 2010, doi: 10.1109/TIE.2009.2026761.

[32] F. Wu, B. Li, and H. B. Gooi, "Principle and control of modified cascaded NPC-GCI with variable topology ability to enhance european efficiency," IEEE Trans. Ind. Electron., vol. 64, no. 2, pp. 1214–1221, Feb. 2017, doi: 10.1109/TIE.2016.2615035.

[33] H. Lin et al., "A Simplified 3-D NLM-Based SVPWM Technique with Voltage-Balancing Capability for 3LNPC Cascaded Multilevel Converter," IEEE Trans. Power Electron., vol. 35, no. 4, pp. 3506–3518, Apr. 2020, doi: 10.1109/TPEL.2019.2938606.

[34] X. Wei et al., "Parallel open-circuit fault diagnosis method of a cascaded full-bridge npc inverter with model predictive control," IEEE Trans. Ind. Electron., vol. 68, no. 10, pp. 10180–10192, Oct. 2021, doi: 10.1109/TIE.2020.3028801.

[35] A. Rahouma, J. C. Balda, and R. Adapa, "A Medium-Voltage SiC Flying Capacitor Converter Design for 25-kV Distribution Systems," Proc. 2021 IEEE 12th Int. Symp. Power Electron. Distrib. Gener. Syst. PEDG 2021, Jun. 2021, doi: 10.1109/PEDG51384.2021.9494228.

[36] W. Choi et al., "Reviews on grid-connected inverter, utility-scaled battery energy storage system, and vehicle-to-grid application - Challenges and opportunities," in 2017 IEEE Transportation and Electrification Conference and Expo, ITEC 2017, 2017, pp. 203–210, doi: 10.1109/ITEC.2017.7993272.

[37] R. A. Rana, S. A. Patel, A. Muthusamy, C. W. Lee, and H. J. Kim, "Review of multilevel voltage source inverter topologies and analysis of harmonics distortions in FC-MLI," Electronics (Switzerland), vol. 8,

no. 11. MDPI AG, p. 1329, Nov. 01, 2019, doi: 10.3390/electronics8111329.

[38] "FF1800XTR17T2P5BPSA1 Infineon Technologies | Discrete Semiconductor Products | DigiKey." https://www.digikey.com/en/products/detail/infineon-technologies/FF1800XTR17T2P5BPSA1/22157839 (accessed May 06, 2024).

[39] "FF450R33T3E3BPSA1 Infineon Technologies | Discrete Semiconductor Products | DigiKey Marketplace." https://www.digikey.com/en/products/detail/infineon-technologies/FF450R33T3E3BPSA1/9829584 (accessed May 06, 2024).

[40] A. Bolotnikov et al., "Overview of 1.2kV - 2.2kV SiC MOSFETs targeted for industrial power conversion applications," in Conference Proceedings - IEEE Applied Power Electronics Conference and Exposition - APEC, May 2015, vol. 2015-May, no. May, pp. 2445–2452, doi: 10.1109/APEC.2015.7104691.

[41] J. Hayes et al., "Dynamic Characterization of Next Generation Medium Voltage (3.3 kV, 10 kV) Silicon Carbide Power Modules | VDE Conference Publication | IEEE Xplore," 2017, Accessed: Aug. 27, 2021. [Online]. Available: https://ieeexplore.ieee.org/abstract/document/7990668.

[42] J. B. Casady, T. McNutt, D. Girder, and J. Palmour Jeff, "A CREE COMPANY Medium Voltage SiC R&D update," 2016.

[43] A. Reznik, M. G. Simoes, A. Al-Durra, and S. M. Muyeen, "LCL Filter design and performance analysis for grid-interconnected systems," IEEE Trans. Ind. Appl., vol. 50, no. 2, pp. 1225–1232, 2014, doi: 10.1109/TIA.2013.2274612.

[44] T. Lahlou, M. Abdelrahem, S. Valdes, and H. G. Herzog, "Filter design for grid-connected multilevel CHB inverter for battery energy storage systems," 2016 Int. Symp. Power Electron. Electr. Drives, Autom. Motion, SPEEDAM 2016, pp. 831–836, 2016, doi: 10.1109/SPEEDAM.2016.7525972.

[45] A. Rahouma, D. A. Porras, G. G. Oggier, J. C. Balda, and R. Adapa, "Optimal Medium-Voltage Cascaded H-Bridge Converters for High-Power Distribution System Applications," IEEE J. Emerg. Sel. Top. Power Electron., vol. 12, no. 2, pp. 1406–1415, Apr. 2024, doi: 10.1109/JESTPE.2023.3296725.

[46] Y. Jiao and F. C. Lee, "New modulation scheme for three-level active neutral-point-clamped converter with loss and stress reduction," IEEE Trans. Ind. Electron., vol. 62, no. 9, pp. 5468–5479, Sep. 2015, doi: 10.1109/TIE.2015.2405505.

[47] "CAB320M17XM3 1700 V, 3.5 mΩ, Half-Bridge SiC Power Module | Wolfspeed." https://www.wolfspeed.com/products/power/sic-power-modules/xm3-power-module-family/cab320m17xm3/ (accessed Apr. 27, 2024).

[48] Y. Chen et al., "3.3 kV Low-Inductance Full SiC Power Module," in 2023 IEEE Applied Power Electronics Conference and Exposition (APEC), Mar. 2023, pp. 2634–2640, doi: 10.1109/APEC43580.2023.10131290.

[49] "CAB425M12XM3." https://assets.wolfspeed.com/uploads/2020/12/cab425m12xm3.pdf (accessed Jul. 06, 2023).

Analog Control of a 2.5 kW GaN Based CRM PFC with Input Filter Optimization

Naveed Ishraq
Power Electronics and control Engineering
Laboratory (PEACE)
Arizona State University
Mesa, AZ, USA
nishraq@asu.edu

Ayan Mallik
Power Electronics and control Engineering
Laboratory (PEACE)
Arizona State University
Mesa, AZ, USA
amallik3@asu.edu

Abstract— **This paper investigates zero-voltage-switching (ZVS) control of a single-phase GaN-based critical-conduction-mode (CRM) totem-pole power factor correction (TPFC) in fully analog domain. Critical mode operation is used by the totem-pole PFC rectifier in order to achieve both high frequency operation and hence smaller magnetics while also achieving high efficiency due to soft switching. A digitally controlled CRM TPFC faces challenges due to current sensing delays and zero-current detection (ZCD) sensitivity to high di/dt noise, which complicates ZVS control. For designing a fully analog controlled CRM TPFC, a deadtime optimization is performed over the entire line cycle to minimize the cycle averaged CRM turn-on loss. Furthermore, the design methodology and volume optimization of input electromagnetic interference (EMI) filter for CRM totem-pole PFC converter is proposed. Finally, a 2.5kW GaN-based 120Vac/240Vac to 400V dc totem-pole PFC is demonstrated with 98.12% peak efficiency operating between 70 kHz (near peak) and 460 kHz (near zero) switching frequency.**

I. INTRODUCTION

The totem-pole PFC is well known as the simplest topology among all the bridgeless boost PFC structures [1]. However, the totem-pole PFC using silicon MOSFETs suffers from large power loss resulting from the reverse-recovery of the body diode [2]. With the advent of high voltage gallium-nitride (GaN) devices with low on-resistance and zero reverse recovery loss, GaN based continuous conduction mode TPFC has demonstrated high efficiency [3] keeping switching frequency below or around 100kHz. For GaN devices the turn-off loss is much smaller than the turn-on loss [4]. The turn-on loss can be mitigated through ZVS strategy reducing the overall switching loss significantly. Thus, the soft switching capability of GaN devices is utilized to operate CRM TPFC with very high efficiency and power density [5]. Typically, hysteresis control and variable on-time control are implemented digitally for controlling the CRM TPFC [6-7]. However, the hysteresis control requires high-bandwidth current sensing followed by analog to digital conversion which results in sensing delay. Current sensing delay impact is significant in high frequency control which deviates the converter switching frequency, total harmonic distortion (THD) and inductor current from normal operation [8]. The signal processing delay in microcontroller (MCU) causes significant undesired negative current and thus increases the current ripple and conduction loss. In variable on-time control, the ZCD circuit is sensitive to high frequency di/dt or dv/dt noise and hence on the PCB switching network loop inductance, and the control network is prone to anomalous

switching actions when an erroneous ZCD signal is generated due to a disturbance from the switching noise. In addition, the exact calculation of on-time using MCU could be complicated and often requires large execution time when compared to the switching period.

In addition to the totem-pole PFC control, the EMI filter volume of the converter volume is also critical as it consumes almost one-third of the total space [9]. EMI filter is effectively employed for reducing the EMI noise emissions generated due to integration of power electronics converter into the grid. Thus, the unwanted emissions should be suppressed to fulfill noise emission standards, such as CISPR and FCC for frequencies beyond 150 kHz [10]. However, further studies need to address two major issues in the critical mode PFC rectifier. The first issue is existence of a high input current ripple. It requires a large DM-EMI filter with increased converter size and cost. Another major challenge is the variable switching frequency which makes it difficult to analyze and design the DM EMI filter of CRM boost PFC converter. Most of the current research primarily examines EMI noise in constant-frequency CCM PFC converters [11-13]. To overcome these challenges, this paper proposes a fully analog controlled CRM TPFC with ZVS capability. Finally, a 2.5 kW single phase CRM TPFC hardware prototype is developed and tested to validate the efficacy of the analog control strategy in both steady-state and dynamic transient conditions. Moreover, an accurate model of the DM EMI spectrum for the CRM TPFC is obtained using inductor current FFT analysis to estimate the filter attenuation requirement. Finally, a design methodology of an optimal EMI filter considering the variable switching frequency operation of the CRM TPFC within the entire line voltage and load range is proposed.

The key contributions of this paper can be summarized as follows:

a) Design of a fully analog controlled CRM TPFC converter with ZVS capability.

b) Optimization of switch dead time for minimizing cycle-averaged turn ON loss.

c) Extensive analysis of the input current Fast Fourier transform (FFT) model of the CRM PFC converter for conducted EMI spectrum quantification.

d) Multivariable lumped formulation of the DM EMI filter inductor and capacitor volume followed by multi-constraint optimization using Lagrange multiplier method.

979-8-3315-1612-3/25 $31.00 © 2025 IEEE

This paper is organized as follows. In section II, the operation and analog implementation of the CRM TPFC converter are discussed. The optimum dead time generation over the line cycle for minimizing the device turn-on loss is provided in Section III. Section IV discusses the CRM PFC inductor current FFT model for conducted EMI quantification for designing the EMI filter followed by its volume optimization. Finally, the experimental results validating the operation and control of CRM TPFC in analog domain is presented Section V. Section VI puts forward the conclusions with relevant remarks.

II. ANALOG IMPLEMENTATION OF ZVS-ENABLED CRM TPFC CONVERTER

In Fig. 1, the circuit schematic of the TPFC converter is depicted, which consists of the boost inductor L, high switching frequency leg formed by switches S1 and S2, and line frequency leg switches consisting of S_3 and S_4. To better illustrate the full-line-cycle ZVS strategy, Fig. 2 shows the inductor current waveform, gate signals and the state trajectory within one switching cycle of the single-phase CRM PFC converter. For simplicity, the discussion considers the positive half-line cycle of the input voltage and assumes that the input voltage remains nearly constant within one switching period.

As shown in Fig. 2(a), during time interval I, t_{on}, the inductor is charged with S_2 and S_4 on. During time interval II, t_{r1}, S_1 and S_2 are both off. The converter utilizes the resonance between boost inductor L and device junction capacitor to achieve ZVS of S_1. C_{oss1} is discharged from V_o to zero and C_{oss2} is charged from zero to V_o, and the inductor current resonates through its peak value. The inductor current i_L and junction capacitor voltage v_{Coss1} during this interval can be expressed as,

$$i_L(t) = i(t_{on})cos\omega t + \frac{V_{in}}{Z_n}sin\omega t \quad (1)$$

$$v_{coss1}(t) = V_{in} + i(t_{on})Z_n sin\omega t - V_{in}cos\omega t \quad (2)$$

Here, $Z_n = \sqrt{L/2C_{oss}}$, $\omega = 1/\sqrt{2LC_{oss}}$ and $i(t_{on})$ is current at the end of interval I. In time interval III, the inductor is discharged with S_1 on and S_2 off. Typically, when $V_{in} \leq 0.5V_o$, switch S_1 is turned off when inductor current reaches zero and mode IV starts to discharge the C_{oss2} (S_2 junction capacitance) i.e., $V_{DS2} = 0$ for ZVS turn on. However, when $V_{in} \geq 0.5V_o$, switch S_1 is kept on for an extended time t_{ex} after the inductor current zero crossing so that enough inductor energy is stored to discharge C_{oss2} during interval IV. During time interval IV, t_{r2}, S_1 and S_2 are both off. C_{oss2} is discharged from V_o to zero and C_{oss1} is charged from zero to V_o, and the inductor current resonates through its minimum value. The inductor current i_L and junction capacitor voltage v_{Coss1} during this interval can be expressed as

$$i_L(t) = i(0)cos\omega t + \frac{V_o-V_{in}}{Z_n}sin\omega t \quad (3)$$

$$v_{coss2}(t) = V_{in} + (V_o - V_{in})cos\omega t - i(0)Z_n sin\omega t \quad (4)$$

Here, $i(0) = 0$ for $V_{in} \leq 0.5$ and $i(0) = \sqrt{V_o(2V_{in} - V_o)}/Z_n$ for $V_{in} \geq 0.5V_o$. The time t_{r2} required to discharge C_{oss2} from V_o to zero is expressed as, $t_{r2} = \frac{\pi - cos^{-1}(V_{in}/(V_o-V_{in}))}{\omega}$ when V_{in}

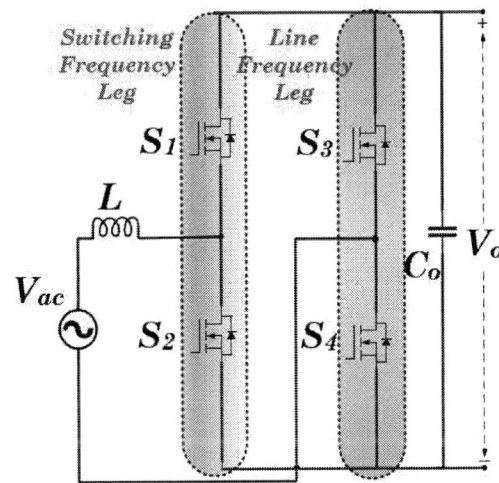

Fig. 1. Circuit diagram of GaN based CRM totem-pole PFC converter.

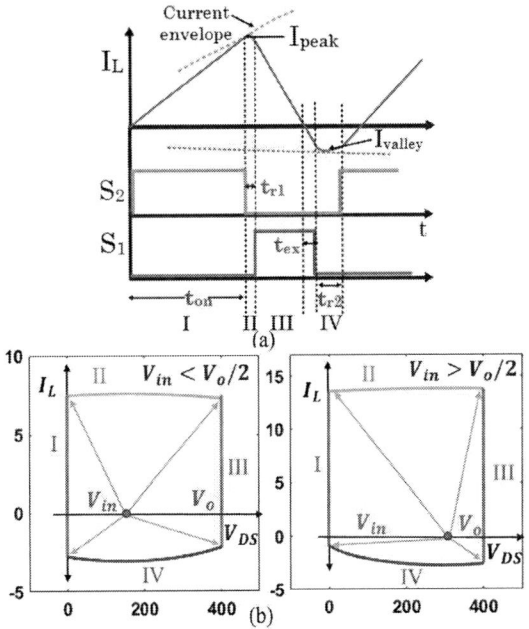

Fig. 2. Operation of CRM TPFC within one switching cycle. (a) Inductor current and gate signals, (b) State plane trajectories of inductor current and drain-to-source voltage.

$\leq 0.5V_o$ and $t_{r2} = \frac{\pi - cos^{-1}((V_o-V_{in})/V_{in})}{\omega}$ $V_{in} \geq 0.5V_o$. However, for analog implementation of CRM PFC control, the valley currents are used to compare with the inductor current in this work to avoid the complex analog circuitry required to generate expressions of $i(0)$. The valley currents can be expressed as,

$$i_{valley} = \begin{cases} \frac{V_o - V_{in}}{Z_n} & V_{in} \leq 0.5V_o \\ \frac{V_{in}}{Z_n} & V_{in} \geq 0.5V_o \end{cases} \quad (5)$$

Fig. 2(b) shows the state plane trajectories of inductor current i_L and drain-to-source voltage v_{DS} of switch S_2. It can be observed during interval III that after the inductor current reaches zero, switch S_2 conduction is extended and reaches negative for both

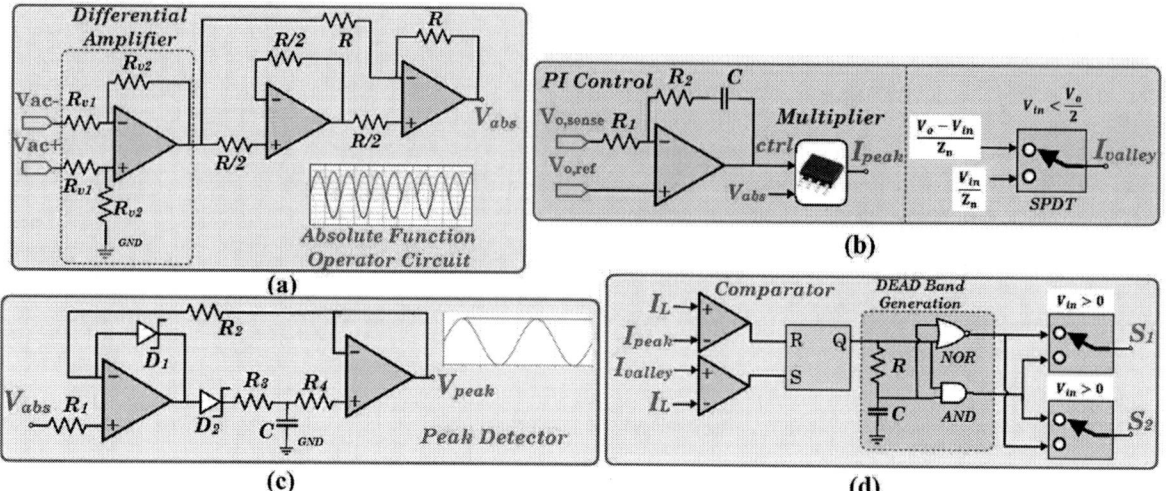

(a)

(b)

(c)

(d)

Fig. 3. Analog implementation of CRM PFC. (a) Input voltage absolute circuit, (b) Peak and valley current generation, (c) Peak detector circuit, (d) SR latch action and deadband generation.

cases. During interval IV the v_{DS} reduces from V_o to zero and the time required for discharging C_{oss2} during this interval is,

$$t_{r2} = \begin{cases} \dfrac{\dfrac{3\pi}{4} - cos^{-1}\left(\dfrac{V_{in}}{\sqrt{2}(V_o - V_{in})}\right)}{\omega} & V_{in} \leq 0.5V_o \\ \dfrac{\pi}{2\omega} & V_{in} \geq 0.5V_o \end{cases} \quad (6)$$

To implement the CRM control using analog circuits, the input AC voltage V_{AC}, inductor current I_L and output DC link voltage V_o are sensed and routed to several subcircuit blocks comprising operational amplifiers, diodes, logic gates, comparators, multiplier etc. which are assigned to perform certain mathematical functions. Fig. 3(a) shows the absolute value calculator circuit block for rectifying the input voltage. The first stage differential amplifier is used to scale down V_{ac} by 1/100. The second and third stage use only positive bias supplies and twice of 2nd stage output is subtracted from the 1st stage output to rectify the sinusoid signal. The output signal goes to the peak detector circuit block in Fig. 3(c) that utilizes BAT17 Schottky diodes to sense the peak sinusoid values. This peak detector output is amplified by $\frac{V_{o,rated}}{V_{AC,rated}}$ to generate the output voltage reference $V_{o,ref}$ required in the output voltage proportional integral (PI) controller in Fig. 3(b). The resistors and capacitors in the analog PI controller op=amp are calculated appropriately to realize the required K_p and K_i constants. Its output is then multiplied with the sensed input voltage using AD633 multiplier to generate the I_{peak} envelope. In order to generate the I_{valley} envelope, the input voltage and output voltage are used as the inputs for differential amplifier and the gain is determined by Z_n which is selected based on the used boost inductor and time-related effective output capacitance of the GaN switch. The AD619 single pole, double-throw (SPDT) switch is used to separate the valley currents when V_{ac} is above or below $0.5V_o$. In Fig. 3(d), the sensed inductor current I_L with 1 MHz bandwidth is then compared with the peak and valley current envelopes using high speed comparator TLV3603. The comparator outputs are then used to set and reset the SR latch

which controls the switch turn on and turn off. Finally, a deadband generator circuit is employed to produce the gate signals for switches S_1 and S_2 utilizing AND and NOR logic gates. The required discharging time t_{r2} for attaining ZVS condition is controlled by the RC time constant, and its optimization is discussed in the subsequent section. Similarly, a deadband circuit is used for line frequency leg switches S_3 and S_4.

III. Optimum Dead Time Generation and Loss Model for Analog CRM PFC Converter

To utilize the soft switching advantage of the CRM PFC converter, it is crucial to allow enough deadtime to discharge the output capacitance voltage. Other dominant loss mechanisms of the CRM TPFC converter include the device conduction loss, turn-off switching loss, inductor core loss, and winding loss.

A. Optimum Deadtime Generation

The selection of deadtime is crucial for attaining the ZVS of the high frequency switches in CRM TPFC. The deadtime required is dependent on the input voltage, output voltage, boost inductor and the device junction capacitance. The ZVS and non-ZVS phenomenon for a range of deadtimes over a quarter line cycle is depicted in Fig. 4. Notably, the deadtime calculation involves an inverse trigonometric function that

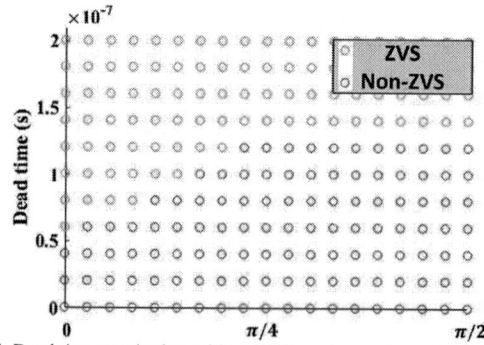

Fig. 4. Dead time required to achieve ZVS over a quarter switching cycle.

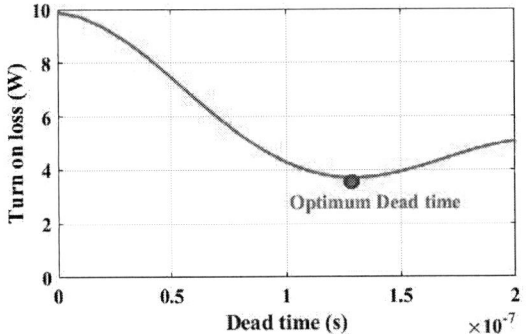

Fig. 5. Optimum dead time for minimizing cycle-averaged turn ON loss..

limits the use of variable deadband in analog control implementation with high accuracy. Hence, the deadtime optimization is performed over the entire line cycle to minimize the cycle averaged CRM turn-on loss which is illustrated in Fig. 5. The instantaneous voltage and current at the switching instants were determined using (3) and (4) for each cycle and the computed switching loss is then averaged over the full line cycle. It can be observed that an optimum deadtime of 127 ns results in the minimum turn-on loss.

B. Loss Modeling

The GaN switches loss include the device turn ON loss, device turn OFF loss, conduction loss, inductor core and winding loss. The switching turn on/off losses are the switching transition losses caused by the device current and voltage overlapping. Since a fixed dead time obtained through minimizing the cycle-averaged turn-on loss is realized through analog implementation, there are certain portions in the line cycle which will observe partial ZVS leading to some amount of device turn-on loss. Mitigating this through full-range ZVS would require adaptive deadtime modulation (as illustrated in Fig. 6) that is a primary limitation of analog control and can be potentially resolved through digital control. The turn ON/OFF losses (P_{on} and P_{off}) can be calculated as,

$$P_{on} = \frac{1}{T_{line}} \int_0^{T_{line}} V_{DS,on} I_{D,on} t_{on} f_s(V_{in}, t) dt \quad (7)$$

$$P_{off} = \frac{1}{T_{line}} \int_0^{T_{line}} V_{DS,off} I_{D,off} t_{off} f_s(V_{in}, t) dt \quad (8)$$

where $f_s(V_{in}, t) = \left(\frac{V_{in,pk}^2}{4LP_o}\right)\left(1 - \frac{V_{in,pk}|sin\omega t|}{V_o}\right)$ is the switching frequency as depicted in Fig. 6, T_{line} is the line cycle period, V_{DS} and I_D are the instantaneous drain to source voltage and drain current respectively which are calculated using (1)–(4). The device conduction loss for the high frequency leg switches and line frequency leg switches in the CRM TPFC are calculated as,

$$P_{c,HF} = \left(\frac{2P_{out}}{\sqrt{3}V_{in}}\sqrt{1 - \frac{8\sqrt{2}V_{in}}{3\pi V_o}}\right)^2 R_{DSon} \quad (9)$$

$$P_{c,LF} = \left(\frac{\sqrt{2}P_{out}}{V_{in,pk}}\right)^2 R_{DSon} \quad (10)$$

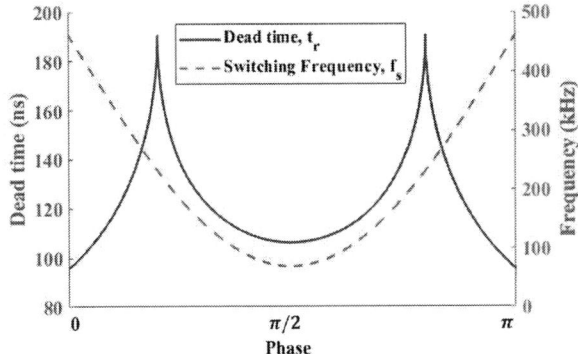

Fig 6. Adaptive dead time and switching frequency variation over the line cycle for 2.5kW CRM TPFC converter operating at V_{in}=240Vac.

The inductor core loss is also an important factor while designing CRM TPFC due to its high inductor current ripple (200% of the average inductor current). A general form of the core loss can be determined from the improved Generalized Steinmetz equation [14] and expressed as follows, which is used in this design for inductor core loss formulation.

$$P_{core} = \frac{1}{T} \int_0^T k_i \left|\frac{dB}{dt}\right|^\alpha (\Delta B)^{\beta-\alpha} dt \quad (11)$$

$$k_i - \frac{k}{(2\pi)^{\alpha-1} \int_0^{2\pi} |cos\theta|^\alpha 2^{\beta-\alpha} d\theta} \quad (12)$$

where α, β and k are Steinmetz coefficients given by the inductor manufacturer datasheet. ΔB is the peak-to-peak magnetic flux density and f_s is the ripple frequency. Due to the higher current ripple in the CRM TPFC compared to CCM PFC converters, higher flux density swing i.e., ΔB will occur.

$$\Delta B = \frac{L\Delta I}{nA} \quad (13)$$

Other losses include the capacitor ESR losses and the copper loss. Based on the developed loss model, the loss breakdown pie chart of the 2.5kW CRM TPFC at full load is shown in Fig. 7. The total loss incurred in the system is 46.6 W. It can be observed that due to the ZVS capability of the analog control CRM TPFC and employment of the optimum dead time loss, the switching turn-on loss.

Power Loss (%) for 2.5kW CRM TPFC

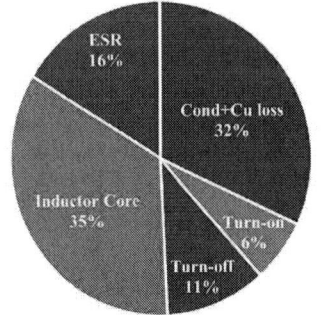

■ Cond+Cu loss ■ Turn-on ■ Turn-off ■ Inductor Core ■ ESR

Fig 7. Loss breakdown of the 2.5 kW CRM totem-pole PFC for V_{in}=240Vac and V_o=400V DC.

IV. Differential Mode EMI Spectrum Quantification for CRM PFC Converter

In order to design the DM EMI filter for the CRM totem-pole PFC converter, the boost inductor current spectrum should be derived. The attenuation required for designing the EMI filter is then determined from DM EMI spectrum. This section presents the analysis of the critical conduction mode inductor current in the frequency domain and discusses the volume optimization of the DM EMI filter.

A. CRM PFC Inductor Current Spectrum Analysis

The CRM PFC converter utilizes the resonance between boost inductor L and MOSFET body capacitor C_{oss} to achieve ZVS. For the CRM totem-pole PFC converter, ZVS can only be realized when the input voltage is lower than half of the output voltage. When the input voltage increases above this bound, ZVS cannot be passively achieved, resulting in partial hard switching loss. To ensure ZVS, additional on-time is kept after the inductor current zero crossing so that the energy stored in the inductor is sufficient to achieve ZVS. To develop the inductor current FFT model, the peak current and valley currents are evaluated over the full line cycle. The minimum current i_{valley} reached by the inductor current by employing the ZVS strategy is given in (5). By assuming that the input voltage remains nearly constant within one switching period and the average inductor current is an average of peak inductor current i_{peak} and valley current i_{valley}, the peak inductor current can be determined as,

$$i_{peak} = \begin{cases} \frac{4P}{V_{pk}}\sin(\omega t) + \frac{V_o - V_{in}}{Z_n} & V_{in} < 0.5V_o \\ \frac{4P}{V_{pk}}\sin(\omega t) + \frac{V_{in}}{Z_n} & V_{in} > 0.5V_o \end{cases} \quad (14)$$

From (14) the conduction on-time of S_2 during the positive inductor current is expressed as follows,

$$t_{on} = \begin{cases} \frac{4LP}{V_{pk}^2} + \sqrt{2LC_{oss}}\left(\frac{V_o - V_{in}}{V_{in}}\right) & V_{in} < 0.5V_o \\ \frac{4LP}{V_{pk}^2} + \sqrt{2LC_{oss}} & V_{in} > 0.5V_o \end{cases} \quad (15)$$

where P is the output power and V_{pk} is the peak of the input voltage. The peak current and valley current envelope of the inductor over a line cycle is shown in Fig. 8. To calculate the critical mode inductor current spectrum in the frequency domain, the first step is to represent mathematically the high-frequency inductor current ripple of Fig. 9. The inductor current is bounded by an envelope developed from i_{peak} and i_{valley} throughout the line cycle. During the kth switching cycle, the

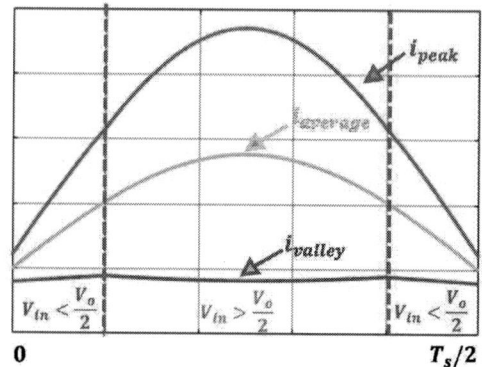

Fig. 8. Inductor current envelope over the line cycle.

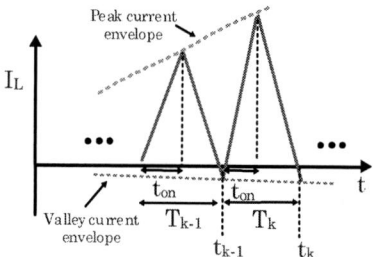

Fig. 9. Switching cycle waveforms of the inductor current.

triangular inductor current $f_+(t)$ and $f_-(t)$ during time interval $t_{on}(k)$ and $t_{off}(k)$ respectively can be expressed as follows,

$$f_+(t) = \frac{i_{pk}(k) - i_{valley}(k-1)}{t_{on}(k)}t - \frac{t(k)}{t_{on}(k)}[i_{pk}(k) - i_{valley}(k-1)] \quad (16)$$

$$f_-(t) = \frac{i_{valley}(k) - i_{pk}(k)}{t_{off}(k)}t - \frac{t(k+1)}{t_{off}(k)}[i_{valley}(k) - i_{pk}(k)] + i_{valley}(k) \quad (17)$$

The inductor current can be represented by the sum of triangular current waveform f(t) at each switching cycle. The Fourier coefficients can be determined from the input inductor current from which the FFT of the input inductor current is developed. Since the inductor current above is half-wave odd symmetric, only the odd harmonic components are present. From the computed Fourier coefficients a_n and b_n as shown in (18) and (19), where n is the harmonic order and T is the line period, the input current spectrum of the CRM PFC converter is obtained. To verify the current spectrum analysis, Fig. 11 illustrates the current spectrum developed from the mathematical model along with the simulation model. The amplitudes of the current harmonics closely match their calculated values.

$$a_n = \frac{4}{T}\left[\frac{A}{t_{on}}\left(\frac{(t_{k-1}\sin(n\omega_o t_{k-1}) - (t_{k-1}+t_{on})\sin(n\omega_o(t_{k-1}+t_{on})))}{n\omega_o}\right) + \frac{(\cos(n\omega_o(t_{k-1}+t_{on})) - \cos(n\omega_o t_{k-1})}{n^2\omega_o^2} + \right.$$
$$\left. \frac{A}{t_{off}}\left(\frac{((t_{k-1}+t_{on})\sin(n\omega_o(t_{k-1}+t_{on})) - t_k\sin(n\omega_o t_k)}{n\omega_o}\right) + \frac{(\cos(n\omega_o(t_k)) - \cos(n\omega_o(t_{k-1}+t_{on})))}{n^2\omega_o^2}\right) + \frac{A}{t_{off}}t_k\left(\frac{\sin(n\omega_o t_k) - \sin(n\omega_o(t_{k-1}+t_{on}))}{n\omega_o}\right)\right] \quad (18)$$

$$b_n = \frac{4}{T}\left[\frac{A}{t_{on}}\left(\frac{(t_{k-1}\cos(n\omega_o t_{k-1}) - (t_{k-1}+t_{on})\cos(n\omega_o(t_{k-1}+t_{on})))}{n\omega_o}\right) + \frac{(\sin(n\omega_o(t_{k-1}+t_{on})) - \sin(n\omega_o t_{k-1})}{n^2\omega_o^2} - \right.$$
$$\left. \frac{A}{t_{off}}\left(\frac{((t_{k-1}+t_{on})\cos(n\omega_o(t_{k-1}+t_{on})) - t_k\cos(n\omega_o t_k)}{n\omega_o}\right) + \frac{(\sin(n\omega_o(t_k)) - \sin(n\omega_o(t_{k-1}+t_{on})))}{n^2\omega_o^2}\right) + \frac{A}{t_{off}}t_k\left(\frac{\cos(n\omega_o(t_{k-1}+t_{on})) - \cos(n\omega_o t_k)}{n\omega_o}\right)\right] \quad (19)$$

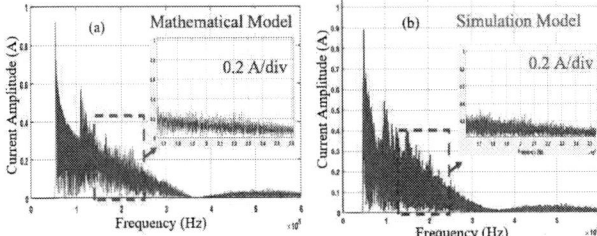

Fig. 10. Inductor current amplitude spectrum for CRM TPFC. (a) Mathematical model, (b) Simulation model.

B. Design and Volume Optimization of the DM EMI Filter

In this section the design of the DM EMI filter with respect to volume is evaluated considering the DM noise level complies with the EMI standards. The DM filter design for the CRM TPFC converter is based upon the selection of LC product to achieve the required attenuation level at a certain design frequency. However, due to the variable switching frequency operation the selection of design frequency is not as straightforward as a constant frequency totem-pole PFC converter [15]. The DM EMI spectrum for V_{in}=240Vac, and L_{boost}=56µH at 2.5kW based on the mathematical model developed in the previous section is shown in Fig. 10. It can be observed that the noise peaks are spread out over the frequency spectrum and selecting the highest noise peak will not guarantee the EMI standard compliance for lower frequencies. This is because the attenuation provided by the LC filter at lower frequencies gets smaller compared to the design frequency f_D (frequency corresponding to the highest noise peak). As a result, the selection of design frequency for the CRM TPFC is reiterated until all the noise peaks are attenuated below the EMI standard. The DM filter corner frequency is determined based on the design frequency f_D and the attenuation requirement and can be expressed as,

$$f_{DM} = \frac{f_D}{10^{\frac{Att_{DM}(f_D)[dB]}{40}}} \qquad (20)$$

Moreover, the attenuation requirement can be fulfilled by numerous other possible combinations of L_{DM} and C_{DM} having the same LC product. This section discusses the selection of optimum L_{DM} and C_{DM} values to minimize the DM filter volume as DM EMI filter contributes a major part of the volume of the CRM TPFC converter due to its large inductor current ripple. For determining the total volume of the two-stage DM EMI filter, the individual DM inductor and capacitor volume needs to be estimated [16]. The filter capacitor volumetric cost function is developed using a regression model that utilize the rated voltage V, capacitance C and stored energy CV^2. The capacitor volume can be approximated as,

$$V_c = K_c C + K'_c V + K''_c CV^2 \qquad (21)$$

Similarly, the volume estimation of the inductors based on toroidal cores is also modeled. The filter inductor volume is proportional the stored inductive energy, $V_L \propto K_L LI^2$. Where K_L is the proportionality between the stored energy and the inductor volume. This factor can be calculated from manufacturer's data for different inductor core dimensions, inductance values, and

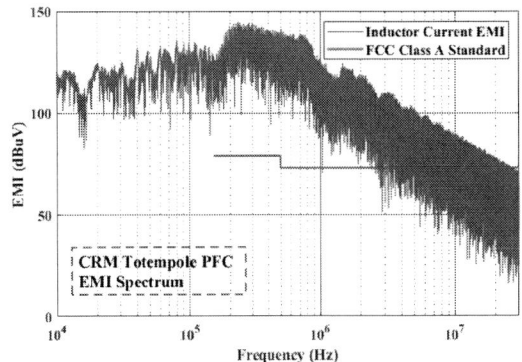

Fig. 11. Mathematical model of the Differential mode (DM) EMI spectrum for CRM TPFC converter without filter.

current ratings. Thus, the filter inductor volume can be approximated as,

$$V_L = K_L L + K'_L I + K''_L LI^2 \qquad (22)$$

The coefficients for the inductor and capacitor volume after employing the regression model are shown in Table I.

TABLE I. COEFFICIENTS FOR PASSIVE VOLUME ESTIMATION

Coefficients for Inductor Volume		Coefficients for Capacitor Volume	
K_L	$0.91 cm^3/mH$	K_c	$2.18 cm^3/\mu F$
K'_L	$0.38 cm^3/A$	K'_c	3e-3 cm^3/V
K''_L	$0.237 cm^3/(mH.A^2)$	K''_c	4.6e-5 $cm^3/(\mu F.V^2)$

The optimum DM EMI filter utilizing the filter inductor and capacitor coefficients can be obtained by minimizing the total volume of the DM filter expressed as,

$$V_{tot} = V_{boost} + V_{LDM} + V_{CDM1} + V_{CDM2} \qquad (23)$$

where V_{LDM} and V_{CDM} are the volumes of each L_{DM1} and each C_{DM1}, respectively, which can be expressed similarly as done in (21) and (22). The boost inductor L_{boost} selection is not selected as part of the volume optimization since it dictates the ZVS criteria of the converter. To attain minimum filter volume, V_{tot} is to be minimized with a constraint of the required level of attenuation as shown,

$$Att_{DM}(f_D) = (2\pi f_D)^4 (L_{boost} C_{DM1} L_{DM} C_{DM2}) \geq Att_{req} \qquad (24)$$

For minimizing (23) while satisfying the constraints given by (24), Lagrange multiplier method was employed to find the optimized volume of the filter inductor and capacitor and can be formulated as,

$$L_{DM} = \sqrt[3]{\frac{Att_{req}}{(2\pi f_D)^4 L_{boost}} \left(\frac{K_c + K''_c V^2}{K_L + K''_L I^2}\right)^2} \qquad (25)$$

$$C_{DM} = \sqrt[3]{\frac{2 Att_{req}}{(2\pi f_D)^4 L_{boost}} \left(\frac{K_L + K''_L I^2}{K_c + K''_c V^2}\right)^2} \qquad (26)$$

This shows that the optimized values of the DM filter stage components depend on the required amount of attenuation (Att_{req}), the rated filter capacitor voltage, and the rated current of the designed filter inductor. Finally, Fig. 12 shows a flowchart demonstrating the steps to design the optimal filter volume based on the analytical EMI noise estimation approach. The attenuation requirement of 63 dB and the design frequency of 210 kHz are selected based on the current spectrum developed

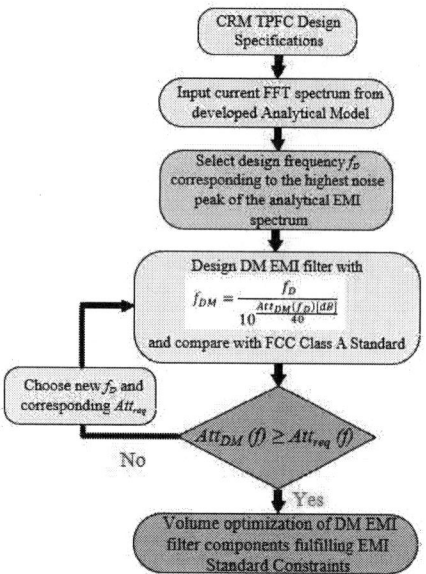

Fig. 12. DM EMI filter design flowchart for CRM TPFC.

from analytical model in Fig. 11. By using the volume optimized DM filter equations, the filter inductor and capacitor are selected. The resulting two-stage DM filter corner frequencies are located at 29.2 kHz and 8.5 kHz. The passive parameters for the CCM and CRM totem-pole PFC converters are depicted in Table II. Due to the variable frequency operation and large peak-to-peak ripple current of the CRM TPFC, its overall passive component volume is larger compared to the CCM TPFC.

TABLE II. PASSIVES VOLUME COMPARISON FOR CCM VS. CRM PFC

Topology	L_{DM} (μH)	$C_{DM1,2}$ (μF)	L_{boost} (μH)	f_s(kHz)	Total AC-side Passive volume (cm³)
Totem-pole CCM PFC	79	0.27	512	160	53.3
CRM TPFC	650.0	0.53	56	70-460	65.6

V. HARDWARE PROTOTYPING AND EXPERIMENTAL RESULTS

To verify the single-phase CRM totem-pole PFC converter operation and ZVS control, a 2.5 kW hardware prototype with an integrated EMI filter (see Fig. 13(a)) is developed and tested. The hardware has a modular architecture consisting of a daughter card for the full-bridge converter and control board for the analog CRM control implementation. The prototype is designed for an input voltage rating of V_{ac}=240Vac at 60Hz and output voltage regulated at V_o=400Vdc. The 650V GaN device GS66516T from GaN systems is used for the full bridge implementation. Fig. 13(b) shows the control board underneath the main power board, that carries out the CRM ZVS control with analog circuits and generate the gate signals for the switches $S_1 - S_4$. Fig. 14 illustrates the key steady state waveforms of the of the CRM TPFC converter for an input voltage of 240Vac rms and 2.5 kW output power. The input current I_{in} of 10.46A rms and with a THD of 4.77% is found to be in phase with the input voltage. Subsequently, an output

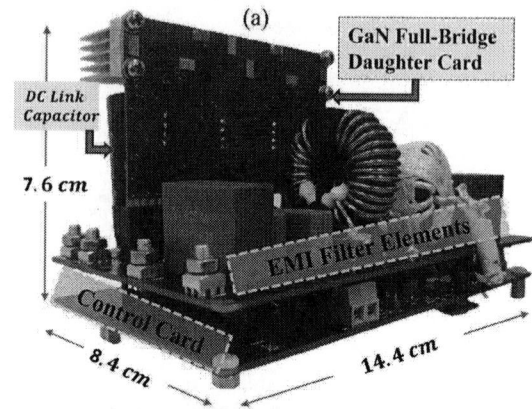

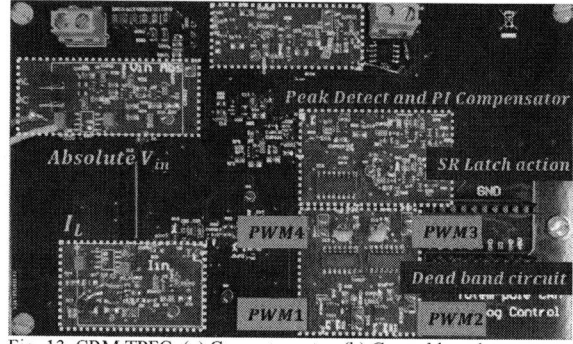

Fig. 13. CRM TPFC. (a) Converter setup (b) Control board.

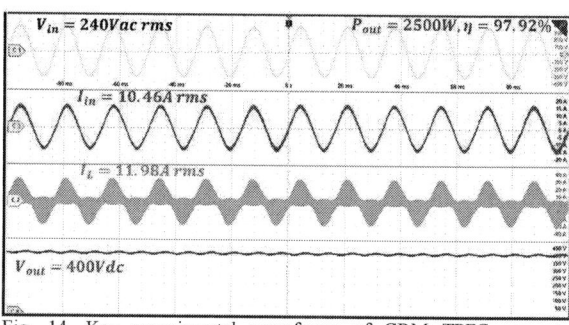

Fig. 14. Key experimental waveforms of CRM TPFC converter operating at 2.5 kW.

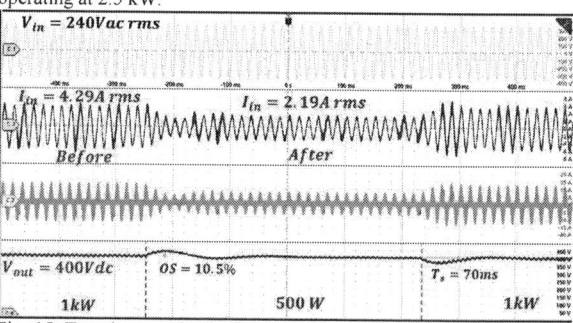

Fig. 15. Transient results during load change from 1 kW to 500 W and again back to 1 kW.

voltage PI controller was designed for $K_p = 1$ and $K_i = 100$. The dynamic response of the output voltage for a load transition from 1 kW to 500W and again back to 1 kW is shown in Fig. 15. The designed compensator regulates the V_o at 400V with an overshoot of 10.5%. and settling time of 70ms. Fig. 16 shows

979-8-3315-1612-3/25 $31.00 © 2025 IEEE

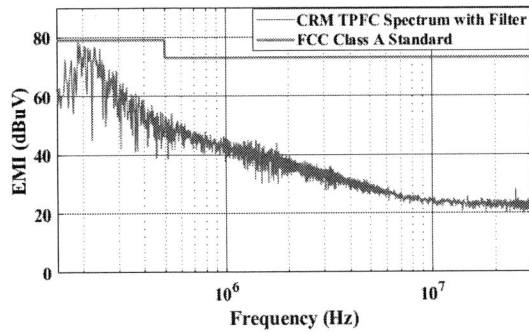

Fig. 16. Experimental DM EMI noise spectrum of CRM TPFC converter operating at 2.5kW for input voltage V_{AC}=240Vac rms.

the experimental DM EMI noise spectrum for $V_{in} = 240Vac$ and 2.5 kW operation ranging between 150 kHz and 30 MHz frequency. The compliance of the FCC Class A standard for the CRM TPFC converter demonstrates the usefulness of the DM EMI filter design methodology proposed in this paper. From the thermal perspective of the converter operating at 2.5kW for 240Vac(rms) the maximum temperature attains 72.4°C without use of any external fans where the ambient temperature is 23°C.

VI. CONCLUSIONS

In this paper, the analog control of a single-phase CRM TPFC rated at 2.5 kW power is investigated. By utilizing the state trajectory analysis over a switching cycle, the ZVS conditions are derived over a full line cycle. The control architecture for the analog circuit implementation is divided into several subcircuit blocks and their functions are discussed. An optimization routine was performed to select the dead time for the high frequency leg switches to minimize the turn-on losses over a full line cycle. The design methodology of DM EMI filter for the CRM TPFC converter is proposed which undergoes variable switching frequency operation. After selection of the DM filter corner frequency, the filter inductor and capacitor volume optimizations are performed which was relatively higher compared to CCM totem-pole PFC converter due to its larger current ripple. A 2.5 kW experimental prototype is developed to verify the analog CRM TPFC control action which attains 97.92% efficiency at rated condition and peak efficiency of 98.12% at 1870W while demonstrating 500W to 1kW both-way load transient.

ACKNOWLEDGEMENT

This material is based upon work supported by the National Science Foundation under Grant Number 2236846.

REFERENCES

[1] L. Huber, Y. Jang, M. M. Jovanovic, "Performance evaluation of bridgeless PFC boost rectifiers," IEEE Trans. Power Electron., vol. 23, no. 3, pp. 1381–1390, May 2008.

[2] B. Su and Z. Lu, "An interleaved totem-pole boost bridgeless rectifier with reduced reverse-recovery problems for power factor correction," IEEE TPEL., vol. 25, no. 6, pp. 1406–1415, Jun. 2010.

[3] Q. Huang and A. Q. Huang, "Review of GaN totem-pole bridgeless PFC," in CPSS Transactions on Power Electronics and Applications, vol. 2, no. 3, pp. 187-196, Sept. 2017.

[4] E. A. Jones, F. F. Wang and D. Costinett, "Review of commercial GaN power devices and GaN converter design challenges," in IEEE Journal of Emerging and Selected Topics in Power Electronics, vol. 4, no. 3, pp.707-719, Sept. 2016.

[5] Z. Liu, F. C. Lee, Q. Li and Y. Yang, "Design of GaN-based MHz totempole PFC rectifier," in IEEE Journal of Emerging and Selected Topics in Power Electronics., vol. 4, no. 3, pp. 799-807, Sept. 2016.

[6] C. Zhao, J. Zhang and X. Wu, "An Improved Variable On-Time Control Strategy for a CRM Flyback PFC Converter," in IEEE Transactions on Power Electronics, vol. 32, no. 2, pp. 915-919, Feb. 2017.

[7] J. -W. Kim, H. -S. Youn and G. -W. Moon, "A Digitally Controlled Critical Mode Boost Power Factor Corrector With Optimized Additional On Time and Reduced Circulating Losses," in IEEE Transactions on Power Electronics, vol. 30, no. 6, pp. 3447-3456, June 2015.

[8] J. Sun, N. N. Strain, D. J. Costinett and L. M. Tolbert, "Analysis of a GaN-Based CRM Totem-Pole PFC Converter Considering Current Sensing Delay," in 2019 IEEE Energy Conversion Congress and Exposition (ECCE), Baltimore, MD, USA, 2019.

[9] K. M. Golmakani and H. Heydari, "*Volume-Optimized DM-EMI Filter Design for a TM-PFC Converter Based on the Worst Condition*," 2022 12th Smart Grid Conference (SGC), Kerman, Iran, Islamic Republic of, 2022, pp. 1-5, doi: 10.1109/SGC58052.2022.9998900.

[10] CISPR 16: 1993 "Specification for radio disturbance and immunity measuring apparatus and methods".

[11] S. Brehaut, "Analysis EMI of a PFC on the band-pass 150 kHz–30 MHz for a reduction of the electromagnetic pollution," in Proc. Appl. Power. Electr. Conf., 2004, pp. 695–700.

[12] L. Rossetto, S. Buso and G. Spiazzi, "Conducted EMI issues in a boost PFC design," *INTELEC - Twentieth International Telecommunications Energy Conference (Cat. No.98CH36263)*, San Francisco, CA, USA, 1998, pp. 188-195, doi: 10.1109/INTLEC.1998.793497.

[13] C. Reece, N. Ishraq and A. Mallik, "Development of a Half-Bridge Cell-Derived All-Inclusive Conducted Emission Noise Model for SiC-based Single Phase Boost PFC Converter," in *IEEE Transactions on Power Electronics*, doi: 10.1109/TPEL.2024.3485035.

[14] K. Venkatachalam, C. R. Sullivan, T. Abdallah and H. Tacca, "Accurate prediction of ferrite core loss with nonsinusoidal waveforms using only Steinmetz parameters," *2002 IEEE Workshop on Computers in Power Electronics, 2002. Proceedings.*, Mayaguez, PR, USA, 2002, pp. 36-41, doi: 10.1109/CIPE.2002.1196712.

[15] P. Rathod, N. Ishraq, A. Chandwani and A. Mallik, "Input Current FFT Model-derived Comprehensive Comparison of Totem-pole PFC and H-Bridge PFC Converter DM EMI Performances," *2023 7th International (CERA)*, Roorkee, India, 2023, pp. 1-6, doi: 10.1109/CERA59325.2023.10455370.

[16] N. Ishraq and A. Mallik, "Design of a 2.5 kW Four-Level Interleaved Flying Capacitor Multilevel Totem-Pole PFC Converter With AC-Side Passive Volume Optimization," in *IEEE Open Journal of Power Electronics*, vol. 5, pp. 214-231, 2024, doi: 10.1109/OJPEL.2024.3359479.

An iTHD and Efficiency Optimized Control Method for Triangular Conduction Mode Totem Pole Bridgeless PFC with Zero Current Detection

Brent McDonald
Texas Instruments
Dallas, TX USA
b-mcdonald@ti.com

Sheng-yang Yu
Texas Instruments
Dallas, TX USA
seanyu@ti.com

Abstract— **A power factor correction (PFC) topology and new control methodology are introduced for high-performance cost-effective PFC. The topology enables the use of cost-effective component materials and zero voltage switching (ZVS) operation over the entire line cycle. In addition, a new zero current detector circuit (ZCD) and algorithm are employed to enhance ZVS and total harmonic distortion (THD) while maintaining cycle-by-cycle control of the current. Proof of concept is demonstrated in a 5 kW prototype with a power density of 120 W/in³, an efficiency of 99.2% and a THD < 5%.**

Keywords—Power factor, PFC, THD, zero voltage switching, ZVS, zero current detection, ZCD

I. INTRODUCTION

The need for cost effective solutions to improve PFC light load and peak efficiency while shrinking passive components is becoming difficult with conventional continuous conduction mode (CCM) control. Complex multi-mode solutions are driving significant research to address these concerns ([1], [2]). Reference [3] discusses a method for improving power density with the use of coupled inductors and a new control method to achieve ZVS, however, the efficiency is below 99 %. In addition, [3] suffers from high conduction loss and high turn-off loss under light load conditions leading to reduced efficiency. The approach in this paper is able to maintain high efficiency down to lighter loads by virtue of the variable frequency control algorithm which reduces the ripple current at light load. Reference [4] discusses a method to maintain ZVS that depends entirely on the accuracy of the calculations, which introduces robustness concerns. This paper will address the efficiency non-optimization and robustness concern of the calculations through the use of a zero-current detector (ZCD). Reference [5] discusses a method for improving THD by using a variable slope ramp for computing the controller on-time with input voltage. The THD achieved in [5] is approximately 2 times higher than the method used in this paper. Reference [5] also does not achieve ZVS at all line voltage conditions making it less desirable for high efficiency applications. Reference [6] discusses a closed loop interleaving strategy for 2 phase critical conduction mode (CrM) PFC. The digital controller discussed in this paper offers a simpler solution than that presented in [6]. Reference [7] proposes a look up table approach that can require significant memory resources. Reference [8] addressed these concerns by introducing a control scheme based on a novel concept of using a zero-voltage detector (ZVD) to enable the

exact solution to the system differential equations. The result was both high efficiency and low THD.

Concerns about the lack of cycle by cycle current control raise questions about how robustly the control can address significant system disturbances. The challenge addressed by this paper is to demonstrate how the high-performance levels achieved in [8] can be maintained while simultaneously controlling the current on a cycle by cycle basis. The paper will present an exact solution to the iTHD and ZVS problem with an integrated zero-current detector to facilitate the required cycle-by-cycle control. The result is an interleaved solution with an efficiency of 99.2 %, an iTHD below 4 % and the added benefit of cycle by cycle current control.

II. PFC TOPOLOGY

Fig. 1 shows the schematic for a 2-phase totem pole (TTPL) bridgeless PFC. Triangular conduction mode (TCM) is utilized due to its ability to achieve ZVS in every switching cycle for all load and line conditions. Fig. 2 shows an alternative topology dubbed integrated triangular conduction mode (iTCM) [9]. This topology has the benefit of preventing the high frequency ripple current from flowing through the input source. The topology also offers the ability to use lower cost materials for the inductor L_g by splitting the high frequency and low frequency ripple current content into two different devices (L_g and L_b). Since L_b carries all the high frequency current a potentially lower cost structure can be used for L_g. The control algorithm discussed in the next section will work equally well for both iTCM and TCM topologies.

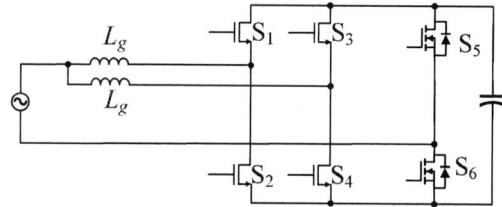

Fig. 1. 2-Phase TTPL bridgeless TCM PFC Topology

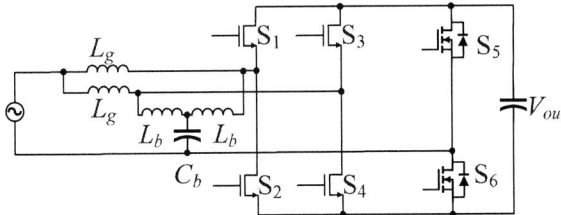

Fig. 2. 2-Phase TTPL bridgeless iTCM PFC Topology

III. CONTROL

The FET timing parameters are defined in Table I.

TABLE I: TIMING PARAMETER DEFINITIONS

Timing Parameter	Normalized Parameter	Definition
t_{cf}	θ_1	On-time of the control FET. This is the FET that defines the duty cycle for the converter. During the positive ½ cycle this is the low side FET. During the negative ½ cycle this is the high side FET.
t_{rp}	θ_2	Dead time between the control FET turnoff and synchronous rectifier turnon
t_{sr}	θ_3	On-time of the SR. During the positive ½ cycle this is the high side FET. During the negative ½ cycle this is the low side FET.
$t_{sr,ext}$	$\theta_{3,ext}$	On-time of the SR when the inductor current is negative during the positive ½ cycle. On-time of the SR when the inductor current is positive during the negative ½ cycle.
t_{rv}	θ_4	Dead time between the synchronous rectifier turnoff and the control FET turnon

[8] uses the state plane to derive an exact solution to the differential equations governing the ZVS transition and the average inductor current through L_g shown in Fig. 1. The solution is made possible by setting the dead time between the synchronous rectifier (SR) turn off and the control field effect transistor (CF) turn on event exactly equal to ¼ of a resonant period $\left(\frac{1}{2 \cdot \pi \cdot \sqrt{2 \cdot L_g \cdot C_{oss}}}\right)$ or $\frac{\pi}{2}$. This solution will only work if it is coupled with an additional feedback element. In the case of [8] one such feedback element is the ZVS switching status of the CF dubbed the ZVD signal. Because the ZVD signal arrives after the switch event has occurred, it can't be used to control the FETs and current on a cycle by cycle basis.

The approach outlined in this paper will utilize a ZCD signal from the SR to achieve similar performance with added advantage of cycle by cycle control.

Fig. 3 and Fig. 4 illustrates the control mechanism during the positive ½ cycle. The SR, the high side FET, turns on after the dead time following the turn off of the CF (the low side FET). The SR stays on until it encounters the high side ZCD signal. After the control sees the ZCD signal it holds the SR on for an additional duration of time equal to $t_{sr,ext}$.

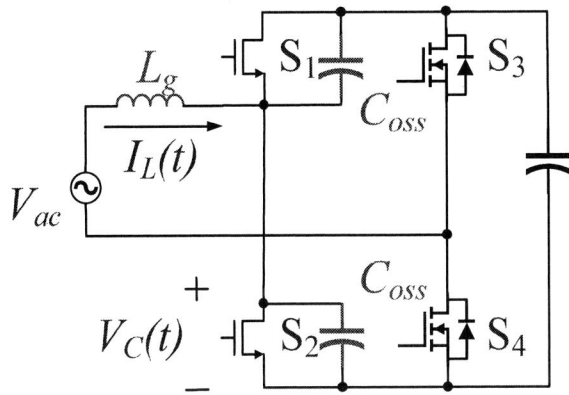

Fig. 3. ZCD Timing waveform schematic

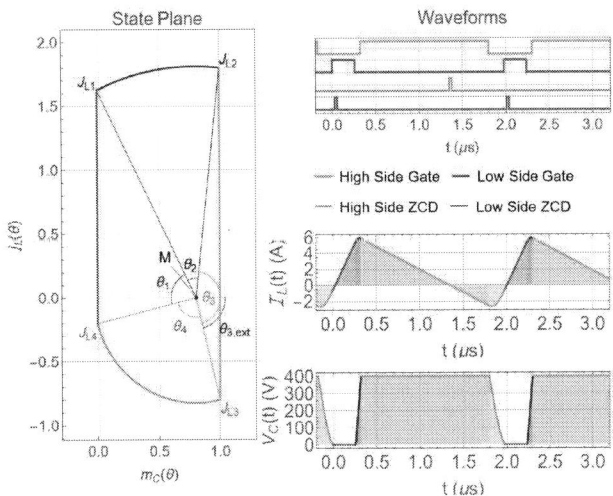

Fig. 4. ZCD Timing waveforms & State Plane

The normalized version of $t_{sr,ext}$ can be solved for explicitly as $\theta_{3,ext}$ by recognizing that the transition from J_{L2} to J_{L3} occurs via a constant voltage with a normalized value of 1. This means that the normalized current $j_l(t)$ will ramp according to equation (1).

$$j_l(t) = \theta_{3,ext} \cdot m_c(t) \tag{1}$$

Solving equation (1) for $\theta_{3,ext} = \frac{J_{L3}}{1-M}$. Fig. 5 shows the equations for the complete system solution and how to calculate the normalized values. It's worth emphasizing that this is an exact solution to the differential equations and enables both high efficiency and low iTHD. Details on the derivation of these equations can be found in [8].

979-8-3315-1612-3/25 $31.00 © 2025 IEEE

System Equations	Normalization
$J_{L1} = \sqrt{(1-M)^2 + \dfrac{4(1-M)M\pi J_{LR}}{F}}$	$V_{base} = V_{out}$
$J_{L2} = \sqrt{-1 + 2M + J_{L1}^2}$	$I_{base} = \dfrac{V_{base}}{Z_0}$
$J_{L3} = M$	$Z_0 = \sqrt{\dfrac{L}{C}}$
$J_{L4} = 1 - M$	$\omega_0 = \dfrac{1}{\sqrt{LC}}$
$\theta_1 = \dfrac{J_{L1}+J_{L4}}{M}$	$\theta = \omega_0 t$
$\theta_2 = \tan^{-1}\left(\dfrac{M}{J_{L1}}\right) + \tan^{-1}\left(\dfrac{1-M}{J_{L2}}\right)$	$F = \dfrac{2\pi f_s}{\omega_0}$
$\theta_{3,ext} = \dfrac{J_{L3}}{1-M}$	$M = \dfrac{V_{in}}{V_{base}}$
$\theta_4 = \dfrac{\pi}{2}$	$m_c(t) = \dfrac{V_C(t)}{V_{base}}$
	$j_L(t) = \dfrac{I_L(t)}{I_{base}}$

Fig. 5. ZCD solution and normalization

In practice there is a delay from the time the inductor current crosses 0 A to the time that the SR is turned off. This delay (t_d) will result in increased iTHD if not accounted for. In order to minimize the impact of t_d on the iTHD, the FET timing parameters are adjusted according to equations (2) and (3).

$$t_{sr,ext} = t_{sr,ext,calc} - t_d \qquad (2)$$

$$t_{cf} = \begin{cases} t_{cf,calc} - t_{sr,ext,calc} \cdot \dfrac{1-M}{M} & t_{sr,ext,calc} < 0 \\ t_{cf,calc} & otherwise \end{cases} \qquad (3)$$

In addition a current loop is added to ensure any additional steady state errors are eliminated due to delay variations not accounted for in (2) and (3). This adjustment is illustrated in Fig. 6.

As discussed in [8], it is also necessary to estimate the frequency of operation in order for the algorithm to work fast and precisely. [8] introduces a feedforward term to accomplish this. In the algorithm of this paper, the timing period of the ZCD signal is sufficiently accurate estimate of the frequency to achieve the ZVS behavior and low iTHD required by the end equipment. This timing measurement simplifies the math needed to be done by the microcontroller in [8]. The frequency adjustment is also illustrated in Fig. 6.

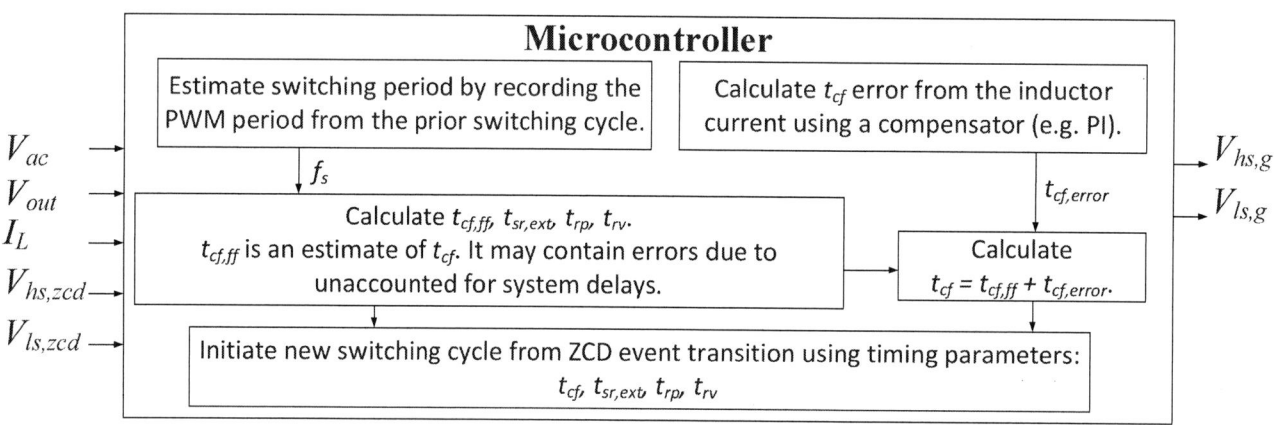

Fig. 6. Control algorithm

IV. RESULTS

Fig. 7 shows the prototype – PMP23475: Variable Frequency, ZVS, GaN-Based 5-kW Two-Phase Totem-Pole PFC Reference Design with Zero Current Detection – used to demonstrate the performance of the new control method. Fig. 8 and Fig. 9 illustrates the cycle by cycle control of the new algorithm under a DC input. The design utilizes GaN switches with integrated ZCD functionality, LMG3527R030. The integration offers a high-performance cost-effective alternative to discrete circuitry. The figure shows the measured timing waveforms for the system with $V_{in,dc} = 330$ V, $V_{out} = 400$ V and $I_{out} = 2$ A. The plot shows the switch node voltage between the source of the high side FET and the drain of the low side FET. This waveform shows resonant transitions that are commensurate with what is expected with ZVS. The figure also shows the output of the ZCD detector asserting itself a small delay time after the measured zero crossing of the current. The SR is also shown held on for an additional length of time ($t_{sr,ext}$) after the zero crossing.

Fig. 7. Prototype converter

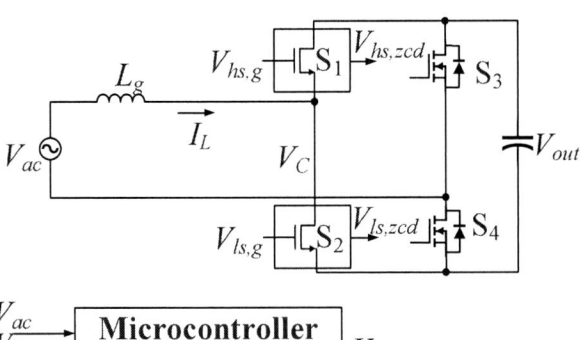

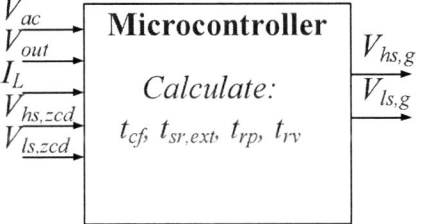

Fig. 8. Switch timing block diagram

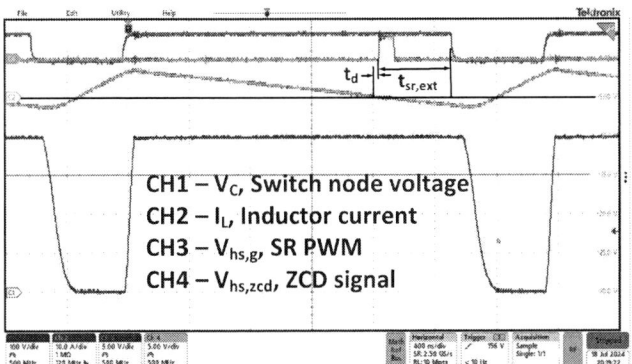

CH1 – V_C, Switch node voltage
CH2 – I_L, Inductor current
CH3 – V_hs,g, SR PWM
CH4 – V_hs,zcd, ZCD signal

Fig. 9. Switch timing waveforms

Fig. 10 shows measured 2 phase operation under AC input when $V_{in} = \frac{1}{2} V_{out}$. The top set of plots show the inductor currents of both phases and the bottom plots show the switch node voltages. Operation is occurring during the negative half cycle. Fig. 11 and Fig. 12 show 2 phase operation when V_{in} is

above ½ V_{out} and below ½ V_{out}. In all cases ZVS is overserved and the phases are operation with 180 ° of phase shift.

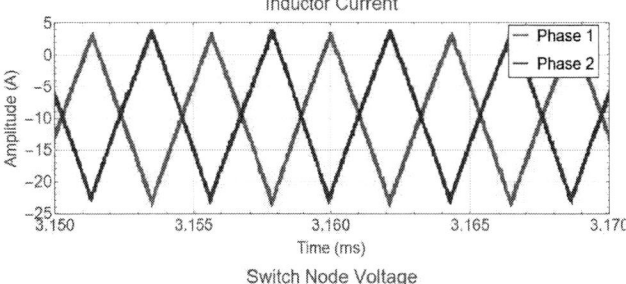

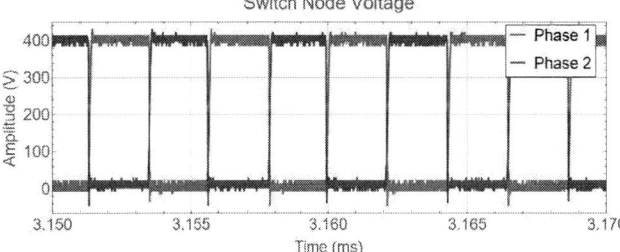

Fig. 10. 2 phase operation, $V_{in} = V_{out}/2$, $V_{in} = 230$ V$_{rms}$, $V_{out} = 400$ V, Power = 5 kW

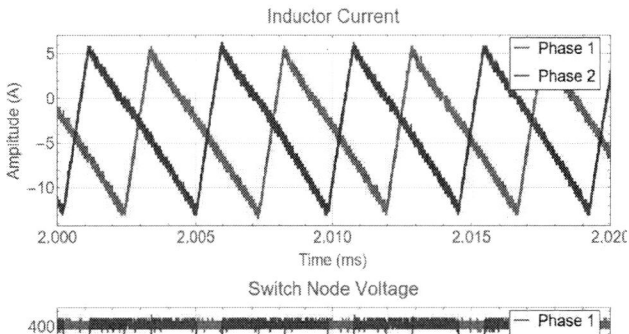

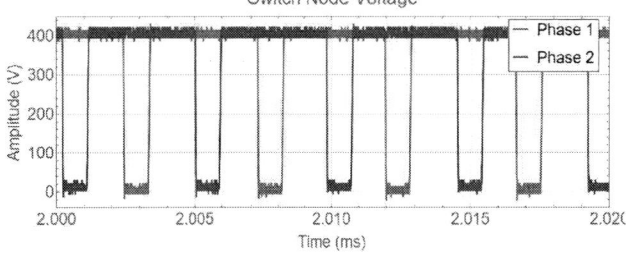

Fig. 11. 2 phase operation, $V_{in} > V_{out}/2$, $V_{in} = 230$ V$_{rms}$, $V_{out} = 400$ V, Power = 5 kW

979-8-3315-1612-3/25 $31.00 © 2025 IEEE

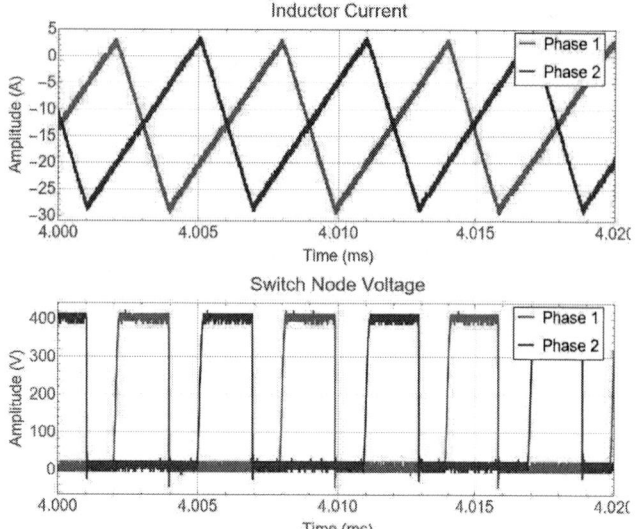

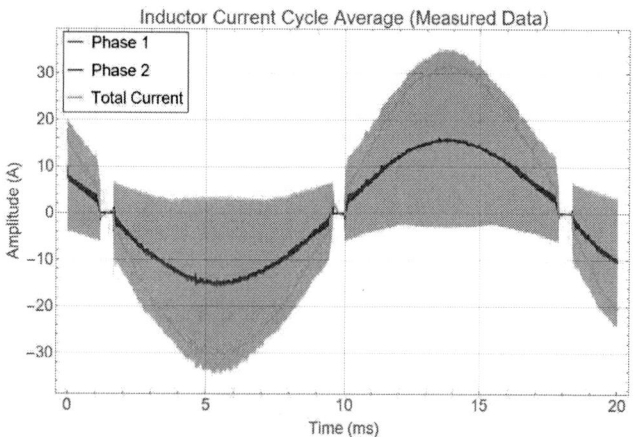

Fig. 12. 2 phase operation, $V_{in} < V_{out}/2$, $V_{in} = 230$ V$_{rms}$, $V_{out} = 400$ V, Power = 5 kW

Fig. 13 shows the ripple current envelopes of the inductor current along with the average of those currents. The total average current is also included in the figure. The system is running at full power with 2 phases.

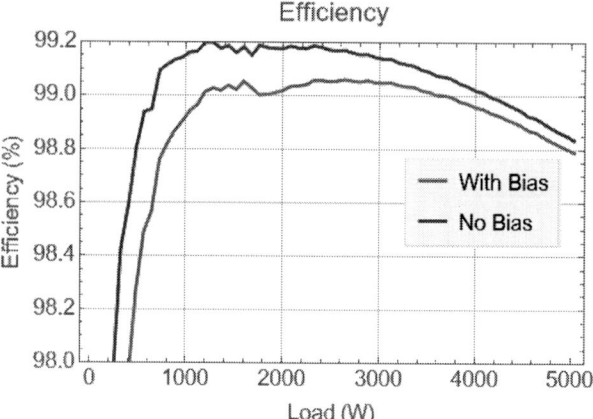

Fig. 14. Efficiency with phase shedding

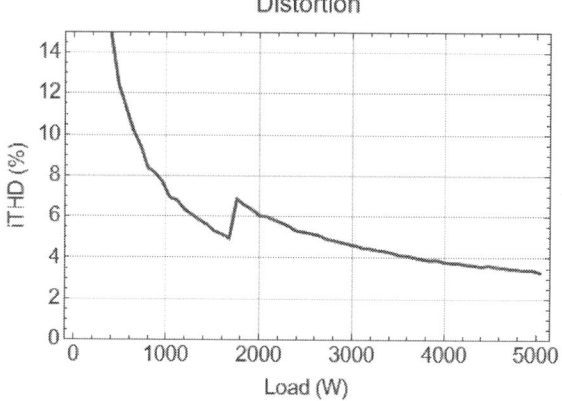

Fig. 15. iTHD with phase shedding

V. CONCLUSIONS

This paper presented a new algorithm that utilizes an exact solution to the differential equations governing the ZVS behavior and iTHD. The solution achieves 99.2% efficiency and less than 4% iTHD while maintaining cycle by cycle control over the current waveform.

Fig. 13. Ripple current envelope and cycle averages, $V_{in} = 230$ V$_{rms}$, $V_{out} = 400$ V, Power = 5 kW

Fig. 14 shows the measured efficiency with and without bias power. Bias power includes energy delivered to the gate drivers and control circuitry. Fig. 15 shows the iTHD of the new algorithm. The peak efficiency measured is 99.2 % and the full load iTHD is less than 4%. Phase shedding/adding occurs around 1.8 kW.

REFERENCES

[1] R. Fernandes and O. Trescases, "A Multimode 1-MHz PFC Front End With Digital Peak Current Modulation," in IEEE Transactions on Power Electronics, vol. 31, no. 8, pp. 5694-5708, Aug. 2016, doi: 10.1109/TPEL.2015.2499194.

[2] S. F. Lim and A. M. Khambadkone, "A multimode digital control scheme for boost PFC with higher efficiency and power factor at light load," 2012 Twenty-Seventh Annual IEEE Applied Power Electronics Conference and Exposition (APEC), 2012, pp. 291-298, doi: 10.1109/APEC.2012.6165833.

[3] L. Huang, W. Yao, Z. Lu, "Interleaved totem-pole bridgeless PFC rectifier with ZVS and low input current ripple," in Proc. IEEE Energy Convers. Congr. Expo., 2015, pp. 166-171.

[4] Liu, Zhengyang (2017-06-08). Characterization and Application of Wide-Band-Gap Devices for High Frequency Power Conversion. Virginia Polytechnic Institute and State University, Blacksburg, VA. http://hdl.handle.net/10919/77959

[5] J. W. Kim, S. M. Choi, and K. T. Kim, "Variable on-time control of the critical conduction mode boost power factor correction converter to improve zero-crossing distortion," in Proc. IEEE Power Electronics and Drive Systems Conf. (PEDS), Nov. 2005, pp. 1542-1546.

[6] X. Xu and A. Huang, "A novel closed loop interleaving strategy of multiphase critical mode boost PFC converters," in Proc. IEEE Appl. Power Electron. Conf., 2008, pp. 1033-1038.

[7] D. Neumayr, D. Bortis, E. Hatipoglu, J. W. Kolar and G. Deboy, "Novel efficiency-Optimal Frequency Modulation for high power density DC/AC converter systems," 2017 IEEE 3rd International Future Energy Electronics Conference and ECCE Asia (IFEEC 2017 - ECCE Asia), 2017, pp. 834-839, doi: 10.1109/IFEEC.2017.7992148.

[8] B. Majmunović, B. A. McDonald, S. -Y. Yu and J. Strydom, "90°-Valley Unified Controller for Zero-Voltage-Switching Quasi-Square-Wave (ZVS-QSW) Boost Converter," in IEEE Transactions on Power Electronics, vol. 39, no. 6, pp. 6930-6940, June 2024, doi: 10.1109/TPEL.2024.3378168.

[9] D. Rothmund, D. Bortis, J. Huber, D. Biadene and J. W. Kolar, "10kV SiC-based bidirectional soft-switching single-phase AC/DC converter concept for medium-voltage Solid-State Transformers," 2017 IEEE 8th International Symposium on Power Electronics for Distributed Generation Systems (PEDG), 2017, pp. 1-8, doi: 10.1109/PEDG.2017.7972488.

Resonance Current Suppression for AC-DC Active-Clamp Flyback Converter by Triangular Current Mode

Yasuo Uchida
Electrical, Electronics, and Information Engineering
Nagaoka University of Technology
Niigata, Japan
s203126@stn.nagaokaut.ac.jp

Hiroki Watanabe
Electrical, Electronics, and Information Engineering
Nagaoka University of Technology
Niigata, Japan
hwatanabe@vos.nagaokaut.ac.jp

Jun-ichi Itoh
Electrical, Electronics, and Information Engineering
Nagaoka University of Technology
Niigata, Japan
itoh@vos.nagaokaut.ac.jp

Abstract—This paper proposes the conduction loss reduction control method under Zero Voltage Switching (ZVS) operation of an AC-DC active-clamp flyback (ACF) converter for USB-Power Delivery. The proposed control method reduces the resonance current in the active clamp circuit by the reduction on-time of the clamp switch, while Triangular Current Mode control achieves ZVS to reduce the switching loss of the ACF converter. Moreover, the proposed control method provides the Power Factor Correction (PFC) capability without an additional PFC circuit because the proposed control method controls the input current by adjusting the on-time of the primary-side switch of the flyback converter. The detailed operation mode of both the conventional control method and the proposed control method is described. Experimental results of the 100-W active-clamp flyback converter demonstrate that the proposed method reduces the RMS current in the active-clamp circuit by 52.7% compared with the conventional control method.

Keywords—Flyback converter, Triangular Current Mode, Zero Voltage Switching

I. Introduction

AC adapters and USB-Power Delivery (USB-PD) have been widely used for home appliance applications such as laptops and smartphones. A flyback converter is the popular circuit topology for their application because of its uncomplicated configuration, galvanic isolation between AC and low voltage DC side, and low cost. This application requires high-efficiency power conversion to have a high power density and extension of the power range of USB-PD.

For the high-efficiency power conversion, the control methods to achieve the soft switching for the flyback converter have been studied. The quasi-resonant (QR) control is the candidate of the soft-switching control method because it provides valley switching to reduce switching losses [1-6]. In the QR control, the drain-source voltage of the primary-side switch of the flyback converter fluctuates during the zero-current period by the resonance between the parasitic parameters and self-inductance of the transformer. The QR control ensures that the flyback converter achieves valley switching by adjusting the turn-on timing of the switching device to the valley of the drain-source voltage. However, the QR control does not achieve the complete ZVS because the input resonance parameters decide the valley voltage. As a result, the switching losses still occur.

Active-clamp flyback (ACF) converter achieves both ZVS and surge voltage suppression [7-12]. The ACF converter utilizes the resonance between the leakage inductance and the clamp capacitor to discharge the parasitic capacitor before turned-on of the semiconductor switch. However, the conduction and the copper losses increase due to the large resonance current.

The RMS current reduction method has been considered to reduce the conduction and the copper losses of the ACF converter [13]. This method uses two transformers and two secondary-side diodes with a parallel connection. However, this method requires additional components for the parallel connection.

This paper proposes the Triangular Current Mode (TCM) control for the ACF converter. The contribution of the paper is efficiency improvement owing to the reduction of the conduction and the copper losses. The proposed control method has a negative magnetizing-current period by adjusting the switching period of the secondary-side synchronous rectifier to discharge the parasitic capacitor. As a result, the ACF converter achieves ZVS without the resonance current in the active-clamp circuit. Moreover, the on-state of the clamp switch is set shorter than the one of the synchronous rectifier to reduce the resonance current.

This paper is organized as follows; First, the configuration of the ACF converter is shown. Secondly, the feature of the conventional control method is described, and the benefit of the proposed control method is described. Next, the on-state of the primary-side switch to achieve PFC is calculated. Finally, the fundamental operation and each evaluation are confirmed by experimental results.

II. Circuit Configuration

Fig. 1 shows the isolated AC-DC converter based on the ACF converter. The input side of the converter connects to the

979-8-3315-1612-3/25 $31.00 © 2025 IEEE

AC voltage source, and the output side connects to the DC voltage source. This circuit consists of the synchronous rectifier and the ACF converter.

The synchronous rectifier improves the conversion efficiency because the on-resistance of the MOSFET is lower than the one of the diode. The synchronous rectifier switches synchronously with the grid frequency to full-wave rectify. In addition, The small capacitor is applied to the DC-link capacitor. Thus, the flyback converter controls the current of the AC voltage source to achieve Power Factor Correction (PFC) because the DC-link capacitor current is small, and the flyback converter provides isolation. In addition, The MOSFET is applied to the secondary side of the flyback converter to achieve the primary-side switch S_1 ZVS by conducting the negative-magnetizing current. The shunt resistor is applied to the secondary-side circuit to detect the magnetizing current i_{Lm}. The active-clamp circuit suppresses surge voltage.

III. CONTROL METHOD

A. Conventional control method

Fig. 2 shows each waveform of the ACF converter with the conventional control method. The conventional control method discharges the parasitic capacitor of the primary-side switch C_{ds_S1} by the resonance current between the leakage inductance L_1 and the clamp capacitor C_C to achieve ZVS for the primary-side switch S_1. This chapter describes the details of each mode.

<Mode 1>

In this mode, the primary-side switch S_1 turns on. The active-clamp switch S_C keeps off state. The input voltage v_{in} is applied to the leakage inductance L_1 and the magnetizing inductance L_m of the transformer. The magnetizing current i_{Lm} linearly increases during this mode and flows into the primary side of the ACF converter.

<Mode 2>

This mode starts when the primary-side switch S_1 turns off. In this mode, the leakage current i_{L1} continues to flow. The parasitic capacitor of the primary-side switch C_{ds_S1} is charged by the leakage current i_{L1}. When charging the parasitic capacitor of the primary-side switch C_{ds_S1} is completed, the parallel diode of the clamp switch S_C and the secondary-side diode are turned on. The resonance current between the leakage inductance L_1 and the clamp capacitor C_C occurs. In addition, the magnetizing current i_{Lm} flows into the secondary side of the ACF converter. Thus, the drain-source voltage of the clamp switch v_{ds_SC} decreases to 0 V because the leakage current i_{L1} discharges the parasitic capacitor of the clamp switch C_{ds_SC}. In this mode, The magnetizing current i_{Lm} linearly decreases because the output voltage V_{out} is applied to the magnetizing inductance L_m of the transformer.

<Mode 3>

This mode starts when the clamp switch S_C turns on. In this mode, the clamp switch S_C achieves ZVS. The resonance current and the magnetizing current i_{Lm} continue to flow. When this

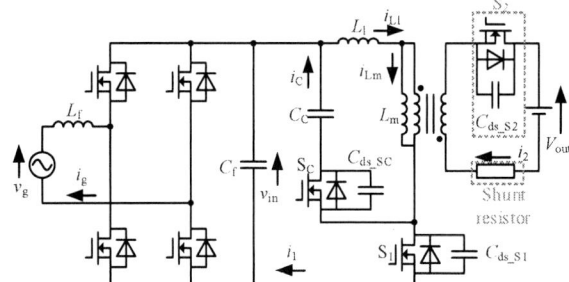

Fig. 1. Circuit configuration of AC-DC ACF converter.

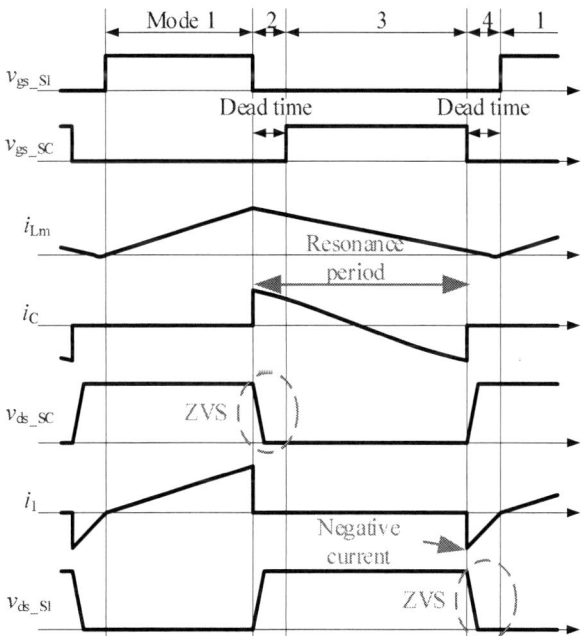

Fig. 2. Key waveforms of ACF converter with conventional control method. Primary-side switch S_1 achieves ZVS by resonance current. Clamp switch S_C also achieve ZVS.

mode is finished, the clamp capacitor current i_C is negative by the resonance current.

<Mode 4>

This mode starts when the clamp switch S_C turns off. During this mode, the leakage current i_{L1} continues to flow. The parasitic capacitor of the clamp switch C_{ds_SC} is charged by the leakage current i_{L1}. When charging of the parasitic capacitor of the clamp switch C_{ds_SC} is completed, the parallel diode of the primary-side switch S_1 is turned on. Thus, the drain-source voltage of the primary-side switch v_{ds_S1} decreases to 0 V because the leakage current i_{L1} discharges the parasitic capacitor of the primary-side switch C_{ds_S1}.

When next mode 1 starts, the primary-side switch S_1 achieves ZVS.

The conventional control method provides that all switches of the ACF converter achieve ZVS. However, the conduction and the copper losses increase due to the large resonance current between the leakage inductance L_1 and the clamp capacitor C_C.

B. Proposed control method

Fig. 3 shows each waveform of the ACF converter with the proposed control method. The proposed control method discharges the parasitic capacitor of the primary-side switch C_{ds_S1} by the negative magnetizing current to achieve ZVS for the primary-side switch S_1. This chapter describes the details of each mode.

<Mode 1>

In this mode, the primary-side switch S_1 turns on for the time of the primary-side switch T_1. The secondary-side switch S_2 and the active-clamp switch S_C keeps off state. The input voltage v_{in} is applied to the leakage inductance L_l and the magnetizing inductance L_m of the transformer. The magnetizing current i_{Lm} linearly increases during this mode and flows into the primary side of the ACF converter.

<Mode 2>

This mode starts when the primary-side switch S_1 turns off. In this mode, the leakage current i_{Ll} continues to flow. The parasitic capacitor of the primary-side switch C_{ds_S1} is charged by the leakage current i_{Ll}. When charging the parasitic capacitor of the primary-side switch C_{ds_S1} is completed, the parallel diode of the secondary-side switch S_2 and the clamp switch S_C are turned on. The resonance current between the leakage inductance L_l and the clamp capacitor C_C occurs. Thus, the drain-source voltage of the clamp switch v_{ds_SC} decreases to 0 V because the resonance current discharges the parasitic capacitor of the clamp switch C_{ds_SC}. In addition, the magnetizing current i_{Lm} flows into the secondary side of the ACF converter. Thus, the drain-source voltage of the secondary-side switch v_{ds_S2} decreases to 0 V because the magnetizing current i_{Lm} discharges the parasitic capacitor of the secondary-side switch C_{ds_S2}. In this mode, The magnetizing current i_{Lm} linearly decreases because the output voltage V_{out} is applied to the magnetizing inductance L_m of the transformer.

<Mode 3>

This mode starts when the secondary-side switch S_2 turns on for the time of the secondary-side switch T_2, and the clamp switch S_C turns on for the time of the clamp switch T_C. In this mode, the secondary-side switch S_2 and the clamp switch S_C achieve ZVS. The resonance current and the magnetizing current i_{Lm} continue to flow. When this mode is finished, the clamp capacitor current i_C is negative by the resonance current.

<Mode 4>

This mode starts when the clamp switch S_C turns off. In this mode, The resonance current between the leakage inductance L_l and the clamp capacitor C_C does not flow. Thus, the proposed control method reduces the conduction time of the clamp switch S_C to reduce the RMS current of the clamp capacitor current i_C. The leakage current i_{Ll} continues to flow into the parallel diode of the primary-side switch S_1 until the leakage current i_{Ll} reaches 0 A. In addition, The magnetizing current i_{Lm} continues to flow into the secondary side of the ACF converter after the magnetizing current i_{Lm} crosses 0 A to achieve ZVS for the primary-side switch S_1. When the magnetizing current reaches the bottom, that is enough value to achieve ZVS for the primary-side switch S_1, this mode ends.

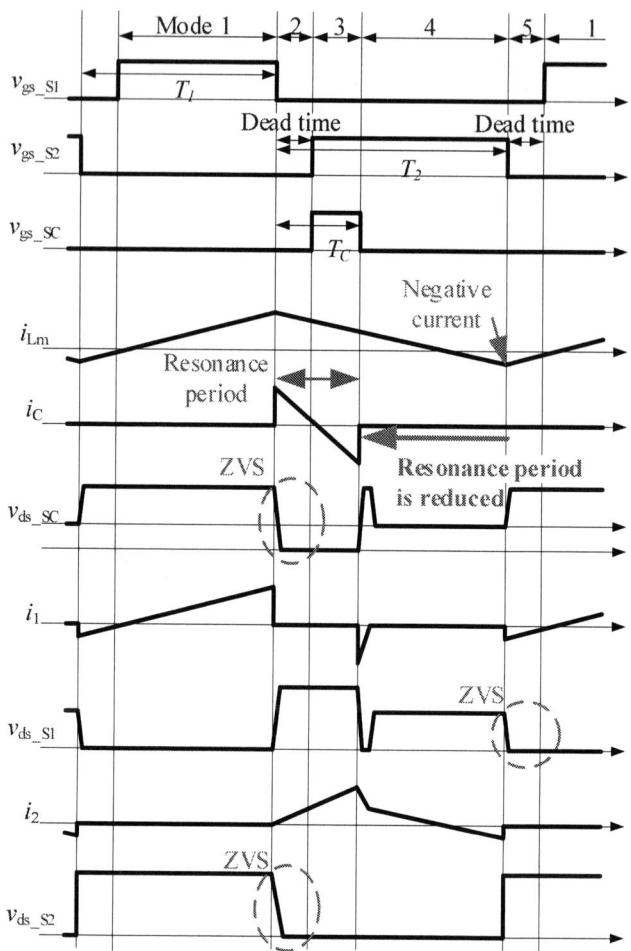

Fig. 3. Key waveforms of ACF converter with proposed control method. Primary-side switch S_1 achieves ZVS by negative magnetizing current. Secondary-side switch S_2 and active clamp switch S_C also achieve ZVS.

<Mode 5>

This mode starts when the secondary-side switch S_2 turns off. During this mode, the magnetizing current i_{Lm} continues to flow into the parallel diode of S_1. Thus, the drain-source voltage of the primary-side switch v_{ds_S1} decreases to 0 V because the magnetizing current i_{Lm} discharges the parasitic capacitor of the primary-side switch C_{ds_S1}.

When next mode 1 starts, the primary-side switch S_1 achieves ZVS.

The proposed control method provides that all ACF converter switches achieve ZVS. In addition, the proposed control method reduces the RMS current of the clamp capacitor current i_C by the reduction conduction time of the resonance current between the leakage inductance L_l and the clamp capacitor C_C.

C. Bottom current detection

Fig. 4 shows the magnetizing current i_{Lm} waveforms without parasitic parameters. According to Fig. 4, the magnetizing current i_{Lm} reaches the bottom while the magnetizing current i_{Lm} flows into the secondary side of the ACF converter. Thus, the proposed control method detects the secondary-side current i_2 by the shunt resistor.

Fig. 5 shows the zero-crossing detection circuit composed of the differential amplifier, buffer amplifier, LPF, Hysteresis comparator, and isolator. According to Fig. 5, the shunt resistor and the amplifier detect the secondary-side current i_2. After that, the hysteresis comparator detects the secondary-side current i_2 polarity. The ACF converter with the proposed control method requires isolation of the detected value of the secondary-side current i_2 because the primary side and the secondary side of the ACF converter are isolated. However, the propagation delay of isolation of analog detection value by an isolation amplifier is long. Thus, the proposed control method detects zero-crossing of the secondary-side current i_2.

The proposed control method estimates the secondary-side current based on the current drop time of the magnetizing current i_{Lm} from zero to bottom current I_{bot}. According to Fig. 4, the relationship between the current drop time T_{2bot} and the bottom current I_{bot} is expressed as

$$T_{2bot} = \frac{L_m}{NV_{out}} I_{bot} \tag{1}$$

where L_m is the magnetizing inductance of the transformer, V_{out} is the output voltage of the ACF converter, and N is the turn ratio of the transformer.

D. On-time calculation of the proposed control method

The proposed control method obtains the full-wave rectified average current of the primary-side current i_{1avg} to provide PFC. According to Fig. 4, the average primary-side current i_{1avg} is expressed as

$$i_{1avg}(t) = \frac{1}{2T_{sw}} \left(I_{peak} \times (T_1 - T_{1bot}) - I_{bot} T_{1bot} \right) \tag{2}$$

where T_{sw} is the switching period, I_{peak} is the peak of the magnetizing current, T_1 is the time of the primary-side switch, T_{1bot} is the reset time of the magnetizing current from the bottom current to zero. In addition, T_{1bot}, T_{sw}, and I_{peak} are expressed as

$$T_{1bot} = \frac{L_m}{|v_{in}(t)|} I_{bot} \tag{3}$$

$$T_{sw} = \frac{|v_{in}(t)| + NV_{out}}{NV_{out}} T_1 \tag{4}$$

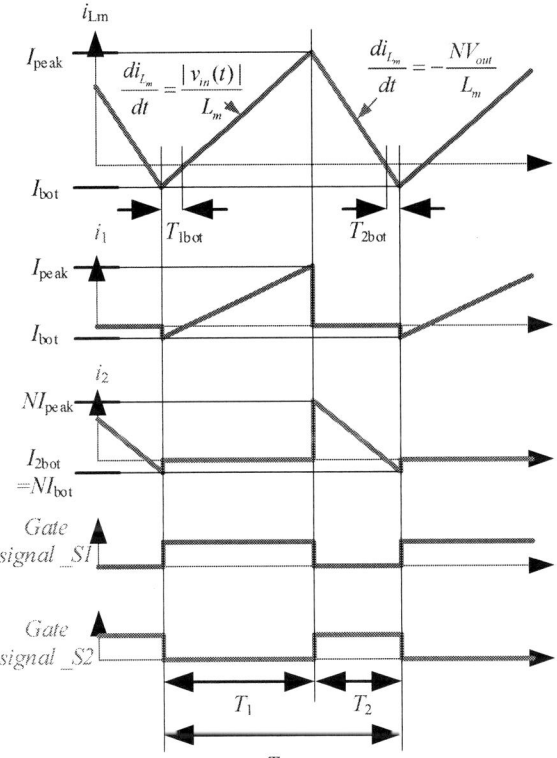

Fig. 4. Waveforms of Magnetizing current of flyback converter.

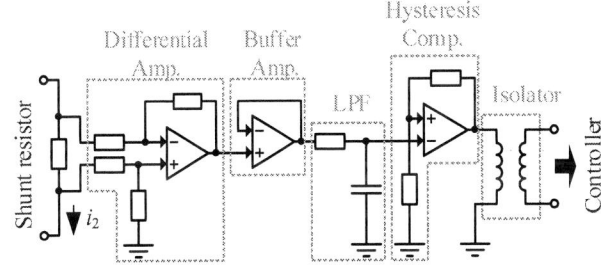

Fig. 5. Circuit configuration of zero-crossing detection circuit. This circuit detects zero-crossing of magnetizing current to control bottom current I_{bot}.

$$I_{peak} = \frac{|v_{in}(t)|}{L_m} T_1 - I_{bot} \tag{5}$$

where v_{in} is the input voltage. According to (2) ~ (5), the ratio of the average current of the primary-side current to the input voltage $K_{iin}(t)$ is expressed as

$$K_{iin}(t) = \frac{i_{1avg}}{|v_{in}(t)|} = \frac{NV_{out}\left(T_1 - 2I_{bot}L_m \dfrac{1}{|v_{in}(t)|}\right)}{2L_m\left(|v_{in}(t)| + NV_{out}\right)} \tag{6}$$

$K_{iin}(t)$ is constant in order to provide PFC. The following equation expresses $K_{iin}(t)$, assuming the input power P and the RMS input voltage V_{in}.

$$K_{iin}(t) = \frac{P}{V_{in}^2} \qquad (7)$$

According to (6) and (7), at the time of the primary-side switch T_1 is expressed as

$$T_1 = \frac{2PL_m\left(\left|v_{in}(t)\right| + NV_{out}\right)}{NV_{out}V_{in}^2} + 2I_{bot}L_m\frac{1}{\left|v_{in}(t)\right|} \qquad (8)$$

According to the slope of the magnetizing current i_{Lm}, the time of the secondary-side switch, T_2 is expressed as

$$T_2 = \frac{\left|v_{in}(t)\right|}{NV_{out}}T_1 \qquad (9)$$

Finally, at the time of the clamp switch, T_C is expressed as

$$T_C = a \times \frac{\left|v_{in}(t)\right|}{NV_{out}}T_1 \qquad (10)$$

where a is the ratio of the on-time of the clamp switch to the one of the secondary-side switches. a is typically between 0.0 and 1.0 depending on the drain-source voltage allowance of the switches because the drain-source voltage of the primary-side switch v_{ds_S1} and the clamp switch v_{ds_SC} increase due to reduce discharging time of the clamp capacitor C_C when a is small. On the other hand, the reduction of the RMS current in the active clamp circuit is expected. The optimization of the a will be considered for the reduction of the conduction loss in another paper.

IV. EXPERIMENTAL RESULTS

Table 1 shows experimental parameters. This chapter shows experimental results using Fig. 1 to confirm ZVS and PFC. Note that, the AC voltage source is applied to the input side, and the DC voltage source is applied to the output side. In addition, the ratio of the on-time of the clamp switch to the one of the secondary-side switch a is 0.2. The prototype circuit demonstrates the validity of the proposed control method.

Fig. 6 shows the input waveforms of the ACF converter with the proposed control method. According to Fig 7, the proposed control method provides high-power factor operation. The sinusoidal input current THD is obtained as an experimental result at 5.46%.

Fig. 7 shows each waveform of the ACF converter with the proposed control method when the primary-side switch S_1 turns on under the condition that the grid voltage is the peak value. According to Fig. 7, the primary-side switch of the ACF converter achieves ZVS by the negative primary-side current i_1.

Table 1: Experimental parameters.

Symbol	Quantity	Value
f_{sw}	Switching frequency	30kHz ~ 60kHz
P_{out}	Output power	100 W
V_{in}	Input voltage	100 V
V_{out}	Output voltage	24 V
L_m	Magnetizing inductance	306 μH
L_l	Leakage inductance	24.7 μH
N	Turns ratio	5
C_C	Clamp capacitor	2.0 μF
L_f	Filter inductance	2.13 mH
C_f	Filter capacitor	440 nF

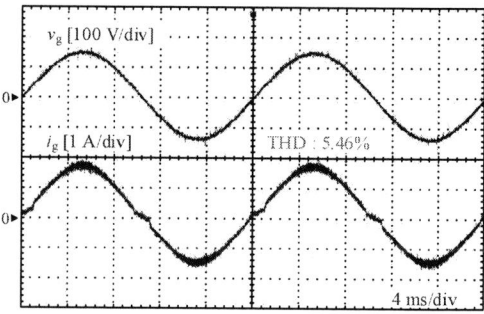

Figure 6: Input waveforms of ACF converter with proposed control method. the proposed control method provides PFC.

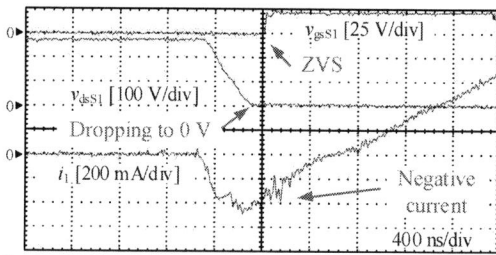

Fig. 7. Switching waveforms of primary-side switch S_1 of ACF converter with proposed control method. ZVS is achieved by TCM control.

Fig. 8 shows each current waveform when the input voltage v_{in} is 100 Vdc. Fig. 8(a) shows the current waveforms of the ACF converter with the conventional control method. According to Fig. 8(a), its resonance current continues to flow while the secondary-side current i_2 flows. Fig. 8(b) shows the current waveforms of the ACF converter with the proposed control method. According to Fig. 8(b), the conduction time of the clamp capacitor current i_C of the proposed control method is shorter than that of the conventional control method. As experimental results, the proposed control method reduces the RMS current of the clamp capacitor current i_C by 52.7%.

Fig. 9 shows the conduction loss of each switch of the ACF converter. The conduction loss is calculated from the RMS current and the on-resistance of the switches referred to in the datasheet. According to Fig. 9, the proposed control method

979-8-3315-1612-3/25 $31.00 © 2025 IEEE   52

reduces the total conduction loss by 5.60%. In the flyback converter, A switch with small on-resistance is applied to the secondary-side switch S_2 because the drain-source voltage of the secondary-side switch v_{ds_S2} is lower than the one of the primary-side switch v_{ds_S1} and the clamp switch v_{ds_SC}. Thus, the reduction in the total conduction loss is improved because the on-resistance of the clamp switch S_C, which large RMS current flows, is small.

V. CONCLUSION

This paper discussed the control method for a flyback converter to reduce the conduction and the copper losses under ZVS operation. The proposed control method reduces the RMS current by decreasing the conduction time of the active-clamp circuit, and the switches of the ACF converter achieve ZVS by TCM control. In addition, the proposed control method provides PFC. As experimental results, the proposed control method provides that the AC-DC ACF converter achieves PFC, the switches achieve ZVS, and the RMS current of the clamp capacitor current i_C is reduced by 52.7%. In future work, the TCM control without a shunt resistor will be considered in order to implement of the efficiency.

REFERENCES

[1] C. Wang, S. Xu, W. Shen, S. Lu and W. Sun, "A Single-Switched High-Switching-Frequency Quasi-Resonant Flyback Converter," IEEE Transactions on Power Electronics, vol. 34, no. 9, pp. 8785-8786, Sep. 2019.

[2] J. Zhang, H. Zeng, and X. Wu, "An Adaptive Blanking Time Control Scheme for an Audible Noise-Free Quasi-Resonant Flyback Converter," IEEE Transactions on Power Electronics, vol. 26, no. 10, pp. 2735-2742, Oct. 2011.

[3] X. Wu, Z. Wang and J. Zhang, "Design Considerations for Dual-Output Quasi-Resonant Flyback LED Driver With Current-Sharing Transformer," IEEE Transactions on Power Electronics, vol. 28, no. 10, pp. 4820-4830, Dec. 2012.

[4] C. Zhao, X. Xie, H. Dong and S. Liu, "Improved Synchronous Rectifier Driving Strategy for Primary-Side Regulated (PSR) Flyback Converter in Light-Load Mode," IEEE Transactions on Power Electronics, vol. 29, no. 12, pp. 6506-6517, Dec. 2014.

[5] C-H. Min and J-I. Ha, "Inner Supply Data Transmission in Quasi-Resonant Flyback Converters for Li-Ion Battery Applications Using Multiplexing Mode," IEEE Transactions on Power Electronics, vol. 34, no. 1, pp. 64-73, Jan. 2019.

[6] Y-C. Kang, C-C. Chiu, M. Lin, C-P. Yeh, J-M. Lin and K-H. Chen, "Quasiresonant Control With aDynamicFrequency Selector and Constant Current Startup Technique for 92% Peak Efficiency and 85% Light-Load Efficiency Flyback Converter," IEEE Transactions on Power Electronics, vol. 29, no. 9, pp. 4959-4969, Sep. 2014.

[7] R. Watson, F.C. Lee and G.C. Hua, "Utilization of an active-clamp circuit to achieve soft switching in flyback converters," IEEE Transactions on Power Electronics, vol. 11, no. 1, pp. 162-169, Jan. 1996.

[8] W. Meng, L. Li and F. Zhang, "Soft-switching Resonant Active Clamp Flyback Converter based-on GaN HEMTs for MHz High Step-up Applications," in 2021 IEEE Workshop on Wide Bandgap Power Devices and Applications in Asia, 2021, pp. 57-62.

[9] F-Z. Lin, T-J. Liang, K-H. Chen and K-F. Liao, "Primary-Side-Controlled AC-DC Single-Stage Active Clamp Flyback Converter," in 2023 IEEE Energy Conversion Congress and Exposition, 1997, pp. 3407-3414.

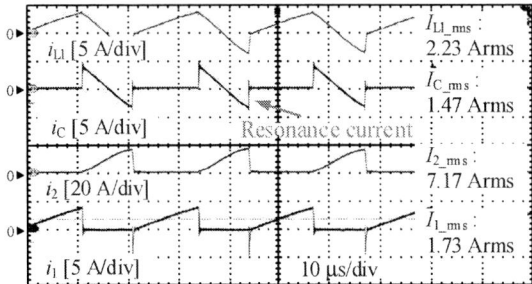

(a) With conventional control method.

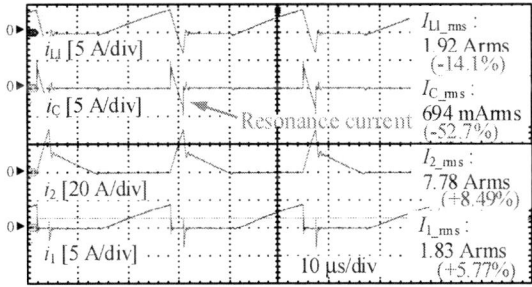

(b) With proposed control method.

Figure 8: Current wavefoms of ACF converter. The proposed control method reduces conduction time of resonance current. It reduced RMS current of i_C by 52.7%.

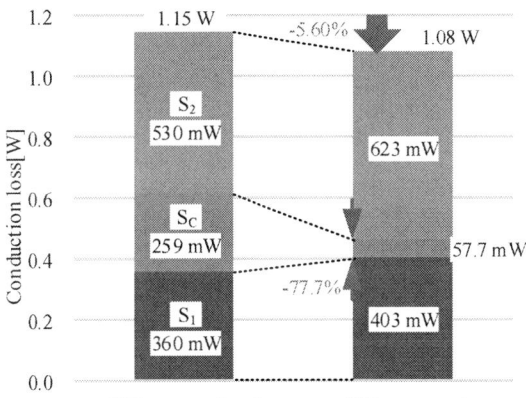

Fig. 9. Conduction loss comparison. Proposed control method reduces the total conduction loss by 5.60%.

[10] Y. Yao, C. Wang, D. Sun and C. Sheng, "A High Precision PSR Constant Voltage Control Method for Active-Clamp Flyback Converter," IEEE Transactions on Industrial Electronics, (early access).

[11] Y. -M. Liu and L. -K. Chang, "Single-Stage Soft-Switching AC–DC Converter With Input-Current Shaping for Universal Line Applications," IEEE Transactions on Industrial Electronics, vol. 56, no. 2, pp. 467-479, Feb. 2009

[12] L, Xue and J. Zhang, "Highly Efficient Secondary-Resonant Active Clamp Flyback Converter," IEEE Transactions on Industrial Electronics, vol. 65, no. 2, pp. 1235-1243, Feb. 2018.

[13] Y-K, Lo and J-Y. Lin, "Active-Clamping ZVS Flyback Converter Employing Two Transformers," IEEE Transactions on Power Electronics, vol. 22, no. 6, pp. 2416-2423, Nov. 2007.

A Universal DC Fast Charging Front-End with Optimized Film Capacitor Design

Sayan Paul*, Anurag Singh*, Tejas Bhuse†, Trent Martin†, Hien Nguyen†, Inder Vedula†,
Nikola Milivojević†, Dragan Maksimović* and Luca Corradini*

*Colorado Power Electronics Center
Department of Electrical, Computer and Energy Engineering
University of Colorado, Boulder, CO 80309, USA
email: {sayan.paul, anurag.singh-1, maksimov, luca.corradini}@colorado.edu

† Freewire Technologies
1811 Pike Rd #2-D, Longmont, CO 80501
email: {tbhuse, tmartin, hnguyen, ikumar, nmilivojevic}@freewiretech.com

Abstract—This paper proposes a three-leg universal charging front-end capable of both single-phase (1ϕ) and three-phase (3ϕ) connectivity and operation depending on the deployment configuration. Hardware re-utilization of both active and passive components across 1ϕ and 3ϕ deployment modes results in a versatile, high power density charging unit. The paper discloses the design and volume/loss/cost optimization methodology for the unit's dc link and buffering capacitors in film technology for a $800\,\mathrm{V}$ dc bus, $50\,\mathrm{kW}$-3ϕ / $33\,\mathrm{kW}$-1ϕ case study. Experimental operation is reported for a scaled-down $10\,\mathrm{kW}$ unit prototype.

Index Terms—Universal charger, active power filter, volume optimization

I. INTRODUCTION

Off-board *universal* chargers, capable of *both* single phase (1ϕ) and three phase (3ϕ) connectivity depending on grid availability at the deployment site, are becoming increasingly attractive, especially in the U.S. because of the widespread availability of 1ϕ as well as 3ϕ low-voltage (LV) grid. When coupled with integrated energy storage [1], they can service a variety of charging locations, including those affected by weak grid characteristics [2], while maintaining a limited power rating. On-board universal chargers are also extremely attractive as they enable level-2 charging from the ac grid in different locations worldwide [3]. Lower development and maintenance costs, and increased modularity, are additional advantages of universal chargers. Furthermore, bidirectional capability can be used to offer vehicle-to-grid (V2G) functionality.

This paper proposes the universal charging front-end shown in Fig. 1a. With selectors in the '3ϕ' position, the unit connects to the low-voltage (LV) 208 V-480 V 3ϕ grid, and the resulting three-phase rectifier is modulated as a unity power factor corrector (PFC). With selectors in the '1ϕ' position, on the other hand, the unit connects to the LV 240 V 1ϕ grid. In this configuration, legs A and B operate as a single-phase PFC, whereas leg C is re-utilized as an active power filter (APF) unit rerouting second-harmonic ($2\omega_o$) fluctuating power generated by the PFC to a highly optimized, low-volume, wide voltage swing capacitive energy buffer C_{buf}

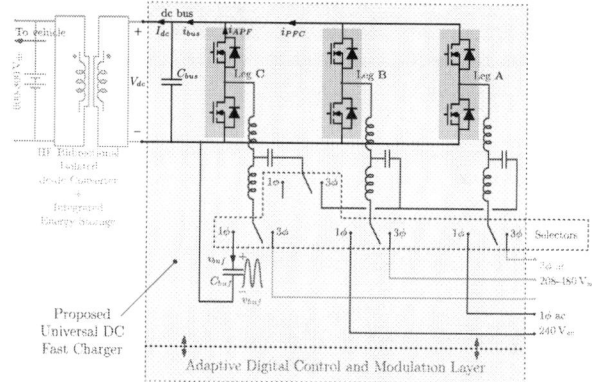

(a)

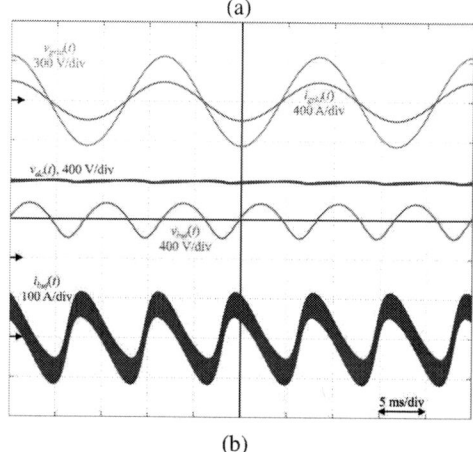

(b)

Fig. 1: (a) Topology of the proposed $3\phi/1\phi$ universal charger, and (b) simulated $33\,\mathrm{kW}$-1ϕ operation with the optimized film capacitor design: (from top to bottom) grid voltage and current, bus voltage v_{dc}, buffer voltage v_{buf} and current i_{buf}.

[4]–[6]. The operating waveforms of the system in $33\,\mathrm{kW}$-1ϕ configuration are exemplified in Fig. 1b, in which the charger operates at unity power factor per action of legs A and B, while the APF (leg C) provides the fluctuating

979-8-3315-1612-3/25 $31.00 © 2025 IEEE

power required for 1ϕ operation, resulting in a very low-ripple bus voltage v_{dc} and a wide-swing buffer voltage v_{buf}. Of several APF topologies [3], [5], [7], [8], the buck-type shunt is employed in this work because of its favorable characteristics for power-dense applications [5], [6]. Compared with other universal charging solutions, the proposed topology offers a higher degree of hardware re-utilization, reduces the number of switches and passive components, and eliminates the need for over-designing the LCL filter [3], [9].

This paper presents a design procedure and results for volume/loss/price assessment and optimization of the APF and dc link capacitors C_{buf} and C_{bus} across the film vs. CeraLink® technology space, as well as across the 1ϕ vs. 3ϕ operating modes. Design considerations are carried out around a case study charger rated $50\,\text{kW-}3\phi$ / $33\,\text{kW-}1\phi$, and with a $800\,\text{V}$ dc bus for compatibility with emerging/next generation battery charging needs.

The paper is organized as follows. Section II provides the quantitative analysis required for deciding the specifications for C_{buf} and C_{bus} design. A brief description of different capacitor technologies, their limitations, and considerations for the design is given in Section III followed by volume optimization and comparative volume/loss/cost analysis of C_{buf} and C_{bus} designs with film and CeraLink® technologies. Section IV verifies the proposed design through experiment, simulation, and hardware-in-the-loop emulation. Finally, the paper is concluded in Section V.

II. DERIVING SPECIFICATIONS FOR OPTIMAL BUFFER AND BUS CAPACITORS DESIGN

A. Quasi steady-state analysis of active power filter in 1ϕ configuration and specifications of buffer capacitor

In unity power factor (UPF) operation, the grid voltage, v_g,

$$v_g = V_{pk}\sin\omega_o t, \quad (1)$$

and current, $\bar{i}_g$,

$$\bar{i}_g = I_{pk}\sin\omega_o t \quad (2)$$

result in a rectified PFC stage current having a dc component, $I_{dc} = \frac{P_{dc}}{V_{dc}}$, where $P_{dc} \triangleq \frac{V_{pk}I_{pk}}{2}$, and a second harmonic ($2\omega_o$) component, where ω_o is the fundamental frequency in radians per second, and V_{dc} is the dc bus voltage. Throughout the paper, a bar over a variable denotes the average over one switching period, $T_s \triangleq 1/f_s$, and f_s is the switching frequency.

It is desired that APF leg injects a current $\bar{i}_{APF}$ into the dc bus node that cancels out the $2\omega_o$ component of $\bar{i}_{PFC}$; resulting in a bus current i_{bus} of Fig. 1a free of line-frequency ripple, and consequently more relaxed design constraints for the dc bus capacitor C_{bus}. Under the assumption that APF input current control is working with zero steady-state error, $\bar{i}_{APF}$ should satisfy

$$\bar{i}_{APF} = I_{dc}\cos 2\omega_o t. \quad (3)$$

With a quasi-steady-state approximation, the voltage drop across the inductor in series with C_{buf} is assumed to be zero.

Henceforth, the expression of the required duty cycle of the top switch of leg C, d_C, to synthesize the input current of APF as (3) can be derived as [6],

$$d_C(t) = \sqrt{\left(\frac{V_{buf,RMS}}{V_{dc}}\right)^2 - \frac{1}{2}\left(\frac{C_{buf,min}}{C_{buf}}\right)\sin\left(2\omega_o t\right)}. \quad (4)$$

Here, $C_{buf,min}$ is the minimum C_{buf} value required for this operation [4]–[6],

$$C_{buf,min} = \frac{2P_{dc}}{V_{dc}^2\omega_o}. \quad (5)$$

$V_{buf,RMS}$ is the line-cycle RMS value of $\bar{v}_{buf}$. Once $C_{buf} > C_{buf,min}$ is selected, the constraint $0 \le d_C(t) \le 1$ should be satisfied for all $\omega_o t \in [0, \pi]$, which leads to a feasible range for $V_{buf,RMS}$,

$$\sqrt{\frac{C_{buf,min}}{2C_{buf}}} \le \frac{V_{buf,RMS}}{V_{dc}} \le \sqrt{1 - \frac{C_{buf,min}}{2C_{buf}}}. \quad (6)$$

Observe that at minimum C_{buf}, there is only one $V_{buf,RMS}$ value at which the APF can operate, that is $V_{dc}/\sqrt{2}$.

The peak buffer capacitor voltage $V_{buf,pk}$ can be determined from (4),

$$V_{buf,pk} \approx V_{dc}\sqrt{\left(\frac{V_{buf,RMS}}{V_{dc}}\right)^2 + \frac{1}{2}\left(\frac{C_{buf,min}}{C_{buf}}\right)} \quad (7)$$

Due to the operation of the APF, the current flowing through the buffer capacitor, $\bar{i}_{buf}$, is

$$\bar{i}_{buf} = \frac{-\bar{i}_{APF}}{d_C}, \quad (8)$$

where $\bar{i}_{APF}$ and d_C can be substituted from (3) and (4), respectively. The RMS value of $\bar{i}_{buf}$, denoted by $I_{buf,RMS}$, can be evaluated numerically for any pair of $C_{buf}/C_{buf,min}$ and the corresponding choice of $V_{buf,RMS}/V_{dc}$ within its range (6).

Along with the dc voltage and RMS current values, C_{bus} design with film capacitors also needs an estimate of the low-frequency voltage ripple across $\bar{v}_{buf}$ [10]. A more detailed discussion regarding this is given in Section III-A. For the periodic $\bar{v}_{buf}$ having $2\omega_o$ and its multiple harmonic components, the peak-to-peak low-frequency voltage ripple of $\bar{v}_{buf}$ divided by $2\sqrt{2}$ is the quantity of our interest here, which is denoted by $V_{buf,ac}$. The peak-to-peak ripple voltage $\bar{v}_{buf}$, denoted by $\Delta\bar{v}_{buf}$, can be evaluated as below,

$$\Delta\bar{v}_{buf} = \frac{1}{\omega_o C_{buf}}\int_{\theta_1}^{\theta_2} \bar{i}_{buf}(\omega_o t)d(\omega_o t), \quad (9)$$

where the above integration is performed on an interval $\omega_o t \in [\theta_1, \theta_2]$, over which $\bar{i}_{buf}$ is positive. Note that $\bar{i}_{buf}$ is periodic with the fundamental frequency of $2\omega_o$. Thereafter, $V_{buf,ac}$ is calculated as

$$V_{buf,ac} = \frac{\Delta\bar{v}_{buf}}{2\sqrt{2}}. \quad (10)$$

Observe that the choice of $V_{buf,RMS}$ within its feasible range (6), influences the $V_{buf,pk}$, $I_{buf,RMS}$, and $V_{buf,ac}$. To

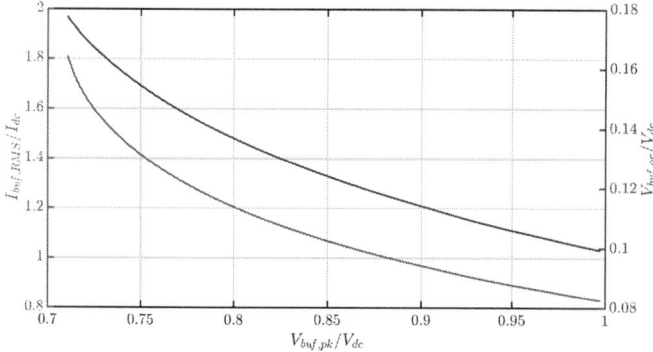

Fig. 2: Normalized $I_{buf,RMS}$ and $V_{buf,ac}$ as a function of $V_{buf,pk}$ for $C_{buf}/C_{buf,min} = 2$.

illustrate this, Fig. 2 shows their variations in normalized scale for one choice of C_{buf}, that is, $C_{buf}/C_{buf,min} = 2$. As $V_{buf,RMS}$ increases, $V_{buf,pk}$ also increases but both $I_{buf,RMS}$ and $V_{buf,ac}$ decrease monotonically. The same trends are observed for all the other C_{buf} values. The variations of the maximum and minimum values of the above three quantities with C_{buf} are shown in Fig. 3 on a normalized scale.

Traditionally, buffer capacitors are designed for their minimum value, i.e., $C_{buf} = C_{buf,min}$, [4], [6]. However, the volume of a capacitor depends not only on its capacitance but also on the dc voltage and the RMS current ratings. For film capacitors, the peak-to-peak ac voltage ripple is also important. Therefore, *the minimum volume for C_{buf} does not necessarily correspond to $C_{buf} = C_{buf,min}$.* We saw in the above analysis that with different choices of $C_{buf}/C_{buf,min}$ and $V_{buf,RMS}/V_{dc}$, one can change $V_{buf,pk}$, $I_{buf,RMS}$, and $V_{buf,ac}$. To find the optimal volume of the buffer capacitor across a range of design choices, this work proposes to consider the set of $(C_{buf}, V_{buf,pk}, I_{buf,RMS})$ or $(C_{buf}, V_{buf,pk}, V_{buf,ac})$, as applicable for different capacitor technologies, within the ranges as indicated in Fig. 3.

B. Design Specifications for the DC-link Capacitor

The dc-bus capacitor, C_{bus}, should have a dc voltage rating greater than V_{dc}. With the assumption that the $2\omega_o$ component of the rectifier current i_{PFC} in 1ϕ operation is entirely removed by the APF, the purpose of C_{bus} is to filter the high-frequency ripple current in the dc-bus for both the 3ϕ and 1ϕ operations. If $I_{ph,RMS}$ is the RMS value of the 60 Hz grid current, the maximum line-cycle RMS value due to the switching ripples in 3ϕ rectifier, denoted by $\tilde{I}_{bus,3\phi}$, operating at UPF with conventional space-vector modulation, is $0.65 \times I_{ph,RMS}$, [11]. The ripple RMS current for 1ϕ operation, $\tilde{I}_{bus,1\phi}$, including both rectifier and APF units is derived below.

To derive the line-cycle RMS of the dc-link switching ripples for 1ϕ operation, denoted by $\tilde{I}_{PFC,1\phi}$, following assumptions are made- 1) the phase current is purely sinusoidal with no switching ripple, 2) the rectifier is operating at UPF, 3) the 1ϕ inverter is modulated with unipolar modulation scheme where the duty ratios of the top switches of phase legs A and

B are $\frac{1}{2} + \frac{\bar{v}_{AB}}{2V_{dc}}$ and $\frac{1}{2} - \frac{\bar{v}_{AB}}{2V_{dc}}$, respectively. The expression of $\tilde{I}_{PFC,1\phi}$ is provided below,

$$\tilde{I}_{PFC,1\phi} = I_{ph,RMS}\sqrt{\frac{8M}{3\pi} - \frac{3M^2}{4}}, \qquad (11)$$

where $M \triangleq \frac{V_{pk}}{V_{dc}}$. The maximum value of $\tilde{I}_{PFC,1\phi}$, $\tilde{I}^*_{PFC,1\phi}$, is given below, which occurs at $M^* = \frac{16}{9\pi} \approx 0.57$.

$$\tilde{I}^*_{PFC,1\phi} = \frac{8}{3\sqrt{3}\pi}I_{ph,RMS} \approx 0.49I_{ph,RMS} \qquad (12)$$

To find the ripple current RMS solely due to the APF operation, denoted by $\tilde{I}_{APF,1\phi}$, one needs to first find the ripple current over a switching period, T_s. Fig. 4 shows the i_{buf} and i_{APF} waveforms over a switching cycle. The RMS ripple in i_{APF} during a switching period, $\tilde{i}_{APF}$, can be calculated for the trapezoidal waveform as

$$\tilde{i}^2_{APF} = d_C\left(\bar{i}^2_{buf} + \frac{\Delta i^2_{buf}}{12}\right) - \bar{i}^2_{APF}$$

$$= \underbrace{d_C(1 - d_C)\bar{i}^2_{buf}}_{\text{First Part}} + \underbrace{d_C\frac{\Delta i^2_{buf}}{12}}_{\text{Second Part}}, \qquad (13)$$

where $\Delta i_{buf} = \frac{V_{dc}T_s}{L_f}d_C(1 - d_C)$, L_f is the value of the filter inductor in series with C_{buf}. Note, $\bar{i}_{buf} = -I_{dc}\cos(2\omega_o t)/d_C$. In (13), $\tilde{i}_{APF}$ is decomposed into two parts- for a given pair of $C_{buf}/C_{buf,min}$ and $V_{buf,RMS}/V_{dc}$, the first and second parts are directly proportional to I_{dc} and $\frac{V_{dc}T_s}{L_f}$, respectively. The line-cycle RMS ripple current due to APF operation, $\tilde{I}_{APF,1\phi}$, is determined as (14). Here, $\tilde{I}_{APF,1}$ and $\tilde{I}_{APF,2}$ are the RMS values over the line cycle of the first and second parts, respectively,

$$\tilde{I}^2_{APF,1\phi} = \frac{1}{\pi}\int_0^\pi \tilde{i}^2_{APF}(\omega_o t)d(\omega_o t) = \tilde{I}^2_{APF,1} + \tilde{I}^2_{APF,2} \qquad (14)$$

Observe that both $\tilde{I}_{APF,1}$ and $\tilde{I}_{APF,2}$ vary with the choice of $C_{buf}/C_{buf,min}$ and its corresponding $V_{buf,RMS}/V_{dc}$ within the range of (6). As the integration of (14), after substituting d_C from (4), cannot be expressed in terms of the elementary functions, $\tilde{I}_{APF,1}$ and $\tilde{I}_{APF,2}$ are calculated numerically for any given pair of $C_{buf}/C_{buf,min}$ and $V_{buf,RMS}/V_{dc}$. Fig. 5a shows the normalized plots of $\tilde{I}_{APF,1}$ and $\tilde{I}_{APF,2}$ as functions of $V_{buf,RMS}$ within its range for $C_{buf}/C_{buf,min} = 2$. Similarly, the maximum and minimum values of $\tilde{I}_{APF,1}$ and $\tilde{I}_{APF,2}$ can be found for all possible ratios $C_{buf}/C_{buf,min}$, which are shown in Fig. 5b and 5c, respectively.

With the assumption that the harmonic components of the ripple currents of the APF and PFC units are orthogonal, the total ripple RMS current of C_{bus} in the 1ϕ configuration is determined as follows,

$$\tilde{I}_{bus,1\phi} = \sqrt{\tilde{I}^2_{PFC,1\phi} + \tilde{I}^2_{APF,1\phi}}. \qquad (15)$$

If the PFC is modulated with a triangular carrier of frequency f_s, the ripple components of i_{PFC} will have sidebands

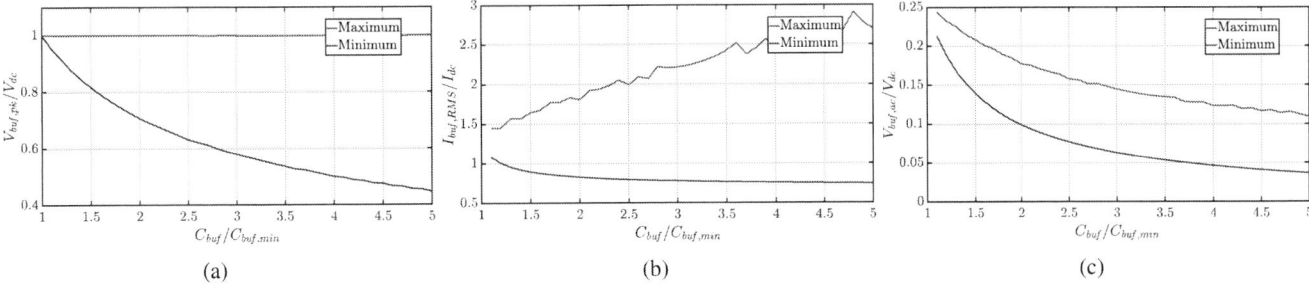

Fig. 3: Normalized plots show the variations of the maximum and minimum values of three quantities as a function of $C_{buf}/C_{buf,min}$- (a) $V_{buf,pk}$, (b) $I_{buf,RMS}$, (c) $V_{buf,ac}$.

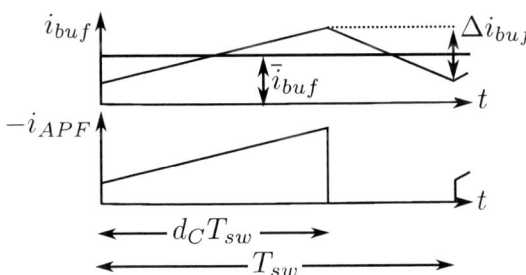

Fig. 4: i_{buf} and i_{APF} over a switching cycle.

around $2f_s$ and its multiples. Whereas modulation of APF unit with the same triangular carrier of frequency f_s will result in having sidebands of i_{APF} around f_s and its multiples. The expression of (15) becomes more accurate when the switching frequencies of the carriers are chosen such that their least common multiple is several order higher than them individually (e.g., 25 kHz and 28 kHz).

The current rating of C_{bus}, $\tilde{I}_{bus}$, will be the maximum among $\tilde{I}_{bus,1\phi}$ and $\tilde{I}_{bus,3\phi}$ so that it can support both 1ϕ and 3ϕ operation, i.e.,

$$\tilde{I}_{bus} = \max\left(\tilde{I}_{bus,1\phi}, \tilde{I}_{bus,3\phi}\right). \qquad (16)$$

Assuming that the entire $\tilde{I}_{bus}$ is concentrated at f_s, the minimum C_{bus} value which is needed to limit the peak-to-peak voltage ripple in the dc bus within $\Delta V_{dc}\%$ is

$$C_{bus} \geq \left(\frac{\sqrt{2}\tilde{I}_{bus}}{\pi f_s V_{dc}}\right)\frac{100}{\Delta V_{dc}(\%)} \qquad (17)$$

III. Volume Optimizations of Buffer and DC Bus Capacitors

The volumes of C_{buf} and C_{bus} are minimized for the target specifications of the universal charger mentioned in Table I. Note that the worst-case $I_{ph,RMS}$, in which the bus capacitor is designed, for both the 1ϕ and 3ϕ configurations, are the same. A brief description of different capacitor technologies, their limitations and considerations for the design follows the two subsections on volume minimization of C_{buf} and C_{bus}.

TABLE I: Target specifications of the universal charger

$P_{dc,3\phi}$ / $P_{dc,1\phi}$	50 kW / 33 kW
V_{dc}	800 V
$I_{ph,RMS}$	140 A
ω_o	$2\pi \times 60$ rad/sec
f_s	25 kHz
ΔV_{dc}	5 %

A. Capacitor Technologies and Considerations

Electrolytic, film, multilayer ceramic and CeraLink® are the four types of capacitor to consider for the design. Electrolytic capacitor has limited lifetime and reliability issues. It is observed that the costs of ceramic arrangements for the design of C_{buf} and C_{bus} are prohibitively high. Therefore, both electrolytic and ceramic capacitors are excluded from the design. The capacitance values of the CeraLink® capacitor is a function of temperature, dc bias voltage, and ac ripple voltage- here the nominal capacitance values at $25\,^\circ$C are considered for volume optimization. Generally, the frequency vs. RMS current rating plot of the CeraLink® capacitors is given at high frequency (> 10 kHz). Although this plot can be referenced for the C_{bus} design, C_{buf} requires the RMS current at the $2\omega_o$ frequency, 120 Hz here. To evaluate it, one needs to: 1) find the ESR values at two frequencies – $2\omega_o$ and high frequency, suppose F_r – where the RMS rating is given; 2) find the corresponding $2\omega_o$ RMS current so that the product $I^2 \times$ ESR remains the same at both $2\omega_o$ and F_r. For some capacitors, such as the TDK CeraLink® B58035U series, the ESR vs. frequency plots are given for > 1 kHz on the log-log scale. The plot is extended to estimate the ESR value at the $2\omega_o$ frequency.

The above approach of finding the RMS current at $2\omega_o$ frequency considering a loss-limited design is not entirely adequate for film capacitors, because the current rating of the such components below 1 kHz is limited by the *corona discharge effect* [10]. For example, the 900 V, 20 μF B32716H9206K000 dc-link film capacitor has a current rating of 15 A at 10 kHz, but its RMS ac voltage rating is limited to 127 V, resulting in an RMS current of 120 Hz as $127 \times (2\pi \times 120) \times 20 \times 10^{-6} = 1.92$ A. Because of the corona discharge limit, *dc-link film capacitors* are the preferred choice for C_{bus} – in which only high-frequency switching ripple current needs to be handled

979-8-3315-1612-3/25 $31.00 © 2025 IEEE

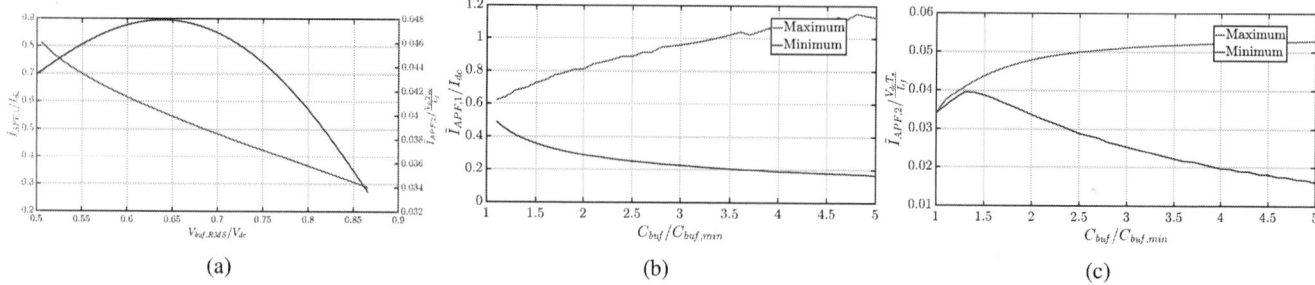

Fig. 5: (a) $\tilde{I}_{APF,1}$ and $\tilde{I}_{APF,2}$ as functions of $V_{buf,RMS}$ for $C_{buf}/C_{buf,min} = 2$, (b) variations of maximum and minimum $\tilde{I}_{APF,1}$ with C_{buf}, (c) variations of maximum and minimum $\tilde{I}_{APF,2}$ with C_{buf}.

– but not for C_{buf}, in which a strong 120 Hz component is present. For this reason, *ac film capacitors* are considered for C_{buf} because of their relatively higher ac rating. For the same reason, it is also legitimate to consider the voltage rating $V_{buf,ac}$, rather than $I_{buf,RMS}$, when designing C_{buf} with film capacitors.

B. Volume Minimization of the Buffer Capacitor

For the target specifications, Table I, one has $C_{buf,min} \approx 274\,\mu F$. A numerical search algorithm is run to determine the parallel combination of any one type of capacitor, from an exhaustive database of commercially available components, such that the combination results in minimum volume for each C_{buf} value between $300\,\mu F$ to $1.5\,mF$ with a $10\,\mu F$ search resolution. The components database consists of the following sets, as available on the distributor websites like *Mouser* and *DigiKey*:

1) Seventy-two ac film capacitors with *i)* dc voltage rating 450 V and above, *ii)* ac voltage rating 230 V RMS and above, and *iii)* capacitance value ranging from $2\,\mu F$ to $60\,\mu F$. These capacitors belong to the series Vishay MKP1847C and MKP1847H, TDK B33331V7, B32354S, B32355C4, B32332I6, and B33331I6.

2) Nine flex-assembly CeraLink® capacitors from TDK B58035U series and with dc voltage ratings of 500 V, 700 V, and 900 V.

The algorithm objectives are set to comply with *i)* the required C_{buf} value, *ii)* the required $V_{buf,pk}$ value, and *iii)* the required RMS current $I_{buf,RMS}$ (applicable for MLCC and CeraLink®) or ac voltage rating $V_{buf,ac}$ (applicable to ac film). The search algorithm facilitates to vary $V_{buf,RMS}/V_{dc}$ factor within its range, (6), for each C_{buf} value to vary the corresponding $V_{buf,pk}$, $I_{buf,RMS}$, and $V_{buf,ac}$, as discussed in Section II-A and their ranges are shown in Fig. 3 in normalized scale. The algorithm also accounts for a safety margin of 10% in addition to the required $V_{buf,pk}$ and $V_{buf,ac}$ to search for the voltage ratings of the capacitors.

Table II lists the minimum volume obtained for each type of capacitor and the corresponding ESR loss, cost, and manufacturer part numbers. Observe that for both technologies, the minimum volume is obtained with the choice $C_{buf} > C_{buf,min}$, which also allows one to keep the dc voltage rating

lower than $V_{dc} = 800\,V$ even after considering the 10% margin in addition to the required $V_{buf,pk}$ and $V_{buf,ac}$. The minimum volume with ac film choice is achieved when eighteen $35\,\mu F$ capacitors are paralleled to realize $C_{buf} = 630\,\mu F$ and the corresponding volume, lost and price are $1835\,cm^3$, 6 W, and \$ 316, respectively. Whereas, four hundred and seventy CeraLink® B58035U7105M062, $1\,\mu F$ capacitors need to be paralleled to attain the minimum volume of $190\,cm^3$ with the corresponding loss and price of 240 W, and \$ 3,581. The choice of $V_{buf,RMS}/V_{dc} = 0.55$ in both cases resulted in the optimal volume.

The results of the search are also compared in (volume, loss) and (volume, price) planes shown in Fig. 6. To attain a comparatively lower volume – suppose below $300\,cm^3$ – CeraLink® is the go-to choice, but the minimum price for this arrangement is \$ 3,000. The estimated loss with the CeraLink® arrangement is also higher as compared to film technology. While the best-total-volume ac film solution is comparable in loss and price with other film solutions (below 10 W and approx. \$ 300), the volume is way higher than what can be achieved by CeraLink®. Considering the price and loss advantages of film capacitor, B32332I6356J082 is finally down-selected for this application.

C. Volume minimization of the DC bus Capacitor

As shown in Fig. 1a, both the converter side and grid side inductors are connected in series with C_{buf} during APF operation. With our choice of $30\,\mu H$ inductors on each side to meet the grid standards of 5% total-harmonic-distortion (THD), the resulting $L_f = 60\,\mu H$ for APF operation. $\tilde{I}_{APF,1}$ and $\tilde{I}_{APF,2}$ of (14) are first evaluated for $V_{buf,RMS}/V_{dc} = 0.55$, as obtained from C_{buf} volume optimization, and the values are 28.3 A and 25.3 A, respectively. Thereafter, $\tilde{I}_{bus,1\phi} = 77.4\,A$ is obtained from (12), (14), and (15). For 3ϕ operation, the required $\tilde{I}_{bus,3\phi}$ is 90.1 A. Therefore, the current rating of C_{bus} is 90.1 A and the capacitor value should be higher than $40\,\mu F$ to limit ΔV_{dc} within 5%. The voltage rating of C_{bus} is considered as 900 V with a safety margin of 100 V over V_{dc}.

A design search algorithm is then run to comply with the above specifications and attain the minimum volume arrangement. The component database for C_{bus} design, as available on the distributor website, is as follows:

979-8-3315-1612-3/25 \$31.00 © 2025 IEEE

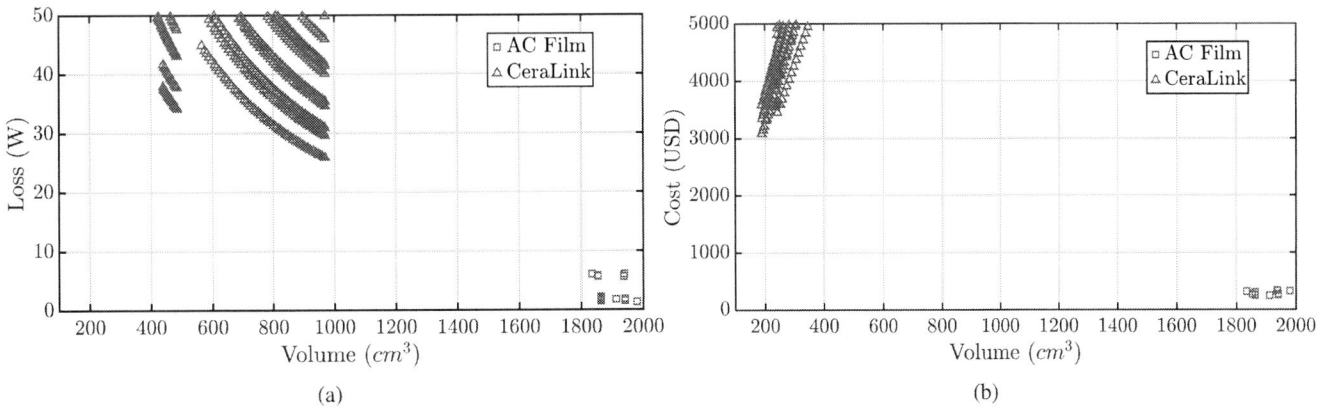

Fig. 6: Comparison of C_{buf} design with different capacitor technologies in (a) volume-loss plane, and (b) volume-price plane.

TABLE II: Summary of best-total-volume and corresponding total cost and loss of film and CeraLink® capacitor designs (price indications are based on 100-unit quotes on distributor sites).

		Part No.	No. of caps	Volume	Loss	Cost
C_{buf}	AC Film	B32332I6356J082 (35 µF, 450 V AC RMS, 630 V DC)	18	1835 cm³	6 W	$ 316
	CeraLink®	B58035U7105M062 (1 µF, 700 V DC)	470	190 cm³	240 W	$ 3581
C_{bus}	DC-link Film	B32776G9156K000 (15 µF, 900 V)	3	99.5 cm³	24.6 W	$ 15.7
	CeraLink®	B58043I9563M052 (33 nF effective value, 900 V)	1213	55.3 cm³	21.8 W	$ 1147.5

1) Thirty dc-link film capacitors ranging from 1 µF to 30 µF belonging to Vishay MKP1848C, MKP1848, and MKP1848Se series, TDK B32776 series, and Kemet C4AU series.

2) Five TDK CeraLink® capacitors – B58031I9254M062, B58043I9563M052 – and three capacitors from the B58035U flex assembly series. Effective capacitance values are used, instead of nominal values, as they are more suitable for dc-link with small high-frequency ac ripple [12].

The summary of best-total-volume arrangements for the above technologies and their volume, loss, and cost are provided in Table II. The comparison in volume-loss and volume-cost planes is shown in Fig. 7. As can be seen, the minimum volume attained by CeraLink® is approximately half that of the dc-link film solution. They incur an almost similar loss, see Fig. 7a, but the price of CeraLink® is much higher than that of the dc link film arrangement. Fig. 7b also shows that the minimum cost to meet the C_{bus} specifications with CeraLink® is approx. $ 900. Based on the above observations, the 3×15 µF, 900 V dc-link film capacitor design with B32776G9156K000 represents the optimal choice for C_{bus}.

IV. SIMULATIONS, HARDWARE-IN-THE-LOOP TESTS AND EXPERIMENTAL RESULTS

The proposed design is verified at full power through simulation in MATLAB® Simulink/PLECS, hardware-in-the-loop (HIL) emulation in Typhoon HIL 604, and experiment in a scaled-down 10 kW prototype (Fig. 8). Both the simulation and the HIL setup consider the finally down-selected C_{buf} (630 µF) and C_{bus} (40 µF) values. The power rating of the hardware prototype is scaled down by five, i.e., 10 kW in 3ϕ and 6.7 kW in 1ϕ. With this reduced power, $C_{buf,min} \approx$

55 µF. A similar volume optimization is performed for 6.7 kW operation in 1ϕ and 5×30 µF (MKP1847C630275Y5, 275 V AC, 600 V DC) is selected as C_{buf}. The corresponding $V_{buf,RMS}/V_{dc} = 0.447$. The optimal volume of C_{bus} in the reduced power is realized by 2×12.5 µF C4AUOBW5125M3FJ 900 V dc-link capacitors. A modified L_f value of 150 µH is used in the experiment keeping the switching frequency same as 25 kHz to meet the grid THD requirement.

Fig. 1b and Fig. 9a show v_g, i_g, v_{buf}, and i_{buf} waveforms of 1ϕ operation, obtained from the simulated and HIL, respectively, for the target specifications mentioned in Table I with 240 V ac grid voltage RMS. The peak voltage attained by v_{buf} is 570 V, and the peak-to-peak ripple voltage is 355 V, that is $V_{buf,ac} = 355/2\sqrt{2} = 125.5$ V. Observe that both of these voltages are within the allowable limit for the design choice made with ac film capacitor, Table II. Fig. 10a shows i_{PFC}, i_{APF} and the current flowing through C_{bus}, denoted by i_{Cbus}. The spectra of i_{PFC} and i_{APF} from DC to 900 Hz are shown in Fig. 10b and 10c, respectively. Observe that the DC (40.3 A) and 120 Hz peak (40.5 A) values are almost equal in i_{PFC}; and the 120 Hz component is compensated by the same frequency component of i_{APF} (40.7 A). As a result, 120 Hz component flowing through C_{bus} is negligible (< 0.05 A). The RMS values of i_{Cbus} obtained from simulation and HIL setup are 71.1 A and 75.1 A, which are close to the predicted $\tilde{I}_{bus,1\phi}$ in Section III-C, 77.4 A. The peak-to-peak voltage ripple across C_{bus} obtained in HIL emulation is 70 V, slightly higher than the original specification. This mismatch might occur due to the simplified assumption that the complete ripple RMS is concentrated at f_s. An accurate analysis of the dc-link voltage ripple of the universal charger is considered as a potential future work.

Experimental waveforms of the system in 1ϕ configuration

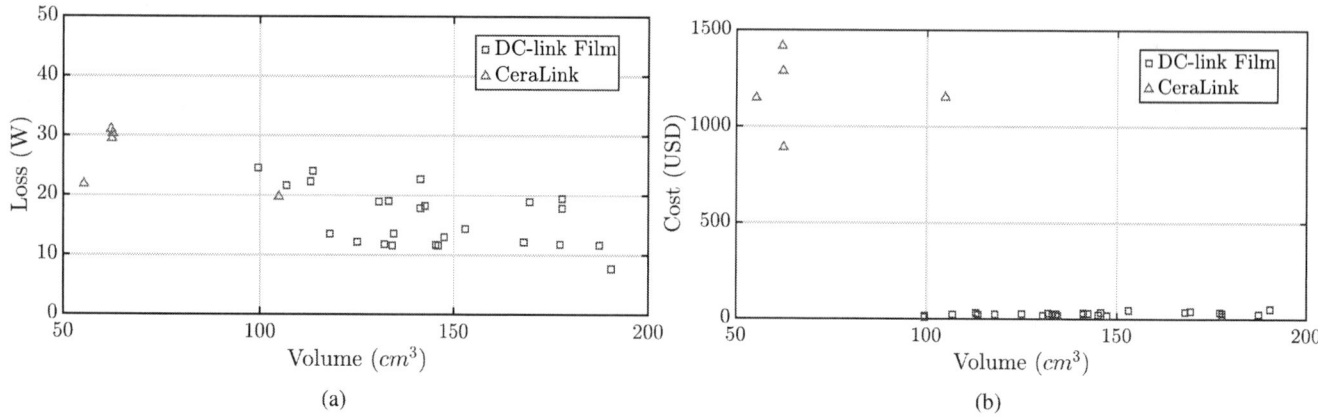

Fig. 7: Comparison of C_{bus} design with different capacitor technologies in (a) volume-loss plane, and (b) volume-price plane.

Fig. 8: Scaled-down 10 kW experimental prototype.

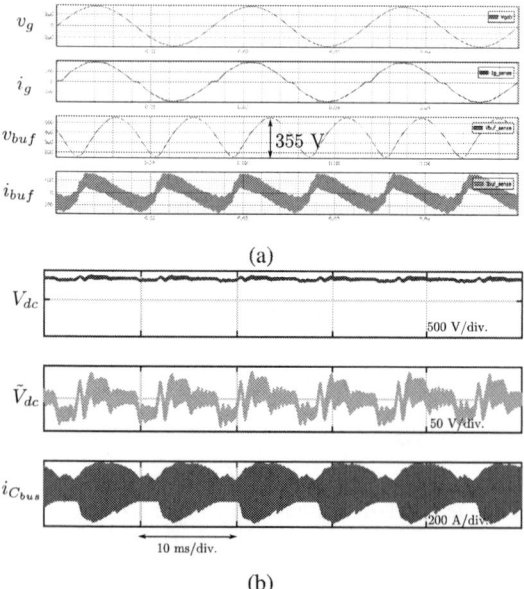

Fig. 9: Hardware-in-the-loop results of 1ϕ operation for $V_{dc} = 800$ V, 240 V ac RMS, and 33 kW-(a) v_g, i_g, v_{buf}, and i_{buf}, (b) V_{dc}, ripple in dc-link ($\tilde{V}_{dc}$), and i_{Cbus}.

are reported in Fig. 11. Fig. 11a and 11b show the v_g, i_g, v_{buf} and I_{dc} waveforms, without and with APF unit, respectively, under rectifier operation with closed-loop control of the grid current at $V_{dc} = 400$ V, $v_g = 150$ V RMS, 1.6 kW. The active filtering brings down the $2\omega_o$ (120 Hz) component of I_{dc} from 3.98 A peak to 0.31 A peak. Along with v_g, i_g, and v_{buf}, the i_{buf} waveform is shown in Fig. 11c which corresponds to inverter operation at $V_{dc} = 600$ V, 180 V ac RMS, and 3.5 kW. Note, the nature of i_{buf} is similar to what was observed in simulation, Fig. 1b, and HIL emulation, Fig. 9a.

V. CONCLUSIONS

This paper proposes a universal dc fast charging front-end capable of both 1ϕ and 3ϕ connectivity and bidirectional operation. The converter, composed of three half-bridge legs operating on a common dc bus, works as a three-phase rectifier when in 3ϕ mode. In 1ϕ operation, on the other hand, two legs operate as a single phase PFC, while the third leg is re-utilized as an APF unit which filters the $2\omega_o$ fluctuating power component using a highly optimized capacitive energy buffer. In the paper, design optimization and results for the buffer and bus capacitors for a 50 kW-3ϕ, 33 kW-1ϕ universal charger are presented, which lead to highly optimized filters entirely based on the film technology. AC film capacitors are selected for energy buffering functions over dc-link film components because of the corona discharge effect limit of the latter, which affects their ac ripple voltage rating. On the other hand, dc-link film technology gives a compact solution at a low cost for the bus capacitor design whose main purpose is to filter the switching ripple in the dc-bus. The paper also reveals that the minimum volume of the buffer capacitor does not always correspond to the conventional design with a minimum capacitance, as it also requires the full dc-bus voltage rating of the capacitor. Therefore, an analytical design formulation is proposed to take into account the dc and ac voltage ratings, and RMS current ratings along with the required capacitance for the volume minimization. The proposed design is verified at full rated power through simulation and hardware-in-the-

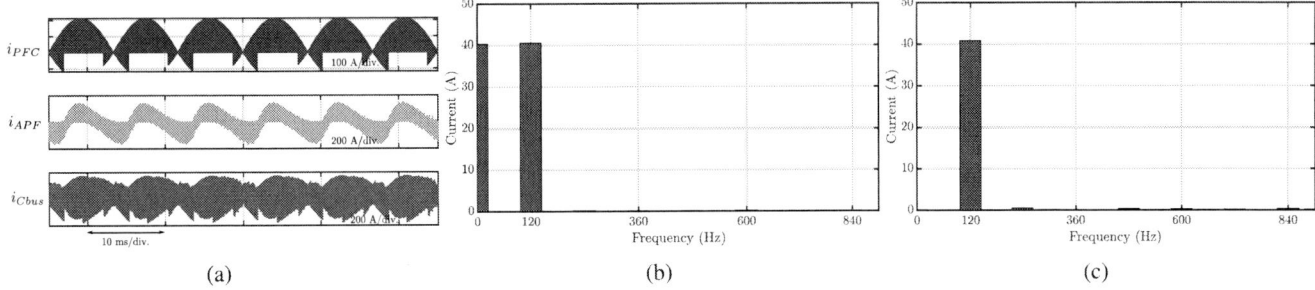

Fig. 10: Simulation results of 1ϕ operation for $V_{dc} = 800\,\text{V}$, 240 V ac RMS, and 33 kW (a) i_{PFC}, i_{APF}, and i_{Cbus}, (b) spectrum of i_{PFC}, (c) spectrum of i_{APF}.

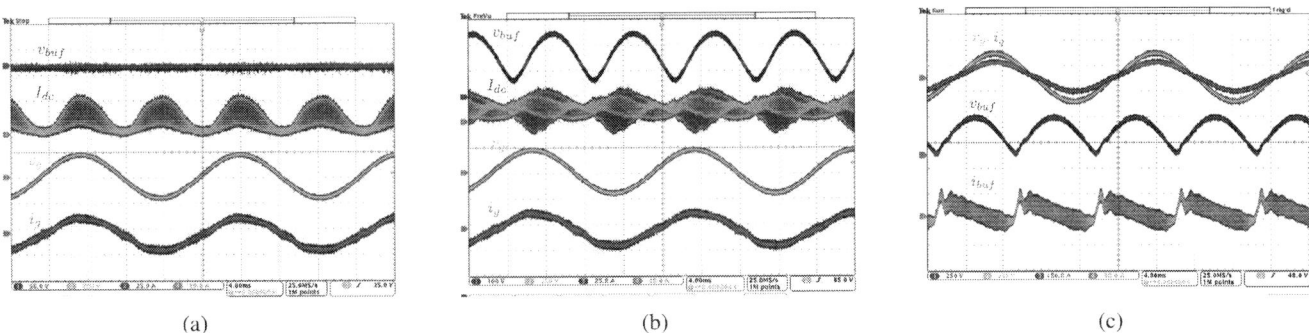

Fig. 11: Experimental results in 1ϕ operation. (a)-(b): Rectifier operation with inner current control at $V_{dc} = 400\,\text{V}$, $v_g = 150\,\text{V}$ RMS, 1.6 kW (a) without and (b) with APF operating; (c) Inverter operation with resistive load at $V_{dc} = 600\,\text{V}$, 180 V ac RMS, 3.5 kW, with APF enabled.

loop emulation. Experimental results are presented on a scaled-down 10 kW unit.

REFERENCES

[1] H. Tu, H. Feng, S. Srdic, and S. Lukic, "Extreme fast charging of electric vehicles: A technology overview," *IEEE Transactions on Transportation Electrification*, vol. 5, no. 4, pp. 861–878, 2019.

[2] M. M. Mahfouz and R. Iravani, "A supervisory control for resilient operation of the battery-enabled dc fast charging station and the grid," *IEEE Transactions on Power Delivery*, vol. 36, no. 4, pp. 2532–2541, 2021.

[3] D. Menzi, S. Weihe, J. A. Anderson, M. Kasper, J. Huber, and J. W. Kolar, "Novel s-link enabling ultra-compact and ultra-efficient three-phase and single-phase operable on-board ev chargers," in *Proc. 24th IEEE Workshop on Control and Modeling for Power Electronics (COMPEL)*, 2023, pp. 1–7.

[4] R. Wang, F. Wang, D. Boroyevich, R. Burgos, R. Lai, P. Ning, and K. Rajashekara, "A high power density single-phase pwm rectifier with active ripple energy storage," *IEEE Transactions on Power Electronics*, vol. 26, no. 5, pp. 1430–1443, 2011.

[5] D. Neumayr, D. Bortis, and J. W. Kolar, "Ultra-compact power pulsation buffer for single-phase dc/ac converter systems," in *Proc. 8th IEEE International Power Electronics and Motion Control Conference (IPEMC-ECCE Asia)*, 2016, pp. 2732–2741.

[6] H. Sarnago and O. Lucía, "High power density on-board charger featuring power pulsating buffer," *IEEE Open Journal of Power Electronics*, vol. 5, pp. 162–170, 2024.

[7] R. Ghosh, M.-x. Wang, S. Mudiyula, U. Mhaskar, R. Mitova, D. Reilly, and D. Klikic, "Industrial approach to design a 2-kva inverter for google little box challenge," *IEEE Transactions on Industrial Electronics*, vol. 65, no. 7, pp. 5539–5549, 2018.

[8] A. Mukhopadhyay and V. John, "Solid-state tuning restorer for second-harmonic lc filter in single-phase converters," *IEEE Transactions on Industry Applications*, vol. 60, no. 1, pp. 658–671, 2024.

[9] H. Zhao, Y. Shen, W. Ying, S. S. Ghosh, M. R. Ahmed, and T. Long, "A single- and three-phase grid compatible converter for electric vehicle on-board chargers," *IEEE Trans. Power Electron.*, vol. 35, no. 7, pp. 7545–7562, 2020.

[10] TDK, "Film capacitors, general technical information," 2018. [Online]. Available: https://www.tdk-electronics.tdk.com/download/530754/480aeb04c789e45ef5bb9681513474ba/pdf-generaltechnicalinformation.pdf

[11] J. Kolar and S. Round, "Analytical calculation of the rms current stress on the dc-link capacitor of voltage-pwm converter systems," *IEE Proceedings - Electric Power Applications*, vol. 153, pp. 535–543(8), July 2006. [Online]. Available: https://digital-library.theiet.org/content/journals/10.1049/ip-epa_20050458

[12] TDK, "Ceralink capacitors technical guide," 2024. [Online]. Available: https://www.tdk-electronics.tdk.com/download/2973492/6259e5169d3c406278919d23314defd5/ceralink-technical-guide.pdf

Power Characterization of a 1200-V/800-V 22-kVA 30-kHz Unity-Gain Dual-Active-Bridge Converter Prototype

Radhika Sarda*, Abishek Sethupandi[†], Madasamy Palavesha Thevar[†], Howe Li Yeo[†], Praveenkumar Palani[†],
Vaisambhayana B. Sriram[‡] and Anshuman Tripathi[‡]

*Energy Research Institute at NTU, Interdisciplinary Graduate Programme, Nanyang Technological University, Singapore
Email: radhika010@e.ntu.edu.sg

[†]Amperesand, Singapore
Email: abishek@amperesand.io, madasamy@amperesand.io, howe_li@amperesand.io, praveen@amperesand.io

[‡]Energy Research Institute at NTU, Nanyang Technological University, Singapore
Email: vsriram@ntu.edu.sg, antri@ntu.edu.sg

Abstract—**Dual-active-bridge (DAB) converters are crucial for enabling high-frequency operation and providing isolation in applications such as solid-state transformers (SSTs). While unity-gain DAB converters, with input and output voltages matched, are theoretically expected to achieve high efficiency across all power levels with inherent zero-voltage switching (ZVS) ability, the increased stored energies in the device output capacitance (C_{oss}) reduce efficiency at low loading conditions because of incomplete ZVS (iZVS). Triple-phase shift modulation provides an opportunity for efficiency improvement and is widely studied for non-unity gain DAB converters. The calculation of duty cycles for TPS scheme in existing literature cannot be applied directly for unity-gain DAB converters. This article aims to provide a simple formula for calculation of duty cycles to ensure minimum peak current for unity-gain DAB converters. The proposed calculation of duty ratios is evaluated experimentally with a 1200-V/800-V 22-kVA 30-kHz DAB converter prototype offering up to 10% improvement of efficiency at light loads and ensuring ZVS at all power levels.**

Index Terms—**dual-active-bridge, efficiency, incomplete zero-voltage switching, modulation.**

I. INTRODUCTION

Dual-active bridge (DAB) converters are used where high-frequency (HF) operation and isolation requirements are a necessity such as solid-state transformers, microgrids, renewable energy integrations, electric-vehicle chargers, etc [1], [2]. The widespread adoption of DAB converters is attributed to the ease of control, inherent zero-voltage switching (ZVS), and reduced thermal management requirements [2], [3].

A typical DAB converter is shown in Fig. 1 with v_{in} and v_{out} as the input and output voltage source respectively and with a $n : 1$ turns-ratio HF transformer operating at f Hz providing the isolation [2]. In Fig. 1, L_{ext} is the external inductance to facilitate power flow P through the DAB converter. Among the different input-output voltage ratio of DAB converters, unity-gain DAB converters, i.e., $v_{in} = nv_{out}$, inherently ensures ZVS at different load levels using a simple single-phase modulation (SPS) technique. For DAB converters with non-unity voltage gain ratio, advanced modulation

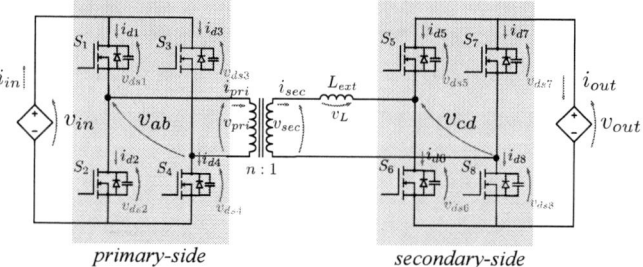

Fig. 1. A typical dual-active bridge with high-frequency transformer.

techniques such as dual-phase shift (DPS), extended DPS and triple-phase shift (TPS) modulation techniques have been widely explored [1], [4]–[10]. The modulation techniques differs from each other on the basis of the duty cycles of the HF converter voltage (v_{ab} and v_{cd} in Fig. 1) waveforms, represented as d_1 and d_2 for the primary and secondary-side bridges respectively. In SPS, the HF converter voltage across the transformer is maintained with the duty-ratio as 0.5 (i.e., $d_1 = d_2 = 0.5$) and a phase-shift between them denoted as ϕ. In DPS or extended DPS, one of the duty cycle is varied such that, $0 \leq d_1 \leq 0.5$, $d_2 = 0.5$ or $d_1 = 0.5$, $0 \leq d_2 \leq 0.5$. In TPS, both d_1 and d_2 are controlled such that $0 \leq d_1 \leq 0.5$ and $0 \leq d_2 \leq 0.5$. Note that, the choice of modulation technique impacts the peak and root-mean-squared values of the current through the devices and the HF magnetics and the active power delivery to the output. Thus, hybrid modulation techniques are also employed based on different power requirement [11].

In ideal situations, the necessary condition for ZVS is when the device (S_i, $i \in \{1, 2, ...8\}$) turns on at t_{Si}^{th} instant with the respective current (i_{di}) flowing in the negative direction (i.e., through the anti-parallel diode of the device), i.e., $i_{di}|_{t_{Si}\uparrow} < 0$. However, with presence of device output capacitance C_{oss} the modified requirements for ensuring ZVS are (i) sufficient minimum current in the negative direction, i.e., $i_{d1}|_{t_{S1}\uparrow} \leq$

TABLE I
DIFFERENT DAB PROTOTYPES.

Literature	Power ratings	Modulation	Low load efficiency	Remarks
[4]	750-V/750-V 100-kW 16-kHz 1:1	SPS, DPS	93% @ 10% load	$d_2 = f(P, f, L_{ext}, \phi, v_{in} - nv_{out})$, reduces to SPS for unity-gain DAB converters
[5][6]	50-V/400-V 1.5-kW 60-kHz 1:8	SPS, TPS	82.5% (SPS), 94% (TPS) @ 10% load	Duty cycle calculation not discussed for unity-gain DAB converters
[1]	260-V/200-V 1-kW 20-kHz 1.1:1	SPS, DPS/EPS	91% (SPS), 92.5% (DPS/EPS) @ 20% load	$d_1 = f(P, f, L_{ext}, \phi, v_{in} - nv_{out})$, reduces to SPS for unity-gain DAB converters
[7]	(80V-50V)/100-V 320-W 10-kHz 1:1	SPS, DPS, TPS	60% (SPS), 80% (DPS), 94% (TPS) @ 20% load	$d_1, d_2 = f(P, f, L_{ext}, \phi, v_{in} - nv_{out})$, reduces to SPS for unity-gain DAB converters
[8]	Specifications not given	SPS, DPS, TPS	78% (SPS), 88% (DPS), 92% (TPS) @ 25% load	Particle Swarm Optimization-based look-up table approach, not easily scalable, unity-gain DAB converters not discussed
[9]	100-V/100-V 500W 2.5 kHz (no transformer)	SPS	84% (SPS), 93% (TPS) @ 12.5% load	Does not consider effect of iZVS, SPS for unity-gain DAB converters
[10]	(150-V-300-V)/120-V 1500-W 40-kHz	SPS	80% (SPS) @ 16% load	Does not consider effect of iZVS, SPS for unity-gain DAB converters

$-i_{L-min}$ and (ii) sufficient time for discharging process to complete in form of deadtime, i.e., $|t_{S2\downarrow} - t_{S1\uparrow}| \approx t_{ZVS}$ (S_1 and S_2 are considered here, for instance). The stored energy in device output capacitance increases square-fold which gets dissipated in form of heat in case of incomplete ZVS (iZVS), which becomes a concern for high-voltage applications, substantially impacting reliability of the switches [12]–[14]. Despite theoretically claiming high efficiency at all power levels for unity-gain DAB converters, the experimental results of have shown otherwise owing to iZVS. The results of some of the developed high-voltage high-power DAB prototypes with their modulation technique is summarized in Table I.

It can be inferred from Table I that most prototypes of DAB converters have reported ways to improve efficiency but are either based on non-unity voltage gain ratios or lookup table based which cannot be easily scaled. Moreover, TPS modulation techniques derived does not provide direct duty ratios applicable for unity-gain DAB converters. The duty ratio calculation formula for TPS merges to SPS when the term $(v_{in} - nv_{out})$ approaches zero, unable to resolve the ZVS problem at light loads for unity-gain DAB converters. In view

of this drawback, this paper examines modulation techniques for unity-gain DAB converters which ensures ZVS operation, suitable for high-power and high-voltage applications where iZVS range uses up a significant portion of the power due to limited current. This article presents a simple analytical and intuitive formula for deriving TPS duty ratios for unity-gain DAB converters, which could be easily scaled to different unity-gain DAB converter systems.

This paper is structured as follows. Section II provides the background of the incomplete ZVS in a DAB. Section III analyses different modulation technique for unity-gain DAB converters. Section IV presents the loss breakdown of the system in consideration. Section V provides the efficiency results of a developed DAB prototype. Section VI concludes the paper.

II. REQUIREMENTS FOR ZVS

The process of ZVS is detailed in [12]–[14]. The key requirements are stated here: (i) minimum inductor current (i_{L-min}) to ensure sufficient energy to discharge or charge C_{oss} and (ii) sufficient dead-time (t_d) to complete the process of discharging or charging the corresponding device output

979-8-3315-1612-3/25 $31.00 © 2025 IEEE

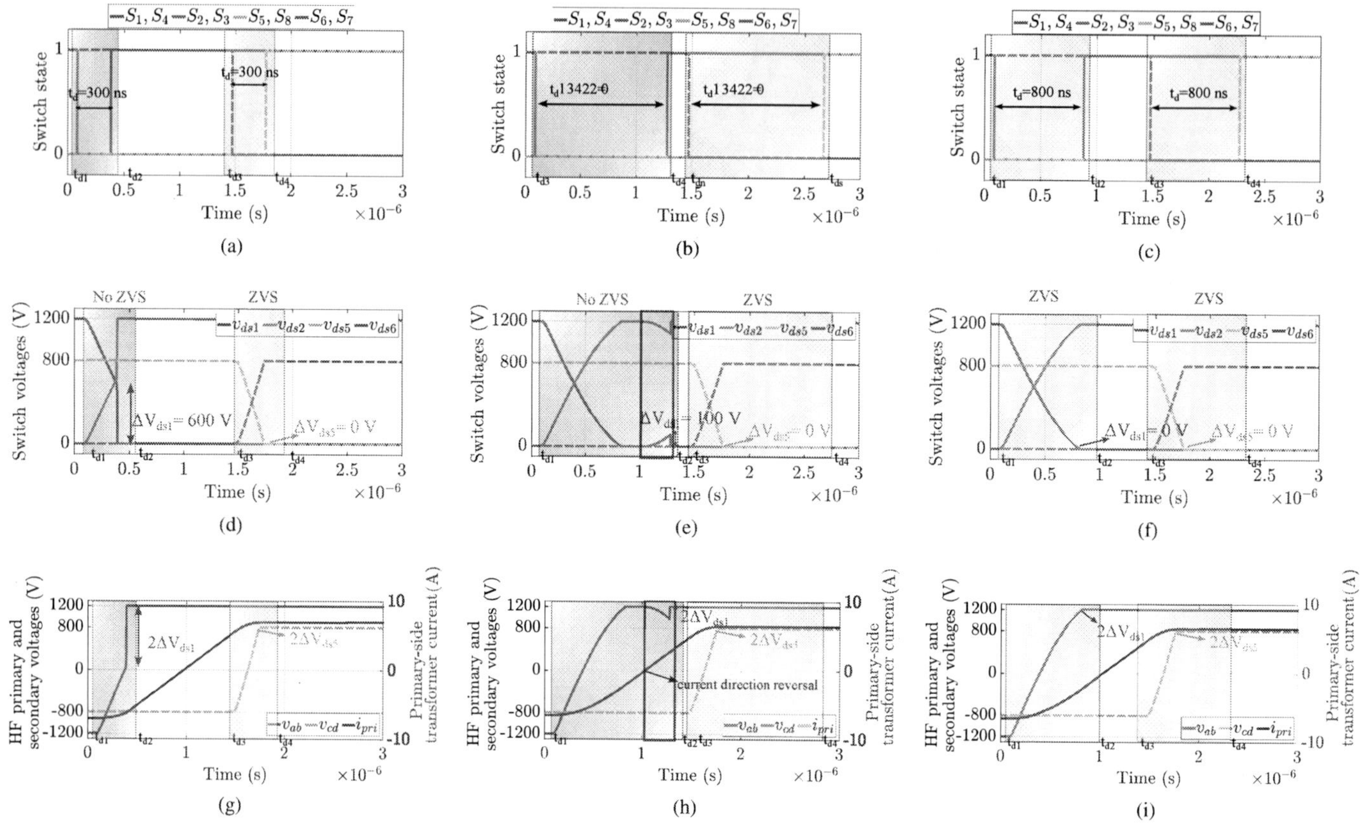

Fig. 2. Simulation waveforms highlighting incomplete ZVS of a DAB converter with device output capacitances, during different deadtimes. From top to bottom, switching states (S_1-S_8) in the top row, corresponding drain-source voltages (v_{dsi}, $i \in 1, 2, 5, 6$) in the middle row and HF converter voltage waveforms (v_{ab} and v_{cd}) and primary-side transformer current (i_{pri}) in the bottom row are represented. From left to right, simulations are performed with: (a) t_d=300 ns, (b) t_d=1200 ns and (c) t_d=800 ns.

capacitances. Fig. 2 shows the HF voltage waveforms obtained in simulations of a 1200-V/800-V 22-kVA 30-kHz DAB converter with the device capacitances (C_{oss} =1.5 nF, arbitrarily chosen) for different conditions of (i_{L-min}, t_d). Specifically, from top to bottom, switching states (S_1-S_8) in the top row, corresponding drain-source voltages (v_{dsi}, $i \in 1, 2, 5, 6$) in the middle row and HF converter voltage waveforms (v_{ab} and v_{cd}) and primary-side transformer current (i_{pri}) in the bottom row are represented. From left to right, the simulations are performed for different deadtimes ($t_d \in \{300$ ns, 1200 ns, 800 ns$\}$) for the same i_{L-min} = 6A.

In Fig. 2(a), (d) and (g), with the chosen deadtime i.e., $t_d = 300$ ns, the device output capacitances are not discharged completely because the condition (ii) is violated. The iZVS process can be easily identified with the abrupt voltage change at the high-frequency converter voltage waveform in Fig. 2(g), denoted as ΔV_{ds1}. Similarly, for the case with t_d =1200 ns, in Fig. 2(b), (e) and (h), iZVS is observed due to violation of condition (ii). As the chosen deadtime is too high, the high-frequency transformer current (i_{pri}) reverses in direction recharging the device capacitances C_{oss}. The iZVS process in Fig. 2(h) is identified similar to Fig. 2(g) with the presence of ΔV_{ds1}. However, when both the conditions for ZVS are satisfied, as seen in Fig. 2(c), (f) and (i) for case with $t_d = 800$

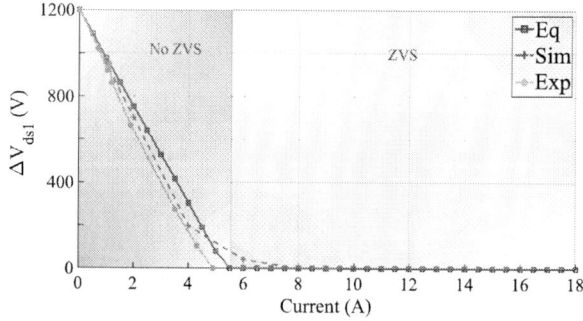

Fig. 3. Abrupt voltage changes indicating iZVS obtained from HF converter voltage waveforms.

ns, no abrupt voltage changes in form of ΔV_{ds} is visible on the high-frequency converter voltage waveforms.

To summarize the conclusions from Fig. 2, satisfying conditions (i) and (ii) are important in ensuring ZVS operation of the DAB converter.

A. Calculation of i_{L-min}

Minimum energy required to be stored in the external inductor, i.e., $L_{ext}i_{L-min}^2/2$ for ZVS operation, is determined by

TABLE II

COMPARISON OF MODULATION TECHNIQUES CONSIDERED FOR UNITY-GAIN DUAL-ACTIVE-BRIDGE CONVERTERS.

SINGLE-PHASE SHIFT	TPS OVERLAPPING	TPS NON-OVERLAPPING
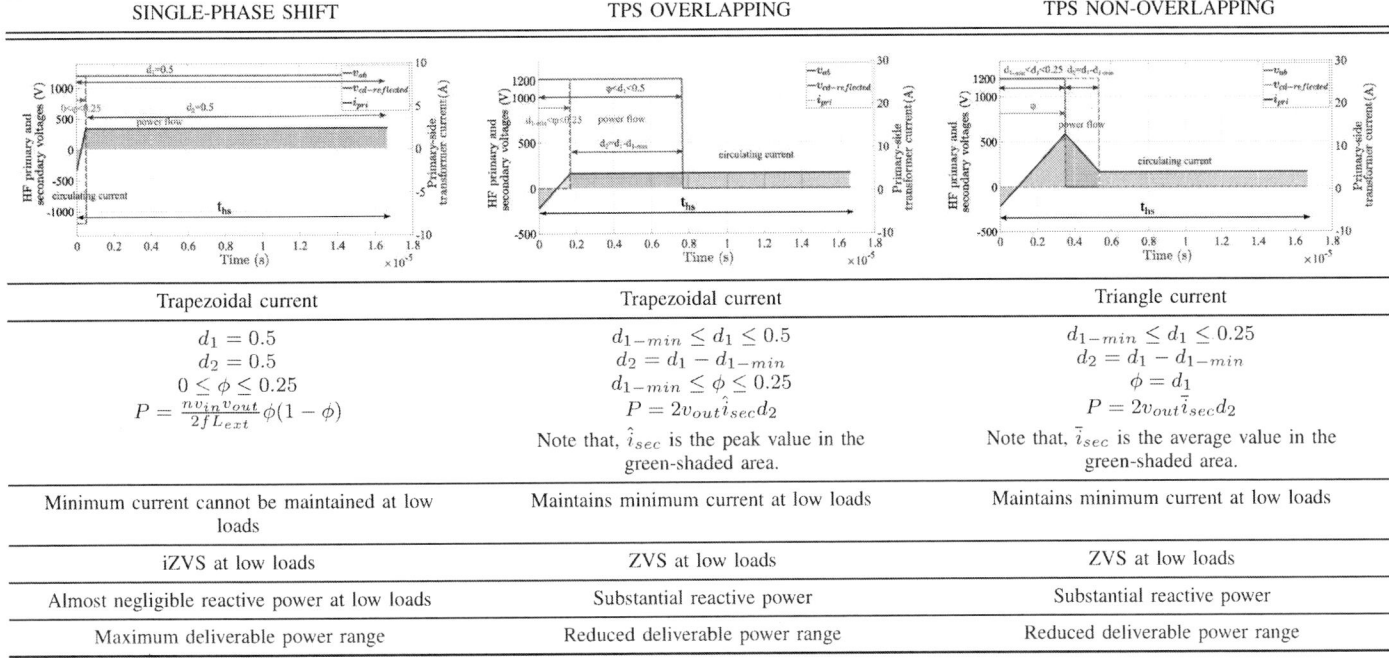		
Trapezoidal current	Trapezoidal current	Triangle current
$d_1 = 0.5$ $d_2 = 0.5$ $0 \leq \phi \leq 0.25$ $P = \frac{n v_{in} v_{out}}{2 f L_{ext}} \phi (1 - \phi)$	$d_{1-min} \leq d_1 \leq 0.5$ $d_2 = d_1 - d_{1-min}$ $d_{1-min} \leq \phi \leq 0.25$ $P = 2 v_{out} \hat{i}_{sec} d_2$ Note that, $\hat{i}_{sec}$ is the peak value in the green-shaded area.	$d_{1-min} \leq d_1 \leq 0.25$ $d_2 = d_1 - d_{1-min}$ $\phi = d_1$ $P = 2 v_{out} \bar{i}_{sec} d_2$ Note that, $\bar{i}_{sec}$ is the average value in the green-shaded area.
Minimum current cannot be maintained at low loads	Maintains minimum current at low loads	Maintains minimum current at low loads
iZVS at low loads	ZVS at low loads	ZVS at low loads
Almost negligible reactive power at low loads	Substantial reactive power	Substantial reactive power
Maximum deliverable power range	Reduced deliverable power range	Reduced deliverable power range

the charge to be discharged from the device output capacitance C_{oss}. However, C_{oss} is a nonlinear function of the device voltage, i.e., $C_{oss} = f(V_{ds})$ and hence, the constant term provided in device datasheet cannot be used directly. Thus, a charge equivalent device capacitance $C_q(V_{ds})$ is obtained using $C_{oss} - V_{ds}$ curve from the datasheet, and given by,

$$C_q(V_{ds}) = \frac{\int_0^{V_{ds}} C_{oss}(v) dv}{V_{ds}} \qquad (1)$$

is better suited for ZVS analysis. Following this, the minimum energy required in the external inductor is required to satisfy,

$$\frac{1}{2} L_{ext} i_L^2 \geq 2 C_q v_{in}^2. \qquad (2)$$

Note in (2), i_{L-min} needs to be sufficient for four times C_q, owing to four devices in full-bridge configuration. Thus,

$$i_{L-min} = \sqrt{\frac{4 C_q v_{in}^2}{L_{ext}}}. \qquad (3)$$

B. Calculation of t_d

Note that, a resonance circuit with L_{ext} and C_q is formed during the deadtime t_d. Thus, the maximum change in voltage (resonant peaks observed in Fig. 2(i) in the simulation study presented) before the current i_{pri} changes direction is observed at $1/4^{th}$ of the resonance frequency, providing a suitable solution for deadtime calculation as [15],

$$t_d = \frac{\pi}{2} \sqrt{L_{ext} C_q}. \qquad (4)$$

C. Calculation of ΔV_{ds}

Solution for ΔV_{ds} is obtained by approximation method as described subsequently. During t_d interval, L_{ext}-C_q resonant circuit is formed. Using averaged current $\bar{i}_{pri}$ over t_d interval,

$$C_q \frac{dv}{dt} = \bar{i}_{pri}, \qquad (5)$$

where dv represents the change occurred in voltage across C_q during dt time. Here, $dt = t_d$, thus,

$$dv = \bar{i}_{pri} \frac{dt}{C_q}. \qquad (6)$$

Substituting in (6), the initial conditions, yields,

$$\Delta V_{ds} = \frac{1}{4} \left(2 v_{in} - \frac{i_{pri}(t_{d1}) + i_{pri}(t_{d2})}{2} \frac{t_d}{C_q} \right), \qquad (7)$$

where $i_{pri}(t_{d1})$ and $i_{pri}(t_{d2})$ are the values of HF transformer current at time instant of starting and finishing of deadtime interval. The values of $i_{pri}(t_{d1})$ and $i_{pri}(t_{d2})$ is obtained by solving the L_{ext}-C_q resonant circuit, which is omitted here for sake of brevity. The solution obtained from (7) is in close agreement with the simulation and experimental results, as shown in Fig. 3.

III. TPS MODULATION STRATEGIES FOR UNITY-GAIN DAB CONVERTERS

There are different operation modes to generate high-frequency (HF) converter voltage waveforms, however, only two different TPS modulation techniques applicable for unity gain are discussed and shown in Table II. The two modulation

979-8-3315-1612-3/25 $31.00 © 2025 IEEE

Fig. 4. Experimental setup of a 1200-V/800-V 30-kHz 22-kVA DAB converter.

TABLE III
EXPERIMENTAL SETUP PARAMETERS.

System Parameters	Definitions	Values
L_{ext}	DAB inductance	250 μH
v_{in}	Input dc-link voltage	1200 V
v_{out}	Output dc-link voltage	800 V
n	DAB transformer turns ratio	3:2
f	DAB switching frequency	50 kHz
L_m	Magnetizing inductance	8 mH
$S_1 - S_4$	Primary-side switches	Infineon F412MR20W3M1_B11
$S_5 - S_8$	Secondary-side switches	ROHM BSM180D12P2E002

technique are classified as TPS overlapping and TPS non-overlapping modulation technique. The term overlapping indicates overlapping of high-frequency converter voltage waveforms (i.e., v_{ab} and v_{cd}) during a half-switching period, and otherwise for non-overlapping modulation technique.

In comparison with the SPS modulation, both the TPS modulation techniques ensure minimum current in the external inductor as defined in (3), even at light load conditions. However, TPS overlapping modulation technique preserves the trapezoidal current waveform through the HF magnetics, whereas TPS non-overlapping modulation technique drives triangular current. Note that the green area under the curve as shaded is responsible for the power delivery, whereas the brown shaded area indicates circulating power. Owing to trapezoidal-shaped current, among TPS modulation techniques, overlapping modulation technique provides lower circulating current and lower difference between the peak and RMS values of the high-frequency transformer current.

The duty cycles can be designed with a specific control objective, i.e., $\{\phi, d_{1-min}\}$ to ensure the peak current requirement (i_{L-min}) is met and d_2 is adjusted to ensure delivery of the required power. To ensure minimum current at the switching instant, the minimum duty cycle of the primary-side full-bridge converter voltage is given by,

$$ d_{1-min} = \frac{i_{L-min} L_{ext} f}{v_{in}}. \tag{8} $$

IV. EXPERIMENTAL RESULTS

The experimental setup of a 1200-V/800-V 30-kHz 22-kVA DAB converter with parameters listed in Table III is shown in Fig. 4. The primary-side devices are Infineon F4-12MR20W3M1_B11 modules and the secondary-side devices are ROHM BSM180D12P2E002. The high-frequency transformer is custom-built by 9-stacked UU101/114.2/25.4 ferrite core operating at 30-kHz meeting 80-kV insulation requirement. The gate signals are provided through the optical fibre cables from Myway MWPE4-IPFPGA24. The experiments are carried up to a power flow of 9 kW, to emphasize on iZVS operation at low loading conditions. Beyond 9 kW of power

flow, SPS is inherently suitable. A fixed deadtime of 800 ns is used throughout the experiments.

Fig. 5 displays the HF converter voltage waveforms (v_{ab} and v_{cd}) with SPS modulation technique at 0.5 kW, 1.5 kW, 4 kW and 6 kW, from left to right. The presence of ΔV_{ds} on the high-frequency converter voltage v_{ab} at light loads, specifically in Fig. 5(a), (b) and (c) indicates iZVS operation. In contrast, Fig. 5(d) meets the requirement for ZVS with i_{L-min}, thus having negligible ΔV_{ds} on v_{ab}. Note that, from the experimental results presented in Fig. 5(a), (b) and (c), the resonant peak occurs at approximately 650 ns. With the known inductance value L_{ext}, (4) could be used to recalculate the effective capacitance in the system undergoing energy exchange with the external inductor during the deadtime. Thus, from the results, C_q is approximately equal to 800 pF.

In contrast, Fig. 6 shows HF converter voltage waveforms with TPS-overlapping modulation technique at 0.5 kW, 1.5 kW and 4 kW from left to right. Unlike Fig. 5(a)-(c) with SPS modulation technique, Fig. 6(a)-(c) with TPS-overlapping modulation technique indicates ZVS operation as the rising or falling edges of v_{ab} doesn't show the presence of abrupt voltage changes ΔV_{ds}.

Fig. 7 presents the efficiency results, demonstrating improved light load efficiency (for power less than 6 kW) with TPS-overlapping mode in comparison with the SPS modulation technique. Note that, for operation beyond 6 kW, SPS modulation technique is suitable. At 3% loading conditions, 10% improvement of efficiency is observed with TPS-overlapping modulation technique in comparison to the SPS modulation technique. Although the overall improvement in efficiency is modest due to presence of circulating currents, the wide-band-gap devices benefit from reduced switching losses and thus, reduced junction temperature increment.

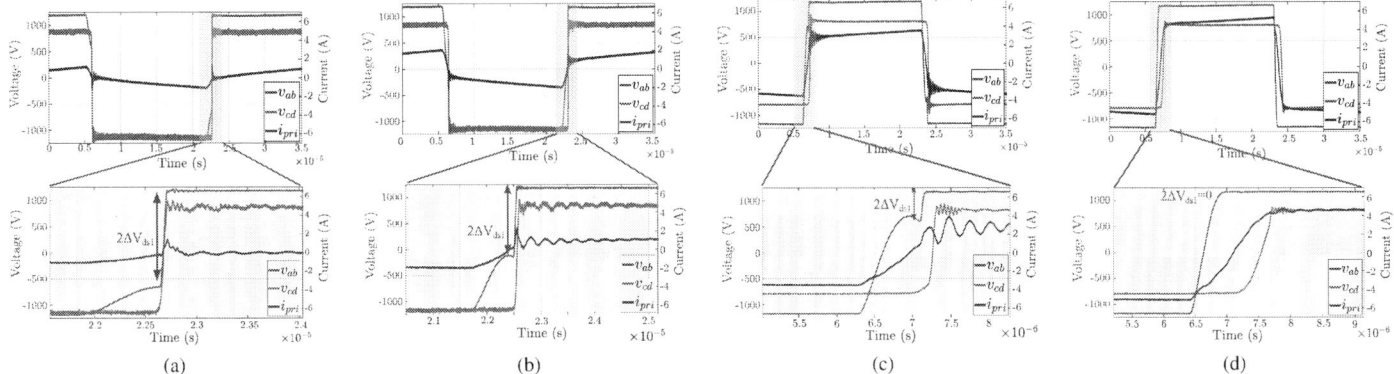

Fig. 5. Experimental waveforms showing high-frequency converter voltages (v_{ab} and v_{cd}) and primary-side high-frequency transformer current (i_{pri}) with SPS modulation technique for different loading conditions: (a) 0.5 kW, (b) 1.5 kW, (c) 4 kW and (d) 6 kW. Figs. 5(a)-(c) highlights iZVS with abrupt voltage changes on v_{ab} at low loading conditions and Fig. 5(d) highlights ZVS operation readily achievable using SPS modulation technique at high loading conditions with a smooth transition during rising edge of v_{ab}.

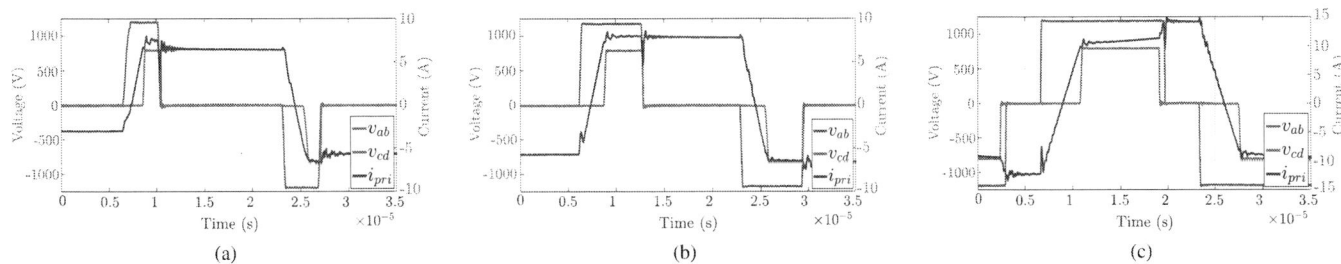

Fig. 6. Experimental waveforms showing high-frequency converter voltages (v_{ab} and v_{cd}) and primary-side high-frequency transformer current (i_{pri}) with TPS-overlapping modulation technique for different loading conditions: (a) 0.5 kW, (b) 1.5 kW and (c) 4 kW. Absence of any abrupt voltage changes on v_{ab} indicates ZVS operation.

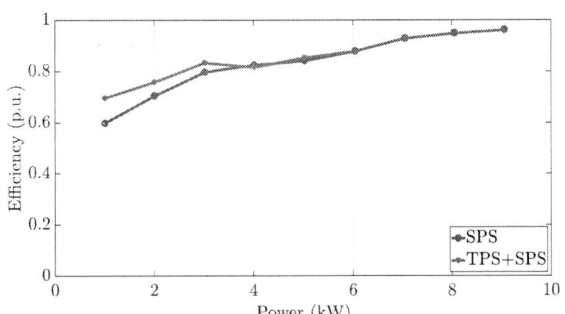

Fig. 7. Comparison of efficiency of the DAB converter between SPS and TPS-overlapping modulation technique. Beyond 6 kW, SPS modulation technique is used, thus the efficiency curves coincide.

V. CONCLUSION

This paper investigates incomplete zero voltage switching (ZVS) in unity-gain dual-active-bridge (DAB) converters under light-load conditions, with simulations incorporating device output capacitance. Two triple-phase shift modulation strategies are proposed to ensure minimum inductor current, thereby achieving ZVS at light loads. The duty cycle calculations are versatile and can be adapted to various unity-gain

DAB converter designs. Experimental results demonstrate that the proposed triple-phase shift modulation improves efficiency by up to 10% at light loads compared to conventional single-phase shift modulation.

ACKNOWLEDGMENT

This work was supported in part by the Energy Research Institute, Nanyang Technological University, Singapore. The authors extend their gratitude to the institute for its financial support.

REFERENCES

[1] B. Zhao, Q. Song, W. Liu, G. Liu, and Y. Zhao, "Universal high-frequency-link characterization and practical fundamental-optimal strategy for dual-active-bridge dc-dc converter under pwm plus phase-shift control," *IEEE Trans. Power Electron.*, vol. 30, no. 12, pp. 6488–6494, Dec. 2015.

[2] B. Zhao, Q. Song, W. Liu, and Y. Sun, "Overview of dual-active-bridge isolated bidirectional dc-dc converter for high-frequency-link power-conversion system," *IEEE Trans. Power Electron.*, vol. 29, no. 8, pp. 4091–4106, 2014.

[3] S. Shao, L. Chen, Z. Shan, F. Gao, H. Chen, D. Sha, and T. Dragicevic, "Modeling and advanced control of dual-active-bridge dc-dc converters: A review," *IEEE Trans. Power Electron.*, vol. 37, no. 2, pp. 1524–1547, Feb. 2022.

[4] R. Haneda and H. Akagi, "Power-loss characterization and reduction of the 750-v 100-kw 16-khz dual-active-bridge converter with buck and

979-8-3315-1612-3/25 $31.00 © 2025 IEEE

boost mode," *IEEE Trans. Ind. Appl.*, vol. 58, no. 1, pp. 541–553, Jan. 2022.

[5] S. Pistollato, T. Caldognetto, P. Mattavelli, and P. Magnone, "Triple-phase shift modulation for dual active bridge based on simplified switching loss model," in *Proc. 2019 AEIT International Annual Conference (AEIT)*. IEEE, Sept. 2019, pp. 1–6.

[6] S. Pistollato, N. Zanatta, T. Caldognetto, and P. Mattavelli, "A low complexity algorithm for efficiency optimization of dual active bridge converters," *IEEE Open J. Power Electron.*, vol. 2, pp. 18–32, 2021.

[7] Y. Deng, S. Yin, J. Chen, and W. Song, "A comprehensive steady-state performance optimization method of dual active bridge dc-dc converters based onwith triple-phase-shift modulation," in *Proc. 2022 IEEE International Power Electronics and Application Conference and Exposition (PEAC)*, Nov. 2022, pp. 221–225.

[8] G. Jean-Pierre, N. Altin, A. El Shafei, and A. Nasiri, "Overall efficiency improvement of a dual active bridge converter based on triple phase-shift control," *Energies*, vol. 15, no. 19, p. 6933, Sept. 2022.

[9] O. M. Hebala, A. A. Aboushady, K. H. Ahmed, and I. Abdelsalam, "Generic closed-loop controller for power regulation in dual active bridge dc-dc converter with current stress minimization," *IEEE Trans. Ind. Electron.*, vol. 66, no. 6, pp. 4468–4478, Jun. 2019.

[10] S. Shao, M. Jiang, W. Ye, Y. Li, J. Zhang, and K. Sheng, "Optimal phase-shift control to minimize reactive power for a dual active bridge dc-dc converter," *IEEE Trans. Power Electron.*, vol. 34, no. 10, pp. 10 193–10 205, Jan. 2019.

[11] F. Krismer and J. W. Kolar, "Efficiency-optimized high-current dual active bridge converter for automotive applications," *IEEE Transactions on Industrial Electronics*, vol. 59, no. 7, pp. 2745–2760, 2012.

[12] Y. Yan, H. Gui, and H. Bai, "Complete zvs analysis in dual active bridge," *IEEE Trans. Power Electron.*, vol. 36, no. 2, pp. 1247–1252, 2021.

[13] M. Kasper, R. M. Burkart, G. Deboy, and J. W. Kolar, "Zvs of power mosfets revisited," *IEEE Trans. Power Electron.*, vol. 31, no. 12, pp. 8063–8067, 2016.

[14] H. L. Yeo, R. Sarda, and V. Sriram, "Device loss characterization procedure for dual-active bridge under light load conditions," in *Proc. 2021 IEEE 12th Energy Conversion Congress & Exposition - Asia (ECCE-Asia)*, 2021, pp. 967–974.

[15] H. Akagi, T. Yamagishi, N. M. L. Tan, S.-i. Kinouchi, Y. Miyazaki, and M. Koyama, "Power-loss breakdown of a 750-v 100-kw 20-khz bidirectional isolated dc-dc converter using sic-mosfet/sbd dual modules," *IEEE Trans. Ind. App.*, vol. 51, no. 1, pp. 420–428, 2015.

Design of Fully Soft-Switched Semi-Dual Active DC-DC Converter for Battery Charging Application

Siva Prabhakar
Department of Electrical Engineering
Indian Institute of Technology Bombay
Mumbai, India
sivaprabhakar123@gmail.com

Shiladri Chakraborty
Department of Electrical Engineering
Indian Institute of Technology Bombay
Mumbai, India
shiladri@ee.iitb.ac.in

Sandeep Anand
Department of Electrical Engineering
Indian Institute of Technology Bombay
Mumbai, India
sa@ee.iitb.ac.in

Abstract—**Semi-dual active bridge converter (SDABC), a uni-directional variant of dual active bridge converter, is popular for battery charging applications. However, it suffers an efficiency dip at different charging instants of the constant current-constant voltage (CC-CV) charging profile, especially at low voltage-medium power conditions of CC mode and high voltage-low power conditions of CV mode. These dips are due to the loss of soft switching at these charging instants. To address this, a design-based approach focusing on transformer turns ratio and leakage inductance is proposed in this digest. The proposed design methodology achieves zero-voltage switching (ZVS) under low voltage-medium power conditions by properly selecting leakage inductance and under high voltage-low power conditions by carefully choosing the turns ratio. In addition, the discontinuous conduction property of SDABC is explored to reduce transformer circulating current at light loads, thereby enhancing efficiency by reducing conduction loss. Analytical results demonstrate that the proposed design enables soft switching across the CC-CV profile and improves efficiency. Validation with a 1.5 kW laboratory-developed hardware prototype of OBC charging 96 V battery shows a 61 % efficiency improvement at the start of CC mode and an improvement of 66 % at 30 % loading at CV mode with the proposed design compared to the conventional design.**

Index Terms—**Isolated DC-DC converter, onboard charger, semi-dual active bridge, zero voltage switching.**

I. INTRODUCTION

Isolated DC-DC converters are widely used in low-power applications like onboard chargers (OBCs) for light electric vehicles (LEVs), data centers, and aerospace applications. The semi-dual active bridge converter (SDABC), shown in Fig. 1, is a cost-effective choice for these applications requiring unidirectional power flow requirements [1]–[6]. However, these applications face wide voltage and current variations, leading to variable load conditions across the load profile for the majority of the duration. Therefore, it is necessary to maintain better efficiency across the entire load profile [7]–[9].

SDABC shown in Fig. 1, with a half-bridge configuration on the primary side and a semi-active bridge configuration on the secondary side of the transformer, is widely used for low-power applications [7]. It typically uses single phase shift (SPS) control for managing the power flow [7], [8]. The duty cycle of switches is maintained at 50 % and the phase shift between the active switch legs is varied to control power flow in SPS control. Conventionally designed SDABC with SPS control loses zero voltage switching (ZVS) and

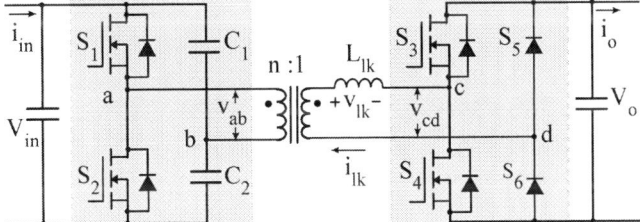

Fig. 1: Semi-dual active bridge converter (SDABC) with half-bridge configuration on the primary side and semi-active bridge configuration with a diode leg and switch leg on the secondary side of the transformer.

has large circulating currents at part load conditions [7], [8]. A hardware-based solution to improve the ZVS range by combining several inductors using additional mechanical or controllable switches to obtain variable inductance values is employed in [10], [11]. However, this involves using extra lossy switches and bulkier magnetic elements. A software-based approach employing variable switching frequency is introduced in [8], [12], [13]. However, a variable frequency results in complex filter designs and may lead to bulkier filters. Another software-based approach using fixed frequency and adaptive dead time control is used in [14]–[16]. However, this requires precise information regarding the switching behavior of active devices and involves complex control. Control complexity is reduced by employing a full-bridge configuration on the primary side and introducing primary side phase-shift variation in addition to SPS control to improve the light-load efficiency in [17]. However, this is an over-design for low-power applications due to the use of a full bridge configuration on the primary side of the transformer [18], [19]. This limits the feasibility of adopting the control strategy suggested in [17]. A hardware-based approach of selecting the transformer parameter to enhance the efficiency is suggested in [7]. However, this approach results in higher conduction loss at heavy load conditions.

In this paper, a hardware-based approach that focuses on selecting transformer turns ratio and leakage inductance is introduced to improve the ZVS range and part load efficiency of SDABC under SPS control. The approach also ensures minimum conduction loss at heavy load conditions. Section

979-8-3315-1612-3/25 $31.00 © 2025 IEEE

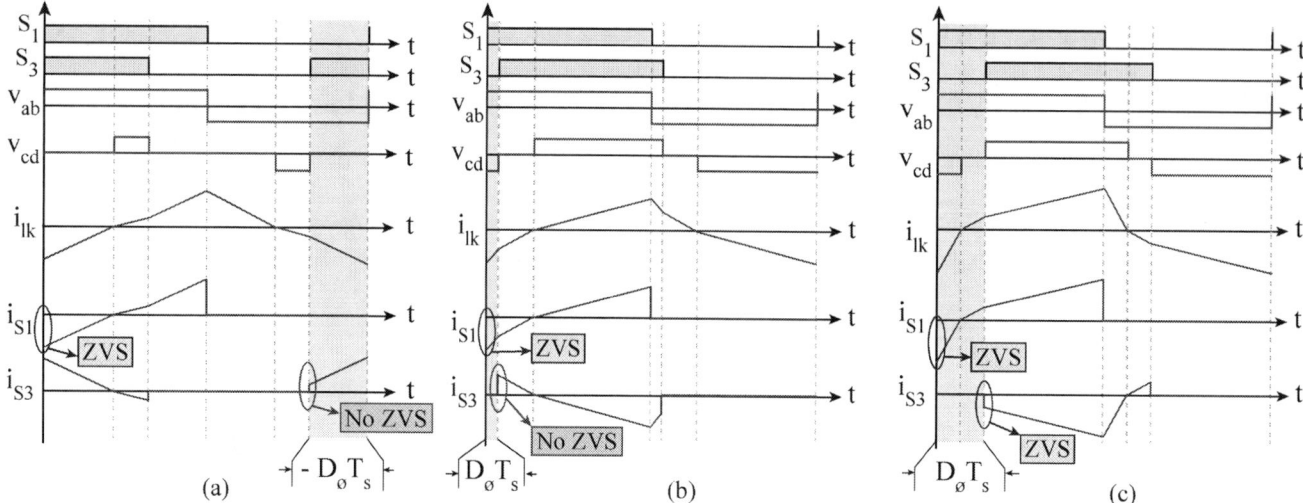

Fig. 2: Representative waveforms of the SDABC for step-down operation i.e. $M(= \frac{V_{in}}{2nV_o}) > 1$ (a) mode-1 $(-0.75T_s < D_\phi T_s < 0)$ (b) mode-2 $(0 < D_\phi T_s < D_\phi * T_s)$ (c) mode-3 $(D_\phi * T_s < D_\phi T_s < 0.5T_s)$. Modes 1 and 2 are suitable for low and medium power ranges, respectively. These modes fail to satisfy the directional ZVS turn-on condition for the secondary-side switches, as can be noticed from the non-negative values of the switch current. However, mode–3 satisfies ZVS turn–on, and it is desirable to extend the operating range of this mode.

II explores the different operating modes of SDABC and discusses the shortcomings of a conventional design approach when employed for a battery charging application. A design methodology proposed to ensure soft switching over the entire charging profile is introduced in section III. In addition, the methodology helps reduce conduction loss at part load conditions. The analytically obtained efficiency values and experimental results for a 1.5 kW laboratory-developed hardware prototype are provided to validate the proposed design methodology in section IV. Section V concludes the paper.

II. SEMI-DUAL ACTIVE BRIDGE: CONVENTIONAL DESIGN AND ITS LIMITATIONS

This section analyses the basic operating modes of SDABC. It also reviews a conventional design approach and highlights its limitations when used for a battery charging application.

A. SDABC operating modes and design parameters

The power transfer in SDABC shown in Fig. 1 depends on the transformer winding current and AC node voltages of the bridge converter. The transformer winding current is the same as the current through the leakage inductance (L_{lk}), denoted as i_{lk}. L_{lk} and turns ratio (n) of the transformer are two critical design parameters of SDABC, as variation in n and L_{lk} influence i_{lk}. This current, i_{lk} is decided by the voltage appearing across L_{lk} (v_{lk}) and the value of L_{lk}. v_{lk} varies with variation in n. The slope of i_{lk} during each transition interval depends on voltage, v_{lk}, and the value of L_{lk}. The AC node voltages of the bridge converter could be varied by changing the duty cycle (D) of the active switch legs and the phase shift (D_ϕ) between the primary and secondary side active switch legs. Thus, for a fixed design, the power transfer in SDABC is controlled by either varying D or D_ϕ. In this paper, SPS control is used, where D of all active switches are fixed at 50 %, and D_ϕ is used to regulate the power flow.

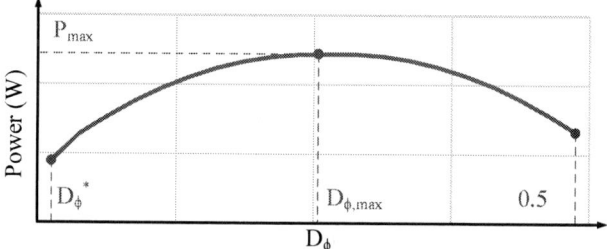

Fig. 3: Variation of power with variation in phase shift for mode-3 operation. The power increases initially with the increase in D_ϕ. However, after $D_{\phi,max}$ power decreases as the current freewheeling dominates to reduce power.

For the SDABC shown in Fig. 1, the voltage conversion ratio is given by $M = \frac{V_{in}}{2nV_o}$, where V_{in} is the input voltage applied to SDABC and V_o is the battery voltage. SDABC performs voltage step-down operation if $M > 1$ and voltage step-up operation if $M < 1$. Thus, SDABC performs voltage step-down or step-up based on the operating voltages and n. For a fixed design, variation in D_ϕ results in different operating modes. Fig. 2 shows the voltage and current waveforms of three key operating modes of SDABC while performing voltage step-down. S_1 and S_3 are the gate signals applied to switches S_1 and S_3 respectively. $D_\phi T_s$ is the phase delay introduced between the switching of S_1 and S_3. Here, T_s is the time period of one switching cycle. v_{ab} and v_{cd} are the voltage across the primary and secondary sides of the transformer. For a half-bridge configuration on the primary side of the transformer, the AC node voltage is half of the input voltage and the polarity is decided by the switching of primary side active switches. Thus, v_{ab} is either $\frac{V_{in}}{2}$ when S_1 is on or $\frac{-V_{in}}{2}$ when S_2 is on. v_{cd} is the AC node voltage of the semi-active bridge with a diode leg and switch leg. Therefore, v_{cd} is either V_o, 0 or $-V_o$ depending on the switching of

979-8-3315-1612-3/25 $31.00 © 2025 IEEE

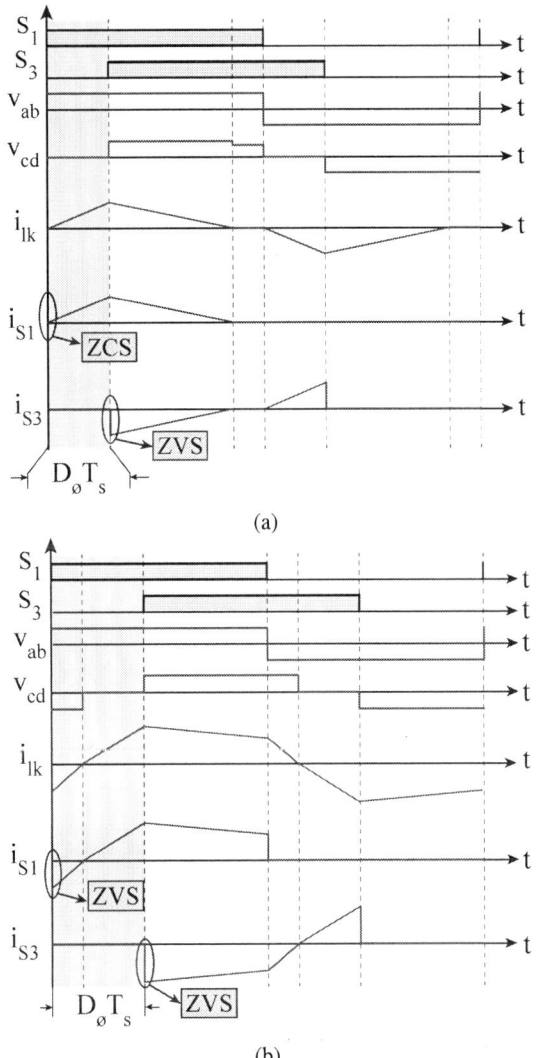

(a)

(b)

Fig. 4: Representative waveforms of SDABC for step-up operation i.e., $M(= \frac{V_{in}}{2nV_o}) < 1$ (a) mode-4 $(0 < D_\phi T_s < D_c T_s)$, suitable for light load (b) mode-5 $(D_c T_s < D_\phi T_s < 0.5T_s)$, suitable for medium and heavy load. Both modes satisfy ZVS turn–on conditions. Thus, step-down operation is desirable for all ranges of CV mode.

switches S_3 and S_4, and the current through the transformer winding. The diodes in the bridge leg of the secondary side bridge conduct based on the direction of current through the bridge leg. i_{S1} and i_{S3} are the currents through switches S_1 and S_3 respectively.

A negative value of switch currents at the switch turn-on instant satisfies the directional ZVS turn-on condition. S_1 satisfies the directional ZVS turn-on condition across all three modes, whereas S_3 satisfies the same only for mode-3. S_3 undergoes hard switching for mode-1 and mode-2 operation. Modes 1, 2, and 3 cover the entire load range for $M > 1$ operation. Mode-1 is defined over the range $-0.75T_s < D_\phi T_s < 0$, mode-2 over the range $0 < D_\phi T_s < D_\phi * T_s$, while mode-3 over the range $D_\phi * T_s < D_\phi T_s < 0.5T_s$. Thus, SDABC goes into different operating modes based on

D_ϕ values. Power transferred to the load is directly related to D_ϕ. Thus, mode-3 with a higher D_ϕ range is suitable for heavy load operation. However, for SDABC, there is an inherent current freewheeling interval on the secondary side as shown by the zero voltage duration of v_{cd}. During this interval, power is not transferred to the load and instead circulates within the power devices and through the transformer windings. This freewheeling interval increases with an increase in phase shift during mode-3 operation. Thus, the power that increases initially decreases after a particular D_ϕ value given by $D_{\phi,max}$. This is the value of D_ϕ at which maximum power transfer (P_{max}) happens for a given design of SDABC. Fig. 3 shows the variation of power with D_ϕ during mode-3 operation. If SDABC is designed to operate in mode-3 at rated conditions, then at medium and light load conditions SDABC operates in modes 2 and 1 respectively, due to variation of D_ϕ values to meet power flow requirements. In mode-1 operation, in addition to hard switching behaviour, the freewheeling duration is high, resulting in more conduction loss and poor efficiency.

Similarly, Fig. 4 shows the representative waveforms of two key operating modes of SDABC covering the entire load range, when performing voltage step-up operation. Unlike $M > 1$ operation, which requires three modes for full load range operation, $M < 1$ operation requires only two operating modes. Thus, medium load requirements could be met by either mode-4 or mode-5, depending on operating voltages and n. Mode-5 is similar to mode-3 in terms of voltage and current waveform shape. Thus, mode-5 is suitable for medium to heavy load conditions. However, for the same power transfer requirements, mode-5 undergoes more current freewheeling than mode-3 [7]. Mode-4 corresponds to low values of phase shift and is suitable for operation in light to medium load conditions. SDABC goes into a discontinuous conduction state during mode-4 operation, allowing S_1 to turn on with zero current. In this mode, i_{lk} remains at zero value instead of freewheeling through the devices and transformer winding, resulting in lesser conduction loss. Thus, for $M < 1$ operation, S_1 satisfies zero current switching (ZCS) turn-on condition for mode-4 operation. Also, mode-4 satisfies directional ZVS turn-on conditions for S_1 as shown in Fig. 4 (a). Mode-5 satisfies the directional ZVS turn-on condition for both switches S_1 and S_3. Thus, for $M < 1$ conditions, primary and secondary side switches undergo soft switching across the two operating modes. However, SDABC could operate in either $M > 1$ or $M < 1$ condition depending on the operating voltages. Thus, care should be taken while designing SDABC to make it operate in either modes 3, 4, or 5 across the load range to ensure ZVS and better efficiency.

B. Conventional Design Approach of SDABC

The conventional design approach selects the values of n and L_{lk} that give minimum conduction loss at P_{max} condition of the charging profile [20]. Fig. 5 (a) shows a sample constant current-constant voltage (CC-CV) charging profile considered for the study. Fig. 5 (b) shows the variation in power across

979-8-3315-1612-3/25 $31.00 © 2025 IEEE

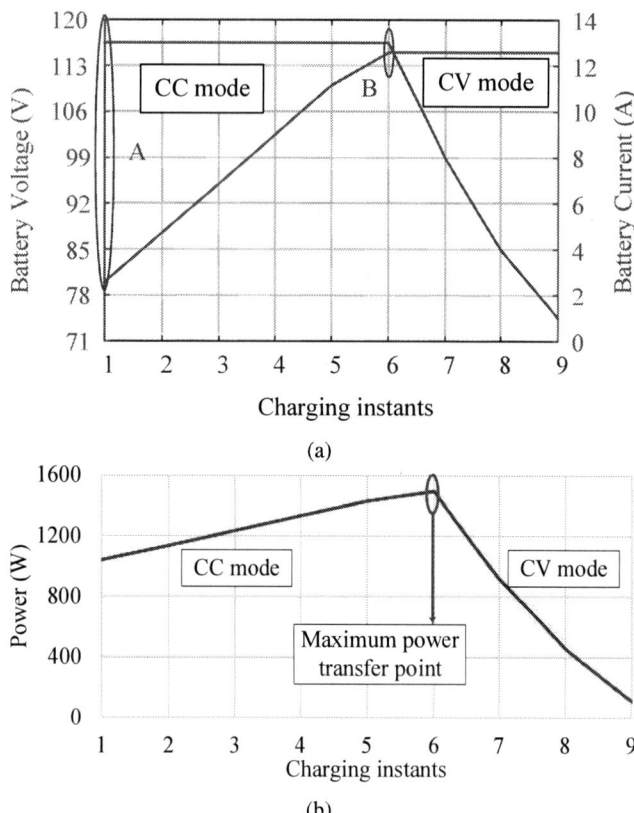

(a)

(b)

Fig. 5: (a) Typical CC-CV charging profile showing instants A and B. Conventional design is done at B and results in mostly $M > 1$ operation across the load range. The proposed design uses instant A for selecting L_{lk} to extend the mode-3 range. n value is selected using instant B to ensure step-up operation across the CV range. (b) Variation in power with different charging instants for the selected CC-CV charging profile.

the charging instants for the same profile. For this charging profile, P_{max} happens at the transition from CC to CV mode. During CC mode, power varies from a medium value to P_{max}, whereas during CV mode power varies from P_{max} to a very low value of power towards the end of CV mode.

For SDABC, based on operating voltages, either mode-3 or mode-5 is suitable for heavy load operation. However, employing the conventional design approach to this charging profile results in SDABC operation in mode-3 at P_{max}. The low freewheeling interval for the same power level for mode-3 operation compared to mode-5 operation is aiding for this particular selection. n value selection results $M > 1$ operation at P_{max} and across the entire CC-CV charging range. This leads to SDABC operation in mode-2 and mode-1 at low voltage-medium power conditions of CC mode and high voltage-low power conditions of CV mode, respectively, resulting in hard switching at these conditions. Hence, conventionally designed SDABC may operate in modes 1, 2, and 3 at low, medium, and high power conditions of the charging profile. However, it is preferred that SDABC operate in modes 3, 4, or 5 over the entire load range to ensure soft switching and lower conduction loss. Thus, design modification is required

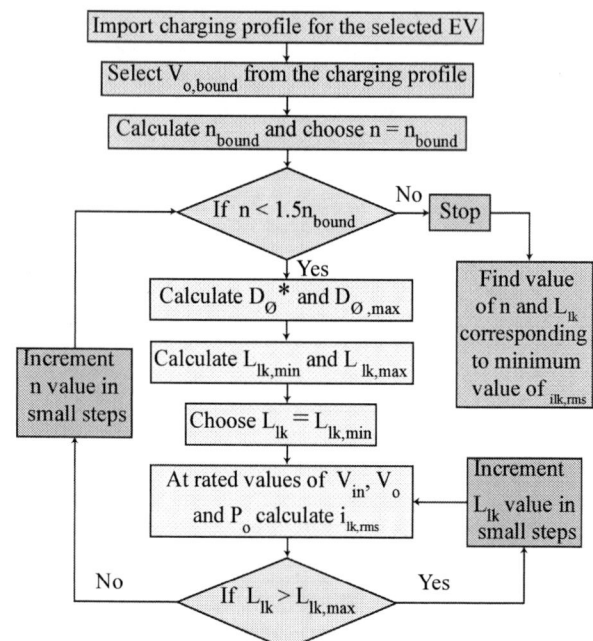

Fig. 6: Flow chart of the proposed design methodology which varies n and L_{lk} within specified limits to make SDABC operate in modes 3, 4, and 5. This ensures ZVS over the entire load range and discontinuous conduction at light load conditions of CV mode to reduce circulating current.

to ensure SDABC operation in either of these modes.

III. PROPOSED DESIGN METHODOLOGY

Fig. 6 illustrates the flow chart of the proposed design methodology used for selecting n and L_{lk} values to ensure soft switching across the charging profile. The proposed design approach aims to make SDABC operate in modes 3, 4, and 5 across the charging profile. V_{in} applied to SDABC is maintained a constant by the rectifier stage of OBC. First, the battery voltage ($V_{o,bound}$) at P_{max} is identified. n value corresponding to $V_{o,bound}$ is calculated as $n_{bound} = \frac{V_{in}}{(2nV_{o,bound})}$. The lower limit for n, n_{min} is set to n_{bound} and the upper limit for n, n_{max} is set to $1.5\,n_{bound}$. Varying n within this range ensures $n > n_{bound}$ for any voltage greater than or equal to $V_{o,bound}$. This ensures SDABC operation with $M < 1$ condition in the entire CV range. Thus, SDABC operates in either mode-4 or mode-5 in the entire CV range. Mode-4 operation at low power ranges of CV mode helps SDABC achieve ZVS and reduced conduction loss. If the n value is selected to be n_{bound}, then SDABC performs voltage step-down operation for the entire CC range.

Another design parameter available for the designer is L_{lk}. A low value of L_{lk} results in steeper current slopes and more power transfer. Thus, for a low value of L_{lk}, a small value of D_ϕ is sufficient to transfer a given amount of power to the load. Therefore, in this approach, the L_{lk} limits are set based on D_ϕ values. For each n value, D_ϕ^* and $D_{\phi,max}$ are calculated. D_ϕ^* is the value of phase shift at the boundary between mode-2 and mode-3 and $D_{\phi,max}$ is the value of phase shift at which maximum power can be transferred for

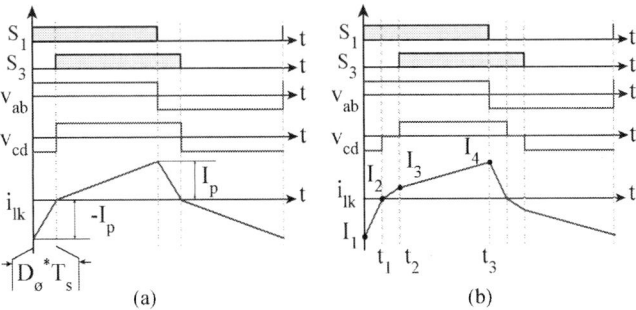

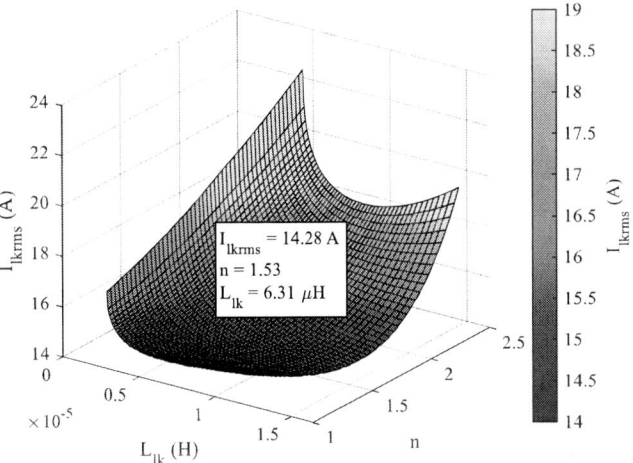

Fig. 7: Representative waveforms of SDABC (a) at the boundary between mode-2 and mode-3 (b) mode-3 operation, highlighting the corner values of i_{lk}. D_ϕ^* corresponds to D_ϕ value at boundary between mode-2 and mode-3. $D_{\phi,max}$ occurs at mode-3 for $M > 1$ and mode-5 for $M < 1$ condition.

Fig. 8: Variation of $I_{lk,rms}$ with variation in n and L_{lk} when employing conventional design strategy for the selected charging profile. n and L_{lk} are varied within their specified limits as per the conventional approach, and the combination of n and L_{lk} with results in a minimum value of $I_{lk,rms}$ is selected. The selected values of n and L_{lk} and the corresponding $I_{lk,rms}$ are highlighted.

a given design of SDABC. D_ϕ^* is calculated at the lowest boundary voltage of CC mode. Now, $L_{lk,min}$ is selected corresponding to this. This ensures that even if $n = n_{bound}$ after design, SDABC operates in mode-3 over the entire CC range. $L_{lk,max}$ is set corresponding to $D_{\phi,max}$. $D_{\phi,max}$ is calculated at $V_{o,bound}$, thereby ensuring that SDABC could transfer the maximum power for the selected charging profile if $L_{lk} = L_{lk,max}$. This is the maximum value of L_{lk} which could transfer the specified P_{max} corresponding to particular operating voltages and n. For each combination of n and L_{lk}, the RMS value of i_{lk} ($i_{lk,rms}$) is calculated at P_{max}. From all $i_{lk,rms}$ values, the value of n and L_{lk} that yields the minimum $i_{lk,rms}$ is selected. This selection ensures that conduction loss is minimal at P_{max}. In addition, restricting n and L_{lk} within these limits allows SDABC operation in modes 3, 4, and 5 across the charging profile. The detailed derivations of D_ϕ^* and $D_{\phi,max}$ are provided below.

1) D_ϕ values corresponding to L_{lk} limits: Fig. 7 (a) shows the representative waveforms of SDABC at the boundary between mode-2 and mode-3. During this condition, $D_\phi = D_\phi^*$, which is the lowest value of D_ϕ which ensures mode-3 operation. The peak value of i_{lk} (I_p) at this boundary condition is given by

$$I_p = (\frac{V_{in}}{2nL_{lk}} + \frac{V_o}{L_{lk}})D_\phi^* T_s = (\frac{V_{in}}{2nL_{lk}} - \frac{V_o}{L_{lk}})(D - D_\phi^*)T_s. \quad (1)$$

On solving the above equation D_ϕ^* could be obtained as

$$D_\phi^* = D(\frac{1}{2} - \frac{nV_o}{V_{in}}). \quad (2)$$

As discussed earlier in Section II A, P_{max} happens at mode-3 or mode-5 operations for $M > 1$ and $M < 1$ conditions respectively. Since mode-3 and mode-5 operation has similar voltage and current shape, for estimating $D_{\phi,max}$ mode-3 operation could be considered. The input power, P_{in} is given by $P_{in} = V_{in}I_{in}$. Here, I_{in} is the average value of the input current. For a converter with a half-bridge configuration on the primary side of the transformer, I_{in} is half of the average value of primary side leakage inductance current over the period DT_s duration. The primary side leakage inductance current

could be obtained by reflecting i_{lk} waveform to the primary side. The corner points of i_{lk} are given by

$$I_1 = I_2 - (\frac{\frac{V_{in}}{2n} + V_o}{L_{lk}})t_1, \quad I_3 = I_2 + \frac{V_{in}}{2nL_{lk}}(t_2 - t_1),$$
$$I_2 = 0, \quad I_4 = I_3 + (\frac{\frac{V_{in}}{2n} - V_o}{L_{lk}})(t_3 - t_2). \quad (3)$$

Here, t_1 is the first zero crossing instant of i_{lk} in a switching cycle, t_2 is $D_\phi T_s$ and t_3 is DT_s. On solving (3), t_1 is given by

$$t_1 = (\frac{V_{in}D - 2nV_oD + 2nV_oD_\phi}{2V_{in} + 2nV_o})T_s \quad (4)$$

Using these corner points and time instants the average value of primary side leakage inductance current is estimated. P is obtained as

$$P = \frac{V_{in}}{2nT_s}(I_3(t_2 - t_1) + I_4(t_3 - t_2 - t_1)) \quad (5)$$

$D_{\phi,max}$ is obtained by equating the derivative of P with respect to D_ϕ to zero as

$$D_{\phi,max} = \frac{D(V_{in}^2 + 2nV_oV_{in} + 4n^2V_o^2)}{2V_o^2 + 4nV_oV_{in} + 4n^2V_o^2}. \quad (6)$$

From (2) and (6), it is evident that D_ϕ^* and $D_{\phi,max}$ depends only on the operating voltages and n.

IV. EXPERIMENTAL RESULTS AND DISCUSSIONS

Fig. 5 shows the charging profile considered for the experimental study. The conventional and proposed design methodologies were employed for this charging profile. Fig. 8 and Fig. 10 show the variation of $I_{lk,rms}$ with variation in n and L_{lk} when employing conventional and proposed design methodologies for this charging profile. n and L_{lk} are varied within their specified limits as per both approaches. The

979-8-3315-1612-3/25 $31.00 © 2025 IEEE

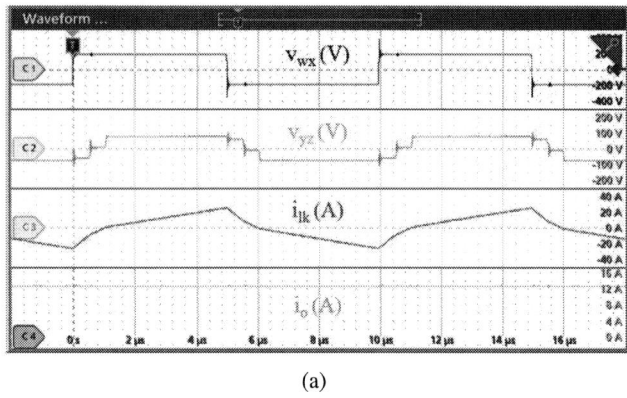 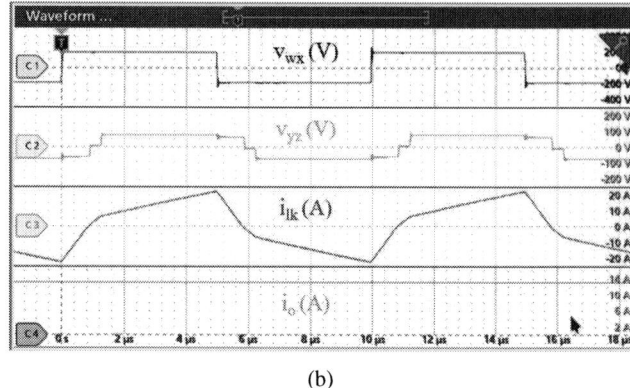

(a)   (b)

Fig. 9: Experimental waveforms of SDABC showing v_{wx} (V), v_{yz} (V), i_{lk} (A) and i_o (A) during the start of CC mode (V_o = 80 V, I_o = 13 A) (a) conventional design - mode-2 operation (no ZVS for secondary side switches) (b) proposed design - mode-3 (ZVS for all switches). The proposed design achieves 32 % efficiency improvement compared to conventional design.

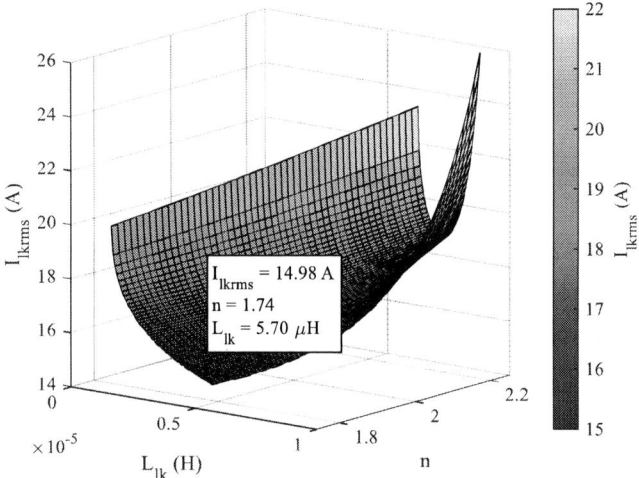

Fig. 10: Variation of $I_{lk,rms}$ with variation in n and L_{lk} when employing proposed design methodology for the selected charging profile. n and L_{lk} is varied within the modified limits as per the proposed approach to ensure SDABC operation in modes 3, 4, and 5 alone. The combination of n and L_{lk} with results in a minimum value of $I_{lk,rms}$ is selected. The selected values of n and L_{lk} and the corresponding $I_{lk,rms}$ are highlighted.

Fig. 11: Analytically obtained efficiency and ZVS status at different charging instants for conventional and proposed design approaches. The efficiency profile is flat for the proposed approach due to ZVS occurrence across the charging instants.

combination of n and L_{lk} with results in a minimum value of $I_{lk,rms}$ is selected for both cases. The values of n and L_{lk} with the conventional approach are 1.53 and 6.31 μH. n_{bound} for the selected profile is 1.739. With the particular selection of n value with the conventional approach, SDABC performs voltage step-down during the entire charging profile. Similarly, with the proposed approach, n and L_{lk} values are 1.74 and 5.7 μH respectively. The selected n value is just above n_{bound} for the charging profile. Conduction loss is expected to increase with an increase in n, since v_{lk} value increases with an increase in n value, resulting in a steeper slope and more RMS current. Thus, the n value near the lower limit (here, n_{bound}) results in lower conduction loss, justifying the selection. This n value ensures $M < 1$ operation across the entire CV range and $M > 1$ across the CC range. The particular selection of L_{lk} from values greater than $L_{lk,min}$

also ensures mode-3 operation over the entire CC range.

Fig. 11 compares the analytically obtained efficiency values and soft switching status at various voltages and current points within the selected CC-CV charging profile. The conventional design fails to maintain soft switching near the beginning of the CC mode and toward the end of the CV mode with the particular selection of n and L_{lk}, leading to reduced efficiency. In contrast, the proposed design ensures soft switching throughout the charging profile, significantly improving efficiency. In addition, SDABC operation in a discontinuous conduction state near the light load condition of CV mode results in a reduction of conduction loss. With the proposed design, SDABC achieves a relatively flat efficiency profile, as illustrated in Fig. 11.

Table I shows the specifications of important parameters of SDABC considered for the experimental study. The experimental setup developed in the laboratory for validation

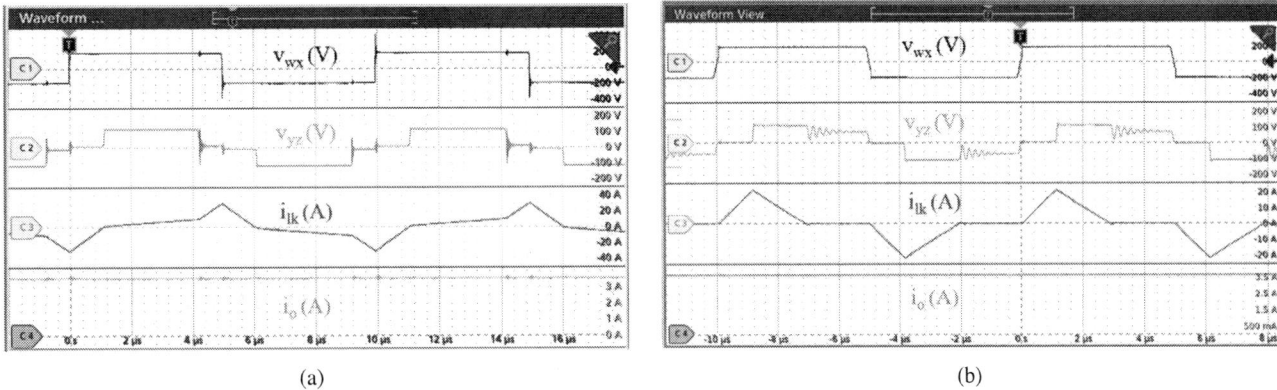

(a)   (b)

Fig. 12: Experimental waveforms of SDABC showing v_{wx} (V), v_{yz} (V), i_{lk} (A) and i_o (A) at 30 % loading during CV mode (V_o = 115 V, I_o = 3.5 A) (a) conventional design - mode-1 operation (no ZVS for secondary side switches) (b) proposed design - mode-4 (ZVS for all switches). The proposed design achieves 67 % efficiency improvement compared to conventional design.

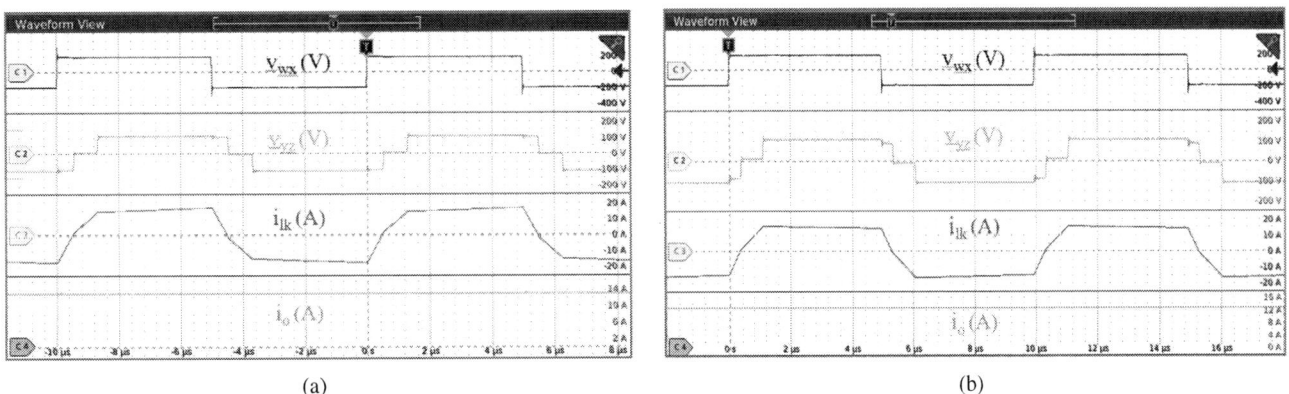

(a)   (b)

Fig. 13: Experimental waveforms of SDABC showing v_{wx} (V), v_{yz} (V), i_{lk} (A) and i_o (A) at maximum power transfer condition (V_o = 115 V, I_o = 13 A) (a) conventional design - mode-3 operation (ZVS for all switches) (b) proposed design - mode-5 (ZVS for all switches). The efficiency values are nearly the same with both approaches at this loading condition.

TABLE I: Key parameters considered for the experimental study

Parameters	Values
Input voltage (V_{in})	400 V
Nominal Battery voltage	96 V
Maximum battery voltage ($V_{o,bound}$)	115 V
Maximum battery charging current	13 A
Power at CC-CV boundary (P_{max})	1500 W
Switching frequency	100 kHz
n for conventional design	1.54
L_{lk} for conventional design	6.27 μH
n for proposed design	1.74
L_{lk} for proposed design	5.7 μH

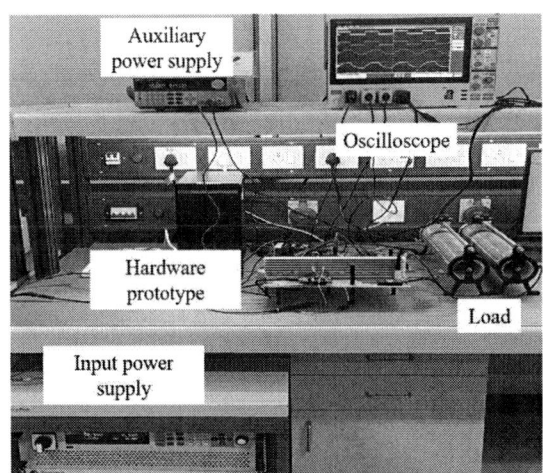

Fig. 14: Experimental setup developed in the laboratory.

is shown in Fig. 14. Fig. 9 (a) and Fig. 9 (b) show the experimental waveforms of SDABC showing v_{wx}, v_{yz}, i_{lk} and i_o at the start of CC mode (V_o = 80 V and I_o = 13 A) with conventional and proposed design respectively. As shown in Fig. 9 (a), SDABC operates in mode-2 with conventional design. The proposed design results in SDABC operation in mode-3 as shown in Fig. 9 (b), ensuring ZVS for all devices. At this loading condition, the conventional approach results in an efficiency of 66.3 % while the proposed design achieves an efficiency of 87.1 %. Similarly, Fig. 12 (a) and Fig. 12

(b) illustrate the same experimental waveforms of SDABC with conventional and proposed designs respectively towards the end of CV mode (V_o = 115 V and I_o = 4 A). The proposed design allows circuit operation in mode-4 to ensure better efficiency. However, with conventional design SDABC operates in mode-1. Efficiency is 58 % with the conventional approach and 97 % with the proposed approach at this loading condition. Finally, Fig. 13 (a) and Fig. 13 (b) show the same experimental waveforms of SDABC with conventional and proposed designs respectively at P_{max} condition (V_o = 115 V and I_o = 13 A). The conventional design results in SDABC operation in mode-3 and the proposed design in mode-5. Although conduction losses are expected to be higher for mode-5 operation than mode-3 operation, the reduction in the turn-off current results in similar efficiency values for this case. At full load condition, the efficiency with the conventional approach is 94 % while with the proposed approach, the efficiency is 93.7 %.

V. CONCLUSIONS

Conventionally designed SDABC fails to meet soft switching criteria at various regions of the CC-CV charging profile, especially at the start of CC mode and at light load conditions of CV mode. Transformer turns ratio (n) and leakage inductance (L_{lk}) are two important design parameters of SDABC. In this paper, a design methodology for selecting these parameters is devised to improve the soft switching range and the efficiency across the charging profile. The proposed design methodology achieves better efficiency at the start of CC mode by enabling a relatively flat current that satisfies directional ZVS turn-on conditions to flow through the devices. At light load conditions of CV mode, the discontinuous conduction mode during step-up operation is utilized to ensure ZVS and minimize conduction loss. With the proposed design, the efficiency at the start of CC mode (low voltage medium power condition) is 87.1 % compared to 66.3 % with conventional design. Furthermore, there is a 66% improvement in efficiency at 30 % loading in CV mode.

REFERENCES

[1] S. Kulasekaran and R. Ayyanar, "Analysis, Design, and Experimental Results of the Semidual-Active-Bridge Converter," in *IEEE Transactions on Power Electronics*, vol. 29, no. 10, pp. 5136-5147, Oct. 2014.

[2] D. Sha, J. Zhang and T. Sun, "Multimode Control Strategy for SiC mosfets Based Semi-Dual Active Bridge DC–DC Converter," in *IEEE Transactions on Power Electronics*, vol. 34, no. 6, pp. 5476-5486, June 2019

[3] Junming Zhang, Fan Zhang, Xiaogao Xie, Dezhi Jiao and Zhaoming Qian, "A novel ZVS DC/DC converter for high power applications," in *IEEE Transactions on Power Electronics*, vol. 19, no. 2, pp. 420-429, March 2004.

[4] T. Mishima and M. Nakaoka, "Practical Evaluations of a ZVS-PWM DC–DC Converter With Secondary-Side Phase-Shifting Active Rectifier," in *IEEE Transactions on Power Electronics*, vol. 26, no. 12, pp. 3896-3907, Dec. 2011.

[5] M. A. H. Rafi and J. Bauman, "Optimal Control of Semi-Dual Active Bridge DC/DC Converter With Wide Voltage Gain in a Fast-Charging Station With Battery Energy Storage," in IEEE Transactions on Transportation Electrification, vol. 8, no. 3, pp. 3164-3176, Sept. 2022.

[6] P. Ma, D. Sha and K. Song, "A Single-Stage Semi Dual-Active-Bridge AC–DC Converter With Seamless Mode Transition and Wide Soft-Switching Range," in *IEEE Transactions on Industrial Electronics*, vol. 70, no. 2, pp. 1387-1397, Feb. 2023.

[7] S. Prabhakar, N. Deshmukh, S. Anand, S. Chakraborty, M. Deo and P. Chaudhary, "Design Methodology to Improve Efficiency of Semi-dual Active Bridge Converter," *2024 IEEE Transportation Electrification Conference and Expo (ITEC)*, Chicago, IL, USA, 2024, pp. 1-6.

[8] N. Deshmukh et al., "Extended High Efficiency Operation of Semi - Active Half-Bridge DC-DC Converter," *2024 IEEE International Communications Energy Conference (INTELEC)*, Bengaluru, India, 2024, pp. 1-6.

[9] N. Deshmukh et al., "Diode Emulation Control for Efficiency Improvement of Dual Active Bridge Converters," *2024 IEEE International Communications Energy Conference (INTELEC)*, Bengaluru, India, 2024, pp. 1-6.

[10] H. Higa and J. -i. Itoh, "Extension of zero-voltage-switching range in dual active bridge converter by switched auxiliary inductance," *2017 IEEE Energy Conversion Congress and Exposition (ECCE)*, Cincinnati, OH, USA, 2017, pp. 5324-5331.

[11] H. Xie, Y. Zhang and F. Cai, "Variable inductance control of DAB converter to reduce reactive power," *2023 IEEE 6th International Electrical and Energy Conference (CIEEC)*, Hefei, China, 2023, pp. 1360-1364

[12] A. Cárcamo, A. Fernandez-Hernandez, F. Gonzalez-Hernando, A. Vázquez, and A. Rodríguez, "Variable switching frequency for zvs over wide voltage range in dual active bridge," *Electronics*, vol. 13, no. 10, 2024. [Online]. Available: https://www.mdpi.com/2079-9292/13/10/1800.

[13] H. Higa and J. -i. Itoh, "Extension of zero-voltage-switching range in dual active bridge converter by switched auxiliary inductance," *2017 IEEE Energy Conversion Congress and Exposition (ECCE)*, Cincinnati, OH, USA, 2017, pp. 5324-5331

[14] J. A. Santiago-González, D. M. Otten and D. J. Perreault, "Light Load Efficiency Improvements in Dual Active Bridge Converters via Dead time Control," *2018 IEEE 19th Workshop on Control and Modeling for Power Electronics (COMPEL)*, Padua, Italy, 2018, pp. 1-7.

[15] B. Ulrich, F. Ohler, F. Schenzle and T. Walter, "A Single Stage Dual Active Half-Bridge Single Phase Solid-State Transformer With Wide Input-Range," *2024 IEEE Applied Power Electronics Conference and Exposition (APEC)*, Long Beach, CA, USA, 2024, pp. 493-500.

[16] J. Li, Z. Chen, Z. Shen, P. Mattavelli, J. Liu and D. Boroyevich, "An adaptive dead-time control scheme for high-switching-frequency dual-active-bridge converter," 2012 Twenty-Seventh Annual IEEE Applied Power Electronics Conference and Exposition (APEC), Orlando, FL, USA, 2012, pp. 1355-1361

[17] M. Lu, X. Li and G. Chen, "A Hybrid Control of a Semi-dual Active-Bridge DC–DC Converter With Minimum Current Stress," in *IEEE Transactions on Power Electronics*, vol. 35, no. 3, pp. 3085-3096, March 2020.

[18] Hui Li, Fang Zheng Peng and J. S. Lawler, "A natural ZVS medium-power bidirectional DC-DC converter with minimum number of devices," in *IEEE Transactions on Industry Applications*, vol. 39, no. 2, pp. 525-535, March-April 2003.

[19] K. R and R. Kalpana, "An Isolated Dual-Input Half-Bridge DC–DC Boost Converter With Reduced Circulating Power Between Input Ports," in IEEE Canadian Journal of Electrical and Computer Engineering, vol. 45, no. 1, pp. 68-76, winter 2022.

[20] B. Sahoo, S. Chakraborty and S. Sarangi, "Cost-Performance Pareto Analysis of DAB-based DC-DC Converter Topologies for LEV Chargers," *2023 IEEE Energy Conversion Congress and Exposition (ECCE)*, Nashville, TN, USA, 2023, pp. 2095-2102.

A ZCS-ZVS Strategy For Low Impedance Dual Active Bridges in MHz range

1st Pushkar Saraf
Electrical and Computer Engineering
The University of Texas at Austin
Austin, United States
pushkarsaraf@utexas.edu

2nd Michael Solomentsev
Electrical and Computer Engineering
The University of Texas at Austin
Austin, United States
mys432@utexas.edu

2rd Alex Hanson
Electrical and Computer Engineering
The University of Texas at Austin
Austin, United States
ajhanson@utexas.edu

Abstract—At low voltage-to-current ratios and high frequencies, the dual active bridge can suffer significant turn-off losses on the high-current side and typical design rules can lead to inductances on the same order as parasitics, introducing significant unpredictability and loss. In this paper, we analyze the circuit behavior when the parasitic inductance of switching loops approaches the leakage inductance and propose operating the dual-active (half) bridge in boundary conduction mode, where the inductor current returns to zero before switching. This control achieves ZCS on the high-current side and ZVS on the high-voltage side, mitigating the effect of parasitic inductance and reducing switching loss. Operation over a wide power range is provided by burst-mode control. Experimental results verify the efficacy of the proposed strategy on a 1MHz 40W DAHB GAN converter which achieves 600 W/in^3 power density and 93% efficiency.

Index Terms—Dual Active Bridge, Burst Mode, ZCS, ZVS

I. INTRODUCTION

The dual active bridge (and its variants such as the dual active half bridge (DAHB)) is a widely used isolated topology for its low rms current, built-in bidirectional power flow, and built-in synchronous rectification [1]–[3]. The DAB belongs to a small subset of bi-directional dc-dc converters that can achieve zero-voltage switching (zvs) on all four switches [4], one of the key attractions for the DAB. Soft switching allows the DAB to operate at high switching frequencies, even into the MHz range.

Several control strategies have been developed to optimize the operation of dual-active-bridge (DAB) converters. These strategies include single-phase-shift (SPS), dual-phase-shift (DPS), and triple-phase-shift (TPS) control, all of which operate at a constant switching frequency. SPS offers simplicity and ease of implementation but with a limited soft-switching range. In contrast, DPS and TPS expand the soft-switching region and improve overall efficiency at the cost of increased computational and control complexity, as highlighted in various studies [5]–[8]. Frequency modulation has also been used to ensure the minimum necessary current to maintain ZVS over the entire operating range [9].

Under light-load conditions, the current through the leakage inductance of the dual-active bridge (DAB) decreases. This reduction in inductor current results in incomplete discharge

Research reported in this publication was supported by BMW.

of the switch's C_{oss} capacitance, leading to partial zero-voltage switching (ZVS), which is undesirable. To maintain ZVS across the entire operating range, it is essential to ensure that a minimum current flows through the leakage inductance, even at lighter loads. Burst mode control addresses this by intermittently switching the converter on and off. During the "on" phase, the converter delivers sufficient energy to maintain ZVS on all switches. The average of the "on" and "off" times effectively simulates a light-load condition while preventing ZVS degradation, ensuring stable performance even at reduced power levels [10]–[12].

While soft switching is useful in general, zero-voltage turn-on has been a greater preoccupation of existing DAB work than zero-current turn-off. However, under certain operating conditions (such as low voltages and high currents), turn-off losses and parasitic inductance become more significant than turn-on losses and parasitic capacitance [13]. In these conditions, the $1/2LI^2$ energy lost in parasitic inductance each cycle is substantial. In addition, the high currents combined with substantial parasitic inductance can lead to significant waveform distortion due to the Ldi/dt voltage across parasitic inductances. This distortion in the switch node waveforms results in non-ideal operation of the converter, contributing to even higher losses and reduced efficiency. To mitigate this, one reported strategy for Dual Active Bridges (DABs) ensures ZVS on one primary leg and ZCS on the other three legs of the converter to minimize switching loss [13]–[17]. However, the gating signals for each leg of primary/secondary are not symmetrical and this increases the complexity of the control, which is especially limiting at higher frequencies and for lower power cost-limited converters. Furthermore, this strategy cannot be applied to low-power variants of DAB such as the dual active half bridge.

To overcome the above challenges, we propose a boundary conduction mode control strategy [18] to achieve ZCS on the high-current side and ZVS on the high-voltage side of DABs or DAHBs. The strategy can be applied to all variants of DAB. To ensure high efficiency over a wide power range, burst mode control is used to maintain the ZCS-ZVS condition over the entire power range. The proposed control is particularly useful for converters in which one side has very low voltage-to-current ratios and the operating frequency is high.

979-8-3315-1612-3/25 $31.00 © 2025 IEEE

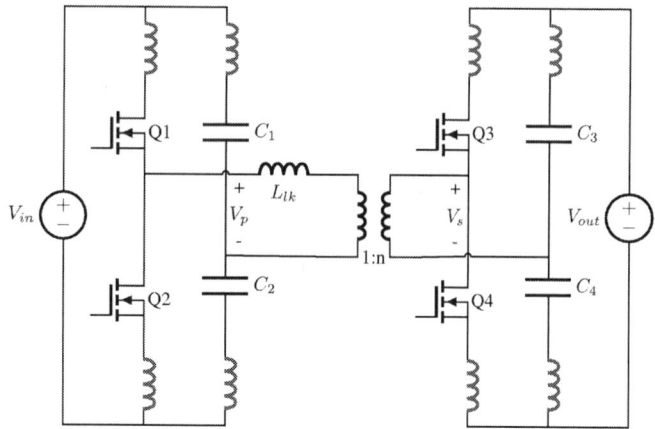

Fig. 1: Schematic of dual active half bridge, parasitic inductances are highlighted in red

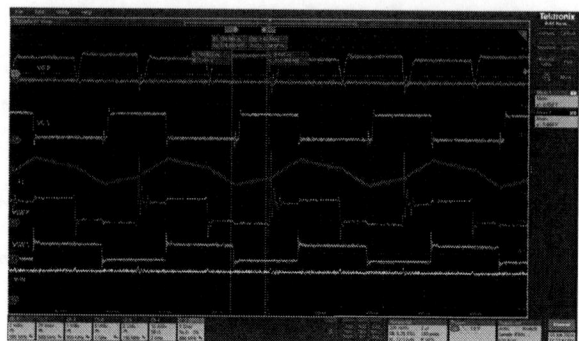

Fig. 2: Heavy load operation of Si Fet Prototype DAHB

First, we examine how parasitic inductance distorts switch node waveforms in low-impedance dual-active half-bridge (DAHB) converters. In the following section, we propose a control strategy combining boundary-conduction mode and burst mode to ensure zero-current switching (ZCS) and zero-voltage switching (ZVS) across the entire operating range. Finally, we present experimental results to demonstrate the improved efficiency and ideal converter waveforms achieved with the proposed control strategy.

II. LOW IMPEDANCE IN DAHBs

A dual active half bridge circuit consists of one half-bridge on each side of the transformer as shown in Fig. 1. Each half bridge is usually operated at 50% duty cycle which leads to quasi-square wave ac voltages across the leakage inductor L_{lk} of the transformer (V_p and V_s respectively at the switching node for primary and secondary leg). The phase shift ϕ between the two switch node voltages is a degree of freedom that allows the current through the leakage inductor to increase, which overall controls output power. In a conventionally-designed converter, the leakage inductor current is therefore trapezoidal (see Fig. 3a). The output power can be then derived as [4] -

$$P_o = \frac{V_{in}V_{out}}{8nf_sL_{lk}}\frac{\phi}{\pi}(1 - \frac{\phi}{\pi}) \tag{1}$$

where n is the turns ratio of the transformer, f_s is the switching frequency, P_o is the output power, and V_{in} & V_{out} are the input and output voltages of the converter.

Usually, the converter is designed to operate at rated power $P_{o,max}$ at or under the maximum phase shift ($\phi_{max} = \pi/2$). So, the target leakage inductance for the converter can be calculated as

$$L_{lk} = \frac{V_{in}V_{out}}{32nf_sP_{o,max}} \tag{2}$$

Typical design rules call for the transformer ratio to be as close to the input-to-output voltage ratio as possible to minimize the rms current and increase the zvs range. Thus,

using ($P_{o,max} = V_{out}I_{o,max}$) and the approximation ($V_{out} \approx nV_{in}$, $I_{in} \approx nI_{out}$), we can write down the required inductance as

$$L_{lk} = \frac{1}{32f_s}\frac{V_{in}}{I_{in,max}} \tag{3}$$

The required inductance clearly depends on the frequency and port voltage-to-current ratio. For applications with low voltage and high current and/or high frequency, the required leakage inductance could be on the same order as typical parasitic inductances in, e.g., switching loops ($\sim$10s nH).

Using the design philosophy discussed earlier, a 1 MHz, 40 W SiFET Dual Active Half-Bridge (DAHB) converter with a leakage inductance in the nanohenry (nH) range has been designed. Figure 2 shows the operation of the converter using single-phase shift control with a 60-degree phase shift. From the figure, it is evident that the primary voltage is much lower than the secondary voltage, approximately four times lower, meaning the primary side half-bridge carries four times the current of the secondary side. Ideally, the converter's switch node waveforms and leakage inductor current should resemble the waveform shown in Figure 3a.

However, in practice, we observe that the primary side switch node voltage exhibits a shelf (V_{shelf}) before reaching the expected V_{in} value when the primary half-bridge is on. This behavior can be attributed to the distribution of the input voltage across the inductive voltage divider consisting of the parasitic inductance (L_p highlighted in red in Fig. 1) and leakage inductance (L_{lk}) of the converter. Additionally, when the primary half-bridge switches on, the current through L_p induces ringing at the switch node, which reaches almost twice the input voltage.

When the primary half-bridge is turned on and the secondary side is turned off, the primary side loop consists of $V_{in} - Q1 - L_{lk} - C_2$ as shown in Fig. 1. Assuming that each half of the switching loop, consisting of Q1 and C_2, introduces an equal parasitic inductance L_p, we can derive the following equation:

$$\frac{L_p}{L_p + L_{lk}} = \frac{V_{in} - V_{shelf}}{V_{in}} \tag{4}$$

where, V_{shelf} is the plateau the V_p reaches before the secondary voltage turns on.

As the current on secondary is low (or, equivalently, the leakage inductance reflected to the secondary is much larger), we can assume that the parasitic inductance on the secondary has a much smaller effect on the voltage and current waveforms of the circuit. So, we can write down the inductor equation to calculate the parasitic inductance on the primary side of the circuit as

$$V_{in} - V_{shelf} = \frac{L_p I_{L_{lk}}}{\Delta t} \qquad (5)$$

Here, the input voltage is 2V and the V_{shelf} is 1.6V and inductor current is 6A for a Δt of 100ns. Substituting these values into (5), we obtain a parasitic inductance of 1.6nH. Using the value of L_p, V_{shelf} and V_{in} in (4), we get the leakage inductance to be 8.33nH. We can see that the parasitic inductance in a loop ($2L_p \approx 3.2$ nH) is around 40% of the leakage inductance. This parasitic inductance must be taken into account when designing low-leakage inductance DAHBs, as it sets a lower bound to the total inductance and therefore an upper bound for power. It also corresponds to substantial turn-off loss.

While the more common objective in DAB design is to achieve ZVS for all switches, ZVS is may not reduce loss as much as ZCS for low-voltage high-current applications and ZVS during a switching instance will not reduce the distortion associated with parasitic inductances that are on the same order as the transformer leakage. ZCS on the low-voltage side would be better suited to mitigate these challenges.

III. CONTROL STRATEGY

As previously discussed, a simple and widely used control technique for operating a DAHB converter is SPS control. In SPS, both half-bridges operate at a 50% duty cycle, and the phase shift between them determines the output current of the converter. The DAHB typically operates in one of three modes, depending on the output current. A commonly observed mode is the heavy-load operation, where the inductor current transitions from an initial negative value to a positive value within a half-switching cycle, as illustrated in Figure 3a.

Under heavy-load conditions, the current direction at each switching instant facilitates the discharge of the half-bridge's C_{oss} capacitance, enabling Zero-Voltage Switching (ZVS) and minimizing turn-on losses for all switches in the DAHB. However, achieving ZVS requires the current to exceed a minimum threshold. If the current falls below this limit, only partial ZVS occurs, leading to increased turn-on losses across the switches.

In contrast, under light-load conditions (see Figure 3c), the current direction during the primary half-bridge switching instants is reversed. As a result, the current unable to discharge the C_{oss} capacitance on the primary side, preventing ZVS. Additionally, the peak current may fall below the minimum threshold required for ZVS, leading to only partial ZVS on the secondary half-bridge as well.

Consequently, conventional control strategies aim to operate the DAHB converter in the heavy-load mode for as much of

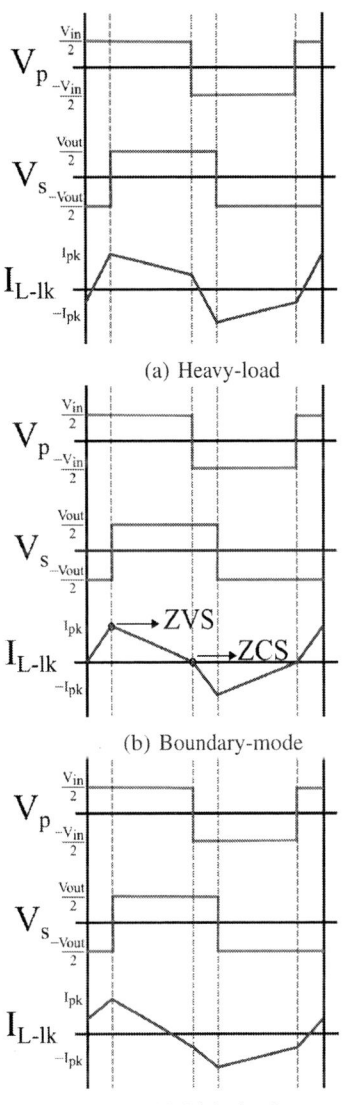

(a) Heavy-load

(b) Boundary-mode

(c) Light-load

Fig. 3: switch-node voltage and inductor current waveforms of a dual-active half bridge in case $V_s > nV_p$

the power range as possible to maximize ZVS and minimize switching losses.

A. Boundary Mode

Under specific operating conditions, the current can enter boundary conduction mode, as shown in Figure 3b. In this mode, the current starts and ends at zero within a half switching cycle. Thus, the current waveform transitions from a trapezoidal to a triangular shape.

In boundary mode, the primary leg (in this example) does not achieve Zero-Voltage Switching (ZVS) because the current is near zero at the switching instant, falling below the minimum current threshold required for ZVS. However, it does achieve Zero-Current Switching (ZCS). While turn-on switching losses persist, the turn-off switching losses associated with the switch parasitic inductance are significantly reduced.

Since the primary leg operates with low voltage and high current, the turn-on loss due to the C_{oss} capacitance is relatively minor compared to the turn-off losses caused by the parasitic inductance, which carries substantial current. As a result, boundary conduction mode can effectively minimize total switching losses under these conditions.

Without loss of generality, let us assume that the primary side voltage (V_p) is less than (V_s/n). So, when the primary-leg is turned on and the secondary leg is turned off, we can write down the increase in inductor current as

$$\Delta I_{L_{lk}} = \frac{\left(V_{in} + \frac{V_{out}}{n}\right)}{2} \frac{\phi T_s}{L_{lk}} \tag{6}$$

similarly, when the primary-leg turns off because the current falls to zero in boundary mode, we get

$$\Delta I_{L_{lk}} = \frac{\left(V_{in} - \frac{V_{out}}{n}\right)}{2} \frac{(1-\phi)T_s}{L_{lk}} \tag{7}$$

using (6) and (7), we can find the condition for Boundary-mode operation as

$$\left(V_{in} + \frac{V_{out}}{n}\right)\phi = \left(V_{in} - \frac{V_{out}}{n}\right)(1-\phi) \tag{8}$$

where T_s is the switching period of the converter. Simplifying (8), we get

$$\frac{\phi}{\pi} = \frac{1}{2}\left(1 - \frac{V_{out}}{nV_{in}}\right) \tag{9}$$

So, for a specific input-to-output voltage ratio and turns ratio, there is a single phase shift that results in boundary mode which achieves ZCS on the high-current side and ZVS on the high-voltage side. Once the turns ratio and phase shift are fixed, the output power that can be pushed through the converter can be determined using (1).

The advantages of boundary mode can be seen in Fig. 4, where $V_{p_{sw}}$ is the high current (primary side in this example) switching node waveform. In Fig. 4a at a non-boundary-mode operating point, the ringing on the switch node is almost equal the input voltage $V_{in} = 6.7V$ due to the significant parasitic inductance relative to the designed leakage inductance. When boundary mode is implemented as in Fig. 4b, the overshoot is just 1.6V and the waveform is much more clean.

B. Control Design for Efficient Power Transfer

In boundary-mode operation, for a given specification of output power, switching frequency, and input-to-output voltage ratio, there are two primary degrees of freedom: leakage inductance and phase shift. To maximize efficiency, it is ideal to minimize both leakage inductance and phase shift. This concept can be illustrated by analyzing the positive half-cycle of the inductor current in DAHB converter.

During the rising segment of the inductor current, the secondary-side half-bridge voltage (V_s) is equal to $V_{out}/2$. Despite the current flowing into the load, the converter effectively draws power out of the output load. This segment contributes to the root mean square (RMS) current but reduces the net power transferred to the load. In contrast, during the falling segment of the inductor current, V_s is positive, and the output

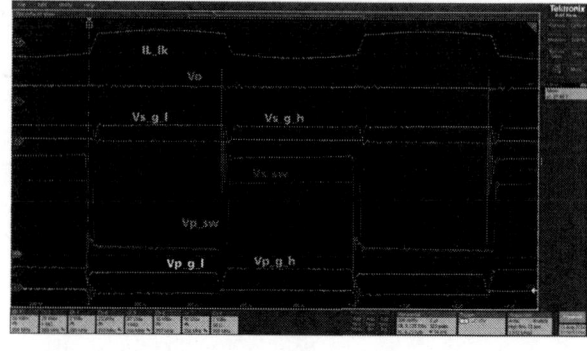

(a)

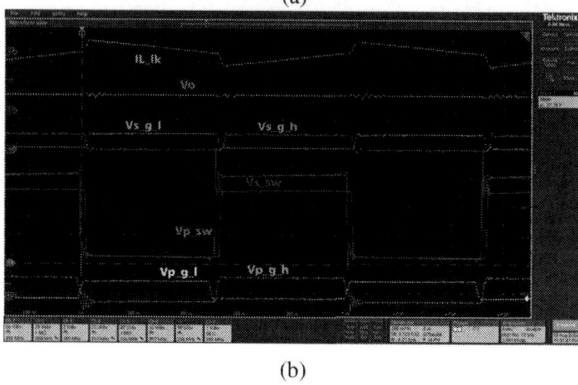

(b)

Fig. 4: Single-Phase shift operation of DAHB (a) No ZCS on primary (b) Boundary-mode operation with ZCS on primary

current is also positive. In this phase, power flows into the load, increasing the net power transferred.

For a fixed switching frequency, input-to-output voltage ratio, and transformer turns ratio, the same output power can be achieved through a combination of leakage inductance and phase shift. However, a longer phase shift and higher leakage inductance extend the inductor charging cycle for the same power transfer. This results in higher RMS current and increased conduction losses, reducing overall efficiency.

Therefore, the ideal boundary-mode operation achieves efficient power transfer by minimizing both phase shift and leakage inductance. This approach ensures reduced conduction losses, optimized performance, and maximum power transfer efficiency.

C. Burst mode operation for wide power range

Using SPS control, the phase shift between the two half-bridges determines the amount of power transferred to the output load. To ensure ZCS on the primary side and ZVS on the secondary side, SPS is operated with a fixed phase shift at the rated output power. Adjusting the power by modifying the phase shift would result in departure from boundary mode operation, causing the loss of ZCS on the high-current side. To maintain these optimal switching conditions, burst-mode control is employed for power adjustment, ensuring proper waveforms (ZVS and ZCS) during bursts.

For closed-loop control, hysteretic burst mode control is implemented. In this strategy, a hysteresis band around the

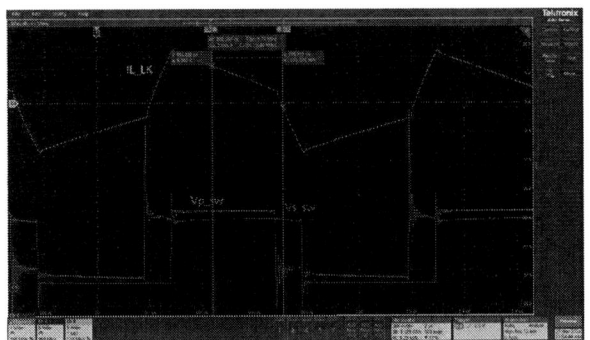

Fig. 5: Heavy load operation of GaN Fet Prototype DAHB

TABLE I: Specifications for DAHB Converter

Dual Active Half Bridge	
Rated Power	40W
Rated Input Voltage	6.7V
Rated Output Voltage	27V
Switching Frequency (f_{sw})	1MHz
Transformer Ratio (1:n)	1:4
Planar Transformer Core	Ferrite 9579130302
Leakage Inductance	8.33nH

output voltage is used to determine the converter's on and off states during burst mode operation. When the output voltage drops to the lower threshold of the hysteresis band, the controller initiates a burst of boundary-mode pulses. These pulses continue until the output capacitor charges to the upper limit of the hysteresis band, at which point the controller halts the burst pulses. The process repeats as the output voltage fluctuates within the hysteresis band, ensuring efficient regulation while preserving optimal switching characteristics.

IV. EXPERIMENTAL RESULTS

Experimental prototype converters using Si FETs and GaN FETs have been designed to validate the proposed ZVS-ZCS burst mode strategy. The converter is engineered to operate within a voltage conversion range of 6.7V to 27V, with a power rating of 40W and a switching frequency of 1 MHz. Additional specifications of the converters are provided in Table I. Components of the Si and GaN Fet prototype are given in Table II. They operate in boundary condition mode with an exceptionally low phase shift of 0.36° to achieve the full power rating. The converters leakage inductance is just 8.33nH, which is comparable to the parasitic inductance within the circuit.

To compare the two converter designs, the GaN Fet converter is also operated at heavy load 2V to 12V 6A condition to calculate the parasitic inductance in its switching loop as shown in Fig. 5. Using the analysis from section II, we obtain the parasitic inductance of 0.4nH which is 1/4th the value of Si based converter prototype.

TABLE II: Components of DAHB Converter

	Components	
	Si Prototype	GaN Prototype
Primary-side Fets	BSC015NE2LS5I	EPC2067
Secondary-side Fets	BSC029N025SG	EPC2055
Gate Driver	UCC27212	LMG1205

Boundary-condition mode operation of the converters can be seen in Fig. 7. The current is triangular ($I_{L_{lk}}$ in Fig. 7) and the primary switching instances ($V_{p_{gl}}$ and $V_{p_{gh}}$ in Fig. 7) happen during zero inductor current. At peak of the inductor current, secondary side (see $V_{p_{gl}}$ and $V_{p_{gh}}$) is switched with deadtime to ensure ZVS. The waveforms show that even at such a high switching frequency the primary half bridge switch node which has 6A flowing through it has really low ringing on V_p.

Burst mode has been used to operate at lower power levels. An example at 80% of full load is shown in Fig. 8a. Fig. 8b shows the zoomed in view of the burst mode controller which shows boundary mode operation as expected. The hysteresis limits set for the controller are 26.9V and 27.1V.

Based on the proposed control, the converters operate efficiently across nearly the entire power range. The SiFET converter is operated in open-loop burst mode control and achieves a peak efficiency of 93.9% at full load. The GaNFET converter achieves a peak efficiency of 92.9% at full power (100% burst). As the burst pulse decreases, efficiency slightly drops due to startup and end effects, but it remains close to full-load efficiency across the entire operating range, from 10% to 100% load. The GaN prototype has a power density of 600 W/in^3 (see Fig. 6b). While the Si prototype uses the same transformer and passive elements, the FETs and their packages are physically larger, which reduces the power density to 145W/in^3 (see Fig. 6a).

The burst mode control, which introduces startup and end effects to maintain efficiency, requires longer bursts on the order of milliseconds. This extended off-period, during which the output voltage must remain constant, necessitates the use of electrolytic capacitors with a total capacitance of 1.8 mF to support the converter. This electrolytic capacitance has not been included in the power density calculation. This extended burst will not be necessary when current sensing is introduced, which will allow the converter to start and end bursts precisely within a single cycle, permitting bursts of just a few switching cycles.

V. CONCLUSION

A boundary-mode control strategy has been proposed to achieve Zero Current Switching (ZCS) on the high-current side and Zero Voltage Switching (ZVS) on the high-voltage side of a low-impedance DAHB circuit. This strategy effectively mitigates the impact of parasitic inductance, which is nearly equal to the circuit's leakage inductance. Boundary-mode not only improves the switch node waveforms but also optimizes

(a)

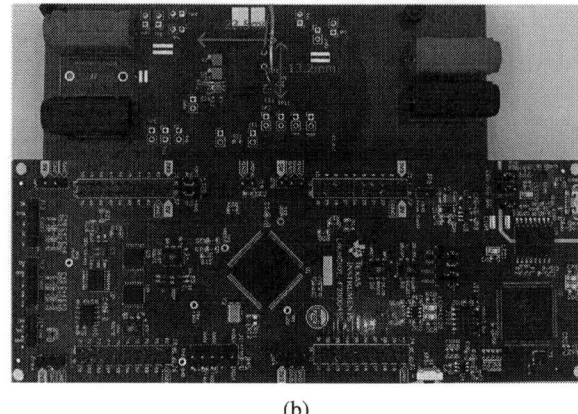
(b)

Fig. 6: Experimental prototypes with active area highlighted in red (a) Si FET DAHB prototype (b) GaNFet DAHB prototype

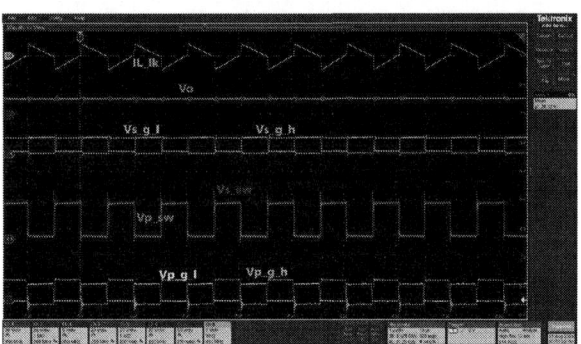

Fig. 7: Boundary Mode Operation of DAHB, ZCS on primary-side and ZVS on secondary-side switching leg

the converter's performance, achieving a peak efficiency of 93.9% at rated power. To maintain these advantages under lighter loads, a burst-mode is implemented. With this control strategy, the converter consistently achieves an overall efficiency exceeding 90% across its entire operating range.

REFERENCES

[1] P. Saraf, C. Deng, and A. Hanson, "Describing function analysis of nonlinear decentralized control scheme for modular power sharing," in *2022 IEEE 23rd Workshop on Control and Modeling for Power Electronics (COMPEL)*, 2022, pp. 1–8.

[2] P. Saraf and A. Hanson, "Decentralized power sharing of ipop buck converters," in *2023 IEEE 24th Workshop on Control and Modeling for Power Electronics (COMPEL)*, 2023, pp. 1–7.

[3] M. Solomentsev and A. J. Hanson, "Highly-scalable differential power processing architecture for on-vehicle photovoltaics," in *2023 IEEE 24th Workshop on Control and Modeling for Power Electronics (COMPEL)*, 2023, pp. 1–7.

[4] P. He and A. Khaligh, "Comprehensive analyses and comparison of 1 kw isolated dc–dc converters for bidirectional ev charging systems," *IEEE Transactions on Transportation Electrification*, vol. 3, no. 1, pp. 147–156, 2017.

[5] H. Bai and C. Mi, "Eliminate reactive power and increase system efficiency of isolated bidirectional dual-active-bridge dc–dc converters using novel dual-phase-shift control," *IEEE Transactions on Power Electronics*, vol. 23, no. 6, pp. 2905–2914, 2008.

[6] B. Zhao, Q. Yu, and W. Sun, "Extended-phase-shift control of isolated bidirectional dc–dc converter for power distribution in microgrid," *IEEE Transactions on Power Electronics*, vol. 27, no. 11, pp. 4667–4680, 2012.

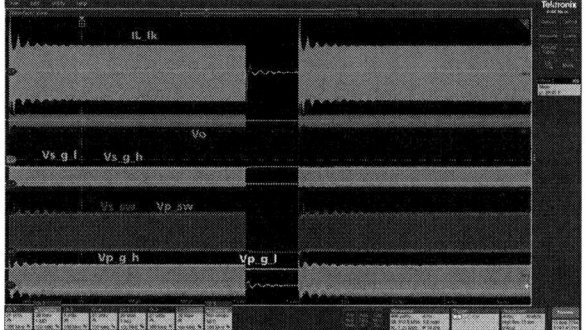
(a)

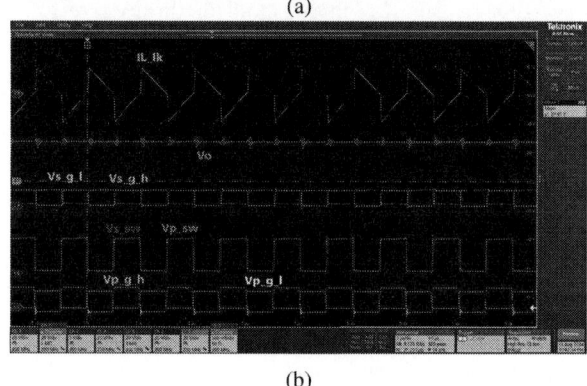
(b)

Fig. 8: Burst mode operation of DAHB: (a) 80% burst (b) zoomed-in view showing boundary mode operation on a cycle-by-cycle basis

[7] N. Hou and Y. W. Li, "Overview and comparison of modulation and control strategies for a nonresonant single-phase dual-active-bridge dc–dc converter," *IEEE Transactions on Power Electronics*, vol. 35, no. 3, pp. 3148–3172, 2020.

[8] J. Zhang, Y. Tang, W. Hu, Z. Zhang, J. Li, and Z. Chen, "Minimum current stress operation of dual active half-bridge converter using triple phase shift control for renewable energy applications," *Energy Reports*, vol. 8, pp. 547–553, 2022, the 8th International Conference on Energy and Environment Research –"Developing the World in 2021 with Clean and Safe Energy". [Online]. Available: https://www.sciencedirect.com/science/article/pii/S2352484722000683

[9] J. Hiltunen, V. Väisänen, R. Juntunen, and P. Silventoinen, "Variable-

979-8-3315-1612-3/25 $31.00 © 2025 IEEE

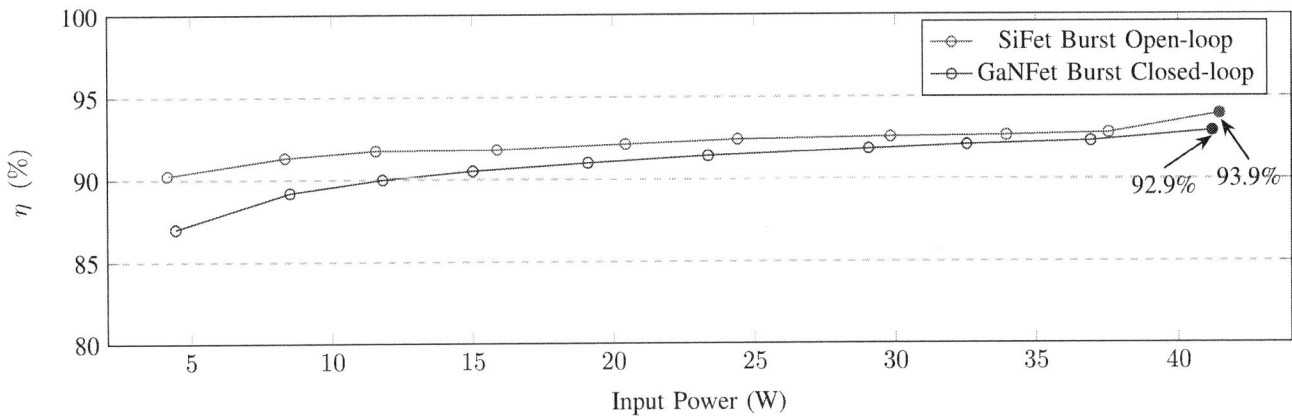

Fig. 9: Measured efficiency curve of the experimental prototype converter.

frequency phase shift modulation of a dual active bridge converter," *IEEE Transactions on Power Electronics*, vol. 30, no. 12, pp. 7138–7148, 2015.

[10] G. G. Oggier and M. Ordonez, "High-efficiency dab converter using switching sequences and burst mode," *IEEE Transactions on Power Electronics*, vol. 31, no. 3, pp. 2069–2082, 2016.

[11] S. Junglas and J. Maas, "Hv dab using loss-minimizing burst sequences for feeding de transducers," in *2021 23rd European Conference on Power Electronics and Applications (EPE'21 ECCE Europe)*, 2021, pp. P.1–P.10.

[12] M. Yaqoob, G. Torrico, and W. Shuqin, "A multi-mode control based asymmetrical dual-active-bridge series-resonant dc-dc converter (dab-src)," in *2022 24th European Conference on Power Electronics and Applications (EPE'22 ECCE Europe)*, 2022, pp. 1–9.

[13] F. Krismer, S. Round, and J. W. Kolar, "Performance optimization of a high current dual active bridge with a wide operating voltage range," in *2006 37th IEEE Power Electronics Specialists Conference*, 2006, pp. 1–7.

[14] H. Chan, K. Cheng, and D. Sutanto, "An extended load range zcs-zvs bi-directional phase-shifted dc-dc converter," in *2000 Eighth International Conference on Power Electronics and Variable Speed Drives (IEE Conf. Publ. No. 475)*, 2000, pp. 74–79.

[15] O. Omotoso, O. Kiselychnyk, R. McMahon, P. Mawby, and P. James, "Analysis of the peak and rms current ratios in a dual active bridge converter with triangular modulation," in *2023 IEEE 2nd Industrial Electronics Society Annual On-Line Conference (ONCON)*, 2023, pp. 1–6.

[16] O. Omotoso, O. Kiselychnyk, R. McMahon, P. Mawby, P. James, and M. Mawby, "The challenges of determining the optimum transformation ratio for a triangular modulated dual active bridge converter," in *2022 IEEE Workshop on Wide Bandgap Power Devices and Applications in Europe (WiPDA Europe)*, 2022, pp. 1–5.

[17] F. Krismer and J. W. Kolar, "Efficiency-optimized high-current dual active bridge converter for automotive applications," *IEEE Transactions on Industrial Electronics*, vol. 59, no. 7, pp. 2745–2760, 2012.

[18] C. C. Mi, H. Bai, C. T. Wang, and S. Gargies, "Operation, design and control of dual h-bridge-based isolated bidirectional dc-dc converter," *Iet Power Electronics*, vol. 1, pp. 507–517, 2008.

Gap in pagination due to withheld paper.

Pages 84-89

A 6.6 kW Highly Efficient Reconfigurable Dual Active Bridge Converter Designed Using Planar Transformer, SiC-FETs and Monolithic Bidirectional Devices

Reza Barzegarkhoo
Chair of Power Electronics
Kiel University
Kiel, Germany
rbar@tf.uni-kiel.de

Fabian Groon
Chair of Power Electronics
Kiel University
Kiel, Germany
fagr@tf.uni-kiel.de

Arkadeb Sengupta
Chair of Power Electronics
Kiel University
Kiel, Germany
arsa@tf.uni-kiel.de

Marco Liserre
Chair of Power Electronics
Kiel University
Kiel, Germany
ml@tf.uni-kiel.de

Abstract—**Efficient performance of a standard dual active bridge (DAB) converter with symmetric full bridges in primary and secondary sides of a galvanic transformer and within a limited range of voltage conversion gain has been extensively proven in both industry and academia. However, when a battery charger system for energy storage or automotive applications operates at a lower power rating, or whenever a wider voltage conversion range is required (200 V to 800 V), such efficient performance is compromised due to the loss of zero voltage switching (ZVS). The concept of reconfigurable DAB converter is developed in this work to alleviate such problem. The proposed topology consists of a T-cell branch and a half-bridge cell in both the primary and secondary sides, which can be reconfigured as a full- or half-bridge circuit. By incorporation of SiC-FETs and a recently developed monolithic bidirectional GaN device in the proposed reconfigurable DAB converter, four different configurations are extracted aiming to preserve the ZVS operation with a high overall efficiency for a wide range of input/output voltage regulation. A planar transformer (PT) design using Pareto-Front optimization is also carried out for the galvanic isolation. Working principle of the proposed concept is discussed, while the design guidance of the PT, and the ZVS range analysis are given in details. A 6.6 kW/50 kHz prototype with a 98% peak efficiency for 200 V to 800 V input voltage and 400 V output voltage is fabricated and its detailed experimental results in each of the operation modes are presented.**

Index Terms—**Bidirectional dual active bridge (DAB), Reconfigurable DAB topology, Monolithic bidirectional GaN device, Planar transformer design.**

I. INTRODUCTION

Dual active bridge (DAB) dc-dc converter with a galvanic isolation, soft-switched performance, and symmetrical full-

This work was supported by the Priority Programme "Energy Efficient Power Electronics 'GaNius' (DFG SPP 2312)".

bridge (FB) design is recognized as an efficient power electronics system for applications requiring bidirectional power flow, e.g., dc-microgrid, solid-state transformer, automotive battery chargers, and energy storage systems [1]. Even though exhibiting an appealing overall efficiency, its performance with unity turn ratio of the high frequency (HF) transformer is degraded at low power as the current passing through the HF transformer is not that sufficient to charge/discharge the parasitic capacitance of the semiconductors to achieve a complete zero voltage switching (ZVS) [2]. Moreover, at this case, the circulating current becomes severely high at low power causing excessive conduction losses within a reactive network [3].

Apart from such problems, regulation of the input/output dc voltages of a standard DAB converter with unity turn ratio of HF transformer over a wide range of the voltages, i.e., 200 V to 800 V, which is a typical range for most of the battery-packs of electric vehicle systems, is not too straight forward as the efficiency is again reduced due to loss of ZVS at large voltage conversion gain [4].

To address the low efficiency of the standard DAB converter at the low power and to compromise its performance for both 400 V and 800 V standard voltages, it is recommended to change the configuration from FB design to half-bridge (HB) one [5]. In this case, the single-phase shift (SPS) modulation applying to the secondary-side of the DAB converter can still possesses more freedom degree to change over a wider range of the input/output dc voltages [6].

As for its galvanic isolation, the concept of planar transformer (PT) is also appealing as it offers a good potential for volume and weight reduction of passive elements leading to

979-8-3315-1612-3/25 $31.00 © 2025 IEEE

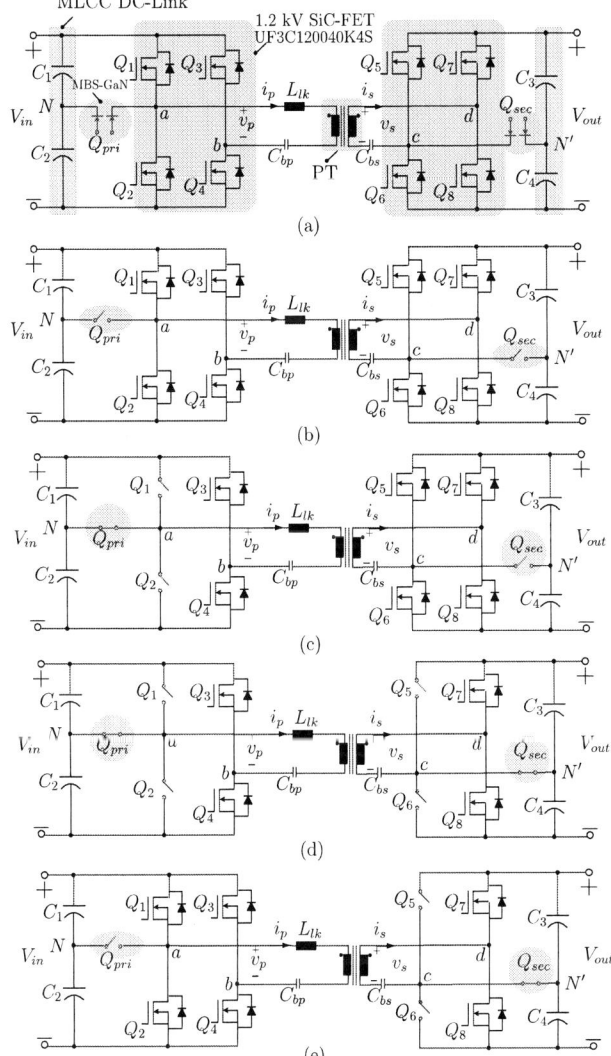

Fig. 1: (a) The proposed reconfigurable DAB converter, and its operation mode based on (b) FB-FB, (c) HB-FB, (d) HB-HB, and (e) FB-HB configurations.

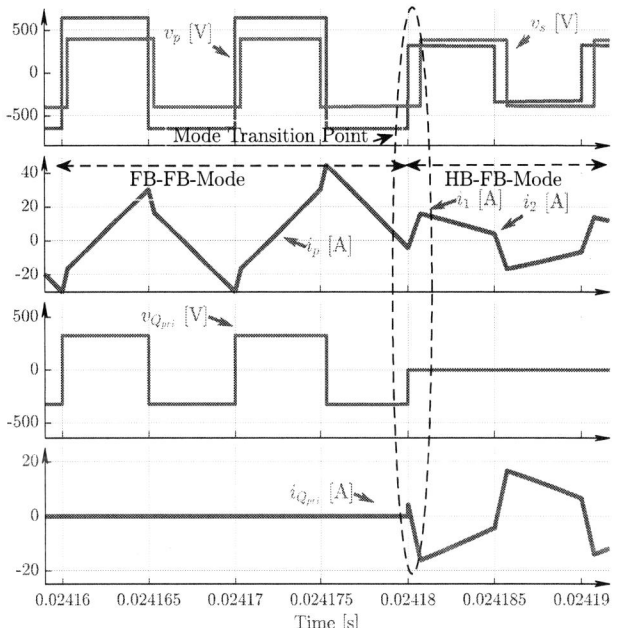

Fig. 2: Typical waveforms of the proposed reconfigurable DAB converter over a wide range of the input dc voltage changes, i.e., a dynamic transition from FB-FB to HB-FB mode, showing the output voltages of the primary and secondary sides of the PT as well as the current passing through the PT, the voltage across Q_{pri}, and the current passing through Q_{pri}.

achieve better overall power density with a predictable parasitics [7]. Apart from providing better thermal performance due to its large surface area in comparison to litz-wire-based transformers, an efficient PT can also facilitate a predictable leakage inductance, which is a propitious feature for ZVS operation of the DAB or resonant converters [8]. Hence, with the help of an efficient PT design within a given area, and through incorporation of advanced semiconductors, i.e., low loss SiC-FETs and GaN devices, new highly efficient topology for the DAB can be developed.

The aim of this article is to investigate the operation of a highly efficient reconfigurable DAB converter, which can change the configuration of the primary and secondary sides of the HF transformer to FB- or HB-based by means of 1.2 kV SiC FETs and a recently developed monolithic bidirectional

switch (MBS)-GaN device. The resultant reconfigurable DAB converter can realize four different operation modes to be used in various conditions, i.e., low power and higher range of voltage conversion gain. Using a Pareto-front optimization, the design consideration of the PT, and also the entire power converter are included to achieve higher power density and overall efficiency for a 6.6 kW system. Moreover, the ZVS range analysis are developed to realize the behavior of the proposed converter at different power level and wide input voltage variation among each of the operation modes. The experimental results show more than 97% overall efficiency at 50 kHz effective switching frequency for all four combinations of the operation modes, while the range of applied voltage is 200 to 800 V for the input and fixed 400 V for the output side.

II. WORKING PRINCIPLES OF THE PROPOSED RECONFIGURABLE DAB CONVERTER

The overall structure of the proposed reconfigurable DAB converter is shown in Fig. 1(a), where in both sides of an PT, a symmetric T-cell plus a half-bridge leg is employed. Multilayer ceramic capacitors (MLCC) have also been used in both sides to provide the dc-link voltages for the input and output. As wide voltage conversion gain, i.e., from 200 V to 800 V in a bidirectional power flow is the prime focus, 1.2 kV SiC-FETs for the bridge devices are chosen, while as for better power density and lower conduction loss, the newly developed four-quadrant device named as MBS-GaN is used for Q_{pri} and

Q_{sec} [9],[10]. Similar to the standard DAB converter, SPS modulation is used to derive the gate switching pulses.

As can be seen from Fig. 1(b), by deactivating the MBS-GaN devices in both sides, the converter can work as a standard DAB converter with a FB-FB-based configuration. Herein, blocking a bipolar stress voltage equals to half value of the input/output voltage is the main challenge of the MBS-GaN. Hence, if the MBS-GaN is replaced with two discrete-based devices in a common-drain-based configuration, then this task would be more challenging as the source of the devices is floated. Although, there is a smooth power flow with a ZVS operation for the legs' devices at large power within unity voltage conversion gain ratio in this mode, the challenge of DAB converter still remains in place as when 800 V input voltage is available and 400 V output is desired, the resonant network at the unity ratio of the HF transfomer would be severely reactive and the efficiency is declined. Concerning Fig. 2(c), the proposed converter can work in a HB-FB-based mode to address this challenge with a wider range of phase-shift adjustment. Hence, by activating the Q_{pri}, and deactivating Q_{sec} within the same SPS modulation, still the highest possible efficiency can be attained.

The proposed reconfigurable DAB converter can also be operated within FB-HB and HB-HB modes as shown in Fig. 1 (d) and (e), respectively thanks to the flexibility of the MBS-GaNs used in both sides. Hence, in a case of lower power rating requirement, the efficiency of the converter can be still kept as high as possible within these two modes as the ZVS operation of the devices in both sides is still achievable. Herein, for the case of FB-HB mode, the converter is operated at the boost mode, i.e., 400 V (200 V) input voltage to 800 V (400 V) output, while using the same SPS modulation with a ZVS operation. On the contrary, the same unity gain similar to the FB-FB mode can be kept within HB-HB mode as well, while higher overall efficiency at the low power is achieved.

Utilization of a single chip device as MBS-GaN instead of two commercially available wide-band-gap devices in a back-to-back condition can also help the converter to possess lower conduction losses during the HB mode of operation [10].

To better visualize the performance of the proposed reconfigurable DAB converter under a wide variation of the input dc voltage, Fig. 2 can be considered. Herein, a typical waveform of v_p and v_s as the primary and secondary voltages of the PT, besides i_p as the current passing through the PT, and $v_{Q_{pri}}$ as the blocking voltage across the MBS-GaN device, with $i_{Q_{pri}}$ as its passing through current are shown, from top to bottom, respectively. As can be seen, when a large voltage conversion gain, d, is required during the variation in the input dc voltage, and whenever the converter is being operated as its FB-FB mode, the peak of i_p becomes high, and the converter tends to lose its ZVS operation due to lossy reactive network within SPS modulation. In this case, the MBS-GaN is blocking a stress voltage, equals to half value of the input voltage in both polarities.

Contemporary, due to flexibility of the proposed reconfigurable DAB converter, the same amount of power to the output

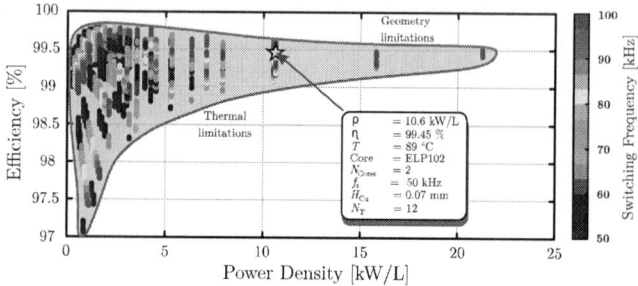

Fig. 3: Pareto-Front optimization for incorporation of an efficient PT in the proposed reconfigurable DAB converter.

side with the same value of the output voltage can still be provided as the converter can change its mode of operation to HB-FB mode during any large transition in the input dc voltage. As can be seen from Fig. 2, within a larger value of phase shift, φ, provided in HB-FB mode, the nature of i_p becomes smooth providing ZVS operation for both primary and secondary sides devices. In this mode, the MBS-GaN device must be able to pass the same amount of current as i_p.

Herein, to calculate the losses across active and passive elements, definition of i_1, and i_2, shown in Fig. 2 as the upward and downward peak values of the i_p is required. Hence, considering f_{sw} as the switching frequency of the converter, a nominal base current as per (1) can be defined [11]:

$$I_b = \frac{V_p}{2\pi f_{sw} L_{lk}} \tag{1}$$

while, the leakage inductance value within a unity turn ratio of primary and secondary sides of the PT is given as:

$$L_{lk} = \frac{V_p V_s \varphi (\pi - \varphi)}{2\pi^2 f_{sw} P} \tag{2}$$

where, P is the operating power of the converter. As per (1) and (2), the amplitudes of the primary- and secondary-side voltages, v_p and v_s, are denoted by the quantities, V_p and V_s, respectively.

Therefore, taking (3) and (4) into account, i_1, and i_2 can be defined as:

$$i_1 = 0.5(2\varphi - (1 - d)\pi) I_b \tag{3}$$

$$i_2 = 0.5(2d\varphi + (1 - d)\pi) I_b \tag{4}$$

Hence, concerning I_m as the magnetizing current of the PT, the rms currents passing through the primary and secondary side of the PT can be obtained as [11]:

$$I_{rms,pr} = \sqrt{\frac{1}{3}(i_1^2 + i_2^2 + (1 - (2\frac{\varphi}{\pi})i_2 i_1))} + I_m \tag{5}$$

$$I_{rms,soc} = \sqrt{\frac{1}{3}(i_1^2 + i_2^2 + (1 - (2\frac{\varphi}{\pi})i_2 i_1))} \tag{6}$$

979-8-3315-1612-3/25 $31.00 © 2025 IEEE

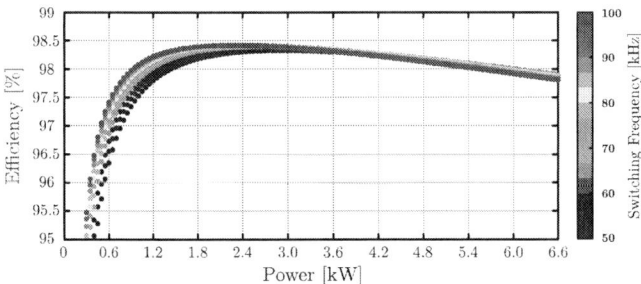

Fig. 4: Pareto-Front result for the entire reconfigurable DAB converter at the FB-FB mode considering the switching/conduction losses of the devices as well as the core and winding losses of the designed PT and the leakage inductor, L_{lk}.

III. PARETO-FRONT OPTIMIZATION FOR THE APPLIED PT

To achieve the highest efficiency/power density, many optimization approaches can be found in the literature to optimize the PT in the proposed reconfigurable DAB converter. Herein, a comprehensive Pareto-Front optimization is used as it is a multi-objective optimization and includes many variables needed for optimization [2]. As the converter operates in different configurations, i.e., FB-FB, FB-HB, HB-FB, and HB-HB modes, the possible worst-case scenario must be identified for the PT design. According the circuit feature of the proposed reconfigurable DAB converter, the FB-FB mode is considered for the PT design, as the nominal power of the converter is at this mode. The design criteria for the PT are unity ratio for the primary and secondary side, 50 kHz switching frequency, and 6.6 kW rated power. As the output voltage of the proposed converter is supposed to be fixed at 400 V, this value is considered for the design of PT. As for the efficiency analysis, the core and winding losses of the PT are taken into account. Hence, the core losses can be estimated by the Steinmetz equation for a square wave signal as [12]:

$$P_{\text{v,Core}} = C_{\text{m}} f_{\text{sw}}^{\alpha} B_{\text{max}}^{\beta}. \tag{7}$$

where α, β, and C_m are the material constants of the selected core. Alternatively, the maximum flux density, B_{max}, is calculated by the applied primary voltage, v_p, across the PT, the number of turns, N_T, on the primary and the cross-section area of the core, A_E, as:

$$B_{\text{max}} = \frac{v_{\text{p}}}{4 N_T f_{\text{sw}} A_{\text{E}}}. \tag{8}$$

On the other hand, winding losses consist of the DC and AC resistances. The DC resistance of the PT comes from the mean length track, MLT, of the winding traces, the number of turns, the width of the traces, L_{Cu}, the electrical resistivty of copper, ρ, and the height of the traces, H_{Cu} as:

$$R_{\text{DC}} = \frac{\rho \cdot MLT}{L_{\text{Cu}} H_{\text{Cu}}}. \tag{9}$$

Taking (9) into account, the AC resistance increases the effective resistance of the windings as skin-effect and its prox-

imity influence the resistance of the windings. This resistance can be estimated by (10) as [13]:

$$\frac{R_{\text{AC}}}{R_{\text{DC}}} = \frac{\xi}{2} \left[(H_{\text{ext}} - H_{\text{int}})^2 \frac{\sinh(2\xi) + \sin(2\xi)}{\cosh(2\xi) - \cos(2\xi)} \right.$$
$$\left. + 2 H_{\text{ext}} H_{\text{int}} \frac{\sinh(\xi) - \sin(\xi)}{\cosh(\xi) + \cos(\xi)} \right]. \tag{10}$$

where, ξ is the ratio between the copper thickness, and the skin depth, while, H_{ext} and H_{int} represent the magnetic field strength at the top and bottom side of the layer stack, respectively. Winding losses are then calculated based on the rms current passing through the primary and secondary side of the PT, $I_{\text{rms,pr}}$ and $I_{\text{rms,sec}}$, as:

$$P_{\text{Cu}} = R_{\text{AC}} I_{\text{i}}^2. \tag{11}$$

Considering the described core and winding losses, the Pareto optimization is carried out. The maximum temperature of the design is limited to 125 °C. All acceptable solutions are plotted in Fig. 3 and the selected design is highlighted. Based on this Pareto-Front, semi-interleaved windings for the PT are considered, which can significantly reduce the PT winding losses and its leakage inductance [8]. Based on the measurement, the final PT induces around 2 uH leakage inductance at 50 kHz switching frequency.

Concerning the above analysis as for the PT and regarding the switching and conduction losses of the chosen SiC-FETs, i.e., UF3C120040K4S, a Pareto-Front result for the entire converter is given as shown in Fig. 4. Herein, 400 V-to-400 V voltage conversion is considered as for the FB-FB mode of the proposed converter, while a fixed value of φ equals to 18.5 degrees within different values of L_{lk} as per (2) for each operating point, i.e., switching frequency and power level, is taken into account. As can be deduced from Fig. 4, achieving $\geq 97.5\%$ efficiency at high power is always attainable even with lower rating of the switching frequency, i.e., 50 kHz. The same analysis for the HB-FB mode of the proposed reconfigurable DAB converter can also be done when 800 V-to-400 V is an objective. As for lower power rating, HB-HB and FB-HB modes as well can be considered taking the conduction losses of the MBS-GaN devices used in the proposed converter into account.

IV. ZVS RANGE ANALYSIS

The ZVS range of the proposed reconfigurable DAB converter is derived in terms of the output voltage and operating power level. The ZVS condition based on the current polarity is employed in this paper. The ZVS conditions are then obtained in terms of the operating power, P, for the primary-side bridge as:

$$\sqrt{1 - \frac{8 f_{sw} L_{lk} P}{V_p V_s}} \leq \frac{V_p}{V_s}, \tag{12}$$

and for the secondary-side bridge as:

$$\sqrt{1 - \frac{8 f_{sw} L_{lk} P}{V_p V_s}} \leq \frac{V_s}{V_p}. \tag{13}$$

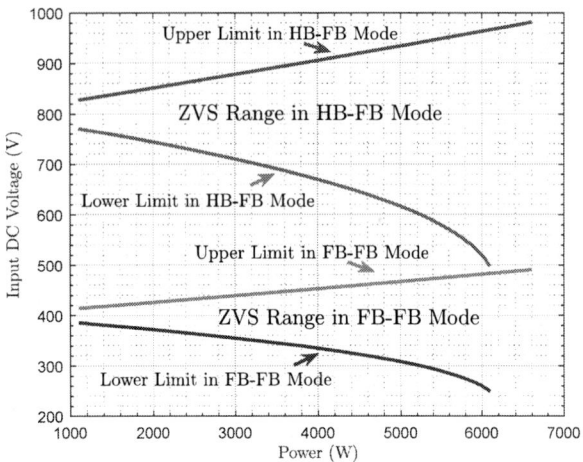

Fig. 5: Analytical ZVS range limits of the proposed reconfigurable DAB converter within its HB-FB to FB-FB modes based on current polarity condition when the output voltage of the converter is fixed at 400 V, at 50 kHz switching frequency and $L_{lk} = 25uH$.

Since the bridges on each side are reconfigurable, the ac voltage amplitudes, V_p and V_s, are related to the corresponding dc voltages, V_{in} and V_{out}, using the factors, k_1 and k_2, respectively, as (14) and (15).

$$V_p = k_1 V_{in} \tag{14}$$

$$\text{where, } k_1 = \begin{cases} 0.5, & \text{primary-side in HB mode} \\ 1, & \text{primary-side in FB mode} \end{cases}$$

$$V_s = k_2 V_{out} \tag{15}$$

$$\text{where, } k_2 = \begin{cases} 0.5, & \text{secondary-side in HB mode} \\ 1, & \text{secondary-side in FB mode} \end{cases}$$

To obtain the ZVS range in terms of the terminal voltages, (14) and (15) are substituted into (12) and (13). The relations describing the ZVS of the primary- and secondary-side bridges are then obtained as (16) and (17), respectively.

$$\sqrt{1 - \frac{8 f_{sw} L_{lk} P}{k_1 k_2 V_{in} V_{out}}} \leq \frac{k_1 V_{in}}{k_2 V_{out}} \tag{16}$$

$$\sqrt{1 - \frac{8 f_{sw} L_{lk} P}{k_1 k_2 V_{in} V_{out}}} \leq \frac{k_2 V_{out}}{k_1 V_{in}} \tag{17}$$

The equality condition is imposed on the primary-side ZVS relation in (16), and the resulting expression rearranged, to obtain the relation describing the lower limit of ZVS, as (18).

$$V_{in}^3 - \frac{k_2^2 V_{out}^2}{k_1^2} V_{in} + \frac{8 f_{sw} L_{lk} P k_2 V_{out}}{k_1^3} = 0 \tag{18}$$

Similarly, imposing equality on the secondary-side ZVS relation in (17), and the rearranging, the relation describing the upperlimit of ZVS is obtained as (19).

$$V_{in}^2 - \frac{8 f_{sw} L_{lk} P}{k_1 k_2 V_{out}} - \frac{k_2^2 V_{out}^2}{k_1^2} = 0 \tag{19}$$

TABLE I: Main parameters used for the experimental prototype

Element	Type and Description
Switching Frequency/Rated Power	50 kHz/ 6.6 kW
SiC-FETs	1.2 kV/UF3C120040K4S
MBS-GaN	±650V/TP65F060WS
MLCCs	10×220nF/GRM55DR7LW224KW1L
V_{in}/V_{out}	200-800V/400V
Microprocessor	DSP-TMS320F28379D
C_{bp}/C_{bs}	8×470nF/2220Y1K00474KXS2
L_{lk}	25uH/E42/21/20 N87
Gate Drivers	UCC21520
Isolated dc/dc ICs	MGJ1D051505MPC
Sensors	AMC3330

Based on (18), and (19), and considering 50 kHz switching frequency with $L_{lk} = 25uH$ for a 400 V given output voltage, the ZVS limits of the proposed converter over a wide range of input dc voltage changes within FB-FB and HB-FB modes are illustrated in Fig. 5. As can be seen, at larger rate of output power, i.e., 5.5 kW, during the FB-FB mode, the upper and lower limits of the input dc voltage with an ZVS, which can keep the efficiency still high, is around 480 V and 300 V, respectively, to stabilize a 400 V output voltage. As for the HB-FB mode and at the same amount of power, this limit can be even larger from 950 V to 550 V to get a fixed output voltage at 400 V with a ZVS operation and high efficiency. Apparently, owing to the fixed value of L_{lk}, the transient from 480 V input dc voltage, i.e., the upper limit in the FB-FB mode, to 550 V i.e., the lower limit in the HB-FB mode needs further attention as due to lose of ZVS, the efficiency would be dropped and electro magnetic interference (EMI) noises might be propagated.

V. EXPERIMENTAL RESULTS

To verify the performance of the proposed reconfigurable DAB converter in all the operations modes, a 6.6 kW/50 kHz ($5 \times 32 \times 30$ cm^3) prototype shown in Fig. 6(a) is developed, while the used parameters and components in the experiments are tabulated in Table I. Apart from the designed PT for 400 V at 50 kHz switching frequency, and 1.2 kV SiC-FETs from Qorvo, MBS-GaN devices manufactured by Transphorm and introduced as four-quadrant switch are used in the prototype. These first-generation four-quadrant devices are cascode-based with an integrated common-drain design. The first generation of this device has been comprised of two low-voltage normally-off Si devices plus a high-voltage normally-on GaN device, while providing 16 A rms continuous current and ±650 V voltage blocking capability with an equivalent on-state resistance of 70mΩ. As these are cascode devices, the same -5 V to 15 V gate-driver ICs, i.e., UCC21520 as used for SiC-FETs, have been employed for the MBS-GaN devices but with recommended gate-resistances as shown in Fig. 6(b). On the contrary, the MBS-GaN devices offer two gate-source leads; hence, two isolated dc supplies, i.e., MGJ1D051505MP, with a dual channel driver per each gate-source lead have been considered. Contemporary, the value of L_{lk} for the proposed reconfigurable DAB converter is chosen as 25 uH

979-8-3315-1612-3/25 $31.00 © 2025 IEEE

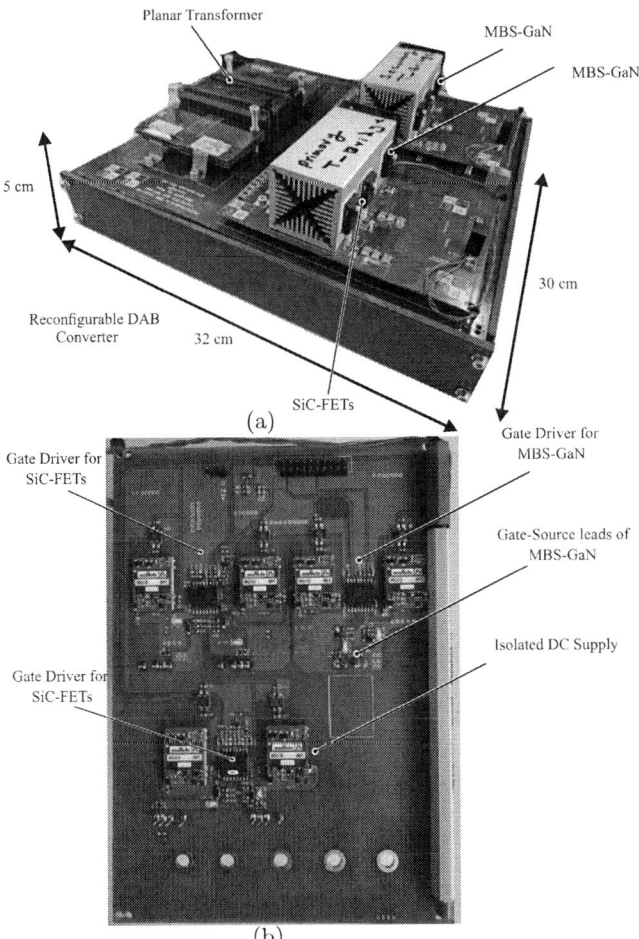

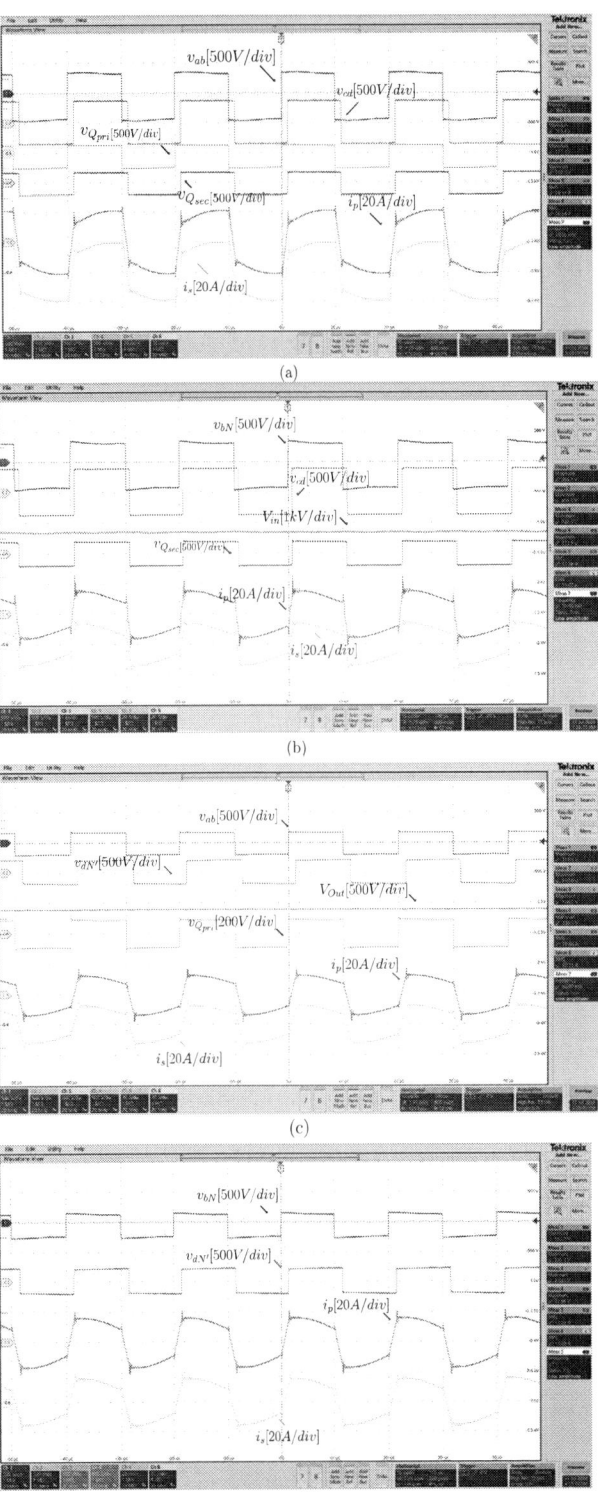

Fig. 6: (a) 6.6 kW fabricated prototype for the proposed reconfigurable DAB converter (Output capacitors, DSP with measurement sensors, and L_{lk} are all included in bottom side of the PCB), and (b) a close shot for the driver side of one bridge including SiC-FETs and MBS-GaN.

as an extreme case in the FB-FB mode based on (2) with a custom-oriented design, while the gate switching pulses are provided through a DSB within the SPS modulation technique. The value of dc-link MLCCs have also been designed based on the recommendation in [11].

Alternatively, to avoid saturation issue of the PT in case of unbalanced currents caused by mismatched PWMs or other asymmetries in the system, the dc blocking capacitors, C_{bp}, and C_{bs} are used in the design. The value of these HF capacitors must be designed to handle the rms currents on both sides of the PT as per (20) [11]:

$$C_{\mathrm{bp}} = C_{\mathrm{bs}} = \frac{100}{4\pi^2 f_{\mathrm{sw}}^2 L_{lk}} \qquad (20)$$

As for the first experiment, a FB-FB based configuration of the primary and secondary sides of the PT [see Fig. 1(b)] is considered, while the rated power is 6.6 kW and both the input and output voltage are fixed at 400 V. As can be seen from Fig. 7(a), both the MBS-GaN devices in both sides of

Fig. 7: Experimental results of the proposed reconfigurable DAB converter at (a) FB-FB Mode [6.3 kW/400 V to 400 V], (b) HB-FB Mode [4.4 kW/760 V to 400 V], (c) FB-HB Mode [2 kW/200 V to 400 V], and (d) HB-HB Mode [2.5 kW/400 V to 400 V].

the PT are blocking a stress voltage of ±200V, while the rms current passing through the PT is around 18 A. Fig. 7 (b) can

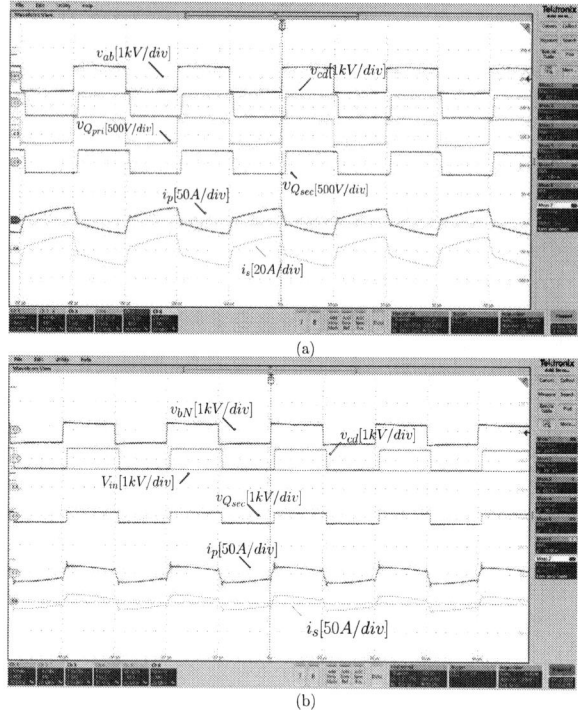

Fig. 8: Experimental results of the proposed reconfigurable DAB converter at (a) FB-FB Mode [5.8 kW/460 V to 400 V], and (b) HB-FB Mode [3.8 kW/680 V to 400 V].

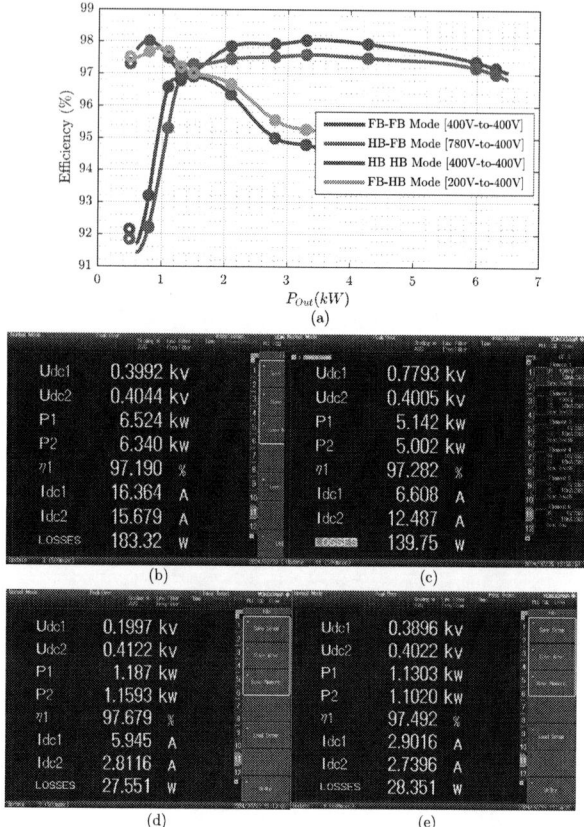

Fig. 9: (a) Overall efficiency results over a wide range of output power, and the detailed efficiency result for the (b) FB-FB-based mode, (c) HB-FB-based mode, (d) FB-HB-based mode, and (e) HB-HB-based mode.

also confirm the correct operation of the proposed converter in the HB-FB mode [see Fig. 2 (c)] at 4.4 kW output power. Herein the input dc voltage is around 760 V, while the output voltage is fixed at 400 V. As can be seen, the MBS-GaN in the primary side is passing around 12 A current with a 40°C maximum device temperature as per the heatsink/cooling system designed for the proposed converter.

Fig. 7 (c) and (d) also show the detailed waveforms of the proposed converter at FB-HB (2 kW) and HB-HB (2.5 kW) modes, respectively, while the input voltage is set at 200 V, and 400 V, respectively, and the output voltage is kept fixed at 400 V. Concurrently, to validate the ZVS analysis illustrated in Fig. 5, the performance of the proposed converter at different input voltages is also evaluated and shown in Fig. 8(a) and (b). Herein, as for the FB-FB mode, the input dc voltage is set at 460 V, while a fixed 400 V at the output at 5.8 kW output power is desired. The experimental result of this test is shown in Fig. 8(a), while due to the ZVS action of the devices at both sides of the PT based on Fig. 5, around 97.2% efficiency got obtained in practice. As can be seen from the shape of i_p, the converter is almost in its marginal range to lose the ZVS at this mode. The same observation as for the HB-FB mode can be seen from Fig. 8(b), where the input dc voltage is set at around 680 V, while the output voltage is fixed at 400 V and the output power is 3.8 kW. Confirming the analysis shown in Fig. 5, the converter can still operate at its ZVS range, while giving around 97.1% overall efficiency based on the measurement.

The efficiency results of the proposed reconfigurable DAB converter under a wide range of the output voltage are measured using the power analyzer WT1800 from Yokogawa and is shown for each of the operation modes in Fig. 9(a). As can be seen, owing to flexibility of the proposed structure in the FB-HB and HB-HB modes, the overall efficiency even at the low power is more than 96%, while the peak efficiency at 3.3 kW power/400 V-to-400 V input-output voltage condition in FB-FB mode is 98.1%. Details of the efficiency results over different input/output voltages and power can be seen from Fig. 9(b)-(e). The efficiency of the proposed structure in HB-HB and FB-HB modes at larger power is declined as the MBS-GaN devices become severely lossy; however, it might be improved either by the use of discrete-based bidirectional devices or by utilization of future generation of this device.

Finally, to monitor the thermal behavior of the MBS-GaN devices at both sides of the PT, the converter has been kept running continuously for five minutes at HB-FB and HB-HB modes. As has been captured by the thermal cam and shown in Fig. 10(a), the maximum temperature of the MBS-GaN at HB-FB mode at 3.8 kW power and under 770 V input dc voltage to 400 V fixed output voltage, is around 43.6°C, while the SiC-FETs temperature is about 38°C due to their much lower ON-

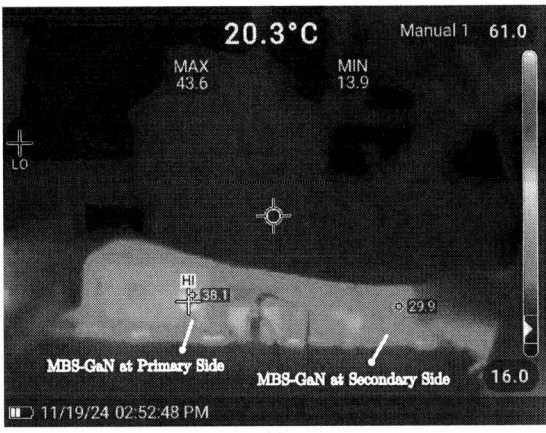

(a)

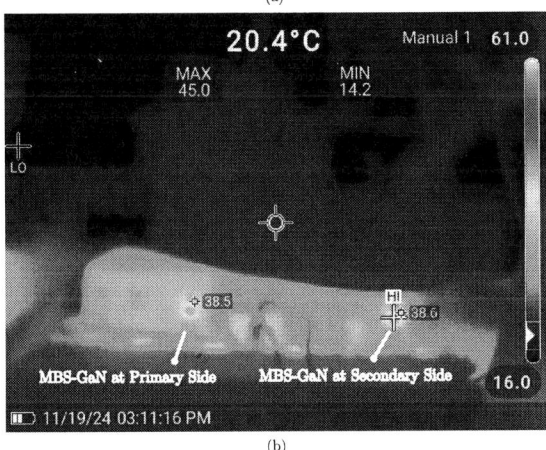

(b)

Fig. 10: MBS-GaN devices temperature in primary and secondary side of the PT after five minutes continuous running at (a) HB-FB mode [3.8 kW/770 V to 400 V], and (b) HB-HB mode [2 kW/400 V to 430 V]

state resistance. As per HB-HB mode and when both MBS-GaN devices at two sides of the PT are ON at 2 kW output power and 400 V-to-430 V input-output voltage condition, their maximum temperature after five minutes becomes around 45°C as per Fig.10(b). Herein, as for the HB-FB mode, the measured efficiency was around 97.5% and at HB-HB mode, it was about 96.45%, which can both confirm high overall efficiency of the proposed reconfigurable DAB converter under the ZVS range discussed in Section IV. The performance of the proposed converter at the transient case study, when changing its mode of operation, will be the focus of next studies.

VI. CONCLUSION

A highly efficient reconfigurable DAB converter using an PT with hybrid utilization of SiC and GaN wide-band-gap devices is proposed in this paper. As the four-quadrant device used in the proposed topology is always ON when an HB mode in one or two sides of the PT is needed, utilization of the MBS-GaN device provided by Transphorm Company

can be helpful to attain high overall efficiency. Four different operation modes based on the input voltage value are extracted from the proposed reconfigurable DAB converter. Thanks to these, the converter maintains high overall efficiency even at the low power or under the wide voltage conversion gain condition. Working principle, design guidance of the PT, ZVS range analysis of entire the converter and several experimental results using a 6.6 kW prototype have been presented confirming more than 97% overall efficiency over a wide range of operating conditions.

VII. ACKNOWLEDGMENT

Based on Material Transfer Agreement of "Four Quadrant Switches made with ARPA-E funding," between Chair of Power Electronics in Kiel University, Germany and Transphorm INC, USA, authors would like to acknowledge the provider as for MBS-GaN device used in this research.

REFERENCES

[1] L. Li, G. Xu, D. Sha, Y. Liu, Y. Sun and M. Su, "Review of Dual-Active-Bridge Converters With Topological Modifications," in IEEE Transactions on Power Electronics, vol. 38, no. 7, pp. 9046-9076, July 2023.

[2] R. M. Burkart and J. W. Kolar, "Comparative η– ρ– σ Pareto Optimization of Si and SiC Multilevel Dual-Active-Bridge Topologies With Wide Input Voltage Range," in IEEE Transactions on Power Electronics, vol. 32, no. 7, pp. 5258-5270, July 2017.

[3] S. Shao, H. Chen, X. Wu, J. Zhang and K. Sheng, "Circulating Current and ZVS-on of a Dual Active Bridge DC-DC Converter: A Review," in IEEE Access, vol. 7, pp. 50561-50572, 2019.

[4] H. Bai and C. Mi, "Eliminate Reactive Power and Increase System Efficiency of Isolated Bidirectional Dual-Active-Bridge DC–DC Converters Using Novel Dual-Phase-Shift Control," in IEEE Transactions on Power Electronics, vol. 23, no. 6, pp. 2905-2914, Nov. 2008.

[5] H. Higa, S. Takuma, K. Orikawa and J. -i. Itoh, "Dual active bridge DC-DC converter using both full and half bridge topologies to achieve high efficiency for wide load," 2015 IEEE Energy Conversion Congress and Exposition (ECCE), Montreal, QC, Canada, 2015, pp. 6344-6351.

[6] N. Hou and Y. W. Li, "Overview and Comparison of Modulation and Control Strategies for a Nonresonant Single-Phase Dual-Active-Bridge DC–DC Converter," in IEEE Transactions on Power Electronics, vol. 35, no. 3, pp. 3148-3172, March 2020.

[7] Z. Ouyang, O. C. Thomsen and M. A. E. Andersen, "Optimal Design and Tradeoff Analysis of Planar Transformer in High-Power DC–DC Converters," in IEEE Transactions on Industrial Electronics, vol. 59, no. 7, pp. 2800-2810, July 2012.

[8] F. Groon, H. Beiranvand, G. Can and M. Liserre, "Circular Economy Oriented and Reconfigurable Planar Transformer Design for Isolated DC/DC Converters," PCIM Europe 2024; International Exhibition and Conference for Power Electronics, Intelligent Motion, Renewable Energy and Energy Management, Nürnberg, Germany, 2024, pp. 3375-3384.

[9] https://www.transphormusa.com/en/news/arpaefourquadrantswitch.

[10] R. Barzegarkhoo, T. Pereira and M. Liserre, "Unlocking the Opportunities of Bidirectional GaN Devices in Isolated DC/DC Converters and Multilevel Inverters," CIPS 2024; 13th International Conference on Integrated Power Electronics Systems, Düsseldorf, Germany, 2024, pp. 15-24.

[11] "Design Guide: TIDA-010054 Bidirectional, Dual Active Bridge Reference Design for Level 3 Electric Vehicle Charging Stations." (2019).

[12] C. P. Steinmetz, "On the Law of Hysteresis," in Transactions of the American Institute of Electrical Engineers, vol. IX, no. 1, pp. 1-64, Jan. 1892.

[13] I. Villar, "Multiphysical characterization of medium-frequency power electronic transformers," Ph.D dissertation, EPFL, p. 234, 20.

Interleaved Switched-Inductor-Based SIPO Partial Power Converter Module for Battery Management Systems

Fengwang Lu
Department of Electrical Engineering
City University of Hong Kong, Hong Kong SAR, China
Email: fluwang@foxmail.com

Henry Shu-Hung Chung
Department of Electrical Engineering
City University of Hong Kong, Hong Kong SAR, China
Email: eeshc@cityu.edu.hk

Abstract—**Given the diverse operating voltage range and different operating modes of the battery management system (BMS) within renewable energy systems and DC buses, there is a demand for a highly efficient up/down bidirectional converter. A modular-based and series-connected partial power converter (S-PPC) using an interleaved switched-inductor series-input-parallel-output (SIPO) structure is proposed. Each module is a dual active clamped flyback-derived (DACF) bidirectional DC-DC converter without output capacitor. A 2.4kW, 800V / 600V prototype with the efficiency of 98.79% has been built and evaluated. The experimental results are favorably compared with theoretical predictions.**

I. INTRODUCTION

Electric vehicles (EVs) require a wide voltage range bidirectional DC-DC converter to enable efficient energy flow between DC buses —commonly referred to as the 800V high-voltage platform, with an operational range of 550V to 900V— and the high-voltage battery management system (BMS). This is essential to accommodate the broad voltage variations associated with battery charging and discharging characteristics. Conventional PWM bidirectional DC-DC converters, such as buck/boost converters [1], are commonly used in medium- and low-power applications due to their wide voltage range. By avoiding the hard switching and device voltage stresses associated with buck/boost converters, CLLC [2], [3] and DAB [4] converters, utilizing the merits of zero-voltage-switching (ZVS) capability and high power density, have been widely studied for high-power applications. However, the significant device current stresses resulting from high power levels in applications like electric vehicles (EVs) introduce new challenges. These include increased component costs due to the low $R_{ds\text{-}on}$ requirements of MOSFETs, the use of high-frequency MLCCs, and more complex thermal management. This is particularly evident in the selection of high-current MOSFETs and resonant capacitors in CLLC converters.

To address these limitations, partial power processing (PPP) converters are gaining attraction. PPP converters only process a portion of the input power, transferring it to the output while maintaining load regulation with reduced power losses. Two primary classifications of PPP converters exist based on their connection scheme: differential power processing (DPP) converters [5] and S-PPCs [6]. Differential PPP converters primarily address current imbalances in string PVs [7], while series-connected PPP converters act as voltage regulators between the DC bus and various sources [8]. This paper will present a SIPO partial power converter module, which is a DACF bidirectional DC-DC converter without output capacitor. A 2.4kW prototype has been built and evaluated.

II. OPERATING PRINCIPLE OF THE PROPOSED ARCHITECTURE

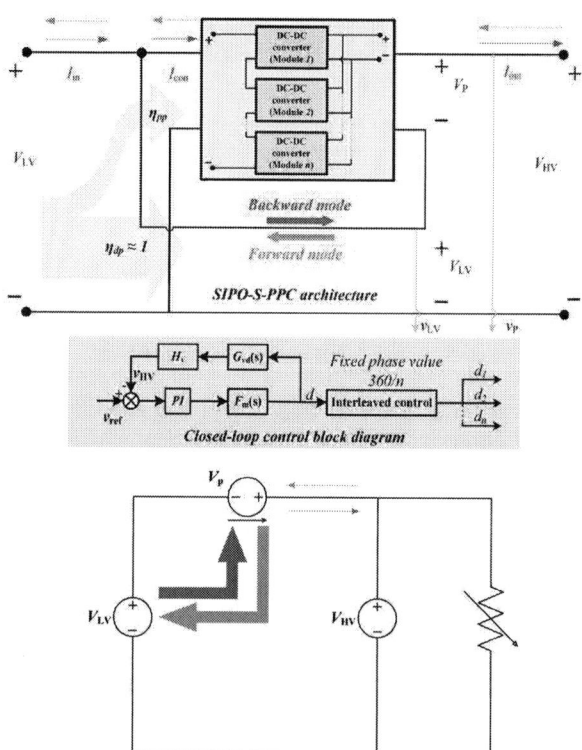

Fig. 1. Proposed SIPO-S-PPC schematic with bidirectional power flow

---

This work was supported by the Green Tech Fund from the Hong Kong Special Administrative Region, China, under Project GTF202020166.

Each module is a bidirectional DACF converter for its merits of wide voltage gain range, ZVS operation under heavy load and straightforward control. The modules are under interleaved phase-shifted control to reduce output current ripple. Since the structure and operation of the converter are symmetrical between the high-voltage and low-voltage sides, the power conversion operation from high-voltage to low-voltage side is illustrated in this digest.

Fig. 1 shows the architecture of a BMS charged by or discharged to the DC bus via the proposed SIPO-S-PPC architecture, which consists of N isolated bidirectional DC-DC converter modules. Their high-voltage ports are connected in series, while their low-voltage ports are connected in parallel. This configuration can reduce the voltage stress on semiconductors and extend the voltage gain range without requiring transformer of high turns ratios. Hence, such a configuration facilitates the use of low voltage devices and high efficiency. The total efficiency of the SIPO-S-PPC, η_{total}, can be shown to be

$$\eta_{total} = \frac{V_{HV} \cdot I_o}{V_{LV} \cdot I_{in}} = \frac{I_o + I_{con} \cdot \eta_{pp}}{I_o + I_{con}} = \frac{I_o + I_{con} \cdot \eta\left(I_o / n\right)}{I_o + I_{con}} \quad (1)$$

where η_{pp} is the efficiency of SIPO structure and $\eta\left(I_o/n\right)$ denotes the efficiency of the DACF converter at I_o/n load. We should focus on two key strategies: minimizing direct power flow and selecting a DC-DC converter with superior efficiency performance at light load. Using a common duty ratio control method with a PI controller in the voltage loop is a practical approach to simplifying the controller design for power electronics applications. Ignoring the minor power difference during basic modules, this method allows for effective regulation of the output voltage while streamlining control structure.

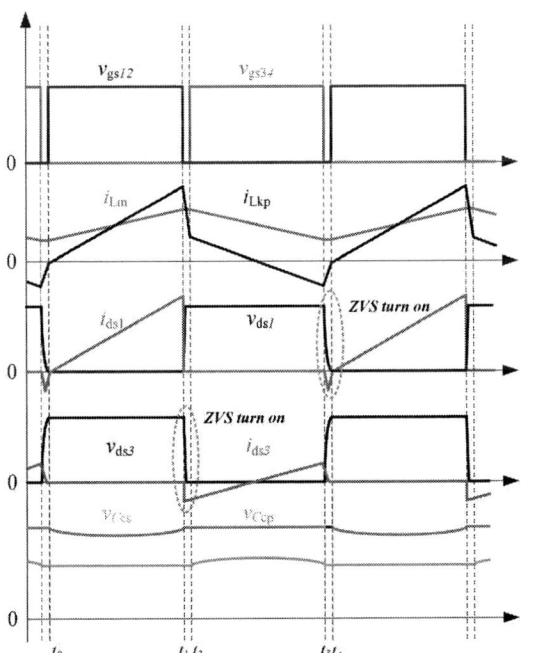

Fig. 2. Key waveforms.

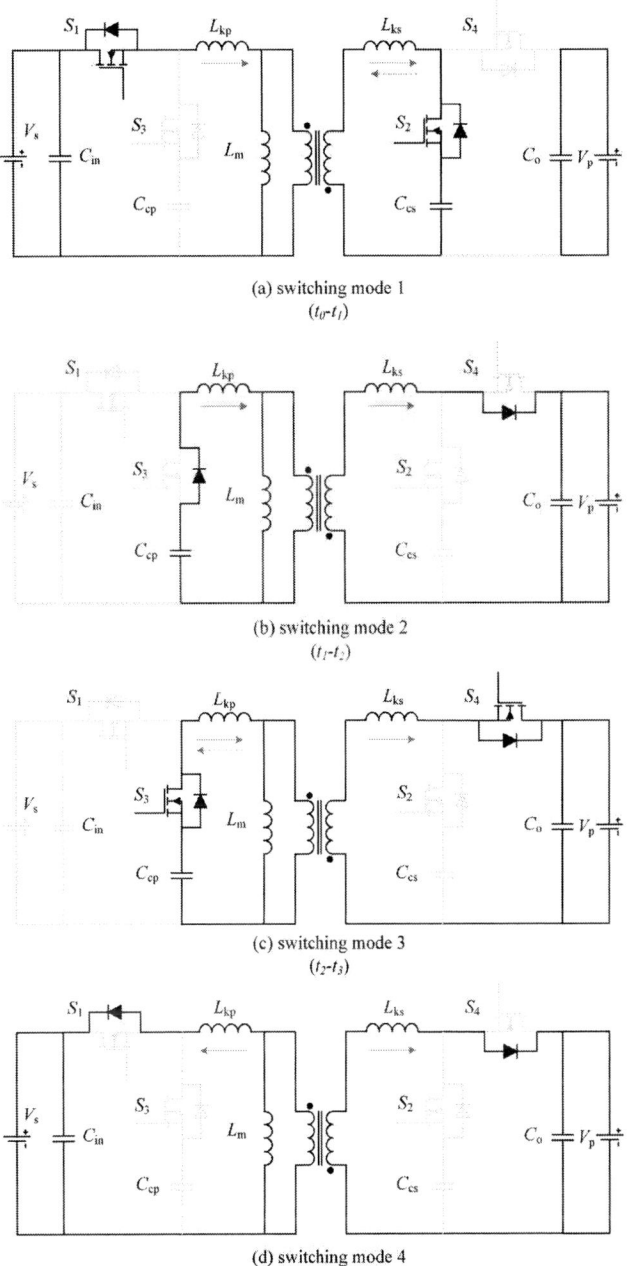

(a) switching mode 1
$(t_0 \text{-} t_1)$

(b) switching mode 2
$(t_1 \text{-} t_2)$

(c) switching mode 3
$(t_2 \text{-} t_3)$

(d) switching mode 4
$t_3 \text{-} t_4$

Fig. 3. operating mode of the proposed converter

As shown in Figs. 2 and 3, active clamping circuits with S_3 and C_{Cp} (forward mode) and S_2 and C_{Cs} (backward mode) recycle the energy stored in the leakage inductances L_{kp} and L_{ks}. Apart from the inherent zero-voltage switching (ZVS) at turn on in the active clamping circuit, the resonance between parasitic capacitance and leakage inductance also enables the ZVS of S_1 at turn on that further reduces switching loss.

III. STEADY-STATE ANALYSIS OF THE DACF CONVERTER

To simplify the theoretical analysis, all components are assumed to be ideal, and the effect of the leakage inductances is assumed to be negligible. This is justified by the proposed converter operating in CCM mode with a relatively large magnetic inductance, which dominates the energy transfer characteristics. The steady-state characteristics of DACF converter can be derived by applying the voltage-second balance principle across coupled inductors L_m. Because of the interleaved phase-shifted common duty control, the voltage transfer ratio is related to the V_{HV} and V_{LV} by the following equation:

$$\begin{cases} V_{HV} = \dfrac{3-2D}{3-3D} V_{LV} \\ D = \dfrac{3V_{HV} - 3V_{LV}}{3V_{HV} - 2V_{LV}} \end{cases} \quad (2)$$

IV. EXPERIMENTAL RESULTS AND COMPARISON

TABLE I
PARAMENTS AND SPECIFICATIONS OF THE PROTOTYPE

Parameter	Specification
LV V_{LV}	570-700 V
HV V_{HV}	800 V
Switching Frequency	100 kHz
SiC MOSFET	GC2M0080120D1
Clamping Capacitor	5uF
Magnetic devices	704uH/1.62uH
Turns Ratio	1:1

To validate the converter design (based on Table I parameters), a 2.4kW prototype with a controller card was built (as shown in Fig. 4). As shown in Fig. 5, using the DACF converter as basic unit (V_{LV} =600V, V_{HV} =800V, V_p =200V), the

TABLE II
COMPARISON WITH OTHER COUNTERPARTS

Topologies	[9]	[10]	[11]	[12]	[13]	Proposed
N_ MOSFETs	12	6	5	6	6	12
N_ diodes	6	2	2	2	4	0
N_ transformers	2, n=1	1, n=3	2, n=3,5.6	1	1, n=2	3, n=1
N_ inductors	8	1	0	1	1	0
N_ capacitors	15	5	5	2	4	9
Soft switching	Yes	Yes	Yes	Yes	No	Yes
Voltage stress of devices	V_{in} /8	V_{in}	-	V_{in} /2	V_{in} /(1-D)	V_{in} /(3-3D)
Voltage gain, n=1	-	-	-	$\pm 1/2D*_{\Delta v}$	$\pm D/(1-D) *_{\Delta v}$	$D/3(1-D)*_{\Delta v}$
DOF	3	2	2	2	2	3
Input voltage regulation range	1-1.6 kV	32-40 V	355-410 V	187-253 V	370 V	570-700 V
Output voltage	100 V	400 V	350 V	220 V	288-400 V	800 V
Energy transfer direction	Uni-	Uni-	Uni-	Bidirectional	Uni-	Bidirectional
Power level	1000 W	200 W	2000 W	750 W	2700 W	2400 W
Peak efficiency	97.3%	95.3%	99.4%	99.066%	98%	98.79%

where D is the duty cycle of switches S_1 and S_2. The voltage stress across switches S_1-S_4 can be written by

$$U_{S1} = U_{S2} = U_{S3} = U_{S4} = \frac{1}{3-3D} V_{LV}. \quad (3)$$

In order to achieve the ZVS operation of S_1, the dead time t_{dead} and output current should satisfy the condition: 1) The resonant period should be less than 0.25T; 2) The energy stored in the leakage inductance (E_{Lk}) should be greater than the energy stored in the parasitic capacitance (E_C).

$$\begin{cases} t_{dead} \leq 0.5\pi\sqrt{L_{Kp} \cdot C_{ds}} \\ E_{Lk} \geq E_C \end{cases} \quad (4)$$

$$\begin{cases} t_{dead} \leq 0.5\pi\sqrt{L_{Kp} \cdot C_{ds}} \\ I_o \geq \sqrt{\dfrac{C_{ds}}{L_{Kp}}} \cdot \dfrac{V_{LV}}{6} \end{cases} \quad (5)$$

SIPO-S-PPC can achieve high efficiency (peak efficiency 98.79%) due to the SIPO and PPP structure. Besides, while the prototype demonstrates good overall power density, the substantial volume occupied by the film capacitors negatively impacts its effective power density in high voltage platforms.

Fig. 4. Photo of the prototype.

A. Comparison with other S-PPCs

Table II shows the comparison of the proposed topology with other counterparts such as the high step-up converter with buck converter in [10], which perform well in power density. Compared to its counterparts, the proposed SIPO-S-PPC has more degrees of freedom (DOF) in voltage gain than others, meaning that it is determined not only by the duty cycle of each converter unit but also by the turns of the transformer and the number of basic units in the SIPO structure. Meanwhile, the flexible SIPO structure facilitates the feasibility of extending the power level by adding or reducing the number of DACF converter units. The current pulse or ripple from the DACF converter can also be alleviated by the interleaved phase-shifted control method, which is beneficial for extending the life of batteries and capacitors. Additionally, lower device stress due to the special SIPO structure contributes to the application of low $R_{ds\text{-}on}$ devices like GaN MOSFETs and improves reliability. The function of bidirectional conversion also enables applications in high-voltage battery management systems (BMS).

B. Experimental results

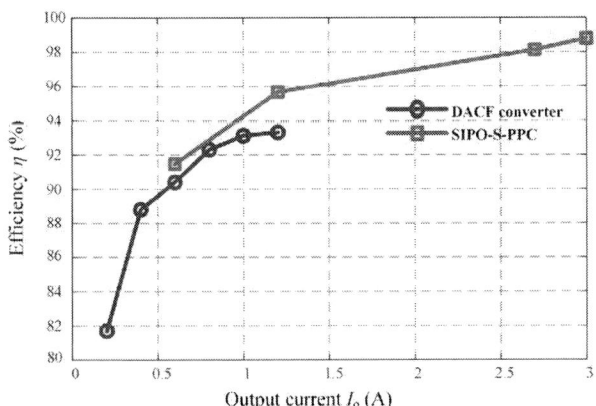

Fig. 5. Measured efficiency under different load conditions.

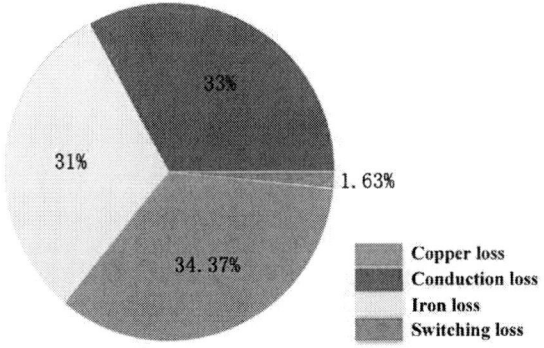

Fig. 6. Loss breakdown results of the SIPO-S-PPC.

It can be seen from Fig. 6 that conduction losses, which include copper losses in magnetic devices and conduction losses in MOSFETs, dominate the theoretical total loss (approximately 27 W) at a 2.4 kW load with an output voltage of 800 V. ZVS operation, except for the MOSFETs in the secondary clamping circuits of each basic unit, helps to alleviate switching losses. In

order to design for larger leakage inductance, we need to use a greater number of turns and a specialized winding method.

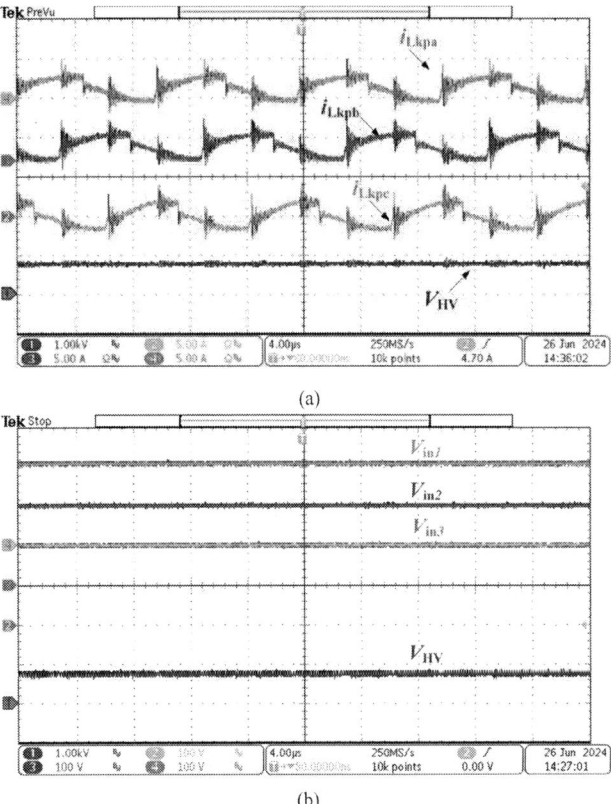

Fig. 7. Critical waveforms of the SIPO-S-PPC. (a) Currents across leakage inductance. (b) Input voltages across input capacitors.

In Fig. 7(b), the voltage differences among the input voltages of each module are minimal, as also shown in Fig. 7(a). This indicates that the common duty ratio control method realizes the average power distribution and regulates effectively in SIPO-S-PPC. The critical waveforms of the DACF converter are shown in Fig. 8, where in good agreement with theoretical analysis in Fig. 2.

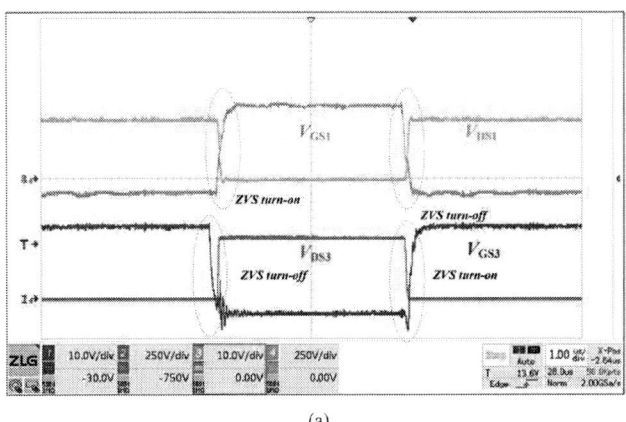

(a)

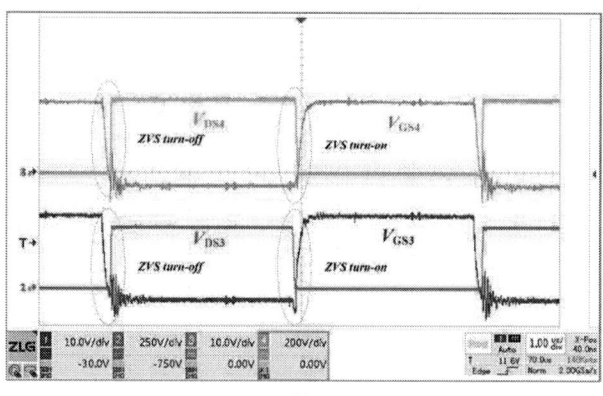

(b)

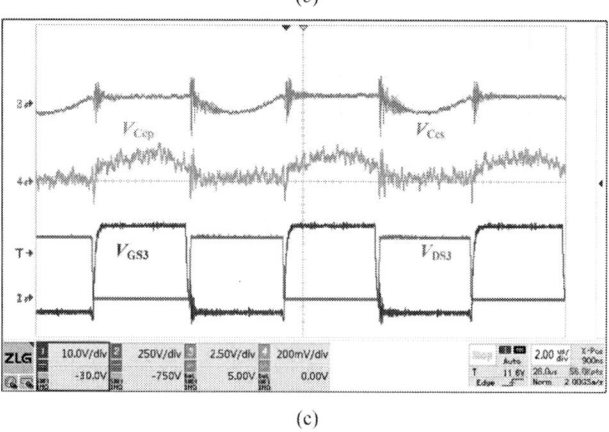

(c)

Fig. 8. Critical waveforms of the DACF converter. (a)Voltage waveforms of S1 and S2. (b) Voltage waveforms of S2 and S3. (c) Voltage ripple waveforms of clamping capacitor.

V. CONCLUSION

This paper mainly introduces the operating principle of the proposed SIPO-S-PPC architecture, which exhibits the following merits: 1) versatile and extensible SIPO-S-PPC architecture; 2) bidirectional power flow; 3) high conversion efficiency; 4) simple control; 5) low current ripple.

REFERENCES

[1] M. Safayatullah, M. T. Elrais, S. Ghosh, R. Rezaii and I. Batarseh, "A Comprehensive Review of Power Converter Topologies and Control Methods for Electric Vehicle Fast Charging Applications," IEEE Access, vol. 10, pp. 40753-40793, 2022. *(references)*

[2] Y. -F. Wang, B. Chen, Y. Hou, Z. Meng and Y. Yang, "Analysis and Design of a 1-MHz Bidirectional Multi-CLLC Resonant DC–DC Converter with GaN Devices," IEEE Transactions on Industrial Electronics, vol. 67, no. 2, pp. 1425-1434, Feb. 2020.

[3] H. Li et al., "Bidirectional Control with Fitting Model-Based Synchronous Rectification and Input Ripple Current Feedforward for SiC Bidirectional CLLC EV Charger," IEEE Transactions on Industrial Electronics, vol. 70, no. 9, pp. 9136-9146, Sept. 2023.

[4] X. Chen, G. Xu, H. Han, D. Liu, Y. Sun and M. Su, "Light-Load Efficiency Enhancement of High-Frequency Dual-Active-Bridge Converter Under SPS Control," IEEE Transactions on Industrial Electronics, vol. 68, no. 12, pp. 12941-12946, Dec. 2021.

[5] N. G. F. dos Santos, J. R. R. Zientarski and M. L. d. S. Martins, "A Review of Series-Connected Partial Power Converters for DC–DC Applications," IEEE Journal of Emerging and Selected Topics in Power Electronics, vol. 10, no. 6, pp. 7825-7838, Dec. 2022.

[6] N. Hou, L. Ding, P. Gunawardena, T. Wang, Y. Zhang and Y. W. Li, "A Partial Power Processing Structure Embedding Renewable Energy Source and Energy Storage Element for Islanded DC Microgrid," IEEE Transactions on Power Electronics, vol. 38, no. 3, pp. 4027-4039, March 2023.

[7] P. S. Shenoy, K. A. Kim, B. B. Johnson and P. T. Krein, "Differential Power Processing for Increased Energy Production and Reliability of Photovoltaic Systems," IEEE Transactions on Power Electronics, vol. 28, no. 6, pp. 2968-2979, June 2013.

[8] J. R. R. Zientarski, M. L. da Silva Martins, J. R. Pinheiro and H. L. Hey, "Evaluation of Power Processing in Series-Connected Partial-Power Converters," IEEE Journal of Emerging and Selected Topics in Power Electronics, vol. 7, no. 1, pp. 343-352, March 2019.

[9] R. Guan et al., "A High Step-Down Partial Power Processing Switched-Capacitor Converter for Wide Input Voltage Range Medium Voltage DC Applications," IEEE Transactions on Power Electronics, vol. 38, no. 10, pp. 12265-12277, Oct. 2023.

[10] X. Sang, Y. Wang, S. Gao, Y. Guan and D. Xu, "Analysis and Design of a Partial Power Processing Architecture for High Step-Up Applications," IEEE Transactions on Power Electronics, vol. 38, no. 7, pp. 8654-8665, July 2023.

[11] O. Abdel-Rahim, A. Chub, H. M. Maheri, A. Blinov and D. Vinnikov, "High-Performance Buck-Boost Partial Power Quasi-Z-Source Series Resonance Converter," IEEE Access, vol. 10, pp. 130177-130189, 2022.

[12] J. R. R. Zientarski, M. L. d. S. Martins, J. R. Pinheiro and H. L. Hey, "Series-Connected Partial-Power Converters Applied to PV Systems: A Design Approach Based on Step-Up/Down Voltage Regulation Range," IEEE Transactions on Power Electronics, vol. 33, no. 9, pp. 7622-7633, Sept. 2018.

[13] A. Diab-Marzouk and O. Trescases, "SiC-Based Bidirectional Ćuk Converter With Differential Power Processing and MPPT for a Solar Powered Aircraft," IEEE Transactions on Transportation Electrification, vol. 1, no. 4, pp. 369-381, Dec. 2015.

Single Sensor-Based Fault Localization and Detection in GaN Three-Phase Dual Active Bridge Converters

Satyam Sa, Yi Han, Cheng Feng Wang, and Olivier Trescases

The Edward S. Rogers Sr. Department of Electrical & Computer Engineering, University of Toronto, Canada

E-mail: satyam.sa@mail.utoronto.ca

Abstract—**This paper presents a novel approach for single-switch fault detection and localization in three-phase dual active bridge (DAB) converters. The proposed control scheme leverages an existing current sense resistor that is typically used for output current regulation, to extract the necessary $\alpha\beta$ parameters for detecting and localizing open-circuit faults (OCFs). Additionally, secondary-side short-circuit faults (SCFs) are detected and localized without the need for additional sensors. Simulated results demonstrate successful detection and localization of both OCFs and SCFs during 400V operation. The proposed method is validated on a 400V, 6.6kW liquid-cooled converter equipped with 650V GaN devices, operating at a switching frequency of 300 kHz and achieving a peak efficiency of 96.6% at 6.5 kW. Experimental results confirm the successful detection and localization of single-switch OCFs within four switching cycles.**

I. INTRODUCTION

Given the tight volume constraints for electric vehicle (EV) sub-systems and the safety-critical nature of automotive power electronics, onboard chargers (OBCs) must be power-dense, highly efficient, and reliable. Conventional OBCs typically consist of an ac-dc stage followed by a dc-dc stage. A three-phase (TP) dual active bridge (DAB) converter enables higher power density and reliability for the dc-dc stage when compared with single-phase designs. This topology minimizes volume by leveraging flux cancellation to shrink the magnetic core. The input and output filter sizes can also be reduced through ripple cancellation, enhancing overall OBC density [1]. The TP-DAB also offers better fault tolerance, allowing the system to continue operating at a reduced capacity by phase shedding in the event of a phase failure, thus preventing a complete shutdown [2].

Gallium nitride (GaN) devices enable higher-frequency operation compared to silicon (Si) and silicon carbide (SiC), thereby reducing converter volume and enhancing on-board charger (OBC) density. GaN devices [3] offer a 13X lower $R_{on}Q_g$ figure of merit for 400 V applications compared to Si [4]. Nevertheless, challenges in fault tolerance and reliability persist, particularly with key gate-drive issues such as false turn-on due to low threshold voltages, narrow voltage margins, and increased dv/dt and di/dt effects. These issues arise from low parasitic capacitance and high slew rates [5]–[7], collectively contributing to higher failure rates.

A 2018 survey found that 76% of failures in converter systems are attributed to semiconductor devices (57%) and gate drivers (19%) [8]. Short-circuit faults (SCFs) and open-circuit faults (OCFs) in power devices are major contributors to these failures. In GaN-based TP-DAB converters, the in-

creased number of semiconductor failure points compared to single-phase designs makes a reliable fault detection and localization algorithm essential for maintaining system reliability and mitigating the risks associated with added complexity and a higher switch count. SCFs, often caused by gate or source metal shorts [9], lead to overcurrent conditions, requiring rapid detection mechanisms [10], [11]. OCFs, typically resulting from bond wire lifts or gate driver failures [9], are influenced by converter topology, necessitating topology-specific fault detection methods. Various OCF detection and localization strategies for TP-DAB topologies have been explored in the literature [12]–[15].

In [12], OCF detection and localization are achieved using six current sensors to measure phase currents on both the primary and secondary transformer sides. These measurements determine the magnetizing current for each phase, and by analyzing its polarity along with the phase currents, OCFs can be accurately located; however, the reliance on multiple sensors makes this method relatively costly. An alternative approach in [13] uses six voltage sensors to monitor switch node voltages for OCF detection and localization, which, while effective, adds complexity, cost, and potential noise issues from switch-node sensing. The detection algorithm in [14] requires only three current sensors and applies a variable threshold for each phase, reducing the sensor count, but its dependence on finely tuned thresholds increases noise susceptibility. In [15], an OCF detection and localization algorithm in the $\alpha\beta$ frame requires only two sensors, simplifying the detection process.

While these approaches address OCF faults, none of them can address SCFs without additional sensing, as noted in [11]. In [10], SCFs are detected by monitoring the voltage across local power decoupling capacitors and using a de-saturation circuit for protection. However, this requires voltage sensors at each of the six half-bridges in a TP-DAB converter, greatly increasing the sensor count. Alternatively, [11] proposes a GaN device with a sense-HEMT for SCF detection. While this approach achieves SCF detection within 36 ns, the 400 V GaN IC technology is still under development, and currently, only limited commercial GaN devices offer integrated sense-HEMTs.

This work presents a novel scheme for detecting OCFs and secondary side SCFs in TP-DAB converters by utilizing an existing current sense resistor, typically used for output current regulation. The proposed fault detection method is experimentally validated on a 400-V, 6.6-kW liquid-cooled GaN-based converter operating at 300 kHz. The system architecture is described in Section II, with the fault detection and localization algorithm in Section III, followed by experimental results in Section IV, and conclusions in Section V.

II. SYSTEM ARCHITECTURE

A TP-DAB converter system comprises two three-phase active bridges connected by three inductors and a three-phase transformer, as depicted in Fig. 1. Additionally, the system includes a current sense resistor, R_{sns}, to regulate the output current, as discussed in [16], [17]. The power transferred by the TP-DAB converter, P, is determined by the phase shift between the primary and secondary sides, along with the transformer configuration. For a YΔ transformer configuration, the power transfer for TP-DAB is given by

$$
P = \begin{cases} \frac{nV_{link}V_{batt}}{2\pi f_{sw}L}\phi & , \quad 0 \le \phi \le \frac{\pi}{6} \\ \frac{nV_{link}V_{batt}}{2\pi f_{sw}L}\left[\frac{3}{2}\left(\phi - \frac{\phi^2}{\pi}\right) - \frac{\pi}{24}\right] & , \quad \frac{\pi}{6} \le \phi \le \frac{\pi}{2} \end{cases} \quad (1)
$$

where

ϕ = phase difference between V_{AN} on the primary side and V_{ab} on the secondary side

L = leakage inductance used in the design

f_{sw} = switching frequency of the converter

V_{link} = power-factor-correction stage output

V_{batt} = battery voltage

The output current delivered to the battery is primarily dc, as the filter capacitor, C_{filter}, smooths out the high-frequency current content that would otherwise carry fault signatures during a failure event. As a result, the battery current is unsuitable for fast fault detection. In the proposed method, the existing R_{sns} is re-positioned to sense the current I_{sns} before the bulk output filter capacitor, as illustrated in Fig. 1. This allows for the detection and localization of OCFs at all switch locations, as well as secondary-side SCFs, without affecting output current regulation.

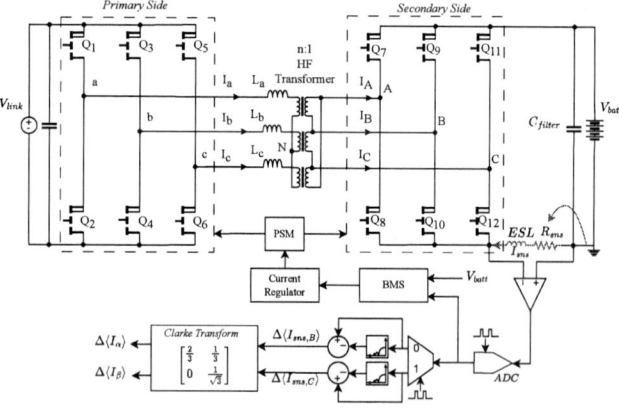

Fig. 1. TP-DAB converter with fault detection and localization scheme.

Under ideal conditions, I_{sns} represents the active rectification of the phase currents based on the states of the active switches. This results in a ripple at six times the switching frequency, f_{sw}, due to the switching combinations, as shown in Fig. 2. In practical implementations, however, C_{filter} consists of a main bus capacitor, C_{bus}, and high-frequency decoupling capacitors, C_{HF}, at each phase leg. The use of C_{HF} as close as possible to power devices minimizes the power loop inductance, L_{PL}, as illustrated in Fig. 3.

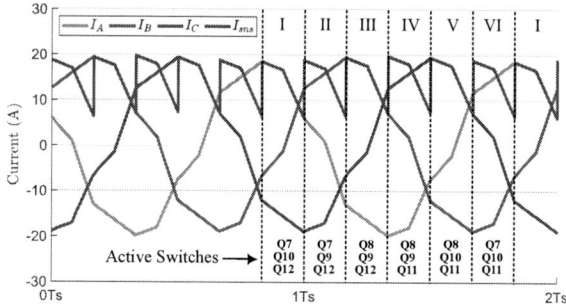

Fig. 2. Operating currents of the TP-DAB. Without the presence of C_{HF}, I_{sns} is a combination of the three phase currents, I_{a-c}, rectified, as dictated by the switch states.

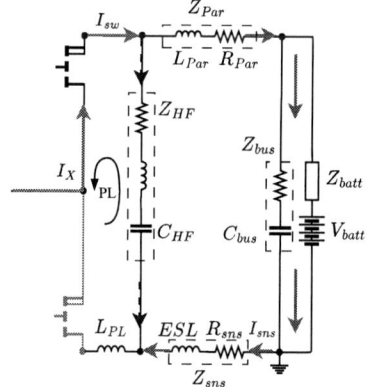

Fig. 3. Distribution of the phase current among the filtering elements and the output node. Finite element analysis on the PCB gives $L_{par} = 50\,\text{nH}$, $R_{sns} = 5\,\text{m}\Omega$, $ESL = 1\,\text{nH}$, C_{HF} is a ceramic capacitor, and C_{bus} is an electrolytic capacitor with appropriately derated values based on the datasheet.

The addition of C_{HF} alters the ideal I_{sns} waveform which both smooths and delays the actual I_{sns} relative to the ideal I_{sns}, as shown in Fig. 4. Based on the more accurate model depicted in Fig. 3, I_{sns} can be expressed by

$$
I_{sns}(s) = I_{sw}(s) \cdot Z_{tf}(s) \quad (2)
$$

where Z_{tf} is the bridge to sense current transfer function

$$
Z_{tf}(s) = \frac{Z_{HF}}{Z_{HF} + Z_{par} + (Z_{batt} \parallel Z_{filter}) + Z_{sns}} \quad (3)
$$

Z_{HF} = Impedance of high frequency bypass capacitors

Z_{par} = Impedance of parasitic inductance

Z_{batt} = Impedance of battery

Z_{filter} = Impedance of bus capacitors

Z_{sns} = Impedance of current sense resistor

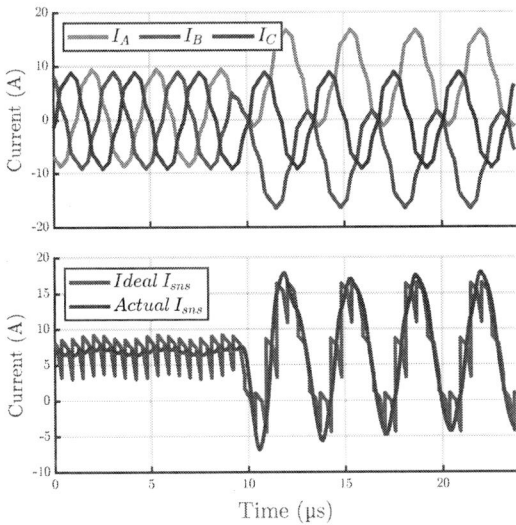

Fig. 4. Current waveform before and after OCF operation.

III. PROPOSED FAULT-DIAGNOSIS SOLUTION

A. Open-Circuit Fault Detection and Localization

An open-circuit fault in any single switch (Q_1–Q_{12}) introduces a dc component in the phase currents [14], which is reflected in I_{sns}, as shown in Fig. 4. By utilizing I_{sns} and the switching configurations depicted in Fig. 5(a) and Fig. 5(b), the dc component in the secondary phase B (I_B) and C (I_C) currents can be extracted. When switches Q_8, Q_9, and Q_{12} are on, I_{sns} can be expressed as in (4); similarly, when switches Q_8, Q_{10}, and Q_{11} are on, I_{sns} follows (5), as shown in Fig. 5(a) and Fig. 5(b), respectively.

$$I_{sns} = I_B \cdot Z_{tf} \qquad (4)$$

$$I_{sns} = I_C \cdot Z_{tf} \qquad (5)$$

As discussed in [15], the dc component information from I_B and, I_C is sufficient to accurately locate the OCF in a TP-DAB converter with a YΔ configuration. Therefore, the necessary $\alpha\beta$ parameters for fault detection and localization can be extracted from I_{sns}.

In the proposed algorithm, I_{sns} is sampled during these periods with a precisely pre-calibrated delay from the gating signal, ensuring accurate data capture.

$$\begin{aligned} I_{sns,B}[n] &= I_{sns}(nt_{s,B} + \delta) \\ I_{sns,C}[n] &= I_{sns}(nt_{s,C} + \delta) \end{aligned} \qquad (6)$$

where $t_{s,B}$ and $t_{s,C}$ represents the sampling instances of phase B and C, respectively. The delay, δ, compensates for the phase shift introduced by Z_{par}, Z_{HF}, Z_{bus} and Z_{batt} at $6f_{sw}$, which is approximately equal to 250 ns based on the estimated component values. To increase the robustness to component variations, the dc shift in I_B and I_C represented by $\Delta\langle I_{sns,B}\rangle[n]$ and $\Delta\langle I_{sns,C}\rangle[n]$, is determined by subtracting the moving average from the instantaneous sampled value

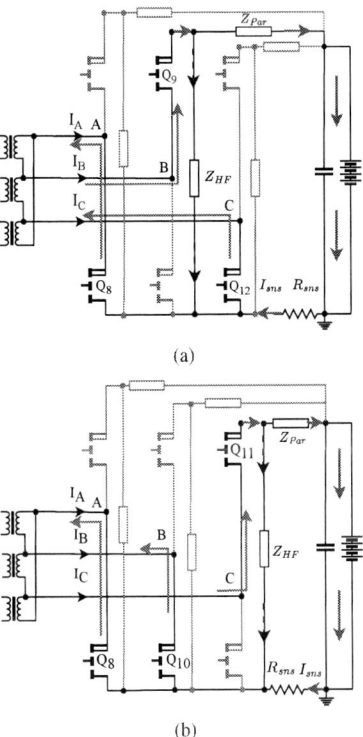

Fig. 5. (a) Positive rectification of phase B current. (b) Positive rectification of phase C current.

according to (7), where m is the moving average window size. The $\Delta\langle I_\alpha\rangle$ and $\Delta\langle I_\beta\rangle$ parameters are calculated from $\Delta\langle I_{sns,B}\rangle[n]$ and $\Delta\langle I_{sns,C}\rangle[n]$ as given by (8).

$$\begin{aligned} \Delta\langle I_{sns,B}\rangle[n] &= I_{sns,B}[n] - \frac{1}{m}\sum_{k=1}^{m} I_{sns,B}[n-k] \\ \Delta\langle I_{sns,C}\rangle[n] &= I_{sns,C}[n] - \frac{1}{m}\sum_{k=1}^{m} I_{sns,C}[n-k] \end{aligned} \qquad (7)$$

$$\begin{bmatrix} \Delta\langle I_\alpha\rangle \\ \Delta\langle I_\beta\rangle \end{bmatrix} = \frac{1}{n}\begin{bmatrix} \frac{2}{3} & \frac{1}{3} \\ 0 & \frac{1}{\sqrt{3}} \end{bmatrix}\begin{bmatrix} \Delta\langle I_{sns,B}\rangle \\ \Delta\langle I_{sns,C}\rangle \end{bmatrix} \qquad (8)$$

Using (7) and (8), the offsets in $\Delta\langle I_\alpha\rangle$ and $\Delta\langle I_\beta\rangle$ are calculated and compared to the threshold, I_{th}, to identify the faulty switch, as shown in Table I. The simulated response at 400 V in Fig. 6, 7, and 8 confirms successful detection of OCFs at Q_1, Q_6, and Q_7 respectively.

B. Short-Circuit Fault Detection and Localization

A short-circuit fault (SCF) can occur under two main scenarios. The first scenario is when the switch is already conducting, but due to a fault, it fails to turn off as expected. The second scenario occurs when the switch is in the off state, but unexpectedly turns on because of a high dv/dt.

To determine the time required to detect a short-circuit (SC) fault, the ability of the device to withstand SC pulses was tested using the setup shown in Fig. 9(a) [18]. Three 650-V-rated GaN devices were evaluated for their capability to endure a 10 μs SC event at 400 V. The corresponding short-circuit waveforms for each device are displayed in Fig. 10.

TABLE I
Fault Localization Criteria

Fault Location	Q_1	Q_2	Q_3	Q_4	Q_5	Q_6	Q_7	Q_8	Q_9	Q_{10}	Q_{11}	Q_{12}	*
$\Delta\langle I_\alpha\rangle > I_{th}$	1	0	0	1	0	1	0	1	1	0	0	0	0
$\Delta\langle I_\alpha\rangle < -I_{th}$	0	1	1	0	1	0	1	0	0	1	0	0	0
$\Delta\langle I_\beta\rangle > I_{th}$	0	0	1	0	0	1	0	1	0	1	1	0	0
$\Delta\langle I_\beta\rangle < -I_{th}$	0	0	0	1	1	0	1	0	1	0	0	1	0
$\lvert\Delta\langle I_\alpha\rangle\rvert > \lvert\Delta\langle I_\beta\rangle\rvert$	1	1	0	0	0	0	1	1	1	1	0	0	-

*Fault Free Operation

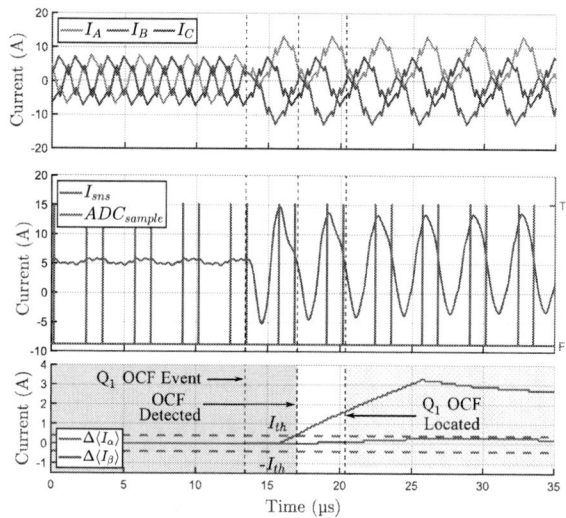

Fig. 6. Simulated phase current, I_{sns}, ADC sample signals, and $\alpha\beta$ currents in the TP-DAB topology during OCF at Q_1.

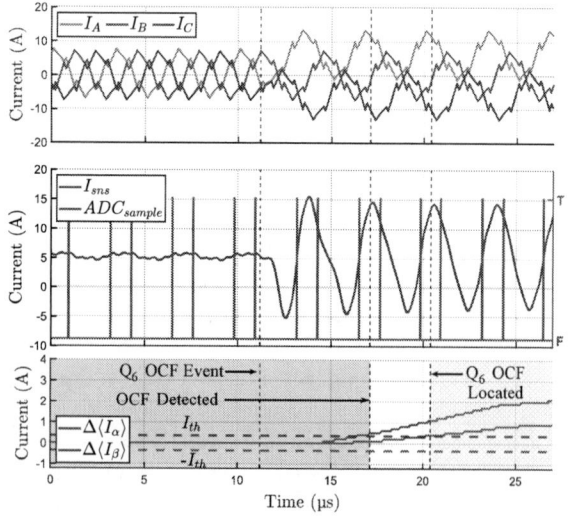

Fig. 7. Simulated phase current, I_{sns}, ADC sample signals, and $\alpha\beta$ currents in the TP-DAB topology during OCF at Q_6.

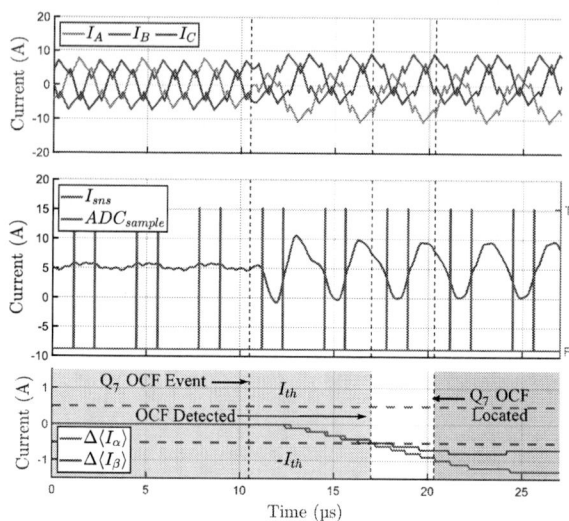

Fig. 8. Simulated phase current, I_{sns}, ADC sample signals, and $\alpha\beta$ currents in the TP-DAB topology during OCF at Q_7.

The results, with the assumption that each device represents the typical characteristics of its family, indicate that larger GaN devices have longer short-circuit withstand time, primarily due to their increased thermal capacity and lower current density. Both factors contribute to the ability to handle high current levels for a longer period before failure.

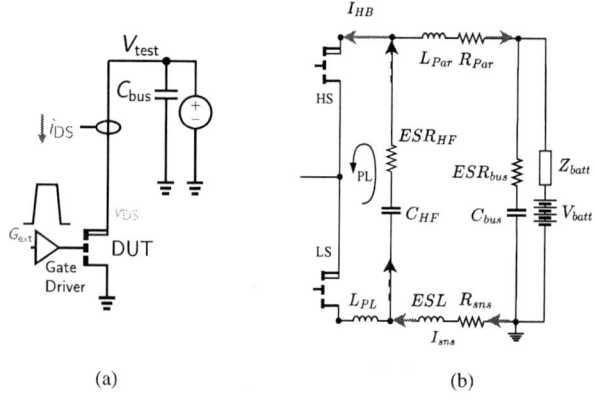

Fig. 9. (a) SC test circuit [18] and (b) SC simulation setup for half bridge.

To validate the proposed algorithm and analyze the dynamic behavior of I_{HB} and I_{sns} during an SCF, a half-bridge circuit, shown in Fig. 9(b), was simulated using the GS66516T SPICE model at 400 V. The response of I_{sns} for different parasitic inductance values, L_{par}, was compared to the shoot-through current, I_{HB}, as shown in Fig. 11. The initial peak shoot-through current is supplied by C_{HF} and later shared between C_{HF} and C_{bus}, depending on the bridge-to-sense current transfer function given by (3). The Bode plot of Z_{tf} for various parasitic inductance values is shown in Fig. 12. As L_{par} increases, I_{sns} takes longer to decay, the oscillation frequency decreases, and the phase lag increases. Although placing R_{sns} within the power loop, PL, could reduce SCF detection time, it would degrade power loop performance.

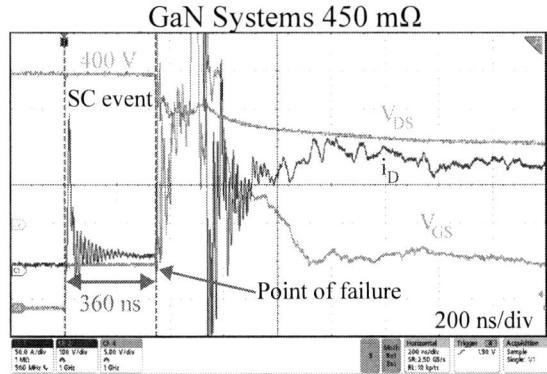

GaN Systems 450 mΩ

GaN Systems 225 mΩ

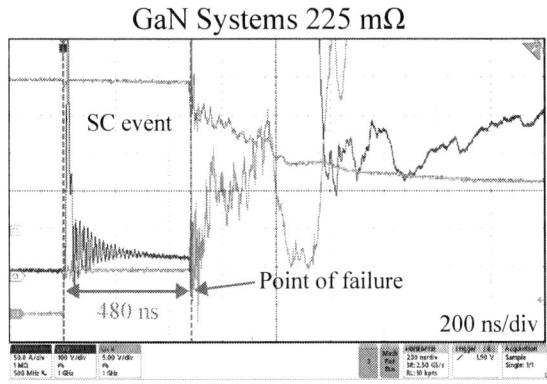

GaN Systems 150 mΩ

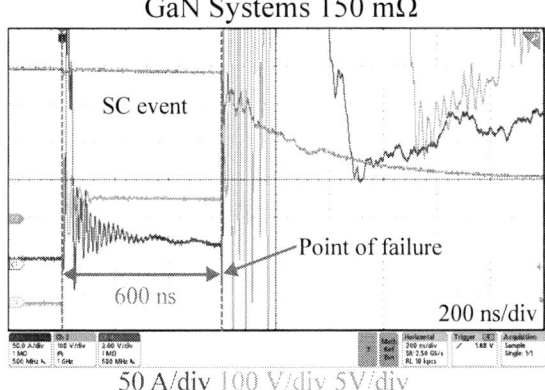

Fig. 10. Measured V_{gs}, V_{ds}, and i_D waveforms for SC events performed at 400V until device failure for three different GaN HEMTs [18].

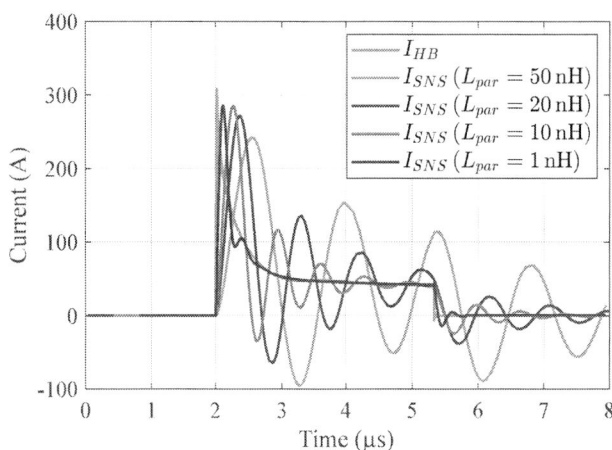

Fig. 11. Simulated dynamic I_{sns} behavior with varying parasitic inductance.

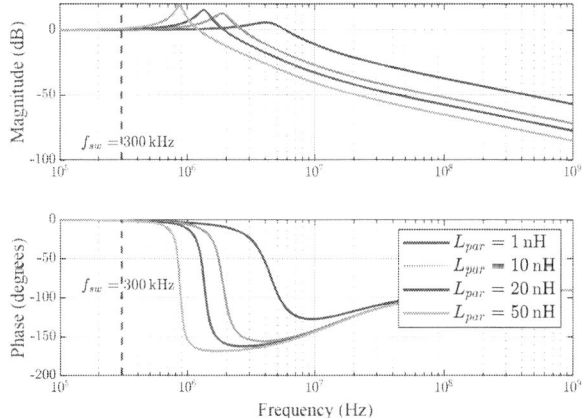

Fig. 12. Bode plot of Z_{tf} with varying parasitic inductance.

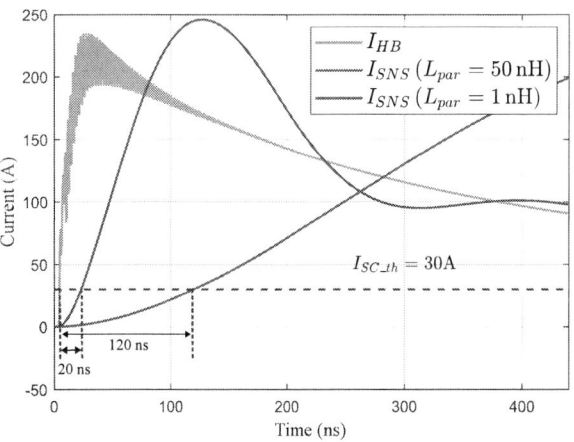

Fig. 13. Simulated short-circuit detection time for $L_{par} = 1$ nH and $L_{par} = 50$ nH, with the short-circuit threshold, $I_{(SC\_th)}$, set to 30 A.

The short-circuit threshold for SCF detection is set to $I_{th} = \min\{$device current rating, $1.5 \times$peak operating current$\}$. During normal operation of the designed TP-DAB converter at 6.6 kW, the maximum peak current experienced by the devices on the secondary side is 20 A. The GS66516T GaN devices are rated for 60 A. Therefore, the short-circuit threshold, $I_{(SC\_th)}$, for SCF detection is set to 30 A. Based on simulations of the proposed method, the SCF detection time is 20 ns for a parasitic inductance L_{par} of 1 nH, and 120 ns for $L_{par} = 50$ nH, as shown in Fig. 13. The simulated detection time is less than the minimum SC withstand time reported in Fig. 10, proving the effectiveness of the proposed method.

Since only one switch transition occurs during state changes,

as illustrated in Fig. 2, SCFs can be localized easily. For example, if a fault is detected in one of the secondary-side switches during switch state I, causing it to remain permanently on, and an SCF is observed at the beginning of switch state IV, the fault can be attributed to switch Q_{12}. Similarly, if the SCF is observed at the beginning of switch state III, the fault can be traced to switch Q_7. By analyzing transitions between switch states and noting that only one switch changes state at a time, the precise location of the SCF can be identified. With timely detection and localization of the switch failing as an SCF, the proposed algorithm prevents any cascading faults from occurring.

The proposed algorithm is able to detect SCFs only on the secondary side. To enable the detection of SCFs on the primary side, an additional current sense resistor, R_{sns}, must be placed on the primary side in a similar configuration to that shown in Fig. 1.

IV. EXPERIMENTAL RESULTS

A 400-V, 6.6-kW liquid-cooled experimental prototype using 650-V GaN devices, switching at 300 kHz, was implemented to validate the proposed algorithm, as shown in Fig. 14, with key features listed in Table II

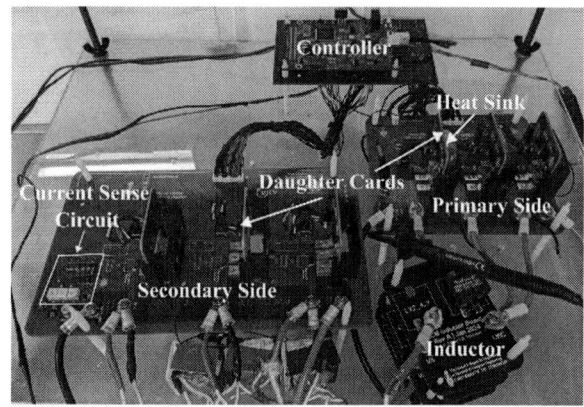

Fig. 14. 400-V, 6.6-kW, liquid cooled TP-DAB converter prototype.

TABLE II
TP-DAB CONVERTER SPECIFICATION

Parameter	Specification
Link Voltage	360-440 V
Battery Voltage	320-450 V
Peak Power	6.6 kW
Switching frequency, f_{sw}	300 kHz
GaN FET	GS66516T, $R_{DS(on)} = 25\,m\Omega$
Gate drive	+4/-4V
Shunt Current Sensor	5 mΩ
Parallel ADC	AD9200, 10 bits
Differential Op-Amp	AD8129

Three custom GaN half-bridge modules are used in the prototype and are mounted on motherboards for both the primary and secondary sides. Top-side cooled GaN devices (GS66516T) are employed, positioned on the module's back side. Each GaN module is cooled using an off-the-shelf heat sink. Thermal pads with a thermal conductivity of 12.8 W/mK are applied to connect the GaN FETs to the heat sink, providing both thermal conductivity and electrical isolation between the devices and the heat sink. A 3D-printed sleeve is used to ensure stable contact between the heat sink and the GaN module.

The switching waveforms on both the primary and secondary sides, including the three-phase currents on the primary side and the output current, are shown in Fig. 15, demonstrating successful operation at 6.5 kW with a peak efficiency of 96.6%, as illustrated in Fig. 16. The efficiency plot of the converter at 400 V under various load conditions is also shown in Fig. 16.

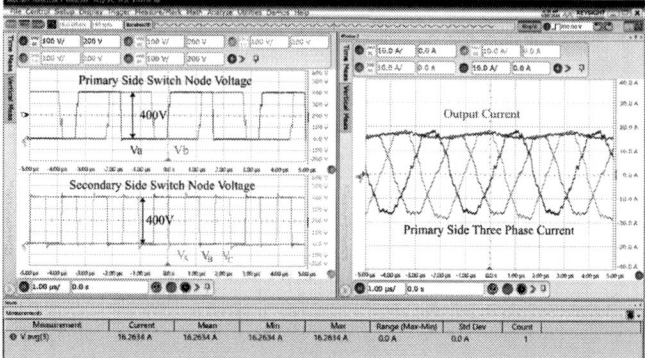

Fig. 15. Switching waveforms on the primary and secondary sides, including three phase currents on the primary side and the output current at 6.5 kW. Voltage scale: 100 V/div, current scale: 10 A/div.

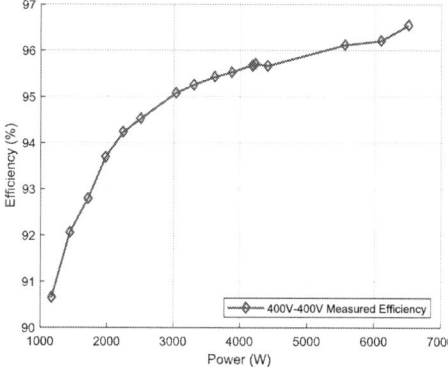

Fig. 16. Efficiency curve of TP-DAB converter operating at 400 V.

A low-side, single-shunt current sensing circuit was implemented on the secondary side of the motherboard, as shown in Fig. 14. This circuit measures I_{sns} on the secondary side, which is then converted by a differential amplifier into an equivalent voltage V_{sns}. A high-bandwidth op-amp was used to minimize any delay added to the detection time by the amplifier. A single-switch OCF event is triggered by pulling down the gating pulses of the selected switches Q_1, Q_6, and Q_7. Following this triggered OCF event, each of the three-phase currents exhibits a unique dc shift depending on the location of the OCF. This shift is reflected in I_{sns} as predicted by the simulation, and subsequently in V_{sns} as experimentally observed in Figs. 17 to 19. The immediate effect of changes in I_{sns} on V_{sns}, demonstrates that the bandwidth of the op-amp does not delay the algorithm's response.

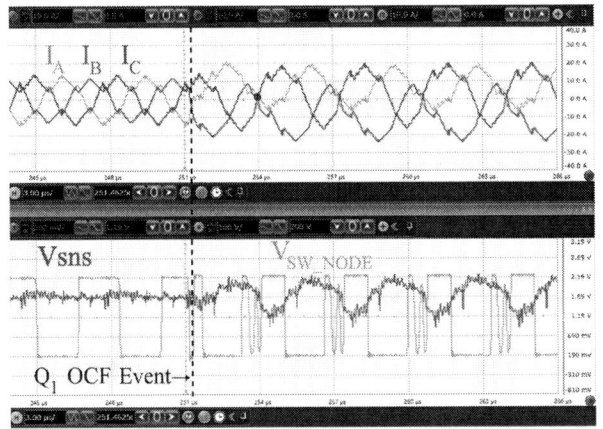

Fig. 17. Measured phase currents, V_{sns}, and switch node voltage of affected leg during OCF at Q_1.

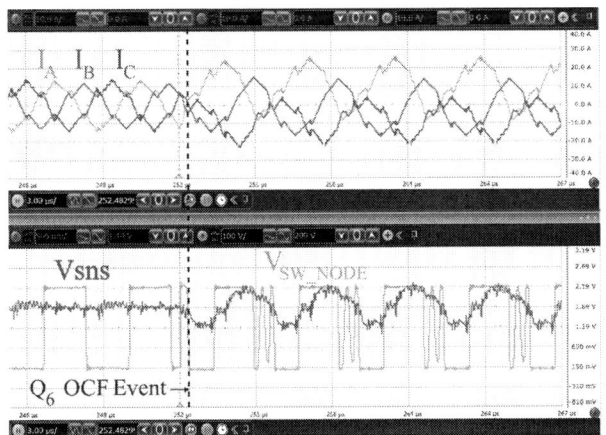

Fig. 18. Measured phase currents, V_{sns}, and switch node voltage of affected leg during OCF at Q_6.

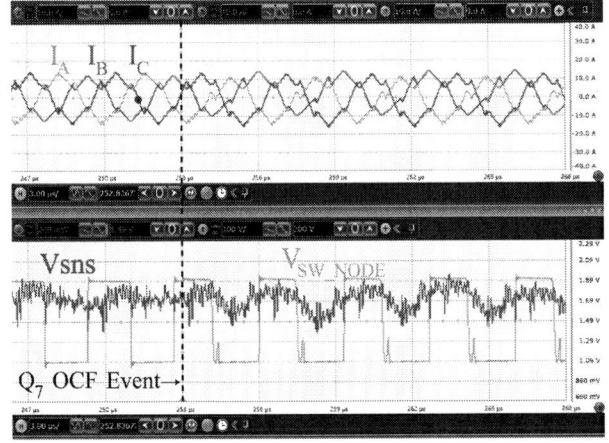

Fig. 19. Measured phase currents, V_{sns}, and switch node voltage of affected leg during OCF at Q_7.

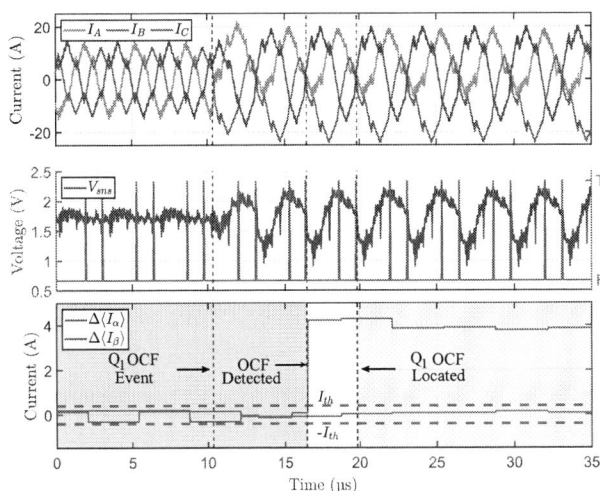

Fig. 20. Measured FPGA internal signals are aligned with the experimentally observed V_{sns} and phase currents during an OCF at Q_1.

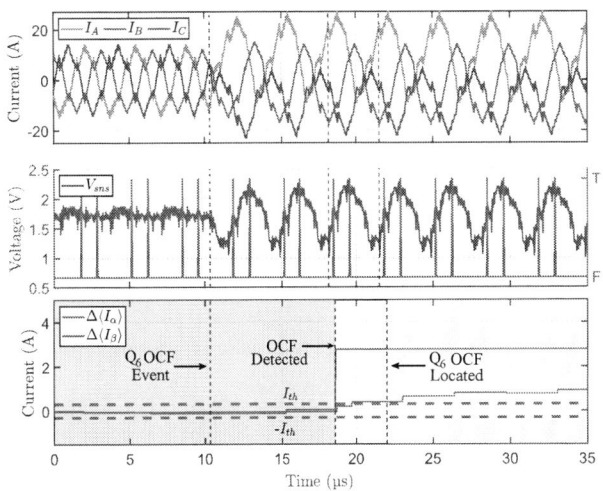

Fig. 21. Measured FPGA internal signals are aligned with the experimentally observed V_{sns} and phase currents during an OCF at Q_6.

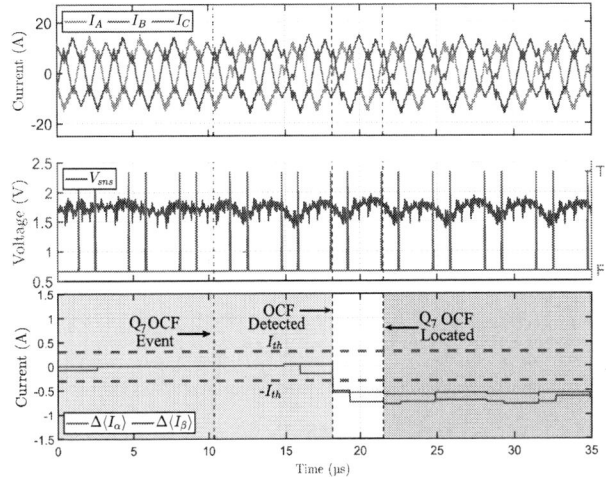

Fig. 22. Measured FPGA internal signals are aligned with the experimentally observed V_{sns} and phase currents during an OCF at Q_7.

V_{sns} is sampled by a parallel ADC and processed by an FPGA to compute $\Delta\langle I_\alpha\rangle$ and $\Delta\langle I_\beta\rangle$, using (7) and (8), as shown in Figs. 20 to 22. A moving average window, $m = 16 = 2^4$, is used to simplify division in the digital domain. This larger m value was required to mitigate the effect of high-frequency noise at 1.8 MHz, 6x the switching frequency. Under fault-free conditions, $\Delta\langle I_\alpha\rangle$ and $\Delta\langle I_\beta\rangle$ exhibit small non-zero values due to system noise, as shown in Figs. 20 to 22. This noise influences the choice of I_{th} and the minimum detection time. The computed values of $\Delta\langle I_\alpha\rangle$ and $\Delta\langle I_\beta\rangle$ are then compared against the threshold $|I_{th}| = 0.4$ to detect and localize the faulty switch based on the criteria in Table I. After fault detection, the algorithm waits for one switching cycle to allow $\Delta\langle I_\alpha\rangle$ and $\Delta\langle I_\beta\rangle$ to stabilize before confirming the fault location, as illustrated in Figs. 20 to 22. The OCF is detected within four switching cycles, demonstrating the effectiveness of the proposed method.

V. Conclusions

A fault-detection and localization algorithm for three-phase dual active bridge converters is presented. The algorithm utilizes the existing shunt current resistor, originally designed for output current control, to detect open-circuit faults (OCFs) and secondary-side short-circuit faults (SCFs). This approach reduces processing requirements and eliminates the need for additional sensors. A tunable delay parameter, δ, is incorporated for calibration, allowing the detection algorithm to account for parasitic-induced delays. The proposed algorithm was tested on a converter that achieved 96.6% efficiency at 6.5 kW and 400 V while switching at 300 kHz. Successful detection of OCFs within four switching cycles demonstrates the method's effectiveness. SPICE simulations of the half-bridge configuration reveal that increased parasitic inductance causes I_{sns} to lag behind the shoot-through current, delaying SCF detection. Based on the estimated parasitic impedance of the design, an SCF detection time of 120 ns was predicted in the simulation. This is faster than the observed short-circuit withstand time of 360 ns for a smaller device compared to the prototype used. The timely detection and localization of OCFs and SCFs help prevent cascading failures, avoiding catastrophic damage to high-power converters.

VI. Acknowledgments

This research was supported by the Natural Sciences and Engineering Research Council (NSERC) of Canada and the Taiwan Semiconductor Manufacturing Company (TSMC). The authors thank Samantha K. Murray for her work on short-circuit analysis, as well as Prof. Wai Tung Ng and his graduate researchers, Jingyuan Liang and Chun Yin Au Yeung, for their assistance with the thermal design of the prototype.

References

[1] C. P. Dick, A. König, and R. W. De Doncker, "Comparison of Three-Phase DC-DC Converters vs. Single-Phase DC-DC Converters," in *2007 7th International Conference on Power Electronics and Drive Systems*, 2007, pp. 217–224.

[2] H. Wouters and W. Martinez, "Bidirectional Onboard Chargers for Electric Vehicles: State-of-the-Art and Future Trends," *IEEE Transactions on Power Electronics*, vol. 39, no. 1, pp. 693–716, 2024.

[3] GaN Systems, "GS-065-008-1-L 650 V. E-mode GaN transistor," https://gansystems.com/wp-content/uploads/2022/07/GS-065-008-1-L-DS-Rev-220712.pdf, 2022, rev. 220712.

[4] Infineon, "IPD65R225C7 650 V CoolMOS C7 Power Device," https://www.infineon.com/cms/en/product/power/mosfet/n-channel/500v-950v/ipd65r225c7/, 2020, rev. 2.1, 2020-05-26.

[5] J. Roberts, J. Styles, and D. Chen, "Integrated gate drivers for e-mode very high power gan transistors," in *2015 IEEE International Workshop on Integrated Power Packaging (IWIPP)*, 2015, pp. 16–19.

[6] O. Trescases, S. K. Murray, W. L. Jiang, and M. S. Zaman, "GaN Power ICs: Reviewing Strengths, Gaps, and Future Directions," in *2020 IEEE International Electron Devices Meeting (IEDM)*, 2020, pp. 27.4.1–27.4.4.

[7] S. Li *et al.*, "Understanding Electrical Parameter Degradations of P-GaN HEMT Under Repetitive Short-Circuit Stresses," *IEEE Transactions on Power Electronics*, vol. 36, no. 11, pp. 12 173–12 176, 2021.

[8] L. Ferreira Costa and M. Liserre, "Failure Analysis of the dc-dc Converter: A Comprehensive Survey of Faults and Solutions for Improving Reliability," *IEEE Power Electronics Magazine*, vol. 5, no. 4, pp. 42–51, 2018.

[9] S. S. Khan and H. Wen, "A Comprehensive Review of Fault Diagnosis and Tolerant Control in DC-DC Converters for DC Microgrids," *IEEE Access*, vol. 9, pp. 80 100–80 127, 2021.

[10] K. Wang *et al.*, "A Reliable Short-Circuit Protection Method with Ultra-Fast Detection for GaN based Gate Injection Transistors," in *2019 IEEE 7th Workshop on Wide Bandgap Power Devices and Applications (WiPDA)*, 2019, pp. 43–46.

[11] W. L. Jiang *et al.*, "An Integrated GaN Overcurrent Protection Circuit for Power HEMTs Using SenseHEMT," *IEEE Transactions on Power Electronics*, vol. 37, no. 8, pp. 9314–9324, Aug 2022.

[12] A. Davoodi *et al.*, "A Novel Transistor Open-Circuit Fault Localization Scheme for Three-Phase Dual Active Bridge," in *2018 Australasian Universities Power Engineering Conference (AUPEC)*, 2018, pp. 1–6.

[13] S. S. Khan and H. Wen, "A Fast and Low-Cost Open-Circuit Fault Detection and Isolation Technique for Three-Phase Dual-Active-Bridge Converters Based on Finite State Machines," *IEEE Transactions on Power Electronics*, vol. 39, no. 2, pp. 2751–2766, 2024.

[14] S. K. Rastogi, S. S. Shah, B. N. Singh, and S. Bhattacharya, "Mode Analysis, Transformer Saturation, and Fault Diagnosis Technique for an Open-Circuit Fault in a Three-Phase DAB Converter," *IEEE Transactions on Power Electronics*, vol. 38, no. 6, pp. 7644–7660, 2023.

[15] S. Sa, Y. Han, S. A. Assadi, M. Shawkat Zaman, and O. Trescases, "Localization of Open-Circuit Faults in GaN-Based Three-Phase Dual Active Bridge Converters with Reduced Sensing Requirements," in *2024 IEEE Applied Power Electronics Conference and Exposition (APEC)*, 2024, pp. 461–467.

[16] H. M. Binqadhi, M. Ismail Hossain, A. Bineshaq, and M. A. Abido, "Fault Current Control of Three-Phase Dual Active Bridge DC–DC Converter Employing Modified Asymmetrical Duty Cycle Control Modulation," *IEEE Access*, vol. 12, pp. 87 348–87 360, 2024.

[17] J. Sun *et al.*, "Improved Model Predictive Control for Three-Phase Dual-Active-Bridge Converters With a Hybrid Modulation," *IEEE Transactions on Power Electronics*, vol. 37, no. 4, pp. 4050–4064, 2022.

[18] S. K. Murray, "Leveraging GaN Integrated Circuits for In-Situ Device Monitoring and High-Density Power Converters," Ph.D. dissertation, University of Toronto, Toronto, Ontario, Canada, November 2024.

Enhanced Cocharge Operation Scheme in Bidirectional PhaseShift Full-Bridge Converters with Eliminated Voltage Overshoot and Reduced Freewheeling Current

Tien-Sheng Li, Minh Ngo, Rolando Burgos, Dong Dong

Center for Power Electronics Systems (CPES), Virginia Tech, Blacksburg, VA 24061, USA

Email: tsli@vt.edu

Abstract—For high power density, high efficiency, wide voltage range, and low current ripple in isolated bidirectional DC/DC converters for battery charging applications, the phase-shift full-bridge (PSFB) converter is employed in this paper. However, the PSFB converter exhibits significant voltage overshoot on the transformer's secondary side, increasing the voltage stress on the secondary-side switches. Additionally, it generates a large freewheeling current, leading to extra conduction losses and larger transformer sizes, which compromise both efficiency and power density. This paper proposes an enhanced operating method to eliminate voltage overshoot while reducing freewheeling current. A simplified model is developed to analyze and predict the voltage overshoot under the proposed operation, offering insights into circuit behavior and design guidelines for bidirectional PSFB converters.

Keywords—DC/DC converter, PSFB converter, voltage overshoot, free-wheeling current

I. INTRODUCTION

With the expansion of applications like electric vehicle charging, fuel cells, and on-board chargers [1]-[5], bidirectional DC/DC converters with wide voltage range and low current ripple are attracting significant attention. Innovations in wide-bandgap materials, such as Silicon Carbide, are advancing efficiency and performance, allowing isolated bidirectional converters to deliver high power density across broad voltage ranges [5]-[15]. These developments make them especially appealing by minimizing system volume and cost while maximizing efficiency.

There are several popular circuit topologies in isolated DC/DC converters, like dual-active bridge (DAB) converters, resonant converters, and phase-shift full-bridge (PSFB) converters [16]-[19]. Among these topologies, PSFB converters have gained significant attention in recent years due to their notable advantages in their simple topology, voltage regulation capability, low switching loss, and low current ripple [20]-[26]. Fig. 1 indicates the circuit topology of PSFB converters, where L_k is defined as the equivalent transformer leakage inductance reflecting on the secondary side, and L_o is the output inductor of the converter.

However, PSFB converters have typically not been used in bidirectional applications due to a major challenge: when power

is delivering from V_{LV} side to V_{HV} side, also known as Boost-type PSFB, a large voltage overshoot on the transformer's secondary side, e.g. v_{sec} in Fig. 1, is generated. This voltage overshoot would cause a severe voltage stress on secondary side switches, which is S_{1-4} in Fig. 1. To observe this phenomenon, define i_t is the transformer current on secondary side, i_o is the output inductor current, and v_{sec} is the transformer voltage on secondary side as shown in Fig. 1. The simulated key waveforms of the PSFB are shown in Fig. 2, under condition listed in Table I, where C_{oss} is defined as the output capacitance of switching devices.

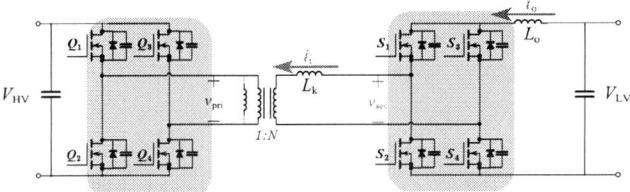

Fig. 1. Phase-Shift Full-Bridge Converter

In Fig. 2, the duration when power is transferring from the LV side to the HV side is defined as Power Transmission Mode (PTM), and the period when current is freewheeling in the circuit is defined as Current Freewheeling Mode (CFM). At t_1, when PTM initiates, a significant voltage overshoot occurs, with the voltage reaching up to 8 times the value of $V_{HV}N$. This phenomenon has been discussed and analyzed in previous literature [20]-[22]. Accordingly, it is caused by the resonance of L_k and C_{oss} of S_{1-4}. In [20], the voltage overshoot has been modelled and explained, where the quantified peak voltage was derived as (1), I_{ini} is defined as the initial current in i_t when the voltage overshoot occurs, e.g. transformer current i_t at t_1.

TABLE I
PARAMETERS OF BOOST MODE PSFB SIMULATION

Variable	Value
Power	60 kW
V_{HV}	1000 V
V_{LV}	200 V
N	0.5
L_k	125 nH
C_{oss} of Q_{1-4}, S_{1-4}	2 nF
f_s	200 kHz

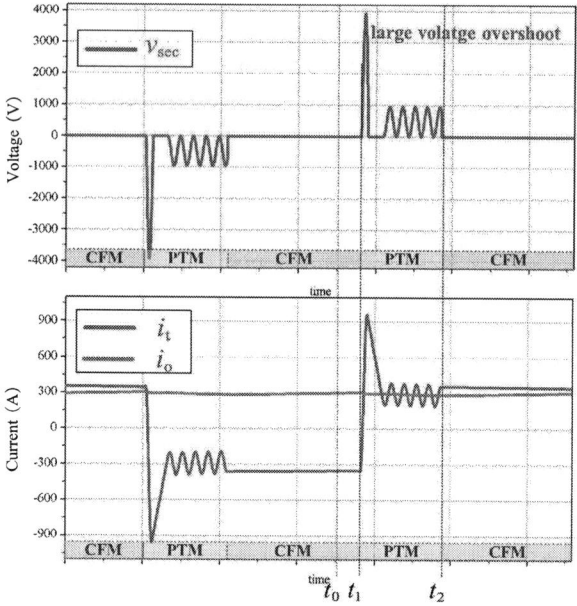

Fig. 2. Key waveforms of Boost mode PSFB operation

$$V_{peak} = V_{HV}N + \sqrt{(-V_{HV}N)^2 + \left(\frac{\sqrt{2C_{oss}L_k}(I_o - I_{ini})}{2C_{oss}}\right)^2} \quad (1)$$

Conventional approaches utilizes passive clamping circuits to eliminate the voltage overshoot [23][24]. However, adding supplementary clamping circuits not only reduces efficiency, also requires additional space for cooling systems, ultimately lowers the power density and circuit efficiency.

In recent years, research has increasingly focused on active control methods to address voltage overshoot [21][22][25][26]. As shown in (1), adjusting I_{ini}, effectively controls voltage overshoot, which is the main focus on prior literature. To eliminated voltage overshoot, design I_{ini} as equation (2), and the voltage overshoot can be eliminated.

$$\begin{cases} I_{ini} = I_o \\ V_{peak} = 2V_{HV}N \end{cases} \quad (2)$$

In literature [21], a precharge technique was proposed to adjust I_{ini}, the direction of i_t is reversed during precharge duration at the beginning of CFM, as the simulation is shown in Fig. 3 (a), under the same condition as Table I. When PTM starts, the circuit operates with current $i_t = I_o$ and the voltage overshoot is eliminated. This technique has been applied in various recent literature as well [25][26].

Another technique called cocharge was proposed in [22]. In order to improve the efficiency, [22] combined precharge and cocharge operation to eliminate voltage overshoot and reduce reverse power that was caused from precharge duration. However, due to the lack of co-charge operation modelling, a small overshoot still exists, as the simulation shown in Fig. 3 (b) with the same condition as Table I. In addition, a significant CFM current was produced comparing to Fig. 3 (a). Observing Fig. 2 and Fig. 3, all techniques show a large CFM transformer current, which results in large circulating current, higher

transformer size and worse circuit efficiency. In this paper, an enhanced operation will be introduced to eliminate voltage overshoot while maintaining a small CFM current.

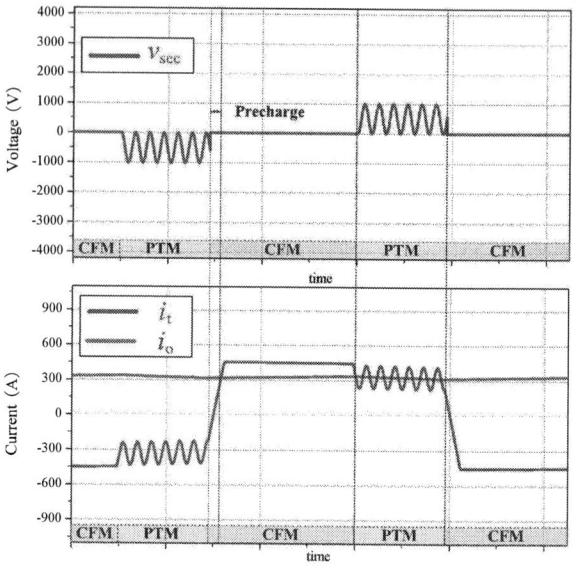

(a)

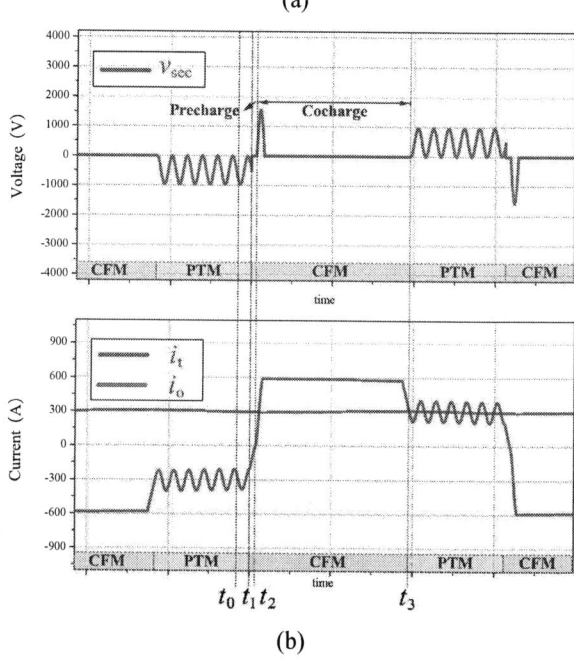

(b)

Fig. 3. Key waveforms of Boost mode PSFB operation with
(a) Precharge operation from [21] (b) Cocharge operation from [22]

II. PRINCIPLE OF COCHARGE OPERATION

The circuit operation proposed in [22] is illustrated in Fig. 4. At t_0, the circuit starts during PTM. At t_1, S_1 and S_4 conduct, grounding the secondary side of transformer. During this precharge phase, $V_{HV}N$ begins charging L_k, reversing i_t till reaches 0A. At t_2, S_2, S_3 and Q_3 turn off while Q_4 turns on, starting cocharge phase. I_o then charges i_t, aligning i_t with I_o till t_3, when PTM resumes. As noted in previous literature, voltage overshoot typically occurs when PTM, which is eliminated in this operation. However, as showing in Fig. 3 (b), despite the voltage overshoot at t_3 is eliminated, a voltage overshoot is generated at t_2, when cocharge duration initiates. In order to fully eliminate the voltage overshoot and properly design the operation, cocharge operation should be modelled.

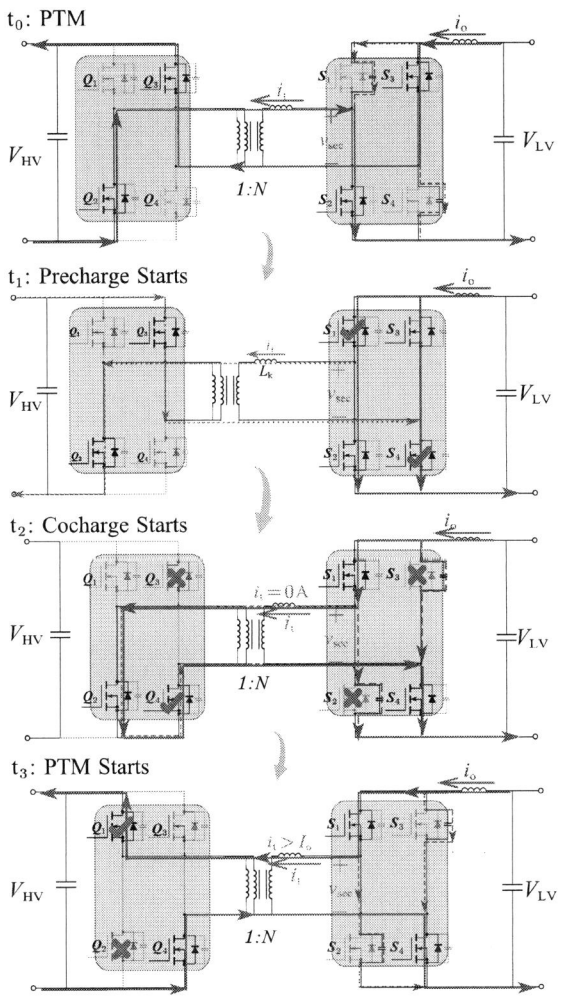

Fig. 4. Circuit operation of cocharge technique in [22]

The voltage overshoot occurs at t_2 when cocharge initiates. Fig. 5 shows the power switch connections, while Fig. 6 (a) depicts an equivalent circuit diagram. In the circuit diagram, D_{S2} and D_{S3} are the body diodes of S_2 and S_3, and $C_{oss\ S2}$ and $C_{oss\ S3}$ are the output capacitance of S_2 and S_3. Given that the output inductor L_o is generally much larger than leakage inductance L_k, i_o can be approximated as a constant current source. The equivalent circuit can be further simplified as Fig. 6 (b), where C_{eq} is the $C_{oss\ S2}$ in parallel with $C_{oss\ S3}$. Inductance L_k resonates with capacitance C_{eq} during cocharge phase. By deriving the time variant equation of v_{sec} (t), the voltage overshoot during cocharge can be explored. The circuit shown in Fig. 6 (b) meets the relation shown in equation (3).

The initial condition of L_k and C_{eq} is provided in equation (4). The initial current existing in L_k is defined as I_{ini}. Since C_{eq} is the

output capacitance of switching devices, voltage is 0 V when devices are turning off.

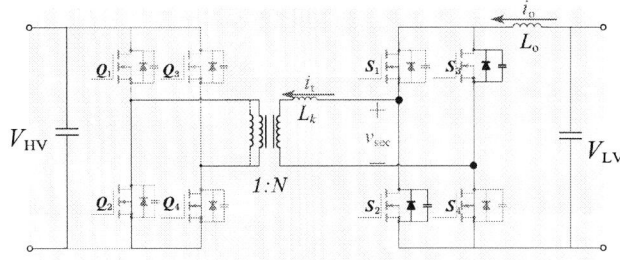

Fig. 5. Circuit diagram during cocharge

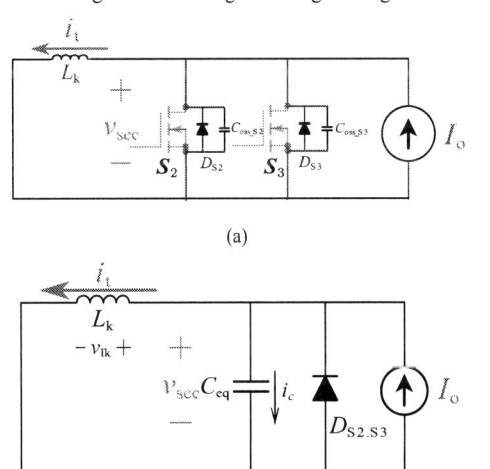

(a)

(b)

Fig. 6. (a) Equivalent circuit diagram during cocharge (b) Equivalent simplified circuit during cocharge

$$\begin{cases} v_{sec} = v_{lk} \\ i_c + i_t = I_o \end{cases} \tag{3}$$

$$\begin{cases} i_t (t = 0^+) = I_{ini} \\ v_{sec} (t = 0^+) = 0 \end{cases} \tag{4}$$

By deriving the second-order differential equation for v_{sec} (t), the results can be shown in (5). To predict the voltage overshoot during cocharge, the peak voltage V_{peak} of equation (5) is derived as (6). It can be found that the peak voltage would be influenced by components C_{oss}, L_k, power condition I_o, and initial current I_{ini}. Similar as original cases discussed in [20], smaller I_{ini} results in more significant V_{peak}. The derived v_{sec} (t) is also depicted in Fig. 7, along with the corresponding i_t (t) and I_o waveforms.

$$v_{sec}(t) = \frac{\sqrt{C_{eq}L_k}\,(I_o - I_{ini})}{C_{eq}} \sin\left(\frac{1}{\sqrt{C_{eq}L_k}}t\right) \tag{5}$$

$$V_{peak} = \frac{\sqrt{2C_{oss}L_k}\,(I_o - I_{ini})}{2C_{oss}} \tag{6}$$

At t_α, resonance starts with current $i_t = I_{ini}$, causing a voltage overshoot. At t_β, resonance voltage reaches its maximum, meanwhile, i_t reaches I_o. At t_γ, voltage reaches 0 and the resonance is clamped by $D_{S2,S3}$. By comparing equation (6) with equation (1), the voltage overshoot during cocharge can be evaluated relative to the original case. It can be found (6) is only a fraction of (1), indicating that the voltage overshoot in cocharge operation is naturally smaller than in the original case.

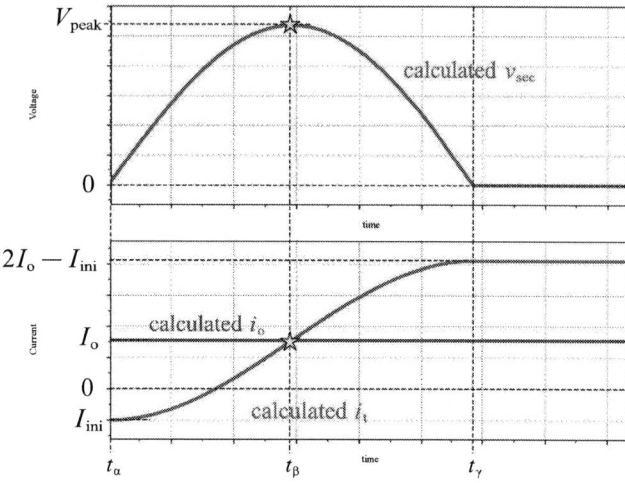

Fig. 7. Modelled waveform of v_{sec} and i_{t}

III. PROPOSED ENHANCED COCHARGE OPERATION

In order to fully eliminate the voltage overshoot, as well as decrease the freewheeling current, this paper proposes an enhanced operation that is able to eliminate the voltage overshoot, as well as decrease the freewheeling current. The operation of a half-cycle is shown as Fig. 8.

The circuit begins in PTM. At t_1, when CFM starts, secondary side is grounded and the circuit enters a precharge phase to adjust the freewheeling current. At t_2, once the designed freewheeling current is reached, the primary side is grounded, and the current begins freewheeling. At t_3, the circuit transitions to a cocharge phase, where L_k and C_{eq} starts resonating as described in the previous section. The freewheeling current between from t_2 to t_3 erves as the initial current for this resonance, represented as I_{ini} in equation (4)-(6). The cocharge phase last till t_4, when the circuit enters PTM again, which forms a half-cycle operation.

By leveraging the nature of cocharge operation, equation (6) is designed to satisfy $V_{\text{peak}} = 2V_{\text{HV}}N$, thereby eliminating the voltage overshoot. Accordingly, the freewheeling current can be determined as shown in equation (7), and this current can be adjusted by designing the precharge duration. From equation (7), it can be observed that when I_o is small (e.g., under low-power operation), $I_{\text{ini}} < 0$ may occur. As discussed in the previous section, a larger I_{ini} results in a smaller voltage overshoot. Therefore, setting $I_{\text{ini}} = 0$ in this case not only eliminates the voltage overshoot but also reduces the CFM current to 0 A, minimizing the circuit's conduction losses.

$$I_{\text{ini}} = I_o - \frac{4V_{\text{HV}}NC_{\text{oss}}}{\sqrt{2L_k C_{\text{oss}}}}, \ I_{\text{ini}} \in [0, I_o] \qquad (7)$$

Comparing the proposed operation to original and precharge case shown in [21], instead of entering PTM after current freewheeling duration, a cocharge duration is added

before PTM, making the simplified circuit model change as indicated in Fig. 9. The difference in the model is the primary reason for the suppression on voltage overshoot from equation (1) to equation (6).

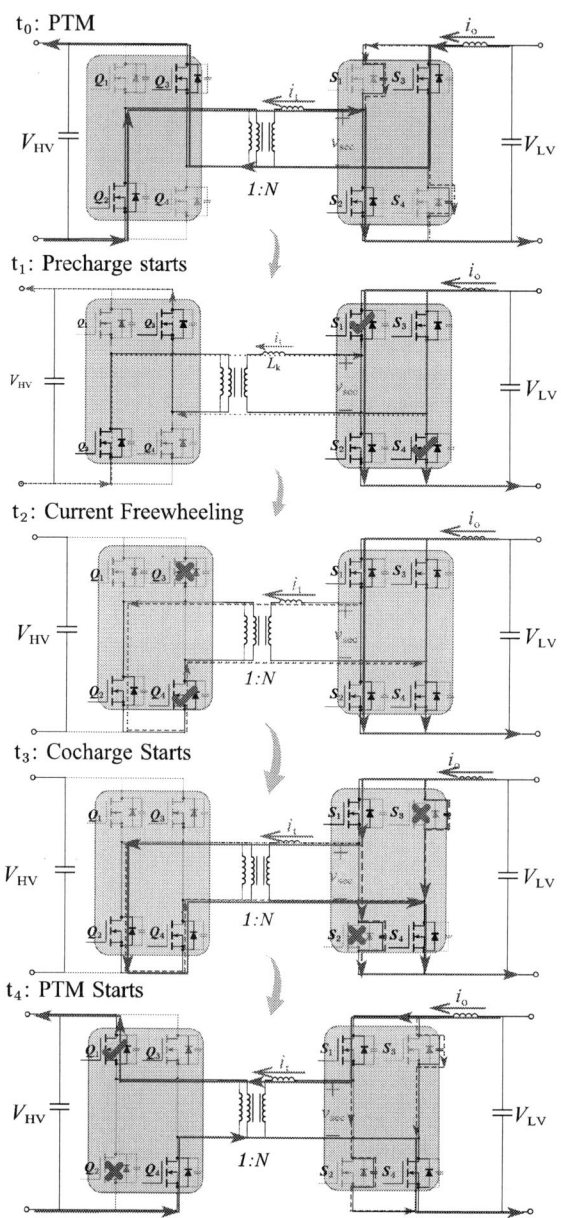

Fig. 8. Circuit operation of the proposed enhanced cocharge operation

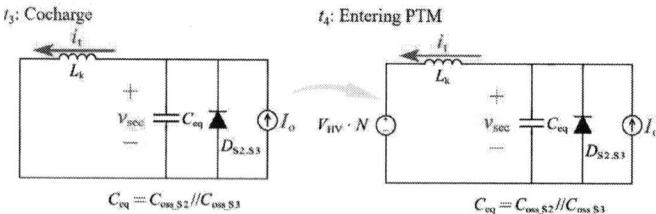

Fig. 9. Equivalent model changing from cocharge to PTM

The simulation of the proposed operation is shown in Fig. 10, with the condition listed in Table I. It can be observed that with the proposed operation, the voltage overshoot is eliminated, while maintaining a much lower CFM current compared to original and any of the previous techniques shown in Fig. 2 and Fig. 3.

Comparing with the cocharge operation present in [22], instead of designing a long cocharge phase during the entire CFM, the proposed operation maintained most of the CFM time operating in freewheeling mode to avoid the high transformer current.

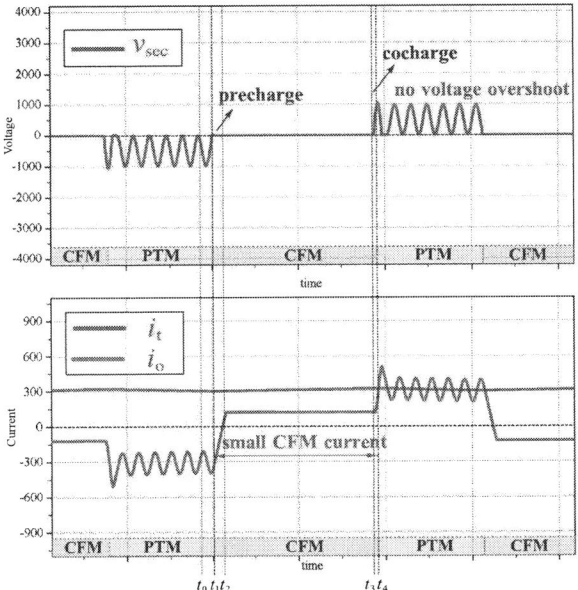

Fig. 10. Key waveforms of Boost mode PSFB operation with proposed enhanced cocharge operation

A comparison of normalized V_{peak} prediction and the root-mean-square (rms) transformer current under 60 kW PSFB design is demonstrated in Fig. 11, where the red bar represents the peak voltage, and the purple bar represents the transformer rms current. The voltage overshoot and transformer current on original case is defined as 1 p.u. With the proposed operation, the voltage overshoot can be eliminated and the transformer current can be reduced for 37 % compared to the original operation.

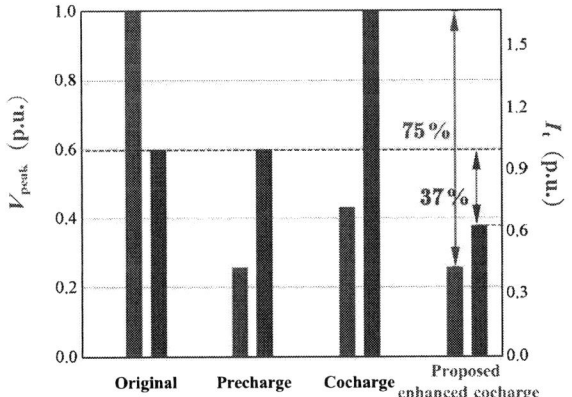

Fig. 11. Normalized comparison between different techniques

Moreover, the normalized V_{peak} and rms I_t comparison under different power conditions are also illustrated in Fig. 12. It can be seen that with the proposed technique, the voltage overshoot is always eliminated, while the rms transformer current is always maintained at low current.

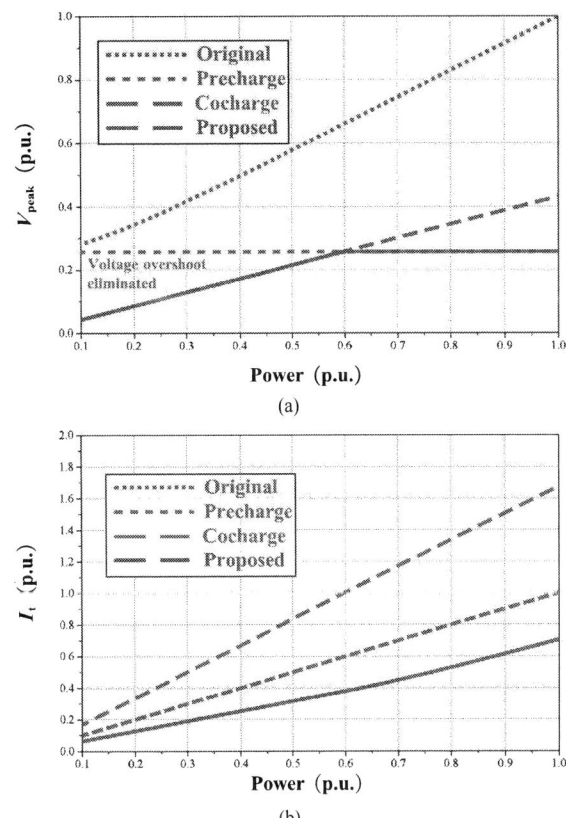

Fig. 12. Normalized comparison between different techniques under different power conditions

IV. EXPERIMENTAL RESULTS

A. Prototype Implementation

A prototype PSFB circuit was developed to validate the proposed analysis and technique, as shown in Fig. 13. The converter's parameters are summarized in Table II. The transformer was designed with a 1:1 turns ratio and a leakage inductance of 4.01 µH, reflected on the secondary side. The output inductor L_o is designed at 200 uH to ensure a low current ripple at LV side.

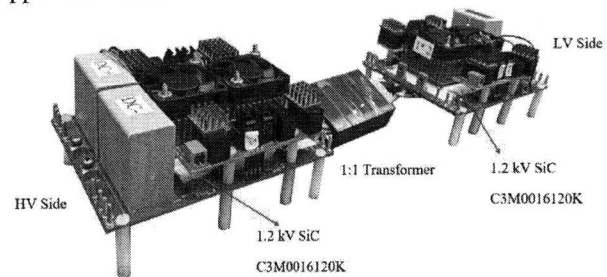

Fig. 13. Prototype circuit of PSFB

Both the HV and LV side use the 1.2 kV SiC C3M0016120 device, with 2 devices in parallel on each position. As state in Fig. 2, the voltage overshoot can be **8** times of $V_{HV}N$, the circuit operates at a low power condition of 265 W to avoid severe voltage overshoot.

TABLE II
PARAMETERS OF HARDWARE CIRCUIT AND EXPERIMENT CONDITIONS

Variable	Value
L_k	4.01 uH
L_m	102 uH
L_o	200 uH
$I_{o(avg)}$	5.5 A
V_{LV}	48 V
V_{HV}	72 V
f_s	100 kHz

B. Experiment Result

The experimental results under conditions listed in Table II are shown in Fig. 15, following the circuit diagram in Fig. 1. Channel 1 – 4 are v_{pri}, v_{sec}, i_t, i_o, as defined in Fig. 1.

The experiment results under original operation are shown in Fig. 14. It can be seen that the voltage overshoot is as large as 602 V while at a 48/ 72V PSFB converter, and a large freewheeling current present in the transformer during CFM.

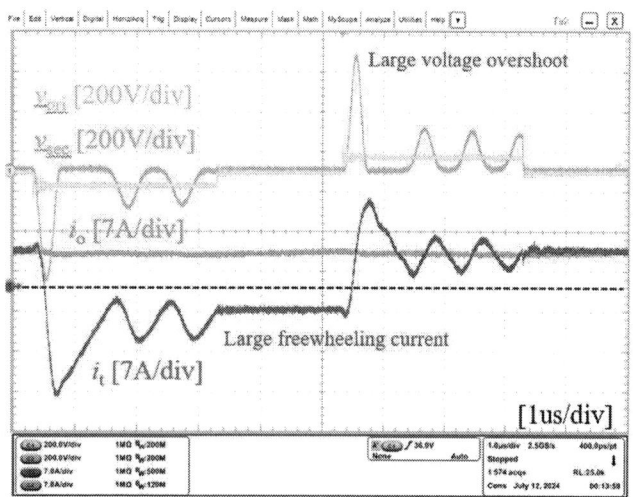

Fig. 14. Experiment waveforms of original operation

Fig. 15 shows the result of applying precharge operation on bidirectional PSFB converter with the same condition as original operation. It can be found that i_t is aligned with i_o at the beginning of CFM, ensuring the condition of equation (2) and voltage overshoot is eliminated.

With the same condition, Fig. 16 shows the result of utilizing the proposed enhanced cocharge operation. It can be seen that the freewheeling current is controlled at 0 A, where the transformer current is minimized. Meanwhile, by observing the time difference between v_{pri} and v_{sec}, it can be found that a cocharge phase is added before PTM. Owing to the nature of cocharge operation, the voltage overshoot is eliminated as well, proving the feasibility and effectiveness of the proposed enhanced cocharge operation.

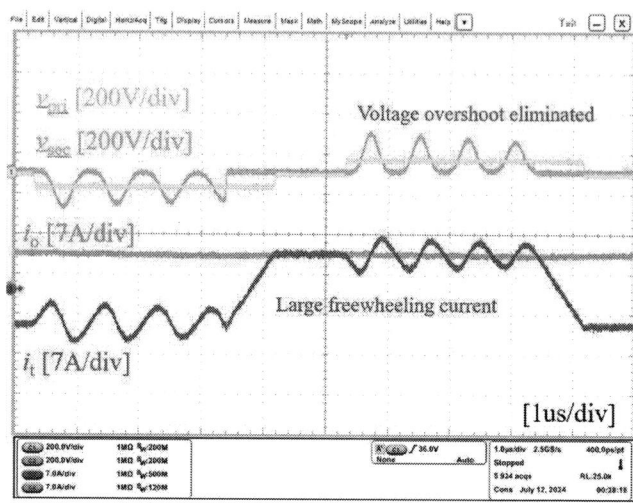

Fig. 15. Experiment waveforms of precharge operation

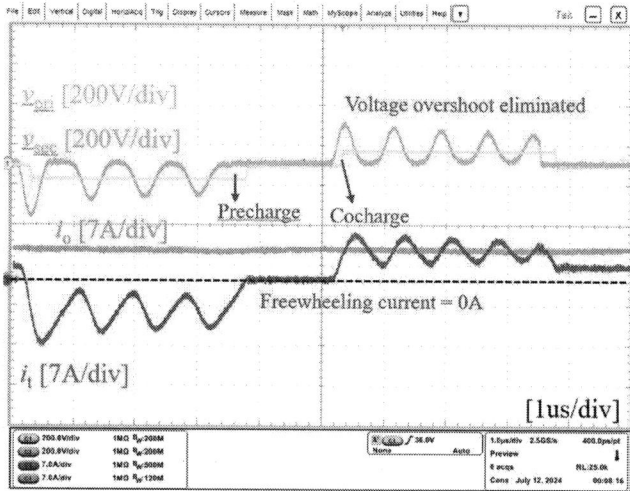

Fig. 16. Experiment waveforms of proposed enhanced cocharge operation

V. CONCLUSION

This paper proposes an enhanced operation for bidirectional PSFB converters to eliminate inherent voltage overshoot, suppress voltage stress on power switching devices, reduce transformer current, and improve both power density and efficiency. A cocharge operation is introduced and analyzed. By leveraging the low peak resonant voltage characteristic of cocharge, the proposed method eliminates voltage overshoot while maintaining a small freewheeling current. With this technique, the transformer RMS current is reduced by 37%, and the peak voltage is suppressed by 75%. A comparison with previous techniques is presented, and the proposed method is validated through hardware experiments.

REFERENCES

[1] T. Yuan, F. Jin and Q. Li, "Analysis and Comparison of Integrated Planar Transformers for 22-kW On-Board Chargers," in IEEE Transactions on Power Electronics, vol. 39, no. 9, pp. 11368-11385, Sept. 2024, doi: 10.1109/TPEL.2024.3410878.

[2] T. Yuan, F. Jin and Q. Li, "A 22-kW On-Board Charger (OBC) with an Integrated Planar Inductor and Transformer," 2024 IEEE Applied Power

979-8-3315-1612-3/25 $31.00 © 2025 IEEE

Electronics Conference and Exposition (APEC), Long Beach, CA, USA, 2024, pp. 1300-1304, doi: 10.1109/APEC48139.2024.10509419.

[3] X. Yu, J. Feng, L. Zhu and Q. Li, "Modelling of A Planar Omnidirectional Wireless Power Transfer System," 2024 IEEE Applied Power Electronics Conference and Exposition (APEC), Long Beach, CA, USA, 2024, pp. 2609-2615, doi: 10.1109/APEC48139.2024.10509107.

[4] Y. -T. Huang, C. -C. Yang, T. -S. Li and Y. -M. Chen, "A Feedforward Voltage Control Strategy for Reducing the Output Voltage Double-Line-Frequency Ripple in Single-Phase AC–DC Converters," in IEEE Journal of Emerging and Selected Topics in Power Electronics, vol. 9, no. 6, pp. 6605-6612, Dec. 2021, doi: 10.1109/JESTPE.2021.3083258.

[5] T. -S. Li, Y. -H. Yang, C. -A. Cheng and Y. -M. Chen, "A Variable DC-Link Voltage Determination Method for Motor Drives with SiC MOSFETs," 2020 IEEE Workshop on Wide Bandgap Power Devices and Applications in Asia (WiPDA Asia), Suita, Japan, 2020, pp. 1-6, doi: 10.1109/WiPDAAsia49671.2020.9360266.

[6] Q. Yang, A. Nabih, R. Zhang, Q. Li and Y. Zhang, "A Converter Based Switching Loss Measurement Method for WBG Device," 2023 IEEE Applied Power Electronics Conference and Exposition (APEC), Orlando, FL, USA, 2023, pp. 8-13, doi: 10.1109/APEC43580.2023.10131509.

[7] X. Yang, J. Liu, B. Wang and G. Zhang, "Pulsed Overcurrent Capability of Power Semiconductor Devices in Solid-State Circuit Breakers: SiC MOSFET vs. Si IGBT," 2022 IEEE Applied Power Electronics Conference and Exposition (APEC), Houston, TX, USA, 2022, pp. 966-973,

[8] X. Yang et al., "Ultrafast Optically Controlled Power Switch: A General Design and Demonstration With 3.3 kV SiC MOSFET," in IEEE Transactions on Electron Devices, doi: 10.1109/TED.2024.3485018.

[9] X. Yang et al., "Evaluation and MHz Converter Application of 1.2-kV Vertical GaN JFET," in IEEE Transactions on Power Electronics, vol. 39, no. 12, pp. 15720-15731, Dec. 2024, doi: 10.1109/TPEL.2024.3445667.

[10] Qiuzhe Yang, Feng Jin, Qiang Li, "An Accurate Temperature-Based Method for Fast Switching Loss Extraction of WBG Device," 2024 IEEE Applied Power Electronics Conference and Exposition (APEC), Los Angeles, CA, USA, 2024, pp. 2183-2187, doi: 10.1109/APEC48139.2024.10509222

[11] C. -W. Chang, M. Spieler, A. EL-Refaie, R. A. Torres, R. Burgos and D. Dong, "A Current Balancing Gate Driver for Dynamic Current Sharing of Paralleled SiC MOSFETs with Kelvin-Source Connection," in IEEE Transactions on Power Electronics, doi: 10.1109/TPEL.2024.3475288.

[12] Qiuzhe Yang, Shuo Wang, Qiang Li, "Modeling and Analysis of The Balance Network for Common Mode EMI Noise Suppression," in Proc. IEEE Energy Convers. Congr. Expo., 2024.

[13] C. -W. Chang et al., "Thermal Consideration and Design for a 200-kW SiC-Based High-Density Three-Phase Inverter in More Electric Aircraft," in IEEE Journal of Emerging and Selected Topics in Power Electronics, vol. 11, no. 6, pp. 5910-5929, Dec. 2023, doi: 10.1109/JESTPE.2023.3308854.

[14] C. -W. Chang, M. Spieler, R. Burgos and D. Dong, "Evaluation and Efficiency Comparison of Soft-Switching ARCP SiC-Based Traction Inverters in Electric Vehicles," 2023 IEEE Applied Power Electronics Conference and Exposition (APEC), Orlando, FL, USA, 2023, pp. 1417-1422, doi: 10.1109/APEC43580.2023.10131452.

[15] C. -A. Cheng, C. -C. Chang, T. -S. Li, Z. -J. Chen and Y. -M. Chen, "Initial Rotor Position Startup Process Emulation Based on Electric Motor Emulator," 2020 IEEE 9th International Power Electronics and Motion Control Conference (IPEMC2020-ECCE Asia), Nanjing, China, 2020, pp. 605-610, doi: 10.1109/IPEMC-ECCEAsia48364.2020.9368013.

[16] F. Jin, A. Nabih, T. Yuan and Q. Li, "A High-Efficiency High-Density Three-Phase CLLC Resonant Converter With a Universally Derived Three-Phase Integrated Transformer for On-Board-Charger Application," in IEEE Transactions on Power Electronics, vol. 39, no. 4, pp. 4350-4366, April 2024, doi: 10.1109/TPEL.2024.3354679.

[17] F. Jin, T. Yuan, A. Nabih and Q. Li, "Efficient Integrated Magnetics With Winding Cancellation Technique to Reduce Common-Mode EMI Noise for a Single-Phase CLLC Converter," in IEEE Transactions on Power Electronics, vol. 39, no. 11, pp. 14758-14774, Nov. 2024, doi: 10.1109/TPEL.2024.3436044.

[18] X. Yu, J. Feng, L. Zhu and Q. Li, "Design and Optimization of a Planar Omnidirectional Wireless Power Transfer System for Consumer Electronics," in IEEE Open Journal of Power Electronics, vol. 5, pp. 311-322, 2024, doi: 10.1109/OJPEL.2024.3360878.

[19] X. Yu, J. Feng and Q. Li, "A Planar Omnidirectional Wireless Power Transfer Platform for Portable Devices," 2023 IEEE Applied Power Electronics Conference and Exposition (APEC), Orlando, FL, USA, 2023, pp. 1654-1661, doi: 10.1109/APEC43580.2023.10131566.

[20] T. -S. Li, M. Ngo, R. Burgos and D. Dong, "Modeling and Analysis of Voltage Overshoot in Bidirectional Phase-Shift Full Bridge Converters," 2024 IEEE Sixth International Conference on DC Microgrids (ICDCM), Columbia, SC, USA, 2024, pp. 1-7, doi: 10.1109/ICDCM60322.2024.10665239.

[21] M. Escudero, D. Meneses, N. Rodriguez and D. P. Morales, "Modulation Scheme for the Bidirectional Operation of the Phase-Shift Full-Bridge Power Converter," in IEEE Transactions on Power Electronics, vol. 35, no. 2, pp. 1377-1391, Feb. 2020, doi: 10.1109/TPEL.2019.2923804.

[22] Y. Yang et al., "An Improved Control Scheme for Reducing Circulating Current and Reverse Power of Bidirectional Phase-Shifted Full-Bridge Converter," in IEEE Transactions on Power Electronics, vol. 37, no. 10, pp. 11620-11635, Oct. 2022, doi: 10.1109/TPEL.2022.3170310.

[23] Song-Yi Lin and Chern-Lin Chen, "Analysis and design for RCD clamped snubber used in output rectifier of phase-shift full-bridge ZVS converters," in IEEE Transactions on Industrial Electronics, vol. 45, no. 2, pp. 358-359, April 1998, doi: 10.1109/41.681236.

[24] C. -Y. Lim, Y. Jeong and G. -W. Moon, "Phase-Shifted Full-Bridge DC–DC Converter With High Efficiency and High Power Density Using Center-Tapped Clamp Circuit for Battery Charging in Electric Vehicles," in IEEE Transactions on Power Electronics, vol. 34, no. 11, pp. 10945-10959, Nov. 2019, doi: 10.1109/TPEL.2019.2899960.

[25] T. -W. Huang, H. -J. Chiu, G. -C. Wang and Y. -C. Chang, "Current-Fed Phase-Shifted Full-Bridge Converter With Secondary Harmonic Current Reduction for Two-Stage Inverter in Energy Storage System," in IEEE Transactions on Power Electronics, vol. 38, no. 9, pp. 10716-10728, Sept. 2023, doi: 10.1109/TPEL.2023.3282911.

[26] R. Maddipudi, N. Kummari and S. Chattopadhyay, "A Secondary-Side Phase-Shifted Bidirectional Soft-Switching DC–DC Converter Employing Step Voltage Switching for Eliminating Device Voltage Overshoot," in IEEE Transactions on Power Electronics, vol. 38, no. 7, pp. 8583-8596, July 2023, doi: 10.1109/TPEL.2023.3263205.

DC Bias Elimination in Isolated DC-DC Converters using Fundamental-Frequency Ripple

Arkadeb Sengupta
Chair of Power Electronics
Kiel University
Kiel, Germany
arsa@tf.uni-kiel.de

Thiago Pereira
Chair of Power Electronics
Kiel University
Kiel, Germany
tp@tf.uni-kiel.de

Marco Liserre
Chair of Power Electronics
Kiel University
Kiel, Germany
ml@tf.uni-kiel.de

Abstract—The performance of an isolated dc-dc converter is degraded by a persistent dc bias in the transformer current. Existing dc bias elimination methods estimate the bias either by directly sampling the terminal voltage ripple, which is susceptible to switching noise and requires precise measurements, or by using additional sensors in the ac network, which add to the complexity and cost of implementation. This paper proposes a dc bias elimination strategy for isolated dc-dc converters based on a band-pass filtering of the terminal voltage ripple. The relation between the dc bias and the amplitude of the fundamental (switching-frequency) component of the ripple is analyzed. Based on this, a dc bias elimination technique is proposed, which suppresses switching noise in the ripple and can be implemented using analog circuitry and low-resolution sensors. The steady-state and transient performance of the proposed method is validated in simulations and experiments on a 150 W dual-active-bridge (DAB) converter.

Index Terms—dc bias elimination, dual active bridge, switching ripple, tuned filter.

I. INTRODUCTION

Transformer-isolated dc-dc converters offer scope for efficient operation with large voltage conversion ratios, in applications such as battery chargers and power supplies for information technology. However, practical asymmetries in manufacturing and implementation may lead to a non-zero volt-second across the transformer, producing a dc bias in the current flowing in the transformer windings and thus, deteriorating converter performance. Accordingly, active methods to eliminate this dc bias are proposed in the literature [1]–[5].

Active bias elimination methods cancel a measured dc bias by applying an opposing dc excitation on the ac network. The dc bias measurement is crucial to this approach. One class of methods measures the dc bias by sensing various ac network quantities, such as the magnetic flux in the transformer core [1], or the transformer winding current [2]. These approaches, while effective, require one or more specialized sensors associated with the transformer windings or core. In contrast, the method proposed in [3] uses the terminal voltage ripple to estimate the dc bias. A variant of this method is proposed for single-sided measurements in [4]. In [5], a low-cost analog implementation for the method proposed in [3] is presented. The method in [5] is validated on a converter with a relatively high ripple amplitude ($1 - 5\%$), which is necessitated by this class of methods. While the terminal voltage (and its ripple)

are accessible quantities, the estimation of the dc bias using the ripple waveform without low-pass filtering, as proposed in [3]–[5], leaves the measurement susceptible to switching noise [6]. This leads to concerns of signal-to-noise ratio, and requires precise sensing and sampling arrangements to ensure the accuracy of measurements [4].

A relation between the dc bias in the transformer current, and the switching-frequency component of the terminal voltage ripple, was identified qualitatively for isolated dc-dc converters in [7], and further exploited to encode information in the ripple. This switching-frequency component of the terminal voltage ripple is readily extracted, and presents a straightforward basis to measure dc bias in the transformer windings. In this paper, a strategy to detect and eliminate dc bias in transformer windings, using this switching-frequency component of terminal voltage ripple, is proposed and validated. The main contributions of the paper are:

(a) A method to detect and quanitfy the dc bias in the transformer current, using the fundamental-frequency ripple, is proposed. The proposed method employs an indirect, filtered measurement of the ripple.

(b) An approach to eliminate the dc bias is outlined based on the the proposed dc bias detection scheme.

(c) The proposed bias elimination strategy is validated in simulations and through experiments on an isolated dual-active-bridge (DAB) converter.

This paper is organized as follows. The analytical relationship between the dc bias current in the reactive network of a full-bridge isolated dc-dc converter, and the ripple at its dc terminals, is derived in Section II. Based on this relationship, a method to estimate the dc bias current, and further, to eliminate it through a leg duty ratio adjustment of the full bridge, is outlined in Section III. The proposed dc bias estimation and elimination approach is validated through circuit simulations and experiments on an isolated dual-active-bridge converter in Section IV. Section V concludes the paper.

II. ANALYTICAL RELATION OF TERMINAL VOLTAGE RIPPLE SPECTRUM TO DC BIAS

A generic bidirectional two-port isolated dc-dc converter is shown in Fig. 1a, with primary- and secondary-side full-bridges, comprising switches S_1-S_4 and S_5-S_8, respectively.

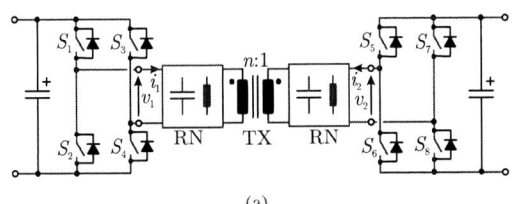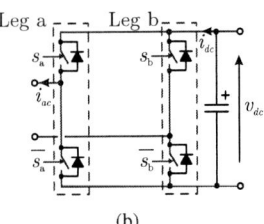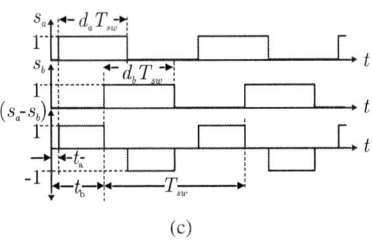

Fig. 1: (a) Circuit schematic of a generic full-bridge isolated 2-port dc-dc converter. (b) Schematic of a full bridge showing the switching functions corresponding to each switch. (c) Representative waveforms of the switching signals.

The ac terminals of each full bridge are connected to an $n:1$ transformer (TX) through a reactive network (RN) on each winding. A dc bias develops in transformer winding only if the configuration of the corresponding resonant network allows dc current to flow. For instance, a series capacitor, e.g., in a series-resonant reactive network, prevents the flow of dc current and hence, the problem of dc bias does not arise. Such resonant network configurations are excluded in the present discussion.

When operated below saturation, the magnetizing inductance of the transformer behaves as a short circuit for dc quantities and thus, decouples the steady-state dc currents in the windings. Hence, for a given transformer winding, the effect of dc bias on the physical quantities of the dc terminal of the corresponding full bridge, is analyzed for a single full bridge, shown in Fig. 1b. The voltage, v_{dc}, and unfiltered current, i_{dc}, at the dc terminal, as well as the ac terminal current, i_{ac}, are shown. The operation of each leg (a, b) of the full bridge is described by a switching function, s_x, as

$$s_x = \begin{cases} 0, & \text{if the top switch of leg } x \text{ is on,} \\ 1, & \text{if the bottom switch of leg } x \text{ is on,} \end{cases} \quad (1)$$
$$\text{where, } x = (a, b).$$

The switching function, $\bar{s}_x$, is defined as the logical complement of the function, s_x, i.e., $\bar{s}_x = 1 - s_x$. A complementary mode of operation, with switching functions, s_x and $\bar{s}_x$, is considered for the switches constituting each leg.

Representative waveforms of the switching pulses are shown in Fig. 1c. The pulses have a frequency, f_{sw}, and the top switch of each leg has a duty ratio, d_x. The phase information of the pulses is represented by a time delay, t_x, between the rising edge of the pulse and an arbitrary temporal origin, as shown in Fig. 1c. The pulses are expressed as the Fourier series,

$$s_x = d_x + \sum_{k \in N} \frac{2}{k\pi} \sin k\pi d_x \sin 2\pi k f_{sw}(t - t_x). \quad (2)$$

The unfiltered dc link current, i_{dc}, is related to the ac link current, i_{ac}, in terms of the switching functions, s_a and s_b, of legs a and b, respectively, as (3).

$$i_{dc} = i_{ac}(s_a - s_b) \quad (3)$$

The difference, $(s_a - s_b)$, of the leg switching functions is evaluated from the expressions for the individual switching functions in (2), as (4).

$$s_a - s_b = (d_a - d_b) + \sum_{k \in N} \frac{2}{k\pi} \sin k\pi d_a \sin 2\pi k f_{sw}(t - t_a)$$
$$- \sum_{k \in N} \frac{2}{k\pi} \sin k\pi d_b \sin 2\pi k f_{sw}(t - t_b) \quad (4)$$

In conventional operating modes of full-bridge isolated dc-dc converters, the leg duty ratios are set to a constant value of 50%, i.e., $d_a = d_b = 0.5$, [8], to extract the maximum possible ac voltage from the full bridge. This consideration simplifies the expression in (4) to that given in (5).

$$s_a - s_b = \sum_{\text{odd } k} s_k \sin(2\pi k f_{sw} t - \phi_k) \quad (5)$$
$$\text{where, } s_k = \frac{4}{k\pi}(-1)^{\frac{k-1}{2}} \sin \pi k f_{sw}(t_b - t_a),$$
$$\text{and, } \phi_k = \pi k f_{sw}(t_a + t_b) - \frac{\pi}{2}$$

The relation in (5) is substituted in (3) to obtain the time-domain expression for the unfiltered dc-link current, i_{dc}, in terms of the ac-link current, i_{ac}, as (6).

$$i_{dc} = i_{ac} \sum_{\text{odd } k} s_k \sin(2\pi k f_{sw} t - \phi_k) \quad (6)$$

The ac terminal current, i_{ac}, is expressed as the sum of a dc bias, i_0, and a quarter-wave symmetric ac component, as

$$i_{ac} = i_0 + \sum_{\text{odd } m} i_m \sin(2\pi m f_{sw} t - \theta_m) \quad (7)$$

The expression of the current, i_{ac}, in (7) is substituted into the relation in (3), to obtain the expression for the unfiltered dc-link current, as (8).

$$i_{dc} = i_0 \sum_{\text{odd } k} s_k \sin(2\pi k f_{sw} t - \phi_k)$$
$$+ \sum_{\text{odd } m} i_m \sin(2\pi m f_{sw} t - \theta_m) \sum_{\text{odd } k} s_k \sin(2\pi k f_{sw} t - \phi_k) \quad (8)$$

The expression in (8) is further simplified to separate the distinct frequency components in the current, i_{dc}, as (9).

979-8-3315-1612-3/25 $31.00 © 2025 IEEE   119

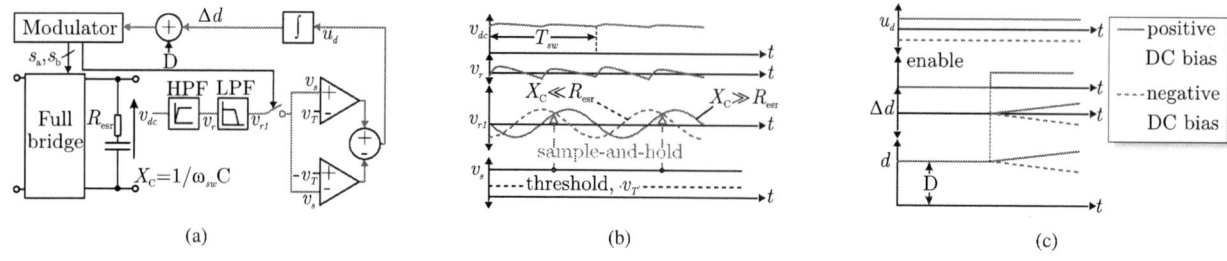

Fig. 2: (a) Schematic of the proposed dc bias elimination scheme. (b) Representative waveforms of the dc bias sensing and conditioning system, and (c) the dc bias compensation process that adjusts the leg duty ratio, d.

$$i_{dc} = \overbrace{I_{dc}}^{\text{dc}} + \overbrace{i_0 \sum_{\text{odd } k} s_k \sin(2\pi k f_{sw} t - \phi_k)}^{\text{ripple, odd switching-frequency harmonics}}$$
$$+ \overbrace{\sum_{\text{odd } m} \sum_{\text{odd } k \neq m} i_m s_k \sin(2\pi m f_{sw} t - \theta_m) \sin(2\pi k f_{sw} t - \phi_k)}^{\text{ripple, even switching-frequency harmonics}}$$
$$(9)$$

As indicated in (9), the current, i_{dc}, comprises ripple components at even and odd harmonics of switching frequency, with the latter being proportional to i_0. In particular, the amplitude, $I_{dc,1}$, of the switching-frequency (i.e., $k = 1$) component ($i_{dc,1}$) of the current, i_{dc}, is obtained from (5) and (9), as

$$I_{dc,1} = \frac{4}{\pi} i_0 \sin \pi f_{sw}(t_b - t_a). \quad (10)$$

Hence, the amplitude of the fundamental (i.e., switching-frequency) component, V_{r1}, of the terminal voltage ripple, is given by the product of the ripple current component in (10) and the equivalent impedance, Z_f, of the combination of the output filter and the load at switching frequency, as

$$V_{r1} = \frac{4}{\pi} i_0 Z_f \sin \pi f_{sw}(t_b - t_a). \quad (11)$$

The design of the output filter ensures that its impedance at switching frequency is negligible compared to the load [9]. Hence, the impedance, Z_f, does not change appreciably during operation, compared to the (known) impedance of the output filter. Additionally, the quantity, $t_b - t_a$, represents the time delay between the switching functions of the two legs of a full bridge and is, hence, a known quantity. Thus, the relationship between the dc bias, i_0, in the ac terminal current, and the amplitude, V_{r1}, of the resulting fundamental-frequency ripple at the dc terminal, is rewritten in terms of a known, but variable, factor, K_r, as

$$V_{r1} = K_r i_0. \quad (12)$$

The proportional relationship between the dc bias current, i_0, in a transformer winding supplying a full-bridge and the amplitude, V_{r1}, of the first-harmonic terminal ripple of the full bridge is described by (12). This fundamental-frequency component of ripple can be extracted, and the remaining components suppressed, using a tuned filter similar to that

detailed in [7]. This result is exploited in the design of the bias elimination scheme described next.

III. PROPOSED DC BIAS ELIMINATION SCHEME

The foregoing section derives a proportional relation, (12), between the magnitude of the bias current in the ac terminal current of a full bridge, and the amplitude of the fundamental-(switching) frequency ripple component of the dc terminal voltage. In the proposed technique, this ripple component is extracted from the measured dc voltage and used to estimate the dc bias. The bias is then cancelled by applying an opposing dc voltage at the ac terminals.

The proposed bias elimination scheme is illustrated in Fig. 2a, and corresponding waveforms are presented in Figs. 2b and 2c. The dc bus voltage, v_{dc}, is passed through a high-pass filter (HPF) to extract the ripple, v_r, and subsequently through a tuned low-pass filter (LPF) to extract the switching-frequency component, v_{r1}, of the ripple. The peak value, V_{r1}, of the switching-frequency ripple is directly related to the dc bias current and is measured by sampling the ripple waveform, v_{r1}. The sample, v_s, is used directly in the bias elimination algorithm. The sampling instant is chosen, by analyzing the phase of the switching-frequency ripple, as follows.

The switching-frequency component of the unfiltered dc current, i_{dc}, has a phase, ϕ_1, given by (9). The phase, ϕ_1, is obtained from (5), as

$$\phi_1 = 2\pi f_{sw} \left(\frac{t_a + t_b}{2} \right) - \frac{\pi}{2}. \quad (13)$$

The peak of the switching-frequency current component occurs at

$$t = \frac{\phi_1 + \pi/2}{2\pi f_{sw}}$$
$$= \frac{t_a + t_b}{2}. \quad (14)$$

Hence, the peak of the switching-frequency current ripple coincides with the temporal mid-point between the rising edges of the switching functions of the two legs of the full bridge, which are shown in Fig. 1c. The peak of the switching-frequency voltage ripple, v_{r1}, which is the measured quantity, is shifted from this temporal instant by an interval determined by the impedance of the output filter.

The magnitude of the output filter impedance influences the amplitude of the peak fundamental frequency ripple, as

979-8-3315-1612-3/25 $31.00 © 2025 IEEE

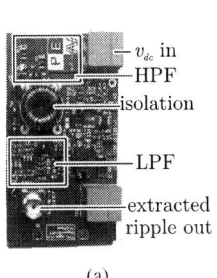 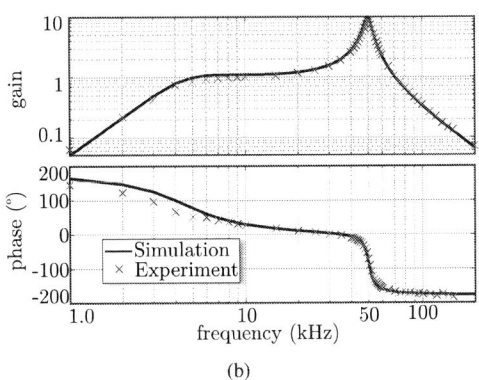 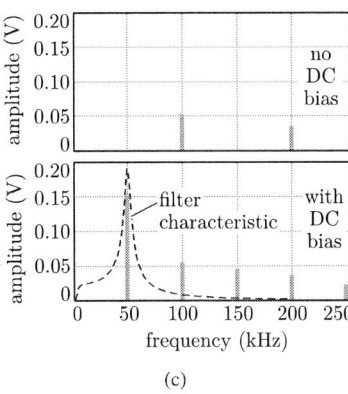

Fig. 3: (a) Fabricated filter for extracting fundamental component of ripple. (b) Simulated and experimental gain and phase responses of the filter. (c) Simulated frequency components in the terminal voltage ripple with and without dc bias in the reactive network. The simulated filter characteristic is overlaid to illustrate the extraction of the ripple component.

evident in (11). The filter, connected to the dc terminals, is explicitly shown in Fig. 2a. The impedance of this filter at switching frequency is formed of a resistive (R_{esr}) and a reactive ($X_C = 1/2\pi f_{sw}C$) component. The phase (α) of the switching-frequency voltage ripple, v_{r1}, relative to the switching-frequency current ripple, $i_{dc,1}$, depends on the relative magnitude of resistance and the reactance, as (15).

$$\alpha = \tan^{-1} 2\pi f_{sw} C R_{esr} \qquad (15)$$

If the angle, α, of the filter impedance is known from measurements, the sampling instant of the switching-frequency ripple is delayed by an equivalent duration. However, when this angle is not known, especially since these values may change with the aging of the filter, the following approximate method is employed to decide the sampling instant.

Two extreme cases, corresponding to $R_{esr} \ll X_C$ and $R_{esr} \gg X_C$, yielding phase delays, $\alpha = 0$ and $\alpha = \pi/2$, respectively, are illustrated in Fig. 2a. The sampling points indicated in Fig. 2a, located half-way between the instants of peak in the two cases, are hence within an angle of $\pi/4$ of the peak. Hence, sampling the extracted ripple, v_{r1}, at these points yields at least $(\sin \pi/4 =) 1/\sqrt{2}$ times the amplitude. Thus, the amplitude of the ripple is measured up to a factor of $1/\sqrt{2}$. The phase lag of the low-pass filter, measured at the switching frequency, is also considered in the adjustment of the sampling instant. For example, the phase near resonance of a tuned second-order low-pass filter, with a high quality factor, is approximately 90°. Hence, the sampling is synchronized at the midpoint of the peak and zero crossing of the quantity, $(s_a - s_b)$. Thus, depending on the impedance of the output filter at switching frequency, the amplitude of the ripple, and hence, the dc bias, is measured to an accuracy of within 30%. If the dc bias is negative, the waveforms have the opposite polarity, and hence peak detection by rectification cannot be employed for amplitude measurements.

The dc bias elimination process, illustrated in Fig. 2a, employs the existing method of imposing a dc excitation of opposing polarity and suitable magnitude on the ac network. The magnitude of this dc excitation depends on the the leg duty ratios, d_a and d_b, of the full bridge legs, a and b, as illustrated in Figs. 1b and 1c. As described in Section II, the nominal values of these duty ratios is 0.5. The duty ratio perturbation, Δd, of a full bridge is defined as (16).

$$\Delta d = d_a - d_b \qquad (16)$$

The applied dc voltage bias, v_b, is related to the dc terminal voltage, V_{dc}, of the full bridge through the duty ratio perturbation, Δd, as (17).

$$v_b = \Delta d V_{dc} \qquad (17)$$

Thus, the measured peak of the ripple is used to adjust the applied duty ratio perturbation, in order to eliminate the bias current, by comparing the sampled ripple, v_s, with a threshold voltage, v_T, or its negated version, $-v_T$. The update rule for the bias elimination is given by (18).

$$\Delta d \leftarrow \begin{cases} \Delta d + \delta, & v_s \geq v_T \\ \Delta d - \delta, & v_s \leq -v_T \\ \Delta d, & \text{otherwise} \end{cases} \qquad (18)$$

The value of the duty ratio adjustment, δ, is obtained from the temporal resolution of the digital control platform where the proposed bias elimination scheme is implemented. Considering a clock frequency, f_{clk}, the duty ratio adjustment, the value of the duty ratio step, δ, is set to the smallest possible value, f_{sw}/f_{clk}. Thus, the minimum possible value of the bias voltage, v_b, is obtained from (17) as $V_{dc}f_{sw}/f_{clk}$.

The corresponding resolution in the dc bias current adjustment depends on the equivalent resistance, r, of the reactive network. Hence, the minimum bias current adjustment possible is equal to the ratio of the minimum bias voltage, and the resistance of the reactive network, i.e., $V_{dc}f_{sw}/(r f_{clk})$.

The minimum dc bias current that the algorithm seeks to adjust is thus set to half of this value, and is substituted in (10). Further, the gain, K_{bpf}, of the band-pass filter, and the factor of $1/\sqrt{2}$ due to the phase at the sampling instant yields the threshold, v_T, as

$$v_T = \frac{\sqrt{2} f_{sw} V_{dc}}{\pi f_{clk} r} K_{bpf} Z_f \sin \pi f_{sw}(t_b - t_a). \qquad (19)$$

979-8-3315-1612-3/25 $31.00 © 2025 IEEE

TABLE I

(A) Circuit Parameters, and (B) Operating Conditions of the DAB.

(a)

Turn ratio, $n:1$	0.87:1
Primary-referred series inductance, L	110 µH
Primary-referred series resistance, r	0.65 Ω
dc bus capacitance, C_{dc}	0.24 mF

(b)

Input voltage, V_g	110 V
Output power, P_o	150 W
Output voltage, V_o	100 V
Switching frequency, f_{sw}	50 kHz

Fig. 4: (a) Experimental DAB prototype along with the control platform and the filter. (b) Simulated, and (c) experimental waveforms of the extracted ripple, v_{r1}, and the transformer secondary current, i_2, of the DAB, showing the elimination of a dc bias with a positive amplitude through the proposed strategy.

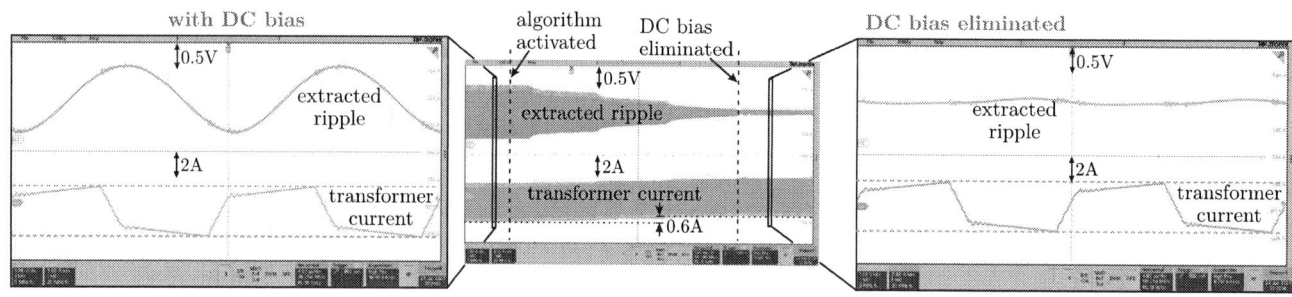

Fig. 5: Experimental waveforms of transformer current (i_2, green) and extracted first-harmonic terminal voltage ripple (v_{r1}, blue), validating the proposed dc bias elimination approach. Corresponding waveforms magnified to the time scale of the switching period show operation prior to (inset, left) and post (inset, right) dc bias elimination.

Finally, the proposed method can eliminate the dc bias of a transformer winding only when the corresponding bridge comprises one or more active devices. For example, the dc bias in the secondary winding of a unidirectional LLC-resonant topology cannot be eliminated using the proposed method.

IV. Simulation and Experimental Results

The proposed dc bias elimination strategy is validated in circuit simulation and experiments on a DAB prototype. The circuit parameters and operating conditions of the DAB are presented in Tables Ia and Ib, respectively.

The extraction of the switching-frequency terminal voltage ripple, which is a key step in the proposed approach, is achieved using the analog filter shown in Fig. 3a. The high-pass characteristic is realized using a passive $R-C$ circuit, while the low-pass characteristic is realized using a second-order Sallen-Key stage. The frequency response of the filter is presented in Fig. 3b. The gain response shows a peak at the switching frequency (50 kHz), with an experimental attenuation of 25× between the fundamental (f_{sw}) and harmonic ($2f_{sw}$ and higher) components. Hence, the extraction

of the fundamental ripple, which depends on the dc bias, is verified. The phase lag introduced by the filter at the switching frequency is approximately 71°.

The operation of the converter is simulated, and the frequency components in the terminal voltage ripple, with and without a ~ 1 A dc bias in the primary winding current of the transformer, are illustrated in Fig. 3c. The introduction of odd-harmonic components in the terminal voltage ripple is thus confirmed. The measured gain response of the designed band-pass filter is overlaid on the plot to illustrate the extraction of the fundamental ripple component.

The DAB prototype is shown in Fig. 4a. The elimination of a bias current, through the proposed algorithm, is studied for the operating point tabulated in Table Ib. Simulated and experimental waveforms of the extracted ripple, v_{r1}, and the transformer secondary current, i_2, are presented in Figs. 4b and 4c, respectively. The dc bias is eliminated in steps, corresponding to the repeated update of the duty ratio perturbation described by (18). A dc bias of 0.6 A is eliminated in about 5 ms, thus validating the proposed strategy.

The proportional relation between the amplitude of the dc

979-8-3315-1612-3/25 $31.00 © 2025 IEEE

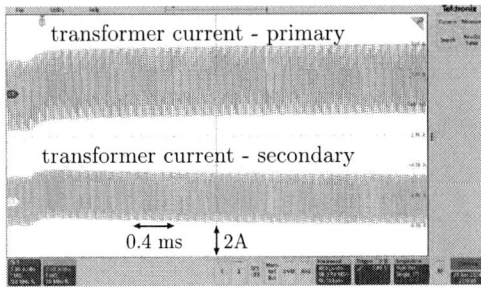

Fig. 6: Waveforms of transformer primary and secondary currents under application of step dc bias on primary winding, showing the decoupling between steady-state dc currents in transformer windings.

bias and the fundamental ripple, at a given operating point, is evident from the respective waveforms. This dependence makes the proposed strategy a precise alternative to existing ripple-based strategies that operate on switching-cycle deviations in the peaks of the ripple waveform.

The dependence is further illustrated in Fig. 5. For the considered operating condition, the extracted fundamental ripple, v_{r1}, has a $\sim 0.7\,\mathrm{V}$ peak for a $0.6\,\mathrm{A}$ dc bias, while the same quantity has a negligible peak value when the dc bias is removed. Thus, the extracted fundamental ripple employed in the proposed strategy is a reliable indicator of the dc bias in the transformer current.

The proposed method requires the measurement of the ripple at each dc terminal, in order to detect a dc bias at the corresponding ac terminal. The waveforms of transformer primary and secondary current, following the application of a step bias voltage to the primary winding, are presented in Fig. 6 to support this requirement. Following the voltage step, the primary winding current retains a bias, whereas the bias in the current in the secondary winding first develops and then decays. Thus, a dc bias in the primary winding cannot be detected using the terminal ripple of the secondary-side full bridge if the transformer operates well outside saturation.

V. Conclusion

This paper proposes a dc bias elimination strategy for isolated full-bridge dc-dc converters based on a measurement of the switching-frequency harmonic of the voltage ripple at the dc terminals. The ripple is processed using a tuned analog filter and the dc bias is estimated without high-resolution measurements. An algorithm to eliminate the dc bias is developed, and its performance validated in simulations and experiments on a DAB converter. The proposed approach is implemented using only analog conditioning circuitry in conjunction with the digital controller of the converter. The hardware and software simplicity of the proposed strategy makes it applicable to various isolated dc-dc converter topologies.

Acknowledgment

This research has received funding from the European Innovation Council (EIC) under the European Union's Horizon 2020 research and innovation programme, vide grant agreement no. 101057679 within the framework of the project Super-HEART.

References

[1] G. Ortiz et al., "Flux balancing of isolation transformers and application of "the magnetic ear" for closed-loop volt–second compensation," *IEEE Transactions on Power Electronics*, 2014.

[2] G. Qiu et al., "A fluxgate-based current sensor for dc bias elimination in a dual active bridge converter," *IEEE Transactions on Power Electronics*, 2022.

[3] S. M. Kaviri et al., "A digital active dc-eliminating method for dc/dc converters," *IEEE Transactions on Power Electronics*, 2019.

[4] P. Lenzen et al., "Dc-bias elimination in high-frequency dual active bridge dc/dc converters through single-sided measurements," in *2024 IEEE Applied Power Electronics Conference and Exposition (APEC)*, 2024.

[5] P. Lenzen et al., "Dc-bias reduction in high-frequency dual active bridge dc-dc converters through slow dc measurements," in *PCIM Europe 2024; International Exhibition and Conference for Power Electronics, Intelligent Motion, Renewable Energy and Energy Management*, 2024.

[6] R. Cvetanovic et al., "Switching noise propagation and suppression in multisampled power electronics control systems," *IEEE Transactions on Power Electronics*, 2024.

[7] A. Sengupta et al., "A low-complexity load-decoupled talkative power conversion strategy for the dual-active-bridge converter," in *2024 IEEE Energy Conversion Congress and Exposition (ECCE)*, 2024.

[8] A. K. Jain et al., "Pwm control of dual active bridge: Comprehensive analysis and experimental verification," *IEEE Transactions on Power Electronics*, vol. 26, no. 4, pp. 1215–1227, 2011.

[9] M. K. Kazimierczuk et al., "Accurate design of output filter for dc–dc pwm buck converter and derived topologies," *IEEE Transactions on Circuits and Systems I: Regular Papers*, 2023.

Tunable Matching Network with Dual Phase-Switched Impedance Modulation Actuators

Alexander Jurkov
MKS Instruments Inc.
Rochester, NY, USA
alexander.jurkov@mks.com

David Perreault
Massachusetts Institute of
Technology
Cambridge, MA, USA
djperrea@mit.edu

Abstract—Accurate, rapid, and dynamically-controlled impedance matching offers significant advantages for a wide range of present and emerging radio-frequency (RF) power applications. This work develops a solid-state switched-mode tunable matching network (TMN) that enables faster and more precise impedance matching than conventional techniques. The implementation is based on a recently-proposed technique, Phase-Switched Impedance Modulation (PSIM), which entails the switching of passive elements at the RF operating frequency, effectively modulating their impedances. PSIM has shown promise for implementing fast, solid-state TMNs for high-frequency (HF) applications at kilowatt power levels. This work presents the first TMN design with multiple independently-controlled PSIM tuning elements. The prototype TMN, based on a Π-network topology with GaN FETs, matches a wide range of load impedances to 50 Ω, targeting 13.56 MHz inductively-coupled plasma (ICP) processes. Experimental results demonstrate successful operation up to 150 W input power, highlighting the TMN's potential for significant improvements in RF power delivery systems.

Keywords—impedance matching, tunable matching network (TMN), phase-switched impedance modulation (PSIM), radio-frequency (RF) power amplifier, inductively-coupled plasma (ICP)

I. INTRODUCTION

Delivering radio-frequency (RF) power into widely-varying load impedances at kilowatt levels is essential for a wide range of applications [1]-[8]. RF amplifiers and inverters are typically designed to operate efficiently at a fixed load impedance, making significant variations in load impedance particularly challenging. To address this challenge, tunable matching networks (TMNs) are commonly employed between the load and the generator [5]-[10]. TMNs transform the varying load impedance to a fixed input impedance (e.g., 50 Ω) suitable for the amplifier. Conventional TMNs use dynamically tuned adjustable passive components, such as mechanically adjusted variable capacitors or inductors [11] driven by servo motors, reconfigurable capacitor/inductor banks [12], [13], or high-power varactors [14], [15]. Although effective, these methods often result in bulky, costly systems with limited resolution and slow response times.

One recently-proposed solid-state technique showing significant promise in implementing rapidly-tunable reactance elements is phase-switched impedance modulation (PSIM) [16], [17]. PSIM involves switching a capacitor under zero-voltage-switching (ZVS) conditions (using, for e.g., a GaN transistor in parallel with the capacitor), modulating its effective reactance seen at the fundamental frequency by adjusting the switch's duty cycle. This solid-state technique allows for accurate and rapid impedance modulation at high frequencies (tens of MHz), enabling the implementation of TMNs with microsecond tune times and kilowatts of RF power handling capability. Previously demonstrated PSIM-based TMN systems [16], [17] incorporate a single PSIM tunable reactance element while also employing narrow-band modulation of the operating frequency to match a two-dimensional region of load impedances to 50 Ω. However, some applications, such as certain RF plasma processes for semiconductor fabrication, require rapid impedance matching over a wide load range while operating at a fixed frequency. This necessitates the implementation of TMN system with multiple, independent high-speed tunable elements.

This paper presents the first design of a TMN system with two independently-controlled PSIM actuators, targeting high-bandwidth impedance matching over a wide inductive load range associated with inductively-coupled plasma (ICP) processing at 13.56 MHz. The work explores various aspects, including network topology selection, PSIM implementation, TMN control, and PSIM switch synchronization.

II. PHASE-SWITCHED IMPEDANCE MODULATION

Phase-Switched Impedance Modulation (PSIM) modulates impedance by switched-mode modulation of the voltage and current waveforms on a reactive element [16]-[18]. Fig. 1 illustrates a parallel combination of a capacitor C_0 and an ideal switch driven by a sinusoidal current source. When the switch is on (q is high), the capacitor voltage v_C is clamped to zero, and all current i_C flows through the switch. When the switch is off (q is low), v_C is determined by the current through C_0. By adjusting the phase of the switch control signal q with respect to the current i_C, zero-voltage switching (ZVS) can be achieved at both turn-on and turn-off, enabling efficient high-frequency operation.

Fourier analysis shows that the fundamental component of the capacitor voltage under ZVS lags the current by 90°, indicating the switched capacitor network behaves as a capacitor at the switching frequency. Adjusting the switch duty cycle controls the peak capacitor voltage and the magnitude ratio of the fundamental components of v_C and i_C, thus modulating the apparent capacitance C_{EFF} of the network. Increasing the switch duty cycle from 0% to 100% modulates the effective capacitance from C_0 to infinity (short circuit). In practice, the modulation range may be limited by switch losses and switching harmonics [16].

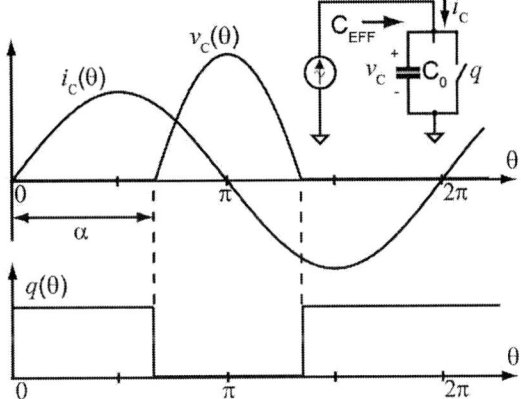

Fig. 1. Conceptual operation of a phase-switched capacitor and its voltage and current waveforms. The effective capacitance C_{EFF} at the switching frequency can be controlled by modulating the switch duty-cycle, related to the switch conduction angle α [16].

III. TUNABLE MATCHING NETWORK DESIGN

The TMN prototype presented here is designed to match a power amplifier (PA) with a 50 Ω source impedance to a load impedance Z_L with 1 Ω to 40 Ω resistive and 10 Ω to 40 Ω reactive range. This system targets operation at 13.56 MHz with up to 150 W of input power. The ability to provide rapid impedance matching at such inductive loads and power levels is of great importance to generating and driving inductively-coupled plasma (ICP) for small-scale semiconductor manufacturing equipment [7].

The design is based on a Π-network with fixed series branch reactance X and PSIM-based tunable capacitive shunt elements C_1 and C_2, as shown in Fig. 2. This network topology allows for implementation of both PSIM elements with ground-referenced switches thus simplifying their driving and the overall system construction. The typical load impedance matching range one can achieve with this TMN design is illustrated in Fig. 3 for a given tuning range of capacitors C_1 and C_2. The minimum $C_{1,MIN}$ and maximum $C_{1,MAX}$ values of C_1 determine the maximum and minimum conductance bounds of the matching range respectively. On the other hand, the tuning range of C_2 determines the achievable susceptive range of the load, with $C_{2,MIN}$ and $C_{2,MAX}$ corresponding to the maximum and minimum susceptance bounds respectively.

It can be shown that irrespective of the tuning range of C_1 and C_2, the ideally-lossless Π-network in Fig. 2 can only match loads with conductance component G_L less than a certain maximum allowed conductance $G_{MAX} = R_S/X^2$. Fig. 3 illustrates this restricted range on the Smith chart with the red cross-hatched conductance circle. Decreasing the minimum bound $C_{1,MIN}$ of the C_1 tuning range expands the TMN's matching range towards higher conductances until the matching

range reaches the restricted conductance circle G_{MAX}. Further decrease in $C_{1,MIN}$ beyond this point results in *overlapping* of the matching range, i.e. the impedance transformation the TMN provides is no longer unique and the additional tuning range of C_1 does not contribute any further expansion of the matching range of the TMN.

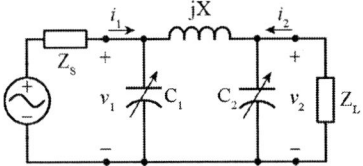

Fig. 2. Schematic of a TMN design example based on a Π-network matching a load impedance Z_L to a source impedance Z_S with tunable shunt capacitors C_1 and C_2.

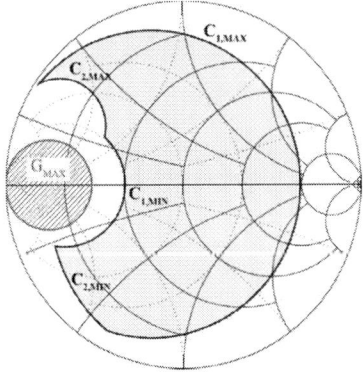

Fig. 3. Typical load impedance matching range (shaded in yellow) for a given tuning range $[C_{1,MIN}; C_{1,MAX}]$ and $[C_{2,MIN}; C_{2,MAX}]$ of capacitors C_1 and C_2, respectively for the Π-network-based TMN in Fig. 2. The Smith chart is normalized to Z_S.

As an initial step in the design of the TMN, we first determine the series branch reactance X and the required tuning range $[C_{1,MIN}; C_{1,MAX}]$ and $[C_{2,MIN}; C_{2,MAX}]$ for the two shunt capacitors to achieve the desired impedance matching range. The tuning range of the Π-network shunt capacitors can be described in terms of the minimum capacitance required, termed here *base capacitance*, along with its maximum modulation factor necessary to cover the full load impedance matching range. (Modulation factor for a tunable capacitor is defined here as the ratio of its capacitance to its base capacitance). Fig. 4 shows the minimum capacitance and required maximum modulation factor for the two shunt capacitors C_1 and C_2 in the Π-network of Fig. 2 versus the series branch reactance X necessary to achieve the target load impedance range. (Detailed network design equations are provided and discussed in [19].)

Note from Fig. 4 that as the reactance X decreases, the base value of both capacitors increases. This dependence arises from the fact that the size of the restricted range for a Π-network is inversely proportional to X. Smaller branch reactance results in narrower restricted region (larger G_{MAX}) and requires larger shunt capacitances to achieve the same impedance matching range. On the other hand, increasing the branch reactance X, one trades modulation in the load-side capacitor C_2 for modulation in the input-side capacitor C_1 (see Fig. 4).

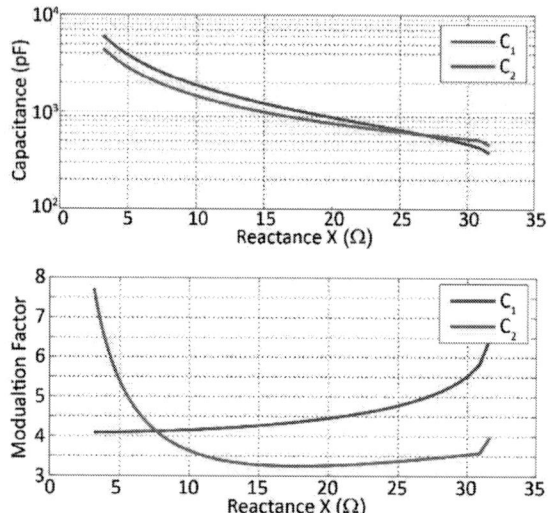

Fig. 4. Minimum capacitance (top) and maximum capacitance modulation factor (bottom) versus branch reactance X for an example TMN design based on the Π-network of Fig. 2 for a 1 Ω-40 Ω resistive and 10 Ω-40 Ω reactive load range.

When implementing tunable capacitors with PSIM, one generally aims to minimize the required capacitance modulation factor, as high modulation factors result in higher voltage stress on the switching devices, larger PSIM losses, increased waveform distortion and unwanted harmonic content in the input / output of the TMN.

It can be seen from Fig. 4 that for the prototype design considered here, the required modulation factor for the load-side capacitor C_2 is minimum when X is approximately 15 Ω - 20 Ω. For this design, we select $X \approx 15$ Ω, resulting in a modulation factor of approximately 3.3 for C_2 and 4.3 for C_1. Plasma generation and driving applications often have strict requirements on the amount of harmonic content that can be injected into the plasma and is typically limited to less than - 20dBc. In the TMN prototype considered here, where the tunable shunt elements are implemented with PSIM, the harmonic content injected into the load is predominantly determined by the amount of modulation in the load-side PSIM element. Hence, in order to minimize harmonic content injected into the load, we tradeoff lower modulation in the load-side capacitor for slightly higher one in the input-side capacitor.

TMN implementation based on the Π-network topology of Fig. 2 with fixed series branch reactance and tunable PSIM-based shunt capacitive elements is shown in Fig. 6. Each PSIM element includes a switch Q_1 (or Q_2) and an external capacitance C_{P1} (or C_{P2}), which absorbs the device capacitance and provides the necessary PSIM capacitance tuning range. As mentioned, this architecture favors PSIM implementation by using ground-referenced switches, considerably simplifying

their gate-drive circuitry. In the TMN prototype design considered here, we implement both PSIM elements with 650V GaN devices (GS66516T, GaN Systems). At 200W of input power, this leaves about 300 V of margin on the input-side PSIM switch assuming a 50 Ω TMN input impedance. To prevent over-stressing the output-side PSIM switch, one must appropriately limit output power depending on the particular load impedance.

The series branch components, C_2 and L_2 in Fig. 6 are selected to realize the desired reactance X from Fig. 4 (approximately 15 Ω at 13.56 MHz for this design) while also limiting injection of switching harmonics from one PSIM element to the other. Input and output filters, consisting of L_1, C_1, L_3, and C_3 are series-resonant at the operating frequency of 13.56 MHz, and limit harmonic injection from the PSIM elements to the power amplifier and load, respectively. Additionally, C_1, C_2, and C_3 provide DC blocking between the generator, the load, and the PSIM elements.

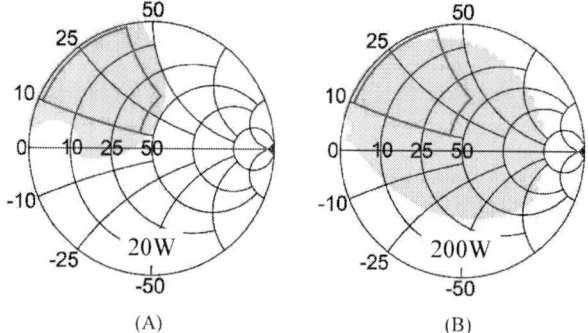

(A)   (B)

Fig. 5. Simulated load matching range (yellow) for the PSIM-based TMN design of Fig. 6 at 20W (A) and 200 W (B) input power with $L_1 = 1.46$ μH, $L_2 = 193$ nH, $L_3 = 1.83$ μH, $C_1 = 93.9$ pF, $C_2 = 10$ nF, $C_3 = 75.2$ pF, $C_{P1} = 400$ pF, $C_{P2} = 900$ pF, and GS66516T, 650V GaN devices for Q_1 and Q_2. The effective capacitance of both PSIM elements is modulated up to 4x the respective base capacitance. The desired operating load range is plotted with a red contour.

Due to the highly-nonlinear characteristic of the FET output capacitance, the PSIM base capacitance can dramatically decrease with increasing switch voltage and power level. This variation can cause significant modulation of the TMN matching range with operating power, especially when the external capacitance is small relative to the device capacitance. To ensure that the desired matching range is achieved over the entire power range of interest, C_{P1} and C_{P2} must be appropriately selected. The expected load impedance matching range for the TMN of Fig. 6 with $C_{P1} = 400$ pF and $C_{P2} = 900$ pF is shown in Fig. 5 for 20 W and 200 W of TMN input power and up to 4x modulation of both PSIM elements. As can be seen from Fig. 5, at low power levels, the matching range tends to cover mostly inductive loads and extends to loads with very small resistive components. However, ss the power level increases, the PSIM base capacitance decreases, and the matching range tends to shift towards capacitive loads with higher resistive components.

979-8-3315-1612-3/25 $31.00 © 2025 IEEE

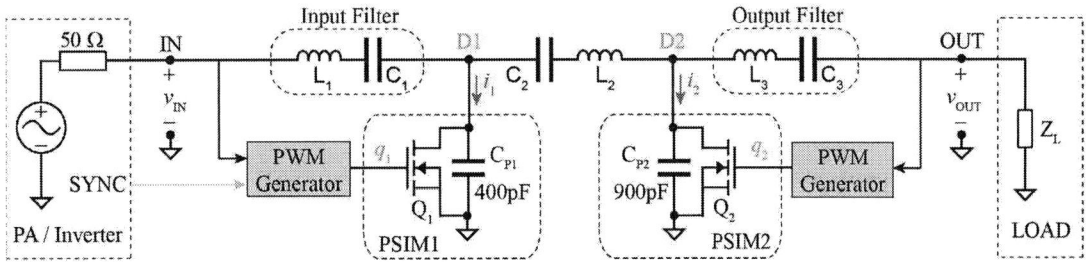

Fig. 6. Schematic of the pi-network TMN prototype with two PSIM elements matching a load Z_L to a power amplifier with 50 Ω output impedance with L_1 = 1.46 µH, L_2 = 193 nH, L_3 = 1.83 µH, C_1 = 93.9 pF, C_2 = 10 nF, C_3 = 75.2 pF, C_{P1} = 400 pF, C_{P2} = 900 pF, and GS66516T, 650V GaN devices for Q_1 and Q_2. The gate-drive signals q_1 and q_2 for the PSIM switches are synchronized to the network's input and output voltages respectively.

IV. TMN CONTROL AND SWITCH SYNCHRONIZATION

The impedance transformation between the output and input ports of the TMN of Fig. 6 is controlled by adjusting the effective PSIM capacitances, which in turn is achieved by appropriately controlling the conduction angle of switches Q_1 and Q_2. To illustrate the operation of the TMN and provide some insight into the control of the PSIM elements, consider the example (see Fig. 7) of transforming an inductive load Z_L = 10+j50 Ω connected to the output port of the network in Fig. 6 to a 50 Ω resistance seen looking into the TMN input port.

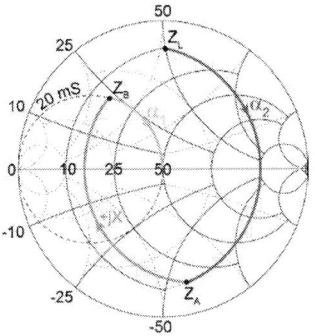

Fig. 7. Example impedance matching trajectory in the Smith chart (A) for the Π-network TMN design (B) matching an inductive load Z_L = 10+j50 Ω to 50 Ω by appropriately selecting the switch conduction angles α_1 and α_2 of the input and output side PSIM elements respectively.

Such an impedance transformation corresponds to matching Z_L to an rf power amplifier with a 50 Ω output impedance. The Smith chart in Fig. 7 shows the load Z_L and plots the trajectory of impedance transformation introduced by each component of the TMN as one traverses from the output port to the input port of the network. Since the output filter L_3-C_3 is series resonant at 13.56 MHz, at this operating frequency it has no transformation effect on the load impedance. The output-side PSIM element effectively behaves as a tunable shunt capacitor, and as Fig. 7 shows, increasing the conduction angle α_2 (related to the switch duty cycle) of Q_2 causes the impedance Z_A to traverse on a constant conductance contour. The series reactance branch formed by L_2 and C_2 causes a fixed reactive offset of $+jX$ to Z_A and transforms the impedance to Z_B. Note that if one properly selects the conduction angle of Q_2 to provide the right amount of susceptance, one can adjust Z_A so that after the $+jX$ reactive

offset, Z_B lies on the constant conductance circle of 20 mS. Finally, by appropriately selecting the conduction angle α_1 of the input-side PSIM switch Q_1, one can introduce enough shunt susceptance to Z_B to bring the input impedance of the TMN to 50 Ω. Similarly to the output filter, the input filter formed by L_1 and C_1 is resonant at 13.56 MHz and has little effect on the input impedance at this frequency. In general, one can think of the Q_2 switch duty cycle as a handle to adjust the TMN input conductance, while the Q_1 switch duty-cycle controls the TMN input admittance.

A. PSIM Switch Synchronization

The ability to synthesize the appropriate gate waveform for driving the PSIM switches is of crucial importance to the TMN operation. As Fig. 1 suggests, for a particular PSIM element, the gate drive signal q should be synchronized to the current i_C flowing through the switched-capacitor network. Furthermore, one must be able to accurately control the conduction angle of the switch. In essence, this requires one to generate a PWM gate waveform with a variable duty cycle synchronized to the switched capacitor current i_C. The resolution with which one can vary the duty cycle determines the resolution with which the effective capacitance can be modulated and hence sets the limit on the overall TMN impedance matching resolution. Accurate phase measurement of RF currents is often challenging; therefore, synchronizing the gate drive signals directly to the switched-capacitor current waveform can be problematic. As an alternative approach, in the TMN design presented here, we synchronize the gate drive signals of the input- and output-side PSIM elements with the TMN's input- and output-port voltages, v_{IN} and v_{OUT}, respectively.

To illustrate this approach, consider the TMN design shown in Fig. 6. Since the TMN's input filter is series-resonant at the 13.56 MHz nominal operating frequency, the fundamental frequency components of v_{IN} and the drain voltage v_{D1} are in phase at this frequency. Furthermore, since the phase-switched capacitor network PSIM1 behaves effectively as a capacitor, the fundamental frequency component of its current i_1 leads that of v_{D1} by 90°. Hence, by synchronizing the gate drive signal directly to the TMN's input voltage (and by appropriately phase-shifting it), one is able to effectively synchronize the switching of the PSIM1 capacitor to its current. Analogous approach is used to synchronize the switching of the output-side PSIM

979-8-3315-1612-3/25 $31.00 © 2025 IEEE   127

element. Since the output filter is series-resonant at the operating frequency, the fundamental component of the current i_2 leads that of the output voltage v_{OUT} by 90°, and so by synchronizing the gate-drive signal q_2 to the output port voltage and appropriately phase-shifting it, one effectively achieves indirect synchronization between i_2 and q_2. This is the underlying principle of the proposed gate-drive synchronization approach.

Since the Π-network TMN presented here does not rely frequency tuning, if the input and output filters are properly tuned for resonance at the operating frequency, the phase shift between the fundamental components of the PSIM currents and the network's input- and output-port voltages is approximately constant and equal to 90° over the entire matching range of the TMN. This greatly simplifies the control of the network especially in situations where relatively large filter characteristic impedances may be required to deal with harmonic content.

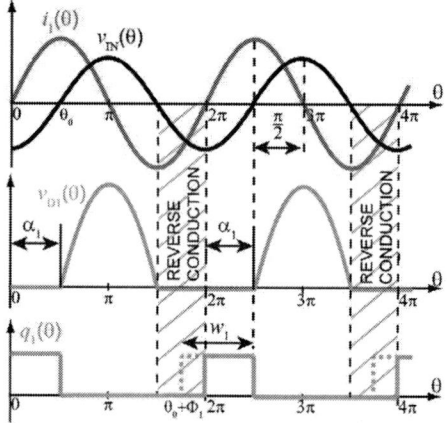

Fig. 8. Current and voltage waveforms for the input-side PSIM element in the Π-network TMN of Fig. 6. The switch gate waveforms q_1 is phase-locked to the TMN input voltage v_{IN}, which lags the PSIM current i_1 by approximately $\Theta_0 = 90°$. By controlling the switch duty cycle one can adjust the effective capacitance of the switched capacitor.

As Fig. 8 illustrates in more detail, the gate-drive signal q_1 for the input-side PSIM switch having variable pulse width w_1 is phase-locked to the TMN's input voltage v_{IN}, i.e. a phase shift Φ_1 is maintained between the rising edge of q_1 and the negative-to-positive v_{IN} transition. (The PLL-based gate drive signal generation circuit employed here is capable of independently controlling both Φ_1 and w_1 over nearly a 360° range.) Note that if the switch is not turned on immediately after the drain voltage v_{D1} rings back down to zero, the switch goes into reverse conduction mode clamping the drain voltage near zero (since at this instant the current i_1 is negative). Assuming the switched-capacitor current i_1 in Fig. 8 is purely sinusoidal, it can be shown that the duration of the reverse conduction mode is also α_1. ZVS operation is guaranteed, provided that the switch is turned on anytime during this reverse conduction interval. This allows one to use the body-diode conduction mode effectively as a ZVS *safety margin* by alleviating the requirement for precise switch turn-on. Analogous control strategy is applied to the output-side PSIM element, where the gate-drive signal q_2 is synchronized to the output port voltage v_{OUT}.

B. PWM Generation

In order to modulate the effective capacitance of the PSIM elements in the Π-network TMN and control its impedance transformation ratio, as illustrated by Fig. 8, one is required to generate pulse-width modulated (PWM) gate-drive waveforms with dynamically controlled angular pulse width w and phase shift Φ with respect to a reference signal. Here we use a PLL-based approach.

The phase-locked PWM circuit comprises a cascade of two PLL modules (see Fig. 9). Each PLL module is designed to generate an output signal at its OUT terminal such that the signal fed back to its IN terminal is frequency-locked to the reference signal at its REF terminal and is phase-shifted from it by a certain amount. This phase shift is digitally controlled by a microcontroller and can be adjusted from -180° to 180° with a 10-bit resolution. The resolution is a result of the particular implementation of the PLL module and can be easily increased.

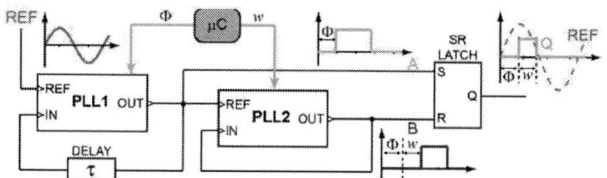

Fig. 9. Block diagram of the circuit used to generate a PWM waveform with dynamically controllable pulse width w and phase shift Φ with respect to a reference signal REF comprising a cascade of two phase-locked loop modules PLL1 and PLL2.

To illustrate the operation of the PWM generation circuitry, consider Fig. 9 and assume for a moment that the time delay element in the feedback path of PLL1 is zero. The output signal A of PLL1 is frequency-locked to the RF input and phase-shifted from it by Φ. For this particular implementation of PLL1, a phase shift of Φ between the reference and the output signals implies that the rising edge of the output signal pulse lags the negative-to-positive transition in the reference signal by Φ. In turn, the output of PLL1 serves as the input of PLL2, whose output signal B is phase-shifted by w from signal A. For the SR latch implementation in Fig. 9, the output Q is set high by a rising edge of signal A (set input S) and is cleared by a rising edge of signal B (reset input R). Thus, signals A and B are combined through the SR latch to produce the signal Q with variable angular pulse width w and phase shift Φ between its rising edge and the negative-to-positive transition of the RF signal. Note that by selecting the time delay τ in the feedback path of PLL1 to match the delay of the SR latch, one can eliminate the dependence on frequency of the phase shift between Q and the *REF* signal. Since, the TMN prototype discussed here does not rely on frequency tuning to provide impedance matching between the load and the source, accurate estimation and tuning of the delay element τ in the feedback loop of PLL1 is not critical, and in fact, it can be entirely omitted. The gate-waveform generation module in Fig. 9 is capable of generating a variable duty-cycle gate drive which can be adjusted from 10 % to 90 % with 0.1 % resolution while maintaining synchronization to the REF signal even if operating

979-8-3315-1612-3/25 $31.00 © 2025 IEEE

frequency changes. In the Π-network TMN prototype of Fig. 6, two separate PWM generators based on Fig. 9 are used to generate the gate-drive signals q_1 and q_2 for the input- and output-side PSIM elements.

V. TMN SYSTEM IMPLEMENTATION

The complete TMN system comprises a liquid-cooled PSIM module, PWM generation module, an input filter, and an output filter in an aluminum enclosure as shown in Fig. 10.

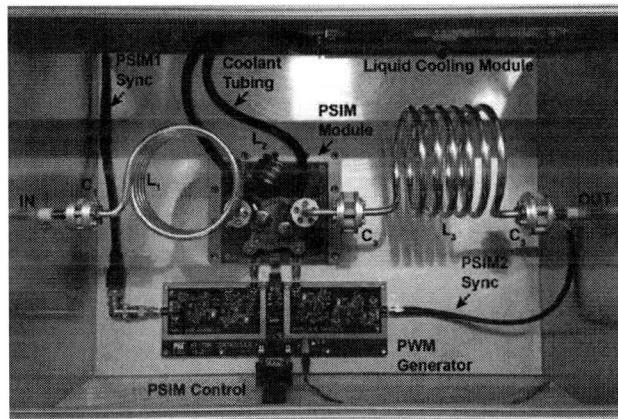

Fig. 10. Full TMN system assembly comprising the PWM Generation module, the PSIM module and the input and output filters inside of its aluminum enclosure (12" x 10" x 18") along with the liquid cooling system for the PSIM devices.

The PSIM module, integrated on a single FR4 PCB, implements the Π-network portion of the TMN design from Fig. 6, with transistors Q_1 and Q_2 forming the input- and output-side PSIM shunt elements, respectively, and L_2–C_2 providing the series branch reactance X. Here we use 650 V GaN transistors (GS66516T, GaN Systems) for both Q_1 and Q_2 resulting in voltage margin of approximately 300 V for the input-side switch and 100 V for the output-side switch over most of the desired load matching range for a TMN input power of up to 200 W. These GaN devices offer small gate resistance of 0.34 Ω and gate-to-source capacitance of 520 pF. To ensure very tight gate-drive loops and reduce voltage ringing on the gate during transistor switching, the gate drivers are included on the PSIM module PCB and are placed on the bottom layer of the board, immediately below the transistors. The gate drivers used here (MAX5048C, Maxim Integrated) have a typical pull-up and pull-down output resistances of 0.8 Ω and 0.3 Ω, respectively, and a propagation delay of less than 8 ns. When paired with the GS66516T GaN devices and powered from a 6 V regulated bus, these gate drivers allow for fast turn-on and turn-off of less than 5 ns, which is sufficient for PSIM operation at 13.56 MHz. Externally-added shunt capacitors C_{P1} and C_{P2} are selected to provide the necessary PSIM base capacitance. In this prototype, C_{P1} and C_{P2} are realized respectively with 430 pF and 1000 pF type C0G ceramic capacitors (see Table I), resulting in a total charge-equivalent PSIM base capacitance of approximately 760 pF for the input-side PSIM element and 1330 pF for the output-side PSIM element for a 350 V peak drain-to-source voltage.

The GaN FETs are attached to the back of a liquid-cooled 40mm x 40mm cold plate (PLT-UN40F, Koolance). Coolant is circulated through the cold-plate approximately at the rate of 1 L/min using a standalone, off-the-shelf liquid cooling module (CL-W0052-01, Thermaltake).

The series branch inductance L_2 in Fig. 6 is approximately 365 nH and is constructed out of 3.5 turns of copper tubing with 0.188" diameter, 0.230" winding pitch and 1" coil inner diameter (air core). Its quality factor is measured to be approximately 300 at 13.56 MHz. The capacitance C_2, also implemented with type C0G ceramic capacitors, is selected to be 750 pF so as to resonate a portion of the L_2 inductance and provide the desired 15 Ω net series branch reactance X.

The input and output filter inductors L_1 (1.89μH) and L_3 (2.26 μH) are constructed from 4.5 and 6 turns of 0.25" soft-tempered, silver-platted copper tubing with a 3" inner coil diameter and have 400 mil and 550 mil winding pitch respectively (see Table I). The quality factor of L_1 and L_3 is measured to be approximately 850 and 960 respectively at 13.56 MHz.

The PWM generation module in Fig. 10 comprises two separate PWM generators – one for each of the PSIM gate drivers. Both PWM generators are based on the cascaded PLL design of Fig. 9 and are capable of synthesizing PWM waveforms with dynamically-controlled duty cycle and phase shift with respect to a reference signal. The input- and output-side PWM generators are synchronized to the TMN input- and output-port voltages, respectively. One can adjust the duty-cycle and phase of the PWM signals over nearly a 10 % - 90% and ±180° range respectively with a 10-bit resolution via an RS-232 serial communication link with an on-board microcontroller (PIC18F26K22, Microchip).

TABLE I.   VALUES AND IMPLEMENTATION DETAILS FOR KEY COMPONENTS IN THE TMN SYSTEM DESIGN OF FIG. 6.

Component	Value	Implementation Description
L_1	1.89 μH	4.5 turns of 0.25" copper tubing with 0.3 mil silver plating, 400 mil winding pitch, 3" ID, air-core (Q ≈ 852 at 13.56 MHz)
L_2	365 nH	3.5 turns of 0.188" copper tubing, 230 mil winding pitch, 1" ID, air-core (Q ≈ 300 at 13.56 MHz)
L_3	2.26 μH	6 turns of 0.25" copper tubing with 0.3 mil silver platting, 550 mil winding pitch, 3" ID, air-core (Q ≈ 960 at 13.56MHz)
C_1	77.5 pF	9x 7.5 pF, 2.5 kV 2325 C0G ceramic 1x 10pF, 2.5 kV 2325 C0G ceramic
C_2	750 pF	C_{2A} in series with C_{2B} C_{2A}: 5x 300 pF, 630V 1111 C0G ceramic C_{2B}: 5x 300 pF, 630V 1111 C0G ceramic
C_3	60 pF	C_{3A} in series with C_{3B} C_{3A}: 10x 12 pF, 7.2 kV 3838 C0G ceramic C_{3B}: 10x 12 pF, 7.2 kV 3838 C0G ceramic
C_{P1}	430 pF	10x 43 pF, 1kV, 1111 C0G ceramic
C_{P2}	1000 pF	10x 100 pF, 1kV, 1111 C0G ceramic
Q_1, Q_2		GS66516T, GaN Systems

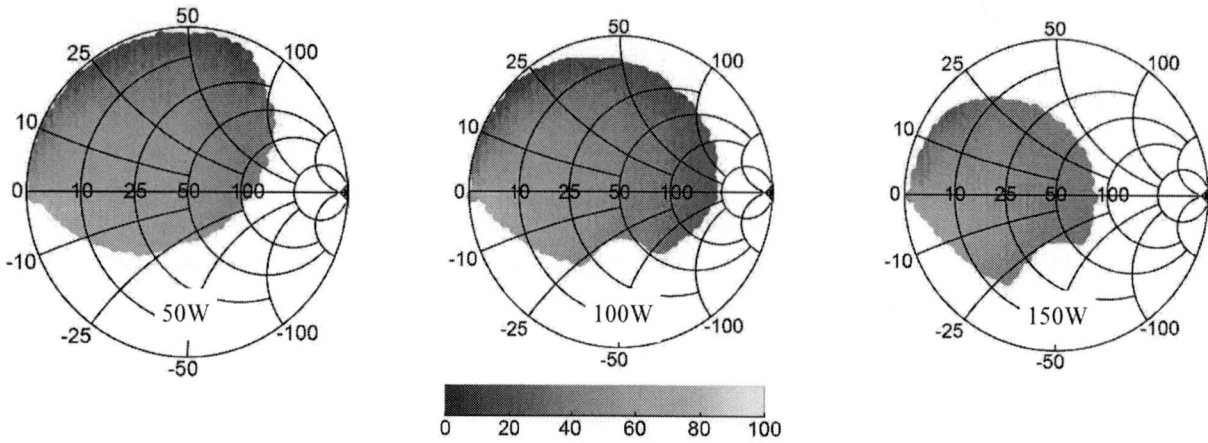

Fig. 11. Measured power efficiency (%) over the load impedance matching range of the TMN prototype for 50W, 100W and 150W of input power.

VI. TMN PERFORMANCE AND MATCHING RANGE

The dual-PSIM TMN's capability to match a wide range of load impedances to 50 Ω, typical in inductively-coupled plasma processes, is demonstrated experimentally. Fig. 11 shows the TMN system efficiency across its impedance matching range for input power levels of 50 W, 100 W, and 200 W. The highest efficiency is observed for small resistive loads with minimal reactive components. In these cases, the PSIM transistors operate at small conduction angles or remain off, reducing power dissipation in the FETs primarily to C_{OSS} losses. Maximum efficiency is observed for a 1.8 - j0.5 Ω load corresponding to both PSIM switches being off for the entire duration of the RF cycle.

As can be seen from Fig. 11, the power efficiency drops quickly for load impedances with large reactive and small resistive components due to the necessity for large PSIM capacitance modulation and FET conduction angles. Under these conditions, most of the power dissipation in the PSIM FETs is attributed to conduction losses. Furthermore, as power level increases, for a given load impedance, the peak drain-to-source voltage across the PSIM FETs increases. This in turn leads to a decrease in their base capacitance requiring larger switch duty cycle resulting in even larger FET conduction losses.

The evaluation considers load impedances that do not exceed 300 V on the GaN FETs or dissipate more than 40 W per transistor, hence the effective narrowing of the tuning range with increasing power level. Measurements of the TMN efficiency and the voltage stress $V_{D1,PK}$ and $V_{D2,PK}$ on the input- and output-side PSIM devices respectively are listed in Table II for matching various load impedances to 50 Ω at 13.56 MHz and 100 W of TMN input power. The conduction angles α_1^* and α_2^* in Table II correspond to the total conduction angles of the PSIM transistors Q_1 and Q_2 respectively and include both forward and reverse device conduction.

TABLE II. MEASURED TMN SYSTEM EFFICIENCY (η) AND PEAK VOLTAGE STRESS ON THE SWITCHING DEVICES $V_{D1,PK}$, AND $V_{D2,PK}$, FOR MATCHING VARIOUS LOAD IMPEDANCES TO 50 Ω AT 13.56 MHz AND APPROXIMATELY 100 W OF TMN INPUT POWER ALONG WITH THE CORRESPONDING SWITCH CONDUCTION ANGLES α_1^* AND α_2^*.

LOAD	α_1^* (°)	α_2^* (°)	$V_{D1,PK}$ (V)	$V_{D2,PK}$ (V)	P_{IN} (W)	P_{OUT} (W)	η (%)
1+j10	184	148	170	200	108	20	18.5
1.5+j25	191	125	196	260	103	12	11.7
1.8-j0.5	0	0	216	70	98	80	81.6
5+j10	152	148	186	170	101	55	54.5
5+j40	203	117	192	272	101	20	19.8
20+j10	148	121	204	172	109	78	71.6
20+j40	187	129	202	238	108	48	44.4
40+j10	187	121	198	192	109	72	66.1
40+j40	191	98	195	220	101	54	53.5

Fig. 12 shows an oscilloscope screenshot of the TMN input and output port voltage waveforms along with the drain-to-source voltages v_{D1}, v_{D2} and gate-drive waveforms q_1, q_2 for both PSIM switches corresponding to matching a 5 + j40 Ω load impedance to 50 Ω at 13.56 MHz and 100 W of input power. As can be seen from Fig. 12, the TMN output port voltage appears to be almost purely-sinusoidal as is the case with matching most load impedances over the range shown in Fig. 11. Furthermore, Fig. 12 shows that the switching of the PSIM devices is appropriately synchronized to the respective PSIM currents resulting in ZVS operation of the switches.

As we demonstrate here, the PSIM-based TMN prototype is indeed able to match a wide range of load impedances by appropriately controlling the conduction angle of the two PSIM switches and in turn modulating the effective capacitance of the PSIM elements. We demonstrate that the switching of both PSIM devices can be successfully synchronized to the respective currents through the PSIM elements allowing for ZVS operation of the transistors over the entire matching range of the TMN. Evaluating the dynamic performance of this prototype is the subject of future work.

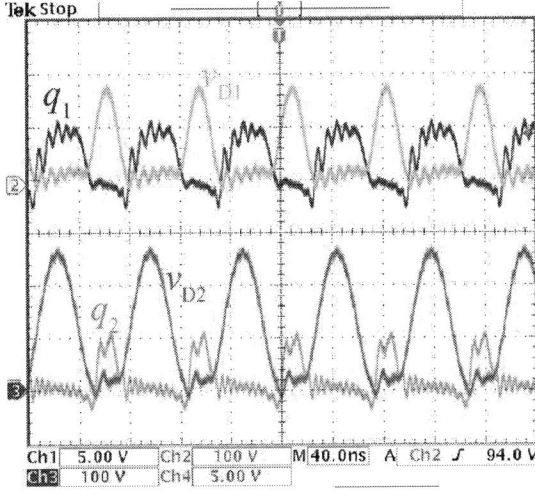

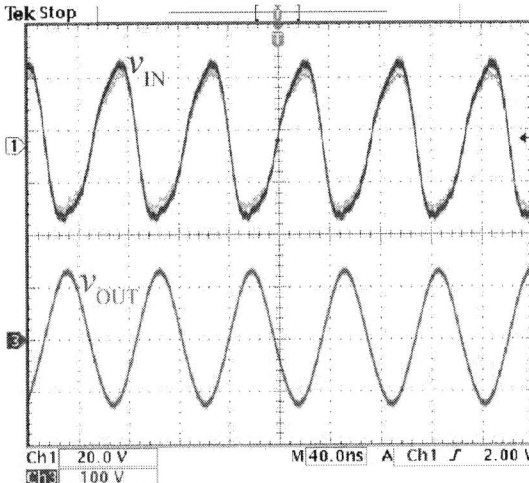

Fig. 12. Measured voltage waveforms corresponding to matching a 5+j40 Ω load impedance to 50 Ω at 100 W of TMN input power: q_1, v_{D1} – gate-drive and drain-to-source voltage respectively for the input-side PSIM switch, q_2, v_{D2} – gate-drive and drain-to-source voltage respectively for the output-side PSIM switch, v_{IN} – TMN input port voltage, v_{OUT} – TMN output port voltage.

VII. CONCLUSION

This work presents the first TMN design featuring multiple PSIM tuning elements, utilizing Π-network topology with two shunt PSIM-based capacitive tuning elements operating at 13.56 MHz and targeting inductively-coupled plasma applications. The TMN demonstrates its ability to match a wide range of load impedances at a fixed operating frequency, and up to 150 W of input RF power. This architecture enables rapid impedance tuning without the need for dynamic adjustment of the operating frequency, offering significant potential for improving efficiency in RF power delivery systems.

REFERENCES

[1] S. Sohn, J. T. Vaughan and A. Gopinath, "Auto-tuning of the RF transmission line coil for high-fields magnetic resonance imaging (MRI) systems," *2011 IEEE MTT-S International Microwave Symposium*, Baltimore, MD, 2011, pp. 1-4.

[2] A. Abuelhaija, K. Solbach and A. Buck, "Power amplifier for magnetic resonance imaging using unconventional Cartesian feedback loop," *2015 German Microwave Conference*, Nuremberg, 2015, pp. 119-122.

[3] F. H. Raab *et al.*, "Power amplifiers and transmitters for RF and microwave," in *IEEE Transactions on Microwave Theory and Techniques*, vol. 50, no. 3, pp. 814-826, March 2002.

[4] B. Regensburger *et al.*, "High-performance large air-gap capacitive wireless power transfer system for electric vehicle charging," *2017 IEEE Transportation Electrification Conference and Expo (ITEC)*, Chicago, IL, 2017, pp. 638-643.

[5] S. Sinha, A. Kumar and K. K. Afridi, "Improved design optimization of efficient matching networks for capacitive wireless power transfer systems," *2018 IEEE Applied Power Electronics Conference and Exposition (APEC)*, San Antonio, TX, 2018, pp. 3167-3173.

[6] E. Waffenschmidt, "Dynamic resonant matching method for a wireless power transmission receiver," *IEEE Transactions on Power Electronics*, vol. 30, no. 11, pp. 6070-6077, Nov. 2015.

[7] A. Al Bastami *et al.*, "Dynamic Matching System for Radio-Frequency Plasma Generation," in *IEEE Transactions on Power Electronics*, vol. 33, no. 3, pp. 1940-1951, March 2018.

[8] S. Voronin and C. Vallée, "50 Years of Reactive Ion Etching in Microelectronics," *IEEE Transactions on Materials for Electron Devices*, vol. 1, pp. 49-63, 2024

[9] Y. Lim, H. Tang, S. Lim and J. Park, "An Adaptive Impedance-Matching Network Based on a Novel Capacitor Matrix for Wireless Power Transfer," in *IEEE Transactions on Power Electronics*, vol. 29, no. 8, pp. 4403-4413, Aug. 2014.

[10] G. J. J. Winands, A. J. M. Pemen, E. J. M. van Heesch, Z. Liu and K. Yan, "Matching a pulsed power modulator to a corona plasma reactor," *2007 IEEE International Pulsed Power Conference*, pp. 587-590, June 2007.

[11] G. Bacelli, J. V. Ringwood, and P. Iordanov, "Impedance matching controller for an inductively coupled plasma chamber: L-type matching network automatic controller," *4th International Conference on Information in Control, Automation and Robotics*, Angers, France, 2007.

[12] B. Chae, J. Min, Y. Suh, J. Kim and H. Kim, "Impedance Matching Scheme of Electrical Variable Capacitors Using SiC MOSFET for 13.56MHz RF Plasma Systems," *2019 IEEE Energy Conversion Congress and Exposition (ECCE)*, Baltimore, MD, USA, 2019, pp. 392-398

[13] J. Min, B. Chae, Y. Suh, J. Kim and H. Kim, "Next-Generation Variable Capacitors to Reduce Capacitance Variable Time Using SiC MOSFETs and p-i-n Diodes in 13.56-MHz RF Plasma Systems," in *IEEE Journal of Emerging and Selected Topics in Power Electronics*, vol. 10, no. 2, pp. 1353-1362, April 2022

[14] F. C. W. Po, E. de Foucauld, D. Morche, P. Vincent, and E. Kherherve, "A novel method for synthesizing an automatic matching network and its control unit," *IEEE Transactions on Circuits and Systems I*, vol. 58, no. 9, pp. 2225-2236, Sept. 2011.

[15] M. T. Arnous, Z. Zhang, S. E. Barbin and G. Boeck, "Characterization of high voltage varactors for load modulation of GaN-HEMT power amplifier," *2015 17th International Conference on Transparent Optical Networks (ICTON)*, Budapest, Hungary, 2015, pp. 1-4

[16] A. S. Jurkov, A. Radomski and D. J. Perreault, "Tunable Matching Networks Based on Phase-Switched Impedance Modulation," *in IEEE Transactions on Power Electronics*, vol. 35, no. 10, pp. 10150-10167, Oct. 2020

[17] A. Al Bastami, A. Jurkov, D. Otten, D. T. Nguyen, A. Radomski and D. J. Perreault, "A 1.5 kW Radio-Frequency Tunable Matching Network Based on Phase-Switched Impedance Modulation," *in IEEE Open Journal of Power Electronics*, vol. 1, pp. 124-138, 2020.

[18] W. Gu, and K. Harada, "A new method to regulate resonant converters," IEEE Transactions on Power Electronics, Vol. 3, No. 4, Oct. 1988.

[19] A. S. Jurkov, *Techniques for Efficient Radio Frequency Power Conversion*, Ph.D. dissertation, Dept. of Electrical Engineering and Computer Science, Massachusetts Institute of Technology, Cambridge, MA, USA, 2019. Available: https://hdl.handle.net/1721.1/122558

Soft-Switched Pulsed Bias Plasma Supply System

Julia Estrin*, Alexander Jurkov*†, David J. Perreault*
*Massachusetts Institute of Technology
{jestrin, djperrea}@mit.edu
†MKS Instruments, Inc.
{alexander.jurkov}@mksinst.com

Abstract—**Radio Frequency (RF) generators are essential as bias voltage sources in plasma-enhanced semiconductor manufacturing processes. Employing pulsed waveforms to generate plasma offers significant improvements in manufacturing precision. However, producing these waveforms is challenging due to the need for high voltages (in the kV range), high frequencies (hundreds of kHz to low MHz), precise timing, and broadband frequency content. Traditional methods to generate these waveforms are constrained by semiconductor voltage ratings, resulting in either low-voltage outputs or complex circuits to achieve higher pulse voltages. This work introduces a simple and compact method for generating pulsed bias voltages for plasma processing, enabling reduced loss compared to the more complex systems previously described in the literature. The approach synthesizes the pulsed waveform at a low, convenient voltage and then uses an auto-transformer to step up the voltage to the desired level. A coaxial cable-based transformer having low leakage inductance is developed to provide scaling with sufficient fidelity across a wide frequency range. Zero voltage switching (ZVS) is achieved on all devices, ensuring low-loss operation. The proposed system is validated through a lab bench prototype that generates pulses of 2.1 kV at 400 kHz. The proposed system further offers both adjustable pulse duty ratio and slew rate, providing enhanced control and versatility for various applications.**

Index Terms—**Capacitively coupled plasma (CCP), pulse power generation, soft switching.**

I. INTRODUCTION

Future semiconductor manufacturing technologies must enable further miniaturization and increased production throughput of integrated circuits [1]. Essential to these advancements is the improved control of plasma-enhanced semiconductor manufacturing processes, such as plasma etching and plasma-enhanced chemical vapor deposition (PECVD). Plasma generation involves applying an RF power source across gases within a plasma chamber, as illustrated for a capacitively coupled plasma (CCP) in Figure 1a. This power source, known as the bias voltage, facilitates energetic ion bombardment of the substrate (wafer) and is a primary control mechanism for the ion energy [2]. A CCP system can be modeled as a series resistor-capacitor branch with a sheath region represented by a capacitive element, C_{sheath}, and a resistive element R_{plasma}, which sets the real output power [3].

Traditionally, plasma generation employs a sinusoidal RF signal, with modulation of ion energies through adjustments in the bias waveform's amplitude and frequency. However, this approach results in a broadband ion energy distribution function (IEDF) within the CCP, which limits etch feature profiles, etch selectivity, and the quality of PECVD films [5],

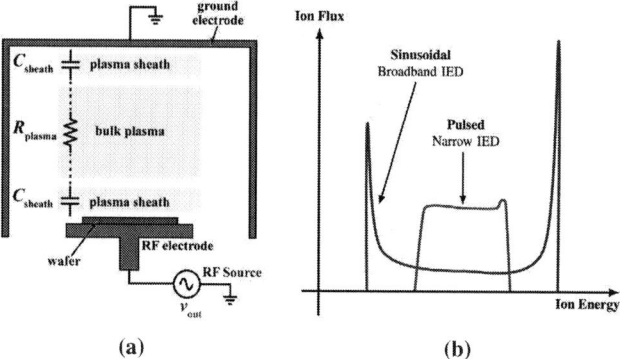

Fig. 1: (a) A representation of a capacitively coupled plasma chamber; (b) IEDFs for a traditional sinusoidal waveform and pulsed waveform [4].

[6]. In contrast, using a pulsed RF signal — a focal point of research in recent years [1] — produces a narrower IEDF, as seen in Figure 1b, enhancing etch selectivity and overall performance. Its output power is modulated via the pulse duty cycle and amplitude [5].

Generating pulsed bias waveforms for capacitively coupled plasma presents significant challenges. It requires high-frequency operation and high voltages that often exceed typical device ratings. Several techniques have been proposed for generating these waveforms. A straightforward method involves feeding the desired waveform into a linear amplifier, such as the Class-A amplifier used in [4]. However, this technique suffers from poor efficiency.

Another approach is to directly synthesize the pulse from a high-voltage dc source using current and voltage source inverters. Various techniques have been demonstrated in both literature and patents. [7] and [8] use a switch-mode inverter with multiple variable dc sources to set the output voltage. A half-bridge inverter structure to generate a positive pulse across the CCP load is patented in [9]. In [10], a current-source type pulse generator is developed, which uses series-stacked devices to maintain a high input voltage.

A multi-level inverter structure can be used to generate pulses with larger amplitudes for a given semiconductor device voltage rating. These topologies enable the use of low-voltage-rated devices but require more complex control [11]–[13]. Both [14] and [15] utilize a neutral-point clamped type inverter structure to accommodate a larger input dc supply voltage,

979-8-3315-1612-3/25 $31.00 © 2025 IEEE

thereby generating higher amplitude pulses.

Series stacking of devices and multi-level inverters are the dominant techniques to produce higher voltage pulses. However, both methods suffer from high system complexity, component count, and costs. Therefore, a need exists for a simple, compact, and inexpensive approach for generating bias pulsed waveforms.

This work presents the Bias Pulser, a pulsed bias waveform generator that is simpler and more compact than the methods previously described in the literature and which enables low-loss zero-voltage switching of the devices to be achieved. The pulsed waveform is shaped at a convenient, low voltage with a soft-switched inverter and then stepped up via an auto-transformer. This technique enables the use of lower-voltage semiconductor switching devices and leverages low-leakage magnetics design techniques to realize an auto-transformer with reduced output voltage ringing and electromagnetic interference (EMI).

Section II presents the proposed system topology, detailing the circuit operation and component selection. In Section III, a bench-top prototype of the Bias Pulser is developed to drive an equivalent model nominal plasma load. Section IV showcases experimental waveforms generated by the Bias Pulser system, accompanied by performance analysis. Finally, Section V concludes the paper.

II. PROPOSED BIAS PULSER SYSTEM

The Bias Pulser circuit design consists of a half-bridge inverter used to generate the pulsed waveform across the primary of the auto-transformer, with the output of the auto-transformer connected to the CCP load. The inverter's output voltage capability is determined by the maximum blocking voltage of the MOSFETs and the auto-transformer turns ratio. Low loss is achieved by leveraging the transformer's magnetizing current for zero-voltage switching across all devices.

A. Topology and Control

The circuit topology of the proposed Bias Pulser is depicted in Figure 2. The half-bridge inverter, consisting of MOSFETs Q_1 and Q_2 — each shown with their inherent body diode and output capacitance — produces a bipolar pulsed waveform at the node labeled v_x. The high and low side amplitudes of the pulse are defined by the dc voltage sources V_1 and V_2, respectively. If desired, either V_1 or V_2 can be replaced by a clamping capacitor. The node v_x is located on the primary side of a (tapped) auto-transformer winding. Given that this application - the generation of capacitively coupled plasma - does not require galvanic isolation, the use of an auto-transformer provides additional gain for a given amount of winding loss.

The auto-transformer is designed to have a specific magnetizing inductance L_μ (as seen from the primary winding), which facilitates soft switching of the MOSFET devices. The relative values of input voltage sources V_1 and V_2 are selected to maintain volt-second balance over the magnetizing inductance; balance is maintained automatically if one of the

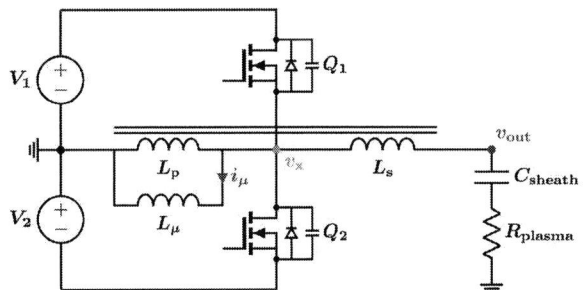

Fig. 2: Bias Pulser circuit topology.

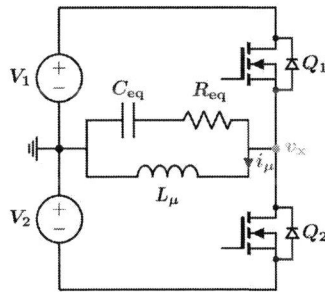

Fig. 3: Simplified ac equivalent circuit model for the Bias Pulser.

input supplies is replaced by a clamping capacitor. The slew rate of the pulsed waveform is dictated by the dead time required to achieve ZVS. It is easily modified by tuning the equivalent value of L_μ to suit varying application needs.

B. Circuit Analysis

To streamline the circuit analysis, a simplified circuit model is presented in Figure 3. The simplification involves reflecting the RC equivalent plasma load to the primary side of the transformer, according to the transformer's turns ratio, and re-referencing the MOSFET device capacitances C_{Q1} and C_{Q2} to ground. The turns ratio N for a step-up auto-transformer is defined as $(N_p + N_s)/N_p$, where N_p is the number of turns on the primary winding and N_s is the number of turns on the secondary winding. Thus, the reflected load impedance is $C_{\text{sheath,ref}} = N^2 C_{\text{sheath}}$ and $R_{\text{plasma,ref}} = R_{\text{eq}} = (1/N^2)R_{\text{plasma}}$.

As the circuit analysis focuses on the dynamic behavior of the Bias Pulser circuit, we can examine the circuit through an ac perspective, with dc voltage sources replaced by short circuits. In this configuration, the MOSFETs' output capacitances C_{Q1} and C_{Q2} appear also in parallel with the magnetizing inductance.

In many applications, the plasma resistance is negligible compared to the impedance of the capacitive sheath and is thus excluded from the analysis. The equivalent system capacitance C_{eq} can then be described by the parallel combination of $C_{\text{sheath,ref}}$, C_{Q1}, and C_{Q2}.

The operation of the Bias Pulser, using the simplified circuit shown in Figure 3, comprises four distinct stages that repeat periodically. The high and low side switches operate alternately, separated by a period of dead time. This dead time

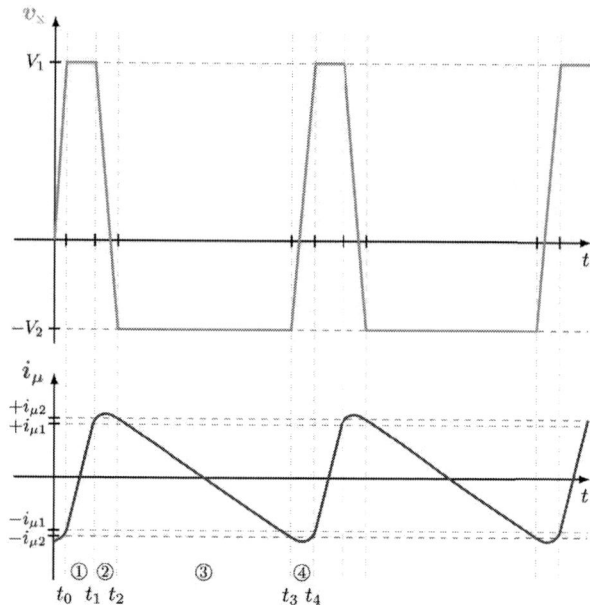

Fig. 4: Relative inverter switching node and magnetizing inductor current waveforms.

	Node Voltage (v_x)	Inductor Current (i_μ)
Stage ①	V_1	$\frac{V_1}{L_\mu}t - i_{\mu 1}$
Stage ②	$A_1 \sin(\omega_0 t) + B_1 \cos(\omega_0 t)$	$\frac{1}{\omega_0 L_\mu}[B_1 \sin(\omega_0 t) - A_1 \cos(\omega_0 t)] + C_1$
Stage ③	$-V_2$	$-\frac{V_2}{L_\mu}t + i_{\mu 2} + \frac{V_2}{L_\mu}t_2$
Stage ④	$A_2 \sin(\omega_0 t) + B_2 \cos(\omega_0 t)$	$\frac{1}{\omega_0 L_\mu}[B_2 \sin(\omega_0 t) - A_2 \cos(\omega_0 t)] + C_2$

$$A_1 = \frac{V_1 \cos(\omega_0 t_2) + V_2 \sin(\omega_0 t_1)}{\sin(\omega_0 (t_1 - t_2))} \quad A_2 = -\frac{V_1 \cos(\omega_0 t_3) + V_2 \cos(\omega_0 t_4)}{\sin(\omega_0 (t_3 - t_4))}$$

$$B_1 = -\frac{V_1 \sin(\omega_0 t_2) + V_2 \cos(\omega_0 t_1)}{\sin(\omega_0 (t_1 - t_2))} \quad B_2 = \frac{V_1 \sin(\omega_0 t_3) + V_2 \sin(\omega_0 t_4)}{\sin(\omega_0 (t_3 - t_4))}$$

$$C_1 = i_{\mu 1} + \frac{A_1 \cos(\omega_0 t_1) - B_1 \sin(\omega_0 t_1)}{\omega_0 L_\mu} \quad C_2 = -i_{\mu 2} + \frac{A_2 \cos(\omega_0 t_3) - B_2 \sin(\omega_0 t_3)}{\omega_0 L_\mu}$$

TABLE I: Bias pulse generator circuit equations for pulsed waveform voltage and inductor current, where $\omega_0 = \frac{1}{\sqrt{L_\mu C_{eq}}}$.

occurs when both switches are turned off or when clamping occurs through the MOSFET body diode. The corresponding relative waveforms for the node v_x and the magnetizing inductor current i_μ are illustrated in Figure 4.

Stage ① $(t_0 - t_1)$: This stage begins with switch Q_1 on, while Q_2 is off, clamping the node v_x to the top power supply V_1. The switch Q_1 remains on until the rising inductor current i_μ reaches the reference current $+i_{\mu 1}$, signaling the start of Stage 2.

Stage ② $(t_1 - t_2)$: This stage begins with Q_1 turning off. The inductor current begins to flow through the capacitor C_{eq}, which ensures a ZVS turn-off as the voltage across Q_1 begins to rise in a resonant manner. Both switches Q_1 and Q_2 remain off for the remainder of the stage. The inductor L_μ continues to resonate with the capacitor C_{eq} until the output voltage v_x drops to the negative rail $-V_2$. Subsequently, the body diode of switch Q_2 begins conducting and clamps v_x to the bottom supply. The switch Q_2 can then be turned on while the diode

is conducting, facilitating a zero voltage turn-on and leading into Stage 3.

Stage ③ $(t_2 - t_3)$: Mirroring Stage 1, Stage 3 involves switch Q_2 remaining on, with v_x clamped to the negative rail voltage $-V_2$. The inductor current i_μ decreases until it reaches the reference current $-i_{\mu 2}$, signaling the start of Stage 3.

Stage ④ $(t_3 - t_4)$: This stage is analogous to Stage 2. It begins with switch Q_2 turning off, which requires the inductor current to flow through the capacitor. This ensures the voltage across Q_2 rises in a resonant manner, enabling ZVS turn-off. With both switches remaining off, the voltage v_x rises until it reaches V_1. This forces the body diode of Q_1 to begin conducting, clamping v_x to the positive rail. As the body diode conducts, Q_1 is turned on with ZVS turn-on, restarting the cycle.

The semiconductor manufacturing application dictates the duration of each operational stage. Table I provides the equations describing these waveforms. User-defined parameters include the duration of each operational stage, the equivalent capacitance C_{eq}, and the voltage level of one of the two input power supplies.

The value of the magnetizing inductance L_μ is particularly critical, as it directly influences pulse rise and fall times. To achieve a faster slew rate, a smaller L_μ is required; however, it must still be large enough to facilitate ZVS of all devices. The method for selecting L_μ is explored in Section II-C.

Additionally, L_μ must maintain a volt-second balance throughout each cycle to avoid core saturation. Therefore, the ratio between the two dc input power supplies V_1 and V_2 is carefully chosen to ensure that the time integral of the pulse waveform remains zero over a complete cycle.

C. Component Value Selection

The Bias Pulser's ability to adjust pulse duty ratio and slew rate relies on precise selection of key component values, particularly the magnetizing inductance L_μ. This inductance is crucial not only for achieving the desired slew rate but also for ensuring ZVS. Additionally, the ratio between the input power supplies must satisfy the volt-second balance across L_μ. A system of equations, derived from the circuit dynamics detailed in Table I and the volt-second balance in Equation 1, is used to determine these values.

The volt-second balance equation for the magnetizing inductance L_μ, based on the equation in Table I, is as follows:

$$
\begin{aligned}
0 = {}& V_2(t_2 - t_3) - V_1(t_0 - t_1) + \\
& \frac{1}{\omega_0}\Big[A_1\big(\cos(\omega_0 t_1) - \cos(\omega_0 t_2)\big) + A_2\big(\cos(\omega_0 t_3) - \cos(\omega_0 t_4)\big) \\
& - B_1\big(\sin(\omega_0 t_1) - \sin(\omega_0 t_2)\big) - B_2\big(\sin(\omega_0 t_3) - \sin(\omega_0 t_4)\big)\Big]
\end{aligned}
\tag{1}
$$

Equation 2 defines the necessary system of equations to appropriately determine each component's value in the Bias Pulser system. A practical approach to solving these equations involves setting V_1, C_{eq}, t_0, t_1, t_2, and t_4 as known variables. The value of C_{eq} is determined by the CCP load, the transformer turns ratio, and the output capacitances of the

MOSFETs. The timing parameters are chosen based on the desired pulse shape.

$$
\begin{cases}
\omega_0 &= 1/\sqrt{L_\mu C_{\text{eq}}} \\
i_{\mu 1} &= \frac{V_1}{L_\mu} t_1 - i_{\mu 1} \\
i_{\mu 2} &= \frac{1}{\omega_0 L_\mu}[B_1 \sin(\omega_0 t_2) - A_1 \cos(\omega_0 t_2)] + C_1 \\
-i_{\mu 2} &= -\frac{V_2}{L_\mu} t_3 + i_{\mu 2} + \frac{V_2}{L_\mu} t_2 \\
-i_{\mu 1} &= \frac{1}{\omega_0 L_\mu}[B_2 \sin(\omega_0 t_4) - A_2 \cos(\omega_0 t_4)] + C_2 \\
0 &= V_2(t_2 - t_3) - V_1(t_0 - t_1) + \\
& \quad \frac{1}{\omega_0}\Big[A_1\big(\cos(\omega_0 t_1) - \cos(\omega_0 t_2)\big) + A_2\big(\cos(\omega_0 t_3) - \cos(\omega_0 t_4)\big) \\
& \quad -B_1\big(\sin(\omega_0 t_1) - \sin(\omega_0 t_2)\big) - B_2\big(\sin(\omega_0 t_3) - \sin(\omega_0 t_4)\big)\Big]
\end{cases}
\tag{2}
$$

The remaining unknowns are L_u, V_2, t_3, $i_{\mu 1}$, $i_{\mu 2}$, and ω_0. The values of $i_{\mu 1}$ and $i_{\mu 2}$ are important for determining the maximum magnetizing current, which is a key factor in transformer design. The calculated value of V_2 establishes a relative ratio between V_1 and V_2 that satisfies the volt-second balance equation. In practice, the sum of V_1 and V_2 must remain within the blocking voltage capabilities of the MOSFETs, and adjustments may be required to ensure compatibility.

Due to the complexity of the non-linear equations derived from the circuit's operation, analytical solutions are not feasible. Therefore, computer-aided numerical solvers, specifically 'lsqnonlin' available in MATLAB, are used to find solutions.

To correct any discrepancies in the selected values of V_1 and V_2, a dc blocking capacitor can be placed in series with either supply or directly with the primary transformer winding. The capacitor will charge to maintain volt-seconds balance across L_μ in periodic steady state operation. The capacitor must be sized to store sufficient energy and withstand the maximum voltage required.

Alternatively, one of the two dc power supplies can be entirely substituted with a capacitor. For example, the bottom supply V_2 is replaced by a capacitor while the top supply V_1 remains. Conversely, the bottom supply can be retained while replacing the top supply with a capacitor. L_μ will charge the capacitor to the voltage necessary to achieve volt-second balance. The capacitor must be large enough to store the required energy for a full cycle, clamping the voltage as needed. It may be beneficial to add a damping leg (e.g., a resistor or a resistor-capacitor series combination) in parallel with this capacitor to mitigate any ac oscillations, as discussed in the context of filter design in [16].

III. BIAS PULSER DESIGN

A laboratory prototype was designed and constructed based on load and operating specifications for a nominal plasma load, with the design framework outlined in Table II. These parameters guided the selection of the remaining circuit components, transformer gain, and inverter control configurations.

A. Inverter Design

To generate high-voltage pulses while minimizing the required transformer gain and its associated losses, the inverter employs silicon carbide (SiC) devices. 1200 V SiC devices were chosen for their affordability and availability. However,

Parameter	Symbol	Value
Output Voltage (peak-to-peak)	V_{out}	2.1 [kV]
Pulse Frequency	f_{sw}	400 [kHz]
Target Duty Ratio	D	10 - 30 [%]
Pulse Rise/Fall Time	t_{d}	100 - 300 [ns]
Equivalent Plasma Load	R_{plasma}	1 [Ω]
Equivalent Sheath Load	C_{sheath}	300 [pF]

TABLE II: Design specifications for Bias Pulser bench prototype.

for reliability, these devices are derated to approximately 60% of their maximum blocking voltage, resulting in an allowed dc bus voltage of 700 V. Consequently, a transformer gain of 3 is required to achieve 2.1 kV pulses. The key components selected for the Bias Pulser prototype are listed in Table III. A Kepco KLP 600-4-1.2k programmable dc power supply is used for V_1. V_2 is replaced with a $10\,\mu\text{F}$ capacitor in parallel with a $3\,\text{k}\Omega$ resistor.

	Description	Manufacturer	Manufacturer Part
Inverter	SiC MOSFETs	STMicroelectronics	SCT020H120G3AG
	Gate Driver	Infineon	1ED3124MU12HXUMA1
	Gate Isolated PSU	Recom Power	R2S-2405/R2S-2415
Plasma Load	Capacitors	KYOCERA AVX	27 x 1206GXXXXXAT
	Resistors	Bourns Inc.	10 x CRM2512-FX-10R0ELF

TABLE III: Key components used for the Bias Pulser inverter and equivalent plasma load prototypes.

Considerable effort was directed at optimizing the inverter's commutation loop to minimize additional leakage inductance in this high-frequency design, which would otherwise lead to increased ringing on the pulsed waveform and elevated EMI. Efforts to reduce the commutation loop's length were inspired by the design study in [17]. The selected layout for the half-bridge inverter, which uses a single TO-263 packaged SiC MOSFET per switch position, places the two devices in line with a return path directly below on an adjacent circuit board layer, as shown in Figure 5. Due to the soft-switched nature of the design, minimal cooling is required, so top-side surface-mount heat sinks are used.

B. Transformer Design

The Bias Pulser employs a 1:3 auto-transformer to step up the 700 V peak-to-peak pulses generated by the inverter to 2.1 kV peak-to-peak across the plasma load. A critical challenge in using a transformer for this application is ensuring that it can accurately scale the pulsed voltage with high fidelity across the broad frequency spectrum of the pulsed waveform. To this end, the transformer's design is focused on minimizing parasitic effects that could disrupt waveform synthesis, particularly aiming to reduce leakage inductance, which can contribute to poor scaling, voltage ringing, and increased EMI.

Minimizing transformer leakage inductance requires limiting the area where leakage fields can form between the primary and secondary windings, while also adhering to voltage breakdown requirements. A practical approach to achieve this is through the use of a coaxial cable, where one conductor

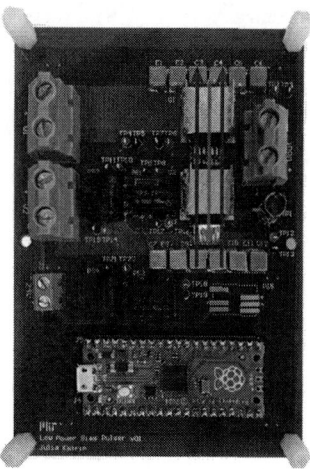

Fig. 5: Bias Pulser inverter PCB, with arrows highlighting the commutation loop.

acts as the primary winding and the other as the secondary winding.

To achieve a gain larger than 1, multiple sections of coaxial cable are employed. The primary winding is formed by connecting one conductor from each section in parallel, while the secondary winding is constructed by series connecting the other conductors from each section. For the implementation highlighted in this work, the outer conductors of the coaxial cables are connected in parallel to form the primary winding, while the inner conductors are connected in series as the secondary winding, as illustrated in Figure 6.

The only unlinked flux between the windings is confined to the space between the inner and outer conductor. Additionally, the coaxial cable provides consistent spacing between the two conductors along the entire cable's length, enabling a simple conversion of the geometrically defined distributed inductance of Equation 3 into lumped parameters that can be included in circuit analysis and simulation:

$$L'_{\text{coax}} = \frac{\mu}{2\pi} \ln\left(\frac{b}{a}\right), \qquad (3)$$

where L'_{coax} is the distributed inductance of the coaxial cable per unit length, μ is the permeability of the insulating material, a is the inner conductor radius, and b is outer conductor inner radius. Therefore, the transformer's leakage inductance is the distributed inductance of the cable multiplied by the cable length. Adjusting the cable geometry allows for controllable and predictable leakage inductance.

The 1:3 auto-transformer used for this design uses custom-made coaxial cables and an EI core configuration with specifications outlined in Table IV and shown in Figure 7. Based on Equation 3, the theoretical primary side leakage inductance for this single-turn, two-cable auto-transformer is approximately 5 nH. However, due to the additional external connections needed for the coaxial cables, the actual leakage inductance increases to 45 nH.

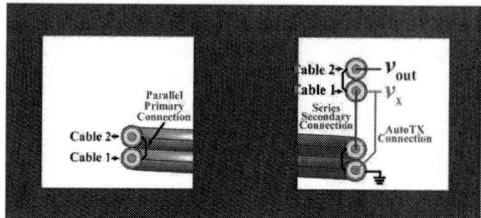

(a) Connections required to configure the coaxial cable as an auto-transformer, with a single-turn primary winding formed by the parallel connection of the outer conductors of the two cable sections and a secondary winding formed by the series connection of the inner conductors.

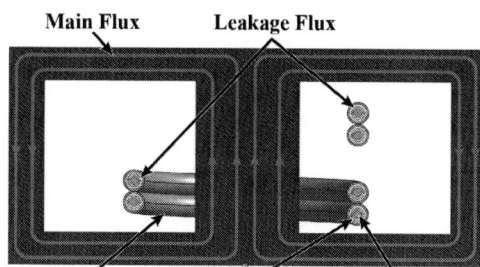

(b) Flux paths within the coaxial cable transformer, showing that leakage flux is confined to the region between the inner and outer conductors.

Fig. 6: A 1:3 auto-transformer constructed using 2 coaxial cable sections.

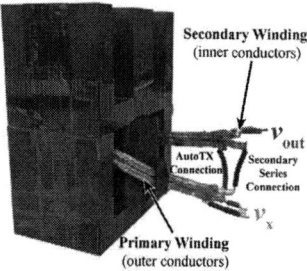

Fig. 7: The constructed 1:3 auto-transformer with specifications detailed in Table IV. The top E core is flipped to be used as a I core section.

	Parameter	Part Description
Construction	Inner Conductor	Litz - 3000 strand/46 AWG
	Insulation	10 mil thick FEP heat shrink
	Outer Conductor	3 mm Copper Braid
	EI Core	4 x TDK 100/60/28 N97

	Parameter	Symbol	Value
Specifications	Turn Ratio	N	1 : 3
	Primary Turn Count	N_p	1
	Magnetizing Inductance	L_μ	$1 - 10\,\mu\text{H}$
	Leakage Inductance	L_{lk}	45 nH
	Insulation Level		10 kV HiPot

TABLE IV: 1:3 autotransformer components and key specifications.

L_μ	$D = 20\%$	$D = 25\%$
$7\ \mu H$ $l_{\text{gap}} = 0.03\ \text{mm}$	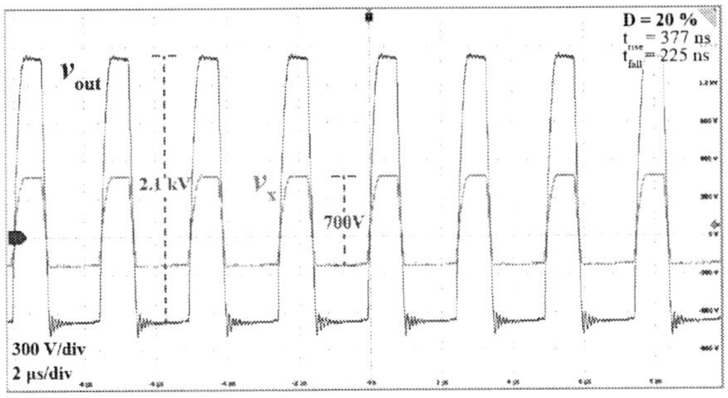	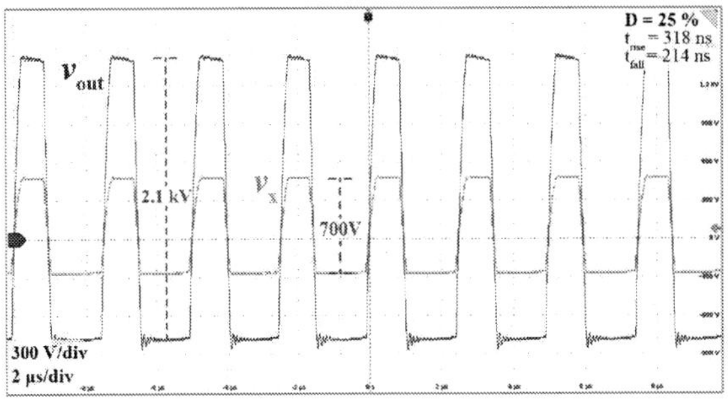
$5.5\ \mu H$ $l_{\text{gap}} = 0.06\ \text{mm}$	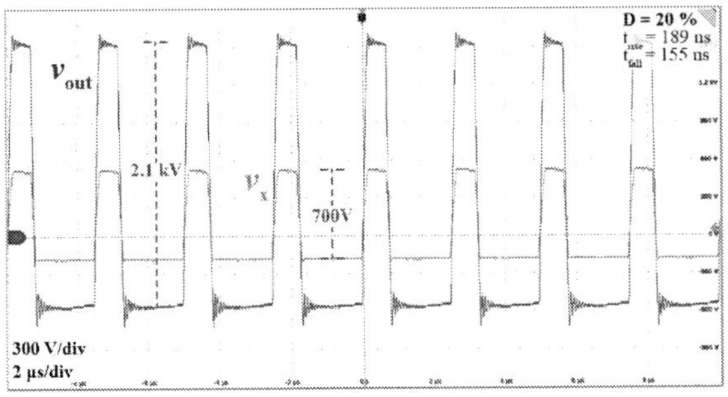	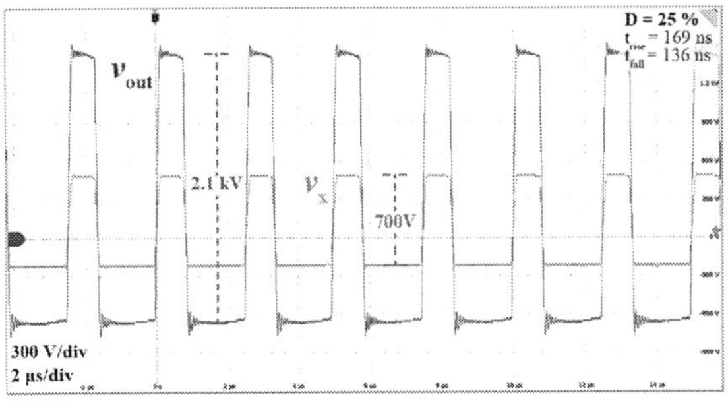
$4.5\ \mu H$ $l_{\text{gap}} = 0.09\ \text{mm}$	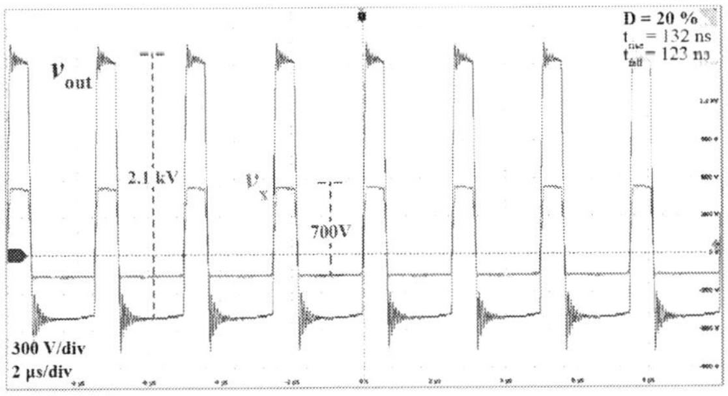	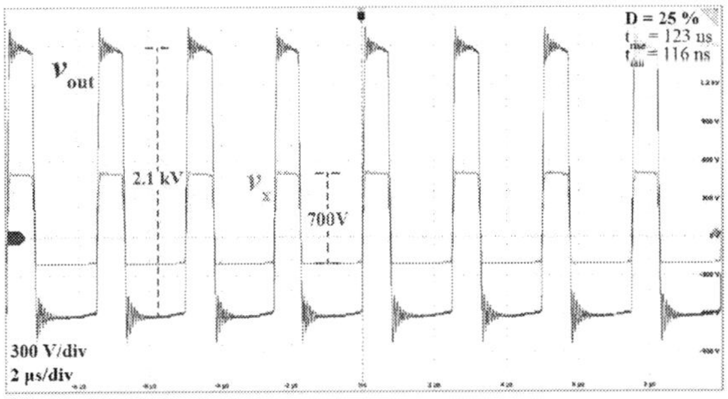

TABLE V: Oscilloscope screenshots of exemplar experimental waveforms for Bias Pulser, performing with various duty ratios and slew rates. The orange waveforms are measured at the primary of the auto-transformer and the blue waveforms are seen across the load.

IV. EXPERIMENTAL RESULTS

The Bias Pulser was operated to meet the design specifications outlined in Table II. For benchtop testing, a representative equivalent plasma load model of $300\ \text{pF}$ and $1\ \Omega$ was used, as shown in Figure 8.

Fig. 8: Equivalent RC plasma load used for benchtop testing, where $C_{\text{sheath}} = 301\ \text{pF}$ and $R_{\text{plasma}} = 1.01\ \Omega$.

The duty ratio is modulated through open-loop control of the MOSFETs' gate drive signals, and, if necessary, adjustments of the input power supplies to ensure the transformer receives the maximum voltage allowable by the inverter (700 V rail-to-rail). This work defines the duty ratio as the pulse high time divided by the period, t_{high}/T. The slew rate is controlled by tuning of the magnetizing inductance L_μ. For these experiments, L_μ is set by adjusting the core gap length. If dynamic tuning of the equivalent magnetizing inductance is required, a switched bank of inductors could be connected in parallel with L_μ.

Table V provides example waveforms with duty ratios of 20% and 25%. Three values of L_μ - $7\ \mu H$, $5.5\ \mu H$, and $4.5\ \mu H$ - are used with rise and fall times of approximately $300\ \text{ns}$, $150\ \text{ns}$, and $125\ \text{ns}$, respectively.

The Bias Pulser maintained ZVS throughout its entire operating range, evidenced by smooth transitions between the top and bottom dc voltage rails. The output waveforms exhibit appropriate gain with high fidelity and minimal ringing. For a given duty ratio, the slew rate of the pulse rise and fall time can be adjusted by tuning the value of L_μ. A smaller L_μ

allows for a higher slew rate of the pulse waveform, such that any reasonable slew rate can be achieved. A practical lower limit is set on the value of L_μ such that zero voltage switching is not compromised. Looking across a range of duty ratios for a single value L_μ, the slew rates slightly vary. This is due to the inherent coupling between the duty ratio and slew rate.

V. CONCLUSION

Using high-voltage pulsed waveforms to generate capacitively coupled plasma represents a transformative advancement for semiconductor manufacturing processes like plasma etching and plasma-enhanced chemical vapor deposition. These waveforms enable precise control over plasma energy, significantly enhancing manufacturing outcomes. However, traditional methods for generating these waveforms typically involve cumbersome, complex, and costly equipment. This paper introduces the Bias Pulser, an innovative approach to pulse generation utilizing a soft-switched half-bridge inverter operating at a low voltage, paired with a low-leakage-inductance auto-transformer. This method facilitates the generation of high-voltage pulsed waveforms for plasma applications with a compact, low-loss design.

The careful design of the Bias Pulser's PCB layout and associated magnetics allows for flexible, low-loss performance. The half-bridge inverter achieves zero voltage switching by utilizing the soft-switching current provided by the transformer's magnetizing inductance. The transformer design uses coaxial cables as windings, which confine leakage flux fields to the insulation between the inner and outer conductors, reducing transformer leakage inductance. Additionally, the system allows adjustable duty ratios and slew rates, managed through inverter control and adjustments to the magnetizing inductance.

Validation of this design was conducted using a laboratory bench prototype that drove an equivalent resistor-capacitor plasma load model ($C_{\text{sheath}} = 300\,\text{pF}$, $R_{\text{plasma}} = 1\,\Omega$). The prototype successfully generated pulsed waveforms with a peak-to-peak voltage of 2.1 kV at a frequency of 400 kHz. Demonstrations of these waveforms, featuring variable duty ratios and slew rates, effectively showcase the robust performance of the circuit and magnetic design.

REFERENCES

[1] V. M. Donnelly and A. Kornblit, "Plasma etching: Yesterday, today, and tomorrow," *Journal of Vacuum Science & Technology A*, vol. 31, no. 5, p. 050825, Sep. 2013. [Online]. Available: https://doi.org/10.1116/1.4819316

[2] S. Voronin and C. Vallée, "50 Years of Reactive Ion Etching in Microelectronics," *IEEE Transactions on Materials for Electron Devices*, pp. 1–15, 2024. [Online]. Available: https://ieeexplore.ieee.org/document/10579058/

[3] F. F. Chen and J. P. Chang, *Lecture Notes on Principles of Plasma Processing*. Boston, MA: Springer US, 2003, ch. Introduction to plasma sources, pp. 25–30. [Online]. Available: http://link.springer.com/10.1007/978-1-4615-0181-7

[4] S.-B. Wang and A. E. Wendt, "Control of ion energy distribution at substrates during plasma processing," *Journal of Applied Physics*, vol. 88, no. 2, pp. 643–646, Jul. 2000. [Online]. Available: https://pubs.aip.org/jap/article/88/2/643/488595/Control-of-ion-energy-distribution-at-substrates

[5] A. Wendt and S.-B. Wang, "Method And Apparatus for Plasma Processing with Control of On Energy Distribution at the Substrates," U.S. Patent US Patent 6,201,208 B1, Mar., 2001.

[6] K. Luu and A. Radomski, "Piecewise RF Power Systems and Methods for Supplying Pre - Distorted RF Bias Voltage Signals to an Electrode in a Processing Chamber," U.S. Patent US Patent 10,396,601 B2, Aug., 2019.

[7] Q. Yu, E. Lemmen, K. Wijnands, and B. Vermulst, "A Switched-Mode Power Amplifier for Ion Energy Control In Plasma Etching," in *2020 22nd European Conference on Power Electronics and Applications (EPE'20 ECCE Europe)*. Lyon, France: IEEE, Sep. 2020, pp. P.1–P.8. [Online]. Available: https://ieeexplore.ieee.org/document/9215859/

[8] ——, "Accurate Ion Energy Control in Plasma Processing by Switched-Mode Power Converter," in *2022 International Power Electronics Conference (IPEC-Himeji 2022- ECCE Asia)*. Himeji, Japan: IEEE, May 2022, pp. 498–505. [Online]. Available: https://ieeexplore.ieee.org/document/9806909/

[9] V. Brouk, R. Heckman, and D. J. Hoffman, "System, method and apparatus for controlling ion energy distribution," US Patent US9 287 086B2, Mar., 2016. [Online]. Available: https://patents.google.com/patent/US9287086B2/un

[10] B. Chae, J. Min, Y. Suh, H. Kim, and H. Kim, "Current-Source-Type Pulse Current Generator With Reduced Waveform Distortion for Capacitively Coupled Plasma Systems," *IEEE Transactions on Industry Applications*, vol. 57, no. 3, pp. 2578–2590, May 2021. [Online]. Available: https://ieeexplore.ieee.org/document/9345414/

[11] J. Rodriguez, Jih-Sheng Lai, and Fang Zheng Peng, "Multilevel inverters: a survey of topologies, controls, and applications," *IEEE Transactions on Industrial Electronics*, vol. 49, no. 4, pp. 724–738, Aug. 2002. [Online]. Available: http://ieeexplore.ieee.org/document/1021296/

[12] J. Rodriguez, L. Franquelo, S. Kouro, J. Leon, R. Portillo, M. Prats, and M. Perez, "Multilevel Converters: An Enabling Technology for High-Power Applications," *Proceedings of the IEEE*, vol. 97, no. 11, pp. 1786–1817, Nov. 2009. [Online]. Available: http://ieeexplore.ieee.org/document/5290111/

[13] H. Akagi, "Multilevel Converters: Fundamental Circuits and Systems," *Proceedings of the IEEE*, vol. 105, no. 11, pp. 2048–2065, Nov. 2017. [Online]. Available: http://ieeexplore.ieee.org/document/7900354/

[14] H.-j. Kim, H.-B. Kim, J.-H. Kim, J.-h. Kim, C.-H. Park, and Y.-H. Choi, "Voltage generator, voltage waveform generator, semiconductor device manufacturing apparatus, voltage waveform generation method, and semiconductor device manufacturing method," US Patent US10 516 388B1, Dec., 2019. [Online]. Available: https://patents.google.com/patent/US10516388B1/en?oq=US10516388B1

[15] J.-S. Kim, J.-K. Han, J. Kim, and G.-W. Moon, "A New High-Power Density Tailored Waveform Modulator for Plasma Processing," *IEEE Transactions on Industrial Electronics*, pp. 1–11, 2024. [Online]. Available: https://ieeexplore.ieee.org/document/10444697/

[16] J. G. Kassakian, D. J. Perreault, G. C. Verghese, and M. F. Schlecht, "26 Electromagnetic Interference and Filtering," in *Principles of Power Electronics*, 2nd ed. Cambridge University Press, Aug. 2023, pp. 771–792. [Online]. Available: https://www.cambridge.org/highereducation/product/9781009023894/book

[17] E. Persson, "Optimizing PCB Layout for HV GaN Power Transistors," *IEEE Power Electronics Magazine*, vol. 10, no. 2, pp. 65–78, Jun. 2023. [Online]. Available: https://ieeexplore.ieee.org/document/10167536/

979-8-3315-1612-3/25 $31.00 © 2025 IEEE

Analysis and Design of a Cyclo-Active-Bridge Inverter for Single-Stage Three-Phase Grid Interface

Mian Liao$^\diamond$, Tanuj Sen$^\diamond$, Yang Wu†, and Minjie Chen$^\diamond$
$^\diamond$*Princeton University, Princeton, NJ, United States*
†*ABB Corporate Research, Västerås, Sweden, United States*
Email: {mianl, minjie}@princeton.edu

Abstract—This paper presents the analysis and design of a cyclo-active-bridge (CAB) inverter for single-stage three-phase grid interface. The CAB inverter is an integration of three single-phase dual-active-bridge (DAB) converters and three cyclo converters, featuring individual cyclo-phase-shift modulation and single-stage power processing for high efficiency and fast dynamics. The topology utilizes a multi-port high-frequency ac transformer link to eliminate bulky dc-link capacitors and filter inductors while achieving isolation and bidirectional power flow for the system. The dc-side of the inverter comprises three half-bridge circuits whose switched nodes are connected in a delta configuration. The ac-side of the inverter includes three half-bridge cycloconverters, each providing independent ac outputs. The output power of each phase is regulated through phase-shifting the primary and secondary sides with decoupled power flow in the three phases, enabling a simplified control scheme and enhanced system flexibility. An experimental prototype with a 48-V dc input, three-phase 240-Vrms ac outputs, and 600-W total output power has been developed and tested. Experimental results verified the functions and efficacy of the CAB inverter.

Index Terms—three-phase inverter, high-frequency AC transformer link, dual-active-bridge (DAB) converter, cyclo converter, phase-shift modulation, zero-voltage-switching (ZVS)

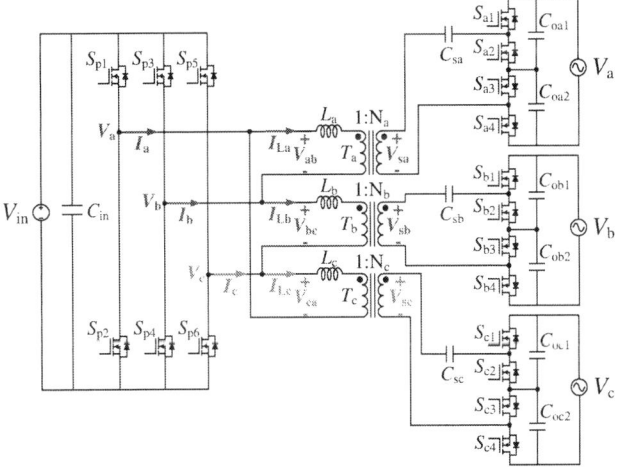

Fig. 1. Topology for three-phase CAB inverter. The primary side consists of three half-bridge circuits and three series inductors, similar to a three-phase DAB converter. The secondary side contains three independent half-bridge cycloconverters for three isolated outputs. Each cycloconverter is linked to the primary side with a high-frequency solid-state transformer. It opens opportunities for unbalanced operation at output with different voltage amplitudes or frequencies.

I. INTRODUCTION

The rapid digital transformation driven by data centers and AI cloud computing, alongside the accelerating adoption of electric vehicles (EVs), is reshaping global energy consumption patterns [1]. These technologies, while enabling smarter, more connected societies, demand unprecedented levels of energy. This surge in energy demand has placed excessive stress on the electric grid, underscoring the need for efficient, sustainable, and decentralized energy solutions. Distributed energy resources (DERs), such as solar farms, are emerging as pivotal technologies to meet these demands [2]. They can generate clean and abundant low-voltage dc power, offering localized energy resilience, and providing ideal solutions to support the energy needs of modern computing infrastructure.

For the modern grid with high DER penetrations, high-performance inverters are utilized to ensure high efficiency for dc-ac power conversion, offer fast transient response for load change, and contribute to the improved grid resiliency [3]. To interface a low-voltage dc from a solar farm with the high-voltage ac grid, the conventional approach usually employs a two-stage inverter topology [4]. The first stage performs dc-dc conversion, followed by dc-ac conversion in the second stage. While effective, this traditional design has several limitations. It requires more switches and components, leading to increased power loss and reduced power density. Additionally, the intermediate stage necessitates the use of a bulky filter capacitor to stabilize the dc link. Due to the high capacitance requirements, electrolytic capacitors are typically used, but these components have some notable drawbacks, including a short operational lifespan, considerable physical size, and poor high-frequency transient response [5], [6].

To address the limitations of multi-stage inverters, single-stage inverters with a high-frequency ac (HVAC) transformer-link have been widely explored for grid-tied applications [7], [8]. Contrary to traditional multi-stage designs, HFAC inverters eliminate the bulky dc-link capacitors and filter inductors, resulting in a more compact system and improved transient response [9]. Among various HFAC topologies, dual-active-bridge (DAB) based dc-ac inverters [10], [11] are attractive due to their inherent advantages. These include bidirectional power flow capability, galvanic isolation, and zero-voltage-switching (ZVS) over a wide operation range [12], [13], making them a compelling choice for HFAC inverters.

In DAB applications, single phase shift (SPS) modulation is

979-8-3315-1612-3/25 $31.00 © 2025 IEEE

commonly used to regulate the output voltage and current of a DAB converter [12]. This straightforward control technique is widely favored for its simplicity and ease of implementation. Power flow magnitude and direction are controlled by adjusting the phase shift between the switched-node square wave voltages of the primary and secondary sides. For enhanced performance, advanced techniques such as extended phase shift (EPS) [14], double phase shift (DPS) [15], and triple phase shift (TPS) [16] can be employed. These methods generate quasi-square wave voltages on both sides, reducing inductor current, minimizing conduction losses, and expanding the soft-switching operating range. However, they come with the trade-off of increased control complexity and the need for additional switches on both sides [17], [18].

This paper presents a single-stage three-phase cyclo-active-bridge (CAB) inverter as shown in Fig. 1, which comprises a three-phase switching unit on the primary side and several cycloconverters on the secondary side. The delta configuration on the primary side integrates the transformers and ensures the decoupled power flow control of each phase. The topology offers ZVS opportunities for both primary and secondary switches across a wide operation range. A feedback controller similar to [19], [20] was implemented to regulate the three-phase ac outputs [21]. A compact CAB inverter prototype with planar magnetics has been developed to verify the effectiveness of the key principles.

II. OPERATING PRINCIPLES

The CAB inverter integrates a three-phase DAB converter with three independent cycloconverters. The DAB configuration offers several advantages, including bidirectional power flow and galvanic isolation facilitated by a high-frequency transformer. It also supports zero-voltage-switching (ZVS) across a wide operating range, enhancing the overall efficiency. The widely used phase shift control method for the DAB converter enables rapid and precise regulation of the output voltages across a wide range.

The cycloconverter converts the high-frequency ac output generated by the switching devices to the lower frequency grid voltages. In this design, the high-frequency ac input corresponds to the device switching frequency, while the low-frequency ac output aligns with the 60 Hz grid frequency. Each cycloconverter consists of four switches, a dc blocking capacitor, and an output capacitor.

Fig. 2 illustrates the key waveforms for the proposed CAB inverter. All switches operate at a 50% duty cycle, and the phase shift between each leg on the primary side is maintained at 120°, resulting in three-level phase-to-phase voltages V_{ab}, V_{bc}, and V_{ca}. The cycloconverter can generate both positive and negative output voltages. In the positive cycle, the bottom two switches of the cycloconverter are continuously on, while the top two switches function as a half-bridge rectifier. Conversely, during the negative cycle, the top two switches are continuously on, and the bottom two switches serve as the rectifier. In the illustrated waveforms, the system is operating in the positive cycle for all phases, hence the signals for switches S_{a3}, S_{a4}, S_{b3}, S_{b4}, S_{c3}, and S_{c4} are maintained high. For Phase A,

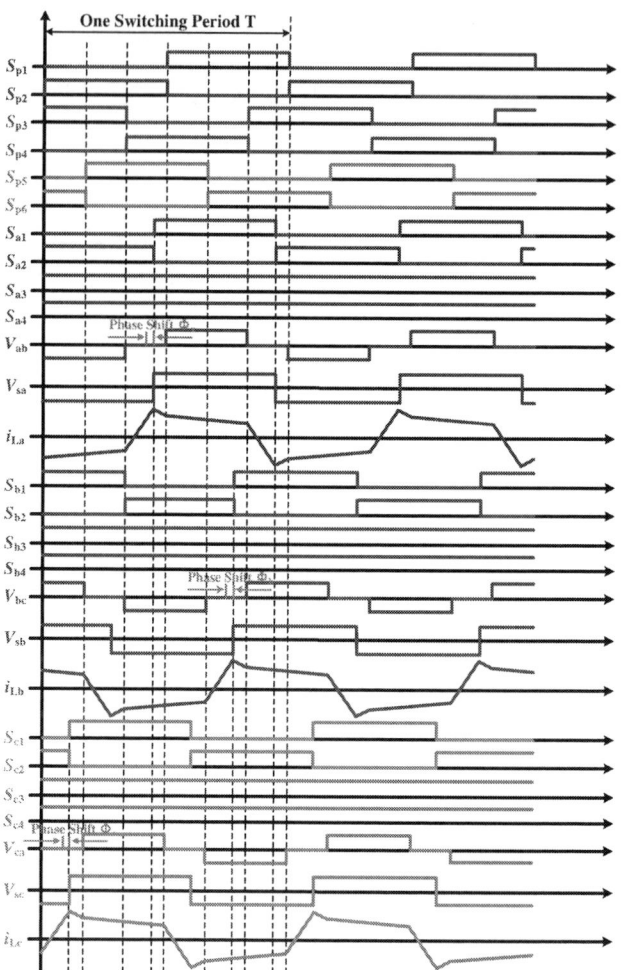

Fig. 2. Key waveforms for the proposed three-phase inverter. SPS modulation scheme is used to regulate the output voltage of each phase. Gate signals of switches and switched-node voltages are shown here.

the phase shift ϕ_a for controlling the output power is defined as the phase difference between the fundamental components of V_{ab} and V_{sa}. The reference point for the phase shift is set as 30° ahead of the turn-on action of the primary side's top switches. When the phase shift is set to 0°, no power is delivered to the secondary side. For Phase B, the phase shift ϕ_b is defined in the same way between V_{bc} and V_{sb}. For Phase C, the phase shift ϕ_c is defined in the same way between V_{ca} and V_{sc}. The microprocessor generates these phase shifts, allowing the output power for each phase to be controlled solely by the corresponding phase shift between inverter and rectifiers. This design feature enables decoupled power flow control, resulting in a flexible and robust system.

III. POWER FLOW ANALYSIS FOR THE CAB INVERTER

Since the power flow of each phase is decoupled, the three-phase topology can be simplified to a single-phase model for analyzing the relationship between the phase shift and output power. Fig. 3 (a) illustrates the equivalent circuit for Phase A. On the primary side, a full bridge operates with

979-8-3315-1612-3/25 $31.00 © 2025 IEEE   140

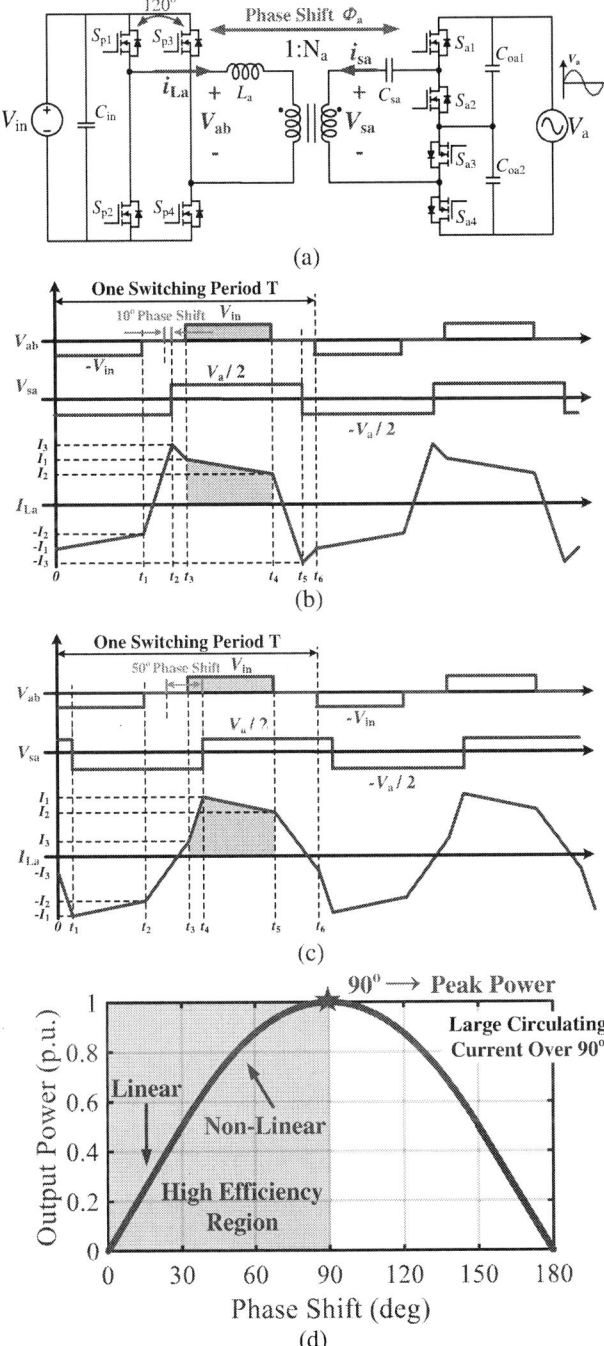

(a)

(b)

(c)

(d)

Fig. 3. (a) Equivalent circuit for a single-phase operation. (b) Key waveforms for 10°(< 30°) phase shift (c) Key waveforms for 50°(> 30°) phase shift.

50% duty cycle and 120° phase shifts between each leg, creating three-level voltage V_{ab}. On the secondary side, the half-bridge cycloconverter produces two-level voltage V_{sa}. The system operates under two conditions: 1) when the phase shift ϕ_a ranges from 0° to 30°, as shown in Fig. 3 (b); and 2) when the phase shift ϕ_a ranges from 30° to 90°, as shown in Fig. 3 (c). $t_1 \sim t_6$ represents the transition time between different switching sequences. The output power reaches its maximum at 90° phase shift. Beyond this point, the system

primarily generates reactive power which limits the overall system efficiency. The series inductance is represented as L_a and its current is I_{La}. The switching period is T, and its corresponding switching frequency is f_{sw}. The turns ratio is N_a. The phase duty cycle is defined as $D_a = \phi_a/180°$. To calculate the power flow of one phase in a three-phase CAB inverter, two operating conditions are analyzed separately.

A. Power Flow Analysis Under $0 \leq \phi_a \leq 30°$

Fig. 4 illustrates the mode analysis under condition 1, focusing specifically on the positive half cycle of the output ac waveforms. In this case, the secondary side switches S_{a1} and S_{a2} operates as a half-bridge circuit, while switches S_{a3} and S_{a4} remains continuously on. Consequently, the output capacitance C_{oa2} is bypassed. In contrast, during the negative half cycle, S_{a3} and S_{a4} function as a half-bridge circuit, with S_{a1} and S_{a2} continuously on.

In Fig. 4, Modes 1 to 6 represent time periods from $t_0 \sim t_6$, respectively. The change in inductor current for each mode can be determined based on the voltage across the inductor. Since Mode 1 to 3 and Mode 4 to 6 are symmetrical, analyzing only Modes 1 to 3 is sufficient to explain the full cycle operation. These relationships are expressed by the following equations:

$$
\begin{cases}
\dfrac{V_a}{2N_a} - V_{in} = L_a \cdot \dfrac{I_1 - I_2}{T/3} \\[2mm]
\dfrac{V_a}{2N_a} = L_a \cdot \dfrac{I_2 + I_3}{(1/12 + D_a/2)T} \\[2mm]
\dfrac{V_a}{2N_a} = L_a \cdot \dfrac{I_3 - I_1}{(1/12 - D_a/2)T}
\end{cases}
\tag{1}
$$

where I_1, I_2, and I_3 are inductor currents during the switching transitions between Modes 1, 2 and 3, as shown in Fig. 3 (a).

The inductor currents can be expressed as:

$$
\begin{cases}
I_1 = \dfrac{(3D_a + 1)V_a - 2N_aV_{in}}{12N_aL_a}T \\[2mm]
I_2 = \dfrac{(3D_a - 1)V_a + 2N_aV_{in}}{12N_aL_a}T \\[2mm]
I_3 = \dfrac{3V_a - 4N_aV_{in}}{24N_aL_a}T
\end{cases}
\tag{2}
$$

The output power can be calculated with the input voltage and current during one cycle:

$$
P_{out} = \frac{\int_0^T V_{in}I_{La}(t)dt}{T}
\tag{3}
$$

Then the output power can be represented with the input voltage V_{in} and output voltage V_a:

$$
P_{out} = \frac{D_aV_{in}V_a}{6f_{sw}N_aL_a}
\tag{4}
$$

The relationship between the input voltage and the output voltage can be rewritten as:

$$
\frac{V_{out}}{V_{in}} = \frac{D_aR_{out}V_a}{6f_{sw}N_aL_a}
\tag{5}
$$

where R_{out} is the equivalent output resistance on the ac side.

979-8-3315-1612-3/25 $31.00 © 2025 IEEE   141

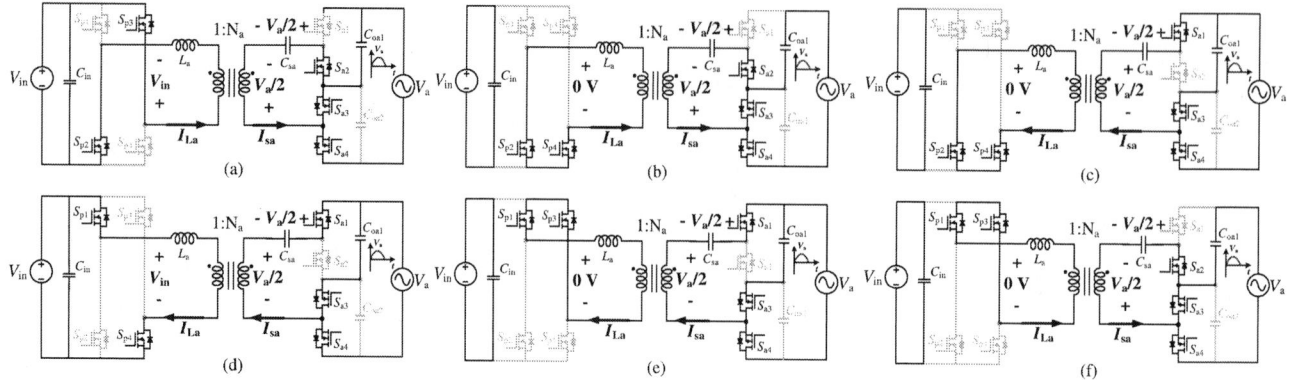

Fig. 4. For Condition 1 ($0 \leq \phi_a \leq 30°$), operating mode analysis during positive half cycle of the output ac waveform: (a) Mode 1; (b) Mode 2; (c) Mode 3; (d) Mode 4; (e) Mode 5; (f) Mode 6.

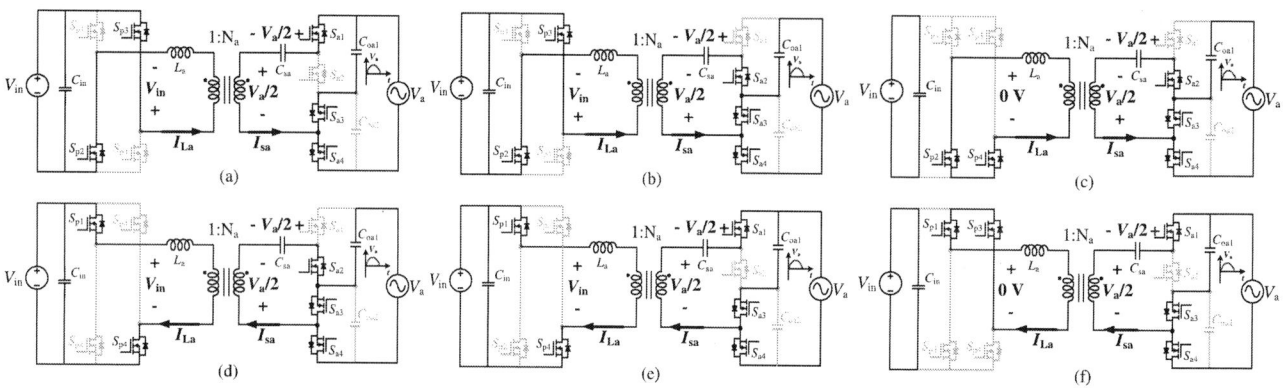

Fig. 5. For Condition 2 ($30° < \phi_a \leq 90°$), operating mode analysis during positive half cycle of the output ac waveform: (a) Mode 1; (b) Mode 2; (c) Mode 3; (d) Mode 4; (e) Mode 5; (f) Mode 6.

B. Power Flow Analysis Under $30° < \phi_a \leq 90°$

Fig. 5 illustrates the mode analysis under condition 2, focusing specifically on the positive half cycle of the output ac waveforms. In this case, the secondary side switches S_{a1} and S_{a2} operates as a half-bridge circuit, while switches S_{a3} and S_{a4} remains continuously on. Consequently, the output capacitance C_{oa2} is bypassed. In contrast, during the negative half cycle, S_{a3} and S_{a4} function as a half-bridge circuit, with S_{a1} and S_{a2} continuously on.

Modes 1 to 6 represent time periods from $t_0 \sim t_6$, respectively. The change in inductor current for each mode can be determined based on the voltage across the inductor. Similarly, analyzing only Modes 1 to 3 is sufficient to explain the full cycle operation due to symmetry:

$$\begin{cases} \dfrac{V_a}{2N_a} - V_{in} = L_a \cdot \dfrac{I_1 - I_2}{(5/12 - D_a/2)T} \\[2mm] \dfrac{V_a}{2N_a} = L_a \cdot \dfrac{I_2 + I_3}{T/6} \\[2mm] \dfrac{V_a}{2N_a} + V_{in} = L_a \cdot \dfrac{I_1 - I_3}{(D_a/2 - 1/12)T} \end{cases} \quad , \quad (6)$$

where I_1, I_2, and I_3 are inductor currents during the switching transitions shown in Fig. 3 (a).

The inductor currents can be expressed as:

$$\begin{cases} I_1 = \dfrac{V_a + (4D_a - 2)N_a V_{in}}{8N_a L_a}T \\[2mm] I_2 = \dfrac{(3D_a - 4)V_a + 2N_a V_{in}}{12N_a L_a}T \\[2mm] I_3 = \dfrac{(1 - 3D_a)V_a - 2N_a V_{in}}{12N_a L_a}T \end{cases} . \quad (7)$$

The output power can be calculated with the input voltage and current during one cycle:

$$P_{out} = \dfrac{\int_0^T V_{in} I_{La}(t)dt}{T}. \quad (8)$$

Then the output power can be represented with the input voltage V_{in} and output voltage V_a:

$$P_{out} = \dfrac{V_{in} V_{out}}{4N_a L_a f_{sw}}\left[\dfrac{2}{3}D_a - (D_a - \dfrac{1}{6})^2\right]. \quad (9)$$

The relationship between the input voltage and the output voltage can be rewritten as:

$$\dfrac{V_{out}}{V_{in}} = \dfrac{R_{out}}{4N_a L_a f_{sw}}\left[\dfrac{2}{3}D_a - (D_a - \dfrac{1}{6})^2\right]. \quad (10)$$

where R_{out} is the equivalent output resistance on the ac side.

In condition 1), the phase shift controls the output power

TABLE I
PROTOTYPE PARAMETERS

Parameter	Description	Value
V_{in}	DC input voltage	48 V_{dc}
V_{out}	AC output voltage	240 V_{rms}
f_{sw}	switching frequency	200 kHz
$L_a/L_b/L_c$	series inductor	5 μH
$N_a/N_b/N_c$	turns ratio	3:12
$C_{sa}/C_{sb}/C_{sc}$	DC blocking capacitor	3 μF
$C_{oa1}/C_{ob1}/C_{oc1}$	output filter capacitor	3 μF
$C_{oa2}/C_{ob2}/C_{oc2}$	output filter capacitor	3 μF

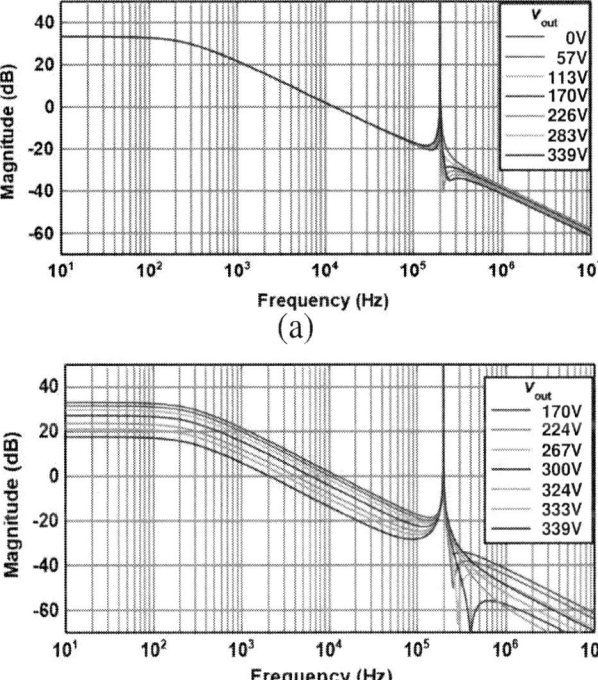

Fig. 6. Magnitude response of the per-phase small-signal transfer function $G_{v_{out}\phi a}(s)$ of the CAB inverter for different output voltages: (a) in the linear operating condition; (b) in the non-linear operating condition. The gain of the system varies with the output voltage.

linearly. In condition 2), the phase shift controls the output power non-linearly but monotonically. The maximum power is achieved in condition 2 when $\phi_a = 90°$ or $D_a = 1/2$ as shown in Fig. 3 (d). When ϕ_a sweeps from 90° to 180°, the circulating current greatly increases but the output power gradually drops. To reduce the conduction loss and maintain high efficiency, it is desirable to operate the system within the range from 0° to 90°. The same principles and equations apply to Phase B and Phase C, with their respective phase shifts.

C. Closed-Loop Power Flow Control

In order to generate continuous ac output waveforms, the output voltages needs to closely and rapidly follow a sinusoidal reference signal. To gain a deeper understanding of the system's behavior, it is essential to derive its transfer function. Through small-signal analysis, the system transfer function can be expressed as follows:

$$G_{vout\phi a}(s) = \frac{1}{D}(2V_{out}R_{out}\omega_s - $$
$$\pi N_a L_a R_{out}(s^2 + \omega_s^2)(I_{pc}\cos\phi_a + I_{ps}\sin\phi_a) + \quad (11)$$
$$4V_{in}\sin\frac{\pi}{3}N_a R_{out}\lambda_a(\omega_s\sin\phi_a + s\cos\phi_a).$$

where I_{pc} and I_{ps} represent the coefficients of the cosine and sine terms, respectively, for the fundamental frequency component of the inductor current. Parameter D is defined as:

$$D = 2R_{out}s + L_a N_a^2 \pi^2 (s^2 + \omega_s^2)(C_{oa1}R_{out}s + 1). \quad (12)$$

Parameter λ serves as an adjustment factor to account for the influence of high-frequency harmonic components. Under condition 1, corresponding to linear operation, the parameter λ is defined as:

$$\lambda = \frac{\pi^2}{12\sin\frac{\pi}{3}\sin\phi_a} - \frac{\cos\phi_a}{\sin\phi_a}. \quad (13)$$

Under condition 2, corresponding to non-linear operation, the parameter λ is defined as:

$$\lambda = \frac{\pi(\pi - 2\phi_a)}{8\sin\frac{\pi}{3}\sin\phi_a} - \frac{\cos\phi_a}{\sin\phi_a}. \quad (14)$$

Fig. 6 (a) shows the magnitude response of the system under linear operating conditions while Fig. 6 (b) shows the magnitude response of the system under non-linear operating conditions. The gain of the system varies with the output voltage in both conditions. The detailed derivation for the transfer function and the design for the closed-loop controller can be found in [21].

IV. ZERO VOLTAGE SWITCHING

Achieving ZVS, especially on the high-voltage side, is important for reducing the turn-on loss of all switches in a high-frequency design. The turn-on current of switches must be negative to ensure the discharging of parasitic capacitors during dead time. For switches on the primary side, for example S_{p1}, the turn-on current is $I_{p1} = I_a = I_{La} - I_{Lc}$ as shown in Fig. 1. It depends on the currents flowing through both Phase A and Phase C, thus determined by both phase shift ϕ_a and ϕ_c. The circuit parameters used in this study are listed in Table. I. Fig. 7 illustrates the turn-on currents of switch S_{p1} under varying phase shift combinations of ϕ_a and ϕ_c, as well as different load conditions. The turn-on current I_{p1} remains negative across all operating scenarios, ensuring that S_{p1} consistently achievs ZVS during its turn-on period. By carefully optimizing the circuit parameters, ZVS can be guaranteed for all primary-side switches across different phase shift combinations of ϕ_a, ϕ_b, and ϕ_c.

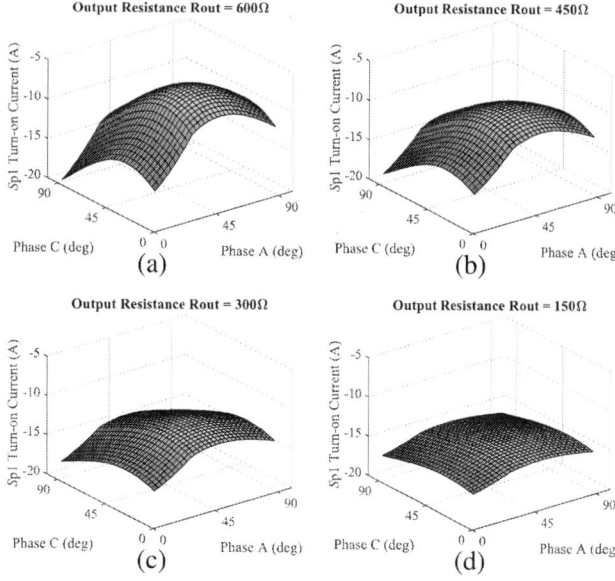

Fig. 7. Turn-on currents of switch S_{p1} with different phase shift combination of ϕ_a and ϕ_c, as well as different load conditions: (a) when $R_{out} = 600$ Ω; (b) when $R_{out} = 450$ Ω; (c) when $R_{out} = 300$ Ω; (d) when $R_{out} = 150$ Ω.

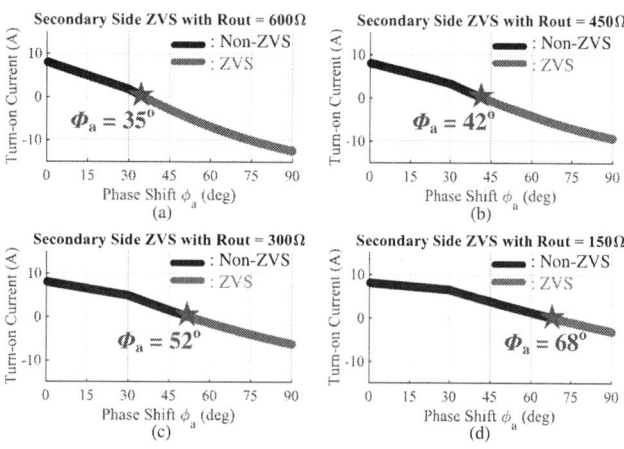

Fig. 8. Turn-on currents of switch S_{a1} with different phase shifts ϕ_a, as well as different load conditions: (a) when $R_{out} = 600$ Ω; (b) when $R_{out} = 450$ Ω; (c) when $R_{out} = 300$ Ω; (d) when $R_{out} = 150$ Ω.

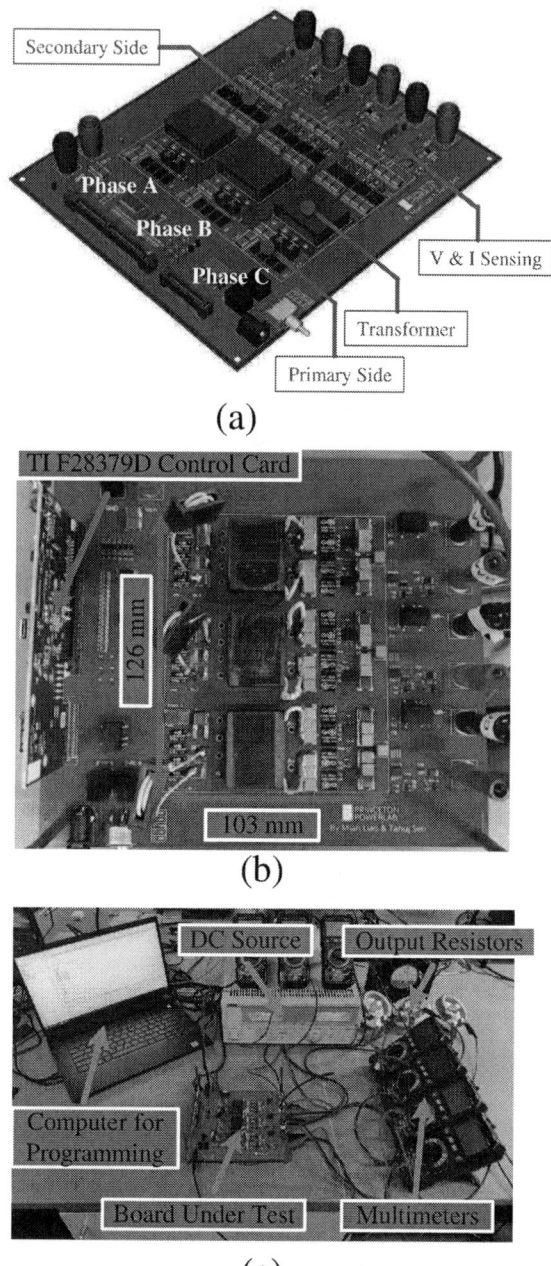

Fig. 9. (a) 3D View of the prototype for the three-phase CAB inverter. (b) Top side view of the prototype, highlighting the power stage dimension of 126 mm × 103 mm. (c) Experimental setup used for testing the proposed three-phase inverter.

For switches on the secondary side, the turn-on current of switch S_{a1} is given by $I_{a1} = I_{sa}$ as shown in Fig. 3 (a). This current is only determined by the primary side phase current I_{La} and the phase shift ϕ_a. Fig. 8 demonstrates the turn-on currents of S_{a1} under varying phase shifts of ϕ_a and different load conditions. The critical points between ZVS and non-ZVS regions are marked with blue stars in the Fig, 8. ZVS on the secondary side is achievable when the phase shift exceeds a critical threshold. Addiitonally, as the load resistance and output power increase, the system achieves a wider range of ZVS under phase shift modulation. In conclusion, the design is well-suited for high-voltage and high-power applications, as ZVS can be reliably achieved under these operating conditions.

V. EXPERIMENTAL VERIFICATION

A three-phase inverter was designed and tested to track voltage reference at 60 Hz. The parameters designed for hardware implementation are shown in Table I. The system is powered by a 48 V dc input, producing three-phase 240 Vrms outputs. The pulse-width-modulation (PWM) signals operate at a switching frequency of 200 kHz, generated by the Texas Instruments F28379D control card. On the primary side, 100 V GaN Systems GS61008Ps are utilized as switching devices. To

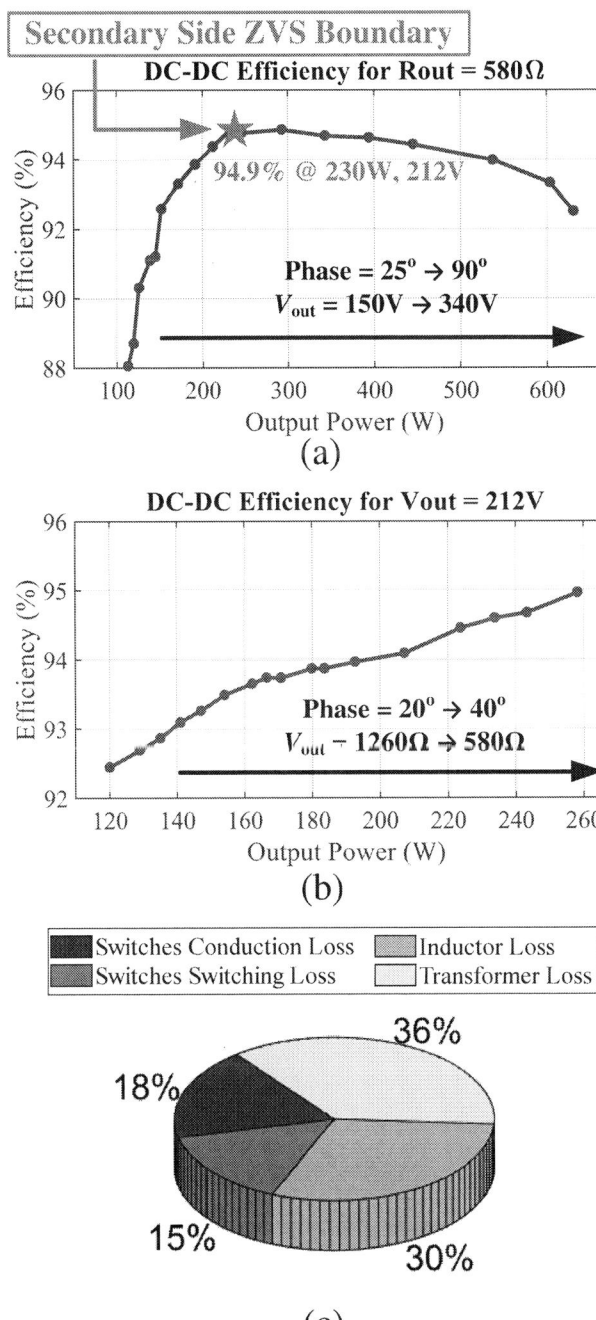

Fig. 10. (a) Dc-dc efficiency with different phase shifts and different output voltages under 580 Ω output resistance. Gating losses are included. (b) Dc-dc efficiency with a consistent output voltage by adjusting phase shifts and output resistance. Gating losses are included. (c) Loss breakdown at the secondary side's ZVS critical point under 580 Ω output resistance.

reduce turn-on resistance and conduction losses, each switch consists of two parallel devices. On the secondary side, 650V GaN Systems GS66508Bs are utilized as the switching device, handling peak output voltages of $240\sqrt{2}$ V (339 V). Both sides employ the isolated half-bridge gate driver 2EDF7275K. The isolated power supply XP ISE0505A is used on the secondary side to support gate drivers and sensors with floating grounds. A series inductance of 5 μH is chosen to satisfy the required

output power and ensure ZVS operation on the primary side. It is designed with planar magnetics Ferroxcube ER23/3.6/13 as the core and litz wires as the winding. The system's designed parameters are consistent across all three phases, achieving symmetrical operation and uniform output power. The high-frequency transformer, built with planar magnetics Ferroxcube ER32/6/25 in 3C92 material. The turns ratio is designed as 3:12 for stepping up the voltage. PCB windings are used on the primary side, while Litz wires are employed on the secondary side. The dc blocking capacitor is selected as 3 μF on the secondary side. This capacitance is sufficiently large to prevent dc current from saturating the magnetic core without affecting the characteristics of the system. Additionally, a 3 μF output filter capacitor is used to minimize output voltage ripple while maintaining the dynamic performance.

Fig. 9 (a) shows a 3D view of the prototype for the three-phase CAB inverter, showcasing the symmetrical design of phases A, B, and C on the board. The high-frequency solid-state transformers are centrally located between the primary and secondary sides. Voltage and current sensors are placed between the secondary side and the outputs to provide feedback for closed-loop control. Fig. 9 (b) shows the top view of the prototype. The power stage dimension is 126 mm × 103 mm, featuring a small size and low profile. Fig. 9 (c) illustrates the experimental setup used for testing the prototype.

Fig. 10 (a) shows the measured dc-dc efficiency of the three-phase CAB inverter with 48 V input and three 580 Ω output load resistance. For efficiency measurement, the phase shifts are varied from 25° to 90°, resulting in 150 V - 350 V output voltages and 110 W - 630 W output power. The 94.9% peak efficiency under 230 W output power is achieved at the boundary between ZVS and non-ZVS operation for the secondary side. The efficiency at the 630 W full load is about 92.5%. Fig. 10 (b) presents the measured dc-dc efficiency with a fixed output voltage of 212 V. To maintain this output voltage, the phase shift and output power are adjusted accordingly in open-loop. The results indicate that efficiency improves as output power increases. Gate drive losses and other auxiliary losses are not included in the efficiency measurements.

Fig. 10 (c) provides a breakdown of power losses at 230 W output power, 212 V output voltage, and 580 Ω output resistance, which corresponds to the critical point for achieving ZVS on the secondary side. At this operating condition, power loss are attributed to turn-off switching loss (15%) and conduction loss (18%) of the switches on both the primary and secondary sides. Additionally, significant portions of the power loss originate from inductor (30%) and transformer (36%). Further optimization of magnetic components can enhance overall efficiency.

Fig. 11 (a) shows the switched-node voltages V_{ab} and V_{sa}, and the inductor current I_{La} of Phase A with 48 V – 200 V dc-dc operation. ZVS is achieved for both primary and secondary sides, ensuring minimized turn-on voltage spikes. Fig. 11 (b) shows the closed-loop control waveforms for three-phase ac output voltages with 60 Hz and 48 V peak voltage. A PID feedback controller is used to reduce total harmonic distortion (THD) of the output voltage. Details about the controller implementation and small signal models are provided in [21].

979-8-3315-1612-3/25 $31.00 © 2025 IEEE

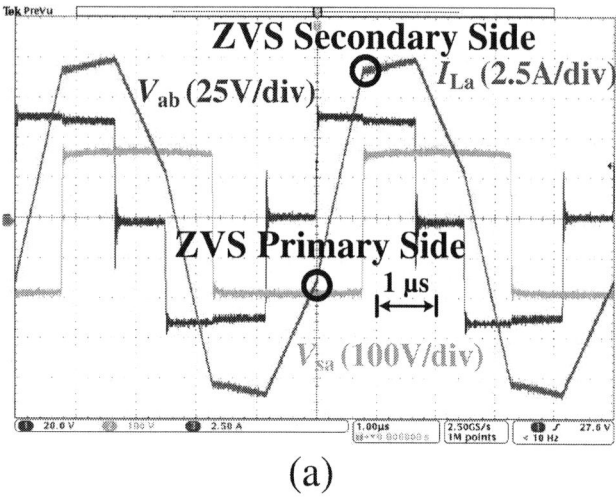

(a)

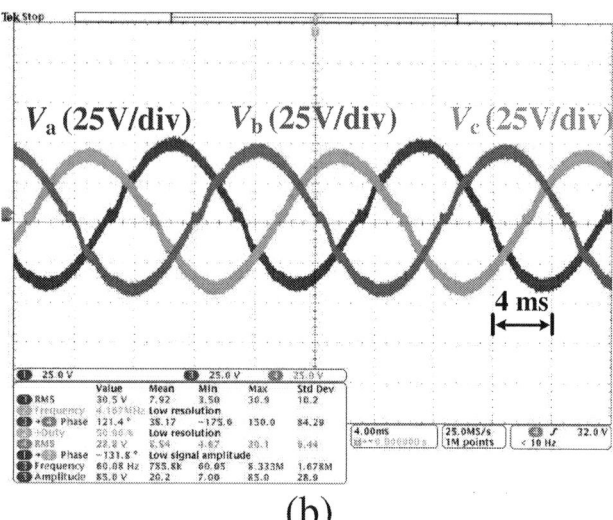

(b)

Fig. 11. (a) Waveforms for the switched-node voltages V_{ab} and V_{sa}, and the inductor current I_{La} in Phase A. Both the primary side and the secondary side achieve ZVS operation. (b) Three-phase output voltage waveforms with 60 Hz and 48 V peak voltage.

VI. CONCLUSIONS

This paper presents the theory, design methods, and hardware implementation of the cycle-active-bridge converter - a single-stage three-phase inverter. The operating principles of the inverter and the decoupled power flow control are illustrated. The single-phase circuit is analyzed and the power flow equations are calculated under different operating conditions. ZVS conditions are discussed for both the primary side and the secondary side across various operation ranges. A hardware prototype was designed and built with planar magnetics and compact system size. Experimental results are shown to validate the three-phase operation and high dc-dc efficiency of the proposed inverter.

ACKNOWLEDGEMENTS

This work was supported by NSF ASCENT Grant Award Number #2328241 and the NJEDA Wind Institute Fellowship.

REFERENCES

[1] Y. C. Lee and A. Y. Zomaya, "Energy efficient utilization of resources in cloud computing systems," *The Journal of Supercomputing*, vol. 60, no. 2, pp. 268–280, 2010.

[2] J. Huang, C. Jiang, and R. Xu, "A review on distributed energy resources and microgrid," *Renewable and Sustainable Energy Reviews*, vol. 12, no. 9, pp. 2472–2483, 2008.

[3] N. Pogaku, M. Prodanovic, and T. C. Green, "Modeling, analysis and testing of autonomous operation of an inverter-based microgrid," *IEEE Transactions on Power Electronics*, vol. 22, no. 2, pp. 613–625, 2007.

[4] G. Ding, F. Gao, H. Tian, C. Ma, M. Chen, G. He, and Y. Liu, "Adaptive dc-link voltage control of two-stage photovoltaic inverter during low voltage ride-through operation," *IEEE Transactions on Power Electronics*, vol. 31, no. 6, pp. 4182–4194, 2016.

[5] M. Chen, K. K. Afridi, and D. J. Perreault, "Stacked switched capacitor energy buffer architecture," *IEEE Transactions on Power Electronics*, vol. 28, no. 11, pp. 5183–5195, 2013.

[6] ——, "A multilevel energy buffer and voltage modulator for grid-interfaced microinverters," *IEEE Transactions on Power Electronics*, vol. 30, no. 3, pp. 1203–1219, 2015.

[7] A. Trubitsyn, B. J. Pierquet, A. K. Hayman, G. E. Gamache, C. R. Sullivan, and D. J. Perreault, "High-efficiency inverter for photovoltaic applications," *2010 IEEE Energy Conversion Congress and Exposition*, pp. 2803–2810, 2010.

[8] D. Das, N. Weise, K. Basu, R. Baranwal, and N. Mohan, "A bidirectional soft-switched dab-based single-stage three-phase ac–dc converter for v2g application," *IEEE Transactions on Transportation Electrification*, vol. 5, no. 1, pp. 186–199, 2019.

[9] A. K. Bhattacharjee and I. Batarseh, "Sinusoidally modulated ac-link microinverter based on dual-active-bridge topology," *IEEE Transactions on Industry Applications*, vol. 56, no. 1, pp. 422–435, 2020.

[10] M. Wang, S. Guo, Q. Huang, W. Yu, and A. Q. Huang, "An isolated bidirectional single-stage dc–ac converter using wide-band-gap devices with a novel carrier-based unipolar modulation technique under synchronous rectification," *IEEE Transactions on Power Electronics*, vol. 32, no. 3, pp. 1832–1843, 2017.

[11] P. Morsali, S. Dey, A. Mallik, and A. Akturk, "Switching modulation optimization for efficiency maximization in a single-stage series resonant dab-based dc–ac converter," *IEEE Journal of Emerging and Selected Topics in Power Electronics*, vol. 11, no. 5, pp. 5454–5469, 2023.

[12] R. De Doncker, D. Divan, and M. Kheraluwala, "A three-phase soft-switched high-power-density dc/dc converter for high-power applications," *IEEE Transactions on Industry Applications*, vol. 27, no. 1, pp. 63–73, 1991.

[13] M. Liao, H. Li, P. Wang, T. Sen, Y. Chen, and M. Chen, "Machine learning methods for feedforward power flow control of multi-active-bridge converters," *IEEE Transactions on Power Electronics*, vol. 38, no. 2, pp. 1692–1707, 2023.

[14] B. Zhao, Q. Yu, and W. Sun, "Extended-phase-shift control of isolated bidirectional dc–dc converter for power distribution in microgrid," *IEEE Transactions on Power Electronics*, vol. 27, no. 11, pp. 4667–4680, 2012.

[15] B. Zhao, Q. Song, W. Liu, and W. Sun, "Current-stress-optimized switching strategy of isolated bidirectional dc–dc converter with dual-phase-shift control," *IEEE Transactions on Industrial Electronics*, vol. 60, no. 10, pp. 4458–4467, 2013.

[16] N. Hou, W. Song, and M. Wu, "Minimum-current-stress scheme of dual active bridge dc–dc converter with unified phase-shift control," *IEEE Transactions on Power Electronics*, vol. 31, no. 12, pp. 8552–8561, 2016.

[17] H. Qin and J. W. Kimball, "Generalized average modeling of dual active bridge dc–dc converter," *IEEE Transactions on Power Electronics*, vol. 27, no. 4, pp. 2078–2084, 2012.

[18] S. S. Shah and S. Bhattacharya, "A simple unified model for generic operation of dual active bridge converter," *IEEE Transactions on Industrial Electronics*, vol. 66, no. 5, pp. 3486–3495, 2019.

[19] H. Qin and J. W. Kimball, "Closed-loop control of dc–dc dual-active-bridge converters driving single-phase inverters," *IEEE Transactions on Power Electronics*, vol. 29, no. 2, pp. 1006–1017, 2014.

[20] K. Takagi and H. Fujita, "Dynamic control and performance of a dual-active-bridge dc–dc converter," *IEEE Transactions on Power Electronics*, vol. 33, no. 9, pp. 7858–7866, 2018.

[21] T. Sen, M. Liao, Y. Wu, and M. Chen, "Modeling and control of a cyclo-active-bridge inverter for single-stage three-phase grid interface," *2025 IEEE Applied Power Electronics Conference and Exposition (APEC)*, 2025.

Modular Nanosecond Pulse Generator Leveraging GaN and SiC for Versatility and Performance.

P. Briz, *Student Member, IEEE*, H. Sarnago, *Senior Member, IEEE*,
O. Lucía, *Senior Member, IEEE*
Dept. of Electrical Engineering and Communications
Universidad de Zaragoza / I3A
Zaragoza, Spain
pbriz@unizar.es

Abstract—This paper introduces a novel, modular nanosecond pulsed electric field generator designed to address some of the critical challenges of the generation of high-voltage pulses with short durations, down to a few ns, while maintaining a low output impedance, relaxing characterization and impedance matching of the load requirements. The proposed system architecture, which features a multilevel topology, a digital control platform, and high performance SiC power switches with custom GaN drivers, enables the generation of high-voltage and high-current pulses that achieve switching times of hundreds of picoseconds. By leveraging the capabilities of GaN and SiC transistors, the generator provides precise control over the main pulse characteristics: amplitude, dv/dt and duration. A custom-designed GaN-based discrete driver is implemented to achieve the desired switching speed and control over pulse duration. The modular design enhances flexibility and scalability. Experimental validation confirms the ability of the generator to produce nanosecond-long pulses with peak voltages up to 3.6 kV, demonstrating its potential to significantly advance electroporation research and technology.

Index Terms—Nanosecond electroporation, high frequency and power LiDAR drivers, pulsed electric field, nanosecond high voltage pulses

I. Introduction

The application of nanosecond pulsed electric field (nsPEF) is generating significant interest in both experimental and industrial areas, e.g.: gene transfer, medical applications (where cardiac tissue and tumor ablation stand out) [1]–[3], food processing [4], [5], or biomass drying [6]. By applying a pulsed electric field (PEF) to cells, its transmembrane potential increases, and a phenomenon known as electroporation (EP) occurs. This phenomenon is based on an increase of the cell membrane permeability through the application of high-intensity electric fields. Specifically, in contrast to conventional EP that uses µs long pulses, one of the sought-after effects of the application of nsPEF is the permeabilization of intracellular membranes while minimizing the effect on the extracellular membrane or minimizing the lethal effects of EP [7].

The voltage requirements in these applications can range from a few kV/cm to hundreds of kV/cm, with voltage requirements increasing dramatically as the pulse frequency

This work was partly supported by Projects PID2022-136621OB-I00, PDC2021-120898-I00, TED2021-129274B-I00 and ISCIII PI21/00440, co-funded by MCIN/AEI/10.13039/501100011033 and EU through FEDER and NextGenerationEU/PRTR programs, and by the DGA-FSE, and by the DGA under PhD grant.

(a)

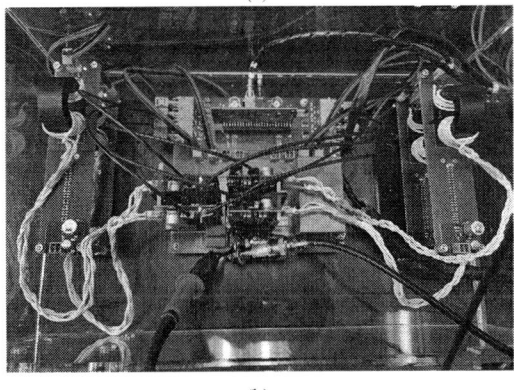

(b)

Fig. 1: (a) Full prototype render. Power modules connected to the main board where control logic and load are connected. (b) Full working prototype. (b) Power module PCBs interconnected by means of the main board. Note the isolated power supplies for each module on each side of the picture, and the FPGA board where control signals are generated. On the bottom of the picture, the high precision, high bandwidth shunt resistor, can be seen. The load is a Gene Pulser® / Micropulser™ electroporation cuvette with a yeast strain suspension.

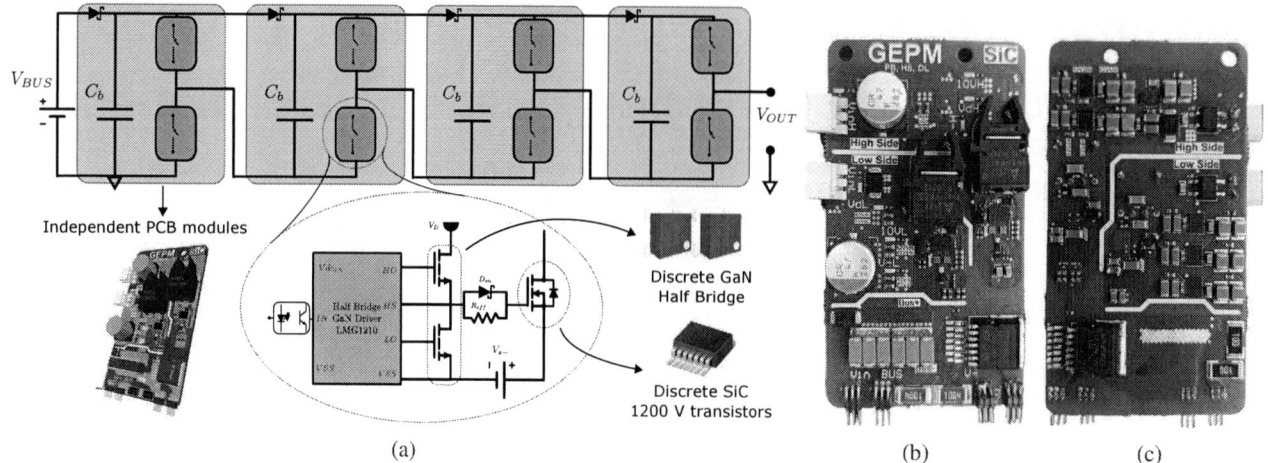

Fig. 2: (a) Main circuit diagram of the converter, with the description of the most relevant parts. The circuit consists of series-connected half-bridge inverters. Each half-bridge stage is in an independent printed circuit board (PCB) power module, with its own isolated control signals (optical isolation) and voltage supplies (galvanic isolation). The main power switches are silicon carbide (SiC) transistors, and the drivers consist of a discrete implementation of a gallium nitride (GaN) half-bridge. (b), (c) Half bridge power module and drivers, top and bottom respectively.

increases [8], [9].One of the greatest technological challenges in the field of EP is the generation of nanosecond pulses, as it requires power converters capable of delivering high levels of voltage and current in very short times. If the impedance of the target is known and constant, there are resonant converters that can achieve the desired results [10]–[14]. However, the evolution of the impedance is complex, depends on the applied electric field, and decreases during EP and nsPEF treatments [15], [16]. Moreover, in many applications, the impedance of the target depends on non-controllable factors and cannot be modified. In this sense, a critical aspect is the independence of the applied electric field and the impedance of the tissue or medium in which it is to be applied. Recently, several non-resonant designs are being developed in order to improve the load dependence and control versatility [13], [14], [17]–[19]. This paper proposes the design and implementation of a prototype, shown in Fig. 1, that allows the application of unipolar pulse width down to $10\,\text{ns}$, with amplitudes and currents up to $3.6\,\text{kV}$ and $250\,\text{A}$, respectively, with a minimum output impedance, a controllable $\mathrm{d}v/\mathrm{d}t$, and an adjustable pulse frequency from $10\,\text{mHz}$ up to $50\,\text{MHz}$, continuously or in bursts. This represents a relevant step forward for different specific nsPEF applications [20], [21].

II. PROPOSED CONVERTER

The proposed converter's modular design enhances its versatility and scalability. The control board's ability to sequentially activate modules expands waveform generation capabilities, including precise control over $\mathrm{d}v/\mathrm{d}t$. Fiber optic transceivers ensure optical isolation of control signals, while galvanic isolation in auxiliary power supplies minimizes coupling capacitance. This meticulous design is crucial, considering the rapid switching times and high voltage, which can lead to

$\mathrm{d}v/\mathrm{d}t$ exceeding $1000\,\text{V/ns}$. This can potentially cause electromagnetic compatibility (EMC) issues in some environments, so the sequential activation of the modules can help reducing electromagnetic interferences (EMIs).

As depicted in Fig. 2a, the proposed multilevel topology comprises series-connected half-bridge inverters. Each module contributes to a fraction of the total output voltage, enabling efficient power distribution and precise voltage regulation.

A. Driver design

To minimize switching time and increase control capabilities over transistor activation, the power device drivers have been implemented using discrete components. The design is based on a half-bridge topology with fast, low-voltage EPC2214 GaN devices, aiming to deliver a peak current up to $30\,\text{A}$ to the gate of the power transistors, minimizing switching times. A voltage source is added between the low side of the driver and the power transistor source to obtain a negative gate voltage during the inactive period.

This reduces the possibility of activation due to the Miller effect, and further decreases switching times. Moreover, the rise and fall times of the driver output are only a few hundreds of picoseconds, being able to withstand fast-rising pulsed currents up to $47\,\text{A}$ and voltages up to $60\,\text{V}$ (despite this, in this implementation, the gate voltage is kept in the recommended device ratings to ensure reliability).

III. IMPLEMENTATION AND EXPERIMENTAL RESULTS

The proposed implementation utilizes $1200\,\text{V}$, $17\,\text{m}\Omega$ SiC Infineon transistors (IMBG120R017M2H) as power devices. These transistors, housed in a low-inductance TO-267-7 package, can handle peak currents exceeding $200\,\text{A}$. While GaN transistors were initially considered due to their rapid switching speeds, concerns regarding reliability and preliminary test results led to their exclusion.

Recent advancements in trench SiC power devices have yielded impressive switching speeds, approaching those of GaN, while offering superior blocking voltage capabilities and enhanced noise immunity [22]. To fully exploit the potential of these power devices and ensure stable, reliable activation for sub-nanosecond pulses, a custom gate driver employing low-voltage GaN transistors has been implemented.

This driver is capable of delivering 47 A peak current, under 700 ps rise / fall time and a minimum input pulse of 4 ns. The GaN half-bridge employed as a driver is driven by a Texas Instruments high frequency GaN half-bridge driver, LMG1210, with a single PWM input and resistor-controlled dead-time. Fig. 1b shows the implemented prototype, including the power modules (Figs. 2b and 2c), a main board to interconnect the modules, galvanically isolated power supplies for each voltage domain, that also hosts the digital FPGA PCB where the control logic is implemented. This implementation focuses on minimizing inductance in two places: first, in the gate-loop of the devices, because otherwise it may limit the peak current flowing into the main transistors and, moreover, cause ringing issues. Secondly, in the main power loop, to achieve a minimum output impedance, ensuring this way that all the output voltage reaches the load and energy transfer is maximized. In Fig. 3d, the voltage on the gate of the power transistor is shown.

A. Measurement setup

To measure the main waveforms of the generator, the following equipment was used:

- Teledyne LeCroy® 500 MHz, 6 kV passive voltage probe, PPE-6kV-A.
- Teledyne LeCroy® 500 MHz, 400 V passive volgate probe, PP026.
- Teledyne LeCroy® WaveSurfer, 1 GHz, 12-bit oscilloscope.
- Powertek ® 25 mΩ shunt resistor, 1200 MHz bandwidth, 50 Ω output, SDN-414-025, connected directly to the scope by a 50 Ω cable, to measure output current.

B. Experimental results

Finally, the generator evaluation is performed using both biological targets and resistive loads. The biological loads are yeast strain suspension of *Saccharomyces cerevisiae SafAle™ S-04* in a 2 S/cm conductivity buffer. Waveforms with those loads are shown in Fig. 3a and 3c. Waveforms associated to the resistive loads are shown in Fig. 3b and 3d. This generator allowed to develope a study that lays the groundwork for optimizing PEF parameters [23].

In every case, the switching times of the converter are sufficiently low to achieve a controlled pulse duration, but 10 ns is the limit case, especially when dealing with high currents (Fig. 3b). Here, a 15 Ω resistive load was used, so the resulting current is 240 A for a pulse amplitude of 3.6 kV. This current saturated the scope input so it is not represented on the screen.

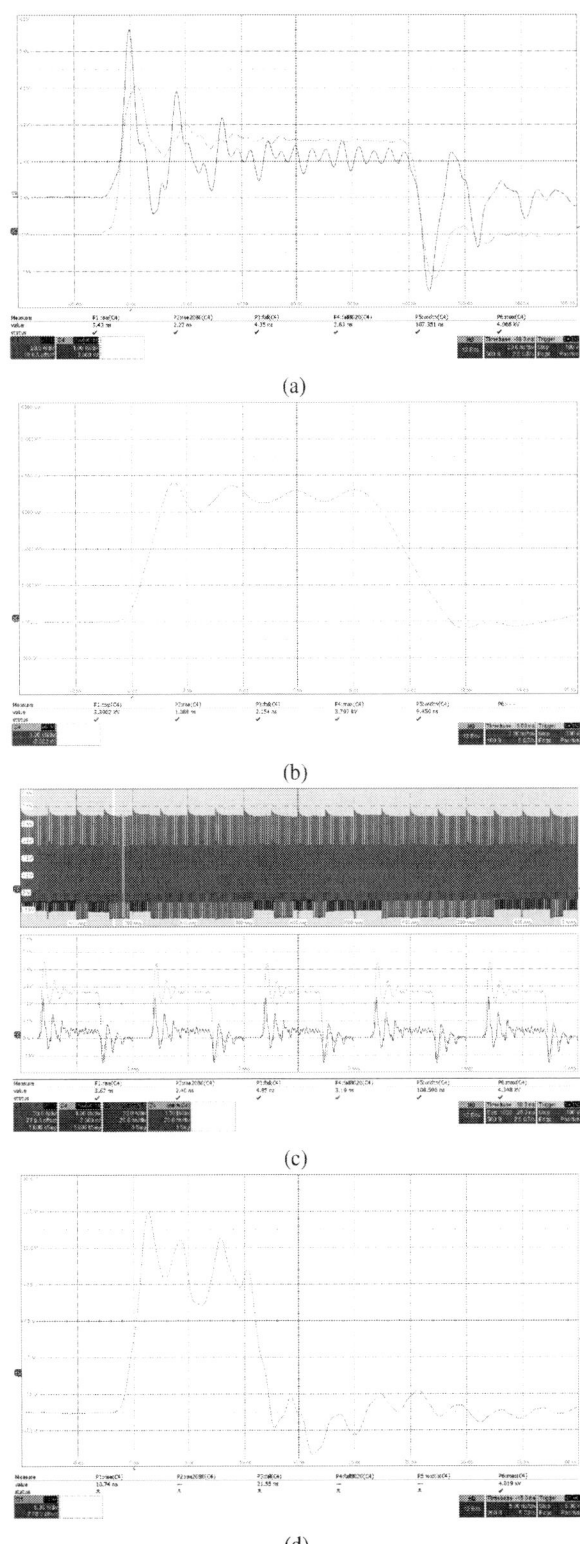

(a)

(b)

(c)

(d)

Fig. 3: (a) Single 100 ns long pulse with and amplitude of 3 kV and about 20 A of current, applied to a buffer containing yeast cells. (b) Single 10 ns long pulse with an amplitude of 3.6 kV, applied to a 15 Ω resistive load. (c) 100 ns long pulses applied to a buffer containing yeast cells in 1 kHz bursts of 100 pulses. (d) driver output voltage.

IV. CONCLUSIONS

This work proposes a modular, solid-state, digitally controlled high voltage pulse generator, aimed at enabling experimentation procedures based on nsPEF with a broad range of load impedances. Its design focuses on the versatility of pulsed patterns and the minimization of output impedance. The implementation of the proposed design results in a generator that allows to evaluate its performance using real laboratory loads, as well as resistive loads. Thanks to the discrete implementation of GaN-based drivers, the maximum performance of the SiC transistors can be exploited, making this converter useful for generation of nsPEF with durations equal to or greater than $10\,\mathrm{ns}$, with the ability to apply up to $3.6\,\mathrm{kV}$, and frequencies up to $50\,\mathrm{MHz}$. A future prototype focused on further reducing the output impedance could decrease the minimum output pulse down to $4\,\mathrm{ns}$, the minimum for the driver to perform with a strong repeatability. The modular design is scalable, and therefore the range of applications of the proposed design is broader than others found in the literature.

REFERENCES

[1] A. Sugrue, E. Maor, A. Ivorra, V. Vaidya, C. Witt, S. Kapa, and S. Asirvatham, "Irreversible electroporation for the treatment of cardiac arrhythmias," *Expert Review of Cardiovascular Therapy*, vol. 16, pp. 349–360, 5 2018.

[2] T. Kotnik, W. Frey, M. Sack, S. Haberl Meglič, M. Peterka, and D. Miklavčič, "Electroporation-based applications in biotechnology," *Trends in Biotechnology*, vol. 33, pp. 480–488, 8 2015.

[3] W. Szlasa, O. Michel, N. Sauer, V. Novickij, D. Lewandowski, P. Kasperkiewicz, M. Tarek, J. Saczko, and J. Kulbacka, "Nanosecond pulsed electric field suppresses growth and reduces multi-drug resistance effect in pancreatic cancer," *Scientific Reports*, vol. 13, p. 351, 1 2023.

[4] G. B. Pintarelli, C. T. S. Ramos, J. R. d. Silva, M. J. Rossi, and D. O. H. Suzuki, "Sensing of Yeast Inactivation by Electroporation," *IEEE Sensors Journal*, vol. 21, pp. 12027–12035, 5 2021.

[5] C. Zhang, X. Lyu, R. N. Arshad, R. M. Aadil, Y. Tong, W. Zhao, and R. Yang, "Pulsed electric field as a promising technology for solid foods processing: A review," *Food Chemistry*, vol. 403, p. 134367, 2023.

[6] R. P. Joshi, A. L. Garner, and R. Sundararajan, "Review of Developments in Bioelectrics as an Application of Pulsed Power Technology," *IEEE Transactions on Plasma Science*, vol. 51, pp. 1682–1717, 7 2023.

[7] O. G. M. Khan and A. H. El-Hag, "Biological Cell Electroporation Using Nanosecond Electrical Pulses," *JOURNAL OF MEDICAL IMAGING AND HEALTH INFORMATICS*, vol. 1, no. 3, pp. 278–283, 2011.

[8] R. Sundararajan, "Nanosecond Electroporation: Another Look," *MOLECULAR BIOTECHNOLOGY*, vol. 41, no. 1, pp. 69–82, 2009.

[9] J. A. Costa, P. X. d. Oliveira, L. S. Pereira, J. Rodrigues, and D. O. H. Suzuki, "Sensitivity Analysis of a Nuclear Electroporation Model—A Theoretical Study," *IEEE Transactions on Dielectrics and Electrical Insulation*, vol. 28, no. 6, pp. 1850–1858, 2021.

[10] X. Cheng, S. Chen, Y. Lv, H. Chen, and B. M. Novac, "An Improved High Voltage Pulse Generator With Few Nanoseconds Based on the Synergy of DOS and LTD Topologies for Supra Electroporation," *IEEE Transactions on Industrial Electronics*, vol. 70, pp. 7855–7866, 8 2023.

[11] G.-p. Wang, F. Li, X. Jin, and F.-l. Song, "A cascade nanosecond pulse generator based on two-stage DSRDs," in *2020 IEEE International Conference on High Voltage Engineering and Application (ICHVE)*, pp. 1–4, IEEE, 9 2020.

[12] F. Cao, D. Jiang, Y. Liu, Y. Tian, X. Ran, Y. Long, T. Ito, X. Hu, G. Weng, H. Akiyama, and S. Chen, "Subnanosecond Marx Generators for Picosecond Gain-Switched Laser Diodes," *IEEE Photonics Journal*, vol. 16, pp. 1–8, 2 2024.

[13] P. Butkus, A. Murauskas, S. Tolvaišienė, and V. Novickij, "Concepts and Capabilities of In-House Built Nanosecond Pulsed Electric Field (nsPEF) Generators for Electroporation: State of Art," *Applied Sciences*, vol. 10, p. 4244, 6 2020.

[14] P. Leveque and D. Arnaud-Cormos, "Generators and applicators for nanosecond pulsed electric field," in *2012 6th European Conference on Antennas and Propagation (EUCAP)*, pp. 351–355, IEEE, 3 2012.

[15] B. Lopez-Alonso, H. Sarnago, O. Lucia, P. Briz, and J. M. Burdio, "Real-Time Impedance Monitoring During Electroporation Processes in Vegetal Tissue Using a High-Performance Generator," *Sensors*, vol. 20, no. 11, 2020.

[16] A. Bernardis, M. Bullo, L. G. Campana, P. Di Barba, F. Dughiero, M. Forzan, M. E. Mognaschi, P. Sgarbossa, and E. Sieni, "Electric field computation and measurements in the electroporation of inhomogeneous samples," *Open Physics*, vol. 15, pp. 790–796, 12 2017.

[17] I. Davies, C. Merla, A. Tanori, A. Zambotti, J. Bishop, C. Palego, and C. P. Hancock, "Push-Pull Configuration of High Power MOSFETs for Generation of Nanosecond Pulses for Electropermeabilization of Isolated Cancer Stem Cells," in *2018 48th European Microwave Conference (EuMC)*, pp. 866–869, IEEE, 9 2018.

[18] X. Rao, X. Chen, J. Zhou, L. Sun, and J. Liu, "A Digital Controlled Pulse Generator for a Possible Tumor Therapy Combining Irreversible Electroporation With Nanosecond Pulse Stimulation," *IEEE Transactions on Biomedical Circuits and Systems*, vol. 14, pp. 595–605, 2 2020.

[19] C.-Y. Ho, T.-J. Liang, K.-H. Chen, and K.-F. Liao, "Design and Implementation of Cascoded Dual-Half-Bridge Resonant Converter with GaN E-HEMT for High Input Voltage Applications," in *2024 IEEE Applied Power Electronics Conference and Exposition (APEC)*, pp. 114–121, IEEE, 2 2024.

[20] P. Lamberti, S. Romeo, A. Sannino, L. Zeni, and O. Zeni, "The Role of Pulse Repetition Rate in nsPEF-Induced Electroporation: A Biological and Numerical Investigation," *IEEE Transactions on Biomedical Engineering*, vol. 62, pp. 2234–2243, 9 2015.

[21] H. Liu, C. Yao, Y. Zhao, X. Chen, S. Dong, L. Wang, and R. V. Davalos, "In Vitro Experimental and Numerical Studies on the Preferential Ablation of Chemo-Resistant Tumor Cells Induced by High-Voltage Nanosecond Pulsed Electric Fields," *IEEE Transactions on Biomedical Engineering*, vol. 68, no. 8, pp. 2400–2411, 2021.

[22] A. I. Emon, Mustafeez-ul-Hassan, A. B. Mirza, J. Kaplun, S. S. Vala, and F. Luo, "A Review of High-Speed GaN Power Modules: State of the Art, Challenges, and Solutions," *IEEE Journal of Emerging and Selected Topics in Power Electronics*, vol. 11, pp. 2707–2729, 6 2023.

[23] P. Briz, A. Berzosa, B. Lopez-Alonso, J. Marin-Sanchez, C. Calvo-Chueca, H. Sarnago, O. Lucia, and J. Raso, "Comparative Study of the Effects of Nanosecond and Microsecond Pulsed Electric Fields on Saccharomyces cerevisiae," in *5th World Congress on Electroporation and Pulsed Electric Fields in Biology, Medicine, and Food & Environmental Technologies* (C. Merla, F. Apollonio, and S. Mahnič-Kalamiza, eds.), (Rome), pp. 283–284, ISEBTT, 9 2024.

A Variable Frequency Technique for EMI and Efficiency Improvements in High-Level Count Flying Capacitor Multilevel Converters

Francesca Giardine, Sahana Krishnan, Logan Horowitz, Robert C. N. Pilawa-Podgurski
Email: {fgiardine, sahana_krishnan, logan_h_horowitz, pilawa}@berkeley.edu
University of California, Berkeley, CA, USA

Abstract—**This work assesses the use of a variable switching frequency scheme specifically designed for a flying capacitor multilevel (FCML) converter in dc-ac operation to spread out the peaks of conducted electromagnetic interference (EMI) and improve efficiency. Ripple constraints of the flying capacitor voltages and inductor current set the basis for a deterministic switching frequency pattern, which can also be considered a targeted dithering scheme. Furthermore, by leveraging the constraints set by the passive components of the designed prototype and reducing the switching frequency, switching losses are mitigated and an increase in efficiency is gained. Hardware results are provided for a 2 kW 8-level FCML inverter and showcase improved conducted EMI peaks and efficiency for the high performance, high-level count prototype.**

Index Terms—**dc-ac converters, multilevel converters, modulation, hybrid switched-capacitor converters**

I. INTRODUCTION

Growing calls for electric transportation, flight electrification, and renewable energy integration necessitate efficient and power-dense converters [1]–[3]. The flying capacitor multilevel (FCML) converter [4] is an increasingly popular solution, in part due to its superior volumetric and gravimetric power densities which are achieved via the use of high figure-of-merit switches and energy-dense capacitors [5]–[7]. For an FCML converter operated in dc-ac mode, the passive components are typically rated for the peak voltage and current ripples which occur at peak power in the ac cycle. However, these ripples are variable across a single ac cycle as they have a duty cycle dependency which is far more pronounced than for conventional, 2-level designs. As the switching frequency and converter components are rated for the worst-case ripple conditions, the passive components are not fully utilized throughout the ac cycle.

The use of high figure-of-merit gallium nitride high-electron-mobility transistors (GaN HEMTs) enables increasingly high switching frequencies for multilevel converters [8], [9]. An important consideration is the generation of unwanted conducted electromagnetic interference (EMI) and how this propagates in the system as the frequency increases [5], [10]–[12]. The input side of an FCML converter operated as a step-down inverter has a current flow with a fundamental harmonic at the switching frequency f_{sw}, much like the 2-level buck converter equivalent. This poses possible EMI challenges at

a lower frequency [1] rather than at the multiplied effective frequency at the ac output of the inverter [5], [13], which can have detrimental effects on the equipment attached to the dc bus [3], [14], [15]. Beyond the inclusion of a passive EMI filter at the converter input port, there are modulation techniques that can be used to improve the conducted emissions at harmonics of the switching-frequency fundamental by spreading out the characteristic peaks [16]–[18]. For these techniques, however, the passive components must be sufficiently sized for the lowest frequency used in the modulation scheme, which may further penalize power density.

This work analyzes and demonstrates in hardware how a technique previously used to improve the efficiency of an FCML inverter can be leveraged more strategically to improve EMI performance as well [19]. In the previous work [19], the observed efficiency improvement was accomplished by reducing the switching frequency in low-ripple regions of the ac cycle for a grid-frequency inverter. With respect to EMI mitigation, the ripple constraints here create feasible operating conditions that make it possible to generate a deterministic switching pattern sequence that effectively functions as a spread-spectrum frequency modulation technique. In this work, we assess the efficacy of this approach for a higher level count converter, which is less frequently characterized for EMI. Hardware results are provided for a high performance 8-level FCML inverter with GaN switches. Improved measured conducted EMI at the input of the converter and improved efficiency results demonstrate the merits of this technique.

This paper continues as follows: Section II explores the general guidelines for developing the variable switching frequency scheme and some practical implementation considerations, Section II-C presents a hardware validation with substantial EMI and efficiency improvements compared to conventional techniques, and Section IV provides concluding remarks.

II. OVERVIEW OF THEORY AND CONTROL

A. Conventional Operation

The schematic for an 8-level FCML inverter with a split-bus input is shown in Fig. 1. A general N-level FCML has $N-1$ switch pairs and $N-2$ flying capacitors. Each flying capacitor has a dc voltage offset of $v_{c,k} = V_{in} \cdot k/(N-1)$, where $k \in [1, N-2]$, while the switches must block a dc voltage of $\frac{V_{in}}{N-1}$.

979-8-3315-1612-3/25 $31.00 © 2025 IEEE

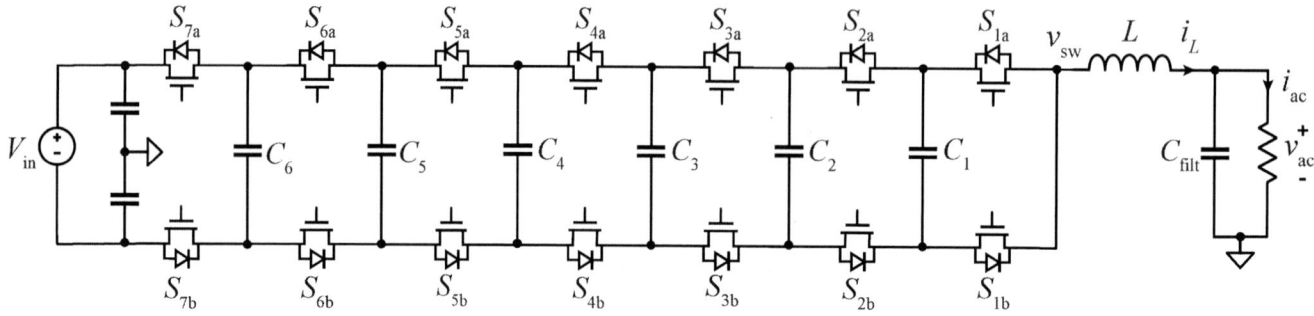

Fig. 1: Simplified schematic for an 8-level FCML converter with a split input bus.

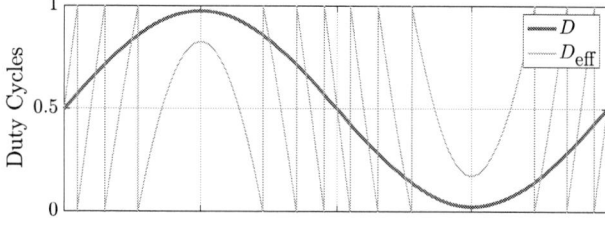

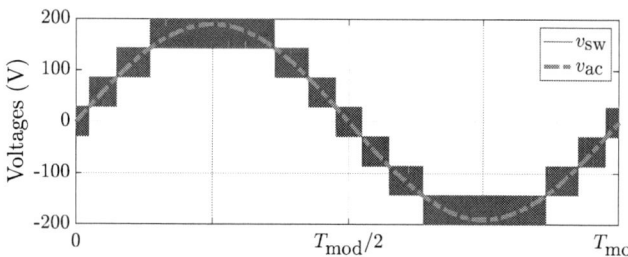

Fig. 2: a) Example sinusoidal duty cycle and the corresponding effective duty cycle D_{eff} for an 8-level FCML inverter. b) Switch node voltage waveform v_{sw} and filtered ac voltage v_{ac} for the split-bus converter.

Under conventional operation with phase-shifted pulse-width modulation (PS-PWM) [20], the switching frequency observed at the switch node v_{sw} is $f_{\text{eff}} = (N-1) \cdot f_{\text{sw}}$.

In the split-bus configuration with triangular PWM carriers bounded by 0 and 1, the duty cycle used for an inverter may be generated as

$$D = \frac{1}{2} + M \sin(2\pi f_{\text{mod}} t) \qquad (1)$$

where M is the modulation depth and f_{mod} is the ac modulation frequency [21]. The generation of this duty cycle assumes that $D = \frac{1}{2}$ corresponds to an output voltage of $0\,\text{V}$ due to the split input bus.

The effective duty cycle observed at the switch node v_{sw} is given by $D_{\text{eff}} = D \cdot (N-1) - floor(D \cdot (N-1))$. The corresponding effective duty cycle is highlighted in Fig. 2, alongside waveforms for the switch-node voltage v_{sw} and the ac output voltage v_{ac}. This duty cycle influences the volt-second balance observed across the inductor, which results in the current ripple described by (2).

$$\Delta i_{L,\text{pp}} = \frac{V_{\text{in}} D_{\text{eff}}(1 - D_{\text{eff}})}{L f_{\text{sw}} (N-1)^2}. \qquad (2)$$

The other major passive components in the FCML are the flying capacitors, which have a load-dependent peak-to-peak voltage ripple given by (3).

$$\Delta v_{C,\text{pp}} = \begin{cases} \frac{|i_{\text{ac}}| \cdot D_{\text{eff}}}{(N-1) \cdot f_{\text{sw}} \cdot C_{\text{fly}}} & D \leq \frac{1}{N-1} \\ \frac{|i_{\text{ac}}|}{(N-1) \cdot f_{\text{sw}} \cdot C_{\text{fly}}} & \frac{1}{N-1} < D < \frac{N-2}{N-1} \\ \frac{|i_{\text{ac}}| \cdot (1 - D_{\text{eff}})}{(N-1) \cdot f_{\text{sw}} \cdot C_{\text{fly}}} & \frac{N-2}{N-1} \leq D \end{cases} \qquad (3)$$

Typically, it is desirable to limit the capacitor voltage ripple to a small fraction of its dc value, so as not to increase the voltage stress on the adjacent transistors. In this work, the voltage ripple allowed on the flying capacitors is set for the load current at full power, and is not adjusted for varying load currents.

B. Variable Frequency Operation

As can be seen from (2) and (3), the current and voltage ripples in the FCML are both dependent on switching frequency, and are both functions of where in the ac cycle the converter is operating. The expression for inductor current ripple in (2) is rearranged as a time-varying function of switching frequency $f_{\text{sw}, i_{Lpp}}(t)$, where $\Delta i_{L,\text{pp}}$ is set to its fixed maximum allowable value $\Delta i_{L,\text{pp, max}}$. In the same way, the piecewise expression in (3) is rearranged to be in terms of $\Delta v_{C,\text{max}}$ to generate another time-varying lower limit $f_{\text{sw}, v_{Cpp}}(t)$. To generate the switching frequency envelope, the frequency may be calculated as

$$f_{\text{sw}}(t) = \max[f_{\text{sw}, i_{Lpp}}(t), f_{\text{sw}, v_{Cpp}}(t)], \\ f_{\text{sw, min}} \leq f_{\text{sw}}(t) \leq f_{\text{sw, max}}. \qquad (4)$$

The absolute upper limit $f_{\text{sw, max}}$ is set as the nominal operating frequency for which $\Delta i_{L,\text{pp, max}}$ and $\Delta v_{C,\text{max}}$ are rated. The absolute lower limit, $f_{\text{sw, min}}$, may be set to be high enough to avoid exciting resonance in the LC filter at the output, as described in [19]. In this work however, the lower limit $f_{\text{sw, min}}$ is further constrained by both level count and modulation frequency, as described in Section II-C.

979-8-3315-1612-3/25 $31.00 © 2025 IEEE

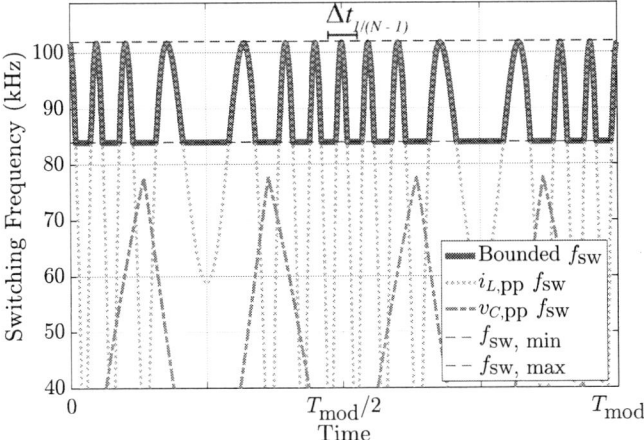

Fig. 3: Example continuous variable switching frequency limits for an 8-level FCML inverter with a nominal switching frequency of 102 kHz, and low modulation frequency f_{mod}.

Fig. 3 shows the ideal allowable switching frequency region across a full ac cycle for the 8-level converter described in this work. The bounded switching frequency waveform creates a duty cycle generated reference value for the updating of each switching frequency instance.

C. Considerations for Hardware Implementation

This work further considers the role and limitations imposed by the modulation frequency (f_{mod}) and by the level count in generating the possible sequence of switching frequency events. For a low modulation frequency such as the common grid frequencies (50-60 Hz), updating the period after each switching instance results in a frequency sequence with very granular steps across the ac cycle.

However, as the period length T_{mod} decreases with higher modulation frequencies, the number of possible switching

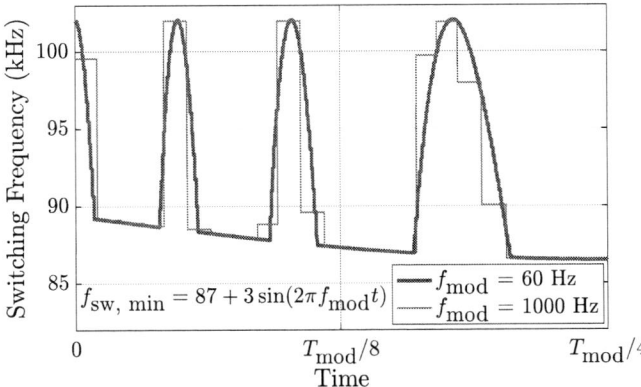

Fig. 4: Discrete frequency envelopes from simulation for both 60 Hz and 1 kHz modulation frequencies showing the impact of sizing f_{mod} for an 8-level converter. Note that here, the lower limit for $f_{\mathrm{sw, min}}$ is set as a sinusoid with a frequency of f_{mod}, to prevent the switching frequency from saturating at a hard limit for much of the ac cycle.

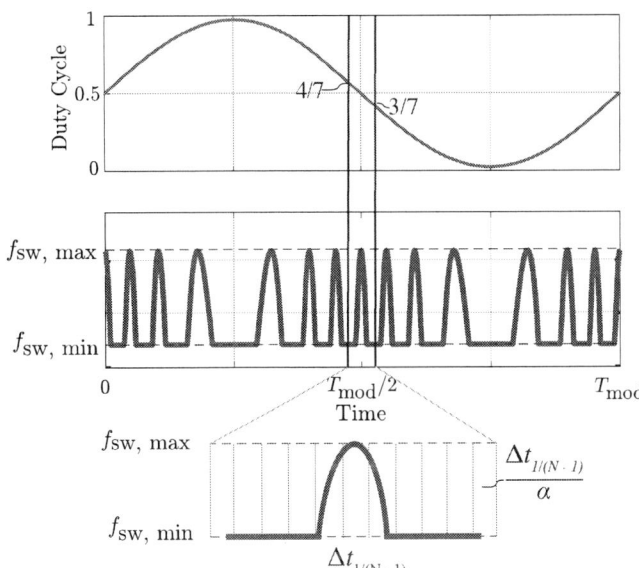

Fig. 5: Plot showing how $\Delta t_{1/(N-1)}$ relates to the sinusoidal duty cycle, and how during the shortest sub-duration a minimum number of steps is set by $f_{\mathrm{sw, min}}$.

frequency updates across the ac cycle is diminished for the same band of switching frequencies. For a higher modulation frequency, depending on the sampling instant, this might result in a dithering scheme that appears pseudo-random.

The simulated discrete switching frequency steps for different modulation frequencies are shown for a portion of T_{mod} in Fig. 4. For the 60 Hz grid frequency, the switching frequency waveform is smooth, as is updated at a rate much faster than this modulation frequency. For a modulation frequency of

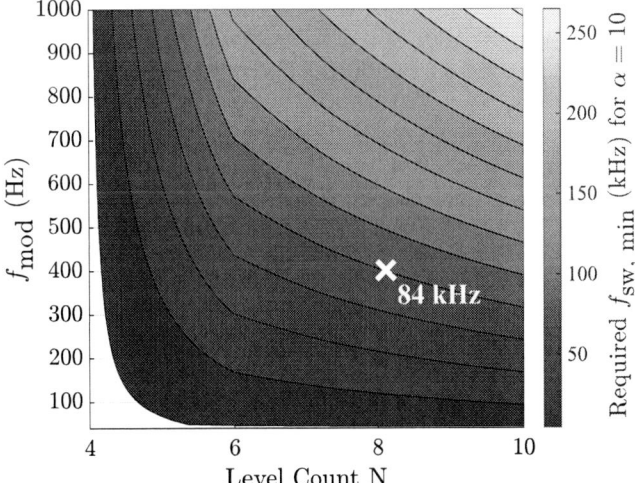

Fig. 6: Contour plot showing how $f_{\mathrm{sw, min}}$ scales across both level count N and ac modulation frequency f_{mod} to ensure that at least 10 steps ($\alpha \geq 10$) are generated across the brief sub-duration defined by $\Delta t_{1/(N-1)}$. For the 8-level FCML with $f_{\mathrm{mod}} = 400\,\mathrm{Hz}$, this is computed to be at least $84\,\mathrm{kHz}$.

1 kHz, the frequency may jump from the maximum switching frequency to the minimum in some portions of the ac cycle.

To better understand the impacts of level count and modulation frequency on the ideal discrete switching frequency sequence, the shortest sub-duration between nominal duty ratios within the ac-cycle is considered. For simplicity, this is computed in the continuous time domain based on the duty cycle in (1). For an even level split-bus FCML, this sub-duration is between $D = \frac{1}{2} + \frac{1}{2(N-1)}$ and $D = \frac{1}{2} - \frac{1}{2(N-1)}$. For an odd level count, this sub-duration occurs between $D = \frac{1}{2} + \frac{1}{N-1}$ and $D = \frac{1}{2}$. The time duration for these two cases is computed as $\Delta t_{1/(N-1)}$, and is given in (5).

$$\Delta t_{1/(N-1)} = \begin{cases} \dfrac{\arcsin(\frac{1}{2M(N-1)})}{\pi f_{\text{mod}}} & N, \text{even} \\[3mm] \dfrac{\arcsin(\frac{1}{M(N-1)})}{2\pi f_{\text{mod}}} & N, \text{odd} \end{cases} \quad (5)$$

To guarantee a user-specified minimum number of steps during this time window in order to achieve a more continuous frequency envelope in hardware, a minimum required switching frequency cutoff is obtained as

$$f_{\text{cutoff}} \geq \alpha \frac{1}{\Delta t_{1/(N-1)}} \quad (6)$$

where α is the minimum number of discrete steps in $\Delta t_{1/(N-1)}$.

Fig. 5 shows how this sub-duration $\Delta t_{1/(N-1)}$ relates to the duty cycle for the 8-level converter examined in this work. The sub-duration is visually divided into a minimum number of possible steps. The upper limits for switching frequency are

determined in this case by the peak inductor current ripple.

In Fig. 6, we examine how an increase in modulation frequency makes it challenging to set a low switching frequency limit in order to enable the variable switching frequency scheme. For an 8-level inverter with $f_{\text{mod}} = 400\,\text{Hz}$, we see that in order to guarantee at least 10 steps in $\Delta t_{1/(N-1)}$, a switching frequency of at least $84\,\text{kHz}$ is required.

III. HARDWARE IMPLEMENTATION

The hardware prototype and pre-compliant benchtop EMI setup used to validate the switching technique is shown in Fig. 7. Applicable hardware parameters are shown in Table I. This converter was designed with minimal parasitic inductances in the power path, which helps to reduce conducted EMI [22]. The variable switching frequency technique is implemented using a TI C2000 ControlCard, and the duty cycle and switching frequency are concurrently updated from their corresponding shadow registers via a synchronization event set by the PWM carrier signal. This ensures that there are no violations of passive ripple constraints despite the frequency hopping behavior of the frequency envelope.

Measured converter waveforms under the variable switching frequency scheme for V_{in}, v_{sw}, and v_{ac} are showcased in Fig. 8. No difference in flying capacitor voltage balancing was observed in the switch node voltage waveform v_{sw} for both fixed and variable switching frequency operation.

Efficiency measurements were recorded using a Keysight IntegraVision PA2201A. An efficiency sweep in Fig. 9 shows improved efficiencies across the load range for the variable switching frequency technique. A peak efficiency of

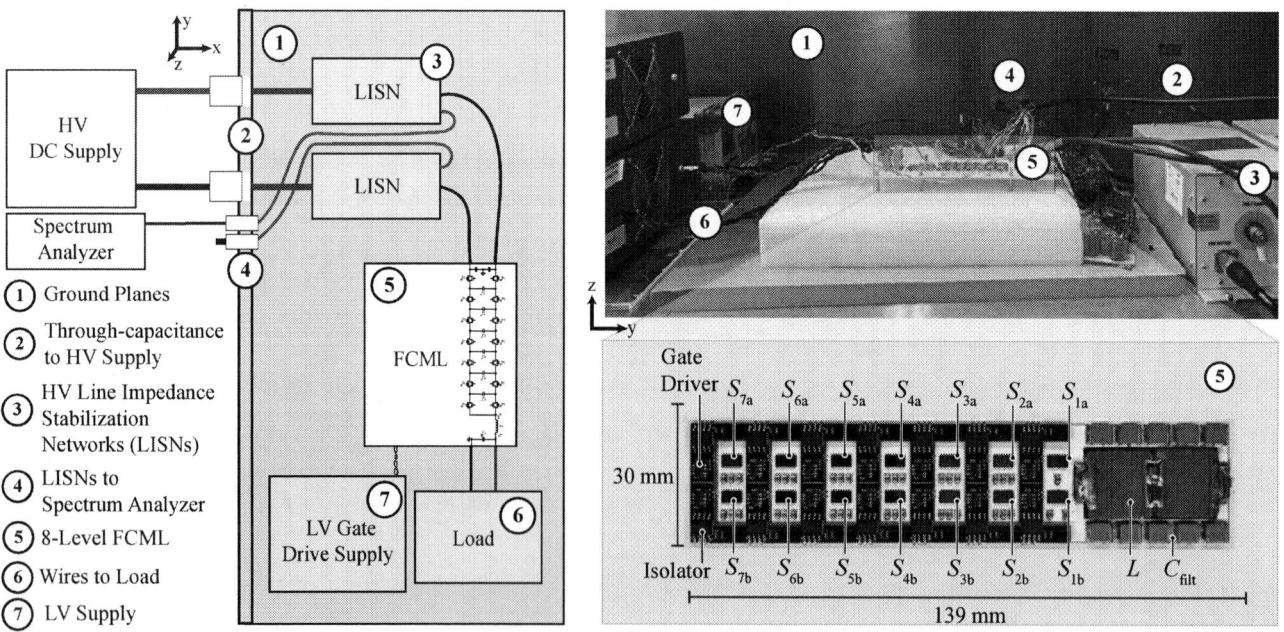

Fig. 7: On the left, a diagram of an aerial view of the benchtop EMI setup shows the connection of the line impedance stabilization networks (LISNs) to the HV dc converter bus. Labels are provided for each of the major EMI setup components. On the right, a close-up photograph of the test setup shows the converter connected to both the HV LISNs, and the ac resistive load. An inset shows the front of the high performance hardware prototype.

979-8-3315-1612-3/25 $31.00 © 2025 IEEE

Table I: Converter Operating Conditions

Parameter	Value
V_{in}	400 V
f_{mod}	400 Hz - 600 Hz
f_{sw}	84 kHz - 102 kHz
v_{ac}	135 V_{rms}
P_{out}	350 - 2000 W
$\Delta v_{C_{pp,\,max}}$	10.5 V
$\Delta i_{L_{pp,\,max}}$	3.0 A
Derated C_{fly}	$\approx$ 3 µF, paralleled 2.2 µF
	450 V, C5750X6S225K250KA
GaN HEMT	100 V, 1 mΩ, EPC 2361
L	6.6 µH, 2× IHLP6767GZER3R3M01
C_{filt}	6 × 150 nF
	250 V, TDK 5750C0G2E154J230KE
Microcontroller	TI C2000 TMDSCNCD28379D

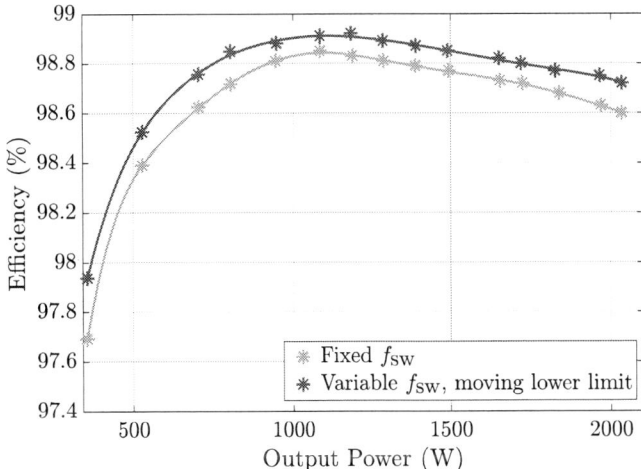

Fig. 9: Plot of efficiencies for both fixed frequency and variable frequency operation for f_{mod} = 600 Hz.

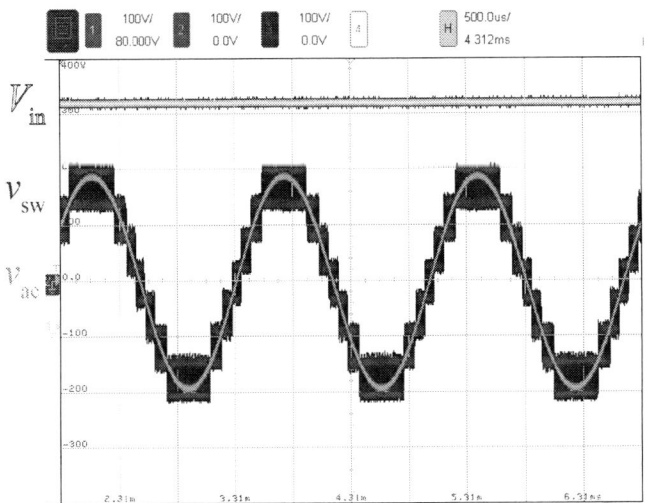

Fig. 8: Measured waveforms of the input voltage V_{in}, the switch-node voltage v_{sw}, and the ac output voltage v_{ac} at 700 W and f_{mod} = 600Hz.

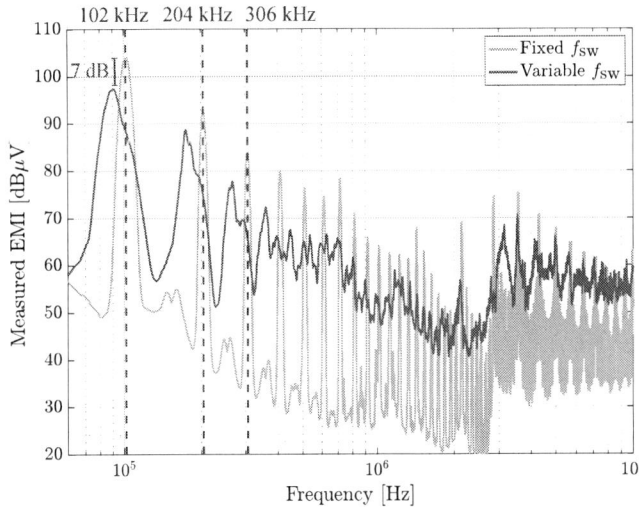

Fig. 10: Measured input EMI for f_{mod} = 400 Hz.

98.85 % was observed at 1080 W for the fixed frequency case, compared to 98.9 % at 1190 W for the variable switching frequency operation. Substantial efficiency improvements are observed and may be primarily attributed to an improvement in overlap losses, as the particular GaN HEMTs used featured very low on-state resistances.

EMI measurements are taken for both the fixed frequency case and the variable switching frequency case at 700 W. CISPR average measurements are provided in this work to capture the impact of the variable switching frequency scheme. Substantial EMI improvements are shown in Fig. 10 for the f_{mod} = 400 Hz case. For fixed frequency operation, the peak of the fundamental is at the 102 kHz switching frequency, and the subsequent peaks are at harmonics of this frequency. The application of the variable switching frequency scheme reduces the peak of the fundamental by 7 dBµV. A spread

sub-peak forms around 91.2 kHz in this case. For operational conditions that allow for a larger difference between $f_{sw,\,min}$ and $f_{sw,\,max}$, this spreading would be more substantial.

Additional measurements taken at f_{mod} = 600 Hz validate that this scheme is still effective at higher modulation frequencies, when there are fewer switching frequency updates per ac cycle. As discussed in Section II-C, fewer discrete time steps are available for the same switching frequency limits. As shown in Fig. 11, the increasingly discretized switching frequency envelope for the 600 Hz case remains an effective mechanism to reduce the EMI peaks, with improvements on the order of 5 dBµV. Like the 400 Hz case, the sub-peak occurs at 91.9 kHz, showing consistency between both modulation frequencies. This congruity suggests that this technique can be applied at higher modulation frequencies for the same limits of $f_{sw,\,min}$ and $f_{sw,\,max}$ without loss in functionality.

979-8-3315-1612-3/25 $31.00 © 2025 IEEE

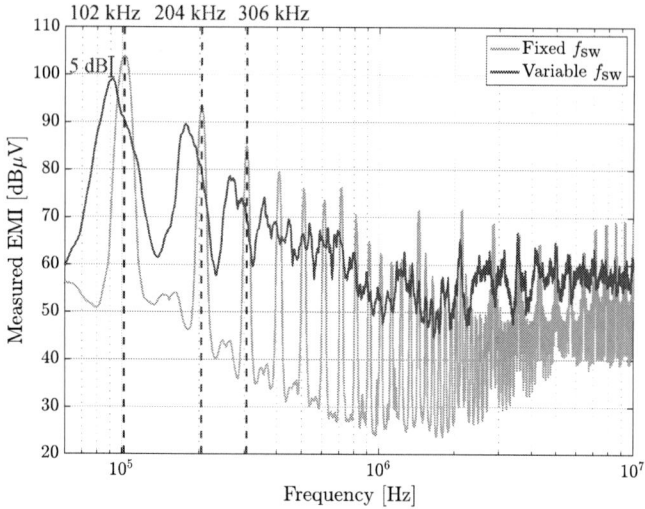

Fig. 11: Measured input EMI for $f_{\mathrm{mod}} = 600\,\mathrm{Hz}$.

IV. CONCLUSION

This work demonstrates a variable frequency switching scheme that can be used for peak EMI spreading in dc-ac FCML converters. The approach is generalized to handle a range of modulation frequencies and level counts such that a deterministic and sufficiently granular pattern is generated. Hardware results showcase a substantial reduction in input-side EMI peaks for an 8-level FCML at different ac modulation frequencies. In outlining the possible switching frequency conditions that the converter is rated for, this work opens up the possibility of creating other FCML converter specific spread-spectrum frequency routines. Furthermore, the efficiency of the high-performance prototype is increased, as the average switching frequency in this case is reduced.

V. ACKNOWLEDGMENT

This material is based upon work supported by the National Science Foundation Graduate Research Fellowship Program under Grant No. DGE 2146752. Any opinions, findings, and conclusions or recommendations expressed in this material are those of the authors and do not necessarily reflect the views of the National Science Foundation. The authors acknowledge financial support from the Berkeley Power and Energy Center (BPEC).

REFERENCES

[1] N. Pallo, R. S. Bayliss, and R. C. N. Pilawa-Podgurski, "A multi-phase segmented drive comprising arrayed flying capacitor multi-level modules," in *2021 IEEE Applied Power Electronics Conference and Exposition (APEC)*, 2021, pp. 192–199.

[2] B. Pasquet, S. Vinnac, J.-M. Blaquière, T. Meynard, and S. Sanchez, "Design and efficiency measurement of a sub-unit for a 20kw dc-dc multiphase power converter," in *2023 IEEE International Conference on Electrical Systems for Aircraft, Railway, Ship Propulsion and Road Vehicles International Transportation Electrification Conference (ESARS-ITEC)*, 2023, pp. 1–6.

[3] D. Hamza, M. Qiu, and P. Jain, "Interface impedance consideration in the design of an input emi filter for grid-tied pv micro-inverter," in *IECON 2011 - 37th Annual Conference of the IEEE Industrial Electronics Society*, 2011, pp. 1390–1395.

[4] T. Meynard and H. Foch, "Multi-level conversion: high voltage choppers and voltage-source inverters," in *PESC '92 Record. 23rd Annual IEEE Power Electronics Specialists Conference*, 1992, pp. 397–403 vol.1.

[5] J. Azurza Anderson, G. Zulauf, P. Papamanolis, S. Hobi, S. Mirić, and J. W. Kolar, "Three levels are not enough: Scaling laws for multilevel converters in ac/dc applications," *IEEE Transactions on Power Electronics*, vol. 36, no. 4, pp. 3967–3986, 2021.

[6] O. Lorenz and J. Sanchez, "Ultra low-profile flying capacitor 7-level 3kw pfc with optimized high frequency layout and active balancing using 100v gan," in *2024 IEEE Applied Power Electronics Conference and Exposition (APEC)*, 2024, pp. 22–28.

[7] N. C. Brooks, J. Zou, S. Coday, T. Ge, N. M. Ellis, and R. C. N. Pilawa-Podgurski, "On the size and weight of passive components: Scaling trends for high-density power converter designs," *IEEE Transactions on Power Electronics*, pp. 1–19, 2024.

[8] T. Modeer, N. Pallo, T. Foulkes, C. B. Barth, and R. C. N. Pilawa-Podgurski, "Design of a gan-based interleaved nine-level flying capacitor multilevel inverter for electric aircraft applications," *IEEE Transactions on Power Electronics*, vol. 35, no. 11, pp. 12 153–12 165, 2020.

[9] M. T. Elrais, M. Safayatullah, and I. Batarseh, "Generalized architecture of a gan-based modular multiport multilevel flying capacitor converter," *IEEE Transactions on Power Electronics*, vol. 38, no. 8, pp. 9818–9838, 2023.

[10] *Vehicles, boats and internal combustion engines – Radio disturbance characteristics – Limits and methods of measurement for the protection of on-board receivers*, International Special Committee on Radio Interference Std., Rev. 5.0, 2021.

[11] R. Hartwig, A. Hensler, T. Ellinger, S. Ag, and T. Ilmenau, "Emi filter for a three-phase 800 khz nine-level flying capacitor gan multilevel inverter," in *2021 23rd European Conference on Power Electronics and Applications (EPE'21 ECCE Europe)*, 2021, pp. P.1–P.10.

[12] Q. Huang, Q. Ma, P. Liu, A. Q. Huang, and M. de Rooij, "3kw four-level flying capacitor totem-pole bridgeless pfc rectifier with 200v gan devices," in *2019 IEEE Energy Conversion Congress and Exposition (ECCE)*, 2019, pp. 81–88.

[13] N. Ishraq and A. Mallik, "Design of a 2.5 kw four-level interleaved flying capacitor multilevel totem-pole pfc converter with ac-side passive volume optimization," *IEEE Open Journal of Power Electronics*, vol. 5, pp. 214–231, 2024.

[14] B. Soleymani and S. Eren, "Impacts of high-frequency harmonics of input current on a multi-string full-bridge solar inverter," in *2023 25th European Conference on Power Electronics and Applications (EPE'23 ECCE Europe)*, 2023, pp. 1–6.

[15] J. Yu, K. Lee, J. Cho, and B. Choi, "The study on emi characteristics under various operational conditions in dc-dc converter for electric vehicle," in *2024 IEEE 15th International Symposium on Power Electronics for Distributed Generation Systems (PEDG)*, 2024, pp. 1–4.

[16] K. Mainali and R. Oruganti, "Conducted emi mitigation techniques for switch-mode power converters: A survey," *IEEE Transactions on Power Electronics*, vol. 25, no. 9, pp. 2344–2356, 2010.

[17] R. Gamoudi, D. Elhak Chariag, and L. Sbita, "A review of spread-spectrum-based pwm techniques—a novel fast digital implementation," *IEEE Transactions on Power Electronics*, vol. 33, no. 12, pp. 10 292–10 307, 2018.

[18] Q. Li, X. Zhao, D. Jiang, and J. Chen, "Voltage ripple control of flying capacitor three-level inverter with variable switching frequency pspwm," *IEEE Transactions on Industrial Electronics*, vol. 69, no. 4, pp. 3313–3323, 2022.

[19] F. Giardine, Y. Wu, and R. C. N. Pilawa-Podgurski, "A variable switching frequency control technique for dc-ac flying capacitor multilevel converters to improve efficiency and inductor utilization," in *2024 IEEE Energy Conversion Congress and Exposition (ECCE)*, 2024.

[20] B. McGrath and D. Holmes, "Multicarrier pwm strategies for multilevel inverters," *IEEE Transactions on Industrial Electronics*, vol. 49, no. 4, pp. 858–867, 2002.

[21] M. G. Taul, N. Pallo, A. Stillwell, and R. C. N. Pilawa-Podgurski, "Theoretical analysis and experimental validation of flying-capacitor multilevel converters under short-circuit fault conditions," *IEEE Transactions on Power Electronics*, vol. 36, no. 11, pp. 12 292–12 308, 2021.

[22] L. Horowitz and R. C. Pilawa-Podgurski, "High power density flying capacitor multilevel inverter for electric aircraft with a stacked pcb interleaved hybrid commutation loop design," in *2023 IEEE Applied Power Electronics Conference and Exposition (APEC)*, 2023, pp. 1065–1069.

Analysis and Implementation of Minimum-Sensor Capacitor Voltage Estimators for Flying Capacitor Multilevel Converters

S. Tahmid Mahbub, Rahul K. Iyer, Ivan Z. Petric, Robert C. N. Pilawa-Podgurski

Dept. of Electrical Engineering and Computer Sciences, University of California, Berkeley, USA
Email: {tahmid, rkiyer, pilawa}@berkeley.edu, ivan5ric@ieee.org

Abstract—This work studies the estimation of capacitor voltages in the Flying Capacitor Multilevel (FCML) converter via direct measurements of only the input voltage, switched node voltage, and a minimum number of additional voltage sensors. Prior work developing capacitor voltage estimators for FCML converters has demonstrated that capacitor voltages are unobservable over specific ranges of duty ratios that depend on the converter level count. This work derives the minimum number of differential capacitor voltage sensors required to ensure capacitor voltage observability over the entire conversion ratio range and discusses where these sensors must be placed. Experimental results employing industry-standard digital signal processor hardware verify capacitor voltage observability with the proposed approach and demonstrate improved converter performance compared to the dithering solution previously proposed for retaining observability.

I. INTRODUCTION

The Flying Capacitor Multilevel (FCML) [1] converter enables simultaneous use of smaller magnetic components and high-figure-of-merit low-voltage devices in designs, and has been recently explored in applications demanding highly efficient and compact power conversion solutions [2]–[6].

A generalized N-level FCML converter, shown in Fig. 1, is made up of $n_c = N - 1$ series-connected complementary switch pairs and $M = N - 2$ flying capacitors. In steady-state operation, the converter's switching signals are generated with symmetric Phase-Shifted Pulse-Width Modulation (PS-PWM) shown in Fig. 2, yielding a switched node voltage with effective frequency $f_{\mathrm{eff}} = (N-1)f_s$ as shown in Fig. 3a. The reduced instantaneous switched node voltage and its higher fundamental frequency enable the use of an $(N-1)^2$ lower filter inductance L, compared to a 2-level converter, to meet a current ripple specification. Furthermore, the nominal balanced flying capacitor voltages are given by $v_{c,k} = \frac{k}{n_c}V_{\mathrm{in}}$ where $k = \{0, ..., n_c - 1\}$ is an integer. This corresponds to an equal nominal blocking voltage of $\frac{1}{n_c}V_{\mathrm{in}}$ across each switch in the converter, enabling the use of low-voltage switches with high figure-of-merit in place of high-voltage switches rated for the full supply voltage [7], [8].

In a FCML converter operated with PS-PWM, the capacitor voltages reach their nominal steady-state values via a "natural balancing" phenomenon characterized in [9]–[11]. Imbalance in capacitor voltages results in the inductor current ripple

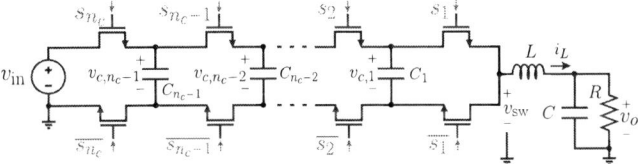

Fig. 1: Schematic of an N-level FCML converter showing $n_c = N - 1$ complementary switch pairs and $M = N - 2$ flying capacitors.

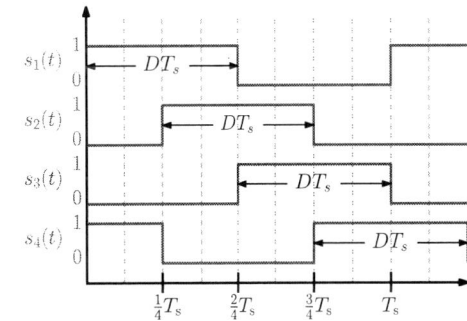

Fig. 2: Switching signals for a 5-level FCML converter generated with duty ratio $D = 0.5$ under PS-PWM.

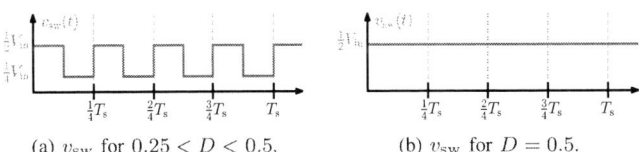

(a) v_{sw} for $0.25 < D < 0.5$.   (b) v_{sw} for $D = 0.5$.

Fig. 3: Switched node voltage v_{sw} under PS-PWM at different duty ratios D with balanced flying capacitor voltages for a 5-level FCML converter.

deviating from its nominal symmetric triangular shape, which drives nonzero average capacitor currents that serve to rebalance the system. The natural balancing dynamics are typically slow and underdamped. In converter start-up and shutdown scenarios, the oscillatory line transient response results in unequal voltage stresses across the power devices in the converter, and may cause switch blocking voltages to surpass the device ratings.

Recent work has investigated closed-loop "active balancing" control to regulate the flying capacitor voltages to their balanced fractions of the supply voltage [12]–[19]. However, most active balancing requires measurement of the capacitor voltages, which is typically accomplished with differential

979-8-3315-1612-3/25 $31.00 © 2025 IEEE

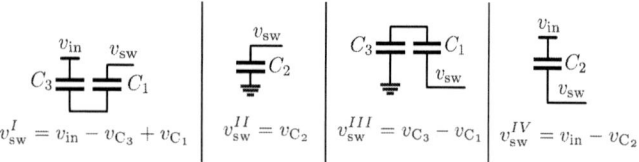

Fig. 4: Flying capacitor connections with $D = 0.5$ for the 5-level FCML converter. Each phase corresponds to a quarter of the switching period T_s.

voltage measurement circuits. As the required hardware in such schemes scales linearly with the converter level count N, this approach is cost-prohibitive for higher level-count designs and can significantly increase the overall system volume. Thus, recent research [12], [20]–[22] has investigated estimation of capacitor voltages with a reduced number of sensors by using only ground-referenced measurements of the supply input and the switched node voltage v_{sw}.

This work presents a systematic framework for: 1) identifying which capacitor voltages cannot be estimated solely with input and switched node measurements, and 2) subsequently identifying where additional sensor(s) must be placed to recover observability of all capacitor voltages, under the PS-PWM scheme. The proposed approach addresses cost-effective sensing for reliable control of flying capacitor voltages—a key obstacle to industry adoption of FCML converters. The results of this work are especially applicable to high level-count multilevel inverters in which the conversion ratio crosses several regions where capacitor voltages are unobservable in each line cycle.

Section II discusses the capacitor estimation problem, considering the relationship between the switched node voltage, the input voltage and the flying capacitor voltages. Section III analyzes the positioning of the minimum number of additional differential voltage sensors that must be placed to observe all capacitor voltages. Section IV compares the performance of the proposed minimum-sensor design with previously reported schemes for both a 5-level and a 7-level FCML converter.

II. THE CAPACITOR VOLTAGE ESTIMATION PROBLEM

The switched node voltage can be written as a combination of the M flying capacitor voltages and the input voltage in each switching phase. Under the assumption that the capacitor voltage dynamics are slow compared to the switching period, this relationship is described [22] in the form shown in (1). In this representation, $[v_{sw}]$ and $[v_{in}]$ are column vectors describing the values of the switched node and input voltages for each switching phase, $[v_c]$ is a column vector describing each flying capacitor's voltage, $[C]$ is a matrix that captures the polarity of capacitor connections to the switched node in each switching phase, and $[W]$ is a diagonal matrix describing whether the input voltage is connected to the switched node in a particular switching phase.

$$\underset{n_{ph} \times 1}{[v_{sw}]} = \underset{n_{ph} \times M}{[C]} \cdot \underset{M \times 1}{[v_c]} + \underset{n_{ph} \times n_{ph}}{[W]} \cdot \underset{n_{ph} \times 1}{[v_{in}]} \quad (1)$$

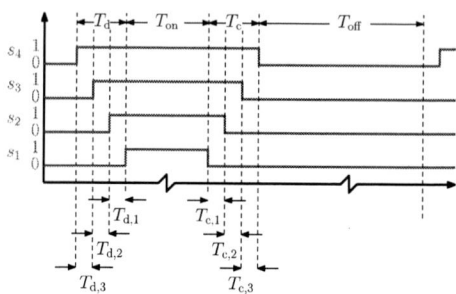

Fig. 5: Switching signals for a 5-level FCML converter generated with Quasi-Two-Level (Q2L) Modulation when inductor current has a negative valley and a positive peak. The capacitors are individually discharged during the $T_{d,1}$, $T_{d,2}$ and $T_{d,3}$ intervals through the negative inductor current. The capacitors are individually charged during the $T_{c,1}$, $T_{c,2}$ and $T_{c,3}$ intervals through the positive inductor current.

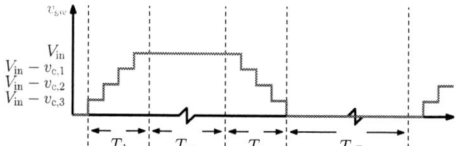

Fig. 6: Example waveform for the switched node voltage v_{sw} with Q2L switching. $T_{on} \gg T_d$, $T_{on} \gg T_c$.

A. Modulation Scheme

The relationship described in (1) depends on the modulation scheme being used. Fig. 2 and 4 show the switching signals and corresponding flying capacitor connections respectively for a 5-level FCML converter operating under PS-PWM with a 50% duty ratio. At this operating point, (1) is expanded as

$$\underbrace{\begin{bmatrix} v_{sw}^I \\ v_{sw}^{II} \\ v_{sw}^{III} \\ v_{sw}^{IV} \end{bmatrix}}_{[v_{sw}]} = \underbrace{\begin{bmatrix} 1 & 0 & -1 \\ 0 & 1 & 0 \\ -1 & 0 & 1 \\ 0 & -1 & 0 \end{bmatrix}}_{[C]} \underbrace{\begin{bmatrix} v_{C_1} \\ v_{C_2} \\ v_{C_3} \end{bmatrix}}_{[v_c]} + \underbrace{\begin{bmatrix} 1 & 0 & 0 & 0 \\ 0 & 0 & 0 & 0 \\ 0 & 0 & 0 & 0 \\ 0 & 0 & 0 & 1 \end{bmatrix}}_{[W]} \underbrace{\begin{bmatrix} v_{in}^I \\ v_{in}^{II} \\ v_{in}^{III} \\ v_{in}^{IV} \end{bmatrix}}_{[v_{in}]}.$$

(2)

If the matrix $[C]$ has full-column-rank, the unique minimum-norm solution for the flying capacitor voltages can either be determined exactly via the pseudoinverse [23] $[v_c] = [C]^\dagger ([v_{sw}] - [W][v_{in}])$ or approximately via iterative methods such as that detailed in [22], [24]. The issue of capacitor voltage observability—whether capacitor voltages can be measured using only the switched node and input voltage measurements—therefore arises when $[C]$ is not full-column-rank, as detailed in [20]–[22].

When using an alternate modulation scheme, (2) is rewritten with different values of $[C]$ and $[W]$ depending on the sequence of switching states that appears over the switching period. For example, with the quasi-2-level (Q2L) modulation scheme [15]–[18], at most one capacitor is connected to the switched node in any switching phase, which guarantees observability of all capacitor voltages. To ensure the switched node voltage measurements are sufficiently far from switching transitions, the durations of the states when capacitors are connected must be sufficiently long.

Switching signals and an example switched node voltage waveform under the Q2L scheme are shown in Fig. 5 and

Fig. 6 respectively. For the rising edge of the switched node voltage under the Q2L scheme (during the T_d interval in Fig. 6), the system is described by

$$\underbrace{\begin{bmatrix} v_{\mathrm{sw}}^I \\ v_{\mathrm{sw}}^{II} \\ v_{\mathrm{sw}}^{III} \end{bmatrix}}_{[v_{\mathrm{sw}}]} = \underbrace{\begin{bmatrix} 0 & 0 & -1 \\ 0 & -1 & 0 \\ -1 & 0 & 0 \end{bmatrix}}_{[C]} \underbrace{\begin{bmatrix} v_{C_1} \\ v_{C_2} \\ v_{C_3} \end{bmatrix}}_{[v_c]} + \underbrace{\begin{bmatrix} 1 & 0 & 0 \\ 0 & 1 & 0 \\ 0 & 0 & 1 \end{bmatrix}}_{[W]} \underbrace{\begin{bmatrix} v_{\mathrm{in}}^I \\ v_{\mathrm{in}}^{II} \\ v_{\mathrm{in}}^{III} \end{bmatrix}}_{[v_{\mathrm{in}}]}.$$

(3)

$[C]$ is full-column-rank, implying observable capacitor voltages. Despite the observability advantage, the Q2L scheme relinquishes the $(N-1)^2$ reduction of filter inductance as compared to PS-PWM, which remains the most commonly used modulation scheme for the FCML converter. This work focuses on minimum-sensor estimation specifically under the PS-PWM scheme.

B. System Observability and Unobservable States

Under the PS-PWM scheme, matrix $[C]$ is rank-deficient when the duty ratio can be expressed as $D = \frac{m}{n_c}$, where $m \in \mathbb{N}$ and m and n_c are not co-prime [20], [21]. In practical implementations, the unobservable region is a range of duty ratios around these points as short-duration switching phases cannot be reliably sampled without coupling noise from the switching transitions into the switched node voltage measurements. At these conversion ratios, the total number of switching phases in a period is n_c. Therefore the rank deficiency, defined as $\rho_d := n_c - 1 - \mathrm{rank}([C])$, indicates the number of unobservable capacitor voltages. To recover observability at non-co-prime duty ratios, the number of differential capacitor voltage sensors required, N_s, is also given by the rank deficiency

$$N_s = \rho_d = n_c - 1 - \mathrm{rank}([C]) \tag{4}$$

The unobservable capacitor voltages can be identified from non-zero entries in the basis vectors spanning the null space of $[C]$. From the rank-nullity theorem, the number of basis vectors spanning the null space is equal to ρ_d. As an example, the null-space of $[C]$ from the 5-level converter example of (2) is spanned by

$$\{[n_1]\} = \left\{ \begin{bmatrix} 1 & 0 & 1 \end{bmatrix}^T \right\}. \tag{5}$$

v_{C_1} and v_{C_3} can be expressed in terms of a common-mode and differential-mode value. The common-mode value is defined as $(v_{C_1} + v_{C_3})/2$ and the differential mode value is defined as $v_{C_3} - v_{C_1}$. The nullspace basis vector $[n_1]$, in (5), indicates that a common offset applied to v_{C_1} and v_{C_3} is indeterminate from the switched node voltage measurements and only the differential mode value can be determined. This is readily verified from the switched node equations in Fig. 4. Thus, these two capacitor voltages cannot be uniquely estimated without additional information.

For a 7-level FCML ($n_c = 6$), duty ratios corresponding to $2/6$ ($m = 2$), $3/6$ ($m = 3$) and $4/6$ ($m = 4$) yield switched node voltage equations where certain capacitor voltages are unobservable. When operating at $D = 3/6$, the null-space basis vectors of the corresponding $[C]$ matrix are

$$\{[n_1],[n_2]\} = \left\{ \begin{bmatrix} 1 & 0 & 0 & 1 & 0 \end{bmatrix}^T, \begin{bmatrix} 0 & 1 & 0 & 0 & 1 \end{bmatrix}^T \right\}. \tag{6}$$

When operating at $D = 2/6$ and $D = 4/6$, the null-space basis vector is

$$\{[n_1]\} - \left\{ \begin{bmatrix} 1 & 0 & 1 & 0 & 1 \end{bmatrix}^T \right\} \tag{7}$$

Following the arguments made for the 5-level converter, the capacitor voltages cannot be uniquely estimated without additional sensors. Notably, the null-space basis vectors in (6) are shifted versions of each other due to the circulant nature of the capacitor connections under PS-PWM.

III. SENSOR PLACEMENT STRATEGY

The rank deficiency of the system's $[C]$ matrix, ρ_d, dictates the number of additional sensors needed to uniquely estimate all flying capacitor voltages at the corresponding duty ratio. Table I specifies the minimum number of flying capacitor voltage sensors to enable estimation across operating duty ratios for FCML converters under PS-PWM, up to 13 levels ($n_c = 12$). The duty ratio is given by m/n_c.

As described in Section II-B, the unobservable capacitor voltages are identified by non-zero entries in the null-space basis vectors. In the 5-level case, adding a sensor to either v_{C_1} or v_{C_3} enables estimation of all flying capacitor voltages uniquely. In this approach, only 3 sensors are needed for the switched node voltage, the input voltage and either flying capacitor, with only the flying capacitor voltage requiring a differential sensor. This is two fewer than the total number of sensors required when measuring all flying capacitor voltages directly. For higher level count converters, the difference between the minimum-sensor design and a design sensing all capacitors is more significant.

In the case of measuring v_{C_1} directly, the additional measurement information can be captured by the augmented system described in (8), where $v_{C_1,\mathrm{meas}}$ indicates the directly measured capacitor voltage. Note that appending the elemen-

TABLE I
MINIMUM ADDITIONAL SENSOR COUNT

n_c / m	2	3	4	5	6	7	8	9	10	11	12
1	0	0	0	0	0	0	0	0	0	0	0
2		0	1	0	1	0	1	0	1	0	1
3			0	0	2	0	0	2	0	0	2
4				0	1	0	3	0	1	0	3
5					0	0	0	0	4	0	0
6						0	1	2	1	0	5
7							0	0	0	0	0
8								0	1	0	3
9									0	0	2
10										0	1
11											0

tary row vector $[1 \; 0 \; 0]$ to the $[C]$ matrix results in the full-column-rank *augmented* matrix $[C_{\text{aug}}]$.

$$
\begin{bmatrix}
v_{\text{sw}}^{I} \\
v_{\text{sw}}^{II} \\
v_{\text{sw}}^{III} \\
v_{\text{sw}}^{IV} \\
\hline
v_{C_1,\text{meas}}
\end{bmatrix}
=
\begin{bmatrix}
1 & 0 & -1 \\
0 & 1 & 0 \\
-1 & 0 & 1 \\
0 & -1 & 0 \\
\hline
1 & 0 & 0
\end{bmatrix}
\begin{bmatrix}
v_{C_1} \\
v_{C_2} \\
v_{C_3}
\end{bmatrix}
+
$$

$$
\underbrace{\phantom{[v_{\text{sw}}]}}_{[v_{\text{sw}}]}
\qquad
\underbrace{\phantom{[C_{\text{aug}}]}}_{[C_{\text{aug}}]}
\qquad
\underbrace{}_{[v_c]}
\tag{8}
$$

$$
\begin{bmatrix}
1 & 0 & 0 & 0 \\
0 & 0 & 0 & 0 \\
0 & 0 & 0 & 0 \\
0 & 0 & 0 & 1 \\
\hline
0 & 0 & 0 & 0
\end{bmatrix}
\begin{bmatrix}
v_{\text{in}}^{I} \\
v_{\text{in}}^{II} \\
v_{\text{in}}^{III} \\
v_{\text{in}}^{IV}
\end{bmatrix}.
$$

$$
\underbrace{\phantom{[W_{\text{aug}}]}}_{[W_{\text{aug}}]}
\qquad
\underbrace{\phantom{[v_{\text{in}}]}}_{[v_{\text{in}}]}
$$

For the 7-level converter, the null-space basis vectors corresponding to operation at $D = 3/6$, given in (6), indicate that one sensor must measure either v_{C_1} or v_{C_4} and a second sensor must measure v_{C_2} or v_{C_5}. The null-space basis vector corresponding to operation at $D = 2/6$ and $D = 4/6$, given in (7), indicates that one sensor must measure either v_{C_1}, v_{C_3}, or v_{C_5}. To preserve observability across all duty ratios, the 7-level converter has several possible combinations of additional sensors. Of particular interest is the addition of sensors for v_{C_1} and v_{C_2}. This minimum-sensor addition places the sensors at the lowest voltage capacitors and ensures capacitor voltage observability over the entire conversion ratio range.

Elementary row vectors corresponding to the measurement of v_{C_1} and v_{C_2}—$[1 \; 0 \; 0 \; 0 \; 0]$ and $[0 \; 1 \; 0 \; 0 \; 0]$— can be appended to the $[C]$ matrix corresponding to the 7-level FCML converter to obtain the corresponding full column-rank *augmented* matrix $[C_{\text{aug}}]$ similar to the 5-level case in (8).

In the general case, the minimum number of sensors can be placed sequentially from C_1, C_2 and onwards, or from C_M, C_{M-1} and so on. Placing sensors at the lowest voltage capacitors is advantageous as lower common-mode voltages require less attenuation on resistive dividers, which is helpful for reducing measurement noise, and allows for smaller routing clearances in the sensing circuits.

As a further example, an 11-level FCML ($n_c = 10$) converter may be considered. Table I indicates that 4 flying capacitor voltage sensors are needed. Adding sensors from either end—$\{v_{C_1}, v_{C_2}, v_{C_3}, v_{C_4}\}$ or $\{v_{C_9}, v_{C_8}, v_{C_7}, v_{C_6}\}$—will enable recovering observability across all duty ratios.

IV. EXPERIMENTAL CHARACTERIZATION

The proposed estimator is implemented on a Texas Instruments TMS320F28379D DSP. Assuming that the dynamics of capacitor voltages are slow compared to the rate at which they appear in the estimator equations, they are estimated using Richardson's iterative method [24]. This work extends the estimator design of [22] by incorporating information from the additional minimum sensors. The outputs of the iterative solver corresponding to the estimated capacitor voltages are augmented by the measured differential capacitor voltage readings as in (8) for the 5-level case. This can be implemented

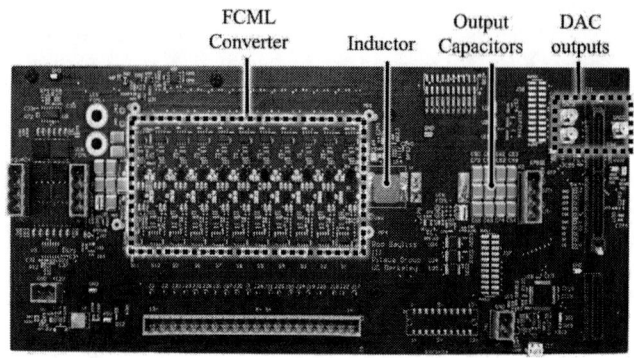

Fig. 7: FCML converter prototype used in this work. $L = 4.7\,\mu\text{H}$, $C_{\text{fly}} = 8.8\,\mu\text{F}$, $f_{\text{sw,eff}} = 200\,\text{kHz}$.

TABLE II

CONVERTER SPECIFICATIONS

Converter Parameter	Value
Effective Switching Frequency	200 kHz
Inductance	4.7 μH
Flying Capacitance	4x 2.2 μF
Switches	EPC2302
Gate Driver	LM5114
Isolator	ADUM5240
Cap. Voltage Sensor	AD8429ARZ-R7

by skipping the iterative computation for each capacitor that has an attached sensor and using the capacitor voltage directly.

The estimator is verified on the prototype shown in Fig. 7. The hardware is a 12-level converter with key parameters given in Table II, with the ability to configure to a lower level count by shorting consecutive switching cells. To verify minimum-sensor operation, the estimator is verified on both a 5-level and a 7-level converter. To maintain similar harmonic content in the inductor current without changing the inductance, the effective switching frequency, f_{eff}, was kept constant between the 5-level and 7-level tests.

The digital-to-analog converters (DACs) on the DSP were used to output the estimated capacitor voltages so that they may be measured on an oscilloscope along with the real measured flying capacitor voltages. The signals are scaled on the oscilloscope for direct comparison with measured capacitor voltages. Experiments were carried out with natural balancing under the PS-PWM scheme.

A. Estimation for a 5-Level Converter

The 12-level converter is reconfigured as a 5-level FCML converter by shorting seven switching cells. The converter is operated at the unobservable nominal duty ratio $D = 0.5$.

Results are first shown in Fig. 8 for the converter operating under an input voltage transient typical of a startup scenario highlighting the inability to correctly estimate v_{C_1} and v_{C_3} without the additional voltage sensor on C_1. Fig. 9 illustrates the erroneous estimation in steady-state without additional sensor information for C_1. The estimated voltages snap to the correct values once the sensor information is used for capacitor C_1. In these tests, v_{C_2} is unaffected by the use of the sensor since its voltage is an observable state. In contrast, Fig. 10

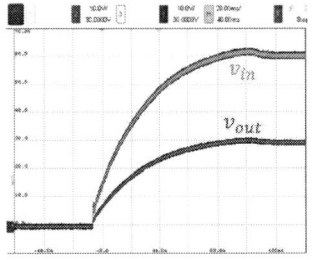

(a) Input and output voltages during startup from 0 V to 60 V with $D = 0.5$.

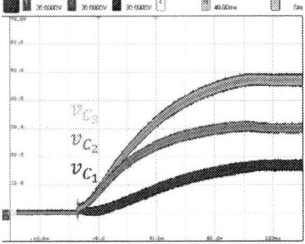

(b) Measured flying capacitor voltages during startup.

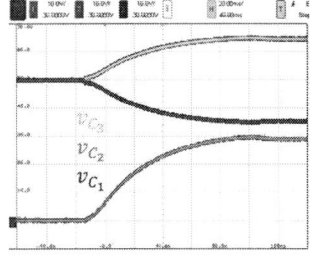

(c) Estimated capacitor voltages with no additional capacitor voltage sensors.

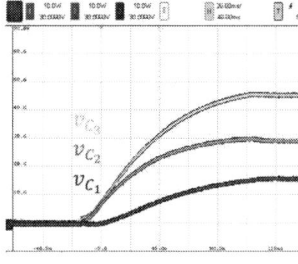

(d) Estimated capacitor voltages with additional sensor for C_1.

Fig. 8: Supply startup test 0 V to 60 V with $D = 0.5$ for the 5-level FCML converter. At this operating point, v_{C_1} and v_{C_3} cannot be correctly estimated from just the input and switched node measurement. The addition of a sensor for C_1 enables correct estimation of all flying capacitor voltages.

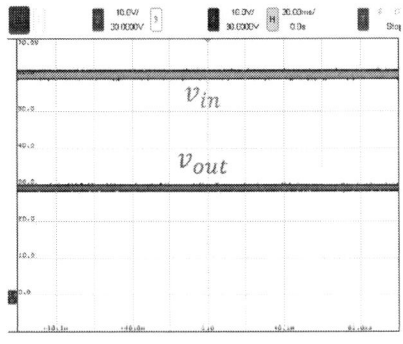

(a) Steady state input (60 V) and output (30 V) voltages with $D = 0.5$.

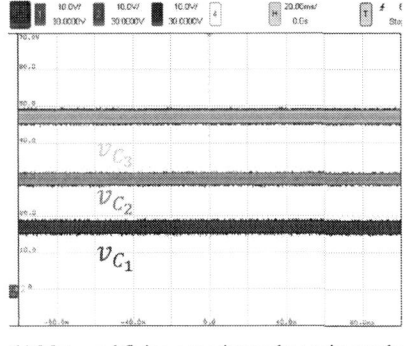

(b) Measured flying capacitor voltages in steady state operation.

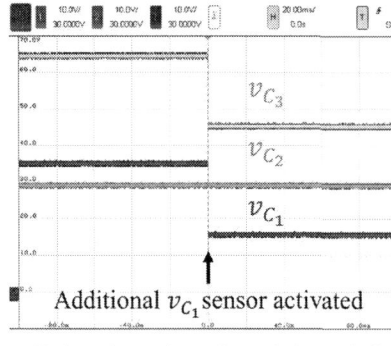

(c) Estimated capacitor voltages before and after additional sensor for C_1 is used.

Fig. 9: Steady-state estimation test for the 5-level FCML converter. Before the v_{C_1} sensor is used, only v_{C_2} is correctly estimated. Once the sensor is used, v_{C_1} and v_{C_3} snap to the correct estimates that match measurements. v_{C_2} is unaffected by the use of the sensor.

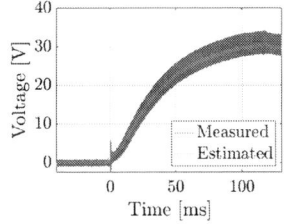

(a) Differential mode voltage.

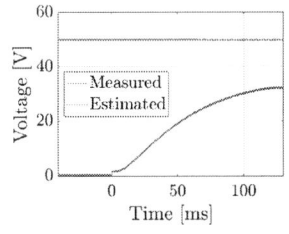

(b) Common mode voltage.

Fig. 10: Cap voltages v_{C_1} and v_{C_3}, during input startup from 0 V to 60 V, for the 5-level FCML converter in Section IV-A. The differential mode voltage is defined as $v_{C_3} - v_{C_1}$. The common mode voltage is defined as $\left(v_{C_3} + v_{C_1} \right) /2$. The estimated differential voltages match the measured values. The common mode voltage cannot be correctly estimated without additional sensors.

depicts the incorrect common-mode voltage estimation for v_{C_1} and v_{C_3} without the use of the additional sensor.

Prior work [22] has demonstrated the use of a dithering scheme, operating at observable duty ratios $D = 0.485$ and 0.515 for an effective conversion ratio of 50%. Fig. 11 illustrates the result of a similar dithering implementation and highlights the increased ripple on the flying capacitor voltages as well as on the supply line voltage. The estimator proposed in this work does not suffer from the increased ripple observed with dithering.

B. Estimation for a 7-Level Converter

The 12-level converter is reconfigured as a 7-level FCML converter by shorting five switching cells. For clarity of the presented experimental data, the measured and estimated capacitor voltages from multiple oscilloscopes are combined and plotted on a single set of axes via post-processing in MATLAB.

Results are shown in Fig. 12 for the converter operating under an input voltage transient typical of a startup scenario. The converter is operating at an unobservable duty ratio of $D = 0.5$. This experiment illustrates the inability to correctly estimate v_{C_1}, v_{C_2}, v_{C_4} and v_{C_5} without the additional voltage sensors on C_1 and C_2.

As indicated from (6), only the difference between v_{C_1} and v_{C_4} can be computed uniquely but not their common mode voltage. Similarly, only the difference between v_{C_2} and v_{C_5} can be computed uniquely but not their common mode voltage. This is observed in Fig. 12b. As illustrated in Fig. 10, the startup transient is coupled into both the common mode and differential mode voltages. Since the estimated values in Fig. 12b do not capture the correct common mode voltage as described above, they arrive at incorrect steady-state values before the measured capacitor voltages have reached steady-state. The parameter v_{C_3} is correctly estimated without the need for additional sensors. The addition of sensors for C_1 and C_2 enables correct estimation of all flying capacitor voltages.

979-8-3315-1612-3/25 $31.00 © 2025 IEEE

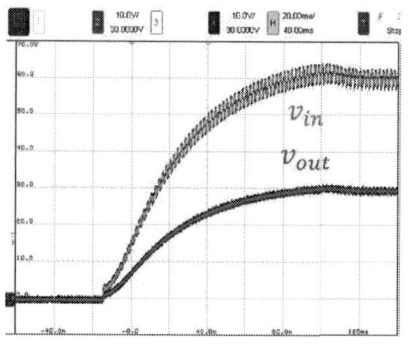
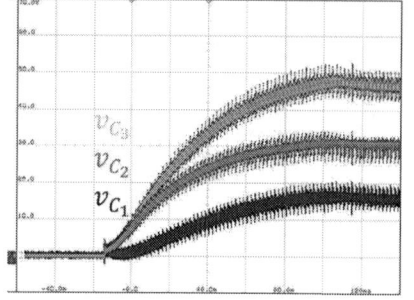
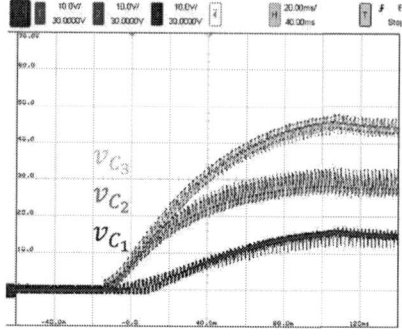

(a) Input and output voltages during startup to 60 V with $D = 0.5 \pm 0.015$ dithering.

(b) Measured flying capacitor voltages during startup with dithering.

(c) Estimated capacitor voltages during startup with dithering.

Fig. 11: Dithering test for the 5-level FCML converter without using sensor for C_1. Dithering around $D = 0.5$ enables estimating v_{C_1} and v_{C_3} at an otherwise unobservable operating point. However, the impact of dithering propagates to all flying capacitor voltages and the input voltage.

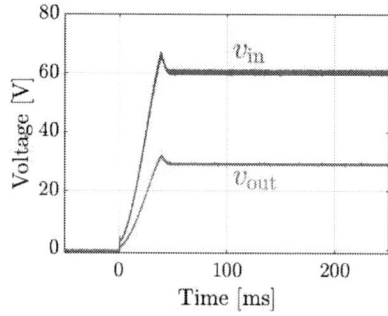
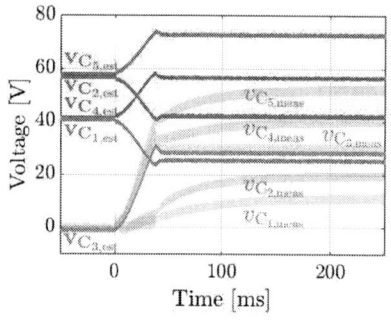
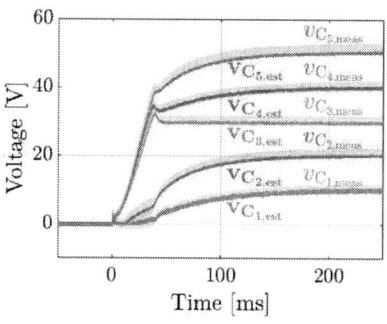

(a) Input and output voltages during startup from 0 V to 60 V with D = 0.5.

(b) Capacitor voltages with no additional capacitor voltage sensors.

(c) Capacitor voltages with additional sensor for C_1 and C_2.

Fig. 12: Supply startup test 0 V to 60 V with $D = 0.5$ for the 7-level FCML converter. Lighter shaded colors—annotated as $v_{C_{x,meas}}$—represent measured capacitor voltages; darker shaded colors—annotated as $v_{C_{x,est}}$—represent voltages estimated by the DSP. At this operating point, v_{C_1}, v_{C_2}, v_{C_4} and v_{C_5} cannot be correctly estimated from just the input and switched node measurements. The addition of sensors for C_1 for C_2 enables correct estimation of all flying capacitor voltages.

V. CONCLUSIONS

This work analyzes and demonstrates an implementation of minimum-sensor capacitor voltage estimation for FCML converters over the entire conversion ratio range. The presented framework for a general N-level converter demonstrates the ability to place sensors at the lowest capacitor voltages under PS-PWM. The estimator is implemented on a commercial digital signal processor and demonstrates excellent tracking of the capacitor voltages for practical 5-level and 7-level converters. Future work will investigate implementation of the estimator alongside closed-loop active balancing control in FCML converters with a small number of sensors, enabling reduction in converter volume and cost.

ACKNOWLEDGMENT

The information, data, or work presented herein was funded in part by the Advanced Research Projects Agency-Energy (ARPA-E), U.S. Department of Energy, under Award Number DE-AR0001609 in the OPEN 2021 program monitored by Dr. Olga Blum Spahn. The views and opinions of authors expressed herein do not necessarily state or reflect those of the United States Government or any agency thereof.

REFERENCES

[1] T. Meynard and H. Foch, "Multi-level conversion: high voltage choppers and voltage-source inverters," in *PESC '92 Record. 23rd Annual IEEE Power Electronics Specialists Conference*, 1992, pp. 397–403 vol.1.

[2] Y. Lei, C. Barth, S. Qin, W.-C. Liu, I. Moon, A. Stillwell, D. Chou, T. Foulkes, Z. Ye, Z. Liao, and R. C. N. Pilawa-Podgurski, "A 2-kw single-phase seven-level flying capacitor multilevel inverter with an active energy buffer," *IEEE Transactions on Power Electronics*, vol. 32, no. 11, pp. 8570–8581, 2017.

[3] T. Modeer, N. Pallo, T. Foulkes, C. B. Barth, and R. C. N. Pilawa-Podgurski, "Design of a gan-based interleaved nine-level flying capacitor multilevel inverter for electric aircraft applications," *IEEE Transactions on Power Electronics*, vol. 35, no. 11, pp. 12 153–12 165, 2020.

[4] L. Horowitz and R. C. Pilawa-Podgurski, "A 14-level fcml inverter for electric vehicles with optimal capacitors achieving 175 kw/kg and 380 kw/l power density," in *2024 IEEE Applied Power Electronics Conference and Exposition (APEC)*, 2024, pp. 1009–1013.

[5] O. Lorenz and J. Sanchez, "Ultra low-profile flying capacitor 7-level 3kw pfc with optimized high frequency layout and active balancing using 100v gan," in *2024 IEEE Applied Power Electronics Conference and Exposition (APEC)*, 2024, pp. 22–28.

[6] Q. Huang, Q. Ma, P. Liu, A. Q. Huang, and M. A. de Rooij, "99% efficient 2.5-kw four-level flying capacitor multilevel gan totem-pole pfc," *IEEE Journal of Emerging and Selected Topics in Power Electronics*, vol. 9, no. 5, pp. 5795–5806, 2021.

[7] J. Azurza Anderson, G. Zulauf, J. W. Kolar, and G. Deboy, "New figure-of-merit combining semiconductor and multi-level converter properties," *IEEE Open Journal of Power Electronics*, vol. 1, pp. 322–338, 2020.

979-8-3315-1612-3/25 $31.00 © 2025 IEEE

[8] J. Azurza Anderson, G. Zulauf, P. Papamanolis, S. Hobi, S. Mirić, and J. W. Kolar, "Three levels are not enough: Scaling laws for multilevel converters in ac/dc applications," *IEEE Transactions on Power Electronics*, vol. 36, no. 4, pp. 3967–3986, 2021.

[9] T. Meynard, M. Fadel, and N. Aouda, "Modeling of multilevel converters," *IEEE Transactions on Industrial Electronics*, vol. 44, no. 3, pp. 356–364, 1997.

[10] R. H. Wilkinson, T. A. Meynard, and H. du Toit Mouton, "Natural balance of multicell converters: The general case," *IEEE Transactions on Power Electronics*, vol. 21, no. 6, pp. 1658–1666, 2006.

[11] Z. Xia, B. L. Dobbins, and J. T. Stauth, "Natural balancing of flying capacitor multilevel converters at nominal conversion ratios," in *2019 20th Workshop on Control and Modeling for Power Electronics (COMPEL)*, 2019, pp. 1–8.

[12] G. Farivar, A. M. Y. M. Ghias, B. Hredzak, J. Pou, and V. G. Agelidis, "Capacitor voltages measurement and balancing in flying capacitor multilevel converters utilizing a single voltage sensor," *IEEE Transactions on Power Electronics*, vol. 32, no. 10, pp. 8115–8123, 2017.

[13] C.-Y. Lu, D.-H. Lin, and H.-C. Chen, "Decoupled design of voltage regulating and balancing controls for four-level flying capacitor converter," *IEEE Transactions on Industrial Electronics*, vol. 68, no. 12, pp. 12 152–12 161, 2021.

[14] R. K. Iyer, I. Z. Petric, R. S. Bayliss, N. C. Brooks, and R. C. N. Pilawa-Podgurski, "A high-bandwidth parallel active balancing controller for current-controlled flying capacitor multilevel converters," *IEEE Transactions on Power Electronics*, pp. 1–15, 2024.

[15] R. K. Iyer, S. T. Mahbub, and R. C. N. Pilawa-Podgurski, "Dynamical modeling and control of the flying capacitor multilevel converter under quasi-two-level switching for active balancing in light-load conditions," in *2024 IEEE Workshop on Control and Modeling for Power Electronics (COMPEL)*, 2024, pp. 1–7.

[16] R. K. Iyer, S. T. Mahbub, and R. C. Pilawa-Podgurski, "Quasi- two-level switching for active balancing of flying capacitor multilevel converters under light-load conditions," in *2024 IEEE International Communications Energy Conference (INTELEC)*, 2024, pp. 1–7.

[17] P. Czyz, P. Papamanolis, F. Trunas Bruguera, T. Guillod, F. Krismer, V. Lazarevic, J. Huber, and J. W. Kolar, "Load-independent voltage balancing of multi-level flying capacitor converters in quasi-2-level operation," *Electronics*, vol. 10, no. 19, 2021. [Online]. Available: https://www.mdpi.com/2079-9292/10/19/2414

[18] S. Mersche, D. Bernet, and M. Hiller, "Quasi-two-level flying-capacitor-converter for medium voltage grid applications," in *2019 IEEE Energy Conversion Congress and Exposition (ECCE)*, 2019, pp. 3666–3673.

[19] M. Khazraei, H. Sepahvand, K. A. Corzine, and M. Ferdowsi, "Active capacitor voltage balancing in single-phase flying-capacitor multilevel power converters," *IEEE Transactions on Industrial Electronics*, vol. 59, no. 2, pp. 769–778, 2012.

[20] Z. Xia, B. L. Dobbins, J. S. Rentmeister, and J. T. Stauth, "State space analysis of flying capacitor multilevel dc-dc converters for capacitor voltage estimation," in *2019 IEEE Applied Power Electronics Conference and Exposition (APEC)*, 2019, pp. 50–57.

[21] Z. Xia, K. Datta, and J. T. Stauth, "State-space modeling and control of flying-capacitor multilevel dc–dc converters," *IEEE Transactions on Power Electronics*, vol. 38, no. 10, pp. 12 288–12 303, 2023.

[22] I. Z. Petric, R. K. Iyer, N. C. Brooks, and R. C. N. Pilawa-Podgurski, "A real-time estimator for capacitor voltages in the flying capacitor multilevel converter," in *2022 IEEE 23rd Workshop on Control and Modeling for Power Electronics (COMPEL)*, 2022, pp. 1–8.

[23] A. Ben-Israel and T. N. E. Greville, *Generalized Inverses: Theory and Applications*. New York, NY: Springer, 2003.

[24] L. F. Richardson, "The approximate arithmetical solution by finite differences of physical problems involving differential equations, with an application to the stresses in a masonry dam," *Philosophical Transactions of the Royal Society of London. Series A, Containing Papers of a Mathematical or Physical Character*, vol. 210, pp. 307–357, 1911. [Online]. Available: http://www.jstor.org/stable/90994

979-8-3315-1612-3/25 $31.00 © 2025 IEEE

Single-Stage Bidirectional High-Frequency Link DC to Three-Phase AC (4-Wire) Grid-Tied Microinverter

Aniruddh Marellapudi[†], Satish Belkhode[*], Joseph Benzaquen[†], and Deepak Divan[†]

[†]Center for Distributed Energy, Georgia Institute of Technology, Atlanta, GA, USA
[*]Department of Electrical Engineering, Indian Institute of Technology Roorkee, Roorkee, India
am123@gatech.edu

Abstract — **This work presents a novel, single-stage, inertia-less isolated converter (IIC) designed to interface low-voltage PV and battery storage to the three-phase AC grid at 480 VAC. This approach addresses key challenges with multi-stage DC/AC topologies conventionally used in grid-connected PV and energy storage applications. Specifically, elimination of the intermediate DC capacitor improves power density, reliability, and system cost. Further, single stage operation enabled by the proposed pulse density modulation (PDM) scheme simplifies plant control, providing high-bandwidth, closed loop control of the inverter output voltages without PWM dead-time and dwell-time compensation loops. The proposed PDM scheme manages volt-second and amp-second balance on the transformer, mitigating injection of DC flux on both the primary and secondary windings. Transformer parasitic elements are utilized to enable zero voltage switching (ZVS) of the AC bridge devices. This work provides a detailed analysis of the DC to 3-phase AC IIC topology, exploring the impact of critical parasitic elements and commutation of the AC bridge devices. A 3-phase, 2 kW 125 VDC to 480 VAC unit has been built and tested, validating ZVS operation of the AC bridge devices. Projected system efficiency at full load is >97.3%.**

Keywords—single-stage DC/AC converter, bidirectional, three-phase microinverter, ZVS soft-switching, pulse density modulation, grid-interactive converter, ultra-low leakage transformer

I. INTRODUCTION

Rapidly declining costs of PV solar, energy storage, and power converters have driven exponential growth in the deployment of distributed energy resources (DERs), all of which require a DC to AC interface, often with bidirectional power flow capability [1], [2]. In many applications, the need is for a 1-10 kW microinverter building block, which needs to convert low voltage (LV) DC to the AC grid voltage, typically using high-frequency (HF) isolation with high efficiency [3], [4]. As penetrations of DERs rise, the requirements of microinverter systems are evolving to include three-phase, 4-wire operation for flexibility under imbalanced conditions, capability to support AC microgrids or direct grid connections in grid-following and grid-forming modes, and the ability to scale from tens of kilowatts to megawatts using modular building blocks [5], [6].

The traditional approach to realizing bidirectional power flow control is with a high-frequency isolated DC/DC converter, such as a dual active bridge (DAB) [7] or series resonant converter (CLLC) converter followed by a conventional DC/AC inverter [8], [9], [10], [11]. The DAB stage uses a finite leakage inductor, L_{lkg}, with power transfer achieved through the control of phase angle difference between the two high-frequency AC waveforms across L_{lkg}. This involves careful selection of L_{lkg} and closed-loop control over the required phase shift for the desired

power transfer [12]. Control of the DAB converter is highly sensitive to minute angle differences, especially with low values of leakage inductance. The inverter operates with a PWM control strategy, typically at a lower frequency to reduce switching loss. An intermediate DC link capacitor is needed between the two stages to absorb the high-frequency current ripple, and to decouple the two power stages by maintaining low-ripple DC voltage on the intermediate DC bus. Given the large capacitance requirement, the DC link capacitor is often realized with an electrolytic capacitor, severely limiting reliability and power density [13].

To improve power density in isolated DC/AC conversion, several single-stage techniques have been proposed. Recently, single-stage DC/AC conversion has been achieved with a DC side full bridge and AC side cycloconverter separated by high-frequency isolation, but once again using a finite leakage inductor [14]. The principle of operation for this topology mirrors that of the DAB. However, the phase shift operation with the cycloconverter involves closed-loop control with complex real-time computation required for accurate power transfer. Further, this configuration utilizes the indirect capacitive filter stages on AC-side, leading to relatively slower dynamics due to the inertia of the power transfer mechanism.

A second category of solutions achieves direct single-stage DC/AC conversion without the requirement of indirect decoupling capacitive stage through use of the single-stage high-frequency link isolated configuration [15], [16], [17]. This single-stage high-frequency link converter is operated with pulse width modulation (PWM) on either DC or AC side bridge as reported in [16] and [17], respectively. In each switching cycle, utilization of the high frequency link voltage is modulated using a zero state, whose duration is computed to achieve a sinusoidal PWM type voltage on the output filter inductor. One major challenge of this approach arises from the mismatch in the applied active state and zero state durations over the switching cycle leading to DC flux in the transformer core, causing increased losses and potential saturation of the transformer, and limiting the overall power handling capability of the converter. Furthermore, as these approaches still utilize a low but finite leakage inductance, management of trapped leakage energy requires complex circuitry to avoid significant losses and increased device voltage stress.

A third approach for single-stage DC/AC conversion is with the Soft-Switching Solid-State-Transformer (S4T), which utilizes a flyback principle to achieve single-stage isolated power transfer [1]. As a reduced DC link converter, the energy stored in the magnetizing inductance in the S4T is >10X lower than that stored in the DC link capacitor of a similarly rated voltage source inverter (VSI) [18]. Nevertheless, the S4T still

979-8-3315-1612-3/25 $31.00 © 2025 IEEE

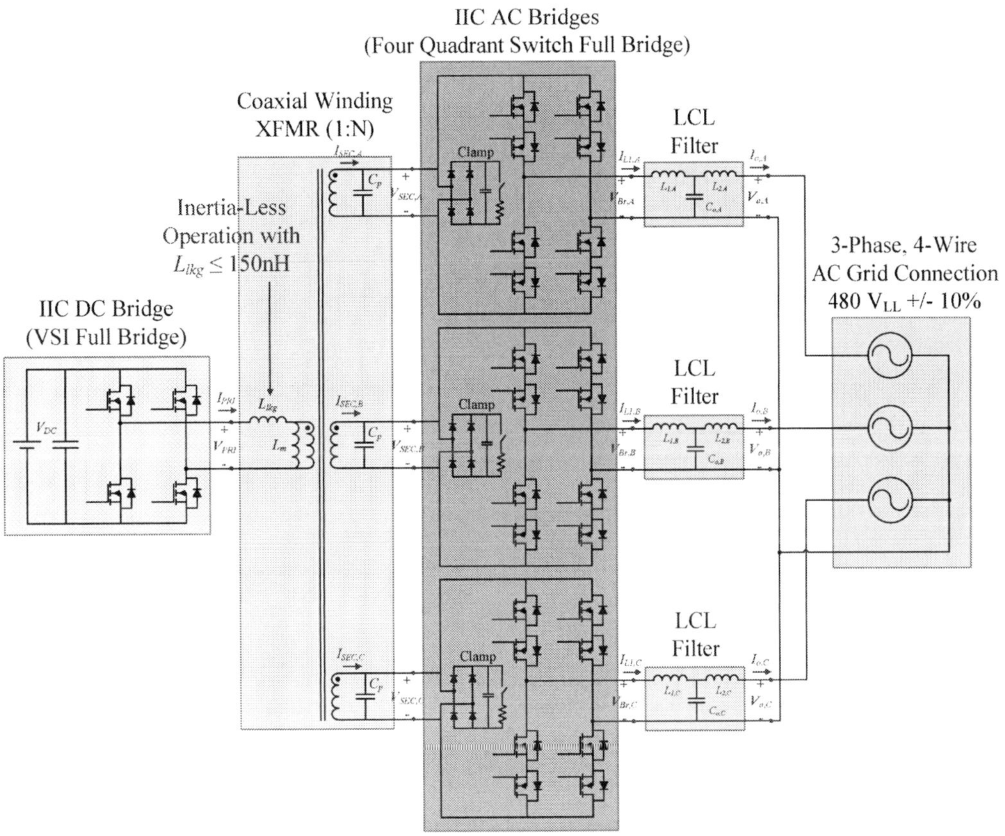

Fig. 1. Topology of the DC to 3-phase AC IIC microinverter consisting of a VSI full-bridge (DC bridge), three four-quadrant-switch full bridges (AC bridges), and a four-winding ultra-low leakage inductance transformer.

comprises finite 'inertia,' and the control and topology complexity stemming from the resonant circuitry greatly increases system complexity, challenging ubiquitous usage in low-cost 1-10 kW microinverters.

Thus, although some of the proposed technologies realize single-stage DC/AC power conversion, they either require a complex control mechanism or comprise additional energy storage elements that represent 'inertia' and can slow down the overall controllability and dynamic performance of the plant. Particularly for cost-sensitive applications such as low-cost residential solar microinverters, energy access portals, storage-integrated home energy management systems, and EV chargers, existing approaches have been constrained by the cost and size impact of these storage elements. To address the aforementioned challenges, this paper presents an isolated single-stage bidirectional DC to three-phase AC converter, where energy is transferred across an ultra-low leakage high frequency transformer, resulting in an inertialess isolated converter (IIC). The IIC concept was initially presented in a single-phase configuration in [19], but several aspects warranted further exploration and refinement. The IIC topology eliminates the need for an intermediate DC capacitor stage, improving power density and reliability while also reducing the converter cost. Paired with a unique discrete pulse modulation (DPM) strategy, the converter achieves zero DC flux in the transformer core and high-bandwidth closed loop control of the AC output voltages without the need for complex real time vector duration calculations. The rest of the paper presents a detailed discussion of the topology, principle of operation, modulation and control, and experimental results in both single-phase and three-phase applications.

II. TOPOLOGY

This work presents a single-stage, bidirectional DC to 3-phase AC microinverter based on the inertialess isolated converter (IIC) topology that addresses several limitations of state-of-the-art approaches [14], [15], [16], [17]. As shown in Fig. 1, the DC to 3-phase IIC topology comprises a voltage-sourced DC full bridge, an ultra-low leakage high frequency transformer, and three four-quadrant-switch (FQS) based full bridges whose outputs are connected to LCL filters and to the grid. Unlike single-stage DC to AC DAB-based configurations where a finite leakage inductance is used as the energy transfer element, the IIC features nearly zero leakage inductance, forcing the voltages on the DC and AC bridges to be essentially equal. Thus, there is no energy stored in the path from the DC bridge to the output of AC bridge, producing an inertialess converter.

Ultra-low leakage inductance in the transformer is achieved with the coaxially winding transformer (CWT) [20]. This reduces the complexity of the circuitry required to manage the leakage energy. Nonetheless, a small amount of leakage energy, which is stored in the practically constructed CWT can be dissipated using the simple clamping circuit as shown in Fig. 1. The clamping circuit also helps in limiting the high-frequency ringing overvoltage on the transformer secondary, clamping the stress on the AC bridge devices.

III. PRINCIPLE OF OPERATION

The key waveforms showing the principle of operation of the IIC are given in Fig. 2 and Fig 3. The DC bridge is operated to produce a high-frequency ($F_{SW} = 1/T_{SW}$) quasi-square wave, with equal positive and negative vector durations, and a short 'zero-voltage' duration as shown in Fig 3. This quasi-square wave voltage serves as the high frequency link voltage, corresponding to the DC link voltage in a standard voltage source inverter, however with the added constraint of volt-second balance on the transformer to prevent saturation. With an ultra-low leakage transformer such as a CWT, the quasi-square wave is reflected with nearly zero phase-shift across the turns ratio to generate a high voltage (HV) quasi-square-wave on all three secondary windings, $V_{SEC,n}$. Through application of flipping logic, each AC bridge then unfolds its transformer secondary voltage, $V_{SEC,n}$, to produce a cycle of two positive pulses or two negative pulses on its bridge output, $V_{BR,n}$, making the system amenable to high-bandwidth discrete pulse density modulation (PDM) strategies. Critically, the DC bridge is fed a quasi-square wave so that volt-second balance is achieved on the transformer primary, while the commutation of the AC bridge achieves AC side control objectives and ensures amp-second balance on the secondary.

Given that the IIC features no intermediate energy storage elements, the voltage of each secondary bridge, $V_{SEC,n}$, is ideally equal to the quasi-square wave reflected through the transformer turns ratio. However, two critical parasitic elements must be considered. First, the finite leakage inductance of the transformer, L_{lkg}, seen from the DC side is around 100 nH. Second, the parasitic capacitance of the AC bridge, C_p, composed of the capacitance of the AC bridge PCB plane structure in addition to the composite C_{oss} of the AC bridge devices, is in the order of 100-1000 pF. When combined, these parasitic elements cause a MHz-order LC oscillation, with resonant frequency and characteristic impedance defined in (1) and (2), on $V_{SEC,n}$ as the DC bridge switches states from the 'zero state' to the positive (or negative vector). To avoid overvoltage across the AC side switches, a minimal clamping circuit, consisting of a diode bridge and a capacitor, is added across each AC side transformer connection.

$$F_{res} = \frac{1}{2\pi\sqrt{(L_{lkg} \times N^2)C_p}} \tag{1}$$

$$Z_0 = \sqrt{\frac{(L_{lkg} \times N^2)}{C_p}} \tag{2}$$

Interestingly, commutation of the AC bridge 4QS devices can be designed to utilize L_{lkg} and C_p to enable ZVS turn-on and turn-off. At the end of the positive half cycle of the quasi-square wave, the DC bridge goes into overlap mode while the AC bridge devices remain gated on (State 2), causing V_{Cp} to decrease towards 0 V. When V_{Cp} reaches 0 V, the AC bridge is

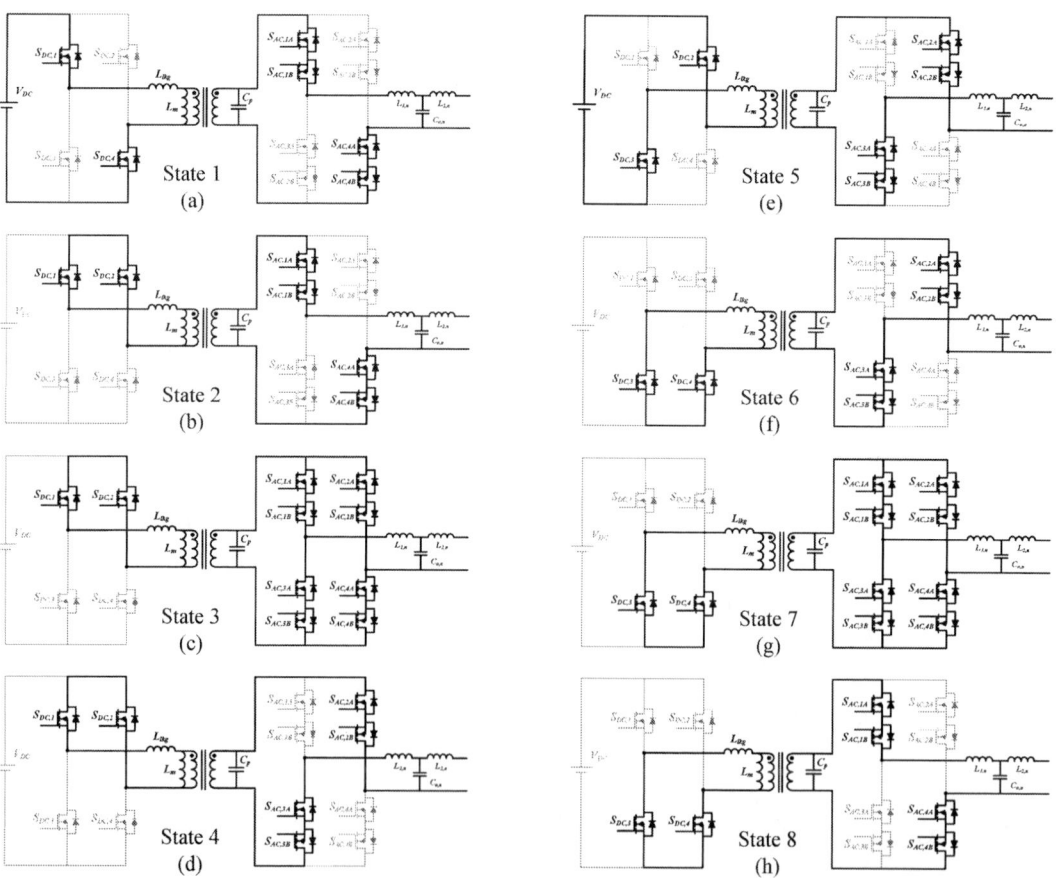

Fig. 2. Operating states over one switching cycle: (a) state 1 – positive half cycle of quasi-square wave; (b) state 2 – zero state 1, in which the transformer secondary capacitance discharges to 0 V, providing ZVS turn on conditions for the incoming AC bridge devices; (c) state 3 – zero state 2, overlap period in which all AC bridge devices are on; (d) state 4 – ZVS turn-on of AC bridge devices prior to start of negative half cycle of quasi-square wave (driven by DC bridge); (e)-(h) correspond to (a)-(d) for the negative half cycle of the quasi-square wave.

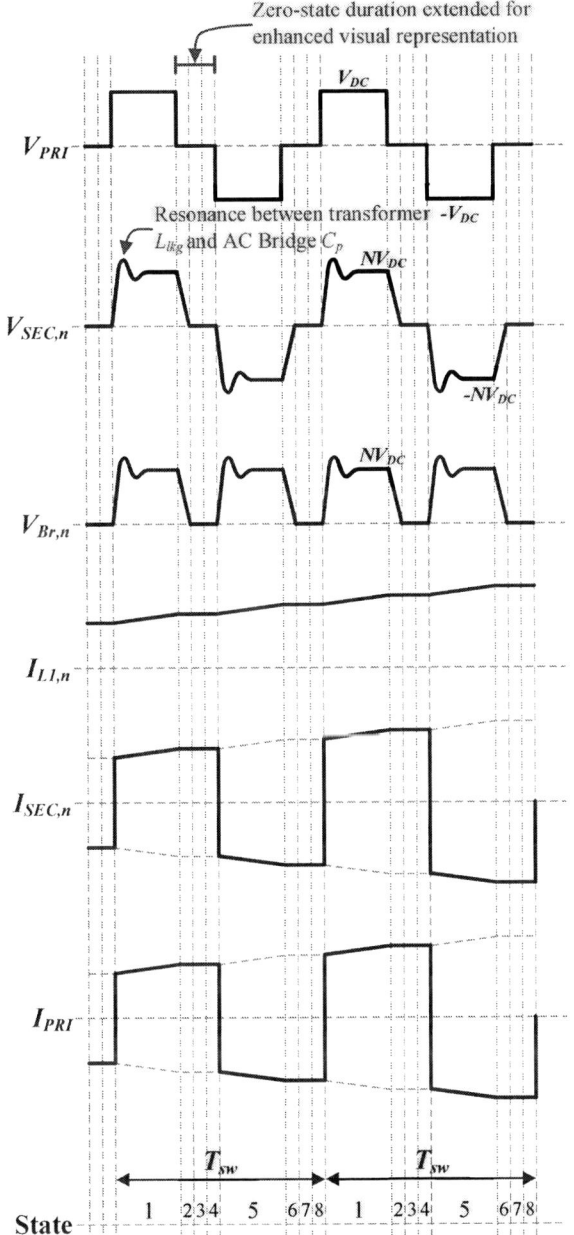

Fig. 3. Conceptual waveforms of the IIC under the proposed discrete pulse modulation. (state 1 – 8 correspond to Fig. 2 (a) – (h)).

commanded into overlap mode (State 3), thus enabling ZVS turn-on of the two incoming FQS devices. Similarly, immediately before the start of the negative half cycle of the quasi-square wave (State 4), the outgoing AC bridge devices are gated off again under ZVS conditions. This process is repeated for the negative half cycle, ensuring that all AC bridge FQS devices are commutated through the 'zero state,' providing the conditions for ZVS turn-on and turn-off of all AC bridge devices. This process, enabled by the proposed PDM method, eliminates switching losses in AC bridge devices and limits dV/dt and EMI on the bridge output voltages, $V_{Br,n}$, while also maintaining volt-second balance on the transformer primary and providing a simple actuation scheme.

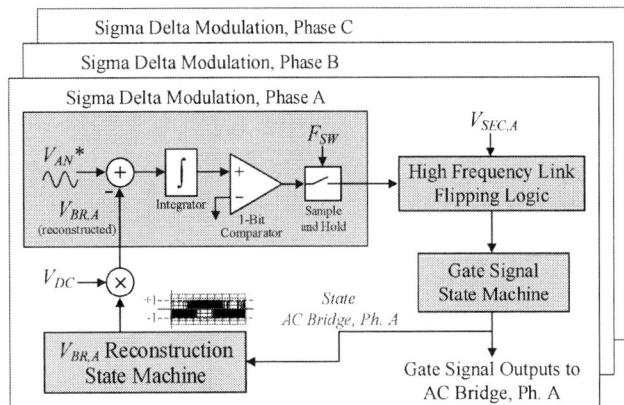

Fig. 4. Three-phase sigma-delta modulation scheme for the DC to three-phase IIC microinverter.

IV. THREE-PHASE AC SIGMA-DELTA MODULATION

While several PWM schemes have been proposed to synthesize low-frequency AC outputs from isolated high-frequency links, these methods suffer from potential injection of DC or low-frequency flux into the transformer stemming from pulse duration mismatches [16], [17]. PWM schemes rely on application of variable-length AC bridge pulses whose durations are computed to achieve sinusoidal PWM output voltages. In real-world application, dead-times and dwell-times must be applied to ensure safe commutation, adding to pulse duration non-idealities and necessitating inner compensation loops. In addition, real time computation of vector durations to achieve net zero flux may result in complex control and an increased number of switching events per high-frequency cycle.

An alternate approach exists, eliminating the challenges of existing PWM schemes in single-stage high-frequency link converters. The proposed sigma-delta pulse density modulation scheme, shown in Fig. 4 and based on the approach described in [21], enables the delivery of an integral number of discrete pulses to each AC bridge output, $V_{Br,n}$, such that fine control of the output voltage spectral components is realized. In the IIC, sigma-delta modulation is paired with flipping logic, which enables application of discrete pulses with appropriate polarity on the AC bridge output while ensuring volt-second balance of the transformer with the primary-side quasi-square wave.

The sigma-delta PDM modulation method necessarily acts on the instantaneous error between the phase voltage reference, V_{AN}^*, and the AC bridge output voltage, $V_{Br,n}$, before connecting to any passive element, providing direct high-bandwidth, closed-loop control of the inverter output voltages. Each AC bridge output voltage, a high-voltage signal with components at the switching frequency, F_{SW}, can either be sensed directly with low-noise analog circuitry, or reconstructed using the applied switching states of the AC bridges in the controller itself. The integrated error output between the reference voltage and the applied discrete pulses is passed to the AC bridge in the form of switching states to reduce the error and track the reference. The flipping logic impresses the right polarity on the output while ensuring that DC flux is not present. A suitably designed LCL filter stage is connected to interface with the three-phase AC grid at 480 V_{LL}. In summary, the proposed PDM scheme achieves high-bandwidth, closed-loop, and independent control of the three-phase output voltages while simultaneously ensuring

TABLE I. SIMULATION AND EXPERIMENTAL PROTOTYPE PARAMETERS

Parameter	Value
Power Converter General Specifications	P_{rated} = 2 kW, V_{DC} = 125 V, F_{SW} = 40 kHz, $V_{o,LL}$ = 480 V_{LL} +/- 10%, 60 Hz, $\eta_{2kW} \geq$ 97.3%
Transformer	Turns Ratio N = 1:5, L_m = 350 µH, L_{lkg} = 150 nH
LCL Filter	$L_{1,n}$ = 18 mH, $L_{2,n}$ = 1.5 mH, $C_{o,n}$ = 7 µF
AC Bridge Parasitic Capacitance	C_p = 500 pF
DC Bridge Power Devices	IPB068N20NM6ATMA1, 200 V Si MOSFET
AC Bridge Power Devices	NTBG070N120M3S, 1.2 kV SiC MOSFET
Clamp	Diode: C4D02120E-TR, 1.2 kV SiC Schottky C_{clamp} = 1 µF

Fig. 5. Experimental prototype of the 2 kW, 125 VDC to 480 V_{LL} three-phase AC IIC microinverter, showing the DC and AC bridges and the CWT.

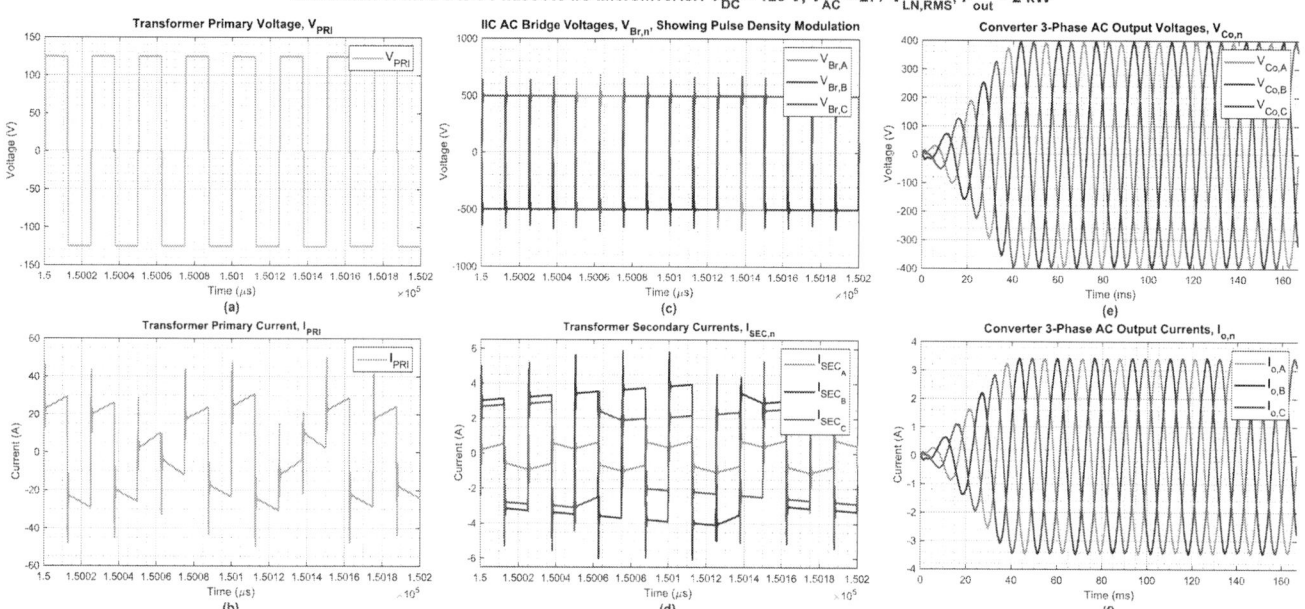

Fig. 6. Simulation results of the DC to three-phase AC IIC microinverter at 480 V_{LL} and 2 kW, showing the discrete pulse modulation scheme.

transformer volt-second balance and eliminating the challenges of partial pulses that arise in PWM modulation schemes. Thus, the proposed PDM scheme provides one of the simplest and most robust control strategies for this family of single-stage high-frequency link, bidirectional DC/AC converters.

V. SIMULATION RESULTS

The design and operation of the 2 kW DC to three-phase AC IIC microinverter was then extensively verified through simulation. Key system parameters of both the simulation and the experimental prototype are given in Table I, and the built experimental prototype is shown in Fig. 5. Fig. 6 provides simulation results of the microinverter system operating with 2 kW of power transformer from a 125 V DC bus to a three-phase AC grid connection at 480 V_{LL}.

Simulation studies of the DC to three-phase AC IIC microinverter confirmed several critical design elements. Specifically, as seen in Fig. 6(e) and Fig. 6(f), all three converter output voltages are ramped from 0 to 277 V_{RMS}, and all three converter output currents are ramped from 0 to 2.4 A_{RMS}. Fig. 6(a) – (d) depict critical converter waveforms across eight 40

kHz switching cycles beginning at t = 150 ms. Fig. 6(a) shows the DC bridge quasi-square wave voltage, with a total zero-state duration of 1 µs per 40 kHz cycle (T_{SW} = 25 µs). The quasi-square wave is reflected across all three transformer secondary windings with minimal phase-shift due to the ultra-low leakage inductance of the transformer. Fig. 6(c) shows the three AC bridge output voltages, $V_{BR,n}$, validating that each AC bridge unfolds its high-frequency link voltage, $V_{SEC,n}$, to apply a sequence of two positive pulses or two negative pulses onto the phase output according to the sigma-delta error signal. The transformer secondary winding currents are given in Fig. 6(d), showing the flipping logic and elimination of DC flux. As can be seen from Fig. 6(b), I_{PRI} results from the sum of the secondary winding currents, $I_{SEC,n}$, reflected through the transformer turns ratio, as described in (3).

$$I_{PRI} = N \times (I_{SEC,A} + I_{SEC,B} + I_{SEC,C}) \quad (3)$$

The impact of transformer and AC bridge parasitic elements L_{lkg} and C_p are seen in Fig. 6(b), (c), and (d). As the DC bridge transitions from a zero-state to either the positive or negative segment of the quasi-square wave, a damped MHz-order LC oscillation is observed on all transformer secondary voltages.

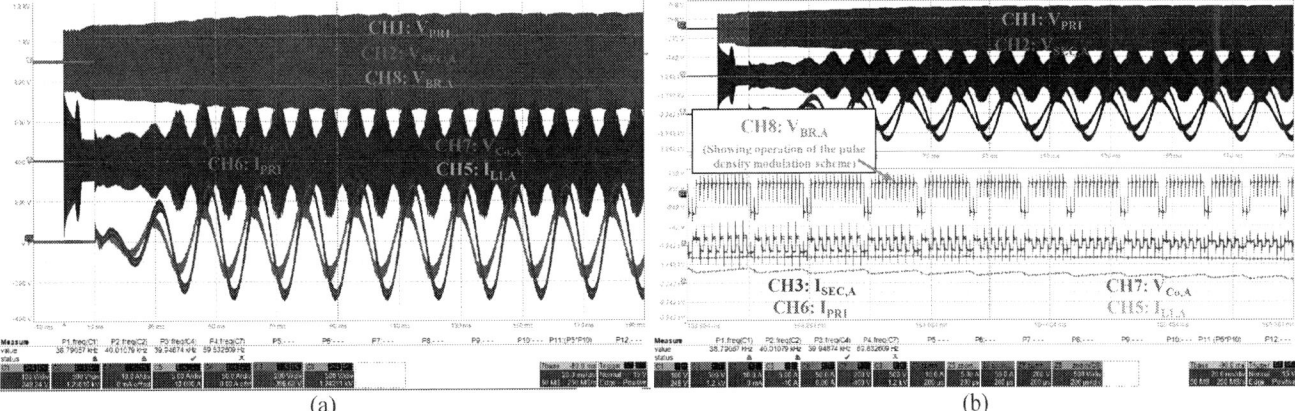

(a)   (b)

Fig. 7. (a) Experimental results showing steady state AC operation at 200 V_{RMS}, 60 Hz, (b) and zoomed-in waveform validating operation of the pulse density modulation scheme. Overvoltage on the AC bridge is well controlled by the minimal clamp circuit.

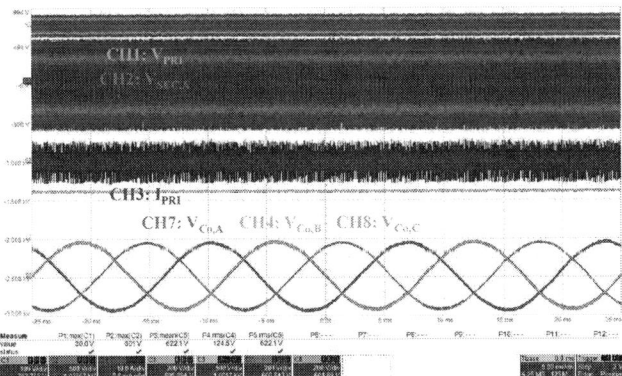

Fig. 8. Experimental waveform of the 2 kW, DC to 3-phase AC IIC microinverter, validating three-phase AC operation.

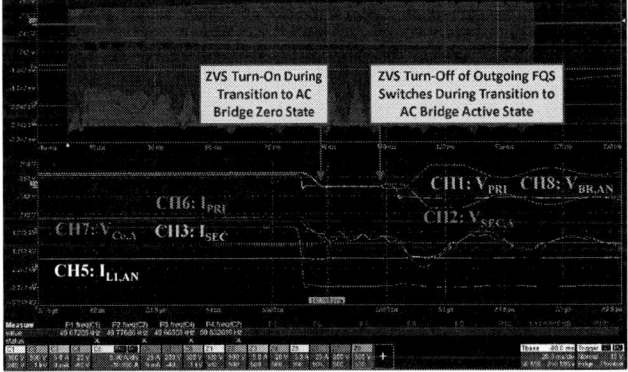

Fig. 9. Experimental waveform showing ZVS transitions of AC bridge devices. dV/dt of the AC bridge devices is limited by transformer L_{lkg} and C_p.

The resonant frequency and characteristic impedance were determined to be 3.6 MHz and 87 Ω using (1) and (2), respectively. Importantly, high-frequency impedance stemming from the physical construction of the transformer windings and the PCB layout provide natural damping of this LC oscillation. Action of the minimal clamp circuit, as described in [18], further mitigates the overvoltage stress caused by this LC oscillation.

Simulation studies confirmed stable operation of the DC to three-phase AC IIC microinverter across varying load levels, load power factors, and modulation indices. Multiple LV DC sources can be interfaced to the 125 V DC bus using standard DC/DC converters, enabling fully bidirectional multiport operation. As this work serves to describe the topology and converter level design considerations of the three-phase IIC microinverter, study of high-level grid interactive control under dynamic conditions will be presented in future work.

VI. EXPERIMENTAL RESULTS

After concluding simulation studies, converter operation was experimentally validated under single-phase and three-phase operating conditions. Fig. 7 provides experimental results from steady-state single-phase AC operation in which the AC reference voltage, V_{AN}^*, was ramped from 0 V to 200 V_{RMS} over two 60 Hz line cycles. The zoomed waveform from Fig. 7(b), taken near the peak of the AC reference, validates operation of the proposed sigma-delta pulse density modulation scheme.

Given the single-phase testing condition, a twice-line-frequency pulsating current was seen in the transformer primary and secondary currents, I_{PRI} and $I_{SEC,A}$, in Fig. 7. However, as seen in the three-phase testing result from Fig. 8, I_{PRI} shows a stable DC envelope as expected. Under three-phase operation, while each AC bridge draws a twice-line-frequency pulsating power, the transformer primary experiences the reflected sum of secondary currents, minimizing the DC bus capacitance requirement. In addition, as a single-stage DC/AC high-frequency link converter, the IIC eliminates the need for an intermediate capacitive energy storage element on the AC bridges and eliminates the finite leakage inductance in DAB-type configurations, further increasing power density and reliability.

Fig. 9 provides experimental validation of the ZVS soft-switching performance of the AC bridge devices. Specifically, while the switching transients of the DC bridge quasi-square wave (V_{PRI}) exhibit dV/dt stress of 5 kV/μs, the transformer secondary voltage shows reduced dV/dt stress of 3 kV/μs due to the action of L_{lkg} and C_p. The AC bridge then unfolds this dV/dt-limited high-frequency link voltage onto the phase output, reducing EMI injection. The ZVS soft-switching of the AC bridge devices in the IIC topology mitigates key challenges in the integration of SiC MOSFETs whose inherent dV/dt capability exceeds 25 kV/μs in hard-switching applications [22], [23]. Lastly, table II details all converter loss components at full load (2 kW), showing a full load system efficiency of >97.3%.

979-8-3315-1612-3/25 $31.00 © 2025 IEEE   169

TABLE II. BREAKDOWN OF LOSSES IN THE 2 KW 480 V$_{LL}$ IIC MICROINVERTER

DC Bus Voltage	Converter Output Power	DC Bridge Conduction Loss	DC Bridge Switching Loss	AC Bridge Clamp Loss	AC Bridge Conduction Loss	XFMR Total Loss	LCL Filter Loss	Gate Driver + Control Power	Total Loss	Full Load System Efficiency
125 V	2 kW	10.9 W	17.3 W	2.0 W	4.9 W	10.0 W	5.0 W	5.0 W	55.1 W	97.3 %

VII. CONCLUSIONS AND FUTURE WORK

A novel, single-stage, inertialess isolated converter (IIC) is proposed to provide a high-efficiency, bidirectional, and easily controllable interface between LV DC resources, such as PV and batteries, and the AC grid at 480 V$_{LL}$. The IIC topology together with the proposed sigma-delta PDM modulation scheme achieve a simple, low-cost, and bidirectional DC to three-phase AC conversion with essentially zero DC flux over every switching cycle. The presented experimental results demonstrate the inertialess operation of the system under DC to three-phase AC operation, and confirm management of critical parasitic elements and ZVS soft-switching of the AC bridge devices, critical for high conversion efficiency and low EMI. Projected efficiency from three-phase operation at 2 kW is >97.3%.

Future work will address optimization of the modulation scheme using unipolar sigma-delta modulation to minimize LCL filter size, and active filter configurations for robust grid interactions. In addition, future work will investigate multi-inverter systems built around IIC microinverters in various grid-interactive scenarios. Due to the absence of inner dead-time and dwell-time compensation loops, it is predicted that the IIC paired with the sigma-delta PDM modulation scheme provides a robust, high-bandwidth, and easily controllable plant for massively scalable systems.

REFERENCES

[1] Z. An, X. Han, V. R. Chowdhury, J. Benzaquen, R. Kandula, and D. Divan, "A tri-port current-source soft-switching medium-voltage string inverter for large-scale solar-plus-storage farms," *IEEE Trans. Power Electron.*, vol. 37, no. 11, pp. 13808–13823, Nov. 2022.

[2] J. M. Carrasco *et al.*, "Power-Electronic Systems for the Grid Integration of Renewable Energy Sources: A Survey," in *IEEE Transactions on Industrial Electronics*, vol. 53, no. 4, pp. 1002-1016, June 2006.

[3] S. B. Kjaer, J. K. Pedersen and F. Blaabjerg, "A review of single-phase grid-connected inverters for photovoltaic modules," in *IEEE Transactions on Industry Applications*, vol. 41, no. 5, pp. 1292-1306, Sept.-Oct. 2005.

[4] D. Cao, S. Jiang, F. Z. Peng and Y. Li, "Low cost transformer isolated boost half-bridge micro-inverter for single-phase grid-connected photovoltaic system," *2012 Twenty-Seventh Annual IEEE Applied Power Electronics Conference and Exposition (APEC)*, Orlando, FL, USA, 2012.

[5] Y. Lin *et al.*, "Research Roadmap on Grid-Forming Inverters," National Renewable Energy Laboratory, NREL/TP-5D00-73476, Nov. 2020. [Online]. Available: https://www.nrel.gov/docs/fy21osti/73476.pdf.

[6] W. Du *et al.*, "Modeling of Grid-Forming and Grid-Following Inverters for Dynamic Simulation of Large-Scale Distribution Systems," *IEEE Trans. Power Delivery*, vol. 36, no. 4, pp. 2035–2045, Aug. 2021.

[7] M. N. Kheraluwala, R. W. Gascoigne, D. M. Divan, and E. D. Baumann, "Performance characterization of a high-power dual active bridge DC-to-DC converter," *IEEE Trans. Ind. Appl.*, vol. 28, no. 6, pp. 1294–1301, Nov.–Dec. 1992.

[8] R. Hao, S. Belkhode, J. Benzaquen, and D. Divan, "Multimode Control of HF Link Universal Minimal Converters - Part I: Principles of Operation," *Proc. 2023 IEEE Energy Conversion Congress and Exposition (ECCE)*, Nashville, TN, USA, 2023, pp. 2872–2878.

[9] B. Majmunović *et al.*, "1 kV, 10-kW SiC-Based Quadruple Active Bridge DCX Stage in a DC to Three-Phase AC Module for Medium-Voltage Grid Integration," *IEEE Trans. Power Electron.*, vol. 37, no. 12, pp. 14631–14646, Dec. 2022.

[10] T. Chen, R. Yu, and A. Q. Huang, "A Bidirectional Isolated Dual-Phase-Shift Variable-Frequency Series Resonant Dual-Active-Bridge GaN AC–DC Converter," *IEEE Trans. Ind. Electron.*, vol. 70, no. 4, pp. 3315–3325, Apr. 2023.

[11] H. Wu, X. Tang, J. Zhao, and Y. Xing, "An Isolated Bidirectional Microinverter Based on Voltage-in-Phase PWM-Controlled Resonant Converter," *IEEE Trans. Power Electron.*, vol. 36, no. 1, pp. 562–570, Jan. 2021.

[12] S. Shao *et al.*, "Modeling and Advanced Control of Dual-Active-Bridge DC–DC Converters: A Review," *IEEE Trans. Power Electron.*, vol. 37, no. 2, pp. 1524–1547, Feb. 2022.

[13] M. G. Varzaneh, W. Emar, M. Iranshahi, N. Kamali-Omidi and A. S. Panah, "DC Link Capacitors Selection and Arrangement Procedure in High Power Inverters: A General Review," *2023 2nd International Engineering Conference on Electrical, Energy, and Artificial Intelligence (EICEEAI)*, Zarqa, Jordan, 2023.

[14] L. Schrittwieser, M. Leibl, and J. W. Kolar, "99% Efficient isolated three-phase matrix-type DAB buck–boost PFC rectifier," *IEEE Trans. Power Electron.*, vol. 35, no. 1, pp. 138–157, Jan. 2020.

[15] N. Kummari, S. Chakraborty, and S. Chattopadhyay, "An isolated high-frequency link microinverter operated with secondary-side modulation for efficiency improvement," *IEEE Trans. Power Electron.*, vol. 33, no. 3, pp. 2187–2200, Mar. 2018.

[16] S. Muroyama, T. Aoki, and K. Yotsumoto, "A control method for a high frequency link inverter using cycloconverter techniques," *Proc. Eleventh Int. Telecommun. Energy Conf.*, Florence, Italy, 1989, pp. 19.1/1–19.1/6 vol. 2.

[17] S. Norrga, "Experimental Study of a Soft-Switched Isolated Bidirectional AC–DC Converter Without Auxiliary Circuit," *IEEE Trans. Power Electron.*, vol. 21, no. 6, pp. 1580–1587, Nov. 2006.

[18] D. Divan, R. P. Kandula, and M. J. Mauger, "The Case for Soft Switching in Four-Quadrant Power Converters," *IEEE J. Emerging Sel. Topics Power Electron.*, vol. 9, no. 6, pp. 6545–6560, Dec. 2021.

[19] S. Belkhode, N. Prabhu, J. Benzaquen, and D. Divan, "Single-Stage Bidirectional Inertia-less Isolated DC/AC Converter," *Proc. 2024 IEEE Appl. Power Electron. Conf. Exposition (APEC)*, Long Beach, CA, USA, 2024, pp. 348–353.

[20] M. Rauls, D. Novotny, and D. Divan, "Design considerations for high-frequency coaxial winding power transformers," *IEEE Trans. Ind. Appl.*, vol. 29, no. 2, pp. 375–381, Mar.–Apr. 1993.

[21] G. Luckjiff, I. Dobson, and D. Divan, "Interpolative sigma delta modulators for high frequency power electronic applications," *Proc. PESC '95 - Power Electron. Specialist Conf.*, Atlanta, GA, USA, 1995, pp. 444–449 vol. 1.

[22] N. Oswald, P. Anthony, N. McNeill, and B. H. Stark, "An Experimental Investigation of the Tradeoff between Switching Losses and EMI Generation With Hard-Switched All-Si, Si-SiC, and All-SiC Device Combinations," *IEEE Trans. Power Electron.*, vol. 29, no. 5, pp. 2393–2407, May 2014.

[23] L. Zhang, X. Yuan, X. Wu, C. Shi, J. Zhang, and Y. Zhang, "Performance Evaluation of High-Power SiC MOSFET Modules in Comparison to Si IGBT Modules," *IEEE Trans. Power Electron.*, vol. 34, no. 2, pp. 1181–1196, Feb. 2019.

Analysis and Design of a Constant Current LCC Class-E Inverter

Ju Gao
School of Integrated Circuits
Peking University
Beijing, China
gaoju@stu.pku.edu.cn

Ziheng Liu
School of Integrated Circuits
Peking University
Beijing, China
zihengliu@stu.pku.edu.cn

Jiayin He
School of Integrated Circuits
Peking University
Beijing, China
hejiayin@stu.pku.edu.cn

Hongjie Peng
School of Integrated Circuits
Peking University
Beijing, China
2301111840@stu.pku.edu.cn

Chengkang Ao
School of Integrated Circuits
Peking University
Beijing, China
2301111809@stu.pku.edu.cn

Jinyan Wang
School of Integrated Circuits
Peking University
Beijing, China
wangjinyan@pku.edu.cn

Abstract—This paper presents a Class-E inverter that provides constant output current and zero-voltage switching (ZVS) over wide load range by integrating the input switch unit with the output network. Both the duty cycle and switching frequency remain constant even with load variations. Utilizing frequency-domain analysis, the new inverter topology can simplify the design procedure of the constant current realization and the load-independent ZVS. A prototype operating at 6.78 MHz, with a 10-V input and 0.8 A output, was developed to validate the design. Both LTspice simulations and experimental results demonstrated ZVS and a constant AC output current across load resistances of 0.5, 40 and 80 Ω, achieving a load range 4 times greater than existing technologies.

Keywords—Class E inverter, ZVS, constant current, load independent

I. INTRODUCTION

The pursuit of miniaturized power supply devices has intensified the demand for high-frequency switch power conversion circuits, which offer the dual benefits of reduced component size and increased system efficiency [1]-[2]. Power sources are applied in high-tech applications, encompassing wireless power transfer, gate drivers, and battery charging systems [3]-[4]. The Class-E converter design stands out in this context, offering a compelling solution with its single-switch operation and inherent ZVS feature [5]. The evolution of Class-E inverters has been driven by the need to achieve high efficiency and soft-switching across variable load conditions. Traditional Class-E inverters achieve high efficiency at a specific load, but their efficiency declines as the load deviates from the optimal point due to hard switching outside of these conditions [6]. To address this, various techniques have been employed to maintain ZVS with load change, including duty cycle modulation and the use of additional transformers or matching networks [7]. Original Class-E inverters, with infinite

input inductance, offered 90% efficiency at minimum resistance but saw a decline to 65% at ten times that resistance [8]. The finite Lin in Class-E inverters can be minimized to create a resonant current, resulting in smaller size and better transient performance [9]. It has been employed to provide a stable output voltage across a spectrum of load conditions, leveraging ZVS irrespective of the magnitude of input inductance Lin [10]. However, the challenge of sustaining a uniform output current from a pure Class E configuration remains elusive [8]. The output network typically features a series resonant circuit, but converting a constant voltage source to a current source requires more passive components [11]. A ZVS Class E inverter has been proposed [12] to ensure a consistent AC current output, but the complexity of time-domain analysis escalates with the intricacy of the circuit architecture. This study aims to design a Class-E inverter with finite input inductance to maintain constant current and ZVS across a broad load range.

In this study, we propose a load-independent Class E inverter designed via the frequency-domain approach, ensuring the maintenance of ZVS by optimizing the resonant network. It achieves a constant current output across a wide range of load resistances without the need for additional components or feedback mechanisms. The detailed design procedure for constant current mode and ZVS realization is discussed in Section II. Due to the implementation of the LCC output network, the inverter enhanced performance for constant current operation, with analysis of its key waveforms included. Employing the frequency-domain method, the design offers an efficient approach to constant current inverter construction. In Section III, the methodology has been demonstrated to be effective across a wide range of loads, with the efficacy of this approach validated in both simulation and experimental prototypes. To verify the proposed topology, a prototype with a 10V input and 0.6A output was simulated and then measured.

II. OPERATION PRINCIPLE OF TOPOLOGY

The schematic representation of the proposed constant-current ZVS Class E inverter is presented in Figure 1(a), which is composed of an input switch unit (V_{in}, S, L_{in} and C_{in}), as well as an output unit (L_1, C_1 and C_2). The analysis begins with the resonant input unit featuring a single switch, which successfully maintains a constant output voltage. The switch is activated by a square wave input with a 50% duty cycle denoted as f_s. A detailed analysis of the ZVS condition for the switch is conducted using frequency-domain equations. The output unit is engineered to transform the voltage source into a current source and efficiently filters out the harmonic components of the voltage. The operating frequency is designed to ensure load independence. Consequently, the circuit sustains a stable current despite variations in the load.

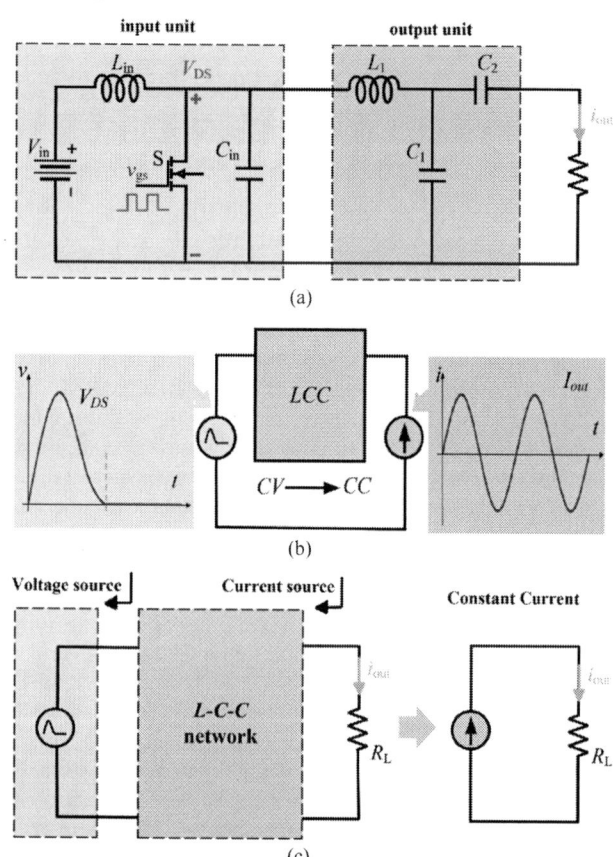

Figure 1: Schematic of the Class-E inverter (a) and the process of formulating its equivalent circuit (b) for a constant output current (c).

A. Input Unit and ZVS

The non-linear input unit converts the DC input voltage V_{in} to the half-sine AC waveform. Because the inductor L_{in} and capacitor C_{in} resonate in a series configuration with the input unit, the V_{DS} can be switched to be half-sine waveform with any frequency. In accordance with Kirchhoff's voltage law, the expression for V_{DS} during the OFF state ($0<\theta<2\pi D$) is derived as follows:

$$\begin{cases} V_{DS} = V_{in} - L_{in}C_{in}\dfrac{dV_{DS}^2}{dt^2} \\ V_{DS} = [A\sin(\lambda\omega t) - \cos(\lambda\omega t) + 1]V_{in} \cdot \\ A = \dfrac{\lambda\pi + \sin(\lambda\pi)}{1 - \cos(\lambda\pi)}, \lambda = \dfrac{1}{\omega\sqrt{L_{in}C_{in}}} \end{cases} \quad (1)$$

To obtain the exact switching-period of the half-sine V_{DS} waveform showed as the yellow curve in Fig. 2(a), the zero points of the V_{DS} expression (t_1 and t_2) can be solved with a switching duty cycle, as follows:

$$t_1 = \frac{2n\pi}{\lambda\omega}, \ t_2 = \frac{2\varphi + n\pi + \pi/2}{\lambda\omega}, \ n = 0,1,2,\dots, \quad (2)$$

$$\varphi = \arctan(1/A). \quad (3)$$

Then the critical duty cycle D that makes V_{DS} just drop to zero at $\theta = 2\pi D$ is obtained.

B. Output Unit and Constant Current Output

To ensure that the output current amplitude remains independent of the load resistance RL, the design of the passive resonant network is contingent upon the nature of the power source. As illustrated in Fig. 1(c), the output unit, which incorporates an L-C-C network, is meticulously crafted to maintain a constant current output and to facilitate the transformation of a voltage source into a current source. The correlation between the fundamental voltage of V_{DS} (V_{DS_1}) and the output current I_{out} is articulated through the derivation:

$$V_{DS\_1} = (1 - \omega^2 L_1 C_1)R_L \cdot I_{out} + \frac{1 - \omega^2 L_1 C_1 - \omega^2 L_1 C_2}{\omega C_2} \cdot I_{out} \quad (4)$$

To ensure a load-independent constant current output, the operating frequency should be selected at the point where the derivation of I_{out} with respect to R_L equals zero, as determined by the following condition:

$$\omega = \sqrt{\frac{1}{L_1 C_1}} \cdot \quad (5)$$

The output network is designed to selectively extract the fundamental frequency components from the output waveform while effectively suppressing the higher-order harmonics. This design allows the output current to be nearly sinusoidal in shape, maintaining a consistent current gain across a wide range of load resistances R_L from 0.5 Ω to 200 Ω as shown in Fig. 2(b). Furthermore, current gain at a frequency of 13.56 MHz is 24 dB lower than at 6.78 MHz that reduces the second harmonic frequency.

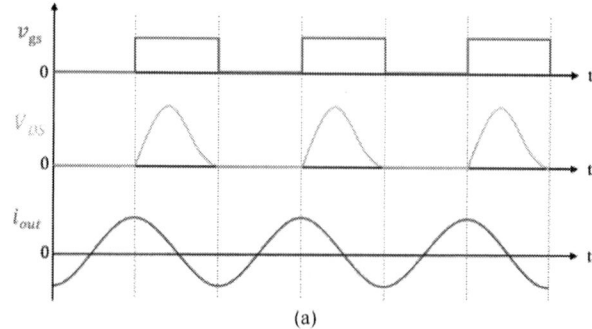

(a)

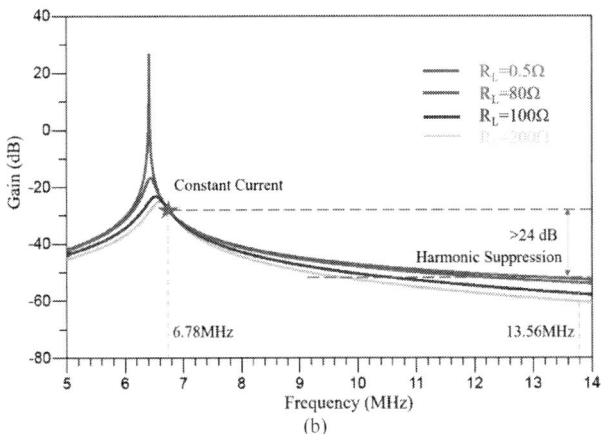

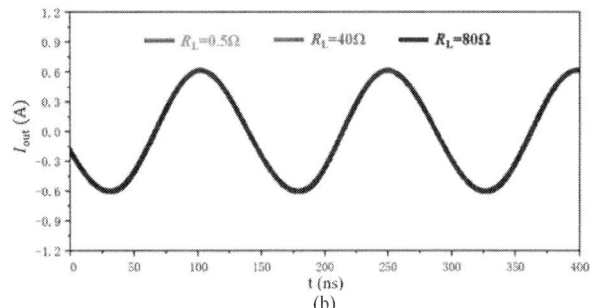

(b)

Fig. 3 Simulated ZVS (a) and constant current output (b) in different load resistance.

Figure 2: (a) Key waveforms of the Class E inverter. (b) Small-signal current gain of the L-C-C unit.

III. SIMULATION AND EXPERIMENT

The constant current Class-E inverter is validated through a prototype with V_{in} = 10V, I_m = 0.6A, and R_{Lmax} = 80Ω. It operates at a 6.78 MHz switching frequency with 50% duty cycle. The components L_{in} = 200nH and C_{in} = 6nF are selected to ensure ZVS condition. The output resonant unit, consisting of L_1 = 560nH, C_1 = 1nF, and C_2 = 0.1nF, is designed to maintain a constant output current. The other parameters of the circuit were designed in the order and with the equations listed in Table I.

TABLE I.   PARAMETERS OF THE DESIGNED CLASS E INVERTER

Parameters	Value	Theory
V_{in}	10 V	Based on (1)
D	50 %	duty cycle
f_s	6.78 MHz	switching frequency
L_{in}	200 nF	$\sqrt{L_{in}/C_{in}}$ and $\sqrt{L_{in}C_{in}}$
C_{in}	6 nF	
L_1	560 nF	$\omega^2 = 1/L_1C_1$
C_1	1 nF	
C_2	100 pF	Based on (4)

Fig. 3 presents simulation waveforms with a duty cycle of D = 0.5 and varying load resistances R, specifically at 0.5, 40, and 80Ω. The achievement of superior load-independent ZVS is demonstrated in Fig. 3(a). Constant current output is shown in Fig. 3(b), indicating its successful implementation of load-independent operation.

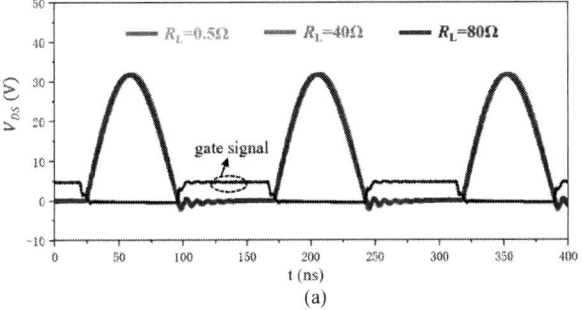

(a)

The presented topology of Class-E inverter targeting ZVS and a stable current output across a broad load spectrum has been experimentally validated as depicted in Fig. 4(a). The power switch is an Infineon-GaN System GS66508B which is driven by the TI UCC27614 gate driver. A RIGOL DP2031 provides a 10V input, while a Keysight 81150A function generator supplies the switching pulse at 6.78 MHz with a 50% duty cycle. The load is varied using 100W power resistors. The experimental setup and the inverter circuit board are displayed in Fig. 4. Table II provides a detailed list of the key components.

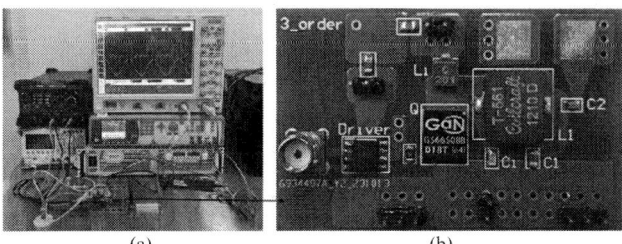

(a)   (b)

Figure 4: (a)Photograph of the measured prototype and (b) top view of the inverter.

TABLE II.   BILL OF MATERIALS

Parameters	Value	Manufacturer	Model Number
Driver	30 V, 10A	TI	UCC27614DR
S	650V, 30A	Infineon	GS66508B
R_{Lmax}	100W, 0.5Ω	PAKAN	RX24 0.5RJ
R_{Lmax}	100W, 80Ω	PAKAN	RX24 80RJ
C_{in}	6.2 nF	KEMET	C0805C622JAGACAUTO
L_{in}	200 nF	Coilcraft	XEL4030-201MEC
C_1	1 nF	Murata	GCM21B5C2J102JX03L
L_1	560 nF	Coilcraft	MVR1251T-561MLC
C_2	100 pF	Murata	GCM21A5C2J101FX01D

The performance of circuit board under varying loads was evaluated in Fig. 5. The voltage across the power switch V_{DS} is observed to drop to zero before each turn-on instance, indicating the ZVS under load variation showed in Fig. 5(a). Employing a Keysight 1147B probe, the output current was consistently measured at around 0.6A across a load resistance span from 0.5Ω to 80Ω, thereby confirming the excellent load-independent performance.

979-8-3315-1612-3/25 $31.00 © 2025 IEEE

TABLE III. COMPARISON OF THIS WORK AND RECENTLY CONSTANT CURRENT INVERTER.

Index	TPE 2017[8]	TPE 2018 [10]	TPE 2021 [12]	TIE 2021 [11]	**This work**
Topology	Class EF	Class EF	Class E	Class E	**Class E**
Switch frequency	6.78MHz	13.56 MHz	1 MHz	6.78MHz	**6.78MHz**
constant current range	5 - 100Ω (20x)	0.56 - 6Ω (10x)	1 - 20Ω (40x)	1 - 8.6Ω (8.6x)	$0.5 - 80\Omega$ (160x)
Passive Components	9	7	5	7	**6**
Load-independent ZVS	NO	YES	YES	YES	**YES**

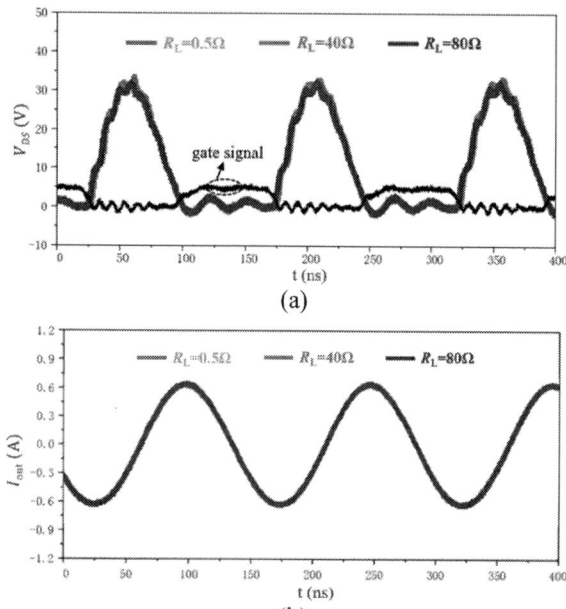

Figure 5: Experimental (a) ZVS and (b) constant current output in different load resistance.

Table III presents a detailed comparative analysis of the proposed circuit alongside existing resonant inverters with load-independent characteristics. It is noted that finite input inductor is used for Class E topology and an infinite inductor is used for Class EF topologies. In terms of the load-independent operating range, existing designs demonstrate load independence from 3 to 40 times the minimum load resistance, while the proposed circuit has been validated for a load range exceeding 160 times. The design method simplifies the process of designing constant current inverter while ensuring the desired operating performance of the proposed circuit.

IV. CONCLUSIONS

In this paper, a design of constant-current ZVS *LCC* Class E inverter over a wide load range was proposed and verified. The design process of load independent ZVS inverter is streamlined by employing a frequence-domain method. The new topology effectively maintains a constant output current over 160 times load range while achieving ZVS conditions. Validation has been successfully conducted through simulations and experimental tests.

ACKNOWLEDGMENT

This work is supported by the National Key Research and Development Program of China under Grant 2023YFB3905703 and the National Natural Science Foundation of China under Grant No. U2241220. The authors would like to appreciate the help from the Dongke Semiconductor (Anhui) Company.

REFERENCES

[1]. J. Millán, P. Godignon, X. Perpiñá, A. Pérez-Tomás and J. Rebollo, "A Survey of Wide Bandgap Power Semiconductor Devices, "IEEE Trans. Power Electron., vol. 29, no. 5, pp. 2155-2163, May 2014.

[2]. N. O. Sokal and A. D. Sokal, "Class E-A new class of high-efficiency tuned single-ended switching power amplifiers," IEEE J. Solid-State Circuits, vol. 10, no. 3, pp. 168-176, June 1975.

[3]. I. A. Mashhadi, E. Ovaysi, E. Adib and H. Farzanehfard, "A Novel Current-Source Gate Driver for Ultra-Low-Voltage Applications," IEEE Trans. Ind. Electron., vol. 63, no. 8, pp. 4796-4804, Aug. 2016.

[4]. W. Zhang and C. C. Mi, "Compensation Topologies of High-Power Wireless Power Transfer Systems," IEEE Trans. Veh. Technol., vol. 65, no. 6, pp. 4768-4778, June 2016.

[5]. L. Roslaniec, A. S. Jurkov, A. A. Bastami and D. J. Perreault, "Design of Single-Switch Inverters for Variable Resistance/Load Modulation Operation," IEEE Trans. Power Electron., vol. 30, no. 6, pp. 3200-3214, June 2015.

[6]. L. Gu, W. Liang and J. Rivas-Davila, "A multi-resonant gate driver for Very-High-Frequency (VHF) resonant converters," *2017 IEEE 18th Workshop on Control and Modeling for Power Electronics (COMPEL)*, Stanford, CA, USA, 2017, pp. 1-7.

[7]. A. Ayachit, F. Corti, A. Reatti and M. K. Kazimierczuk, "Zero-Voltage Switching Operation of Transformer Class-E Inverter at Any Coupling Coefficient," in *IEEE Transactions on Industrial Electronics*, vol. 66, no. 3, pp. 1809-1819, March 2019.

[8]. S. Liu, M. Liu, S. Yang, C. Ma, and X. Zhu, "A novel design methodology for high-efficiency current-mode and voltage-mode class-E power amplifiers in wireless power transfer systems," IEEE Trans. Power Electron., vol. 32, no. 6, pp. 4514–4523, Jun. 2017.

[9]. S. Park and J. Rivas-Davila, "Duty Cycle and Frequency Modulations in Class-E DC–DC Converters for a Wide Range of Input and Output Voltages," in *IEEE Transactions on Power Electronics*, vol. 33, no. 12, pp. 10524-10538, Dec. 2018.

[10]. S. Aldhaher, D. C. Yates and P. D. Mitcheson, "Load-Independent Class E/EF Inverters and Rectifiers for MHz-Switching Applications," IEEE Trans. Power Electron., vol. 33, no. 10, pp. 8270-8287, Oct. 2018.

[11]. L. Zhang and K. Ngo, "A Constant-Current ZVS Class-E Inverter With Finite Input Inductance," IEEE Trans. Ind. Electron., vol. 68, no. 8, pp. 7693-7696, Aug. 2021.

[12]. T. Sensui,"Load-Independent Class E Zero-Voltage-Switching Parallel Resonant Inverter," IEEE Trans. Power Electron., vol. 36, no. 11, pp. 12805-12818, Nov. 2021.

Series Connected Class-E Push-Pull Converters using GaN HEMT for High-Efficiency RF Generators in Float Zone Silicon Production

Faheem Ahmad
AAU Energy
Aalborg University
Aalborg, Denmark
faah@energy.aau.dk

Thore Stig Aunsborg
AAU Energy
Aalborg University
Aalborg, Denmark
tsu@energy.aau.dk

Jannick Kjær Jørgensen
AAU Energy
Aalborg University
Aalborg, Denmark
jkj@energy.aau.dk

Stig Munk-Nielsen
AAU Energy
Aalborg University
Aalborg, Denmark
smn@energy.aau.dk

Abstract—Ultra-pure Float Zone silicon production uses induction heating to melt poly-crystal silicon ingot into mono-crystal structure. Radio-frequency (RF) generators used for induction heating operate at around 2.4 MHz. The generators currently used by the industry are based on vacuum-tube triode and thus suffer from low efficiency. Wide bandgap devices such as gallium-nitride (GaN) can achieve the high switching frequency however the power levels required by such RF generators cannot be achieved by a single high frequency converter. Series connection of Class-E push-pull converters in a multi-cell architecture is a way to achieve higher power, however challenge arises when multiple Class-E push-pull converters operating at 2.4 MHz are connected in series leads to instability in the system. In this paper the authors have identified this instability mechanism as the parallel resonance between GaN HEMT output capacitance and external capacitor and presents a solution to mitigate the issue. Finally, the paper demonstrates three Class-E push-pull converters operating in series. Output of 4.1 kW at 98% efficiency is achieved.

Index Terms—Class-E push-pull, gallium nitride, induction heating, multi-cell, RF generator

I. INTRODUCTION

In industries that require induction heating at high power (several kW) and high frequency (several MHz) such as in production of ultra-pure Float Zone (FZ) silicon, radio frequency (RF) generators employing vacuum tube triodes are used. But the efficiency of such RF generators are limited to 60% [1]. Thus modern semiconductor material technology such as GaN HEMT devices can be used to achieve the several MHz switching frequency requirement. However, the high power requirement in several kW is still evading the GaN HEMT technology due to limited voltage rating of 650 V. To achieve higher power than what is achievable by a single converter, applications involving ultra-high frequency (UHF) and very-high frequency (VHF) utilize power combiners to couple power from several converters, also referred

This work is supported by the Greenheat project which is sponsored by Innovation Fund Denmark, and is a collaboration between Topsil Globalwafers A/S, Kallesoe Machinery A/S, and Aalborg University. The Greenheat project is stated to run for three years from 2023-2026 and aims to develop and demonstrate two full-scale prototypes of radio frequency generators that replaces existing vacuum tube triode based RF generators in the industry.

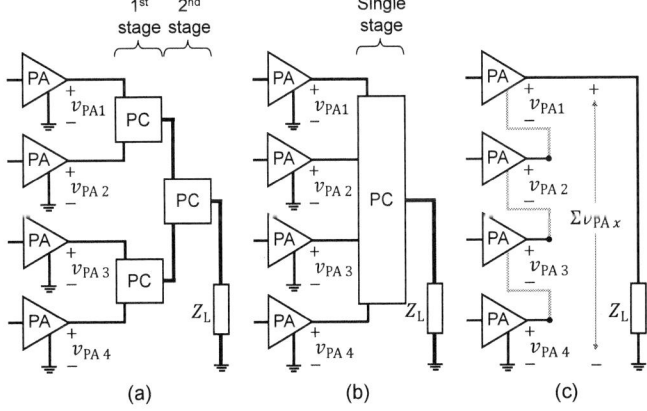

Fig. 1. Different strategies to connect multiple power amplifers (PA) (a) staggered power combiner (PC) [2], (b) single-stage PC [3], and (c) series connection of PA presented in this paper [4].

as power amplifiers as presented in [2]. Taking this idea of utilizing power combiners, sub-UHF applications have also been presented based on single-stage power combiners [3]. The concept of utilizing power combiners (PC) to couple power from several power amplifiers (PA) have been shown in Fig. 1(a) and (b). In Fig. 1(c) a third strategy to connect several power amplifier in series is presented in this paper. The authors first proposed this series connection of PA operating at several MHz that does not utilize PC in [4]. In this paper the authors have presented series connection of three Class-E push-pull converter cells to achieve 1.1 kV (pk) from 650 V GaN HEMT based RF generator and an output power of 4.1 kW at 98% efficiency.

In section II, two Class-E push-pull converter cells are connected in series to demonstrate the proposed methodology. The section also presents the challenge faced due to the external capacitor, added to linearize the non-linear behavior of GaN HEMT's output capacitance (C_{OSS}). Section III elaborates the parallel resonance originating due to the external capacitor and the implementation of damping resistor. The section also

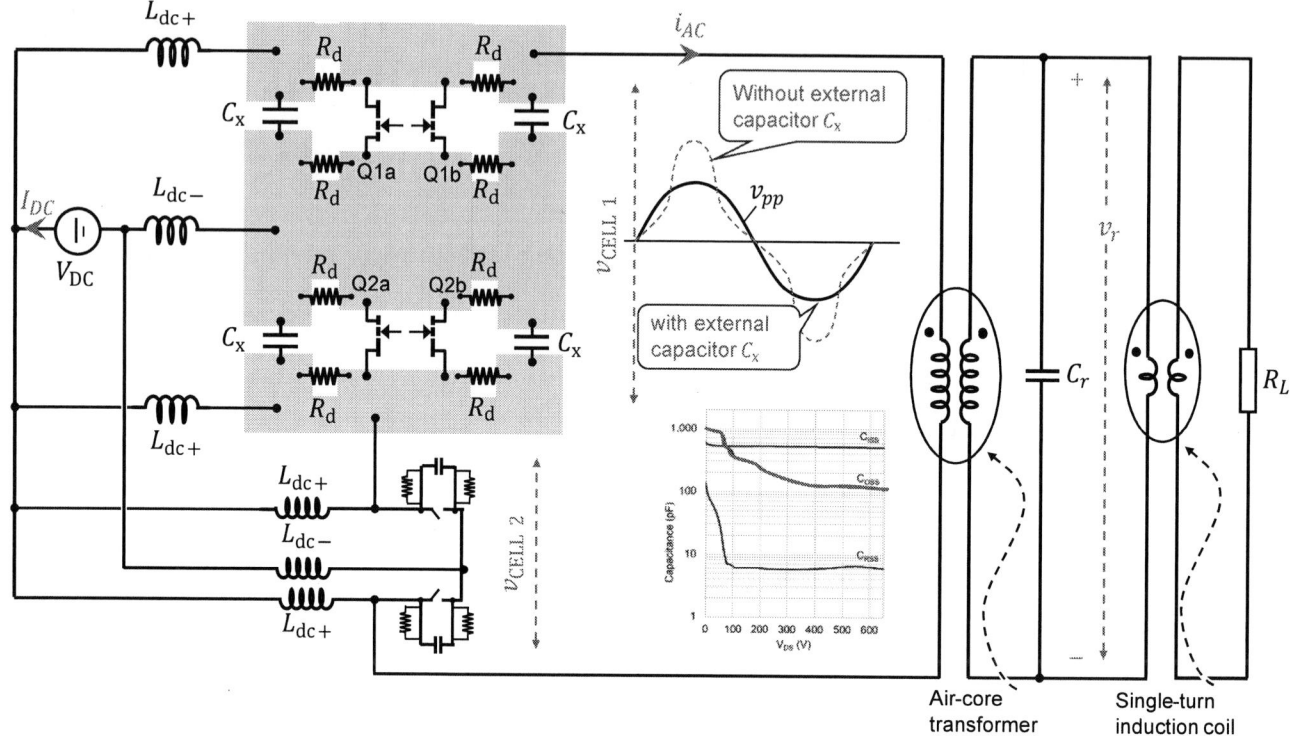

Fig. 2. Schematic of series connected two Class-E push-pull converter with resonant tank load.

shows method to find the optimum value of damping resistor. Finally section IV presents three cell operation results.

II. SERIES CONNECTION OF CLASS-E PUSH-PULL CONVERTER CELLS

Fig. 2 shows two Class-E push-pull cells in series connection. One of the cells is shown containing details of the PCB layout with copper layer (yellow polygons). The second cell is shown only as a schematic for simplicity. On the input side they are in parallel connection to the DC supply via DC choke inductors (L_{dc}). Standard Class-E push-pull has DC choke inductors only on the push-pull arm (L_{dc+}) as is shown in Fig. 2. However the presented topology has an additional choke inductor (L_{dc-}) on the DC ground as well to provide high impedance between the cells on its DC side. To achieve high current capacity each of the switch in Class-E push-pull topology is implemented via two paralleled 650 V, 60 A GaN HEMT devices [5]. This is shown in detail for the first cell while not indicated for the second cell in the simplified schematic view. In Class-E operation the output capacitance (C_{OSS}) of the semiconductor switch resonates with the load. However, the non-linear characteristics of the C_{OSS} leads to high voltage stress across the device [6]. Thus an external capacitor (C_X) is placed in parallel to both GaN HEMT as shown in Fig. 2. The external capacitor C_X, dampens out the non-linear behavior of C_{OSS}. Indicatory waveforms, with and without the external capacitor as well as the non-linear behavior of C_{OSS} is embedded in Fig. 2. Fig. 2 also shows the load used for testing in this paper. It comprises a

resonant tank system connected to the RF generator via an air-core transformer. The resonant tank system and the air-core transformer has been previously presented in [7]. The resonant tank presents a resistive impedance of 100 Ω at its natural frequency of around 2.4 MHz.

Finally based on the proposed strategy of series connection, two cell tests are conducted. Output waveforms for the two cells (v_{CELL1}, and v_{CELL2} respectively) are presented in Fig. 3. The waveforms contain two results, with and without the damping resistance (R_d). In the zoomed section of the results it is seen that the cell voltage (v_{CELL}) has a high frequency ripple of approximately 55 MHz, riding on top of the fundamental component without the damping the resistance (R_d). The ripple is seen to be more prominent for cell 1 when there is no damping resistor as compared to cell 2. However the ripple amplitude is clearly higher for both the cells when R_d is removed. It is important to point out that the cell voltages have a peak of about 100 V in Fig. 3. This is because as the DC input voltage is increased for higher v_{CELL}, the ripple content amplitude keeps on growing. This has sometimes led to the system becoming unstable. Whereas when R_d is added the high frequency oscillations have been dampened significantly. Thus the addition of damping resistance (R_d) in parallel to the PCB copper trace between the GaN HEMT and external capacitor (C_X) as shown in Fig. 2 has played a key role in realizing a stable, series connected, several MHz Class-E push-pull power amplifiers. The value of R_d used during the experiment shown in Fig. 3 are 2 Ω.

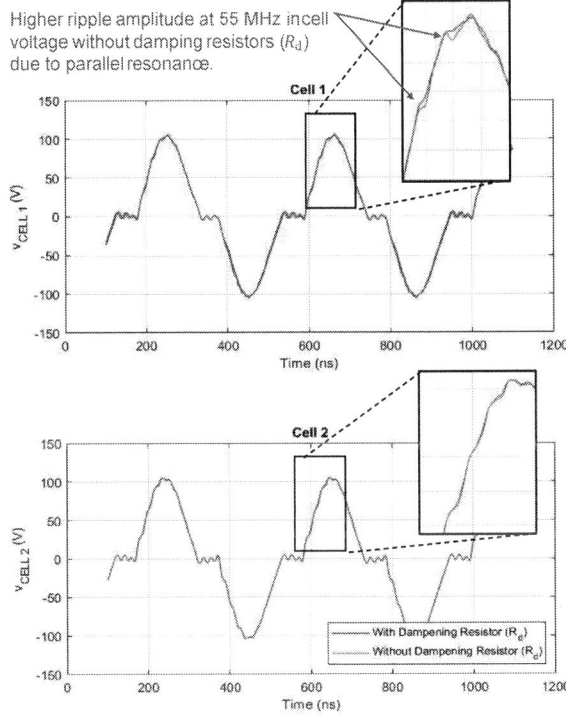

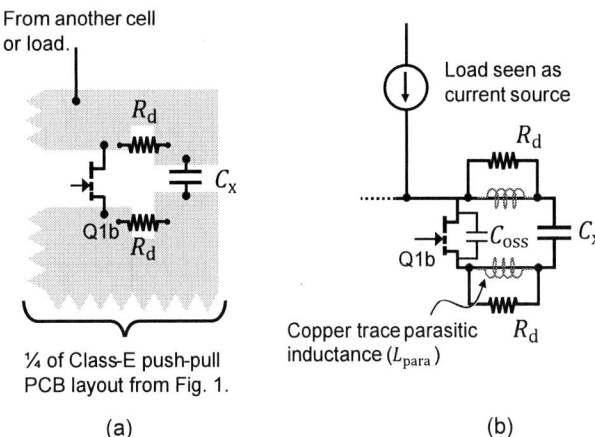

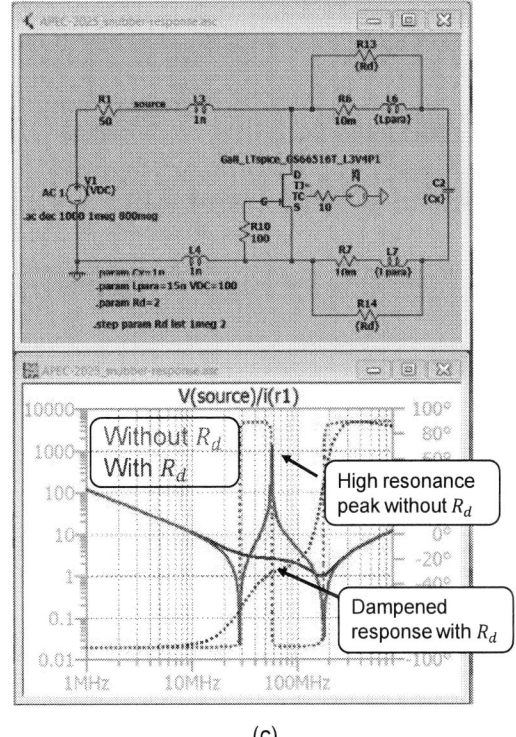

Fig. 3. High frequency ripple seen in Class-E push-pull cells output voltage without the damping resistor R_d.

Fig. 4. (a) A quarter portion of PCB for analysis, (b) equivalent schematic, and (c) parallel resonance due to external capacitor C_X and damping by R_d.

The explanation for the high frequency ripple content in absence of damping resistance R_d is given in next section along with the optimum value of R_d.

III. PARALLEL RESONANCE DUE TO EXTERNAL CAPACITOR (C_X) AND METHOD TO DAMPEN USING RESISTANCE (R_D)

To analyze the parallel resonance due to external capacitor (C_X) a quarter of the PCB layout shown in Fig. 2 is analyzed (due to symmetry) and shown in Fig. 4(a). The equivalent schematic for which is presented in Fig. 4(b). The copper trace has been replaced with parasitic inductance (L_{para}). And the connection from another cell or load is represented as a current source. GaN HEMT's C_{OSS} is also explicitly shown here. Fig. 4(c) shows the AC analysis of the schematic with and without the damping resistor R_d. In the impedance vs frequency waveforms, a high resonance peak at about 55 MHz is seen due to the parallel resonance between the C_{OSS}, C_X and the trace parasitic inductance L_{para} when R_d is removed. The values for C_X and L_{para} are taken to be 1 nF and 15 nH respectively. Since C_{OSS} has voltage dependent behavior, the AC signal source in simulation has been initiated with 100 V DC. This leads equivalent output capacitance ($C_{OSS(eq)}$) to be approximately 650 pF according to (1).

$$C_{OSS(eq)} = \frac{1}{V_{DS}} \int_0^{V_{DS}} C_{OSS}(v)\,dv \tag{1}$$

In Fig. 4(c) when damping resistance R_d is added to the circuit the high resonance peak is completely damped. The high resonance peak is responsible for higher amplitude of the ripple voltage for the two cell test waveforms shown in Fig. 3. This high frequency ripple leads to additional losses in the GaN HEMT devices and sometimes can lead to instability in the multi-cell RF generator. Therefore the damping resistor R_d has been a major finding during this work to achieve stable operation in multi-cell series connection of Class-E push-pull resonant converters.

The equivalent schematic of Fig. 4(c) is used to develop transfer function to analyze it further. The transfer function

979-8-3315-1612-3/25 $31.00 © 2025 IEEE

for the two system, without damping resistor R_d ($T(s)$) and with R_d ($T_d(s)$) are given as (2) and (3) respectively,

$$T(s) = \frac{s^2 L_{para} C_x + 1}{s(s^2 L_{para} C_{OSS(eq)} C_x + C_{OSS(eq)} + C_x)} \quad (2)$$

$$T_d(s) = \frac{s^2 R_d L_{para} C_x + s L_{para} + R_d}{s(s^2 R_d L_{para} C_{OSS(eq)} C_x + (s L_{para} + R_d)(C_{OSS(eq)} + C_x))} \quad (3)$$

Fig. 5 shows the step response for the two systems. It is seen that $T(s)$ is an oscillatory system. However when R_d is added, $T_d(s)$ shows a damped response. And as the values for R_d is decreased from 8 Ω to 2 Ω, the damping increases.

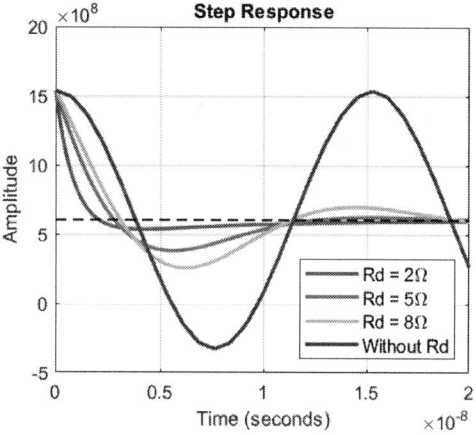

Fig. 5. Step response for the two systems, $T(s)$ (without R_d) and $T_d(s)$ (with R_d) for different values of R_d.

Based on the characteristic equation of $T_d(s)$ the system damping co-efficient (ζ) is given in (4). As the equivalent output capacitance ($C_{OSS(eq)}$) is the only varying component in (4) therefore the optimum value of R_d which is able to achieve critically damped system for all values of $C_{OSS(eq)}$ is calculated by keeping $\zeta = 1$ to find optimum R_d at different V_{DS}.

$$\zeta = \frac{1}{2R_d} \sqrt{\frac{L_{para}(C_{OSS(eq)} + C_x)}{C_{OSS(eq)} C_x}} \quad (4)$$

Fig. 6 illustrates R_d value for achieving critical damping across various V_{DS} levels. The primary y-axis of Fig. 6 represents both C_{OSS} and $C_{OSS(eq)}$ for 650 V, 60 A GaN HEMT device used in the Class-E push-pull topology presented in this paper. The secondary y-axis shows the optimum R_d value, which increases with the drain-source voltage. For the values of L_{para} and C_X presented in this paper, the optimal R_d ranges from 2.7 Ω to 4.5 Ω. Thus, a value of $R_d = 2$ Ω was selected for the work presented in this paper. Fig. 5 also demonstrates that with $R_d = 2$ Ω, oscillations are fully damped, ensuring stable operation for the series connected multi-cell Class-E push-pull converters operating at 2.4 MHz.

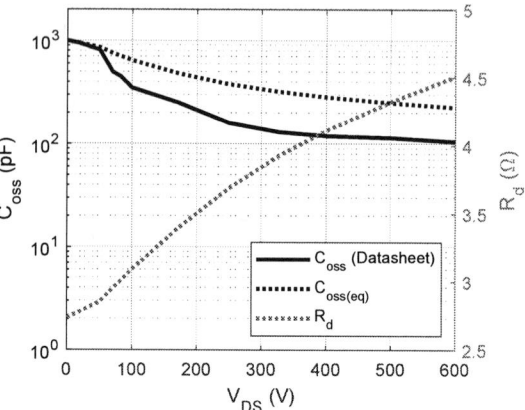

Fig. 6. Output capacitance (C_{OSS}) and equivalent output capacitance ($C_{OSS(eq)}$) at different V_{DS}. Secondary axis shows the optimum R_d value for a critically damped system.

IV. THREE-CELL EXPERIMENTAL RESULTS

Based on the analysis and finding presented thus far. The damping resistor R_d has been successful in allowing higher voltage operation that what was previously possible. Thus another cell is added for a three cell Class-E push-pull RF generator for high power test as shown in Fig. 7. The three cells are connected via standard 2 mm thick copper wires. One end of cell 1 and cell 3 are connected to the air-core transformer which is connected to the resonant tank. The experimental setup in Fig. 7 also shows measurement probes. Three differential probes (HVD3206) are connected to the three cells to measure cell voltage ($v_{CELL1, 2, and 3}$). Another differential probe (ADP305) is connected to load resonant tank to measure v_r as indicated in Fig. 2. To measure various currents, a Pearson current transformer (CT 3972) is used to measure RF generator output current (i_{AC}) and a Teledyne clamp meter (CP150) for input DC current (i_{DC}). These measured signals are also indicated in Fig. 2.

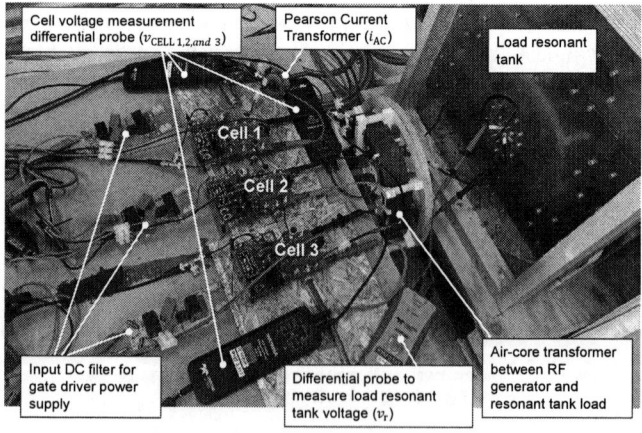

Fig. 7. Experimental setup for three cell series connected Class-E push-pull RF generator with resonant tank load.

Fig. 8 shows the voltage and current waveforms of the measured signals. Fig. 8(a) shows cell output voltage ($v_{CELL1, 2, and 3}$). The peak value for v_{CELL} is 358 V(pk) which leads to a total of 1.1 kV(pk) at air-core transformer primary shown. The oscilloscope math function F2 shows the total RF generator voltage in Fig. 8(a). The second set of waveforms in Fig. 8(b) shows DC input current (i_{DC}) of 51 A at 82 V V_{DC}. Thus the input power to the three cell Class-E push-pull RF generator is 4.18 kW. The RF generator output current (i_{AC}) is 32 A(pk). The resonant tank voltage (v_r) is 930 V(pk) which gives an RMS value of 640 V as shown by oscilloscopes self-calculation indicated in Fig. 8(b). And therefore the active output power dissipated in load is $\frac{640^2}{100}$ or 4.1 kW.

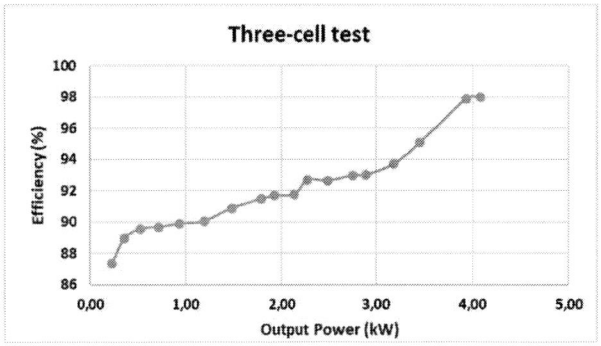

Fig. 9. Overall DC-AC efficiency results shows a maximum efficiency of 98% at 4.1 kW output power.

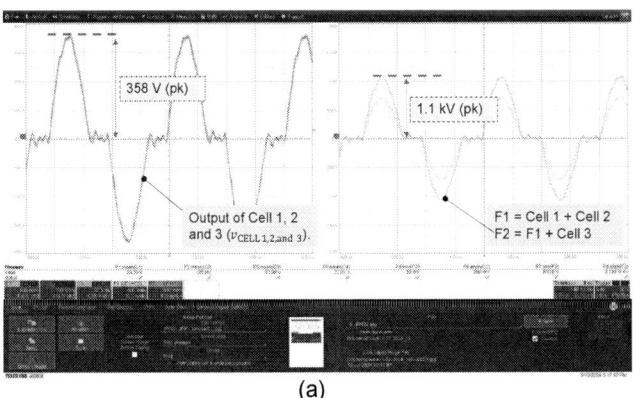

(a)

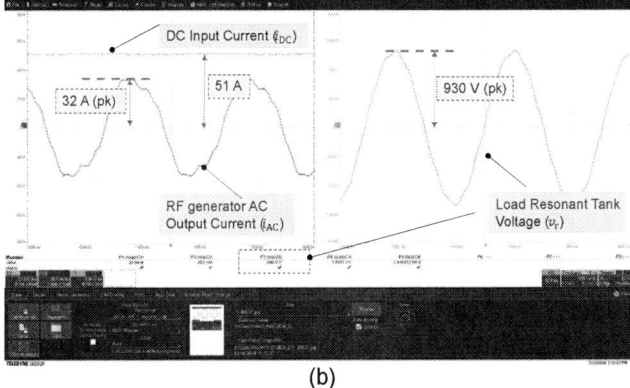

(b)

Fig. 8. Experimental waveform for three-cell tests at 4.1 kW output power. (a) Cell output voltages for three cells $v_{CELL 1, 2, and 3}$ and total RF generator voltage F2 = v_{CELL1} + v_{CELL2} + v_{CELL3}. (b) The RF generator input DC current (i_{DC}) and output RF current (i_{AC}). And finally the resonant tank voltage v_r.

Measurements were taken throughout the test procedure which leads to the efficiency curve as presented in Fig. 9. At low power the efficiency is below 90 %. However as the output power is increased the efficiency goes above 96 % at 4 kW of output power and reaches a maximum of 98 % at 4.1 kW of output power.

V. CONCLUSION

In this paper series connected Class-E push-pull resonant converters are presented to achieve high power at 2.4 MHz

of operating frequency. The series connection method of three Class-E push-pull topology is the first, ever presented as per authors knowledge. However the concept of series connecting several resonating converters operating at multi-MHz frequency is not limited to current source topology of Class-E. The authors have presented similar results with a voltage source topology Class-PN [4]. Thus series connection of resonating converters is a viable solution for higher power without the need of power combiners at several MHz or operating frequency. Nevertheless, it is important to point out the fact that before the addition of R_d in Class-E push-pull topology, it had been impossible to achieve such high operating voltages with three cells as presented in this paper. This is because as voltage was increased, the cells would enter instability and collapse the output voltage leading to damaged GaN HEMT devices.

REFERENCES

[1] T. L. Wilson, "Radio-Frequency Dielectric Heating in Industry," *Electric Power Research Institute, California, USA*, 3 1987.

[2] A. Jain, P. R. Hannurkar, D. K. Sharma, A. K. Gupta, A. K. Tiwari, M. Lad, R. Kumar, P. D. Gupta, and S. K. Pathak, "Design and characterization of 50 kW solid-state RF amplifier," *International Journal of Microwave and Wireless Technologies*, vol. 4, no. 6, p. 595–603, 2012.

[3] S. Jeon and D. Rutledge, "A 2.7-kW, 29-MHz Class-E/F$_{odd}$ Amplifier with a Distributed Active Transformer," in *IEEE MTT-S International Microwave Symposium Digest, 2005.*, 2005, pp. 1927–1930.

[4] F. Ahmad, T. S. Aunsborg, A. B. Jørgensen, and S. Munk-Nielsen, "From vacuum tubes to modern semiconductors: Opportunities for industrial heating industry," *IEEE Transactions on Industry Applications*, vol. 60, no. 4, pp. 6488–6498, 2024.

[5] GaN Systems, "GS66516T - Top-side cooled 650V E-mode GaN transistor," 2024. [Online]. Available: https://gansystems.com/wp-content/uploads/2021/10/GS66516T-DS-Rev-210727.pdf

[6] M. J. Chudobiak, "The use of parasitic nonlinear capacitors in class E amplifiers," *IEEE Transactions on Circuits and Systems I: Fundamental Theory and Applications*, vol. 41, no. 12, pp. 941–944, 1994.

[7] T. S. Aunsborg, S. B. Duun, S. Munk-Nielsen, and C. Uhrenfeldt, "Development of a Current Source Resonant Inverter for High Current MHz Induction Heating," *IET power electronics*, vol. 15, no. 1, p. 1–10, 2022.

STATE OF THE ART 1.7kV LATERAL GaN HEMTs, AN ALTERNATIVE TO SiC

Karthick Murukesan[1], Robert Yang[1], Kamal Varadarajan[1] Sorin Georgescu[1], Doug Kang[1]

[1]Power integrations, 5245 Hellyer Avenue San Jose, CA USA 95138

Email: karthick.murukesan@power.com

Abstract— **In this work, we demonstrate industry's first, true 1.7kV -rated lateral GaN HEMT with state of the art switching and transient overvoltage capability suitable for automotive and industrial applications. Based on the PowiGaN™ technology platform of Power Integrations, the 1.7kV GaN HEMT shows stable off-state leakage beyond 2.5kV with a typical breakdown voltage of 3kV. Reliable switching performance up to peak V_{DS} of 1.45kV with stable R_{DSON} has been demonstrated in lateral GaN HEMTs for the first time in this work. Further, reliability qualification data through 1000Hrs of HTRB, THBT and DHTOL stress tests is presented demonstrating the field readiness of the 1.7kV GaN HEMT. Performance of the 1.7kV GaN HEMT in a power conversion application was evaluated in a 60W isolated flyback power supply and results comparable to an equivalent SiC power switch were observed for key performance metrics (efficiency, thermals) under a wide range of input and load conditions. Our work demonstrates the potential of 1.7kV lateral GaN HEMT as a cost-effective alternative to SiC in high voltage applications.**

Keywords—1.7kV GaN HEMT, Reliability, SiC

I. INTRODUCTION

Wide bandgap semiconductor (WBG) by virtue of its high critical electric field and superior intrinsic properties enable high frequency, high density power converters. SiC and GaN are the leading WBG semiconductors which have had market success in power electronic applications. GaN with epitaxially engineered layers can form a high mobility 2-dimensional electron gas (2DEG) without external potential enabling normally on lateral High Electron Mobility Transistors (HEMTs). For reliable, safe power switching operations normally OFF HEMTs are desirable. Cascoded HEMTs, MISHEMTs, p-GaN gate HEMTS are the commonly used normally OFF HEMT architectures. GaN, though promising, has a dynamic component of R_{DSON} due to its intrinsic traps which makes the reliability, manufacturability, and converter design challenging. With improved epitaxy, device design improvements and reliable gate driver design the challenges have been addressed reasonably well in the past decade. With this GaN HEMTs are a commercial success in <650V power switching applications, specifically in charger/wall socket applications. Beyond 650 V, only two manufacturers have released lateral GaN HEMTs with a rated voltage of 900 V [1][2]. Owing to this, applications needing WBG power devices rated 1200V and beyond are constrained to use vertical SiC devices. However, cost is still a significant barrier to SiC adoption[3]. GaN on the other hand can be a cheaper alternative to SiC in high voltage applications if reliable high voltage switching can be realized. With automotive EVs transitioning from 400V to 800-1000V battery bus to enable rapid charging and with the elimination of 12/48V battery, there are a slew of new low power high voltage (600V-1.2kV) DC-DC, AC-DC

applications which could be catered by GaN. With this application scenario, there have been demonstrations of 1.2kV[4][5][6][7], 1.7kV[8], 3kV[9] and 6.5kV[10] rated lateral GaN HEMTs targeting automotive and industrial applications. However, none of these works demonstrated a) true switching capability meeting industry standard 80% derating or b) reliability test results in accordance with JEDEC standards. Both the above-mentioned milestones are critical for successful field deployment of high voltage lateral GaN HEMTs. In this work we have developed and qualified robust 1.7kV cascode GaN with demonstrated switching capability of ~ 1.45kV (85% of rated voltage) providing the necessary switching margin for high voltage applications. A high voltage 60W isolated flyback power supply unit was built with the 1.7kV GaN cascode device and benchmarked against an equivalent SiC power switch to demonstrate lateral GaN HEMT technology's ability to be a cost-effective alternative to SiC in high power conversion applications.

II. DEVICE CHARACTERISTICS, RELIABLITY AND POWER SUPPLY EVALUATION

Normally ON 1.7kV lateral GaN HEMTs are engineered utilizing proprietary PowiGaN™ technology platform. Normally off operation is achieved by connecting the normally ON GaN HEMT in series with a low voltage silicon MOSFET in cascode configuration as in Fig.1. By employing proprietary techniques, a higher breakdown voltage, higher true switching voltage margin is achieved. This state-of-the-art GaN cascode die is designed area scalable to target R_{DSON} ranging from 50mOhm – 2.5Ohm to cover a wide range of potential applications.

A. Device electrical characteristics

The pulsed output characteristics of a typical 380mOhm GaN cascode device is shown in Fig. 2. illustrating high current capability. The 1.7kV GaN cascode device has a R_{DSON} temperature coefficient of 1.5 at Tj=100°C (Fig. 3.) and its behaviour over typical operating range is comparable with a cascoded SiC power switch with similar Rdson [11]. Off state characteristics of a typical 1.7kV GaN cascode are shown in Fig. 4. illustrating stable leakage beyond 2.5kV with a breakdown of 3kV. This ensures that the device has excellent transient overvoltage capability compared to SiC devices with similar voltage rating. For GaN HEMT power devices stable high voltage switching is of paramount importance to enable reliable power supply operation. The GaN cascode device switches reliably up to 1.45kV in a continuous mode with typical switching waveforms as shown in. Fig. 5. The transient voltage capability of the GaN cascode allows it to switch through momentary increase in supply voltages up to 1.7kV without suffering a hard failure. This capability is

979-8-3315-1612-3/25 $31.00 © 2025 IEEE

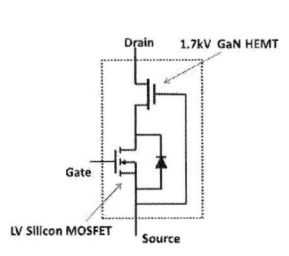

Fig. 1. Circuit schematic of the 1.7kV GaN cascode device

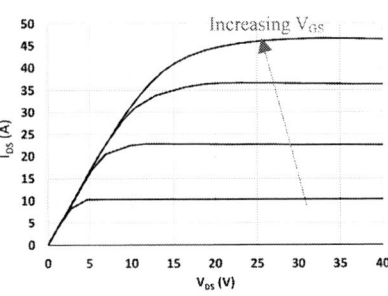

Fig. 2. Output characteristics of a 380mOhm, 1.7kV GaN cascode device illustrating high current capablity

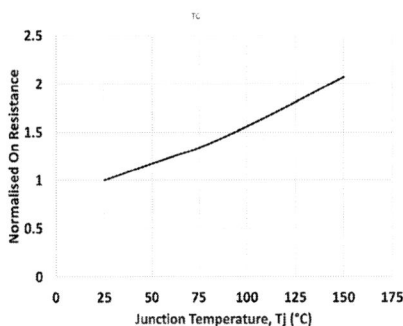

Fig. 3. Normalized on resistance with respect to temperature of the 1.7kV GaN cascode device

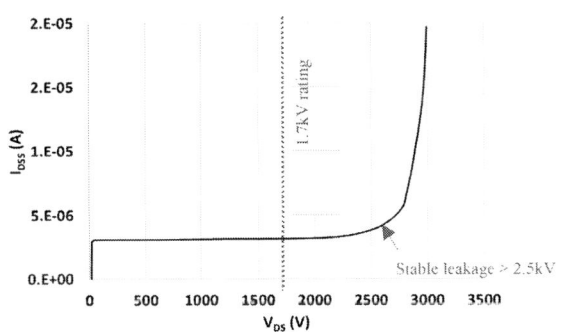

Fig. 4. Typical off state characteristics of 1.7kV GaN cascode device demonstrating stable leakage beyond 2.5kV and a breakdown of 3kV

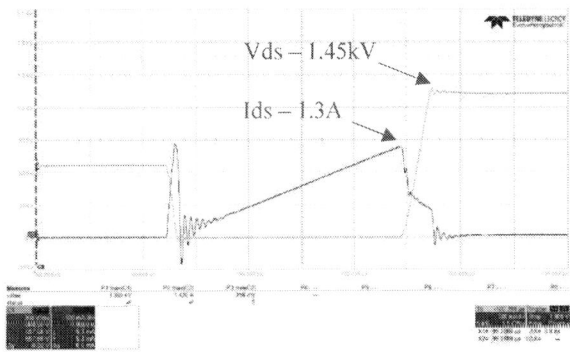

Fig. 5. Waveform illustrating switching with peak VDS of 1.45kV

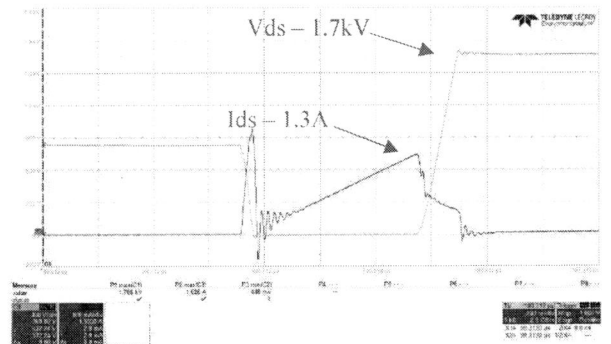

Fig. 6. Waveform illustrating switching with peak VDS of 1.7kV representing surge conditions

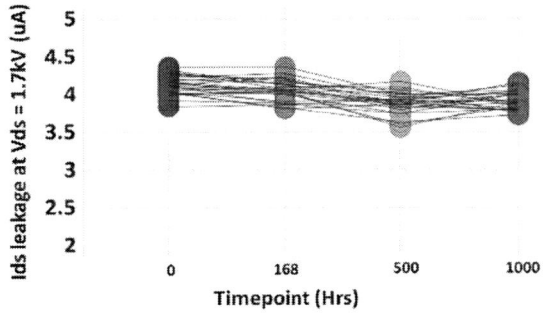

Fig. 7. Stable off state drain leakage through 1000Hrs., of HTRB stress at 1.36kV/150°C

demonstrated by reliably switching at a peak V_{DS} of 1700 V without any issues as shown in Fig. 6.

B. Reliability Evaluation

Industry standard reliability evaluation procedures[12], [13] as detailed in Table1 are performed to qualify the 1.7kV GaN cascode device for high voltage, high reliability applications. Clean results were obtained after 1000Hrs of HTRB and THBT at 1.36kV and 800V, respectively.

HTRB	DHTOL	THBT
Pass, 1.36kV 1000 Hrs., 150°C	Pass,1.45kV, 1000 Hrs., 125 °C	Pass, 800V 1000 Hrs., 85% H*

Table 1: Reliability qualification summary (*H-humidity)

The drain to source off state leakage current during the HTRB shows excellent stability as shown in Fig.7. demonstrating the device robustness. In addition to the static OFF state high voltage reliability tests, switching reliability at the product level was evaluated using a flyback switcher IC with the 1.7kV GaN cascode as power switch (Fig. 2.). Dynamic High Temperature Operating Life (DHTOL) test is performed as specified in JEP 180[13], a JEDEC guideline for GaN switching reliability evaluation. A custom test bed as shown in Fig. 8. was developed to evaluate DHTOL on multiple units in parallel at 125°C for 1000Hrs switched at 1.45kV at target application frequency. During DHTOL, at multiple timepoints device performance metrics are sampled.

Fig. 8. Custom test-bed used to perform DHTOL on multiple flyback switcher ICs with 1.7kV GaN at 125°C

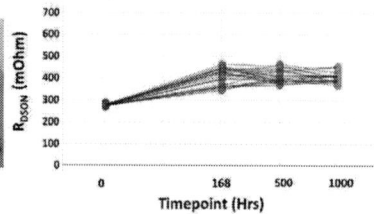

Fig. 9. Stable $R_{DS(ON)}$ with limited shift observed during 1000 Hrs., of DHTOL

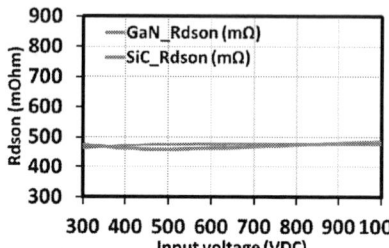

Fig. 10. R_{DSON} of the GaN cascode and SiC switch measured in-situ in the power supply at various Vin

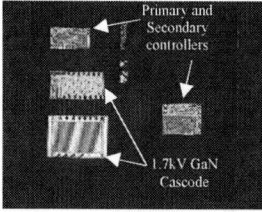

Fig. 11. Flyback switcher IC incorporating GaN cascode device incl. primary, secondary controllers

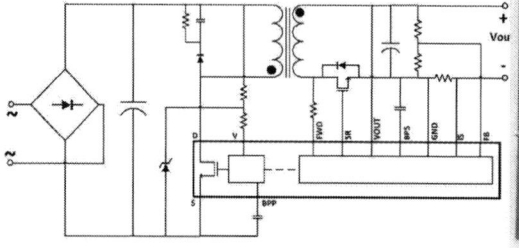

Fig. 12. Power supply design schematic where the flyback switcher IC utilizing 1.7kV GaN cascode or 1.7kV SiC is evaluated

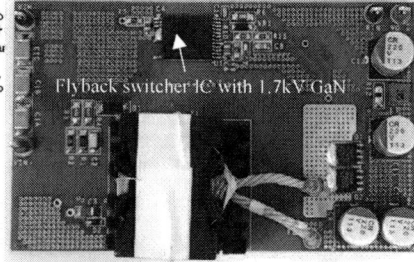

Fig.13. Top-side photograph of the power supply design used in the evaluation.

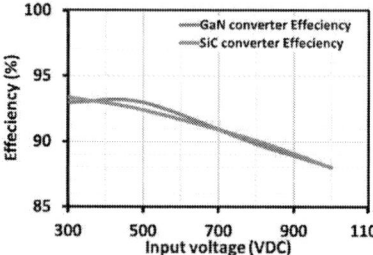

Fig. 14. Efficiency comparison over a range of input voltage (300-1000V) at full load

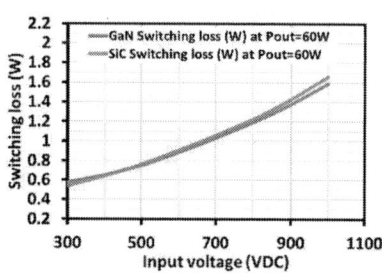

Fig. 15. Switching loss comparison over a range of input voltage (300-1000V) at full load

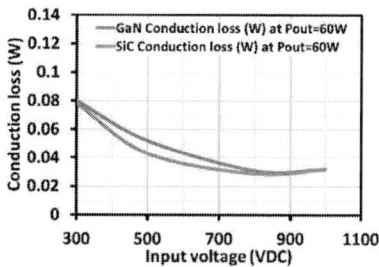

Fig. 16. Conduction loss comparison over a range of input voltage (300-1000V) at full load

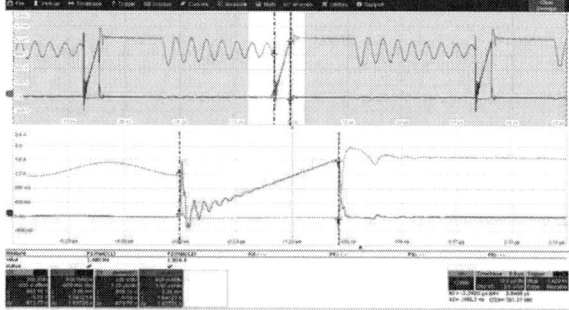

Fig. 17. Power supply switching waveforms while steady state thermals were measured

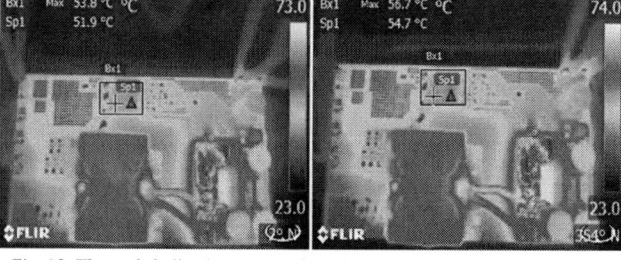

Fig. 18. Thermals indicating comparable temperature rise during power supply operation (illustrated in Fig.17) for both SiC (left) and GaN HEMT (right)

$R_{DS(ON)}$ of the 1.7kV GaN cascode monitored over the course of 1000 hrs. of DHTOL stress at 1450 V / 2.3 A / 125°C as shown in Fig. 9. shows stable performance from the earliest timepoint with a limited shift <30%. In comparison, dynamic $R_{DS(ON)}$ increase ranging between 40% and 50% is observed even at a 600V switching condition on reported 1200V GaN FET [6] making it unsuitable for applications that require a true 1200V rating. The DHTOL is performed at 85% derating (1.45kV) against standard 80% (1.36kV) providing an additional margin at device level.

C. Power supply evaluation and benchmarking with SiC

The performance of the 1.7kV GaN cascode device in a power conversion application was evaluated after replacing a 1.7kV SiC switch with comparable Rdson (Fig. 10) in an integrated flyback switcher IC belonging to the InnoSwitch™3 EP product family[14]. The IC incorporated the high-voltage primary-side switch, a primary-side controller, and a secondary-side controller for synchronous rectification as shown in Fig. 11. and was evaluated in an automotive flyback power supply design using a 60 W, 24 V / 2.5A test set-up described in RDR-919Q[15]. The schematic

979-8-3315-1612-3/25 $31.00 © 2025 IEEE

for the power supply design, detailed in RDR-919Q, is shown in Fig. 12., along with the top-side photograph of the circuit board used in the evaluation (Fig. 13.). The efficiency of the flyback converter with the SiC switch and the GaN cascode device was found to closely match across a wide range of input voltages (300-1000V) at load conditions as shown in Fig.14. The average efficiency of GaN and SiC based converters are ~90% at Vin of 800VDC. The conduction and switching losses measured over a complete switching cycle show comparable results for both GaN Cascode and SiC over the entire input voltage range as shown in Figs. 15 and 16. Comparable thermal performance was observed after 2 hours of stable power supply operation in both cases with temperature rise below 30°C with respect to the ambient. Maximum temperature of 56.7°C and 53.8°C (Fig. 18.) were observed with GaN and SiC respectively after switching the power supply for 2 hours at Vin=1000VDC as shown in Fig. 17.

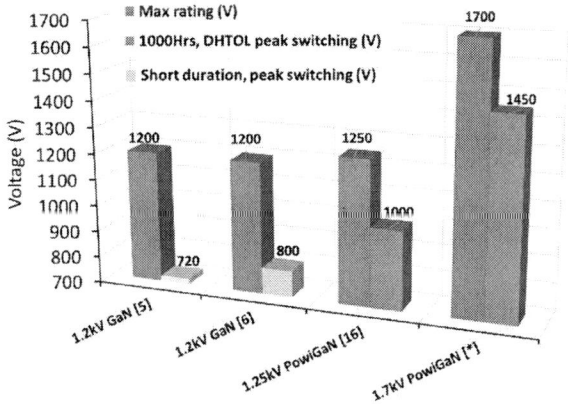

Fig. 19. Benchmarking lateral high voltage GaN technologies rated >1kV comparing switching capability, *-this work.

III. CONCLUSION

In this work we have successfully demonstrated the industry's first 1.7 kV normally off lateral GaN cascode HEMT with true 1.45kV switching capability. The performance of the 1.7kV GaN cascode device in a power conversion application was evaluated and compared against an equivalent SiC power switch. Matching results were observed for key performance metrics (efficiency, thermals) under a wide range of input and load conditions. With 75% additional breakdown voltage margin, reliable 1000hrs DHTOL switching at 1.45kV with stable dynamic Rdson, the reported 1.7kV lateral GaN HEMTs are state of the art, as benchmarked with other lateral GaN HEMTs[16] (Fig.19).

REFERENCES

[1] Transphorm: 900V GaN FETs; 2024., https://www.transphormusa.com/en/products/900V.

[2] Di Paolo Emilio M. Power Integrations launches 900V GaN flyback switcher ICs. Power Electronics News. 2023., https://www.powerelectronicsnews.com/power-integrations-launches-900v-gan-flyback-switcher-ics/.

[3] Ahmad M. Silicon carbide's wafer cost conundrum and the way forward. EDN Asia. 2023.,

"https://www.ednasia.com/silicon-carbides-wafer-cost-conundrum-and-the-way-forward/."

[4] S. Kumar, K. Geens, A. Vohra, D. Wellekens, D. Cingu, E. Fabris, T. Cosnier, H. Hahn, B. Bakeroot, N. Posthuma, R. Langer, and S. Decoutere, "1.2 kV enhancement-mode p-GaN gate HEMTs on 200 mm engineered substrates," *IEEE Electron Device Letters*, pp. 1–1, 2024, doi: 10.1109/LED.2024.3361164.

[5] G. Gupta, M. Kanamura, B. Swenson, C. Neufeld, T. Hosoda, P. Parikh, R. Lal, and U. Mishra, "1200V GaN Switches on Sapphire: A low-cost, high-performance platform for EV and industrial applications," in *Technical Digest - International Electron Devices Meeting, IEDM*, Institute of Electrical and Electronics Engineers Inc., 2022, pp. 3521–3524. doi: 10.1109/IEDM45625.2022.10019381.

[6] S. Li, Y. Ma, W. Lu, M. Li, L. Wang, Z. Zhang, T. Zhu, Y. Li, J. Wei, L. Zhang, S. Liu, and W. Sun, "1200V E-mode GaN Monolithic Integration Platform on Sapphire with Ultra-thin Buffer Technology."

[7] Y. Du, H. Yan, P. Luo, X. Tan, E. M. S. Narayanan, H. Kawai, S. Yagi, and H. Narui, "Investigation on Shift in Threshold Voltages of 1.2 kV GaN Polarization Superjunction (PSJ) HFETs," *IEEE Trans Electron Devices*, vol. 70, no. 1, pp. 178–184, Jan. 2023, doi: 10.1109/TED.2022.3225695.

[8] X. Li, J. Wang, J. Zhang, Z. Han, S. You, L. Chen, L. Wang, Z. Li, W. Yang, J. Chang, Z. Liu, and Y. Hao, "1700 V High-Performance GaN HEMTs on 6-inch Sapphire With 1.5 μm Thin Buffer," *IEEE Electron Device Letters*, vol. 45, no. 1, pp. 84–87, Jan. 2024, doi: 10.1109/LED.2023.3335393.

[9] N. E. K. H. Y. S. & N. H. Sheikhan A, " Evaluation of a 3 kV Polarization Superjunction GaN HEMT," *PCIM Europe Conference Proceedings, Vol. , Jun. 2024, pp. 549–556.

[10] J. Cui, J. Wei, M. Wang, Y. Wu, J. Yang, T. Li, J. Yu, H. Yang, X. Yang, J. Wang, X. Liu, D. Ueda, and B. Shen, "6500-V E-mode Active-Passivation p-GaN Gate HEMT with Ultralow Dynamic R on," in *2023 International Electron Devices Meeting (IEDM)*, IEEE, Dec. 2023, pp. 1–4. doi: 10.1109/IEDM45741.2023.10413742.

[11] 1700V SiC JFETs, https://www.qorvo.com/products/power-solutions/sic-jfets.

[12] JEP 198 JEDEC Standard, Guideline for HTRB, 2023.

[13] 180 JEDEC Standard, Guideline for switching Reliablity Evaluation for GaN Ver 1.0. 2023.

[14] 250V GaN Switcher IC. Power Electronics News. 2023. DI Paolo Emilio M. Power Integrations Releases Innovative 1, https://www.powerelectronicsnews.com/power-integrations-releases-innovative-1250v-gan-switcher-ic/

[15] Power Integrations Applicaitons Engineering Department. Reference Design Report for a 60W INN3949CQ. 2023,

https://www.power.com/sites/default/files/document s/rdr-919q_60w_high_input_voltage_psu_automotive_in noswitch3-aq-1700v_sic.pdf.

[16] K. Varadarajan, A. Ankoudinov, R. Yang, A. Kudymov, B. Shankar, K. Murukesan, S. Georgescu, J. Rongavilla, and D. Kang, "Reaching Beyond 1200 V: Lateral GaN HEMTs for High-Reliability EV and Industrial Applications," in *PCIM Europe Conference Proceedings*, Mesago PCIM GmbH, 2024, pp. 1596–1598. doi: 10.30420/566262217.

Modeling and Characterization of Current and Future 1.2 kV Wide Bandgap Semiconductor-based MOSFETs

Sushanta Gautam[1], Austin M. Szczublewski[1], Samuel K. Atwimah[1], Aidan P. Fox[1],
William M. Collings[1], Tolen Nelson[1], Daniel G. Georgiev[1], Raghav Khanna[1],
Andrew D. Koehler[2], and Karl D. Hobart[2]

[1]*EECS Dept., The University of Toledo, Toledo, OH, USA*
[2]*U.S. Naval Research Laboratory, Washington, DC, USA*
Sushanta.Gautam@rockets.utoledo.edu

Abstract—This paper presents the design and simulation of a 1.2 kV vertical GaN based MOSFET. First, a physics-based TCAD model of the proposed vertical GaN MOSFET structure is discussed. The proposed MOSFET structure employs hybrid edge termination to achieve high-breakdown with low on-resistance. The feasibility of the proposed device is demonstrated through static characterization, and a circuit-simulation model of the GaN MOSFET is extracted from the TCAD design. Next, an experimental Double Pulse Test (DPT) is performed on similarly-rated commercially available Si and SiC MOSFETs. An empirically validated circuit simulation model of the test-stand is then used to compare the simulated switching performance of the theoretical GaN MOSFET with that of the experimentally measured SiC and Si MOSFETs. The results show that the low technology readiness level GaN MOSFET achieved superior performance to the Si MOSFET, while demonstrating comparable performance to the SiC MOSFET. Thus, this paper can be used to further guide and improve both the breakdown capability and switching characteristics of future vertical GaN MOSFETs.

Index Terms—gallium nitride; wide band gap semiconductors; TCAD modeling; switching characterization; behavioral modeling; vertical GaN MOSFET; hybrid edge termination; Double Pulse Test; SaberRD simulation

I. INTRODUCTION

The semiconductor industry has grown substantially in recent years. This expansion has been largely fueled by the increasing demand for efficient and faster switching devices, particularly in high-power and high-frequency applications. Among the various advancements in this field, wide bandgap (WBG) semiconductor devices, especially those based on gallium nitride (GaN), have gained considerable attention. GaN devices are highly sought after due to their superior properties, such as faster switching speeds and reduced conduction losses at high voltages. These advantages stem from GaN's wide bandgap of 3.4 eV, large electric field strength, high electron mobility (2000 cm^2/V-s), low carrier lifetimes, and high saturation velocity (6.63 x 10^{6}cm/s) [1], [2], [3].

This work was supported in part by Dana Inc. and the U.S. Office of Naval Research under grant number N00014-21-1-2832, and was approved for public release under DCN 543-2240-24. Distribution Statement A. Approved for Public Release. Distribution Unlimited.

In recent years, the maturation of GaN technology has significantly impacted the performance and efficiency of power electronic systems. One notable impact is evident in lateral GaN-based High Electron Mobility Transistors (HEMTs), which have proven highly effective in low-voltage applications (under 1000V). However, the high voltage capability of lateral GaN HEMTs is restricted by the trade-off between extending the gate-drain spacing and increasing the device surface area [4]. Extending the gate-drain spacing to accommodate higher voltages necessitates proportional increases in the device's physical dimensions, potentially resulting in impractically large chip sizes for certain applications. Similarly, enhancing current handling capabilities requires enlarging the device's surface area. A significant amount of current in lateral devices flows near the interface, causing non-uniform heat generation. This concentration of heat near the surface poses substantial challenges for heat dissipation, especially at high currents [5].

Nevertheless, recent advancements in material processing techniques have enabled the development of thick, low-doped drift layers capable of supporting higher voltages [6]. This progress has paved the way for designing vertical architectures in GaN-based devices. The development of vertical architectures has highlighted the capacity of GaN devices to fully leverage their high voltage potential [7]. Vertical GaN devices provide several distinct advantages over their lateral counterparts. For example, vertical devices can achieve high voltage-handling without necessitating a proportional increase in chip area, thereby maintaining a compact form factor. Moreover, vertical architectures enable more effective heat dissipation through more uniform distribution within the bulk of the device, leading to improved thermal management. Additionally, these devices offer greater current handling capacity and mitigate numerous surface-related issues that typically affect lateral designs. Furthermore, vertical GaN devices demonstrate better scalability, making them suitable for a wider range of applications [8], [9].

This paper focuses on the design and simulation of a 1.2 kV GaN-based planar MOSFET, which is formed on an epitaxially

979-8-3315-1612-3/25 $31.00 © 2025 IEEE

grown p-n–n+ wafer. Technology Computer Aided Design (TCAD) is used to develop the physics-based GaN MOSFET model. The study also includes empirical characterization of commercially available 1.2 kV silicon (Si) and silicon carbide (SiC) MOSFETs. These devices are evaluated for their transient performance using a double pulse test (DPT) setup. The results of these tests are then used to validate a circuit simulation model of the empirical DPT testbed. Furthermore, a behavioral model of the 1.2 kV GaN MOSFET is extracted from TCAD and instantiated in circuit simulation to compare its dynamic performance with that of empirically validated models of the commercially available Si and SiC MOSFETs. The findings of this research suggest that the low-technological readiness level (TRL) GaN MOSFET exhibits performance that is comparable to the commercially available SiC MOSFET and superior to the Si MOSFET.

II. TCAD MODELING OF GaN MOSFET

Vertical GaN planar MOSFETs, formed on epitaxially grown p-n-n+ wafers, are investigated through physics-based TCAD Sentaurus (Synopsys) simulation to understand their electrical performance and reliability under various conditions. Fig. 1 shows the considered structural composition and layout of the proposed MOSFET. The n-type plug, known as the JFET region, and the source deposition region are created through ion implanted Si counter-doping on top of a grown p/n⁻/n⁺ power diode stack. This counter-doping technique is used because it avoids selective p-type doping in GaN, via either Mg implantation or etch and regrowth, and mitigates issues related to the activation of p-type dopants [10].

The structure of the MOSFET shown in Fig. 1 is that of a vertical planar MOSFET, with a drift layer thickness of 8 μm and a Si doping concentration of 2×10^{16} cm⁻³, and a p-layer thickness of 0.37 μm with a Mg doping concentration of 1×10^{18} cm⁻³. Optimization of the structure, as detailed in [11] and [12], leads to a low specific on-resistance of 0.26 mΩ*cm², and a breakdown voltage of approximately 1.2 kV. The inversion channel width is set to 1 μm, and the widths of the source and n-plug JFET gaps are minimized to 1 μm, which is the smallest dimension achievable by proposed fabrication capabilities.

As the n-type JFET-plug and source deposition regions are proposed to be formed via ion implantation, a strategy was developed to minimize the implantation tail, which can adversely affect the performance of the device. This challenge was effectively addressed by designing an implant schedule that would provide a desired box profile of the net doping. Specifically, a 9-implant schedule was devised, optimizing the implantation parameters to achieve the desired doping profile. As illustrated in Fig. 2, each individual implantation step is modeled using a Pearson distribution to describe the distribution of dopants [13]. The parameters of each step, energy and dose, are optimized to provide a net implantation profile such that the entirety of the counter-doped region is n-type while minimizing the variation from a target net doping of 5×10^{16} cm⁻³ as well as minimizing the implantation tail

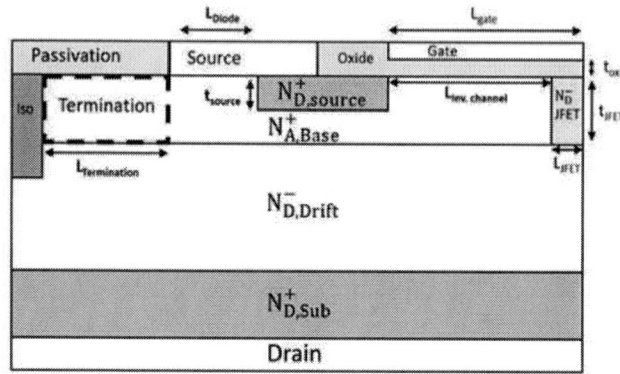

Fig. 1. 2-D half cross-sectional view of the GaN MOSFET structure.

into the drift layer. This carefully designed implant profile results in a maximum net n-type doping concentration of 2×10^{17} cm⁻³ and effectively minimizes the implantation tail to approximately 60 nm, which is critical for achieving high breakdown voltage.

To manage the electric fields at the periphery of the device and prevent premature breakdown, a planar edge termination technique known as hybrid edge termination was implemented [14]. This innovative technique utilizes nitrogen implants to form a high-resistive region, which mitigates peripheral field effects, as detailed in [12]. By creating this high-resistive region, the hybrid edge termination technique ensures that the electric field is distributed more uniformly, thereby reducing electric field crowding and the associated risks of premature breakdown.

As a result, this approach not only enhances the overall performance and reliability of the device but also contributes to its longevity under high-voltage conditions. A notable

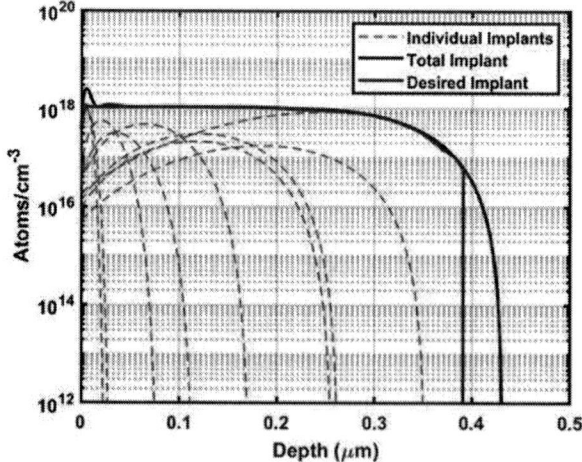

Fig. 2. Plug Implant design. Dashed red lines are the individual implants. The solid black line is the sum of the implants. The solid blue line is the desired profile.

979-8-3315-1612-3/25 $31.00 © 2025 IEEE

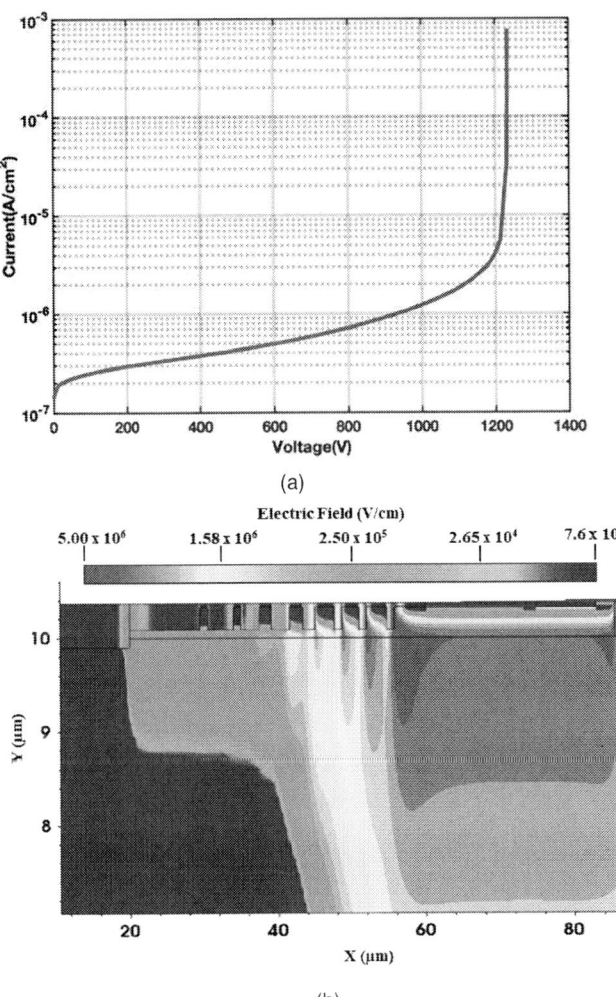

Fig. 3. Breakdown characteristics of 1.2 kV GaN MOSFET. (a) I-V curve. (b) Electric field distribution.

the behavior of the transistor. This level of detail, while beneficial for precision, significantly increases the computational resources and time needed to perform simulations, making it less practical for large-scale or rapid analyses.

Behavioral models use equivalent circuits and mathematical modeling to provide relatively swifter simulation speed, at the expense of not having any physical meaning [16]. These models abstract the physical details and instead focus on capturing the functional behavior of the device through simplified representations [17]. This allows for quicker simulations, which is advantageous when dealing with extensive circuit designs or when rapid iterative testing is required. However, the lack of physical grounding of behavioral models can sometimes lead to less accurate predictions under certain conditions, limiting their applicability in scenarios where high fidelity is crucial. Despite this, behavioral models are highly valuable in preliminary design phases and in large-scale simulations where the primary goal is to understand overall system performance rather than detailed physical phenomena. The speed and efficiency of these models facilitate rapid prototyping and iterative development.

To allow for a fair comparison to commercially available devices, the nominal characteristics of the theoretical GaN model were scaled. This was achieved by examining the on-resistance of the selected 1.2 kV SiC MOSFET [18], which is measured at 350 mΩ, and utilizing the well-established Baliga Figure of Merit (BFOM) [19], [20], which characterizes a device's specific on-resistance as a function of breakdown voltage. An approximated area of the commercially available SiC MOSFET was surmised by assuming the device was at the theoretical BFOM limit of SiC. Thusly, an area scaling factor for the GaN MOSFET was derived to scale the nominal TCAD data to a size comparable to the SiC MOSFET. The derived scaling factor was then implemented into the TCAD simulation in order to produce characteristic curves that would enable a rigorous comparison. Next, a behavioral model of the GaN MOSFET was implemented in the SaberRD environment. This behavioral model aims to replicate the real-world performance of a scaled GaN MOSFET under various operating conditions. Consequently, Fig. 4 presents the scaled behavioral model of the GaN MOSFET, illustrating modeled IV and CV characteristics. Here, the normalized TCAD curves are extracted and imported into circuit-simulation to implement the behavioral model. Thus, the TCAD model serves as the "target" or "measured" value for the behavioral model to replicate. The scaling process ensures that the comparison between the GaN MOSFET and the SiC and Si counterparts is based on equivalent performance metrics, thereby providing a level playing field for assessing their respective advantages and limitations. As seen in Fig. 4, good agreement is obtained between the behavioral model and the targeted TCAD model so that an accurate assessment of the GaN's MOSFET's dynamic performance can be obtained through circuit simulation. Next, data-sheet based behavioral models of 1.2 kV Si [21] and SiC [18] power MOSFETs were meticulously developed within the Synopsys SaberRD environment. Empirical IV and

achievement of this technique is the attainment of a breakdown voltage of approximately 1200 V, as shown by the simulated reverse I-V and electric field distribution at breakdown in Fig. 3 [12]. As will be demonstrated later, the model is also capable of simulating other standard static behaviors, such as Forward current-voltage (IV) characteristics, in addition to capacitance-voltage (CV) characteristics. These characteristics can be used to instantiate a circuit simulation model of the theoretical GaN MOSFET, which can be compared with empirically validated models of commercially available and similarly-rated Si and SiC MOSFETs.

III. BEHAVIORAL MODEL DEVELOPMENT

There are a variety of transistor modeling techniques ranging from purely physics-based, to purely behavioral, to hybrid physics and behavioral models [15]. High precision models can be realized by using physics-based models, which are based on the physical structure and typically require detailed information about the device geometry, material properties, and complex mathematical equations to accurately represent

979-8-3315-1612-3/25 $31.00 © 2025 IEEE

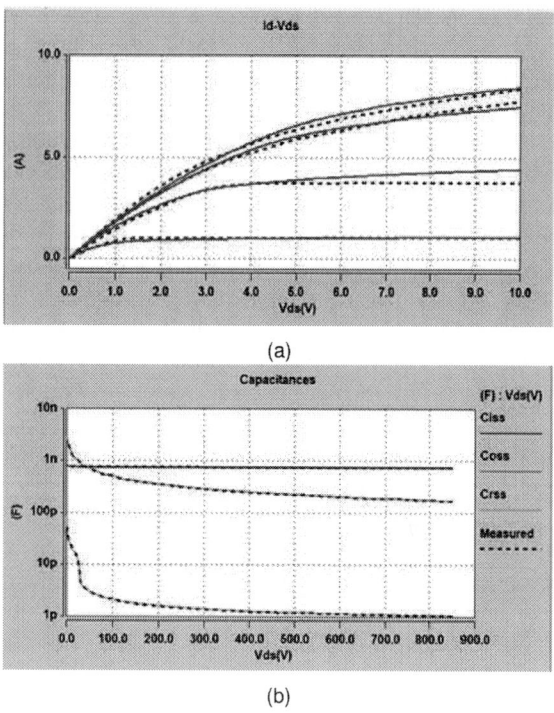

Fig. 4. TCAD based behavioral models of comparably scaled GaN MOSFET. Solid lines are the behavioral model in SaberRD, and dotted lines are the TCAD target. (a) Forward Curves. (b) Device Capacitances.

CV device characteristics from the respective data sheets, were employed as benchmarks to construct these behavioral models for both the Si and SiC MOSFETs. These models are visually represented in Fig. 5 and Fig. 6, respectively, showing good agreement with their experimental counterparts.

The solid line is the trend of the behavioral model, and the dashed line is the graph from the data-sheet. By aligning the behavioral model trends with the data-sheet graphs, the accuracy of the models can be validated, ensuring that they reliably replicate the performance of the actual devices under various operating conditions. This validation step is crucial for confirming that the developed models can be effectively used for further simulations and analysis in the design and optimization of power electronic systems.

IV. DPT RESULTS

An experimental DPT test-stand was prototyped to empirically evaluate the switching performance of both the Si and SiC MOSFETs. The DPT setup is illustrated in Fig. 7. A detailed schematic of the DPT is shown in Fig. 7(a), while Fig. 7(b) provides a photograph of the actual experimental setup. The test parameters include a bus voltage of 600 V, which is half the rated-value of the MOSFETs under-study, and a load inductance of 1.6 mH, resulting in a steady-state load current of 2 A. A SiC power Schottky diode [22], was used as the freewheeling diode. The behavioral model of the Si was then simulated in a model of the DPT using the

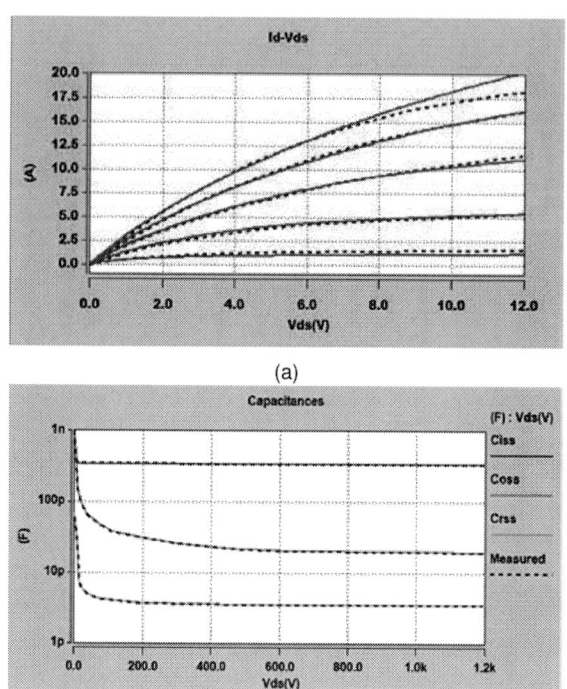

Fig. 5. Datasheet based behavioral models of SiC MOSFET. (a) Forward Curves. (b) Device Capacitances.

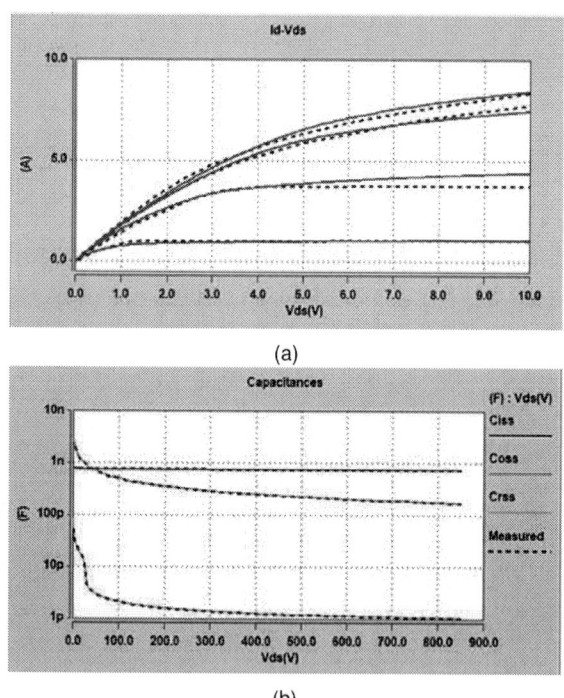

Fig. 6. Datasheet based behavioral models of Si MOSFET. (a) Forward Curves. (b) Device Capacitances.

Synopsys SaberRD environment. An optimization procedure, as previously reported in [23], was used in SaberRD to tune

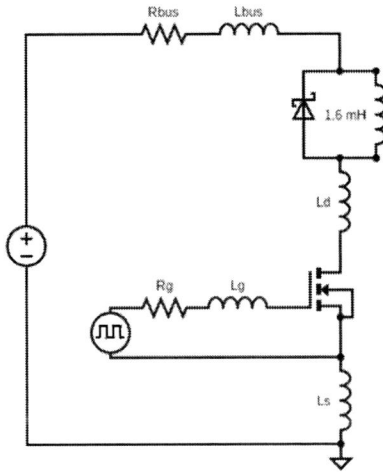

(a)

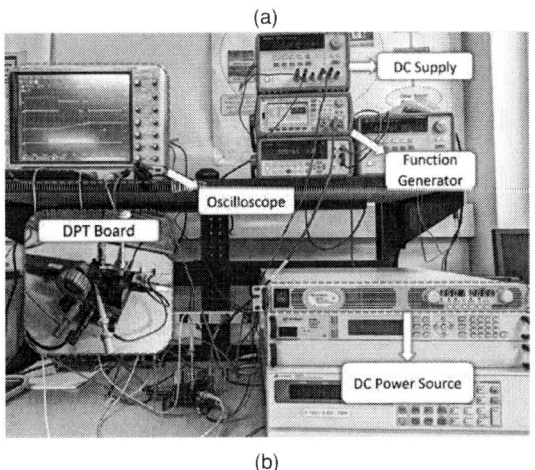

(b)

Fig. 7. Double Pulse Test setup. (a) Schematic. (b) Experimental Setup.

the parasitic parameters of the test-stand in order to obtain an accurate representation of the experimental system. This is an essential consideration for characterization of fast-switching GaN devices, where the parasitic components can severely impact device behavior, particularly under the influence of extremely high spectral content [23]. By properly accounting for these parasitics, the simulated DPT could be empirically validated and used as a basis on which the transient performance of the theoretical 1.2 kV GaN MOSFET can be projected. The components selected for optimization include the bus-inductance and resistance, as well as the per-terminal leg inductances for the SMD package used for both the Si and SiC MOSFETs. A comparison of experimental (red) and simulated (blue) switching waveforms at 600 V and 2 A, is shown in Fig. 8 for the Si MOSFET. Turn-off behavior is shown in Fig. 8(a), and turn-on behavior is shown in Fig. 8(b). Good agreement is obtained between measured and simulated performance. The extracted values from the tuning

algorithm are provided in Table I. L_{bus} and R_{bus} account for bus impedance and C_j accounts for the additional junction capacitance which is connected in parallel to the free wheeling diode. The optimized values are all practical and within reason. To further validate the test-stand, the tuned values of the parasitics from Table I were used with an orthogonal dataset, of 600 V and 4 A, as shown in Fig. 9. Fig. 9(a) and Fig. 9(b) illustrate comparison of experimental (red) and simulated (blue) waveforms for turn-off and turn-on instances of the Si MOSFET for the orthogonal dataset. In this orthogonal dataset simulation, there is no tuning. Rather, the extracted parameters from Table I are held at their optimized values to determine the extent of their predictive power. The good agreement shown in Fig. 8 and Fig. 9, indicates that the circuit simulation model of the empirical DPT test-stand is valid and can be used to predict the performance of the theoretical 1.2 kV GaN MOSFET. Similar undertakings were performed for the SiC MOSFET, but will not be shown here for the sake of space.

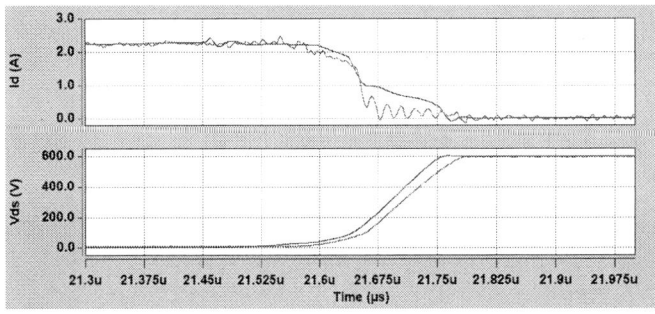

(a)

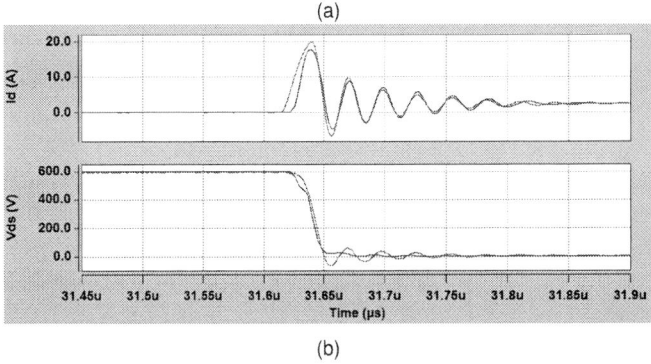

(b)

Fig. 8. Si MOSFET experiment (red) and simulation (blue). (a) Turn off. (b) Turn on.

The empirically validated model of the DPT test-stand was then used to simulate the 1.2 kV GaN behavioral model, using the optimized parasitic values from Table I. Turn-off and turn-on waveforms for the theoretical GaN MOSFET are shown in Fig. 10. The simulated switching losses were calculated and were then compared to the experimentally measured switching losses of the Si and SiC MOSFETs.

Table II summarizes the switching losses comparison between experimentally tested Si and SiC MOSFETs against the simulated GaN MOSFET. For total switching loss, the low-TRL GaN MOSFET shows comparable performance to

979-8-3315-1612-3/25 $31.00 © 2025 IEEE

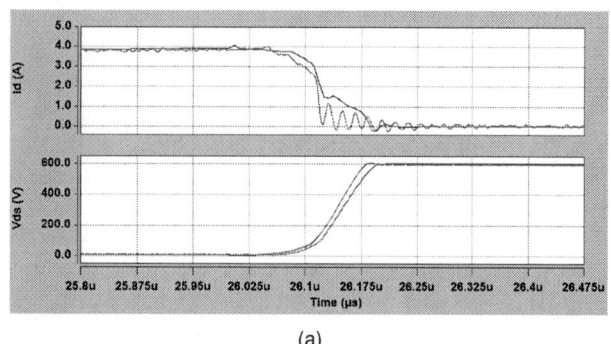

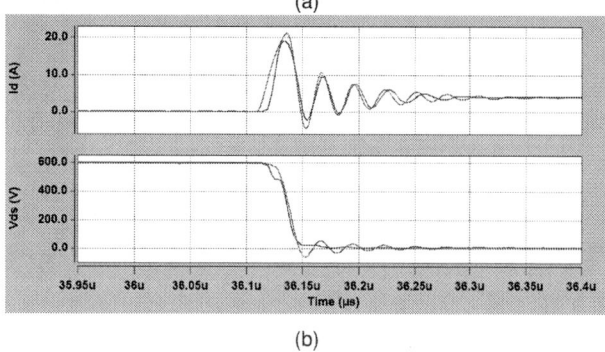

Fig. 9. Orthogonal dataset for Si MOSFET experiment (red) and simulation (blue). (a) Turn off. (b) Turn on.

TABLE I: Tuned Parameters in SaberRD.

Parameter	Value
L_{bus}	40.381 nH
R_{bus}	0.1971 Ω
C_j	200.232 pF
L_{source}	3.022 nH
L_{drain}	7.16 nH
L_{gate}	5.068 nH

the commercially available SiC MOSFET, and significantly improved performance over the Si MOSFET. It is instructive to note that the 1.2 kV GaN MOSFET is still in the development phase with a very low TRL, and thus these results merit further investigation into the potential of GaN devices for future high voltage power electronics. Although the thermal management capabilities of SiC are better than those of GaN, theoretical projections for the voltage-blocking capability, switching capability, and conduction capability of GaN are superior to both Si and SiC [24]. Thus, further development into GaN MOSFET technology so that it can realize its full performance entitlement, would allow for a well-founded comparison against high-TRL SiC MOSFETs. In particular, a fully-optimized GaN MOSFET would in theory outperform a similarly rated SiC MOSFET at room temperature. However, a more compelling comparison could be undertaken in harsher operating environments.

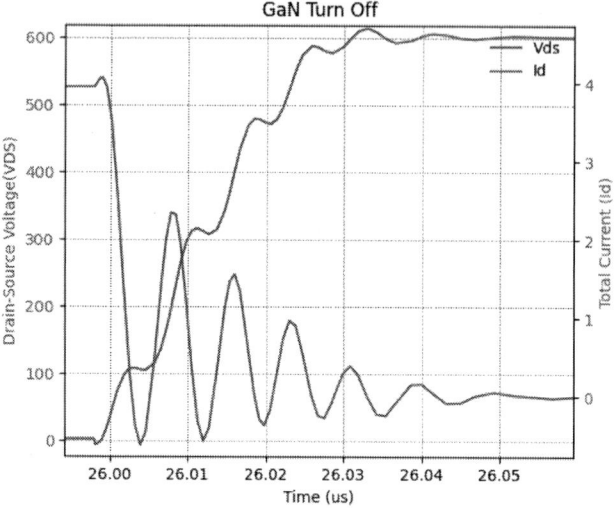

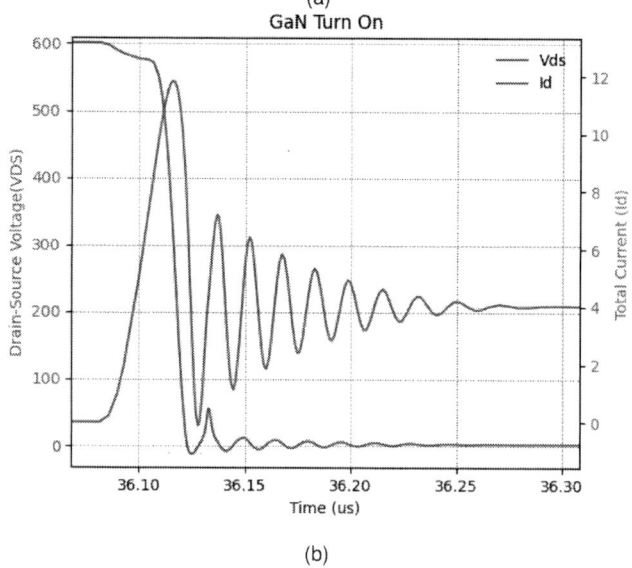

Fig. 10. Simulated switching waveforms for GaN with V_{ds} (red) and I_d (blue). (a) Turn off. (b) Turn on.

TABLE II: Switching Losses Comparison.

Material	Turn-on Losses (μJ)	Turn-off Losses (μJ)
GaN MOSFET	68.27	6.0849
SiC MOSFET	80.29	6.427
Si MOSFET	134.64	11.971

V. CONCLUSION

This work provides a comparison of the switching performance between commercially available 1.2 kV Si and SiC MOSFETs and a 1.2 kV theoretical GaN MOSFET. First, datasheet based behavioral modeling of Si and SiC MOSFETs was conducted in SaberRD, followed by designing a vertical GaN planar MOSFET in physics-based TCAD Sentaurus (Synopsys) environment. Switching performances of Si and

SiC MOSFETs were characterized using an experimental DPT test-stand. These characterizations were used to develop an empirically validated model of the DPT test-stand. A behavioral model of the GaN MOSFET was then developed and implemented in the empirically validated model of the DPT test-stand. It was found that despite being a low TRL device, the prospect of the GaN MOSFET in high voltage power electronics is promising, as it showed comparable performance to SiC and significantly better performance over Si. Furthermore, the TCAD model presented in this paper demonstrates the feasibility of high-voltage vertical GaN MOSFETs and provide a basis on which the technology can be further developed and optimized.

ACKNOWLEDGMENT

The authors gratefully acknowledge the support of Dana Inc. and the U.S. Office of Naval Research (grant number N00014-21-1-2832 and approved under DCN # 543-2240-24), at the direction of Captain LJ Petersen, USN (Ret). Distribution Statement A. Approved for Public Release. Distribution unlimited.

REFERENCES

[1] E. A. Jones, F. F. Wang, and D. Costinett, "Review of commercial gan power devices and gan-based converter design challenges," *IEEE journal of emerging and selected topics in power electronics*, vol. 4, no. 3, pp. 707–719, 2016.

[2] Z. Guo, C. Hitchcock, and T.-S. P. Chow, "Comparative performance evaluation of lateral and vertical GaN high-voltage power field-effect transistors," *Japanese Journal of Applied Physics*, vol. 58, no. SC, p. SCCD09, may 2019. [Online]. Available: https://dx.doi.org/10.7567/1347-4065/ab1123

[3] S. N. Mohammad, A. A. Salvador, and H. Morkoc, "Emerging gallium nitride based devices," *Proceedings of the IEEE*, vol. 83, no. 10, pp. 1306–1355, 1995.

[4] S. K. Atwimah, T. Nelson, P. Pandey, A. P. Fox, D. G. Georgiev, A. G. Jacobs, A. D. Koehler, K. D. Hobart, T. J. Anderson, and R. Khanna, "Modeling framework to compare high voltage vertical gan pn and merged pn-schottky diodes," in *2024 IEEE Applied Power Electronics Conference and Exposition (APEC)*. IEEE, 2024, Conference Proceedings, pp. 2663–2669.

[5] M. Haziq, S. Falina, A. A. Manaf, H. Kawarada, and M. Syamsul, "Challenges and opportunities for high-power and high-frequency algan/gan high-electron-mobility transistor (hemt) applications: A review," *Micromachines*, vol. 13, no. 12, 2022. [Online]. Available: https://www.mdpi.com/2072-666X/13/12/2133

[6] Z. Yang, Y. Ma, M. Porter, H. Gong, Z. Du, H. Wang, Y. Luo, L. Wang, and Y. Zhang, "Breakdown voltage and leakage current of the nonuniformly activated lightly doped p-gan," *IEEE Transactions on Electron Devices*, vol. 71, no. 9, pp. 5589–5596, 2024.

[7] M. Sun, Y. Zhang, X. Gao, and T. Palacios, "High-performance gan vertical fin power transistors on bulk gan substrates," *IEEE Electron Device Letters*, vol. 38, no. 4, pp. 509–512, 2017.

[8] Y. Ma, H. Wang, S. Chen, and C. Liu, "Gan vertical mosfets with monolithically integrated freewheeling merged pn-schottky diodes (mps-mos) for 1.2-kv applications," *IEEE Transactions on Electron Devices*, vol. 71, no. 8, pp. 4570–4577, 2024.

[9] D. Ji, B. Ercan, and S. Chowdhury, "Experimental determination of velocity-field characteristic of holes in gan," *IEEE Electron Device Letters*, vol. 41, no. 1, pp. 23–25, 2019.

[10] T. Narita, H. Yoshida, K. Tomita, K. Kataoka, H. Sakurai, M. Horita, M. Bockowski, N. Ikarashi, J. Suda, T. Kachi *et al.*, "Progress on and challenges of p-type formation for gan power devices," *Journal of Applied Physics*, vol. 128, no. 9, 2020.

[11] T. Nelson, P. Pandey, D. G. Georgiev, M. R. Hontz, A. D. Koehler, K. D. Hobart, T. J. Anderson, A. Ildefonso, and R. Khanna, "Hybrid edge termination in vertical gan: Approximating beveled edge termination via discrete implantations," *IEEE Transactions on Electron Devices*, vol. 69, no. 12, pp. 6940–6947, 2022.

[12] T. Nelson, "Examination of hybrid edge termination on vertical gan planar mosfets," in *Proceedings of the 2024 GOMACTech*, 2024, Conference Proceedings, p. 81.

[13] D. G. Ashworth, R. Oven, and B. Mundin, "Representation of ion implantation profiles by pearson frequency distribution curves," *Journal of Physics D: Applied Physics*, vol. 23, pp. 870–876, July 1990.

[14] P. Pandey, T. M. Nelson, M. R. Hontz, D. G. Georgiev, R. Khanna, A. G. Jacobs, J. S. Lundh, J. C. Gallagher, A. D. Koehler, and K. D. Hobart, "Hybrid edge termination for high-voltage vertical gan devices: Empirical validation and robust processing tolerance," *IEEE Transactions on Electron Devices*, 2024.

[15] H. A. Mantooth, K. Peng, E. Santi, and J. L. Hudgins, "Modeling of wide bandgap power semiconductor devices—part i," *IEEE Transactions on Electron Devices*, vol. 62, no. 2, pp. 423–433, 2014.

[16] M. R. Ahmed, R. Todd, and A. J. Forsyth, "Predicting sic mosfet behavior under hard-switching, soft-switching, and false turn-on conditions," *IEEE Transactions on Industrial Electronics*, vol. 64, no. 11, pp. 9001–9011, 2017.

[17] Z. Duan, T. Fan, X. Wen, and D. Zhang, "Improved sic power mosfet model considering nonlinear junction capacitances," *IEEE Transactions on Power Electronics*, vol. 33, no. 3, pp. 2509–2517, 2017.

[18] Wolfspeed, "C3m0350120j - 1200v, 7.2 a silicon carbide mosfet," 2024. [Online]. Available: https://www.wolfspeed.com/products/power/sic-mosfets/1200v-silicon-carbide-mosfets/c3m0350120j/

[19] B. J. Baliga, "Gallium nitride devices for power electronic applications," *Semiconductor Science and Technology*, vol. 28, no. 7, p. 074011, jun 2013. [Online]. Available: https://dx.doi.org/10.1088/0268-1242/28/7/074011

[20] J. Tsao, S. Chowdhury, M. Hollis, D. Jena, N. Johnson, K. Jones, R. Kaplar, S. Rajan, C. Van de Walle, and E. Bellotti, "Ultrawide-bandgap semiconductors: research opportunities and challenges," *Advanced Electronic Materials*, vol. 4, no. 1, p. 1600501, 2018.

[21] STMicroelectronics, "Sth12n120k5-2ag - n-channel 1200 v, 7 a power mosfet," 2024. [Online]. Available: https://www.st.com/en/power-transistors/sth12n120k5-2ag.html

[22] ——, "Stpsc20h12g2-tr - sic power schottky diode," 2024. [Online]. Available: https://www.st.com/en/diodes-and-rectifiers/stpsc20h12g2-tr.html

[23] W. Collings, T. Nelson, A. Sellers, R. Khanna, A. Courtay, S. Jimenez, and A. Lemmon, "Optimization algorithms for dynamic tuning of wide bandgap semiconductor device models," in *2021 IEEE Applied Power Electronics Conference and Exposition (APEC)*. IEEE, 2021, Conference Proceedings, pp. 2427–2433.

[24] M. Buffolo, D. Favero, A. Marcuzzi, C. De Santi, G. Meneghesso, E. Zanoni, and M. Meneghini, "Review and outlook on gan and sic power devices: industrial state-of-the-art, applications, and perspectives," *IEEE Transactions on Electron Devices*, 2024.

2.5-kV 6.4-ns 100-kHz Repetitive GaN Marx Generator

Ruize Sun*†, Ci Pan*, Wanjun Chen* and Bo Zhang*

Email: rzsun@uestc.edu.cn

*University of Electronic Science and Technology of China, Chengdu 610054, China

†Institute of Electronic and Information Engineering of UESTC in Guangdong, Dongguan 523808, China

Abstract—**Repetitive Marx generator based on GaN HEMTs is proposed and experimentally verified. The self-triggering GaN Marx generator prototypes are realized and only one gate driver ICs is required. With a DC input ranging from 100 to 500 V, the proposed GaN Marx generators consistently produce output voltage pulses at repetition frequencies between 10 and 100 kHz, achieving a voltage conversion efficiency (VCE)—defined as the ratio of output voltage to the product of the number of stages and input voltage—of 100%. The 5-stage Marx generator utilizing GaN HEMTs delivers pulses with a peak voltage of 2.5 kV, a rise time of 6.4 ns, and a repetition rate of 100 kHz. This overall performance surpasses that of previously reported Marx generators using Si avalanche transistors, SiC MOSFETs, and GaN HEMTs. Therefore, the proposed GaN HEMT-based Marx generator presents a highly promising solution for applications requiring highly repetitive pulsed power supplies.**

Keywords—GaN HEMT, Marx generator, self-trigger

I. INTRODUCTION

Traditional Marx generators typically use semiconductor switches like silicon (Si) IGBTs, MOSFETs, and avalanche transistors (ATs). The multiple-stage Marx generators with 9.6-kV pulses at 5 kHz [1] or with achieving rise times of 50 ns [2] and 3.4 ns [3] have been realized, however, the circuits required external drivers and faced synchronization issues with limited voltage conversion efficiency (VCE) and repetition rates.

Wide-bandgap devices like GaN HEMTs and SiC MOSFETs offer advantages such as high switching frequency and low on-resistance, making them ideal for pulsed power applications. A 4-stage GaN Marx generator achieved 2-kV pulses with a 15-ns rise time [4] though each switch required an external driver. Other designs using SiC MOSFETs had VCEs below 80% and repetition rates of only 10 kHz.

This paper presents a novel Marx generator using self-triggering GaN HEMTs. This design achieves 2.5-kV pulses with a 6.4-ns rise time, 100-kHz repetition rate, and nearly 100% VCE, addressing previous challenges in synchronization and efficiency.

II. GaN REPETITIVE MARX GENERATOR

Figure 1 depicts the schematic of the proposed repetitive Marx generator using self-triggering GaN HEMTs. The design features five stages, each with a GaN HEMT (Q_1-Q_5) and energy-storage capacitor (C_1-C_5).

The DC power supply V_{DC}, inductor L_0, and resistor R_0 simulate the charging circuit. Diodes D_0 to D_9 manage

current flow, while capacitors C_{CSi} (i=2-5) control charge distribution between energy-storage capacitors and parasitic input capacitances. Gate resistors R_{Gi} (i=2-5) reduce gate voltage oscillations. Only one gate driver is needed for Q_1.

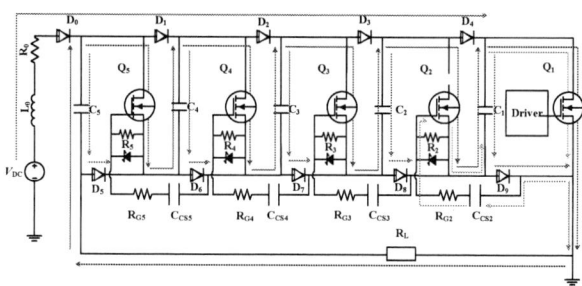

Figure 1: Schematic of the proposed repetitive Marx generator with charge-control self-triggering GaN HEMTs. Thin solid and dashed line shades are added, respectively, to explain the OFF-state and ON-state operation. Thick line shade denotes the self-triggering path in turn-on operation.

LTspice simulations were conducted to validate the proposed GaN HEMT Marx generator. Figure 2 shows voltage waveforms for varying stages and V_{DC} values. In Figure 2(a), V_{GS} of Q_1 toggles between 0 and 5 V, controlled by the driver IC, while V_{GS} for Q_2 to Q_5 sequentially rises. Figure 2(b) displays the corresponding V_{DS} of Q_1 to Q_5, which drops to 0 in sequence. Figure 2(c) shows V_{OUT} for stages 2 to 5 with a 1 kΩ load, where the pulse rise time increases with more stages due to cumulative delays. V_{OUT} matches the product of the stage number and V_{DC}, reflecting a *VCE* of 100%. Figure 2(d) confirms V_{OUT} is proportional to V_{DC} (200 to 500 V) with a consistent rise time.

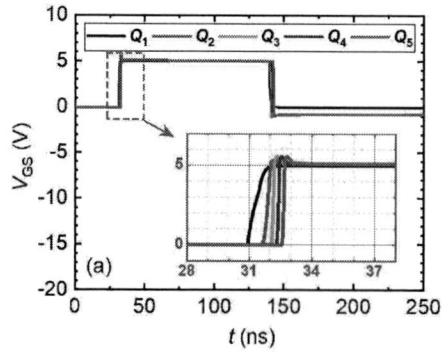

979-8-3315-1612-3/25 $31.00 © 2025 IEEE

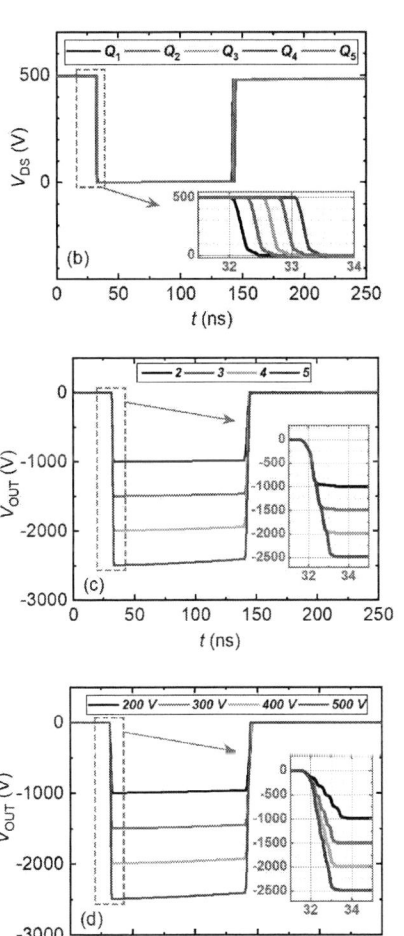

Figure 2: Simulated waveforms of (a) V_{GS} and (b) V_{DS} for Q_1 to Q_5, (c) V_{OUT} for 2-5 stage generator with inset of rising edges with V_{DC} of 500 V, (d) V_{OUT} for a 5-stage generator with V_{DC} of 200 /300 /400 /500 V, where R_L is 1 kΩ.

III. RESULTS AND DISCUSSION

Figure 3 presents the photo of the workbench and prototype of the proposed 5-stage repetitive Marx generator with GaN HEMTs. The gate driver IC used is the ADuM4121 from Analog Devices, and the GaN HEMTs are the 650-V/29-A INN650D080BS from Innoscience. The prototype, measuring 8.0 cm × 11.5 cm, features diodes D_0 to D_9 positioned on the back for compactness.

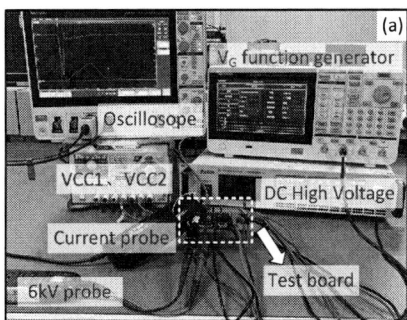

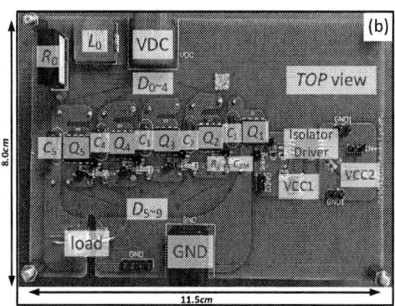

Figure 3: Photograph of (a) the workbench of test circuit, (b) top views of the PCB test board.

Figure 4(a) shows the typical output voltage V_{OUT} waveforms for a 3-stage Marx generator with GaN HEMTs at V_{DC} values from 100 to 500 V and a load resistance R_L of 1 kΩ. V_{OUT} reaches $3 \times V_{DC}$ with a VCE of 100% in about 5 ns, remains stable for the 100-ns on-time of Q_1, and then drops to zero. Figure 4(b) displays the gate-source voltage V_{GS} waveforms for Q_1 to Q_3 at V_{DC} of 500 V. The overshoot and ringing in V_{GS} due to fast rise times are within the safe-operating area of the GaN HEMTs, with delays among V_{GS} signals of only a few nanoseconds. Figure 4(c) illustrates V_{OUT} waveforms for generators with 1 to 5 stages at V_{DC} of 500 V. The self-triggering circuit effectively addresses driver synchronization issues, enabling rapid triggering of GaN HEMTs and achieving fast voltage pulses with a VCE of 100%.

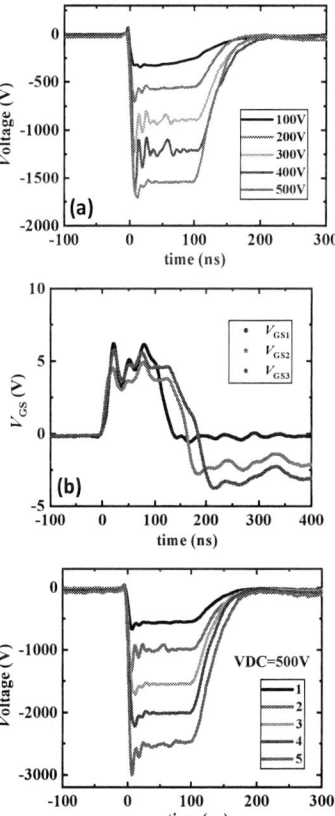

Figure 4: Measured waveforms for Marx generator with GaN HEMTs and R_L of 1 kΩ: (a) V_{OUT} in a 3-stage generator; (b) V_{GSi} in a 3-stage generator; (c) V_{OUT} with V_{DC} of 500 V in a 1 to 5 stage generator

979-8-3315-1612-3/25 $31.00 © 2025 IEEE

Repetitive pulsed power output characteristics were measured using gas discharge tubes (GDTs) to verify the maximum pulsed current capability of the GaN HEMTs. Figure 5 displays the typical V_{OUT} and output current I_{OUT} waveforms for the proposed 5-stage Marx generator with GaN HEMTs. At V_{DC} of 500 V, V_{OUT} reaches a peak value of 2.5 kV with a rise time of 6.4 ns. The output current peaks at 56.1 A, which is close to the 58-A maximum pulsed current rating of the INN650D080BS.

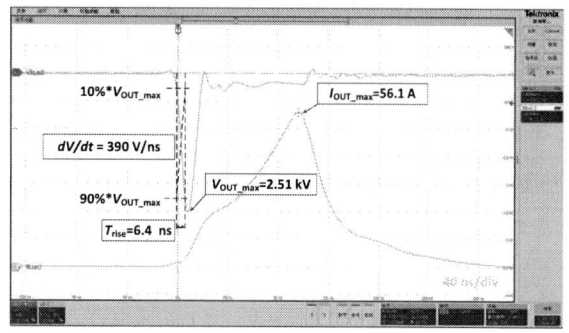

Figure. 5 Measured typical waveforms of (a) V_{OUT} and (b) I_{OUT} of the 5-stage GaN Marx generator.

Figure 6 shows the measured V_{OUT} waveforms of the 5-stage GaN Marx generator at 10 and 100 kHz. Stable and clean voltage pulses with a 2.5 kV peak are consistently achieved.

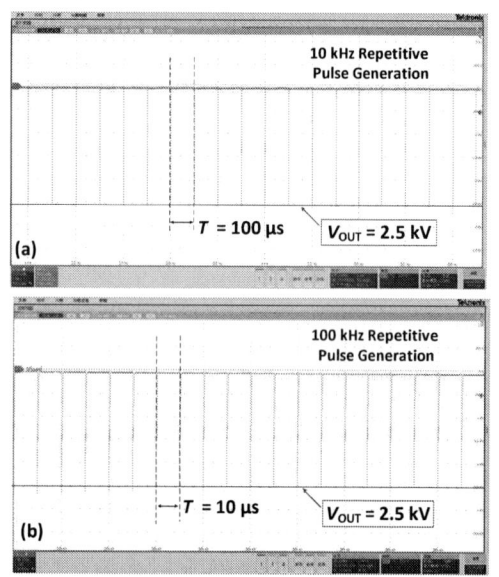

Figure. 6 Measured repetitive V_{OUT} waveforms of the 5-stage Marx generator at f of (a) 10 kHz and (b) 100 kHz.

Repetitive pulsed output at 100 kHz was measured for the proposed Marx generators with 5 stages. Stable output voltage pulses with peak values proportional to the number of stages, indicating a VCE of 100%. This confirms the feasibility and reliability of multi-stage Marx generators with GaN HEMTs operating at 100 kHz.

To compare the Marx generator with the different switches, Table I lists the key parameters from previous studies. It can be seen that using GaN HEMT devices as switches in a Marx circuit enables a stable output with a repetition frequency of up to 100 kHz and can achieve nanosecond-level rising time at higher voltage. This work has realized the compact nanosecond HV pulse generator based on GaN HEMT with overall improved performance.

TABLE I BENCHMARK OF REPETITIVE MARX GENERATORS

Ref.	Switch	Gate drive	Stage/ device	V_{OUT} (kV)	t_{rise} (ns)	VCE	f (kHz)
This work	*GaN HEMT*	*Self-trigger*	*5 / 5*	*2.5*	*6.4*	*~100%*	*100*
[2]	GaN HEMT	All-drive	4 / 4	2	15	~100%	single
[4]	SiC MOS	Self-trigger	20/ 20	15.3	45	95.6%	10
[5]	Si AT	Self-trigger	5 / 20	2.5	3.4	73%	15
[6]	Si AT	Self-trigger	20 / 20	1.65	0.23	30.5%	100
				0.73	0.23	13.5%	700

IV. CONCLUSION

The proposed 5-stage compact GaN Marx generator achieves a 2.5 kV peak, 6.4 ns risetime, and 100 kHz repetition rate with nearly 100% VCE, all using just one gate driver IC. In Table I, it demonstrates the lowest risetime, highest repetition rate, and highest VCE compared to reported SiC and GaN Marx generators, outperforming Marx generators with Si ATs for repetitive pulsed power supplies. This 100 kHz GaN Marx generators reduces the need for multiple driver ICs, easing synchronization challenges, and demonstrate enhanced repetitive output with 100% VCE, highlighting the potential for repetitive pulsed power applications.

ACKNOWLEDGMENT

This work was supported in part by the Natural Science Foundation of Sichuan under Grant 2024NSFSC1409 and in part by the Guangdong Basic and Applied Basic Research Foundation under Grant 2024A1515030102.

REFERENCES

[1] Z. Zhou, Z. Li, J. Rao, S. Jiang and T. Sakugawa, "A High-Performance Drive Circuit for All Solid-State Marx Generator," *IEEE Trans. on Plasma Sci.*, vol. 44, no. 11, pp. 2779-2784, Nov. 2016.

[2] X. Wang, Q. Huang, L. Xiong, L. Xu, Q. Chen and Q. Xiong, "A Compact All-Solid-State Repetitive Pulsed Power Modulator Based on Marx Generator and Pulse Transformer," *IEEE Trans. on Plasma Sci.*, vol. 46, no. 6, pp. 2072-2078, June 2018.

[3] J. Rao, W. Zhang, S. Jiang and Z. Li, "Nanosecond pulse generator based on cascaded avalanche transistors and Marx circuits," *IEEE Trans. Dielectr. Electr. Insul.*, vol. 26, no. 2, pp. 374-380, April 2019.

[4] Y. Chen, J. Du, L. Feng, J. Xiao, J. Zhang and Y. He, "Comparative Study on Driving Switching Characteristics of GaN-FET and SiC-MOSFET in Transient High Voltage Pulse Discharge Circuit," *2020 IEEE Int. Conf. on High Voltage Engineering and Application*, Beijing, China, 2020.

[5] S. Shen, J. Yan, G. Sun and W. Ding, "Improved Auxiliary Triggering Topology for High-Power Nanosecond Pulse Generators Based on Avalanche Transistors," *IEEE Trans. on Power Electron.*, vol. 36, no. 12, pp. 13634-13644, Dec. 2021.

[6] J. Chen, X. Du, Q. Luo, X. Zhang, P. Sun and L. Zhou, "A Review of Switching Oscillations of Wide Bandgap Semiconductor Devices," *IEEE Trans. on Power Electron.*, vol. 35, no. 12, pp. 13182-13199, Dec. 2020.

Novel Dual Output LDO Architecture
in 650-V GaN Technology for Power ICs

Plinio Bau, Thanh Hai Phung,
Wise-Integration, Grenoble, France,
Email: plinio.bau@wise-integration.com

Deniz Aygun, Bart Coomans, Mike Wens
MinDCet NV, Leuven, Belgium

Abstract—**This paper presents a novel topology for a dual output linear regulator (LDO) for integrated functions in GaN technology. The proposed topology uses an alternative topology compared to the existing designs in literature and can generate a 6 V and 12 V regulated output voltage for monolithic gate drivers in 650-V normally-off GaN power transistors. This work presents the design of the circuit, the measurements and a comparison with existing solutions presented in other published papers.**

Keywords— linear regulator, low dropout regulator, LDO, Gallium nitride; HEMTs; monolithic integrated circuits; power integrated circuits.

I. INTRODUCTION

Monolithic integration of the gate driver and the power transistor on the same GaN-on-Si chip allows higher performance and power density, by eliminating parasitic components in the gate drive circuitry and consequently allowing for faster and reliable switching behavior. The first circuit to be integrated, offering the most significant benefit, is the gate driver. To ensure on-chip power supply regulation for the gate driver, it is recommended to implement a linear regulator circuit within the same GaN die.

In the most widely used industrial technology in the market for GaN transistors, the optimal gate voltage is 6 V. At this level, the device's lifetime is maximized, with components designed to operate continuously for over 10 years. To achieve a 6 V gate drive voltage and enable rapid switching for minimized switching losses, a higher power supply rail is usually required.

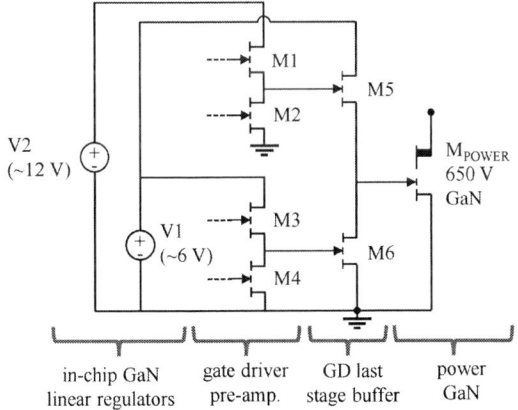

Fig. 1. Monolithically integrated linear regulators avoid area consuming bootstrap capacitor in gate drivers supply and allows the pull up transistor (M5) of the last stage buffer be fully activated during the switching on transition.

This higher supply rail is necessary to activate the gate of the high-side transistor (M5) in the final buffer stage of the gate driver, as shown in Fig. 1.

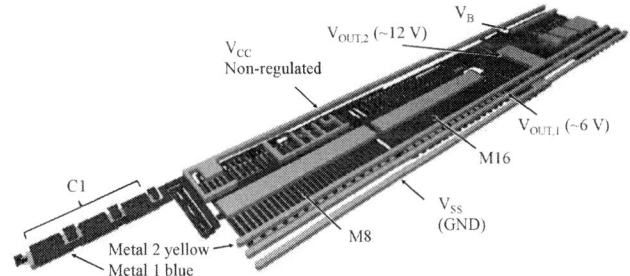

Fig. 2. A 3D representation of the fabricated dual output LDO in GaN-on-Si technology.

There are two main methods [1] to produce the power supply rail for the gate of the pull up transistor (M5 in Fig. 1). One method is by using a bootstrap capacitor that requires a significant area on the chip. Another topology for gate driver for GaN technology uses dual power supply rails of around 6 V and 12 V for the output buffer stage. Having a 12 V linear regulator allows the gate of the pull up transistor (M5) to be fully switched on. In this case the sizing of M5 requires less surface for the shortest gate rise time possible.

Most logic circuits within a gate driver can be powered by a 6 V supply rail. Therefore, a linear regulator should be designed to supply a higher current capacity at this voltage level, while requiring significantly lower current capability for nodes operating at 12 V.

The proposed architecture presented in this work demonstrates fast switching behavior in the nanosecond range, which is characteristic of GaN transistors. While some AC simulations were conducted, power supply rejection (PSR) is beyond the scope of this study. It is worth noting that applying CMOS layout best practices is also crucial for GaN-based architectures. Fig. 2 shows a 3D representation of the layout of the proposed linear regulator, where symmetric transistor placement and dummy structures are incorporated to optimize performance.

A crucial aspect of circuit integration is the effective management of power sequences. To ensure stable operation and protect the components, the desired startup and shutdown sequence of the linear regulator (LDO) is illustrated in Fig. 3.
The 6 V output is buffered using an off-chip decoupling capacitor, while the 12 V output does not have an externally accessible pad. As charging a large off-chip capacitor requires relatively more time, the LDO design must ensure that the

12 V output does not reach its final value before the 6 V output. This design approach prevents exceeding the safe operating limits of the gate terminals of transistors. Therefore, the topology is specifically designed to guarantee that the 12 V output becomes operational only after the 6 V output.

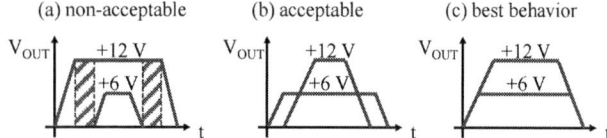

Fig. 3. Illustration of the expected power-up sequence behavior to prevent exceeding 6 V at the gate of any GaN transistor. Waveforms (b) and (c) depict safe operating conditions that do not risk damage to the GaN transistor gate. In contrast, waveform (a) illustrates a scenario that could compromise reliability.

II. DESIGN

A. Voltage Reference

The proposed linear regulator topology employs a stacked structure, resembling that of a voltage reference circuit. To facilitate a better understanding of the linear regulator, the operation of a voltage reference circuit with a similar structure is presented first.

Significant variation in threshold voltage (V_{TH}) and saturation current ($i_{DS,SAT}$) are typical in small-sized devices in emerging GaN process [2]. Fig. 4 (a) shows a current source implemented with a depletion-mode GaN (dGaN) transistor and a resistor in its drain. The transfer characteristics obtained from this circuit is illustrated in Fig. 4 (b).

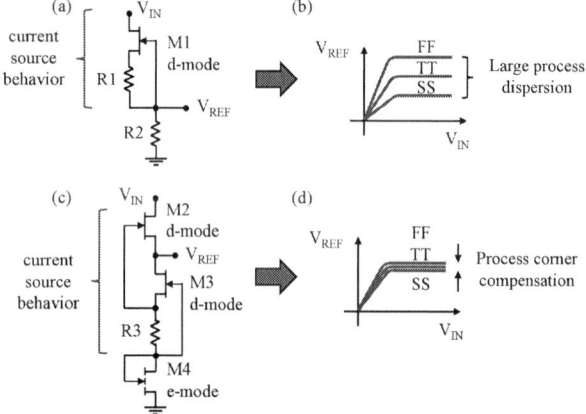

Fig. 4. (a) Current source circuit with a resistive load, showing (b) significant variation in output voltage. By employing a specific (c) stacked structures, (d) compensation for process variation can be achieved.

By stacking two d-mode (normally-on) GaN transistor with one enhancement-mode (normally-off), a compensation for process variation is obtained. Fig. 4 illustrates this effect. Fig. 5 demonstrates the behavior of threshold voltage (V_{TH}) variation as a function of the process corner position. It is possible to observe that the same physical mechanisms that increase the threshold value of an eGaN transistor, also increase it for the dGaN transistor.

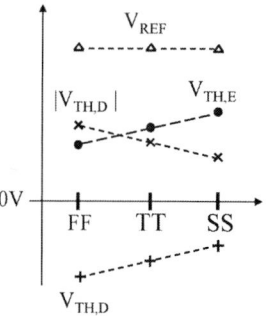

Fig. 5. A representation of the process compensation effect achieved by stacking eGaN and dGaN transistors. This configuration balances the threshold variations between the eGaN and dGaN transistors, resulting in improved process compensation.

However, the absolute value of the threshold voltage for dGaN transistors decreases. By combining these effects, the resulting output voltage (V_{REF}) variation remains constant across all process corner values.

The proposed voltage reference topology, shown in Fig. 4 (c), employs a stacked structure to bias the gate of the depletion-mode GaN transistor M2, which serves as a pass element in the design. The dGaN transistor M3, along with the eGaN transistor M4, forms a stack that creates a voltage drop across the threshold (V_{TH}), exhibiting self-regulatory behavior for M2. The advantages of this structure in terms of process, voltage, and temperature (PVT) compensation have been detailed in previous publications by the same authors [3]. Equation (1) and Fig. 4 (c) display the output voltage of around 3 V when the input voltage is above this value.

$$V_{REF}= V_{TH,E} + 2 \times |V_{TH,D}| \qquad (1)$$

B. Linear Regulator

A linear regulator can be obtained using a similar structure as the voltage reference previously explained. By doubling the number of transistors, using two eGaN and four dGaN transistors instead of one eGaN and two dGaN, the output voltage increases from approximately 3 V to 6 V. The output voltage of the linear regulator, when the input voltage exceeds 6 V, is given by eq. (2).

$$V_{OUT,1}= 2 \times V_{TH,E} + 4 \times |V_{TH,D}| \qquad (2)$$

To achieve a 12 V supply, the same structure is stacked on top of the 6 V stack, effectively doubling the output voltage.

C. Schematic of the First Version

The schematic of the first version of the dual-output linear regulator is shown in Fig. 6. A similar structure is used for each output. Transistor M7 is configured to act as a capacitor, improving the dynamic behavior of the circuit. Transistor M8 and resistor R2 form a current source, sinking 200 µA to the ground (V_{SS}) node. This current source is matched with a biasing branch, composed of transistors M10 through M14, ensuring symmetric operation between the two outputs.

979-8-3315-1612-3/25 $31.00 © 2025 IEEE

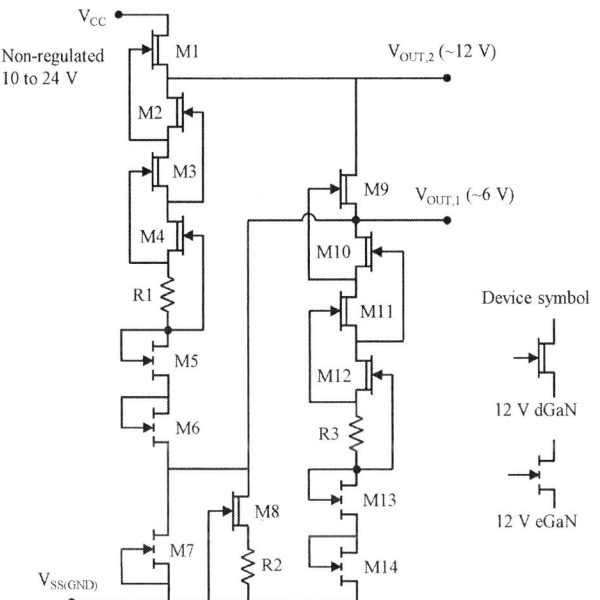

Fig. 6. Schematic of a dual-output linear regulator. A similar structure is used for both outputs.

This topology offers the advantage of maintaining both outputs at the same level when the input voltage is below 6 V. Additionally, the process compensation effect is achieved with a relatively simple design using a minimal number of devices.

An issue was observed in the prototypes of this circuit: higher current consumption by the system-on-chip (SoC) occurred when the switching voltage exceeded 150 V and the switching frequency surpassed 400 kHz. While the static current consumption remained below 10 mA at low voltage and frequency, it increased to up to 20 mA under high-voltage, high-frequency conditions. Simulations identified the root cause as frequency instability in the linear regulator, where the 12 V output exhibited oscillations exceeding 1 V peak-to-peak under certain conditions.

D. Schematic of the Second Version

To address the issues in the first version, the following improvements were implemented:
- Current density in specific metal paths was enhanced.
- A second transistor was added to the main transistors M1 and M9 to mitigate gate leakage effects.
- A trimming system with two pads for two fuses was introduced.
- Instability was resolved by adding RC filters.

The circuit elements R1, R4, C1 and C2 were critical for improving stability. These components eliminated the oscillations observed in simulations on the 12 V output, thereby reducing circuit consumption under high-frequency operation.

The schematic of the second version of the linear regulator is presented in Fig. 7. Dummy structures are omitted for simplicity.

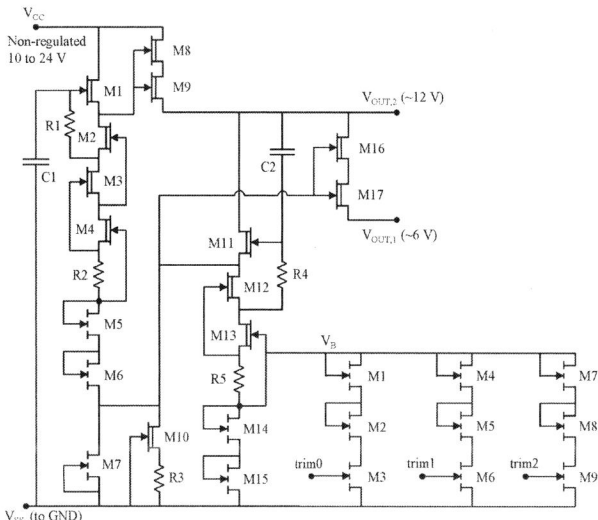

Fig. 7. Schematic of the second version of the linear regulator.

E. Trimming System

The schematic of the trimming circuit is presented in Fig. 8. The fuse element, labeled "fuse0" in the schematic, consists of a narrow metal path with a total resistance of 38.5 Ω, designed to fuse with currents below 100 mA. The binary-to-thermometer code block measures 112 μm × 201 μm.

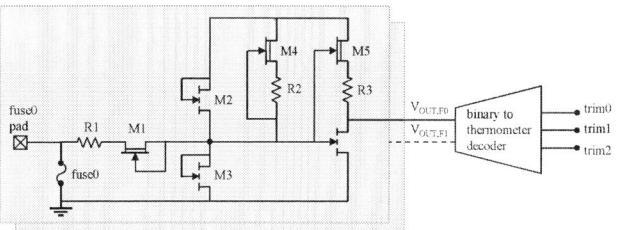

Fig. 8. Trimming circuit used to minimize process variation in the regulated output.

A trimming table is provided in **Table I**, detailing the available trim settings. The number of trimming bits is limited due to the significant area required by the fuse pads, and the design prioritizes the minimum configuration necessary to reduce process variation within design specifications.

TABLE I.

TRIMMING TABLE

	Fuse0	Fuse1	Process corner	Difference in output
trim0	-	fused	SS	+200 mV
trim1	fused	fused	~SS	+100 mV
trim2	-	-	TT	0 mV
trim3	fused	-	FF	-100 mV

The default trim setting is "trim2". If desired, the 6 V output can be adjusted by fusing either or both fuses according to the trimming table. These adjustments will result in a corresponding increase or decrease in the 12 V output voltage as well.

III. SIMULATED VALUES

A. Simulated Values for the First Version

Simulations of the two outputs of the first version of the linear regulator is presented in Fig 9. For the 12 V output, a load resistance of 25 kΩ was used. For the 6 V output, load resistances (R_L) ranging from 620 Ω to 3 kΩ were simulated. These values correspond to a maximum load of 10 mA and a minimum designed load of 2 mA, respectively.

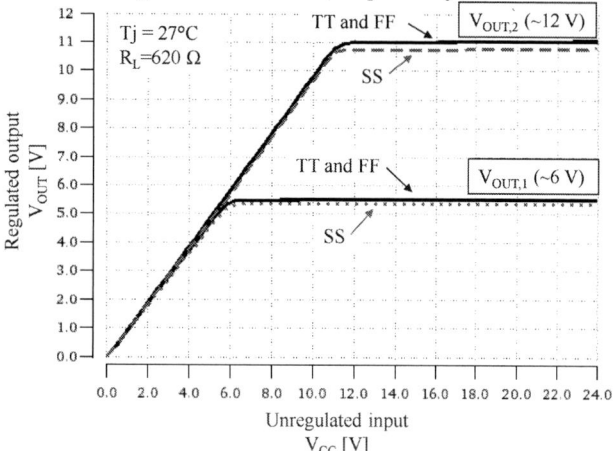

Fig. 9. The static transfer characteristics with simulated values obtained using Cadence™.

It can be observed that the corner values for fast (FF) and typical (TT) process corners are nearly identical, while the slow (SS) corner also remains close. This performance consistency is achieved despite the significant variations in threshold voltage and saturation current typically seen in low-voltage GaN transistors within this technology.

A thermal simulation of the 6 V output is presented in Fig. 10. Thermal compensation is achieved by appropriately sizing transistors M13 and M14, with larger dimensions selected to minimize the thermal coefficient of the circuit, resulting in improved thermal stability.

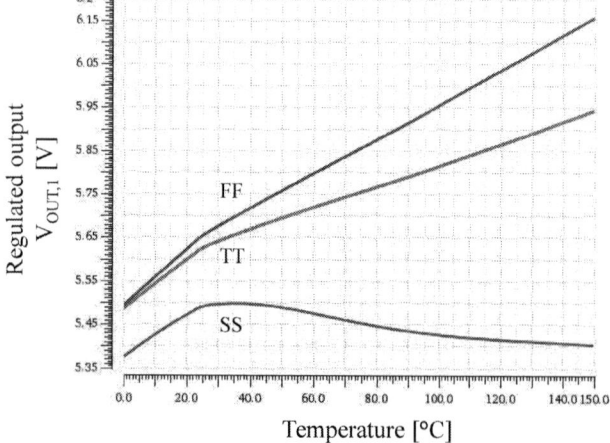

Fig. 10. Simulation of the thermal drift of the 6 V output. The sizing of the transistor is made to obtain maximum 6.2 V in the worst case. The three different curves are for the process corners values.

B. Simulated Values for the Second Version

The simulation results for the outputs of the second version of the linear regulator are presented in Fig. 11. For this simulation, the load currents were 633 μA for the 12 V output and 353 μA for the 6 V output, corresponding to the design specifications for low load conditions. These values represent the power supply consumption of a gate driver and an entire system-on-chip (SoC) that includes additional analog functions.

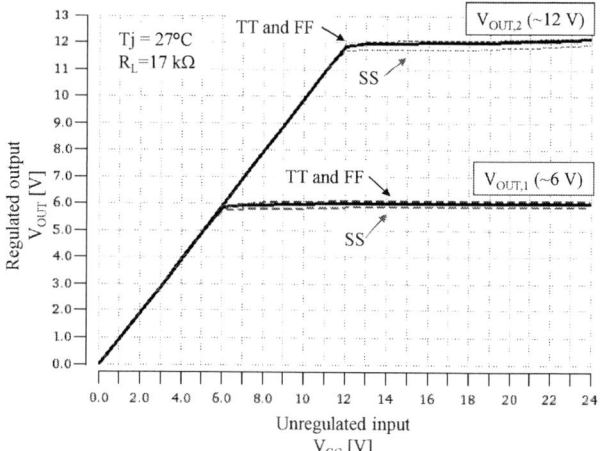

Fig. 11. The static transfer characteristics with simulated values in Cadence™. The solid line is for the typical (TT) fabrication corner and dashed lines showing the process corner values for slow (SS) and fast (FF) corners.

For the second version of the dual-output LDO, when the non-regulated input is 20 V, the typical output values are 6.10 V and 12.11 V for the low- and high-voltage outputs, respectively. At the SS process corner, the values are 5.96 V and 11.88 V, while at the FF corner, the outputs are 6.16 V and 12.17 V.

To address discrepancies between measured and simulated values observed in the first version, the output voltages of the second version were increased by 300 mV. This adjustment ensures better alignment with the expected performance during fabrication and testing.

IV. MEASURED VALUES

The first version of the circuit was fully characterized, while the second version has only been evaluated at the wafer level. Complete characterization of the second version was not available at the time of this publication.

A. Measured Values for the First Version

Measurement results for the first version are presented in Fig. 12. These values were obtained from 10 different devices fabricated within the same lot. The measurement was performed with an external load current of 1 mA, in addition to an internal load of 1 mA from the system-on-chip (SoC), resulting in a total load of 2 mA for the LDO. The maximum input voltage used in the tests was 24 V, although some tests were conducted at higher voltages to explore the performance limits. Fig. 13 presents similar data to Fig. 12, with the key difference being a higher load current of 10 mA.

979-8-3315-1612-3/25 $31.00 © 2025 IEEE

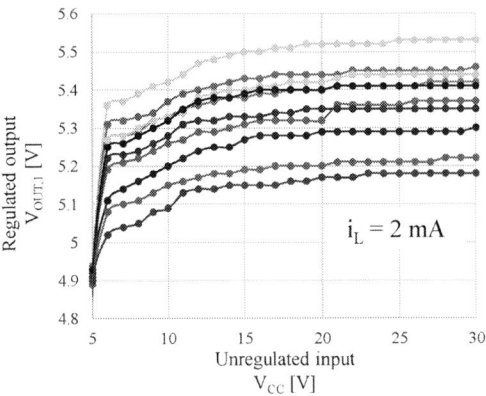

Fig. 12. The measured transfer characteristics (V_{OUT} vs V_{IN}) for one output of the linear regulator. The different curves are for 10 different packaged parts.

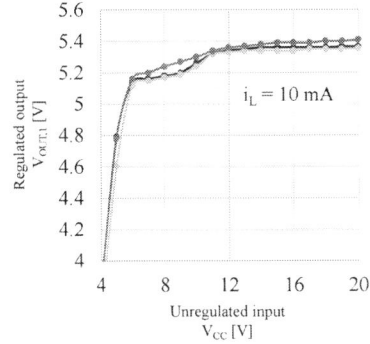

Fig. 13. The measured output for 3 different packaged parts and load current of 10 mA.

The measured temperature drift, as shown in Fig. 14, closely aligns with the simulated trend across temperature variations (Fig. 10).

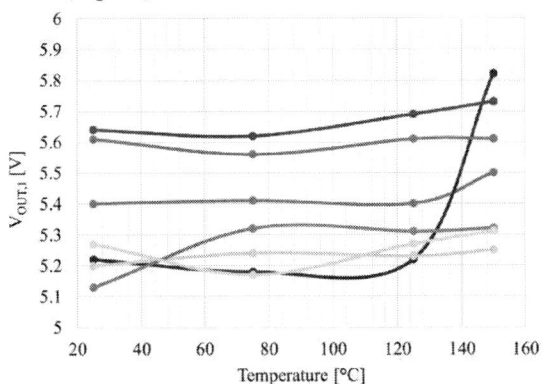

Fig. 14. The measured thermal drift of the 6 V output of the linear regulator for 7 different packaged devices with same design.

Table II compares the post-layout (PEX) simulated values with the average measured values for 10 devices. This comparison is based on the first version of the design, as measurements for the second version with packaged dies were not available at the time of publication.

TABLE II.
COMPARING MEASURED VALUES TO PARASITIC EXTRACTED (POST-LAYOUT) SIMULATION FOR THE FIRST VERSION OF THE LDO.

Parameter	Condition	PEX Simulated	Measured Average
$V_{OUT,1}$	V_{CC}=20 V I_L=1 mA	5.66 V	5.35 V
$V_{OUT,2}$	V_{CC}=20 V I_L=1 mA	11.32 V	10.70 V
Line Regulation	I_L=1 mA	0.18 %	2.1 %
Temp. Variation	I_L=1 mA, 25°C-150°C	1.66 %	2.6 %
Load Regulation	V_{CC}=20 V, I_L=1-10 mA	2.82 %	3.9%
Current Cons.	R_L=100kΩ, 25°C-150°C	411 µA	437 µA

It can be observed that the measured values for the first version of the LDO are approximately 300 mV lower than the simulated values. Similarly, the 12 V output shows a difference of 600 mV. Other performance parameters appear to align closely with the simulated values, with minimal deviation.

B. Measured Values for the Second Version

Wafer measurements for the second version of the GaN LDO are presented in Fig. 15. It is important to note that wafer tests are affected by light, and thus, the results should not be directly used to validate the circuit's performance. Instead, they should serve as a reference for comparison, with due consideration of the test conditions.

The LDO's load during these measurements corresponds to the quiescent current consumption of the system-on-chip (SoC), which is 1.38 mA. Additionally, Fig. 16 presents the transfer characteristics under an external load of 552 Ω. This load resistance corresponds to an output current of 10 mA.

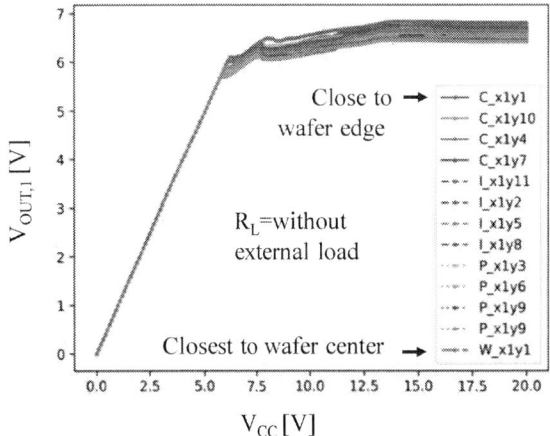

Fig. 15. The measured output characteristics at wafer probe test without external load. The different curves are for different dies in the same wafer. The dies were chosen equidistant from each other from the center to the edge of the wafer.

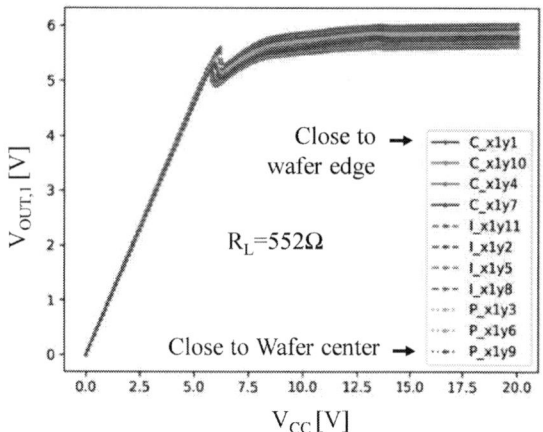

Fig. 16. The measured output characteristics at wafer probe test. This is measured with an external load resistance of 552 Ω.

To evaluate process variation across the wafer, the output voltage of the 6 V LDO is plotted as a function of the approximate distance from the wafer's center. The die closest to the center is plotted at the origin and dies at equidistant points from the center to the edge of the wafer are depicted in Fig. 17. No strong correlation is observed between the LDO output voltage and the distance from the wafer center.

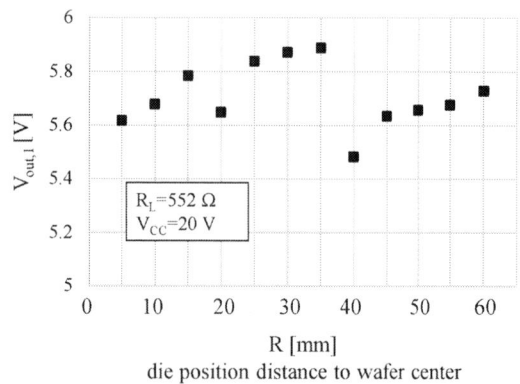

Fig. 17. The measured output characteristics at wafer probe test as a function of distance from the center of the wafer.

Another test involved loading the circuit beyond the maximum current rating of 10 mA to observe the voltage drop at the 6 V output. Table III presents the measured values for the die closest to the wafer center when the non-regulated input voltage is set to 20 V. These measurements show an accepted performance for high values of load current.

TABLE III.

MEASURING MAXIMUM LOAD CURRENT

RL	Iin	Iout	VOUT,1
552 Ω	12 mA	10 mA	5.74 V
266 Ω	22 mA	20 mA	5.5 V
180 Ω	30 mA	28 mA	5.2 V
86 Ω	57 mA	54.6 mA	4.7 V

V. DISCUSSION OF RESULTS

Table IV compares this work with previously published work. In this table the first version of the circuit is reported because the measurements for the second version were not ready before the publishing of this paper.

TABLE IV.

COMPARING THIS WORK TO THE PUBLISHED STATE-OF-THE-ART

Parameter	Unit	Ref. [4]	Ref. [5]	This work
Year	NA	2021	2022	2025
Process	NA	GaN	GaN	GaN
V_{IN}	[V]	10-24	10-15	10-24
V_{OUT}	[V]	6	5	**6 and 12**
I_{Load}	[mA]	NA	0.25-100	1-10
$I_{Quiescent}$	[μA]	NA	2000	**411**
Area	[mm²]	0.31	NA	**0.06**

The article published in [4] does not report the current consumption of the LDO circuit alone. Publication [5] lacks information regarding the layout area, while other publications have reported similar work using CMOS technology [6-8].

For the second version of the linear regulator, the area has been increased from 0.06 mm² to 0.078 mm² to improve uniformity of the current density distribution and thus reliability of the circuit.

The transfer characteristics and line regulation are within acceptable performance limits. However, the average output value of output number one (the 6 V output) under no-load conditions was observed to be too high for the second version of the linear regulator. This could potentially damage the gate of devices rated for 6 V. The unexpected increase is likely related to a reduction in gate leakage due to the stacking of the pass devices. Further testing with packaged dies is required for the second version.

VI. CONCLUSION

Compared to previously published LDOs in GaN technology, the proposed solution offers an alternative topology that can output both 6 V and 12 V, making it particularly suitable for GaN gate drivers. A lower current consumption and area compared to the published state-of-the-art is also achieved, offering a better solution for mass production applications.

ACKNOWLEDGMENT

The authors would like to thank the integrated circuit designer engineer Sebastian Gaviria-Duque for the technical contributions that consisted in the main idea, first simulations and layouts. Also, for the characterization team at Wise-integration that performed most of the measures.

REFERENCES

[1] Basler, Michael. Extended monolithic integration levels for highly functional GaN power ICs. Dissertation, Universität Freiburg, 2022.

[2] S. K. Murray *et al.*, "On-Chip Dynamic Gate-Voltage Waveform Sampling in a 200-V GaN-on-SOI Power IC," in *IEEE Journal of*

Emerging and Selected Topics in Power Electronics, vol. 10, no. 6, pp. 7150-7161, Dec. 2022

[3] P. Bau, *et al.*, "Voltage Reference and Zero Current Detector Monolithically Integrated on p-GaN Technology Designed for Process Corners Compensation," *2023 IEEE Applied Power Electronics Conference and Exposition (APEC)*, Orlando, FL, USA, 2023.

[4] Y. -Y. Kao *et al.*, "Fully Integrated GaN-on-Silicon Gate Driver and GaN Switch With Temperature-Compensated Fast Turn-on Technique for Achieving Switching Frequency of 50 MHz and Slew Rate of 118.3 V/Ns," in *IEEE Journal of Solid-State Circuits*, vol. 56, no. 12, pp. 3619-3627, Dec. 2021.

[5] P. Wang *et al.*, "A Monolithic GaN LDO Based on 12 V/0.5 μm GaN-on-Si Power Technology Achieving 20 ns Settling Time and 22 MHz UGF," *2022 IEEE 16th International Conference on Solid-State & Integrated Circuit Technology (ICSICT)*, Nangjing, China, 2022.

[6] L. Wang, D. Zhou, N. He, Y. Xu, X. He and Z. Chen, "Design of Transient Enhanced LDO Circuit for GaN HEMT Gate Driver," *2021 6th International Conference on Integrated Circuits and Microsystems (ICICM)*, Nanjing, China, pp. 40-44, 2021.

[7] C. Desai, D. Mandal, B. Bakkaloglu and S. Kiaei, "A 1.66 mV FOM Output Cap-Less LDO With Current-Reused Dynamic Biasing and 20 ns Settling Time," in *IEEE Solid-State Circuits Letters*, vol. 1, no. 2, pp. 50-53, Feb. 2018.

[8] Y. Lu, W. -H. Ki and C. Patrick Yue, "An NMOS-LDO Regulated Switched-Capacitor DC–DC Converter With Fast-Response Adaptive-Phase Digital Control," in *IEEE Transactions on Power Electronics*, vol. 31, no. 2, pp. 1294-1303, Feb. 2016.

[9] P. Bau, *et al.*, "Three different designs of Integrated Gate Drivers in 650-V e-mode GaN Power IC" *2024 IEEE Workshop on Wide Bandgap Power Devices and Applications in Europe (WiPDA Europe)*, Cardiff, United Kingdom, pp. 1-6, Sep. 2024.

979-8-3315-1612-3/25 $31.00 © 2025 IEEE

Impact of Substrate Bias on the Stability of Bidirectional GaN HEMT in Hard- and Soft-Switching

Qihao Song
Center for Power Electronics
Systems (CPES)
Virginia Tech
Blacksburg, USA
qihao95@vt.edu

Hongchang Cui
Center for Power Electronics
Systems (CPES)
Virginia Tech
Blacksburg, USA
chc121@vt.edu

Qiang Li
Center for Power Electronics
Systems (CPES)
Virginia Tech
Blacksburg, USA
lqvt@vt.edu

Yuhao Zhang
Department of Electrical and
Electronic Engineering
The University of Hong Kong
Hong Kong, China
yuhzhang@hku.hk

Abstract—**Many power electronics circuits for DC/AC conversion require power semiconductors with bidirectional voltage-blocking capability. Such a bidirectional device is typically achieved by connecting two unidirectional devices in series. The emerging monolithic bidirectional GaN-on-Si HEMTs (MBD-HEMT) offer an attractive alternative by replacing these two devices with a single-device solution. Despite the promise, substrate management in GaN MBD-HEMTs remains a critical challenge. The substrate of GaN MBD-HEMTs is usually floating, resulting in a constantly changing substrate bias that can be either positive or negative during the hard- or soft-switching. This issue can be resolved by adding a substrate management circuit, which, however, increases device cost and complexity. To optimize the substrate management for GaN MBD-HEMT, the impact of dynamic substrate bias on device parametric stability needs to be thoroughly understood. This work addresses this gap by *in situ* characterizations of dynamic on-resistance (dyR_{ON}) and threshold voltage (V_{th}) in a GaN HEMT with the floating substrate under both hard- and soft-switching conditions. A relatively large dyR_{ON} shift is found when the substrate is floating or shorted to drain during the hard switching, whilst the dynamic V_{th} shift is minimal in these cases. Instead, the back-gating effect, attributed to the changing substrate bias, i.e., negative during hard-switching and positive during soft-switching, is suspected to be the primary cause. These results can provide practical guidance for optimizing the stability of MBD-HEMT.**

Keywords—*GaN-on-Si HEMT, monolithic power integration, hard-switching, soft-switching, stability, DC-AC Converter*

I. INTRODUCTION

Leveraging the unique device structure, GaN power devices allow for new functionalities that are distinct from what Si and SiC can offer, such as monolithic integration for high-voltage devices [1], [2], [3], [4], [5], [6]. However, the commercialization and application of these devices are hindered by their stability, reliability, and robustness, which could be more complex than conventional discrete devices [7].

Bi-directional AC switches are currently utilized in power electronics circuits like DC/AC multi-level T-type inverters [9], current source inverters [10], AC/AC matrix inverters [11], and AC/DC bridgeless PFCs. Monolithically integrated bidirectional GaN-on-Si HEMTs (MBD-HEMT) have recently gained extensive traction from the industry and academia, as they can allow the minimization of parasitic stray inductance, as well as achieving at least four times lower specific on-resistance (R_{ON}) compared to the conventional bidirectional device implementation by two anti-series connection of two unidirectional devices [12], [13], [14], [15], [16], [17]. However, the Si substrate termination is a key concern for the stability of such bidirectional GaN devices built on Si substrates [18], [19], [20]. For commercial unidirectional GaN HEMTs, the substrate is typically shorted to the low-potential side, i.e., source. However, such a substrate-source connection is impractical for the monolithic bi-directional GaN HEMT, as the low-potential side alternates.

Many bidirectional GaN-on-Si HEMTs still utilize floating substrates [15], [21]. However, the floating Si substrate has a changing potential during device switching, which is likely to deteriorate the performance of the device [22]. Although this issue can be solved by adding an extra circuit to switch the substrate potential [23], it adds an extra cost and design complexity. Recent studies have revealed that a floating substrate can significantly impact the high-frequency soft-switching loss [24] and dynamic R_{ON} (dyR_{ON}) [22] of the GaN-on-Si HEMTs. Despite some prior studies on the impact of

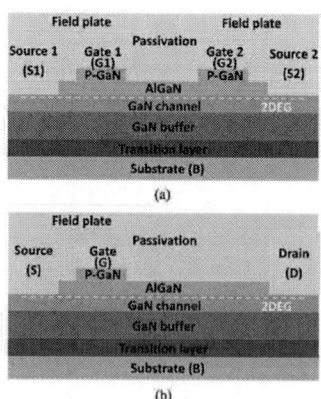

Fig. 1. Schematics of (a) a monolithic bidirectional GaN HEMT and (b) a unidirectional GaN HEMT.

This work is supported by the Center for Power Electronics Systems (CPES) Power Management Consortium.

979-8-3315-1612-3/25 $31.00 © 2025 IEEE

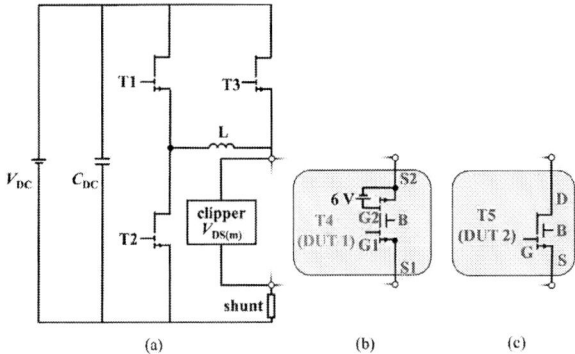

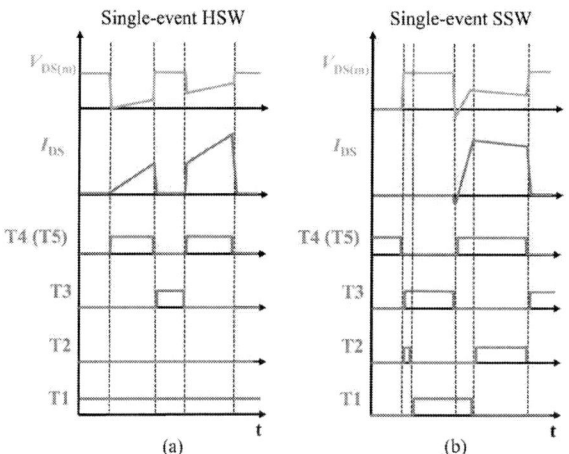

Fig. 2. Schematic of (a) the full-bridge-based DPT setup, (b) bidirectional GaN HEMT with one gate turned ON, and (c) a unidirectional GaN HEMT.

Fig. 4. Switching sequences to achieve (a) HSW and (b) SSW for the DUTs in the half-bridge DPT.

Fig. 3. A photo of the test setup.

substrate potential on dyR_{ON} of bidirectional GaN HEMT [18], the underlying physical mechanisms remain unclear. Particularly, it is unclear if such dyR_{ON} is due to the dynamic shift of threshold voltage (V_{th}), and if these shifts differ in hard- and soft-switching conditions.

This work begins by examining a monolithic bidirectional GaN HEMT, characterizing the substrate bias waveforms under both soft- and hard-switching conditions. Building on these observations, we conducted the stability studies using a unidirectional GaN-on-Si HEMT with a floating substrate, leveraging its simpler structure and driving requirements while maintaining a similar substrate bias condition. Through *in situ* characterization of the dyR_{ON} and dynamic V_{th} (dyV_{th}), this research reveals the impact of a floating substrate on device stability under soft- and hard-switching operations, which provide practical guidance for substrate management optimization for monolithic bidirectional GaN-on-Si HEMT.

II. DUT AND TEST SETUP

A. Device Under Test (DUT)

Two devices under test (DUTs) are tested in this work. DUT 1 is an industrial monolithic bidirectional 650-V 25-mΩ GaN HEMT, while DUT 2 is an industrial 650-V 50-mΩ GaN HEMT with a floating substrate. The schematics of these two DUTs are shown in Fig. 1(a) and (b), respectively. Both DUTs are fabricated on Si substrates and utilize Schottky-type P-gates for normally-OFF operation. The bidirectional GaN HEMTs consist of two gates (G1 and G2), two source electrodes (S1 and S2),

and one shared drift region that can block voltage bidirectionally, like a common-drain configuration. The unidirectional GaN HEMTs consist of a single gate (G), drain (D), and source (S). Both DUTs have a controllable substrate electrode (B).

Note that for the general device characterization purpose, for many power electronics applications, the bidirectional GaN HEMT can be equivalent to a unidirectional GaN HEMT by permanently turning ON one of the gates [18], [23]. This is because, in many converter operations, e.g., a T-type converter, one of the gates in the bidirectional GaN HEMT is always ON, and the other gate is actively switching, which behaves similarly to a unidirectional GaN HEMT. For example, for the bidirectional GaN HEMT in Fig. 1(a), considering that S2 blocks high voltage, G2 is ON, and S1 actively switches, the bidirectional device can be simplified to be equivalent to a unidirectional GaN HEMT shown in Fig. 1(b) for many applications.

B. Test Setup

The DUTs are dynamically characterized in a full-bridge-based double-pulse test (DPT) platform, as shown in Fig. 2, similar to the one proposed in [25]. The power supply V_{DC} delivers up to 400 V bus voltage in this test, in parallel with the 100-µF bus capacitors C_{DC}. Load inductor L of 220 µH is utilized. Power switches T1, T2, and T3 are realized by GaN HEMTs (GS66516T). Either T4 (DUT 1) or T5 (DUT 2) are connected to the circuit. For DUT 1, one side is constantly turned ON by applying an isolated 6-V voltage supply between G2 and S2. For both DUTs, the gate-to-source voltages (V_{G1-S1} and V_{GS}) are monitored using a Tektronix TPP1000 probe; the drain/source-to-source voltages (V_{S2-S1} and V_{DS}) and substrate-to-source voltages (V_{B-S1} and V_{BS}) are measured by Tektronix THDP0200 probes; the drain/source-to-source currents (I_{S2-S1} and I_{DS}) are sensed by a commercial 0.1-Ω co-axial shunt resistor (SSDN-414-10). In addition, to extract the dyR_{ON}, a commercial voltage clipper is parallel to the DUTs to accurately measure the DUTs' ON-state voltage ($V_{DS(m)}$); the dyR_{ON} measurement using the clipper has been described in [26], [27], [28], [29], [30]. Fig. 3 shows a photo of the test setup. Fig. 4 (a) and (b) show the switching sequences to achieve hard-switching (HSW) and soft-

979-8-3315-1612-3/25 $31.00 © 2025 IEEE   203

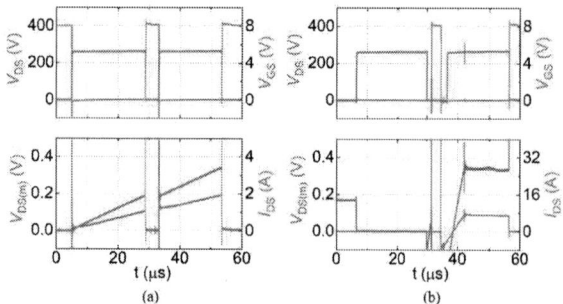

Fig. 5. Test waveforms of DUT 2 under (a) HSW and (b) SSW.

switching (SSW) of DUT and the consequent waveforms in an exemplar 400-V/3.5-A HSW and 400-V/6-A SSW are shown in Fig. 5(a) and (b).

C. Dynamic R_{ON} and V_{th} Extraction

The dyR_{ON} and dyV_{th} measurements are performed on DUT 2. As illustrated in Fig. 6(a), the dyR_{ON} and V_{th} are extracted from the test waveform from the DUT 2' second turn-ON transit and fully-ON period, respectively. DyR_{ON} is calculated by Ohm's law from $V_{DS(m)}$ and I_{DS} waveforms, while dyV_{th} is extracted by reconstructing the dynamic I_{DS}-V_{GS} characteristics using the I_{DS} and V_{GS} data extracted from the switching waveforms, similar to the method proposed in [31]. DyR_{ON} and dyV_{th} are extracted from DPT with a gate driving resistances (R_G) of 500 Ω. The relatively large R_G used for dyV_{th} extraction enables the reconstruction of dynamic transfer characteristics with sufficient data resolution and less sensitivity to the possible temporal misalignment of the voltage and current signals [31]. Fig. 6(b) illustrates a dyV_{th} extraction at V_{DC} of 400 V. V_{GS} and I_{DS} data from the waveform are mapped to the same curve, forming dynamic transfer characteristics, where dyV_{th} is extracted at a 10-mA crossing point.

Fig. 7 shows the normalized R_{ON}, i.e., dyR_{ON} / static R_{ON}, measured for the DUT 2 and a Si MOSFET (IPBE65R050CFD7A). The Si MOSFET is dyR_{ON}-free and

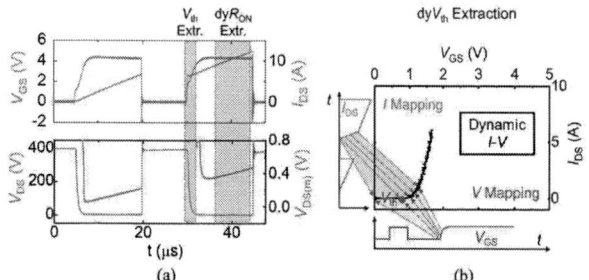

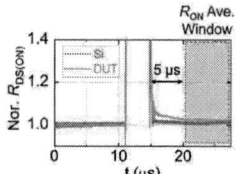

Fig. 6. (a) A typical DPT waveform used for V_{th} and dyR_{ON} extraction with an R_G of 500 Ω. (b) DyV_{th} extraction from DPT waveform.

Fig. 7. Calibration of the dyR_{ON} measurement of DUT 2 and Si MOSFET.

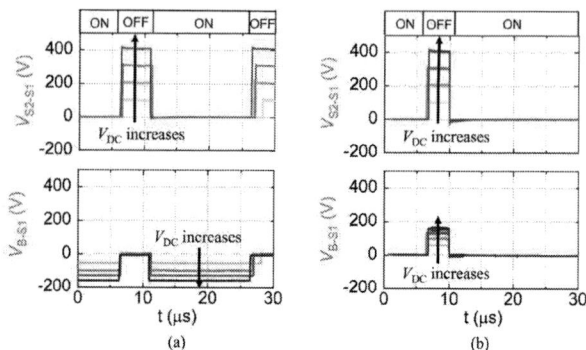

Fig. 8. V_{S2-S1} and V_{B-S1} waveform of the DUT 1 in DPT with different V_{DC} under (a) HSW and (b) SSW.

unitized for calibration to eliminate any possible offset of the clipper circuit that may cause an error. The dyR_{ON} of the DUT is extracted by averaging the data points across a 10-μs window after 5 μs when the DUT fully turns ON.

III. TEST RESULTS

A. Substrate Voltage under Hard- and Soft-Switching

The substrate voltage of both DUTs is monitored during both HSW and SSW tests, with that of DUT 1 demonstrated in Fig. 8 (a) and (b), respectively, as an example. The V_{S2-S1} and V_{B-S1} waveforms are recorded at V_{DC} of 100 V, 200 V, 300 V, and 400 V. Several noteworthy observations can be drawn from the results. First, V_{B-S1} can be either positive or negative during switching. Under hard switching, V_{B-S1} is zero when DUT 1 is OFF and drops to a negative value after DUT turns ON due to the B-S1 parasitic capacitance discharging. Conversely, under soft-switching, V_{B-S1} is zero when DUT 1 is ON and rises to a positive value after DUT 1 turns OFF. However, a common trend observed in both cases is that V_{B-S1} decreases after DUT 1 is ON.

Identical substrate behavior is observed in the unidirectional GaN HEMT with a floating substrate (i.e., DUT 2). This further confirms that the substrate bias of bidirectional GaN HEMT is similar to that of a unidirectional device with G2 kept ON.

B. Dynamic R_{ON} and V_{th} Shifts

Due to its simpler structure and driving requirements, DUT 2 is utilized to characterize the dyR_{ON} and dyV_{th} in the HSW and SSW. Note that dyV_{th} of GaN HEMTs with a floating substrate has been seldom reported previously.

Three substrate terminations are investigated: the substrate is connected to the source (common source), the substrate is

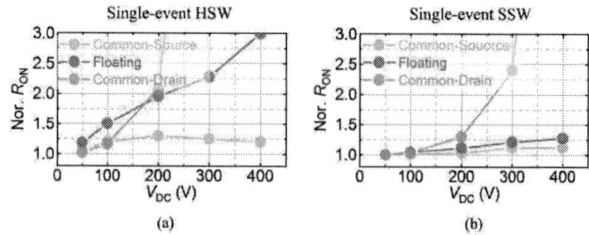

Fig. 9. DyR_{ON} with different substrate connections under (a) HSW and (b) SSW. The dyR_{ON} is normalized to the static R_{ON} value.

979-8-3315-1612-3/25 $31.00 © 2025 IEEE

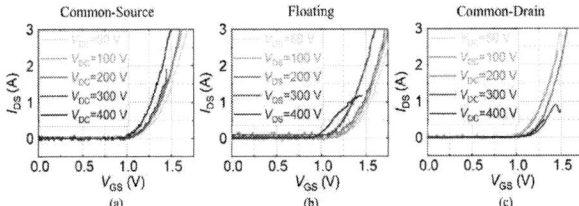

Fig. 10. DyR_{ON} measured with (a) common-source, (b) floating, and (c) common-drain substrates.

floating, and the substrate is shorted to the drain (common drain). Fig. 9 (a) and (b) show the normalized R_{ON} of DUT 2 as a function of V_{DC} with these three terminations under HSW and SSW, respectively. For consistency, the peak I_{DS} is maintained at 6 A in all test conditions. The results reveal that the common-source configuration provides the best DUT performance, achieving the smallest dyR_{ON} among the three substrate terminations. The floating-substrate case exhibits a higher dyR_{ON} than the common-source configuration during HSW but is comparable to the common-source case in SSW. In contrast, the common-drain configuration significantly deteriorates device performance, with the dyR_{ON} of DUT 2 increasing drastically beyond the measurement range when V_{DC} exceeds approximately 200 V to 300 V.

To understand if such dyR_{ON} shift is due to the V_{th} shift, dyV_{th} is extracted with three substrate terminations, as presented in Fig. 10. Overall, DUT 2 exhibits a small dyV_{th} shift across all terminations, e.g., less than 0.25 V. For common-source and floating substrates, dyV_{th} both show negative shifts as V_{DC} increases, which can be attributed to the elevated potential of the p-GaN layer [31]. Such negative dyV_{th} shifts cannot fully account for the observed dyR_{ON} increase since R_{ON} is supposed to decrease with a smaller V_{th} at the same V_{GS}. In contrast, for common-drain connections, V_{th} increases as V_{DC} goes up. However, the V_{th} shifts remain small, e.g., ~ 0.15 V, and still cannot best explain the drastic dyR_{ON} shifts observed with common-drain substrate connections.

IV. DISCUSSIONS AND PHYSICAL EXPLANATIONS

Since dyV_{th} shifts are small under both positive and negative V_{BS}, the back-gating effect is likely the primary cause of the significant dyR_{ON} increases. As illustrated in Fig. 11(a), under positive V_{BS}, electrons can be injected from the Ohmic contact or the 2DEG and subsequently captured by initially empty traps in the buffer layer, leading to the accumulation of negative charges in the buffer stack. Conversely, under negative V_{BS}, the initially empty acceptor traps in the buffer layer can become ionized, potentially by capturing electrons injected from the Si substrate [Fig. 11(b)]. In both scenarios, whether V_{BS} is positive or negative, the buffer layer accumulates a net negative charge. This negative charge partially depletes the 2DEG, resulting in a significant increase in dyR_{ON}. On the other hand, such a net buffer charge, and thus dyR_{ON}, are expected to be low for a source-connected substrate, which is consistent with the test results in Fig. 9.

V. CONCLUSIONS

This work investigates the substrate potential behavior of a monolithic bidirectional GaN HEMT during HSW and SSW

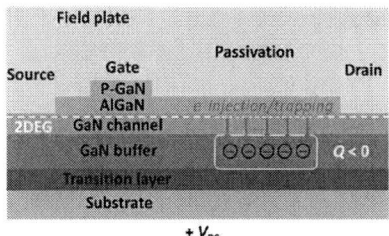

(a)

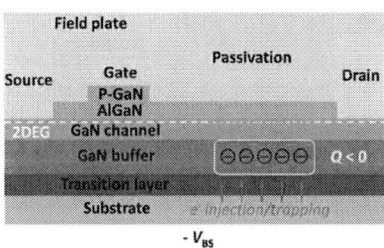

(b)

Fig. 11. Trapping behavior of DUT with (a) positive and (b) negative V_{BS}.

conditions, as well as the impact of positive/negative substrate bias on dyR_{ON} and dyV_{th}. During HSW, the substrate potential is zero when the device is blocking and drops to a negative value when the device turns ON. In contrast, during SSW, the substrate potential is zero when the device is ON but rises to a positive value when the device turns OFF.

Subsequently, a unidirectional GaN HEMT with a tunable substrate electrode is employed to investigate the impact of varying substrate potentials on GaN HEMT stability. Substrate connections, including common-source, common-drain, and floating configurations, are studied. For the first time, both dyR_{ON} and dyV_{th} of a GaN power HEMT with different substrate terminations are *in situ* monitored simultaneously in a full-bridge-based DPT. It is found that the common-source substrate gives the best dyR_{ON} performance while the common-drain substrate connections induce the highest dyR_{ON}. V_{th} is found to have a relatively small shift across all three substrate connections and cannot account for the large dyR_{ON} shifts with floating and common-drain substrates in HSW. The root cause is attributed to a net negative charge in the buffer layer induced by negative and positive substrate bias during HSW and SSW, respectively, which partially depletes the 2DEG and increases dyR_{ON}. These findings offer guidance for the optimized substrate management design of monolithic bidirectional GaN HEMTs.

REFERENCES

[1] Y. Zhang, F. Udrea, and H. Wang, "Multidimensional device architectures for efficient power electronics," *Nat Electron*, vol. 5, no. 11, Art. no. 11, Nov. 2022, doi: 10.1038/s41928-022-00860-5.

[2] M. Xiao *et al.*, "Multi-Channel Monolithic-Cascode HEMT (MC2-HEMT): A New GaN Power Switch up to 10 kV," in *2021 IEEE International Electron Devices Meeting (IEDM)*, Dec. 2021, p. 5.5.1-5.5.4. doi: 10.1109/IEDM19574.2021.9720714.

[3] H. Xu, G. Tang, J. Wei, Z. Zheng, and K. J. Chen, "Monolithic Integration of Gate Driver and Protection Modules With P-GaN Gate Power HEMTs," *IEEE Transactions on Industrial Electronics*, vol. 69, no. 7, pp. 6784–6793, Jul. 2022, doi: 10.1109/TIE.2021.3102387.

[4] R. Sun, J. Lai, W. Chen, and B. Zhang, "GaN Power Integration for High Frequency and High Efficiency Power Applications: A Review," *IEEE*

Access, vol. 8, pp. 15529–15542, 2020, doi: 10.1109/ACCESS.2020.2967027.

[5] X. Li *et al.*, "200 V Enhancement-Mode p-GaN HEMTs Fabricated on 200 mm GaN-on-SOI With Trench Isolation for Monolithic Integration," *IEEE Electron Device Letters*, vol. 38, no. 7, pp. 918–921, Jul. 2017, doi: 10.1109/LED.2017.2703304.

[6] M. Basler, N. Deneke, S. Mönch, R. Reiner, B. Wicht, and R. Quay, "Monolithically Integrated GaN Gate Drivers– A Design Guide," *IEEE Open Journal of Power Electronics*, vol. 4, pp. 487–497, 2023, doi: 10.1109/OJPEL.2023.3290190.

[7] J. P. Kozak *et al.*, "Stability, Reliability, and Robustness of GaN Power Devices: A Review," *IEEE Transactions on Power Electronics*, vol. 38, no. 7, pp. 8442–8471, Jul. 2023, doi: 10.1109/TPEL.2023.3266365.

[8] E. Gurpinar and A. Castellazzi, "Single-Phase T-Type Inverter Performance Benchmark Using Si IGBTs, SiC MOSFETs, and GaN HEMTs," *IEEE Transactions on Power Electronics*, vol. 31, no. 10, pp. 7148–7160, Oct. 2016, doi: 10.1109/TPEL.2015.2506400.

[9] H. Dai, R. A. Torres, J. Gossmann, W. Lee, T. M. Jahns, and B. Sarlioglu, "A Seven-Switch Current-Source Inverter Using Wide Bandgap Dual-Gate Bidirectional Switches," *IEEE Transactions on Industry Applications*, vol. 58, no. 3, pp. 3721–3737, May 2022, doi: 10.1109/TIA.2022.3149461.

[10] P. Guerriero, S. Orcioni, I. Matacena, and S. Daliento, "A GaN based bidirectional switch for matrix converter applications," in *2020 International Symposium on Power Electronics, Electrical Drives, Automation and Motion (SPEEDAM)*, Jun. 2020, pp. 375–380. doi: 10.1109/SPEEDAM48782.2020.9161876.

[11] T. Hirota, K. Inomata, D. Yoshimi, and M. Higuchi, "Nine Switches Matrix Converter Using Bi-directional GaN Device," in *2018 International Power Electronics Conference (IPEC-Niigata 2018 -ECCE Asia)*, May 2018, pp. 3952–3957. doi: 10.23919/IPEC.2018.8507399.

[12] T. Morita *et al.*, "650 V 3.1 mΩcm2 GaN-based monolithic bidirectional switch using normally-off gate injection transistor," in *2007 IEEE International Electron Devices Meeting*, Dec. 2007, pp. 865–868. doi: 10.1109/IEDM.2007.4419086.

[13] M. Wolf, O. Hilt, and J. Würfl, "Gate Control Scheme of Monolithically Integrated Normally OFF Bidirectional 600-V GaN HFETs," *IEEE Transactions on Electron Devices*, vol. 65, no. 9, pp. 3878–3883, Sep. 2018, doi: 10.1109/TED.2018.2857848.

[14] H. Wang *et al.*, "Experimental Demonstration of Monolithic Bidirectional Switch With Anti-Paralleled Reverse Blocking p-GaN HEMTs," *IEEE Electron Device Letters*, vol. 42, no. 9, pp. 1264–1267, Sep. 2021, doi: 10.1109/LED.2021.3098040.

[15] G. Baratella *et al.*, "Monolithic 650-V Dual-Gate p-GaN Bidirectional Switch," *IEEE Transactions on Electron Devices*, vol. 71, no. 11, pp. 6904–6909, Nov. 2024, doi: 10.1109/TED.2024.3456077.

[16] M. T. Alam, J. Chen, R. Bai, S. S. Pasayat, and C. Gupta, "High-Voltage (>1.2 kV) AlGaN/GaN Monolithic Bidirectional HEMTs With Low On-Resistance (2.54 mΩ · cm2)," *IEEE Transactions on Electron Devices*, vol. 71, no. 1, pp. 733–738, Jan. 2024, doi: 10.1109/TED.2023.3330133.

[17] J. Huber and J. W. Kolar, "Monolithic Bidirectional Power Transistors," *IEEE Power Electronics Magazine*, vol. 10, no. 1, pp. 28–38, Mar. 2023, doi: 10.1109/MPEL.2023.3234747.

[18] C. Kuring *et al.*, "Impact of Substrate Termination on Dynamic On-State Characteristics of a Normally-off Monolithically Integrated Bidirectional GaN HEMT," in *2019 IEEE Energy Conversion Congress and*

Exposition (ECCE), Baltimore, MD, USA: IEEE, Sep. 2019, pp. 824–831. doi: 10.1109/ECCE.2019.8912793.

[19] S. A. Albahrani *et al.*, "Modeling of the Impact of the Substrate Voltage on the Capacitances of GaN-on-Si HEMTs," *IEEE Transactions on Electron Devices*, vol. 66, no. 12, pp. 5103–5110, Dec. 2019, doi: 10.1109/TED.2019.2948828.

[20] S. Moench *et al.*, "Asymmetrical Substrate-Biasing Effects at up to 350V Operation of Symmetrical Monolithic Normally-Off GaN-on-Si Half-Bridges," in *2019 IEEE 7th Workshop on Wide Bandgap Power Devices and Applications (WiPDA)*, Oct. 2019, pp. 28–35. doi: 10.1109/WiPDA46397.2019.8998934.

[21] C. Kuring, O. Hilt, J. Böcker, M. Wolf, S. Dieckerhoff, and J. Würfl, "Novel monolithically integrated bidirectional GaN HEMT," in *2018 IEEE Energy Conversion Congress and Exposition (ECCE)*, Sep. 2018, pp. 876–883. doi: 10.1109/ECCE.2018.8557741.

[22] S. Yang, C. Zhou, S. Han, J. Wei, K. Sheng, and K. J. Chen, "Impact of Substrate Bias Polarity on Buffer-Related Current Collapse in AlGaN/GaN-on-Si Power Devices," *IEEE Trans. Electron Devices*, vol. 64, no. 12, pp. 5048–5056, Dec. 2017, doi: 10.1109/TED.2017.2764527.

[23] C. Kuring *et al.*, "Active substrate termination of discrete and monolithic bidirectional GaN HEMTs in a T-type inverter," in *2022 24th European Conference on Power Electronics and Applications (EPE'22 ECCE Europe)*, Sep. 2022, p. P.1-P.11.

[24] Q. Song, A. Briga, V. Veprinsky, R. Volkov, Q. Li, and Y. Zhang, "Minimizing Output Capacitance Loss in GaN Power HEMT," *IEEE Transactions on Power Electronics*, vol. 39, no. 8, pp. 9120–9126, Aug. 2024, doi: 10.1109/TPEL.2024.3399237.

[25] T. Zhao, R. Burgos, and J. Xu, "Dynamic ON-Resistance Characterization of GaN HEMT Under Soft-Switching Condition," in *2021 IEEE 8th Workshop on Wide Bandgap Power Devices and Applications (WiPDA)*, Nov. 2021, pp. 246–249. doi: 10.1109/WiPDA49284.2021.9645123.

[26] X. Yang *et al.*, "Dynamic RON Free 1.2-kV Vertical GaN JFET," *IEEE Transactions on Electron Devices*, vol. 71, no. 1, pp. 720–726, Jan. 2024, doi: 10.1109/TED.2023.3338140.

[27] X. Yang *et al.*, "Evaluation and MHz Converter Application of 1.2-kV Vertical GaN JFET," *IEEE Transactions on Power Electronics*, pp. 1–11, 2024, doi: 10.1109/TPEL.2024.3445667.

[28] L. Rossetto and G. Spiazzi, "A Fast ON-State Voltage Measurement Circuit for Power Devices Characterization," *IEEE Transactions on Power Electronics*, vol. 37, no. 5, pp. 4926–4930, May 2022, doi: 10.1109/TPEL.2021.3129613.

[29] T. Cappello, A. Santarelli, and C. Florian, "Dynamic RON Characterization Technique for the Evaluation of Thermal and Off-State Voltage Stress of GaN Switches," *IEEE Transactions on Power Electronics*, vol. 33, no. 4, pp. 3386–3398, Apr. 2018, doi: 10.1109/TPEL.2017.2710281.

[30] R. Li, X. Wu, S. Yang, and K. Sheng, "Dynamic on-State Resistance Test and Evaluation of GaN Power Devices Under Hard- and Soft-Switching Conditions by Double and Multiple Pulses," *IEEE Transactions on Power Electronics*, vol. 34, no. 2, pp. 1044–1053, Feb. 2019, doi: 10.1109/TPEL.2018.2844302.

[31] X. Lu *et al.*, "Impact of V_{th} Instability of Schottky-Type p-GaN Gate HEMTs on Switching Behaviors," *IEEE Transactions on Power Electronics*, vol. 39, no. 9, pp. 11625–11636, Sep. 2024, doi: 10.1109/TPEL.2024.3405320.

Characterization of LED Driven GaN-based Photoconductive Switches

Samuel K. Atwimah[1], Tolen M. Nelson[1], Geoffrey M. Foster[2], Daniel G. Georgiev[1], Andrew D. Koehler[2], Alan G. Jacobs[2], Karl D. Hobart[2], Michael R. Hontz[3], and Raghav Khanna[1]

[1]EECS Dept., The University of Toledo, Toledo, OH, USA
[2]U.S. Naval Research Laboratory, Washington, DC, USA
[3]Naval Surface Warfare Center Philadelphia Division: Philadelphia, PA, USA
Samuel.Atwimah@utoledo.edu

Abstract—This paper presents the characterization of a carbon-doped gallium nitride (GaN) photoconductive semiconductor switch (PCSS). A novel method was devised to study the devices' dynamic behaviour using a Cascaded Double Pulse Test (C-DPT), which utilizes commercial off-the-shelf (COTS) near-ultraviolet LEDs as the optical trigger. The C-DPT setup enabled evaluation of the turn-on and turn-off dynamics of lateral GaN PCSS devices, thereby demonstrating proof-of-concept for the proposed dynamic characterization method. While carbon doping has proven effective in achieving high resistivity and reducing leakage currents, the study identifies challenges posed by deep-level carbon traps, which significantly prolong the turn-off time. Additionally, the results highlight the important role of optical intensity in driving the turn-on of the PCSS. These findings emphasize the need to explore alternative impurities and high-power UV LEDs for improved switching performance. This work provides valuable insights into optimizing and characterizing GaN PCSS devices for advanced power electronics applications.

Index Terms—photoconductive semiconductor switch; gallium nitride; double pulse test; III-V semiconductors; wide bandgap semiconductors; COTS LED

I. INTRODUCTION

Wide bandgap semiconductors, such as gallium nitride (GaN) have revolutionized power electronics by addressing the growing demand for high-efficiency, high-voltage devices in modern power systems. With its wide bandgap of $\sim 3.4\,\text{eV}$, high critical electric field of 3 MV/cm, and electron mobility of 1800 cm²/V·s [1]. GaN is particularly critical as industries strive to electrify applications requiring power converters that can switch voltages between 600 V and 20 kV, balancing high power density with operational efficiency.

In the realm of wide bandgap devices, vertical GaN structures such as JFETs and MOSFETs are under development to meet the growing demands for high-voltage and high-speed power applications [2], [3]. However, photoconductive semiconductor switches (PCSS) have emerged as a highly promising alternative. By relying on high-resistivity materials and optical triggering mechanisms, PCSS devices address several limitations inherent to conventional unipolar GaN devices.

This work was supported in part by the U.S. Office of Naval Research under grant number N00014-21-1-2832, and was approved for public release under DCN 543-2214-24. Distribution Statement A. Approved for Public Release. Distribution Unlimited.

Traditional unipolar devices face a fundamental tradeoff between on-resistance and breakdown voltage. In these devices, the drift layer is critical for voltage blocking but contributes significantly to on-state resistance, particularly in high-voltage systems that require a thicker or more lightly doped drift layer. This tradeoff results in increased conduction losses and poses challenges for achieving efficient high-voltage designs [4].

PCSS devices utilize optical control mechanisms to generate a high density of free carriers across the resistive layer, effectively lowering resistivity during operation. Their voltage-blocking capability is determined by the gap size between electrodes [5], allowing high-voltage operation without significantly increasing the material's on-resistance. This unique attribute of decoupling voltage-handling capability from conduction loss positions PCSS as a compelling alternative to conventional 1-D devices, offering significant advantages for advanced power applications.

Furthermore, the optical triggering mechanism in PCSS eliminates common failure modes such as false turn-on effects caused by high dV/dt [6], [7], gate dielectric breakdown from voltage overshoot, and excessive electromagnetic interference (EMI) [8], all of which are often observed in fast-switching GaN devices. These issues often necessitate compromises in device performance, where switching speeds are intentionally reduced to prevent damage [9]. By isolating the input port of the device from its output, PCSS devices overcome these constraints, enabling optimal performance without sacrificing reliability or safety.

PCSS devices operate in two optical modes, intrinsic and extrinsic [10], and two electrical modes, linear and nonlinear [5]. The nonlinear mode occurs when the PCSS is operated near the breakdown voltage. In this mode, carriers generated by the incident light are accelerated by the electric field with enough energy to ionize additional carriers via impact ionization. This causes filaments of electron-hole plasma to appear, locking on the device until the current is cutoff. Although this results in a nonlinear current-voltage (I-V) characteristic, the lock-on is undesirable for continuous power electronics, and the locality of the filaments results in reliability concerns. The linear mode occurs when the PCSS is operated suitably below the breakdown voltage such that there is insignificant impact-ionization to cause filament generation. In this mode, each incident photon generally results in one electron-hole pair (ignoring any multi-photon absorption). When the light source is terminated, rather than locking in the on-state due to sustained impact ionization, the carriers eventually all recombine, and the layer returns to its semi-insulating nature [11]. The intrinsic mode occurs when above-bandgap energy light is used to induce band-to-band generation of carriers, while the extrinsic mode utilizes the impurity photovoltaic (IPV) effect [12], [13], where sub-bandgap energy light generates carriers from defect levels within the bandgap.

Among the materials explored for PCSS devices, carbon-doped GaN has shown great promise due to its ability to achieve high resistivity and reduced leakage currents, making it a prime candidate for high-voltage applications [10]. These characteristics are particularly advantageous in the linear mode, where maintaining semi-insulating behavior is important. While much research has focused on optimizing the material properties and fabrication techniques of carbon-doped GaN PCSS [14], a detailed understanding of the dynamic behavior of these devices remains limited.

This paper presents a study of the static and dynamic performance of carbon-doped GaN lateral PCSS devices and proposes a cost-effective method for their characterization. Unlike traditional PCSS setups that rely on expensive lasers for optical triggering [8], [10], [15] —this work demonstrates the use of commercial off-the-shelf (COTS) near-ultraviolet LEDs as optical triggers. A novel Cascaded Double Pulse Test (C-DPT) is proposed, which consists of a low-voltage and high-voltage DPT. The low-voltage DPT drives the COTs LED, which is strategically positioned over the high-voltage DPT, containing the PCSS as the device under study, in order to characterize its dynamic behavior. This approach significantly reduces costs and complexity while maintaining reliable device activation.

The fabricated lateral carbon-doped GaN PCSS devices exhibit high blocking voltages and are operated in linear mode. Through detailed analysis, the static and dynamic characteristics of these devices are explored, emphasizing their potential application in high-performance power electronics.

II. FABRICATION AND OPERATION

Fig. 1 illustrates the overall structure of the PCSS devices under study. A 4 μm GaN layer was grown on a silicon carbide (SiC) substrate via metal organic chemical vapor deposition (MOCVD), incorporating a carbon doping concentration of $1.5 \times 10^{17} \text{cm}^{-3}$. Ti/Al/Ni/Au contacts were deposited using a lift-off technique, followed by an 800°C rapid anneal. Three sets of PCSS devices with varying electrode gaps—100 μm, 30 μm, and 20 μm—were fabricated.

The carbon doping level and its associated deep acceptor level in GaN ($\sim 3.1 \, \text{eV}$ below the conduction band) are particularly relevant, as they contribute to the semi-insulating properties and photoconductive response of the device. These properties were characterized using photoionization spectroscopy. Fig. 2 shows the difference between the on and off-current of

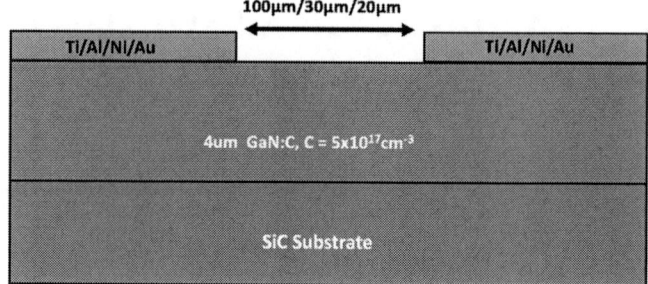

Fig. 1. Structure of fabricated PCSS devices.

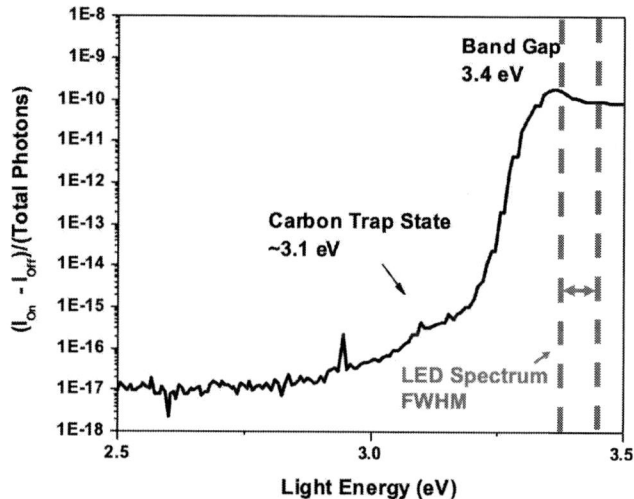

Fig. 2. Measured monochromatic photo-ionization data of PCSS.

the device, as a ratio to the total number of photons applied. To obtain the data in Fig. 2, a parameter analyzer [16] was used to measure the current under an applied voltage and optical triggering was provided by an external monochromatic light source, which was directed across the electrode gap of the PCSS. The illumination wavelengths were varied during testing to evaluate the spectral photo-response. Illumination at energies at and above the GaN bandgap energy ($\sim 3.4 \, \text{eV}$) resulted in a strong photo-response, with the on-current exceeding the off-current by several orders of magnitude, as seen in Fig. 2, indicating a strong intrinsic mode photo-response due to the direct bandgap nature of GaN. Sub-bandgap illumination ($> 3.25 \, \text{eV}$) also induced a significant photocurrent, indicating contributions from carbon-related defect levels, consistent with the IPV effect, and indication of a strong extrinsic mode operation in this region. This effect, which involves excitation of carriers from defect levels to band edges, plays a significant role in the photoconductive properties of carbon-doped GaN PCSS [17].

III. STATIC CHARACTERIZATION

To measure the reverse I-V characteristics, a high voltage power supply [18] was used and the measurements were taken in a vacuum probe station. These measurements resulted in

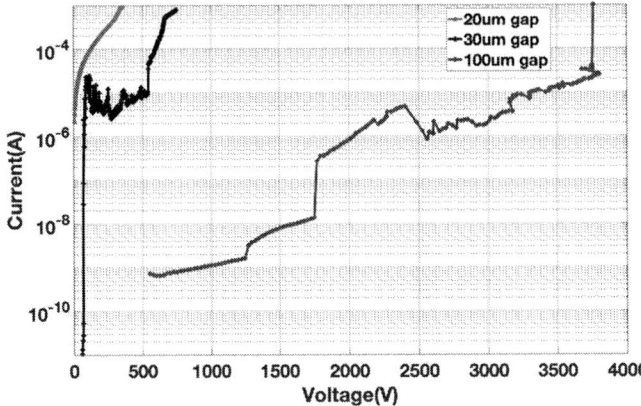

Fig. 3. Measured reverse I-V curves of fabricated PCSS devices.

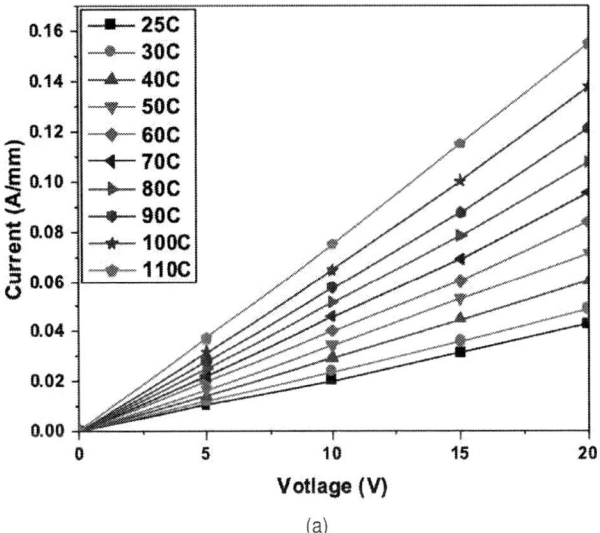

(a)

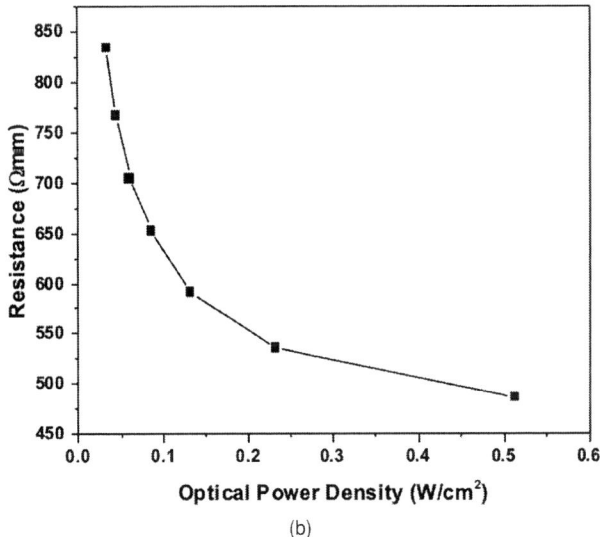

(b)

Fig. 5. Measured forward static characteristics of 100um gap PCSS. (a) I-V curve at different temperatures. (b) Resistance vs. Optical power density curve.

breakdown voltages of approximately 3.8 kV for the 100 μm device, 745 V for the 30 μm device, and 350 V for the 20 μm device. Fig. 3 highlights the impact of gap size on device performance and demonstrates the scalability of carbon-doped GaN PCSS for high-voltage applications.

To investigate the performance of the devices, a COTS LED [19] with a peak wavelength of 365 nm and full width at half maximum (FWHM) of ∼ 10 nm was utilized to optically trigger the 100 μm gap device. Fig. 2 shows the FWHM of the LED spectrum (red dashed lines) overlaid on the measured monochromatic photoionization data of the PCSS, demonstrating that the LED provides incident light in the region of peak photo-response. The LED is driven by an LED driver that provides a low ripple current and precise control of the LED's optical output. The LED was mounted on a separate board positioned above the PCSS enabling controlled illumination across the device gap, as shown in Fig. 4. The voltage across the device was swept from 0 to 20 V and its static DC characteristics were analyzed.

As shown in Fig. 5(a), the effect of temperature on the static

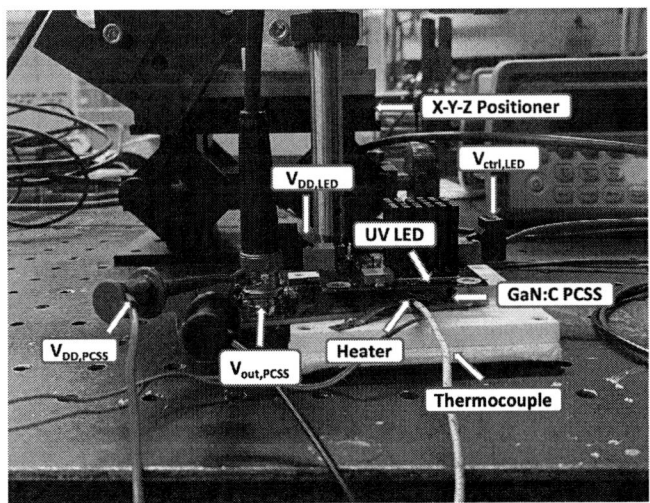

Fig. 4. Experimental setup for forward static characterization.

characteristics was significant, with higher temperatures leading to lower resistance. This is attributed to bandgap narrowing at elevated temperatures, which increases the number of band-to-band direct transitions, thereby enhancing the photocurrent. Additionally, while keeping the PCSS biased at 10 V, the optical intensity was varied by adjusting the distance between the LED and the device. The on-state resistance of the PCSS was observed to decrease with increasing optical intensity, as shown in Fig. 5(b). This behavior is due to the increased generation of free carriers at higher optical intensities, which reduces the on-resistance of the device [17].

IV. DYNAMIC CHARACTERIZATION

A Cascaded Double Pulse Test (C-DPT) is proposed and used to study the switching characteristics of the GaN de-

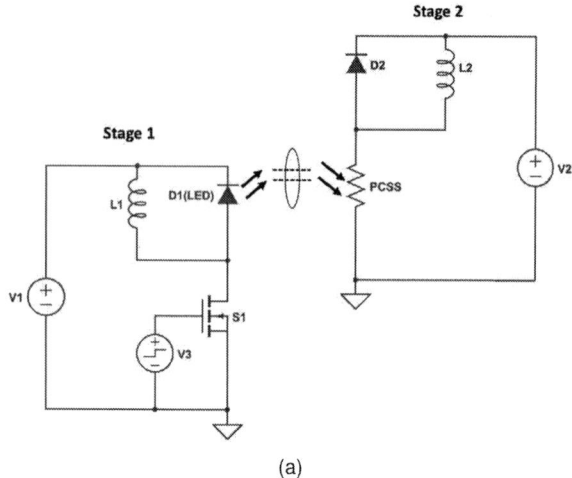

(a)

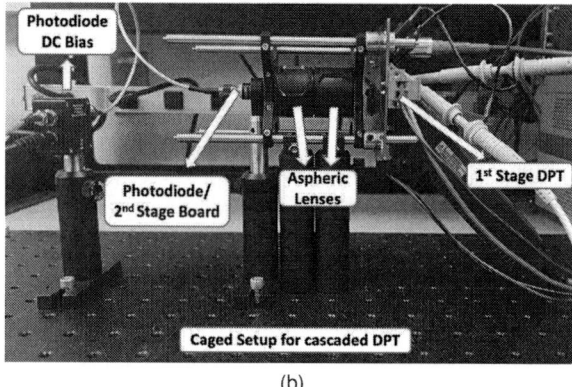

(b)

Fig. 6. Cascaded Double Pulse Test. (a) Schematic. (b) Experimental setup with photodiode in place of second stage DPT for optical power characterization.

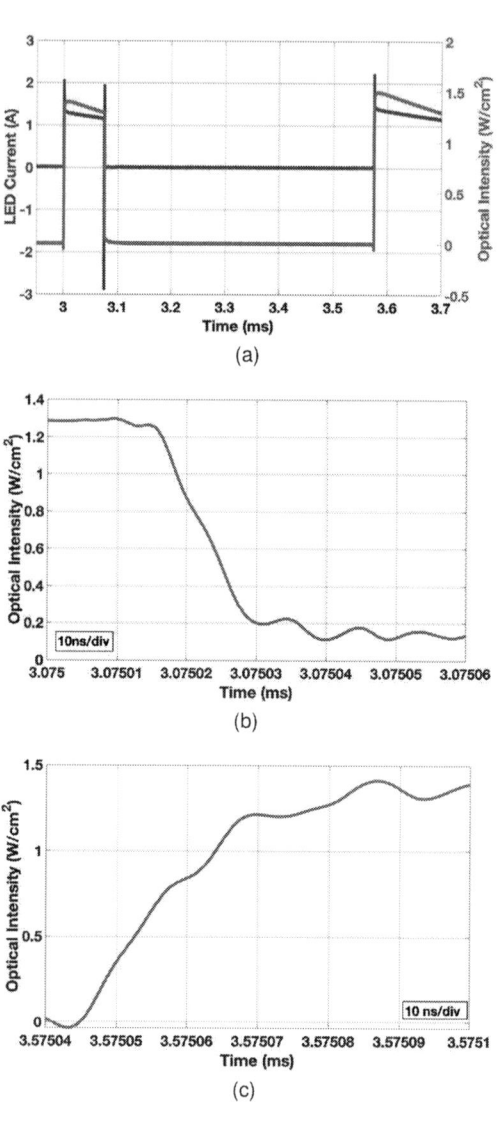

(a)

(b)

(c)

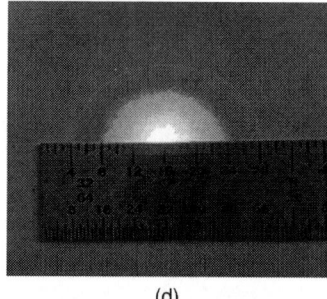

(d)

Fig. 7. Stage 1. (a) Measured LED current and optical power. (b) Measured optical intensity turn off edge. (c) Measured optical intensity turn on edge. (d) Measured spot size.

vices. The C-DPT allows characterization of both the driving LED and PCSS switching losses. The C-DPT arrangement is provided in Fig. 6, with the schematic provided in Fig. 6(a), and a photograph of the experimental setup with a photodiode in place of the second stage DPT for characterization of the optical power in Fig. 6(b). The C-DPT consists of two stages, a low-voltage driving side, and a high-voltage output side. The two stages are configured and oriented so that the LED, functioning as the freewheeling diode on the low-voltage driving side, can provide optical power to the PCSS, contained on the high-voltage output side. The low-voltage DPT drives the LED; the optical output of the LED is then used to control the high-voltage output DPT and characterize the PCSS switching dynamics. This setup allows for the characterization of the entire PCSS system, including the PCSS and the optical driver. It enables the testing of fast edge rates of the LED without the need for a specialized power supply.

A. Stage 1

To verify that the high-power near-UV LED can switch fast enough for MHz operation, the LED switching in the low-voltage DPT, was examined. The low-voltage DPT, hereto termed as *Stage 1*, uses a low-voltage GaN high electron mo-

bility transistor (HEMT) [20] as the switch, a 4 mH inductor as the load inductor, and an LED [21], as the freewheeling diode. A pair of UV fused silica aspheric lenses [22] with focal lengths of 20 mm and a numerical aperture of 0.65 are used to collect and focus the light emitted by the LED. The entire setup is housed in a compact caged configuration, with the

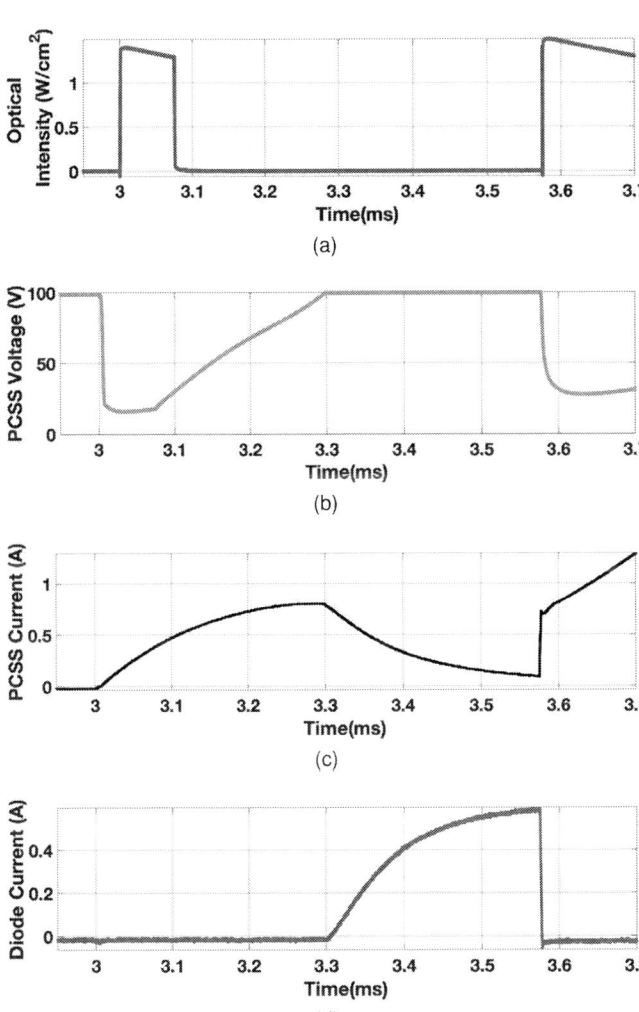

Fig. 8. Switching waveforms for the second stage of C-DPT with a 20 μm gap and 14 mm width PCSS. (a) Optical intensity. (b) PCSS Voltage. (c) PCSS Current.(d) Freewheeling Diode Current.

LED centered in the middle of the board, as shown in Fig. 6(b). Using a fast photodiode [23] approximately 2.7 mm away from the edge of the lens, switching waveforms of the optical signal from the LED were measured. The results are presented in Fig. 7, which includes: (a) the LED current and the corresponding optical power detected by the photodiode, (b) the turn-on edge, (c) the turn-off edge, and (d) the measured spot size of the optical signal, which is ∼ 2.8 mm. The optical turn-on time was measured at 35 ns, while the optical turn-off time was measured at 25 ns.

B. Stage 2

In the second stage of the Cascaded Double Pulse Test (C-DPT), hereto termed *Stage 2*, the PCSS operates in a high-voltage configuration. The test circuit for Stage 2 includes a 1.2 kV SiC Schottky diode [24] as the freewheeling diode, an 8 mH inductor, and a PCSS, which serves as the active switch and features a 20 μm gap, a 14 mm width, and an interdigitated finger structure. To ensure consistent optical intensity delivery,

the Stage 2 board was positioned such that the PCSS was placed in the same location as the photodiode in Stage 1.

The switching characteristics of the 20 μm gap PCSS are shown in Fig. 8, where the PCSS voltage and current waveforms illustrate the device's response during switching. The waveforms provided in Fig. 8 demonstrate the proof-of-concept feasibility of the proposed C-DPT for characterizing PCSS devices. However, the results also highlight areas for improving the performance of the PCSS device. As seen in Fig. 8, during the turn-off phase of the PCSS, a slow turn-off was observed, the voltage switches in ∼ 2.0 ms while the current lags the voltage. The current initially increases as the inductor charges through the still conducting PCSS until it reaches sufficient resitivity to achieve freewheeling of the diode; then it slowly decreases as its resitivity is gradually restored to the static blocking state. This is directly linked to the interaction between free carriers and carbon-related deep-level traps in the GaN. Because carbon doping introduces deep acceptor levels approximately ∼ 3.1 eV below the conduction band, sub-bandgap light from the LED excites trapped electrons into the conduction band, changing the occupancy of the traps. During turn-off, when the optical signal is removed, the empty traps must re-capture the excess electrons to restore the semi-insulating characteristic. The electron capture rate from the conduction band to an empty trap state is given by the following equation [25]:

$$C_n^c = \sigma_n v_{th}^n n, \tag{1}$$

where C_n^C represents the electron capture rate, σ_n is the electron thermal capture cross-section of the trap, v_{th}^n is the thermal velocity of electrons and n is the free electron concentration.

The capture rate C_n^C is dependent on the capture cross-section of the carbon level. The turn-off time could be improved by finding alternative dopants with increased capture cross-sections while still providing high photoconductivity and semi-insulating properties [such as iron (Fe) or co-doping carbon (C) and Fe [26], [27]]. Another approach is to use optical quenching via a lower energy optical signal to restore semi-insulating nature in which electrons in the valence band are excited to the unoccupied defect level via a second optical source during turn-off [28], [29].

Furthermore, in larger area devices, this effect is amplified, leading to even more pronounced delays in the turn-off process. Fig. 9 compares the turn-on waveforms of the device with a 20 μm gap and 14 mm width to a device with a 30 μm gap and 151.2 mm width. The smaller device exhibited a higher on-resistance of ∼ 60 Ω, while the larger device achieved a much lower on-resistance of about ∼ 3.5 Ω. However, the larger device could not be fully characterized due to having a similar turn-off time constant. As a result, it never reaches a sufficiently high on-resistance for the freewheeling diode to conduct, causing the inductor to continue charging.

Fig. 10 shows the turn-on behavior of the 20 μm gap PCSS (smaller device) under varying optical power conditions. The

979-8-3315-1612-3/25 $31.00 © 2025 IEEE

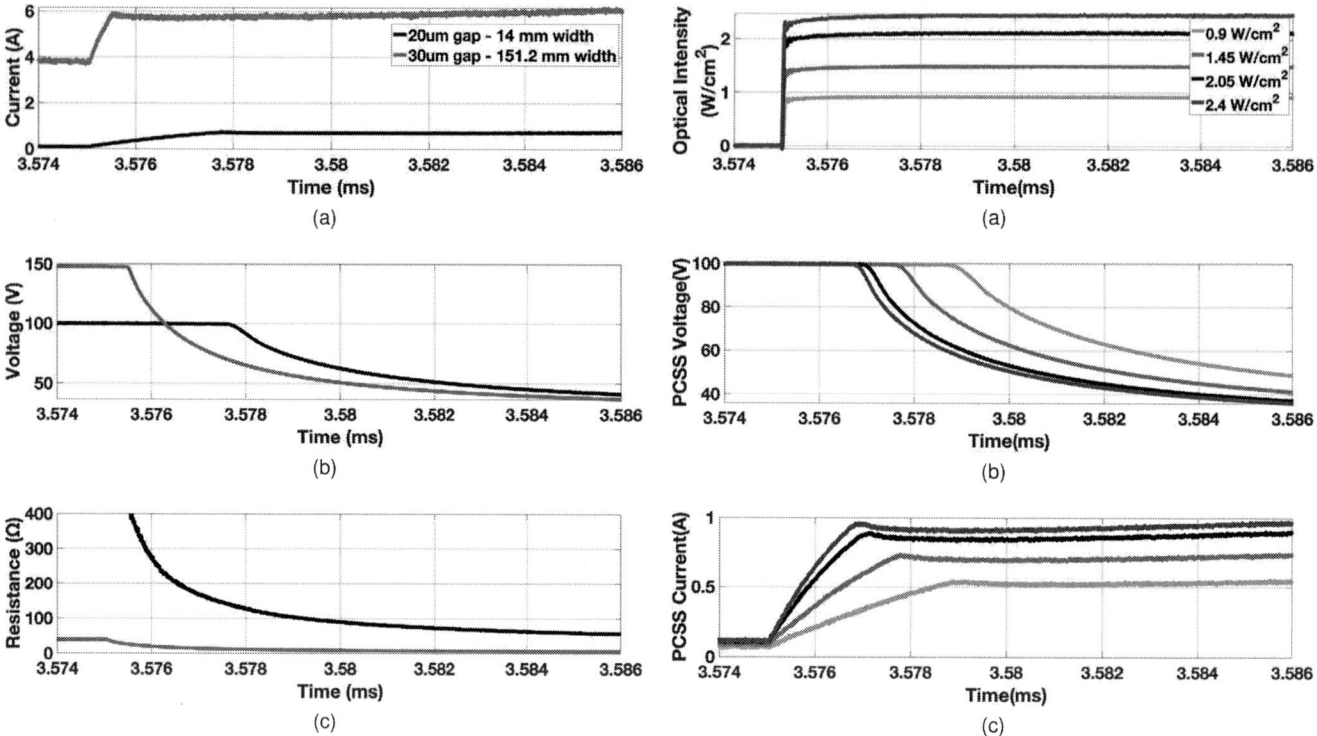

Fig. 9. Measured turn-on switching waveforms comparison of: 14 mm width vs. 151.2 mm width PCSS devices (a) Current. (b) Voltage. (c) Resistance.

Fig. 10. Turn-on waveforms of a 20 μm gap and 14 mm PCSS with varied optical intensity. (a) Optical intensity. (b) PCSS Voltage. (c) PCSS Current.

turn-on is observed to occur within $\sim 2.0\,\mu s$ at an initial optical intensity of $\sim 1.4\,\text{W/cm}^2$, as indicated by the PCSS current waveform. However, the voltage response lags behind the current and takes longer, with a fall time $> 6.0\,\mu s$. The analysis of optical intensity variation revealed a strong dependence of the turn-on rate on optical intensity: higher optical intensity resulted in a faster turn-on, while lower optical intensity significantly slowed the process. This occurred despite the LED's optical power transient during turn-on showing no significant dependence on intensity.

The turn-on phase of the PCSS is driven by both the band-to-band excitation from the above-bandgap LED spectrum and electron excitation from the carbon trap levels to the conduction band from the sub-bandgap LED spectrum, which is directly influenced by the photon flux generated by the LED. The optical electron emission rate from an occupied acceptor level into the conduction band (g_{nt}^{opt}) is described as [12]:

$$g_{nt}^{opt} = \int_{\lambda_g}^{\lambda_{n,\max}} \sigma_n^{\text{opt}}(x,\lambda)\phi_{\text{ph}}(x,\lambda)\,d\lambda, \qquad (2)$$

where $\phi_{\text{ph}}(x,\lambda)$ represents the photon flux, and $\sigma_n^{\text{opt}}(x,\lambda)$ is the optical emission cross-section for electrons. The integral accounts for all photons within the energy range capable of interacting with the trap levels to facilitate carrier excitation. As the optical intensity increases, the photon flux $\phi_{\text{ph}}(x,\lambda)$ also increases, enhancing g_{nt}^{opt}. This leads to a faster emission rate, thereby accelerating the turn-on process.

These results highlight the critical role of optical intensity in driving the turn-on dynamics of the PCSS. Achieving faster switching requires higher photon flux, emphasizing the importance of exploring higher power UV LEDs, high power laser diodes, and improving the optical focusing setup. Developing LEDs capable of delivering greater optical power could significantly enhance the performance of PCSS devices in high-speed applications.

V. Conclusion

This study investigated the dynamic behavior of carbon-doped GaN photoconductive semiconductor switches (PCSS) using a Cascaded Double Pulse Test (C-DPT) setup. The C-DPT provided a simple and cost-effective method to evaluate the turn-on and turn-off dynamics of PCSS devices driven by COTS LEDs. Carbon doping was shown to be effective in achieving high resistivity and reducing DC leakage currents, making these devices well-suited for high-voltage applications. The turn-on process was found to be driven by the optical intensity delivered to the PCSS, with higher photon flux leading to faster switching. However, the study also revealed significant limitations associated with deep-level carbon traps, which extend the turn-off process due to their impact on carrier capture dynamics. These traps hinder the device's ability to achieve rapid switching, posing a challenge for high-speed switching applications.

The C-DPT setup proved advantageous for its ability to characterize the dynamic performance of both the PCSS and its optical driver. This methodology is particularly valuable for advancing the development of photoconductive devices by enabling precise comparisons across varying optical intensity levels and device configurations.

To address the challenges identified in this work, future research should explore alternative PCSS architectures, such as vertical structures, which could offer improved electrical performance. Additionally, investigating alternative impurities, such as iron, may maintain the high resistivity achieved with carbon while mitigating the effects of deep-level trapping. Reducing turn-off delays through impurity engineering would significantly enhance the suitability of GaN PCSS devices for high-speed applications. Additionally, advancements in LED technology are critical. High-power UV LEDs capable of delivering greater optical power could drive faster turn-on rates and improve overall device performance, making GaN PCSS devices more competitive in demanding power electronics environments.

By addressing these material and optical challenges, the performance of GaN PCSS devices can be further optimized to meet the needs of high-speed and high-voltage power electronics. The combination of material advancements and innovative test methodologies like the C-DPT can accelerate progress toward scalable and efficient solutions for next-generation power systems.

ACKNOWLEDGMENT

The authors gratefully acknowledge the support of the U.S. Office of Naval Research (grant number N00014-21-1-2832 and approved under DCN # 543-2214-24), at the direction of Captain LJ Petersen, USN (Ret). Distribution Statement A. Approved for Public Release. Distribution unlimited.

REFERENCES

[1] E. A. Jones, F. F. Wang, and D. Costinett, "Review of commercial gan power devices and gan-based converter design challenges," *IEEE journal of emerging and selected topics in power electronics*, vol. 4, no. 3, pp. 707–719, 2016.

[2] J. Liu, M. Xiao, Y. Zhang, S. Pidaparthi, H. Cui, A. Edwards, L. Baubutr, W. Meier, C. Coles, and C. Drowley, "1.2 kv vertical gan fin jfets with robust avalanche and fast switching capabilities," in *2020 IEEE International Electron Devices Meeting (IEDM)*. IEEE, 2020, pp. 23–2.

[3] C. Langpoklakpam, A.-C. Liu, Y.-K. Hsiao, C.-H. Lin, and H.-C. Kuo, "Vertical gan mosfet power devices," *Micromachines*, vol. 14, no. 10, p. 1937, 2023.

[4] B. J. Baliga, *Fundamentals of power semiconductor devices*. Springer Science & Business Media, 2010.

[5] E. Majda-Zdancewicz, M. Suproniuk, M. Pawłowski, and M. Wierzbowski, "Current state of photoconductive semiconductor switch engineering," *Opto-Electronics Review*, vol. 26, no. 2, pp. 92–102, 2018.

[6] N. Perera, K. Ledins, S. W. Fung, L. Efthymiou, K. Mukherjee, J. Findlay, and P. Comiskey, "Evaluation of gan hemt dv/dt immunity and dv/dt induced false turn-on energy loss," in *2024 IEEE Applied Power Electronics Conference and Exposition (APEC)*. IEEE, 2024, pp. 729–736.

[7] W. Li, K. Nomoto, K. Lee, S. Islam, Z. Hu, M. Zhu, X. Gao, M. Pilla, D. Jena, and H. G. Xing, "Development of gan vertical trench-mosfet with mbe regrown channel," *IEEE Transactions on Electron Devices*, vol. 65, no. 6, pp. 2558–2564, 2018.

[8] J. Leach, R. Metzger, E. Preble, and K. Evans, "High voltage bulk gan-based photoconductive switches for pulsed power applications," in *Gallium Nitride Materials and Devices VIII*, vol. 8625. SPIE, 2013, pp. 294–300.

[9] D. Han, S. Li, W. Lee, W. Choi, and B. Sarlioglu, "Trade-off between switching loss and common mode emi generation of gan devices-analysis and solution," in *2017 IEEE Applied Power Electronics Conference and Exposition (APEC)*. IEEE, 2017, pp. 843–847.

[10] A. D. Koehler, T. J. Anderson, A. Khachatrian, A. Nath, M. J. Tadjer, S. P. Buchner, K. D. Hobart, and F. J. Kub, "High voltage gan lateral photoconductive semiconductor switches," *ECS Journal of Solid State Science and Technology*, vol. 6, no. 11, p. S3099, 2017.

[11] G. M. Loubriel, F. J. Zutavern, A. G. Baca, H. Hjalmarson, T. A. Plut, W. D. Helgeson, M. W. O'Malley, M. H. Ruebush, and D. J. Brown, "Photoconductive semiconductor switches," *IEEE Transactions on Plasma Science*, vol. 25, no. 2, pp. 124–130, 1997.

[12] J. Yuan, H. Shen, F. Zhong, and X. Deng, "Impurity photovoltaic effect in magnesium-doped silicon solar cells with two energy levels," *physica status solidi (a)*, vol. 209, no. 5, pp. 1002–1006, 2012.

[13] J. Verschraegen, S. Khelifi, M. Burgelman, and A. Belghachi, "Numerical modeling of the impurity photovoltaic effect (ipv) in scaps," in *21st European Photovoltaic Solar Energy Conference*, vol. 396. Citeseer, 2006.

[14] T. J. Anderson, A. D. Koehler, L. E. Luna, J. C. Gallagher, J. K. Hite, M. A. Mastro, A. G. Jacobs, B. Feigelson, K. D. Hobart, and F. J. Kub, "Process development for gan-based photoconductive semiconductor switches (pcss)," in *Electrochemical Society Meeting Abstracts aimes2018*, no. 34. The Electrochemical Society, Inc., 2018, pp. 1155–1155.

[15] B. Heshmat, H. Pahlevaninezhad, Y. Pang, M. Masnadi-Shirazi, R. Burton Lewis, T. Tiedje, R. Gordon, and T. E. Darcie, "Nanoplasmonic terahertz photoconductive switch on gaas," *Nano letters*, vol. 12, no. 12, pp. 6255–6259, 2012.

[16] Tektronix, "Keithley 4200A-SCS Parameter Analyzer," Accessed on: Sept. 30, 2024. [Online]. Available: https://www.tek.com/en/products/keithley/4200a-scs-parameter-analyzer

[17] G. M. Foster *et al.*, "Characterization of optically modulated semi-insulating gan photoconductive semiconductor switches," in *Proc. CS-MANTECH 2023*, May 2023.

[18] Spellman, "Bertan 225 High Voltage Power Supply," Accessed on: Oct. 30, 2024. [Online]. Available: https://www.spellmanhv.com/-/media/en/Products/225.pdf

[19] Inolux, "IN-C68QACTMU2," [Online], 2024, Accessed: Aug. 7, 2024. [Online]. Available: https://www.inolux-corp.com/details.php?i=256

[20] Efficient Power Conversion, "EPC8004," [Online], 2024, Accessed: Aug. 7, 2024. [Online]. Available: https://epc-co.com/epc/products/gan-fets-and-ics/epc8004

[21] Violumas, "VS7272C45L9-365," [Online], 2024, Accessed: Oct. 7, 2024. [Online]. Available: https://violumas.com/products/high-power-7272-series/

[22] Thor Labs, "ASL2520-UV," [Online], 2024, Accessed: Aug. 7, 2024. [Online]. Available: https://www.thorlabs.com/thorproduct.cfm?partnumber=ASL2520-UV

[23] ——, "SM05PD3A," [Online], 2024, Accessed: Aug. 7, 2024. [Online]. Available: https://www.thorlabs.com/thorproduct.cfm?partnumber=SM05PD3A

[24] STMicroelectronics, "STPSC20H12G2-TR," Accessed on: Nov. 06, 2024. [Online]. Available: https://www.st.com/en/diodes-and-rectifiers/stpsc20h12g2-tr.html

[25] S. M. Sze, Y. Li, and K. K. Ng, *Physics of semiconductor devices*. John wiley & sons, 2021.

[26] V. Meyers, D. Mauch, J. Mankowski, J. Dickens, and A. Neuber, "Characterization of the optical properties of gan: Fe for high voltage photoconductive switch applications," in *2015 IEEE Pulsed Power Conference (PPC)*. IEEE, 2015, pp. 1–4.

[27] Y. Chen, H. Lu, D. Chen, F. Ren, R. Zhang, and Y. Zheng, "High-voltage photoconductive semiconductor switches fabricated on semi-insulating hvpe gan: Fe template," *physica status solidi (c)*, vol. 13, no. 5-6, pp. 374–377, 2016.

[28] Z. Huang, D. Mott, P. Shu, R. Zhang, J. Chen, and D. Wickenden, "Optical quenching of photoconductivity in gan photoconductors," *Journal of applied physics*, vol. 82, no. 5, pp. 2707–2709, 1997.

[29] T.-Y. Lin, H.-C. Yang, and Y. Chen, "Optical quenching of the photoconductivity in n-type gan," *Journal of Applied Physics*, vol. 87, no. 7, pp. 3404–3408, 2000.

Development and Validation of Repetitive Transient Gate Overvoltage Rating for GaN HEMTs

Ricardo Garcia
Efficient Power Conversion
El Segundo, USA
Ricardo.Garcia@epc-co.com

Angel Espinoza
Efficient Power Conversion
El Segundo, USA
Angel.Espinoza@epc-co.com

Siddhesh Gajare
Efficient Power Conversion
El Segundo, USA
Siddhesh.Gajare@epc-co.com

Shengke Zhang
Efficient Power Conversion
El Segundo, USA
Shengke.Zhang@epc-co.com

Abstract—In this work, time-dependent accelerated gate reliability testing is conducted on GaN high-electron-mobility-transistors (HEMTs), featuring Schottky type pGaN gate. By understanding the underlying failure mechanism, a physics-based lifetime model is developed to project gate lifetime. A 1% duty cycle-based repetitive transient gate overvoltage rating is developed by scaling the projected lifetime at 7 V_{GS} continuous bias to the 10-year expected mission lifespan. To validate the proposed 1% duty cycle factor (DC_{Factor}), an inductive switching circuit is implemented to emulate the transient gate overvoltage ringing that is commonly observed in GaN-based applications. A suite of GaN HEMTs were tested to over one trillion overvoltage pulses with a peak V_{GS} of 7 V at 25°C and 125°C, where neither catastrophic device failure nor significant parameter shifts were found. Not only did the test results confirm the validity of the 1% duty cycle-based repetitive transient overvoltage rating, but it also demonstrated the excellent gate overvoltage robustness of GaN HEMTs.

Keywords—GaN, Repetitive Transient Overvoltage, Reliability Robustness, Duty Cycle, Lifetime, high electron mobility transistor (HEMTs).

I. Introduction

Gallium nitride (GaN) high-electron-mobility transistors (HEMTs) have attracted significant attention from the power electronics industry owing to the superior power conversion efficiency. Therefore, GaN HEMTs have been utilized in increasingly more advanced applications such as DC-DC converters in datacenters for artificial intelligence (AI) applications, light detection and ranging (LiDAR) for cars, and motor drives for humanoid robots, drones, and power tools [1-3]. Schottky-type pGaN gate is one of the most popular gate implementations among commercially available enhancement-mode GaN HEMTs.

Understanding the fundamental failure mechanism responsible for the pGaN gate breakdown has been the focus of multiple accelerated gate reliability studies [3-10], where the applied gate bias significantly exceeded the maximum rated gate voltage by datasheet specification. However, previous studies did not offer a tangible guideline to translate the projected lifetime based on the accelerated gate wearout study to a repetitive transient gate overvoltage stress condition that is more applicable to real-world switching applications. GaN-based converters typically feature high switching frequency and high slew rates, which is more likely to lead to repetitive gate overvoltage ringing during the turn-on transients. Therefore, it

is highly desired to specify a cycle-by-cycle transient gate rating that can address the gate overvoltage ringing concerns frequently raised by the users.

In this work, a duty cycle-based repetitive transient gate overvoltage rating of 7 V gate-source bias (V_{GS}) is first developed by comparing the projected lifetime under continuous DC bias against the 10-year mission lifespan. To validate the proposed repetitive gate overvoltage specification, an inductive switching test circuit is implemented to stress the gate of GaN HEMTs with ~3 MHz high switching frequency. The resulting resonance-like transient gate voltage waveform produced by the test setup is representative of the gate overvoltage ringing that is commonly observed during the turn-on transients in high-speed switching applications. Next, a wide range of GaN HEMTs were tested by this inductive switching test circuit to over one trillion pulses at room temperature (25°C) and high temperature (125°C), where the gate voltage waveforms were monitored continuously. Static electrical characteristics measurements of the GaN HEMTs were carried out before and after the one trillion pulses stress and showed insignificant parametric shifts, providing the validity of the repetitive transient gate overvoltage specification proposed.

II. Lifetime Model Development

A. Developing Lifetime Model

Many researchers, including the authors, have established that under accelerated forward V_{GS}, hot electron bombardment at the TiN gate metal/pGaN interface causes impact ionization and electron-hole multiplication process, which leads to the eventual dielectric breakdown between the gate and the source metal field plate [2-10]. By understanding the underlying gate wearout mechanism, a physics-based lifetime model is developed, as shown in (1):

$$MTTF \propto exp[(b/V_{GS})^m] \qquad (1)$$

where b is a temperature dependent coefficient. In this work, b is treated as a constant because the voltage-dependent gate reliability testing and derivation was performed at room temperature, 25°C. m is an exponent that is typically ranging from 1 to 2 [11-14]. m=1.6 is used in this work.

Accelerated time-dependent gate reliability testing was conducted on EPC2212 [19] at four different static DC V_{GS} of 8 V, 8.5 V, 9 V and 9.5 V, well exceeding the maximum gate-

979-8-3315-1612-3/25 $31.00 © 2025 IEEE

source voltage rating of 6 V. Fig. 1 shows that the lifetime model provides a good fit to the measured mean-time-to-fail (MTTF), showing the applicability and validity of the gate lifetime model. When projecting the MTTF fit to 100 ppm (part per million) failure rate, the predicted lifetime at 7 V_{GS} is estimated to be 3.3 x 10^6 seconds. Comparing the lifetime at 7 V to a typical mission lifespan of 10 years (3.1 x 10^8 seconds) yields approximately 1%. The 1% total lifetime can then be translated to a duty cycle factor (DC_{Factor}) in every switching cycle as shown in Fig. 2. Fig. 2 illustrates a simplified switching gate waveform, where T_S represents the switching period, which is the inverse of the switching frequency, and T_O is the duration of the transient overvoltage ringing. Thus, the DC_{Factor} can be established as the ratio between T_O and T_S, which suggests that GaN HEMTs shall be able to sustain a repetitive 7 V_{GS} overvoltage spike with a duration of 1% of each switching period with a low failure rate.

Here is an example to demonstrate how to use the 1% DC_{Factor} specification to evaluate the reliability risk in a design. If a converter operates at 1 MHz switching frequency (T_S=1 µs), a repetitive overvoltage spike is found in the gate turn-on transients due to unoptimized gate loop inductance, where the spike has a peak V_{GS} of 7 V with a time interval of 8 ns above 6 V_{GS}. Dividing 8 ns with 1 µs of T_S yields 0.8%, which is less than 1% DC_{Factor}. Therefore, a much lower than 100 ppm failure rate after 10 years of continuous operation is expected.

III. VALIDATION

A. Experiment Setup

In real-world switching applications, the transient overvoltage ringing waveform typically shows a sinusoidal shape with resonance-like characteristics. To validate the 1% DC_{Factor} based overvoltage specification, an inductive switching test system was developed to emulate an overvoltage ringing phenomenon. A previous study published [24] by the authors has shown that gate breakdown failures caused by the dynamic switching circuit shared identical electrical and physical failure characteristics as the DC static stress. In addition, the Weibull slope parameter from dynamic switching testing with a peak V_{GS} of 10 V was found consistent with the static gate stress with a continuous V_{GS} of 10 V. All evidence suggests that the dynamic

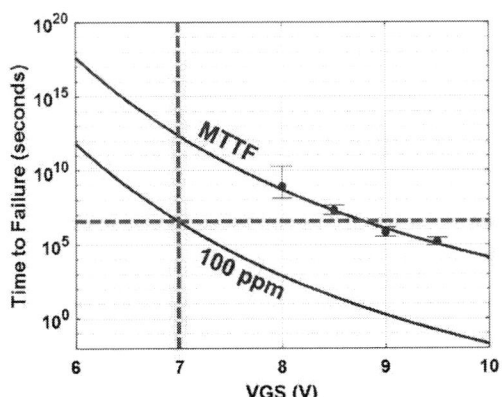

Fig. 1.   Time to failure vs. gate bias at 25°C of EPC2212 [19] and the solid lines indicate the impact ionization lifetime model.

gate overvoltage switching testing circuit is stressing the same failure mechanism as the DC static stress.

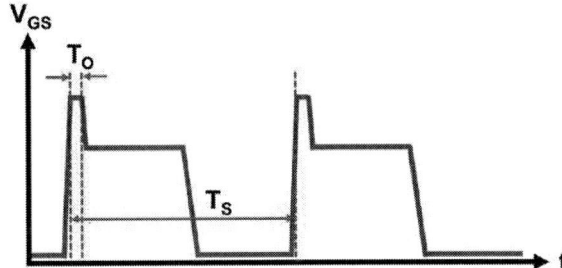

Fig. 2.   Illustration of the 1% DCFactor overvoltage specification, which is defined by the ratio between T_O (overvoltage duration) and T_S (switching period)

Fig. 3 (a) shows the testing block diagram circuit. Fig. 3 (c) shows a picture of the testing board setup comprised of five main components, a DC source V_{IN}, the inductor (L), gate driver for FET Q, and device under test (DUT). The DUTs are assembled onto test coupons with an edge connector design, which is also compatible with the pre-stress and post-stress static electrical characteristics measurement. L is used to emulate the gate loop parasitic inductance typically found in application. FET Q services as a controllable charging path for the inductor, which is driven by the gate driver. The general working principles of the inductive gate testing circuit are consistent with the unclamped inductive switching circuit [4, 17, 18]. The circuit operation can be split into two main phases, phase 1 being the charging phase and phase 2 being the transient phase.

During Phase 1 (the charging phase), FET Q in Fig. 3 (b) is turned ON, during which the input voltage (V_{IN}) charges the inductor L and the inductor current begins to rise, as demonstrated in Phase 1 of Fig. 3 (b). The inductor current (I) can be represented as V_{IN} times the ON time (t) of FET Q divided by L, which is denoted by (2)

$$I = V_{IN} \times t / L \qquad (2)$$

where t is the FET Q turn-on time to charge the inductor, which is denoted by the turn-on duty cycle (3)

$$DC_Q = t/T_S \qquad (3)$$

where T_S is the switching period that is the inverse of switching frequency. Switching frequency of 2.6 – 3 MHz was used for the inductive switching testing. DC_Q was adjusted individually to ensure a consistent time interval of ~70 ns for the V_{GS} greater than 6 V.

It's noted that during phase 1 the V_{GS} that is applied on the DUT is equivalent to the voltage drop across the FET Q, which is approximately 0 V. In Phase 2 (the transient phase), the FET Q turns off, and a half-sinusoidal transient gate voltage waveform is generated due to the resonance between L and the parasitic capacitances, as shown in Phase 2 of Fig. 3 (b). The primary resonant components are the inductor, L and the total loop capacitance, which can be modeled by the output capacitance (C_{OSS}) of FET Q and the input capacitance (C_{ISS}) of the DUT in parallel connection. To test the DUTs at higher temperatures, a resistive heating system is developed, featuring a proportional-integral-derivative (PID) control system, as

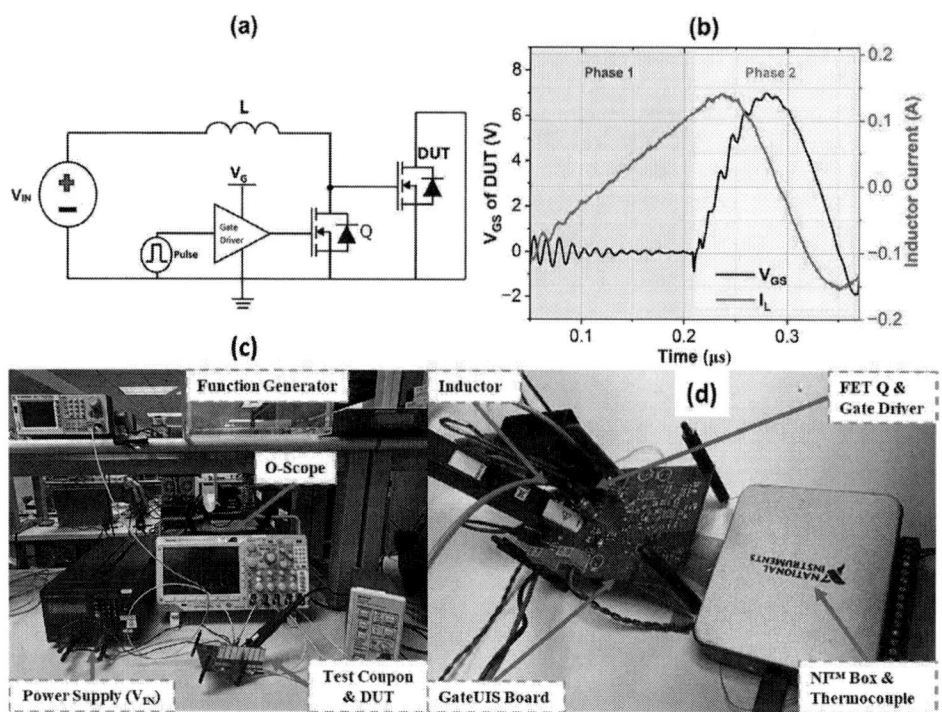

Fig. 3. (a) a simplified schematic of the inductive gate switching test system; (b) the measured inductor current and VGS waveforms with a peak voltage of 7 V; (c) picture of the test setup; (d) picture of the temperature forcing setup

shown in Fig. 3 (d). A thermo-couple is first directly attached to the backside of the DUT to monitor the DUT's case temperature continuously. Next, a TO-247 power resistor is mounted on top of the thermo-couple through a thermal pad on the backside of the DUT to accurately control the device junction temperature by leveraging the low junction-to-case thermal resistance of EPC's GaN HEMTs, where the thermo-couple provides a feedback loop to the controlling power resistor. The heating system is controlled by a NI 6001 device [20] in conjunction with a LabVIEW [21] program.

B. Results and Discussion

Four different GaN HEMT products and three parts per product with a drain-source (V_{DS}) rating from 50 V to 200 V were tested with a peak V_{GS} of 7 V to a trillion pulses at 25°C ambient temperature. All devices selected have gate-source voltage rating from -4 to 6 V. Fig. 4 shows the evolution of threshold voltage (V_{th}) and on-state resistance ($R_{DS(ON)}$) of a representative device from each product. Device characterization was conducted prior to testing and after reaching a trillion cycles. As shown in Fig. 4, the post-stress measurements are well below the datasheet limit of each product. The post-stress parametric values were normalized to the pre-test measurements to quantify the parametric shift, where +/-8% was found for V_{th} and +/-5% for $R_{DS(ON)}$. Therefore, the parametric shift observed in Fig. 4 is likely the result of measurement variation and should not be considered as significant parametric shift caused by the gate overvoltage stress.

The dynamic inductive switching test circuit shown in Fig. 3 (a) is a zero-voltage switching (ZVS) setup, where the drain and source terminals are tied together to ground during the experiment. Cheng et al [10] reported longer gate lifetime under a resistive hard switching test circuit with a drain bias than that measured without a drain bias. In addition, a longer gate lifetime was found with an increase of drain bias when subjected to resistive hard switching test. This is explained by TCAD simulation results, which show that the off-state drain bias uplifts the conduction band of the AlGaN barrier layer, reducing the electron injection rate and therefore slowing down the gate wearout. Therefore, the projected gate lifetime based on the ZVS and the DC testing method should be applicable to the hard switching stress conditions.

In addition, 125°C repetitive gate overvoltage testing was performed on two representative products of a 50 V_{DS}-rated (EPC2057 [22]) and one 200 V_{DS}-rated (EPC2307 [23]) for another trillion pulses with a peak V_{GS} of 7 V. Fig. 5 shows insignificant shift in the static parameter measurements after an additional trillion cycles of stress at 125°C, indicating temperature is not a significant acceleration factor. This finding is also consistent with the results reported by other literature [4-6], in which the gate lifetime was found to be improving at higher temperatures. Thus, the test results under repetitive transient gate overvoltage stress suggest that the Schottky-type pGaN gate implemented in EPC's GaN HEMTs products are capable of 7 V transient overvoltage rating, demonstrating the robustness of the gates of GaN HEMTs.

A total of 15 GaN DUTs from four different products were subjected to over 1 trillion 7 V_{GS} spikes, which correlates to a total of 15 trillions pulses. Since every device was tested with a consistent time interval of ~70 ns with V_{GS}>6 V, the total stress time is calculated to be approximately 1.1 x 10^6 seconds, which

979-8-3315-1612-3/25 $31.00 © 2025 IEEE   216

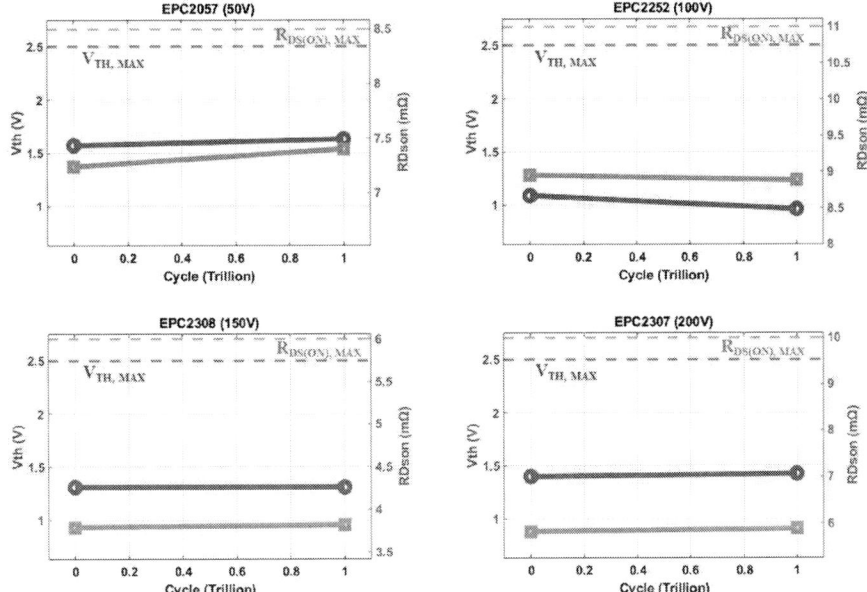

Fig. 4. Parametric comparison of pre- and post- stress (1-trillion pulses with peak voltage of 7 V) of four representative GaN HEMT products.

is about one third of the projected gate lifetime at 7 V static gate bias. At the end of the 15 trillion pulses, no observable parameter shift was measured, suggesting there is still significant margin in lifetime before the GaN DUTs would show any measurable parametric degradation.

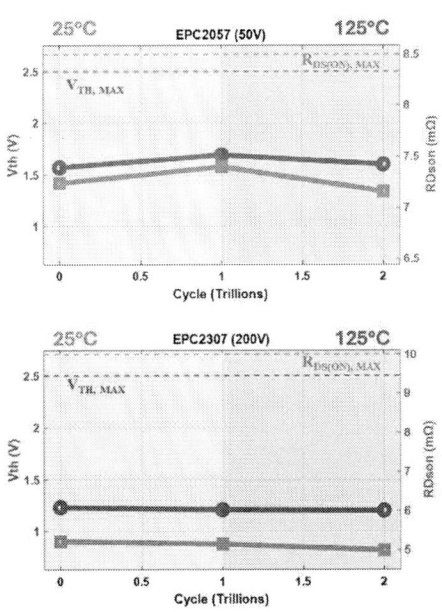

Fig. 5. Parametric comparison of pre- and post-stress (2-trillion pulses with peak voltage of 7 V at 25°C (blue shaded) and 125°C (orange shaded) forced heating) of two representative GaN HEMT products

IV. CONCLUSION

This work proposed a duty cycle-based repetitive transient gate overvoltage rating at 7 V_{GS} that is 1 V higher than 6 $V_{GS,Max}$ of GaN HEMTs, providing significantly greater margin (~2 V) from the recommended gate operating voltage. This specification is validated by testing over 15 trillion cycles of transient gate overvoltage testing with a peak voltage of 7 V at 25°C and 125°C across a broad range of GaN HEMT products, showing insignificant parametric shift. This work demonstrates the validity and applicability of the proposed 1% duty cycle based transient gate overvoltage rating also showing the excellent long term gate reliability and robustness.

REFERENCES

[1] A. Lidow et al, GaN Transistors for Efficient Power Conversion. USA: John Wiley & Sons, 2019.

[2] A. Lidow, GaN Power Devices and Applications. CA, USA: Power Conversion Publications, 2022.

[3] S. Gajare et al, "GaN reliability and lifetime projections: phase 16," EPC, El Segundo, CA, USA, 2024. A. Jones, B. Smith, and C. Maxwell, "Title of their journal paper," IEEE Transactions on Power Electronics, vol. 17, no. 1, pp. 45-55, Jan. 1995.

[4] B. Wang et al., "Gate Robustness and Reliability of P-Gate GaN HEMT Evaluated by a Circuit Method," IEEE Transactions on Power Electronics, vol. 39, no. 5, pp. 5576-5589, May 2024.

[5] B. Wang et al, "Dynamic Gate Breakdown of p-Gate GaN HEMTs in Inductive Power Switching," IEEE Electron Device Lett., vol. 44, no. 2, pp. 217-220, Feb. 2023.

[6] J. He et al, "Frequency- and Temperature-Dependent Gate Reliability of Schottky-Type p-GaN layer HEMTs," IEEE Trans. Electron Devices, vol. 66, no. 8, pp. 3453-3458, Aug. 2019.

[7] A. N. Tallarco et al, "Investigation of the pGaN Gate Breakdown in Forward-biased GaN-based Power HEMTs," IEEE Electron Device Lett., vol. 38, no. 1, pp. 99–102, Jan. 2017.

[8] M. Tapajna et al, "Gate reliability investigation in normally-off p-type-GaN cap/AlGaN/GaN HEMTs under forward bias stress," IEEE Electron Device Lett., vol. 37, no. 4, pp. 385–388, Apr. 2016.

[9] T. Wu et al, "Forward bias gate breakdown mechanism in enhancement-mode pGaN gate AlGaN/GaN high-electron mobility transistors," IEEE Electron Device Lett., vol. 36, no. 10, pp. 1001–1003, Aug. 2015.

[10] Y. Cheng et al, "Gate Reliability of Schottky-Type p-GaN Gate HEMTs Under AC Positive Gate Bias Stress with a Switching Drain Bias," IEEE Electron Device Lett, vol. 43, no. 9, pp. 1404-1407, Sept. 2022.

[11] T. L. Ooi, P. L. Cheang, A. H. You and Y. K. Chan, "Mean multiplication gain and excess noise factor of GaN and Al0.45Ga0.55N avalanche photodiodes," Eur. Phys. J. Appl. Phys., vol. 92, no. 1, Oct. 2020, Art. no. 10301.

[12] L. Cao et al, "Temperature Dependence of Electron and Hole Impact Ionization Coefficients in GaN," IEEE Trans. Electron Devices, vol. 68, no. 3, pp. 1228-1234, March 2021.

[13] L Cao et al, "Experimental characterization of impact ionization coefficients for electrons and holes in GaN grown on bulk GaN substrates," Appl. Phys. Lett., vol. 112, no. 26, Jun. 2018.

[14] Y. Okuto and C. R. Crowell, "Threshold Energy Effect on Avalanche Breakdown Voltage in Semiconductor Junctions," Solid-State Electron., vol. 18, no. 2, pp. 161-168, Feb. 1975.].

[15] M. Young, The Technical Writer's Handbook. Mill Valley, CA: University Science, 1989.

[16] T.-S. Yeoh, R. S. Nair, and S.-J. Hu, "Mos transistor gate oxide breakdown stress dependence and their related models," in Proc. 5th Int. Symp. Phys. Failure Anal. Integr. Circuits, Nov./Dec. 1995, pp. 127–131.

[17] J. Liu et al., "1.2-kV Vertical GaN Fin-JFETs: High-Temperature Characteristics and Avalanche Capability," IEEE Trans. Electron Devices, vol. 68, no. 4, pp. 2025–2032, Apr. 2021

[18] Q. Song, R. Zhang, J. P. Kozak, J. Liu, Q. Li, and Y. Zhang, "Robustness of Cascode GaN HEMTs in Unclamped Inductive Switching," IEEE Trans. Power Electron., vol. 37, no. 4, pp. 4148–4160, Apr. 2022

[19] Efficient Power Conversion Corporation, "EPC2212 – Enhancement-mode power transistor," EPC2204 datasheet. [Online]. Available: https://epc-co.com/epc/Portals/0/epc/documents/datasheets/EPC2212_datasheet.pdf

[20] National Instruments 6001 Box Specifications [Online]. Available: https://www.ni.com/docs/en-US/bundle/usb-6001-specs/resource/374369a.pdf

[21] National Instruments LabView Program Specifications [Online]. Available: https://www.ni.com

[22] Efficient Power Conversion Corporation, "EPC2057 – Enhancement-mode power transistor," EPC2057 datasheet. [Online]. Available: https://epc-co.com/epc/Portals/0/epc/documents/datasheets/EPC2057_datasheet.pdf

[23] Efficient Power Conversion Corporation, "EPC2307 – Enhancement-mode power transistor," EPC2307 datasheet. [Online]. Available: https://epc-co.com/epc/Portals/0/epc/documents/datasheets/EPC2307_datasheet.pd

[24] A. Espinoza, R. Garcia, S. Gajare, and S. Zhang, " Common Wearout Mechanism of pGaN Gate in GaN HEMTs under DC and Inductive Switching Test Methods," 2021 IEEE 1th Workshop on Wide Bandgap Power Devices & Applications (WiPDA), Dayton, Ohio, USA, 2024

Junction Temperature Monitoring of GaN HEMT by Using On-Resistance with Voltage Clamp and Current Shunt

Xiao Wang	Mingrui Zou	Jiakun Gong	Yulei Wang	Zheng Zeng
Chongqing University	Chongqing University	Chongqing University	Chongqing University	Chongqing University
Chongqing, China	Chongqing, China	Chongqing, China	Chongqing, China	Chongqing, China
20206100@cqu.edu.cn	zoumingrui@cqu.edu.cn	gonjiakun@cqu.edu.cn	yulei_wang@cqu.edu.cn	zengerzheng@cqu.edu.cn

Abstract—**Due to the unique device cell structure and power packaging of the GaN HEMT device, the online junction temperature monitoring approach of the GaN HEMT is a remaining challenge. In this paper, based on the temperature-sensitive electrical parameter on-resistance, a novel on-state resistance measurement circuit is proposed to observe the online junction temperature of the GaN HEMT. To accurately measure the on-state voltage, the proposed clamp circuit is carefully designed to remove the bias voltage of the clamp diode. Besides, the high bandwidth shunt is utilized to detect the drain current, which enables the online condition monitoring of the on-resistance. Furthermore, to ensure the validation of the proposed clamp circuit, comprehensive experiments are presented to demonstrate the junction temperature monitoring capability. The proposed circuit can be easily integrated into the gate driver or the power packaging of the GaN HEMT.**

Keywords—***GaN HEMT, junction temperature monitoring, on-resistance, voltage clamp, current shunt***

I. INTRODUCTION

Due to the higher switching frequency, lower on-resistance, reduced switching loss, and affordable device cost, the GaN HEMT is considered as a promising power device for industrial applications, compared to the Si counterpart [1]. Meanwhile, due to the deteriorated reliability and durability of the GaN HEMT, the online junction temperature monitoring of the GaN HEMT is a key challenge for thermal management and reliability enhancement [2]. However, limited by the compact power packaging and unique device specification, the online and accurate junction temperature monitoring is the remaining challenge for the GaN HEMT.

Some research has focused on the temperature-sensitive electrical parameter (TSEP) of the GaN HEMT for junction temperature monitoring, owing to low invasiveness of the approach. For instance, the short-circuit current, influenced by the junction temperature, is a specific TSEP [3]. Nevertheless, due to the short-circuit capability of the GaN HEMT, it may result in damages and fails for the online junction temperature monitoring. Moreover, although some TSEPs, like threshold voltage, are safe solutions [4], the shift and inaccuracy issues

make them unsuitable for the GaN HEMT application. Besides, regarding the turn-on delay for junction temperature monitoring [5], the ultra-high switching speed of the GaN HEMT may lead to difficulties in measurement routine and circuit design. In general, due to the uncertainty of the operation scheme and electric circuit, the available TSEPs are hard to be effectively integrated and online employed.

To fulfill the research gap, a novel on-state resistance measurement circuit for online junction temperature monitoring of the GaN HEMT is proposed in this paper. In Section II, the electric circuit to measure the on-resistance of GaN HEMT is introduced, along with an explanation of the detection circuit scheme for on-state voltage and drain current. In Section III, the test rig is exhibited to calibrate the map between on-resistance and junction temperature, and the experimental steps are explained. Besides, the preliminary relationship between on-resistance and junction temperature is obtained to verify the feasibility of the proposed method. In addition, the functionality and error of the proposed circuit will be further validated through online junction temperature monitoring.

II. PRINCIPLE AND DESIGN OF JUNCTION TEMPERATURE MONITORING CIRCUIT FOR GAN HEMT

A. Basic Operation Principle of Junction Temperature Monitoring Based on On-Resistance TSEP

The on-resistance R_{dson} is proven to be correlated with the junction temperature T_j of the GaN HEMT, which can be calculated by using the on-state voltage drop V_{dson} and the drain current I_d [6], as presented in Fig. 1.

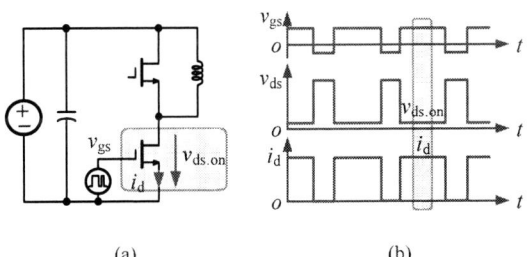

(a)   (b)

Graduate Research and Innovation Foundation of Chongqing, China (Grant No. CYB240023).

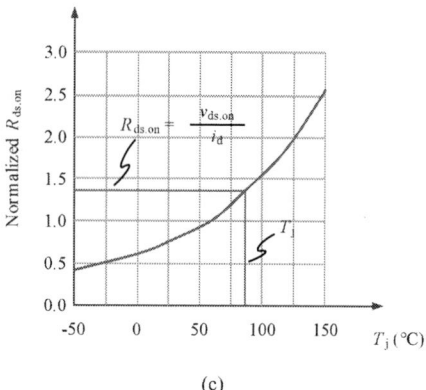

(c)

Fig. 1. Configuration of temperature-sensitive on-resistance of GaN HEMT. (a) Schematic circuit, (b) switching waveform, and (c) measured map between junction temperature and on-resistance.

As a result, the junction temperature of the GaN HEMT can be monitored online based on the measured map between the T_j and R_{dson}.

B. Proposed Circuit for Junction Temperature Monitoring Based on On-Resistance TSEP

1) Clamp Circuit for On-State Voltage Measurement

In order to accurately measure the low voltage drop in the on-state, and to withstand the high blocking voltage in the off-state, the clamp circuit is critical to measure the on-state voltage. However, the available clamp circuits are presented and compared, as shown in TABLE I and Fig. 2. It is found that integrability and general accuracy remain the primary challenges of clamp circuits. Regarding general accuracy, the voltage drift suppression of clamp diode in some circuits is limited, impeding their ability to measure clamping voltage accurately under general conditions. Different operating currents and temperatures can cause variations in the forward voltage across the clamp diode. Additionally, specific clamp circuit need the assistance of additional sampling circuits to acquire accurate voltage signals, which portability needs to be improved. Regarding integrability, the number of active components, component selection requirements, and the complexity of circuit restrict the integrability level. These limitations and challenges further impedes the application of clamp circuits for online monitoring. In this paper, a novel clamp circuit based on Schottky diode, and op-amp bias is proposed to eliminate voltage drift and improve measurement accuracy, as shown in Fig. 2(f).

TABLE I. COMPARISON OF VARIOUS CLAMP CIRCUITS

	Accuracy	Integrability	Sensitivity	Portability	Drift Suppression	Noise influence	Other
MOSFET [7]	★★	★ (1 active devices, extra driver)	★★★ (0.045~0.51% K^{-1})	★ (Extra signal)	No	☆	Temperature drift and parasitic parameters of MOSFET
Zener [8]	★	★★ (1 active device)	★★ (0.3% K^{-1})	★★★	No	☆☆ (15mV Nosie)	Diode partial voltage requires calibration
Passive Decoupling [9]	★★	★★★ (0 active device)	★★ (0.3% K^{-1})	★★	No	☆	Diode current limitation
Dual-diode Compensation [10], [11]	★★★	★★ (2 active devices)	★★★ (0.3~0.5% K^{-1})	★★★	Yes	☆☆	Two diodes require thermally coupled
Current Mirror [6]	★★★	★ (≥ 2 active devices)	★ (0.08% K^{-1})	★★	Yes	☆☆☆	Current mirror needs to be thermally coupled with two diodes
Proposed Circuit	★★★	★★ (2 active devices)	★★★★ (0.58~0.72% K^{-1})	★★★	Yes	☆☆	Constant current source ensures the two-diode thermally coupled

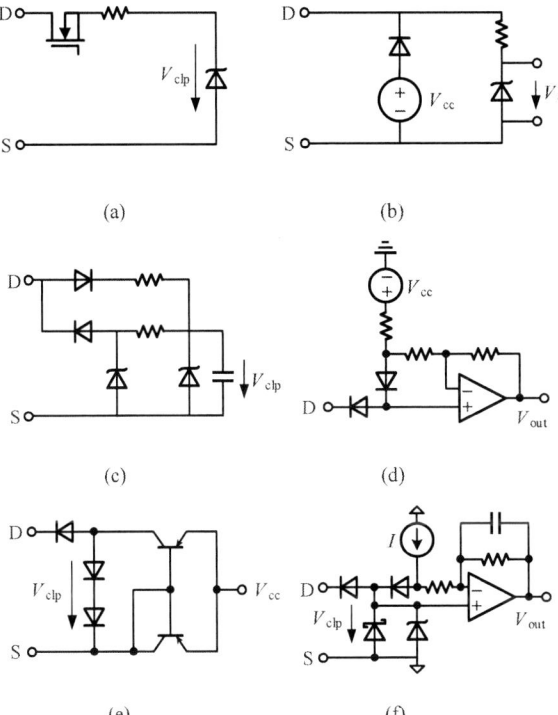

Fig. 2. Schematic clamp circuits for junction temperature monitoring. (a) MOSFET clamp circuit, (b) Zener clamp circuit, (c) Passive Decoupling clamp circuit, (d) Dual-diode Compensation clamp circuit, (e) Current Mirror clamp circuit and (f) Proposed clamp circuit.

2) Shunt Circuit for Drain Current Measurement

To accurately measure the drain current, the current shunt with low resistance and high bandwidth is utilized. Due to the presence of uncertain DC bias, the Rogowski coil is not suitable. Additionally, the hall element is not reasonable due to its limited bandwidth and bulk size. In contrast, the current shunt could be easily integrated in the PCB board and minimized parasitic inductance when introduced. In this paper, the current shunt has been designed based on the impedance matching criteria to meet the low parasitic and high bandwidth requirements [12], as shown in Fig. 3. A resistance of $R_L=100m\Omega$ is selected to satisfy the impedance matching requirements with the characteristic impedance Z_0 of the current shunt that satisfies the integration size constraints. This design ensures a 3 GHz measurement bandwidth and near-zero parasitic inductance. Furthermore, the resistive current probe guarantees synchronization of the current signal, avoiding delays in current signal measurement.

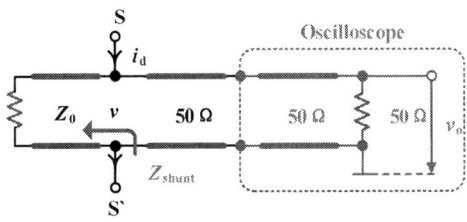

Fig. 3. Schematic diagram of current shunt with impedance matching

3) Integrated Circuit and Operation Principle of On-Resistance Measurement

The proposed on-resistance measurement with clamp circuit and current shunt is demonstrated, as shown in Fig. 4(a). The online junction temperature conditioning principle by using the TESP on-resistance is characterized, as depicted in Fig. 4(b).

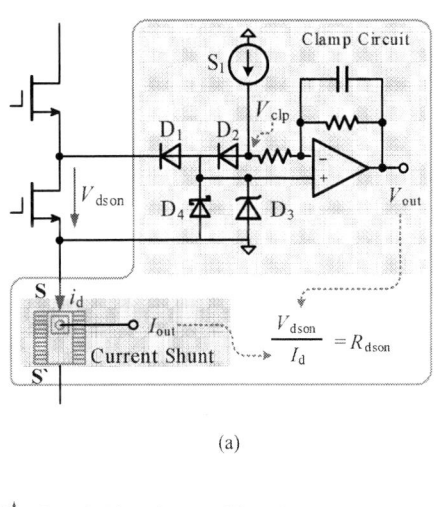

(a)

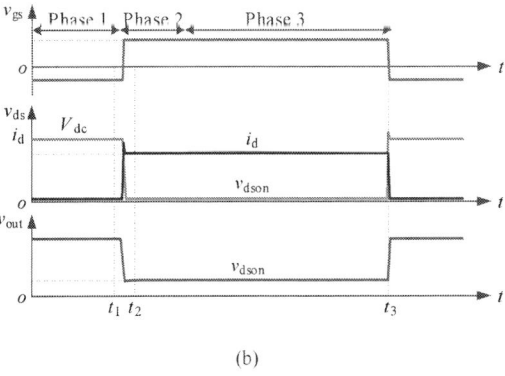

(b)

Fig. 4. On-resistance measurement with clamp circuit and current shunt. (a) Schematic diagram and (b) operation waveforms.

The proposed circuit is composed of the following parts: the Schottky diodes D_1 and D_2 of the same package used as high voltage devices. The current source S_1 that generates a constant current. The clamp diode D_3 that provides a stable clamping voltage when DUT is off. In addition, the Schottky diode D_4 that rapidly charges the parasitic capacitance of D_1 when DUT is transitioning on. The op-amp with a high gain bandwidth, a low input distortion voltage and low input temperature drift. Two resistors of the same resistance value are linked to the op-amp negative feedback port to achieve voltage drift compensation. The capacitor is used to improve the high-frequency characteristics of the clamp circuit.

The detailed operation is described as follows. In phase 1: during the off-state, the output of clamp circuit V_{out} equals to the voltage drop of clamp diode D_3. In phase 2: during the on-state, the parasitic capacitance of the clamp diode is charged by Schottky diode D_4 so that V_{clp} can quickly follow the actual value of V_{dson}. In phase 3: When the V_{clp} reaches the steady-state, the

979-8-3315-1612-3/25 $31.00 © 2025 IEEE

leakage currents in the clamp diode D_3 and Schottky diode D_4 are negligible. The current source S_1 supplies a constant current of 10 mA to ensure that the currents flowing through the two Schottky diodes, D_1 and D_2, are nearly equal. Because D_1 and D_2 are in the same package, located close enough together, and flow the same amount of current, they are thermally coupled and thus have the same voltage drop, V_{dio}. As a result, the on-state voltage can be expressed as

$$V_{\mathrm{dson}} = V_{\mathrm{clp}} - 2V_{\mathrm{dio}}. \tag{1}$$

Due to the pseudo-broken of the op-amp, the voltage drops of the resistor V_{res} and the clamp diode V_{dio} are equal. Therefore, the output voltage of the clamp circuit is the on-state voltage of the GaN HEMT, which can be written as

$$V_{\mathrm{out}} = V_{\mathrm{clp}} - 2V_{\mathrm{res}} = V_{\mathrm{clp}} - 2V_{\mathrm{dio}} = V_{\mathrm{dson}}. \tag{2}$$

Moreover, once the drain current I_d is measured by the shunt. The R_{dson} of the GaN HEMT can be obtained in conditions of different junction temperatures, drain-source voltages, and drain currents.

III. EXPERIMENTAL VALIDATION

A. online junction temperature map

To calibrate the $R_{\mathrm{dson}} - T_j$ map of the GaN HEMT, the test rig, including the hotplate and thermocouple, is set up, as shown in Fig. 5. The 60 A / 650 V GaN HEMT from CloudSemi Inc. with TOLT packaging is employed. The thermocouple is built in the hotplate to reduce the error in junction temperature calibration. Meanwhile, the proposed clamp circuit and current shut are integrated in the PCB board with the gate driver.

To observe the map between T_j and R_{dson}, the GaN HEMT is heated and held among 30 to 120℃. During the steady-state, the junction temperature approximates the case temperature, which is monitored by the thermocouple. Besides, the on-state voltage and drain current are recorded by the clamp circuit and current shunt. Similarly, the junction temperature map in the conditions of different voltage (100 to 400 V) and current (10 to 60 A) are measured, as depicted in Fig. 6(a) and(b). Therefore, the calculated $R_{\mathrm{dson}} - T_j$ map can be obtained, as demonstrated in Fig. 6(c). In practice, according to the observed map, the online junction temperature of the GaN HEMT can be estimated.

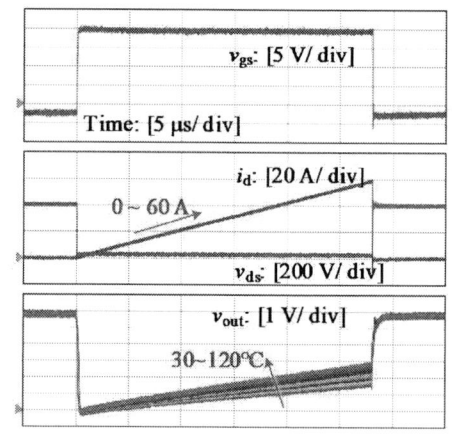

Fig. 5. Configuration of test rig for junction temperature calibration. (a) Schematic diagram and (b) experimental platform.

(a)

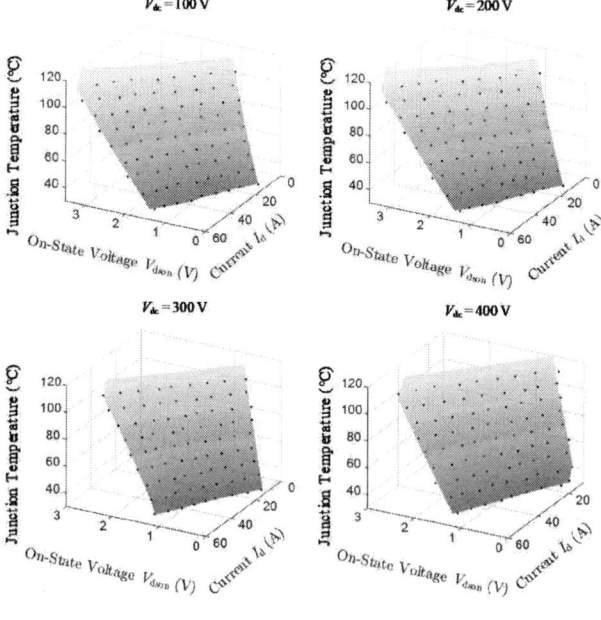

(b)

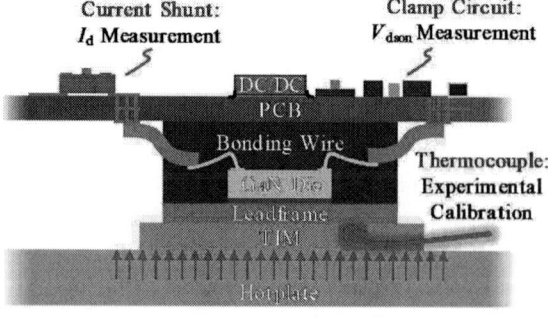

(a)

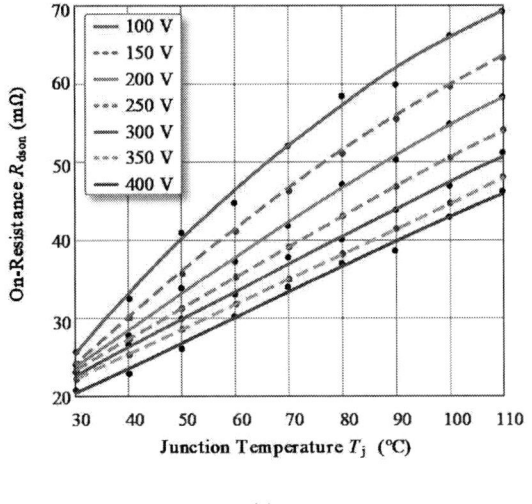

(c)

Fig. 6. Experimental results of junction temperature map. (a) Measured switching waveforms under different temperatures, (b) observed temperature map under different voltage and current, and (c) junction temperature map between R_{dson} and T_j.

B. TSEP sensitivity

The sensitivity of the voltage clamp circuit under general operating conditions is considered. With $V_{\text{dc}}=200\text{V}$ and $I_d=50\text{A}$, the junction temperature is slowly changed from $T_j=30°\text{C}$ to $T_j=110°\text{C}$. The curve of on-state voltage drop with junction temperature is shown in Fig. 7. A quadratic polynomial fit curve for V_{dson} and T_j is extracted.

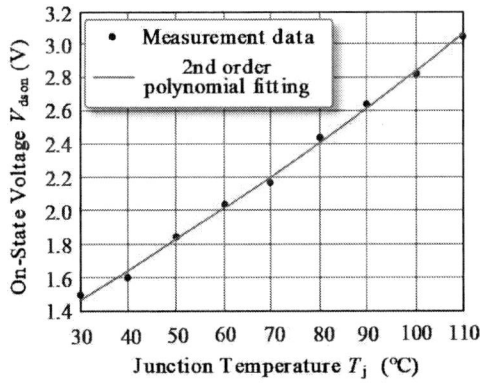

Fig. 7. Measured map between junction temperature T_j and on-state voltage V_{dson} at 200 V and 50 A.

Based on the relative sensitivity [13] and the slope of the quadratic polynomial fit curve, the TSEP sensitivity of the clamp at different temperatures can be expressed as

$$S = \frac{|s_v|}{val_{\max}}, \qquad (3)$$

where $val_{\max}$ is the maximum test voltage allowed in the clamp circuit's operating temperature range, s_v is the absolute

sensitivity of the on-state voltage at different temperatures. Based on the calculations, the TESP sensitivity of the measurement circuit was calculated to be in the range of 0.58%-0.72% K^{-1}, which proves that the clamp circuit satisfies the sensitivity requirements for TSEP acquisition.

C. online junction temperature monitoring verification

The accuracy of the junction temperature monitoring circuit is evaluated under general operating conditions. With $V_{\text{dc}}=250\text{V}$, the map of the measured on-resistance and the on-resistance predicted by the model in Fig. 6(c) corresponding to different temperatures is shown in Fig. 8(a). It can be found that the predicted on-resistance increases with the junction temperature, which is consistent with the trend of the actual on-resistance change. The residuals of the measured on-resistance and the model-predicted on-resistance versus temperature are shown in Fig. 8(b). All the residuals are randomly distributed near the zero line and form a band pattern parallel to the zero line, which proves that the predictive model reaches the linearity condition and the variance uniformity. At 40°C, the occurrence of outliers can be attributed to lower junction temperatures corresponding to reduced clamping voltages. This condition renders the system more susceptible to external noise. Nevertheless, this outlier remains within acceptable limits. The mean square error of the measured on-resistance and the on-resistance predicted by the model is 6.54×10^{-7}, which indicates that the error between the predicted and actual values is small, and the model demonstrates a high predictive capability.

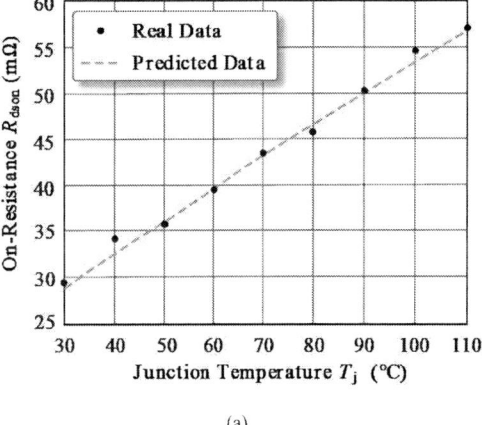

(a)

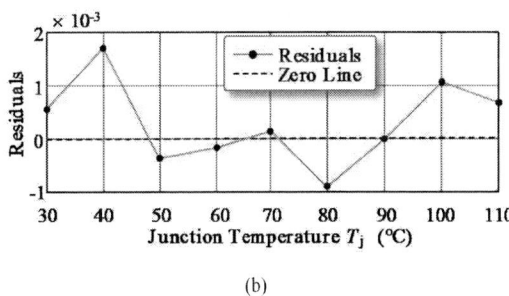

(b)

Fig. 8. Experimental results of verification at 250 V. (a) junction temperature map between the test on-resistance and the predicted on-resistance, and (b) residuals between the test on-resistance and the predicted on-resistance.

IV. CONCLUSIONS

To achieve online junction temperature monitoring of the GaN HEMT, the temperature-sensitive on-resistance is demonstrated and confirmed in this paper. To accurately observe the on-resistance in real time, the new clamp circuit and current shunt are proposed to measure the on-state voltage and drain current of the GaN HEMT. Comprehensive experiments are presented to ensure the feasibility and validation of the proposed junction temperature conditioning approach. Furthermore, junction temperature monitoring circuits will be considered for use in multi-chip situations and in package modules.

ACKNOWLEDGMENT

This work was supported by the Graduate Research and Innovation Foundation of Chongqing, China (Grant No. CYB240023).

REFERENCES

[1] E. A. Jones, F. F. Wang, and D. Costinett, "Review of commercial GaN power devices and GaN-based converter design challenges," *IEEE J. Emerg. Select. Top. Power Electron.*, vol. 4, no. 3, pp. 707-719, Sep. 2016.

[2] J. P. Kozak, R. Zhang, M. Porter, Q. Song, J. Liu, B. Wang, R. Wang, W. Saito, and Y. Zhang, "Stability, reliability, and robustness of GaN power devices: A review," *IEEE Trans. Power Electron.*, vol. 38, no. 7, pp. 8442-8471, Jul. 2023.

[3] L. Ceccarelli, R. Wu, and F. Iannuzzo, "Evaluating IGBT temperature evolution during short circuit operations using a TSEP-based method," *Microelectron. Reliab.*, vol. 100-101, p. 113423, 2019.

[4] X. Jiang, J. Wang, H. Yu, J. Chen, Z. Zeng, X. Yang, and Z. J. Shen, "Online junction temperature measurement for SiC MOSFET based on dynamic threshold voltage extraction," *IEEE Trans. Power Electron.*, vol. 36, no. 4, pp. 3757-3768, Apr. 2021.

[5] Z. Lu and F. Iannuzzo, "Online junction temperature extraction for cascode GaN devices based on turn-on delay," *IEEE Trans. Power Electron.*, vol. 39, no. 8, pp. 10250-10260, Aug. 2024.

[6] H. Qin, J. Peng, Z. Zhang, F. Zhang, X. Zhao, and Z. Xu, "Junction temperature prediction method of GaN HEMT power devices based on accurate on-voltage testing," *Energy Reports*, vol 9, pp. 389-395, 2023.

[7] K. Muñoz Barón, K. Sharma, M. Nitzsche and I. Kallfass, "Accounting for Acquisition Circuit Temperature in Accurate Online Junction Temperature Estimation," in *Proc. EPE'21 ECCE Europe*, Ghent, Belgium, pp. 1-8, 2021.

[8] P. Asimakopoulos, K. Papastergiou, T. Thiringer, M. Bongiorno and G. Le Godec, "On Vce Method: In Situ Temperature Estimation and Aging Detection of High-Current IGBT Modules Used in Magnet Power Supplies for Particle Accelerators," *IEEE Trans. Ind. Electron.*, vol. 66, no. 1, pp. 551-560, Jan. 2019.

[9] N. Badawi, O. Hilt, E. Bahat-Treidel, J. Böcker, J. Würfl and S. Dieckerhoff, "Investigation of the Dynamic On-State Resistance of 600 V Normally-Off and Normally-On GaN HEMTs," *IEEE Trans. Ind. Electron.*, vol. 52, no. 6, pp. 4955-4964, Nov.-Dec. 2016.

[10] S. Bęczkowski, P. Ghimre, A. R. de Vega, S. Munk-Nielsen, B. Rannestad and P. Thøgersen, "Online Vce measurement method for wearout monitoring of high power IGBT modules," in *Proc. EPE'13 ECCE Europe*, Lille, France, pp. 1-7, 2013.

[11] M. Guacci, D. Bortis and J. W. Kolar, "On-state voltage measurement of fast switching power semiconductors," *CPSS Trans. Power Electron. Appl.*, vol. 3, no. 2, pp. 163-176, Jun. 2018.

[12] Y. Wang et al., "Miniaturized Current Shunt With High Bandwidth and Low Parasitics for High-Integrated Applications: Electro-Thermal Considerations and Co-Design," *IEEE Trans. Power Electron.*, vol. 39, no. 12, pp. 15732-15747, Dec. 2024.

[13] Y. Avenas, L. Dupont and Z. Khatir, "Temperature Measurement of Power Semiconductor Devices by Thermo-Sensitive Electrical Parameters—A Review," *IEEE Trans. Power Electron.*, vol. 27, no. 6, pp. 3081-3092, Jun. 2012.

False Turn-on Failure and Protection of p-gate GaN HEMT in MHz Class-E Resonant Inverter

Ziheng Liu
School of Integrated Circuits
Peking University
Beijing, China
zihengliu@stu.pku.edu.cn

Ju Gao
School of Integrated Circuits
Peking University
Beijing, China
2201111467@stu.pku.edu.cn

Hongjie Peng
School of Integrated Circuits
Peking University
Beijing, China
2301111840@stu.pku.edu.cn

Jiayin He
School of Integrated Circuits
Peking University
Beijing, China
hejiayin@stu.pku.edu.cn

Jinyan Wang
School of Integrated Circuits
Peking University
Beijing, China
wangjinyan@pku.edu.cn

Maojun Wang
School of Integrated Circuits
Peking University
Beijing, China
mjwang@pku.edu.cn

Abstract—Featuring simple structure and natural soft-switching, the single-switch resonant power inverters usually operate at several MHz frequency. However, gate-node voltage rings caused by parasitic inductors challenge the device security and limit the operating voltage. To address this problem, this paper investigates the mechanism and protection of the false turn-on (FTN) problem as the p-gate gallium nitride (GaN) HEMT turning off. After FTN, the gate-control capability losses and source-node shows large leakage. Two important causes of gate-node voltage ring are high dv/dt noise on drain-node, resonance between device junction capacitors and parasitic inductors. Through circuit modelling, parameter design and experimental verification, reducing the Miller current amplitude in the gate-to-drain capacitor (C_{GD}) can suppress the gate-node voltage ring, by using external small passive LC network.

Keywords—Resonant inverter, p-gate GaN, Miller current, false turn-on

I. INTRODUCTION

As an excellent wide bandgap semiconductor, enhancement-mode p-gate GaN power devices have been widely used in power modules [1]. Thanks to high electron mobility and low on-resistance, p-GaN devices have pushed switch-mode converter into MHz frequency with high efficiency. Derived from RF amplifier, Class-E topologies have been proven to realize DC-AC and DC-DC conversion at HF to VHF bands [2] [3]. However, for board-level circuits, higher switching frequency leads to more serious voltage oscillation, risking damaging GaN devices. Especially, Class-E resonant topology features large voltage stress and the power drain-node shows large dv/dt during the power switch turning off, then Miller current is flowing through Miller capacitor C_{GD} and AC-voltage occurs on the gate-node [4]. Compared to Si power devices, p-gate GaN devices have lower threshold voltage, so FTN problem is huge concern in MHz Class-E power conversion

applications. In addition, the gate threshold voltage of the p-gate GaN device has been found to be negatively shifted at high voltage stress (-0.5V shifted when the drain-to-source voltage rises to over 100V) [5] [6], which increases the risk of damage. Therefore, it is essential to investigate engineering approach to prevent FTN damage. In bridge type switching circuits, the FTN phenomenon has been modelled considering parasitic inductor/capacitor [7]. To mitigate the effect of parasitic inductor, external resistance (includes ferrite bead) [8] and multiloop layout optimization [9] are added to the gate-node so that the high-frequency oscillation energy in parasitic inductors can be reduced. Several active circuit-based approaches have been proposed to help mitigate gate-node voltage oscillation, applying negative bias [10] and connecting active conductive channel [11] [12]. However, these methods limit the switching speed and increasing system control complexity respectively, they do not suit for MHz switching applications.

Therefore, to avoid the above short-comes, we demonstrate a protection approach away from false turn-on problem for MHz Class-E power inverter. Different from most current methods which always focus on suppressing the energy stored in parasitic inductors, the energy delivered in Miller capacitor C_{GD} is partly separated to another resonant tank according to this work. Thanks to the single-switch structure of Class-E topology, the external resonant network can participate in shaping the drain-to-source voltage with other LC components. Only small passive LC network is connected to the drain-port, greatly reducing the amplitude of Miller current. The proposed method is presented with a four-order Class-E inverter, where the Miller current is proven to be affected by the dv/dt of drain-to-source voltage and the Miller capacitance. The mechanism that the external LC resonant tank mitigates the Miller current is also analyzed. Finally, both the prototypes of original Class-E inverter and the modified Class-E inverter are demonstrated by hardwire.

979-8-3315-1612-3/25 $31.00 © 2025 IEEE

II. FALSE TURN-ON PROBLEM

P-gate GaN device has layer-by-layer structure and planar conductive channel, as shown in Fig. 1. The two-dimensional electron gas (2DEG) formed by the polarization effect between the AlGaN barrier layer and the GaN channel layer. Three electrodes (gate, drain and source) are fabricated at the top of the heterojunction platform. For FTN analysis, it is important to determine the junction capacitances of the power device. The electrodes and 2DEG region form capacitor C_{GD}, C_{DS}, which are all voltage-dependent capacitors. The device-under-test (DUT) of this work is given to a set of capacitance measurements, including input capacitor C_{iss} ($C_{GS}+C_{GD}$), output capacitor C_{oss} (C_{DS}) and Miller capacitor C_{GD}. As shown in Fig. 2, C_{GD} decreases with the depletion of 2DEG region. It is noted that the value of C_{GD} is as large as 20pF~50pF when the drain voltage stress is lower than 50V. During the turning-off period of power switch, such Miller capacitance will lead to large Miller current under high dv/dt.

Considering parasitic inductor L_d, drain-node voltage with high dv/dt will generate high frequency voltage oscillation during turning off. Then the oscillation current flows to C_{GD} and C_{DS}, the current in C_{GD} branch leads to voltage ring on gate-node under the induction of L_g and R_g. Fig. 3 gives the waveforms of node-voltages with oscillation ripples; FTN occurs as the amplitude of v_{gs} exceeds threshold voltage V_{th}. If FTN occurs at off-state, the conductive channel will experience thermal breakdown due to the simultaneous presence of high electric field and high current. Different to the gate overvoltage [13] protection, conventional voltage-clamp approaches are useless for FTN because the gate voltage is at low state. As a common solution, using external resistance module to suppress the resonant energy has been proposed in existing literatures. However, the adding of external resistance causes large RLC delay time, which do not suit for MHz applications. As a replacement, reducing the Miller capacitance by applying other resonant elements can also mitigate Miller current and will not affect the switching speed. Therefore, this work aims to realize small Miller current at high dv/dt by adjusting the Miller capacitance.

III. FTN IN CLASS-E POWER INVERTER

For comprehensively present the FTN phenomenon and mechanism in MHz p-gate GaN switching circuit, a 6.72MHz Class-E power inverter is used for analysis and demonstration. Both the circuit waveform behavior and internal behavior of p-gate GaN device are discussed in this section.

A. Circuit Waveform Behavior

A 6.72MHz Class-E inverter is designed to demonstrate the FTN phenomenon. The circuit diagram of the inverter is shown in Fig. 4(a), where the gate driving stage is connected to the ground because only the off-state is considered for FTN analysis (A square waveform generator should be connected before R_g normally). The L_1, C_1 are used to invert the input DC-voltage V_{IN} to be half-sine AC voltage v_d. The L_r, C_r, L_2 and C_2 form output resonator, transferring the fundamental voltage

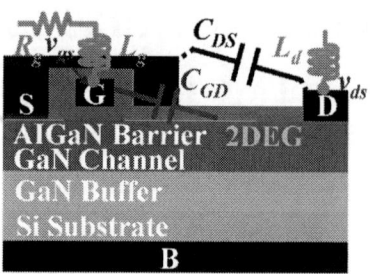

Fig. 1. Schematic of the commercial p-gate GaN device. The parasitic components and Miller capacitor are included.

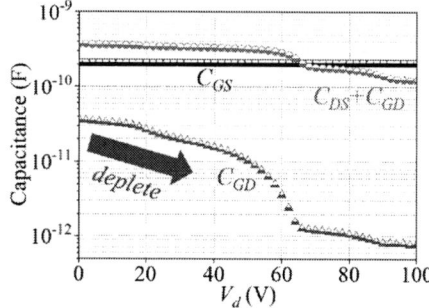

Fig. 2. Measured plots of junction capacitances of a commercial p-gate GaN device (Infineon: GS66506T [14]).

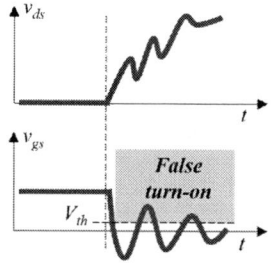

Fig. 3. Theoretical v_{ds} and v_{gs} waveforms when FTN occurs.

component of v_d to the load R_L. Because the output resonator always shows high impedance at high-order harmonic frequencies, the resonant current i_r can be assumed to be purely sinusoidal current, as follows:

$$i_r = I_0 \sin(\theta + \varphi). \qquad (1)$$

To obtain the slew rate at the beginning of off-state, the expression of drain voltage v_d should be derived firstly. By adopting Kirchhoff's Law, the circuit in Fig. 4(a) can be modelled as follows:

$$KVL: V_{IN} = L_1 \frac{di_{L_1}}{dt} + v_d, \quad KCL: i_{L_1} = C_1 \frac{dv_d}{dt} + i_r, \qquad (2)$$

$$v_d = V_{IN} - \frac{1}{\lambda_1^2 \omega_0^2} \frac{dv_d^2}{dt^2} - \omega_0 L_1 I_0 \cos(\theta + \varphi), \qquad (3)$$

$$\lambda_1 = \frac{1}{\omega_0 \sqrt{L_1(C_1 + C_{DS})}}, \qquad (4)$$

where ω_0 refers to the switching angular frequency. Substituting these two conditions to (4): v_d is zero as power switch just turns off; the average value of v_d in one period is V_{IN}. Then the v_d is solved as follows:

979-8-3315-1612-3/25 $31.00 © 2025 IEEE

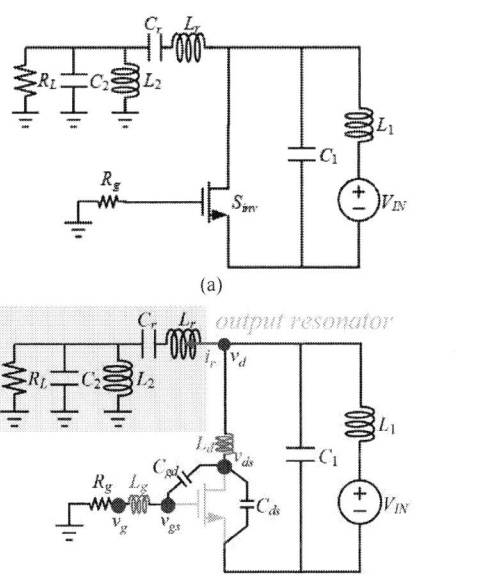

(a)

output resonator

(b)

Fig. 4. Circuits diagrams of the original MHz Class-E power inverter (a) without detailed parasitic elements and (b) with detailed parasitic elements.

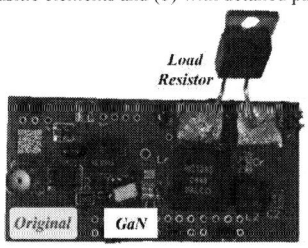

Load Resistor

Original   GaN

Fig. 5. Top-view of the original inverter prototype (The DUT has been damaged due to FTN).

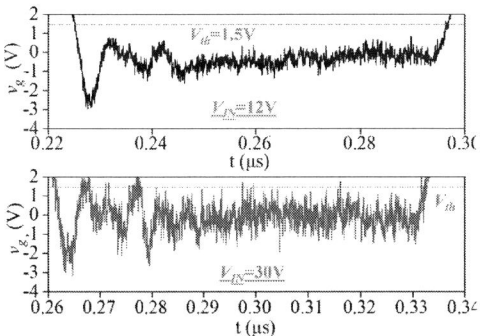

Fig. 6. Measured v_g waveforms of the original inverter prototype under V_{IN}=12V and V_{IN}=30V.

$$v_d = c_1 \cos(\lambda_1 \theta) + c_2 \sin(\lambda_1 \theta) + \frac{\lambda_1^3 I_0^2 \omega_0 L_1}{\lambda_1^2 - 1} \cos(\theta + \varphi) + V_{IN}, \quad (5)$$

$$c_1 = -\frac{\lambda_1^3 I_0^2 \omega_0 L_1}{\lambda_1^2 - 1} \cos \varphi - V_{IN}, \quad (6)$$

$$c_2 = \frac{(-c_1)\sin(\lambda_1 \pi) + \frac{2\lambda_1^4 I_0^2 \omega_0 L_1}{\lambda_1^2 - 1}\sin \varphi + \lambda_1 V_{IN} \pi}{1 - \cos(\lambda_1 \pi)}. \quad (7)$$

Then, the dv/dt of v_d at the beginning time of off-state is written as followed:

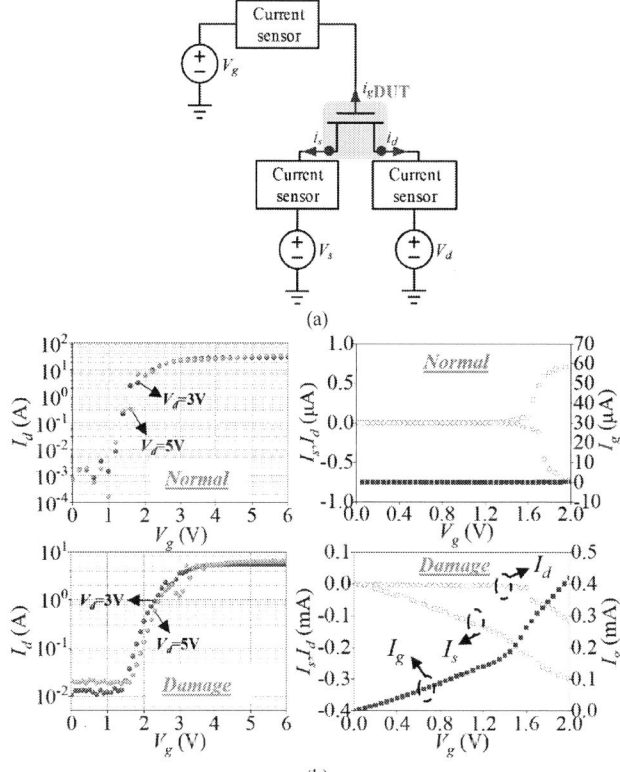

(a)

(b)

Fig. 7. (a) Static test platform of the DUT. (b) Transfer characteristic, static leakage current of normal commercial p-gate GaN device and damaged device.

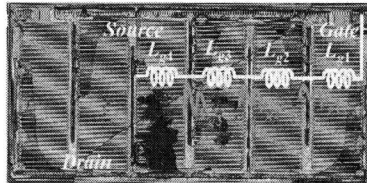

Fig. 8. Micrograph of the chip of damaged p-gate GaN device (Infineon: GS66506T [14]).

$$(dv_d/dt)\big|_{\theta=0} = c_2 \lambda_1 \cos(\lambda_1 \theta) - \frac{\lambda_1^3 I_0^2 \omega_0 L_1}{\lambda_1^2 - 1}\sin \varphi. \quad (8)$$

It is obvious that the dv/dt is positive correlation with V_{IN}. At given output power and switching frequency, the value of dv/dt can be calculated according to (8). In this work, the maximum of dv/dt is 3V/ns to 6V/ns at 12V input and 30V input respectively. As shown in Fig. 4(b), the conductive loop (including L_d, C_{GD}, L_g and R_g) will be formed at off-state and generate Miller current. Fig. 5 shows the test board, and the measured waveforms are shown in Fig. 6. As V_{IN} increases from 12V to 30V, the oscillation voltage amplitude of v_g becomes about 2 times, exceeding the threshold voltage of the commercial p-gate GaN device (Infineon: GS66506T). As shown in Fig. 5, the p-gate GaN device detaches from the welding position after FTN occurs. It should be noted that the FTN damage will seriously limit the max output power of MHz Class-E power inverter.

979-8-3315-1612-3/25 $31.00 © 2025 IEEE   227

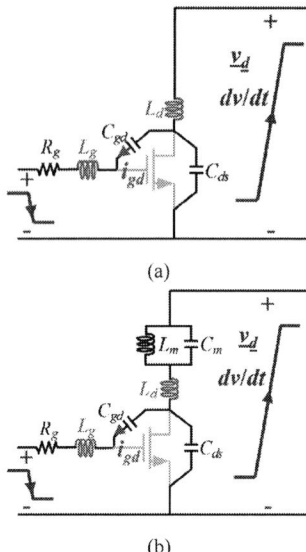

(a)

(b)

Fig. 9. Analytical circuit model of the switching stage (a) without external L_m-C_m resonant tank, (b) with external L_m-C_m resonant tank.

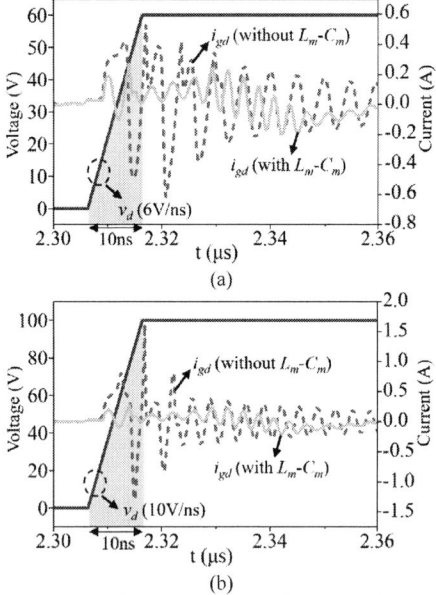

(a)

(b)

Fig. 10. Simulated Miller current i_{gd} under (a) dv/dt=6V/ns and (b) dv/dt=10V/ns. (An external C_{GD}=20pF is used to test Miller current)

B. Device Behavior

The damaged device is characterized with B1505 platform, as shown in Fig. 7(a). As shown in Fig. 7(b), the saturation output current becomes 10 times lower after FTN problem, the leakage current flows between gate-node and source-node. This commercial p-gate GaN device is composed of several finger-like structures, and the reason for this phenomenon may be that some of the finger-like device structures have been burned out. To verify this point, the damaged device is taken for microscopic observation. Fig. 3(c) shows the top-view of damaged chip, the burn areas can be observed to be larger as far away from gate-node, which is caused by the distribution effect of parasitic inductors. Although the damaged device still has

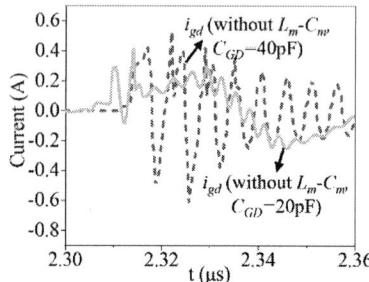

Fig. 11. Simulated Miller current at external C_{GD}=40pF and external C_{GD}=20pF.

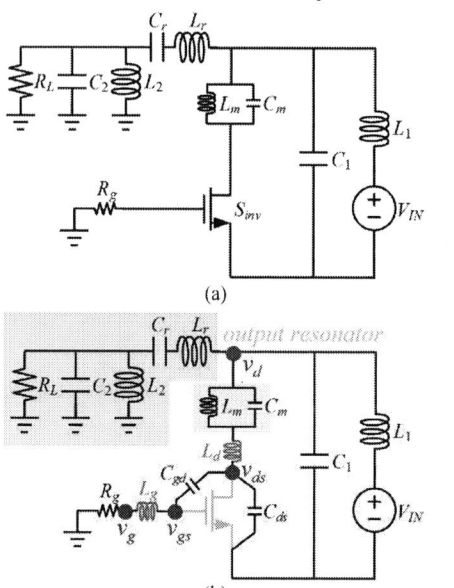

(a)

(b)

Fig. 12. Circuits diagrams of the modified MHz Class-E power inverter (a) without detailed parasitic elements and (b) with detailed parasitic elements.

gate control capability, it will soon enter the saturation region during operation.

IV. MODIFIED CLASS-E POWER INVERTER

As shown in Fig. 9(a), the Miller current i_{gd} is generated from the resonant of L_d-C_{GD}-L_g-R_g. A possible solution to suppress the gate ring is to mitigate the resonant current in C_{GD}. Excluding additive gate resistance and active circuits, a LC network will help to absorb high-frequency AC-current and do not affect switching speed. A parallel resonant tank L_m-C_m is added to the drain-node of power switch, as shown in Fig. 9(b). Considering network L_m-C_m, the impedance of it (Z_e) is written as follows:

$$Z_e = j\omega L_m \left\| \frac{1}{j\omega C_m} = \frac{1}{j\omega \left[\frac{(\omega/\omega_m)^2 - 1}{L_m} \right]}, \quad (9)$$

where the ω_m refers to the resonant frequency of L_m-C_m. The L_m-C_m can be equivalent to a capacitor when the operating frequency exceeds ω_m. indicating that the total capacitance of L_m-C_m and C_{GD} must be smaller than original C_{GD}. Then, it means that the value of C_{GD} has decreased and Miller current will be naturally mitigated.

979-8-3315-1612-3/25 $31.00 © 2025 IEEE   228

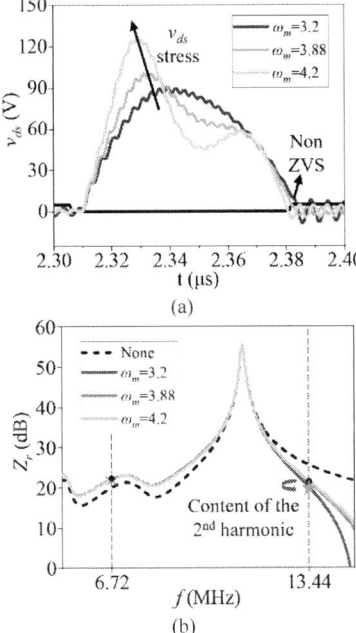

Fig. 13. Simulated (a) v_{ds} waveforms and (b) Z_r impedance characteristic of the modified inverter.

To verify the above analysis, a set of simulations of the Miller current are made, as shown in Fig. 10. The results show that the amplitude of Miller current is successfully suppressed after adding L_m-C_m network under dv/dt of 6V/ns up to 10V/ns. It is noted that the drain voltage v_d is pulse voltage in the simulation, indicating the proposed modification can be applied in both resonant and bridge-type switching power conversion. In order to verify the decisive role of C_{GD} value in suppressing Miller current, the simulations using external drain-to-gate capacitance 20pF and 40pF are set. As shown in Fig. 11, without L_m-C_m network, the Miller current can be mitigated by using smaller external drain-to-gate capacitance.

To verify this statement, a modified 6.72MHz Class-E inverter is designed and fabricated. Fig. 12 gives the circuit diagrams of the modified Class-E power inverter. The circuit parameters should be carefully designed with three goals: Zero-voltage-switching (ZVS) realization, low v_d stress and proper oscillation suppression ability. For L_m-C_m network, a key design parameter is the resonant frequency ω_m. Normally, the resonant frequency should be larger than switching frequency so that the network will show high impedance at higher oscillation frequency. To properly determine the ω_m value, the effects on the v_{ds} waveform is focused. Fig. 13(a) plots the v_{ds} waveforms under different ω_m value and Fig. 13(b) shows the corresponding characteristic impedance seen from the drain-node of power switch. At small ω_m value, the 2nd harmonic voltage component is suppressed and vds waveform risks losing ZVS performance. On the contrary, too large ω_m value will lead to more 2nd harmonic voltage component and greatly increase the v_{ds} stress. Consequently, the ω_m value is chosen to be 3.88.

Then, the L_m and C_m are determined as 65nH and 570pF respectively, resonating at 3.88 times of ω_0. The other circuit

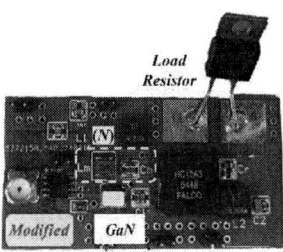

Fig. 14. Top-view of the modified inverter prototype.

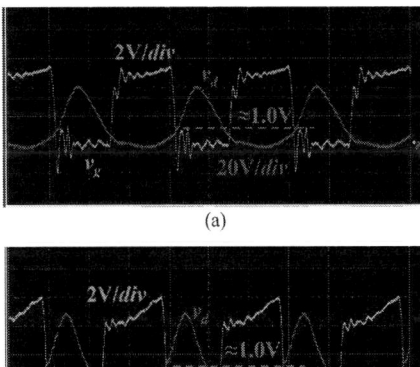

Fig. 15. Measured v_g, v_d waveforms of the modified inverter prototype under V_{IN}=12V and V_{IN}=30V.

parameters are: L_1=150nH, C_1=1.8nF, L_r=800nH, C_r=736pF, L_2=150nH, C_2=373pF, R_L=20Ω. Compared to the original test board, the same gate driver, gate resistor, p-gate GaN device and driving stage layout are used in the modified test board. Fig. 14 shows the test board of the modified Class-E inverter. At 12V and 30V input, the measured v_g and v_d is shown in Fig. 15. In the modified test board, the gate rings maintain lower than 1.2V under V_{IN}=12V and V_{IN}=30V, FTN do not occur anymore. The results prove that the proposed LC network successfully suppress the Miller current in C_{GD}, regardless of high dv/dt. Fig. 16(a) and Fig. 16(b) show the detailed v_g waveform comparison during on-state and off-state respectively. Obviously, either turning-on speed or the turning-off speed are the same before and after adding L_m-C_m network.

V. CONCLUSIONS

In this work, the mechanism and protection approach of FTN problem in MHz Class-E power inverter are discussed. As drain-node shows higher dv/dt, the Miller current flowing in C_{GD} increases and leads to larger gate-node oscillation. Due to low threshold voltage of p-gate GaN device, FTN will damage the device soon, conductive channel forms between the gate-node and source-node. Adding LC filter network on drain-node help to suppress the gate rings because the amplitude of resonant current is reduced, which has been verified with hard-wire experiment. Thanks to the single-switch structure of Class-E power inverter, the external L_m-C_m can serve as part of the resonant network and shape the drain-to-source voltage. Compared to traditional protection methods, such as adding

979-8-3315-1612-3/25 $31.00 © 2025 IEEE

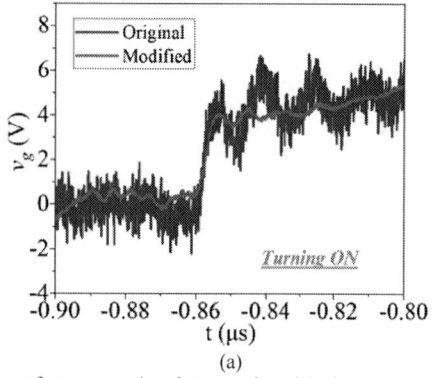 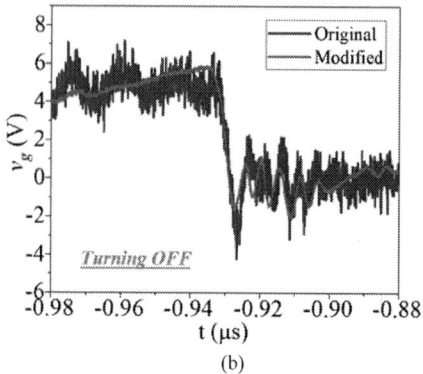

(a)　　　　　　　　　　　　　　　　　(b)

Fig. 16. The v_g waveform comparison between the original prototype and modified prototype. (a) Turning on period, (b) turning off period.

EMI filter on gate-node, increasing gate resistance and connecting active filter circuit, the proposed method does not affect switching speed and increase the complexity of control stage. This work has the potential to serve as a solution for improving the reliability of p-gate GaN power devices in the next generation of very-high-frequency power conversions.

ACKNOWLEDGMENT

This work is supported by the National Key Research and Development Program of China under Grant 2023YFB3905703 and the National Natural Science Foundation of China under Grant No. U2241220. The authors would like to appreciate the help from the Dongke Semiconductor (Anhui) Company.

REFERENCES

[1]. G. Eason, A. I. Emon, Mustafeez-ul-Hassan, A. B. Mirza, J. Kaplun, S. S. Vala and F. Luo, "A Review of High-Speed GaN Power Modules: State of the Art, Challenges, and Solutions," *IEEE J. Emerg. Sel. Topics Power Electron.*, vol. 11, no. 3, pp. 2707-2729, June 2023.

[2]. J. M. Rivas, R. S. Wahby, J. S. Shafran and D. J. Perreault, "New Architectures for Radio-Frequency DC–DC Power Conversion," *IEEE Trans. Power Electron.*, vol. 21, no. 2, pp. 380-393, March 2006.

[3]. Z. Liu, F. Meng, K. Ma and K. S. Yeo, "Current Harmonics Analysis and Design for Load-Independent ZVS Single-Switch Resonant DC/DC Converter," *IEEE Trans. Power Electron.*, vol. 37, no. 9, pp. 10877-10888, Sept. 2022.

[4]. A. Lemmon, M. Mazzola, J. Gafford and C. Parker, "Instability in Half-Bridge Circuits Switched with Wide Band-Gap Transistors," *IEEE Trans. Power Electron.*, vol. 29, no. 5, pp. 2380-2392, May 2014.

[5]. M. Nuo, Y. Wu, J. Yang, Y. Hao, M. Wang and J. Wei, "Time-Resolved Extraction of Negatively Shifted Threshold Voltage in Schottky-Type p-GaN Gate HEMT Biased at High VDS," *IEEE Trans. Electron Devices*, vol. 70, no. 7, pp. 3462-3467, July 2023.

[6]. H. Xu, J. Wei, R. Xie, Z. Zheng, J. He and K. J. Chen, "Incorporating the Dynamic Threshold Voltage Into the SPICE Model of Schottky-Type p-GaN Gate Power HEMTs," *IEEE Trans. Power Electron.*, vol. 36, no. 5, pp. 5904-5914, May 2021.

[7]. X. Long, Z. Jun, L. Pu, D. Chen and W. Liang, "Analysis and Suppression of High Speed Dv/Dt Induced False Turn-on in GaN HEMT Phase-Leg Topology," *IEEE Access*, vol. 9, pp. 45259-45269, 2021.

[8]. T. Liu, T. T. Y. Wong and Z. J. Shen, "A Survey on Switching Oscillations in Power Converters," *IEEE J. Emerg. Sel. Topics Power Electron.*, vol. 8, no. 1, pp. 893-908, March 2020.

[9]. K. Wang, L. Wang, X. Yang, W. Chen and H. Li, "A Multiloop Method for Minimization of Parasitic Inductance in GaN-Based High-Frequency DC–DC Converter," in *IEEE Transactions on Power Electronics*, vol. 32, no. 6, pp. 4728-4740, June 2017.

[10]. H. Zhou, M. Priestley, J. Fletcher and K. Sun, "A Gate Driver with a Negative Turn Off Bias Voltage for GaN HEMTs," *2020 IEEE 9th International Power Electronics and Motion Control Conference (IPEMC2020-ECCE Asia)*, Nanjing, China, 2020, pp. 1083-1086.

[11]. Z. Zhang, F. Wang, L. M. Tolbert and B. J. Blalock, "Active Gate Driver for Crosstalk Suppression of SiC Devices in a Phase-Leg Configuration," *IEEE Trans. Power Electron.*, vol. 29, no. 4, pp. 1986-1997, April 2014.

[12]. J. Shu, J. Sun, Z. Zheng and K. J. Chen, "A Gate Driver with a Low-Voltage GaN HEMT for False Turn-on Suppression and Gate Reliability Enhancement of SiC MOSFETs," *2024 IEEE Applied Power Electronics Conference and Exposition (APEC)*, Long Beach, CA, USA, 2024, pp. 724-728.

[13]. B. Wang *et al.*, "Gate Robustness and Reliability of P-Gate GaN HEMT Evaluated by a Circuit Method," *IEEE Trans. Power Electron.*, vol. 39, no. 5, pp. 5576-5589, May 2024.

[14]. GS66506T. Datasheet: https://cn.gansystems.com/wp-content/uploads/2020/03/GS66506T-DS-Rev-200227.pdf.

Heat Extraction from Ferrite Cores Using Metallic Laminations

Alyssa Brown, Duy T. Nguyen, Alex J. Hanson
University of Texas at Austin
2501 Speedway, Austin, TX 78712, USA
{rileyalyssa3, daniel.nguyen2304, ajhanson}@utexas.edu

Abstract—**Designers of power magnetic devices face thermal constraints at all power and frequency levels and must optimize core size to ensure proper operating temperatures. Ferrite cores, commonly used for medium to high frequency designs, are poor thermal conductors which complicates effective heat extraction when cooling is only available on the surface. We propose that the addition of metallic laminations (i.e. metallic cooling planes) within the core can improve heat extraction without adverse effects to the performance of the magnetic component. We demonstrate this conclusion with an analytic solution of the expected eddy current loss in the plate(s) and a calculation of the expected impact on the overall component's loss which is largely independent of the details of the component design and applies across broad frequency ranges. Through FEA and experiment, we demonstrate improved natural convection of a prototype transformer with a metallic lamination in the center of ferrite core sets. We first demonstrate the improvement in heat extraction and the negligible loss of the plate using Ansys Maxwell and Icepak simulations. We further validate the proposed approach by testing a prototype transformer in a 2 kW, 150 kHz LLC converter.**

I. INTRODUCTION

Magnetic components are often the largest and among the lossiest components in power converters, even as elevated frequencies and optimized designs have improved their performance [1]. The size of magnetic components, for a given frequency, is usually limited by their operating temperature – smaller components experience more loss and have less surface area from which to extract heat [2]. Therefore, loss reduction and heat extraction (the emphasis of this work) for magnetic components are critical to improving power converter density. Heat extraction is particularly important for the rapidly-growing category of applications operating in the hundreds of kHz or MHz regimes where ferrites become the necessary core choice to manage core loss. The poor thermal conductivity of ferrite (1 to 5 $\frac{W}{mK}$ [3] versus aluminum which is ~200 $\frac{W}{mK}$) prevents effective heat extraction, particularly from the inner parts of the core. This challenge is exacerbated in high-power applications, such as solid state transformers [4]–[6], electric ship power distribution [7], and renewable energy [8], which increasingly adopt high frequencies and ferrite cores of substantial size.

Researchers have taken many avenues in pursuit of better thermal performance for high power transformers. One avenue involves optimizing the insulation of the conductors that is required for medium and high voltage operation [9]. For dry-type transformers, advancements in potting materials/methods

using epoxy or epoxy composites [10]–[12], including partial potting structures [13], have shown improved thermal performance while also providing the required insulation. Bobbins have been improved through the use of 3D printing that allow airflow/cooling in-between the windings [14], [15]. [16] showed that the thermal performance of a transformer can be improved by simply replacing the bobbin with a thin insulator at the cost of a more cumbersome assembly process. Improvements in the insulation and bobbin of transformers are especially valuable and low cost since both items are typically necessary for high power transformers.

Another avenue for improvement involves the addition of materials or active cooling devices, which is the direction taken in this paper. Although these have the potential to add weight, cost, or volume, with optimal design their improvements to the thermal performance of a transformer can outweigh these costs. Many high power converters use forced air or liquid cooling to keep operating temperatures within range [17]. These same methods can be expanded specifically to the magnetic components. [18] and [19] discuss the design of optimized air-cooling methods for planar transformers which show significant improvement in the operating temperatures while maintaining a good power density. Liquid/gas cooling methods using water or refrigerants were developed in [20] and [21] to specifically remove heat from within a transformer's windings. Immersion of high power transformers in oil is common practice to improve cooling and isolation and the efficacy of this practice was investigated and modeled in [22] and [23]. In [24], a hybrid air and liquid cooling design method was developed to optimize the design of a high power, medium frequency, and medium voltage transformer.

Heat pipes are also widely used to aid in the conduction of heat from within the windings [25], [26]. [27] uses copper foil in tandem with some heat pipes as a light-weight means of extracting heat from within both the windings and core. In a similar vain, the use of air channels in-between stacked core pieces in [28] improved the heat extraction and increased the airflow velocity within the high power transformer.

We propose that the addition of metallic cooling planes, shaped much like laminations in line-frequency steel transformers [29], can greatly improve heat extraction from the interior of ferrite cores. As long as the cooling plane is oriented parallel to the direction of magnetic flux, minimal induced eddy currents and loss are expected. The reluctance model

---

979-8-3315-1612-3/25 $31.00 © 2025 IEEE

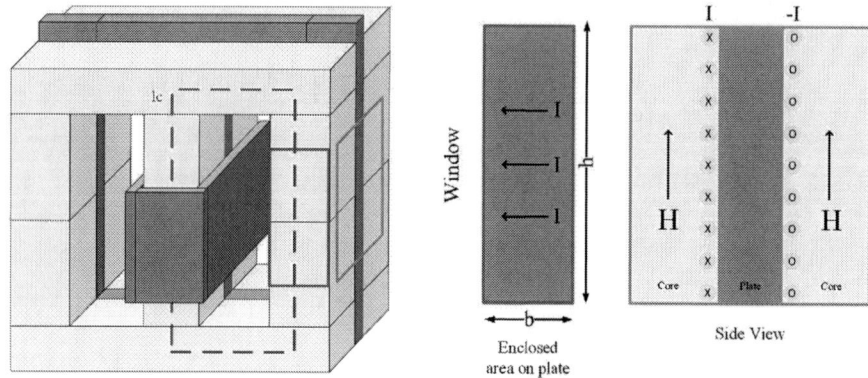

Fig. 1: Diagram of proposed transformer with internal cooling plate. The red enclosed area is on the surface of the plate with a height of h and a width of b. The green enclosed area is on the side of the transformer and it illustrates the eddy current (yellow circles) induced in the plate from the H field in the core.

of the transformer is only slightly changed, and therefore the impact on the transformer design procedure is also small. Little work has been published on the addition of heat conducting elements to the inside of a transformer's core. In [30], cooling planes made of aluminum nitride ceramic are added in-between stacked U cores and are claimed to add no additional eddy current losses while improving the hot spot of the transformer by 22.6% in simulation but there is no comparison made experimentally. Similarly, [31] and [32] use aluminum plates, which we use as well, within the core to help with heat conduction. However, the losses in the cooling system are not thoroughly analyzed and since so many plates are used at once, it is difficult to tell which placement leads to the best improvement.

In summary, while many prior works have examined methods to cool magnetic components, few have separated the core into pieces in order to extract heat from the interior. The few that have attempted it have reported isolated results without systematic investigation.

In this work, we demonstrate improved thermal performance by including a cooling plane within the core of a 2 kW 150 kHz transformer with negligible additional loss. In simulation and experiments, we compare our design to a control transformer that uses thermal management on the outer surfaces of the core. We begin in section II with a theoretical discussion on the expected losses induced in the metallic cooling plane and derive an equation based on the ratio of core loss to loss in the plate to gain perspective on the effects of frequency, core material properties, and number of plates used in the transformer. In section III, we use FEA electromagnetic and thermal simulations to verify both the induced loss and the thermal benefit of the plane. We further validate this experimentally in section IV using a 2 kW 400 V to 800 V LLC converter that is akin to on-board chargers found in electric and hybrid vehicles.

II. LOSS WITHIN THE COOLING PLANE(S)

Consider a magnetic component with a winding wrapped around an E core, as shown in the diagram in Fig. 1. A plate (shown in blue in the diagram) is inserted in such a way that it splits the core in half and the magnetic flux is parallel to the plate. Taking a look at the section of the core enclosed in the red rectangle, we can calculate the loss due to the eddy current in this part of the plate. It is given by $2I^2 R = 2I^2 \rho \frac{b}{A}$, where I is the current on one surface of the plate, b is the path length that the current must flow through, and A is the cross sectional area that the current sees. The cross sectional area, A, is equal to $h\delta$, where h is the "height" of the cross sectional area and the skin depth $\delta = \sqrt{\frac{2\rho}{\omega\mu_0}}$ is the "width." The current is $I = K \times h$, where K is the surface current density (A/m). The magnitude of K is equal to the magnitude of H just outside the plate, which is $B/\mu = B/\mu_r \mu_0$. For simplicity, b, which is the distance from the outside of the core to the window, is approximately $\sqrt{A_c/2}$, where A_c is the cross sectional area of the center post of the core and typically E core legs (which is where the enclosed area is defined) have a cross sectional area of $A_c/2$. Assembling these pieces yields

$$\text{Loss for this section} = 2(Kh)^2 \times \rho \frac{\sqrt{A_c/2}}{h\delta} \quad (1)$$

$$= \sqrt{2}(Kh)^2 \times \rho \frac{\sqrt{A_c}}{h\delta} \quad (2)$$

This is the loss for the chosen section of the plate. To extend this to the full geometry, we can sum the loss due to plate sections, where $\Sigma h = 2l_c$, the core length. Thus we can substitute $h = 2l_c$ into the equation which results in the following for the power loss in the whole plate:

$$\text{Total loss} = \frac{2\sqrt{2}H^2\sqrt{A_c}\rho l_c}{\delta} \quad (3)$$

$$= \frac{2\sqrt{2}B^2\sqrt{A_c}l_c\rho}{\mu^2\delta} \quad (4)$$

979-8-3315-1612-3/25 $31.00 © 2025 IEEE

TABLE I: Self inductances and losses of the simulated transformers.

	Primary (μH)	Secondary 1 (mH)	Secondary 2 (mH)	Avg. Core Loss (W)	Avg. Copper Loss (W)	Avg. Solid Loss (mW)
With Plane	179.58	2.85	2.89	1.69	2.71	19.1
Without Plane	176.26	2.80	2.84	1.60	2.71	0

We can further generalize this loss by finding the loss *in the plate* per unit volume *of core*, which is given by

$$\text{Loss per unit volume} = \frac{2\sqrt{2}B^2\rho}{\mu^2\delta\sqrt{A_c}} \tag{5}$$

$$= \frac{2\sqrt{2}\rho}{\mu^2\sqrt{\frac{2\rho}{\omega\mu_0}}\sqrt{A_c}}B^2 \tag{6}$$

$$= \sqrt{\frac{\omega\mu_0\rho}{A_c}}\frac{2B^2}{\mu^2} \tag{7}$$

$$= \sqrt{\frac{\omega\rho}{A_c\mu_0^3}}\frac{2B^2}{\mu_r^2} \tag{8}$$

For a well-designed magnetic component, the B field is likely to be $\hat{B}$, which is the excitation that yields a maximum acceptable core loss density, P_v (generally 200-500 mW/cm^3) [33].

A good quantity for comparison between materials, then, would be a ratio of the total plate loss to the total core loss, which is the same as the ratio of the core loss per unit volume and the plate loss per unit volume (of core). We would like for this loss ratio to be much less than 1:

$$1 \gg \frac{\text{Loss in Plate}}{\text{Loss in Core}} = \sqrt{\frac{\omega\rho}{A_c\mu_0^3}}\frac{2\hat{B}^2}{P_v\mu_r^2} \tag{9}$$

$$= \sqrt{\frac{8\pi\rho}{A_c\mu_0^3 f^3}\frac{\mathcal{F}^2}{\mu_r^2 P_v}} \tag{10}$$

Where $\mathcal{F}$ is the standard performance factor of the material defined as $\mathcal{F} = \hat{B}f$. Consider Fair-Rite 67 at 10 MHz with $\hat{B} = 10$mT, $P_v = 200$ mW/cm^3, $A_c = 0.2$ cm^2 and $\mu_r = 40$ with an aluminum plate ($\rho = 3.2 \cdot 10^{-8}$ $\Omega\cdot$m). Plugging these values into (10) gives 0.14, which indeed is less than 1, even with such a generous core loss density. Therefore, we can conclude that the plate in this scenario is likely to yield losses that are ~7x smaller than core loss.

Frequency has an interesting impact on the result. While frequency explicitly appears in the numerator in equation (9), we know that $\hat{B}$ itself is a decreasing function of frequency for a given material. Equation (10) reveals that, for constant performance factor, elevated frequency actually helps our cause. This is contrary to the usual assumption of higher losses at high frequency due to eddy currents. The reason for this is largely because the skin depth shrinks as the square root of frequency, but the *allowable* B field tends to shrink rapidly with frequency, and $|K| = |H| = |B|/\mu$. The loss ratio defined in (10) is plotted in Fig. 2 against frequency for different core materials whose performance factors are given in [33]–[35] with $P_v = 200$ mW/cm^3 and $A_c = 0.2$ cm^2. This plot shows that the loss ratio is always smaller than 1, but it suggests that plates have the least amount of loss in

the lower frequency range and may become too lossy in the higher frequency regime. We can extend this analysis to a wide application space, since (9) is a function of known material parameters and only one size parameter. If we know $\hat{B}$ and μ_r of the best-performing materials across frequency, we can make broad statements about the utility of such cooling plates.

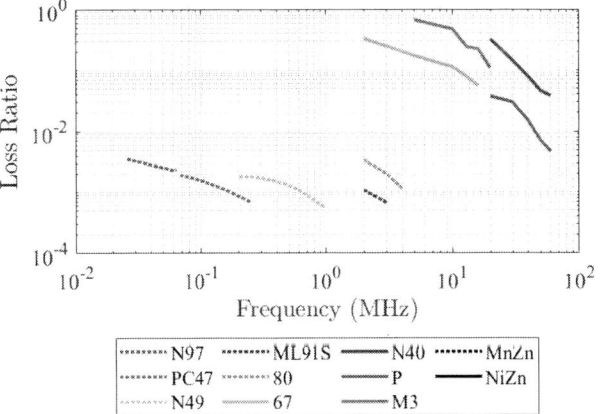

Fig. 2: Plot of (10) versus frequency for high $\mathcal{F}$ materials with $P_v = 200$ mW/cm^3 and $A_c = 0.2$ cm^2. Since all materials have a loss ratio < 1, this means the addition of aluminum cooling plates could be viable for a variety of operating frequencies. To further improve the high frequency loss ratios, one could allow for a larger P_v.

This calculation also reveals a strong dependence on μ_r. For example, consider N97 at 40 kHz, which has a performance factor of ~12,900 HzT, or a value of $\hat{B} = \frac{12,900\text{Hz T}}{40,000\text{Hz}} = 0.3101$T. This test point is convenient, as $\hat{B}$ is fairly close to the saturation flux density (at any lower of a frequency, the equations may need to be modified to apply B_{sat} as the limit). N97 has an initial permeability $\mu_r = 2300$. With this data and setting $P_v = 200$ mW/cm^3 and $A_c = 0.2$ cm^2 as before, (9) yields a ratio of 0.003, even smaller than the high-frequency case with Fair-Rite 67. This is because, while $\hat{B}$ increases as frequency is lowered, μ_r also increases substantially substantial (the ratio in (9) is proportional to $1/\mu_r^2$ which is $2300^2 = 5.29 \times 10^6$ for N97 versus $40^2 = 1.6 \times 10^3$ for 67). This is also exemplified in Fig. 2, where MnZn materials, which typically have $\mu_r > 500$, are predicted to experience significantly less loss as opposed to the NiZn materials with $\mu_r < 100$.

Larger core sizes tend to recommend the use of a cooling plate, as the added loss of a single plate is a weak function of core size, while core loss scales linearly with volume for

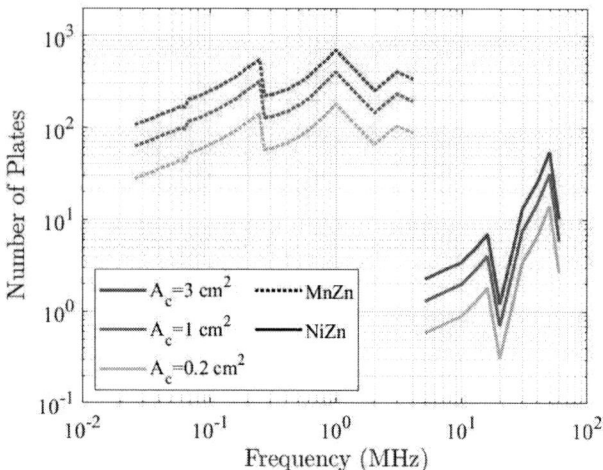

Fig. 3: The number (N) of allowable aluminum plates with an increase of 10% in loss budget for high $\mathcal{F}$ materials with different core sizes (A_c) with $P_v = 200$ mW/cm^3. This shows that larger core sizes can use more plates and that lower frequency materials are able to use significant number of plates with little additional loss.

constant B-field excitation. The advantage of interior cooling is also particularly pronounced for large cores.

If multiple plates are desired, we can multiply the result by N. Then, if we set an acceptable "added-loss" budget, we can determine the allowable N. Using the two data points from before as an example, we find that if we allow core loss (actual core loss plus added plate loss) to increase by 10%, then Fair Rite 67 at 10 MHz would almost allow one plate, while N97 at 40 kHz permits nearly 39 plates. We have plotted this same scenario for the highest $\mathcal{F}$ materials with different core sizes in Fig. 3 with $P_v = 200$ mW/cm^3. As expected, larger core sizes allow for the use of more cooling plates but the limit on the thermal improvement versus the number of plates is something that needs to be further explored. Nevertheless, we can see that the loss in the plate(s) no matter the core size, core material, or number of plates remains low as compared to the core loss, especially for lower frequency materials that are more common at high power (which is the target application of this thermal management technique).

III. SIMULATION RESULTS

A. Electromagnetic Simulations

To verify that the addition of a cooling plane within the transformer does not induce significant loss or alteration to the magnetic properties of the transformer, we performed a transient simulation using Ansys Maxwell. We designed a 8:32:32 center tapped transformer that is made of two sets of N87 E 42/21/20 cores combined to form one large core with a gap in the center post of ~0.2 mm.

The primary winding is litz wire of 435 strands of 40 AWG wire (~14 AWG equivalent) and each secondary is litz wire of

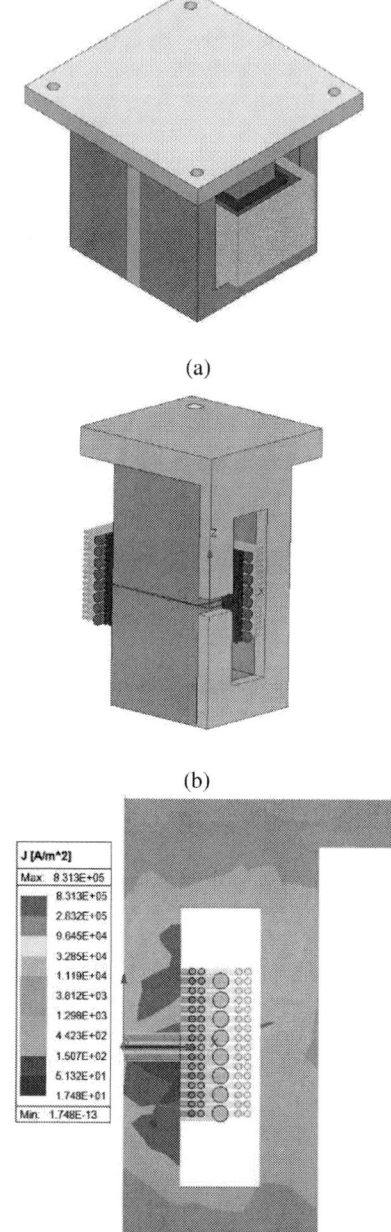

Fig. 4: (a) Full model of transformer. (b) Center view of one fourth of the simulation model of the designed transformer. (c) J field plot on the cooling plane showing small amounts of induced eddy currents.

420 strands of 48 AWG (~22 AWG equivalent). The cooling plane is a custom plate made of aluminum that is designed to fit the dimensions of the core. It is 5 mm thick and is the same height as an E core set so that it can be neatly sandwiched between the two core sets, as shown in Fig. 4a. The center post of the plate is cut to keep it away from the core gap since the fringing fields could induce a significant

amount of eddy current in the plate. The center view of the quarter model of the transformer is shown in Fig. 4b; the whole transformer is modeled by mirroring the quarter model across the XZ and YZ planes. A symmetry multiplier is applied in the quarter model for all of the calculations performed by Ansys. In Ansys Maxwell, the core loss model for material N87 has been applied. The aluminum plate had eddy effects applied in order to see the induced currents within the plate. The windings are excited using current waveforms that were generated using the LTspice simulation of the designed LLC converter.

The J field on the surface of the plate is plotted in Fig. 4c and shows that little current is induced in the cooling plane. Significant eddy currents are only induced near the gaps. The gap is only in the bottom E-core piece, which is why there is more current density on the bottom side of the cut in the gap. The maximum values are lower than typical winding current densities (500 A/cm^2), and we further recall that these currents exist only within ∼1 skin depth of the surface.

Table 1 shows the average values of the simulated copper and core loss for the whole transformer. The average solid loss is listed as well but it is only applicable for the plate transformer since it is the loss from the eddy currents induced in the aluminum plate. The total loss in the transformer increases to only 1.025 times its value without the plate, further verifying that the cooling plane does not compromise the performance of the transformer.

We also investigated the effects of the plane on the inductance of the transformer. Extracted data in Table 1 shows that the cooling plane only causes a slight variation between the simulated inductances, which indicates that the magnetic performance is very similar.

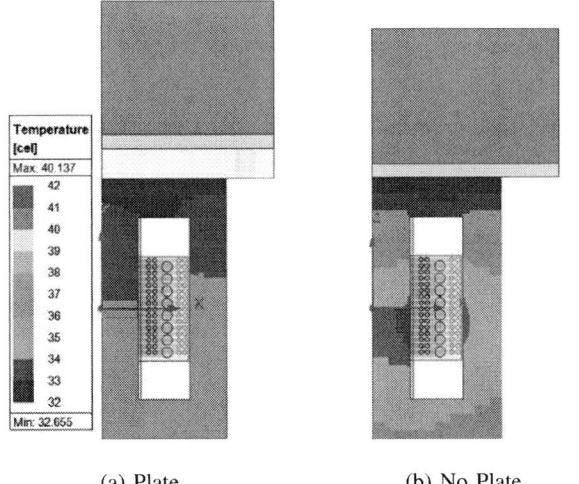

(a) Plate   (b) No Plate

Fig. 5: Plots of simulated temperature fields in Ansys Icepak for a cross section of the proposed transformer with (a) and without (b) the added cooling plane.

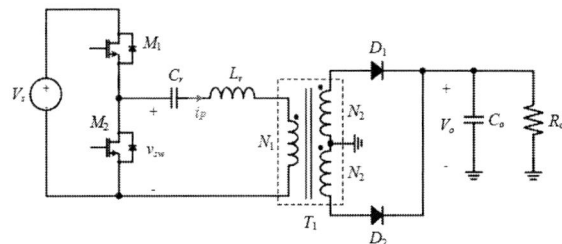

Fig. 6: Schematic of the 2 kW LLC converter in which the transformers are tested.

B. Thermal Simulations

Coupling the electromagnetic losses from the transient simulation of the transformer to Ansys Icepak allowed for thermal analysis of the transformer under full load condition. A heat sink was added to the top of both transformers and they were simulated to steady state using natural convection in ambient conditions. The results of this thermal simulation are pictured in Fig. 5 where the surface temperature fields for the inner core faces of the transformers are included. These images highlight that the cooling plane creates ∼ 5 − 6°C or a ∼ 12% improvement in the hot spot temperature of the core. The temperature of the core is also more uniform as heat from within the core has a more direct path to the heat sink.

IV. EXPERIMENTAL VALIDATION

The center-tapped transformer under analysis was designed to operate in a 2 kW 400 V to 800 V LLC converter which is similar to on-board chargers in EVs. The converter operates in open loop at 150 kHz and is used to drive the transformer so that its temperature can be measured in-situ. The schematic of the converter is featured in Fig. 6 and the prototype of the transformer is pictured in Fig. 7. As mentioned previously, the cooling plane is sandwiched between two sets of E 42/21/20 cores. The cooling plane is a 5 mm thick piece of aluminum that was cut using a CNC machine to achieve clean window cutouts for the windings and a smooth finish for mating with the core pieces. The windings are wrapped concentrically around the center post and are isolated from each other and the core using two layers of Kapton tape between each interface. The core sets are held in place by a 3D printed bobbin and with Kapton tape on the outside. Both transformers have the same heat sink attached to the top of the structures. The plate has threaded screw holes that allow for the heat sink to be securely fastened. To provide compression to the transformer without a plate, a zip-tie is used.

A thermocouple probe is placed inside the winding window, near the cooling plane on the center post of the primary side core set. This probe is connected to an EXTECH TM5000 thermocouple datalogger which is used to log the temperature of the center of the core during converter operation. The transformer is tested in ambient conditions as it is isolated from the circuitry which uses forced convection as shown in Fig. 9. Both trials ran until steady state temperature of the

TABLE II: Converter efficiency and self inductances of prototype transformer.

	Converter Efficiency	Primary (μH)	Secondary 1 (mH)	Secondary 2 (mH)
With plane	93.97%	176.95	2.78	2.76
Without plane	93.98%	178.1	2.81	2.79

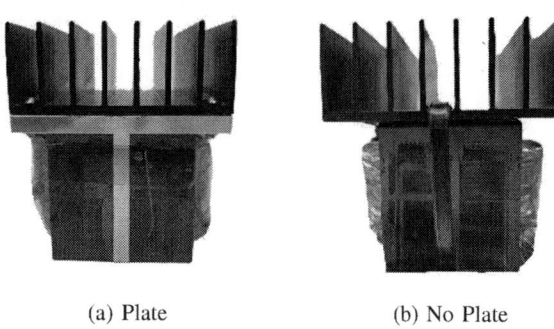

(a) Plate   (b) No Plate

Fig. 7: Transformers under test with added heat sink.

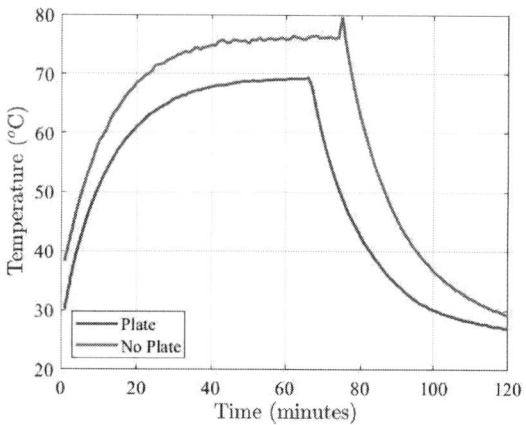

Fig. 8: Measured temperature of core vs. time showing an improvement in operating temperature with the addition of the cooling plane.

transformer was reached. As shown in Fig. 8, the addition of the cooling plane does improve the ΔT of the transformer by about 10% (7°C) once steady state operation is achieved. This result aligns well with the 12% difference found between steady state temperatures in the full power simulation of the transformers, however, the absolute temperatures of the cores in simulation are much cooler than the experimental values. Since this is a slightly more copper loss dominated design, the

Fig. 9: Experimental setup of the LLC converter operating at 2 kW with the transformer isolated from the rest of the circuitry.

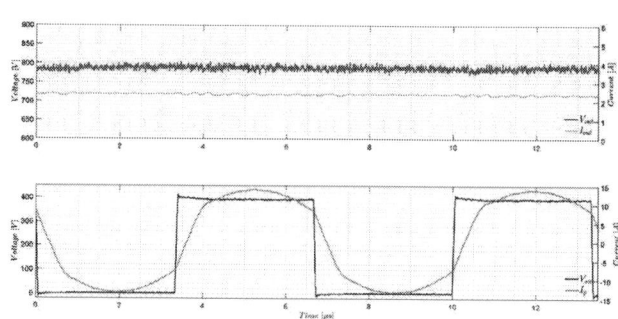

Fig. 10: Experimental waveforms from the LLC converter operating at 2 kW.

windings heat up more than the core but the effect of this heat on the temperature of the core is not as evident in simulation as it is in experimental measurements. Thus the temperature difference is likely due to the trapped heat from the windings increasing the measured temperature of the core. Nevertheless, the percent improvement in core temperature is very similar for both simulation and experimental results.

The slight peak in the no plate temperature is likely from EMI noise while turning the input voltage down by hand. It does not appear in the plate version because the datalogger did not capture the peak while the converter was being turned off.

The waveforms of the LLC converter operating at 2 kW are provided in Fig. 10. The switch node voltage and primary current are slightly out of phase to achieve ZVS. The phase angle and dead times required for ZVS is relatively constant with or without the cooling plane in the transformer despite the slight change in primary inductance as shown in Table II. The efficiency of each trial and of each transformer is also provided in Table II which are calculated from the measured input and output powers taken at steady state temperature from a Tektronix PA3000 power analyzer that was connected to the converter. The small variation of 0.01% demonstrates that the cooling plane has little effect on the efficiency of the system.

V. Conclusion

In this work, we demonstrate the benefit of adding a metallic cooling plane for the heat extraction from the interior of ferrite cores. This is verified with both electromagnetic/thermal simulations and with experiments using a 2 kW LLC converter and showed a $\geq$ 10%. This is the first work that aims to systematize the use of cooling plates within transformers to improve their thermal behavior with negligible effects on their operation.

REFERENCES

[1] D. J. Perreault, J. Hu, J. M. Rivas, Y. Han, O. Leitermann, R. C. Pilawa-Podgurski, A. Sagneri, and C. R. Sullivan, "Opportunities and challenges in very high frequency power conversion," in *2009 Twenty-Fourth Annual IEEE Applied Power Electronics Conference and Exposition*, 2009, pp. 1–14.

[2] J. G. Kassakian, D. J. Perreault, G. C. Verghese, and M. F. Schlecht, *Principles of Power Electronics*, 2nd ed. Cambridge University Press, 2023.

[3] *Ferrite Summary*, TDK Corporation, December 2021. [Online]. Available: product.tdk.com/en/system/files/dam/doc/product /ferrite/ferrite/ferrite-core/catalog/ferrite_summary_en.pdf

[4] M. Liserre, L. Costa, Z. Guo, D. D'Amato, S. S. Queiroz, and A. Huang, "Last developments and new technologies in solid-state transformer," in *2024 IEEE 15th International Symposium on Power Electronics for Distributed Generation Systems (PEDG)*, 2024, pp. 1–8.

[5] J. E. Huber and J. W. Kolar, "Volume/weight/cost comparison of a 1mva 10 kv/400 v solid-state against a conventional low-frequency distribution transformer," in *2014 IEEE Energy Conversion Congress and Exposition (ECCE)*, 2014, pp. 4545–4552.

[6] X. She, A. Q. Huang, and R. Burgos, "Review of solid-state transformer technologies and their application in power distribution systems," *IEEE Journal of Emerging and Selected Topics in Power Electronics*, vol. 1, no. 3, pp. 186–198, 2013.

[7] N. Rajagopal, R. Raju, T. Moaz, and C. DiMarino, "Design of a high-frequency transformer and 1.7 kv switching-cells for an integrated power electronics building block (ipebb)," in *2021 IEEE Electric Ship Technologies Symposium (ESTS)*, 2021, pp. 1–8.

[8] J. Carrasco, L. Franquelo, J. Bialasiewicz, E. Galvan, R. PortilloGuisado, M. Prats, J. Leon, and N. Moreno-Alfonso, "Power-electronic systems for the grid integration of renewable energy sources: A survey," *IEEE Transactions on Industrial Electronics*, vol. 53, no. 4, pp. 1002–1016, 2006.

[9] Q. Chen, R. Raju, D. Dong, and M. Agamy, "High frequency transformer insulation in medium voltage sic enabled air-cooled solid-state transformers," in *2018 IEEE Energy Conversion Congress and Exposition (ECCE)*, 2018, pp. 2436–2443.

[10] Z. Li, E. Hsieh, Q. Li, and F. Lee, "High-frequency transformer design with medium-voltage insulation for resonant converter in solid-state transformer," *IEEE transactions on power electronics*, 2023.

[11] Z. Li, E. Hsieh, Q. Li, and F. C. Lee, "High-frequency transformer design with medium-voltage insulation for resonant converter in solid-state transformer," *IEEE Transactions on Power Electronics*, vol. 38, no. 8, pp. 9917–9932, 2023.

[12] H. Wang, Z. Guo, S. M. Tayebi, X. Zhao, Q. Huang, R. Yu, Q. Yang, Y. Li, and A. Q. Huang, "Thermal design consideration of medium voltage high frequency transformers," in *2020 IEEE Applied Power Electronics Conference and Exposition (APEC)*, 2020, pp. 2721–2726.

[13] C. Chen, Z. Guo, and A. Q. Huang, "A 10kw/200khz pcb-winding transformer with high insulation voltage for solid-state transformer applications," in *2024 IEEE Applied Power Electronics Conference and Exposition (APEC)*, 2024, pp. 869–874.

[14] Z. Guo, S. Rajendran, J. Tangudu, Y. Khakpour, S. Taylor, L. Xing, Y. Xu, X. Feng, and A. Q. Huang, "A novel high insulation 100 kw medium frequency transformer," *IEEE Transactions on Power Electronics*, vol. 38, no. 1, pp. 112–117, 2023.

[15] Z. Guo, R. Yu, W. Xu, X. Feng, and A. Q. Huang, "Design and optimization of a 200-kw medium-frequency transformer for medium-voltage sic pv inverters," *IEEE Transactions on Power Electronics*, vol. 36, no. 9, pp. 10 548–10 560, 2021.

[16] A. Dey, N. Shafiei, R. Khandekar, W. Eberle, and R. Li, "Improving thermal performance of high frequency power transformers using bobbinless transformer design," in *2020 19th IEEE Intersociety Conference on Thermal and Thermomechanical Phenomena in Electronic Systems (ITherm)*, 2020, pp. 291–297.

[17] C.-W. Chang, X. Zhao, R. Phukan, R. Burgos, S. Uicich, P. Asfaux, and D. Dong, "Thermal consideration and design for a 200-kw sic-based high-density three-phase inverter in more electric aircraft," *IEEE Journal of Emerging and Selected Topics in Power Electronics*, vol. 11, no. 6, pp. 5910–5929, 2023.

[18] M. Ngo, Y. Cao, D. Dong, R. Burgos, K. Nguyen, and A. Ismail, "Forced air-cooling thermal design methodology for high-density, high-frequency, and high-power planar transformers in 1u applications," *IEEE Journal of Emerging and Selected Topics in Power Electronics*, vol. 11, no. 2, pp. 2015–2028, 2023.

[19] Y. Ruan, Y. Cao, D. Dong, and Q. Li, "A high-efficiency modular air-cooling method for pcb winding with the additive manufacturing," in *2023 IEEE Applied Power Electronics Conference and Exposition (APEC)*, 2023, pp. 449–455.

[20] L. Heinemann, "An actively cooled high power, high frequency transformer with high insulation capability," in *APEC. Seventeenth Annual IEEE Applied Power Electronics Conference and Exposition (Cat. No.02CH37335)*, vol. 1, 2002, pp. 352–357 vol.1.

[21] K. Li, L. Kui, X. Xie, X. Xie, X. Xie, K. He, H. Kai, Y. Lei, L. Yao, Z. C. Feng, F. Zhaozan, F. Zhaozan, T. Chen, and T. Chen, "Thermal design of a 2-phase flow cooled medium-frequency 140kva transformer for railway applications," *Vehicle Power and Propulsion Conference*, 2020.

[22] S. Tenbohlen, N. Schmidt, C. Breuer, S. Khandan, and R. Lebreton, "Investigation of thermal behavior of an oil-directed cooled transformer winding," *IEEE Transactions on Power Delivery*, vol. 33, no. 3, pp. 1091–1098, 2018.

[23] L. Chen, K. Zhou, D. Li, J. Zhao, X. Gao, R. Tong, and H. Du, "Temperature characteristics analysis of transformer cooling systems under variable loads and inter-turn short-circuit faults," in *2023 International Conference on Power System Technology (PowerCon)*, 2023, pp. 1–4.

[24] S. Beheshtaein, S. Beheshtaein, A. Alshafei, A. Alshafei, G. Jean-Pierre, G. Jean-Pierre, N. Altin, N. Altin, M. Khayamy, M. Khayamy, R. Cuzner, R. Cuzner, A. Nasiri, A. Nasiri, and A. Nasiri, "An optimal design of a hybrid liquid/air cooling system for high power, medium frequency, and medium voltage solid-state transformer," *International Symposium on Power Electronics for Distributed Generation Systems*, 2021.

[25] J. Biela and J. W. Kolar, "Cooling concepts for high power density magnetic devices," in *2007 Power Conversion Conference - Nagoya*, 2007, pp. 1–8.

[26] M. Pavlovsky, S. de Haan, and J. Ferreira, "Design for better thermal management in high-power high-frequency transformers," in *Fourtieth IAS Annual Meeting. Conference Record of the 2005 Industry Applications Conference, 2005.*, vol. 4, 2005, pp. 2615–2621 Vol. 4.

[27] R. A. Friedemann, F. Krismer, and J. W. Kolar, "Design of a minimum weight dual active bridge converter for an airborne wind turbine system," in *2012 Twenty-Seventh Annual IEEE Applied Power Electronics Conference and Exposition (APEC)*, 2012, pp. 509–516.

[28] P. Czyz, T. Guillod, D. Zhang, F. Krismer, J. Huber, R. Färber, C. M. Franck, and J. W. Kolar, "Analysis of the performance limits of 166 kw/7 kv air- and magnetic-core medium-voltage medium-frequency transformers for 1:1-dcx applications," *IEEE Journal of Emerging and Selected Topics in Power Electronics*, vol. 10, no. 3, pp. 2989–3012, 2022.

[29] R. W. Erickson and D. Maksimovic, *Fundamentals of Power Electronics*, 3rd ed. Springer Cham, 2020.

[30] Z. Guo, C. Chen, R. Yu, and A. Q. Huang, "A high power 200kw medium frequency transformer with improved thermal management," in *2023 IEEE Energy Conversion Congress and Exposition (ECCE)*, 2023, pp. 5587–5591.

[31] M. Pavlovsky, S. W. H. de Haan, and J. A. Ferreira, "Reaching high power density in multikilowatt dc–dc converters with galvanic isolation," *IEEE Transactions on Power Electronics*, vol. 24, no. 3, pp. 603–612, 2009.

[32] G. Ortiz, M. Leibl, J. W. Kolar, and O. Apeldoorn, "Medium frequency transformers for solid-state-transformer applications — design and experimental verification," in *2013 IEEE 10th International Conference on Power Electronics and Drive Systems (PEDS)*, 2013, pp. 1285–1290.

[33] A. J. Hanson, J. A. Belk, S. Lim, C. R. Sullivan, and D. J. Perreault, "Measurements and performance factor comparisons of magnetic materials at high frequency," *IEEE Transactions on Power Electronics*, vol. 31, no. 11, pp. 7909–7925, 2016.

[34] Y. Han, G. Cheung, A. Li, C. R. Sullivan, and D. J. Perreault, "Evaluation of magnetic materials for very high frequency power applications," *IEEE Transactions on Power Electronics*, vol. 27, no. 1, pp. 425–435, 2012.

[35] M. Solomentsev and A. J. Hanson, "At what frequencies should air-core magnetics be used?" *IEEE Transactions on Power Electronics*, 2022.

Folded Flex-PCB Winding Planar Transformer for High-Frequency Isolated DC-DC Converters

Soundhariya G. Soundararajan
Dept. of Electrical Engineering
KU Leuven- EnergyVille
Leuven, Belgium
gssoundhariya1999@gmail.com

Hans Wouters
Dept. of Electrical Engineering
KU Leuven - EnergyVille
Leuven, Belgium
hans.wouters@kuleuven.be

Wout Vanderwegen
Dept. of Electrical Engineering
KU Leuven - EnergyVille
Leuven, Belgium
wout.vanderwegen@kuleuven.be

Wilmar Martinez
Dept. of Electrical Engineering
KU Leuven - EnergyVille
Leuven, Belgium
wilmar.martinez@kuleuven.be

Abstract—**In modern high-frequency DC-DC converters such as electric vehicle (EV) chargers, transformers can commonly contribute to ~45% of the losses and a third of the box volume. Meanwhile, the magnetics can strain such converters' thermal management and electromagnetic interference (EMI), especially when targeting high power densities. This underlines the need for innovative solutions in magnetic design. One such innovation arises from recent advancements in flexible printed circuit board (PCB) manufacturing, allowing highly interleaved windings with sufficient copper thickness while eliminating termination losses between PCB boards. Therefore, this paper assesses flex-PCB designs for high-frequency transformers in EV chargers, addressing challenges and opportunities to conventional PCB-winding designs. The key modelling and trade-offs are detailed, considering the impact of parasitic capacitance, core loss, winding loss, leakage inductance, and thermal management for different configurations. From the analytical and simulation results, a design candidate with low transformer loss and parasitic effect is selected and experimentally characterised for its implementation in a 500 kHz CLLC converter.**

Index Terms—**Flex-PCB Transformer, Planar Magnetics, Design Optimisation, DC-DC converter, CLLC converter, EV Charging, On-Board Chargers**

I. INTRODUCTION

The automotive industry imposes strict demands for on-board chargers with high energy efficiency, compactness, and EMI management. Advancements in high-frequency wide bandgap transistors and soft-switching converter topologies have shifted the bottleneck for improved performance towards magnetic devices [1]. In response, novel winding techniques, optimisation algorithms, interleaving methods, integration strategies, and core structures have pushed the magnetic performance forward [2]–[5]. Planar transformers with printed circuit board (PCB) windings have generally emanated as a solution for low-profile design, good thermal dissipation,

This work was supported in part by the European Union's Horizon Europe under Grant No. 101056857 (PowerDrive) and FWO Research Foundation Flanders under Grant No. 1SHE524N.

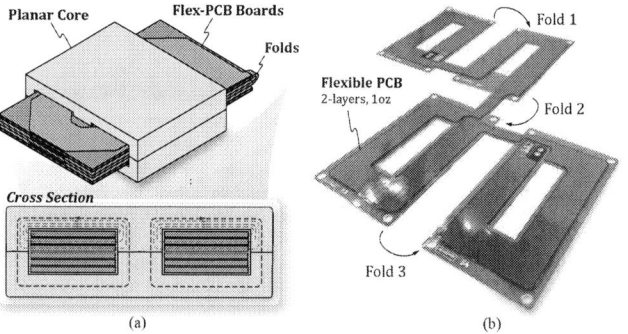

Fig. 1. (a) Structure of flex-PCB transformer (b) Flex-PCB geometry and proposed folding structure

controllable parasitics, and highly manufacturable designs. Furthermore, they enable interleaved designs, significantly reducing the AC proximity losses. However, board-to-board terminations significantly increase AC losses while being difficult to model, and parasitic capacitances between windings can limit the high-frequency performance of such devices.

Recent developments in flexible PCB manufacturing provide novel design opportunities. An illustration of a flex-PCB transformer is illustrated in **Fig. 1a**, where multiple layers of PCBs are strategically stacked around the magnetic core to create primary and secondary windings. These windings are depicted in **Fig. 1b**, and can be folded from a single sheet of flexible PCB material. Dielectric mediums such as FR-4 and polyamide, which provide robust high-voltage insulation, are utilised as substrate materials. Flex-PCBs offer compelling possibilities over traditional rigid PCBs, such as up to 80 % material savings, more design flexibility, and a significant drop in termination losses [6]. While previous research on flex-PCB transformers investigated better folding techniques, designs, and material minimisation [7], [8], their implementations in high-power converters remain limited. Being in their prelimi-

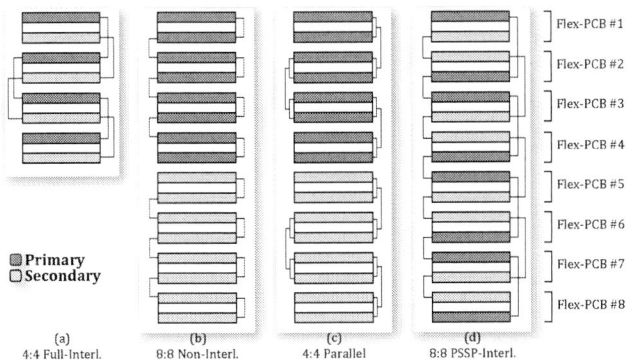

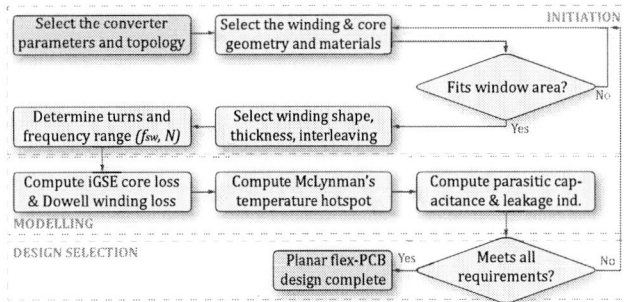

Fig. 3. Planar transformer design flowchart

Fig. 2. Flex-PCB transformers with different winding configurations: (a) 4:4 full interleaving (*FL1*), (b) 8:8 non-interleaving (*NL1*), (c) 4:4 parallel-connected non-interleaved (*P1*), (d) 8:8 PSSP-interleaving (*PI1*)

nary research stage, flex-PCB transformers open a vast window of opportunity to explore in terms of design and application.

This paper presents an in-depth analysis of flex-PCB transformers featuring novel interleaved winding configurations designed for DC-DC converters specifically tailored for EV charging systems to enhance power density and overall efficiency. The paper is structured as follows. **Section II** introduces the various flex-PCB transformer models selected for this study. **Section III** delves into the modelling of the transformer's magnetic, parasitic, loss, and thermal parameters. A thorough investigation of these transformer models and their associated design trade-offs is discussed in **Section IV**, supported by finite element analysis (FEA) of the chosen model. **Section V** showcases the experimental results obtained from the prototype testing, including a comparative analysis between the proposed flex-PCB transformer model and conventional rigid PCB transformers.

II. OPTIMISED FLEX-PCB TRANSFORMER DESIGN

Considering current manufacturing capabilities in commercial flex-PCB boards, four possible implementations of flex-PCB transformers with distinct winding configurations and copper thicknesses are selected, cf. Fig. 2. These designs are selected for the targeted 4-kW CLLC converter operable at a frequency range of 200 kHz to 800 kHz. An ELP64/10/50 core of 3F36 material known for its low core loss properties is chosen for the transformer based on the required power and frequency level. Based on the core geometry, a PCB copper width of 16 mm is used to limit the average current density. Overall, these parameters are selected to strike a balance between minimal winding loss and compact size while adhering to thermal and isolation requirements. The edges of the PCB copper traces are partially curved to prevent current crowding. The folding pattern of the windings with different interconnections is presented in **Fig.1b** and are mounted on the EI core, cf. **Fig.1a**. Standard FR-4 is selected as the substrate material due to its cost-effectiveness and high breakdown voltage of 20 kV/mm.

Several winding configurations can be established using a 2-layer flex-PCB board as depicted in **Fig. 2**. In the proposed four models, the fully-interleaved (*FL1*) 4:4 transformer of **Fig. 2a** and PSSP-interleaving (*PI1*) of **Fig. 2d** achieve a minimal mmf-profile. Meanwhile, the PSSP-interleaving also reduces the parasitic capacitance [2]–[5], [7], [8]. The 8:8 non-interleaved (*NL1*) structure of **Fig. 2b** can be constructed using two separate flex-PCB boards, similar to the 4:4 parallel (*P1*)-connected transformer of **Fig. 2c**, of which the parallel connected layers nearly double the current carrying capability. **Fig. 3** illustrates the transformer design procedure, featuring minimisation of the losses based on the selection of the number of turns, frequency, and winding configuration.

III. TRANSFORMER MODELLING

Transformer parameters reflect the transformer's nonlinearity and frequency-dependent characteristics. Hence, transformer parameter characterisation is crucial for the converter's reliable operation within a suitable range of frequencies and system excitation levels. The mathematical equations for estimating transformer design parameters are explained below.

A. Leakage Inductance

In general, leakage inductance is an undesirable characteristic affecting the system's stability. However, in a resonant converter, minimal leakage inductance is necessary to achieve a soft switching mechanism. This inductance allows the energy stored in the flex-PCB layers to provide the required current needed to charge and discharge the converter switches during dead time, minimising the commutation losses [9] and enhancing the converter efficiency. Equation (1) determines the leakage inductance needed for the soft switching operation of the DC-DC converter. It captures the dependency of leakage inductance on the proximity and skin effect for diverse winding geometries and interleaving configurations.

$$
L_k = \frac{n l_t}{2w} \sqrt{\frac{\mu_0\, \rho_{cu}\, \eta_p}{\pi f_{sw}}} \left[\frac{\sinh 2\xi + \sin 2\xi}{\cosh 2\xi - \cos 2\xi} \right.
$$
$$
\left. + \frac{2(m^2 - 1)}{3} \frac{\sinh \xi - \sin \xi}{\cosh \xi + \cos \xi} \right] \quad (1)
$$

Here, n is the number of winding turns, l_t is the mean turn length of the winding, μ_0 is the permeability of free space,

ρ_{cu} is the copper resistivity, η_p is the porosity factor of the windings, w is the width of the copper trace. m is the relative magneto-motive force (MMF) increase per layer given by the equation (2).

$$m = \frac{mmf(h)}{mmf(h) - mmf(0)} \qquad (2)$$

The eddy current generated in the winding is dependent on the skin depth δ and the height of the PCB trace h, which is given as a ratio $\xi = h/\delta$ and the skin depth is calculated using equation (3).

$$\delta = \sqrt{\frac{\rho}{\pi \mu_0 f_{sw}}} \qquad (3)$$

B. Magnetising Inductance

Like leakage inductance, magnetising inductance is vital for soft switching operation. A low magnetising inductance enables an extended soft switching region for a wider operating range and power handling capability, as seen in [10]. The circulatory current induced by the concerned inductance in the flex-PCB layers lowers the switching losses but increases the conduction losses marginally [11]. Therefore, the desired magnetising inductance is selected based on the trade-off between the conduction loss and the voltage variation range [12] while the soft switching operation is ensured for the operating frequency range as seen in equation (4).

$$L_m \leq \frac{t_d}{8(C_{os} + C_{stray})f_{sw}} \qquad (4)$$

Here, t_d is the dead time of the switch. C_{os} and C_{stray} are the output capacitance of the switches and stray capacitance of the transformer, respectively. The desired magnetising inductance is achieved by altering the air gap between the transformer cores.

C. Parasitic Capacitance

The parallel arrangement between the winding layers of the flex-PCB transformer induces a reprehensible parasitic capacitance in the device, which results in voltage spikes and current oscillations affecting the system's behaviour. Besides, these need to be discharged during the dead time to achieve zero voltage switching turn-on. Equation (5)-(11) computes the parasitic capacitance of planar transformers based on the inter-winding and intra-winding capacitance for different interleaving structures. The capacitance between two flex-PCB C_o is given in the equation (5) where the two PCB layers have an overlapping area S, separated by a distance d. ϵ_r is the relative permittivity of the dielectric medium and ϵ_o is the permittivity of free space.

$$C_o = \epsilon_0 \epsilon_r \frac{S}{d} \qquad (5)$$

The turn-to-turn capacitance C_d in a single layer of flex-PCB is calculated using equation (6) where n_t is the number of turns in each PCB layer.

$$C_d = \frac{(n_t + 1)(2n_t + 1)}{6n_t} C_o \qquad (6)$$

The intra-winding capacitance C_{ii} for each set of primary and secondary winding is estimated for different types of winding structures using equation (7) where n_{ii} is the number of PCB layers with the same winding.

$$\begin{aligned} C_{ii} &= \frac{4(n_{ii} - 1)}{n_{ii}^2} C_d \qquad n_{ii} > 1 \\ C_{ii} &= \frac{4}{n_{ii}^2} C_d \qquad n_{ii} = 1 \end{aligned} \qquad (7)$$

Equation (8) computes the inter-winding capacitance C_{ij} between different types of winding layers.

$$C_{ij} = n_{ij} C_d \qquad (8)$$

Here, n_{ij} is the number of PCB layers with different windings. The primary capacitance C_p and secondary capacitance C_s are calculated using the equations (9) and (10) [13].

$$C_p = C_{pp} + (1 - k)C_{ps} \qquad (9)$$

$$C_s = k^2 C_{ss} - k(1 - k)C_{ps} \qquad (10)$$

Here, k is the transformer turns ratio. C_{pp}, C_{ss} and C_{ps} are the primary intra-winding, secondary intra-winding and inter-winding capacitance. The overall stray capacitance of the transformer C_{str} is an aggregate of the primary and secondary capacitance as provided in equation (11).

$$C_{str} = C_p + C_s \qquad (11)$$

D. Core Loss

The Improvised Generalised Steinmetz Equation (iGSE) for arbitrary waveforms is adopted to compute core loss [14]. The equation (12) defines the empirical relation between the core loss per unit volume P_v, magnetic flux density B_m, switching frequency f_{sw} and time t.

$$P_v = \frac{1}{T} \int_0^t k_i \left| \frac{dB}{dt} \right|^\alpha \Delta^{\beta - \alpha} dt \qquad (12)$$

$$k_i = \frac{k}{(2\pi)^{\alpha - 1} 2\pi |\cos \theta|^\alpha 2^{\beta - \alpha} d\theta} \qquad (13)$$

Here, k, α and β are Steinmetz parameters which are obtained from the curve fitting graph in the ferrite core data sheet.

E. Winding Loss

Dowell's equation for winding loss of flat wires is restructured and applied for PCB winding where the orthogonality between proximity and skin effect is considered for different geometries and winding structures [15]. Equation (14) presents the ratio of AC resistance to DC resistance.

$$F_r = \frac{\xi}{2}\left[\frac{\sinh\xi + \sin\xi}{\cosh\xi - \cos\xi} + 2(m^2 - 1)\frac{\sinh\xi - \sin\xi}{\cosh\xi + \cos\xi}\right] \quad (14)$$

Here, F_r is the resistance ratio in which the first term is based on the skin effect factor and the second term represents the proximity effect factor. The DC and AC winding resistance R_{dc} and R_{ac} are presented in equation (15).

$$R_{dc} = \frac{n_l \, \rho_{cu} \, l_t}{h \, w} \quad \text{and} \quad R_{ac} = R_{dc} \, F_r \quad (15)$$

The total winding loss P_{cu} is calculated by equation (16).

$$P_{cu} = I_{out} \, R_{ac}^2 \quad (16)$$

Here, I_{out} is the output current. The total magnetic loss of a flex-PCB transformer P_{tloss} is the summation of the core loss and winding loss of the transformer.

F. Temperature Rise

Thermal management is a critical design challenge that ensures the transformer's reliability, longevity, and performance. In general, transformers have an operating temperature ranging from 40 °C to 130 °C. FR-4 and polyamide PCBs have temperature tolerance up to 150 °C and 300 °C respectively [16]. The McLyman thermal model is used for flex-PCB transformer thermal modelling [17]. This model offers a good balance between accuracy and computational complexity. The temperature rise in the transformer is calculated using the equation (17).

$$\Delta T = 450 \left(\frac{P_{tloss}}{A_e}\right)^{0.826} \quad (17)$$

Here, ΔT is the temperature rise, and A_e is the effective area for heat dissipation in cubic centimetres. From equation (18), the thermal resistance of the flex-PCB transformer R_{th} is estimated at approximately.

$$R_{th} = \left(\frac{450}{P_{tloss}^{0.174}}\right)\left(\frac{1}{A_e}\right)^{0.826} \quad (18)$$

IV. Design Trade-off Evaluation

The proposed four transformer models are analytically validated using the established mathematical models for frequencies ranging from 200 kHz to 1 MHz and copper thickness of 0.5 oz and 1 oz for the selected CLLC converter. **Figs. 4a-e** presents the core loss, winding loss, leakage inductance, parasitic capacitance, and temperature for the proposed four flex-PCB transformer models at 500 kHz and 1 oz winding thickness. **Fig. 5** displays the overall magnetic loss for different frequencies and copper thickness. Based on these analytical

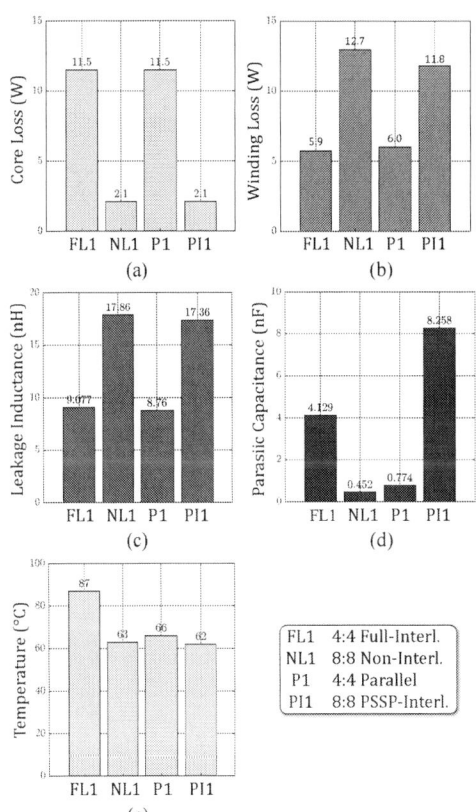

Fig. 4. Flex-PCB transformer results comparison for the four models at 500 kHz and 1 oz winding (a) Core loss (b) Winding loss (c) Leakage inductance (d) Parasitic capacitance (e) Temperature

findings, this section delves into the intricate design considerations to be made in the design of a flex-PCB transformer. While most of these are similar to the design of a rigid PCB transformer, **Section V-C** draws their comparison.

A. Number of Winding Turns

A critical aspect of transformer design is the selection of a number of winding turns. A higher number of winding turns increases the winding loss for the models; see 8:8 *NL1* and 8:8 *PI1* in **Fig. 4b**. This is attributed to the increase in the effective length of conduction, which proportionally increases the DC resistance of the winding. Also, with a higher number of flex-PCB conducting layers, the proximity factor elevates, resulting in proximity losses. Similarly, the leakage inductance is higher for these cases, as seen in **Fig. 4c**, which is because of poor flux leakage between the primary and secondary windings. On the contrary, the core loss is low for these models, as shown in **Fig. 4a**. With a higher number of turns in the winding, the inductance of the winding increases as $L \propto n^2$. This leads to a drop in the magnetic field strength per unit current, which weakens the AC magnetic flux density. This automatically reduces the core loss as $P_v \propto B_m$. Thus, an optimum balance between the core and winding loss should be made to keep the overall transformer loss low. This can be seen in **Fig. 5** where 8:8 NI1 - 1 oz and 8:8 *PI1* - 1 oz with eight turns have

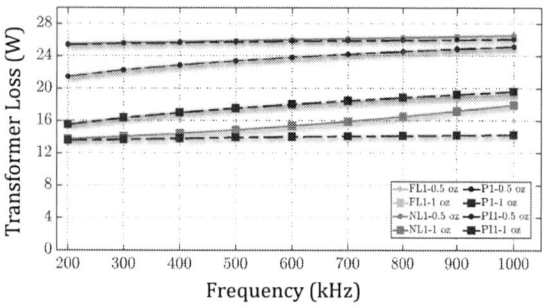

Fig. 5. Flex-PCB transformer total magnetic loss comparison for different frequencies and copper thickness

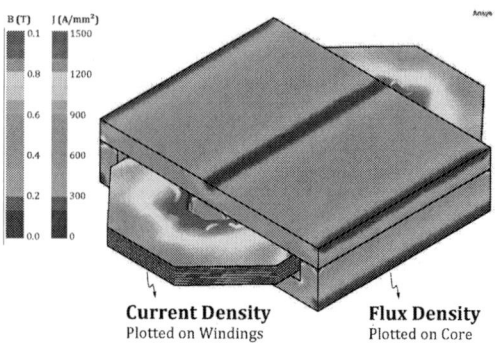

Current Density
Plotted on Windings

Flux Density
Plotted on Core

Fig. 6. Ansys Maxwell eddy current simulation results for *PI1* structure at 500 kHz

low transformer loss compared to the other two models with 4 turns each. Although these models have low transformer loss, the leakage inductance is high, which can be reduced by altering the winding configuration.

B. Winding Configuration

Another pivotal challenge in transformer design is choosing an appropriate winding technique. The interleaving flex-PCB transformer models - 4:4 *FI1* and 8:8 *PI1* have low leakage inductance and winding loss in comparison with their counterparts with the same number of winding turns. These models have improved flux leakage cancellation and a uniform magnetic field distribution between the windings. However, the stray capacitance for these two models rises significantly in the ratio of 5 and 18 in contrast with their counterparts. This upsurge is due to the continuous interaction between the adjacent layers of the primary and secondary windings. These windings, being close to one another, get easily affected by each other's electric field and store opposite electric charges, which leads to an increase in their capacitance level. The 4:4 *P1* winding model- 1 oz has a low stray capacitance of 0.774 nH and provides better thermal handling capacity, but it causes material wastage and higher manufacturing cost, as seen in **Figs. 4de**. On the contrary, the 8:8 *NL1* - 1 oz model has the least parasitic capacitance of 0.452 nH because of less interaction between the primary and secondary winding, which minimizes the proximity effect. However, it has a high transformer loss compared to its counterpart. This parasitic component is an unsettling component that negatively impacts the system's behaviour. Hence, a transformer model with the least parasitic component is an ideal choice for selection. Here, the non-interleaving winding configuration is a good choice when operated at a frequency where the transformer loss converges $\pm$ 2.5 % with an interleaving arrangement as seen in **Fig. 5** between frequencies 200 kHz to 1 MHz.

C. Copper Thickness

The thickness of copper windings in the flex-PCB transformer is a key factor for minimising the parasitic component and power loss. With thick copper wires as seen in **Fig. 4c** and **Fig. 5**, the cross-sectional area of the winding increases, which inversely reduces the resistance of the windings (R_{dc}

$\propto 1/A_w$), resulting in lower copper losses as seen in **Fig. 5**. Further, with thick copper layers, the dielectric between the PCB layers is also thicker, which weakens the electric field's ability to store electric charges and reduces parasitic capacitance. Finally, thicker copper traces allow for slimmer turns to be used, reducing the capacitance between adjacent turns. Overall, thick copper traces offer significant merits. However, there is a constraint in terms of the commercial availability of copper traces with a thickness above 1 oz for flexible PCBs.

D. Switching Frequency

CLLC converters are prominently known for operating at high frequencies and have higher switching and transformer loss. It is necessary to find an optimum switching frequency that will equip these losses and improve system performance. In flex-PCB transformers, at low frequencies, the transformer loss is less due to the minimal impact of skin and proximity effect, but there is a high leakage inductance between frequencies 200 kHz to 600 kHz because of weak magnetic field strength. Inevitably, a trade-off needs to be made between the leakage inductance and transformer loss. A frequency of 500 ~ 600 kHz is suitable for efficient converter operation, as depicted in **Fig. 5**.

Based on the analytical results and design trade-offs discussed, notably, the *PI1*-structure using PSSP-interleaving in an 8:8 transformer achieves the lowest overall loss (13.9 W) for 1 oz copper thickness while reducing the parasitic capacitance compared to an 8:8 fully-interleaved structure. Moreover, the excessive core losses of 4:4 transformers would strain the thermal management, indicating an important limitation in flex-PCB design. Namely, while the number of turns achievable in a single PCB layer is increased, the total achievable number of turns is still limited. Thus, their field of application leans towards high-frequency converters, in which the high frequency reduces the volt seconds of the excitation.

The high capacitance of the PSSP-interleaving is overcome by the non-interleaved *NL1*-winding, which obtains a very low inter-winding capacitance. However, the power loss for this model increases significantly, apprising the selection of an appropriate operating frequency. Thus, the 8:8 transformer

979-8-3315-1612-3/25 $31.00 © 2025 IEEE

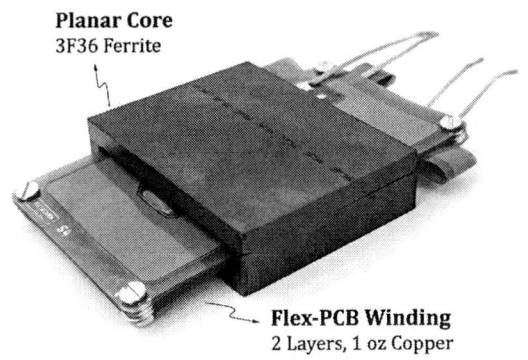

Planar Core
3F36 Ferrite

Flex-PCB Winding
2 Layers, 1 oz Copper

Fig. 7. Flex-PCB transformer prototype featuring PSSP-interleaved windings structure and a 3F36 planar EI-core

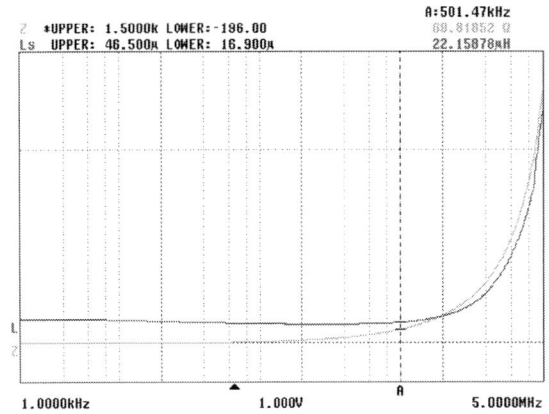

Fig. 8. Experimental characterisation of the inductances measured with a Hioki impedance analyzer up to 5 MHz.

with PSSP-interleaving having a copper thickness of 1 oz and operating at a frequency of 500 kHz is selected and validated in Ansys Maxwell in **Fig. 6**. The FEA simulation confirms a uniform magnetic field distribution across the core and an acceptable current distribution in the windings.

V. EXPERIMENTAL VALIDATION

This section presents the prototype of the selected transformer 8:8 PSSP-interleaving structure for high-frequency EV charging applications. The construction, experimental characterization and scope of this novel transformer design are elucidated below.

A. Transformer Prototype

The transformer prototype for the PSSP-winding structure, i.e., 8:8 *PII*, is built, cf. **Fig. 7**. Planar E64/10/50 and complementary I core of 3F36 material suitable for high-frequency application are combined to form the transformer core. The eight primary and secondary turns are built in the desired pattern around the core, using two discrete flex-PCB boards. Each board has four 2-layers within which the winding turns are distributed, as depicted in **Fig. 1b**. The PCB layers are designed to have minimal termination losses and smooth thermal dissipation. The spacing between the layers around the transformer core is adjusted using screws; see **Fig. 7**.

B. Experimental Characterisation

The experimental characterisation of the inductances for the 8:8 *PII* structure is presented in the **Fig. 8** and **Fig. 9**. A Hioki IM3570 Impedance Analyzer was used to measure the transformer inductance and impedance. The results show stable leakage and magnetizing inductance over a frequency range of 100 Hz to 500 kHz, as illustrated in **Fig. 8**. At the operating frequency of 500 kHz, the primary and secondary self-inductance measure is 22.1 µH, while the total leakage inductance equals 0.19 µH for a magnetising inductance of 22.0 µH. This indicates the need for an additional series of resonant inductors for use in resonant-type isolated dc-dc converters.

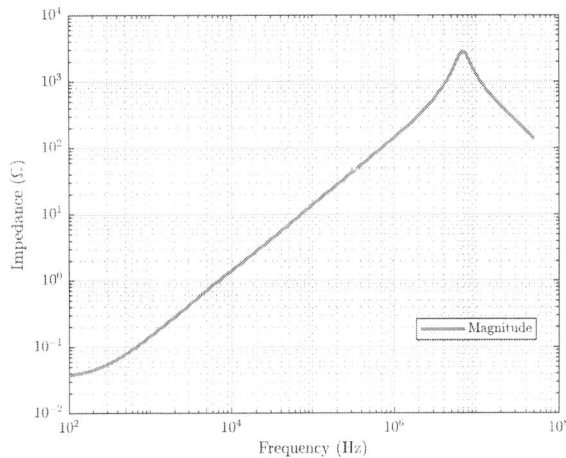

Fig. 9. Experimental characterisation of the open-circuit impedance, indicating the first self-resonant frequency at 6.98 MHz when measured with a Bode 100 network analyzer.

Due to the high resonant frequencies, which lie even outside the measurement range of up to 50 MHz, the parasitic capacitances cannot be accurately measured according to [18]. The self-resonant frequency induced by the primary intra-winding inductance and the primary self-inductance is observed at 6.98 MHz as seen in **Fig. 9**. The measurement is performed using an Omikron Lab Bode 100 Network Analyzer where the secondary winding is open-circuited, according to [18].

The experimentally measured inductances and analytically determined parasitic capacitances are implemented in a circuit simulation model of the converter in Plexim PLECS. Herein, the zero voltage turn-on of the transistors can be confirmed as preliminary validation before converter implementation. As shown in **Fig. 10**, zero voltage switching can indeed be achieved, even with consideration of the transformer and transistor parasitic capacitances.c

979-8-3315-1612-3/25 $31.00 © 2025 IEEE

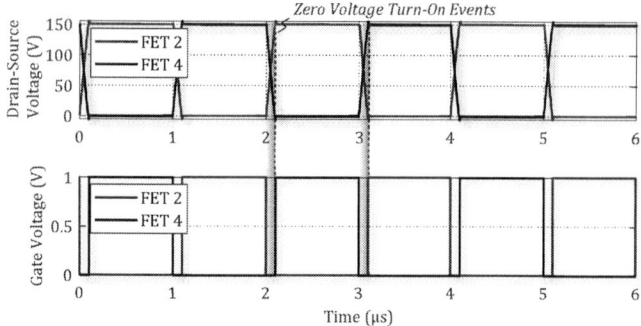

Fig. 10. Circuit simulation results confirming zero voltage switching at full-power 500 kHz operation with voltages and currents in the primary and secondary side resonant tanks indicating resonant operation.

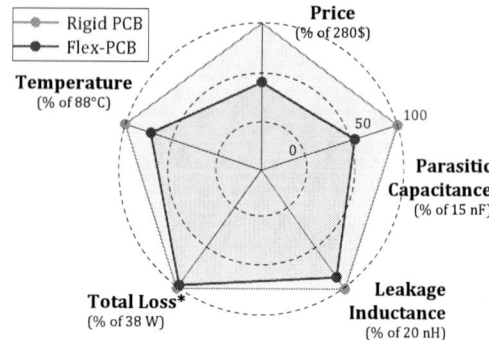

Fig. 11. Comparative analysis of the proposed transformer versus a rigid PCB reference design

C. Flexible vs Rigid PCB Transformer

Based on these findings, the 8:8 *PI1* flex-PCB transformer is compared with a conventional rigid 8:8 *PI1* transformer design and their results are summarised in **Fig. 11**. The rigid transformer consists of four PCB boards, each with four layers. One of the most notable features of flex-PCBs is their low parasitic capacitance, despite having thinner dielectric layers. This is due to the possibility of increasing the air-gap spacing between the PCB films in each board. While thicker rigid PCBs are also available, this inadvertently increases the cost significantly, especially when requiring sufficient copper thickness. Additionally, the flex-PCB design allows for tighter coupling between winding, resulting in lowered leakage inductance. It is worth noting, however, that flex-PCB transformers with numerous winding layers may experience increased leakage inductance due to the cumulative effect of air gaps on the porosity factor [19], [20].

Flex-PCB transformers shine when it comes to thermal management, as their design facilitates better heat dissipation and improved airflow. This is especially true when relying on air-cooling of the windings, which is typically the case as expensive potting materials and manufacturing processes counteract the benefit of PCB windings. Therefore, even with similar resistive losses, the thermally-limited transformer design is facilitated, and a more compact transformer could be possible. Besides, by eliminating the terminations between boards, termination losses are reduced, and fewer production steps are required. Still, the flex-PCB is to be connected to the converter's motherboard. Future research could explore more complex models incorporating vias and terminations for more accurate approximations [21].

For relatively low-power applications, flex-PCB transformers offer an economical solution compared to rigid PCBs that require boards with a high number of layers or multiple PCBs. However, as power requirements increase, so does the cost of these transformers. Additionally, their delicate structure poses challenges for widespread adoption in high-power industrial applications [22], especially as creepage and clearance become a greater constraint. Based on these findings, however, flex-PCB transformer designs can be further explored for relatively low-power onboard charging applications where their unique advantages can be fully leveraged.

VI. CONCLUSION

In response to modern challenges in automotive converters, this paper analyses and validates novel flex-PCB transformer designs for CLLC resonant converters in bidirectional EV chargers. Flex-PCBs, known for their cost-effectiveness and design flexibility, were optimized to enhance power density, efficiency, and manufacturability. Among the investigated configurations, the 8:8 transformer with PSSP-interleaving (*PI1*) proved the most promising, offering low overall loss and overcoming challenges due to parasitic capacitance. Experimental results confirmed that the *PI1* structure maintains stability up to 1 MHz. The proposed flex-PCB design shows superior performance and cost-effectiveness compared to conventional PCB designs, though limitations remain in low-frequency and high-current scenarios due to fewer winding layers and lower copper weight. Future work will focus on the further implementation and integration of the flex-PCB transformer into the charger and evaluating the long-term reliability of flex-PCB transformers in demanding automotive environments.

REFERENCES

[1] H. Wouters and W. Martinez, "Bidirectional Onboard Chargers for Electric Vehicles: State-of-the-Art and Future Trends," in IEEE Transactions on Power Electronics, vol. 39, no. 1, pp. 693-716, Jan. 2024, doi: 10.1109/TPEL.2023.3319996.

[2] H. Wouters, H. Pervaiz, T. Geboers, Y. Zuo, W.-R. Lin, and W. Martinez, "Interleaved PCB winding planar transformer for electric vehicle charging CLLC converters," in 2024 IEEE Applied Power Electronics Conference and Exposition (APEC), pp. 3216–3223, [Online]. Available: https://ieeexplore.ieee.org/document/10509410

[3] C. Sullivan, "Cost-constrained selection of strand diameter and number in a litz-wire transformer winding," vol. 16, no. 2, pp. 281–288.

[4] B. Li, Q. Li, and F. C. Lee, "High-frequency PCB winding transformer with integrated inductors for a bidirectional resonant converter," vol. 34, no. 7, pp. 6123–6135.

[5] Y. Zhang, Y. Chen, D. Xu, K. Mino, and Y. Okuma, "Utilizing flexible printed circuit board (FPCB) to realize passives integration in LLC resonant converter," in 2008 Twenty-Third Annual IEEE Applied Power Electronics Conference and Exposition, pp. 1465–1471, [Online]. Available: https://ieeexplore.ieee.org/document/4522917

[6] Buttay, Cyril, et al. "Application of the PCB-embedding technology in power electronics–State of the art and proposed development." 2018 Second International Symposium on 3D Power Electronics Integration and Manufacturing (3D-PEIM). IEEE, 2018.

[7] G. K. Y. Ho, C. Zhang, B. M. H. Pong, and S. Y. R. Hui, "Modelling and analysis of the bendable transformer," vol. 31, no. 9, pp. 6450–6460, conference Name: IEEE Transactions on Power Electronics. [Online]. Available: https://ieeexplore.ieee.org/document/7328745

[8] Z. Ouyang, O. C. Thomsen, and M. A. E. Andersen, "Optimal design and tradeoff analysis of planar transformer in high-power DC–DC converters," vol. 59, no. 7, pp. 2800–2810.

[9] A. Chandwani, and A. Mallik,"Parametric modeling and characterisation of leakage-integrated planar transformer for CLLC DC-DC converter," IEEE Transactions on Magnetics, 58(6):1–8, 2022.

[10] W. G. Hurley and, W. H. Wolfle, "Transformers and inductors for power electronics: Theory, design and applications." John Wiley and Sons, 2013.

[11] B. Lu, W. Liu, Y. Liang, F. C. Lee, and J. D. Van Wyk.,"Optimal design methodology for LLC resonant converter'" In Twenty-First Annual IEEE Applied Power Electronics Conference and Exposition, 2006. APEC'06., pages 6–pp. IEEE, 2006.

[12] Zhang, Zhengda, et al. "High-efficiency high-power-density CLLC resonant converter with low-stray-capacitance and well-heat-dissipated planar transformer for EV on-board charger." IEEE Transactions on Power Electronics 35.10 (2020): 10831-10851.

[13] L. Deng, P. Wang, X. Li, H. Xiao, and T. Peng, "Investigation on the parasitic capacitance of high frequency and high voltage transformers of multi-section windings," vol. 8, pp. 14 065–14 073, conference Name: IEEE Access. [Online]. Available: https://ieeexplore.ieee.org/document/8959179

[14] J. Muhlethaler, J. Biela, J. W. Kolar, and A. Ecklebe, "Improved core-loss calculation for magnetic components employed in power electronic systems," vol. 27, no. 2, pp. 964–973.

[15] P. L. Dowell, "Effects of eddy currents in transformer windings," vol. 113, no. 8, pp. 1387–1394, publisher: IET Digital Library.

[16]] Anonymous, "Polyimide pcb: A comprehensive guide for beginners," https://hilelectronic.com/polyimide-pcb/, 2024.

[17] C. W. T. McLyman, "Transformer and inductor design handbook," 3rd ed., ser. Electrical and computer engineering. Marcel Dekker, no. 121.

[18] C. Liu, L. Qi, X. Cui and X. Wei, "Experimental Extraction of Parasitic Capacitances for High-Frequency Transformers," in IEEE Transactions on Power Electronics, vol. 32, no. 6, pp. 4157-4167, June 2017, doi: 10.1109/TPEL.2016.2597498

[19] Whitman, Daniel J. "The Effect of Winding Curvature and Core Permeability on the Power Losses and Leakage Inductance of High-Frequency Transformers." (2021).

[20] Tria, Lew Andrew. Modelling and design of planar PCB transformers for high-frequency DC-DC converters. Diss. UNSW Sydney, 2017.

[21] M. K. Ranjram, P. Acosta and D. J. Perreault, "Design Considerations for Planar Magnetic Terminations," 2019 20th Workshop on Control and Modeling for Power Electronics (COMPEL), Toronto, ON, Canada, 2019, pp. 1-8, doi: 10.1109/COMPEL.2019.8769642

[22] De Jong, Erik CW, Braham JA Ferreira, and Pavol Bauer. "Toward the next level of PCB usage in power electronic converters." IEEE Transactions on Power Electronics 23.6 (2008): 3151-3163.

979-8-3315-1612-3/25 $31.00 © 2025 IEEE

Winding Strategy Analysis and Optimization for High-Current Matrix Transformer

Bima Nugraha Sanusi, Pinhe Wang, Michael A.E. Andersen, and Ziwei Ouyang

Department of Electrical and Photonics Engineering

Technical University of Denmark (DTU)

Kongens Lyngby, Denmark

bnusa@dtu.dk, piwa@dtu.dk, maea@dtu.dk, ziou@dtu.dk

Abstract—**Matrix transformer exhibits characteristic that can create low winding loss in high current application. This paper outlines the modeling, analysis, and optimization of winding strategy in the proposed matrix transformer structure. The impact of multiple interleaving structure, the number of winding layers, and copper volume to loss are analyzed. From 2D FEM analysis, the interleaving structure with the lowest AC resistance ratio is found. However, the 3D FEM analysis shows an increased AC resistance, due to the not captured proximity effect in part of the winding window. This fact is often overlooked in matrix transformer design and can change the design result. The analysis is verified in a DCX prototype achieving 91.3% efficiency and 488A output current capability.**

Index Terms—**matrix transformer, high current, DC transformer, winding design**

I. BACKGROUND

The New York Times has reported that, by 2027 AI servers could use between 85 to 134 TWh annually [1]. Most power converter in the data center will have a step-down function, which is natural because of the very low operating voltage of modern processors (below 1V). Several state-of-the-art solutions include switched capacitors topology [2], PWM switched inductors [3], and series resonant converter [4]. Among them, the LLC resonant converter is highly suitable for DC transformer (DCX) application, e.g. from 400V to 48V [5], or from 48V to 1.8V [6]. This topology allows for a high efficiency design, thanks to the primary switch zero voltage switching (ZVS) and secondary rectifier zero current switching (ZCS) operation. Furthermore, a minimum footprint can also be achieved, thanks to the matrix transformer implementation.

Matrix transformer characteristics matches well with the high step-down ratio and high output current requirement. It enables single turn implementation in both secondary and primary windings, rendering a low winding loss. It is often used in LLC resonant converter with full wave rectifier [4], [7] to limit the RMS current going through the secondary winding. These previous works optimized the transformer design with regard to core loss versus winding loss trade-off. However, there are several parameters in the winding design which have not been analyzed. Besides, a new design is presented in this work to create a higher turns ratio in the transformer, without sacrificing the area.

This work was supported by the European Research Council (ERC) Consolidator Grant under project H3PMAG.

Therefore, this paper aims to analyze and optimize the winding arrangement in a matrix transformer with high output current. A new transformer structure is proposed to suit the high step-down ratio of the DCX converter. The transformer design will be described and modeled in detail. In particular, multiple interleaving structure, the number of layers and, the copper volume impact to winding loss will be compared. The 2D and 3D FEM simulation will be used to find the optimal winding structure and calculate the AC resistance. Next, a measurement based on a selected case will verify the result. Finally, a design optimization is performed based on the simulation result and a prototype is built for verification.

II. A HIGH-CURRENT MATRIX TRANSFORMER

The matrix transformer is applied to the series resonant converter in Fig. 1. The converter utilizes half-bridge inverter on the primary and full-wave rectifier on the secondary side. It is designed to achieve a voltage step-down ratio (V_o/V_i) of 48. Therefore the transformer will have a 24:1 turns ratio with multiple parallel phases (N_{phs}) on the secondary, as shown by the blue boxes.

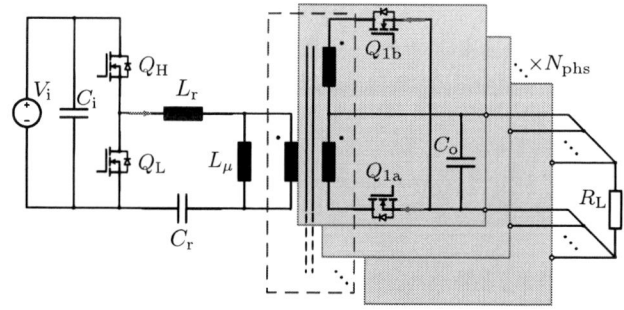

Fig. 1: Series resonant converter where the high-current matrix transformer (dashed line box) is used. Parallel phases on the secondary are shown by the blue boxes.

The new high-current matrix transformer structure is described in Figure 2. The core structure is shown in Fig. 2(a) along with its dimensions. There are 2 identical cores stacked vertically. A single unit consists of one center leg with the size of $C_l \times C_m$, 12 cylindrical side legs with the diameter of D_c, the top and bottom plates with the thickness of C_t. An airgap of l_g can be inserted to control the magnetizing inductance.

979-8-3315-1612-3/25 $31.00 © 2025 IEEE

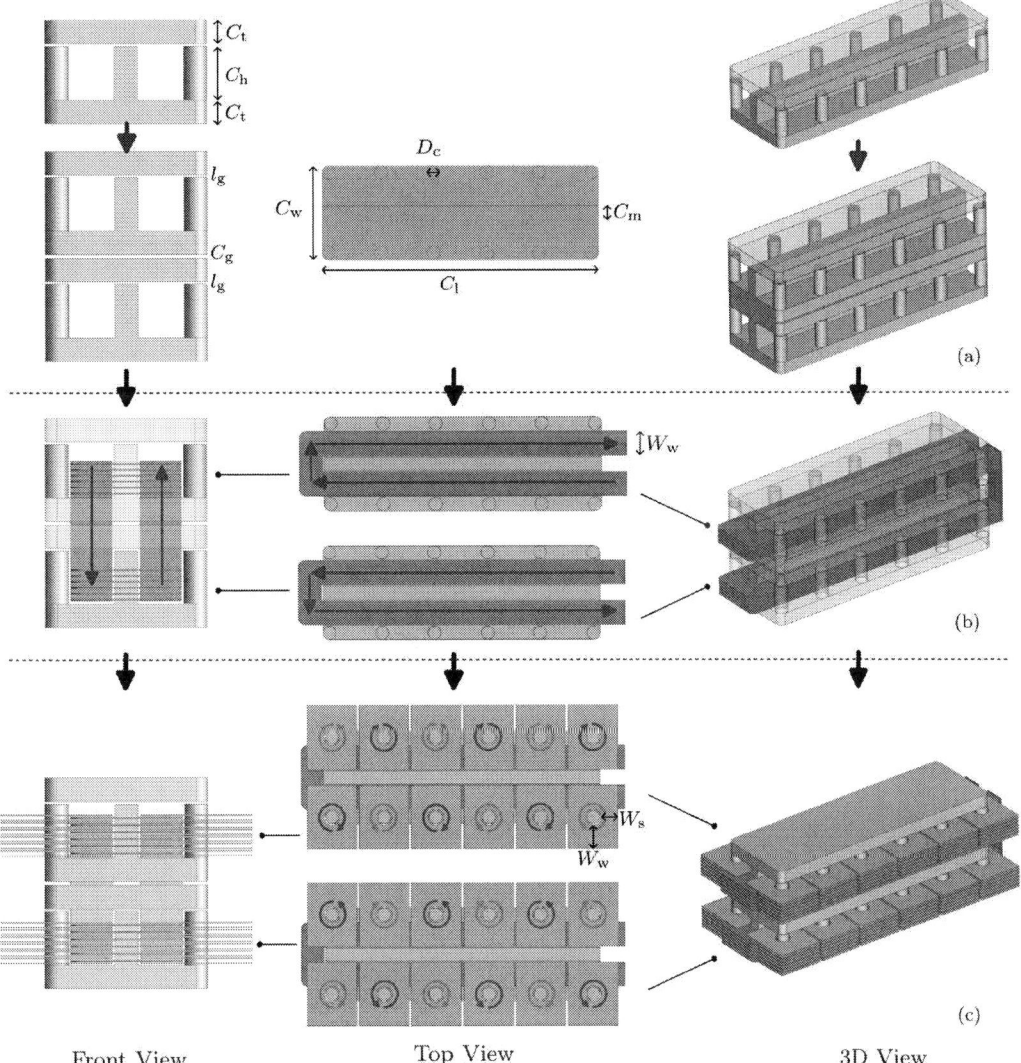

Front View   Top View   3D View

Fig. 2: The high-current matrix transformer structure: (a) magnetic core structure and dimension, (b) primary winding is added to the core structure, (c) secondary windings are added to the structure. The total achieved turns ratio, when seen as a single transformer, is $n_t = 24$. The front, top and isometric 3D view are shown here.

The gap C_g between the 2 cores is used to isolate the operation of both cores.

The two transformer core unit are connected by a single primary winding, as shown in Fig. 2(b). There is only one turn for each core. The winding from the top core is connected to the bottom core's by the vertical connection at the back of the transformer, as can be seen on the right hand side of Fig. 2(b). The red arrows in the top and front view pictures show the primary winding direction. This method offers an attractive way to create vertical power delivery structure, using the third dimension to minimize converter footprint.

Next, the transformer secondary winding is integrated into the structure. Figure 2(c) displays the final structure. The secondary windings are wound around the cylindrical side legs and each leg will create one center tap rectifier phase.

It means there are 2 windings with opposite polarity around each side leg. Furthermore, there will be 12 phases for 1 core unit, and 24 phases for the complete matrix transformer. The top view picture shows the polarity of the secondary windings. During the positive half-cycle, if the primary current flows following the red arrows in Fig. 2(b), the secondary windings with blue arrows will conduct. The other half of the center tap will conduct during the other half-cycle.

There are 2 variables that define the windings width: W_w and W_s. The primary winding is only controlled by W_w while, the secondary is controlled by W_w and W_s. These variables are defined in Eqn. 1 and Eqn. 2. As one of the design criterion, the transformer area (A_{xfmr}) is defined as Eqn. 3.

$$W_w = \frac{C_w - C_m - 2 \cdot D_c}{2} \qquad (1)$$

979-8-3315-1612-3/25 $31.00 © 2025 IEEE   247

$$W_s = \frac{C_l - 6 \cdot D_c}{5} \tag{2}$$

$$A_{xfmr} = (C_l + 2.W_w) \cdot (C_w + 2.W_w) \tag{3}$$

III. WINDING ANALYSIS BASED ON 2D FEM

A. Interleaving Structure Impact

The first parameter to be analyzed is the impact of interleaving structure to winding resistance. Winding interleaving is one of the most effective strategy to mitigate the high AC resistance [8]. There are 4 considered structures in this work and they are shown in Fig. 3, where 2 parallel layers for each winding are shown in the figure for illustration. More layers can be added and the pattern repeats. These cross sections are the matrix transformer's cross section between the center leg and one of side legs. Secondary A and B each represent half of the center tap secondary.

In this 2D analysis, the assumed dimensions are listed in Tab. I, where W_t is the copper thickness and W_i is the insulation thickness between two layers. There is also a 0.2 mm distance between the copper and the core, accounting for the manufacturing limitation.

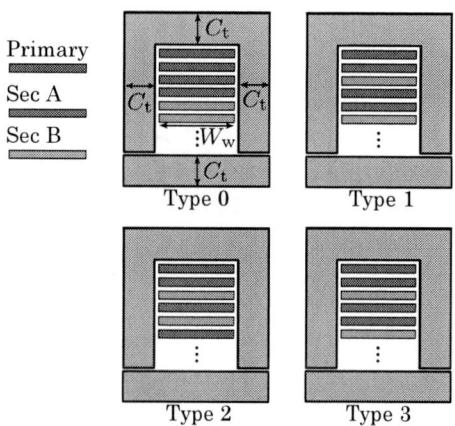

Fig. 3: Interleaved winding structure considered for analysis. Only 2 parallel layers for each winding are shown here.

TABLE I: Parameters for Interleaving Structure Analysis

Parameter	Value
W_w	3.6 mm
W_t	100 μm
W_i	100 μm
C_t	1.5 mm

The AC to DC resistance ratio is presented in Fig. 4, for frequency of up to 1 MHz. Interleaving type 0 is not shown in the figure due to its much higher value ($R_{AC}/R_{DC} > 5$). For the primary winding, interleaving type 1 (T1) gives the highest ratio at $f_{AC} < 360$ kHz but, above that type 2 (T2) generates higher resistance. Type 3 (T3) interleaving has the lowest R_{AC}, with an almost constant value up to 1 MHz. For

the secondary windings, T2 interleaving also gives the highest R_{AC}, starting at $f_{AC} > 100$ kHz. Similarly, T3 interleaving has the best performance for secondary windings. At 500 kHz the R_{AC} is only about 10% higher than R_{DC} and the impact is almost identical for both secondary A and B.

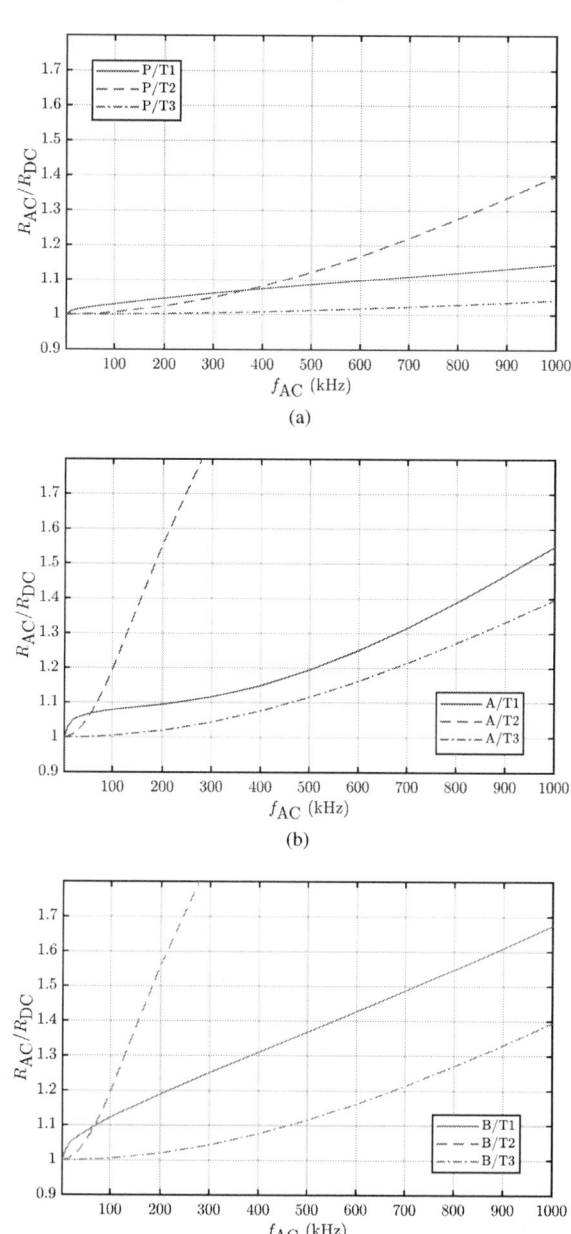

Fig. 4: AC to DC resistance ratio for different windings: (a) primary winding, (b) secondary A winding, (c) secondary B winding. T1/T2/T3 represents different interleaving structure.

B. Number of Layers Impact

The next parameter to be investigated is the number of parallel layers for each winding. Since PCB winding implementation will be used, only even number of parallel layers is considered, i.e. 2, 4, and 6 layers per winding. An example to

979-8-3315-1612-3/25 $31.00 © 2025 IEEE   248

illustrate the meaning of this parallel layers is shown in Fig. 5. *6L* means 6 parallel layers for each winding. The result of the resistance calculation is shown in Fig. 6 for different winding, all using T3 interleaving structure. Secondary winding B has exactly the same resistance as secondary winding A so, only one is plotted here. Looking at the result, it is clear that the number of layers does not influence the R_{AC}/R_{DC} value. The same winding dimension as in the previous section is also considered here.

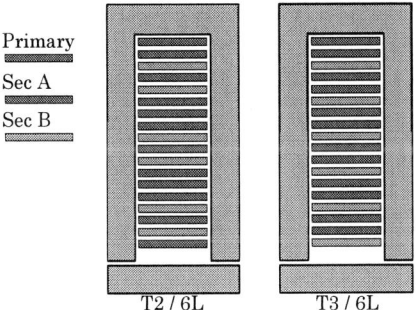

Fig. 5: Example of interleaved winding structure: T2/6L means interleaving Type 2 with 6 parallel layers, T3/6L means interleaving Type 3 with 6 paralel layers.

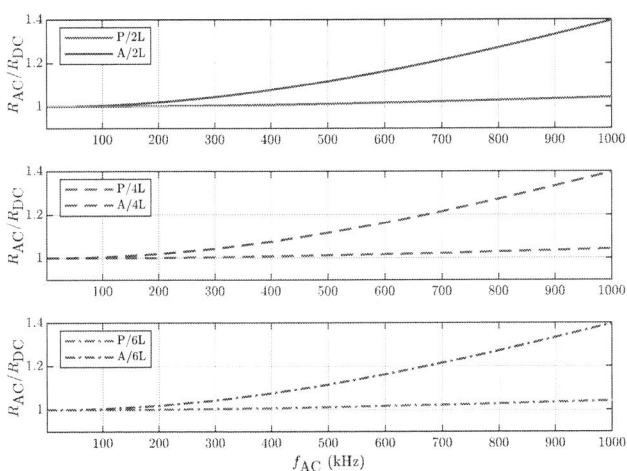

Fig. 6: AC to DC resistance ratio for different number of parallel layers. P represents primary winding and A represents secondary A winding. 2L/4L/6L represents the number of parallel layers.

IV. 3D EFFECT ON WINDING INTERLEAVING

A. Current Distribution

3D FEM simulation is built to refine the result from previous 2D analysis. The geometry can be seen in Fig. 2 and symmetry can be leveraged to reduce the model size. For instance, it would be enough to only simulate one cell of the transformer which consists of only 1 core leg and 1 secondary phase. The considered operating frequency here is 500 kHz.

The current density in the winding cross section is shown in Fig. 7 and Fig. 8. Fig. 7 shows the result from 2D simulation,

which can be considered as the ideal case when full MMF cancellation heppens. Meanwhile, Fig. 8 shows the result from 3D simulation, which shows how the actual current density distribution is. Four cross sections are shown in Fig. 8 and only the section between center leg and side leg experience the interleaving effect. This can be seen from the more uniform current distribution. On the other 3 cross sections, a severe proximity effect can be observed. This will consequently increase the AC resistance significantly.

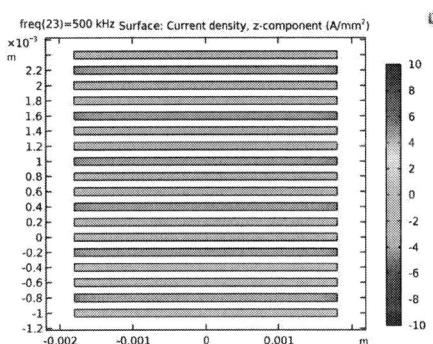

Fig. 7: Current density distribution from 2D simulation. T3 interleaving and 6 parallel layers per winding is considered.

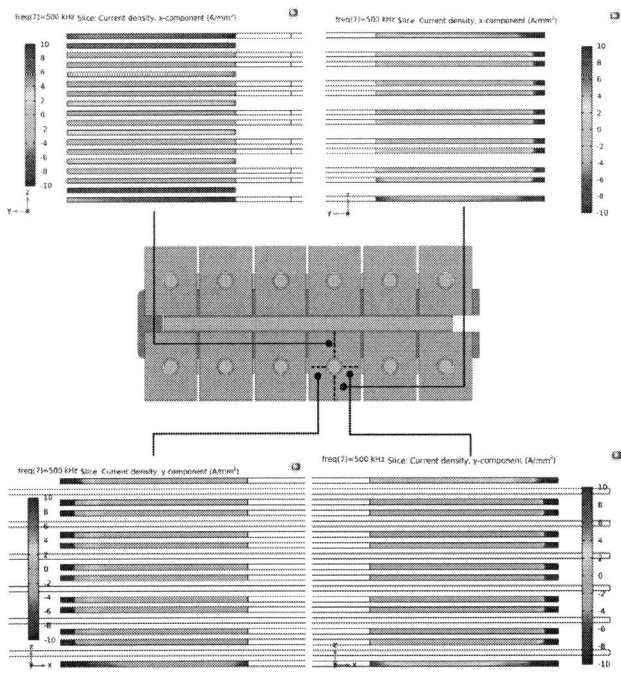

Fig. 8: Current density distribution from 3D simulation. T3 interleaving and 6 parallel layers per winding is considered. 4 different cross sections are shown.

B. AC Resistance

For the sake of brevity, only the results from T3 interleaving and 6 parallel layers will be analyzed here. Although, from previous analysis, it can be seen that this combination also has the lowest resistance. The comparison between 2D and 3D simulation results is presented in Fig.9.

The 3D simulation shows a significantly higher R_{AC}/R_{DC} for all three windings. For the secondary windings (A and B), the ratio is about 10x higher than the 2D result, even with the best interleaving structure according to 2D analysis. For the primary winding, the difference is less dramatic, e.g. 3x higher at 500 kHz. This discrepancy between 2D and 3D is caused by the sections of windings which do not have primary/secondary interleaving effect. In particular, a big part of the secondary winding does not overlap with the primary. Meanwhile, the primary winding has more area which overlaps with the secondary, hence the discrepancy is smaller. This effect is often ignored in the design process [4], [9], [10], even though it makes the result inaccurate.

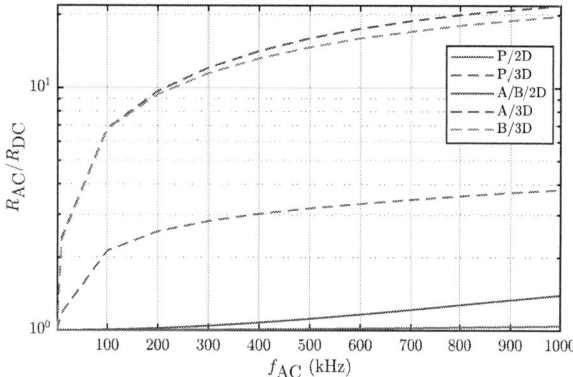

Fig. 9: Comparison of AC to DC resistance ratio between 2D and 3D simulation result

V. OPTIMIZATION AND VERIFICATION

A. Core and Winding Loss Model

The core loss can be calculated by splitting the core into 3 sections: the side legs (Vol_{leg}) where the secondary windings are wound, the center legs (Vol_{cen}) where the primary is wound, and the top/bottom plate (Vol_{plt}) of each core unit. Each section has different cross section (A_{Fe}), flux density (ΔB), and volume. Due to the non-sinusoidal flux waveform, *iGSE* is used to calculate the core loss density ($P_{\text{v,iGSE}}$). The selected core material in this work is DMR50 from DMEGC. The adopted approach is akin to the method in [11].

There are 3 components in the total winding loss (P_{Cu}), which are reflected in Eqn. 5. The first is the primary winding loss, which only includes AC component. The second is the AC component of secondary winding loss. $I_{\text{sec,AC}}$ here accounts only for half of the center tap winding, hence there are A and B components in the equation. Thirdly, there is the DC component of the secondary winding. The DC resistance ($R_{\text{DC,P}}$, $R_{\text{DC,A}}$, $R_{\text{DC,B}}$) depend on the geometry parameters. The AC to DC resistance ratio ($f_{\text{AC/DC}}(f_{\text{s}})$) is derived from the 3D simulation result in Fig. 9 by means of curve fitting.

The rms current calculation follows the formulas derived in [12]. Meanwhile, N_{ph} is the number of parallel phases on the secondary.

$$
\begin{aligned}
P_{\text{Fe}} = 2 \times (&Vol_{\text{leg}} \cdot P_{\text{v,iGSE}}(\Delta B_{\text{leg}}) + \\
&Vol_{\text{cen}} \cdot P_{\text{v,iGSE}}(\Delta B_{\text{cen}}) + Vol_{\text{plt}} \cdot P_{\text{v,iGSE}}(\Delta B_{\text{plt}}))
\end{aligned} \quad (4)
$$

$$
\begin{aligned}
P_{\text{Cu}} = I_{\text{pri,rms}}{}^2 \cdot f_{\text{AC/DC,P}}(f_{\text{s}}) \cdot R_{\text{DC,P}} + \\
(I_{\text{sec,AC}}{}^2 \cdot f_{\text{AC/DC,A}}(f_{\text{s}}) \cdot R_{\text{DC,A}} + I_{\text{sec,AC}}{}^2 \cdot f_{\text{AC/DC,B}}(f_{\text{s}}) \cdot R_{\text{DC,B}}) \cdot N_{\text{ph}} + \\
(I_{\text{sec,DC}}{}^2 \cdot R_{\text{DC,A}} + I_{\text{sec,DC}}{}^2 \cdot R_{\text{DC,B}}) \cdot N_{\text{ph}} \quad (5)
\end{aligned}
$$

B. Design Optimization

To verify the previous analysis, the high-current matrix transformer with 24:1 turns ratio is built. Interleaving structure T3 and 6 parallel layers per winding are used, which means 18 layers PCB technology is adopted. The switching frequency is pre-selected to be 500 kHz in this work, after considering the employed core material and power switches performance. Copper thickness of 100 μm is also used.

The converter design space is explored, particularly focusing on the transformer design. The considered output current at 1V is 500A. The explored design parameters are listed in Tab. II, which will probe the trade-off between loss, and transformer area. The performance space result is shown in Fig. 10. The total loss (P_{Loss}) includes the core loss and winding loss. The transformer area (A_{xfmr}) definition was given in Eqn. 3. Additionally, the copper volume is shown as the color axis. It can be seen that more copper does not always mean lower loss. The performance space demonstrates the theoretical limit of the DCX design with the current technology. The implemented prototype's real performance is also indicated in Fig. 10.

TABLE II: Design Space Parameters

Parameter	Value	Selected
D_c	[2 ... 6] mm	2 mm
C_m	[2 ... 6] mm	2 mm
C_l	[32 ... 86] mm	42 mm
W_w	[2 ... 13] mm	3.6 mm

C. Verification and Measurement

One of the key features in this converter is the vertical DC output connection which enables vertical power delivery to the load. This feature can be seen in the final prototype assembly presented in Fig. 11. The DC bus bars are inserted through the two PCBs. The boxed dimension here excludes the signal connectors and debugging area.

To verify the transformer analysis, the resistance is measured and compared with the 3D analysis result in Fig.12. The resistance value here is already the sum of all windings resistance, hence the value lies between the P and A/B curve in Fig. 9. Copper plates are used to short the secondary windings during the measurement. The R_{AC}/R_{DC} ratio shows a matching result. In the frequency of interest range (100 kHz to 1 MHz), the difference lies under 20%, compared to the measurement result.

979-8-3315-1612-3/25 $31.00 © 2025 IEEE

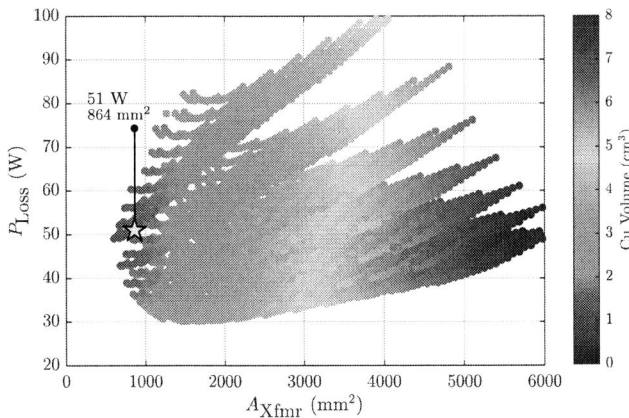

Fig. 10: Design space exploration result parameterized by total loss (P_{Loss}), transformer area (A_{xfmr}), and copper volume. The implemented prototype's real performance is indicated by the star symbol.

In terms of performance, a peak efficiency of 91.3% is achieved at 190A output. The full load efficiency is 88.6% at 488A output, without gate driver loss. At this point, 51W of loss is dissipated in the transformer with 40.6W coming from the secondary winding AC resistance. The converter is operated at 540 kHz, which is slightly above the resonant point, thereby giving some margin to ensure that ZVS is achieved, but creating higher AC resistance.

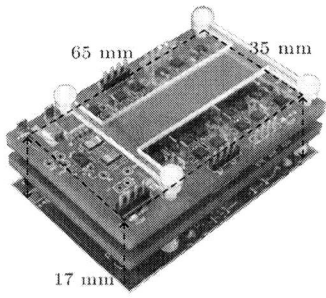

Fig. 11: Final prototype and its dimension. Vertical copper bars enable the vertical power delivery.

TABLE III: DCX Converter Performance

V_i (V)	V_o (V)	I_{out} (A)	Density (W/in^3)	Efficiency (%)
48.25	0.954	488	197	88.6

VI. CONCLUSION

The design of a new high-current matrix transformer is introduced in this paper. It is suitable for a high-current DCX converter with high voltage step-down ratio. In this work, the transformer is used in an LLC converter to achieve 48:1 step-down ratio with 500A output. The winding analysis and optimization steps are also outlined here. From 2D FEM analysis, the interleaving structure with the lowest $R_{\text{AC}}/R_{\text{DC}}$ ratio is chosen. Meanwhile, the 3D FEM analysis shows a significant increase in $R_{\text{AC}}/R_{\text{DC}}$, even after employing the

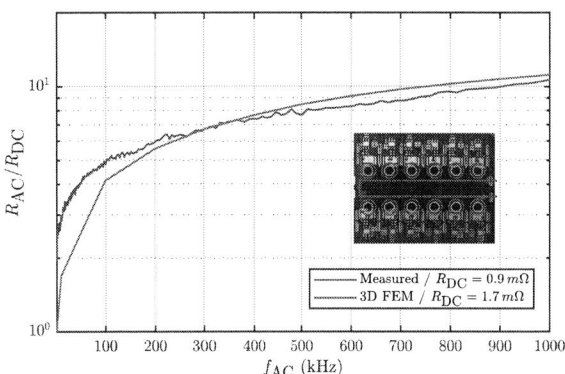

Fig. 12: Measured vs. 3D FEM simulated resistance ratio $R_{\text{AC}}/R_{\text{DC}}$. The value includes the sum of primary and secondary winding resistance.

best interleaving strategy. This fact must be taken into account during the design as it can radically change the result. Finally, a prototype is built and the analysis result is verified.

REFERENCES

[1] The New York Times, "A.I. Could Soon Need as Much Electricity as an Entire Country," October, 2023. [Online]. Available: https://www.nytimes.com/2023/10/10/climate/ai-could-soon-need-as-much-electricity-as-an-entire-country.html

[2] N. M. Ellis, Y. Zhu, and R. C. Pilawa-Podgurski, "Gallium nitride-based 48v-to-1v point-of-load (pol) converter for aerospace telecommunications and computing applications," in *2024 IEEE Applied Power Electronics Conference and Exposition (APEC)*, 2024, pp. 1384–1388.

[3] Y. Elasser, J. Baek, K. Radhakrishnan, H. Gan, J. P. Douglas, H. K. Krishnamurthy, X. Li, S. Jiang, V. De, C. R. Sullivan, and M. Chen, "Mini-lego cpu voltage regulator," *IEEE Transactions on Power Electronics*, vol. 39, no. 3, pp. 3391–3410, 2024.

[4] X. Ren, J. Zhang, Y. Jiang, X. Li, and T. Long, "A 48-to-1v llc dc transformer," in *2023 IEEE 24th Workshop on Control and Modeling for Power Electronics (COMPEL)*, 2023, pp. 1–5.

[5] A. Nabih and Q. Li, "Design of 98.8% efficient 400-to-48-v *llc* converter with optimized matrix transformer and matrix inductor," *IEEE Transactions on Power Electronics*, vol. 38, no. 6, pp. 7207–7225, 2023.

[6] P. R. Prakash, A. Nabih, Y. Liang, S. Kudva, M. Mosa, C. T. Gray, and Q. Li, "A 2400 w/in3 1.8 v bus converter enabling vertical power delivery for next-generation processors," in *2024 IEEE Applied Power Electronics Conference and Exposition (APEC)*, 2024, pp. 910–917.

[7] M. Li, C. Wang, Z. Ouyang, and M. A. E. Andersen, "Optimal design of a matrix planar transformer in an llc resonant converter for data center applications," *IEEE Journal of Emerging and Selected Topics in Power Electronics*, vol. 11, no. 2, pp. 1778–1787, 2023.

[8] H. Wouters, H. Pervaiz, T. Geboers, Y. Zuo, W.-R. Lin, and W. Martinez, "Interleaved pcb winding planar transformer for electric vehicle charging cllc converters," in *2024 IEEE Applied Power Electronics Conference and Exposition (APEC)*, 2024, pp. 3216–3223.

[9] Y. Cai, M. H. Ahmed, Q. Li, and F. C. Lee, "Optimal design of megahertz llc converter for 48-v bus converter application," *IEEE Journal of Emerging and Selected Topics in Power Electronics*, vol. 8, no. 1, pp. 495–505, 2020.

[10] C. Wang, M. Li, Z. Ouyang, T.-G. Zsurzsan, and M. A. Andersen, "Pentacentra transformer for multiphase llc converter in high-current data center application," *IEEE Transactions on Power Electronics*, vol. 39, no. 1, pp. 1150–1161, 2024.

[11] B. N. Sanusi and Z. Ouyang, "Integrated inductor design for a highly compact embedded battery charger," *IEEE Transactions on Power Electronics*, vol. 37, no. 8, pp. 8873–8885, 2022.

[12] W. Zhang, F. Wang, D. J. Costinett, L. M. Tolbert, and B. J. Blalock, "Investigation of gallium nitride devices in high-frequency llc resonant converters," *IEEE Transactions on Power Electronics*, vol. 32, no. 1, pp. 571–583, 2017.

Investigation on Impact of Transformer Parasitic Capacitance on Standby Power Consumption in Power Converters

Kamran Kamran[13], Andrea Russo[1], Federica Cammarata[2], Claudia Malannino[1], S. Yuri Ciardo[1], and Ziwei Ouyang[3]
Email: {kamran.sogulraja, andrea.russo01, federica.cammarata, claudia.malannino, yuri.ciardo}@st.com, ziou@dtu.dk
[1]Advanced Power Electronics, System Research and Applications (SRA), STMicroelectronics S.r.L. Catania, Italy.
[2]STI2GaN Solutions, APMS, STMicroelectronics S.r.L. Catania, Italy.
[3]Department of Electrical and Photonics Engineering, Technical University of Denmark, Kongens Lyngby, Denmark

Abstract— **The significance of reducing no-load or standby power dissipation is increasingly vital in various isolated power converter applications to save energy and meet stringent energy efficiency standards. While multiple factors contribute to this dissipation, the transformer stands out as a key contributor. But unlike transformer leakage inductance, parasitic capacitance plays a pivotal role, especially in scenarios involving high input voltage and light or no loads. This study delves into the impact of parasitic capacitance on standby power consumption through an analysis of two iterations of a custom-designed planar transformer for 400V to 15V/65W DC-DC converter employing a fly-back converter topology. The findings provide valuable insights for optimizing transformer design to reduce energy loss during standby and lightly loaded operations. This optimization not only enhances energy efficiency but also ensures adherence to stringent no-load energy conservation regulations.**

Keywords—Standby Power, no-load Power, Parasitic Capacitance, Transformer, Flyback.

I. INTRODUCTION

In the global context, a significant amount of energy is dissipated when electronic devices remain in standby or idle modes, awaiting activation signal to operate under normal load conditions. Research indicates that approximately 10 to 15% of residential electricity is consumed annually solely to sustain devices working in idle, no-load, or standby modes [1] [2]. As the global shift towards "Green Energy" gains momentum, minimizing power wastage becomes imperative. Consequently, energy conservation regulatory bodies worldwide continue to enforce much stringent limits on standby power consumption for power converters in various application areas and power levels. For instance, for a 65W converter, which belongs to the range of 49W to 250W External Power Supplies (EPSs), the maximum no-load power specified by the U.S. Department of Energy (DoE) was 750mW as per the California Energy Commission's (CEC) regulations in 2004. Which subsequently reduced to less than 210mW in the DoE's 2016 Level VI specifications [3]. A similar trend can be observed in the European Code of Conduct (EU CoC) V5 Tier 1, where the EU CoC Tier 2 sets even more stringent limits [4]. Table I provides additional information on the limits set for the no-load power consumption requirements for EPSs designed for different power levels, as mandated by both U.S. and EU standards.

TABLE I. MAXIMUM NO-LOAD POWER LIMITS FOR EXTERNAL POWER SUPPLIES

Output Power	US DoE Limits	EU CoC Limits	
	Level VI specs	*Tier 1*	*Tier 2*
$0.3 \leq 1W$	≤100mW	≤ 100mW	≤ 75mW
$1 \leq 49W$	≤100mW	≤ 100mW	≤ 75mW
$49 \leq 250$	≤210mW	≤ 210mW	≤ 150mW
>250W	≤500mW	≤ 500mW	≤ 500mW

Hence, it is often essential for designers to also consider the no-load power dissipation in addition to prioritizing full-load efficiencies during power converter design. This not only helps limiting the energy wastage but also ensures compliance with regulatory standards for product qualification.

II. BACKGROUND WORK

The prior work on impact of parasitics on power consumption is mainly focused on leakage inductance keeping in view its impact of typically full load efficiencies. The research regarding parasitic capacitance predominantly revolves around its impact on electromagnetic interference (EMI). Instead mainly the work in [4] highlighted this issue, focusing on the impact of self and mutual capacitance on efficiency under lightly loaded conditions. The study was primarily centered on comparisons based on measurements of no load power consumption under various winding arrangements of the transformer. Subsequently, recent studies [6], [7], and [8] further extended the work by developing equivalent models for the network of self and mutual capacitances typically for two-winding transformers. These models aimed to represent the equivalent capacitance effects of all parasitic capacitances, allowing to quantify the energy losses during charge and discharge in different operational modes of the converter, illustrating how these losses contribute to unnecessary energy dissipation from the input power source. With growing importance of the subject, further research endeavors are crucial to model and evaluate the distinct effects of parasitic capacitance on standby input power. Therefore, alongside optimizing winding design, it is imperative to account for the influence of

979-8-3315-1612-3/25 $31.00 © 2025 IEEE

other components and conduct system-level optimizations such as integrating magnetics and optimizing overall layout to achieve the maximum possible reduction in the impact of parasitic capacitance on standby power dissipation.

This work delves deep into the impact of parasitic capacitance, on standby power dissipation. The controller operates in Burst Mode during no load or standby conditions. Therefore, the study models, simulates, and measures the energy flow in the capacitive network in Burst Mode, illustrating the effect of capacitance on current and voltage waveforms. The MOSFET switch node capacitance (C_{oss}) and transformer parasitic capacitance are key contributors to standby losses. To mitigate these contributions, the work implements two strategies: Firstly, designing an optimized planar transformer with reduced parasitic capacitance. Additionally, integrating the transformer directly into the converter board to minimize additional capacitance contributions from connections. Secondly, utilizing a solution that incorporates a Pulse Width Modulation (PWM) controller with integrated Gallium Nitride (GaN) switch to achieve ultra-low drain capacitance ($C_{oss} < 7p$). Overall, the study reveals that excessive parasitic capacitance significantly impacts standby losses, which have been reduced by up to 36% for intended converter by reducing parasitic capacitance.

III. IMPACT OF PARASITIC CAPACITANCE - MODELING AND SIMULATION

Before to evaluate the impact of parasitic capacitance, all its components need to be estimated to approximate the total switch node capacitance C_D. Considering the converter topology i.e. flyback converter with main and auxiliary outputs as shown in Fig. 1a, there are number of self and mutual capacitances associated with transformer primary, secondary and auxiliary windings which contribute to C_D. In addition, the output capacitance of MOSFET (C_{oss}) and the junction capacitance of main and auxiliary output diodes do also contribute corresponding to respective turn ratios. Taking winding connections and turn ratios into account, the total contribution of transformer windings capacitance $C_{p_{eq}}$ referred to primary side can be modeled as given in (1) using three capacitor model for coupled windings [9].

$$C_{P_{eq}}=C_p+N_{ps}^2 C_s+C_{ps}\left(N_{ps}-1\right)^2+N_{pa}^2 C_a+C_{pa}\left(N_{pa}-1\right)^2 +C_{as}N_{ps}^2\left(N_{pa}-1\right)^2 \tag{1}$$

To estimate total equivalent capacitance at drain C_D, the other remaining capacitive elements must also be included considering corresponding turn ratio, such as MOSFET C_{oss} and main and auxiliary output diodes junction capacitances C_{ds} and C_{da}, whose values are already known from manufacturer specifications. Once transformer C_{peq} is estimated referred to primary side, the simplified equivalent circuit is shown in Fig. 1b. Equation (2) gives altogether the total drain capacitance including C_{peq} as the total contribution from all transformer winding.

$$C_D=C_{P_{eq}}+C_{oss}+N_{ps}^2 C_{ds}+N_{pa}^2 C_{da} \tag{2}$$

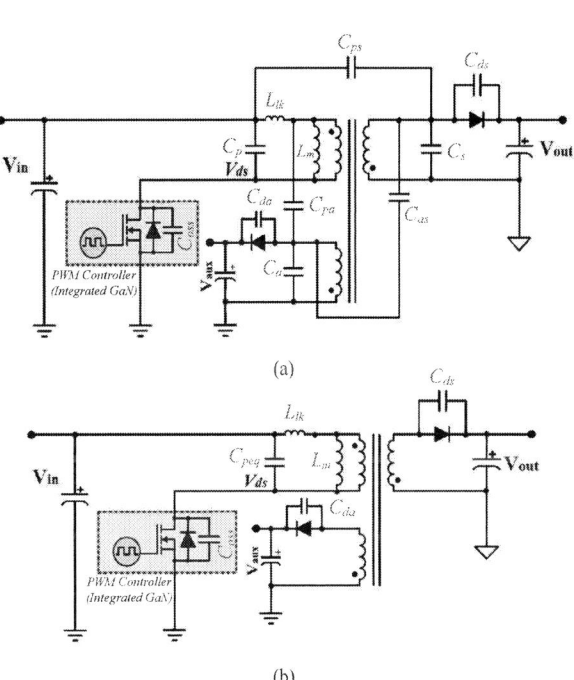

(a)

(b)

Fig. 1. Flyback Converter (a) with all parasitic capacitances with 3-capacitor model (b) with transformer equivalent capacitance referred to primary side.

The PWM controller operates in burst mode in lightly loaded or no load conditions. In burst mode the controller intelligently skips the switching events to reduce the unnecessary switching losses, duly maintaining the required output voltage regulation. Fig. 2 shows the simulation waveforms of standby mode in which secondary output is connected to no load while auxiliary winding supplies minimum standby current to the controller. The switching frequency inside the burst is same 240 kHz but due to burst mode pulse skipping the effective switching frequency is around 6 kHz.

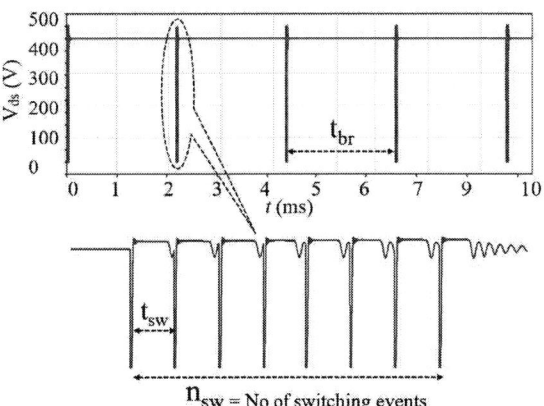

Fig. 2. Simulation waveforms of in standby mode.

Based on the burst mode switching behavior, the effective switching frequency can be given analytically in (3).

$$f_{eff} = \frac{N_{sw}}{N_{sw}t_{sw} + t_{br}} \qquad (3)$$

Where N_{sw} is the total number of switching cycles in one burst event, t_{sw} is the switching time corresponding to the actual switching frequency and t_{br} is the length of burst event as shown in Fig 2.

Parasitic capacitances contribute to the standby losses mainly in two ways, firstly due to its charging and discharging in each switching cycle and secondly due it's impact on switching losses. The cumulative drain capacitance C_D is charged at every turn on event [7] [8]. Knowing the voltage difference across C_D and switching frequency, the total loss incurred at turn on can be found as given in (4).

$$P_{turn_{on}} = \frac{1}{2}(C_D)\left(V_{in} - \frac{V_o}{N}e^{\alpha(t-t_r)}\right)^2 f_{eff} \qquad (4)$$

Where $\alpha = \frac{R_{win}}{2L_m}$ and t_r the resonance time when of L_m and C_D resonate until next turn on event. Equation (4) highlights that the voltage difference across the drain capacitance during each turn-on event is dependent upon the operational behavior of the controller. When the controller operates in Discontinuous Conduction Mode (DCM) due to a light output load, the reflected voltage at the drain may be equal to or lower than Vo/N. This dependency is primarily due to the decaying behavior because of winding resistance R_{win}, the discontinuous mode resonance time (t_r), and the resonance frequency, which are directly affected by value of magnetizing inductance L_m and switch node capacitance C_D. As can be seen from the simulation results depicted in Fig. 3, that the varying parasitic capacitance not only influence the DCM resonance frequency but may also results different levels of drain voltage at the switching events caused by different values of parasitic capacitances resonating with magnetizing inductance L_m.

Additionally, Fig. 3 also demonstrates how the turn-on event can induce a current spike in the primary switch current due to abrupt voltage changes. Notably, an increase in parasitic capacitance leads to a higher peak amplitude of the leading-edge current spike. This effect is further exacerbated by rapid transition time during turn-on events, common in high-frequency applications like those with Gallium Nitride (GaN) solutions, such as the case in this study. The very high leading edge current spikes if not mitigated may also cause other issue such difficulty manage control, especially in current mode controllers and also can cause EMI issues in power supplies.

Secondly the parasitic capacitance negatively influences the switching losses. This contributes mainly to turn-off losses because at turn-on the system operates in deep DCM mode because of light or no-load condition at the output, therefore the switch current has already fallen to zero after preceding turn off event, resulting turn on with zero current switching (ZCS). But at the turn-off the drain voltage and current overlap each other incurring switching loss which can be give in (5).

$$P_{turn_{off}} = \frac{1}{2}V_{ds}(t_{doff} + t_f)I_p f_{eff} \qquad (5)$$

Where t_{doff} and t_f are off delay and fall times respectively and I_p is the peak primary switch current. The unnecessary parasitic capacitance impacts the rate of rise of V_{ds} which consequently increases the overlap between current and voltage. This behavior may be seen in simulation waveform of the V_{ds} in Fig. 3, in which the rise time seems to increase with increasing parasitic capacitance resulting larger overlap between switch current and drain voltage.

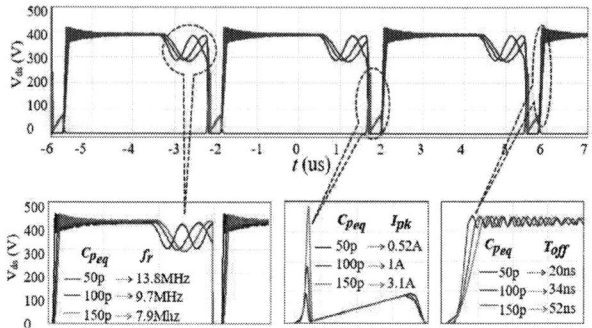

Fig. 3. Simulation waveforms in stand-by mode with varying transformer equivalent capacitance

Therefore, it is essential to carefully consider the impact of parasitic capacitance in power converters, particularly when operating at high switching frequencies and rapid transition times. Additionally, this consideration is vital for applications with a broad input voltage range and for managing power dissipation in scenarios with light or no load.

IV. PARASITIC CAPACITANCE ESTIMATION

It has been established in preceding sections that to estimate total parasitic capacitance, all the self and mutual capacitances among the primary, secondary, and auxiliary windings need to be estimated. These unknown capacitances are approximated using the Finite Element Method (FEM) electrostatic solver [9], [10]. While other analytical methods offer a quicker estimation process, but are prone to inaccuracies when dealing with intricate winding geometries and configurations. On the other hand FEM tools allow to capture actual geometries, and therefore are applicable to almost any kind of intricate windings shapes and configurations. Despite the method used, the estimation of winding capacitance primarily hinges on determining the total stored energy when an electric potential is applied to the windings [9] [10].

For instance, self-capacitance arises from the varying potential between turns within the same winding. However, the electrostatic solver treats all turns within the same winding as a single conductor, resulting in equipotential windings and negligible or no net capacitance. Consequently, each turn within

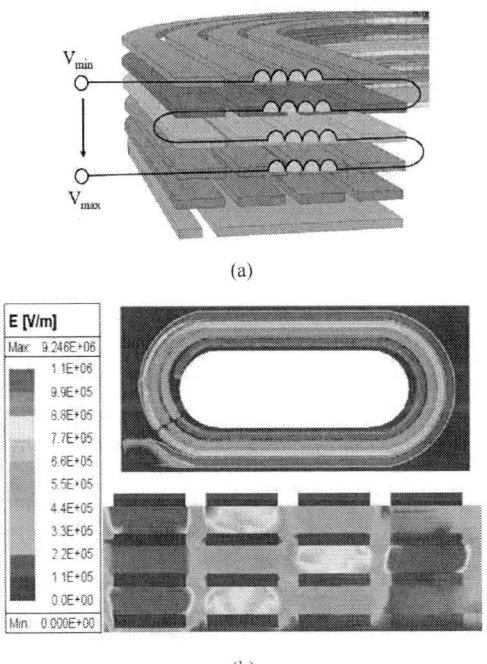

(a)

(b)

Fig. 4. Estimation of Parasitic Capacitance by FEM

the same winding is electrically isolated from the others with a minute gap as shown in Fig 4b, and the total winding voltage is incrementally applied from minimum to maximum on a turn-to-turn basis as described in (6).

$$\frac{n(n-1)-i}{n}V_{in} \quad where\ i=1,2,3.........n \quad (6)$$

Where, n represents the number of turns in the given windings, and V_{in} denotes the total winding voltage applied. Additionally, based on the series connection between the winding layers, the direction of the applied potential is adjusted accordingly, as illustrated in Fig. 4a. Consequently, the electric field distribution varies correspondingly based on the applied potential, as can be seen from the Fig. 4b, between layers the field increases in each layer changing increasing pattern alternatively. The total energy stored in electric field is determined by integrating the distributed field, as outlined in (7) taking into account the relative permeability of printed circuit board material in use.

$$E_t = \iiint_0^v \varepsilon_0\,\varepsilon_r E(t)\partial v \quad (7)$$

Knowing total energy E_t and applied potential V_{in} the total self-capacitance may be estimated by (8).

$$E_t = \frac{1}{2}CV_{in}^2 \quad \rightarrow \quad C = \frac{2E_t}{V_{in}^2} \quad (8)$$

For mutual capacitance, the process is relatively straightforward, which involves applying different voltage potentials to two distinct windings, which again results electric field distribution in between windings. The total mutual capacitance is determined by estimated total energy stored in electric field by using (8). Once the self and mutual capacitances are determined for all three windings, the total parasitic capacitance contribution coming from transformer C_{peq} referred to primary can be determined by (1).

V. PROTOTYPING, MEASUREMENT AND RESULTS

Based on analysis and geometrical optimization of the transformer in previous section the non-interleaved solution is discarded because of very high leakage inductance, and excessive switch node voltage. Therefore, two designs prototype have been developed, one with fully interleaving and other partially interleaved layer arrangements as shown in Fig. 5a. The prototype is developed with both variants of transformer integrated in main converter board as shown in Fig. 5b. Besides this, other components are also carefully selected to contribute minimally the drain capacitance, particularly for the primary switch the integrated GaN has been used having very low junction capacitance of the order of less than 7pf.

The measured V_{ds} and I_d waveform for both variants of transformer integrated in main board are shown in Fig. 6a and 6b. As expected, the difference in terms of DCM ringing frequency and leading-edge current spike in primary current is aligned with simulation results because of presence of different parasitic capacitance. Fully interleaved solution results in lower resonance frequency during DCM ringing while partially interleave solution results in higher frequency, because the different drain capacitance is resonating with L_m in both the cases.

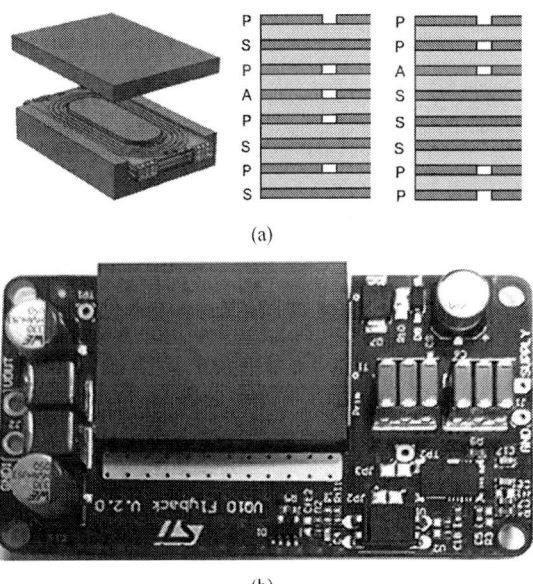

(a)

(b)

Fig. 5. Prototype board with integrated planar transformer

979-8-3315-1612-3/25 $31.00 © 2025 IEEE

During Discontinuous Conduction Mode (DCM), when all the energy is transferred to secondary side the magnetizing inductance L_m resonates with the total drain capacitance C_D. By accurately measuring L_m and resonance time T_r cumulative drain capacitance can be estimated using (9).

$$C_D = \frac{T_r^2}{4\pi^2 L_m} \qquad (9)$$

Where T_r is the period of the resonance frequency of DCM ringing. Using the same converter board for both the cases total drain parasitic capacitance L_m is obtained corresponding to the DCM ringing resonance times as shown in measured V_{ds} waveforms given in Fig. 6a and 6b.

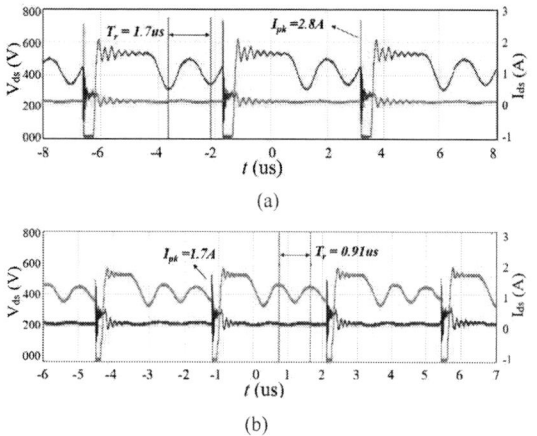

Fig.6. Parasitic Capacitance in flyback converter (a) Fully Interleaved (b) Part. interleaved.

The measured results for both partially interleaved and fully interleaved cases are given in Tab. II. It can be clearly seen that the by making optimization in transformer makes significant reduction in cumulative drain capacitance. For both the cases the converter is operated in standby mode, in which main output is connected to similar output conditions that is no load condition with auxiliary output is delivering minimal standby mode current to the controller IC. There can be seen significant reduction in total input power with respect to decreasing parasitic capacitance.

TABLE II. MEASUREMENT RESULTS FOR BOTH VERSION OF TRANSFORMER

Version	L_m Meas.	L_{lkg} Meas.	C_D (FEM)	C_D Meas.	Standby Power Measured
Fully Interleaved	265uH	0.55uH	254.2pf	276pf	272mW
Part. Interleaved	270uH	1.41uH	68.5pf	79.2pf	171mW

It can be further noticed form measured current waveforms in Fig. 6, with difference in total drain capacitance the turn on current spike in drain current have also been clearly reduced from 2.8A to 1.7A from fully interleaved case to partially interleaved case due to reduction in practice capacitance.

Further, as shown in Fig. 7a and 7b, the measured waveforms in burst mode, observing the value of t_{br} and t_{sw} and number of switching cycles N_{sw} the effective frequency in burst mode is estimated using Eqn. 3. to be 6.1 kHz.

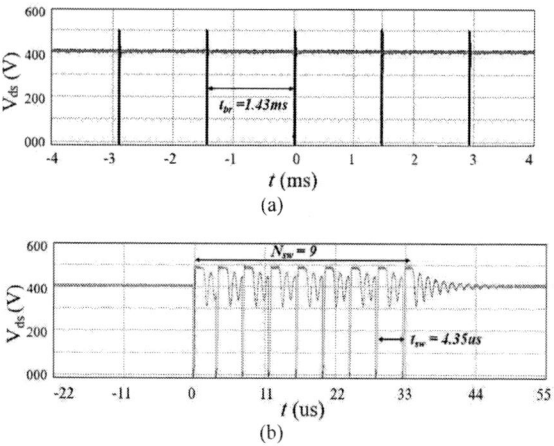

Fig. 7. Switching behavior measured in standby burst mode.

Knowing effective switching frequency, the share of losses of parasitic capacitance and other components of the converter are estimated such as losses coming from controller, snubber and MOSFET, feedback path and other losses of transformer including core and winding loss etc. While plotting total losses distribution for both cases it can be observed from Fig. 8, that significant share of losses is occupied by parasitic capacitance in standby condition, especially in unoptimized transformer. While improved partially interleaved design significantly reduces standby power dissipation because of reduced parasitic capacitance.

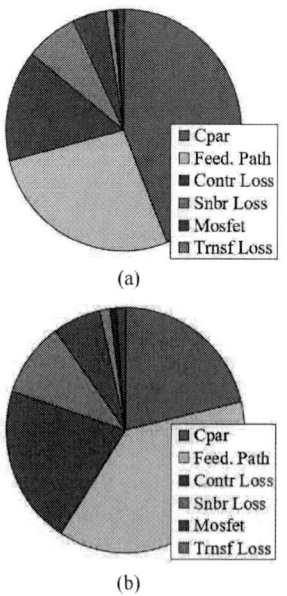

Fig. 8. Standby Losses Distribution (a) Fully Interleaved. (b) Partially Interleaved.

VI. CONCLUSIONS AND FUTURE WORK

The work demonstrates that the efficiency of the power converters under high input voltages and no-load conditions is significantly influenced by unnecessary parasitic capacitance of used magnetic components. While utilizing a planar magnetics solution with a fully interleaved winding arrangement can result in very low leakage inductance, but it may not be optimal for standby or lightly loaded situations due to high parasitic capacitance. On the other hand, an improved partially interleaved solution, striking a good balance between leakage inductance and parasitic capacitance, has demonstrated a reduction in standby power consumption of over 100mW for intended converter. Moving forward, further research should delve deeper into this topic, focusing on developing a comprehensive magnetics design approach that ensures both high efficiency at full load and minimal dissipation at no-load across various converter applications.

REFERENCES

[1] Meier, A., & Siderius "Should the next standby power target be 0-watt?," *ECEEE 2017. Summer Study – Consumption, Efficiency & Limits. European Council for an Energy Efficient Economy*, June 2017.

[2] C. Patro, Y.-S. Cho, A. de Almeida, L. Pagliano, P. Rivire, A. Roscetti, B. Schlomann, and D. Silva, "Stand-by and off mode energy losses in new appliances," in Proc. of *2010 ACEEE Summer study on energy efficiency in buildings*, 2010, pp. 9–282–9–294.

[3] "Level VI energy conservation standard," US Department of Energy (DoE) Directives Invoked on February 10, 2016. Available: *https://www.energy.gov/*

[4] "European Code of Conduct (CoC) and Eco-design 2019/1782 regulations," Invoked on April 01, 2020. Available: *https://www.energystar.gov/*

[5] M. J. Prieto, A. Fernandez, J. M. Diaz, J. M. Lopera and J. Sebastian, "Influence of transformer parasitics in low-power applications," *APEC '99. Fourteenth Annual Applied Power Electronics Conference and Exposition. 1999 Conference Proceedings (Cat. No.99CH36285)*, Dallas, TX, USA, 1999

[6] H. Kewei, L. Jie, H. Xiaolin and F. Ningjun, "Analysis and simulation of the influence of transformer parasitics to low power high voltage output flyback converter," *2008 IEEE International Symposium on Industrial Electronics*, Cambridge, UK, 2008.

[7] D. Leuenberger and J. Biela, "Accurate and computationally efficient modeling of flyback transformer parasitics and their influence on converter losses," *2015 17th European Conference on Power Electronics and Applications (EPE'15 ECCE-Europe)*, Geneva, Switzerland, 2015.

[8] N. Vijaya Kumar and N. Lakshmi Narasamma, "Comparison of Planar Transformer Architectures and Estimation of Parasitics for High Voltage Low Power DC-DC Converter," *2018 IEEE International Conference on Power Electronics, Drives and Energy Systems (PEDES)*, Chennai, India, 2018.

[9] Z. Ouyang, O. C. Thomsen and M. A. E. Andersen, "Optimal Design and Tradeoff Analysis of Planar Transformer in High-Power DC–DC Converters," in *IEEE Transactions on Industrial Electronics*, vol. 59, no. 7, pp. 2800-2810, July 2012

[10] M. B. Shadmand and R. S. Balog, "A finite-element analysis approach to determine the parasitic capacitances of high-frequency multi-winding transformers for photovoltaic inverters," *2013 IEEE Power and Energy Conference at Illinois (PECI)*, Urbana, IL, USA, 2013

PCB-Winding Integrated Transformer for 800-V Dual Active Bridge Converter using 1.2-kV GaN Devices

Hans Wouters
Dept. of Electrical Engineering
KU Leuven - EnergyVille
Leuven, Belgium
hans.wouters@kuleuven.be

Wei-Ren Lin
Dept. of Electrical Engineering
KU Leuven - EnergyVille
Leuven, Belgium
weiren.lin@kuleuven.be

Nicolas Pirson
Dept. of Electrical Engineering
KU Leuven - EnergyVille
Leuven, Belgium
nicolas.pirson@kuleuven.be

Thomas Jochmans
Dept. of Electrical Engineering
KU Leuven - EnergyVille
Leuven, Belgium
thomas.jochmans@kuleuven.be

Yu Zuo
Dept. of Electrical Engineering
KU Leuven - EnergyVille
Leuven, Belgium
yu.zuo@kuleuven.be

Wilmar Martinez
Dept. of Electrical Engineering
KU Leuven - EnergyVille
Leuven, Belgium
wilmar.martinez@kuleuven.be

Abstract—**This paper introduces a novel PCB-winding integrated inductor-transformer for an 800-V Dual Active Bridge (DAB) converter. This isolated dc-dc converter implements 1200-V GaN transistors for their introduction in a next-generation onboard charger. Using 1200-V GaN devices omits the need for complex and expensive multi-level topologies using 650-V GaN devices and challenges the well-established 1200-V SiC devices in the onboard charger application. The newly developed split-PCB integrated inductor-transformer eliminates the series inductor and addresses key design challenges in high-voltage, high-frequency converters, such as parasitic capacitance mitigation. An optimal transformer is achieved through a hybrid optimisation procedure that combines analytical and finite element calculations. Limitations due to parasitic capacitances, which are inadvertently high in PCB winding transformers, are detailed and mitigated through three optimised transformer prototypes. A novel design approach achieves the target leakage inductance by tuning the turns and degree of interleaving in the PCB windings. Furthermore, experiments and simulations provide a comprehensive comparative analysis between the different transformer prototypes. Finally, these are implemented in two distinct proof-of-concept dual active bridge converters with 1200-V GaN transistors, highlighting challenges in implementing these novel technologies. The results underline the potential of PCB-winding integrated magnetics and 1200-V GaN devices towards high-performance yet cost-effective automotive converters.**

Index Terms—**Planar transformer, PCB-winding, integrated magnetics, gallium nitride (GaN), 1200-V GaN, dual active bridge (DAB), parasitic capacitance, onboard charger, dc-dc converter**

I. INTRODUCTION

Developing efficient and reliable power conversion systems is crucial in the competitive landscape of electric vehicle

This work was supported in part by the European Union's Horizon Europe under Grant No. 101056857 (PowerDrive) and FWO Research Foundation Flanders under Grant No. 1SHE524N. Special thanks go out to SMA Magnetics for providing customised ferrite cores.

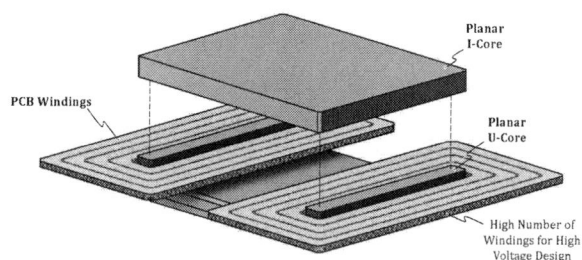

Fig. 1. 3D exploded view of the integrated inductor-transformer using PCB windings and planar ferrite cores, with the PCB substrate invisible to illustrate the copper turns.

(EV) technology, where innovative yet cost-effective solutions provide a competitive advantage. Pushed by fast-charging requirements, the shift towards 800-V powertrains in EVs necessitates the advancement of isolated DC-DC converters capable of handling high voltages while maintaining compact, cost-effective and efficient design [1], [2]. Conventional transistors for OBCs, however, have a 650-V breakdown voltage and necessitate complex multi-level topologies to enable 800-V operation. As such, implementing 1.2-kV Gallium Nitride (GaN) transistors potentially forms a disruptive solution. 1200-V SiC MOSFETs have traditionally been the predominant choice for 800-V battery chargers, which are well-grounded for high-power off-board applications due to their superior thermal performance and high-voltage performance. However, the need for compact and cost-effective onboard chargers rated at *only* 6.6 to 11 kW invites the consideration of GaN transistors as a potentially superior alternative in this market segment. However, 1200-V GaN devices are not yet commercially available, and challenges in cooling and driving such potential devices remain pertinent.

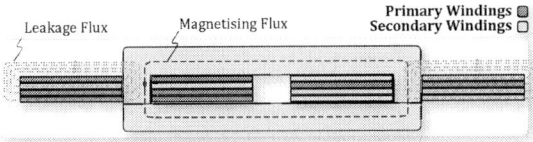

Fig. 2. Cross-section of the UI core with split-PCB winding illustrating the inductor-transformer integration principle.

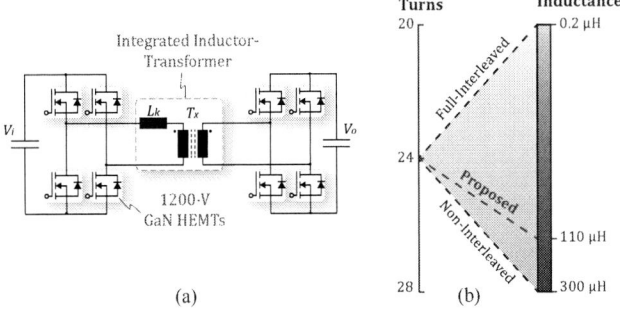

(a)

Fig. 3. (a) Integrated inductor-transformer in the DAB converter, (b) Series-inductor design method based on the amount of interleaving for a certain number of turns

Another disruptive technology entering the automotive market is PCB-winding transformers. Instead of conventional copper-wound devices that often require specialised litz-type windings, the windings can be implemented in printed circuit boards (PCBs), cf. **Fig.1**. Similarly, they drastically reduce costs while being well-suited for high-frequency operations. However, their design for 800-V chargers entails the use of a high number of turns, complicating their design. Furthermore, the inherently high parasitic capacitance challenges the high-frequency operation of GaN devices [3].

Building on these technological advancements, a third innovation in the field is the integration of magnetic components, specifically combining the series inductor and transformer into a single structure. As proven by the magnetic scaling laws [4] and exemplified in various papers [5]–[7], a combined magnetic device, if designed well, can achieve a drastic reduction in magnetic losses and size. Many solutions have been proposed, often requiring an additional flux-conducting path or windings. Alternatively, high-leakage fluxes can be achieved by physically splitting the primary and secondary windings, as depicted in **Fig. 2** [8], [9]. However, their design, parasitic capacitance, thermal management, and the sensitivity of the leakage inductance to the winding geometry still pose major challenges. Described in [8], a CLLC resonant converter with a litz wire integrated inductor-transformer is employed in a 6.6 kW onboard EV charger operating at 55 kHz. It reduces its volume by 37% without sacrificing efficiency. The split-winding design also serves high-power applications, exemplified by the 5 MW medium frequency transformer discussed in [10] and the 125 kW solid-state transformer from [11]. These, however, do not tap into the potential of PCB windings for the implementation of reproducible split-winding integrated magnetics nor into the use of partly interleaved structures.

This paper, therefore, presents a high-frequency, high-voltage integrated planar inductor-transformer using partially interleaved PCB windings with reduced capacitance. The objective is its implementation in the first 1200-V GaN proof-of-concept dual active bridge (DAB) converter, offering a synergy between the high-voltage GaN devices and the planar PCB-winding integrated magnetics. Hence, **Section II** presents the design principle of the split-PCB integrated inductor-transformer, detailing the design parameters, analytical pre-calculations, and finite elements-based parameter selection. This is followed by a study of the parasitic capacitance in **Section III** and mitigation thereof. The experimental results are presented in **Section IV**, followed by conclusions and insight into future work.

II. SPLIT-PCB INTEGRATED INDUCTOR-TRANSFORMER

A. Magnetic Integration Design Principle

The design of the integrated inductor-transformer (cf. **Fig. 3a**) employs a fundamentally simple approach, foregoing additional core elements or windings, and is especially beneficial for PCB windings by doubling the turns per layer. However, the design of the split-PCB winding enlarges the transformer's footprint compared to E-type transformers with windings on the centre leg. Moreover, the inability to incorporate fully interleaved winding structures in split-PCB designs contributes to increased high-frequency losses [12] and, depending on the converter layout, may necessitate shielding to avoid electromagnetic interference (EMI) issues. This leakage flux through the air also implies the need for finite element simulations to consider sufficiently large regions surrounding the component. Finally, this leakage flux is frequency-dependent, challenging their implementation in frequency-modulated resonant converters. These challenges highlight the complex nature of the design and implementation of these split-PCB winding integrated inductor-transformers.

The proposed inductor-transformer design process can be summarised as follows:

1) An analytical pre-calculation determines a feasible range of transformer parameters based on a multi-objective optimisation.
2) Design candidates are selected, for which the envelope of possible leakage inductance is determined using FEA simulation, i.e., the minimum leakage (when fully-interleaved) and maximum leakage (when non-interleaved).
3) The number of interleaved turns is identified that results in a leakage inductance within the target range. This is illustrated in **Fig. 3b**.

B. Optimal Integrated Inductor-Transformer

1) Design Parameters Envelope: A starting point of many high-frequency magnetic devices is the trade-off between

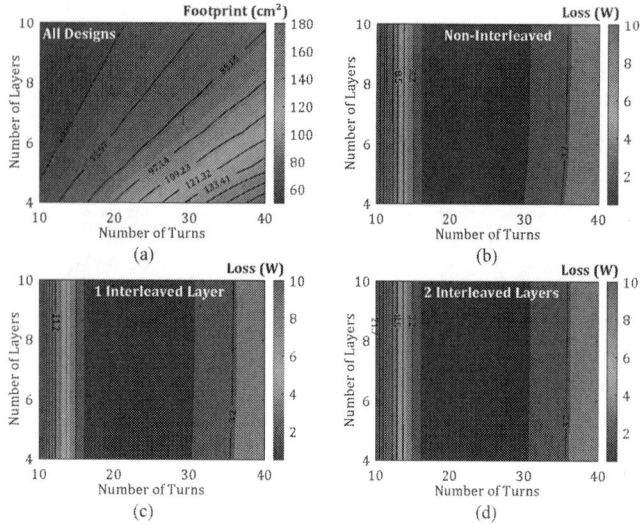

Fig. 4. Contour plots representing the results of the analytical pre-calculations for the transformer design for varying turns and layers (a) Total footprint of all designs, (b) Total magnetic losses for non-interleaved windings, (c) Total magnetic loss for PCB windings with a single interleaved turn, (d) Total magnetic loss when comprising two interleaved turns.

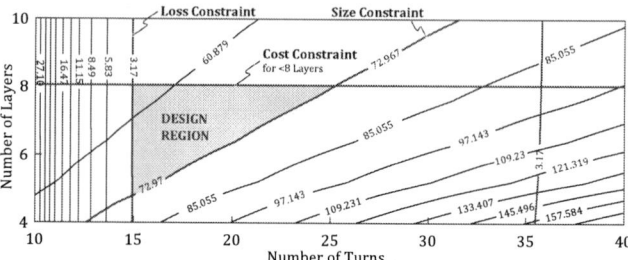

Fig. 5. Combined contour plot of the possible transformer designs for varying number of turns and layers having one interleaved turn, featuring loss and footprint contours and outlining the constraints that confine the design region.

TABLE I
COMPARISON OF DIFFERENT CASES

Design Candidate	A1	A2	A3	B1	B2*
No. of Signal Layers	8	8	8	6	6
No. of Turns per Layer	3	3	3	4	4
Total No. of Turns	24	24	24	24	24
Degree of Interleaving	Non	Medium	High	Medium	Low

winding loss and core loss. Depending on the converter specifications, cooling strategy, and materials used, the desired balance between core and winding losses shifts from a conventional 50-50 distribution towards a more nuanced design that intentionally permits greater losses in one type over the other. The target converter is a high-voltage 800-V isolated dc-dc converter with *only* 1 kW of power, dictating a design limited by high core losses. This explicitly underlines the design choice for the split-PCB design with a high number of turns per layer, adverting the need for an excessively large core. As part of a feasibility assessment, a constant switching frequency of 250 kHz was determined for the GaN devices, which are discrete and consequently experience a greater drive loop inductance. This, combined with a lack of features typical of integrated devices, such as active miller clamps and sensing, creates issues such as high ringing in the gate drive loop and false turn-on of the low-side device when subject to high dv/dt.

2) Analytical Parameter Pre-selection: A pre-selection of the possible transformer parameters is performed using analytical models of the transformer's most prominent losses and constraints. The models below are extensively validated and contextualised in prior work [13]. A copper thickness of 2 oz (70μm) is selected due to its low cost and sub-skin depth thickness at 250 kHz; this applies to harmonics up to 1.2 MHz. Constraining the average current density to 5 A/mm² dictates a track width of 3.6 mm. Both liquid and forced air cooling are investigated, considering the onboard charger application. Furthermore, up to 8-layer PCBs are economically available with a 2 oz copper thickness for all inner and outer layers.

Firstly, the high-frequency conduction losses in the PCB winding are estimated based on Dowell's equation, see Equa-

tion (1). This exposes the dependence of the winding loss on the ratio between skin depth (δ) and track width (h), expressed as $\xi = h/\delta$. Besides, the impact of proximity effects, regulated by interleaved windings, becomes evident by the MMF-factor m. This MMF-factor is determined by the MMF profile for each layer, cf. [14], [15]. However, as a high leakage inductance is desired, only mildly interleaved windings are foreseen. This penalty is accepted because core losses are expected to far outweigh winding losses. As such, the calculations consider only zero, one, or two interleaved layers. Finally, the AC winding losses $P_w = I^2 R_{ac}$ are computed for each combination of interleaving type, number of turns, and number of layers.

$$R_{ac} = R_{dc} \frac{\xi}{2} \left(\frac{sinh(\xi) + sin(\xi)}{cosh(\xi) - cos(\xi)} + (2m-1)^2 \frac{sinh(\xi) - sin(\xi)}{cosh(\xi) + cos(\xi)} \right) \quad (1)$$

The volumetric core losses (P_v) are subsequently estimated using the improved generalised Steinmetz equation (iGSE), as in Equation (2). This takes the square voltage waveforms of the DAB converter into account. Meanwhile, Steinmetz parameters α, β, and k are empirically collected based on the *PowerBrain* database, using a 4-wire core loss measurement setup.

$$P_v = f \int_0^t \frac{k}{2\pi^{\alpha-1} \int_0^{2\pi} |\cos\theta|^\alpha 2^{\beta-\alpha} d\theta} |\frac{dB}{dT}|^\alpha \Delta B^{\beta-\alpha} dt \quad (2)$$

These are combined to estimate the total loss for each possible configuration. Finally, the footprint is calculated considering ample clearances. The breakdown voltage of standard FR-4 dielectric equals 20 kV/mm and 3 kV/mm through air.

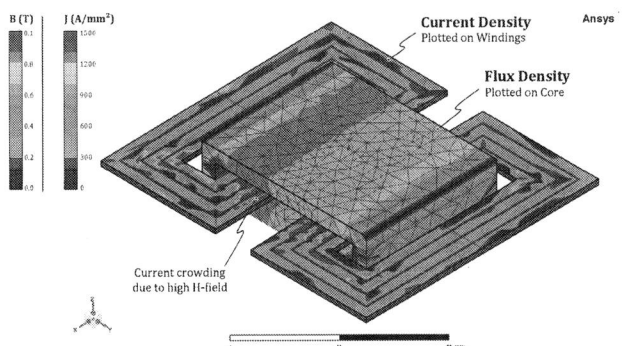

Fig. 6. Ansys Maxwell 3D Eddy Current simulation results at 500 kHz and nominal load, exemplified in a 24:24 transformer using 6 layers, with a single interleaved layer.

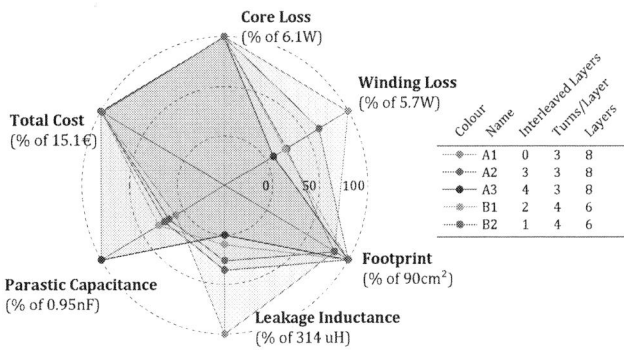

Fig. 7. Spider graph of the resulting KPIs for the different design candidates based on 3D Ansys Maxwell simulations.

The considered clearances are the following: lateral track-to-track is 1 mm (20 kV), minimum pre-preg layer thickness 0.1 mm (2 kV), core layer thickness 0.5 mm (10 kV), via-to-track distances of minimum 1 mm (20 kV), with core-to-track having 1 mm board clearance and 1 mm air (23 kV). Considering the voltage distribution among the windings, which gradually increases, this suffices for 800 V operation.

The results of the analytical pre-calculation are presented in **Fig. 4**, considering the footprint and total loss for different numbers of layers, turns, and interleaved layers. From this, several conclusions are drawn. Firstly, the total loss reaches a minimum value between 15 to 35 turns, with a loss below 3.2 W. Secondly, as expected, the core losses dominate, proven by the minor effect of the interleaving on the total losses (cf. **Figs 4b-d**). This indicates the requirement for effective cooling of the core, alleviating the thermal strain from the poor thermal conduction of PCB windings. Related, the number of layers has a negligible effect on the total loss, which is not the case when AC winding losses dominate. Finally, the footprint, cf. **Fig 4a**, naturally benefits from a higher number of layers.

The trade-off between the footprint and total loss is studied in **Fig. 5**, which combines the contour plot of the footprint with the loss contours of a design with one interleaved layer. Herein, a loss constraint is outlined by the region between 15 and 35 turns. Besides, a cost constraint is imposed that limits the number of layers to eight. Finally, a size constraint defines an approximated design region, as highlighted in **Fig. 5**.

3) FEA Simulation-based Comparison: To perform final parameter selection based on an accurate estimation of the leakage inductance, the expected losses, and related parasitic capacitance, finite elements analysis (FEA) simulations are performed for the highest-ranked design candidates. These are summarised in **Table I**, in which a preference for designs with around 24 turns was implied due to the high leakage requirement. **Fig. 6** exemplifies such a simulation result in Ansys Maxwell 3D, indicating the flux density and current density. The imposed simulation setup and implications are detailed in [13].

The optimisation objectives are minimal cost, total size, and losses while complying with constraints enforced by parasitic

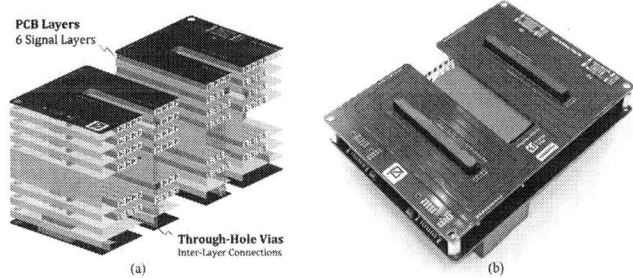

Fig. 8. (a) Isometric section view of the PCB winding stack-up and configuration, (b) PCB winding transformer prototype version 1 featuring 6 layers with 4 turns per layer. Experimental results indicate a parasitic capacitance of 0.9 nF.

capacitance, EMI, and heat dissipation. The results of the 3D Ansys Maxwell simulations are summarised in **Fig 7**. As presented in **Section IIA**, some interleaving is required to achieve the target leakage inductance, which is quantified in these findings. The cost includes only the cost of the PCB and magnetic cores, based on the unit pricing for an order quantity of 100 pieces. Since the ferrite core largely determines the pricing, the layer count only has a minor influence. This, once more, highlights the benefit of PCB windings.

Based on this comprehensive comparison, the 24:24 transformer design *B2* was selected, comprising 6 PCB layers with 4 turns per layer. As illustrated in **Fig. 6a**, the desired integrated leakage inductance of 110 μH is achieved by one interleaved primary and secondary layer placed on the outer layers of the PCB stack-up. Through-hole vias are implemented to connect the different layers, and the availability of an 8-layer board allows for two layers to be utilised for connections inside and outside the transformer.

III. PARASITIC CAPACITANCE REDUCTION

1) High Capacitance Limitation: The proposed PCB-winding inductor-transformer achieves low magnetic loss and a compact footprint but is challenged by high parasitic capacitances. Notably, excessive intra-winding capacitance caused by substantial overlapping copper areas at close proximity causes resonance between the transformer's inductances and its parasitic capacitance (cf. **Section IV**). Reducing the magnetising

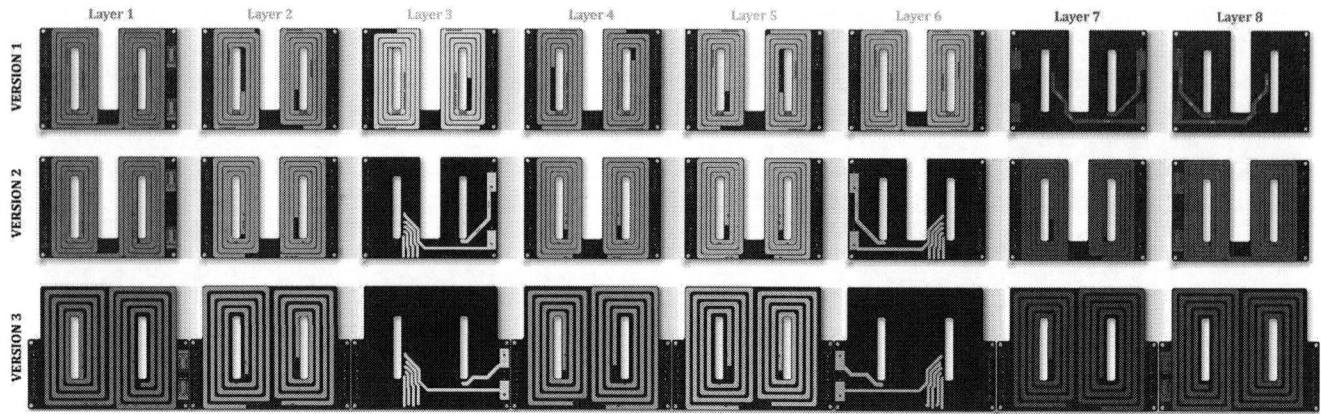

Fig. 9. PCB winding layout of the three transformer versions. Herein, version 2 implements *flyback-type* windings by interconnecting the end of each layer from the inside to the beginning of the next layer at the outside through layers 3 (primary) and 6 (secondary). Version 3 implements a *zig-zag* winding pattern with enlarged lateral spacing between each turn, while windings in adjacent turns align with these clearances.

inductance resolves this, albeit at the cost of excessive DAB circulating currents. Strategies to reduce parasitic capacitance are thus investigated.

These strategies are based on an understanding of how the electrostatic energy is distributed in the transformer and constitutes the parasitic capacitance. As is well known, the capacitance between adjacent layers is expressed as $C_o = \epsilon \frac{S}{d}$, with ϵ the permittivity of the medium, S the overlapping surface area, and d the distance between the planes. However, the total lumped capacitance of a winding then depends on the total energy stored in the field, as dictated by the voltage distribution of the winding ($E \propto U^2$). Note that due to the aspect ratio of the PCB tracks, the lateral turn-to-turn capacitance is negligible compared to the capacitance between overlapping turns of adjacent PCB layers.

2) Flyback-type Windings: The first improvement involves implementing flyback-type windings to achieve a reduced and constant voltage difference between adjacent winding layers, as proposed in [14], [16]. This approach minimises the energy stored in the electrostatic field by altering the interconnections between windings across PCB layers. In traditional designs comprising multiple turns per layer, the most compact solution is to directly connect the end of one layer to the beginning of the next layer using vias (cf. **Fig. 9**. This results in each consecutive layer winding in the opposite lateral direction, e.g., the first layer winds from the inside to the outside, the next layer winds from the outside to the inside, and so on. This change of winding direction causes a high maximum voltage difference between the terminals of adjacent layers of $2V_{in}/N_{layers}$. This configuration is depicted in the layers of version 1 in **Fig. 9**. To address this issue, the flyback-type windings in version 2 modify the interconnection strategy. Herein, layers 3 and 6 are utilised to interconnect the last turn of each layer to the beginning of the next layer to maintains the same winding direction across all layers, e.g., consistently winding from the inside to the outside. Re-routing the connections through these additional layers reduces the maximum voltage difference between adjacent layers, enforc-

Fig. 10. (a) PCB winding transformer prototype version 2 employing flyback-type layers with 10.1% primary capacitance reduction, (b) Version 3 employing flyback-type and misaligned layers with 19.2% primary capacitance reduction.

ing a constant gradient equal to $2V_{in}/N_{layers}$. This reduces the voltage stress between layers and decreases the energy stored in the electrostatic field, reducing the lumped parasitic capacitance.

3) Zig-zag Windings: Another improved winding layout is presented in version 3 of **Fig. 9**. It implements enlarged lateral spacing between PCB turns within a layer. This additional spacing is then used to misalign the turns of the adjacent layers. As a result, a *zig-zag* winding pattern emerges, which, in this instance, reduces the overlapping copper area by a factor of 3.3. Of course, the footprint is significantly penalised by this implementation. Combined with the flyback-type windings, this is implemented in version 3 of the transformer.

IV. EXPERIMENTAL RESULTS

A prototype of each iteration of the transformer is constructed and experimentally characterised. These are implemented in the 800-V GaN dual active bridge proof-of-concept for onboard chargers. This section reports on the results and challenges thereof.

A. Integrated Inductor-Transformers Characterisation

1) Prototyping and Characterisation Method: The version 2 transformer with flyback-type windings and version 3 transformer with zig-zag windings are respectively depicted in **Figs.**

TABLE II
COMPARISON TABLE FOR THE DETAILED TRANSFORMER PARAMETERS IN
VERSION 1 TO 3 PROTOTYPES.

Parameter	Version 1	Version 2	Version 3
Prim. Self-Induct. L_{sp} (μH)	353.5	394.1	314.1
Prim. Leakage Induct. L_{kp} (μH)	76.0	85.6	50.0
Magnetising Induct. L_m (μH)	277.5	308.5	264.1
Prim. Intra-Winding C_p (nF)	0.391	0.352	0.316
Sec. Intra-Winding C_s (nF)	0.391	0.352	0.316
Mutual Capacitance C_{ps} (nF)	0.064	0.109	0.093

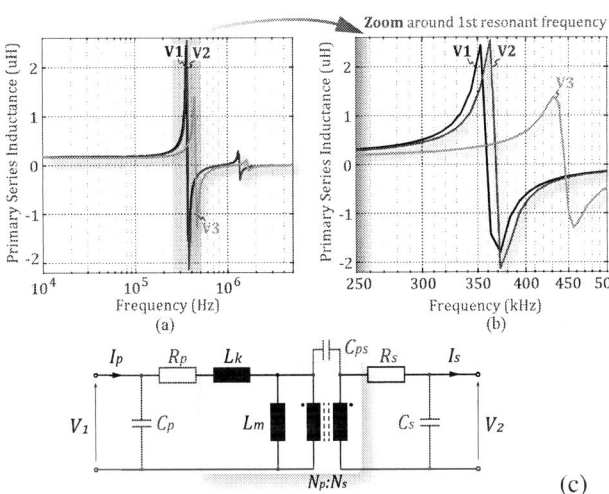

Fig. 11. (a) Open circuit measurements of the transformer prototypes using a Hioki impedance analyser from 100 Hz to 5 MHz, (b) identifying the first self-resonance between magnetising inductance and primary intra-winding capacitance for the three prototypes, (c) Transformer T-model referred to the primary side, including the key parasitic components.

10ab. Each of the prototypes implements the aforementioned 3F36 ferrite UI cores. The magnetising inductance is tuned to 280 μH by an air gap length of approximately 50 μm. Additional polyimide Kapton tape was applied between the PCB and the core, and the PCB windings were placed away from the air gap to alleviate fringing losses due to stray fields surrounding the air gaps. Experimental characterisation of the transformer prototypes is conducted using a Hioki IM3570 impedance analyser (100 Hz~5 MHz) and an Omikron Lab Bode 100 network analyser (1 Hz~50 MHz) according to [17].

2) Transformer Model Characterisation: The characterisation based on the transformer model in **Fig. 11c** is summarised in **Table II**. Experimentally, the primary parasitic capacitance of version 1 is measured at 0.39 nF. The second version with flyback-type windings achieves a 10.1 % reduction of intra-winding capacitance, as the primary and secondary parasitic capacitances are reduced to 0.35 nF. Finally, the zig-zag windings of version 3 further reduce the primary self-capacitance to 0.31 nF, a 19.2 % reduction compared to version 1. To compare the primary self-inductance of the three versions, the magnetising inductance of each was tuned to 180 μH. **Fig. 11** shows the subsequent primary self-inductance measurements. The first self-resonant frequency, approximately between the primary self-inductance and primary intra-winding capacitance, is respectively 353 kHz, 363 kHz, and 433 kHz for versions 1 to 3. It is noteworthy that neither enhancement achieved the degree of improvement anticipated based on their theoretical calculations and modelling. This could be attributed to the electrostatic field skewing even in between misaligned planes (as exemplified in **Fig. 10b**) and the need for a more complex capacitance model that includes the interaction between each of the traces. Nevertheless, version 3 provides ample distance between the operating and fundamental frequencies. This underlines the challenge in PCB winding design to achieve a high magnetising inductance while operating at high frequency, where self-resonance with the intra-winding capacitance may occur.

3) Further Capacitance Reduction: Based on these findings, other suggestions can be made to further reduce the parasitic capacitance. A pragmatic solution is the use of low-permittivity PCB substrates. Standard FR-4 materials, such as the one used in these prototypes, have a relative permittivity ranging between 3.8~4.8. Commercial PCB manufacturers often provide high-frequency PCB solutions with specialised

low-dielectric constant (Dk) materials, in which a Dk as low as 2.0 can be reached. Naturally, the parasitic capacitance will reduce accordingly. However, the cost of such PCBs is 10 times higher than that of standard FR-4 materials in the case of the presented PCBs. Further capacitance reduction can be achieved using higher copper thickness in the PCB. Designers can choose to penalise the AC resistance by being aware of the skin depth at their operating frequency to reduce the required width of the copper traces. For instance, the fundamental frequency of this 250 kHz DAB could allow for 3 oz (105 μm) copper tracks to be used. Hence, the width of the traces can be reduced by 33%. This, similar to the zig-zag windings, reduces the overlapping copper areas. Finally, thicker PCB stack-ups are also available, which increase the distance between adjacent PCB layers. Since the capacitance between planes is inversely proportional to the distance separating them, the total capacitance could be reduced significantly. Again, the constraint, however, is the increased cost, especially when combined with copper thicknesses above 1 oz.

B. Dual Active Bridge Converter

1) 1200-V GaN Device Implementation Challenges: This work implements one of the first power converters using 1200-V GaN high-electron-mobility transistors (HEMT), namely the GPIHV30DFN HEMTs from GaNPower with a 65 mΩ on-resistance. Due to their superior electrical properties, GaN HEMTs are excellent candidates for power switches in high-speed operations up to hundreds of kilo hertz and mega hertz. The implementation of discrete 1200 V GaN devices in 800 V converters is not without challenges. For 250 kHz and 800 V, the slew rates are already sufficiently high to induce significant levels of noise in the drive loop of the transistors. This is because the high slew rate at the drain induces a current through the gate-drain capacitance C_{GD}, creating a voltage

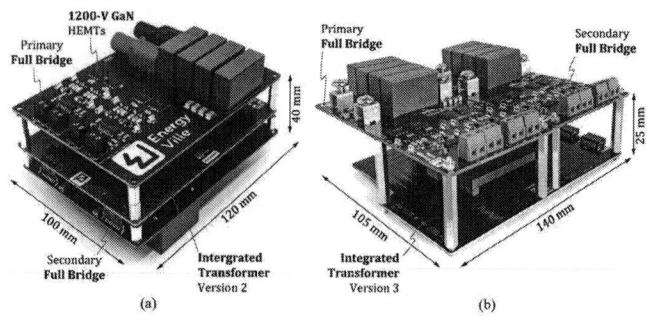

Fig. 12. DAB converter prototype with 1200-V GaN transistors, (a) 3-layer converter with version 1-2 transformer, (b) 2-layer converter with version 3 transformer.

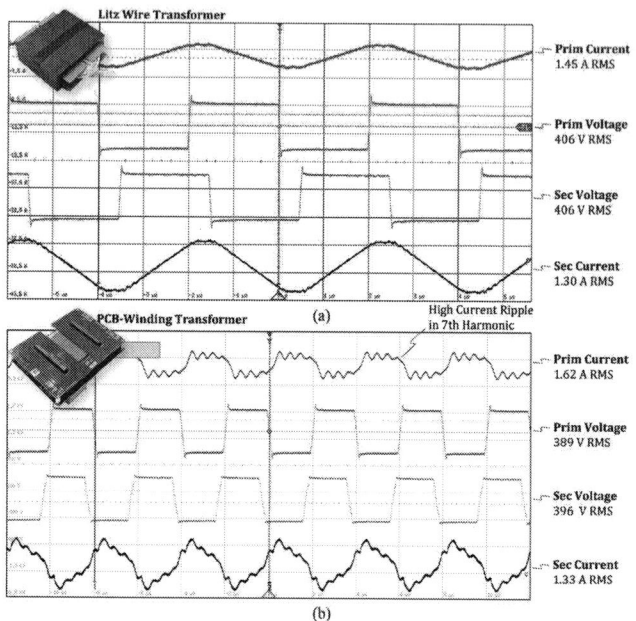

Fig. 13. Dual active bridge converter preliminary experimental waveforms with 1200-V GaN devices operating at 400 V, 250 kHz, with (a) Reference litz wire transformer and inductor, (b) Integrated PCB-winding transformer, depicting high harmonic content and resonance.

spike relative to the C_{GD}/C_{GS} ratio, being quite high in GaN HEMTs. This, combined with their low threshold voltage (1.3 V), provides low margins for avoiding false turn-on. This can be addressed by using a negative off-voltage, but this inadvertently increases reverse conduction losses. Integrated devices mitigate this through, e.g., a miller clamp. A second challenge is related to the inductance in the drive loop, which has to be kept low to reduce spiking and ringing. This requires the gate loop length to be minimised, which can be achieved more effectively by devices with integrated drivers. The third major issue is thermal extraction. The GPIHV30DFN has a DFN8x8 package, which is among the packages typical for GaN HEMTs. Their size of 64 mm² is quite small compared to conventional SiC and Si power transistors, impeding the extraction of heat. The DAB in this paper is cooled through vias under thermal pads of GaN HEMTs to the bottom layer of the PCB, where a heatsink is attached. In a later iteration, the HEMTs were placed on copper coin inserts, directly connecting copper to the thermal source pad.

2) DAB Converter Prototyping: The single-phase dual active bridge topology is selected for its soft-switching capabilities and phase-shift control. **Figs. 12ab** show the dual active bridge converter prototypes. First, a stacked prototype is constructed in a box volume of 100×120×40 mm³. Having the converter split into three boards, forced-air cooling from the side of the board can effectively dissipate the heat. A second packaging strategy was implemented using the version 3 transformer due to its larger footprint. A flatter 2-layer structure is designed comprising a single board for both full bridges. This air-cooled DAB features a box volume of 140×105× 25 mm³. Board-to-board connections are done using Phoenix Contact FR 1,27 series connectors. A liquid-cooled single-PCB converter is foreseen in future work for improved cooling capability at increased power.

3) Primary Self-Resonance Limitation: The first challenge faced when implementing PCB-winding transformers in the DAB is primary self-resonance. Compared to preliminary DAB operating waveforms at 250 kHz using a conventional litz-wire transformer (cf. **Fig. 13a**), the PCB-winding prototypes have an order of 10 magnitude higher intra-winding capacitance. This is illustrated by the waveforms of **Fig.**

13b of the version 2 DAB prototype. This necessitated the reduction of the intra-winding capacitance, as discussed in detail in previous sections, to enable similar operations with planar devices. Furthermore, an efficiency of only 89.1 % was achieved using a reference litz wire transformer and inductor, confirming the potential of the newly designed PCB-winding integrated inductor-transformer.

4) Leakage Resonance Limitation: As shown in **Fig. 14a**, the square-wave voltage of the DAB converter contains odd harmonics (3rd, 5th, 7th, etc.). The resonant tank formed by the transformer's inductances and parasitic capacitances may cause resonances at specific harmonics, e.g. the seventh. This is confirmed in circuit simulations in Plexim PLECS upon implementing the full transformer model of **Fig. 11c** in a DAB converter. The resulting Fourier series, in **Fig. 14a**, confirm the impact of the leakage and magnetising inductance values, as e.g. the use of a higher leakage of 150 µH further eliminates this harmonic, see **Fig. 14b**. **Fig. 15** shows an improved DAB implementation with increased leakage inductance, confirming that this resonance is successfully eliminated. Herein, the operating frequency is even increased to 500 kHz to demonstrate the high-frequency operating capabilities of the GaN HEMTs, which will be subjected to further investigation.

V. CONCLUSIONS

This paper presents a novel 800-V dual active bridge converter prototype for electric vehicle onboard chargers, successfully implementing state-of-the-art 1.2-kV GaN devices and novel magnetics. The design introduced a split-winding PCB-integrated inductor-transformer that effectively integrates

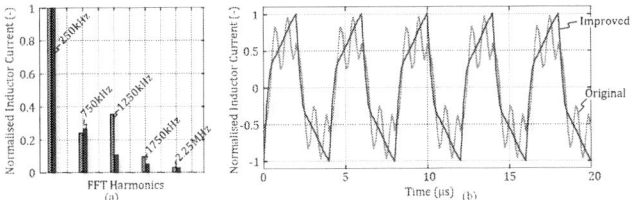

Fig. 14. Experimentally characterised transformer parameters implemented in a circuit simulation of the DAB converter in PLECS, (a) Normalised Fourier transform results, (b) Normalised current waveform in steady-state.

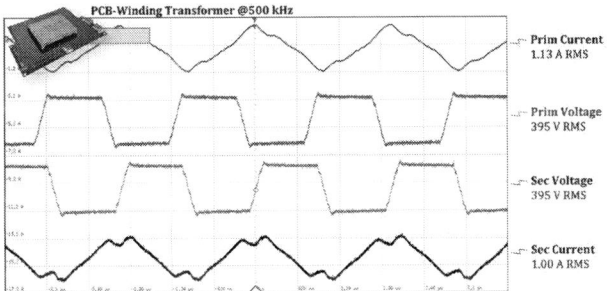

Fig. 15. Dual active bridge converter experimental waveforms with 1200-V GaN devices operating at 400 V, 500 kHz with the version 3 integrated PCB-winding transformer.

Fig. 16. PCB integration of the DAB and proposed inductor-transformer.

the series inductor, achieving a cost-effective, compact, and efficient solution while addressing key challenges such as parasitic capacitance. The transformer design is optimised through a combination of analytical optimisations and extensive finite element analysis. The implementation of flyback-type and zig-zag winding layouts demonstrated significant reductions in parasitic capacitance, enhancing the transformer's high-frequency performance. Experimental results validated the feasibility of the proposed design while highlighting implementation challenges due to resonances and the use of discrete GaN devices. These are quantified and addressed, with 400-V operation performed at 250 kHz as well as 500 kHz.

Future research vectors are to further reduce parasitic capacitance through the use of low-permittivity PCB substrates and thicker PCBs. Furthermore, a liquid-cooled prototype that integrates the converter and transformer PCBs is being developed to improve thermal management and manufacturability while eliminating high-frequency terminations, with a first prototype depicted in **Fig. 16**. This research underscores the potential of both 1.2-kV GaN devices and PCB-winding integrated magnetics to develop compact, efficient, and cost-effective high-voltage converters for EV applications. The advancements made herein contribute to the progression toward next-generation onboard chargers that meet the increasing demands for cost, efficiency and power density in electric vehicles.

REFERENCES

[1] H. Wouters and W. Martinez, "Bidirectional onboard chargers for electric vehicles: State-of-the-art and future trends," *IEEE Transactions on Power Electronics*, vol. 39, no. 1, pp. 693–716, 2024. [Online]. Available: https://ieeexplore.ieee.org/document/10265141

[2] A. Khaligh and M. D'Antonio, "Global trends in high-power on-board chargers for electric vehicles," *IEEE Trans. Veh. Technol.*, vol. 68, no. 4, pp. 3306–3324.

[3] G. Andrioli, M. Pajnić, S. Calligaro, and R. Petrella, "Exploiting the depth: Design, analysis, and implementation of high-power-density high-frequency transformers for one rack unit CLLC DCX converters," in *2023 IEEE Applied Power Electronics Conference and Exposition (APEC)*, pp. 3255–3262.

[4] C. R. Sullivan, B. A. Reese, A. L. F. Stein, and P. A. Kyaw, "On size and magnetics: Why small efficient power inductors are rare," in *2016 3D-PEIM*, pp. 1–23.

[5] A. Nabih and Q. Li, "A method to embed resonant inductor into PCB matrix transformer for high-density resonant converters," *IEEE Transactions on Power Electronics*, vol. 39, no. 2, pp. 2385–2400.

[6] Y. Liu, H. Wu, J. Zou, Y. Tai, and Z. Ge, "CLL resonant converter with secondary side resonant inductor and integrated magnetics," *IEEE Transactions on Power Electronics*, vol. 36, no. 10, pp. 11 316–11 325.

[7] S.-Y. Yu, C. Hsiao, and J. Weng, "A high frequency CLLLC bi-directional series resonant converter DAB using an integrated PCB winding transformer," in *2020 IEEE Applied Power Electronics Conference and Exposition (APEC)*, pp. 1074–1080.

[8] J. Yang, X. Wu, G. Liu, D. Ping, and Z. Deng, "Modeling and design of integrated inductor and transformer considering superposed flux density in on-board-charger," in *2020 IEEE Applied Power Electronics Conference and Exposition (APEC)*, pp. 879–884, ISSN: 2470-6647.

[9] Y. Park, S. Chakraborty, and A. Khaligh, "DAB converter for EV on-board chargers using bare-die SiC MOSFETs and leakage-integrated planar transformer," *IEEE Trans. Transp. Electrification*, pp. 1–1.

[10] M. Kaymak, M. E. Fincan, and R. W. De Doncker, "Core-type transformer design method with integrated series inductance for DC-DC converters," in *ICPE 2019 - ECCE Asia*, pp. 1581–1587.

[11] E. S. Lee, J. H. Park, M. Y. Kim, and J. S. Lee, "High efficiency integrated transformer design in DAB converters for solid-state transformers," *IEEE Transactions on Vehicular Technology*, vol. 71, no. 7, pp. 7147–7160.

[12] B. Li, Q. Li, F. C. Lee, Z. Liu, and Y. Yang, "A high-efficiency high-density wide-bandgap device-based bidirectional on-board charger," *IEEE J. Emerg. Sel. Top. Power Electron.*, vol. 6, no. 3, pp. 1627–1636.

[13] H. Wouters, H. Pervaiz, T. Geboers, Y. Zuo, W.-R. Lin, and W. Martinez, "Interleaved PCB winding planar transformer for electric vehicle charging CLLC converters," in *2024 IEEE Applied Power Electronics Conference and Exposition (APEC)*, pp. 3216–3223. [Online]. Available: https://ieeexplore.ieee.org/document/10509410

[14] Z. Ouyang, O. C. Thomsen, and M. A. E. Andersen, "Optimal design and tradeoff analysis of planar transformer in high-power DC–DC converters," *IEEE Trans. Ind. Electron.*, vol. 59, no. 7, pp. 2800–2810.

[15] J. Ferreira, "Improved analytical modeling of conductive losses in magnetic components," *IEEE Trans. Power Electron.*, vol. 9, no. 1, pp. 127–131.

[16] J. Biela and J. Kolar, "Using transformer parasitics for resonant converters - a review of the calculation of the stray capacitance of transformers," in *2005 IAS Annual Meeting*, vol. 3, pp. 1868–1875 Vol. 3.

[17] C. Liu, L. Qi, X. Cui, and X. Wei, "Experimental extraction of parasitic capacitances for high-frequency transformers," *IEEE Transactions on Power Electronics*, vol. 32, no. 6, pp. 4157–4167.

Comparative Assessment of Inductance Modeling for PCB-based Circular Spiral Coils in Inductive Power Transfer Systems

Gaia Petrillo, Drazen Dujic

Power Electronics Laboratory - PEL
École Polytechnique Fédérale de Lausanne - EPFL
Lausanne CH-1015, Switzerland
gaia.petrillo@epfl.ch, drazen.dujic@epfl.ch

Abstract—**To improve the efficiency of inductive power transfer (IPT) systems, it is beneficial to maximize the power transfer by enhancing the quality factor of coils and the coupling between them. Therefore, it is crucial to rely on accurate, flexible, and practical models to perform a successful design optimization of the coils and to improve their quality factor. This paper focuses on the inductance model selection and compares different modeling methods reported in the literature in terms of accuracy, implementation ease, and flexibility to address design variations. The results of the analyzed methods are compared with 2D and 3D FEM simulations and measurements on prototyped coils with different characteristics. The analysis and measurements provide a direct comparison and allow the selection of suitable models for subsequent design optimization.**

Index Terms—**inductance, modeling, coupling coils, IPT**

I. INTRODUCTION

In recent years, Inductive Power Transfer (IPT) has become a commercial reality for many different applications from consumer electronics to medical devices, EV charging, and industrial robots, and it is still the focus of intense research in many new applications such as gate drivers, auxiliary power supplies, drones and autonomous vehicles [1]. Nevertheless, despite their convenience, the efficiency of IPT systems is still low when compared to wired alternatives, which limits their practical wider applicability. Different aspects of an IPT system can be optimized to improve overall system efficiency. One of those is the inductive coupling link efficiency, which depends on the quality factor (Q) of the coils and their coupling coefficient (k). While in many applications k is constraint due to the distance requirements between transmitter and receiver, Q can be improved for any distance and inductance requirement through a thoughtful optimization of the coil design.

While different technologies can be used for IPT coils, this paper focuses only on PCB coils. PCBs offer many advantages such as low cost, ease of manufacturability, possibility of batch fabrication, which is interesting for modular design, durability, stability of parameters, and therefore repeatability and reliability, design flexibility, and suitability for high-frequency operating systems [2].

Not many studies can be found in literature on the optimization of PCB-based IPT coils. Others have investigated the optimization of PCB-based magnetic components, but parameters are chosen based on Finite Elements Analysis (FEA), as in [3]. This solution is very computationally expensive, and it would be more convenient to identify a reliable analytical approach to perform the Q optimization. To define Q of an IPT coil, the models for inductance (L) and resistance (R) are needed. In this work, the focus is only on the comparative assessment of the inductance models. While the approach for accurate L modeling is well defined and assessed in case of spiral coils with circular cross section, as presented in [4], a well consolidate approach can not be found in literature in case of rectangular cross section coils, which is the case of PCB-based designs.

This paper compares several commonly used inductance models, thoroughly considering them in terms of accuracy, ease of implementation, flexibility to geometry variations. All the considered models are compared against the 2D and 3D FEM simulations and measurements on a variety or manufactured coils. The aim of this comparative analysis is to assess the applicability of the available models to the framework of design optimization of PCB-based coils for IPT. Nevertheless, the study methodology is valid also for different applications of planar magnetics.

Multiple coils were manufactured, as support for the validation of the models on of PCB based coils, whose inductance value is in the range of fractions to few μH and whose footprint can be contained in an area of 12x12 cm, all the design present uniform pitch and width.

II. COIL MODELING APPROACHES

In this section, four modeling methods for inductance, available in literature are discussed highlighting their major characteristics. It may be helpful to refer to the Appendix in order to better understand the global picture and the relation between the different models. The definition of geometrical parameters used in the modeling equation are given in Fig. 1.

All modeling methods, expect for the Method 1, evaluate the total coil inductance as the sum of the contribution of the

979-8-3315-1612-3/25 $31.00 © 2025 IEEE

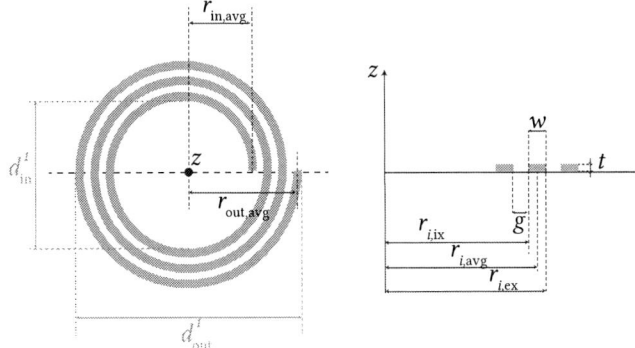

Fig. 1. One Layer winding of PCB coil, top and section view. Track parameters: w - width, t - thickness, g - gap, N - number of turns, $r_{in,avg}$ - most internal turn final radius (in the center of w), $r_{out,avg}$ - most external turn starting radius (in the center of w), $r_{i,ix}$ - internal radius at the start of any turn, $r_{i,avg}$ - mid-w radius at the start of any turn, and $r_{i,ex}$ - external radius at the start of any turn

self inductance of each turn (L_i), and the one of the mutual inductance between each couple of turns ($M_{i,j}$). Calculation of mutual inductance is simplified by the fact that $M_{i,j} = M_{j,i}$ for any i,j. General expression for total coil inductance is given in (1)

$$L_{tot} = \sum_{i=1}^{N} L_i + 2 \sum_{i=1}^{N-1} \sum_{j=i+1}^{N} M_{i,j} \qquad (1)$$

A. Method 1

In this method, proposed in [5], a single equation, given in (2), estimates the total coil inductance, already taking into account the contribution of all the turns. It applies for planar circular spiral coils, but in the same article the authors propose a modification of this formula for squared, hexagonal and octagonal shapes. In any case, the assumption is made that the turns have no vertical displacement and that w and g are constant in all the structure.

$$L_{tot} = \frac{\mu_o N^2 d_{avg}}{2} \left(ln\left(\frac{2.46}{\rho}\right) + 0.20 \cdot \rho^2 \right) \qquad (2)$$

$$d_{avg} = \frac{d_{out}^1 + d_{in}^1}{2} \qquad \rho = \frac{d_{out}^1 - d_{in}^1}{d_{out}^1 + d_{in}^1} \qquad (3)$$

Fill ratio (ρ) and average diameter (d_{avg}) definitions are given in (3). It is important to to define d_{out} and d_{in} perpendicular to each other as in the Fig. 1 when implementing this model. As shown in [6] this is an approximate equation based on the current sheet approximation method, and according to the authors, it is valid for $g/w < 3$.

B. Method 2

This method was developed in [6] for on-chip radio frequency inductors and transformers. It is also applied in [7], for PCB coils for IPT. It evaluates independently the different contribution as in (1) and it applies concentric circles approximation. Equation for L_i and $M_{i,j}$, given respectively in (4) and

(5), are obtained from series expansions of Neumann formula and applying current sheet theory (i.e. terms including distance are substituted with mean distance of the finite cross section). According to the authors the approximation in good within 5% for $\rho^{II} < 0.6$ as defined in (6). This method provides all provisions for uneven gaps, but not for uneven track widths or vertical misalignments. In the implementation it must be kept in mind that G is not the gap between two successive turns, but the effective distance edge to edge between the two specific turns considered in the calculation $G = g + (j - i - 1)(w + g)$ with $j > i$.

$$L_i = \frac{\mu 2 r_{i,avg}}{2} \left[ln\left(\frac{2 r_{i,avg}}{w}\right) + 0.9 + 0.05 \frac{w^2}{r_{i,avg}^2} \right] \qquad (4)$$

$$
\begin{aligned}
M_{ij} = {} & \frac{\mu_0 (r_{i,avg} + r_{j,avg})}{2} \left[ln\left(\frac{r_{i,avg} + r_{j,avg}}{w + G}\right) \right. \\
& - 0.6 + 0.7 \left(\frac{w + G}{r_{i,avg} + r_{j,avg}}\right)^2 \\
& \left. + \left(0, 2 + \frac{(r_{i,avg} + r_{j,avg})^2}{12(w + G)^2}\right) \frac{w^2}{(r_{i,avg} + r_{j,avg})^2} \right]
\end{aligned}
$$
$$(5)$$

$$\rho^{II} = \frac{r_{i,avg} - r_{j,avg}}{r_{i,avg} + r_{j,avg}} \qquad (6)$$

C. Method 3

This method is explained in [8] and used in [9] and [10]. It is presented for conductors with circular cross section, but, as it is based on filament current approximation, it can be applied also for rectangular cross-section, where the filament current is located in the center of the cross-section. Also in this case, spiral tracks are approximated with concentric circles and the contribution of self and mutual inductance are evaluated independently, as in (1). The contributions are evaluated considering the definition of inductance (9). Biot-Savart Rule, in (7), gives the magnetic flux density generated in a point Q by an elementary filament current source. In (7) $\vec{dl}$ is the length of the filament current source, I is the current carried by $\vec{dl}$ that is generating the flux, $\vec{q}$ is the unity vector pointing from the current element (source) to the point of interest (Q) where the magnetic field is calculated, and R_Q is the norm of that vector. The total magnetic field generated by a line current source is given by (8). Flux is evaluated integrating the magnetic flux density over the area of the turn, and the inductance of a turn is calculated as in (9). It is important to notice that: θ and x_k are related to the location of the current source, while Q, and therefore r and φ, are related to the turn taken in into account in the area calculation. Equation (9) is applicable to calculate both L_i and M_{ij}. For L_i the turn considered as current source and the one delimiting the area for the flux calculation are the same. Instead, for M_{ij} two different turns are considered as source and as border of the area for flux calculation.

979-8-3315-1612-3/25 $31.00 © 2025 IEEE

$$dB = \frac{\mu_o}{4\pi} \frac{I \, d\vec{l} \times \vec{q}}{R_Q^2} \tag{7}$$

$$\vec{B} = \frac{\mu_o}{4\pi} \int_{line} \frac{I \, d\vec{l} \times \vec{q}}{R_Q^3} = \int_{\theta=0}^{2\pi} \frac{\vec{I}(\theta) \times \vec{q}(x_q, y_q, z_q, x_k, \theta)}{R_Q^3(x_q, y_q, z_q, x_k, \theta)} \tag{8}$$

$$L = \frac{\Phi}{I} = \int_0^{2\pi} \int_0^{R_{turn}} \vec{B}(r \cdot cos\varphi, r \cdot sin\varphi, 0) \times [0; 0; 1] d\varphi dr \tag{9}$$

This method is very versatile as it can take into account uneven width and pitch and vertical misalignment, since, for each turn, only the location of the equivalent filament wire is used in the equation. On the other side, it is very difficult to implement, triple integral needs to be properly discretized and multiple loops are needed to evaluate each contribution.

D. Method 4

This method, presented in [11], is based on circular approximation, and calculates the different contributions separately as in (1). The authors propose to model each circular turn with two filament conductors located in $r^* + \alpha$ and $r^* - \alpha$, to take better into account the width of the track. r^* and α are calculated as in (10) for each turn i. This formula apply only when the track width is bigger than its thickness ($w > t$), which is always true in the study.

$$r^* = r_{i,avg}\left(1 + \frac{t_i^2}{24\,r_{i,avg}^2}\right) \qquad \alpha = \sqrt{\frac{w_i^2 - t_i^2}{12}} \tag{10}$$

Once the radii of the filament conductors are calculated for each track with (10) as in Fig. 2, both self and mutual inductance are calculated using Maxwell mutual inductance equation for a coaxial pair of circular filament conductors shown in (15), this is clarified in (11).

$$L_i = M_{12} \qquad M_{ij} = \frac{M_{13} + M_{14} + M_{23} + M_{24}}{4} \tag{11}$$

This method is very versatile as it can easily take into account uneven gap, width or thickness between the turns. Moreover, it includes vertical displacement, so it is suitable to model any configuration, as long as all the turns are coaxial.

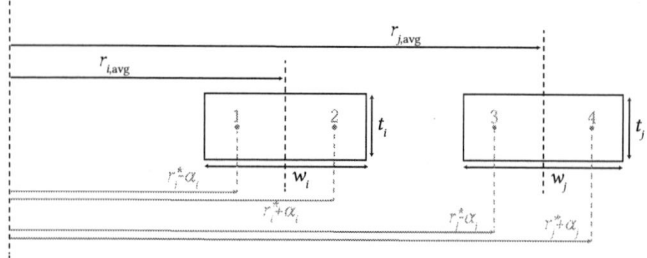

Fig. 2. Cross section of two turns of a coil where filament conductors, used in Method 4, are highlighted

E. Concentric Circles Approximation

As mentioned in previous sections, the majority of the methods approximate the spiral coil with a set of concentric circles. In a spiral though, the radius continuously changes with the angle, and it can be evaluated using the formula for Archimedean spiral in [12], therefore there is a degree of freedom in the value selected as radius ($r_{eq,i}$) for each concentric circle. It is assumed that $r = r_{ex,i}$ is the radius at the external end of a spiral turn, $r = r_{ex,i} - (w_i + g_i)$ is the radius at the internal end of the turn. Since the variation is continuous and uniform, it is a good approximation to use, as an equivalent radius ($r_{eq,i}$) for the concentric circle representing each turn, the one calculated as in (12).

$$r_{eq,i} = r_i - 0.5 \cdot (w_i + g_i) \tag{12}$$

III. Validation on 1-Layer coils

A. Validation approaches

All the presented methods have been implemented in MATLAB, and their accuracy has been validated against FEM simulations and measurements, on a set of coils whose parameters are swiped over a range of interest. Such a range can be defined as follow: r_{out} is smaller than 6 cm, the w is within 2-6 mm, the g between the track is 1-3 mm, the N is up to 5.

FEM simulations are performed in ANSYS, first with MAXWELL 2D (symmetric on the z-axis) and with MAXWELL 3D, using respectively 0.1% and 1% relative error to define the convergence. 3D implementation is simplified by the existence of User Defined Primitives that can be parametrized to perform parametric tests; this implementation can be adapted to multilayer structures but does not allow to uneven gaps or widths. Conversely, accuracy of 3D simulations is largely influenced by the good definition of the boundary conditions and the coil terminations.

In total, 19 1-Layer coils were also manufactured, their design parameters and measured values are reported in Table I. Measurements were performed using Bode 100 Vector Network Analyzer from Omicron Lab with the Impedance adapter B-WIC; the experimental setup is shown in Fig. 3. All the measurement and simulations are performed at 6.78MHz.

B. Validation of the approximation

To verify accuracy of the spiral to circular approximation, some coils were manufactured with circular turns. In the circular design each turn $r_{i,avg}$ is selected to be the same as

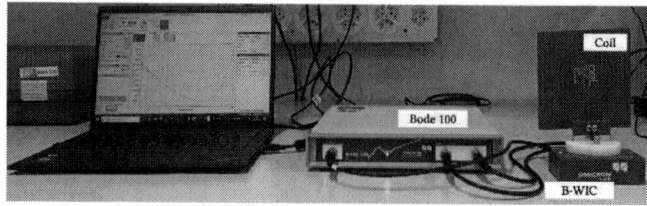

Fig. 3. Inductance Measurement Setup

979-8-3315-1612-3/25 $31.00 © 2025 IEEE

TABLE I
Manufactured 1-Layer coils

Sample	$r_{out,ex}$	N	w	g	$L\ [\mu H]$
01	55	3	2	1	1.92
02	55	3	3	1	1.66
03	55	3	4	1	1.45
04	55	3	6	1	1.14
05	55	3	2	2	1.72
06	55	3	3	2	1.51
07	55	3	4	2	1.33
08	55	3	6	2	1.05
09	55	3	2	3	1.58
10	55	3	3	3	1.38
11	55	3	4	3	1.22
12	55	3	6	3	0.96
13	55	5	2	2	3.51
14	55	5	3	2	2.90
15	55	5	4	2	2.40
16	40	3	3	1	1.02
17	40	3	3	2	0.90
18	40	3	3	3	0.81
19	40	5	3	2	1.53

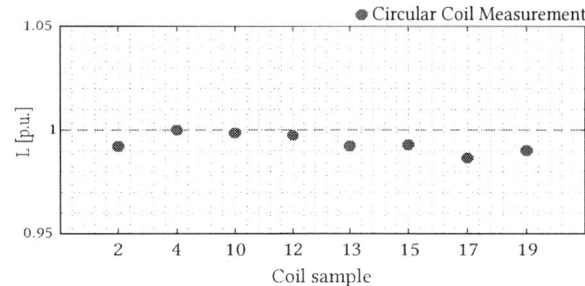

Fig. 5. Difference in measurements between circular and spiral (base) design

the equivalent radius $r_{eq,i}$ from (12). Fig. 4 show an example of spiral and circular design for the same coil parameters, while Fig. 5 shows the value of measured inductance for the circular designs in p.u. considering as base the measured value of the spiral design with same parameters. In all the considered case, the difference between the two designs is below 1.5%, therefore the approach suggested in (12) is considered valid.

C. Discussion on performance of models

The inductance values obtained through the MATLAB implementation of the presented modeling methods, together with the ANSYS simulation results are collected in Fig. 6. All the results are normalized with respect to the experimentally measured values of inductance for each coil, this allows to better illustrate relative error of different methodologies. It

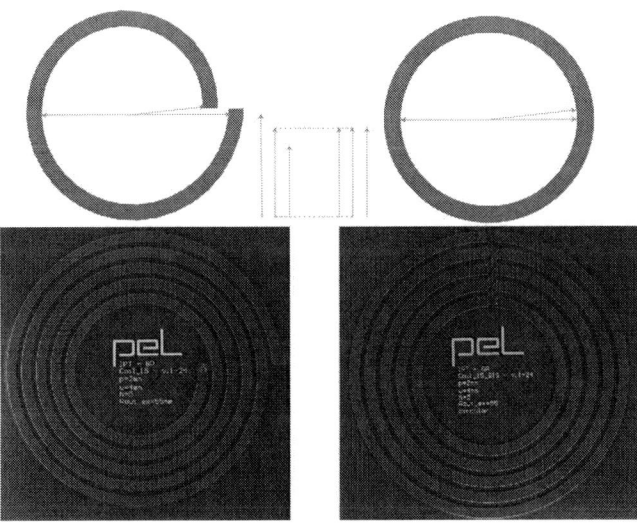

Fig. 4. On the top: equivalence between one turn of a spiral and its circular approximation; On the bottom: Coil 15 in its spiral design and circular approximation according to (12)

emerges that Methods 1-3 and 2D FEM always underestimate the value of inductance. This is a conservative error when the objective is to optimize the quality factor of the coils.

Method 1 is extremely simple and fast to implement and execute, it does not require to use the approximation as in (12), but just to follow the geometric definitions given in Fig. 1. The error is always below 6.3%, as all the cases of interest respect the condition $g/w < 3$. Considering the sampled cases accuracy seem to decrease with increasing w and N. The main drawback is its low flexibility, which makes this method very suitable for fast evaluation of simple coils, but not ideal for the intended optimization study.

Method 2 present similar error to the previous one going up to 6.5%, also in this case the condition on ρ^{II} is always respected in the considered cases. Likewise the previous case, for this method, it emerges that the accuracy decreases when the track width increases. In this implementation the approximation as in (12) was applied. Overall this method is more complex than the previous, and has anyway a limited flexibility since it can only take into account uneven gaps.

Method 3 implementation is hardest since proper integral discretion needs to be ensured. The limit of the integration area for the current carrying filament has been set on the border of the track (half the width away from the center of the coil where the ideal filament current is placed), because of this the best modeling results are achieved using $r_{eq} = r_{ex,avg}$. This method gives excellent results, with an error always below 5% and it can take into account very flexibly any set of concentric coils, but its computational time is extremely high when compared to other analytical models.

Method 4 is also very accurate, with errors staying within 4.5%, and its implementation is simple as elliptic integrals can be calculated with Matlab functions, therefore iteration loops are only needed to consider the different turns and the different filament currents for each turn, this makes this method computationally efficient, ~ 250 times faster than Method 3. In this case, the approximation as in (12) is considered in this implementation. On the negative side, it shows both positive and negative errors, which is not a conservative approach for the quality factor optimization. Also in this case, the error is increasing with the track width, which is unexpected since this method is designed to take into account w.

To summarize results of the comparison, error statistics and

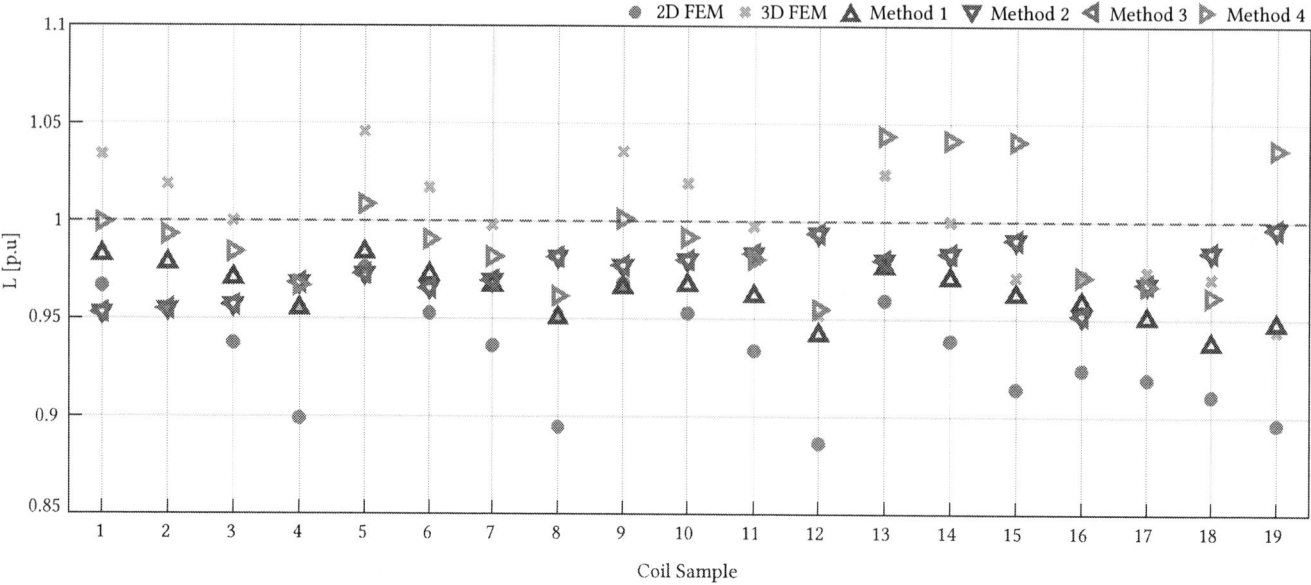

Fig. 6. Comparison of modeling results in p.u. using actual coil's measurements as base

some qualitative indicators are reported in Table II.

IV. CONCLUSIONS AND FUTURE WORK

This paper compares four analytical models for the inductance of circular spiral PCB-based coils. Analytical models flexibility and implementation ease are assessed, and their accuracy is estimated through a comparison with the results of 2D and 3D FEM simulations and with experimental measurements, with the aim of identifying the most suitable approach to be used in IPT design optimization routines. Method 4 is identified as the most flexible and accurate one, preserving very low execution times and it is therefore suggested for optimization studies. The presented results are relevant for the optimization of quality factor of IPT coils, and therefore to improve power transfer efficiency of IPT systems based on PCB. They can also be applied in different domains such as on-chip radio frequency inductors and transformers and other PCB-based planar inductors.

APPENDIX

This appendix gives a short background on well known fundamental electromagnetic equations to provide an understanding of the common framework from which different solutions presented in the modeling methods are derived. Neumann formula is an exact integral formula for mutual inductance of closed filiform conductors, it can be obtained directly from mutual inductance definition using the expression of the vector potential ($\vec{A}$). This equation, reported in (13), is exact and valid for any couple of filiform closed loops regardless of their shape and orientation in space.

$$M_{i,j} = \frac{\mu_o}{4\pi} \oint_{\gamma_j} \oint_{\gamma_i} \frac{\vec{dl_i} \cdot \vec{dl_j}}{r_{ji}} \qquad (13)$$

If we consider a couple of concentric circular filiform wires, (13) can be simplified in cylindrical coordinates, using geometrical equivalences. It can be therefore be rewritten as in (14), where R_i and R_j are the radii of the two circles and d is the vertical distance between the two centers. The resulting equation is still exact.

$$M_{i,j} = \frac{\mu_o}{4\pi} \int_0^{2\pi} d\varphi_i \int_0^{2\pi} d\varphi_j \frac{R_i R_j cos(\varphi_j - \varphi_i)}{\sqrt{R_i^2 + R_j^2 - 2R_i R_j cos(\varphi_j - \varphi_i) + d^2}}$$
$$(14)$$

This specific formulation in cylindrical coordinates of mutual inductance between two concentric circular filament wires, can be rewritten using elliptic integrals as in (15), where $K(k)$ and $E(k)$ are the elliptic integrals of first and second kind and k is the modulus defined as (16). (15) is still and exact integral formula, but its implementation may be simplified in numeric computing environments (e.g. MATLAB has a predefined function to solve elliptic integrals which remove the need to discretize the nested circulations presents in (14)). (15) is also known as Maxwell equation for mutual inductance for a coaxial pair of circular filament conductors.

$$M_{i,j} = \mu_o \sqrt{R_i R_j} \left[\left(\frac{2}{k} - k \right) K(k) - \frac{2}{k} E(k) \right] \qquad (15)$$

TABLE II
SUMMARY OF COMPARATIVE ANALYSIS

Metrics	Analytic Methods				FEM	
	1	**2**	**3**	**4**	**2D**	**3D**
Abs Max Error [%]	6.3	6.6	4.8	4.5	11.4	5.6
Median Error [%]	-3.4	-3.3	-2.3	-0.9	-6.4	-0.2
Implementation ease	✓ ✓	✓	✓	✗	✓ ✓	✓
Flexibility	✗	✓	✓ ✓	✓ ✓	✓ ✓	✓
Computational time	✓	✓	✗	✓	✗ ✗	✗ ✗ ✗

$$k = \frac{2\sqrt{R_i R_j}}{\sqrt{(R_i + R_j)^2 + d^2}} \qquad (16)$$

This equations can be approximated using series expansions to obtain non-integral closed form equations like the one presented in Method 1 and 2.

ACKNOWLEDGMENT

The results presented in this paper are a part of the project "Inductive Low-Power Transfer with High Insulation Capabilities" that has been founded by the Swiss National Science Foundation (SNSF, Grant agreement No. 215183).

REFERENCES

[1] Y. Wang, Z. Sun, Y. Guan, and D. Xu, "Overview of megahertz wireless power transfer," *Proceedings of the IEEE*, vol. 111, no. 5, pp. 528–554, 2023.

[2] A. Islam, "Design and optimization of printed circuit board inductors for wireless power transfer system," *Circuits and Systems*, vol. 04, pp. 237–244, 01 2013.

[3] O. C. Spro, F. Mauseth, and D. Peftitsis, "High-voltage insulation design of coreless, planar pcb transformers for multi-mhz power supplies," *IEEE Transactions on Power Electronics*, vol. 36, no. 8, pp. 8658–8671, 2021.

[4] J. P. K. Sampath, "Design and optimization of wireless power transfer system," PhD Thesis, Nanyang Technological University, Singapore, 2017.

[5] S. Mohan, M. del Mar Hershenson, S. Boyd, and T. Lee, "Simple accurate expressions for planar spiral inductances," *IEEE Journal of Solid-State Circuits*, vol. 34, no. 10, pp. 1419–1424, 1999.

[6] S. S. Mohan, "The design, modeling and optimization of on-chip inductor and transformer circuits," PhD Thesis, Sanford University, 1999.

[7] X. Du, "Modeling and optimization of pcb coils for inductive power transfer," PhD Thesis, EPFL, 2023.

[8] Y. Dou, X. Huang, Z. Ouyang, and M. Andersen, "Modelling and design for position-free inductive couplers with the homogeneous-flux transmitting coil in wireless power transfer systems," *TechRxiv*, 2023.

[9] Y. Dou, "Modelling and design in advanced megahertz range wireless power transfer systems," PhD Thesis, Technical University of Denmark, 2021.

[10] J. Huang, Y. Dou, X. Huang, Z. Zhang, Z. Ouyang, and M. A. Andersen, "Optimization of a 6.78-mhz inductive power transfer system for unmanned aerial vehicles," *IEEE Transactions on Power Electronics*, vol. 38, no. 10, pp. 11 940–11 952, 2023.

[11] M. Noh, T. V. Bui, K. T. Le, and Y.-W. Park, "Analysis of uncertainties in inductance of multi-layered printed-circuit spiral coils," *Sensors*, vol. 22, no. 10, 2022. [Online]. Available: https://www.mdpi.com/1424-8220/22/10/3815

[12] S. R. Khan, S. K. Pavuluri, and M. P. Y. Desmulliez, "Accurate modeling of coil inductance for near-field wireless power transfer," *IEEE Transactions on Microwave Theory and Techniques*, vol. 66, no. 9, pp. 4158–4169, 2018.

Compact Air-Core Inductors for Variable Frequency Soft-Switching in 3 Phase Inverters

Youssef A Fahmy
Dept. of Electrical Engineering
Columbia University
New York, USA
y.fahmy@columbia.edu

Matthias Preindl
Dept. of Electrical Engineering
Columbia University
New York, USA
matthias.preindl@columbia.edu

Abstract—An air-core inductor is designed and experimentally tested for use in a 3-phase 47kW grid-tied soft-switching inverter. The combination of switching frequencies in the hundreds of kiloHertz and peak currents over 100A, necessary for soft-switching over the entire grid cycle, causes significant iron loss in cored inductor designs. The novel use of air-core inductors in this application, with power levels of 8kW per inductor, eliminates these losses, improving overall converter efficiency. A comparison in design and performance with a ferrite-cored inductor is made and the air-core is shown to be $\geq 0.5\%$ more efficient at rated power, with higher savings at partial power. The design reduces weight while increasing volume to maintain inductance. Radiated electromagnetic interference is analyzed in a simulation and is shown to be benign to the circuit's operation in practice.

Index Terms—air-core inductors, magnetics, power electronics, variable frequency

I. INTRODUCTION

Small, efficient power electronics are a key part of the electrification of global energy use. These electronics typically require magnetics, be they inductors, transformers, or both, to provide everything from galvanic isolation, to filtering, to resonance. However, these magnetics can be difficult to design optimally and often require complex simulations to select interacting parameters such as core size, core material, airgap, and winding structure. Losses in particular can be calculated but are not always accurate, especially when the exciting waveforms are non-sinusoidal, high frequency, or high magnitude [1].

Inductors for electric vehicle (EV) related power electronics are increasing in rating from less than one to tens of kilowatts per inductor. At the same time, the push for power density has driven up switching frequencies [2], [3]. However, reducing volume purely by increasing frequencies is limited by losses in magnetic core materials.

This combination of increasing powers and switching frequencies has led to situations of high magnetic fields that are changing rapidly. For example, in a 3 phase grid tied inverter such as those in [4] switching frequencies of over $1MHz$ are achieved variably throughout the grid cycle. The converter, meanwhile, is being pushed to $> 20kW$ per stage. The amount of energy stored in an inductor increases with the square of the current meaning the cores of potential inductors, which suffer

losses that increase non-linearly with both increased fields and frequencies [1], are processing, and losing, much more energy.

One simplification that has been used in other fields is the use of coreless or "air-core" inductors and transformers [2], [5], [6]. These inductors are typically used in high frequency applications, such as radio frequency and chip power supplies, where magnetic materials would otherwise saturate. Some attempts have been made in the power electronics literature for air-core magnetics [7]–[11], but not at current and voltage levels sufficient for EVs.

This paper will explore the performance of air-core inductors under these conditions. They have several advantages that will be exploited in an example circuit: a soft-switching grid-tied inverter. First, the lack of magnetic material means that there is no chance of nonlinear inductance behavior through saturation [12]. For the same reason, there are no core losses. Both the hysteresis loss, caused by nonlinear realignment of magnetic domains, and eddy current loss, caused by induced electric fields in the core itself, are no longer possible. These losses, commonly modeled by variations on the Steinmetz equation [1], [13], are the most complicated to model. The inductor design process is significantly simplified by discounting these equations, which are empirical and accurate over a limited range of operating conditions.

There are several potential drawbacks to air-core inductors that must be addressed before design can begin. Without a core, the magnetic flux is no longer guided and contained; this has several ramifications. The reduced flux density means that the inductance available in a given volume is smaller. However, for the application being considered, an inductance range of just 1-10µH is necessary, which, as will be seen, is easily achievable in a relatively compact size. Although air-core inductors may not be feasible for designs with low switching frequencies or high inductances, they should not be discounted for situations with medium to high frequencies and high powers.

A second potential issue is the increase in copper loss. In order to achieve the same inductance as a cored design, an air-core inductor will require more turns and/or more area. This means an increase in the length of copper, increasing both the DC and frequency-dependent resistance, as well as potentially increasing proximity effects [14]. The implication of this is that the decrease in core losses must be greater than the increase in winding losses so that the net loss is lower overall. For simplicity, this paper will not consider toroidal inductors, only solenoidal, as they tend to require longer lengths of wire for a given inductance.

Lastly, the absence of a core results in increased electromagnetic interference (EMI) caused by the unconfined magnetic field. Most crucially, this EMI must not disrupt the proper actuation of the power electronic circuit from which it is generated. Nominal circuit operation will ensure that there is no feedback loop that may damage the electronics or interfere with control. If it can be shown that electronics that are not additionally hardened against EMI can still operate effectively with an air-core inductor in close proximity, then other EMI concerns can be mitigated with common techniques such as metallic shielding around the converter as a whole.

This paper presents the design and operation of an air-core inductor for use in a softswitching grid tied inverter. Comparisons will be made with a ferrite-cored inductor of similar inductance. The requirements and design are shown in Section II. Next, a simulation is conducted using Ansys Maxwell to examine the magnetic fields in Section III. The experimental results are provided in Section IV. Finally, in Section V, the paper is concluded

II. INDUCTOR REQUIREMENTS AND DESIGN

A. Inductance

The first parameter to choose for the inductor is the required inductance itself. Three identical inductors will be used in the power factor correction (PFC) stage of an EV charger as described in [4]. This circuit topology, shown in Fig. 1, differs from a traditional inverter by the connection of the DC minus to the capacitor star point. This decouples the switching instances of the phases and allows for single phase analysis to describe the system.

The relationship between inductor current ripple ΔI and switching frequency f_{sw} is given by

$$f_{sw} = \frac{D(1-D)V_{DC}}{\Delta I L} \qquad (1)$$

where D is the duty cycle, V_{DC} is the DC bus voltage, and L is the filter inductance to be determined. It is this relationship that will be manipulated to ensure soft-switching.

The inductance value is chosen such that soft-switching will be maintained in all operating conditions under variable switching frequency operation [4]. It will be assumed that the DC bus value is a constant and set such that the duty cycle range, which depends on the ratio of the bus to the grid voltage magnitude, is roughly between $[0.1, 0.9]$. This will prevent the main control parameter, D, from saturating.

The ripple needed to soft-switch is determined by

$$\Delta I = \alpha(I_{grid} + I_{Lim}) \qquad (2)$$

where α is a parameter set at 2, I_{grid} is the sinusoidal grid current, and I_{Lim} is a parameter set such that there is enough excess ripple to soft-switch the MOSFETs. The value of

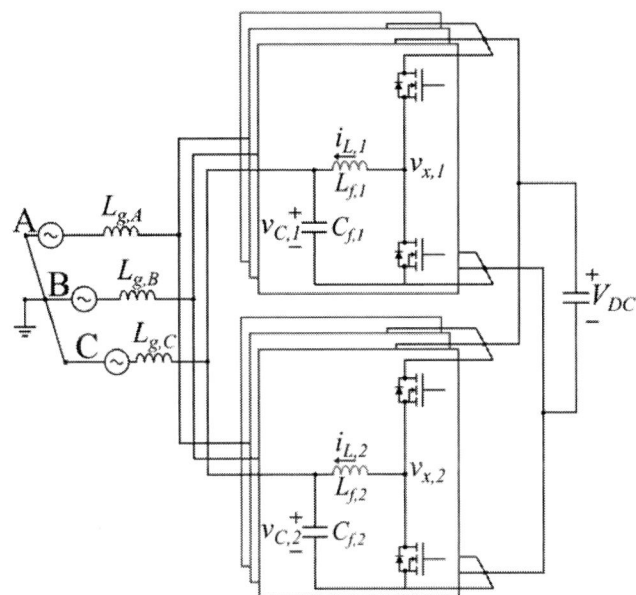

Fig. 1: Circuit topology of the EV charger stage with the designed inductor highlighted in red.

I_{grid} will be determined by the specifications of the converter under consideration. For example, if using a European 400V grid at 50Hz, to achieve 25kW the grid current needs to be approximately 51 A_{pk}.

To determine the necessary inductance (1) can be solved for L and, using the above values, the maximum ripple, which for real power occurs at the peaks of grid voltage, can be compared to the inductance. The lower bound on the minimum switching frequency is set by the top of the audible range. Meanwhile, switching frequencies are limited from above by duty cycle distortions that arise when rise, fall, and dead times become a significant portion of the pulse time, around 600kHz.

A value of 8-10μH operating with a target bus of 800V provides a good margin for operation in the 40-50kHz range at maximum current in a 25kW converter. Inductance values that are smaller than calculated using this procedure will simply provide excess current ripple. Although this is not necessarily optimal for overall efficiency, it will ensure soft-switching operation throughout the grid cycle. Having a target range instead of a specific value will allow for inductance variation that inevitably arises when manufacturing magnetics.

B. Air-Core Design

The inductance of an air-core inductor is described by the empirical Wheeler formula [15]. It is traditionally given by

$$L = \frac{a^2 n^2}{9a + 10b} \qquad (3)$$

where L is the inductance in μH, a is the radius, and b is the length of the inductor, both given in inches. The formula assumes a single winding layer, does not consider spacing between windings ($p = d$), and is accurate for $b > 0.8a$. To simplify analysis, these assumptions will be adopted for the

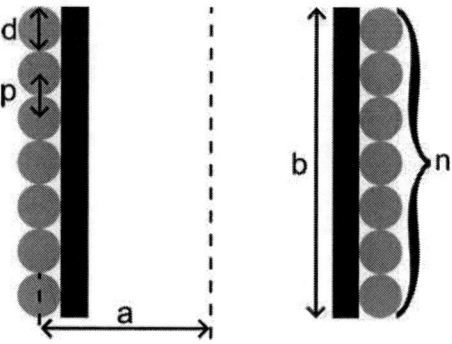

Fig. 2: Key design parameters of an air-core inductor shown in cross section.

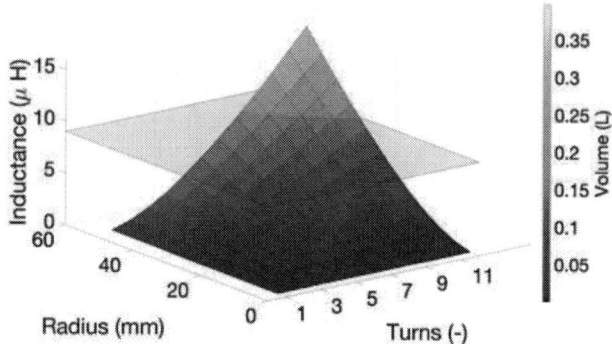

Fig. 3: The optimized Wheeler equation showing L vs. a and n, along with the desired $9\mu H$ of plane of inductance.

remainder of the air-core design. A sketch of the design with relevant parameters is shown in Fig. 2.

The air-core is designed with the primary goal of maximizing efficiency. Without a core, the only source of loss is the windings themselves. Resistive losses can be modeled by

$$P_L = I^2 R(\omega) \qquad (4)$$

where I is the current in the conductor and $R(\omega)$ is the frequency dependent resistance. Three major phenomena are known to affect $R(\omega)$ and therefore the resistive losses. The simplest is the DC component that is purely based on the geometry and type of conductor.

There are two frequency-dependent effects that increase the resistance of a wire when carrying alternating currents. First, in the skin effect, alternating magnetic fields generate eddy currents that force the conducting current into an annular "skin" of depth $\delta = \sqrt{\frac{2\rho}{\omega\mu}}$ where μ is the permeability of the material and ρ is its resistivity. Second, the proximity effect describes the tendency of currents in adjacent wires to bundle, again because of eddy currents. Both effects are summarized in the Dowell equation [16], which describes the change in resistance as frequencies increase.

These effects are commonly mitigated through the use of many small, individually isolated wires that are bundled to form Litz wire. To reduce the proximity effect, the strands are twisted within the wire, exposing each wire to different electromagnetic fields. The use of this wire, while more expensive, greatly reduces the frequency dependent effects in the range of interest, thus improving efficiency.

The equations describing these mechanisms can be formally incorporated into an optimization problem to solve for the minimally resistive inductor at a given inductance as is done in [17]. However, a much simpler approach will be taken here that will result in only a minor deviation from the previously calculated optimum.

The idea is to maximize the inductance of an air-core inductor for a given wire length assuming a given wire diameter. All of the loss mechanisms described above are directly dependent on the length of the conductor. By minimizing this length for a given inductance, all forms of loss are mitigated.

To do this, the substitutions $a = \frac{w}{2\pi n}$ and $b = nd$ are made into (3), resulting in

$$L = \frac{w^2 n}{9(2\pi w) + 10d(2\pi n)^2}, \qquad (5)$$

where w is the total length of wire.

Setting the derivative of L with respect to the number of turns equal to zero results in a point of maximum inductance and the value of n that achieves it. Using this value and some algebraic manipulations, an optimal "shape ratio", the ratio of inductor length to diameter, of $b/2a = 0.45$ is derived. This is close to the ratio of $b/2a = 0.408$ in [17] and the air-core rule of thumb of "radius equal to length". It is also notable that the optimal point found is in a broad, shallow minimum, meaning that small deviations from the reported optimal shape ratio will not result in large resistance changes.

The behavior of (3) with a fixed shape ratio of 0.45, is shown in Fig. 3. The target inductance of $9\mu H$ is given by the gray plane. Only a full turn of wire encloses area and provides inductance quantifiable by the Wheeler equation. Therefore, the resulting inductance changes in steps as the integer number of turns increases. To further narrow the inductance selection, the coloring in Fig. 3 is based on the total volume of the air-core cylinder. Solenoids with more turns but smaller radii have a lower volume and thus are more desirable.

The final air-core inductor used was chosen with a value of $9.2\mu H$ and has a radius of 42mm, a height of 40mm, and 10 turns. A small change (2mm) in the effective radius due to the thickness of the wire is taken into account. The Litz wire is equivalent to 10AWG and supports frequencies up to 800kHz, well within the operating range of this converter.

C. Cored Inductor Comparison

The cored inductor against which the air-core will be compared was designed following the efficiency-volume optimization process described in [4]. The result is a compact design with two paralleled E42/21/20 3F36 Ferroxcube cores, a 6.2mm airgap, and 8 paralleled turns of Litz wire. It has an inductance of $8.3\mu H$. A MnZn ferrite core was chosen because of its availability and its combination of high permeability

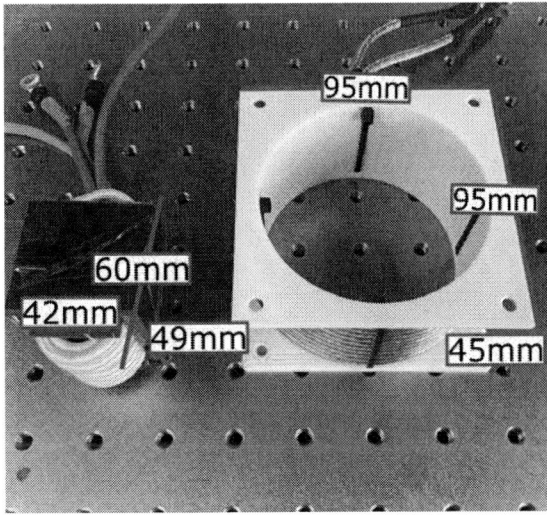

Fig. 4: Physical prototypes of the cored and air-core inductors.

and low loss. Other materials, such as carbonyl iron and NiZn ferrite, have lower permeability and are intended for higher frequency operation than will be used here. The air- and ferrite-core inductors being compared are shown in Fig. 4.

In addition to the copper loss mechanisms described above, the cored inductor experiences iron losses. These come in two forms, induced eddy current and hysteresis losses. The structure of the ferrite means that the hysteresis losses dominate. Both of these magnetic losses are modeled by the Steinmetz equation, a curve fitting equation that relates the power loss density to the frequency and magnetic flux density as in

$$P_v = kf^aB^b. \tag{6}$$

The equation, while widely used, is not accurate for non-sinusoidal excitations. This can be partially compensated for using the improved generalized Steinmetz equation [13], but even this does not account for DC bias currents. As will be shown, the currents seen by the inductors are sawtooth waveforms that vary in magnitude, frequency, and bias, making them particularly difficult to model using any form of the Steinmetz equation.

Overall, the volume of the cored inductor is 0.123L while the air-core is 0.255L, or roughly double. The weight, meanwhile, is reduced from 359g to 248g or approximately 30% thanks to the removal of the core and despite the increase in wire.

III. SIMULATION

To compare the magnetic properties of the two inductors, both were simulated in Ansys Maxwell. The inductors were excited with a $50kHz$, $150A_{Pk}$ current that represents the highest peak ripple currents experienced by the inductors during operation.

Fig. 5a shows the distribution of the magnetic flux density in the EE core inductor as a result of the applied excitation. As can be seen, despite the airgap, which is necessary to forestall

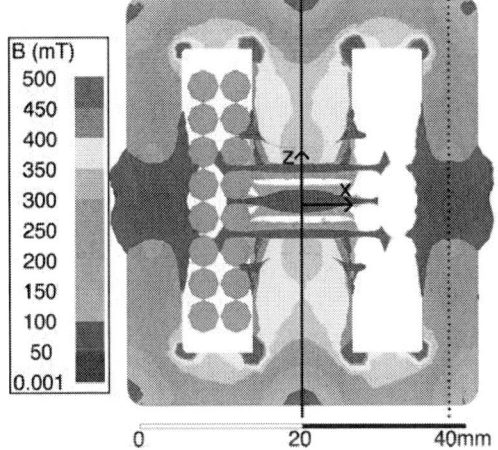

(a) Cored inductor simulation of magnetic flux density

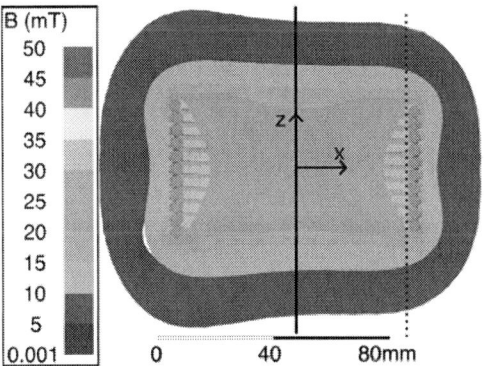

(b) Air-Core inductor simulation of magnetic flux density

Fig. 5: The solid and dashed lines represent the central and offset axes plotted in Fig. 6. Note the differences in B-field and length scales.

saturation of the core, the magnetic field density within the core is high, with parts exceeding $500mT$. The increased flux at the corners, caused by fringing fields, represents locations of high loss, which limit the operation of the core. This is at the worst-case grid operating point in terms of magnetic flux and will change with the phase of the grid.

Further increasing the airgap beyond the 6.2mm (already large) used here would reduce the flux density in the core, reducing inductance and loss. In this regime, the practicality of the core as a guide for flux becomes less evident. As the airgap increases, the effect of the core on the inductor decreases and a cored inductor approaches an air-core.

In order to maintain practical use with a core, it is typical to increase the core volume rather than the airgap. However, increasing the volume to sufficiently reduce the magnetic flux density to prevent high loss density would bring it to a volume comparable to that of the air-core inductor. At which point, the air-core core inductor should be considered equally with the cored inductor in terms of inductance and power density.

The lack of core results in a lower overall magnetic flux density, as seen in the 10x lower scale of Fig. 5 as compared

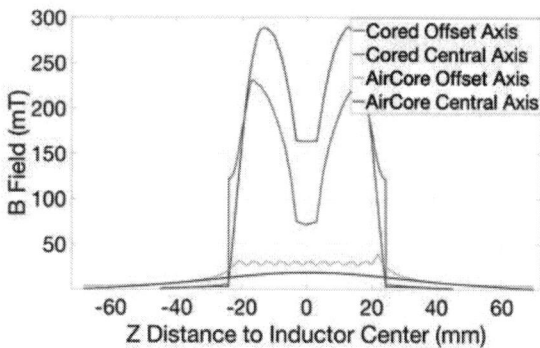

Fig. 6: Magnetic flux density comparison between cored and coreless inductors along two slices in the z direction.

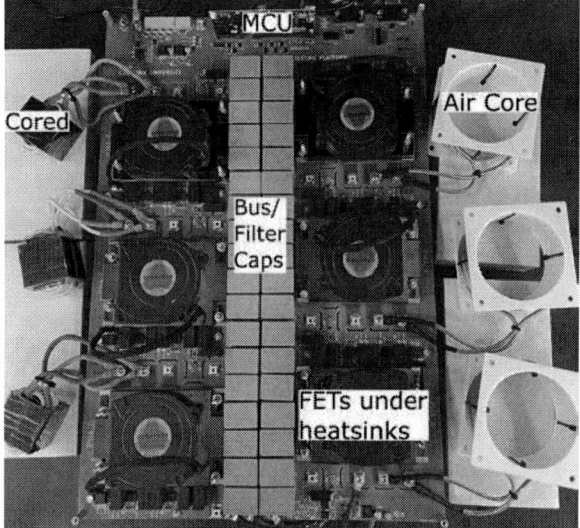

Fig. 7: Software-defined power electronics testing platform, showing the PCB and attached inductors.

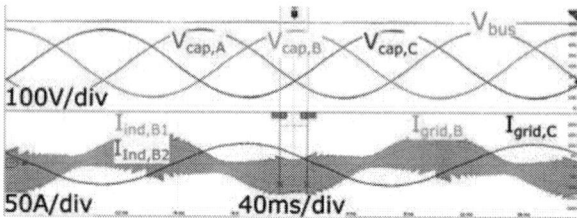

(a) Results using the air-core inductor showing the current ripple and resulting grid current while operating at $47kW$.

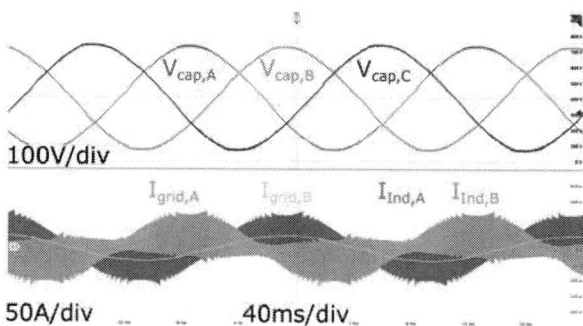

(b) Results using the cored inductor at 20kW, non-interleaved.

Fig. 8: Capacitor and bus voltages and grid and inductor currents during operation.

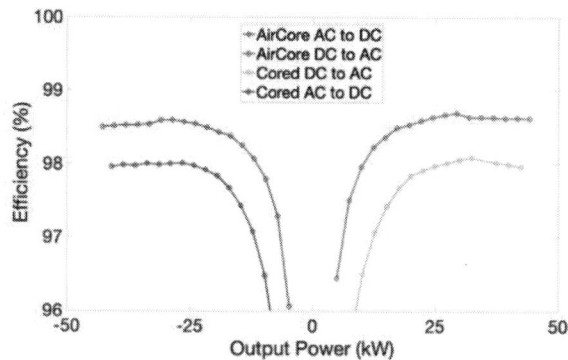

Fig. 9: Efficiency comparison between air-core and cored inductor setups as measured with a power analyzer.

to Fig. 5a. However, the magnetic field also drops off more slowly. A comparison of slices of the magnetic fields of the two inductors on the same scale is shown in Fig. 6. Two slices are taken, one along the central axis of each inductor for easy comparison and the other along an offset axis. The cored inductor's offset axis was chosen in the middle of one of the side prongs, while the air-core's is directly adjacent to the wires, where the magnetic fields are expected to be strongest. As expected, the peak of the cored inductor is much higher but is better contained. However, while the B-field of the air-core persists for twice as long it does so at relatively low levels.

IV. EXPERIMENTAL SETUP AND RESULTS

In order to compare the performance of the two inductors, both are placed into the same circuit and operated under the same voltage and current conditions. The circuit consists of two parallel, interleaved three phase PFC stages, as in Fig. 1, made to interface with the European grid of $400V_{LL}$.

The software-defined power electronics testing platform used for experimental verification is shown in Fig. 7. The PCB has six half-bridge subcircuits with a common DC bus that can be configured as desired by attaching different magnetics and adjusting the control. For compactness and comparison purposes the figure shows three cored and three coreless inductors attached. However, when testing the interleaved circuitry, all six half bridges would have either air-core or cored inductors.

As can be seen, the inductors are placed adjacent to the PCB and, despite the proximity, the EMI generated in both cases is not enough to interfere with the proper operation of the circuit. This includes the digital gate-drive signals, analog sensor signals, and the micro-controller. A theoretical prototype could be packaged with a shielded case, which would block EMI from propagating and affecting any other components.

Oscilloscope waveforms using the air-core inductors are

shown in Fig. 8a while those for the cored inductor are in Fig. 8. Both show capacitor voltages, grid currents, and inductor currents. The air-core results of Fig. 8a in particular show interleaved results with the inductor currents overlaid but 180 degrees out of phase to partially cancel the ripple. The cored waveforms are shown to demonstrate the similarity in operation of the two inductors and the flexible nature of the setup, in which just one half of the converter can be used.

Fig. 9 shows the difference in efficiency as power flow increases both to and from the battery. As can be seen, the air-core significantly outperforms the cored inductor across the entire range. The percent increase in savings is largest at lower powers but is still significant at high powers. This could be because of increased savings at higher frequencies, typical at lower powers. As an example of absolute savings, the air-core reduces loss by 235W at an output power of 40kW.

V. CONCLUSION

This work shows the feasibility of air-core inductors in power electronics designs for variable-frequency converters at power levels required by EV chargers. Their use under high currents and frequencies simultaneously has not been explored previously in the literature. A simple design principle, the ratio of height to radius, is presented to create efficient air-core inductors at a desired inductance.

The air-core inductors are tested in hardware under variable switching frequencies and compared with a cored inductor. They have been shown to increase converter efficiency by 0.5% at rated power and even more at lower powers where higher frequencies are needed. Compared with the cored design, the volume and extent of the magnetic field are doubled while the weight of the air-core is decreased by 30%. Future work may include comparisons with lower permeability cores and methods to decrease volume. In addition, boundary conditions for the use of air-core as a substitute for cored inductors based on frequency and current magnitudes will be investigated.

REFERENCES

[1] T. Guillod, J. S. Lee, H. Li, S. Wang, M. Chen, and C. R. Sullivan, "Calculation of ferrite core losses with arbitrary waveforms using the composite waveform hypothesis," in *IEEE Applied Power Electronics Conference and Exposition (APEC)*, Mar. 2023.

[2] M. Solomentsev and A. J. Hanson, "At what frequencies should air-core magnetics be used?" *IEEE Trans. on Power Electronics*, vol. 38, no. 3, pp. 3546–3558, Mar. 2023.

[3] D. J. Perreault, J. Hu, J. M. Rivas, *et al.*, "Opportunities and challenges in very high frequency power conversion," in *IEEE Applied Power Electronics Conference and Exposition*, Washington, DC: IEEE, Feb. 2009.

[4] M. Jahnes, L. Zhou, Y. Fahmy, and M. Preindl, "A peak 1.2-MHz, >99.5% efficient, and >10-kW/l power dense soft-switched inverter for EV fast charging applications," *IEEE Trans. on Transportation Electrification*, vol. 10, no. 3, pp. 5520–5532, Sep. 2024.

[5] A. Mediano and F. J. Ortega-Gonzalez, "Class-e amplifiers and applications at MF, HF, and VHF: Examples and applications," *IEEE Microwave Magazine*, vol. 19, no. 5, pp. 42–53, 2018.

[6] C. R. Sullivan, D. V. Harburg, J. Qiu, C. G. Levey, and D. Yao, "Integrating magnetics for on-chip power: A perspective," *IEEE Trans. on Power Electronics*, vol. 28, no. 9, pp. 4342–4353, Sep. 2013.

[7] R. T. Naayagi and A. J. Forsyth, "Design of high frequency air-core inductor for DAB converter," in *IEEE International Conference on Power Electronics, Drives and Energy Systems (PEDES)*, Dec. 2012.

[8] U. Pratik and Z. Pantic, "Design of variable air-core coupled co-axial solenoidal inductors," in *IEEE Energy Conversion Congress and Exposition (ECCE)*, Oct. 2022.

[9] W. Liang, L. Raymond, and J. Rivas, "3-d-printed air-core inductors for high-frequency power converters," *IEEE Trans. on Power Electronics*, vol. 31, no. 1, pp. 52–64, Jan. 2016.

[10] E. Asahina, M. Fukuoka, I. Masuda, A. Nagai, K. Maeda, and M. Ishitobi, "Structure of air-core power inductor with high energy density and low copper loss," *IEEE Trans. on Magnetics*, vol. 59, no. 11, Nov. 2023.

[11] C. D. Meyer, S. S. Bedair, B. C. Morgan, and D. P. Arnold, "High-inductance-density, air-core, power inductors, and transformers designed for operation at 100–500 MHz," *IEEE Trans. on Magnetics*, vol. 46, no. 6, pp. 2236–2239, Jun. 2010.

[12] R. A. Salas and J. Pleite, "Simulation of the saturation and air-gap effects in a POT ferrite core with a 2-d finite element model," *IEEE Transactions on Magnetics*, vol. 47, no. 10, pp. 4135–4138, Oct. 2011.

[13] K. Venkatachalam, C. Sullivan, T. Abdallah, and H. Tacca, "Accurate prediction of ferrite core loss with nonsinusoidal waveforms using only steinmetz parameters," in *IEEE Workshop on Computers in Power Electronics*, Jun. 2002, pp. 36–41.

[14] P. Murgatroyd, "The optimal form for coreless inductors," *IEEE Trans. on Magnetics*, vol. 25, no. 3, pp. 2670–2677, May 1989.

[15] H. Wheeler, "Simple inductance formulas for radio coils," *Proc. of the Institute of Radio Engineers*, vol. 16, no. 10, pp. 1398–1400, Oct. 1928.

[16] P. Dowell, "Effects of eddy currents in transformer windings," *Proc. Inst. Electr. Eng. UK*, vol. 113, no. 8, p. 1387, 1966.

[17] T. Ibuchi and T. Funaki, "A study on copper loss minimization of air-core reactor for high frequency switching power converter," in *IEEE International Symposium on Power Electronics for Distributed Generation Systems (PEDG)*, Jul. 2013.

Simulation and Experimental Research on Cooling Performance of Fully-Immersed Evaporative Cooling High-Frequency Transformer

Zhanlei Liu
State Key Lab of Electrical Insulation and Power Equipment
Xi'an Jiaotong University
Xi'an, China
lzl0283@stu.xjtu.edu.cn

Lingyu Zhu
State Key Lab of Electrical Insulation and Power Equipment
Xi'an Jiaotong University
Xi'an, China
zhuly1026@xjtu.edu.cn

Yuntian Gao
State Key Lab of Electrical Insulation and Power Equipment
Xi'an Jiaotong University
Xi'an, China
2203211439@stu.xjtu.edu.cn

Yongliang Dang
State Key Lab of Electrical Insulation and Power Equipment
Xi'an Jiaotong University
Xi'an, China
dyl877759724@stu.xjtu.edu.cn

Cao Zhan
Center for Power Electronics Systems
Virginia Tech
Blacksburg, VA, USA
caozhan@vt.edu

Shengchang Ji
State Key Lab of Electrical Insulation and Power Equipment
Xi'an Jiaotong University
Xi'an, China
jsc@xjtu.edu.cn

Abstract—**Efficient and reliable cooling scheme design is non-trivial for high-power high-frequency transformer (HFT). Traditional air cooling HFT faces the problem of low heat dissipation efficiency and high temperature rise. To tackle this problem, fully-immersed evaporative cooling technology is applied in the cooling of HFT in this paper. A fully-immersed evaporative cooling HFT prototype and two air cooling HFT prototypes are made. The power losses of HFT are calculated. A finite element (FE) simulation model for the fully-immersed evaporative cooling HFT is established to calculate the steady state temperature distribution. A HFT temperature test platform is built to measure HFT temperature rise. The proposed FE model can accurately calculate the steady-state temperature of fully-immersed evaporative cooling HFT. In addition, the temperatures of evaporative cooling HFT and air cooling HFTs are compared. Experimental results show that evaporative cooling technology can significantly decrease the steady-state temperature rise of HFT and improve the uniformity of HFT temperature distribution.**

Keywords—**High-frequency transformer, temperature rise, temperature distribution, fully-immersed evaporative cooling, finite element simulation.**

I. INTRODUCTION

High-frequency transformer (HFT) is an important component in isolated DC-DC converters, having wide applications in DC power conversion occasions such as traction power supplies [1], PV inverters [2], charging stations [3] and medium-voltage DC interconnections [4]. With the emergence of high-voltage and high-power switching devices [5]-[7], HFTs are developing towards higher voltage level and higher power density [8], [9]. Operating temperature is an important design objective and assessment indicator of HFT since overheating in HFT can cause insulation degradation and failure [10], [11]. The small size and high power loss density increase heat load density and heat dissipation difficulty of HFT. To ensure that the operating temperature of HFT under rating condition is below maximum allowable temperature, efficient and reliable heat dissipation designs are very essential.

Natural convection cooling [12] is the simplest cooling method, yet having the disadvantage of low cooling capacity and is not suitable for the cooling of high-power HFTs.

Forced convection cooling by adding fans is the most commonly used cooling method for high-power electrical and electronic devices [13], [14], including high-power HFTs [11], [15] since forced convection cooling has higher cooling capacity than natural convection cooling. However, the fans will introduce additional costs and noises. In addition, for epoxy resin potting HFTs, forced convection cooling cannot quickly and efficiently take away the heat generated inside the HFT, leading to temperature hotspots inside the core and windings. Watering cooling is another efficient cooling method for high-power HFTs [16]. However, the water cooling heat dissipation devices are expensive, complex and of low reliability, limiting its applications in high-power HFTs with high insulation and high power density requirements.

Evaporative cooling utilizes the phase change latent heat of working fluid to cool the devices. In addition to the advantages of high heat transfer coefficient, evaporative cooling technique can reduce the cooling device energy consumption and noises, showcasing advantageous application prospects in the cooling of high-power density devices [17]. In [18], two-phase cooling method is used in the cooling of power electronic devices. In [19], a U-shape evaporative cooling radiator is designed for cooling of HFT. The hotspot temperature is lowered and the temperature distribution uniformity is improved. However, now there lacks a comprehensive analysis of cooling performance of fully-immersed evaporative cooling (FIMEC) HFT.

This paper gives both simulation and experimental analysis on cooling performance of FIMEC HFT. In Section II, the core and winding losses of HFT are calculated. In Section III, the simulation model of FIMEC HFT is introduced. In Section IV, the simulation and experiment results of FIMEC HFT temperature distribution are analyzed. The temperature distributions of fan-cooling HFTs and FIMEC HFT are compared. Section V concludes this paper.

II. POWER LOSS CALCULATION OF HFT

A. Introduction of HFT protptypes

In this paper, three HFT prototypes are made for investigations and comparisons. The core and winding

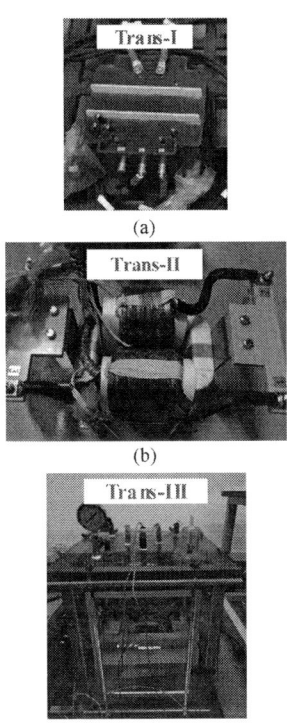

(a)

(b)

(c)

Fig. 1. Three HFT prototypes used in this paper. (a) Epoxy resin potting HFT with fan cooling. (b) Non-epoxy resin potted HFT with fan cooling. (c) FIMEC HFT. These three HFT prototypes have the same core and winding structures.

structures of these three HFT prototypes are the same. The first HFT prototype (Trans-I) is an epoxy resin potting HFT with fan cooling, as illustrated in Fig. 1(a). Eighteen thermocouples are embedded in the HFT to measure the operation temperatures at different locations inside the HFT. The second HFT prototype (Trans-II) is a non-epoxy resin potted HFT with fan cooling. Sixteen thermocouples are embedded in the HFT to measure the operation temperatures at different locations inside the HFT. The third HFT prototype (Trans-III) is a FIMEC HFT, which includes Trans-II, a sealed container and coolant. The container is made by polycarbonate and organic glass boards for clear observation of internal boiling phenomenon. Aluminum plates are not selected to avoid induced eddy currents. A barometer is installed on the container to monitor the internal air pressure. Several terminals are installed on the surface of the container to connect the inside and outside wires. A water-cooling coil pipe is installed inside the container to condense the internal vapor. The coolant is placed in the container and HFT is immersed in the coolant. Fluorinert electronic liquid FC-72 from 3M™ is selected as the coolant in this paper. The boiling point of FC-72 is 56°C and the vaporization latent heat is 88 kJ/kg, making FC-72 suitable for cooling of HFT.

The HFT prototypes are made by nanocrystalline core and square litz wire windings. The structures of the core and windings are shown in Fig. 2. The HFT core is made by stacking two 40mm-thickness nanocrystalline cores. The HFT windings are concentric windings wound on two core limbs. Both the primary and secondary windings are six turns per layer and two layers on each core limb. The

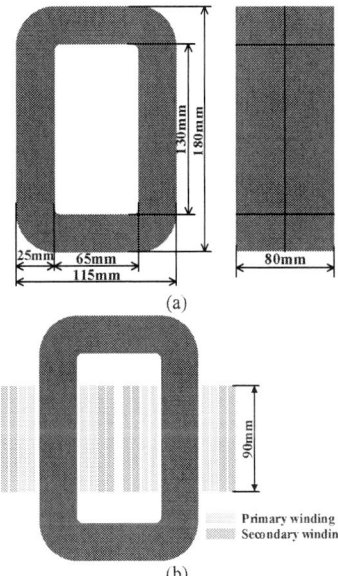

(a)

Primary winding
Secondary winding

(b)

Fig. 2. Structures of (a) nanocrystalline core and (b) litz wire windings of the HFT prototypes selected in this paper.

section size of the square litz wire is 6mm*12mm and the height of the litz wire winding is 90mm.

B. Core loss calculation

Steinmetz equation is the most popular core loss calculation method. The Original Steinmetz equation (OSE) [20] can only be applied to calculate core loss under sinusoidal excitation. To extend the applicability of core loss calculation method to non-sinusoidal excitations, various improved Steinmetz equations are proposed, including Modified Steinmetz equation (MSE) [21], Generalized Steinmetz Equation (GSE) [22], Improved Generalized Steinmetz Equation (IGSE) [23], Waveform Coefficient Steinmetz Equation (WcSE) [24], etc. Among these improved core loss calculation methods, IGSE is the most popular one for its high calculation accuracy. IGSE is based on the hypothesis that the total core loss in a magnetization cycle can be calculated by the summation of losses of major and minor hysteresis loops. The instantaneous core loss depends on the magnetization rate dB/dt as well as peak-to-peak magnetic flux density $\triangle B$. The time-average core loss in a magnetization cycle can be expressed as

$$P_V = \frac{1}{T}\int_0^T k_i \left|\frac{dB}{dt}\right|^\alpha (\triangle B)^{\beta-\alpha} dt \qquad (1)$$

where k, α and β are Steinmetz coefficients and

$$k_i = \frac{k}{2^{\beta+1}\pi^{\alpha-1}(0.2761+\dfrac{1.7061}{\alpha+1.354})} \qquad (2)$$

A core loss test platform is established as shown in Fig. 3. A small toroidal nanocrystalline core is selected as test sample. A power amplifier is utilized to provide excitations to the core. An oven is used to control the core temperature during test process. The measured core losses under sinusoidal excitation at different frequencies and flux density

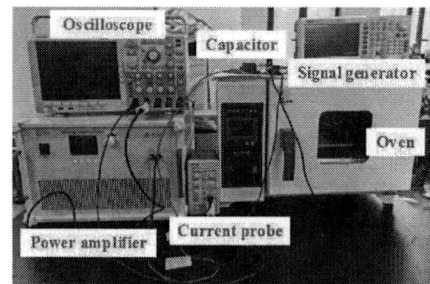

Fig. 3. Core loss test platform.

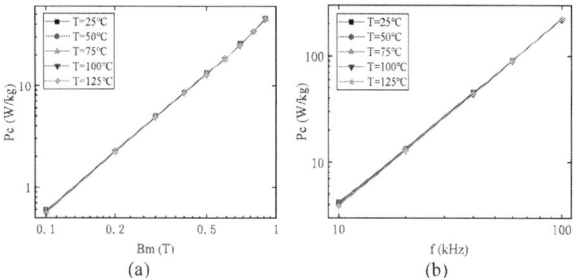

Fig. 4. Core loss measurement results. (a) Variations of core loss with flux density amplitude. (b) Variations of core loss with frequency.

amplitudes are shown in Fig. 4. It can be seen that core loss increases linearly with frequency and flux density amplitude in logarithmic coordinate. In addition, temperature has little impact on core loss.

Applying equation (1), the core loss of HFT prototypes is calculated to be 18.4W at rated voltage of 700V.

C. Winding loss calculation

The winding current waveform of HFT is typically non-sinusoidal wave. According to the Fourier decomposition method, winding loss under non-sinusoidal current can be calculated by the summation of winding loss under all harmonics.

$$P_w = \sum I_n^2 R_{ac,n} \tag{3}$$

where I_n is the RMS of winding n-th harmonic current, $R_{ac,n}$ is the winding ac resistance at n-th harmonic frequency.

An impedance analyzer is utilized to measure the winding ac resistance at different frequencies. The winding ac resistance measurement platform is shown in Fig. 5. The winding ac resistance measurement result is shown in Fig. 6. In a DAB converter with single-phase-shift (SPS) modulation, the winding current is dependent on the square voltage amplitude, leakage inductance of HFT and phase-shift angle ϕ. Under 10° phase-shift angles, winding loss is calculated to be 128.5W.

In summary, the core and winding losses of HFT under rated load condition are shown in Table. I.

III. SIMULATION MODEL OF FIMEC HFT

A. Boiling heat transfer coefficient formulas

FIMEC technology immerses the HFT in evaporative cooling coolant and takes away the heat of HFT by vaporization latent heat of the coolant. The working principle of FIMEC HFT is illustrated in Fig. 7. Under rated load condition, the coolant is in saturated nuclear pool boiling

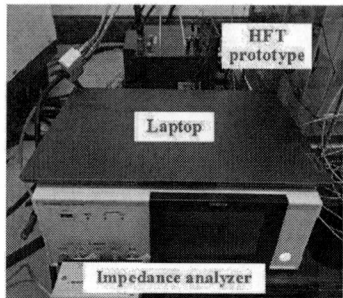

Fig. 5. Winding ac resistance measurement by an impedance analyzer.

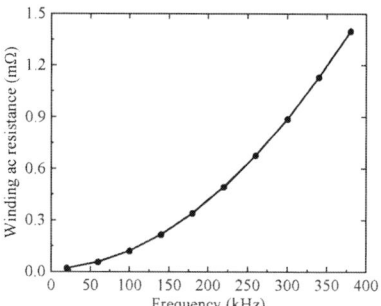

Fig. 6. Winding ac resistance measurement result.

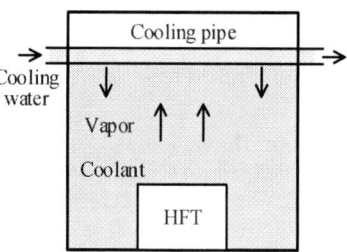

Fig. 7. Principle of FIMEC HFT.

TABLE. I. CORE AND WINDING LOSSES OF HFT UNDER RATED LOAD CONDITION

Rated load condition	Core loss (W)	Winding loss (W)
U=700V, ϕ=10°	18.4	128.5

state. The phase transition process of coolant can remove a large amount of heat, which is stored in the coolant vapor. The vapor will condense on the wall of cooling pipe and exchange the heat into cooling water.

The physical process and mechanism of saturated nuclear pool boiling are very complex. Theoretical derivation on the heat transfer coefficient of saturated nuclear pool boiling is infeasible. Many experimental correlation formulas have been developed to estimate the heat transfer coefficient of saturated nuclear pool boiling, including *Rohsenow* correlation [25], *Cooper* correlation [26] and *Imura* correlation [27]. The *Rohsenow* nuclear pool boiling heat transfer coefficient correlation can be expressed as

$$h_R = \left(\frac{q_{nb}''}{i_{lv}}\right)^{1-r}\left[\mu_l\sqrt{\frac{\sigma}{g(\rho_l-\rho_v)}}\right]^r\frac{c_{p,l}}{C_{sf}}\mathrm{Pr}_l^{-s} \tag{4}$$

where $\mathrm{Pr}_l = c_l\mu_l/k_l$, Pr_l is Prandtl number of liquid, $c_{p,l}$ is specific heat of liquid, μ_l is dynamic viscosity of the liquid, k_l is thermal conductivity of liquid, i_{lv} is latent heat of vaporization, g is gravitational acceleration, ρ_l and ρ_V are

979-8-3315-1612-3/25 $31.00 © 2025 IEEE

densities of liquid and vapor, σ is liquid-vapor surface tension, C_{sf} is surface fluid factor, $r=1/3$, s is the exponent of Prandtl number, $s=1$ for water and $s=1.7$ for other fluids.

The *Cooper* nuclear pool boiling heat transfer coefficient correlation can be expressed as

$$h_{Coop} = 55(q_{nb}^{''})^{0.67}(P/P_{crit})^{0.12-0.2\log R_{a,p}}(-\log(P/P_{crit}))^{0.55}M_{mol}^{-0.5}$$

(5)

where P is working pressure, P_{crit} is the critical pressure, R_{crit} is the average roughness expressed in μm. If unknown, a default value of $R_{a,p}=1\mu$m should be taken. M_{mol} is the relative molecular weight.

The *Imura* nuclear pool boiling heat transfer coefficient correlation can be expressed as

$$h_{Imu} = 0.32\left(\frac{\rho_l^{0.65}k_l^{0.3}c_{p,l}^{0.7}g^{0.2}}{\rho_v^{0.25}i_{lv}^{0.4}\mu_l^{0.1}}\right)\left(\frac{P_v}{P_{atm}}\right)^{0.3}q_{nb}^{''\,0.4} \quad (6)$$

where P_v is vapor pressure and P_{atm} is atmospheric pressure.

The *Rohsenow* correlation is the mostly widely used nuclear boiling heat transfer coefficient correlation. It has good accuracy for most nuclear boiling cases. The surface fluid factor C_{sf} should be adjusted for different fluid/surface material combinations. The *Cooper* correlation is also a widely used empirical correlation for nuclear boiling heat transfer coefficient. The surface roughness $R_{a,p}$ should be adjusted for different material surface. The *Imura* correlation is especially proposed for two-phase closed thermosyphons and can achieve good accuracy at high fluid filling ratios. The comparison of heat transfer coefficients calculated by these three correlations are shown in Fig. 8.

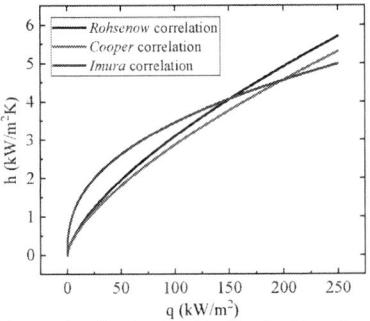

Fig. 8. Calculation results of nuclear pool boiling heat transfer coefficients.

B. Finite element model of FIMEC HFT

A finite element model (FEM) of FIMEC HFT is established in COMSOL Multiphysics. To improve the computation speed of FEM, the irrelevant components such as the sealed container are neglected. In addition, the nanocrystalline core, primary winding and secondary winding are modeled as a solid respectively by homogenization modeling method. The equivalent anisotropic thermal conductivities are utilized in the modeling of core and windings. The core loss and winding loss are taken as heat sources and applied in FEM. The Heat Transfer in Solids and Fluids module in COMSOL Multiphysics is utilized to simulate temperature field. Since simulating the generation, movement and vanishment of bubbles is non-trivial, the Nuclear Boiling Heat Flux boundary condition is applied to the HFT surface and *Rohsenow* correlation is used to calculate the steady state temperature of FIMEC HFT.

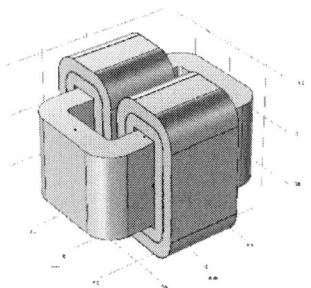

Fig. 9. Finite element model of FIMEC HFT neglecting the container and coolant.

IV. SIMULATION AND EXPERIMENTAL RESULTS

A. Simulation result of FIMEC HFT temperature

The temperature distribution of FIMEC HFT under rated load condition is shown in Fig. 10. The maximum temperature of FIMEC HFT is about 60.9°C, located in the primary windings. The minimum temperature of FIMEC HFT is about 58°C, located in the nanocrystalline core. The maximum temperature difference in the FIMEC HFT is within 3°C. Thus, FIMEC technology can not only decrease HFT hotspot temperature, but also improve HFT temperature distribution uniformity.

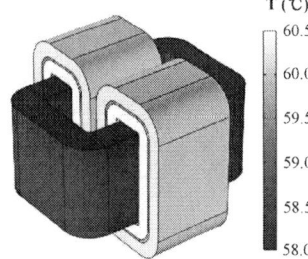

Fig. 10. Temperature distribution of FIMEC HFT under rated load condition.

The variations of winding temperature with phase-shift angle are shown in Fig. 11. With the increases of phase-shift angle or winding loss, winding temperatures also increase. When phase-shift angle increases from 5° to 10°, winding temperature increments are merely about 2°. Thus, FIMEC technology can effectively decrease HFT temperature and reduce temperature fluctuations during load variation.

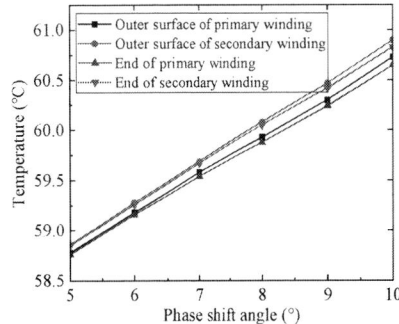

Fig. 11. Variations of winding temperature with phase-shift angle.

B. Experimental result of FIMEC HFT temperature

To experimentally verify the effectiveness of FIMEC technology in cooling of HFT, a FIMEC HFT prototype as shown in Fig. 1(c) is made. A HFT temperature measurement platform is established as shown in Fig. 12. The HFT test power supply is used to provide rated load excitation condition. The multi-channel temperature tester is used to record the temperature rise curve of HFT. Several K-type thermocouples are embedded inside the HFT to measure the internal temperature. The locations of the embedded thermocouples are shown in Fig. 13. The curves of temperature rise under rated load condition at different locations of the HFT are shown in Fig. 14. It takes about 3 hours for the FIMEC HFT to reach nuclear boiling state. Since the boiling of coolant requires an overheating of several degrees, typically 6°C for FC-72 used in this paper, a temperature drop phenomenon can be observed after boiling in the temperature rise curve. Then it takes about 1 additional hour for the FIMEC HFT to reach the final thermal steady state. In thermal steady state, the boiling phenomenon can be clearly observed as shown in Fig. 15.

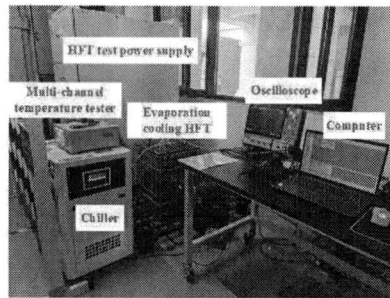

Fig. 12. HFT temperature measurement platform.

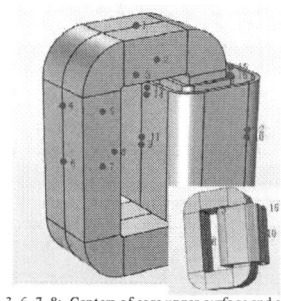

1, 2, 3, 6, 7, 8: Centers of core upper surface and side surface;
4, 5: Core surfaces near winding end;
9, 10: Centers of primary winding outer surface;
11, 12: Centers of secondary winding outer surface;
13, 14, 15, 16: Ends of primary and secondary windings

Fig. 13. Locations of embedded thermocouples inside the HFT.

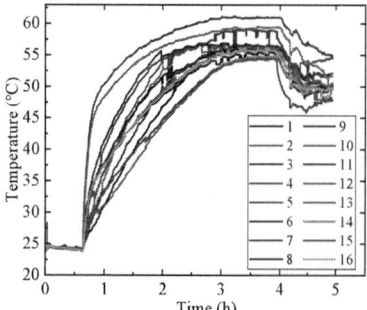

Fig. 14. Temperature rise curve of the HFT.

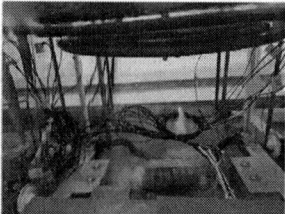

Fig. 15. Boiling phenomenon in the FIMEC HFT.

The comparison of simulation and measurement results of HFT steady state temperatures is shown in Fig. 16. From the temperature measurement results, the maximum temperature is about 55.0°C, located in the outer surface of primary windings. The minimum temperature is about 47.8°C, located in the secondary winding ends. The temperature distribution in the FIMEC HFT is relatively uniform and the maximum temperature difference is within 8°C. The maximum errors between temperature simulation and measurement results are about 10°C. On the one hand, the measured temperatures may deviate from the actual values due to the immersion of thermocouples and electromagnetic interference from the excitation voltage of HFT. On the other hand, the inappropriate selection of parameters in *Rohsenow* correlation can reduce the temperature calculation accuracies.

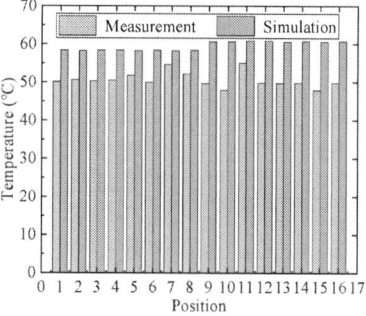

Fig. 16. Comparison of simulation and measurement results of HFT steady state temperatures.

C. Comparison of temperatures between three HFT prototypes

In this paper, finite element models of the other two HFT prototypes are established. Fig. 17(a) shows the temperature distribution of epoxy resin potting HFT with fan cooling (Trans-I). The maximum temperature is over 125°C, located in the primary windings. The secondary windings are also over 125°C. The core surrounded by the primary windings is over 125°C and the upper and lower cores not surrounded by the primary windings is only 90°C. Fig. 17(b) shows the temperature distribution of non-epoxy resin potted HFT with fan cooling (Trans-II). The maximum temperature is about 110°C, located in the primary windings in core window. The secondary windings temperature out of core window is below 100°C. The core temperature in the windward side is very low as about 40°C. The core temperature in the leeward side can reach 60 to 70°C.

979-8-3315-1612-3/25 $31.00 © 2025 IEEE

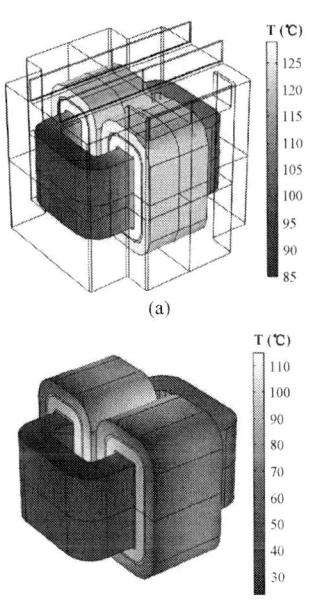

(a)

(b)

Fig. 17. Temperature distribution of (a) epoxy resin potting HFT with fan cooling and (b) Non epoxy resin potted HFT with fan cooling.

Fig. 18 shows the comparison of steady state temperatures of these three HFT prototypes. Due to the low thermal conductivity of epoxy resin material, the epoxy resin potting HFT with fan cooling (Trans-I) has the maximum hotspot temperature and average temperature, being 130.0°C and 108.2°C. In addition, the maximum temperature difference is about 57.5°C. For non-epoxy resin potted HFT with fan cooling (Trans-II), the hotspot temperature and average temperature are 106.4°C and 77.9°C, reduced by about 20°C and 30°C respectively compared with epoxy resin potting HFT (Trans-I). The maximum temperature difference is about 54.2°C. The FIMEC technology can significantly reduce the HFT hotspot temperature and average temperature to 55.0°C and 50.6°C. The maximum temperature difference is only about 7.2°C. The comparison of hotspot temperature, average temperature and maximum temperature difference of these three HFT prototypes are summarized in Table. II.

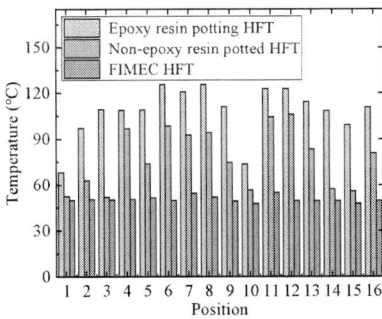

Fig. 18. Comparison of steady state temperatures of these three HFT prototypes.

TABLE. II. HOTSPOT TEMPERATURE, AVERAGE TEMPERATURE AND MAXIMUM TEMPERATURE DIFFERENCE FOR THE THREE HFT PROTOTYPES

HFT prototypes	Trans-I	Trans-I	Trans-III
Hotspot temperature (°C)	130.0	106.4	55.0
Average temperature (°C)	108.2	77.9	50.6
Maximum temperature difference (°C)	57.5	54.2	7.2

V. CONCLUSION

This paper conducted simulation and experimental research on cooling performance of FIMEC HFT. The main conclusions are listed as follows.

1) A finite element simulation model of FIMEC HFT is established. The maximum error between calculated and measured temperatures is within 10%. This simulation model can accurately calculate the temperature field of FIMEC HFT.

2) The steady-state temperature distribution of FIMEC HFT is uniform. The maximum temperature difference is within 8°C. With the increase of power loss, the temperature rise of FIMEC HFT is small. FIMEC technology can prevent sharp temperature rise.

3) Compared with epoxy resin potting and non-epoxy resin potted HFT with fan cooling, FIMEC technology can significantly decrease the steady-state temperature and improve the temperature distribution uniformity. The maximum temperature is reduced from 130.0°C and 106.4°C to 55.0°C and the maximum temperature difference is reduced from 57.5°C and 54.2°C to 7.2°C.

REFERENCES

[1] C. Gu, Z. Zheng, L. Xu, K. Wang and Y. Li, "Modeling and Control of a Multiport Power Electronic Transformer (PET) for Electric Traction Applications," *IEEE Trans. Power Electron.*, vol. 31, no. 2, pp. 915-927, Feb. 2016.

[2] T. Liu et al., "Design and Implementation of High Efficiency Control Scheme of Dual Active Bridge Based 10 kV/1 MW Solid State Transformer for PV Application," *IEEE Trans. Power Electron.*, vol. 34, no. 5, pp. 4223-4238, May 2019

[3] M. Vasiladiotis and A. Rufer, "A Modular Multiport Power Electronic Transformer With Integrated Split Battery Energy Storage for Versatile Ultrafast EV Charging Stations," *IEEE Trans. Ind. Electron.*, vol. 62, no. 5, pp. 3213-3222, May 2015.

[4] D. Dong, M. Agamy, J. Z. Bebic, Q. Chen and G. Mandrusiak, "A Modular SiC High-Frequency Solid-State Transformer for Medium-Voltage Applications: Design, Implementation, and Testing," *IEEE J. Emerg. Sel. Topics Power Electron.*, vol. 7, no. 2, pp. 768-778, June 2019.

[5] W. Sung and B. J. Baliga, "On Developing One-Chip Integration of 1.2 kV SiC MOSFET and JBS Diode (JBSFET)," *IEEE Trans. Ind. Electron.*, vol. 64, no. 10, pp. 8206-8212, Oct. 2017.

[6] C. Zhan et al., "A Novel Sensor-Reduction Condition Monitoring Approach for MMC Submodule IGBTs Based on Statistics of Inferred On-State Voltage," *IEEE J. Emerg. Sel. Topics Power Electron.*, vol. 12, no. 1, pp. 1068-1077, Feb. 2024.

[7] C. Zhan et al., "Intelligent Condition Monitoring of Multiple Thermal Degradation of IGBT Modules Based on Case Temperature Matrix," *IEEE Trans. Power Electron.*, vol. 39, no. 10, pp. 12490-12501, Oct. 2024.

[8] T. Yuan, F. Jin, Z. Li, C. Zhao and Q. Li, "Design of an Integrated Transformer With Parallel Windings for a 30-kW LLC Resonant Converter," *IEEE Trans. Power Electron.*, vol. 38, no. 11, pp. 14317-14333, Nov. 2023.

[9] T. Yuan, F. Jin and Q. Li, "Analysis and Comparison of Integrated Planar Transformers for 22-kW On-Board Chargers," *IEEE Trans. Power Electron.*, vol. 39, no. 9, pp. 11368-11385, Sept. 2024.

[10] H. Wang et al., "Thermal Design Consideration of Medium Voltage High Frequency Transformers," *2020 IEEE Applied Power Electronics Conference and Exposition (APEC)*, New Orleans, LA, USA, 2020, pp. 2721-2726.

[11] M. Ngo, Y. Cao, D. Dong, R. Burgos, K. Nguyen and A. Ismail, "Forced Air-Cooling Thermal Design Methodology for High-Density, High-Frequency, and High-Power Planar Transformers in 1U

Applications," *IEEE J. Emerg. Sel. Topics Power Electron.*, vol. 11, no. 2, pp. 2015-2028, April 2023.

[12] P. S. Nasab, R. Perini, A. Di Gerlando, G. M. Foglia and M. Moallem, "Analytical Thermal Model of Natural-Convection Cooling in Axial Flux Machines," *IEEE Trans. Ind. Electron.*, vol. 67, no. 4, pp. 2711-2721, April 2020.

[13] D. Christen, M. Stojadinovic and J. Biela, "Energy Efficient Heat Sink Design: Natural Versus Forced Convection Cooling," *IEEE Trans. Power Electron.*, vol. 32, no. 11, pp. 8693-8704, Nov. 2017.

[14] M. Kopeć, R. Olbrycht, P. Gamorski and M. Kałuża, "The Influence of Air Humidity on Convective Cooling Conditions of Electronic Devices," *IEEE Trans. Ind. Electron.*, vol. 65, no. 12, pp. 9717-9727, Dec. 2018.

[15] M. Mogorovic and D. Dujic, "100 kW, 10 kHz Medium-Frequency Transformer Design Optimization and Experimental Verification," *IEEE Trans. Power Electron.*, vol. 34, no. 2, pp. 1696-1708, Feb. 2019.

[16] M. Leibl, G. Ortiz and J. W. Kolar, "Design and Experimental Analysis of a Medium-Frequency Transformer for Solid-State Transformer Applications," *IEEE J. Emerg. Sel. Topics Power Electron.*, vol. 5, no. 1, pp. 110-123, March 2017.

[17] L. Ruan, G. Gu, X. Tian and J. Yuan, "The comparison of cooling effect between evaporative cooling method and inner water cooling method for the large hydro generator," *2007 International Conference on Electrical Machines and Systems (ICEMS)*, Seoul, Korea (South), 2007, pp. 989-992.

[18] J. B. Campbell, L. M. Tolbert, C. W. Ayers, B. Ozpineci and K. T. Lowe, "Two-Phase Cooling Method Using the R134a Refrigerant to Cool Power Electronic Devices," *IEEE Trans. Ind. Appl.*, vol. 43, no. 3, pp. 648-656, May-june 2007.

[19] Y. Zhangbin, K. Huang, Z. Sixiang, B. Xiong and L. Daijun, "Optimal design of U-shaped evaporative cooling radiator for high frequency transformer," *2022 25th International Conference on Electrical Machines and Systems (ICEMS)*, Chiang Mai, Thailand, 2022, pp. 1-5.

[20] C. P. Steinmetz, "On the law of hysteresis," *AIEE Trans.*, vol. 9, pp. 3–64, 1892.

[21] J. Reinert, A. Brockmeyer, and R. W. A. A. De Doncker, "Calculation of losses in ferro- and ferrimagnetic materials based on the modified Steinmetz equation," *IEEE Trans. Ind. Appl.*, vol. 37, no. 4, pp. 1055-1061, July-Aug. 2001.

[22] Jieli Li, T. Abdallah and C. R. Sullivan, "Improved calculation of core loss with nonsinusoidal waveforms," *Proc. Conf. Rec. IEEE Ind. Appl. Conf. 36th IAS Annu. Meeting*, 2001, vol. 4, pp. 2203–2210.

[23] K. Venkatachalam, C. R. Sullivan, T. Abdallah, and H. Tacca, "Accurate prediction of ferrite core loss with nonsinusoidal waveforms using only Steinmetz parameters," *Proc. IEEE Workshop Comput. Power Electron.*, 2002, pp. 36–41.

[24] W. Shen, F. Wang, D. Boroyevich and C. W. Tipton, "Loss Characterization and Calculation of Nanocrystalline Cores for High-Frequency Magnetics Applications," *IEEE Trans. Power Electron.*, vol. 23, no. 1, pp. 475-484, Jan. 2008.

[25] W.M. Rohsenow, "A Method of Correlating Heat Transfer Data for Surface Boiling of Liquids," *MIT Division of Industrial Cooperation*, Cambridge, Mass, 1951.

[26] M.G. Cooper, "Saturation nucleate pool boiling—a simple correlation," *Inst. Chem. Eng. Symp. Ser.* 86 (2) (1984) 785–793.

[27] H. Imura, H. Kusuda, J.I. Ogata, et al., "Heat transfer in two-phase closed-type thermosyphons," *JSME Transactions* 45 (1979) 712–722.

High-Efficiency PCB-Embeddable Inductor for Vertical Power IVR Applications

Youssef Kandeel, Liang Ye, John Flannery,
Cian Ó Mathúna, Ranajit Sai, Séamus O'Driscoll
Tyndall National Institute
University College Cork
Cork, Ireland
youssef.kandeel@tyndall.ie, liang.ye@tyndall.ie,
John.flennary@tyndall.ie, cian.omathuna@tyndall.ie,
ranajit.sai@tyndall.ie, seamus.odriscoll@tyndall.ie

Takayuki Tsuchida, Naoya Terauchi, Sumiaki Kishimoto,
Toshio Hiraoka, Masanori Nagano
TAIYO YUDEN Co. Ltd.
Tokyo, Japan
ta-tsuchida@jty.yuden.co.jp, n-terauchi@jty.yuden.co.jp,
s-kishimoto@jty.yuden.co.jp, thiraoka@jty.yuden.co.jp,
mnagano@jty.yuden.co.jp

Abstract - **This paper presents a substrate/PCB embeddable inductor device for Integrated Voltage Regulator (IVR) application. The inductor is fabricated with two new 100 MHz pre-release magnetic materials to suit highly integrated IVR for vertical power. The measured characteristics of the magnetic materials are presented and used in simulation models for the inductor device. The inductor is designed according to high-volume manufacturing capabilities with a target nominal height equal to the PCB core thickness of 0.3 mm, so that it can be embedded within the application PCB core or the SoC package substrate core and facilitate vertical power delivery. The fabricated inductor achieved 1.54 *nH/mΩ* and simulations indicate that it will achieve 97.5% inductor efficiency for V_{IN} = 1.8 V, V_{OUT} = 0.9 V, F_{SW} = 100 MHz and I_{DC} = 2 A.**

Index Terms – **Embedded inductor, integrated inductor, integrated voltage regulator (IVR), DC-DC converter, power supply in package (PSiP, PwrSiP), vertical power.**

I. INTRODUCTION

The demand in the high-performance computing (HPC) market for more efficient and highly integrated solutions motivates research into improving voltage regulators and their components' performance. High-performance microprocessors utilize multiphase voltage regulator topologies to divide the total load current between a large number of phases, with an inductor per phase or inverse-inductor-coupled phases arrangement, [1][2][3]. Noting that inductors usually comprise the largest portion of the power converter volume, motivates the need to make inductor devices smaller, embeddable and capable of higher frequency operation. Design trade-offs require improvements in the inductor device's performance, particularly minimizing the inductor's loss, and parasitic capacitances, both shunt loading on the converter's switching node and series capacitive injections to the output, while meeting the converter electrical specifications and packaging capabilities. This requires a magnetic material that maintains its permeability at frequencies much higher than the switching frequency and low loss under large signal excitation, even if this comes at the cost of reducing the material's permeability [4], and thereby increasing the inductor device size.

There are different magnetic materials proposed in the literature for very high frequency (50-100 MHz) embeddable inductors, such as Intel's iron-alloy-based magnetic

microparticle material [5], Panasonic HBS1 [6], Intel's composite magnetic material [7], and multilayer amorphous CoZrTa thin-film [8], while others have employed air-core, as in [1][2][9].

IVR inductor devices may be coupled or non-coupled and may come in different structures such as solenoid, toroid, spiral, racetrack, strip-line, and coaxial. The coupled-inductor aspect is not part of the scope of this paper.

To deliver a PCB embeddable inductor suitable for IVR applications, this paper presents two new, pre-release, magnetic materials developed by TAIYO YUDEN Co. Ltd. to provide very low loss at 100 MHz operation. These materials are implemented in a strip-line prototype inductor structure, which has lower feed-through capacitance than the multi-turn structure.

There are PCB technologies that can embed such low-height devices within the PCB core, such as [10], and which can suit vertical power delivery [11].

This paper is structured as follows. Section II presents the application scope and strip-line inductor background. The measured properties of TAIYO YUDEN's magnetic materials are presented in Section III. The fabricated devices' small signal characterizations and modelling are presented in Section IV. The devices' performances, under typical IVR specifications, are compared with the relevant literature in Section V. Finally, conclusions are discussed in Section VI.

II. STRIP-LINE INDUCTOR DESIGN

100-MHz IVRs usually operate at relatively low input voltage, such as 1.8 V, and their common application is to step down the input voltage to a lower level, such as the 0.6 V core voltage, for an advanced geometry CMOS SoC. Hence, they require an output filter inductor with wide-bandwidth inductance but with low shunt and feed-through capacitances, because of large voltage slew rates on the converter's switching node. Representative converter specifications assumed for this study are detailed in TABLE I, i.e. a typical step-down IVR (buck) specification. The frequency is assumed a fixed 100 MHz; hence, the inductor is required to maintain an inductive characteristic until significantly greater than 1 GHz to filter switching harmonics effectively and prevent switching node

979-8-3315-1612-3/25 $31.00 © 2025 IEEE

slew rate feedthrough to the output capacitor's ESL, i.e. the self-resonance frequency (SRF) of the inductor is required to be higher than 1 GHz.

For IVR applications, an inductance range of 2.5 to 10 nH is appropriate. For the study purpose, the output current peak-to-peak ripple (ΔI_{PK-PK}) in TABLE I is calculated at the corresponding inductance value, however, actual applications may have other different current ripple requirements.

The strip-line inductor considered consists of a copper conductor single strip with a rectangular cross-section as shown in Fig. 1. The conductor is surrounded by a magnetic material, which is assumed to have the same thickness above, below and on either side of the conductor as detailed in Fig. 1(b). The strip-line structure is chosen for this study as its parasitic capacitance is generally much less than multiturn structures, hence, it offers greater bandwidth.

TABLE I
IVR TARGET SPECIFICATIONS

Symbol	Quantity	Value			Unit
V_{IN}	Input voltage	1.8			V
V_{OUT}	Output voltage	0.9			V
F_{SW}	Switching frequency	100			MHz
I_{DC}	Output DC current	2			A
L	Inductance	2.5	5	10	nH
ΔI_{PK-PK}	Output current ripple	1.8	0.9	0.45	A

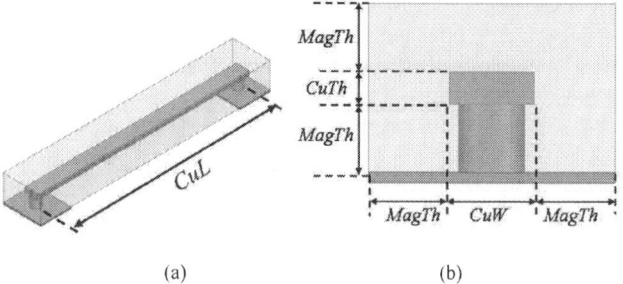

(a)   (b)

Fig. 1. Strip-line inductor model: (a) 3D structure, (b) cross section view.

The inductance of the strip-line inductor L_{SL} is calculated as $L_{SL} = L_{Air} + L_{Mag}$, where L_{Air} is the air-core inductance calculated in (1), and L_{Mag} is the magnetic core inductance calculated in (2).

$$L_{Air} = \frac{\mu_0 Cu_L}{4\pi}\left(0.5 + \ln\frac{2Cu_L}{Cu_W + Cu_{Th}} + \frac{Cu_W + Cu_{Th}}{3Cu_L}\right) \quad (1)$$

$$L_{mag} = \frac{\mu_0 \mu_r Cu_L Mag_{Th}}{2(Cu_W + Mag_{Th}) + 2(Cu_{Th} + Mag_{Th})} \quad (2)$$

where μ_0 is the air permeability $\mu_0 = 4\pi 10^{-7}$, μ_r is the magnetic material's relative permeability, Cu_L, Cu_W and Cu_{Th} are the copper trace's length, width and thickness respectively, and Mag_{Th} is the magnetic material thickness.

The strip-line inductor's DC resistance (R_{DC}) is calculated as $R_{DC} = \frac{\rho Cu_L}{Cu_W Cu_{Th}}$, where ρ is the copper conductivity.

The design parameters are tuned initially to achieve the electrical specifications in TABLE I and to minimize the resistance (R_{DC}), within the inductor's fabrication process capabilities. The starting point of the inductor dimensions tuning is based on the manufacturing technology capabilities, i.e., $Cu_W \geq 100\ \mu m$, $Cu_{Th} \geq 60\ \mu m$, $Mag_{Th} \geq 100\ \mu m$. First, Cu_{Th} is fixed at a certain value, then Mag_{Th} is chosen to maintain the device height at 0.3 mm and then varying Cu_W and Cu_L to meet the electrical specifications. The mathematically synthesized design is then modelled in the Finite Element (FEA) Simulator.

III. MAGNETIC MATERIAL PROPERTIES

A. Small signal complex permeability

This study uses the pre-release magnetic materials, TY-M5 and TY-M6 developed by TAIYO YUDEN to suit IVR applications. The materials' small signal properties were measured on a toroid sample using an impedance analyzer model No E4990A with test fixture 16454A. The measured complex permeability (μ' and μ'') vs frequency characteristics for the materials are shown in Fig. 2 and Fig. 3, respectively. The sample was measured from 1 MHz to 3 GHz, within the impedance analyser's capability. The measured data shows ferromagnetic resonance (FMR) around 2 GHz.

Permeability spectra are noisy at low frequencies (~ 1-5 MHz) due to equipment limitations while measuring very small impedances. Permeability extraction is not trusted around the material's FMR or the device resonance frequency, which occur close to the upper range capability of the analyser. To overcome these two measurement limitations and to model the device to 10 GHz, the materials' measured permeability data are modified, through smoothing the permeability data, up to 10 MHz and through substitution with LLG formula-based [12] and modelled material data between approximately 500 MHz and 10 GHz. The measured and modified data are shown in Fig. 2 and Fig. 3 for TY-M5 and TY-M6 materials, respectively. Note that, LLG formula extrapolation, between 500 MHz and 10 GHz is used to give insight into material effects on inductor device behaviour, as it is understood that the mentioned materials do not have useful magnetic characteristics beyond approximately 1 GHz. The modified μ'' of TY-M6 has a relatively low peak value as a result of using a second-order LLG equation and reducing the μ'' fitting error at frequencies less than 500 MHz. The magnetic material's permittivity and the device's physical structure contribute to the device's high-frequency capacitances and the standard LLG models contribute a negative permeability above FMR. The capacitances and the transition to negative permeability, both contribute to phase loss in the range 1 − 10 GHz. The modified permeability vs frequency data are entered in the materials models for the device Finite Element Analysis (FEA) simulations.

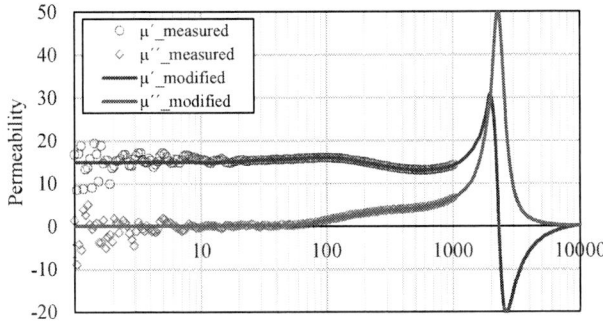

Fig. 2. TY-M5 material small signal complex permeability, data points for measured values, and solid lines for the modified data which is smoothed at low frequency and modelled by the LLG formula from 800 MHz to 10 GHz.

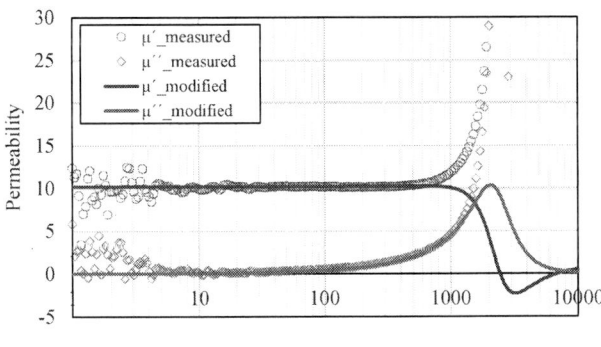

Fig. 3. TY-M6 material small signal complex permeability, data points for measured values, and solid lines for the modified data which is smoothed at low frequency and modelled by the LLG formula from 500 MHz to 10 GHz.

B. Relative permittivity

The materials' measured relative permittivities vs frequency are shown in Fig. 4, and are necessary for the FEA model to predict the devices' self-resonance frequencies (SRF) and determine their applicable bandwidths. The permittivity was measured on materials samples with plated electrodes and using an impedance analyzer with the test fixture 16034E from 100 Hz to 40 MHz. Higher frequency data was extrapolated using a power function. TY-M5 shows nearly double the permittivity of TY-M6, however, the impact of this difference is small with strip-line devices compared with multiturn devices.

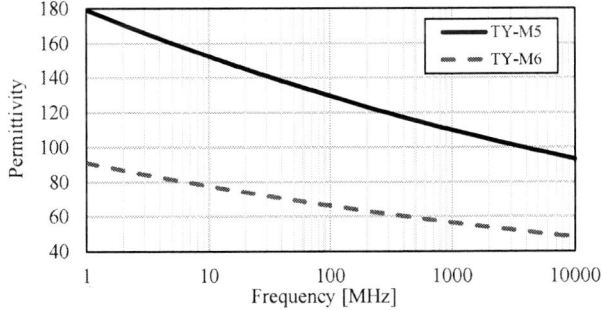

Fig. 4. TY-M5 and TY-M6 measured relative permittivity vs frequency.

C. B-H magnetizing curves

Fig. 5 (a) presents the measured B-H magnetizing characteristics of the materials, which are used to calculate the normalized permeabilities in Fig. 5 (b). Fig. 5 (b) indicates the saturation characteristics of the magnetic materials, e.g. normalized permeabilities drop to 70% at flux densities of approximately 0.8 and 0.7 T for TY-M5 and TY-M6 materials, respectively.

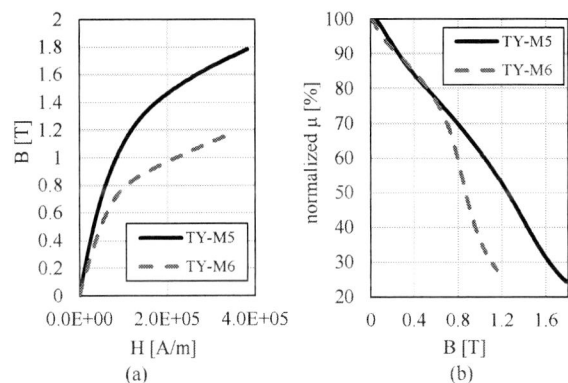

Fig. 5. TY-M5 and TY-M6 (a) measured B-H curve, and (b) normalized permeability.

D. Specific magnetic core loss

The TY-M5 and TY-M6 materials' loss characteristics in Fig. 6 are calculated using Steinmetz coefficients extracted from large-signal measurements. The Steinmetz coefficients apply over a frequency range of 1-50 MHz and flux density of 1-100 mT, however, they are extrapolated to predict the core loss at 100 MHz.

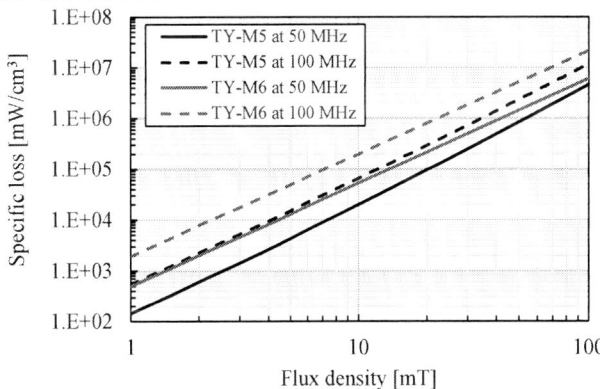

Fig. 6. Steinmetz formula curves for TY-M5 and TY-M6 materials specific loss, extracted from the measured data.

Core loss measurements at 100 MHz will be performed for future work. The TY-M5 material in Fig. 6 shows significantly lower loss than TY-M6 at 50 and 100 MHz, however, their loss characteristics seem to cross over at higher flux densities, > 100 mT. Hence, the inductor requirements and excitation levels will determine which material would perform better.

IV. DEVICE PERFORMANCE

A. Fabricated devices

Due to time limitations, only one device size was fabricated, with each material type, for this this publication.

A sample fabricated prototype device, shown in Fig. 7, has dimensions of 2.5 x 0.4 x 0.35 mm. The height is higher than the target design value, but the technology can achieve 0.3 mm and there is a roadmap to even thinner devices and so will be embeddable in the core of the SoC package substrate or in the application PCB.

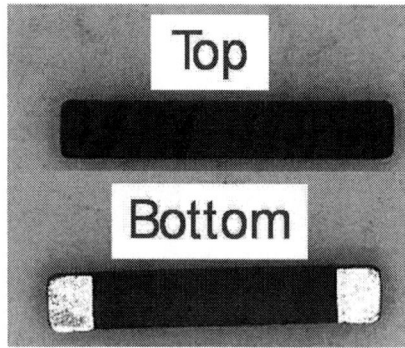

Fig. 7. Picture of a sample of the fabricated inductor.

B. Device FEA modelling and characterization

The inductor devices were modelled and simulated in Ansys software with the properties of TY-M5 and TY-M6 materials presented in the previous section. Ansys HFSS was used to simulate the devices' small signal impedance characteristics vs frequency, and Ansys Maxwell was used to simulate the devices' large signal losses at 100 MHz.

The measured R_{DC} for the fabricated devices with TY-M5 and TY-M6 materials are 6.4 and 6.8 mΩ, respectively. The inductor devices were assembled on a test board and characterized under small signal excitation using an impedance analyzer from 1 MHz to 1 GHz. The test board's impedance was "de-embedded" from the measured data.

Both the de-embedded measured data and the FEA model results are shown in Fig. 8 and Fig. 9 for both theTY-M5 and TY-M6 based devices, respectively.

The small-signal measured and FEA-modelled inductance characteristics are shown in Fig. 8(a) and Fig. 9(a). There are flat characteristics to 100 MHz and show inductive characteristics (positive phase) to greater than 2GHz. At 100 MHz, the devices had measured inductances of 9.84 and 7.75 nH with TY-M5 and TY-M6, respectively. The measured inductances are 1.7-2 nH higher than the FEA-modelled results and are understood to be as a result of higher magnetic permeabilities in the fabricated devices, than in the toroidal test samples.

The devices' phase characteristics in Fig. 8(b) and Fig. 9(b) show phase transition through zero at greater than 1 GHz, however, this is due to the magnetic materials' FMRs and permittivities, not due to the device structure's geometry-based capacitances.

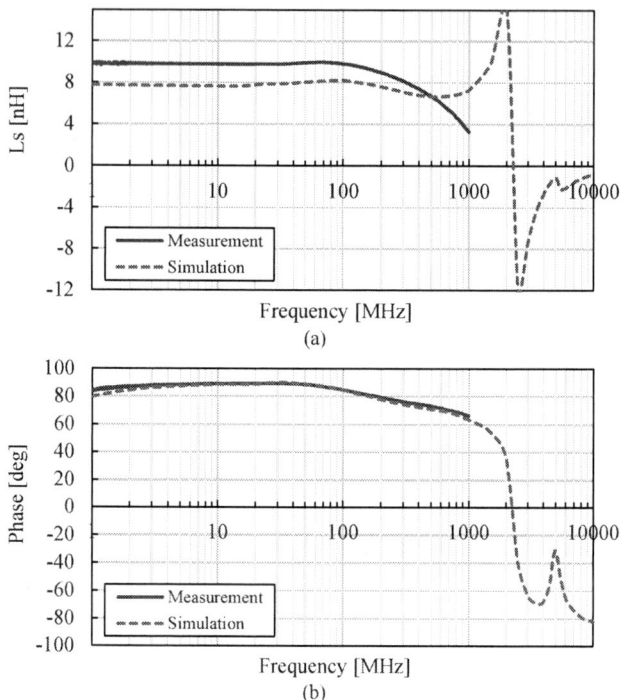

Fig. 8. Impedance analyzer measured data (solid line) and FEA model results (dashed line) of the strip-line inductor with TY-M5 small signal properties: (a) Inductance vs frequency, (b) Phase vs frequency.

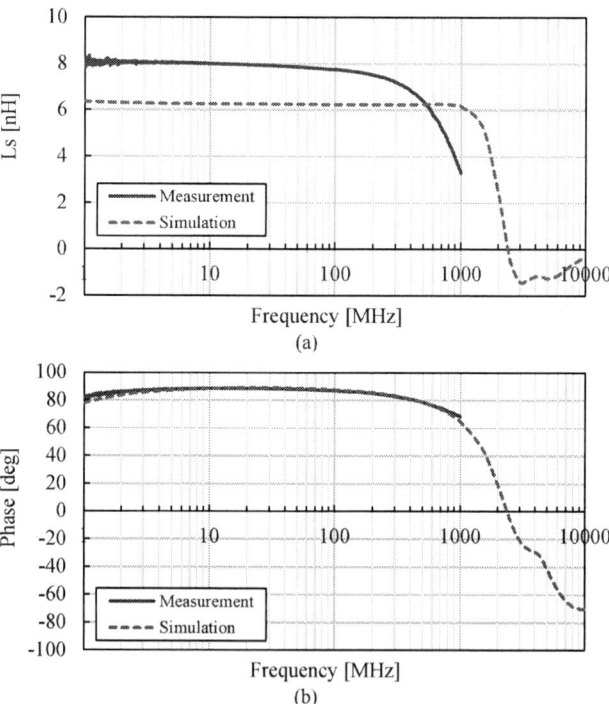

Fig. 9. Impedance analyzer measured data (solid line) and FEA model results (dashed line) of the strip-line inductor with TY-M6 small signal properties: (a) Inductance vs frequency, (b) Phase vs frequency.

V. Discussion and Comparison

To evaluate the fabricated inductors, the inductors' performance parameters are compared in TABLE II against some of the relevant published magnetic material-based devices. Lower density air-core inductors are not considered relevant for this comparison. Each reference in TABLE II considered different converter specifications, but the inductances and AC resistances were extracted at 100 MHz to achieve comparative FOMs. However, the inductor losses and efficiencies for the reference paper inductors are not presented as they would not be valid for comparison. The material losses for the Taiyo Yuden TY-M5 and TY-M6 based devices' R_{AC} losses at 100 MHz were extracted from the FEA simulations with the materials' large–signal loss properties as presented in Fig. 6.

In terms of the measured L/R_{DC} parameter, the inductor manufactured with TY-M5 material would achieve 6.2 and 7.4 times the values for the devices in [5] and [7], respectively, and the inductor manufactured with TY-M6 material would achieve 4.6 and 5.5 times the values for the devices in [5] and [7], respectively. In terms of the $L/Area$ parameter, the inductor manufactured with TY-M5 material achieves 1.6 times those for the devices in [5] and [7], and the inductor manufactured with TY-M6 material would achieves 1.3 and 1.2 times those for the devices in [5] and [7], respectively. Inductors manufactured with TY-M5 and TY-M6 materials have calculated efficiencies of 97.5 and 96.6 %, respectively. This means that the TY-M5 and TY-M6 properties will enable the devices to achieve high inductance density and high efficiency.

TABLE II
INDUCTOR DEVICES COMPARISON AT SPECIFICATIONS IN TABLE I

Reference	[7], Intel 2021	[5], Intel 2020	[13], IBM 2013	This work TY-M5	This work TY-M6
L at 100 MHz [nH]	2.5	3•	5	9.84▲	7.75▲
R_{DC} [mΩ]	12	12•	270	6.4▲	6.8▲
R_{AC} at 100 MHz [mΩ]	47.6■	104.7■	2900	283.8▼	348.1▼
B_{AC} [mT]				10.2▼	6.5▼
Footprint area [mm²]	0.4	0.5•	0.245	1	1
Volume [mm³]	-	0.09■	-	0.35	0.35
L/R_{DC} [nH/mΩ]	0.208	0.25•	0.019	1.54	1.14
$L/Area$ [nH/mm²]	6.3	6	20.4	9.84	7.75
$L/Volume$ [nH/mm³]	-	33.3	-	28.1	22.2
ΔI_{PK-PK} [A]♦				0.46	0.58
R_{DC} loss [mW]♦				26	28
R_{AC} loss [mW]♦				20	39
Total loss [mW]♦				46	67
Inductor efficiency [%]				97.5	96.4

▲ Measured values, as presented in Section IV.B.
▼ Based on device FEA simulation with the material large signal loss properties at 100 MHz, presented in Fig. 6, and current amplitude of 0.45 A.
• Values reported by [7].
■ Estimated based on published small signal data.
♦ Calculated according to specifications in TABLE I.

VI. Conclusion

This paper presents the design parameters achievable for a PCB/substrate embeddable strip-line inductor for IVR applications based on TAIYO YUDEN's manufacturing technology and its newly developed TY-M5 and TY-M6 materials for 50 to 100+ MHz IVR applications. The paper presents permeability, permittivity, B-H magnetization and core loss characterizations for the TY-M5 and TY-M6 pre-release materials. The measured materials' characteristics were used in the FEA simulation models to estimate the device bandwidths and ensure good high-frequency performances. The fabricated inductor's total height is 0.35 mm, so that it could be embedded horizontally within the PCB-Core layer of the PCB or package-substrate of the z-dimension stack-up, to enable vertical power delivery. The fabricated inductor with TY-M5 material achieved 1.54 nH/mΩ in a size of 0.35 mm³; and calculated inductor efficiency of 97.5% at the converter specification of V_{IN} = 1.8 V, V_{OUT} = 0.9 V, F_{SW} = 100 MHz and I_{DC} = 2 A. Further characterizations of the fabricated inductors, thermal characterizations of the magnetic materials, and other device design formats will be investigated in future work. Mechanical stress, thermal expansion coefficient mismatch with substrate and inductor device contact plating design considerations have not been addressed in this work, but nevertheless, it is concluded that exceptional magnetic material based inductor performance can be achieved in a very low profile and suitable for use in 100 MHz package-integrated VR.

References

[1] E. A. Burton et al., "FIVR - Fully Integrated Voltage Regulators on 4th Generation Intel® Core™ SoCs," in IEEE Applied Power Electronics Conference and Exposition (APEC), 2014, pp. 432–439. doi: 10.1109/APEC.2014.6803344.

[2] K. Bharath and S. Venkataraman, "Power Delivery Design and Analysis of 14nm Multicore Server CPUs with Integrated Voltage Regulators," in IEEE Electronic Components and Technology Conference (ECTC), 2016, pp. 368–373. doi: 10.1109/ECTC.2016.322.

[3] Y. Kandeel, S. O'Driscoll, C. O'Mathuna, and M. Duffy, "Optimum Phase Count in a 5.4-W Multiphase Buck Converter Based on Output Filter Component Energies," IEEE Trans. Power Electron., vol. 38, no. 4, pp. 4909–4920, Apr. 2023, doi: 10.1109/TPEL.2022.3226708.

[4] K. Radhakrishnan, M. Swaminathan, and B. K. Bhattacharyya, "Power Delivery for High-Performance Microprocessors - Challenges, Solutions, and Future Trends," IEEE Trans. Components, Packag. Manuf. Technol., vol. 11, no. 4, pp. 655–671, Apr. 2021, doi: 10.1109/TCPMT.2021.3065690.

[5] M. Sankarasubramanian et al., "Magnetic Inductor Arrays for Intel® Fully Integrated Voltage Regulator (FIVR) on 10th generation Intel® Core™ SoCs," in IEEE Electronic Components and Technology Conference (ECTC), 2020, pp. 399–404. doi: 10.1109/ECTC32862.2020.00071.

[6] C. A. Barros et al., "Proposed Inductor Power Loss Metric and Novel Embedded Toroidal Inductor for Integrated Voltage Regulators," IEEE Trans. Components, Packag. Manuf. Technol., vol. 11, no. 11, pp. 1935–1947, Nov. 2021, doi: 10.1109/TCPMT.2021.3116942.

[7] K. Bharath et al., "Integrated Voltage Regulator Efficiency Improvement using Coaxial Magnetic Composite Core Inductors," in Electronic Components and Technology Conference, 2021, pp. 1286–1292. doi: 10.1109/ECTC32696.2021.00208.

[8] S. Raju et al., "Thin-Film Magnetic Inductors on Silicon for Integrated Power Converters," in IECON 2020 The 46th Annual

Conference of the IEEE Industrial Electronics Society, 2020, pp. 2292–2295. doi: 10.1109/IECON43393.2020.9255388.

[9] Y. Ding, X. Fang, R. Wu, and J. K. O. Sin, "Fan-Out-Package-Embedded Coupled Inductors for Integrated Voltage Conversion," in *International Symposium on Power Semiconductor Devices and ICs (ISPSD)*, 2020, pp. 356–359. doi: 10.1109/ISPSD46842.2020.9170128.

[10] M. Morianz and H. Stahr, "Embedded power electronics on the way to be launched," in *European Microelectronics Packaging Conference (EMPC)*, Friedrichshafen, Germany, 2016. [Online]. Available: https://ieeexplore.ieee.org/document/7390738

[11] S. Krishnakumar and I. Partin-Vaisband, "Vertical Power Delivery for Emerging Packaging and Integration Platforms - Power Conversion and Distribution," in *EEE 36th International System-on-Chip Conference (SOCC)*, 2023. doi: 10.1109/SOCC58585.2023.10256973.

[12] J. Xu, B. Dai, Y. Ren, Y. Wang, and X. Huang, "Electromagnetic and microwave properties of NiFe/NiFeO multilayer thin films," *J. Mater. Sci. Mater. Electron.*, vol. 26, no. 5, pp. 2931–2936, 2015, doi: 10.1007/s10854-015-2779-8.

[13] N. Sturcken *et al.*, "A 2.5D Integrated Voltage Regulator Using Coupled-Magnetic-Core Inductors on Silicon Interposer," *IEEE J. Solid-State Circuits*, vol. 48, no. 1, pp. 244–254, 2013, doi: 10.1109/JSSC.2012.2221237.

An Adaptive Zero Current Switching Control Technique for Multi-Resonant Switched-Capacitor Converters

Haifah B. Sambo, Rose A. Abramson, Sahana Krishnan and Robert C. N. Pilawa-Podgurski

Department of Electrical Engineering and Computer Sciences
University of California, Berkeley
Email: {hsambo, rose_abramson, sahana_krishnan, pilawa}@berkeley.edu

Abstract—Combining energy dense capacitors with efficient switch utilization, resonant switched-capacitor (ReSC) converters have gained popularity due to their ability to simultaneously achieve high efficiencies and high power densities. Multi-phase ReSC converters, here referring to topologies that operate with more than two operating circuit states, are especially attractive as they can achieve high conversion ratios with a fewer number of capacitors and switches than conventional two-phase ReSC converters. However, the topological complexity of these converters, combined with real-world circuit non-idealities, make the phase timings required to operate at resonance challenging to estimate with precision. Here, we present an adaptive control technique that can dynamically track and converge to zero current switching (ZCS) operation to achieve high performance, and facilitate larger scale adoption of these topologies. The concept was validated on a high-performance 48-to-6 V hardware prototype, where the proposed active control technique achieved nearly 20% reduction in power loss at the highest tested load (30 A) when compared with conventional open-loop control.

I. INTRODUCTION

With the advent of high-performance computing, the power demand of processors and large-scale computing systems has grown exponentially [1]. To reduce the power delivery network (PDN) distribution losses in data center server racks, industry has widely adopted a two-stage power conversion approach wherein the 48 V bus is first stepped down to an intermediate bus voltage, and then stepped down to the point-of-load (PoL) voltages required by various storage and computing units. While an intermediate bus voltage of 12 V [2] is the most common, research suggests that lower intermediate bus voltages (between 4 - 8 V) can enable more efficient operation of the voltage regulator modules (VRMs) in the second-stage and facilitate vertical power delivery [3]–[5], both of which can further reduce PDN losses. This architecture, although desirable, requires innovative high-conversion ratio circuit topologies in the first-stage that can densely and efficiently process increasing amounts of currents.

By leveraging energy-dense class II dielectric multi-layer ceramic capacitors (MLCCs) for energy transfer [6] and small resonant inductors for soft-charging capabilities [7], [8], resonant switched-capacitor (ReSC) converters have achieved state-of-the-art power densities and efficiencies for fixed-ratio 48 V step-down in data centers and computing applications in general [9]–[18]. Moreover, most ReSC converters can operate

with zero current switching (ZCS) to eliminate the voltage-current (V-I) overlap losses of the transistors [19].

To minimize total active and passive component count, previously developed high-conversion-ratio 48 V intermediate bus converters utilize cascaded switched-capacitor stages requiring multi-phase, multi-resonant operation [11], [20]–[22]. The complexity of these converters, combined with circuit non-idealities (including parasitics, component tolerances, class II MLCC derating and aging, inductor soft saturation) render the timing required to operate at resonance for each phase challenging to estimate with precision. Therefore, using conventional open-loop control techniques, complete ZCS operation cannot be achieved without manual phase tuning, hindering widespread adoption.

This work proposes a feedback control technique that can dynamically track soft switching conditions and achieve multi-resonant ZCS operation via sensing of the inductor switch node voltage. The concept was validated on a high performance 48-to-6 V ReSC cascaded series-parallel (CaSP) converter [22], and achieved nearly 20% power loss reduction at the highest tested load (30 A) when compared with open-loop control.

This work expands on a previous ZCS autotuning technique proposed in [23] and is the first demonstration of this feedback control scheme on a high-conversion-ratio, multi-phase and multi-resonant ReSC prototype, proving its generalizable nature. Moreover, previous methods of sensing used a comparator and fixed voltage reference for the feedback loop [23], which can cause slight inaccuracies in the point of convergence at heavy loads. In this work, a more suitable analog sensing circuitry based on slope (i.e. $\frac{dv}{dt}$) detection [24] is used for increased accuracy across passive component, line, and load variations.

The remainder of the paper is organized as follows: Section II gives an overview of the structure and operation of the 8-to-1 CaSP converter. Section III analyzes the dead-time dynamics that are leveraged to detect incomplete ZCS conditions and describes the switch node voltage sensing circuit implemented. Finally, Section IV provides experimental validation of the proposed adaptive ZCS controller which achieves nearly 20% power loss reduction at the highest tested load (30 A) when compared with open-loop control.

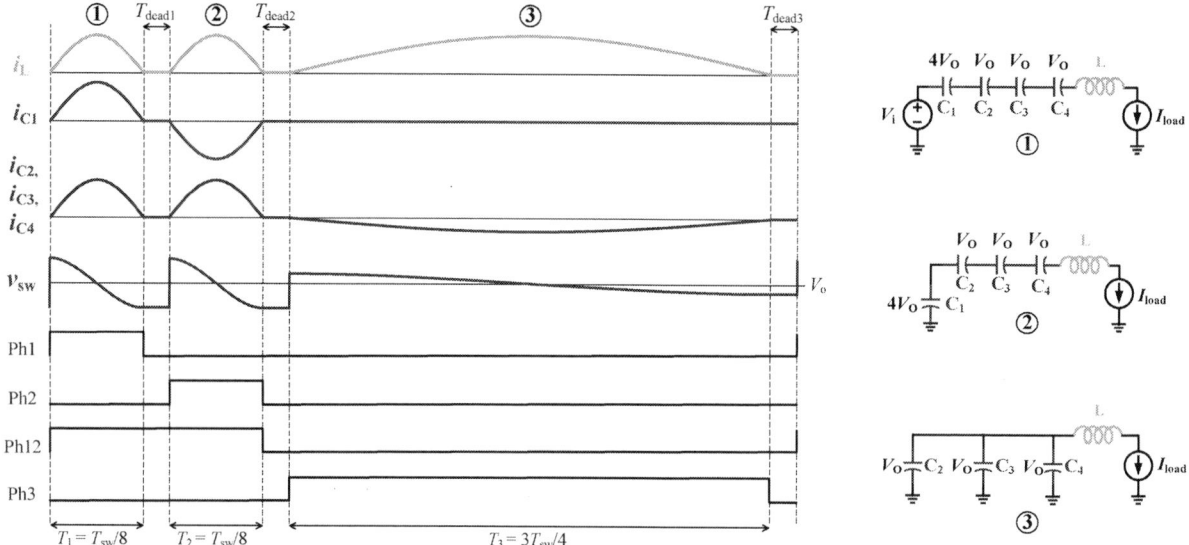

Fig. 1: Schematic drawing of the 8-to-1 cascaded series-parallel (CaSP) ReSC converter.

Fig. 2: Key waveforms of the 8-to-1 cascaded series-parallel (CaSP) converter under ideal ZCS conditions, requiring multiphase, multi-resonant operation. The deadtimes, T_{dead1}, T_{dead2} and T_{dead3}, typically on the order of tens of nanoseconds, are expanded in this figure for better visibility.

II. OPERATING PRINCIPLES

The schematic drawing of the CaSP converter employed in this work is shown in Fig. 1. It comprises a 2-to-1 switched-capacitor stage cascaded with a 4-to-1 series-parallel stage. By having more than two operating circuit states (here referred to as multi-phase operation), this topology is able to achieve an overall 8-to-1 conversion ratio with a lower number of components than many other two-phase ReSC converters, such as the 8-to-1 series-parallel [11], [25].

The gate signals for the converter, the inductor current (i_L), the flying capacitor currents (i_{C1} - i_{C4}) and the switch node voltage (v_{sw}) under ideal ZCS operation are shown in Fig. 2. Fig. 2 also illustrates the three operating phases of the converter, separated by deadtimes. During Phases 1 and 2, the front-end stage of the converter achieves a 2-to-1 step down.

During Phase 3, the converter operates with a 4-to-1 parallel mode operation, resulting in a continuous output voltage (V_o) of 6 V for a 48 V input (V_i).

As shown in Fig. 2, the equivalent capacitance during Phases 1 and 2 is equal to the series combination of all flying capacitors C_1 - C_4, while in Phase 3, it is equal to the parallel combination of flying capacitors C_2 - C_4. As a result, the resonant LC tank formed in Phases 1 and 2 is different than that of Phase 3, resulting in multi-resonant operation – i.e., different resonance timings across phases. With multi-resonant converters, ZCS can still be achieved by setting each phase duration to half the resonant period of its equivalent LC tank.

To ensure soft-charging operation [26] of the flying capacitors, C_2 - C_4 must be sized equally:

979-8-3315-1612-3/25 $31.00 © 2025 IEEE

$$C_2 = C_3 = C_4 = C \qquad (1)$$

Therefore, the equivalent capacitance of the LC tank formed during Phase 3 is:

$$C_2 + C_3 + C_4 = 3C, \qquad (2)$$

and the duration of Phase 3 in ZCS operation is:

$$T_3 = \pi\sqrt{3LC}. \qquad (3)$$

Although soft-charging is not contingent on the capacitance of C_1 [27], an optimum value can be chosen to achieve equal peak inductor currents across all phases. To satisfy the charge balance requirement of C_1, the charge transferred by the inductor during Phase 1 must be equal to the charge transferred by the inductor during Phase 2. Additionally, to satisfy the charge balance requirements of C_2 - C_4, the charge transferred by the inductor during Phases 1 and 2 combined must be equal to $\frac{1}{3}$ of the charge transferred by the inductor during Phase 3.

Given the equivalent capacitance of the LC tank formed during Phases 1 and 2 is:

$$\frac{1}{\frac{1}{C_1} + \frac{1}{C_2} + \frac{1}{C_3} + \frac{1}{C_4}} = \frac{1}{\frac{1}{C_1} + \frac{3}{C}}, \qquad (4)$$

the duration of Phases 1 and 2 must satisfy the following equations to achieve equal peak inductor currents across all phases:

$$T_1 = T_2 = \pi\sqrt{L\frac{1}{\frac{3}{C} + \frac{1}{C_1}}}, \qquad (5)$$

and

$$T_1 = T_2 = \frac{T_3}{6} = \pi\sqrt{\frac{LC}{12}}. \qquad (6)$$

Combining (5) and (6), the optimum value of C_1 can be solved:

$$C_1 = \frac{C_2}{9} = \frac{C_3}{9} = \frac{C_4}{9} = \frac{C}{9} \qquad (7)$$

Theoretically, the phase durations T_1, T_2 and T_3 can be set according to (3) and (6) to operate the CaSP converter with ZCS. In practice, circuits non-idealities such as parasitics, component tolerances, class II MLCC derating and aging, inductor soft saturation, line and load variations render resonance timing challenging to estimate with precision. Consequently, open-loop control techniques often result in incomplete ZCS conditions.

III. SWITCH NODE VOLTAGE SENSING AND FEEDBACK CONTROL

With incomplete ZCS, the inductor current is either positive or negative at switching transitions. Fig. 2 depicts ideal and

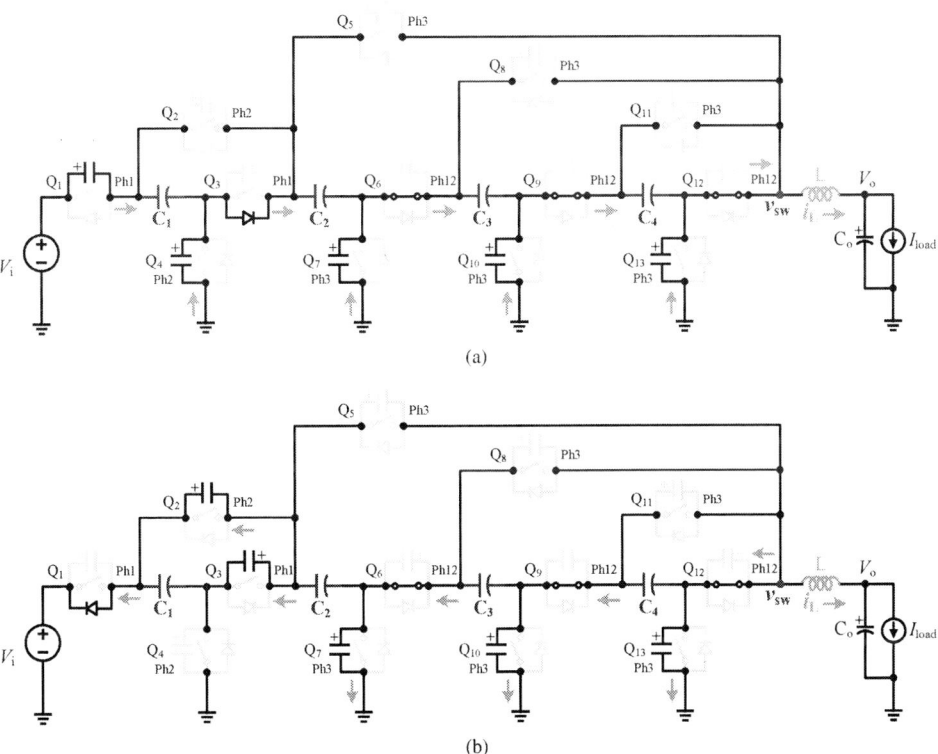

Fig. 3: Equivalent circuit of the 8-to-1 cascaded series parallel (CaSP) converter during T_{dead1}. (a) Positive turn-off inductor current. (b) Negative turn-off inductor current.

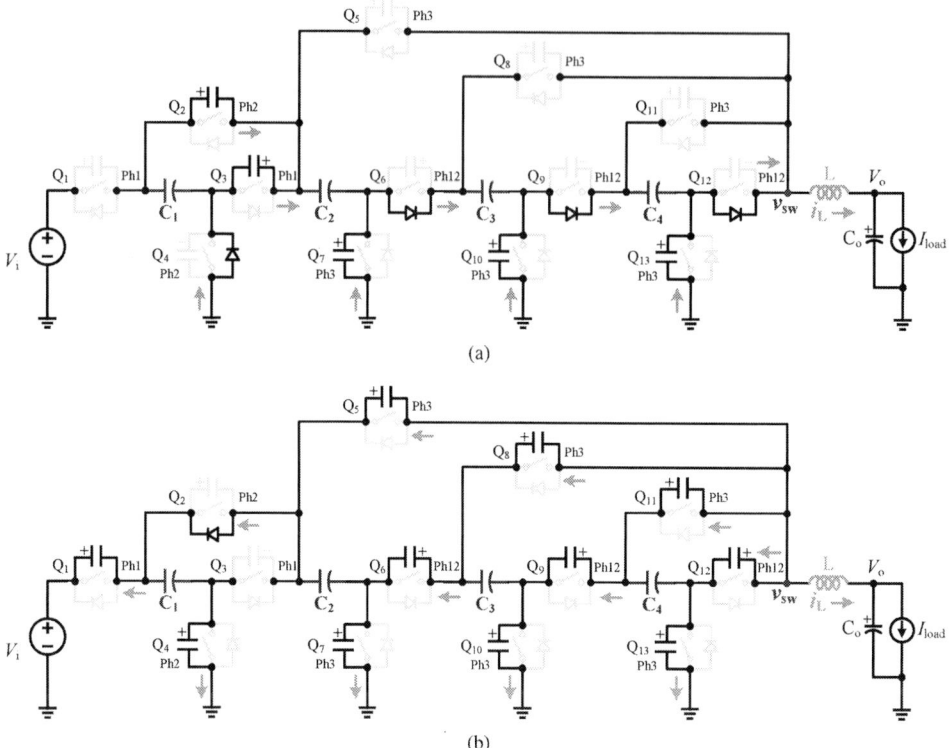

(a)

(b)

Fig. 4: Equivalent circuit of the 8-to-1 cascaded series parallel (CaSP) converter during T_{dead2}. (a) Positive turn-off inductor current. (b) Negative turn-off inductor current.

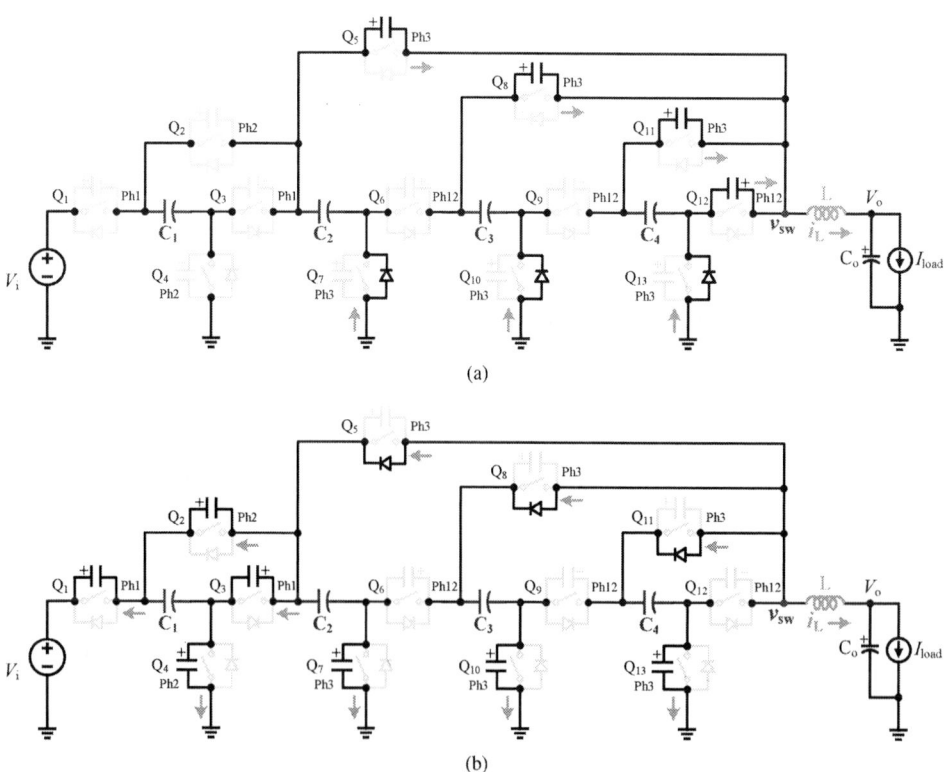

(a)

(b)

Fig. 5: Equivalent circuit of the 8-to-1 cascaded series parallel (CaSP) converter during T_{dead3}. (a) Positive turn-off inductor current. (b) Negative turn-off inductor current.

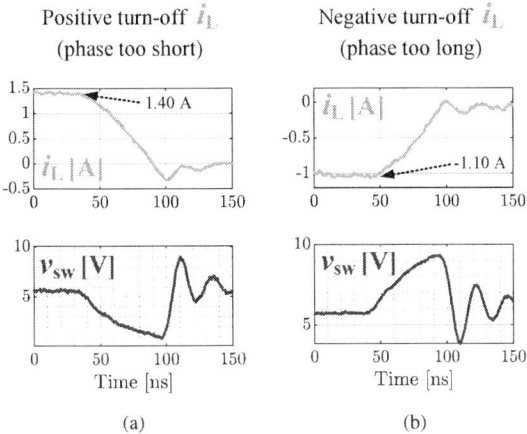

Positive turn-off i_L
(phase too short)

Negative turn-off i_L
(phase too long)

(a)

(b)

Fig. 6: Experimental waveforms of the inductor current (i_L) and the switch node voltage (v_{sw}) during the deadtime. Waveforms confirm that (a) v_{sw} decreases during the deadtime if the turn-off i_L (1.40 A) is positive, and (b) v_{sw} increases during the deadtime if the turn-off i_L (-1.10 A) is negative.

complete ZCS conditions, where the inductor current (i_L) is equal to 0 A during the deadtimes T_{dead1}, T_{dead2}, and T_{dead3}. As a result, the switch node voltage (v_{sw}) remains constant and equal to the value observed at the end of the previous phase. However, in the case of incomplete ZCS, the residual i_L can discharge or charge the C_{oss} of certain switches during T_{dead1}, T_{dead2}, and T_{dead3}. Consequently, a decrease or increase of v_{sw} will be observed.

The equivalent circuits of the converter under incomplete ZCS conditions during T_{dead1}, T_{dead2}, and T_{dead3} are shown in Fig. 3, Fig. 4, and Fig. 5 respectively. Fig. 3a, Fig. 4a, and Fig. 5a correspond to the case of positive turn-off i_L, indicating

the preceding phase was too short to achieve ZCS. Fig. 3b, Fig. 4b, and Fig. 5b correspond to the case of negative turn-off i_L, indicating the preceding phase was too long to achieve ZCS.

With positive turn-off i_L, the C_{oss} of the switches connected to the switch node will discharge during the deadtime as shown in Fig. 3a, Fig. 4a and Fig. 5a, causing a decrease of v_{sw}. On the other hand, with negative turn-off i_L, the C_{oss} of the switches connected to the switch node will charge during the deadtime as shown in Fig. 3b, Fig. 4b and Fig. 5b, causing an increase of v_{sw}. This behavior is illustrated in the experimental waveforms for i_L and v_{sw} shown in Fig. 6. The slope of the measured v_{sw} during the deadtime is indicative of the polarity of the turn-off current in the previous phase, and therefore, whether its duration was too short or long. By sensing v_{sw} during deadtimes, incomplete ZCS conditions can be detected and used to actively adjust each phase duration to achieve precise ZCS at every switching transition.

A schematic drawing of the high-bandwidth and high-speed circuit used to sense voltage deviations on v_{sw} during the nanosecond-scale deadtimes is shown in Fig. 8a. The first stage of the sensing circuit is a scaling amplifier which replicates the resistor-divided v_{sw} waveform at its output such that:

$$v_{amp} = \frac{R_2}{R_1} \cdot v_{sw} \tag{8}$$

The second stage is a differentiator amplifier whose output is determined by the slope of v_{sw}:

$$v_{slope} = V_{bias} - R_d C_d \cdot \frac{dv_{amp}}{dt} \tag{9}$$

Finally, the comparator stage outputs the digital signal feeding into the controller, here implemented using an FPGA. If v_{sw} decreases during the deadtime, indicating a positive turn-off i_L, the signal v_{comp} is triggered and the corresponding

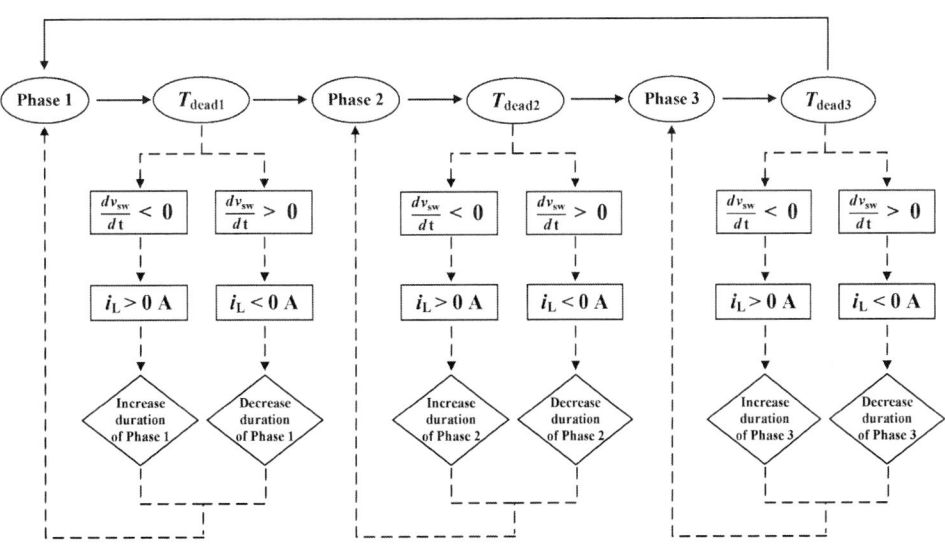

Fig. 7: Flowchart of the proposed control scheme.

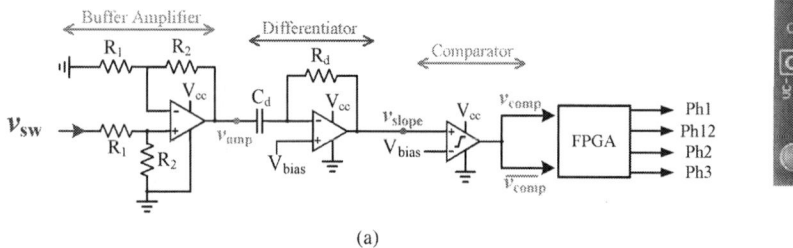
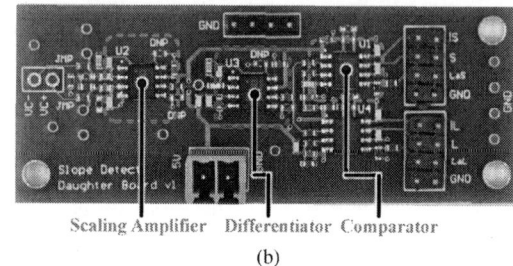

(a)

(b)

Fig. 8: High-bandwidth analog sensing circuit used to detect v_{sw} deviations during the deadtime, indicative of incomplete ZCS conditions. (a) Schematic drawing. (b) Hardware implementation as daughterboard.

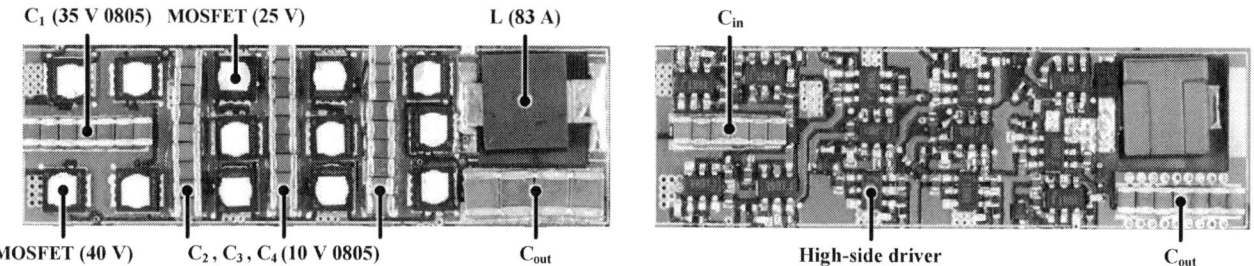

Fig. 9: Hardware prototype of the tested 48-to-6-V cascaded series parallel (CaSP) converter.

TABLE I: List of the active and passive components used in the hardware prototypes.

Component	Part number	Parameters
Switch Q_1-Q_4,Q_5, Q_7	Infineon IQE013N04LM6CG	40 V, 1.5 mΩ
Switch Q_6, Q_8-Q_{13}	Infineon IQE006NE2LM5CG	25 V, 0.65 mΩ
Flying capacitor C_1	TDK C2012X5R1V226M125AC	X5R, 35 V, 22 μF*$\times$14 (in parallel)
Flying capacitor C_2-C_4	Murrata GRM21BR61A476ME15L	X5R, 10 V, 47 μF*$\times$18 (in parallel)
Inductor L	Pulse FP0805R1-R06-R	83 A, 58 nH, 0.17 mΩ
Input capacitor C_{in}	TDK C2012X7S2A105M125AB	X7S, 100 V, 1 μF*$\times$6 (in parallel)
Output capacitor C_{out}	Murata GRM21BR61A476ME15L	X7S, 10 V, 47 μF*$\times$7 (in parallel)
	KEMET C1210C107M4PAC7800	X5R, 16 V, 100 μF*$\times$4 (in parallel)
Gate driver	Analog Devices LTC4440-5	High-side gate driver, 80 V
Bootstrap diode	ON Semiconductor NSR0340V2T1G	Schottky diode, 40 V
High-bandwidth amplifier	Analog Devices AD8091	110 MHz bandwidth, single-supply
High-speed comparator	Analog Devices AD8611	4 ns propagation delay, single-supply

* The capacitance listed in this table is the nominal value before dc bias derating.

phase will be increased gradually in the next cycles until the turn-off i_L reaches 0 A. If v_{sw} increases during the deadtime, indicating a negative turn-off i_L, the signal $\overline{v_{comp}}$ is triggered and the corresponding phase will be decreased gradually until complete ZCS is achieved. The flowchart of the proposed control scheme is summarized in Fig 7. In this application, sensing the switch-node voltage to implement the control scheme is preferred to directly sensing the inductor current, given current sensing typically incurs larger packages, lower accuracy, higher losses or costs than voltage sensing.

IV. EXPERIMENTAL VALIDATION

The switch node sensing circuit is implemented as a daughterboard as shown in Fig 8b. The amplifier AD8091 is used for the scaling and differentiator stages, and the comparator AD8611 is used for high-speed analog-to-digital conversion.

The proposed active ZCS control technique was verified on the 48-to-6 V CaSP converter prototype shown in Fig 9. The switches, flying capacitors and resonant inductor are on the top side of the board, and the gate driver circuitry is on the bottom side. A comprehensive list of the components used for the daughterboard and the power stage are shown in Table I. Operating parameters of the converter are listed in Table II.

Fig. 10 displays the key measured waveforms that verify the convergence process towards multi-resonant ZCS operation. The initial switching frequency and duty cycle for the convergence test are selected to generate incomplete ZCS operation with negative turn-off inductor currents. After convergence of the controller, the inductor current waveform demonstrates that multi-resonant ZCS operation is precisely achieved.

The full-system efficiency of the hardware prototype is shown in Fig. 11, for up to 30 A of average output current.

979-8-3315-1612-3/25 $31.00 © 2025 IEEE

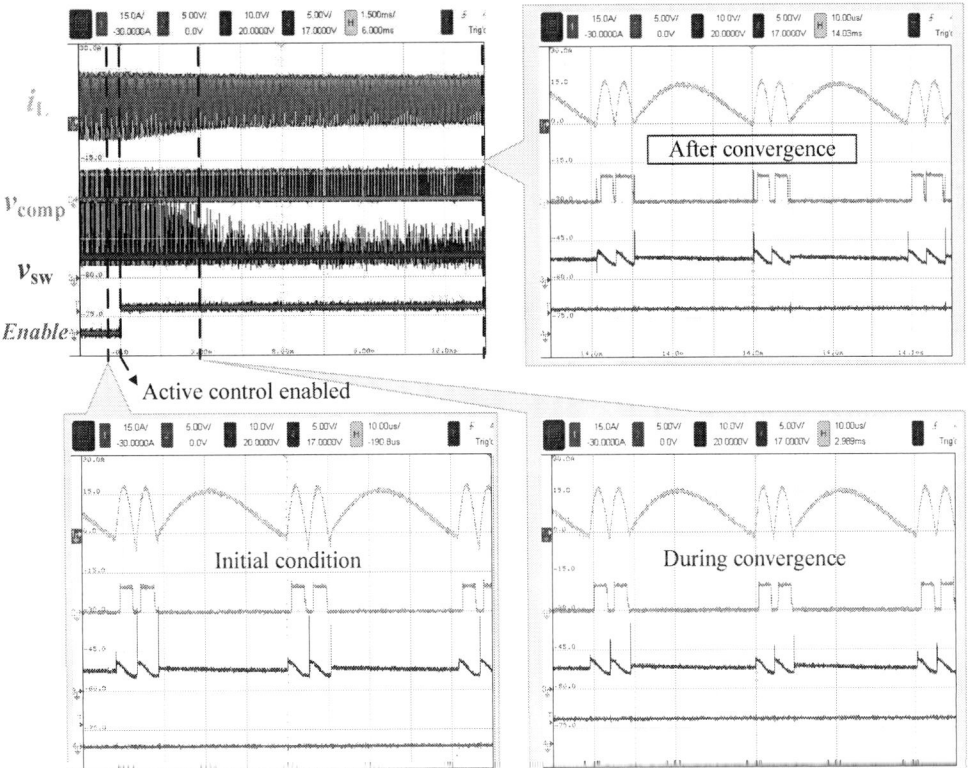

Fig. 10: Measured inductor current i_L, comparator output v_{comp}, switch-node voltage v_{sw}, and active control *Enable* signal demonstrating the convergence towards multi-resonant ZCS operation at 10.0 A of average output current.

TABLE II: Operating parameters of the 48-to-6 V CaSP Converter.

Parameter	Nominal Value	Range
Input Voltage	48 V	40 - 60 V
Output Voltage	6 V	5 - 7.5 V
Maximum Load Current	30 A	
Maximum Power	180 W	
Resonant Frequency	25 kHz [†]	20 - 30kHz

[†] The resonant frequency, seemingly low, is typical for this topology [11], [21], [22] and reflects the capacity of ReSC converters to reach high power densities at low switching frequencies.

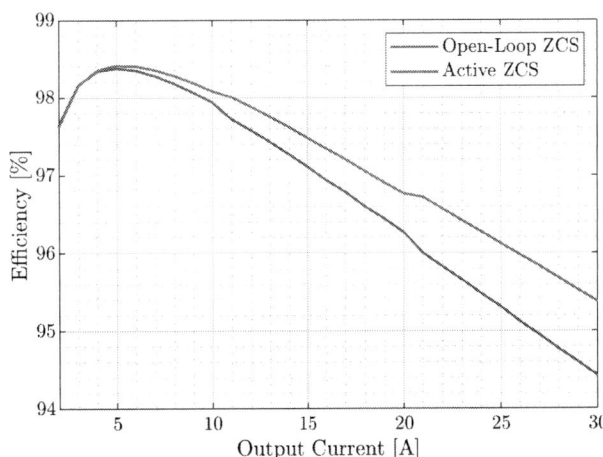

Fig. 11: Full system efficiency comparison of the hardware prototype.

Two cases are compared: open-loop ZCS and active ZCS. With open-loop ZCS, the phase durations are constant and estimated according to (3) and (5), accounting for dc bias derating of the flying capacitors and parasitic output inductance. With active ZCS, the phase durations are adjusted by the feedback controller. Efficiency measurements were obtained using a high-precision WT5000 Yokogawa power analyzer. With the proposed active control technique, nearly 20% reduction in power loss was achieved at the highest tested load when compared with open-loop control.

V. CONCLUSION

This work demonstrates an adaptive ZCS control scheme applied to a 8-to-1 multi-phase and multi-resonant cascaded series-paralell (CaSP) converter, proving the generalizable nature of this technique and its applicability to a wide range of ReSC converters. Relying on high-bandwidth and high-speed slope detection of the switch node voltage (v_{sw}) during deadtimes, the feedback controller can operate across a wider range of operating conditions with improved accuracy compared to previous works which relied on sensing of specific dc voltage levels. The technique is tested on a high-density, high-efficiency 48-to-6 V hardware prototype, operating with a nominal switching frequency of 25 kHz and up to 30 A of average output current. At full load, the closed-loop control was able to achieve nearly 20% reduction in total losses.

VI. Acknowledgment

This material is based upon work supported by the National Science Foundation Graduate Research Fellowship Program under Grant No. DGE 2146752. Any opinions, findings, and conclusions or recommendations expressed in this material are those of the authors and do not necessarily reflect the views of the National Science Foundation.

The authors acknowledge financial support from the Berkeley Power and Energy Center (BPEC).

References

[1] K. Radhakrishnan, M. Swaminathan, and B. K. Bhattacharyya, "Power delivery for high-performance microprocessors—challenges, solutions, and future trends," vol. 11, no. 4, 2021, pp. 655–671.

[2] P. Sandri, "Increasing Hyperscale Data Center Efficiency: A Better Way to Manage 54-V48-V-to-Point-of-Load Direct Conversion," *IEEE Power Electronics Magazine*, vol. 4, no. 4, pp. 58–64, 2017.

[3] M. H. Ahmed, F. C. Lee, Q. Li, M. de Rooij, and D. Reusch, "GaN Based High-Density Unregulated 48 V to x V LLC Converters with ??? 98% Efficiency for Future Data Centers," in *PCIM Europe 2019; International Exhibition and Conference for Power Electronics, Intelligent Motion, Renewable Energy and Energy Management*, 2019, pp. 1–8.

[4] S. Lu, H. Ma, J. Yi, X. Li, Y. Pan, and J. Xu, "A novel two-stage solution for low-power 48-V voltage regulator module (VRM) using high step-down hybrid LLC with integrated planar transformer," *International Journal of Circuit Theory and Applications*, vol. 52, pp. 1714–1732, 11 2023.

[5] H. Gan, S. Jiang, S. Teng, S. Yamamoto, V. Chivukula, B. Edwards, C. Chung, J. Chen, M. Mohideen, G. Sizikov, and X. Li, "Vertical Power Delivery for 1000 Amps Machine Learning ASICs," in *2024 IEEE Applied Power Electronics Conference and Exposition (APEC)*, 2024, pp. 906–909.

[6] N. C. Brooks, J. Zou, S. Coday, T. Ge, N. M. Ellis, and R. C. N. Pilawa-Podgurski, "On the Size and Weight of Passive Components: Scaling Trends for High-Density Power Converter Designs," *IEEE Transactions on Power Electronics*, vol. 39, no. 7, pp. 8459–8477, 2024.

[7] Y. Yeung, K. Cheng, S. Ho, K. Law, and D. Sutanto, "Unified Analysis of Switched-Capacitor Resonant Converters," *IEEE Transactions on Industrial Electronics*, vol. 51, no. 4, pp. 864–873, 2004.

[8] M. Shoyama, T. Naka, and T. Ninomiya, "Resonant switched capacitor converter with high efficiency," in *2004 IEEE 35th Annual Power Electronics Specialists Conference (IEEE Cat. No.04CH37551)*, vol. 5, 2004, pp. 3780–3786 Vol.5.

[9] S. Jiang, S. Saggini, C. Nan, X. Li, C. Chung, and M. Yazdani, "Switched Tank Converters," *IEEE Transactions on Power Electronics*, vol. 34, no. 6, pp. 5048–5062, 2019.

[10] S. Webb and Y.-F. Liu, "A Zero Inductor-Voltage 48V to 12V/70A Converter for Data Centers with 99.1% Peak Efficiency and 2.5kW/in3 Power Density," in *2020 IEEE Applied Power Electronics Conference and Exposition (APEC)*, 2020, pp. 1858–1865.

[11] Z. Ye, R. A. Abramson, T. Ge, and R. C. Pilawa-Podgurski, "Multi-Resonant Switched-Capacitor Converter: Achieving High Conversion Ratio With Reduced Component Number," *IEEE Open Journal of Power Electronics*, vol. 3, pp. 492–507, 2022.

[12] R. Rizzolatti, C. Rainer, S. Saggini, and M. Ursino, "High Density Hybrid Switched Capacitor Converter for Data-Center Application," in *2021 IEEE Applied Power Electronics Conference and Exposition (APEC)*, 2021, pp. 1288–1293.

[13] Z. Tian, Y. Guan, W. Wang, and D. Xu, "Research on High Efficiency and High Density 48 V-5 V Multi-Resonant Switched Capacitor Converter," *CPSS Transactions on Power Electronics and Applications*, vol. 7, no. 3, pp. 229–238, 2022.

[14] A. Dago, M. Leoncini, S. Saggini, S. Levantino, and M. Ghioni, "Hybrid Resonant Switched-Capacitor Converter for 48–3.4 V Direct Conversion," *IEEE Transactions on Power Electronics*, vol. 37, no. 11, pp. 12 998–13 002, 2022.

[15] M. Qiu, M. Wei, X. Liu, H. Meng, and D. Cao, "A Matrix Auto-transformer Switched-Capacitor Converter for Data Center Application," *IEEE Transactions on Power Electronics*, vol. 38, no. 12, pp. 14 982–14 999, 2023.

[16] H. Wu, Y. Zhang, and Z. Li, "Hybrid Resonant Converter-Based 8:1 Bus Converter With 3.5 kW/in3 and 98.6%-Efficient for 48 V Data-Center Power Systems," *IEEE Transactions on Power Electronics*, vol. 39, no. 1, pp. 36–41, 2024.

[17] Y. Han, W. Hu, Y. Yuan, Z. Zhang, and F. Blaabjerg, "Multiratio-Multiresonance Switched-Capacitor Step-Down Converter for DC Bus Voltage Conversion," *IEEE Transactions on Power Electronics*, vol. 38, no. 12, pp. 16 180–16 195, 2023.

[18] Y. Guan, X. Li, W. Wang, B. Li, W. Yang, Y. Wang, and D. Xu, "A High-Performance 3:1 Conversion Ratio DC–DC Converter: Analysis Method and Modular Adoption," *IEEE Transactions on Power Electronics*, vol. 39, no. 4, pp. 4412–4425, 2024.

[19] D. Cao and F. Z. Peng, "A family of zero current switching switched-capacitor dc-dc converters," in *2010 Twenty-Fifth Annual IEEE Applied Power Electronics Conference and Exposition (APEC)*, 2010, pp. 1365–1372.

[20] W. Xie, S. Li, Y. Zheng, K. M. Smedley, J. Wang, Y. Ji, and J. Yu, "A Family of Step-Up Series–Parallel Dual Resonant Switched-Capacitor Converters With Wide Regulation Range," *IEEE Transactions on Power Electronics*, vol. 35, no. 3, pp. 2724–2736, 2020.

[21] R. A. Abramson, Z. Ye, and R. C. Pilawa-Podgurski, "A High Performance 48-to-8 V Multi-Resonant Switched-Capacitor Converter for Data Center Applications," in *2020 22nd European Conference on Power Electronics and Applications (EPE'20 ECCE Europe)*, 2020, pp. 1–10.

[22] R. A. Abramson, Z. Ye, T. Ge, and R. C. Pilawa-Podgurski, "A High Performance 48-to-6 V Multi-Resonant Cascaded Series-Parallel (CaSP) Switched-Capacitor Converter," in *2021 IEEE Applied Power Electronics Conference and Exposition (APEC)*, 2021, pp. 1328–1334.

[23] H. B. Sambo, Y. Zhu, T. Ge, N. M. Ellis, and R. C. N. Pilawa-Podgurski, "Autotuning of Resonant Switched-Capacitor Converters for Zero Current Switching and Terminal Capacitance Reduction," in *2023 IEEE Applied Power Electronics Conference and Exposition (APEC)*, 2023, pp. 1217–1224.

[24] R. A. Abramson, S. Krishnan, M. E. Blackwell, and R. C. N. Pilawa-Podgurski, "An Active Split-Phase Control Technique for Hybrid Switched-Capacitor Converters Using Capacitor Voltage Discontinuity Detection," in *2023 IEEE 24th Workshop on Control and Modeling for Power Electronics (COMPEL)*, 2023, pp. 1–7.

[25] M. Makowski, "Realizability conditions and bounds on synthesis of switched-capacitor DC-DC voltage multiplier circuits," *IEEE Transactions on Circuits and Systems I: Fundamental Theory and Applications*, vol. 44, no. 8, pp. 684–691, 1997.

[26] R. C. Pilawa-Podgurski, D. M. Giuliano, and D. J. Perreault, "Merged two-stage power converter architecture with soft charging switched-capacitor energy transfer," in *2008 IEEE Power Electronics Specialists Conference*, 2008, pp. 4008–4015.

[27] N. Patle, R. A. Abramson, R. K. Iyer, and R. C. N. Pilawa-Podgurski, "A Ripple-Equivalent Circuit-Based Method for Analyzing Soft-Charging Operation in Hybrid Switched-Capacitor Converters," in *2024 ECCE IEEE Energy Conversion Congress and Expo*, 2024.

Small-Signal Analysis and External Ramp Design for Multiphase Current-Mode Constant On-Time Control with Phase Overlapping

Sundaramoorthy Sridhar, Qiang Li

Center for Power Electronic Systems
The Bradley Department of Electrical and Computer Engineering
Virginia Tech, Blacksburg, VA, USA
sundaramoorthy@vt.edu

Abstract—High-current processor core voltage regulators (VRs) use current-mode constant on-time (COT) control to simplify phase interleaving and achieve high control bandwidth. Reducing VR bus voltage and increasing phase count results in steady-state phase overlapping within their operating duty range. With phase overlap, a smaller external ramp reduces the inner current loop stability margin and introduces large magnitude peaks in control-to-output response. These peaks could lead to double crossover, reduce gain margin, and make high-bandwidth outer voltage loop unstable in phase overlapping regions. Increasing the ramp damps these peaks and improves the gain margin but reduces the phase margin at the target bandwidth. Hence, this paper simplifies infinite-order describing function model and provides external ramp design guidelines to attain desired stability margin at the target bandwidth. The simplified model and small-signal analysis are validated using SIMPLIS simulation and experiments from a six-phase current-mode COT buck converter.

Index Terms—Multiphase current-mode COT control, phase overlapping, small-signal model, external ramp, design guideline

I. INTRODUCTION

The increasing current demands and decreasing voltage tolerance bands (TOBs) for today's multi-core CPUs and GPUs introduce new trends in multiphase buck VR design. First, with processor currents reaching 1000A [1], [2], the VR phase count is drastically increased due to inductor size and DrMOS current constraints [1]. For example, reference [1] shows that the phase count today reaches as high as 32. Second, the intermediate bus voltage (or VR input voltage) is reduced to increase duty cycle, push switching frequency and achieve high power density without sacrificing efficiency. The conventional 12V input [3] is now reduced to as low as 3.3 V for integrated VRs [4], [5].

To regulate processor core voltage within TOBs, symmetrically interleaved multiphase current-mode COT VRs (shown in Fig. 1) is widely used [6]. Compared to constant frequency VRs, current-mode COT VRs have smaller switching action delay and achieve higher control bandwidth [7], [8]. Also, they eliminate phase-locked loops (PLLs) and simplify phase interleaving [9], [10]. If this VR has N phases then COT pulses of two or more phases do not overlap at steady-state provided duty cycle $0 < D < 1/N$. Hence, this duty range is referred as "no-phase overlapping" region. However, if duty cycle $1/N < D < 1$ then COT pulses of two or more phases overlap at steady-state. Hence, the duty range $1/N < D < 1$ is

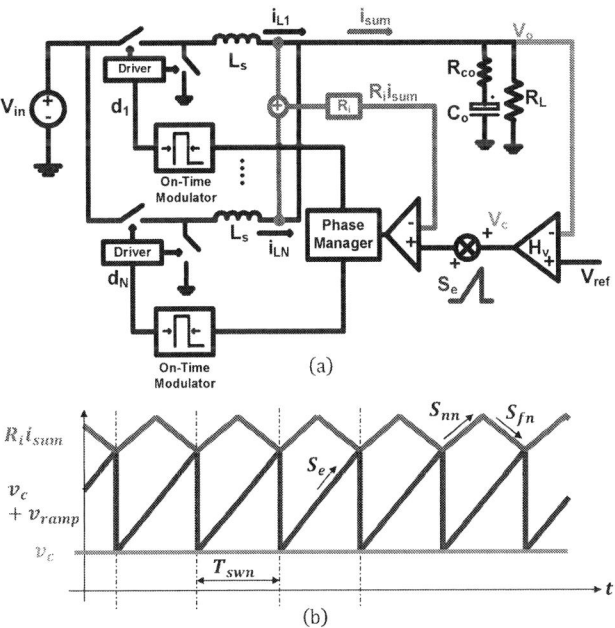

Fig. 1. Multiphase current-mode COT control based on pulse distribution method. (a) Circuit diagram and (b) steady-state modulator waveforms.

simply referred as "phase overlapping region". As mentioned before, the increasing phase count and reducing VR bus voltage make steady-state phase overlapping inevitable in practical multiphase COT buck VRs. For example, a sixteen-phase interleaved VR used for 12V to 1V conversion must operate with phase overlap because its duty cycle $D = 1/12 > 1/16$. In single-phase current-mode COT control, the current loop is stable for all duty cycles without the external ramp [7], [11]. However, the total current loop of multiphase current-mode COT control (shown in Fig. 1) becomes unstable when its duty cycle $D > 1/N$ (or with phase overlapping) [12]. Hence, multiphase current-mode COT VRs need external ramp to preserve stability and improve noise immunity.

To study external ramp effects on outer voltage loop stability margin, Tian et al. [9] provided an equivalent circuit model for multiphase current-mode COT control. Later, Li et al. [13] extended Tian's analysis [9] to multiphase buck converters with coupled inductors. The models from [9] and [13] were derived based on the assumption that the control waveforms for multiphase and single-phase current-mode COT

979-8-3315-1612-3/25 $31.00 © 2025 IEEE

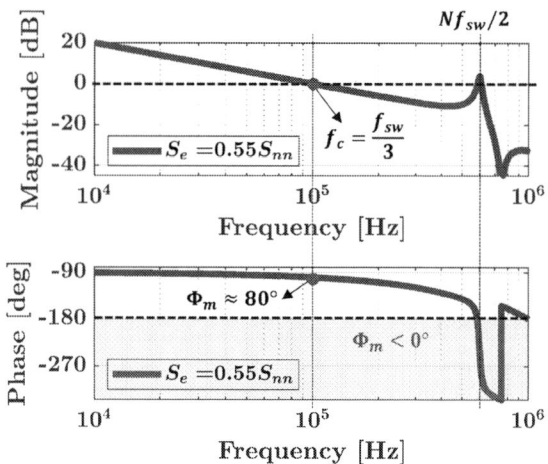

Fig. 2. Outer voltage loop gain of a four phase COT buck converter with two-phase overlap ($D = 0.4$) and $S_e = 1.1S_{ec}$ where critical ramp $S_{ec} = S_{nn}/2$. Circuit parameters: $V_{in} = 10$V; $f_{sw} = 300$ kHz; $L_s = 150$ nH; $R_i = 5$ mΩ.

buck converters have identical modulation principle. However, total current feedback signal exhibits COT modulation only when $D < 1/N$ for an N phase buck VR. When duty cycle $D > 1/N$, phase interleaving forces the total current to exhibit constrained variable on-time and off-time modulation instead of COT modulation. Hence, the approach of extending single-phase COT control models for multiphase design is valid only when duty cycle $D < 1/N$.

Recently, reference [14] provided a general multiphase current-mode COT control model valid in both no-phase overlapping ($D < 1/N$) and phase-overlapping region ($D > 1/N$). This model predicted the inner total current loop instability for $D > 1/N$ (or with phase overlap) and provided the critical ramp for its stability [12]. However, it is possible that the system can become unstable while closing outer voltage loop even with a stable inner current loop [15]–[17]. For example, Fig. 2 shows the voltage loop gain in two-phase overlapping region with external ramp slightly greater than the critical ramp for inner current loop. In this scenario, current loop is stable but outer voltage loop becomes unstable because of double crossover at half the effective switching frequency, i.e., $f_{swn}/2 = Nf_{sw}/2$. Hence, external ramp must be carefully chosen for high-bandwidth designs. This paper first simplifies the exact control-to-output transfer function from [14] using pade approximation for the two-phase overlapping case. Then, relationships between the external ramp, outer voltage loop bandwidth, gain margin, and phase margin are derived. Finally, design guideline is proposed using these relationships to achieve the desired gain and phase margin at the target bandwidth.

II. DESCRIBING FUNCTION MODEL

After closing inner total current loop, the overall outer voltage loop gain can be written as:

$$T_v(s) = H_v(s)G_{vc}(s) \qquad (1)$$

where $G_{vc}(s)$ is the control-to-output transfer function. The exact control-to-output transfer function based on describing function (DF) method [14] is given below:

$$G_{vc}(s) = \frac{\hat{v}_o(s)}{\hat{v}_c(s)} = \frac{Y_{vc}(s)Z_{eq}(s)}{1 - Y_{vo}(s)Z_{eq}(s)} \qquad (2)$$

where output filter impedance $Z_{eq} = R_L||(R_{co}+1/sC_o)$. The control DF, $Y_{vc}(s)$, and output DF, $Y_{vo}(s)$, are given below:

$$Y_{vc}(s) = \frac{\hat{i}_{sum}(s)}{\hat{v}_c(s)} = \frac{V_{in}f_{swn}(1 - e^{-sT_{on}})}{sL_s\Delta(s)}, \qquad (3a)$$

$$Y_{vo}(s) = \frac{\hat{i}_{sum}(s)}{\hat{v}_o(s)} = \left(\frac{N}{sL_s}\right)(R_iY_{vc}(s) - 1). \qquad (3b)$$

The characteristic polynomial $\Delta(s)$ is given by:

$$\Delta(s) = S_{fn}+S_e\left(1 - e^{-sT_{swn}}\right)+(S_{nn}+S_{fn})\sum_{u=1}^{M-1}e^{-usT_{swn}}. \qquad (4)$$

Referring to Fig. 1(b) and (4), $T_{swn} = 1/f_{swn}$ denotes the total current period (or) effective switching period. The terms $S_{nn} = R_i(MV_{in} - NV_o)/L_s$ and $S_{fn} = R_i(NV_o - (M-1)V_{in})/L_s$ denotes the sensed total current rising slope and falling slope, respectively. Also, M denotes the number of phase overlapped. If duty cycle $(M-1)/N < D < M/N$ where $M = 2, 3, \cdots, N$ then COT pulses of M phases overlap in one effective switching period. Hence, the duty range $(M-1)/N < D < M/N$ can be called "M-phase overlapping region". For no-phase overlapping case (or $D < 1/N$), small-signal analysis and compensation design guideline were discussed in [9]. From a practical standpoint, today's commercial controllers cannot overlap large number of phases at steady-state because of duty cycle constraints imposed by integrated circuit (IC) under-voltage lock-out (UVLO). Hence, this paper discusses the outer voltage-loop gain design for two-phase overlapping case in detail.

III. SMALL-SIGNAL ANALYSIS IN TWO-PHASE OVERLAPPING REGION

A. Control-to-output transfer function

With two-phase overlapping ($1/N < D < 2/N$), one can reduce the control DF, $Y_{vc}(s)$, and the output DF, $Y_{vo}(s)$, using modified Pade approximation [12] as follows:

$$Y_{vc}(s) \approx \frac{1}{R_i}\left(\frac{1 + \dfrac{s}{Q_1\omega_1} + \dfrac{s^2}{\omega_1^2}}{1 + \dfrac{s}{Q_{se}\omega_1} + \dfrac{s^2}{\omega_1^2}}\right)\frac{1}{1 + \dfrac{s}{Q_1\omega_o} + \dfrac{s^2}{\omega_o^2}} \qquad (5a)$$

$$Y_{vo}(s) \approx \frac{T_{sw}}{2L_s}\left(1 - ND - \frac{2}{\pi Q_{se}}\right)\bigg/\left(1 + \frac{s}{Q_{se}\omega_{se}}\right) \qquad (5b)$$

Here, the double pole and zero locations due to COT and external ramp in (5a) are $\omega_o = \pi/T_{on}$ and $\omega_1 = \pi/T_{swn}$. The Q-factors of the pole pairs in (5a) are:

$$Q_{se} = \frac{2}{\pi}\left(\frac{2S_{fn} + S_{nn}}{2S_e - S_{nn}}\right) \qquad (6)$$

$$Q_1 = 2/\pi \qquad (7)$$

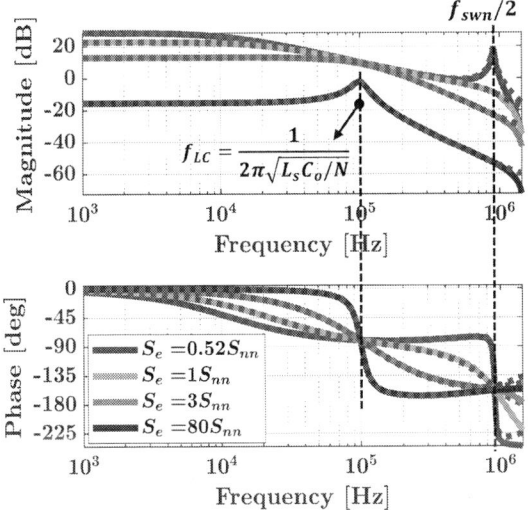

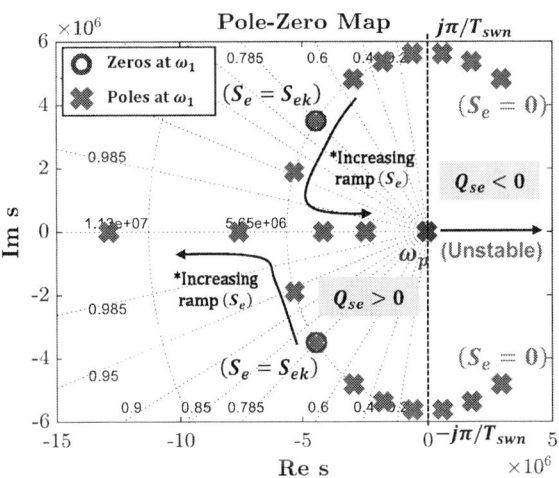

Fig. 4. Pole-zero map of control-to-output voltage transfer function in two-phase overlapping region ($1/N < D < 2/N$) with increasing external ramp.

Fig. 3. Control-to-output response in two-phase overlapping region ($1/N < D < 2/N$) with different external ramp. Circuit parameters: V_{in} = 10V; N = 6; $D = 0.2$; f_{sw} = 300 kHz; L_s = 150 nH; R_i = 5 mΩ; R_{co} = 1.4 mΩ; C_o = 100 μF; R_L = 200 mΩ.

Fig. 3 shows the bode plots of control-to-output transfer function with different ramp slopes. In Fig. 3, red dash lines and solid lines shows the response obtained by substituting reduced order (5) and the infinite order results (3) in (2), respectively. The magnitude plots show large resonant peaks at half the steady-state effective switching frequency ($f_{swn}/2 = Nf_{sw}/2$) with smaller ramp slopes. This behavior is very similar to the control-to-output response of peak-current controlled buck converters operating with duty cycle close to 0.5 without external ramp [15], [16]. Referring to Fig. 3, larger external ramp slopes over-damps the system and reduces the high-frequency control-to-output transfer function phase. Hence, $Q_{se} < 0.5$ is not recommended in practical designs. For $Q_{se} > 0.5$, one can simplify the describing function result (2) using (5) as:

$$G_{vc}(s) \approx K_c \frac{1 + sR_{co}C_o}{1 + s/\omega_p} Y_{vc}(s). \quad (8)$$

where the dc gain and the low-frequency pole in (8) are

$$K_c = \frac{R_L/R_i}{1 - R_Lk_2/R_i}, \quad (9a)$$

$$\omega_p = \frac{1}{R_LC_o}\frac{1 - R_Lk_2/R_i}{1 + R_{co}/R_L - R_{co}k_2/R_i}, \quad (9b)$$

$$k_2 \approx \frac{R_iT_{sw}}{2L_s}\left(1 - ND - \frac{2}{\pi Q_{se}}\right). \quad (9c)$$

B. Effect of external ramp on pole-zero movement

Fig. 4 shows the poles and zero locations of transfer function (5a) with increasing external ramp excluding the COT double pole at $\omega_o = \pi/T_{on}$. Without external ramp, i.e., $S_e = 0$, the high-frequency poles at $\omega_1 = \pi/T_{swn}$ would have negative Q-factor as shown in (10).

$$Q_{se} = -2(2S_{fn} + S_{nn})/(S_{nn}\pi) < 0 \quad (10)$$

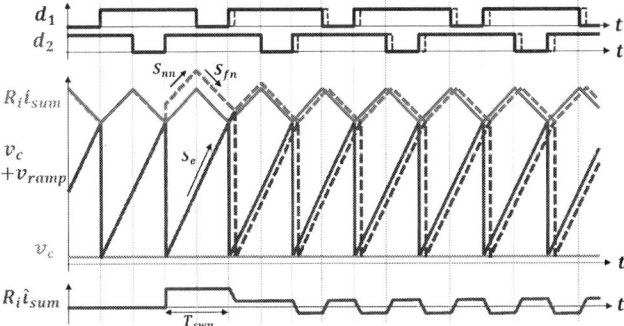

Fig. 5. Perturbed total current waveform in two-phase overlapping region ($1/N < D < 2/N$) with $S_e = S_{nn} + S_{fn}$. Here, a two-phase buck converter with two-phase overlap is taken as an example. After one effective switching cycle, the total current reaches steady-state similar to single-phase current-mode COT control without external ramp.

Hence, total current loop becomes unstable without external ramp when duty cycle $1/N < D < 2/N$. As external ramp increases, the poles will start moving towards left-half plane. When $S_e = S_{nn}/2$, $Q_{se} = \infty$ and the system poles lie on the imaginary axis. When $S_e > S_{nn}/2$, $Q_{se} > 0$, the system poles will have a negative real part and total current loop will become stable. Hence, the critical ramp for inner current loop with two-phase overlap is $S_{ec} = S_{nn}/2$ [12].

Referring to Fig. 4 and (5a), if $Q_{se} = Q_1 = 2/\pi$ then the double pole and double zero at $\omega_1 = \pi/T_{swn}$ in transfer function (5a) will cancel each other and

$$Y_{vc}(s) = \frac{\hat{i}_{sum}(s)}{\hat{v}_c(s)} \approx \frac{1}{R_i}\frac{1}{1 + sT_{on}/2 + s^2T_{on}^2/\pi^2}. \quad (11)$$

The transfer function (11) matches the control-to-inductor current transfer function of single-phase current-mode COT control without external ramp [7]. Since multiphase and single-phase COT control dynamics becomes identical with $Q_{se} = 2/\pi$, the external ramp (S_e) for which $Q_{se} = 2/\pi$ can be considered as optimal external ramp for two-phase overlapping

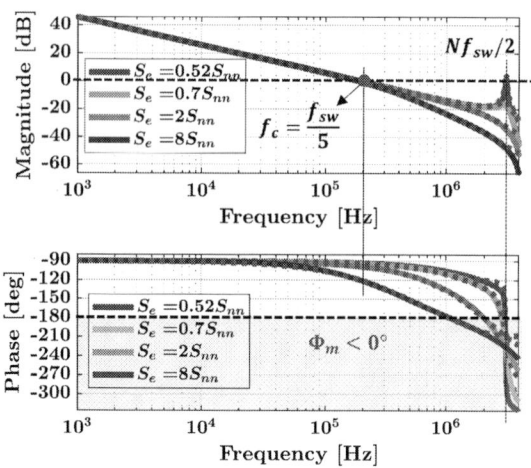

Fig. 6. Outer voltage loop gain (T_v) in two-phase overlapping region ($1/N < D < 2/N$) for six-phase converter with different external ramp and fixed bandwidth. Circuit parameters: $V_{in} = 8$V; $N = 6$; $D = 0.25$; $f_{sw} = 1$ MHz; $L_s = 150$ nH; $R_i = 5$ mΩ; $R_{co} = 1.4$ mΩ; $C_o = 100$ μF; $R_L = 200$ mΩ.

region. Using (6), this optimal external ramp was found to be:

$$S_{ek} = S_{nn} + S_{fn} \tag{12}$$

Fig. 5 shows the time-domain response of total current for small perturbations with external ramp $S_e = S_{ek}$. The optimal ramp S_{ek} eliminates the external ramp dynamic in the control-to-total current transfer function (3a). Hence, the total current error reaches steady-state within one effective switching cycle and exhibits dynamics identical to single-phase current-mode COT control without an external ramp.

C. Outer Voltage Loop Gain Design

As shown in Fig. 3 and Fig. 4, the quality factor Q_{se} can be made smaller by increasing external ramp and the high-frequency poles that originate after closing total current loop can be damped. Thus, eliminating a potential instability with high-bandwidth designs. However, introducing too much external ramp reduces the control-to-output response phase at high frequencies (above one third of switching frequency). Generally, current-mode controllers use Type-II compensation of the form (13) for output voltage regulation. These compensators cannot provide phase boost at high-frequency. Consequently, the maximum phase margin that can be achieved using Type-II compensator is reduced when a large external ramp is used. Hence, external ramp design must consider outer voltage loop stability margin requirements.

$$H_v(s) = \frac{H_{v0}}{s}\left(\frac{1 + s/\omega_{zc}}{1 + s/\omega_{pc}}\right) \tag{13}$$

Referring to (1), (8), and (13), the compensator zero (ω_{zc}) and the compensator pole (ω_{pc}) are typically chosen to cancel the power-stage pole (ω_p) and capacitor equivalent series resistance (ESR) zero ($\omega_{esr} = 1/R_{co}C_o$) in (8), respectively. After performing this cancellation, the gain $H_{v0} = \omega_c/K_c$ can be adjusted to achieve the desired voltage-loop bandwidth

$\omega_c = 2\pi f_c$. For $Q_{se} > 1$, sensitivity of k_2 and hence the low-frequency pole (ω_p) and dc gain (K_c) of transfer function (8) to external ramp variation is negligible. Hence, one can approximate high frequency outer voltage loop gain for two-phase overlapping case as:

$$T_v(s) \approx \frac{\omega_c}{s} \frac{1 + \dfrac{s}{Q_1\omega_1} + \dfrac{s^2}{\omega_1^2}}{\left(1 + \dfrac{s}{Q_{se}\omega_1} + \dfrac{s^2}{\omega_1^2}\right)\left(1 + \dfrac{s}{Q_1\omega_o} + \dfrac{s^2}{\omega_o^2}\right)} \tag{14}$$

Fig. 6 shows the outer voltage loop gain $T_v(s)$ for different ramp slopes by fixing bandwidth at $f_c = f_{sw}/5$. In Fig. 6, solid lines and red dash lines shows the comparison between exact loop gain (1) and approximate loop gain (14). Smaller external ramp introduces double crossover in the loop gain $T_v(s)$, makes the gain margin negative, and leads to an unstable outer voltage loop. However, larger ramp slopes decrease the phase margin at the target bandwidth. Hence, an external ramp range exists for a given bandwidth within which both gain and phase margin requirements can be met simultaneously. To determine this range, this section simplifies (14) and studies the gain and phase margin dependence on external ramp for given bandwidth.

1) External Ramp Effect on Gain Margin: The voltage loop gain margin can be defined as:

$$G_m \text{(in dB)} = -20\log_{10}|T_v(j2\pi f_{pc}^v)| \tag{15}$$

where $f_{pc}^v = \frac{\omega_1}{2\pi} = Nf_{sw}/2$ denotes the phase crossover frequency of loop gain $T_v(s)$. Substituting $f_{pc}^v = Nf_{sw}/2$ in (14), the magnitude of loop gain $T_v(s)$ at its phase crossover frequency was found to be:

$$|T_v(j2\pi f_{pc}^v)| \approx \left(\frac{\pi}{N^3D^2}\right)\left(\frac{f_c}{f_{sw}}\right)Q_{se}. \tag{16}$$

Substituting (16) in (15), one can relate Q_{se} to the gain margin (G_m) for a given bandwidth (f_c) as follows:

$$Q_{se}(f_c, G_m) = \left(\frac{f_{sw}}{f_c}\right)\left(\frac{N^3D^2}{\pi}\right)10^{-G_m\text{(in dB)}/20} \tag{17}$$

Using (6), one can rewrite (17) in terms of external ramp slope as follows:

$$S_e = \frac{S_{nn}}{2} + \left(\frac{f_c}{f_{sw}}\right)\left(\frac{2S_{fn} + S_{nn}}{N^3D^2}\right)10^{(G_m/20)} \tag{18}$$

Equation (18) is the first design equation which relates gain margin and external ramp for given bandwidth. Fig. 7 plots this equation for different bandwidth. As shown in Fig. 7, increasing ramp slope increases the gain margin of the system. By substituting $G_m = 0$ in (18), the critical external ramp with closed voltage loop (S_{ec}^{vc}) can be obtained as:

$$S_{ec}^{vc} = \frac{S_{nn}}{2} + \left(\frac{f_c}{f_{sw}}\right)\left(\frac{2S_{fn} + S_{nn}}{N^3D^2}\right) \tag{19}$$

Referring to (19), if bandwidth f_c is much lower than switching frequency (f_{sw}) then critical ramp with closed voltage loop $S_{ec}^{vc} \approx S_{nn}/2$ and becomes same as critical ramp for inner current loop.

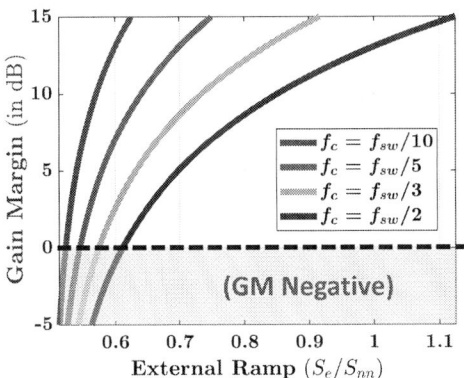

Fig. 7. External ramp vs. gain margin (GM) for different control bandwidth for a six-phase converter operating with $D = 0.25$.

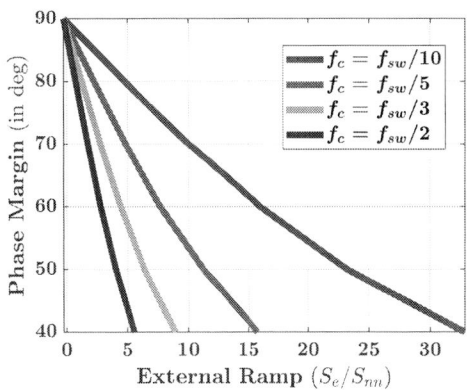

Fig. 8. External ramp vs. phase margin for different control bandwidth for a six-phase converter operating with $D = 0.25$.

2) External Ramp Effect on Phase Margin: The outer voltage loop gain phase margin can be defined as follows:

$$\Phi_m = 180° + \angle T_v(j\omega_c) \quad (20)$$

If bandwidth $f_{sw}/10 \leq f_c \leq f_{sw}/2$ then using (14) and (20), we can relate the phase margin with bandwidth (f_c) and the quality factor (Q_{se}) as follows:

$$Q_{se}(f_c, \Phi_m) = \frac{2f_c}{Nf_{sw}} \left(1 - \frac{4f_c^2}{N^2 f_{sw}^2}\right)^1 \tan(\Phi_m - \Phi_{uc}) \quad (21)$$

where the phase lag at target bandwidth Φ_{uc} equals:

$$\Phi_{uc} = \tan^{-1} \frac{\pi f_c/f_{sw}}{N\left(1 - \frac{4f_c^2}{N^2 f_{sw}^2}\right)} - \tan^{-1} \frac{\pi f_c T_{on}}{1 - 4f_c^2 T_{on}^2} \quad (22)$$

Using (21) and (6), the external ramp and phase margin can be related as follows:

$$S_e = \frac{S_{nn}}{2} + \frac{1}{\pi}\left(\frac{2S_{fn} + S_{nn}}{Q_{se}(f_c, \Phi_m)}\right) \quad (23)$$

Equation (23) is the second design equation which relates phase margin and external ramp for given bandwidth. Fig. 8 plots (23) and shows the dependence of phase margin on external ramp for different bandwidth. As seen in Fig. 6 and Fig. 8, adding more external ramp reduces the phase margin at target bandwidth.

D. External Ramp Design Example

This section presents a design example based on the equations (18) and (23) by taking a six-phase converter operating with two-phase overlap. The other circuit parameters used are: $V_{in} = 8V$; $N = 6$; $D = 0.25$; $f_{sw} = 1$ MHz; $L_s = 150$ nH; $R_i = 5$ mΩ; $R_{co} = 1.4$ mΩ; $C_o = 100$ μF; $R_L = 200$ mΩ. Assume that a bandwidth $f_c = f_{sw}/3$, gain margin $G_m \geq G_m^{min} = 10$dB and phase margin $\Phi_m \geq \Phi_m^{min} = 60°$ are required for outer voltage regulation loop. From here, the external ramp design procedure for two-phase overlapping region involve two steps:

1) *Determine the minimum ramp that meets the gain margin requirement (G_m):*

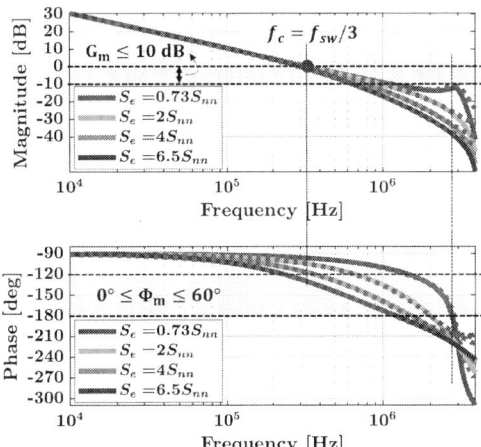

Fig. 9. Verification of ramp design equations with bandwidth $f_c = f_{sw}/3$ for a six-phase converter.

For given bandwidth (f_c), the gain margin increases with an increase in external ramp as shown in (18) and Fig. 7. Hence, gain margin requirement sets the minimum value of external ramp needed. According to (18), minimum ramp needed can be determined as:

$$S_e^{min} = \frac{S_{nn}}{2} + \frac{f_c}{f_{sw}}\left(\frac{2S_{fn} + S_{nn}}{N^3 D^2}\right) 10^{(G_m^{min}/20)}. \quad (24)$$

For the example considered here, the minimum ramp needed equals $S_e^{min} = 0.73 S_{nn}$ according to (24).

2) *Determine the maximum ramp that meets phase margin requirement (Φ_m):*

For a given bandwidth (f_c), the phase margin decreases with an increase in external ramp as shown in (23) and Fig. 8. Hence, phase margin requirement sets the maximum value of external ramp needed. According to (23), maximum ramp needed can be determined as:

$$S_e^{max} = \frac{S_{nn}}{2} + \frac{1}{\pi}\left(\frac{2S_{fn} + S_{nn}}{Q_{se}(f_c, \Phi_m^{min})}\right) \quad (25)$$

For the example considered here, the maximum ramp that can be used without violating phase margin requirement is $S_e^{max} =$

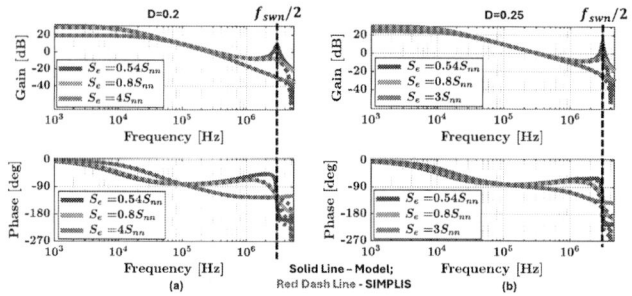

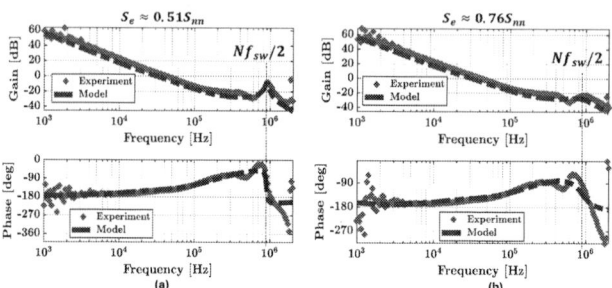

Fig. 10. SIMPLIS simulation verification of control-to-output transfer function (8) with two different duty cycles within two-phase overlapping region for different ramp. Circuit parameters: V_{in} = 8V; N = 6; f_{sw} = 1 MHz; L_s = 150 nH; R_i = 5 mΩ; R_{co} = 1.4 mΩ; C_o = 100 μF; R_L = 200 mΩ.

Fig. 12. Experimental verification of outer voltage loop gain for two-phase overlapping case. (a) Loop gain with ramp $S_e = 0.51S_{nn}$ and (b) Loop gain with ramp $S_e = 0.76S_{nn}$.

As seen in Fig. 12, the model and experimental results match well above switching frequency $f_{sw} = 300$kHz.

For external ramp $S_e = 0.51S_{nn}$ (close to critical value $S_{ec} = 0.5S_{nn}$), the double pole Q-factor, i.e., $Q_{se} \approx 45.8 >> 1$. Hence, Fig. 12(a) shows a sharp phase drop and large magnitude peak at $f_1 = Nf_{sw}/2 = 900$ kHz verifying double pole existence at that frequency. Fig. 12(b) shows the outer voltage loop gain with a larger ramp $S_e = 0.76S_{nn}$. As for the smaller ramp, the model matches the experiments well. In Fig. 12(b), the double pole is well-damped, and there is no magnitude peak at $f_1 = 900$ kHz because $Q_{se} \approx 1.64$. Referring to (5a), the system has a double zero and double pole at the frequency $f_1 = Nf_{sw}/2$ with two-phase overlap. At $f_1 = Nf_{sw}/2$, the double zero provides a magnitude dip equal to $-20\log_{10}(Q_1)$ and double pole introduces a magnitude peak equal to $20\log_{10}(Q_{se})$. Hence, the net dB magnitude at $f_1 = Nf_{sw}/2$ approximately equals to $20\log_{10}(Q_{se})$ - $20\log_{10}(Q_1)$. This is why even a double pole Q-factor, which is slightly greater than 1, does not show significant peaking in Fig. 12(b).

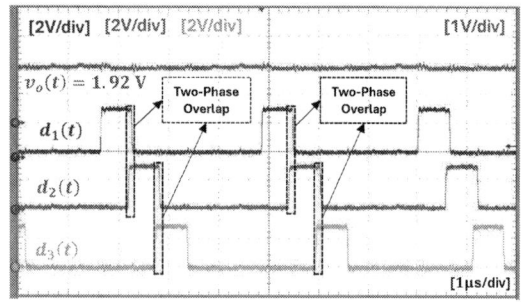

Fig. 11. Steady-State PWM waveforms of first three-phases of the six-phase converter used for experiments under two-phase overlapping conditions.

$4.4S_{nn}$ according to (25). For a given bandwidth (f_c), a ramp $S_e^{min} < S_e < S_e^{max}$ ensures that gain and phase margin of voltage loop are greater than minimum required gain margin (G_m^{min}) and phase margin (Φ_m^{min}), respectively.

IV. SIMULATION AND EXPERIMENTAL VERIFICATION

Using SIMPLIS software, Fig. 10 verifies the simplified model (8) for two-phase overlapping case. The proposed model was also verified experimentally using the Texas Instrument's demo board for the TPS53667 controller [18]. TPS53667 controller can support six phases and it is based on multiphase current-mode COT control architecture shown in Fig. 1. The experimental parameters are: V_{in} = 10V; V_o = 1.92V; L_s = 150nH; f_{sw} = 300kHz. For V_{in} = 10V and V_o = 1.92V, duty cycle D = 0.192. Also, $0.1667 < D = 0.192 < 0.333$ and two subsequent phases overlap at steady-state as shown in Fig. 11. The output capacitor network has twenty 100μF ceramic capacitors and four 470μF/5mΩ electrolytic capacitors.

To expose the possible peaking identified at $f_{swn}/2 = Nf_{sw}/2 = 900$kHz in section III-C, outer voltage loop gain was experimentally measured by fixing the bandwidth at $f_c \approx 40$kHz and varying the external ramp. Fig. 12(a) and Fig. 12(b) shows the outer loop gain with two-different ramp slopes $S_e = 0.51S_{nn}$ and $S_e = 0.76S_{nn}$, respectively. The model response in Fig. 12 was calculated by multiplying approximate control-to-output transfer function (8), including composite capacitor effect, with outer loop compensator transfer function.

CONCLUSION

This paper investigates the small-signal behavior of multiphase current-mode COT control with phase overlapping. Phase overlapping introduces large resonance peaks in control-to-output response with smaller ramp slopes. These peaks could introduce double crossover in voltage loop gain. Hence, high-bandwidth multiphase COT VRs can become unstable even with a stable inner current loop in phase overlapping regions. For a given bandwidth, increasing ramp slopes improve gain margin by damping these peaks but decreases phase margin. Hence, this work simplifies the infinite-order control-to-output transfer function and provides external ramp design equations which can be used to meet gain and phase margin specifications simultaneously. The simplified transfer function was validated using SIMPLIS simulation and experimental results. Future work will generalize the results for two-phase overlapping case to arbitrary number of phase overlapping.

ACKNOWLEDGMENT

This work was supported by Power Management Consortium (PMC) in Center for Power Electronics Systems (CPES), Virginia Tech.

979-8-3315-1612-3/25 $31.00 © 2025 IEEE

APPENDIX I: OPTIMAL RAMP DERIVATION FOR TWO-PHASE OVERLAPPING REGION WITH SAWTOOTH RAMP

Section III B derived the optimal ramp for two-phase overlapping case using the rational transfer function approximation for infinite-order $Y_{vc}(s)$. Here, we show another way to derive optimal ramp directly from infinite-order $Y_{vc}(s)$ given by (3a). For two-phase overlapping case, $M = 2$ and equation (3a) reduces as follows:

$$Y_{vc}(s) = \frac{V_{in}}{sL_s} \frac{f_{swn}(1 - e^{-sT_{on}})}{S_e + S_{fn} + (S_{nn} + S_{fn} - S_e)e^{-sT_{swn}}} \quad (26)$$

Referring to (26), if $S_e = S_{nn} + S_{fn}$ then the coefficient of $e^{-sT_{swn}}$ in denominator becomes zero and hence eliminating the dynamic due to external ramp in system natural response. Substituting $S_e = S_{nn} + S_{fn}$, and the expressions for S_{nn} and S_{fn} given in section II in (26), we get

$$Y_{vc}(s) = \frac{V_{in}}{sL_s} \frac{f_{sw}(1 - e^{-sT_{on}})}{S_f} \quad (27)$$

where $S_f = R_i V_o/L_s$ denotes the falling slope of each phase inductor current. Equation (27) is identical to control-to-inductor current describing function for single-phase COT control without external ramp provided by J. Li [7].

APPENDIX II: INSTABILITY WITH DELAYED SAWTOOTH RAMP AND PHASE OVERLAP

To reduce jitter in control waveform, improve noise immunity and ensure stability, multiphase COT controllers implement external ramp compensation in two different ways. The first implementation is a pure sawtooth ramp signal, i.e., the ramp is active throughout a total current period and it is reset at the end of a total current cycle, as shown in Fig. 1(b). Fig. 13 shows the second implementation where external ramp is deactivated during total current ramp up time (or rise time) and activated only during the total current ramp down time (or falling time) [9], [13]. The ramp implementation shown in Fig. 13 can be called as 'delayed' sawtooth ramp because the linear rising phase of ramp signal is delayed by the total current rise time in every total current cycle.

Compared to the delayed ramp, the pure sawtooth ramp is more popular in commercial multiphase COT control products. Therefore, the main section of the paper studied in detail the multiphase COT control dynamics with pure sawtooth external ramp compensation. For no phase overlapping case (or duty cycle $D < 1/N$), the multiphase current-mode COT VRs have identical small-signal transfer functions with pure sawtooth and delayed sawtooth ramp. This is because each phase COT time equals the total current ramp up time (or rise time) in every total cycle for duty cycle $D < 1/N$. However, with phase overlap or duty cycle $D > 1/N$, the total current rise time not be constant after introducing perturbation [14]. Hence, the delayed ramp and pure sawtooth ramp have different small-signal models and stability properties with phase overlap. This appendix will show that delayed sawtooth ramp shown in Fig. 13 cannot stabilize the multiphase current-mode COT VRs with phase overlap (or duty cycle $D > 1/N$).

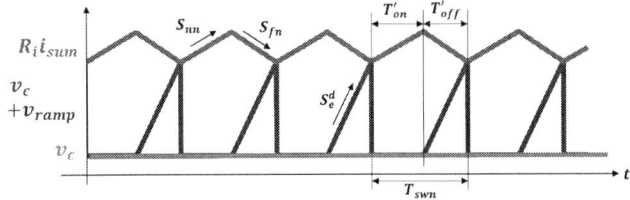

Fig. 13. Modulator waveforms with delayed sawtooth ramp compensation.

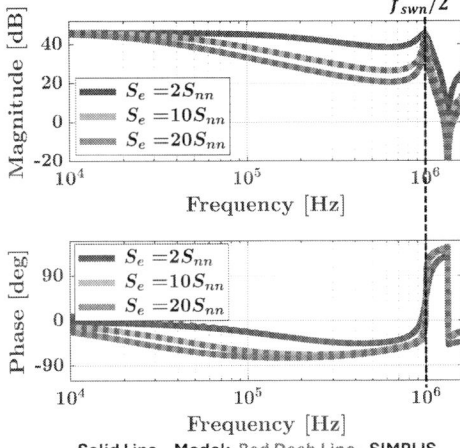

Solid Line – Model; Red Dash Line - SIMPLIS

Fig. 14. Control-to-total current response with delayed sawtooth ramp in two-phase overlapping region. The magnitude response shows large resonant peaks and phase increases around the frequency $f_{swn}/2 = 1/(2T_{swn})$ indicating the presence of right-half plane poles. Circuit parameters: $V_{in} = 8$V; $N = 4$; $D = 0.375$; $f_{sw} = 500$ kHz; $L_s = 150$ nH and $R_i = 5$ mΩ.

The poles of the control-to-total current transfer function, i.e., v_c-to-i_{sum} transfer function, determine the inner total current loop stability. The control-to-total current transfer function for delayed sawtooth ramp, valid for duty range $0 < D < 1$, was derived using describing function method presented in [14] and shown below.

$$\frac{\hat{i}_{sum}(s)}{\hat{v}_c(s)} = \frac{V_{in}f_{swn}(1 - e^{-sT_{on}})}{sL_s\Delta_d(z)}. \quad (28)$$

The characteristic polynomial with delayed ramp

$$\Delta_d(z) = S_{fn} + S_e^d\left(1 - z^{-M}\right) + (S_{nn} + S_{fn})\sum_{u=1}^{M-1} z^{-u} \quad (29)$$

where $z = e^{sT_{swn}}$. The characteristic polynomial $\Delta_d(z)$ could introduce unstable poles in closed-loop transfer function (28). Hence, the location of the roots of $\Delta_d(z)$ in the z-plane determines the total current loop stability with delayed sawtooth ramp. To prove that delayed ramp cannot stabilize the system with phase overlap, it is sufficient to show that $\Delta_d(z)$ has at least one root outside unit circle for all $S_e^d > 0$ in phase overlapping regions (or $M \geq 2$). Here, we verify this statement for two-phase overlapping case.

For the duty cycle $1/N < D < 2/N$, two-phases overlap $(M = 2)$ and the characteristic polynomial for delayed sawtooth ramp (29) reduces as follows:

$$\Delta_d(z) = S_{fn} + S_e^d\left(1 - z^{-2}\right) + (S_{nn} + S_{fn})z^{-1}. \quad (30)$$

979-8-3315-1612-3/25 $31.00 © 2025 IEEE

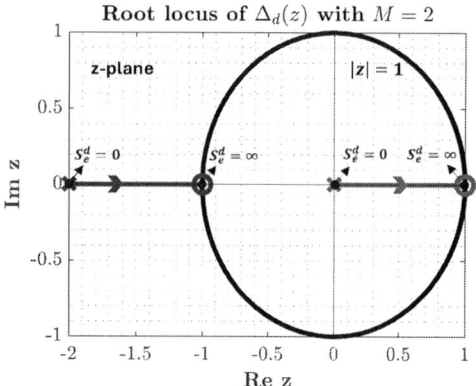

Fig. 15. Root locus of $\Delta_d(z)$ in two-phase overlapping region as delayed ramp slope S_e^d varies from 0 to infinity.

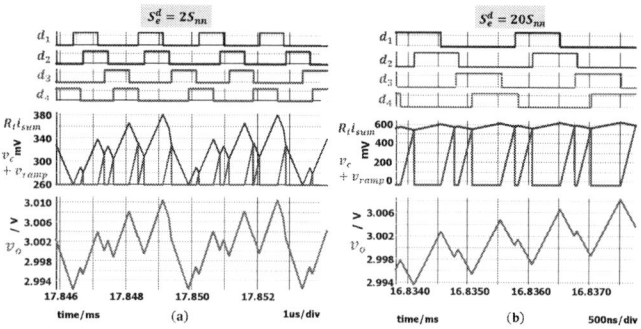

Fig. 16. Simulation verification of instability in two-phase overlapping region of a four-phase current-mode COT controlled buck with delayed sawtooth ramp compensation. Circuit parameters: V_{in} = 8V; N = 4; D = 0.375; f_{sw} = 500 kHz; L_s = 150 nH; R_i = 5 mΩ.

According to Jury's criterion [19], [20], the first two necessary conditions for a second-order discrete-time polynomial

$$F(z) = a_0 z^{-2} + a_1 z^{-1} + a_2 \quad \text{where} \quad a_2 > 0$$

to have all its roots inside unit circle are:

$$F(1) > 0, \tag{31a}$$
$$F(-1) > 0. \tag{31b}$$

Substituting $z = 1$ and $z = -1$ in (30), we get

$$\Delta_d(1) = S_{nn} + 2S_{fn} > 0, \tag{32}$$
$$\Delta_d(-1) = -S_{nn} < 0. \tag{33}$$

The sensed total current slopes, i.e., S_{nn} and S_{fn}, are always positive within two-phase overlapping region. Hence, the second necessary condition in Jury's stability test is not satisfied. Consequently, the system always has one closed-loop pole outside unit circle in two-phase overlapping region. Fig. 15 shows the root locus of $\Delta_d(z)$ for two-phase overlapping case by varying delayed ramp slope S_e^d. As shown using Jury's stability test, one root of the $\Delta_d(z)$ always lies outside unit circle. Fig. 16 verifies the instability in two-phase overlapping region with delayed sawtooth ramp using SIMPLIS simulation. Although the results are shown only for two-phase overlapping case, it can be proved using Nyquist analysis that delayed

ramp cannot stabilize the system for any number of phase overlapping.

REFERENCES

[1] W. Huang, D. Clavette, S. Zhou, and M. Rodrigues, "A 32-phase 1200-ampere dc/dc converter for data center and artificial intelligence systems," in *2021 IEEE Applied Power Electronics Conference and Exposition (APEC)*, 2021, pp. 2017–2023.

[2] H. Gan, S. Jiang, S. Teng, S. Yamamoto, V. Chivukula, B. Edwards, C. Chung, J. Chen, M. Mohideen, G. Sizikov, and X. Li, "Vertical power delivery for 1000 amps machine learning asics," in *2024 IEEE Applied Power Electronics Conference and Exposition (APEC)*, 2024, pp. 906–909.

[3] J. Williams, *Step-Down Switching Regulators*, Analog Devices, 1989. [Online]. Available: https://www.analog.com/media/en/technical-documentation/application-notes/an35f.pdf

[4] K. Radhakrishnan, M. Swaminathan, and B. K. Bhattacharyya, "Power delivery for high-performance microprocessors—challenges, solutions, and future trends," *IEEE Transactions on Components, Packaging and Manufacturing Technology*, vol. 11, no. 4, pp. 655–671, 2021.

[5] *EP71xx IVR Series*, Empower Semicondutor, 2024. [Online]. Available: https://www.empowersemi.com/wp-content/uploads/2024/07/EP71xx-IVR-Series-Product-Brief-Rev2p0.pdf

[6] C.-S. Cheng, J.-R. Huang, and C.-S. Li, "Circuit and method for constant on-time control for an interleaved multiphase voltage regulator," U.S. Patent 8 159 197 B2, Apr. 2012.

[7] J. Li and F. C. Lee, "New modeling approach and equivalent circuit representation for current-mode control," *IEEE Transactions on Power Electronics*, vol. 25, no. 5, pp. 1218–1230, 2010.

[8] X. Duan and A. Q. Huang, "Current-mode variable-frequency control architecture for high-current low-voltage dc–dc converters [letters]," *IEEE Transactions on Power Electronics*, vol. 21, no. 4, pp. 1133–1137, 2006.

[9] S. Tian, "Equivalent circuit model of high frequency pwm and resonant converters," Ph.D. dissertation, Virginia Polytechnic Institute and State University, 2015.

[10] C.-J. Tsai, H.-H. Chen, and C.-J. Chen, "A phase interpolated dual-phase adaptive on-time controlled buck converter," *IEEE Transactions on Circuits and Systems I: Regular Papers*, vol. 71, no. 11, pp. 5155–5165, 2024.

[11] N. Yan, X. Ruan, and X. Li, "A general approach to sampled-data modeling for ripple-based control—part ii: Constant on-time and constant off-time control," *IEEE Transactions on Power Electronics*, vol. 37, no. 6, pp. 6385–6396, 2022.

[12] S. Sridhar and Q. Li, "Multiphase constant on-time control with phase overlapping–part ii: Stability analysis," *IEEE Transactions on Power Electronics*, vol. 39, no. 3, pp. 3156–3174, 2024.

[13] C. Li, L. Wang, M. Fu, and H. Wang, "Small-signal modeling of multi-phase trans-inductor voltage regulator modules in datacenter applications," in *2024 IEEE Applied Power Electronics Conference and Exposition (APEC)*, 2024, pp. 633–638.

[14] S. Sridhar and Q. Li, "Multiphase constant on-time control with phase overlapping—part i: Small-signal model," *IEEE Transactions on Power Electronics*, vol. 39, no. 6, pp. 6703–6720, 2024.

[15] R. Ridley, "A new, continuous-time model for current-mode control (power convertors)," *IEEE Transactions on Power Electronics*, vol. 6, no. 2, pp. 271–280, 1991.

[16] F. Tan and R. Middlebrook, "A unified model for current-programmed converters," *IEEE Transactions on Power Electronics*, vol. 10, no. 4, pp. 397–408, 1995.

[17] W.-C. Liu, C.-H. Cheng, P. P. Mercier, and C. C. Mi, "Small-signal analysis and design of constant on-time controlled buck converters with duty-cycle-independent quality factors," *IEEE Transactions on Power Electronics*, vol. 38, no. 7, pp. 8379–8393, 2023.

[18] *TPS53667 datasheet*, Texas Instruments, 2017. [Online]. Available: https://www.ti.com/lit/ds/symlink/tps53667.pdf

[19] E. I. Jury, "A simplified stability criterion for linear discrete systems," *Proceedings of the IRE*, vol. 50, no. 6, pp. 1493–1500, 1962.

[20] G. F. Franklin, J. D. Powell, M. L. Workman *et al.*, *Digital control of dynamic systems*. MA: American Mathematical Society, 1998.

Multiphase Constant-On-Time Minimum-Deviation Controller for Modern Processors

1st Duo Li, 2nd Gianluca Roberts, 4th Aleksandar Prodic
Department of Edward S. Rogers Sr. Department of Electrical and Computer Engineering
University of Toronto
Toronto, Canada
jasonliduo98@gmail.com, gianluca.roberts@mail.utoronto.ca, prodic@ece.utoronto.ca

3rd Alan Wu
Intel Corporation
North York, Canada
almtbwu@gmail.com

Abstract—In Voltage Regulator Module (VRM) applications, rapid dynamic load conditions often lead to significant output voltage deviation and negatively impact the dynamic efficiency of converters optimized for static efficiency. This paper proposes a novel multiphase Constant-On-Time Minimum Deviation (COT Min Dev) Controller to mitigate dynamic losses and reduce converter sizes. This marks the first application of a Minimum Deviation Controller in a multiphase system, validated on a 4-phase buck converter with a 12 V input and 1 V output, achieving a reduction in output voltage deviation from 280 mV to 80 mV during a 10 A - 90 A load step. Additionally, it was validated on a 3-level 4-phase buck converter prototype, where the output voltage deviation was reduced from 280 mV to 125 mV for the same load step, with input voltage initial drop eliminated. The experimental results demonstrate effective phase balancing, enhanced transient performance, and a reduction in converter size.

Index Terms—Minimum Deviation Controller, flying capacitor, multilevel multiphase converter, Voltage Regulator Module (VRM)

I. INTRODUCTION

The Constant-On-Time (COT) Controller [1]–[4] is one of the most popular controllers for modern Voltage Regulator Modules (VRMs). However, it has been increasingly challenged by the more rapidly fluctuating workloads, particularly in Graphics Processing Unit (GPU) applications. These rapidly changing conditions necessitate faster controllers to prevent large output voltage deviations, which negatively impact efficiency and stability.

The Minimum Deviation Controller [5], known for minimizing output voltage deviation by detecting both the rising and falling zero-crossing points of the output capacitor current and extending operation for D/2 and D'/2, has so far only been applied in single-phase applications. A direct implementation of the Minimum Deviation Controller in multiphase applications leads to phase balancing issues. Without effective phase sharing, certain phases may face a risk of reaching the inductor saturation point. At this point, the Minimum Deviation Controller becomes ineffective, resulting in reduced efficiency and potential instability.

To solve this problem, this paper introduces a novel multiphase Constant-On-Time Minimum Deviation (COT Min Dev) Controller. This controller is well-suited for conventional multiphase buck converters (Fig. 1), providing effective phase balancing, reducing output capacitor size, and improving dynamic efficiency, thereby achieving a lower environmental impact compared to a traditional Constant-On-Time (COT) controller [6].

Furthermore, the controller is adapted for use in multilevel multiphase buck converters (Fig. 9) [7], leveraging the growing interest in flying capacitor topologies [8]–[15]. This faster multiphase controller enables the reduction of both input bulk capacitor size and output capacitor size while minimizing dynamic losses.

II. CONSTANT-ON-TIME MINIMUM DEVIATION FOR MULTIPHASE BUCK CONVERTER

A. Open Loop Operation Of 4-phase Buck Converter

A conventional 4-phase buck converter was selected for this study due to its compatibility with the power requirements of our industry partner's product. Fig. 2 illustrates the open-loop operation of a conventional 4-phase buck converter. During the off time, all synchronous rectifiers conduct (state I in Fig. 1). During the on time, phases are turned on in sequential order and only one conducts at the time (state A-D in Fig. 1). The on time and off time are determined by the switching frequency and duty ratio just like the single-phase buck converter.

B. Phase Balancing Issue

A straight approach to implementing the Multiphase Minimum Deviation Controller would be to detect the rising and falling Zero-Crossing (ZC) points of the output capacitor current i_C and extend the operation for $D/2$ and $D'/2$ respectively, where D' is $1 - D$, as it is done in a single phase implementation [16]. Instead of firing all pulses with one phase, the operation rotates among the 4 phases. Note that the peaks in the i_C waveform come from different phases. However, this approach leads to phase balancing issues during both steady-state and transient operations.

During steady-state operation, phase balancing issues arise due to variations in PCB trace lengths from the output of each phase to the i_C sensing point, leading to differing delays in the i_C waveform peaks. This discrepancy results in unequal on-times across the phases. As illustrated in Fig. 3, phases 1 and 3 are located further from the i_C sensing point to accommodate the footprint of the 2-phase coupled inductors. This leads to

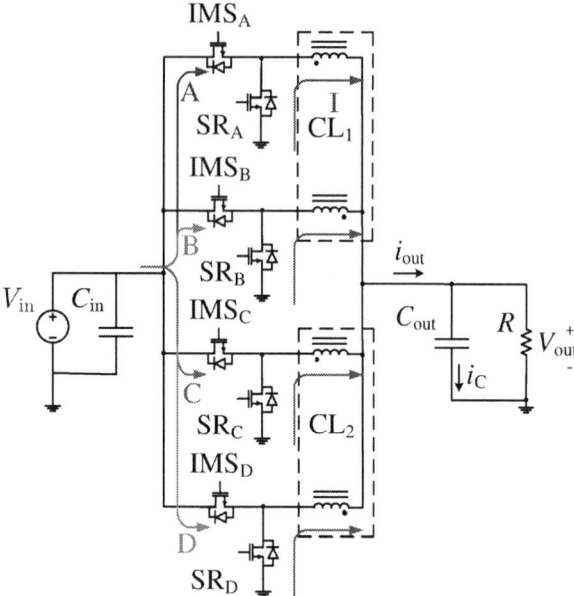

Fig. 1. Current flow diagram illustrating four on-time states (A–D), during which the total inductor current rises sequentially, and the off-time state (I), during which all synchronous rectifiers conduct. In each on-time state, only one phase is active, and one of the four states occurs after the off-time in a rotational manner. For simplicity, the input wire resistance and inductance are not shown here.

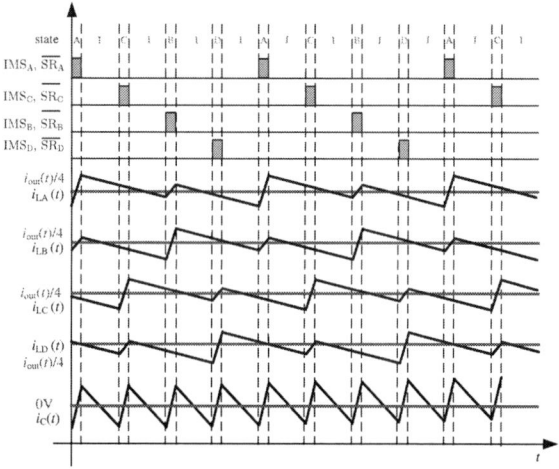

Fig. 2. Gating signals of converter with associated inductor current for each phase and total inductor current.

longer delays in their peaks, which causes longer on-times and higher currents compared to phases 2 and 4 (Fig. 4). The current balancing issue also occurs during transient response. When a transient event occurs, all 4 phases are turned on. However, when the total current of all phases reaches the new load value, the phases at that point have different values, leading to imbalanced current distribution. The experiment was conducted with an effective switching frequency of 1 MHz (a per-phase switching frequency of 250 kHz). While the severity

of the phase balancing problem may be reduced with lower switching frequencies or an optimized PCB layout, it is unavoidable when employing a straightforward implementation approach.

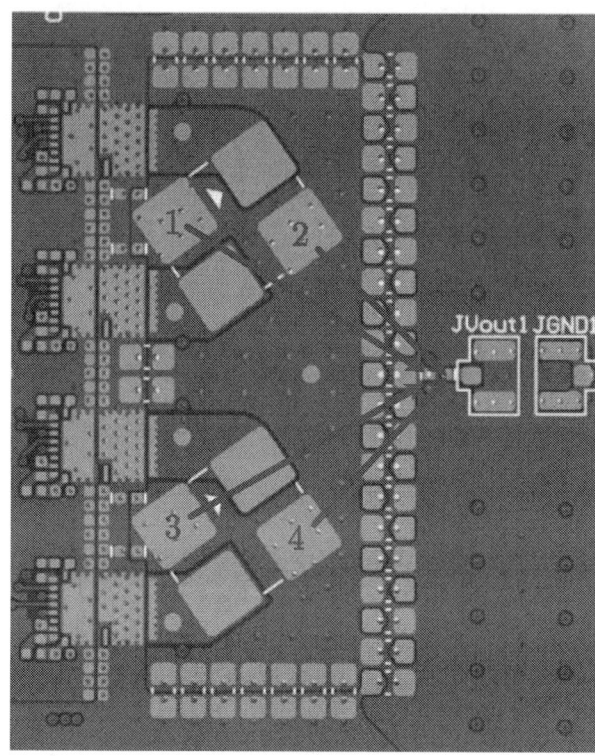

Fig. 3. PCB layout of the 4-phase buck converter with 2-phase coupled inductors. The couple inductor footprints were placed in a diagonal way, resulting in differences in distances from each phase to the differentiator sense point (labeled JVout1).

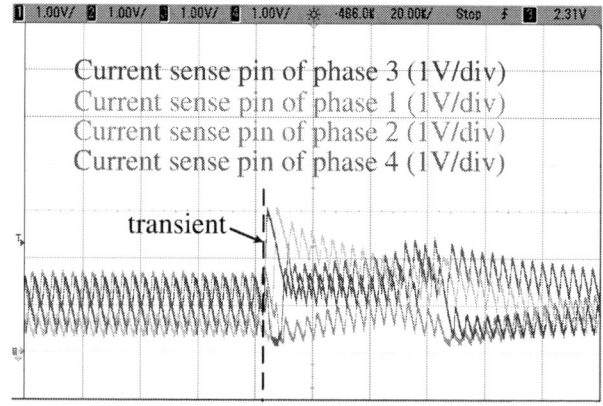

Fig. 4. Current sense pins of all phases of 4-phase buck converter with the straight implementation of Multiphase Minimum Deviation Controller at 10 - 50 A, 1/8 duty, 600 Hz load.

C. Principle of Operation

To address the phase balancing issues, the COT Min Dev Controller was introduced. During the on-time, the controller functions as a COT Controller, by issuing the same on-time duration (D) for each phase (Fig. 6). During the off-time, it operates as a Minimum Deviation Controller, extending the duration for $D'/2$ after detecting the falling zero-crossing point on the output capacitor current i_C. By doing this, the controller utilizes information about the inductor current, to improve transient performance while ensuring phase balancing. Even when the zero-crossing points of each phase experience different delays due to PCB trace length variations, the controller ensures that, over a complete switching cycle, each phase has the same total on-time. Consequently, all phases contribute an equal amount of current to the load, maintaining balanced operation. During transients, the controller also provides stability, effective phase balancing, and improved performance (Fig. 5).

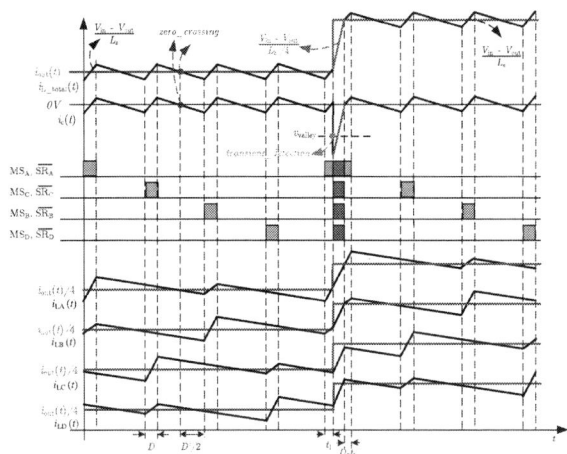

Fig. 6. Gating signals of the 4-phase converter during a light-to-heavy load step, illustrating the transient detection process and associated inductor currents for each phase, total inductor current and differentiated output voltage.

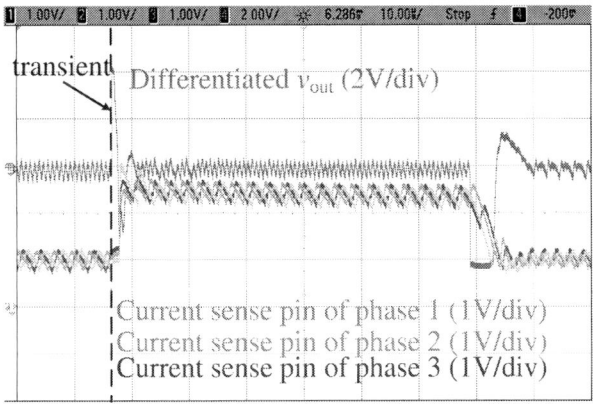

Fig. 5. Current sense pin of phase 1 (green), phase 2 (yellow), phase 3 (purple) and differentiated v_{out} (red) of 4-phase buck converter with COT Min Dev Controller at 10 - 150 A, 1/8 duty, 2 kHz load.

This method combines the best features of both Constant-On-Time and Minimum Deviation Controllers, operating with variable frequency. The upper frequency is bounded by the width of the constant-on-time pulses, while a maximum off-time counter limits the lower frequency. If the falling zero-crossing point is not detected within the maximum off-time duration, the controller progresses to the next state, the off-time second half (Fig. 8), ensuring continuous operation. During the off-time second half state, the zero-crossing signal should remain high. If it becomes low unexpectedly, the state reverts to the off-time first half. Once the zero-crossing signal is high again, the controller re-enters the off-time second half state, thereby maintaining continuous operation. This implementation has lower hardware requirements, compared to the approach of per-phase current sensing.

Transient is detected when i_c experience sudden drops and becomes lower than a valley value (Fig. 6). The valley value is typically set in the range of 3 to 5 times the i_c ripple. During transient, all 4 phases will be turned on, ensuring the fastest recovery rate. After the transient, a stored counter value is used to ensure that the phase interrupted by the transient will continue to finish the remaining time ($D - t_1$) after transient. Therefore, the phase balancing is maintained even after transient.

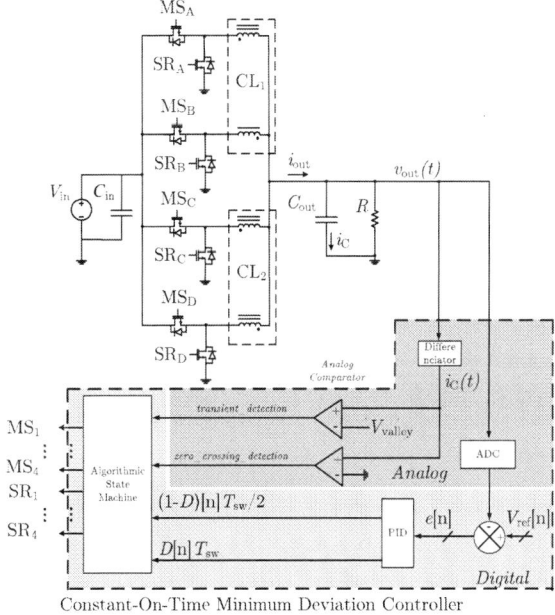

Fig. 7. Diagram of 4-phase buck converter and the implementation of its complementary COT Min Dev Controller.

The controller consists of both digital and analog circuits (Fig. 7). The analog circuits are employed for fast zero-crossing and transient detection, utilizing information from the output capacitor current (i_c). The output capacitor current is derived by differentiating the output voltage (v_{out}).

The digital logic is responsible for implementing the PID compensator to adjust the duty cycle D and generating the gating signals required for the converter's operation.

D. Couple Inductor

This controller can also benefit from the use of coupled inductors. When one of the phases is on (connected to V_{in}),

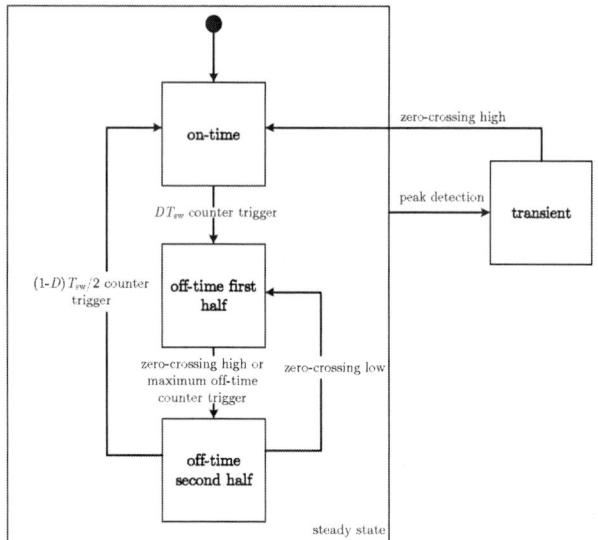

Fig. 8. The Algorithmic State Machine of the COT Min Dev Controller shown in Fig.7.

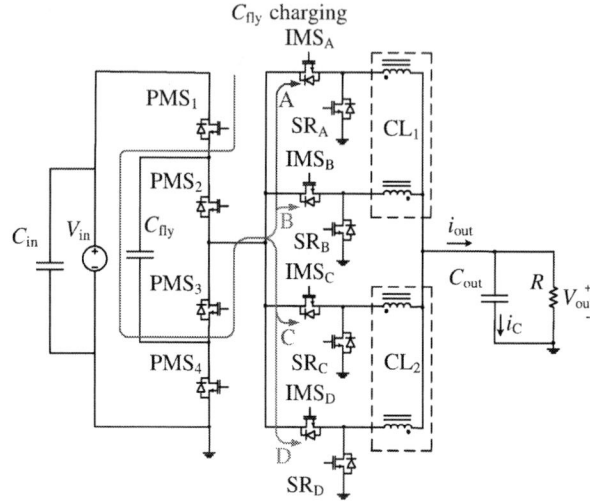

Fig. 9. Current flow diagram for 3-level 4-phase buck converter, with four different on-time states during the C_{fly} charging period. In each state, only one phase is active.

the other phases are off (connected to ground). The effective inductance is L_s, which is the sum of $L_m + L_k$, where L_s is the self inductance, L_m is the magnetizing inductance, and L_k is the leakage inductance. The phases that are off will experience a rise as well due to the negative coupling. This mechanism reduces current ripple and inductor volume during steady state. During transient, all 4 phases are turned on. If discrete inductors are used, the effective inductance becomes $L/4$ instead of L. In a coupled inductor configuration, when all phases are turned on, the mutual inductance L_m is effectively eliminated because the magnetic fields generated by each phase in the coupled inductor structure support each other. Instead of opposing each other (as in steady state with alternate phases on and off), they combine constructively. Therefore, effective inductance becomes $L_k/4$ instead of $L_s/4$ (L_m is eliminated), further increasing slew rate and, therefore, the speed of voltage recovery.

III. CONSTANT-ON-TIME MINIMUM DEVIATION FOR MULTILEVEL MULTIPHASE BUCK CONVERTER

A. Open Loop Operation Of 3-level 4-phase Buck Converter

In the realm of power delivery for high-performance applications, the implementation of three-level multiphase buck converters has proven to be an important advancement [7]. Central to this topology is the inclusion of a Flying Capacitor (C_{fly}), nominally rated at 6V. This critical component effectively halves the switch node voltage swing from 12V to 6V, thereby reducing the voltage stress on the components, enhancing the overall efficiency of the system.

During start-up, the flying capacitor (Fig. 9) is charged to a voltage of 6 V. This operation ensures that the downstream buck converters receive an input voltage of 6 V. The mechanism of operation for these downstream buck stages closely mirrors that of a conventional buck converter, wherein the

Intermediate Main Switches (IMS) and the Synchronous Rectifier (SR) are sequentially activated based on the predetermined duty ratio.

Fig. 11 shows how the three-level multiphase buck converter operates: the cycle begins with the charging state of the flying capacitor (C_{fly}) (Fig. 9). In this stage, every downstream buck converter sees 6 V at the phase node, and functions in a sequential manner, adjusting its SR and IMS according to their respective duty cycles. After 5 cycles, the system enters the C_{fly} discharging state where the 12 V input is disconnected and C_{fly} provides 6 V directly to the buck converters (Fig. 10). Each phase continues to operate in turns. After another 5 cycles, the system reverts to the charging state, reconnecting the 12 V input to recharge C_{fly} to 6 V, and the cycle restarts. Natural phase balancing can be achieved if the Primary Main Switch (PMS) operates at a frequency that is an odd fraction of the switching frequency of the downstream buck converters. This operating scheme also ensures natural voltage balancing at the flying capacitor. Additionally, the PMS employs Zero Current Switching (ZCS) to minimize switching losses.

During the off-time, all Synchronous Rectifiers (SRs) conduct similarly to their operation in a 4-phase buck converter (state I in Fig. 1), allowing the inductor current to decrease.

B. Principle of Operation

The novel COT Min Dev Controller adapted well to this 3-level topology. In steady state, the downstream IMS and SR function similarly to the conventional 4-phase buck converter described, utilizing the falling zero-crossing detection and COT pulses. The key difference lies in the flying capacitor C_{fly}, which is switched at 1/10 of the IMS frequency.

During transients, the behavior of the lower stream IMS mirrors that of multiphase converters (Fig. 12), with all four phases turning on simultaneously until the inductor current reaches the zero-crossing point, leveraging the advantages

979-8-3315-1612-3/25 $31.00 © 2025 IEEE

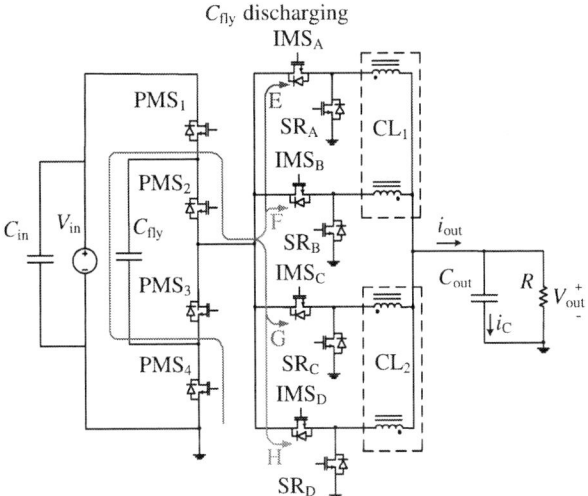

Fig. 10. Current flow diagram for 3-level 4-phase buck converter, with four different on-time states during the C_{fly} discharging period.

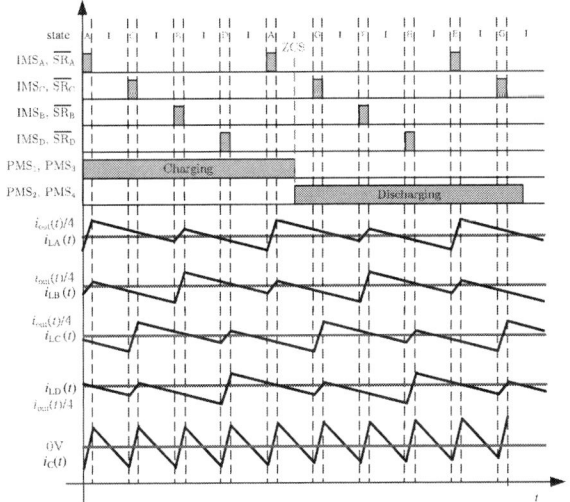

Fig. 11. Gating signals of 3-level 4-phase converter and associated inductor current for each phase and total inductor current [7].

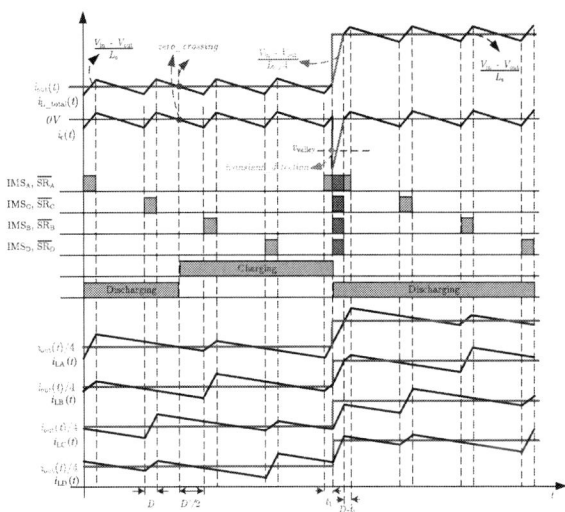

Fig. 12. Gating signals of 3-level 4-phase converter during a light-to-heavy load step with associated inductor current for each phase, total current ripple and differentiated output voltage.

of coupled inductors. For the PMS, each transient initiates the start of a discharging period, causing C_{fly} to experience a voltage deviation. This mechanism effectively decouples the input capacitor from the load during transients, thereby reducing the required input capacitor size.

This approach is preferred because there are regulations for V_{out} and V_{in} deviation but not for V_{cfly} deviation. Conventionally, for consumer GPU applications, input capacitors account for the majority of the total volume in the converter. This is essential to prevent significant input voltage drops during transients, which could lead to power supplies shutting down.

This method works also thanks to the inductor, when inductors are connected to the flying capacitor, there is no non-adiabatic charging loss. This innovative strategy of utilizing

C_{fly} to reduce the reliance on input bulk capacitors during transients is first presented in this paper.

The primary drawback of this approach is that the output capacitor needs to be doubled, as the slew rate is halved compared to connecting the phase node to the 12 V input, a method known as the C_{fly} bypass technique. However, the use of the COT Min Dev Controller mitigates this by enabling a faster transient response and reducing the overall output capacitance requirement.

The controller implementation for 3-level 4-phase buck converter is very similar to one for conventional multiphase buck converter, except that the additional combinational logic for PMS (Fig. 13).

IV. EXPERIMENTAL RESULTS

A. Constant-On-Time Minimum Deviation Controller for Multiphase Converters

The concept is verified with a 3-level 4-phase buck converter prototype with 8 layers (Fig. 14), list of main components is shown in Table I. For experiments conducted on the 4-phase buck converter, PMS1 and PMS2 are turned on, while PMS3 and PMS4 are turned off.

The transient behavior of the COT Min Dev Controller is compared with the conventional COT Controller on the 4-phase buck converter, using identical output capacitors (9×47 µF and 2×22 µF). The output voltage deviation is significantly reduced, from 280 mV to 80 mV, for a 10 A - 90 A load step when using the COT Min Dev Controller (Fig. 15 and Fig. 16). It is important to note that during transient events, the COT Min Dev Controller turns on all phases simultaneously to maximize the transient response rate.

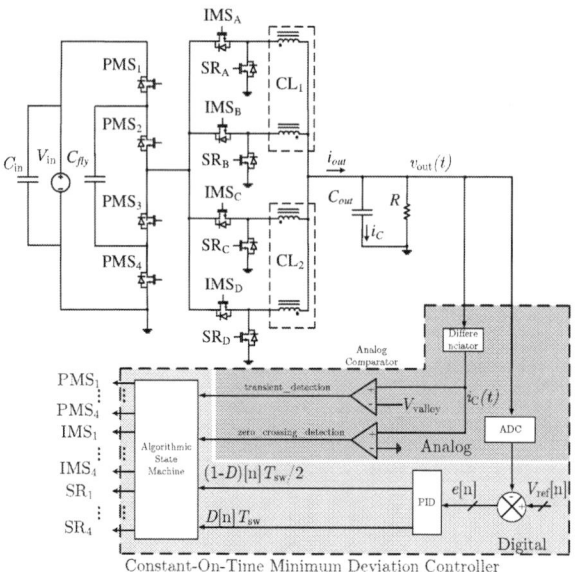

Fig. 13. Diagram of 3-level 4-phase buck converter and its complementary COT Min Dev Controller. For simplicity, the input wire resistance and inductance are not shown here.

TABLE I
LIST OF MAIN COMPONENTS USED IN THE EXPERIMENTAL PROTOTYPE

Component	Quantity & Main Parameters
Inductor	Intel's internal part (120 nH, 80 A)
Flying Capacitor	31x 22 µF / 16 V
Output Capacitor	9x 47 µF / 6.3 V, 4x 22 µF / 4 V
Input Capacitor	21x 1 µF / 16 V, 4x 560 µF / 16 V
Switches	4x MPS MP87006 (IMS) 4x BSZ031NE2LS5ATMA1 (PMS1) 5x BSZ031NE2LS5ATMA1 (PMS3) 5x IQE006NE2LM5 (PMS2) 4x IQE006NE2LM5 (PMS4)
Gate Drivers	4x LTC4440

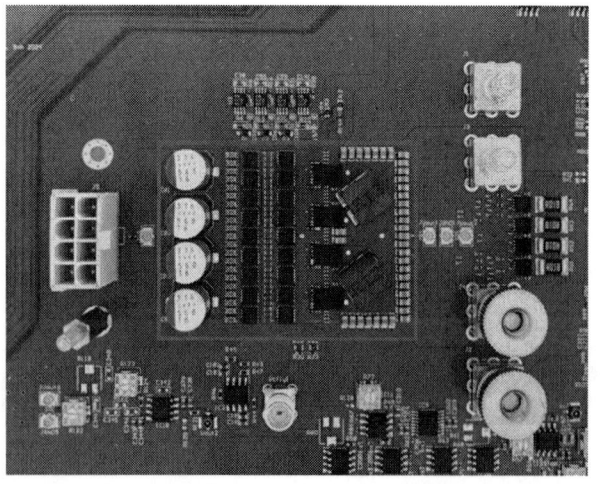

Fig. 14. Photo of assembled PCB of the 3-level 4-phase converters. The converter part (in red square) is 54 mm × 38 mm.

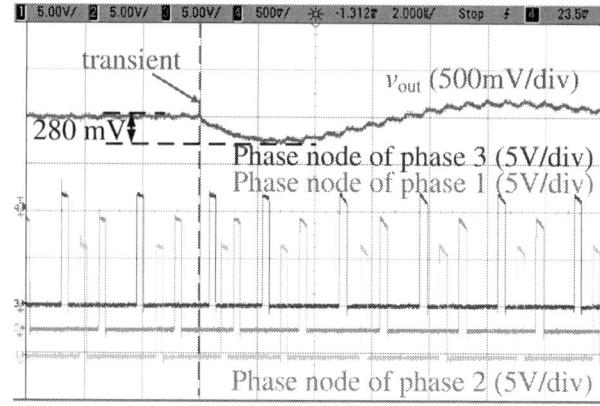

Fig. 15. Experimental output voltage deviation for load step 10 A - 90 A for 12 V - 1 V 4 phase buck converter with conventional COT Controller. Effective switching frequency is 1.7 MHz. Per-phase switching frequency is 417 kHz.

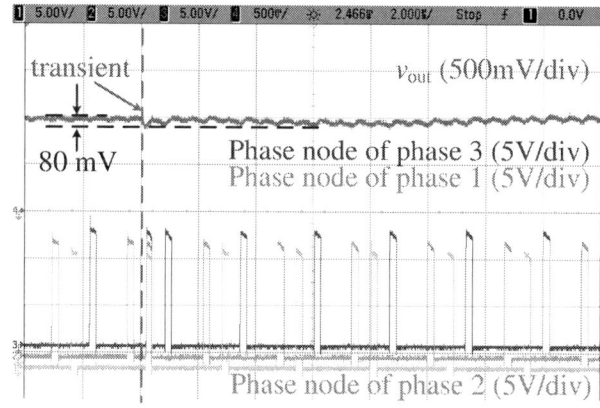

Fig. 16. Experimental output voltage deviation for load step 10 A - 90 A for 12 V - 1 V 4 phase buck converter with COT Min Dev Controller. Effective switching frequency is 1.7 MHz. Per-phase switching frequency is 417 kHz.

B. Constant-On-Time Minimum Deviation Controller for Multilevel Multiphase Converters

Fig. 17 and Fig. 18 illustrate the differences in input voltage drops during transient events occurring in the C_{fly} charging state and discharging states. A high D_0 signal indicates a charging period, while a low D_0 signal indicates a discharging period. It is observed that there is no instantaneous voltage drop at the input capacitor (C_{in}) if a transient occurs during the discharging period (Fig. 18). However, there is a larger input voltage ripple when the flying capacitor is connected to the input again in the subsequent charging period.

After a transient during a discharging period, it is observed that v_{cfly} drops below its nominal value of 6 V (Fig. 18). In the subsequent charging period, a higher voltage is seen at the phase node, allowing more charges to flow into the flying capacitor. As a result, v_{cfly} gradually rises back to its nominal value through a process known as the natural voltage balancing of C_{fly}.

Output voltage deviation is also compared between the conventional COT Controller and the COT Min Dev Controller

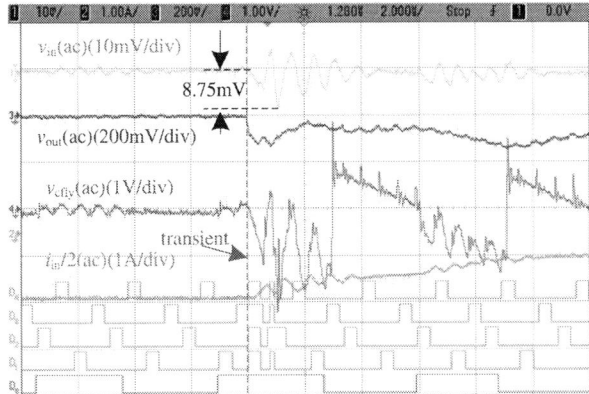

Fig. 17. Experimental $v_{in}(ac)$, $v_{cfly}(ac)$, $v_{out}(ac)$, $i_{in}/2(ac)$, IMS$_A$-IMS$_D$ gating signals (D$_1$-D$_4$) and PMS$_1$ gating signal (D$_0$), of a 3-level 4-phase buck converter with the COT Min Dev Controller at 10 A - 160 A load step. Transient occurs in C_{fly} charging period. Please note that v_{out} will recover to the nominal voltage over time because it takes time for the PID compensator to adjust the duty ratio in response to the heavy load.

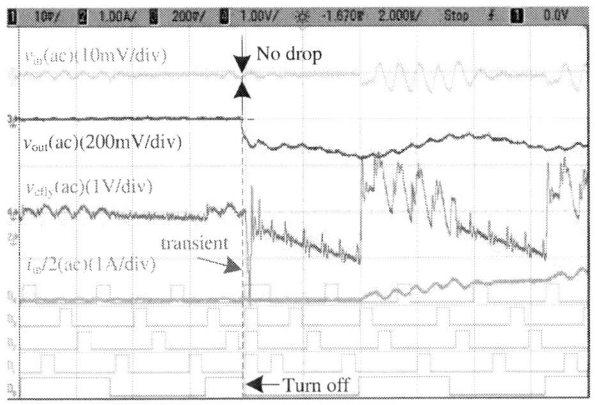

Fig. 18. Experimental $v_{in}(ac)$, $v_{cfly}(ac)$, $v_{out}(ac)$, $i_{in}/2(ac)$, IMS$_A$-IMS$_D$ gating signals (D$_1$-D$_4$) and PMS$_1$ gating signal (D$_0$), of a 3-level 4-phase buck converter with the COT Min Dev Controller at 10 A - 160 A load step. Transient occurs in C_{fly} discharging period. Please note that v_{out} will recover to the nominal voltage over time because it takes time for the PID compensator to adjust the duty ratio in response to the heavy load.

on the 3-level 4-phase buck converter (Fig. 19 and Fig. 20). It is observed that the output voltage deviation is drastically reduced from 280 mV to 125 mV.

C. Volume Reduction

For 4-phase buck converters, experiments show that reducing an additional 200 mV of output voltage deviation requires increasing the output capacitance by a factor of at least five. This adjustment alters the overall volume and area distribution of the converter (Fig. 21 and Fig. 22), assuming the same output voltage deviation. The conventional buck converter used for comparison is based on an existing industry design provided by our industry partner. It is important to note that two extremes exist in this design space: one where the output capacitance remains constant to achieve smaller voltage deviation, and another, where the voltage deviation remains

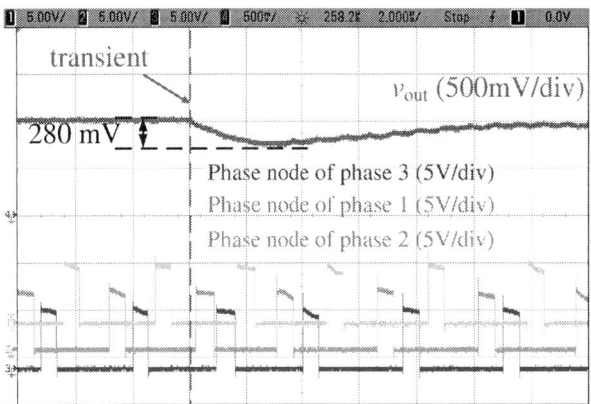

Fig. 19. Experimental output voltage deviation for load step 10 A - 90 A for 12 V - 1 V 3-level 4-phase buck converter with conventional COT Controller. Effective switching frequency is 1.7 MHz. Per-phase switching frequency is 417 kHz.

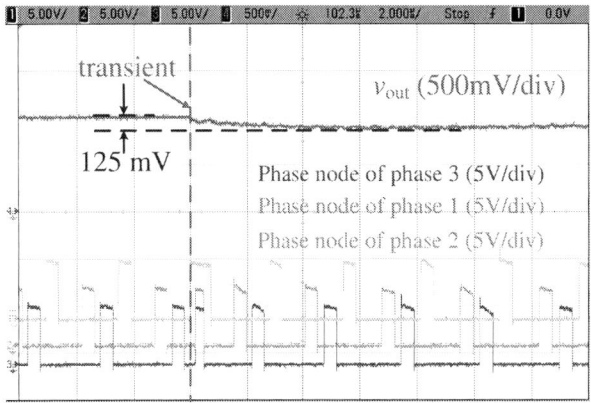

Fig. 20. Experimental output voltage deviation for load step 10 A - 90 A for 12 V - 1 V 3-level 4-phase buck converter with COT Min Dev Controller. Effective switching frequency is 1.7 MHz. Per-phase switching frequency is 417 kHz. Please note that v_{out} will recover to the nominal voltage over time because it takes time for the PID compensator to adjust the duty ratio in response to the heavy load.

constant to achieve a smaller output capacitance. In practice, a balance between these trade-offs can be selected [17].

For 3-level 4-phase buck converters with COT Min Dev Controller, the volume reduction is even more pronounced compared to conventional 4-phase buck converter with COT Controller. When comparing the required capacitance for a COT controller to achieve the same output voltage deviation, the output capacitor size can be reduced approximately to 1/5 of the original. However, with the flying capacitor discharging state in the 3-level topology, the slew rate is reduced by half during transient. This results in an output voltage deviation of 125 mV for the 3-level topology, nearly doubling the 80 mV seen in the 2-level topology. Consequently, the output capacitor needs to be doubled compared to the 2-level case, meaning it can only be reduced to 2/5 of the original size, given the same deviation.

On the other hand, the input capacitor can be reduced by half because the input voltage initial drop during transients

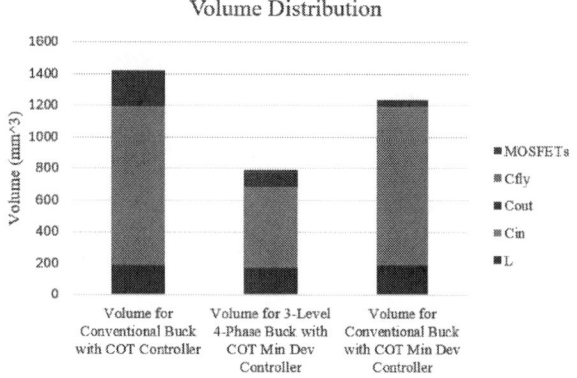

Fig. 21. Converter volume distribution of 4-phase buck converter with conventional COT Controller, 3-level 4-phase buck converter with COT Min Dev Controller and 4-phase buck converter with COT Min Dev Controller.

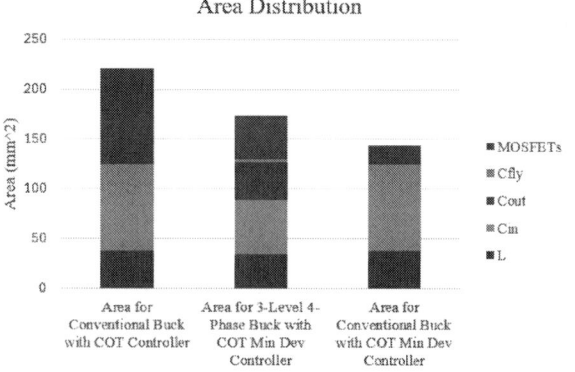

Fig. 22. Converter area distribution of 4-phase buck converter with conventional COT Controller, 3-level 4-phase buck converter with COT Min Dev Controller and 4-phase buck converter with COT Min Dev Controller. Note that the MOSFET area shown includes the packaging size of the discrete MOSFETs, not just the silicon area.

is eliminated. However, half of the input capacitance is still required because the input voltage experiences greater ripple during the subsequent charging state.

For the 3-level converter, reducing the phase node voltage from 12V to 6V allows the inductor value to be reduced to 10/11 of its original value, even with a larger duty ratio.

This 3-level topology, combined with the COT Min Dev Controller, enables significant reductions in overall size, even considering the additional MOSFETs and flying capacitors.

V. CONCLUSION

In conclusion, this paper presents a novel multiphase Constant-On-Time Minimum Deviation Controller. Previous Min Dev Controllers were limited to single-phase applications. The unique transient techniques of the controller successfully minimized v_{out} deviation, resulting C_{out} reduction. The implementation on the COT Min Dev Controller with the 3-level topology eliminates initial v_{in} drop during transients and significantly reduces input capacitor size. Furthermore, the versatility of this controller allows for broader applications

across systems with any number of phases. In configurations exceeding 6 phases for a 3-level topology, or 12 phases for a 2-level topology, the overlap of Pulse Width Modulation (PWM) signals may interfere with zero-crossing detection of the output capacitor current. In such cases, phase synchronization can be employed as a potential solution.

REFERENCES

[1] J. Sun, "Characterization and performance comparison of ripple-based control methods for voltage regulator modules," *IEEE Transactions on Power Electronics*, vol. 21, no. 2, pp. 346–353, 2006.

[2] C. Song, "Accuracy analysis of constant-on current-mode dc–dc converters for powering microprocessors," in *Proc. IEEE APEC*, 2009, pp. 97–101.

[3] R. Redl and J. Sun, "Ripple-based control of switching regulators—an overview," *IEEE Transactions on Power Electronics*, vol. 24, no. 12, pp. 2669–2680, 2009.

[4] Monolithic Power Systems (MPS), "The past and present of cot control: Article: Mps," https://www.monolithicpower.com/en/learning/resources/the-past-and-present-of-cot-control, accessed: Aug. 2, 2024.

[5] L. Lu, T. Moiannou, and A. Prodić, "Single mode near minimum deviation controller for multi-level flying capacitor converters," in *2019 IEEE Applied Power Electronics Conference and Exposition (APEC)*, 2019, pp. 1751–1757.

[6] C. Thompson, "Ai is an energy hog. this is what it means for climate change," https://www.technologyreview.com/2024/05/23/1092777/ai-is-an-energy-hog-this-is-what-it-means-for-climate-change/, 2024, accessed: 2024-09-24.

[7] G. Roberts, N. Vukadinović, and A. Prodić, "A multi-level, multi-phase buck converter with shared flying capacitor for vrm applications," in *2018 IEEE Applied Power Electronics Conference and Exposition (APEC)*, 2018, pp. 68–72.

[8] T. A. Meynard and H. Foch, "Multilevel conversion: high voltage choppers and voltage-source inverters," in *Power Electronics Specialists Conference, 1992. PESC '92 Record., 23rd Annual IEEE*, 1992, pp. 397–403 vol.1.

[9] ——, "Multilevel converters and derived topologies for high power conversion," in *Proceedings of IECON '95 - 21st Annual Conference on IEEE Industrial Electronics*, vol. 1, 1995, pp. 21–26 vol.1.

[10] W. Kim, D. M. Brooks, and G. Y. Wei, "A fully-integrated 3-level dc/dc converter for nanosecond-scale dvs with fast shunt regulation," in *2011 IEEE International Solid-State Circuit Conference*, 2011.

[11] Y. Lei, W.-C. Liu, and R. C. N. Pilawa-Podgurski, "An analytical method to evaluate and design hybrid switched-capacitor and multilevel converters," *IEEE Transactions on Power Electronics*, vol. 33, no. 3, pp. 2227–2240, Mar. 2018.

[12] J. S. Rentmeister, C. Schaef, B. X. Foo, and J. T. Stauth, "A flying capacitor multilevel converter with sampled valley-current detection for multi-mode operation and capacitor voltage balancing," in *2016 IEEE Energy Conversion Congress and Exposition (ECCE)*, Oct. 2016, pp. 1–8.

[13] Y. Jang, M. Jovanović, and Y. Panov, "Multiphase buck converters with extended duty cycle," in *2006 IEEE Applied Power Electronics Conference and Exposition (APEC)*, Mar. 2006, pp. 38–44.

[14] K. Nishijima, K. Harada, T. Nakano, T. Nabeshima, and T. Sato, "Analysis of double step-down two-phase buck converter for VRM," in *INTELEC 05 - Twenty-Seventh International Telecommunications Conference*, Sep. 2005, pp. 497–502.

[15] K. Abe, K. Nishijima, K. Harada, T. Nakano, T. Nabeshima, and T. Sato, "A novel three-phase buck converter with bootstrap driver circuit," in *2007 IEEE Power Electronics Specialists Conference*, Jun. 2007, pp. 1864–1871.

[16] L. Lu, D. Li, and A. Prodić, "Absolute minimum deviation controller for multi-level flying capacitor direct energy transfer converters," in *2020 IEEE Applied Power Electronics Conference and Exposition (APEC)*, 2020, pp. 305–311.

[17] P.-L. Wong, F. C. Lee, X. Zhou, and J. Chen, "Vrm transient study and output filter design for future processors," in *Industrial Electronics Society, 1998. IECON'98. Proceedings of the 24th Annual Conference of the IEEE*, vol. 1. IEEE, 1998, pp. 410–415.

Closed-loop Control of A Dual-side Series/Parallel Piezoelectric-Resonator-based DC-DC Converter

Wen-Chin B. Liu
Dept. of ECE
University of California San Diego
La Jolla, USA
brianliu@ucsd.edu

Gaël Pillonnet
University Grenoble Alpes
CEA-Leti
Grenoble, France
gael.pillonnet@cea.fr

Patrick P. Mercier
Dept. of ECE
University of California San Diego
La Jolla, USA
pmercier@ucsd.edu

Abstract—This work introduces a control strategy for hybrid piezoelectric-resonator-base (PR-based) DC-DC converters, specifically focusing on the Dual-side Series/Parallel PR-based (DSPPR) converter. By analyzing the out-of-phase operation waveforms, which occur when the switching frequency is misaligned with the steady-state PR resonant frequency, a finite-state-machine-based phase-locked loop (PLL) is established to track the optimal PR operation, thereby maximizing efficiency performance. Moreover, incorporating adaptive synchronous rectification facilitates the effective operation of a hybrid PR-based converter while enhancing overall efficiency. A voltage regulation and loss minimization loop with PI compensations is also introduced, enabling fully closed-loop operation with precise regulation capability. In addition, a frontside switched capacitor voltage balancing loop is designed to regulate the capacitor voltage. A discrete prototype of the dual-side series/parallel PR-based converter is built with a (20/0.2)mm PZT radial vibration mode PR to verify the proposed control strategy. It should be noted that the proposed control strategy is adaptable to various other hybrid PR-based converter designs.

Index Terms—Piezoelectric resonator, switched capacitor, hybrid converter, closed-loop control.

I. INTRODUCTION

Driven by the need for compact, high-performance DC-DC converters, various materials have been investigated as alternatives to magnetic devices [1]. Among these, piezoelectric resonators (PRs) garner significant attention for their thin planar form factor, cost-effectiveness, linear frequency scaling, and promising efficiency and power density. As a result, numerous PR-based converters have been reported, demonstrating impressive performance across diverse voltage ranges [2], [3]. Meanwhile, [4]–[6] demonstrates switched-capacitor hybrid PR-based converters that strive to improve efficiency at low voltage conversion ratios (VCRs), defined as V_o/V_{in}. Additionally, [7] showcases a hybrid PR-based converter realized through integrated circuit (IC) technology, with enhanced efficiency at even lower VCRs (VCRs ≤ 0.1).

However, most of the reported converters are verified through open-loop testing, as the closed-loop dynamics of PRs within the system are complex [2]–[8]. This omission stems from the need for intricate and precise timing control in PR-based converters. To ensure optimal performance of a PR-based converter, a controller is required to manage frequency, pulse width, dead time, and phase shift modulations, presenting significant challenges.

A few prior works have explored closed-loop strategies for PR-based converters. For example, [9] employs pulse-frequency modulation (PFM) for regulation, yet lacks consideration of optimal timing, which may result in zero-voltage switching (ZVS) losses under variable load conditions. [10] introduced a fixed-frequency control to avoid the spurious modes effects. [11] achieves optimal operation with regulation; however, this control is case-specific, and its sense-based nature may be susceptible to noise. [12] addresses noise immunity concerns with a static closed-loop control that aims for optimal operation. Despite these developments, there remains a lack of control strategies specifically designed for hybrid PR-based converters. Moreover, previous works have not addressed the out-of-phase/frequency behavior, a crucial factor for understanding system dynamics.

This paper introduces a control strategy for hybrid PR-based converters [4]–[7], which leverages a phase-locking concept through a "sense and predict" mechanism that examines the out-of-phase/frequency behavior of PR waveforms. This work specifically focuses on the dual-side series/parallel PR-based converter [7] due to its topological complexity compared to [4]–[6]. Importantly, the proposed control method exhibits adaptability, enabling seamless adaptation to the aforementioned PR-based converters, as they share similar operational characteristics, specifically the six-stage switching sequences, $\{V_{in} - V_o, zero, V_o\}$ defined in [2]. To validate the proposed strategy, a discrete prototype featuring a (20/0.2)mm PZT radial vibration mode PR is constructed, with closed-loop operation managed by a digital controller.

II. DUAL-SIDE SERIES/PARALLEL PIEZOELECTRIC-RESONATOR-BASED DC-DC CONVERTER

Fig. 1 shows the Dual-side Series/Parallel PR-based (DSPPR) DC-DC converter, which merges two 2:1 series/parallel switched capacitors (SCs) on the frontside and backside of the baseline PR converter [2]. Leveraging the merged SCs, the converter achieves optimal efficiency at an 8:1 VCR [7] and the regulation is realized through the dedicated operational phase arrangement of the PR. The PR is characterized by the Butterworth Van-Dyke (BVD) model

---

979-8-3315-1612-3/25 $31.00 © 2025 IEEE   315

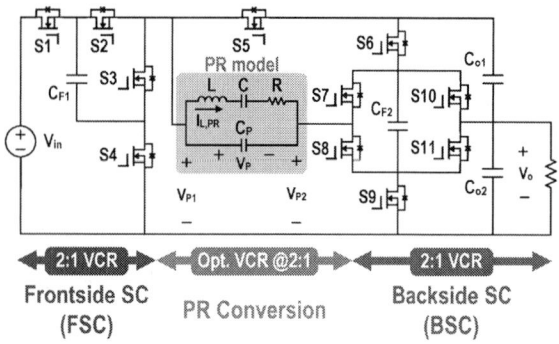

Fig. 1. Dual-side Series/Parallel PR-based (DSPPR) converter [7]

[13], which incorporates the mechanical vibration components, R, L, and C, and the static piezoelectric capacitor, C_P.

The operation waveforms of the DSPPR converter are depicted in Fig. 2. The converter operates in seven distinct phases, with the frontside SC (FSC) in series/parallel interleaving modes between cycles and the backside SC (BSC) in series/parallel configurations within each cycle. In this work, the equivalent PR input/output voltage is denoted as $V_{in,PR}$, and $V_{o,PR}$, corresponding to values of $V_{in}/2$ and $2V_o$, respectively. Here, the PR can be configured into 3 distinct stages: 1) an open stage, 2) a connected stage, and 3) a shorted stage, which are used for ZVS operation, energy transfer, and current free-wheeling, respectively.

In each cycle, the phases proceed as follows:

- **Phase 1S/1P**: The PR is opened. and $i_{L,PR}$ softly discharges C_P, leading to a voltage rise in the negative PR node, V_{P2}, until it reaches $V_{o,PR}$.
- **Phase 2S/2P**: Switches S6, S7, and S11 are enabled with zero-voltage switching (ZVS), connecting the PR to both the input and output to deliver energy and simultaneously charge the PR.
- **Phase 3**: Either switches S1 and S3 or S2 and S4 are deactivated, returning the PR to an open state. C_P undergoes a soft discharge, where the positive PR node, V_{P1}, gradually falls to $V_{o,PR}$.
- **Phase 4**: ZVS turn-on of switch S5 creates a freewheeling path, allowing $i_{L,PR}$ to circulate until its polarity changes.
- **Phase 5**: In this opened PR state, C_P is softly charged by $i_{L,PR}$ where V_{P2} gradually falls 0.
- **Phase 6**: Switches S8, S9, and S10 are turned on with ZVS, allowing the PR to release stored energy to the output in this connected state.
- **Phase 7**: The PR is once again in the opened state, where C_P is softly charged by $i_{L,PR}$, causing V_{P1} to rise until $V_{P1} = V_{in,PR}$, initiating a new cycle.

III. FEEDBACK LOOPS AND CONTROL STRATEGY

A. Control logic of driving signals

According to Fig. 2, the accurate operating frequency and the on/off timings are crucial for achieving optimal performance in a PR-based converter including ZVS and soft-

charging. Here, Fig. 2 illustrates the driving logic based on the operation waveforms, where t_0 to t_6 represent the on/off timing of the driving signals.

Given the variable frequency nature of PR operation, the operating frequency needs to be phase-locked with the PR resonance. This synchronization allows the switches to be turned on/off accordingly, enabling voltage regulation, loss minimization, and ZVS. Therefore, phase-locked control becomes imperative. Once the frequency and phase are established, the subsequent driving logic can be determined as follows and the timing control summary is illustrated in Table I.

- **Signal (S1,3), (S2,4)**: Turned on at the beginning of a cycle, t_0, and turned off based on the voltage regulation loop at t_2. Here, the duty cycles of (S1,3) and (S2,4), $D_{S1,3}$, and $D_{S2,4}$, range across phases 1 and 2 with phase 2 being the only phase connected to the input. Hence, $D_{S1,3}$ and $D_{S2,4}$ can be used to control the amount of energy sourced from the input, towards achieving output voltage regulation. The voltage regulation capability can be verified by solving the voltage conversion ratio of the PR using conservation of charge and energy (CoC and CoE) [2], as shown in (1). Here, q_n denotes the charge in the n-th phase. It can be found that the portion of q_2 determines the voltage gain and, therefore, the output voltage.

$$0 \leq \frac{V_{o,PR}}{V_{in,PR}} = \frac{q_2}{2q_2 + q_4} \leq \frac{1}{2}. \tag{1}$$

- **Signal (S5)**: Turned on after a delay time, DT_{S5}, when V_{P1} falls to $V_{o,PR}$ at t_3, and turned off according to the loss minimization loop at t_6. Here, the turn-off timing is related to the phase 7 duration, during which $i_{L,PR}$ charges C_P. That is, at a given frequency, the duty cycle of S5, D_{S5}, determines the phase 7 duration. To ensure precise charging time in phase 7 and prevent additional losses at t_0, the turn-off timing of S5 is regulated by a feedback compensator within the loss minimization loop.
- **Signal (S6,7,11)**: Turned on after a delay time, DT_{S6-11}, when V_{P2} rises to $V_{o,PR}$ at t_1, and turned off at half cycle, t_4, where $i_{L,PR}$ reverts its polarity.
- **Signal (S8,9,10)**: The driving signal (S8,9,10) mirrors the signal (S6,7,11) but operates with a 180-degree out-of-phase configuration, which turns on/off at t_5 and t_0 respectively. Upon closer examination of the delay times, DT_{S6-11}, for (S8,9,10) and (S6,7,11), the duration of phases 1 and 5, respectively, a pattern emerges. Throughout phases 1 and 5, $i_{L,PR}$ charges and discharges C_P with an identical current amplitude and a voltage difference of ($V_{o,PR}$). Furthermore, their deactivation aligns precisely with the zero-crossing of $i_{L,PR}$. Thus, signal (S8,9,10) replicates signal (S6,7,11) with a half-cycle phase shift.

B. Phase-locked loop

Before designing the phase-locked mechanism for a PR-based converter, it is important to understand how the PR behaves during out-of-phase operation, which can be achieved

979-8-3315-1612-3/25 $31.00 © 2025 IEEE

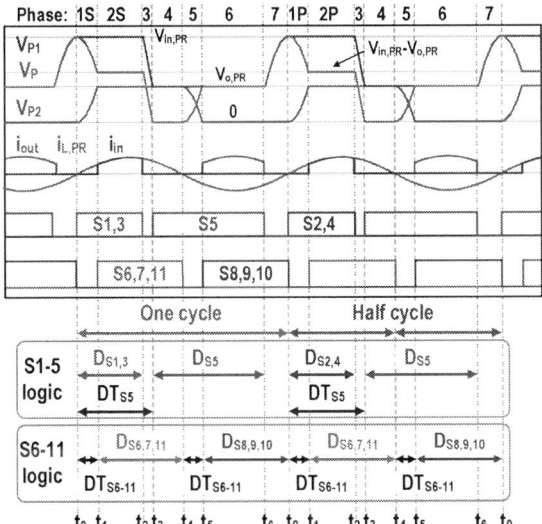

Fig. 2. Operation waveforms and control logic of the DSPPR converter

TABLE I
TIMING CONTROL OF EACH DRIVING SIGNAL

Signal	Turn-on control	Turn-off control
(S1,3)/ (S2,4)	PLL (t_0)	Voltage regulation loop (t_2)
S5	Comparator (t_3)	Loss minimization loop (t_6)
(S6,7,11)	Comparator (t_1)	PLL (t_4)
(S8,9,10)	Comparator (t_5)	PLL (t_0)

by inspecting V_P, V_{P1}, and V_{P2} at the zero-crossing of $i_{L,PR}$. Given the intricate nature of V_P measurement in practice, the analysis of out-of-phase behavior is performed through the examination of V_{P1} and V_{P2}. Here, the beginning of a cycle is defined as t_0, as illustrated in Fig. 2, where $i_{L,PR}$ experiences zero-crossing in the ideal scenario.

1) PR positive node analysis, V_{P1}: The ideal and most common out-of-phase V_{P1} waveforms are shown in Fig. 3(a)-(f). Here, two distinct sampling points, denoted as Pre_{smp1} and Pre_{smp2}, are assigned for pre-sampling immediately before a new cycle. According to the pre-sample results in Fig. 3, the relationship can be partially identified, as shown in Table II.

For instance, in (b)-(d) of Fig. 3, when Pre_{smp1} is greater than Pre_{smp2}, f_{sw} is smaller than f_{PR} because the polarity of $i_{L,PR}$ reverses before the new cycle, discharging the V_{P1} node, and causing a decrease in both V_{P1} and Pre_{smp2}.

In an ideal case, V_{P1} gradually rises to $V_{in,PR}$ before the subsequent switches are activated, thus entering the next phase. However, deviations in the switching frequency, f_{sw}, from the ideal PR operating frequency, f_{PR}, may result in V_{P1} not precisely equaling $V_{in,PR}$ at t_0, as demonstrated in (b)-(f) of Fig.3. For instance, in (b)-(d) of Fig. 3, when Pre_{smp1} is greater than Pre_{smp2}, f_sw is smaller than f_{PR} because the polarity of $i_{L,PR}$ reverses before the new cycle, discharging V_{P1} node, and causing a decrease in both V_{P1} and Pre_{smp2}. It is worth noting that, based on Table II, the relationship between f_{sw} and f_{PR} cannot be solely determined by V_{P1} waveforms since t_6 turn-off timing also plays a role in the

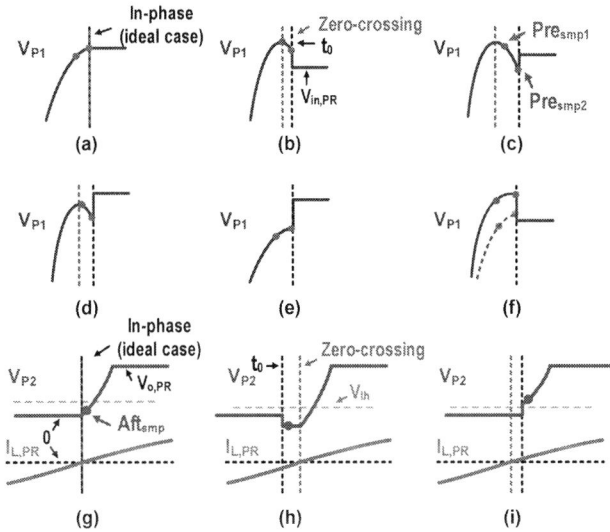

Fig. 3. PR positive/negative node, V_{P1}, V_{P2}, waveforms in proximity to t_0

TABLE II
V_{P1} AND V_{P2} IN/OUT-OF-PHASE WAVEFORMS SUMMARY

Singal	Case	Observation	Conclusion
V_{P1}	(a)	$Pre_{smp2} \approx V_{in,PR} > Pre_{smp1}$	$f_{sw} \approx f_{FR}$
	(b)	$Pre_{smp1} > Pre_{smp2} > V_{in,PR}$	$f_{sw} < f_{PR}$
	(c)	$Pre_{smp1} > Vin,PR > Pre_{smp2}$	$f_{sw} < f_{PR}$
	(d)	$Vin,PR > Pre_{smp1} > Pre_{smp2}$	$f_{sw} < f_{PR}$
	(e)	$V_{in,PR} > Pre_{smp2} > Pre_{smp1}$	$f_{sw} > f_{PR}$ or $t_6 > t_{6,ideal}$
	(f)	$Pre_{smp2} > Pre_{smp1} > V_{in,PR}$ or $Pre_{smp2} > V_{in,PR} > Pre_{smp1}$	$t_6 < t_{6,ideal}$
V_{P2}	(g)	$V_{th} > Aft_{smp} > 0$	$f_{sw} \approx f_{PR}$
	(h)	$V_{th} > 0 \geq Aft_{smp}$	$f_{sw} > f_{PR}$
	(i)	$Aft_{smp} > V_{th} > 0$	$f_{sw} < f_{PR}$ or t_6 timing

final value of V_{P1} at t_0, i.e., case (e) and (f).

2) PR negative node analysis, V_{P2}: Apart from the V_{P1} waveform, the V_{P2} waveform near t_0 appears to be another powerful indicator. Similar to V_{P1}, the V_{P2} waveform also provides insights into frequency deviations, as shown in Fig. 3(g)-(h). Here, an after-sampling point, Aft_{smp}, is assigned to capture V_{P2} after the initiation of a new cycle, and a threshold voltage, V_{th}, is used to determine the frequency deviation.

In Fig. 3(g)-(h), three scenarios are illustrated: (g) $f_{sw} \approx f_{PR}$, (h) $f_{sw} > f_{PR}$, and (i) $f_{sw} < f_{PR}$ where the conclusions are presented in Table II. In case (g), the ideal case, Aft_{smp} is a small positive value due to the soft charging of C_P with a small current at the very beginning of the new cycle. In case (h), Aft_{smp} shows a voltage drop, a consequence of the body-diode conduction initiated before the zero-crossing, where a current path is established from the ground to the PR input through the body diodes of S8 and S9, as well as the path of switches (S1,3) or (S2,4). In case (i), $i_{L,PR}$ reverses before a new cycle, causing the discharging of C_P and, hence, presenting a voltage step at t_0. It is noteworthy that the t_6 turn-off timing also affects the voltage step in case (i).

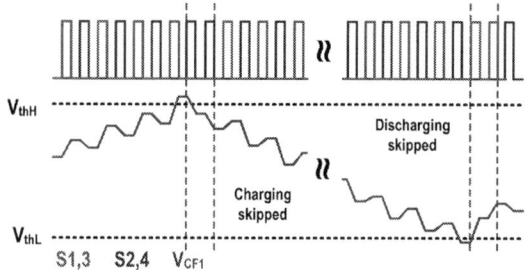

Fig. 4. Frontside switched capacitor (FSC) balance loop control

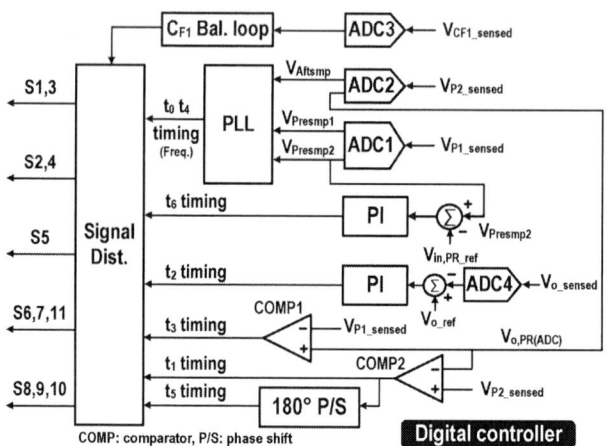

Fig. 5. Block diagram for the closed-loop control

3) Phase-locked loop implementation: Analyzing the V_{P1} and V_{P2} waveforms near t_0 provides insights into frequency adjustments, as inferred from Table II. The frequency modulation can be executed based on (a-d), (g), and (h), facilitated by (e), (f), and (i), given that (a-d), (g), and (h) are deterministic cases, while (e), (f), and (i) are indeterministic cases influenced by the t_6 timing. For example, upon detecting case (b), the controller can increase f_{sw} to synchronize with f_{PR}; on the other hand, identifying case (h) allows the controller to decrease f_{sw} to align with f_{PR}. That is, according to Table II, a state machine within a digital controller can be designed to achieve frequency/phase alignment where (a) and (g) occur as $f_{sw} \approx f_{PR}$, (b)-(d) are used to identify $f_{sw} < f_{PR}$ (pre-sample points), and (h) is used to detect $f_{sw} > f_{PR}$ (after-sample point).

C. Voltage regulation and loss minimization loops

1) Voltage regulation loop: The voltage regulation loop involves a feedback mechanism that senses the output voltage and feeds the sensed signal into a proportional-integral (PI) controller with a reference voltage, generating duty cycle commands. The commands are subsequently used to control switches S1-4, which, in turn, implicitly control energy sourced from the input in phase 2, as shown in Fig. 2.

2) Loss minimization loop: The loss minimization loop constitutes another feedback operation designed to precisely align V_P with $V_{in,PR}$ at t_0 by leveraging the sampled result, Pre_{smp2}, from the phase-locked loop. This feedback loop feeds Pre_{smp2} to another PI compensator along with the $V_{in,PR}$ reference, generating an off-time command for t_6 in Fig. 2. This off-time governs the soft-charging duration of C_P in phase 7 and eventually influences the final V_P value at t_0.

D. Frontside switched capacitor balance Loop

The frontside switched capacitor (FSC) operates in a manner similar to that of the three-level buck converter [14]–[17]. In this configuration, the FSC voltage (V_{CF1}) is not self-balanced, which leads to voltage drift due to mismatches in parameters such as on-resistance and turn-on duration between series and parallel modes. To address the voltage drift and ensure optimal performance, a "skip-and-repeat" balance loop, utilizing hysteresis thresholds, is employed, as shown in Fig. 4. Here, the hysteresis thresholds, V_{thH} and V_{thL}, are set to activate the balance loop control when V_{CF1} reaches

these thresholds. Specifically, the series mode (charging mode) is skipped, while the parallel mode (discharging mode) is repeated when $V_{CF1} \geq V_{thH}$, and the opposite action is taken when $V_{CF1} \leq V_{thL}$.

E. Comparator sensing synchronization

As shown in Table I, the turn-on timings of switches S5 and S6-11 are determined by the results of comparator sensing. The turn-on timing of switch S5 is generated by comparing the sensed $V_{o,PR}$ and V_{P1} signals, occurring when V_{P1} falls to $V_{o,PR}$. For switches S6-11, the timing is based on the comparison between the sensed $V_{o,PR}$ and V_{P2} signals, triggering when V_{P2} rises to $V_{o,PR}$ as switches S6, S7, and S11 are ready to turn on. The turn-on timing for switches S8, S9, and S10 is phase-shifted by 180° relative to the comparison result, as described in Section III. Consequently, only one comparator output is required to turn on the backside switches S6-11. Additionally, the $V_{o,PR}$ sensing value can be obtained by leveraging the V_{P2} sensing point with a dedicated sensing interval between t_1 and t_4, effectively reducing circuit complexity.

Unlike traditional sense-and-react comparator-based control, the control timings in this design are not directly applied as driving signals. Instead, the timings are fed back to the controller, where the sensing result in this cycle, x_n, is compared with the result in the previous cycle, x_{n-1}, with an additional timing margin, ≈ 33ns in this design, and then determines the turn-on timing for the next cycle. Although body conduction loss is introduced by the timing margin, this approach ensures operational integrity and enhances noise immunity, effectively preventing false turn-on of the switches. Consequently, a sense-process-and-react behavior is implemented in this design.

F. Block diagram for closed-loop control

The block diagram depicting the finalized closed-loop control is shown in Fig. 5. Here, a digital controller, TMS320F28397D, is utilized in this work, whose analog comparators share the same pins as the ADC blocks, reducing

979-8-3315-1612-3/25 $31.00 © 2025 IEEE

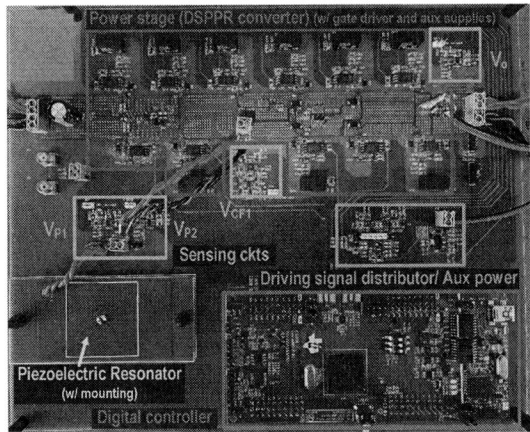

Fig. 6. Circuit implementation for closed-loop test

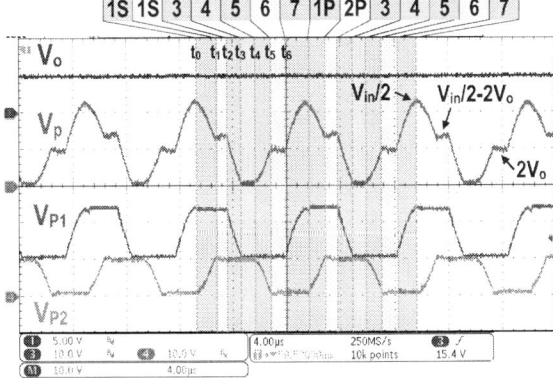

Fig. 7. Closed-loop steady-state waveforms at 48/5V, 0.1A

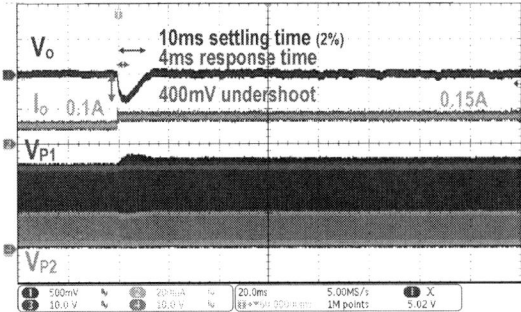

Fig. 8. Load step-up dynamics at 48/5V, 0.1-0.15A

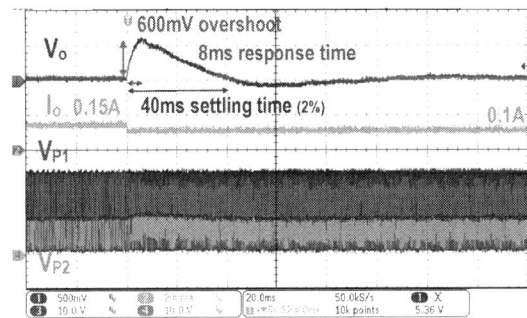

Fig. 9. Load step-down dynamics at 48/5V, 0.15-0.1A

system complexity with four feedbacks: V_{CF1}, V_{P1}, V_{P2}, and V_o. In Fig. 5, the comparators, $COMP1$, and $COMP2$, share the pins with $ADC1$ and $ADC2$ respectively. Additionally, the reference voltage, $V_{o,PR(ADC)}$, for the comparators can be obtained from oversampling of V_{P2} in $ADC2$. Ultimately, a fully closed-loop system with reduced complexity is designed.

IV. IMPLEMENTATIONS AND EXPERIMENTAL RESULTS

The DSPPR converter prototype is built, as shown in Fig. 6, with a digital controller, F28379D, operating at 150 MHz clock frequency. A disk-shaped PIC 181 PR with a size of (20/0.2) mm in radius and thickness is applied to this design, exhibiting resonant and anti-resonant frequencies of 113 and 129 kHz, respectively, with a quality factor of ≈ 700, which is defined as $(\sqrt{L/C})/R$. EPC2241s are used as active switches. The output capacitors and flying capacitors are $\approx 10\mu$F MLCCs.

The steady-state waveforms with closed-loop control enabled are shown in Fig. 7, with each corresponding phase and timing annotated. The measured waveforms are similar to the ideal waveforms, as shown in Fig. 2, demonstrating soft-charging of C_P and quasi-ZVS. Due to the inherent non-idealities, such as sensing resolution, offsets, and processing delays, ZVS is not fully achieved, though it is close.

Transient results are demonstrated in Fig. 8 and Fig. 9 with a load step from 0.1A to 0.15A and a 0.1A/μs slew rate. Here,

the undershoot voltage, response time, and 2% settling time are 400mV, 4ms, and 10ms, respectively, in the load step-up, and 600mV, 8ms, and 40ms, respectively, in the load step-down.

The FSC balance waveforms are shown in Fig. 10, revealing that when the balancing loop is disabled, there is a voltage offset due to parameter mismatch. On the other hand, when the balancing loop is enabled, the series-mode-skipped control is activated, repeating an additional discharging operation to realign V_{CF1} to $V_{in}/2$, which is 24V in the test scenario.

The efficiency of the DSPPR converter was measured across a range from 36-60V to 4-6V (VCR < 0.125) at various load currents. A comparison was conducted with previously reported topologies, including the Baseline PR [2], Backside Series/Parallel PR-based (BSPPR) [6], and Frontside Series/Parallel PR-based (FSPPR) [4] converters, as shown in Fig. 11. For a fair comparison, efficiency curves were obtained using the same power and control design components

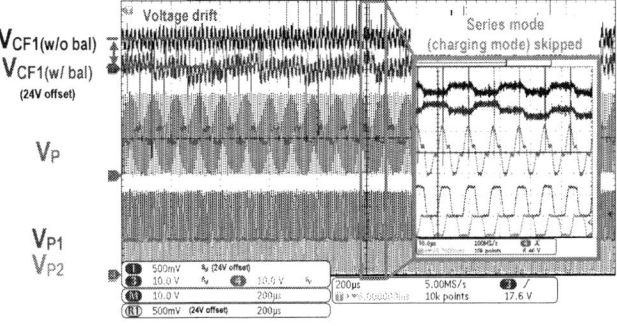

Fig. 10. Frontside switched capacitor balance waveforms at 48/5V

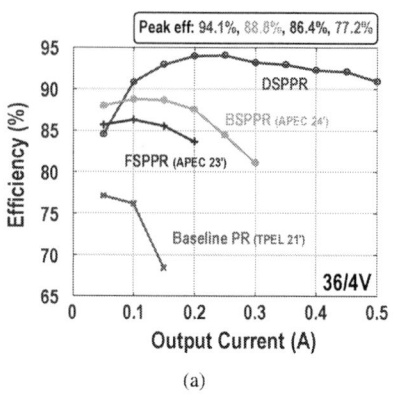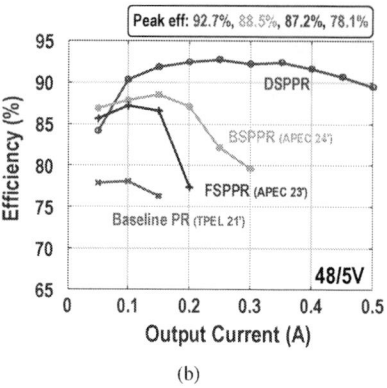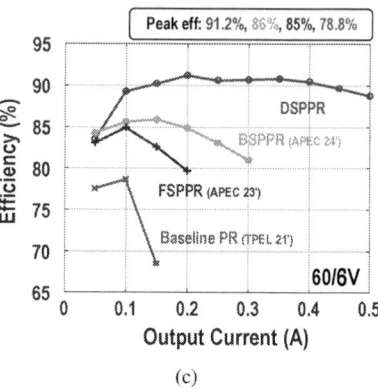

Fig. 11. Efficiency comparison at (a) 36V-to-4V, (b) 48V-to-5V, (c) 60V-to-6V

for each topology, as depicted in Fig. 6. Results indicate that DSPPR achieves better performance at low VCR conditions, with peak efficiencies of 94.1% at 36V-to-4V and 0.2A output, showcasing an improvement of approximately 5% to 17% over other topologies. Additionally, output capability is enhanced by the dual-side SC strategy, lowering the PR resonant current and, as a result, maximizing the utilization of the PR devices, which is typically limited by their maximum current density. However, DSPPR shows slightly lower efficiency under light-load conditions due to the additional power FETs in the power loop.

V. CONCLUSIONS

This paper introduces a new control strategy for hybrid PR-based converters. Each control signal timing is illustrated in detail, and accordingly, the driving strategy is developed. By examining the out-of-phase waveforms of the PR, a state machine is designed for PLL control. In addition, the regulation and loss minimization loops are implemented to achieve ideal operation and regulation. Finally, a prototype is built featuring a disk-shaped PIC 181 PR with dimensions of (20/0.2)mm in radius and thickness, where the closed-loop control is realized by a digital controller. The measurement results validate the proposed control strategy and demonstrate the ability to regulate output. A comparison of efficiency performance is also conducted along with previously reported topologies.

ACKNOWLEDGEMENT

This work was supported in part by the Power Management Center (PMIC) an NSF I/UCRC, award number 2052809

REFERENCES

[1] P. A. Kyaw, A. L. F. Stein, and C. R. Sullivan, "Fundamental examination of multiple potential passive component technologies for future power electronics," *IEEE Transactions on Power Electronics*, vol. 33, no. 12, pp. 10 708–10 722, 2018.

[2] J. D. Boles, J. J. Piel, and D. J. Perreault, "Enumeration and analysis of dc-dc converter implementations based on piezoelectric resonators," *IEEE Transactions on Power Electronics*, vol. 36, no. 1, pp. 129–145, 2021.

[3] M. Touhami, G. Despesse, and F. Costa, "A new topology of dc-dc converter based on piezoelectric resonator," *IEEE Transactions on Power Electronics*, vol. 37, no. 6, pp. 6986–7000, 2022.

[4] W.-C. B. Liu and P. P. Mercier, "A series/parallel magnetic-less step-down converter based on piezoelectric resonators," in *2023 IEEE Applied Power Electronics Conference and Exposition (APEC)*, 2023, pp. 484–489.

[5] Q. Li, Y. Hou, and K. K. Afridi, "Merged switched-capacitor piezoelectric-resonator based dc-dc converter with high voltage conversion ratio," in *2023 IEEE 24th Workshop on Control and Modeling for Power Electronics (COMPEL)*, 2023, pp. 1–8.

[6] W.-C. B. Liu and P. P. Mercier, "A merged backside series/parallel hybrid piezoelectric-resonator-based dc-dc converter," in *2024 IEEE Applied Power Electronics Conference and Exposition (APEC)*, 2024.

[7] W.-C. B. Liu, G. Pillonnet, and P. P. Mercier, "An integrated dual-side series/parallel piezoelectric resonator-based dc–dc converter," *IEEE Journal of Solid-State Circuits*, pp. 1–13, 2024.

[8] B. Pollet, G. Despesse, and F. Costa, "A new non-isolated low-power inductorless piezoelectric dc-dc converter," *IEEE Transactions on Power Electronics*, vol. 34, no. 11, pp. 11 002–11 013, 2019.

[9] G.-S. Seo, J.-W. Shin, and B.-H. Cho, "A magnetic component-less series resonant converter using a piezoelectric transducer for low profile application," in *The 2010 International Power Electronics Conference - ECCE ASIA -*, 2010, pp. 2810–2814.

[10] E. Stolt, W. D. Braun, L. Gu, J. Segovia-Fernandez, S. Chakraborty, R. Lu, and J. Rivas-Davila, "Fixed-frequency control of piezoelectric resonator dc-dc converters for spurious mode avoidance," *IEEE Open Journal of Power Electronics*, vol. 2, pp. 582–590, 2021.

[11] M. Touhami, G. Despesse, F. Costa, and B. Pollet, "Implementation of control strategy for step-down dc-dc converter based on piezoelectric resonator," in *2020 22nd European Conference on Power Electronics and Applications (EPE'20 ECCE Europe)*, 2020, pp. 1–9.

[12] J. J. Piel, J. D. Boles, J. H. Lang, and D. J. Perreault, "Feedback control for a piezoelectric-resonator-based dc-dc power converter," in *2021 IEEE 22nd Workshop on Control and Modelling of Power Electronics (COMPEL)*, 2021, pp. 1–8.

[13] K. Van Dyke, "The piezo-electric resonator and its equivalent network," *Proceedings of the Institute of Radio Engineers*, vol. 16, no. 6, pp. 742–764, 1928.

[14] T. Meynard and H. Foch, "Multi-level conversion: high voltage choppers and voltage-source inverters," in *PESC '92 Record. 23rd Annual IEEE Power Electronics Specialists Conference*, 1992, pp. 397–403 vol.1.

[15] Y. Liu, A. Kumar, D. Maksimovic, and K. K. Afridi, "A high-power-density high-efficiency three-level buck converter for cellphone battery charging applications," in *2018 IEEE Energy Conversion Congress and Exposition (ECCE)*, 2018, pp. 5265–5270.

[16] Z. Li, Z. Xue, C. Liang, Y. Zhang, M. Duan, S. Zhao, X. Liu, Z. Guo, and L. Geng, "A single-input dual-output three-level buck converter for soc applications," in *2022 IEEE International Conference on Integrated Circuits, Technologies and Applications (ICTA)*, 2022, pp. 127–128.

[17] S. Udomkaew, K. Sengsui, W. Saksiri, R. Gavagsaz-Ghoachani, M. Phattanasak, and S. Pierfederici, "Current and capacitor-voltage balancing controls of three-level buck converter," in *2023 Research, Invention, and Innovation Congress: Innovative Electricals and Electronics (RI2C)*, 2023, pp. 360–364.

979-8-3315-1612-3/25 $31.00 © 2025 IEEE

High-Bandwidth Embedded Rogowski Coil on Multilayer Substrate with Minimal Contribution to Power Loop Inductance

Takahiro Okamoto
Graduate School of Natural Science and Technology
Okayama University
Okayama, Japan
pnc26471@s.okayama-u.ac.jp

Masataka Ishihara
Graduate School of Natural Science and Technology
Okayama University
Okayama, Japan
masataka.ishihara@ec.okayama-u.ac.jp

Kazuhiro Umetani
Graduate School of NaturalScience and Technology
Okayama University
Okayama, Japan
umetani@okayama-u.ac.jp

Eiji Hiraki
Graduate School of NaturalScience and Technology
Okayama University
Okayama, Japan
hiraki@okayama-u.ac.jp

Abstract— **Power devices using GaN-HEMTs require high-bandwidth current sensors for fast switching current sensing, switch device protection and system control. PCB-embedded Rogowski coils are suitable for high frequency current sensing because they are electrically non-contact, have no magnetic core, and can be compactly mounted in the system. However, in conventional Rogowski coils, it is necessary to pass the current path through the center of the coil. This causes an extension of the current path, and consequently, an increase in parasitic inductance in the power loop of GaN-HEMT devices. In this paper, we propose a Rogowski coil consisting of a multilayer printed-circuit-board (PCB) substrate, which pick-up coil embedded in a laminated bus bar. The pick-up coil is also designed with microstrip lines to allow measurement of high-frequency current. Experimental results confirm that the bus bar of the proposed Rogowski coil exhibited low parasitic inductance and can detect high-frequency currents.**

Keywords—Current sensor, Parasitic Inductance, Laminated bus bar, Microstrip line, Impedance matching

I. INTRODUCTION

The next-generation semiconductor GaN-HEMT, which has emerged in recent years, achieves low on-resistance and high-speed switching. Consequently, inverters using GaN-HEMT can miniaturize surrounding passive components through increased drive frequency, which has led to extensive research and development of compact, high power density inverters and converters to date [1]–[3]. In such power modules employing GaN-HEMT, a current sensor in the switching current measurement module is required for device protection and system control. Additionally, since the short-circuit withstand capability of GaN-HEMTs is limited [4], a high-bandwidth current sensor capable of instantaneously detecting abnormal currents is essential.

As shown in Fig. 1, the Rogowski coil consists of a coil winding and an integration circuit. The basic current measurement principle of the Rogowski coil is that the

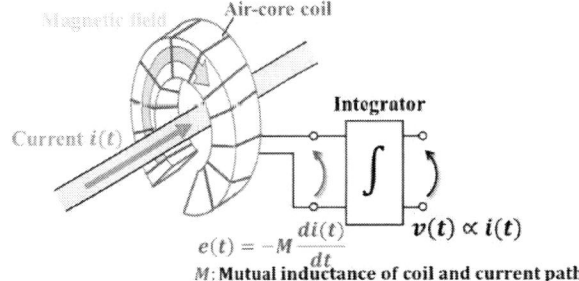

$$e(t) = -M\frac{di(t)}{dt} \qquad v(t) \propto i(t)$$

M : Mutual inductance of coil and current path

Fig. 1. Structure of Rogowski Coil

alternating magnetic field generated by the current links with the coil, inducing a voltage across the coil windings proportional to the derivative of the current according to Faraday's law of electromagnetic induction. This induced voltage is then integrated by the integrator, allowing the current to be detected. The Rogowski coil is electrically non-contact with the current and does not use a magnetic core, making it widely used as a current sensor for power devices, In recent years, compact Rogowski coils designed on PCB substrates have been actively researched and developed [5]–[8].

However, the conventional Rogowski coil structure shown in Fig. 1 presents two challenges. The first is the low bandwidth of the coil winding. To accurately detect fast-rising currents, such as those generated by GaN-HEMTs, it is difficult to achieve accurate measurement with existing Rogowski coils, which have a bandwidth of around 100 MHz. The second challenge is the increase in parasitic inductance associated with the implementation of the Rogowski coil. In conventional Rogowski coil structures, the current path must pass through the center of the coil, requiring the current path to be routed out and reconnected when implementing the Rogowski coil in a power module. This increases the length of the current path and, consequently, the parasitic inductance along the path. An increase in parasitic inductance in the power loop of power devices can cause significant switching surges in semiconductor

979-8-3315-1612-3/25 $31.00 © 2025 IEEE
321

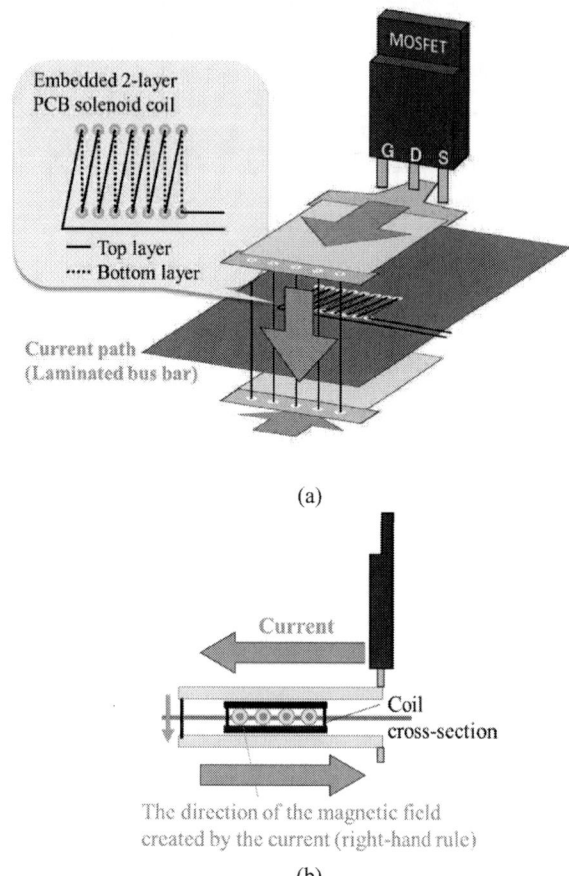

(a)

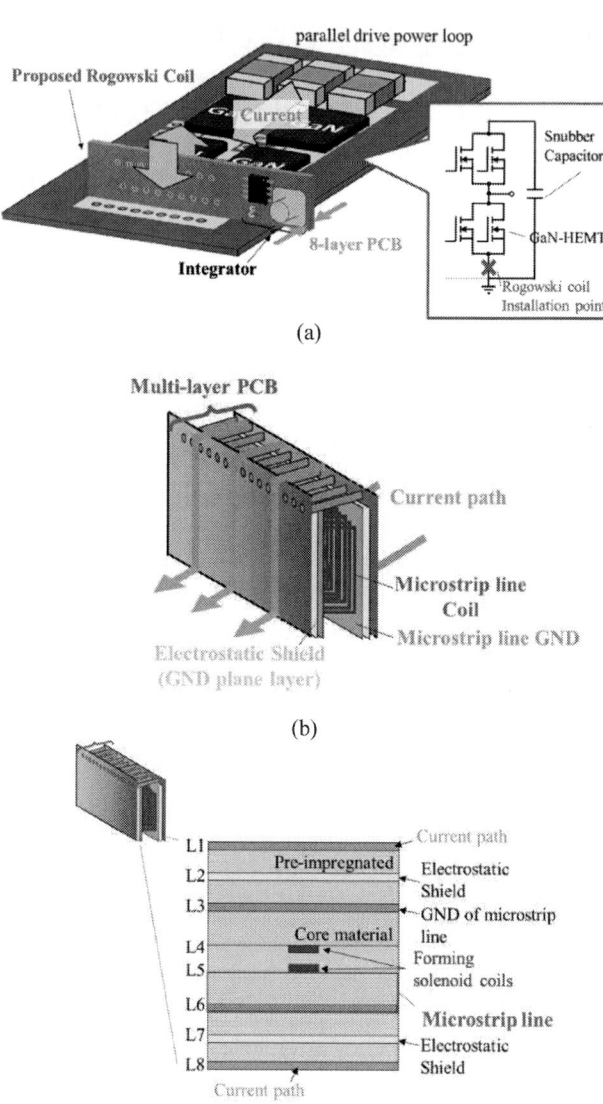

(a)

(b)

Fig. 2. Coil structure of Embedded Rogowski Coil (a), (b) side view

devices, posing a risk of device and system failure. Additionally, the need to route the current path through the coil center limits the width of the current path to the coil diameter, making it difficult to secure sufficient current path width for handling large currents.

To address the first issue of low bandwidth in the coil winding, the study in [9] designs the Rogowski coil winding as a microstrip transmission line with characteristic impedance, with both ends terminated by resistors for impedance matching. This approach suppresses the high-frequency reflection of voltage and current, characteristic of high-frequency regions, thereby preventing the degradation of the coil winding's frequency response. To address the second issue of increased parasitic inductance, an embedded Rogowski coil structure, as shown in Fig. 2, has been proposed [10]–[12]. As shown in Fig. 2(a), the current flows in opposing directions in a laminated structure, with the coil winding embedded inside as a solenoid coil formed within a two-layer substrate. According to Fig. 2(b), the magnetic field generated by the opposing currents links with the embedded coil winding in the same direction, inducing a voltage across the coil winding proportional to the derivative of the current, enabling current detection based on the same principle as conventional Rogowski coils. Meanwhile, the magnetic fields generated externally by the opposing currents cancel each other out, thereby minimizing the increase in parasitic inductance caused by the implementation of the Rogowski coil [13].

Fig. 3. Schematic diagram of the proposed embedded Rogowski Coil. (a) Example of implementation in a GaN-HEMT power module. (b) Schematic diagram. (c) 8-layer substrate configuration.

In this study, a high-bandwidth embedded Rogowski coil, constructed using a multi-layer PCB, is proposed to simultaneously address both challenges. The proposed Rogowski coil minimizes the increase in parasitic inductance, and through impedance matching, enhances the bandwidth of the coil winding. The effectiveness of the proposed Rogowski coil is verified by evaluating both the suppression of parasitic inductance and the improved high-bandwidth performance of the coil winding.

II. THE PROPOSED COIL STRUCTURE OF ROGOWSKI COIL

Fig. 3 shows a schematic of the proposed embedded Rogowski coil. The proposed Rogowski coil consists of an 8-layer PCB, and, as shown in Fig. 3(a), it is mounted vertically relative to the horizontal power loop current path of the GaN-HEMT parallel-driven power module. Since the surface layer of the proposed structure serves as a laminated current path, it minimizes the increase in parasitic inductance caused by the installation of the Rogowski coil. As shown in Fig. 3(c), the coil winding is formed as a solenoid-type coil using the signal lines of the microstrip line on layers 4 and 5. The ground (GND) of the microstrip line utilizes the adjacent layers, 3 and 6, respectively. Thus, the coil winding functions as a transmission line with characteristic impedance, enabling impedance matching at both ends of the coil winding. This suppresses voltage and current reflections during high-frequency current measurement and increases the bandwidth of the coil winding. To prevent noise caused by electrostatic coupling of high-frequency currents flowing on the surface layer and interference from external electromagnetic noise, electrostatic shielding layers connected to GND are implemented on layers 2 and 7 of the microstrip line's GND layer. Since the electrostatic shielding layers must be connected to GND, the Rogowski coil needs to be installed at the low-side GaN-HEMT source terminal (GND).

III. LAYOUT OF THE PROPOSED ROGOWSKI COIL

A. The current path

As shown in Fig. 5(a) and Fig. 5(e), the current path is arranged on the 1st and 8th layers. The current path on the 1st layer is connected to 50 Ω from the signal line of the SMA edge connector. It forms an opposing trace that connects to the GND of the SMA edge connector on the lower side of the layout on the 8th layer through a via on the upper side of the layout. The 50 Ω resistor ensures impedance matching with the characteristic impedance of the measurement equipment when measuring the transfer characteristics of the Rogowski coil. The path width of the current route is designed to correspond to the switching current path width of a parallel-driven GaN-HEMT power module.

B. Microstrip line coil

The coil winding is formed as a solenoid coil using the inner 4th and 5th layers, as shown in Fig. 5(d). The 4th and 5th layers are connected via buried via holes. The coil winding has an inter-winding spacing at least 10 times the width of each line, and the 3rd and 6th layers are used as the ground for the microstrip line, enabling the design as a microstrip line with characteristic impedance, as shown in Equation (1) [13]. The GND layers of the 3rd and 6th microstrip lines are connected by vias in the central part of Fig. 5(c). This approach eliminates paths for eddy currents to flow via the vias with respect to the magnetic flux linked to the coil winding, thus stabilizing the microstrip line's GND. Where ε_r is the relative permittivity of the dielectric, w is the line width, t is the copper thickness, and h is the layer spacing, designed to achieve a characteristic impedance of 47 Ω.

$$Z_0 = \frac{87}{\sqrt{\varepsilon_r + 1.41}} \ln\left(\frac{5.98h}{0.8w + t}\right) \quad (1)$$

Additionally, the self-inductance of the coil winding is approximately 75nH, as calculated using equation (2), where μ_0 is the permeability of free space and l_{track} is the length of the coil winding.

$$L_s = \frac{\mu_0 l_{track}}{2\pi} \ln\left(\frac{5.98h}{0.8w + t}\right) \quad (2)$$

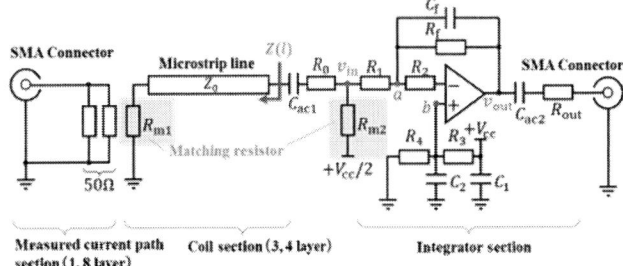

Fig. 4. Proposed structure characteristic evaluation circuit

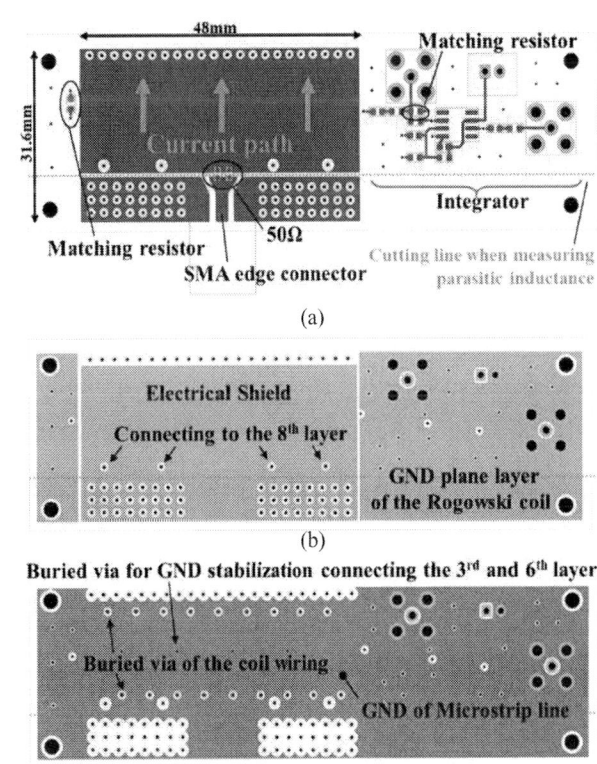

(a)

(b)

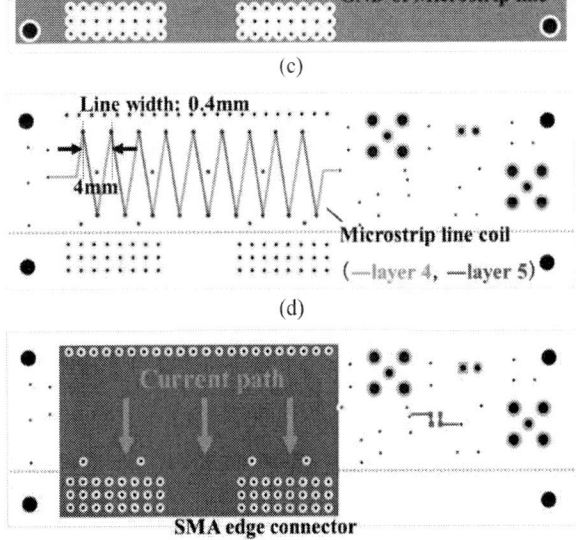

(c)

(d)

(e)

Fig. 5. PCB layout of the proposed Rogowski coil. (a)layer 1 (b) layer 2, 7 (c) layer 3, 6 (d) layer 4, 5 (e) layer 8

C. Electrical Shield

A GND plane electrostatic shield is provided on layers 2 and 7, where the coil winding is embedded. This shield prevents interference from noise caused by electrostatic coupling of high-frequency currents on the surface layer and external electromagnetic noise from affecting the GND of the microstrip line on layers 3 and 6. Since the electrostatic shield must be connected to GND, the proposed Rogowski coil must be attached to the source terminal of the low-side GaN-HEMT in the inverter. To establish the opposing current path, the electrostatic shield is connected to the 8th layer via a through-hole on the lower side of the layout in Fig. 5(b).

D. Integrator

The integrator, as shown in Fig. 4, is designed as an inverting integrator using an operational amplifier (THS3201 from Instruments). The transfer function is shown in equation (3), and the cutoff frequency of the integrator is shown in equation (4). The integrator's low-frequency gain is set to $R_f/R_1 = 10$, with a cutoff frequency of $f_c = 700[\text{kHz}]$. The component values for the integrator are listed in Table 1. The circuit layout utilizes layers 1 and 8, with all inner layers designated as GND planes.

$$G(j\omega) = -\frac{R_f}{R_1}\frac{1}{1 + j\omega C_f R_f} \tag{3}$$

$$f_c = \frac{1}{2\pi C_f R_f} \tag{4}$$

The capacitors C_{ac1} and C_{ac2} act as coupling capacitors, blocking DC current, and can be considered shorted in the AC equivalent circuit. The resistor R_2 is used to prevent oscillations in the current-feedback operational amplifier and is connected to the inverting input terminal of the op-amp. The resistor R_{out} sets the output impedance of the integrator, matching it to the impedance of the subsequent stage. Meanwhile, the input impedance of the integrator can be expressed by Equation (5). Due to the virtual short of the op-amp, points a and b maintain equal potential, effectively making points an equivalent to GND in the AC equivalent circuit. Thus, the input impedance of the integrator can be represented as the combined impedance of R_1 and R_{m2}, ensuring impedance matches with the characteristic impedance of the coil winding in the previous stage.

$$Z_{in} = \frac{R_1 \cdot R_{m2}}{R_1 + R_{m2}} \tag{5}$$

IV. EVALUATION

A. Evaluation of Parasitic Inductance in the Current Path

The surface layout of the measurement board for assessing the parasitic inductance of the measured current path is shown in Fig. 6. The backside of the measurement board is a ground plane, where current flows directly beneath the surface current and is connected to the SMB connector's ground pin through via holes. By comparing the parasitic inductance in the current path with the Rogowski coil attached at the mounting position shown in Fig. 6 and without the coil (connected with copper foil instead), the parasitic inductance of the Rogowski coil itself is determined. This parasitic inductance is calculated by connecting a capacitor in parallel with the current path parasitic inductance at the base of the SMB connector and deriving it from the parallel resonance point of the impedance characteristic.

TABLE I. PARAMETERS OF THE INTEGRATORS

Symbols	Values
R_1	$110[\Omega]$
R_2	$510[\Omega]$
R_3, R_4	$10[k\Omega]$
R_f	$560[\Omega]$
R_{out}	$51[\Omega]$
C_1	$0.1[\mu F]$
C_2	$4.7[\mu F]$
C_f	$390[pF]$
C_{ac1}, C_{ac2}	$4.7[\mu F]$

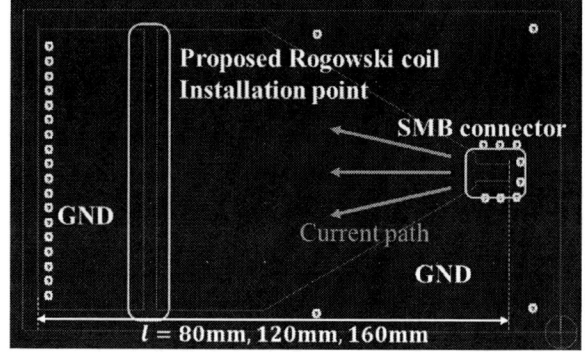

Fig. 6. Top layout for parasitic inductance measurement

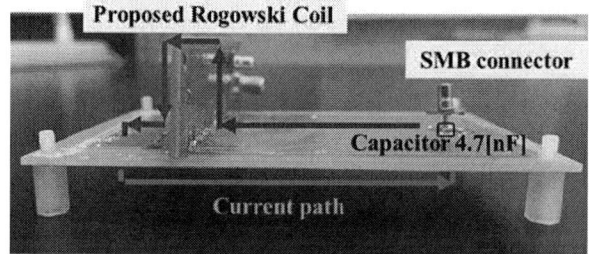

Fig. 7. Side view of parasitic inductance measurement board

Since the capacitor includes parasitic elements, the precise parasitic inductance is extracted by fitting the parallel resonance characteristics obtained in circuit simulation—considering the equivalent circuit model with parasitic components of the capacitor—to the measured resonance characteristics.

A side view of the measurement board with the Rogowski coil attached is shown in Fig. 7. An impedance analyzer (IM7581, HIOKI) was used for impedance measurement, and a 4.7 nF capacitor (KEMET) was employed. The equivalent circuit model accounting for parasitic components of the capacitor was obtained from the KEMET website. The current path length was varied to 80[mm], 120[mm], and 160[mm], and impedance characteristics for each were measured. Fig. 8 shows the relationship between the current path length and the parasitic inductance with and without the Rogowski coil attached. From Fig. 8, the differential parasitic inductance of the current path indicates that the parasitic inductance of the proposed Rogowski coil is consistently very low, remaining below 1[nH].

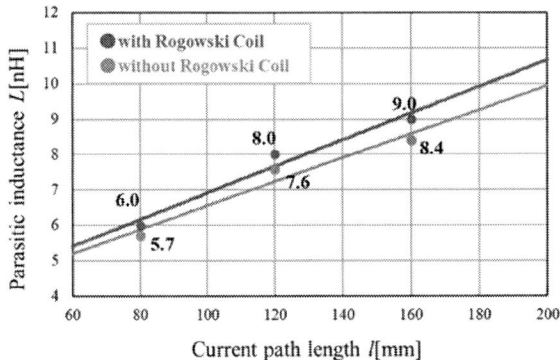

Fig. 8. Parasitic inductance characteristics

B. Transfer Characterisitc of the Proposed Rogowski Coil

First, the characteristic impedance of the designed microstrip line is derived. Equation (6) shows the relationship between the characteristic impedance and the coil input impedance $z(\lambda/4)$ when the wavelength is one-fourth. Based on Equation (6), the characteristic impedance of the microstrip line can be determined by connecting an arbitrary resistor R_{m1} at the start of the coil and measuring the coil's input impedance at one-fourth wavelength.

$$Z(\lambda/4) = Z_0 \frac{R_{m1} + jZ_0 \tan(\pi/2)}{Z_0 + jR_{m1} \tan(\pi/2)} = \frac{Z_0{}^2}{R_{m1}} \quad (6)$$

Fig. 9 shows the frequency characteristics of the coil input impedance when 150 Ω is connected to the starting end of the coil. The impedance was measured using an impedance analyzer (HIOKI, IM7581). According to Equation (6), at the frequency where the wavelength is one-fourth, the input impedance consists only of a real component, displaying resonance characteristics. Therefore, the impedance at one-fourth wavelength corresponds to the peak value in the frequency

response. Based on Equation (6), the characteristic impedance was found to be 37.6[Ω]. Compared to the design value of 47 [Ω], the derived value is smaller, likely due to the prepreg thickness between layers 3-4 and 5-6 of the microstrip line being thinner than the nominal value.

Next, we measure the impedance matching and transmission characteristics of the proposed Rogowski coil structure at the coil's start and end points, both in matched and mismatched conditions. Fig. 10 shows the circuit diagram for measuring the transmission characteristics, using a vector network analyzer (Keysight, E5061B). In the case of impedance matching at the coil's start, a 36 Ω resistor is connected, and in the mismatched case, a 0 Ω resistor is used. At the coil's end, a 130 Ω resistor is connected in parallel when matching the impedance, ensuring that the combined impedance with the network analyzer's 50 Ω coaxial cable matches the characteristic impedance of the coil windings. In the mismatched case, a 1 kΩ resistor is added in

series. Fig. 11 shows the transmission characteristics of the coil winding in both matched and mismatched conditions. Fig. 11 shows that the gain increases with a slope of +20[dB/dec] from 1MHz to about 100MHz during impedance matching and mismatching, respectively, and that a voltage proportional to the derivative of the current under test is induced between the coil windings. The phase characteristic shows -90 degrees at a few

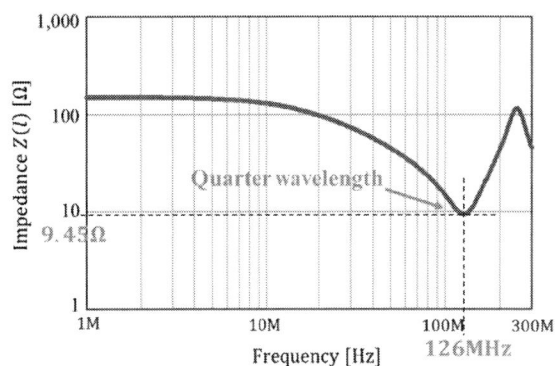

Fig. 9. Input impedance characteristics

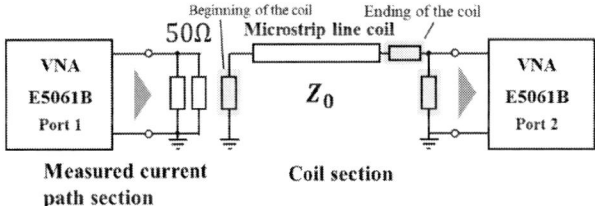

Fig. 10. Circuit for measuring coil transfer characteristics

MHz but begins to change significantly at tens of MHz This is because, at these frequencies, the wavelength of the measured current becomes non-negligible compared to the coil winding length, causing the voltage induced in the coil to not remain constant along the coil, resulting in phase shifts. Fig. 11(a) shows that impedance matching at the beginning of the coil gives transfer characteristics that suppress the quarter-wavelength resonance characteristics. Additionally, matching the impedance at the coil's end increases the gain of the output characteristic. This is because matching the coil's end ensures that the induced voltage is transmitted without reflection between the coil and the integrator, maximizing the power transfer.

The transmission characteristics of the proposed Rogowski coil, from the measured current path to the output of the integrator circuit, are shown in Figure 12. In this configuration, both the start and end points of the coil (including the input impedance of the integrator circuit) are matched to the characteristic impedance of the coil. The circuit constants for the integrator are shown in Table 1. Fig. 12 shows that the gain characteristics of the Rogowski coil are flat and constant up to 100MHz. This flat characteristic is equivalent to the combined gain of the coil and the integrator circuit, indicating that the Rogowski coil outputs a voltage that is a scaled version of the measured current. However, beyond 100 MHz, the frequency response exhibits characteristics that combine the resonance of the coil and the oscillatory behavior that may arise from the integrator circuit. In this frequency range, the measured current cannot be detected accurately. Regarding the phase characteristic of the Rogowski coil, it is found that, like the gain, the phase is a combination of the phase characteristics of the coil and the integrator circuit. This means that the phase between the measured current and the voltage output from the Rogowski coil is not constant. This variation in phase is likely due to the propagation of the induced voltage along the microstrip coil, and compensatory circuitry for phase shift needs to be considered.

979-8-3315-1612-3/25 $31.00 © 2025 IEEE

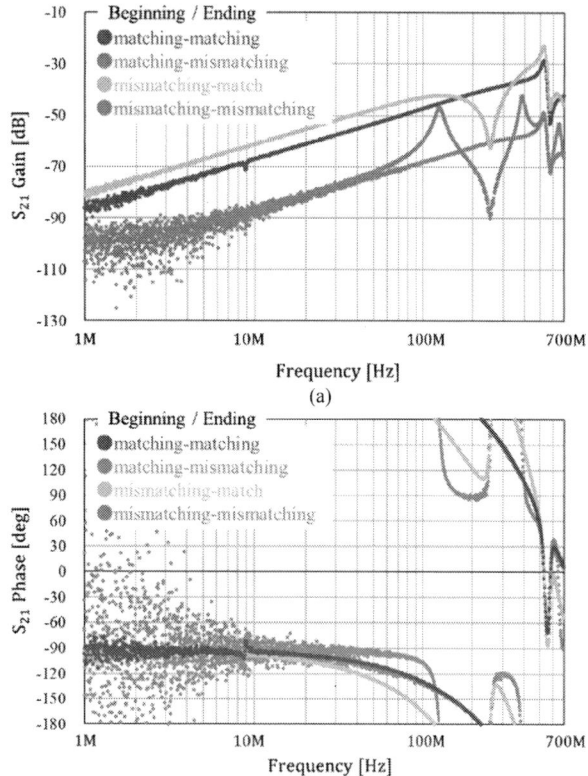

Fig. 11. Coil transfer characteristic. (a) Gain characteristic. (b) Phase characteristics

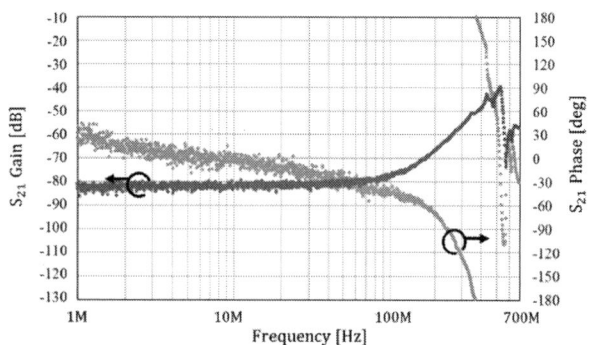

Fig. 12. Transfer characteristic of the proposed Rogowski coil

Furthermore, the bandwidth of the proposed Rogowski coil structure is limited by the integrator circuit, which restricts its effective range to around 100MHz. Improving the bandwidth of the integrator circuit will be an important future challenge.

V. CONCLUSION

In this study, a new embedded Rogowski coil with an electrostatic shield composed of multilayer substrates is proposed as a current sensor for GaN-HEMT power modules, which does not contribute to an increase in parasitic inductance. From the parasitic inductance evaluation, it was confirmed that the parasitic inductance of the current path under test is small, less than 1[nH], and that the increase in parasitic inductance caused by the extension of the current path can be minimized.

Experiments also confirmed the effectiveness of the impedance matching in the proposed Rogowski coil and the ability to accurately detect the current under test, based on the gain characteristics of the frequency response measurement of the coil. However, the bandwidth limitation of the integrating circuit remained an issue. Therefore, improving the bandwidth of the integrating circuit is a future issue.

REFERENCES

[1] N. G. M. Thao, K. Naruse and K. Fujisaki, "Reduction of Harmonics and Inverter Temperature in Experimental GaN-based Motor Drive System at High Frequencies Using LC Filter," 2022 IEEE Ninth International Conference on Communications and Electronics (ICCE), Nha Trang, Vietnam, 2022, pp. 507-512

[2] D. Nehmer, M. Hepp, W. Wondrak and M. -M. Bakran, "Switching a eMode GaN HEMT under conditions of an inverter module for electrical vehicles (EV)," 2023 25th European Conference on Power Electronics and Applications (EPE'23 ECCE Europe), Aalborg, Denmark, 2023I. S. Jacobs and C. P. Bean, "Fine particles, thin films and exchange anisotropy," in Magnetism, vol. III, G. T. Rado and H. Suhl, Eds. New York: Academic, 1963, pp. 271–350.

[3] P. Han, P. Liu, Q. Huang, Z. Chen and A. Q. Huang, "A 650 V, 2.1 mohm GaN Half-bridge Power Module for 400V EV Traction Inverter Application," 2022 IEEE Energy Conversion Congress and Exposition (ECCE), Detroit, MI, USA, 2022, pp. 1-6

[4] D. Bisi et al., "Short-Circuit Capability with GaN HEMTs : Invited," 2022 IEEE International Reliability Physics Symposium (IRPS), Dallas, TX, USA, 2022, pp. 1-7

[5] P T Nandh Kishore, Sumit Kumar Pramanick, and Soumya Shubhra Nag, "Development of a PCB Embedded High Bandwidth Coil Based Current Sensor Suitable for Characterizing GaN Devices", 2023 11th International Conference on Power Electronics and ECCE Asia (ICPE 2023 - ECCE Asia), Jeju Island, Korea, 2023, pp99-104

[6] H. Li, Z. Xin, X. Li, J. Chen, P. C. Loh and F. Blaabjerg, "Extended Wide-Bandwidth Rogowski Current Sensor With PCB Coil and Electronic Characteristic Shaper," in IEEE Transactions on Power Electronics, vol. 36, no. 1, pp. 29-33, Jan. 2021

[7] Y. Shi, Z. Xin, P. C. Loh and F. Blaabjerg, "A Review of Traditional Helical to Recent Miniaturized Printed Circuit Board Rogowski Coils for Power-Electronic Applications," in IEEE Transactions on Power Electronics, vol. 35, no. 11, pp. 12207-12222, Nov. 2020

[8] T. Tao, Z. Zhao, W. Ma, Q. Pan and A. Hu, "Design of PCB Rogowski Coil and Analysis of Anti-interference Property," in IEEE Transactions on Electromagnetic Compatibility, vol. 58, no. 2, pp. 344-355, April 2016

[9] Y. Wang et al., "Transmission Line Rogowski Coil: Isolated Current Sensor With Bandwidth Exceeding 3 GHz for Wide-Bandgap Device," in IEEE Transactions on Power Electronics, vol. 38, no. 11, pp. 13599-13605, Nov. 2023

[10] L. Zhao, J. D. van Wyk and W. G. Odendaal, "Planar embedded pick-up coil sensor for power electronic modules," Nineteenth Annual IEEE Applied Power Electronics Conference and Exposition, 2004. APEC '04., Anaheim, CA, USA, 2004, pp. 945-951 vol.2

[11] Y. Xue, J. Lu, Z. Wang, L. M. Tolbert, B. J. Blalock and F. Wang, "A compact planar Rogowski coil current sensor for active current balancing of parallel-connected Silicon Carbide MOSFETs," 2014 IEEE Energy Conversion Congress and Exposition (ECCE), Pittsburgh, PA, USA, 2014, pp. 4685-4690

[12] Y. Kuwabara, K. Wada, J. -M. Guichon, J. -L. Schanen and J. Roudet, "Implementation and Performance of a Current Sensor for a Laminated Bus Bar," in IEEE Transactions on Industry Applications, vol. 54, no. 3, pp. 2579-2587, May-June 2018

[13] A. Sagehashi, K. Kusaka, K. Orikawa, J. -i. Itoh and A. Momma, "Pattern design criteria of main circuit using printed circuit boards for parasitic inductance reduction," 2014 16th International Power Electronics and Motion Control Conference and Exposition, Antalya, pp. 569-574, Turkey, 2014

979-8-3315-1612-3/25 $31.00 © 2025 IEEE

Operating and Switching Frequency Circulating Current Control in Paralleled High Power Adjustable Speed Drives with Common DC Link

Kevin Lee[1] Zhihao Song[2] Wenxi Yao[2] Bo Wei[2]

Eaton[1]
Menomonee Falls, WI, USA

Zhejiang University[2]
Hangzhou, China

Abstract—There can be two types of circulating currents in paralleled high power adjustable speed drives (ASDs) with common DC link: (a). It varies with motor operating speed such as at 50Hz, caused by unbalanced output inductors and hardware variation; (b). It occurs at inverter switching frequency, associated with pulse width modulation (PWM) carrier shift between paralleled ASDs. This paper models their equivalent circuit, analyzes the effects, and introduces unified controller solutions mitigating effectively both types of circulation currents. Simulation results on two 500HP (370kW), 590A ASDs with induction machine (IM) are obtained to verify theoretical derivations, both are further validated experimentally using two 3.6kW paralleled ASDs including rectifiers, DC chokes, inverters, output inductors, and a three-phase RL load.

I. INTRODUCTION

Paralleling high-power adjustable speed drives (ASD) can achieve modularity and redundancy to scale up power rating reliably. Due to variations of parameters between paralleled ASDs such as semiconductor rise and fall time, pulse width modulation (PWM) phase shift, deadtime, output inductor tolerances [1], zero sequence and/or cross circulating currents exist and may cause overstresses of ASDs, degrade system reliability and power quality. Since the circulating currents flow between paralleled inverters instead of into the motor, system current carrying capacity can be compromised with unbalanced loss distribution.

There have been various published methods for minimizing circulating currents. The relationship between common mode voltage and circulating current is described in [2], a discrete PWM method is identified to minimize the peak and rms values of the circulating current. A concept to suppress both cross and zero sequence circulating currents applying distributed control strategy is proposed in [3]. Analysis and reduction of high frequency circulating current caused by carrier phase shift in paralleled inverters are studied in [4]. Compared to conventional coupled inductors, [5] introduces a novel filter structure to suppress circulating currents based on the sequence of PWM voltage harmonics, achieving 33% reduction in inductor weight and size. An averaged model for predicting zero sequence current dynamics in two paralleled three-phase boost rectifiers is developed in [6]. Under unbalanced grid voltage and unbalanced filter parameters, a modified PI-quasi-resonant and the feedforward controller is proposed in [7] to mitigate the zero-sequence circulating current (ZSCC) resolving the shortcomings of conventional proportional and integral (PI) current controllers. To reduce the ZSCC and common mode voltage (CMV) simultaneously, a carrier-based pulse width modulation (CBPWM) method is proposed in [8]. An analysis is carried out for circulating current during the inductors mismatch condition in [9], and a CMV offset signal is also proposed to limit the flow of low frequency component of the circulating current. A deadbeat control method is realized in [10] by dwell time adjustment of small vectors in each space vector modulation. Another deadbeat control strategy for n-paralleled inverters is described in [11] by adjusting the duty cycle of the small vectors in each PWM sampling period, thus it limits the ZSCC peak current amplitude. Based on the generalized model, a deadbeat control solution is proposed in [12] for circulating current suppression with the advantages of easy to implement and removal of inter-module communication requirement. Since accurate inductance parameters are necessary in the deadbeat control strategy, an online inductance identification method based on a model reference adaptive system is presented in [13] to compensate for the influence of inaccurate inductance parameter on the circulating current suppression and current control. From the circulating current model and relationship between each harmonic component and phase difference of the carrier wave, a method for restraining circulating current is discussed in [14], which is based on phase adjustment of triangular carrier wave. The impact of carrier phase difference on high frequency circulating current is analyzed and verified in [15], where a closed loop control approach is proposed to suppress the switching frequency circulating current caused by carrier phase difference. In addition, the deadbeat and closed loop control are combined to suppress the low and high frequency circulating currents simultaneously to improve the overall performance of circulating current suppression. However, a unified and efficient solution for mitigating circulating currents at both motor operating speed and PWM frequency of paralleled high power ASDs is limited in literatures. This paper contributes to fill the gap and provides practical engineering perspectives.

In Section II, the modeling of paralleled high power ASDs with common DC link is established. The equivalent circuit in synchronous reference frame is presented. Circulating current mitigation solutions at both motor operating speed and ASD inverter switching frequency are proposed in Section III. Analytical results are verified through simulation studies in Section IV using two 500HP (370kW), 590A ASDs and an induction machine (IM) platform. The experimental setup of using two 3.6kW paralleled ASDs including AC-DC, and DC-AC stages with DC chokes and output inductors, and a three-phase RL load demonstrates the effectiveness of the proposed circulation current mitigation solutions, even under 30% output inductance variation.

II. SYSTEM MODELING OF PARALLELED HIGH POWER ASDS WITH COMMON DC LINK

A. Definitions of Circulating Currents

Fig. 1 shows two paralleled ASDs with a common DC link. The inductor of each inverter output phase is connected in parallel at the IM terminal. The common DC link makes the rectifier and inverter approximately decoupled, so this paper primarily analyzes the inverter side circulating currents, which do not pass through the rectifier. However, abc-axis circulating currents contain operating frequency components, which are usually transformed into αβo-axis in stationary reference frame or dqo-axis synchronous reference frame circulating currents for simpler analysis. Two examples of the transformed circulating currents are illustrated in Fig. 2 and Fig. 3. With 3 upper and 3 lower switches turned on in each inverter in Fig. 2, the zero-sequence circulating current can be controlled through the o-axis voltage. When there is a PWM carrier phase shift, Phase-A circulating current in Fig. 3 occurs between two inverter phases, which can be controlled by αβ-axis or dq-axis inverter voltages. In this paper, due to unbalanced output inductors and asynchronous PWM switching, the paralleled ASD modeling effective circulating current mitigation solutions will be presented from theory to practice.

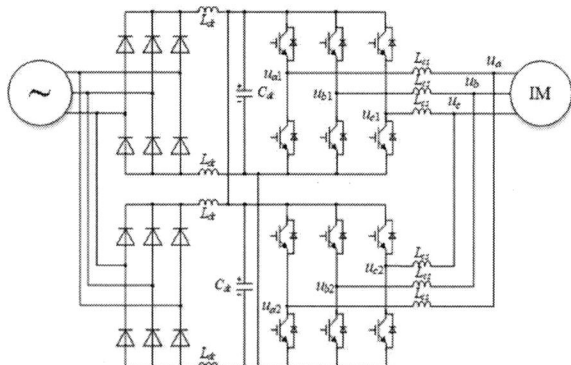

Figure 1: Two paralleled ASDs with a common DC link.

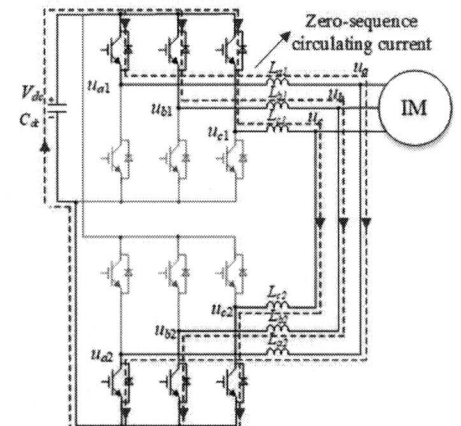

Figure 2: Zero sequence circulating current of paralleled ASDs.

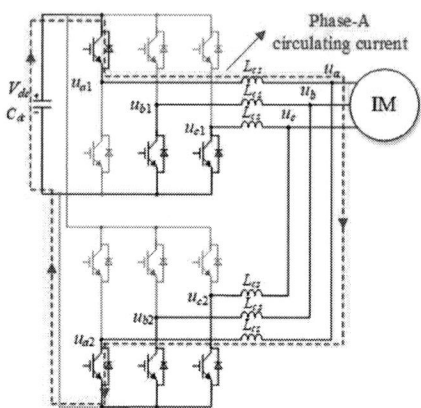

Figure 3: Phase-A circulating current path of paralleled ASDs.

B. System Modeling Descriptions of Paralleled ASDs with Common DC link

To achieve the IM rotor flux field-oriented control (FOC), the voltage equations are transformed to the dqo-axis quantities of (1) in synchronous reference frame, with corresponding equivalent circuit in Fig. 4, where $u_{d_kN}, u_{q_kN}, u_{o_kN}$ and $u_{dN_m}, u_{qN_m}, u_{oN_m}$ represent dqo-axis kth ASD output voltages and IM terminal voltages, L_{cs} and R_{cs} are current sharing inductance and its equivalent resistance. It is worth noting that the virtual midpoint potential of the dq-axis on the inverter side is the same as the motor midpoint potential, i.e., $u_N = u_{Nm}$.

$$
\begin{bmatrix} u_{d_kN} \\ u_{q_kN} \\ u_{o_kN} - u_{N_kN} \end{bmatrix} = \begin{bmatrix} u_{dN_m} \\ u_{qN_m} \\ u_{oN_m} - u_{N_kN_m} \end{bmatrix} +
$$
$$
\begin{bmatrix} sL_{cs} + R_{cs} & -\omega_e L_{cs} & 0 \\ \omega_e L_{cs} & sL_{cs} + R_{cs} & 0 \\ 0 & 0 & sL_{cs} + R_{cs} \end{bmatrix} \begin{bmatrix} i_{dk} \\ i_{qk} \\ i_{ok} \end{bmatrix} k = 1,2 \quad (1)
$$

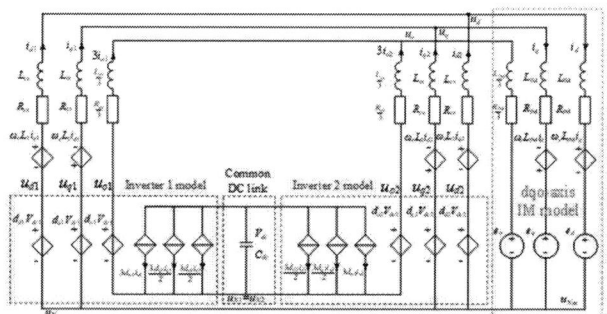

Figure 4: Paralleled ASDs model in synchronous reference frame.

C. Motor Control Structure of Paralleled ASDs with Common DC Link

The IM dq-axis voltage equation satisfies:

$$
u_{dqNm} = R_{IM}i_{dq} + (s + j\omega_e)L_{IM}i_{dq} + e_{dq} \quad (2)
$$

where the equivalent IM stator resistance is $R_{IM} = R_r L_m^2 / L_r^2 + R_s$, equivalent IM inductance is $L_{IM} = L_s - L_m^2 / L_r$ and equivalent back electromotive force is $e_{dq} = \frac{L_m}{L_r}\left(j\omega_r - \frac{1}{t_r}\right)\psi_{rdq}$.

When paralleling two ASDs, the sum of their output currents is seen by the motor, and the average output voltage is defined in (3). The IM current can be controlled using the classic complex vector PI controller, as shown in Fig. 5. Its transfer function $G_{c\_dq}^{avg}(s)$ satisfies (4), where ω_c is the designed crossover frequency.

$$
\begin{aligned}
u_{dq}^{avg} &= \frac{(u_{d1N} + u_{d2N})}{2} = (sL_{cs} + R_{cs})i_{dq}/2 + u_{dqNm} \\
&= [R_{IM} + R_{cs}/2 + (s + j\omega_e)(L_{IM} + L_{cs}/2)]i_{dq} + e_{dq}
\end{aligned} \tag{3}
$$

$$
G_{c\_dq}^{avg}(s) = \frac{\omega_c}{s}[R_{IM} + \frac{R_{cs}}{2} + (s + j\omega_e)(L_{IM} + \frac{L_{cs}}{2})] \tag{4}
$$

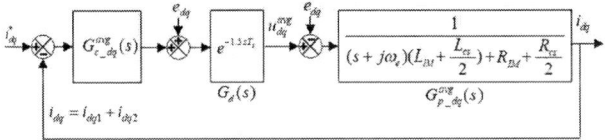

Figure 5: IM current control block diagram of paralleled ASDs.

III. CIRCULATING CURRENT MITIGATION IN ASDs WITH COMMON DC LINK

A. Circulating Current Mitigation Solution at IM Operating Speed

Since the circulating currents do not flow through the IM, it can be controlled separately. The average inverter voltage u_{dq}^{avg} and inverter voltage difference $u_{\alpha\beta o}^{diff}$ are used to control the IM three-phase currents and circulating current respectively. The dq-axis inverter voltage difference u_{dq}^{diff} is defined in (5), where the inverter current difference is $i_{dq}^{diff} = (i_{dq1} - i_{dq2})/2$. Then, the dq-axis quantities are converted to their αβ-axis equivalents. the inverter current difference $i_{\alpha\beta}^{diff}$ can be controlled to zero using a complex proportional, integral, and resonant (PIR) controller, as shown in Fig. 6.

$$
u_{dq}^{diff} = \frac{(u_{d1N} - u_{d2N})}{2} = [(s + j\omega_e)L_{cs} + R_{cs}]i_{dq}^{diff} \tag{5}
$$

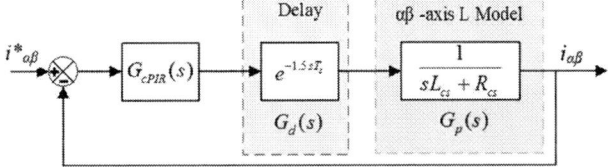

Figure 6: Inverter current difference control of paralleled ASDs.

In the αβ-axis, both the positive and negative sequence components can be regulated through a PR controller in (6):

$$
G_{cPR}(s) = K_p + \frac{K_r s}{s^2 + \omega_e^2} \tag{6}
$$

The open loop transfer function is expressed in (7):

$$
\begin{aligned}
G_o(s) &\approx G_{cPR}(s)G_d(s)G_p(s) \\
&= \frac{e^{-1.5sT_s}}{s + R_{cs}/L_{cs}}(K_p + \frac{K_r s}{s^2 + \omega_e^2})
\end{aligned} \tag{7}
$$

Where ω_e, K_r are resonant frequency and gain, K_p is proportional gain, T_s is sampling time. The current controller

bandwidth can be selected to be 1/20th-1/10th of the switching frequency with a phase margin of 65°. With a switching frequency at 4kHz, the Bode plot of (7) is shown in Fig. 7, where the gain margin is 10.5dB at 667Hz and the phase margin is 62.7° at 200Hz.

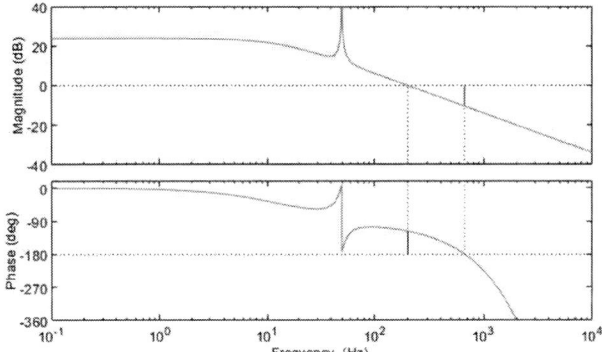

Figure 7: Bode diagram of the PR current open loop controller.

Unlike the dq-axis circulating current, the o-axis circulating current flows through the common DC link. The o-axis inverter voltage difference Δu_{Lo} is derived in (8).

$$
\begin{aligned}
\Delta u_{Lo} &= \frac{u_{Lo1} - u_{Lo2}}{2} \\
&- \frac{L_{m1} + L_{m2}}{2}\frac{di_{o1}}{dt} + R_{cs}i_{o1} \\
&+ \frac{U_{Lo1} - U_{Lo2}}{2}
\end{aligned} \tag{8}
$$

Where u_{Lo1}, u_{Lo2} are zero sequence voltage of three-phase output inductors, U_{Lo1}, U_{Lo2} are fundamental components of u_{Lo1}, u_{Lo2}. L_{m1}, L_{m2} are average inductances of inverters 1 and 2 output inductors. Fig. 8 defines the o-axis circulating current control block diagram.

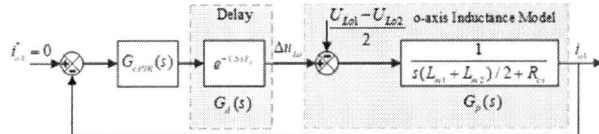

Figure 8: o-axis circulating current control of paralleled ASDs.

Assuming $L_{cs} = (L_{m1}+L_{m2})/2$, the o-axis open loop transfer function is presented in (9):

$$
\begin{aligned}
G_o(s) &= G_{cPIR}(s)G_d(s)G_p(s) \\
&= \frac{e^{-1.5sT_s}}{sL_{cs} + R_{cs}}(\frac{sK_p + K_i}{s} \\
&+ \frac{K_r s}{s^2 + \omega_e^2})
\end{aligned} \tag{9}
$$

Where the PIR controller transfer function satisfies (10):

$$
\begin{aligned}
G_{cPIR}(s) &= K_p + \frac{K_i}{s} + \frac{K_r s}{s^2 + \omega_e^2} \\
&= \frac{sK_p + K_i}{s} + \frac{K_r s}{s^2 + \omega_e^2}
\end{aligned} \tag{10}
$$

The Bode plot of (9) is shown in Fig. 9, where the gain margin and phase margin are set as: 10.5dB at 667Hz and the phase margin is 62.9° at 200Hz respectively.

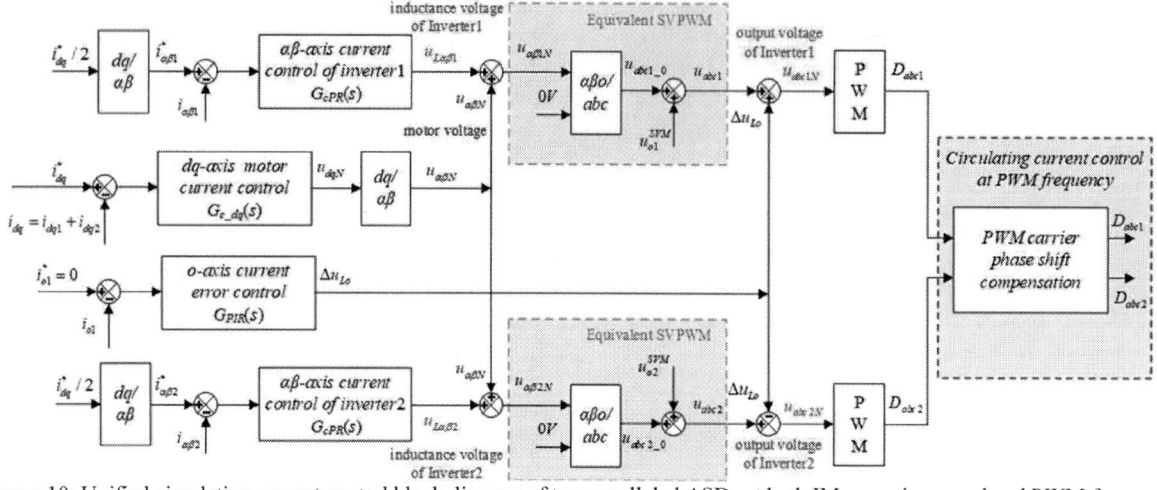

Figure 10: Unified circulating current control block diagram of two paralleled ASDs at both IM operating speed and PWM frequency.

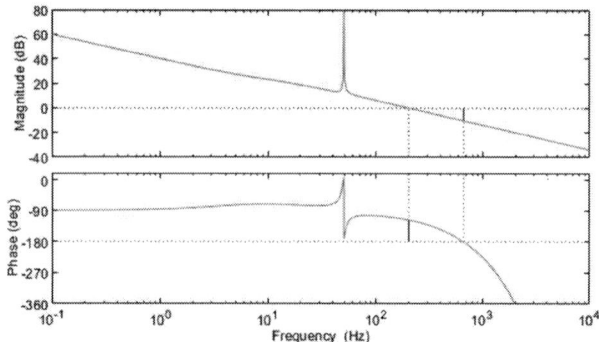

Figure 9: Bode diagram of o-axis PIR current open loop controller.

Combining Fig. 5, Fig. 6 and Fig. 8, the overall system control structure for mitigating circulation currents at IM operating speed is proposed, as shown in the left side of PWM blocks in Fig. 10.

B. Circulating Current Mitigation Solution at ASD Inverter Switching Frequency

As depicted in Fig. 11, this higher frequency cross current is caused by asynchronous PWM switching events in paralleled ASDs, especially due to any carrier phase shift. Assuming equal duty cycle between the two ASDs, and the PWM carrier Tri_2 of inverter 2 lags Tri_1 of inverter 1 by T_d, each phase circulating current is captured at the PWM peak and valley. D_x is duty cycle, g_{x1}, g_{x2} are gate commands of phase x of the two inverters, where x can be phases of a, b, or c. u_x^{diff}, i_x^{diff} are phase x voltage difference and circulating current between the two inverters.

Between t_1 and t_2, $u_x^{diff} = -V_{dc}/2$, and between t_3 and t_4, $u_x^{diff} = V_{dc}/2$. In addition, $t_2 - t_1 = t_4 - t_3 = T_d/2$. I_{xTP}^{diff}, I_{xTV}^{diff} are phase x circulating currents captured at the PWM carrier peak and valley, their amplitudes are defined in (11). Due to the PWM carrier phase shift between two inverters, the high frequency peak to peak zero sequence circulation current is expressed in (12).

$$|I_{xTP}^{diff}| = |I_{xTV}^{diff}| = \frac{V_{dc}T_d}{4L_{cs}} \qquad (11)$$

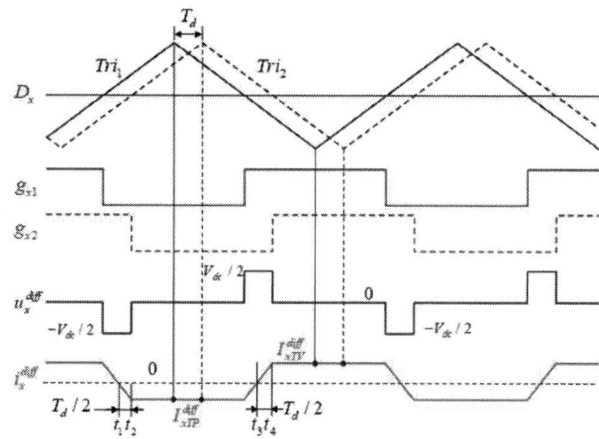

Figure 11: PWM phase shift between two paralleled ASDs.

$$I_{xm}^{diff} = |I_{xTP}^{diff} - I_{xTV}^{diff}| = \frac{V_{dc}T_d}{2L_{cs}} \qquad (12)$$

It can be observed from (12) that I_{xm}^{diff} increases linearly with increasing PWM carrier delay of T_d. For instance, at V_{dc} = 540V, switching frequency f_{sw} = 500Hz, carrier phase shift delay T_d = 10μs, and output inductance L_{cm} = 12.3μH, I_{xm}^{diff} = 219A. Refer to Fig. 11, when the PWM carrier Tri_2 of inverter 2 lags Tri_1 of inverter 1, $I_{xTP}^{diff} < 0$ at the PWM carrier peak, and $I_{xTV}^{diff} > 0$ at the PWM carrier valley. Therefore, $I_{xTP}^{diff} - I_{xTV}^{diff} < 0$. To reduce the circulating current caused by T_d, the 1st inverter carrier Tri_1 should be made lagging by T_d, where T_d is defined in (13).

$$T_d = -\frac{2L_{cs}}{V_{dc}}\left(I_{xTP}^{diff} - I_{xTV}^{diff}\right) \qquad (13)$$

Based on the above principle, the proposed PWM carrier phase shift compensation method is illustrated in Fig. 12, where T_{d0} is the time that the 2nd inverter carrier Tri_2 lags behind Tri_1 before carrier phase correction, ΔT_{dsyn} is the time that 1st inverter carrier Tri_1 lags behind after each phase correction, $T_{dsyn} = \sum \Delta T_{dsyn}$ is the total lag time of Tri_1 compared to its original carrier after the phase correction, $T_d = T_{d0} - T_{dsyn}$ is the time that Tri_2 lags behind Tri_1 after the carrier phase correction and K is the adjustable gain. The system closed loop transfer function satisfies (14). When $K = 1$, the carrier phase correction achieves a deadbeat control, which is fast in response but poor in robustness. Therefore, the gain K is set to a value smaller than 1 to reduce the transient problems that may be caused by too fast carrier phase adjustment. Furthermore, the carrier phase correction does not need to run every control cycle. It can be run at a lower frequency, e.g., once per fundamental cycle.

$$G_c = \frac{Kz^{-1}/(1-z^{-1})}{1+Kz^{-1}/(1-z^{-1})} = \frac{K}{z-1+K} \quad (14)$$

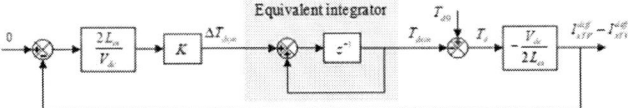

Figure 12: Carrier phase correction method of paralleled ASDs.

IV. SIMULATION RESULTS

The full digital time domain simulation platform has been established in Matlab/Simulink on a 750kW, 380V, 4-pole IM load with two paralleled ASDs at 500HP (370kW), 590A each.

The 1st scenario is to show the current sharing characteristics with unbalanced output inductors of $L_{a1} = 14.7\mu H$, $L_{a2} = 9.8\mu H$ among two paralleled ASDs, and without applying the proposed solutions. The IM operates at 80% rated speed and 100% load, Fig. 13 illustrates the simulation results of phase A currents with unequal peak current distribution of $i_{a1} = 750A$, $i_{a2} = 1050A$. The red trace represents the total current to the IM, the pink trace is phase A circulating current at the IM operating speed. The overstressed ASD can lead to degraded system efficiency and reliability.

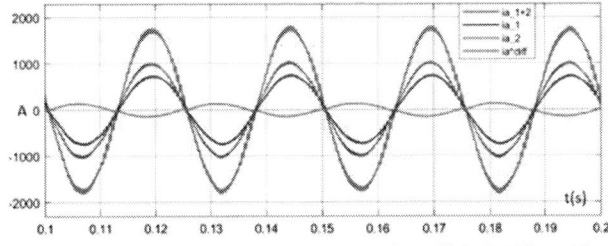

Figure 13. Simulated phase A currents of paralleled ASDs with unequal output inductors.

The 2nd scenario has the same condition as in the analytical case using (12), where the 1st inverter PWM carrier Tri_1 lags the 2nd inverter Tri_2 by $T_d = 10\mu s$, $V_{dc} = 540V$, and $L_{cs} = 12.3\mu H$. Two ASDs operate at 80% load and 80% rated IM speed. Again, before applying the proposed solutions, Fig. 14 illustrates the simulation results of phase A currents and the high frequency circulating current due to the PWM carrier phase shift. The simulation result shows that $|i_{a1} - i_{a2}|_{max}/2 = I_{am}^{diff} = 219A$, which agrees with the analytical finding in Section III.B.

Fig. 15 demonstrates the circulating current mitigation effectiveness with the initial PWM carrier delay of $T_d = 10\mu s$, and ~30% ASD output inductors unbalance. There are four segments in transition from (1) to (4) to illustrate the sequence.

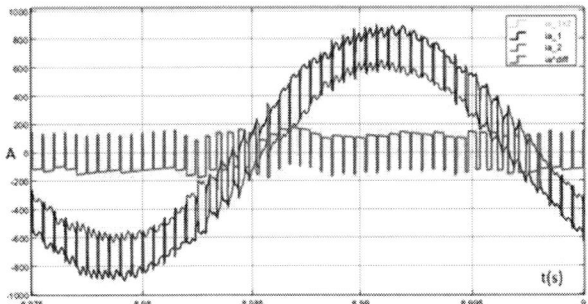

Figure 14. Simulated phase A currents and circulating current of paralleled ASDs with PWM carrier phase shift.

The IM operates at 80% load and 80% rated speed: (1) No mitigation solutions are applied from 0s to 0.1s; (2) Apply the circulating current mitigation solution at IM operating speed in Section III.A from 0.1s to 0.2s; (3) Apply both (2) and the circulating current mitigation solution at ASD inverter switching frequency as described in Section III.B from 0.2s to 0.3s; (4) Apply the dynamic 100% load reversal at 0.3s. In segment (2), it can be observed that the two ASD phase A output currents share the load equally after 0.1s, the high frequency circulating current is still significant in the pink trace. In segment (3), the circulating current at switching frequency is attenuated to nearly zero after 0.2s. Even upon and after the dynamic step load reversal in segment (4), the two inverters share currents as designed, demonstrating the effectiveness of the circulating current compensation strategies.

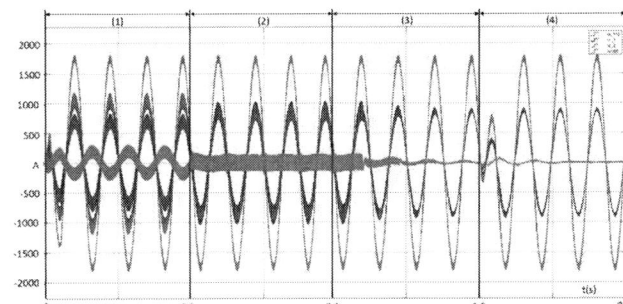

Figure 15. Simulated circulation current mitigation effectiveness of the proposed methods in steady state and dynamic conditions.

V. EXPERIMENTAL RESULTS

Since the circulating current is independent of the IM load, the RL load is adopted for the performance validation. The experimental platform in Fig. 16 has been established, including a TMS320F28377D microprocessor, two 3.6kW paralleled inverters, and a three-phase variable RL load. The input three-phase voltage is 380V, 50Hz with a DC link voltage of 540V and a switching frequency at 2kHz.

Fig. 17 records the measured inverter output currents (pink and blue), the total output current (red) and the circulating current (yellow). The unbalanced output inductors of the two paralleled inverters are configured as 2mH and 1.4mH respectively, with an initial PWM carrier phase shift of $T_d = 10\mu s$. Without any circulating current mitigation, the two inverters' phase A currents are unbalanced, the total peak current is 2A and the circulation current peak reaches 1.2A.

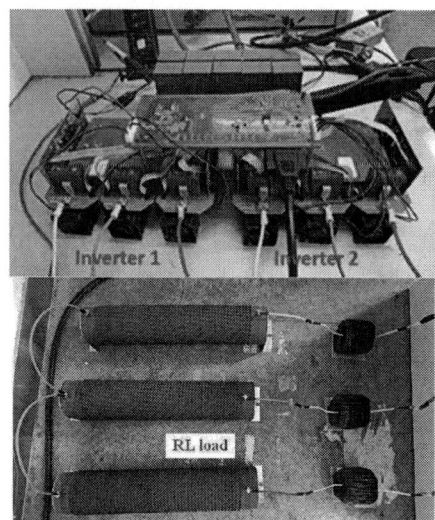

Figure 16: Experimental setup of two paralleled inverters and three-phase RL load.

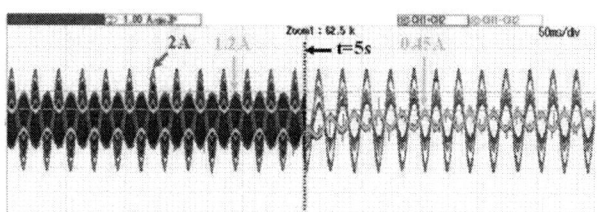

Figure 17: Experimental results of high-frequency circulation current mitigation effectiveness.

In the case where two inverters have different output inductances with unequal PWM phase shift, the circulating current (yellow) has both low frequency and high frequency components as shown in Fig. 17. When the carrier phase correction method is applied at 5s, the circulating current amplitude (yellow) is reduced from 1.2A to 0.45A, validating the effectiveness of the proposed PWM carrier phase correction method as described in III.B. However, the circulating current at the IM operating frequency caused by output inductors mismatch still exists. Then, Fig. 18 captures the low-frequency circulation current mitigation results. At 8s, after applying circulating current control solutions at the IM operating frequency of 50Hz as described in III.A, the circulating current amplitude (yellow) gradually decreases from 0.45A to 0.1A while the total output current (red) maintains 2A with two phase A currents sharing the load equally, validating the effectiveness of the unified circulating current control strategy proposed in Fig. 10.

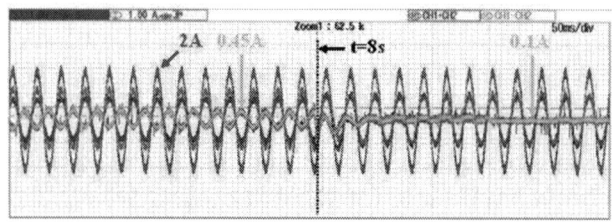

Figure 18: Experimental results of the IM operating frequency circulation current mitigation effectiveness.

VI. CONCLUSIONS

Circulating currents caused by unbalanced output inductors and asynchronous PWM angle shifts can overstress one or more of paralleled ASDs with common DC link, which may lead to system efficiency and reliability degradation. In this paper, the equivalent circuit of such a system is modeled and analyzed in dq synchronous reference frame. A unified compensation strategy is proposed for mitigating circulating currents at both IM operating speed and PWM switching frequency including a PWM carrier phase shift correction algorithm in the paralleled ASDs. The IM control system is independent of circulating current mitigation and is implemented in dq synchronous reference frame. The phase circulating current component related to the IM operating frequency, as well as the zero-sequence component are processed in αβ stationary reference frame. Furthermore, a PWM carrier phase shift compensation strategy based on phase circulating current sampling is proposed for the switching frequency circulating current mitigation. Theoretical findings are verified in simulation on two 500HP (370kW), 590A ASDs platform with a 750kW, 380V, 4-pole IM and validated experimentally using two 3.6kW paralleled ASDs including rectifiers, DC chokes, inverters, output inductors, and a three-phase RL load, fully demonstrating the mitigation effectiveness, even under 30% output inductance variation and a PWM phase delay of 10μs.

REFERENCES

[1] T. Itkonen, J. Luukko, A. Sankala, T. Laakkonen, and R. PöllÄnen, "Modeling and analysis of the dead-time effects in parallel PWM two-level three-phase voltage-source inverters," IEEE Transactions on Power Electronics, vol. 24, no. 11, pp. 2446–2455, 2009.

[2] R. Maheshwari, G. Gohil, L. Bede, and S. Munk-Nielsen, "Analysis and modelling of circulating current in two parallel-connected inverters," IET Power Electronics, vol. 8, no. 7, pp. 1273–1283, 2015.

[3] B. Wei, J. M. Guerrero, J. C. Vasquez, and X. Guo, "A circulating-current suppression method for parallel-connected voltage-source inverters with common dc and ac buses," IEEE Transactions on Industry Applications, vol. 53, no. 4, pp. 3758–3769, 2017.

[4] X. Zhang, W. Li, Y. Xiao, G. Wang, and D. Xu, "Analysis and suppression of circulating current caused by carrier phase difference in parallel voltage source inverters with SVPWM," IEEE Transactions on Power Electronics, vol. 33, no. 12, pp. 11007–11020, 2018.

[5] S. Ohn, H-S. Jung, D. Boroyevich, and Seung-Ki Sul, "A novel filter structure to suppress circulating currents based on the sequence of sideband harmonics for high-power interleaved motor-drive systems," IEEE Transactions on Power Electronics, vol. 35, no. 1, pp. 853–866, 2020.

[6] Z. Ye, D. Boroyevich, J. Choi, and F. C. Lee, "Control of circulating current in two parallel three-phase boost rectifiers," IEEE Transactions on Power Electronics, vol. 17, no. 5, pp. 609–615, 2002.

[7] H. Chen, and X. Xing, "Circulating current analysis and suppression for module grid-connected inverters under unbalanced conditions," IEEE Access, vol. 6, pp. 69120–69129, 2018.

[8] X. Xing, X. Li, C. Qin, J. Chen, and C. Zhang, "An optimized zero-sequence voltage injection method for eliminating circulating current and reducing common mode voltage of parallel-connected three-level converters," IEEE Transactions on Industrial Electronics, vol. 67, no. 8, pp. 6583–6596, 2020.

[9] K. Shukla, and R. Maheshwari, "Circulating current suppression in parallel interleaved 2L VSIs using modified CM offset based method during inductors mismatch condition," IEEE Transactions on Industry Applications, vol. 57, no. 3, pp. 3143–3153, 2021.

[10] X. Xing, Z. Zhang, C. Zhang, J. He, and A. Chen, "Space vector modulation for circulating current suppression using deadbeat control strategy in parallel three-level neutral-clamped inverters," IEEE Transactions on Industrial Electronics, vol. 64, no. 2, pp. 977–987, 2017.

[11] X. Xing, C. Zhang, A. Chen, H. Geng, and C. Qin, "Deadbeat control strategy for circulating current suppression in multi-paralleled three-level inverters," IEEE Transactions on Industrial Electronics, vol. 65, no. 8, pp. 6239–6249, 2018.

[12] X. Liu, T. Liu, A. Chen, X. Xing, and C. Zhang, "Circulating current suppression for paralleled three-level T-type inverters with online inductance identification," IEEE Transactions on Industry Applications, vol. 57, no. 5, pp. 5052–5062, 2021.

[13] J. Liu, X. Sun, B. Ren, and Q. Zhang, "Dynamic circulating current suppression method for multiple hybrid power parallel grid-connected inverters with model reference adaptive system," IEEE Transactions on Industrial Electronics, vol. 69, no. 5, pp. 4364–4375, 2022.

[14] L. Jian-bao, L. Hua, and Q. Xinxin, "Study on restraint of circulating current in parallel inverters system with SPWM modulation by adjusting phases of triangular carrier waves," 2nd International Symposium on Instrumentation and Measurement, Sensor Network and Automation (IMSNA), pp. 477–480, 2013.

[15] Z. Xueguang, W. Li, Y. Xiao, G. Wang, and D. Xu, "Analysis and suppression of circulating current caused by carrier phase difference in parallel voltage source inverters with SVPWM," IEEE Transactions on Power Electronics, vol. 33, no. 12, pp. 11007–11020, 2018.

979-8-3315-1612-3/25 $31.00 © 2025 IEEE

Mixed-Signal Sliding Mode Controller for Non-Inverting Buck-Boost Photovoltaic DC Optimizers

Anurag Singh, Sayan Paul, Dragan Maksimović and Luca Corradini
Department of Electrical, Computer and Energy Engineering
University of Colorado, Boulder, CO 80309
Email: {anurag.singh-1, sayan.paul, maksimov, luca.corradini}@colorado.edu

Abstract—**Photovoltaic (PV) dc optimizers require input voltage regulation over a wide range to maximize power extracted from PV panels. As such, seamless reference tracking and robust stability are required under all operating conditions. This work introduces a mixed-signal sliding mode controller for non-inverting Buck-Boost (NIBB) dc optimizers to address these challenges. The controller, implemented around a low-cost DSP platform, achieves seamless reference tracking in pass-through mode – where the input voltage is close to the output voltage – while retaining the reduced inductor and switch current stresses typical of multi-mode NIBB modulations. Design guidelines for the implementation are outlined, and experimental results validate the effectiveness of the proposed controller on a 600 W NIBB converter.**

Index Terms—**sliding mode control, dc optimizers**

I. INTRODUCTION

In residential photovoltaic (PV) systems, PV panels can be interfaced with a grid-tied inverter through series-stacked *dc optimizers* [1], [2], as shown in Fig. 1. Each dc optimizer performs Maximum Power Point Tracking (MPPT) by varying the voltage at the PV panel terminals. The input voltage V_g of the optimizer varies over an extensive range depending on the panel's irradiance, operating temperature, and conditions such as partial shading, while the inverter regulates the dc bus voltage. The non-inverting Buck-Boost (NIBB) converter, with its topological versatility and capability for step-up and step-down operation, is often used to implement dc optimizers.

Due to the practical limitations of pulse width modulators, multi-mode controllers for NIBB topologies invariably face

challenges around the so-called *pass-through region* when the conversion ratio $M = V_o/V_g$ is close to 1. This typically leads to pulse skipping, chattering of controlled state variables, and potential instability [3]–[7]. This work proposes a *Mixed-Signal Sliding Mode Controller* (MS-SMC) capable of achieving smooth operation around the pass-through region while retaining the advantages of the reduced inductor and switch current stress commonly seen in multi-mode NIBB modulations. The controller architecture, depicted in Fig. 2, is constructed around the mixed-signal resources of a low-cost commercial DSP platform [8]. A fast, analog-based, but digitally programmable sliding mode control layer implements the cycle-by-cycle control of the NIBB. Here, the sliding surface function $\sigma(t)$ is analog-compared to generate the basic switching events for the converter, while a fast digital logic implemented using the platform's Configurable Logic Block (CLB) provides the additional processing required for the applied gate signals in different operating modes. A slower digital layer, controlled by the DSP, programs the sliding surface and controller set-point and provides higher-level functions such as MPPT, communication, telemetry, and other ancillary features. By combining analog and digital resources in a versatile and programmable manner, the proposed archi-

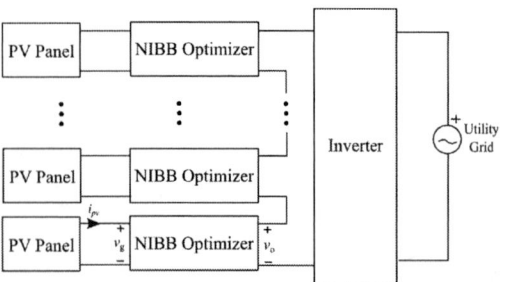

Fig. 1: Photovoltaic system based on several series-connected dc optimizer modules interfacing with a grid-tied inverter.

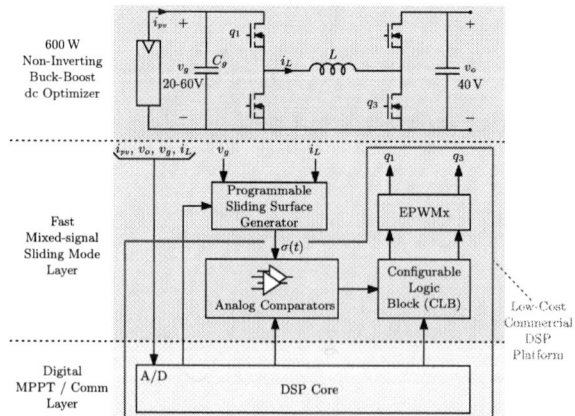

Fig. 2: High-level depiction of the proposed mixed-signal sliding mode controller, with general case study specifications.

979-8-3315-1612-3/25 $31.00 © 2025 IEEE

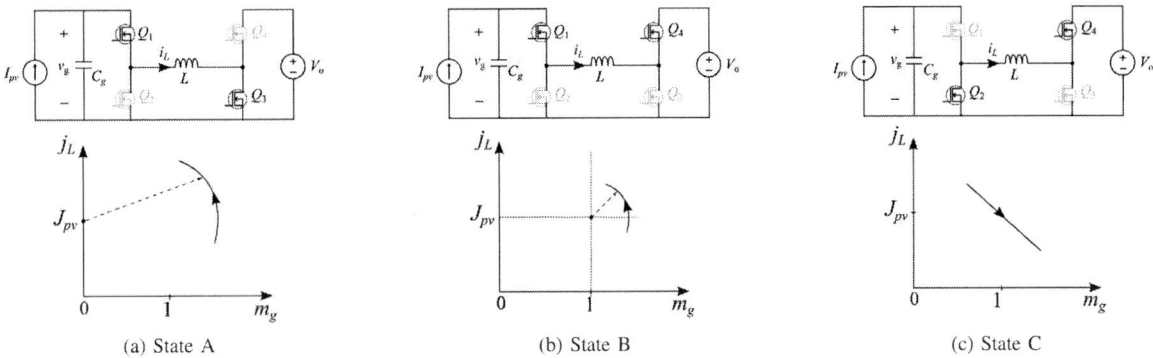

Fig. 3: The three switching states of the NIBB converter with accompanying normalized state-plane trajectories.

tecture reduces the computational latency and attains a faster switching rate in a low-cost implementation when compared to its purely digital counterparts [9]–[11].

The rest of the paper proceeds as follows: Section II describes the operation and general implementation of the proposed mixed-signal sliding mode controller, while Section III outlines the design methodology adopted for the specific NIBB dc optimizer case study considered in this work. Section IV illustrates the experimental validation of the proposed implementation on a $600\,\text{W}$ NIBB converter prototype operated as a dc optimizer.

II. PROPOSED MIXED-SIGNAL SLIDING MODE CONTROL

In all generality, the instantaneous dynamic equations of the converter can be expressed as

$$\frac{di_L}{dt} = \frac{1}{L}\left(v_g q_1 - V_o(1 - q_3)\right) \tag{1a}$$

$$\frac{dv_g}{dt} = \frac{1}{C_g}\left(I_{pv} - \frac{v_g}{R_g} - i_L q_1\right), \tag{1b}$$

with q_1 and q_3 as the gate signals for Q_1 and Q_3, respectively.

There are a total of four possible converter states depending on the switches q_1 and q_3 switch states. The three states involved in direct power transfer are highlighted in Fig. 3. Note that the axis of the plots are normalized with $V_{base} = V_o$ and $I_{base} = V_{base}/R_0$, where $R_0 = \sqrt{L/C_g}$ is the characteristic impedance. State trajectories associated with the resonant states A and B can be drawn as a family of circles centered at $(0, J_{pv})$ and $(1, J_{pv})$, respectively, while state C generates a linear trajectory with slope $-1/J_{pv}$. With this normalization, the large-signal behavior of the converter can be conveniently studied using purely geometrical arguments [12], [13].

A. Operation of the Proposed Sliding Mode Controller

The sliding surface equation is defined as

$$\sigma \triangleq \boldsymbol{K}\left(\boldsymbol{x} - \boldsymbol{x}_{ref}\right) = 0, \tag{2}$$

where $\boldsymbol{K} \triangleq [K_1\ K_2]$, $\boldsymbol{x} \triangleq [v_g\ i_L]^T$, and where $\boldsymbol{x}_{ref} \triangleq [v_{g,ref}\ i_{L,ref}]^T$ is the controller set-point [14]–[17]. The $i_{L,ref}$ set-point is obtained as a low-pass version of the sensed

inductor current. For given $v_{g,ref}$, the operating $i_{L,ref}$ can be found geometrically as the point where the trajectories have opposing tangents when crossing the sliding surface ($\sigma = 0$) for a two-state modulation. The proposed controller operates across three main operating regions:

- *Boost region*, defined as $v_g < V_{\text{g,bst-pt}} < V_o$;
- *Buck region*, defined as $V_o < V_{\text{g,bck-pt}} < v_g$;
- *Pass-through region*, defined as $V_{\text{g,bst-pt}} < v_g < V_{\text{g,bck-pt}}$.

where $V_{\text{g,bst-pt}}$ and $V_{\text{g,bck-pt}}$ denote the pass-through region boundaries on the boost and the buck sides, respectively.

The normalized state-plane trajectories in the three modes of operations are depicted in Fig. 4. In boost mode (Fig. 4(a)), the converter alternates between states A and B, resulting in the steady-state trajectory translating along the $j_L = J_{PV}$ line; in buck mode (Fig. 4(c)), on the other hand, the converter alternates between states B and C, causing the steady-state trajectory to shift along $j_L = J_{PV}m_g$.

Within the pass-through region (Fig. 4(b)), states A and C terminate when $\sigma(t)$ crosses thresholds H_1 and H_2, respectively. However, the two distinct instances of state B – color-coded red and green in Fig. 4(b) – are each terminated by separate digital counters, c_1 and c_2, upon reaching thresholds $t_{c1} \in [0, T_{c1}]$ and $t_{c2} \in [0, T_{c2}]$. Use of counters c_1 and c_2 in pass-through mode is illustrated in greater detail in Fig. 5. As V_g increases, t_{c1} rises while t_{c2} falls, shifting dominance from the green to the red instance of state B. The counter thresholds for these two instances of state B adjust as follows,

$$t_{c1} = \alpha \cdot T_{c1}, \quad t_{c2} = (1 - \alpha) \cdot T_{c2}, \tag{3}$$

where $\alpha = \left(V_{\text{g,ref}} - V_{\text{g,bst-pt}}\right)/\left(V_{\text{g,bck-pt}} - V_{\text{g,bst-pt}}\right)$.

State-plane analysis of each operational mode reveals that the converter's trajectory consistently forms a closed figure and the average value of the controlled variable, specifically v_g, is determined by the geometry of this closed trajectory. The approach here relies on the principle that, as the operating point transitions from the boost region, through the pass-through region, and into the buck region, the shape of the closed trajectory should evolve gradually, thereby avoiding abrupt changes in the average value. Furthermore, the proposed modulation scheme ensures that the converter operates

979-8-3315-1612-3/25 $31.00 © 2025 IEEE

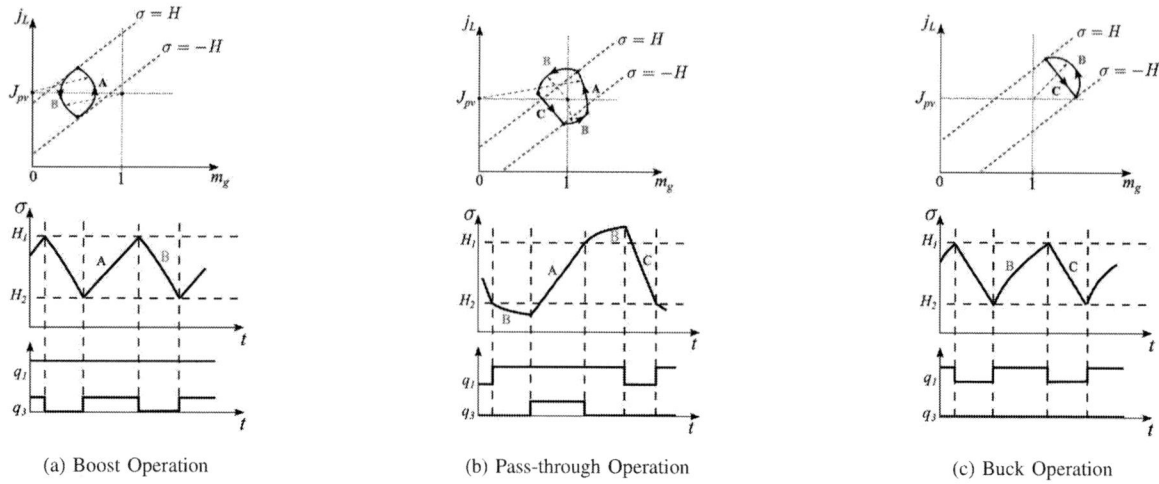

Fig. 4: Normalized state plane trajectory for the three modes of operation.

at a significantly reduced $i_{L,\text{ref}}$ compared to standard buck-boost modulation, as illustrated in Fig. 6.

Interestingly, the counter-based implementation discussed above is an extension of the conventional, hysteresis-based SMC, with the key difference that the hysteresis is a level-based threshold, whereas the counters are time-based thresholds. Counter-termination events $c_1 = t_{c1}$ and $c_2 = t_{c2}$ act as additional inputs to the SMC controller, making switching between the three states possible in the pass-through region.

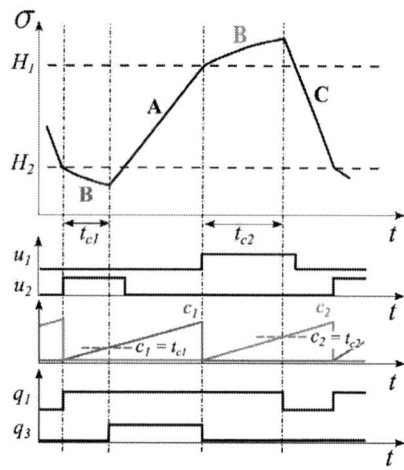

Fig. 5: Modulation scheme in pass-through operation.

B. Hardware Implementation

Implementation of the mixed-signal controller of Fig. 2 is conducted around Texas Instruments LAUNCHXL F280049C development board [8]. A detailed diagram of the controller architecture is shown in Fig. 7. The sliding surface function $\sigma(t)$ is generated in the analog domain via external instrumentation and summing amplifiers. A digital potentiometer functions as an adjustable gain resistor in the instrumentation amplifier

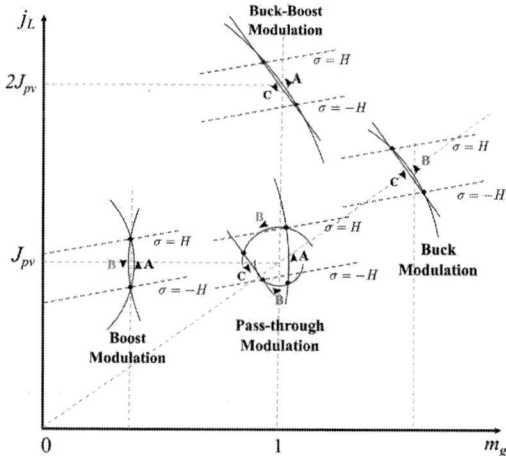

Fig. 6: Comparison of the proposed pass-through modulation with the traditional buck-boost modulation. States A and B are circular arcs centered at $(0, J_{pv})$ and $(1, J_{pv})$, respectively.

stage. It is programmable via SPI by the microcontroller, enabling modification of the sliding surface coefficients K_1 and K_2.

The analog signal $\sigma(t)$ is fed to the onboard CMPSS (Comparator Sub-system), where each module includes two comparators and internal DACs for hysteresis thresholds H_1 and H_2. The CMPSS also performs digital filtering to mitigate switching noise from the sensed $\sigma(t)$, ensuring stable operation. In the subsequent CLB (Configurable Logic Block) stage, sequential logic uses the comparator outputs u_1 and u_2, along with internal counter termination events c_1 and c_2, to generate gate logic signals q_1 and q_3. These logic signals are then fed to the EPWM module, which adds the required dead times to produce the final gate signals $q_{1,\text{PWM}}$ and $q_{3,\text{PWM}}$. The q_1 and q_3 signals are also fed to the ECAP module to measure switching frequency in real-time, allowing for on-the-

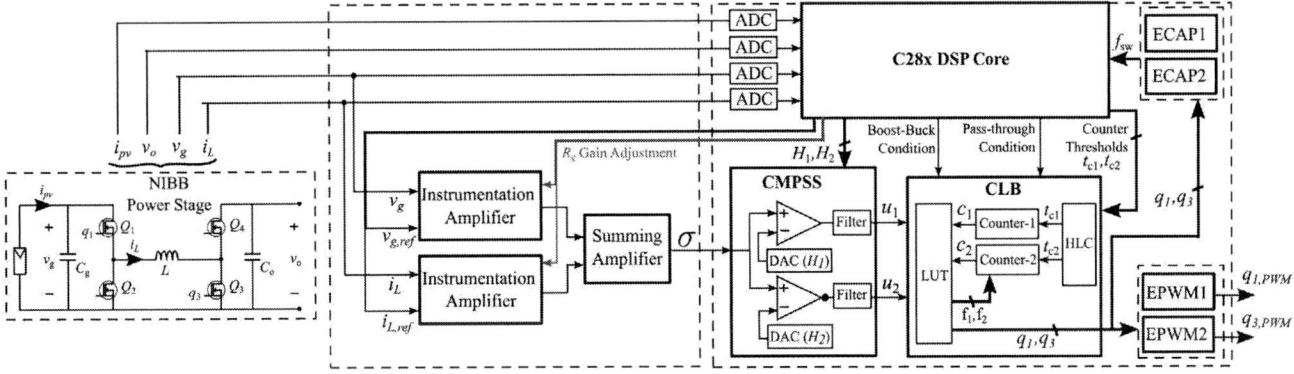

Fig. 7: Hardware implementation of the mixed-signal sliding mode controller using resources of a low–cost DSP platform [8].

fly adjustment of hysteresis thresholds to specifically limit the maximum switching frequency in boost and buck modes. The implementation described above provides complete run-time control over all parameters used in the sliding mode controller, leading to enhanced system programmability over traditional SMC implementations and enabling support of diverse options for advanced modulation schemes.

C. Generation of the gate logic signals

The CLB has three central units utilized in the implementation: the Look-up Tables (LUTs), Counters, and High-Level Controller (HLC). The sequential logic that generates the gate logic signals q_1 and q_3 is depicted as a finite state machine in Fig. 8a. The sequential logic for gate logic signals can be derived for the pass-through region (4a), boost region (4b), and buck region (4c). The counter enable flags, f_1 and f_2, corresponding to counter-1 and counter-2 (4d), respectively, activate the appropriate counter during state B based on the context of the preceding state. This ensures that the two instances of state B are terminated after the appropriate durations.

$$\begin{cases} q_1' = q_1 \cdot q_3 + \overline{q_1} \cdot \overline{q_3} \cdot u_2 + q_1 \cdot \overline{q_3} \cdot (c_2 + \overline{u_1} \cdot f_1) \\ q_3' = q_1 \cdot q_3 \cdot \overline{u_1} + q_1 \cdot \overline{q_3} \cdot (c_1 + u_2 \cdot f_2) \end{cases} \quad (4a)$$

$$\begin{cases} q_1' = 1 \\ q_3' = q_1 \cdot q_3 \cdot \overline{u_1} + q_1 \cdot \overline{q_3} \cdot u_2 \end{cases} \quad (4b)$$

$$\begin{cases} q_1' = q_1 \cdot \overline{q_3} \cdot \overline{u_1} + \overline{q_1} \cdot \overline{q_3} \cdot u_2 \\ q_3' = 0 \end{cases} \quad (4c)$$

$$\begin{cases} f_1' = \overline{f_2} \cdot (\overline{u_1} + u_2) + \overline{u_1} \cdot u_2 \\ f_2' = \overline{f_1} \cdot (u_1 + \overline{u_2}) + u_1 \cdot \overline{u_2} \end{cases} \quad (4d)$$

III. DESIGN GUIDELINES

In the subsequent sections, we explore the selection of the hysteresis levels (H_{min}, H_{max}) and the maximum counter times (T_{c1}, T_{c2}) based on a given specification of sliding surface coefficients: $K_1/K_2 = 0.8$, pass-through region limits $(V_{g,\text{bst-pt}} = 35\,\text{V}, V_{g,\text{bck-pt}} = 45\,\text{V})$, and f_{sw} limits: $(100\,\text{kHz} - 200\,\text{kHz})$. It is important to note that these limits, here taken

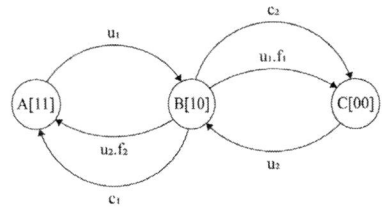

(a) Pass-through Region

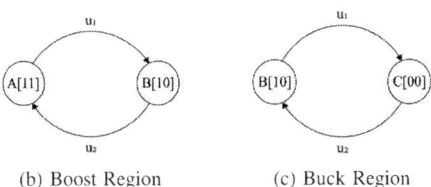

(b) Boost Region   (c) Buck Region

Fig. 8: Finite State Machine (FSM) illustrating state transitions, with each state uniquely defined by the gate logic values $[q_1, q_3]$.

TABLE I: Summary of Design Parameters

Parameter	Value	
	Initial Design	Final Design
K_1'/K_2'	0.8	0.83
f_{sw}	100 kHz to 200 kHz	80 kHz to 200 kHz
$V_{g,\text{bst-pt}}, V_{g,\text{bck-pt}}$	(35 V, 45 V)	(35 V, 45 V)
H_{min}	0.4	0.4
T_{c1}, T_{c2}	$(6.5\mu s, 8\mu s)$	$(7\mu s, 9\mu s)$

as inputs to the design procedure, involve several trade-offs including converter efficiency as well as reachability/existence requirements for state B. As such, they should be subsequently refined based on simulation and/or experimental evaluation.

A. Sliding Surface Coefficients

Let the sensor gains for v_g and i_L be $G_{v_g,\text{sns}}$ and $G_{i_L,\text{sns}}$ respectively. Similarly, assume the instrumentation amplifier gains to be $G_{v_g,\text{IA}}$ and $G_{i_L,\text{IA}}$ along each path. The imple-

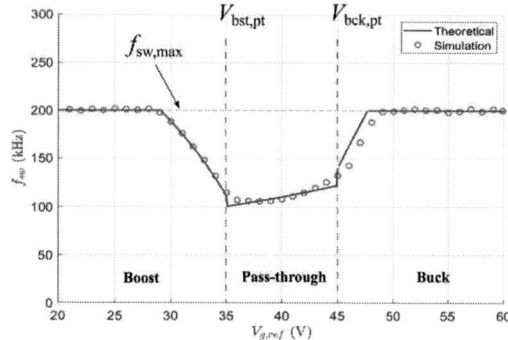

Fig. 9: Theoretically estimated vs. simulated switching frequency (f_{sw}) variation across boost, buck, and pass-through regions, based on the approximate expressions reported in the Appendix, including the maximum switching rate limitation.

mented coefficients can be written as

$$K_1' = G_{v_g,\text{sns}} \cdot G_{v_g,\text{IA}} \tag{5a}$$

$$K_2' = G_{i_L,\text{sns}} \cdot G_{i_L,\text{IA}} \tag{5b}$$

The modified sliding surface equation then becomes $\sigma = K_1'(v_{g,\text{sns}} - v_{g,ref}') + K_2'(i_{L,\text{sns}} - i_{L,ref}')$ where the reference values $v_{g,ref}' = G_{v_g,\text{sns}} \cdot v_{g,ref}$ and $i_{L,ref}' = G_{i_L,\text{sns}} \cdot i_{L,ref}$ are sensor gain adjusted values provided by the microcontroller's DAC. The designer can choose the $G_{v_g,\text{IA}}$ and $G_{i_L,\text{IA}}$ such that the the implemented gain ratio K_1'/K_2' equals K_1/K_2.

B. Choosing the hysteresis levels

Across the pass-through region, hysteresis thresholds are kept fixed at H_{min}, which can be evaluated from (6) for a given choice of minimum switching frequency $f_{sw,min}$ at the pass-through region boundaries.

$$f_{sw,min} = \min\left\{ \left(\frac{1}{T_{bst}}\right)_{V_{bst,pt}}, \left(\frac{1}{T_{bck}}\right)_{V_{bck,pt}} \right\} \tag{6}$$

Here, $T_{bst}(H, V_{g,\text{ref}}) = t_{A,\text{bst}} + t_{B,\text{bst}}$ and $T_{bck}(H, V_{g,\text{ref}}) = t_{B,\text{bck}} + t_{C,\text{bck}}$ represent the steady-state periods for boost and buck operations, respectively. These periods are functions of the hysteresis levels (H) and the input voltage reference ($V_{g,\text{ref}}$) (see Appendix).

C. Determination of Counter Thresholds

The counter thresholds $t_{c1} \in [0, T_{c1}]$ and $t_{c2} \in [0, T_{c2}]$ vary with the reference voltage $V_{g,\text{ref}}$ within the pass-through region defined by boundaries $V_{g,\text{bst-pt}}$ and $V_{g,\text{bck-pt}}$. The maximum counter thresholds T_{c1} and T_{c2} are chosen to keep the duration of state B duration continuous across these boundaries.

$$T_{c1} = t_{B,bck}|_{V_{g,bck-pt}}, \quad T_{c2} = t_{B,bst}|_{V_{g,bst-pt}}, \tag{7}$$

where the expressions for $t_{B,bck}$ and $t_{B,bst}$ are provided in the Appendix.

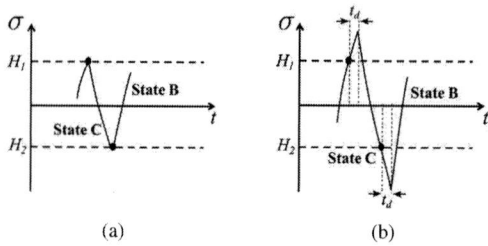

Fig. 10: Effect of delays on σ: (a) shows transitions without delays, and (b) shows transitions with delays.

D. Limiting maximum switching frequency

Fig. 9 illustrates the variation of the switching rate across the input voltage range. As seen in the plot, the switching frequency is minimum around the pass-through region, and tends to rapidly increase as soon as the converter enters buck or boost mode of operation. Limiting the switching frequency is usually desirable to reduce switching losses and maintain efficiency. For this reason, a switching frequency limitation mechanism is embedded in the proposed controller, which dynamically adjusts the hysteresis band H to prevent the switching rate from exceeding $f_{sw,max} = 200\,\text{kHz}$. This control can be implemented using an ECAP module to measure the clock cycles between consecutive rising edges of gate signals q_1 or q_3, providing real-time switching frequency estimates. The hysteresis bandwidth H is then dynamically adjusted based on the measured frequency f_{meas} relative to the maximum frequency $f_{sw,max}$, governed by periodic adjustments as in (8).

$$H = \begin{cases} \min(H + \Delta H, H_{\max}), & \text{if } f_{meas} > f_{sw,max} \\ \max(H - \Delta H, H_{\min}), & \text{if } f_{meas} < f_{sw,max} \\ H, & \text{otherwise} \end{cases} \tag{8}$$

where ΔH is the step adjustment for H, and $H_{\min}$ and $H_{\max}$ represent the minimum and maximum hysteresis values, respectively. $H_{\min}$ is determined earlier in Section III-B on selecting appropriate hysteresis levels, while $H_{\max}$ is constrained by the reference voltage of the internal DACs within the CMPSS sub-module.

As seen in Fig. 9, the switching frequency limitation kicks in when the converter is deep into the buck or boost modes of operation, effectively limiting f_{sw}. Near and inside the pass-through region, on the other hand, the switching rate is not constrained and is estimated using the expressions reported in the Appendix. Fig. 9 compares the theoretically expected f_{sw} variation with the simulated one, confirming the effectiveness of both the switching rate limitation mechanisms, and the good accuracy of the provided expressions, valid for $f_{sw} < f_{sw,max}$.

E. Effect of delays on the state durations

The design methodology is based on the knowledge of state durations in boost and buck modes. In practical implementa-

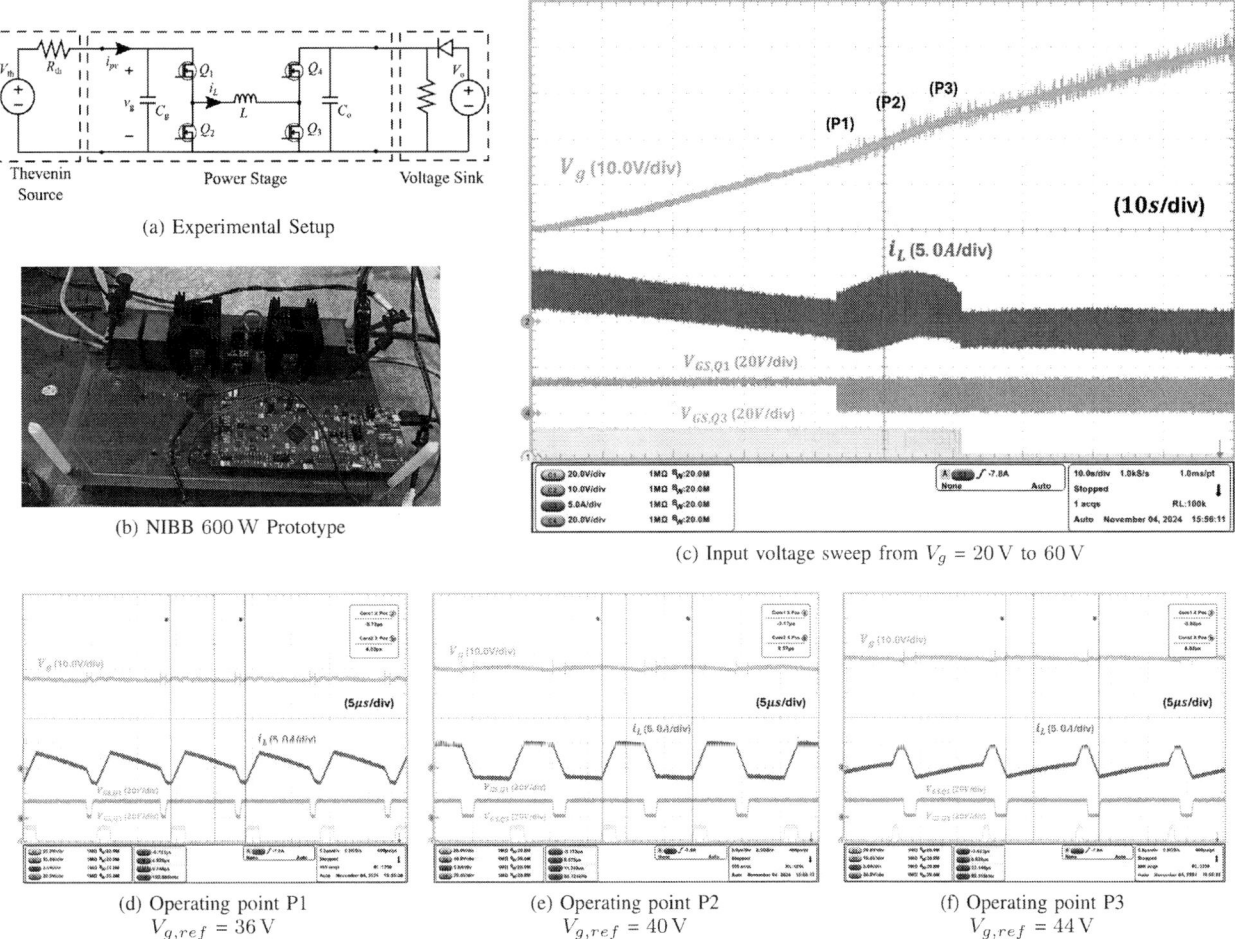

(a) Experimental Setup

(b) NIBB 600 W Prototype

(c) Input voltage sweep from $V_g = 20\,\text{V}$ to $60\,\text{V}$

(d) Operating point P1
$V_{g,ref} = 36\,\text{V}$

(e) Operating point P2
$V_{g,ref} = 40\,\text{V}$

(f) Operating point P3
$V_{g,ref} = 44\,\text{V}$

Fig. 11: Experimental results for input voltage sweep with input Thévenin source $V_{g,th} = 104\,\text{V}$ and $R_{g,th} = 4.5\,\Omega$ (set up for $P_{max} = 600\,\text{W}$ at $V_g = 52\,\text{V}$), and fixed output voltage $V_o = 40\,\text{V}$.

tions, factors such as dead times, digital filtering, and propagation delays cause the actual steady-state periods to deviate from theoretical estimates. The magnitude of these deviations depends on the specific constituent states within each mode and the operating point. Fig. 10 illustrates the impact of delays by comparing two scenarios: one with delays and one without. During each comparison event, the *extended state* exceeds the hysteresis level by $\Delta\sigma$ during t_d, requiring the *return state* to compensate for the overshoot. Consequently, this results in deviations in the switching period, which becomes more pronounced when the *extended state* changes faster than the *return state*. Adjustments to the initial design parameters are therefore necessary to mitigate these effects. For reference, Table I summarizes the finalized design parameters for the case study discussed.

IV. EXPERIMENTAL RESULTS

The NIBB experimental prototype is shown in Fig. 11a, and a list of main components used to implement the proposed mixed-signal controller is reported in Table II.

In the following tests, the converter operates off a Thévenin input source set for a maximum power of 600 W at $V_g = 52\,\text{V}$,

TABLE II: Component Values for NIBB Converter

Component	Value
Input Capacitor (C_g)	20 μF
Inductor (L)	10 μH
Switching MOSFETs (Q_1 to Q_4)	FDP083N15A
Gate Driver	1EDB8275FXUMA1
Current Sense Amplifier	INA241A4 (100V/V)
Voltage Sense Amplifier	ADA4851-2Y
Instrumentation Amplifier	INA849
Digital Potentiometer	MAX5394LATA+
Control Platform	TI F280049C
Switching Frequency	80-200 kHz
Power Ratings	600W max, 20-60V input

and with V_o fixed at 40 V. An input voltage sweep between 20 V and 60 V is illustrated in Fig. 11c. The converter transitions smoothly through the pass-through region – here defined by a $\pm5\,\text{V}$ window around V_o – with a limited excursion of the average inductor current. Throughout the sweep, the converter switching frequency ranges between 80 kHz and

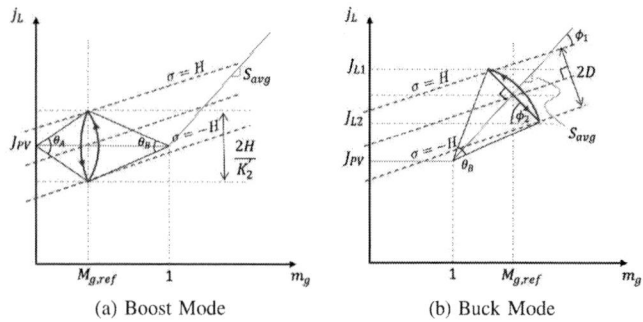

(a) Boost Mode　　　(b) Buck Mode

Fig. 12: Typical steady-state state plane trajectories.

200 kHz. Fig. 11(d)–(f) present a sequence of three steady-state operating points, illustrating how the q_3 pulse shifts from left to right as V_g increases. This shift reflects changes in t_{c1} and t_{c2}—the thresholds of counter-1 and counter-2, respectively—that determine the duration of the two instances of state B.

V. CONCLUSION

This work presents a mixed-signal sliding mode controller for NIBB-based PV dc optimizers. The approach enables seamless tracking of the input voltage reference throughout the converter's operating range. It provides an advanced modulation in the pass-through region, which reduces the inductor and switch current stresses. The controller is implemented around a low-cost commercial DSP platform and uses mixed-signal resources to get configurable parameters and minimal delay from the sensing to output. Experimental results on a 600 W NIBB converter operated as a dc optimizer are presented to validate the proposed controller architecture.

APPENDIX A
THEORETICAL ESTIMATION OF STATE DURATIONS

The first step in calculating the state durations in the boost and buck modes is to translate the sliding surface equation (2) onto the normalized state plane (m_g vs j_L),

$$\sigma = K_{1,\text{norm}}(m_g - M_{g,\text{ref}}) + K_{2,\text{norm}}(j_L - J_{L,\text{ref}}), \quad (9)$$

where $K_{1,\text{norm}} = K_1 V_{\text{base}}$ and $K_{2,\text{norm}} = K_2 I_{\text{base}}$.

A. Boost Operation

Referring to Fig. 12(a), in boost operation, the converter alternates between states A and B. The vertical distance between the two switching instants on the state plane is given by $2H/K_{2,\text{norm}}$. The durations of states A and B are derived as shown in (10).

$$t_{A,\text{bst}} = \frac{\theta_A}{\omega_0} = 2\sqrt{LC_g}\tan^{-1}\left(\frac{H/K_{2,\text{norm}}}{M_{g,\text{ref}}}\right), \quad (10a)$$

$$t_{B,\text{bst}} = \frac{\theta_B}{\omega_0} = 2\sqrt{LC_g}\tan^{-1}\left(\frac{H/K_{2,\text{norm}}}{1 - M_{g,\text{ref}}}\right). \quad (10b)$$

B. Buck Operation

In buck operation, as shown in Fig. 12(b), the converter alternates between states B and C. The slope of the state C trajectory is given by:

$$\frac{dj_L}{dm_g} = -\frac{1}{J_{PV}} = \tan(\pi - \phi_2). \quad (11)$$

Using input-output power balance, it can be shown that the reference inductor current is related to the photovoltaic current by $J_{L,\text{ref}} = J_{PV} M_{g,\text{ref}}$. For $M_{g,\text{ref}} > 1$, the slope of the ideal average inductor current trajectory becomes:

$$S_{\text{avg}} = \frac{J_{L,\text{ref}} - J_{PV}}{M_{g,\text{ref}} - 1} = J_{PV}. \quad (12)$$

This implies that the state C trajectory is perpendicular to the ideal average trajectory, $j_{pv} = J_{PV} m_g$, shown as a line on the state plane. Let ϕ_1 denote the angle between the average $J_{L,\text{ref}}$ trajectory (line $j_{pv} = J_{PV} m_g$) and the sliding surface (defined by $K_{1,\text{norm}}$ and $K_{2,\text{norm}}$). The angle ϕ_1 is expressed as:

$$\phi_1 = \tan^{-1}\left(-\frac{K_{1,\text{norm}}}{K_{2,\text{norm}}}\right) - \tan^{-1}(J_{PV}). \quad (13)$$

The perpendicular distance between the switching boundaries $\sigma = H$ and $\sigma = -H$ is defined as $2D$, where:

$$D = \frac{H/K_{2,\text{norm}}}{\sqrt{1 + \left(\frac{K_{1,\text{norm}}}{K_{2,\text{norm}}}\right)^2}}. \quad (14)$$

Assuming that the average trajectory $(M_{g,\text{ref}}, J_{L,\text{ref}})$ lies on the state C trajectory, and noting that $(J_{L1} - J_{L2}) = 2D\sec(\phi_1)\sin(\phi_2)$, the durations of states B and C are given in (15).

$$t_{B,\text{bck}} = 2\sqrt{LC_g}\tan^{-1}\left(\frac{D\sec(\phi_1)}{\sqrt{(M_{g,\text{ref}} - 1)^2 + (J_{L,\text{ref}} - J_{PV})^2}}\right), \quad (15a)$$

$$t_{C,\text{bck}} = 2D\sqrt{LC_g}\sec(\phi_1)\sin(\phi_2). \quad (15b)$$

C. Pass-through Operation

For estimating the steady-state period in this mode, the following simplifying assumptions are made: during the two instances of state B, the change $\Delta\sigma \approx 0$, and states A and C exhibit constant $d\sigma/dt$,

$$t_{A,\text{pt}} = \frac{2H}{\dot{\sigma}_A}, \quad t_{C,\text{pt}} = \frac{2H}{\dot{\sigma}_C}, \quad (16)$$

where $\dot{\sigma}_A$ and $\dot{\sigma}_C$ represent $d\sigma/dt$ for states A and C, respectively, evaluated at the equilibrium point $(M_{g,\text{ref}}, J_{L,\text{ref}})$. The durations of the two instances of state B are determined by the counter thresholds t_{c1} and t_{c2}, as defined in (3).

REFERENCES

[1] G. Walker and P. Sernia, "Cascaded dc-dc converter connection of photovoltaic modules," in *Proc. 33rd IEEE Power Electronics Specialists Conference (PESC)*, vol. 1, 2002, pp. 24–29.

[2] L. Linares, R. W. Erickson, S. MacAlpine, and M. Brandemuehl, "Improved energy capture in series string photovoltaics via smart distributed power electronics," in *Proc. 24th IEEE Applied Power Electronics Conference and Exposition (APEC)*, 2009, pp. 904–910.

[3] R. Paul and D. Maksimović, "Analysis of PWM nonlinearity in non-inverting buck-boost power converters," in *Proc. 39th IEEE Power Electronics Specialists Conference (PESC)*, 2008, pp. 3741–3747.

[4] P.-C. Huang, W.-Q. Wu, H.-H. Ho, and K.-H. Chen, "High efficiency and smooth transition buck-boost converter for extending battery life in portable devices," in *Proc. 1st IEEE Energy Conversion Congress and Exposition (ECCE)*, 2009, pp. 2869–2872.

[5] C. Restrepo, T. Konjedic, J. Calvente, and R. Giral, "Hysteretic transition method for avoiding the dead-zone effect and subharmonics in a non-inverting buck–boost converter," *IEEE Trans. Power Electron.*, vol. 30, no. 6, pp. 3418–3430, 2015.

[6] N. Zhang, G. Zhang, and K. W. See, "Systematic derivation of dead-zone elimination strategies for the noninverting synchronous buck–boost converter," *IEEE Trans. Power Electron.*, vol. 33, no. 4, pp. 3497–3508, 2018.

[7] C.-H. Tsai, Y.-S. Tsai, and H.-C. Liu, "A stable mode-transition technique for a digitally controlled non-inverting buck–boost dc–dc converter," *IEEE Trans. Ind. Electron.*, vol. 62, no. 1, pp. 475–483, 2015.

[8] *TMS320F28004x Real-Time Microcontrollers*, Texas Instruments, 2023, rev. G. [Online]. Available: https://www.ti.com/lit/gpn/tms320f280049c

[9] S. Narula, L. Corradini, and D. Maksimović, "Unified sliding-mode control of non-inverting buck-boost converters," in *Proc. 24th IEEE Workshop on Control and Modeling for Power Electronics (COMPEL)*, 2023, pp. 1–7.

[10] J. Celikovic, P. Cavallini, S. Abedinpour, and D. Maksimović, "Minimum-deviation transient response in non-inverting buck-boost dc-dc converters," in *Proc. 21st IEEE Workshop on Control and Modeling for Power Electronics (COMPEL)*, 2020, pp. 1–8.

[11] M. Oppenheimer, I. Husain, M. Elbuluk, and J. De Abreu Garcia, "Sliding mode control of the Cuk converter," in *Proc. 27th IEEE Power Electronics Specialists Conference (PESC)*, vol. 2, 1996, pp. 1519–1526.

[12] R. Munzert and P. Krein, "Issues in boundary control [of power convertors]," in *PESC Record. 27th Annual IEEE Power Electronics Specialists Conference*, vol. 1, 1996, pp. 810–816 vol.1.

[13] M. Greuel, R. Muyshondt, and P. Krein, "Design approaches to boundary controllers," in *PESC97. Record 28th Annual IEEE Power Electronics Specialists Conference. Formerly Power Conditioning Specialists Conference 1970-71. Power Processing and Electronic Specialists Conference 1972*, vol. 1, 1997, pp. 672–678 vol.1.

[14] V. Utkin, "Variable structure systems with sliding modes," *IEEE Transactions on Automatic Control*, vol. 22, no. 2, pp. 212–222, 1977.

[15] H. Sira-Ramirez, "Sliding motions in bilinear switched networks," *IEEE Transactions on Circuits and Systems*, vol. 34, no. 8, pp. 919–933, 1987.

[16] P. Mattavelli, L. Rossetto, G. Spiazzi, and P. Tenti, "General-purpose sliding-mode controller for dc/dc converter applications," in *Proc. 24th IEEE Power Electronics Specialist Conference (PESC)*, 1993, pp. 609–615.

[17] R. Venkatramanan, "Sliding mode control of power converters," Ph.D. dissertation, California Institute of Technology, Pasadena, California, May 1986.

A Current Sensorless Output Voltage Tracking Controller-Observer for a Boost Inverter Using Feedback Linearization

1st Ion Leandro dos Santos
Dept. of Electrical and Electronics Eng.
Federal University of Santa Catarina
Florianopolis, Brazil
ionleandrods@gmail.com

2nd Tailan Orlando
Dept. of Electrical and Electronics Eng.
Federal University of Santa Catarina
Florianopolis, Brazil
tailan_orlando@hotmail.com

3rd Yohannes Amilcar Tekle Scherer
Dept. of Systems and Automation Eng.
Federal University of Santa Catarina
Florianopolis, Brazil
yohascherer@gmail.com

4th Telles Brunelli Lazzarin
Dept. of Electrical and Electronics Eng.
Federal University of Santa Catarina
Florianopolis, Brazil
telles@inep.ufsc.br

5th Hector Bessa Silveira
Dept. of Systems and Automation Eng.
Federal University of Santa Catarina
Florianopolis, Brazil
hector.silveira@ufsc.br

Abstract—This paper presents a new feedback linearization-based controller-observer for output voltage tracking of a Boost inverter. The nonlinear inductor current estimator proposed has a much lower computational cost than an Extended Kalman Filter, for instance, and it is shown to possess a kind of separation principle with global convergence properties. Experimental results were achieved by emulating a 250 W Boost inverter on a HIL (Hardware-In-the-Loop) platform and implementing the controller-observer on a microcontroller to corroborate the proposed control strategy. The transient performance of the inverter was significantly improved according to previous works of the authors, along with a steady-state output voltage THD reduction from 11.57% to 0.44% at rated power. These results show the viability of the proposed approach for various applications and its potential for further performance and efficiency enhancement.

Index Terms—Sensorless, Voltage Tracking, Controller-observer, Feedback Linearization.

I. INTRODUCTION

With the expansion of Power Electronics applications in recent decades, inverters (or dc-ac converters) have become increasingly prevalent in energy processing systems. Due to the broad range of applications, numerous inverter topologies have been developed. In [1], a novel Boost inverter (Fig. 1(a)) was proposed, offering several advantageous features, such as reduced number of switches, step-up transformerless characteristic, and a common-grounded circuit. However, in off-grid applications, the nonlinear characteristics inherent to Boost-based inverters lead to natural harmonic distortion in the output voltage. To mitigate this, [1] introduced a static linearization technique to improve the Total Harmonic Distortion (THD) of the inverter's output voltage. Despite these improvements, the static linearization did not account for the dynamic requirements imposed by operational perturbations

and failed to achieve an output voltage THD below 10% under steady-state conditions.

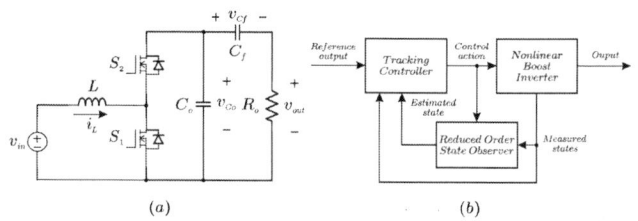

Fig. 1. (a) Boost Inverter [1] - (b) Structure of the proposed tracking controller-observer using a nonlinear reduced-order state estimator.

To enhance performance and improve converter robustness across all operating ranges, closed-loop operation using linear control methods was initially tested. However, these controllers did not achieve satisfactory transient performance; they struggled to start the inverter from an idle state without high THD and failed to meet performance requirements under perturbations. Consequently, this paper proposes a nonlinear feedback control approach (Fig. 1(b)) to improve the converter's dynamic performance, as also achieved in [2]–[5].

Regarding control, it is well-known that linear controllers designed based upon the linearized dynamics of nonlinear systems usually only ensure local asymptotic convergence properties in closed-loop [6]. Hence, herein, it has relied on feedback linearization techniques (see e.g. [17], [7]) in order to achieve global ones for the tracking error of a sinusoidal reference output for the output voltage $v_{out}(t) = v_{Co}(t) - v_{Cb}(t)$ of the considered (nonlinear) Boost inverter. Differently from [3]–[5], we stabilize the (linear) tracking error dynamics directly, so that tracking controller design is more straightforward. Nonetheless, feedback linearization typically requires

979-8-3315-1612-3/25 $31.00 © 2025 IEEE

measurement of all state variables (in our case, three sensors would be required for measuring $v_{Co}(t)$, $v_{Cb}(t)$ and $i_L(t)$), which can significantly increase the overall cost of controller implementation. In terms of state estimation for nonlinear systems, the Extended Kalman Filter (EKF) is widely used in practice. However: it only guarantees local asymptotic stability of the estimation error; the separation principle is not valid in general; and computational costs may inhibit its application when the sampling period is relatively small and/or the system is of high order, since time-varying matrix computations with inversion are required [8]–[10]. Therefore, we propose a new feedback linearization based nonlinear tracking controller-observer for the referred Boost inverter in which a kind of separation principle holds with global convergence properties and only a single scalar nonlinear time-invariant state equation is needed to estimate the unmeasured inductor current $i_L(t)$, thus significantly reducing computational costs with respect to EKF. The obtained experimental results show that it is robust to considerably high load disturbances.

This work presents the nonlinear average model of the considered Boost inverter, the design of the referred new current sensorless output voltage tracking controller-observer, along with preliminary experimental results validating the proposed control strategy applied to the Boost inverter. The obtained improvements in THD reduction and dynamic performance suggest potential applications ranging from off-grid systems, such as UPS, to on-grid systems.

II. AVERAGE MODEL AND CONTROL PROBLEM

Power converters have a switched behavior that can make its analysis rather complex due its variable structure dynamics. In order to obtain an approximate smooth model that considerably reduces the complexity involved in system analysis as well as controller design, the average model of a power converter has been used [12], [13]. A formalization under some reasonable hypothesis for the validity of such (smooth) average models for switched converters have been discussed in [14], [15], which in turn is analogous to standard averaging techniques for smooth dynamical system (see e.g. [7]) and, loosely speaking, consisting in "averaging out" the fast switching dynamics over the slow dynamics of the system.

A. Average model

The Boost inverter considered in this paper exhibits two topological states, as illustrated in Fig. 2, according to the behavior of the switches S_1 and S_2.

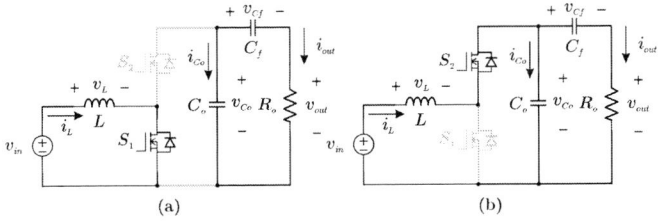

Fig. 2. (a) First topological state - (b) Second topological state

For each of theses topological states there is an associated set of differential equations that describes the corresponding dynamics of the converter. For the first topological state, we have

$$\begin{cases} \dot{i}_L(t) = \dfrac{1}{L}V_{in}, \\[2mm] \dot{v}_{Co}(t) = -\dfrac{1}{R_oC_o}v_{Co}(t) + \dfrac{1}{R_oC_o}v_{Cb}(t), \\[2mm] \dot{v}_{Cb}(t) = \dfrac{1}{R_oC_b}v_{Co}(t) - \dfrac{1}{R_oC_b}v_{Cb}(t), \end{cases}$$

which can be rewritten in the form $\dot{x}(t) = A_1x(t) + B_1V_{in}$:

$$\frac{d}{dt}\begin{bmatrix} i_L(t) \\ v_{Co}(t) \\ v_{Cb}(t) \end{bmatrix} = \begin{bmatrix} 0 & 0 & 0 \\ 0 & -\frac{1}{R_oC_o} & \frac{1}{R_oC_o} \\ 0 & \frac{1}{R_oC_b} & -\frac{1}{R_oC_b} \end{bmatrix} \begin{bmatrix} i_L(t) \\ v_{Co}(t) \\ v_{Cb}(t) \end{bmatrix} + \begin{bmatrix} \frac{1}{L} \\ 0 \\ 0 \end{bmatrix} V_{in}. \tag{1}$$

And, for the second topological state,

$$\begin{cases} \dot{i}_L(t) = \dfrac{1}{L}V_{in} - \dfrac{1}{L}v_{Co}(t), \\[2mm] \dot{v}_{Co}(t) = \dfrac{1}{C_o}i_L - \dfrac{1}{R_oC_o}v_{Co}(t) + \dfrac{1}{R_oC_o}v_{Cb}(t), \\[2mm] \dot{v}_{Cb}(t) = \dfrac{1}{R_oC_b}v_{Co}(t) - \dfrac{1}{R_oC_b}v_{Cb}(t), \end{cases}$$

so that it can rewritten as $\dot{x}(t) = A_2x(t) + B_2V_{in}$:

$$\frac{d}{dt}\begin{bmatrix} i_L(t) \\ v_{Co}(t) \\ v_{Cb}(t) \end{bmatrix} = \begin{bmatrix} 0 & -\frac{1}{L} & 0 \\ \frac{1}{C_o} & -\frac{1}{R_oC_o} & \frac{1}{R_oC_o} \\ 0 & \frac{1}{R_oC_b} & -\frac{1}{R_oC_b} \end{bmatrix} \begin{bmatrix} i_L(t) \\ v_{Co}(t) \\ v_{Cb}(t) \end{bmatrix} + \begin{bmatrix} \frac{1}{L} \\ 0 \\ 0 \end{bmatrix} V_{in}. \tag{2}$$

According to [12], [15], as long as the switching frequency is sufficiently high, the (smooth) average model is given by

$$\dot{x}(t) = A_{avg}(d(t))x(t) + B_{avg}(d(t))V_{in}, \tag{3}$$

where

$$\begin{aligned} A_{avg}(d(t)) &= A_1d(t) + A_2\big(1 - d(t)\big), \\ B_{avg}(d(t)) &= B_1d(t) + B_2\big(1 - d(t)\big), \end{aligned} \tag{4}$$

$d(t) \in [0,1]$ is the duty cycle defined as

$$d(t) = \frac{T_1(t)}{T_s},$$

meaning that the converter remains in the first topological state during the time interval $[t, t + T_1(t)]$ with $0 < T_1(t) \le T_s$ and $T_s > 0$ is the switching period of the converter. When $T_1(t) = 0$, the converter is then in the second topological state at t. Thus, from (1)–(4), we obtain the average model for the converter:

$$\begin{cases} \dot{i}_L(t) = \dfrac{1}{L}V_{in} - \dfrac{1}{L}v_{Co}(t)u(t), \\[2mm] \dot{v}_{Co}(t) = \dfrac{1}{C_o}i_L(t)u(t) + \dfrac{1}{R_oC_o}\Big(v_{Cb}(t) - v_{Co}(t)\Big), \\[2mm] \dot{v}_{Cb}(t) = \dfrac{1}{R_oC_b}\Big(v_{Co}(t) - v_{Cb}(t)\Big), \\[2mm] v_{out}(t) = v_{Co}(t) - v_{Cb}(t), \end{cases} \tag{5}$$

TABLE I
DESIGN SPECIFICATION

Parameter	Value
Rated power (P_{out})	250 W
Rated input voltage (V_{in})	100 V
Peak output voltage value ($V_{out,pk}$)	155 V
Output voltage (rms) (V_{out})	110 V
Switching frequency (f_s)	100 kHz
Output voltage frequency (f_0)	60 Hz
Input current ripple (ΔI_L)	20% of $I_{L,pk}$
Output voltage ripple (ΔV_{out})	2.5% of $V_{out,pk}$
L inductance	275 μH
C_o capacitance	2.2 μF
C_b capacitance	500 μF
R_o resistance	48.4 Ω

where $v_{out}(t) = v_{Co}(t) - v_{Cb}(t)$ is the considered output voltage to be controlled, and $u(t) = 1 - d(t) \in [0, 1]$ is chosen as the control input for simplicity.

B. Average model validation

In order to validate de average model and continue throughout its control problem, a simulation was carried out. The design specification used both in model validation and control-observer approach is presented in Table I.

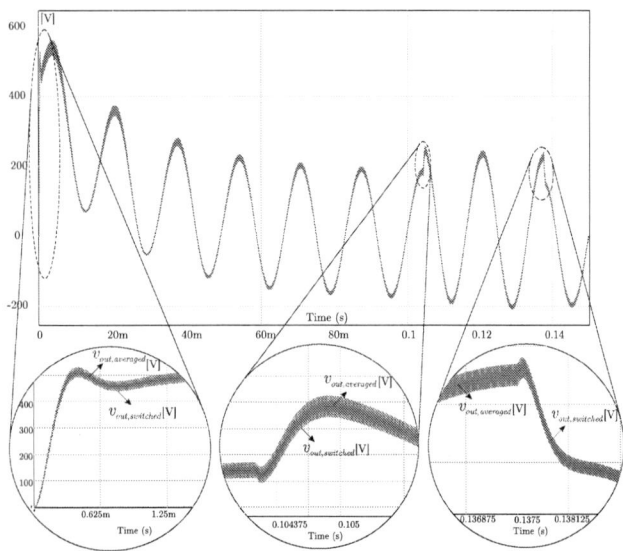

Fig. 3. Curves of average model and switched inverter during open-loop system initialization and two $u(t)$ amplitude steps of $+20\%$ and -20% obtained through simulation for average model validation ($v_{out,averaged}$ (red) and $v_{out,switched}$ (green))

A simulation was carried out so that the average model could be validated. In this simulation, the inverter and its average model were submitted to same initial conditions ($i_L(0) = 0$, $v_{Co}(0) = 0$, and $v_{Cf}(0) = 0$) and an input $u(t) = 1 - d(t) = k(0.625 + 0.33\sin(2\pi 60t))$ such that $k = 1$, while the parameters used can be seen in Table I. For both switched and average model, the linearizing function λ presented in [1] was used to guarantee a sufficient low THD. In the instant $t_1 = 0.1041667\ s$ ($6 + 1/4$ cycles

after initialization), k is changed to 1.2 and in the instant $t_2 = 0.1375\ s$ ($8 + 1/4$ cycles after initialization) $k = 1$ again. The response of the switched converter and its average model can be seen in Fig. 3. As one can see, the average model represented the low frequency dynamics of the switched inverter.

C. Control Problem

Consider the Boost inverter presented in Fig. 1. We assume throughout the paper that $v_{Co}(t)$ and $v_{Cb}(t)$ are measured. Consider the reference output trajectory $v_{ref}(t) = 155\sin(2\pi 60t)$, $t \geq 0$ for the output voltage $v_{out}(t) = v_{Co}(t) - v_{Cb}(t)$. The control problem here treated is to design a tracking controller-observer based on the average model (5), i.e. determine a control law $u = \alpha(i_L, v_{Co}, v_{Cb}, t)$ such that the closed-loop tracking error $e(t) = v_{out}(t) - v_{ref}(t)$ is such that $e(t) \cong 0$ for all sufficiently large $t \geq 0$ and any initial conditions $v_{Co}(0) \geq 0$, $v_{Cb}(0) \geq 0$, $i_L(0)$, $\widehat{i_L}(0)$, where $\widehat{i_L}$ is an estimate for i_L in order to reduce the number of required sensors. We also assume that the closed-loop control satisfies $0 < u_{min} \leq u(t) \leq 1$ for all $t \geq 0$, which is a reasonable hypothesis in practice.

III. TRACKING CONTROLLER-OBSERVER DESIGN

We begin by designing a tracking feedback controller for (5) based on feedback linearization techniques assuming that $i_L(t)$ is measured, i.e. $\widehat{i_L} = i_L$. It is straightforward to see from the last three equations of (5) that

$$\dot{v}_{out}(t) = \frac{1}{C_o}i_L(t)u(t) - \frac{C_o + C_b}{R_o C_o C_b}v_{out}(t). \tag{6}$$

By construction, the feedback linearization control law

$$u(t) = \frac{C_o}{i_L(t)}\left[\frac{C_o + C_b}{R_o C_o C_b}v_{out}(t) + w\right],$$

where $w \in \mathbb{R}$ is the new control input, linearizes the dynamics of $v_{out}(t)$ as

$$\dot{v}_{out}(t) = w.$$

Let $e(t) = v_{out}(t) - v_{ref}(t)$ be the output voltage tracking error. Hence, choosing

$$w = \dot{v}_{ref}(t) - k\left(v_{out}(t) - v_{ref}(t)\right),$$

where $k > 0$ is the control gain, ensures that

$$\dot{e}(t) + ke(t) = 0,$$

and thus $e(t) = e(0)\exp(-kt) \to 0$ as $t \to \infty$. It should be noted that $i_L(t)$ cannot vanish for such control law to be well-defined. However, the inductor current $i_L(t)$ must oscillate around zero when $v_{out}(t) \equiv v_{ref}(t)$. Therefore, the control law above cannot be implemented in practice. This motivates us to propose two feedback linearization loops: an inner and an outer one.

We first show the design of the outer loop. Based on the first equation of (5), it is clear that

$$u(t) = -\frac{L}{v_{Co}(t)}\left[-\frac{V_{in}}{L} + \dot{i}_{Lref}(t) - k_1\Big(i_L(t) - i_{Lref}(t)\Big)\right],$$
$$(7)$$

where $k_1 > 0$ is a control gain and $i_{Lref}(t)$ is the reference inductor current determined by the inner loop given shortly, implies

$$\dot{\xi}(t) + k_1\xi(t) = 0,$$

where $\xi(t) = i_L(t) - i_{Lref}(t)$ is the inductor current tracking error. Hence, $\xi(t) = \xi(0)\exp(-k_1 t) \to 0$ as $t \to \infty$. However, $\dot{i}_{Lref}(t)$ in (7) is not explicitly known. For simplicity, we temporarily consider in the sequel that $\dot{i}_{Lref}(t) \equiv 0$ and that $i_L(t) \equiv i_{Lref}(t)$. Therefore, the control law for the outer loop becomes

$$u(t) = -\frac{L}{v_{Co}(t)}\left[-\frac{V_{in}}{L} - k_1\Big(i_L(t) - i_{Lref}(t)\Big)\right]. \quad (8)$$

We point out that since $u(t) \in [0,1]$ for $t \geq 0$ in (5), it is implicitly assumed in the expression above that $u(t)$ saturates at 0 (lower limit) and at 1 (upper limit). In particular, $u(t) = 1$ when $v_{Co}(t) = 0$.

Now, as for the inner loop, from (6) and the hypothesis that $i_L(t) \equiv i_{Lref}(t)$, we define $i_{Lref}(t)$ as

$$i_{Lref}(t) = \frac{C_o}{u(t)}\left[\frac{C_o + C_b}{R_o C_o C_b}v_{out}(t)\right.$$
$$\left. + \dot{v}_{ref}(t) - k_2\Big(v_{out}(t) - v_{ref}(t)\Big)\right], \quad (9)$$

where $k_2 > 0$ is another control gain, since, by construction, it ensures that

$$\dot{e}(t) + k_2 e(t) = 0,$$

and hence $e(t) = e(0)\exp(-k_2 t) \to 0$ as $t \to \infty$.

Assume that (5) in closed-loop with (8) and (9) is such that $\dot{i}_{lref}(t) \cong 0$ for all sufficiently large $t \geq 0$. Note that (8) can be rewritten as

$$u(t) = -\frac{L}{v_{Co}(t)}\left[-\frac{V_{in}}{L} \pm \dot{i}_{Lref}(t) - k_1\Big(i_L(t) - i_{Lref}(t)\Big)\right],$$

so that $\dot{\xi}(t) + k_1\xi(t) = -\dot{i}_{Lref}(t)$. Therefore, $\xi(t) = i_L(t) - i_{Lref}(t) \cong 0$ for all sufficiently large $t \geq 0$. This follows from well-known results in linear systems theory (see e.g. [16]). Consequently, using (6) and (9), we obtain that $\dot{e}(t) + k_2 e(t) = \xi(t)u(t)/C_0$. Hence, $e(t) \cong 0$ for sufficiently large $t \geq 0$. We thus conclude that (8) and (9) is a tracking feedback controller that solves the considered control problem for (5) (as long as $u(t)$ does not saturate for all sufficiently large $t \geq 0$). However, a current sensor for $i_L(t)$ is required. In order to overcome such drawback, we propose in the sequel a nonlinear state observer for $i_L(t)$ based on the measurements of $v_{Co}(t)$, $v_{Cb}(t)$ and the control $u(t)$. We remark that our nonlinear observer was inspired by the standard (linear) reduced-order observers for linear systems (see e.g. [8]).

Let $y = [v_{Co} \quad v_{Cb}]' \in \mathbb{R}^2$ (column vector) be the measured output, where $'$ denotes vector transpose, and let

$P = [q \quad r] \in \mathbb{R}^2$ (row vector) be an adjustable observer gain with $q > 0$. From the first equation of (5), we define the following nonlinear state observer for $i_L(t)$:

$$\dot{\widehat{i}}_L(t) = \frac{1}{L}V_{in} - \frac{1}{L}v_{Co}(t)u(t) + P\big(\dot{y}(t) - \widehat{\dot{y}}(t)\big), \quad (10)$$

where

$$\widehat{\dot{y}}(t) = \begin{bmatrix} \dfrac{v_{Cb}(t) - v_{Co}(t)}{R_o C_o} + \dfrac{\widehat{i}_L(t)}{C_o}u(t) \\[2ex] \dfrac{v_{Co}(t) - v_{Cb}(t)}{R_o C_b} \end{bmatrix}$$

is an estimate of

$$\dot{y}(t) = \begin{bmatrix} \dot{v}_{Co}(t) \\ \dot{v}_{Cb}(t) \end{bmatrix} = \begin{bmatrix} \dfrac{v_{Cb}(t) - v_{Co}(t)}{R_o C_o} + \dfrac{i_L(t)}{C_o}u(t) \\[2ex] \dfrac{v_{Co}(t) - v_{Cb}(t)}{R_o C_b} \end{bmatrix}$$

by the second and third equations of (5). Hence,

$$\dot{\widehat{i}}_L(t) = \frac{1}{L}V_{in} - \frac{1}{L}v_{Co}(t)u(t) + \frac{q}{C_o}\Big(i_L(t) - \widehat{i}_L(t)\Big)u(t),$$

and thus

$$\dot{\psi}(t) = -\frac{q}{C_o}\psi(t)u(t),$$

where $\psi(t) = i_L(t) - \widehat{i}_L(t)$ is the estimation error. Consequently,

$$(\dot{\psi^2})(t) = -\frac{2q}{C_o}\psi^2(t)u(t) \leq -\frac{2q u_{min}}{C_o}\psi^2(t)$$

(recall the assumption that $0 < u_{min} \leq u(t) \leq 1$ for every $t \geq 0$), and therefore $|\psi(t)| \leq |\psi(0)|\exp(-q u_{min}t/C_o)$ for all $t \geq 0$ by the Comparison Lemma (see [5]). This means that $\psi = 0$ is globally exponentially stable. Now, to overcome the dependence of $\dot{y}$ in (10), we take $r = 0$ and consider

$$\bar{i}_L(t) \triangleq \widehat{i}_L(t) - Py(t) = \widehat{i}_L(t) - qv_{Co}(t),$$

so that the nonlinear reduced-order state observer (10) is equivalent to

$$\dot{\bar{i}}_L(t) = \frac{1}{L}V_{in} - \frac{1}{L}v_{Co}(t)u(t) - \frac{q}{C_o}\left[\frac{1}{R_o}\Big(v_{Cb}(t) - v_{Co}(t)\Big)\right.$$
$$\left. + \Big(\bar{i}_L(t) + qv_{Co}(t)\Big)u(t)\right],$$
$$\widehat{i}_L(t) = \bar{i}_L(t) + qv_{Co}(t), \quad (11)$$

where $q > 0$ is an adjustable gain. Based once again on well-known results in linear systems theory, we conclude that the designed nonlinear tracking controller-observer that solves our control problem is given by (8), (9) and (11) with $\widehat{i}_L(t)$ substituted for $i_L(t)$ in (8), that is:

$$\dot{\overline{i}}_L(t) = \frac{1}{L}V_{in} - \frac{1}{L}v_{Co}(t)u(t) - \frac{q}{C_o}\left[\frac{1}{R_o}\Big(v_{Cb}(t) - v_{Co}(t)\Big)\right.$$
$$\left. + \Big(\overline{i}_L(t) + qv_{Co}(t)\Big)u(t)\right],$$

$$\widehat{i}_L(t) = \overline{i}_L(t) + qv_{Co}(t),$$

$$u(t) = \frac{1}{v_{Co}(t)}\left[V_{in} + Lk_1\Big(\widehat{i}_L(t) - i_{Lref}(t)\Big)\right], \qquad (12)$$

$$i_{Lref}(t) = \frac{C_o}{u(t)}\left[\frac{C_o + C_b}{R_o C_o C_b}v_{out}(t)\right.$$
$$\left. + \dot{v}_{ref}(t) - k_2\Big(v_{out}(t) - v_{ref}(t)\Big)\right],$$

where $k_1 > 0$, $k_2 > 0$, $q > 0$ are adjustable gains. We point out that a kind of global separation principle has thus been achieved in a nonlinear setting.

Fig. 4 shows the block diagram corresponding to the proposed tracking controller-observer applied to the considered Boost inverter. Robustness issues are assessed in the experimental results presented in the next section.

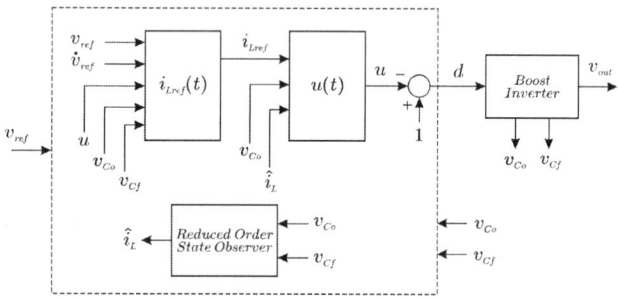

Fig. 4. Block diagram illustrating the proposed nonlinear tracking controller-observer.

IV. Experimental Results

To verify the proposed tracking controller-observer for the Boost inverter, the converter's switched model was implemented using the Typhoon HIL 404, while the control strategy was deployed on the F28069M LaunchPad™. The Typhoon HIL 404 emulated the converter in real-time with a 200 ns step size, generating the analog signals read by the F28069M, which in turn controlled and commanded the converter switches. The system specifications are the same as those given in Table I. The controller-observer gains used are $k_1 = 62500$, $k_2 = 25$, and $q = 0.022$. The observer differential equation discretization was done using Euler's Method with a sample frequency of 100 kHz, which was the same sample frequency used by all ADC of the F28069M microntroller. The experimental setup is shown in Fig. 5.

Fig. 6 was capture through oscilloscope showing the switched behavior of the inverter for the sake of validation that the Typhoon HiL was emulating the Boost Inverter as expected. It can be seen the low frequency envelope and the detailed switched behavior of the inverter. The maximum

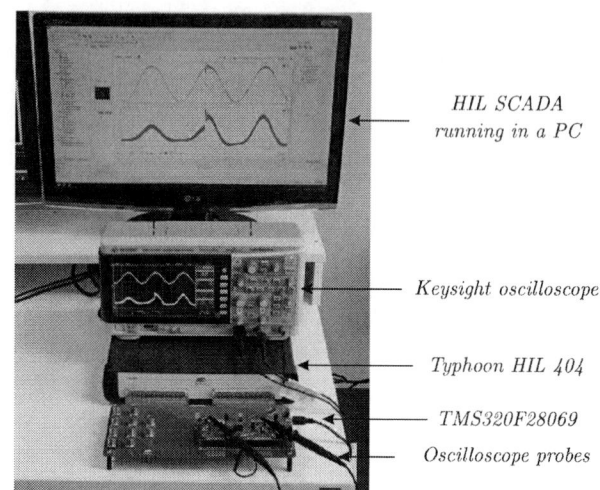

Fig. 5. Experimental HIL setup for experimental validation.

measured voltage in the switches S_1 and S_2 was approximately 450 V.

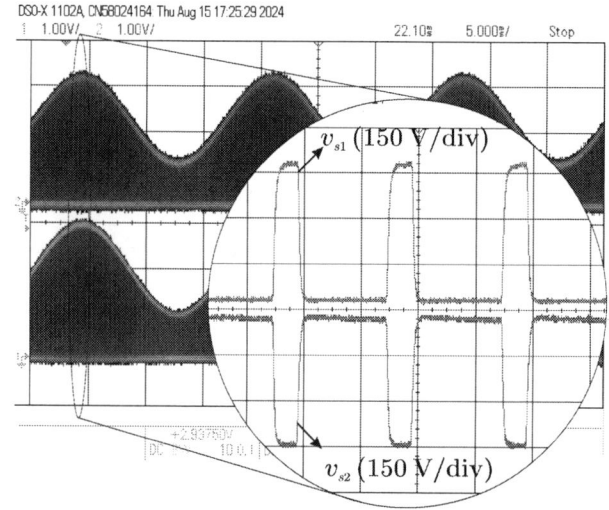

Fig. 6. Curves of v_{s1} (150 V/div) and v_{s2} (150 V/div) experimentally obtained through HIL emulation.

The start-up of the system were tested and it is presented in Fig. 7. The state of the system in instant of initialization was rated input voltage and rated load and with all the discussed observer and control system activated. The inverter observer-based tracking system has presented a startup behavior that does not show any overshoot in both input current i_L and output voltage v_{out}. There were no steady-state error

The steady-state rated operation of the inverter and its observer-based tracking system were validated through Fig. 8. The system has presented no amplitude or phase error reaching peak output voltage of 155 V. In this figure, one can see the input voltage and its characteristic inverter output voltage. This result presented an 0.44% output voltage THD.

With the aim of testing the dynamic response of the system

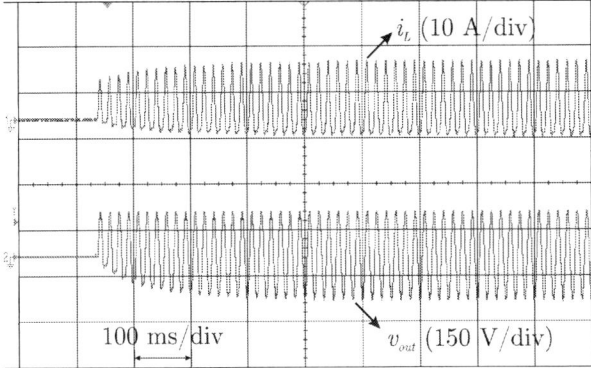

Fig. 7. Curves of i_L (10 A/div) and v_{out} (150 V/div) experimentally obtained through HIL emulation during startup.

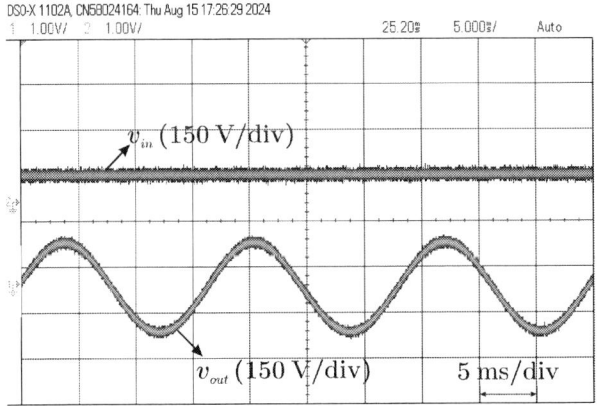

Fig. 8. Curves of v_{in} (150 V/div) and v_{out} (150 V/div) experimentally obtained through HIL emulation.

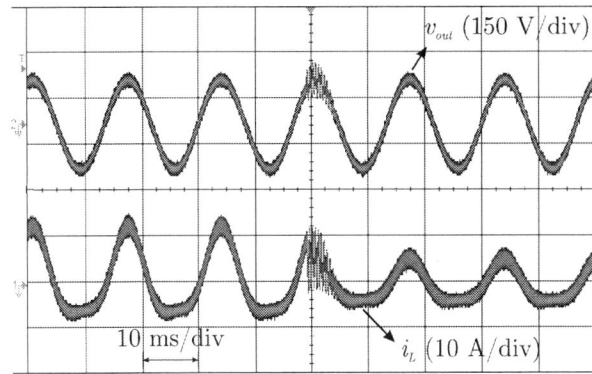

Fig. 9. Behavior of v_{out} (150 V/div) and i_L (10 A/div) for a load step-down from 100% to 50%.

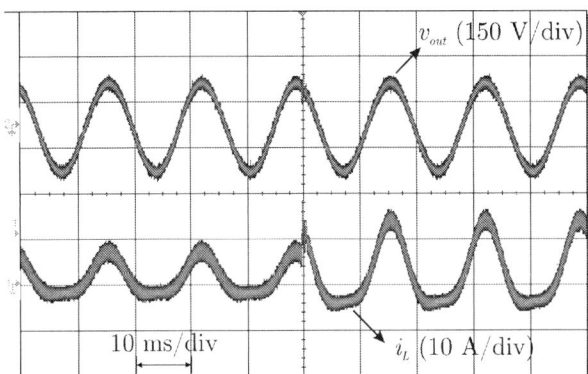

Fig. 10. Behavior of v_{out} (150 V/div) and i_L (10 A/div) for a load step-up from 50% to 100%.

in front of load disturbances, the closed-loop observer-based control system was subjected to a decrease in load from 100% to 50% of its rated power. This step disturbance was implemented by increasing the load resistance from 48.4 Ω to 96.8 Ω in a time close to the moment of maximum energy accumulated in the system (when the output voltage reaches its peak) after the system was in steady-state behavior. As can be seen in Fig. 9, its transient response exhibits characteristics of a system with reduced damping, which corroborate the physical interpretation of a larger resistance load. Moreover, the output voltage and input current settling time was approximately 5 ms (less than a third of operation cycle).

In a similar way, a load step from 50% to 100% was applied. For this load disturbance, the load resistance was decreased from 96.8 Ω to 48.4 Ω at a time close to the peak of the output voltage. The settling time was approximately 3 ms with no steady-state error. This result can be seen in Fig. 10.

Finally, the THD was measured in steady-state for some load values with purpose of verify experimentally the output voltage quality applied to the load. As a results, a THD vs. output power curve was traced and it is presented in Fig. 11. The curve obtained shows that, with load increasing, the output voltage THD decreases. This result is supported by the fact that

with the systems move toward its nominal operational point, its parameters come closer to the controller and observer used parameters.

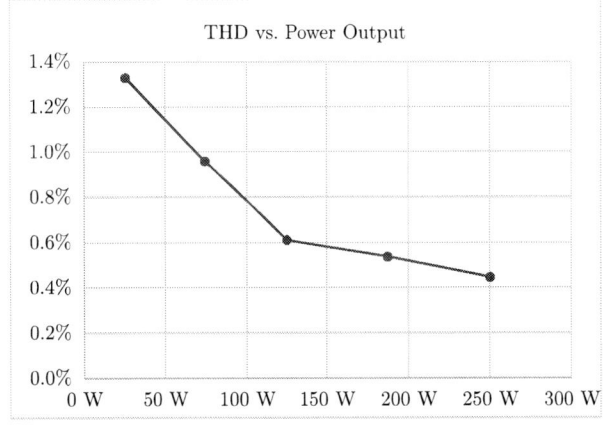

Fig. 11. THD vs. Power output curve obtained experimentally.

Lastly, to validate the robustness of the observer against inverter parameter variations, a test was conducted by varying the inductance L in the inverter, while the controller-observer system maintained the rated value of L as presented in Table I.

Initially, a 10% parameter variation was applied, demonstrating the robustness of the observer, though no significant differences were observed in the resulting curves. Subsequently, the inverter inductance L was reduced to 50% of its rated value, and a load step-up was applied, increasing from 50% to 100% of its nominal value, to examine the effect of parameter variation under load disturbances. The test results showed that the controller-observer system is sufficiently robust to handle parameter variations of up to 50% in L, with a rapid response to load disturbances. These results are shown in Fig. 12.

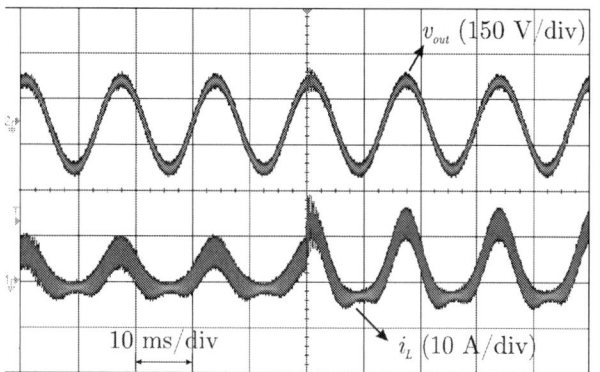

Fig. 12. Behavior of v_{out} (150 V/div) and i_L (10 A/div) for a load step-up from 50% to 100% with a parametric variation of 50% in the inductance value of L.

V. CONCLUSIONS AND FUTURE WORK

We presented a new current sensorless output voltage tracking controller-observer for a Boost inverter that exhibits a kind of global separation principle. The proposed reduced-order state observer for the inductor current has a much lower computational cost than other approaches (e.g. EKF). Experiments with a Typhoon HIL 404 and F28069M LaunchPadTM significantly reduced the steady-state output voltage THD from 11.57% to 0.44% at rated power compared to previous work of the authors. Furthermore, disturbance tests showed improved transient performance and a test with parameter variation was carried out, validating the robustness of the controller-observer system. Moreover, the system has presented a start-up capability with no overshoot that can be useful for inverter stand-alone operations. In terms of future work, preliminary simulation results have indicated that the robust tracking controller-observer proposed in this paper can be successfully adapted to the case where the considered Boost inverter is connected to the grid.

ACKNOWLEDGMENT

This work was partially supported by National Council for Scientific and Technological Development (CNPq) and National Fund for Scientific and Technological Development (FNDCT) of the Ministry of Science, Technology and Innovations (MCTI), Brazil. Process number 406978/2021-2 and 407826/2022-0.

REFERENCES

[1] I. L. Dos Santos and T. B. Lazzarin, "Two Switches Common Grounded Transformerless Step-Up and Step-Down Inverter," 2024 IEEE Applied Power Electronics Conference and Exposition (APEC), Long Beach, CA, USA, 2024, pp. 600-605, doi: 10.1109/APEC48139.2024.10509240.

[2] S. Yang, Y. Tang, Z. Xu, M. Zagrodnik, G. Amit and P. Wang, "Feedback linearization based current control strategy for modular multilevel converters," 2017 IEEE Applied Power Electronics Conference and Exposition (APEC), Tampa, FL, USA, 2017, pp. 659-665, doi: 10.1109/APEC.2017.7930764..

[3] C. A. Busada, R. A. Fantino and J. A. Solsona, "High-Performance Control and Power Decoupling of a Grid-Tied Differential Boost Inverter," in IEEE Transactions on Power Electronics, vol. 39, no. 4, pp. 4042-4049, April 2024, doi: 10.1109/TPEL.2023.3348434.

[4] M. G. Judewicz, S. A. González, E. M. Gelos, J. R. Fischer and D. O. Carrica, "Exact Feedback Linearization Control of Three-Level Boost Converters," in IEEE Transactions on Industrial Electronics, vol. 70, no. 2, pp. 1916-1926, Feb. 2023, doi: 10.1109/TIE.2022.3158015.

[5] S. Gomez Jorge, J. A. Solsona and C. A. Busada, "Nonlinear Control of a Two-Stage Single-Phase DC–AC Converter," in IEEE Journal of Emerging and Selected Topics in Industrial Electronics, vol. 3, no. 4, pp. 1038-1045, Oct. 2022, doi: 10.1109/JESTIE.2022.3151003.

[6] J. Huang, Nonlinear Output Regulation: Theory and Applications, SIAM, 2004.

[7] H. Khalil, Nonlinear Systems, 3rd edition, Prentice Hall, 2002.

[8] H. Kwakernaak and R. Sivan, Linear Optimal Control Systems, John Wiley & Sons, 1972.

[9] T. Karvonen, Stability of Linear and Non-Linear Kalman Filters, Master's Dissertation, Helsinki University, 2014.

[10] M. Grewal and A. Andrews, Kalman Filtering: Theory and Practice Using MATLAB, John Wiley & Sons, 2008.

[11] N. Mohan, T. M. Undeland, and W. P. Robbins, "Power Electronics: Converters, Applications, and Design", 3rd ed. Hoboken, NJ: Wiley, 2003.

[12] R. D. Middlebrook and S. Cuk, "A general unified approach to modelling switching-converter power stages," 1976 IEEE Power Electronics Specialists Conference, Cleveland, OH, USA, 1976, pp. 18-34, doi: 10.1109/PESC.1976.7072895.

[13] B. Lehman and R. M. Bass, "Extensions of averaging theory for power electronic systems," in IEEE Transactions on Power Electronics, vol. 11, no. 4, pp. 542-553, July 1996, doi: 10.1109/63.506119.

[14] S. Meo and L. Toscano, "Some New Results on the Averaging Theory Approach for the Analysis of Power Electronic Converters," in IEEE Transactions on Industrial Electronics, vol. 65, no. 12, pp. 9367-9377, Dec. 2018, doi: 10.1109/TIE.2018.2821620.

[15] S. R. Sanders, J. M. Noworolski, X. Z. Liu and G. C. Verghese, "Generalized averaging method for power conversion circuits," in IEEE Transactions on Power Electronics, vol. 6, no. 2, pp. 251-259, April 1991, doi: 10.1109/63.76811.

[16] C. T. Chen, Linear System Theory and Design, 3rd edition, Oxford University Press, 1999

[17] A. Isidori, Nonlinear Control Systems, 3rd edition, Springer-Verlag, 1995.

Modeling and Control of a Cyclo-Active-Bridge Inverter for Single-Stage Three-Phase Grid Interface

Tanuj Sen$^\diamond$, Mian Liao$^\diamond$, Yang Wu†, and Minjie Chen$^\diamond$

$^\diamond$*Princeton University, Princeton, NJ, United States*
†*ABB Corporate Research, Västerås, Sweden, United States*
Email: {tsen, minjie}@princeton.edu

Abstract—This paper presents the small-signal modeling and control of a cyclo-active-bridge (CAB) inverter for single-stage three-phase grid interface, featuring single-stage decoupled three-phase power flow control. The strengths of the CAB inverter and its control architecture include: (1) single-stage dc-ac power conversion; (2) highly integrated magnetic components; and (3) three independently controlled output phases. It allows for both balanced and unbalanced 3-Φ operation for grid forming and energy routing applications. The First Harmonic Approximation (FHA) method, along with the application of a correction factor, is used for finding an accurate small-signal model of the inverter. This small-signal model is then used to design a per-phase controller to control the per-phase output of the inverter as an ac output. A hardware prototype of this inverter is designed and tested in closed-loop to generate a 3-Φ ac output at 60 Hz.

Index Terms—Cyclo-active-bridge inverter, three-phase, small-signal model, integrated magnetics, first harmonic approximation, correction factor, balanced operation, unbalanced operation

I. INTRODUCTION

WITH the rapid surge in energy demand as well as the growing need for switching to environmentally sustainable means of energy generation, the dependence on renewable sources of energy like solar or wind energy is ever increasing. These energy sources can be used to feed power directly into the existing grid infrastructure or can be used independently to meet the energy needs of small localized areas. Power electronic inverters are playing a critical role in interfacing these future energy systems either to the grid or for localized microgrid applications [1].

Conventional inverter designs usually involve single stage as well as multistage configurations with a low-pass filter at the output [2]–[4]. These inverters can be configured to operate either as a Grid-Forming Inverter (GFM), which directly control the grid voltage and frequency and provide a reference for the entire grid [5], or they can be used as Grid-Following Inverters (GFL), which primary feed power into the grid [6]. In this work, we present a 3-Φ cyclo-active-bridge (CAB) inverter which borrows on the concept of the dual active bridge (DAB) dc-dc converter. The primary side is composed of a three-phase dc-ac inverter with a cyclo-converter stage on the secondary side, connected by a three-phase transformer, as shown in Fig. 1. The topology is similar to the I3DAB dc-dc converter presented in [7]. The power flow can be controlled by changing the phase shift between the primary and secondary side voltages, similar to a DAB. This technique

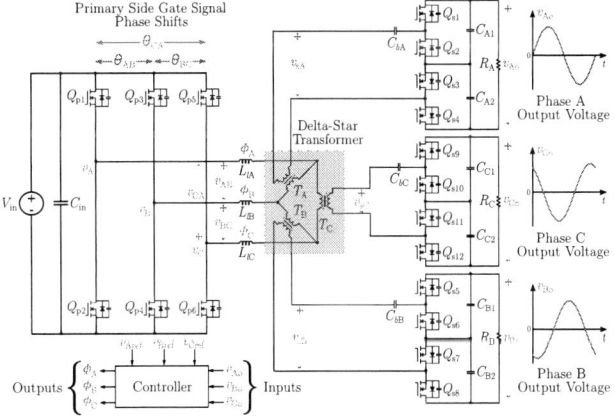

Fig. 1. Topology of the three-phase cyclo-active-bridge (CAB) inverter, with a four-switch cyclo-converter on each phase of the secondary side. Each phase generates a sinusoidal output voltage, while the power delivered by each phase can be controlled by the phase shift (ϕ_N) between the primary side and secondary side voltages of the inverter. A dedicated controller is used to shape the sinusoidal voltage at the output of each phase. The transformer used is an integrated three-phase delta-star configuration transformer.

is used to shape the 60 Hz ac output, eliminating the need for a low-frequency output filter, and enabling more flexible 3-Φ power flow control. The advantage of this topology is that the three output ports are completely decoupled from each other in their operation, which allows for 1) single-stage power processing; 2) balanced and unbalanced 3-Φ operation, leading to a high control flexibility; 3) fast power response due to the high-frequency-link; and 4) direct phase-to-phase power flow without passing through the dc-port.

To control the operation of the CAB inverter as either a GFM or GFL inverter, a dedicated controller needs to be designed, which requires the extraction of the small-signal continuous-time model of the inverter. Previous research has extensively explored methods to derive the continuous-time model of the DAB converter [1], [8]–[10], which is useful for regulating the output voltage of the converter. [11], [12] use variations of the first harmonic approximation method to extract the small-signal converter model. In dc-dc applications, proportional-integral (PI) controllers are often used for controlling the converter output voltage. Proportional resonant (PR) controllers [13] find utility in dc-ac applications as they introduce infinite gain at the fundamental ac frequency. They can also be used in combination with a PI controller, as in [14], to handle both dc and ac components at the output.

In this paper, the 3-Φ CAB inverter is introduced in detail in Section II. Due to its decoupled phase operation, it is treated as a single-phase dc-ac inverter and the First Harmonic Approximation (FHA) technique is employed for extracting its small-signal continuous-time model. The correction factor approach used in [12] is also employed to improve the accuracy of the extracted model, as shown in Section III. This model of the CAB inverter, at various output voltage levels, is then collectively used to design a suitable ac output voltage controller in Section IV, which is capable of generating a balanced or unbalanced 3-Φ ac output at the three decoupled output ports of the inverter. The operation of the designed closed-loop system is verified through both simulations and experimental results in Section V.

II. THREE PHASE CYCLO-ACTIVE-BRIDGE INVERTER: OVERVIEW

The Cyclo-Active-Bridge (CAB) inverter topology is shown in Fig. 1. This inverter takes a dc voltage as an input and converts it into three separate ac outputs, with galvanic isolation provided between the input and output by the three-phase transformer. The transformer also helps to provide a more flexible voltage conversion ratio between the input and the output. The operating principles and mechanism of the CAB inverter are closely derived from the operation of the half-bridge cyclo-converter based single phase dc-ac DAB inverter presented in [15], [16], and the 3-Φ dc-dc I3DAB converter topology introduced in [7]. Similar to as mentioned in [7] for the I3DAB topology, the primary side of the CAB inverter employs a three-phase half-bridge inverter, with the three phases connected in the delta configuration. The three phase switching signals of the primary side are equally phase shifted by 120° to generate a symmetrical three-level primary side voltage. The secondary side is composed of three separate phases, However, unlike [7], where each secondary side phase is composed of a full-bridge rectifier to generate a dc output voltage, the CAB inverter makes use of a cyclo-converter for each phase, similar to the implementation shown in [15]. This cyclo-converter allows for an ac output voltage at each of the three output phases. The frequency and amplitude of this ac ouptut voltage can be varied by properly controlling the value of the phase shift ϕ between the primary side and secondary side voltages. An advantage of the CAB inverter, derived from the I3DAB converter proposed in [7] due to their similar construction, is that the three output ports of such an inverter are isolated from each other and operate independently. This implies that the output voltage and power delivered by each phase can be independently controlled through the per-phase phase shift between the primary and secondary side transformer winding voltages, with no influence from the other two phases. Hence, the proposed 3-Φ inverter can be considered as three separated single phase dc-ac cyclo-inverters, with Fig. 2 highlighting the structure and the voltage and current waveforms of one of the three phases.

Due to the delta-configuration three-phase inverter on the primary side of the CAB inverter, with the per-phase switching

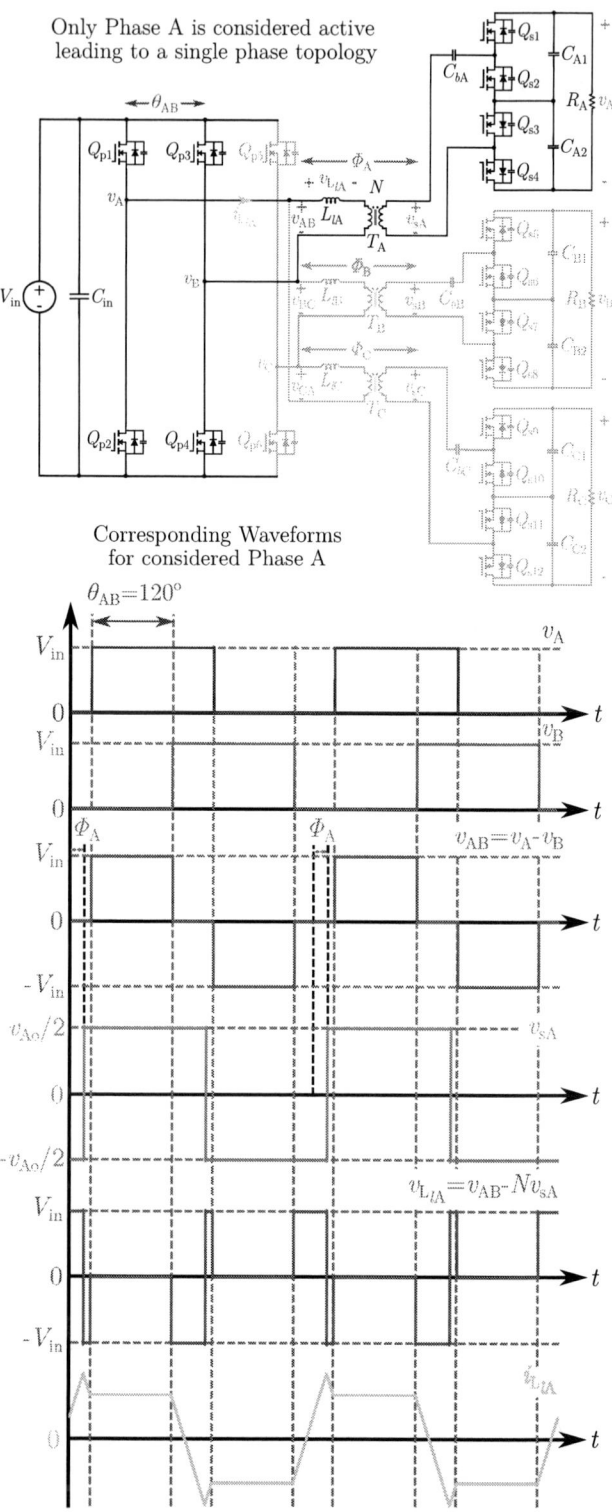

Fig. 2. The circuit diagram and the corresponding voltage and current waveforms in the linear mode of operation, with one of the output phases (Phase A) active. The primary side voltage v_{AB} is a three-level voltage formed by the square-wave switch node voltages of phase A and B, which are phase-shifted by $\theta_{AB}=120°$. The secondary side voltage v_{sA} is phase shifted (ϕ_A) with respect to v_{AB}, which determines the voltage drop L_{llA} across the inductor L_{lA}, which in turn determines the current flowing through the inductor.

signals having a 120° phase shift between them, the per-phase primary side voltage is a symmetrical three-level voltage, formed by the phase shifted switch-node voltages of the two associated primary side half-bridge legs. On the other hand, the secondary side voltage of the CAB inverter is effectively a two-level square wave voltage with a peak amplitude of half of the output voltage. The phase shift per-phase phase shift ϕ between the primary and secondary side voltages is determined from the mid-point of the zero-level of the primary side voltage, as shown in Fig. 2. Due to the specific shapes of the primary and secondary side voltages, the per-phase operation of the CAB inverter can be divided into two separate modes or regimes, based on the value of the phase shift ϕ between the primary and secondary side voltage for each phase. These two modes are called the linear and non-linear mode of operation. In the linear mode of operation, which applies for the range $0 \leq |\phi| \leq \frac{\pi}{6}$, the per-phase power and the output voltage equations for the CAB inverter are as follows:

$$P_{\text{out}} = \frac{V_{\text{in}} V_{\text{out}}}{6\pi N L f_{\text{sw}}} \phi, \quad \forall \quad 0 \leq |\phi| \leq \frac{\pi}{6}. \quad (1)$$

$$V_{\text{out}} = \frac{V_{\text{in}} R_{\text{out}}}{6\pi N L f_{\text{sw}}} \phi. \quad (2)$$

where V_{in} and V_{out} are the input and output voltages, respectively. L is the phase leakage inductance, N refers to the transformer turns ratio, f_{sw} is the switching frequency. Due to the linear relationship between the output power and the phase shift ϕ, this regime of operation is called the linear mode. Note that this linear relationship exists only when the output voltage remains invariant. Furthermore, the linear relationship between the output voltage and ϕ is maintained only for constant load operation. On the other hand, for the range $\frac{\pi}{6} \leq |\phi| \leq \frac{\pi}{2}$, the per-phase power and output voltage with respect to ϕ are given as:

$$P_{\text{out}} = \frac{V_{\text{in}} V_{\text{out}}}{4\pi N L f_{\text{sw}}} \left(\phi - \frac{\phi^2}{\pi} - \frac{\pi}{36} \right), \quad \forall \quad \frac{\pi}{6} \leq |\phi| \leq \frac{\pi}{2}. \quad (3)$$

$$V_{\text{out}} = \frac{V_{\text{in}} R_{\text{out}}}{4\pi N L f_{\text{sw}}} \left(\phi - \frac{\phi^2}{\pi} - \frac{\pi}{36} \right). \quad (4)$$

which highlights the non-linear relationship between P_{out}, V_{out} and ϕ, hence the name non-linear mode for this regime of operation. The per-phase maximum power is attained at $\phi = \pi/2$, with the power curve being symmetrical around $\phi = \pi/2$, as shown in Fig. 3 (considering constant output voltage). Negative power flow can be achieved using negative values of ϕ. Using these relations, the CAB inverter can be operated to produce an ac voltage at the output, by varying the phase shift ϕ accordingly. More details about the operation of the CAB inverter can be found in [17].

III. SMALL SIGNAL MODELING OF THE THREE-PHASE CYCLO-ACTIVE-BRIDGE INVERTER

The extraction of the small-signal model of the CAB inverter is of great importance in order to design a suitable controller for controlling the output voltage to be ac. Since the three phases of the CAB inverter are completely decoupled from each other, the small-signal model can be developed on a

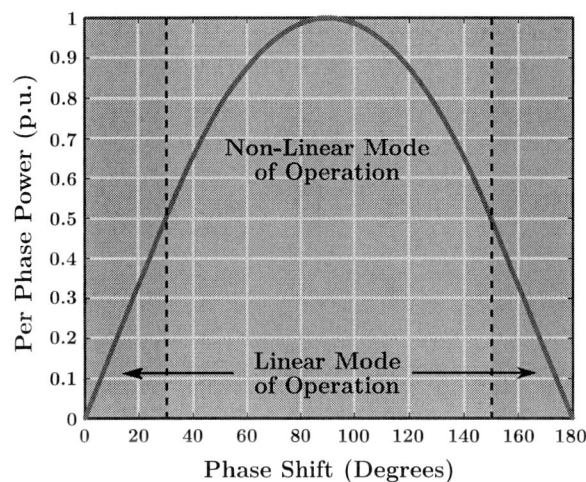

Fig. 3. Variation of the per-phase power (p.u.) with respect to the phase shift ϕ between the primary and secondary side voltages of the CAB inverter, assuming a constant output voltage.

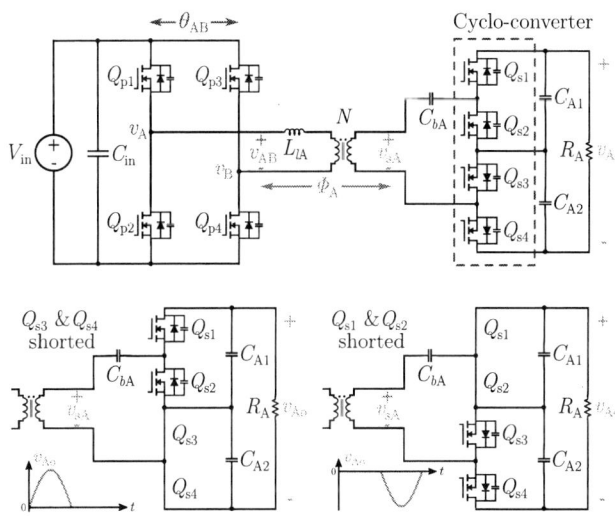

Fig. 4. A simplified single-phase model of the proposed 3-Φ CAB inverter, as the three phases are decoupled from each other. The operation principle of the cyclo-converter on the secondary side inverter is shown. The upper half-bridge operates conventionally while the lower half bridge is shorted when the output voltage is positive and vice-versa.

per-phase basis, with each phase being treated as an equivalent single-phase DAB converter, as shown in Fig. 4. In this paper, we use the First Harmonic Approximation (FHA) technique to develop the small-signal model, along with the the use of the correction factor concept introduced in [12], to account for the higher order harmonics of the current and voltage which are neglected in the FHA. Moreover, due to the differing modes of operation of the CAB inverter with respect to ϕ, the small-signal model needs to be developed separately for the linear mode and non-linear mode to correctly account for the varying system dynamics. We begin by considering the small-signal model for the linear mode and then extend the approach for the non-linear mode as well.

The per-phase primary and secondary side switched-mode voltages are shown in Fig. 2. The fundamental component, at the angular switching frequency of the inverter ω_s, of the primary and secondary side voltages, V_P and V_S, and the current through the inductor i_L are given as follows:

$$V_P = \frac{4V_{in}\sin\left(\frac{\pi}{3}\right)}{\pi}\sin\omega_s t. \tag{5}$$

$$V_S = \frac{4v_o}{2\pi N}\sin(\omega_s t - \phi) = v_s\sin\omega_s t + v_c\cos\omega_s t. \tag{6}$$

$$\frac{di_L}{dt} = \frac{V_P - V_S}{L}. \tag{7}$$

Note that V_S has a factor of the turns ratio of the transformer N as it is referred to the primary side of the inverter. From (5) and (6), it is evident that the fundamental components of the primary and secondary side voltages are also phase shifted by ϕ. As a result, V_S has a sine (v_s) and a cosine (v_c) component, which also leads to the inductor current i_L to be composed of a sine (i_{Ls}) and a cosine (i_{Lc}) component. Thus, the fundamental component of the inductor current and its differential become:

$$i_L = i_{Lc}\cos\omega_s t + i_{Ls}\sin\omega_s t. \tag{8}$$

$$\frac{di_L}{dt} = \left(\frac{di_{Lc}}{dt} + \omega_s i_{Ls}\right)\cos\omega_s t + \left(\frac{di_{Ls}}{dt} - \omega_s i_{Lc}\right)\sin\omega_s t. \tag{9}$$

The current flowing into the inverter output capacitor can be computed as the product of the inductor current referred to the secondary side and the switching signal of the cyclo-converter. Since the switching signal is practically a square wave with a duty ratio of 50%, phase shifted by ϕ, its fundamental component representation is:

$$s_2 = \frac{1}{2} - \frac{2}{\pi}\sin\phi\cos\omega_s t + \frac{2}{\pi}\cos\phi\sin\omega_s t. \tag{10}$$

When multiplied with the secondary side current, we get:

$$s_2 \cdot \frac{i_L}{N} = -\frac{i_{Lc}}{N\pi}\sin\phi + \frac{i_{Ls}}{N\pi}\cos\phi + \text{harmonic terms}. \tag{11}$$

Since the output capacitor voltage is practically dc, we consider only the zeroth-order components of currents and the voltages, leading to the following governing equation for the output voltage:

$$C_O\frac{dv_o}{dt} = -\frac{i_{Lc}}{N\pi}\sin\phi + \frac{i_{Ls}}{N\pi}\cos\phi + \frac{v_o}{R_{out}}. \tag{12}$$

Considering i_{Ls}, i_{Lc} and v_o to be the state variables of the system, we arrive at the following state equations:

$$\frac{di_{Ls}}{dt} = \frac{4V_{in}}{\pi L}\sin\frac{\pi}{3} - \frac{4v_o}{2\pi NL}\cos\phi + \omega_s i_{Lc} \tag{13}$$

$$\frac{di_{Lc}}{dt} = \frac{4v_o}{2\pi NL}\sin\phi - \omega_s i_{Ls} \tag{14}$$

$$\frac{dv_o}{dt} = -\frac{i_{Lc}}{C_O N\pi}\sin\phi + \frac{i_{Ls}}{C_O N\pi}\cos\phi + \frac{v_o}{R_{out}C_O}. \tag{15}$$

These equations only take the fundamental components of the voltages and the current into account. However, by neglecting the effect of the higher order harmonics, the small-signal model would lack the desired accuracy and would

fail to model the variation of the system transfer function with a changing output voltage. Therefore, we make use of a correction factor γ for the state variables, introduced in [12], to account for the effect of these additional harmonics. The power transfer between the primary and secondary side, considering the two fundamental frequency sources V_P and V_S, connected by the inductance L is given as:

$$P_{out(1)} = \frac{4V_{in}v_o}{\pi^2\omega_s NL}\sin\left(\frac{\pi}{3}\right)\sin\phi. \tag{16}$$

while the actual power flow in the linear-mode is given by (1). Therefore, dividing (1) by (16), we get the required correction factor γ for v_o:

$$\gamma = \frac{P_{out}}{P_{out(1)}} = \frac{\pi^2\phi}{12\sin\left(\frac{\pi}{3}\right)\sin\phi}. \tag{17}$$

$$v_{o,mod} = \gamma v_o. \tag{18}$$

In a similar manner, the correction factor for i_{Ls}, i_{Lc} can be determined. The total active power delivered by the fundamental component of the inductor current depends solely on i_{Ls} and is given as:

$$P_{out(1)} = \frac{1}{2}\frac{4V_{in}i_{Ls}}{\pi\omega_s L}\sin\left(\frac{\pi}{3}\right)\sin\phi. \tag{19}$$

Based on (19), we find the correction factor for i_{Ls} to be γ as well, leading to the following corrected state variable relation:

$$i_{Ls,mod} = \gamma i_{Ls}. \tag{20}$$

Lastly, the corrected form of i_{Lc} is calculated by comparing the values of the total reactive power to the reactive power due to the fundamental component of the current, in a manner similar to that shown in [12], leading to the following relation:

$$i_{Lc,mod} = \gamma i_{Lc} + (\gamma - 1)\frac{4V_{in}}{\pi\omega_s L}. \tag{21}$$

Using the aforementioned relations, the modified state equations can be formulated, as shown below:

$$\frac{di_{Ls,mod}}{dt} = \frac{4V_{in}}{\pi L}\sin\frac{\pi}{3} - \frac{4V_{out,mod}}{2\pi NL}\cos\phi + \omega_s i_{Lc,mod}. \tag{22}$$

$$\frac{di_{Lc,mod}}{dt} = \frac{4v_{o,mod}}{2\pi NL}\sin\phi - \omega_s i_{Ls,mod} + \frac{d\frac{4(\gamma-1)V_{in}}{\omega_s\pi L}\sin\left(\frac{\pi}{3}\right)}{dt}. \tag{23}$$

$$\frac{dv_{o,mod}}{dt} = -\frac{i_{Lc,mod}}{C_O N\pi}\sin\phi + \frac{i_{Ls,mod}}{C_O N\pi}\cos\phi + \frac{v_{o,mod}}{R_{out}C_O} + \frac{4(\gamma-1)V_{in}}{\omega_s\pi^2 LC_O}\sin\left(\frac{\pi}{3}\right)\sin\phi. \tag{24}$$

Upon perturbing the state-space variables and ϕ in the modified state-space equations, we arrive at the following state-space matrices for the small signal model.

979-8-3315-1612-3/25 $31.00 © 2025 IEEE

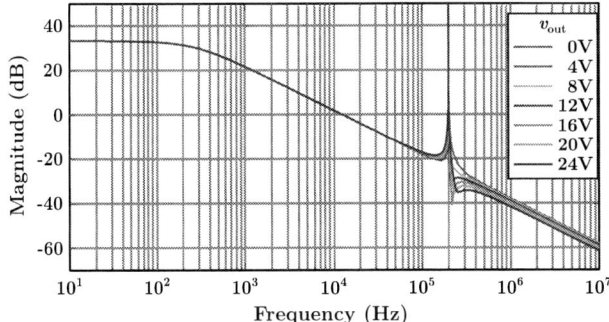

Fig. 5. Magnitude response of the per-phase small-signal transfer function $G_{v_o\phi}(s)$ of the CAB inverter for different output voltages, in the linear mode.

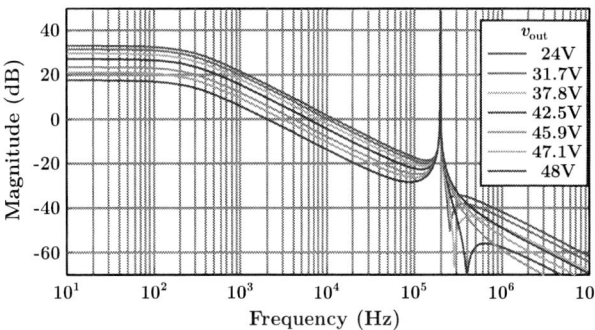

Fig. 6. Magnitude response of the per-phase small-signal transfer function $G_{v_o\phi}(s)$ of the CAB inverter for different output voltages, in the non-linear mode. The gain of the system varies with the output voltage.

$$A = \begin{bmatrix} 0 & \omega_s & \frac{-4\cos\phi}{2\pi NL} \\ -\omega_s & 0 & \frac{4\sin\phi}{2\pi NL} \\ \frac{\cos\phi}{N\pi C_O} & \frac{-\sin\phi}{N\pi C_O} & \frac{-1}{R_{out}C_O} \end{bmatrix} \quad (25)$$

$$B = \begin{bmatrix} \frac{-4V_{out}\sin\phi}{2\pi NL} \\ \frac{-4V_{out}\cos\phi}{2\pi NL} + \frac{4V_{in}}{\omega_s\pi L}\sin\left(\frac{\pi}{3}\right)\lambda\frac{d}{dt} \\ -\frac{i_{Lc}}{C_O N\pi}\cos\phi - \frac{i_{Ls}}{C_O N\pi}\sin\phi + \frac{4V_{in}\lambda}{\omega_s\pi^2 NLC_O}\sin\left(\frac{\pi}{3}\right)\sin\phi \end{bmatrix} \quad (26)$$

$$C = \begin{bmatrix} 0 & 0 & 1 \end{bmatrix} \quad (27)$$

for the state-space matrix $x = [i_{Ls,mod}\ i_{Lc,mod}\ v_{o,mod}]^T$ and the input matrix $x = [\phi]$. The parameter λ in B is defined as:

$$\lambda = \frac{\pi^2}{12\sin\frac{\pi}{3}\sin\phi} - \frac{\cos\phi}{\sin\phi}. \quad (28)$$

Therefore, the small-signal transfer function relating the output voltage v_o to the phase shift ϕ for the linear mode of operation for the CAB inverter can be computed through:

$$G_{v_o\phi}(s) = C(s \cdot I - A)^{-1}B. \quad (29)$$

The bode plot showing the magnitude response of the each phase of the CAB inverter system, when operating in the linear mode, is computed and shown in Fig. 5. For the magnitude response, the following per-phase inverter parameters were selected: a switching frequency f_{sw} of 200 kHz to generate a 48 V ac voltage at the output terminals of each phase, with a phase inductance L of 5 μH, a transformer turns ratio N of 1.33, an input voltage of 48 V, while C_{load} and R_{load} are 24 μF and 23 Ω, respectively. The per-phase magnitude response is shown for varying output voltage values, as is the case for ac operation. However, the voltage variation is only taken from 0 V to 24 V, as this is the range of linear-mode of operation. From the magnitude response, it is evident that the gain of the system at lower frequencies remains constant with respect to the output voltage, which is a characteristic of the linear-

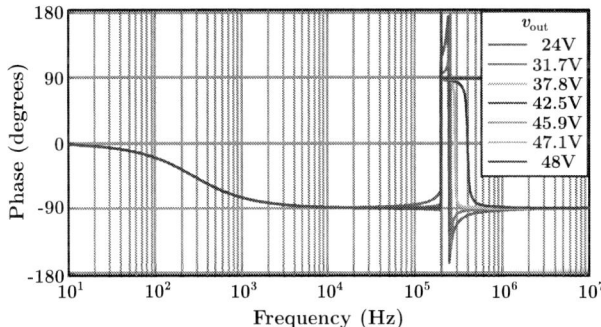

Fig. 7. Phase response of the per-phase small-signal transfer function $G_{v_o\phi}(s)$ of the CAB inverter for different output voltages, in the non linear mode of operation. The phase of the system shows a spike at the switching frequency.

mode of operation. Since the considered system is undamped, a large resonance peak appears at the switching frequency.

In a similar manner, the per-phase system transfer function for the CAB inverter can be derived for the non-linear mode of operation. All the state equations have the same form as the linear-mode. However, due to the per-phase power equation being different for the non-linear mode (3), the value of the correction factor is different for this mode:

$$\gamma = \frac{P_{out}}{P_{out(1)}} = \frac{\pi^2\left(\phi - \frac{\phi^2}{\pi} - \frac{\pi}{36}\right)}{8\sin\left(\frac{\pi}{3}\right)\sin\phi}. \quad (30)$$

The resulting state-space matrices for the non-linear mode have the same form as the linear-mode matrices as given in (25), (26) and (27). However, the value of λ in this small-signal model is different, due to the different value of the correction factor γ in this mode of operation. The non-linear mode value of λ is given as:

$$\lambda = \frac{\pi(\pi - 2\phi)}{8\sin\frac{\pi}{3}\sin\phi} - \frac{\cos\phi}{\sin\phi}. \quad (31)$$

Using this value of λ in the state-space matrices and then applying these matrices to (29), the per-phase transfer function $G_{v_o\phi}(s)$ for the non-linear mode of operation can be determined. The bode magnitude plot for $G_{v_o\phi}(s)$ for varying output voltages v_o in the non-linear range is shown in Fig. 6. In

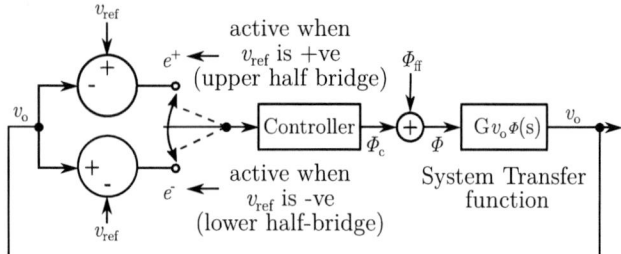

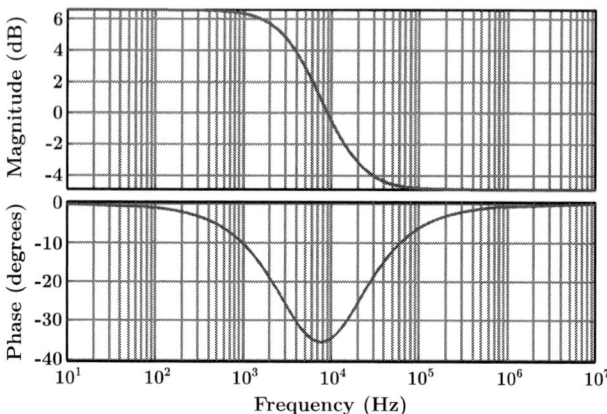

Fig. 8. Block diagram of the per phase closed loop system with the controller. The polarity of the processed error signal depends on the polarity of the reference signal, which also influences which half of the cyclo-converter is active. The controller takes this error as an input and converts it into the required phase shift between the primary and secondary side voltages.

Fig. 9. Magnitude and phase response of the designed per-phase controller $G_{ovc}(s)$ for the CAB inverter. The lag compensator introduces a phase dip around the desired crossover frequency of $f_{sw}/20 = 10$ kHz, which allows the system to have a phase margin of around 60° for all output voltages.

the non-linear mode of operation, the gain of the system varies with the output voltage. This is due to λ having a factor of the phase-shift ϕ in it, and the value of λ has a direct correlation with the gain of the system. Hence, as ϕ is varied to change the output voltage, the gain of the system changes.

The phase variation of the system over frequency of operation, for output voltages in the non-linear mode of operation, is shown in Fig. 7. The phase variation is independent of any changes in the output voltage and only shows a jump at the switching frequency. The phase variation shows the same characteristic for the linear mode of operation as well. The phase information is of great importance for designing a suitable closed-loop controller for the CAB inverter, as is shown in the following section.

IV. CONTROLLER DESIGN FOR THE CAB INVERTER

Figure 4 shows the operation of the cyclo-converter on the secondary side (per-phase) of the CAB inverter to generate an ac voltage at the output, by varying the phase shift ϕ accordingly. To generate both positive and negative voltages at the output, the voltage reference needs to be positive for one half of the desired ac voltage period and negative for the other half. Thus, the error signal that is fed into the controller needs to have the proper polarity, as is shown in Fig. 8. Since the frequency of the output ac voltage is many orders of magnitude lower than the switching frequency, the CAB inverter can effectively be treated as a dc-dc converter. Therefore, the magnitude and phase responses shown in the previous section can be used collectively to design a per-phase controller, which can modulate the output voltage of the phase as an ac voltage of required amplitude and frequency.

Due to the variable gain of the system in the non-linear mode, the crossover frequency will be different for different output voltages. In this case, we choose the system magnitude response with the highest gain ($v_{out} = 24$ V) as the basis in order to compute the required gain of the controller, so that the crossover frequency of the modulated system shifts to $f_{sw}/20$. This would cause the crossover frequency of the modulated system to be lower than $f_{sw}/20$ at other output voltages, which effectively attenuates the impact of switching frequency disturbances on the closed loop system.

An important aspect for controller design is to ensure suitable gain and phase margin for the system. To design the

controller, a phase margin of 60° is chosen at the desired crossover frequency. A lag compensator is designed to this effect. The pole and zero of the lag compensator are chosen such that the phase margin of the system at different output voltages remains in the vicinity of the 60° target. The designed per-phase controller, therefore, has the following form, with the mentioned values:

$$G_{ovc}(s) = K_p \left(\frac{s + \omega_z}{s + \omega_p} \right) = \frac{1}{1.754} \left(\frac{s + 2\pi \cdot 15 \ kHz}{s + 2\pi \cdot 4 \ kHz} \right) \quad (32)$$

The magnitude and phase response of the designed controller are shown in Fig. 9. The designed controller responds to any small signal variations in the output voltage. The phase shift value generated by the controller, together with a feedforward value of the phase shift, which accounts for the large signal variation of the output voltage when operating to produce an ac output, regulates the output voltage of each phase, as shown in Fig. 8.

V. SIMULATIONS AND EXPERIMENTAL VERIFICATION

The operation of the 3-Φ cyclo-active-bridge circuit, as shown in Fig. 1, was simulated in PLECS, with the circuit parameter values mentioned previously in Section III. The switching frequency for the inverter was selected as 200 kHz, similar to the value taken in the small-signal model design. In order to get an ac output at the three phases of the CAB inverter, the designed per-phase controller was employed, operating each phase in a closed loop, independent of the other phases. This closed-loop controlled inverter is first simulated to generate a 3-Φ 48 V balanced ac output at 60 Hz for an input voltage of 48 V dc. The simulation results for the balanced case are provided in Fig. 10. The simulation results show that the designed per-phase controller is able to shape the output voltage of each phase of the inverter into the desired sinusoidal voltage with an amplitude of 48 V and a phase shift of 120° between each of the three phases, resulting in a balanced 3-Φ output. To highlight the capability of the designed closed-loop

979-8-3315-1612-3/25 $31.00 © 2025 IEEE

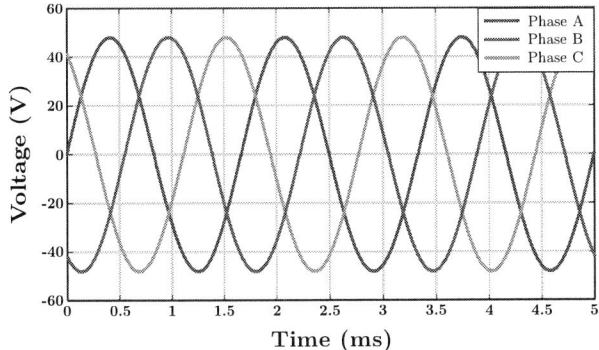

Fig. 10. Simulation results showing the desired operation of the 3-Φ output voltages of the inverter in the closed loop. The designed per phase controller creates a per phase sinusoidal voltage with an amplitude of 48 V at 60 Hz, with a phase shift of 120° between them.

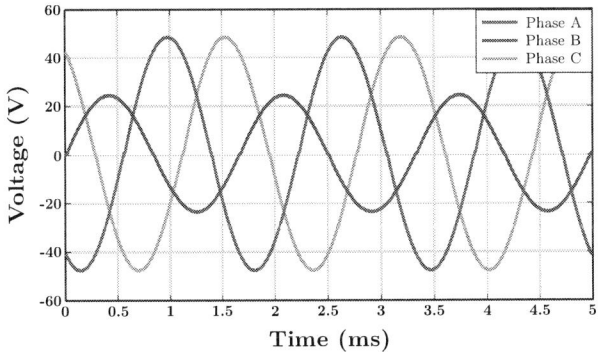

Fig. 11. Simulation results showing the unbalanced closed loop operation of the 3-Φ inverter. The controller for Phase A holds its output voltage at 24 V 60 Hz, while Phase B and Phase C output voltages are at 48 V and 60 Hz. The phase shift between the phases is maintained at 120°.

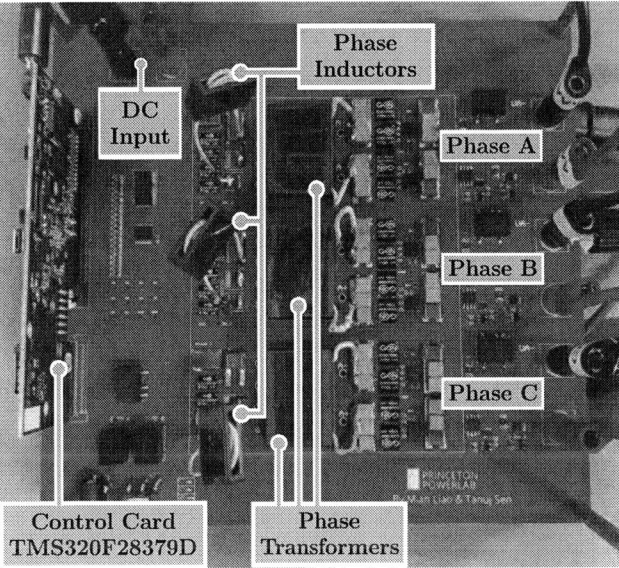

Fig. 12. Hardware prototype of the cyclo-converter based 3-Φ CAB inverter. The primary side terminals of the transformers are routed on the PCB and are connected in a delta configuration, while the secondary side terminals are formed using litz wires, connected separately to three independent secondary side phases without balancing needs.

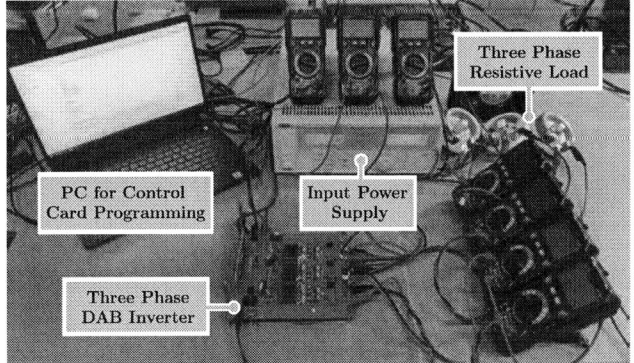

Fig. 13. Complete hardware setup of the cyclo-converter based 3-Φ CAB inverter, with the three output phases of the inverter connected three equal resistive loads of 50 Ω each, simulating a balanced load at the output. The inverter is provided a dc supply of 48 V at the input. The control card TMS320F28379D implements the closed loop control to produce a 3-Φ ac voltage at the output.

inverter to operate with an unbalanced 3-Φ output, the inverter is driven with three sinusoidal voltage references which are 120° shifted with respect to each other. However, the reference for Phase A is set to have an amplitude of 24 V, while Phase B and Phase C reference amplitudes are set as 48 V. The unbalanced operation simulation results, with Phase A voltage having half the amplitude of Phase B and Phase C voltages, in accordance with the provided references, are shown in Fig. 11.

The designed inverter circuit was also implemented on hardware, with the experimental prototype shown in Fig. 12. Each switch on the primary side of the inverter was realized with two 100 V/90 A GS61008P GaN MOSFETs in parallel, while the switches on the secondary side make use of the 650 V/30 A GS66508B GaN MOSFETs provided by GaN systems. The phase inductance of 5 μH was designed with a 3F46 planar ER core with litz wire windings, while the transformer makes use of a 3C92 planar ER core with PCB windings for the primary side and litz wire for the secondary side windings. The TMS320F28379D control card provided by Texas Instruments is used for generating the gate signals and for implementing the closed loop control. The three output phases of the inverter are connected to three separate resistive loads. The full dc-rated 300 W 3-Φ experimental setup is shown in Fig. 13.

Similar to the simulation results, the experimental results for the inverter being controlled to provide a balanced 3-Φ output voltage at 48 V and 60 Hz are shown in Fig. 14. At every zero crossing of the per phase output voltage, there is a slight discontinuity which appears due to change in the active half-bridge of the cyclo-converter on the secondary side, as depicted in Fig. 4 previously. Additionally, the unbalanced operation of the inverter was also tested through experiments, the results of which are provided in Fig. 15. In this case, Phase A is completely turned off, while Phase B and Phase C are controlled to produce a sinusoidal output voltage with an amplitude of 48 V at 60 Hz. Even in this case, the discontinuity at the zero crossing of the phase voltages, due to the bridge

979-8-3315-1612-3/25 $31.00 © 2025 IEEE

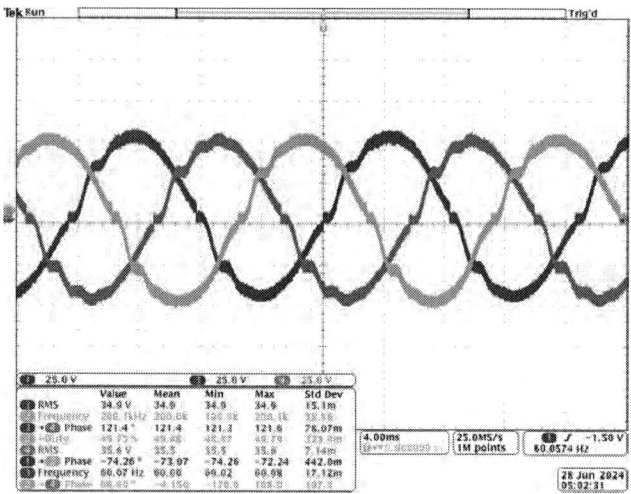

Fig. 14. Experimental results showing the 3-Φ output voltages of the inverter, with the per phase controller set to create a per phase 60 Hz sinusoidal voltage with an amplitude of 48 V. The three phases are shifted of 120° among each other.

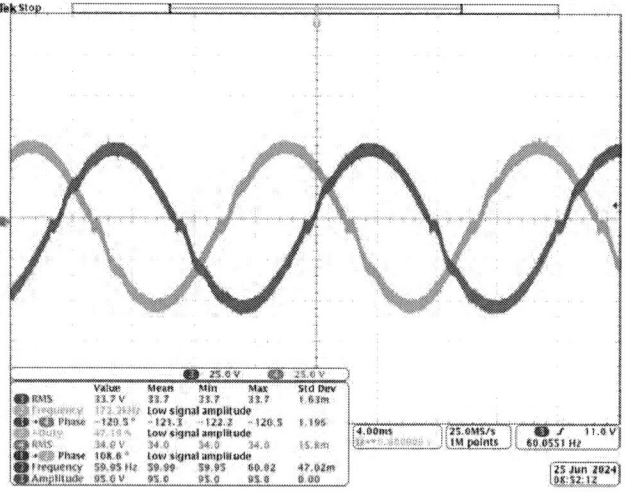

Fig. 15. Experimental results showing the unbalanced 3-Φ output voltages of the inverter, with the per phase controllers for Phase B and Phase C set to create a per phase 60 Hz sinusoidal voltage with an amplitude of 48 V, while Phase A is completely turned off. The phase shift between Phase B and Phase C output voltages is still maintained at 120°.

switching action of the cyclo-converter, is observable. The experimental results verify the versatility of the designed inverter in producing both balanced and unbalanced 3-Φ ac output voltages.

VI. CONCLUSION

The small-signal modeling using the First Harmonic Approximation technique and control design of the single-stage 3-Φ cyclo-active-bridge inverter is presented in this paper. The three phases of the inverter are isolated from each other, thus, this inverter can be used for both balanced and unbalanced 3-Φ operation. Moreover, due to this decoupled nature, a dedicated per phase controller can be designed for each of the three phases, allowing each phase to be controlled similar to a 1-Φ inverter. A prototype of the CAB inverter is designed on

hardware, using integrated magnetics and its operation is tested at a frequency of 200 kHz. Simulations as well as experimental measurements show that the closed loop operation of the proposed inverter topology, using the per-phase controller designed for the CAB inverter is indeed capable of generating both balanced and unbalanced 3-Φ ac output voltages at 60 Hz.

ACKNOWLEDGEMENTS

This work was supported by NSF ASCENT Grant Award Number #2328241 and the NJEDA Wind Institute Fellowship.

REFERENCES

[1] J. Yang, S. Guenter, G. Buticchi, C. Gu, M. Liserre, and P. Wheeler, "On the impedance and stability analysis of dual-active-bridge-based input-series output-parallel converters in dc systems," *IEEE Transactions on Power Electronics*, vol. 38, no. 8, pp. 10 344–10 358, 2023.

[2] S. Anttila, J. S. Döhler, J. G. Oliveira, and C. Boström, "Grid forming inverters: A review of the state of the art of key elements for microgrid operation," *Energies*, vol. 15, no. 15, 2022. [Online]. Available: https://www.mdpi.com/1996-1073/15/15/5517

[3] P. Morsali, S. Dey, A. Mallik, and A. Akturk, "Switching modulation optimization for efficiency maximization in a single-stage series resonant dab-based dc–ac converter," *IEEE Journal of Emerging and Selected Topics in Power Electronics*, vol. 11, no. 5, pp. 5454–5469, 2023.

[4] S. Jain and V. Agarwal, "A single-stage grid connected inverter topology for solar pv systems with maximum power point tracking," *IEEE Transactions on Power Electronics*, vol. 22, no. 5, pp. 1928–1940, 2007.

[5] L. Kong, Y. Xue, L. Qiao, and F. Wang, "Control design of passive grid-forming inverters in port-hamiltonian framework," *IEEE Transactions on Power Electronics*, vol. 39, no. 1, pp. 332–345, 2024.

[6] D. Pattabiraman, R. H. Lasseter., and T. M. Jahns, "Comparison of grid following and grid forming control for a high inverter penetration power system," in *2018 IEEE Power Energy Society General Meeting (PESGM)*, 2018, pp. 1–5.

[7] J. Böhler, F. Krismer, T. Sen, and J. W. Kolar, "Optimized modulation of a four-port isolated dc–dc converter formed by integration of three dual active bridge converter stages," in *2018 IEEE International Telecommunications Energy Conference (INTELEC)*, 2018, pp. 1–8.

[8] S. S. Shah and S. Bhattacharya, "A simple unified model for generic operation of dual active bridge converter," *IEEE Transactions on Industrial Electronics*, vol. 66, no. 5, pp. 3486–3495, 2019.

[9] O. M. Hebala, A. A. Aboushady, K. H. Ahmed, S. Burgess, and R. Prabhu, "Generalized small-signal modelling of dual active bridge dc/dc converter," in *2018 7th International Conference on Renewable Energy Research and Applications (ICRERA)*, 2018, pp. 914–919.

[10] M. Safayatullah and I. Batarseh, "Small signal model of dual active bridge converter for multi-phase shift modulation," in *2020 IEEE Energy Conversion Congress and Exposition (ECCE)*, 2020, pp. 5960–5965.

[11] H. Qin and J. W. Kimball, "Generalized average modeling of dual active bridge dc–dc converter," *IEEE Transactions on Power Electronics*, vol. 27, no. 4, pp. 2078–2084, 2012.

[12] S. S. Shah and S. Bhattacharya, "Large small signal modeling of dual active bridge converter using improved first harmonic approximation," in *2017 IEEE Applied Power Electronics Conference and Exposition (APEC)*, 2017, pp. 1175–1182.

[13] H. Cha, T.-K. Vu, and J.-E. Kim, "Design and control of proportional-resonant controller based photovoltaic power conditioning system," in *2009 IEEE Energy Conversion Congress and Exposition*, 2009, pp. 2198–2205.

[14] H. Qin and J. W. Kimball, "Closed-loop control of dc–dc dual-active-bridge converters driving single-phase inverters," *IEEE Transactions on Power Electronics*, vol. 29, no. 2, pp. 1006–1017, 2014.

[15] A. Trubitsyn, B. J. Pierquet, A. K. Hayman, G. E. Gamache, C. R. Sullivan, and D. J. Perreault, "High-efficiency inverter for photovoltaic applications," in *2010 IEEE Energy Conversion Congress and Exposition*, 2010, pp. 2803–2810.

[16] B. J. Pierquet and D. J. Perreault, "A single-phase photovoltaic inverter topology with a series-connected energy buffer," *IEEE Transactions on Power Electronics*, vol. 28, no. 10, pp. 4603–4611, 2013.

[17] M. Liao, T. Sen, Y. Wu, and M. Chen, "Analysis and design of a cyclo-active-bridge inverter for single-stage three-phase grid interface," in *2025 IEEE Applied Power Electronics Conference and Exposition (APEC)*, 2025.

979-8-3315-1612-3/25 $31.00 © 2025 IEEE

Turn-On Transient Modeling of 10 kV SiC MOSFET Half-Bridge Power Module in LTspice

Nianzun Qi
AAU Energy
Aalborg University
Aalborg, Denmark
nqi@energy.aau.dk

Jannick Kjær Jørgensen
AAU Energy
Aalborg University
Aalborg, Denmark
jkj@energy.aau.dk

Gao Liu
AAU Energy
Aalborg University
Aalborg, Denmark
gaol@energy.aau.dk

Zhixing Yan
AAU Energy
Aalborg University
Aalborg, Denmark
zhya@energy.aau.dk

Morten Rahr Nielsen
AAU Energy
Aalborg University
Aalborg, Denmark
mrni@energy.aau.dk

Asger Bjørn Jørgensen
AAU Energy
Aalborg University
Aalborg, Denmark
abj@energy.aau.dk

Hongbo Zhao
AAU Energy
Aalborg University
Aalborg, Denmark
hzh@energy.aau.dk

Stig Munk-Nielsen
AAU Energy
Aalborg University
Aalborg, Denmark
smn@energy.aau.dk

Abstract—Medium-voltage SiC MOSFETs (> 3.3 kV) feature high breakdown voltage and fast switching speed. During the turn-on transient of 10 kV SiC MOSFETs, drain-source voltage drops from 6000 V (rated voltage) to 0 V within hundreds of nanoseconds. In this paper, an improved device modeling approach is proposed to describe the turn-on transients of medium-voltage SiC MOSFETs, with the consideration on the short-channel effect, the threshold voltage hysteresis effect, and the soft-transient characteristics. A custom-packaged power module in a half-bridge configuration, where each switch position contains a single 10 kV SiC MOSFET die (QPM3-10000-0300), is modeled and simulated in LTspice as the case study. Compared with the double pulse test experiments at several operating points, the proposed simulation model for 10 kV SiC MOSFET module can provide more accurate dv/dt, di/dt behavior compared to the traditional model with only static characterization.

Index Terms—Medium voltage, 10 kV, SiC MOSFET, simulation model, LTspice.

I. Introduction

Medium voltage (MV) silicon-carbide (SiC) MOSFETs are the promising techniques because of their high switching speed, low switching loss, and high junction temperature characteristics compared to MV Silicon devices [1]–[3]. Fig. 1 demonstrates the 10 kV SiC MOSFET power modules using the power module packaging structure [4]. To understand the module switching behaviors, further assisting the external circuitry design of applications, an accurate and efficient simulation model is indispensable [5].

Extensive studies have been conducted on the SiC MOSFET models for time domain simulations, which mainly focus on finding the suitable equations, developing the algorithm

This work is co-supported by the MV-BASIC project and the MVolt project. The MV-BASIC project is co-funded by AAU Energy of Aalborg University, together with industrial project partners KK Wind Solutions and Siemens Gamesa. MVolt project is co-funded by AAU Energy of Aalborg University, Innovation Fund Denmark, Siemens Gamesa Renewable Energy, Vestas Wind Systems, and KK Wind Solutions.

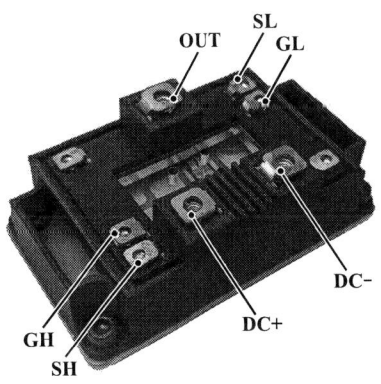

Fig. 1: Custom packaged medium-voltage half-bridge power module [4].

to extract the model parameters, and improving the model convergence [6]. However, the unique SiC-based device characteristics, such as short channel effect, threshold voltage hysteresis effect, and soft transition between the linear region and saturation region are also critical for the switching modeling, which is overlooked during the static I-V characterization.

The I-V characteristics from the datasheet is normally measured by the power device analyzer, which serves as the modeling database [7]. It provides low v_{ds} region information (< 60 V) limited by the power level of measurement equipment. However, the narrow scope of the I-V characteristic, relative to the 10 kV devices, is far from sufficient to account for the physical effects occurring during dynamic switching transients on a nanosecond scale.

The semi-physical model proposed in [8] simplifies the Shichman-Hodges' physical equations by setting channel-length modulation parameter (λ) to 0 to avoid the overestimation. The simulated double pulse test results meet well with

$$i_{ch} = \begin{cases} 0, & v_{gs} < V_{th}, \\ (k_{p1} + k_{p2}v_{gs})\left(v_{gs} - V_{th} - \dfrac{v_{ch}}{2}\right)v_{ch}\left(1 + \lambda v_{ch}\right), & v_{ch} < v_{gs} - V_{th}, \\ \dfrac{(k_{p1} + k_{p2}v_{gs})}{2}\left(v_{gs} - V_{th}\right)^2\left(1 + \lambda v_{ch}\right), & v_{ch} \geq v_{gs} - V_{th}. \end{cases} \tag{1}$$

the experiments at low current working conditions. But it can cause significant deviations in the estimated rate of change in current (di/dt) and voltage (dv/dt) between the simulation results and experimental observations with increasing load current [9]. [10] adopts single pulse tests to obtain saturation region data but still lacks an analysis of the threshold voltage hysteresis effect. It is also possible to consider all the effects in a series of analytical equations to get real-time solutions for the circuit variables, which require numerous mathematical procedures and are not user-friendly [11][12].

In this paper, an improved channel modeling method for 10 kV SiC MOSFET will be proposed. Compared to the traditional modeling method, the static I-V characteristics will be combined with the measured I-V characteristics in high v_{ds} region, being used to extract the parameters in the channel equations. The short channel effect, threshold voltage hysterisis effect, and soft transition effect are both taken into consideration in the channel equaitons. The accuracy of the developed device model is validated by comparing the switching transients in LTspice simulation results and experimental waveforms at several switching conditions.

II. PROBLEM FORMULATION

Fig. 2 shows the circuit model for the 10 kV SiC MOSFET half bridge power module. The module model comprises both the 10 kV SiC MOSFET model and the parasitic model from the module packaging. The chip model introduce the non-linear characteristics of the MOSFETs. As shown in the Fig. 3, the SiC MOSFET die model includes channel current (i_{ch}), three intrinsic capacitors (C_{gd}, C_{gs}, C_{ds}), drift resistance (R_{drift}), body diode (Diode), and the internal gate resistance (R_{gin}). The modeling of channel current describes the non-linear variations in current with respect to the gate-source voltage (v_{gs}) and drain-source voltage (v_{ds}).

In the existed SPICE models for SiC MOSFET, the modified level-1 MOSFET model is widely adopted to describe the voltage controlled current source characteristics [8]. Three simple analytical equations define the cutoff, linear region, and the saturation region, shown in Eq. (1). There are five parameters that characterize the model: channel coefficient (k_{p1}, k_{p2}), threshold voltage (V_{th}), and channel-length modulation constant (λ), and drift resistance (R_{drift}). Based on the static I-V characteristics derived from either the data sheet or the curve tracer, the parameters can be extracted step by step [7].

The static I-V data for the 10 kV SiC MOSFET module is shown in Fig. 4, which was measured by the Keysight B1506A device analyzer. The V_{th} is around 4.1 V. The R_{drift} can be

determined at the vertical intercept of extrapolated R_{ds} versus $(v_{gs} - V_{th})^{-1}$ as shown in Fig. 5. It is worth noticing that the drift region of 10 kV SiC MOSFET cell is long to withstand higher voltage, resulting the larger R_{drift} (249 $m\Omega$) compared to low-voltage devices. Then, excluding the voltage drop on the R_{drift}, the other three parameters k_{p1}, k_{p2}, and λ can be further extracted, which is -0.1178, 0.1189, and 0.0066.

The intrinsic capacitances (C_{gs}, C_{dg}, and C_{ds}) are also crucial for the switching transient modeling. This paper focus

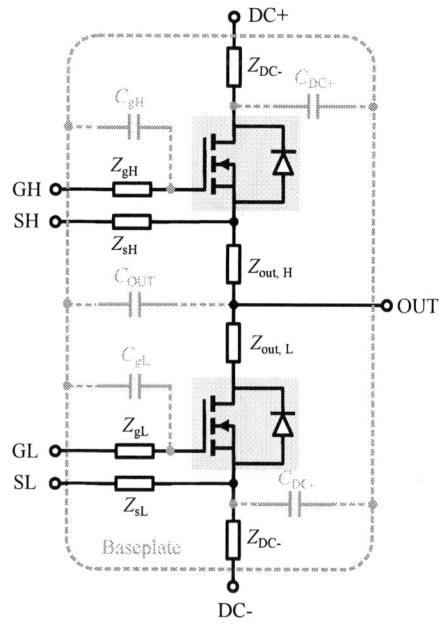

Fig. 2: Circuit model of 10 kV SiC MOSFET power module considering the packaging parasitics.

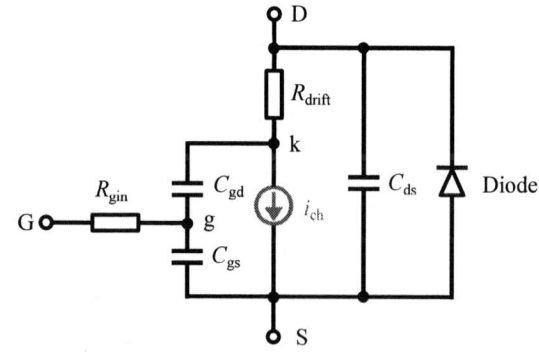

Fig. 3: Subcircuit model of 10 kV SiC MOSFET die in the power module.

979-8-3315-1612-3/25 $31.00 © 2025 IEEE   358

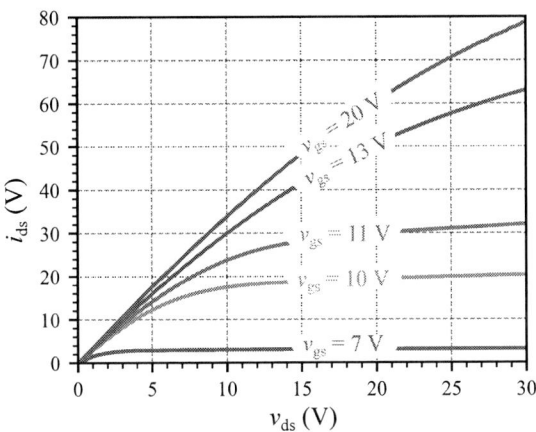

Fig. 4: *I-V* characteristics of 10 kV SiC MOSFET power module.

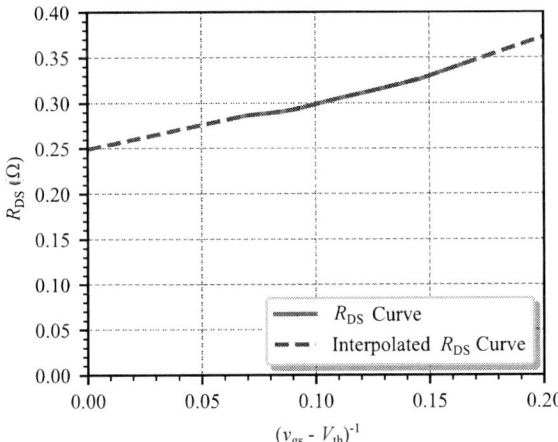

Fig. 5: Relationship of R_{ds} versus $(v_{gs} - V_{th})^{-1}$.

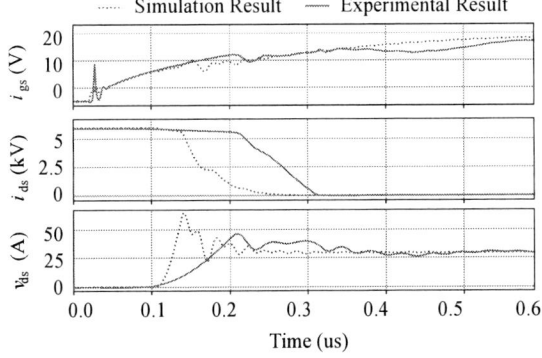

Fig. 6: Double pulse test results comparison at 6000 V / 30 A using the traditional modeling method.

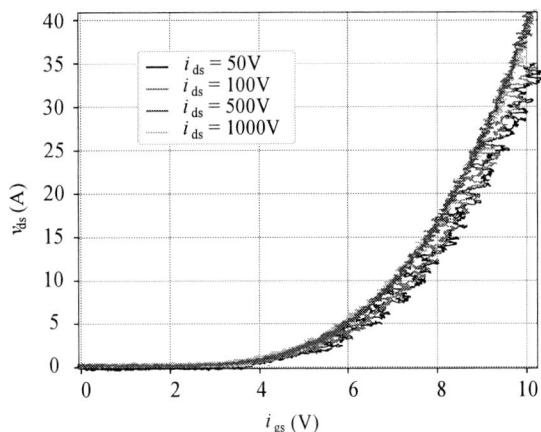

Fig. 7: Measurements of the transfer characteristics i_{ch} (v_{gs}) for different values of v_{ds} ranging from v_{ds} = 50 V to v_{ds} = 1 kV.

packaging can be found in [13].

Integrated the traditional channel model with the other circuit parts in LTspice, the simulated turn on results is shown in Fig. 6. The deviation in the current rise stage is obvious. Simulated di/dt (1.2 A/ns) is three times larger than the measured result (0.4 A/ns), which indicates that higher transconductance (g_m) is estimated at the v_{ds} = 6 kV.

It can be attributed to the larger fitted parameter λ (0.0066), derived from the static characteristic in the low v_{ds} region (< 30 V). For the 10 kV SiC MOSFET, the high sensitivity to the λ value makes this typical modeling approach less robust. Therefore, it is worthwhile to investigate a more suitable modeling approach for the 10 kV SiC MOSFET.

III. PROPOSED METHOD

In this section, the improved channel current model for 10 kV SiC MOSFET is introduced with detailed parameters explanation. The i_{ds}-v_{gs} relationship in the current rise stage is captured and transferred into the part of dynamic *I-V* characterization. With the data of both low and high v_{ds} region, key parameters in the improved model can be fitted through Curve Fitting Toolbox in MATLAB.

A. Model description

The hyperbolic tangent function $\tanh()$ is adopted to model the channel current, as shown in (2) and (3). The $\tanh()$ function is a continuous function that exhibits better convergence performance in circuit simulators, compared to the piecewise functions shown in Eq. (1)[14].

$$i_{ch} = k_1 \times \tanh(k_2 \times v_{ks}) + k_3 \times v_{ks} \qquad (2)$$

$$k_1, k_2, k_3 = f(v_{gs} - v_{th.shift}) \qquad (3)$$

$$v_{ks} = v_{ds} - i_{ch} \cdot R_{drift} \qquad (4)$$

By adjusting the parameters k_1, k_2, and k_3, each output curve under different v_{gs} can be effectively described using

on the channel modeling part, so the capacitance part will not be discussed. The value for the parasitic from the module

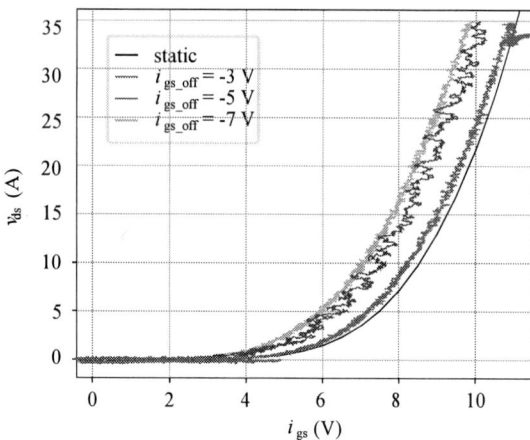

Fig. 8: Measurements of the transfer characteristics i_{ch} (v_{gs}) for different values of turn off gate voltage ($v_{gs,off}$) ranging from $v_{gs,off} = 0$ V to $v_{gs,off} = -7$ V.

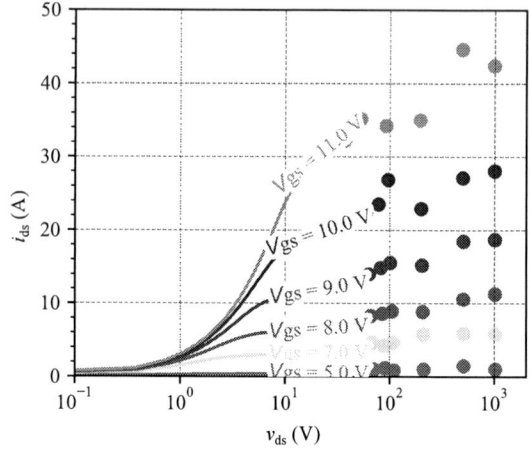

Fig. 9: The full picture of I-V characteristic of 10 kV SiC MOSFET.

Eq. (3). The values of k_1 to k_3 are the functions of v_{gs}. The k_1 determines the saturation current level, k_2 defines the slope of the linear region.

The last term $k_3 \times v_{ks}$ accounts for the short-channel effect. The short-channel effect is due to the drain-induced barrier lowering effect, which is more observable in SiC MOSFET due to its shorter channel length. It is indicated by an increase in i_{ch} with v_{ds} in the saturation region. As shown in Fig. 7, with the same gate-source voltage, the current is increasing with the drain-source voltage.

The threshold voltage hysteresis effect, induced by the the defect states at the SiC / SiO2 interface, is significant and non-negligible. It can lead to dynamically lower v_{th} before the miller region during the turn on transients. In the Eq. (3), the $v_{th,shift}$ is introduced to represent the impact. As shown in Fig. 8, the threshold voltage is decreasing as the turn off gate voltage $v_{gs,off}$ becomes more negative.

B. Dynamic I-V characterization

The static I-V characterization refer to the tests conducted by the parametric device analyzer by the Keysight B1506a. The measurements involve a low DC-bias voltage sweep applied on the device gate and drain for tens of microseconds.

However, there are two limitations for static characterization. Firstly, the niche measurement product is usually limited to tens of kilowatts. Either the high voltage low current range or low voltage high current region can be measured. Secondly, the effect of the interface defect is underestimated due to the inconsistent time constant. The time constant of the interface defect is much higher than the quasi-static measurements.

In this paper, the dynamic I-V characterization is conducted, as the supplement test. Double pulse tests with larger gate resistance are used to obtain the dynamic I-V characteristics of 10 kV SiC MOSFET. The current rise stage (di/dt phase) of the second turn-on switching transient in DPTs is extracted to the dynamic transfer curves [15]. In this stage, the required gate-source voltage (v_{gs}) equals v_{th} at the beginning

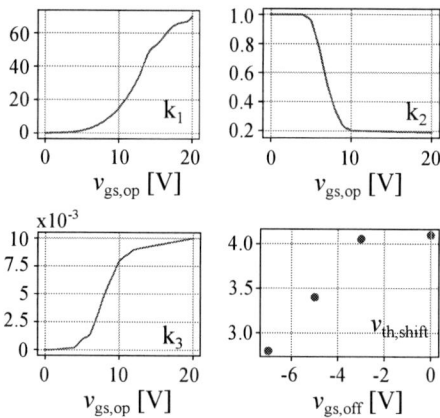

Fig. 10: The k_1, k_2, k_3, and $v_{th,shift}$ with the change of v_{gs} and $v_{gs,off}$.

and reaches v_{miller} at the end while the v_{ds} remain as high as the DC voltage. A larger gate resistance (220 Ω) is chosen to decrease the voltage drop on v_{ds} and v_{gs} due to the loop parasitic inductance [16]. Then, the full picture of the I-V characterization combined with the static and dynamic is shown in Fig. 9. Based on the I-V data, the k_1, k_2, k_3, and $v_{th,shift}$ is extracted as shown in Fig. 10.

LTspice is used for circuit simulations in this paper. The simulation model of 10 kV SiC MOSFET power module is constructed as shown in Fig. 11. The subcircuit of 10 kV SiC MOSFETs are both on the high side and low side, which are shown in Fig. 12.

IV. EXPERIMENTAL VERIFICATION

A. Double pulse test platform

To validate the proposed model, the double pulse test (DPT) platform shown in Fig. 13 is implemented. All the components and equipment used in the setup is illustrated in Table I.

979-8-3315-1612-3/25 $31.00 © 2025 IEEE

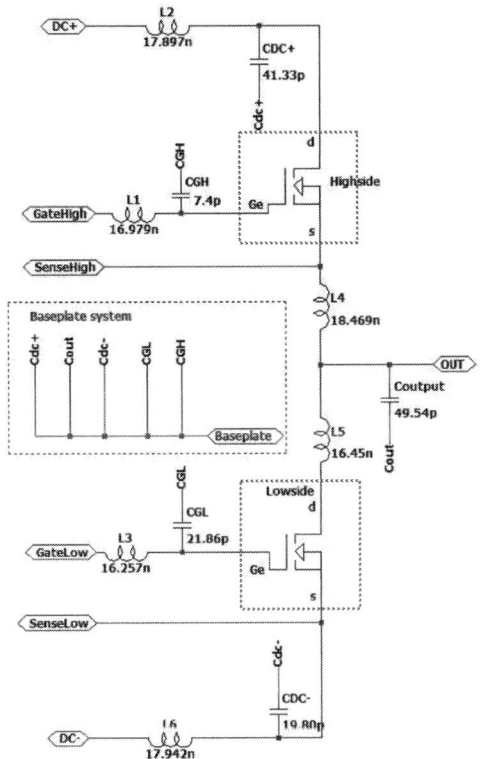

Fig. 11: Simulation model of the 10 kV half-bridge SiC MOSFET power module in LTspice, with parasitic inductances (L_1–L_6) and parasitic capacitances (C_{DC+}, C_{GH}, C_{output}, C_{GL}, and C_{DC-}) coupling to the grounded baseplate.

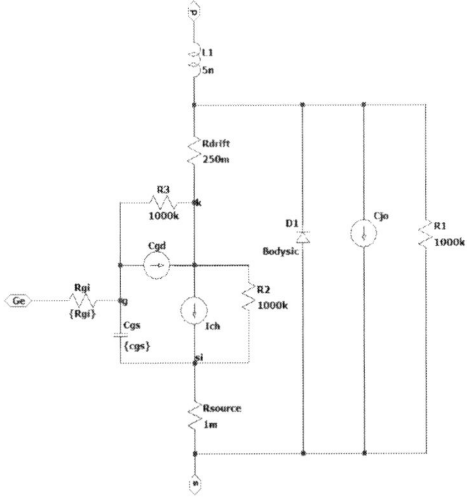

Fig. 12: Subcircuit of the 10 kV SiC MOSFET die, which corresponds to the detailed internal model of a single switch in the half-bridge module shown in Fig. 11.

B. Accuracy discussion

The turn-on transient of low-side DPT simulations with the proposed model are compared with the experimental waveforms, as shown in Fig. 14. The load current is maintained at 30 A and the DC voltage changes from 2 kV to 6 kV.

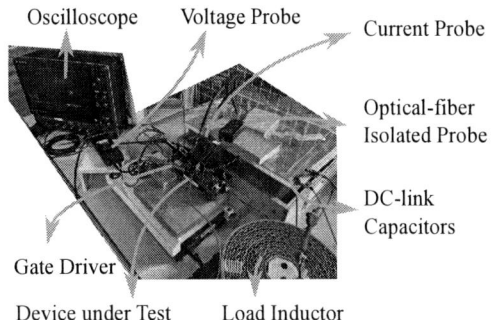

Fig. 13: Experimental double pulse test platform.

TABLE I: Equipment and test conditions for the DPT experiments.

Equipment/Device	Key Parameters
Magna power supply	XR 6000-1.0
Oscilloscope	WaveRunner 8000
Voltage probe for v_{ds}	HVD3605A
Voltage probe for v_{gs}	Micsig MOIP03P
Current probe for i_{ds}	Pearson 7427A
DC-link capacitors	$4 \times 50\,\mu F$ / $5\,kV$
Load inductor	47 mH
Gate voltage V_{GG} / V_{EE}	20 V / -5 V
Load current I_{Load}	30 A
DC-link voltage V_{DC}	2 kV - 6 kV
Junction temperature T_j	25°C
Gate resistance R_g	20 Ω

TABLE II: Error Analysis of the proposed model for 10 kV SiC MOSFET module.

Quantity	DC-link voltage	Experiments	Simulations	Error
	2 kV	34 V/ns	44 V/ns	29%
dv/dt	4 kV	44 V/ns	56 V/ns	27%
	6 kV	56 V/ns	65 V/ns	16%
	2 kV	0.45 A/ns	0.55 A/ns	22%
di/dt	4 kV	0.44 A/ns	0.53 A/ns	20%
	6 kV	0.40 A/ns	0.43 A/ns	7.5%

For the v_{gs} at turn on transient, the simulated waveforms are identical to the experimental waveforms with limited errors. The simulated miller voltage is closer to the measured value (11 V) compared to the traditional model in Fig. 6 (around 9.9 V). The measured voltage spike at the beginning is due to the charging and discharging on the parasitic capacitance of the gate driver board in the fast speed, which is not included in the simulations.

979-8-3315-1612-3/25 $31.00 © 2025 IEEE

······ Simulation Result   —— Experimental Result

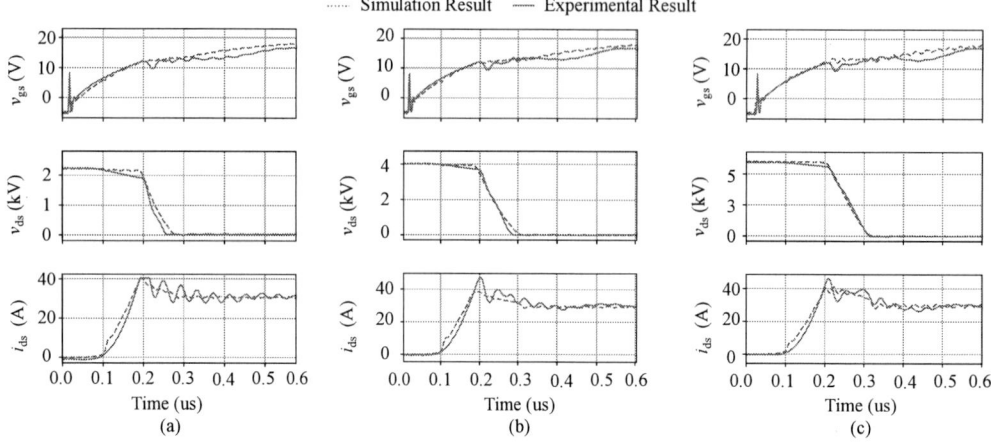

Fig. 14: Comparison between experimental results and simulation results (a) 2000 V / 30 A, (b) 4000 V / 30 A, and (c) 6000 V / 30 A.

For the v_{ds} and i_{ds}, the simulated slew rates align well with experiments. With $R_g = 20\ \Omega$, 10 kV SiC MOSFET is switching at high dv/dt (40 to 60 V/ns) but at a relatively low di/dt (around 0.5 A/ns). Error analysis is shown in Table. II, which shows better transient prediction of proposed model compared to the conventional modeling method in Section II.

V. CONCLUSION

The static and dynamic I-V characterizations were conducted to build a 10 kV SiC MOSFET die model, taking into account physical effects such as short-channel effects, threshold voltage hysteresis, and soft-transition behavior. A close match between simulated and measured turn-on switching transients under various operating conditions validates the model's dynamic performance with improved accuracy.

REFERENCES

[1] B. F. Kjærsgaard, G. Liu, M. R. Nielsen, R. Wang, D. N. Dalal, *et al.*, "Parasitic Capacitive Couplings in Medium Voltage Power Electronic Systems: An Overview," *IEEE Transactions on Power Electronics*, vol. 38, no. 8, pp. 9793–9817, 2023.

[2] J. Millán, P. Godignon, X. Perpiñà, A. Pérez-Tomás, and J. Rebollo, "A Survey of Wide Bandgap Power Semiconductor Devices," *IEEE Transactions on Power Electronics*, vol. 29, no. 5, pp. 2155–2163, 2014.

[3] B. J. Baliga, "Silicon Carbide Power Devices: Progress and Future Outlook," *IEEE Journal of Emerging and Selected Topics in Power Electronics*, vol. 11, no. 3, pp. 2400–2411, 2023.

[4] A. B. Jørgensen, N. Christensen, D. N. Dalal, S. D. Sønderskov, S. Bęczkowski, *et al.*, "Reduction of Parasitic Capacitance in 10 kV SiC MOSFET Power Modules Using 3D FEM," in *2017 19th European Conference on Power Electronics and Applications (EPE'17 ECCE Europe)*, 2017, pp. 1–8.

[5] P. Sochor, A. Huerner, Q. Sun, and R. Elpelt, "Characteristics of SiC MOSFET Compact Models Suitable for Virtual Prototyping of Power Electronic Circuits," in *2023 11th International Conference on Power Electronics and ECCE Asia (ICPE 2023 - ECCE Asia)*, 2023, pp. 112–119.

[6] B. W. Nelson, A. N. Lemmon, B. T. DeBoi, M. M. Hossain, H. A. Mantooth, *et al.*, "Computational Efficiency Analysis of SiC MOSFET Models in SPICE: Static Behavior," *IEEE Open Journal of Power Electronics*, vol. 1, pp. 499–512, 2020.

[7] P. Yang, W. Ming, and J. Liang, "A Step-by-step Modelling Approach for SiC Half-bridge Modules Considering Temperature Characteristics," in *2020 IEEE Energy Conversion Congress and Exposition (ECCE)*, 2020, pp. 2827–2834.

[8] J. K. Jorgensen, N. Christensen, D. N. Dalal, A. Bjorn Jorgensen, H. Zhao, *et al.*, "Loss Prediction of Medium Voltage Power Modules: Trade-offs between Accuracy and Complexity," in *2019 IEEE Energy Conversion Congress and Exposition (ECCE)*, 2019, pp. 4102–4108.

[9] Z. Dong, X. Wu, H. Xu, N. Ren, and K. Sheng, "Accurate Analytical Switching-On Loss Model of SiC MOSFET Considering Dynamic Transfer Characteristic and Qgd," *IEEE Transactions on Power Electronics*, vol. 35, no. 11, pp. 12 264–12 273, 2020.

[10] R. Chen, M. Lin, X. Huang, F. Wang, and L. M. Tolbert, "An Improved Turn-on Switching Transient Model of 10 kV SiC MOSFET," in *Proc. IEEE Applied Power Electronics Conference and Exposition (APEC)*, 2022, pp. 1–8.

[11] G. L. Rødal, Y. V. Pushpalatha, D. A. Philipps, and D. Peftitsis, "Capacitance Variations and Gate Voltage Hysteresis Effects on the Turn-ON Switching Transients Modeling of High-Voltage SiC MOSFETs," *IEEE Transactions on Power Electronics*, vol. 38, no. 5, pp. 6128–6142, 2023.

[12] N. Wang, J. Zhang, and F. Deng, "Improved SiC MOSFET Model Considering Channel Dynamics of Transfer Characteristics," *IEEE Transactions on Power Electronics*, vol. 38, no. 1, pp. 460–471, 2023.

[13] B. F. Kjærsgaard, T. S. Aunsborg, J. K. Jørgensen, D. N. Dalal, M. Takahashi, *et al.*, "Loss Imbalance in SiC Half-Bridge Power Module," in *2023 IEEE 6th International Electrical and Energy Conference (CIEEC)*, 2023, pp. 3247–3253.

[14] J. Nakashima, T. Horiguchi, Y. Mukunoki, M. Hagiwara, T. Urakabe, and S. Harada, "Automated Flexible Modeling for Various Full-SiC Power Modules," *IEEE Transactions on Power Electronics*, vol. 38, no. 5, pp. 6094–6107, 2023.

[15] P. Hofstetter, R. W. Maier, and M.-M. Bakran, "Influence of the Threshold Voltage Hysteresis and the Drain Induced Barrier Lowering on the Dynamic Transfer Characteristic of SiC Power MOSFETs," in *2019 IEEE Applied Power Electronics Conference and Exposition (APEC)*, 2019, pp. 944–950.

[16] H. Sakairi, T. Yanagi, H. Otake, N. Kuroda, and H. Tanigawa, "Measurement Methodology for Accurate Modeling of SiC MOSFET Switching Behavior Over Wide Voltage and Current Ranges," *IEEE Transactions on Power Electronics*, vol. 33, no. 9, pp. 7314–7325, 2018.

979-8-3315-1612-3/25 $31.00 © 2025 IEEE

A Compact, Automated Sawyer-Tower System for Characterization of the High-Frequency, Soft-Switching C_{oss} Loss of Wide Bandgap Devices

Katherine Liang
Electrical Engineering
Stanford University
Stanford, CA 94305
katliang@stanford.edu

Malachi Hornbuckle
Electrical Engineering
Stanford University
Stanford, CA 94305
malachih@stanford.edu

Juan Rivas-Davila
Electrical Engineering
Stanford University
Stanford, CA 94305
jmrivas@stanford.edu

Abstract—**This paper presents a novel implementation of the Sawyer-Tower system for characterizing the large-signal C_{oss} loss of power semiconductor devices, enhanced to be cost-effective, compact, and fully automated. The system integrates a hybrid class D converter, a control and signal generation system, and programmable test equipment to measure C_{oss} losses based on user inputs to a Python script. The resulting characterization values show strong alignment with previous literature, but with a system that demonstrates more than tenfold reduction in cost, time, and space. This system significantly enhances accessibility for rapid, large-scale characterization of C_{oss} loss across various drain waveform parameters, including frequency, voltage amplitude, $\frac{dV}{dt}$, and waveform shape.**

I. INTRODUCTION

The advancement of wide bandgap (WBG) devices has played a part in enabling efficient power converter operation at multi-MHz frequencies under soft-switching conditions [1]–[6], minimizing switching losses through zero-voltage switching (ZVS). While ZVS effectively reduces switching losses, significant losses still occur in the output capacitance (C_{oss}) of devices [7]. When a device in a converter utilizing ZVS is in its off-state, its C_{oss} is charged up to its peak drain-source voltage and discharged down to zero once per switching cycle. However, some charges are trapped in the device during this process [8], and loss is generated via charge-voltage hysteresis. Investigations into this loss mechanism in materials such as Gallium Nitride (GaN) and Silicon Carbide (SiC) have been undertaken [7]–[10]. Since this loss mechanism occurs once per switching cycle, characterizing large-signal C_{oss} loss at high frequencies is critical for the optimized design of power converters at multi-MHz frequencies and beyond.

The Sawyer-Tower circuit is widely used for characterizing C_{oss} loss [11]–[13], as shown in Fig. 1. In the circuit, the device-under-test (DUT) remains in its off-state to extract its large-signal C_{oss} behaviour. Traditionally, a signal generator and power amplifier are used to apply the required high-voltage and high-frequency waveform across the device to emulate the waveform across the device during operation in

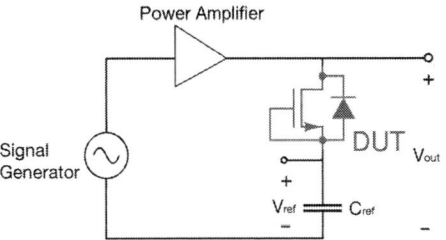

Fig. 1. Schematic of traditional Sawyer-Tower circuit.

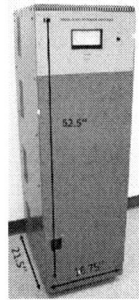

Fig. 2. Image of traditional RF power amplifier with dimensions.

a power converter [7]. However, these components are large (as shown in Fig. 2) and expensive. Furthermore, the characterization process requires an expert user to tune the system, sometimes even having to design a new board for each DUT [7]. This lengthy process poses a significant barrier for the widespread adoption of large-signal C_{oss} characterization. To address these challenges, this paper introduces an automated Sawyer-Tower system that leverages a hybrid class D converter and control subsystem to measure C_{oss} losses in power devices.

II. SYSTEM ARCHITECTURE

The proposed system, as shown in Fig. 3, can be compartmentalized into 3 subsystems — a hybrid class D converter, a signal generation circuit, and a control algorithm.

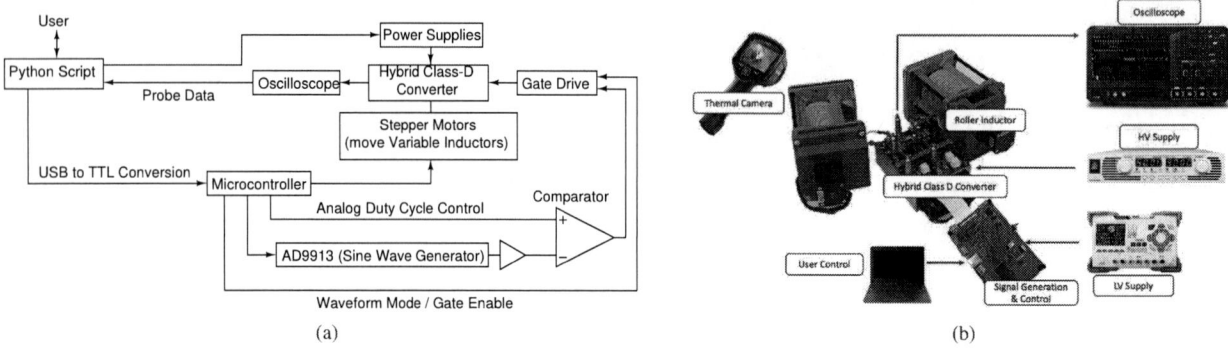

Fig. 3. (a) Block diagram demonstrating control sequence, and (b) experimental setup.

A. Hybrid Class D Converter

A hybrid class D converter, as shown in Fig. 4 and detailed in [14], was designed to emulate the function of the power amplifier in the traditional Sawyer-Tower system. While the hybrid class D converter is not able to amplify arbitrary voltage waveforms like the amplifier in Fig. 2, it applies waveforms across the DUT that mimic those seen by the device in power converters. The hybrid class D converter is smaller and cheaper than the amplifier in Fig. 2 because it is specifically designed to generate the selection of waveforms commonly seen by switching devices in high frequency converters. The most common waveform applied across devices is approximately sinusoidal, such as in class E amplifiers [15]. Sinusoidal waveforms can be applied to the DUT by operating the hybrid converter in current mode, as shown in Fig. 4(c). Additionally, various power converter topologies employ trapezoidal modulation to reduce switching losses, such as in modular multilevel dc-dc converters [16] and dual active bridge dc-dc converters [17] for high-frequency applications. The hybrid class D converter can operate in voltage mode to apply a trapezoidal waveform across the DUT, as shown in Fig. 4(b).

A key element of note in the hybrid class D converter schematic is the variable inductor L_R, which translates to two roller inductors rotated by stepper motors in the physical system. In trapezoidal mode, L_R adjusts $\frac{dV}{dt}$, while in sinusoidal mode, L_R resonates with the switch node capacitance to generate a sinusoidal waveform across the DUT.

B. Signal Generation Circuit

To drive the switches in the topology shown in Fig. 4, a signal generation circuit was designed. A top-level overview of this subsystem is shown in Fig. 5. This subsystem utilizes a direct digital synthesizer (AD9913) to generate a low-amplitude sine wave, which is subsequently amplified using a wideband operational amplifier (OPA855). The amplified sine wave is then compared to a dc reference voltage from the microcontroller with a high-speed comparator (LTC6752) to generate a square wave output. The reference voltage,

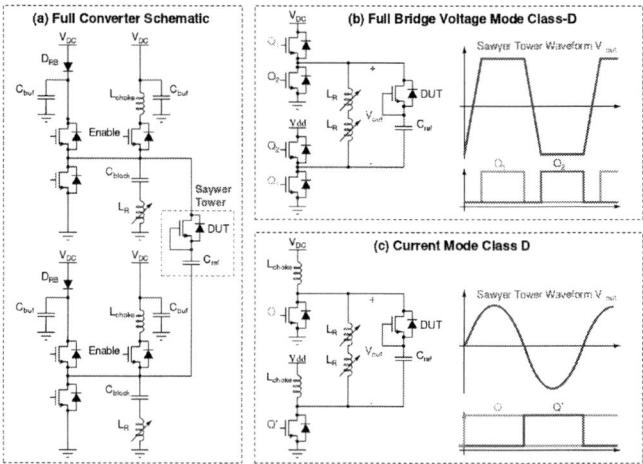

Fig. 4. (a) Schematic of hybrid class D converter, (b) voltage-mode operation, and (c) current-mode operation [14].

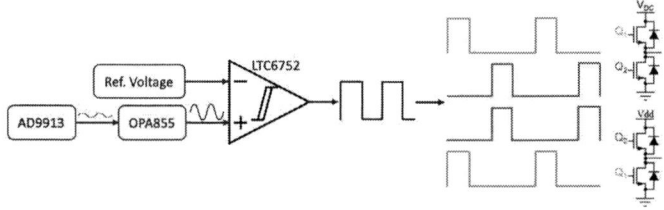

Fig. 5. Overview of signal generation circuit.

determined by the control algorithm, varies based on the desired waveform to be applied across the DUT. The resulting square wave from the comparator is used to generate two pairs of complementary gate drive signals (illustrated in blue and red in Fig. 5), which control the four switches of the hybrid converter. The physical implementation of this subsystem is pictured as the green circuit board in Fig. 3(b).

979-8-3315-1612-3/25 $31.00 © 2025 IEEE

C. Control Algorithm

A control algorithm integrates the previous two subsystems, as shown in Fig. 3(a). The Python script prompts users to input desired values for switching frequency, drain-to-source voltage, waveform mode (i.e. trapezoidal or sinusoidal), and $\frac{dV}{dt}$ if the system is operating in trapezoid mode. The onboard microcontroller (ATMEGA328P) then relays the frequency value to the sine wave generator to generate gating signals at this specified frequency. The drain-to-source voltage parameter is communicated to a programmable power supply via the PyVISA library, which is a Python package used to control test equipment through the Virtual Instrument Software Architecture (VISA). Additionally, the user-specified waveform mode is converted to a digital logic level by the microcontroller, which is then fed into the signal generation circuit to toggle between trapezoid and sinusoid modes of operation.

In trapezoid mode, the user-specified $\frac{dV}{dt}$ value is used to tune the two roller inductors through an iterative process. The inductor tuning algorithm begins by extracting the current $\frac{dV}{dt}$ value from the oscilloscope trace and comparing it to the target value. If the desired $\frac{dV}{dt}$ value is not yet achieved, the microcontroller commands the stepper motors to adjust the inductors, and this process repeats until the user-specified $\frac{dV}{dt}$ value is achieved, as detailed in the following section.

III. Test Procedure

Before operating at points of interest, the system undergoes a series of initialization procedures, including calibrating the roller inductors, adjusting the duty cycle, and deskewing the oscilloscope traces to ensure measurement accuracy.

The flow of an operating point run is detailed in a block diagram shown in Fig. 6, and the key equations of the algorithm are explained as follows.

Prior to testing with the DUT, system calibration is performed by running the desired operating points with a high-Q capacitor in place of the DUT. This capacitor is selected to approximately match the average nonlinear C_{oss} present in the DUT. To approximate initial conditions, resonant capacitances at the switch node of the hybrid converter are estimated based on equations derived in [14]. As shown in Fig. 4(b), the capacitance seen by the switch node in voltage mode operation, $C_{trapezoid}$, is given by:

$$C_{trapezoid} = 2\left(\frac{C_{ref}C_{ideal}}{C_{ref} + C_{ideal}} + C_Q \right) \tag{1}$$

where C_{ideal} is the value of the high-Q capacitor in place of the DUT, C_{ref} is the value of the capacitor that is in series with the DUT, and C_Q accounts for the capacitance seen at the switch node from non-DUT transistors in the system, as presented in manufacturer datasheets.

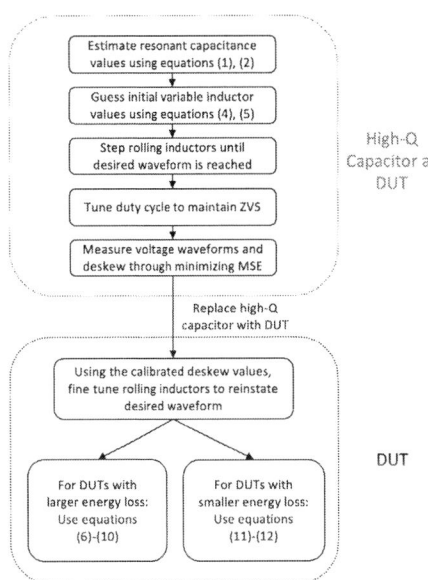

Fig. 6. Block diagram demonstrating flow of each operating point run.

In current mode, the switch node capacitance $C_{sinusoid}$ is:

$$C_{sinusoid} = \frac{C_{ref}C_{ideal}}{C_{ref} + C_{ideal}} + C_Q \tag{2}$$

Using these capacitances, initial duty cycle and inductance values are determined. For voltage mode operation, the duty cycle D is calculated using the user-defined $\frac{dV}{dt}$.

The required inductance $L_{R,trapezoid}$ in voltage mode is initially approximated with the equation from [14]:

$$L_{R,trapezoid} = \frac{D(1-D)\left(\frac{T_s}{2}\right)^2}{4C_{trapezoid}} \tag{3}$$

where $T_s = \frac{1}{f_s}$, with f_s being the switching frequency.

For current mode operation, the inductance $L_{R,sinusoid}$ is given by:

$$L_{R,sinusoid} = \frac{1}{\omega_s^2 C_{sinusoid}} \tag{4}$$

where $\omega_s = 2\pi f_s$.

The system then measures the resulting waveform and adjusts the inductors until the desired waveform is achieved. To tune to a specific $\frac{dV}{dt}$ in voltage mode, the sensitivity of inductance to step changes is evaluated by calculating $\Delta\frac{dV}{dt}$ per step of the roller inductor. Adjustments are made based on the calculated increment per step of the roller inductor:

$$\text{Number of Steps} = \frac{\left(\frac{dV}{dt}\right)_{desired} - \left(\frac{dV}{dt}\right)_{actual}}{\left(\Delta\frac{dV}{dt}\right)_{per\ step}} \tag{5}$$

In current mode, the algorithm locates corners in the waveform and fine tunes the variable inductance values until the waveform's rising and falling periods match the desired period.

Then, to fine tune the duty cycle in both voltage and current mode, the duty cycle value is incrementally increased until current output from the high voltage power supply is minimized, thereby maintaining ZVS of the hybrid converter.

Once the system parameters have been tuned to produce the user-defined waveform, data is extracted based on the Sawyer-Tower method. To characterize the amount of energy dissipated from the DUT in its off-state, three waveforms in the system are measured: the positive and negative nodes of V_{out} as well as the voltage at the top of C_{ref}. The definition of these nodes is illustrated in Fig. 1, while the measured waveforms are shown in Fig. 7.

Then, the following post-processing steps are used to extract the dissipated energy (E_{diss}) from the measured waveforms:

$$V_{ref} = V_{ref}^{+} - V_{out}^{-} \tag{6}$$

$$Q_{oss} = V_{ref} C_{ref} \tag{7}$$

$$V_{out} = V_{out}^{+} - V_{out}^{-} \tag{8}$$

$$V_{DUT} = V_{out} - V_{ref} \tag{9}$$

$$E_{diss} = \oint_{T_s} V_{DUT} \cdot dQ_{oss} \tag{10}$$

Applying (6)-(10) yields the Sawyer-Tower voltage waveforms shown in Fig. 8 and the corresponding charge-voltage (Q-V) plot shown in Fig. 9. The waveforms in Fig. 8 illustrate the post-processed data utilized to construct the Q-V plot. The charging and discharging curves in the Q-V plot show hysteresis, which translates to E_{diss}.

Since the precise time delay of these waveforms relative to each other is critical for extracting E_{diss}, care must be taken to accurately deskew the three traces. Assuming the high-Q capacitor dissipates a negligible amount of energy, the node voltages at its terminals should be aligned in time. Therefore, using the waveform for V_{ref} as a reference, the second trace (i.e. the positive node of V_{out}) is deskewed to achieve maximal overlap with the reference trace (through minimizing the mean-square-error between the two normalized traces). The third trace (i.e. the negative node of V_{out}) is then deskewed to ensure approximately zero E_{diss} since the high-Q capacitor should dissipate significantly less energy than a DUT.

After calibration, the high-Q capacitor is replaced with the DUT, and if needed, further fine-tuning of the inductors and gating signal duty cycle (following the same algorithm as previously described) is performed to reinstate the desired waveform. Finally, the oscilloscope trace data is processed by integrating the Q-V curve using (6)-(10). This method of Q-V

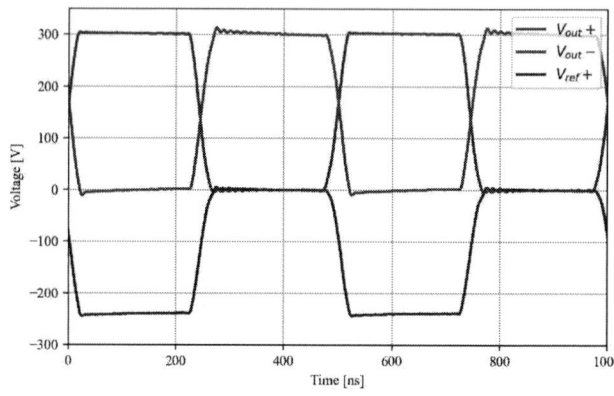

(a)

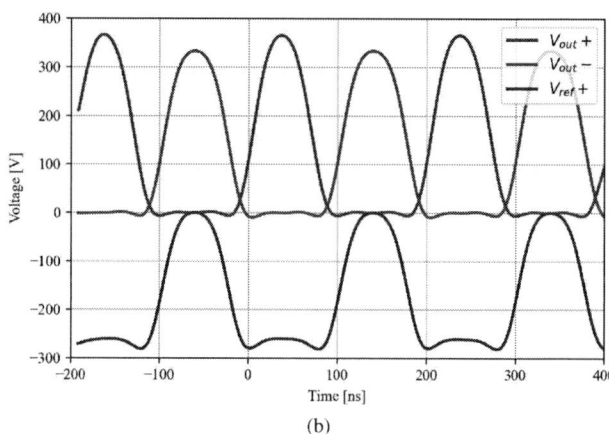

(b)

Fig. 7. Measured waveforms of the positive node of V_{out} ($V_{out}+$), negative node of V_{out} ($V_{out}-$), and the positive node of V_{ref} ($V_{ref}+$) for the system operating in: (a) voltage mode, (b) current mode.

curve integration is suitable for devices with larger energy loss. However, due to the oscilloscope's limited sampling rate and the oscillations present in the waveforms, accurately determining skew values is challenging. Therefore, for devices with extremely small energy loss (i.e. on the order of sub-µJ), thermal measurements of the DUT can be used to calculate power dissipation (P_{diss}) and E_{diss} once the device has reached thermal equilibrium during system operation using the following equations:

$$P_{diss} = \frac{T_{DUT} - T_{ambient}}{R_{th}} \tag{11}$$

$$E_{diss} = \frac{P_{diss}}{f_{sw}} \tag{12}$$

where T_{DUT} is the equilibrium device temperature, $T_{ambient}$ is the ambient temperature, and R_{th} is the device's thermal resistance.

IV. EXPERIMENTAL RESULTS

The compact and automated system was used to characterize three devices, as summarized in Table I.

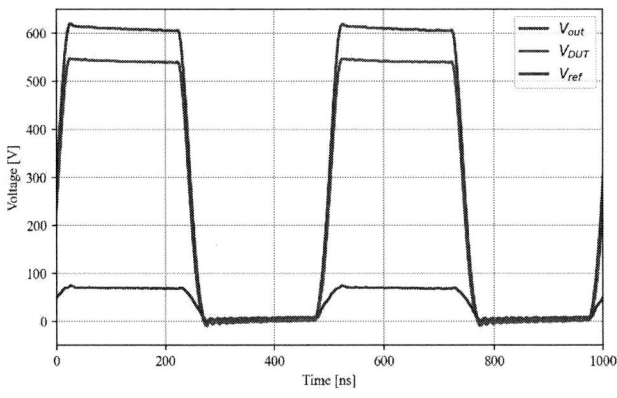

(a)

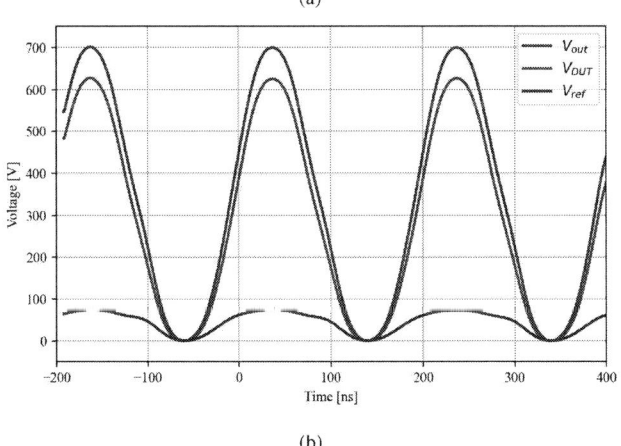

(b)

Fig. 8. Post-processed Sawyer-Tower waveforms with data from Fig. 7 for the system operating in: (a) voltage mode, (b) current mode.

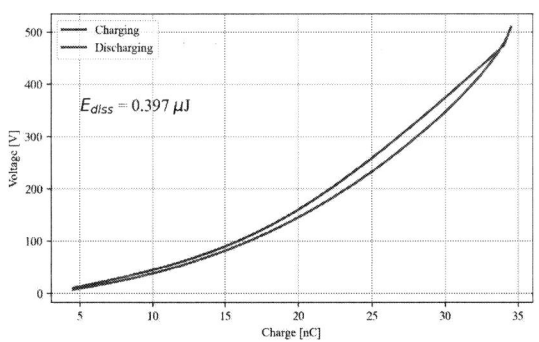

Fig. 9. Q-V plot demonstrating energy dissipation of GS66504B under 2 MHz trapezoidal wave.

TABLE I
PARAMETERS OF SELECTED TEST DEVICES FROM MANUFACTURER DATASHEETS

Component	Technology	V_{DS} [V]	I_D [A]	Reported C_{oss} [pF]
GS66508B	GaN	650	30	65
NTBG025N065S	SiC	650	46	133
IPB60R040C7	Si SJ	600	73	85

Experimental results for the system are presented for up to switching frequencies of 10 MHz. Higher frequencies can be supported up to 30 MHz. Note that L_R in the hybrid converter is inversely proportional to f_s^2, so at higher frequencies, lower inductance values are needed. However, the roller inductors in the system impose a practical limit on the minimum achievable inductance, which constrains the system's maximum operating frequency.

The results of characterizing the energy dissipation of the devices listed in Table I are presented in Fig. 10. Fig. 10(a), 10(b), and 10(c) depict E_{diss} measurements with trapezoidal waveforms applied across DUTs as $\frac{dV}{dt}$ is varied; Fig. 10(d), 10(e), and 10(f) depict E_{diss} with sinusoidal waveforms applied across DUTs as V_{DS} is varied. The measured data shows trends in E_{diss} with respect to $\frac{dV}{dt}$, V_{DS} amplitude (which is noted in Fig. 10 as V_{pp} on DUT), and frequency in the tested devices.

In Si superjunction (SJ) devices, output capacitance losses arise primarily due to charge trapping within the n- and p-pillars during depletion [18]. When a trapezoidal waveform with a higher $\frac{dV}{dt}$ is applied, the rapidly changing electric field leaves minimal time for charges in the pillars to redistribute, which results in incomplete depletion of carriers, leading to higher output capacitance loss. A similar effect is observed at higher switching frequencies, where rapid charge injection and removal reduce the time available to fully extract charges, thereby increasing losses. Additionally, applying a higher V_{DS} across the device increases the electric field across the drift region, causing more charges to accumulate in the pillars and enhancing the likelihood of charge trapping. Fig. 10(c) and Fig. 10(f) reveal that Si SJ devices exhibit dissipation energies comparable in magnitude to those of SiC and GaN devices, despite operating at a significantly lower V_{DS}. This is attributed to the substantial nonlinearity in the Si SJ output capacitance, which is more than 100 times greater than the value reported in the datasheet at bias voltages under 25 V.

For SiC devices, E_{diss} trends can be explained by incomplete ionization [19]. In SiC, dopants do not fully ionize at the device's operating temperature, resulting in carrier concentrations that are much lower than the doping concentrations, known as incomplete ionization. Lower ionized dopant concentrations lead to higher resistivity, and thereby higher losses. As $\frac{dV}{dt}$ and V_{DS} increase, the finite ionization time constant limits the carriers' ability to respond quickly [19], leading to increased resistivity and losses. Regarding E_{diss} trends with changing frequency, it is noteworthy in Fig. 10(e) that energy dissipation remains similar between switching frequencies of 2.5 MHz and 5 MHz. This result aligns with findings in [19], which indicate that C_{oss} losses in SiC devices can saturate at high frequencies.

In GaN-on-Si HEMT devices, we speculate output capacitance loss to be due to a combination of Si substrate

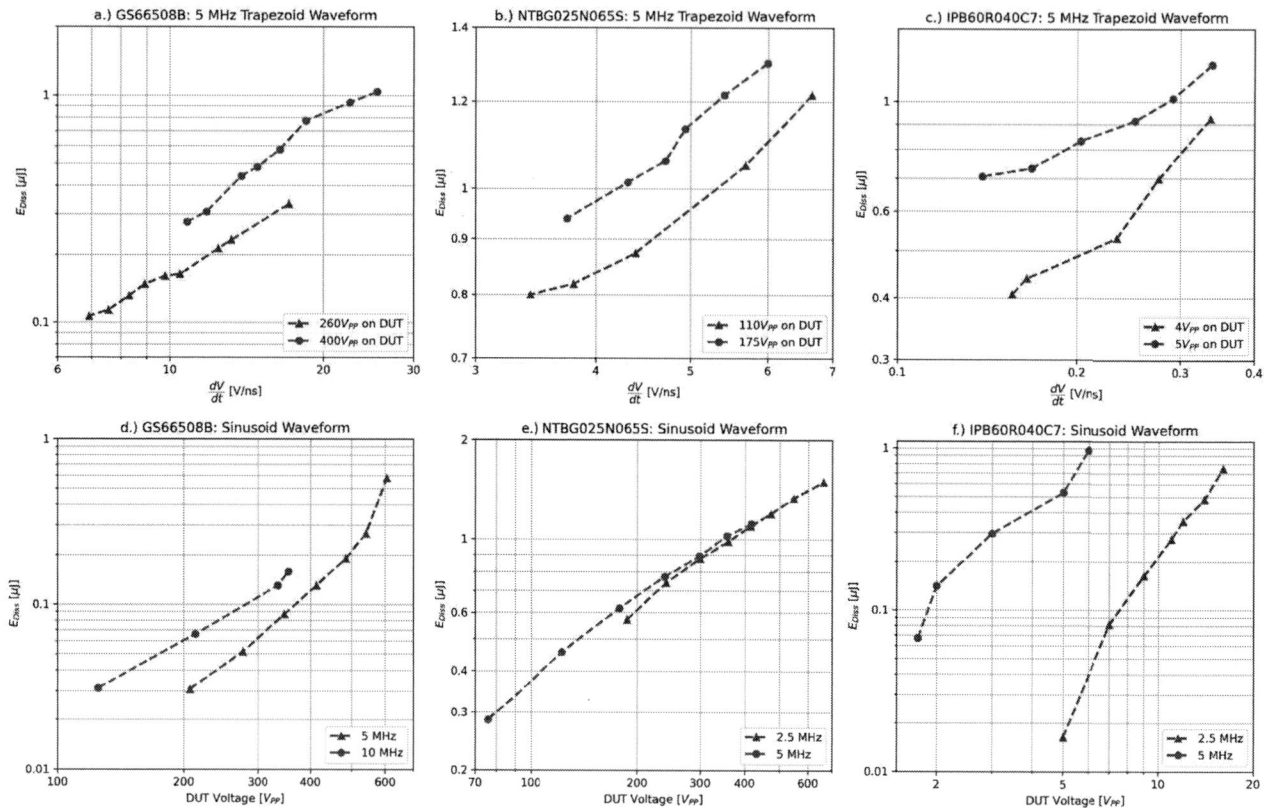

Fig. 10. E_{diss} vs. $\frac{dV}{dt}$ with 5 MHz trapezoidal excitation applied to (a) GS66508B, (b) NTBG025N065S, (c) IPB60R040C7; E_{diss} vs. DUT voltage with sinusoidal excitation applied to (d) GS66508B, (e) NTBG025N065S, and (f) IPB60R040C7.

loss and interface state trapping [8], [20]. However, due to the propriety processes used by device manufacturers, it is difficult to pinpoint the exact causes. For Si substrate loss, the displacement current (I_{dis}) through the Si substrate of the device has been shown to increase with $\frac{dV}{dt}$ and frequency [8]. Thus, at elevated $\frac{dV}{dt}$ and switching frequencies, $I_{dis}^2 R_{substrate}$ losses become more significant, where $R_{substrate}$ is the resistance of the device's Si substrate which varies largely between different fabrication technologies. Additionally, the GaN stack in GaN-on-Si HEMTs includes a buffer layer between the GaN and Si layers to alleviate stress from lattice mismatch and thermal expansion rate mismatch, where trap sites exist. As discovered in [8], applying a high voltage across the device will accelerate charge in the channel, and the kinetic energy of such charges can penetrate the buffer layer and become trapped, further contributing to output capacitance losses.

V. SYSTEM COMPARISON

The integrated system presents improvements in size, cost, and ease of operation. The power amplifier in the traditional setup occupied a considerable volume of 21.5" x 16.75" x 52.5", as shown in Fig. 2, whereas the proposed system is significantly more compact, with dimensions of 18" x

15" x 4". The material cost of the new system, including the two variable inductors, the hybrid class D converter system, and the control and signal generation system, total to $1,300. This is greater than a tenfold improvement in cost of equipment alone, since the power amplifier used in the traditional Sawyer-Tower system costs over $40,000. Furthermore, the traditional test system required the design of new boards for each DUT, and it took up to 8 hours of manual work to test one device. By contrast, the automated setup of the new system can test 10 operating points of a device in 1 hour with minimal manual intervention.

To validate that the performance of the proposed system matches that of the traditional Sawyer-Tower system, data was measured with the GS66504B device for comparison to data from [7] using the $\frac{dV}{dt}$ normalization procedure from that work. As shown in Fig. 11, the trends and E_{diss} values from both systems are well-aligned. Slightly elevated values in the proposed system are attributed to waveform oscillations arising from parasitic inductances in the hybrid class D converter board.

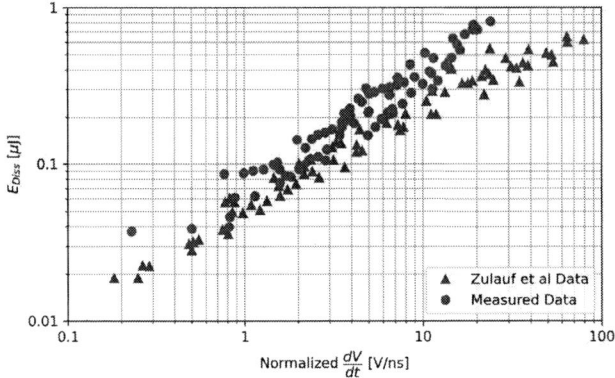

Fig. 11. Comparison of E_{diss} versus normalized $\frac{dV}{dt}$ for GS66504B.

VI. CONCLUSION

The proposed Sawyer-Tower test system effectively substitutes the large and costly equipment required for traditional test systems with a hybrid class D converter and a signal generation circuit. Furthermore, the implemented control architecture fully automates the system, making large-signal C_{oss} characterization more accessible. GaN, SiC, and Si devices are characterized and show strong alignment with anticipated trends as well as with data from previous literature to validate this compact and automated measurement system.

ACKNOWLEDGMENT

The authors would like to thank PowerAmerica for their support of this work.

APPENDIX

In an effort to increase the accessibility of C_{oss} loss characterization, we have uploaded the code required for system setup, control, and automation to a public GitHub repository: github.com/SUPER-Lab-Stanford/automated-sawyer-tower

REFERENCES

[1] Y. Liu, Z. Ouyang, and M. A. E. Andersen, "Review of soft-switching high-frequency GaN-based single-phase bridgeless rectifier," in *2021 IEEE Workshop on Wide Bandgap Power Devices and Applications in Asia (WiPDA Asia)*, 2021, pp. 411–416.

[2] Z. Liu, "Characterization and application of wide-band-gap devices for high frequency power conversion." Ph.D. dissertation, Virginia Tech, Blacksburg VA, 2017.

[3] P. Korta, L. V. Iyer, and N. C. Kar, "Soft–switching EV traction inverter exploiting full potential of wide bandgap devices," in *IECON 2020 The 46th Annual Conference of the IEEE Industrial Electronics Society*, 2020, pp. 4703–4708.

[4] B. Agrawal, L. Zhou, A. Emadi, and M. Preindl, "Variable-frequency critical soft-switching of wide-bandgap devices for efficient high-frequency nonisolated dc-dc converters," *IEEE Transactions on Vehicular Technology*, vol. 69, no. 6, pp. 6094–6106, 2020.

[5] K. Li, P. Evans, and M. Johnson, "SiC and GaN power transistors switching energy evaluation in hard and soft switching conditions," in *2016 IEEE 4th Workshop on Wide Bandgap Power Devices and Applications (WiPDA)*, 2016, pp. 123–128.

[6] X. Han, R. P. Kandula, K. Kandasamy, D. Divan, and M. Saeedifard, "Soft-switching characterization of 3.3 kV reverse-blocking SiC devices." in *2018 IEEE 6th Workshop on Wide Bandgap Power Devices and Applications (WiPDA)*, 2018, pp. 185–191.

[7] G. Zulauf, S. Park, W. Liang, K. N. Surakitbovorn, and J. Rivas-Davila, "Coss losses in 600 V GaN power semiconductors in soft-switched, high- and very-high-frequency power converters," *IEEE Transactions on Power Electronics*, vol. 33, no. 12, pp. 10 748–10 763, 2018.

[8] J. Zhuang, G. Zulauf, J. Roig, J. D. Plummer, and J. Rivas-Davila, "An investigation into the causes of Coss losses in GaN-on-Si HEMTs," in *2019 20th Workshop on Control and Modeling for Power Electronics (COMPEL)*, 2019, pp. 1–7.

[9] Q. Song, Q. Li, and Y. Zhang, "Output capacitance loss in wide-bandgap and superjunction power transistors: Impact of switching voltage and current," in *2023 IEEE 10th Workshop on Wide Bandgap Power Devices Applications (WiPDA)*, 2023, pp. 1–4.

[10] X. Yang, Q. Song, R. Zhang, B. Wang, S. Pidaparthi, A. Walker, C. Drowley, and Y. Zhang, "Evaluation of dynamic Ron, Coss loss, and short-circuit ruggedness of 650V and 1200V industrial vertical GaN JFETs," in *2024 36th International Symposium on Power Semiconductor Devices and ICs (ISPSD)*, 2024, pp. 283–286.

[11] B. Kohlhepp, D. Kuebrich, and T. Duerbaum, "Experimental study of the Coss-losses occurring during ZVS transitions – emphasis on low and high voltage GaN-HEMTs," in *CIPS 2020; 11th International Conference on Integrated Power Electronics Systems*, 2020, pp. 1–6.

[12] J. Fedison and M. Harrison, "Coss hysteresis in advanced superjunction MOSFETs," in *2016 IEEE Applied Power Electronics Conference and Exposition (APEC)*, 2016, pp. 247–252.

[13] M. Samizadeh Nikoo, A. Jafari, N. Perera, and E. Matioli, "Measurement of large-signal Coss and Coss losses of transistors based on nonlinear resonance," *IEEE Transactions on Power Electronics*, vol. 35, no. 3, pp. 2242–2246, 2020.

[14] M. Hornbuckle, S. Abrego, K. Liang, S. Davidova, Z. Tong, and J. Rivas-Davila, "Voltage waveform generation for sawyer-tower Coss loss measurements using a hybrid power converter," in *PCIM Europe 2024; International Exhibition and Conference for Power Electronics, Intelligent Motion, Renewable Energy and Energy Management*, 2024, pp. 2719–2724.

[15] M. Hayati, A. Lotfi, M. K. Kazimierczuk, and H. Sekiya, "Generalized design considerations and analysis of class-E amplifier for sinusoidal and square input voltage waveforms," *IEEE Transactions on Industrial Electronics*, vol. 62, no. 1, pp. 211–220, 2015.

[16] S. Yin, Z. Zeng, S. Debnath, and M. Saeedifard, "Modeling and ZVS operation of the isolated modular multilevel DC–DC converter with a unified trapezoidal wave modulation," *IEEE Transactions on Power Electronics*, vol. 39, no. 7, pp. 8306–8322, 2024.

[17] D. Goldmann, S. Schramm, and H.-G. Herzog, "Triangular and trapezoidal modulation for dual active bridge DC-DC converters with fast switching semiconductors," in *2019 21st European Conference on Power Electronics and Applications (EPE '19 ECCE Europe)*, 2019, pp. P.1–P.10.

[18] J. Roig and F. Bauwens, "Origin of anomalous Coss hysteresis in resonant converters with superjunction FETs," *IEEE Transactions on Electron Devices*, vol. 62, no. 9, pp. 3092–3094, 2015.

[19] Z. Tong, J. Roig-Guitart, T. Neyer, J. D. Plummer, and J. M. Rivas-Davila, "Origins of soft-switching Coss losses in SiC power MOSFETs and diodes for resonant converter applications," *IEEE Journal of Emerging and Selected Topics in Power Electronics*, vol. 9, no. 4, pp. 4082–4095, 2021.

[20] M. Guacci, M. Heller, D. Neumayr, D. Bortis, J. W. Kolar, G. Deboy, C. Ostermaier, and O. Häberlen, "On the origin of the Coss-losses in soft-switching GaN-on-Si power HEMTs," *IEEE Journal of Emerging and Selected Topics in Power Electronics*, vol. 7, no. 2, pp. 679–694, 2019.

Enhancing Behind-the-Meter Visibility of grid edge PV Systems and Electric Vehicle Charging Loads through Integration of Compact Low-Cost Sensors

Mehrnaz Madadi
Electrical Engineering department
North Carolina State University
Raleigh, US
mmadadi@ncsu.edu

Paul Ohodnicki
Mechanical and Material Science department
University of Pittsburgh
Pittsburgh, US
PRO8@pitt.edu

Subhashish Bhattacharya
Electrical Engineering department
North Carolina State University
Raleigh, US
sbhatta4@ncsu.edu

Abstract—The growing adoption of solar panels, battery storage, and Electric Vehicle (EV) chargers is driving a profound transformation in the distribution grid system. However, limited visibility into Behind-the-Meter (BTM) Distributed Energy Resources (DERs) generation restricts the full integration of these resources as grid assets. This lack of insight hampers efforts to enhance grid reliability, resilience, reduce operational costs, and expand photovoltaic (PV) hosting capacity. By leveraging existing data from distributed inverters, which are already monitored by controllers, it is possible to improve BTM PV generation visibility without incurring additional hardware costs. Furthermore, the introduction of a compact, low-cost bolt-on sensor can further enhance visibility into BTM PV generation. This paper presents a Power Hardware-in-the-Loop (PHIL) testbed designed to evaluate the feasibility of the "virtual inverter sensing node" and "bolt-on sensor" concepts. These concepts aim to improve BTM visibility by collecting data from BTM DERs, loads such as EV charging stations, and micro-Phasor Measurement Units (μ PMU) within a distribution feeder. The paper demonstrates how real-time sensor data can be processed into actionable measurements within both the physical and virtual sensor interfaces using the PHIL testbed. Additionally, the study compares the benefits of cloud-based data aggregation with the use of μ PMU as data aggregators, as demonstrated through experimental testing.

Index Terms—smart sensor, Behind the meter, visibility, Electrical Vehicle, bolt-on sensor, Power Hardware In the loop, micro Phasor Measurement unit.

I. INTRODUCTION

Distribution systems (DSs) are increasingly incorporating behind-the-meter (BTM) distributed energy resources (DERs) such as solar photovoltaics (PV), battery storage, and electric vehicles (EVs), alongside flexible buildings with controllable loads. However, many of these DER interfaces are grid-following, which complicates coordination between prosumers and network operators. Additionally, maintaining voltage and frequency stability becomes challenging with the rising penetration of renewable energy, dynamic prosumer actions, and low system inertia, which can destabilize the grid [1].

As BTM DERs become more widespread, their combined

Funded by: US Department of Energy Solar Energy Technology Office under award # DE-0009632

impact can be both positive and negative. High penetration of these resources may lead to issues like increased component congestion and voltage fluctuations within the system. Furthermore, since many BTM DERs are not utility-owned, their adoption can distort load forecasting, leading to inaccuracies in phase imbalance detection and demand response analysis. These inaccuracies may result in flawed balancing plans and actions, further complicating grid management.

A case study in New York State [2] highlighted the potential of BTM DERs to provide grid services through a DER Management System (DERMS). The study found that DERs with frequency-watt droop response could help maintain grid stability as thermal synchronous generators are replaced by renewable energy sources. Additionally, BTM DERs were shown to provide frequency regulation services similar to those of large utility-scale generation, reinforcing their potential to support grid stability despite the challenges posed by their integration. Limited visibility into BTM DERs restricts the full utilization of their potential as grid assets, negatively impacting grid reliability and efficiency. One significant challenge for utilities is the lack of visibility into the charging patterns of Electric Vehicles (EVs) within distribution networks. Often, utilities do not have access to dedicated sensors monitoring EV charging in residential areas, which complicates their ability to effectively plan and manage power distribution systems. As EV adoption grows, this lack of insight further underscores the need for improved data integration and more sophisticated grid management strategies.

The global deployment of smart meters presents a valuable opportunity for utilities to gather data that enhances their understanding of EV usage within their networks. However, since EVs are generally connected behind the meter, utilities typically lack dedicated meters to track EV charging, meaning they can only monitor the total load recorded by the smart meters. To address this gap, [3] proposes a non-intrusive, training-free approach for detecting BTM EV charging events using advanced metering infrastructure (AMI) data, such as that from smart meters. This technique analyzes data on charging start times, durations, and power levels to probabilistically estimate the charging behavior of customers over a year of meter data. These inferred charging patterns are then applied to evaluate the effects of EV charging on the IEEE-8500 test feeder, providing valuable insights into the broader impacts of EV integration into the grid.

In [4], a model has been developed to simulate the charging demand profiles of plug-in electric vehicles (PEVs) alongside real-world data from three distinct home groups with varying distributed energy resource (DER) configurations. A valley-filling smart charging algorithm has been examined to optimize the charging process for electric vehicles, enhance the utilization of solar photovoltaic (PV) systems, lower emissions, and alleviate peak demand. Results indicate that smart charging not only helps extend the lifespan of transformers, but also aids in reducing greenhouse gas (GHG) emissions. Although the deployment of PEVs contributes to lower GHG emissions, integrating DERs and smart charging solutions further reduces emissions. This combination of DERs and smart charging is shown to counteract the negative impacts of PEVs on the distribution grid, promote PV adoption, and contribute to reductions in both emissions and peak demand.

The study in [5] addresses the challenge of disaggregating Behind-the-Meter (BTM) electric vehicle (EV) load traces from smart meter data. It identifies three interrelated sub-problems based on the typical characteristics of EV charging profiles: (a) detecting the presence of BTM EVs, (b) estimating the EV charging rate, and (c) identifying the EV charging periods. A unified iterative algorithmic framework is developed to address all three sub-problems. Notably, the proposed algorithms do not rely on prior knowledge of the actual EV load traces, instead estimating the BTM EV load traces in an "unsupervised" manner. This method is designed to operate with 15-minute interval data, a typical resolution in utilities using smart meters.

Most studies focus on utilizing data already collected by smart sensors, with few exploring the design of smart Behind-the-Meter (BTM) sensors and the collection and analysis of BTM data. [6] proposes embedding a grid-edge distributed energy resource (DER) chip that hosts a Distributed Energy Resource Management System (DERMS) algorithm into next-generation smart meters. To bridge the gap between electric utilities and DERs behind the meter, the development of smart, low-cost, compact BTM sensors is essential [7].

This paper explores methods for leveraging existing data from DER inverter controllers as virtual sensing nodes, introducing bolt-on sensors and various data collection strategies to enhance monitoring and data analysis of BTM renewable energy sources and dynamic loads.

II. SMART LOW-COST SENSORS

To address the challenge of limited visibility into behind-the-meter renewable energy generation or load consumption, two economical sensors have been introduced. Both sensors communicate via radio frequency (RF) signals, eliminating the need for an internet connection. This makes them particularly well-suited for rural areas, where monitoring of generation

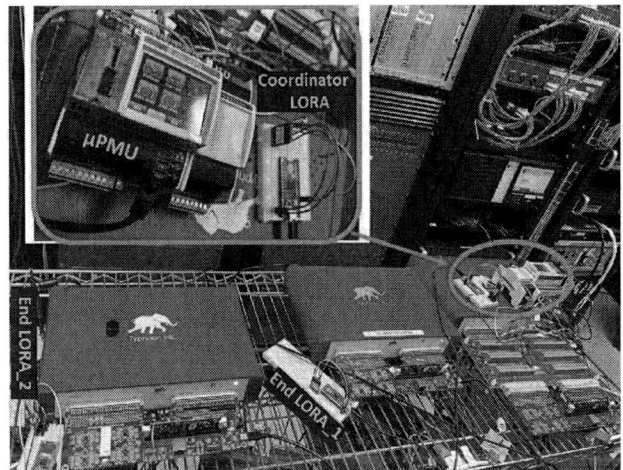

Fig. 1. The Experimental testbed to validate the concept of virtual sensors

or load consumption can be difficult due to lack of internet access.

A. Virtual Sensor Nodes

As previously discussed, one of the main factors contributing to blind spots in the visibility of grid-edge resources is the high cost of advanced measurement and sensing devices. Solar inverter controllers already collect data on both the DC and AC sides to ensure protection and optimize operation. The core concept behind virtual sensors is to explore how to leverage the existing data from Distributed Energy Resource (DER) inverter controllers as virtual sensing nodes. By implementing efficient data management and communication pathways, utilities can gain enhanced visibility at minimal additional cost, improving real-time monitoring of DER performance.

However, with the current 15-minute sampling interval used by commercial inverters, achieving effective behind-the-meter (BTM) visibility is not feasible. To provide adequate visibility of BTM photovoltaic (PV) generation, the sampling interval should be reduced to at least 3 minutes. Depending on the data collection objectives, the sampling interval in virtual sensors could vary, potentially reaching as low as 10 microseconds for more precise measurements.

B. Bolt-on Sensor

To enhance the visibility of behind-the-meter (BTM) solar generation and address data privacy concerns associated with commercial inverters, a new data acquisition device: the bolt-on sensor was developed (Shown in fig.2). This sensor is installed at the output of commercial inverters, where the photovoltaic (PV) system connects to the grid, and measures both voltage and current. The data is then transmitted via a LoRa gateway to an Amazon Web Services (AWS) account. Figure 3 shows the bolt-on sensor and its wireless communication capabilities.

This non-intrusive sensor is easy to install and can be connected directly at the integration point with the grid utility, effectively mitigating data privacy issues. The voltage rating

Fig. 2. The bolt-on sensor

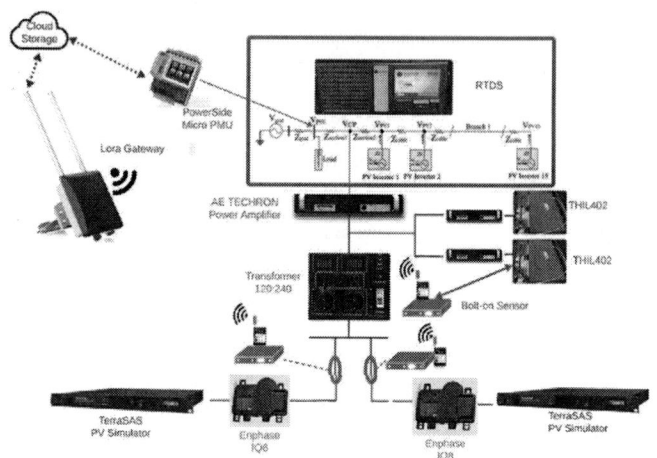

Fig. 3. The Experimental testbed integrating multiple inverters and mPMUs from different vendors

is 250V, and the current rating is 50A. Furthermore, the cost of the sensor is under 150. Figure 1 illustrates the installation of multiple bolt-on sensors at the output of Enphase microinverters, with the measured data being sent to cloud storage (AWS in our case) via integrated LoRa modules.

III. DATA AGGREGATORS FOR BTM DATA COLLECTION

Although the virtual and bolt-on sensors are fundamentally different from each other, the data collected from both can be aggregated either through a physical data aggregator or a cloud-based platform. Each data aggregator has its own advantages and disadvantages. The choice between them depends on factors such as the required data location and cost considerations.

A. μPMU as a physical data aggregator

A μPMU (micro Phasor Measurement Unit) is a device that captures high-precision voltage phase angles, or synchrophasors, in distribution feeders, with the capability to store and transmit data in real-time. One of the main benefits of integrating low-cost sensors with μPMUs is their ability to facilitate communication with utility distribution systems. This integration improves the data by adding context through algorithms that incorporate time synchronization and GPS information. The data collected by the μPMU can be monitored online via the device's IP address. A commercial μPMU from Powerside was installed at the Point of Common Coupling (PCC) on the feeder in the PHIL testbed, as shown in Fig. 4. While the μPMU measures the PCC voltage and currents, its four analog inputs can also be utilized to collect data from distribution feeder inverters. The first input will be reserved for identifying the node whose data is being monitored, while the remaining three analog inputs can be used to transmit the active power (P), reactive power (Q), and voltage (V) of a single PV inverter, or the active power generated by three PV inverters ([P1, P2, P3]). If the BTM data consists of the P, Q, and V of a single inverter and all nodes connected to the coordinator are end nodes, the maximum number of inverters whose [P, Q, V] data can be monitored within a 3-minute period, with a 2-second delay, is 90. While using a μPMU as a data aggregator can be cost-effective—since a single device

can monitor and collect data from up to 90 inverters ([P, Q, V]), there are several drawbacks that need to be addressed. Firstly, the μPMU supports only single-channel communication, meaning only one end node can communicate with the coordinator at a time, which can reduce data transmission efficiency. Additionally, minor changes in the spreading factor or frequency can adversely affect the reliability of communication. The system also has limited data storage capacity, which could pose challenges for long-term data collection and analysis. Moreover, there is a 2-3 second delay in viewing monitoring data on the μPMU's web interface, which makes real-time analysis difficult. Accessing the web interface requires being physically close to the μPMU, which is not ideal for remote monitoring. Finally, the system lacks the capability for real-time analysis or advanced computations on the stored data, limiting its use for dynamic grid management.

B. Cloud-based data aggregator

An alternative approach avoids using the μPMU for low-cost sensor communication and instead relies on an off-the-shelf LoRaWAN gateway. This method effectively addresses the challenges outlined in III-A. The proposed system offers several benefits: it enables multi-channel, multi-node parallel communication with up to 8 channels on the selected chipset, adapts to changes in the spreading factor or frequency, and provides a significantly extended range, enhancing coverage and connectivity. Data received can be stored both on the coordinator node and in cloud storage within seconds. This data can then be utilized to create highly customizable web interfaces and perform in-depth analyses. Since the data is stored in the cloud, it is accessible from anywhere with an internet connection, allowing for more comprehensive analysis and enabling bi-directional communication. While the cost may be higher in this case, the reliability is improved, as data collection is independent of the μPMU. Therefore, any failure of the μPMU will not affect the data collection process.

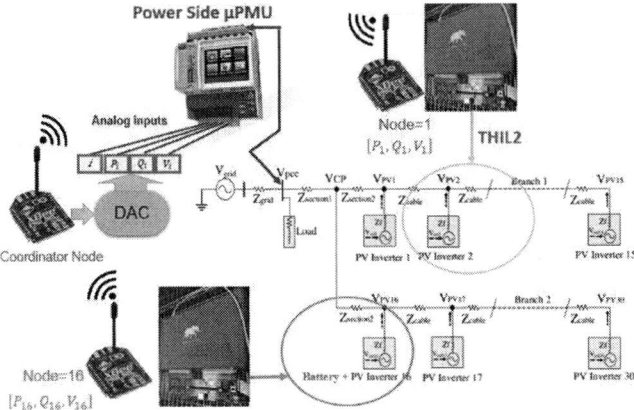

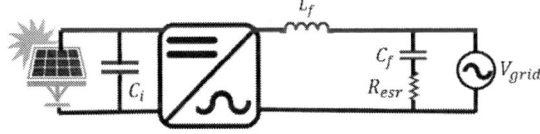

Fig. 4. Proposed RF-module-based communication with μPMU as a data aggregator

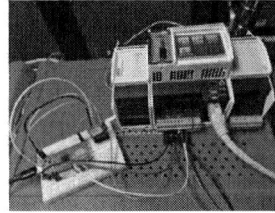

Fig. 5. Block diagram of simulated PV system in Typhoon HIL

IV. EXPERIMENTAL TESTBED

Multiple Power-Hardware-in-the-Loop (PHIL) simulation testbeds were developed to verify and evaluate the performance of virtual sensor nodes, bolt-on sensors, and the μPMU serving as a measurement device for capturing voltage and current data from the PCC node and also as a data aggregator for collecting information, highlighting the benefits of using a LoRa gateway and incorporating virtual data aggregators into the system.

A. Virtual sensor nodes and μPMU as a data aggregator

To validate the concept of virtual sensors, the PV and battery systems were simulated using Typhoon Hardware-in-the-Loop (THIL) 402 devices. The collected BTM data is transmitted through a wireless communication network based on radio frequency (RF) and aggregated by the Micro Phasor Measurement Unit (μPMU) at the Point of Common Coupling (PCC). Fig.4 illustrates the proposed communication network designed to collect BTM data from various inverters, with the μPMU serving as a data aggregator for monitoring. As shown, the data from each inverter can be monitored via the LoRa modules, utilizing the four analog inputs of the μpmu.
Two PV systems were simulated with Typhoon HIL (THIL) devices, operating at a unity power factor. The block diagram of the simulated PV systems is shown in Fig.5. Data on active power (P), Node voltage (V) and the frequency (f) for each node were sent to the coordinator node connected to the μPMU.for each node measured in Typhoon devices SCADA environment, is shown in Fig.7 and Fig.9. The transmitted data will be stored on the internal storage of the μPMU or could be monitored in real-time using the IP address of the device and the related web interface shown in (Fig. 8 and Fig. 10).

a. LORA module 1 and THIL at node 1

b. LORA module 2 and THIL at node 2

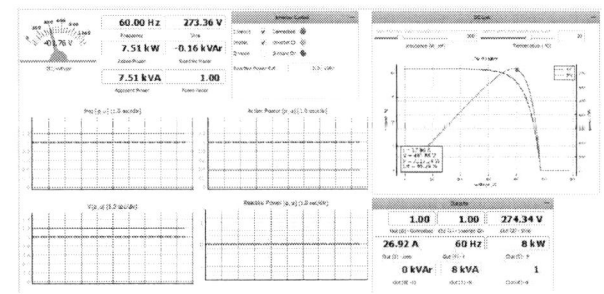

c. LORA module 3, raspberry pi, and μPMU as a data aggregator at the coordinator node

Fig. 6. The experimental testbed to prove the virtual sensor concept with μPMU as a data aggregator

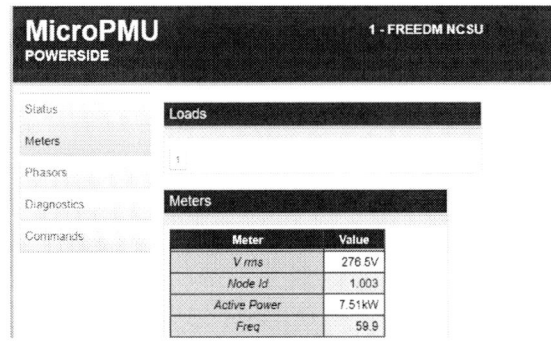

Fig. 7. Measured P, V, f at node 1 of THIL1

Fig. 8. Measured P, V, f and node ID (1) shown at μPMU web interface

The installed PV capacity is $15kW$ at $1000W/m^2$ solar irradiation and $20°C$. It has been assumed that while at node 1, the solar irradiation is $500W/m^2$ and therefore, the generated power is approximately $7.5kW$ (Fig.7), at node 2, the PV system is generating $15kW$ at $1000W/m^2$ solar irradiation and $20°C$ Fig.9. The controller at each node is Raspberry pi 4 Model B that transmits each node's data to the coordinator raspberry pi SC15184 4 Model B and it is sent to the 4 analog inputs of the μPMU with a two-second delay. The whole test bed is shown in Fig.6.

Tables I and II compare the data measured within the

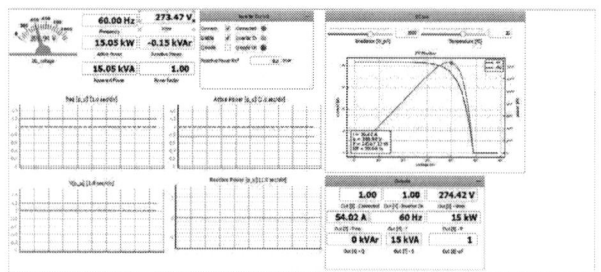

Fig. 9. Measured P, V, f at node 2 of THIL2

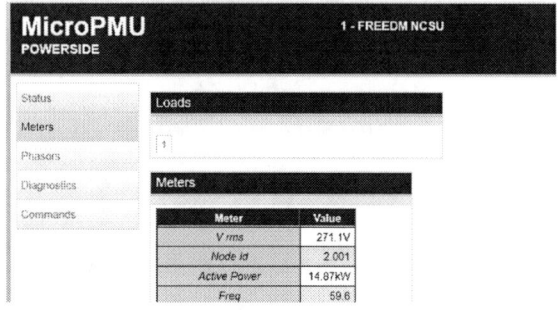

Fig. 10. Measured P, V, f and node ID (2) shown at μPMU web interface

TABLE I
VIRTUAL SENSOR AT NODE 1

Node1	μPMU	THIL	Error%
V(V)	276.6	273.36	1.18
P(kW	7.51	7.51	0
f(Hz)	60	59.9	0.16

TABLE II
VIRTUAL SENSOR AT NODE 2

Node2	μPMU	THIL	Error%
V(V)	271.7	273.47	0.65
P(kW	15.05	14.87	1.19
f(Hz)	60	59.5	0.83

Typhoon devices' SCADA environment with the data received by the μPMU's web interface, showing an error margin of less than %2. The transmitted data is stored in the μPMU's internal storage, and can be monitored with a 2-3 second delay via the device's IP address and web interface.

B. Bolt-on sensor and cloud-based data aggregator

Fig. 11 shows the schematic of the real feeder simulated in the RTDS. The node where the μpmu is connected to measure voltage, current, and phase angle is highlighted on the feeder. The node where the external load is connected for bolt-on sensor performance validation is represented by a current source. Fig.12 shows the experimental testbed set up in the RTDS lab.

In the first test scenario, a resistor load of $R_1 = 100\Omega$ is connected in parallel with $R_2 = 68\Omega$ resulting in an equivalent resistance of $R_{eq} = 41\Omega$. These two resistors are then placed in series with two 15Ω resistors, giving a total external load of 71Ω. A bolt-on sensor is used to measure the voltage and current across the total external load. The

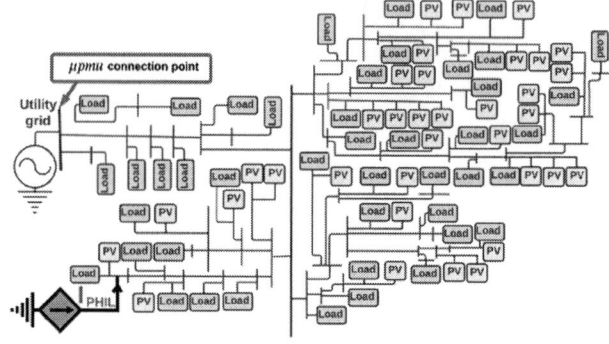

Fig. 11. The Schematic of the simulated feeder in RTDS

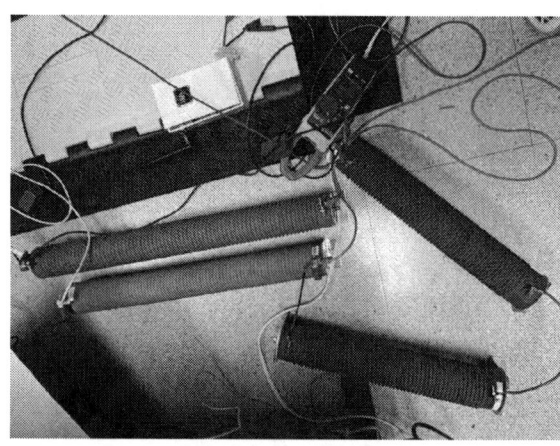

Fig. 12. The experimental testbed for Bolt-on sensor performance evaluation

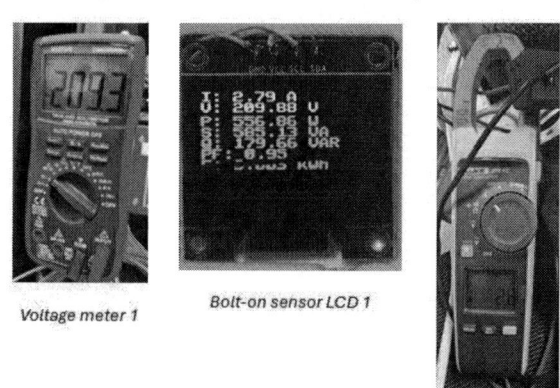

Fig. 13. Comparison of bolt-on sensor readings with voltage and current meter measurements for test 1

external voltage applied to the load is $V_{ac} = 209v$ and the load current is $I_{load} = 2.9A$.

In the second test scenario, the resistor load of $R_1 = 100\Omega$ is disconnected, resulting in an equivalent resistance of $R_{eq} = 98\Omega$. While the voltage applied to the load remains at $V_{ac} = 209v$, the load current deceases to $I_{load} = 2.1A$. Fig.13 and Fig.14 Show the measurements taken by the bolt-on sensor, displayed on its LCD, compared to the voltage and current readings from a commercial voltmeter and ammeter

Voltage meter 2

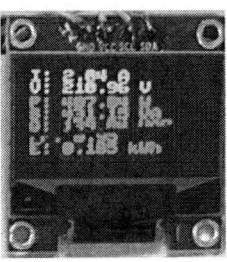
Bolt-on sensor LCD 2

Current meter 2

Fig. 14. Comparison of bolt-on sensor readings with voltage and current meter measurements for test 2

in the both test cases. The data taken by the bolt-on sensor is transmitted to be collected on the Amazon Web Service (AWS) for further analysis.

C. Electrical Vehicle load integration and the μPMU measurements

The distribution feeder characteristics are as follows: the line-to-line voltage is $138kV$, and the phase-to-line voltage is $80kV$. The total PV generation connected is $0.36MW$, with 36 sources, each rated at $0.01MW$. Additionally, 51 loads are connected, with a total rated load of $0.691MW$. The μPMU is connected to the analog output of the RTDS to measure the grid voltage, current, and phase angle, replicating its connection points in the real distribution feeder.

The NREL EVI-Pro Lite API tool was used to extract data for integrating the load profiles of 1,000 Level 1 and 1,000 Level 2 EV charging stations in Raleigh, NC. For the Level 1 charging points, operating at 120V, the load connected is 52 kW per phase, with three single-phase transformers rated at 75 kVA, 7.2/0.12 kV ratio. For the Level 2 charging points, operating at 227V, the load connected is also 52 kW per phase, with three single-phase transformers rated at 75 kVA, 7.2/0.277 kV ratio.

The NREL EVI-Pro Lite API is a tool provided by the National Renewable Energy Laboratory (NREL) that allows users to access and interact with data related to electric vehicle infrastructure and energy consumption. Specifically, EVI-Pro Lite is a web-based application designed to assist planners, engineers, and policymakers in modeling and evaluating the deployment of electric vehicle (EV) charging infrastructure.

Figure 15 shows the phase B voltage, while Figure 16 displays the phase B current at the node where the Level 1 EV charging station load was connected. The data for voltage corresponds to the timestamp when all EVs, after being fully charged, were disconnected from the grid. The data for Current corresponds to the timestamp when half of the EVs (500 EVs), after being fully charged, were disconnected from the grid.

Fig. 17 shows the grid voltage phasors, the current phasors, and the active and reactive power measured at the grid in the RTDS, while all EVs are charging and the PV systems are

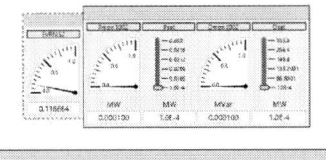

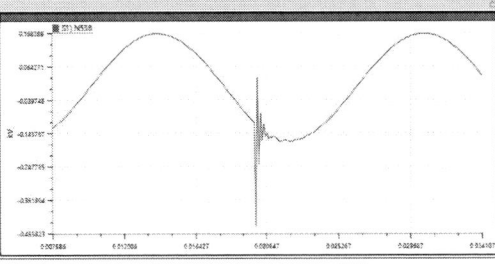

Fig. 15. The Phase b voltage at the point of level 1 EV charging stations connection

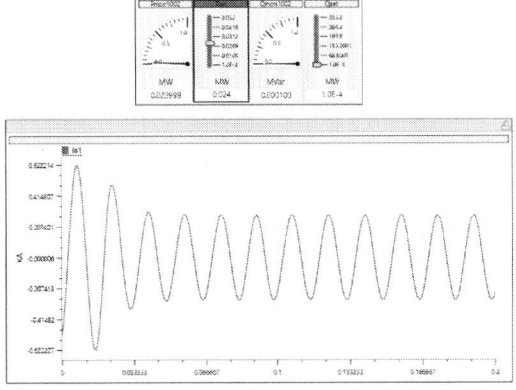

Fig. 16. The Phase b current at the point of level 1 EV charging stations connection

generating power at their full rated capacity. Fig. 18 shows the same data corresponding to when the PVs are not generating such as after sunset.

Fig.19 shows a snapshot of the μpmu web-interface and fig.20 illustrates the voltage and current phasors of the grid measured by μpmu, shown on its web-interface. The three phase voltages, currents, active, and reactive powers are shown while all EVs are charging and the PV systems are generating power at their full rated capacity.

Fig.21 shows a snapshot of the μpmu web-interface and fig.22 illustrates the voltage and current phasors of the grid measured by μpmu, shown on its web-interface. The three phase voltages, currents, active, and reactive powers are shown while all EVs are charging but the PVs are not generating such as after sunset.

V. CONCLUSION

Two low-cost sensors—a compact bolt-on sensor and a virtual sensor node—were introduced and thoroughly evaluated for their effectiveness in measuring Behind-the-Meter (BTM)

979-8-3315-1612-3/25 $31.00 © 2025 IEEE

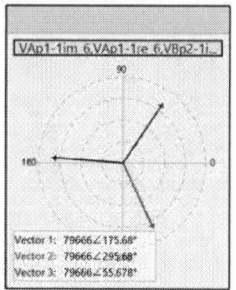
a. V phasors with PV generation in RTDS

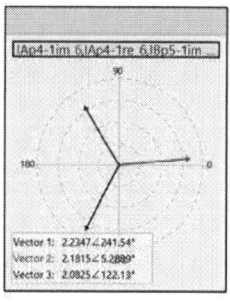
b. I phasors with PV generation in RTDS

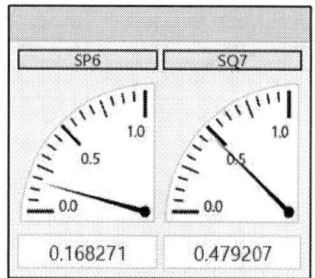
c. The active (P) and reactive (Q) power measured at the grid with PV generation in RTDS

Fig. 17. The voltage, current phasors and the active (P) (Mw) and reactive (Q) (Mvar) power measured at the grid in RTDS with PV generation at full EV load

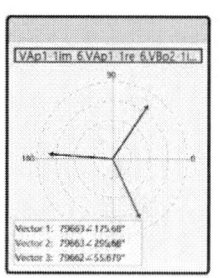
a. V phasors without PV generation in RTDS

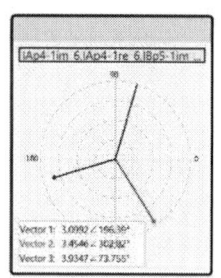
b. I phasors without PV generation in RTDS

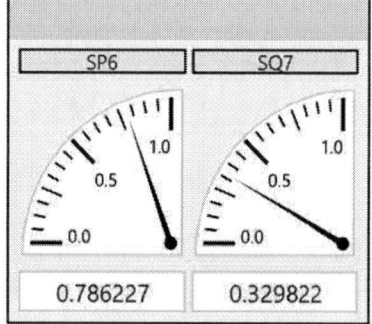
c. The active (P) and reactive (Q) power measured at the grid without PV generation in RTDS

Fig. 18. The voltage, current phasors and the active (P) (Mw) and reactive (Q) power (Mvar) measured at the grid in RTDS without PV generation at full EV load

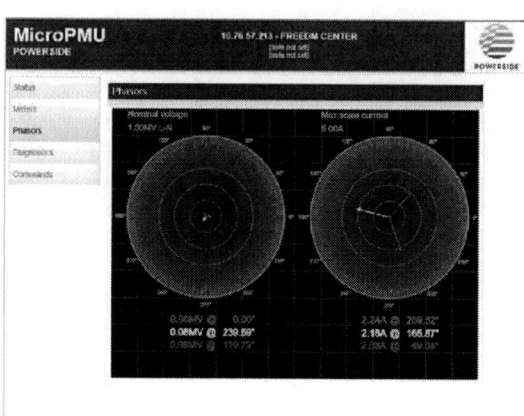

Fig. 19. The μpmu web interface output with PV generation at full EV load

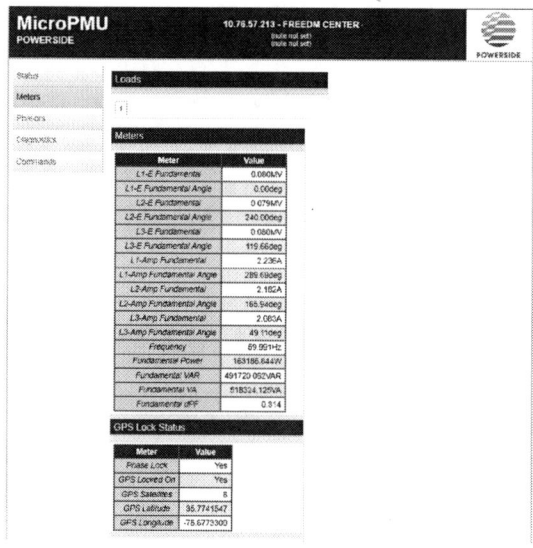

Fig. 20. The μpmu web interface voltage and current phasors with PV generation at full EV load

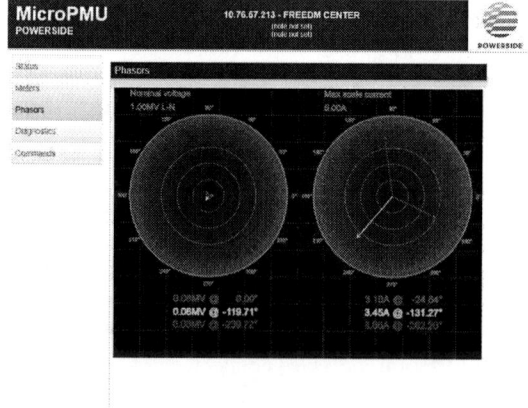

Fig. 21. The μpmu web interface output without PV generation at full EV load

photovoltaic (PV) generation and loads, such as Electric Ve-

hicle (EV) charging stations. This assessment was conducted using a Power Hardware-in-the-Loop (PHIL) testbed, which

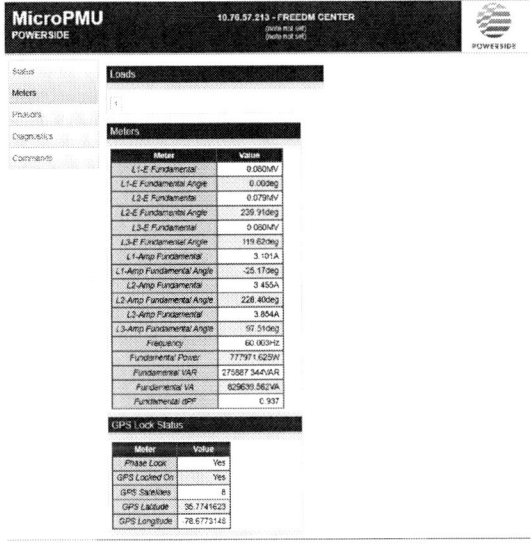

Fig. 22. The μpmu web interface voltage and current phasors without PV generation at full EV load

integrated a commercial micro-Phasor Measurement Unit (μ PMU) alongside the proposed sensors to enable real-time benchmarking of BTM visibility. The results demonstrate that integrating bolt-on and virtual node sensors provides more comprehensive data on BTM renewable energy sources (RES) and EV charging stations, significantly enhancing the accuracy of BTM disaggregation algorithms. When combined with existing methods, this integration offers a cost-effective and reliable solution for improving BTM visibility. This approach paves the way for the widespread adoption of Distributed Energy Resource Management Systems (DERMS), as many utilities are planning to invest in advanced metering infrastructure in the near future.

Acknowledgment

The authors gratefully acknowledge the financial support of the US Department of Energy Solar Energy Technology Office under award #DE-0009632.

References

[1] A. Srivastava et al., "Distribution System Behind-the-Meter DERs: Estimation, Uncertainty Quantification, and Control," in IEEE Transactions on Power Systems, 2024, pp.1-16.

[2] H. Hooshyar et al., "Grid Services by Behind-the-Meter Distributed Energy Resources: NY State Grid Case Study," 2023 IEEE Power & Energy Society General Meeting (PESGM), Orlando, FL, USA, 2023, pp. 1-5.

[3] Feng Li, Élodie Campeau, Ilhan Kocar, Antoine Lesage-Landry, "Inferring electric vehicle charging patterns from smart meter data for impact studies", Electric Power Systems Research, Volume 235, 2024.

[4] B. Hudson, G. Razeghi, S. Samuelsen, "Mitigating impacts associated with a high-penetration of plug-in electric vehicles on local residential smart grid infrastructure", Journal of Power Sources, Volume 593, 2024.

[5] K. Pu and Y. Zhao, "Behind-the-Meter Disaggregation of Residential Electric Vehicle Charging Load," 2022 IEEE International Conference on Communications, Control, and Computing Technologies for Smart Grids (SmartGridComm), Singapore, Singapore, 2022, pp. 426-431.

[6] J. Comden, J. Wang, S. Ganguly, S. Forsyth, R. Gomez and A. Bernstein, "Hardware-in-the-Loop Evaluation of Grid-Edge DER Chip Integration Into Next-Generation Smart Meters," 2023 IEEE International Conference on Communications, Control, and Computing Technologies for Smart Grids (SmartGridComm), Glasgow, United Kingdom, 2023, pp. 1-6.

[7] M. Madadi, R. Beddingfield, P. Ohodnicki and S. Bhattacharya, "Utilizing Smart Inverter Virtual-Sensor Nodes for Enhanced Behind-the-Meter Visibility in High PV Penetration Distribution Feeders," 2023 IEEE Energy Conversion Congress and Exposition (ECCE), Nashville, TN, USA, 2023, pp. 610-616.

Supercapacitor based TMS pulse generator design-Experimental results versus MATLAB MOSFET simulation model

Soniya Raju
School of science
university of Waikato
Hamilton, New Zealand
sr231@students.waikato.ac.nz

Nihal Kularatna
School of Engineering
University of Waikato
Hamilton, New Zealand
nihalkul@waikato.ac.nz

Marcus Wilson
School of Science
University of Waikato
Hamilton, New Zealand
marcus.wilson@waikato.ac.nz

Alistair Steyn-Ross
School of Engineering
University of Waikato
Hamilton, New Zealand
alistair.steyn-ross@waikato.ac.nz

Abstract—**Supercapacitors are ideal for short-term energy storage due to their high instantaneous power capability. Unlike traditional high-voltage DC power supplies typically used in transcranial magnetic stimulation (TMS) applications, a pre-charged supercapacitor module can serve as a cost-effective alternative for building a pulse generator with adjustable waveform capabilities. This approach utilizes a supercapacitor module with a capacitor bank rated at a few hundred farads and 10 V, combined with a step-up transformer-based converter and medium-voltage MOSFETs to meet TMS requirements. The output of this pulse generator, including the effects of varying gate voltages, can be accurately predicted using a robust model. This paper presents the design of a MOSFET model, incorporating a newly derived equation that effectively captures the subthreshold, near-threshold, and above-threshold regions of the MOSFET operation. Experimental results are provided and compared with MATLAB model simulations, demonstrating the model's accuracy and effectiveness.**

Index Terms—**Electrical stimulation, Magnetic stimulation, Pulse circuits, Supercapacitors, Transcranial magnetic stimulation (TMS)**

I. INTRODUCTION

High-voltage and high-power pulse generators are essential for applications such as transcranial magnetic stimulation (TMS) [1–6], lightning surge simulators, and fence energizers [7]. Traditional designs, combining high-voltage DC sources with pulse-shaping circuits and high-voltage semiconductor switches, face significant challenges due to safety isolation requirements and high component costs [1, 6, 8–13]. Despite innovations like solid-state Marx generators and Tesla transformers aimed at enhancing efficiency, these systems remain complex, expensive, and require multiple stages and components. Most existing pulse generators also lack provisions for pulse shaping and demand high-voltage supplies or high-voltage capacitors, further complicating their use [14–21]. Additionally, TMS pulse generators, especially those for small animals, are underexplored, with limited dedicated circuits and simulation models. A supercapacitor-based pulse generator [22] offers a promising solution, leveraging higher energy density, rapid charging and discharging capabilities, and longer lifetimes. This approach could provide more efficient, cost-effective, and precise pulse control, advancing TMS research significantly.

Existing TMS circuitry uses MOSFET as switches, but they have the potential to shape coil current more precisely. This study introduces a model that predicts coil current by varying the gate applied to the MOSFET, incorporating a novel MOSFET drain current equation derived from real-time experimental results. Emphasizing MOSFET operation at and near the threshold region is critical for achieving high voltage across the TMS coil during switch-off. The model overcomes challenges such as limited-accurate data available from MOSFET datasheets.

II. SUPERCAPACITOR BASED PULSE GENERATOR (SCPG)

In a TMS pulse generator applications, the pulse generator is expected to drive certain amount of energy into the TMS coil and this may be in the range of few tens of joules to few hundred joules. With this requirement coming as primary requirement, we see that the commercially available supercapacitors (SC) could be easily configured into a pre-stored energy module to deliver multiple test pulses in one shot when required without using a continuously rated high-voltage DC power supply. The cost of this approach was found to be much less here due to the elimination of a high-voltage DC power supply.

The design approach for this TMS pulse generator leverages the capability of a charged supercapacitor to release a very high instantaneous current, which is predominantly controlled by the load side impedance rather than the SC's internal resistance. Considering a small supercapacitor module composed of 4 series-connected 380 F cells, this module has a short-circuit current capability of $\frac{10}{4\,\text{ESR}}$, where the equivalent series resistance (ESR) of a single cell is approximately 3.2 mΩ. In cases where a larger storage module is required, using 3000 F cells with an ESR of 0.23 mΩ provides 14 times greater current capability. This straightforward practical consideration indicates that the primary loop current is mainly governed by

979-8-3315-1612-3/25 $31.00 © 2025 IEEE   378

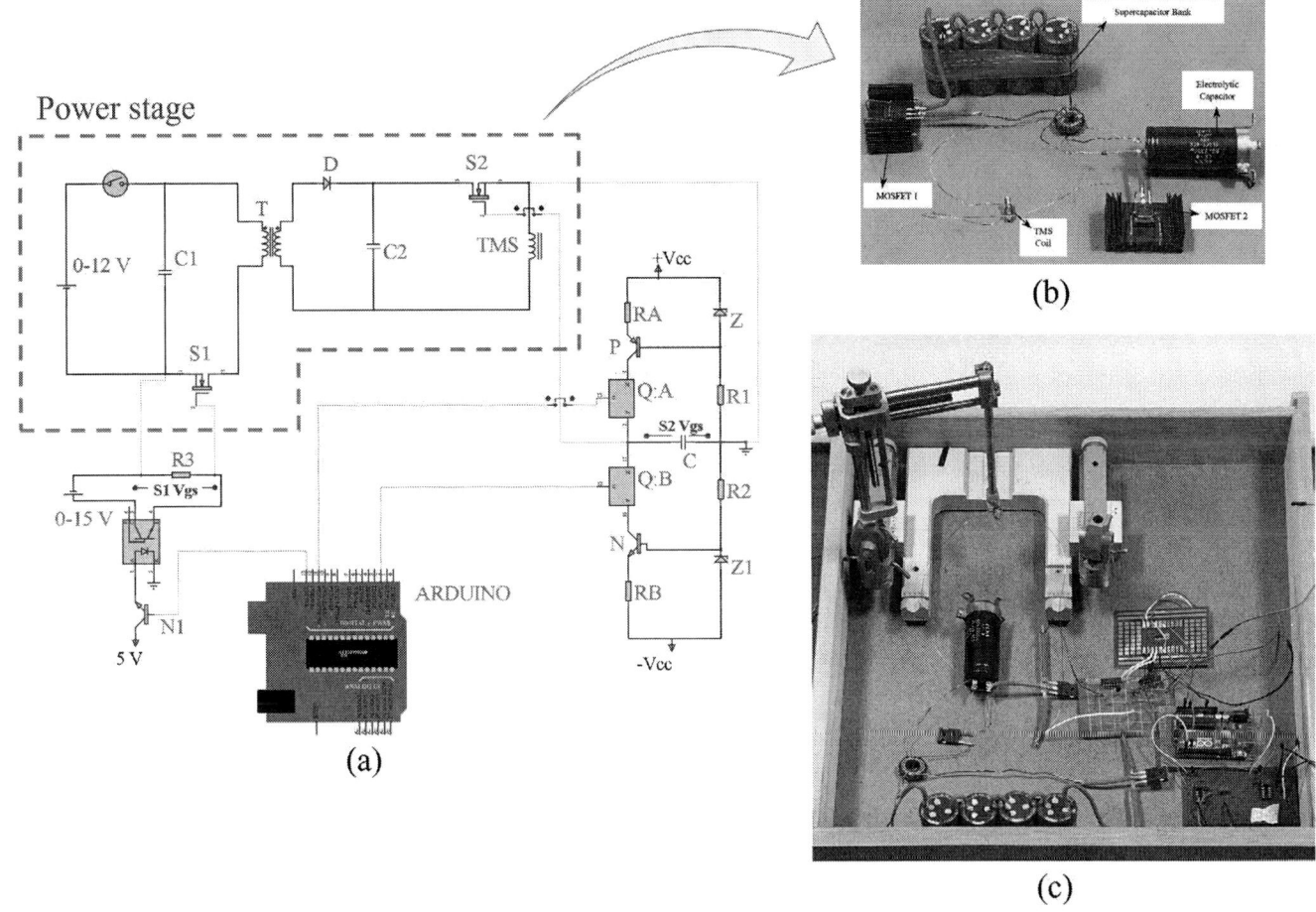

Fig. 1: Proof-of-concept of prototype- (a) Overall system showing external drivers to S1 from Arduino processor board and triangle voltage waveform generator for S2; (b) experimental power stage; (c) Full circuit prototype setup

the resistance and inductance parameters of the primary side of the step-up transformer.

Our team has successfully developed a supercapacitor-based pulse generator (SCPG) [22] as shown in Fig. 1, capable of producing a magnetic field of approximately 400 mT suitable for small animal brain excitation. This prototype can shape the output TMS current and voltage, crucially without needing additional external snubber or hardware setups like the existing topologies [6, 11]. The triangular gate input waveform, with its adjustable slope, enables a slower turn-on and turn-off than using MOSFET as switch, impacting the current and voltage pulses in the TMS coil system. Increasing the symmetry of the triangular waveform to 80% (80:20 rise-fall) resulted in a steeper fall slope as shown in Fig. 2, higher rate of change of current (di/dt), and a narrower pulse width of the voltage and current pulses. The developed MATLAB model of the MOSFET accurately predicts these outcomes, making it essential for future work in optimizing pulse shapes for TMS applications.

Using a 1 V input SC supply, we achieved a 250 V output TMS voltage pulse. For higher peak voltage of the output pulses we can proportionally increase the SC bank voltage. This can be further increased by adjusting the input triangle

TABLE I: List of components and the associated costs

Label	Component	Nominal rating	Cost (USD)
C1	Electrolytic-capacitor	330 μF, 450 V	7
C2	Supercapacitor	380 F, 3 V, 3.2 mΩ	12
S1	Charging switch	40 V, 1.35 mΩ, 350 A	5
S2	Discharging switch	650 V, 24 mΩ, 120 A	24
D	High-frequency diode	1.2 kV, 75 A	5
T	Step-up transformer	1:60 Turns ratio	2
A	ARDUINO UNO	5 V, 14 digital I/O pin	30
	Others		20
		Total cost	105

gate voltage symmetry. The results shown in Fig. 2 are for a 1 V input, which already produced a significant magnetic field (400 mT). Higher voltage outputs and pulse shaping are possible with change in optimal input gate voltage which can be determined through simulation models. Table I provides a breakdown of the cost of each component used

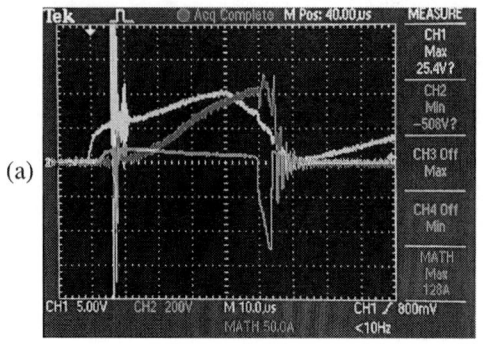 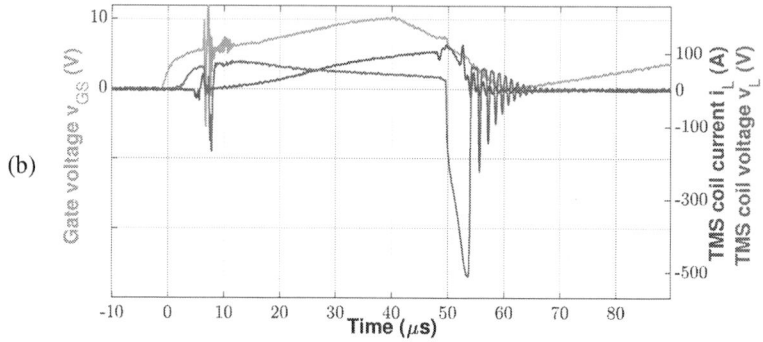

Fig. 2: (a) Oscilloscope measurement for gate voltage rise-fall symmetry set to 80:20. Note pronounced increase in peak coil voltage and emergence of ~1 MHz oscillations arising from reduced gate-voltage fall time.; (b) MATLAB capture of oscilloscope traces shown in (a)

in the prototype, with the total amount reaching 105 USD, highlighting the affordability and cost-effectiveness of the design. The total cost of 105 USD demonstrates that the design is economically viable for research or small-scale manufacturing purposes.

One key advantage of this design is that the series-connected supercapacitor module can be pre-charged using any suitable laboratory power supply. This allows the SC module to deliver repeated pulse sets for a given experiment without requiring continuous connection to a DC power supply. Additionally, we have developed a new MOSFET equation that fills a gap in the literature, providing a model that closely aligns with experimental data.

III. NEED FOR NEW EQUATION

In our application, we require the MOSFET (S2 in Fig. 1) to operate at and near threshold point since the high voltage across the TMS coil is achieved at the time of switch off (when gate voltage goes below threshold). Surprisingly, this region has not been thoroughly analyzed by researchers. Analytical solutions such as the parallel plate charge control model and the unified charge control model exist [23–25], yet they lack precise analytical solutions for the sub-threshold or near-threshold region.

For many applications, the approximate solution for drain current used from these models is:

$$i_{\mathrm{D}} = 2\, i_{\mathrm{o}} \ln\left[1 + \frac{1}{2} \exp\left(\frac{v_{\mathrm{GS}} - v_{\mathrm{T}}}{\eta\, v_{\mathrm{Th}}}\right)\right] \qquad (1)$$

where i_0 is the initial current (at threshold), applied gate voltage (v_{GS}), threshold voltage (v_{T}), thermal voltage (v_{Th}), and a sub-threshold ideality factor (η) is used. This solution is suitable for both above and below the threshold but not near the threshold [26–28]. All the analytical approaches available are designed for lower power and small-signal MOSFETs, whereas we are using a power MOSFET. There is no specific analytical calculation data available for power MOSFETs. We initially tried to use Equation (1) to fit the data points of the

measured drain current versus gate voltage, but failed to do so as shown in Fig. 3(a). Thus we modified the drain current equation as a function of gate voltage only as:

$$i_{\mathrm{D}} = \frac{i_{\mathrm{o}}}{\ln 2} \ln\left(1 + \left[\exp\left(\frac{v_{\mathrm{GS}} - v_{\mathrm{T}}}{\eta\, v_{\mathrm{Th}}}\right)\right]^p\right) \qquad (2)$$

where we have introduced a new parameter p to describe the transition from below-threshold to above-threshold as we need the current curve to change from exponential to linear. Figure 3(a) illustrates the failed curve fit using Eq. 1, while Fig. 3(b) demonstrates the successful fit achieved by employing the modified equation (2), which aligns much better with the data points (black dotted points).

The reason for the modification becomes apparent by analyzing Eq. (2) in three different regions.

The Eq. 2 can be further expressed in terms of a variable X:

$$i_{\mathrm{D}} = \frac{i_{\mathrm{o}}}{\ln 2} \ln[1 + X^p], \quad \text{where} \quad X = \exp\left(\frac{v_{\mathrm{GS}} - v_{\mathrm{T}}}{\eta\, v_{\mathrm{Th}}}\right). \qquad (3)$$

A. Subthreshold

In the subthreshold region ($v_{\mathrm{GS}} < v_{\mathrm{T}}$), where $X \ll 1$, we use $p = 1$ (we have given the maximum value for the p since subthreshold region characteristics is of more exponential behavior):

$$i_{\mathrm{D}} = \frac{i_{\mathrm{o}}}{\ln 2} \ln[1 + X], \qquad (4)$$

and using

$$\ln[1 + X] \approx X \quad \text{for small X}$$

we obtain

$$i_{\mathrm{D}} \approx \frac{i_{\mathrm{o}}}{\ln 2} \exp\left(\frac{v_{\mathrm{GS}} - v_{\mathrm{T}}}{\eta\, v_{\mathrm{Th}}}\right). \qquad (5)$$

This equation resembles an exponential growth similar to general MOSFET characteristics [29].

979-8-3315-1612-3/25 $31.00 © 2025 IEEE

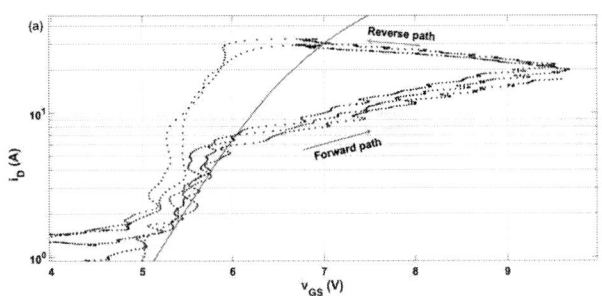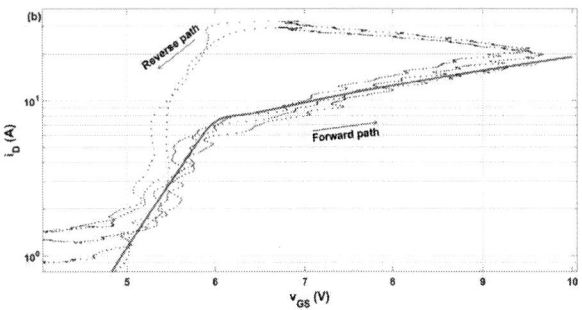

Fig. 3: (a) Current curve fit (red line) of the data points (black dots) from experiment using Eq. (1) which failed to fit; (b) current curve fitting done using modified equation Eq. (2), which was a successful fit with the forward data points

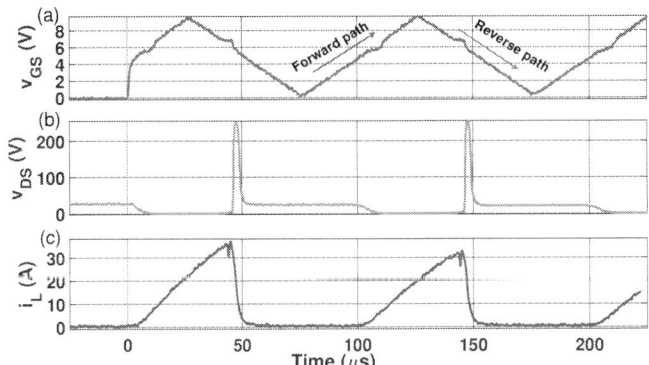

Fig. 4: Circuit parameters measured during TMS coil excitation- (a) Triangular control voltage of MOSFET 2 gate, 50% symmetry gate voltage with forward path (0 to 10 V) and reverse path (10 V to 0) marked (initial 25 μs of the triangular wave is not captured accurately due to the oscilloscope triggering configuration); (b) measured drain-source voltage and (c) coil current from the experimental setup used for the mosfet model

B. Above-threshold

In the above-threshold region ($v_{\text{GS}} > v_{\text{T}}$), where $X \gg 1$ and we use $p \approx 0.2$ (the value of p reduces from below-threshold to above-threshold as we need the current curve to change from exponential to linear), i_{D} becomes a linear equation:

$$i_{\text{D}} = \frac{i_{\text{o}}}{\ln 2} \ln[1 + X^p] \qquad (6)$$

and using

$$\ln[1 + X^p] \approx \ln X^p \quad \text{for large } X^p$$

we obtain

$$i_{\text{D}} = \frac{i_{\text{o}}}{\ln 2} \ln[X^p] = \frac{i_{\text{o}}}{\ln 2} \ln\left[\exp\left(\frac{v_{\text{GS}} - v_{\text{T}}}{\eta\, v_{\text{Th}}}\right)\right]^p$$
$$= p \frac{i_{\text{o}}}{\ln 2} \left(\frac{v_{\text{GS}} - v_{\text{T}}}{\eta\, v_{\text{Th}}}\right) \qquad (7)$$

The equation (2) requires a gate-voltage dependent variable p to capture the changes occurring from subthreshold to above-threshold behaviors. Typically, small-signal MOSFETs exhibit an η value ranging from 1 to 2. However, in this analysis, we selected the η value to achieve a more accurate fit. In this context, η is treated as a fitting parameter. Thus, choosing η values greater than 1 is justified to attain a better alignment with the experimental data, particularly given the specific behavior of power MOSFETs.

C. At-threshold

At threshold, when $v_{\text{GS}} = v_T$, $X^p = 1$ for any value of p

$$i_{\text{D}} = \frac{i_{\text{o}}}{\ln 2} \ln[1 + X^p] = i_{\text{o}} \qquad (8)$$

The modified term ($\ln 2$) was added in Eq. (2), so that at threshold condition i_{D} should be equal to i_0.

The crucial aspect of our application revolves around understanding MOSFET characteristics during turn-off, particularly the voltage across and current through the inductor as the MOSFET transitions to its non-conducting state. This necessitates a robust equation that accurately models behavior below, near and above threshold. Through analysis of experimental data, we derived the new equation (Eq. 2) capable of effectively modeling high-power MOSFET behavior, thus addressing the limitations of existing analytical approaches.

By modifying the equation to depend solely on gate voltage, we achieved a successful fit that aligns with all operational regions. This method treats the MOSFET as a variable resistance controlled by gate-source voltage, simplifying analysis compared to switch-based models. It indirectly accounts for the influence of drain-source voltage on resistance, enhancing model accuracy.

IV. Development of Matlab model

The TMS excitation circuit (RLC) is the last stage of the prototype, illustrated in Fig. 1. The objective is to develop a model that solves the differential equation of the RLC circuit. since

$$v_{\text{C}} = v_{\text{L}} + (r_{\text{L}} + r_{\text{C}} + r_{\text{DS}})i_{\text{L}} \qquad (9)$$

979-8-3315-1612-3/25 $31.00 © 2025 IEEE   381

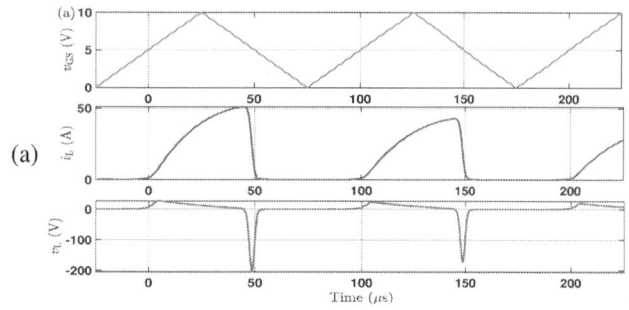

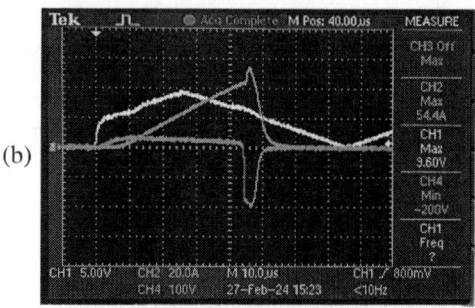

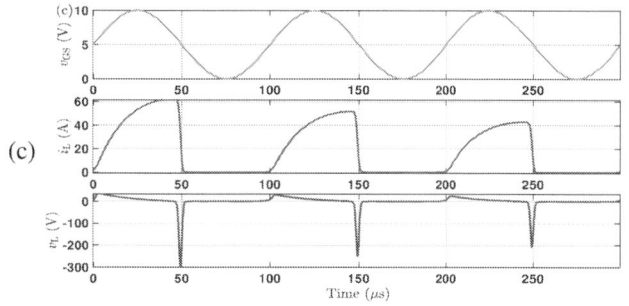

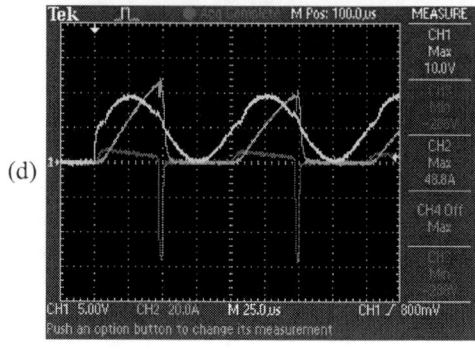

Fig. 5: (a) `ode15s` Solver solution of the differential equation with triangle gate voltage. The TMS voltage pulse has an amplitude of 200 V and current pulse with 50 A, which is almost identical to the oscilloscope output; (b) Oscilloscope measurement of MOSFET gate voltage (yellow), TMS coil voltage (green) and TMS coil current (blue). Gate voltage rise-fall symmetry is set at 50:50 (c) `ode15s` Solver solution of the differential equation with sinusoidal gate voltage input; (d) Oscilloscope measurement of sinusoidal gate voltage (yellow), TMS coil voltage (magenta) and TMS coil current (blue)

where r_L (internal resistance of the TMS coil), r_C (ESR of the capacitor C2), and L (inductance of the TMS coil) are known from experimental measurements. Given the measured values of v_C (C2 capacitor voltage) and v_L (inductor voltage), we can determine $i_L(t)$. Our approach is as follows:

1) Development of the drain-source current equation, $i_{L_{fit}} = i_D$ as a function of v_{GS}: As in Eq. (2).
2) Computing the resistance of the MOSFET: Curve fitting is performed by plotting the measured v_{DS} versus v_{GS} (from Fig. 4) and deriving an equation for $v_{DS_{fit}}$ to fit the data points. Curve fitting involves creating a mathematical function or curve that best approximates the series of data points obtained from plotting v_{DS} versus v_{GS}. With the $v_{DS_{fit}}$ equation in hand, the $r_{DS_{fit}}$ of the MOSFET can be determined by:

$$r_{DS_{fit}} = \frac{v_{DS_{fit}}(v_{GS})}{i_{L_{fit}}(v_{GS})} \qquad (10)$$

where

$$v_{DS_{fit}} = \frac{1}{2}\left(D_1 + D_2 + (D_2 - D_1)\tanh\left(\frac{v_{GS} - v_T}{\eta_1\, v_{TH}}\right)\right) \qquad (11)$$

Equation (2) is modified to read as:

$$i_{L_{fit}} = \frac{i_o}{\ln 2}\ln\left(1 + \left[\exp\left(\frac{v_{GS} - v_T}{\eta_2\, v_{TH}}\right)\right]^{p_{fit}}\right) \qquad (12)$$

where the exponent $p \to p_{fit}$ is now a tanh function of MOSFET gate voltage,

$$p_{fit} = \frac{1}{2}\left(p_1 + p_2 + (p_2 - p_1)\tanh\left(\frac{v_{GS} - v_T}{\eta_3\, v_{TH}}\right)\right) \qquad (13)$$

3) MATLAB model: Now we can use this $r_{DS_{fit}}$ in the differential Equation. (9) as:

$$\frac{di_L}{dt} = \frac{1}{L}\left[v_C - (r_L + r_C + r_{DS_{fit}}(v_{GS}))i_L\right] \qquad (14)$$

and solve it for a given $v_{GS}(t)$ profile. The circuit's performance hinges on the MOSFET's turn-off behavior, and relying solely on the on-resistance provided in datasheets or MOSFET models in simulation software like SPICE is inadequate. This model addresses the gap in the literature where a universal drain-source current equation applicable across all regions is missing, leading to the necessity of developing a new equation.

V. VERIFICATION

The MATLAB model for our supercapacitor-based pulse generator (SCPG) was validated using the `ode15s` solver, aligning closely with experimental data. Figure 5(a) and 5(b) compares simulation and experimental results for a 50% symmetry gate voltage, demonstrating consistency. The MOSFET behavior was accurately captured through curve fitting. A 10 kHz, 10 V sinusoidal gate voltage (Fig. 5(c)) showed simulation results matching prototype tested results based on the oscilloscope outputs (Fig. 5(d)), affirming the model's accuracy in predicting TMS behavior under varied conditions.

VI. CONCLUSION

In conclusion, the developed prototype is a low-cost supercapacitor based pulse generator (SCPG) for TMS with pulse shaping capabilities, achieved without the need for any external snubber or additional wave-shaping circuit. The developed MATLAB model for our SCPG demonstrates robust accuracy in predicting TMS current and voltage waveforms under various input conditions. Through careful validation against experimental data, including sinusoidal and triangular gate voltages, the model reliably captures MOSFET behavior. This tool enhances understanding and optimization of pulse generator performance, offering valuable insights for future research and applications in neurostimulation and related fields.

ACKNOWLEDGMENT

The authors gratefully acknowledge the University of Waikato Doctoral Scholarship for research support, as well as the Neurological Foundation and the Hackett Memorial Trust for providing research funding and a travel grant.

REFERENCES

[1] A. V. Peterchev, R. Jalinous, and S. H. Lisanby, "A transcranial magnetic stimulator inducing near-rectangular pulses with controllable pulse width (cTMS)," *IEEE Transactions on Biomedical Engineering*, vol. 55, no. 1, pp. 257–266, 2007.

[2] A. V. Peterchev, Z.-D. Deng, and S. M. Goetz, "Advances in transcranial magnetic stimulation technology," *Brain stimulation: methodologies and interventions*, pp. 165–189, 2015.

[3] J. Selvaraj, P. Rastogi, N. Prabhu Gaunkar, R. L. Hadimani, and M. Mina, "Transcranial magnetic stimulation: Design of a stimulator and a focused coil for the application of small animals," *IEEE Transactions on Magnetics*, vol. 54, no. 11, pp. 1–5, 2018.

[4] F. A. Khokhar, L. J. Voss, D. A. Steyn-Ross, and M. T. Wilson, "Design and demonstration in vitro of a mouse-specific transcranial magnetic stimulation coil," *IEEE Transactions on Magnetics*, vol. 57, no. 7, pp. 1–11, 2021.

[5] A. D. Tang, A. S. Lowe, A. R. Garrett, R. Woodward, W. Bennett, A. J. Canty, M. I. Garry, M. R. Hinder, J. J. Summers, R. Gersner *et al.*, "Construction and evaluation of rodent-specific rTMS coils," *Frontiers in neural circuits*, vol. 10, p. 47, 2016.

[6] A. V. Peterchev and D. L. Murphy, "Controllable pulse parameter transcranial magnetic stimulator with enhanced pulse shaping," in *2013 6th International IEEE/EMBS Conference on Neural Engineering (NER)*. IEEE, 2013, pp. 121–124.

[7] "Electric fencing basics," am.gallagher.com/en-US/Solutions/Case-Study-Listings/Electric-Fencing-Basics, accessed: 2022-02-15.

[8] J. Mankowski and M. Kristiansen, "A review of short pulse generator technology," *IEEE Transactions on plasma science*, vol. 28, no. 1, pp. 102–108, 2000.

[9] A. Elserougi, S. Ahmed, and A. Massoud, "A boost converter-based ringing circuit with high-voltage gain for unipolar pulse generation," *IEEE Transactions on Dielectrics and Electrical Insulation*, vol. 23, no. 4, pp. 2088–2094, 2016.

[10] X. Zan, D. R. Torres, R. Kheirollahi, X. Lu, S. Zheng, F. Lu, and A.-T. Avestruz, "Medium voltage pulse power generator for accurate current interruption," *IEEE Transactions on Industrial Electronics*, vol. 70, no. 4, pp. 3604–3615, 2023.

[11] S. M. Goetz, M. Pfaeffl, J. Huber, M. Singer, R. Marquardt, and T. Weyh, "Circuit topology and control principle for a first magnetic stimulator with fully controllable waveform," in *2012 Annual International Conference of the IEEE Engineering in Medicine and Biology Society*. IEEE, 2012, pp. 4700–4703.

[12] L. M. Koponen and A. V. Peterchev, "Transcranial magnetic stimulation: principles and applications," *Neural Engineering*, pp. 245–270, 2020.

[13] K. Ali, K. Wendt, M. M. Sorkhabi, M. Benjaber, T. Denison, and D. J. Rogers, "xTMS: A pulse generator for exploring transcranial magnetic stimulation therapies," in *2023 IEEE Applied Power Electronics Conference and Exposition (APEC)*, 2023, pp. 1875–1880.

[14] L. Pang, T. Long, K. He, Y. Huang, and Q. Zhang, "A compact series-connected sic mosfets module and its application in high voltage nanosecond pulse generator," *IEEE Transactions on Industrial Electronics*, vol. 66, no. 12, pp. 9238–9247, 2019.

[15] W. Zeng, C. Yao, S. Dong, Y. Wang, J. Ma, Y. He, and L. Yu, "Self-triggering high-frequency nanosecond pulse generator," *IEEE Transactions on Power Electronics*, vol. 35, no. 8, pp. 8002–8012, 2020.

[16] Y. Liu, R. Fan, X. Zhang, Z. Tu, and J. Zhang, "Bipolar high voltage pulse generator without h-bridge based on cascade of positive and negative marx generators," *IEEE Transactions on Dielectrics and Electrical Insulation*, vol. 26, no. 2, pp. 476–483, 2019.

[17] L. Cheng, K. Mei, Z. Chen, W. Jia, Y. Wang, H. Wang, L. Xie, S. Shen, and W. Ding, "High-voltage repetitive nanosecond pulse generator utilizing power synthesis of modified avalanche transistorized marx circuits," *IEEE Transactions on Instrumentation and Measurement*, vol. 71, pp. 1–16, 2022.

[18] J. Yan, S. Shen, and W. Ding, "High-power nanosecond pulse generators with improved reliability by adopting auxiliary triggering topology," *IEEE Transactions on*

Power Electronics, vol. 35, no. 2, pp. 1353–1364, 2020.

[19] L. Cheng, Z. Chen, H. Wang, F. Guo, G. Wu, L. Xie, J. Xiao, Y. Wang, S. Shen, and W. Ding, "A novel avalanche transistor-based nanosecond pulse generator with a wide working range and high reliability," *IEEE Transactions on Instrumentation and Measurement*, vol. 70, pp. 1–14, 2021.

[20] Y. Zhao, W. Xie, J. Jiang, L. Chen, S. Feng, M. Wang, and Z. Wang, "Replacement of marx generator by tesla transformer for pulsed power system reliability improvement," *IEEE Transactions on Plasma Science*, vol. 47, no. 1, pp. 574–580, 2019.

[21] L. Li, M. Ning, C. Dehuai, L. Lun, K. Qiang, L. Mingjia, C. Yong, and P. Yuan, "Study on double resonant performance of air-core spiral tesla transformer applied in repetitive pulsed operation," *IEEE Transactions on Dielectrics and Electrical Insulation*, vol. 22, no. 4, pp. 1916–1922, 2015.

[22] S. Raju, N. Kularatna, and M. Wilson, "Supercapacitor based adjustable high power pulse generator for medical research applications," in *IECON 2023-49th Annual Conference of the IEEE Industrial Electronics Society*. IEEE, 2023, pp. 1–6.

[23] N. D. Arora, *MOSFET models for VLSI circuit simulation: theory and practice*. Springer Science & Business Media, 2012.

[24] J. Brews, W. Fichtner, E. Nicollian, and S. Sze, "Generalized guide for mosfet miniaturization," in *1979 International Electron Devices Meeting*. IEEE, 1979, pp. 10–13.

[25] Y. Byun, K. Lee, and M. Shur, "Unified charge control model and subthreshold current in heterostructure field-effect transistors," *IEEE Electron Device Letters*, vol. 11, no. 1, pp. 50–53, 1990.

[26] T. Ytterdal, Y. Cheng, and T. A. Fjeldly, *Device modeling for analog and RF CMOS circuit design*. John Wiley & Sons, 2003.

[27] T. A. Fjeldly and M. Shur, "Threshold voltage modeling and the subthreshold regime of operation of short-channel mosfets," *IEEE Transactions on Electron Devices*, vol. 40, no. 1, pp. 137–145, 1993.

[28] T. A. Fjeldly, M. Shur, and T. Ytterdal, *Introduction to device modeling and circuit simulation*. John Wiley & Sons, Inc., 1997.

[29] P. van der Meer, A. van Staveren, and A. H. van Roermund, *Low-power deep sub-micron CMOS logic: sub-threshold current reduction*. Springer Science & Business Media, 2004, vol. 841.

Application of Artificial Intelligence for Modeling SiC Power MOSFETs

Fredo Chavez[1a], Danial Bavi[2a], Sourabh Khandelwal[3a]

[a]Macquarie University, Australia

[1]fredo.chavez@mq.edu.au, [2]danial.bavi@hdr.mq.edu.au, [3]sourabh.khandelwal@mq.edu.au,

Abstract—An accurate and flexible neural network(NN)-based I-V model for SiC MOSFET is created in this paper. The technique starts with the generating TCAD simulation of close to 200 SiC MOSFETs with variations in key physical parameters. Compact model formulations and their parameters are automatically generated from the variations in the TCAD simulations using machine learning techniques. The NN-based compact model is then trained using 78k data points using the TCAD simulated data. The unphysical behavior of NN-based compact models having a negative g_m and g_D, causing convergence issues during circuit simulation is also tackled for the first time. The fully trained NN-based compact model results show good accuracy with the fitting I-V, g_m and g_D for multiple devices. The developed NN-based is then converted to a standard Verilog-A file which can be imported into a circuit simulator.

Keywords — SiC MOSFET, machine learning, neural network (NN), parameter generation

I. Introduction

Silicon Carbide (SiC) power MOSFETs is one of the key technologies for power electronics [1], [2]. To create more powerful and efficient electronic designs, a fast and accurate compact model is needed. Several types of compact models have been developed for SiC MOSFETs [3]–[5] in recent years, including a neural network(NN) model [6] capable of modeling a single device. Recent developments on neural network(NN)-based compact models offer fast development time, excellent accuracy and fast simulation speed [7]–[11], and can greatly improve existing SiC MOSFET models.

The paper expands the capability of using NN for modeling SiC MOSFET. This work presents a technique to develop an accurate and flexible NN-based compact model for SiC MOSFETs. Unlike in [6], the proposed technique automatically creates compact model parameters from the variations in the I-V characteristics capable of modeling different devices after one-time training, avoiding the time-consuming training process for each device. Also, the non-physical behavior of having a negative transconductance (g_m) and output conductance (g_D) produced by NN-based compact models, which can cause convergence issues during circuit simulation, in multiple devices has been tackled for the first time. The developed NN-based compact model is tested to model commercial SiC Power MOSFETs.

The rest of the paper is organized as follows: Section II shows the development of the NN-based SiC compact model, which includes the generation of the dataset, parameter generation, and the training of the NN. Section III analyzes the trained NN, accuracy evaluation, model deployment and testing. Finally, concluding remarks are presented in Section IV.

II. SiC NN-based Compact Model Development

The traditional approach for developing compact models involves mathematical and behavioral modeling of the transistor. This process requires a deep understanding of device physics to account for all physical and non-physical behavior of devices, and can take from several months to years. The use of artificial intelligence, or NN, has significantly reduced the development time for these compact models. NN-based compact models have the advantage of modeling complex non-linear behavior with very high accuracy, and efficient model equations but require a large amount of training data [9]. These advantages of NN-based compact models come at the cost of interpretability as NN models retain their 'black box' nature [12].

The development of the SiC NN-based compact model follows the workflow shown in Figure 1. The presented technique starts with the generation of the training data using TCAD simulation. Then, compact model parameters are generated from the variations in the TCAD devices. Afterwards, a NN is trained from the generated TCAD data and generated compact model parameters. Finally, the trained NN model is converted to a standard Verilog-a file and then tested on the TCAD simulations and commercial SiC MOSFET device.

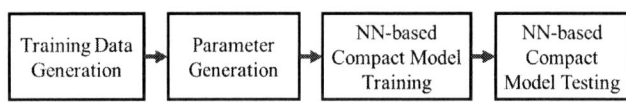

Fig. 1. SiC MOSFET NN-based compact model development workflow.

A. Training Data Generation

A nominal SiC MOSFET TCAD structure is created in Silvaco. Key physical parameters that affect I-V device performance, such as work function, mobility, and oxide thickness, are uniformly varied to create multiple devices with different I-V characteristics. TCAD parameters that may cause a negative g_m and g_D under extreme cases, like self-heating and carrier velocity saturation, are turned off, as non-physical negative g_m and g_D can lead to non-convergence issues. Nearly 200 SiC devices are created from the variations in the physical parameters. For each device, the gate voltage V_G is swept from

979-8-3315-1612-3/25 $31.00 © 2025 IEEE

0 to 18V, with a step size of 1V, while the drain voltage V_D is also swept from 0 to 20V, with a step size of 1V. The drain current I_D simulations comprise the training dataset for the NN-based SiC compact model. The training data is split into 90% training and 10% validation.

B. Parameter Generation

To generate the parameters for every SiC device, we have adapted the technique presented in [11]. Unlike in [11], we included the transconductance g_m and output conductance g_D to improve the fitting accuracy for these electrical characteristics later during the training process. The g_m and g_D for every device are numerically obtained from the TCAD I_D simulations of every device. The I_D, g_m, and g_D are then normalized and then used to fit a *principal component analysis* (PCA) model to generate new parameters for every device, as shown in Figure 2(a). The PCA model is created in using an open-source machine learning library called scikit-learn [13]. The PCA model projects the variation in the normalized I_D, g_m, and g_D data to a lower dimensional space to create the compact model parameters P_n.

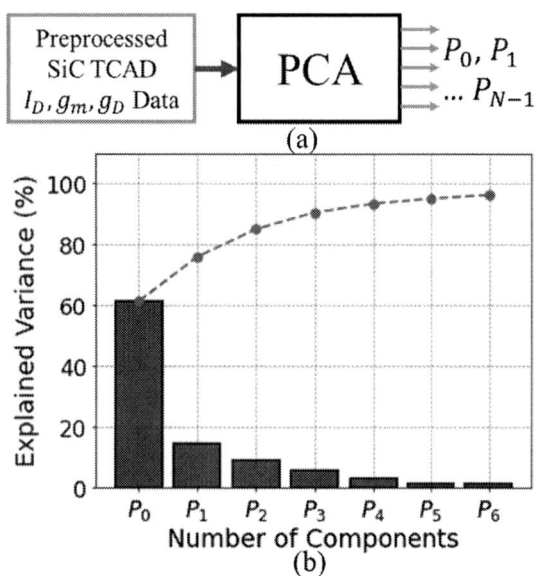

Fig. 2. (a)Automatic parameter generation from variations in SiC TCAD simulation. (b) A Pareto chart showing how each component contributes to the explained variance in the I_D, g_m, and g_D

In PCA model fitting, the total number of parameters generated by PCA N can be controlled. Figure 2 (b) shows a Pareto chart of how each parameter contributes to capturing the variation in the I_D, g_m , and g_D data. For example, P_0 explains nearly 60% of the variance in the data, while P_1 explains nearly 15% of the variance. The higher the N, the higher the explained variance in the dataset, and also leads to higher accuracy later when the NN-based compact model is trained. Higher N is not always desirable, as it can also generate parameters derived from the noise present in the

dataset that can lead to overfitting later during NN training. For this application, we have chosen to keep 7 parameters, which account for 95% of the variation in the I_D, g_m and g_D.

C. Neural-Network Structure and Training

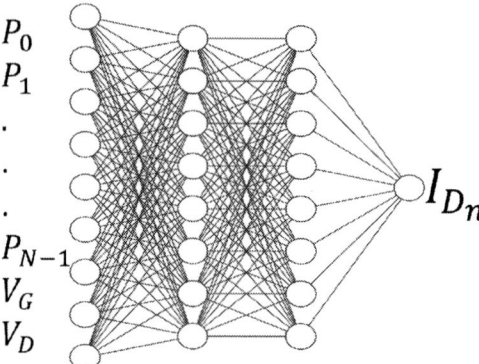

Fig. 3. The structure of the SiC NN-based compact model.

The compact model parameters P_n, voltages V_G and V_D are then used to train the NN-based compact model shown in Fig. 3. Before training, the I_D are normalized to produce I_{D_n}, which is the output of the NN. The structure is *Multilayer Perceptron Neural Network* (MLPNN), which has two hidden layers with 8 neurons each. Each neuron in the hidden layer uses the 'softplus' activation function, as it produces smooth derivatives and can help reduce the occurrence of negative g_m and g_D.

The training of the NN minimizes the loss function, which involves iteration of forward propagation, calculation of the loss function, and backpropagation until the desired loss is achieved. In forward propagation, the input data passes through the layers and neurons of the NN to calculate the predicted normalized drain current $I_{D_n,pred}$. The $I_{D_n,pred}$ and the true normalized drain current $I_{D_n,true}$ are used to calculate the loss function. The loss is then propagated backwards to calculate the gradients of the loss to adjust the weights and biases of every neuron. The optimization algorithm used during training is Adam.

The loss function is defined in (1). The terms in (2) are included to improve the fitting accuracy in I_D, g_m and g_D, and are calculated within the training data available, or within the V_G and V_D bias step of the TCAD simulated data, which is 1V. We have noticed that if we only use the terms (2) for the calculation of the loss function, we were able to model the devices accurately, but the predictions made by the trained NN produces negative g_m and g_D even though this does not occur in the training data. We also found out that common training techniques to further reduce the loss like training the NN longer, using a deeper and more complex NN structure is not enough to remove this issue in the training data. We then introduce the terms in (3) to impose a soft penalty to reduce the occurrence of negative g_m and g_D. The terms in (3) are calculated on a finer V_G and V_D step size of 0.2V to enhance the robustness of the NN at the cost of longer training time.

$$loss = A + B + C + D + E + F + G \qquad (1)$$

$$
\begin{aligned}
A &= w_1 \cdot MSE(I_{D_{n,true}} - I_{D_{n,pred}}) \\
B &= w_2 \cdot MSE(g_{m_{n,true}} - g_{m_{n,pred}}) \\
C &= w_3 \cdot MSE(g_{D_{n,true}} - g_{D_{n,pred}}) \\
D &= w_4 \cdot MSE(\frac{\partial g_{m_{n,true}}}{\partial V_G} - \frac{\partial g_{m_{n,pred}}}{\partial V_G}) \\
E &= w_5 \cdot MSE(\frac{\partial g_{D_{n,true}}}{\partial V_D} - \frac{\partial g_{D_{n,pred}}}{\partial V_D})
\end{aligned}
\qquad (2)
$$

$$
\begin{aligned}
F &= -w_6 \cdot Mean(g_{m_{n,pred,negative}}) \\
G &= -w_7 \cdot Mean(g_{D_{n,pred,negative}})
\end{aligned}
\qquad (3)
$$

The NN is created and trained using an open-source DL library called PyTorch [14]. The initial learning rate is set to 1e-4 but further reduced later during training. The batch size is set to 399, which is the product of the total bias points in V_G and V_D, to allow the calculation of the second-and third-derivatives of I_D for every device during training.

III. RESULTS ANALYSIS

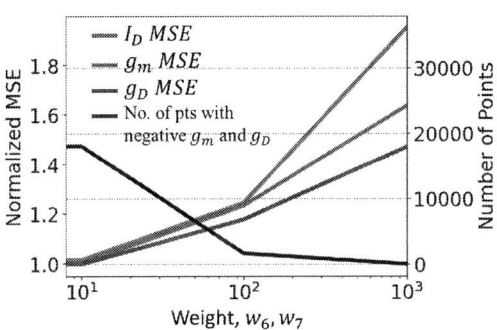

Fig. 4. The I_D, g_m, and g_D MSE analysis when w_6 and w_7 are increased to decrease the occurrence of negative g_m, and g_D to zero.

To prevent convergence issues for all the trained TCAD devices, we experimented with the weights w_6 and w_7 in (3) while holding w_1 to w_5 in (2) to 1. The NN is trained for 20 thousand epochs for different values of w_6 and w_7 and recorded the lowest calculated loss during training. A zero occurrence of negative g_m and g_D is recorded for all of the devices included in the training when w_6 and w_7 are at 1000, as shown in Figure 4. Consequently, the mean-squared error (MSE) in I_D, g_m and g_D can increase by more than 1.5 times when the w_6 and w_7 are set to zero. This shows that the weights in the loss function can be optimized for accuracy while minimizing or eliminating the occurrence of negative g_m and g_D. Furthermore, The trained NN model was able to model all of the devices in the training and validation dataset

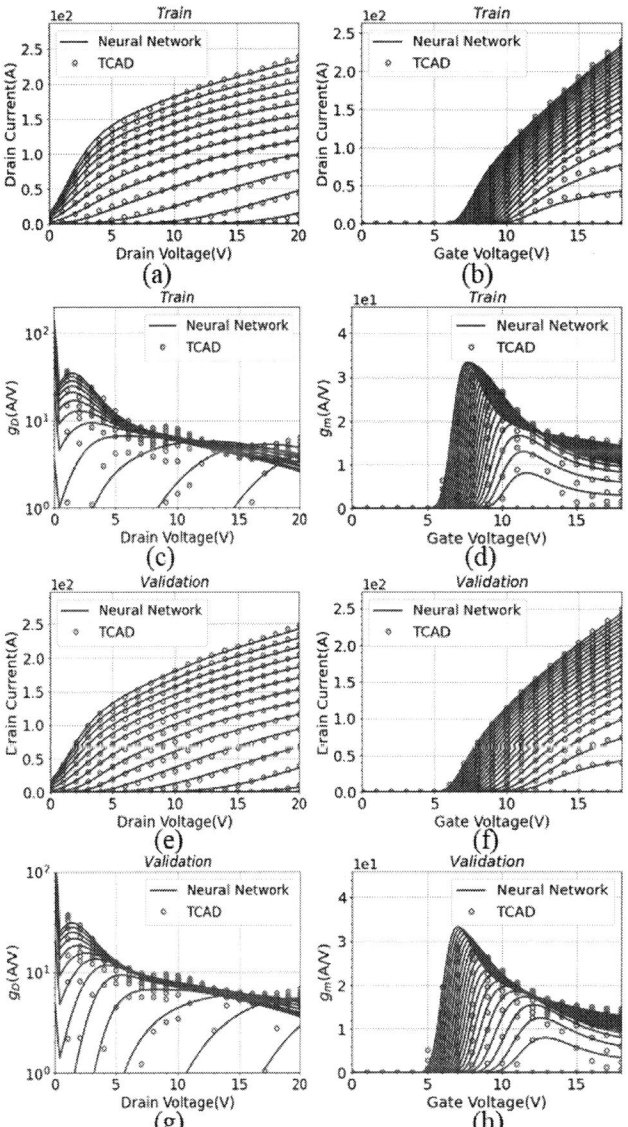

Fig. 5. The fitting of output characteristics, transfer characteristics, g_m, and g_D with TCAD simulated data in the training (a to d) and validation dataset (e to h)

with excellent accuracy. The fitting of output characteristics, transfer characteristics, g_m and g_D in the TCAD simulated data is shown in Figure 5.

After training, the NN is then converted to a standard Verilog-A file that can be imported to a standard circuit simulator. Figure 6 shows the schematic of the imported Verilog-A file in Cadence AWR for I-V simulation. We then tested the NN-based compact model to fit the output characteristics of commercial SiC MOSFET devices. This is done by tuning the generated compact model parameters P_0 to P_6. Figures 7(b) and (c) show a good accuracy in fitting a 750V, 27mΩ, and a 650V, 6.7mΩ SiC MOSFET devices. The accuracy of fitting these devices is mainly limited by the data used during training.

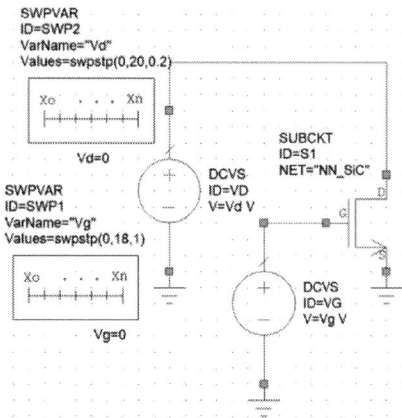

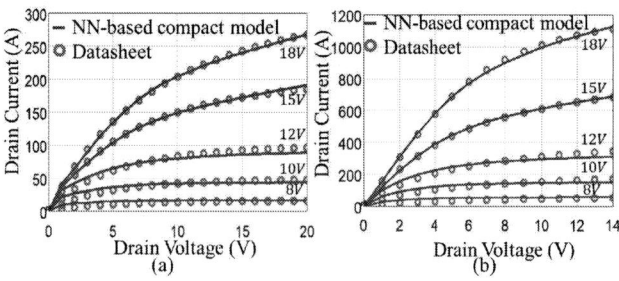

Fig. 6. The NN-based SiC compact model inside a circuit simulator

Fig. 7. A (a) 750V, $27m\Omega$, and a (b) 650V, $6.7m\Omega$ SiC MOSFET is modeled using the developed NN-based compact model.

To further enhance this accuracy, increasing the scope of the possible variations in the I-V characteristics should be done, such as increasing the variations in the physical parameters in TCAD simulation, increasing the sweep of V_G and V_D bias points, and increasing the total number of devices included during training. The limits of the accuracy of this approach have yet to be explored.

This work has shown the promising capability of creating a flexible and accurate NN-based compact model for SiC MOSFET. The loss function is redefined and new terms were introduced to account for the non-physical behavior of multiple devices in NN-based compact models. Even though the presented technique is demonstrated for SiC MOSFET, it can also be applied to other types of devices. The development of the CV model and switching tests of the NN-based compact model will be investigated in the future.

IV. Conclusion

An NN-based compact model for SiC MOSFET is created in this paper. The proposed technique can create a flexible and accurate NN-based compact model for different SiC MOSFET devices. The unphysical behavior in the NN-based compact model, and the issue of convergence due to negative g_m and g_D for multiple devices have been tackled for the first time. The modified loss function can prevent the non-physical negative g_m and g_D for all devices at a relatively small cost of fitting

accuracy. The trained NN can be converted to a standard Verilog-A file that can be imported to a standard simulator.

References

[1] S. Hazra, A. De, L. Cheng, J. Palmour, M. Schupbach, B. A. Hull, S. Allen, and S. Bhattacharya, "High switching performance of 1700-v, 50-a sic power mosfet over si igbt/bimosfet for advanced power conversion applications," *IEEE transactions on Power Electronics*, vol. 31, no. 7, pp. 4742–4754, 2015.

[2] B. Shi, A. I. Ramones, Y. Liu, H. Wang, Y. Li, S. Pischinger, and J. Andert, "A review of silicon carbide mosfets in electrified vehicles: Application, challenges, and future development," *IET Power Electronics*, vol. 16, no. 12, pp. 2103–2120, 2023.

[3] Y. Mukunoki, K. Konno, T. Matsuo, T. Horiguchi, A. Nishizawa, M. Kuzumoto, M. Hagiwara, and H. Akagi, "An improved compact model for a silicon-carbide mosfet and its application to accurate circuit simulation," *IEEE Transactions on Power Electronics*, vol. 33, no. 11, pp. 9834–9842, 2018.

[4] A. U. Rashid, M. M. Hossain, A. I. Emon, and H. A. Mantooth, "Datasheet-driven compact model of silicon carbide power mosfet including third-quadrant behavior," *IEEE Transactions on Power Electronics*, vol. 36, no. 10, pp. 11 748–11 762, 2021.

[5] A. S. Kashyap, C.-P. Chen, and V. Tilak, "Compact modeling of silicon carbide lateral mosfets for extreme environment integrated circuits," in *2011 International Semiconductor Device Research Symposium (ISDRS)*. IEEE, 2011, pp. 1–2.

[6] Y. H. Lee, M. Zhang, P. J. Niu, P. F. Ning, L. Liu, and S. S. Lee, "Simplified silicon carbide mosfet model based on neural network," in *Materials Science Forum*, vol. 954. Trans Tech Publ, 2019, pp. 163–169.

[7] C.-T. Tung and C. Hu, "Neural network-based bsim transistor model framework: Currents, charges, variability, and circuit simulation," *IEEE Transactions on Electron Devices*, vol. 70, no. 4, pp. 2157–2160, 2023.

[8] W. Dai, Y. Li, B. Peng, L. Zhang, R. Wang, and R. Huang, "Benchmarking artificial neural network models for design technology co-optimization," in *2023 International Symposium of Electronics Design Automation (ISEDA)*. IEEE, 2023, pp. 423–427.

[9] J. Wang, Y.-H. Kim, J. Ryu, C. Jeong, W. Choi, and D. Kim, "Artificial neural network-based compact modeling methodology for advanced transistors," *IEEE Transactions on Electron Devices*, vol. 68, no. 3, pp. 1318–1325, 2021.

[10] L. Zhang and M. Chan, "Artificial neural network design for compact modeling of generic transistors," *Journal of Computational Electronics*, vol. 16, pp. 825–832, 2017.

[11] F. Chavez, D. Bavi, N. C. Miller, and S. Khandelwal, "A neural network-based manufacturing variability modeling of gan hemts," in *2024 IEEE 36th International Conference on Microelectronic Test Structures (ICMTS)*. IEEE, 2024, pp. 1–4.

[12] C. Park, H. Cho, and J. Lee, "Enhancing interpretability of neural compact models: Towards reliable device modeling," *IEEE Journal of the Electron Devices Society*, 2024.

[13] F. Pedregosa, G. Varoquaux, A. Gramfort, V. Michel, B. Thirion, O. Grisel, M. Blondel, P. Prettenhofer, R. Weiss, V. Dubourg, J. Vanderplas, A. Passos, D. Cournapeau, M. Brucher, M. Perrot, and E. Duchesnay, "Scikit-learn: Machine learning in Python," *Journal of Machine Learning Research*, vol. 12, pp. 2825–2830, 2011.

[14] A. Paszke, S. Gross, F. Massa, A. Lerer, J. Bradbury, G. Chanan, T. Killeen, Z. Lin, N. Gimelshein, L. Antiga, A. Desmaison, A. Kopf, E. Yang, Z. DeVito, M. Raison, A. Tejani, S. Chilamkurthy, B. Steiner, L. Fang, J. Bai, and S. Chintala, "Pytorch: An imperative style, high-performance deep learning library," 2019, version 1.3.1, Softplus activation available as part of the nn module. [Online]. Available: https://pytorch.org/

Multi-Objective Design Automation in Power Electronics Using Bayesian Optimization Techniques

Tung-Tan Nguyen
Department of Electronics and
Computer Engineering
National Taiwan University of
Science and Technology
Taipei, Taiwan
nttung.ntust@gmail.com

Man-Hay Pong
Department of Electronics and
Computer Engineering
National Taiwan University of
Science and Technology
Taipei, Taiwan
mhp@mail.ntust.edu.tw

Huang-Jen Chiu
Department of Electronics and
Computer Engineering
National Taiwan University of
Science and Technology
Taipei, Taiwan
hjchiu@mail.ntust.edu.tw

Abstract— **This paper presents a design automation framework for power converters, integrating finite element analysis (FEA) simulations with Bayesian optimization (BO) to enhance design precision and decision-making. The methodology conducts multi-objective optimization, balancing efficiency and cost, while accounting for discrete design variables such as specific semiconductor device options. Detailed FEA simulations are embedded within the optimization loop to accurately model and refine magnetic components, ensuring practical and high-performance outcomes. Engineers are provided with a Pareto front (PF) of optimized solutions, enabling an objective selection of the most suitable design. The approach is validated through a 10-kW interleaved boost converter (IBC), demonstrating its capability to deliver cost-effective, high-efficiency solutions tailored for advanced power converter converters.**

Keywords—Bayesian optimization, design automation, FEA, IBC, multi-objective design.

I. INTRODUCTION

The increasing necessity for clean energy has led to significant advancements in photovoltaic (PV) systems [1]. This trend has heightened the demand for design automation of PV converters, aiming to reduce time-to-market and development costs. Using artificial intelligence (AI) or machine learning (ML) optimization to relieve design engineers from repetitive trial-and-error design steps is highly desirable for achieving optimal efficiency and performance. Additionally, calculating accurate losses in magnetic components requires detailed simulations with FEA software like ANSYS, which can be time-consuming and expensive. Therefore, minimizing the number of iterations is crucial for the FEA-included optimization process. Our prior work [2], [3] has successfully employed single-objective BO combined with FEA simulation to enhance efficiency. In practice, multiple objectives, such as efficiency and cost, need to be considered. To address this, a Pareto Front (PF) should be generated to enable engineers to make informed decisions. Traditional methods for creating a PF through parameter sweeps demand substantial computing resources. This paper addresses this issue by proposing a design automation routine that uses multi-objective BO to efficiently generate PF.

This work is supported by the National Science and Technology Council in Taiwan with project number: NSTC 113-2221-E-011 -108 -MY2.

In [4], the non-dominated sorting genetic algorithm II (NSGA-II) addresses design optimization challenges by balancing efficiency, cost, and reliability across various design variables but struggles with complex PFs. Approach [5] extends this to multi-converter optimization for PV systems, it employs an idealized waveform and performs parameter sweeps; however, this approach neglects the influence of various design inputs on the waveform, which can ultimately impact the final outcome. The grid elitist multi-objective genetic algorithm (GEMOGA) in [6] addresses NSGA-II shortcomings but relies on single-condition modeling, leading to potential inaccuracies, it also faces high computational demand because of numerous optimization iterations. This paper builds on the multi-objective BO process from [7], combining multi-objective BO and FEA simulations to overcome the mentioned challenges. It accelerates prototyping while balancing multiple objectives, the effectiveness of method is demonstrated through a 10kW dc-dc stage in a PV inverter system in Fig. 1 (a).

II. INTERLEAVED BOOST CONVERTER

A. Overview of Interleaved Boost Converter

The DC-DC boost converter is favored for its high efficiency and simplicity in PV systems, but it struggles with

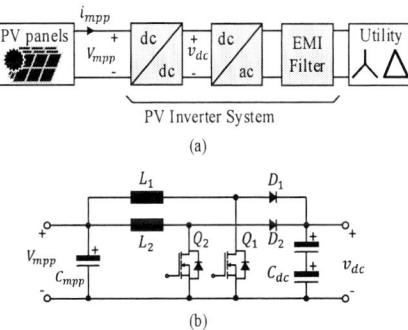

Fig. 1. The PV inverter system and a typical dc-dc IBC topology. (a) The PV inverter system which is connected to PV panels at input and directly connected to utility at output. (b) the two-phase IBC topology as a typical dc-dc stage in Fig 1 (a).

TABLE I. IBC Specification and Constraints

Parameters	Value	Unit
MPP input voltage (V_{mpp})	200-500; nominal 300	V
IBC output voltage (v_{dc})	350-800; nominal 600	V
Rated output power (P_{out})	10	kW
Inductor current ripple max ($\Delta I_{L,max}$)	50	%

TABLE II. CEC Eficiency's Weighted Factors

Parameters	Description	Value
w_1	Weighted factor of 10% load condition	0.04
w_2	Weighted factor of 20% load condition	0.05
w_3	Weighted factor of 30% load condition	0.12
w_4	Weighted factor of 50% load condition	0.21
w_5	Weighted factor of 75% load condition	0.53
w_6	Weighted factor of 100% load condition	0.05

high input current and output voltage ripples. Interleaved boost converters (IBC), depicted in Fig. 1 (b), address these issues by distributing the current across multiple paths, thereby reducing ripple and the size of passive components [8], [9]. The two-phase IBC employs the 180-degree phase shift operation, which helps to lower conduction losses and boost overall efficiency. These characteristics make IBCs highly effective, offering reduced switching losses, and superior performance compared to conventional boost converters. In this work, high-power IBC prototypes are built using the proposed design automation method.

B. IBC Specifications

The IBC serves two crucial functions: implementing the maximum power point tracking (MPPT) for PV inverters and enhancing the string voltage to a higher level. In terms of system flexibility, the maximum power point (MPP) input voltage typically varies across a wide range, enabling the PV inverter to adjust to string voltage variations. Table I illustrates that the selected MPP voltage for the IBC ranges from 200 V to 500 V. A nominal value, $v_{mpp(nom)}$, of 300 V, is suitable as an input voltage for a standard string configuration in residential PV systems. Regarding variable sunlight intensity, assuming the MPPT function operates efficiently, the MPP voltage usually fluctuates within a narrow range for a fixed string configuration. However, the MPP current significantly depends on irradiance conditions. Therefore, this paper designs the IBC optimally at the nominal MPP voltage, $v_{mpp(nom)}$, while also taking different load conditions into account.

Table I also shows that the IBC output voltage (v_{dc}) ranges from 350 V to 800 V, which corresponds to the dc-link voltage of PV inverters. This wide range of dc-link voltage makes it possible for inverters to be modulated corresponding to different grid systems, such as three-phase 120/208 Vac, 220/380 Vac, or even 277/480 Vac systems. For optimization, the paper selects the three-phase 220/380 Vac as the typical grid system. Thus, the nominal output voltage of the IBC, $v_{dc(nom)}$, is typically set with a buffer of 10% to 15% above the peak line-to-line RMS voltage, $V_{LL(rms)}$, and it can be calculated as follows:

$$v_{dc(nom)} = V_{LL(rms)} \times \sqrt{2} \times [1.10 - 1.15] \qquad (1)$$

In this paper, $v_{dc(nom)}$ of 600 V is selected to be the operating point for IBC design optimization. Henceforth, values of 300V input and 600V output voltages are the operating condition to be optimized for IBC converter and the loading conditions are described more detailed in Section III.

III. Proposed Design Automation Routine

A. Objective functions

A multi-objective optimization problem [7] can be defined by a pair (Ω, f), where $\Omega \subseteq \mathbb{R}^n$ represents the search space and n denotes its dimensionality. The function f is an m-dimensional real-valued vector function over Ω, expressed as f: $\Omega \to \mathbb{Y}$, where $\mathbb{Y} \subseteq \mathbb{R}^m$ is the objective space. The n components of x $\in \Omega$ are known as design variables. Solving the multi-objective optimization problem involves finding a set of values $x^* \subset \Omega$, known as the Pareto-front (PF) set. In practice, the objectives often conflict with each other, so the PF signifies a compromise among these conflicting objectives. As efficiency and cost of converter are always important objectives, this paper aims to identify optimal solutions that balance the California Energy Commission (CEC) efficiency [10] and the converter cost of an IBC. Thus, the objective function is defined as f = [η_{CEC}; C_{conv}].

$$\eta_{CEC} = \sum_{i=1}^{6} w_i \eta_i \; ; \; C_{conv} = \sum_{j=1}^{k} q_j c_j \qquad (2)$$

Where η_{CEC} is the CEC efficiency, w_i and η_i are the weighted factors (Table II) and efficiencies at various conditions respectively; C_{conv} is the total cost of IBC, q_j and c_j are the quantity and cost of each kind of component in IBC respectively.

B. Design variables

The selection of design variables is a critical step in any power converter design optimization problem. However, the development of material technologies presents a vast array of options for power components, such as switching devices, inductor cores, capacitors, and more. Consequently, designers often face challenges in selecting the right components for power converter design, involving many trial-and-error steps. Therefore, this paper incorporates component selection into the design automation process to help designers overcome these challenges.

The switches devices including SiC MOSFETs, Diodes and inductors' core materials of IBC are treated as local design variables that obviously do not affect the waveforms of converter. Moreover, various capacitors are also considered among the design variables due to their considerable cost implications. A set of components, Table III, is selected as the search domain according to the IBC specification data. This method guarantees that the search domain effectively covers the converter specifications, while preventing the inclusion of overly broad and unnecessary component options.

979-8-3315-1612-3/25 $31.00 © 2025 IEEE

TABLE III. COMPONENT SELECTION

SiC MOSFETs	Cost (p.u*)	$R_{ds(on)}$ (mΩ)	C_{oss} (pF)	---
1. IMZA120R007M1H	1.823	7.00	420	---
2. IMZ120R030M1H	0.473	30.0	116	---
3. C3M0016120K	2.575	16.0	230	---
4. SCTWA70N120G2V-4	0.904	30.0	176	---
5. C3M0040120K	0.613	53.5	103	---
6. SCTWA60N120G2-4	0.797	52.0	113	---

SiC diodes	Cost (p.u*)	V_F** (V)	R_d** (mΩ)	---
1. C4D20120D	0.462	1.6	75.00	---
2. STPSC30H12C	0.282	1.55	42.92	---
3. C4D30120D	0.693	1.73	45.46	---
4. IDW40G120C5B	0.400	1.47	24.53	---
5. IDW30G120C5B	0.322	1.47	32.62	---
6. STPSC40H12C	0.434	1.55	32.60	---

Inductor cores	Cost (p.u*)	k_c**	α**	β**
1. KAM184-075A	4	10.718	1.377	2.239
2. KH184-060A	5.2	02.748	1.485	2.169
3. KS184-075A-HF	1.2	11.758	1.378	2.078
4. KNF184-075A	2	15.984	1.407	2.159

Electrolytic capacitors	Cost (p.u*)	Capacitor -C (μF)	ESR (mΩ)	---
1. MAL229960471E3	0.858	470	300	---
2. ALH82A471DL550	0.309	470	261	---
3. EKMS551VSN471MA60S	0.496	470	564	---
4. ALA7DA471EE550	0.323	470	375	---

Notes:

* 1 p.u = 1000 NTD (the cost based on online sources and can be varied by the time being)

** V_F, R_d is diode forward voltage and series resistance respectively; k_c, α, β represents inductor core loss density data $p_v = k_c f^\alpha B^\beta$

Consequently, the search domain remains manageable and focused on the optimal choices. In Table III, key parameters of selected components are extracted from datasheets, and as cost is one objective to optimize, the unit costs are also listed with data from the same online site.

Additionally, to do more comprehensive optimization and ensure whole converter high performance, the switching frequency, f_{sw}, and the inductance, L, will be used as the global design variables which influence the converter's waveforms. The f_{sw} has been selected from 40 kHz to 120 kHz, offers a wide range of switching frequency for both switching losses and inductor losses trade-off consideration. Moreover, once f_{sw} range is established, the inductance range can be derived based on the inductor current ripple constraint outlined in Table I. For reducing searching domain while keep details of each design options, the switching frequency range and inductance range are divided into different options by step sizes; each f_{sw} step size is set at 16 kHz while each L step size is set at 50 μH, therefore the f_{sw} and L options are listed in (3) and (4) as follows:

$$f_{sw} = [40 \text{ kHz, } 56 \text{ kHz, } ..., 120 \text{ kHz}] \quad (3)$$

$$L = [180 \text{ μH, } 230 \text{ μH, } ..., 730 \text{ μH }] \quad (4)$$

Finally, the comprehensive design variable vector which includes both electronic components – local design variables and key design parameters – global design variables for IBC can be expressed in (5) as follows:

$$x = [\text{switch; diode; capacitor; core type; L; } f_{sw}] \quad (5)$$

It is important to emphasize that the paper introduces the design automation method aimed at effectively producing a set of design vectors x that balance the trade-off between CEC efficiency and cost of IBC.

C. Bayesian optimization

Bayesian optimization (BO) is a method for efficiently finding the best solution to complex problems where each observation is costly [11]. It uses a probabilistic model, typically

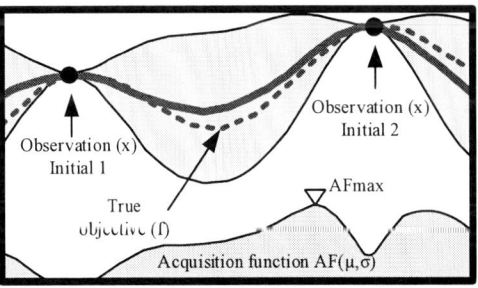

a) 1st iteration

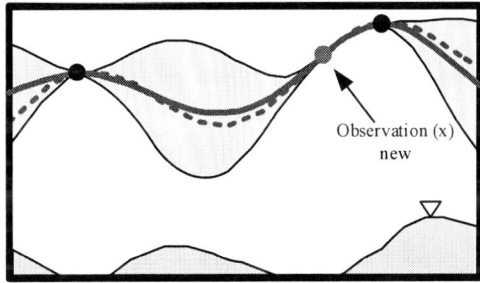

b) 2nd iteration

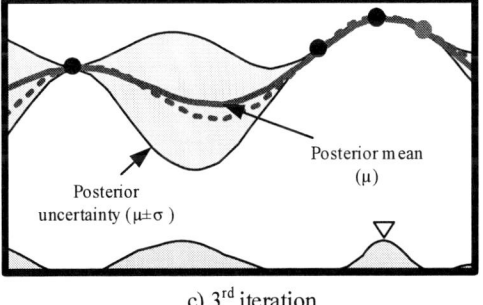

c) 3rd iteration

Fig. 2. The illustration of Bayesian optimization process a) the 1st iteration starts with two initial observations, the probabilistic model has been built, then provides mean and uncertainty, acquisition function is calculated to provide the next observation point (at AFmax). b), c) are the 2nd, 3rd iterations respectively. Observation data increases, curve fitting get improved, posterior mean is closer to true objective.

Gaussian process – GP, to predict the performance of different solutions and includes an acquisition function (AF) to decide which solution to try next in Fig. 2. The AF balances exploring new possibilities and exploiting known good one, helping to quickly find high-quality solutions with minimal observations.

However, one of the challenge when using BO is that GP cannot directly handle discrete electronic components. To address this, a new design variable vector, x_p, is proposed in (6), which includes key parameters for each component rather than the components themselves. This approach enables the GP to use Euclidean distance-based kernel function [12] to calculate the correlation of any pair of design variable vectors and then predict the posterior mean and uncertainty, and finally determine new potential observation point based on maximum of AF for next iteration.

$$x_p = [R_{ds(on)}; C_{oss}; V_F; R_d; C; ESR; k_c; \alpha; \beta; L; f_{sw}] \quad (6)$$

The typical kernel function, Matérn kernel, is applied in this paper. For example, when smoothness parameter $\vartheta = 2.5$ and the characteristic length-scale $l = 1$, the Matérn kernel is expressed as follows:

$$k_{\vartheta=2.5, l=1}(x, x') = \left(1 + \sqrt{5}r + \frac{5}{3}r^2\right) e^{-\sqrt{5}r} \quad (7)$$

where $r^2 = (x - x')\text{diag}(\theta)(x - x')^T$, and $\text{diag}(\theta)$ is a diagonal matrix with positive length-scale vector θ as diagonal values.

The value r denotes the distance between two vectors, scaled by a positive length-scale vector θ. The intuition behind this is two points, x and x', are considered more similar when smaller r value observed then results in a larger covariance. The positive length-scale vector θ is hyperparameter that can be optimized during fitting process by maximizing the log-likelihood [13].

In [7], multi-objective BO had been proposed and its effectiveness was also proven through four benchmark functions, the method quickly converged to pareto-front set within only 30 iterations. Leveraging the less iteration benefit, both circuit and FEA simulation can be incorporated into optimization process for overcoming both efficient losses estimation and compromise solutions searching challenges.

D. Design automation routine

The new multi-objective BO-based design automation routine has been proposed in Fig. 3. There are two key loops in the flowchart:

- The inner loop utilizes both circuit and Ansys simulations to determine IBC efficiencies under various loads, based on the CEC efficiency loading conditions outlined in Table II. The circuit simulation is responsible

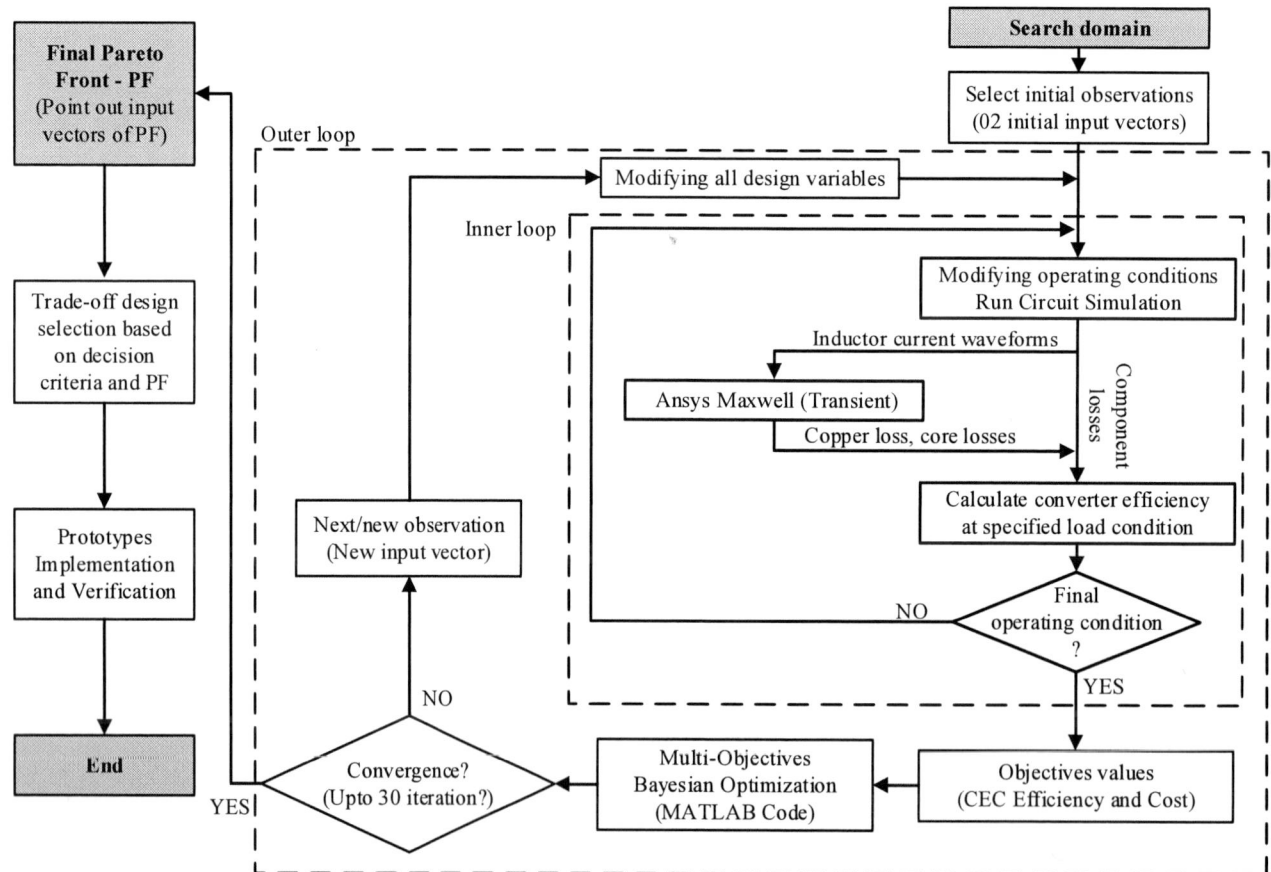

Fig. 3. Multi-objective Bayesian optimization-based design automation routine

for providing IBC component losses, excluding copper and inductor core losses. Ansys Maxwell 3D transient simulation has been used to estimate the total inductor losses by applying excitation signals from the circuit simulation. This method offers more practical results compared to theoretical calculations. The inner loop is repeated for each CEC efficiency loading condition. Therefore, the IBC efficiencies at specific operating conditions can be determined.

- The core of the outer loop is the multi-objective BO algorithm implemented in MATLAB, responsible for suggesting the next potential observation point based on GP results of CEC efficiencies and IBC costs from previous observations. Once the next observation point is determined, the outer loop modifies all design parameters to feed into the inner loop. This routine is repeated for 30 iterations to find the optimal Pareto front set, exploring total cases (vectors) of design variables.

It is worth to emphasize that the routine starts with two initial observations, x_{int1} and x_{int2} in Table IV, selected randomly from search domain, the initial function evaluation is vital for starting BO so that inner loop is used for achieving efficiencies of these initial observations. Once convergence or the maximum number of iterations is reached, the final PF set is easily generated. Based on the decision criteria of designer, several potential design options could be selected for final implementation and verification.

IV. OPTIMIZATION RESULT AND VERIFICATION

The proposed design automation routine quickly converges to the Pareto front set which shows the trade-off between CEC efficiency and cost in Fig. 4 (a). In this paper, the optimal solution point, red square – "1", has been selected because of two following reasons:

- The PF set is categorized into two regions. The first region incorporates IBC costs lower than 9.2 p.u., while the second region includes IBC costs exceeding 10 p.u. A notable increase in efficiency data is observed from the first region to the second region. Therefore, selecting a design option within the second region is advantageous for achieving higher converter efficiency.

- In the second region, the PF trend reveals a relatively flat slope in terms of CEC efficiency despite rising costs. This suggests that the efficiency gains of the converter are insignificant with increasing expenses. Consequently, opting for a design with higher costs in this region is not justifiable.

The Fig. 4 shows the measured data of selected prototype's efficiency which illustrate that measured data are very close to estimated results from design automation routine. However, additional prototype corresponding with black square - "2" or verified point in Fig. 4 (a) has been built for more measurement verification; the measured efficiencies depicted as green stars match well with PF values. Table IV summarizes that the measured CEC efficiency and estimated CEC efficiency of selected prototype are 98.95 % and 99.10 % respectively while the measured CEC efficiency and estimated CEC efficiency of

TABLE IV. INITIAL OBSERVATIONS AND SELECTED DESIGN OPTION

Domain	Parameter	x_{init1}	x_{init2}	$x_{selected}$	$x_{verified}$
Design space	MOSFET's index	2	3	5	5
	Diode's index	1	5	6	2
	Capacitor's index	1	4	3	3
	Core's index	2	1	3	3
	Inductance (µH)	280	680	380	530
	Frequency (kHz)	72	120	40	104
Objective space	Estimated cost (p.u)	---	---	10.33	09.15
	Real cost (p.u)	22.57	17.67	10.44	08.89
	Cost error (%)	---	---	-1.05	+2.84
	Estimated CEC eff (%)	98.73	98.83	99.10	98.54
	Measured CEC eff (%)	---	---	98.85	98.58
	Efficiency error (%)	---	---	+0.25	-0.04

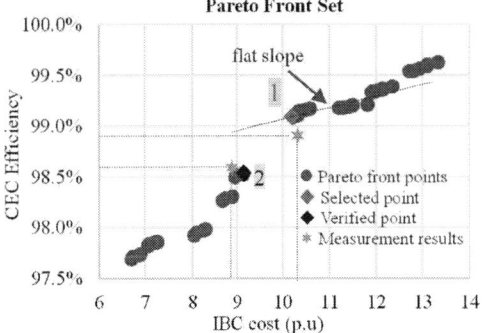

(a) PF set obtained by design automation routine

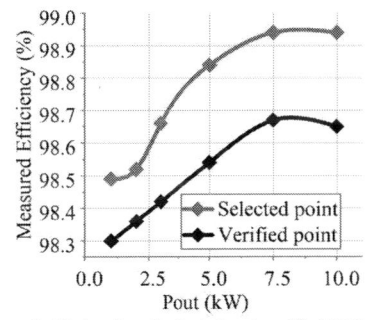

(b) Measured efficiencies of selected and verified IBC prototypes

(c) Efficiency measurement of selected prototype at different CEC efficiency's load conditions

Fig. 4. The automation design optimization results and efficiency measurement data.

verified prototype are 95.58 % and 98.54 % respectively. Furthermore, the real IBC cost and estimated IBC cost of selected prototype are 10.44 p.u and 10.33 p.u respectively while the real IBC cost and estimated IBC cost of verified prototype are 8.89 p.u and 9.15 p.u respectively. The results reports that the surrogate model by GP has good accuracy, low error on both cost and efficiency estimation.

Consequently, this work effectively minimizes the number of required prototypes and accelerates the prototyping process. The entire design process takes 45 computing hours, 42% of the time goes to ANSYS and circuit simulation, the remaining 58% to MATLAB optimization and establishment of Pareto Front.

V. CONCLUSIONS

The paper successfully introduces the BO-based and FEA-based design-automation method to achieve compromise solutions for multi-objective problems. The approach efficiently balances the CEC efficiency and cost for an IBC. The optimization process incorporates the selection of semiconductor devices and magnetic cores, effectively addressing the challenges posed by discrete input variables. Two 10-kW IBC prototypes from Parato front based on engineer's selecting criteria have been built for verification after 30 iterations and 45 computing hours of optimizing process. This design approach could be easily extended and applied to various power electronics converters in the same manner as described in this work.

REFERENCES

[1] F. Blaabjerg, D. M. Lonel, "Renewable Energy Devices and Systems with Simulation in MATLAB and ANSYS," *CRC press, Taylor & Francis group,* 2017

[2] N. T. Tung, M. -H. Pong and H. -J. Chiu, "Magnetics Design Optimization for LLC Converter employing Machine Learning," *2023 11th International Conference on Power Electronics and ECCE Asia (ICPE 2023 - ECCE Asia),* Jeju Island, Korea, Republic of, 2023, pp. 1219-1224.

[3] N. T. Tung, M. -H. Pong and H. -J. Chiu, "Design Automation of Power Converters by Machine Learning with Prior Parameter Boundaries," *The 44th Republic of China Electric Power Engineering Symposium and the 20th Taiwan Power Electronics Symposium,* Taipei, Taiwan, 2023.

[4] G. Adinol, G. Graditi, P. Siano, and A. Piccolo, "Multiobjective optimal design of photovoltaic synchronous boost converters assessing efficiency, reliability, and cost savings," *IEEE Trans. Ind. Information.,* vol. 11, no. 5, pp. 10381048, Oct. 2015. doi: 10.1109/TII.2015.2462805.

[5] R. M. Burkart and J. W. Kolar, "Comparative life cycle cost analysis of Si and SiC PV converter systems based on advanced η-ρ-σ multiobjective optimization techniques," *IEEE Trans. Power Electron.,* vol. 32, no. 6, pp. 4344–4358, Jun. 2017.

[6] T. Delaforge and S. Mariethoz, "Design Automation of Power Electronic Converters a Grid Elitist Multiobjective Genetic Algorithm," in *Proc. IEEE Appl. Power Electron. Conf. Expo.,* 2020, pp. 2892–2899.

[7] P. P. Galuzio, E. H. de Vasconcelos Segundo, L. dos Santos Coelho, and V. C. Mariani, "MOBOpt — multi objective Bayesian optimization," *SoftwareX,* vol. 12, p. 100520, 2020, ISSN: 2352-7110. DOI: https://doi.org/10.1016/j.softx.2020.100520. [Online]. Available: http://www.sciencedirect.com/science/article/pii/ S2352711020300911.

[8] S. K. Singh, A. Haque, "Performance evaluation of MPPT using boost converters for solar photovoltaic system," *2015 Annual IEEE India Conference (INDICON),* DOI: 10.1109/INDICON.2015.7443516.

[9] A. R. Krishnan, S. Mohammed S and S. Manafudeen, "Comparison of P&O MPPT Based Solar PV System with Interleaved Boost Converter," *2019 2nd International Conference on Intelligent Computing, Instrumentation and Control Technologies (ICICICT),* 2019, pp. 1370-1376.

[10] "European or CEC Efficiency,", [online] Available: https://www.pvsyst.com/help/inverter_euroeff.htm.

[11] P. I. Frazier, "A tutorial on Bayesian optimization", *arXiv:1807.02811,* 2018, [online] Available: https://arxiv.org/abs/1807.02811.

[12] C. E. Rasmussen & C. K. I. Williams, "Gaussian Processes for Machine Learning," the MIT Press, 2006, ISBN 026218253X, 2006 Massachusetts Institute of Technology, [online] Available: www.GaussianProcess.org/gpml.

[13] B. Shahriari, K. Swersky, Z. Wang, R. P. Adams, and N. de Freitas,"Taking the human out of the loop: A review of Bayesian optimization," *Proc. IEEE,* vol. 104, no. 1, pp. 148–175, Jan. 2016.

Reduced Order Thermal Modelling of Multi-chip Silicon Carbide Power Modules

Aamir Rafiq
Wolfspeed, Inc, UK
aamir.rafiq@wolfspeed.com

Blake Nelson
Wolfspeed, Inc, Fayetteville, USA
blake.nelson@wolfspeed.com

Marshal Olimmah
Wolfspeed, Inc, Fayetteville, USA
marshal.olimmah@wolfspeed.com

Abstract— **Multi-chip silicon carbide (SiC) power modules are gaining in popularity due to high current demand in automotive applications. To ensure reliable operation of the power module, thermal coupling between adjacent dies needs to be considered when estimating junction temperature of the power device. Finite-element (FE) simulations can accurately estimate the effect of these mutual thermal interactions on the junction temperature of the power device. However, the increased complexity and computational cost of FE modelling makes it unsuitable for power converter simulations needing prolonged runtime. This paper presents a reduced-order modelling (ROM) approach for thermal analysis of power modules based on vector fitting in ANSYS Twin Builder. The model is validated through static as well as dynamic measurements on the power module. Lastly, the efficacy of the model to simulate mutual thermal coupling effects is demonstrated with PLECS simulations.**

Keywords—silicon-carbide, thermal modelling, reduced order model.

I. Introduction

Increasing demand for high current SiC power modules with ultra-low parasitic inductance necessitates packaging of more dies within close proximity inside the power module. This causes thermal interactions among the dies and, therefore, undesired increase in the junction temperature of the power device [1].

Traditional Cauer network based thermal models can be modified to include thermal coupling effects [1-3]. Li et. al [1] demonstrate a thermal impedance matrix approach for modelling the thermal interactions between adjacent dies in a power module. The non-diagonal elements of the thermal impedance matrix represent the thermal coupling between the dies, and this effect is modeled in the cauer network by including cross coupled thermal impedances. Ref [2],[3] demonstrate similar approaches for modelling the thermal interactions between adjacent dies. However, all these techniques require modelling of additional thermal impedance paths in the network, thereby increasing the complexity of the modelling process. Moreover, the additional impedance paths rely on the specific module geometry and need to be reworked for different module variants.

FE simulations are useful for accurately estimating the effect of thermal coupling on the device junction temperature. These simulations solve the heat transfer equation, while considering the boundary conditions as well as the material properties of the power module. While these simulations are highly useful for characterizing the thermal performance of the power module, e.g., determining module R_{TH} and Z_{TH}, they are not suitable for analyzing thermal performance over long time durations. Often there is a need to look at the power module thermal performance over the whole drive cycle in

electric vehicle (EV) applications needing several hours of run-time.

Achieving the accuracy of FE simulations over long run times is a challenge with traditional thermal modelling approaches. This paper proposes thermal modelling of a Wolfspeed WolfPACK power module by constructing its state-space thermal model with Ansys Twin Builder. The state-space model is grounded in high fidelity thermal simulations of the power module carried out in Ansys ICEPAK. Through vector fitting, the Twin Builder constructs a state space model with a reduced order suitable for use with commonly used circuit simulators while at the same time preserving the modelling detail of the underlying FEA model.

II. Reduced Order Modelling

Ignoring radiative effects, the heat transfer equation can be expressed as:

$$\rho c_p \frac{\partial T}{\partial t} = \nabla(k \nabla T) + h(T - T_{ambient}) \tag{1}$$

where ρ is the density, and c is the specific heat capacity, h is the heat transfer coefficient of the media wherein the solution of the equation is sought. Finite element software, such as ANSYS ICEPAK, then discretizes the object into nodes, and the relationship between the temperature at the nodes and the material properties is represented in discrete form as

$$[C_T]\{\dot{T}\} + [K]\{T\} = \{F\} \tag{2}$$

where, $[C_T]$ and $[K]$ are the specific heat capacity and thermal conductivity between the nodes in the FEA model. $\{T\}$ represents the nodal temperatures; and the vector $\{F\}$ is the heat flux boundary conditions applied to the model. ICEPAK solves the discretized heat equation to determine the temperature at various nodes of the model in response to the heat flux inputs to the dies.

Equation (2) can be transformed in a state-space form to obtain the relationship between the temperature of the dies in response to power inputs. The full-order system matrix can then be reduced with the help of Krylov subspace methods to get a reduced order thermal model of the power module [4-5]. An alternate approach, used in this paper, consists of vector fitting the temperature response of the dies subjected to step

Fig. 1. CCB032M12FM3 WolfPACK™ Power Module

979-8-3315-1612-3/25 $31.00 © 2025 IEEE

power inputs at each position. This is possible since for a Linear Time-Invariant (LTI) system, the output can be completely characterized by its impulse or step response.

Development of thermal reduced-order model (ROM) for the Wolfspeed CCB032M12FM3 power module, shown in Fig. 1, is demonstrated with a two-stage process. The FM3 power module contains six switches forming a three-phase inverter, with each switch position consisting of a single die. Firstly, the junction temperature response is generated with a power step input to each die/switch position. The resulting temperature response is captured for each switch position in the module. Fig. 2 shows the temperature distribution across all dies in a Wolfspeed FM power module when FET 1 is subjected to a 100 W step power input in an ICEPAK steady-state thermal simulation. Secondly, after repeating the process for the remaining die positions, all the collected responses, generated with ICEPAK, are then vector fitted with ANSYS Twin Builder to develop a ROM of the power module in the simplified state-space model:

$$\dot{x} = [A]x + [B]u$$
$$y = [C]x + [D]u \tag{3}$$

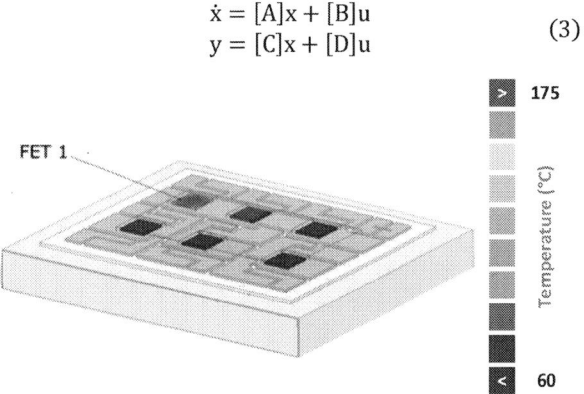

Fig. 2. Steady-state temperature distribution across the dies owing to a 100 W power injection through FET 1

Where, 'u' is the power step input and 'y' is the temperature output. 'x' represents the internal states of the system which don't possess any physical meaning. As shown in Fig. 3, the inputs to the state-space system are the power inputs to the dies/switches in the module, and the output is configured to be the junction temperature of various dies/switches inside the module. The obtained lower order matrices (A, B, C, and D) are then directly used with circuit simulators, e.g. PLECS, for determining the thermal behavior of the power module.

An overview of the process for obtaining the system matrices A,B,C, and D through vector fitting is provided now with a two input -two output system example in Fig. 4.

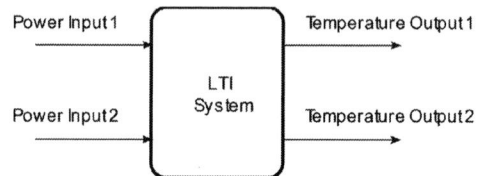

Fig. 4. A two-input two-output LTI system

The temperature output of this system can be represented as the convolution of input signals with the impulse response of the system as

$$y_1(t) = x_1(t) * h_{11}(t) + x_2(t) * h_{12}(t)$$
$$y_2(t) = x_1(t) * h_{21}(t) + x_2(t) * h_{22}(t) \tag{4}$$

Where, h_{ij} is the impulse response of the system at output position i due to input at position j. Keeping $x_2(t) = 0$ and applying only $x_1(t)$ gives the following output:

$$y_1(t) = x_1(t) * h_{11}(t)$$
$$y_2(t) = x_1(t) * h_{21}(t) \tag{5}$$

Similarly, keeping the input $x_1(t)=0$ and only applying $x_2(t)$ as an input to the system gives the following output:

$$y_1(t) = x_2(t) * h_{12}(t)$$
$$y_2(t) = x_2(t) * h_{22}(t) \tag{6}$$

The convolution of the impulse response and the input to the system is mathematically expressed as,

$$y(t) = \int_{-\infty}^{+\infty} h(t - \tau)u(\tau)d\tau \tag{7}$$

Since solving the convolution integral is not straighforward, and is diffcuilt to implement with most circuit simulators, a frequency domain represnetation of the system is more useful for finding the impulse resposne of the system from the avaiable system meausrments [6]. Applying Laplace transform to both sides of the equation above, a frequency domain representation of equation (7) is obtained as

$$Y(s) = H(s)U(s), \tag{8}$$

Where, $s=\sigma+j\omega$ is the complex frequency. The vector fitting problem can be expressed now. If there are k measuremnts of the transfer fucntion

$$H_k = H(j\omega_k)k = 1, \ldots, \bar{k}, \tag{9}$$

Determine the fucntion H(s) that approximates the measurments

$$\tilde{H}(j\omega_k) \simeq H_k \forall k = 1, \ldots, \bar{k} \tag{10}$$

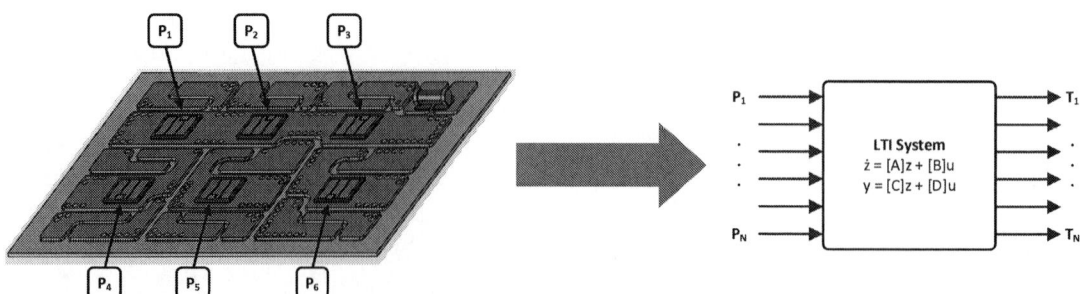

Fig. 3. (a) CCB032M12FM3 power module and (b) state-space model captures the details of the FEA model in a simplified representation

$\tilde{H}(s)$ can be expressed as a ratio of two polynomials as

$$\tilde{H}(s) = \frac{n(s)}{d(s)}. \quad (11)$$

The polynomials $n(s)$ and $d(s)$ are chosen to be partial fractions with the Gustavsen and Semlyen algorithm [7] as below

$$n^{(i)} = c_0^{(i)} + \sum_{n=1}^{\bar{n}} \frac{c_n^{(i)}}{s - p_n^{(0)}},$$

$$d^{(i)} = 1 + \sum_{n=1}^{\bar{n}} \frac{d_n^{(i)}}{s - p_n^{(0)}}, \quad (12)$$

Selecting the poles (p_n) as well as the parameters c_n and d_n for these polynimials is governed by the methodology of the particular vector fitting algorithms used, and is beyond the scope of this paper. Finally, the derived rational function $\tilde{H}(s)$ needs to be reconstructed such that the transfer fucntion between the input and output is the same as represented with A,B, C and D paremeters of the system.

$$\tilde{H}(s) = D + C(sI_N - A)^{-1}B, \quad (13)$$

Since there are infinite realizations of the system which can satisfy above equation, specific realizations like Gilbert's [8] can be utilized for vector fitting the system to a state space realization. For this paper, Ansys Twin Builder performed the task of extracting $\tilde{H}(s)$ from the temperature responses obtained via Ansys ICEPAK. The A, B, C, and D parameters of the ROM are then fitted in the state-space block in PLECS as shown in Fig. 5.

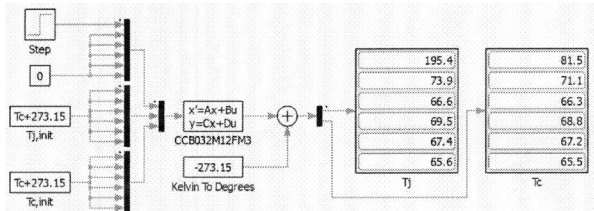

Fig. 5. State-space parameters of the extracted ROM of the power module imported in PLECS

As shown in Fig. 5, the A, B, C, and D parameters, representing the thermal ROM of the power module are plugged in the state space block available from the PLECS library. The number of inputs and outputs of the block will be configured during the setup of the ROM in Ansys. On the left side of the block, the top six signals represent the power input to the module for each of the six dies. The next twelve signals are used to input the initial temperature for the six junctions and the six case nodes of the power module. Generally, the initial temperature is equal to the temperature of the cooling liquid used in the Ansys ICEPAK simulation.

On the right side of the state-space block, the first six signals are the junction temperature outputs, and the next six represent the case temperature outputs of the model. As will be demonstrated later, since both the estimates for the junction and the case temperature are available from the state-space block output, the PLECS simulation can be used to estimate the thermal resistance from junction to case ($R_{TH,JH}$) as well as transient thermal impedance from junction to case ($Z_{TH,JH}$) of the power module. The case temperature probe in ROM is based on the probe configured in the ICEPAK simulation, and it corresponds to the thermocouple placement below the

power module junction in accordance with AQG-324. $R_{TH,JH}$ is calculated as

$$R_{TH,JH} = \frac{T_{J,Steady-State} - T_{H,Steady-State}}{P_D}. \quad (14)$$

Where $T_{J,Steadt-State}$ is the steady-state junction temperature and $T_{H,Steady-State}$ is the steady-state case temperature available from the thermal ROM model. P_D is the power input through the switch. Similarly, $Z_{TH,JH}$ can be extracted with the developed ROM model in Fig. 5.

For switch position 1, the die's transient temperature response when subjected to a step input of 100 W is generated with ICEPAK as well as the state-space thermal model in PLECS. The results, compared in Fig. 6, can validate the methodology of mapping the FEA based thermal model in ICEAPK to a single state-space block in PLECS.

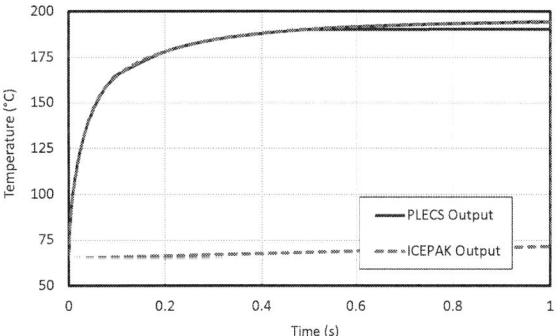

Fig. 6. Transient temperature response from ICEPAK and PLECS: A comparison

III. STATIC AND DYNAMIC TEST RESULTS

To demonstrate the efficacy of the reduced-order modelling approach, a Simcenter Micred T3STER system is used to capture the steady-state temperature across the dies when a power input is applied to each die. Fig. 7 shows the steady-state temperature at each die position measured in response to a power input of 60 W. Similarly, a power input of 60 W is fed to the PLECS state-space thermal model (shown in Fig. 5), and the steady-state temperature captured from the simulation is shown in Fig. 8. Comparing the test results Fig. 7 to the thermal results from PLECS in Fig. 8, the simulation can estimate the temperature within ~6% of the measured test results.

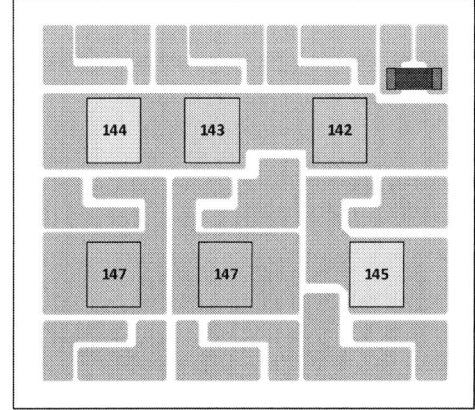

Fig. 7. Junction temperature measurements per die position

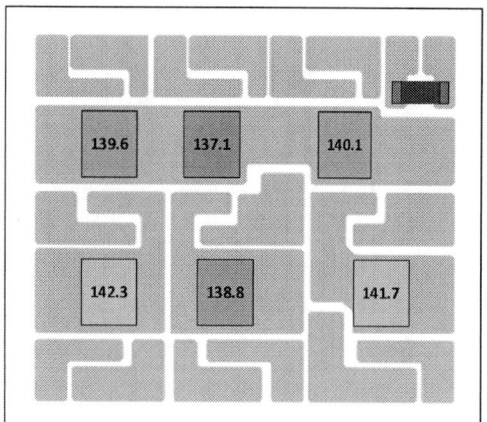

Fig. 8. Estimated junction temperature with PLECS state-space model with only one switched excited at a time

Lastly, all the dies/switch positions are simultaneously fed together with a power input of 60 W in the PLECS model, and the measured steady-state temperature is plotted for each position in Fig. 9. The effect of mutual thermal coupling is observed in Fig. 9 owing to the higher junction temperatures estimated with the state-space model in PLECS.

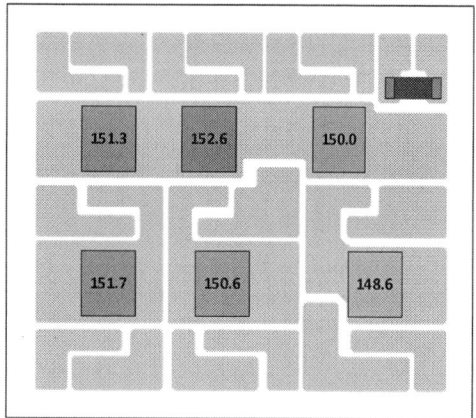

Fig. 9. Estimated junction temperature with the PLECS state-space model with all the switches excited simultaneously

With the static measurements validated, the next objective is to demonstrate the efficacy of the state-space model to be able to capture the thermal dynamics of the power module. The transient thermal impedance from junction to heatsink, $Z_{TH,JH}$ is obtained from T3ESTER and is also estimated with the simulation. The simulation result in Fig. 10 shows excellent correlation with the T3STER measurements, therefore, demonstrating the ability of the simulation to capture the thermal dynamics of the module accurately.

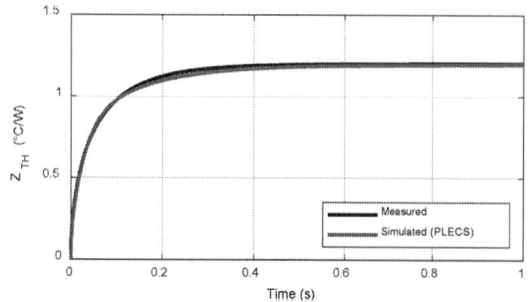

Fig. 10. $Z_{TH,JH}$ measurements from T3STER compared with $Z_{TH,JH}$ extracted with the PLECS state-space model

IV. COMPARISON OF ROM WITH CAUER MODEL

With both the static as well as the dynamic test results validated, the state-space model can now be deployed to demonstrate advantages of state-space thermal modelling of the power module over the traditional Cauer network models in an inverter application.

A two-level inverter test stand utilizing the averaged switch model is developed in PLECS. The device losses are implemented with loss lookup tables. The temperature estimated with the thermal models is fed back iteratively to the loss calculation models for more accurate loss modelling. The operating parameters of the inverter are listed in Table. 1 below.

TABLE 1. INVERTER OPERATING PARAMETERS

Parameter	Value
DC Link Voltage	800V
Output Phase Current	30A
Switching Frequency	20 kHz
Fundamental Frequency	300 Hz
Gate Resistance	20Ω
Coolant Temperature	60°C
Modulation scheme	SVPWM

To compare the state-space with the traditional Cauer network model, both the thermal models were included in the PLECS inverter model. The values of the Cauer network model parameters for the Wolfspeed power modules are available in [9].

The junction temperature response for all the six dies inside the FM power module is captured with the PLECS simulation and is shown in Fig. 11(a). For the Cauer network model, since the model is identical for all three phases, the temperature response is identical for the top three as well as the bottom three devices as shown in Fig. 11(b). Comparing the junction temperature response from the PLECS thermal model with the Cauer network thermal model, the following observations are made:

Firstly, the peak temperature observed with the state-space thermal model is about 174°C, which is higher than the 151°C observed with the traditional Cauer network model. This can be attributed to the effect of mutual thermal coupling which is modeled with the state-space model. The traditional Cauer network model fails to capture the effect of mutual thermal coupling between the dies inside the power module. Since the maximum permissible limit for the junction temperature is only 175°C, the state-space thermal modelling can, therefore, help ensure the device operates within it safe-operating area (SOA) for various use cases.

Secondly, the temperature swing observed with the Cauer model is about 20°C. On the other hand, the swing can be as low as 13°C with the state-space model. Both the aforementioned parameters, the peak T_J as well as the temperature swing, heavily influence the lifetime and reliability of the power device. Therefore, thermal state-space model can be useful for estimating the lifetime of the power module with actual use cases with more accuracy.

Even for the dies inside the module, there is a variation of about 4°C between the dies '4' and '6' in the bottom switch positions as well as dies '3' and '2' in the top switch positions. Such variation in the junction temperature is not captured with the traditional Cauer network. Lastly, the thermal state-space model, owing to its accurate representation of $Z_{TH,JH}$, gives a more accurate dynamic junction temperature response, which is observed to be rising slowly than the temperature response observed from the Cauer network model.

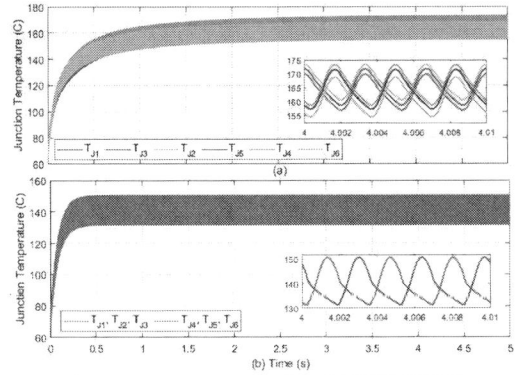

Fig. 13. Junction temperature response of the CCB032M12FM3 power module in an inverter application

V. DRIVE CYCLE ANALYSIS

One key advantage of PLECS and its minimization of simulation complexity is the capability to run extremely long simulations of complex systems. The Worldwide harmonized Light-duty vehicles Test Cycle (WLTC) Class 3b, for example, has a 30-minute-long duration (or 1800 seconds), and other drive cycles can be multiple hours long. Attempting to simulate such a long duration in highly detailed solvers such as ANSYS is highly impractical. Even SPICE based solvers struggle with minute long simulations, especially when considering closed loop interaction between electrical and thermal networks.

PLECS, however, is suitable for the long durations simulations, although the high RAM utilization of such as simulation should be considered (as paging to disk can dramatically decrease system performance). One consideration is that each scope in PLECS will save each trace at each time step through the simulation and should therefore be used sparingly. In longer drive cycles, each trace of each scope can easily consume more than a gigabyte of memory. Similarly, any calculations that are not mandatory can be eliminated to improve simulation speed and reduce RAM usage.

Fig. 12 shows a high-level overview of the system simulation developed to test the CCB032M12FM3 model across a full drive cycle. A space-vector PWM controller was used to create the gate signals, and three ideal current sources were used for the load. The inputs to both the controller and

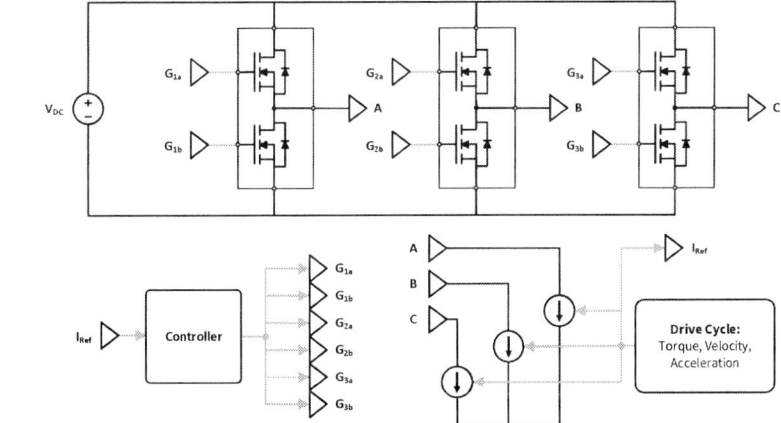

Fig. 11. Drive cycle PLECS simulation

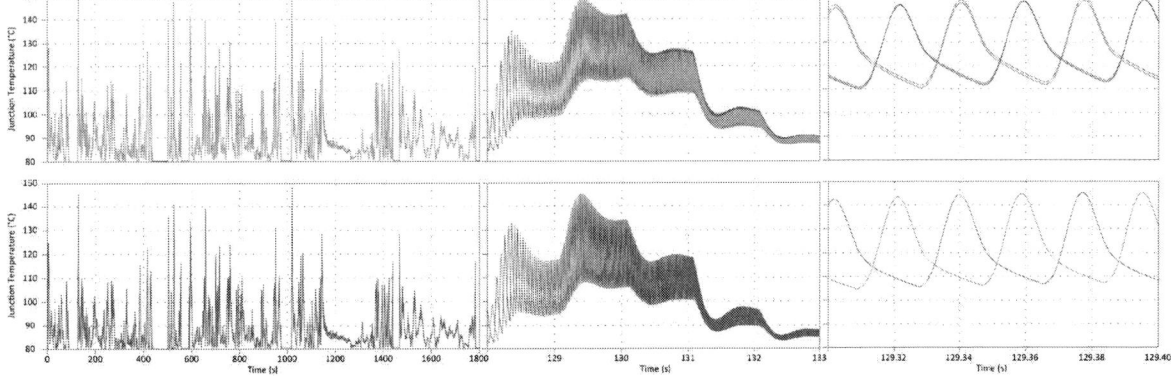

Fig. 12. Drive Cycle simulated results, for State-Space matrix model (top), and traditional model (bottom)

load were read from a csv file, which defined power factor, reference current, DC voltage, modulation factor, and output frequency at 10 ms timesteps.

Fig. 13 shows the simulated junction temperature of the state-space matrix model and compares it to the junction temperature calculated by a traditional Cauer network approach. For the traditional model, only high side and low side are calculated as the model is not detailed enough to differentiate the temperature delta between the phases. The module rarely reaches steady state in the drive cycle studied as the current setpoint typically changes on the order of half a second. Because of the transient thermal impedance of the two models are quite close below one second, the junction temperatures predicted by the two models are much closer in this simulation than observed in the steady state result (shown in Fig. 11), showing a delta of ~5 °C instead of 23°C. This indicates that, although the cross-heating effects are quite significant in the context of steady-state performance, they may be less critical in some real-world applications due to their relatively large time constant. Simulations such as the PLECS model described in this section are necessary to differentiate between applications highly dependent on cross-heating, and those where the effect of cross-heating is more modest.

VI. CONCLUSION

This paper presents the reduced order thermal matrix model of multi-chip SiC power module. The model is obtained by vector fitting the step power response data, obtained from thermal simulations, with Ansys Twin Builder. The static as well as dynamic performance of the model is validated through measurements on the power module. Lastly, the ability of the ROM to model mutual thermal coupling effects is demonstrated with an inverter simulation in PLECS. Future work on the ROM will investigate the effect of ROM system order on the temperature estimation accuracy and study its tradeoff with the simulation speed in inverter applications.

REFERENCES

[1] H. Li *et al.*, "Improved thermal couple impedance model and thermal analysis of multi-chip paralleled IGBT module," *2015 IEEE Energy Conversion Congress and Exposition (ECCE)*, Montreal, QC, Canada, 2015, pp. 3748-3753.

[2] A. S. Bahman, K. Ma and F. Blaabjerg, "A Lumped Thermal Model Including Thermal Coupling and Thermal Boundary Conditions for High-Power IGBT Modules," in *IEEE Transactions on Power Electronics*, vol. 33, no. 3, pp. 2518-2530, March 2018.

[3] H. Wang *et al.*, "A Thermal Network Model for Multichip Power Modules Enabling to Characterize the Thermal Coupling Effects," in *IEEE Transactions on Power Electronics*, vol. 39, no. 5, pp. 6225-6245, May 2024.

[4] C. Entzminger, W. Qiao, L. Qu and J. L. Hudgins, "A High-Accuracy, Low-Order Thermal Model of SiC MOSFET Power Modules Extracted from Finite Element Analysis via Model Order Reduction," *2019 IEEE Energy Conversion Congress and Exposition (ECCE)*, Baltimore, MD, USA, 2019, pp. 4950-4954.

[5] H. B. Aissia, J. Jay, S. Xin and R. Knikker, "Thermal Reduced Order Model for an Electronic Power Module," *2018 24rd International Workshop on Thermal Investigations of ICs and Systems (THERMINIC)*, Stockholm, Sweden, 2018, pp. 1-4.

[6] Piero Triverio, "Vector Fitting," 2019. Available online: https://arxiv.org/pdf/1908.08977

[7] B. Gustavsen and A. Semlyen, "Rational approximation of frequency domain responses by vector fitting," in IEEE Transactions on Power Delivery, vol. 14, no. 3, pp. 1052-1061, July 1999, doi: 10.1109/61.772353.

[8] E. G. Gilbert, "Controllability and observability of multivariable control systems", *SIAM J. Contr.*, vol. 1, pp. 128-151, 1963.

[9] https://www.wolfspeed.com/tools-and-support/power/ltspice-and-plecs-models/

Design and Evaluation of Dual-Resolver Emulation for Control System Verification in Aerospace Actuation Applications

Tomas Sadilek, Julian Opificius, Jason Wright, Alec Leslie, Jeremie Tuzizila, Cesar Alzate,
Hunter Burnett, Joshua Atkinson, and Justin Stricula
Curtiss-Wright Controls, Shelby, NC, USA

Abstract—Unlike much of industrial power electronics, aerospace actuation comes with the requirement of additional design rigor and exhaustive validation and verification of firmware and hardware. The elevated degree of design assurance stems from the safety criticality of the application. Much of the design - power electronics, firmware, and mechanical subsystems - needs to be done in parallel. The integrity with which a full aerospace application control system can be verified under dynamic operating conditions while the rest of the system is being designed or constructed can be improved by precise emulation of critical sensors such as shaft resolver. In this paper, we propose and evaluate a low-cost dual-resolver emulator for the control of aircraft doors.

I. INTRODUCTION

Two commonly considered sensing solutions for measuring accurate rotor position in rotary electric machines are the encoder and resolver [1]. Encoders are typically opto-electrical devices, which allow the drive system to accurately determine the rotor position. However, opto-electrical encoders are rarely used in rugged applications, e.g. high-vibration or temperature range, such as aerospace, due to concerns with reliability. In such applications, resolvers are preferred. There are several kinds of resolvers, though the typical resolver is effectively a three-winding rotating transformer with a primary excitation winding and two secondary windings, which are placed in an orthogonal fashion with respect to each other, such that the magnetic flux linking each secondary winding to the primary winding is a function of the transformer rotation (**Fig. 1**). Typical resolver waveforms are shown in **Fig. 2** - the voltage envelope across each secondary winding is the sine and cosine function of the rotor angle. Thus, the rotor angle can be easily determined based on the amplitude ratio and phase angles of these two voltages. The frequency of the envelope of the sine

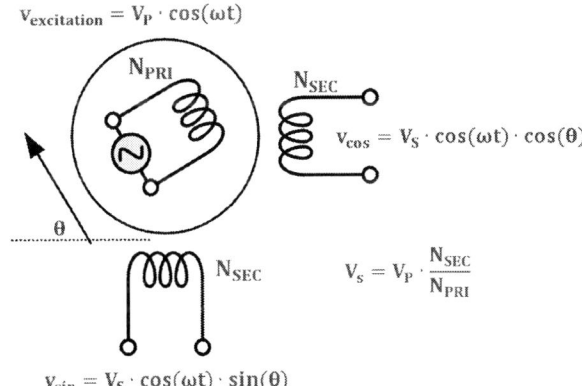

Fig. 1. Notional resolver structure.

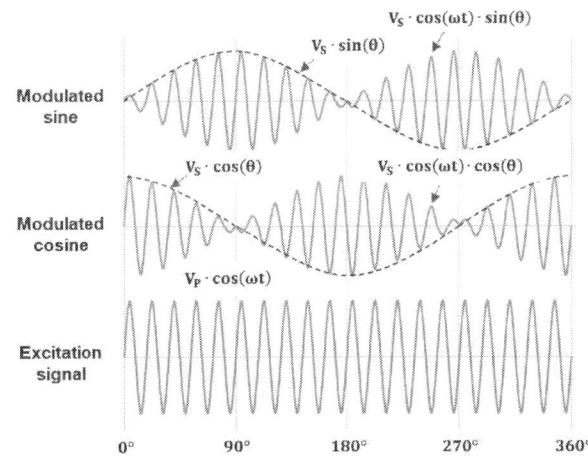

Fig. 2. Fundamental resolver waveforms.

and cosine signals is equal to the mechanical frequency of the shaft multiplied by the pole-pair factor of the resolver. The sine and cosine envelopes are modulated by a high-frequency excitation signal. The high excitation frequency of the resolver allows reasonably small magnetic core size while avoiding core saturation, and permits accurate position and velocity demodulation at both standstill and high speeds.

With regards to the position sensor interface to the motor drive, encoders provide a simple digital electrical interface

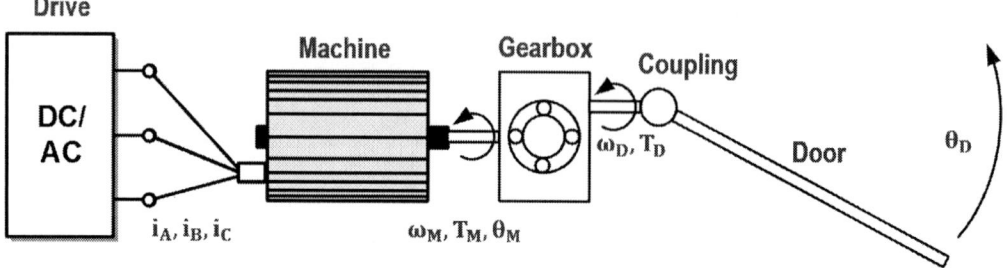

Fig. 3. Simplified door actuation system with two resolvers.

consisting of quadrature-encoded signals **A**, **B**, and **I** pulse (**I** - index, sometimes denoted as **NM** - north marker). Interfacing with a digital sensor is very simple - just several wires in a harness are needed, and digital isolation that increases noise immunity is easy to achieve.

On the other hand, the resolver interface is much more complex and nuanced - resolvers need to be excited at a specified frequency and with a specified voltage. Further, since resolvers are effectively transformers with finite magnetizing inductance, the current draw can range from several mA to a hundred mA. Resolvers also come with a custom voltage transformation ratio - values between 1:1 and 4:1 are typical. For these reasons, interfacing with a resolver is not straightforward. Modern hardware-in-the-loop (HIL) platforms, such as the PLECS RT Box [2], offer the option to interface both with digital encoders and also with resolvers. However, these platforms are meant to emulate the power electronics drive and connect to a single resolver - not the other way around, where it is the resolver that is being emulated.

Unlike most terrestrial industrial industry segments, aerospace actuation is highly demanding with regard to system reliability, imposing particular demands on firmware design, which needs extremely rigorous and time-consuming DO-178 and DO-254 processes [3]. For this, and other practical reasons, such as geographical team dislocation or hardware availability, it is beneficial to have the requisite hardware for control system development and testing emulated to the greatest extent possible.

Commercial equipment that emulates resolvers does exist [4] but the cost (>$25,000) may be prohibitive for some applications, especially if multiple teams require such functionality. Some previous work on resolver emulation was presented by Analog Devices [5], though a number of

analog circuits were used in the approach and only limited details on the design were provided. Furthermore, some commercial resolver emulator equipment is limited in the excitation frequency (<10 kHz), which in turn renders the resolver emulator unusable for physically small resolvers, which require higher excitation frequency due to the low number of turns.

II. EXAMPLE - AIRCRAFT DOOR ACTUATION

Interestingly, the typical power levels in the aerospace industry, excluding propulsion, are quite low - in the kW range. At the same time, the voltage levels of drives and therefore the corresponding electric machines are tied to the 3-ph 115 Vrms [6] or 230 Vrms [7] AC distribution buses. As a result, a well-designed drive/motor combo therefore features high speed and low torque. The actuation of doors or control surfaces is therefore typically made possible by employing gearbox assemblies that, as a general rule, tend to significantly amplify torque. Consequently, the gearbox output speed is proportionally reduced.

Equations governing a fixed-ratio mechanical gearbox (such as the one in **Fig. 3**) are shown below:

$$\omega_D = k_{gearbox}\omega_M \tag{1}$$

$$T_D = \frac{\eta_{gearbox}}{k_{gearbox}}T_M \tag{2}$$

Where T_D is the gearbox output torque, T_M is the gearbox input torque, $k_{gearbox}$ is the gear ratio, and $\eta_{gearbox}$ is the gearbox efficiency.

In typical aerospace applications, $k_{gearbox}$ is less than one since machine torque needs to be amplified. Gearbox efficiency $\eta_{gearbox}$ is highly variable, depending on gear type, gear ratio, tooth geometry, material properties, tooth contact surface treatment, and internal temperature.

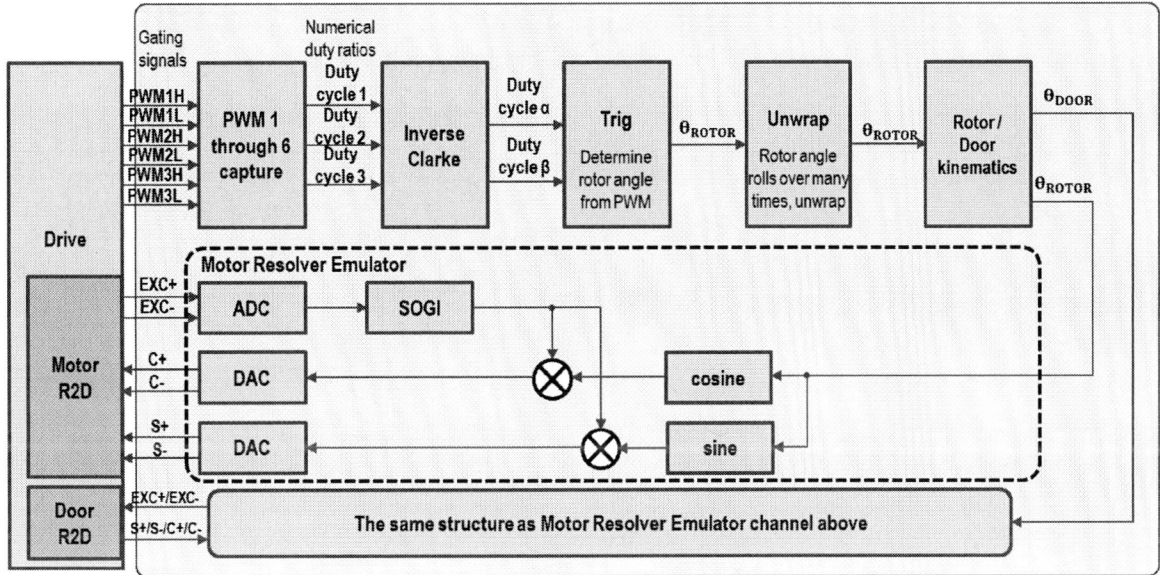

Fig. 4. Dual-resolver emulator structure.

Since the actuation motor shaft typically spins at high speed and needs to be geared down significantly to provide sufficient torque for the application, the motor-mounted resolver used for controlling the stator magnetic field (field-oriented control) accumulates high total count very rapidly, and is subject to rapid and continuous rollover, and is thus not an optimal source of end-effector position. A separate, absolute position or angle measuring device is therefore preferred.

As an example of actuation system health monitoring, a linearity cross-check between the high-speed motor displacement, scaled by the gear ratio, and the actuated surface position, is performed to determine whether the mechanical linkage, which includes the gearbox and other elements, is functional. **Fig. 3** shows a simplified system described thus far. The electric motor speed is ω_M, torque is T_M, and shaft angle is θ_M. The gearbox output speed and torque values are ω_D and T_D, respectively. The door angle, as measured at the coupling element, is θ_D.

III. RESOLVER EMULATOR DESIGN

As mentioned above, oftentimes it is not practical to evaluate the system behavior only once the system is fully built. Thus, the two resolvers and the mechanical system (**Fig. 3** - gearbox, coupling, door) require emulation.

The proposed structure of the dual-resolver interface is shown in **Fig. 4**. In this simplified approach, the high-level motor drive and system control firmware is tested for correct functionality by driving open-loop PWM signals to indicate the desired rotor angle. These PWM signals are captured and filtered by the resolver emulator and the angle is determined using the inverse Clarke transformation.

Note that when verifying systems under DO-178 [3] and DO-254 [8] processes, the test platform must be representative of the final target environment, so actual PWM drive signals, which would be sent to the power stage for current amplification, are used to convey the desired motor shaft angle to the resolver emulators, rather than a simple digital data stream, which would otherwise be far simpler. Any specially furnished serial communication equipment would also not be part of the final target environment, further impacting the integrity of the "representative target environment."

Using this approach to capture the desired motor position, the motor, shaft, and load dynamics are neglected. That is, the decoded PWM angle is immediately reflected onto the rotor shaft angle. This means that the motor dynamics, both electrical, such as winding current, and mechanical, such as the moment of inertia, are not taken into consideration. The idea behind this emulation approach is to test the complete control system bar the fundamental motor phase current regulation algorithm, which is typically FOC, and the motor velocity feedback control. While that might seem like significant simplification, it needs to be noted that in this scenario, only the high-level control system needs to be verified - what happens if the gearbox breaks during a door closing or door opening sequence, what happens when a resolver wire gets disconnected, etc. Low-level control system design (current loop, velocity loop) would be verified during a different stage of the system design.

An additional level of fidelity would be achieved by using a commercial or custom HIL platform to emulate the machine by capturing the PWM pulses, determining the winding currents with regards to winding inductance and bEMF, and feeding the emulated phase currents measurement signals back to the drive analog inputs to close the phase current regulation loop. For system-level behavior, however, these fast phase current dynamics are of diminished interest. As a result, in this emulator incarnation, it is assumed that the motor position is determined by the angle decoded from the gating signals following the simple formulae: $v_{phA} = d_{top-switch-phase-A} \cdot V_{DC}$, where $d_{top-switch-phase-A}$ is the low-pass filtered gating signal of the top switch in phase A, and V_{DC} is the nominal DC link voltage. The gate signals of all three phase legs are captured to properly determine the output voltage vector angle.

As shown in **Fig. 4**, the proposed solution is nearly fully digital. That is, the excitation signal is captured and digitized by an analog-to-digital converter (ADC), and its angle is extracted using a second-order generalized integrator (SOGI) [9] structure. The excitation signal is then multiplied by the sine and cosine values corresponding to the motor angle. That is, the resulting output signals are $M_{sin} = V_S \cdot cos(\theta_{EXC}) \cdot sin(\theta_{MOTOR})$ and $M_{cos} = V_S \cdot cos(\theta_{EXC}) \cdot cos(\theta_{MOTOR})$, consistent with **Fig. 2**.

The popular and inexpensive TI LaunchPad LAUNCHXL-F28379D is used as the resolver emulator base. While it has many features, such as dual CPU, dual CLA (control law accelerator), and various communication buses, the primary feature required for the operation of the resolver emulator is the very fast digital-to-analog converter (DAC), which can be updated at about 1 MHz frequency. A high update rate is crucial for the resolver emulator since the emulator needs to output the sine and cosine of the relatively slow-changing position signal multiplied by the reconstructed excitation signal. Should the excitation signal frequency be 20 kHz (the typical maximum), at least 20 samples per period would be required to produce reasonably smooth waveforms.

Other approaches might include the use of multiplying

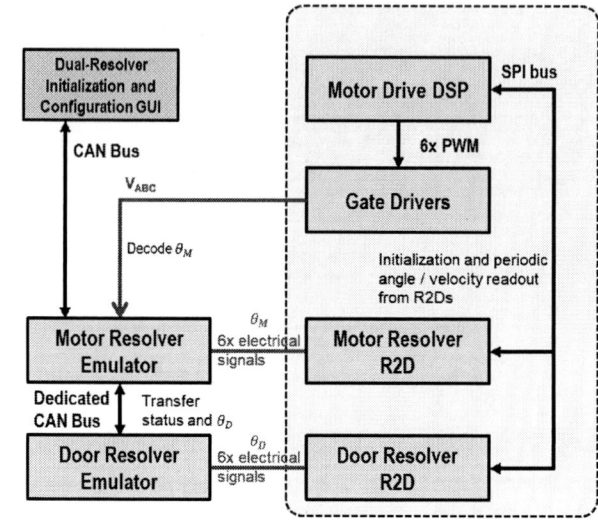

Fig. 5. Notional structure of how the proposed dual-resolver emulator is used to verify the high-level control system.

DACs, though there is the potential issue of signal scaling in the analog domain. Thus, a fully digital and reconfigurable solution for the dual-resolver emulator is preferred.

It needs to be noted that the update rate of the proposed resolver emulator is not an integer multiple of the resolver excitation frequency. The reasoning behind this important detail is that it avoids the incursion of beat-frequency (aliasing) effects. For example, should the resolver excitation frequency be 10 kHz, the emulator update frequency should not be 200 or 300 kHz, but rather 205 or 305 kHz, or other values resulting in no or insignificant beat-frequency effects.

Referring to **Fig. 4** and **Fig. 5**, our example scenario has two resolver emulator channels, because of the previously described need for separate feedback devices for the high-speed velocity control of the motor, and for the precision position feedback. Initially, both resolver emulators were to be placed inside the same LaunchPad. However, due to the limitation of only two fast DAC channels being available, augmentation with external DACs would have been required. Alas, readily available external DACs could not meet the propagation latency requirement, which is the delay between the computation of new analog value and the DAC analog output latching the updated value. For this reason, a second LaunchPad was used as the second resolver emulator channel. The second channel has the same code structure and DAC setup as the first resolver emulator. It also samples the excitation signal of the second resolver-to-digital (R2D) IC using its own ADC. However, the position angle setpoint comes from the first resolver emulator channel

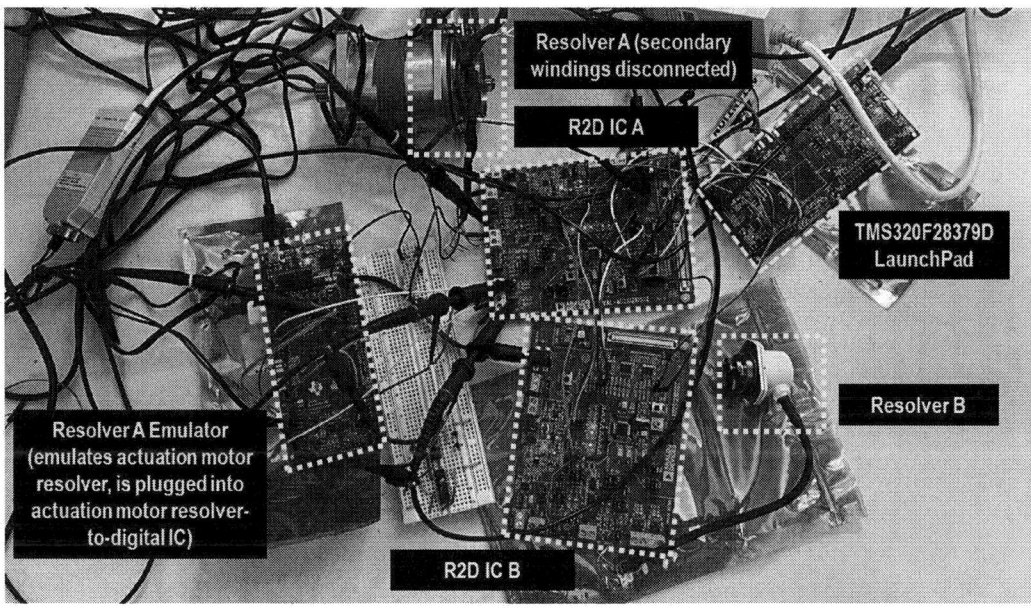

Fig. 6. Experimental setup showing two R2D IC demo cards connected to a drive control DSP (full-power motor drive is currently being designed). Further, the LaunchPad card on the left successfully emulates Resolver A - notice the wires connected to the R2D IC A evaluation board. In this picture, only resolver A is being emulated - the physical resolver is disconnected. Resolver B is still connected to its respective R2D card. In the final experimental configuration, which is not shown in this paper, there are two Resolver Emulator cards, each connected to an R2D circuit card. The two Resolver Emulators cards communicate with each other using a fast CAN bus.

via a fast, dedicated CAN bus with only two devices. The 2nd angle (actuation surface angle θ_D) is determined in the first emulator and transferred to the second emulator with about a 2-ms delay. The CAN bus update rate is 1 kHz.

IV. EXPERIMENTAL RESULTS

Fig. 6 shows the control system as well as the resolver emulator prototype, in its nascent flying-lead form. Two resolver-to-digital (R2D) cards from Analog Devices (AD2S1210) [10] are connected to a target DSP (TI TMS320F28379D LaunchPad in the top-right corner) [11] that facilitates the initialization of the two R2D and subsequent periodic readout of resolver position and velocity signals. A 10-MHz serial link (SPI) is used for both R2D ICs. The required readout time for position and velocity signals of the drive motor and the position signal of the door position signal is ≈ 15 μs.

Fig. 7 shows the resolver excitation signal at the analog input of the resolver emulator. This signal is sampled at a reasonably high rate of ~ 400 kHz. The intent is to execute the required calculations as fast as possible without violating the CPU timing constraints. The sampling and calculation rate of 400 kHz provides an oversampling factor of 20 should the excitation signal be 20 kHz. A second-order generalized integrator is used to filter out the incoming excitation signal

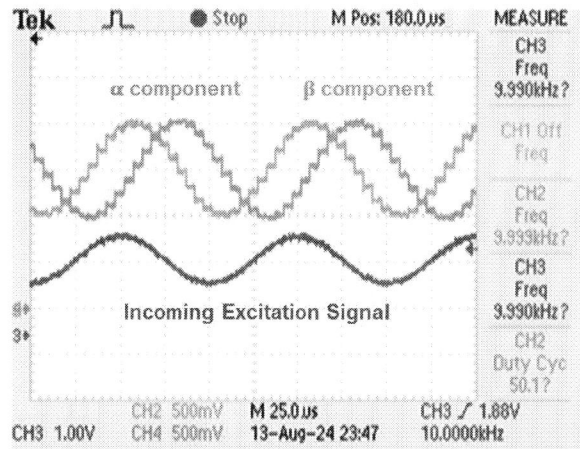

Fig. 7. Incoming excitation signal as well as α and β components obtained from SOGI. The DAC update rate in this figure is 200 kHz, though the final update rate was 405 kHz.

and provide a set of orthogonal output signals. The α and β components are shown in **Fig. 7** as well. Generally, the SOGI structure is used for 1-ph PLL applications for 50/60 Hz grid voltages. In this case, the tracked signal is 10 kHz. Proper scaling of SOGI gains results in correct behavior as shown in **Fig. 7**. Ideally, the α component would be in phase with the incoming signal. Despite the excitation frequency being known and constant, it is still fairly high. The sampling, computation, and DAC delays can be clearly seen in the figure. However, since the excitation frequency is constant,

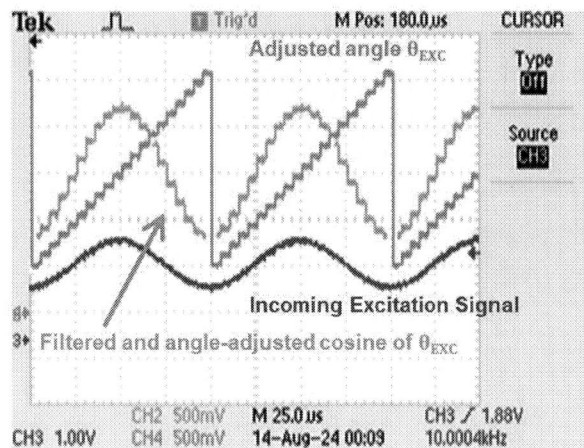

Fig. 8. Incoming excitation signal and adjusted angle θ_{EXC} as well as resulting modulating signal $\cos(\theta_{EXC})$.

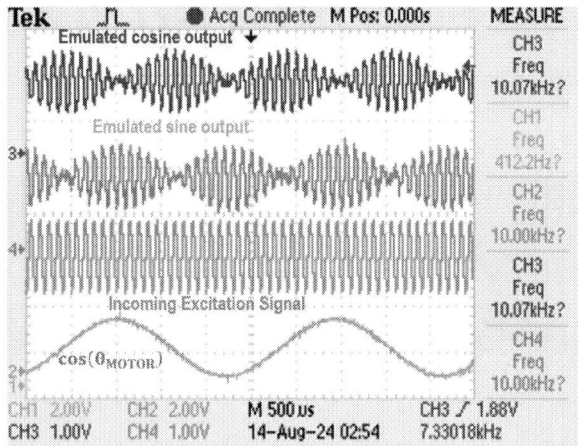

Fig. 9. From top to bottom: emulated cosine output, emulated sine output, incoming excitation signal, and cosine of motor angle.

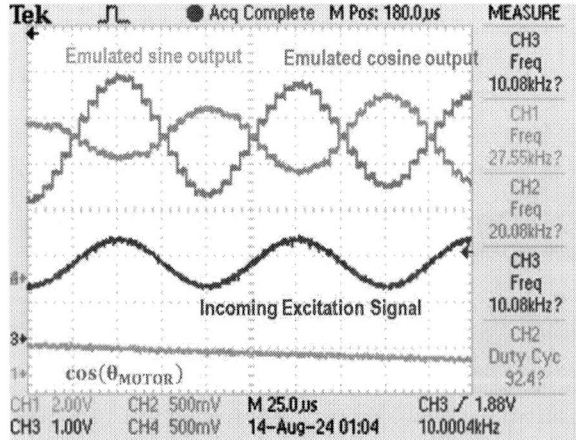

Fig. 10. From top to bottom: emulated cosine output, emulated sine output, incoming excitation signal, and cosine of motor angle. This is a zoomed-in version of the previous figure.

the angle obtained from the α and β components can be easily corrected by adding an offset, as shown in **Fig. 8**. Finally, **Fig. 9** (zoom: **Fig. 10**) shows the experimental waveforms with the following settings: $f_{MOTOR} = 413$ Hz, $f_{EXC} = 10$ kHz. These waveforms are successfully decoded by the R2D IC (AD2S1210), thus verifying the resolver emulation concept. With this configuration, it is possible to emulate various system behavior without requiring access to a full-system bench or even partial power electronics hardware.

Fig. 11 shows the combined data as sent to a control computer by both the motor drive, which obtains the velocity and position data from its R2D converter connected to a resolver emulator, and the resolver emulator, which provides the position signal to the motor drive. In this particular test case, the resolver emulator shaft velocity is suddenly decreased from 14,300 rpm to 10% of that value, 1,430 rpm. This step response represents a scenario with more extreme dynamics than possible with a physical system. Nonetheless, it tests the limits of the motor drive system - can it follow such resolver behavior? The resulting velocity tracking was found to be within the spec of the used R2D chip.

Furthermore, an important test that combined two resolver emulators was performed - as shown in **Fig. 3**, one emulator is connected to the motor shaft and is used to drive the motor (provides ω_M and θ_M. A second resolver measures the door angle (θ_D). **Fig. 12** shows the anticipated normal system behavior - motor angle and door angle increment in a proportional fashion, confirming no mechanical linkage issue. In this particular example, a total effective gear ratio of 12 is used; in practical physical systems, this ratio would be in the thousands. On the other hand, **Fig. 13** shows abnormal system behavior - motor angle and door angle increment in a proportional fashion until a gearbox failure occurs, which leads to the engagement of a mechanical brake and stopping the motion of the door, while the electrical motor and gearbox input shaft is still rotating. This fault can be easily injected using the dual-resolver emulator but is nearly impossible to properly test with the actual mechanical hardware.

It should be noted that the hardware setup shown in **Fig. 6** is not yet final - all the flying wires are to be replaced with a base PCB, resulting in a more professionally-looking and ruggedized test fixture. However, the motherboard PCB will

979-8-3315-1612-3/25 $31.00 © 2025 IEEE

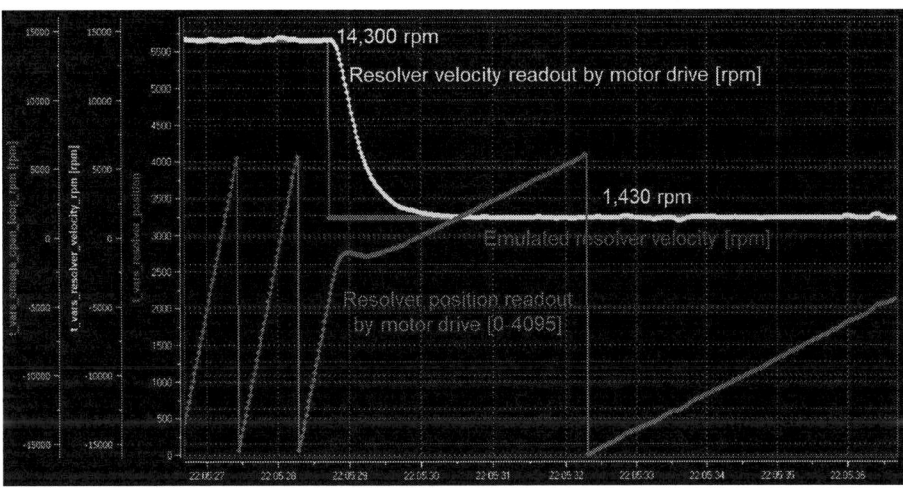

Fig. 11. In this particular test case, the resolver emulator shaft velocity undergoes a step change from 14,300 rpm to 10% of that value, 1,430 rpm. This is clearly a non-physical scenario, but it tests the limits of the motor drive system.

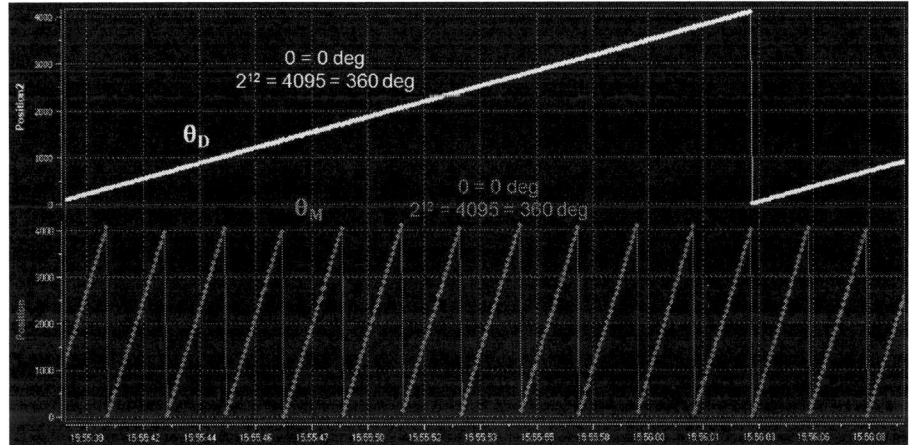

Fig. 12. Normal system behavior - motor angle and door angle increment in a proportional fashion, confirming no mechanical linkage issue. In this particular example, a total effective gear ratio of 12 is used; in practical physical systems, this ratio would be in the thousands.

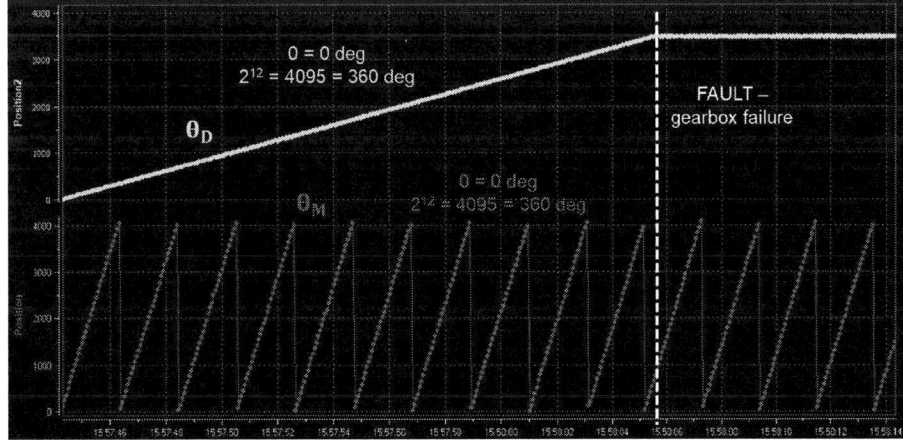

Fig. 13. Abnormal system behavior - motor angle and door angle increment in a proportional fashion until a gearbox failure occurs, which leads to the engagement of a mechanical brake and stopping the motion of the door, while the electrical motor and gearbox input shaft is still rotating. In this particular example, a total effective gear ratio of 12 is used; in practical physical systems, this ratio would be in the thousands.

979-8-3315-1612-3/25 $31.00 © 2025 IEEE

be manufactured after the submission deadline of this paper.

V. CONCLUSIONS

As electromechanical actuation systems continue to displace mechanical and hydraulic systems in aerospace applications, cost-effective, practical yet rigorous design verification of those digital solutions is increasingly important. It has been shown that the proposed dual-resolver emulator approach can aid a design team with the validation and verification of mission-critical high-level behavior encoded in firmware, from nominal operation, to difficult-to-test faults such as gearbox or coupling damage, without the need for actual hardware, thus developing confidence in the system as the design matures. This allows verification to commence earlier in the product life cycle, enabling earlier detection of design errors, and minimizing their impact on development time and cost. Furthermore, there are no personal safety concerns with this verification method. The nearly fully digital, low-cost solution (<$500) has been tested experimentally in a dual-channel configuration - possible system fault modes have been investigated and a fault in the mechanical subsystem was emulated.

REFERENCES

[1] "Encoder versus Resolver. What the difference between them?." https://eltra-encoder.eu/news/resolver-vs-encoder. [Accessed: 2024-08-14].

[2] "RT Box - the HIL Platform for Power Electronics." https://www.plexim.com/products/rt_box. Accessed: 2024-08-14.

[3] "DO-178C - Info on this software standard from MathWorks." https://www.mathworks.com/solutions/aerospace-defense/standards/do-178.html. Accessed: 2024-08-14.

[4] "5330A Synchro/Resolver Simulator, Programmable." https://www.naii.com/model/5330A. Accessed: 2024-08-14.

[5] "High Accuracy Resolver Simulator System with Fault Injection Function." https://www.analog.com/en/resources/analog-dialogue/articles/high-accuracy-resolver-simulator-system-with-fault-injection-function.html. Accessed: 2024-08-14.

[6] G. Gong, M. L. Heldwein, U. Drofenik, J. Minibock, K. Mino, and J. W. Kolar, "Comparative evaluation of three-phase high-power-factor ac-dc converter concepts for application in future more electric aircraft," *IEEE Transactions on Industrial Electronics*, vol. 52, no. 3, pp. 727–737, 2005.

[7] M. Hartmann, J. Miniboeck, H. Ertl, and J. W. Kolar, "A three-phase delta switch rectifier for use in modern aircraft," *IEEE Transactions on Industrial Electronics*, vol. 59, no. 9, pp. 3635–3647, 2011.

[8] "DO-254 - Design Assurance Guidance for Airborne Electronic Hardware." https://www.rtca.org/training/do-254-training/. Accessed: 2024-08-14.

[9] M. Ciobotaru, R. Teodorescu, and F. Blaabjerg, "A new single-phase pll structure based on second order generalized integrator," in *2006 37th IEEE Power Electronics Specialists Conference*, pp. 1–6, 2006.

[10] "AD2S1210 - Variable Resolution, 10-bit to 16-bit R/D Converter with Reference Oscillator." https://www.analog.com/en/products/ad2s1210.html. Accessed: 2024-08-14.

[11] "LAUNCHXL-F28379D." https://www.ti.com/tool/LAUNCHXL-F28379D. Accessed: 2024-08-14.

Un-terminated Blackbox Modeling for Electric Machines

Xinliang Yang
CPES
Virginia Tech
Arlington, VA, USA
xinliangy@vt.edu

Vladimir Mitrovic
CPES
Virginia Tech
Arlington, VA, USA
vlmitr@vt.edu

Qing Lin
CPES
Virginia Tech
Blacksburg, VA, USA
qingl19@vt.edu

Rolando Burgos
CPES
Virginia Tech
Blacksburg, VA, USA
rolando@vt.edu

Abstract—With the increasing penetration of power electronics in modern power systems, such as more electrified aircraft and ships, small-signal impedance-based design and stability evaluation have become widely accepted at the system level. This method is preferred because it does not require detailed information about the physical components and control loops, allowing for "blackbox" characterization. Typically, impedance measurement is conducted at the interfaces among subsystems and encompasses all dynamics from upstream or downstream, necessitating time-consuming re-measurements after any system variation. To overcome the need for re-measurements and maintain intellectual property for the original equipment manufacturer (OEM) through the blackbox approach, the un-terminated terminal behavior model (TBM) is introduced in this paper. This model pertains solely to the devices under test, remaining unaffected by the source side's output impedance and the load side's input impedance. In the rotor flux-orientated d-q domain, this paper presents analysis and procedures for source-load impedance decoupling to measure the un-terminated TBM for electric machines. Experimental tests were conducted on an interior permanent magnet synchronous machine to identify its blackbox un-terminated TBM. An induction machine serves as a torque perturbation source emulating loads or a governor, and a power hardware-in-the-loop (PHIL) system provides electrical power supply with voltage perturbation.

Index Terms—un-terminated terminal behavior model (TBM), electric machine, impedance measurement, blackbox modeling, power hardware-in-the-loop (PHIL).

I. INTRODUCTION

With the increasing penetration of power electronics equipment and more frequent system variations induced by circuit relays or breakers [1] during energy optimization management or faults, respectively, it is nearly impossible to maintain a comprehensive and detailed model of the entire power system. This reality underscores the infeasibility of using state space equation-based methods for stability evaluation. Instead, the impedance-based method proves advantageous as it does not require transparency of either physical components or controllers [2]. Through on-site frequency sweeping at one of the electrical or mechanical interfaces, as illustrated in Fig. 1, the total dynamics of upstream and downstream components are captured as multi-port network containing frequency responses or immittance (impedance or admittance). Based on the extracted aggregated immittance model, subsequent activities such as impedance matching [3], fault diagnosis [4]

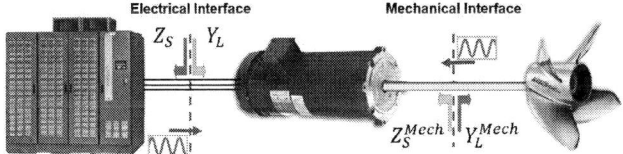

Fig. 1. Typical small signal impedance measurement progress at electrical and mechanical interfaces.

and islanding detection [5], [6] can be conducted. Additionally, the general Nyquist stability criterion may be applied to the minor loop ratio involving these immittances [7]. Such aggregated Thevenin or Norton equivalent representations can be achieved through analytical modeling, simulation, or on-site experimental measurements. From large networks, such as solid-state transformers [8] comprising numerous components, to multi-port systems exhibiting only terminal behavior, the substantial simplification relies not only on the assumption of "mild linear" dynamics under small disturbances but also on the complete observability from terminal variables to inner state variables in linear networks. This approach is further utilized in node tearing technology for accelerating EMT simulations [9].

The utilization of multi-port network in power electronics could be traced back to several decades ago. The concept of two-port network equivalence for DC-DC converters was initially introduced by Cho [10], highlighting a generalized representation through terminal behavior that includes two-wire inputs and outputs. In the domain of three-phase systems, Hiti's research [11] pioneered the three-port network terminal behavior model (TBM) for converters. Within the realm of electric machines, when mechanical variables are converted into their electrical equivalents based on the foundational physical equations, it is possible to establish a three-port network [4], [12].

Although analytical modeling can be used to determine the immittance characteristics of a multi-port network, conflicting interests between system integrators, who prioritize system transparency, and OEMs, who are concerned about protecting intellectual property, pose significant challenges in obtaining the necessary parameters of physical

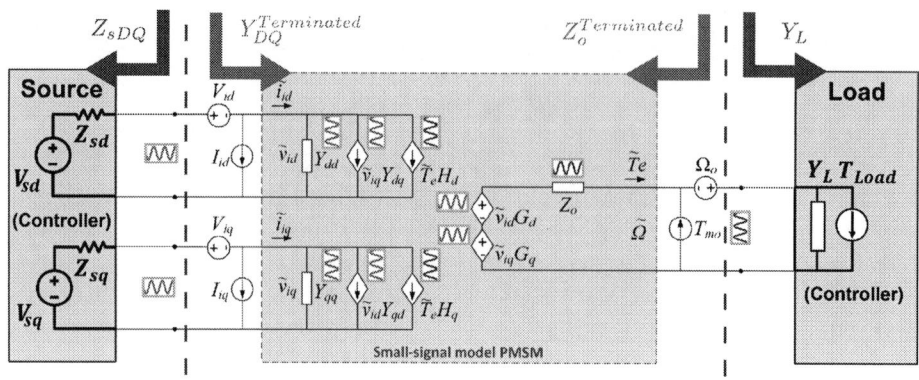

Fig. 2. Small signal equivalent circuit of the IPMSM in dq-axis.

systems and internal controllers. Consequently, the demand for impedance measurement equipment has increased, enabling system integrators to perform small-signal perturbation tests at the interfaces between subsystems to derive operating-point-specific impedance values [13]. However, since these measured impedances inherently capture the aggregated dynamics of all downstream or upstream components—commonly referred to as terminated Thévenin or Norton TBM—such measurements must be repeated whenever subsystems are added to or removed from the overall system. Examples include the disconnection of renewable energy sources during low-demand periods or the replacement of generators. These repeated measurements are both time-consuming and resource-intensive. To address these challenges and enhance efficiency, this paper introduces a method for measuring and utilizing un-terminated TBM. Using an interior permanent magnet synchronous machine (IPMSM) as a case study, experiments are conducted to demonstrate the complete procedures and validate their effectiveness.

The rest of this paper would be organized as follows. Section II introduce the so-called terminated and un-terminated TBMs for electric machines in rotor flux orientated d-q domain. And their characteristcs and relation would be analyzed, which motivates demanding of the un-terminated TBM. Section III provides the extraction method for the un-terminated TBM with simulation verification. And Section IV would demonstrate the experimental results with power hardware-in-the-loop (PHIL) equipments. Section V contains conclusion and consideration of following-up researches.

II. ELECTRIC MACHINE TERMINATED AND UN-TERMINATED TERMINAL BEHAVIOR MODEL

A. General Modeling of Permanent Magnet Synchronous Machines

In this paper, the dominant electromagnetic-mechanical dynamics of electric machines at the fundamental frequency are analyzed without direct consideration of thermal impact [14] which could reflect as electric and magnetic parameter variation. To simplify the representation of electric machine dynamics, the fundamental wave model is typically developed

within a rotating reference frame. This modeling approach eliminates the time-varying components inherent in the three-phase stationary reference frame, thereby facilitating more straightforward analysis. Using the Park transformation, the stator variables are mapped into the d-q reference frame, where the d-axis is aligned with the rotor flux. For an IPMSM, the governing differential equations in the d-q domain are expressed as follows:

$$
\begin{aligned}
v_d &= R_s i_d + L_d \frac{di_d}{dt} - P\omega_m i_q L_q, \\
v_q &= R_s i_q + L_q \frac{di_q}{dt} + P\omega_m (i_d L_d + \lambda_{PM}), \\
T_e &= \frac{3}{2} P \left(i_q (i_d L_d + \lambda_{PM}) - i_d i_q L_q \right), \\
T_e &= T_L + B\omega_m + J \frac{d\omega_m}{dt},
\end{aligned}
\tag{1}
$$

where items with subscript d or q corresponds to d or q axis component, L_d and L_q are d-q axis inductance with different value due to inherent IPMSM characteristics, T_e and T_L are electromagnetic torque and mechanical load torque, J is total inertia for both the machine of interest and the mechanical load, B is vicious friction coefficient, ω_m is rotor angular speed in rad/s, λ_{pm} is the permanent flux linkage magnitude.

Nonlinearities in the system arise from the coupling between state variables and model inputs, exemplified by terms such as $\omega_m i_q$ and $\omega_m i_d$. These nonlinear interactions significantly increase the complexity of system analysis and control. Nevertheless, by assuming the system operates within a "quasi-linear" regime around a specific operating point, local linearization techniques can be applied. This approach approximates the system dynamics as a linear time-invariant (LTI) model, thereby simplifying the analysis and enabling the application of conventional linear control methods.

By using first order linearization, the resulting linearized

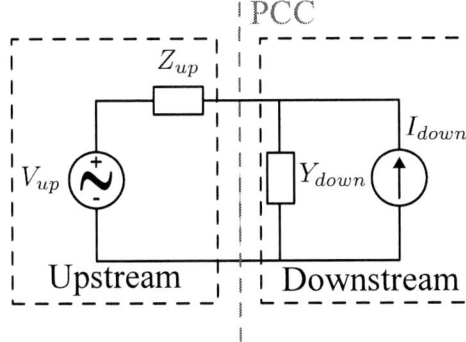

Fig. 3. Impedance representation of an interconnected upstream-downstream system separated at the point of common connection (PCC).

state-space model can be expressed as:

$$
\begin{bmatrix} \frac{di_d}{dt} \\ \frac{di_q}{dt} \\ \frac{d\omega_m}{dt} \end{bmatrix} = \begin{bmatrix} \frac{1}{L_d} & 0 & 0 \\ 0 & \frac{1}{L_q} & 0 \\ 0 & 0 & -\frac{1}{J} \end{bmatrix} \begin{bmatrix} v_d \\ v_q \\ t_L \end{bmatrix}
$$

$$
+ \begin{bmatrix} -\frac{R_s}{L_d} & \frac{L_q P\omega_{m0}}{L_d} & \frac{L_q P I_{q0}}{L_d} \\ -\frac{L_d P\omega_{m0}}{L_q} & -\frac{R_s}{L_q} & -\frac{L_d P I_{d0} + P\lambda_{pm}}{L_q} \\ \frac{3P(I_{sd}-I_{sq})I_{q0}}{2J} & \frac{3P[(I_{sd}-I_{sq})I_{d0}+\lambda_{pm}]}{2J} & -\frac{B}{J} \end{bmatrix} \quad (2)
$$

$$
\cdot \begin{bmatrix} i_d \\ i_q \\ \omega_m \end{bmatrix}
$$

where Laplace transformation ($y(s) = [C(sI - A)^{-1}B + D]x(s)$, A is state matrix, B is input matrix, C is identity output matrix, and D is zero feedthrough matrix) could help to describe the system in s domain as:

$$
\begin{bmatrix} i_d(s) \\ i_q(s) \\ \omega_m(s) \end{bmatrix} = \begin{bmatrix} Y_{dd}(s) & Y_{dq}(s) & H_d(s) \\ Y_{qd}(s) & Y_{qq}(s) & H_q(s) \\ G_d(s) & G_q(s) & -Z_o(s) \end{bmatrix} \cdot \begin{bmatrix} v_d(s) \\ v_q(s) \\ t_L(s) \end{bmatrix} \quad (3)
$$

$$
\rightarrow \begin{bmatrix} i_{dq}(s) \\ \omega_m(s) \end{bmatrix} = \begin{bmatrix} Y_{DQ}(s) & H_{dq}(s) \\ G_{dq}(s) & -Z_o(s) \end{bmatrix} \cdot \begin{bmatrix} v_{dq}(s) \\ t_L(s) \end{bmatrix} \quad (4)
$$

where there are 9 impedance or transfer functions relate three inputs and three outputs. And all of them are only related to the parameters of electric machines without impact from the upstream motor drives and downstream mechanical loads.

B. Terminated and Un-terminated TBMs

Following the same governing equations, the mechanical aspects of the electric machine can be represented through an analogous equivalent circuit. In this mapping, mechanical variables torque and speed are substituted with their electrical counterparts—current and voltage, respectively. Consequently, the total equivalent circuit for an IPMSM is illustrated in Fig. 2. This circuit features two electrical terminals and one mechanical terminal. The upstream electrical components, including the dynamics of the motor drive and controller, are modeled using a Thevenin equivalent, which comprises series-connected output impedance Z_{sDQ} and associated voltage

sources. Similarly, the downstream mechanical components, encompassing the mechanical load and any pertinent controller dynamics, are modeled via a Norton equivalent that includes the admittance Y_L and current sources. Un-terminated TBM is the 3 by 3 impedance matrix T(s), as described in (3), which delineates the relationship between model inputs and outputs, independent of the influences from Z_{sDQ} and Y_L. On the machine side, terminated TBM is the impedance scalar $Z_o^{Terminated}$ at mechanical terminal and the 2 by 2 impedance matrix $Y_{DQ}^{Terminated}$ at electrical terminal.

When there is small amplitude perturbation, due to the non-ideality of the electrical source (Z_{sDQ}) and mechanical load (Y_L), the direct ratio between specific frequency components of state variables and model inputs, in other words the terminated TBM, always contains electromechanical coupling. Analytical expression for such coupling could be derived by substituting $\omega_m(s)Y_L(s) = t_L(s)$ and $v_{dq}(s) = Z_{sDQ}(s)i_{dq}(s)$ into (3), resulting in follows:

$$
\begin{aligned}
Y_{DQ}^{Terminated} &= Y_{DQ} + H_{dq}(1 + Z_oY_L)^{-1}Y_L G_{dq}, \\
Z_o^{Terminated} &= Z_o + G_{dq}(1 + Z_{sDQ}Y_{DQ})^{-1}Z_{sDQ}H_{dq}.
\end{aligned} \quad (5)
$$

where non-zero source-load (upstream-downstream) impedance Z_{sDQ} and Y_L boosts the coupling.

C. Demanding of The Un-terminated TBM

In impedance-based approaches commonly employed at interfaces or points of common connection (PCC), interconnected upstream and downstream subsystems are characterized by their respective input and output impedances, as depicted in Fig. 3. The voltage at the PCC can be determined by the following equation:

$$
V_{PCC} = (I_{down}Z_{up} + V_{up})(I + Z_{up}Y_{down})^{-1} \quad (6)
$$

where I is identity matrix or scalar 1 depending on the amount of system dimension, Y_{down} is the downstream input admittance. There needs an assumption that the upstream voltage is stable if the downstream is removed, and the downstream current can be stable with zero upstream impedance Z_{up}. Then the overall system stability thus hinges on whether the impedance ratio $Z_{up}Y_{down}$ meets the Nyquist criterion or not.

For IPMSM that bridge electrical and mechanical parts, the stability assessments can be executed at both electrical and mechanical interfaces. Here, the input admittance at the electrical interface corresponds to the electrical terminated TBM $Y_{DQ}^{Terminated}$, while the output impedance at the mechanical interface is represented by the mechanically terminated TBM $Z_o^{Terminated}$. In other words, for system integrators, terminated TBMs are preceding basis for general Nyquist stability criteria evaluation at interfaces, which come from un-terminated TBMs and source-load impedance, as shown in (5). Therefore, given Z_{sDQ} and Y_L, the time-consuming impedance measurement could be circumvented if un-terminated TBM could be provided by OEMs. Furthermore, this methodology eliminates the necessity to disclose parametric information about the system and its control strategies, thereby safeguarding the intellectual property of OEMs.

979-8-3315-1612-3/25 $31.00 © 2025 IEEE

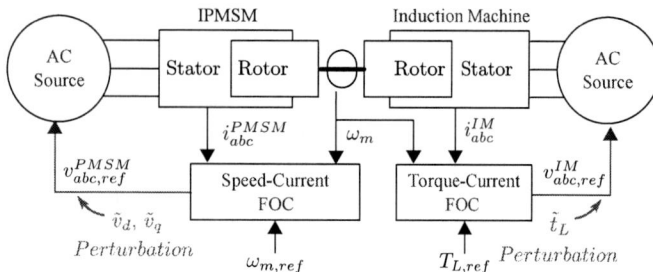

Fig. 4. Un-terminated TBM extraction schematics.

TABLE I
SIMULATION PARAMETERS FOR IPMSM AND IM

Parameter	IPMSM	IM
Pole Pairs (P)	3	2
Stator Resistance (R_s)	0.0636 Ω	2.3 Ω
Stator Inductance	$L_d, L_q = 6.1, 17$ mH	10 mH
Rotor Resistance (R_r)	N/A	1.4 Ω
Rotor Inductance (L_r)	N/A	19.8 mH
Stator-rotor Mutual Inductance (L_m)	N/A	352 mH
Inertia (J)	0.0067 kg·m²	0.0152 kg·m²
Friction Coefficient (B)	0.0002 N·m·s	0.0002 N·m·s
Inner Loop Bandwidth	500 Hz	500 Hz
Outer Loop Bandwidth	50 Hz	50 Hz
Speed Estimation Bandwidth	100 Hz	100 Hz

This paper emphasizes the time efficiency gained in removing impedance re-measurement for electric machine replacement scenarios. Indeed, this methodology can be advantageous to any component integration within the power system network, provided that their un-terminated TBM is available prior to system integration.

III. EXTRACTION OF TERMINATED AND UN-TERMINATED TBMS

A. Steady State Operating Points Manipulation

Figure 4 illustrates the control setup for model extraction, where an additional induction machine (IM) serves as an active mechanical load emulator. After soft starting [15], both machines operate under traditional rotor-flux-oriented field-oriented control (FOC), featuring high-bandwidth inner current control loops. The outer control loop of the IPMSM regulates speed, while that of the IM regulates torque. Through adjusting the reference speed and torque, the desired operating point, represented by $[I_d, I_q, \Omega_m]^T$, could be indirectly manipulated.

B. Perturbation Injection and Signal Post Processing

Perturbation injection based on frequency sweeping is typically achieved using an additional impedance measurement unit (IMU) [16], which can provide shunt current or series voltage injection. In this paper, the focus is solely on the machine given the assumption that Z_{sDQ} and Y_L are known for calculation of un-terminated TBM $T(s)$. Consequently, the motor drive can serve as the injection facility at the electrical

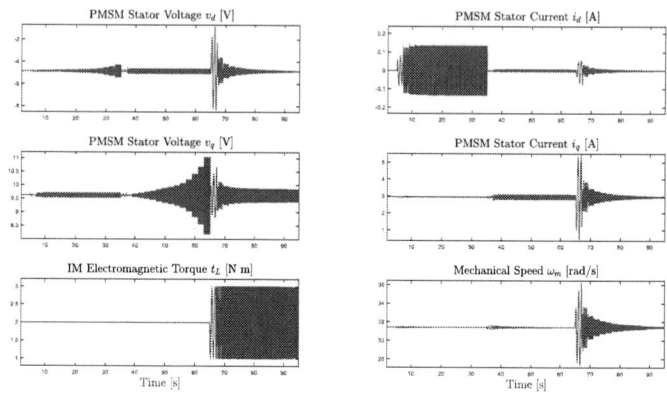

Fig. 5. At operating point $[I_d, I_q, \Omega_m]^T = [0 \text{ A}, 2.94 \text{ A}, 300 \text{ rpm}]^T$, waveform of state variables and model inputs of balanced IPMSM during perturbation.

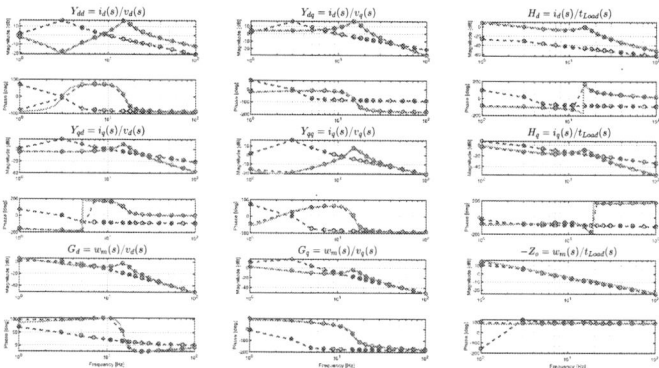

Fig. 6. At operating point $[I_d, I_q, \Omega_m]^T = [0 \text{ A}, 2.94 \text{ A}, 300 \text{ rpm}]^T$, balanced IPMSM impedance comparison among extracted terminated TBM (blue), analytical un-terminated TBM (orange), and extracted un-terminated TBM (red).

interface, while the IM can provide torque perturbation. The potentially unexpected harmonics coming from dead time need compensation or mitigation [17]

In three phase systems without the neutral line, typically there are four impedance describing electrical terminal behaviors, like the 2 by 2 matrix $Y_{DQ}^{Terminated}$, which requires one more perturbation to avoid rank deficient problem. For the three-port network representing three phase rectifiers/inverters or electric machines in this paper, three linear independent perturbations at the same frequency are needed for 3 by 3 unknown impedance matrix as follows:

$$
T(s) = \begin{bmatrix} Y_{dd}(s) & Y_{dq}(s) & H_d(s) \\ Y_{qd}(s) & Y_{qq}(s) & H_q(s) \\ G_d(s) & G_q(s) & -Z_o(s) \end{bmatrix}
$$
$$
= \begin{bmatrix} i_{d1}(s) & i_{d2}(s) & i_{d3}(s) \\ i_{q1}(s) & i_{q2}(s) & i_{q3}(s) \\ \omega_{m1}(s) & \omega_{m2}(s) & \omega_{m3}(s) \end{bmatrix} \begin{bmatrix} v_{d1}(s) & v_{d2}(s) & v_{d3}(s) \\ v_{q1}(s) & v_{q2}(s) & v_{q3}(s) \\ t_{L1}(s) & t_{L2}(s) & t_{L3}(s) \end{bmatrix}^{-1}
$$
(7)

where the subscripts 1, 2 and 3 represent corresponding state variables or model inputs during v_d, v_q and t_L perturbation, respectively. As long as the linear independence is proactively designed, it is feasible to perturb one of the model inputs

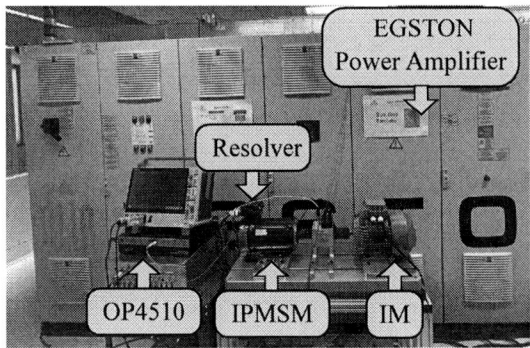

Fig. 7. Experimental Setup.

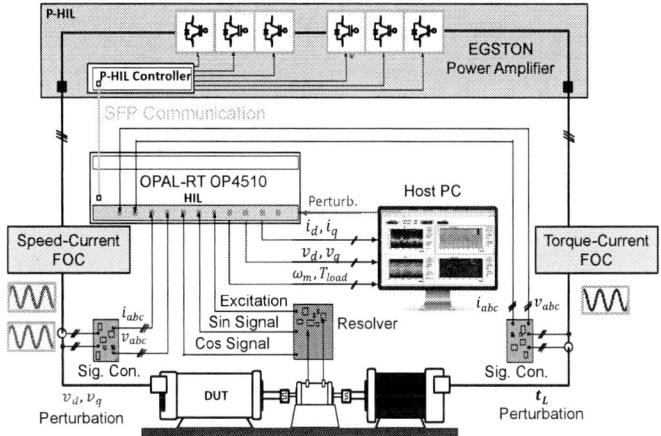

Fig. 8. Experimental diagram for power flow and signal flow.

three times consecutively. However, to circumvent any issues of dependency, the most straightforward approach is to perturb these three model inputs independently.

For selecting the perturbation amplitude, typically 5% of the rated value is chosen. During perturbations involving the motor drive or IM, the control loops of both machines remain closed, which implies that disturbances introduced through the feedforward path in the control loop are significantly attenuated. This attenuation is particularly pronounced in the low-frequency range, where the loop gain is relatively high. Consequently, larger perturbations is required, with the extent of increase depending on the designed magnitude of the loop gain.

C. Simulation Verification

In Matlab&Simulink environment, the simulation corresponding Fig. 4 is constructed with system parameters and critical controller indexes listed in Table. I.

By setting reference speed of IPMSM as 5 Hz and the reference torque of IM as $T_L = 2$ N·m, the steady state operating point $[I_d, I_q, \Omega_m]^T = [0\ \text{A}, 2.94\ \text{A}, 300\ \text{rpm}]^T$ is reached. Fig. 5 illustrates the time-domain waveforms of the state variables and model inputs, where the first five seconds correspond to steady-state operation. The inertia of the electric machine often behaves as a low-pass filter, attenuating higher-frequency dynamics. Therefore, stability evaluation focuses primarily on the low-frequency range. To capture the frequency response, a logarithmic frequency sweep is performed from 1 to 100 Hz, with fifteen evenly distributed frequencies selected for injection. Each frequency injection lasts for 2 s, and the three model inputs are perturbed sequentially across the frequency range, resulting in a total perturbation duration of approximately 90 s

As anticipated based on prior analysis, all state variables and model inputs exhibit oscillatory behavior during single model input perturbation due to the immittance characteristics of the non-ideal motor drive and mechanical loads. Following the perturbation, frequency component extraction is carried out using the Fast Fourier Transform (FFT). The FFT window length is set to match the duration of a single frequency perturbation, which is 2 s. The Fourier coefficients of the frequency components for all model inputs and state variables

are subsequently processed through (7) to derive the un-terminated TBM $T(s)$.

The comparison among extracted terminated TBM (blue), analytical un-terminated TBM (orange), and extracted un-terminated TBM (red) are shown in Fig. 6. The high matching rate between analytical and extracted un-terminated TBM show the effectiveness of the proposed method.

IV. EXPERIMENTAL DEMONSTRATION

A. Experimental Setup

The complete experimental setup is depicted in Fig. 7, while the power flow and signal flow diagram is illustrated in Fig. 8. Instead of utilizing embedded systems such as DSP or FPGA, the controller functionality is implemented on a hardware-in-the-loop (HIL) platform, the OP4510, which enables direct development within the Simulink environment. The motor drive operation is achieved by a high-bandwidth power amplifier, the COMPISO SYSTEM UNIT (CSU) 2000 from EGSTON [18]. This amplifier functions as six independently controllable voltage sources, receiving control commands directly from the OP4510. The interior permanent magnet synchronous machine (IPMSM), model Pactorq E145E3C6N0A8010 [19], and the induction machine (IM), model Marathon R322B [20], are both operated under standard rotor flux-oriented control (RFOC). The rotor speed and position are estimated using a phase-locked loop (PLL) algorithm applied to resolver signals. Electrical measurements are performed using internal sensors in the CSU 2000, and the data is transmitted to the OP4510 via a small form-factor pluggable (SFP) channel.

Since the total inertia J represents the combined inertia of both the machine and the load during the modeling process, it is important to note that the load torque t_L is estimated indirectly by analyzing the stator current of the IM to determine the electromagnetic torque, rather than employing a torque meter positioned between the two machines. To make full use of the real time computing ability, the sampling period for control is set as 50 μs . And the memory limitation of OP4510

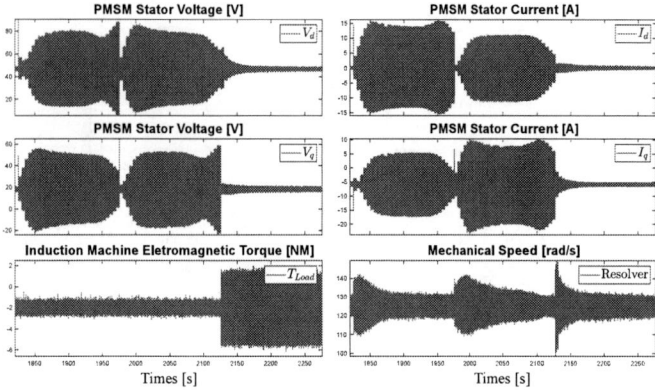

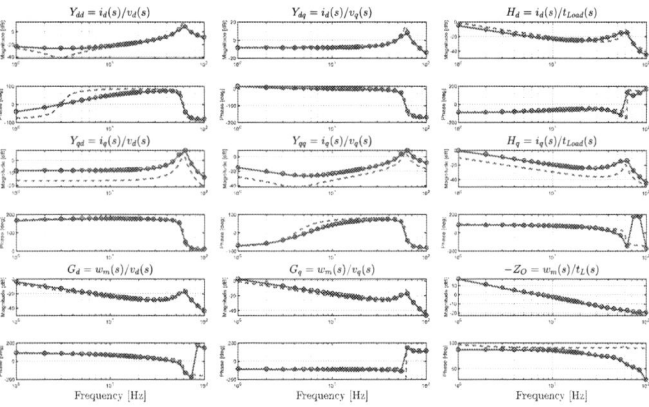

Fig. 9. Experiment at operating point $[I_d, I_q, \Omega_m]^T = [0\ \text{A}, -6\ \text{A}, 1200\ \text{rpm}]^T$, waveform of state variables and model inputs of the IPMSM Pactorq E145E3C6N0A8010 during perturbation.

Fig. 10. At operating point $[I_d, I_q, \Omega_m]^T = [0\ \text{A}, -6\ \text{A}, 1200\ \text{rpm}]^T$, impedance comparison between analytical un-terminated TBM (orange) and extracted un-terminated TBM (blue).

requires trade off among data collection sampling rate, sampling length, and un-terminated TBM modeling accuracy. The selected data sample period is $500\ \mu s$ second with decimation factor of 10 comapred with the sampling period for control.

B. Experimental Results

Setting the reference torque of IM as $T_L = -2\ \text{N} \cdot \text{m}$, Fig. 9 shows the monitored signals during consecutive perturbations. After the system reaches the steady state at $[I_d, I_q, \Omega_m]^T = [0\ \text{A}, -6\ \text{A}, 1200\ \text{rpm}]^T$, consecutive perturbation of v_{dq} and t_L are activated. Thirty evenly distributed frequencies from 1 to 100 Hz on a logarithmical scale are selected. Each perturbation lasts for 5 seconds, and FFT is performed on each 5-second to extract components of each state variable and model inputs at the particular frequencies. The extracted un-terminated TBMs T(s) is shown in Fig. 10. The system parameters used for analytical T(s) come from data sheets and parameter measurement procedures following IEEE standard 1812-2023 [21]. High matchiness between the analytical T(s) and experiment-based T(s) demonstrates effectiveness of the introduced procedures for extraction of un-terminated TBMs.

The subtle deviation can be attributed to several primary factors:

1) Parameter Mismatch: Due to the nonlinear hysteresis characteristics of magnetization or temperature variations, the d-q axis inductances (L_{dq}) and stator resistance (R_s) deviate from the values specified in the datasheet and those obtained through measurements.

2) Limited Bandwidth of Speed-Angle Estimation: The bandwidth of the speed-angle estimation is approximately 40 Hz, which is significantly lower than the desired 100 Hz to suppress noise. This limitation introduces a substantial phase lag in the extracted open-loop impedance (Z_o) at higher frequency ranges.

3) Non-Ideal Mechanical Connection: The mechanical coupling between the two machines is not ideal, exhibiting damping and spring-like effects caused by the rigid coupling between the machines.

V. CONCLUSION

To address the challenges of repeated measurements and to protect intellectual property through a black-box approach, this paper introduces the un-terminated TBM as an effective interface between system integrators and OEMs. A detailed methodology for extracting the un-terminated TBM is presented for IPMSMs, with the approach being extendable to a wide range of components within power system networks. Experimental results demonstrate strong agreement with analytical predictions based on system parameters derived from datasheets and measurements. In addition to resolving the issues discussed in the previous section to enhance model extraction accuracy, future work will focus on stability evaluation. Furthermore, the general applicability of the proposed method will be validated for other equipment, such as grid-tied converters and various types of electric machines.

REFERENCES

[1] H. Liu, J. Zhou, T. Zhao and X. Xu, "Si IGBT and SiC MOSFET Hybrid Switch-Based Solid State Circuit Breaker for DC Applications," 2022 IEEE Energy Conversion Congress and Exposition (ECCE), Detroit, MI, USA, 2022, pp. 1-6.

[2] J. Sun, "Small-Signal Methods for AC Distributed Power Systems–A Review," in IEEE Transactions on Power Electronics, vol. 24, no. 11, pp. 2545-2554, Nov. 2009.

[3] H. Xiao, H. Peng, H. Sun, Y. Zhao, X. Liu and C. Jiang, "Automatic Impedance Matching With Dual Time-Scale P&O in Fully Self-Powered Electromagnetic Vibration Energy Harvesting," in IEEE Transactions on Power Electronics, vol. 39, no. 3, pp. 3377-3390, March 2024.

[4] X. Yang, Q. Lin, V. Mitrovic and R. Burgos, "Extraction and Evaluation of Decoupled Electrical Terminal-Behavior Model for Machine Parameter Variation and Fault Detection," IECON 2024- 50th Annual Conference of the IEEE Industrial Electronics Society, Chicago, IL, USA, 2022, in press.

[5] H. Chen, Q. Huang, W. Li, X. Xiang, H. Luo and X. He, "An Impedance-Based Islanding Detection Method for DC Microgrids with multiple DGs," 2020 4th International Conference on HVDC (HVDC), Xi'an, China, 2020, pp. 1025-1030.

[6] H. Chen et al., "An impedance-based islanding detection method for DC microgrid with multiple distributed generators," in CSEE Journal of Power and Energy Systems.

[7] Q. Lin, B. Wen, R. Burgos, X. Li, Q. Wang and X. Li, "D-Q Impedance Modeling and Stability Analysis of a Three-Phase Four-Wire System With Single-Phase Loads," in IEEE Transactions on Power Electronics, vol. 38, no. 9, pp. 11169-11182, Sept. 2023.

[8] H. Cao, L. Du, F. Guo, Z. Ma and Y. Zhao, "A Triple Active Bridge (TAB) Based Solid-State Transformer (SST) for DC Fast Charging Systems: Architecture and Control Strategy," 2023 IEEE Energy Conversion Congress and Exposition (ECCE), Nashville, TN, USA, 2023, pp. 855-860.

[9] K. Strunz and E. Carlson, "Nested Fast and Simultaneous Solution for Time-Domain Simulation of Integrative Power-Electric and Electronic Systems," in IEEE Transactions on Power Delivery, vol. 22, no. 1, pp. 277-287, Jan. 2007.

[10] B. H. Cho and F. C. Y. Lee, "Modeling and analysis of spacecraft power systems," in IEEE Transactions on Power Electronics, vol. 3, no. 1, pp. 44-54, Jan. 1988.

[11] S. Hiti, D. Boroyevich and C. Cuadros, "Small-signal modeling and control of three-phase PWM converters," Proceedings of 1994 IEEE Industry Applications Society Annual Meeting, Denver, CO, USA, 1994, pp. 1143-1150 vol.2.

[12] D. Boroyevich, I. Cvetković, D. Dong, R. Burgos, F. Wang and F. Lee, "Future electronic power distribution systems a contemplative view," 2010 12th International Conference on Optimization of Electrical and Electronic Equipment, Brasov, Romania, 2010, pp. 1369-1380.

[13] J. Huang, K. A. Corzine and M. Belkhayat, "Small-Signal Impedance Measurement of Power-Electronics-Based AC Power Systems Using Line-to-Line Current Injection," in IEEE Transactions on Power Electronics, vol. 24, no. 2, pp. 445-455, Feb. 2009.

[14] R. Ilka et al., "Multi-Physics Modeling of Power Electronic Converters with Liquid Immersion Cooling," 2023 IEEE Energy Conversion Congress and Exposition (ECCE), Nashville, TN, USA, 2023, pp. 4698-4704.

[15] J. Zhou, H. Liu, T. Zhao, X. Xu and Y. Wang, "SiC Bidirectional Solid-State Circuit Breaker with Soft-Start Function for Motor Control Center," 2023 IEEE Applied Power Electronics Conference and Exposition (APEC) Orlando, FL, USA, 2023, pp. 2307-2312.

[16] Z. Shen, M. Jaksic, P. Mattavelli, D. Boroyevich, J. Verhulst and M. Belkhayat, "Design and implementation of three-phase AC impedance measurement unit (IMU) with series and shunt injection," 2013 Twenty-Eighth Annual IEEE Applied Power Electronics Conference and Exposition (APEC), Long Beach, CA, USA, 2013, pp. 2674-2681,

[17] Ji, Yi, Yong Yang, Jiale Zhou, Hao Ding, Xiaoqiang Guo, and Sanjeevikumar Padmanaban. 2019. "Control Strategies of Mitigating Deadtime Effect on Power Converters: An Overview" Electronics 8, no. 2: 196.

[18] "CSU 200 Power Amplifier," EGSTON Power Electronics, Oct. 03, 2024. https://www.egstonpower.com/portfolio/csu200/

[19] POWERTEC, " Brushless Motor Pactorq E Series Manual,", 2012. [Online]. Available: https://www.powertecmotors.com/wp-content/uploads/PacTorq-Motor-Manual-w_E252-Addendum.pdf.

[20] Marathon, "Specification Sheet," R322B, Jan, 2019. [Online]. Available: https://documents.mrosupply.com/product_documents/59/18/5918058/arathon_TCA2P22AE211GAA009_Specification_Sheet_uPCox7i.pdf.

[21] "IEEE Guide for Testing Permanent Magnet Machines," in IEEE Std 1812-2023 (Revision of IEEE Std 1812-2014) , vol., no., pp.1-88, 11 Dec. 2023.

7.2 kW GaN-Based DAB Converter with 37 kW/L Power Density and High Efficiency

Esmaeil Jalalabadi and Xiaoyu Wang, *Senior Member IEEE*
Dept. of Electronics, Carleton University, Ottawa, Canada
Email: esmaeiljalalabadi@cmail.carleton.ca,
xiaoyuwang3@cunet.carleton.ca

Jaksa Rubinic, Yang Jiao, Lucas Lu
Infineon Technologies AG, Ottawa, Canada
Email: jaksa.rubinic@infineon.com,
yang.jiao@infineon.com, lucas.lu@infineon.com

Abstract- **This paper presents a GaN-based high power density and isolated DC-DC converter with dual active bridge topology for DC/DC stage of a bidirectional 400V EV on board charger (OBC). Gallium Nitride (GaN) technology in top side cooled package is employed to achieve higher switching frequency with more compact passive components solutions. The paper introduces the design parameters for the DAB circuit with wide range output, variable frequency controller as well as two high frequency planar transformer designs to achieve 125 kW/L power density. System level loss breakdown and circuit protection scheme and implementation are also provided. The 7.2kW DAB prototype has been implemented and tested in constant voltage (CV) mode, proving exceptional power density of 37 kW/L with 98.8% peak efficiency.**

Index Terms— **Dual Active Bridge (DAB), On Board Charger (OBC), Power Density, wide-bandgap devices, Isolated DC-DC Converter, Fault Protection, Digital Control, Variable frequency.**

I. INTRODUCTION

Battery chargers are pivotal in the advancement of EVs, influencing both charging time and battery longevity through their parameters. An efficient and reliable battery charger must be cost-effective, compact, lightweight, and capable of high output power [1]. Lightweight solutions are critical in the automotive sector due to stringent size constraints for onboard chargers (OBCs) [2]. Additionally, galvanic isolation is often necessary for safety and voltage adaptation between the grid and the DC side [3], with high-frequency (HF) transformers helping to minimize size and weight [4]. Among DC-DC topologies, the Dual Active Bridge (DAB) converter stands out for its bidirectional power flow, galvanic isolation, straightforward control, minimal components, and high efficiency. This converter uses two active full bridges connected by an HF transformer, achieving bidirectional power transfer with minimal energy storage. It is ideal for battery energy storage systems, EV charging converter, and renewable energy interfaces where efficient power conversion and isolation are crucial.

Figure 1 illustrates the proposed single-phase bidirectional OBC using interleaved Totem-pole PFC and full-bridge DAB topology. The interleaved GaN-based PFC stage has been already developed and more details can be found in [5].

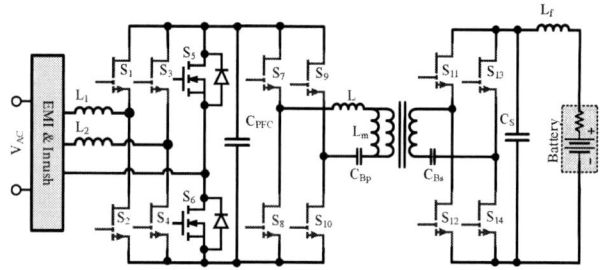

Fig. 1 Simplified structure of the proposed single-phase bidirectional onboard charger (OBC)

Power transfer in a DAB converter is determined by the voltage amplitude, switching frequency, and phase-shift between the voltages across the transformers. By controlling the gate signals of the two full bridges, one can control the power flow's direction and magnitude. High-frequency operation not only reduces passive components' size and weight, but also improves the converter's dynamic response [6-8].

Research in the literature has extensively focused on improving various performance aspects of DAB converters, including efficiency, power transfer capability, and dynamic performance. These improvements can be broadly categorized into two approaches: modulation-based methods and topology modification techniques. Modulation methods aim to optimize the voltage applied to the leakage inductor by controlling the phase shift angle and switching frequency [7-11].

In this paper, the following design objectives are considered:

➢ Achieving high power density with optimizing transformer design considering losses, parameter consistency and thermal management.
➢ Minimizing power loop and gate loop inductances to adapt for high switching speed of GaN devices
➢ Implementing robust control strategies to achieve good performance over the whole EV battery charging profile
➢ Reliable protections for transformer & GaN devices
➢ High efficiency and thermal management

II. DAB CIRCUIT PARAMETER DESIGN

TABLE I
DAB System and Circuit Parameters

Input Voltage (DC+AC)	DC = 385v (355~425v) AC =< 15v 120Hz (90~130) Hz
Output Voltage	240v~450v
Max Power	7200 W
Switching Frequency F_{sw}	150kHz ~1 MHz
Transformer Leakage Inductor (L)	5µH
Transformer Magnetizing Inductance (L_m)	750 µH
Switches	8 × IFX GaN (R_{DSon} = 20mΩ, C_{OSS} = 150pF)
Dead-time	50 ns
Transformer Blocking Capacitors (C_{Bp}, C_{Bs})	3 µF
Input and output capacitors (C_P, C_S)	100 µF
Output filter (L_f)	170nH

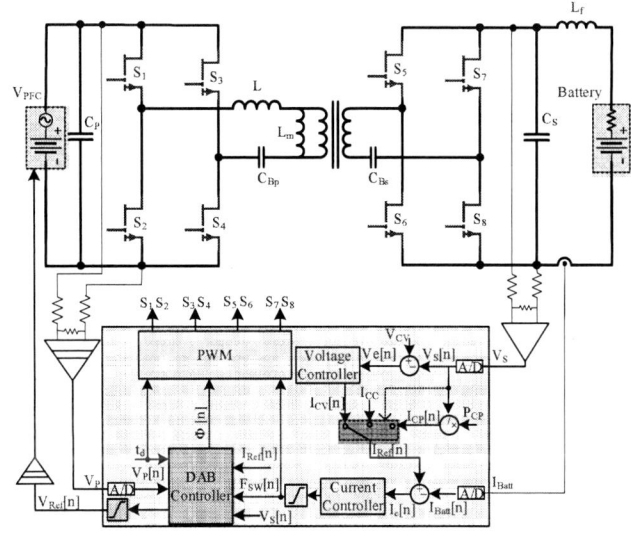

Fig. 2. Overall control scheme of the proposed approach

The optimized design parameters and component selection is critical to maintain a high efficiency performance for a wide range of input/output voltage and power. The circuit parameters and specifications for the proposed 7.2 kW DAB converter including, input/output voltage range, transformer turns ratio and leakage inductance, switching frequency range, input/output capacitors, and battery side filter in a 400V on-board charger (OBC) application are detailed in Table I. The parameters will serve as the basis for simulation and will be instrumental in illustrating the proposed design and control methodology. The design procedure corresponding to these values will be elaborated upon in the subsequent sections.

OVERALL CONTROL SCHEME

The derived power transfer equations highlight that power can be effectively controlled by manipulating the primary DC link voltage (i.e., PFC output voltage), switching frequency, and phase shift, φ_{ZVS}. Therefore, in the proposed control strategy, the switching frequency and φ_{ZVS} are selected as the primary control variables to enable a fast response to power demands. Meanwhile, the average of the PFC DC link voltage is used as a secondary control variable to maximize the practical ZVS region of the DAB and to improve efficiency.

The overall block diagram of the proposed DAB controller is shown in Figure 2. In this design, the PFC output voltage is modeled as a combination of DC and AC voltage sources. The DC reference value for the PFC is generated by the proposed controller based on the output reference current, output voltage, and the derived power transfer equations. The AC voltage source for the PFC is modeled as follows:

$$\tilde{V}_{PFC}(t) = \frac{P_{out}}{C_O \omega_g} \sin\left(2\omega_g t\right) \tag{1}$$

Where ω_g is the grid frequency, C_O is the total capacitance of the PFC output capacitor banks, and P_{out} is the PFC output power or the DAB input power. C_O is set to 1.47 mF as specified in [5].

The current controller ensures that the reference current is accurately regulated with its output being the switching frequency. This arrangement compensates for AC variations in the input voltage by adjusting the switching frequency. Higher switching frequencies result in lower output current.

The voltage controller is primarily used for the Constant Voltage (CV) charging mode, and its output is the reference output current. The proposed controller also supports Constant Current (CC) and Constant Power (CP) modes, which can be activated as needed.

The DAB controller calculates and sets the PFC reference voltage and φ_{ZVS} in real time. The PWM module represents the digital PWM peripherals of the MCU, which generate the final PWM signal sequence for the corresponding gate drivers. The MCU is placed on the secondary side, sharing the same ground as the battery side. Proper isolation is provided for primary voltage measurement and PFC voltage command/signals, as illustrated in Figure 2.

The key control variables considered are the input voltage, ranging from 340V to 440V, and the switching frequency, which varies between 150 kHz and 1 MHz. However, in light load conditions, there is a noticeable loss of ZVS, either for the secondary switches or the primary switches, depending on whether the converter is operating in buck/boost or

979-8-3315-1612-3/25 $31.00 © 2025 IEEE   417

charging/discharging mode. This non-ZVS region indicates that ZVS benefits diminish under lighter loads, which could lead to reduced efficiency.

III. ZERO VOLTAGE SWITCHING (ZVS) CRITERIA

The mathematical formulation for the proposed control is developed. The introduced control variable φ_{ZVS} ensures soft switching transitions and the minimum practical current with the required dead-time. To avoid losing ZVS during zero current crossing, φ_{ZVS} must satisfy the dead-time period constraint as follows:

$$\varphi_{zvs} \geq \frac{t_d}{T_{sw}} \tag{2}$$

Figure 3 illustrates the dead-time region, and the minimum required φ_{zvs} for maintaining ZVS operation in the proposed approach during zero crossings.

Let us define $V_s' = V_s / n$ where V_s' is the secondary voltage of the transformer transferred to the primary side, and n is the transformer turn ratio. The corresponding transition switching current I_{zvs}, during the zoomed interval in Figure 3 is calculated as follows:

$$I_{zvs} = \frac{\varphi_{zvs}\left(V_p + V_s'\right)T_{sw}}{4L} \tag{3}$$

Where T_{sw} is the switching period. Utilizing the energy equation, the minimum value of I_{zvs} required to ensure the practicality of the ZVS approach at the transition current can be derived. This minimum current is constrained by the switch output capacitance and maximum voltage, and is expressed as follows:

$$I_{zvs} \geq 2V_p\sqrt{\frac{C_{oss}}{L}} \tag{4}$$

To ensure ZVS for all switches in the minimum $|\varphi_{zvs}|$ is determined by combining equations (2) and (3), resulting in the following condition:

$$|\varphi_{zvs}| \geq \frac{8V_p F_{sw}\sqrt{LC_{oss}}}{\left(V_p + V_s'\right)} \tag{5}$$

Therefore, φ_{zvs} must satisfy both equations (1) and (4), leading to the final requirement:

$$|\varphi_{zvs}| \geq F_{sw} \times max\left\{\frac{8V_p\sqrt{LC_{oss}}}{\left(V_p + V_s'\right)}, t_d\right\} \tag{6}$$

This condition ensures that the ZVS is maintained for all switches, thereby enhancing the overall efficiency of the converter.

IV. Design Procedure

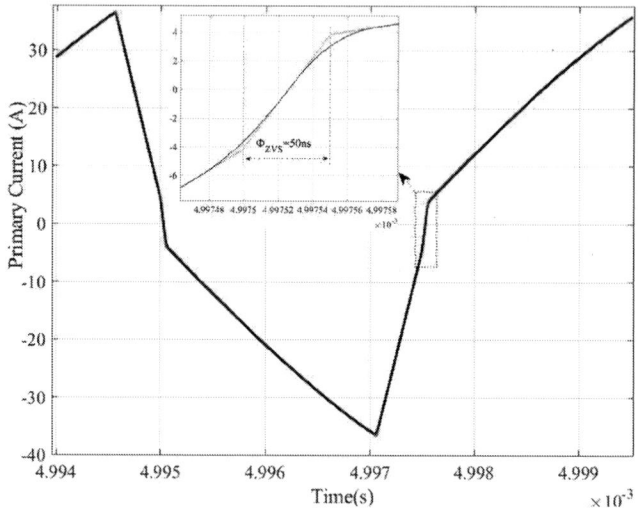

Fig. 3. Minimum phase shift and dead time for ZVS criteria.

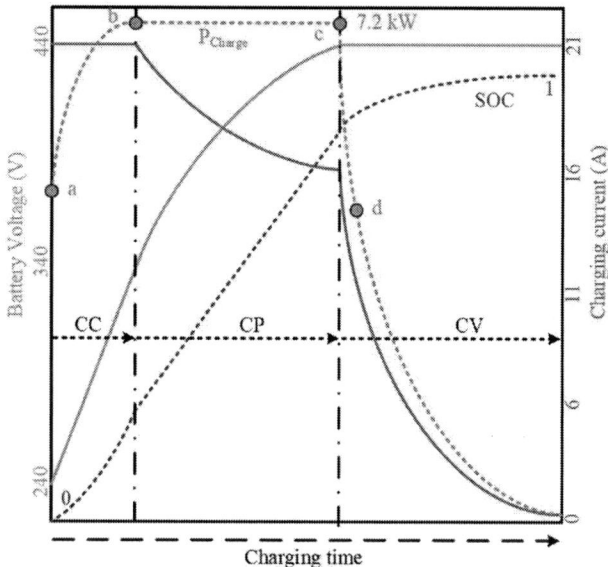

Fig. 4. Charging profile of a typical 400v EV battery.

A typical 400V EV battery charging curve for a 7.2 kW OBC is illustrated in Figure 4. The curve shows the battery voltage and corresponding charging modes required for the reliable operation of the EV battery. This charging profile is considered crucial in the design of the Dual Active Bridge (DAB) circuit parameters, as the system is expected to operate primarily in Grid-to-Vehicle (G2V) charging mode.

The design procedure will ensure that the DAB converter meets the necessary requirements for efficient and reliable operation throughout the charging process.

To meet the operating profile of the battery and the PFC output voltage range, the transformer turns ratio is selected based on the midpoint of the PFC output voltage range under nominal working conditions. This midpoint is approximately 385V, and the battery voltage at half charge is around 390V. Consequently, a unity turns ratio (*n=1*) is considered to simplify the design and ensure efficient power transfer.

A. Magnetizing Inductance and Peak Flux Density

Next, the transformer parameters, such as magnetizing inductance and peak flux density, need to be calculated based on the DAB circuit parameters. These parameters are crucial for ensuring that the transformer operates efficiently without saturation and within the desired switching frequency range.

The magnetizing inductance (L_m) and peak flux density (B_p) are determined using the following equations:

$$L_m = \frac{N^2 \mu A_c}{l_c} \tag{7}$$

$$B_p = \frac{\mu N I_{P_m}}{l_c} \tag{8}$$

Where N is the number of turns on the primary side of the transformer, μ is the magnetic permeability of the core, A_c is the core area, l_c is the core length, and I_{P_m} is the peak magnetizing current which is calculated as:

$$I_{P_m} = \frac{V_{dc} T_{sw}}{4 L_m} = \frac{V_{dc}}{4 L_m F_{sw}} \tag{9}$$

V_{dc} is the DC link voltage, T_{sw} is switching period, and F_{sw} is the switching frequency.

B. Core Material and Switching Frequency

Given the requirement for high power density, the core material N97 is selected, which has a saturation flux density of about 400mT. To avoid core saturation and high losses, the peak flux density B_p is limited to 200mT. The minimum switching frequency required to meet this criterion is calculated as:

$$F_{sw} \geq \frac{V_{dc_{max}}}{4 \times 0.2 \times N A_c} \tag{10}$$

A larger transformer size ($N A_c$), allows for a lower minimum switching frequency.

EELP cores, as in Figure 5, are chosen for their high power density and effective heat dissipation. Using three EELP 38/8/25 cores in parallel increases the core area.

With a 1:1 turn ratio for PCB winding design, the number of turns for the planar transformer is set to 8:8 as in Figure 6. The minimum switching frequency is calculated as:

$$F_{sw} \geq \frac{440}{4 \times 0.2 \times 8 \times 3 \times 0.000194} \rightarrow F_{sw} \geq 120 \ kHz \tag{11}$$

Leakage inductance is calculated to meet power delivery and ZVS requirements under full load conditions. With a decided

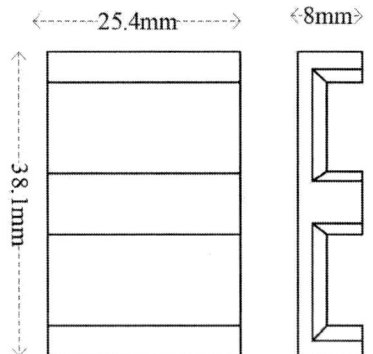

Fig. 5. Selected magnetic cores for HF transformer design.

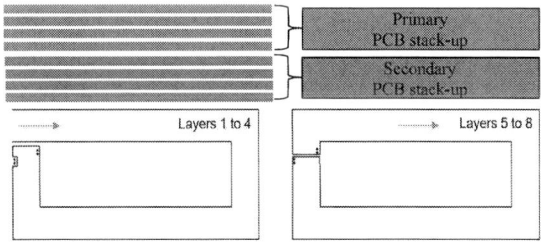

Fig. 6. Proposed HF transformer PCB windings and stack-up.

switching frequency of 150kHz and a worst-case battery voltage of 440V, the leakage inductance is determined to be 5μH. Finally, the switches are selected based on the maximum current (35A), peak voltage (440V), and maximum switching frequency (1MHz). Infineon 650V/20mΩ CoolGaN® device with 20mΩ R_{DSon}, 60A maximum current, and 650V maximum drain-source voltage, are chosen for both primary and secondary switches.

The design approach ensures the converter meets the performance, efficiency, and power density requirements for the application. Figure 7 shows the loss breakdown simulation results for different operation points, *a* to *d* as in Figure 4 using Mathcad.

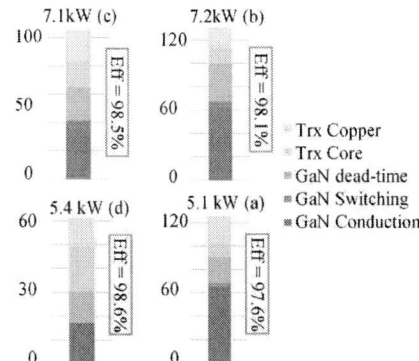

Fig. 7. Loss breakdown simulation at different charging points.

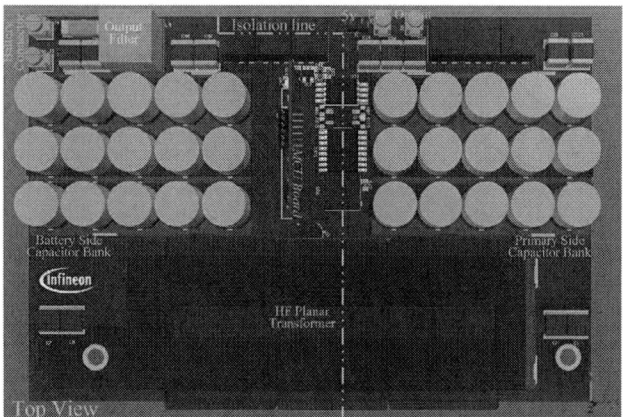

Fig. 8. Top view of the designed DAB converter.

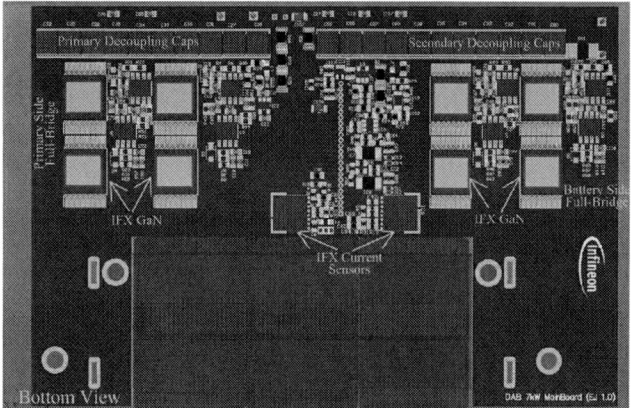

Fig. 9. Bottom view of the designed DAB converter.

V. DAB Prototype

Top and bottom side 3D views of designed DAB converter with brief components description is illustrated in Figures 8 and 9. Figure 10 shows the realized hardware of DAB 7.2kW and the corresponding dimensions. The prototype proves outstanding 37 kW/L power density.

An alternative winding design using Litz wire with AWG 15, 8:9 turns, for the winding area with winding factor of 0.7, is developed providing 25mΩ DC resistance, and satisfying the transformer design values and dimensions. This transformer design is shown in Figure 11.

VI. Experimental Results

The proposed design based on wire windings is shown in Figure 12 for the test performance. Wire winding design is chosen as the PCB winding transformer does not show an acceptable performance due to higher winding capacitance and AC resistance. Also, wire winding design is much better than PCB winding from manufacturability and cost point of views.

Fig. 10. Implemented 7.2kW DAB converter.

Fig. 11. High-density wire-winding DAB converter transformer design

Fig. 12. Tested design of 7.2kW DC-DC DAB converter

The overall dimension of these two designs is identical, and it will not affect the overall converter power density.

Figures 13 to 15 show the DAB converter waveforms in ZVS regions for start-up, light load at 1-MHz, and full load at 150-kHz switching frequency, respectively.

Black waveform shows the primary side voltage, the blue waveform shows the secondary side voltage, the red waveform shows the transformer primary side current, and the green waveform shows the battery side current. Start-up test results show a smooth and stable charging current.

The proposed design and controller can work under ZVS at about 2.8 kW power at high frequency (1 MHz) with above 97% efficiency and good thermal performance.

Figure 16 shows the in-progress test setup in half load at maximum efficiency of 98.83%.

Thermal performance at full-load conditions is shown in Figure 17. Some efficiency test results at different load conditions are shown in Table II. The proposed compact converter design proves above 98.4% efficiency for a wide range of power (2.2 kW to 7.2kW).

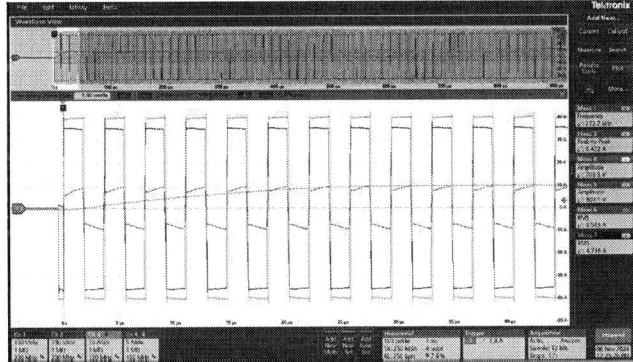

Fig. 13. Start-up under ZVS for DC-DC DAB converter

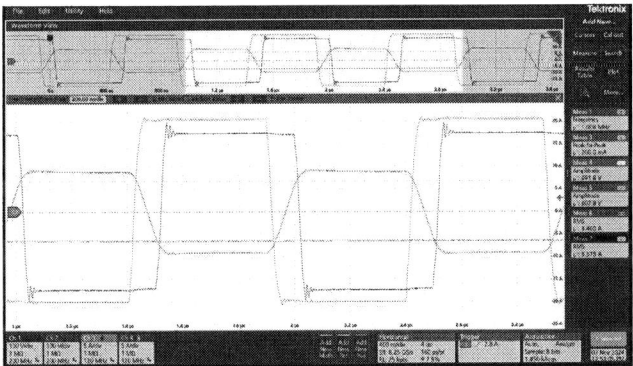

Fig. 14. Light load at one MHz switching frequency under ZVS

TABLE III
EFFICIENCY RESULTS AT $V_P = 355V$ AND $V_S = 410V$

	@ 150 kHz	@ 300 kHz	@ 500 kHz	@ 1 MHz
2.2 kW	98.7	98.3	98.0	97.1
3.6 kW	98.8	98.0	97.6	96.3
5.8 kW	98.5	97.7	97.1	
7.2 kW	98.4			

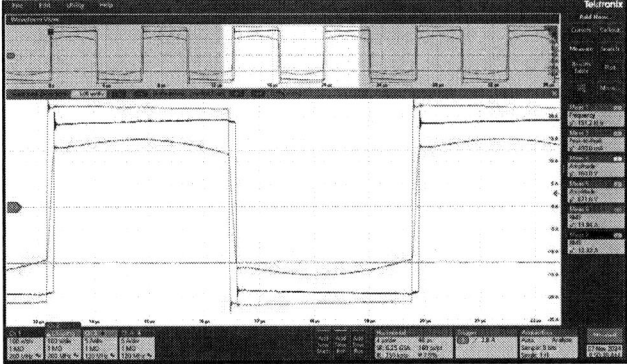

Fig. 15. Full load at 150 kHz switching frequency under ZVS

Fig. 16. Peak efficiency test setup at about half load charging mode

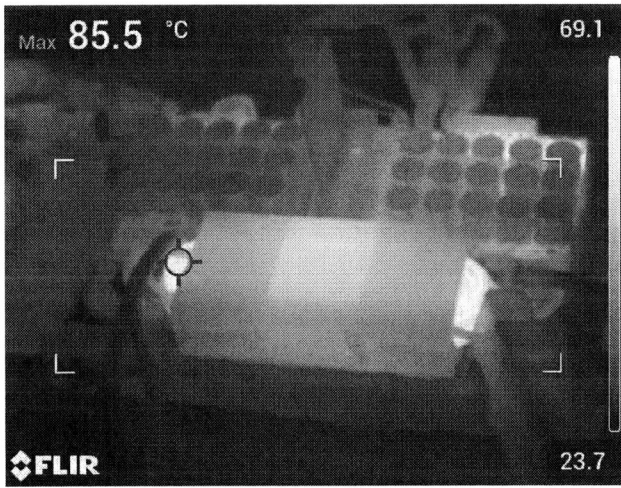

Fig. 17. Thermal performance under full load charging mode.

VII. Conclusions

This paper presented a high-power-density, isolated DC-DC converter using GaN technology for EV on-board chargers. The proposed 7.2 kW dual active bridge (DAB) converter achieved exceptional power density of 37 kW/L with outstanding 98.8% peak efficiency. Key features include wide-range output capability, variable frequency control, and high-frequency planar transformer designs. The prototype demonstrated excellent performance across various operating conditions, including ZVS operation at light and full loads. These results highlight the potential of GaN-based converters for compact, efficient EV charging solutions.

Future work may focus on further optimizing the control strategy for ultra-light load conditions and exploring its applicability in other power conversion scenarios.

ACKNOWLEDGMENT

This research was made possible through the generous funding provided by the MITACS program. The authors gratefully acknowledge MITACS for their financial support, which enabled the collaboration between academia and industry, fostering innovation in power electronics.

The authors also extend their appreciation to our colleagues and the technical staff at our institution and Infineon Technologies Canada Inc. for their assistance and support during the experimental phase of this research.

REFERENCES

[1] Yilmaz, M.; Krein, P.T. Review of battery charger topologies, charging power levels, and infrastructure for plug-in electric and hybrid vehicles. IEEE Trans. Power Electron. 2013, 28, 2151–2169.

[2] Zhang, S.; Zhang, C.; Xiong, R.; Zhou, W. Study on the optimal charging strategy for lithium-ion batteries used in electric vehicles. Energies 2014, 7, 6783–6797.

[3] J. Lu et al., "A Modular-Designed Three-Phase High-Efficiency High-Power-Density EV Battery Charger Using Dual/Triple-Phase-Shift Control," in IEEE Transactions on Power Electronics, vol. 33, no. 9, pp. 8091-8100, Sept. 2018, doi: 10.1109/TPEL.2017.2769661.

[4] A. KhakparvarYazdi and S. Ali Khajehoddin, "Low-Profile Fractional Planar Transformer Based on a Novel Infinite-shape pcb winding For 5kW Dual Active Bridge Converter," 2024 IEEE Applied Power Electronics Conference and Exposition (APEC), Long Beach, CA, USA, 2024, pp. 838-845, doi: 10.1109/APEC48139.2024.10509476.

[5] E. Jalalabadi et al., "GaN-Based Bidirectional 6.6kW Interleaved Totem-Pole PFC with 13 kW/L Power Density and High Efficiency," PCIM Europe 2024; International Exhibition and Conference for Power Electronics, Intelligent Motion, Renewable Energy and Energy Management, Nuremberg, Germany, 2024.

[6] H. van Hoek, M. Neubert, and R. W. De Doncker, "Enhanced modulation strategy for a three-phase dual active bridge—Boosting efficiency of an electric vehicle converter," IEEE Trans. Power Electron., vol. 28, no. 12, pp. 5499–5507, Dec. 2013.

[7] F. Krismer, "Modeling and optimization of bidirectional dual active bridge DC-DC converter topologies," Ph.D. dissertation, Eidgenössische Technische Hochschule, ETH Zürich, Zürich, Switzerland, 2010.

[8] S. Wei, Z. Zhao, K. Li, L. Yuan, and W. Wen, "Deadbeat current controller for bidirectional dual-active-bridge converter using an enhanced SPS modulation method," IEEE Trans. Power Electron., vol. 36, no. 2, pp. 1274–1279, Feb. 2021. control," IEEE Trans. Power Electron., vol. 35, no. 9, pp. 9886–9903, Sep. 2020.

[9] B. Zhao, Q. Song, and W. Liu, "Power characterization of isolated bidirectional dual-active-bridge DC–DC converter with dual-phase-shift control," IEEE Trans. Power Electron., vol. 27, no. 9, pp. 4172–4176, Sep. 2012.

[10] Z. Guo and X. Han, "Control strategy of AC-DC converter based on dual active bridge with minimum current stress and soft switching," IEEE Trans. Power Electron., vol. 37, no. 9, pp. 10178–10189, Sep. 2022.

A Novel Interleaving Method for High Power Integrated Electric Vehicle Charger with Three-Phase Permanent Magnet Synchronous Motor

Ryota Tanaka
Research Division
Nissan Motor Co., Ltd.
Kanagawa, Japan
ryota-tanaka@mail.nissan.co.jp

Toshihiro Kai
Research Division
Nissan Motor Co., Ltd.
Kanagawa, Japan
to-kai@mail.nissan.co.jp

Kenta Takishima
Research Division
Nissan Motor Co., Ltd.
Kanagawa, Japan
kenta-takishima@mail.nissan.co.jp

Yoshiyuki Nagai
Research Division
Nissan Motor Co., Ltd.
Kanagawa, Japan
y-nagai@mail.nissan.co.jp

Tetsuya Hayashi
Research Division
Nissan Motor Co., Ltd.
Kanagawa, Japan
t-hayashi@mail.nissan.co.jp

Kantaro Yoshimoto
Department of Robotics and Mechatronics
Tokyo Denki University
Tokyo, Japan
kantaro@mail.dendai.ac.jp

Abstract— The voltage of electric vehicle (EV) batteries has shifted from the conventional 400 V class to the 800 V class, necessitating boost converters to enable charging 800 V class batteries with 400 V class chargers. To reduce the additional costs associated with these boost converters, integrated chargers utilizing a traction motor and inverter have been proposed. Interleaving strategies that employ three-phase windings or only two out of the three phases have been suggested to effectively suppress battery charging current ripple. However, these strategies face challenges such as increased magnetic losses due to interleaving and higher root-mean-square current per phase when conduction is restricted to two phases. These challenges result in temperature rises in permanent magnet synchronous motors, thereby limiting both output and the duration of fast charging. This paper proposes a new interleaving method that uses three-phase windings while suppressing magnetic losses for high-power charging with permanent magnet synchronous motors. The effectiveness of this method is confirmed through finite element analysis and experimental results.

Keywords— *Integrated charger, Step-up converter, DC charging, Electric vehicle, Permanent magnet synchronous motor, magnetic loss reduction, Interleaving*

I. INTRODUCTION

In recent years, there has been a trend toward increasing the capacity of electric vehicle (EV) batteries to extend driving range. Fast charging of high-capacity batteries can be achieved by raising the voltage of EV batteries from the conventional 400 V class to the 800 V class, thereby enhancing the charging output. However, installing a boost converter in EV to charge 800 V class batteries with 400 V class chargers can result in higher equipment costs. To mitigate these challenges without significant cost escalation, integrated chargers that utilize a drive motor and an inverter have been proposed [1]–[10]. The integrated charger can utilize the motor windings, switching

devices of the inverter, smoothing capacitors, harnesses, cooling systems, current sensors, and controller, allowing it to accommodate a high output while maintaining a high cost increases. Nevertheless, the inductance of the motor windings, which are not designed for charging, and the switching frequency of the inverter may not sufficiently suppress the charging current ripple. Charging current ripple can potentially cause overdischarging and overcharging of the battery, which is why standards (IEC 61851-23) specify the allowable limits.

To address the limited inductance of motor windings, strategies such as interleaving with three-phase windings [3]–[5] or using only two of the three phases [1][2] have been suggested to effectively suppress charging current ripple. However, these approaches face challenges, including increased magnetic losses due to interleaving, leading to temperature rises in motor magnets, and heating of motor windings due to higher root-mean-square current per phase when conduction is restricted to two phases. These temperature increases limit both the output and the duration of fast charging. This paper proposes a novel switching method that reduces magnetic losses while energizing all three phases. In previous researches [1]–[5], symmetric phase-shift interleaving has been used, where the switching phases of each UVW phase are evenly shifted according to the number of coils being energized. While symmetric interleaving can minimize the fundamental harmonic component of the charging current ripple, it causes an increase in magnetic losses due to the periodic variation in the direction of the magnetic flux generated by the motor. The proposed method involves designing the phase shifts of interleaving to eliminate the component of magnetic flux which increases the magnetic loss. This paper provides results from finite element analysis (FEA) and experiments that demonstrate the magnetic loss reduction effects of the proposed method.

II. INTERLEAVING FOR MAGNETIC LOSS REDUCTION

This section theoretically explains the reason why applying conventional symmetric interleaving to integrated chargers using motor windings leads to increased magnetic losses. Next, based on the theory, the design method for the phase shift of interleaving to reduce the magnetic loss is proposed.

A. Symmetric phase-shift interleaving of inverter switching

Symmetric interleaving of inverter switching reduces the fundamental frequency ripple in the charging current [1]–[5]. Losses owing to the current ripple generated in the motor intensify because of the d-axis current ripple motor described by a direct–quadrature (dq) coordinate system increases rotor magnet eddy current losses, which warrants attention [11]. The circuit diagram of the three-phase permanent magnet (PM) synchronous motor and inverter is shown in Fig. 1. Each phase of the inverter has complementary upper and lower arm switches, which can be in one of two states: either the upper arm is on or off. Consequently, the three phases of the inverter can adopt one of eight possible switching states, S0–S7, as detailed in Table I. Based on the switching state, the UVW phase voltages v_u, v_v, and v_w take on values of either the inverter voltage E_{dc} or 0. The dq-axes voltages v_d and v_q listed in Table I were obtained via a coordinate transformation expressed as

$$\begin{bmatrix} v_d \\ v_q \end{bmatrix} =$$

$$\sqrt{\frac{2}{3}} \begin{bmatrix} \cos\theta & \cos\left(\theta - \frac{2}{3}\pi\right) & \cos\left(\theta + \frac{2}{3}\pi\right) \\ -\sin\theta & -\sin\left(\theta - \frac{2}{3}\pi\right) & -\sin\left(\theta + \frac{2}{3}\pi\right) \\ \frac{1}{\sqrt{2}} & \frac{1}{\sqrt{2}} & \frac{1}{\sqrt{2}} \end{bmatrix} \begin{bmatrix} v_u \\ v_v \\ v_w \end{bmatrix} \quad (1)$$

In interleaving switching, when the on-duty D is 50% and the switching phases Φ_u, Φ_v, Φ_w are shifted by one-third of a

switching period, the switching changes at intervals of one-sixth of the period, as depicted in Fig. 2 (a). The relationship between v_d, v_q, and currents i_d and i_q is expressed as follows:

$$\begin{bmatrix} v_d \\ v_q \end{bmatrix} = \begin{bmatrix} R + pL_d & -L_q\omega \\ L_d\omega & R + pL_q \end{bmatrix} \begin{bmatrix} i_d \\ i_q \end{bmatrix} + \begin{bmatrix} 0 \\ \psi_a\omega \end{bmatrix} \quad (2)$$

Here, L_d and L_q denote the inductances of the motor windings along the d-axis and q-axis, respectively. R is the winding resistance, ω denotes the motor's electrical angular velocity, and ψ_a stands for the magnetic flux of the permanent magnet. For instance, with L_d = 100 μH, L_q = 300 μH, a switching frequency f_{sw} = 10 kHz, and sufficiently low values of R and ω, the current components i_d and i_q exhibit the pattern shown in Fig. 2(b). In motors where $L_d < L_q$, a relatively large d-axis current ripple can occur, resulting in increased rotor magnetic losses. Because cooling of rotor magnets is difficult, they are prone to demagnetization due to elevated temperatures. Therefore, minimizing rotor magnetic losses is crucial to enhance high-power output.

B. The phase shift design of interleaving for magnetic loss reduction

A novel method for suppressing rotor magnetic losses while interleaving the three phases is proposed in this study. To minimize eddy current losses in the rotor magnet, the d-axis current ripple Δi_d must be reduced to zero. As indicated by (2), maintaining v_d at zero prevents the occurrence of Δi_d. Table I shows that when the rotor angle $\theta = \pi/2$, $v_d = 0$ V in switching states S1 and S4 (in addition to S0 and S7). Therefore, by employing a switching method that alternates between S1 and S4, Δi_d can be eliminated. The transitions between switching states in this method are illustrated in Fig. 3(a). One of the three phases, specifically the u phase, is shifted by half a switching period, making it an interleaving phase where the upper arms of two phases alternate. The i_d and i_q waveforms for this scenario are shown in Fig. 3(b). Similar switching can be achieved by selecting S2 and S5 or S3 and S6 at $\pi/2 + \pi/3$ or $\pi/2 - \pi/3$, respectively, to ensure Δi_d is zero. However, the i_q peak of the proposed interleaving method reaches 36 A, which is 15% higher than that of conventional interleaving. This increase in i_q peak may result in greater motor vibrations and noise; however, the fundamental and harmonic frequencies remain the same or exceed the inverter switching frequency. For instance, a 10 kHz frequency falls within a range unlikely to cause human discomfort.

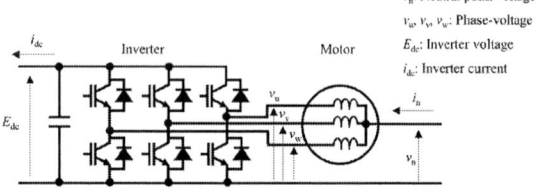

v_n: Neutral point voltage
v_u, v_v, v_w: Phase-voltage
E_{dc}: Inverter voltage
i_{dc}: Inverter current

Fig. 1. Schematic of integrated EV charger with traction motor/inverter

TABLE I. LIST OF SWITCHING STATES AND VOLTAGES FOR UVW–PHASE AND DQ–AXES

State	U	V	W	v_u	v_v	v_w	v_d	v_q	v_0
S0	0	0	0	0	0	0	0	0	$-\sqrt{3}v_n$
S1	1	0	0	E_{dc}	0	0	$\sqrt{2/3}E_{dc}\cos\theta$	$-\sqrt{2/3}E_{dc}\sin\theta$	$E_{dc}/\sqrt{3} - \sqrt{3}v_n$
S2	1	1	0	E_{dc}	E_{dc}	0	$\sqrt{2/3}E_{dc}\cos(\theta - \pi/3)$	$-\sqrt{2/3}E_{dc}\sin(\theta - \pi/3)$	$2E_{dc}/\sqrt{3} - \sqrt{3}v_n$
S3	0	1	0	0	E_{dc}	0	$-\sqrt{2/3}E_{dc}\cos(\theta + \pi/3)$	$\sqrt{2/3}E_{dc}\sin(\theta + \pi/3)$	$E_{dc}/\sqrt{3} - \sqrt{3}v_n$
S4	0	1	1	0	E_{dc}	E_{dc}	$-\sqrt{2/3}E_{dc}\cos\theta$	$\sqrt{2/3}E_{dc}\sin\theta$	$2E_{dc}/\sqrt{3} - \sqrt{3}v_n$
S5	0	0	1	0	0	E_{dc}	$-\sqrt{2/3}E_{dc}\cos(\theta - \pi/3)$	$\sqrt{2/3}E_{dc}\sin(\theta - \pi/3)$	$E_{dc}/\sqrt{3} - \sqrt{3}v_n$
S6	1	0	1	E_{dc}	0	E_{dc}	$\sqrt{2/3}E_{dc}\cos(\theta + \pi/3)$	$-\sqrt{2/3}E_{dc}\sin(\theta + \pi/3)$	$2E_{dc}/\sqrt{3} - \sqrt{3}v_n$
S7	1	1	1	E_{dc}	E_{dc}	E_{dc}	0	0	$\sqrt{3}(E_{dc} - v_n)$

979-8-3315-1612-3/25 $31.00 © 2025 IEEE

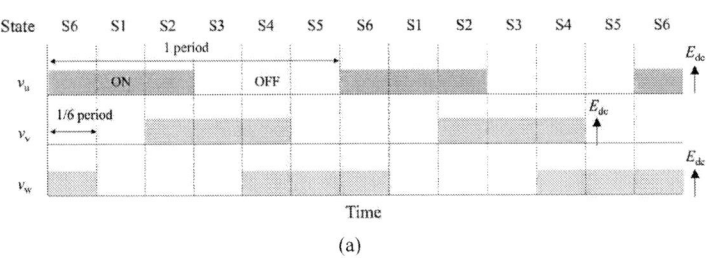
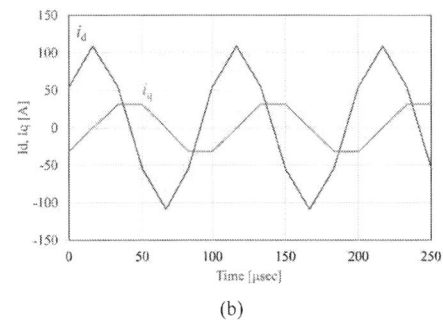

(a)

(b)

Fig. 2. Conventional interleaving. Duty = 50%, $(\Phi_u, \Phi_v, \Phi_w) = (0°, 120°, 240°)$. (a) Transition of the switching state and phase voltages. (b) dq-axes current.

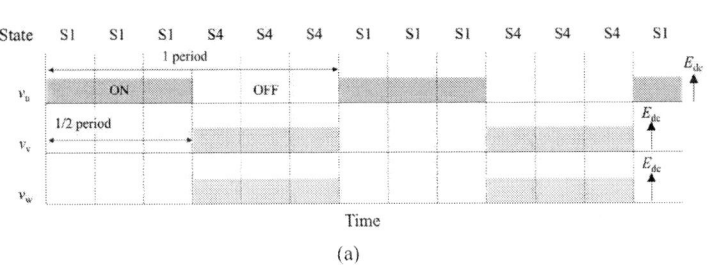
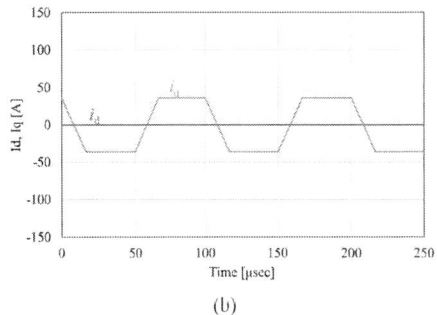

(a)

(b)

Fig. 3. Proposed interleaving. Duty = 50%, $(\Phi_u, \Phi_v, \Phi_w) = (0°, 180°, 180°)$. (a) Transition of the switching state and phase voltages. (b) dq-axes current.

III. FINITE ELEMENT ANALYSIS OF MOTOR MAGNETIC LOSS

An electromagnetic field analysis using FEA was conducted to verify the effectiveness of the proposed method in reducing magnetic losses. The analysis focused on the drive motor of the Nissan LEAF [12], with the analysis conditions detailed in Table II. The relationship between the rotor electrical angle and the breakdown of iron losses, specifically the stator core loss, rotor core loss, and magnetic loss, when applying both the conventional and proposed methods is illustrated in Fig. 4. At a rotor angle of $\theta = \pi/2$ with the proposed method, magnetic loss is almost completely suppressed, which is expected to prevent output derating due to elevated magnet temperatures. As mentioned in the previous section, by selecting the switching states according to the rotor angle, Δi_d can be reduced to zero with a periodicity of $\pi/3$ electrical angle. Therefore, if the rotor angle during charging can be adjusted by $\pi/3$ electrical angle, the proposed method can effectively always suppress the magnetic loss. In contrast, when examining the magnet losses associated with the conventional method, no clear dependence on the electrical angle is observed. Even at $\theta = 0$, where the magnet loss is minimized, approximately 400 W of magnetic loss still occurs. This indicates that the proposed method effectively reduces magnet losses compared to the conventional method.

Here, it was observed that as the rotor angle approaches $\theta = \pi/2$ with the proposed method, the magnetic losses decrease while rotor core losses tend to increase. The magnetic flux distribution and magnetic loss distribution at a rotor angle of θ

TABLE II. FINITE ELEMENT ANALYSIS CONDITIONS

Parameters	Symbol	Value
Neutral point voltage	v_n	450 V
Inverter voltage	E_{dc}	800 V
Inverter current	i_n	245 A
Switching frequency	f_{sw}	10 kHz
Electrical rotor angle	θ	$0-\pi/2$

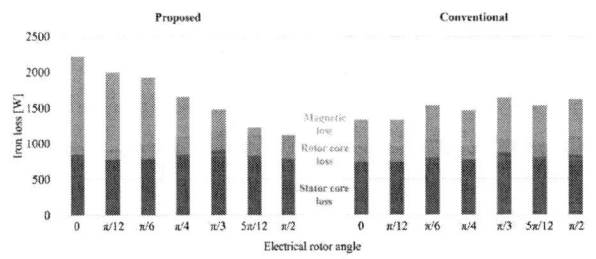

Fig. 4. Breakdown of iron loss analysis for conventional and proposed interleaving methods

= 0 are shown in Fig. 5. the magnetic flux induced from the stator flows perpendicularly into the magnets close to the rotor surface, resulting in the generation of eddy currents. Conversely, at $\theta = \pi/2$, the stator magnetic flux flows into the rotor core, circumventing the rotor magnets, which is the cause of the increased rotor core losses, as shown in Fig. 6.

979-8-3315-1612-3/25 $31.00 © 2025 IEEE

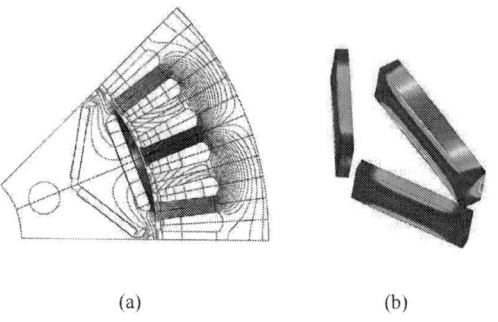

(a)   (b)

Fig. 5.   Rotor angles $\theta = 0$ with the proposed interleaving. (a) Magnetic flux distribution in the motor core and magnet. (b) Loss distribution in the magnet.

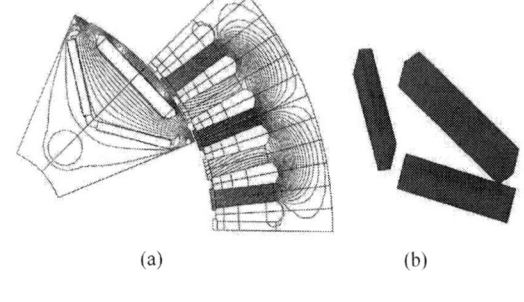

(a)   (b)

Fig. 6.   Rotor angles $\theta = \pi/2$ with the proposed interleaving. (a) Magnetic flux distribution in the motor core and magnet. (b) Loss distribution in the magnet.

IV.   EXPERIMENTAL RESULTS

An experiment was conducted to verify the effect of magnetic loss reduction using the proposed method on the temperature rise in an actual EV traction motor. The circuit diagram of the experimental setup is shown in Fig. 7, and the test conditions are detailed in Table III. Owing to facility limitations, a step-down experiment was conducted using v_1 of 400 V with an output limit of 80 kW, alongside v_2 of 320 V. The motor used in the experiment was the drive motor of the Nissan LEAF [12], and the inverter incorporated Si IGBT module (FS600R07A2E3 by Infineon). The IGBTs in the UVW phases were switched with the same duty cycle across all three phases to maintain a neutral-point voltage v_2 of 320 V. The switching phase for the UVW phase is $(\Phi_u, \Phi_v, \Phi_w) = (0°, 120°,$

240°) in the conventional method, and $(\Phi_u, \Phi_v, \Phi_w) = (0°, 180°, 180°)$ in the proposed method. A continuous load variation test was conducted, with the output increasing stepwise from 5 to 80 kW.

The experimentally obtained waveforms of the UVW phase currents i_u, i_v, i_w and i_d, i_q for both the conventional and proposed interleaving methods are shown in Fig. 8 and Fig. 9, respectively. The proposed control method reduced Δi_d (Fig. 9(b)) by half compared to the conventional three-phase interleaving Δi_d (Fig. 9(a)). The temperature of the rotor magnets, measured using a thermocouple inserted into the rotor magnets via slip rings, confirmed that the temperature rise in the rotor magnets was halved with the proposed control (Fig. 10).

TABLE III.   EXPERIMENTAL CONDITIONS

Parameters	Symbol	Value
Inverter voltage	v_1	400 V
Neutral point voltage	v_2	320 V
Switching frequency	f_{sw}	10 kHz
Inverter capacitance	C_1	1277 uF
Neutral point capacitoance	C_2	1277 uF
Electrical rotor angle	θ	$\pi/2$

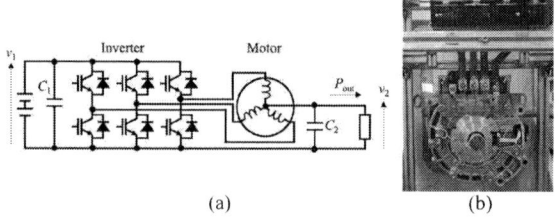

(a)   (b)

Fig. 7.   Experimental bench. (a) Circuit diagram. (b) Test motor.

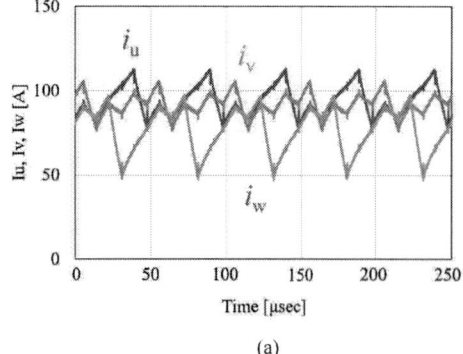

(a)

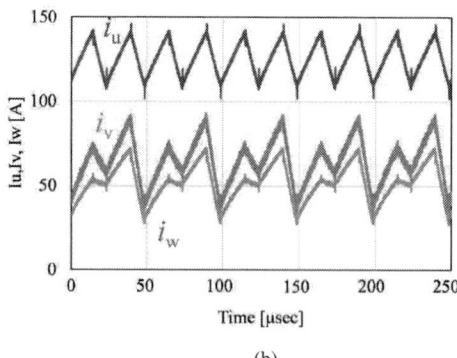

(b)

Fig. 8.   Waveforms of phase current. (a) Conventional interleaving. (b) Proposed interleaving

979-8-3315-1612-3/25 $31.00 © 2025 IEEE

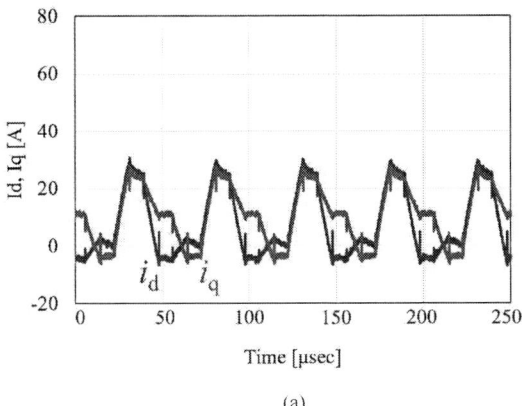

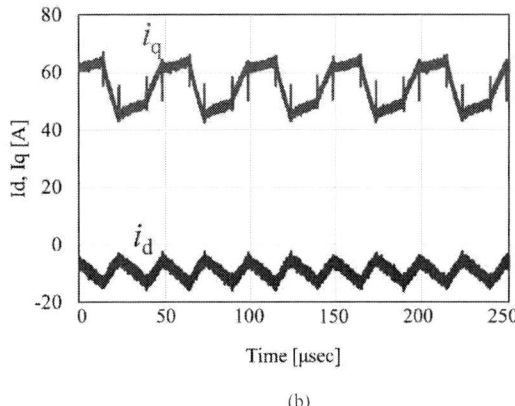

(a)   (b)

Fig. 9. Waveforms of dq-axes current. (a) Conventional interleaving. (b) Proposed interleaving.

In this experiment, the three-phase current was unbalanced under the proposed control (Fig. 8(b)) due to the open-loop voltage control, leading to relatively lower current values in the V and W phases. Consequently, a comparison of the measurements from the temperature sensor installed in the V-phase coil indicated that the proposed method resulted in a lower coil temperature (Fig. 10). However, if the U-phase coil temperature had been measured, it is likely that a higher temperature would have been observed. Implementing feedback control could improve the balance of the three-phase current. Typically, EV motor drive systems are equipped with current sensors that allow for the measurement of the currents in each UVW phase, which makes it easy to implement feedback control for current balancing.

V. CONCLUSIONS

This study developed a novel interleaving method designed to simultaneously suppress the temperature rise of rotor magnets and motor windings for integrated chargers that utilize a drive motor and inverter. By mitigating the temperature rise, fast charging becomes feasible even after high-speed driving, when the temperatures of the magnets and windings approach their upper limits. The effectiveness of this method was validated through both FEA and experimental testing. The FEA results indicated that magnetic losses generated by conventional three-phase interleaving can be reduced to nearly zero. Additionally, the experimental results demonstrated that the temperature increase in the magnets could be halved. When implementing an unbalanced three-phase interleaving method like the one proposed, incorporating current-balance control for the three phases becomes even more critical.

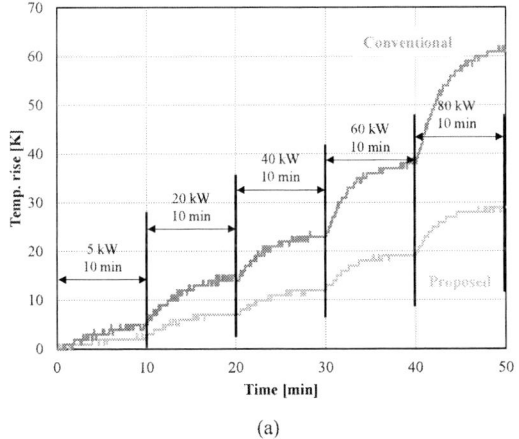

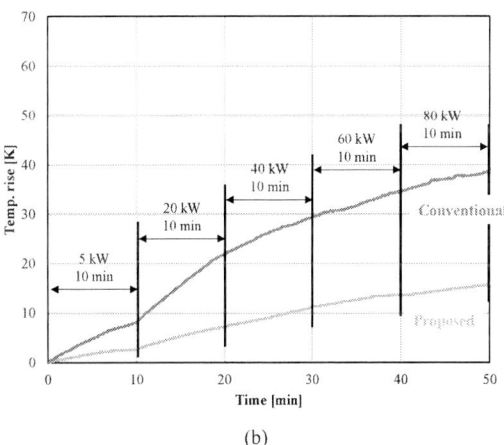

(a)   (b)

Fig. 10. Transition of motor temperature for conventional and proposed interleaving. (a) Rotor magnet temperature. (b) Stator coil (V-pahse) temperature.

REFERENCES

[1] J. Yang, G. C. Lim, C. Hwang and J. -I. Ha, "Two-Phase Interleaved DC Charging Method of Integrated Charger for Efficiency Enhancement," 2024 IEEE Applied Power Electronics Conference and Exposition (APEC), Long Beach, CA, USA, 2024, pp. 801-806.

[2] H. Bao, W. Yao and Y. Li, "Design and Implementation of a Motor Phase-access Integrated Boost Charging Scheme for Electric Vehicle," 2022 IEEE International Power Electronics and Application Conference and Exposition (PEAC), Guangzhou,Guangdong, China, 2022, pp. 211-216.

[3] C. Viana and P. W. Lehn, "A Drivetrain Integrated DC Fast Charger With Buck and Boost Functionality and Simultaneous Drive/Charge Capability," IEEE Transactions on Transportation Electrification, vol. 5, no. 4, pp. 903-911, Dec. 2019.

[4] I. Subotic, E. Levi and N. Bodo, "A Fast On-Board Integrated Battery Charger for EVs Using an Asymmetrical Six-Phase Machine," 2014 IEEE Vehicle Power and Propulsion Conference (VPPC), Coimbra, Portugal, 2014, pp. 1-6.

[5] Y. Xiao, C. Liu and F. Yu, "An Integrated On-Board EV Charger with Safe Charging Operation for Three-Phase IPM Motor," IEEE Transactions on Industrial Electronics, vol. 66, no. 10, pp. 7551-7560, Oct. 2019.

[6] Y. Xiao, C. Liu and F. Yu, "An Effective Charging-Torque Elimination Method for Six-Phase Integrated On-Board EV Chargers," IEEE Transactions on Power Electronics, vol. 35, no. 3, pp. 2776-2786, March 2020.

[7] M. S. Diab, A. A. Elserougi, A. S. Abdel-Khalik, A. M. Massoud and S. Ahmed, "A Nine-Switch-Converter-Based Integrated Motor Drive and Battery Charger System for EVs Using Symmetrical Six-Phase Machines," IEEE Transactions on Industrial Electronics, vol. 63, no. 9, pp. 5326-5335, Sept. 2016.

[8] I. Subotic, N. Bodo and E. Levi, "An EV Drive-Train With Integrated Fast Charging Capability," IEEE Transactions on Power Electronics, vol. 31, no. 2, pp. 1461-1471, Feb. 2016.

[9] C. Lai, K. L. V. Iyer, K. Mukherjee and N. C. Kar, "Analysis of Electromagnetic Torque and Effective Winding Inductance in a Surface-Mounted PMSM During Integrated Battery Charging Operation," IEEE Transactions on Magnetics, vol. 51, no. 11, pp. 1-4, Nov. 2015.

[10] Y. Ito and H. Haga, "Controlling Torque Ripple Vibration in a Single-Phase PFC Using the Neutral Point of an IPMSM," IEEJ Journal of Industry Applications, vol. 11, no. 11, pp. 815-821, Dec. 2022.

[11] N. Limsuwan, T. Kato, C. Y. Yu, J. Tamura, D. Reigosa, K. Akatsu, R. D. Lorenz, "Secondary resistive losses with high-frequency injection-based self-sensing in IPM machines," 2011 IEEE Energy Conversion Congress and Exposition, Phoenix, AZ, USA, 2011, pp. 622-629.

[12] S. Oki, Y. Sato, "Nissan LEAF and e-POWER: Evolution of Motors and Inverters," IEEJ Journal of Industry Applications, vol. 13, no. 1, pp. 8-16, Jan. 2024.

A Three-phase CLLC Resonant Converter with Integrated Planar Magnetics for 22-kW On-board Chargers

Tianlong Yuan
Center for Power Electronics Systems (CPES)
Blacksburg, USA
tianlong@vt.edu

Zhangwei Xiang
Center for Power Electronics Systems (CPES)
Blacksburg, USA
xiangzhangwei@vt.edu

Abdelrahman Ali
Center for Power Electronics Systems (CPES)
Blacksburg, USA
mahgoub@vt.edu

Feng Jin
Delta
Raleigh, USA
feng.jin@deltaww.com

Qiang Li
Center for Power Electronics Systems (CPES)
Blacksburg, USA
lqvt@vt.edu

Wendell DA-CUNHA-ALVES
Valeo
Cergy, France
wendell.da-cunha-alves@valeo.com

Xiaoshan Liu
Valeo
Cergy, France
xiaoshan.liu@valeo.com

Abstract—In response to the growing demand for higher power and reduced volume, 22-kW on-board chargers incorporating integrated planar magnetics are increasingly gaining traction. Among the preferred solutions for the high-power DC-DC conversion stage are three-phase CLLC converters, valued for their inherent isolation capabilities and exceptional efficiency. Despite their advantages, several challenges persist under this evolving paradigm. The first challenge pertains to integrating the resonant inductance directly into the transformer to achieve higher power density. This integration is critical for optimizing space and performance. Secondly, the elevated power levels introduce significant current stress, necessitating innovative approaches to manage and mitigate this stress effectively. This paper delves into the integration of planar inductors and transformers, focusing specifically on winding structures designed to minimize winding losses. Through comprehensive analysis and development, we present a novel three-phase planar transformer with built-in resonant inductance. This transformative design is tested and validated within a 22-kW CLLC resonant converter. The empirical results demonstrate the converter's impressive capabilities, achieving a power density of 30.7 kW/L and a peak efficiency of 98.2%. These findings underscore the efficacy of planar devices and transformers in enhancing performance metrics.

Keywords—Wide-band-gap devices, Planar magnetics, Resonant converter

I. INTRODUCTION

The advent of wide-bandgap (WBG) semiconductor devices [1]-[4] has significantly advanced power conversion technologies, enabling converters to achieve higher operating frequencies, improved efficiency, and enhanced power density [5]-[9]. These advancements have catalyzed the global adoption of green energy solutions, driving their integration across diverse applications [10]-[12].

However, these technological strides also introduce complex challenges to converter design. Key issues include thermal management, electromagnetic interference (EMI) mitigation, and the optimization of magnetic components, all of which require innovative approaches and rigorous engineering solutions [13]-[22].

DC-DC converters play a important role in voltage regulation and isolation, which are essential for the ongoing development and optimization of On-Board Chargers (OBCs) in electric vehicles (EVs) [23]-[26]. Among the various topologies, resonant converters have gained prominence due to their ability to achieve zero voltage switching (ZVS) under all operating conditions, thereby enhancing efficiency and reliability [27]-[32].

Transformers are a critical component in resonant converter design, as they contribute significantly to the total loss and volume of the system. Consequently, extensive research has been conducted to optimize magnetic design and performance [33]-[34]. For instance, studies on winding optimization have focused on minimizing winding losses and improving current distribution, yielding significant advancements [35]-[37].

979-8-3315-1612-3/25 $31.00 © 2025 IEEE

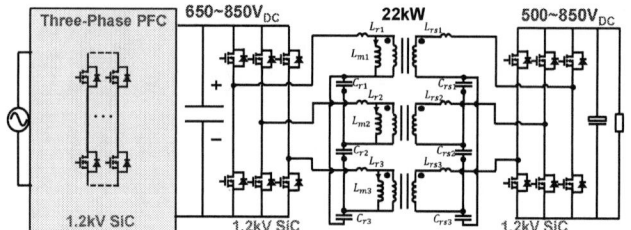

Fig 1: Three-phase 22-kW CLLC resonant converter topology.

A notable innovation in transformer design is the integration of resonant inductance into the transformer structure. This approach substantially increases the overall power density while maintaining a reasonable total transformer loss [37]-[46]. Research on integrated transformers has included the development of 11-kW designs [38]-[39], where various structures were compared, and detailed analyses of flux distribution across core plates were performed. Additionally, a three-phase integrated transformer was proposed in [40] for a 16.5-kW CLLC resonant converter. This work highlighted the advantages of three-phase designs in high-power applications.

However, it is worth noting that the converter in [40] was specifically designed for single-phase Power Factor Correction (PFC) as the first stage. When scaling up to three-phase PFC for 22-kW OBCs, several new challenges arise. The bus voltage in three-phase systems is inherently higher, necessitating a greater amount of integrated leakage inductance for effective regulation. Unfortunately, leakage inductance does not exhibit linear control characteristics so far, complicating the design process. Additionally, as the power level increases or more leakage inductance is integrated, the winding loss can surpass the core loss, significantly impacting the efficiency and performance of the converter.

This paper addresses and resolves these challenges by introducing a more practical method for linearly controlling leakage inductance. Various integration methods are meticulously compared to identify the most effective approach. Furthermore, we present a detailed winding structure designed to minimize winding loss, thereby enhancing overall efficiency.

To validate these concepts, a 22-kW three-phase CLLC resonant converter prototype was developed. The prototype's topology is illustrated in Fig. 1, and its specifications are detailed in Table I. Through rigorous testing, the prototype demonstrated the feasibility and effectiveness of the proposed solutions, marking a significant advancement in the design and implementation of high-power, efficient OBCs for electric vehicles.

Table I - Converter Parameters

Parameters	Value	Parameters	Value
P_{rated}	22 kW	n	1
f_o	250 kHz	L_m	20 μH
V_{in}	650-850 V	L_r	4.4 μH
V_{out}	500-850 V	$nV_{\text{out}}/V_{\text{in}}$	0.77~1

II. UNIFIED TRANSFORMER INTEGRATION METHOD

A general three-phase planar integrated transformer structure is illustrated in Fig. 2, with its corresponding equivalent circuit model depicted in Fig. 3. In this configuration, Leg A and Leg C function as an integrated transformer, while Leg B operates as a pure transformer. The magnetizing and leakage inductances, denoted in Fig. 3, are calculated using the expressions in (1), (2) and (3). Here, L_{ma} and L_{mb} represent the magnetizing inductances for Legs A, C, and B, respectively, while L_{kp} and L_{ks} correspond to the leakage inductances. N_1, N_2, N_{tx1} and N_{tx2} denote the turn numbers on each leg, and R_{g1} and R_{gtx} represent the air gap reluctances.

In this design, Legs A and C are designated as the integrated legs, while Leg B is the transformer leg. The L_n value, for the entire transformer structure is subsequently calculated using (4).

$$L_{ma} = \frac{2N_1 N_2}{R_{g1}} \quad (1)$$

$$L_{mb} = \frac{N_{tx1} N_{tx2}}{R_{gtx}} \quad (2)$$

$$L_{kp} = L_{ks} = \frac{(N_1 - N_2)^2}{R_{g1}} \quad (3)$$

$$L_n = \frac{2N_1 N_2 R_{gtx} + N_{tx1} N_{tx2} R_{g1}}{(N_1 - N_2)^2 R_{gtx}} \quad (4)$$

One potential candidate structure meeting the design specifications, referred to as Candidate I, is depicted in Fig. 4. The leakage inductance integrated, L_{kp} and L_{ks} are both $4.4\mu H$. The magnetizing inductance, L_{ma}, is $6.6\mu H$. The magnetizing

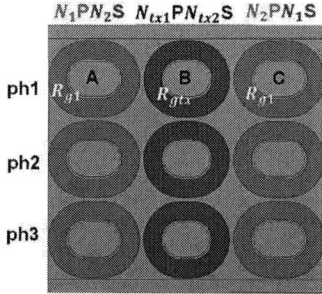

Fig 2: Proposed universal integrated transformer.

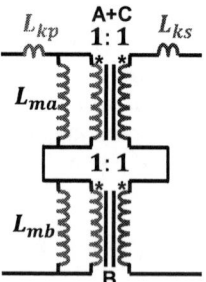

Fig 3: Transformer equivalent circuit model.

979-8-3315-1612-3/25 $31.00 © 2025 IEEE

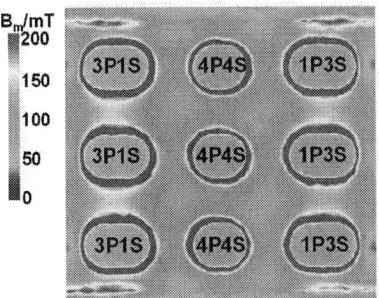

Fig 4: Flux density simulation and core structure for candidate I.

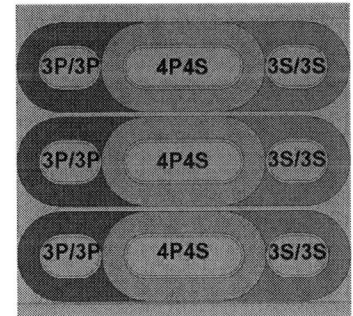

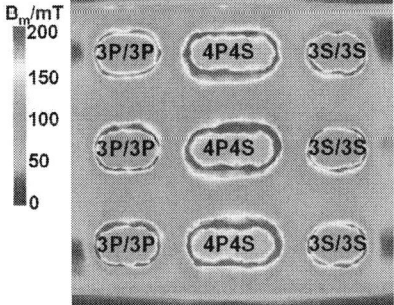

Fig 5: Transformer structure and its simulated flux distribution result.

inductance generated by the pure transformer, L_{mb}, is 13.4 μH. The loss distribution of such transformer is shown in Table II..

The loss distribution for this transformer design is summarized in Table II, providing the efficiency and performance information of the proposed structure.

However, the winding loss in Candidate I is observed to be 50% higher than the core loss. To address this issue, Candidate II, featuring fewer turns, was proposed and is illustrated in Fig. 5. Parallel windings are introduced for leg A and leg C to reduce the current stress on the inductors. This design is also encompassed within the equivalent circuit model. If N_2 is set to zero, Legs A and C transition to functioning purely as inductors. In this configuration, the primary and secondary windings can be integrated, which helps to further minimize the transformer footprint.

In Candidate II, as shown in Fig. 5, the integrated leakage inductances, L_{kp} and L_{ks}, remain at 4.4 μH, while the magnetizing inductance increases to 20 μH. The loss distribution

for this transformer is presented in Table II. While Candidate II demonstrates a smaller footprint, it incurs a higher overall loss compared to Candidate I.

Nevertheless, the design of Candidate II has significant potential for further optimization, particularly by focusing on reducing winding loss, which could enhance its performance and efficiency while maintaining its compact size.

Table II - Transformer loss comparison

Parameters	Cand. I	Cand. II
Core loss	110W	116W
Winding loss	160W	162W
Total loss	270W	278W
Footprint	15000mm^2	14200mm^2

III. TRANSFORMER OPTIMIZATION

A unified resonant inductance integration model is proposed in the previous section. With the integration model, two integration candidates are proposed and compared. For high power applications, the current stress is high. So, fewer turns number seem to be a better option. However, in table II, candidate II has fewer turns number but more winding loss. It is because the parallel windings on the inductors are directly in parallel, so the current distribution is uneven, resulting in even higher winding loss than candidate I. This chapter focuses on solving these problems by optimizing the winding structure and the transformer.

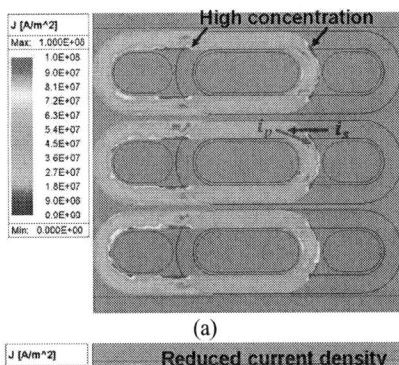

(a)

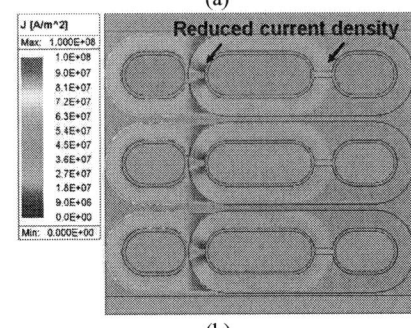

(b)

Fig 6: Winding structure optimization: (a) structure with current concentration; (b) optimized structure with smaller current density.

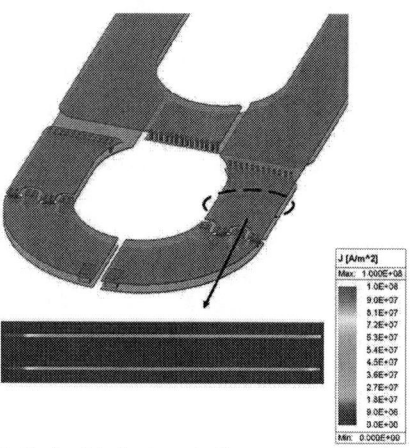

Fig 7: Optimized inductor winding structure and its current distribution.

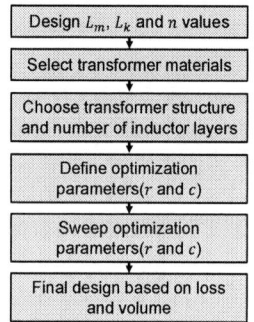

Fig 8: Transformer design and optimization process.

A. Winding structure optimization

As illustrated in Fig. 6(a), high current concentration areas are observed on the windings, primarily caused by the proximity effect. This phenomenon leads to increased winding loss, as the secondary side current flows in the direction indicated by the blue arrow, causing the primary current to concentrate in the highlighted region.

To mitigate this issue, an optimized winding design, shown in Fig. 6(b), was implemented. This optimization reduces the winding loss by 6.4% while significantly improving the current distribution. Specifically, the current density in the previously concentrated region is substantially reduced, resulting in enhanced efficiency and better thermal performance.

B. Twist Winding Design for Inductors

To minimize the winding loss in the inductors, parallel windings are employed. However, to ensure effective performance, it is essential to analyze the current distribution to guarantee proper current sharing among the parallel windings. To achieve this, the windings are twisted, as depicted in Fig. 8. This twisted winding design helps equalize the magnetic strength integration along the parallel windings, effectively eliminating the current imbalances caused by variations across different layers.

Simulation results confirm that this approach achieves perfect current sharing. Additionally, interleaved vias are incorporated to further reduce the winding loss associated with the vias, enhancing the overall efficiency.

For the hardware implementation, two four-layer 4oz PCBs are utilized. The daughter board is soldered directly onto the main board, forming a compact and efficient structure.

C. Transformer design and optimization

The optimization process is outlined in the flow chart shown in Fig. 8. The process begins with evaluating device losses over a range of deadtimes under nominal working conditions [47]-[52]. Based on the analysis, a deadtime of 130 ns is selected, ensuring a maximum magnetizing inductance of 20 μH to achieve ZVS.

The transformer core material is then chosen to be DMR96A, selected for its relatively low core loss (P_v) as shown in Fig. 9 and the manufacturer's capability to produce cores with larger dimensions, which align with the design requirements for efficiency and scalability.

Next, the transformer's winding width and core radius are defined as the key optimization parameters, as these dimensions critically influence performance and compactness. Finite Element Analysis (FEA) simulations are conducted to calculate the total magnetic losses for various combinations of these parameters.

Finally, a trade-off analysis between loss and footprint is performed, and the final design parameters are selected at the intersections where both criteria are optimized as shown in Fig. 10. For different radius and winding width, there is a total loss information. The black straight lines are the points with the same footprint. Each line is tangent to the loss contour. So, for each loss contour, there is a point with the smallest footprint. Finally,

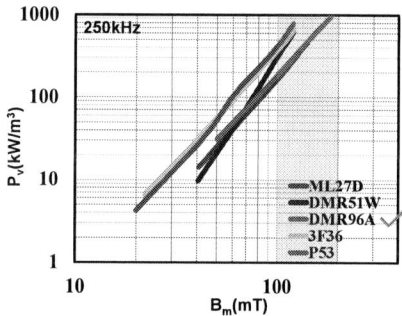

Fig 9: Transformer core material comparison.

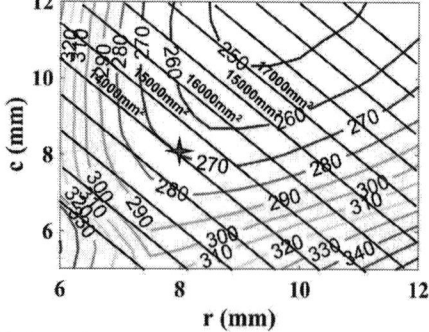

Fig 10: Transformer loss and footprint trade off.

Fig 11: 22-kW OBC DCDC stage prototype.

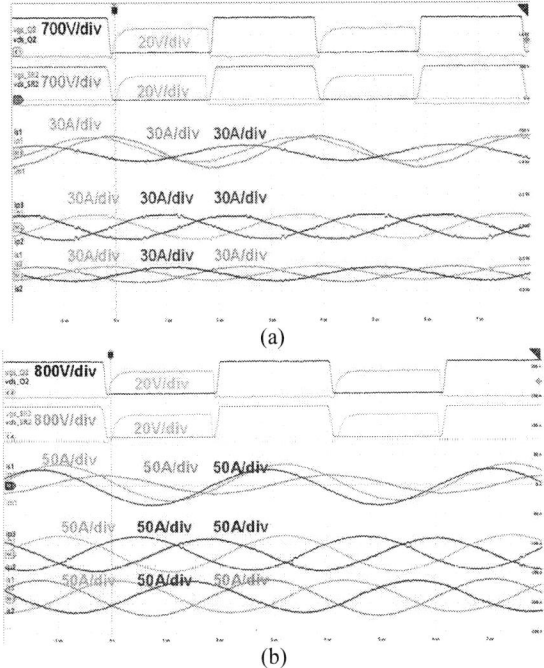

Fig 12: Testing results: (a) $V_{in} = 750V; V_o = 750V; P_o = 4.4\ kW$; (b) $V_{in} = 750V; V_o = 750V; P_o = 22\ kW$;

the radius is selected to be 8 mm, and the winding width is 8.5 mm. Transformer information is shown in Table III.

Table III -Transformer loss distribution

Core loss	Winding loss	Total loss	Footprint
114W	152W	268W	14200mm²

IV. HARDWARE PROTOTYPE AND ITS PERFORMANCE

A 22-kW hardware prototype was constructed, as shown in Fig. 11. The selected power devices are C3M0016120K, and the gate drivers used in the design are UCC21750. The transformer core material is DMR96A, chosen for its favorable magnetic properties and compatibility with the design specifications.

The tested waveforms under various operating conditions are presented in Fig. 12, confirming that ZVS is achieved across all conditions. The system efficiency, illustrated in Fig. 13, demonstrates a peak efficiency of 98.2%, achieved with the proposed integrated planar transformer. Additionally, the design achieves an impressive power density of 30.7 kW/L.

The loss distribution under nominal operating conditions is detailed in Fig. 14, providing insights into the efficiency and

thermal performance of the prototype. The comparison between this hardware to the other publications is shown in Table IV. The efficiency of the prototype is lower because more leakage inductance is integrated, but the highest power density is realized among all the candidates thanks to the optimization in the core structure and the winding structure.

V. CONCLUSION

This paper focuses on the comparison and optimization of three-phase integrated transformers. A practical approach to integrating resonant inductance using a linear control methodology is proposed, enhancing both functionality and efficiency. The optimized winding structure demonstrates a significant reduction in current density, while the twisted winding design ensures effective current sharing among parallel windings.

To validate the proposed concepts, a 22-kW hardware prototype was developed. The prototype achieves a peak efficiency of 98.2% and an impressive power density of 30.7 kW/L, confirming the effectiveness of the design innovations.

ACKNOWLEDGMENT

The authors would like to thank Valeo for sponsoring this project. The authors would like to thank DMEGC for donating the customized core samples.

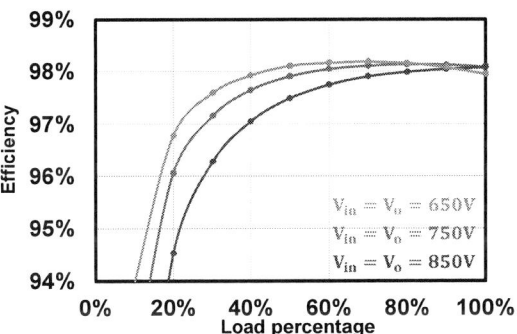

Fig 13: Efficiency curves under different test conditions.

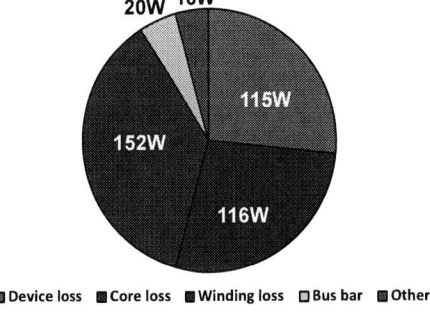

Fig 14: Loss distribution at 750V, 22kW.

TABLE IV
PERFORMANCE COMPARISON

Parameter	[18]	[19]	[20]	[21]	This work
Input voltage / V	$650 \sim 850$	$550 \sim 850$	$500 \sim 850$	$550 \sim 850$	$650 \sim 850$
Output voltage / V	$550 \sim 850$	$450 \sim 850$	$450 \sim 850$	$450 \sim 850$	$500 \sim 850$
Max. power / kW	22	11	11	16.5	22
Resonant Freq. / kHz	250	250	250	250	250
Peak efficiency	98.5%	98.5%	98.4%	98.8%	98.2%
Power density / kW/L	11.6	25.6	25.6	30.4	30.7
Magnetic Components	1	1	1	1	1

REFERENCES

[1] J. J. Kim, J. -H. Park, S. Sabri, B. Fetzer, B. Hull and S. -H. Ryu, "Investigation into Relationship of the Switching Performance and Short-Circuit Withstand Time on 1.2 kV 4H-SiC Power MOSFETs," *2024 36th International Symposium on Power Semiconductor Devices and ICs (ISPSD)*, Bremen, Germany, 2024, pp. 148-151, doi: 10.1109/ISPSD59661.2024.10579672.

[2] X. Yang *et al.*, "Evaluation of Dynamic RON, Coss Loss, and Short-Circuit Ruggedness of 650V and 1200V Industrial Vertical GaN JFETs," *2024 36th International Symposium on Power Semiconductor Devices and ICs (ISPSD)*, Bremen, Germany, 2024, pp. 283-286, doi: 10.1109/ISPSD59661.2024.10579632.

[3] T. Neyer, M. Domeij, H. Das and S. Sunkari, "Is there a perfect SiC MosFETs Device on an imperfect crystal?," *2021 IEEE International Reliability Physics Symposium (IRPS)*, Monterey, CA, USA, 2021, pp. 1-6, doi: 10.1109/IRPS46558.2021.9405098.

[4] X. Yang, J. Liu, B. Wang and G. Zhang, "Pulsed Overcurrent Capability of Power Semiconductor Devices in Solid-State Circuit Breakers: SiC MOSFET vs. Si IGBT," *2022 IEEE Applied Power Electronics Conference and Exposition (APEC)*, Houston, TX, USA, 2022, pp. 966-973, doi: 10.1109/APEC43599.2022.9773378.

[5] Y. -T. Huang, C. -C. Yang, T. -S. Li and Y. -M. Chen, "A Feedforward Voltage Control Strategy for Reducing the Output Voltage Double-Line-Frequency Ripple in Single-Phase AC–DC Converters," in IEEE Journal of Emerging and Selected Topics in Power Electronics, vol. 9, no. 6, pp. 6605-6612, Dec. 2021, doi: 10.1109/JESTPE.2021.3083258.

[6] Y. Jiang et al., "Efficiency Optimization of WPTS Based on Variable Angle Phase Shift Control for EV charging," 2020 IEEE 9th International Power Electronics and Motion Control Conference (IPEMC2020-ECCE Asia), Nanjing, China, 2020, pp. 747-752, doi: 10.1109/IPEMC-ECCEAsia48364.2020.9368148.

[7] V. R. Vakacharla, A. K. Rathore, R. K. Singh and S. K. Mishra, "Analysis and Design of Current-fed LCL Series Resonant Converter with Capacitive Doubler," *2020 IEEE Industry Applications Society Annual Meeting*, Detroit, MI, USA, 2020, pp. 1-6, doi: 10.1109/IAS44978.2020.9334840.

[8] Wu, M., Li, S., Zhou, X., Yu, X.: A synchronization method based on current orthogonal decomposition considering harmonic components for wireless power transfer system. IET Power Electron. 1–9 (2024). https://doi.org/10.1049/pel2.12702.

[9] C. -W. Chang, M. Spieler, R. Burgos and D. Dong, "Evaluation and Efficiency Comparison of Soft-Switching ARCP SiC-Based Traction Inverters in Electric Vehicles," 2023 IEEE Applied Power Electronics Conference and Exposition (APEC), Orlando, FL, USA, 2023, pp. 1417-1422, doi: 10.1109/APEC43580.2023.10131452.

[10] Y. Li et al., "A Universal Parameter Design Method of Resonant Coils Under Multiple Boundary Constrains for Wireless Power Transfer Systems," 2023 IEEE Energy Conversion Congress and Exposition (ECCE), Nashville, TN, USA, 2023, pp. 6489-6496, doi: 10.1109/ECCE53617.2023.10362919.

[11] M. Wu, X. Yu, X. Yang, W. Chen and L. Wang, "An Efficiency Optimization Method for the Multiple Coils WPT System Against the Pad Misalignment," in IEEE Transactions on Transportation Electrification, doi: 10.1109/TTE.2024.3410676.

[12] Y. Wei, Q. Luo and A. Mantooth, "A Hybrid Half-bridge LLC Resonant Converter and Phase Shifted Full-bridge Converter for High Step-up Application," *2020 IEEE Workshop on Wide Bandgap Power Devices and Applications in Asia (WiPDA Asia)*, Suita, Japan, 2020, pp. 1-6, doi: 10.1109/WiPDAAsia49671.2020.9360292.

[13] V. R. Vakacharla, A. K. Rathore, R. K. Singh and S. K. Mishra, "Fixed-Frequency Zero Voltage Switching CurrentFed (L) (LC) Resonant DC-DC Converter," *2020 IEEE Transportation Electrification Conference & Expo (ITEC)*, Chicago, IL, USA, 2020, pp. 1000-1005, doi: 10.1109/ITEC48692.2020.9161723.

[14] C. -W. Chang, M. Spieler, A. EL-Refaie, R. A. Torres, R. Burgos and D. Dong, "A Current Balancing Gate Driver for Dynamic Current Sharing of Paralleled SiC MOSFETs with Kelvin-Source Connection," in IEEE Transactions on Power Electronics, doi: 10.1109/TPEL.2024.3475288.

[15] C. -A. Cheng, C. -C. Chang, T. -S. Li, Z. -J. Chen and Y. -M. Chen, "Initial Rotor Position Startup Process Emulation Based on Electric Motor Emulator," 2020 IEEE 9th International Power Electronics and Motion Control Conference (IPEMC2020-ECCE Asia), Nanjing, China, 2020, pp. 605-610, doi: 10.1109/IPEMC-ECCEAsia48364.2020.9368013.

[16] T. O. Olowu, H. Jafari and A. Sarwat, "Voltage-Controlled Series Resonant DC-DC Converter for Solid State Transformer Applications," *2021 IEEE Transportation Electrification Conference & Expo (ITEC)*, Chicago, IL, USA, 2021, pp. 237-241, doi: 10.1109/ITEC51675.2021.9490128.

[17] C. -W. Chang et al., "Thermal Consideration and Design for a 200-kW SiC-Based High-Density Three-Phase Inverter in More Electric Aircraft," in IEEE Journal of Emerging and Selected Topics in Power Electronics, vol. 11, no. 6, pp. 5910-5929, Dec. 2023, doi: 10.1109/JESTPE.2023.3308854.

[18] T. -S. Li, Y. -H. Yang, C. -A. Cheng and Y. -M. Chen, "A Variable DC-Link Voltage Determination Method for Motor Drives with SiC MOSFETs," 2020 IEEE Workshop on Wide Bandgap Power Devices and Applications in Asia (WiPDA Asia), Suita, Japan, 2020, pp. 1-6, doi: 10.1109/WiPDAAsia49671.2020.9360266.

[19] Qiuzhe Yang, Shuo Wang, Qiang Li, "Modeling and Analysis of The Balance Network for Common Mode EMI Noise Suppression," in Proc. IEEE Energy Convers. Congr. Expo., 2024.

[20] N. Jia, X. Tian, L. Xue, H. Bai, L. M. Tolbert and H. Cui, "Integrated Common-Mode Filter for GaN Power Module With Improved High-Frequency EMI Performance," in IEEE Transactions on Power Electronics, vol. 38, no. 6, pp. 6897-6901, June 2023, doi: 10.1109/TPEL.2023.3248092.

[21] J. Cao, X. Zhang, P. Rao, S. Zhou, F. Zhou and Q. Zhang, "Design of Three-Phase Delta-Delta LLC Resonant Converter," *2020 IEEE Vehicle*

Power and Propulsion Conference (VPPC), Gijon, Spain, 2020, pp. 1-5, doi: 10.1109/VPPC49601.2020.9330918.

[22] N. Jia, X. Tian, L. Xue, H. Bai, L. M. Tolbert and H. Cui, "In-Package Common-Mode Filter for GaN Power Module with Improved Radiated EMI Performance," 2022 IEEE Applied Power Electronics Conference and Exposition (APEC), Houston, TX, USA, 2022, pp. 974-979, doi: 10.1109/APEC43599.2022.9773764.

[23] T. -S. Li, M. Ngo, R. Burgos and D. Dong, "Modeling and Analysis of Voltage Overshoot in Bidirectional Phase-Shift Full Bridge Converters," 2024 IEEE Sixth International Conference on DC Microgrids (ICDCM), Columbia, SC, USA, 2024, pp. 1-7, doi: 10.1109/ICDCM60322.2024.10665239.

[24] C. Li, X. Yu, Y. Li, H. Dang, J. Liu and S. Du, "A Variable Frequency Modulation Strategy for Current-Fed Dual-Active-Bridge Converter to Expand ZVS Range," 2024 IEEE Applied Power Electronics Conference and Exposition (APEC), Long Beach, CA, USA, 2024, pp. 468-473, doi: 10.1109/APEC48139.2024.10509332.

[25] L. Zhu et al., "Charging Pad as the Transformer: Integration of On-board Charger, Auxiliary Power Module and Wireless Charger for Electric Vehicles," 2023 IEEE Energy Conversion Congress and Exposition (ECCE), Nashville, TN, USA, 2023, pp. 1718-1724, doi: 10.1109/ECCE53617.2023.10362363.

[26] L. Zhu, H. Bai and A. Brown, "Model and Control of a Current-Fed Dual Active Bridge Based Ultrawide-Voltage-Range Auxiliary Power Module for 400 V/800 V Electric Vehicles," in *IEEE Transactions on Power Electronics*, vol. 39, no. 3, pp. 3263-3276, March 2024.

[27] Y. Cao, K. Ngo and D. Dong, "Resonant Commutation Electronic-Embedded DC Transformer (RC-EET DCX) With Quasi-Trapezoidal Current and Natural Current Sharing," in *IEEE Transactions on Power Electronics*, vol. 39, no. 8, pp. 9736-9751, Aug. 2024.

[28] G. Yu, T. Yuan, X. Chen and Q. Li, " A Three phase 22kW Soft Switching Based Two-stage Onboard Charger," 2024 IEEE Energy Conversion Congress and Exposition (ECCE), Phoenix, AZ, USA, 2024.

[29] F. Jin, C. Zhao, T. Yuan and Q. Li, "Rotation Control of Synchronize Rectifier to Improve Thermal Performance of LLC Converter under Boost Mode Operation," 2024 IEEE Applied Power Electronics Conference and Exposition (APEC), Long Beach, CA, USA, 2024, pp. 1727-1733, doi: 10.1109/APEC48139.2024.10509185.

[30] P. R. Prakash et al., "A 2400 W/in3 1.8 V Bus Converter Enabling Vertical Power Delivery for Next-Generation Processors," *2024 IEEE Applied Power Electronics Conference and Exposition (APEC)*, Long Beach, CA, USA, 2024, pp. 910-917, doi: 10.1109/APEC48139.2024.10509453.

[31] T. Yuan, F. Jin and Q. Li, "A 22-kW On-Board Charger (OBC) with an Integrated Planar Inductor and Transformer," 2024 IEEE Applied Power Electronics Conference and Exposition (APEC), Long Beach, CA, USA, 2024, pp. 1300-1304, doi: 10.1109/APEC48139.2024.10509419.

[32] Y. Liang, P. R. Prakash, A. Nabih and Q. Li, "Design Optimization of a 3.3 V Bus Converter for Vertical Power Delivery in Next-Generation Processors," *2024 IEEE Energy Conversion Congress and Exposition (ECCE)*, Phoenix, AZ, USA, 2024.

[33] C. Tu, K. Ngo and X. Yu, "Diamond-Window Resonant Inductor with Significant AC Flux," 2023 IEEE Applied Power Electronics Conference and Exposition (APEC), Orlando, FL, USA, 2023, pp. 405-412, doi: 10.1109/APEC43580.2023.10131237.

[34] M. Wu et al., "A Compact Coupler With Integrated Multiple Decoupled Coils for Wireless Power Transfer System and its Anti-Misalignment Control," in IEEE Transactions on Power Electronics, vol. 37, no. 10, pp. 12814-12827, Oct. 2022, doi: 10.1109/TPEL.2022.3166888.

[35] T. Yuan, F. Jin, Z. Li and Q. Li, "Current Sharing Analysis of a High Power Transformer with Parallel Windings," 2023 IEEE Applied Power Electronics Conference and Exposition (APEC), Orlando, FL, USA, 2023, pp. 1551-1556, doi: 10.1109/APEC43580.2023.10131421.

[36] M. Wu et al., "Modeling of Litz-Wire DD Coil With Ferrite Core for Wireless Power Transfer System," in IEEE Transactions on Power Electronics, vol. 38, no. 5, pp. 6653-6669, May 2023, doi: 10.1109/TPEL.2022.3222228.

[37] T. Yuan, F. Jin and Q. Li, "Analysis and Comparison of Integrated Planar Transformers for 22-kW On-Board Chargers," in IEEE Transactions on Power Electronics, vol. 39, no. 9, pp. 11368-11385, Sept. 2024, doi: 10.1109/TPEL.2024.3410878.

[38] F. Jin, T. Yuan, A. Nabih and Q. Li, "Efficient Integrated Magnetics With Winding Cancellation Technique to Reduce Common-Mode EMI Noise for a Single-Phase CLLC Converter," in IEEE Transactions on Power Electronics, vol. 39, no. 11, pp. 14758-14774, Nov. 2024, doi: 10.1109/TPEL.2024.3436044.

[39] T. Yuan, F. Jin, G. Yu and Q. Li, " Analysis and Comparison of Integrated Planar Transformers for CLLC Resonant Converters," 2024 IEEE Energy Conversion Congress and Exposition (ECCE), Phoenix, AZ, USA, 2024.

[40] F. Jin, A. Nabih, T. Yuan and Q. Li, "A High-Efficiency High-Density Three-Phase CLLC Resonant Converter With a Universally Derived Three-Phase Integrated Transformer for On-Board-Charger Application," in IEEE Transactions on Power Electronics, vol. 39, no. 4, pp. 4350-4366, April 2024, doi: 10.1109/TPEL.2024.3354679.

[41] T. Yuan, F. Jin, Z. Li, C. Zhao and Q. Li, "Design of an Integrated Transformer With Parallel Windings for a 30-kW LLC Resonant Converter," in IEEE Transactions on Power Electronics, vol. 38, no. 11, pp. 14317-14333, Nov. 2023, doi: 10.1109/TPEL.2023.3291954.

[42] T. Yuan, F. Jin, Z. Li and Q. Li, "High Frequency High Power Integrated Transformer Design for Resonant Converters with SiC Devices," 2022 IEEE 9th Workshop on Wide Bandgap Power Devices & Applications (WiPDA), Redondo Beach, CA, USA, 2022, pp. 170-175, doi: 10.1109/WiPDA56483.2022.9955265.

[43] T. Yuan, F. Jin and Q. Li, "Parasitic Capacitance Analysis of Integrated Transformers with Parallel Windings," 2024 IEEE Energy Conversion Congress and Exposition (ECCE), Phoenix, AZ, USA, 2024.

[44] F. Jin, T. Yuan, A. Nabih, Z. Li and Q. Li, "Efficient Integrated Magnetics with Winding Cancellation Technique to Reduce Common-Mode EMI Noise for A Single Phase CLLC Converter," 2024 IEEE Energy Conversion Congress and Exposition (ECCE), Phoenix, AZ, USA, 2024.

[45] J. Cao, X. Zhang, P. Rao, S. Zhou, F. Zhou and Q. Zhang, "Design of Three-Phase Delta-Delta LLC Resonant Converter," *2020 IEEE Vehicle Power and Propulsion Conference (VPPC)*, Gijon, Spain, 2020, pp. 1-5, doi: 10.1109/VPPC49601.2020.9330918.

[46] Y. Cao et al., "3.5 kW/in3 Planar Coupled Inductor Design and Optimization for a 50 kW 3-level Four-Switch Buck-Boost (3L-FSBB) Converter," *2022 IEEE Energy Conversion Congress and Exposition (ECCE)*, Detroit, MI, USA, 2022, pp. 1-8, doi: 10.1109/ECCE50734.2022.9947382.

[47] Q. Yang, A. Nabih, R. Zhang, Q. Li and Y. Zhang, "A Converter Based Switching Loss Measurement Method for WBG Device," 2023 IEEE Applied Power Electronics Conference and Exposition (APEC), Orlando, FL, USA, 2023, pp. 8-13, doi: 10.1109/APEC43580.2023.10131509.

[48] D. Tochigi, K. Takashima, T. Isobe, H. Tadano and M. Tomoyuki, "Experimental Verification of a Model of Switching Transients Considering Device Parasitic Capacitance for the Loss Estimation of Soft-switching Power Converters," *2019 IEEE 4th International Future Energy Electronics Conference (IFEEC)*, Singapore, 2019, pp. 1-7, doi: 10.1109/IFEEC47410.2019.9015017.

[49] D. Goldmann, S. Mayer, S. Schramm and H. -G. Herzog, "Comparing Switching and Conduction Losses of Uni- and Bidirectional SiC Semiconductor Switches for AC Applications," *2021 23rd European Conference on Power Electronics and Applications (EPE'21 ECCE Europe)*, Ghent, Belgium, 2021, pp. P.1-P.9, doi: 10.23919/EPE21ECCEEurope50061.2021.9570496.

[50] J. Lin, K. Ma and Y. Zhu, "Statistics-based Switching Loss Characterization of Power Semiconductor Device," *2020 IEEE Energy Conversion Congress and Exposition (ECCE)*, Detroit, MI, USA, 2020, pp. 3811-3814, doi: 10.1109/ECCE44975.2020.9236230.

[51] K. Li, P. Evans and M. Johnson, "SiC and GaN power transistors switching energy evaluation in hard and soft switching conditions," *2016 IEEE 4th Workshop on Wide Bandgap Power Devices and Applications (WiPDA)*, Fayetteville, AR, USA, 2016, pp. 123-128, doi: 10.1109/WiPDA.2016.7799922.

[52] W. Hua, L. Peng, T. Liu, C. Chen, X. Liu and Y. Kang, "A Datasheet-Based Loss Model and Efficiency Analysis for Vienna Rectifiers Using SiC Power Devices," *2020 IEEE 1st China International Youth Conference on Electrical Engineering (CIYCEE)*, Wuhan, China, 2020, pp. 1-6, doi: 10.1109/CIYCEE49808.2020.9332615.

Reconfigurable LLC Resonant Converter for Wide Voltage Range and Reduced Voltage Stress in DC-Connected EV Charging Stations

Yu Zuo
Dept. of Electrical Engineering
KU Leuven - EnergyVille
Leuven, Belgium
yu.zuo@kuleuven.be

Xiaobing Shen
Dept. of Electrical Engineering
KU Leuven - EnergyVille
Leuven, Belgium
xiaobing.shen@kuleuven.be

Bangli Du
Dept. of Electrical Engineering
KU Leuven - EnergyVille
Leuven, Belgium
bangli.du@kuleuven.be

Qingcheng Sui
Dept. of Electrical Engineering
KU Leuven - EnergyVille
Leuven, Belgium
qingcheng.sui@ kuleuven.be

Tim Geboers
Dept. of Electrical Engineering
KU Leuven - EnergyVille
Leuven, Belgium
tim.geboers1@kuleuven.be

Wilmar Martinez
Dept. of Electrical Engineering
KU Leuven - EnergyVille
Leuven, Belgium
wilmar.martinez@ kuleuven.be

Abstract— **The LLC resonant converter features its high efficiency and compact design but struggles with achieving a wide voltage gain. To address this challenge, multi-stage circuits, hybrid control strategies, and multi-mode operations are utilized, though these solutions increase the system's complexity. This paper presents an innovative LLC resonant converter with a stacked bridge structure, employing simple PFM and secondary turn-on/off control. By incorporating a stacked half-bridge primary and a reconfigurable multi-voltage rectifier, the proposed design extends the output voltage range within a narrow switching frequency range while reducing voltage stress on both the primary and secondary sides by half. The reconfigurable rectifier operates as a double-voltage rectifier at low output voltages and a quadruple-voltage rectifier at high output voltages. The converter adapts to varying voltage requirements, making it suitable for high voltage applications like on-board charging and DC fast charging. Experimental validation with a prototype operating at 400V input and 200-800V output at 500W demonstrates effective soft-switching across the voltage range and confirms the feasibility of the design.**

Keywords— *LLC resonant converter, reconfigurable rectifications, wide voltage range, low voltage stress, EV charging*

I. INTRODUCTION

The trend in electric vehicle (EV) development is towards higher charging voltages to enhance charging speed, reduce system losses and heat, optimize battery management, and improve power and range [1-2]. Models like the Kia EV6, BYD Seal, and Audi A6 Avant are adopting high-voltage charging. Consequently, DC fast charging stations must handle higher and a wide range of voltages (250-400V and 600-900V) to meet diverse vehicle requirements, necessitating DC-DC converters with high voltage output and wide-range regulation [3-4]. Similar requirements also apply to photovoltaic energy systems, where high voltage output and broad-range regulation are

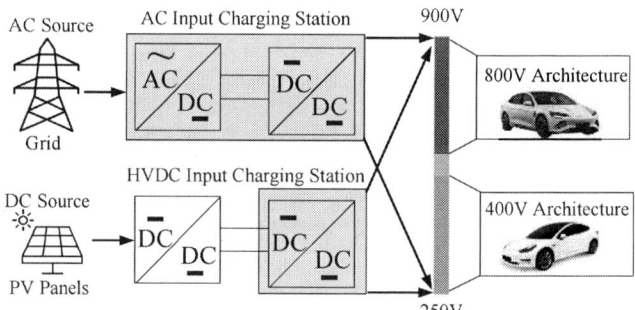

Fig.1. DC Fast Charging Configurations

essential for efficient operation [5-8]. To achieve a wide output voltage range in resonant DC-DC converters, several methods can be employed: multi-stage circuits utilizing two-stage [9-11] or series-parallel configurations [12-15], hybrid control strategies employing modulations such as Pulse Frequency Modulation (PFM), Pulse Width Modulation (PWM), and phase-shift modulations [16-19], and multi-mode operation incorporating active and passive components [20-22].

Key research in this area includes: [23] presents a quasi-two-stage buck-LC converter. The LC resonant stage operates in DC transformer mode, while the buck stage achieves voltage regulation. It operates in four modes using PWM to maintain high efficiency. [24] proposes a reconfigurable circuit with two full-bridge modules on the secondary side. By controlling auxiliary switches, the converter switches between input-parallel output-series for high voltage and input-parallel output-parallel for low voltage, using triple-phase shift modulation. [25] introduces an LLC resonant converter with two transformers and three bridge legs, employing a pulse-width plus pulse-frequency modulation strategy. It reconfigures the

979-8-3315-1612-3/25 $31.00 © 2025 IEEE

rectifier into full-bridge, hybrid voltage-multiplier, and voltage-multiplier modes. [26] proposes an H5-bridge-based asymmetric LLC resonant converter using PFM. Adjusting the H5-bridge switch pattern allows six operation modes with varying voltage gains. [27] discusses an LLC resonant converter with stacked half-bridges. The inverter and rectifier each have two modes, reconfiguring modulation schemes during bidirectional operation, resulting in three voltage-gain ranges with a narrow modulation frequency. These innovations enhance the flexibility and universality of DC fast charging statios, catering to various EV voltage levels.

This paper proposes an LLC resonant converter with a stacked half-bridge primary and a reconfigurable rectifier, extending the output voltage range within a narrow switching frequency while halving the voltage stress on both sides. The advantages and contributions of this paper include: 1) The reconfigurable rectifier operates across a wide voltage range, functioning as a double-voltage rectifier at low output voltages and a quadruple-voltage rectifier at high output voltages. This design not only extends the output voltage range but also narrows the operating frequency range, which benefits circuit design and optimization. 2) It achieves soft-switching across the entire output voltage range, even with PFM modulation and straightforward secondary-side on-off control. 3) The stacked bridge structure with low voltage stress is well-suited for applications requiring high input and high output voltages, such as DC fast charging stations connected to a high-voltage bus supply and EV charging. The structure of this article is as follows: Section II explains the proposed DC-DC converter, its working principle and design flowchart. Then, Section III presents the experimental results and analysis. Finally, conclusion ends the paper.

II. PROPSED TOPOLOGY AND THEORETICAL ANALYSIS

A. Proposed Topology Discription

The stacked-bridge LLC resonant converter proposed in this paper is shown in Fig.2. It consists of stacked half-bridge inverter on the primary side and a reconfigurable rectifier on the secondary side. Stacked half bridge structure is made up of four active switches (denoted as S_1, S_2, S_3, S_4), which are arranged in a series of half-bridges, along with two DC-linked capacitors, C_1 and C_2, to stabilize the input voltage and generate an intermediate voltage level. Accordingly, the voltage stress on the primary switches is reduced to half of the input voltage. This reduction in voltage stress on the active switches makes the inverter structure suitable for high input voltage scenarios. Its resonant tank circuit incorporates a resonant inductor L_r, a resonant capacitor C_r, a high-frequency transformer with a n:1 turns ratio, and a magnetizing inductance L_m. The switching frequency is small than series resonant frequency f_r. It helps achieve zero-voltage switching (ZVS) on the primary side and zero-current switching (ZCS) on the secondary side.

$$f_r = \frac{1}{2\pi\sqrt{L_r C_r}} \quad (1)$$

This reconfigurable rectifier combines two types of voltage-doubling rectification circuits to provide flexible operation modes suited to varying voltage requirements. The first type of voltage-doubling rectifier is a full-wave rectifier, consisting of

two capacitors and two diodes (denoted C_{b1}, C_{b2}, D_{r1}, and D_{r2}). This configuration charges the output capacitor alternately during the positive and negative half cycles through the transformer. The second type of voltage-doubling rectifier is half-wave rectifier. In this arrangement, component C_{b3} works with the transformer to charge the output capacitor during the positive half cycle, while the transformer charges capacitor C_{b3} during the negative half cycle. In the proposed converter's rectifier section, a combination of stacked diode bridges (denoted D_1, D_2, D_3, D_4) and four voltage-doubling capacitors (C_{d1}, C_{d2}, and C_{d3}, C_{d4}) is utilized. Additionally, two auxiliary

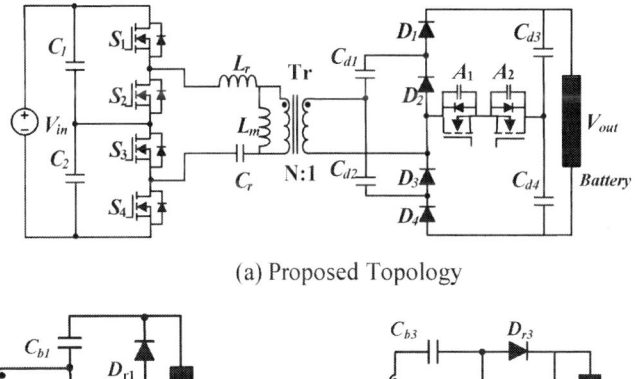

(a) Proposed Topology

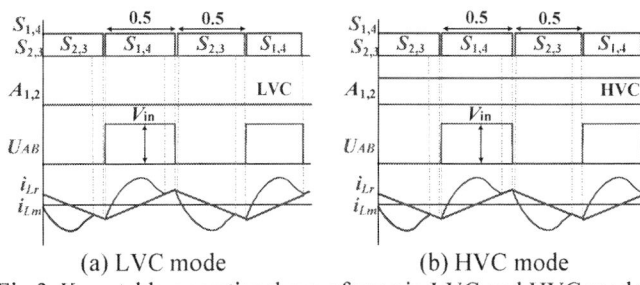

(b) Voltage-doubling rectifier A

(c) Voltage-doubling rectifier B

Fig.2. The proposed LLC resonant DC-DC converter

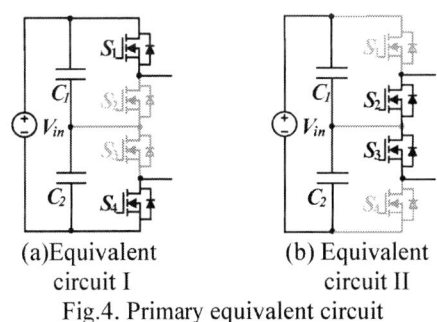

(a) LVC mode

(b) HVC mode

Fig.3. Key stable operational waveforms in LVC and HVC modes

(a)Equivalent circuit I

(b) Equivalent circuit II

Fig.4. Primary equivalent circuit

switches, A_1 and A_2, are incorporated to decide the type of rectifications. When these two switches are turned off, the rectifier functions as a voltage-doubling rectifier. When the switches are turned on, it operates as a quadruple-voltage rectifier.

B. Working Principle

For a 400V EV charging architecture and an 800V architecture, this paper treats 400V as the Low Voltage Charging Mode (LVC) with a voltage range of 200-400V, while 800V is considered the High Voltage Charging Mode (HVC) with a range of 400-800V. Both modes utilize Pulse Frequency Modulation (PFM) to regulate the voltage range. The steady-state waveforms under different charging modes are shown in Fig. 3.

In both charging modes, the control method on the primary side remains the same: switches S_1 and S_4 have the same on and off times, with dead time neglected, and a duty cycle of 0.5. The drive signals for switches S_2 and S_3 are complementary to those of S_1 and S_4, which means that S_2 and S_3 have identical drive signals. Within one switching period, ignoring switching transition times, dead time, and current commutation, the inverter primarily operates in two working modes. When switches S_1 and S_4 are both conducting, the inverter output voltage equals the input voltage;when switches S_2 and S_3 are conducting, the inverter output voltage is zero. The equivalent circuit of the inverter is shown in Fig. 4.

In Low Voltage Charging (LVC) mode, the transformer's positive voltage, combined with the charge from capacitor C_{d1}, supplies energy to the output capacitors C_{d3} and C_{d4} in series, effectively increasing the output voltage, as illustrated in Fig 5(a). Simultaneously, capacitor C_{d2} is directly charged by the transformer to maintain stable energy transfer. When the transformer voltage reverses and becomes negative, capacitor C_{d2} releases its stored energy to output capacitors and load, while capacitor C_{d1} recharges in the opposite direction, as depicted in Fig. 5(b). This configuration enables the output voltage amplitude to be twice that of capacitor C_{d1} or C_{d2} in LVC mode.

In High Voltage Charging (HVC) mode, the operation shifts to allow the output capacitors C_{d3} and C_{d4} to be charged alternately during the positive and negative half-cycles, respectively. Specifically, during the positive half-cycle, the transformer and capacitor C_{d1} together charge the output capacitor C_{d3}, while in the negative half-cycle, the transformer and capacitor C_{d2} charge the output capacitor C_{d4}. This configuration results in an output voltage that is four times the voltage of either capacitor C_{d1} or C_{d2}.

III. TOPOLOGY DESIGN CONSIDERATIONS AND CONTROL

A. Units esign Considerations

The design of the converter parameters primarily involves selecting the resonant components L_r, L_m, and C_r, as well as determining the transformer turns ratio n and capacitor values, as outlined in the Design Algorithm. It is worth noting that the expression for the equivalent output resistance of the resonant tank differs between LVC and HVC modes, as shown below.

TABLE I. SWITCHING AND VOLTAGE CONDITION UNDER DIFFERENT CHARGING TYPE

Charging type	Turning-on switches	Turning-off switches	V_{AB}	V_{out}/V_{cd1}
LVC mode	S_1,S_4	S_2,S_3,A_1,A_2	V_{in}	2
	S_2,S_3	S_1,S_4,A_1,A_2	0	2
HVC mode	S_1,S_4,A_1,A_2	S_2,S_3	V_{in}	4
	S_2,S_3,A_1,A_2	S_1,S_4	0	4

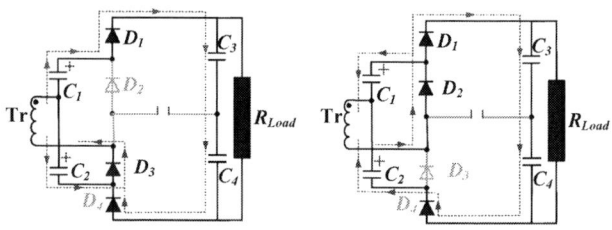

(a)Equivalent circuit I   (b) Equivalent circuit II
Fig.5. Secondary equivalent circuits in LVC mode

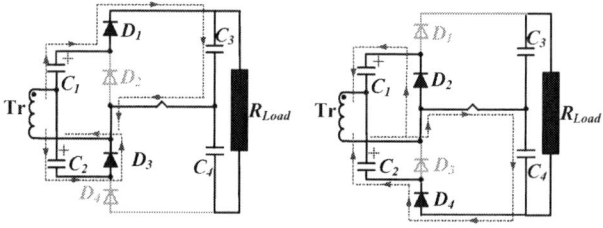

(a)Equivalent circuit I   (b) Equivalent circuit II
Fig.6. Secondary equivalent circuits in HVC mode

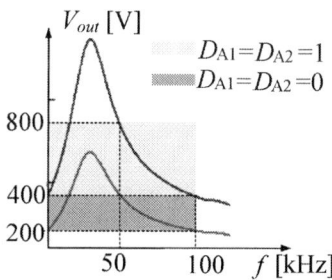

Fig.7. Output voltage curves in LVC and HVC modes

$$R_{eq\_LVR} = \frac{2n^2 R_{Load}}{\pi^2} \quad (2)$$

$$R_{eq\_HVR} = \frac{n^2 R_{Load}}{2\pi^2} \quad (3)$$

Aside from the resonant capacitor C_r, the input and secondary-side capacitors in this study have well-defined polarities. Consequently, their capacitance values are selected following standard design principles for DC filter capacitors.

$$C = \frac{I}{\Delta V f_s} \qquad (4)$$

Where I represents either the input current on the primary side or the output current on the secondary side, and ΔVout denotes the allowable output voltage ripple. f_s is the switching frequency, with the design calculations based on the minimum switching frequency.

Fig 7 shows the frequency response curves of the output voltage Vout in different operating modes. The yellow region represents the state in which secondary-side switches A_1 and A_2 are always on, indicating operation in High Voltage Charging (HVC) mode. Conversely, the blue region represents the state in which the secondary-side switches remain off, corresponding to Low Voltage Charging (LVC) mode. Obviously, due to the two distinct charging modes, the switching frequency range can be reduced by half compared to non-reconfigurable circuits.

IV. EXPERIMENAL RESULTS

A 500W prototype was constructed. The parameters of the prototype are listed in Table II. In testing, the input voltage was fixed at a constant 400V, while the output voltage varied from 200V to 800V. Specifically, 200-400V for LVC mode and 400-800V for HVC mode. The switching frequency ranged from 50kHz to 100kHz with an output power of 500W.

The experimental results include a minimum input voltage in LVC mode with a switching frequency of 100kHz, as shown in Fig. 8. In HVC mode, the output voltage reaches 495V with a switching frequency of 70kHz, as illustrated in Fig. 9.

In Low Voltage Charging (LVC) mode, a stable voltage balance is maintained across the secondary capacitors C_{d1} and Cd2 both of which hold a steady voltage of 100V, as shown in Fig. 8(a). Similarly, the series output capacitors C_{d3} and C_{d4} each stabilize at 100V, as depicted in Fig. 8(b). Consequently, the voltage ratio between the output voltage and each secondary capacitor C_{d1} or C_{d2} is 2:1. This balanced configuration facilitates soft-switching in the primary switches, with Zero-Voltage Switching (ZVS) achieved for the primary switch S_2 as an example, as illustrated in Fig. 8(c).

In High Voltage Charging (HVC) mode, voltage balance is similarly achieved across both the secondary and series output capacitors. In this mode, the secondary capacitors C_{d1} and C_{d2} maintain a stable voltage of 124V with equal amplitudes, as shown in Fig. 9(a). The series output capacitors C_{d3} and C_{d4} each stabilize at 250V, as illustrated in Fig. 9(b). Unlike the LVC mode, the voltage ratio between the output voltage and each secondary capacitor C_{d1} or C_{d2} is now changed from 2:1 to 4:1. Soft-switching is also achieved in HVC mode, minimizing switching losses and ensuring efficient operation, as demonstrated in Fig. 9.

This consistent voltage balance and soft-switching capability across both LVC and HVC modes enhance the converter's efficiency and reliability, reducing switching stress and optimizing power transfer across a wide output voltage range. Additionally, the peak efficiency reaches 93.9% when output voltage is 400V and its corresponding loss distribution is shown in Fig.10. Among them, the diode losses account for the largest portion. In the future, by incorporating synchronous rectification,

TABLE II. CONVERTER PARAMETERS

Components	Parameters
Input voltage range (V_{in})	400V
Output voltage range (V_{out})	200-800V
Transformer turns ratio (n)	2:1
Magnetizing inductance (L_m)	160μH
Resonant inductance (L_r)	25.3μH
Resonant capacitance (C_r)	100nF
Switching frequency range	50-100kHz
Capacitors (C_1, C_2)	10μF
Capacitors (C_{d1}, C_{d2})	4.7μF
Capacitors (C_{d3}, C_{d4})	10μF
Output Power (P_{out})	500W
Deadtime between switches	200ns

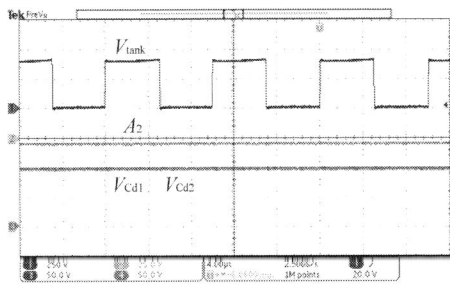

(a) Voltage balance between secondary capacitors

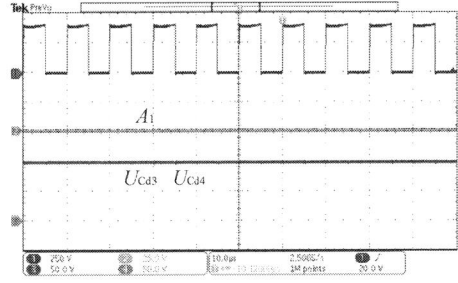

(b) Voltage balance between output capacitors

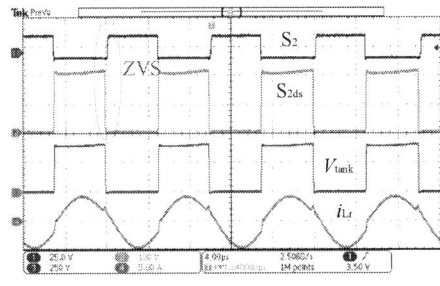

(c) soft-switching results

Fig.8 Experimental waveforms in LVC mode

the whole system efficiency will be further improved. Overall, the proposed converter is suitable for high voltage applications, providing half the voltage stress on both sides. It also achieves a wide output voltage range while maintaining soft-switching capabilities.

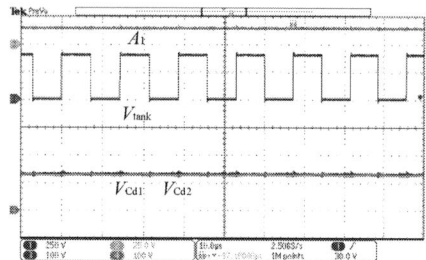

(a) Voltage balance between secondary capacitors

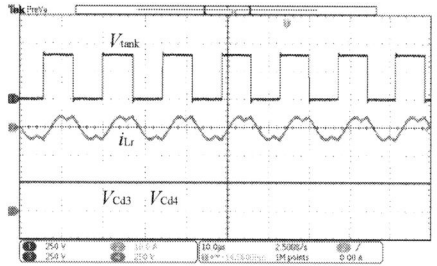

(b) Voltage balance between output capacitors

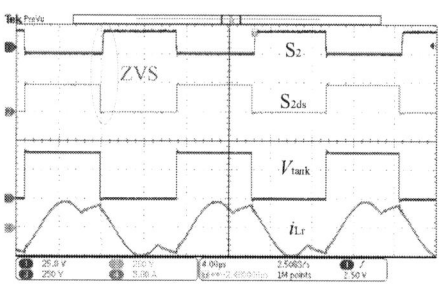

(c) soft-switching results

Fig.9 Experimental waveforms in HVC mode

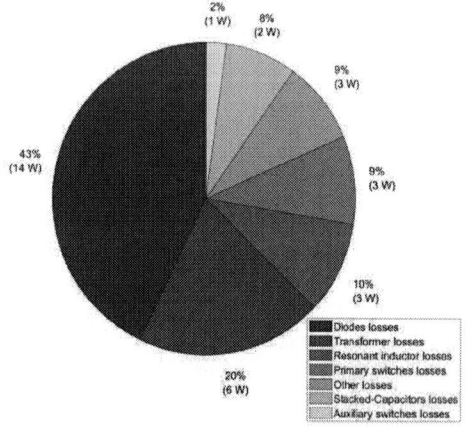

V. CONCLUSION

In this paper, an innovative LLC resonant converter featuring a stacked half-bridge in the primary side and a reconfigurable multi-voltage rectifier is presented. The proposed method effectively extends the output voltage range through PFM control and reconfigurable rectifications, reducing voltage stress on both sides by half. Experimental results demonstrates a prototype operating with a 400V input and an output range of 200-800V at a rated power of 500W with peak efficiency up to 93.9%, achieving soft-switching across the entire range. This converter is particularly suited for high voltage applications like DC fast charging stations and EV charging, offering a versatile and efficient solution.

REFERENCES

[1] C. Suarez and W. Martinez, "Fast and Ultra-Fast Charging for Battery Electric Vehicles – A Review," 2019 IEEE Energy Conversion Congress and Exposition (ECCE), Baltimore, MD, USA, 2019, pp. 569-575, doi: 10.1109/ECCE.2019.8912594.

[2] N. Hou, K. Qin, R. Wei, Y. Zhang and Y. W. Li, "Control Techniques With Low Computational Burden for a DAB-Based Two-Stage DC–DC EV Charging Converter System," in IEEE Transactions on Power Electronics, vol. 39, no. 9, pp. 10865-10875, Sept. 2024, doi: 10.1109/TPEL.2024.3405193.

[3] Y. Wei, S. Zhang, J. Liu and H. A. Mantooth, "Hybrid PFM and PWM Modulation Strategy for Stacked Structure LLC Resonant Converter With Wide Input Voltage Range Application," in IEEE Transactions on Industry Applications, vol. 60, no. 3, pp. 4038-4053, May-June 2024, doi: 10.1109/TIA.2024.3355382.

[4] M. Li, Z. Ouyang, M. A. E. Andersen and B. Zhao, "Self-Driven Gate Driver for LLC Synchronous Rectification," in IEEE Transactions on Power Electronics, vol. 36, no. 1, pp. 56-60, Jan. 2021, doi: 10.1109/TPEL.2020.3003417.

[5] K. Kanathipan and J. Lam, "High-Frequency Interlinking High-Voltage Gain PV Converter Modules With Embedded Power Balancing Technique for DC-Distributed System," in IEEE Journal of Emerging and Selected Topics in Power Electronics, vol. 12, no. 2, pp. 1769-1781, April 2024, doi: 10.1109/JESTPE.2024.3369534.

[6] Y. Zuo, X. Shen, B. Du, D. B. Cobaleda, H. Wouters and W. Martinez, "Fixed-frequency Dual PWM Interleaved Boost LLC Resonant Converter for A Wide Input Voltage for Photovoltaic Applications," 2024 IEEE 10th International Power Electronics and Motion Control Conference (IPEMC2024-ECCE Asia), Chengdu, China, 2024, pp. 4172-4177, doi: 10.1109/IPEMC-ECCEAsia60879.2024.10567987.

[7] K. Kanathipan and J. Lam, "A High Voltage Gain Isolated PV Micro-Converter With a Single-Voltage Maximum Power Point Tracking Control Loop for DC Micro-Grid Systems," in IEEE Journal of Emerging and Selected Topics in Industrial Electronics, vol. 3, no. 3, pp. 755-765, July 2022, doi: 10.1109/JESTIE.2021.3130473.

[8] Y. Zuo, D. B. Cobaleda, X. Shen and W. Martinez, "High Step-up Ratio Interleaved Boost L-LLC Resonant Converter with PWM and PFM Control for Wide Input and Output Voltage Range," 2024 IEEE Applied Power Electronics Conference and Exposition (APEC), Long Beach, CA, USA, 2024, pp. 1396-1402, doi: 10.1109/APEC48139.2024.10509079.

[9] W. Wang, Y. Liu, J. Zhao, P. Zhang and P. C. Loh, "A Dynamic Control Method for Buck + LLC Cascaded Converter With a Wide Input Voltage Range," in IEEE Transactions on Power Electronics, vol. 38, no. 2, pp. 1522-1534, Feb. 2023, doi: 10.1109/TPEL.2022.3208872.

[10] V. K. Goyal and A. Shukla, "Two-Stage Hybrid Isolated DC–DC Boost Converter for High Power and Wide Input Voltage Range Applications," in IEEE Transactions on Industrial Electronics, vol. 69, no. 7, pp. 6751-6763, July 2022, doi: 10.1109/TIE.2021.3099245.

[11] Z. Hou, S. C. Kao, D. Jiao and J. -S. Lai, "Variable Turns-Ratio Matrix Transformer based LLC Converter for Two-Stage Electric Vehicle Auxiliary Power Module Applications," 2023 IEEE Energy Conversion Congress and Exposition (ECCE), Nashville, TN, USA, 2023, pp. 5859-5865, doi:

10.1109/ECCE53617.2023.10362741.

[12] Z. Hou, S. C. Kao, J. -S. Lai, Z. Xu, C. Chen and C. -L. Wang, "A Cost-Effective Winding Structure On Modular Matrix Transformer LLC Application," 2022 IEEE Energy Conversion Congress and Exposition (ECCE), Detroit, MI, USA, 2022, pp. 1-7, doi: 10.1109/ECCE50734.2022.9948049.

[13] R. Wei and Y. R. Li, "Optimized Design and Fast-Dynamic Control for ISOP-Connected Hybrid CLLC-DAB System With Partial Power Processing Property," in IEEE Transactions on Power Electronics, vol. 39, no. 7, pp. 8844-8857, July 2024, doi: 10.1109/TPEL.2024.3389588.

[14] Z. Wei, H. Wang, Y. Lu, D. Shu, G. Ning and M. Fu, "Bidirectional Constant Current String-to-Cell Battery Equalizer Based on L2C3 Resonant Topology," in IEEE Transactions on Power Electronics, vol. 38, no. 1, pp. 666-677, Jan. 2023, doi: 10.1109/TPEL.2022.3205440.

[15] C. Wang, M. Li, Z. Ouyang, T. -G. Zsurzsan and M. A. E. Andersen, "Pentacentra Transformer for Multiphase LLC Converter in High-Current Data Center Application," in IEEE Transactions on Power Electronics, vol. 39, no. 1, pp. 1150-1161, Jan. 2024, doi: 10.1109/TPEL.2023.3323298.

[16] M. -H. Park, X. Zhang, Y. Jeong and G. -W. Moon, "A High Efficiency Boost Preregulator Merging With an Asymmetric LLC Standby Converter in DC Power Distribution System for Data Center," in IEEE Transactions on Power Electronics, vol. 39, no. 8, pp. 9804-9813, Aug. 2024, doi: 10.1109/TPEL.2024.3379382.

[17] Q. Zhao, Y. Gao, H. Ding, Z. Wu, X. Li and D. Wang, "Bidirectional Hybrid DC–DC Resonant Converter With Wide Voltage Gain Range," in IEEE Transactions on Industrial Electronics, doi: 10.1109/TIE.2024.3468726.

[18] A. Awasthi, S. Bagawade and P. K. Jain, "Analysis of a Hybrid Variable-Frequency-Duty-Cycle-Modulated Low-Q LLC Resonant Converter for Improving the Light Load Efficiency for a Wide Input Voltage Range," in IEEE Transactions on Power Electronics, vol. 36, no. 7, pp. 8476-8493, July 2021, doi: 10.1109/TPEL.2020.3046560.

[19] M. Pahlevani, S. Pan and P. Jain, "A Hybrid Phase-Shift Modulation Technique for DC/DC Converters With a Wide Range of Operating Conditions," in IEEE Transactions on Industrial Electronics, vol. 63, no. 12, pp. 7498-7510, Dec. 2016, doi: 10.1109/TIE.2016.2593679.

[20] Q. Zhao, J. Zhang, C. Fu, Y. Chen and Q. Yang, "A Structure-Reconfigurable LLC Resonant Converter With Wide Gain Range," in IEEE Journal of Emerging and Selected Topics in Power Electronics, vol. 11, no. 4, pp. 4057-4067, Aug. 2023, doi: 10.1109/JESTPE.2023.3281929.

[21] C. -H. Jo and D. -H. Kim, "Reconfigurable LLC Resonant Converter for Bidirectional Electric-Vehicle Chargers," in IEEE Transactions on Power Electronics, vol. 38, no. 12, pp. 15168-15172, Dec. 2023, doi: 10.1109/TPEL.2023.3319504.

[22] Y. Zuo, X. Pan, J. Zhu, J. Ye and Y. Wang, "A Bidirectional Isolated LLC Resonant Converter with Configurable Structure for Wide Output Voltage Range Applications," 2020 IEEE 9th International Power Electronics and Motion Control Conference (IPEMC2020-ECCE Asia), Nanjing, China, 2020, pp. 1716-1721, doi: 10.1109/IPEMC-ECCEAsia48364.2020.9367645.

[23] Z. Lu, M. Su, G. Xu, L. Li, Y. Liu and X. Chen, "Integrated Quasi-Two-Stage Buck-LC Converter With Wide Voltage Gain Range," in IEEE Transactions on Power Electronics, vol. 39, no. 8, pp. 9827-9838, Aug. 2024, doi: 10.1109/TPEL.2024.3390942. [A]

[24] O. Zayed, A. Elezab, A. Abuelnaga and M. Narimani, "A Dual-Active Bridge Converter With a Wide Output Voltage Range (200–1000 V) for Ultrafast DC-Connected EV Charging Stations," in IEEE Transactions on Transportation Electrification, vol. 9, no. 3, pp. 3731-3741, Sept. 2023, doi: 10.1109/TTE.2022.3232560. [B]

[25] X. Tang, Y. Xing, H. Wu and J. Zhao, "An Improved LLC Resonant Converter With Reconfigurable Hybrid Voltage Multiplier and PWM-Plus-PFM Hybrid Control for Wide Output Range Applications," in IEEE Transactions on Power Electronics, vol. 35, no. 1, pp. 185-197, Jan. 2020, doi: 10.1109/TPEL.2019.2914945.[C]

[26] C. Li, M. Zhou and H. Wang, "An H5-Bridge-Based Asymmetric LLC Resonant Converter With an Ultrawide Output Voltage Range," in IEEE Transactions on Industrial Electronics, vol. 67, no. 11, pp. 9503-9514, Nov. 2020, doi: 10.1109/TIE.2019.2952778.[D]

[27] Y. Zuo, X. Pan and C. Wang, "A Reconfigurable Bidirectional Isolated LLC Resonant Converter For Ultra-Wide Voltage-Gain Range Applications," in IEEE Transactions on Industrial Electronics, vol. 69, no. 6, pp. 5713-5723, June 2022, doi: 10.1109/TIE.2021.3088355.

Design and Control of GaN based Three-Phase/ Single-Phase Combo Three-Level Flying Capacitor PFC for OBC Applications

Nidhi Haryani
Milan Jovanovic Power Electroncis Laboratory (MPEL)
Raleigh, NC, USA

Laszlo Huber
Milan Jovanovic Power Electroncis Laboratory (MPEL)
Raleigh, NC, USA

Anup Anurag
Milan Jovanovic Power Electroncis Laboratory (MPEL)
Raleigh, NC, USA

Juan Ruiz
Milan Jovanovic Power Electroncis Laboratory (MPEL)
Raleigh, NC, USA

Peter Barbosa
Milan Jovanovic Power Electroncis Laboratory (MPEL)
Raleigh, NC, USA

Abstract— **As power demand for on-board chargers (OBC) is increasing to meet the fast-charging requirements, 800 V DC bus voltage is being adopted to reduce the copper losses as compared to 400 V DC bus voltage for PFC in OBCs. For higher voltage operation, multi-level converter topologies employ the benefit of reducing device stress and common-mode voltage. Hence a GaN based three-level flying capacitor converter is presented in this paper for 800 V DC bus voltage OBC. Since flying capacitor converter has the benefit of multiplying the inductor current frequency as compared to the switching frequency thus reducing filter size, it becomes attractive for high density OBC applications. The design of the converter, specifically the GaN based flying capacitor phase leg considering symmetrical gate loop and power loop design for all the devices is discussed in detail. It is critical to achieve symmetry in gate loop and power loops as a small asymmetry can cause unbalance in flying capacitor voltage owing to the small rise and fall times in GaN devices. Also, an active feedback method for balancing of flying capacitor voltage is presented which works on the principle of adjusting duty ratios to charge/ discharge the flying capacitor to achieve good balancing.**

Keywords— *Flying Capacitor Converter, Power Factor Corrector (PFC), Active Voltage Balancing, Sinusoidal Pulse Width Modulation (SPWM), Discontinuous Pulse Width Modulation (DPWM), Continuous Conduction Mode (CCM), Neutral Point Clamped (NPC)*

I. INTRODUCTION

For high power (>10 kW) charging systems, increasing the PFC DC bus voltage to >600 V is imminent owing to the high copper losses at 400 V. Three-phase two-level Si/SiC devices-based converters are the most common choice for 800 V bulk voltage as two-level converters are simple to design and control. However as multi-level converter topologies employ the benefit of reducing device stress and reduced common-mode voltage, they are also a potential candidate for EV chargers operating at 800 V.

The most common bidirectional multi-level topologies are Active Neutral Point Clamped (ANPC), Diode Neutral Point Clamped (DNPC), Cascaded H-bridge (CHB), T-type and Flying capacitor converter [1]&[3]. For an N-level converter, each device will block a voltage of $V_{DC}/(N-1)$. Though T-type topologies don't give the benefit of reducing the blocking voltage for all devices [2]. The flying capacitor converter has an added advantage of multiplying the inductor current frequency (f_{iL}) by N-1 times the switching frequency (f_{sw}) [3],

$$f_{iL}=(N-1)*f_{sw} \qquad (1)$$

This leads to the benefit of reducing the size of the inductor and EMI filter while keeping the switching losses lower. A three-phase three-level flying capacitor PFC is shown in Fig. 1 with the possibility of operating in both three-phase and single-phase modes. The circuit can operate as three-phase AC/DC bidirectional converter or a single-phase PFC with S_N and S_P acting as unfolder leg depending on the position of the relay. S_N and S_P comprise the unfolder leg with the switches switching at line frequency.

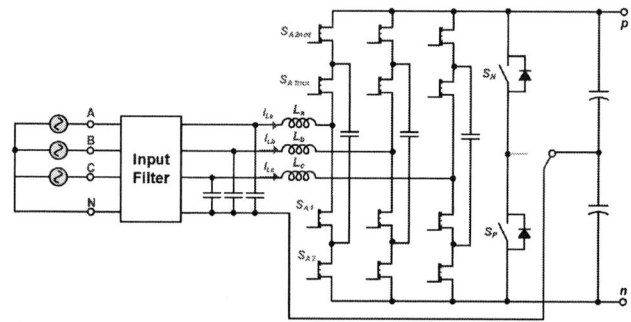

Fig. 1. A three-phase/ single phase combo three-level flying capacitor converter with GaN devices for the three phases in three-phase mode and for the high frequency leg in single-phase mode

The detailed switching operation in three-phase CCM SPWM and the frequency multiplication effect is discussed in Section II along with the proposed active feedback method for flying capacitor voltage balancing. The controller design is also discussed comprehensively in section II. The detailed design of the converter including GaN based three-level phase leg and inductors is discussed in section III. The experimental results in three-phase four-wire (3φ-4W) as well as three-phase three-wire (3φ-3W) configuration are discussed in Section IV.

II. FREQUENCY DOUBLING & ACTIVE VOLTAGE BALANCING OF FLYING CAPACITOR VOLTAGE

A. Basic Operation

As shown in Fig. 1, when phase neutral is connected to DC bus mid-point, each phase can operate as an independent half bridge single-phase converter as shown in Fig. 2(a). For a balanced symmetrical operation, the flying capacitor must be charged to $V_{dc}/2$, i.e. $V_{Cfl}=V_{dc}/2$. The two pairs of complementary devices, S_{A1} and S_{A1not} and S_{A2} and S_{A2not} are operated as two interleaved pair of switches with 180° phase shift between the two carrier signals as shown in Fig. 2(b). The detailed switching operation for $0<V_{AN}<V_{dc/2}$ is shown in Fig. 2(b).

In the beginning of the switching cycle, when S_{A1} and S_{A2not} are on, the voltage across the inductor v_L is V_{AN}, hence the inductor current increases linearly. When S_{A1} is turned off at the end of DT_s (where T_s is the switching period), S_{A1not} is turned on and S_{A2not} is still on, v_L is equal to $V_{AN}-V_{dc}/2$, hence the inductor current decreases. Further, at the end of $T_s/2$, S_{A2not} is turned off and S_{A2} is turned on. Thus v_L becomes equal to V_{AN} again. Similarly, at the end of $T_s/2+DT_s$, S_{A2} is turned off, S_{A1not} and S_{A2not} conduct and v_L becomes equal to $V_{AN}-V_{dc}/2$. Thus effectively, the inductor voltage and current frequency (f_{iL}) doubles as compared to the switching frequency (f_s). From inductor volt-sec balance,

$$D_A=0.5-V_{AN}/V_{dc} \qquad (2)$$

For SPWM operation, duty ratio for other phases can be calculated like phase A.

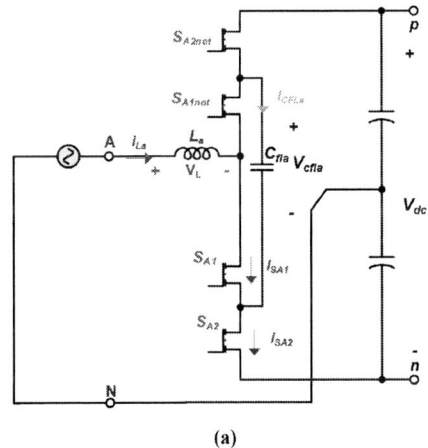

(a)

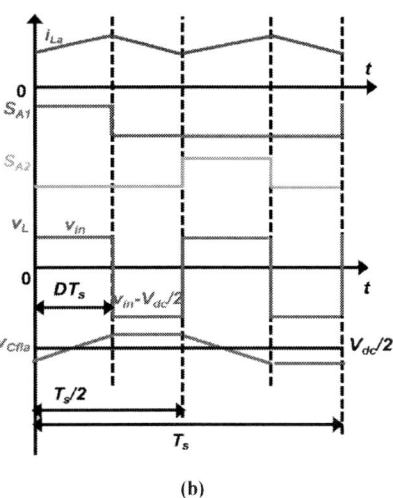

(b)

Fig. 2(a) Equivalent circuit of Phase A during three-phase operation. (b) Gating signals and inductor current during one switching cycle of phase A.

B. Active Balancing of Flying Capacitor Voltage

Phase-shifted PWM operation between S_{A1} and S_{A2} leads to natural balancing of flying capacitor voltage [4]-[6] as the flying capacitor is charged by the difference between i_{SA1} and i_{SA2} (Fig. 2(a)). Here the symbols are representative of phase A, however the theory is valid for all the phases.

$$i_{CFLa}=i_{SA2}-i_{SA1}=(d_{SA2}-d_{SA1})i_{La} \qquad (3)$$

For perfect symmetric operation, $d_{SA2}=d_{SA1}$, however practically due to asymmetric delays in gate signals or asymmetric gate loop and power loop design, the flying capacitor voltage can deviate from $V_{DC}/2$ which can lead to higher voltage stress on one set of devices and loss of frequency multiplication effect. Hence for robust operation a feedback controller is added [4].

The error between reference flying capacitor voltage ($V_{DC}/2$) and measured V_{CFL} (after filtering) is fed into a P or PI controller (as shown in Fig. 3) and the output v_{BAL} multiplied by the sign(i_L) is added to the duty ratios of S_{A2} and subtracted from the duty ratio of S_{A1} (as shown in Fig. 4).

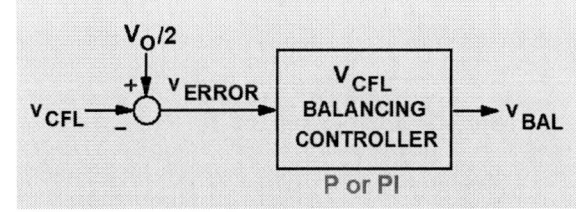

Fig. 3. Feedback control for active balancing of flying capacitor voltage.

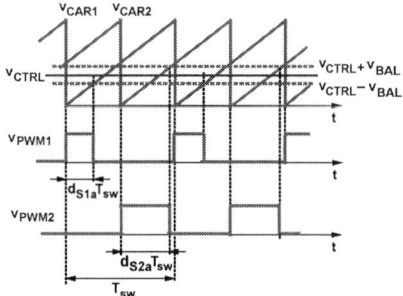

Fig. 4. Duty ratio adjustment with balancing controller output being added/ subtracted to the current controller output.

Usually, in DC-DC FCML converters, the low-pass filter bandwidth for FC voltage sensing and hence the FC voltage balancing loop is quite low (<100 Hz). This design works well for DC-DC case with the exception for light-load conditions as in light load, the inductor current is low and there's low current to charge/ discharge the flying capacitor according to equation (3). However, in an AC/DC converter, the inductor current goes to zero twice every line cycle and hence the flying capacitor voltage goes close to the unbalanced value every time the current crosses zero as shown in Fig. 5(a). The simulation results shown in Fig. 5(a) are for V_{phrms}= 230 V, V_{dc}= 800 V, P_o= 11 kW, L= 65 μH, f_s=65 kHz, C_{fl}= 6.58 μF, low pass filter (LPF) bandwidth= 100 Hz, K_p=0.0001 with active balancing enabled at 0.1 s. The unbalance is created in simulation by using 183.6° phase shift between S_{A1} and S_{A2}.

If K_p is increased further, the system becomes unstable because of low gain margin. To increase the controller speed, the LPF bandwidth is increased to 10 kHz with the same value of K_p, the results are shown in Fig. 5(b). There's still significant 120 Hz ripple, the speed of the controller needs to be increased further, hence K_p can be increased at this high bandwidth and the system will have enough gain margin to remain stable. The simulation results shown in Fig. 5(c) are for LPF bandwidth= 10 kHz, K_p=0.01 with active balancing enabled at 0.1 s.

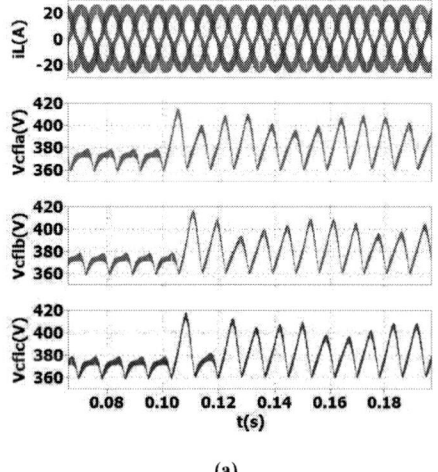

(a)

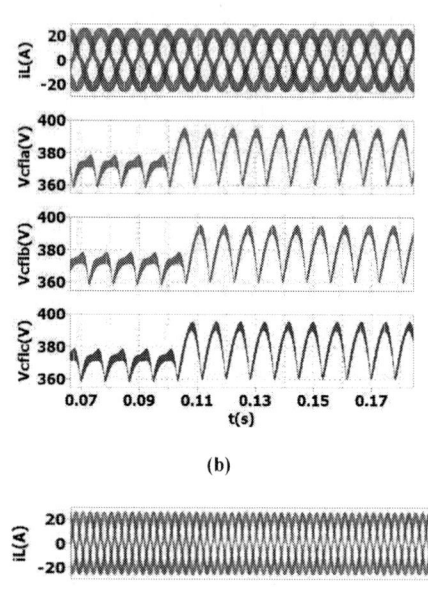

(b)

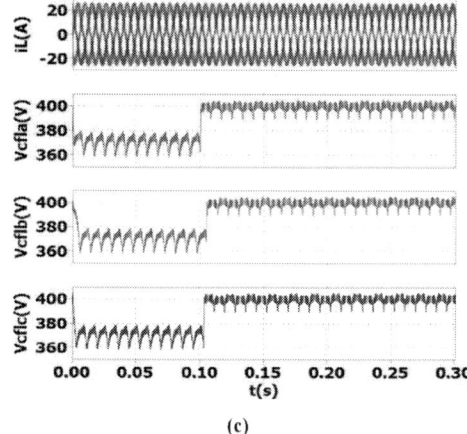

(c)

Fig. 5. Simulation results with active balancing enabled at 0.1s, the unbalance being introduced by injecting a >180° phase shift between S_{A1} and S_{A2} for(a) LPF bandwidth= 100 Hz, K_p=0.0001 (b) LPF bandwidth= 10 kHz, K_p=0.0001 and (c) LPF bandwidth= 10 kHz, K_p=0.01

III. HARDWARE DESIGN

The devices used in the high frequency phase legs are 2*GS66516T (blocking voltage: 650 V) in parallel, the flying capacitor is designed for a ripple of ±5 V at 800 V V_{dc}, 11 kW P_o. The total value of the flying cap is 6.58 μF, the combination is realized by 14*0.47 μF ceramic capacitors. The GaN based flying capacitors phase leg and inductor design are discussed in detail below.

A. GaN based Flying Capacitor Phase Leg Design

Owing to the fast transients in GaN devices, it is necessary to design small gate loops and power loops [7]-[8] to avoid any overvoltage on the gate as well as the drain. It is also very important to design the gate loop and power loop completely symmetrical for the two GaN devices in parallel to prevent any current mismatch between the devices as well as any oscillation on the gate voltage during transient [7]-[8]. Secondly, it is of utmost importance that the two sets of complementary devices have symmetrical gate loop and power loop as a mismatch in these critical loops design will cause an unbalance in the flying capacitor voltage. The small power loop design with decoupling caps right next to the devices is shown in Fig. 6. The top layer (shown in Fig. 6(a)) has the decoupling caps for the outer pair of devices while the bottom layer has the decoupling caps for the inner pair of complimentary devices (as shown in Fig. 6(b)). The gate drivers are isolated comprising of two stages: an isolator + gate driver. The non-isolated gate drivers are placed on the bottom layer right underneath the devices resulting in a very small gate loop as shown in Fig. 6(b).

The flying capacitors are mounted on a secondary daughter card underneath the main board (as shown in Fig. 7) to optimize the power loops and gate loops as discussed above.

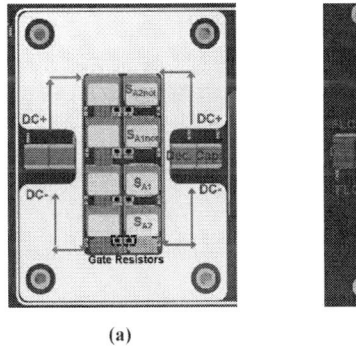 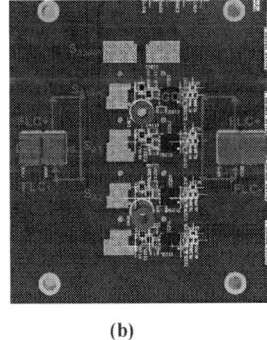

(a)   (b)

Fig. 6 . Layout of devices in one phase leg: (a) Top layer with decoupling caps for outer devices (b) Bottom Layer with decoupling caps for inner devices

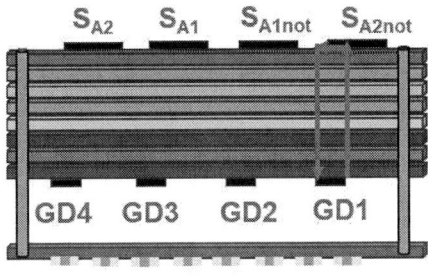

Fig. 7 . Side view of the board structure with highlighted gate loop and flying capacitor daughter card

B. Inductor Design

The inductance is designed for 40% ripple at peak value of phase voltage at V_{phrms}= 230 V, V_{dc}= 800 V, P_o= 11 kW, f_s= 65 kHz, f_{iL}= 130 kHz

$$L=V_{phmax}*D*T_s/(0.4*I_{peak}) \qquad (4)$$

Where D is given by (2)

Thus the value of inductance is 65 µH. The inductors are built with high B_{Sat} nanocrystalline material (F3CC), the smallest commercially available core is selected. The litz wire chosen is 0.06 mm/400 strands achieving a current density of 14 A/mm^2. For loss optimization, B_{max} is varied from 0.2-1 T as shown in Fig. 8. The total inductor loss is optimized for V_{phrms}= 230 V, V_{dc}= 800 V, P_o= 11 kW and it is minimum at B_{max}= 0.24 T, n=28 and l_g= 5.9 mm. The core and winding losses are calculated based on Steinmetz [10] and Sullivan's

[11] equation respectively. For optimum design, core loss, P_{cb}=3.1 W/phase and winding loss, P_{wb}=3.43 W/phase at V_{phrms}= 230 V, V_{dc}= 800 V, P_o= 11 kW, f_s= 65 kHz, f_{iL}= 130 kHz.

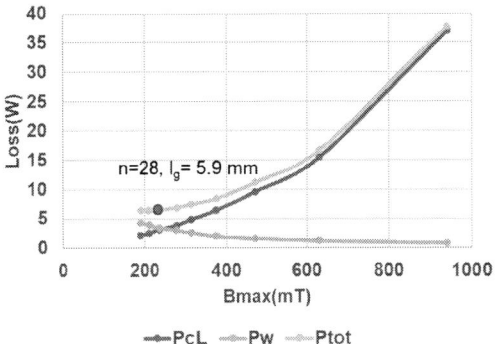

Fig. 8. B_{max} sweep for inductor loss optimization

IV. EXPERIMENTAL VERIFICATION

The hardware setup with inductors and AC filter capacitors is shown in Fig. 9. The circuit is a 3-ph/1-ph combo as required in OBC applications. The slow leg consists of SiC devices. The flying capacitors daughter card is mounted on top of the mother board is shown in Fig. 10. The details of the components are shown in Table I.

TABLE I.   COMPONENTS

Component	Description
GaN devices	2*GS66516T (650 V, 60 A) in parallel/ switch
Slow leg devcies for 1-ph operation	2*NTH4L014N120M3P (1.2 kV, 127 A) in parallel/ switch
Flying capcitors	14*C2220C474KCRACTU (MLCC - SMD/SMT 500V 0.47uF X7R 2220)/ phase
DC bus bulk capacitors	8*ALA7DA391DF550 (Aluminum Electrolytic Capacitors - 550V 390 µF)

Fig. 9. HW setup with inductors and filter capacitors

979-8-3315-1612-3/25 $31.00 © 2025 IEEE

Fig. 10. Bottom side of the board showing flying capacitor daughter card

The key experimental waveforms for 3ϕ-4W (AC neutral connected to DC bus mid-point) configuration at V_{phrms}= 230 V, V_{dc}= 800 V, P_o= 11 kW, L= 65 µH, f_s= 65 kHz, C_{fl}= 6.58 µF are shown in Fig. 11. The flying capacitor voltage is very close to 400 V (as shown in the figure) for all the three phases, thus achieving good balancing. The measured flying capacitor voltage for phase A, B and C is 404.7 V, 400.8 and 404.7 V respectively.

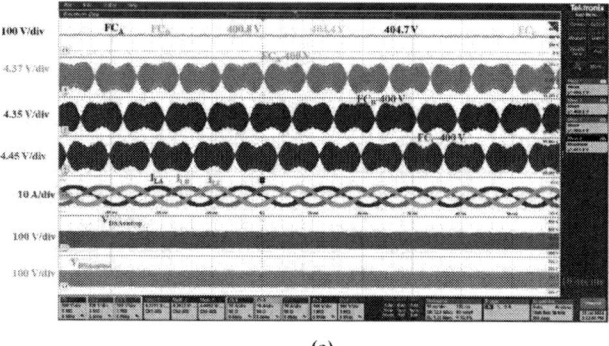

(a)

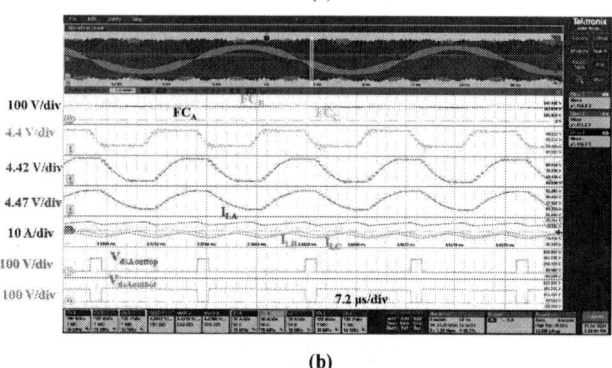

(b)

Fig. 11. Key experimental waveforms at V_{phrms}= 230 V, V_{dc}= 800 V, P_o= 11 kW, f_s= 65 kHz for a 3ϕ-4W system: (a) Multiple line cycles and (b) Zoom-in waveform. From top to bottom: 1. Flying capacitor voltages for phase A, B and C 2. Phase A flying capacitor voltage ripple 3. Phase B flying capacitor voltage ripple 4. Phase C flying capacitor voltage ripple 5. Phase A.B and C inductor currents. 6. $V_{DS,SA1not}$ and 7. $V_{DS,SA1}$

The key experimental waveforms for 3ϕ-3W (AC neutral not connected to DC bus mid-point) configuration at V_{phrms}= 230 V, V_{dc}= 800 V, P_o= 11 kW, L= 65 µH, f_s=65 kHz, C_{fl}= 6.58 µF are shown in Fig. 12. The flying capacitor voltage is very close to 400 V (as shown in the figure) for this case also. The measured flying capacitor voltage for phase A, B and C is 408.7 V, 399.8 and 407.4 V respectively. The ripple shape is slightly asymmetric as compared to the 3ϕ-4W system; however, the scale of ripple measurement is also small (~5 V/div). This is because in a 3ϕ-4W system, each phase runs as an independent single-phase converter and the effect of any asymmetry caused due to gate driver and controller delays or different device properties is more easily contained.

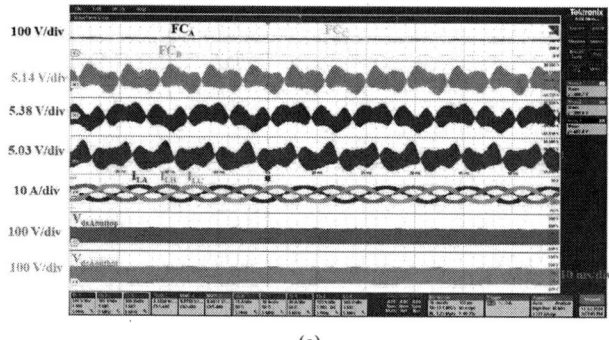

(a)

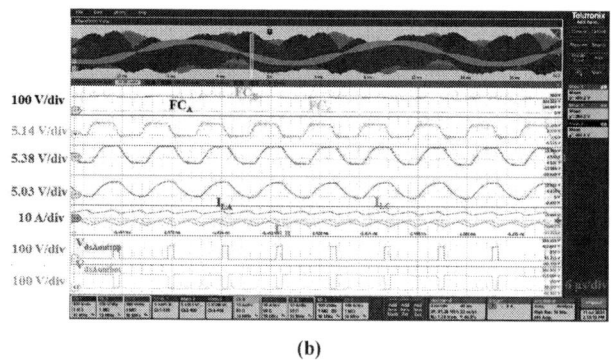

(b)

Fig. 12. Key experimental waveforms at V_{phrms}= 230 V, V_{dc}= 800 V, P_o= 11 kW, f_s= 65 kHz for a 3ϕ-3W system: (a) Multiple line cycles and (b) Zoom-in waveform. From top to bottom: 1. Flying capacitor voltages for phase A, B and C 2. Phase A flying capacitor voltage ripple 3. Phase B flying capacitor voltage ripple 4. Phase C flying capacitor voltage ripple 5. Phase A.B and C inductor currents. 6. $V_{DS,SA1not}$ and 7. $V_{DS,SA1}$

The thermal image of the converter at V_{phrms}= 230 V, V_{dc}= 800 V, P_o= 11 kW, L= 65 µH, f_s=65 kHz, C_{fl}= 6.58 µF is shown in Fig. 13. The temperatures marked are directly beneath the GaN devices on the board, the temperatures are slightly uneven due to uneven air flow, the coolest phase has more air coming on it. The airflow distribution will be improved in the future. For all the three phases the maximum temperatures on the board are below 45 °C, the junction temperatures can be estimated to be ~10-15 °C higher. Thus power rating of this hardware can be increased comfortably.

979-8-3315-1612-3/25 $31.00 © 2025 IEEE

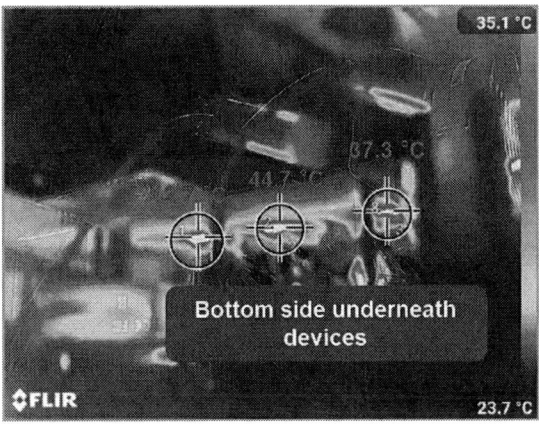

Fig. 13. Thermal camera image of the converter at V_{phrms}= 230 V, V_{dc}= 800 V, P_o= 11 kW, , f_s= 65 kHz

The measured efficiency vs load for V_{phrms}= 230 V, V_{dc}= 800 V, L= 65 µH, f_s=65 kHz is shown in Fig. 14, the peak efficiency achieved is 98.35% at P_o= 11 kW. As compared to state-of-the-art two-level SiC converter operating in CCM [9], the efficiency achieved is ~0.8% higher at the same power level while the switching frequency and inductor current frequency in [9] are <50 kHz. Further, efficiency can be improved in CCM SPWM by optimizing the inductor design and reducing dead-times and gate resistors.

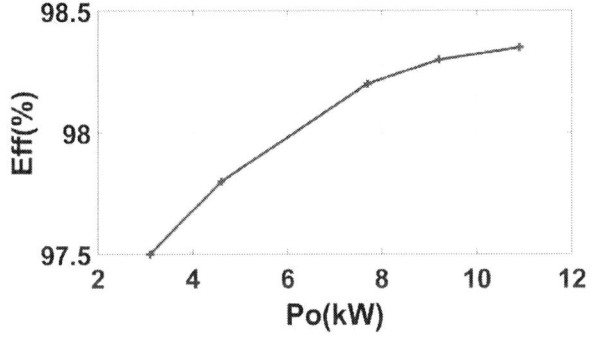

Fig. 14. Measured efficiency vs load at V_{phrms}= 230 V, V_{dc}= 800 V, L= 65 µH, f_s=65 kHz

The loss breakdown based on a model for V_{phrms}= 230 V, V_{dc}= 800 V, P_o= 11 kW, L= 65 µH, f_s=65 kHz, C_{fl}= 6.58 µF are shown in Fig. 15. The detailed component losses are shown in Table II. The switch losses are calculated based on the double pulse test (DPT) data presented in [7]. The calculated efficiency is 98.7% which is close to the measured efficiency. The final inductor losses came out to be higher than projected in section III as the bobbin area and perimeter is much higher than the core geometry. Thus the efficiency can be improved by optimizing the inductor considering these revised parameters. The efficiency can be further improved by DPWM technique such that the phase carrying maximum current is clamped. Based on calculations this can help increase the efficiency by ~0.3%.

TABLE II.   LOSS BREAKDOWN BASED ON CALCULAITON

Component	Loss (W)
GaN conduction (Pcond)	36 (T_j= 65°C)
GaN turn-on (Pturnon)	55.77
GaN turn-off (Pturnoff)	1.6
Inductor core (PcL)	6.33
Inductor winding (PwL)	34.5
GaN dead-time conduction (PcondDT)	2.7
Total	140.4
Projected Efficiency (%)	98.72

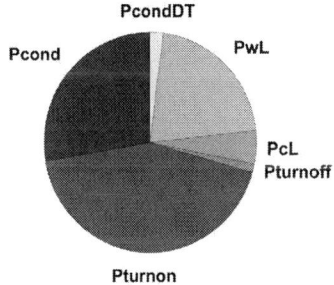

Fig. 15. Loss breakdown based on loss model at V_{phrms}= 230 V, V_{dc}= 800 V, P_o= 11 kW, L= 65 µH, f_s=65 kHz

V. CONCLUSIONS

A three-phase GaN based flying capacitor three-level converter is demonstrated at 800 V DC bus voltage and 11 kW output power. The switching operation is discussed in detail showing that the inductor current frequency is double the switching frequency. A robust active balancing method based on flying capacitor voltage measurement is proposed, the details of the controller design are also discussed. It is shown that the flying capacitor balancing controller has to be much faster than the output voltage controller owing to the drop of the voltage towards its unbalanced value near AC current zero crossing.

The three-level GaN phase leg design is also discussed in detail stating the importance of small gate loop and power loop and symmetry between the two sets of complementary switches to achieve good balancing for flying capacitor voltage. Experimental results are shown for three-phase four-wire as well as three-phase three-wire configuration. The measured efficiency is 98.35% at V_{phrms}= 230 V, V_{dc}= 800 V, P_o= 11 kW, L= 65 µH, f_s=65 kHz, f_{iL}= 130 kHz with SPWM which is 0.8% higher than state-of-the-art SiC based two-level hard-switched AC/DC converters operating at the same power level while at a much lower inductor current frequency.

979-8-3315-1612-3/25 $31.00 © 2025 IEEE

REFERENCES

[1] L. G. Franquelo, J. I. Leon and E. Dominguez, "Recent advances in high-power industrial applications," *2010 IEEE International Symposium on Industrial Electronics*, Bari, Italy, 2010, pp. 5-10.

[2] D. Leuenberger and J. Biela, "Triangular Current Mode Operation of a Three Phase Interleaved T-Type Inverter for Photovoltaic Systems", *Power Conversion and Intelligent Motion Europe*, May 2012.

[3] Jih-Sheng Lai and Fang Zheng Peng, "Multilevel converters-a new breed of power converters," *IAS '95. Conference Record of the 1995 IEEE Industry Applications Conference Thirtieth IAS Annual Meeting*, Orlando, FL, USA, 1995, pp. 2348-2356 vol.3.

[4] A. M. Y. M. Ghias, J. Pou, M. Ciobotaru and V. G. Agelidis, "Voltage balancing method for the multilevel flying capacitor converter using phase-shifted PWM," *2012 IEEE International Conference on Power and Energy (PECon)*, Kota Kinabalu, Malaysia, 2012

[5] Q. Ma, Q. Huang and A. Q. Huang, "Dual-Loop High Speed Voltage Balancing Control for High Frequency Four-level GaN Totem-Pole PFC With Small Flying Capacitors," *2020 IEEE Energy Conversion Congress and Exposition (ECCE)*, Detroit, MI, USA, 2020, pp. 6218-6225

[6] M. Khazraei, H. Sepahvand, K. A. Corzine and M. Ferdowsi, "Active Capacitor Voltage Balancing in Single-Phase Flying-Capacitor Multilevel Power Converters," in *IEEE Transactions on Industrial Electronics*, vol. 59, no. 2, pp. 769-778, Feb. 2012.

[7] B. Sun, R. Burgos, N. Haryani, S. Bala and J. Xu, "Design, characteristics and application of pluggable low-inductance switching power cell of paralleled GaN HEMTs," *IECON 2017 - 43rd Annual Conference of the IEEE Industrial Electronics Society*, Beijing, China, 2017, pp. 1077-1082.

[8] N. Haryani, J. Wang and R. Burgos, "Paralleling 650 V/ 60 A GaN HEMTs for high power high efficiency applications," *2017 IEEE Energy Conversion Congress and Exposition (ECCE)*, Cincinnati, OH, USA, 2017, pp. 3663-3668.

[9] Wolfspeed Inc., "CRD-22AD12N: 22 kW Bi-Directional Active Front End (AFE)", User Guide PRD-02282.

[10] E. C. Snelling, Soft Ferrites, Properties and Applications, 2nd ed. London, U.K.: Butterworths, 1988.

[11] C. R. Sullivan, "Computationally efficient winding loss calculation with multiple windings, arbitrary waveforms, and two-dimensional or three-dimensional field geometry," in IEEE Transactions on Power Electronics, vol. 16, no. 1, pp. 142-150, Jan. 2001.

Optimization Strategy for Battery Electric Vehicle (BEV) DC Fast Charging (FC) in Cold Environments

Seif Sarofim, Cheng Feng Wang, Satyam Sa, Avram Kachura, Isaac Muscat and Olivier Trescases
The Edward S. Rogers Sr. Department of Electrical & Computer Engineering, University of Toronto, Canada
E-mail: seif.sarofim@mail.utoronto.ca

Abstract—**BEV fast charging in cold weather is severely limited by slow kinetics, high impedance, and degradation concerns in the lithium battery. This paper presents an offline open-loop optimization approach that considers the electrical and thermal behavior of the battery, including its thermal management system. This scalable approach is experimentally validated at cold temperature in a custom thermal chamber on a single Tesla Model 3 battery module using a Battery Assisted DC Fast Charger with four parallel-connected 25 kW DC-DC converters.**

I. INTRODUCTION

In 2022, 330,000 fast chargers were added globally [1]. These chargers are crucial for equitable EV adoption, offering an alternative to home charging in dense urban areas [1]. As governments shift incentives from vehicle subsidies to public charging infrastructure [1], maximizing charger utilization and the average charging power across varying conditions is critical. Vehicles only achieve peak charging power under specific conditions — low state of charge (SOC) and high battery temperature, which must be actively managed to avoid thermal runaway. Fast charging in cold weather presents challenges, as reduced battery performance and increased degradation risk are significant concerns. Lithium plating, the dominant aging mechanism below 25°C [2], is especially problematic during fast charging [3]. In [4], a 30-minute longer charge time and up to 30% lower end SOC was observed at 0°C compared to 25°C ambient. In this work, the charging performance of a 2020 Tesla Model 3 Standard Range Plus (SR+), shown in Fig. 1(a), was measured at various temperatures. The maximum charge power was 60 kW at battery temperatures of 10°C and below, compared to 146 kW in optimal conditions, as shown in Fig. 2. The vehicle uses waste heat from the drive inverter for battery conditioning when stationary. The Model 3 SR+ battery pack consists of four identical liquid-cooled modules, shown in Fig. 1(b) with a custom battery management system (BMS).

Common battery charging strategies such as CC-CV and CP-CV, based on heuristic approaches, offer simple implementations for embedded systems, however, they struggle to adapt to varying conditions due to the extensive time and resources required for long-term battery cycle life experiments, especially at the pack level and across different temperatures [5]–[7]. Electro-chemical models provide detailed insights into internal cell parameters and degradation but are computationally intensive, challenging to

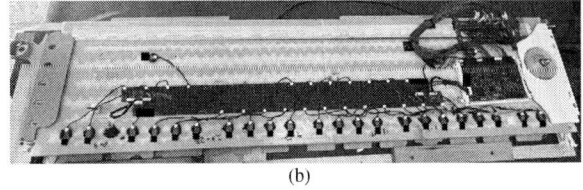

Fig. 1. (a) 2020 Tesla Model 3 SR+ at a FC station. (a) Model 3 battery module with a custom BMS and temperature sensors.

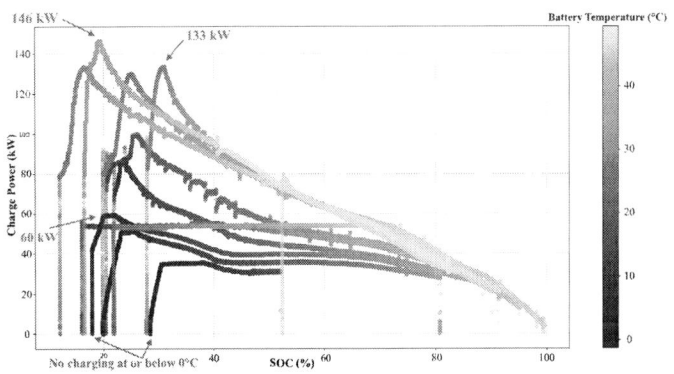

Fig. 2. Measured Model 3 charge power according to minimum battery pack temperature and SOC.

characterize, and deviate at high current densities [8]. Among the over 50 works on health-aware fast charging reviewed in [6], only two considered more than a single cell and often relied on three-electrode cells, which are not commercially viable. Comprehensive reviews of BEV fast charging approaches identified the following gaps in literature: 1) experimental validation of fast charge approaches scaled to the battery pack level, 2) low temperature charging studies, 3) combined modeling of the thermal and electrical response

---

979-8-3315-1612-3/25 $31.00 © 2025 IEEE

of the battery including battery thermal management system (BTMS), 4) development of derating methods according to battery state of health (SOH), 5) consideration of the cell-to-cell thermal and electrical imbalance on fast charging strategies and aging [5], [6], [8]–[10].

II. OPTIMIZATION STRATEGY

The proposed optimization strategy, shown in Fig. 3, takes the following inputs: cell and coolant initial temperature, $T_{cell,0}$ and $T_{coolant,0}$, ambient temperature, $T_{ambient}$, initial and target SOC, SOC_0 and SOC_r, respectively. Subject to constraints on terminal voltage, V_{term}, cell temperature, T_{cell}, charge current magnitude and rate of change, the optimization outputs a charge curve with the minimal charge time, J_t, which includes charge current, I_{chrg}, battery HVAC system power, P_{HVAC}, and coolant flow rate, $\dot{V}(t)$. The algorithm runs offline, outside the vehicle, and is therefore not constrained by the vehicle's computational resources. Cloud computing infrastructure can thus be leveraged to simulate a variety of temperature and SOC conditions. The resulting charge curves would be compiled into a lookup table (LUT) and deployed onto the vehicle, either on the charge controller or battery management system. An accurate electro-thermal model is essential for this open-loop scheme. Fig. 4 presents a model tuning methodology that uses vehicles telematics. Model tuning is beyond the scope of this work.

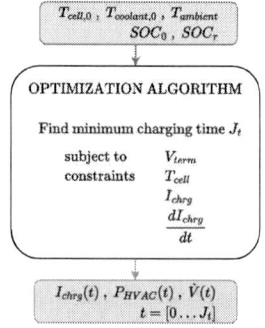

Fig. 3. Proposed optimization strategy.

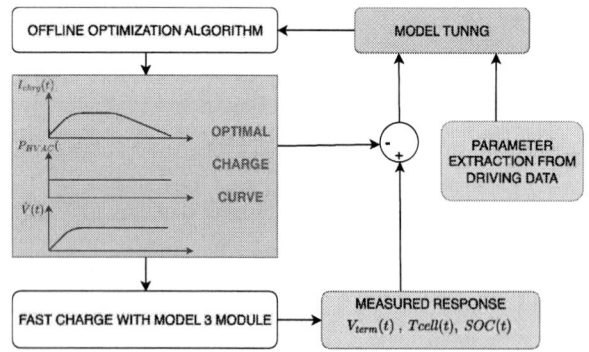

Fig. 4. Proposed model tuning framework based on vehicle measurements.

The battery model consists of an electrical and thermal model run sequentially as shown in Fig. 5. The electrical model is a 1R-ECM, a Thevenin equivalent circuit model (ECM) with an SOC-dependent voltage source in series with a non-linear resistor dependent on temperature and SOC [11], [12]. The model assumes that all cells are balanced and identical; no variation in impedance, capacity, uniform aging and current distribution. These assumption have been made to simply the model and reduce computational complexity. The reduced-order numerical model approach from [13] was used to model a subset of the battery pack, where the thermal system was converted into an electrical analogous circuit. The four Model 3 battery modules are cooled in parallel, each of which contains seven parallel cooling manifolds. As such, the coolant flow rate and heater power were assumed to be divided equally among them, and only one cooling manifold was modeled. A parametric study was conducted to select an optimal time-step that allowed for minimized computational time and sufficient simulation accuracy.

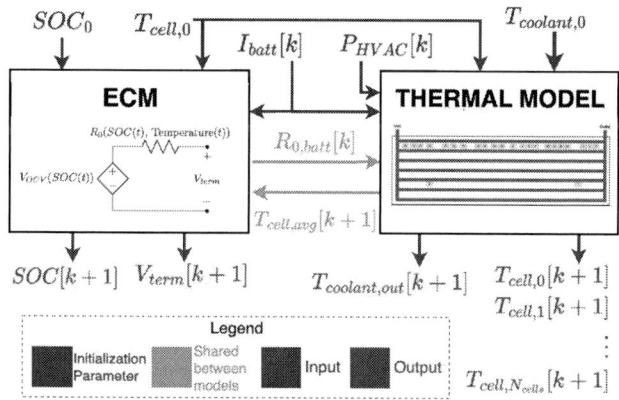

Fig. 5. Modeling approach: interaction between the electrical and thermal battery model.

The proposed global optimization strategy was benchmarked against a greedy algorithm that maximizes SOC rise at each time step. In contrast, the global algorithm optimizes the entire charging sequence, accounting for constraints over time. The global optimization leverages the two-layer approach presented in [14], where charge-time minimization and charge current selection are decoupled into outer and inner layers, respectively. This work expands on [14] by modeling the battery thermal management system, impedance temperature and SOC dependence, limiting degradation, allowing a variable optimization problem size and not constraining battery self-heating, which is beneficial in cold weather. The outer loop employs forward (increasing charge time) and backward (decreasing charge time) searches, as shown in Fig. 6(a). SOC error, defined by (1) where SOC_{end} is the SOC at the end of the charge duration, is a monotonic function of J_t. The forward search uses a coarse time step of five minutes, $t_{step,coarse}$; once a charge time meeting the SOC error threshold, ε_{SOC}, is found, a backward search with a fine time step of 60 seconds, $t_{step,fine}$, begins. The backward search will exit if no charge

979-8-3315-1612-3/25 $31.00 © 2025 IEEE

time meets ε_{SOC} or if the previous forward search time is reached. The forward search begins at the minimum charge time, $J_{t,min}$, and continues until the maximum charge time, $J_{t,max}$, is reached, failing if SOC_{error} cannot be met.

The inner loop determines charge current for a given charge time, J_t, using a convex cost function solved with the interior point method, as shown in Fig. 6(b). The optimization time step, Δt, is fixed at 10 seconds, and the number of current steps, N, is defined by (2). A third optimization layer limits lithium plating at low temperatures using a maximum current LUT based on real-world data.

$$SOC_{error} = |SOC_{\text{end}} - SOC_{\text{r}}| \quad (1)$$

$$N = \frac{J_t}{\Delta t} = \frac{J_t}{10} \quad (2)$$

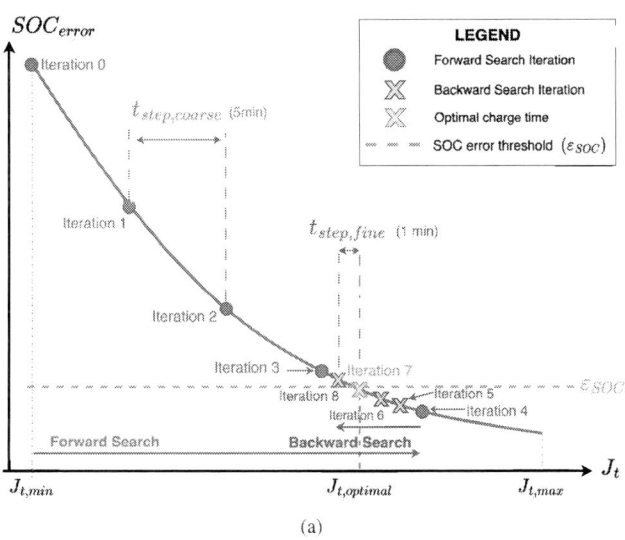

(a)

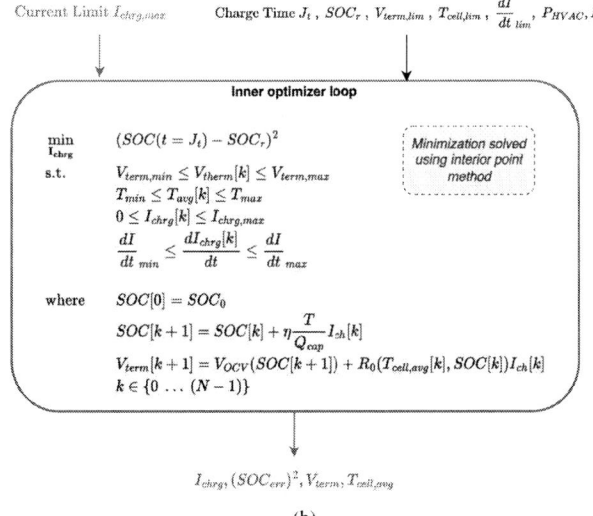

(b)

Fig. 6. (a) Example outer loop search where the optimal charge time, $J_{t,optimal}$, is found after nine iterations. (b) Inner optimization loop.

III. SIMULATION RESULTS

In this study, all simulations were executed on a compact 0.8L computer equipped with an AMD Ryzen 9 7940HS mobile processor (8 cores, 16 threads) and 32GB of RAM. The greedy algorithm averaged a completion time of 3 minutes, while the global algorithm took 12 hours on average. The global algorithm duration is dominated by inner optimization loop iterations, particularly during the backward search of the outer loop near the optimal charge time, $J_{t,optimal}$. The `scipy` and `numpy` Python libraries called by the optimizers did not utilize the AVX-512 instructions supported by the processor, which could potentially enhance performance.

The charge curves from the greedy and global algorithms are shown in Fig. 7(a) and (b), respectively. The greedy algorithm, which maximized the current at each time-step, failed to reach the 80% SOC target while maintaining cell temperatures below 40°C. It led to rapid temperature rise and required significant current reduction, resulting in a terminal SOC of 72% due to breaching temperature constraints, as shown in Fig. 7(a). In contrast, the global optimizer met the target SOC and maintained temperature limits by using a constant current beyond the lithium plating limit, as shown in Fig. 7(b). The simulations show that in cold weather, lithium plating, rather than impedance or cell temperature, is the primary performance limiter, and maximizing HVAC heating power is most beneficial.

IV. OPTIMIZATION SIMULATIONS SWEEP

The performance of the greedy and global algorithms were analyzed in simulation across thirty-six possible scenarios, encompassing nine combinations of starting and ending SOCs, as given in TABLE I, and four sets of temperature conditions, as detailed in TABLE II. While all optimization scenarios were executed for the greedy algorithm, only a subset were run using the global algorithm due to computational time limitations.

TABLE I
OPTIMIZATION SWEEP SOC RANGE

Start SOC (%)	Target End SOC (%)
10%	50%
20%	70%
30%	80%

TABLE II
OPTIMIZATION SWEEP TEMPERATURE CONDITIONS

Case	$T_{cell,avg,0}$ (°C)	$T_{coolant,0}$ (°C)	T_a (°C)	$T_{cell,lim}$ (°C)
A	1.50	2.25	-5.0	55.0
B	3.00	6.00	-5.0	40.0
C	14.00	18.00	10.0	55.0
D	25.00	30.00	10.0	45.0

The results of the greedy and global optimizers converged under temperature conditions A and C but diverged for conditions B and D. As discussed in Section II, the global optimizer's backward search time step is fixed at 60 seconds,

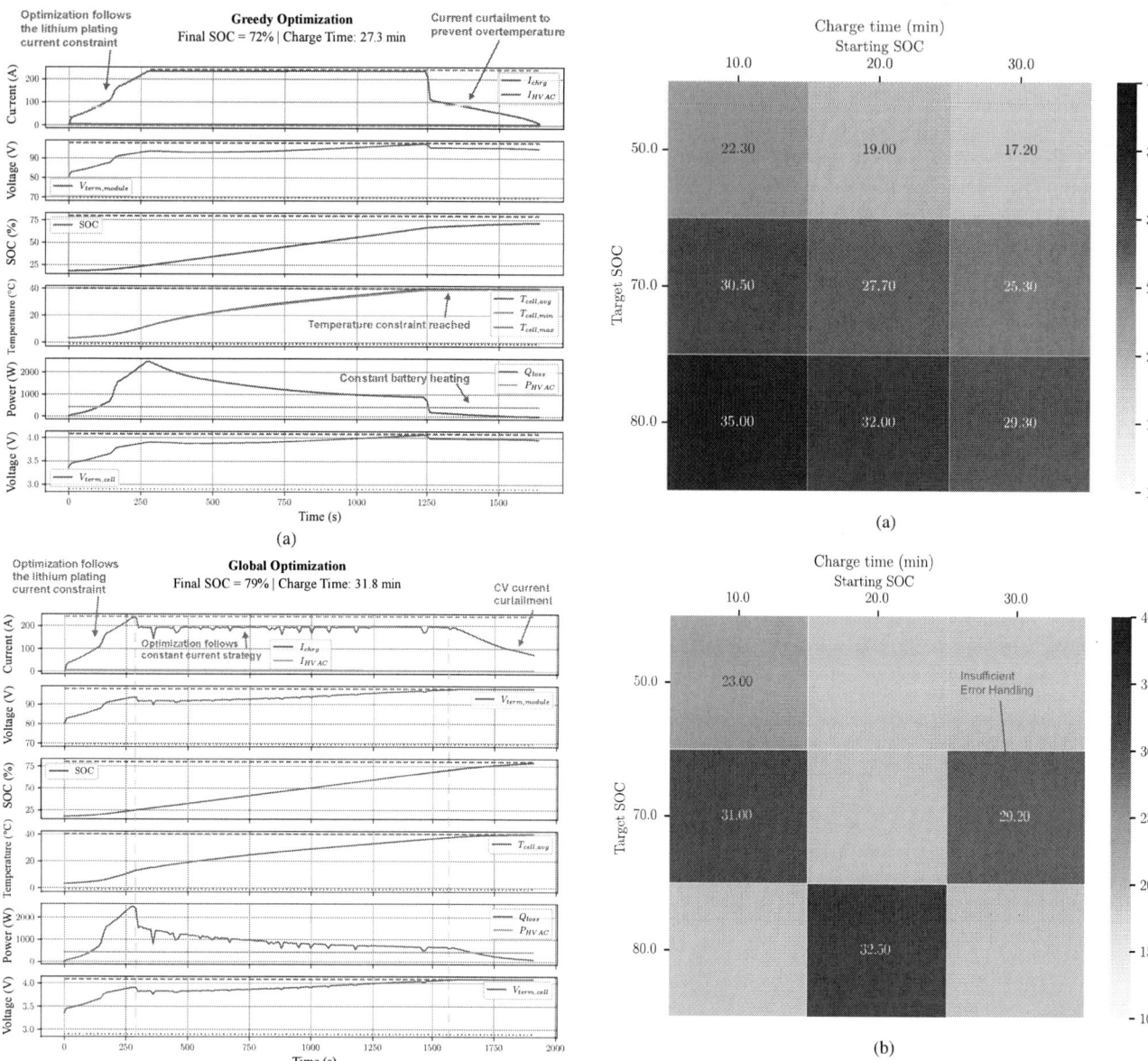

Fig. 7. Greedy (a) and global (b) optimization simulation results for a module fast charge from 18% to 80% SOC, with $T_{\text{cell,avg,0}} = 3°C$, $T_{\text{ambient}} = -3°C$, and maximum station current $I_{\text{max}} = 240A$. Greedy is unable to meet the 80% target SOC without exceeding 40°C.

Fig. 8. Greedy (a) and global (b) optimizer charge time for Case A, $T_{\text{cell,avg,0}} = 1.5°C$. The global optimization charge durations are within one minute of the greedy optimization durations. The cases shown in grey were not run.

while the greedy optimizer's time step is 10 seconds. Consequently, the global optimizer converged toward slightly longer charge durations, within one minute of the greedy optimizer's charge duration, as shown in Fig. 8. This constraint was introduced to reduce run-time, as each inner loop iteration required an average of 90 to 120 minutes.

In temperature condition D (25°C), the greedy algorithm maximized the current at the onset of fast charging, resulting in significant current curtailment to prevent exceeding temperature constraints. This curtailment led to an increase in charge time, as indicated by the triangular matrices annotated in Fig. 9(a). In some cases, the charge times were

longer than those in temperature condition A (1.5°C), shown in Fig. 8(a). By optimizing for the entire duration rather than individual time steps, the global algorithm selected a charge current that resulted in a more balanced temperature profile throughout the charge cycle. The global algorithm's charge times are shown in Fig. 9(b). The global algorithm decreased the charge time by 1 minute and enhanced the $P_{\text{chrg,avg}}$ by 1kW when charging from 10% to 50% SOC, and cut the charge time by approximately 3 minutes and increased $P_{\text{chrg,avg}}$ by 2kW when charging from 20% to 80% SOC. This outcome highlights the limitations of the greedy approach when design constraints are applied. Meeting temperature constraints poses a trade-off for vehicle OEMs, as allowing higher battery temperatures requires a more robust cooling system, which increases both cost and weight.

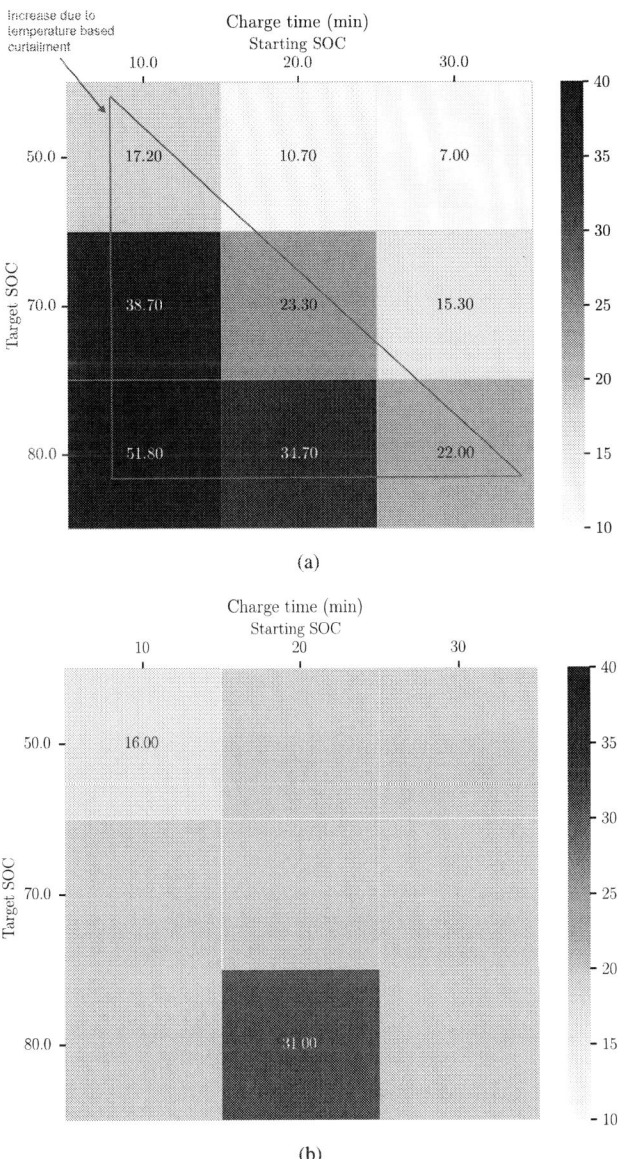

Fig. 9. Greedy (a) and global (b) optimizer charge time for Case D, $T_{\text{cell,avg,0}} = 25°C$. The cases shown in grey were not run.

The battery SOC at the end of the charge duration for the greedy and global algorithms in Case B are shown in Fig. 10. They highlight a distinct advantage for the global algorithm. As discussed in Section III, the greedy algorithm failed to meet the target end SOC for the scenarios depicted by the lower triangular matrix in Fig. 10(a). It significantly reduced the current to limit temperature rise and maintain cell temperatures below 40°C. The global algorithm selected a lower constant current maintained throughout the majority of the charge without breaching the temperature constraint. Notably, in cases with a 50% target end SOC, the shorter charge duration allowed for the maximization of the current without breaching the temperature limit.

The sensitivity of the global algorithm to minimum charge time bound estimates and error handling was notable. A failed inner optimization loop, such as exceeding the allowed number of function evaluations, led the algorithm to converge to a longer charge time, as annotated in Fig. 8(b). A large initial guess for the minimum charge time bound could cause the optimization to overlook the optimal charge time.

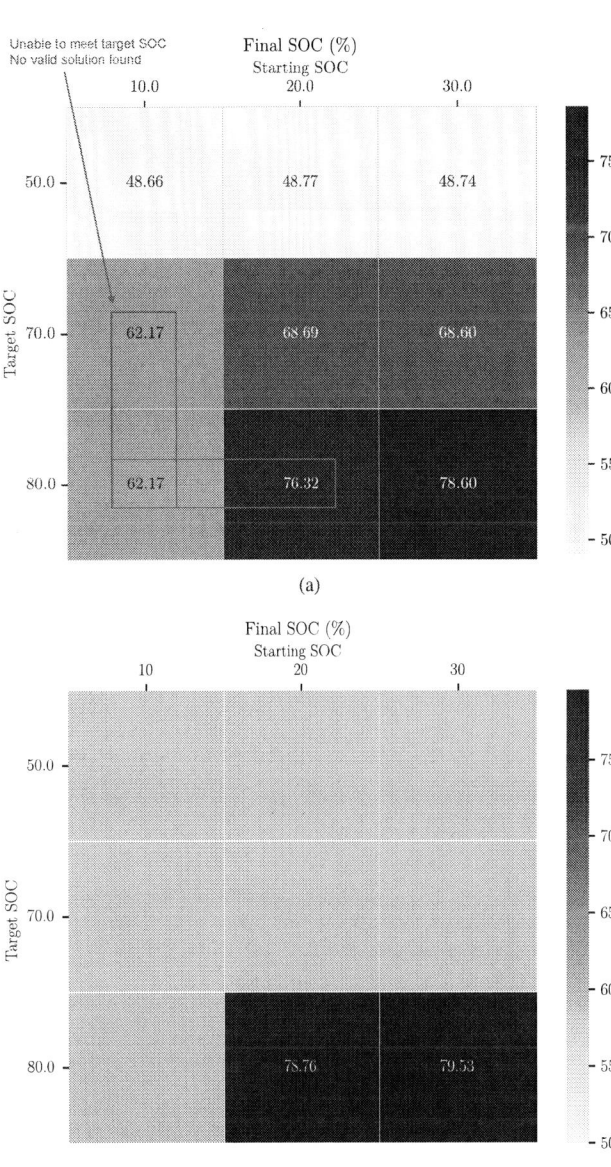

Fig. 10. Greedy (a) and global (b) final SOC for Case B, $T_{\text{cell,avg,0}} = 3°C$. In three starting SOC conditions, greedy is unable to meet the 80% target SOC without exceeding 40°C. The cases shown in grey were not run.

V. EXPERIMENTAL VALIDATION

The simulation results were validated using a Tesla Model 3 battery module (13.2 kWh, 88.8V nominal), shown in Fig. 1(b). The charging and thermal control design is shown in Fig. 11(a) and its implementation in Fig. 11(c). The charge curve generated by the offline optimization was loaded into a GUI which relayed the charge current requests to the custom Battery Assisted DC Fast Charger (BA-DCFC), shown in Fig. 11(b). The stationary battery, consisting of 12 second life Tesla Model S modules in series, powers four 25 kW isolated bi-directional DC-DC converters in parallel. A custom thermal chamber was fabricated to emulate the vehicle battery enclosure and ambient temperatures. During charging, the battery was heated with an inline heater. The current was curtailed if it exceeded the maximum allowed at the measured battery temperature.

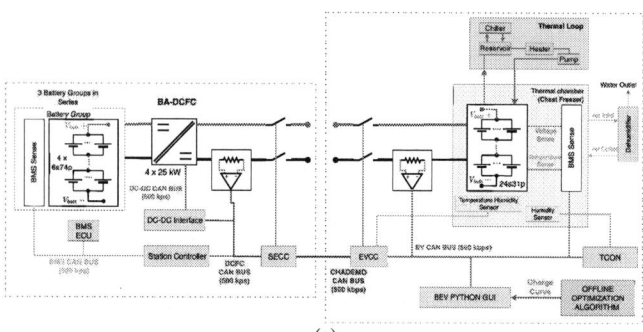

(a)

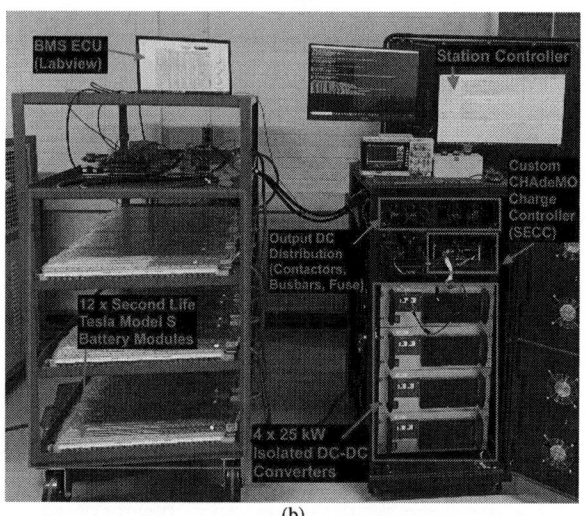

(b)

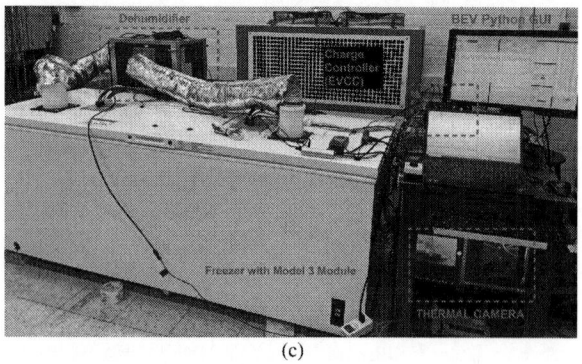

(c)

Fig. 11. (a) Experimental test setup. (b) 100 kW BA-DCFC implementation. (c) Model 3 module validation setup including thermal chamber.

The global optimized profile in Fig. 7(b) was imposed on the battery, as illustrated in Fig. 12(a). A peak charging power of 18.9 kW (approx. 195A at 97V) was achieved. The experimental results were significantly affected by discrepancies in the thermal model. At the charge on-set the current was curtailed according to the lithium plating constraint because the module's temperature rise was overestimated, as shown in Fig. 12(b). A temperature difference of approximately 10°C was noted for most of the charging process, however, this discrepancy should not be interpreted as a measure of model accuracy since it presupposes a higher charging current. Similarly, the measured battery module terminal voltage was approximately 1.5V higher than simulated, as a result of the lower temperature hence higher impedance. This lead to a breach of the simulation terminal voltage constraint and converters entered CV mode sooner than in simulation, at $t = 1400s$ compared to $t = 2000s$, as shown in Fig. 12(a) and 12(c).

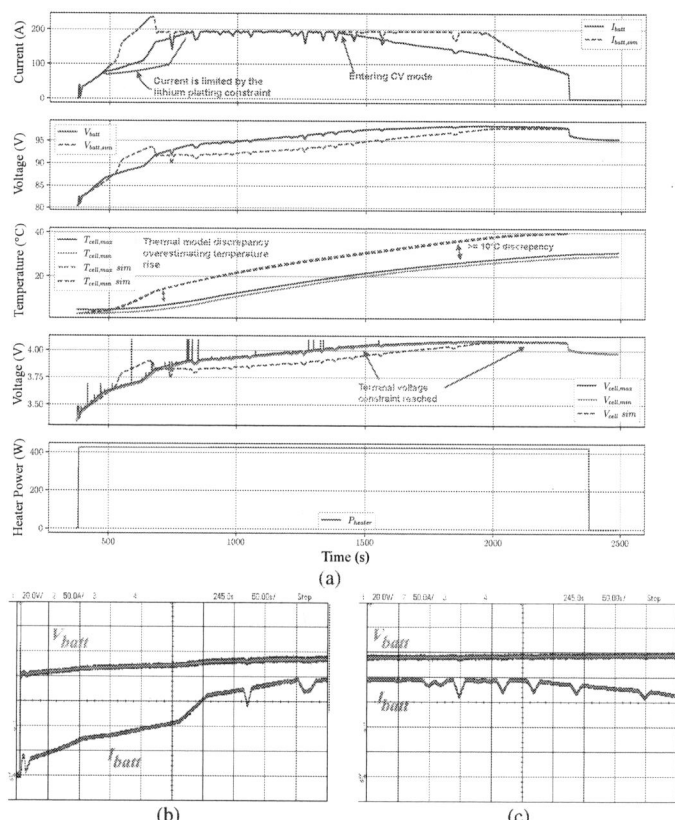

(a)

(b)　　　　　　(c)

Fig. 12. (a) Measured fast charge experimental result compared to the expected simulation result. (b) Measured current ramp-up, limited by the lithium plating constraint, and module voltage. (c) Measured current curtailment when entering CV.

VI. MODEL DISCREPANCIES

To quantify model discrepancies, two simulations were conducted based on the experimental measurements: an ECM-only simulation with measured temperature input into the electrical model, aimed at validating the ECM model and temperature sensing accuracy, and a full electro-thermal model simulation using measured current, flow rate, and HVAC power.

979-8-3315-1612-3/25 $31.00 © 2025 IEEE

The electrical model accurately tracked the terminal voltage response, however, deviations appeared as SOC and temperature increased, reaching a peak voltage discrepancy of 1.4V, as shown in Fig. 13(a). The module SOC and lumped resistance extracted from measurements using a hybrid pulse power characterization (HPPC) test [15] closely matched the simulation results. As such, the voltage discrepancy was attributed to the first and second order capacitive battery impedance elements and underscores the limitations of the 1R-ECM model compared to more complex models such as 1RC-ECM and 2RC-ECM. The in-situ Battery Electrochemical Impedance Spectroscopy (EIS) methodology presented in [12] enables the extraction of an accurate battery ECM across various temperatures. This methodology could be leveraged to enhance the optimization's electrical model by implementing a 2RC-ECM instead of a 1R-ECM.

The thermal model discrepancy narrowed, with an approximate 6°C difference at the end of the charge, though the temperature profile remained significantly divergent, maintaining a 10°C difference overall, as shown in Fig. 13(b). The thermal model deviation is primarily attributed to ambient temperature variation within the custom thermal chamber and the impedance distribution within the thermal model.

VII. CONCLUSIONS

An offline fast charge optimization methodology considering electrical, thermal, and aging behavior, including the battery thermal management system, is proposed. It relies on existing commercially available batteries and is compatible with FC standards. The scalable methodology can adapt to various battery configurations, cell chemistries, and thermal management systems, facilitating iterative development of FC strategies for both current and future vehicle platforms. This work fills a gap in the literature by experimentally validating FC strategies at the module scale in cold temperatures. Model discrepancies inherent to this open-loop approach underscore the necessity for a systematic model-tuning method in both the electrical and thermal domains. The 1R-ECM model was found to be insufficient for fast charging as it underestimated the terminal voltage, and could be enhanced through EIS [12].

Self-tuning would enhance scalability by adapting charging strategies to the vehicle's aging profile, addressing a gap in the current literature. The proposed methodology would also benefit from a local closed-loop feedback mechanism running on the vehicle that responds to model disturbances and dynamic station limitations (e.g. maximum available current) by leveraging a large set of precomputed charge paths from the offline optimization.

As shown in Fig. 14, vehicle manufacturers can leverage the proposed offline optimization strategy, hosted on cloud computing infrastructure, to simulate various charging scenarios, battery configurations, chemistries, and thermal management systems. Pre-computed charging paths are

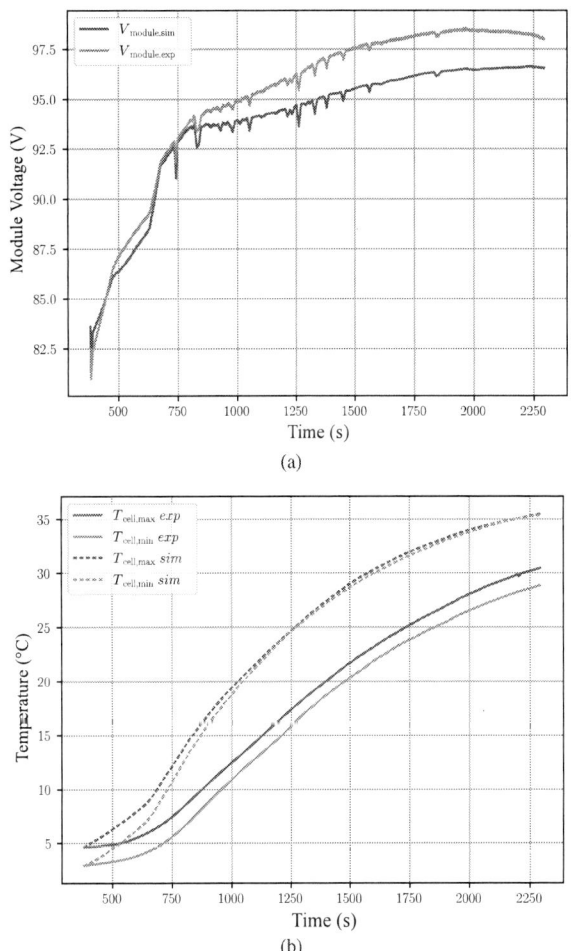

Fig. 13. (a) Module terminal voltage ECM simulation result based on measured fast charge current and cell temperature. (b) Cell temperature simulation result from thermal model using experimental data as input.

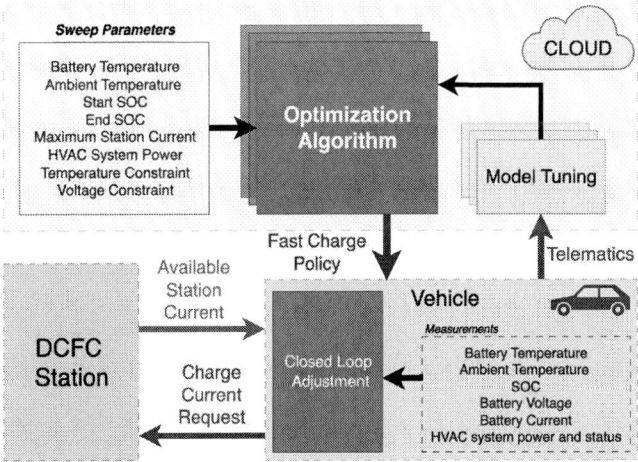

Fig. 14. Proposed FC optimization strategy leveraging cloud infrastructure and on-vehicle closed loop feedback. Telematics including EIS measurements would be used to improve model accuracy.

transmitted to EV fleets and dynamically adjusted during charging via an on-vehicle closed-loop controller. The optimization process is continuously refined using a battery model updated with telematics data, including EIS measurements.

VIII. ACKNOWLEDGMENTS

This work was supported by the Litens Automotive Partnership, and the Natural Sciences and Engineering Research Council of Canada (NSERC).

REFERENCES

[1] International Energy Agency, "Global EV Outlook 2023: Catching up with Climate Ambitions," International Energy Agency, Tech. Rep., Apr. 2023.

[2] T. Waldmann, M. Wilka, M. Kasper, M. Fleischhammer, and M. Wohlfahrt-Mehrens, "Temperature dependent ageing mechanisms in Lithium-ion batteries – A Post-Mortem study," *Journal of Power Sources*, vol. 262, pp. 129–135, Sep. 2014.

[3] T. Waldmann, B.-I. Hogg, M. Kasper, S. Grolleau, C. G. Couceiro, K. Trad, B. P. Matadi, and M. Wohlfahrt-Mehrens, "Interplay of Operational Parameters on Lithium Deposition in Lithium-Ion Cells: Systematic Measurements with Reconstructed 3-Electrode Pouch Full Cells," *Journal of The Electrochemical Society*, vol. 163, no. 7, p. A1232, Apr. 2016.

[4] Y. Motoaki, W. Yi, and S. Salisbury, "Empirical analysis of electric vehicle fast charging under cold temperatures," *Energy Policy*, vol. 122, pp. 162–168, Nov. 2018.

[5] Y. Gao, X. Zhang, Q. Cheng, B. Guo, and J. Yang, "Classification and Review of the Charging Strategies for Commercial Lithium-Ion Batteries," *IEEE Access*, vol. 7, pp. 43 511–43 524, 2019.

[6] N. Wassiliadis, J. Schneider, A. Frank, L. Wildfeuer, X. Lin, A. Jossen, and M. Lienkamp, "Review of fast charging strategies for lithium-ion battery systems and their applicability for battery electric vehicles," *Journal of Energy Storage*, vol. 44, p. 103306, Dec. 2021.

[7] N. Tian, H. Fang, and Y. Wang, "Real-Time Optimal Lithium-Ion Battery Charging Based on Explicit Model Predictive Control," *IEEE Transactions on Industrial Informatics*, vol. 17, no. 2, pp. 1318–1330, Feb. 2021.

[8] A. Tomaszewska, Z. Chu, X. Feng, S. O'Kane, X. Liu, J. Chen, C. Ji, E. Endler, R. Li, L. Liu, Y. Li, S. Zheng, S. Vetterlein, M. Gao, J. Du, M. Parkes, M. Ouyang, M. Marinescu, G. Offer, and B. Wu, "Lithium-ion battery fast charging: A review," *eTransportation*, vol. 1, p. 100011, Aug. 2019.

[9] P. Makeen, H. A. Ghali, and S. Memon, "A Review of Various Fast Charging Power and Thermal Protocols for Electric Vehicles Represented by Lithium-Ion Battery Systems," *Future Transportation*, vol. 2, no. 1, pp. 281–299, Mar. 2022.

[10] N. Ghaeminezhad and M. Monfared, "Charging control strategies for lithium-ion battery packs: Review and recent developments," *IET Power Electronics*, vol. 15, no. 5, pp. 349–367, 2022.

[11] W. Waag, S. Käbitz, and D. U. Sauer, "Experimental investigation of the lithium-ion battery impedance characteristic at various conditions and aging states and its influence on the application," *Applied Energy*, vol. 102, pp. 885–897, Feb. 2013.

[12] Z. Gong, A. Kachura, S. A. Assadi, N. Cusimano, J. Piruzza, J. Xu, and O. Trescases, "An EV-Scale Demonstration of In-Situ Battery Electrochemical Impedance Spectroscopy and BMS-Limited Pack Performance Analysis," *IEEE Transactions on Industrial Electronics*, pp. 1–10, 2022.

[13] C. Escobar, Z. Gong, C. Da Silva, O. Trescases, and C. H. Amon, "Effect of Cell-to-Cell Thermal Imbalance and Cooling Strategy on Electric Vehicle Battery Performance and Longevity," in *2022 21st IEEE Intersociety Conference on Thermal and Thermomechanical Phenomena in Electronic Systems (iTherm)*, May 2022, pp. 1–9.

[14] Q. Ouyang, R. Ma, Z. Wu, and Z. Wang, "Optimal Fast Charging Control for Lithium-ion Batteries," *IFAC-PapersOnLine*, vol. 53, no. 2, pp. 12 435–12 439, Jan. 2020.

[15] J. P. Christophersen, "Battery Test Manual For Electric Vehicles, Revision 3," Idaho National Lab. (INL), Idaho Falls, ID (United States), Tech. Rep. INL/EXT-15-34184, Jun. 2015.

979-8-3315-1612-3/25 $31.00 © 2025 IEEE

DC-Link Voltage Reduction with Synergetic Common-Mode Voltage Control of Single-Phase Two-Stage Non-Isolated EV Chargers

Dongsu Lee
Electrical and Computer Engineering Department
Seoul National University
Seoul, Korea
babypighill0@snu.ac.kr

Juwon Lee
Electrical and Computer Engineering Department
Seoul National University
Seoul, Korea
wronskian@snu.ac.kr

Jung-Ik Ha
Electrical and Computer Engineering Department
Seoul National University
Seoul, Korea
jungikha@snu.ac.kr

Abstract— **As interest in climate issues grows, research on electric vehicles (EVs) and their charging technologies is advancing. Non-isolated topologies have been proposed to improve EV charger efficiency and power density. However, removing the galvanic isolation stage increases leakage currents, potentially causing electric shocks to users. These currents, generated by low-frequency (LF) common-mode voltage (CMV), are difficult to attenuate with EMI filters. Injecting LF CMV into the power converter can reduce leakage currents but requires higher DC-link voltage, increasing losses and reducing efficiency. This paper proposes a two-stage non-isolated EV charger with an AC/DC power factor correction (PFC) converter and a DC/DC three-level regulator. The PFC and regulator synergistically share the CMV to lower the DC-link voltage while maintaining leakage current reduction. The proposed method enables operation at lower DC-link voltage and higher modulation index (MI) regions. Experimental validation using a prototype confirms its effectiveness.**

Keywords—**Common-Mode Voltage, Leakage Current, Non-Isolated EV Charger, Synergetic Control, Touch Current, Transformerless, Virtual Grounding Control.**

I. INTRODUCTION

Recently, global interest in climate issues related to low-carbon and carbon-neutral initiatives has been rising. Electric vehicles (EVs) present an attractive alternative to internal combustion engine vehicles, as they utilize clean energy and produce no gas emissions, making them an effective solution for complying with environmental regulations. By 2035, it is projected that half of all vehicles sold worldwide will be electric, and the number of global public EV charging points is expected to increase by four times compared to 4 million in 2023 [1]. Most EV chargers include galvanic isolation stages to comply with common-mode (CM) current standards, even though regulations such as IEC 61851 and UL 2954 do not explicitly mandate insulation stages. To ensure user safety, a 60 Hz transformer in the low-frequency (LF) range or a solid-state transformer in the high-frequency (HF) range is used [2]. However, incorporating transformers introduces additional energy conversion stages, which makes the system larger and more complex, resulting in the reduction of power density and degraded efficiency [3]. To address these disadvantages, removing galvanic isolation is considered recently; however, this allows CM leakage current to flow within the system, potentially resulting in touch currents that could cause electric shocks to users. A common approach to ensure safety in the absence of galvanic isolation is the use of residual current devices (RCDs). In single-phase systems, an RCD measures the sum of currents flowing through the two lines, which

corresponds to the leakage current flowing through the ground, and disconnects the circuit when this sum reaches a trip level, deviating from zero. According to IEC 61851-1, the leakage current trip level for RCDs installed in EV chargers is specified as AC 30 mA, and the regulation for human touch current is set at an RMS value of 3.5 mA [4].

In a single-phase grid, line-to-neutral (LN) configuration is widely used where one line from a three-phase grid and the neutral line is utilized to form a single-phase system. A single-phase LN system continuously applies a common-mode voltage (CMV) equal to half of the grid voltage across the two lines, resulting in low-frequency (LF) leakage currents flowing within the non-isolated system [5]. When an RCD is employed in a single-phase LN system, the leakage current flowing to the ground can easily reach the trip level, leading to frequent and unintended RCD trips, known as nuisance tripping. For example, in a 220V LN single-phase system with a total y-capacitor value C_y of 1 μF, the amplitude of the LF leakage current is approximately 59 mA, which exceeds the RCD's trip threshold, causing nuisance tripping. To address this issue by reducing the system's leakage current, the use of EMI filters or modifications to the total y-capacitor, C_y, could be explored. However, electromagnetic interference (EMI) filters are primarily designed to attenuate HF CM currents in the switching frequency range, and significant resources are required to attenuate LF CM currents within the grid frequency range. The y-capacitance, connected between the power system and the protective earth (PE), is used to suppress EMI and reduce both HF noise and leakage currents. Reducing the value of the y-capacitance can decrease the magnitude of leakage currents in the low-frequency range. However, the total parallel y-capacitance of an EV charger is constrained by the circuit configurations of other electrical devices connected to the output power line, and its design cannot be arbitrarily modified. IEC 61851-23 specifies a maximum total parallel y-capacitance, C_y, of 1 μF [6].

To reduce LF CM leakage currents without modifying the value of the y-capacitance or adding EMI filters, the concept of controlling CMV in power converters is first introduced in [7]. This approach has been applied to various applications, including grid interfaces [7], PV systems [8], and EV chargers [9], [10]. In EV charger applications, when rectifying a single-phase AC grid with a power factor correction (PFC) converter, an additional step-down stage is required to satisfy the wide output voltage range required by EV batteries. Additionally, there is some controversy as to whether power ripple occurring at twice the grid frequency due to single-phase grid rectification degrades battery charging performance [11]. To

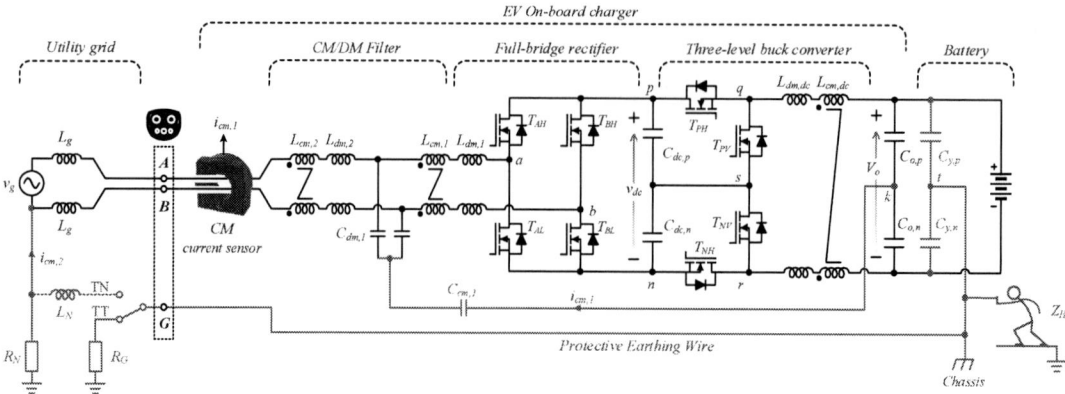

Fig. 1. Circuit diagram of a single-phase two-stage non-isolated converter connected to an LN grid system. It consists of a full-bridge PFC and a three-level stacked buck converter as the regulator, with the neutral points of the input filter and output filter capacitors connected.

meet the low voltage requirements of EV batteries and eliminate power ripple, a two-stage converter, consisting of a PFC (AC/DC stage) cascaded with a step-down regulator (DC/DC stage), is implemented. As seen in [5], [7], [8], [9], [10], [12], [13], [14], active CMV control requires a higher DC-link voltage because the voltage at the converter's poles is determined by the sum of the differential-mode (DM) voltage reference at the converter's input and output and the CM controller's voltage reference. In such two-stage converters, when input and output voltage conditions are the same, an increase in DC-link voltage leads to a higher PFC boost ratio and a lower regulator step-down ratio. Higher DC-link voltage magnifies switching losses and current ripple, thereby increasing inductor core and conduction losses, which degrades efficiency [15]. Therefore, reducing DC-link voltage while controlling CMV to reduce leakage current is critical for the operation of single-phase non-isolated EV chargers. This paper proposes a control method for a two-stage, non-isolated EV charger consisting of a PFC and a regulator in a cascaded structure. The synergetic CMV control method injects LF CMV to reduce leakage current while synergistically sharing CMV control between the PFC and regulator to lower the DC-link voltage. The effectiveness of the synergetic CMV control method is validated through experiments conducted on a prototype.

II. LINE FREQUENCY LEAKAGE CURRENT CONTROL

A. Common-Mode Equivalent Circuit Modeling

The circuit diagram of a two-stage non-isolated EV charger that rectifies a single-phase LN grid to supply DC voltage is shown in Fig. 1. The circuit consists of a PFC and a regulator in a cascaded configuration. The PFC is implemented as a full-bridge active rectifier, while the regulator is designed as a three-level stacked buck converter. A buck converter consisting of a single half-bridge cannot simultaneously control both DMV and CMV. Therefore, the regulator is configured using two half-bridge buck converters in a stacked arrangement rather than as a full-bridge to reduce switching losses and voltage stress. Input filter is configured to mitigate HF EMI noise. An inner loop is formed by connecting the input filter's neutral point to the DC output neutral point, which helps to localize HF CMV. This configuration, known as a floating filter, effectively reduces HF EMI noise generated by the converter's pulse-width modulation (PWM) [16]. The charging connector is connected to the utility grid's power lines (A, B) and the grounding point

(G). Two different grounding schemes were analyzed: TT (Terra-Terra) with $R_N = 10\ \Omega$ and $R_G = 100\ \Omega$ and TN (Terra-Neutral) with $R_N = R_G = 0$ [3]. Here, R_N represents the electrode impedance of the utility grid, while R_G is the grounding impedance of the local ground. The vehicle chassis, a metallic part accessible to the user, is connected to the grounding point G of the charging connector through a protective earthing wire and is also connected to the output side of the charger via the y-capacitance. The charger's y-capacitances, $C_{y,p}$, and $C_{y,n}$, are each less than 500 nF. To simulate human contact scenarios, a human body impedance model, represented by Z_H, is constructed in accordance with IEC 60990, as shown in Fig. 6. The two ends of the impedance model are connected to the vehicle chassis and ground, respectively [17]. The entire circuit is represented by the simplified circuit shown in Fig. 2.

In this circuit, the differential-mode voltage (DMV) and CMV of the grid are denoted as $v_{dm,g}$ and $v_{cm,g}$, respectively. The magnitude of $v_{dm,g}$ is equal to the nominal grid voltage, v_g. In a single-phase LN grid system, one end of the power line is grounded, resulting in the magnitude of $v_{cm,g}$ being half of v_g. The stray inductance of the power line is represented as L_g, and in circuits configured with a TN grounding system, the neutral line's stray inductance is denoted as L_n. The virtual neutral point m of the DC-link voltage corresponds to the midpoint of the potentials p and n and is expressed mathematically as

$$v_m = \frac{v_p + v_n}{2}. \tag{1}$$

If the potentials of the input port a and b of the PFC, as seen from the virtual neutral point m of the DC-link, are denoted as v_{am} and v_{bm}, respectively, the CMV of a and b as seen from m, $v_{cm,A}$, is expressed as

$$v_{cm,A} = \frac{v_{am} + v_{bm}}{2}. \tag{2}$$

Similarly, the DMV of a and b, $v_{dm,A}$, corresponds to the DMV between the input ports a and b of the PFC and is expressed as

$$v_{dm,A} = v_{am} - v_{bm}. \tag{3}$$

If the potentials of the output ports q and r of the regulator, as seen from the virtual neutral point m of the DC-link, are denoted as v_{qm} and v_{rm}, respectively, the CMV of q and r as seen from m, $v_{cm,D}$, is expressed as

979-8-3315-1612-3/25 $31.00 © 2025 IEEE

$$v_{cm,D} = \frac{v_{qm} + v_{rm}}{2}. \tag{4}$$

Similarly, the DMV of q and r, $v_{dm,D}$, corresponds to the DMV between the output ports q and r of the regulator and is expressed as

$$v_{dm,D} = v_{qm} - v_{rm}. \tag{5}$$

The overall CMV of the power converter, $v_{cm,conv}$, with CMV injection is expressed as

$$v_{cm,conv} = v_{cm,A} - v_{cm,D}. \tag{6}$$

L_{dm1} and L_{dm2} represent the DM inductance of the input EMI filter, while $i_{dm,A2}$ and $i_{dm,A1}$ are the DM currents flowing in the DM loop of the input filter. $L_{dm,dc}$ denotes the DM inductance component of the regulator, and $i_{dm,D}$ is the DM current flowing in the output DM loop. i_{cm1} represents the CM current flowing in the inner loop of the floating EMI filter, while i_{cm2} is the CM current flowing in the ground loop. Notably, i_{cm2} corresponds to the leakage current of the circuit which generates touch current through the human body, potentially causing electric shock.

Since the touch current of the body is worse in the TT grounding scheme than in the TN, only the TT grounding scheme is assumed in the following discussion [3]. When the circuit in Fig. 2 is configured with a TT grounding system, the CM equivalent circuit is as shown in Fig. 3. The HF CMV generated by the switching operation of the power converter induces HF leakage current through the ground loop, which is attenuated by the floating EMI filter. However, the LF CM leakage current caused by the grid's CM voltage, $v_{cm,g}$, cannot be attenuated by the EMI filter and flows through the ground loop. By short-circuiting the filter inductance, which is neglected in the LF range, and highlighting only the components with dominant impedance, the simplified CM equivalent circuit is shown in Fig. 4.

B. Low-Frequency Leakage Current Control

Using this circuit, the transfer function $G_{cp}(s)$ from the converter's CMV, $v_{cm,conv}$, injected through LF CMV control to the ground loop leakage current, $i_{cm,2}$, is approximated in the LF range as

$$G_{cp}(s) = \frac{i_{cm,2}}{v_{cm,conv}} \approx \frac{1}{\frac{1}{sC_y} + R_N + R_G \parallel Z_H}. \tag{7}$$

$$G_{cc}(s) = G_{cp}^{-1}(s)F_{cc}(s) \tag{8}$$

The input of the CMV controller designed to reduce LF CM leakage current is the CM leakage current, $i_{cm,2}$, measured using a CM leakage current sensor. The output is the converter's CMV reference, $v_{cm,conv}^*$. The transfer function of the controller, $G_{cc}(s)$, which is described as

$$G_{cc}(s) = G_{cp}^{-1}(s)F_{cc}(s), \tag{9}$$

is designed based on [10], using the resonant controller $F_{cc}(s)$, which is described as

$$F_{cc}(s) = \sum_{k=odd} K_r(k) \frac{2\omega_{rc}s}{s^2 + 2\omega_{rc}s + (k\omega_g)^2}, \tag{10}$$

and the plant transfer function $G_{cp}(s)$. Here, ω_g denotes the grid frequency, ω_{rc} represents the damping coefficient,

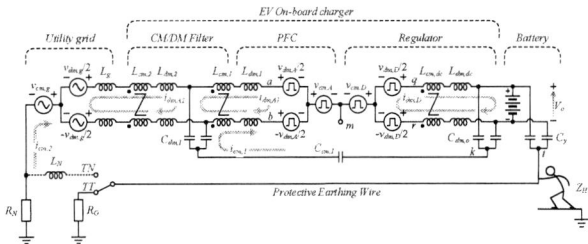

Fig. 2. Simplified circuit of a single-phase two-stage non-isolated converter.

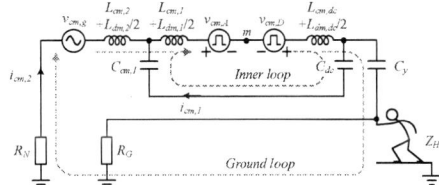

Fig. 3. Common-mode equivalent circuit of a single-phase two-stage non-isolated converter.

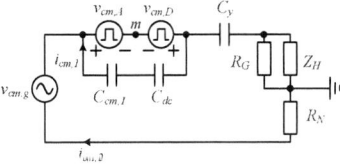

Fig. 4. Simplified common-mode equivalent circuit of a single-phase two-stage non-isolated converter, showing only the dominant components in the line-frequency range

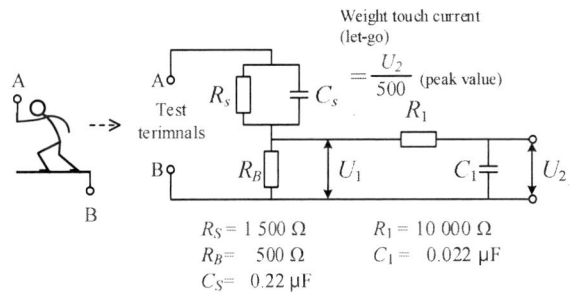

Fig. 6. Circuit diagram of the human body impedance model defined in IEC 60990. The touch current is calculated as $U_2/500$.

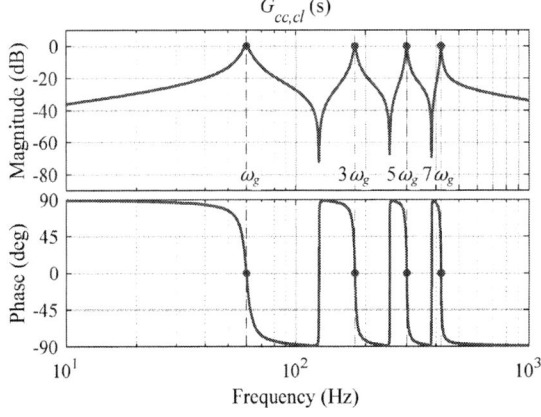

Fig. 5. Bode diagram of the closed-loop transfer function of leakage current controller.

979-8-3315-1612-3/25 $31.00 © 2025 IEEE   459

and $K_r(k)$ refers to the gain of the resonant controller at the k-th harmonic. The plant transfer function $G_{cp}(s)$ used for the controller design is assumed to have the same values as those of the actual plant parameters. The bode plot of the closed-loop transfer function, $G_{cc,cl}$, of the designed CMV controller is shown in Fig. 5. The dotted lines indicate the fundamental and harmonic frequencies of the grid, where the magnitude of the closed-loop transfer function is 1, and the phase is 0 degrees. The CMV controller reduces the LF leakage current caused by the grid CMV in a single-phase LN system and generates the CMV reference for this purpose.

III. SYNERGETIC COMMON-MODE VOLTAGE CONTROL

The CMV reference generated by the controller is actively applied by the power converter, which consists of a PFC and a regulator connected in a two-stage configuration. This application is broadly categorized into three methods. In the first method, the PFC primarily applies the CMV reference, while in the second method, the regulator applies the CMV reference. In the third method, the CMV sharing method, the CMV is synergistically distributed between the PFC and the regulator. In all three methods, the magnitude of the total CMV generated by the cascaded two-stage power converter is identical, resulting in the same LF leakage current reduction performance. However, the minimum voltage range of the DC-link, located between the PFC and the regulator, varies depending on the method. If the input and output voltages are the same, increasing the DC-link voltage leads to higher switching losses, inductor core losses, and conduction losses in the converter, thereby reducing efficiency. Therefore, minimizing the DC-link voltage is crucial. Consequently, this paper first explains the methods where either the PFC or the regulator primarily applies the CMV reference and finally describes a synergetic control method, where the PFC and the regulator appropriately share the CMV application to reduce the minimum DC-link voltage.

A. PFC-Priority Common-Mode Voltage Control

Since the PFC primarily applies the CMV reference, the total CMV applied by the converter, $v_{cm,conv}$, is equal to the CMV applied by the PFC, $v_{cm,A}$. In this case, the regulator outputs the constant DC voltage by regulating the power ripple generated during the rectification of the single-phase AC input by the PFC. The regulator does not apply any CMV voltage, meaning $v_{cm,D}$ is 0. In a single-phase LN grid system, one end of the power line is grounded, resulting in the magnitude of $v_{cm,g}$ being half of the grid voltage, v_g. The CMV reference output by the LF leakage current reduction controller is approximately equal to $v_{cm,g}$, assuming the voltage drop across the CM inductance is negligible. Thus, the CMV applied by the PFC and the regulator is approximated as

$$v_{cm,A} = v_{cm,conv} \approx \frac{v_g}{2}, \qquad v_{cm,D} = 0. \quad (11)$$

The input to the PFC is a single-phase AC system. Using the circuit shown in Fig. 2, the differential input voltage is expressed as

$$v_{dm,A} = v_{dm,g} - 2L_{dm,2}\frac{di_{dm,A2}}{dt} - 2L_{dm,1}\frac{di_{dm,A1}}{dt}. \quad (12)$$

If the voltage drop across the DM inductance is significantly smaller than the grid DM voltage, the PFC DM voltage, $v_{dm,A}$, is approximated as

$$v_{dm,A} \approx v_g. \quad (13)$$

The regulator's differential output is DC voltage. Using the circuit shown in Fig. 2, the output voltage is expressed as

$$v_{dm,D} = V_o + 2L_{dm,dc}\frac{di_{dm,D}}{dt}. \quad (14)$$

If the voltage drop across the DM inductance is significantly smaller than the output DM voltage, the regulator DM voltage, $v_{dm,D}$, is approximated as

$$v_{dm,D} \approx V_o. \quad (15)$$

When the PFC rectifies the input voltage and current in phase alignment, and the regulator maintains constant output voltage and current, the instantaneous DC-link voltage, v_{dc}, exhibits ripple at twice the system frequency. Let V_{dc} represent the average value of v_{dc} and V_r represent the amplitude of the second harmonic ripple component. Then, the voltages v_{pm} and v_{nm}, as seen from the virtual neutral point m for the terminals p and n, satisfy

$$v_{pm} > \frac{V_{dc} - V_r}{2}, \qquad v_{nm} < -\frac{V_{dc} - V_r}{2}. \quad (16)$$

At this point, the voltages at the input and output ports of the converter fall within the DC-link voltage range, satisfying

$$v_{nm} < v_{am}, v_{bm}, v_{qm}, v_{rm} < v_{pm}. \quad (17)$$

This relation is rewritten as follows:

$$-\frac{V_{dc} - V_r}{2} < \min(v_{am}, v_{bm}, v_{qm}, v_{rm}), \quad (18)$$

$$\max(v_{am}, v_{bm}, v_{qm}, v_{rm}) < \frac{V_{dc} - V_r}{2}. \quad (19)$$

The lower bound of V_{dc} is given by:

$$V_{dc} > V_r + \max(2V_g, V_o). \quad (20)$$

This expression is derived from the relationship between the input and output pole voltages (3)-(6) of the converter. Here, V_g represents the amplitude of the grid voltage, v_g. If V_r is sufficiently small compared to V_{dc}, V_r is approximated as

$$V_r \approx \frac{P_{in}}{2\omega_g C_{dc} V_{dc}}. \quad (21)$$

In this approximation, P_{in} denotes the average input power of the power converter, C_{dc} represents the total series capacitance of the DC-link, and ω_g is the angular velocity of the input system [18].

B. Regulator-Priority Common-Mode Voltage Control

Since the regulator primarily applies the CMV reference, the total CMV applied by the converter, $v_{cm,conv}$, is equal to the CMV applied by the regulator, $v_{cm,D}$. In this case, the regulator injects the LF CMV as well as regulates constant DMV. The operation of the LF leakage current reduction controller remains unchanged, and thus, the CMV reference output by the controller is equal to $v_{cm,g}$. Therefore, the CMV applied by the PFC and the regulator is approximated as

$$v_{cm,A} = 0, \qquad v_{cm,D} = -v_{cm,conv} \approx -\frac{v_g}{2}. \quad (22)$$

The differential input voltage of the PFC and the output voltage of the regulator are identical to those in the PFC-priority case and similarly be approximated as (13) and (15). The lower bound of V_{dc} is given by:

$$V_{dc} > V_r + \max(V_g, V_o + V_g). \qquad (23)$$

C. Synergetic Common-Mode Voltage Control

Based on (20) and (23), when V_o is smaller than V_g, the lower bound of the DC-link voltage is smaller in the case of regulator-priority for CMV sharing. Conversely, when V_o is larger than V_g, the lower bound of the DC-link voltage is smaller in the case of PFC-priority for CMV sharing. Therefore, by appropriately sharing the CMV reference between the PFC and the regulator based on V_o, the lower bound of the DC-link voltage is further reduced. Assume that both the PFC and the regulator apply CMV and that the total CMV applied by the converter, $v_{cm,conv}$, is shared in the ratio of k to $(1-k)$. Here, k is a constant between 0 and 1 and varies depending on V_o. The operation of the LF CMV controller remains unchanged, and thus, the CMV reference output by the controller is equal to $v_{cm,g}$. Accordingly, the CMV applied by the PFC and the regulator is approximated as

$$v_{cm,A} = k v_{cm,conv} \approx \frac{k v_g}{2},$$
$$v_{cm,D} = -(1-k) v_{cm,conv} \approx -\frac{(1-k) v_g}{2}. \qquad (24)$$

The differential input voltage of the PFC and the output voltage of the regulator are identical to those in the PFC-priority case and can similarly be approximated as (13) and (15). The lower bound of the DC-link voltage is given by:

$$V_{dc} > V_r + \max\left((1+k)V_g, V_o + (1-k)V_g\right). \qquad (25)$$

The constant k that minimizes the right-hand side of the equation is determined as

$$k = \min\left(\frac{V_o}{2V_g}, 1\right). \qquad (26)$$

With this relation, the lower bound of the V_{dc} is rewritten as

$$V_{dc} > V_r + \max\left(\frac{V_o}{2} + V_g, V_o\right). \qquad (27)$$

The lower bound of V_{dc} as a function of V_o is visualized in Fig. 7, based on (20), (23), and (27). When the CMV is controlled synergistically, where the PFC and regulator share the CMV application, the converter can operate at a lower DC-link voltage compared to the PFC-priority or regulator-priority CMV control method. In this case, the constant k, which determines the degree of CMV sharing, is calculated using (26). The output value of the LF CMV controller, $v_{cm,conv}^*$, is multiplied by k and $-(1-k)$, respectively, and these values are assigned as the CMV references for the PFC and regulator.

IV. Experiment Result

The system control block diagrams for each CMV application method are shown in Fig. 8. The overall controller consists of the DM controllers for the PFC and regulator, as well as the controller for reducing LF leakage current. These controllers, respectively, output DM and CM voltage references, which are used to calculate the port voltages of the PFC and regulator, v_{am}, v_{bm}, v_{qm}, and v_{rm}, through (2)-(6). Fig. 8 (a) illustrates the PFC-priority CMV application method, where the CMV applied by the PFC is determined through (11). Fig. 8 (b) represents the regulator-priority CMV

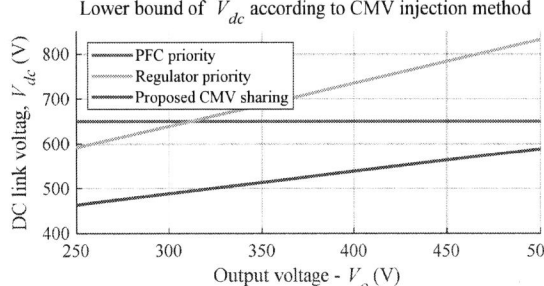

Fig. 7. The lower bound of the DC-link voltage as a function of the output

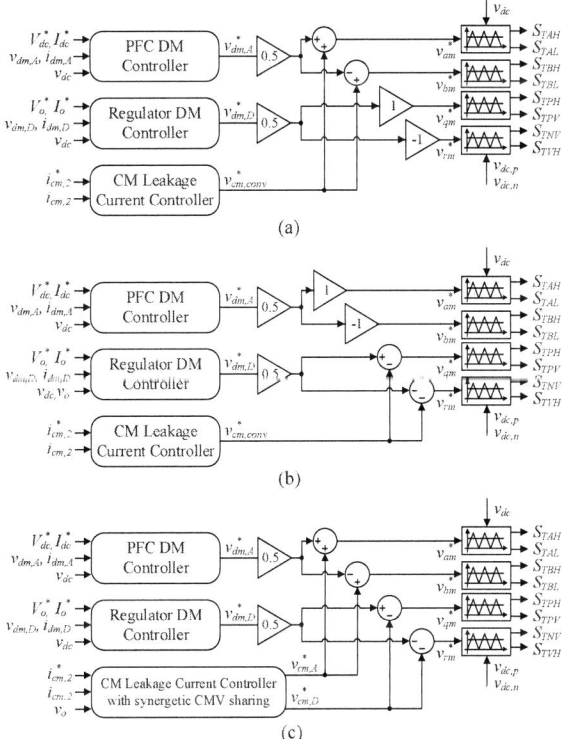

Fig. 8. Control block diagrams for CMV reference generation methods: (a) PFC-priority CMV control, (b) Regulator-priority CMV control, (c) Synergetic CMV control.

application method, where the CMV applied by the regulator is determined through (22). Fig. 8 (c) shows the shared CMV application method, where the CMV applied by the PFC and regulator is determined through (24). In the shared method, the constant k, required to calculate the CMV sharing ratio between the PFC and regulator, is given by (26). The control algorithm is implemented using an STM32 microcontroller unit (MCU).

An experimental setup is developed to validate the effectiveness of the synergetic CMV control method. The PFC is implemented as a two-level full-bridge rectifier, and the regulator is implemented as a three-level stacked buck converter, both operating at a switching frequency of 20 kHz. A third-order LCL EMI filter is employed to attenuate the HF CMV caused by the power converter's PWM. The output side of the converter is connected to the protective earth, the same as the vehicle chassis, through y-capacitance. The y-capacitances, $C_{y,p}$ and $C_{y,n}$, are 330 nF each, resulting in a total parallel y-capacitance, C_y, of 660 nF. The PFC's input

979-8-3315-1612-3/25 $31.00 © 2025 IEEE   461

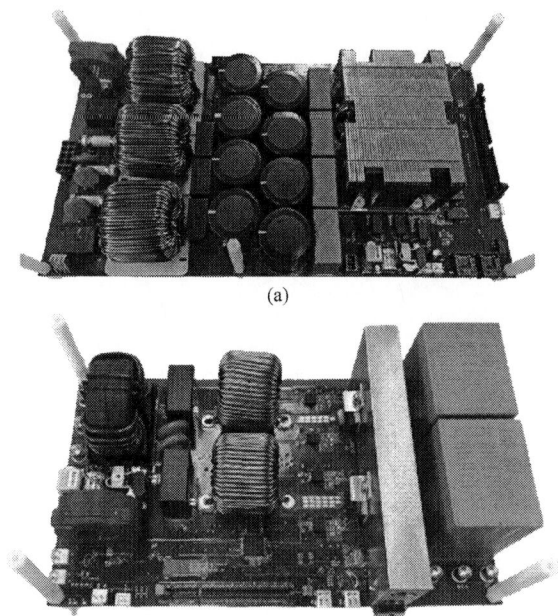

(a)

(b)

Fig. 9. Photos of the implemented two-stage non-isolated converter prototype: (a) full-bridge PFC, (b) stacked three-level buck converter

voltage is 220 V from a single-phase LN grid system, with an output voltage range of 250 V to 500 V. The target EV battery voltage is 400 V. The single-phase LN grid system is implemented with TT grounding, where R_N and R_G are 10 Ω and 2000 Ω, respectively, to simulate the worst-case scenario. Fig. 9 (a) shows the prototype of the PFC, and Fig. 9 (b) shows the prototype of the regulator. The LF leakage current is measured using a common-mode current sensor. A CTSR-0.3 sensor features a bandwidth exceeding the grid frequency and mA-level resolution. CMV reference information is exchanged between the PFC and regulator through CAN communication. The PWM switching signals are implemented on the MCU using triangular modulation.

The experimental waveforms are presented in Fig. 10. Fig. (10) (a), (b), and (c) illustrate the DC-link voltage reduction achieved through synergetic CMV control for converter output voltages of 250 V, 375 V, and 500 V, respectively. The graphs show the grid voltage (v_g), DC-link voltage (V_{dc}), output voltage (V_o), and leakage current ($i_{cm,2}$). In region A of Fig. (10) (a), before leakage current control is applied, the leakage current, $i_{cm,2}$, shows that the EMI filter does not attenuate the LF component. As a result, $i_{cm,2}$ reaches a peak of 36 mA, which can cause nuisance tripping of the RCD. By applying PFC-priority CMV control, the LF leakage current is reduced. The LF component of $i_{cm,2}$ converges to 0 within two cycles. In region B, where the LF leakage current has been mitigated through CMV control, the RMS values of $i_{cm,2}$ and the human body touch current are 1.1 mA. Thus, the LF leakage current control satisfies both the leakage current and touch current regulations specified in IEC 61851-1. However, in the case of PFC-priority CMV application, the DC-link voltage operates at 640 V, including the control margin. When the synergetic CMV sharing method is used, where the PFC and regulator share the CMV application, the DC-link voltage is controlled to be greater than 436 V and operates at 473 V, including the control margin. Compared to the PFC-priority or regulator-priority CMV control methods, the synergetic

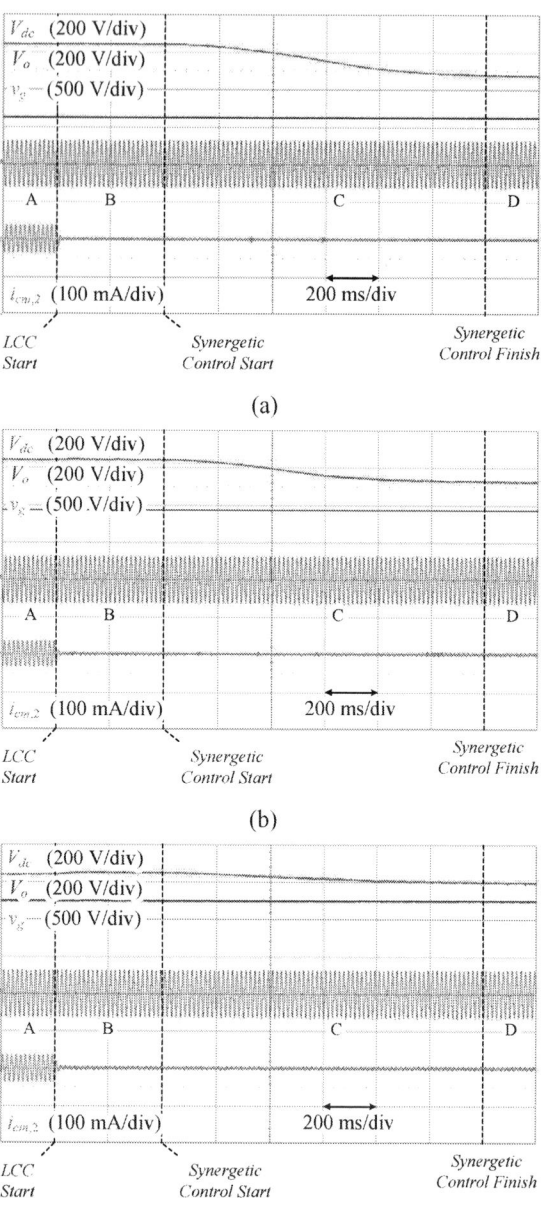

(a)

(b)

(c)

Fig. 10. Experimental results demonstrating the reduction of DC-link voltage using synergetic CMV control according to different output voltages: (a) $V_o = 250$ V, (b) $V_o = 375$ V, (c) $V_o = 500$ V.

control method reduces the DC-link voltage by 167 V. In region D, where the LF leakage current has been mitigated through control, the RMS values of $i_{cm,2}$ and the human body touch current are 1.1 mA. For an output voltage of 375 V in Fig. 10 (b), the leakage current $i_{cm,2}$ in region A reaches a maximum of 36 mA. In region B, the RMS values of $i_{cm,2}$ and the human body touch current are 1.2 mA. Using synergetic CMV control, the DC-link voltage in region D can be reduced from 640 V to 523 V, while the RMS values of $i_{cm,2}$ and the human body touch current in region D are 1.3 mA. For an output voltage of 500 V in Fig. 10 (c), the leakage current $i_{cm,2}$ in region A reaches a maximum of 35 mA. In region B, the RMS values of $i_{cm,2}$ and the human body touch current are 1.5 mA. Using synergetic CMV control, the DC-link voltage in region D can be reduced from 640 V to 523 V, while the

RMS values of $i_{cm,2}$ and the human body touch current in region D are 1.6 mA. Thus, within the target voltage range of 250 V to 500 V, the system is operated at a lower DC-link voltage using synergetic CMV control while maintaining the performance of the CMV controller compared to the PFC-priority and CMV control methods. Furthermore, the system satisfies both the leakage current and touch current regulations specified in IEC 61851-1 throughout the operating range.

V. CONCLUSION

Single-phase non-isolated EV chargers face significant challenges with leakage current due to the absence of galvanic isolation, leading to issues such as nuisance tripping of RCDs in the LF range and electric shocks to users caused by touch current. This paper proposes a control method for a two-stage, non-isolated EV charger comprising a cascaded structure of a PFC and a regulator. The method reduces leakage current by injecting LF CMV while simultaneously lowering the DC-link voltage through synergetic CMV sharing between the PFC and the regulator. As a result, the synergetic CMV control method enables the system to operate at a lower DC-link voltage while maintaining input and output conditions as well as leakage current reduction performance. The method complies with the leakage current and touch current regulations of IEC 61851-1.

ACKNOWLEDGMENT

This work was supported in part by the BK21 FOUR Program of the Education and Research Program for Future ICT Pioneers, Seoul National University, in part by the Seoul National University Electric Power Research Institute, and in part by the National Research Foundation of Korea (NRF) grant funded by the Korea Government (MSIT) (RS-2024-00354687).

REFERENCES

[1] "Global EV Outlook 2024 – Analysis," IEA. Accessed: Aug. 15, 2024. [Online]. Available: https://www.iea.org/reports/global-ev-outlook-2024

[2] H. Tu, H. Feng, S. Srdic, and S. Lukic, "Extreme Fast Charging of Electric Vehicles: A Technology Overview," *IEEE Trans. Transp. Electrific.*, vol. 5, no. 4, pp. 861–878, Dec. 2019.

[3] J. Wang *et al.*, "Nonisolated Electric Vehicle Chargers: Their Current Status and Future Challenges," *IEEE Electrific. Mag.*, vol. 9, no. 2, pp. 23–33, Jun. 2021.

[4] *Electric vehicle conductive charging system - Part 1: General requirements*, IEC Std. 61851-1, 2017.

[5] T. R. Oliveira, W. W. A. G. Silva, S. I. Seleme, and P. F. Donoso-Garcia, "PLL-Based Feed-Forward Control to Attenuate Low-Frequency Common-Mode Voltages in Transformerless LVDC Systems," *IEEE Trans. on Ind. Applicat.*, vol. 55, no. 3, pp. 3151–3159, May 2019.

[6] *Electric vehicle conductive charging system - Part 23: DC electric vehicle supply equipment*, IEC Std. 61851-23, 2014.

[7] D. Dong, F. Luo, D. Boroyevich, and P. Mattavelli, "Leakage Current Reduction in a Single-Phase Bidirectional AC–DC Full-Bridge Inverter," *IEEE Transactions on Power Electronics*, vol. 27, no. 10, pp. 4281–4291, Oct. 2012.

[8] S. B. Santra, A. Acharya, T. R. Choudhury, B. Nayak, and C. K. Panigrahi, "A modified carrier-based PWM technique for minimization of leakage current in transformer less single-phase grid-tied PV system," *Electr Eng*, vol. 103, no. 1, pp. 447–461, Feb. 2021.

[9] D. Zhang, D. Cao, J. Huber, J. Everts, and J. W. Kolar, "Nonisolated Three-Phase Current DC-Link Buck–Boost EV Charger With Virtual Output Midpoint Grounding and Ground Current Control," *IEEE Trans. Transp. Electrific.*, vol. 10, no. 1, pp. 1398–1413, Mar. 2024.

[10] J. Lee, D. Lee, S. Lee, and J.-I. Ha, "Direct Leakage Current Control Method of Single-Phase Non-Isolated EV Charger," in *2024 IEEE 10th International Power Electronics and Motion Control Conference (IPEMC2024-ECCE Asia)*, May 2024, pp. 5053–5058.

[11] S. Bala, T. Tengnér, P. Rosenfeld, and F. Delince, "The effect of low frequency current ripple on the performance of a Lithium Iron Phosphate (LFP) battery energy storage system," in *2012 IEEE Energy Conversion Congress and Exposition (ECCE)*, Sep. 2012, pp. 3485–3492.

[12] D. Dong, F. Luo, X. Zhang, D. Boroyevich, and P. Mattavelli, "Grid-Interface Bidirectional Converter for Residential DC Distribution Systems—Part 2: AC and DC Interface Design With Passive Components Minimization," *IEEE Trans. Power Electron.*, vol. 28, no. 4, pp. 1667–1679, Apr. 2013.

[13] F. Chen, R. Burgos, D. Boroyevich, and X. Zhang, "Low-Frequency Common-Mode Voltage Control for Systems Interconnected With Power Converters," *IEEE Transactions on Industrial Electronics*, vol. 64, no. 1, pp. 873–882, Jan. 2017.

[14] F. Chen, R. Burgos, and D. Boroyevich, "A Bidirectional High-Efficiency Transformerless Converter With Common-Mode Decoupling for the Interconnection of AC and DC Grids," *IEEE Transactions on Power Electronics*, vol. 34, no. 2, pp. 1317–1333, Feb. 2019.

[15] J. M. S. Callegari, A. F. Cupertino, V. de N. Ferreira, and H. A. Pereira, "Minimum DC-Link Voltage Control for Efficiency and Reliability Improvement in PV Inverters," *IEEE Transactions on Power Electronics*, vol. 36, no. 5, pp. 5512–5520, May 2021.

[16] D. A. Rendusara and P. N. Enjeti, "An improved inverter output filter configuration reduces common and differential modes dv/dt at the motor terminals in PWM drive systems," *IEEE Transactions on Power Electronics*, vol. 13, no. 6, pp. 1135–1143, Jan. 1998.

[17] *Methods of measurement of touch current and protective conductor current*, IEC Std. 60990, 2016.

[18] Y. Liu, Y. Sun, M. Su, M. Zhou, Q. Zhu, and X. Li, "A Single-Phase PFC Rectifier With Wide Output Voltage and Low-Frequency Ripple Power Decoupling," *IEEE Transactions on Power Electronics*, vol. 33, no. 6, pp. 5076–5086, Jun. 2018.

DC-DC Converter Architecture for Fast Electric Vehicle (EV) Battery Charging Applications

Shibaji Basu, *Student Member, IEEE*,   Arjun Ivimey, *Member, IEEE*,   Praveen Jain, *Fellow, IEEE*
shibaji.basu@queensu.ca, arjun.ivimey@queensu.ca, praveen.jain@queensu.ca
ECE, Queen's University
Kingston, Canada

Abstract—This paper introduces an innovative DC-DC converter architecture designed for fast Electric Vehicle (EV) battery charging. The proposed system utilizes a novel multi-primary, single secondary adder transformer. Multiple resonant full-bridge inverters supply the transformer's primaries, and the combined secondary voltage is rectified by an active bridge to power the EV battery. This input-parallel-output-series architecture promotes modularity and ensures inherent current sharing. The converter achieves a wide output voltage range (250-920V) and a peak power output of 180kW while maintaining soft switching transitions across the entire operating voltage range. Detailed mathematical modeling is demonstrated for system design and selection of temporal characteristics under various load conditions. To validate the concept, a scaled-down prototype with a voltage range of 47V-210V and power range of 0.35kW-1kW, demonstrating peak efficiency of 97.2% was developed.

Index Terms—Fast EV battery charging, Adder Architecture

I. INTRODUCTION

UNPRECEDENTED global climate change, driven by decades of reliance on fossil fuels like coal and petroleum, has compelled governments to address environmental concerns. A key strategy for a greener future is reducing emissions through transport sector electrification. Advances in battery technology, along with government incentives and rebates, are accelerating the adoption of EVs [1].

Fast EV battery charging technology is a major hindrance to widespread adoption of EVs. With rising power demands on EV chargers, engineers are advancing power converter topologies to achieve higher peak and overall efficiency. A comprehensive review [2] highlights the need for converter designs that offer a wide output voltage range and deliver high power levels (175 kW−350 kW) [3], while maintaining high efficiency and soft-switching capabilities [2]. In spite of the popularity of traditional phase-shift (PS) PWM topologies [4], there is an increasing trend toward cascaded [5], partial power processing [7], and reconfigurable [8], [9] converter configurations for EV battery charging. A 50kW fast charger featuring a parallel-connected diode bridge and filters is presented in [4]. However, alongside the constraints of a limited ZVS range and filter size [6], [4] lacks details on ensuring proper current sharing between parallel-connected modules, which may be affected by filter or device tolerances.

A cascaded topology combining an unregulated resonant converter with a non-isolated buck converter is proposed in [5] to achieve a wide output voltage range (50–650 V) for battery loads. However, the inclusion of additional passives, high-voltage switches, and diodes in the buck converter increases system cost. Furthermore, the hard-switched operation of the buck converter at higher power levels (20 kW) and high voltages (700 V) raises potential reliability concerns. The partial power (PP) processing [7] DC-DC converter configuration demonstrates remarkable peak efficiency of 98.8%. However, the converter has a limited output voltage range (314 - 450V) and it's conception requires a complex transformer design to ensure current sharing in the secondary winding. Re-configurable power converter configurations, as discussed in [8] and [9], require additional components such as four-quadrant semiconductor switches [8] or mechanical relays [9] to achieve high efficiency and soft-switching over a wide output voltage range. The DC-DC converter in [8] is limited by the inclusion of extra diode pairs and a complex transformer design. While [9] achieves a wide voltage range (150–950 V) and a full-load efficiency of 97%, its reliance on extra DC-blocking capacitors and mechanical switches for implementing desirable features across the voltage range adds complexity to the design.

A DC-DC converter employing an adder transformer configuration is presented in [10]. The presented architecture achieves a peak efficiency of 95.5%, while ensuring soft-switching for most of the switches across the entire output voltage range (42–223 V). [10] demonstrates how the phase and duty cycle modulation technique effectively mitigates tolerance effects in multiple series-connected resonant converter modules within an adder transformer-based DC-DC converter architecture. The adder transformer-based DC-DC converter's ability to mitigate resonant component tolerance in multi-module systems, as shown in [10], underpins the converter architecture proposed in this paper. Innovative approaches−[11] using coupled inductor network and [12] using current sharing control (CSC) have been and continue to be developed to facilitate effective current sharing among multiple DC-DC converter modules, enhancing reliability in high-current applications. Unlike the previous configuration [10], the present conception consists of multiple (n) $m:1$ toroid core transformers with a single common secondary turn, as shown in Fig.1. Each of the primary turns are fed individually by full-bridge series resonant inverters. The output voltage of the inverters are 'added' by the multi-level transformer and subsequently rectified by an active bridge in the secondary to produce the requisite DC voltage at the battery load.

979-8-3315-1612-3/25 $31.00 © 2025 IEEE

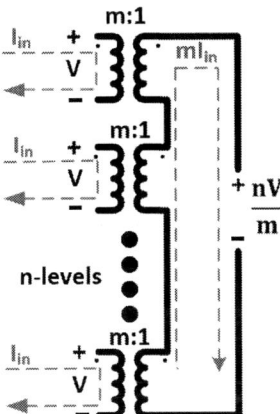

Fig. 1: Adder Transformer Configuration: n-levels with a turns ratio of $m:1$ per-level

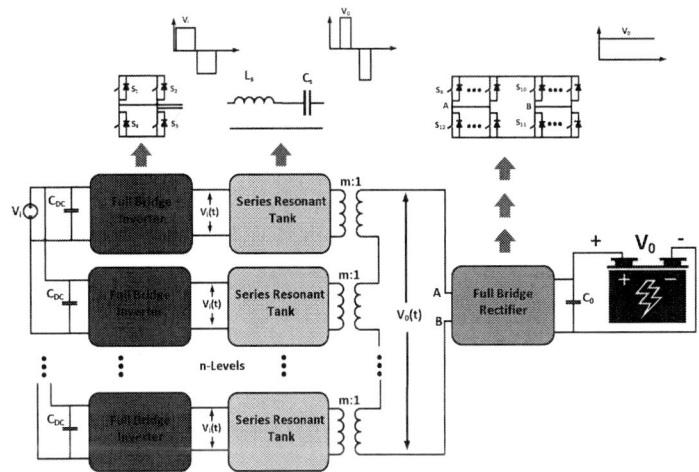

Fig. 2: Proposed n-Level DC-DC Converter Architecture with $m:1$ Adder Transformer per Level

The step-down transformer configuration ensures high current (180-400A) at the load side. This novel approach guarantees that the same current flows through multiple high-frequency resonant inverters connected to the transformer's primary side, eliminating the need for additional hardware [11] or complex control techniques [12]. By employing phase shift modulation (PSM) with variable switching frequency (97kHz - 100kHz) and duty cycle (0.4, 0.45, 0.47), a wide output voltage range is achieved without requiring cascaded [5] or reconfigurable [9] converter configurations. The adopted modulation scheme [10] ensures soft-switching transitions for all active bridge switches across the full operating voltage range (250V - 950V). A detailed modal analysis and design methodology for a 180 kW EV battery charger achieving peak efficiency of 98.21% is presented. Soft-switching conditions, referred as Design Check points are determined for all operating conditions. A 1 kW ow scale experimental prototype is built to validate the theoretical proposition. The prototype operates at a peak efficiency of 97.2% and undergoes soft-switching transition for all switches for every operating conditions.

The layout of the paper is as follows - DC-DC converter architecture, modal analysis and operating principles are outlined in Section II, while design methodology is detailed in Section III. Section IV describes the magnetics design. Experimental results are presented in Section V. Conclusions from the work are drawn in Section VI.

II. DC-DC CONVERTER ARCHITECTURE, MODAL ANALYSIS AND OPERATING PRINCIPLES

A. Proposed DC-DC Converter Specifications

The adder transformer as shown in Fig. 1 is the focal point of the converter architecture conception. The total converter schematic is shown in Fig.2. It comprises of n-numbered full-bridge high frequency resonant inverters feeding n, $m:1$ toroid core transformer primary. A buck based PFC rectifier is assumed to be providing the input DC voltage (400V).

B. Principle of Operation and Modal Analysis

Typical per-cycle operation of the converter with time period T is shown in Fig.3. G1 - G8 represent the gate-drive logic input signals to switches S1 - S8. $v_i(t)$ and $v_0(t)$ represent the inverter output and rectifier input voltage with phase angle ϕ_2 between them. $i_i(t)$ and $v_{CS}(t)$ represent the resonant current and capacitor voltage respectively. The full bridge inverters on the primary side of the transformer operate with zero phase-shift between the levels and at fixed 50% duty cycle for all operating points. The common secondary, as shown in Fig. 4 to Fig. 5, along with the equivalent circuit in Fig. 6, ensures uniform current flow through all the resonant inverters, eliminating the need for complex current balancing techniques. Despite tolerance of multiple resonant components, the converter operating conditions are decided by a net resonant inductance (L_{eq}) and capacitance (C_{eq}). L_{eq} and C_{eq} for the n-level proposed DC-DC converter are shown in (1)

$$L_{eq} = \frac{nL_s}{m^2}, \quad C_{eq} = \frac{m^2C_s}{n} \qquad (1)$$

The active full-bridge rectifier operates with a variable phase shift and a reduced, adjustable duty cycle. To meet the fast charger's current demands, multiple parallel switches are used per leg of each half-bridge on the secondary side, as shown in Fig. 4 [2]. This configuration increases the capacitive (C_{OSS}) load, prompting an investigation into the dead-time dynamics on the secondary side. Neglecting the primary side inverter dead time, the operation of the converter in symmetrical positive half-cycle can be explained by five sub-intervals. For the sake of brevity, only two sub-intervals - Energy Exchange Interval and Transition Interval 1a are described in this paper. Time domain analysis of the converter is performed in the mentioned intervals with the assumption of ideal devices, components, source characteristics and at steady-state conditions. The analysis provides insights into converter operation, forms the basis for converter component design and operating point specifications over a wide load range. The circuit configuration for each interval are shown in Fig.4 to Fig.5.

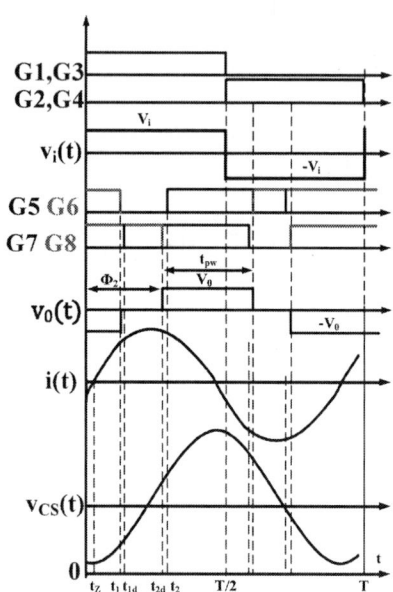

Fig. 3: Ideal Operating Waveforms of Proposed DC-DC Converter

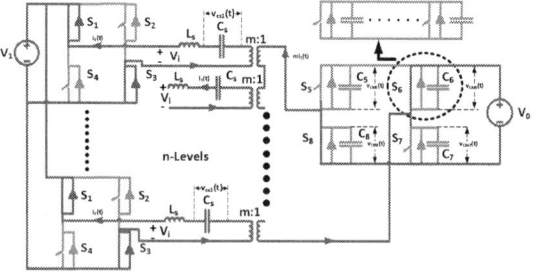

Fig. 4: Energy Exchange Interval

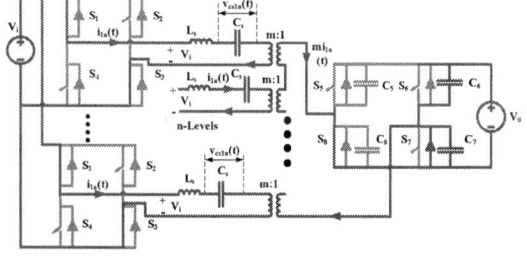

Fig. 5: Transition Interval 1a

Energy Exchange Interval [0 to t_1]: As depicted in Fig.4 switches S1, S3 of all the primary modules and switches S6, S8 of secondary side are ON. Resonant current ($i_1(t)$) and capacitor voltage ($v_{cs1}(t)$) are given in (2) and (3). $v_{CM5}(t)$ to $v_{CM8}(t)$ represent the secondary side parallel switch capacitance voltage. Per-level resonant inductance and capacitance are given by L_s and C_s. V_i and V_0 are the input

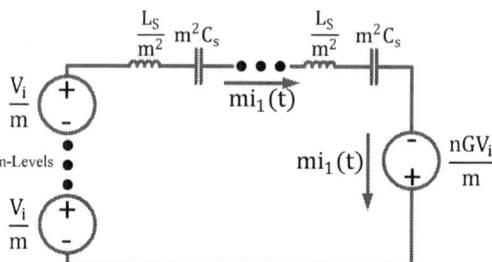

Fig. 6: Simplified Equivalent Circuit for Energy Exchange Interval

and output DC voltage respectively for the n-level converter. The solutions to the state variables are given in (2) to (7).

In (2) and (3), variable G is the per-level voltage gain of the converter. I_0 and V_{CS0} are resonant current and capacitor voltage at $t = 0$. ω_R and Z_R are the resonant frequency and characteristic impedance respectively. The final values of the state-variables in this interval are obtained by putting $t = t_1$ in (2) to (7) and are denoted by I_{t1}, V_{CSt1}, V_{CM5t1}, V_{CM6t1}, V_{CM7t1}, V_{CM8t1} respectively. Resonant current zero crossing in this interval establishes the first Design Check point (detailed in Section III) of the converter.

$$i_1(t) = I_0 \cos(\omega_R t) - \frac{V_{CS0}\sin(\omega_R t)}{Z_R}$$
$$+ \frac{nV_i \sin(\omega_R t)(Gm + 1)}{Z_R}, \qquad (2)$$
$$v_{CS1}(t) = V_{CS0}\cos(\omega_R t)$$
$$+ nV_i(1 - \cos(\omega_R t))(Gm + 1)$$
$$+ I_0 Z_R \sin(\omega_R t) \qquad (3)$$
$$v_{CM5}(t) = nGV_i \qquad (4)$$
$$v_{CM6}(t) = nGV_i \qquad (5)$$
$$v_{CM7}(t) = 0 \qquad (6)$$
$$v_{CM8}(t) = 0 \qquad (7)$$

Transition Interval 1a [t_1 to t_{1d}]: This interval marks the transition between the energy exchange interval and charging interval. Time t_{1d} is the instant at which switch S7 output capacitance C_7 discharges to zero and C_6 of switch S6 charges to V_0. The evaluation of t_{1d} provides estimate of minimum dead time t_{dd1} necessary between switches of the leading rectifier leg and establishes the second Design Check point for ensuring soft-switch turn ON of switch S7 for the full operating voltage range. Switch S6 turns OFF, while switches S1 and S3 of all primary modules are ON. The reflected switch output capacitance is assumed to have a constant value of C_m. The temporal behavior of switch parallel capacitance voltage, (8) and (9), during interval 1a, undergoing transition are used to determine dead time and hence are the only state variable

979-8-3315-1612-3/25 $31.00 © 2025 IEEE

presented.

$$v_{CM61a}(t) = \frac{(V_{CM6t1}(2C_m + C_s cos(\omega_{RD}(t-t_1))))}{(2C_m + C_s)}$$
$$+ \frac{(C_s V_{CSt1}(cos(\omega_{RD}(t-t_1)) - 1))}{(2C_m + C_s)}$$
$$+ \frac{(C_s V_i n(1 - cos(\omega_{RD}(t-t_1)))(Gm+1))}{(2C_m + C_s)}$$
$$+ I_{t1} C_{R2} Z_{R2} sin(\omega_{RD}(t-t_1)) \quad (8)$$

$$v_{CM71a}(t) = V_{CM7t1}$$
$$- \frac{(C_s V_{CM7t1}(cos(\omega_{RD}(t-t_1)) - 1))}{(2C_m + C_s)}$$
$$- \frac{(C_s V_{CSt1}(cos(\omega_{RD}(t-t_1)) - 1))}{(2C_m + C_s)}$$
$$- \frac{(C_s V_i n(1 - cos(\omega_{RD}(t-t_1)))(Gm+1))}{(2C_m + C_s)}$$
$$- C_{R2} Z_{R2} I_{t1} sin(\omega_{RD}(t-t_1)) \quad (9)$$

where,

$$\omega_{RD} = \sqrt{\frac{1}{L_s C_{eq}}} \quad C_{eq} = \frac{2C_m C_s}{C_m + C_s}, \quad C_m = \frac{nC_i}{m^2}$$
$$C_{R2} = \sqrt{\frac{2C_s}{2(2C_m + C_s)}}, \quad Z_{R2} = \sqrt{\frac{L_s}{C_m}} \quad (10)$$

The final values of the state-variables in Interval 1a are obtained by putting $t = t_{1d}$ in (8) to (9). The following condition is true in steady state.

$$I_0 = -I_{T2}, \quad V_{CS0} = -V_{CST2} \quad (11)$$

In (11), the I_{T2} and V_{CST2} refer to steady state value of resonant current and capacitor voltage at $t = \frac{T}{2}$. Without considering losses in the converter, the relationship between input and output power for an n-level converter can be stated as:

$$nV_i I_{avg} = \frac{(mnGV_i)^2}{nm^2 R_L} \quad (12)$$

In the above equation, R_L is the load resistance per level emulating a battery load and the average current I_{avg} is given in (13). For ease of computation, the dynamics of transition intervals are ignored.

$$I_{avg} = \frac{2}{T}(\int_0^{t_1} i_1(t)dt + \int_{t_1}^{t_2} i_2(t)dt + \int_{t_2}^{\frac{T}{2}} i_{T2}(t)dt) \quad (13)$$

The zero-crossing time t_z ($\phi_z = 2\pi f_s t_z$) of the resonant current is determined from (2) and is stated in (14).

$$t_z = \frac{1}{\omega_R} tan^{-1}(\frac{Z_R I_0}{n(Gm+1)V_i - V_{CS0}}) \quad (14)$$

III. CONVERTER DESIGN METHODOLOGY

A. Design Approach & Required Specifications

The intent of modal analysis is to evaluate the resonant components (L_s and C_s) and determine suitable operating specifications, such as phase angle and duty cycle, for a given

switching frequency across various load and output voltage conditions. Additionally, a key objective is to assess the design check points (as outlined in the previous section) to ensure soft-switching for all rectifier switches under all operating conditions. Design Check Point I (DCP-I), corresponding to the current zero crossing, while Design Check Point II (DCP-II) relating to transition interval 1a are elaborated upon in this section. A higher number of input levels (n) combined with a transformer step-down ratio (m) reduces the current rating of primary-side switches and minimizes resonant component size and specifications. A 20-level converter with step down turns ratio of 10 : 1, having output of 46V per-level would meet the proposed DC-DC converter specifications with available switches [13]. The parameter 'k' and load resistance R for the n-level converter in Table II is defined in (15).

$$k = \frac{\omega_R}{2\pi f_s (= \omega_s)}, \quad R = nR_L \quad (15)$$

TABLE I: Initial Converter Parameters for Design

Parameters	Values
V_i	400V
G_c	1.15
n	20
m	10
$R_{(variation)}$	0.375Ω - 5.11Ω
$P_{L(variation)}$	60kW - 180kW
$f_{s(initial)}$	100kHz
$d_{initial}$	0.4
$k_{variation}$	0.8, 0.85, 0.9, 0.92, 0.95
$\phi_{2(initial)}$	50.4°
$\phi_z(R_L = 5.11Ω)$	9°

ZVS angle for resonant current was stipulated at 9° which facilitated soft-switching while keeping the circulating current in check. (11), (12) and (14) are solved simultaneously with the information provided in Table I for evaluating L_s (C_s), I_0, V_{CS0} and G for V_{out} of 920V at R of 5.11Ω. The choice of ϕ_2 arose from required G value of 1.15. Consequently, the derived values are used to evaluate resonant current (RMS and peak values), peak inductor voltage, peak capacitor voltage, transition times t_{1d} and t_{2d}. Modal analysis predicts the phase angle ϕ_2, duty cycle d and switching frequency f_s for the chosen converter parameters required for other load conditions. The transition time t_{1d} is estimated from (8).

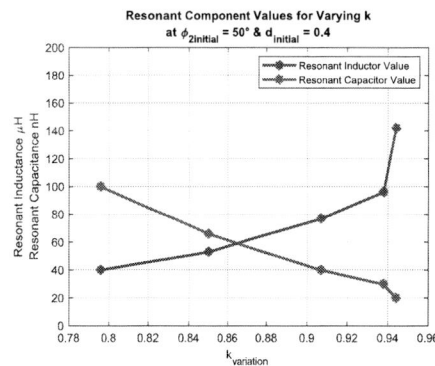

Fig. 7: Resonant Component Values with Varying 'k' Value

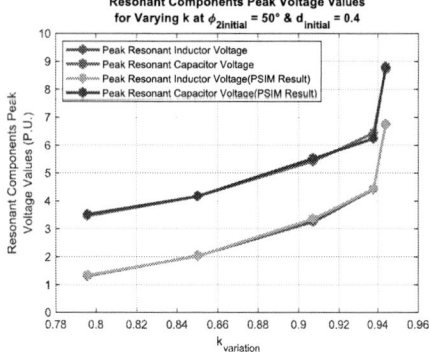

Fig. 8: Resonant Component Voltage Drop Values (in p.u.) with Varying 'k' Value

B. Design Check Point I

The zero crossing of the resonant current forms an important Design Check point for all operating conditions. It is the necessary condition with chosen converter parameters not only for inverter switches, but also to ensure soft-switching transition of switches in the leading rectifier leg (Switch S6 and S7) for all operating load conditions. The condition is stated in (16).

$$\frac{2\pi(dT + t_2 - \frac{T}{2})}{T} > \phi_Z \qquad (16)$$

C. Design Check Point II

Considering Design Check point I is satisfied, Design Check Point II deals with the dead time t_{dd1} required to ensure soft switching in the leading rectifier leg for all operating conditions for chosen converter parameters and load conditions. Check Point II conditions are stated in (17), (18) and (19).

$$t_{dd1} > t_{1d} - t_1 \qquad (17)$$

$$v_{CM61a}(t = t_{1d}) \approx V_0 \qquad (18)$$

$$v_{CM71a}(t = t_{1d}) \approx 0 \qquad (19)$$

Parameter k (15) variation as indicated in Table I results in five possible resonant tank (L_s and C_s) combination. The resonant tank circuit values at initial chosen phase angle ϕ_2 and duty-cycle d is shown in Fig.7. Using base quantities (20), resonant capacitor and inductor peak voltage variation is shown in Fig.8.

$$V_{base} = V_i, \quad I_{base} = \frac{\max(I_{out})}{m} \qquad (20)$$

Higher per-unit (p.u.) drop of resonant component voltage for $k = 0.92$ and $k = 0.95$, as shown in Fig.8 led to exclusion of resonant component combination from further design considerations. Preliminary chosen resonant component combinations are: L_s - 40μH, C_s - 100nF; L_s - 53μH, C_s - 65nF; L_s - 77μH, C_s - 40nF.

Operating condition sweep (250 - 950V) of the preliminary chosen resonant component combination. Operating parameters - ϕ_2, f_s and d obtained from modal analysis for different loading conditions must satisfy Design Check Points I and

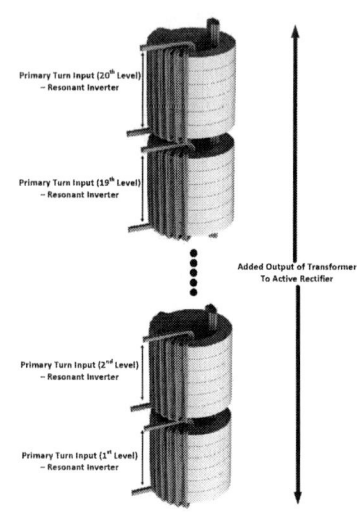

Fig. 9: 20-Level Adder Transformer Configuration

II. The final resonant converter design is selected based on peak and overall efficiency, resonant component voltage drop (p.u.), and semiconductor switch turn-off current. Section IV provides a detailed evaluation of full voltage range efficiency and operating specifications, culminating in the selection of the most suitable design for the fast EV battery charger.

D. Transformer Design

The converter conception is inspired from pulsed power applications [14] but with the modification of increased primary turns for application compatibility. The highest peak current ($\sim$70A) and primary input voltage ($\sim$460V) at $t_{pw} = 4\mu$s, B_{pk} of 0.05T are used for the transformer design. The cross-sectional area and the window area are evaluated from (21) and (22) [15].

$$A_P = V_0 t_{pw}/(m\Delta B(= 2B_{pk})) \qquad (21)$$

$$A_W = \frac{mA_{PW}n_{PW} + A_{SW}n_{SW}}{K_U} \qquad (22)$$

where, A_{PW}, n_{PW}, A_{SW} and n_{SW} are the area, number of wires in parallel for primary and secondary turns respectively. A_W is the transformer window area and K_U is the winding factor, typically assigned a value of 0.3. The transformer has three and twenty-one 10 AWG wire in parallel for the primary winding and and secondary winding. The cross-sectional and window area were evaluated as 2100 sq.mm and 2300sq.mm respectively. To meet the window and cross-section specifications, six stacked TX107/65/18 [16] toroid cores are used in the adder transformer configuration shown in Fig.9. The magnetizing inductance L_m per unit height is given by (23).

$$L_m = \frac{\mu_0 \mu_r m^2 \ln\left(\frac{R_2}{R_1}\right)}{2\pi} \qquad (23)$$

where, μ_0, μ_r, R_1, R_2 are free space permeability, material permeability, internal and external radius of the toroid core.

979-8-3315-1612-3/25 $31.00 © 2025 IEEE

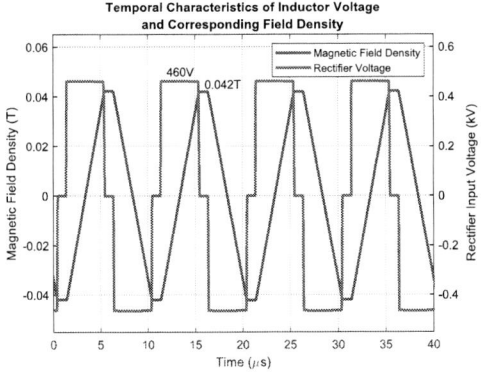

Fig. 10: Temporal Characteristics of Transformer Primary Voltage and Corresponding Flux Density at Highest Voltage Operating Condition

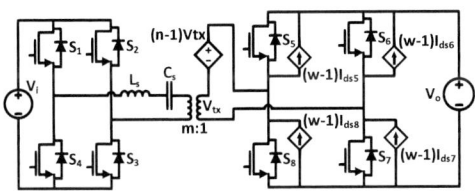

Fig. 11: QSpice Simulation Circuit

The temporal profile of the magnetic field density, for the highest output voltage condition, is obtained from the transformer primary voltage as shown in Fig.10. The temporal characteristics of magnetic flux density were utilized to evaluate transformer core losses ($\frac{mW}{cm^3}$) employing the improved generalized Steinmetz equation (iGSE) [17]. Transformer losses (core loss resistance R_c) for all operating conditions (different battery voltages) were evaluated by the same procedure. Primary and secondary winding losses were evaluated following the procedure shown in [15] considering wire configuration as previously mentioned.

IV. DC-DC CONVERTER SIMULATION RESULTS

A. QSPICE Simulation Results

The proposed 20-level DC-DC Converter was modeled as a single inverter module with the corresponding transformer segment and a voltage dependant voltage source in series with the secondary winding. The voltage dependant voltage source multiplies the output voltage of the transformer by n prior to being fed to the rectifier. A current dependant current source is placed in parallel with each of the rectifier switches to model the effects of multiple parallel MOSFETs in the rectifier circuit. The inverter and rectifier switches were modelled using built in 650V and 1200V SiCFET models provided by Qorvo [13]. The simulation circuit is shown in Fig.11. Efficiency for all operating conditions for the preliminary resonant component combinations were evaluated in QSPICE as shown in Fig.12. The 40μH and 100nF resonant component combination shows the highest peak and overall efficiency.

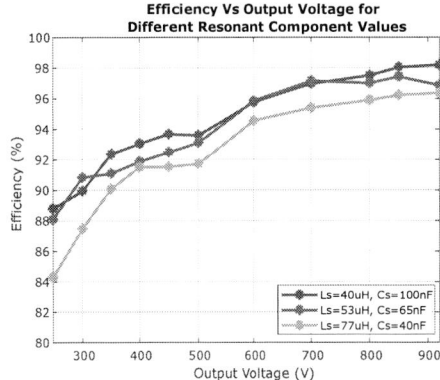

Fig. 12: QSpice Effciency Results

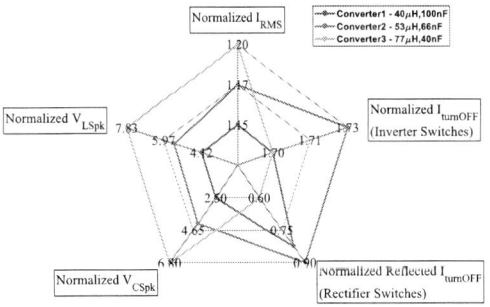

Fig. 13: Performance Comparison Between Preliminary Choice of DC-DC Converters

TABLE II: Simulation Results

V/I	d(f)	G/stage	ϕ_2	I_0	V_{CS0}
[V]/[A]	[Hz]		[Degs]	[A]	[V]
920/180	0.4 [100k]	0.115	49 (49)	-9.14 (-8.3)	-517 (-526)
250/400	0.47 [97k]	0.03125	92 (92)	-68 (-67)	-322 (-333)

TABLE III: Simulation Results

[V]/[A]	V_{CSpk}	V_{Lpk}	$t_0[t_1]$	t_{dd1}
[V]/[I]	[V]	[V]	ns	ns
920/180	0.53k(0.53k)	1.39k(1.39k)	260 [385]	142 (110)
250/400	1.07k(1.07k)	1.60k(1.60k)	2020 [2300]	16 (20)

B. PSIM Simulation Results

The comparison of worst case critical parameters like I_{rms} and V_{CSpk}, presented in Fig. 13, highlights the superiority of the 40 μH and 100 nF resonant component combination. Combined with its superior overall efficiency, this selection finalizes the design with 40 μH and 100 nF as the optimal resonant components. Circuit parameters at a few operating points for the chosen resonant converter have been tabulated in Tables III and IV. Simulated values obtained in PSIM for a few parameters - I_0, V_{CS0}, V_{CSpk}, V_{Lpk}, ϕ_2 and t_{dd1} are shown in parenthesis for comparison. As observed in Table III, $t_0 < t_1$ - Design Check Point I condition is satisfied for all the operating points. Based on values evaluated for t_{dd1} a dead time of eight degrees for all the converter half-bridges should satisfy Design Check Point II. The close agreement

TABLE IV: Experimental Prototype System Attributes

Attribute	Value/Description
k	∼0.8
L_s/level	76μH (EE type with E56/24/19)
C_s/level	55nF (KEMET Film Capacitors)
V_i/level	120V
R	7 - 48Ω
n	2
m	$\frac{4}{3}$
V_{out}	47 - 210V
P_{out}	0.33 - 1kW
Primary Switches	UJ3C065030K3S (650V,43mΩ, SiCFET)
Secondary Switches	STW55NM60ND (600V,60mΩ,MOSFET)

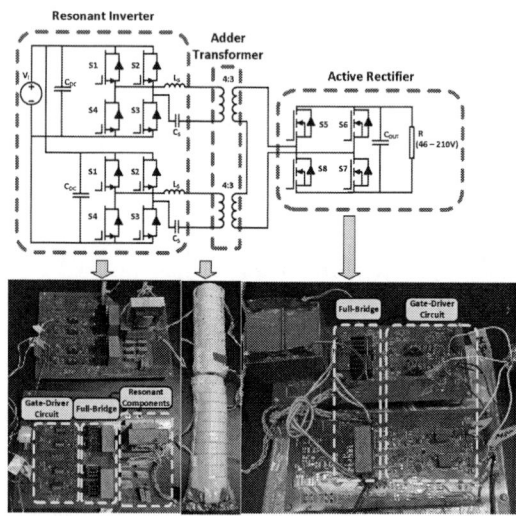

Fig. 14: Prototype DC-DC Converter Experimental Setup

with PSIM simulation results validates the presented modal analysis. Phase angle (ϕ_2), duty-cycle (d) and switching frequency (f_s) variation of 47°, 0.07 and $3kHz$ are converter requirements for obtaining full load voltage variation while maintaining soft-switching transition for all switches for all operating conditions.

V. EXPERIMENTAL RESULTS

A. Experimental Setup

A scaled down 1 kW experimental prototype with two inverter modules was implemented to validate the proposed architecture. Keeping parity with the theoretical design proposition, k was chosen to be ∼ 0.8 and switching frequency to be 100 kHz. Available TX51/32/19 [16] toroid magnetic cores prompted the transformer ratio to be scaled down by a factor of 7.5 ($N_p = 4$ and $N_s = 3$). Each level of adder transformer was constructed by stacking twelve TX51/32/19 toroid magnetic cores achieving a magnetizing inductance of 853 μH and a leakage inductance of 10 μH. Considering the choices in k, f_s, n, R and m, the modal analysis evaluation resulted in resonant tank circuit configuration to be L_s - 76μH and C_s - 55nF. The experimental prototype features are given below in Table IV. Each of the modules was provided with 120V DC input. The experimental prototype setup is shown in Fig.14.

B. Experimental Waveforms & Analytical Results

TABLE V: Experimental Prototype Operating Conditions & Circuit Parameters

V_{out}	R	ϕ_2	$f_s(d)$	I_0	V_{CS0}
[V]	[Ω]	[Degs]	[kHz]()	[A]	[V]
47 (48)	7	81	99k (0.47)	-8.39 (-8.73)	-73 (-63)
109 (112)	18	92	100k (0.4)	-8 (-7.63)	-133 (-137)
210 (214)	48	59	100k (0.4)	-2.4 (-2.3)	-173 (-179)

TABLE VI: Experimental Prototype Operating Conditions & Circuit Parameters

V_{out}	V_{CSpk}	I_{rms}	t_Z	$Eff.(\eta)$
[V]	[V]	[A]	[s]	[%]
47 (48)	238 (237.8)	5.62 (5.8)	2μ (2.02μ)	89.86
109 (112)	250 (239)	5.83 (5.8)	1.53μ (1.45μ)	94.65
210 (214)	191 (182)	4.49 (4.3)	0.39μ (0.5μ)	97.22

Inverter output voltage, rectifier input voltage and resonant inductor current waveforms are shown in this section. Fig.15 and Fig.16 show the experimental waveforms at load

conditions of $R = 7\Omega$ and $R = 48\Omega$. Design Check points I - II, for soft switching conditions are satisfied for all the experimentally observed operating points. A phase angle variation of 38°, duty cycle variation of 0.07 and switching frequency variation of 1kHz is necessary for the DC-DC converter prototype to achieve full gain variation while maintaining soft-switching transitions for all switches. Load condition $R = 26\Omega$ was chosen to show ZVS conditions for leading rectifier bridge switches. Experimental results

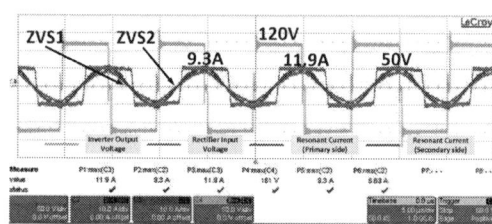

Fig. 15: Experimental Inverter Output Voltage, Rectifier Input Voltage, Resonant Current (primary and secondary) for load $R = 7\Omega$. ZVS1 and ZVS2 indicate zero-voltage-switching for rectifier top and bottom switches of leading rectifier leg.

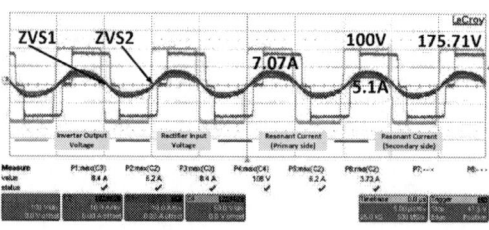

Fig. 16: Experimental Inverter Output Voltage, Rectifier Input Voltage, Resonant Current (primary and secondary) for load $R = 48\Omega$.

of the proposed DC-DC Converter operating conditions for full (scaled down) voltage range of operation is tabulated in

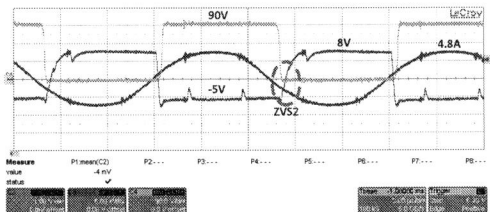

Fig. 17: ZVS Condition for Lower Switch of Leading Rectifier Leg - S7 at load $R = 26\Omega$,

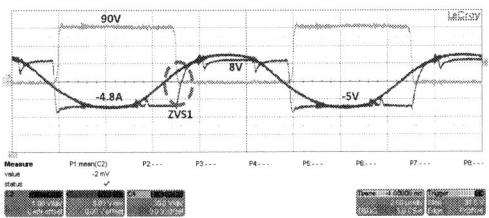

Fig. 18: ZVS Condition for Upper Switch of Leading Rectifier leg - S6 at load $R = 26\Omega$,

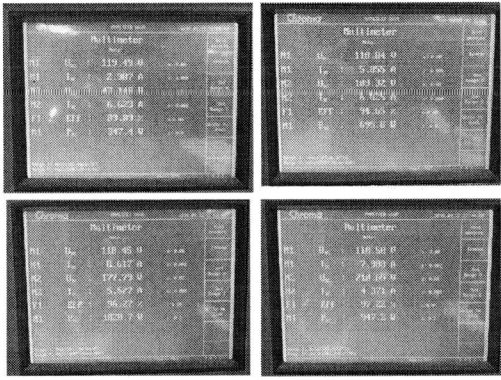

Fig. 19: Prototype DC-DC Converter Efficiency Over Wide Output Voltage Range Measured With Chroma 6630 Power Analyzer. Efficiency (clockwise from top) at $R = 7\Omega$, $R = 18\Omega$, $R = 32\Omega$, $R = 48\Omega$

Tables V and VI at 120V DC input per level. The operating conditions and converter circuit parameters evaluated from the modal analysis (within parentheses) are compared with experimental results. Close correlation with the experimental results emphasizes the suitability of the proposed converter architecture in wide voltage EV battery charging applications and re-iterates the veracity of the presented mathematical analysis. Experimental prototype efficiency over wide output voltage is shown in Fig.19.

VI. CONCLUSION

A new adder transformer based DC-DC converter architecture has been presented with wide output voltage range without the use of cascaded or complicated re-configurable circuits. The unique multiple primary, single common secondary, step-down adder transformer configuration ensures equal current through all the resonant components, thereby avoiding complex current balancing techniques. The steady-state analysis

of the converter has been performed and performance characteristics presented. It has been shown that converter maintains soft-switching for the full-voltage range. The proof-of-concept has been experimentally verified. The 1 kW prototype shows a peak efficiency of 97.2% at 210V DC output voltage.

REFERENCES

[1] Ronanki, D.; Kelkar, A.; Williamson, S.S. Extreme Fast Charging Technology—Prospects to Enhance Sustainable Electric Transportation. Energies 2019, 12, 3721. https://doi.org/10.3390/en12193721

[2] H. Tu, H. Feng, S. Srdic and S. Lukic, "Extreme Fast Charging of Electric Vehicles: A Technology Overview," IEEE Transactions on Transportation Electrification, vol. 5, no. 4, pp. 861-878, Dec. 2019, doi: 10.1109/TTE.2019.2958709.

[3] ABB Electric Vehicle Infrastructure - Terra HP high power charging. Website Reference: https://www.evchargesolutions.com/ABB-Terra-HP-175-DC-Fast-Charging-Station-p/abb-ter-hp-175-dcfc.htm

[4] Yoon, Hm., Kim, Jh. and Song, Eh., "Design of a novel 50 kW fast charger for electric vehicles", J. Cent. South Univ. 20, 372–377 (2013). https://doi.org/10.1007/s11771-013-1497-8

[5] Woo-Seok Lee, Jin-Hak Kim, Jun-Young Lee and Il-Oun Lee, "Design of an isolated DC/DC topology with high efficiency of over 97% for EV fast chargers." IEEE Transactions on Vehicular Technology Vol.68, No.12, pp. 11725-11737, December 2019.

[6] M. Pahlevani, S. Eren, A. Bakhshai and P. Jain, "A Series–Parallel Current-Driven Full-Bridge DC/DC Converter," IEEE Transactions on Power Electronics, vol. 31, no. 2, pp. 1275-1293, Feb. 2016, doi: 10.1109/TPEL.2015.2417773.

[7] Y. Cao, M. Ngo, N. Yan, D. Dong, R. Burgos and A. Ismail, "Design and Implementation of an 18 kW 500 kHz 98.8% Efficiency High Density Battery Charger With Partial Power Processing," IEEE Journal of Emerging and Selected Topics in Power Electronics, vol. 10, no. 6, pp. 7963-7975, Dec. 2022, doi: 10.1109/JESTPE.2021.3108717.

[8] D. Shu and H. Wang, "An Ultrawide Output Range LLC Resonant Converter Based on Adjustable Turns Ratio Transformer and Reconfigurable Bridge," IEEE Transactions on Industrial Electronics, vol. 68, no. 8, pp. 7115-7124, Aug. 2021, doi:10.1109/TIE.2020.3009588.

[9] S. Mukherjee, J. M. Ruiz and P. Barbosa, "A High Power Density Wide Range DC–DC Converter for Universal Electric Vehicle Charging," IEEE Transactions on Power Electronics, vol. 38, no. 2, pp. 1998-2012, Feb. 2023, doi: 10.1109/TPEL.2022.3217092.

[10] S. Basu and P. Jain, "Resonant Based DC-DC Converter for Fast EV Battery Charging Applications Using Novel Adder Architecture - Analysis, Design & Resonant Components Tolerance Effects," 2023 IEEE Conference on Power Electronics and Renewable Energy (CPERE), Luxor, Egypt, 2023, pp. 1-7, doi: 10.1109/CPERE56564.2023.10119598.

[11] X. Zhang, S. Pan and P. Jain, "A Discrete Coupled Multiphase Interleaved LLC Converter with Symmetrical Components Analysis," IEEE Transactions on Power Electronics, doi: 10.1109/TPEL.2023.3279822.

[12] Z. Ye, P. K. Jain and P. C . Sen, "A Full-Bridge Resonant Inverter With Modified Phase-Shift Modulation for High-Frequency AC Power Distribution Systems," IEEE Transactions on Industrial Electronics, vol. 54, no. 5, pp. 2831-2845, Oct. 2007, doi: 10.1109/TIE.2007.896030.

[13] Website reference: https://www.qorvo.com/products/discrete-transistors/sic-fets

[14] Shibaji Basu, Sunil Swamy, Samir K. Sahoo, R.K. Rajawat, "Development of 22 kV, 1 kHz rep-rated solid-state Pulser based on Linear Transformer Driver topology", Journal of Instrumentation Vol.9, No.12, December 2014.

[15] Colonel W.T. Mclyman, Title: *Transformer and Inductor Design Handbook*, Edition(Third), Idyllwild, California, U.S.A: Marcel Dekker, Inc., 2004, 533 pages.

[16] Website reference: https://www.ferroxcube.com/en-global/download/index/product_catalog

[17] K. Venkatachalam, C. R. Sullivan, T. Abdallah and H. Tacca, "Accurate prediction of ferrite core loss with nonsinusoidal waveforms using only Steinmetz parameters," 2002 IEEE Workshop on Computers in Power Electronics, 2002. Proceedings., Mayaguez, PR, USA, 2002, pp. 36-41, doi: 10.1109/CIPE.2002.1196712

Fast Simulator for the Estimation of Inverter DC-link Temperature in e-Drives Subjected to Highly Variable Working Cycles

Simone Giuffrida
Dipartimento Energia "G. Ferraris"
Politecnico di Torino
Torino, 10129, Italy
simone.giuffrida@polito.it

Fabio Mandrile
Dipartimento Energia "G. Ferraris"
Politecnico di Torino
Torino, 10129, Italy
fabio.mandrile@polito.it

Radu Bojoi
Dipartimento Energia "G. Ferraris"
Politecnico di Torino
Torino, 10129, Italy
radu.bojoi@polito.it

Abstract—Accurate estimation of losses and operating temperatures is extremely important for the design of traction inverters and its validation for specific operating conditions. Although several methods are already available in the literature to estimate the semiconductor power losses, the power losses and temperature evaluation of inverter DC-link capacitors are less explored. Therefore, this paper presents a fast simulation model of traction eDrive to provide accurate estimation of DC-link loss and temeperature, as well as DC link voltage ripple evaluation. The proposed model is implemented in Simulink and consists of both a loss model based on loss maps and a circuital model to estimate the dc-link voltage ripple. This model is very fast and therefore it is very useful when the eDrive is subject to long and highly variable driving cycles. Moreover, the model accounts for the influence of the main parts of the powertrain and allows a proper verification of the selected DC link capacitor.

Index Terms—traction inverter, dc-link, simulation

I. INTRODUCTION

The assessment of power losses and thermal behavior represents a critical challenge for the design of inverters for electric vehicles that are inherently subjected to the highly variable working cycles. In the literature, available papers deal with the estimation of the losses and junction temperature of traction power modules over a certain load cycle [1], [2]. However, these models do not focus on the DC-link capacitors that have to be included in a complex simulation model that includes the influence of the main part of the traction system. As a fundamental element of traction inverters, the DC-link capacitor bank typically represents a significant portion of the converter physical volume and mass [3]. Current state-of-art design practices [4] typically size the DC-link based on peak stress conditions (maximum current ripple scenarios), inevitably resulting in conservative oversizing of the capacitor bank for applications with highly variable working cycles. Although this conventional methodology, which is widely used in industrial applications where the load is quite constant, ensures operational safety margins, it leads to suboptimal resource utilization and larger component costs. Prior to prototype validation, designers require sophisticated simulation tools to evaluate the performance boundaries of proposed DC-link

design. Various analytical frameworks for evaluating capacitor thermal cycling under predetermined operational profiles [5]–[8]. However, these investigations predominantly emphasize lifespan prediction rather than the sizing process. The design of the capacitor bank is typically done without considering its integration in the powertrain model and it often fails capturing the dynamic interaction with it. Previous investigations have explored DC-link sizing through mathematical formulations [9]. Although these analytical approaches provide fundamental behavioral aspects of DC-link capacitors, they do not suit well for long and variable load cycles. This research introduces an innovative rapid simulation framework for evaluating DC-link performance within the context of electric traction systems. In contrast to conventional analytical methodologies [9], the proposed approach utilizes pre-computed powertrain loss maps to enhance to reduce the simulation time, while preserving accuracy. Moreover, the simulation architecture incorporates the DC-link model within a more complete powertrain simulation, enabling thorough analysis of subsystem interactions. This framework encompasses both electrical and thermal domains, delivering precise estimation of power dissipation, thermal aspects, and voltage fluctuations under dynamic operating conditions.

The paper is organized as follows. Section I introduces the need of dc-link model to be integrated in the entire powertrain simulation. Section II presents the powertrain simulation framework. Section III presents the proposed DC-link design methodology, with the explanation of the loss map generation algorithm, the thermal parameter extraction procedures and the obtained simulation results.

II. TRACTION E-DRIVE MODEL

The proposed traction model is shown in Fig.1 and it consists of four main blocks: vehicle longitudinal dynamic model, eMotor, energy source and inverter. According to a specific driving cycle that imposes the speed and acceleration, the vehicle model provides to the eMotor the operating electrical torque T_e and rotational speed n_{mot}. The motor model provides to the inverter model the phase peak voltage

979-8-3315-1612-3/25 $31.00 © 2025 IEEE

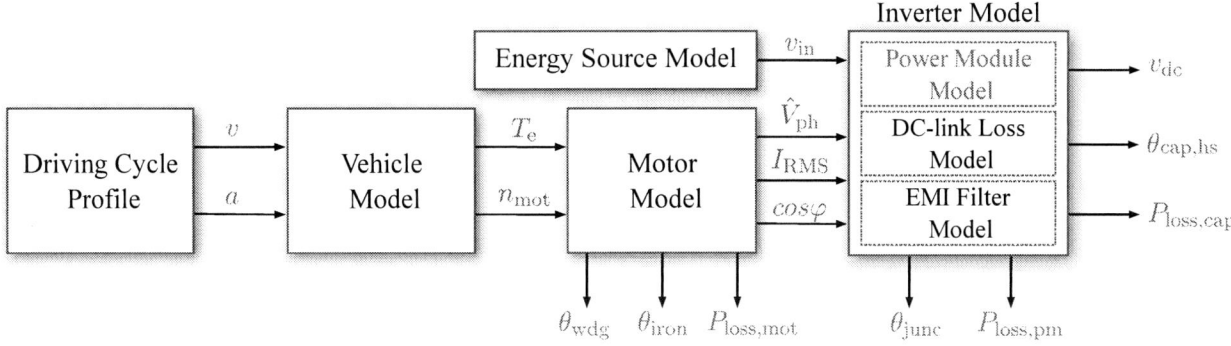

Fig. 1: Full traction e-Drive model.

$\hat{V}_{\text{ph}}$, motor RMS current I_{RMS} and power factor $cos\varphi$ to the inverter model. This block also provides the motor power losses $P_{\text{loss,mot}}$, the core mean temperature θ_{iron} and the winding mean temperature θ_{wdg} information, however, these variables are not considered in this work since they are not relevant for the DC-link design. The input voltage v_{in} is provides by the energy source model (battery). The three-phase traction inverter model receives as input the information coming from the energy source and motor models to compute the mean DC-link voltage v_{dc}, the hot-spot temperature of the capacitor bank $\theta_{\text{cap,hs}}$, the power losses $P_{\text{loss,cap}}$, the semiconductor junction temperature θ_{junc} and the semiconductor power losses $P_{\text{loss,pm}}$. The inverter model can be divided in three main components: the power module, the DC-link and the EMI filter. For this work, only the DC-link and EMI filter models are considered, while the energy source is modeled for simplicity as a constant voltage source.

In Tab.I are reported the main nominal specification of the selected inverter used in the next sections.

TABLE I
Inverter Main Nominal Specifications

Parameter	Description	Value
S	Apparent Power	150 kVA
I_{n}	RMS Phase Current	150 A
V_{dc}	DC-link Voltage	400 V
f_{sw}	Switching Frequency	30 kHz

A. Motor Model

The motor model is based on flux maps (current-to-flux linkage relationship) in the rotor (dq) frame, as in (1). These maps can be generated either through Finite Element Analysis (FEA) when the motor design is known, or through experimental measurements [10].

$$\begin{cases} \lambda_{\text{d}} = f(i_{\text{d}}, i_{\text{q}}) \\ \lambda_{\text{q}} = f(i_{\text{d}}, i_{\text{q}}) \end{cases} \quad (1)$$

The flux maps elaboration enables prediction of motor performance computing the motor control trajectories: Maximum Torque per Ampere (MTPA) below base speed and Maximum Torque per Volt (MTPV) for flux-weakening operation. The Maximum Torque per Speed (MTPS) is also computed starting from the flux maps, it represents the best torque production versus speed for a certain DC-link voltage and inverter current limit.

The motor model consists of multiple 2D Look-Up-Tables (LUTs) that take torque and speed as inputs and provide peak voltage and RMS phase current and power factor as outputs, as reported in Fig.2. A key advantage of this approach is its independence from specific torque control methods, as it relies directly on the machine's actual capabilities derived from flux map processing. This feature enables immediate voltage/current requirement determination without complex real-time calculations. For this work, the motor model uses experimentally obtained flux maps of a Brusa HSM1 traction motor.

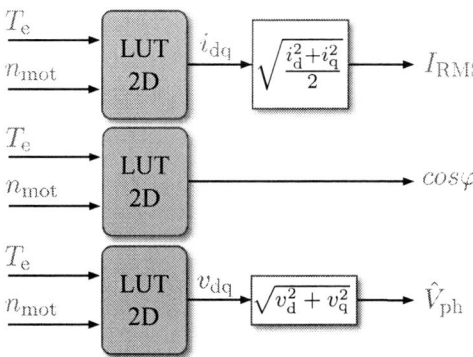

Fig. 2: Motor model main internal blocks.

B. DC-link Model

The DC-link model is part of the inverter block and it consists of two main parts: the loss model and the circuital model. The aim of the loss model block is to estimate the dc-bank losses (Fig.3a) and the hot-spot temperature. The inverter

979-8-3315-1612-3/25 $31.00 © 2025 IEEE   473

model is based on a 4-dimensional (4D) loss map taking as input the motor model outputs to compute the losses which are sent to the DC-link thermal model to compute $\theta_{\mathrm{cap,hs}}$, then the estimated hot-spot temperature is sent as feedback to the loss model to adjust the power losses according to the capacitor temperature. In general, this characteristic is related to the capacitor Equivalent Series Resistance (ESR) variation in temperature dependently on the capacitor technology [11]. Moreover, the thermal model also take as input θ_{amb} representing the ambient temperature where the DC-link is located. The modulation index M is computed starting from the DC-link voltage and phase peak voltage quantities.

The DC-link circuital model (Fig.3b) estimates the capacitor voltage variation under variable load conditions. Indeed, due to the limited dynamic response of the energy source [12], [13] and the decoupling behavior (low pass filter) of the EMI stage, the instantaneous load changes are supplied by the DC-link bank leading to a DC-link voltage variation. Therefore, the aim of this block is to check that the DC-link capacitance is sufficiently high to effectively provide the required instantaneous power and that the maximum and minimum voltage limits are not exceeded.

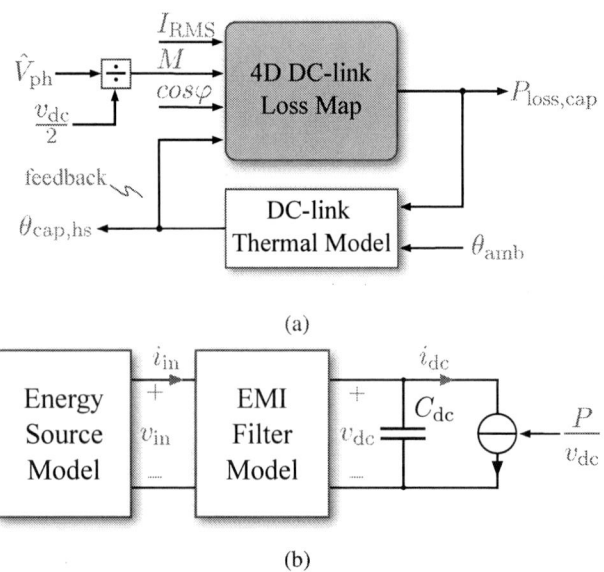

(a)

(b)

Fig. 3: (a) DC-link full 4D loss map model, (b) simplified circuital model for mean DC-link voltage estimation under variable load.

III. PROPOSED DC-LINK DESIGN METHODOLOGY

The proposed DC-link design methodology is suitable for highly variable load aiming to highlight the advantages compared to the traditional design approach. For this reason, Table II reportes the main specifications, provided by datasheet, of the C4AQLBW5500M3JK discrete film capacitor [14]. The $V_{\mathrm{dc,n}}$ is the rated DC voltage, C^* is the nominal capacitance, I_{RMS}^* is the maximum continuative RMS current stress at the testing frequency f^*, R_{ESR}^* is the ESR at f^* and R_{th} is

the hot-spot to ambient thermal resistance, while θ_{rated} is the nominal continuative operating temperature of the capacitor (Table II).

TABLE II
Selected Capacitor Main Datasheet Specifications

$V_{\mathbf{dc,n}}$	$\mathbf{C}^*$	$\mathbf{I}_{\mathbf{RMS}}^*$	$\mathbf{R}_{\mathbf{ESR}}^*$	$\mathbf{R}_{\mathbf{th}}$	$\theta_{\mathbf{rated}}$	**Material**
500V	50μF	22.8A	2.6mΩ	18°C/W	85°C	PP

A. Traditional Design Approach

Usually, the design of DC-link capacitors must address two critical requirements: managing maximum RMS current stress and controlling peak-to-peak voltage ripple [4]. The RMS current influences power losses and capacitor temperature rise, while voltage ripple affects semiconductor device peak voltage and converter operation quality.

The RMS current flowing into the DC-link capacitor analytically expressed by neglecting the phase output switching ripple [9]. As example, Fig.4a shows the RMS current stress in the DC-link in the $M - \varphi$ plane at nominal load current.

However, the peak-to-peak charge ripple $\Delta Q_{\mathrm{C_{dc},pp,max}}$ requires numerical calculation as it is not straightforward to obtain an analytical formulation. The $\Delta Q_{\mathrm{C_{dc},pp,max}}$ stress in the $M - \varphi$ plane is reported in Fig.4b. As previously mentioned, the minimum required DC-link capacitance C_{dc} is obtained, calculated by the maximum of two values as reported in (2).

$$C_{\mathrm{min}} = \max[C_{\mathrm{dc},\Delta V_{\mathrm{pp}}}, C_{\mathrm{dc},I_{\mathrm{RMS}}}] \qquad (2)$$

The voltage ripple-based minimum capacitance ($C_{\mathrm{dc},\Delta V_{\mathrm{pp}}}$) is calculated in (3) directly from the ratio between the worst case ripple and the maximum allowed peak-to-peak voltage ripple ($\Delta V_{\mathrm{dc,pp,max}}$). In Fig.4c is reported the minimum required capacitance as a function of $\Delta V_{\mathrm{dc,pp,max}}$, assuming as worst case a worst case peak-to-peak voltage ripple equal to the 5% of the nominal DC-link voltage resulting a $C_{\mathrm{dc},\Delta V_{\mathrm{pp}}}$ of 90μF.

$$C_{\mathrm{dc},\Delta V_{\mathrm{pp}}} = \frac{\Delta Q_{\mathrm{C_{dc},pp,max}}}{\Delta V_{\mathrm{dc,pp,max}}} \qquad (3)$$

The current stress-based minimum capacitance $C_{\mathrm{dc},I_{\mathrm{RMS}}}$ calculation is reported in (6) as the the product of the selected capacitor capacitance and the ration between the total DC-link RMS current stress $I_{\mathrm{C_{dc},RMS,max}}$ and the individual capacitor RMS current stress $I_{\mathrm{RMS,cap}}$ considered located all at the switching frequency. This last value is obtained by manipulating (4) and (5) to maintain the same temperature difference ($\Delta\theta$) for the different frequency conditions. The capacitor equivalent series resistance at the switching frequency $R_{\mathrm{ESR}}(f_{\mathrm{sw}}$ is obtained from the ESR vs frequency curve provided by datasheet (Fig.5). Applying (6) it results $C_{\mathrm{dc},I_{\mathrm{RMS}}} = 233\mu$F.

979-8-3315-1612-3/25 $31.00 © 2025 IEEE

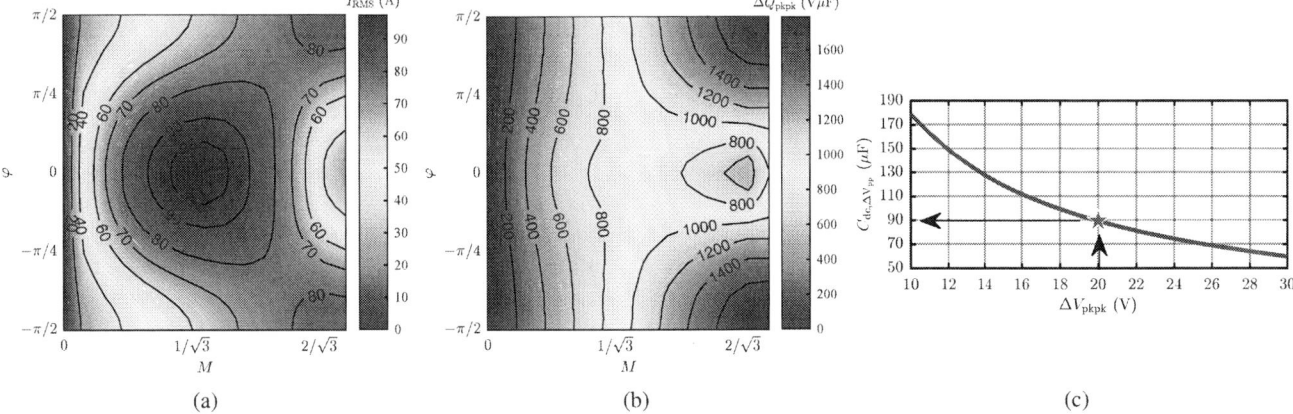

Fig. 4: (a) Peak-to-peak charger ripple map in the M-φ plane, (b) DC-link RMS current map in the M-φ plane, (c) Minimum capacitance vs peak-to-peak voltage ripple.

$$\Delta\theta = P_{\text{loss}} \cdot R_{\text{th}} \tag{4}$$

$$P_{\text{loss}} = R_{\text{ESR}} \cdot I_{\text{RMS,cap}}^2 \tag{5}$$

$$C_{\text{dc},I_{\text{RMS}}} = C^* \frac{I_{C_{\text{dc}},\text{RMS,max}}}{I_{\text{RMS}}^*} \sqrt{\frac{R_{\text{ESR}}(f_{\text{sw}})}{R_{\text{ESR}}^*(f^*)}} \tag{6}$$

Therefore, to satisfy (2) the number of selected capacitors in parallel n_{p} is equal to 5, resulting an overall installed DC-link capacitance (C_{dc}) of 250μF.

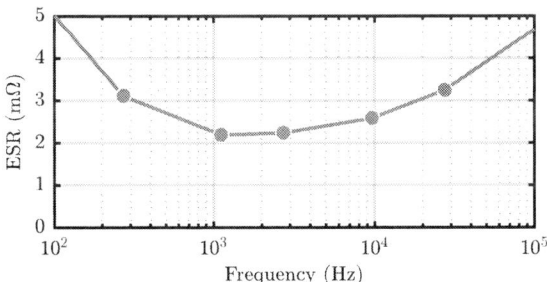

Fig. 5: ESR vs frequency variation of the selected C4AQLBW5500M3JK capacitor.

B. Proposed Design for Highly Variables Working Cycles

The previously presented traditional design approach consists of meeting both the voltage ripple and the thermal current constraints, considering a continuative RMS current stress on the DC-link at nominal load. However, if we take into account only the rated RMS current stress for highly variable working cycles, DC-link oversize will result due to the limited current capability of the capacitors.

The proposed DC-link design methodology takes as a starting point the minimum capacitance value to meet the peak-to-peak voltage ripple limit 3) and then checks the capacitor thermal constraint ($\theta_{\text{cap,hs}} < \theta_{\text{rated}}$), considering a realistic

working cycle and taking into account the influence of the main part of the powertrain. If the thermal limit is not met, it is possible to add just the extra amount of capacitor in parallel to ensure a safe DC-link operation.

Therefore, to just satisfy the peak-to-peak voltage ripple constraint, only two selected capacitors are required in parallel, obtaining $C_{\text{dc}} = 100\mu$F. A comparison between traditional and proposed design methods is provided in Fig. 6 for weight, volume, cost and losses, assuming the same thermal limits over the working cycle. It is possible to notice that there is a large decrease of the weight, volume and cost at the expense of higher losses. In any case, it is mandatory to ensure that these losses do not exceed over the working cycle the capacitor thermal limit.

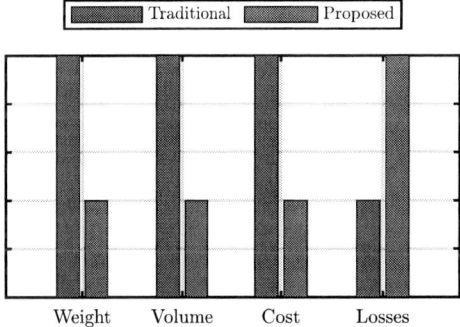

Fig. 6: Comparison for DC-link bank weight, volume, cost and power losses between the traditional and the proposed design approach.

C. DC-link Loss Map Generation

To ensure a fast powertrain simulation, a DC-link loss model based map is implemented, as described in Sec.II-B. The map creation process is described in Fig.7. The first step is to define the input parameters containing the capacitor characteristics and all the operating point which constitute the map. For each operating point, the capacitor DC current i_{dc} is extracted and the Fast Fourier Transform (FFT) is computed. The FFT current values are used, combined with the ESR vs frequency capacitor characteristic (Fig.5), to compute the power losses according to (7). Then, the power loss value is loaded in the map and a new operating point is selected. This iterative procedure ends when the power losses are computed for all the operating point defined in input.

$$P_{\text{loss}} = \frac{n_{\text{p}}}{2} \sum_{k=1}^{n} R_{\text{ESR}}(f_{\text{k}}, \theta_{\text{cap,hs}}) \left(\frac{\hat{I}_{\text{C}}(f_{\text{k}}, M, cos\varphi)}{n_{\text{p}}} \right)^2 \quad (7)$$

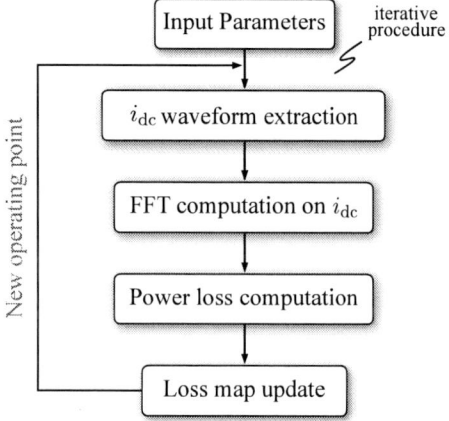

Fig. 7: DC-link map creation algorithm.

D. DC-link Thermal Parameter Estimation

The DC-link thermal model block consists of a first order Cauer network. The equivalent dc-link thermal resistance $R_{\text{th,eq}}$ is obtained in (8) by dividing the single capacitor thermal resistance R_{th} with n_{p}. Then, a corrective factor k_{exc} is applied to take into account the reduced available exchange area A_{exc} compared to the total one A_{tot} due to the thermal interaction between close capacitors ($k_{\text{exc}} = A_{\text{exc}}/A_{\text{tot}} \approx 0.8$). The computed $R_{\text{th,eq}}$ is equal to 10.8°C/W.

$$R_{\text{th,eq}} = \frac{R_{\text{th}}}{n_{\text{p}}} \cdot k_{\text{exc}} \quad (8)$$

The formula to estimate the dc-link equivalent capacitance $C_{\text{th,eq}}$ is reported by (9), considering only the dielectric material of the capacitor. The specific heat capacity (c_{p}) of the dielectric material is assumed equal to 1100 J/(kgK) [15] while the material weight (W_{mat}) is assumed to be as 80% of the total one, resulting in $C_{\text{th,eq}} = 139$ J/kg.

$$C_{\text{th,eq}} = c_{\text{p}} \cdot W_{\text{mat}} \quad (9)$$

E. Simulation Results

The proposed model (Fig .1) has been simulated in Matlab/Simulink for a WLTC driving cycle, as shown in Fig.8. The considered vehicle model has the following data: mass m = 2000 kg, friction model $F_{\text{road}} = a + bv + c\dot{v} + mgsin(\alpha)$,

with $\alpha = 0°$, $a = 196.2$ N, $b = 0$ Ns/m, $c = 0.536$ Ns^2/m^2, gear ratio $\tau = 10$, gear efficiency $\eta = 94\%$.

The voltage source model is simply implemented considering a constant voltage source equal to 400V. The EMI filter model consists of an equivalent differential mode inductance $L_{\text{dm}} = 50\mu$F and a resistance $R_{\text{dm}} = 20$mΩ. The DC-link power loss map consists of 1728 points ($12 \times 12 \times 12$) and does not depend on the hot-spot DC-link temperature as the selected film capacitor does not have an ESR temperature dependency.

The main simulation results are shown in Fig.9, with focus on the output quantities of the vehicle, motor and DC-link models. It is possible to notice that the capacitance value is sufficiently high to take charge of the instantaneous power oscillation required by the load, generating an acceptable mean dc-link voltage oscillation of few volts (Fig.9f). Moreover, in Fig.9h it is reported the hot-spot temperature pattern over the driving cycle which satisfy the capacitors thermal limit thus confirming the feasibility of the designed DC-link.

The total simulation time is 1.94 s, without including map computation, with a computer equipped with Intel I9 processor. Although the loss map computation is not optimized and it takes approximately 3 minutes, however it is computed once for each DC-link design and can be used for all the next driving cycle simulations.

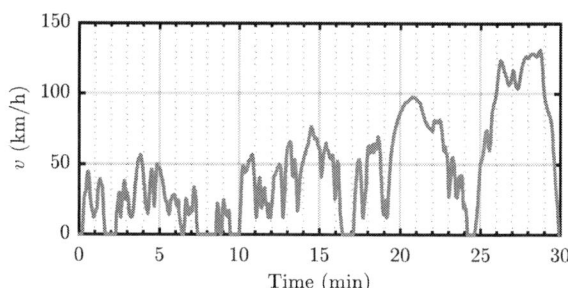

Fig. 8: Selected WLTC driving cycle.

IV. CONCLUSION

This paper has presented a comprehensive simulation framework for fast evaluation of DC-link performance in electric vehicle applications. The proposed model combines loss mapping and circuital analysis to accurately predict the DC-link behavior under variable operating conditions, while maintaining good computational efficiency.

The novel design methodology is specifically tailored for highly variable working cycles demonstrates substantial improvements over traditional approaches, achieving significant reductions in weight, volume, and cost while ensuring safe thermal operation. The proposed methodology addresses a critical gap in existing DC-link design approaches by providing a practical tool for evaluating capacitor performance within complete powertrain operation. Results from WLTC driving cycle simulations validate both thermal compliance and acceptable voltage ripple characteristics with the optimized design, while achieving simulation times of just 1.94

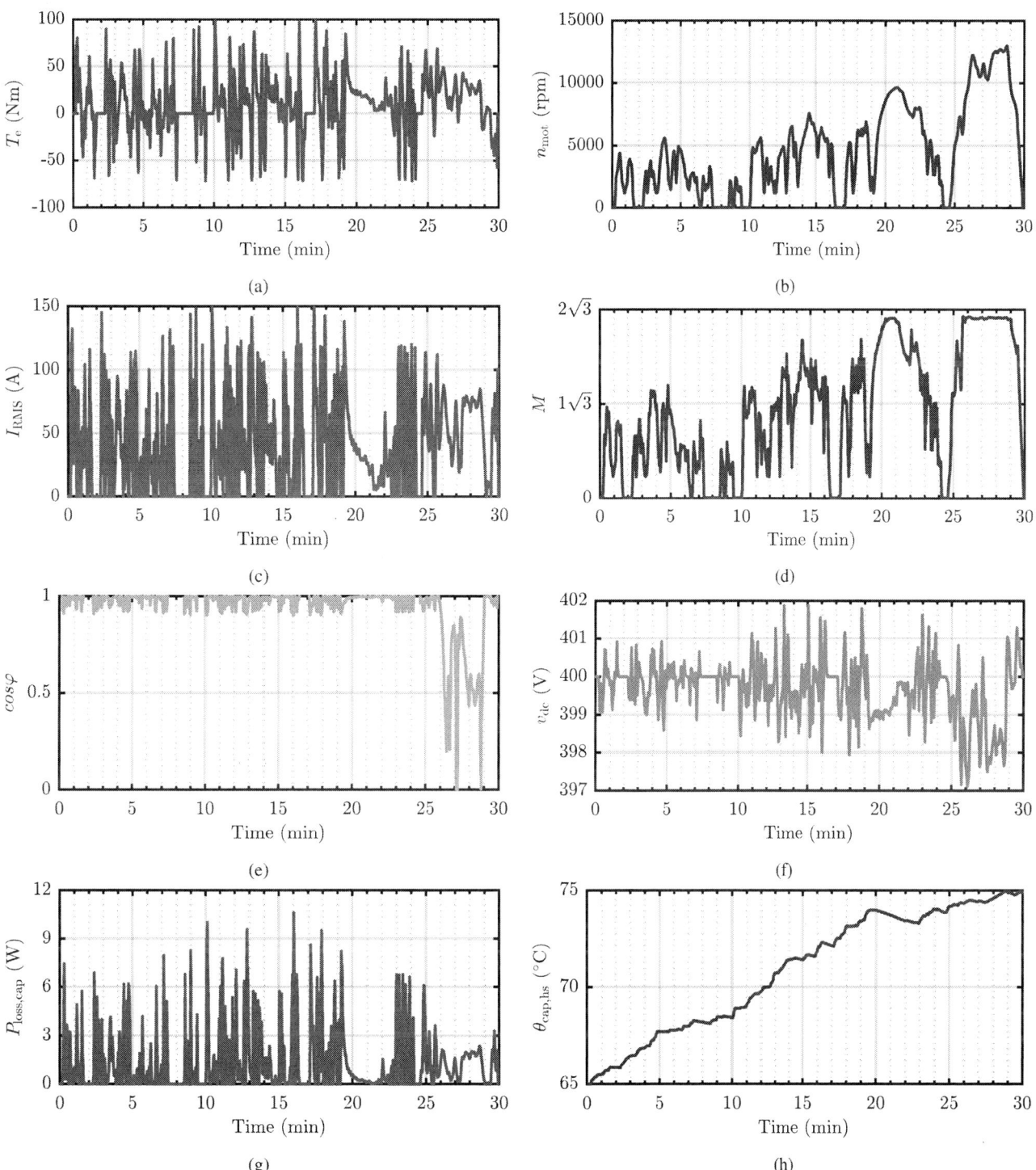

Fig. 9: Driving cycle simulation results: (a) motor electric torque, (b) motor mechanical speed, (c) motor RMS currents, (d) modulation index, (e) power factor, (f) DC-link voltage, (g) DC-link power losses and (h) DC-link hot-spot temperature.

s for complete cycle analysis. Future work will focus on extending the model to include additional capacitor technologies, implementing more sophisticated thermal models, and incorporating reliability prediction capabilities.

REFERENCES

[1] S. Giuffrida, L. Tolosano, F. Mandrile, C. Romano, M. Tranchero, and R. Bojoi, "Fast simulator with inverter temperature estimation for traction edrives in vehicles subjected to driving cycles," in *PCIM Europe 2024; International Exhibition and Conference for Power Electronics, Intelligent Motion, Renewable Energy and Energy Management*, 2024, pp. 248–253.

[2] H. Ye, K. Yang, H. Ge, P. Magne, and A. Emadi, "A drive cycle based electro-thermal analysis of traction inverters," in *2014 IEEE Transportation Electrification Conference and Expo (ITEC)*, 2014, pp. 1–6.

[3] D. Cittanti, F. Stella, E. Vico, C. Liu, J. Shen, G. Xiu, and R. Bojoi, "Analysis, design, and experimental assessment of a high power density ceramic dc-link capacitor for a 800 v 550 kva electric vehicle drive inverter," *IEEE Transactions on Industry Applications*, vol. 59, no. 6, pp. 7078–7091, 2023.

[4] H. Wen, W. Xiao, X. Wen, and P. Armstrong, "Analysis and evaluation of dc-link capacitors for high-power-density electric vehicle drive systems," *IEEE Transactions on Vehicular Technology*, vol. 61, no. 7, pp. 2950–2964, 2012.

[5] A. Sangwongwanich, Y. Shen, A. Chub, E. Liivik, D. Vinnikov, H. Wang, and F. Blaabjerg, "Mission profile-based accelerated testing of dc-link capacitors in photovoltaic inverters," in *2019 IEEE Applied Power Electronics Conference and Exposition (APEC)*, 2019, pp. 2833–2840.

[6] H. Wang, P. Davari, H. Wang, D. Kumar, F. Zare, and F. Blaabjerg, "Lifetime estimation of dc-link capacitors in adjustable speed drives under grid voltage unbalances," *IEEE Transactions on Power Electronics*, vol. 34, no. 5, pp. 4064–4078, 2019.

[7] H. Wang and H. Wang, "Capacitive dc links in power electronic systems-reliability and circuit design," *Chinese Journal of Electrical Engineering*, vol. 4, no. 3, pp. 29–36, 2018.

[8] D. Zhou, H. Wang, and F. Blaabjerg, "Mission profile based system-level reliability analysis of dc/dc converters for a backup power application," *IEEE Transactions on Power Electronics*, vol. 33, no. 9, pp. 8030–8039, 2018.

[9] J. Kolar, T. Wolbank, and M. Schrodl, "Analytical calculation of the rms current stress on the dc link capacitor of voltage dc link pwm converter systems," in *1999. Ninth International Conference on Electrical Machines and Drives (Conf. Publ. No. 468)*, 1999, pp. 81–89.

[10] E. Armando, R. I. Bojoi, P. Guglielmi, G. Pellegrino, and M. Pastorelli, "Experimental identification of the magnetic model of synchronous machines," *IEEE Transactions on Industry Applications*, vol. 49, no. 5, pp. 2116–2125, 2013.

[11] S. Chowdhury, E. Gurpinar, and B. Ozpineci, "Capacitor technologies: Characterization, selection, and packaging for next-generation power electronics applications," *IEEE Transactions on Transportation Electrification*, vol. 8, no. 2, pp. 2710–2720, 2022.

[12] H. Qu and Z. Ye, "Comparison of dynamic response characteristics of typical energy storage technologies for suppressing wind power fluctuation," *Sustainability*, vol. 15, p. 2437, 01 2023.

[13] W. Jiang and B. Fahimi, "Active current sharing and source management in fuel cell–battery hybrid power system," *IEEE Transactions on Industrial Electronics*, vol. 57, no. 2, pp. 752–761, 2010.

[14] "Kemet c4aqlbw5500m3jk datasheet," https://search.kemet.com/component-documentation.

[15] B. Weidenfeller, M. Höfer, and F. R. Schilling, "Thermal conductivity, thermal diffusivity, and specific heat capacity of particle filled polypropylene," *Composites Part A: Applied Science and Manufacturing*, vol. 35, no. 4, pp. 423–429, 2004. [Online]. Available: https://www.sciencedirect.com/science/article/pii/S1359835X03003440

A Monolithic Regulated 160 MHz Resonant DC-DC Converter

Giacomo Ripamonti*, Stefano Michelis*, Georgios Bantemits*, Pablo Daniel Antoszczuk*, Khalil Khalife*, Nils Hans Van Der Blij*, Sokratis Koseoglou*, Mattia Balutto⊎, Francesco Driussi†, Stefano Saggini†

∗ CERN - European Organization for Nuclear Research
Espl. des Particules 1, 1211 Meyrin, Switzerland

⊎ Politecnico di Milano
Piazza Leonardo da Vinci 32, 20133 Milano, Italy

† DPIA - University of Udine
Via delle Scienze 208, 33100 Udine, Italy

Email: giacomo.ripamonti@cern.ch, stefano.michelis@cern.ch, georgios.bantemits@cern.ch,
pablo.daniel.antoszczuk@cern.ch, khalil.khalife@cern.ch, nils.hans.van.der.blij@cern.ch,
sokratis.koseoglou@cern.ch, mattia.balutto@polimi.it, francesco.driussi@uniud.it, stefano.saggini@uniud.it

Abstract—This paper presents a monolithic dual phase 2:1 DC-DC converter featuring a 2.0-2.5 V input range, employing quasi-resonant regulation and using LC resonant tanks with coupled inductors. The output voltage can be regulated in the 0.8 – 1 V range by regulating the ON-time of the switches, while the switching frequency corresponds to the resonant frequency of the LC tanks (160 MHz). The usage of air-core inductors and exclusively MOM capacitors leads to a cost-effective implementation. The proposed converter has been validated through a 3.45 mm² (2.15 mm x 1.6 mm) fully integrated prototype implemented in a 28 nm CMOS process. Experimental results demonstrate a load current capability of 225 mA and a peak full load efficiency of 77.9 % at 2.5 V input and 1.0 V output. Performance analysis reveals comparable efficiency, efficiency enhancement factor (EEF) and power density with the current state of the art. The presented converter can be used as the basis for a family of modular monolithic DC-DC converter, as more power stages can be stacked to increase the conversion ratio.

I. INTRODUCTION

MOORE'S law has been a fundamental driver in the scaling of semiconductor technology, enabling continuous advancements in the development of low-power, high-performance electronic systems. This scaling has allowed the electronics industry to push the boundaries of miniaturization, but it faces substantial challenges due to the inherent physical limitations of passive components, especially at smaller nodes [1], [2]. On the other hand, power management integrated circuits (PMICs), which use passive components as energy storage elements, have become essential in managing and optimizing power consumption across a wide range of applications, from Internet of Things devices to wearable electronics and beyond [3]–[5]. Inductor-based DC-DC converters are often preferred for their high efficiency and ability to handle substantial power loads. However, integrating inductors on chip remains a significant challenge due to the bulky nature

of these components, which hinder further miniaturization of systems [6]–[8]. As a result, there has been an increased focus on exploring alternative power converter topologies to overcome the limitations of inductor-based designs [9]–[12].

Fully integrated voltage regulators (FIVRs), for instance, have emerged as a viable solution, allowing for improved control over voltage domains directly on-chip, which is beneficial for dynamic voltage scaling (DVS) in modern electronics [13], [14].

Switched-capacitor (SC) converters have become an important alternative to inductor-based converters, relying on more compact on-chip capacitors rather than large inductors. However, SC converters face efficiency limitations due to the charge redistribution mechanism [15]. To address these constraints, hybrid and resonant switched-capacitor (ReSC) converters have been developed [16], [17]. They leverage a combination of SC and inductive elements to enhance efficiency compared to their SC counterparts [18].

Additionally, innovative approaches, such as auto-oscillator structures, have also been explored to further optimize power conversion efficiency in a variety of applications [19].

This paper introduces a monolithic 2:1 dual-phase ReSC DC-DC converter. This converter is designed in a 28 nm CMOS technology, using exclusively low-voltage transistors rated at 0.9 *V*, metal-oxide-metal (MOM) capacitors, and air-core metal-trace coupled inductor. The proposed design is intended to employ standard manufacturing processes, reducing costs. Finally, this converter can serve as a base stackable cell for building DC-DC converters with higher conversion ratios.

This paper is organized as follows: Section II introduces the power stage and its operation, while Section III details the implementation techniques adopted for the power stage. Section IV presents the closed-loop control and the startup procedure, while Section V details the experimental results of

---

979-8-3315-1612-3/25 $31.00 © 2025 IEEE

the developed prototype.

II. MULTIPHASE DC-DC POWER STAGE

Fig. 1 illustrates the power stage of the dual-phase DC-DC converter, a similar structure to [18]. The two operational

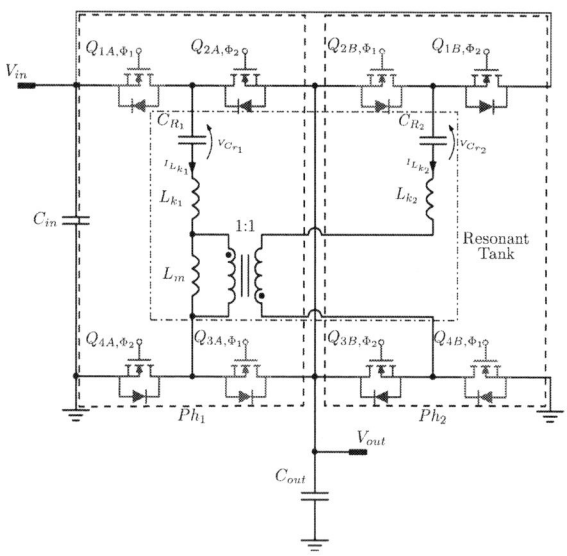

Fig. 1: Power stage of the dual-phase DC-DC converter. The red and blue transistors are respectively driven by driving signals Φ_1 and Φ_2.

phases, A and B, operate in complementary mode (they are phase-shifted by 180 degrees), enhancing the on-chip input and output capacitance utilization and reducing input and output voltage ripple. The implementation described in this work features a two-phase powertrain that employs a resonant tank. The tank is composed of a two-phase coupled inductor and two resonant capacitors, realized with metal-oxide-metal (MOM) capacitors. The usage of MIM capacitors as in [18] would require an additional mask for fabrication, therefore the adoption of MOM elements leads to a more cost-effective solution. The air-core coupled inductor can be implemented in a standard CMOS process, without the need for a dedicated thin-film magnetic core, which would boost the performance but leads to additional costs. Each interleaved phase has a nominal 2:1 conversion ratio and operates with two switching phases per period. As presented in [20], one can describe the relationship between each linked flux (λ_1, λ_2) and the respective tank current ($I_{L_{k1}}$, $I_{L_{k2}}$) by the following inductance matrix:

$$\begin{bmatrix} \lambda_1 \\ \lambda_2 \end{bmatrix} = \mathbf{L} \begin{bmatrix} I_{L_{k1}} \\ I_{L_{k2}} \end{bmatrix} \xrightarrow{where} \mathbf{L} = \begin{bmatrix} L_{11} & L_{12} \\ L_{21} & L_{22} \end{bmatrix} \quad (1)$$

As in [18], interleaved operation results in identical inductor currents with opposite polarities, enabling positive coupling and thus higher effective resonant inductance ($L_{r,i}$).

$$L_{r,i} = L_{ii}(1+k) = L_{k,i} + 2L_m, \quad (2)$$

where $k = \frac{L_{12}}{\sqrt{L_{11}L_{22}}} \approx \frac{L_m}{L_m + L_k}$ is the coupling factor, L_m the magnetizing inductance, $L_{k,i}$ is the i-winding leakage inductance, $i \in [1,2]$. Consequently, assuming the input

voltage and the output voltage of the cell constant, the resonant frequency of each resonant tank can be expressed as:

$$\begin{cases} f_{r1} = \frac{1}{2\pi\sqrt{L_{r1}C_{r1}}} = \frac{1}{2\pi\sqrt{(L_{k,1}+2L_m)C_{r1}}} \\ f_{r2} = \frac{1}{2\pi\sqrt{L_{r2}C_{r2}}} = \frac{1}{2\pi\sqrt{(L_{k,2}+2L_m)C_{r2}}}. \end{cases} \quad (3)$$

Assuming a high quality factor for both the resonant tanks, the tank currents can be approximated with a sinusoidal waveform (as depicted in Fig. 2).

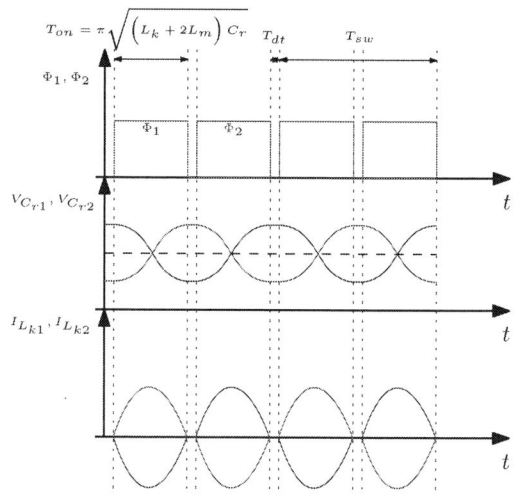

Fig. 2: Power stage waveforms.

This resonant tank implementation improves quality factor and chip utilized area compared to uncoupled two-phase designs. Conversely, keeping a constant available area the resonant frequency (corresponding to the switching frequency) can be reduced by employing a coupled inductor, thus reducing the frequency-dependent losses and boosting the efficiency.

III. POWER STAGE IMPLEMENTATION

Due to the exclusive use of $0.9\ V$-rated devices in this implementation, each CMOS half-bridge introduced in Fig. 1 is constructed by stacking two low-voltage transistors: the main switch (connected to the rail) and a clamping device with a fixed gate voltage. By using this technique, each half-bridge is able to withstand a voltage of $1.5\ V$. The half-bridges constituted by Q_3 and Q_4 are referred to ground and can be therefore driven by gate drivers powered by a $0.9\ V$ line. Such $0.9\ V$ domain can be generated from V_{in} by a linear regulator. After the end of the start-up phase, the output voltage of the converter can be used to power the gate drivers of Q_3 and Q_4, boosting the efficiency.

Fig. 3 provides instead a representation of the architecture employed to drive the floating half bridges (Q_1, Q_2). This schematic highlights the key functional blocks of the proposed design, including two low dropout regulators (LDOs) that generate the $0.9\ V$ power supplies of the gate drivers, a pair of capacitive level shifters and the gate drivers. All the nMOS devices are isolated from the substrate by a deep nwell connected to V_{in} (the nwell-to-substrate junction is capable to withstand the full input voltage). Furthermore, Fig. 3 depicts

979-8-3315-1612-3/25 $31.00 © 2025 IEEE

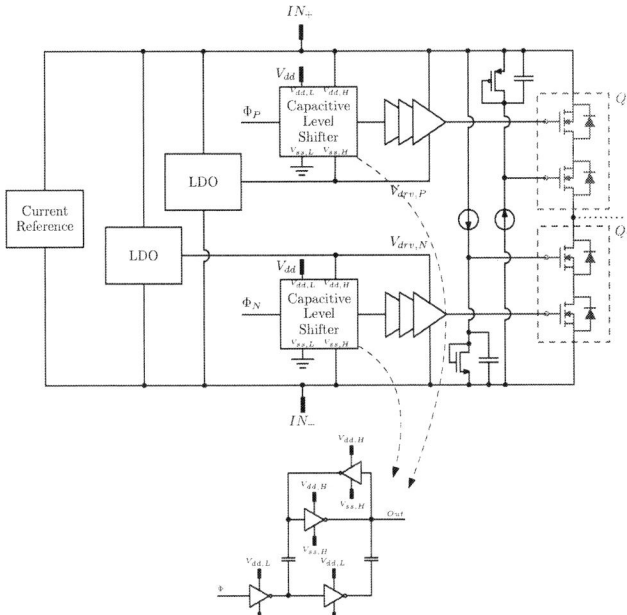

Fig. 3: Block diagram of the driving scheme adopted for the floating half-bridges.

the internal architecture of the capacitive level shifter, which allows the adoption of exclusively $0.9\ V$-rated devices.

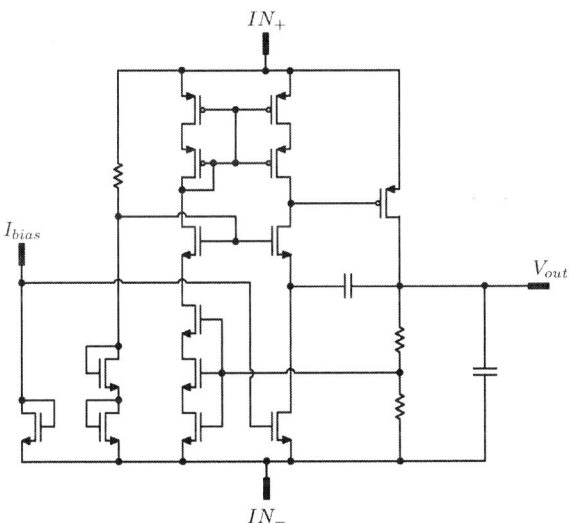

Fig. 4: Bandgap-less LDO.

The schematic of the LDO used to power the gate driver of the nMOS switch is presented in Fig. 4. The output voltage of the LDOs (V_{out} in Fig. 4 and $V_{drv,N}$ in Fig. 3) is proportional to a beta multiplier current reference, which avoids the usage of a dedicated bulkier bandgap voltage reference circuit in the power stage. The V_{gs} of nMOS devices where the current from the beta multiplier reference flows acts as the reference voltage, allowing to obtain an output voltage between $0.82\ V$ and $0.98\ V$ in all PVT and temperature (from $-30\ °C$ to $80\ °C$) corners, ensuring long term reliability for the $0.9\ V$ core transistors. The dual schematic is used for the LDO

providing VSS (called $V_{drv,P}$ in Fig. 3) to the driver of the pMOS switch.

The resonant tank comprises a two-phase magnetically coupled inductor (represented by L_{k1}, L_{k2}, L_m in Fig. 1) and a pair of MOM capacitor-based resonant capacitors ($C_r 1$, C_{r2}). The careful optimization of these passive elements is fundamental in achieving the desired frequency response, target inductance, and optimized total series resistance. Specifically, the coupled inductor has been obtained parallelizing all available thick metals. Additionally, the resonant capacitors have been placed below the inductor, avoiding any metal loop in the routing to suppress eddy currents. The LC tank has been simulated and refined using electromagnetic simulations.

The presented power stage does not rely on any FET with voltage rating larger than $0.9\ V$, while only the capacitors in the level shifter and the nwell-to-substrate junction must withstand the input voltage V_{in}. As long as such two elements are within the specified voltage rating, multiple power stages can be stacked to obtain an higher conversion ratio.

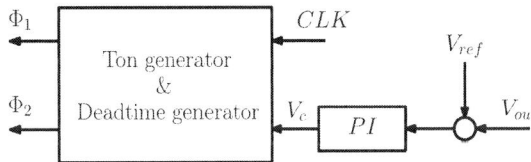

Fig. 5: Block diagram of the control circuit.

IV. CLOSED-LOOP CONTROL AND STARTUP

A control circuit, represent in Fig 5, has been developed to control the output voltage in the $0.8 - 1.0\ V$ range. It adapts the ON-time of the switches, while keeping the switching frequency fixed and equal to the resonant frequency of the coupled LC tank. The control loop is relatively simple: it is composed of a PI controller, a Ton generator and a deadtime generator, and it uses an input clock whose frequency corresponds to the resonant frequency. Compared to [18], the proposed control leads to a faster transient response and a reduced output ripple for light loads. It does not guarantee zero-current switching (ZCS), but the related switching losses can be contained in a monolithic implementation, as the switching speed can be large.

A soft start procedure that ensures an acceptable voltage stress for all used device has been developed: in particular, an LDO is placed between the output of the power stage and the output of the converter. A clamping circuit ensures that the output of the power stage never drops more than $1.5\ V$ below V_{in}, guaranteeing that all the devices stay in a safe operating area. The LDO is initially off, and when the input voltage is raised the output of the power stage is initialized to $V_{in} - 1.5\ V$ by the clamping circuit. During the startup, the power stage starts regulating the voltage at its output to the desired value, while the output voltage of the converter is gradually ramped up by the LDO. At the end of the startup, the pass device of the LDO is fully turned on, and the sensing point of the output voltage for the control is moved from the output of the power stage to the output of the converter.

979-8-3315-1612-3/25 $31.00 © 2025 IEEE

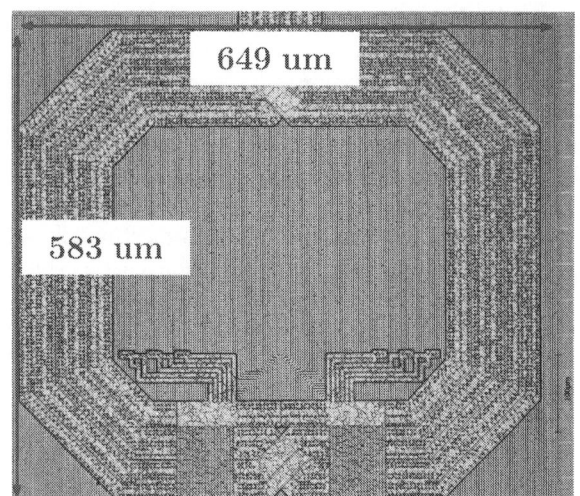

Fig. 6: Micrograph of the custom coupled air core inductor.

V. EXPERIMENTAL RESULTS

To validate the proposed architecture, a prototype converter was fabricated in a commercial 28-nm CMOS process. Fig. 7 shows the die micrograph of the proposed converter. The total die area is $3.45\ mm^2$ ($2.15\ mm$ x $1.6\ mm$), with the power stage occupying $2.15\ mm$ x $1.17\ mm$ and the control circuitry measuring $2.15\ mm$ x $0.43\ mm$. The resonant tank is composed by the coupled inductor ($L_{11} = L_{22} = 1.89\ nH$ with a coupling factor of $k = 0.567$ and series resistance of $501\ m\Omega$) and two $335\ pF$ MOM resonant capacitors, resulting in a resonant frequency (corresponding to the switching frequency) of $160\ MHz$. The custom coupled air core inductors are shown in 6.

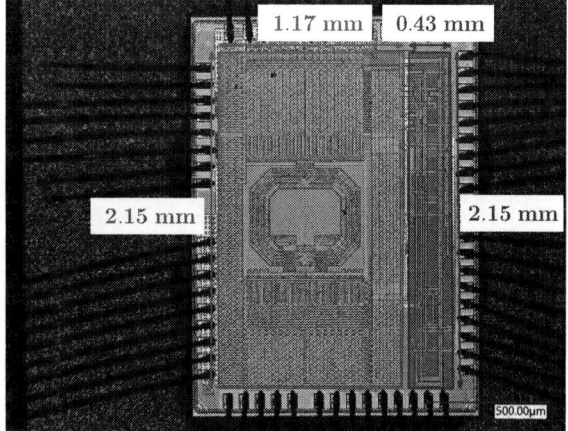

Fig. 7: Micrograph of the DC-DC converter ASIC. The power stage is highlighted in red, while the control circuitry is highlighted in blue.

Fig. 8 and Fig. 9 depict the efficiency of the converter in multiple operating points ($V_{out} = 0.8$–$1\ V$, $V_{in} = 2.1$–$2.5\ V$): the closer is the conversion ratio $\frac{V_{out}}{V_{in}}$ to 0.5, the larger is the efficiency for the same load, since the ON-time of the switches approaches half a resonance period (full resonant operation).

The maximum load current instead increases if the conversion ratio moves further from 0.5, as the resonant tank voltage excitation becomes larger.

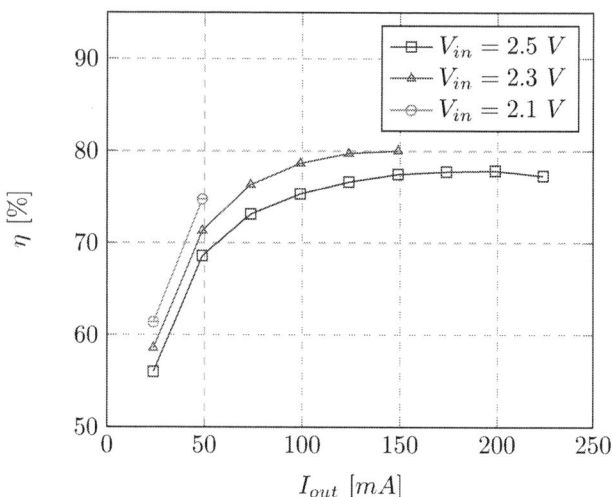

Fig. 8: Efficiency plot for $V_{out} = 1\ V$.

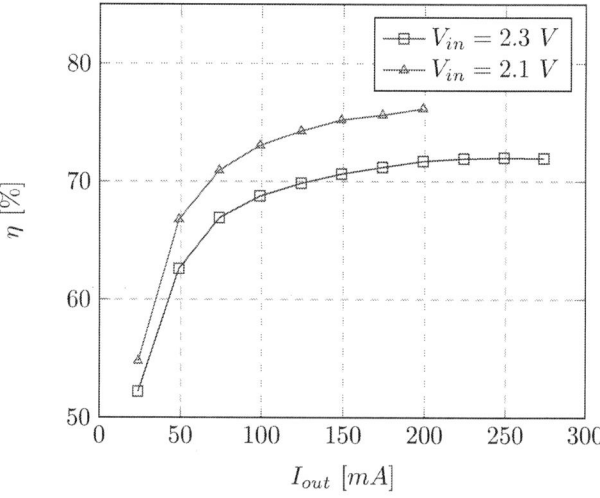

Fig. 9: Efficiency plot for $V_{out} = 0.8\ V$.

Fig. 10 illustrates the output voltage response during a load step, while Fig. 11 depicts the output voltage response during a load release transient. The control loop proves to be stable. Furthermore, V_{out} deviates from its steady-state value by less than $100\ mV$ and returns to steady state in $\approx 300\ ns$, demonstrating the fast transient response of the control loop.

Fig. 12 illustrates the output voltage waveforms of the DC-DC converter and of the power stage during the startup phase, demonstrating the desired soft start operation: the output of the power stage is initialized by a clamping circuit, ensuring that the voltage stress of the floating half bridges is kept lower than $1.5\ V$. After enabling the switching, the output LDO gradually increases V_{out}, which correctly converges to the output voltage of the switching power stage after approximately $50\ \mu s$.

Table I reports instead the comparison with the state-of-art of fully integrated converters, showing comparable efficiency,

979-8-3315-1612-3/25 $31.00 © 2025 IEEE

	[21]	[22]	[23]	[24]	[18]	[19]	This work
Technology	14 nm	65 nm	90 nm	65 nm	180 nm	180 nm	28 nm
Topology	Buck	3LB	SC	SIC	ReSC	EM-coupled class-D LC	ReSC
Conversion ratio	1.33:1	2.6:1	2:1	1.33:1	2:1	2:1	2:1
Interleaved phases	1	2	9	1	2	4	2
f_{sw} [MHz]	70	-	70	45	47.5	1250	160
V_{in} [V]	1.6	1.8	3.0-3.6	1.2	22.4-4.4	1-3.6	2-2.5
V_{out} [V]	1.2	0.7	1.3-1.5	0.6-0.9	1.0-2.2	0.4-1.6	0.8-1.0
C_{in} [nF]	-	-	-	-	7.0	0	1.0
C_{fly} - C_{res} [nF]	-	2 x 1.5	2	1.72	2 x 1.7	2 x 0.23	0.34
C_{out} [nF]	10	1.9	3.2	3.1	7	0.0044	6.7
L [nH]	2.5 package	2 x 1.5	-	0.85	7.7 coupled	7.8 coupled	3 coupled
Area [mm^2]	2.16	2.09	3.24	0.65	8.93	1.61	3.45
η_{peak} [%]	88	64	77	78	85.5	67	77.9
Power density [$\frac{W}{mm^2}$]	0.28	0.067	0.022	0.55	0.053	0.21	0.065
EEF [%]	15	39.2	43.9	4	45.5	39	48.7

TABLE I: Comparison with state-of-the-art fully-integrated DC-DC converters.

power density and efficiency enhancement factor ($EEF = 1 - \frac{\eta_{LDO}}{\eta_{DC-DC}}$, where η_{LDO} and η_{DC-DC} are respectively the efficiencies of an LDO and of a DC-DC converter with the same conversion ratio) with the current state of the art.

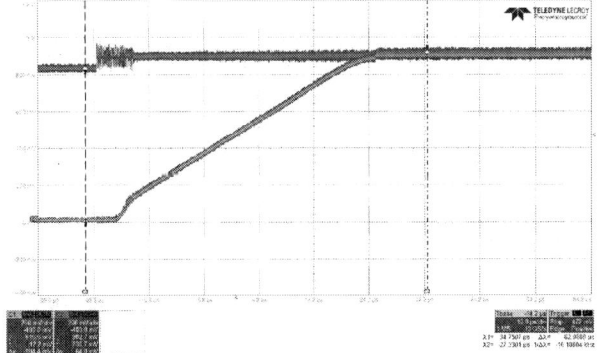

Fig. 12: V_{out} of the converter (yellow trace) and output voltage of power stage (purple trace) during startup, $V_{in} = 2.3\ V$.

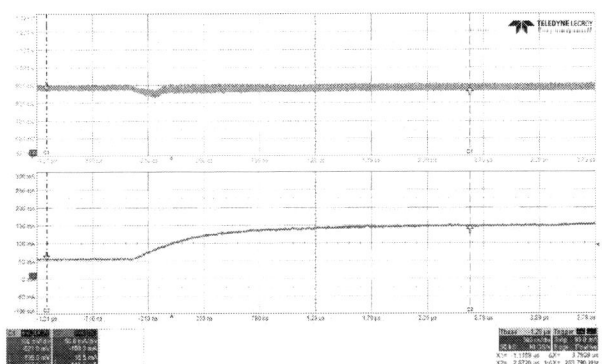

Fig. 10: V_{out} transient (yellow trace) during a load step (purple trace), $V_{in} = 2.3\ V$.

VI. CONCLUSION

In this paper, a monolithic 2:1, 2.5 V-input, regulated dual-phase DC-DC converter implemented using 0.9 V-rated MOS-FETs and metal-oxide-metal (MOM) capacitors is proposed. The usage of MOM capacitors and air-core inductors allows a more cost-effective manufacturing compared to monolithic converters employing MIM capacitors and thin-film magnetic cores.

A dedicated soft-start procedure is included to ensure a controlled power-up sequence, preventing inrush currents and voltage stress on the devices. The regulation scheme has been carefully designed and simulated, achieving expected performance under a variety of load conditions.

Experimental measurements demonstrate that the converter achieves comparable metrics to the current state of the art, including power conversion efficiency, power density, and efficiency enhancement factor (EEF).

The modular architecture of the proposed converter presents significant opportunities for further exploration and optimization: in particular, multiple power stages can be stacked to achieve higher conversion ratios compared to the presented 2:1 ratio. This scalability can potentially broaden the applicability

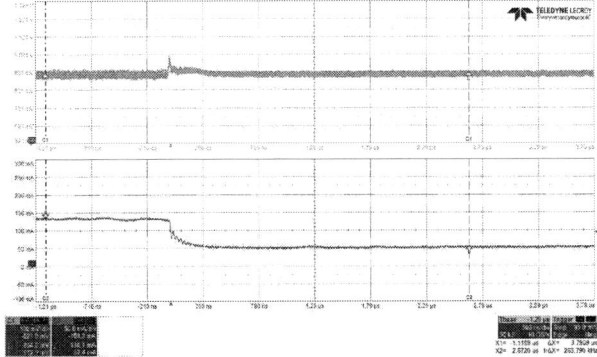

Fig. 11: V_{out} transient (yellow trace) during a load release (purple trace), $V_{in} = 2.3\ V$.

of the presented monolithic converter in areas such as high-performance computing and IoT devices.

REFERENCES

[1] W. M. Holt, "1.1 moore's law: A path going forward," in *2016 IEEE International Solid-State Circuits Conference (ISSCC)*, 2016, pp. 8–13.

[2] J. T. Stauth, "Pathways to mm-scale dc-dc converters: Trends, opportunities, and limitations," in *2018 IEEE Custom Integrated Circuits Conference (CICC)*, 2018, pp. 1–8.

[3] M. Steyaert, F. Tavernier, H. Meyvaert, A. Sarafianos, and N. Butzen, "When hardware is free, power is expensive! is integrated power management the solution?" in *ESSCIRC Conference 2015 - 41st European Solid-State Circuits Conference (ESSCIRC)*, 2015, pp. 26–34.

[4] D. Blaauw, D. Sylvester, P. Dutta, Y. Lee, I. Lee, S. Bang, Y. Kim, G. Kim, P. Pannuto, Y.-S. Kuo, D. Yoon, W. Jung, Z. Foo, Y.-P. Chen, S. Oh, S. Jeong, and M. Choi, "Iot design space challenges: Circuits and systems," in *2014 Symposium on VLSI Technology (VLSI-Technology): Digest of Technical Papers*, 2014, pp. 1–2.

[5] M. Yip, R. Jin, H. H. Nakajima, K. M. Stankovic, and A. P. Chandrakasan, "18.2 a fully-implantable cochlear implant soc with piezoelectric middle-ear sensor and energy-efficient stimulation in 0.18m hvcmos," in *2014 IEEE International Solid-State Circuits Conference Digest of Technical Papers (ISSCC)*, 2014, pp. 312–313.

[6] N. Tang, B. Nguyen, R. Molavi, S. Mirabbasi, Y. Tang, P. Zhang, J. Kim, P. P. Pande, and D. Heo, "Fully integrated buck converter with fourth-order low-pass filter," *IEEE Transactions on Power Electronics*, vol. 32, no. 5, pp. 3700–3707, 2017.

[7] M. Kar, A. Singh, A. Rajan, V. De, and S. Mukhopadhyay, "An all-digital fully integrated inductive buck regulator with a 250-mhz multi-sampled compensator and a lightweight auto-tuner in 130-nm cmos," *IEEE Journal of Solid-State Circuits*, vol. 52, no. 7, pp. 1825–1835, 2017.

[8] N. Sturcken, E. O'Sullivan, N. Wang, P. Herget, B. Webb, L. Romankiw, M. Petracca, R. Davies, R. Fontana, G. Decad, I. Kymissis, A. Peterchev, L. Carloni, W. Gallagher, and K. Shepard, "A 2.5d integrated voltage regulator using coupled-magnetic-core inductors on silicon interposer delivering 10.8a/mm2," in *2012 IEEE International Solid-State Circuits Conference*, 2012, pp. 400–402.

[9] S. Abedinpour, B. Bakkaloglu, and S. Kiaei, "A multistage interleaved synchronous buck converter with integrated output filter in 0.18 μm sige process," *IEEE Transactions on Power Electronics*, vol. 22, no. 6, pp. 2164–2175, 2007.

[10] Y. K. Ramadass, A. A. Fayed, and A. P. Chandrakasan, "A fully-integrated switched-capacitor step-down dc-dc converter with digital capacitance modulation in 45 nm cmos," *IEEE Journal of Solid-State Circuits*, vol. 45, no. 12, pp. 2557–2565, 2010.

[11] L. Chang, R. K. Montoye, B. L. Ji, A. J. Weger, K. G. Stawiasz, and R. H. Dennard, "A fully-integrated switched-capacitor 21 voltage converter with regulation capability and 90

[12] W. Kim, D. M. Brooks, and G.-Y. Wei, "A fully-integrated 3-level dc/dc converter for nanosecond-scale dvs with fast shunt regulation," in *2011 IEEE International Solid-State Circuits Conference*, 2011, pp. 268–270.

[13] S. Amin, H. Krishnamurthy, H. Do, C. Alvarez, M. Hill, K. Radhakrishnan, V. De, S. Weng, K. Ravichandran, J. Tschanz, W. Gomes, and J. Douglas, "A 5.4v-vin, 9.3a/mm2 10mhz buck ivr chiplet in 55nm bcd featuring self-timed bootstrap and same-cycle zvs control," in *2024 IEEE Symposium on VLSI Technology and Circuits (VLSI Technology and Circuits)*, 2024, pp. 1–2.

[14] E. A. Burton, G. Schrom, F. Paillet, J. Douglas, W. J. Lambert, K. Radhakrishnan, and M. J. Hill, "Fivr — fully integrated voltage regulators on 4th generation intel® core™ socs," in *2014 IEEE Applied Power Electronics Conference and Exposition - APEC 2014*, 2014, pp. 432–439.

[15] C. Tse, S. Wong, and M. Chow, "On lossless switched-capacitor power converters," *IEEE Transactions on Power Electronics*, vol. 10, no. 3, pp. 286–291, 1995.

[16] G. Ripamonti, M. Ursino, S. Saggini, S. Michelis, and F. Faccio, "Regulated resonant switched-capacitor point-of-load converter architecture and modeling," *IEEE Transactions on Power Electronics*, vol. 36, no. 4, pp. 4815–4827, 2021.

[17] A. Dago, M. Leoncini, S. Saggini, A. Brunero, A. Gasparini, O. Zambetti, S. Levantino, and M. Ghioni, "A high-power-density quasi-resonant switched-capacitor dc–dc converter with single semiperiod tank current modulation," *IEEE Transactions on Power Electronics*, vol. 39, no. 2, pp. 2100–2114, 2024.

[18] P. H. McLaughlin, Z. Xia, and J. T. Stauth, "A monolithic resonant switched-capacitor voltage regulator with dual-phase merged-lc resonator," *IEEE Journal of Solid-State Circuits*, vol. 55, no. 12, pp. 3179–3188, 2020.

[19] A. Novello, G. Atzeni, J. Künzli, G. Cristiano, M. Coustans, and T. Jang, "A 1.25-ghz fully integrated dc–dc converter using electromagnetically coupled class-d lc oscillators," *IEEE Journal of Solid-State Circuits*, vol. 56, no. 12, pp. 3639–3654, 2021.

[20] M. Chen and C. R. Sullivan, "Unified models for coupled inductors applied to multiphase pwm converters," *IEEE Transactions on Power Electronics*, vol. 36, no. 12, pp. 14 155–14 174, 2021.

[21] C. Schaef, N. Desai, H. Krishnamurthy, S. Weng, H. Do, W. Lambert, K. Radhakrishnan, K. Ravichandran, J. Tschanz, and V. De, "8.5 a fully integrated voltage regulator in 14nm cmos with package-embedded air-core inductor featuring self-trimmed, digitally controlled variable on-time discontinuous conduction mode operation," in *2019 IEEE International Solid-State Circuits Conference - (ISSCC)*, 2019, pp. 154–156.

[22] W. Godycki, B. Sun, and A. Apsel, "Part-time resonant switching for light load efficiency improvement of a 3-level fully integrated buck converter," in *ESSCIRC 2014 - 40th European Solid State Circuits Conference (ESSCIRC)*, 2014, pp. 163–166.

[23] T. M. Van Breussegem and M. S. J. Steyaert, "Monolithic capacitive dc-dc converter with single boundary–multiphase control and voltage domain stacking in 90 nm cmos," *IEEE Journal of Solid-State Circuits*, vol. 46, no. 7, pp. 1715–1727, 2011.

[24] N. Tang, W. Hong, B. Nguyen, Z. Zhou, J.-H. Kim, and D. Heo, "Fully integrated switched-inductor-capacitor voltage regulator with 0.82-a/mm2 peak current density and 78

Reconfigurable Trans-Inductor Voltage Regulator with Improved Light Load Efficiency in Data Center Applications

Ziyao Wang, Zehui Li and Haoyu Wang
School of Information Science and Technology
ShanghaiTech University, Shanghai, China
Shanghai Engineering Research Center of Energy Efficient and Custom AI IC
wanghy@shanghaitech.edu.cn

Abstract—In data centers, the trans-inductor voltage regulator (TLVR) is emerging as a promising topology for point-of-load converters, designed to deliver high-current, fast-transient power to the XPU. Since XPUs typically operate under light load conditions, enhancing TLVR light load efficiency is essential. This paper proposes a novel reconfigurable TLVR design that incorporates a 4-quadrant switch within the secondary side coupling loop of the trans-inductor. At heavy load, the configuration retains the characteristics of a conventional TLVR. While at light load, the TLVR is reconfigured to decouple the interleaved phases, thus improving light load efficiency. A 12V/1.8V, 500kHz, 80A 4-phase TLVR with this secondary switch was designed and tested, with experimental results demonstrating an average light load efficiency improvement of 2.9%.

Index Terms—data center, light load efficiency, TLVR, voltage regulator modules.

I. INTRODUCTION

With the rapid development of cloud computing, big data, and artificial intelligence, the energy consumption of data centers is booming. It uses about 2% of the global electricity supply and will grow to 4% ~ 8% by 2030 [1]. However, the load point XPUs operate in idle mode for over 89% of the time, with power consumption at less than 10% of full load capacity [2]. Therefore, enhancing light load efficiency is crucial for energy saving in data centers.

In point-of-load applications, there is a high demand for the ability to handle large currents and fast current slew rates. The trans-inductor regulator (TLVR) topology is a ideal candidate, offering reduced conduction losses, output current ripple cancellation, and improved dynamic response [3] [4]. However, as a coupled inductor buck-derived converter, TLVR suffers from low light-load efficiency due to the interphase coupling and the high step-down conversion ratio. This coupling introduces additional driving and conduction losses [5]. Moreover, high step-down buck converters face challenges such as low duty cycle and hard switching, which further degrade the efficiency [6]. Narrow duty cycles increase output voltage sensitivity, complicate control and gate drive design, and lead to higher RMS current, exacerbating conduction losses [7]. Switching losses are increased, and the benefits of interleaving are diminished [8].

Several techniques have been proposed to enhance efficiency, particularly to extend their duty cycles. Cascading converters offer a straightforward method to achieve a higher duty cycle but increase the component count and control complexity, which ultimately reduces efficiency [9]. In [10], quadratic topologies combining series converters are proposed to reduce switch count. However, they still suffer from high voltage and current stress on the switches, which impacts overall performance. In [11] [12], another series capacitor-based technique is proposed to extend the duty cycle, reduce voltage stress on MOSFETs, and ensure current balance across phases. However, this approach faces challenges such as hard switching of MOSFETs and increased circuit complexity due to a larger number of components. To mitigate switching losses, series-resonant structures are investigated in series capacitor converters [13]. However, hard switching in high-side MOSFETs and zero current switching (ZCS) in low-side MOSFETs remain challenges, with zero voltage switching (ZVS) required to eliminate the switching losses.

Other methods address issues associated with coupled inductors. In [5], synchronized control of two coupled inductor phases is proposed. It eliminates power loss from dual zero current touching, but the current ripple increases. In [14], a tapped-inductor-based converter is studied to extend the duty cycle and reduce switching stress. However, leakage inductance in the tapped inductor often resonates with switches' parasitic capacitance which limits the overall performance. In [15], lossless snubber and active clamp circuit are used to reduce ringing and achieve soft switching, but this adds complexity. Another drawback of coupled inductor converters is the pulsating output current in [14] [15], which leads to excessive equivalent series resistance (ESR) losses in the output capacitance and reduces its lifespan [16]. In [17], a new coupled inductor structure with shorter winding paths is proposed to suppress the magnetic loss. Still, the asymmetry in the multi-phase structure results in unequal phase-coupling coefficients, leading to unbalanced phase ripple current cancellation and larger output capacitors. Additionally, nonlinear inductors have been suggested to enhance light load efficiency in [18] [19]. By increasing inductance, current ripple-related

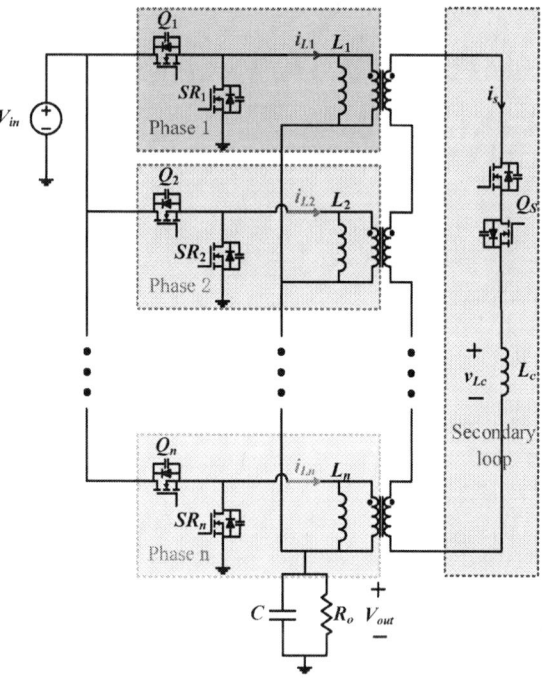

Fig. 1. Schematic of the proposed reconfigurable TLVR.

losses at light load can be minimized, although the magnetic design becomes more complex.

In this paper, we propose a novel reconfigurable TLVR designed to enhance light load efficiency. By incorporating a four-quadrant switch into the coupling loop on the secondary side of the trans-inductor, our design facilitates dynamic phase coupling control. During heavy loads, the switch is turned on to support conventional TLVR operation, ensuring a rapid transient response. Under light load conditions, the switch deactivates, decoupling the TLVR phases. This strategy not only optimizes phase shedding, but also increases the equivalent steady-state inductance, thereby reducing current ripple and associated losses, leading to improved light load efficiency. Compared to existing approaches, our solution boasts a simplified structure, ease of implementation, straightforward control, and excellent scalability, making it suitable for high-power applications.

The remainder of this paper is structured as follows. Section II outlines the reconfigurable TLVR and explains its operation principles. Section III delves into the methodology for improving light load efficiency. Section IV presents a loss analysis model to demonstrate the effects of our proposed solution on overall system losses. Experimental results are provided in Section V. Lastly, Section VI concludes the paper.

II. PROPOSED RECONFIGURABLE TLVR

The schematic of the proposed reconfigurable TLVR is shown in Fig. 1. The primary side consists of an interleaved multi-phase buck converter, where each phase includes a

half-bridge and a transformer. The primary winding is connected between the half-bridge switching point and the output, while all secondary windings are connected in series with a compensating inductor L_c, and an additional switch Q_s. This design builds on the conventional TLVR framework by adding a reconfigurable component: a four-quadrant switch Q_s, positioned in the coupling loop on the secondary side of the trans-inductor. In practice, Q_s is implemented using two back-to-back MOSFETs to control phase coupling. The operational principle of the reconfigurable TLVR is illustrated in Fig. 2. Depending on the load condition, the converter offers two operation modes as follows.

A. Heavy-load Mode

At heavy loads, the circuit operates in continuous conduction mode (CCM). The secondary switch is turned on to maintain phase coupling, preserving the characteristics of a conventional TLVR. For a 4-phase example, when any phase on the primary side is activated, the voltage across the secondary inductor L_c is $V_{in} - 4V_{out}$; when all phases are deactivated, it becomes $-4V_{out}$. This voltage determines the secondary current i_{Lc}. The TLVR inductor current i_L can be viewed as the superposition of i_{Lc} and the inductor current of uncoupled buck, as illustrated in Fig. 3.

B. Light-load Mode

At light load, the secondary switch is turned off, disconnecting the secondary coupling circuit. Without the influence of secondary current, the converter is reconfigured to function as a multi-phase buck converter, inheriting its characteristics. This mode effectively reduces power loss. A detailed analysis is provided in Section III.

C. Effect of Switch

The equivalent circuit in the secondary loop is shown in Fig. 4, assuming that the leakage inductance of the coupled inductor is negligible.

In heavy-load mode, Q_s is turned on and can be treated as a resistor R_{eq}, representing the on-resistance of the two MOSFETs. This additional resistance contributes to conduction losses. To evaluate its effect on overall efficiency, we first calculate the secondary current. Using the previously mentioned voltage, the effective value of the secondary current is derived as follows:

$$I_{s-heavy} = \frac{(1-4D)V_{out}}{2\sqrt{3}L_c f_s} \quad (1)$$

As shown in (1), $I_{s-heavy}$ is independent of the load. When the output current increases, the additional losses introduced by the secondary switch remain constant, represented by $I_{s-heavy}^2 R_{eq}$. However, since the calculated value of $I_{s-heavy}$ is much smaller than the heavy-load current, this extra conduction loss accounts for less than 1% of the primary-side MOSFET losses, making it negligible in terms of overall performance.

In light-load mode, we aim to reduce the secondary current to zero to fully decouple the phases. However, when Q_s is

979-8-3315-1612-3/25 $31.00 © 2025 IEEE

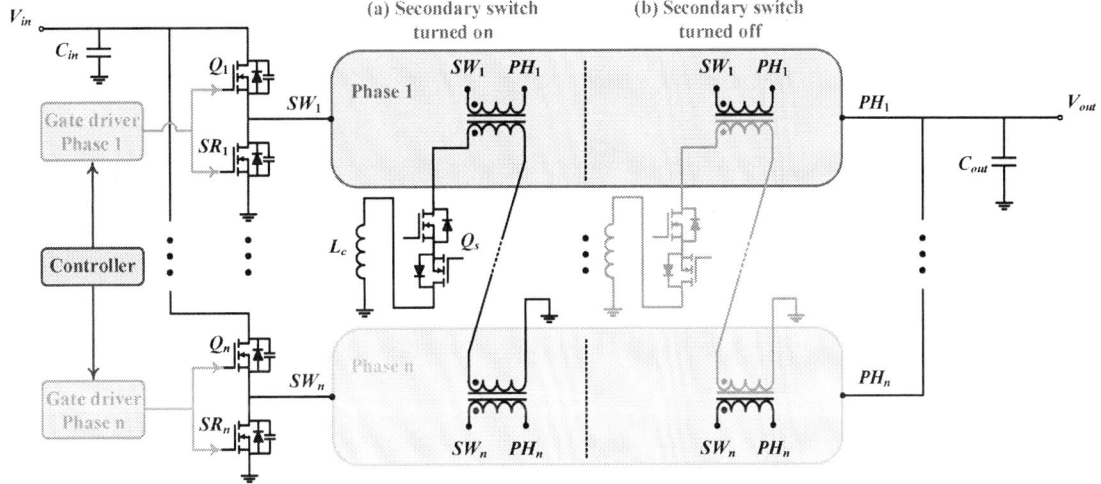

Fig. 2. Operation principle of reconfigurable TLVR.

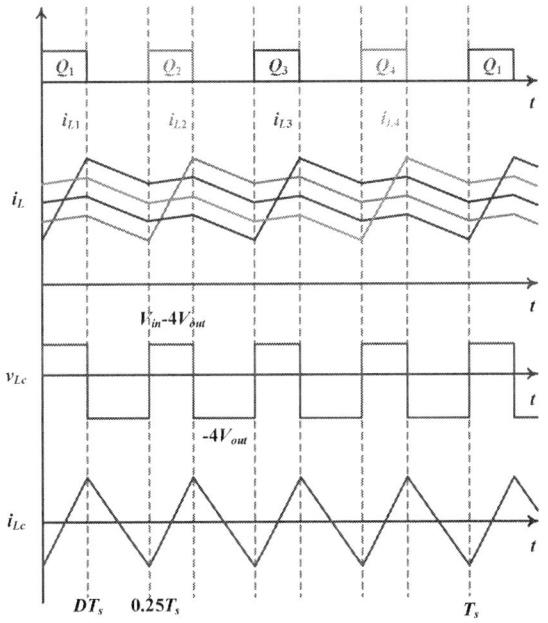

Fig. 3. Main converter waveforms for the proposed reconfigurable TLVR in heavy-load mode.

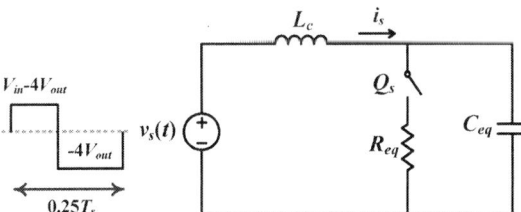

Fig. 4. Equivalent circuit in the secondary loop.

To ensure this resonant current does not affect the primary current, we can select MOSFETs such that the output capacitance satisfies $I_{s-light} < 0.01A$, which is much smaller than the primary current. In this case, we can treat the resonant current as negligible, considering the secondary loop as an open circuit.

III. LIGHT LOAD EFFICIENCY IMPROVEMENT METHOD

A. Enhanced Phase Shedding

In multi-phase Buck converters, phase shedding is commonly used to improve light load efficiency by reducing switching loss, core loss, and driving loss through a reduction in the number of active phases [20] [21]. However, in conventional TLVR, the efficiency gains from phase shedding are limited due to phase coupling. When one phase is ON (with its upper MOSFET turned on) and the redundant phases are OFF (with their upper MOSFETs turned off), a coupled current still flows in the OFF phases by the conduction of secondary loop, as shown in Fig. 5. This current introduces additional losses. If the synchronous rectifier (SR) channel is off, the coupled current flows through the SR body diode, causing

turned off, the output capacitors of the MOSFETs (represented as C_{eq} in Fig. 4) remain in the secondary loop, potentially causing a resonant current. The impedance in circuit can be represented as $j\omega L_c + 1/j\omega C_{eq}$. The effective value of the resonant current can be calculated as

$$I_{s-light} = \frac{V_{Lc}}{|\omega L_c - \frac{1}{\omega C_{eq}}|} \qquad (2)$$

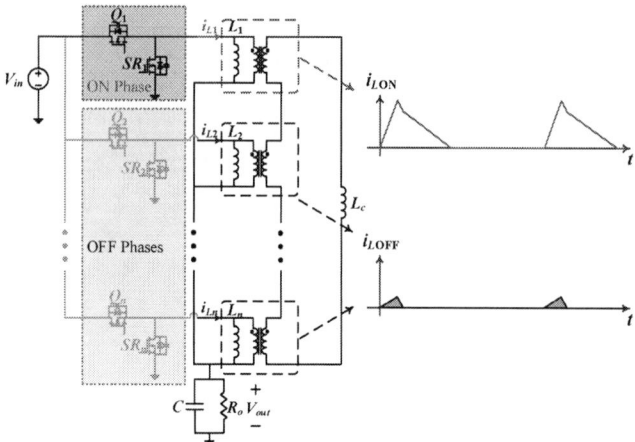

Fig. 5. Inductor current waveforms of TLVR with phase shedding.

extra conduction losses. If the SR channel is on, driving losses increase. The proposed reconfigurable TLVR addresses this issue by turning off the secondary switch, which decouples the phases and eliminates the coupling current in the OFF phases, thereby improving light load efficiency.

B. Increased Equivalent Steady-state Inductance

Decoupling the phases increases the equivalent steady-state inductance of the circuit, which helps reduce ripple-related losses and improve efficiency. The steady-state equivalent inductance of an n-phase coupled inductor is introduced in [22] and shown in (3).

$$L_{ss} = \frac{(L_{self} - M)(L_{self} + (n-1)M)}{L_{self} + (\frac{n-1}{1-D} - 1)M} \tag{3}$$

where L_{self} is the self-inductance, and M is the mutual inductance coefficient, which can be calculated as:

$$L_{self} = (1 - \frac{k^2}{n + \beta})L_{nc} \tag{4}$$

$$M = -\frac{k^2}{n + \beta}L_{nc} \tag{5}$$

Here L_{nc} represents the inductance of each phase when the coupled winding loop is open, $L_{nc} = L_m + L_k$, with L_m is the magnetic inductance, L_k is the leakage inductance. k is the coupling coefficient, $k = L_m/L_{nc}$, and $\beta = L_c/L_{nc}$. According to (3)–(5), it is evident that the mutual inductance coefficient M directly influences the steady-state inductance L_{ss}. When phases are coupled, ($M \neq 0$), L_{self} can be calculated using (4). When phases are decoupled ($M = 0$), L_{self} equals L_{nc}, which is larger than the L_{self} calculated from (4). This results in an increased L_{ss}, and a larger L_{ss} leads to reduced current ripple, thereby lowering conduction losses.

IV. POWER LOSS ANALYSIS AT LIGHT LOAD

A. Core Loss

In this paper, we decouple the secondary loop and simultaneously enable phase shedding at light load. As a result, the secondary-side current approaches zero, and only the magnetic loss of the ON phase and the primary copper loss need to be considered. The magnetic flux variation can be derived as

$$\Delta B = \frac{L \Delta i}{2n A_e} \tag{6}$$

where n is the winding turns and A_e is the effective cross-sectional area of the magnetic core. The transformer core loss can be expressed as

$$P_{core} = V_c k f_s^x \Delta B^y \tag{7}$$

where V_c is the core volume, k, x, y are coefficients determined by the core material, which can be derived in dataset. The total inductor loss can is the sum of core loss and winding loss, expressed as

$$P_{inductor} = P_{core} + I_{rms}^2 R_{winding} \tag{8}$$

where I_{rms} is the effective value of load current, and $R_{winding}$ is resistance of the primary winding.

B. ON-phase MOSFET Losses

The ON-phase MOSFET losses constitute the primary portion of light load losses. These include switching losses (P_{sw}), driving losses (P_{dri}), and the conduction losses (P_{cond}) in the upper and lower MOSFETs. The switching losses are given by:

$$P_{sw} = \frac{1}{2}V_{ds}I_{ds}(t_{on} + t_{off})f_s \tag{9}$$

where t_{on} and t_{off} are the turn-on and turn-off times of the MOSFETs, which can be approximated by the following equations [2]:

$$t_{on} = R_G C_{iss} \ln \frac{V_{gs} - V_{th}}{V_{gs} - V_{gp}} + R_G C_{gd} \frac{V_{ds}}{V_{gs} - V_{gp}} \tag{10}$$

$$t_{off} = R_G C_{iss} \ln \frac{V_{gp}}{V_{th}} + R_G C_{gd} \frac{\Delta Q_{ds}}{\Delta V_{ds}} \frac{V_{ds}}{V_{gp}} \tag{11}$$

The driving losses are expressed as:

$$P_{dri} = Q_g V_{gs} f_s \tag{12}$$

Finally, the conduction losses can be calculated as:

$$P_{cond} = \int_0^{T_s} I_{ds-on}^2 dt \times R_{ds-on} f_s \tag{13}$$

where R_G is the gate resistance, C_{iss} is the input capacitance, V_{gs} is the gate-source voltage, V_{gp} is the gate plateau voltage, V_{th} is the gate threshold voltage, C_{gd} is the gate-drain capacitance, and Q_g is the gate charge. ΔQ_{ds} and ΔV_{ds} represent the changes in the drain-source charge and voltage, respectively.

979-8-3315-1612-3/25 $31.00 © 2025 IEEE

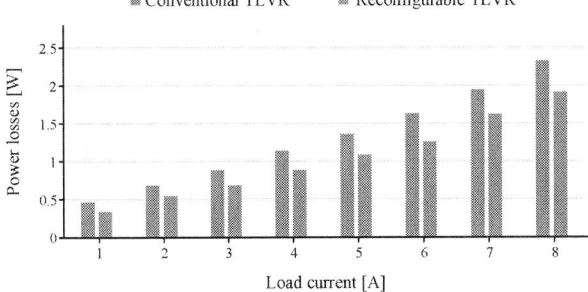

Fig. 6. Comparison of light load losses between the conventional TLVR and the reconfigurable TLVR.

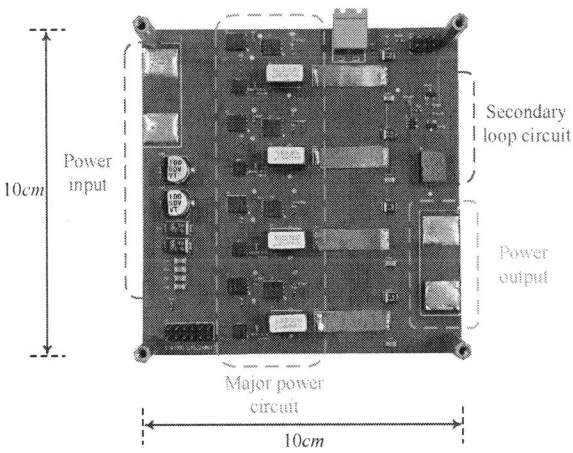

Fig. 7. Prototype of the reconfigurable TLVR.

C. OFF-phase MOSFET Losses

For the OFF phases, losses occur when no driving signal is provided, allowing the MOSFET's body diode to conduct the coupled current. The losses in this case include the conduction loss $P_{cond-BD}$ and the recovery loss P_{rr} of the body diode. The conduction loss in the body diode is given by:

$$P_{cond-BD} = V_F \cdot I_D + I_D^2 \cdot R_F \tag{14}$$

where V_F is the forward voltage drop of the body diode, I_D is the current through the body diode, and R_F is the resistance to the body diode. The recovery loss, due to the reverse recovery charge of the body diode, is:

$$P_{rr} = Q_{rr} V_{ds} f_s \tag{15}$$

where Q_{rr} is the reverse recovery charge of the body diode, and V_{ds} is the drain-source voltage.

In summary, the total losses as a function of load current for both the reconfigurable TLVR and the conventional TLVR are compared in Fig. 6. The reconfigurable TLVR reduces losses and improves light load efficiency when the secondary switch is turned off.

TABLE I
COMPONENTS AND CIRCUIT PARAMETERS

Parameter	Value
Input Voltage	12V
Output Voltage	1.8V
Load Current(full load)	80A
Phase Number	4
Switching Frequency	500kHz
Controller	TMS320F28335
Coupled Inductor	PGL6215.201HLT
Primary-side MOSFET	BSC0802LS
Primary-side Driver	2EDL8024G
Secondary-side MOSFET	DMN2053
Secondary-side Driver	1EDN7512B

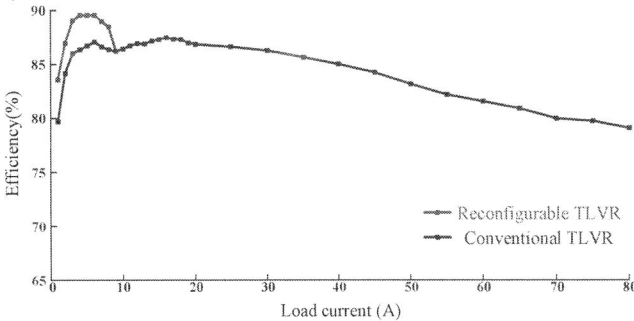

Fig. 8. Efficiency curves for reconfigurable TLVR and conventional TLVR.

V. EXPERIMENTAL RESULTS

To validate the efficiency improvement of the proposed scheme compared to the conventional TLVR, a 12V/1.8V, 500kHz, 80A 4-phase reconfigurable TLVR is designed and tested, as shown in Fig. 7. The critical parameters are listed in Table I. Both the magnetizing inductance L_m and the compensation inductance L_c are 200nH. From (3) - (5), L_{nc} is calculated as 205nH and L_{ss} is 180.7nH. Conventional TLVR operation is simulated by keeping the secondary switch turned on. The light load is defined as less than 10% of the full load, corresponding to output currents below 8A. To better observe the efficiency comparison at light load, additional test points are taken when the load current is below 20A. The efficiency curves for both configurations are plotted in Fig. 8.

As shown in Fig. 8, the reconfigurable TLVR exhibits higher efficiency than the conventional TLVR. As the load current decreases to below 8A, the efficiency of the conventional TLVR behaves as shown by the blue curve, due to phase shedding. In contrast, the reconfigurable TLVR achieves a higher efficiency, as represented by the red curve. This agrees well with the theoretical analysis. The full load efficiency is approximately 80%. The peak efficiency occurs at an output current of 4A, reaching 89.6%. The experimental waveforms for this condition are shown in Fig. 9.

Fig. 9 shows a comparison of the inductor currents for the reconfigurable TLVR and the conventional TLVR. As seen in the figure, when the converter operates at light load and phase shedding is employed, induced current flows in the OFF phases of the conventional TLVR. With the proposed method,

979-8-3315-1612-3/25 $31.00 © 2025 IEEE

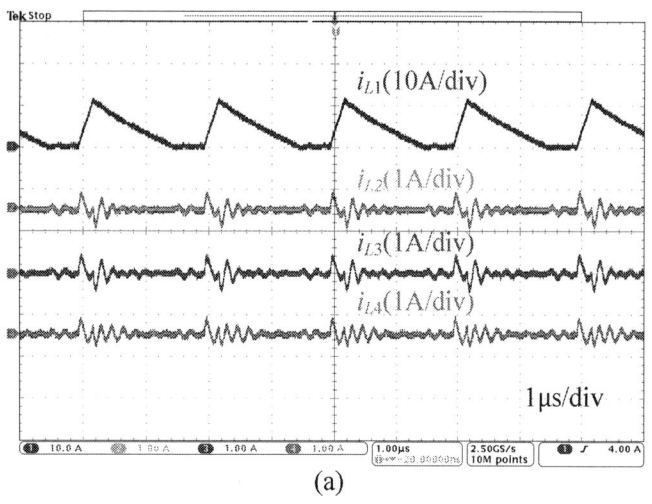

(a)

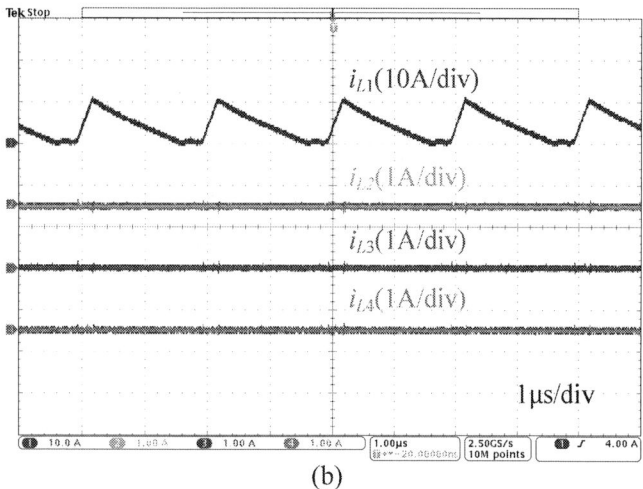

(b)

Fig. 9. Experiment waveforms at 4A output current (phase shedding):(a) Conventional TLVR. (b) Reconfigurable TLVR with secondary switch turned off.

this induced current is almost eliminated, resulting in reduced current ripple. Consequently, the light load efficiency improves from 86.4% to 89.6%. For each point between 1A and 8A of output current, the average efficiency improvement is 2.9%.

VI. CONCLUSION

This paper proposes a reconfigurable method to enhance the light load efficiency of TLVRs. By incorporating an additional secondary switch to control the coupling between phases, the proposed scheme maintains the high transient response of conventional TLVRs under heavy loads while significantly improving efficiency at light loads. Experimental results confirm that phase decoupling enhances the effectiveness of phase shedding, reduces current ripple, and achieves an average light load efficiency improvement of 2.9%. This method offers a promising approach for improving the overall performance of TLVRs in light load conditions.

REFERENCES

[1] Supermicro, "Top ten best practices for a green data center," https://www.supermicro.com/white_paper/Top_10_Best_Practices_for_a_Green_Data_Center.pdf, 2022, online; accessed October 17, 2024.

[2] J. Liang and H. Wang, "Light load efficiency boost technique for switched tank converters based on hybrid zvs-zcs control," in *Proc. Int. Power Electron. Conf. (IPEC-ECCE Asia)*. IEEE, May 2022, pp. 2231–2235.

[3] S. Jiang, X. Li, M. Yazdani, and C. Chung, "Driving 48v technology innovations forward - hybrid converters and trans-inductor voltage regulator (tlvr)," in *Proc. IEEE Appl. Power Electron. Conf. Expo. (APEC)*, New Orleans, LA, Mar. 2020.

[4] N. Zhang, C. Zhan, G. Ye, C. Chen, X. Li, and J. Yi, "Analysis of multi-phase trans-inductor voltage regulator with fast transient response for large load current applications," in *Proc. IEEE Int. Symp. Circuits Syst. (ISCAS)*, May 2021, pp. 1–5.

[5] J. Sun, M. Xu, Y. Ren, and F. C. Lee, "Light-load efficiency improvement for buck voltage regulators," *IEEE Trans. Power Electron.*, vol. 24, no. 3, pp. 742–751, Mar. 2009.

[6] J. Liang, L. Wang, M. Fu, J. Liang, and H. Wang, "Overview of voltage regulator modules in 48 v bus-based data center power systems," *CPSS Trans. Power Electron. Appl.*, vol. 7, no. 3, pp. 283–299, 2022.

[7] I. A. Mashhadi, B. Soleymani, E. Adib, and H. Farzanehfard, "A dual-switch discontinuous current-source gate driver for a narrow on-time buck converter," *IEEE Trans. Power Electron.*, vol. 33, no. 5, pp. 4215–4223, 2017.

[8] M. Biswas, S. Majhi, and H. B. Nemade, "A high step-down dc-dc converter with reduced inductor current ripple and low voltage stress," *IEEE Trans. Ind. Appl.*, vol. 57, no. 2, pp. 1559–1571, 2020.

[9] Y. Ren, M. Xu, K. Yao, and F. C. Lee, "Two-stage approach for 12-v vr," *IEEE Trans. Power Electron.*, vol. 19, no. 6, pp. 1498–1506, 2004.

[10] M. Veerachary, "Two-switch semiquadratic buck converter," *IEEE Trans. Ind. Electron.*, vol. 64, no. 2, pp. 1185–1194, 2016.

[11] L. Wang, C. Li, J. Liang, and H. Wang, "A multi-phase series capacitor trans-inductor voltage regulator with high switching frequency and fast dynamic response," in *Proc. IEEE Appl. Power Electron. Conf. Expo. (APEC)*, Orlando, FL, Mar. 2023.

[12] C. Li, L. Wang, G. Zheng, M. Fu, and H. Wang, "Small-signal modeling and loop analysis of ultrafast series capacitor trans-inductor voltage regulator with constant on-time control," *IEEE Trans. Power Electron.*, in press, DOI: 10.1109 / TPEL. 2024.3488734.

[13] C. Tu, R. Chen, and K. D. Ngo, "Steady-state analysis of series-resonator buck converter," *IEEE Trans. Power Electron.*, vol. 37, no. 10, pp. 12 327–12 335, 2022.

[14] D. A. Grant, Y. Darroman, and J. Suter, "Synthesis of tapped-inductor switched-mode converters," *IEEE Trans. Power Electron.*, vol. 22, no. 5, pp. 1964–1969, 2007.

[15] T. Yao, M. Ban, and Y. Liu, "A high conversion ratio converter based on tapped-series capacitor circuit," *IEEE Trans. Power Electron.*, 2024.

[16] B. Soleymani, O. Bagheri, E. Adib, and S. Eren, "A zvs high step-down converter with reduced component count and low ripple output current," *IEEE Open J. Power Electron.*, 2024.

[17] Y. Dong, Y. Yang, F. C. Lee, and M. Xu, "The short winding path coupled inductor voltage regulators," in *Proc. IEEE Appl. Power Electron. Conf. Expo. (APEC)*, Austin, TX, USA, Feb. 2008, pp. 1446–1452.

[18] J. Kaiser and T. Duerbaum, "An overview of saturable inductors: Applications to power supplies," *IEEE Trans. Power Electron.*, vol. 36, no. 9, pp. 10 766–10 775, 2021.

[19] F. Li, L. Wang, B. Wu, L. Yu, and K. Wang, "A novel planar nonlinear coupled inductor for improving light and intermediate load efficiency of dc/dc converters," *IEEE J. Emerg. Sel. Topics Power Electron.*, vol. 11, no. 2, pp. 2004–2014, Apr. 2023.

[20] M. A. Alharbi, A. M. Alcaide, M. Dahidah, P. Montero-Robina, S. Ethni, V. Pickert, and J. I. Leon, "Rotating phase shedding for interleaved dc–dc converter-based evs fast dc chargers," *IEEE Trans. Power Electron.*, vol. 38, no. 2, pp. 1901–1909, 2022.

[21] J.-T. Su and C.-W. Liu, "A novel phase-shedding control scheme for improved light load efficiency of multiphase interleaved dc–dc converters," *IEEE Trans. Power Electron.*, vol. 28, no. 10, pp. 4742–4752, Oct. 2013.

[22] F. Zhu and Q. Li, "Coupled inductors with an adaptive coupling coefficient for multiphase voltage regulators," *IEEE Trans. Power Electron.*, vol. 38, no. 1, pp. 739–749, Jan. 2023.

Fully Integrated Voltage Regulators (FIVRs) with Package In-situ Coupled CoaxMIL Inductor for High Power Density Microprocessor Applications

[†‡]Jaeil Baek, [‡]Beomseok Choi, [‡]Siddharth Kulasekaran, [‡]Huong Do, [‡]Brandon Marin, [‡]Jose Chavarria, [‡]Leigh Wojewoda, [‡]Kaladhar Radhakrishnan

[†] *Korea Advanced Institute of Science & Technology (KAIST)*, [‡]*Intel Corporation*
Email: jaeil.baek@kaist.ac.kr, {beomseok.choi, kaladhar.radhakrishnan}@intel.com

Abstract—CoaxMIL inductors are the latest package in-situ inductors with a high ratio of inductance to resistance for fully integrated voltage regulators (FIVRs). However, it is challenging to achieve both high efficiency and high density with CoaxMIL due to the fundamental performance trade-off of uncoupled inductors. This paper introduces a coupled package integrated coaxial magnetic inductor (CoaxMIL-C), which relieves the trade-off of CoaxMIL. By adding phase coupling to CoaxMIL structures, the proposed CoaxMIL-C reduces the inductor size by 33% and improves the transient performance while maintaining the efficiency of conventional CoaxMIL inductors. Performance is verified with lab measurements and validated with a FIVR test bench with external inductor pads. CoaxMIL-C provides 24 A/mm^2 maximum density and enables the FIVR to achieve 89% peak efficiency at 1.75 V to 0.75 V conversion.

Index Terms—coupled inductors, fully integrated voltage, microprocessors, package in-situ inductors, voltage regulators

I. INTRODUCTION

The power consumption of data centers is growing exponentially to meet the insatiable demand for AI-based workloads. System architects optimize the design at the rack scale and maximize the amount of computing within a rack. This requires power delivery solutions to provide a large amount of power in a relatively small area. Today's data center GPUs primarily rely on two-stage lateral power delivery solutions with voltage regulators placed along the periphery of the GPU package [1], [2] (Fig. 1(a)). This has the benefit of keeping the overall height of the system low. However, its long lateral current path causes higher I^2R losses with increasing current demand. One workaround is a shift to vertical power delivery [3]–[5] (Fig. 1(b)). In this structure, the second-stage VRs are moved to the backside of the motherboard(MB), which greatly reduces the current path and the form factor of the motherboard. However, as the current demand increases, there may not be enough room on the backside of the MB to fit both the VRs and the output capacitors. Furthermore, the increased overall z-height of the system can limit the number of GPUs accommodated in a rack.

Integrated Voltage Regulator (IVR) based design is one of the most promising candidates for high-power microprocessor applications. Fig. 1(c) shows a typical IVR-based design where the first stage is modified to convert input voltage into 1.8

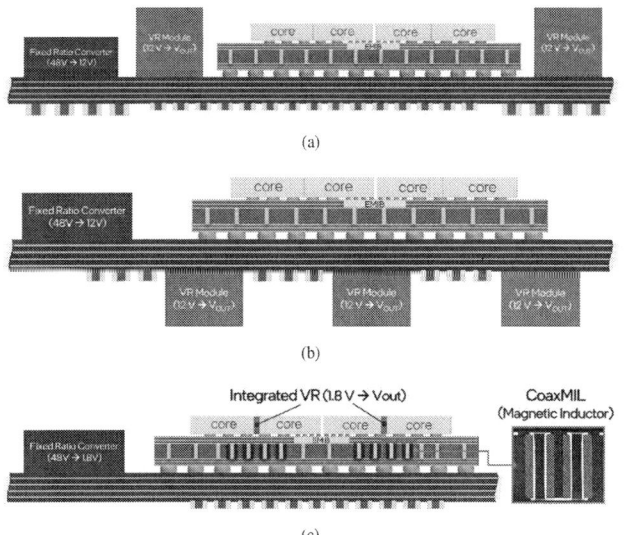

Fig. 1: Power delivery architectures for high-performance computing devices. (a) Lateral power delivery architecture with fixed ratio converter and voltage regulators. Power is laterally delivered to the microprocessor on the motherboard. (b) Vertical power delivery architecture with voltage regulators on the backside of the MB. Z-height becomes a constraint. (c) IVR-based vertical power delivery architecture. It achieves a low Z-height and reduced current from the MB to the package. Package in-situ CoaxMIL inductors can be used.

V [6]–[9], and the second-stage IVR is integrated into the package and die [10]–[12]. The reduction of I^2R losses in the MB and improvement of 48 V to 1.8 V converters [6]–[9] make this solution more attractive. In addition, such a conversion scheme achieves a low Z-height like the lateral solution. While this simplifies the power delivery design requirements on the board, the IVRs have to manage a lot of complexities, such as large currents, fast transient, and voltage regulation within relatively small areas. The inductor has played a critical role in the performance of any buck-based IVRs [12]–[28]. It has a significant impact on the efficiency and transient performance of IVRs. Furthermore, the overall power density of the IVR is largely determined by the inductor. In this paper, a new coupled coaxial inductor is proposed, which can be implemented in the core of the package substrate and provide the best-in-class current density while enabling good efficiency and transient response characteristics.

979-8-3315-1612-3/25 $31.00 © 2025 IEEE

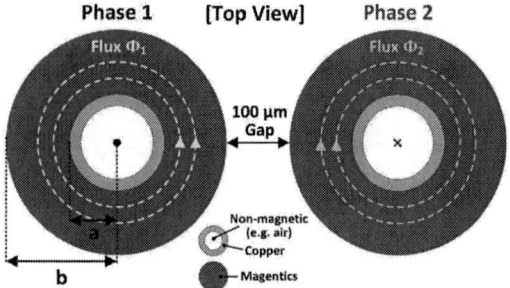

Fig. 2: Two uncoupled CoaxMIL inductors. A 100 μm non-magnetic gap (spacing) between the two magnetics makes the flux of each inductor decoupled and two inductors independent.

Fig. 3: Two-phase coupled CoaxMIL inductor. Two magnetics are merged, which removes the 100 μm gap and creates a new design parameter d, which relates to the coupling coefficient.

II. PACKAGE IN-SITU COUPLED COAXMIL INDUCTORS

Various package in-situ inductors have been proposed for IVRs. There are two types of inductors: 1) Air Core Inductors (ACI) and 2) Magnetic inductors. ACIs have been widely used due to their good manufacturability, requiring only package copper layout design to implement the inductor [12]–[14]. However, ACIs are difficult to scale to a small form factor since one has to choose between lower inductance or higher resistance within a fixed area, which results in lower efficiency of IVRs. On the other hand, the use of magnetics improves the ratio of inductance to resistance value, resulting in efficient IVRs. Thin film magnetic inductors utilizing multiple laminated layers of ferromagnetic alloys such as CoZrTa (CZT) and composite magnetic materials have been studied [17]–[20]. However, scaling thin film inductors has been challenging due to limitations in their manufacturing complexity and cost. Also, PCB and package integrated inductors are actively studied [22]–[28]. However in a package substrate, there are more challenges added due to smaller form-factor demand and isolation from high-speed IO. To address these challenges, substrate-integrated magnetic core inductors, named CoaxMIL inductors, were proposed [22], [23].

A. Uncoupled CoaxMIL Inductors

The CoaxMIL inductor consists of a copper-plated thru-hole (PTH) and a magnetic material encapsulating it, as shown in Fig. 2 [22], [23]. The current flowing through an inner conductor (copper) generates circular loops of magnetic flux. The magnetic flux is stronger when it is closer to the inner conductor and decreases when it is closer to the outer edge of the magnetics. The inductance of a coaxial structure, shown in Fig. 2, can be calculated using a formula derived from the logarithmic form of Bessel functions:

$$L = h \cdot \ln(\frac{b}{a}) \cdot \mu_r \cdot \frac{\mu_0}{2\pi} \qquad (1)$$

where h is the height of the coaxial structure, a is the radius of inner copper, b is the radius of the magnetic material, μ_r is the relative permeability of magnetic material, and μ_0 is the permeability of free space. Since parameters a and h are determined by the design rules of products, b and μ_r are usually allowable design parameters of CoaxMIL inductors. Due to the clearance design rule between magnetics, there must

be a minimum space (e.g., 100 μm) between two CoaxMILs. Therefore, multiple CoaxMIL inductors designed adjacent to each other are seen as uncoupled structures because the non-magnetic spacing between CoaxMIL inductors decouples the fluxes of the inductors. Hence, each inductor can be seen as independent and has only one inductance value for steady-state and transient operations. This results in a performance trade-off between the efficiency, transient, and power density of FIVRs.

B. Coupled CoaxMIL Inductors (CoaxMIL-C)

Fig. 3 shows the proposed coupled CoaxMIL (CoaxMIL-C) inductor. Unlike the uncoupled CoaxMIL inductor shown in Fig. 2, the proposed CoaxMIL-C inductor fills the 100 μm gap between two CoaxMIL inductors with magnetic material symmetrically. This magnetic-filled gap allows the flux of each inductor to be coupled. In the proposed CoaxMIL-C inductor, a magnetic material encapsulates two PTHs as coupled windings. The two PTHs have opposite current directions to implement a negative coupling between two inductors. By doing this, the proposed CoaxMIL-C has one more critical design parameter (d) on top of the existing design parameters of uncoupled CoaxMIL inductors [23], which increases the design freedom of inductors and controls the coupling coefficient of the proposed CoaxMIL-C. To understand the impact of the design parameter d, the EM simulation (ANSYS EDT) was conducted by varying d under the fixed h (1 mm), a (75 μm), and b (200 μm) conditions. Fig. 4 shows the impact of d on the footprint, coupling coefficient, and effective inductances of a two-phase CoaxMIL-C inductor. For the comparison with uncoupled CoaxMIL inductors, simulation results of the CoaxMIL 2.0 inductor designed in [23], which has the same a and h but has 225 μm of larger b and 550 μm of d, are also included. From Figs. 4(a) and (b), as d decreases, the size of CoaxMIL-C decreases while increasing the coupling coefficient. Since the uncoupled CoaxMIL 2.0 inductor has a larger b than CoaxMIL-C, it has a larger footprint than CoaxMIL-C even at the same d condition. From Fig. 4(c), due to stronger coupling, as d decreases, per-phase steady-state inductance L_{pss} increases, reducing the inductor ripple current, and per-phase transient inductance L_{ptr} decreases, so improve both efficiency and transient performance of FIVRs [29]. Since

979-8-3315-1612-3/25 $31.00 © 2025 IEEE

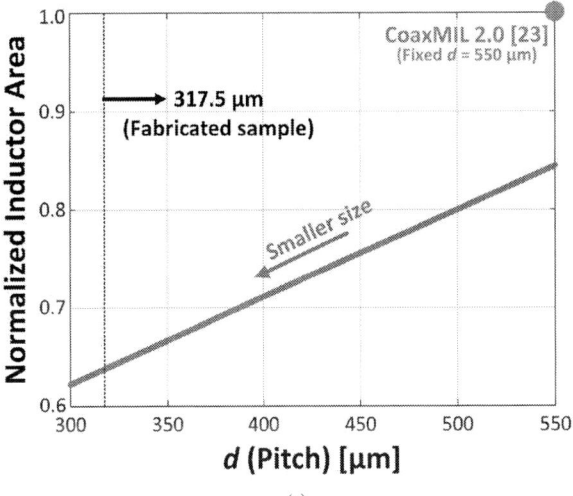

(a)

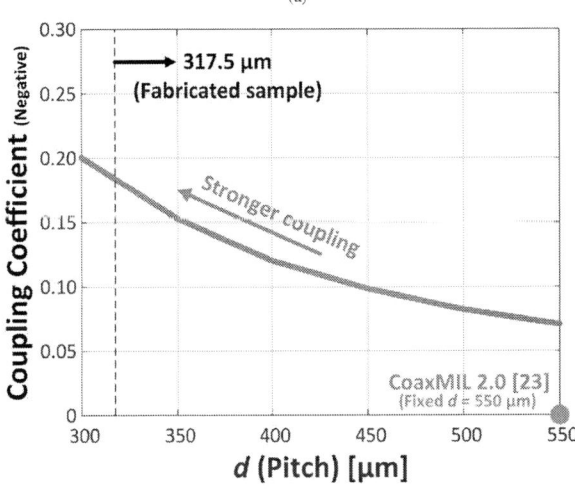

(b)

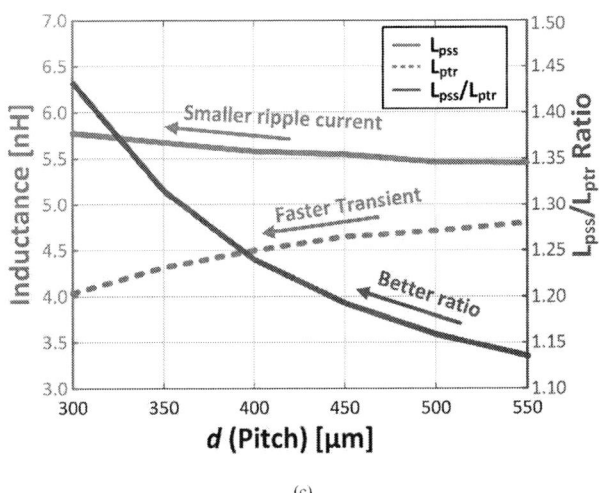

(c)

Fig. 4: Finite-element simulation results of a two-phase CoaxMIL-C with varied d and fixed h, a, and b. (a) Impact on XY footprint. (b) Impact on Coupling coefficient. (c) Impact on L_{pss} and L_{ptr}. Due to manufacturing constraints, the prototype was fabricated with 317.5 μm of d.

the CoaxMI 2.0 is uncoupled and has fixed one inductance of 4.8 nH, its L_{pss} to L_{ptr} ratio is 1.

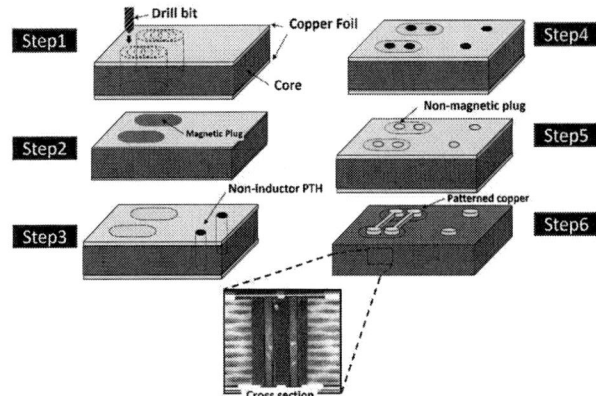

Fig. 5: Detailed steps describing the coupled CoaxMIL process flow - Step 1: Create the cavity for magnetic plug using multiple mechanical through-hole (MTH) shots. Step 2: Plug cavities with a magnetic plug, then grind back the excess plug. Step 3: Plate copper to cover and protect magnetic plug from harsh subsequent steps, then drill plated through-holes (PTHs) that are used for non-inductor functions. Step 4: Clean the non-inductor PTHs then drill the coaxial inner PTH and use high-pressure water cleaning. Step 5: PTH wall copper plating for both indcutor and non-inductor PTHs, then non-magnetic plug. Step 6: Final copper plating and subtractive etching to form surface copper pattern.

III. PROTOTYPE AND MEASURED RESULTS OF COAXMIL-C

Coupled CoaxMIL prototypes were fabricated in an organic substrate core comprised of re-inforced glass cloth in epoxy resin with 317.5 μm PTH pitch (d). Even though smaller d is desired for stronger coupling and smaller size, manufacturing and other design rule constraints limit scaling. The manufacturing process is almost identical to the prior CoaxMILs [22], [23], which does not disrupt the typical package manufacturing flow. Details and additional fabrication processes for CoaxMIL-C are discussed in the following part. The fabricated prototype comprises a two-phase negatively coupled configuration with four terminals (Fig. 6). In this configuration, one single-phase inductor consists of two PTH passes, which is required in the fully integrated voltage regulator configuration [12]. The diameter of PTH is 150 μm with 25 μm thick copper plating. All other design parameters are the same as the conditions used in Section II-B.

A. Fabrication Process

Coupled CoaxMIL inductors were fabricated in a very similar way to previous work [22] [23]. The main difference between previous works is the formation of a larger cavity to house the magnetic plug material, as shown in Step 1 of Fig. 5. Subsequent steps are identical to existing CoaxMIL inductors. The formation of plated through-holes (PTHs) for non-inductor is done separately to ensure that the magnetic material is not damaged by harsh chemical etching steps.

B. Magnetic Plug Material

The key functional component of the coaxial inductor is an epoxy composite with magnetic fillers that serve as the magnetic core; we call this a magnetic plug or magnetic

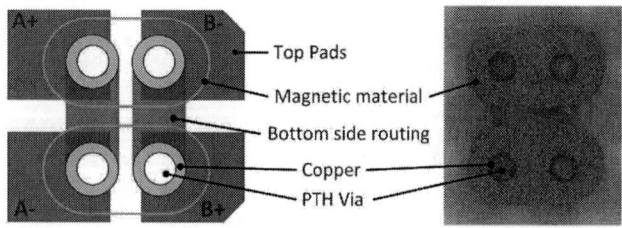

Fig. 6: Layout view of the Coupled CoaxMIL structure measured(left) and the corresponding cross-section picture(right). (A+ & A-) and (B+ & B-) form a pair of inductors. Two CoaxMIL inductors are negatively coupled. The XY footprint of the prototype is 0.41 mm^2.

material. These epoxy composites offer multiple benefits, such as high mechanical strength, low cure shrinkage, stability with thermal cycling, and good chemical resistance. Additionally, epoxy-based materials are preferred for their known compatibility with existing substrate material sets and processes. The magnetic plug used in this work is identical to the material used in the CoaxMIL 2.0 inductor [23].

C. Passive Characteristics

Passive characteristics of CoaxMIL-C, such as inductance, resistance, and saturation current, are measured using a Vector Network Analyzer (VNA) and digital multimeters. The measurement method used is identical to what was used for the substrate-based ACI [13] and CoaxMIL [22], [23]. Fig. 7(a) shows the measured inductances according to frequency. L_{11} and L_{22} are self-inductances, and L_{12} is the mutual inductance. The polarity of L_{12} is inverted in the plot for better readability. An insignificant 4% (0.20 nH) difference between L_{11} and L_{22} at 50 MHz is observed due to manufacturing tolerance - PTH position, copper thickness, uneven magnetic material pasting, magnetic material position, etc. CoaxMIL-C shows similar self-inductance to CoaxMIL 2.0 even though it has a smaller magnetic radius (200 μm) than CoaxMIL 2.0 (225 μm), thanks to the magnetic-filled gap. Due to the negative coupling, L_{12} is -0.84 nH at 50MHZ, which results in (-)0.18 coupling coefficient, 3.98 nH of L_{trans}, and 5.38 nH of L_{pss} at 1.75 V input voltage and 0.75 V output voltage conditions [29]. For reference, CoaxMIL 2.0 has 4.6 nH self-inductance. Fig. 7(b) shows the measured resistance. CoaxMIL-C has similar resistances at 50 MHz but shows slightly higher DC resistance compared to CoaxMIL 2.0 due to design and manufacturing variances of the samples measured. Fig. 7(c) shows the inductance with different DC bias currents. It shows that CoaxMIL-C has a soft-saturation characteristic, and the inductance drops by 30% at 10 A bias current condition, where the current density of CoaxMIL-C is achieved as 24.4 A/mm^2.

D. Active Characteristics

FIVR test bench and package were designed and fabricated to compare the performance of CoaxMIL-C with previous CoaxMIL inductors, as shown in Fig. 8. Half-bridge switches and metal-insulated-metal (MIM) capacitors of the FIVR are integrated into die and placed under a heat spreader. In

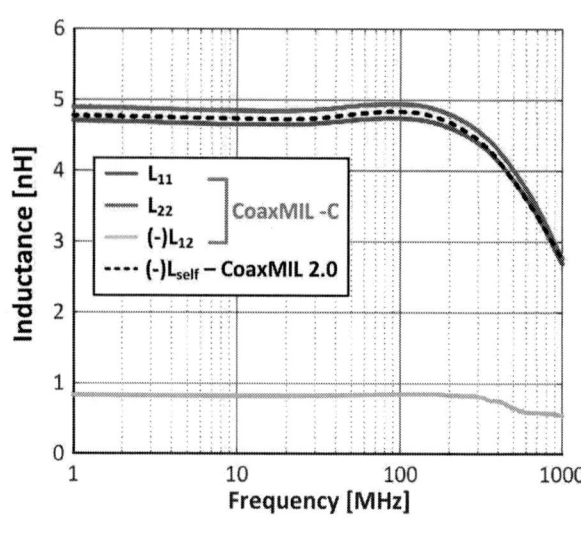

(a)

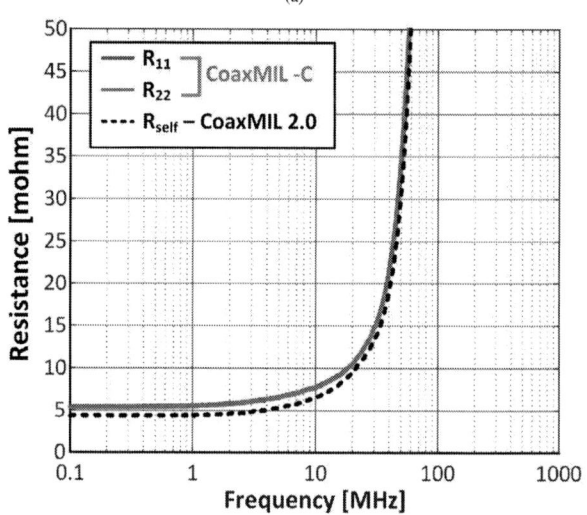

(b)

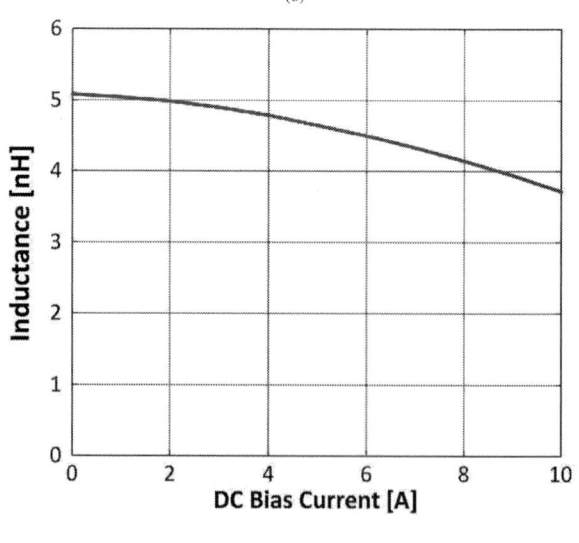

(c)

Fig. 7: Measured passive characterization of CoaxMIL-C. (a) Inductance vs frequency. (b) Resistance vs frequency. (c) Current saturation. The sign of cross-terms L_{12} is inverted for better readability.

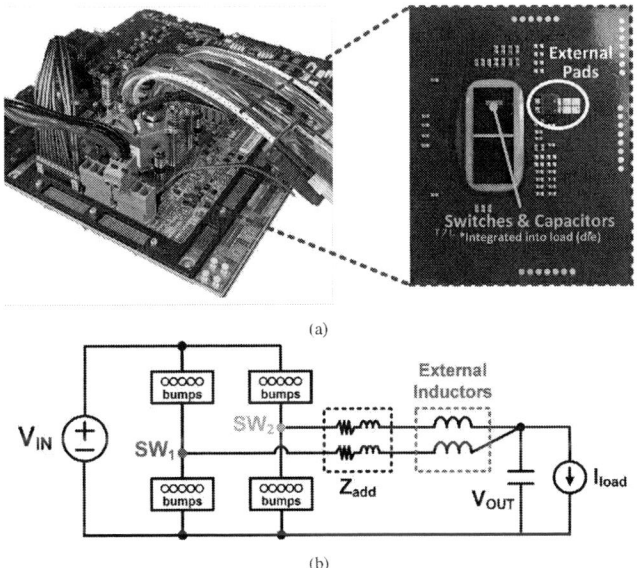

(a)

(b)

Fig. 8: Photo and simplified circuit diagram of the test bench. (a) FIVR testbench(left) and the FIVR test package (right) with external pads to evaluate inductors. Switches and capacitors of the FIVR are integrated into a load (die). On top of the die, a heat sink is placed to control the temperature of the die. (b) Simplified circuit diagram. Due to the external pads, the test bench includes additional parasitic impedance.

the FIVR package, there are external pads where different prototypes of inductors can be placed for fair comparison. The external pad introduces additional parasitic inductance and resistance. The additional impedance, Z_{add}, can be extracted through 3D package model extraction and lab measurements. The simulation and measurement results show around 0.25 nH parasitic inductance and 22 mΩ AC resistance at 50MHz and 4.5 mΩ parasitic DC resistance. In real products, since CoaxMIL inductors are placed in the package core, real products have smaller resistances than the current prototype setup. The test board includes precision control of variables such as output current, input voltage, output voltage, and temperature to compare inductor performance accurately in different conditions. This active measurement setup allows for non-linear characteristics of the inductor to be evaluated as in a complete VR system.

The efficiency of the 2-phase FIVR with CoaxMIL-C is evaluated for multiple output voltages with a fixed 1.75 V input voltage and 50 MHz switching frequency. The results are shown in Fig. 9(a). All auxiliary losses, such as gate driver loss and control circuitry loss, are included in the efficiency. FIVR with CoaxMIl-C achieves over 90% peak efficiency at output voltages above 0.9 V. Although the CoaxMIL-C is much smaller than CoaxMIL 2.0 [23], it enables comparable efficiency with CoaxMIL 2.0 due to its higher steady-state inductance (L_{pss}). Fig. 9(b) and (c) show results for open-loop and closed-loop transient performance under a 5 A load step change for both CoaxMIL-C and CoaxMIL 2.0 inductors. The result of the open-loop load transient confirms that the L_{trans} of CoaxMIL-C is smaller than CoaxMIL 2.0 because it can be clearly seen that CoaxMIL-C shows faster recovery over CoaxMIL 2.0. After optimizing the closed-loop controller for each inductor, CoaxMIL-C improves the voltage droop

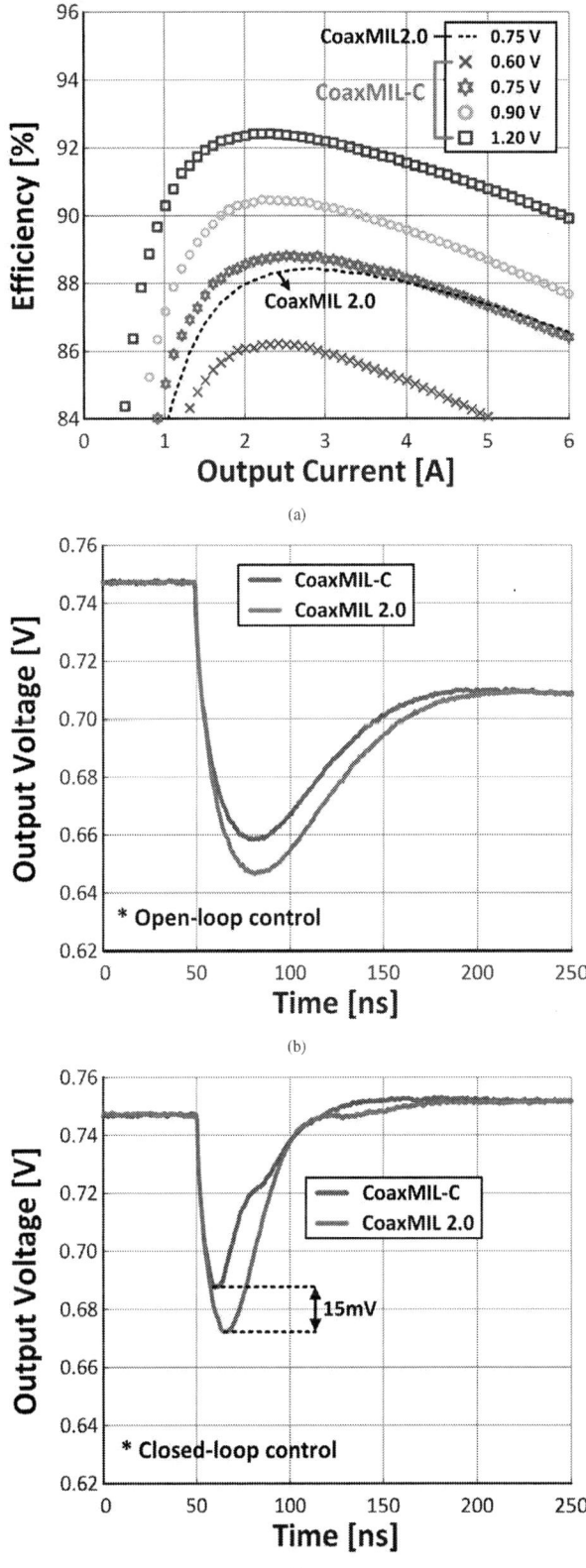

(a)

(b)

(c)

Fig. 9: Measured active characteristics. (a) Efficiency at 1.75 V input and different output voltage. (b) Transient waveforms with a 5 A load current step without closed-loop control. (c) Transient with a 5 A load current step with closed-loop control.

979-8-3315-1612-3/25 $31.00 © 2025 IEEE

TABLE I: Performance Comparison of CoaxMIL-C Against Inductor Technologies for IVRs.

Year	Note	Design	Type	L_{self}	L_{trans}	Q_{peak}	I_{max} [†]	Footprint[*]	Area Density
This Work	CoaxMIL-C	In Package	Magnetics	4.8 nH	3.9 nH	50 @ 50MHz	10 A	0.41 mm^2	24.4 A/mm^2
2016	ACI [13]	In Package	Air Core	1.2 nH	0.9 nH	24 @ 140MHz	8 A	1.68 mm^2	4.7 A/mm^2
2020	MCI [17]	On Package	Magnetics	5.2 nH	1.7 nH	8 @ 80MHz	0.9 A[‡]	0.35 mm^2	2.5 A/mm^2
2021	CoaxMIL 1.0 [22]	In Package	Magnetics	2.3 nH	2.3 nH	65 @ 75MHz	8 A	0.5 mm^2	16 A/mm^2
2024	CoaxMIL 2.0 [23]	In Package	Magnetics	4.6 nH	4.6 nH	60 @ 30MHz	10 A	0.61 mm^2	16.4 A/mm^2

† Lower value between 1) Maximum continuous copper current under 120 °C for 5 years and 2) Bias current for a 30% inductance drop.
‡ Saturation limited current (90% inductance drop).
* Footprint includes space for clearance between inductors and other adjacent coppers.

performance by 15mV thanks to lower L_{trans} of 3.9 nH of CoaxMIL-C over $L_{self} = L_{trans}$ of 4.6 nH of CoaxMIL 2.0.

E. Performance Comparison

Table I compares key characteristics of CoaxMIL-C with other state-of-the-art integrated inductors for IVRs. Magnetic-based inductors have higher inductance than the air-core inductor. Even though the MCI inductor achieves the highest inductance, it has very low I_{max} due to the low current capability of high permeability magnetic material. CoaxMIL-C achieves extremely high current density (A/mm^2) compared to existing inductor technologies due to its small footprint.

IV. CONCLUSIONS

This paper presents a new package in-situ inductor technology, CoaxMIL-C, for high current density fully integrated voltage regulators (FIVRs). By adding coupling to the CoaxMIL inductor structure, CoaxMIL-C reduces the form factor significantly compared to CoaxMIL2.0 while maintaining FIVR efficiency and improving voltage drop mitigation performance. CoaxMIL-C is fabricated in a very similar way to the existing CoaxMIL inductors. CoaxMIL-C achieves a current denstiy of 24.4 A/mm^2 and enables the FIVR to obtain over 90% peak efficiency at 1.75 V to over 0.9 V conversions while providing good transient performance.

REFERENCES

[1] X. Lyu et al., "Composite Modular Power Delivery Architecture for Next-Gen 48V Data Center Applications," *2018 1st Workshop on Wide Bandgap Power Devices and Applications in Asia (WiPDA Asia)*, Xi'an, China, 2018, pp. 343-350.

[2] M. H. Ahmed et al., "48-V Voltage Regulator Module With PCB Winding Matrix Transformer for Future Data Centers," *IEEE Transactions on Industrial Electronics*, vol. 64, no. 12, pp. 9302-9310, Dec. 2017.

[3] H. -P. Le et al., "Vertical Power Delivery and Heterogeneous Integration for High-Performance Computing," *2023 IEEE BiCMOS and Compound Semiconductor Integrated Circuits and Technology Symposium (BCICTS)*, Monterey, CA, USA, 2023, pp. 32-35.

[4] M. Ursino, S. Saggini, S. Jiang and C. Nan, "Vertical Power Flow 48 V to PoL VRM Architectures With Hybrid Pre-Regulator," *IEEE Journal of Emerging and Selected Topics in Power Electronics*, vol. 11, no. 5, pp. 4907-4917, Oct. 2023.

[5] H. Gan et al., "Vertical Power Delivery for 1000 Amps Machine Learning ASICs," *2024 IEEE Applied Power Electronics Conference and Exposition (APEC)*, Long Beach, CA, USA, 2024, pp. 906-909.

[6] S. Jiang et al., "Driving 48V technology innovations forward-hybrid converters and trans-inductor voltage regulator (TLVR)", Industry Session of *2020 IEEE Applied Power Electronics Conference and Exposition (APEC)*, Mar. 2020.

[7] J. Baek et al., "Vertical Stacked LEGO-PoL CPU Voltage Regulator," *IEEE Trans. on Power Electron.*, vol.37, no.6, pp. 6305-6322, Jun. 2022.

[8] T, Ge, R. Abramson, Z. Ye and R. C. N. Pilawa-Podgurski, "Core Size Scaling Law of Two-Phase Coupled Inductors - Demonstration in a 48-to-1.8 V Hybrid Switched-Capacitor MLB-PoL Converter," *IEEE Applied Power Electronics Conference and Exposition (APEC)*, Houston, TX, 2022, pp. 1500-1505.

[9] P. R. Prakash, A. Nabih, Y. Liang, S. Kudva, M. Mosa, C. T. Gray and Q. Li, "A 2400 W/in^3 1.8 V Bus Converter Enabling Vertical Power Delivery for Next-Generation Processors," *IEEE Applied Power Electronics Conference and Exposition (APEC)*, Long Beach, CA, 2024, pp. 910-917.

[10] Ferric, Fe1736 Preliminary Product Brief [Online]. Available: https://www.ferric.com/fe1736brief.

[11] Infineon, TDM22544D & TDM22545D daul-phase power modules [Online]. Available: https://www.infineon.com/cms/en/product/promopages/power-modules/.

[12] E. A. Burton, G. Schrom, F. Paillet, J. Douglas, W. J. Lambert, K. Radhakrishnan and M. J. Hill, "FIVR – Fully Integrated Voltage Regulators on 4th Generation Intel(R) Core^{TM} SoCs," *IEEE Applied Power Electronics Conference and Exposition (APEC)*, Fort Worth, TX, 2014, pp. 432-439.

[13] W. J. Lambert, M. J. Hill, K. Radhakrishnan, L. Wojewoda and A. E. Augustine, "Package Inductors for Intel Fully Integrated Voltage Regulators," *IEEE Transactions on Component, Packaging and Manufacturing Technology*, vol. 6, no. 1, pp. 3-11, January 2016.

[14] H. Lin et al., "System Optimization: High-Frequency Buck Converter With 3-D In-Package Air-Core Inductor," *IEEE Transactions on Component, Packaging and Manufacturing Technology*, vol. 12, no. 3, pp. 401-409, March. 2022.

[15] M. Sankarasubramanian et al., "Magnetic Inductor Arrays for Intel(R) Fully Integrated Voltage Regulator (FIVR) on 10th generation Intel(R) Core^{TM} SoCs," *IEEE Electronics Components and Technology Conference (ECTC)*, Orlando, FL, 2020, pp. 399-404.

[16] K. Tang, Y. Ji, and L. Cheng, "A 6-Phase Integrated Voltage Regulator With Multi-Phase Transient Optimization Technique in 28-nm CMOS Process," *IEEE Transactions on Circuits and Systems II: Express Briefs*, vol. 71, no. 10, pp. 4596-4600, Oct. 2024.

[17] W. J. Lambert, M. J. Hill, K. P. O'Brien, K. Radhakrishnan and P. Fischer, "Study of Thin-Film Magnetic Inductors Applied to Integrated Voltage Regulators," *IEEE Transactions on Power Electronics*, vol. 35, no. 6, pp. 6208-6220, June 2020.

[18] Y. He, Z. Zhong, H. Zhang and F. Bai, "On-Chip Coupled Solenoid Inductors for Integrated Power Conversion," *IEEE Transactions on Electron Devices*, vol. 68, no. 12, pp. 6292-6295, December 2021.

[19] N. Wang et al., "A Novel Thin Film Cascade Matrix Coupled Inductor for Integrated Voltage Regulators," *IEEE Transactions on Industrial Electronics*, vol. 36, no. 12, pp. 13349-13354, Dec. 2021.

[20] N. Sturcken et al., "A 2.5D Integrated Voltage Regulator Using Coupled-Magnetic-Core Inductors on Silicon Interposer," *IEEE Journal of Solid-State Circuits*, vol. 48, no. 1, pp. 244-254, Jan. 2013.

[21] T. Sun et al., "Substrate-Embedded Low-Resistance Solenoid Inductors for Integrated Voltage Regulators," *IEEE Transactions on Component,*

Packaging and Manufacturing Technology, vol. 10, no. 1, pp. 134-141, Jan. 2020.

[22] K. Bharath et al., "Integrated Voltage Regulator Efficiency Improvement using Coaxial Magnetic Composite Core Inductors," *IEEE Electronics Components and Technology Conference (ECTC)*, San Diego, CA, 2021, pp. 1286-1292.

[23] B. Choi, J. Baek, B. C. Marin, S. Qu, S. Kulasekaran, J. I. Chavarria, L. E. Wojewoda and K. Radhakrishnan, "CoaxMIL 2.0 - Next Generation Coaxial Magnetic Integrated Inductors for Higher Efficiency Fully Integrated Voltage Regulator," *IEEE Electronics Components and Technology Conference (ECTC)*, Denver, CO, 2024, pp. 1044-1047.

[24] F. Zhu and Q. Li, "A Novel PCB-Embedded Coupled Inductor Structure for a 20-MHz Integrated Voltage Regulator," *IEEE Journal of Emerging and Selected Topics in Power Electronics*, vol. 10, no. 6, pp. 7452-7463, December. 2022.

[25] J. Qiu et al., "A toroidal power inductor using radial-anisotropy thin-film magnetic material based on a hybrid fabrication process," *2013 Twenty-Eighth Annual IEEE Applied Power Electronics Conference and Exposition (APEC)*, Long Beach, CA, USA, 2013, pp. 1660-1667.

[26] T. Fukuoka et al., "An 86% Efficiency, 20MHz, 3D-Integrated Buck Converter with Magnetic Core Inductor Embedded in Interposer Fabricated by Epoxy/Magnetic-Filler Composite Build-Up Sheet," *2019 IEEE Applied Power Electronics Conference and Exposition (APEC)*, Anaheim, CA, USA, 2019, pp. 1561-1566.

[27] Y. Kondo et al., "Embedded planar power inductor technology for package-level DC power grid," *2015 International Conference on Electronics Packaging and iMAPS All Asia Conference (ICEP-IAAC)*, Kyoto, Japan, 2015, pp. 814-817.

[28] J. Li, V. Tseng, Z. Xiao, and H. Xie, "A High-Q In-Silicon Power Inductor Designed for Wafer-Level Integration of Compact DC-DC Converters," *IEEE Transactions on Industrial Electronics*, vol. 32, no. 5, pp. 3858-3867, May 2017.

[29] M. Chen and C. R. Sullivan, "Unified Models for Coupled Inductors Applied to Multiphase PWM Converters,"*IEEE Transactions on Power Electronics*, vol. 36, no. 12, pp. 14 155–14 174, 2021.

Multiphase Lateral Flux Indirect Coupled Inductor for Vertical Power Delivery Voltage Regulator Module

Adhistira M. Naradhipa and Qiang Li

Center for Power Electronics Systems (CPES), Virginia Polytechnic Institute and State University, Blacksburg, VA, USA
adhistira@vt.edu

Abstract—**The rising trend of artificial intelligence (AI) usage in many applications requires high-performance processors, demanding power to an unprecedented level. Recently, in the 48 V two-stage conversion system, the vertical power delivery (VPD) solution is sought, where the second stage is placed directly underneath the processor to remove the "last inch" power loss found in the lateral power delivery (LPD) architecture. However, the VPD solution gives a strict size requirement for the voltage regulator module (VRM), where the bottleneck is usually the magnetic component. In addition, high-performance processors require a fast transient response. This article proposed a new air gap-less powder-core-based multiphase integrated lateral flux negative coupled inductor structure with a footprint of only 100 mm^2 and a height of 2.9 mm, enabling a high-density and fast-transient VRM solution. The negative coupling is achieved electrically through the coupled winding, which enables symmetrical N-phase coupling and straight-phase winding, resulting in an extremely small DCR. The proposed inductor is experimentally tested at up to 300 A (75 A/phase) to prove its high current handling capability, achieving a high 3 A/mm^2 current density.**

Index Terms—**indirect coupled inductor, lateral flux, trans-inductor voltage regulator, vertical power delivery.**

I. INTRODUCTION

With the surge of artificial (AI) and machine learning (ML) applications, the demand for high-performance processors (XPU) is significantly increasing. Powering the high-performance XPU has significant challenges due to their high current and fluctuating power demand with fast current slew rate [1]. Recently, academia and industry have been looking at the two-stage solution with the vertical power delivery (VPD) architecture [2]-[4], as shown in Fig. 1. The first stage step down 48 V input to intermediate bus voltage [5]-[8]. The second-stage voltage regulator module (VRM) is placed directly underneath the XPU to remove the "last inch" power loss found in the conventional lateral power delivery (LPD) architecture. However, designing a highly efficient, low-profile VRM, including the output capacitor layer, that could fit within the XPU footprint and achieve fast transient response is critical in the VPD architecture, where the bottleneck usually comes from the magnetic component.

Negative-coupled inductor technology has a non-linear inductance characteristic, known to have small transient-related inductance and large current ripple-related inductance [9]-[13]. Therefore, a VRM with a negative-coupled inductor can achieve a fast transient response without sacrificing efficiency.

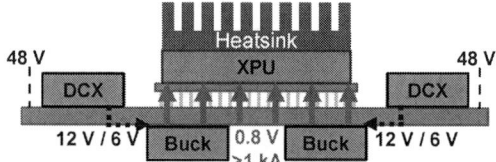

Fig. 1. High-performance processor with vertical power delivery architecture.

Moreover, the benefit of the negative-coupled inductor becomes more prominent when more phases are coupled [13].

Among many negative coupled inductor structures, the lateral flux structure achieves the highest inductance density [14]-[15]. Thus, it suits small-sized, low-profile inductors for sandwiched VRM structures in VPD solutions. The lateral flux inductor structure in [14]-[15] achieves negative coupling through direct magnetic flux interaction, named a direct coupled inductor (DCL), which can reduce the magnetic size due to the DC flux cancellation. However, symmetrically coupling high phase numbers of the VRMs using DCL is difficult. Furthermore, the structure in [14] requires more than a single turn winding to realize the negative coupling, which hinders implementation for high-current applications due to the high winding resistance.

Recently, indirect coupled inductor (ICL) technology has become a popular solution, with the name of trans-inductor voltage regulator (TLVR), to achieve symmetrical N-phase negative coupled inductor while keeping single-turn phase winding [16]-[25]. However, to the author's knowledge, no ICL structure has been developed leveraging the lateral flux concept.

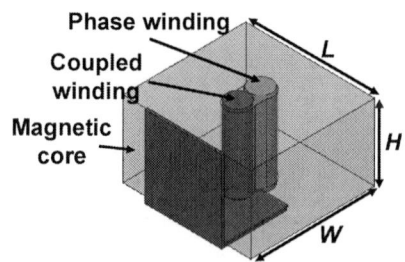

Fig. 2. Proposed unit-cell of the lateral flux indirect coupled inductor.

979-8-3315-1612-3/25 $31.00 © 2025 IEEE

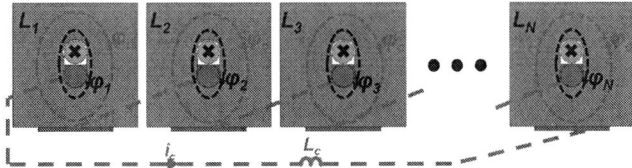

Fig. 3. Multiphase indirect coupled inductor using the proposed lateral flux unit cell structure.

This article proposed a new powder-core-based lateral flux ICL structure. The powder core has a significantly higher saturation flux than the ferrite core, enabling air-gapless design. Therefore, a small-sized inductor with reduced manufacturing difficulty and inductance tolerance between the products is achievable. The single-straight phase winding enables an extremely small DC resistance (DCR), which is suitable for the high current requirement of VPD. Moreover, the proposed inductor can be easily combined, creating a multiphase-integrated inductor suitable for multiphase VRM assembly. The proposed four-phase integrated inductor is designed with a 100 mm² footprint and a height of only 2.9 mm, which is tested using a multiphase buck converter to prove the negative coupling operation and high current handling capability up to 300 A, achieving a high current density of 3 A/mm².

II. PROPOSED LATERAL FLUX INDIRECT NEGATIVE COUPLED INDUCTOR

A. Basic Structure and Operation

The proposed inductor's unit cell consists of the magnetic core, phase winding, and coupled winding, as shown in Fig. 2. The phase winding connects the switching node to the output voltage node of the VRM, carrying each VRM's phase current. As the inductor in the 3D VRM structure is sandwiched between the active and output capacitor layers, the straight phase winding of the proposed structure enables an extremely small winding DCR. Therefore, the proposed structure is suitable for minimizing the conduction loss in high dc-current applications. The coupled winding wraps around the magnetic core vertically to capture the magnetic flux generated by the phase winding, which is lateral to the magnetic core footprint.

Referring to Fig. 3, the phase flux, ϕ_N, is created when a current flows in the respective phase winding, which is lateral to the inductor footprint. Then, the ϕ_N will induce a voltage in the coupled winding. Because the coupled winding creates a closed loop circuit, the current i_c flows in the coupled winding loop, inducing coupled winding flux ϕ_c in each inductor. Therefore, all inductor phases will see the flux change in one of them through ϕ_c, which makes the inductor operate as a negative coupled inductor [18]. The inductor L_c in the coupled winding loop controls the ICL coupling coefficient, which can be realized by an additional inductor or coupled winding trace inductance.

B. Proposed Inductor Unit-Cell Design

The footprint of the commercial two-phase sandwiched VRM structure is 9 mm x 10 mm, with a height of around 8

TABLE I
PROPOSED LATERAL FLUX COUPLED INDUCTOR DESIGN TARGET

Parameter	Value
Length L	5 mm
Width W	5 mm
Height H	< 3 mm
Non-coupled inductance L_{nc}	≥ 60 nH at 0 A
Phase winding DCR DCR_{ph}	≤ 0.1 $\mu\Omega$
Coupled winding DCR DCR_{cw}	≤ 0.8 $\mu\Omega$
Peak current capability I_{pk}	75 A/phase

mm without the output capacitor layer [26]-[27]. Assuming 45 mm² is the VRM footprint for each phase, TABLE I shows the design target of the proposed inductor. The inductor footprint is set to be 25 mm², smaller than the total VRM footprint, to allow some space for power and signal pins between the output capacitor and active layers.

There are two essential inductances for the VRM performance using the ICL: the steady-state ripple and transient inductance, L_{ss} and L_{tr}, respectively. The L_{ss} and L_{tr} depend on the non-coupled inductance value, L_{nc}, which is the inductance value of each phase when the coupled winding loop is open, and coupling coefficient α, as expressed in (1) and (2). The n is the number of coupled phases, D is the duty cycle, j is the number of top switch on-time overlap. As seen from (3), in ICL, the α can be controlled through inductor L_c.

$$L_{ss} = L_{nc} \frac{(n-1)(n-|\alpha|)}{n-1-[n-2j-2+\frac{j(j+1)}{nD}+\frac{nD(n-2j-1)+j(j+1)}{nD}]|\alpha|} \quad (1)$$

$$L_{tr} = L_{nc} \frac{(n-1)(1-|\alpha|)}{n-1+|\alpha|} \quad (2)$$

$$|\alpha| = \frac{(n-1)}{n-1+\frac{L_c}{L_{nc}}} \quad (3)$$

Usually, the L_{tr} is pre-determined for a given processor, which can be achieved by the combination of L_{nc} and $|\alpha|$. Looking at (1), a higher L_{ss} can be achieved with higher L_{nc} for a given $|\alpha|$. Therefore, maximizing the L_{nc} is desirable to improve the VRM efficiency. After the L_{nc} is maximized, the L_c can control $|\alpha|$ and achieve the target L_{tr}.

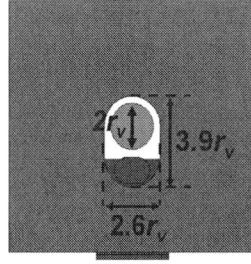

Fig. 4. Window area of the proposed lateral flux indirect coupled inductor.

As the proposed inductor uses an air-gap-less powder-core material, only the magnetic core dimension can change the

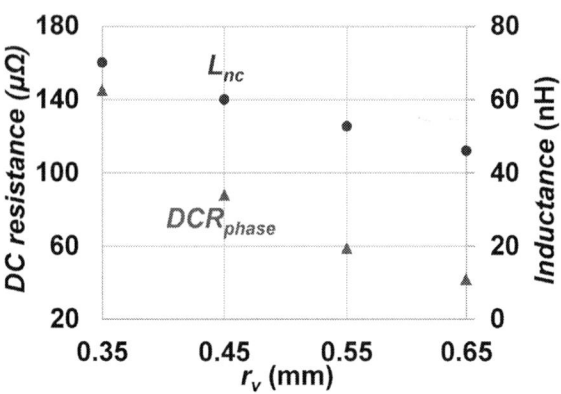

Fig. 5. Impact of r_v to the DCR_{ph} and the L_{nc} at zero dc-bias with $L = 5$ mm, $W = 5$ mm, $H = 3$ mm, and $\mu_r = 80$.

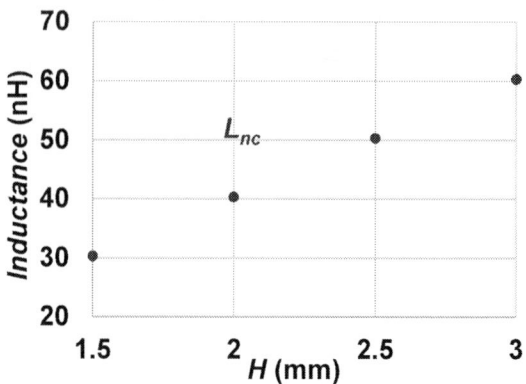

Fig. 7. Impact of H to the L_{nc} at zero dc-bias with $L = 5$ mm, $W = 5$ mm, $r_v = 0.45$ mm, and $\mu_r = 80$.

inductance value at a given dc current. The first step is to design the phase winding to achieve the targeted DCR_{ph} by sweeping the phase winding radius, r_v. The r_v not only impacts the DCR_{ph}, but also the L_{nc} because it determines the magnetic core hole size for both phase and coupled windings, as illustrated in Fig. 4. Because the phase winding structure is only a simple straight cylinder, the DCR_{ph} can be easily calculated using (4). The L_{nc} value is calculated using a 3D FEA simulation tool, and the magnetic core initial permeability is 80.

$$DCR_{ph} = \rho \frac{H}{\pi r_v^2} \qquad (4)$$

Figure 5 shows that a larger r_v results in a smaller DCR_{ph} and L_{nc}. The L_{nc} reduces as r_v increases because a larger r_v gives less core volume. The r_v of 0.45 mm is selected to achieve an extremely small DCR_{ph} of 87 $\mu\Omega$ while keeping enough L_{nc} value. With the selected r_v, the window area dimension can be calculated, and the coupled winding can be designed. Figure 6 shows the coupled winding dimensions to fit within the given window area and have a 0.3 mm distance between the phase and the coupled winding. Due to the more complicated coupled winding structure, the DCR_{cw} value is simulated using the 3D FEA simulation tool. With a given dimension, as shown in Fig. 6, the DCR_{cw} is still below the maximum value given in TABLE I.

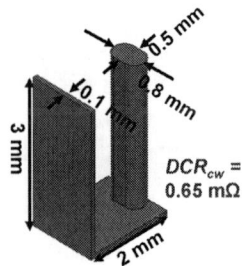

Fig. 6. Coupled winding dimension and the simulated DCR_{cw} value.

With a given window area and the inductor footprint, only the inductor height H will now impact the inductance value. Figure 7 shows the L_{nc} at different core heights H. As shown in Fig. 7, smaller H reduces the L_{nc} because the core volume reduces with a smaller H value. Note that in Fig. 7, the L_{nc} is directly proportional to the H value. The proportionality between the L_{nc} and H is because the inductance density does not change with H, which is the unique characteristic of the lateral flux inductor [14]. The H of 3 mm is selected to achieve the target L_{nc}.

III. PROPOSED INTEGRATED LATERAL FLUX INDIRECT NEGATIVE COUPLED INDUCTOR

For higher-power processors, having four or more VRs within a single module is desirable to reduce the assembly and manufacturing effort [19]. Moreover, the assembly efforts and cost can be further reduced if the inductors for different phases can be integrated into a single magnetic core.

This article further proposed a multiphase integrated lateral flux coupled inductor. The idea is to connect several magnetic core sides of the proposed lateral flux indirect coupled inductor unit-cell with no coupled winding, which is possible due to the simplicity of the unit-cell structure. Figure 8 shows the integration method for the proposed lateral flux ICL. Figure 8(a) shows two lateral flux ICLs, which can be combined into a two-phase integrated lateral flux ICL, as shown in Fig. 8(b). Furthermore, the four-phase integrated lateral flux ICL, as shown in Fig. 8(c), can be created by combining four unit-cells ICL into a single magnetic core. The integration method can be further extended to create an N-phase integrated lateral flux ICL. From another point-of-view, N number of phase and coupled winding pairs can be inserted into a single magnetic core to create an integrated lateral flux ICL. Therefore, the solution is easily scalable.

Figure 9 shows the L_{nc} comparison between the four-phase integrated version and the discrete version, in which the integrated version has a higher L_{nc}. The increase in L_{nc} is due to the larger magnetic core area seen by each phase winding in the integrated version. Therefore, integration reduces the

979-8-3315-1612-3/25 $31.00 © 2025 IEEE

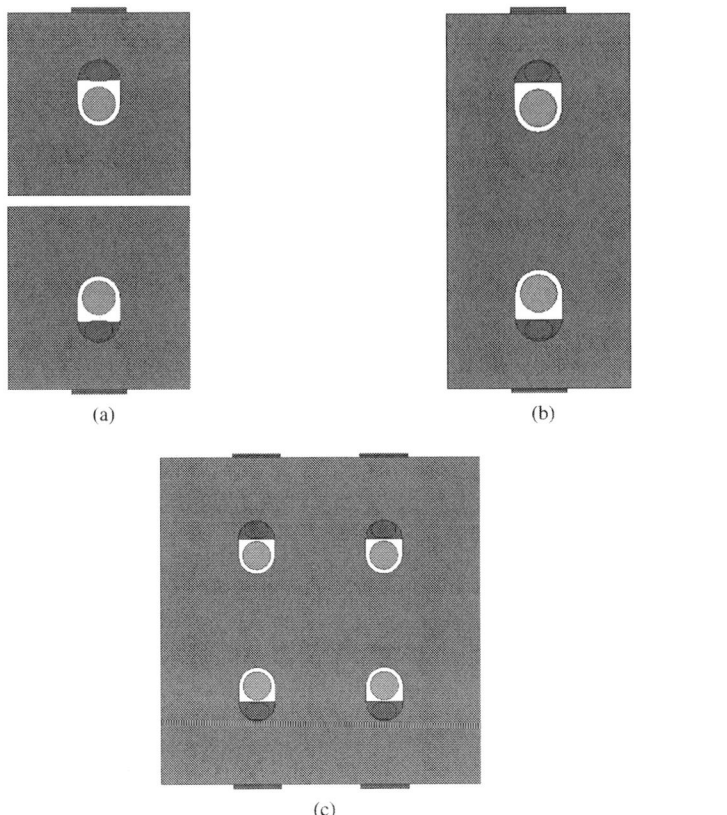

(a)　　　　　　(b)

(c)

Fig. 8. Proposed magnetic core integration of the lateral flux coupled inductor: (a) two separate inductor (b) two-phase integrated (c) four-phase integrated.

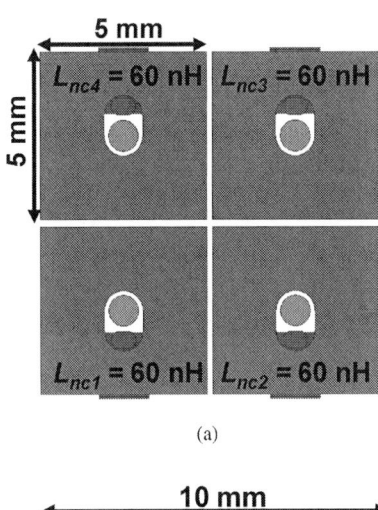

(a)

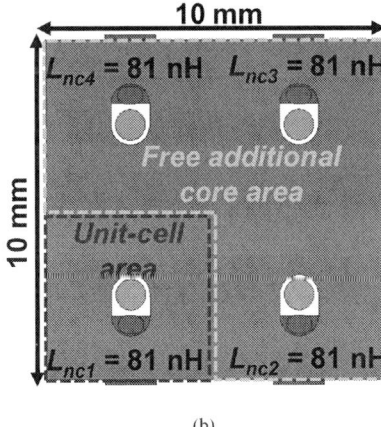

(b)

Fig. 9. Proposed magnetic core integration of the lateral flux coupled inductor: (a) four unit-cell (b) four-phase integrated.

manufacturing effort and brings the added benefit of increasing the L_{nc} value by 35% from 60 nH of the unit cell to 81 nH of the four-phase integrated version without increasing the total footprint.

IV. EXPERIMENTAL RESULTS

The four-phase integrated lateral flux coupled inductor is fabricated using a powder core material from Microgate with dimensions 10 mm x 10 mm x 2.9 mm, as shown in Fig. 10. From Fig. 10, it can be seen that no air gap exists in the proposed inductor.

The TI TPS536C7EVM-051 evaluation board is used to test the proposed inductor. Figure 11 shows the proposed four-phase integrated indirect coupled inductor with the associated phase and coupled winding on the buck converter evaluation board. Firstly, to measure the L_{nc}, the coupled winding is open, and the inductor is tested under different DC loading conditions with an input voltage of 10.8 V, output voltage of 0.8 V, and 800 kHz switching frequency. The inductor current waveform is measured using the Rogowski coil CWT06 from Power Electronics Measurement. Figure 12 shows the i_{L1} and v_{L1} waveforms, in which the inductance is calculated using (5). The L_{nc} is achieved by subtracting the result from (5) by the extra inductance, around 15 nH, from the long phase winding. As shown in Fig. 13, the calculated L_{nc} from

measurement is around 81 nH at 0 A, which matches well with the simulation results. Furthermore, the proposed inductor still has L_{nc} of 52 nH/phase at 300 A. Thus, the proposed inductor has a high current density of 3 A/mm^2 with a height of only 2.9 mm.

$$L_{nc} = v_L \frac{\Delta t}{\Delta i_L} \qquad (5)$$

Figure 14(a) shows the phase current waveforms at a full continuous current of 160 A when the coupled winding loop is closed. Using Fig. 14(a), the L_{ss} and L_{tr} can be calculated following the method in [12]. The calculated L_{ss} and L_{tr} from measurement results in Fig. 14(a) are 66 nH and 28.4 nH, respectively, which includes the extra leakage inductance from the long phase winding. Figure 14(b) shows the phase and coupled winding current waveforms. As shown in Fig. 14(b), the coupled winding current has a frequency four times the switching frequency. This proves that the proposed inductor operates as an indirect negative coupled inductor.

In addition to the measured current waveforms, Fig. 15 shows the transient performance between open and closed coupled windings. Figure 15(a) shows the output voltage overshoot and undershoot with a 160 A load current step

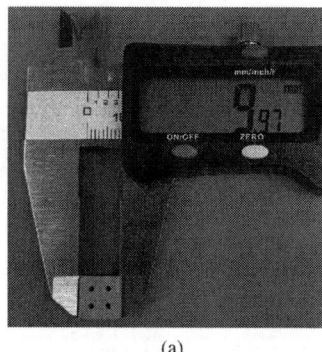

(a)

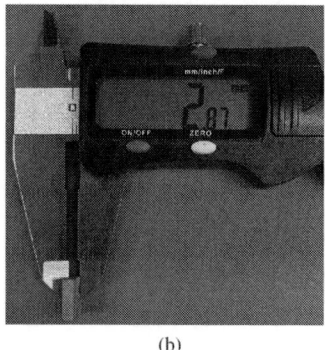

(b)

Fig. 10. Fabricated powder magnetic core of the proposed four-phase integrated lateral flux indirect coupled inductor:(a) top-view (b) side-view.

Fig. 11. The proposed four-phase integrated lateral flux indirect coupled inductor inductor in the buck converter evaluation board.

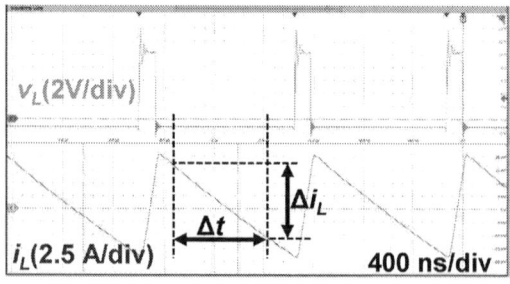

Fig. 12. Measured phase current and voltage waveforms of the proposed inductor with open coupled winding at $v_{in} = 10.8$ V, $v_o = 0.8$ V, $i_{Load} = 75$ A/phase, and $f_s = 800$ kHz.

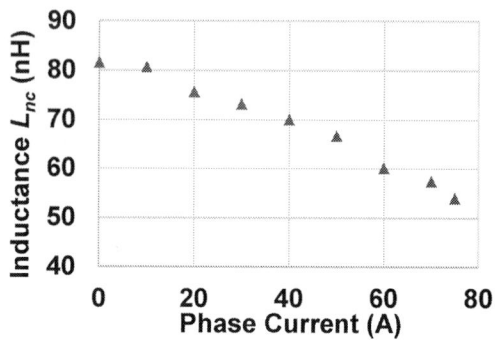

Fig. 13. Calculated inductance L_{nc} at different phase current.

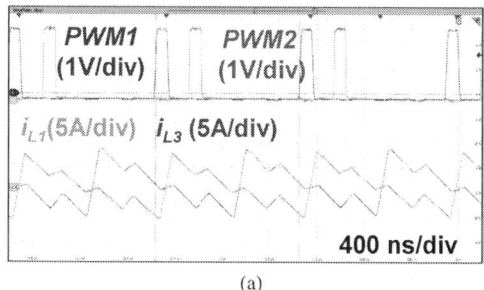

(a)

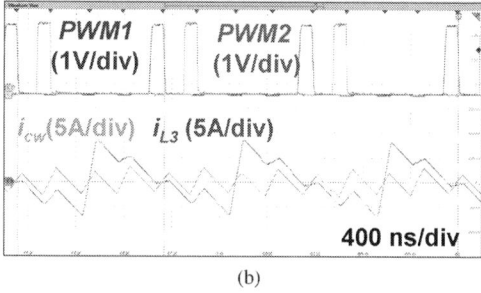

(b)

Fig. 14. Measured current waveforms of the proposed inductor:(a) two-phase winding currents, (b) phase and coupled winding currents at $v_{in} = 10.8$ V, $v_o = 0.8$ V, $i_{Load} = 160$ A/phase, and $f_s = 800$ kHz.

and when the coupled winding is open, in which the L_{nc} determines the transient performance. On the other hand, Fig. 15(b) shows a smaller output voltage overshoot V_{os} and a faster settling time t_{set} are achieved with a closed secondary winding because the transient performance is determined by the small L_{tr}. Although the voltage undershoots V_{us} during the step-up transient of both cases have a similar value, the settling time of the closed coupled winding case is faster. The same value of V_{us} is because the output capacitor parasitics determines the first output voltage dip, which is the same in both cases.

V. CONCLUSIONS

A new multiphase lateral flux indirect coupled inductor structure is proposed, achieving an extremely small DCR with a single-straight phase winding. The proposed inductor uses a powder-core material, which enables an air gap-less design and achieves a high current density of 3 A/mm² with a height

979-8-3315-1612-3/25 $31.00 © 2025 IEEE

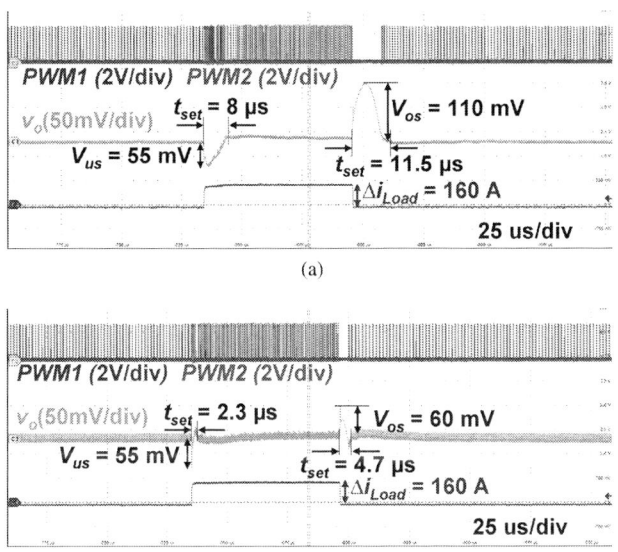

Fig. 15. Measured transient response at v_{in} = 10.8 V, v_o = 0.8 V, i_{Load} = 0-160 A, and C_o = 5.2 mF with: (a) open and (b) closed coupled winding.

of only 2.9 mm. The coupled winding is added to make the inductor operate as a negative coupled inductor using the ICL concept, which eases the scalability to couple with any number of other phases symmetrically. In conclusion, this article proposes a new lateral flux indirect coupled inductor structure suitable for high current vertical power delivery VRM.

ACKNOWLEDGMENT

The authors would like to thank Google LLC for supporting this work.

This work was conducted using core samples donated in kind by Microgate.

REFERENCES

[1] Y. Fard, S. S. D. Naidu, H. D. Tamdem and B. Vafakhah, "Trans- Inductors versus Discrete Inductors in Multiphase Voltage Regulators: An Analytical and Experimental Comparative Study," *2023 IEEE Applied Power Electronics Conference and Exposition (APEC)*, Orlando, FL, USA, 2023.

[2] "VERTICAL POWER DELIVERY STRUCTURE FOR ULTRA-HIGH CURRENT APPLICATIONS", *Technical Disclosure Commons*, (September 26, 2018) [Online]. Available: https://www.tdcommons.org/dpubs_series/1534.

[3] "Multi-rail Integration in Vertical Power Delivery", *Technical Disclosure Commons*, (November 18, 2020) [Online]. Available: https://www.tdcommons.org/dpubs_series/3786.

[4] H. Gan et al., "Vertical Power Delivery for 1000 Amps Machine Learning ASICs," *2024 IEEE Applied Power Electronics Conference and Exposition (APEC)*, Long Beach, CA, USA, 2024.

[5] P. R. Prakash et al., "A 2400 W/in^3 1.8 V Bus Converter Enabling Vertical Power Delivery for Next-Generation Processors," *2024 IEEE Applied Power Electronics Conference and Exposition (APEC)*, Long Beach, CA, USA, 2024.

[6] X. Xu and Q. Li., "Symmetric Series-Capacitor Buck in 48V-to-12V Regulated Conversion for High-Performance Server Boards," *2024 IEEE Energy Conversion Congress and Exposition (ECCE)*, Phoenix, AZ, USA, 2024.

[7] A. Dago, M. Balutto, S. Saggini, M. Leoncini, S. Levantino and M. Ghioni, "Hybrid Resonant Switched Tank Converters for High Step-Down Voltage Conversion," in *IEEE Transactions on Power Electronics*, vol. 39, no. 11, pp. 14838-14851, Nov. 2024.

[8] Z. Ye, Y. Lei and R. C. N. Pilawa-Podgurski, " A 48-to-12 V cascaded resonant switched-capacitor converter for data centers with 99% peak efficiency and 2500 W/in 3 power density" in *2019 IEEE Applied Power Electronics Conference and Exposition (APEC)*, Anaheim, CA, USA, 2019.

[9] P.L. Wong, P. Xu, P. Yang, and F. C. Lee, "Performance improvements of interleaving VRMs with coupling inductors," in *IEEE Transactions on Power Electronics*, vol. 16, no. 4, pp. 499–507, Jul. 2001, doi: 10.1109/63.931059.

[10] J. Li, A. Stratakos, A. Schultz, and C. R. Sullivan, "Using coupled inductors to enhance transient performance of multi-phase buck converters," in *Proc. IEEE Appl. Power Electron. Conf.*, 2004, pp. 1289–1293.

[11] W. Huang and B. Lehman, "Inversely Coupled Inductors With Small Volume and Reduced Power Loss for Switching Converters," in *IEEE Transactions on Power Electronics*, vol. 38, no. 6, pp. 6779-6783, June 2023, doi: 10.1109/TPEL.2023.3241883.

[12] A. M. Naradhipa, F. Zhu and Q. Li, "Ultra-Low-Profile Twisted Core Inductor for Vertical Power Delivery Voltage Regulator," *2024 IEEE Applied Power Electronics Conference and Exposition (APEC)*, Long Beach, CA, USA, 2024.

[13] Y. Dong, "Investigation of multiphase coupled-inductor buck converters in Point-of-Load applications," *Ph.D. dissertation, Virginia Polytechnic Institute and State University*, Blacksburg, VA, USA, 2009.

[14] F. C. Lee and Q. Li, "High-Frequency Integrated Point-of-Load Converters: Overview," in *IEEE Transactions on Power Electronics*, vol. 28, no. 9, pp. 4127-4136, Sept. 2013, doi: 10.1109/TPEL.2013.2238954.

[15] J. Baek, Y. Elasser and M. Chen, "MIPS: Multiphase Integrated Planar Symmetric Coupled Inductor for Ultrathin VRM," in *IEEE Transactions on Power Electronics*, vol. 38, no. 5, pp. 5609-5614, May 2023, doi: 10.1109/TPEL.2023.3236152.

[16] F. Zhu and Q. Li, "Coupled Inductors With an Adaptive Coupling Coefficient for Multiphase Voltage Regulators," in *IEEE Transactions on Power Electronics*, vol. 38, no. 1, pp. 739-749, Jan. 2023, doi: 10.1109/TPEL.2022.3203855.

[17] S. Krishnamurthy, D. Wiest and Y. Zhou, "Trans-Inductor Voltage Regulator (TLVR): Circuit Operation, Power Magnetic Construction, Efficiency and Cost Trade-Offs," in *PCIM Europe 2022; International Exhibition and Conference for Power Electronics, Intelligent Motion, Renewable Energy and Energy Management*, Nuremberg, Germany, 2022, pp. 1-6, doi: 10.30420/565822053.

[18] M. Xu, F. C. Lee, and Y. Ying, "Coupled-inductor multi-phase buck converters," *U.S. Patent US7791321B2*, Sep. 7, 2010.

[19] S. Jiang, et. al., "Next TLVR Innovations: Topologies, Magnetics and Control," in *Proc. APEC Ind. Session*, Feb. 27, 2024.

[20] H. Shao, T. Zhao, D. Fu, D. Huang and J. Zhou, "Analytic Model and Design Procedure of the Single-Secondary Trans-Inductor Voltage Regulator," *2021 IEEE Energy Conversion Congress and Exposition (ECCE)*, Vancouver, BC, Canada, 2021.

[21] C. Li, L. Wang, M. Fu and H. Wang, "Small-Signal Modeling of Multi-Phase Trans-Inductor Voltage Regulator Modules in Datacenter Applications," *2024 IEEE Applied Power Electronics Conference and Exposition (APEC)*, Long Beach, CA, USA, 2024.

[22] P. Kumar, J. Tippetts, S. S. Deepak Naidu and P. Brusco, "Efficiency Impact of Phase Firing Order in Dual-Sided Power Entry with Trans-Inductor Voltage Regulators (TLVR)," *2024 IEEE Applied Power Electronics Conference and Exposition (APEC)*, Long Beach, CA, USA, 2024.

[23] A. Ikriannikov and D. Yao, "Converters with Multiphase Magnetics: TLVR vs CL and the Novel Optimized Structure," *PCIM Europe 2023; International Exhibition and Conference for Power Electronics, Intelligent Motion, Renewable Energy and Energy Management*, Nuremberg, Germany, 2023.

[24] J. Amanor-Boadu, R. Rice, A. Shuma, R. Bazaz and H. Tamdem, "Pre and post silicon server platform transient performance using transinductor voltage regulator," *2023 IEEE 41st VLSI Test Symposium (VTS)*, San Diego, CA, USA, 2023.

[25] N. Zhang, C. Zhan, G. Ye, C. Chen, X. Li and J. Yi, "Analysis of Multi-Phase Trans-Inductor Voltage Regulator with Fast Transient Response for Large Load Current Applications," *2021 IEEE International Symposium on Circuits and Systems (ISCAS)*, Daegu, Korea, 2021.

[26] "MPC22163-130", *Monolithic Power Systems, [Online].* Available: www.monolithicpower.com/en/products/power-management/data-center/mpc22163-130.html.

[27] "BMR511", *FLEX, [Online].* Available: flexpowermodules.com/products/bmr511?model=BMR511X044%2F002.

A High Density Three-Level Quadratic Buck Hybrid Converter for 48V-to-PoL Conversion

Kejia Wang
Dept. of Electrical and Computer Eng.
Iowa State University
Ames, USA
k6wang@iastate.edu

Si Yuan Sim
Dept. of Electrical and Computer Eng.
Iowa State University
Ames, USA
simsy@iastate.edu

Yin Quen Choong
Dept. of Electrical and Computer Eng.
Iowa State University
Ames, USA
choong@iastate.edu

Xin Zhang
IBM T.J. Watson Research Center
Yorktown Heights, USA
xzhang@ibm.com

SriHarsh Pakala
NXP
Chandler, USA
sriharsh.pakala@nxp.com

Cheng Huang
Dept. of Electrical and Computer Eng.
Iowa State University
Ames, USA
chengh@iastate.edu

Abstract— **This paper presents a new hybrid topology for 48V point-of-load (PoL) DC-DC converters. The proposed three-level quadratic buck converter (3LQB) consists of a three-level buck converter at the front-end merged with an inductor-first buck converter to achieve the required high ratio conversion with only 6 switches. The three-level buck stage allows for the reduction of the inductor value, while the inductor-first buck stage takes advantage of parallel paths through two inductors to deliver more power, with phase interleaving that results in smaller ripples at the output. These two factors combined contribute to improvement of efficiency and density, while the quadratic duty-cycle to voltage conversion relation provides the benefits of duty-cycle extension needed for high-ratio voltage conversion. The measured peak efficiency for a 48V to 1-3.3V conversion using GaN switches is 89.4%, 91.1% and 93.5%, and the power density is calculated to be 1kW/in³, 1.5kW/in³, and 3.3kW/in³ for an output of 1V, 1.5V and 3.3V, respectively, with a max load current of 45A.**

Keywords—Hybrid converter, DC-DC converter, point-of-load converter, 48V, direct step down, efficiency

I. INTRODUCTION

As 48V power delivery networks have become increasingly common in data centers and automotive applications, many different high ratio DC-DC converter topologies that can achieve 48V-to-PoL conversion have been introduced [1-11]. In particular, hybrid converters with switched capacitors and inductors have received significant attention due to their potential in realizing high power density as compared to transformer-based designs, which rely on bulky transformers. However, many hybrid converters utilize a switch capacitor network to first step-down the input voltage and require the use of many (8-16) switches [1-10], which results in more components thus increasing cost, failing points, and complicated gate driving circuits. Reducing the step-down ratio of the switch capacitor network can reduce switch count but this increases the voltage swing and inductor size of the inductive stage, which degrades power efficiency and density. The

This work is supported by the Semiconductor Research Corporation under Project 2810.080.

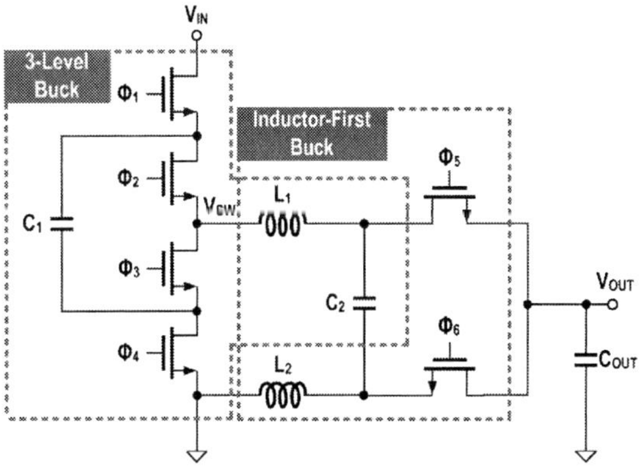

Fig. 1. Block diagram of the proposed three-level quadratic buck.

superimposed quadratic buck (SQB) converter that was recently introduced relaxes the duty cycle requirements due to the quadratic conversion ratio and requires only 4 switches in total to achieve the large conversion ratio [11]. However, the inductor in the SQB connects directly to the high-voltage input, thus a large inductor is required to reduce the current ripple and AC loss in the core of the inductor, as well as limiting the output voltage ripple. In addition, the switching node also swings between 48V and ground, causing more voltage stress on the switches and more losses during the transition. The efficiency of the converter is also affected as lower-voltage rating devices with a better FoM cannot be used for reliable performance. Therefore, this work attempts to improve the power density of the topology while using the minimum number of switches.

II. THE PROPOSED 3LQB

The proposed three-level quadratic buck (3LQB) converter consists of a three-level buck converter [12-13] merged with an inductor-first buck stage [14], as shown in Fig. 1. By having a three-level front-end, the volt-second product across the inductor is dropped by half, and the effective frequency is twice that of a conventional buck converter. These two factors allow

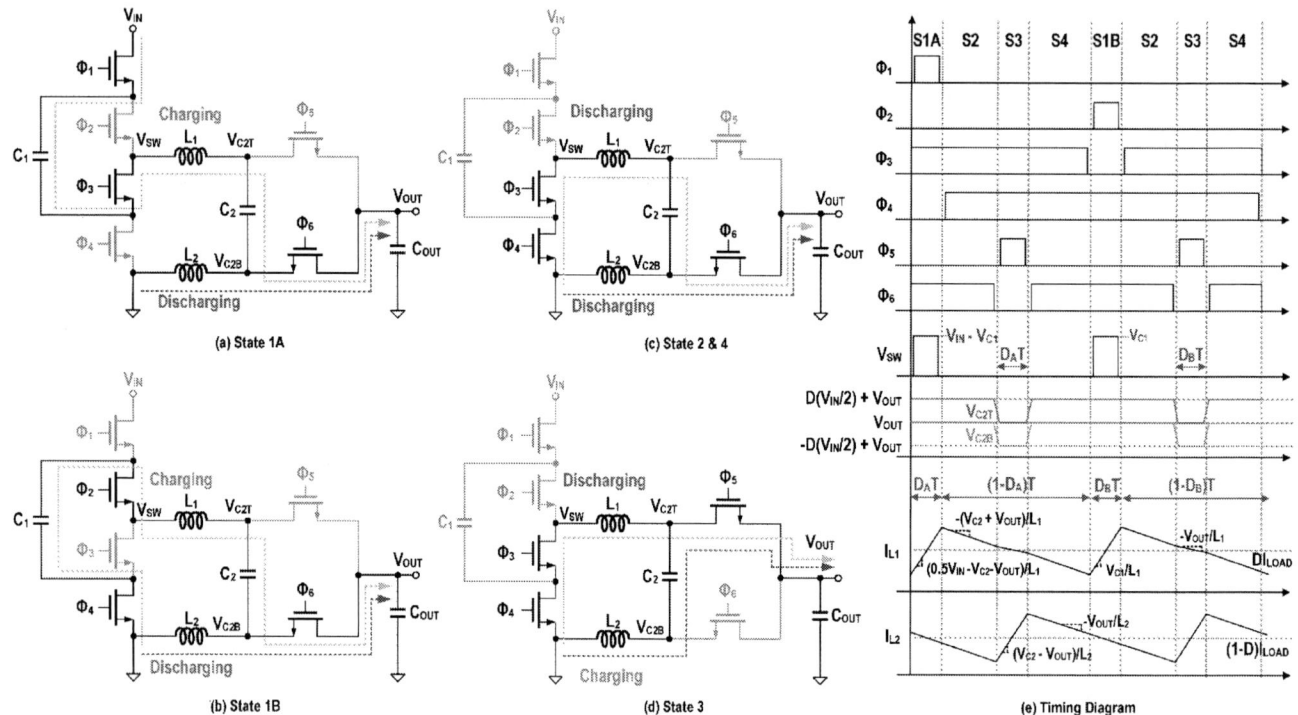

Fig. 2. Operation principle of the proposed three-level quadratic buck with (a – d) the four operational states and (e) timing diagram.

the inductor size to be reduced with the addition of only 2 switches and 1 flying capacitor, also achieving a more relaxed duty cycle. In addition, the flying capacitor also clamps the voltages across the switches in the three-level buck to half V_{IN}, allowing the use of lower voltage rated switches for realizing higher efficiency.

A. Operation Principle

The timing diagram illustrating the states of operation for the proposed design is depicted in Fig. 2. As shown in the diagram, the four unique states for the proper operation of the proposed design and the order of the states are as follows:

$$S1A \rightarrow S2 \rightarrow S3 \rightarrow S4 \rightarrow S1B \rightarrow S2 \rightarrow S3 \rightarrow S4 \quad (1)$$

A single full period, $2T$, for the entire switching operation consists of states $S1A/B$ once and $S2/3/4$ twice but the effective switching frequency (f_{SW}) of the entire converter is doubled. Once the converter has reached its steady state, the operation for each state is as follows: 1) In state 1A ($S1A$, Fig. 2a), the inductor L_1 is charged through the flying capacitor C_1 from V_{IN} and the capacitor voltage increases. Inductor L_2 is discharging to the output during this state. 2) Then, in state 2 ($S2$, Fig. 2c), both inductors are discharged. 3) Next, in state 3 ($S3$, Fig. 2d), inductor L_2 is being charged when the voltage of the bottom plate of C_2 is switched to $V_{OUT} - V_{C2}$ whereas inductor L_1 is still being discharged but the slope becomes smaller. 4) The converter then moves to the discharging state again in state 4 ($S4$) which is the same as state 2 ($S2$). 5) Once the operation has completed a half cycle, the next operation is state 1B ($S1B$, Fig. 2b) where inductor L_1 is being charged again but this time the

flying capacitor C_1 is being discharged into the inductor, so the voltage across it starts decreasing. The process of charging and discharging of C_1 ensures that C_1 is balanced to half of V_{IN} which is essential to utilizing lower voltage rated switches for achieving higher efficiency. 6) Lastly, the states $S2/3/4$ are repeated to complete one full switching cycle. Assuming that the flying capacitor C_1 voltage is balanced and that the duty cycles, D_A and D_B, are matched for every on-time where the inductors are charging such that $D_A=D_B=D$, the volt-second balance equations can be written for the two inductors:

$$\frac{\frac{1}{2}V_{IN}-V_{C2}-V_{OUT}}{L_1}DT = \frac{V_{C2}+V_{OUT}}{L_1}(1-2D)T + \frac{V_{OUT}}{L_1}DT \quad (2)$$

$$\frac{V_{C2}-V_{OUT}}{L_2}DT = \frac{V_{OUT}}{L_2}(1-D)T \quad (3)$$

Solving (3) for V_{OUT} allows the equation to be simplified to:

$$DV_{C2} = V_{OUT} \quad (4)$$

Substituting (4) into (2) allows for the voltage conversion ratio of the entire converter to be derived:

$$V_{OUT} = D^2 * \frac{V_{IN}}{2} \quad (5)$$

It can be observed that the conversion ratio is quadratic which is beneficial to the high-ratio conversion such as 48V to 1/3.3V as it relaxes the duty cycle requirement. This also leads to better switch utilization at the inductor-first stage which conducts the full output load current.

Due to the addition of the three-level buck front end, the converter has a maximum duty cycle of 0.5, with the proposed switching scheme to ensure C_1 is balanced [13]. However, even with this duty cycle limitation, for a 48V input, the converter is able to generate output voltages from 1V up to 6V which covers the range of supply voltages for most digital systems.

B. Inductor Current Analysis

The ratio of the average current and the ripple current through the two inductors do not remain the same when the duty cycle changes for a different output voltage. This is caused by the quadratic duty cycle, and the relationship for the average current is defined as:

$$I_{L2} = \left(\frac{1}{D} - 1\right) I_{L1} \qquad (6)$$

The ripple current for both inductors is then defined to be:

$$\Delta I_{L1} = \frac{\frac{1}{2}V_{IN} - V_{C2} - V_{OUT}}{L_1} DT \qquad (7)$$

$$\Delta I_{L2} = \frac{V_{C2} - V_{OUT}}{L_2} DT \qquad (8)$$

Because the two inductor currents are 180° out-of-phase, the smallest output voltage ripple can be achieved by equalizing the ripple current from both inductors. By equating (7) and (8), it can be found that asymmetrical inductors are required to achieve that condition, and this tends to lead to the inductance of L_1 being larger than L_2 [11]. On the other hand, as long as the output voltage ripple is small enough, the condition of $\Delta I_{L1} = \Delta I_{L2}$ does not have to be guaranteed, and inductance different from this condition can be chosen for efficiency optimization, which will be the actual case in this design and efficiency measurement. Furthermore, from (6) it is observed that for a D less than 0.5, the average current through L_2 is always larger than L_1, while to have proper inductor current ripple cancellation effect, L_2 is typically much smaller than L_1. This is beneficial to relax the trade-offs when selecting L_2, especially when the conversion ratio is large. Even though the two inductors are typically not the same, and the current ratio between the two inductors is not consistent at various output voltages, the topology has inherent inductor current balancing. With small ripple approximation, the ampere-second balance equation can be used to compute the total charge of C_2 for one half-cycle, which is from $S1A$ to $S4$:

$$\int_{T_0}^{T_0 + T_S} I_{C2}\, dt = I_{L1}(D_A) + I_{L1}(1 - 2D_A) - I_{L2}(D_A) \qquad (9)$$

Therefore, if the on-time duty cycle, D_A, is matched for $S1A$ and $S3$, (9) can be simplified and proves that the inductor current relationship in (6) will always hold true. The same equation can be derived for the second half-cycle which may have a different on-time duty cycle, D_B, in order to compensate for any mismatch in the balancing of the flying capacitor C_1 for the three-level buck stage.

III. MEASUREMENT RESULTS

A prototype using 6 GaN power transistors was developed and measured to verify the performance of the proposed design (Fig. 3), and the parts utilized are listed in Table 1. The key switch node waveforms are depicted in Fig. 4. Due to the flying capacitor C_1 clamping the maximum voltage in the three-level buck stage to half V_{IN}, the switches used are only rated for 40V

TABLE I.   CIRCUIT COMPONENT AND PARAMETERS

Item	Design Selection
$Q_1 - Q_4$	EPC2055 (40V, EPC)
$Q_5 - Q_6$	EPC2067 (40V, EPC)
L_1	XGL6060-472 (4.7µH, Coilcraft)
L_2	XGL6060-471 (0.47µH, Coilcraft)
C_1	2.2µF*8 (50V, Murata)
C_2	10µF*5 (25V, Murata)
C_{OUT}	47µF*6 (10V, Murata)
Gate Driver	UCC27511ADBVR (TI)
Isolators	ISO7710FQDRQ1 (TI)
Schottky Diodes	PMEG6010CEJ, PMEG2030CER

Fig. 3. PCB prototype of the 3LQB with additional capacitors for C_1 & C_2 placed at the bottom along with the Schottky diodes.

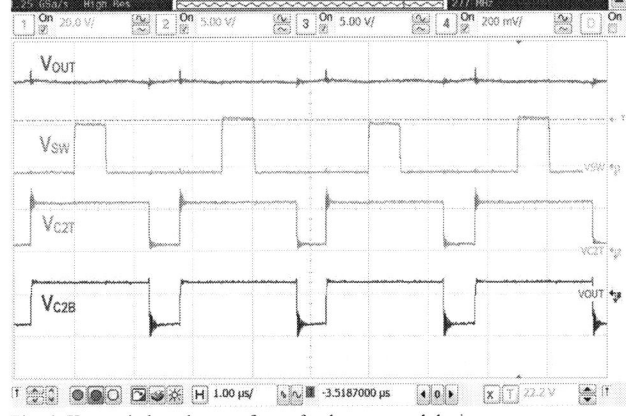

Fig. 4. Key switch node waveforms for the proposed design.

for better performance despite a 48V input voltage. Schottky diodes are also used to assist in the reverse current conduction for Q_3, Q_4 and Q_6 during the dead-time, because the voltage drop across the GaN device during any reverse current conduction is much larger compared with a conventional Silicon device, which can contribute to increased switching loss and dominate when the switching frequency is high [15]. As discussed earlier, since the output voltage ripple is acceptable, the selected inductors for this particular design are off of the optimum ratio calculated by (7) and (8), prioritizing reduced AC loss and higher efficiency over ripple cancellation. The

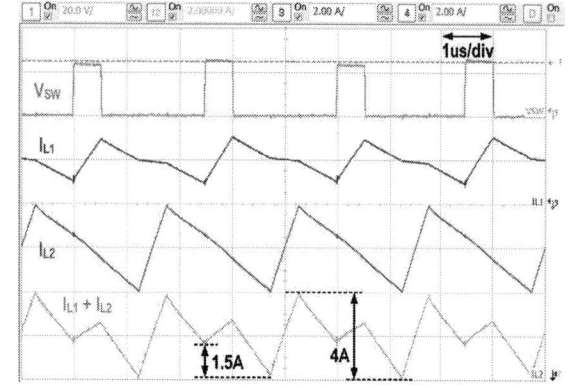

(a) $L_1 = 4.7\mu H$ and $L_2 = 0.47\mu H$, non-optimum inductor ripple cancellation.

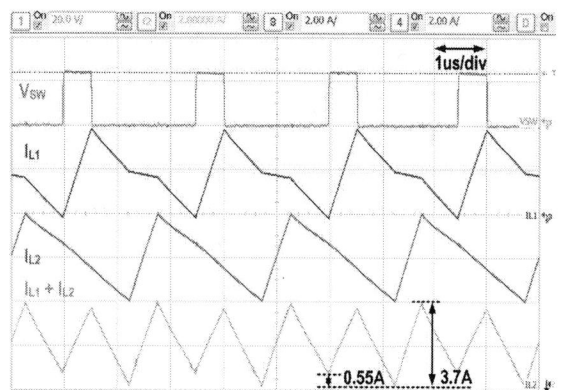

(b) $L_1 = 2.2\mu H$ and $L_2 = 0.47 \mu H$, optimum inductor ripple cancellation.

Fig. 5. Measured inductor current waveforms with (a) non-optimum ripple cancellation and slightly larger ripple and (b) optimum ripple cancellation.

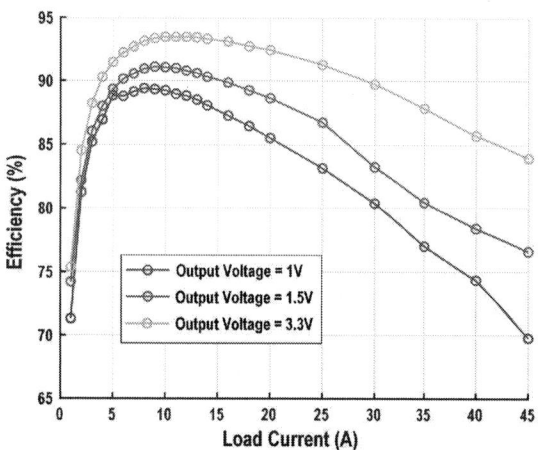

Fig. 6. Efficiency at different output voltages with $L_1 = 4.7\mu H$ and $L_2 = 0.47\mu H$.

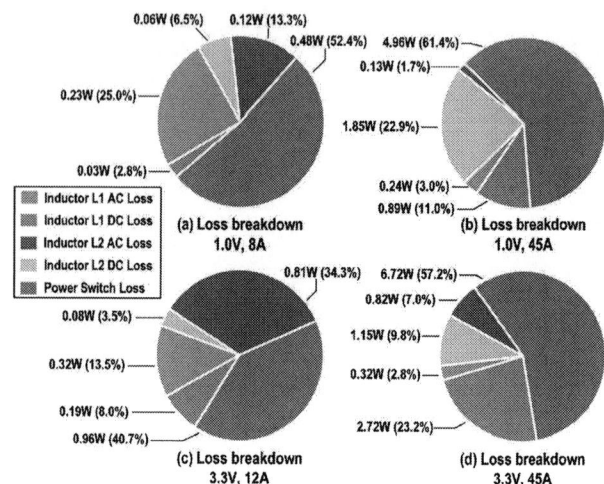

Fig. 7. Simulated power loss breakdown. Inductor AC losses include winding and core losses; power switch losses include all losses related to power switches.

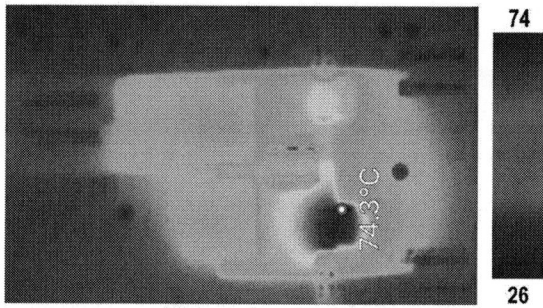

Fig. 8. Thermal image at 30A, 1V output and air-cooling only.

inductor current waveforms for the inductors used in this design and the other with near-optimum ratio are shown in Fig. 5. As shown in the summed inductor current waveforms, although the non-optimum inductor ratio (Fig. 5a) has imperfect ripple cancellation, the difference is minor enough not to affect the output voltage ripple, while the efficiency difference is relatively more significant.

The effective switching frequency of the converter is set to 400kHz, and the output voltage is measured from 1V to 3.3V.

The switching frequency can be increased to further reduce the size of the output inductors and increase power density, however, with trade-offs in efficiency and the dead-time tuning, which is realized in a low-end FPGA in this work. Fig. 6 shows the measured efficiency for various output voltages with the peak efficiency of the 3LQB being 93.5% at 3.3V output voltage and 89.4% at 1V along with the breakdown of the power losses in Fig. 7. The efficiency is maintained above 70% across the entire targeted current range of 45A, dipping slightly below 70% at the 1V max load condition. As a result, the converter achieves a power density of 3,317 W/in³ for a 3.3V output and 1,005 W/in³ for a 1V output. The loading range and power density could be further extended with better thermal design.

A thermal image of the converter is shown in Fig. 8 with only air-cooling at thermal equilibrium. For a higher load current, the temperature of Q_6 starts to heat up rapidly due to the increase of conduction loss and requires the use of heat-sinks to reduce thermal resistance and dissipate heat. Due to the low component count of the topology, only a single device is conducting in a given phase as compared to other topologies which rely on multiple phases and devices to conduct the load current. The performance of the converter could be improved by using a larger device such as EPC2066 for Q_6 which has a smaller on-resistance and larger package to dissipate more heat. Table II compares this work with prior hybrid converter

TABLE II.   COMPARISON TABLE WITH PREVIOUS WORK

	APEC 19 [1]	ECCE 21 [2]	TPE 24 [3]	TIA 23 [4]	ECCE 22 [5]	APEC 24 [6]	APEC 23 [11]	This Work
Topology	MPMIH	DIH	SDIH	DPHD	CADIHD	SDIH	SQB	3LQB
Num. of Switches	9	8	14	10	9	16	4	6
Num. of Inductors	3	2	2	1	2	2	2	2
Num. of Capacitors	5	5	10	6	4	10	1	2
Switching Freq.	300 kHz	750 kHz	750 kHz	250 kHz	500 kHz	1 MHz	300 kHz	400 kHz
V_{IN} (V)	24 – 54	48	48	36 – 65	40 – 65	48	48	48
V_{OUT} (V)	1 – 2	1 – 3	1 – 3	1 – 2	1	1	1	1 – 5
Max I_{OUT} (A)	40A	70A	60A @ 3V 65A @ 2V 115A @ 1V	32A	80A	50A	20A	45A
Peak Efficiency	94.6% @ 2V 92.2% @ 1V	93.8% @ 3V 92.2% @ 2V 87.5% @ 1V	91.4% @ 3V 89.4% @ 2V 83.5% @ 1V	92.7% @ 2V 90.6% @ 1V	94.3% @ 1V	91.9% @ 1V	94.5% @ 1V	93.5% @ 3.3V 91.1% @ 1.5V 89.4% @ 1V
Power Density* (W/in³)	425 @ 2V 213 @ 1V	892 @ 3V 715 @ 2V 422 @ 1V	2,464 @ 3V 1,807 @ 2V 1,588 @ 1V	481 @ 2V 241 @ 1V	636 @ 1V	2,020# @ 1V	73 @ 1V	3,317 @ 3.3V 1,508 @ 1.5V 1,005 @ 1V

* Calculated with maximum power over total power stage component volume.

Calculated with maximum power over box volume.

designs. For a fair comparison, only measurements at the output voltage range of 1V to 3.3V are provided as other topologies did not measure with a 5V output. Compared to other state-of-the-art hybrid PoL converters, the proposed 3LQB design achieves a decent efficiency and high power density while utilizing the fewest number of switches and passive components.

IV. CONCLUSION AND FUTURE WORK

This work presents a new hybrid topology for 48V-to-PoL converters with a three-level stage followed by an inductor-first stage which reduces the size of the output inductors that usually dominate the size of a PoL converter. The design is able to achieve 48V to 1-5V conversion, 93.5% peak efficiency, and 3.3-kW/in³ maximum density while using only 6 switches and few passive components. Future work includes developing a close-loop controller with small-signal analysis to achieve a fast and stable voltage regulation.

ACKNOWLEDGEMENTS

Kejia Wang and Si Yuan Sim contributed equally to this paper.

REFERENCES

[1] R. Das, G. -S. Seo, D. Maksimovic and H. -P. Le, "An 80-W 94.6%-Efficient Multi-Phase Multi-Inductor Hybrid Converter," *IEEE Applied Power Electronics Conference and Exposition (APEC)*, Anaheim, CA, USA, 2019, pp. 25-29.

[2] N. M. Ellis and R. C. N. Pilawa-Podgurski, "Modified Split-Phase Switching with Improved Fly Capacitor Utilization in a 48V-to-POL Dual Inductor Hybrid-Dickson Converter," *IEEE Energy Conversion Congress and Exposition (ECCE)*, Vancouver, BC, Canada, 2021, pp. 1735-1740.

[3] N. M. Ellis, R. A. Abramson, R. Mahony and R. C. N. Pilawa-Podgurski, "The Symmetric Dual Inductor Hybrid Converter for Direct 48V-to-PoL Conversion," *IEEE Transactions on Power Electronics*, vol. 39, no. 6, pp. 7278-7289, June 2024

[4] C. Chen, J. Liu and H. Lee, "Dual-Path Hybrid Dickson Converter for High-Ratio Conversions in Point-of-Load Applications," *IEEE Transactions on Industry Applications*, vol. 59, no. 6, pp. 6914-6925, Nov.-Dec, 2023

[5] W. Han, C. Chen, J. Liu and H. Lee, "An 80A 48V-Input Capacitor-Assisted Dual-Inductor Hybrid Dickson Converter for Large-Conversion Ratio Applications," *IEEE Energy Conversion Congress and Exposition (ECCE)*, Detroit, MI, USA, 2022, pp. 1-5.

[6] N. M. Ellis, Y. Zhu and R. C. N. Pilawa-Podgurski, "Gallium Nitride-based 48V-to-1V Point-of-Load (PoL) Converter for Aerospace Telecommunications and Computing Applications," *IEEE Applied Power Electronics Conference and Exposition (APEC)*, Long Beach, CA, USA, 2024, pp. 1384-1388.

[7] M. Halamicek, T. McRae and A. Prodić, "Cross-Coupled Series-Capacitor Quadruple Step-Down Buck Converter," *2020 IEEE Applied Power Electronics Conference and Exposition (APEC)*, New Orleans, LA, USA, 2020, pp. 1-6.

[8] J. W. Kwak and D. Brian Ma, "A Conduction-Loss-Conscious 4-Level Power Converter with Tri-Path Synchronous Rectification for High Step-Down DC-DC Conversion," *2024 IEEE Applied Power Electronics Conference and Exposition (APEC)*, Long Beach, CA, USA, 2024

[9] Y. Zhu, J. Zou and R. C. N. Pilawa-Podgurski, "A 1500-A/48-V-to-1-V Switching Bus Converter for Next-Generation Ultra-High-Power Processors," *IEEE Transactions on Power Electronics*, vol. 39, no. 9, pp. 11340-11355, Sept. 2024

[10] Y. Elasser et al., "Mini-LEGO CPU Voltage Regulator," *IEEE Transactions on Power Electronics*, vol. 39, no. 3, pp. 3391-3410, March 2024

[11] J. W. Kwak and D. B. Ma, "Superimposed Quadratic Buck Converter for High-Efficiency Direct 48V/1V Applications," *IEEE Applied Power Electronics Conference and Exposition (APEC)*, Orlando, FL, USA, 2023, pp. 2253-2259.

[12] D. Reusch, F. C. Lee and M. Xu, "Three level buck converter with control and soft startup," *IEEE Energy Conversion Congress and Exposition (ECCE)*, San Jose, CA, USA, 2009, pp. 31-35.

[13] V. Yousefzadeh, E. Alarcon and D. Maksimovic, "Three-level buck converter for envelope tracking applications," *IEEE Transactions on Power Electronics*, vol. 21, no. 2, pp. 549-552, March 2006

[14] A. Abdulslam and P. P. Mercier, "A Continuous-Input-Current Passive-Stacked Third-Order Buck Converter Achieving 0.7W/mm2 Power Density and 94% Peak Efficiency," *IEEE International Solid-State Circuits Conference (ISSCC)*, San Francisco, CA, USA, 2019, pp. 148-150.

[15] Y. -Y. Kao et al., "A Monolithic GaN-Based Driver and GaN Power HEMT with Diode-Emulated GaN Technique for 50MHz Operation and Sub-0.2ns Deadtime Control," *IEEE International Solid-State Circuits Conference (ISSCC)*, San Francisco, CA, USA, 2022, pp. 228-230.

Air-LEGO: A Magnetic-Free Ultra-Thin 24V-to-1V 120A VRM with Air-Coupled Inductors

Haoran Li°, Wenliang Zeng°, Youssef Elasser†, and Minjie Chen°

°*Princeton University, Princeton NJ, United States*
†*Nvidia Research, Durham NC, United States*
Email: {haoranli, wz6581, minjie}@princeton.edu

Abstract—This paper presents a magnetic-free ultra-thin 24 V-to-1 V 120 A linear-extendable group-operated point-of-load (LEGO-PoL) voltage regulation module (VRM) with air-coupled inductors (Air-LEGO). The thickness of the Air-LEGO converter is very low (about 3 mm), facilitating compact packaging, relaxed thermal constraints, and magnetic-free integration. The module operates by initially stepping down the 24 V input voltage using three series-stacked 2:1 switched-capacitor submodules, resulting in an average voltage of 4 V on the virtual intermediate buses. This virtual bus voltage is then further stepped down to below 1 V through three two-phase buck submodules. Each of these six phases is capable of delivering a peak current of 20 A and regulating the output voltage. Buck submodules are implemented with air-coupled inductors, where designs based on both PCB and Litz-wire windings are explored and compared. Experimental results highlight the viability and potential of incorporating air-coupled inductors into VRM designs, while also addressing associated drawbacks and essential design considerations.

Index Terms—voltage regulation module, point-of-load converter, air-core inductor, coupled inductor.

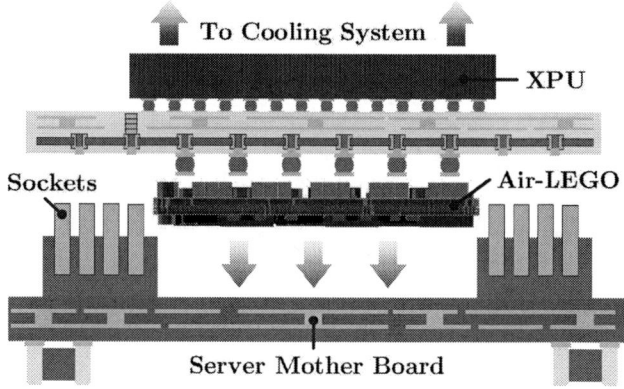

Fig. 1. Ultra-thin Air-LEGO VRM embedded into a CPU or GPU package that fits within the constrained space for high-level compact integration. The air-coupled inductor enables an overall system height lower than 3 mm while effectively mitigating magnetic thermal constraints.

I. INTRODUCTION

RAPID advancement of high-performance computing systems, such as CPUs and GPUs, has significantly amplified the demand for compact, efficient, and reliable power delivery solutions. These technological strides necessitate voltage regulation modules (VRMs) that can handle higher currents and ensure stable power delivery within the constrained spaces of modern electronic devices, as depicted in Fig. 1. As CPUs and GPUs continue to grow in computational power, their associated power consumption scales accordingly, creating a substantial challenge for VRMs to deliver adequate power without compromising critical factors such as spatial efficiency, energy efficiency, and signal integrity [1], [2].

Typically, VRMs have relied on magnetic core-based inductors to meet the requirements for voltage regulation and current handling capabilities. While these inductors have proven effective, they come with inherent limitations, including core losses, thermal dissipation requirements, and substantial volume, all of which constrain the overall efficiency and miniaturization potential of VRMs [3], [4]. Additionally, the need for a magnetic core also hinders high density integration, advanced packaging, and effective thermal management [5]–[9]. Converters utilizing magnetic core inductors often face challenges due to the temperature sensitivity of the core materials. Unlike semiconductor devices, which can typically withstand higher operating temperatures (up to a few hundred °C), magnetic cores are much more temperature-sensitive (less than 100°C). Key parameters such as permeability and core losses can vary substantially with temperature changes [10], and the performance of magnetic cores may degrade or even fail under elevated temperatures. These thermal limitations impose constraints on the integration, reliability, and optimization of VRM designs.

In contrast, air-core inductors eliminate the these temperature-related dependencies entirely. Without the need for magnetic materials, air-core designs offer greater thermal resilience, expanding the design space and enhancing the adaptability of converters to various thermal environments. This freedom not only relaxes the temperature constraints of magnetic cores but also enables more robust and optimized VRM architectures, providing improved scalability and performance for modern high-power applications. Without satiation, core loss, and frequency limit, air-core magnetics can be pushed to operate at very high frequencies while maintaining high system performance [11], [12]. Multi-phase air-core magnetics, e.g., the origami inductors developed in [13], can benefit from multiphase coupling for flux cancellation, ripple reduction, and improved transient [14].

This paper systematically investigates the potential of using air-coupled inductors in ultra-thin VRM designs, which bypasses the limitations imposed by magnetic cores. The proposed Air-LEGO converter leverages air-coupled inductors

979-8-3315-1612-3/25 $31.00 © 2025 IEEE

to achieve a low profile of just 3 mm. Air-core inductors offer promising opportunity in the MHz frequency range [11]. The two-phase or multiphase coupling of these inductors fundamentally further reduce the energy storage needed in the inductors, ensuring a fast dynamic response [13], [14].

Key contributions of this paper include analysis, simulation, and experimental validation of the Air-LEGO architecture, with a focus on demonstrating the feasibility, benefits, and challenges associated with air-coupled inductor-based VRMs in high step-down applications. The rest of the paper is organized as follows: Section II introduces the circuit topology and operation principles of Air-LEGO architecture; Section III demonstrates the design consideration and simulation results of air-coupled inductors; Section IV presents the experimental results of the Air-LEGO prototype, highlighting the feasibility and associated drawbacks of using air-coupled inductors. Finally, Section V concludes the paper.

II. LEGO ARCHITECTURE AND OPERATION PRINCIPLES

The Air-LEGO VRM design builds upon the Linear-Extendable Group Operated (LEGO) architecture [5], [6], as illustrated in Fig. 2. The LEGO architecture features a series-input and parallel-output configuration, making it particularly well-suited for high-current power delivery applications that require large voltage conversion ratios. The LEGO architecture includes two primary building blocks: an initial step-down stage employing series-stacked 2:1 switched-capacitor (SC) units and a secondary step-down stage utilizing multi-phase buck units, together with multi-phase coupled inductors.

In the first stage of this Air-LEGO design, an input voltage of 24 V is stepped down to three intermediate bus voltages of 4 V each using a set of three series-stacked 2:1 switched-capacitor submodules. This series-stacked arrangement not only ensures balanced voltage distribution across all submodules but also minimizes voltage stress on individual components, thereby enhancing reliability and enabling the use of lower-voltage-rated devices.

The second stage further converts the intermediate bus voltage of 4 V to the target output voltage, such as 1 V, using three multi-phase buck converter submodules. Each submodule features two interleaved phases, operating at a duty cycle of approximately 25%, which significantly improves current-handling capability while effectively reducing output voltage ripple. The interleaved operation helps distribute thermal dissipation and reduces electromagnetic interference (EMI), contributing to the overall system's efficiency and reliability. In this design, each phase is capable of delivering up to 20 A, culminating in a total output current capacity of 120 A.

Moreover, the modular nature of the LEGO architecture allows for easy customization and scalability, which is critical for addressing the diverse power demands of modern computing systems. Additional series-stacked or parallel submodules can be incorporated to accommodate varying voltage conversion ratios or to support even higher current demands.

Figure 3 depicts the switching strategy of LEGO architecture. The 2:1 switched-capacitor unit operates with fixed complementary 50% duty ratio gate signals (ϕ_1 and ϕ_2) at a

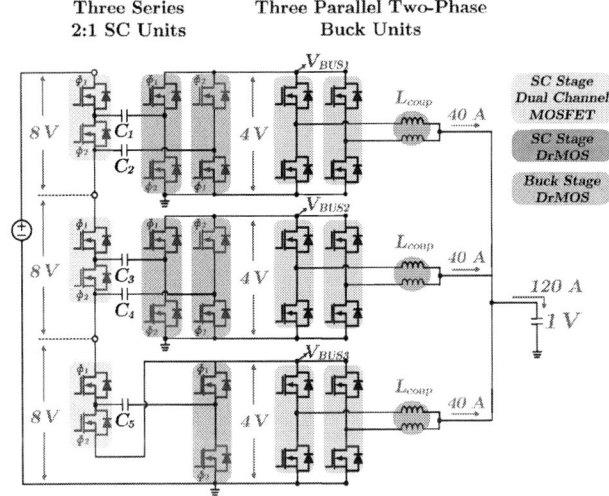

Fig. 2. Principles of LEGO architecture, consisting of three series-stacked switched-capacitor submodules and three paralleled multi-phase buck submodules with coupled inductors. LEGO architecture provides large voltage conversion ratio and high current capacity with reduced switch stress for high current VRM applications.

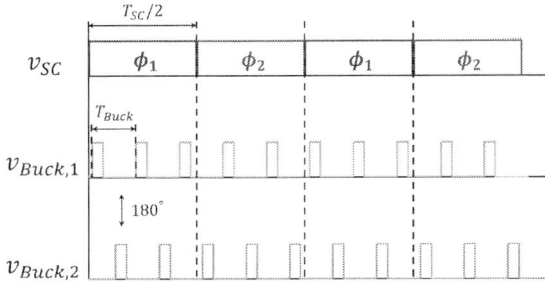

Fig. 3. Theoretical gate drive modulation strategy of LEGO architecture. The switched-capacitor unit is driven by low frequency 50% duty ratio square waves with 180° phase difference. The two-phase buck unit is driven by a higher frequency sequence with desired duty ratio, where two phases are 180° interleaved. Soft-charging of capacitors, zero-current switching of SC unit, and phase current balancing through phase rotation are achieved.

relatively low switching frequency (i.e., a few hundred kHz). This low-frequency operation reduces switching losses, enhancing overall efficiency. Meanwhile, the buck stage utilizes interleaved high-frequency operation, which reduces output current ripple and provides a high control bandwidth for fast transient and precise regulation.

A unique feature of this architecture is the functional integration of the two stages. Without the need for a resonant inductor in the switched-capacitor stage or a large dc decoupling capacitor between the stages, the switched-capacitor and buck units are inherently merged. The buck units act as controlled current sources to soft-charge the switched capacitors. The capacitors in the switched-capacitor units are continuously charged and discharged by the current sourced from the buck stage. Furthermore, by coordinating the switching sequences, the buck inductor currents only flow through the switched-capacitor units when the high-side buck switches are active.

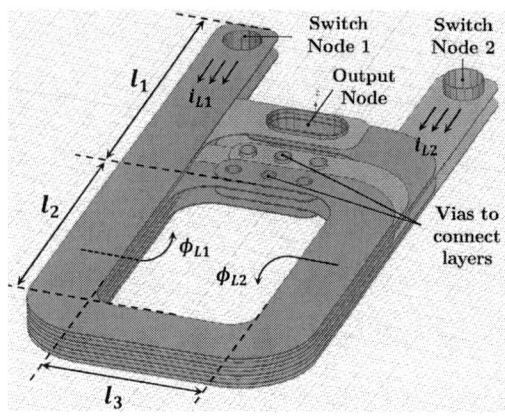

Fig. 4. Basic geometry the two-phase air-coupled inductor. Three parameters need to be determined, where the coupling factor between two phases is mostly determined by l_2 and l_3, while l_1 provides additional self-inductance.

This enables zero-current switching (ZCS) in the switched-capacitor units by aligning their switching transitions with the free-wheeling state of the buck units, significantly reducing switching losses.

The LEGO architecture also incorporates mechanisms to ensure automatic current balancing among phases. Current balancing is achieved by properly coordinating the switching frequencies of the switched-capacitor and buck stages. Since the absence of a large dc decoupling capacitor leads higher ripple in the virtual intermediate bus voltages, there is potential for phase current mismatch in the buck units. To address this, a passive phase rotation scheme [5] is implemented. When the buck switching frequency is chosen as:

$$f_{Buck} = (2k+1) \cdot f_{SC}, \quad k = 1, 2, 3... \tag{1}$$

where f_{Buck} and f_{SC} are the switching frequencies of the buck stage and switched-capacitor stage, respectively, an odd number of buck switching events (e.g., five, as shown in Fig. 3) occurs within one switching cycle of the switched-capacitor stage. This odd-switching frequency alignment causes the buck phases to rotate through different switch positions, with each phase taking its turn to experience the highest virtual bus voltage ripple. This ensures balanced average input voltages across all buck switches and effectively mitigates phase current mismatches, resulting in uniform current distribution across the entire system.

III. AIR-COUPLED INDUCTOR DESIGN

The design of air-coupled inductors for the Air-LEGO converter requires optimization of the inductor winding geometry to balance self-inductance, mutual inductance, and winding resistance. Figure 4 depicts the fundamental geometry of the 2-phase air-coupled inductor with 2 turns each phase, focusing on three critical dimensions: l_1, l_2, and l_3. These dimensions directly impact the electromagnetic properties of the inductor. The overlap region defined by $l_2 \times l_3$ plays a key role in determining the coupling factor between the two phases. A higher coupling factor enhances energy transfer and ripple cancellation but may increase parasitic effects. Conversely,

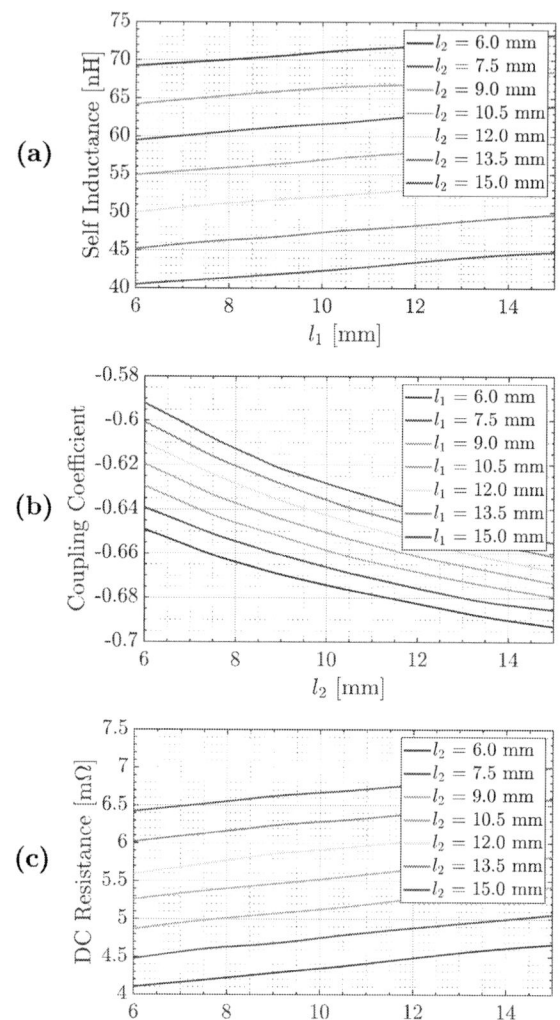

Fig. 5. Simulation results with parameter sweeping for different l_1 and l_2 combinations: (a) self inductance per phase; (b) coupling coefficient between phases; and (c) Dc resistance per phase.

the l_1 portion, which remains largely uncoupled, provides additional self-inductance, offering an extra degree of freedom to determine the characteristics of the inductor. The dimension l_3 is largely constrained by the physical footprint of the power devices and cannot be significantly adjusted without affecting the overall layout of the VRM. This makes the proper selection of l_1 and l_2 critical to achieving an optimal trade-off between performance and efficiency. A comprehensive parameter sweep simulation was conducted, evaluating the interplay between self-inductance, mutual inductance, and winding resistance under various configurations. Figure 5 presents the simulation results for different combinations of l_1 and l_2.

Based on the simulation results, l_1 was set to 12 mm and l_2 to 8 mm, values that offer a balanced design capable of meeting the desired system performance metrics. This configuration ensures adequate self-inductance, sufficient coupling, and reasonable winding resistance.

For PCB winding implementations, interleaving is a critical

979-8-3315-1612-3/25 $31.00 © 2025 IEEE

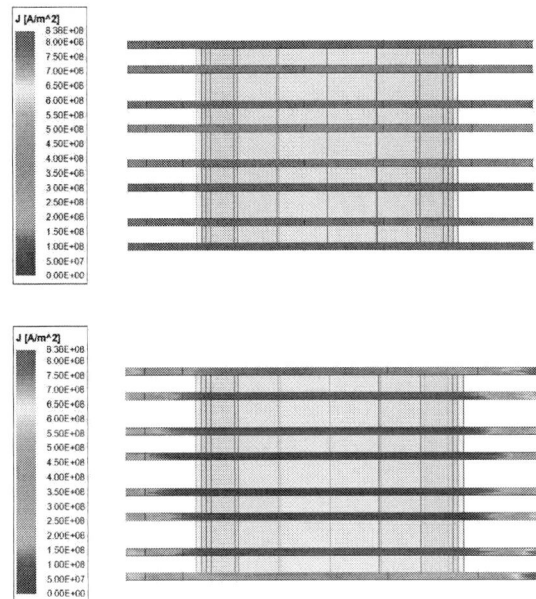

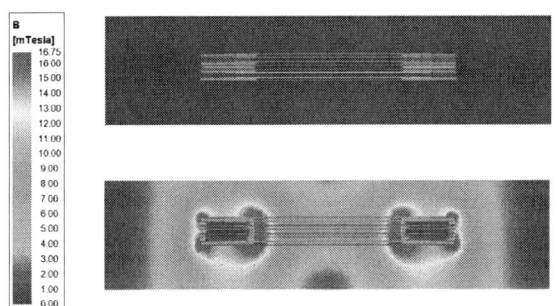

Fig. 6. Simulated current density distributions for interleaved structure (top) and non-interleaved structure (bottom), where interleaving greatly mitigates the skin effect and proximity effect, and reduces the winding ac resistance.

Fig. 7. Simulated flux density distributions for interleaved structure (top) and non-interleaved structure (bottom), where interleaving effectively minimizes fringing flux and confines it around the winding plane.

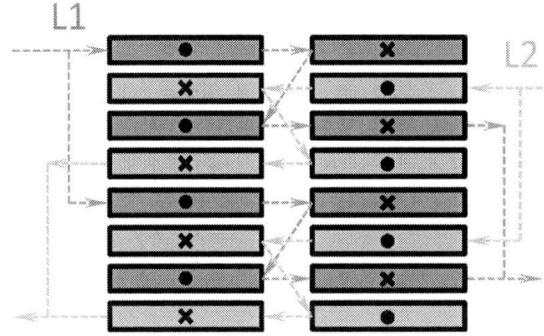

Fig. 8. The interleaved winding structure in this design, where each phase contains two sets of parallel windings, and the two phases are fully interleaved in an 8-layer PCB stack.

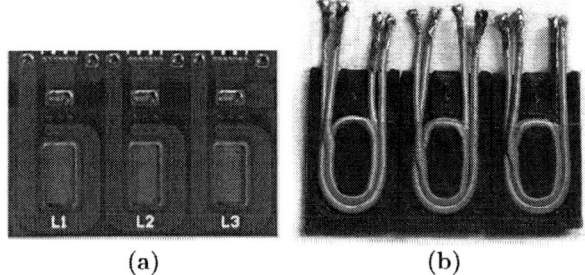

Fig. 9. Pictures of the air-coupled inductors: (a) PCB traces as windings; (b) Litz-wire as windings.

TABLE I
COMPARISON OF PCB-WINDING INDUCTOR AND LITZ-WIRE INDUCTOR.

Winding Type	R_{dc} [mΩ]	R_{ac}[mΩ] @2MHz	L_s [nH]	L_m [nH]
PCB Traces	4.46	17.9	47.56	-27.83
Litz-Wires	3.51	11.8	60.21	-34.85

technique to mitigate losses caused by the skin effect and proximity effect, which are especially pronounced at high switching frequencies [15]. Figure 6 illustrates the simulated current distributions for examples of interleaved and non-interleaved structures in an 8-layer PCB stack, highlighting the impact of interleaving on reducing ac resistance (R_{ac}). In the non-interleaved structure, the skin effect and proximity effect lead to uneven current distribution across the conductor, resulting in a simulated R_{ac} that is 2.25 times higher than in the interleaved design. Layer interleaving ensures that current flows more uniformly, reducing resistive losses and improving the efficiency of the inductor.

Additionally, interleaving minimizes the effects of fringing flux. A common concerns with air-core designs is the associated electromagnetic interference (EMI). There is potentially high-frequency stray magnetic fields dispersing into the surrounding space, interfering with the operation of nearby devices. However, interleaving provides an effective solution to this issue. As shown in Figure 7, interleaving significantly reduces fringing flux by arranging the windings so that opposing magnetic fields largely cancel each other. This cancellation ensures that the majority of the flux remains confined near the plane of the windings, mitigating the potential EMI risks, improving the signal integrity, and greatly enhancing the compatibility of the air-core inductors with other devices in densely packed systems. This capability is particularly important in compact VRM designs, where maintaining electromagnetic compatibility is critical for ensuring stable and reliable operation. Through proper interleaving, the air-coupled inductors achieve desired performance without compromising the system's EMI characteristics. Additional shielding sheets can be attached to the air-coupled inductor to further mitigate the EMI.

Figure 8 presents the selected interleaving structure, where each phase comprises two parallel winding sets, and the two

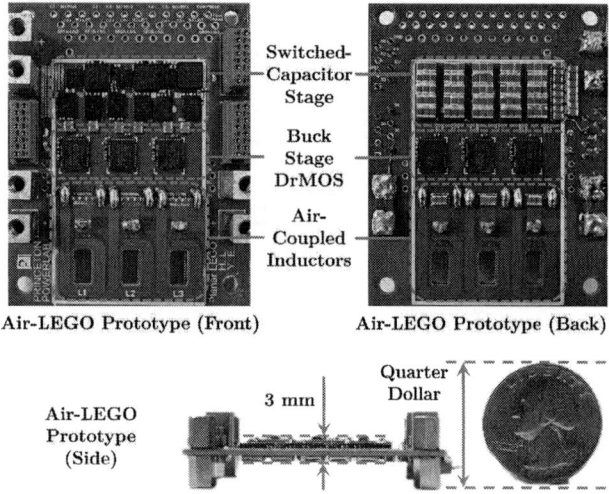

Switched-Capacitor Stage

Buck Stage DrMOS

Air-Coupled Inductors

Air-LEGO Prototype (Front)   Air-LEGO Prototype (Back)

Air-LEGO Prototype (Side)   3 mm   Quarter Dollar

Fig. 10. Pictures of the Air-LEGO prototype (with PCB-trace inductors) from front, back, and side views, achieving a low profile of 3 mm.

phases are fully interleaved to optimize performance. This structure helps to reduce both dc and ac resistance, and mitigates the EMI as aforementioned. The impact of parasitic capacitance caused by this fully interleaved structure is not obviously observed in the experiment.

For Litz wire implementations, a similar geometric configuration and size are adopted to match the PCB windings, ensuring similar electromagnetic characteristics. The wire gauge of the Litz wire is selected to closely match the total thickness of the PCB windings, allowing for a comparable low-profile design while retaining the advantages of reduced ac resistance. To maintain uniform parameters and achieve effective coupling, a bobbin and frame are specifically designed for the Litz wire inductors. These components ensure the windings remain tightly packed and precisely aligned, optimizing their electromagnetic properties.

Figure 9 showcases the completed inductors for both PCB and Litz wire implementations. The PCB-winding inductors are implemented with two 8-layer PCBs in parallel, where the copper thickness is 1 oz each and the total thickness is 2.1 mm. The Litz-wire inductors are implemented with Litz wires consisting of 300 strands of 46 AWG wires, resulting in a similar total thickness of 2.1 mm after being pressed. Table I provides a comparative analysis of their key performance metrics. The results indicate that while PCB windings offer superior integration, easier manufacturing, and lower profile, Litz wire inductors demonstrate slightly better efficiency at higher frequencies due to their inherent ability to mitigate the skin effect more effectively. Overall, the choice between PCB and Litz wire implementations depends on the specific application requirements. PCB windings are preferable for compact, high-density designs where integration is paramount, while Litz wire inductors are better suited for scenarios requiring maximum efficiency at ultra-high frequencies. Both approaches underscore the importance of optimizing winding design to achieve the best trade-offs between efficiency, EMI, and manufacturability in modern VRM systems.

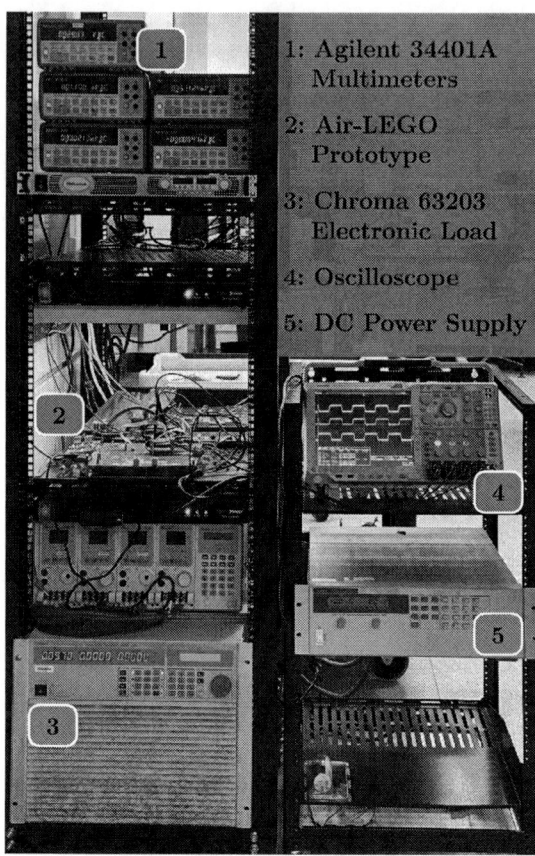

1: Agilent 34401A Multimeters

2: Air-LEGO Prototype

3: Chroma 63203 Electronic Load

4: Oscilloscope

5: DC Power Supply

Fig. 11. Pictures of the testing platform, including dc power supplies, electronic loads, digital multimeters with current shunts for power measurement, and an oscilloscope.

IV. EXPERIMENTAL RESULTS

To validate the strengths and performance of the Air-LEGO converter, a fully operational prototype was constructed and thoroughly tested. The floating switches in the switched-capacitor (SC) stage are implemented using Onsemi NT-TFD2D8N03P1E dual-channel MOSFETs. These switches are driven by Infineon 2EDF7275K isolated gate drivers, and a TI UCC27212 charge pump is employed for bootstrapping, providing the necessary gate drive voltage for the floating switches. The remaining switches in the SC stage are implemented with Onsemi FDMF3039 DrMOS devices, chosen for their integrated design and high efficiency. All switches in the buck stage utilize Infineon TDA21490 DrMOS modules, which are optimized for high switching frequencies and low conduction losses. The switching frequencies of the SC-stage and the buck stage are 400 kHz and 2 MHz, respectively.

Figure 10 provides a photograph of the prototype featuring PCB-winding inductors. The image highlights the compact integration of the air-coupled inductors, which significantly contribute to the converter's ultra-thin profile. The air-core design not only eliminates the thermal constraints associated with magnetic cores but also enables a high degree of design flexibility. The overall prototype demonstrates the feasibility of combining air-coupled inductors with the LEGO architecture

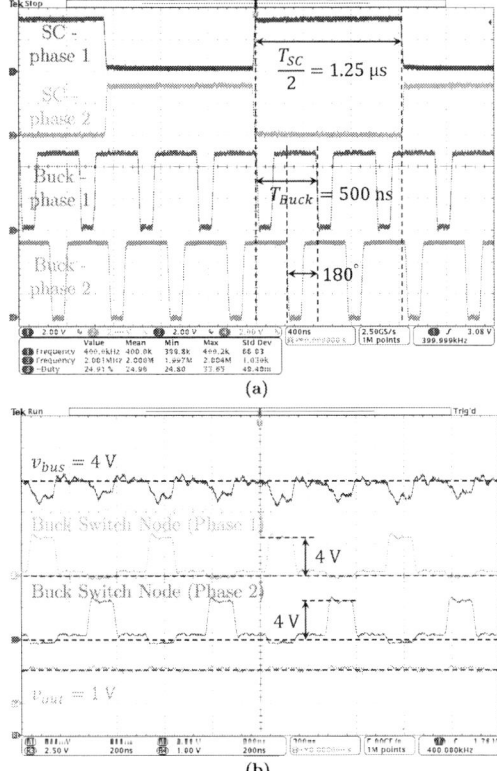

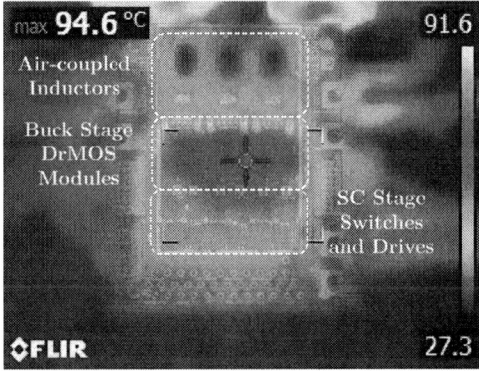

Fig. 12. Key experimental waveforms of the Air-LEGO prototype: (a) switched-capacitor stage 400 kHz gate drive signals, with 50% duty ratio and 180° phase shift; 2 MHz two-phase gate drive signals for one of the buck submodules, with 180° phase interleaving. (b) 4 V intermediate bus voltage, two-phase interleaved buck stage switch node voltage, and 1 V output voltage.

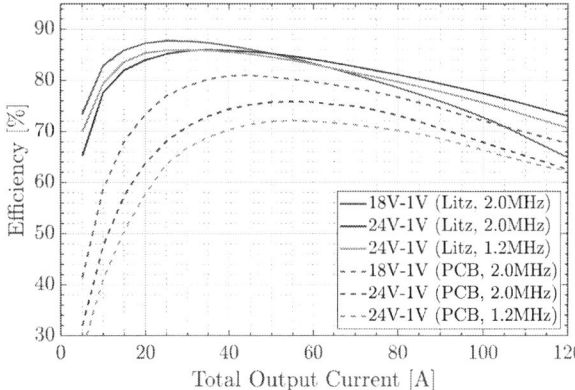

Fig. 14. Measured efficiency of the Air-LEGO prototype with either the Litz-wire inductors or PCB winding inductors, at different input voltages and different buck stage switching frequencies, excluding consumption of gate drives. Prototype with Litz-wire air-coupled inductors demonstrates higher efficiency.

Fig. 13. Thermal image of the Air-LEGO prototype with PCB windings at 24-to-1 V operation and 120 A output current. The DrMOS modules in the buck stage shows the highest temperature.

to achieve a low-profile VRM design. Notably, a second prototype was also constructed using aforementioned Litz-wire inductors, while keeping the rest of the design identical. The interfaces of the Litz wire-based inductors are fully compatible with the PCB-winding version, ensuring consistency in testing and performance comparison.

A series of tests were conducted to evaluate the performance of the prototype. Figure 11 presents the testing platform, which

includes a dc power supply, an electronic load, digital multimeters with current shunts for accurate power measurement, and an oscilloscope for signal monitoring. During the tests, the Air-LEGO prototype was cooled using a single dc fan without the addition of a heat sink, demonstrating its inherent thermal efficiency. In fact, the low-profile and compact design of the Air-LEGO converter, enabled by the flat air-coupled inductor structure, offers significant advantages for thermal management. This thin geometry makes it exceptionally easy to integrate a heat sink directly onto the system, providing effective cooling for both the semiconductor devices and the inductors. The potential to combine heat dissipation for these components further simplifies the overall thermal management and ensures enhanced system reliability, especially under high-power operating conditions.

Key experimental waveforms are shown in Fig. 12, illustrating the switching behavior and the typical operation of the Air-LEGO prototype. Figure 13 provides a thermal image of the Air-LEGO prototype with PCB windings while delivering a full load output current of 120 A at a 1 V output voltage. In this case, the hottest components are the buck stage switches, which handle the majority of the power dissipation.

To evaluate the performance of the Air-LEGO prototype under various operating conditions, efficiency tests were conducted across different input voltages and buck stage switching frequencies. Figure 14 presents the experimentally measured efficiency curves, highlighting the system's behavior under these varying conditions.

Efficiency measurements highlight the differences between the PCB winding and Litz wire winding implementations. With 24 V input voltage, the prototype with PCB windings achieved a peak efficiency of 75.8%, while the Litz wire windings demonstrated a significantly higher peak efficiency of 85.9%, benefiting from reduced ac resistance due to the optimized wire geometry. At full load, the efficiencies were 73.1% and 62.5%, respectively. More specifically, the efficiency test results reveal several valuable insights and conclusions worth discussing in detail:

979-8-3315-1612-3/25 $31.00 © 2025 IEEE   515

- **Different air-coupled inductor implementations:** Efficiency test results consistently show that Litz-wire inductors achieve noticeably higher efficiency compared to PCB-winding inductors. This efficiency advantage is primarily attributed to differences in resistive losses within the windings. As shown in Table I, Litz wire inductors exhibit significantly lower dc and ac resistance despite having nearly identical geometric dimensions to their PCB winding counterparts. The key reason for this difference lies in the limitations of the PCB-based design. With the overall thickness constrained, the copper layers in PCB windings are inherently thin, resulting in much higher winding resistance, particularly in terms of dc resistance. Additionally, Litz wire inductors offer superior mitigation of skin effect and proximity effect, which are significant at the operating frequencies of the Air-LEGO converter. This advantage enables Litz wire to achieve lower ac resistance as well, further reducing the resistive loss. On the other hand, the trade-off between the two implementation methods lies in their manufacturability and consistency. PCB windings are better suited for standardized production processes, offering excellent uniformity and repeatability in manufacturing. In contrast, Litz wire inductors, which rely on manual or semi-automated winding processes, are more challenging to produce with the same level of consistency, making scalability more difficult. Additionally, differences in the inductance values between the two types of inductors also contribute to variations in the overall efficiency of the converter. However, this impact is relatively minor compared to the influence of resistive losses.

- **Different buck stage switching frequencies:** The switching frequency of the buck stage significantly affects the efficiency of the converter, primarily because it operates in a hard-switching mode. Typically, lower switching frequencies are preferred to reduce switching losses and improve efficiency. However, in the test of Air-LEGO prototype, it was observed that operating at 1.2 MHz did not result in higher efficiency as expected; instead, it led to a measurable decrease in efficiency compared to the 2 MHz case. This is mostly due to the use of air-coupled inductors. Unlike conventional magnetic-core inductors, air-core inductors provide significantly lower inductance values. Consequently, the circuit experiences much larger current ripple, which is highly sensitive to switching frequency. At 1.2 MHz, the reduction in switching losses achieved by lowering the frequency is outweighed by the increase in resistive losses caused by the larger current ripple. This increased ripple not only amplifies conduction losses in the inductor windings but also impacts other components, such as switches and capacitors, leading to an overall decrease in efficiency. These findings align with the general understanding that air-core inductors are better suited for operation at relatively high switching frequencies. At higher frequencies, conventional magnetic-core inductors often suffer from substantial core losses, whereas air-core inductors maintain high efficiency. Additionally, higher frequencies help reduce current ripple, thereby improving overall performance. The tests also showed that at 1.2 MHz, there was a slight efficiency improvement under light-load conditions. In such scenarios, overall losses are dominated by switching losses rather than conduction losses, making the reduction in switching frequency beneficial.

- **Different input voltages:** Generally, larger voltage conversion ratios lead to lower efficiency, and our test results broadly align with this observation. Higher input voltages increase the switching losses in both the SC stage and the buck stage, while also increasing voltage ripple across the capacitors in the SC stage, further contributing to higher losses. In the current prototype, the switching devices in both the SC stage and the buck stage were initially selected for a 48 V input voltage. However, achieving a 48 V to 1 V conversion requires the buck stage to operate with an extremely low duty ratio, such as 12.5%. This significantly amplifies current ripple, resulting in excessive conduction and ripple-related losses. As a result, the rated input voltage for the prototype was ultimately set to 24 V or 18 V to balance performance and efficiency. Despite this adjustment, the overrating of the switching devices for lower voltages introduces additional inefficiencies. The higher $R_{DS(on)}$ and C_{oss} of the selected devices, due to the higher voltage rating, lead to increased conduction and switching losses, respectively. These factors lead to the overall low efficiency observed in the Air-LEGO prototype, particularly under full-load conditions. An interesting behavior was observed when the input voltage was set to 18 V. As the load current increases, the efficiency at 18 V begins to decline significantly. This is primarily due to the need for a progressively higher duty ratio to maintain a constant output voltage. At higher duty ratios, the SC stage loses the Zero-Current Switching (ZCS) condition depicted in Fig. 3, and transitions into hard-switching operation. This change drastically increases switching losses, causing a noticeable drop in overall efficiency.

Moreover, the loss breakdown is calculated for the Air-LEGO prototype under the conditions of a 24 V input voltage and a 2 MHz buck stage switching frequency, as shown in Fig. 15. The results reveal that the buck stage DrMOS devices contribute the largest portion of the total power loss, consistent with the observations from the thermal image in Fig. 13. The primary sources of these losses are the $R_{DS(on)}$ and C_{oss} characteristics of the over-rated devices, as well as the high switching frequency.

The winding resistive losses of the air-coupled inductors also account for a significant portion of the total losses. These losses are driven by the relatively small inductance values of the air-core inductors, which, combined with the small duty ratio, result in large current ripples. Additionally, the inherent large dc and ac resistances of the air-core inductors also contribute substantially to the overall loss.

Another observation is the increase in SC stage losses beyond approximately 60 A of output current. As the duty ratio increases, the SC stage loses its soft-switching (ZCS) conditions, transitioning into hard-switching operation. This

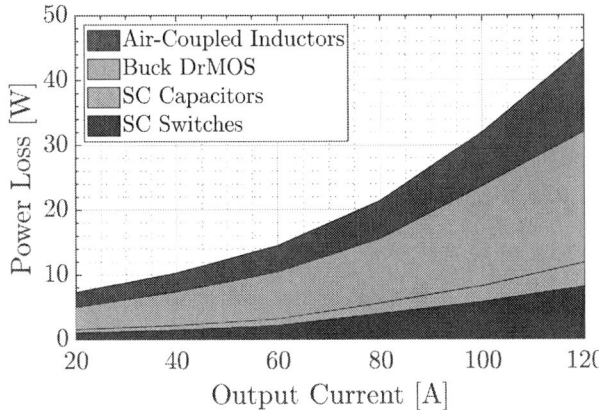

Fig. 15. Calculated loss breakdown of the Air-LEGO prototype during 24-V-to-1-V operation at 2 MHz with PCB-winding inductors. The switched capacitor stage loss is split into switching and conduction losses due to the switches, losses due to the capacitors. In the buck stage, the loss is divided into the device switching and conduction losses and the winding from the air-coupled inductors.

shift introduces additional switching losses, further reducing efficiency at higher load currents.

These insights provide valuable guidance for the design of the next version of Air-LEGO prototype. Key improvements include the better selection of switching devices optimized for the operating frequencies, rated voltages, and currents of the converter. Additionally, the air-core inductor design needs further optimization to increase the inductance value while minimizing both dc and ac resistances, all while maintaining the desired low-profile and thin form factor. These refinements are critical for achieving better efficiency and overall performance in future iterations.

V. Conclusions

This paper demonstrates the feasibility of incorporating air-coupled inductors into the VRM design, enabling ultra-low-profile and compact integration. By eliminating the use of magnetic cores, the design removes the temperature limitations associated with magnetic materials, significantly expanding the possibilities for VRM integration and simplifying thermal management. The absence of core losses also allows for more efficient cooling strategies, making the Air-LEGO architecture highly adaptable to advanced power delivery requirements.

Through comparative experiments, Litz wire windings were shown to outperform PCB windings, achieving superior efficiency due to their lower ac and dc resistance at the selected buck-stage switching frequency of 2 MHz and the designed geometry. Furthermore, proper interleaving of the windings not only helps to reduce ac winding losses but also effectively addresses EMI concerns by confining magnetic flux and minimizing stray fields. These improvements highlight the necessity of careful winding design in optimizing air-core inductor performance.

However, the prototype revealed some efficiency limitations. Both the SC stage and the buck stage experienced significant losses in their switching devices, and additional losses arose from the winding resistance of the air-coupled inductors. These losses underscore one of the inherent challenges of air-core inductor-based VRMs, where the trade-off between efficiency and compactness must be carefully managed.

Acknowledgements

This work was supported by Semiconductor Research Corporation (SRC) and Princeton Andlinger Center for Energy and the Environment.

References

[1] K. Radhakrishnan, M. Swaminathan, and B. K. Bhattacharyya, "Power delivery for high-performance microprocessors—challenges, solutions, and future trends," *IEEE Transactions on Components, Packaging and Manufacturing Technology*, vol. 11, no. 4, pp. 655–671, 2021.

[2] M. Chen, S. Jiang, J. A. Cobos, and B. Lehman, "Design considerations for 48-v vrm: Architecture, magnetics, and performance tradeoffs," in *2023 Fourth International Symposium on 3D Power Electronics Integration and Manufacturing (3D-PEIM)*, 2023, pp. 1–9.

[3] C. R. Sullivan, B. A. Reese, A. L. F. Stein, and P. A. Kyaw, "On size and magnetics: Why small efficient power inductors are rare," in *2016 International Symposium on 3D Power Electronics Integration and Manufacturing (3D-PEIM)*, 2016, pp. 1–23.

[4] B. Choi, J. Baek, B. C. Marin, S. Qu, S. Kulasekaran, J. I. Chavarria, L. E. Wojewoda, and K. Radhakrishnan, "Coaxmil 2.0 – next generation coaxial magnetic integrated inductors for higher efficiency fully integrated voltage regulator," in *2024 IEEE 74th Electronic Components and Technology Conference (ECTC)*, 2024, pp. 1044–1047.

[5] J. Baek, Y. Elasser, K. Radhakrishnan, H. Gan, J. P. Douglas, H. K. Krishnamurthy, X. Li, S. Jiang, C. R. Sullivan, and M. Chen, "Vertical stacked lego-pol cpu voltage regulator," *IEEE Transactions on Power Electronics*, vol. 37, no. 6, pp. 6305–6322, 2022.

[6] Y. Elasser, J. Baek, K. Radhakrishnan, H. Gan, J. P. Douglas, H. K. Krishnamurthy, X. Li, S. Jiang, V. De, C. R. Sullivan, and M. Chen, "Mini-lego cpu voltage regulator," *IEEE Transactions on Power Electronics*, vol. 39, no. 3, pp. 3391–3410, 2024.

[7] P. Wang, Y. Chen, G. Szczeszynski, S. Allen, D. M. Giuliano, and M. Chen, "Msc-pol: Hybrid gan–si multistacked switched-capacitor 48-v pwrsip vrm for chiplets," *IEEE Transactions on Power Electronics*, vol. 38, no. 10, pp. 12 815–12 833, 2023.

[8] Y. Elasser, H. Li, P. Wang, J. Baek, K. Radhakrishnan, S. Jiang, H. Gan, X. Zhang, D. Giuliano, and M. Chen, "Circuits and magnetics co-design for ultra-thin vertical power delivery: A snapshot review," *MRS Advances*, vol. 9, no. 1, pp. 12–24, Feb 2024. [Online]. Available: https://doi.org/10.1557/s43580-023-00724-w

[9] N. M. Ellis, R. A. Abramson, R. Mahony, and R. C. N. Pilawa-Podgurski, "The symmetric dual inductor hybrid converter for direct 48v-to-pol conversion," *IEEE Transactions on Power Electronics*, vol. 39, no. 6, pp. 7278–7289, 2024.

[10] D. Serrano, H. Li, S. Wang, T. Guillod, M. Luo, V. Bansal, N. K. Jha, Y. Chen, C. R. Sullivan, and M. Chen, "Why magnet: Quantifying the complexity of modeling power magnetic material characteristics," *IEEE Transactions on Power Electronics*, vol. 38, no. 11, pp. 14 292–14 316, 2023.

[11] D. J. Perreault, J. Hu, J. M. Rivas, Y. Han, O. Leitermann, R. C. Pilawa-Podgurski, A. Sagneri, and C. R. Sullivan, "Opportunities and challenges in very high frequency power conversion," in *2009 Twenty-Fourth Annual IEEE Applied Power Electronics Conference and Exposition*, 2009, pp. 1–14.

[12] E. A. Burton, G. Schrom, F. Paillet, J. Douglas, W. J. Lambert, K. Radhakrishnan, and M. J. Hill, "Fivr — fully integrated voltage regulators on 4th generation intel® core™ socs," in *2014 IEEE Applied Power Electronics Conference and Exposition - APEC 2014*, 2014, pp. 432–439.

[13] T. Sen, Y. Elasser, and M. Chen, "Origami inductor: Foldable 3-d polyhedron multiphase air-coupled inductors with flux cancellation and faster transient," *IEEE Transactions on Power Electronics*, vol. 39, no. 6, pp. 7312–7328, 2024.

[14] M. Chen and C. R. Sullivan, "Unified models for coupled inductors applied to multiphase pwm converters," *IEEE Transactions on Power Electronics*, vol. 36, no. 12, pp. 14 155–14 174, 2021.

[15] M. Chen, M. Araghchini, K. K. Afridi, J. H. Lang, C. R. Sullivan, and D. J. Perreault, "A systematic approach to modeling impedances and current distribution in planar magnetics," *IEEE Transactions on Power Electronics*, vol. 31, no. 1, pp. 560–580, 2016.

A 15A 48V-Input Dual-Path Hybrid Dickson Converter with 6 mm³ Low Saturation Current Inductors for Point-of-Load Conversion

[1]Hua Chen, [1]Young-Seok Noh, [2]Minxiang Gong, [2]Vivek De, [1]Arijit Raychowdhury

[1]School of Electrical and Computer Engineering, Georgia Institute of Technology, Atlanta, GA, USA
[2]Intel Labs, Hillsboro, OR, USA

hchen364@gatech.edu, nys2000v@gatech.edu, minxiang.gong@intel.com, vivek.de@intel.com,
arijit.raychowdhury@ece.gatech.edu

Abstract—This paper presents a dual-path inductor current hybrid Dickson converter for achieving high step-down ratio and high current density. The proposed topology combines a 4:1 switched-capacitor (SC) Dickson network and two inductive dual-path networks, which efficiently decrease power switch voltages and reduce average inductor currents by 40%. The proposed topology lowers the current flow at two inductors compare to the conventional hybrid Dickson topology, which results in high power density over 1,000A/in³ with 200mΩ, 6mm³ inductors, five flying capacitors, and twelve gallium nitride (GaN) switches. The proposed converter is demonstrated with 89.1% and 82.7% peak efficiency for 48V-4V and 48V-1.7V conversion ratios, respectively.

Index Terms—Hybrid Dickson converter, Dual-path circuit, merged converter, GaN devices, area-efficient inductor, high power density

I. INTRODUCTION

To optimize high-performance computing, AI processors demand large amount of current and draw attention to power delivery network (PDN) distribution losses [1]. Therefore, the high efficiency and compact size of the last stage high step-down DC-DC converter for data centers remains highly interesting. Recently, the multi-level hybrid converter, which overcomes the extreme short on-times and reduces switching voltage tolerances, has achieved high efficiency [2]-[4]. However, the bottleneck of the converter design is the huge inductor size. According to [2]-[4], inductors occupy more than 50% of the converter area, which is not suitable for a system-on-chip (SoC) or integrated voltage regulator (IVR) system. Fig. 1(a) shows the miniaturization trend of the surface-mount device (SMD) inductors. With the scaling down of inductor size, the direct-current resistance (DCR) increases in Fig.1(b) and saturation current decreases in Fig. 1(c) [5]-[7]. Even though [8] and [9] achieve a high peak efficiency with large SMD inductors, efficiency degraded severely at larger output load currents due to inductor DCR loss. With the small SMD inductors, the efficiency degradation becomes even more significant. To mitigate both the larger conduction loss and current handling issues on inductors, this paper introduces a dual-

This work was supported by Intel Custom Funding Grant.

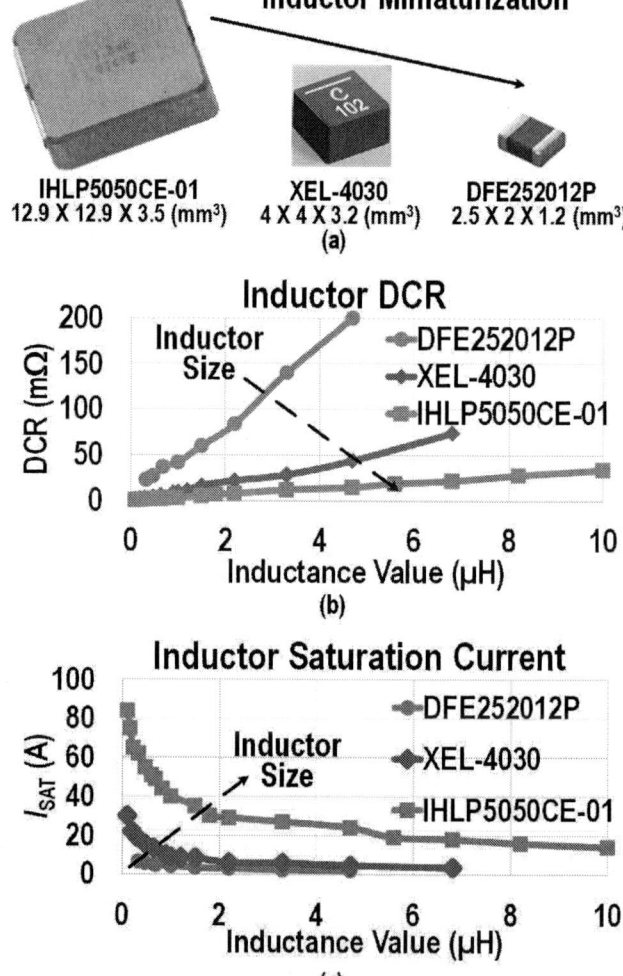

Fig. 1: Motivation: (a) SMD inductor sizes, (b) inductor DCR survey, (c) inductor saturation current survey.

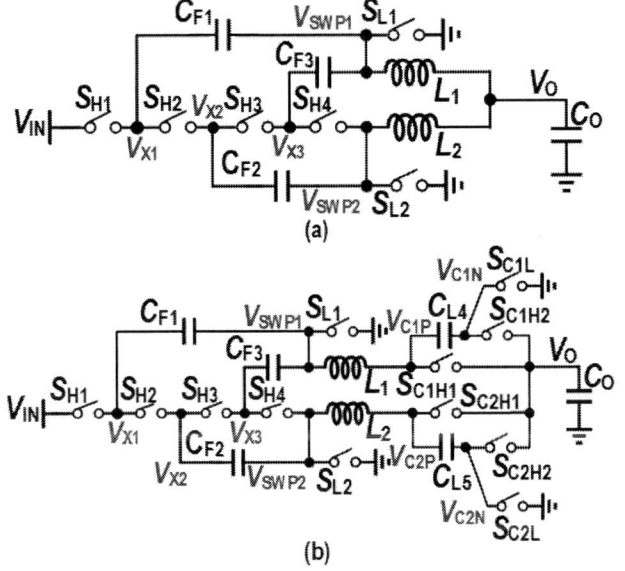

Fig. 2: Topology comparison: (a) conventional two-phase hybrid Dickson converter vs (b) proposed dual-path two-phase hybrid Dickson converter.

path hybrid Dickson converter, which combines the multi-level hybrid Dickson in Fig. 2(a) and dual-path current method to achieve high efficiency at wide load current handling for a 48V-1.7V conversion ratio with area-efficient inductors that only consumed 25-50% of total active converter area in Fig. 2(b).

II. PROPOSED DUAL-PATH HYBRID DICKSON (DPHD) CONVERTER

Fig. 2(b) shows the proposed topology combining the high-side hybrid Dickson converter [2], [8] with low-side inductor dual path [10], [11]. At the high side, four hybrid Dickson switches (S_{H1} - S_{H4}) and three flying capacitors (C_{F1} - C_{F3}) bring down the switching node voltages (V_{SWP1}, V_{SWP2}) by 4X and quadruple duty ratios (D_1, D_2) by 4X along with the switches S_{L1} and S_{L2}. Inductors, L_1 and L_2, are used to enable a two-phase interleaved operation that supports a higher load current. At the low side, two inductor current dual-path circuits consist of S_{C1L}, S_{C1H1}, S_{C1H2}, C_{L4} along with L_1 and S_{C2L}, S_{C2H1}, S_{C2H2}, C_{L5} along with L_2. C_{L4} is parallel with L_1 in θ_{1A}, and C_{L5} is parallel with L_2 in θ_{2A}. Since the output current flows through flying capacitors, C_{L4} and C_{L5}, the dual-path circuit structure releases the current stress from two inductor L_1 and L_2. Therefore, the proposed topology efficiently controls the current flows, which improve the load current with low saturation current inductors.

Fig. 3 shows the circuit operations to enable interleaving modes and dual-path currents, following the sequence θ_{1A}-θ_{1B}-θ_{2A}-θ_{2B}. Fig. 4 shows the circuit operation in details for all key switching and flying capacitor voltages and current nodes at inductors.

At high side, in the mode θ_{1A}, V_{IN} connects with C_{F1}, C_{F2} connects with C_{F3}, C_{L5} connects between L_2 and V_0, and C_{L4}

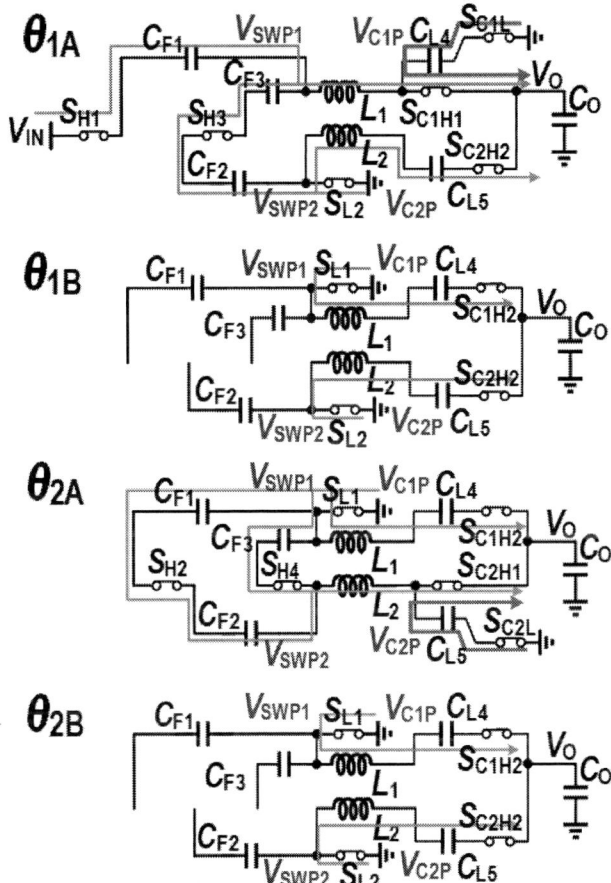

Fig. 3: Operating sequence of proposed dual-path hybrid Dickson converter.

connects between L_1 and ground (GND). C_{F1}, C_{F3}, C_{L5}, and L_1 are all charging, while C_{F2}, C_{L4}, and L_2 are all discharging. In the same way, in the mode θ_{2A}, C_{F1} connects with C_{F2}, C_{L4} connects between L_1 and GND, and C_{L5} connects between L_2 and ground (GND). C_{F1}, C_{F3}, C_{L5}, and L_1 are all discharging, while C_{F2}, C_{L4}, and L_2 are all charging. According to the structure of the two-phase hybrid Dickson converter, the high-side conversion ratio is described by equation (1) and (2). At low side, C_{L4} is parallel with V_O in mode θ_{1A}. After a short charge exchange between C_{L4} and C_O, C_{L4} equals to C_O, and C_{L4} supply the load current along with inductors. In the modes θ_{1B}, θ_{2A}, and θ_{2B}, C_{L4} is in series with C_O and L_1, V_{C1P} quickly reaches to $2V_O$, inductor current flows through L_1 equals the capacitor current C_{L4}. Similarly, in the mode θ_{2A}, C_{L5} is parallel with V_O. C_{L5} supplies current to the load, in θ_{1A}, θ_{1B}, and θ_{2B}, C_{L5} is in series with V_O and L_2. According to the inductor volt-second balance, the low-side conversion ratio is given in equation (3). Consequently, the total conversion ratio is given in equation (4). With the dual-path circuit as above, the inductor current reduction ratio equation is derived by charge-second balance of low-side capacitors as equation (5), which unlike the conventional hybrid Dickson converter, inductors provide 100% of the load according to Fig. 5.

979-8-3315-1612-3/25 $31.00 © 2025 IEEE   519

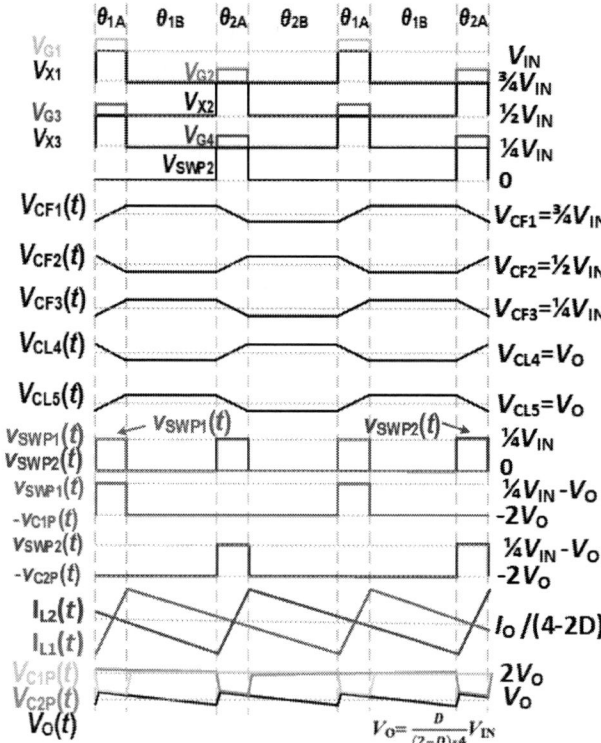

Fig. 4: Circuit Operation of the proposed dual-path hybrid Dickson converter.

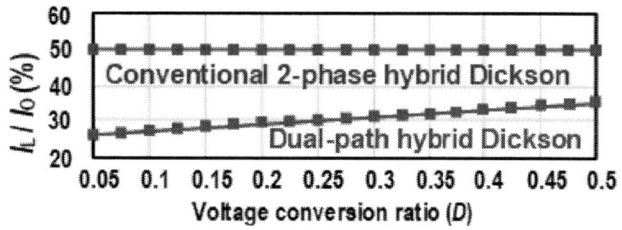

Fig. 5: Inductor current reduction ratio comparison across voltage conversion ratio.

$$v_{SWP1}(t) \approx V_{C1} - V_{C2} = V_{C3} = \frac{1}{4} V_{IN} \quad (1)$$

$$v_{SWP2}(t) \approx V_{IN} - V_{C1} = V_{C2} - V_{C3} = \frac{1}{4} V_{IN} \quad (2)$$

$$\frac{V_O}{v_{SWP1}(t)} = \frac{V_O}{v_{SWP2}(t)} \approx \frac{D}{2-D} \quad (3)$$

$$\frac{V_O}{V_{IN}} \approx \frac{D}{(2-D)*4} \quad (4)$$

$$\frac{I_L}{I_O} \approx \frac{1}{4-2D} \quad (5)$$

III. NEW BOOTSTRAPPED CIRCUITS FOR HIGH-SIDE FLOATING GATE DRIVERS

Fig. 6(a) shows the new bootstrapped circuits of turning on the high-side switches (S_{H1}, S_{H2}, S_{H3}). Each bootstrapped circuit is implemented with a Zener diode, a pre charged capacitor, a Schottky diode, and a gate driver chip (TI LMG1210). As

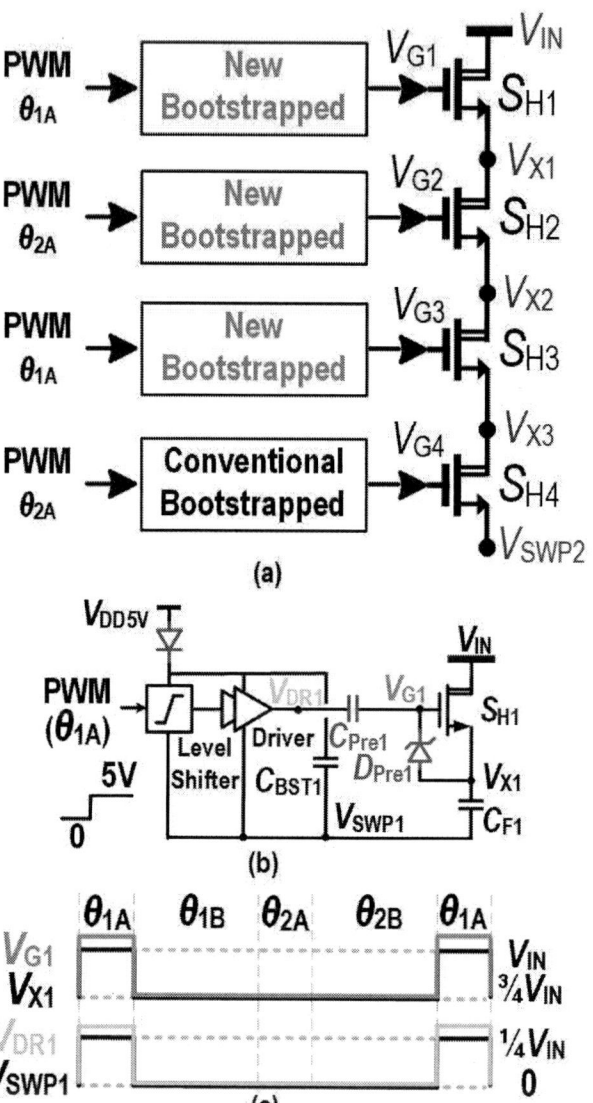

Fig. 6: (a) The refined bootstrapped circuit for high-side GaN power switches, (b) the bootstrapped circuit configuration for GaN S_{H1}.

shown in Fig. 6(b), C_{F1} always maintains the constant voltage difference between V_{X1} and V_{SWP1}, C_{Pre1} holds the voltage different between V_{DR1} and G_1, the bootstrapped circuit only requires one V_{DD} supply to turn on S_{H1} - S_{H4}. The pre-charged capacitor C_{Pre1} and clamped-diode D_{Pre1} not only set the gate voltage boundaries for the high-side switch, but also clamp the maximum voltage between the gate and source for S_{H1}. In Fig. 6(c), the level shifter and gate driver circuit remain in the low voltage domain, and textitC_{Pre1} clamps V_{G1} to high voltage domain. C_{Pre1} pumps V_{G1} higher to fully turn on S_{H1}. The light of this bootstrapped circuit comes from [12]. This method is commonly applied to all different high-side N-switches by a shared V_{DD}. As a result, the entire converter system only requires one V_{DD}.

IV. EXPERIMENTAL RESULTS

The performance of the proposed dual-path hybrid Dickson converter has been verified using a prototype board, as dipicted in Fig. 7, with a 48V input voltage 25.5W output power. The power stage area is 37.8mm X 24.8mm. Active components include GaN power switches, power inductors, and flying capacitors. The total chip coils (DFE252012P) area is $36mm^2$. The component selections and passive design considerations are listed in Table I.

Fig. 7: Dual-path hybrid Dickson prototype.

TABLE I
CIRCUIT COMPONENTS AND PARAMETERS

Item	Design Selection
S_{H1}, S_{H2}, S_{H3}, S_{H4}	EPC2014C, 40V, 16mΩ, EPC
S_{L1}, S_{L2}, S_{C1H1}, S_{C1H2}, S_{C2H1}, S_{C2H2}, S_{C1L}, S_{C2L}	EPC2023, 30V, 1.45mΩ, EPC
L_1, L_2	4.7μH, DFE252012P
C_{F1}, C_{F2}, C_{F3}	1μF, 0603, 50V, MLCC, TDK
C_{L4}, C_{L5}	4.7μF, 0603, 16V, MLCC, TDK
C_O	4.7μF (x2), 0402, 16V, MLCC, Samsung
$D_{ZBST1-8}$, D_{ZH1-3}	Diode Zener (x11), 4.7V, 100mW, CDZVT2R4.7B
D_{BST1-8}	Diode Schottky (x8), 60V, 100mW, RB520CM-60T2R
Gate Drivers	LMG1210 (x8), TI
Controller	LAUNCHXL-F28379D

Measured waveforms are included in Fig. 8. Fig. 8(a) shows measured high-side switching node. The front-end hybrid Dickson circuit converts 48V V_{IN} to two naturally balanced 12V switching nodes V_{SWP1} and V_{SWP2}, The balanced inductor currents L_1 and L_2 are displayed in Fig. 8(b). Fig. 8(c) shows the back-end low-side switching voltage waveforms, which indicates the capacitors C_{L4} and C_{L5} supplies the current to the load in a dual-path circuit configuration. Fig. 8(d) provides the gate timing diagram for GaN S_{H1}, so the bootstrapped circuit properly turn S_{H1} on and off with 19ns rising delay and 24ns falling delay. Fig.9 shows the balanced switching voltage nodes (V_{SWP1} and V_{SWP2}) and inductor currents under 6A load. The two inductors only supply about half of the total current at 6A load. The dual-path circuit helps to reduce the current on inductors, and reduce the inductor DCR loss. Fig. 10 plots 89.1% and 82.7% peak efficiency at 48V-4V and 48V-1.7V, respectively. Fig. 11 plots inductor current reduction ratio is

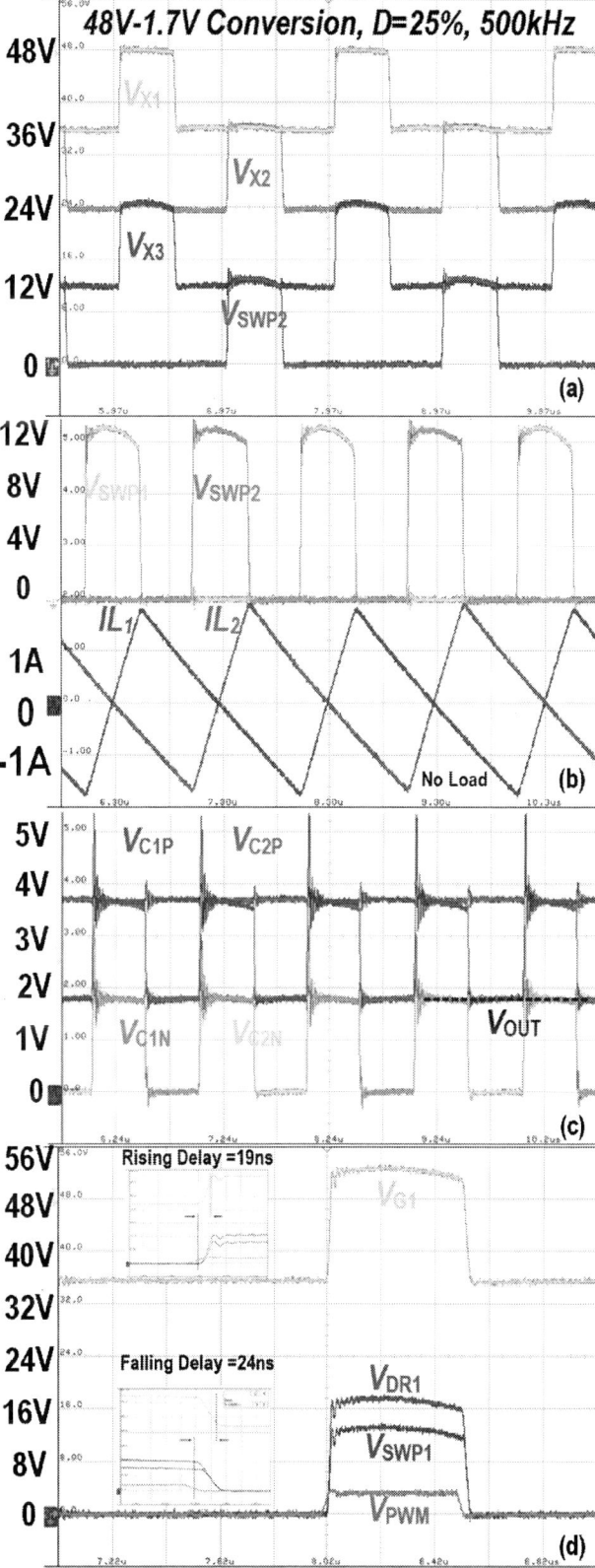

Fig. 8: Measured waveforms at no load: (a) high-side switching nodes, (b) inductor currents, (c) low-side switching nodes, and (d) S_{H1} gate timing.

50-60% across the wide load range, and the measurement is close to the simulation. Fig. 12 shows the estimated loss breakdown comparison. Switch FoM loss (device conduction loss and switching loss), inductor DCR conduction loss, and other loss are all considered. At 1A light load, conventional hybrid Dickson without the dual-path has higher efficiency. However, as the load current increases, the dual-path circuit saves more from inductor DCR loss. Two converters have almost the same loss and efficiency at 8A. After that, the dual-path hybrid Dickson converter becomes more efficient. At 15A load, this proposed converter reduces 22.1% inductor DCR conduction loss and saves 20% of the total loss compared to the conventional hybrid Dickson converter.

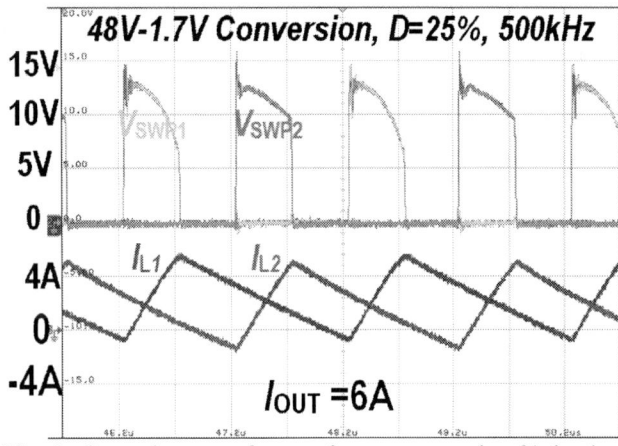

Fig. 9: Operation waveforms of prototype under 6A load.

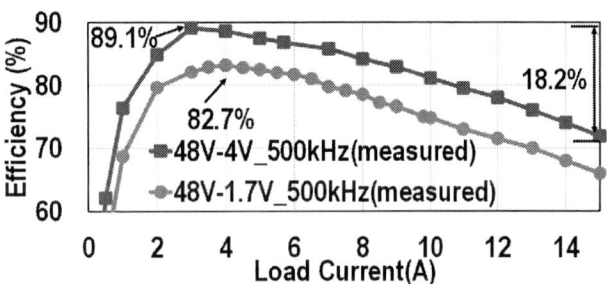

Fig. 10: Measured Efficiency at 48V-4V and 48V-1.7V Conversions.

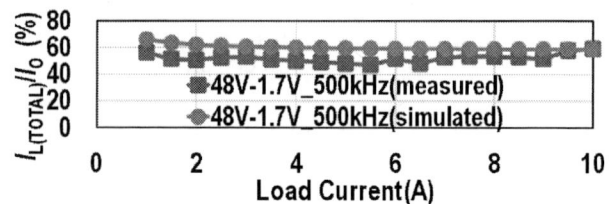

Fig. 11: Inductor current reduction ratio for 48V-1.7V conversion.

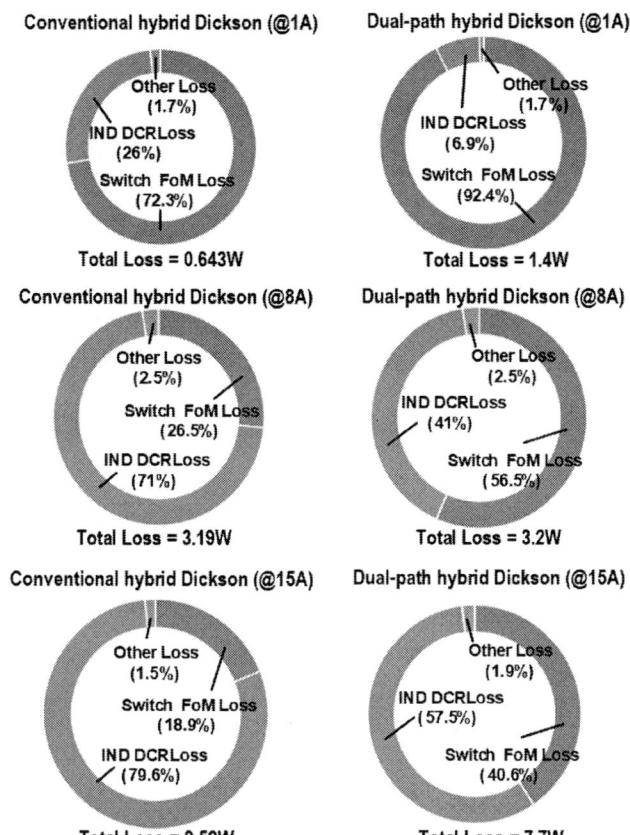

Fig. 12: Estimated loss breakdown comparison across 1-15A load.

TABLE II
COMPARISON WITH STATE-OF-THE ART CONVERTERS

	ECCE 2018 [2]	APEC 2019 [8]	ISSCC 2020 [3]	TPE 2023 [4]	Proposed
Technology	Discrete	Discrete	180nm BCD	Discrete	Discrete
Topology	Dual-phase Dual-inductor	Dual-phase Multi-inductor	Tri-state Double Step Down	Hybrid Dickson CADI	Hybrid Dickson Dual Path
Input Voltage V_{IN}	48V	48V	24V/12V	48V	48V
Output Voltage V_O	1-5V	1-5V	1V	1-2V	1-4V
Max Load Current I_{MAX}	10A	100A	3A	80A	15A
Power Switches	GaN	GaN	On-chip Si	GaN	GaN
Frequency f_{SW}	300kHz	333kHz	0.1-1MHz	500kHz	0.5-1MHz
Inductor	2x1.5μH	4x1μH	2x0.56μH	1x0.26μH	2x1.5μH (equivalent)
# of Flying Capacitor	3	3	2	4	5
Peak Efficiency	93%@48-1V,300kHz	90.9%@48-1V,333kHz	88.3%@24-1V,100kHz	94.3%@48-1V,500kHz	89.1%@48-4V,500kHz
Inductors Volume (mm^3)	1164.9	2329.7	102.4	1394	12-36
Current Density	/	440A/in^3	/	636A/in^3	1,113A/in^3@12A

*Current density for component area only (without C_O and C_{BST}).

V. Conclusions

Compared with the state-of-art converters in Table II, the proposed dual-path hybrid Dickson converter demonstrated the capability to maintain high current density, 1,113A/in^3, at a 48V/4V high step-down DC-DC converter PCB design utilizing 6mm^3 200mΩ DCR inductors. The proposed dual-path hybrid Dickson converter not only mitigates the inductor DCR conduction loss but also enhances the current handling in small power inductors. Future improvements are to include efficient integrating techniques such as on-chip switches, on-chip gate drivers, and on-chip controller to further enhance the power efficiency and power density.

References

[1] "The need for new AI processor power delivery," [Online]. Available: https://www.vicorpower.com/industries-and-innovations/computing.

[2] G. -S. Seo, R. Das and H. -P. Le, "A 95%-Efficient 48V-to-1V/10A VRM Hybrid Converter Using Interleaved Dual Inductors," 2018 IEEE Energy Conversion Congress and Exposition (ECCE), Portland, OR, USA, 2018, pp. 3825-3830, doi: 10.1109/ECCE.2018.8557715.

[3] K. Wei, Y. Ramadass and D. B. Ma, "A Direct 12V/24V-to-1V 3W 91.2%-Efficiency Tri-State DSD Power Converter with Online VCF Rebalancing and In-Situ Precharge Rate Regulation," 2020 IEEE International Solid-State Circuits Conference - (ISSCC), San Francisco, CA, USA, 2020, pp. 190-192, doi: 10.1109/ISSCC19947.2020.9063087.

[4] C. Chen, J. Liu and H. Lee, "Dual-Path Hybrid Dickson Converter for High Ratio Conversions in Point-of-Load Applications," in IEEE Transactions on Industry Applications, vol. 59, no. 6, pp. 6914-6925, Nov.-Dec. 2023, doi: 10.1109/TIA.2023.3295774.

[5] IHLP-5050CE-01 – IHLP® Commercial Inductors, High Saturation Series", [Online]. Available: https://www.vishay.com/docs/34105/ihlp-5050ce-01.pdf.

[6] "XEL4030 – Shielded Power Inductors XEL4030", [Online]. Available: https://www.coilcraft.com/getmedia/8245f050-f190-4295-8c41-7c03d662ee3d/xel4030.pdf.

[7] "DFE252012P - CHIP COIL (CHIP INDUCTOR) for consummer equipment", [Online]. Available: https://www.murata.com/en-global/products/productdetail?partno=DFE252012P-1R0M%23.

[8] R. Das and H. -P. Le, "A Regulated 48V-to-1V/100A 90.9%-Efficient Hybrid Converter for POL Applications in Data Centers and Telecommunication Systems," 2019 IEEE Applied Power Electronics Conference and Exposition (APEC), Anaheim, CA, USA, 2019, pp. 1997-2001, doi: 10.1109/APEC.2019.8722246.

[9] Z. Ye, R. A. Abramson and R. C. N. Pilawa-Podgurski, "A 48-to-6 V Multi-Resonant-Doubler Switched-Capacitor Converter for Data Center Applications," 2020 IEEE Applied Power Electronics Conference and Exposition (APEC), New Orleans, LA, USA, 2020, pp. 475-481, doi: 10.1109/APEC39645.2020.9124384.

[10] Y. Huh, S. -W. Hong and G. -H. Cho, "A Hybrid Structure Dual-Path Step-Down Converter With 96.2% Peak Efficiency Using 250-mΩ Large-DCR Inductor," in IEEE Journal of Solid-State Circuits, vol. 54, no. 4, pp. 959-967, April 2019, doi: 10.1109/JSSC.2018.2882526.

[11] J. -Y. Ko, Y. Huh, M. -W. Ko, G. -G. Kang, G. -H. Cho and H. -S. Kim, "A 4.5V-Input 0.3-to-1.7V-Output Step-Down Always-Dual-Path DC-DC Converter Achieving 91.5%-Efficiency with 250mΩ-DCR Inductor for Low-Voltage SoCs," 2021 Symposium on VLSI Circuits, Kyoto, Japan, 2021, pp. 1-2, doi: 10.23919/VLSICircuits52068.2021.9492478.

[12] L. Pham-Nguyen, V. -Q. Nguyen, D. -M. Nguyen, H. -D. Han, K. -H. Nguyen and H. -P. Le, "A 14-W 94%-Efficient Hybrid DC-DC Converter with Advanced Bootstrap Gate Drivers for Smart Home LED Applications," 2018 IEEE Energy Conversion Congress and Exposition (ECCE), Portland, OR, USA, 2018, pp. 4744-4749, doi: 10.1109/ECCE.2018.8558150.

An Ultra-fast Control Strategy and Pre-current-balancing Measures Prepared for Rapid Transients in Constant On-time Controllers

Yijie Qian
National ASIC System
Engineering Research Center
Southeast University
Nanjing, China
qianyj@seu.edu.cn

Yuan Gao
National ASIC System
Engineering Research Center
Southeast University
Nanjing, China
220226035@seu.edu.cn

Wenze Shu
National ASIC System
Engineering Research Center
Southeast University
Nanjing, China
xduswz@163.com

Lingyun Li
National ASIC System
Engineering Research Center
Southeast University
Nanjing, China
leelingy@163.com

Shen Xu
National ASIC System
Engineering Research Center
Southeast University
Nanjing, China
xus@seu.edu.cn

Weifeng Sun
National ASIC System
Engineering Research Center
Southeast University
Nanjing, China
swffrog@seu.edu.cn

Abstract—**The paper presents an ultra-fast control strategy during step-up transients in constant on-time (COT) controllers. This control strategy significantly improves the magnitude of undershoot over the traditional COT control method. In addition, the pre-current-balancing measures can quickly balance the current deviation caused by the high dynamic strategy without additional sampling circuits so that the current remains balanced during rapid load steps to avoid triggering overcurrent protection. The control strategy is tested on a 16-phase Buck converter, and the results show that it dramatically improves the undershoot and maintains the current balance.**

Keywords—Multi-phase Buck Converters, Dynamic Response, Current Balancing

I. INTRODUCTION

Multi-phase interleaved Buck converters with constant-on-time control (COT) are widely used in processor power supplies. Among products using COT control, the pulse distribution method, as shown in Fig. 1, is popular because of its high reliability and low cost due to its simple structure and the elimination of a phase-locked loop in phase distribution.

COT control was previously considered to have a better dynamic response in step-up scenarios compared to traditional PID control, which in most cases is true. However, with the development of processors that demand higher current exceeds 1000A and transient speed exceeds 1000A/µs [1], the limitations of COT control, especially those based on the pulse distribution method, begin to show and lead to a deterioration of the dynamic response to load step-up scenario. The main factor contributing to this is that the controller needs to manage more than a dozen of phases since the average current and the tolerable peak current of a single phase in a multi-phase Buck converter are limited to about 50A and 100A, respectively, due to restrictions in the

This work is supported by the National Natural Science Foundation of China (62171122).

silicon-based MOSFET process, power losses, cooling method, etc. Combined with constraints such as the need for a minimum off-time for each phase based on safety considerations, COT controllers cannot achieve the excellent dynamic response that can be achieved when the number of phases is low.

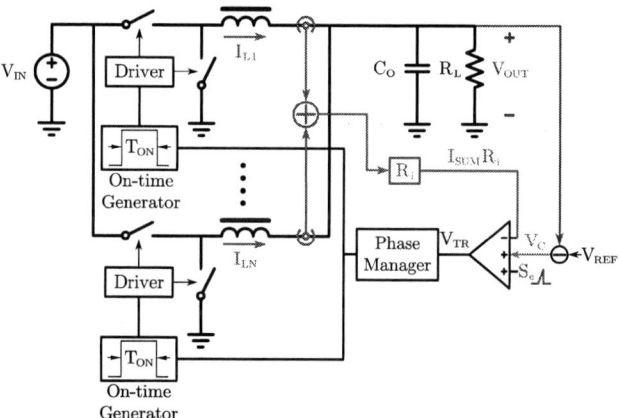

Fig. 1. Conventional current mode constant-on-time (COT) control structure based on the pulse distribution method.

This problem has been recognized by manufacturers, and several measures have been implemented. In steady state, as shown in Fig. 2(a), the current mode constant-on-time (CMCOT) control structure generates pulse signal V_{TR} to control the PWM by comparing the inductor current signal $I_{SUM}R_i$ and the voltage error signal V_C superimposed on the ramp compensation signal S_e. When a significant step-up transient event occurs, the signal V_C+S_e changes rapidly and exceeds the inductor current signal due to the voltage drop, causing the signal V_{TR} to change from the usual pulse to a high level, as shown in Fig. 2(b). The controller needs a particular control strategy when a high level occurs, or it will not be able to suppress the undershoot. In [2] and [3], the PWM will turn on sequentially at a pre-configured time interval called blanking time, regardless of the pulse. However, to maintain the current balancing during transient, the blanking time must be long enough to satisfy the phase interleaving that the dynamic response still gets worse with more phases. In [4] and [5], a more radical measure was introduced, whereby all phases of the PWM would turn on when the output voltage undershoot exceeded a defined threshold. There is a high risk that this measure will result in the current in several of the phases approaching or exceeding the tolerance, which will ultimately lead to the triggering of the failsafe, shutdown, or even damage to the system.

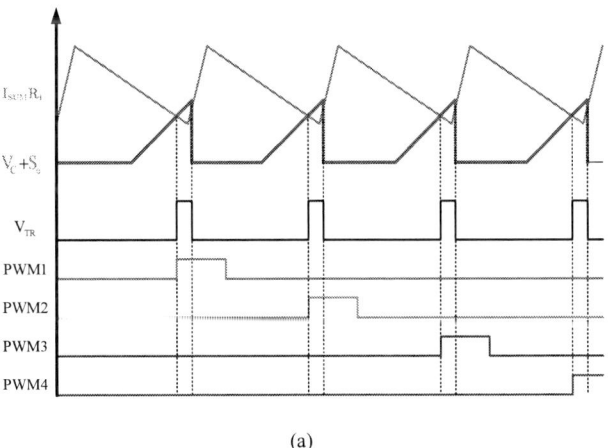

(a)

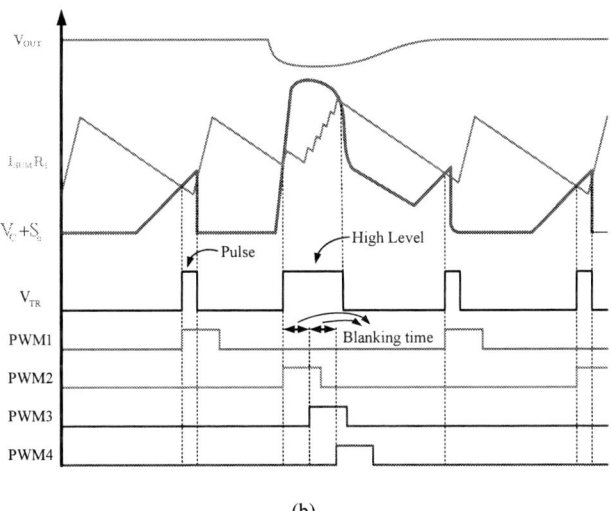

(b)

Fig. 2. Critical waveforms of conventional CMCOT control structure (a). during steady state. (b). during a step-up transient.

This paper proposes a nonlinear control algorithm during transients used in COT controllers based on the pulse distribution method. Compared to the existing algorithms, the proposed control algorithm can break the current limitations of minimum turn-off time and blanking time while maintaining the current balance during and after transients. No additional or higher performance sampling circuits are required, allowing the

pulse distribution method's advantages of simple structure and low cost to be retained.

II. THE EFFECT OF PHASE NUMBER ON STEP-UP DYNAMIC RESPONSE

The effect of the increase in the number of phases on the dynamic response is first analyzed to better design the control algorithm. When a significant step-up transient occurs, the V_{TR} signal changes from the usual pulse signal to a high level. In order for the converter to work correctly during the period of the high level, a time interval called the blanking time t_{blank} is set, and if the length of the high level exceeds t_{blank}, then the controller will allow the next phase to turn on in sequence. The value of t_{blank} needs to be set slightly wider than the width of the pulse signal at the steady state of V_{TR} to prevent false triggering, which is mainly determined by the performance of the comparator used in the controller.

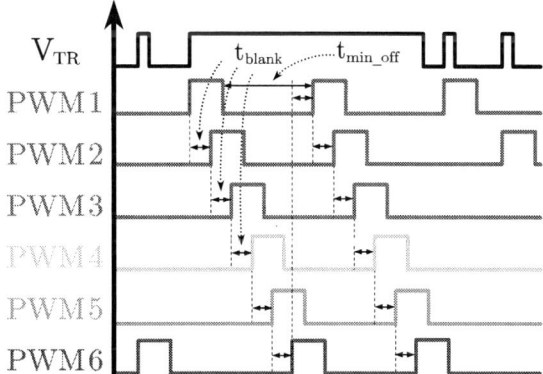

Fig. 3. Waveforms of conventional nonlinear control during a step-up transient.

As shown in Fig. 3, in order to keep the order staggered to ensure current balance, the blanking time t_{blank} and minimum off-time t_{min_off} need to satisfy $N \cdot t_{blank} \geq t_{min\_off} + T_{ON}$. Where N represents the phase number and T_{ON} is the on-time of each phase. The speed of the current injected to the output during a significant step-up transient ΔI_{inject} can be expressed as:

$$\Delta I_{inject} = N \frac{T_{ON} \frac{V_{IN} - V_{OUT}}{L_N} - (N-1) t_{blank} \frac{V_{OUT}}{L_N}}{T_{ON} + (N-1) t_{blank}} \tag{1}$$
$$\cong \frac{V_{IN}}{L_N} - N \frac{V_{OUT}}{L_N}$$

Where L_N represents the inductance of each phase.

From (1), it can be seen that ΔI_{inject} decreases with N increases, worsening dynamic response. The intuitive explanation is that when the current in one phase increases, the currents in all other N-1 phases decrease, slowing down the rate of current change.

This effect can also be explained in terms of the operating frequency of the converter since the frequency variation is the main reason for the high dynamic response of COT controllers. The maximum operating frequency of each phase f_{SW_max} during a step-up transient can be expressed as:

$$f_{SW\_max} = \frac{1}{T_{ON} + t_{min\_off}} \leqslant \frac{1}{N \cdot t_{blank}} \quad (2)$$

The range of the frequency variation can be seen to be in rapid decline with number of phases increasing, which significantly limits the dynamic response. Therefore, it is necessary to design a specific control algorithm for multi-phase converters so that the minimum turn-off time is no longer limited by the number of phases without destroying the current balance of each phase.

III. THE PROPOSED CONTROL ALGORITHM

The proposed control algorithm eliminates the constraint of minimum off-time t_{min_off} on blanking time t_{blank} by splitting the blanking time into two parameters: t_{blank_long} and t_{blank_short}, as shown in Fig. 4. The value of t_{blank_long} is the same as t_{blank} while the value of t_{blank_short} is much shorter. During transients, when the pulse changes into a high-level signal, the first two phases will be turned on at a relatively longer time interval of tblank_long so that the converter can handle small load steps as before. Afterward, if the high level remains, indicating that a moderate or significant load jump has occurred, then the PWM will be turned on sequentially in a much shorter time interval of t_{blank_short}. After the first on-time is finished, all the phases will be turned on again after the set minimum off-time instead of following the blanking time indication. The value of t_{min_off} and t_{blank_short} can be set relatively shorter because they are no longer bound to each other, significantly improving dynamic response.

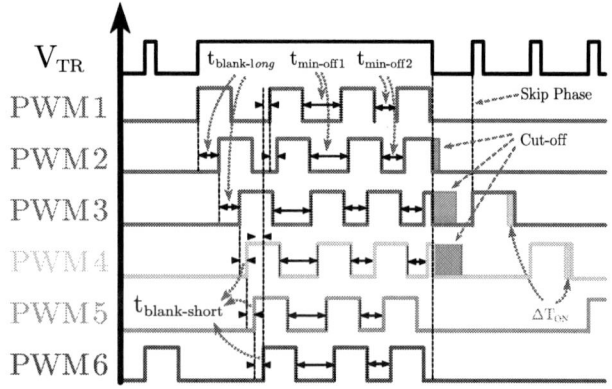

Fig. 4. Waveforms of proposed control algorithm during a step-up transient.

Of course, this modification may result in injecting too much current in a short period of time, with the undershoot followed immediately by a slight overshoot due to the output voltage change lagging behind the capacitor current. Two measures have been introduced to reduce the incidental overshoot. First, the minimum turn-off time is no longer a fixed value, the first turn-off time is longer, and then the turn-off time becomes progressively shorter until it reaches the minimum value. This allows the rate of current injection to increase smoothly, avoiding injecting too much current when a moderate load step occurs. Secondly, when the high-level signal ends, the PWM being turned on will be immediately cut off, preventing additional current injection, which will not help reduce undershoot.

The control strategy and measures described above may lead to significant differences in current between phases, and the low bandwidth of the current-sharing loop previously used in COT controllers is unable to balance the currents in time. Therefore, it is necessary to introduce pre-current-balancing measures. The control strategy leads to unbalancing currents mainly because some phases are turned on more often than others during transient. The proposed control algorithm handles this problem by counting how many times each phase is turned on during the high-level signal, and when the high-level signal ends and the pulse is generated again, the phase manager will skip over the phases that are turned on more than others until all the phases are turned on the same number of times. As for the cut-off measure, the controller will record the remaining time that the phase being cut-off needs to be turned on and slightly increase its turn-on time T_{ON} by a short time ΔT_{ON} each time in steady state until the recorded time is compensated for, thus balancing the current without affecting the output voltage.

IV. EXPERIMENTAL VALIDATION

A 16-phase Buck converter prototype was constructed to validate the proposed control algorithm, as shown in Fig. 5, with its specifications detailed in The converter employs a digital-analog hybrid control scheme. Specifically, current signals, voltage error signals, and slope compensation signals are modulated through analog circuits and then fed into a comparator. The comparator generates the V_{TR} signal, which is processed by an FPGA. The FPGA digitally processes the signals and outputs precise PWM signals for each phase, ensuring high performance and flexibility.

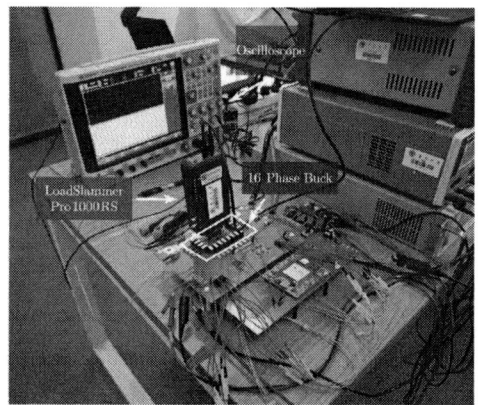

Fig. 5. The 16-phase Buck converter prototype and test setup.

TABLE I. THE SPECIFICATION OF THE PROTOTYPE

parameters/Components		Values
Input Voltage V_{IN}		12V
Output Voltage V_{OUT}		0.98V
Phase Number N		16
Switching Frequency		1MHz
Inductor L_N		150nH
Output Capacitance C_O		6.5mF
Blanking time	t_{blank}	35ns
	t_{blank_long}	35ns
	t_{blank_short}	5ns
t_{min_off}		$15\text{ns} \rightarrow 10\text{ns} \rightarrow 5\text{ns}$

To evaluate the dynamic performance, transient response tests were performed using the LoadSlammer Pro 1000RS from ProGrAnalog. Arbitrarily selected phase current waveforms were recorded to verify current balancing. Phase current signals were sampled from the IOUT port of the MP86957 (an integrated high-side/low-side FET and Driver chip from Monolithic Power Systems), whose current sensing accuracy is within ±2% as specified in its datasheet. The prototype's transient response to a single load step of 0–250 A with a slew rate of 1250 A/µs is shown in Fig. 6.

With the proposed control algorithm, the current in each phase rises extremely faster than the conventional control method, reducing the voltage undershoot from 152mV to 78mV. Furthermore, the pre-current-balancing mechanism effectively redistributes phase currents, achieving balance within ten switching cycles after the transient event ends.

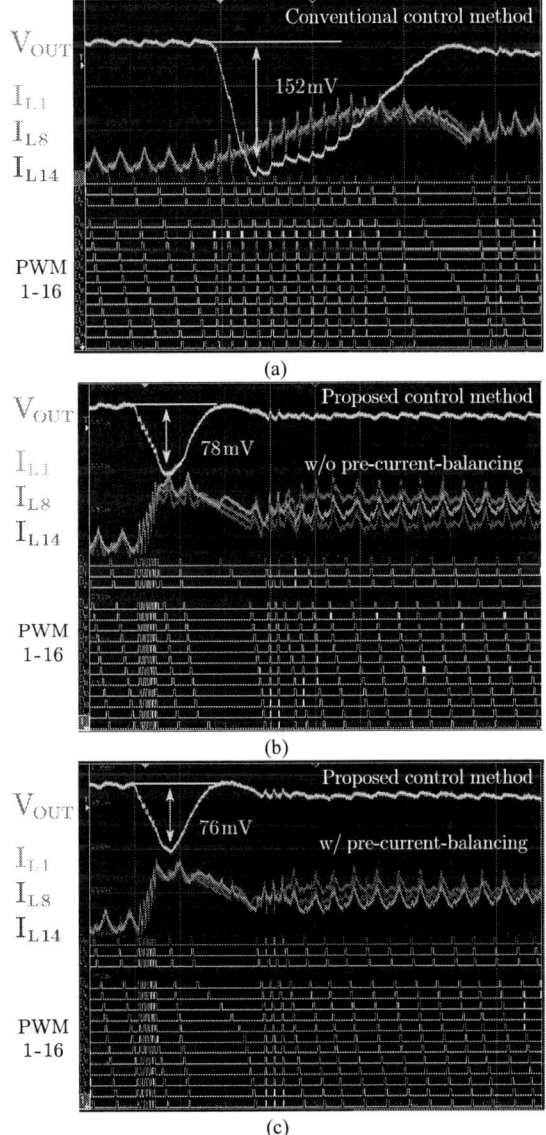

Fig. 6. Dynamic response for a load step of 0-250A (a)Conventional control method. (b)Proposed control method without pre-current-balancing. (c) Proposed control method with pre-current-balancing.

Additional tests were conducted under repetitive 250 A load steps at varying frequencies, ranging from 1 kHz to 1 MHz. The results, shown in Figs. 7–13, demonstrate that the proposed control algorithm maintains consistent current balancing across all phases, even under demanding transient conditions.

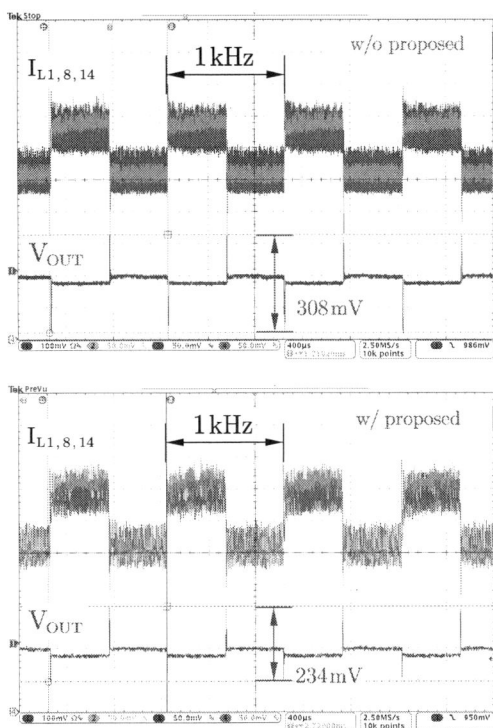

Fig. 7. Dynamic response for 1kHz 250A load steps.

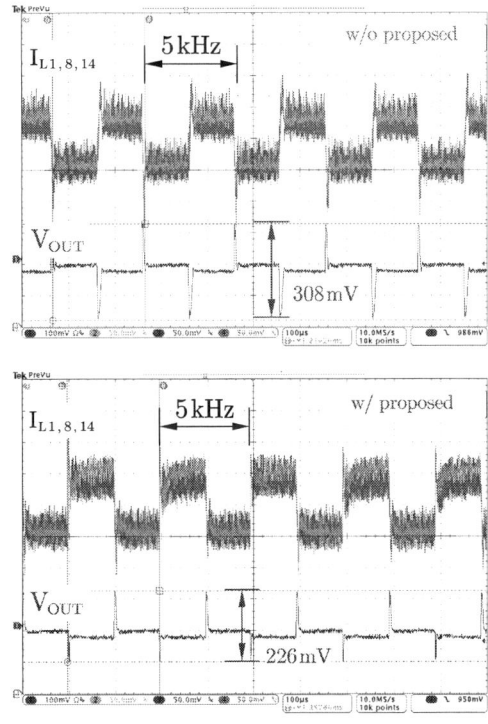

Fig. 8. Dynamic response for 5kHz 250A load steps.

979-8-3315-1612-3/25 $31.00 © 2025 IEEE

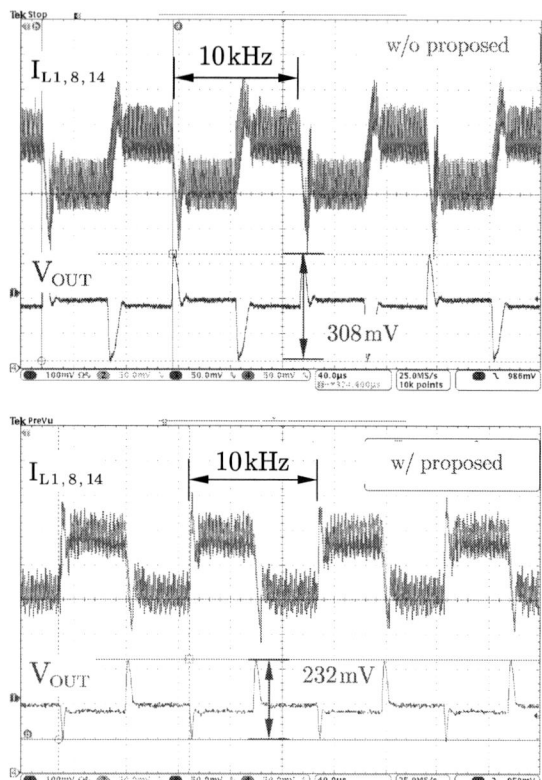

Fig. 9. Dynamic response for 10kHz 250A load steps.

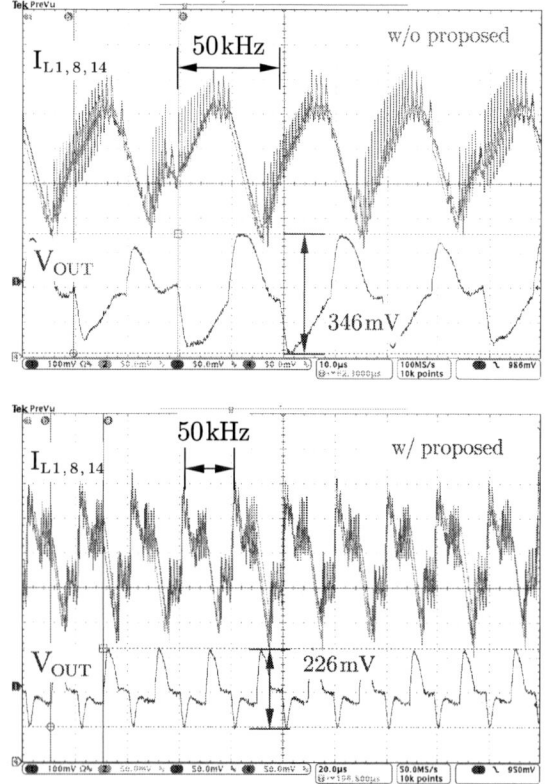

Fig. 10. Dynamic response for 50kHz 250A load steps.

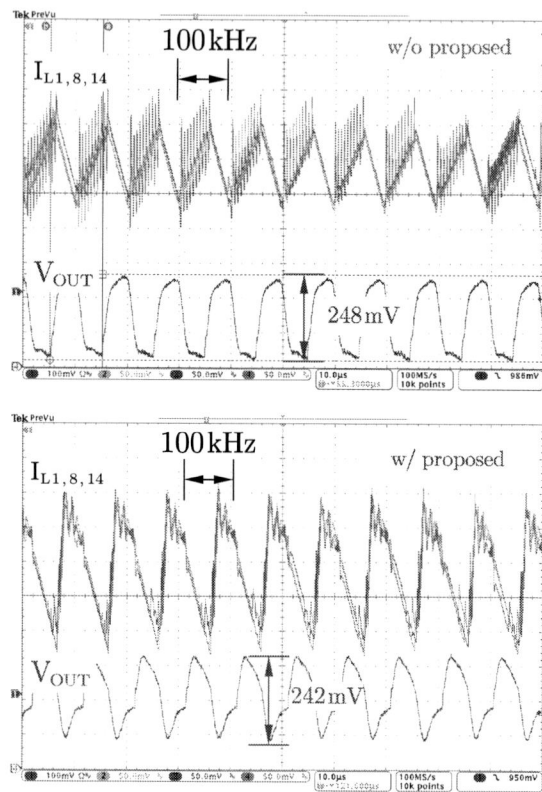

Fig. 11. Dynamic response for 100kHz 250A load steps.

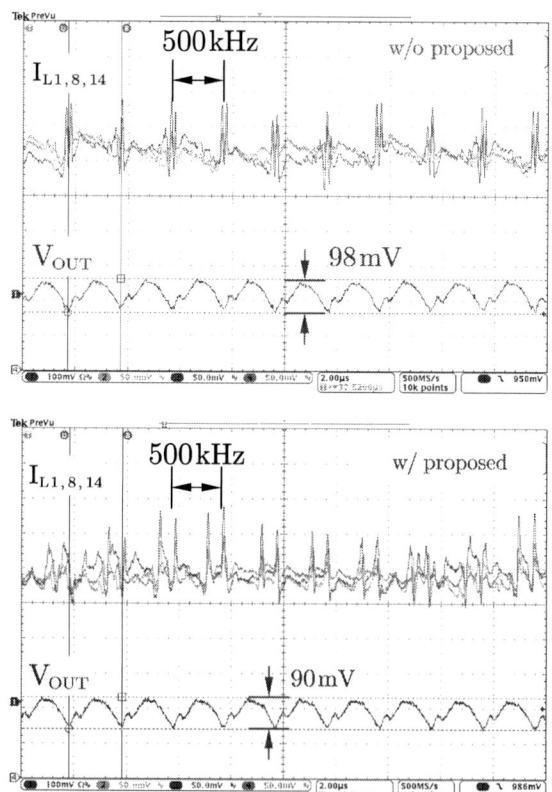

Fig. 12. Dynamic response for 500kHz 250A load steps.

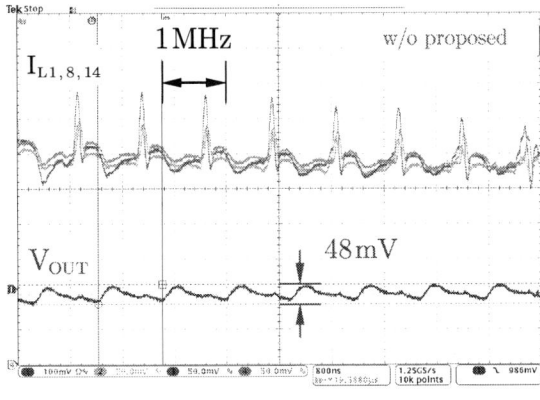

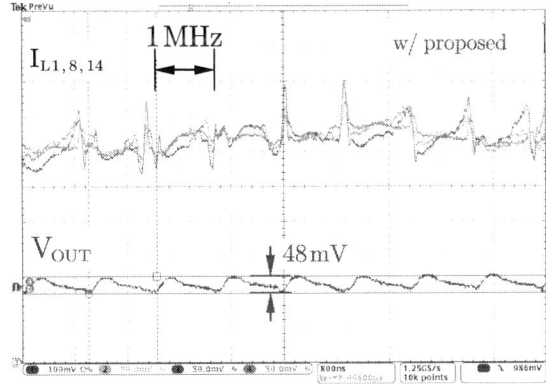

Fig. 13. Dynamic response for 1MHz 250A load steps.

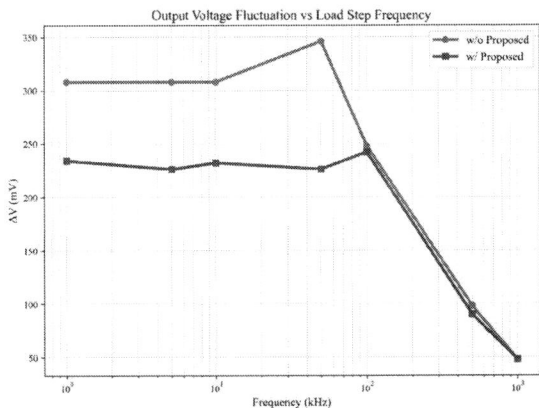

Fig. 14. Comparison of output voltage ripple (ΔV) under different load step frequencies for the conventional control method (w/o Proposed) and the proposed control method (w/ Proposed).

A comparative analysis of output voltage fluctuation (ΔV) between the conventional control method and the proposed control method is summarized in Fig. 14. The results demonstrate that the proposed algorithm significantly reduces voltage fluctuation during repeating load steps, particularly in the low-to-mid frequency range (1 kHz to 100 kHz), where the voltage fluctuation is reduced from a maximum of 346 mV to 242 mV.

In the high-frequency range (above 100 kHz), the load step frequency exceeds the bandwidth of the converter, limiting the ability of the control loop to respond effectively. In this region, the dynamic response behavior is primarily influenced by the characteristics of the power delivery network (PDN), including the impedance of the output capacitors and parasitic elements. As a result, the voltage fluctuation becomes similar between the conventional and proposed methods. Nevertheless, the proposed method achieves consistently lower fluctuation across all frequencies compared to the conventional approach, with the high-frequency fluctuations being inherently more minor than those observed in the low-frequency range.

V. Conclusion

This paper presented a nonlinear control algorithm for multi-phase Buck converters, designed to enhance transient response during step-up load transitions. The proposed algorithm achieves ultra-fast dynamic response while simultaneously balancing phase currents. Experimental results demonstrate that the control method reduces voltage undershoot by up to 49% and rebalances phase currents within ten switching cycles, all without the need for additional current-sensing circuitry. Moreover, the proposed method proves its stability and efficacy across a wide frequency spectrum, effectively suppressing voltage fluctuations during load transients. Future work will focus on extending the proposed algorithm by incorporating phase-shedding mechanisms, enabling enhanced efficiency and control under light-load conditions.

Acknowledgment

This work was supported by the National Natural Science Foundation of China under Grant 62171122. The authors would also like to express their gratitude to Professor Xinning Liu and Nanjing Low Power IC Technology Institute Co., Ltd. for providing the testing equipment used in this research.

References

[1] X. Lou, Q. Li, F. C. Lee, and M. H. Ahmed, "Modeling and analysis of current mode and v2 controls with adaptive voltage positioning (avp) design," in 2021 IEEE Applied Power Electronics Conference and Exposition (APEC), 2021, pp. 2549–2555.

[2] Dual-Loop, Digital, 16-Phase Controller with PMBus Interface for SVI2/AVSBus, Monolithic Power Systems, MP2882 datasheet, June 2021.

[3] Digital, Multi-Phase PWM Controller With PMBus and PWM-VID, Monolithic Power Systems, MP2888A datasheet, Dec. 2018.

[4] 3-Phase Controller with Triple Integrated Driver for VR12.5 Mobile CPU Core Power Supply, Richtek, RT888D datasheet, Aug. 2015.

[5] Dual-channel, 12-phase step-down digital multi-phase D-CAP+™ controller with VR14 SVID and PMBus, Texas Instruments, TPS536C9T datasheet, Sep. 2023.

Loosely Coupled Trans-Inductor Voltage Regulator (LC-TLVR) Inductor as Compensation Inductor (*Lc*)

Pavan Kumar
Data Center Power Solutions
Intel Corporation, *Hillsboro, OR, USA*
pavan.kumar@intel.com

Arturo Sanchez Hernandez
Data Center Power Solutions
Intel Corporation, *Guadalajara, Mexico*
arturo.sanchez.hernandez@intel.com

Abstract—**Trans-Inductor Voltage Regulators (TLVR) are commonly used to power processors to meet stringent transient requirements. TLVR topology requires a separate compensation inductor (*Lc*) to operate effectively. This paper proposes to effectively control the leakage inductance of the TLVR inductor to eliminate the need for a compensation inductor. In addition, the paper demonstrates that a combination of high & low-leakage TLVR inductors can be effectively used to control the compensation inductance required for various implementation without significantly sacrificing performance while eliminating the need for a separate compensation (*Lc*) inductor.**

Keywords— TLVR, Loosely Coupled TLVR Inductor, Dual-Sided Power Delivery, Firing Order, Ripple Current, Compensation Inductor, AC losses, Efficiency.

I. INTRODUCTION

Trans-Inductor Voltage Regulators (TLVRs) have gained wider acceptance owing to their enhanced transient response characteristics to meet the ever-increasing dynamic power demands of modern System-on-Chip (SOC) applications, resulting in reduced output capacitor requirements compared to their multi-phase buck topologies[1]-[5]. The effectiveness of the TLVR topology results from the fact that all the multi-phase TLVR inductors are coupled electrically via the secondary windings connected in series. A compensation inductor (Lc) is also connected in series with the secondary windings (shown in Fig. 1), to control the ripple current and the transient response of the converter. The compensation inductor (Lc) is essential for the proper operation of the converter. The disadvantage of Lc is that it occupies precious real estate near the processor socket leading to a lower power density of the VR solution. The area of the Lc can be utilized for an extra VR phase if it can be eliminated.

This paper outlines a Loosely Coupled TLVR inductor whose leakage can be controller effectively such that the compensation inductor (Lc) can be eliminated thus freeing up area for an additional VR phase. The paper will demonstrate that the leakage inductance of the TLVR inductor can be controlled/tuned such that the effective secondary inductance as seen by the topology can nearly be the same as that with a compensation inductor thus eliminating the need for a discrete Lc component. The performance of the proposed Loosely Coupled TLVR (LC-TLVR) topology is shown to be the same as that of an implementation with a discrete Lc. This paper will

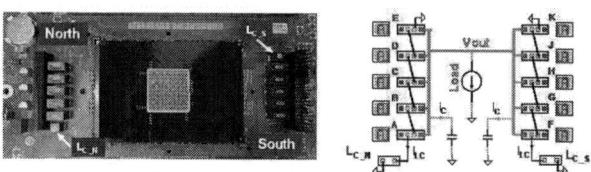

Fig. 1 Typical TLVR topology with Lc & equivalent circuit

also discuss and demonstrate that a combination of low and high-leakage TLVR inductors in the same VR topology can meet the requirements without the need for separate tuning of the leakage or enabling a series of TLVR inductors with different leakage values. A comparison of the performance of the regular TLVR topology with the discrete *Lc*, a LC-TLVR without a discrete *Lc* and a mix-match combination of low-high leakage LC-TLVR will be presented to demonstrate the feasibility of effectively eliminating the compensation inductor for any TLVR topology.

Section II outlines the theory behind the concept of Loosely-Coupled TLVR Inductor and its effect on the ripple currents in the primary & secondary sides of the LC-TLVR inductor. Section III discusses the three case studies that was conducted with the LC-TLVR inductor implementations with simulations and experimental results. Section IV discusses the summary & conclusions while outlining the benefits of the LC-TLVR and its various usages.

II. LOOSELY COUPLED TLVR INDUCTOR (LC-TLVR)

This section outlines the concept of the Loosely-Coupled TLVR Inductor and the theoretical analysis associated with the derivation of the ripple currents in the primary & secondary side of the TLVR inductor.

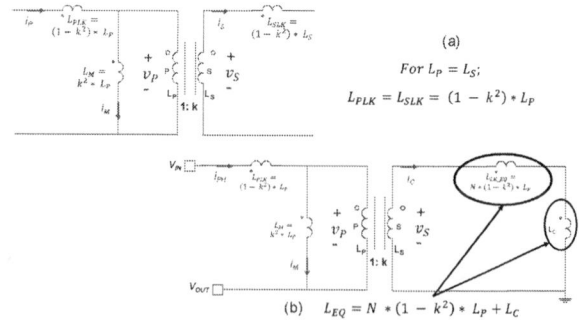

Fig. 2(a) Eq. Circuit of Single TLVR Inductor; (b) Per Phase Equivalent Circuit of an N-phase TLVR Topology

979-8-3315-1612-3/25 $31.00 © 2025 IEEE

Fig. 2(a) shows the equivalent circuit of a single TLVR inductor with primary leakage (L_{PLK}) and secondary leakage (L_{SLK}) respectively, primary magnetizing inductance (L_M) with a coupling factor k between the primary & secondary windings. For a typical TLVR inductor with a turns ratio of unity and the primary & secondary inductance being the same ($L_P = L_S$), the leakage inductances will be the same as given by [6]-[7]:

$$L_{PLK} = L_{SLK} = (1 - k^2) * L_P \tag{1}$$

The per phase equivalent circuit representation of a N-phase VR with TLVR inductors and a discrete compensation inductor (L_C) is shown in Fig. 2(b). It is the seen that the total equivalent secondary inductance L_{EQ} is given by

$$L_{EQ} = N * (1 - k^2) * L_P + L_C \tag{2}$$

The effective secondary inductance is a combination of the total leakage inductances of the N-phases & the compensation inductance L_C as shown. Typically, the leakage inductance is small ($< 5\%$) and hence the secondary inductance (L_{EQ}) is dominated by L_C. If the leakage inductance of the TLVR inductor can be tuned to a higher value (in the range of $8 \sim 15\%$ depending on the requirements of the particular VR) the new effective compensation inductance (L^1_{EQ}) in the secondary needed, can be achieved with the higher leakage TLVR inductors thus eliminating the compensation inductor as shown in Fig. 3(b), where k_1 is the new coupling coefficient ($k_1 < k$)

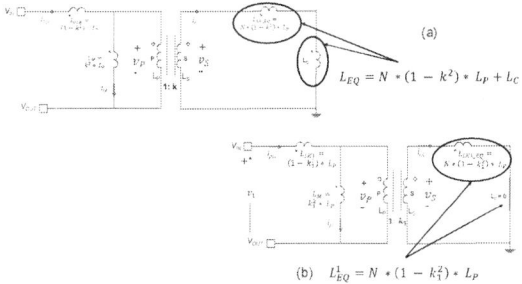

Fig. 3(a) Eq. circuit with discrete Lc; (b) Eq. circuit with higher leakage & Lc eliminated

A Loosely Coupled TLVR (LC-TLVR) inductor can be designed such that the effective secondary inductance as seen by the secondary side can nearly be the same as that with a traditional TLVR design with high coupling coefficient and a discrete L_C. Table I shows the typical electrical parameters of the proposed higher leakage (lower coupling) TLVR inductor in comparison to the original design. The leakage can be adjusted such that the coupling coefficient is appropriately lower compared to the original as needed for a particular VR design

Table I Loosely Coupled TLVR Inductor Design Example

Type	Coil	Inductance @0A [nH]	Tol [%]	Leakage Inductance [nH]	Coupling Factor k_{DS}	DC Resistance [mΩ]	Rated DC Current [Amps] last_20%			Itemp_40C Typ.
							25C	100C	125C	
Original	Primary	100	15	3.4	0.966	0.125	132	113	104	70
	Secondary	100				0.45	132	113	104	40
Loosely Coupled	Primary	100	15	16	0.84	0.125	134	114	107	70
	Secondary	100				0.37	120	102	96	44

Notes (Refer to Appendix I for definition of Leakage & Coupling coefficient) :
1. Coupling coefficient as defined in typical TLVR inductor data sheet (shown above) is $k_{DS} = \{1 - (L_k/L_P)\}$
2. Coupling coefficient as defined in this paper is $k = SQRT\{1 - (L_k/L_P)\}$ since leakage is defined as $L_k = \{1 - k^2\} * L_P$
3. Both definitions are very nearly the same when coupling coefficient in near unity (leakage is small)

depending on various factors (number of phases, transient response, droop & overshoot requirements). It is noted that a series of loosely coupled TLVR inductors can be designed to suit specific applications of the VR although not desirable for a component enabling and supply chain perspective. The choice of coupling factor is mainly influenced by the set of design parameters to be achieved while maintaining a balance between efficiency, space and power density in a particular implementation.

A. VR with Traditional High-Coupling TLVR Inductors & Lc

This section analyzes the traditional TLVR solution with a high-coupling TLVR inductor as a special case of a generic solution. For the purpose of analysis of various cases, the coupling coefficient used in this paper can be defined as:

$k_H \rightarrow$ High coupling $\Rightarrow$ Low leakage inductance ($< 5\%$)

$k_L \rightarrow$ Low coupling $\Rightarrow$ High leakage inductance ($10 \sim 15\%$)

The per phase equivalent circuit of a traditional VR with typical TLVR inductors with low leakage (high coupling coefficient) and a compensation inductance is shown in Fig. 2(b) and Fig. 3(a). The corresponding leakage, magnetizing and the equivalent secondary inductances for this case can be written respectively as

$$L_{Pk_H} = (1 - k_H^2) * L_P \tag{3}$$

$$L_{Mk_H} = k_H^2 * L_P \tag{4}$$

$$L_{EQH} = N * (1 - k_H^2) * L_P + L_C \tag{5}$$

where, L_P is the primary inductance of the TLVR inductor. The peak-peak current in the compensation inductor can now be expressed as (refer to Appendix $(A10)$)

$$\Delta I_{C_{PKPK}} = k_H * \left[\frac{\left\{ V_{IN}(m+1) - N * V_{OUT} \right\} * \left\{ D - \frac{m}{N} \right\}}{L_{EQH} * f_{SW}} \right] * \left\{ \frac{k_H^2 L_P * L_{EQH}}{k_H^4 L_P L_{Pk_H} + k_H^2 L_P * L_{EQH} + L_{Pk_H} L_{EQH}} \right\} \tag{6}$$

The ripple current in the traditional case (high-coupling inductor + Lc) can be rewritten as

$$\Delta I_{Ck_H} = k_H * \left[\frac{\left\{ V_{IN}(m+1) - N * V_{OUT} \right\} * \left\{ D - \frac{m}{N} \right\}}{L_{EQH} * f_{SW}} \right] * A_{EQH} \tag{7}$$

where, A_{EQH} (a dimensionless quantity) is given by:

$$A_{EQH} = \left\{ \frac{k_H^2 L_P * L_{EQH}}{k_H^4 L_P L_{Pk_H} + k_H^2 L_P * L_{EQH} + L_{Pk_H} L_{EQH}} \right\} \tag{8}$$

B. VR with Loosley Coupled TLVR Inductor (no Lc)

For a TLVR topology with all loosely-coupled TLVR inductors (high-leakage & no Lc), the Lc is eliminated. The equivalent circuit will be as shown in Fig. 3(b). In this case, the leakage, magnetizing and equivalent secondary inductances can be written respectively as:

$$L_{Pk_L} = (1 - k_L^2) * L_P \tag{9}$$

$$L_{Mk_L} = k_L^2 * L_P \tag{10}$$

$$L_{EQL} = N * (1 - k_L^2) * L_P \tag{11}$$

It is noted that the secondary inductance as described by (11) does not contain the compensation inductance Lc. The ripple

979-8-3315-1612-3/25 $31.00 © 2025 IEEE

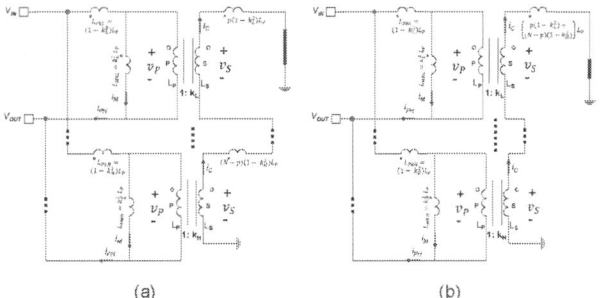

Fig. 4 Eq. Circuit of N-phase TLVR topology with a mix of high & low leakage TLVR inductors

current in secondary for the loosely coupled with all uniform TLVR inductors case can be expressed along the same lines as Eqn. *(7)* & *(8)* as:

$$\Delta I_{Ck_L} = k_L * \left[\frac{\{V_{IN}(m+1) - N*V_{OUT}\}*\{D - \frac{m}{N}\}}{L_{EQL}*f_{SW}} \right] * A_{EQL} \quad (12)$$

where A_{EQL} is given by (similar to *(8)*),

$$A_{EQL} = \left\{ \frac{k_L^2 L_P * L_{EQL}}{k_L^4 L_P L_{Pk_L} + k_L^2 L_P * L_{EQL} + L_{Pk_L} L_{EQL}} \right\} \quad (13)$$

C. Mix of High & Low Leakage TLVR Inductors (no Lc)

The case of an uniform implementation with all low-coupling (high leakage) TLVR inductors can work in most cases (*Case B*). However, the total inductance in the secondary determines not only the ripple current but also the peak voltage overshoot that the VR sees when the load is thrown off. Thus, the total inductance needs to be controlled to achieve this. This is done effectively in the discrete *Lc* case by the choice of *Lc* value. It becomes difficult when using all loosely coupled TLVR inductors unless a series of TLVR inductors with a wide range of leakages are enabled. This is not desirable from component enabling and supply chain perspective. This can be avoided by using a simple combination of high & low leakage TLVR inductors in the same design. Thus the effect of the mix-match of high & low leakage on the ripple and phase currents needs to be analyzed to be able to implement such a design.

Consider a *N*-phase TLVR design with a mix of high & low leakage TLVR inductors with no *Lc*. If "*p*" of the *N* phases have high leakage, while (*N-p*) phases have low leakage. The equivalent circuit can be represented as shown in Fig. **4**. The

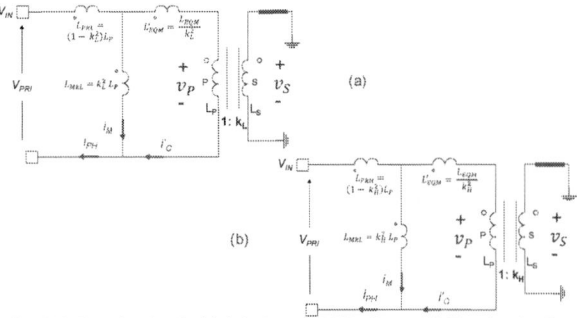

Fig. 5(a) Eq. circuit of a high leakage TLVR phase; (b) Eq. circuit of a low leakage phase

equivalent total secondary inductance (L_{EQM}) and the equivalent coupling coefficient k_M in the mix-match case is given by

$$L_{EQM} = p(1 - k_L^2)L_P + (N - p)(1 - k_H^2)L_P = N(1 - k_M^2)L_P \quad (14)$$

$$k_M = \left[\sqrt{\{k_H^2 - (p/N) * (k_H^2 - k_L^2)\}} \right] \quad (15)$$

The ripple current in the secondary can now be calculated using the same approach outlined in Appendix A. Given that the secondary current is the same for a phase that has either a low or high leakage TLVR inductor (since all secondary windings are connected in series), the ripple current can be expressed as

$$\Delta I_{C_{PKPKM}} = k_L \left[\frac{\left\{ V_{IN}(m+1) - NV_{OUT} \right\}\left\{ D - \frac{m}{N} \right\}}{L_{EQM} * f_{SW}} \right]$$
$$* \left\{ \frac{k_L^2 L_P * L_{EQM}}{(1 - k_L^2)\{k_L^4 L_P^2 + L_P L_{EQM}\} + k_L^2 L_P L_{EQM}} \right\} \quad (16)$$

as derived from an TLVR inductor with low coupling, or

$$\Delta I_{C_{PKPKM}} = k_H \left[\frac{\left\{ V_{IN}(m+1) - NV_{OUT} \right\}\left\{ D - \frac{m}{N} \right\}}{L_{EQM} * f_{SW}} \right]$$
$$* \left\{ \frac{k_H^2 L_P * L_{EQM}}{(1 - k_H^2)\{k_H^4 L_P^2 + L_P L_{EQM}\} + k_H^2 L_P L_{EQM}} \right\} \quad (17)$$

for an equivalent phase with a TLVR inductor with high coupling. Either one of the expressions outlined in *(16)* or *(17)* can be used to determine the ripple current in the secondary. In general, the larger value from *(16)* & *(17)* will be the worst case. It is noted that although the secondary current is the same in all the TLVR inductors, the reflected secondary current to the primary will be different for a given phase and dependent on the coupling factor for that TLVR inductor. Hence, the individual primary ripple current needs to be estimated based on the individual TLVR inductor phase. The equivalent circuit for each of the high and low leakage phases with the corresponding reflected secondary currents are shown in Fig. 5.

For a phase with a high leakage (low coupling) TLVR inductor, the ripple current in the primary is given by

$$\Delta I_{PH\_HL} = \left[\frac{\{V_{IN}(m+1) - N*V_{OUT}\}\{D - \frac{m}{N}\}}{f_{SW}} \right] \left[\frac{(L_{MkL} + L'_{EQM})}{L_{PkL}(L_{MkL} + L'_{EQM}) + L_{MkL}L'_{EQM}} \right] \quad (18)$$

which can be simplified to

$$\Delta I_{PH\_HL} = \left[\frac{\{V_{IN}(m+1) - N*V_{OUT}\}\{D - \frac{m}{N}\}}{L_P * f_{SW}} \right] \left[\frac{k_L^4 L_P + L_{EQM}}{(1 - k_L^2)k_L^4 L_P + L_{EQM}} \right] \quad (19)$$

The corresponding ripple current in the primary for a phase with low leakage (high coupling) TLVR inductor can be expressed as

$$\Delta I_{PH\_LL} = \left[\frac{\{V_{IN}(m+1) - N*V_{OUT}\}\{D - \frac{m}{N}\}}{L_P f_{SW}} \right] \left[\frac{k_H^4 L_P + L_{EQM}}{(1 - k_H^2)k_H^4 L_P + L_{EQM}} \right] \quad (20)$$

For a TLVR inductor that has the same primary inductance (L_P), only the ratio of the leakage inductance to the magnetizing inductance changes for the phase that corresponds to the high & low leakage respectively. The effective phase ripple current due to the multi-phase buck will remain the same in both cases. The reflected current from the secondary will be lower in case of the high leakage (due to low coupling k_L) TLVR inductor

979-8-3315-1612-3/25 $31.00 © 2025 IEEE

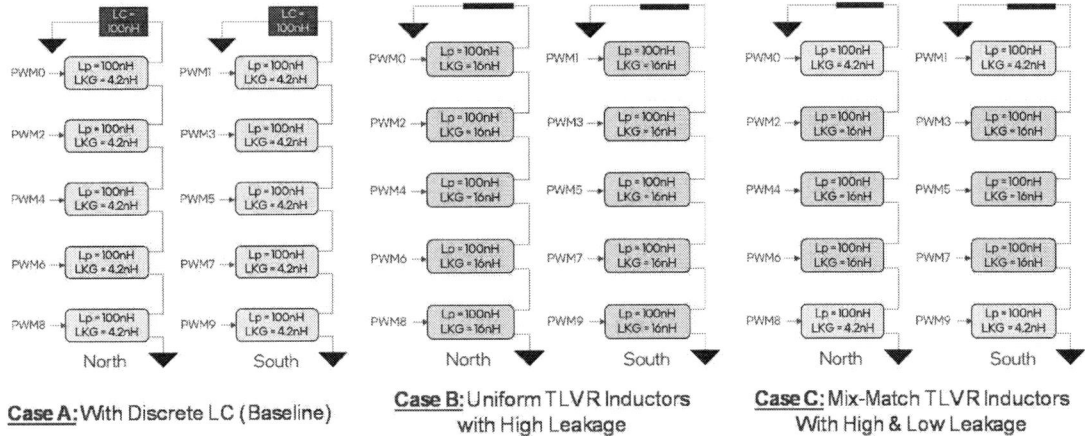

Fig. 6 Case Studies of TLVR inductors with discrete Lc & Proposed Controlled Leakage

while the primary magnetizing current will be higher. In contrast the reflected current is higher in case of the low leakage TLVR phase (due to high coupling k_H) while the primary magnetizing current will be correspondingly lower. Hence, the overall primary ripple current should be similar to that of the uniform case although individual phases have different TLVR inductors from a leakage perspective. This aspect is important and fundamental to maintain the primary current in each phase very nearly the same irrespective of whether the phase has a TLVR inductor with a higher or lower leakage. In addition, this aspect ensures that the primary inductor current information used by VR controllers to perform control functions, current monitoring, fault detection etc. remains largely unaffected. This aspect implies that there should be no substantial differences in control functionality due to the mix-match of the TLVR inductors.

III. LC-TLVR INDUCTOR – CASE STUDY

The theoretical basis for the loosely coupled TLVR inductor was outlined in the previous section. This section describes a case study where the concept of controlled leakage inductance of the TLVR inductor to eliminate the discrete compensation inductor was tested on a 10-phase dual-sided VR.

Fig. 6 shows the 3 case studies that were performed to determine the effectiveness of the higher leakage TLVR inductors. Case A forms the standard baseline configuration and is the basis for comparison. Case B shows a design where all the high leakage TLVR inductors are uniform, and the discrete Lc is

eliminated. Case C shows the mix-match implementation of the high & low leakage TLVR inductors. In this example, two of phases on each side of the VR have been replaced with the high coupling (low leakage inductors) in Case B to illustrate the concept of the design. However, there is no limitation as to combinations possible in this case to achieve the desired equivalent secondary inductance. Various combinations of mix & match are feasible as long as the primary inductance L_P is the same. Although the focus of this paper is on the Loosely coupled TLVR implementation shown in Case B, the novel mix-match topology can provide the best option in many cases (specifically in case where high phase count is needed to deliver higher power) to optimize the performance given the constraints of space while increasing power density.

Table II shows the various parameters as computed for the three cases based on the type of TLVR inductors used for each case. As expected the effective ripple current in the secondary (& hence the primary) increases for cases B & C (since the effective inductance is less). For small increases in current the effect on circuit operation and efficiency of the VR will be minimal. The combination of low coupling (either uniform or mix-match) TLVR inductors will play a key role in determining the performance of the converter.

A. Simulation Results

The individual case studies were simulated with the corresponding values of the TLVR inductor parameters. The focus of the simulations was on cases B & C to determine the

Table II Design Parameters of the VR for individual case study

10-phase, Dual-sided TLVR Topology; Number of phases per side N = 5					
Parameter	Symbol	Units	Case A: Baseline	Case B: Uniform High LKG TLVR	Case C: Mix-Match TLVR Inductors
			L_{LK}=4.2nH; Lc = 100nH	L_{LK}=16nH; Lc=0;	6-ph L_{LK}=16nH; 4-ph L_{LK}=4.2nH; Lc=0
Primary Inductance	L_P	[nH]	100	100	100
Leakage Inductance	L_K	[nH]	4.2	16	Mix
Eff. Coupling Coefficient	k	x	$k_H = \sqrt{1 - \dfrac{L_K}{L_P}}$ $k_H = 0.978$	$k_L = \sqrt{1 - \dfrac{L_K}{L_P}}$ $k_L = 0.916$	$k_M = \left[\sqrt{\{k_H^2 - (p/N)*(k_H^2 - k_L^2)\}}\right]$ $k_M = 0.953$
Eq. Secondary Inductance	L_{EQ}	[nH]	L_{EQH} = 121	L_{EQL} = 80	L_{EQM} = 56.4
Multiplying Factor	A_{EQX}	Ratio	A_{EQH} = 0.9284	A_{EQL} = 0.7361	A_{EQM} = 0.7–0.89
Secondary Ripple Current	$\Delta I_{C\,PRK}$	Ratio	1	~1.12	~2.05

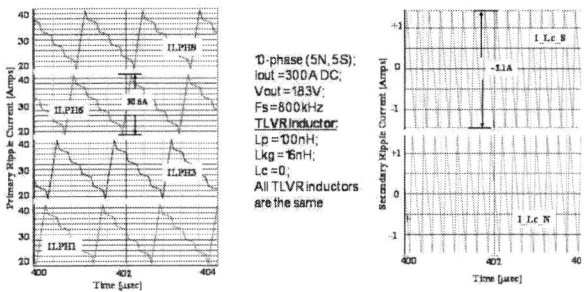

Fig. 7 Case B: Simulated Primary & Secondary Ripple Current

979-8-3315-1612-3/25 $31.00 © 2025 IEEE

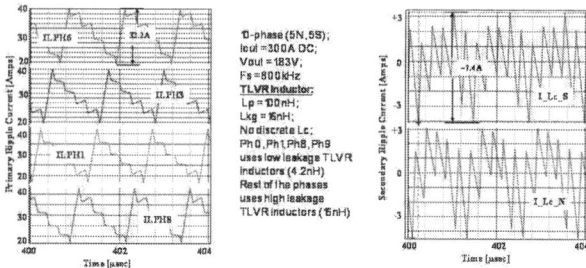

Fig. 8 Case C: Simulated Primary & Secondary Ripple Current

effect of the lower coupling and the mix-match of the TLVR inductors.

Fig. 7 shows the simulated primary & secondary ripple currents for the uniform low coupling TLVR inductors (Case B) whereas Fig. 8 shows the corresponding currents for the mix-match scenario (Case C). As expected, the ripple currents in the secondary are higher, compared to the baseline case (expected nominal secondary ripple current of ~ 2.9A). The mix match case shows the highest ripple (due to low secondary inductance of ~ 56nH compared to 121nH in the baseline). The primary ripple current mismatch in each case is still smaller (~2A). This is because the primary current is dominated by the total primary inductance (which remains the same in all cases) and the reflected current form the secondary is still small. The key observation is that the simulation shows that the currents are within limits and the VR is stable in each case although 2 out of 5 (40%) of the phases are populated with TLVR inductors with lower leakage.

B. Experimental Results

This section discusses the experimental results obtained for the 3 cases on a 10-ph dual-sided TLVR (shown in Fig. 1). Each side (North & South) consists of 5 phases with a Lc inductor in a symmetrical arrangement. The firing order is alternated between each side to minimize ripple currents so that efficiency of the converter is maintained high [8].

The fact that the equivalent secondary inductance in cases B (L_{EQL}) & C (L_{EQM}) are less compared to case A (L_{EQH} -baseline) implies that the secondary ripple current will be higher compared to the baseline. In fact, as computed (shown in Table II with the multiplying factor A_{EQX}) and the simulation results, the ripple current in the secondary should be 10~15% higher in Case B & nearly twice in Case C compared to the baseline. To determine the nature of the ripple current and to see whether it matches the predictions, a small wire loop was introduced into the secondary circuit and the ripple current measured. It is noted that although the loop is small, it does introduce some additional

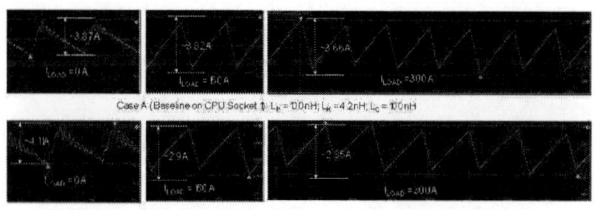

Fig. 9 Secondary Ripple Current Waveforms for Cases A & B

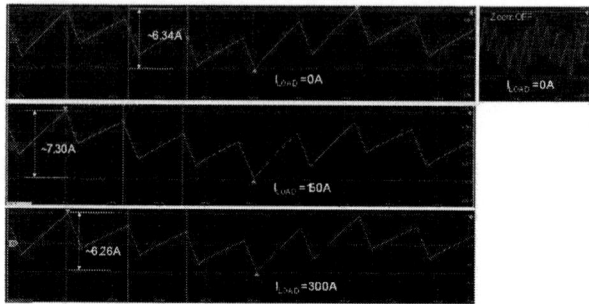

Fig. 10 Secondary Ripple Current for Case C

inductance into the secondary circuit. This causes the actual values of the currents measured to be a little less than the ideal case. However, the relative comparison and behavior should still be representative.

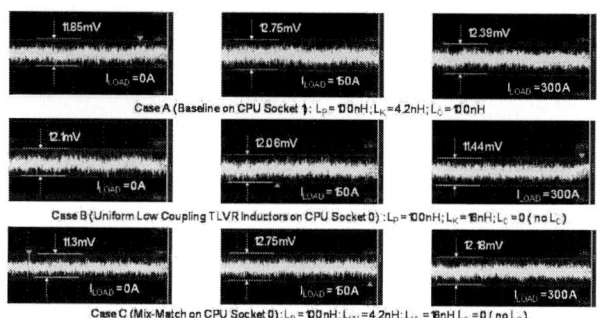

Fig. 11 Output Voltage Ripple Comparison for Cases A, B & C

Fig. 9 shows the measured secondary ripple currents for Cases A & B. It is seen that the ripple difference between Case A & B is small. The individual differences in the magnitudes of the current between the two are attributed to the small variations between the two VRs implemented on two different sockets (which are nearly identical). Fig. 10 shows the measured secondary ripple current for the special case of mix-matched TLVR inductors. As expected, the secondary ripple current is higher (~2x) as predicted by analysis & simulation. The non-zoomed version of the waveform in Fig. 10 matches the simulation result to a large extent & verifies the prediction. In all three cases, the VR operates in a stable manner without any oscillations. The output voltage ripple for each of the cases are shown in Fig. 11. It is seen that the difference in each case (< 1mV) is negligible and well within limits.

To verify the behavior of the converter under various conditions, each case was tested for transient behavior with the VR Test Tool (VRTT). The converter was subjected to transient loading over a frequency range of 1kHz to 1MHz with load duty cycle varied from 10% to 50%. The transient result in each case is shown in Fig. 12. It is seen that in all cases the droop & overshoot are within limits and that the converter is stable. The worst-case values of the droop and the overshoot are summarized in Table III.

The secondary current ripple during steady state in most cases are small and even a 2x increase does not affect the

979-8-3315-1612-3/25 $31.00 © 2025 IEEE

Table III Experimental Results comparison for Case A, B & C

Parameter		Unit		Case A Baseline	Case B: Uniform High LKG TLVR	Case C: Mix-Match TLVR Inductors
				L_{LK}=4.2nH; LC=100nH	no LC; L_{LK}=16nH	6-ph L_{LK}=16nH; 4-ph L_{LK}=4.2nH
				L_{EQ}=121nH	L_{EQ}=80nH	L_{EQ}=56.4nH
Efficiency	at Load = 306 Amps		[%]	90.1	90.1	89.9
Worst Case Transients	Droop (104A-->574A) at Duty Cycle = 50%	Vmin	[Volts]	1.625	1.624	1.63
		Freq	[kHz]	300	140	6
	Overshoot (574A-->51A) at Duty Cycle = 10%	Vmin	[Volts]	1.871	1.868	1.877
		Freq	[kHz]	140	20	500
Ripple	Voltage Ripple	pkpk	[mVolt]	12.75	12.06	12.75
		Load	[Amps]	150	150	150
	Secondary Ripple Current	pkpk	[Amps]	4.18A	4.35	7.54
		Load	[Amps]	0	0	0
Worst Case Secondary Transient Current		pkpk	[Amps]	71.03	72.87	80.75
		Freq	[kHz]	1	1	1
Worst Case Phase Ripple Current		pkpk	[Amps]	20.15	20.33	21.64

operation of the converter. However, the peak currents seen during transients can be very high leading to inductor saturation. Hence, it is important to ascertain the peak ripple currents that occur during transient. Fig. 13 shows the peak current as seen by the secondary winding during a load transient. It is seen that the peak current increase due to the mix-match case is about 13% in this case. This is still within the limits of the inductor saturation current rating and acceptable.

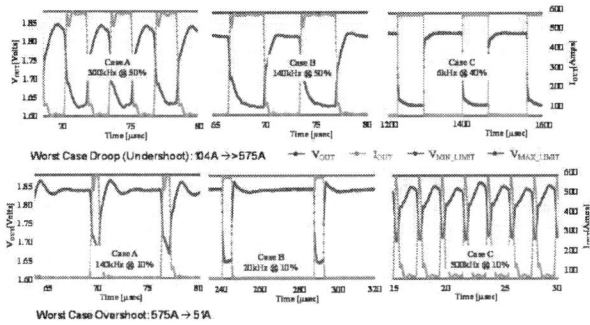

Fig. 12 Transient Results for Cases A, B & C (Left Axis → Voltage, Right Axis → Current)

The secondary ripple current is about 10~15% higher in the case of Case B whereas it is nearly twice that of the baseline in Case C. Hence it is reasonable to expect that the efficiency of the converter will be affected due to this. The efficiency results of the VR for all 3 cases are shown in Fig. 14. It is seen that the efficiency difference between the three cases is in the range of 0.2 ~ 0.3%. Specifically in the range of interest under heavy loads, the efficiency difference is negligible even in the case of mix-match TLVR inductors (Case C).

Thus, it can be reasonably concluded that if the secondary ripple current is not very high (10's of Amps) as seen in the case of asymmetrical firing order [8], the efficiency impact of using the low coupling & a mix-match of low & high coupling TLVR inductors is minimal.

It is noted that in this illustration, the effective secondary inductance was nearly half of the baseline case. This is not intentional from a design perspective but rather a method to demonstrate the feasibility of a mix-match TLVR inductor topology. If the effective secondary inductance can be near to the baseline value, the efficiency impact can be neutralized. In fact, the secondary ripple magnitude can be minimized by adopting a suitable firing order among the phases so that any negative impact of the reduced secondary inductance is overcome. In addition, a balance of reduced secondary inductance against the efficiency is a design criterion that can be used to improve the transient response of the converter. A reduced secondary inductance improves the transient response

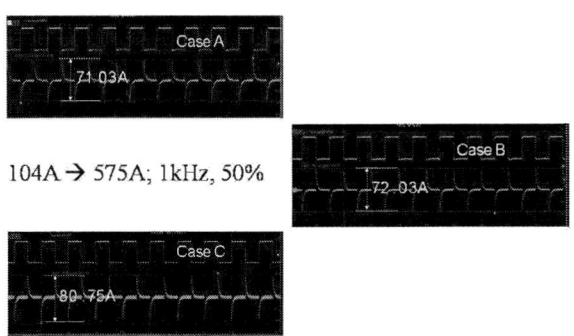

Fig. 13 Peak secondary current during transients

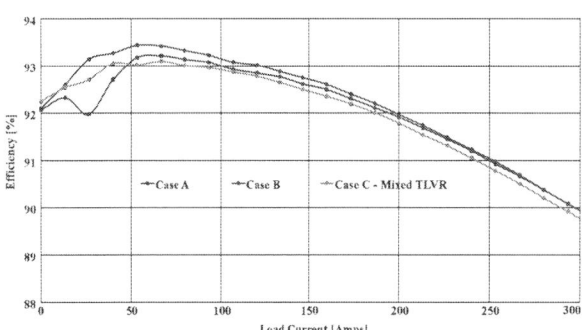

Fig. 14 Efficiency Comparison

while sacrificing efficiency and vice versa. Depending on the specifications, one could settle on a design that balances the two.

The experimental results are summarized in Table III. It is clearly seen that the performance of the Loosely Coupled TLVR design (Case B) is equivalent to that of the traditional TLVR with the discrete Lc inductor. In addition, the mix-match version of the TLVR design is also very similar to that of the other two cases even under the condition that the equivalent secondary inductance is half of that of the baseline and hence the secondary steady state ripple and transient currents are higher. The focus of the mix-match version being that of a technology illustration the performance parameters especially the efficiency can be improved from a design point of view.

IV. SUMMARY & CONCLUSIONS

A Loosely Coupled TLVR (LC-TLVR) Inductor has been proposed in place of the traditional tightly coupled TLVR inductor. The concept of controlling the leakage inductance of the TLVR inductor with a goal to eliminate the discrete compensation inductor Lc to reduce, space and cost while increasing the power density has been explored. A detailed analysis of the concept and the theoretical basis for computing the ripple currents in the new topology has been developed and presented. A LC-TLVR inductor has been developed with a leakage inductance in the range of 10~ 15% of the primary inductance. Proof of concept for the proposed topology has been demonstrated with a 10-phase dual-sided TLVR design by replacing the traditional TLVR topology with the proposed LC-TLVR inductor. A unique mix-match VR topology with a combination of low & high leakage TLVR inductors has been proposed as a feasible solution to tune the equivalent secondary inductance as a design parameter where needed. Detailed simulation and experimental results have been presented to demonstrate the operation of the VR. The impact of the proposed LC-TLVR on the performance of the VR has been shown to be minimal even in the case of the mix-match TLVR inductors.

ACKNOWLEDGMENTS

The authors would like to acknowledge the significant contributions of Mr. Justin Tippetts (formerly with Intel) during the development of the loosely-coupled TLVR inductor concept and the associated inductor specifications. The authors would also like to thank Mr. Bruce Funderburgh (formerly with Intel), Mr. Paul Brusco, Mr. Carlos Javier Camacho Marquez, Mr. Mario Carillo and Mr. Cesar Javier Velazquez Avelar for their contributions in gathering experimental results.

REFERENCES

[1] Shuai Jiang et al., "Driving 48V Technology Innovations Forward – Hybrid Converters and Trans-Inductor Voltage Regulator (TLVR)," IEEE APEC Industry Session 2020.

[2] "Fast multi-phase trans-inductor voltage regulator", Technical Disclosure Commons (May 9, 2019), https://www.tdcommons.org/dpubs_series/2194.

[3] Zhang, Nian, et al., "Analysis of Multi-Phase Trans-Inductor Voltage Regulator with Fast Transient Response for Large Load Current Applications," IEEE International Symposium on Circuits and Systems (ISCAS), 2021.

[4] Amin YF, Satya et. al., "Trans-Inductor vs Discrete Inductors in Multiphase VRs: An Analytical & Experimental Comparative Study", IEEE APEC 2023

[5] "Multiphase Buck Design from start to finish", Application Note slva882b, Texas Instruments Inc.

[6] "TLVR Design Equations", Renesas Electronics Corporation, Nov. 2021.

[7] "Multiphase Buck Converter with TLVR Output Filter", Version 1.2, Infineon Technologies, December 2021.

[8] P. Kumar, J. Tippetts, S. Naidu, P. Brusco, "Efficiency Impact of Phase Firing Order in Dual-Side Power Entry with Trans-Inductor Voltage Regulators (TLVR)", IEEE APEC 2024.

APPENDIX A

LC RIPPLE CURRENT WITH PRIMARY LEAKAGE INCLUDED

Fig. 1(b) shows the per phase equivalent circuit of a N-phase TLVR topology. Since the leakage inductance is small ($< 5\%$) the primary leakage inductance is generally neglected during analysis and the ripple currents computed are simplified. However, given the fact that the LC-TLVR inductors are designed to have a higher leakge (10%~15%) the primary leakage inductance cannot be neglected for the analysis of the LC-TLVR topology. Hence the secondary ripple current needs to be derived with the primary leakage inductance included. **Fig.** 15 shows the definition of leakage, coupling coefficient and the TLVR secondary inductance as transferred to the primary (shown in **Fig.** 1(b)).

The transferred equivalent inductance L'_{EQ} is given by

$$L'_{EQ} = \left(\frac{1}{k^2}\right) * L_{EQ} = \left(\frac{1}{k^2}\right)\{N * (1 - k^2) * L_P + L_C\} \qquad (A1)$$

Since the transferred equivalent inductance L'_{EQ} and the magnetizing inductance L_M are in parallel

$$L_X = \left\{\frac{L_M * L'_{EQ}}{L_M + L'_{EQ}}\right\} \qquad (A2)$$

The total effective inductance as seen by the primary side is

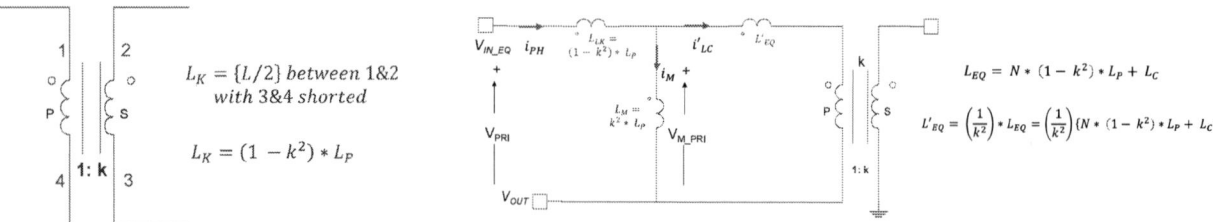

Fig. 15 Definition of Leakage, coupling coefficient & TLVR Secondary Inductance transferred to primary

979-8-3315-1612-3/25 $31.00 © 2025 IEEE   536

$$L_{EFF} = L_{LK} + L_X = L_{LK} + \left\{ \frac{L_M * L'_{EQ}}{L_M + L'_{EQ}} \right\} \tag{A3}$$

The steady state ripple current ($\delta i'_{LC}$) across the transferred equivalent inductance L'_{EQ} is

$$\delta i'_{LC} = \left\{ \frac{V_{M_{PRI}}}{L'_{EQ}} \right\} * \delta t \tag{A4}$$

The voltage across the magnetizing inductance V_{MPRI} is a fraction of the primary voltage V_{PRI} and is defined as

$$V_{M_{PRI}} = \left\{ \frac{V_{PRI} * L_X}{L_{LK} + L_X} \right\} \tag{A5}$$

Substituting for V_{MPRI} from *(A4)* into *(A5)*,

$$\delta i'_{LC} = \left\{ \frac{V_{M_{PRI}}}{L'_{EQ}} \right\} * \delta t = \left\{ \frac{V_{PRI} * L_X}{L_{LK} + L_X} \right\} * \left(\frac{\delta t}{L'_{EQ}} \right) \tag{A6}$$

where, $\delta i'_{LC}$ is the transferred primary current of the secondary steady state ripple current δi_{LC}, which is given by

$$\delta i_{LC} = \Delta I_{C\_PKPK} = \left(\frac{1}{k} \right) \{ \delta i'_{LC} \} = \frac{1}{k} \left\{ \frac{V_{PRI} * L_X}{L_{LK} + L_X} \right\} * \left(\frac{\delta t}{L'_{EQ}} \right)$$
$$= \frac{1}{k} \left\{ \frac{V_{PRI} * L_X}{L_{LK} + L_X} \right\} * \left(\frac{\delta t}{\frac{1}{k^2} L_{EQ}} \right) \tag{A7}$$
$$= k * \left\{ \frac{V_{PRI} * L_X}{L_{LK} + L_X} \right\} * \left(\frac{\delta t}{L_{EQ}} \right)$$

The effective input voltage V_{PRI}, is a function of various design parameters and is given by the expression [x]:

$$V_{PRI} = \left[\frac{\{V_{IN}(m+1) - N * V_{OUT}\} * \left\{D - \frac{m}{N}\right\}}{L_{EQ} * f_{SW}} \right] \tag{A8}$$

where, V_{IN} is the input voltage, V_{OUT} is the output voltage, f_{SW} is the switching frequency, D is the duty cycle & $m = floor$ $(N*D)$. Substituting the relevant expressions into *(A7)*, the steady state ripple current in the compensation inductor can be expressed and simplifed as:

$$\Delta I_{C_{PKPK}} = k * \left\{ \frac{V_{PRI} * L_X}{L_{LK} + L_X} \right\} * \left(\frac{\delta t}{L_{EQ}} \right)$$
$$= k * \left[\frac{\{V_{IN}(m+1) - N * V_{OUT}\} * \left\{D - \frac{m}{N}\right\}}{L_{EQ} * f_{SW}} \right] * \left\{ \frac{L_X}{L_{LK} + L_X} \right\} \tag{A9}$$

$$\Delta I_{C_{PKPK}} = k * \left[\frac{\{V_{IN}(m+1) - N * V_{OUT}\} * \left\{D - \frac{m}{N}\right\}}{L_{EQ} * f_{SW}} \right]$$
$$* \left\{ \frac{k^2 L_P * L_{EQ}}{k^4 L_P L_{LK} + k^2 L_P * L_{EQ} + L_{LK} L_{EQ}} \right\} \tag{A10}$$

If the primary leakage inductance is neglected then $L_{LK} = 0$. This results in the last factor in *(A10)* to be unity. Hence, the steady state ripple current in the Lc for the simplified case can be reduced to

$$\Delta I_{C_{PKPK}} = k * \left[\frac{\{V_{IN}(m+1) - N * V_{OUT}\} * \left\{D - \frac{m}{N}\right\}}{\{L_C + N * (1 - k^2) * L_P\} * f_{SW}} \right] \tag{A11}$$

Novel Complex Permeability Model of Powder Magnetic Materials

Lukas Mueller
Department of Electrical & Computer Engineering
University of Nebraska - Lincoln
Lincoln, USA
lmueller18@unl.edu

James Cox
President
Micrometals, Inc
Anaheim, USA
jcox@micrometals.com

Jun Wang
Department of Electrical & Computer Engineering
University of Nebraska - Lincoln
Lincoln, USA
junwang@unl.edu

Enrique Garcia
Application Engineering
Micrometals, Inc
Anaheim, USA
egarcia@micrometals.com

Abstract—Complex permeability is an important magnetic material parameter in the design of high frequency magnetic components. The complex permeability characteristics of ferrite materials is well known and understood. Currently, there is little information on the complex permeability characteristic of powdered materials available in literature or manufacturer datasheets. Given the potential advantages powdered materials present over ferrites in EMI filters and RF magnetics, it is crucial to model and understand the complex permeability behavior of powder magnetic materials. This paper presents a physical based model to predict the complex permeability of powdered materials over a wide frequency range. The presented model only relies on material data generally provided in powder material datasheets which allows designers to calculate the complex permeability of powder materials over a wide frequency range without requiring additional material information from the manufacturer.

Index Terms—Complex permeability, magnetic materials, very high frequency (VHF), wireless charging

I. INTRODUCTION

Complex permeability is a useful magnetic material parameter in the design of high frequency magnetic components, particularly for EMI filter or resonant inductors. In conjunction with the winding characteristics it can be used to determine the small-signal impedance of inductors or transformers [1]. The complex permeability can also be used to model core loss at small signal excitation, making it helpful in the evaluation of core materials for very frequency applications like inductors used in RF amplifiers or wireless chargers.

The complex permeability characteristics of ferrite materials is well known and understood [2]–[8]. However, there is little information on the complex permeability characteristic of powdered materials available. Given the potential advantages powdered materials have over ferrites in some EMI filters and RF magnetic applications [9]–[11], it is crucial to model and understand the complex permeability characteristics of powder magnetic materials.

Complex Permeability - Definition

The complex series permeability of a magnetic material is defined as:

$$\mu(f) = \mu'(f) - j\mu''(f) \tag{1}$$

with $\mu'(f)$ being the real part of the complex series permeability at a frequency f and $\mu''(f)$ being the imaginary part of the complex series permeability at the frequency f. Using the complex series permeability, the impedance of a wound inductor or transformer due to the core material characteristics can be approximated using:

$$Z_{core}(f) \approx j2\pi f N^2 \frac{A_e}{l_e} \mu(f) \tag{2}$$

which can then be written as:

$$Z_{core}(f) \approx j2\pi f N^2 \frac{A_e}{l_e}\mu' + 2\pi f N^2 \frac{A_e}{l_e}\mu'' \tag{3}$$

The first part of the impedance is imaginary, representing the energy stored in the magnetic core, the second part of the impedance is real, representing losses. The terms of the complex permeability can therefore be used to model the wound magnetic core as an inductive and a resistive element in series. The complex permeability also equates to the dissipation factor (DF) and quality factor (Q) of the magnetic core. The dissipation factor or loss angle of the magnetic material is given as:

$$DF = \tan\delta = \frac{\mu''}{\mu'} \tag{4}$$

The Q of the material is given as:

$$Q_{core} = \frac{1}{\tan\delta} = \frac{\mu'}{\mu''} \tag{5}$$

Complex permeability is a valuable tool to determine the impedance a core material can provide for EMI filters. The complex permeability can also be used to determine the minimum losses or maximum Q an inductor can have at high

979-8-3315-1612-3/25 $31.00 © 2025 IEEE

frequencies, an important metric for high frequency inductor designs. Lastly, the complex permeability information can be used in finite element simulation packages to model the material characteristics better over a wide frequency range. For more information of how to utilize complex permeability in inductor design see [1], [2], [12], [13].

II. LOW FREQUENCY APPROXIMATION

The imaginary part of the complex permeability can be determined from the core loss in the material using the following expression [2]:

$$\mu'' = \frac{P_{core}(f, H_a)}{\pi f H_a^2 \mu_0} \tag{6}$$

with $P_{core}(f, H_a)$ being the core loss density of the material in $\frac{W}{m^3}$ at the frequency f and peak magnetization force H_a (A/m) and μ_0 being the permeability of free space. Core loss is traditionally specified in terms of flux density instead of magnetization force. The equation can then be rearranged to approximate the value based on the flux density [2]:

$$\mu'' \approx \frac{\mu_0 \mu'^2(f) P_{core}(f, B_p)}{\pi f B_p^2} \tag{7}$$

with $P_{core}(f, B_p)$ being the core loss density of the material in $\frac{W}{m^3}$ at the frequency f and peak flux density B_p. This equation is only valid if the Q of the material is high and the real part of the permeability is known and approximately constant.

To determine the imaginary part of the complex permeability it is of paramount importance to use accurate core loss information. The Steinmetz equation is generally used to calculate the core losses in most magnetic materials. However, the standard Steinmetz equation is only an approximate curve-fit which can result in large errors, especially at small signal excitation [14], [15]. Use of the Steinmetz equation to predict the imaginary part of the permeability might therefore lead to large errors.

III. PROPOSED MODEL

Powder cores are produced from pulverized insulated magnetic particles. The insulated particles are mixed with a binder and compressed under high pressure to form a solid magnetic core. Unlike a ferrite, which is homogeneous material, powder cores have magnetic particles suspended in a non-magnetic material, creating a distributed air gap throughout the material. The binder and insulation increases the bulk resistivity of the powder material and decreases the effective bulk permittivity, suppressing bulk eddy current losses and dimensional resonances.

The non-homogeneous composition of powder magnetic materials makes modeling them challenging. An equivalent model needs to be derived specifically for the powder materials. The powder magnetic material is similar to wires in the winding window of a transformer or inductor. Both the wire and the magnetic particles are conductive materials insulated from one another by a non-magnetic, non-conductive material. In the Dowell method, the round wires in a winding are modeled as equivalent conductive sheets [16], [17]. The same principal can be applied to the magnetic particles in powdered core. The magnetic particles can be approximated as and equivalent sheet or laminations of magnetic material. The complex permeability behavior of laminated magnetic sheets is well known and can be used as a foundation for the complex permeability derivation.

Modeling the magnetic particles as magnetic sheets will require knowledge about the materials permeability, conductivity, particle shape and particle shape. While the permeability of the base magnetic material is generally known, there is little information what change in permeability when the material undergoes one of the various pulverization processes. Similarly, the exact conductivity of the magnetic powder is generally unknown [18]. While these values are not known, they can be derived based on other material measurements [18]. To account for this in the derivation of the proposed model, an equivalent skin depth factor F_δ is defined to represent combination of the unknown permeability, conductivity, particle size and particle shape.

The complex permeability of the equivalent magnetic sheets used to model the powder core is then given by [19], [20]:

$$\mu \approx \frac{\mu' e^{\frac{-j\theta_h}{2}} \tanh\left(e^{\frac{-j\theta_h}{2}} \sqrt{j}\sqrt{f} F_\delta\right)}{\sqrt{j}\sqrt{f} F_\delta} \tag{8}$$

where θ_h is the hysteresis phase angle of the base magnetic material.

At high frequencies the complex permeability will be dominated by the eddy current losses, therefore, the hysteresis phase angle θ_h can be neglected and set to zero. The expression then simplifies to:

$$\mu_{material} \approx \frac{\mu' \tanh\left(\sqrt{j}\sqrt{f} F_\delta\right)}{\sqrt{j}\sqrt{f} F_\delta} \tag{9}$$

One key difference is that in the Dowell method the wires have a permeability approximately equal to the surrounding medium. For powder cores, the magnetic particles have a significantly higher permeability than the surrounding medium. The distributed air gap between the particles has to be accounted for when modeling them as a magnetic sheet. At low drive levels, the varying lengths of the distributed air gaps can be modeled as one discrete homogeneous air gap. The complex permeability of a powdered core can then be approximated using:

$$\mu_{core} \approx \frac{\mu_{material}}{F_g + (1 - F_g)\mu_{material}} \tag{10}$$

where F_g is the gapping factor of the material, which varies between zero and one. A value of one would indicate that there is no distributed air gap present, whereas a value of zero would indicate no magnetic material is present in the core.

To solve for the complex permeability, three variables must be determined: the equivalent gapping factor F_g, the relative starting permeability of the base magnetic material $\mu_{material}$ and the equivalent skin depth factor F_δ. As already stated, these parameters have to be derived through other specified material properties.

979-8-3315-1612-3/25 $31.00 © 2025 IEEE

The gapping factor F_g and the relative real starting permeability of the base material are defined through the real starting permeability of the powdered core. The relationship is:

$$F_g = \frac{(\mu'_{core} - 1) \times \mu'_{mat}}{\mu'_{core} \times (\mu'_{mat} - 1)} \quad (11)$$

which can now be normalized by assuming the base magnetic material's real permeability is a factor F_μ times larger than the initial permeability of the pressed powdered core. The relationship is then:

$$F_g = \frac{(\mu'_{core} - 1) \times F_\mu \mu'_{core}}{\mu'_{core} \times (F_\mu \mu'_{core} - 1)} \quad (12)$$

Another parameter is required to solve the system of equations. The drop in relative real permeability can be used to identify the actual fill factor or base material permeability ratio. Real permeability versus frequency data is generally provided by all powder magnetic material suppliers over a limited frequency range, providing sufficient information to solve for complex permeability over a wider frequency range than specified.

To normalize the drop in real permeability, the factor $\Delta F_{L,95-90}$ is used. The factor is defined as:

$$\Delta F_{L,95-90} = \frac{F_{L=90\%} - F_{L=95\%}}{F_{L=90\%}} \quad (13)$$

where $F_{L=90\%}$ is the frequency where the real permeability of the material drops to 90% of its initial value and $F_{L=95\%}$ is the frequency where the real permeability of the material drops to 95% of its initial value.

Lastly, all frequencies are normalized to the set value of $F_{L=90\%}$. This eliminates the need to know the equivalent skin depth of the material, all frequencies can simply be solved in relation to the normalized frequency.

Using the presented dependencies, the factor of F_g can be solved based on $\Delta F_{L,95-90}$ using the chart in Fig. 1. Note that differences in the initial permeability of the pressed powdered core will have a slight effect on the relationship between $\Delta F_{L,95-90}$ and F_μ. However, the influence is greatest with extremely low permeability materials and overall the variation is small.

With the F_μ value identified, the complex permeability can be determined using equation (9) and (10) or using normalized pre-calculated charts. Pre-calculated charts showing the normalized permeability vs. normalized frequency are shown in Fig. 2-6. The extreme values of F_μ are plotted, showing that the real and imaginary permeability characteristics for most practical powdered core materials has to be between these two curves. Of note is that the higher the ratio of base material permeability to the permeability of the pressed core the more gradual the drop off in real permeability with frequency.

IV. EXAMPLE CALCULATION

To highlight the use of the proposed model, an example calculation using a commercially available powder iron material is demonstrated in detail in this section. These calculations can be used as a reference to analyze other powder magnetic

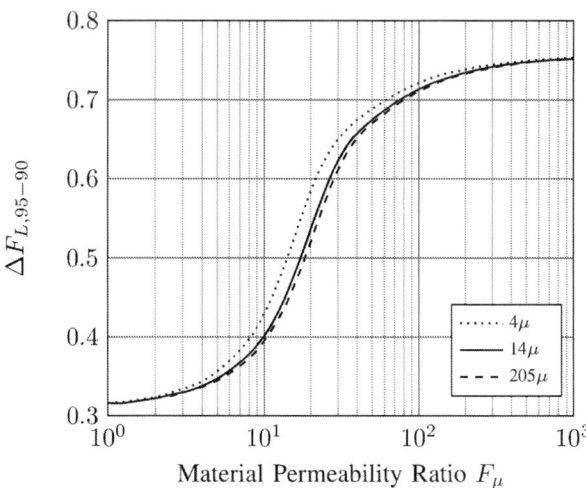

Fig. 1. Dependency between $\Delta F_{L,95-90}$ and F_μ for powdered materials with various initial real relative permeability

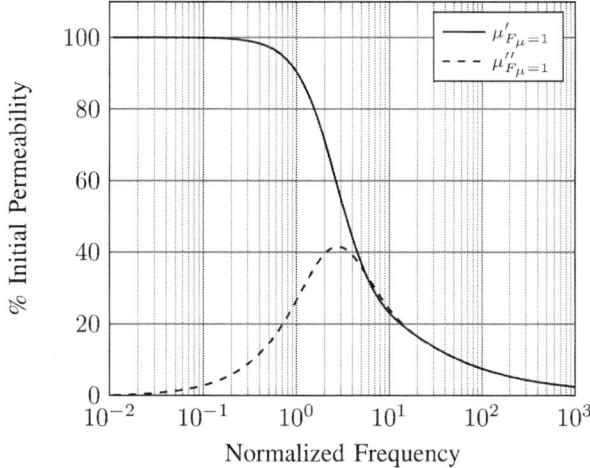

Fig. 2. Complex permeability profiles for powder material with $F_\mu=1$

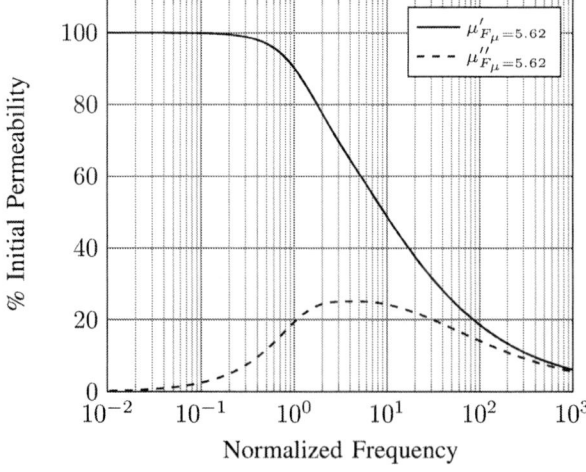

Fig. 3. Complex permeability profiles for powder material with $F_\mu=5.62$

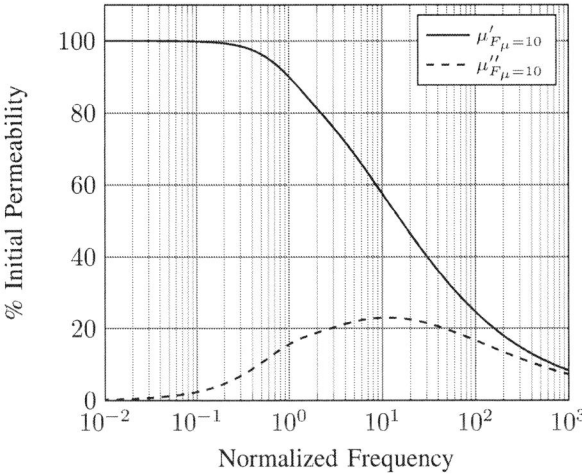

Fig. 4. Complex permeability profiles for powder material with $F_\mu=10$

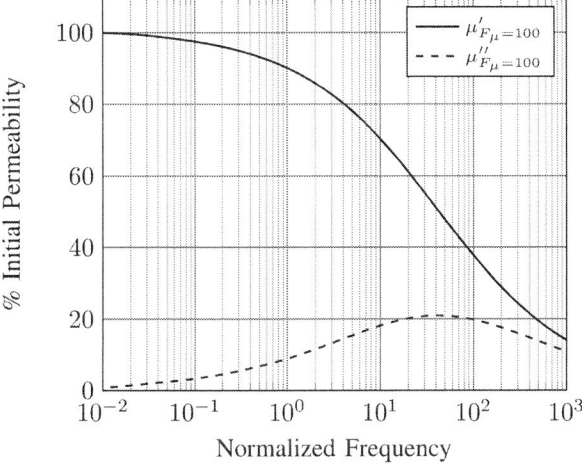

Fig. 5. Complex permeability profiles for powder material with $F_\mu=100$

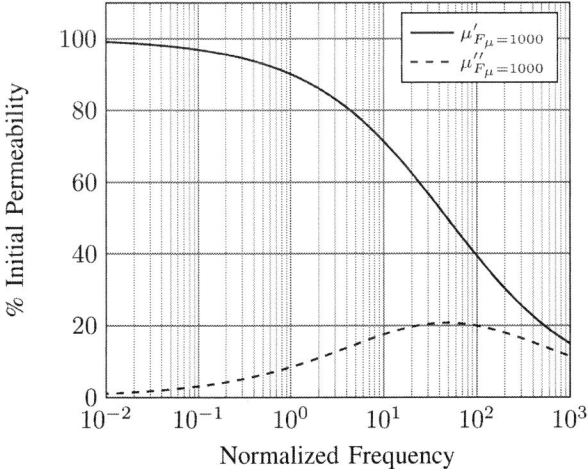

Fig. 6. Complex permeability profiles for powder material with $F_\mu=1000$

material from various suppliers. In this example, Micrometals' Mix 40 is evaluated. The Mix 40 has an initial real relative permeability of 60 and is commonly used in AC line filters or differential model EMI filters. The following steps are performed to calculate the complex permeability of the material.

1) Determine the $F_{L=95\%}$ and $F_{L=90\%}$ frequencies: The frequencies where the initial permeability drops to 95% and 90% need to be determined first. Micrometals provides a curve fit to evaluate real permeability vs. frequency for all its material grades. The equation is:

$$\mu' = \frac{1}{a + bf^c} + d \qquad (14)$$

The fitting parameters for the Mix 40 material are: a=1.86e-2, b=5.98e-8, c=8.23e-1, d=6.64e0 [21]. Other vendors provide similar curve-fits for their materials [22]–[24]. If curve-fits are not available, manufactures generally provide charts showing the real permeability versus frequency behavior over a limited frequency range. If no published information is available on the behavior of real permeability versus frequency, then the frequencies where the inductance drops needs to be evaluated through measurements. As only a 10% drop in real permeability needs to be observed, the required frequency range of the measurement device will be limited, making the measurement feasible using common LCR meters or impedance analyzers.

For the Mix 40, using the curve-fit equation, the value of $F_{L=90\%}$ and $F_{L=95\%}$ were determined to be:

$$F_{L=95\%} = 178500Hz \qquad (15)$$

$$F_{L=90\%} = 410000Hz \qquad (16)$$

The value of $F_{L=90\%}$ for a wide variety of powder magnetic material types is provided [25].

2) Normalize frequencies: All frequencies need to be normalized to the $F_{L=90\%}$ frequency. For the Mix 40, this value was determined to be equal to 410kHz.

3) Find $\Delta F_{L,95-90}$: The value is used to find the approximate equivalent gapping ratio of the material. For the Mix-40 material the value is equals to:

$$\Delta F_{L,95-90} = \frac{410000Hz - 178500Hz}{410000Hz} = 0.5647 \qquad (17)$$

4) Find F_μ: Using the calculated $\Delta F_{L,95-90}$ value, the value of F_μ can be read from Fig. 1. For the Mix 40, based on the chart, the value of F_μ is approximately equal to 25.

5) Calculate full complex permeability spectrum: With the value of F_μ known and the frequency spectrum normalized, the full complex permeability behavior of the material can now be approximated using equation (9) and (10). The non-normalized frequencies are obtained by multiplying the normalized frequency by $F_{L=90\%}$. Instead of solving the design equations, the traces in Fig. 2-6 with the closest F_μ value can be used as an approximation.

The calculated complex permeability of this example is shown in the experimental verification section.

V. EXPERIMENTAL RESULTS

The proposed model was verified by measuring the complex permeability behavior of a 0.38in toroidal core made from Micrometals' Mix 40 powder iron material using a HP4274A and HP4291A. The toroid was wound with 5 turns of 26 AWG wire to minimize the changes in the impedance due to inter-widing capacitance. The calculation from the previous sections were used to generate the theoretical complex permeability behavior. The calculated and measured values are compared to one another in Fig. 7.

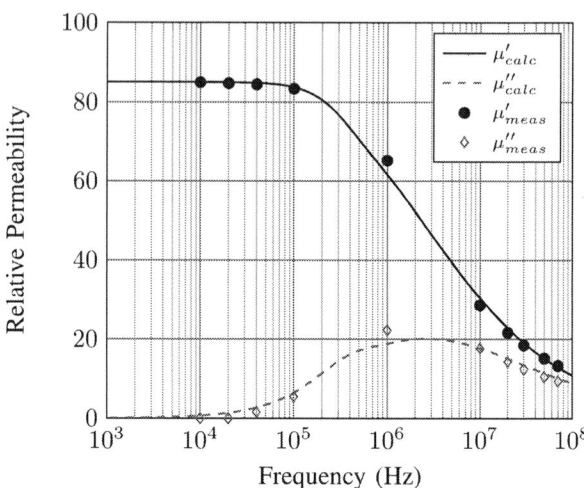

Fig. 8. Comparison between calculated and measured complex permeability of Micrometals' Mix 38 material with $F_\mu \approx 8$

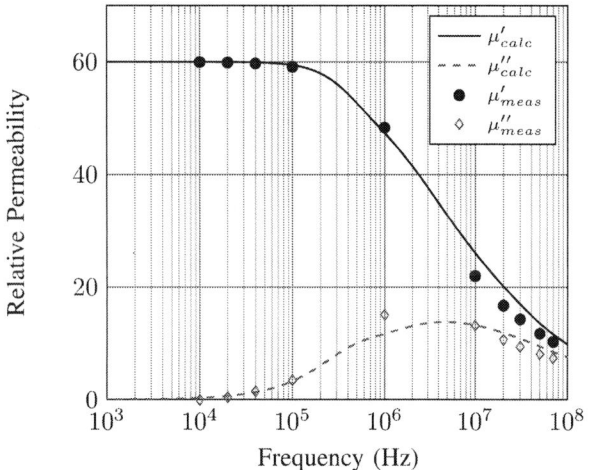

Fig. 7. Comparison between calculated and measured complex permeability of Micrometals' Miz 40 material with $F_\mu \approx 25$

A 0.38in toroidal core made from Micrometals' Mix 38 powder iron material was measured and compared to the model in Fig. 8. The inductor was also wound with 5 turns of 26AWG. The F_μ value was determined based on the real permeability measurements to verify that measuring the real permeability behavior allows predicting the imaginary permeability.

The proposed model matched the measurements well over a wide frequency range, especially considering that only a limited number of datapoints are required to model the complex permeability behavior over a wide frequency range.

VI. LIMITATIONS AND FUTURE WORK

The presented model allows for the evaluation of the complex permeability of powder magnetic materials. The model is straightforward to use and can be utilized using only data commonly supplied by magnetic core manufacturers. However, the ease of use of the model also results in some limitations to the model's accuracy and range of applications.

The model uses the equivalent skin depth factor F_δ to account for the various properties of the magnetic powder, like permeability, conductivity and size. By considering one constant value of F_δ any potential variation in the properties

of the magnetic powder is average over the whole material. While permeability and conductivity tend to be relative constant, the shape and size of the magnetic powder can vary greatly [26], [27]. This variation of the particle size is not considered by the proposed model. As the relationship between complex permeability and F_δ is not linear, effectively averaging the particle size will introduce an error into the model. Carbonyl iron materials will experience the least error due to the generally uniform particle size distribution [18], whereas other materials like Molypermalloy, can have a larger variation is particle size and therefore a larger error. Similarly some high permeability grades of Molypermalloy feature non-spherical particles which will result in anisotropic magnetic behavior [28]. Complex permeability will vary depending on the preferred magnetization direction in these types of materials, an effect not captures by the proposed model. An additional modeling parameter could be added to account for any variations in the characteristics of the magnetic powder. However, the addition of an additional parameter will make the model significantly more complex, making it harder to use and less accessible to practicing engineers.

Another limitation of the proposed model is that it is only valid for a small signal excitation. The distributed air-gap presented in powder materials is considered as a lumped discrete gap in this model. However, in a real powder material the gap is non-linear leading to a gradual partial saturation of sections of the core. Partial saturation of magnetic particles is not considered by the proposed model. High frequency resonant inductors generally operate with low flux densities as core loss would be excessive otherwise, making this limitation less problematic when modeling resonant inductors [29]. However, EMI filter inductors generally operate with significant DC bias currents, causing partial saturation of the core. While previous measurements of powder materials have not shown a significant change in core loss with DC bias [30], this behavior is not fully explored at higher excitation. DC bias operation could therefore impact the complex permeability behavior of powder materials. This is a similar limitation to ferrites, where the change in losses and potentially complex permeability with DC bias requires a more detailed evaluation [15], [30], [31]. A more accurate hysteresis model for powder materials is needed to evaluate the change in complex permeability with DC bias analytically.

Lastly the model does not consider hysteresis losses as it is intended to evaluate materials at high frequencies where obtaining accurate measurements is challenging. The imaginary part of the complex permeability at low frequencies due to hysteresis loss can be approximated using the low frequency approximation shown in this paper.

VII. CONCLUSION

A novel model to calculate the complex permeability of powder magnetic materials is presented here. The model only requires material characteristics that are generally provided by magnetic core producers, making the model easy to use. Experimental results show good agreement between the proposed model and small signal measurements. The model can be used to evaluate the performance of powder magnetic materials at high frequencies for the use in resonant inductors, transformers or EMI filters.

REFERENCES

[1] A. Van den Bossche and V. C. Valchev, *Inductors and Transformers for Power Electronics*. Boca Raton, FL, USA: CRC Press, 2005.

[2] J. Watson, *Applications of Magnetism*. New York, NY, USA: John Wiley & Sons, 1980.

[3] D. Jiles, *Introduction to Magnetism and Magnetic Materials*. Chapman & Hall, 1998.

[4] A. H. Morrish, *The Physical Principals of Magnetism*. Wiley, 1965.

[5] D. Jiles, "Frequency dependence of hysteresis curves in 'non-conducting' magnetic materials," *IEEE Transactions on Magnetics*, vol. 29, no. 6, pp. 3490–3492, 1993.

[6] A. Furuya, Y. Uehara, K. Shimizu, J. Fujisaki, T. Ataka, T. Tanaka, and H. Oshima, "Magnetic field analysis for dimensional resonance in mn–zn ferrite toroidal core and comparison with permeability measurement," *IEEE Transactions on Magnetics*, vol. 53, no. 11, pp. 1–4, 2017.

[7] C. Cuellar, W. Tan, X. Margueron, A. Benabou, and N. Idir, "Measurement method of the complex magnetic permeability of ferrites in high frequency," in *2012 IEEE International Instrumentation and Measurement Technology Conference Proceedings*, 2012, pp. 63–68.

[8] T. Nakamura, "Snoek's limit in high-frequency permeability of polycrystalline ni-zn, mg-zn, and ni-zn-cu spinel ferrites," *Journal of applied physics*, vol. 88, no. 1, pp. 348–353, 2000.

[9] J. Cox, *Iron Powder Cores for High Q Inductors*, Micrometals, Inc, 5615 E. La Palma Ave., Anaheim CA 92807 USA, 1997.

[10] M. L. Heldwein, "Design of minimum volume emc input filters for an ultra compact three-phase pwm rectifier," *Eletronica de Potencia*, vol. 14, pp. 85–96, 05 2009.

[11] L. Mueller, "Analytical evaluation of differential model dc emi filter inductors using material saturation coefficient," in *PCIM Europe 2024; International Exhibition and Conference for Power Electronics, Intelligent Motion, Renewable Energy and Energy Management*, 2024, pp. 2420–2425.

[12] E. C. Snelling, *Soft Ferrites - Properties and Applications*. Mendham: PSMA, 2010.

[13] R. Suárez, M. Tijero, R. Moreno, A. Arriola, and J. M. González, "Influence of complex magnetic permeability on 3-d simulation of mnzn common-mode chokes," in *2023 International Symposium on Electromagnetic Compatibility – EMC Europe*, 2023, pp. 1–6.

[14] C. G. Oliver, "Measurement and modeling of core loss in powder core materials," *ETTC Meeting - TTA Annual meeting*, 2012.

[15] D. Serrano, H. Li, S. Wang, T. Guillod, M. Luo, V. Bansal, N. K. Jha, Y. Chen, C. R. Sullivan, and M. Chen, "Why magnet: Quantifying the complexity of modeling power magnetic material characteristics," *IEEE Transactions on Power Electronics*, vol. 38, no. 11, pp. 14 292–14 316, 2023.

[16] P. Dowell, "Effects of eddy currents in transformer windings," *Proceedings of the Institution of Electrical Engineers*, vol. 113, pp. 1387–1394, 1966. [Online]. Available: https://digital-library.theiet.org/doi/abs/10.1049/piee.1966.0236

[17] R. W. Erickson and D. Maksimovic, *Fundamentals of Power Electronics*, 2nd ed. Springer, 2001.

[18] W. J. Polydoroff, *High-frequency magnetic materials : their characteristics and principal applications*. New York, NY, USA: John Wiley & Sons, 1960.

[19] R. Bozorth, *Ferromagnetism*. Wiley, 1993.

[20] R. Stoll, *The Analysis of Eddy Currents*, ser. International Series of Monographs in Electrical Engineering. Clarendon Press, 1974. [Online]. Available: https://books.google.com/books?id=mojKygAACAAJ

[21] Micrometals. Iron powder products catalog. [Online]. Available: https://www.micrometals.com/design-and-applications/literature/

[22] ChangSungCoperation. Magnetic powder cores. [Online]. Available: https://www.mhw-intl.com/assets/CSC/CSC_Catalog.pdf

[23] Micrometals. Micrometals alloy powder core catalog 2021. [Online]. Available: https://www.micrometals.com/design-and-applications/literature/

[24] Magnetics. 2024 magnetics powder cores catalog. [Online]. Available: https://www.mag-inc.com/Design/Technical-Documents/Powder-Core-Documents

[25] L. Mueller. Tech note 1: Steinmetz coefficients. [Online]. Available: https://www.micrometals.com/design-and-applications/literature/

[26] J. Hua, F. S. Gobber, M. Actis Grande, D. Mortensen, and J. O. Odden, "A numerical modeling framework for predicting the effects of operational parameters on particle size distribution in the gas atomization process for nickel-silicon alloys," *Powder Technology*, vol. 435, p. 119408, 2024. [Online]. Available: https://www.sciencedirect.com/science/article/pii/S0032591024000500

[27] Y. Du, X. Liu, S. Xu, E. Fan, L. Zhao, C. Chen, and Z. Ren, "Numerical simulation of gas atomization and powder flowability for metallic additive manufacturing," *Metals*, vol. 14, no. 10, 2024. [Online]. Available: https://www.mdpi.com/2075-4701/14/10/1124

[28] A. R. Opitz, "Flake magnetic core and method of making same," U.S. Patent 3 255 052, Dec. 9, 1963. [Online]. Available: https://patents.google.com/patent/US3255052A

[29] Y. Han, G. Cheung, A. Li, C. R. Sullivan, and D. J. Perreault, "Evaluation of magnetic materials for very high frequency power applications," in *2008 IEEE Power Electronics Specialists Conference*, 2008, pp. 4270–4276.

[30] J. Muhlethaler, J. Biela, J. W. Kolar, and A. Ecklebe, "Core losses under the dc bias condition based on steinmetz parameters," *IEEE Transactions on Power Electronics*, vol. 27, no. 2, pp. 953–963, 2012.

[31] C. A. Baguley, B. Carsten, and U. K. Madawala, "The effect of dc bias conditions on ferrite core losses," *IEEE Transactions on Magnetics*, vol. 44, no. 2, pp. 246–252, 2008.

979-8-3315-1612-3/25 $31.00 © 2025 IEEE

Design Study Evaluating Impact of Gap Loss on Nanocrystalline Inductor Cores with Experimental Validation

Maurice Sturdivant[1], Brandon Grainger[1], Christopher Bracken[2], Paul Ohodnicki[1,2]

Department of Electrical and Computer Engineering[1]
Department of Mechanical Engineering and Materials Science[2]
University of Pittsburgh
{mds165, bmg10, csb80, pro8} @pitt.edu

Abstract— As electric vehicle (EV) technology matures, the need for compact and efficient power electronics will continue to grow. To increase power density and efficiency, it is ideal to minimize the size and power loss of components, including inductors. A standard inductor design technique is placing an air gap in the magnetic core, allowing designers to tune the inductance and allow higher levels of magnetic field without saturation. The inclusion of gaps can increase the total magnetic core loss beyond those predicted by common models such as the Steinmetz equation, creating the need for accurate gap loss models. In this study, the losses of a nanocrystalline inductor core are experimentally measured and compared to an existing gap loss model, showing significant agreement with predicted values. The model was then integrated into a multi-objective optimization framework to evaluate the fitness of nanocrystalline cores against ferrite cores at different switching frequencies. This work finds that gap losses can increase the total loss of nanocrystalline inductors by up to an order of magnitude as frequency increases, implying that consideration of gap losses is vital to selecting the appropriate core material for design applications.

Keywords— *air gaps, core loss, inductors, loss measurement, nanocrystalline, optimization*

I. INTRODUCTION

With the acceleration of EV development and integration, power electronic systems will be needed to support this transition, with power density and efficiency being key metrics driving the design of these systems. The Department of Energy has set targets to develop converters reaching power density of 100 kW/L and efficiency of 98% as soon as 2025 [1]. A common approach to reducing converter size is to use higher switching frequencies (>50 kHz), but these higher frequencies can increase power losses in the system's components, including inductors. Inductors are also impacted by design characteristics such as the winding, core material, and air gaps. To effectively balance power density and efficiency, component losses must be predicted accurately. Because of the core's sensitivity to these design decisions, this work is focused on inductor core loss.

Magnetic core losses are categorized as classical eddy current loss and hysteresis loss. Classical eddy current loss is the resistive losses caused by magnetic flux-induced eddy currents in the core material. Conversely, hysteresis loss refers to energy consumed by magnetic domain movement in response to the applied magnetic field. Since these losses are caused by the component's magnetizing flux passing through the core, these loss mechanisms will be collectively referred to as magnetizing core loss. The most common core loss model is the Steinmetz Equation, shown in (1), that relates a core material's volumetric loss density p_c to the core, frequency f, peak flux density B_m using the empirically derived coefficients k, α, and β [2].

Including an air gap in the magnetic core can tune the inductance of a component, while also preventing the core from saturating under specific operating conditions. In addition to magnetizing loss, inductor cores begin to experience increases in loss after being cut and gapped [3]. While these additional losses at or near the gap are often called gap losses, they can result from several different mechanisms. Direct gap loss, or increased loss density on the gap-facing surface of the core, is caused by the manufacturing processes used to cut the core. A section of a cut ferrite core is seen in Fig. 1, showing separate bulk and surface regions, denoted by d_b and d_s, respectively [4]. This modeling approach assumes the thickness of the surface region is negligible compared to the bulk region, so that $d_b + 2d_s \approx d_b$.

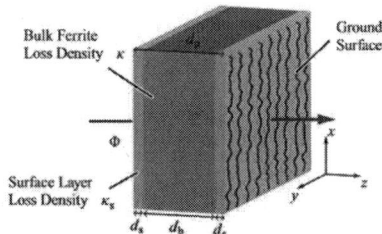

Fig. 1. Ferrite core segment with ground surface. [4]

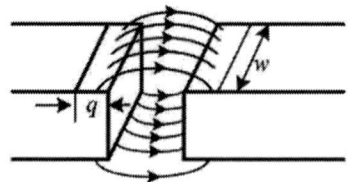

Fig. 2. Fringing flux near air gap. [5]

979-8-3315-1612-3/25 $31.00 © 2025 IEEE

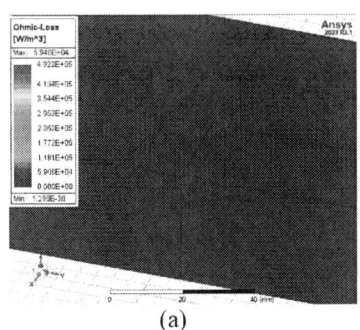 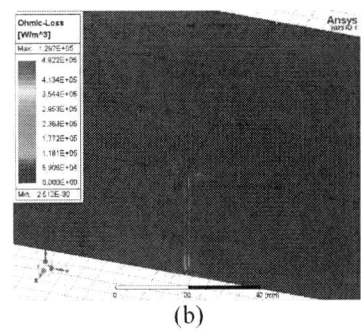 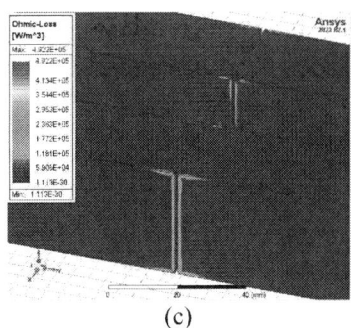

(a)   (b)   (c)

Fig. 3. Ohmic loss plots of core at 25 kHz, 0.1T with (a) 0.25 mm, (b) 0.5 mm, and (c) 1 mm air gaps

To account for direct gap loss on the ground surfaces of a Manganese-Zinc (MnZn) ferrite core, Neumayr et al. uses a separate set of Steinmetz coefficients: k_s, α_s, and β_s [4]. Since the surface regions are assumed to be of negligible thickness, using (2), these coefficients yield a two-dimensional loss density p_{surf}, in units of W/m^2. The total direct gap loss for a ferrite core is the product of this surface loss density and the total ground surface area.

The other source of gap loss is the fringing gap loss. In gapped cores, some of the flux fringes around the gap due to air's lower permeability, as pictured in Fig. 2 [5]. Due to the high electrical conductivity of laminated nanocrystalline core materials, the components of fringing flux normal to its laminations induce additional eddy currents, resulting in localized losses near the edges of the gap. This source of gap loss is assumed to be negligible in ferrites and other highly resistive core materials. Unlike the magnetizing losses and machining-induced direct gap losses, fringing gap loss P_g is also dependent on gap length l_g and lamination width D. Using (3), a model developed by Wang et al. in [3], the gap loss in a nanocrystalline core can be estimated using a numerical constant k_g and coefficients for frequency, flux density, and lamination width.

$$p_c = kf^\alpha B_m^\beta \qquad (1)$$

$$p_{surf} = k_s f^{\alpha_s} B^{\beta_s} \qquad (2)$$

$$P_g = k_g l_g f^{1.72} B_m^2 D^{1.65} \qquad (3)$$

A 3D finite element model of a nanocrystalline inductor core was constructed in Ansys Maxwell, visualizing the fringing flux contributions to eddy current loss in the core. At a fixed frequency of 25 kHz and flux density of 0.1 T, the inductor was simulated with 0.25 mm, 0.5 mm, and 1 mm gaps. Plots of ohmic loss, or eddy current loss in the case of the core, were generated for the core at each gap length, shown in Fig. 3. These plots show higher concentrations of ohmic loss near the air gap and towards the surface of the core laminations. As suggested by (2), these losses also increase significantly with air gap size. This qualitative analysis supports the assumption that the fringing gap losses are a significant source of eddy current loss in gapped nanocrystalline cores.

II. CORE LOSS MEASUREMENT EXPERIMENTS

A. Experiment Procedure

The primary goals of the experiments are to quantitatively observe the impact of increasing gap length on the core loss of a laminated, nanocrystalline inductor and evaluate the ability of the fringing gap loss model to predict these additional gap losses. Core loss can be obtained by directly measuring power dissipated across the inductor, but this approach also includes winding loss. The selected method to measure core loss is the two-winding method [6]. This method uses a primary excitation winding with N_p turns to excite a core under test with magnetic flux and a secondary sense winding with N_s turns to measure the flux in the core. By measuring primary winding current i_p and voltage induced on the secondary winding v_s, Faraday's Law and Ampere's Law can be used to obtain the core's magnetic flux density B and magnetic field intensity H, shown in (4) and (5), [7]. Using (6), B and H can be used to calculate the core loss per cycle, where A_e is effective cross-sectional area of the core and l_m is mean magnetic path length of the inductor.

$$B(t) = \frac{1}{N_s A_e} \int_0^T v_s(\tau) d\tau \qquad (4)$$

$$H(t) = \frac{N_p i_p(t)}{l_m} \qquad (5)$$

$$P_c = A_e l_m \int H dB \qquad (6)$$

A conceptual diagram of the two-winding measurement setup from [7] is shown in Fig. 4. An image of the testbed is pictured in Fig. 5. In this setup, a cut Finemet FT-3TL nanocrystalline core sample was used. The primary winding was placed around the legs of the core and a pair of bobbins was used to divide the six turns around each gap. The secondary winding was placed around the end of the core. The primary was connected to the output of a Keysight 33500B Waveform Generator and AE Techron 7224 RF Amplifier to excite the inductor with sinusoidal currents at selected frequencies. The current and voltage signals of interest were recorded with a Yokogawa WT5000 Power Analyzer.

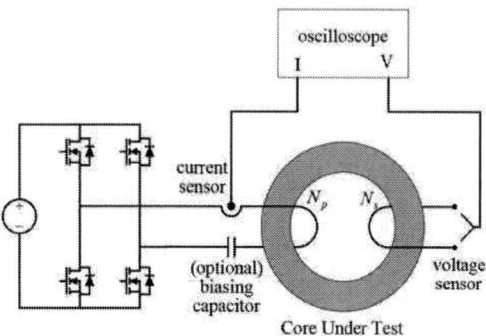

Fig. 4. Conceptual diagram of loss measurement setup. [7]

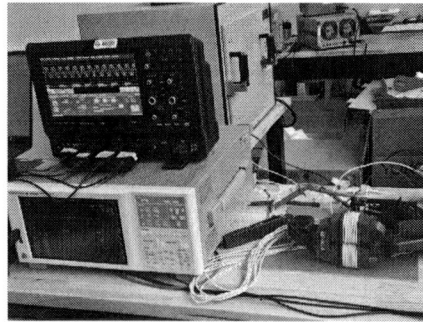

Fig. 5. Core loss measurement testbed

Since core losses are directly related to the frequency of the time-varying fields and magnitude of flux density, the tests were conducted at five frequencies (1 kHz, 5 kHz, 10 kHz, 25 kHz, and 50 kHz), and four target flux densities (0.03 T, 0.05 T, 0.07 T, 0.1 T).

For each measurement, the waveform generator was set to the desired frequency and turned on, exciting the core with a sinusoidal flux excitation. To avoid overloading the amplifier, the current is slowly increased until the desired flux density is reached. The root mean square (RMS) value of the secondary winding voltage measured on the power analyzer and oscilloscope were used to monitor the flux density in the core. When the secondary voltage corresponding to the target flux density is reached, the waveforms from the power analyzer and oscilloscope are recorded.

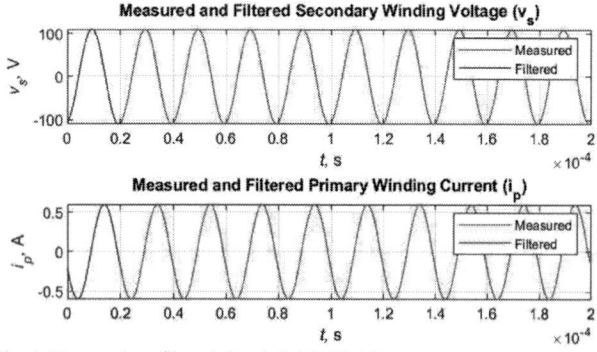

Fig. 6. Measured vs. filtered signals in MATLAB

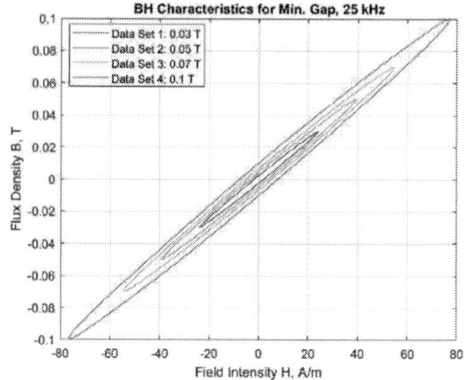

Fig. 7. B-H loops measured from core with minimum gap at 25 kHz

Using least-squares fitting, Fourier series coefficients were fit to the measured signals to filter any noise from them. The waveforms were reconstructed over one period using these coefficients, pictured in Fig. 6. By sampling a single period of the primary current and secondary voltage waveforms and using (4) and (5), the B and H signals were created, allowing the construction of the B-H loops for each test case. A group of B-H loops for a specific gap length and frequency are pictured in Fig. 7. By calculating the area of these loops, the core losses of the sample were obtained using (7).

B. Minimum Gap Baseline Measurements

To validate the setup and measure the impact of gap length on core loss, a baseline of loss values must be established. The baseline was taken by measuring the core's losses with minimum gap size $l_{g_{min}}$, or with a vice holding the cut core halves together and no spacing between them. Since the fringing flux paths are minimized in this configuration, the measured losses are assumed to include only the magnetizing loss and direct gap loss. These minimum gap values are expected to be close to those predicted by (1), using k = 1.633e-05 W/kg, α = 1.57, and β = 2.043, the Steinmetz coefficients for the Finemet core sample.

Fig. 8 shows the experimentally measured core losses for the minimum gap case. These results show significant agreement with the Steinmetz predictions at the higher frequencies tested. At 10 kHz and higher, most of the losses measured are within 20% of the expected Steinmetz losses. Below 10 kHz, the core losses predicted by the Steinmetz equation underestimate the experimentally measured losses. At the 1 kHz test cases, the measured losses were two to three times higher than predicted, resulting in errors between 100% and 200%.

From (4), it is known that at a higher frequency, a larger induced voltage v_s is required to reach the same target flux density. To reach this larger induced voltage, more current is needed to excite the primary winding. Therefore, the tests performed at lower frequencies are operating at lower power levels. This may cause any deviations on the scale of mW present in the setup to become apparent at these lower frequencies.

979-8-3315-1612-3/25 $31.00 © 2025 IEEE 546

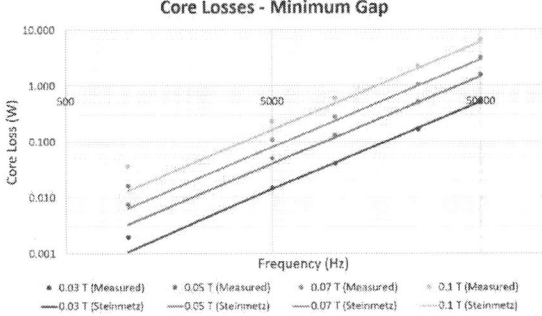

Fig. 8. Minimum gap core loss measurements vs. Steinmetz predictions

C. Calculation of Gap Loss

To estimate gap loss, a set of baseline measurements were obtained before the introduction of a discrete air gap. By incrementing the gap size from the minimum gap case and repeating the loss measurements at the target frequencies and flux densities, the fringing gap loss can be estimated. Any increases in loss from the minimum gap measurements observed as gap length increases are assumed to be the fringing gap loss P_{gap}. Specifically, P_{gap} is the difference between the core loss P_c measured with gap length l_g and core loss measured with $l_{g_{min}}$, taken at the same frequency f and flux density B_m:

$$P_{gap}(l_g, f, B_m) = P_c(l_g, f, B_m) - P_c(l_{g_{min}}, f, B_m) \quad (8)$$

The gap between the core halves was gradually increased using measured sheets of insulation paper and the tests were repeated at gap lengths of 0.127 mm, 0.254 mm, 0.508 mm, and 1.016 mm. After tests were performed, the gap losses were calculated using (8) and compared to values predicted using fringing gap loss model in (2) and the coefficients provided in [3]. Fig. 9 shows plots of the experimental gap losses in response to changing frequency, peak flux density, and gap length, compared to values from the model. The dashed lines represent predicted values that are not accompanied by a corresponding experimental value, as the target flux density could not be reached for those test cases. Specifically, for the 0.508 mm and 1.016 mm gaps at 50 kHz, the current amplifier used was unable to supply enough current to reach 0.1 T.

Overall, the model predicts gap loss with a reasonable degree of accuracy. The gap loss largely behaves as the model predicts, increasing nearly linearly with gap length and quadratically with flux density. However, measured gap losses show a notable deviation from the predictions of the model. Particularly at frequencies of 5 kHz and lower, experimental gap losses are consistently larger than model predictions. The gap loss model was fitted to data from simulations conducted by Wang et al. from 40 kHz to 200 kHz, while frequencies tested in the experiments do not exceed 50 kHz [3].

This deviation in the lower frequency range is therefore reasonable given that the model is being applied to frequencies substantially lower than those used to derive the model. This could also be a result of the variation in magnitude of the measured signals in response to different frequencies, flux densities, and gap lengths. While the largest measured values were as high as 8W, the smallest were on the scale of mW.

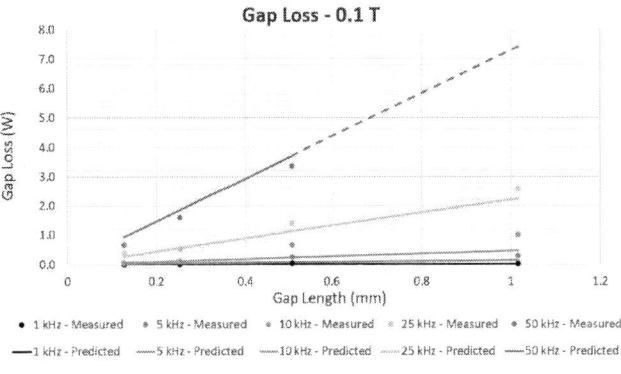

(a)

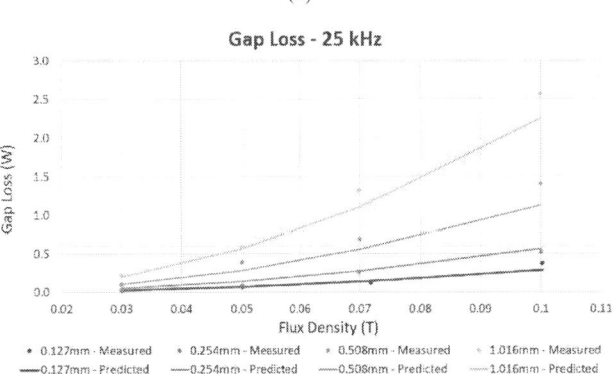

(b)

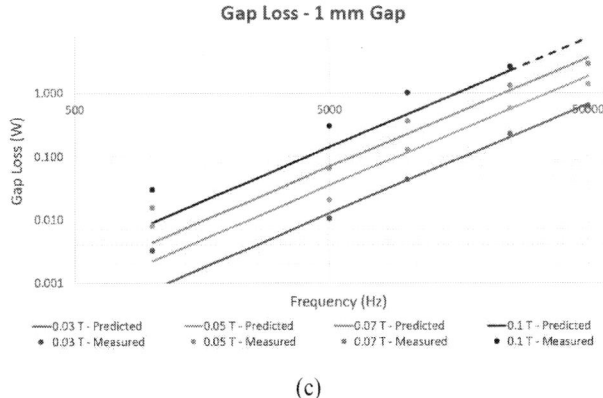

(c)

Fig. 9. Measured Gap Losses vs. (a) Gap Length, (b) Flux Density, and (c) Frequency

III. MULTI-OBJECTIVE OPTIMIZATION

Next, the gap loss models were integrated into a multi-objective optimization tool in MATLAB [8]. This tool employs a genetic algorithm, which generates populations of separate designs. Each design has a set of genes, each corresponding to a characteristic that is used to calculate the design's key metrics and fitness. A fitness function is used to evaluate each member of a population, influencing which characteristics persist into later iterations of the optimization, or generations.

A. Design Space and Constraints

The design space was limited to nanocrystalline cores and Manganese-Zinc ferrite cores, two common classes of core materials. Nanocrystalline cores are known for their high permeability and saturation flux densities, making them attractive options for compact designs in high power applications [9]. Ferrite cores are made of highly resistive metal-oxide ceramic materials, resulting in lower susceptibility to eddy current losses often encountered in high frequency applications [9].

The nanocrystalline core material used permeability data for Finemet FT-3M from the optimization tool and the Steinmetz coefficients for the cut sample of Finemet core used in Section II [8]. These coefficients were selected because the inductor designs are gapped, meaning they would likely resemble that of a cut core. The ferrite core material was assigned permeability data for MN67 and the MnZn ferrite core loss data used by Neumayr in [4,8]. The core material properties used are listed in Table I.

To minimize the impact of AC winding losses on this study, the inductor designs are limited to copper Litz wire conductors and the loss due to skin and proximity effects are neglected. Since the diameter of a Litz wire bundle is larger than that of a solid conductor with equivalent cross-sectional area, an additional packing factor is applied to the conductor diameter [10].

TABLE I.   CORE MATERIAL PROPERTIES AND COEFFICIENTS

Property/Coefficient	Finemet FT-3M	MN67
Relative Permeabilty, μ	50003	26999
Mass Density, ρ (kg/m^3)	7250	4794.4
k (W/kg)	1.633e-05	2.8e-03
α	1.57	1.36
β	2.043	2.77

For the ferrite designs, the gap loss is assumed to be direct gap loss, calculated using (2). The gap loss for nanocrystalline cores was calculated using (3). Pareto optimal fronts were generated, displaying the optimal solutions in the design space, organized by their mass and loss values. While several constraints were used to evaluate designs, those used to bound the key metrics and magnetic properties of the designs are listed in Table II.

TABLE II.   MAIN DESIGN CONSTRAINTS

Metric	Max/Target
Max. Mass	20 kg
Max. Loss	500 W
Max. Flux Density	80% of B_{sat}
Max Temperature Rise	60° C
Target Inductance	12 μH

B. Impact of Gap Loss on Optimal Designs

The chosen application for optimization was a 100 kW,12 μH DC filter inductor, operating at 50 kHz with 110 A of DC current and 56 A current ripple. The goal is to create designs that meet the target inductance while minimizing the mass and loss of inductors created. Each optimization was run for 1000 generations with populations of 1000 designs each.

The optimization was first run considering only winding and magnetizing core loss, the latter being calculated with the Steinmetz Equation and coefficients for each core material. The optimization was then repeated including both magnetizing loss and gap loss. The resulting pareto fronts are pictured in Fig. 10, with nanocrystalline designs marked in blue and ferrite designs in red.

When only magnetizing core losses are considered, Finemet nanocrystalline cores appear to be the optimal choice compared to MnZn ferrite cores, offering the lowest loss and mass of the designs. The pareto optimal front for 12 μH , 50 kHz inductors is dominated by designs with nanocrystalline cores, with the knee occurring near 11 W and 3 kg. A few ferrite core designs offer incremental improvements in losses under 8 W, but this comes at the expense of masses approaching 20 kg. For a 50 kHz switching frequency, the inductors created using nanocrystalline cores offer a better balance between mass and loss than the ferrite cores.

After adding the gap loss models, the nanocrystalline cores still offer the most lightweight designs, but the total losses of the nanocrystalline designs increase significantly. The highest losses increased from approximately 24 W to 120 W, with the designs close to the knee of the curve having losses well above

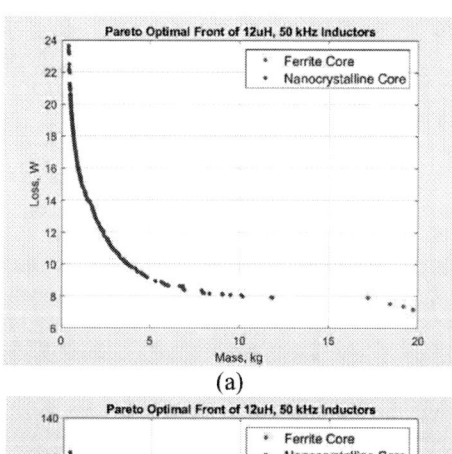

(a)

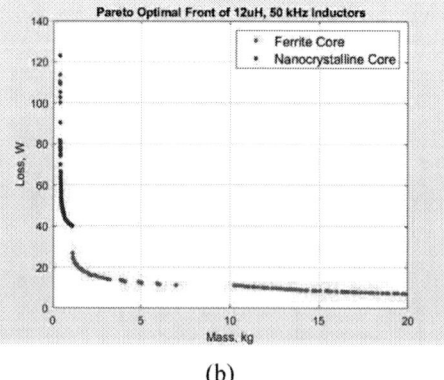

(b)

Fig. 10. Pareto optimal front of 50 kHz inductor designs (a) without gap loss and (b) with gap loss.

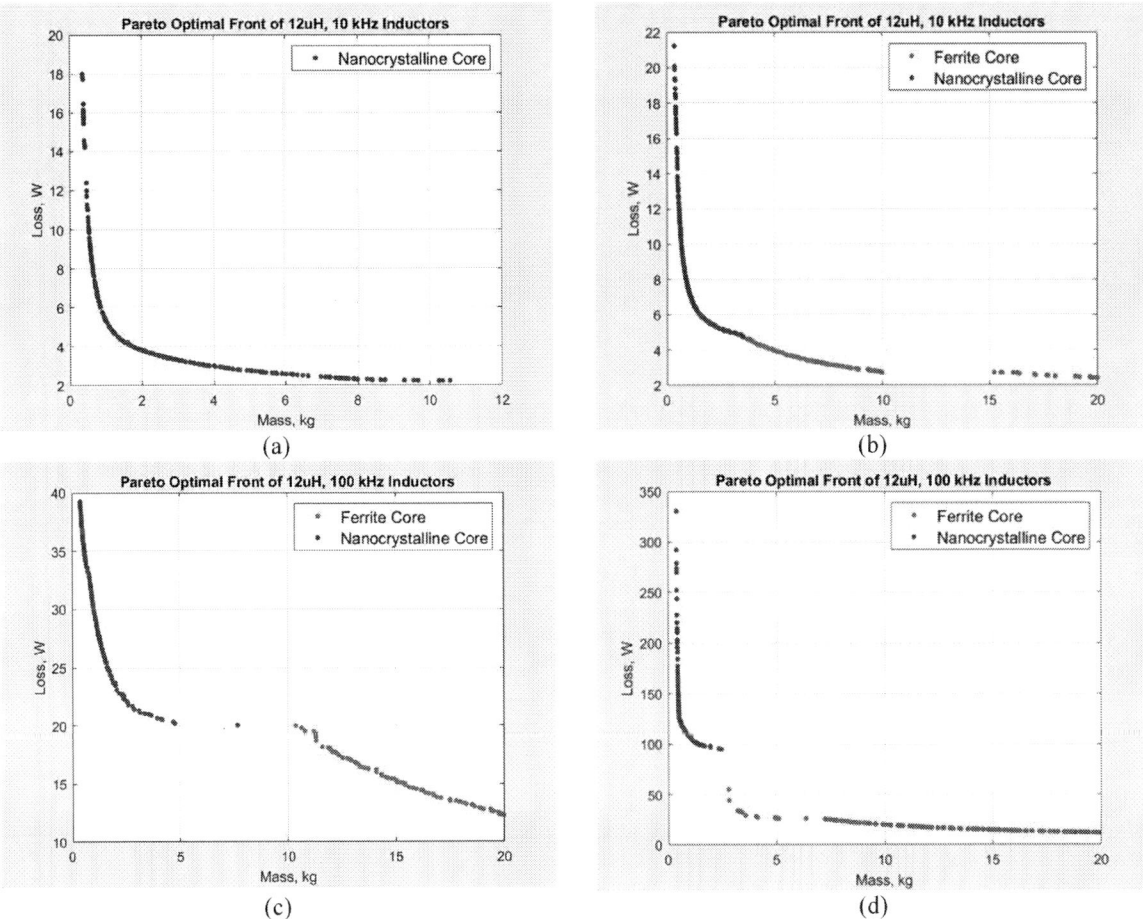

Fig. 11. Pareto optimal fronts of (a) 10 kHz inductor designs without gap loss, (b) 10 kHz designs with gap loss, (c) 100 kHz designs without gap loss, and (d) 100 kHz designs with gap loss

40 W. Ferrite cores become more competitive, providing lower loss than nanocrystalline designs at the cost of higher mass. However, in cases where thermal rise is also a consideration there may be further constraints on the accessible size of nanocrystalline designs which satisfy all design requirements.

While nanocrystalline cores initially appeared to be the optimal core material for the 12 μH inductor designs at 50 kHz, the introduction of gap loss models suggests that both nanocrystalline and ferrite cores provide competitive designs for this application depending upon the application requirements. If smaller mass is strongly preferred for a given design, nanocrystalline cores may be the preferred option. Alternatively, if the losses are a critical consideration, ferrite cores may be the preferred option yielding reduced losses.

C. Tradeoff Frequencies Analysis and Design Comparison

Switching frequency is a key factor in selecting a core material, so it is often useful to identify at which frequencies each material provides more optimal designs. To find the tradeoff frequency between nanocrystalline and ferrite designs for the 12 μH DC filter inductor, the optimizations were repeated at 10 kHz and 100 kHz. The resulting pareto optimal fronts are pictured in Fig. 11.

When gap losses are neglected, nanocrystalline cores offer the best balance between mass and loss at all selected switching frequencies. At 100 kHz, ferrite designs begin to make up a larger portion of the pareto optimal front, suggesting that the designs are nearing a tradeoff point. Many of these ferrite designs still have large masses over 10 kg, twice the mass of nanocrystalline core designs with similar loss. With the gap loss model introduced, the total losses of the nanocrystalline designs increase significantly at all frequencies. While direct gap loss is calculated for ferrite designs, the magnitude is much lower than the fringing gap losses, which quickly become a dominant loss contribution in nanocrystalline inductor designs.

To evaluate the relationship between frequency and the fitness of core materials, one design of each material was selected from each pareto front. A maximum allowable mass was imposed, and the designs closest to this value were selected. For the optimizations neglecting gap loss, this mass was 2.5 kg. A threshold of 5 kg was set for optimizations with gap loss, as the knees of the pareto fronts shift toward higher masses with increasing frequency. The losses of the selected designs were then plotted by frequency in Fig. 12.

These plots align with the trends indicated in the pareto optimal fronts. For gapped inductors, the nanocrystalline

979-8-3315-1612-3/25 $31.00 © 2025 IEEE

designs selected from the pareto fronts neglecting gap loss have lower loss than ferrite designs with similar mass. This incorrectly suggests that nanocrystalline cores are the optimal core material up to 100 kHz, with a possible tradeoff occurring beyond that. This tradeoff point drops to the 10-50 kHz range when gap losses are accounted for, since the losses in nanocrystalline designs overtake the ferrites by 50 kHz and become an order of magnitude larger than the ferrite designs at 100 kHz. In cases where design constraints are thermally limited, the frequency range over which nanocrystalline vs. ferrite designs are preferred can be further modified depending upon specific assumptions of cooling and allowed thermal rise.

D. Nanocrystalline Inductor Design Trends

The optimization work has shown that the inclusion of air gaps has a significant impact on the designs using nanocrystalline inductor cores, increasing the total losses by an order of magnitude in some cases. In addition to locating the tradeoff frequencies between nanocrystalline and ferrite cores, it is also important to identify trends in designs that have been optimized with the gap loss models. Since the goal of the optimization is to minimize mass and loss, nanocrystalline cores will be designed in ways that naturally minimize gap loss.

Two nanocrystalline designs were selected from the 12 μH, 50 kHz pareto optimal fronts: one selected from the designs generated before gap loss models were integrated and one selected after. By normalizing the fitness values for the mass and loss of the 50 kHz inductor designs, an "optimal" design was selected based on the balance achieved between mass and loss. The normalized fitness values of each design were then multiplied to give the designs a third fitness value, impacted by the design's mass and loss. This new fitness value was used to sort the designs, and those with the highest value from each pareto optimal front were selected. The geometries of these designs are pictured in Fig.13, where the core shape is colored in gray and the windings in yellow. The dimensions of the cores are compared in Table III.

After the gap loss model was integrated into the optimization framework, the selected designs near the knee consistently had smaller gap lengths and lamination widths. The core's length, or size along the z-axis, directly corresponds to its lamination width, term D in (3). These trends are expected, as fringing gap loss increases with both gap length and lamination width of the nanocrystalline core. Conversely, these designs tend to have larger core leg widths w_e, impacting size along the x-axis. This appears to be a response to the decreasing core length, as the cross-sectional area A_e, the product of the core length and core leg width, remains similar across designs evaluated near the knee of both pareto fronts. The increased core leg width also leads to a much smaller core window, reducing the area available for conductors. As a result, many of the designs created with gap loss have fewer winding turns.

These trends suggest that using shorter gap lengths and narrower ribbons are ways to mitigate fringing gap losses in nanocrystalline inductor designs. The optimization tending toward these smaller ribbon widths aligns with the work conducted by Calderon-Lopez et al. in [11], which stacked multiple cores with smaller ribbon width to reduce the continuous surface area for the fringing flux-induced eddy currents to form.

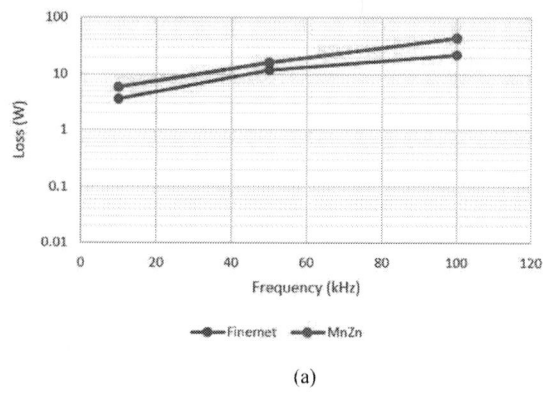

(a)

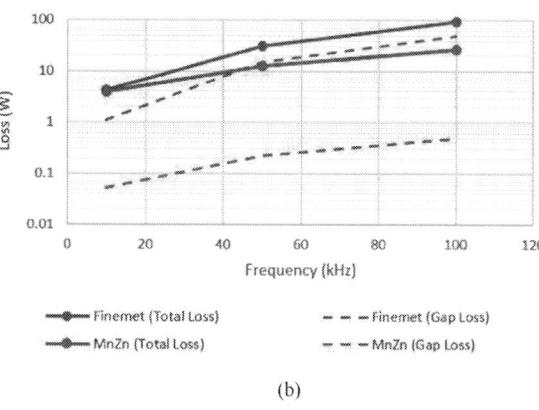

(b)

Fig. 12. Loss Comparison of designs (a) without gap loss and (b) with gap loss as a function of frequency and magnetic material

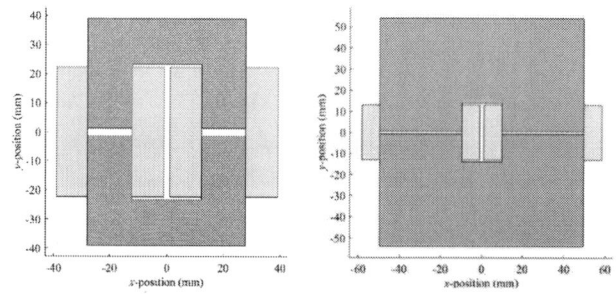

Fig. 13. Geometric Comparison of designs (left) without gap loss and (right) with gap loss

TABLE III.
CORE DIMENSIONS OF SELECTED NANOCRYSTALLINE INDUCTOR DESIGNS

Dimension	Description	No Gap Loss	Gap Loss
l_g	Gap length (mm)	2.22	1.29
D	Lamination width (mm)	32.13	13.43
w_e	U-core end width (mm)	13.43	39.93
A_e	Core cross-sectional area (mm²)	31.65	52.50

IV. Conclusion

In this work, the core losses of a nanocrystalline inductor were experimentally measured in response to increasing gap length and compared to a gap loss model from existing literature. As gap lengths reached 1 mm and frequencies reached 50 kHz, gap losses became a dominant loss contribution, in some cases resulting in core losses twice those predicted by the Steinmetz Equation.

The multi-objective optimization provided further insight on the impact of gap losses on the inductor design process and overall fitness of nanocrystalline cores compared to ferrite cores. For a selected DC filter inductor application, the gap loss becomes dominant in nanocrystalline core designs, often making up more than half of the total inductor losses. When gap loss is neglected, nanocrystalline cores appear to be the optimal material for this application up to 100 kHz. After including gap loss, ferrite core designs become more attractive near 50 kHz due to their lower loss, but still cannot achieve the lower mass that nanocrystalline designs offer. While the high permeability and mass density of nanocrystalline cores make them suitable for use in power magnetic components, accurately predicting loss is critical to reaching the efficiency and power density goals needed to drive EV development.

The content of this digest is based largely on the thesis document found at [12].

Acknowledgments

This work was supported in part by the U.S. Department of Energy's EERE under the Vehicle Technology Office (Award # DE-EE0009870), conducted in collaboration with Eaton Corporation. The work was also supported in part by the Advanced Magnetics for Power and Energy Development industry consortium.

References

[1] US Drive electrical and electronics technical team roadmap | Department of Energy, https://www.energy.gov/eere/vehicles/articles/us-drive-electrical-and-electronics-technical-team-roadmap.

[2] Daniela Rodriguez-Sotelo, Martin A. Rodriguez-Licea, Ismael Araujo-Vargas, Juan Prado-Olivarez, Alejandro Israel Barranco-Gutièrrez, and Francisco J. Perez-Pinal, "Power losses models for magnetic cores: A review". Micromachines, 13, 3 2022.

[3] Yiren Wang, Gerardo Calderon-Lopez, and Andrew J. Forsyth. "High-frequency gap losses in nanocrystalline cores," in *IEEE Transactions on Power Electronics*, vol. 32, no. 6, pp. 4683–4690, June 2017.

[4] D. Neumayr, D. Bortis, J. W. Kolar, S. Hoffmann and E. Hoene, "Origin and quantification of increased core loss in MnZn ferrite plates of a multi-gap inductor," in *CPSS Transactions on Power Electronics and Applications*, vol. 4, no. 1, pp. 72-93, March 2019.

[5] H. Sun, Y. Li, Z. Lin, Z. Wan and H. Liu, "Loss modeling and reluctance quantifying of C-type amorphous core for high-frequency inductors," in *IEEE Journal of Emerging and Selected Topics in Power Electronics*, vol.11,no.1,pp. 691-700, Feb. 2023.

[6] V.J. Thottuvelil, T.G. Wilson, and H.A. Owen. High-frequency measurement techniques for magnetic cores. IEEE Transactions on Power Electronics, 5(1):41–53, January 1990.

[7] S. R. Moon, P. Ohodnicki, K. Byerly and R. Beddingfield, "Soft Magnetic Materials Characterization for Power Electronics Applications and Advanced Data Sheets," *2019 IEEE Energy Conversion Congress and Exposition (ECCE)*, Baltimore, MD, USA, 2019, pp. 6628-6633

[8] S. D. Sudhoff. Power magnetic devices: a multi-objective design approach. IEEE Press, 2014.

[9] R. A. Gomez, D. A. Porras Fernandez, G. G. Oggier, J. C. Balda and Y. Zhao, "Comparison of High-Frequency Ferrite and Nanocrystalline Core Losses Using Identical Geometries," *2022 IEEE 13th International Symposium on Power Electronics for Distributed Generation Systems (PEDG)*, Kiel, Germany, 2022, pp. 1-5

[10] Yang Wang, Sajib Chakraborty, Dai Duong Tran, Thomas Geury, and Omar Hegazy. Design and optimization of the arm inductor for modular multilevel converter. In 2022 International Symposium on Power Electronics, Electrical Drives, Automation and Motion (SPEEDAM), pages 354–359, 2022.

[11] G. Calderon-Lopez, Y. Wang and A. J. Forsyth, "Mitigation of Gap Losses in Nanocrystalline Tape-Wound Cores," in *IEEE Transactions on Power Electronics*, vol. 34, no. 5, pp. 4656-4664, May 2019

[12] Maurice Sturdivant, "Design Study Evaluating Impact of Gap Loss on Inductors with Nanocrystalline Cores", M.S. Thesis, Swanson School of Engineering, University of Pittsburgh, Pittsburgh, 2024. [Online]. Available: http://d-scholarship.pitt.edu.

A Permanent Magnet Variable Inductor for DC Fault Current Limiting Applications

Mark Nations
Electrical and Computer Engineering
North Carolina State University
Raleigh, NC
msnation@ncsu.edu

Subhashish Bhattacharya
Electrical and Computer Engineering
North Carolina State University
Raleigh, NC

Abstract— **A novel permanent magnet variable inductor (PMVI) is proposed for fault-current limiting applications. It is designed to fill the need in hybrid DC circuit breakers (DCCBs) to have low insertion inductance under normal operating conditions, but high inductance under fault conditions to slow down fault current di/dt. Below a designable 'trip' current the magnetic core the device is driven into saturation by permanent magnets. Under a fault condition, the flux induced by the fault current flowing in the inductor winding drives the core out of the saturation region, creating a large increase of inductance. The operating conditions of the permanent magnets are analyzed to prevent partial demagnetization. Electromagnetic analysis is carried out in COMSOL FEA and compared to magnetic equivalent circuit analysis. An augmented MEC branch model is introduced to better capture eddy current effects that have a substantial effect on the PMVI inductance characteristic during high di/dt fault events. A PMVI was designed and built targeting 600V/20A with 50μH inductance at the nominal operating point. This PMVI device achieved 300% peak inductance change during the fault condition.**

Keywords—DC circuit breaker, hybrid circuit breaker, variable inductor, finite element, magnetic equivalent circuit.

I. INTRODUCTION

DC power distribution, including DC microgrids, are seeing increased interest due to potentially higher efficiency and comparably simple integration compared to AC systems. Most modern loads are DC so DC distribution can reduce the number of power conversion stages, reducing cost compared to AC systems. One of the primary barriers to the increased deployment of DC systems is the development of fast and reliable fault protection devices. Creating a resettable circuit breaker for a DC system is greatly complicated by the unavailability of a natural voltage zero crossing as is present in AC systems. The voltage zero crossing provides an inherent mechanism to extinguish the arc that is generated by suddenly opening circuit breaker contacts. DC systems have no such natural zero crossing and therefore arcing can be sustained and destructive. Further, DC systems tend to have low inductance so fault current di/dt can be very high. Therefore, breaker operation speed is of the utmost importance to protect both the downstream system and the breaker itself from excessive current.

Ordinary mechanical breakers tend to be too slow to adequately limit fault current in DC grids [1][2].

Semiconductor-based solid-state circuit breakers have very fast operating speed, but suffer from comparatively high on-state conduction losses because the grid load current always flows in the power semiconductors [3]. Hybrid DC circuit breakers are a class of device containing both a mechanical switch for low loss operation, and a power electronic section to facilitate low stress opening of the mechanical switch. Often, the power electronics section takes the form of a current injector that is used to drive the current in the mechanical switch to zero during an opening event or to generate an artificial voltage zero crossing after the switch has opened [4], [5], [6].

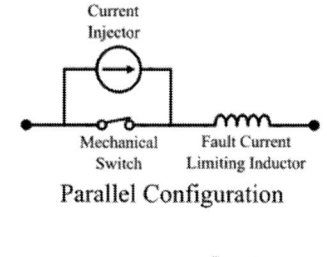

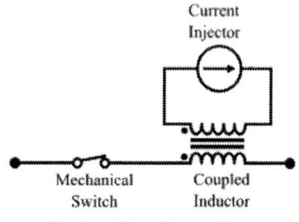

Fig. 1. Simplified parallel and series hybrid DC circuit breaker configurations.

There are various proposed hybrid DC circuit breaker (DCCB) topologies that can be broadly grouped into 'series' and 'parallel' circuits to describe the arrangement of the power electronics path and the mechanical switch conduction path. Nearly all topologies of hybrid circuit breaker use the mechanical switch as the current breaking element, so the operation speed of the breaker is still limited by a mechanical switch. Consequently there has been a great deal of recent work on construction and control of high speed mechanical switches [7], [8], [9]. High speed switches have achieved minimum contact separation in about 1ms which is extremely fast for mechanical switches, but slow enough for high fault di/dt to cause unacceptably high fault current. Therefore, nearly all hybrid circuit breaker designs include an inductor to limit the fault current di/dt. For the parallel circuit configuration this is implemented as a simple inductor at the downstream breaker port. For the series circuit configuration, the current limiting inductor can be combined with a coupled inductor that is also used for current injection. In practice, the current limiting

979-8-3315-1612-3/25 $31.00 © 2025 IEEE

inductor is a large part of the system volume and introduces substantial insertion inductance that is not desirable during normal operation. This paper proposes a novel normally saturated inductor to decouple the normal-operation insertion inductance from the inductance under fault. The proposed permanent magnet variable inductor (PMVI) achieves completely passive operation by utilizing permanent magnets (PMs) to saturate the soft magnetic core of the inductor. The paper discusses different kinds of permanent magnet-based power magnetics in literature and their modes of operation. The next section evaluates the risk of demagnetization of the permanent magnets and mitigation strategies. Then, a modification to magnetic equivalent circuit modeling is introduced to better describe transient events with non-trivial eddy currents. Finally, the construction and experimental evaluation of a small-scale prototype PMVI are shown.

II. BACKGROUND AND PRINCIPLE OF OPERATION

Previous work in permanent magnet biased inductors has focused on increasing power density in linear inductors which operate under unipolar current conditions [10], [11]. For many non-isolated buck-derived converter topologies the current seen by the inductors is unipolar. The approach in these previous works has been to use permanent magnets to bias the magnetic core such that the full linear range of the B-H loop can be utilized by the converter, effectively doubling the DC current rating of the inductor before saturation.

The principle of the PMVI is also unipolar, but that strong permanent magnets are used to bias the magnetic core deeply into saturation. Figure 2A illustrates how the core is biased at zero current. When the DCCB is operating in the nominal range the inductor operates to the left of the B-H corner, and therefore with low flux linkage. When current is increased the core is pushed out of the saturation region and there is a large increase in flux linkage. As current continues to increase the core will re-saturate at the positive side of the B-H loop. For current limiting applications it is often most critical to limit the absolute maximum fault current, so the performance figure of merit is the average inductance in the fault current range (1).

$$\bar{L} = \frac{1}{I_{max} - I_{nom}} \int_{I_{nom}}^{I_{max}} L(i) \, di \qquad (1)$$

To maximize average inductance, it is necessary that the inductor be design as shown in figure 2B, where the inductor begins to re-saturate prior to reaching I_{max}.

III. DESIGN TO MITIGATE DEMAGNETIZATION

When adding permanent magnets to an inductor, it is natural to place them near a single air gap in the magnetic core. If each permanent magnet pole is located near each side of the gap, the magnetic flux flowing in the permanent magnet will

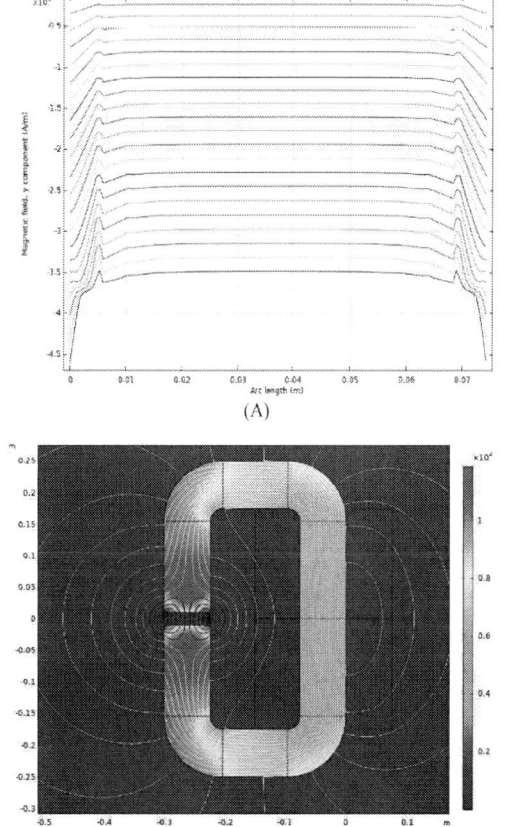

(A)

(B)

Fig. 3. For a PMVI where $A_{PM}=A_{core}$: Demagnetization field across the PM face for various current levels (A), and flux lines at maximum rated current (B). Flux lines linking the core only are white and flux lines linking only the PM are black.

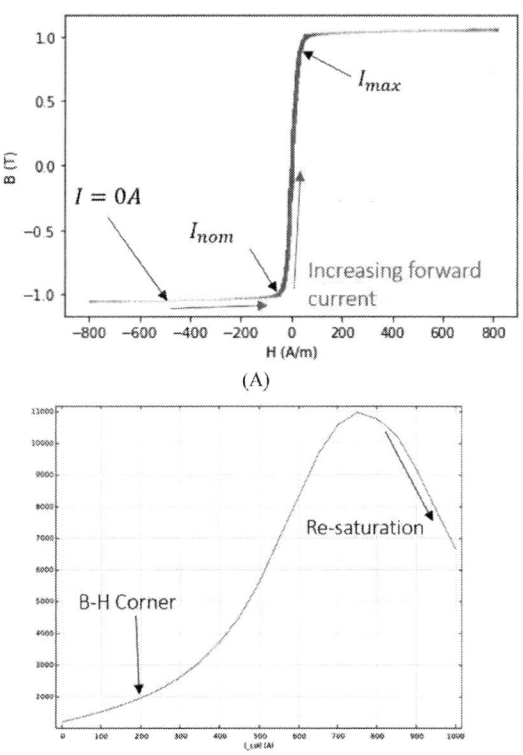

(A)

(B)

Fig. 2. B-H trajectory as a function of PMVI current (A), and simulated inductance-current characteristic using COMSOL FEA (B).

preferentially flow around the high permeance core path, leading to saturation of that path if adequate flux density is sourced. So, a natural location for the permanent magnet to maximize zero-bias saturation of the core is directly in the inductor air gap. However, this location is also subject to the highest demagnetizing field due to induced flux from the fault current. Figure 2A1 and A2 show an example design where a N42 permanent magnet is placed in the of a Metglas 2714 magnetic core subject to maximum winding current. The area of the PM is the same as the area of the core, so flux sourced by the permanent magnet, shown in black, is seen to fringe tightly around the magnet. Figure 1A shows that the demagnetizing field a the magnet surface is much higher at the corners that in the center of the PM. Designing with this area ratio between core and PM results in poor PM utilization. It was determined that the demagnetizing field peaking ate the PM corners can be mitigated by introducing high permeance paths in air around the PM for flux to flow in the maximum fault current condition. In practice this can be most easily accomplished by reducing the PM area to a fraction of the core area. Figure 2B1 and B2 shows the exact same core and current excitation, but the PM area is reduced to 80% of the core area. The additional permeance path around the core gap allows the PM flux to more uniformly return in the fringing path around the gap, greatly reducing the demagnetizing field intensity at the corners. Additionally, it is

clear that the uniformity of the permeability change within the core is also improved.

IV. MAGNETIC EQUIVALENT CIRCUIT ANALYSIS

Analytical analysis of the PMVI is confounded by several factors. For determination of the near-DC inductance vs current characteristic, finite element analysis is effective, but for increased computation speed and integration into optimization it is desirable to obtain an accurate analytical model. Simple permeance calculations are inaccurate because the local magnetic permeability of the PMVI has high spatial variability. This difficulty can be resolved by using magnetic equivalent circuit analysis with carefully selected nodes [12]. The required number of nodes and location of the magnetic nodes is non-obvious due to the unique local saturation characteristics of the PMVI. Electromagnetic simulation using COMSOL FEA was used to inform the MEC node selection. Figure 5 shows the permeability of the PMVI constructed in section V for three different current biases. Ideal MEC nodes have identical

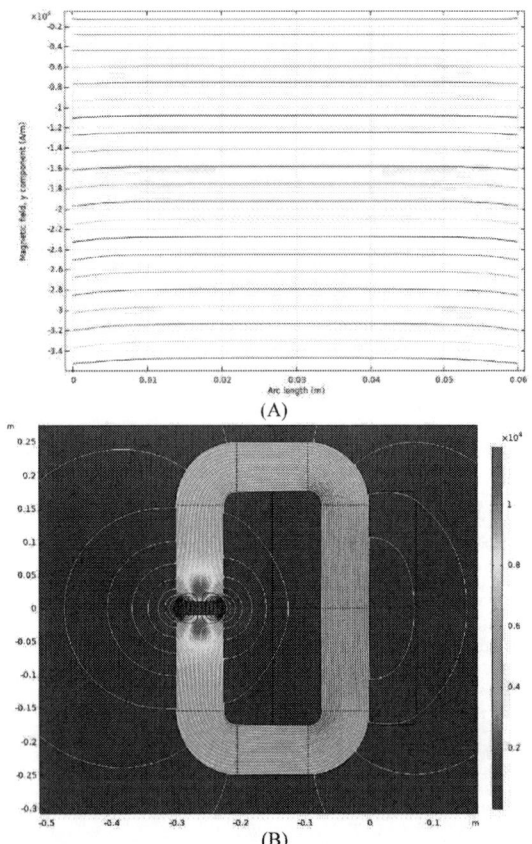

Fig. 4. For a PMVI where $A_{PM}=0.8*A_{core}$: Demagnetization field across the PM face for various current levels (A), and flux lines at maximum rated current (B). Flux lines linking the core only are white and flux lines linking only the PM are black.

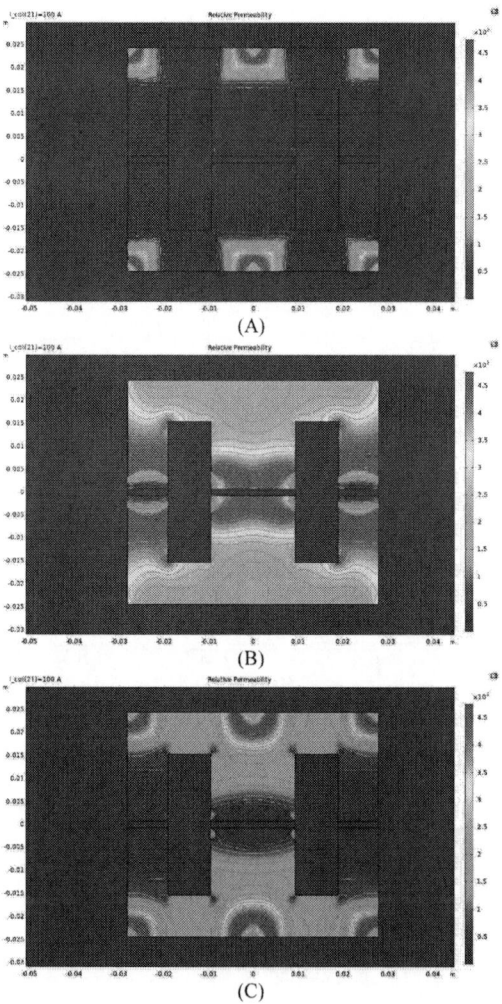

Fig. 5. Magnetic permeability of a PMVI at 0A (A), 50% rated current (B), and 100% rated current (C).

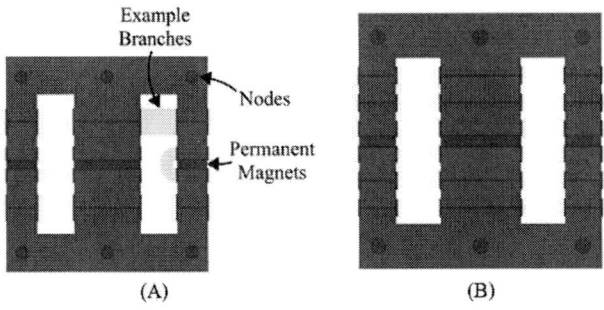

Fig. 6. Initial MEC node model (A), and FEA-informed improved model (B).

magnetic potential across the node and are selected here to additionally split up regions where the gradient of magnetic permeability is large. Observing figure 5 the vertical limbs of the core, and regions near the PMs in particular, have high permeability gradients. These observations were used to derive the MEC node definitions in figure 6. Initial analysis was performed using the model in figure 6A, but the additional granularity provided by the additional nodes in figure 6B improved matching with experimental results by 26%.

Aside from the difficulty modeling the near-DC PMVI inductance, there are additional considerations due to the fast transient operation during a breaker fault. In particular, the event dB/dt is very high compared to what a soft magnetic core or PM may experience during steady state operation of a power converter. The high dB/dt leads to non-trivial eddy currents developing in the PM and/or the soft magnetic core material that substantially changes the effective inductance-current characteristic in operation. Many materials with favorable properties for PMVI performance such as rare earth PMs, amorphous and nanocrystalline soft magnetic cores have substantial electrical conductivity. In magnetic devices and motors rare earth PMs are generally used as stiff flux sources, so they are rarely exposed to high dB/dt events. The PM location in the gap of the PMVI exposes it to a large uniform back MMF thereby causing high dB/dt in the PM. The soft magnetic material is likewise exposed to much higher dB/dt than typical in a steady-state application. Steady-state operation is generally

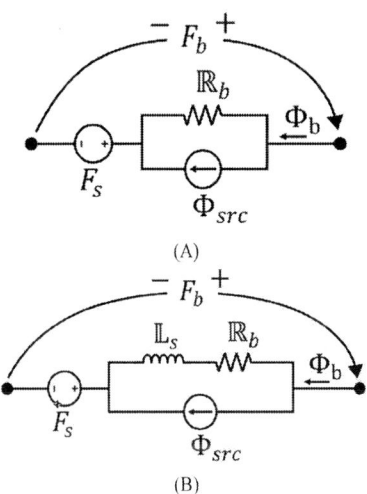

Fig. 7. Traditional MEC branch (A), and modified time-domain branch (B).

thermally limited so the maximum dB/dt is practically limited by magnetic loss. Conversely the PMVI in a hybrid DC circuit breaker operates only in transient so much higher dB/dt are possible. The prototype PMVI (section V) utilized ferrite cores and NbFeB PMs so the permanent magnets are the only components where eddy currents are significant. Figure 7 shows the high dB/dt inductance-current characteristics from COMSOL FEA, the MEC in figure 6B, and experimental results. FEA results match experimental results well, but there is significant error in the MEC curve due to eddy currents in the PMs. The eddy currents always create an opposing MMF to increasing magnetic flux density so in practice the inductance peak is shifted towards higher current, the inductance roll off after saturation is extremely steep, and the average inductance FoM (1) is substantially reduced compared to near-DC operation of the PMVI.

Magnetic equivalent circuit analysis is fundamentally static or quasi-static because the magnetic branch elements depend only on the instantaneous flux values in the branch (for saturable branches). To correctly capture eddy current effects, the ordinary MEC branch must be modified to include time varying elements. The intended behavior is that high dB/dt will generate a back MMF in addition to the MMF due to the branch

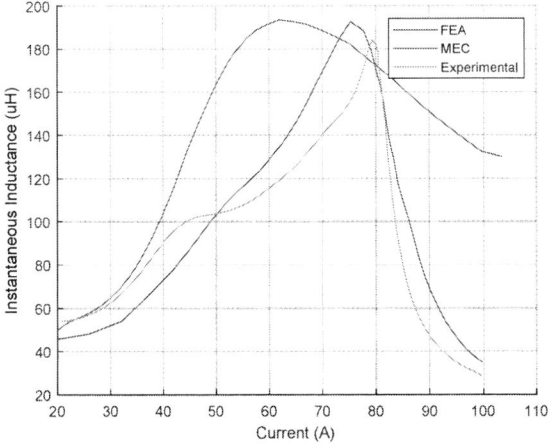

Fig. 8. High dB/dt inductance-current characteristic.

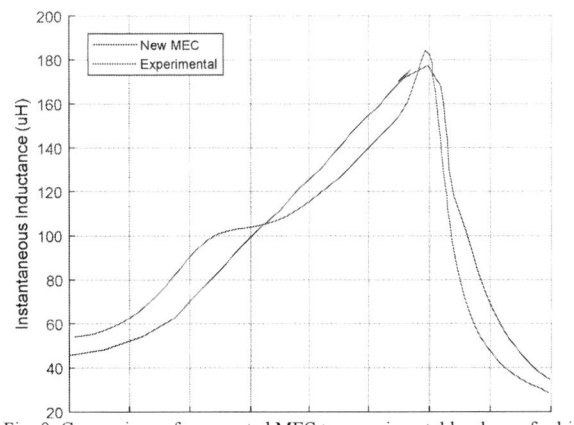

Fig. 9. Comparison of augmented MEC to experimental hardware for high dB/dt fault.

reluctance. For DC fluxes the branch should behave exactly the same as an ordinary MEC branch. The natural component selection for these requirements is a simple inductor. Because magnetic flux is the analogue for current in a MEC, high dΦ/dt will lead to a high back MMF across the inductor. The proposed branch is shown in figure 7.

The MEC is a behavioral model, so it is not necessary to assign physical meaning to the circuit elements, but it is tempting to draw parallels between this MEC inductance and an electrical inductor. There is energy associated with the eddy current, but there is no relationship between the eddy current energy and the MEC inductance. This is easily seen by dimensional analysis. We have defined the inductor, $\mathbb{L}_s$, to behave according to (2). Using (2) we can perform dimensional analysis on $\mathbb{L}_s$ to show that it has units of H^{-1} (3). Therefore, performing ordinary energy and power calculations for inductances using $\mathbb{L}_s$ are highly inappropriate. Indeed, this is why we have elected to use a unique symbol for $\mathbb{L}_s$.

$$F_m = \mathbb{L}_s \cdot \frac{d\Phi}{dt} \tag{2}$$

$$[\mathbb{L}_s] = \frac{s^2 \cdot A^2}{kg \cdot m^2} = H^{-1} \tag{3}$$

Calculating the value of $\mathbb{L}_s$ can be performed by solving the diffusion equation for the eddy current (4). The vector form of this equation (5) is more convenient to use because the various geometric boundary conditions must be used to solve the initial value problem.

$$\nabla^2 H = \sigma \frac{\partial B}{\partial t} \tag{4}$$

$$\frac{\partial^2 B}{\partial x^2} + \frac{\partial^2 B}{\partial y^2} + \frac{\partial^2 B}{\partial z^2} = \sigma \mu \frac{dB}{dt} \tag{5}$$

V. PMVI TEST DEVICE

A test device was designed a built for the specification in Table 1 to represent the required device for a 12kW hybrid DC circuit breaker. As previously mentioned the initial design was carried out largely using COMSOL FEA and the proposed modifications to the MEC analysis were tested against both FEA and experimental results. A worst cast bolted fault at the breaker was assumed for testing purposes so the full rated voltage (600V) was applied directly across the PMVI for the duration of each test. It was assumed that under normal operating conditions the current ripple is no significant, so steady state magnetic loss testing was not performed. Figure 11 shows the constructed inductor using N42 PMs and N87 MnZn ferrite cores. The inductance characteristic, figure 12 shows low current

TABLE I.   PMVI TARGET SPECIFICATIONS

Nominal Current	20A
Peak Fault Current	100A
Inductance Tunability	>50%
Winding Resistance	<5mΩ
Inductance	50μH nom. >100μH average during fault

Fig. 10. 12kW MnZn ferrite-NbFeB PMVI prototype

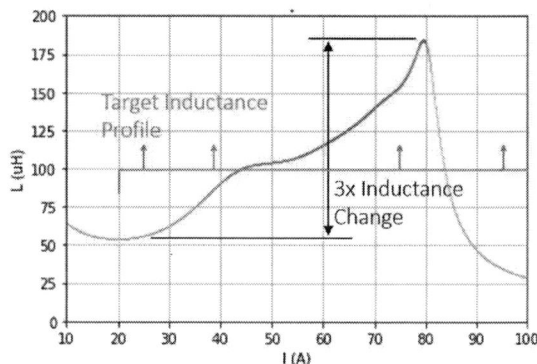

Fig. 11. PMVI inductance-current characteristic during 600V bolted fault.

inductance very close to design specification, with very high peak inductance change of about 300%.

VI. CONCLUSION

In this paper the principle of operation, demagnetization related design, and modeling methods of a novel PMVI device has been discussed as it related to DC fault current limiting applications. Local saturation of the PMVI core requires more and careful placement of nodes for best accuracy. Eddy current effects in rare earth magnets and tape-wound magnetic cores are significant during the fast transients that the PMVI experiences in hybrid DC circuit breaker applications so an augmented MEC branch was introduced to account for time varying effects like eddy currents.

A demonstration device rated nominally for 600V/20A operation with 100A peak overload was designed and built. This PMVI device demonstrated less than 50uH insertion inductance in the rated, and peak inductance approximately 300% higher in the fault region. The PMVI has appealing properties for other applications including magnetic amplifier and pulse stretching networks.

REFERENCES

[1]   D. Min *et al.*, "Thickness-Dependent DC Electrical Breakdown of Polyimide Modulated by Charge Transport and Molecular Displacement," *Polymers*, vol. 10, no. 9, Art. no. 9, Sep. 2018, doi: 10.3390/polym10091012.

[2] C. M. Franck, "HVDC Circuit Breakers: A Review Identifying Future Research Needs," *IEEE Trans. Power Deliv.*, vol. 26, no. 2, pp. 998–1007, Apr. 2011, doi: 10.1109/TPWRD.2010.2095889.

[3] A. Hooshyar and R. Iravani, "Microgrid Protection," *Proc. IEEE*, vol. 105, no. 7, pp. 1332–1353, Jul. 2017, doi: 10.1109/JPROC.2017.2669342.

[4] A. Ray, K. Rajashekara, S. N. Banavath, and S. K. Pramanick, "Coupled Inductor-Based Zero Current Switching Hybrid DC Circuit Breaker Topologies," *IEEE Trans. Ind. Appl.*, vol. 55, no. 5, pp. 5360–5370, Sep. 2019, doi: 10.1109/TIA.2019.2926467.

[5] D. Keshavarzi, E. Farjah, and T. Ghanbarih, "A hybrid DC circuit breaker for DC microgrid based on zero current switching," in *2016 Iranian Conference on Renewable Energy & Distributed Generation (ICREDG)*, Apr. 2016, pp. 45–49. doi: 10.1109/ICREDG.2016.7875917.

[6] X. Pei, O. Cwikowski, A. C. Smith, and M. Barnes, "Design and Experimental Tests of a Superconducting Hybrid DC Circuit Breaker," *IEEE Trans. Appl. Supercond.*, vol. 28, no. 3, pp. 1–5, Apr. 2018, doi: 10.1109/TASC.2018.2793226.

[7] C. Peng, L. Mackey, I. Husain, A. Huang, B. Lequesne, and R. Briggs, "Active damping of ultra-fast mechanical switches for hybrid AC and DC circuit breakers," in *2016 IEEE Energy Conversion Congress and Exposition (ECCE)*, Sep. 2016, pp. 1–8. doi: 10.1109/ECCE.2016.7854816.

[8] C. Peng, I. Husain, and A. Q. Huang, "Evaluation of design variables in Thompson coil based operating mechanisms for ultra-fast opening in hybrid AC and DC circuit breakers," in *2015 IEEE Applied Power Electronics Conference and Exposition (APEC)*, Mar. 2015, pp. 2325–2332. doi: 10.1109/APEC.2015.7104673.

[9] F. Banihashemi, S. Beheshtaein, and R. Cuzner, "Novel Hybrid Circuit Breaker Topology Using a Twin Contact Mechanical Switch," in *2021 9th International Conference on Smart Grid (icSmartGrid)*, Jun. 2021, pp. 132–136. doi: 10.1109/icSmartGrid52357.2021.9551239.

[10] Z. Dang and J. A. Abu Qahouq, "Evaluation of High-Current Toroid Power Inductor With NdFeB Magnet for DC–DC Power Converters," *IEEE Trans. Ind. Electron.*, vol. 62, no. 11, pp. 6868–6876, Nov. 2015, doi: 10.1109/TIE.2015.2436361.

[11] G. M. Shane and S. D. Sudhoff, "Design Paradigm for Permanent-Magnet-Inductor-Based Power Converters," *IEEE Trans. Energy Convers.*, vol. 28, no. 4, pp. 880–893, Dec. 2013, doi: 10.1109/TEC.2013.2274476.

[12] S. D. Sudhoff, G. M. Shane, and H. Suryanarayana, "Magnetic-Equivalent-Circuit-Based Scaling Laws for Low-Frequency Magnetic Devices," *IEEE Trans. Energy Convers.*, vol. 28, no. 3, pp. 746–755, Sep. 2013, doi: 10.1109/TEC.2013.2271976.

Design-Oriented Modeling and Multi-Objective Optimization of Two-Phase Coupled Inductors in Multiphase PWM Converters

Yicheng Zhu, Jiarui Zou, and Robert C. N. Pilawa-Podgurski

Department of Electrical Engineering and Computer Sciences, University of California, Berkeley

Email: {yczhu, jiarui.zou, pilawa}@berkeley.edu

Abstract—**Two-phase coupled inductors can achieve the same steady-state inductance as two discrete inductors with reduced core volume and improved dynamic response, which makes them advantageous in interleaved multiphase PWM converters. However, there is currently no comprehensive and systematic optimization method for two-phase coupled inductors to determine the optimal core dimensions for given design targets. To fill this gap, this paper proposes a design-oriented magnetic circuit model for two-phase coupled inductors based on common-mode and differential-mode decomposition, which offers more physical insights for core geometry design compared to conventional flux analysis. The design of a two-phase coupled inductor is then formulated as a multi-objective optimization problem, aiming to minimize overall volume, power loss, and transient inductance while meeting requirements on steady-state inductance, form factor, and saturation current. The proposed design-oriented model and multi-objective optimization method are validated through comparisons with Monte Carlo simulations using ANSYS and experimental hardware results.**

I. INTRODUCTION

Two-phase coupled inductors have been widely used in interleaved multiphase PWM converters, including applications such as voltage regulator modules (VRMs) [1]–[6], integrated voltage regulators (IVRs) [7], automotive electrical systems [8], and power factor correction (PFC) [9]. Recently, they have also found widespread use in high step-down hybrid switched-capacitor DC-DC converters [10]–[13]. For example, Fig. 1(a) shows the schematic drawing of a two-phase buck converter with a two-phase coupled inductor, CL. This coupled inductor can be electrically modeled with a self-inductance (L) on each phase and a mutual inductance (M) between the two phases, and be physically implemented with a pair of E and I cores and two windings, as shown in Fig. 2(a).

Despite the wide application of two-phase coupled inductors, there is currently no comprehensive and systematic method for designers to optimize core dimensions given specific design targets and constraints. This work attempts to fill this gap with a design-oriented magnetic circuit model based on common-mode (CM) and differential-mode (DM) decomposition and multi-objective optimization. Through CM and DM magnetic flux analysis, the optimal value of a key design parameter, the center-to-side leg width ratio, is derived. Using the proposed model, a two-phase coupled inductor is optimized to minimize overall volume, reduce power loss, and improve transient response. Comparisons with Monte

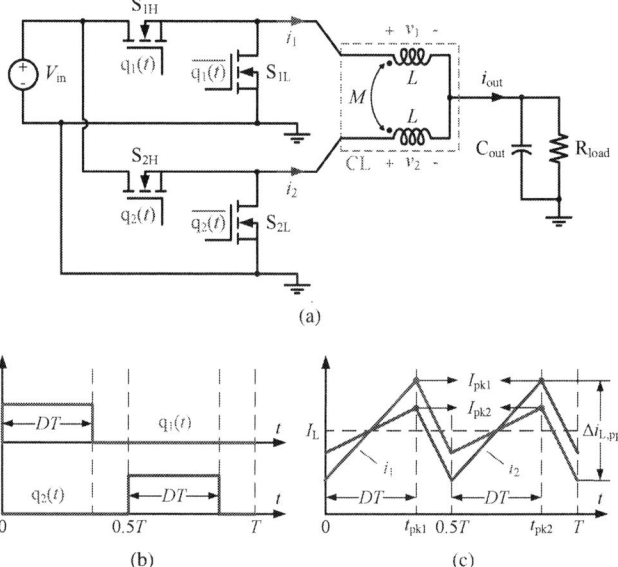

Fig. 1: Two-phase buck converter with a two-phase coupled inductor, CL. (a) Schematic drawing. (b) Two-phase interleaved control signals. (c) Inductor current waveforms.

Carlo simulations using ANSYS and experimental hardware results validate the proposed design-oriented model and multi-objective optimization method.

II. DESIGN-ORIENTED MODELING

As illustrated in the cross-sectional view in Fig. 2(b), the two-phase coupled inductor consists of two side legs and one center leg. Each side leg has an air gap with a length of l_{gs}, and the center leg has an air gap with a length of l_{gc}. The mean magnetic path length on the side and center legs are l_{cs} and l_{cc}, respectively. To ensure a uniform flux density distribution within the magnetic cores, which is typically required to achieve an optimal design, the thicknesses of the I core, w_I, and the base of the E core, w_E, are set equal to the width of the side legs, w_s (i.e., $w_I = w_E = w_s$). Both windings have N turns and the same DC resistance, R_{dc}. The two phases are negatively coupled to achieve DC flux cancellation and transient inductance reduction [1].

979-8-3315-1612-3/25 $31.00 © 2025 IEEE

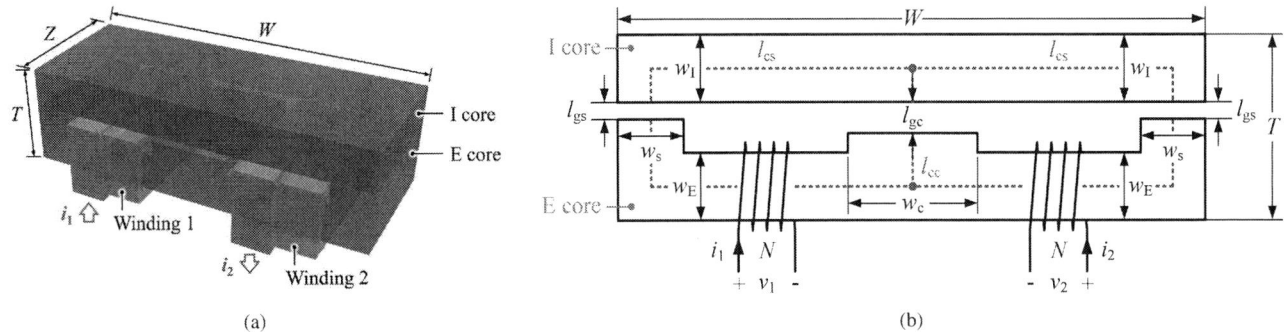

Fig. 2: Two-phase coupled inductor consisting of a pair of E and I cores and two inversely-coupled windings. (a) 3D rendering of the coupled inductor. (b) Dimensions of the cross section. The values of w_I and w_E are set equal to w_s to ensure a uniform flux density distribution within the magnetic cores.

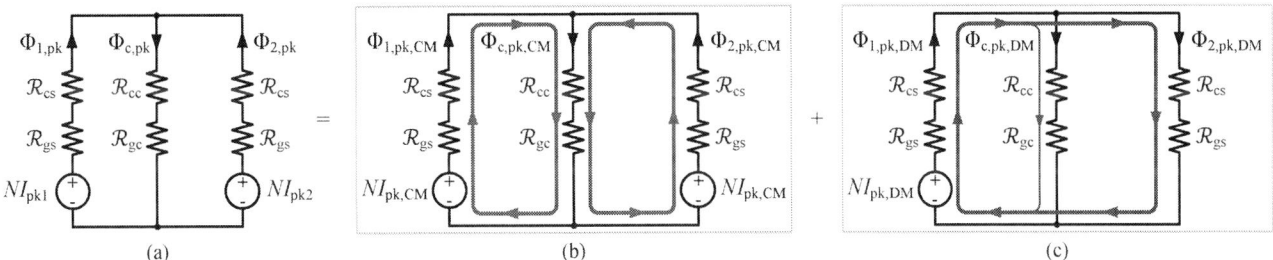

Fig. 3: Magnetic circuit model of the two-phase coupled inductor at t_{pk1}. The magnetic flux through each branch can be decomposed into a CM component (red) and a DM component (blue). (a) Full model. (b) CM model. (c) DM model.

A. Magnetic Circuit Model

Fig. 1(c) shows the coupled inductor current waveforms, i_1 and i_2, in the two-phase buck converter illustrated in Fig. 1(a) under the two-phase interleaved operation shown in Fig. 1(b). As depicted in Fig. 1(c), the coupled inductor currents reach their peak value, I_{pk1}, and sub-peak value, I_{pk2}, at time instants t_{pk1} and t_{pk2}. The magnetic cores are most prone to saturation at t_{pk1} and t_{pk2} when the current in one phase reaches its peak value. Due to the symmetry of the core structure, the magnetic flux distributions at t_{pk1} and t_{pk2} are symmetric about the vertical central axis of the magnetic cores. For simplicity, the magnetic flux analysis in this paper focuses on t_{pk1}, but the same analysis also applies to t_{pk2}.

At t_{pk1}, this coupled inductor can be modeled as the magnetic circuit shown in Fig. 3(a). At this moment, the currents through Winding 1 and Winding 2 are I_{pk1} and I_{pk2}, respectively, which generate the corresponding magnetomotive forces (MMFs) on the side legs of NI_{pk1} and NI_{pk2}. The magnetic reluctances of the side-leg air gap ($\mathcal{R}_{gs}$), the side-leg core ($\mathcal{R}_{cs}$), the center-leg air gap ($\mathcal{R}_{gc}$), and the center-leg core ($\mathcal{R}_{cc}$) can be obtained as

$$\mathcal{R}_{gs} = \frac{l_{gs}}{\mu_0 A_s} \tag{1}$$

$$\mathcal{R}_{gc} = \frac{l_{gc}}{\mu_0 A_c} \tag{2}$$

$$\mathcal{R}_{cs} = \frac{l_{cs}}{\mu_0 \mu_r A_s} \tag{3}$$

$$\mathcal{R}_{cc} = \frac{l_{cc}}{\mu_0 \mu_r A_c}. \tag{4}$$

Parameters A_s and A_c represent the cross-sectional areas of the side and center legs, respectively. $A_s = w_s Z$ and $A_c = w_c Z$, where w_s and w_c are the widths of the side and center legs, respectively, as annotated in Fig. 2(b), and Z is core length, as annotated in Fig. 2(a). The constant μ_0 is the vacuum permeability, and μ_r is the relative permeability of the core material. For simplicity, the magnetic reluctances on the side and center legs can be lumped into two elements, $\mathcal{R}_s$ and $\mathcal{R}_c$, as

$$\mathcal{R}_s = \mathcal{R}_{gs} + \mathcal{R}_{cs} \tag{5}$$

$$\mathcal{R}_c = \mathcal{R}_{gc} + \mathcal{R}_{cc} \tag{6}$$

and related to the coupling coefficient, α, as

$$\alpha \overset{\text{def}}{=} \frac{M}{L} = -\frac{\mathcal{R}_c}{\mathcal{R}_s + \mathcal{R}_c}. \tag{7}$$

A detailed derivation of (7) can be found in [1].

Define the inductor current ripple ratio, α_I, as the ratio of the peak current ripple amplitude to the average inductor current:

$$\alpha_I \overset{\text{def}}{=} \frac{\frac{1}{2}\Delta i_{L,pp}}{I_L} \tag{8}$$

where $\Delta i_{L,pp}$ is the peak-to-peak inductor current ripple and I_L is the per-phase average inductor current, as annotated in Fig. 1(c). By analyzing the inductor current slope during each time interval and solving for the inductor current ripples, the

979-8-3315-1612-3/25 $31.00 © 2025 IEEE

expressions for I_{pk1} and I_{pk2} can be obtained as

$$I_{pk1} = (1 + \alpha_I) I_L \tag{9}$$

$$I_{pk2} = \left[1 - \frac{\alpha_I (D + D'\alpha)}{D' + D\alpha} \right] I_L \tag{10}$$

where D is the duty ratio of the control signals annotated in Fig. 1(b), and $D' = 1 - D$.

B. CM and DM Decomposition

Although the full model in Fig. 3(a) can be used to solve for the magnetic fluxes through each branch, it does not provide sufficient physical insights for core geometry design and optimization. Therefore, this paper aims to establish a design-oriented model for two-phase coupled inductors through CM and DM decomposition. According to the superposition principle, the full model in Fig. 3(a) can be decomposed into the CM model illustrated in Fig. 3(b) and the DM model in Fig. 3(c). At t_{pk1}, since $I_{pk1} > I_{pk2}$, the CM and DM currents, $I_{pk,CM}$ and $I_{pk,DM}$, are defined and obtained as (11) and (12)

$$I_{pk,CM} \stackrel{def}{=} I_{pk2} = \left[1 - \frac{\alpha_I (D + D'\alpha)}{D' + D\alpha} \right] I_L \tag{11}$$

$$I_{pk,DM} \stackrel{def}{=} I_{pk1} - I_{pk2} = \frac{\alpha_I (1 + \alpha)}{D' + D\alpha} I_L. \tag{12}$$

The CM model depicts the magnetic fluxes generated by the identical component ($NI_{pk,CM}$) of the MMFs present on both side legs of the coupled inductor. As shown in red in Fig. 3(b), the CM fluxes come from the side legs and converge in the center leg. Given the same CM MMF and magnetic reluctances on the side legs, the CM fluxes in the side legs are identical and equal to half of the CM flux in the center leg.

The DM model captures the magnetic fluxes excited by the additional component ($NI_{pk,DM}$) of the larger MMF applied to the side legs—specifically, at t_{pk1} in Fig. 1(c), the MMF applied to the left-hand-side (LHS) leg. As shown in blue in Fig. 3(c), the DM flux is generated by the LHS leg and flows into the center leg and the right-hand-side leg.

In multiphase PWM converters, coupled inductors are typically designed with all phases strongly magnetically coupled (i.e., with a coupling coefficient α closer to -1) to leverage the effects of DC flux cancellation and transient inductance reduction. To achieve a strong magnetic coupling, $\mathcal{R}_c$ needs to be much greater than $\mathcal{R}_s$. Therefore, in strong coupling designs, the DM flux primarily circulates around the two side legs, with only a small portion leaking through the center leg, as illustrated in Fig. 3(c).

C. Magnetic Flux Analysis

At t_{pk1}, the magnetic fluxes in the LHS and center legs reach their peak values, which are important to access the saturation limit of the coupled inductor. This subsection analyzes these peak fluxes under given design targets. Based on the CM and DM magnetic circuit models in Fig. 3, the peak CM, DM, and total fluxes in the LHS leg ($\Phi_{1,pk,CM}$, $\Phi_{1,pk,DM}$, and

$\Phi_{1,pk}$) can be obtained and simplified with (7) as

$$\Phi_{1,pk,CM} = \frac{NI_{pk,CM}}{\mathcal{R}_s + 2\mathcal{R}_c} \tag{13}$$

$$\Phi_{1,pk,DM} = \frac{NI_{pk,DM}}{\mathcal{R}_s + \frac{\mathcal{R}_s \mathcal{R}_c}{\mathcal{R}_s + \mathcal{R}_c}} = \frac{NI_{pk,DM}}{(1 - \alpha)\mathcal{R}_s} \tag{14}$$

$$\Phi_{1,pk} = \Phi_{1,pk,CM} + \Phi_{1,pk,DM} \tag{15}$$

and those in the center leg ($\Phi_{c,pk,CM}$, $\Phi_{c,pk,DM}$, and $\Phi_{c,pk}$) can be expressed with $\Phi_{1,pk,CM}$ and $\Phi_{1,pk,DM}$ as

$$\Phi_{c,pk,CM} = 2\Phi_{1,pk,CM} \tag{16}$$

$$\Phi_{c,pk,DM} = \frac{\mathcal{R}_s}{\mathcal{R}_s + \mathcal{R}_c} \Phi_{1,pk,DM} = (1 + \alpha)\Phi_{1,pk,DM} \tag{17}$$

$$\Phi_{c,pk} = \Phi_{c,pk,CM} + \Phi_{c,pk,DM}. \tag{18}$$

In a practical design problem, a coupled inductor typically needs to achieve a minimum steady-state inductance ($L_{ss(min)}$) at a given switching frequency (f_{sw}), output voltage (V_{out}), and duty ratio (D), to meet the peak-to-peak current ripple requirement ($\Delta i_{L,pp}$). This minimum steady-state inductance, $L_{ss(min)}$, can be obtained as

$$L_{ss(min)} = \frac{D'V_{out}}{f_{sw}\Delta i_{L,pp}} = \frac{D'V_{out}}{2\alpha_I f_{sw} I_L}. \tag{19}$$

The steady-state inductance, L_{ss}, of a coupled inductor with a self-inductance of L and a coupling coefficient of α when operating at a duty ratio of D ($D < 0.5$) can be given as

$$L_{ss} = \frac{1 - \alpha^2}{1 + \frac{D}{D'}\alpha} L \tag{20}$$

where L is determined by the magnetic reluctances of the side and center legs, $\mathcal{R}_s$ and $\mathcal{R}_c$, and turns ratio, N, as

$$L = \frac{N^2 (\mathcal{R}_s + \mathcal{R}_c)}{\mathcal{R}_s (\mathcal{R}_s + 2\mathcal{R}_c)}. \tag{21}$$

The minimum self-inductance, L_{min}, required to achieve the minimum steady-state inductance, $L_{ss(min)}$, can be obtained by equating (19) and (20):

$$L_{min} = \frac{(D' + D\alpha)V_{out}}{2\alpha_I(1 - \alpha^2)f_{sw}I_L}. \tag{22}$$

Equating (21) and (22) and rearranging (21) with (7) yields the magnetic reluctances required to achieve the minimum self-inductance, L_{min}:

$$\mathcal{R}_s = \frac{N^2}{(1 - \alpha)L_{min}} \tag{23}$$

$$\mathcal{R}_c = \frac{-\alpha N^2}{(1 - \alpha^2)L_{min}}. \tag{24}$$

The expressions of $\Phi_{1,pk,CM}$ and $\Phi_{1,pk,DM}$ in (13) and (14) can then be simplified by substituting the expressions of $I_{pk,CM}$ and $I_{pk,DM}$ in (11) and (12) and the expressions of

979-8-3315-1612-3/25 $31.00 © 2025 IEEE

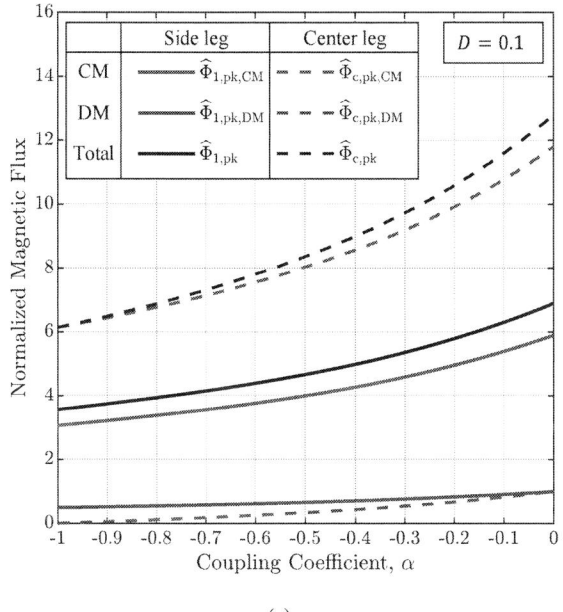
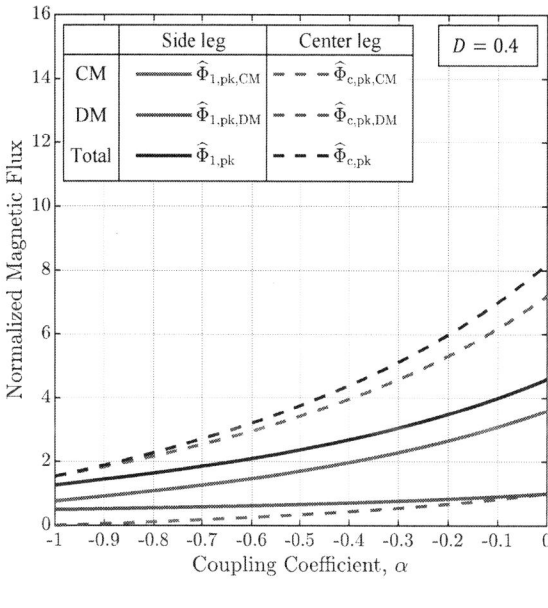

(a) (b)

Fig. 4: Normalized peak CM, DM, and total fluxes through the side and center legs versus coupling coefficient, α. (a) When $D = 0.1$. (b) When $D = 0.4$.

$\mathcal{R}_{\mathrm{s}}$ and $\mathcal{R}_{\mathrm{c}}$ given by (22)-(24):

$$\Phi_{1,\mathrm{pk,CM}} = \left(\frac{\frac{1}{1-\alpha} - D + D'\alpha_{\mathrm{I}}}{\alpha_{\mathrm{I}}} - \frac{1}{1-\alpha} \right) \frac{V_{\mathrm{out}}}{2Nf_{\mathrm{sw}}} \quad (25)$$

$$\Phi_{1,\mathrm{pk,DM}} = \frac{1}{1-\alpha} \cdot \frac{V_{\mathrm{out}}}{2Nf_{\mathrm{sw}}}. \quad (26)$$

To simplify the expressions of $\Phi_{1,\mathrm{pk,CM}}$ and $\Phi_{1,\mathrm{pk,DM}}$ in (25) and (26), define a design factor, k, as

$$k = \frac{\alpha_{\mathrm{I}}}{\frac{1}{1-\alpha} - D + D'\alpha_{\mathrm{I}}}. \quad (27)$$

and a base value for magnetic flux, Φ_{base}, as

$$\Phi_{\mathrm{base}} = \frac{V_{\mathrm{out}}}{2Nf_{\mathrm{sw}}}. \quad (28)$$

Normalizing the peak CM, DM, and total fluxes in (13)-(18) through the side and center legs to Φ_{base} and inserting (27) yields the expressions of normalized fluxes (Φ with hat over):

$$\widehat{\Phi}_{1,\mathrm{pk,CM}} = \frac{\Phi_{1,\mathrm{pk,CM}}}{\Phi_{\mathrm{base}}} = \frac{1}{k} - \frac{1}{1-\alpha} \quad (29)$$

$$\widehat{\Phi}_{1,\mathrm{pk,DM}} = \frac{\Phi_{1,\mathrm{pk,DM}}}{\Phi_{\mathrm{base}}} = \frac{1}{1-\alpha} \quad (30)$$

$$\widehat{\Phi}_{1,\mathrm{pk}} = \frac{\Phi_{1,\mathrm{pk}}}{\Phi_{\mathrm{base}}} = \frac{1}{k} \quad (31)$$

$$\widehat{\Phi}_{\mathrm{c,pk,CM}} = \frac{\Phi_{\mathrm{c,pk,CM}}}{\Phi_{\mathrm{base}}} = \frac{2}{k} - \frac{2}{1-\alpha} \quad (32)$$

$$\widehat{\Phi}_{\mathrm{c,pk,DM}} = \frac{\Phi_{\mathrm{c,pk,DM}}}{\Phi_{\mathrm{base}}} = \frac{1+\alpha}{1-\alpha} \quad (33)$$

$$\widehat{\Phi}_{\mathrm{c,pk}} = \frac{\Phi_{\mathrm{c,pk}}}{\Phi_{\mathrm{base}}} = \frac{2}{k} - 1 \quad (34)$$

Fig. 4 shows the normalized peak CM, DM, and total fluxes in the side and center legs versus coupling coefficient, α, when $D = 0.1$ and $D = 0.4$. As Figs. 4(a) and (b) individually show, strengthening magnetic coupling reduces the total fluxes in the side and center legs due to DC flux cancellation, meaning that the core volume can be reduced while achieving the same design targets. In addition, as α approaches -1, the proportion of the DM flux in the total flux increases. By comparing Figs. 4(a) and (b), it can be observed that with a larger duty ratio (D), the DM flux remains constant while the CM flux decreases, making the DM flux a larger portion of the total flux. Therefore, considering the DM flux is more important for strong coupling designs operating under a large duty ratio.

D. Optimal Center-to-Side Leg Width Ratio, $(w_{\mathrm{c}}/w_{\mathrm{s}})_{\mathrm{opt}}$

A key design parameter for two-phase coupled inductors is the ratio of the center leg width (w_{c}) to the side leg width (w_{s}). To minimize the core volume while achieving the required steady-state inductance and saturation current, this center-to-side leg width ratio ($w_{\mathrm{c}}/w_{\mathrm{s}}$) should be carefully designed. Specifically, it should ensure that the peak flux densities in the side and center legs ($B_{1,\mathrm{pk}}$ and $B_{\mathrm{c,pk}}$) are identical to each other and equal to the saturation flux density (B_{sat}):

$$B_{1,\mathrm{pk}} = B_{\mathrm{c,pk}} = B_{\mathrm{sat}} \quad (35)$$

where

$$B_{1,\mathrm{pk}} = \frac{\Phi_{1,\mathrm{pk}}}{A_{\mathrm{s}}} = \frac{\Phi_{1,\mathrm{pk}}}{w_{\mathrm{s}}Z} \quad (36)$$

$$B_{\mathrm{c,pk}} = \frac{\Phi_{\mathrm{c,pk}}}{A_{\mathrm{c}}} = \frac{\Phi_{\mathrm{c,pk}}}{w_{\mathrm{c}}Z}. \quad (37)$$

979-8-3315-1612-3/25 $31.00 © 2025 IEEE

Equation (35) is a necessary condition for achieving an optimal design, as it ensures equal and full utilization of the side and center legs. Rearranging (35)-(37) provides the relationship between w_c/w_s and the fluxes in the side and center legs:

$$\frac{w_c}{w_s} = \frac{\Phi_{c,pk}}{\Phi_{1,pk}} \qquad (38)$$

In conventional coupled inductor designs, w_c/w_s is determined by considering only the DC flux, which is approximately equal to the CM flux, while neglecting the DM flux. This approximation is based on the fact that the CM flux is typically larger than the DM flux and constitutes the majority of the total flux. Since the CM flux in the center leg is twice that in the side legs, as indicated by (16), w_c/w_s is typically set to 2 in conventional designs [1]:

$$\left(\frac{w_c}{w_s}\right)_{conv} = \frac{\Phi_{c,pk}}{\Phi_{1,pk}} \approx \frac{\Phi_{c,pk,CM}}{\Phi_{1,pk,CM}} = 2 \qquad (39)$$

Although straightforward, this simplified design approach misses the opportunity to optimize the core geometry by considering the CM and DM fluxes simultaneously.

With the DM flux included, the normalized expressions for the peak total fluxes in the side and center legs are given in (31) and (34). According to (38), the optimal center-to-side leg width ratio, $(w_c/w_s)_{opt}$, that ensures identical peak flux densities in the side and center legs can be obtained as

$$\left(\frac{w_c}{w_s}\right)_{opt} = \frac{\Phi_{c,pk}}{\Phi_{1,pk}} = \frac{\widehat{\Phi}_{c,pk}}{\widehat{\Phi}_{1,pk}} = 2 - k. \qquad (40)$$

Fig. 5 shows the values of $(w_c/w_s)_{opt}$ at various α, D, and α_I. As the magnetic coupling strengthens, or as the duty ratio or ripple ratio increases, $(w_c/w_s)_{opt}$ deviates from the conventional rule-of-thumb value of 2 $((w_c/w_s)_{conv})$. Particularly for coupled inductors operating at a duty ratio near 0.5, the influence of DM flux becomes significant and cannot be neglected when determining w_c/w_s.

It is worth noting that (40) assumes balanced currents through the two windings. However, in practice, inductor currents can exhibit some imbalance. For example, multiphase buck converters do not have inherent current balancing capability, and their natural current distribution is determined by the impedance of each phase. Although an active current balance controller, if designed properly, can ensure well-balanced inductor currents during steady-state operation, it cannot completely eliminate the current imbalance during load transients. Furthermore, even in topologies with a natural current balancing mechanism, such as the series-capacitor buck (SCB) converter [14], inductor current imbalance can still occur under heavy-load conditions due to large flying capacitor voltage ripples. Assuming that the peak current I_{pk1} increases by ΔI_{pk1} due to current imbalance, (40) can be modified as

$$\left(\frac{w_c}{w_s}\right)_{opt}' = 2 - k' \qquad (41)$$

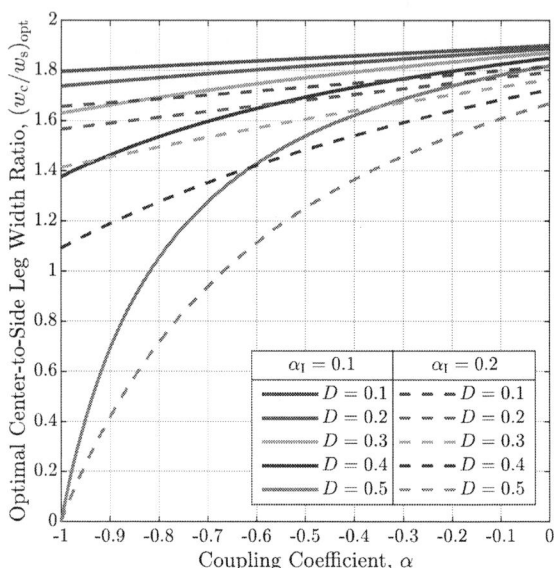

Fig. 5: Optimal center-to-side leg width ratios $((w_c/w_s)_{opt})$ at various coupling coefficients (α), duty ratios (D), and inductor current ripple ratios (α_I). $(w_c/w_s)_{opt}$ deviates from the conventional rule-of-thumb value of 2 as the magnetic coupling strengthens, or as the duty ratio or ripple ratio increases, showing the missed opportunity of core geometry optimization in conventional coupled inductor designs.

where

$$k' = \frac{1 + (1-\alpha)\Delta k}{\frac{1}{k} + \Delta k} \quad \text{and} \quad \Delta k = \frac{D' + D\alpha}{\alpha_I (1-\alpha^2)} \cdot \frac{\Delta I_{pk1}}{I_L}. \quad (42)$$

The derivation of (41) is omitted due to space limitation.

III. MULTI-OBJECTIVE OPTIMIZATION

The three key design objectives of two-phase coupled inductors are minimizing overall volume, reducing power loss, and achieving fast dynamic response. In this section, three metrics are used to quantify these objectives: a) total length, Z_{tot}, as an indicator of overall volume; b) winding DC resistance, R_{dc}, as an indicator of power loss; and c) transient inductance, L_{tr}, as an indicator of dynamic response. With these quantifiable metrics, the design of a two-phase coupled inductor is formulated as a multi-objective optimization problem. In this problem, the coupling coefficient (α) and the width of the side legs (w_s) are treated as two independent design variables, from which other geometric and electrical parameters can be derived.

A. Overall Volume: Total Length, Z_{tot}

In this paper, we assume that the maximum thickness (T_{max}) and maximum width (W_{max}) of the coupled inductor are predetermined based on form factor requirements, while the core length (Z) is treated as a design variable (dimensions annotated in Fig. 2(a)). Since an optimal design should fully utilize the available space, the thickness and width of the coupled inductor are set to their maximum allowable values (i.e., $T = T_{max}$ and $W = W_{max}$). Consequently, the volume of the coupled inductor becomes linearly proportional to its

979-8-3315-1612-3/25 $31.00 © 2025 IEEE

total length (including the cores and windings), Z_{tot}. Thus, Z_{tot} serves as a suitable metric for overall volume.

The cross-sectional area of the side leg, A_{s}, can be obtained by substituting (28), (31), and (35) into (36):

$$A_{\text{s}} = \frac{V_{\text{out}}}{2kNf_{\text{sw}}B_{\text{sat}}}. \qquad (43)$$

where the design factor, k, is given by (27). Therefore, the core length, Z, is given by

$$Z = \frac{A_{\text{s}}}{w_{\text{s}}} = \frac{V_{\text{out}}}{2kNf_{\text{sw}}B_{\text{sat}}w_{\text{s}}}. \qquad (44)$$

Assuming full utilization of the window area for the copper winding, with a core-to-winding clearance of c, the thickness of the winding, T_{Cu}, can be obtained as

$$T_{\text{Cu}} = T_{\text{max}} - 2w_{\text{s}} - 2c. \qquad (45)$$

Therefore, the total length, Z_{tot}, is given by

$$Z_{\text{tot}} = Z + 2(T_{\text{Cu}} + c) = Z + 2T_{\text{max}} - 4w_{\text{s}} - 2c \qquad (46)$$

where Z is given by (44).

B. Power Loss: Winding DC Resistance, R_{dc}

In multiphase PWM converters for VRM applications, inductors typically operate in heavy continuous conduction mode (CCM) under full-load conditions. Given the relatively small ripple compared to the RMS value of the inductor current, magnetic core loss is typically much smaller than winding conduction loss [10]. Consequently, this paper assumes that the coupled inductor loss at full load is dominated by the winding conduction loss. Therefore, the winding DC resistance, R_{dc}, is selected as a metric for power loss.

To minimize R_{dc}, the remainder of this paper assumes that each winding consists of only one turn (i.e., $N = 1$). Similar to the calculation of the winding thickness (T_{Cu}) in (45), the winding width, W_{Cu}, for one-turn inductor designs can be obtained as

$$W_{\text{Cu}} = \frac{1}{2}(W_{\text{max}} - 2w_{\text{s}} - w_{\text{c}}) - 2c \qquad (47)$$

where the width of the center leg, w_{c}, can be calculated from w_{s} and the optimal center-to-side leg width ratio given by (40) under balanced currents or by (41) under unbalanced currents. The length of each one-turn winding can be approximated as

$$L_{\text{Cu}} = \left(Z_{\text{tot}} - 2 \cdot \frac{T_{\text{Cu}}}{2}\right) + 2 \cdot \frac{T_{\text{max}}}{2}$$
$$= Z + 2T_{\text{max}} - 2w_{\text{s}} \qquad (48)$$

where Z is given by (44).

With the geometric parameters (T_{Cu}, W_{Cu}, and L_{Cu}) of the windings derived, R_{dc} can be calculated using the resistivity of copper, ρ_{Cu}, as

$$R_{\text{dc}} = \rho_{\text{Cu}}\frac{L_{\text{Cu}}}{W_{\text{Cu}}T_{\text{Cu}}}. \qquad (49)$$

C. Dynamic Response: Transient Inductance, L_{tr}

Regardless of the control method, the dynamic response of a coupled inductor is fundamentally and physically limited by its per-phase transient inductance, L_{tr}. Therefore, in this work, L_{tr} is chosen as a metric for dynamic response.

With the expression for self-inductance, (L_{min}), from (22), L_{tr} can be obtained as

$$L_{\text{tr}} = (1 + \alpha)L_{\text{min}} = \frac{(D' + D\alpha)V_{\text{out}}}{2\alpha_{\text{I}}(1 - \alpha)f_{\text{sw}}I_{\text{L}}} \qquad (50)$$

D. Multi-Objective Optimization

The expressions for Z_{tot} in (46), R_{dc} in (49), and L_{tr} in (50) are functions of the independent design variables, α and/or w_{s}, while all other quantities are given design parameters or constants. An optimal design should minimize Z_{tot}, R_{dc}, and L_{tr} simultaneously while satisfying the requirements on the minimum steady-state inductance ($L_{\text{ss(min)}}$), form factor (T_{max} and W_{max}), and minimum saturation current ($I_{\text{sat(min)}}$). In summary, the multi-objective optimization of a two-phase coupled inductor can be formulated as

$$\begin{aligned}
\min \quad & Z_{\text{tot}}(\alpha, w_{\text{s}}), \ R_{\text{dc}}(\alpha, w_{\text{s}}), \ L_{\text{tr}}(\alpha) \\
\text{s.t.} \quad & \text{Inductance constraint:} && L_{\text{ss}} \geqslant L_{\text{ss(min)}} \\
& \text{Form factor constraints:} && T \leqslant T_{\text{max}} \\
& && W \leqslant W_{\text{max}} \\
& \text{Saturation limit:} && \max_{I_{\text{L}}=I_{\text{sat(min)}}}(B) \leqslant B_{\text{sat}}.
\end{aligned} \qquad (51)$$

Note that, for simplicity, other natural geometric and electrical constraints are omitted in (51). Furthermore, the choices of setting the number of turns (N) to 1 and limiting the core thickness (T) and width (W) while allowing the core length (Z) to vary are application-specific. While the formulation of the multi-objective optimization problem shown in (51) serves as a specific example, its methodology is generalizable and applicable to a wider range of applications.

IV. SIMULATION AND EXPERIMENTAL VALIDATION

The proposed design-oriented model and optimization method are validated through comparisons with simulation and experimental results. Table I provides the specifications of the example design problem. Figs. 6(a) and (b) present the multi-objective optimization results obtained from a parametric sweep using the proposed model and a Monte Carlo simulation performed in ANSYS, respectively. In Fig. 6(a), the objective metrics (Z_{tot}, R_{dc}, and L_{tr}) were calculated by sweeping the two independent design variables (α and w_{s}) in (51), which was completed within 10 seconds using MATLAB on a desktop computer. By contrast, in Fig. 6(b), the objective metrics have to be extracted through finite element analysis in ANSYS by sweeping w_{s}, $w_{\text{c}}/w_{\text{s}}$, l_{gs}, and l_{gc}, which was automated but required more than 48 hours to complete on the same computer.

The Pareto front of the model-predicted designs in Fig. 6(a) shows good agreement with that of the Monte Carlo simulation results from ANSYS in Fig. 6(b). On the Pareto front, three

TABLE I: Design Specifications

Parameter	Value
Nominal output voltage (V_{out})	1 V
Nominal duty ratio (D)	0.167 ($\frac{1}{6}$)
switching frequency (f_{sw})	200 kHz
Maximum thickness (T_{max})	9 mm
Maximum width (W_{max})	18 mm
Core-to-winding clearance (c)	0.15 mm
Peak-to-peak inductor current ripple ($\Delta i_{L,pp}$)	8 A
Minimum saturation current ($I_{sat(min)}$)	60 A
Additional peak current due to imbalance (ΔI_{pk1})	0.64 A
	(1% of I_{pk1})

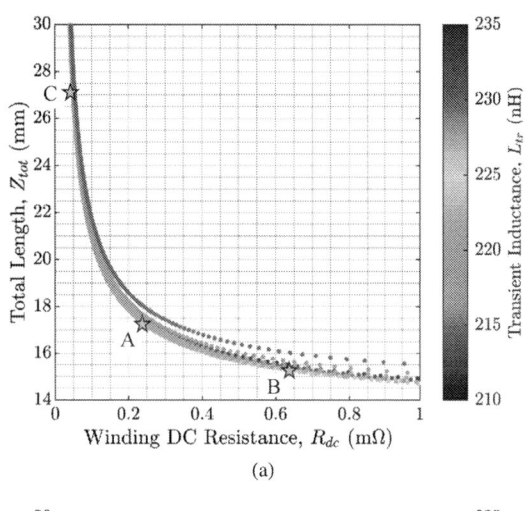

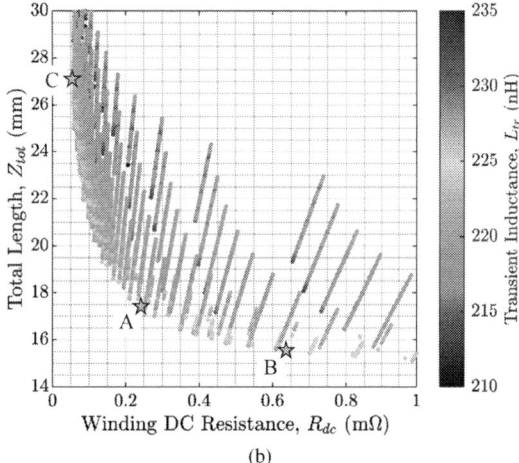

Fig. 6: Multi-objective optimization of a two-phase coupled inductor. (a) Parametric sweep with the proposed design-oriented model. (b) Monte Carlo simulations using ANSYS.

Fig. 7: Photograph of the hardware prototype for model and optimization validation.

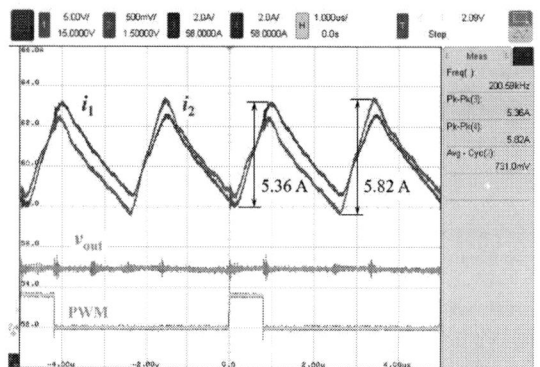

Fig. 8: Measured inductor current waveforms of design A at a load current of 120 A (the required saturation limit).

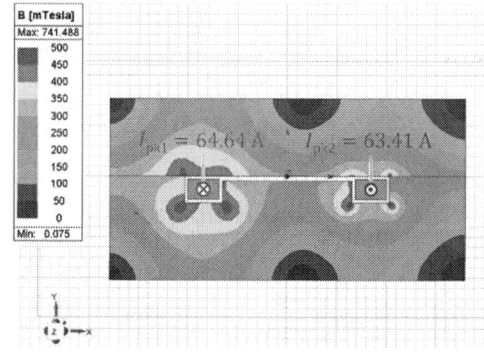

Fig. 9: Cross-sectional magnetic flux density distribution of design A at the required minimum saturation current (60 A per phase). The saturation magnetic flux density of the core material at 100 °C is 410 mT.

designs are selected for experimental validation. Design A is the overall best design that achieves a good balance among the three optimization objectives. Design B is an optimal design that trades off efficiency for a smaller overall volume, while Design C aims to greatly reduce R_{dc} at the cost of a larger inductor size. The three designs are fabricated with custom magnetic cores and copper windings. Fig. 7 presents the photograph of a hardware prototype that contains three SCB converters operating at 200 kHz to test the three coupled in-

ductor designs. Fig. 8 shows the measured current waveforms of design A at a load current of 120 A (the required saturation limit). From the measured inductor current waveforms, the steady-state and transient inductances (L_{ss} and L_{tr}) can be extracted (with the parasitic inductance of the copper jumper for current measurement subtracted). Fig. 9 shows the cross-sectional flux density distribution of design A at the required minimum saturation current (60 A per phase). The saturation magnetic flux density of the core material (DMR95 [15]) at 100 °C is 410 mT. As shown in Fig. 9, the magnetic cores are

TABLE II: Comparison of the electrical and geometric parameters for the three coupled inductors as predicted by the design-oriented model (Model), obtained through Monte Carlo simulations using ANSYS (Sim.), and extracted from experimental results (Exp.)

Design Parameter	Design A			Design B			Design C		
	Model	Sim.	Exp.	Model	Sim.	Exp.	Model	Sim.	Exp.
DC resistance (R_{dc})	0.25 mΩ	0.24 mΩ	0.23 mΩ	0.63 mΩ	0.63 mΩ	0.70 mΩ	0.05 mΩ	0.05 mΩ	0.05 mΩ
Total length (Z_{tot})	17.1 mm	17.4 mm	17.4 mm	15.3 mm	15.3 mm	15.3 mm	27.1 mm	27.2 mm	27.2 mm
Transient inductance (L_{tr})	223 nH	225 nH	211 nH	223 nH	225 nH	211 nH	223 nH	225 nH	211 nH
Steady-state inductance (L_{ss})	521 nH	521 nH	497 nH	521 nH	521 nH	496 nH	521 nH	521 nH	496 nH
Coupling coefficient (α)	-0.91	-0.90	-0.92	-0.91	-0.90	-0.92	-0.91	-0.90	-0.92
Side-leg air gap length (l_{gs})	0.36 mil	0.43 mil	0.30 mil	0.37 mil	0.43 mil	0.30 mil	0.34 mil	0.40 mil	0.30 mil
Center-leg air gap length (l_{gc})	10.1 mil	10.0 mil	9.75 mil	10.1 mil	10.0 mil	9.75 mil	10.1 mil	10.0 mil	9.95 mil
Width of side leg (w_s)	3.84 mm	3.81 mm	3.81 mm	4.09 mm	4.08 mm	4.08 mm	2.77 mm	2.76 mm	2.77 mm
Width of center leg (w_c)	6.44 mm	6.47 mm	6.47 mm	6.85 mm	6.93 mm	6.93 mm	4.65 mm	4.70 mm	4.70 mm
Leg width ratio (w_c/w_s)	1.68	1.70	1.70	1.68	1.70	1.70	1.68	1.70	1.70

on the brink of saturation.

Table II compares the electrical and geometric parameters of the three coupled inductors as predicted by the design-oriented model (Model), obtained through Monte Carlo simulation using ANSYS (Sim.), and extracted from experimental results (Exp.). This comparison shows that the three design objectives of the model-predicted designs align well with those of the simulated and experimental results across all three coupled inductor designs, demonstrating the good accuracy of the design-oriented model and the effectiveness of the multi-objective optimization.

V. CONCLUSIONS

This paper proposes a design-oriented model and multi-objective optimization method for two-phase coupled inductors. Through CM and DM decomposition, this work reveals that, due to the presence of the DM magnetic flux, the optimal center-to-side leg width ratio should be less than 2—the rule-of-thumb value used in conventional designs. Based on the proposed model, the design of two-phase coupled inductors is formulated as a multi-objective optimization problem, aiming for smaller overall volume, lower power loss, and faster dynamic response. Through this optimization method, a Pareto front is determined for an example design problem, from which three designs are selected and implemented for hardware validation. The good agreement between the electric and geometric parameters of the model-predicted designs and those of Monte Carlo simulations and experimental results demonstrates the high accuracy and effectiveness of the proposed model and optimization method.

VI. ACKNOWLEDGEMENTS

The authors acknowledge the financial support from the Berkeley Power and Energy Center (BPEC).

REFERENCES

[1] P.-L. Wong, P. Xu, P. Yang, and F. Lee, "Performance improvements of interleaving VRMs with coupling inductors," *IEEE Transactions on Power Electronics*, vol. 16, no. 4, pp. 499–507, 2001.

[2] J. Li, C. Sullivan, and A. Schultz, "Coupled-inductor design optimization for fast-response low-voltage DC-DC converters," in *2002 IEEE Applied Power Electronics Conference and Exposition (APEC)*, vol. 2, 2002, pp. 817–823.

[3] E. Laboure, A. Cuniere, T. A. Meynard, F. Forest, and E. Sarraute, "A Theoretical Approach to InterCell Transformers, Application to Inter-leaved Converters," *IEEE Transactions on Power Electronics*, vol. 23, no. 1, pp. 464–474, 2008.

[4] Y. Dong, "Investigation of Multiphase Coupled-Inductor Buck Convert-ers in Point-of-Load Applications," Ph.D. dissertation, Virginia Tech, Blacksburg, VA, USA, 2009.

[5] A. Ikriannikov and T. Schmid, "Magnetically coupled buck converters," in *2013 IEEE Energy Conversion Congress and Exposition*, 2013, pp. 4948–4954.

[6] A. M. Naradhipa, F. Zhu, and Q. Li, "Ultra-Low-Profile Twisted Core Inductor for Vertical Power Delivery Voltage Regulator," in *2024 IEEE Applied Power Electronics Conference and Exposition (APEC)*, 2024, pp. 918–924.

[7] F. Zhu and Q. Li, "A Novel PCB-Embedded Coupled Inductor Structure for a 20-MHz Integrated Voltage Regulator," *IEEE Journal of Emerging and Selected Topics in Power Electronics*, vol. 10, no. 6, pp. 7452–7463, 2022.

[8] J. Czogalla, J. Li, and C. Sullivan, "Automotive application of multi-phase coupled-inductor DC-DC converter," in *38th IAS Annual Meeting on Conference Record of the Industry Applications Conference, 2003.*, vol. 3, 2003, pp. 1524–1529 vol.3.

[9] Y. Liu, M. Li, Y. Dou, Z. Ouyang, and M. A. Andersen, "Investigation and Optimization for Planar Coupled Inductor dual-phase interleaved GaN-based Totem-Pole PFC," in *2020 IEEE Applied Power Electronics Conference and Exposition (APEC)*, 2020, pp. 1984–1990.

[10] Y. Zhu, T. Ge, N. M. Ellis, L. Horowitz, and R. C. N. Pilawa-Podgurski, "The Switching Bus Converter: A High-Performance 48-V-to-1-V Architecture with Increased Switched-Capacitor Conversion Ratio," *IEEE Transactions on Power Electronics*, pp. 1–20, 2024.

[11] N. M. Ellis, Y. Zhu, and R. C. Pilawa-Podgurski, "Gallium Nitride-based 48V-to-1V Point-of-Load (PoL) Converter for Aerospace Telecommu-nications and Computing Applications," in *2024 IEEE Applied Power Electronics Conference and Exposition (APEC)*, 2024, pp. 1384–1388.

[12] Y. Zhu, J. Zou, N. M. Ellis, S. Kudva, M. Mosa, C. T. Gray, and R. C. Pilawa-Podgurski, "A Compact 48-V-to-Sub-1-V Switching Bus Converter with 4.7-mm Height for Processor Vertical Power Delivery," in *2024 IEEE Energy Conversion Congress and Exposition (ECCE)*, 2024.

[13] J. Zou, Y. Zhu, N. M. Ellis, L. Horowitz, and R. C. N. Pilawa-Podgurski, "A 48-V-to-1-V Gallium Nitride Switching Bus Converter for Processor Vertical Power Delivery with 2.7 mm Thickness and 3048 W/in^3 Power Density," in *2025 IEEE Applied Power Electronics Conference and Exposition (APEC)*, 2025.

[14] P. S. Shenoy, M. Amaro, J. Morroni, and D. Freeman, "Comparison of a Buck Converter and a Series Capacitor Buck Converter for High-Frequency, High-Conversion-Ratio Voltage Regulators," *IEEE Transactions on Power Electronics*, vol. 31, no. 10, pp. 7006–7015, 2016.

[15] DMEGC, *DMR95 Material Characteristics*, 2014, Accessed: Nov. 1, 2024. [Online]. Available: https://www.dmegc.de/uploads/20230401/DMR95%20Material%20Characteristics.pdf

MagNetX: Extending the MagNet Database for Modeling Power Magnetics in Transient

Hyukjae Kwon◇, Shukai Wang◇, Haoran Li◇, Youssef Elasser‡,
Gyeong-Gu Kang◇, Daniel Zhou◇, Davit Grigoryan◇, and Minjie Chen◇
◇*Princeton University, Princeton, NJ, United States*
‡*Nvidia Research, Durham NC, United States*
Email: {hk1715, sw0123, minjie}@princeton.edu

Abstract—This paper introduces the MagNetX[1] - an extension of the MagNet database to investigate transient hysteresis of magnetic materials. By employing automated data acquisition and measurement systems, detailed transient phenomena are observed and evaluated across a wide range of waveforms and operational conditions. Results reveal significant core loss variations during frequency transitions, and the B-H loop analysis further demonstrates loop area changes during transient phases. The platform enables advanced data-driven modeling of core losses and hysteresis loops over a wide range of amplitudes, frequencies, temperatures, and duty cycles under transient conditions, which are critical for practical circuit design when steady-state models are inadequate, such as modeling the magnetic core behaviors in ac-dc converters, motor drives, or power amplifiers.

Index Terms—power magnetics, core loss, transient, hysteresis loop, machine learning, neural networks

I. INTRODUCTION

Power magnetic components play crucial roles in determining the efficiency, density, cost, and loss of power electronics systems. Among these factors, power loss and material saturation often stand out as significant concerns, impacting overall system design and optimization. The characteristics of the magnetic core materials determine the performance of a magnetic component. However, modeling of magnetic materials across a wide range of operating conditions is still challenging due to the complex inter-dependency of core characteristics on the operating conditions [1]. Numerous factors, such as frequency, dc-bias, excitation waveform, temperature, and duty cycle, influence the loss phenomena, resulting in an incomplete physical understanding of magnetic characteristics.

Steinmetz equation (SE) derived methods (such as iGSE and i²GSE) are among the most widely used models for modeling magnetic core loss [2]–[4]. However, these models demonstrate limitations in accuracy when applied to specific waveform types, potentially leading to malfunctions due to unexpected excessive heat. Data-driven approaches and hybrid data-driven approaches [5], [6], such as neural networks and other machine learning methods, integrate various factors into a unified framework, proving effective in solving multivariable nonlinear regression problems [7], [8]. Data driven techniques have also been adopted to modeling magnetic hysteresis and optimizing inductor design [9]–[11].

[1]MagNetX GitHub Repository: https://github.com/PaulShuk/MagNetX

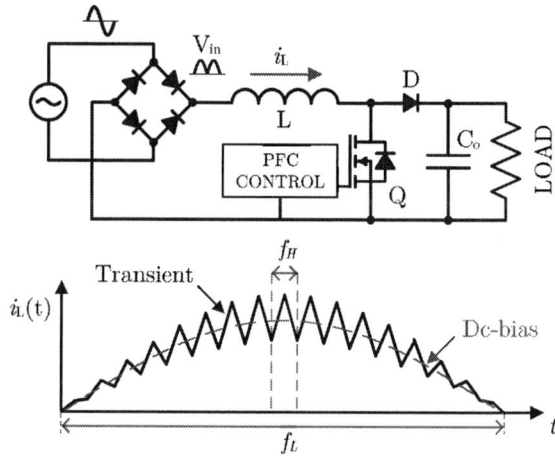

Fig. 1. Transient inductor current waveform with dc-bias in a PFC circuit. The magnetic materials do not operate in periodic steady-state.

Data size and quality are critical for data-driven modeling methods. To overcome these challenges, a large-scale open-source database – the MagNet – has recently been established [12], [13]. This database serves as a comprehensive resource for the data-driven characterization of power magnetics, offering accurate core loss predictions under steady-state conditions with advanced data-driven models such as the neural networks. The current MagNet database primarily addresses steady-state operating conditions. However, transient conditions – such as variations in current, voltage, frequency, and turn-on/off operations – more systematically influence the design process of power converter systems [14], especially in power factor correction (PFC) rectifiers, motor drives, and power amplifiers. Fig. 1 illustrates the inductor current waveform in a PFC circuit, with a high-ripple frequency component and dc-bias. The magnetics do not operate at a fixed frequency, and the excitations never reach a steady state. To model core losses and saturation in PFC inductors, the development of transient modeling is essential. Although the revised iGSE [15] has been applied to PFC core loss analysis, its effectiveness is still limited by data points and frequency range. A large-scale high quality database can greatly advance our understanding about the transient behaviors of power magnetics.

979-8-3315-1612-3/25 $31.00 © 2025 IEEE

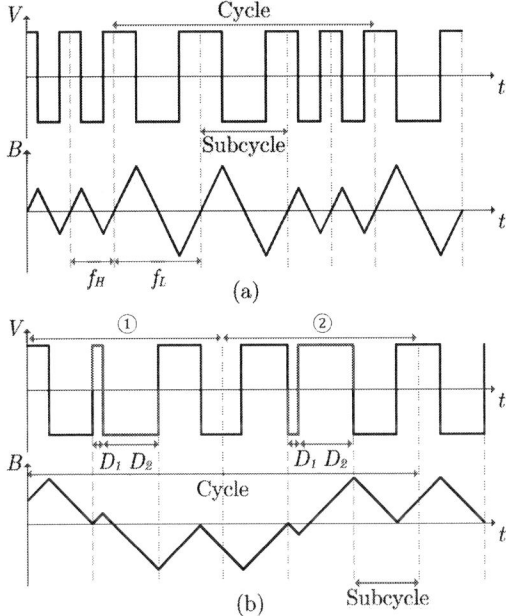

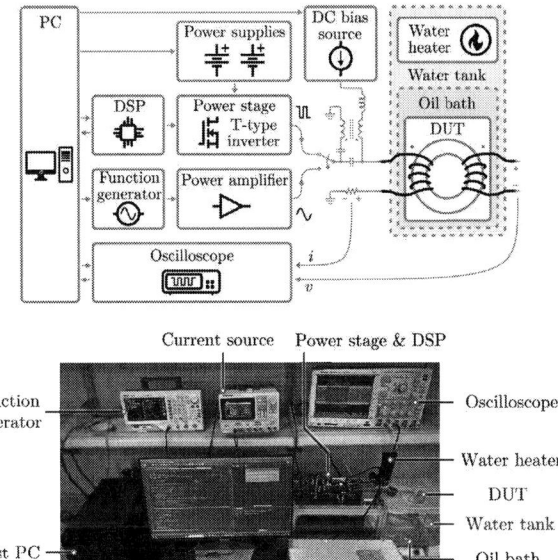

Fig. 2. Step-change voltage and B field waveforms: (a) Triangular excitation with frequency change; (b) Triangular excitation with duty cycle change.

Fig. 3. The MagNetX data acquisition system capable of automatically recording transient hysteresis information for the MagNetX database.

II. DATA ACQUISITION FOR MAGNETX DATABASE

The hysteresis behaviors of magnetic materials are influenced by various factors such as frequency, flux density, waveform shape, temperature, core geometry, and dc-bias. For transient modeling, the memory effects also play significant roles. Data quality determines the model accuracy. A fully automated procedure is necessary to create a large-scale database while ensuring high data quality. To rapidly collect a large amount of transient hysteresis data, step-changes in frequency and varying duty cycles are applied to the magnetic materials and the $v(t)$ and $i(t)$ waveforms are recorded. Fig. 2 (a) shows the ideal step-changed voltage and B field waveforms, assuming no additional effects applied. Two distinct PWM frequency waveforms are employed to observe the transition between them. Each set of cycles must reach a steady-state before shifting to the next frequency condition, thereby minimizing residual effects from the previous waveforms. Fig. 2 (b) shows the voltage with randomly varying duty cycles and the corresponding B field waveforms, which are analyzed to study transient behavior. Both voltage waveforms are designed to avoid the accumulation of dc-bias in the material hysteresis. Details of the setup used to obtain the experimental measurements are provided as follows.

A. MagNetX Data Acquisition System

The two-winding method in [13] was adopted for fast data collection. The dc-bias injection circuitry was not used. This method employs two windings on the device under test (DUT): the primary winding for excitation and current measurement to obtain $H(t)$, and the secondary winding for voltage measurement to obtain $B(t)$. Fig. 3 illustrates an overview of complete automated data acquisition system, which includes the

DUT excited by switch-mode T-type inverter with three-level pulsewidth-modulated (PWM) voltages, voltage and current measurement, and temperature control.

In this design, the triangular flux waveforms are generated using PWM driving signals from a microcontroller (Texas Instruments F38279D controlCARD). The T-type inverter circuit incorporates GaN devices (GS66508B) as the power switches to achieve fast switching. A blocking capacitor is connected in series with DUT to mitigate dc bias current. To ensure precise temperature control and minimize temperature fluctuations during measurements, a large water tank, coupled with an oil bath and a water heater (Anova AN400), is employed to maintain a stable testing environment. Measurements are collected at 25 °C, 50 °C, and 70 °C ambient temperature. A software system controls and coordinates the hardware system, enabling fully automated equipment settings and measuring around 1,000 data points per hour. Additionally, a Python script was developed to generate the desired gating waveforms.

B. Transient Waveform Synthesis

As shown in Fig. 2, step-changes in frequency and random varying duty cycles are applied to observe the transient behaviors. The individual waveform patterns will be referred to as a subcycle, and the complete waveform as a cycle. Fig. 2(a) illustrates triangular waveform with frequency step-changes between high-frequency (f_{H}) and low-frequency (f_{L}) waveforms. To ensure the waveform reaches steady-state before transitioning to another frequency condition, more than 500 subcycles of each frequency waveform are needed. f_{L} ranges from 50 kHz to 500 kHz, while f_{H} is set at twice the value of the low frequency. The PWM excitation voltage waveform is designed to combine two distinct frequency waveforms with

979-8-3315-1612-3/25 $31.00 © 2025 IEEE

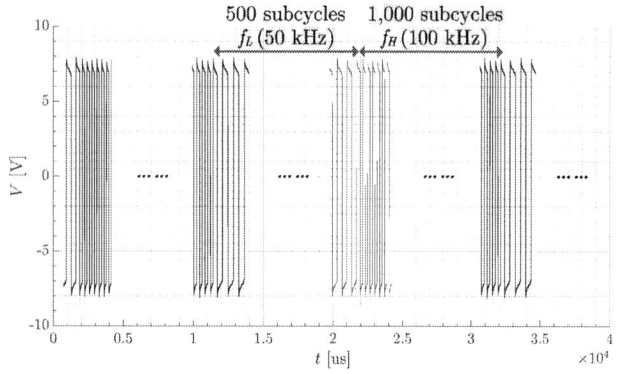

Fig. 4. A frequency step-change excitation: the excitation circuits generate 500 subcycles at 50 kHz, and 1,000 subcycles at 100 kHz.

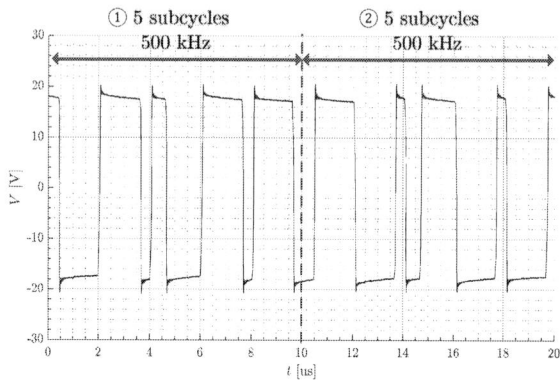

Fig. 5. Example voltage waveform with random duty cycles. The excitation circuit generates two different 5-pulse subcycle patterns every 20 μs.

each cycle of the B field ideally starting and ending at zero, ensuring no dc-bias in the B field.

Similarly, Fig. 2(b) depicts changes in the duty cycle D, which are divided into two segments. The first segment consists of random duty cycles ranging from 0.2 to 0.8, with a step size of 0.1, while the second segment is the inverse of the first, flipped vertically. By combining the two mirrored segments, voltage-time balance is achieved, minimizing the offset in a cycle. Fig. 2(b) illustrates this clearly. In the first segment, D_1 corresponds to the portion of the waveform with positive voltage, while D_2 corresponds to the portion with negative voltage. In the second segment, which is the vertical inverse of the first, the roles of D_1 and D_2 are reversed: D_1 now represents the negative voltage portion, and D_2 represents the positive voltage portion. This inversion ensures that the overall waveform maintains balance, as the sum of positive and negative voltages over time results in zero dc offset and ensuring that the voltage across the inductor core is equal to the exciting voltage. In order to collect and analyze the extensive transient behavior data, waveforms are generated across a frequency range from 50 kHz to 800 kHz. For each frequency, 100 distinct input waveforms are created, each consisting of 100 randomly varying duty subcycles. These excitation waveforms are designed to avoid dc-bias accumulation in the B field over the entire waveform cycle. For rapid transitions to another condition, a lookup table containing information about f_{H}, f_{L}, and the DSP counter compare register values is used to generate the PWM switching signals.

C. Post-Processing of Measurement Signals

An 8-bit oscilloscope (Tektronix DPO4104) is employed to measure the voltage and current waveforms. The total sampling time is adjusted depending on the excitation frequency to capture at least two complete cycles of waveforms. 100,000 samples are saved for each test with bandwidth limited to 20 MHz to avoid excessive switching noise from fast transition. To reduce the data size, the measurements are downsampled from 100,000 to 10,000 samples per test by averaging every ten consecutive samples.

Fig. 4 illustrates an example voltage waveform comprising 1,000 subcycles of $f_H = 100$ kHz and 500 subcycles of

$f_L = 50$ kHz. With a 40 ms oscilloscope window size, the waveform clearly displays two distinct PWM waveforms at different frequencies. Fig. 5 illustrates the example voltage waveform with randomly varying duty cycles. For clarity, 10 subcycles are displayed, rather than the full 100 subcycles. In the following sections of this paper, the waveform with 10 subcycles will be used to illustrate the concepts.

Based on the stored voltage and current measurements, the flux density B is calculated by integrating the voltage, as shown in (1):

$$B(t) = \frac{1}{A_e n_2} \int v_L(t) \, dt. \qquad (1)$$

The magnetic field strength H is derived from the current measurements as follows:

$$H(t) = \frac{n_1}{l_e} i_L(t). \qquad (2)$$

Core loss (P_{loss}) is calculated via voltamperometric method [16] by

$$P_{loss} = \frac{1}{T} \int_{t_0}^{t_0+T} v_L(t) \cdot i_L(t) \, dt. \qquad (3)$$

where v_L and i_L are the measured secondary-side voltage and primary-side current, and n_1 and n_2 are the number of turns of each winding.

III. DATA ANALYSIS ON FREQUENCY CHANGE

Observing the B and H waveforms, as well as the B-H curve, is an effective approach to analyze the transient behavior, as the B and H sequences are strongly correlated and both exhibit memory effects. However, unlike steady-state analysis, which requires only a single cycle of B-H loop because inductor model per cycle remains consistent, observing transient behavior necessitates multiple cycles of waveforms. This results in multiple B-H loops being shown in a single figure. Specifically, more than 200 subcycles are needed to reach a steady state from a transient state in the analysis at f_L=250 kHz for this experiment. Fig. 6 has demonstrated transient characteristics with f_{L1}=75 kHz and D = 0.5 during the transition from $f = f_{L1}$ to $f = f_{H1}$. Fig. 6 (a), (b) and (d) are B, H, and B-H loop, respectively. The cycles

979-8-3315-1612-3/25 $31.00 © 2025 IEEE

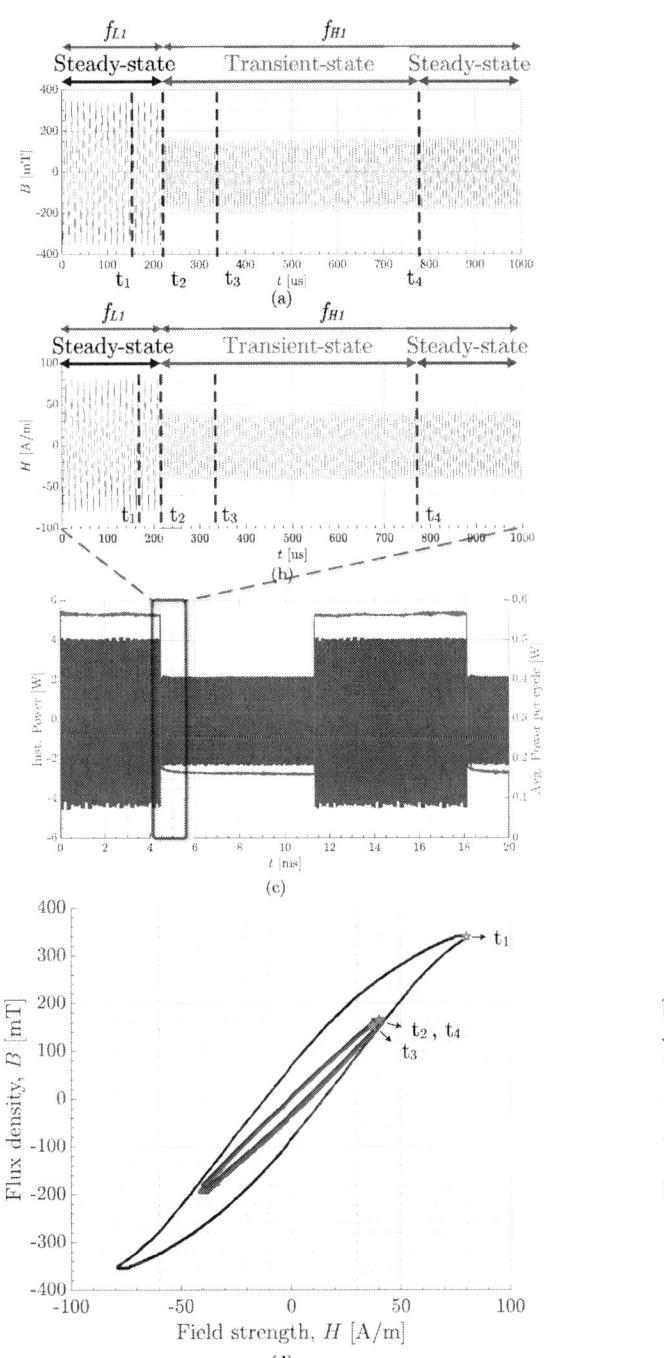

Fig. 6. Transient dynamics at f_{L1}=75 kHz, f_{H1}=150 kHz, and D = 0.5 square wave excitation: (a) Flux density B; (b) Field strength H; (c) B-H loop; (d) Instantaneous power and average power per cycle.

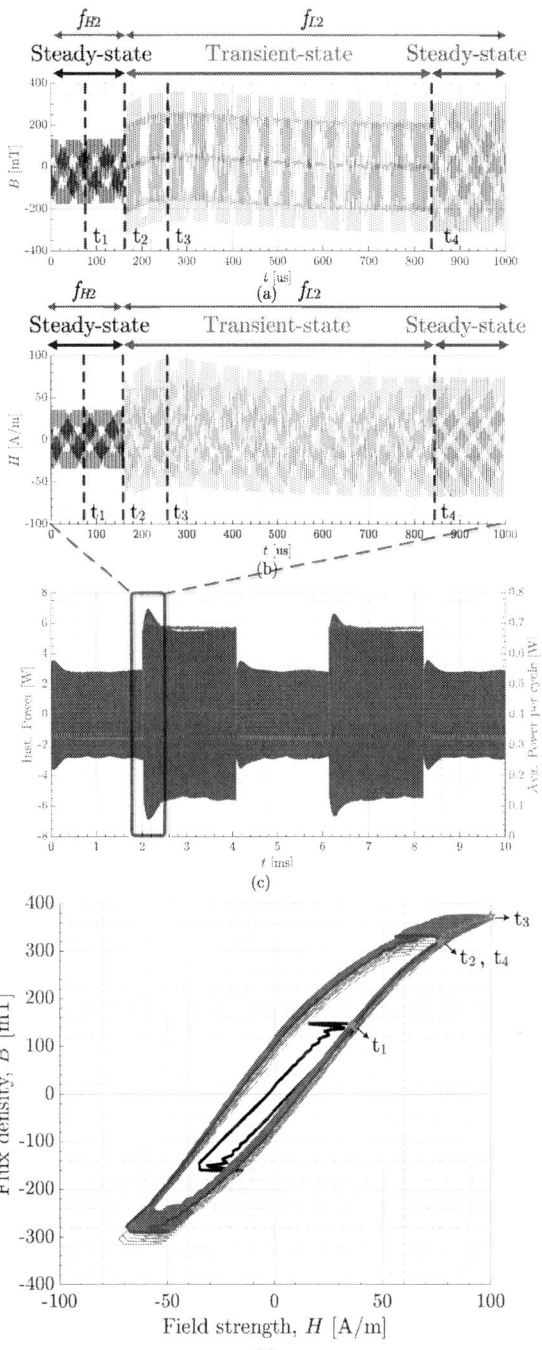

Fig. 7. Transient dynamics at f_{L1}=250 kHz, f_{H1}=500 kHz, and D = 0.5 square wave excitation: (a) Flux density B; (b) Field strength H; (c) B-H loop; (d) Instantaneous power and average power per cycle.

are color-coded to show different states: black represents the steady-state at $f = f_{L1}$, red indicates the transient phase as the frequency changes to $f = f_{H1}$, and blue shows the steady-state at $f = f_{H1}$. Fig. 6 (d) illustrates that during the transient state, the B-H loop initially shifts downward and to the left, before stabilizing back to its steady-state position. Fig. 6 (c) depicts core loss, showing both instantaneous power

and average power loss per cycle. Core loss is lower at $f = f_{H1}$ compared to $f = f_{L1}$, but the difference in core loss is minimal under the same frequency conditions. This observation aligns with the B-H loop results, where the loop area shows little change during transient conditions.

Fig. 7 shows the effect of frequency in triangular waveforms, with f_L adjusted to $f_{L2} = 250$ kHz. It presents a similar

TABLE I
NUMBER OF DATA SEQUENCES CURRENTLY IN THE MAGNETX DATASET

Material	Data Sequence
Ferroxcube 3C90	13,587
Ferroxcube 3C94	9,224
Ferroxcube 3E6	7,407
Ferroxcube 3F4	10,714
Fair-Rite 77	10,726
Fair-Rite 78	9,845
TDK N27	11,456
TDK N30	10,580
TDK N49	7,266
TDK N87	12,313
Total	103,118

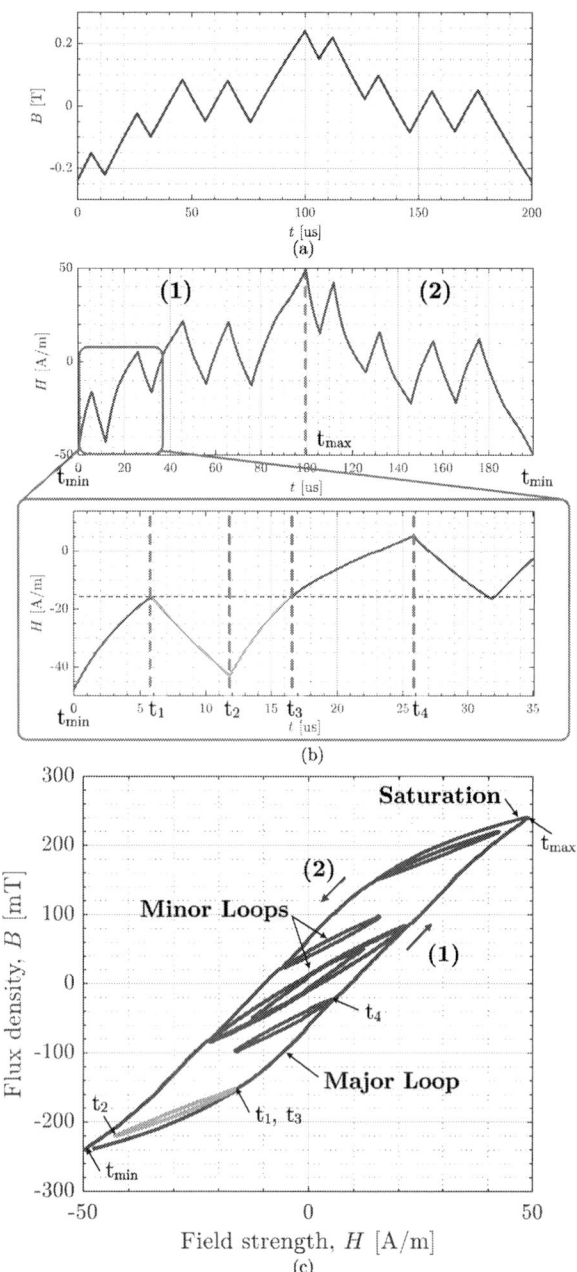

Fig. 8. Transient dynamics at $f_s = 50$ kHz with 10 subcycles: (a) Flux density B; (b) Field strength H; (c) B-H loop.

setup to the previous figure, but captures the transition from $f = f_{H2}$ to $f = f_{L2}$. Fig. 7 (a), (b) and (d) are B, H, and B-H loop at f_{L2}=250 kHz, showing that it takes more cycles to reach steady-state, with the B and H fields exhibiting larger changes compared to Fig. 6. In addition, Fig. 7 (c) illustrates core loss per cycle at f_{L2}=250 kHz, which is noticeably different from Fig. 6 (c). Due to the larger variations in the B and H fields, a peak in core loss is observed at the beginning of the frequency transition. This behavior is also reflected in the B-H loop, where the transient-state cycles (marked in red)

display significant deviations, moving upward and to the right and returning to the steady-state. These deviations result in a noticeable increase in the loop area during the transient phase.

IV. DATA ANALYSIS ON DUTY CYCLE CHANGE

A. Transient Characteristics

In magnetics, previous states influence the subsequent behavior, causing the magnetic response to vary depending on the current values of $B(t)$ and $H(t)$, which correspond to its position in the B-H loop. As a result, even with identical duty cycle sequences, the magnetic behavior can differ based on its position in the loop. This leads to a large number of possible combinations when varying the duty cycles in a voltage waveform. The MagNetX database now contains 100 data sets, each consisting of 100 subcycle sequences, with randomly distributed duty cycles ranging from 0.2 to 0.8 to capture diverse magnetic responses. A total of 103,118 data sequences are collected for 10 materials, with subcycle frequencies f_s ranging from 50 kHz to 800 kHz at three temperatures (25°C, 50°C, and 70°C) as provided in Table I.

Fig. 8 shows an example of the B and H waveforms with randomly varying duty cycles under triangular excitation. The measurements are made using a toroidal core with an outer diameter of 34 mm, an inner diameter of 20.5 mm, and a height of 12.5 mm (designated R34.0 × 20.5 × 12.5 by the manufacturer) of the material N87 by TDK. The waveform consists of 10 subcycles, each driven at a frequency of $f_s = 50$ kHz, yielding an overall cycle frequency of 5 kHz. Since the exciting voltage waveform is designed to be symmetric, the maximum magnetic field, $B_{\max}$, is expected to correspond to a minimum magnetic field value of $-B_{\max}$. Now, the B and H waveforms can be divided into two segments: (1) from the global minimum at $t = t_{\min}$ to the global maximum at $t = t_{\max}$, and (2) from the global maximum at $t = t_{\max}$ to the global minimum at $t = t_{\min}$. These global extrema define the endpoints of the major loop, while each subcycle consists of components that form the minor loop and components that contribute to the major loop. Specifically, the local maxima and minima of each subcycle determine which portions belong to the minor loop and which belong to the major loop. For example, in the first segment (1), the portions where the values of B and H exceed the previous local maxima (considering only those local maxima after the global minimum) correspond

979-8-3315-1612-3/25 $31.00 © 2025 IEEE

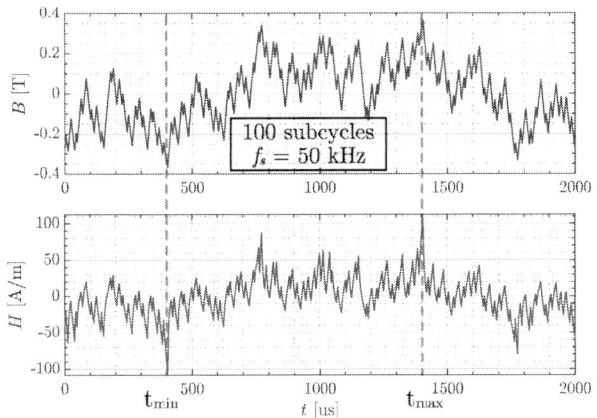

Fig. 9. The example $B(t)$ and $H(t)$ waveforms for a randomly varying duty cycle over 100 subcycles at 50 kHz.

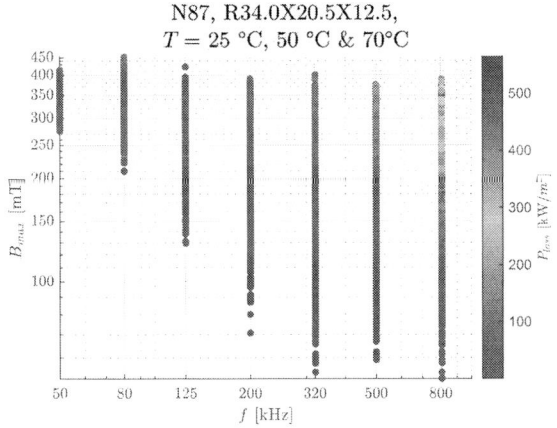

Fig. 10. Core loss under various frequencies, temperatures, and maximum magnetic field values, $B_{\max}$.

to the major loop, such as the intervals from $t_{\min}$ to t_1 and from t_3 to t_4. The remaining portion, such as the interval from t_1 to t_3, corresponds to the minor loops. A similar concept applies to the second segment (2), where the major loop is formed by the portion where the values of B and H drop below the previous local minima. Fig. 9 shows the actual data collected over 100 subcycles at 50 kHz.

B. Core Loss Analysis

Fig. 10 shows the core loss values under various frequencies, temperatures, and maximum magnetic field values, $B_{\max}$. Fewer data points are observed at lower frequencies, such as $f_s = 50$ kHz and $f_s = 80$ kHz. This occurs due to the power stage's voltage range being limited to 5 V to 80 V for the PWM waveforms, which limits the flux density-frequency product for the measurements.

Generally, the core loss increases with higher frequencies and higher $B_{\max}$ values, which is consistent with the Steinmetz equation. However, directly comparing measured core loss results with the Steinmetz equation or determining the Steinmetz

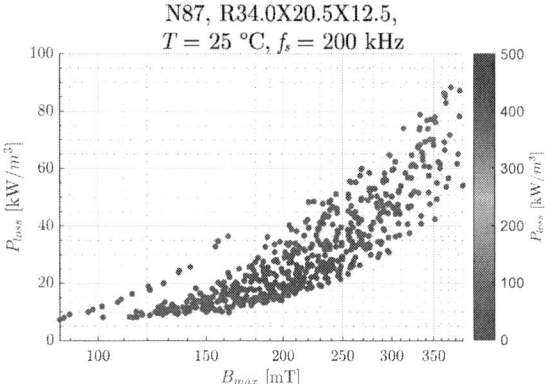

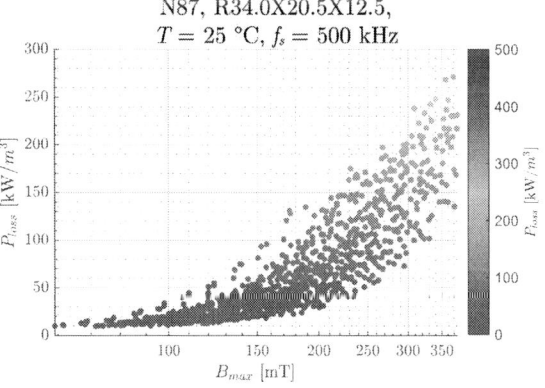

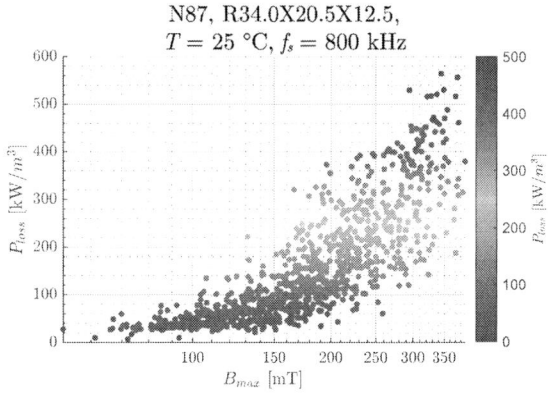

Fig. 11. Core loss at $T = 25$ °C under various frequencies and maximum magnetic field values, $B_{\max}$.

parameters is difficult, as there are multiple data points at the same frequency, temperature, and $B_{\max}$. Although different duty cycle sequences may result in the same $B_{\max}$ and major loop, the minor loops differ, leading to such variations in core loss results. Nevertheless, the overall trend in core loss data can still be observed by varying each parameter, highlighting the complexity of core loss modeling in transient.

Fig. 11 illustrates the core loss as a function of frequency at $T = 25$ °C. As the frequency increases from 200 kHz, 500 kHz, to 800 kHz, the core loss values become more scattered and higher. This is because, at higher frequencies, the capacitive effect in the inductor core becomes more prominent,

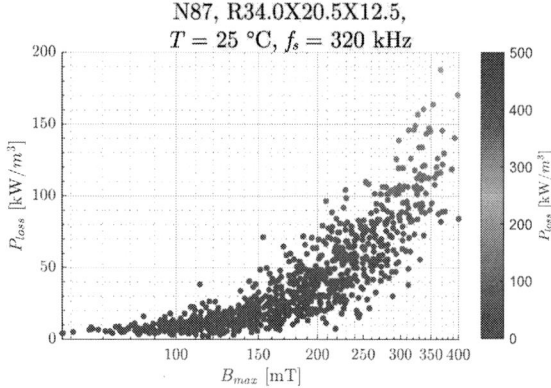

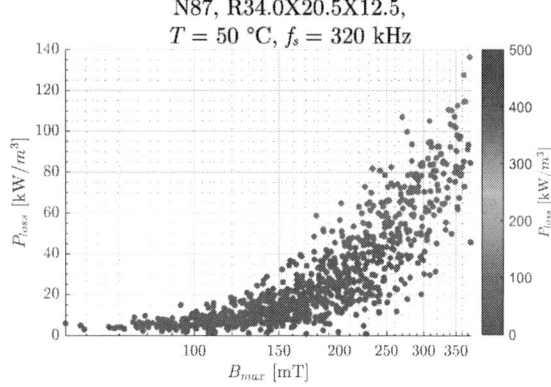

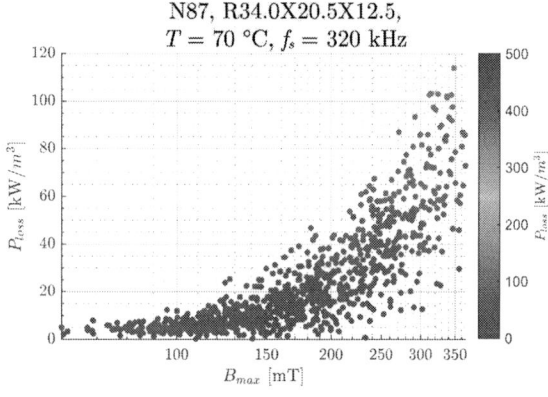

Fig. 12. Core loss at $f_s = 320$ kHz under various temperatures and maximum magnetic field values, $B_{\max}$.

leading to increased ringing and larger B-H loop areas. Fig. 12 shows the core loss as a function of temperature at 320 kHz. In this temperature range, losses decrease as temperature increases, since N87 is known for its minimal losses and high permeability around 90 to 100 °C, which makes it ideal for optimal performance at higher temperatures. However, other requirements, such as reasonable temperature stability or minimal losses at specific temperatures, may also be important. With the extensive data available for both transient and steady-state behavior in MagNetX, users can more rapidly find the core materials that best suits their needs [17], [18].

V. CONCLUSIONS

This paper presents an extended database - MagNetX - as a new development of the MagNet project. The MagNetX database can be used to support analyzing the transient effects in the design and optimization of magnetic components.

ACKNOWLEDGEMENTS

This work was jointly supported by the National Science Foundation under Award #2344664, ITG Electronics, TSMC, pSemi, and the Princeton Andlinger Center for Energy and the Environment.

REFERENCES

[1] D. Serrano, H. Li, S. Wang, T. Guillod, M. Luo, V. Bansal, N. K. Jha, Y. Chen, C. R. Sullivan, and M. Chen, "Why magnet: Quantifying the complexity of modeling power magnetic material characteristics," *IEEE Trans. on Power Electronics*, vol. 38, no. 11, pp. 14 292–14 316, 2023.

[2] C. P. Steinmetz, "On the law of hysteresis," *Transactions of the American Institute of Electrical Engineers*, vol. IX, no. 1, pp. 1–64, 1892.

[3] J. Li, T. Abdallah, and C. Sullivan, "Improved calculation of core loss with nonsinusoidal waveforms," in *Conference Record of the 2001 IEEE Industry Applications Conference. 36th IAS Annual Meeting (Cat. No.01CH37248)*, vol. 4, 2001, pp. 2203–2210 vol.4.

[4] J. Muhlethaler, J. Biela, J. W. Kolar, and A. Ecklebe, "Improved core-loss calculation for magnetic components employed in power electronic systems," *IEEE Transactions on Power Electronics*, vol. 27, no. 2, pp. 964–973, 2012.

[5] H. H. Cui, S. Dulal, S. B. Sohid, G. Gu, and L. M. Tolbert, "Unveiling the microworld inside magnetic materials via circuit models," *IEEE Power Electronics Magazine*, vol. 10, no. 3, pp. 14–22, 2023.

[6] M. Chen *et al.*, "Magnet challenge for data-driven power magnetics modeling," *IEEE Open Journal of Power Electronics*, pp. 1–16, 2024.

[7] Y. LeCun, Y. Bengio, and G. Hinton, "Deep learning," *Nature*, vol. 521, pp. 436–444, 2015.

[8] J. Schmidhuber, "Deep learning in neural networks: An overview," *Neural Networks*, vol. 61, pp. 85–117, Jan 2015.

[9] T. Guillod, P. Papamanolis, and J. W. Kolar, "Artificial neural network (ann) based fast and accurate inductor modeling and design," *IEEE Open Journal of Power Electronics*, vol. 1, pp. 284–299, 2020.

[10] G. Miti and A. J. Moses, "Neural network-based software tool for predicting magnetic performance of strip-wound magnetic cores at medium to high frequency," 2004. [Online]. Available: https://api.semanticscholar.org/CorpusID:108519937

[11] H. Saliah, D. Lowther, and B. Forghani, "Modeling magnetic materials using artificial neural networks," *IEEE Transactions on Magnetics*, vol. 34, no. 5, pp. 3056–3059, 1998.

[12] H. Li, D. Serrano, T. Guillod, E. Dogariu, A. Nadler, S. Wang, M. Luo, V. Bansal, Y. Chen, C. R. Sullivan, and M. Chen, "Magnet: An open-source database for data-driven magnetic core loss modeling," in *2022 IEEE Applied Power Electronics Conference and Exposition (APEC)*, 2022, pp. 588–595.

[13] S. Wang, H. Li, D. Serrano, T. Guillod, J. Li, C. Sullivan, and M. Chen, "A simplified dc-bias injection method for characterizing power magnetics using a voltage mirror transformer," *IEEE Transactions on Power Electronics*, vol. 39, no. 6, pp. 6608–6612, 2024.

[14] H. Cui and K. D. T. Ngo, "Transient core-loss simulation for ferrites with nonuniform field in spice," *IEEE Transactions on Power Electronics*, vol. 34, no. 1, pp. 659–667, 2019.

[15] M. J. Jacoboski, A. de Bastiani Lange, and M. L. Heldwein, "Closed-form solution for core loss calculation in single-phase bridgeless pfc rectifiers based on the igse method," *IEEE Transactions on Power Electronics*, vol. 33, no. 6, pp. 4599–4604, 2018.

[16] M. Mu, Q. Li, D. J. Gilham, F. C. Lee, and K. D. T. Ngo, "New core loss measurement method for high-frequency magnetic materials," *IEEE Transactions on Power Electronics*, vol. 29, no. 8, pp. 4374–4381, 2014.

[17] H. Li, D. Serrano, S. Wang, and M. Chen, "Magnet-ai: Neural network as datasheet for magnetics modeling and material recommendation," *IEEE Transactions on Power Electronics*, vol. 38, no. 12, pp. 15 854–15 869, 2023.

[18] S. Wang, H. Kwon, H. Li, Y. Elasser, G.-G. Kang, D. Zhou, D. Grig-oryan, and M. Chen, "Magnetx: Foundation neural network models for simulating power magnetics in transient," in *2025 IEEE Applied Power Electronics Conference and Exposition*, 2025.

Non-Monotonic Influence of DC Bias on Ferrite Core Loss up to 10 MHz with Sine Wave Excitation

1st Bohua Zhang
Chair of Energy Conversion
TU Dortmund University
Dortmund, Germany
bohua.zhang@tu-dortmund.de

2nd Martin Pfost
Chair of Energy Conversion
TU Dortmund University
Dortmund, Germany
martin.pfost@tu-dortmund.de

Abstract—This study provides a detailed exploration of core losses in MnZn and NiZn ferrite cores under DC bias with sine wave excitation, spanning frequency ranges from 100 kHz to 10 MHz. Contrary to traditional models that predict an increase in core loss with the addition of a DC component, our observations indicate a non-monotonic behavior characterized by distinctive turning points. These turning points vary with operating frequency, peak flux density, and material type, highlighting the complexity and variability in core loss. This discovery not only challenges the prevailing core loss models but also lays the groundwork for optimizing the design of magnetic components through more accurate predictions of loss across various conditions. The capacitive cancellation method was employed to reduce the errors caused by phase discrepancies, ensuring the reliability of our measurements. MnZn cores were used for measurements in the kHz range, while NiZn cores were utilized in the MHz range, each showing consistent trends within their respective frequency domains.

Index Terms—core loss, DC bias, NiZn, MHz range, non-monotonic

I. INTRODUCTION

The design of magnetic components has become increasingly important in modern power electronics due to the rising operating frequencies and growing power demands, which necessitate accurate measuring and modeling of the core loss. In particular, the impact of DC bias remains not fully understood. Consequently, numerous studies have been conducted to investigate the influence of DC bias. Mo *et al.* investigated the impact of DC bias on core losses in ferrites, new loss equations based on experimental data were derived for both hysteresis and eddy-current losses [1]. However, the measuring frequency was limited to 100 kHz. Mu *et al.* conducted a finite element analysis to calculate inductor core losses under DC bias conditions. Unlike the commonly used MnZn ferrites, their study utilized low temperature co-fired ceramic (LTCC) ferrite materials for modeling and experimental validation [2]. Stenglein *et al.* developed a core loss model that accurately predicts core losses under arbitrary excitation with DC bias. Their experimental study focused on MnZn ferrites, with measurements conducted over a frequency range of 100 Hz to 500 kHz [3]. Sanusi *et al.* primarily focused on core losses under square wave excitation with DC bias in MnZn ferrites, operating at frequencies from 500 kHz to 3 MHz. Their research demonstrated that DC bias increases core losses

significantly, and they developed models using Steinmetz Premagnetization Graph (SPG) and artificial neural networks (ANN) to predict these losses [4]. In summary, core losses are typically measured under DC bias using square wave excitation up to 3 MHz, while sine wave excitation is generally limited to 1.5 MHz.

Therefore, this study concentrates on measuring ferrite core losses under DC bias conditions with sine wave excitation at frequencies ranging from 100 kHz to 10 MHz. MnZn materials are used for measurements at frequencies below 1 MHz, while NiZn materials are employed for measurements from 2 MHz to 10 MHz. Finally, this study presents new findings on the relationship between core losses and DC bias at room temperature.

II. CORE LOSS MEASUREMENT SETUP

A. Measurement Method

For a typical ferrite core, the phase angle ϕ between the induced voltage and the exciting current is close to 90°, because the impedance of the magnetizing inductance is typically much larger than the equivalent core loss resistance R_{core}. The classic two-winding method for measuring core losses can introduce a significant error due to its sensitivity to phase discrepancies $\Delta\phi$ [5]. If the core is subjected to sine wave excitation, the relationship between percentage loss error and phase angle can be represented as

$$\frac{\Delta P}{P} = \tan\phi \cdot \Delta\phi \times 100\%. \quad (1)$$

To reduce the error caused by $\Delta\phi$, several methods can be taken into consideration. Mu's inductive cancellation method employs an air-core transformer to completely eliminate the reactive voltage on the secondary side [6]. However, obtaining a well-tuned air-core transformer for the core under test (CUT) is challenging, particularly when L_M is large. Hence, Hou's partial inductive cancellation concept is applied to address this problem by mathematically calculating the error produced by the percentage difference between L_M and L_{AIR} [7]. This approach simplifies the selection of the air-core transformer. However, both the calculation and operation become more complicated. In addition, the measurement at high frequencies is not as accurate as Mu's method.

979-8-3315-1612-3/25 $31.00 © 2025 IEEE

In this study, the capacitive cancellation method is employed, which is particularly suited for measuring core losses with sine wave excitation [8]. High-Q capacitors are utilized to bring the CUT into resonance, thereby reducing the phase angle to near zero and effectively eliminating the total reactive voltage. Fine-tuning is achieved by connecting several capacitors in parallel, which allows for the cumulative addition of their capacitance values to reach the desired total capacitance.

The complete measurement system is illustrated in Fig. 1. Sine wave excitation is produced by an RF power amplifier (model 25A250A from Amplifier Research), while the input small AC signal is generated by an arbitrary wave generator (AWG). Two high voltage differential probes (120 MHz bandwidth) are used to measure V_2 and V_3. The DC bias is generated by the DC power supply, which can provide up to 72 A DC current. Instead of using shunt resistors, which introduce noticeable parasitic inductance at high frequencies and consequently lead to considerable error in measurements, two current clamps (100 MHz bandwidth) are selected for measuring i_{DC} and i_{AC}.

The CUTs are wound in a trifilar arrangement, as shown in Fig. 2. Each winding is designated for DC signal inputs, AC signal inputs, and sensing coils, respectively. The trifilar technique ensures the best coupling, 1:1:1 turns ratio, and minimizes intra-winding capacitance. The number of windings is carefully chosen to ensure impedance matching between the amplifier output (50 Ω) and the transformer, allowing for the maximum utilization of the amplifier power.

To obtain a stable DC signal, two chokes are employed to reduce the interference from AC signals arising from coupling between the windings. Additionally, the differential voltage cancellation technique is implemented by employing two identical CUTs with opposite connected DC input windings, further mitigating AC noise. Fig. 3 demonstrates that AC noise is significantly suppressed in i_{DC} .

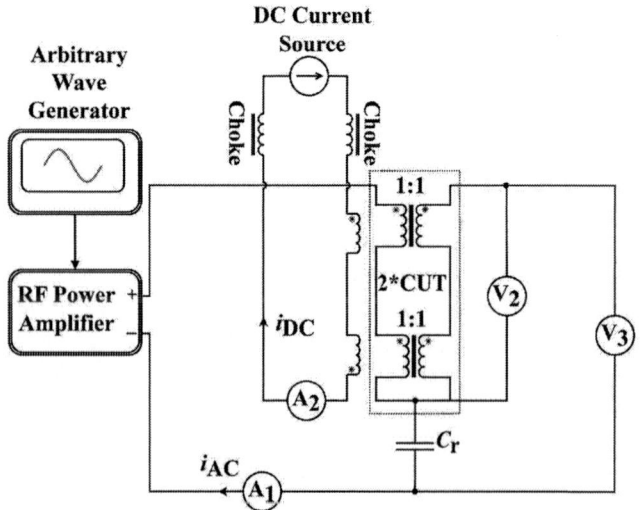

Fig. 1: Schematic of the complete core loss measurement system

Fig. 2: The trifilar-wound CUTs

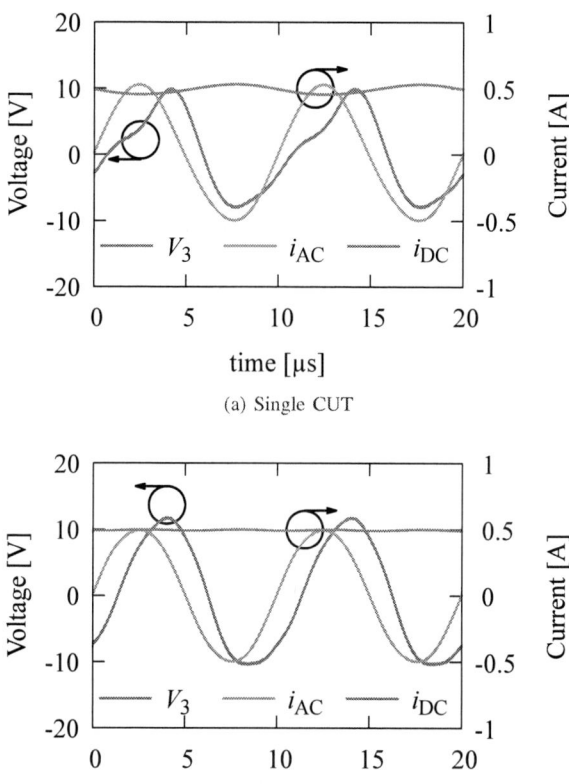

(a) Single CUT

(b) Two CUTs in series

Fig. 3: Typical results for V_3, i_{AC} and i_{DC}, cf. Fig. 1 at $f = 100\,\mathrm{kHz}$, $i_{AC} = 0.5\,\mathrm{A}$, shown for (a) one CUT and (b) two CUTs in series.

To achieve a direct comparison of different materials and geometries, specific core losses P_v are calculated for all CUTs, which can be expressed by integrating V_3 and i_{AC} as

$$P_V = \frac{1}{V_e \cdot T} \int_0^T V_3 \cdot i_{AC}\, dt . \tag{2}$$

979-8-3315-1612-3/25 $31.00 © 2025 IEEE   574

TABLE I: Properties of the CUTs.

Characteristic \ CUT	TDK N87	TDK N49	Fair Rite 61	Fair Rite 67
Base material	MnZn	MnZn	NiZn	NiZn
Permeability μ	2200	1500	125	40
Frequency [MHz]	0.25 - 0.5	0.3 - 1	$\leqslant 25$	$\leqslant 50$
Core shape	R25/15/10	R22/14/8	R33/18/10	R29/19/14
A_e [mm^2]	51.26	32.6	59	69
l_e [mm]	60	54.2	76	73
V_e [mm^3]	3079	1763	4500	5000

The magnetic flux density B, the magnetic field strengths (H_AC and H_DC) can be described by

$$B = \frac{1}{2N \cdot A_\text{e}} \int_0^T V_2 \, dt \tag{3}$$

and

$$H_\text{AC(DC)} = \frac{N \cdot i_\text{AC(DC)}}{l_\text{e}}, \tag{4}$$

where T is the signal period, N is the number of turns, V_e, A_e, l_e are effective core volume, cross-section area and length, respectively.

B. Error Analysis

Besides the error caused by $\Delta\phi$ as discussed above, some other sources of error that affect the accuracy of the experiment are identified below.

- Equivalent series resistance (ESR) of the resonant capacitors lead to considerable error, especially when the equivalent core loss resistance R_core is comparable to the ESR, which can be expressed by

$$\Delta = \frac{ESR}{ESR + R_\text{core}} \times 100\% . \tag{5}$$

To reduce the effects of ESR, ceramic capacitors are used in the measurement, which have low ESR compared to other types of capacitors. Additionally, the desired capacitance value, such as 300 nF, is intentionally achieved by connecting several capacitors in parallel, for example, three 100 nF capacitors, even when a single 300 nF capacitor is available. This configuration significantly reduces the total ESR.
The ESR of an employed capacitor is 200 mΩ, as measured by the impedance analyzer E4990A from Keysight. When three such capacitors are connected in parallel, the total ESR is approximately 70 mΩ. Therefore, the impact of the capacitor ESR can be excluded by

$$P_\text{core} = P_\text{mear} - I_\text{rms}^2 \cdot ESR , \tag{6}$$

where P_mear is the measured core loss in watt, I_rms is the effective current value.
- The experiment is conducted at a controlled temperature of 25°C, achieved using a temperature chamber, therefore reducing the impact of temperature on core losses.

Additionally, all the CUTs have a very small volume, which allows the core losses to consistently remain below a few watts, even under high magnetic flux densities. The excitation period is approximately 1 s, followed by a waiting period of 20 s between tests. These conditions ensure that there is no significant temperature increase during the measurements.
- The propagation delay between the voltage and current probes is measured to be 5.6 ns. The AWG is set to a square wave signal at 10 kHz. Subsequently, the voltage and current are measured using the probes, and the propagation delay is determined by comparing the 50% point on the rising edge of the signal. This delay can be calibrated during data post-processing.

III. Experimental Results

The ferrite materials and core shapes measured in the experiments are summarized in Tab. I. All measured cores were selected to be as similar in size as possible to minimize the error due to volume variations. Initially, the CUTs were investigated using pure sine waves (with i_DC set to zero) and the results were compared with datasheets to verify measurement accuracy. Subsequently, the CUTs were measured under different DC bias conditions.

The measurement results of the four ferrites without DC bias are shown in Fig. 4, demonstrating very good agreement with datasheets. According to (1), the loss error is less than 7% even for a 10° phase discrepancy, when ϕ does not exceed 20°. It should also be noted that the effective permeability μ_eff and reactance increase slightly with higher B. Therefore, the resonant capacitance for a certain frequency is chosen to control ϕ between $-10°$ and $-20°$ at lower B, ensuring accuracy throughout the entire measurement without changing resonant capacitors.

Manufacturers typically provide core loss data only for pure sine signals, necessitating the importance of investigating the impact of DC bias on core losses. The measurement results with DC bias are presented in two distinct ways: at the same frequencies and at the same peak flux densities, as shown in Fig. 5 and Fig. 6.

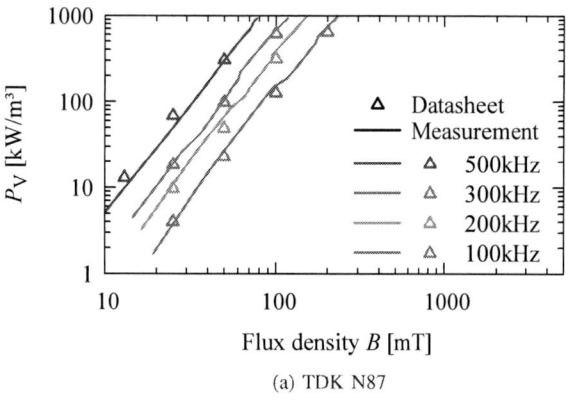

(a) TDK N87

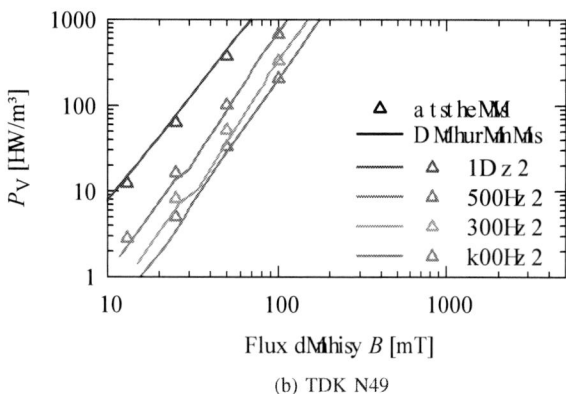

(b) TDK N49

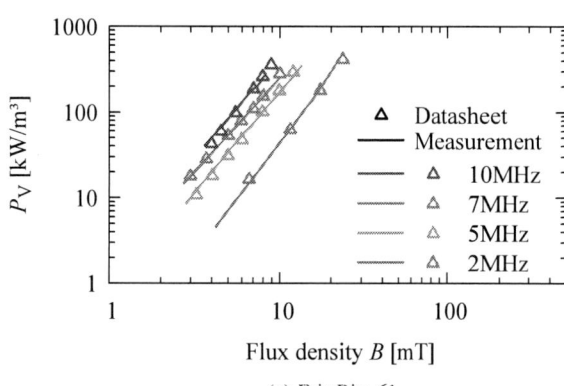

(c) Fair Rite 61

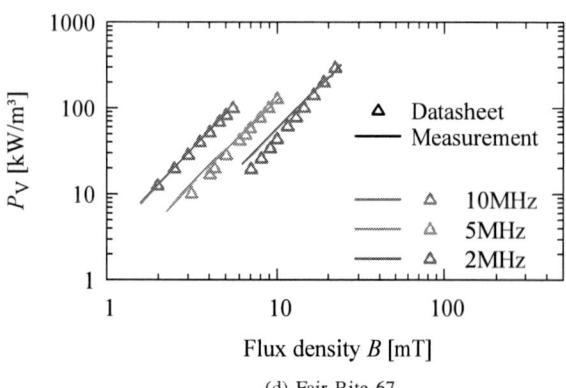

(d) Fair Rite 67

Fig. 4: Comparison of core losses with datasheets.

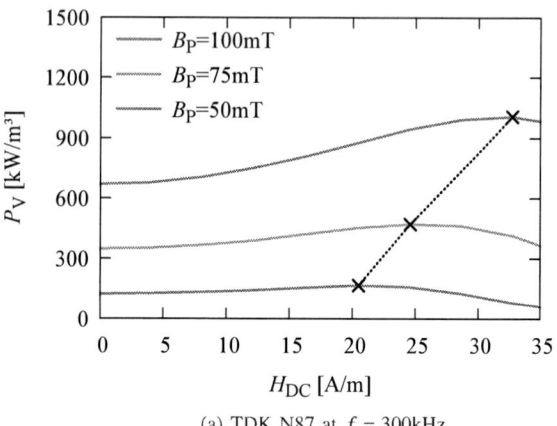

(a) TDK N87 at $f = 300\text{kHz}$

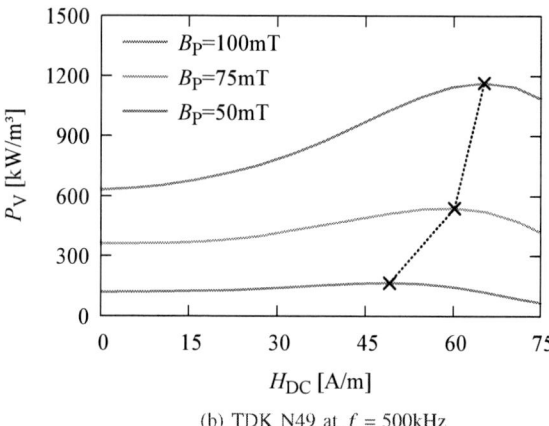

(b) TDK N49 at $f = 500\text{kHz}$

Fig. 5: Core losses versus DC field strength at same frequencies.

Obviously, an increase in H_{DC} does not consistently lead to higher P_v; in fact, the losses begin to decrease after a critical point is reached. This turning point varies with different materials, frequencies f, and peak flux densities B_P, as marked by the black crosses in the figures. Fig. 5 illustrates that the turning point shifts to the right with an increase in B_P at the same frequency. From Fig. 6, similar trends in turning points can be observed at same peak flux densities. These points shift towards higher H_{DC} region as f increases. Notably, they appear at higher levels of H_{DC} for NiZn (low μ) compared to MnZn (high μ) materials.

The increase of the core loss curve under DC bias indicates an expansion in the minor hysteresis loop area. This occurs because the magnetization process becomes harder under certain DC bias conditions due to the local coercive fields impeding domain wall displacements [9]. A possible explanation for the decreasing losses observed at higher DC bias is that the relative incremental permeability μ_Δ degrades with increased DC bias, as calculated by

$$\mu_\Delta = \frac{l_e}{2\mu_0 N^2 A_e} \left(\frac{V_{2,\max}}{2\pi f \cdot i_{AC,\max}} \right). \tag{7}$$

979-8-3315-1612-3/25 $31.00 © 2025 IEEE

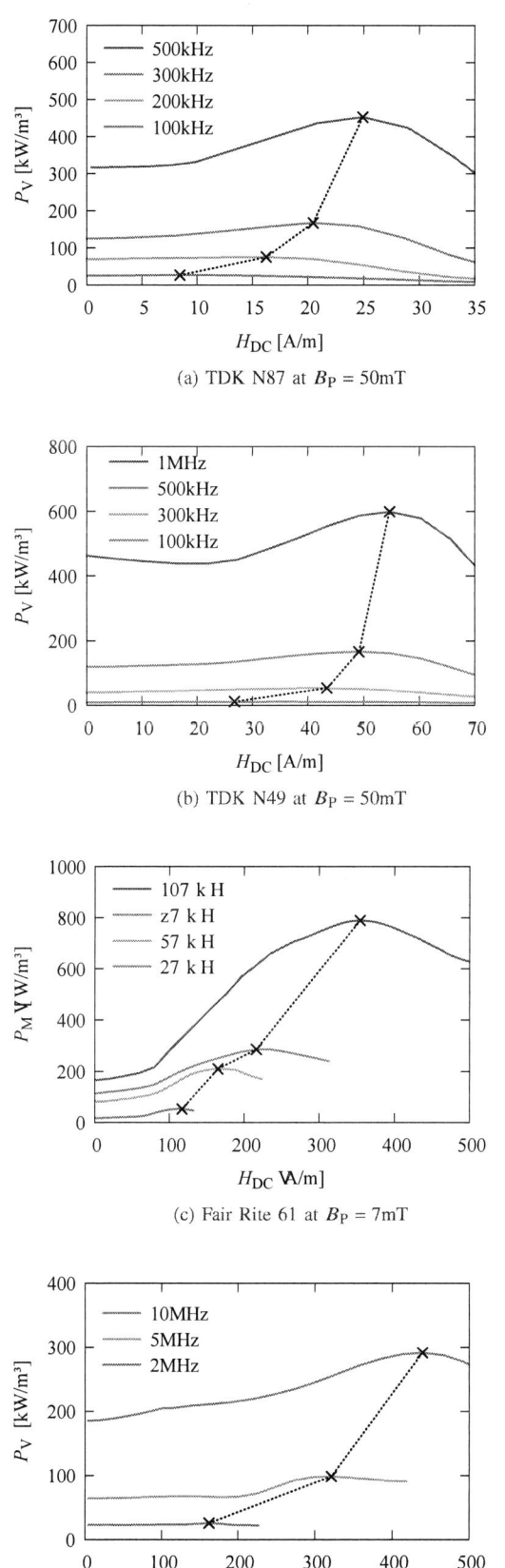

(a) TDK N87 at $B_P = 50$mT

(b) TDK N49 at $B_P = 50$mT

(c) Fair Rite 61 at $B_P = 7$mT

(d) Fair Rite 67 at $B_P = 7$mT

Fig. 6: Core losses versus DC field strength at same peak flux densities.

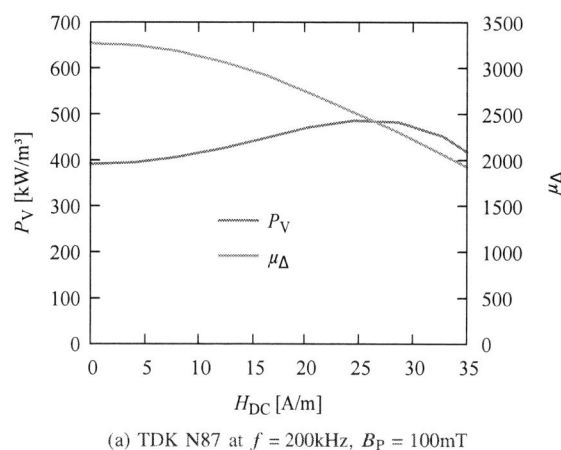

(a) TDK N87 at $f = 200$kHz, $B_P = 100$mT

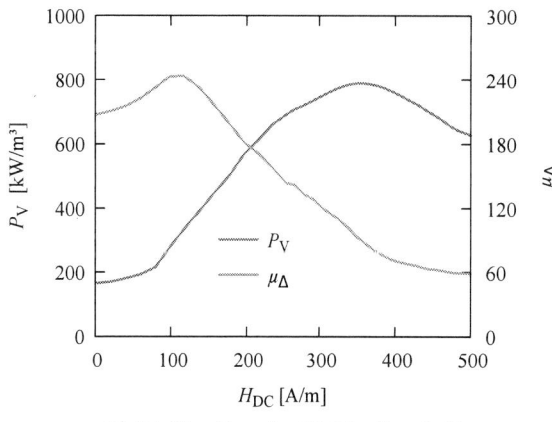

(b) Fair Rite 61 at $f = 10$MHz, $B_P = 7$mT

Fig. 7: Core losses and relative incremental permeability versus DC field strength.

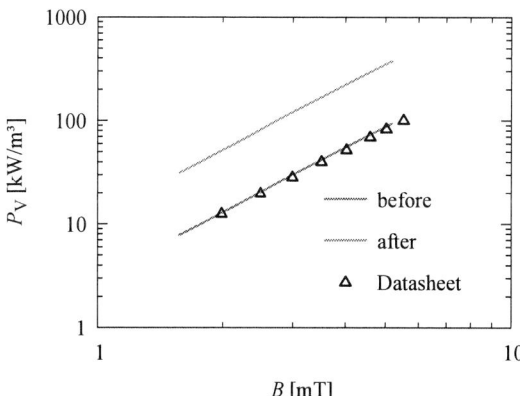

Fig. 8: Change in core loss of Fair Rite 67 before and after the measurement at $H_{DC} > 600$ A/m and $f = 10$ MHz

Fig. 7 shows that μ_Δ decreases with an increase in H_{DC}, indicating more domain walls are aligned to the external magnetic field and the core approaches saturation. It can be inferred that the expansion of the minor loop is constrained by the saturation flux density B_{SAT}.

979-8-3315-1612-3/25 $31.00 © 2025 IEEE   577

Severe damage has been inflicted on NiZn cores due to high magnetic flux densities, as demonstrated in Fig. 8. After the measurement with DC bias, the same ferrite core was measured again with pure sine excitation. This figure reveals an increase in core loss by four to five times, a change that is irreversible. These irreversible deteriorations in the NiZn ferrite cores are mainly due to the increased complexity of domain wall motion, resulting in an irreversible magnetization process and elevated core losses. For a more detailed discussion, see [10].

IV. CONCLUSION

In this study, the impact of DC bias on ferrite cores for both MnZn materials ($\leq 1\,\text{MHz}$) and NiZn materials ($\leq 10\,\text{MHz}$) has been demonstrated. The capacitive cancellation method with differential voltage cancellation technique was applied to the measurements. Turning points in the core loss curves were observed, shifting to the right as both frequency and flux density increased. Additionally, these turning points occur earlier in high-μ materials compared to low-μ materials. Moreover, irreversible core losses change of NiZn materials due to high DC filed strength has been observed.

ACKNOWLEDGMENT

This work was funded by the German Federal Ministry of Education and Research (BMBF), grant 16ME0880. The responsibility for the content of this publication lies with the authors.

REFERENCES

[1] W. K. Mo, D. K. W. Cheng, and Y. S. Lee, "Simple approximations of the DC flux influence on the core loss power electronic ferrites and their use in design of magnetic components," *IEEE Transactions on Industrial Electronics*, vol. 44, no. 6, pp. 788–799, Dec 1997.

[2] M. Mu, F. Zheng, Q. Li, and F. C. Lee, "Finite element analysis of inductor core loss under DC bias condition," in *2012 Twenty-Seventh Annual IEEE Applied Power Electronics Conference and Exposition (APEC)*, Feb 2012, pp. 405–410.

[3] E. Stenglein and T. Duerbaum, "Core loss model for arbitrary excitations with DC bias covering a wide frequency range," *IEEE Transactions on Magnetics*, vol. 57, no. 6, pp. 1–10, June 2021.

[4] B. N. Sanusi, M. Zambach, C. Frandsen, M. Beleggia, A. Michael Jørgensen, and Z. Ouyang, "Investigation and modeling of DC bias impact on core losses at high frequency," *IEEE Transactions on Power Electronics*, vol. 38, no. 6, pp. 7444–7458, June 2023.

[5] J. Zhang, G. Skutt, and F. C. Lee, "Some practical issues related to core loss measurement using impedance analyzer approach," in *Proceedings of 1995 IEEE Applied Power Electronics Conference and Exposition - APEC'95*, vol. 2, 1995, pp. 547–553 vol.2.

[6] M. Mu, F. C. Lee, Q. Li, D. Gilham, and K. D. T. Ngo, "A high frequency core loss measurement method for arbitrary excitations," in *2011 Twenty-Sixth Annual IEEE Applied Power Electronics Conference and Exposition (APEC)*, March 2011, pp. 157–162.

[7] D. Hou, M. Mu, F. C. Lee, and Q. Li, "New high-frequency core loss measurement method with partial cancellation concept," *IEEE Transactions on Power Electronics*, vol. 32, no. 4, pp. 2987–2994, April 2017.

[8] M. Mu, Q. Li, D. Gilham, F. C. Lee, and K. D. T. Ngo, "New core loss measurement method for high frequency magnetic materials," in *2010 IEEE Energy Conversion Congress and Exposition*, Sep. 2010, pp. 4384–4389.

[9] C. A. Baguley, B. Carsten, and U. K. Madawala, "The effect of DC bias conditions on ferrite core losses," *IEEE Transactions on Magnetics*, vol. 44, no. 2, pp. 246–252, Feb 2008.

[10] M. Suzuki, K. Kawano, T. Matsuo, T. Mifune, Y. Uehara, A. Furuya, H. Igarashi, and K. Watanabe, "DC bias effect on the magnetic properties in NiZn ferrite," *Journal of the Japan Society of Powder and Powder Metallurgy*, vol. 61, pp. S245–S247, 03 2014.

Comprehensive SPICE Model for Inductors Considering Magnetic Losses Under DC Bias Current

Yuki Sato
Department of Electrical
Engineering and Electronics,
Aoyama Gakuin University
Sagamihara, JAPAN
yuki-sato@aoyamagakuin.jp

Hirokazu Matsumoto
Department of Electrical
Engineering and Electronics,
Aoyama Gakuin University
Sagamihara, JAPAN
Hmatsumoto@aoyamagakuin.jp

Junichi Kotani
Engineering Division
Panasonic Industry Co., Ltd.
Kadoma, JAPAN
kotani.junichi@jp.panasonic.com

Shohei Tomioka
Engineering Division
Panasonic Industry Co., Ltd.
Kadoma, JAPAN
tomioka.shohei@jp.panasonic.com

Kenichiro Tanaka
Engineering Division
Panasonic Industry Co., Ltd.
Moriguchi, JAPAN
tanaka.kenichiro@jp.panasonic.com

Abstract—**This study presents a SPICE model that accurately represents magnetic loss under DC current conditions in inductors. The proposed circuit enhances the accuracy of frequency response and loss characteristics by introducing a parameter dependent on DC current and a resonance circuit to consider parasitic capacitance. This methodology was validated using a power inductor, revealing precise inductor loss analysis. Additionally, the inductor loss evaluation is compared with the results from a method based on the improved Generalized Steinmetz Equation.**

Keywords—Inductors, magnetic loss, SPICE model.

I. Introduction

Magnetic passive components, such as inductors and transformers, are crucial in power circuits. However, they contribute to significant losses. Therefore, the effective design and analysis of these magnetic passive components are essential for achieving high-efficiency and high-density power electronic circuits.

In the switching power supplies, a triangular wave current flows into the inductors in addition to the DC current I_{dc}. Therefore, it is essential to consider the harmonic components of the triangular wave current and the DC characteristics induced by I_{dc} in the analysis.

A SPICE model (equivalent circuit) integrating Cauer circuits and considering magnetic losses such as hysteresis losses, [1]–[5] has been proven effective in applications across flexible domains. In this circuit, the DC characteristics are inserted into the first-stage inductance of the Cauer circuit, which helps evaluate the circuit behavior and losses due to superimposition. In practical applications, the losses of triangular wave currents have been evaluated through superimposition, demonstrating the possibility of a high-precision evaluation [4]. However, in some instances, the losses may not necessarily match depending on the inductor and excitation conditions.

In this study, we propose a SPICE model that more accurately represents magnetic loss under DC current conditions. We demonstrate its effectiveness by comparing the measured losses under sinusoidal and triangular currents with I_{dc}. Additionally, we applied a method based on the improved generalized Steinmetz equation (iGSE) [6][7] to evaluate the loss under a triangular current with I_{dc}. We then compared these losses with those from the proposed circuit.

II. SPICE Model

A. Conventional SPICE model

First, we provide a detailed description of the conventional SPICE model [4]. To synthesize the conventional SPICE model, the total inductor loss under sinusoidal excitation was assumed to be fitted as

$$P_c = R(f)I_{rms}^2 + Kf^\alpha I_{rms}^\beta, \qquad (1)$$

where I_{rms} is the root-mean-square current, and $R(f)$, K, α, and β are parameters that are determined to fit the losses under sinusoidal excitation. The first term represents the loss related to I_{rms}^2, which corresponds to the copper loss of the coil, including the skin and proximity effects and the eddy current loss in the magnetic material. The second term represents other losses, such as hysteresis and excess loss.

Based on (1), we synthesize the SPICE model shown in Fig. 1. In this model, two Cauer circuits are employed to accurately represent the frequency characteristics of the first and second terms in (1). By extending the SPICE model illustrated in Fig. 1, we could perform loss analysis under arbitrary excitations, such as triangular and square waveforms.

979-8-3315-1612-3/25 $31.00 © 2025 IEEE

To integrate the DC bias characteristic into the SPICE model, the DC characteristic of the inductance ($L\text{-}I_{dc}$ characteristic) was introduced to L_1 and L_{h1}. In fact, the $L\text{-}I_{dc}$ characteristics can be evaluated using commercial instruments and finite element analysis. However, depending on the inductor and excitation conditions, there are cases wherein only $L\text{-}I_{dc}$ to L_1 and L_{h1} yield inaccurate loss-analysis results.

B. Proposed SPICE Model

For a more accurate evaluation of the loss, we propose a new SPICE model, shown in Fig. 2, wherein we modify the multiplier of the current source in the second term of (1) into a function of I_{dc}. This modification is expected to improve the accuracy of loss estimation owing to the enhanced representation capability compared with the conventional SPICE model. Similar to the conventional circuit, the inductances L_1 and L_{h1} in Fig. 2 also possess I_{dc} dependency to express the nonlinearity of the inductor.

Additionally, we introduce a resonant circuit [5] in parallel with the first Cauer circuit to represent the parasitic capacitances that are potentially included in the magnetic device. Since this resonant circuit consists of ladder-type RLC series circuits, multi-resonant characteristics have also been described.

C. Identification of Circuit Parameters

The proposed circuit can simultaneously consider both the magnetic loss under DC bias and the resonant characteristics. Therefore, a wide range of frequency-dependent impedance and inductor loss data under DC bias conditions are required to accurately identify the circuit parameters. In this study, these data were obtained through measurements; however, simulation-based methods such as finite element analysis can also be used to acquire the data.

In the proposed method, we use the following equation to represent the loss with a sinusoidal-wave current when $I_{dc}= 0$ A:

$$P_c = \left(R(f) + K_1 f_1^{\alpha_1}\right)I_{rms}^2 + K_2 f^{\alpha_2} I_{rms}^{\beta}, \qquad (2)$$

where $R(f)$ represents the resistance characteristics measured using a vector network analyzer and an LCR meter. The other parameters, K_1, K_2, α_1, α_2 and β, were identified by fitting the actual loss data at $I_{dc} = 0$A. In this study, we employ a covariance matrix adaptation evolution strategy (CMA-ES) [8] to optimize these parameters.

Upon determining the optimal parameters, the circuit parameters were identified using CMA-ES. In this step, the following optimization problem is solved to synthesize the Cauer circuit for the first term.

$$\boldsymbol{d}_1 = \mathrm{argmin}\left[g_{real}^2(\boldsymbol{d}_1) + g_{imag}^2(\boldsymbol{d}_1)\right], \qquad (3a)$$

$$g_{real}(\boldsymbol{d}_1) = \frac{(R(f) + K_1 f^{\alpha_1}) - \mathrm{Re}[Z_1^{cauer}(\boldsymbol{d}_1)]}{R(f) + K_1 f^{\alpha_1}}, \qquad (3b)$$

$$g_{imag}(\boldsymbol{d}_1) = \frac{X_L(f) - \mathrm{Im}[Z_1^{cauer}(\boldsymbol{d}_1)]}{X_L(f)}, \qquad (3c)$$

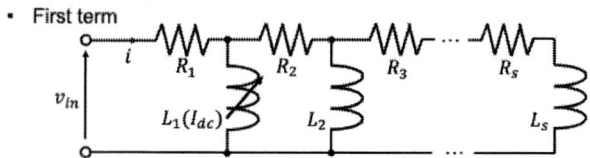

• First term

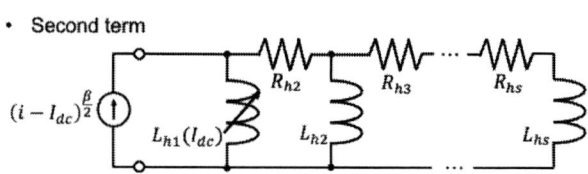

• Second term

Fig. 1 Conventional SPICE model wherein two Cauer circuits are employed to represent the frequency characteritics.

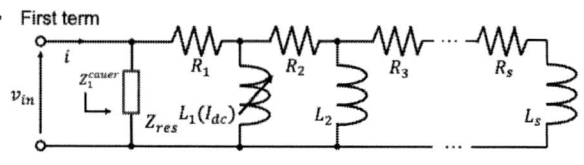

• First term

• Second term

(a) Main circuit

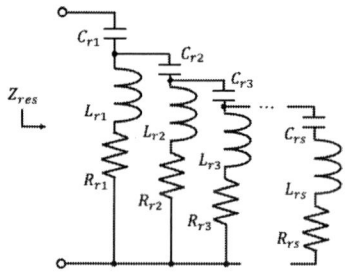

(b) Resonant circuit Z^{res}

Fig. 2 Proposed SPICE model wherein the multiplier β of the current source in the second Cauer circuit is dependent on I_{dc}. Z_{cauer}^1 and Z_{cauer}^2 represent the input impedance of each Caue circuit, respectively.

where $\boldsymbol{d}_1 = [R_1, R_2,\ldots,R_s, L_1, L_2,\ldots,L_s, R_{r1}, R_{r2},\ldots,R_{rs}, L_{r1}. L_{r2},\ldots, L_{rs}, C_{r1}, C_{r2},\ldots,C_{rs}]^t$ and Z_1^{cauer} is the impedance of the Cauer circuit in the first term. $X_L(f)$ is the reactance at frequency f measured by the VNA and LCR meters. To solve (3a) using CMA-ES, we obtain the optimal circuit elements $\boldsymbol{d}$ in the Cauer circuit for the first term.

For the Cauer circuit in the second term, the second source has a multiplier $\beta/2$. Here, we consider the loss under a sinusoidal excitation that is equivalent to the second term in (2).

$$P_2 = \frac{1}{T}\int_0^T R_t |i|^\beta dt = K_2 f^{\alpha_2} I_{rms}^\beta, \qquad (4)$$

where i represents the current passing through the Cauer circuit in the second term. Assuming that the current is $i = \sqrt{2}I_{rms}\cos\omega t$, R_t can be derived as follows:

$$R_t = \frac{2\pi K_2 f^{\alpha_2}}{\int_0^{2\pi}|\sqrt{2}\cos\theta|^\beta d\theta} = \gamma(\beta)K_2 f^{\alpha_2}, \qquad (5)$$

From (5), it can be inferred that coefficient $\gamma(\beta)$ is necessary for the loss equivalency of the second term in (2). Consequently, the optimization problem for determining the circuit parameters for the second term becomes

$$\boldsymbol{d}_2 = \operatorname{argmin}\left[g_{real}^2(\boldsymbol{d}_2) + g_{imag}^2(\boldsymbol{d}_2)\right], \qquad (6a)$$

$$g_{real}(\boldsymbol{d}_2) = \frac{\gamma(\beta)K_2 f^{\alpha_2} - \operatorname{Re}[Z_2^{cauer}(\boldsymbol{d}_2)]}{\gamma(\beta)K_2 f^{\alpha_2}}, \qquad (6b)$$

$$g_{imag}(\boldsymbol{d}_2) = \frac{X_L(f) - \operatorname{Im}[Z_2^{cauer}(\boldsymbol{d}_2)]}{X_L(f)}, \qquad (6c)$$

where $\boldsymbol{d}_2 = [R_{h2},\ldots,R_{hs}, L_{h1}, L_{h2},\ldots,L_{hs}]^t$ and Z_2^{cauer} is the impedance of the Cauer circuit in the second term. Based on the optimized parameters $\boldsymbol{d}_1$ and $\boldsymbol{d}_2$ obtained through the CMA-ES, we can construct the SPICE model under the condition of $I_{dc} = 0$ A.

D. DC Bias Characteristics

After identifying $\boldsymbol{d}_1$ and $\boldsymbol{d}_2$, we introduce the DC bias characteristics. First, the primary inductances L_1 and L_{h1} incorporate the $L\text{-}I_{dc}$ characteristic, which can be measured using commercial instruments. This ensures accurate circuit behavior in power electronic circuits. However, simply introducing DC characteristics into L_1 and L_{h1} may not yield sufficient accuracy for the estimated loss.

To address this issue, multiplier β is modified to $\beta(I_{dc})$, making it a function of I_{dc}. Function $\beta(I_{dc})$ was determined by fitting it to the actual loss measured under DC bias conditions. When determining $\beta(I_{dc})$, the circuit parameters were fixed at $\boldsymbol{d}_1$ and $\boldsymbol{d}_2$, which were optimized under the condition of $I_{dc} = 0$ A.

III. CONSTRUCTION OF SPICE MODEL

A. Test Inductor and Measurement Setup

To validate the effectiveness of the proposed SPICE model, we employ the metal composite-based inductor with dimensions 12.5 x 12.5 x 7.5 mm^3 manufactured by Panasonic Industry Co., Ltd. A bird's-eye view of the inductor is shown in Fig. 3. The frequency characteristics of the impedance measured by the Keysight 4294A are plotted in Fig. 4.

The inductor loss was measured using the capacitive cancellation method [9]–[11], which is suitable for high-frequency loss measurements. Fig. 5 illustrates a schematic of the capacitive cancellation method. In this measurement, a

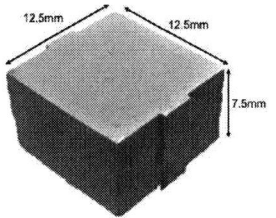

Fig. 3 Test inductor whose size is 12.5 x 12.5 x 7.4 mm^3. The indutor composed of the metal composite material. The inductance is

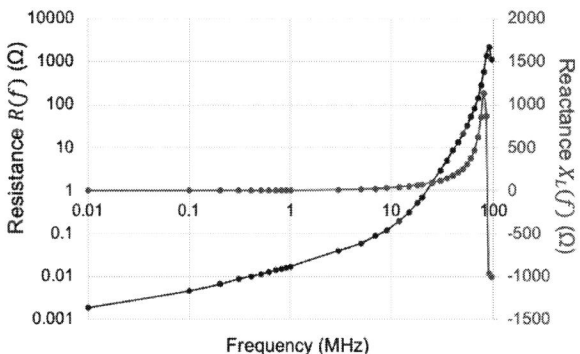

Fig. 4 Frequency-dependent impedance of the test inductor measured by the impedance analyzer (Keysight 4294A), where the black and blue lines represent the real and imaginary parts of the impedance, respectively.

resonant capacitor is connected in series with the inductor (the device under test, DUT) to improve the phase discrepancy between the input current i_1 and excited voltage v_2. At the resonant frequency, the loss can be accurately measured and calculated as

$$P_c = \frac{1}{T}\int_0^T v_3 i_1 dt, \qquad (7)$$

where T denotes the period ($1/f$). In the actual measurement setup, shown in Fig. 6, the matching transformer is directly connected to the amplifier.

B. SPICE Model

First, we identify the optimal parameters in (2) by fitting the actual loss and $R(f)$ data in Fig. 4 without a DC bias. The CMA-ES provided the optimal parameters, which are summarized in Table I. Based on these parameters, we constructed the SPICE model by optimizing (3) and (6). The optimized circuit parameters are summarized in Table II, where R_1 and L_1 correspond to the DC resistance and inductance at low frequencies, respectively. The frequency characteristics of the impedance for each circuit are shown in Fig. 7. The SPICE model yielded highly accurate characteristics that closely matched the resistance obtained from (2).

Next, we introduce this into the DC bias characteristics. As described in Section II, L_1 and L_{h1} incorporate the $L\text{-}I_{dc}$ characteristics. This characteristic was measured by the Kokuyo Electric Co. KC-601 is plotted in Fig. 8, and it exhibits a slight decrease in inductance as I_{dc} increases.

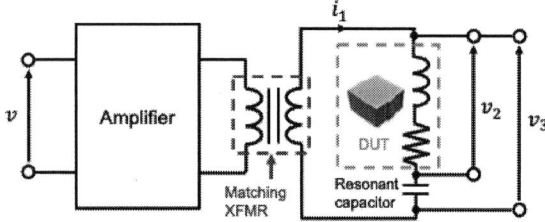

Fig. 5 Schematic of the capacitive cancellation method, in which a capacitor is connected in series with DUT. Loss of the DUT is computed by integrating i_1 and v_3 over a period.

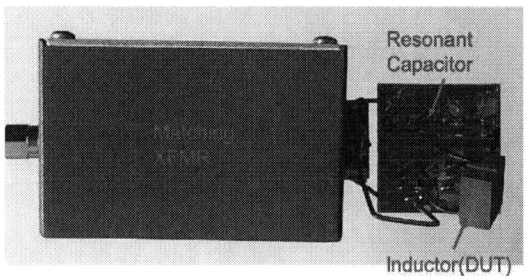

Fig. 6 Measurement setup of the capacitive cancellation method, in which the matching transformer is connected to adjust the impedance discrepancy between the impedance of DUT and amplifier.

Finally, $\beta(I_{dc})$ was determined by fitting it to the actual loss at $I_{dc} = 10, 20, 30$ A, and the resulting curve is illustrated in Fig. 9. This characteristic also exhibits a slight decrease with I_{dc}, indicating a corresponding reduction in loss as I_{dc} increases.

IV. Circuit Analysis Validation

To validate the feasibility of the proposed SPICE model, a circuit analysis in the time domain was performed using LTSpice [12].

A. Sinusoidal Excitation

The response under sinusoidal excitation was analyzed using the proposed SPICE model. Fig. 10 illustrates the analysis results for the case of $I_{dc} = 0$A, where the dots represent the measured results, and the solid line represents the analysis results from the proposed SPICE model. Thus, it can be confirmed that the analysis results closely match the measured results.

Next, a time-domain analysis was performed for the case where $I_{dc} > 0$ A. The analysis results are shown in Fig. 11, where Fig. 11(a), (b), and (c) show the cases with $I_{dc} = 10, 20,$ and 30 A, respectively. The results of the conventional SPICE model, wherein β is not a function of I_{dc}, are also presented. It can be observed that by using the conventional SPICE model, the difference between the analysis and measured results increases with I_{dc}. By contrast, the proposed circuit exhibited a good match with the measurements, confirming the effectiveness of the proposed circuit under sinusoidal excitation.

TABLE I.   Optimized Parameters of (2) for DUT

K_1	α_1	K_2	α_2	β
7.11×10^{-7}	0.682	1.60×10^{-10}	1.29	2.90

TABLE II.   Optimized Circuit Parameters

k	$R_k(\Omega)$	$L_k(\mu H)$	$R_{hk}(\Omega)$	$L_{hk}(\mu H)$
1	0.002	0.555	0.0	0.543
2	4.08	30.2	48.6	372
3	22.3	33.4	249	257
4	465	75.2	989	149
5	5771	50.5	4152	91.4

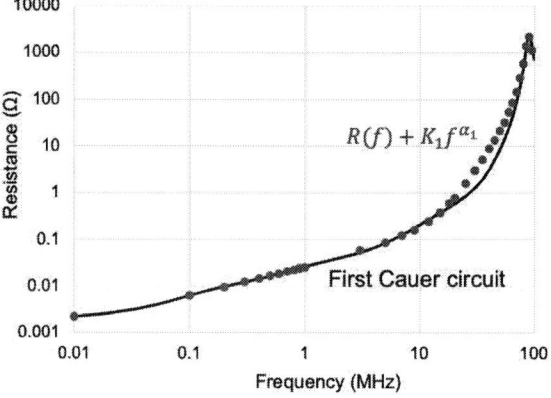

(a)   Comparison of the resisltance related to the fisrt term in (2)

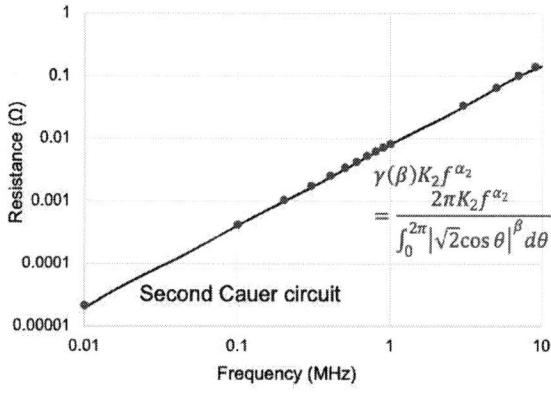

(b)   Comparison of the resistance related to the second term in (2)

Fig. 7 Frequency-dependent of the impedance computed by the constructed Cauer circuit and equation (2), where the black line is computed by the Cauer circuit while the blue dots are computed based on (2) with the optimized parameters.

B. Triangular Excitation

Next, the loss is evaluated using the proposed SPICE model under the triangular current excitation. In fact, for the arbitrary excitation, the improved Generalized Steinmetz equation (iGSE) [6] has been widely used to evaluate the loss of the inductor. Because the iGSE method is extended for the time-domain analysis of the loss from the conventional Steinmetz

979-8-3315-1612-3/25 $31.00 © 2025 IEEE

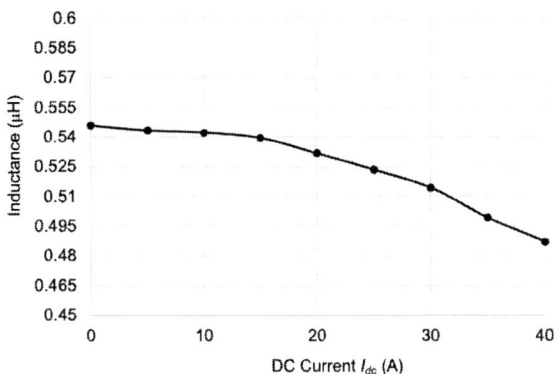

Fig. 8 Inductance characteristc versus DC current I_{dc}, measured by KC-601. The inductance is slightly decreasing with I_{dc}.

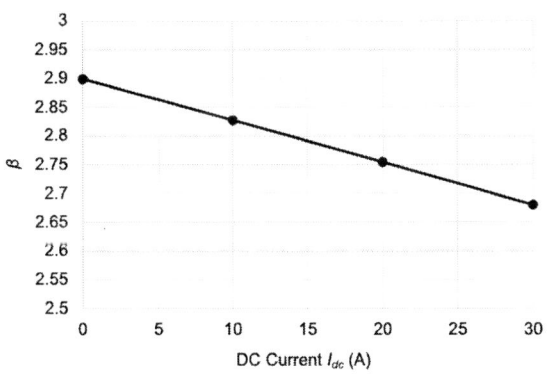

Fig. 9 β against the DC current I_{dc}, which is fitted to the actual loss under the DC bias I_{dc}=10, 20, 30A.

equation, the Steinmetz coefficients are only used to evaluate the loss. Traditionally, iGSE method evaluates the loss according to the magnetic density obtained from finite element analysis. However, in this study, because we already possess the Steinmetz coefficients summarized in Table I, we modified the iGSE to evaluate the inductor to:

$$P_c = \frac{1}{T}\int_0^T k_i \left|\frac{di}{dt}\right|^\alpha \Delta i^{\beta-\alpha}dt, \quad (8a)$$

$$k_i = \frac{K}{(2\pi)^{\alpha-1}\int_0^{2\pi}|\cos\theta|^\alpha 2^{\beta-\alpha}d\theta}, \quad (8b)$$

where K, α, and β correspond to the parameters in (2).

Fig. 12 shows the waveform of the triangular wave applied to the DUT. The measurement with the triangular waveform has been performed with the BH analyzer from Iwatsu Electric Co., Ltd. [13]. The measurement and analysis results for the triangular-wave excitation are shown in Fig. 13. The horizontal and vertical axes in Fig. 12 represent the duty ratio of the triangular wave (T_1/T) and the loss, excluding the DC component, respectively. The current amplitude was set to $\Delta i = 2$ A, and the DC bias currents I_{dc} were (a) 0 A, (b) 10 A, and (c) 30 A. The frequency of each cycle was set to 300 kHz and 500

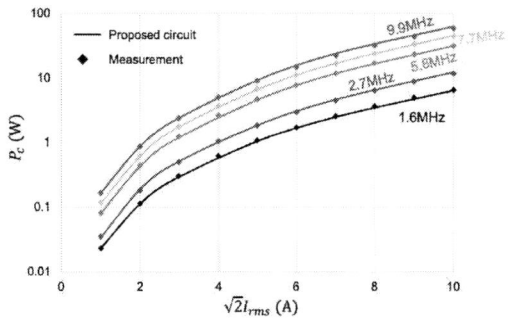

Fig. 10 Comparison of inductor loss under I_{dc}=0A, with lines representing the results from the proposed SPICE model and dots representing measurement results. The different colors indicate various frequencies.

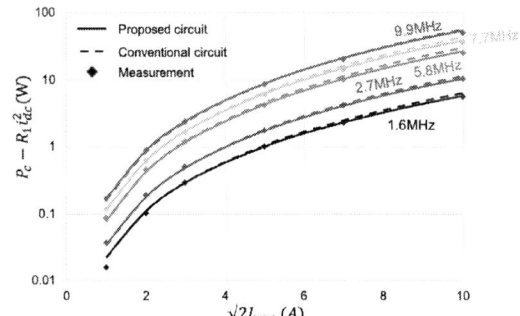

(a) I_{dc}=10A

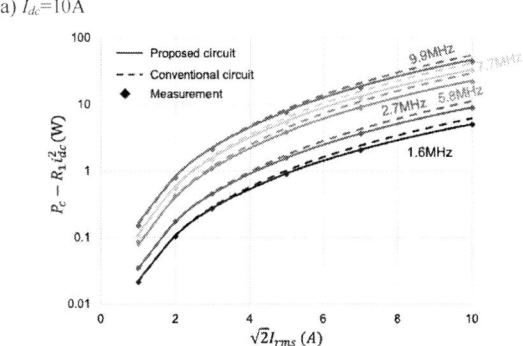

(b) I_{dc}=20A

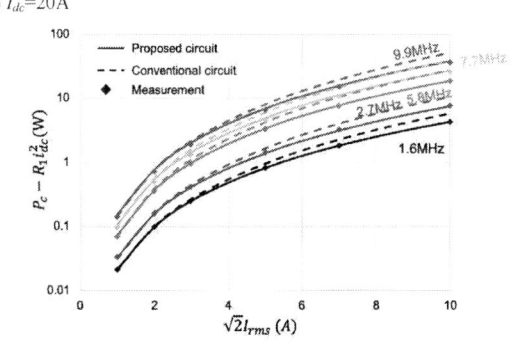

(c) I_{dc}=30A

Fig. 11 Comparison of inductor loss with DC bias current, with lines representing the results from the proposed SPICE model and dots representing measurement results. The different colors indicate various frequencies.

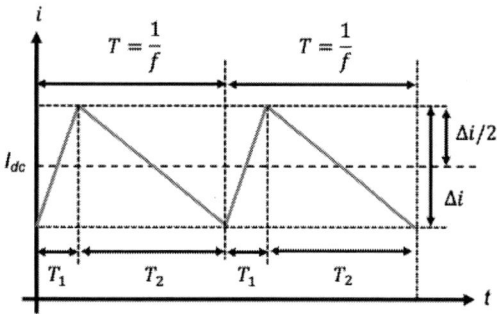

Fig. 12 Waveform of the triangular current which flows into the inductor.

kHz. Fig. 13 confirms that the measurement results closely match the analysis results within the range of the estimation in this study. These results indicate the effectiveness of the proposed method in loss estimation.

Further, it was confirmed that the results of the proposed model matched well with those of the conventional iGSE method. This indicates that the performance of the proposed method is equivalent to that of the commonly used iGSE. However, the iGSE estimates the loss through post-processing after circuit analysis, making it difficult to integrate it directly into SPICE. In contrast, the proposed equivalent circuit can be directly incorporated into the SPICE, allowing the simultaneous evaluation of both converter efficiency and inductor loss, which proves to be a significant advantage.

V. DISCUSSION

In Section IV, the ripple of the input current was fixed at Δi = 2 A. To ensure the comprehensive effectiveness of this method, we varied the amplitude from Δi = 2 A to 8 A at 300kHz with I_{dc}=0A. The current waveforms flowing into the inductor were set to triangular currents with the different duty ratios, as shown in Fig. 12.

Fig. 14 presents the loss results obtained from the measurement, the proposed SPICE model and the method modified based on the iGSE. It was found that the proposed SPICE model and the modified iGSE method closely matched each other across the different amplitudes. On the other hand, the difference in loss between the measurement and the proposed model increases as the ripple current amplitude grows.

To investigate the cause, a breakdown of the loss at a 50% duty ratio is summarized in Fig. 15, where (a) shows the breakdown of the inductor loss and (b) represents the loss ratio between the losses obtained from the 1st and 2nd circuits. The loss is divided into components obtained from the 1st and 2nd circuits, as shown in Fig. 2. The former represents the loss related to I_{rms}^2 which includes DC and AC losses from the coil and eddy current losses in the magnetic material. In contrast, the latter represents the loss due to I_{rms}^β which is considered to include hysteresis and excess losses.

Fig. 15 shows that the loss ratio from the 2nd circuit increases with the current amplitude. This indicates that the proportion of losses related to the magnetic material, such as hysteresis and excess losses, increases with the ripple current amplitude. This suggests that the circuit may not accurately estimate the loss

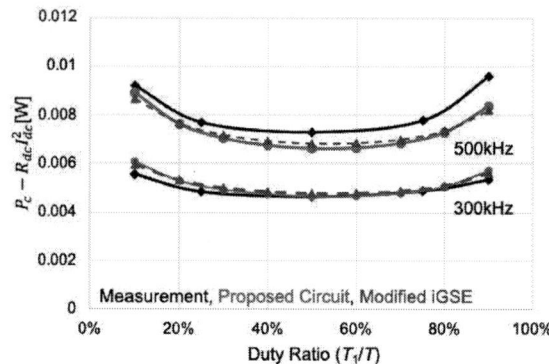

(a) I_{dc}=0A

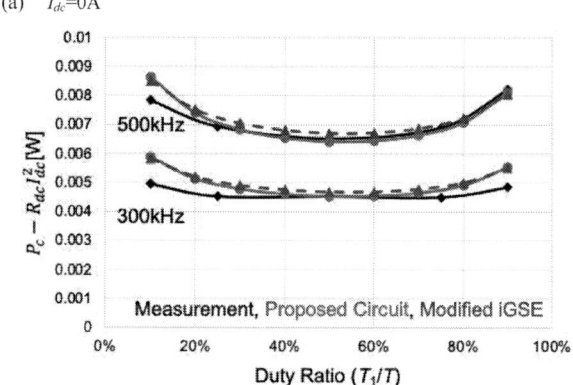

(b) I_{dc}=10A

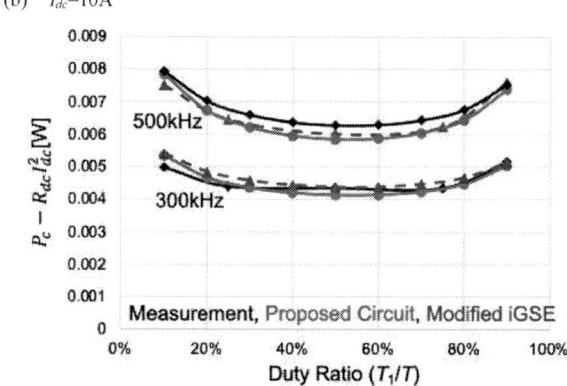

(c) I_{dc}=30A

Fig. 13 Comparison of loss with triangular current at different duty factors. The black, red, and blue lines represent the measurement, the proposed SPICE model, and the modified iGSE, respectively.

when hysteresis and excess losses are dominant. This limitation would also apply when using the iGSE method with a triangular current. Addressing this issue should be a focus of future work.

VI. CONCLUSION

In this study, we developed a comprehensive SPICE model that accurately accounts for the magnetic losses under DC bias conditions in inductors. By introducing a DC-dependent multiplier and a resonant circuit to represent the parasitic capacitance, the proposed model considerably enhanced the accuracy of the frequency response and loss characteristics compared with the conventional model.

Our validation experiments demonstrate that the proposed model closely aligns with the measured loss data for both sinusoidal and triangular current excitations under varying DC bias conditions. Furthermore, the performance of the proposed model was comparable to that of the modified iGSE, which is a commonly used method for loss evaluation. Thus, the model can be directly integrated into SPICE simulations.

ACKNOWLEDGMENT

The authors would like to thank Dr. Yuji Uehara of the Magnetic Device Laboratory Ltd. for valuable support in conducting loss measurements using the capacitive cancellation method.

REFERENCES

[1] W. Cauer, "Die verwirkuchung von wechselstromwiderstiinden vorgeschriebener frequenzabhangigkeit," *Archiv fur Elektrotechnik*, vol. 17, pp. 355–388, 1926.

[2] E. J. Tarasiewicz, A. S. Morched, A. Narang, and E. P. Dick, "Frequency dependent eddy current models for nonlinear iron cores," *IEEE Trans. Power Syst.*, vol. 8, no. 2, pp. 588–597, May 1993.

[3] A. Kameari, H. Ebrahimi, K. Sugahara, Y. Shindo and T. Matsuo, "Cauer Ladder Network Representation of Eddy-Current Fields for Model Order Reduction Using Finite-Element Method," *IEEE Trans. Magn.*, vol. 54, no. 3, pp. 1-4, March 2018, Art no. 7201804, doi: 10.1109/TMAG.2017.2743224.

[4] Y. Sato, K. Kawano, D. Hou, J. Morroni and H. Igarashi, "Cauer-Equivalent Circuit for Inductors Considering Hysteresis Magnetic Properties for SPICE Simulation," *IEEE Trans. Power Electron.*, vol. 35, no. 9, pp. 9661–9668, Sept. 2020, doi: 10.1109/TPEL.2020.2968749.

[5] Y. Sato, K. Kawano, H. Igarashi and H. Matsumoto, "Extended Cauer Equivalent Circuit Model of Inductors: Representing Multi-resonant Characteristics Due to Parasitic Capacitance," *Proceedings of 2023 IEEE Applied Power Electronics Conference and Exposition (APEC)*, Orlando, FL, USA, 2023, pp. 3294-3301, doi: 10.1109/APEC43580.2023.10131598.

[6] K. Venkatachalam, C. R. Sullivan, T. Abdallah and H. Tacca, "Accurate prediction of ferrite core loss with nonsinusoidal waveforms using only Steinmetz parameters," *Proceedings of 2002 IEEE Workshop on Computers in Power Electronics*, Mayaguez, PR, USA, 2002, pp. 36-41, doi: 10.1109/CIPE.2002.1196712.

[7] J. Muhlethaler, J. Biela, J. W. Kolar and A. Ecklebe, "Improved Core-Loss Calculation for Magnetic Components Employed in Power Electronic Systems," *IEEE Trans. Power Electron.*, vol. 27, no. 2, pp. 964-973, Feb. 2012, doi: 10.1109/TPEL.2011.2162252.

[8] Hansen, Nikolaus. "The CMA evolution strategy: A tutorial," arXiv preprint arXiv:1604.00772, 2016.

[9] M. Mu, Q. Li, D. J. Gilham, F. C. Lee and K. D. T. Ngo, "New Core Loss Measurement Method for High-Frequency Magnetic Materials," *IEEE Trans. Power Electron.*, vol. 29, no. 8, pp. 4374–4381, Aug. 2014, doi: 10.1109/TPEL.2013.2286830.

[10] Y. Sato, et. al., "Accuracy Investigation of High-Frequency Core Loss Measurement for Low-Permeability Magnetic Materials," *IEEE Trans. Magn.*, vol. 59, no. 11, pp. 1-5, Nov. 2023, Art no. 6301105, doi: 10.1109/TMAG.2023.3283955.

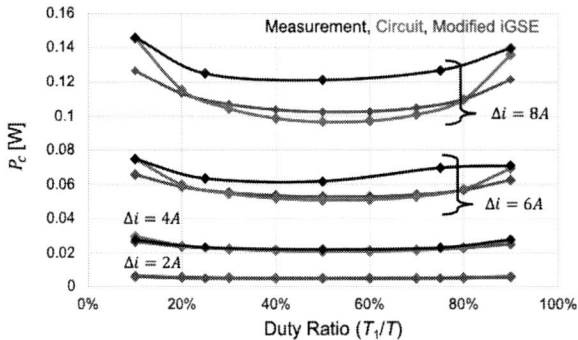

Fig. 14 Comparison of loss with triangular current at different duty factors and amplitudes. The black, red, and blue lines represent the measurement, the proposed SPICE model, and the modified iGSE, respectively.

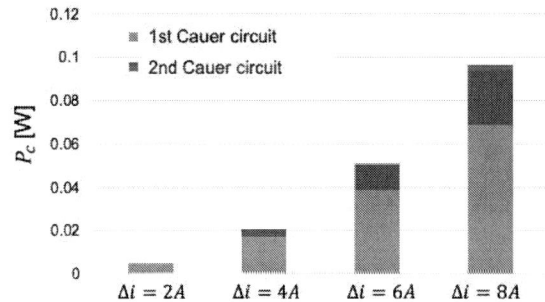

(a) Loss breakdown

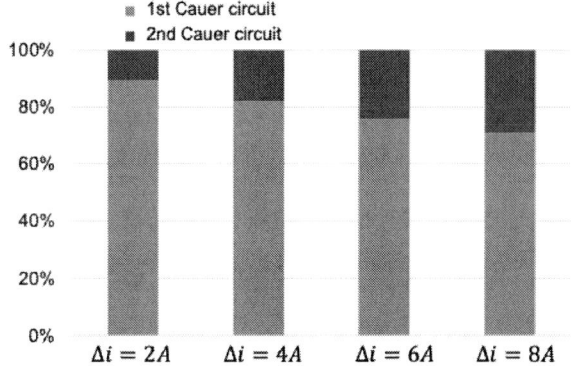

(b) Loss ratio between 1st and 2nd circuit loss

Fig. 15 Breakdown summary of the inductor loss with a different amplitude of the ripple current where the green and purple are obtanied from 1st and 2nd circuits, respectively.

[11] https://www.analog.com/en/resources/design-tools-and-calculators/ltspice-simulator.html

[12] https://www.iwatsu.com/tme/sy/sy8218_top/

AUTHOR INDEX

Abarzadeh, Mostafa 1261
Abbas, Asad 2973
Abotaleb, Youssef 1850
Abrams, Kerry J. 1781
Abramson, Rose A. 291, 2805
Abu-Rub, Omar 3071
Abu-Zaher, Mustafa 2327
Acero, Jesús 2468
Addin, Ali Sharaf 2960
Adeli, Mohammad Hassan 1489
Ademane, Harsha 3133
Adisurja, Ananda Tjakra 1255
Adragna, Claudio 958
Afrasiabi, Seyedeh Nazanin 1279
Afridi, Khurram K. 1640, 1646
Agarwal, Anant 2986
Ahammed, Md Tanvir 2220
Ahmad, Faheem 175
Aider, Youssef 1026
Aiello, Natale 738
Ajmal, Aidha Muhammad 3024
Akamatsu, Keiji 1728
Akter, Tanzila 1844, 2407
Akuta, Hector 761
Alam, Md Didarul 1746
Alam, Muhammad Muneeb 1051, 2569
Alassi, Abdulrahman 3071
Alathamneh, Mohammad 1953
Al-Durra, Ahmed 622, 2871, 3064
Alenezi, Ali 1217
Alexander, Mark 2162
Aleyasin, Seyed Hossein 1408
Ali, Abdelrahman 429
Ali, Jana A. Sheikh 3071
Ali, Kawsar 1529
Alkhatib, Mohamed 1940
Allen, Mark G. 1791
Allgeier, Jan 1919
Allioua, Abdelmoumin 2125
Alou, Pedro 1197
Al-Smadi, Mohammad K. 2779, 2840
Altin, Necmi 1489
Álvarez, Ignacio 3109
Aly, Mokhtar 746, 895, 2290, 2327
Alzahrani, Ahmad 1230
Alzate, Cesar 401
Amano, Yoshiki 3096
Amarathunga, Supun 3030

Amirabadi, Mahshid 1465, 1983
Amitkumar, K. S. 1279
Amler, Adrian 1759, 1767
Amor, Yacine Ayachi 1781
An, Jongchan 3000
Anand, Aniket 1096, 3147
Anand, Sandeep 69
Anantha, Neeraj 1121
Andapally, Bharadwaj Reddy 3119
Andersen, Michael A. E. 246
Anderson, Blake 1850
Ando, Masato 2681
Anekal, Latha 1224
Anjum, Waseah 1148
Antoszczuk, Pablo Daniel 479
Anurag, Anup 9, 442, 1318
Ao, Chengkang 171
Arai, Takamasa 2821
Araki, Hideo 3077
Aravind, G. 1610, 2785
Arduini, Douglas 1159
Asadi, Peyman 2162
Asel, Thaddeus J. 2419
Ashikaga, Toru 2284
Asllani, Besar 2051
Atkinson, Joshua 401
Attanasio, Rosario 3133, 3304
Attukadavil, Jenson Joseph C. 1481
Atwimah, Samuel K. 185, 207
Aunsborg, Thore Stig 175
Avenas, Yvan 1396, 2562
Aygun, Deniz 195
Azzopardi, Stéphane 2718
Bader, Samuel 1681
Bae, Jung-Soo 2228
Bae, Youngmin 3000
Baek, Jaeil 491
Bagci, F. Selin 880
Bahrami-Fard, Milad 930
Bak, Yeongsu 1734
Bakhshai, Alireza 3083
Balakrishnan, Manu 1286
Balamurali, Aiswarya 1096
Balda, Juan C. 27
Balen, Gleisson 1935
Balutto, Mattia 479
Banaie, Amin 1184
Banerjee, Arijit 3089

Bansal, Divyanshu 1610, 2785
Bantemits, Georgios 479
Bao, Mingjun 2628, 2968
Bao, Xiaokun 1143
Barbosa, Peter M. 2002
Barbosa, Peter 9, 442, 1318, 2082, 2296
Barik, Tapas 1184
Barros, Stayner Nóbrega 689
Barzegarkhoo, Reza 90
Basu, Arka 3253
Basu, Shibaji 464
Batard, Christophe 1076
Bau, Plinio 195
Bauer, Pavol 609
Bavi, Danial 385
Bazzi, Ali 2332, 2510
Beckemeyer, Randy 2082
Beig, Abdul R. 2647
Beinarys, Rytis 2009
Belanger, Matthew 2833
Belikov, Juri 1622
Belkhode, Satish 164, 3334
Benson, Mikayla 2413
Bergveld, H.J. 1451
Bertolini, Alessandro 2640
Beura, Kalpana 1940
Beushausen, Steffen 2589
Bezerra, Pedro A.M. 2361
Bhagat, Chinmay 3285, 3291
Bhambay, Rajul 2920
Bhattacharya, Subhashish 370, 552, 1347, 1866
Bhuse, Tejas 1, 54
Biadene, D. 2014
Bian, Fengwei 3312
Bien, Franklin 1629
Blaabjerg, Frede 696, 912, 1501
Blanco, Cristian 1935
Blaquière, Jean-Marc 2718
Blij, Nils Hans Van Der 479
Boby, Mathews 1279
Boisseau, Sébastien 828
Boisson, Guillaume Piquet 1396
Bojoi, Radu 472, 1408
Bolaños, Robert E. 1190
Boles, Jessica D. 1012
Bonanno, Giovanni 1666
Borowy, Bogdan S. 1326
Boroyevich, Dushan 2228
Bosch, Michael 2387
Boutet, Jérôme 828
Bracken, Christopher 544, 3119
Bradford, Paul 2393

Brandão, Danilo I. 1355
Brandão, Dener A. de L. 1355
Briz, P. 147
Briz, Pablo 3109
Brown, Alyssa 231
Brown III, Buck F. 1153
Brückner, Thomas 2960
Bruyere, Paul 2562
Bu, Jiankang 854
Bugade, Vikas 821
Burdío, José-Miguel 2468
Burgos, Rolando 111, 409, 1495, 1551, 2992
Burnett, Hunter 401
Burt, Graeme 3167
Buttay, Cyril 2051
Cairnie, Mark 2228, 2616
Calabretta, Michele 1070
Cammarata, Federica 252
Campbell, Steven 2797
Cao, Hanqing 1810
Cao, Hui 27
Cao, Yue 3036, 3048
Carretero, Claudio 2468
Castro, Alejandro 1427
Catanoso, Matthew 1791
Cattani, Alberto 2640
Cazzaniga, Daniele 958
Cazzitti, Sacha J. 1512
Cerutti, Stefano 738
Cervera, Pedro Alou 788
Cervone, Andrea 1305
Chae, Jongyoon 1899
Chagas, Rafael Bogo Portal 2446
Chakkalakkal, Sreejith 3147
Chakraborty, Shiladri 69
Chambon, Clément 828
Chamorro, Luis Ruiz 788
Chandrasekhar, Nurani 2889
Chang, Che-Wei 1495, 1551, 1564
Chang, Chuan-En 664
Chang, Jun-Yang 16
Chang, Yi-Chun 2143
Chareyron, Mathilde 828
Chatterjee, Bhaskar 1919
Chatterjee, Kallol 821
Chaturvedi, Shivam 2624, 3155
Chaturvedi, V. 1451
Chaudhary, Jai Aditya 3304
Chavarria, Jose 491
Chavez, Fredo 385
Cheema, Muhammad Ali Masood 768
Chellamuthu, Anand 1286

Chen, Cai	2343, 2369, 2426
Chen, Ching-Jan	664, 2131, 2143, 2725, 2735, 2741
Chen, Chun-Yen	938
Chen, Eric	1274
Chen, Guozhu	2059
Chen, Hao	906
Chen, Hongyu	2968
Chen, Hua	518
Chen, Hung-Chi	2687
Chen, Jiahong	1114
Chen, Jiann-Fuh	2043
Chen, Kai-Hui	887
Chen, Kevin J.	1047
Chen, Minjie	139, 349, 510, 566, 1274, 1693, 1882, 2438
Chen, Qiling	2375
Chen, Shih-Gang	938
Chen, Tianxiao	2361
Chen, Ting	2846
Chen, Wanjun	192
Chen, Xi	2628, 2968
Chen, Xingyu	1537, 1741
Chen, Yilun	2535
Cheng, Eric Ka-Wai	3227
Cheng, Jinpeng	1832
Cheng, Kuang-Yao	2157
Cheng, Lin	966, 1687
Cheng, Qi	2236
Cheng, Tzu-Ping	2900
Cheng, Yan	1047
Cheng, Yun-Keng	887
Cheshire, Audrey	682
Chetri, Chandan	3114
Chiu, Huang-Jen	389
Chiu, Jui-Yang	16, 900
Cho, Jaeyong	3187
Choi, Beomseok	491
Choi, Dongho	2311
Choi, Dongmin	1899
Choi, Jinsoo	3187
Choi, Jungwon	1874
Choi, Seokwon	2268
Choi, Seungdeog	943, 951, 1026, 1420, 1858
Choi, Sunghyuk	1659
Choi, Sungjin	3006
Choksi, Kushan	2582
Choo, Vin Loong	1576
Choong, Yin Quen	505
Chowdhury, Vikram Roy	645, 761, 1465, 3059
Chuang, Cheng-Ta	664, 2725
Chung, Henry Shu-Hung	98, 1507, 1582
Ciabattoni, Matteo	1646
Ciardo, S. Yuri	252
Clark, Landon	919
Cobos, Álvaro	1427
Cobos, José A.	1427
Coday, Samantha	971, 2249
Collings, William M.	185
Cong, Yizhou	2986
Contreras-Barrios, René	629
Coomans, Bart	195
Corradini, Luca	1, 54, 334, 2764
Costa, Levy F.	1334, 1341
Costinett, Daniel	3253, 3267
Cox, James	538
Cronin, Jared	2865
Croston, José Andrés Aguilar	2051
Crovetti, Paolo Stefano	738
Cruz, Alfonso	860
Cruz, Mario F.	670
Cui, Hongchang	202
Cui, Wen Tao	1108
Cui, Yujia	2932
Curbow, Austin	1167
D'Amato, Davide	689
Da-Cunha-Alves, Wendell	429
Dai, Hang	3174
Dang, Yongliang	278, 2482
Dannehl, Kai	1774
Dardeer, Mostafa	906
Darvish, Peyman	2453
Das, Shuvangkar Chandra	1184
Datta, Kishalay	1715
Datta, Promit	586
Davari, S. Alireza	2290
De, Vivek	518, 1681
Deboi, Brian	1167
Deboy, Gerald	1444, 2260
Defaz, Samuel	2576, 2582
Dekka, Apparao	2647
Delmar, Aria	1242
Deneke, Niklas	848
Deng, Jianting	3312
Deniz, Erkan	1489
Deppe, Conner	2393
Derbey, Alexis	2562
Desai, Nachiket	1681
Descamps, Anne-Sophie	1076
Deshpande, Ankit Vivek	1459
Dev, Archit	2920
DeVoto, Douglas	1824
Diao, Naizhe	757
Dieckerhoff, Sibylle	2361
DiMarino, Christina	586, 1836, 2228, 2616
Ding, Peiyang	2375

Ding, Wenlong 2713
Divan, Deepak 164, 3334
Do, Huong 491, 1681
Dobakhshari, Sina Salehi 1673
Dominguez, Miguel Alvarez 1640
Dong, Minhai 2075
Dong, ...111, 1495, 1551, 1564, 2992
Driesen, Johan 3124
Driussi, Francesco 479
Dryden, Daniel M. 2419
Du, Bangli 436, 2752
Duan, Bin 2713
Dujic, Drazen 266, 1063, 1305
Dutta, Soham 711
Dworakowski, Piotr 2051
Eguchi, Shinichiro 2828
Ekuewa, Oluwaseun Isaiah 2973
Elasser, Youssef 510, 566, 2438
Elezab, Ahmed 670
El-Fouly, Tarek H.M. 622
Ellis, Nathan M. 2276
Ellis, Philip 1781
El-Refaie, Ayman M. 1551
El-Refaie, Ayman 1230, 1495
El-Saadany, Ehab F. 622
Elsanabary, Ahmed 746
Elshaer, Mohamed 3155
Emadi, Ali 670, 3147
Endo, Shun 2681
Eni, Emanuel 2746
Enjeti, Prasad 727, 1217, 1459, 3054
Enomoto, Jun 3194
Enslin, Johan 1153
Espinar, Alberto 3100
Espinoza, Angel 214
Estrin, Julia 132
Etta, Dheeraj 1640, 1646
Evzelman, Michael 594
Expósito, Alberto Delgado 788, 1803
Fahimi, Babak 930, 3160
Fahmy, Youssef A. 272
Falkenberg, Niklas 2772
Fan, Junchong 1203
Fan, Yucheng 2981
Farantatos, Evangelos 1184
Farivar, Glen G. 1927
Fassi, Youssof 828
Fein, Martin 2348
Feng, Hao 1832
Feng, Kaiyuan 2894
Feng, Wenda 3174
Fernandes, Arnold 1311

Fernandes, Baylon G. 1481
Ferrari, Maximiliano 637
Figueroa, Alejandro 1427
Filho, Braz de J.C. 1355, 1615
Fiore, Michele 1070
Flannery, John 285
Flaten, Paul 682
Forouzesh, Mojtaba 1673, 1892
Forsyth, Andrew J. 1512
Foster, Geoffrey M. 207
Fox, Aidan P. 185
Fox, Matthew 1791
Francés, Airán 868, 3298
Francois, Thomas W. 1311
Frank, Simon 2348
Freeman, Andrew 1242
Fu, Minfan 809, 2846, 3206
Fu, Pengyu 1203, 2986
Fujisaki, Keisuke 1797
Fujita, Jun 1383
Funaki, Tsuyoshi 2813
Funatani, Kenji 2654
Funatsu, Shohei 1237
Furukawa, Akihiko 1383
Gaafar, Mahmoud A. 775, 906, 2327
Gajare, Siddhesh 214
Galamb, Andrew 2527
Gallage, Nirashi Polwaththa 874
Gangadhar, Pratheesh 2920
Gao, Alex 2149
Gao, Ju 171, 225
Gao, Mingze 2992
Gao, Xiang 2846
Gao, Xiaoguang 2070
Gao, Yuan 524, 1034
Gao, Yuntian 278, 2482
Garcia, Enrique 538
García, Pablo 1935
Garcia, Ricardo 214
García, Sofía 3298
García-Espinosa, Antoni 1774
Garza-Arias, Enrique 1459
Gasparini, Alessandro 2640
Gato, Jose 3119
Gautam, Sushanta 185
Gauthier, Jean-Yves 2051
Gauttam, Gaureej 3316
Geboers, Tim 436
Gennaro, Francesco 738
Georgescu, Sorin 180
Georgiev, Daniel G. 185, 207
Gessner, Joerg 1889

Ghanayem, Haneen 1953
Ghartemani, Masoud Karimi 943
Ghitelman, Kolman Puterman 2101
Ghosh, Mohendro Kumar 1326
Ghosh, Prosenjit..2541
Ghosh, Subarto Kumar 1420
Giardine, Francesca 151
Gil, Pablo M. 1701
Ginot, Nicolas 1076
Giuffrida, Simone 472
Gockel, Hendrik..2125
Goetz, Stefan M. 1754, 2846, 3206
Goicoechea, Javier......................................1427
Gomez-Rivera, Luis F. 1774
Gong, Jiakun...219
Gong, Minxiang 518
Gong, Taehyeon 3006
Gong, Xiaowu 1114
Gonzalez, Reynaldo S. 1190
Gonzalez-Castaño, Catalina 629
Goodrich, Dakota...719
Goto, Akiko 1569
Gouy, Louison...1076
Graber, Lukas...................................... 860, 3321
Grainger, Brandon.............................. 544, 1326
Green, Andrew J. 2419
Griepentrog, Gerd 2125
Grigoryan, Davit....................... 566, 1882, 2438
Groon, Fabian 90
Guan, Quanxue..895
Guenther, Robert..1203
Guichon, Jean-Michel.................................2562
Guillod, Thomas 1816
Gunawardena, Pasan 3030
Guo, Heng...2713
Guo, Jiacheng 2375
Guo, Weisheng...3181
Guo, Xiaoqiang...757
Guo, Zhengchen 2703
Guo, Zhongyin 2070
Gurudiwan, Shubhangi 719, 2194
Guthrie, Travis 2162
Gutierrez, Harold 1159
Ha, Jung-Ik 457, 1659, 2268, 2937
Habibolahi, Zahra Sadat 2202
Haddadi, Aboutaleb 1184
Hajisadeghian, Hossein 1666
Halawa, Ali 1473
Hamani, Rachid 1889
Hameed, Aamna Nasir 1673
Hameed, Asad 1972
Han, Yi...103

Hanhart, Michael.......................................2757
Hanna, Rachelle 1396
Hansen, Frederik Lillebæk 2380
Hanson, Alex J. 231, 1121, 2521
Hanson, Alex 77, 2857
Hao, Weijia...2109
Harbi, Ibrahim 895, 2290
Haryani, Nidhi 442
Hasan, Abu Shahir Md Khalid 1844, 2407
Hasan, Md Zakir 1026
Hasan, Syed Imam 1294, 2698
Hassan, Alaaeldien 2327
Hassan, Najam Ul 834
Hassan, Nazmul 1746
Hata, Katsuhiro 1084, 1102, 2284, 2551
Hayashi, Tetsuya 423
He, Bill...3129
He, Binghui..1673
He, JiangBiao 919, 1368
He, Jiayin 171, 225
He, Junlei..3129
He, Xinlong 2066
Heckel, Thomas..1759
Hedenik, Marina 1519
Hedeshi, Hamid Montazeri 2202
Hegde, Anantha...1728
Heinen, Stefan 2757
Heiries, Vincent 828
Heldwein, Marcelo Lobo.............................2446
Hemming, Samuel 670
Heo, Go Woon..1723
Herbert, Edward 2495
Hernandez, Arturo Sanchez..........................530
Herzer, Stefan..1286
Higashiyama, Koji 1728
Hiller, Marc 1919, 2348
Hiraki, Eiji 321, 2654
Hiraoka, Toshio 285
Hirase, Yuko 1946
Hisamochi, Hirofumi..................................1414
Hobart, Karl D.................................... 185, 207
Hoene, Eckart..2361
Hokmabad, Hossein Nourollahi 1622
Hong, Kang...2096
Hontz, Micheal R. 207
Horibe, Masahiro 2821
Hornbuckle, Malachi..........................363, 2241
Horowitz, Logan 151, 2276
Hosani, Khalifa Al............................ 1940, 2871, 3064
Hossain, Md Maksudul 2407
Hossain, Mohammad Safayet.......................1184
Hou, Ting 2375

Hou, Zhengming	1913, 2851
Houska, Brad	3334
Howell, Brandon	2162
Hsieh, Chun-Yu	2735
Hsieh, Hsin-Che	815
Hsu, Jun-Ming	938
Hu, Borong	1439, 2597
Hu, Changsheng	2894
Hu, Changyu	3129
Hu, Jhih-Cheng	2692
Hu, Jiangang	2932
Hu, Shoudong	2764
Huang, Alex Q.	1786
Huang, Cheng	505, 1946
Huang, Hao-Ran	664, 2131
Huang, Ming-Shi	938, 2692
Huang, PengHao	1217
Huang, Peng-Hao	727
Huang, Qinghui	1173, 2603
Huber, J.	2014
Huber, Jonas	1318
Huber, Laszlo	442
Hudgins, Jerry L.	2877
Hudgins, Jerry	1850
Huh, Kum-Kang	3174
Hui, Shu Yuen Ron	3275
Hung, Chien-Chih	16, 900
Hung, Yu-Ting	2735
Huo, Zhenguo	2660
Husain, Iqbal	1746
Husev, Oleksandr	1622, 2173
Hussain, Amir	1990
Hwang, Yun Seong	733
Iannuzzo, Francesco	738, 1070
Ibáñez-Muñoz, Esteban	629
Ibrahim, Ahmed	775
Ibrahim, Eltaib Abdeen D.	775
Ibrahim, Hasan	727, 3054
Ibrahim, Mohamed	670
Ide, Tomoya	1946
Ikriannikov, Alexandr	2149
Iliæ, Milan	2764
Ilka, Reza	1368
Imaeda, Yuta	2431
Imaoka, Jun	2431
Imperiali, Luc	1318
Inokuchi, Seiichiro	2356
Inoue, Shuntaro	782
Irie, Yusuke	2828
Ishido, Ryosuke	1797
Ishihara, Masataka	321, 2654
Ishikura, Yuki	3285, 3291

Ishizuka, Yoichi	2828
Ishraq, Naveed	34, 1135
Islam, Md Khurshedul	943, 951
Islam, Md Majharul	2407
Islam, Nasherul	2059
Islam, Sarwar	1824
Ismail, Ahmed H.	2453
Isobe, Takanori	1946
Ito, Yuki	3248
Itoh, Jun-Ichi	21, 48, 2913
Ivimey, Arjun	464
Iwabuchi, Akio	2828
Iwamoto, Motomitsu	1108
Iyer, Rahul K.	157
Iyer, Vignesh	2764
Jacobs, Alan G.	207
Jafarian, Yousefreza	3083
Jahns, Thomas	3174
Jain, Akshat	658
Jain, Praveen	464, 616, 2953, 3083
Jalakas, Tanel	1622
Jalalabadi, Esmaeil	416
Janabi, Ameer	2597
Jayalath, Sampath	3212
Jeong, Seogyong	834
Jeong, Won Hyo	2937
Jerez, Raiphy	2249
Jha, Kunal	1519
Ji, Shengchang	278, 2482
Ji, Shiqi	2857
Ji, Yichao	966, 1687
Ji, Yingfeng	2889
Jia, Xiaoting	1564
Jiang, C.Q.	795, 3181
Jiang, W.L.	1451
Jiang, Wei	2343, 2426
Jiang, Xi	1114
Jiang, Yang	978
Jiang, Yongbin	3220
Jiao, Dong	1913, 2851
Jiao, Yang	416
Jin, Feng	429
Jin, Liyang	1020, 1564
Jin, Sicong	3018
Jin, Zhiyang	860
Jing, Mengmeng	2713
Jo, Hyeonu	3200
Jo, Hyunkyeong	1629
Jochmans, Thomas	258
Johnson, Brian	711
Johnson, Ken	2510
Jørgensen, Asger Bjørn	357, 1034

Jørgensen, Jannick Kjær 175, 357
Joshi, Kishor 943
Juds, Mark A. 1326, 3119
Jung, Jee-Hoon 834
Jung, Jun-Hyung 689
Jurkov, Alexander 124, 132
Kabashima, Takamune 1728
Kachura, Avram 449, 1905
Kai, Toshihiro 423
Kalathy, Abirami 616, 2953
Kallfass, Ingmar 2387, 3241
Kamalapur, Aakash 2228
Kamran, ... 252
Kanakri, Haitham 2029
Kanathipan, Kajanan 768
Kandeel, Youssef 285
Kang, Byeong-Woo 2948
Kang, Doug. 180
Kang, Eunjin 3012
Kang, Gyeong-Gu 566, 2438
Kang, Seung Hyun 733
Kang, Yong 2066, 2343, 2369, 2426
Kano, Yuko 782
Kanungo, Gautam Dey 821
Kar, Narayan C. 1096
Karanth, Shashank 2746
Karimi-Ghartemani, Masoud 1858
Kataoka, Soya 1237
Katsura, Kenshiro 1299
Kaufmann, Maik 1286
Kawahara, Chihiro 2356
Kawamoto, Keisuke 1569
Kawano, Akihiro 1977
Kelkar, Kapil. 1519
Kennel, Ralph 895
Kerekes, Tamás 738, 3042
Khaburi, Davood Arab 895
Khadka, Purushottam 1040, 2400
Khalid, Saad. 2569
Khalife, Khalil 479
Khan, Faisal 1824
Khan, N. 1451
Khan, Nisar Ahmed 2569
Khan, Shahid Aziz 2624, 3155
Khandelwal, Sourabh 385
Khandla, Dhaval 2920
Khanna, Mudit 854
Khanna, Raghav 185, 207
Khatua, Mausamjeet 1681
Khorasani, Ramin Rahimzadeh 2101
Kim, Byeong-Il 1734
Kim, Chae-Lyn 3200

Kim, Daehyun 3187
Kim, Dong Hwan 1723
Kim, Dongmin 1899
Kim, Han-Gyu 951
Kim, Hongrae 1746
Kim, Hyeon Soo 733
Kim, Jae-Seong 925
Kim, Jaewon 727, 3054
Kim, Jeonghun 761
Kim, Jonghoon 2973, 3000, 3006, 3012
Kim, Jong-Hun 834
Kim, Joon-Seok 1734
Kim, Jungho 1629
Kim, Katherine A. 880
Kim, Minhyeok 3012
Kim, Min-Sik 834
Kim, Myeong-Ho 834
Kim, Namwon 703
Kim, Sung-Oh 2943
Kim, Yura 3006
Kimball, Jonathan W. 1311
Kimpara, Renata 703
Kirtley, James L. 2474
Kisacikoglu, Mithat John 1602
Kishikawa, Ryoko 2821
Kishimoto, Sumiaki 285
Kitano, Junichi 3194
Klidbari, Mohammadreza Khodaparast 2202
Klymenko, Mariia 1590
Knapp, Jeffrey 854
Knappstein, Lukas 2772
Knoll, Jack 2228, 2616
Ko, Bomyeong 3006
Kobayashi, Takumi 3248
Koch, Dominik 2387
Koehler, Andrew D. 185, 207
Koga, Shunsaku 3194
Koga, Takahiro 2828
Kokkonda, Raj Kumar 1347
Kolar, J.W. 2014
Kolar, Johann W. 1318
Kolli, Nithin 1347
Komiyama, Yutaro 3248
Komo, Hideo 2356
Kondo, Hiroki 1102
Kondo, Ryota 2813
Kong, Jiaze 2167
Kong, Jie 2380
Kong, Rui 696
Konishi, Akihiro 3248
Koppolu, Manoj 2920
Korrani, Majid Ghasemi 930, 3160

Kosaka, Takashi	1237, 3096
Koseoglou, Sokratis	479
Kotani, Junichi	579
Kouro, Samir	775, 2327
Kozak, Joseph P.	1211
Kozielski, Kyle	3147
Kragl, Robert	1051
Krishnamoorthy, Harish S.	3316
Krishnamurthy, Harish	1681
Krishnamurthy, Karthik	3129
Krishnan, Sahana	151, 291, 2805
Kritprajun, Paychuda	1184
Ku, Han	900
Kubulus, Pawel Piotr	1034
Kularatna, Nihal	378, 874
Kularatna-Abeywardana, Dulsha	874
Kulasekaran, Siddharth	491
Kumar, Misha	2002, 2082
Kumar, Pavan	530, 3312
Kusaka, Keisuke	3261
Kusunoki, Shigeru	2551
Kutrolli, Uiliam	2332
Kwak, Jin Woong	2541
Kwon, Hyukjae	566, 2438
Kwon, Man Jae	733
Ladhar, Manraj Singh	2322
Laha, Arpan	616, 2953
Lahuerta, Óscar	2468
Lai, Jih-Sheng	815, 1058, 1913, 2851
Lai, Rixin	2138
Lai, Yanwen	1173, 2603
Laird, Ian	2088
Lam, John	768, 2022
Lamar, Diego G.	1701, 1959
Lawniczak, Celine	1129
Lawson, Wayne	1403, 1781
Lazzarin, Telles Brunelli	342
Le, Duc Dung	3155
Le, DucDung	2624
Le, Hoang	2647
Le, Thanh-Long	2718
Leary, Alex M.	2516
Lee, Bonyoung	1629
Lee, Byoung Kuk	733, 1723, 3200
Lee, Byunghun	834
Lee, Chen-Chan	1058
Lee, Dongcheol	3000
Lee, Dong-Choon	1267
Lee, Dongsu	457
Lee, Eun Woo	2311
Lee, Hoi	2236
Lee, Jaea	3012

Lee, Jaehyeong	3006
Lee, Ju-A	3200
Lee, Jun Young	2311
Lee, June-Seok	1734, 2311
Lee, Justin	2138
Lee, Juwon	457
Lee, Kahyun	2937
Lee, Kangbeen	2413
Lee, Kevin	327, 1261, 2907
Lee, Kyo-Beum	925, 2317, 2943, 2948
Lee, Kyungmin	2547, 3281
Lee, Miyoung	3000
Lee, Po-Chang	900
Lee, Seongkyu	3012
Lee, Seunghyun	3012
Lee, Sungjun	3006
Lee, Taewoo	1659
Lee, Ting-Lun	2143
Lee, Wen-Hsuan	2043
Lee, Woongkul	1473, 2413
Lee, Yun-Jin	2317
Lehman, Brad	761, 1465, 1983
Lehmeier, Thomas	1767
Lei, Weihao	1143
Lei, Yiming	3312
Leslie, Alec	401
Leyrer, Thomas	2920
Li, Bing	2932
Li, Chun-I	2741
Li, Duo	307
Li, Haoran	510, 566, 1882, 2438
Li, Heyuan	809, 2846, 3206
Li, Hui	1248, 2075
Li, Jiajun	1590
Li, Lingyun	524
Li, Peidong	3036, 3048
Li, Pengwei	2332
Li, Qiang	202, 299, 429, 498, 1433, 1537, 1557, 1741, 2228, 2488
Li, Ruqi	1159
Li, Sichao	3129
Li, Tien-Sheng	111
Li, Xiang	3312
Li, Xiaoling	1824
Li, Xindong	3212
Li, Xinze	1143
Li, Xuewen	751
Li, Yang	2576
Li, Yanqiao	1590
Li, Yaohua	3220
Li, Yi	1153
Li, Yilei	2035

Li, Yiming	1173	Liu, Wentao	1090
Li, Yuan	761, 1465	Liu, Xiaosen	1544, 2556, 2675
Li, Yunwei	3030	Liu, Xiaoshan	429
Li, Zehui	485	Liu, Y.	1451
Li, Zhenchao	1305	Liu, Yan-Fei	1673, 1892
Lian, Zhina	1090	Liu, Yang	2675, 3321
Liang, Gaowen	1927	Liu, Yifu	3129
Liang, Jingyuan	1108	Liu, Yongjie	1501, 3042
Liang, Katherine	363, 2241	Liu, Yu-Chen	2179
Liang, Tsorng-Juu	887	Liu, Yunting	1179
Liang, Yaogan	1084	Liu, Zeguo	1687
Liao, Hong-Xuan	2692	Liu, Zengyang	2488
Liao, Hsuan	2043	Liu, Zhan	2521
Liao, Kuo Fu	2043	Liu, Zhanlei	278, 2482
Liao, Mian	139, 349, 1882	Liu, Ziheng	171, 225
Libbos, Elie	3089	Locher, Fabrice	2495
Lim, Gyu Cheol	2937	Locke, William	3141
Lim, Je-Yeong	1723	Lodge, Finlay	3167
Lim, Jong-Hun	1723	Logi, Sean	880
Lin, Fanfan	1143	Long, Haihong	651, 2981
Lin, Jesse	1211	Long, Teng	1439, 2597
Lin, Jinshu	2075	Loparo, Kenneth A.	2698
Lin, Lei	2535	Lope, Ignacio	2468
Lin, Qing	409	López, Abraham	1959
Lin, River	1159	Lopez-Torres, Carlos	1774
Lin, Wei-Ren	258	Lu, Che-Yu	1967, 2900
Linares, Daniel Ríos	1375	Lu, Fengwang	98
Liserre, Marco	90, 118, 689, 1148	Lu, Guo-Quan	586, 2228
Liske, Andreas	2348	Lu, Lucas	416
Liu, Baihan	2343, 2426	Lu, Mowei	1754, 2846, 3206
Liu, Caifeng	2066	Lu, Wei	2117
Liu, Chen	3042	Luan, Shaokang	1034
Liu, Chien-Lung	2692	Lucía, O.	147
Liu, Ching-Yao	1058	Lucía, Óscar	3109
Liu, Christopher	1403	Luckett, Benjamin	919
Liu, Chun-Hung	1026	Luise, Claudio	2640
Liu, Gao	357, 1034	Lukic, Srdjan	2527
Liu, Hanbing	3232	Lumod, Phen	1159
Liu, Haoyang	2361	Luo, Fang	2576, 2582
Liu, Hong	2634	Luo, Tianming	2035
Liu, Hongru	2675	Lv, Jianwei	2343, 2369, 2426
Liu, Hualong	1363, 1597	Ma, D. Brian	2541
Liu, Jia	751	Ma, Dingkun	2375
Liu, Jiahong	3042	Ma, Guangji	2070
Liu, Jiaxin	2343, 2369, 2426	Ma, Hangxiao	978
Liu, Jinjie	1114	Ma, Tianlu	3181
Liu, Jinjun	751	Ma, Zhedong	1173, 2603
Liu, Kevin	1274	Ma, Zhiyuan	1786
Liu, Liming	1746	Ma, Zhuxuan	27
Liu, Ming	2521	Maaz, Syed Mohammad	1267
Liu, Sijia	2369, 2426	Mabuchi, Yuuichi	2681
Liu, Wen-Chin B.	315	MacFadyen, Martin	3167

Madadi, Mehrnaz 370
Maddela, Avinash 1715
Maekawa, Sari 2924
Mahbub, S. Tahmid 157
Maheshwari, Anuj 3089
Maji, Sounak 1640
Major, Joshua 1824
Mak, Pui-In 978
Maksimoviæ, Dragan 1, 54, 334, 682, 2764
Malannino, Claudia 252
Mallik, Ayan 34, 1135
Mallik, Ranajay 658
Mandrile, Fabio 472
Manjrekar, Madhav 2883
Mannan, Tahmid Ibne 1420, 1858
Manos, Konstantinos 1274, 1693
Mansour, Mahmoud 719
Mantooth, H. Alan 1844, 2407
Manzoni, Stefano 958
Marcault, Emmanuel 1396
Marellapudi, Aniruddh 164, 3334
Marianne, Julien 828
Marin, Brandon 491
Marquardt, Rainer 2960
Martin, Alexander 1211
Martin, Sébastien 828
Martin, Trent 1, 54
Martinez, Wilmar 238, 258, 436, 2167, 2752, 3124
Martinez-Limia, Alberto 1051
Martins, João R.R.O. 1889
Martins, Rui P. 978
März, Martin 1759, 1767
Mather, Barry 645, 3059
Mathieu, Frédéric 2495
Mathúna, Cian Ó. 285
Matiushkin, Oleksandr 1622, 2173
Matsumori, Hiroaki 1237, 3096
Matsumoto, Hirokazu 579
Matsumoto, Yohei 2681
Matsuo, Takayoshi 2932
Mattavelli, P. 2014
Mattavelli, Paolo 2667
Maureira-Riquelme, Ángel 629
Mauromicale, Giuseppe 1070
Mavencamp, Dan 2157
Mazariegos, Pablo 1427
Mazzer, Simone 1444, 2254, 2260
McDonald, Brent 42
McGrew, Tyler 1557
Mekhilef, Saad 746
Mendes, Arthur 2992
Mercier, Patrick P. 315

Metwly, Mohamed Y. 919
Meyer, Stefan 1034
Miao, Honglei 2059
Michelis, Stefano 479
Milivojeviæ, Nikola 1, 54
Min, Hao 1090
Min, Hyungki 1629
Min, Run 2109
Minato, Yuichiro 2913
Mirafzal, Behrooz 2461
Mirkoviæ, Nikola 788
Mishima, Taichi 3248
Mishra, Santanu K. 2213
Mitcheson, Paul D. 1653
Mitrovic, Vladimir 409
Mitsui, Koji 1299
Miyamae, Masaki 2681
Miyanjou, Kazuki 1977
Miyazaki, Tatsuya 1797
Mo, Liping 795
Mo, Xianghao 1375
Mohammad, Mostak 1635
Mohammadi, Sajjad 2474
Mohseni, Parham 2173
Moniruzzaman, Md 943, 951, 1420
Montejano, Misael 637
Monticone, Francesco 1646
Montoya-Acevedo, Diego 629
Moon, Gun-Woo 1899
Moon, Jinyeong 1473, 2413
Moorthy, Radha Sree Krishna 2797
Morris, Lauryn 1311
Moschopoulos, Gerry 1972
Motoori, Shuichiro 1977
Motto, Eric 2356
Mou, Di 2857
Mou, Shin 2419
Mounesi, Reza 1791
Moursi, Mohamed Shawky El 2871, 3064
Mousavi, Mahdi S. 2290
Mu, Qiang 1388, 2790, 3328
Mu, Wei 2597
Mu, Xuchu 978
Muduli, Utkal Ranjan 1940, 2871, 3064
Mueller, Lukas 538
Muenz, Ulrich 1184
Mühlethaler, Jonas 2495
Mujica, Gabriel 868, 3298
Mukhopadhyay, Anwesha 3267
Mukunoki, Yasushige 2356
Müller, Kilian 3241
Mulumudi, Guru Abhilash 1135

Munk-Nielsen, Stig 175, 357, 1034
Murakami, Haruhiko 1569
Muravleva, Ekaterina 1850
Murillo-Yarce, Duberney 1959
Murray, Samantha K. 1905
Murukesan, Karthick 180
Muscat, Isaac 449
Musolino, Francesco 738
Mustakin, Zaheen 1388, 2790
Na, Woonki 2973, 3000, 3006, 3012
Nabila, Kashfia Tajmim 2877
Nabizadah, Ahmad 3160
Nag, Kumar Joy 990, 997
Nagahara, Teruaki 1569
Nagai, Yoshiyuki 423
Nagano, Masanori 285
Nagar, Anshul 2973
Nagasawa, Shinobu 2610
Nagayoshi, Kenichi 1102, 3096
Nakagaki, Akito 2654
Nakagawa, Shigeki 1797
Nakamura, Hirokazu 1728
Nakamura, Keisuke 1237
Nakano, Satoshi 2551
Nakashima, Junichi 2356
Nakata, Yosuke 1383
Nakata, Yuki 21, 2913
Nam, David 2992
Namadmalan, Alireza 2474
Namba, Akira 1797
Namburi, Krishna 1294
Naradhipa, Adhistira M. 498
Narasimhan, Sneha 1866
Narumanchi, Sreekant 1824
Nasiri, Adel 1489, 1791
Nassaji, Abolfazl 2290
Nassar, Rajaie 586, 2228
Nations, Mark 552
Naval, Sourav 1012
Navarro-Rodríguez, Ángel 1935
Neal, Adam T. 2419
Nelms, R.M. 1953, 2703
Nelson, Blake 395, 1167
Nelson, Tolen M. 207
Nelson, Tolen 185
Ng, Wai Tung 983, 1108
Ngo, Khai D. T. 2228
Ngo, Khai 586
Ngo, Minh 111
Nguyen, Allen T. 840
Nguyen, Calvin 1274
Nguyen, Duy T. 231, 1121

Nguyen, Hien 1, 54
Nguyen, Kien 3248
Nguyen, Tung-Tan 389
Ni, Chuan 2117
Nielsen, Morten Rahr 357
Nikmaram, Behnam 2290
Ning, Guangdong 809
Ning, Guangfu 2096
Ning, Shangxian 2660
Nishijima, Kimihiro 1977
Nishimura, Keigo 3096
Nishio, Haruhiko 1108
Nishizawa, Shin-Ichi 2551
Nitta, Honami 1797
Noesges, Brenton A. 2419
Noguchi, Koichiro 1569
Noh, Young-Seok 518
Norman, Patrick 3167
Notake, Koki 1299, 1414
Núñez, Guillermo 1197
Nuzzo, Jeremy 2387
O'Driscoll, Seamus 285, 2009
Oberdieck, Karl 1051, 2589
Oboreh-Snapps, Oroghene 1311
Ochiai, Yuki 3261
Ohi, Toshi 2821
Ohno, Takashi 21
Ohodnicki, Paul R. 544, 2516, 3119
Ohodnicki, Paul 370, 1326
Okamoto, Takahiro 321
Olalla, David 3100
Olimmah, Marshal 395
Onar, Omer C. 1635
Onishi, Hiroyuki 2431
Onuma, Naoto 2681
Opificius, Julian 401
Orabi, Mohamed 775, 906, 2327
Orlando, Tailan 342
Orr, Allison 1211
Oruganti, V.S.R.Varaprasad 801
Ota, Hiroaki 3248
Ou, Shuyu 1501, 3042
Ouyang, Ziwei 246, 252, 1810
Pahlevani, Majid 616, 2953
Pakala, Sriharsh 505
Palani, Praveenkumar 62
Pallantla, Manikanta 2708
Palmal, Manas 1874
Pan, Ci 192
Pan, Qishan 2207
Panja, Pijush Kanti 821
Paplham, Tyler W. 2516

Parashar, Sanket...1347
Paredes-Camacho, Alejandro.........................1774
Park, Junhyeong..3187
Park, Sung-Bum...3187
Parkhideh, Babak..................................1388, 2790
Parreiras, Thiago M.............................1355, 1615
Pasupuleti, Sai Sushma.................................3316
Patle, Nagesh..2805
Paul, Sayan..1, 54, 334
Paulino, Glaucio H..1274
Pavone, Mario Giuseppe................................738
Peña-Alzola, Rafael..........................1935, 3167
Peng, Fang Z..761
Peng, Hongjie.....................................171, 225
Peng, Xiaochuan..1090
Penof, David..1519
Pereira, Joao..637
Pereira, Lucas.....................................1388, 2790
Pereira, Thiago Antonio..................................118
Peretz, Mor Mordechai....................................594
Pérez, Fernando.........................868, 3298
Pérez, Sara..1197
Perez-Farre, Quirc...1774
Perreault, David J..132
Perreault, David...124
Petriæ, Ivan Z...157
Petriæ, Ivan...2764
Petrillo, Gaia..266
Petucco, Andrea..2667
Pfost, Martin.............. 573, 1129, 1576, 2772
Philippe, Antoine..1396
Phukan, Ripun.....................................2082, 2296
Phung, Thanh Hai...195
Picot-Digoix, Mathis.....................................2718
Piel, Joshua J..2419
Pietrini, Giorgio..670
Pigott, J..1451
Pilawa-Podgurski, Robert C. N................151, 157, 291, 558,
....2276, 2805
Pillonnet, Gaël...315
Pirson, Nicolas...258
Pizzuto, Matteo..1096
Plum, Thomas..1919
Pong, Man-Hay...389
Pool-Mazun, Erick..1459
Popoviæ, Zoya..682
Porras, David A..27
Porter, Matthew.....................................1020, 1564
Pou, Josep..1927
Pourjafar, Saeid...2173
Prabhakar, Siva...69
Pradhan, Rachit...670

Prakash, Surya...1940
Preindl, Matthias.......................272, 1255
Prodiæ, Aleksandar...................307, 990, 997
Punjabi, Shobhana...1159
Qahouq, Jaber A. Abu.....................2779, 2840
Qi, Nianzun......................................357, 1034
Qian, Ting..2117
Qian, Yijie..524
Qiblawey, Yazan..3071
Qin, Yuan..1564
Qin, Zian...609
Qiu, Tian...1040, 2400
Queiroz, Samuel S...............................1334, 1341
Quenette, Vincent..1889
Rabenold, Elizabeth......................................2249
Radhakrishnan, Kaladhar....................491, 1681
Radici, Christian...................1403, 1512, 1781
Rafiq, Aamir...395
Rahman, Md Rashedur.........................943, 951
Rahman, Mohammad Dehan..............1844, 2407
Rahouma, Ahmed...27
Rajagopal, Narayanan...................................1836
Rajpurohit, Chirayu.......................................2764
Raju, Soniya...378
Rallabandi, Vandana...........................1635, 3174
Ram, Achala...2920
Ramasubramanian, Deepak...........................1184
Ramirez, Juan..1211
Ramkumar, S..2708
Ramos, Gabriel V..................................1355, 1615
Ramos, Regina.....................................1197, 1375
Ran, Li...1832
Rana, Dilip...1040
Rana, Mandeep S..2213
Rao, Yifan..1274
Rashid, Syed Saeed...............................1640, 1646
Rathore, Vikas Kumar.....................................594
Raval, Vishwam.......................................727, 3054
Ravichandran, Krishnan.................................1681
Rawat, Shubham..1347
Raychowdhury, Arijit.......................................518
Reddy, Narsimha...3054
Redondo, Alejandro...........................868, 3298
Reinotas, Jurgis...1754
Ren, Linhao..2343
Ren, Sheng..3181
Ren, Xufu...1439
Restrepo, Carlos...629
Rettner, Cornelius..1759
Richardeau, Frédéric.....................................2718
Rikiishi, Yasuhiro..2284
Ripamonti, Giacomo......................................479

Ristic-Smith, Aleksandar 1529
Rivas-Davila, Juan 363, 2241
Rizkalla, Maher 2029
Rizzolatti, Roberto 1444, 2254, 2260
Roberts, Gianluca 307
Rodgers, Aidan 1242
Rodriguez, Ezequiel Ramos 1927
Rodriguez, Fernando 3100
Rodriguez, José 746, 895, 2290, 2327
Rodríguez, Juan 1701, 1959
Rogers, Daniel 1529
Rogers, Michael 1569, 2356
Ronanki, Deepak 2647
Rong, Mingzhe 3220
Rong, Zhenshuai 1439
Rosa, Bruno M.G. 1653
Round, Simon 1803
Roy, Soham 1121
Rubinic, Jaksa 416
Rueß, Manuel 3241
Ruiz, Juan M. 2002
Ruiz, Juan 442, 2082, 2296
Ruoff, Dominik Alexander... 2589
Ruppert, Daniel 1759
Russo, Andrea 252
Ruszczyk, Adam 1803
Sa, Satyam 103, 449
Saberi, Sajad 2840
Sadasivan, Arya 2461
Sadilek, Tomas 401
Sado, Kerry 2833
Saeedifard, Maryam.................................. 2051, 3071, 3321
Saelens, Jonathan 1311
Saggini, Stefano............................... 479, 1444, 2260
Saha, Subrata 1237
Saha, Tarak 3174
Sahoo, Subham 696, 912, 1501
Sai, Ranajit.. 285
Saiga, Kazuma 2551
Saito, Shoji 1569
Saito, Wataru 2551
Sakai, Hiroto 3077
Salari, Omid 3083
Salehi, Maryam 2883
Samanta, Akash.. 3141
Sambo, Haifah B. 291
Sandoval, Rolando 1459
Sangwongwanich, Ariya 738, 1501, 3042
Sanjakdar, Omar 1396, 2562
Santi, Enrico 2833, 2865
Santos, Ion Leandro Dos.. 342
Santos Jr., Euzeli Cipriano Dos...................................... 2029

Sanusi, Bima Nugraha.................................. 246, 1810, 2035
Saraf, Pushkar 77
Sarajian, Ali... 895
Sarda, Radhika 62, 1927
Sarlioglu, Bulent 3174
Sarnago, H. 147
Sarnago, Héctor... 3109
Sarofim, Seif 449
Sati, Shraf Eldin 622
Sato, Yuji 1383
Sato, Yuki 579
Satterlee, Ryan 3133
Satyamsetti, Vijayakrishna 1403
Sauter, Bailey 2764
Sayed-Ahmed, Ahmed 2932
Sba, Baher Abu 2453
Sbabo, Paolo 2667
Schanen, Jean-Luc.. 2562
Scheideler, William 1590
Scherer, Yohannes Amilcar Tekle 342
Sebastián, Javier 1959
Sebata, Kohei 1977
Sekiya, Hiroo 3248
Selvarasu, Uthandi 761, 1465
Sen, Paresh C. 1892
Sen, Tanuj.. 139, 349, 1882
Sengstock, Jonathan 1242
Sengupta, Arkadeb 90, 118, 1148
Seo, Gab-Su 602, 645, 3059
Seo, Seoktae 1629
Sethupandi, Abishek 62
Seugnet, Léo 2718
Shadmand, Mohammad B. 3071
Shafei, Ahmad El 1326
Shah, Shreyas B. 670
Shahbazi, Reza 1179
Shahsavar, Tala Hemmati 1622
Shang, Shuye 3227
Shao, Hang 2138
Shao, Linbo 1020, 1564
Sharma, Mohit 3141
Shen, Andy 3129
Shen, Xiaobing.. 436, 2167
Shi, Guannan 1564
Shillaber, Luke 2597
Shimada, Takae 2681
Shimosako, Shumei 3077
Shin, Se-Un 834
Shivdikar, Saumil 2400
Shoji, Tomokazu 2821
Shrestha, Niranjan 801
Shu, Wenze 524

Siddiquee, Ashraf	1294, 1602
Silveira, Hector Bessa	342
Sim, Dong Hyeon	3200
Sim, Si Yuan	505
Singh, Anurag	1, 54, 334
Singh, Prashant	1026
Singla, Rishabh	3054
Siraj, Ahmed	1040, 2400
Sitta, Alessandro	1070
Smith, John	637
Solecki, Alex	1242
Solomentsev, Michael	77
Son, Gibong	1741
Song, Chen	2075
Song, Keqi	1507
Song, Minwoo	3012
Song, Qihao	202
Song, Xiaoqing	1844, 2407
Song, Yubo	696
Song, Zhihao	327
Sönmez, Ertuðrul	1051
Soundararajan, Soundhariya G.	238
Souri, Naser	2202
Sowers, Elizabeth A.	2419
Sozer, Yilmaz	1294, 1602, 2698
Spiazzi, Giorgio	2667
Spieler, Matthias	1495, 1551
Sridhar, Sundaramoorthy	299
Sriram, Vaisambhayana B.	62, 1927
Srivastava, Shubham	2213
Starke, Michael	637, 703
Stauth, Jason T.	1590, 1715
Steiner, Mark	2356
Stella, Fausto	1408
Steyaert, Bernard	1255
Steyn-Ross, Alistair	378, 874
Stillwell, Andrew	1242
Stokowski, Nicole	1242
Strache, Sebastian	2569
Strathman, Sophia A.	1311
Streit, Jochen	1051
Strezelecki, Ryszard	2173
Stricula, Justin	401
Sturdivant, Maurice	544
Su, Gui-Jia	1635
Su, Mei	2096
Sugie, Hisashi	2730
Sui, Qingcheng	436, 2752
Sukita, Yohei	1102
Sullivan, Charles R.	840, 1816
Sun, Bosheng	1990
Sun, Kai	2488

Sun, Lingwei	1108
Sun, Peiyuan	2375
Sun, Ruize	192
Sun, Weifeng	524
Sun, Xiuhu	3328
Sun, Zhen	3275
Sun, Ziang	2981
Sund, Jade	971
Sune, Joseph Benzaquen	164, 3334
Suntharalingam, Piranavan	670
Suzuki, Asamira	1728
Swaminathan, Madhavan	2101
Sweet, Mark	3167
Swoboda, Philipp	2348
Syed, Hadiuzzaman	1051
Szczublewski, Austin M.	185
Tadakuma, Toshiya	2610
Taguchi, Koichi	2356
Taha, Wesam	3147
Tajima, Shin	782
Takahashi, Keita	2610
Takahashi, Yoshiaki	1414
Takamiya, Makoto	1084, 1102, 2551
Takamura, Yota	1797
Takayama, Naoki	2681
Takeda, Ryo	2821
Takeuchi, Kosuke	21
Takeuchi, Toshiro	2828
Takishima, Kenta	423
Takizawa, Sota	2924
Tan, Matthew	1882
Tan, Siew-Chong	3227
Tanaka, Kenichiro	579
Tanaka, Ryota	423
Tanaka, Shinsaku	2284
Tanaka, Toshiyuki	2828
Tang, Ho-Tin	1507, 1582
Tang, Wenyuan	1363, 1597
Tang, Yi	3220
Tant, Mike	1824
Tariquzzaman, Md.	3036, 3048
Tarutani, Masayoshi	1383
Tatetsu, Riku	1977
Tayebi, Milad	854
Teng, Fei	2527
Teng, Yiyina	757
Terauchi, Naoya	285
Terzija, Vladimir	757
Thacker, Thimothy	2992
Then, Han Wui	1681
Thevar, Madasamy Palavesha	62
Thike, Rajendra	1279

Thirumoorthi, Sathya Rupan............................1866
Thurlbeck, Alastair P. 1602
Tian, Fanghao 2167, 3124
Tian, Jiachen 2327
Tian, Xiaoyang..1754
Tingbari, Vincent Masabiar...............................2973
Tomey, Hala 1211
Tomioka, Shohei 579
Tong, Junhong...1786
Tong, Qiaoling 2109
Torres, Javier..1197
Torres, Renato Amorim......................................1495
Touhami, Mustapha 1012
Tran, Ngoc Ho 2569
Trescases, O. 1451
Trescases, Olivier..................... 103, 449, 676, 1905
Tripathi, Anshuman.................................. 62, 1927
Tsai, Chieh-Ju 664, 2131, 2143, 2725, 2735, 2741
Tschanz, James 1681
Tseng, Chien-Hao 2725
Tsou, Ming-Chang 887
Tsuchida, Takayuki 285
Tuzizila, Jeremie 401
Uchida, Yasuo 48
Uddarraju, Praneeth 1311
Uddin, Muhammad Fasih 2453
Uegaki, Shin 1383
Uematsu, Takeshi 3248
Ulrich, Burkhard 1707, 2303, 2502
Umanand, L. 1610, 2785
Umar, Jamil. 2973
Umar, Muhammad F. 3071
Umetani, Kazuhiro.............................. 321, 2654
Ursino, Mario..................... 1444, 2254, 2260
Uzum, Alper 1294, 2698
Vagnon, Eric 2562
Vanderwegen, Wout...238
Varadarajan, Kamal...180
Vasiæ, Miroslav 788, 1375, 1803
Vedula, Inder....................................... 1, 54
Vergès, Gaël 1905
Vico, Enrico 1408
Vines, Peter 1403, 1512, 1781
Vinnac, Sébastien...2718
Vitale, Gianni 3133, 3304
Vohl, Kenny 2757
Wagner, Tomas..3100
Walters, Andrew..951
Wang, Cheng Feng........................... 103, 449
Wang, Daming 2369
Wang, Haiyan 3312
Wang, Haoyu 485, 983, 1544, 2207, 2556, 2675, 2857

Wang, Hongjie...................................... 719, 2393
Wang, Huai...........................912, 2380, 3042
Wang, Jin...1203, 2986
Wang, Jinyan.....................................171, 225
Wang, Jun.............................538, 1850, 2088
Wang, Kaiyuan....................................3227
Wang, Kejia 505
Wang, Kun...2088
Wang, Kunrong....................................2162
Wang, Laili...2375
Wang, Lei...2162
Wang, Liang 983
Wang, Libing.......................................3174
Wang, Lichong 757
Wang, Linguo.......................................2070
Wang, Lisheng.....................................1248
Wang, Liwei...1153
Wang, Maojun 225
Wang, Meng...3312
Wang, Mengqi.............................2624, 3155
Wang, Pinhe..............................246, 2035
Wang, Qiong 498
Wang, Rudy...9, 1318
Wang, Rui...1063
Wang, Shaozhe.....................................1459
Wang, Shukai...........................566, 1882, 2438
Wang, Shumeng...........................761, 1465
Wang, Shuo 1173, 2603
Wang, Sunqing.....................................3018
Wang, Wei...2692
Wang, Xiao 219
Wang, Xiaohua.....................................3220
Wang, Xiaosheng 795
Wang, Xiaoting....................................3030
Wang, Xiaoyu 416
Wang, Xinlin 809
Wang, Xiongfei....................................1615
Wang, Xuan...1791
Wang, Xuliang...........................1544, 2556, 2675
Wang, Yan...........................1544, 2556, 2675
Wang, Yao...3275
Wang, Yibo...............................795, 3181
Wang, Yicheng.....................................3147
Wang, Yiju...1368
Wang, Yulei 219
Wang, Yunxin.....................................2675
Wang, Zijian...1537
Wang, Ziyao 485
Wang, Zuoshuai....................................3018
Watabe, Kiyoto....................................2551
Watanabe, Hiroki..........................21, 48
Watanabe, Kenichi 1102, 3096

Wehr, Erik	2757
Wei, Anran	1983
Wei, Bo	327
Wei, Jinxiao	1832
Wei, Xing	3042
Wei, Xuanjing	1403
Wei, Yuxin	2713
Weihs, Leon	2757
Weiser, Mathias C.J.	2387, 3241
Weng, Sheldon	1681
Wens, Mike	195
Wheeler, Patrick	895
Wicht, Bernhard	848
Wick, Lukas	2380
Williamson, Sheldon	801, 1224, 2322, 3114, 3141
Wilson, Marcus	378
Winkler, Joseph	848
Wipprecht, Lukas	2303
Wojewoda, Leigh	491
Wong, Andy	2833
Wouters, Hans	238, 258, 2167
Wright, Jason	401
Wu, Alan	307
Wu, Chih-Chiang	2687
Wu, Hsiang-Kai	2687
Wu, Shang-Syun	2179
Wu, Taotao	1090
Wu, Teng	2634, 2660
Wu, Tsai-Fu	16, 900
Wu, Xin	651, 2981
Wu, Xinke	1995
Wu, Yang	139, 349
Wu, Yanqing	1995
Wu, Yingzhe	1248
Wu, Yue	1114, 3220
Wu, Yuxuan	2582
Wunderlich, Andrew	2865
Wunderlich, Ralf	2757
Xi, Zichen	1020
Xia, Xiaoyi	2022
Xiang, Zhangwei	429
Xiao, Junjie	609
Xie, Biyun	919
Xu, Dehong	651, 2894, 2981
Xu, Guo	2096
Xu, Haoran	2109
Xu, Huangsheng	2207
Xu, Limei	2075
Xu, Shen	524
Xu, Wentao	1012
Xu, Wenzhe	2426
Xu, Xinmiao	1433

Xu, Yun	1114
Xu, Ziyang	2521
Xue, Hui	2117
Xue, Yuxiang	1248
Yabuta, Shigenori	2821
Yagielski, John	3174
Yamaguchi, Koji	1299, 1414
Yamamoto, Keisuke	3194
Yamamoto, Masayoshi	2431
Yamanaka, Kimito	1797
Yan, Decheng	3334
Yan, Yiyang	2343, 2426
Yan, Zhaoheng	1114
Yan, Zhixing	357, 1034
Yang, Bowen	602
Yang, Garam	3000
Yang, Hélène T.W. Ma	983
Yang, Juchen	1203
Yang, Liu	2369
Yang, Qichen	860
Yang, Qiuzhe	1537
Yang, Robert	180
Yang, Xin	1020, 1564
Yang, Xingyu	1953
Yang, Xinliang	409
Yang, Yirui	1173, 2603
Yang, Yongheng	3024
Yang, Yun	3227, 3275
Yang, Zineng	1020
Yao, Wenxi	327
Yao, Yuzhou	1203
Yasko, Mohamed	3124
Yato, Shinji	3077
Ye, Liang	285
Ye, Zhengyu	2117
Yeo, Howe Li	62
Yeo, Sungku	2547, 3281
Yi, Lifang	2413
Yi, Zheyuan	2488
Yin, Shan	1248, 2075
Yin, Tianxiang	2535
Yoneyama, Rei	2356
Yoshimoto, Kantaro	423
Yoshimura, Yuto	2654
You, Longxiang	3018
Youssef, Mohamed Z.	3083
Yu, Hao	2343, 2426
Yu, Jingshu	1681
Yu, Ruiyang	854
Yu, Sheng-Han	2131
Yu, Sheng-Yang	42, 1990
Yu, Wensong	2220

Yu, Xiang 1892
Yuan, Hao 2117
Yuan, Huan 3220
Yuan, Jiaqi 670
Yuan, Jingyi 966
Yuan, Song 1114
Yuan, Tianlong 429
Yuan, Tianshu 2375
Yun, Dam 1659, 2268
Zaabi, Omar Al 1940
Zade, Aditya 719, 1004, 2194
Zaitsu, Toshiyuki 1977, 2654
Zaizen, Shohei 2551
Zaman, Mohammad Shawkat 676
Zan, Xin 3232
Zane, Regan 719, 1004, 2194
Zeineldin, Hatem H. 622
Zekorn, Tobias 2757
Zeng, Hank 2138
Zeng, Jia-En 1967
Zeng, Wenliang 510
Zeng, Zheng 219
Zhan, Cao 278, 2482
Zhang, Ben 3181
Zhang, Bing 2070
Zhang, Bo 192
Zhang, Bohua 573
Zhang, Boran 2675
Zhang, Boyi 1850, 2296
Zhang, Cheng 1512, 3212
Zhang, Chenghui 2713
Zhang, Chi 9
Zhang, Desheng 2109
Zhang, Fuxing 2059
Zhang, Haijin 3312
Zhang, Hely 1286
Zhang, Heng 2369
Zhang, Hengbin 1248
Zhang, Hong 2375
Zhang, Honglang 2075
Zhang, Jiazheng 2628, 2968
Zhang, Jincheng 978
Zhang, Jinfeng 1439
Zhang, Li 2535
Zhang, Qingzheng 2894
Zhang, Renjie 3220
Zhang, Shengke 214
Zhang, Shiqi 757
Zhang, Tianyi 2066
Zhang, Weihang 978
Zhang, Xiangrong 3275
Zhang, Xin 505, 1143, 3018

Zhang, Xiong 2343, 2426
Zhang, Yi 2380
Zhang, Yichi 2380
Zhang, Yifan 2343, 2369, 2426
Zhang, Yifu 2746
Zhang, Yingjie 2603
Zhang, Yuanxin 2857
Zhang, Yuhao 202, 1020, 1564
Zhang, Yuxin 2973
Zhang, Zhe 2907
Zhang, Zhenbin 2846, 3206
Zhang, Zheyu 1040, 1153, 2400
Zhang, Zhi Jin 3321
Zhang, Zhining 1203
Zhang, Zichen 1850
Zhao, Delin 3220
Zhao, Fangzhou 1615
Zhao, Hongbo 357, 1034
Zhao, Shuang 2369, 3227
Zhao, Shuofeng 1824
Zhao, Tiefu 1388, 2790, 3328
Zhao, Tuo 1274
Zhao, Wending 1995
Zhao, Yifan 2521, 2846, 3206
Zhao, Yue 27, 2453
Zheng, Zexiang 2343, 2369
Zhou, Daniel H. 1693
Zhou, Daniel 566, 1274, 2438
Zhou, Dao 2380
Zhou, Fei 2541
Zhou, Feng 2624
Zhou, Jiale 1388, 2790, 3328
Zhou, Kunxiao 809
Zhou, Lufan 1803
Zhou, Mingde 2207
Zhou, Wenqi 2589
Zhou, Xigen 2157
Zhou, Yan 1767
Zhou, Yi 651
Zhou, Yuan 2535
Zhou, Yuebin 2369
Zhou, Zongjie 1047
Zhu, Jiaqi 3312
Zhu, Jinli 761, 1465
Zhu, Junjie 2070
Zhu, Lingyu 278, 2482
Zhu, Liyan 1020
Zhu, Yicheng 558, 2276
Zhu, Zhenhai 1995
Zhuo, Fang 2327
Zolfi, Pouya 1230
Zou, Huanghaohe 1786

Zou, Jiaao..2066
Zou, Jiarui...................................... 558, 2276, 2805
Zou, Mingrui...219
Zou, Xudong..2066
Zou, Xuecheng...2109
Zufferli, Kevin 1444, 2260
Zuo, Yu 258, 436, 2167, 2752
Zuo, Zhiling..2070
Zynger-Capaverde, Betina...............................2562

IEEE
445 Hoes Lane
Piscataway, NJ 08854-4141

ISBN 979-8-3315-1612-3